A Textbook of
Organic Chemistry

A Textbook of
Organic Chemistry

V.K. Ahluwalia □ Madhuri Goyal

Narosa Publishing House

New Delhi   Chennai   Mumbai   Calcutta
London

V.K. Ahluwalia
Honorary Visiting Professor
Dr B R Ambedkar Centre for Biomedical Research, and
    former Professor of Chemistry, University of Delhi, Delhi 110 007, India
Former Vice-Chancellor, Manipur University, Imphal, India

Madhuri Goyal
Reader of Chemistry
Deshbandhu College, University of Delhi, Delhi 110 019, India

Copyright © 2000, Narosa Publishing House

NAROSA PUBLISHING HOUSE

6 Community Centre, Panchsheel Park, New Delhi 110 017
22 Daryaganj, Prakash Deep, Delhi Medical Association Road, New Delhi 110 002
35–36 Greams Road, Thousand Lights, Chennai 600 006
306 Shiv Centre, D.B.C. Sector 17, K.U. Bazar P.O., Navi Mumbai 400 705
2F–2G Shivam Chambers, 53 Syed Amir Ali Avenue, Calcutta 700 019
3 Henrietta Street, Covent Garden, London WC2E 8LU, UK

ISBN 81-7319-159-X

Published by N.K. Mehra for Narosa Publishing House, 6 Community Centre,
Panchsheel Park, New Delhi 110 017 and printed at Rajkamal Electric Press,
Delhi 110 040 (India)

Preface

This book is primarily written for undergraduate and graduate students. Our long teaching experience indicates that an individual comprehension of the various topics is improved immeasurable through experimental observations. These aspects have been the guiding aim while writing this book. The text provides facts and theories which relate experimentation to the structure and reactivity of the concerned compounds. The structure and reactivity of an organic compound is related to mechanism of the reaction. Structure of the book and material presented in each chapter will enable students to reach quickly a level of self confidence. Students who wish to study organic chemistry independently will also find it easier to comprehend.

The book containing 48 chapters covers all the topics taught at undergraduate level. The basic principles have been discussed in the beginning (Chapters 1 to 7) which will provide a sound foundation of these principles and concepts. Chapter 2 deals with bonding and properties of organic molecules. The topic has been discussed by taking simple organic molecules as an example but is applicable to complex organic structure which are discussed in Chapters 39 to 48. Classification and nomenclature of organic compounds are given in Chpater 3. This includes the latest IUPAC rules for naming simple mono-, bi- and polyfunctional organic compounds. In Chapter 4, a preliminary treatment to organic reactions and mechanism has been discussed. The detailed discussion of the mechanism of a particular reaction has been incorporated with the class of compound undergoing that reaction which directly hints at the relationship between the structure and properties of that particular class of compound. Chapter 5 on stereochemistry describes the compound in three dimensions i.e., a relationship between the structure of compound (in three dimensions) and their reactivity and related properties. The chapter describes nomenclature of geometrical isomers (E&Z) as well as optical isomers (R&S notations). Chapter 7 on spectroscopy further strengthens the understanding of structure and properties of organic molecules as it is a tool for elucidation of molecular structure with the help of infrared, ultraviolet and NMR. Spectroscopy is an experimental technique that is used and learned in laboratory.

Broadly, basis of first and second year course at undergraduate level in almost all the Indian universities includes the fundamental concepts, structure and reactivity related to aliphatic and aromatic organic molecules. Chapter 8 to 36 discusses various classes of aliphatic and aromatic organic compounds. The reaction like free radical substitution, electrophic and nucleophilic addition and substitution, polymerisation, stereo-specific and stereoselective reactions, tautomerism etc. have been incorporated in appropriate chapters in a student-friendly discussions.

Part of courses at senior undergraduate level are confined to polynuclear hydrocarbons, hetrocyclics and natural products like carbohydrates, proteins, pyrimidines, alkaloids, terpenoides etc. These chapters provide students the practical information and help them appreciate the relevance and pervasiveness of organic chemistry. Polymers, dyes and pesticides have been discussed in Chapters 44 to 48.

In an effort to facilitate orderly learning each chapter has been divided into several section and sub-sections. Unsolved exercises at the end of each chapter will help students in mastering their understanding about the topic. To keep the volume of the book restrictive, numerical and illustrations are provided wherever it was felt necessary.

No work of this volume can proceed without a great deal of day to day help and this one is no exception; it was aided by many persons and authors in this field. To all we express gratitude. Our special thanks are due to our colleagues Sunita Dhigra, Suman Dudeja, Renu Aggarwal, Pooja and Madhu Chopra who were particularly helpful.

I, Madhuri, conveys special thanks to my family, to my parents Late Krishan Kumar Garg and Mrs. Kamlesh Garg, who provided me educational foundation; to my children Amit and Divya, for understanding the time commitment of my work; and most of all to my husband, Rajesh, who not only provided the environment for me to achieve my goals for the work but also wholeheartedly continues to support all my efforts.

I, V.K. Ahluwalia would like to give sincere thanks to my wife Manmohini, for her unwavering patience and support during many evenings and weekends that I have spent in writing and re-writing instead of being with her.

Last but not least we would like to convey our thanks to Prof. Ramesh Chandra, Director, Dr. BR Ambedkar Centre for Biomedical Research, University of Delhi and our colleagues at Department of Chemistry, Deshbandhu College, University of Delhi, for their help.

V.K. AHLUWALIA
MADHURI GOYAL

Contents

1

Purification of Organic Compounds

1.0 INTRODUCTION

The major source of organic compounds are plants and animals along with natural gas, petroleum and coal. The preparation of pure organic compounds from these sources, therefore requires specialised methods for their extraction and purification.

1.1 EXTRACTION OF ORGANIC COMPOUNDS

The organic compounds present in various parts of a plant can be extracted by *Soxhlet apparatus* (Fig. 1.1), which consists of a wide tube provided with a side tube on the left, the siphon tube on the right and a water condenser at the top. The plant material is placed in the wide tube which is attached to a round bottom flask having the solvent.

The plant material like crushed roots, stem, leaves or fruits which is used for the extraction of organic compound is placed in the wide tube and solvent used for extraction is taken in the flask. When the solvent boils, the vapor come in contact with the plant material. When the solvent fills the wide tube, the siphon tube sends it back to the flask. The process is repeated again and again till all the soluble compounds present in the plant source dissolves in solvent. After the extraction is complete the solvent from the flask is distilled and solid compound obtained. It is purified as given in Section 1.2.

1.2 METHODS OF PURIFICATION OF SOLIDS

Solids are purified by a number of methods such as:

(a) *Crystallisation:* A pure solid compound can be separated from its impurities by dissolving it into a suitable solvent at high temperature and collecting its crystals on cooling.

A nearly saturated solution of impure substance is prepared in hot solvent and is heated with bone charcoal which adsorbs the impurities into it followed by filtration while hot or through a hot water funnel as shown in Fig. 1.2. The filtrate

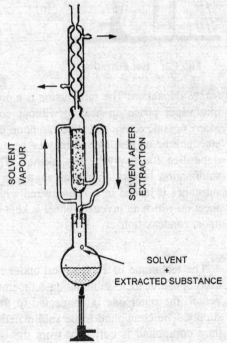

SOLVENT VAPOUR

SOLVENT AFTER EXTRACTION

SOLVENT + EXTRACTED SUBSTANCE

Fig. 1.1: Soxhlet Extractor

on slow cooling crystallises the pure solid compound which is then filtered under reduced pressure using the apparatus shown in Fig. 1.3. The filter pump accelerates the filtration process by suction.

(b)   *Fractional Crystallisation:* Fractional crystallisation is based on the principal of differential solubilities of different compounds in a solvent. A solid mixture having two or three compounds is dissolved at high temperature in an appropriate solvent. The compound having less solubility crystallises out first on cooling leaving behind the other fractions in the solution. This process is repeated several times to get the pure compounds.

Sometimes better results are obtained by using a mixture of two or more solvents for crystallisation. The commonly used solvent mixtures for the crystallisation of organic compounds are:

Alcohol and water
Alcohol and ether
Benzene and petroleum ether
Chloroform and petroleum ether

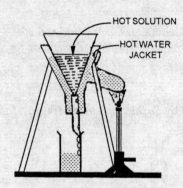

Fig. 1.2:   Hot Filtration

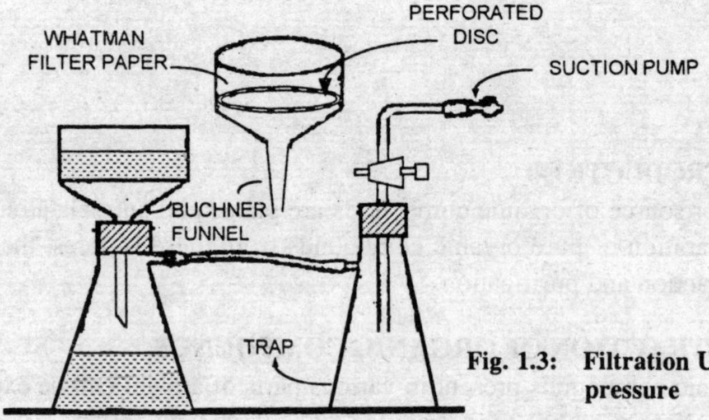

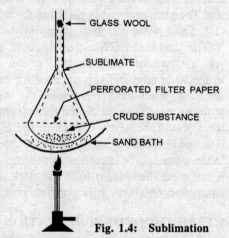

Fig. 1.3:   Filtration Under pressure

(c)   *Sublimation*: The sublimation is a property of solids to pass into vapor phase on heating without coming to liquid phase. Many organic compounds like camphor, naphthalene, anthracene etc. sublime on heating and may be separated from non-volatile impurities effectively by this method. The apparatus used for sublimation is very simple as shown in Fig. 1.4. The impure substance is heated in a dish covered with a perforated asbestos sheet on which an inverted funnel is kept where it gets deposited upon condensation.

Fig. 1.4:   Sublimation

The technique of sublimation under reduced pressure is applicable to only those substances which decompose at their sublimation temperature as shown in Fig. 1.5. The apparatus consists of a double vessel, the outer one is connected to the pump whereas the inner one has continuous flowing water supply. The compound to be sublimated is kept in the outer vessel which is heated from outside. The pure compound is collected from the outer walls of the water cooled inner vessel.

1.3   METHODS OF PURIFICATION OF LIQUIDS

Various methods have been employed for the purification of liquids which are summarised below:

(a)   *Distillation*: Liquids which boil under ordinary conditions of temperature and pressure without decomposition and are associated with non-volatile impurities are purified by simple distillation method in which the liquid vaporises on heating and comes back to its original form by condensation in a water or air condenser. The pure liquid is collected in the receiving flask leaving behind its nonvolatile impurities in the distillation flask.

(b)   *Fractional Distillation*: When the liquid to be separated or purified is a mixture of two or more volatile liquids, a fractionating column is used as shown in Fig. 1.6. The basis of fractional distillation is that the low boiling fraction of a mixture of liquids separates out first followed by the higher boiling fraction one by one.

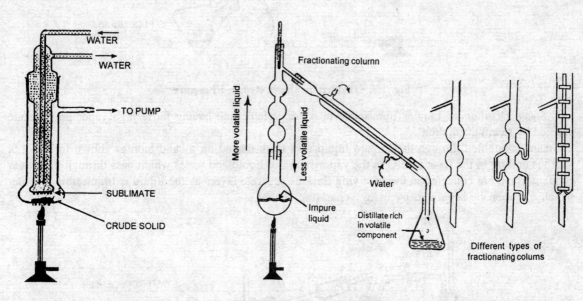

Fig. 1.5:   Sublimation   Fig. 1.6:   Fractional Distillation and Fractionating Columns
                Under Pressure

The vapours of the liquid to be fractionated rise into the column where some of its high boiling fractions condenses back and flows down. The process is repeated again and again till fractions of high volatility go up followed by the others of lower volatility. They are collected separately in the receiving flask.

(c)   *Distillation Under Reduced Pressure*: This method is applicable to those liquids which decompose at their boiling point. The boiling point is reduced by applying pressure.

Distillation under reduced pressure is carried out in an apparatus similar to that for distillation but the receiving adopter is attached to a vacuum pump as shown in Fig. 1.7.

The method is not only useful to avoid decomposition but is at the same time economical too, thus is used in industries.

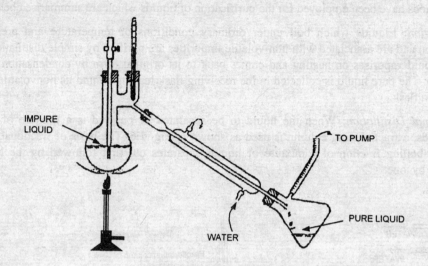

Fig. 1.7: Distillation Under Reduced Pressure

(d) *Steam Distillation*: Liquids immiscible in water, volatile and having fairly high vapor pressure are purified by steam distillation.

Steam is bubbled through the impure liquid in a flask heated on a sand bath as shown in Fig. 1.8. Vigrous boiling in the flask results in the vaporisation of liquid and water which pass through the water condenser and are collected in the receiving flask as separate layers as the liquid is immiscible in water which can then be separated by using separating funnel.

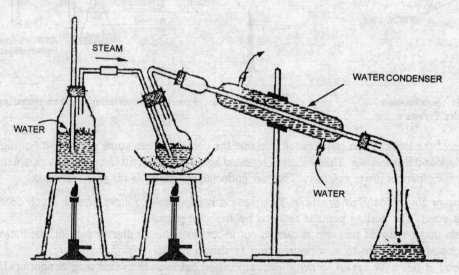

Fig. 1.8: Steam Distillation

(e) *Solvent Extraction*: Liquids can be purified by using appropriate solvent which can selectively dissolve one of the constituents into it. The various constituents are then separated by using separating funnel.

1.4 CHROMATOGRAPHIC METHODS

It is a widely used technique for the isolation, separation, purification and identification of a mixture. It depends on the principle of phase distribution where the different constituents of a mixture are selectively adsorbed on a stationary phase and are then eluted by a mobile phase.

Chromatographic methods have been classified according to the physical state of stationary and mobile phase. The stationary phase can be a solid or a liquid and the mobile phase can be a liquid or a gas. The various chromatographic methods are:

1. Thin layer chromatography or solid-liquid chromatography is an adsorption chromatography in which stationary phase is spread over a glass plate and the liquid is the moving phase.

Similarly, column chromatography, high pressure liquid chromatography and ion exchange involve solid-liquid phase.

2. Partition chromatography or liquid-liquid chromatography involves the separation of two liquid phases, with one of them adsorbed on a solid support.

3. Gas liquid chromatography or gas chromatography, however, has liquid coated over solid as stationary phase and gas the mobile phase.

1.4.1 Thin Layer Chromatography

The thin layer chromatography was developed by Izmailov and Sheaiber in 1938 and accepted in 1958 as a technique for separation and identification of organic compounds. The stationary phase, a solid adsorbent, is spread in the form of a thin layer of about 0.25 mm thickness on a glass plate with a spreader. The commonly used absorbents are silica gel and alumina with a small amount of binder. The sample to be separated or identified is applied at the lower side of the glass plate in the form of symmetrical spots with the help of a tapering capillary or syringe. The coated plate is then placed in a chamber saturated with the solvent as shown in Fig. 1.9.

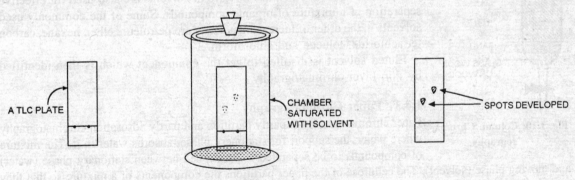

A TLC PLATE   CHAMBER SATURATED WITH SOLVENT   SPOTS DEVELOPED

Fig. 1.9:   Thin Layer Chromatography

The sample to be separated is dissolved in nonpolar solvent and the solvent used for the elution is selected on the basis of nature of the components present in it. When the solvent has moved to a distance of about 10 cm (or less depending upon the size of the TLC plate), the plate is removed from the chamber, solvent front marked and dried. The chromatogram is then developed either by iodine or by reaction with sulphuric acid, potassium permanganate or 2,4-DNP depending upon the nature of compound to be identified. Fluorescent compounds develop the spot by absorption of UV light, when the TLC plate is exposed to UV light.

For a given set of conditions (thickness and nature of adsorbent, solvent etc.), a compound may also be identified by R_f value which is a ratio of distance travelled by the substance and that by the solvent from the point where spotting was done.

$$R_f = \frac{\text{Distance travelled by the substance}}{\text{Distance travelled by the solvent}}$$

1.4.2   Column Chromatography

This technique is commonly used for the separation of a mixture into its components and identifying them by TLC. The stationary phase is solid having alumina, silica gel, cellulose or charcoal as adsorbent. The choice of adsorbent is dependent upon the nature of solvent used for elution and the nature of compounds present in a mixture. The adsorbent must be insoluble in the solvent and must not react with the compounds to be separated.

Cylindrical glass columns of varying sizes are used for the separation of mixtures. The length and width of the column is chosen on the basis of the nature of mixture. The column is packed with glass wool or wool followed by sand and adsorbent as shown in Fig. 1.10.

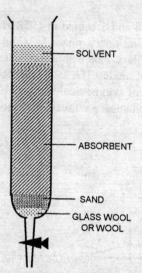

SOLVENT

ABSORBENT

SAND

GLASS WOOL OR WOOL

Fig. 1.10: Column Chromatography

The nature of solvent used for elution (percolation of solvent through the column) depends upon the nature of compounds to be separated. For effective separation, the eluting solvent must be less polar than the components. If the solvent is more polar and strongly adsorbed than the components of the mixture, the separation will not be effective as the components will remain in mobile phase. It is also important to note that the components of the mixture should not be soluble in the solvent used for elution for it will remain permanently adsorbed on the adsorbent. For strongly adsorbed mixtures, polar solvents are used for elution whereas nonpolar solvents are used for weakly adsorbed mixtures. Sometimes mixture of solvents are also used for effective separation of a mixture of organic compounds. Some of the commonly used solvents in the order of increasing polarity are petroleum ether, hexane, carbon tetrachloride, toluene and chloroform.

Eluted solvent is distilled to get the component which is then identified by thin layer chromatography.

1.4.3   Paper Chromatography

Paper chromatography is partly partition and partly adsorption chromatography where paper, the support for stationery phase, absorbs water in it. The mixture of compounds to be separated is partitioned between stationary phase (water) and moving phase (solvent). The cellulose of the paper partitions the components of a mixture so that they migrate differently on its surface undergoing separation. Paper chromatography is used for the separation of mixtures of amino acids and carbohydrates.

(i) *Radial Paper Chromatography*: In this technique a circular Whatmann filter paper number 1 is placed on a petridish containing the solvents. The thin strip or wick is dipped in the solvent. The mixture of compounds to be separated is placed in the centre and the circular paper is covered with another petridish as shown in Fig. 1.11 to prevent evaporation of solvent. The solvent rises through the strip via capillary action and travels through the filter paper along with the mixture of compounds. The compounds form circular bands at different distances from the centre due to differential adsorption and R_f values (Retention factor) calculated as discussed in thin layer chromatography. The R_f value for a compound is dependent upon nature of solvent.

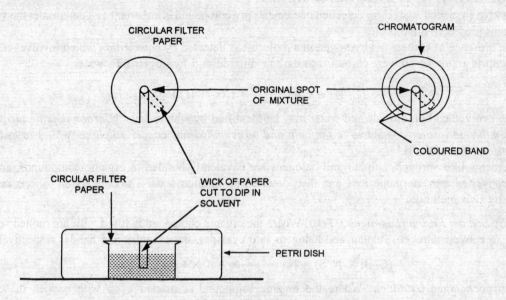

Fig. 1.11:   Radial Paper Chromatography

(ii)   *Ascending and Descending Paper Chromatography*: In this a rectangular paper strip is used for the separation of a mixture of compound as shown in Fig. 1.12. When the paper is suspended into the solvent in such a way that its lower side dips into the solvent, the method is termed as *ascending paper chromatography*. When, the solvent is placed at the top and the upper side of the filter paper dips into it to provide a downward movement of the solvent, the technique is called *descending paper chromatography*.

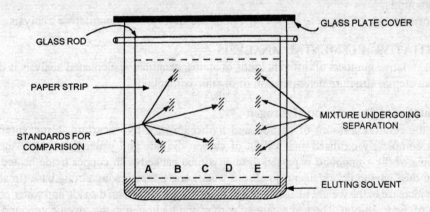

Fig. 1.12:   Ascending Paper Chromatography

If the components of a mixture used for separation do not form colored bands, the paper is either exposed to UV light or sprayed by various developing agents to identify the spots. The R_f values for various compounds may then be calculated by measuring the distance traveled by substance and the solvent.

1.5   QUALITATIVE ELEMENTAL ANALYSIS

Qualitative elemental analysis or detection of elements in a compound is important as it determines the variety of atoms present in a molecule.

The presence of carbon and hydrogen in a molecule is detected by *combustion* which involves heating with cupric oxide to convert carbon into carbon dioxide and hydrogen into water.

$$(C, H) + CuO \xrightarrow{\text{heat}} CO_2 + H_2O + Cu$$

The evolved carbon dioxide and water may be identified by simple tests. [*Carbon dioxide turns lime water milky and water condenses in the bulb and turns anhydrous copper sulphate (white) to hydrated copper sulphate (blue)*].

Elements like nitrogen, sulphur and halogens are covalently bonded in organic compounds, and are first converted into inorganic ions and then detected by familiar tests. The conversion to ions may be done by two methods.

(a)   *By Sodium Fusion (Lassaigne's Test)*: Where the organic compound is fused with hot molten sodium metal to convert nitrogen, sulphur and halogens into cyanides, thiocyanides and halides respectively.

$$(C, H, X, N, S) + Na \longrightarrow NaCN + Na_2S + NaX$$

(b)   *By Schoniger Oxidation*: Where the organic compound is reacted upon with oxygen in sodium hydroxide solution.

$$(C, H, X, N, S) + O_2 \xrightarrow{\text{NaOH}} NaX + NaNO_2 + Na_2SO_3$$

The ions like cyanide, thiocyanide, nitrite, sulphide etc. thus formed may be identified by familiar tests.

By performing the above analysis on toluene, the presence of carbon and hydrogen and in chloroform, carbon, hydrogen and chlorine may be ensured. Further tests show the absence of any other elements in these compounds.

Presence or absence of oxygen, however, cannot be detected by qualitative analysis.

1.6   QUANTITATIVE ELEMENTAL ANALYSIS

To find out the relative numbers of different kinds of atoms, quantitative elemental analysis is done. It is the most important step in structure determination of organic compounds.

1.6.1   Estimation of Carbon and Hydrogen

To find out the relative amount of carbon and hydrogen in a hydrocarbon, a measured amount of compound is completely oxidised and weight of carbon dioxide and water formed are measured. The weighed sample of the compound is passed through a tube packed with copper oxide heated to 600-800° followed by a tube having drying agent (to absorb water) and a tube having strong base (to absorb carbon dioxide). The increase in the weight of each tube gives the weight of carbon dioxide and water produced from the oxidation of hydrocarbon. The percentage of carbon and hydrogen in the given compound may then be calculated as :

Weight of the sample	$= W_1$ g
Weight of carbon dioxide formed	$= W_2$ g
Weight of water formed	$= W_3$ g
Weight of carbon in the sample	$= 12/44 \times W_2$ g
% carbon in the sample	$= 12/44 \times W_2 \times 100/W_1$

Weight of hydrogen in the sample $= 2/18 \times W_3$ g

% hydrogen in the sample $= 2/18 \times W_3 \times 100/W_1$

The direct estimation of oxygen is difficult and it is often calculated by difference.

1.6.2 Estimation of Nitrogen

The covalently bonded nitrogen in organic compounds is estimated by converting it into ionic form.

(a) *Dumas Method*: The organic compound is strongly heated with copper oxide in an atmosphere of carbon dioxide. The nitrogen thus produced is collected in a nitrometer and its volume measured. The carbon dioxide, sulphur dioxide and water formed due to the oxidation of C, H, S etc. are absorbed in strong alkali.

The volume of nitrogen evolved is calculated at STP by substituting the temperature pressure and volume of nitrogen.

Weight of sample $= W_1$ g

Volume of nitrogen at STP $= V$ ml

Since 22.4 liters of gas weighs equal to its molecular weight at STP,

22,400 ml of nitrogen $= 28$ g

V ml of nitrogen $= 28/22,400 \times V$ g

% nitrogen in the sample $= 28/22,400 \times V \times 100/W_1$

(b) *Kjeldahl's Method*: This is the laboratory method for the estimation of nitrogen. When a weighed sample of the compound having nitrogen is digested with concentrated sulphuric acid, potassium sulphate and copper sulphate, ammonium sulphate is produced from which ammonia is liberated by addition of excess of sodium hydroxide. The percentage of nitrogen in the compound is then calculated as follows.

Weight of the sample $= W_1$ g

Volume of 0.1 N H_2SO_4 $= V_1$ ml

Volume of 0.1 N NaOH consumed by remaining acid $= V_2$ ml

Volume of 0.1 N H_2SO_4 used to absorb NH_3 $= (V_1 - V_2)$ ml

$(V_1 - V_2)$ ml of 0.1 N H_2SO_4 $= (V_1 - V_2)$ ml of 0.1 N NH_3

1000 ml of 1 N NH_3 $= 17$ g of NH_3 or 14 g of nitrogen

or 1000 ml of 0.1 N NH_3 $= 1.4$ g of nitrogen

$(V_1 - V_2)$ ml of 0.1N $NH_3 = 1.4/1000 \times (V_1 - V_2)$ g of nitrogen

% nitrogen in the sample $= 1.4/1000 \times (V_1 - V_2) \times 100/W_1$

1.6.3 Estimation of Halogens

Halogens are quantitatively estimated by Carius method in which a weighed sample of the compound having halogen is heated with fuming nitric acid and silver nitrate in a sealed tube. The halogen is converted to silver halide which is weighed and the percentage of halogen (chlorine or bromine) calculated as follows:

Weight of the sample $= W_1$ g

Weight of silver chloride or silver bromide formed $= W_2$ g

143.5 g of AgCl contains 35.5 g of chlorine. Thus, W_2 g of AgCl contains $35.5/143.5 \times W_2$ g of chlorine.

Therefore, % chlorine in the sample $= 35.5/143.5 \times W_2 \times 100/W_1$

Similarly, % bromine in the sample $= 80/188 \times W_2 \times 100/W_1$

where W_2 is the weight of silver bromide produced in the reaction.

1.7   EMPIRICAL FORMULA

With the knowledge of percentage of various elements present in the molecule, we can calculate the empirical formula for the same. The empirical formula expresses the percentage composition or relative number of different kind of elements in a molecule. To calculate the empirical formula,

   (i) Percentage of each element is divided by its atomic weight.

  (ii) Each one is then divided by smallest ratio to get the ratio of quotients.

 (iii) If the ratios obtained are fractional, they are converted into smallest possible whole numbers and this actually is the number of atom of each element.

The above rules may be used to calculate empirical formula as shown in following examples.

Example 1.1: Find the empirical formula of a compound, 0.2801 g of which gave 0.9482 g of CO_2 and 0.1939g H_2O on complete combustion.

$$
\begin{aligned}
\text{Wt Carbon} &= 0.9482 \times 12/44 \\
\text{\% Carbon in 0.2801 g of compound} &= 0.9482/0.2801 \times 12/44 \times 100 = 92.32 \\
\text{Wt Hydrogen} &= 0.1939 \times 2/18 \\
\text{\% Hydrogen in 0.2801 g of compound} &= 0.1939/0.2801 \times 2/18 \times 100 = 7.69
\end{aligned}
$$

To find the empirical formula, divide the percentage composition of the elements by their atomic weights.

$$
\begin{aligned}
\text{Molar ratio for carbon} &= 92.32\ /12 = 7.69 \\
\text{Molar ratio for hydrogen} &= 7.69\ /1 = 7.69
\end{aligned}
$$

Thus, empirical formula of the compound is CH.

1.8   DETERMINATION OF MOLECULAR MASS

Molecular mass accounts for the mass of actual number of different kinds of atoms present in a molecule to give the molecular formula. Molecular mass of a substance in amu (atomic mass unit) is defined as number of times its molecule is heavier than that of hydrogen atom. Molecular mass expressed in grams is termed as gram molecular mass or gram molecule or mol.

Molecular mass is determined by any of the methods discussed below depending upon the nature of compound:

1.8.1   Victor Mayer's Method

Molecular mass of volatile liquids is determined by this method. A known mass of the volatile substance is vaporised in a Victor Mayer's tube and the vapor obtained displaces an equal volume of air in a graduated tube. The volume of vapor collected is measured and reduced to STP using Boyle's Law.

$$
\begin{aligned}
\text{Weight of sample} &= W_1\ \text{g} \\
\text{Volume of vapor at STP} &= V_1\ \text{ml}
\end{aligned}
$$

V_1 ml of vapor (sample) corresponds to W_1 g, thus, 22400 ml corresponds to $W_1/V_1 \times 22400$ g.
Molecular mass of sample $(m) = W_1/V_1 \times 22400$ g

1.8.2   Depression in Freezing Point or Cryoscopic Method

The method is used for nonvolatile solids. A weighed sample of the compound is dissolved in a known quantity of solvent of known freezing point and the freezing point of the mixture is determined. The difference of the two freezing points furnishes the depression in freezing point. The molecular mass is than calculated by following expression.

$$m = \frac{1000 \times K_f \times W_1}{W_2 \times T}$$

where K_f is the molal depression constant, W_1 weight of the sample, W_2 weight of the solvent and T the depression in freezing point

1.8.3   Elevation of Boiling Point or Ebullioscopic Method

The elevation in boiling point (T) is calculated experimentally by dissolving a weighed amount of the sample, W_1 (molecular mass, m) in a known amount of the solvent (W_2), having molal elevation constant K_b. Molecular mass of the sample is given by

$$m = \frac{1000 \times K_b \times W_1}{W_2 \times T}$$

where K_b is the molal elevation constant, W_1 weight of the sample, W_2 weight of the solvent, and T the elevation of boiling point

1.8.4   Silver Salt Method for Acids

Organic acids form sparingly soluble silver salts which upon decomposition form metallic silver residue. A known weight of dried silver salt of organic acid is ignited in a crucible and weight of metallic silver taken.

Let the weight of silver salt be x g and that of silver residue be y g.

Therefore, weight of silver salt that would form 108 g of metallic silver residue, i.e.,

The equivalent weight of salt of acid = $x/y \times 108$ and

The equivalent weight of acid = Equivalent weight of silver salt – Equivalent weight of silver + Equivalent weight of hydrogen = $x/y \times 108 - 108 + 1$ or $(x/y \times 108) - 107$

Molecular weight of acid = Equivalent weight × basicity = $[(x/y \times 108) - 107]n$

The molecular mass of acids and bases can also be determined by volumetric methods. The modern method of molecular mass determination is by mass spectroscopy.

1.9   MOLECULAR FORMULA

Molecular formula expresses the actual number of each kind of atom in a molecule. Having known the percentage composition, empirical formula and molecular mass, we can now calculate the molecular formula for an organic compound.

Example 1.2: 0.18 g of an organic compound on combustion gave 0.264 g carbon dioxide and 0.108 g of water. 6.0 g of the same compound when dissolved in 50 g of water depressed the freezing point of water by 1.24°. Find the molecular formula of the compound (Molal depression constant for water is 1.86).

% carbon	=	$12/44 \times 0.264/0.180 \times 100 = 40$
% hydrogen	=	$2/18 \times 0.108/0.180 \times 100 = 6.6$
% oxygen (by difference) =		$100 - (40 + 6.6) = 53.4$

Molar ratio for carbon   = 40/12 = 3.33
Molar ratio for hydrogen   = 6.6/1 = 6.6
Molar ratio for oxygen   = 53.4/16 = 3.33
Empirical formula   = $C_{3.33}H_{6.6}O_{3.33}$ or CH_2O

$$\text{Molecular mass of compound} = \frac{1000 \times K_b \times W_1}{W_2 \times T} = \frac{1000 \times 1.86 \times 6.0}{50 \times 1.24} = 180$$

$$n = \frac{\text{Molecular mass}}{\text{Empirical formula mass}} = 180/(12 + 2 + 16) \text{ or } 180/30 = 6$$

Molecular formula = $n \times$ Empirical formula = $6 \times CH_2O = C_6H_{12}O_6$

1.10 PROBLEMS

1. What are the general methods of isolation and purification of organic compounds?
2. Comment on the utility of chromatographic methods.
3. Calculate the percentage composition of (carbon and hydrogen) I, II and III from the following data :

	Sample (g)	CO_2 (g)	H_2O (g)
(I)	0.2475	0.4950	0.2025
(II)	0.2813	0.5586	0.0977
(III)	0.4080	0.5984	0.2448

4. An acid salt on quantitative analysis gave 51.4% carbon, 4.3% hydrogen, 12.8% nitrogen, 9.8% sulphur and 7.0 % sodium. Find its empirical formula.
5. A monobasic acid, molecular mass 453, has 79.6% carbon, 13.5% hydrogen and 6.9% oxygen. Find its molecular formula.
6. Find the molecular formula of a dibasic acid having 26.6% carbon, 2.2% hydrogen and its silver salt on ignition gives 71.05% silver.
7. How is nitrogen estimated by Kjeldahl's method?
8. Find the percentage of nitrogen in a compound, 0.2815 g of which gave sufficient nitrogen by Kjeldahl method to neutralise 18.8 cc of N/2 H_2SO_4.
9. Discuss (i) Mixed melting point; (ii) Fractional distillation and (iii) Distillation under reduced pressure
10. How is nitrogen, sulphur and halogens detected in an organic compound?

2
Bonding and Properties of Organic Molecules

2.1 ATOMIC ORBITAL

The understanding of the structure and reactivity of organic compounds begins with the units from which their molecules are constructed i.e. the atoms. We have studies earlier that the atom is made up of certain fundamental particles like electrons, protons and neutrons and subatomic particles like mesons and positrons. The protons and neutrons are present in the nucleus whereas electrons are present in the extra nuclear part which is divided into various orbits. An orbital is a region in space where there is a maximum probability of finding the moving electron.

J.J. Thomson's discovery of electrons in 1897 provided a starting point for our modern electronic theories of atomic structure. According to this discovery an atom is composed of a sphere of positive charge in which electrons are embedded. Lord Rutherford (Nobel Prize in chemistry in 1906) discovered that atoms are mostly space and electrons circulate around a positively charged nucleus in a manner analogous to the rotation of planets around the sun. Niels Bohr (Nobel Prize in physics, 1922) modified the 'planetary model' of Rutherford. He proposed that each electron orbit was associated with a specific energy level. These energy levels are quantised and specific quantum numbers were allotted to each electron orbit. Later in 1929, the physicist Louis De Broglie suggested that electrons have the characters of both particles and waves. His proposal was based on the demonstration that light has properties of wave as well as the momentum associated with particles. In terms of quantum mechanics, we cannot represent an electron as a point in space or a particle moving in a definite pathway. This concept was put forward by Werner Heisenberg. It was termed as 'Heisenberg Uncertainty' principle which states that it is not possible to determine simultaneously the precise position and momentum of an electron.

With the emergence of quantum mechanics, the Bohr's model and the principles of bonding based upon it suffered a serious setback . The present day model of atom considers the electron to be free to move about in the entire volume surrounding the nucleus rather than moving in closed orbits as was believed earlier. The electron moves rapidly towards the nucleus and away from the nucleus in all possible directions, occupying the entire space around it. Thus the moving electron, if visible to the eye, would not appear as a particle but would offer the appearance of diffused spherical cloud formed by its motion (Fig. 2.1). Since the formation of cloud is due to the motion of the electron, it is referred to as electron cloud. The rapidly moving electron would occupy the entire space around the nucleus but at any given time, the highest probability of finding an electron in three dimensional space around the nucleus is referred to as 'Orbital'. An electron can occupy any space within the assigned orbital at a given time, although it tends to occupy certain specific portions of its orbital to a greater extent than other portions.

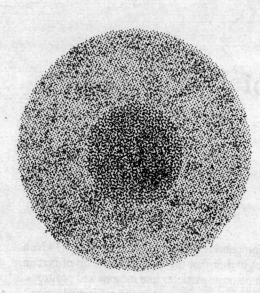

Fig. 2.1:   The Electron Cloud

Quantum Numbers

The atoms of elements heavier than hydrogen contain more than one electron which occupy definite orbitals. These electrons are identified by four quantum numbers:

(a) *The Principal Quantum Numbers (n)* refers to the size of the orbital and the energy of electron in it. The shells corresponding to principal quantum numbers 1, 2, 3, 4, 5, 6 and 7 are termed as K, L, M, N, O, P and Q respectively. Its value can never be zero which would mean that the electron is in the nucleus (distance zero) which is a conclusion contradicted by Rutherford long time back.

(b) *Azimuthal or Subsidiary Quantum Numbers (l)* refers to the shape and direction in which the electron is likely to be found. It has whole number values ranging from zero to $(n-1)$. The region corresponding to azimuthal quantum numbers are called sub-shells and the sub-shells corresponding to $l = 0,1,2,3$ are represented by letters s, p, d, f respectively. The number of sub-shells in a shell is equal to its principal quantum number.

(c) *Magnetic Quantum Numbers (m)* refers to the behavior of the electron in magnetic field as it is an electrically charged particle. Magnetic quantum number can have all integral values from $-l$ to zero to $+l$.

Thus, for 1s electron

$n = 1$ (K shell); $l = 0$ (sub-shell s); $m = 0$; one value of m is represented by s orbital.

For p electrons

$l = 0, 1$; $m = -1, 0, +1$; three values of m are represented by px, py and pz orbitals.

For d electrons

$l = 0, 1, 2$; $m = -2, -1, 0, +1, +2$; five values of m represented by dxy, dxz, dyz, dx^2, dy^2.

For f electrons

$l = 0, 1, 2, 3$; $m = -3, -2, -1, 0, 1, 2, 3$; seven values of m corresponds to f orbitals when placed in magnetic field.

(d) *Spin Quantum Numbers (s)* refers to the behavior of electron in the electrical field. s can either be $+1/2$ or $-1/2$ corresponding to each magnetic quantum number. Thus each p orbital (px, py, pz) will have two electrons corresponding to spin quantum number of $+1/2$ or $-1/2$.

Distribution of electrons (Fig. 2.2) into various shells and sub-shells is governed by following rules:

1. The total number of electrons that can be accommodated in a shell is equal to $2n^2$ where n refers to the principal quantum number. Thus first shell can accommodate two electrons, the second eight, the third 18 and the fourth 32 electrons.

2. The total number of electrons that can be accommodated in a sub-shell is equal to twice the number of orbitals it contain. Thus sub-shell s can have two electrons, sub-shell p can have six electrons since it contains three orbitals (px, py, pz).

3. The maximum number of electrons that can be accommodated in an orbital is two and these two electrons must have their spins opposite in accordance with Pauli's exclusion principal.

4. The electrons occupy the orbital of lowest energy first followed by the orbitals of higher energy. The energy of orbitals of various subshells follows the following sequence:

$$1s < 2s < 2p < 3s < 3p < 4s < 3d < 4p < 5s$$

This is also the order of filling various orbitals.

5. According to Hund's Rule of maximum multiplicity electrons will never pair in degenerate orbitals (orbitals of comparable energy) as long as empty orbitals are available to them, e.g. pairing in p orbital begins with the introduction of fourth electron or in d orbital with sixth electron.

6. Stability to the element is extended with completely filled or exactly half filled sub-shells, e.g. Nitrogen (electronic configuration $1s^2\ 2s^2\ 2p^3$) with its p orbital half filled, is more stable than its neighboring elements oxygen and carbon. The electronic configuration of the first ten elements are given in Table 2.1.

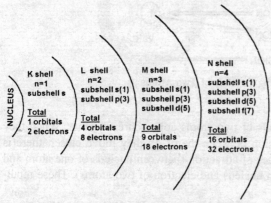

Fig. 2.2:   Distribution of Electrons in Shells and Sub-shells

Table 2.1:   Electronic Configuration

Element	Atomic Number	Configuration
H	1	$1s^1$
He	2	$1s^2$
Li	3	$1s^2\ 2s^1$
Be	4	$1s^2\ 2s^2$
B	5	$1s^2\ 2s^2\ 2px^1$
C	6	$1s^2\ 2s^2\ 2px^1\ 2py^1$
N	7	$1s^2\ 2s^2\ 2px^1\ 2py^1\ 2pz^1$
O	8	$1s^2\ 2s^2\ 2px^2\ 2py^1\ 2pz^1$
F	9	$1s^2\ 2s^2\ 2px^2\ 2py^2\ 2pz^1$
Ne	10	$1s^2\ 2s^2\ 2px^2\ 2py^2\ 2pz^2$

Shapes and Orientation of Orbitals

An orbital is a well defined space where there is a maximum probability of finding a moving electron. In organic chemistry our main concern is s and p orbitals. An s-orbital is spherical in shape (Fig. 2.3) and probability of finding electron in any direction is equal. All s-orbitals are spherical and the only difference between an electron in 1s orbital and 2s orbital is the distance from the nucleus (Fig. 2.4).

The p-orbital is dumb bell shaped (Fig. 2.5) and the lobes of p-orbital are directed along a particular axis. In sub-shell p there are three orbitals of equal energy directed along the three coordinate axis and are named accordingly px, py and pz.

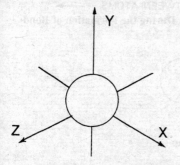

Fig. 2.3:   An s-orbital

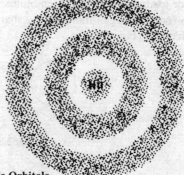

Fig. 2.4:   1s and 2s Orbitals

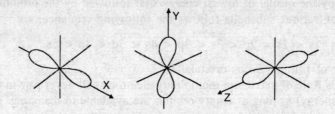

Fig. 2.5:   The p-orbitals

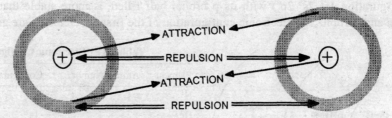

Fig. 2.6:   Forces Between Bonding Orbitals of Two Atoms

2.2   BONDING IN ATOMS

Let us now consider what happens when bonding orbitals of two atoms approach each other. When they are far apart, there are no forces of attraction or repulsion between them. As they move closer, there is overlapping between the electron clouds resulting in forces of attraction (between nucleus of one atom and electron cloud of other) and forces of repulsion (between nucleus and electron of two atoms). These repulsive forces are significant at small distance (Fig. 2.6).

As the two atoms combine to form a bond, there is a lowering in energy due to the attractive forces between the nucleus and electrons. The energy continues to fall till the forces of attraction are dominating over the forces of repulsion. Increase in forces of repulsion increases the energy, whereas the chemical bond between the atoms is formed when energy is at a minimum (Fig. 2.7). Thus, the new arrangement of nuclei and electron is obtained because of bond formation having less energy than the arrangement in isolated atoms. For example, two hydrogen atoms combine to form hydrogen molecule by sharing their electron (Fig. 2.8).

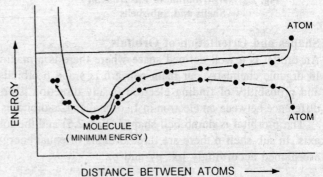

Fig. 2.7:   Energy Changes During the Formation of Bond

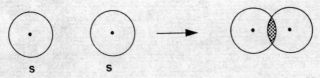

Fig. 2.8:   Bond Formation Due to Orbital Overlap

The electronic configuration of chemically stable inert gases shows that all the orbitals in their atoms are completely filled. On the other hand the atoms of other elements are reactive as they contain half filled orbitals also. During bond formation the half filled orbitals overlap, acquire electron and becomes completely filled. Thus in bond formation only the valence shell electrons and orbitals associated with them are of importance. In valence shell only half filled orbitals take part in bond formation. The half filled orbitals involved in bond formation are referred to as 'Bonding Orbitals'.

Bonding and Anti-Bonding Molecular Orbitals

The electronic configuration of hydrogen atom provided by quantum mechanics serves as a model for other atoms and may be extended for bonding in atoms. According to quantum mechanics the electrons of a chemical bond are the combination of hydrogen like wave functions (ψ_1 and ψ_2) of the two atoms. The new wave function resulted due to linear combination of atomic orbitals (LCAO) is termed as energetically favorable bonding molecular orbital. When electron waves from individual atomic orbitals interact in a destructive manner, an antibonding molecular orbital results with zero probability of electrons between the nuclei.

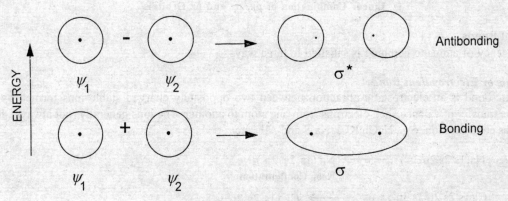

Bonding and Antibonding Molecular Orbitals

It is important to note here, that number of molecular orbitals must always be equal to the number of combining atomic orbitals. The criteria for assigning electrons to molecular orbitals is the same as it was for atomic orbitals.

(i) Molecular orbital with lowest energy is filled first.

(ii) In degenerate molecular orbitals, electron enter singly with same spin before pairing.

(iii) Each molecular orbital can accommodate a maximum of two electron.

Following is the graphical representation of the formation of hydrogen molecule from the linear combination of its 1s atomic orbitals. The bonding molecular orbital (represented as σ), being at a lower energy, is more favorable for electrons than the original 1s orbitals. Therefore the electrons from 1s orbital of two hydrogen atoms are accommodated in bonding molecular orbital whereas antibonding molecular orbital (represented as σ^*) remains empty in ground state.

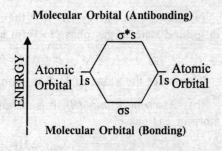

Molecular Orbital (Antibonding)

Molecular Orbital (Bonding)

Graphical Representation of Hydrogen Molecule

Similar graphical representation below shows the combination of px, py and pz orbital forming σ and π bonding and antibonding molecular orbitals.

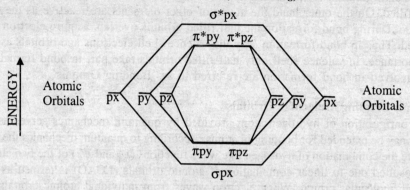

Molecular Orbital (Antibonding)

σ*px

π*py π*pz

ENERGY

Atomic Orbitals   px   py pz   pz py   px   Atomic Orbitals

πpy   πpz

σpx

Molecular Orbital (Bonding)

Linear Combination of px, py and pz Orbitals

Types of Bonds

The tendency of atoms to combine is satisfied in three ways.

1. Ionic or Electrovalent Bond

The ionic bond is an electrostatic attraction between two oppositely charged stable ions formed by the complete transfer of valence shell electrons from one atom to another. The ions generally (not always), have noble gas configuration. e.g., NaCl, KCl, Na_2S and AlF_3 etc.

$$Na(1s^2 2s^2 2p^6 3s^1) \longrightarrow Na^+(1s^2 2s^2 2p^6) + e^-$$
Neon Configuration

$$Cl(1s^2 2s^2 2p^6 3s^2 3p^5) + e^- \longrightarrow Cl^-(1s^2 2s^2 2p^6 3s^2 3p^6)$$
Argon Configuration

$$Na^+ + Cl^- \longrightarrow Na^+Cl^-$$

The formation of ionic bond is related to the ease of formation of cation and anion from the neutral atoms which depend on the following factors:

(a) *Ionisation Energy (I):* It is the minimum energy required to remove an electron from an atom in ground state in gas phase to form an ion.

$$M \longrightarrow M^+ + e^-; \ \Delta H = I$$

The lower the value of I of an atom, the greater will be the ease of formation of cation from it.

(b) *Electron Affinity (E):* It is the amount of energy released when an electron is added to a neutral atom in gas phase.

$$X + e^- \longrightarrow X^-; \ \Delta H = -E$$

The higher the electron affinity of an atom, the easier will be the formation of anion.

2. The Covalent Bond

When bonding orbitals (half filled) of two atoms overlap, the electrons share the same region in the space and a new orbital having two electrons (one electron from each atom) is formed which is called 'Molecular

Orbital' (MO). The molecular orbital is a covalent bond. Like atomic orbital, a molecular orbital can accommodate only two electrons. A covalent bond can be formed if

(a) two combining orbitals are half filled,

(b) bonding orbitals are properly aligned for an effective overlap, and

(c) electrons in the bonding orbitals have opposite spins.

Covalent bonds are of two types:

(i) *Sigma Bond* (σ): It is formed by the linear or end to end overlap of orbitals.

(a) Overlap of two s orbitals (formation of H_2 molecule).

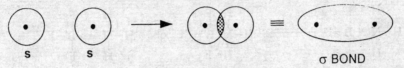

(b) Overlap of s and one of p orbital (e.g. HCl and H_2O).

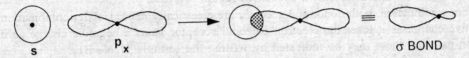

(c) Overlap of two p orbitals (e.g. Cl_2 and F_2).

Sigma bonds are symmetrical around the line drawn between the two nuclei. They are represented by a single line drawn between atoms and electrons involved in sigma bond formation are called sigma electrons.

(ii) *Pi Bond* (π): A π bond is formed by parallel or sideways overlap of p orbitals (e.g., formation of O_2 and N_2).

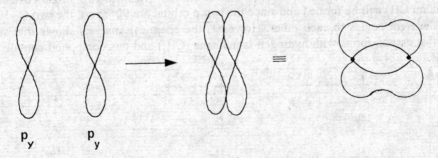

π bond has two lobes; one half of the π bond lies above the plane and other half below the plane of nuclei involved in π bond formation. π bond differ from sigma bond as:

(a) π electrons are loosely held than sigma electrons, so they represent higher energy. As a result π bonds are easily broken and are more reactive than sigma bonds.

(b) Rotation along a π bond is not possible. If any attempt is made to do so, the lobes of p orbital will no longer be planer and result in breaking of π bond.

3. The Coordinate Bond

It is a special type of bond in which both the shared electrons forming the bond are supplied by only one of the two atoms linked together, e.g., when ammonia combines with boron trifluoride, it is the lone pair of electron of nitrogen which is involved in bonding. In boron trifluoride, boron has only six electrons in its valence shell, hence it can accommodate two more electrons to complete its octet. Coordinate bond represented by an arrow pointing away from the atom supplying the lone pair and can be shown as follows:

$$\begin{array}{ccc}
\overset{\displaystyle H}{\underset{\displaystyle H}{H-N:}} + \overset{\displaystyle F}{\underset{\displaystyle F}{B-F}} \longrightarrow & \overset{\displaystyle H}{\underset{\displaystyle H}{H-N:}}\overset{\displaystyle F}{\underset{\displaystyle F}{B-F}} \quad or & \overset{\displaystyle H}{\underset{\displaystyle H}{H-N}}\longrightarrow \overset{\displaystyle F}{\underset{\displaystyle F}{B-F}}
\end{array}$$

Atom that supplies the lone pair is called *donor* and the atom which receives is known as *acceptor*. The coordinate bond is also known as 'dative bond'.

Before combination, both donor and acceptor are electrically neutral. After combination donor gains a positive charge as it loses the electron pair and acceptor gains a negative charge as it receives an electron pair and these may be indicated by writing the formula $H_3 \overset{+}{N} - \overset{-}{B}F_3$.

Once the coordinate bond has been formed, there is no way of distinguishing it from covalent bond, but, since one atom has given a pair of electron, charges are produced in the molecule. Charges may also be produced in the molecule having covalent bonds due to the difference in electronegativity of bonding atoms. For example, $\overset{+}{H}\overset{-}{Cl}$ and $\overset{+}{H}\overset{-}{F}$.

2.3   HYBRIDISATION

The sp³ Hybridisation of Carbon

The electronic configuration of carbon in its ground state is:

$$1s^2\ 2s^2\ 2px^1\ 2py^1\ 2pz^0$$

As energy level diagram it may be represented as in Fig. 2.9. Since there are only two unpaired electrons in 2p orbital of carbon it may be expected that carbon atom will form two single covalent bonds with hydrogen atom. CH_2 will be formed and since the two p orbital are 90° apart, the two C–H bonds would also be at the right angles to each other. However, the chemical analysis shows that simplest stable compound that carbon forms with hydrogen is methane (CH_4) and this compound contain four identical C–H bonds.

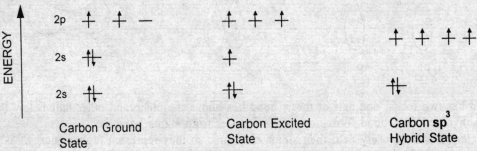

Fig. 2.9:   Carbon in Different States

For the formation of four similar bonds we need four half-filled orbitals of similar energy. Now let us assume that one of the 2s electron is promoted to the empty 2pz orbital. Since 2pz is at a higher energy level than 2s orbital, the process requires energy which is supplied in the form of light or heat. This new state of carbon is called 'exited state' and its electronic configuration can be represented as:

Carbon in exited state $\quad 1s^2\ 2s^1\ 2px^1\ 2py^1\ 2pz^1$

Since there are four unpaired electrons in the valence shell of carbon atom in its exited state, four covalent bonds may be formed by the 1s orbital overlap from four hydrogen atoms resulting in the formation of CH_4 molecule. It is important to note here that out of four C–H bond, one is formed by the overlap of 2s and 1s and other three C–H bonds are formed by the overlap of 2px, 2py and 2pz with 1s. The C–H bond resulting from the overlap of 2s and 1s has no direction whereas the other three C–H bonds are at right angles to each other (see shapes of s and p orbitals in Figs. 2.3 and 2.5). In other words, we can visualise easily that the picture of CH_4 molecule emerging from the above discussion is not true.

Experimental data clearly show that all the four C–H bonds in methane molecule are equivalent and this is possible only when we have all the four half filled orbitals of carbon of equivalent energy. In other words to explain this, the idea of orbital hybridisation was invoked, where one 2s orbital and three 2p orbitals are hybridised together to yield a set of four new orbitals of equivalent energy.

The new orbitals formed are called 'sp^3 hybrid orbitals', since they are formed as a result of combination of one s and 3p orbitals and this process of mixing pure orbitals of different energies to give a set of new orbitals of equivalent energy is called 'Hybridisation' and the carbon is said to be in hybridised state (Fig. 2.9).

Each sp^3 hybrid orbital contain single electron and has 25% s character and 75% p character. sp^3 hybrid orbital as shown in Fig. 2.13 has two unequal lobes separated from each other by a node like in case of p orbitals but there is a difference that in sp^3 hybrid orbital one lobe is very large and other is very small (in p orbital both the lobes are equal in size) so that the electron density is concentrated in one direction making the overlap more effective during bond formation compared to pure orbitals. Hence bond formed by such orbitals is stronger and more stable in comparison to the bonds formed by pure orbitals.

The four sp^3 hybrid orbitals thus formed are equivalent in all respect but differ in their orientation in space as they are directed towards the corners of a regular tetrahedron with the carbon atom located in the center. The angle between any two sp^3 hybrid orbital is 109.5°. The orientation of orbitals in tetrahedral fashion is favored because it provides the maximum possible distance between the half filled sp^3 hybrid orbitals thereby reducing electron-electron repulsion.

With the above discussion, we can summarise that hybridisation is a theoretical concept which is used as realistic modeling of molecular structure that only orbitals of comparable energy undergo hybridisation and that the number of hybrid orbitals generated are always equal to the number of atomic orbitals taken part in hybridisation.

Bonding in Methane (CH_4)

In methane since carbon is attached to four hydrogen atoms through covalent bond as shown in Fig. 2.10, carbon uses one 2s orbital and three 2p orbitals for hybridisation. The four sp^3 hybrid orbitals orient themselves in the corners of a regular tetrahedron with carbon in the center. They overlap with 1s orbital of hydrogen to furnish four C–H bonds which are directed towards the corner of regular tetrahedron. The bond so formed are σ (sigma) bonds and the bond length is the 1.12 Å. The HCH Bond angle in methane is similar to that of a regular tetrahedron, i.e. 109.5°. The data obtained from electron diffraction and spectroscopic studies are also in accordance with the above findings.

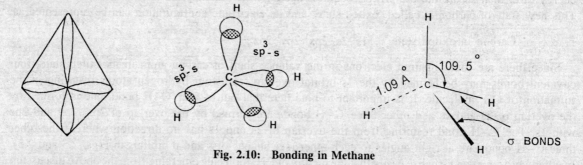

Fig. 2.10:   Bonding in Methane

Bonding in Ethane (C_2H_6)

In ethane molecule we have two carbon atoms attached to each other and with three H-atoms each i.e. again each carbon is bonded to four other atoms or each carbon is sp^3 hybridised.

Here, out of four sp^3 hybrid orbitals of carbon one overlaps with sp^3 hybrid orbital of another carbon resulting in the formation of a sigma (σ) bond of bond length 1.54 Å. Rest of the sigma bonds arising by overlap of sp^3-s orbitals are 1.12 Å apart and bond angles remain same (Fig. 2.11).

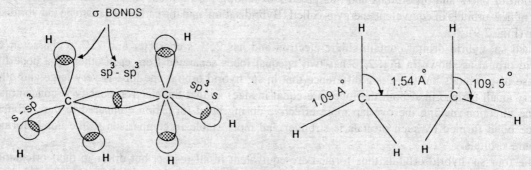

Fig. 2.11:   Bonding in Ethane

The sp^2 Hybridisation of Carbon

When carbon is bonded to three other atoms or groups, e.g., ethylene, benzene, aldehyde or ketone, sp^2 hybridisation results. The ground state and exited state configuration of carbon is given as follows:

Carbon Ground State (GS): $1s^2\ 2s^2\ 2px^1\ 2py^1\ 2pz^0$

Carbon Excited State (ES): $1s^2\ 2s^1\ 2px^1\ 2py^1\ 2pz^1$

Now 2s and only two orbitals from 2p are hybridised together to give three sp^2 hybrid orbitals of equivalent energy. The third orbital, i.e., 2pz of subshell 2p does not take part in hybridisation (Fig. 2.12). Each sp^2 hybrid orbital has 33% s character and 66% p character, thus the shape is almost similar to sp^3 hybrid orbital but is less elongated as shown in Fig. 2.13.

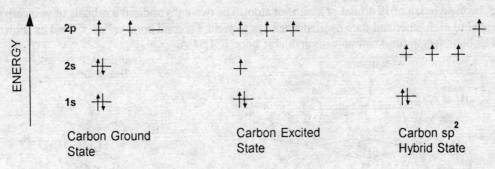

Carbon Ground State

Carbon Excited State

Carbon sp^2 Hybrid State

Fig. 2.12:   Carbon in Different States

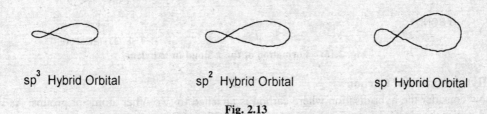

sp^3 Hybrid Orbital

sp^2 Hybrid Orbital

sp Hybrid Orbital

Fig. 2.13

Each sp^2 hybrid orbital formed by hybridisation is identical in shape and energy but differ in its orientation in space with respect to each other. The three sp^2 hybrid orbital lie in the same plane with their axis directed towards the corner of an equilateral triangle and are 120° apart from each other. The unhybridised pz orbital is perpendicular to the plane of sp^2 hybrid orbitals (Fig. 2.14).

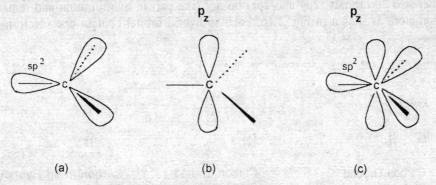

(a)

(b)

(c)

Fig. 2.14:   (a) Orientation of Three sp² Hybrid Orbitals (b) pz Orbital and (c) The Carbon in sp² Hybrid State

Bonding in Ethylene ($H_2C = CH_2$)

Each carbon in ethylene is attached to two hydrogen atoms by a single covalent bond and to other carbon atom by a σ bond and a π bond. Since each carbon is attached to three other atoms, both the carbon atoms in ethylene are in sp^2 hybridised state having three sp^2 hybrid orbitals arranged in the corners of a equilateral triangle in one plane and a pz orbital lying perpendicular to the plane of sp^2 hybrid orbitals. The two such sp^2 hybrid orbitals from each carbon overlap to form a σ bond. Rest of the four sp^2 hybrid orbitals form σ bond by the overlap of 1s orbital of hydrogen atom. The two unhybridised p orbitals of two carbon atoms are parallel to each other and they overlap sideways to yield a second bond called π bond as shown in Fig. 2.15. The bond length for a carbon-carbon double bond is 1.34 Å.

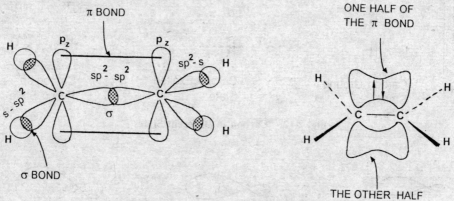

Fig. 2.15:   Formation of the π Bond in Ethylene

The sp Hybridisation of Carbon

Let us now consider the hybridisation where carbon is attached to two other atoms or groups. As we have discussed earlier the electronic configuration of carbon in GS and ES

Carbon Ground State (GS)   : $1s^2\ 2s^2\ 2px^1\ 2py^1\ 2pz^0$
Carbon Excited State (ES) : $1s^2\ 2s^1\ 2px^1\ 2py^1\ 2pz^1$

Now 2s orbital and one of 2p orbital combine together to form two sp hybrid orbitals of equivalent energy. The other two 2p orbitals (2py and 2pz) do not take part in hybridisation and remain as such with each carbon atom as shown in Fig. 2.16. Each sp hybrid orbital contain one electron and lie in

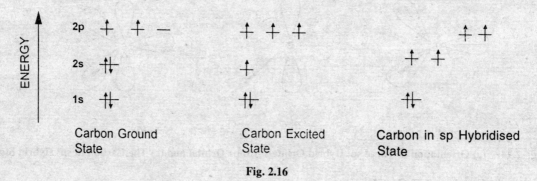

Fig. 2.16

space in a straight line with an angle of 180°. This arrangement allows the sp orbitals to stay as far apart from each other as possible so that electron-electron repulsion in the valence shell is reduced. The unhybridised 2p orbitals occupy the space at right angle to each other along y and z axis respectively as shown in Fig. 2.17.

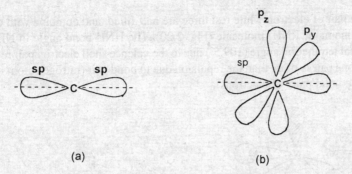

Fig. 2.17: (a) sp Hybrid Orbitals (b) Carbon in sp Hybrid State

Bonding in Acetylene (HC≡CH)

In acetylene each carbon is sp hybridised and is attached to one hydrogen atom and one carbon atom. Out of two sp hybrid orbitals one overlaps with sp hybrid orbital of other carbon forming a σ bond due to sp-sp overlap and the other overlap with 1s orbital of hydrogen to form another σ bond due to sp-s overlap. The two p electrons lying in unhybridised 2p orbitals of each carbon atom along y and z axis overlap sideways giving rise to two π bonds as shown in Fig. 2.18. Out of three bonds shown between the two carbon atoms in acetylene, one is σ bond and two are π bonds. The bond length of carbon-carbon triple bond is 1.20 Å and H–C–C bond angle is 180°.

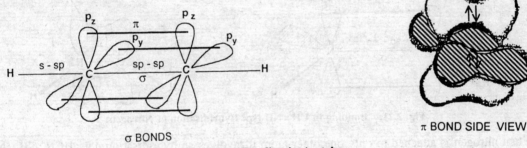

Fig. 2.18:   Bonding in Acetylene

Hybridisation of Nitrogen

To understand the structure of various molecules having nitrogen we shall apply the concept of hybridisation. Fig. 2.19 illustrates the formation of four sp³ hybrid orbitals of nitrogen out of which one

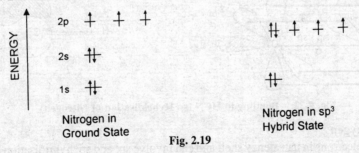

Fig. 2.19

orbital contains a lone pair of electron while rest three are half filled and combine with other atoms to form sigma bonds as in ammonia or RNH_2 molecule (Fig. 2.20). The HNH bond angle in NH_3 is 107°   which is slightly less than normal tetrahedral angle (109.5°) due to the valence shell electron pair repulsion which states that repulsion due to lone pair is more than the repulsion due to bond pair (nitrogen atom in ammonia contains one lone pair and three bond pairs of electrons).

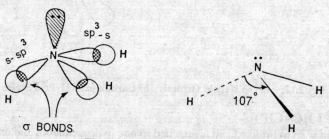

Fig. 2.20:   Bonding in Ammonia

Nitrogen involves sp^2 hybridisation when it is attached to two other atoms or groups, e.g., imines or azo compounds. In case of methylimine ($CH_2=NH$) carbon and nitrogen are in sp^2 hybrid state and have one electron each in their unhybridised p orbital which overlaps sideways to form a π bond as shown in Fig. 2.21.

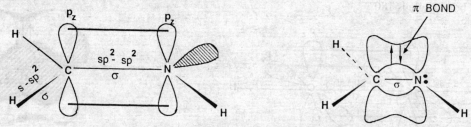

Fig. 2.21:   Bonding in $CH_2=NH$ (sp^2 Hybridisation of Nitrogen)

When nitrogen is attached to only one other atom it involves sp hybridisation e.g., HCN, CH_3CN or N_2. In hydrogen cyanide and methyl cyanide (Fig. 2.22) both carbon and nitrogen involved in bonding are in sp hybrid state and have two π bonds due to the sideways overlap of unhybridised p orbitals as we have discussed in case of acetylene.

Fig. 2.22:   Bonding in HCN (sp Hybridisation of Nitrogen)

Hybridisation of Oxygen

Oxygen atom has six electrons in its valency shell and can involve sp^3 and sp^2 hybridisation as illustrated in the Fig. 2.23. In the molecules of H_2O and CH_3OH, oxygen utilises its sp^3 hybrid orbitals for bonding and they arrange at the corner of a regular tetrahedron (Fig. 2.24). Out of four sp^3 hybrid orbitals of oxygen, two are occupied by lone pair and other two combine with atoms to form a sigma bond. Repulsion due to lone pairs reduces the HOH bond angle to 104° as against a normal tetrahedral bond angle of 109.5°.

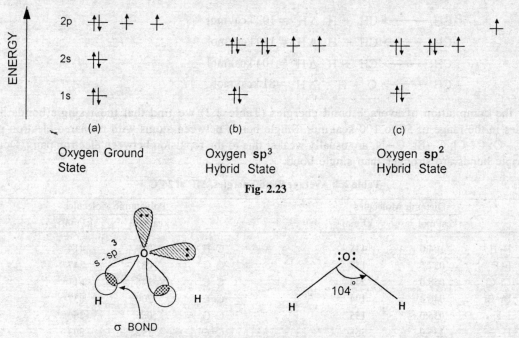

(a)

Oxygen Ground
State

(b)

Oxygen **sp³**
Hybrid State

(c)

Oxygen **sp²**
Hybrid State

Fig. 2.23

Fig. 2.24: Bonding in Water Molecule

The sp² hybridisation of oxygen (Fig 2.23c) is shown in aldehydes and ketones. Both carbon and oxygen are in sp² hybrid state. Fig. 2.25 shows sp² hybridisation in formaldehyde molecule.

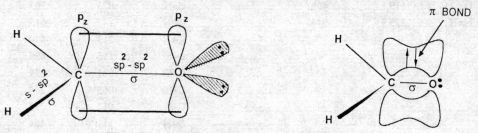

Fig. 2.25: Bonding in Formaldehyde Molecule

2.4 BOND ENERGIES AND BOND LENGTH

Bond Energies

The energy of a chemical bond is a form of potential energy that arises due to attractive forces between atoms. The energy of a bond is specified in two ways:

1. *Average Bond Energy:* It is a measure of average of all similar bond energies in a molecule e.g., the cleavage of four carbon hydrogen bonds in methane to form component atoms requires a total of 397 kcal/mol. Thus the average bond energy of a C–H bond in methane is 397/4 = 99.3 kcal/mol.

$$CH_4 \longrightarrow C + 4H \quad \Delta H° = 397 \text{ kcal/mol}$$

2.   *Bond Dissociation Energy:* It is the energy required to break a specific covalent bond. The energies required to break individual bond in methane are different:

$$CH_4 \longrightarrow CH_3 + H \quad \Delta H^\circ = 102 \text{ kcal/mol}$$

$$CH_3 \longrightarrow CH_2 + H \quad \Delta H^\circ = 110 \text{ kcal/mol}$$

$$CH_2 \longrightarrow CH + H \quad \Delta H^\circ = 104 \text{ kcal/mol}$$

$$CH \longrightarrow C + H \quad \Delta H^\circ = 81 \text{ kcal/mol}$$

By the compilation of average bond energies (Table 2.2) we find that most single bonds have energies in the range of 50 to 100 kcal/mol. Single bonds between atoms with unshared electron pair, e.g., $C-O$; $C-Cl$; $C-Br$; $C-N$, are usually weaker due to the repulsion between electron pairs. Double and triple bonds are stronger than single bond.

Table 2.2: Average Bond Energies, ΔH° at 25°C

Diatomic Molecules			Polyatomic Molecules		
	kcal/mol	kJ/mol		kcal/mol	kJ/mol
H–H	104.0	435	C–H	099	414
F–F	037.5	157	C–C	083	347
Cl–Cl	058.0	243	C=C	146	611
Br–Br	046.3	194	C≡C	200	837
I–I	036.0	155	C–O	086	360
H–F	135.9	568	C=O[1]	192	803
H–Cl	103.1	431	C=O[2]	166	694
H–Br	087.4	365	C=O[3]	176	736
H–I	071.4	299	C=O[4]	179	748
O=O	119.1	498	C–N	073	305
N≡N	225.9	944	C=N	147	615
			C≡N	213	890
			C–F	116	485
			C–Cl	081	339
			C–Br	068	284
			C–I	051	213
			O–H	111	464
			O–O	035	146
			O–Cl	052	218
			O–Br	048	201
			N–H	093	389
			N–N	039	163
			N–O	053	221
			N=N	100	418
			N=O	145	606
			S–H	083	347
			S–S	054	225

1 = carbon dioxide; 2 = formaldehyde; 3 = aldehydes; 4 = ketones

Heat of Reaction

The energy associated with breaking or making of bonds during a chemical reaction is known as *heat of reaction* and is measured in terms of change in enthalpy (ΔH). In terms of thermodynamic quantities there is a relationship between the heat of reaction and standard enthalpy contents of products and reactants. Superscript ($^\circ$) indicates the standard states (25°C temperature and one atmosphere pressure) of reactants and products during enthalpy changes.

$$\Delta H^\circ_{reaction} = \Delta H^\circ_{product} - \Delta H^\circ_{reactant}$$

$$Cl_2 + CH_4 \longrightarrow CH_3Cl + HCl$$

In the above reaction Cl–Cl and C–H bonds are broken and C–Cl and H–Cl bonds are formed. Each conversion is associated with a specific energy change:

Bond Formed	Bond Broken
C–Cl 81 kcal/mol	C–H 99 kcal/mol
H–Cl 103 kcal/mol	Cl–Cl 58 kcal/mol

since breaking requires energy ($\Delta H > 0$) and forming releases energy ($\Delta H < 0$)

$$\Delta H^\circ_{reaction} = (99 + 58) - (81 + 103) = -27 \text{ kcal/mol}.$$

The value of heat of reaction i.e. -27 kcal/mol, indicates that there is a decrease in enthalpy of chemical system if reaction is allowed to take place or we can say that the reaction is 'Exothermic' ($\Delta H < 0$) and heat is given off. The fact that a reaction is exothermic does not guarantee that it will occur as we shall learn this in the Section 4.7, that rate of chemical reactions are governed by energy of activation.

Bond Length

The length of a chemical bond is the result of balance of attractive and repulsive forces operative between two atoms forming bond. Experimentally length of a particular bond is determined by various spectral and diffraction methods and they reflect an average distance between atoms in a molecule which are actually vibrating.

It is interesting to note that bond length vary very little with the change of environment from molecule to molecule. Normal covalent bond length of some common bonds are given in Table 2.3.

Table 2.3: Typical Bond Length in Å

Bond	Bond Length	Bond	Bond Length	Bond	Bond Length
C–C	1.54	C–Cl	1.78	H–H	0.75
C=C	1.34	C≡N	1.15	H–F	0.92
C–N	1.47	C–Br	1.93	N–N	1.094
C≡C	1.20	O–H	0.97	H–Cl	1.27
C–O	1.43	C–I	2.13	Cl–Cl	1.99
C=O	1.23	N–H	1.03	H–Br	1.41
C–F	1.40	C–H	1.12	H–I	1.61
C=N	1.28	S–H	1.35		

There is a correlation between length and strength of a bond between two atoms. The smaller the internuclear distance, the stronger the bond, thus carbon-carbon triple bond is stronger than double bond which in turn is stronger than a single bond, (Table 2.4).

Table 2.4:   Correlation Between Bond Length and Bond Energy

Bond	Length, Å	Energy	
		kcal/mol	kJ/mol
C–H	1.12	99	414
C–C	1.54	83	347
C=C	1.34	146	610
C≡C	1.20	200	836
C–O	1.43	86	359
C=O	1.23	176-192	736-803
C–Cl	1.78	81	339
C–Br	1.93	68	284

2.5   ELECTRONEGATIVITY

Electronegativity of an atom in a molecule is a measure of its power to attract electrons which it is sharing in a covalent bond. In a covalent bond in hydrogen chloride, chlorine has greater power to attract the bonding pair of electron so it is called more electronegative than hydrogen.

Pauling developed a scale of relative electronegativity for most atoms of periodic table. Electronegativity increases when we move from left to right in the periodic table and upward in a column. Table 2.5 shows the electronegativity values of elements according to Pauling scale which provides a reasonable basis for the explanation of properties of organic molecules.

Table 2.5:   Electronegativity Values of Some Elements

H							
2.2							
Li	Be		B	C	N	O	F
1.0	1.6		2.0	2.6	3.0	3.4	4.0
Na	Mg		Al	Si	P	S	Cl
0.9	1.3		1.6	1.9	2.2	2.6	3.2
K	Ca	Zn					Br
0.8	1.0	1.7					3.0
		Cd					I
		1.7					2.7

2.6   POLARITY OF BOND AND DIPOLE MOMENT

A covalent bond between two atoms is formed when their bonding atomic orbitals overlap to give a molecular orbital. The covalent bond thus formed may or may not have polarity depending upon the atoms involved in bond formation e.g. in case of hydrogen molecule or chlorine molecule, they will not have any polarity since the electronegativity of atoms involved in bond formation is similar.

$$H:H \qquad \ddot{\underset{..}{C}}l: \ddot{\underset{..}{C}}l:$$

Now consider the case of covalent bond in hydrogen chloride molecule. Since chlorine atom is more electronegative (3.2) as compared to hydrogen (2.2), the shared electron pair of covalent bond will be shifted towards chlorine atom so that a slight negative charge denoted by δ^- (delta negative) develops on chlorine and positive charge δ^+ (delta positive) will develop on hydrogen atom or in other words we can say that the positive and negative poles have been developed within the molecule so that it will

$$H:\overset{..}{\underset{..}{Cl}}: \quad or \quad H-Cl \quad or \quad \overset{\delta+ \quad \delta-}{H-Cl}$$

now behave as a small magnet and possess a dipole moment. *The dipole moment (μ) is the product of charges and the distance between the charges. Molecules having zero dipole moment are called non-polar molecules.*

$$\underset{\textbf{(Debye units, D)}}{\text{Dipole moment, } \mu} = \underset{\textbf{(Charge, esu)}}{e} \times \underset{\textbf{(Distance, Å)}}{d}$$

As electronic charge is of the order of 10^{-10} esu and distance of the order of 10^{-8} cm, μ is of the order of 10^{-18} esu-cm. This unit is called Debye (D) or $1D = 10^{-18}$ esu-cm.

Dipole moment is a vector quantity and is indicated by an arrow pointing towards the negative end of the covalent bond e.g. $H \rightleftharpoons Cl$. It is important to note that:

(a) In case of molecules containing two or more polar bonds, the dipole moment is the vectorial addition of the dipole moments of the constituent bonds, e.g. dipole moments of methyl chloride is a vectorial addition of dipole moments of three C–H bonds and one C–Cl bond.

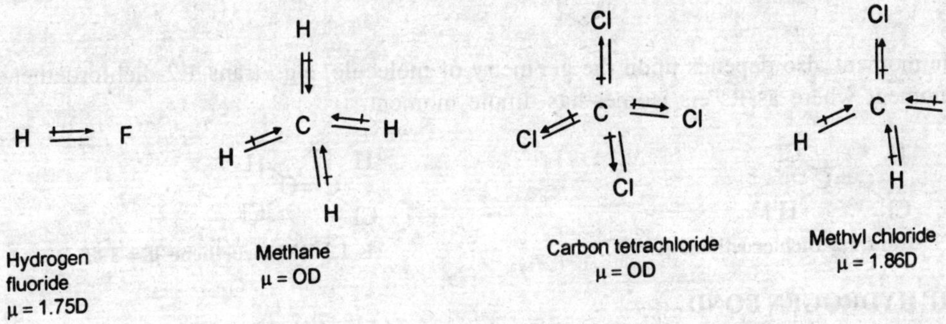

Hydrogen fluoride
$\mu = 1.75D$

Methane
$\mu = OD$

Carbon tetrachloride
$\mu = OD$

Methyl chloride
$\mu = 1.86D$

Fig. 2.26

(b) A symmetrical molecule is non-polar even though it contains polar bonds, e.g., carbon dioxide, methane and carbon tetrachloride are non-polar as their dipole moments are zero (Fig. 2.26).

In the earlier example we have discussed that polar bonds contribute to the dipole moment of molecule. Now, we shall see the role of lone pair of electrons on the polarity of molecules, e.g. the dipole moment of nitrogen trifluoride is ($\mu = 0.24D$) unexpectedly low as the magnetic moment due to unshared pair of electrons opposes the vectorial sum of bond moment. Similar is the case with ammonia and water molecule (Fig. 2.27).

Since dipole moment of carbon dioxide is zero, it has been assigned a linear structure.

$$O \equiv C \equiv O$$

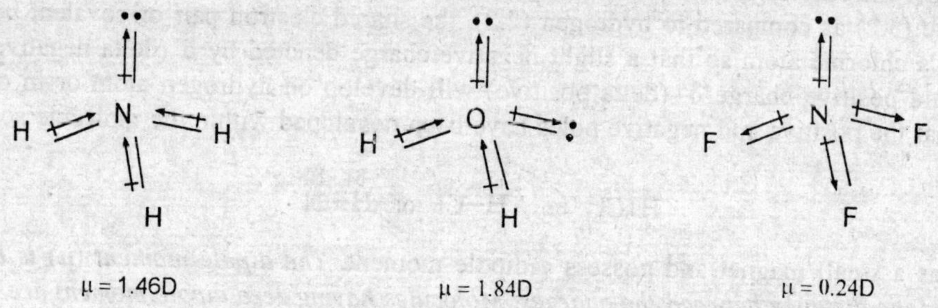

$\mu = 1.46D$ $\mu = 1.84D$ $\mu = 0.24D$

Fig. 2.27:   Dipole Moment of Molecules having Lone Pairs

Dipole moment is also dependent upon state of hybridisation of carbons. Electronegativity of hybridised carbon atoms varies in the following order with highest electronegativity of sp hybridised carbon and so does the dipole moment.

$$sp > sp^2 > sp^3$$

The dipole moment of toluene (0.37D) is due to the difference in electronegativity of methyl carbon (sp^3 hybridised) and carbon of aromatic ring (sp^2 hybridised).

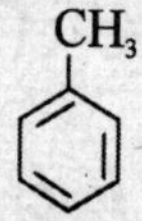

Toluene, $\mu = 0.37D$

Dipole moment also depends upon the geometry of molecule, e.g., trans 1,2-dichloroethene has zero dipole moment where as its cis isomer has dipole moment.

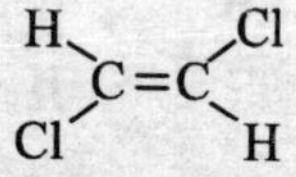

trans 1,2-Dichloroethene $\mu = 0$ D 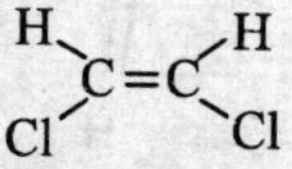   **cis 1,2-Dichloroethene $\mu = 1.85$ D**

2.7   THE HYDROGEN BOND

Compounds having -NH or -OH bonds often exhibit unexpected properties like high boiling points. This is due to the formation of a different type of bond which arises due to the difference in the electronegativity of the two atoms forming the bond (Fig. 2.28). It has been proposed that when hydrogen atom lies between two atoms of high electronegativity, e.g., fluorine, oxygen and nitrogen, it shows a unique property of bond formation between them holding one of the atom with a covalent bond and the other by purely electrostatic force. The electrostatic bond (hydrogen bond) has a strength of about 5-10 kcal/mol which is much less as compared to the strength of a covalent bond. It is represented by a dotted line (Fig. 2.28).

Hydrogen bond arises due to strong dipole-dipole attraction. When hydrogen is attached to a strong electronegative atom by a covalent bond, the bonding pair of electron is drifted towards the other atom exposing the thinly shielded hydrogen nucleus to the electronegative atom of other molecule. This attraction, although, much weaker than the covalent bond is much stronger than a simple dipole-dipole attractions. In case of hydrogen, size is also very important since the small positive charge on hydrogen exerts a large electrostatic attraction. If the atom has a greater radius, the electrostatic forces are weak,

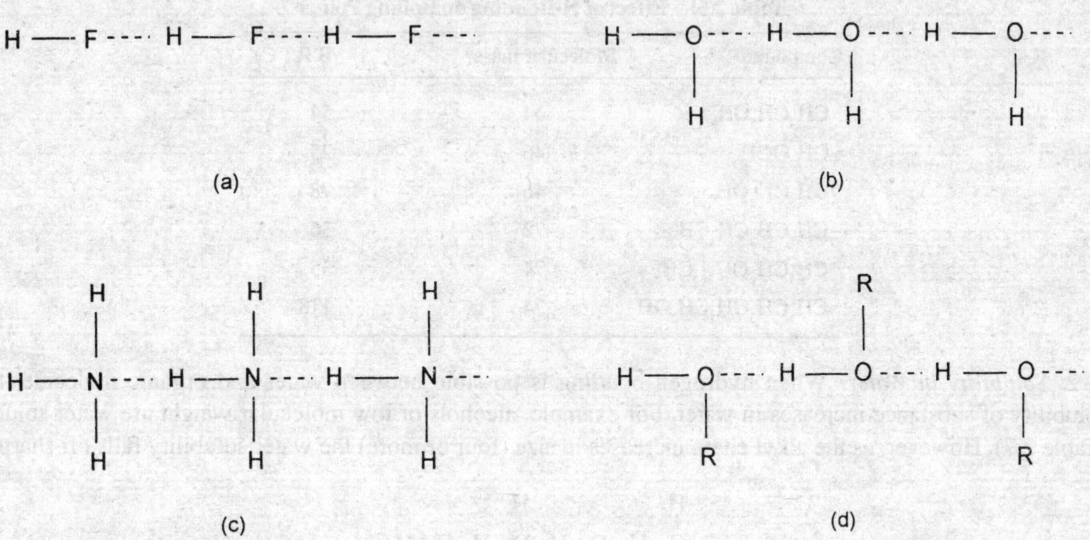

Fig. 2.28: Hydrogen Bonding in (a) Hydrogen Fluoride, (b) Water, (c) Ammonia, and (d) Alcohol

e.g. chlorine having same electronegativity as nitrogen, forms a very weak hydrogen bond as its atomic radius is larger than that of nitrogen. Hydrogen bonds are of two types:

(a) *The Intramolecular Hydrogen Bonding* occurs in compounds like o-nitrophenol where only one molecule is involved in bond formation resulting in ring formation or chelation.

(b) *The Intermolecular Hydrogen Bonding* occurs in compounds like m-nitrophenol or p-nitrophenol, alcohols, carboxylic acids etc. where more than one molecule is involved in bond formation.

O–Nitrophenol   p-Nitrophenol

Carboxylic Acid   Alcohol

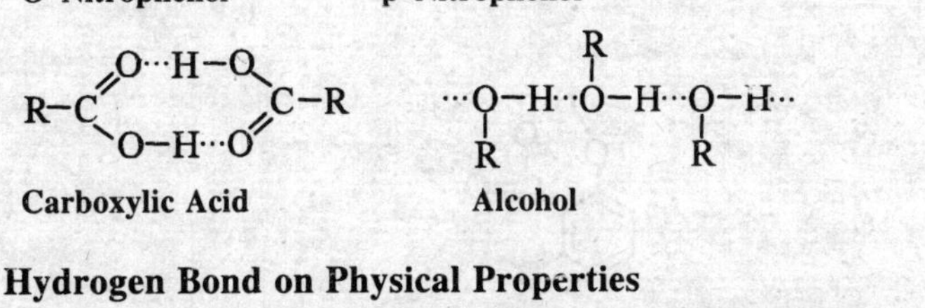

Effect of Hydrogen Bond on Physical Properties

1. *Boiling Point:* It represents the energy required to convert a liquid into gas phase. Compounds having hydrogen bonds will have a higher boiling point than those which do not posses H-bond because more energy is required to break this electrostatic force of attraction. Table 2.6 illustrate that compounds with approximately same molecular weight have significant difference in their boiling points due to hydrogen bonding.

Table 2.6:   Effect of H-Bonding on Boiling Points

Compound	Molecular mass	B.P. (°C)
$CH_3CH_2CH_3$	44	54
CH_3OCH_3	46	25
CH_3CH_2OH	46	78
$CH_3CH_2CH_2CH_3$	72	36
$CH_3CH_2OCH_2CH_3$	74	35
$CH_3CH_2CH_2CH_2OH$	74	118

2. *Solubility in Water:* When hydrogen bonding is possible between water and organic molecule, the solubility of substance increases in water. For example, alcohols of low molecular weight are water soluble (Table 2.7). However, as the alkyl chain increases in size (four or more) the water solubility falls off sharply.

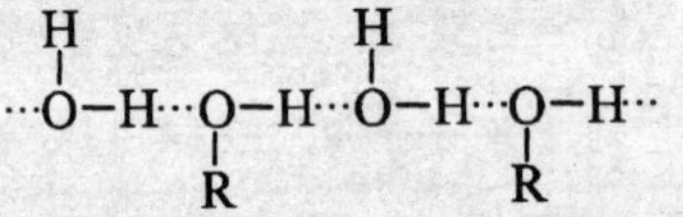

Table 2.7:   Solubility of Alcohol in Water

Alcohol	Solubility (g/100 ml of H_2O)
CH_3OH	Infinite
CH_3CH_2OH	Infinite
$CH_3CH_2CH_2OH$	Infinite
$CH_3CH_2CH_2CH_2OH$	8.0
$CH_3CH_2CH_2CH_2CH_2OH$	2.7
$CH_3CH_2CH_2CH_2CH_2CH_2OH$	0.6

3. *Bond Strength:* o-Hydroxybenzoic acid (salicylic acid) is a stronger acid than its m- and p-isomers due to hydrogen bonding. Similarly 2,6-dihydroxybenzoic acid is highly acidic than benzoic acid. The reason in both the cases is stabilisation of carboxylate anion (formed by the cleavage of O–H bond) by hydrogen bonding.

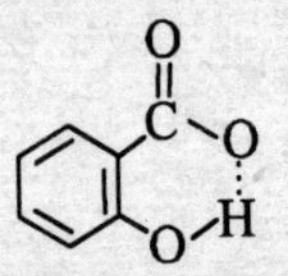

Salicylic Acid Anion

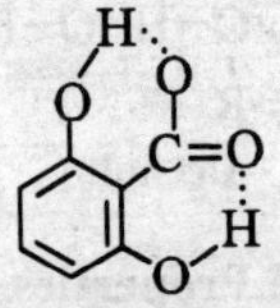

2,6-Dihydroxybenzoic Acid Anion

o-Chlorophenol, on the other hand, is a weaker acid than m- and p-isomer as the acidic proton is involved in intramolecular hydrogen bonding.

Hydrogen bonding affects the properties of various classes of organic compounds and has been discussed in the subsequent chapters of the book.

2.8 ACIDS AND BASES

Acidity and basicity are the two important fundamental properties of organic compounds which are used to understand structure and activity relationship.

Arrhenius (1884) concept defines acid as compounds which ionise in aqueous solution to produce hydrogen ions and bases which ionise to produce hydroxide ions.

Bronsted–Lowry (1923) concept defines acid as a proton donor and base as a proton acceptor. Proton donation and acceptance are reversible reactions as shown below, and are known as deprotonation and protonation respectively. The acid and base in the equation are said to be conjugate of each other; B is the conjugate base (CB) of acid BH^+, and BH^+ is the conjugate acid (CA) of the base B. Since a free proton cannot exist in solution (in sufficient amount so that it can be measured), the acid base reactions are represented as

$$\underset{\text{Base}}{B} + \underset{\text{Proton}}{H^+} \rightleftharpoons \underset{\text{Acid}}{BH^+}$$

$$A_1 + B_2 \rightleftharpoons B_1 + A_2$$

where, $A_1 - B_1$ and $A_2 - B_2$ are conjugate acid base pair, e.g.

Acid(A_1)	+	Base(B_2)	$\rightleftharpoons$	Acid(A_2)	+	Base(B_1)
HCl	+	H_2O	$\rightleftharpoons$	H_3O^+	+	Cl^-
HSO_4^-	+	NH_3	$\rightleftharpoons$	NH_4^+	+	SO_4^{--}
NH_4^+	+	H_2O	$\rightleftharpoons$	H_3O^+	+	NH_3

An acid base reaction always proceeds in the direction of the formation of weak acid and weak base.

$$HA + H_2O \rightleftharpoons H_3O^+ + A^- \tag{2.1}$$

HA here is a strong acid since the equilibrium lies at the right hand side, this is due to the fact that A^- is a weak base and it has very low affinity for proton or in other words we can say that the conjugate base (A^-) of a strong acid (HA) is always a weak base and the conjugate base of a weak acid is always a strong base.

Strength of Acids and Bases

It is a measure of their tendency to lose a proton or to accept a proton respectively and depends on the extent of ionisation. Hence equilibrium constant, K_a of the Reaction 2.1 gives the strength of acid as given below.

$$K_a = \frac{[H_3O^+][A^-]}{[HA][H_2O]}$$

concentration of solvent (H_2O) is constant, thus

$$K_a = \frac{[H_3O^+][A^-]}{[HA]}$$

Similarly, dissociation constant K_b of a reaction gives a quantitative measure of the strength of the base

$$B + H_2O = BH^+ + OH^-$$

$$K_b = \frac{[BH^+][OH^-]}{[B]}$$

The larger the value of K_a or K_b stronger will be the acid or the base.

Acidity of an aqueous solution is measured commonly in terms of pH; the negative logarithm of hydrogen ion activity. The pH scale can be applied only to aqueous medium but most organic compounds are insoluble in water and their reactions are carried out in non-aqueous solvents. Thus to account for their acidic nature a pK_a scale has been adopted as a better measure of acidity for organic compounds. pK_a scale is defined as the negative logarithm of acid dissociation constant.

$$pK_a = -\log(K_a)$$

The pK_a of an acid is actually the pH of an aqueous solution in which the acid is half ionised and is a convenient expression than ionisation constant. The pK_b for a base can similarly be determined and can than be converted into pK_a by using the following relationship:

$$pK_a + pK_b = 14$$

Thus, pK_a is a convenient and generallised method of expression of strength of both acids and bases. The low pK_a value represent strong acid whereas high pK_a value represent a strong base. For example, acetic acid ($pK_a = 4.7$) is a stronger acid than benzoic acid ($pK_a = 6.6$) whereas water ($pK_a = 14.0$) is a stronger base than ethanol ($pK_a = 10.0$).

Lewis (1938) concept defines an acid as a substance that can accept a lone pair of electron and a base as a substance that can donate a pair of electron. For example, in the reaction of boron trifluoride with ammonia, boron has only six electrons in its valency shell and can accept a pair of electron, thus is an acid whereas ammonia can furnish an electron pair for the formation of a covalent bond, thus

$$\underset{\text{Lewis acid}}{F\overset{\displaystyle F}{\underset{\displaystyle F}{-B}}} + :NH_3 \rightleftharpoons F\overset{\displaystyle F}{\underset{\displaystyle F}{-\overset{-}{B}}} :\overset{+}{N}H_3$$

is a base. Diethyl ether (a Lewis base) can also furnish an electron pair to form a covalent bond with boron trifluoride, the Lewis acid.

$$\underset{\text{Lewis acid}}{F\overset{\displaystyle F}{\underset{\displaystyle F}{-B}}} + \ddot{O}(C_2H_5)_2 \rightleftharpoons F\overset{\displaystyle F}{\underset{\displaystyle F}{-\overset{-}{B}}} :\overset{+}{O}(C_2H_5)_2$$

Stannic chloride, aluminum trichloride are Lewis acid since both of them can accept a lone pair of electron.

$$Cl\overset{\displaystyle Cl}{\underset{\displaystyle Cl}{-Al}} + :\ddot{\underset{\cdot\cdot}{Cl}}:^- \rightleftharpoons \left[Cl\overset{\displaystyle Cl}{\underset{\displaystyle Cl}{-Al-Cl}} \right]^-$$

The concepts forwarded by Bronsted and Lewis fits well with bases as electron pair doner (Lewis base) can accept a proton (according to Bronsted definition) easily. For example, $-OH$, HOH, C_2H_5OH and NH_3.

The concept works well with some acids like sulphur trioxide, aluminum and boron trihalides but, does not fit well with hydrochloric acid, HCl (a Bronsted acid), having no vacant orbital to accept an electron pair from a base, thus cannot be labeled as Lewis acid.

2.9 PROBLEMS

1. Draw orbital picture of the following:
   - (a) Methyl chloride (b) Methylene chloride (c) Vinyl chloride (d) Dichloroethene

2. Which is the stronger acid in the following pairs:
   - (a) H_2O and H_3O^+ (b) H_2O and OH^- (c) NH_3 and NH_2^- (d) H_2S and HS^- (e) NH_4^+ and NH_3

3. Explain why:
   - (a) Dipole moment of CCl_4 is zero whereas $CHCl_3$ has dipole moment.
   - (b) Dipole moment of trans-dichloroethene is zero whereas its cis isomer has dipole moment.

4. Predict the direction of dipole moment if any, in the following molecules:
   - (a) $CH_3CH_2CH_2NH_2$ (b) $CCl_3COCH_2CH_3$ (c) CH_3Cl (d) CH_2Cl_2 (e) CH_3COCH_3

5. Explain the following:
   - (a) The H-N-H bond angle of ammonium cation is greater than H-N-H bond angle of ammonia.
   - (b) The H-S-H bond angle in hydrogen sulphide is less than H-P-H bond angle of phosphine.
   - (c) The H-O-H bond angle in water molecule is less than H-N-H bond angle in ammonia.
   - (d) The H-O-H bond angle in water molecule is more than H-S-H bond angle in hydrogen sulphide.

6. Discuss the Lewis concept of acids and bases. Can a proton be regarded as an acid according to both Bronsted and Lewis concept?

7. Arrange the following in the increasing acid/base strength:
   - (a) $ClCH_2COOH$; $CH_3CHClCH_2COOH$; $CH_2ClCH_2CH_2COOH$; CH_3COOH
   - (b) $ClCH_2COOH$; CH_3COOH; C_6H_5COOH; $HCOOH$
   - (c) CH_3NH_2 : $CH_3CH_2CH_2NH_2$; $CH_3CH_2NH_2$; $CH_3CH_2NHCH_3$
   - (d) CH_3NH_2 ; NH_3 ; $CH_3OCH_2NH_2$; $C_6H_5NH_2$

8. Suggest the explanation for the carbon chlorine bond length (Å) in the following compounds:
   - (a) CH_3Cl; 1.78 (b) CH_2Cl_2; 1.77 (c) $CHCl_3$; 1.76
   - (d) CH_2FCl; 1.76 (e) CCl_4; 1.75 (f) F_3CCl; 1.72

9. Which of the following is associated with highest dipole moment and why?
   HF; HCl; HBr ; HI

10. Give reasons:
    - (a) Glycerol is viscous compared to ethanol. (b) Ethanol is freely soluble in water.
    - (c) o-Nitrophenol is steam volatile and not its m- and p-isomer.

11. Write short note on the following:
    - (a) Atomic orbitals (b) Molecular orbitals (c) Electronegativity (d) Bond energy (e) Bond length

3

Classification and Nomenclature of Organic Compounds

This chapter discusses the nomenclature of various classes of organic compounds which we encounter during the study of organic chemistry.

3.1 CLASSES OF ORGANIC COMPOUNDS

Organic compounds have been divided into various categories which are subdivided into classes as follows:

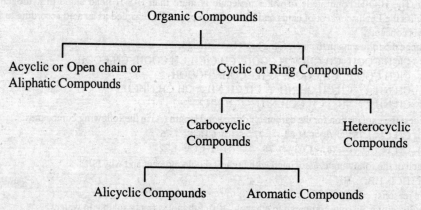

(A) *Acylic Compounds:* Compounds having branched or unbranched open chain of carbon atoms are termed as acyclic or aliphatic compounds. For example,

Methane CH_4

Propane $CH_3CH_2CH_3$ Isobutane $CH_3-\overset{\displaystyle CH_3}{\underset{\displaystyle }{CH}}-CH_3$

(B) *Cyclic or Ring Compounds:* Organic compounds having close chain of carbon atoms are termed cyclic compounds. They can be monocyclic or polycyclic depending upon the number of rings associated with them. They are divided into two categories:

(a) *Carbocyclic Compounds:* Carbocyclic compounds having a ring of carbon atoms are further divided into two categories:

(i) *Aromatic Compounds:* Compounds having a ring of six carbon atoms with alternate double bonds are called aromatic or benzenoid compounds. The carbocyclic compounds can be divided into monocyclic, bicyclic or tricyclic as they have one, two or three rings, for example,

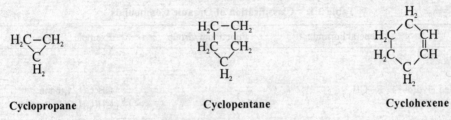

| Benzene (Monocyclic) | Naphthalene (Bicyclic) | Anthracene (Tricyclic) |

(ii) *Alicyclic Compounds:* Alicyclic or Aliphatic cyclic compounds are the compounds having a ring of carbon atoms. They resemble aliphatic hydrocarbons more than cyclic aromatic hydrocarbons.

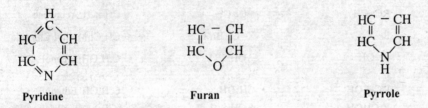

| Cyclopropane | Cyclopentane | Cyclohexene |

(b) *Heterocyclic Compounds:* Compounds having a ring of carbon atoms with one or more heteroatoms like O, N, S are called Heterocyclic compounds.

| Pyridine | Furan | Pyrrole |

Functional Groups and Homologous Series

Hydrocarbons are the backbone of organic compounds as all other compounds are derived from them by substituting one or more atoms of hydrogen by another atom or group (X here may be a halo, nitro, hydroxy

$$R–H \xrightarrow[+X]{-H} R–X$$

or carbonyl group). The properties of compounds 'R – X' depend on the nature of R which may be alkyl or aryl and 'X'. If X remains unchanged in a series of compounds given below, which differ only in number of carbon atoms, we find that there is little change in the chemical properties of compounds although their physical properties like melting point, boiling point or density are affected. For example, the group of compounds given below belong to homologous series of carboxylic acid.

| HCOOH | Methanoic Acid | CH_3CH_2COOH | Propanoic Acid |
| CH_3COOH | Ethanoic Acid | $CH_3CH_2CH_2COOH$ | Butanoic Acid |

The functional group here is carboxyl and each member differs from the member above and below by $-CH_2-$ (methylene). Ethanoic acid, for example has one $-CH_2-$ less than propanoic acid and one $-CH_2-$ more than methanoic acid.

Organic compounds have been classified into a number of classes or homologous series. The characteristics of a homologous series are:

(a) Different members of a homologous series are represented by a general formula. For example, alkanes are represented by C_nH_{2n+2}.

(b) Every member of a homologous series is called homologue and differs from its one higher or one lower member by a $-CH_2-$.

(c) A general method of preparation can be adopted for all the homologues of a series.

(d) All homologues of a series possess similar chemical properties which are characteristics of the functional group of the series.

(e) They show different physical properties due to the difference in their molecular weight.

The concept of functional groups and homologous series has divided organic compounds into various classes which are summarised in Table 3.1.

Table 3.1:   Classification of Organic Compounds

Class	General Formula	Functional Group	Example
1. Hydrocarbons			
a. Saturated Hydrocarbon or Alkanes	$R-CH_3$		CH_3CH_3 Ethane $CH_3CH_2CH_3$ Propane
b. Unsaturated Hydrocarbons			
(i) Alkenes	$RCH=CH_2$	$>C=C<$	$CH_3CH=CH_2$ Propene
(ii) Alkynes	$RC\equiv CH$	$-C\equiv C-$	$CH\equiv CH$ Acetylene
2. Alkyl Halides	$R-X$	$X=Cl, Br, I$	CH_3Cl Methyl Chloride
3. Alcohols a. Monohydric	$R-OH$	$-OH$	C_2H_5OH Ethanol
(i) Primary	RCH_2OH	CH_2OH	C_2H_5OH Ethanol
(ii) Secondary	R_2CHOH	$-OH$	$(CH_3)_2CH-OH$ Isopropyl alcohol
(iii) Tertiary	R_3C-OH	$-OH$	$(CH_3)_3C-OH$ t-butyl alcohol
b. Dihydric alcohol	CH_2OHCH_2OH	OH Group	$CH_2OH.CH_2OH$ Ethylene glycol
c. Trihydric Alcohol	$CH_2OH.CHOH$ CH_2OH	OH Group	CH_2OH $\|$ $CHOH$ $\|$ CH_2OH Glycerol
4. Ethers	$R-O-R'$	$-O-$	$C_2H_5OC_2H_5$ Diethylether
5. Aldehydes	$R-\overset{\text{O}}{\overset{\|}{C}}-H$	$-\overset{\text{O}}{\overset{\|}{C}}-H$	$CH_3\overset{\text{O}}{\overset{\|}{C}}-H$ Acetaldehyde
6. Ketones	$R-\overset{\text{O}}{\overset{\|}{C}}-R$	$-\overset{\text{O}}{\overset{\|}{C}}-$	$CH_3-\overset{\text{O}}{\overset{\|}{C}}-CH_3$ Acetone

(Contd.)

7.	Carboxylic Acid	$R-\overset{\overset{O}{\|}}{C}-OH$	$-\overset{\overset{O}{\|}}{C}-OH$	CH_3COOH Acetic acid
8.	Esters	$R-\overset{\overset{O}{\|}}{C}-OR$	$-\overset{\overset{O}{\|}}{C}-OR$	$CH_3-\overset{\overset{O}{\|}}{C}-OCH_3$ Methylacetate
9.	Acyl Halides	$R-\overset{\overset{O}{\|}}{C}-X$	$-\overset{\overset{O}{\|}}{C}-X$	$CH_3-\overset{\overset{O}{\|}}{C}-Cl$ Acetyl chloride
10.	Amides	$R-\overset{\overset{O}{\|}}{C}-NH_2$	$-\overset{\overset{O}{\|}}{C}-NH_2$	$CH_3-\overset{\overset{O}{\|}}{C}-NH_2$ Acetamide
11.	Anhydrides	$(R-\overset{\overset{O}{\|}}{C}-O)_2O$	$-\overset{\overset{O}{\|}}{C}-O-\overset{\overset{O}{\|}}{C}-$	$(CH_3CO)_2O$ Acetic anhydride
12.	Amines			
	a. Primary	$R-NH_2$	$-NH_2$	$CH_3CH_2NH_2$ Ethylamine
	b. Secondary	R_2-NH	$-NH-$	$CH_3-\overset{\overset{CH_3}{\|}}{N}-H$ Dimethylamine
	c. Tertiary	R_3-N	$-\overset{\overset{\|}{}}{N}-$	$CH_3-\overset{\overset{CH_3}{\|}}{N}-CH_3$ Trimethylamine
13.	Nitrocompounds	$R-NO_2$	$-NO_2$	CH_3NO_2 Nitromethane
14.	Nitriles or or Cyanides	$R-C\equiv N$	$-C\equiv N$	CH_3CN Methyl cyanide (Acetonitrile)
15.	Isonitriles or Isocyanides	$R-N\equiv C$	$-N\equiv C$	C_2H_5NC Ethyl isocyanide (Propioisonitrile)
16.	Sulphonic acid	$R-SO_2OH$	$-SO_2OH$	$C_2H_5SO_3H$ Ethane sulphonic acid
17.	Mercaptans or Thioalcohols	$R-SH$	$-SH$	C_2H_5SH Ethyl mercaptan
18.	Thioethers	$R-S-R$	$-S-$	CH_3SCH_3 Dimethyl Sulphide

3.2   DEVELOPMENT OF SYSTEMATIC NOMENCLATURE

The basis of systematic nomenclature of organic compounds arises through alkane structure. We can name normal straight chain alkanes easily on the basis of number of carbon atoms present in it, like five carbon alkane as pentane; nine carbon nonane; ten carbon decane etc. But alkanes having four carbon atoms or more exist as structural isomers, for example, Butane exists as n-butane and Isobutane.

$$CH_3CH_2CH_2CH_3 \qquad CH_3\!-\!\overset{\overset{\displaystyle CH_3}{|}}{CH}\!-\!CH_3$$

n-Butane   **Isobutane**

and, pentane exists as n-pentane, isopentane and neopentane.

$$CH_3CH_2CH_2CH_2CH_3 \qquad CH_3\!-\!\overset{\overset{\displaystyle CH_3}{|}}{CH}\!-\!CH_2\!-\!CH_3 \qquad CH_3\!-\!\overset{\overset{\displaystyle CH_3}{|}}{\underset{\underset{\displaystyle CH_3}{|}}{C}}\!-\!CH_3$$

n-Pentane   **Isopentane**   **Neopentane**

The number of possible structural isomers increases with the increase of carbon atoms, consequently naming them with trivial names becomes difficult.

Earlier, it was common practice among chemists to name a new compound on the basis of its origin like acetic acid derived its name from vinegar (Latin: *acetum* means vinegar) of which it is a chief constituent. Sometimes the organic compounds are even named after a person like barbituric acid is said to be named after Barbara. Such a system may have certain charms but it is not manageable with the rise in number of compounds.

In 1889, the solution for the above problem was sought by International Chemical Congress by appointing a committee to draw rules for the systematic nomenclature. A report was accepted in 1892 in Geneva but it was found incomplete and in 1930, International Union of Chemistry (IUC) gave a modified report which is also referred to as Liege Rules. This report was further modified by International Union of Pure Applied Chemistry (IUPAC) in the year 1947. The union has issued periodic reports on rules for the systematic nomenclature of organic compounds since that date the most recent of which was published in the year 1979.

3.3   IUPAC NOMENCLATURE

Organic compounds are named by following IUPAC system:

1. The longest chain of carbon atoms is selected.

2. The carbon chain is numbered from the substituents or functional groups. The lowest number is assigned to the carbon atom which is attached to substituent/ functional group.

3. Substituents/ Side chain are indicated by suitable prefixes whereas functional groups are indicated by suitable suffixes.

4. If more than one prefixes are used, they are arranged in alphabetical order.

Despite the fact that present day system adopts IUPAC system of nomenclature, common or trivial names are still used. We shall discuss IUPAC nomenclature of various classes of organic compounds which we come across during the study of aliphatic organic compounds.

Alkanes or Paraffins

1. The longest continuous sequence of carbon atoms is the basis of parent name of an aliphatic hydrocarbon.

The parent name is derived from root word of Greek origin indicating the number of carbon atoms in the longest continuous chain, e.g., a sequence of four atoms is named by using root 'but-' while one of ten carbon atoms is by 'dec-'.

The ending -ane is given to the root word to complete the parent name for an alkane as indicated in Table 3.2.

2. A secondary prefix is employed to designate the structure and position of side chain attached to the parent chain. Hydrocarbon side chain structures are named by using appropriate root name for them with a –yl as the ending. For example, a $CH_3CH_2CH_2$ group is named as propyl (Table 3.3). The position of attachment of side chain group is indicated by numbering the parent chain in such a way so as to give the lowest possible number to the side chain, e.g.,

$$\overset{\displaystyle CH_3}{\underset{|}{CH_3CH}}\,CH_2CH_2CH_3$$

$$CH_3CH_2\overset{\displaystyle CH_3}{\underset{|}{CH}}\,CH\,CH_2CH_2CH_2\overset{\displaystyle CH_3}{\underset{|}{CH}}\,CH_2CH_3 \quad \underset{CH_3}{|}$$

2-Methylpentane (not 4-Methylpentane)   **3,4,8-Trimethyldecane (not 3,7,8-Trimethyldecane)**

3. Multiple side chain groups are labeled with the appropriate Greek multiplying numerical prefix as di, tri, tetra for two, three and four side chains respectively.

$$CH_3CH_2\overset{\displaystyle CH_3}{\underset{\underset{\displaystyle CH_3}{|}}{\overset{|}{C}}}\,CH_2CH_3$$

$$CH_3-\overset{\displaystyle CH_3}{\underset{\underset{\displaystyle CH_3}{\underset{|}{CH_2}}}{\overset{|}{C}}}-CH_2-\overset{\displaystyle CH_3}{\underset{\underset{\displaystyle CH_3}{|}}{\overset{|}{C}}}-CH_3$$

3,3-Dimethylpentane   **2,2,4,4-Tetramethylhexane**
(not 3-Dimethylpentane)   **(not 3,3,5,5-Tetramethylhexane)**

4. When two or more different substituents are attached to parent chain, the names of those groups are arranged in alphabetical order irrespective of their numbers.

$$CH_3CH_2\overset{\displaystyle CH_3}{\underset{|}{CH}}-\overset{\displaystyle CH_2CH_2CH_3}{\underset{|}{CH}}CH_2CH_2CH_3$$

$$CH_3\overset{\displaystyle CH_3}{\underset{|}{CH}}-\overset{\displaystyle CH_2CH_3}{\underset{|}{CH}}CH_2CH_3$$

3-Methyl-4-propylheptane   **3-Ethyl-2-methylpentane**
                                     (not 2-Methyl-3-ethylpentane)

5. To designate the name to aliphatic cyclic compounds the primary prefix cyclo- is placed just before the parent name, e.g., a cyclic four carbon alkane is cyclobutane, a cyclic seven carbon alkane is cycloheptane.

□ **Cyclobutane**   ⬡ **Cycloheptane**

Alkyl substituted cyclic alkanes are named as cycloalkanes rather than cycloalkane derivatives of noncyclic parent. For example:

Methylcyclohexane   **Isopropylcyclohexane**
(not Cyclohexylmethane)   **(not Cyclohexylisopropane)**

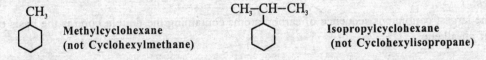

Table 3.2:   Roots and Related Name of Alkanes

Number of Carbons	Root	Formula	Name
1	meth-	CH_4	Methane
2	eth-	CH_3CH_3	Ethane
3	prop-	$CH_3CH_2CH_3$	Propane
4	but-	$CH_3(CH_2)_2CH_3$	Butane
5	pent-	$CH_3(CH_2)_3CH_3$	Pentane
6	hex-	$CH_3(CH_2)_4CH_3$	Hexane
7	hept-	$CH_3(CH_2)_5CH_3$	Heptane
8	oct-	$CH_3(CH_2)_6CH_3$	Octane
9	non-	$CH_3(CH_2)_7CH_3$	Nonane
10	dec-	$CH_3(CH_2)_8CH_3$	Decane
11	undec-	$CH_3(CH_2)_9CH_3$	Undecane
12	dodec-	$CH_3(CH_2)_{10}CH_3$	Dodecane
20	eicos-	$CH_3(CH_2)_{18}CH_3$	Eicosane
30	triacont-	$CH_3(CH_2)_{28}CH_3$	Triacontane
40	tetracont-	$CH_3(CH_2)_{38}CH_3$	Tetracontane
50	pentacont-	$CH_3(CH_2)_{48}CH_3$	Pentacontane

6. When the number of carbon atoms in an alkyl chain exceeds the number of carbon atoms in the ring, The compounds are named as derivatives of noncyclic parent, for example,

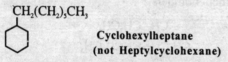

Cyclohexylheptane
(not Heptylcyclohexane)

7. When a side chain possess an additional side chain, the position of secondary side chain is designated by numbering the primary side chain at the point of its attachment to the parent chain. For example,

$$\underset{3\quad2\quad1}{CH_3CH_2CH_2-CHCH_2CH_3}$$

with

$$\overset{1\quad\ 2}{CH_3CHCH_3}$$

3-(1-Methylethyl) hexane
or 3-Isopropylhexane

The trivial names of certain branched alkyl groups listed in Table 3.3 have been retained by IUPAC system. Their IUPAC names have also been incorporated in the table along with their structure

Alkenes:

1. The largest continuous sequence of carbon atoms containing the double bond is the basis of parent name of an alkene.

$$CH_3CH_2CH_2CH=CH_2$$

1-Pentene

$$CH_3CH_2\overset{\textstyle CH_3}{\overset{|}{C}}HCH_2CH=CH_2$$

4-Methyl-1-hexene

2. The numbering of carbon chain is done in such a way so as to give lowest possible number to the double bond even if the side chain alkyl group receive higher number.

Table 3.3:   Trivial and IUPAC Names of Branched Alkyl Group

Trivial Name	Structure	IUPAC Name
Isopropyl	$(CH_3)_2CH-$	1-Methylethyl
Isobutyl	$(CH_3)_2CHCH_2-$	2-Methylpropyl
Sec-Butyl	$CH_3CH_2\overset{\overset{\displaystyle CH_3}{\mid}}{C}H-$	1-Methylpropyl
tert-Butyl	$(CH_3)_3C-$	1,1-Dimethylethyl
Isopentyl	$(CH_3)_2CHCH_2CH_2-$	3-Methylbutyl
Neopentyl	$(CH_3)_3CCH_2-$	2,2-Dimethylpropyl
tert-Pentyl	$CH_3CH_2-\overset{\overset{\displaystyle CH_3}{\mid}}{\underset{\underset{\displaystyle CH_3}{\mid}}{C}}-$	1,1-Dimethylpropyl
Isohexyl	$(CH_3)_2CHCH_2CH_2CH_2-$	4-Methylpentyl

$$\underset{\textbf{4-Methyl-3-heptene (not 4-Methyl-4-heptene)}}{CH_3CH_2CH_2\overset{\overset{\displaystyle CH_3}{\mid}}{C}=CHCH_2CH_3}$$

$$\underset{\textbf{3-Propyl-1-hexene (not 4-Vinylheptane)}}{CH_2{=}CH\overset{\overset{\displaystyle CH_2CH_2CH_3}{\mid}}{C}HCH_2CH_2CH_3}$$

3. If a side chain possess a double bond, it is named as substituent ending in -enyl, e.g., $-CH=CH_2$ side chain (common name vinyl) is derived from C_2 alkane, thus, the name ethenyl is assigned to it. According to IUPAC rules the following three side chains having double bonds may retain their names as shown in Table 3.4.

Table 3.4:   Trivial and IUPAC Names of Unsaturated Side Chain

Trivial Name	Structure	IUPAC Names
Vinyl	$CH_2=CH-$	Ethenyl
Allyl	$CH_2=CHCH_2-$	2-Propenyl
Isopropenyl	$CH_2{=}\overset{\overset{\displaystyle CH_3}{\mid}}{C}-$	1-Methylethenyl

4. Organic compounds having two or three double bonds are called dienes or trienes respectively. If two double bonds are separated by a single bond, the dienes are called conjugated dienes.

$$\underset{\textbf{1,3-Butadiene}}{CH_2{=}CHCH{=}CH_2} \qquad\qquad \underset{\textbf{1,4-Pentadiene}}{CH_2{=}CHCH_2CH{=}CH_2}$$

$$\underset{\textbf{6-Methyl-1,3,5-heptatriene}}{CH_2-\overset{\overset{\displaystyle CH_3}{\mid}}{C}=CHCH{=}CHCH{=}CH_2}$$

Alkynes

1. The longest continuous carbon chain containing triple bond is the basis for the parent name of an alkyne. The ending -yne is added to the root word, e.g., a C_2 alkyne is called ethyne and a C_3 is called propyne.

CH≡CH
Ethyne

$CH_3C≡CH$
Propyne

2. The position of triple bond is shown by numbering the alkyne in such a way so that minimum number is assigned to the triple bond.

$CH_3CH_2C≡CH$
1-Butyne

$\overset{CH_3}{CH_3CHCH_2CH_2C≡CH}$
5-Methyl-1-hexyne

3. If both double and triple bonds are present in the compound then the ending -enyne is given to the root. Lowest possible number is assigned to double or triple bond irrespective of whether -ene or -yne gets the lower number, but when either groups could be assigned the same number, -ene takes priority over -yne and receives the lower number.

CH≡CCH=CHCH₃

3-Penten-1-yne (not 2-Penten-4-yne)

$\overset{CH_3}{CH≡CCHCH=CH_2}$

3-Methyl-1-penten-4-yne (not 3-Methyl-4-penten-1-yne)

Alkyl Halides

1. Alkyl halides are named as substituted hydrocarbons, for example,

CH_3Cl
Chloromethane

$CH_3CH_2CH_2Br$
Bromopropane

2. When double bond, triple bond and substituents are also present along with halogens, they get priority over halogens.

$CH_2=CHCH_2Cl$

3-Chloropropene
(not 1-Chloro-2-propene)

$\overset{Br}{CH_3CHCH=CHCH_3}$

4-Bromo-2-pentene
(not 2-Bromo-3-pentene)

Alcohols and Thiols

1. The longest continuous chain having hydroxy group is the basis for the parent name of an alcohol. The 'e' of corresponding hydrocarbon is dropped, and suffix 'ol' is added. For example,

CH_3CH_2OH
Ethanol

$CH_3CH_2CH_2CH_2OH$
1-Butanol

$\overset{OH}{CH_3CH_2CHCH_2CH_3}$
3-Pentanol

2. Compounds containing two, three or more hydroxy groups are named as -diol, -triol or polyols.

$CH_2OH.CH_2OH$
1,2-Ethanediol

$CH_2OH.CHOH.CH_2OH$
1,2,3-Propanetriol

3. When double bond is also present along with alcoholic group in a compound, hydroxy group takes priority over double bond and the compound is numbered in such a way so as to give

minimum number to the hydroxy group irrespective of whether double bond is or is not a part of the parent chain.

$CH_2=CHCH_2OH$

2-Propen-1-ol

$$CH_3\overset{\overset{\displaystyle CH_3}{|}}{CH}-\overset{\overset{\displaystyle OH}{|}}{CH}CHCH_2CH_2CH_3$$
$$\underset{\displaystyle CH=CH_2}{|}$$

4-Ethenyl-2-methyl-3-octanol or
2-Methyl-4-Vinyl-3-octanol

4. Thiols are sulfur analogs of alcohols and are named by adding suffix -thiol to the parent hydrocarbon. They are also referred to as mercaptans

CH_3CH_2SH

Ethanethiol (Ethylmercaptan)

$CH_3CH_2CH_2SH$

1-Propanethiol (Propylmercaptan)

Ethers and Thioethers

IUPAC has adopted many systems for naming ethers. The preferred practice is to name them as alkoxy alkanes, i.e., alkoxy is treated as substituent on parent hydrocarbon.

$CH_3CH_2OCH(CH_3)_2$

2-Ethoxypropane
(Ethylisopropyl ether)

$\triangleright\!-OCH_3$

Methoxycyclopropane
(Cyclopropylmethyl ether)

Another approach to name ether is to treat oxygen atom as one of the atoms in the parent chain, e.g., ethylpropyl ether can also be named as 3-oxahexane whereas ethyl isopropyl ether is named as 2-Methyl-3-oxapentane.

$\overset{1}{C}H_3\overset{2}{C}H_2\overset{3}{O}\overset{4}{C}H_2\overset{5}{C}H_2\overset{6}{C}H_3$

3-oxahexane (Ethylpropyl ether)

$$\overset{1}{C}H_3\overset{\overset{\displaystyle CH_3}{|}}{\underset{2}{C}H}\overset{3}{O}\overset{4}{C}H_2\overset{5}{C}H_3$$

2-Methyl-3-oxapentane (Ethylisopropyl ether)

Thioethers, the sulfur analogs of ether are named as sulfides or alkylthio derivatives of parent hydrocarbon..

$CH_3SCH_2CH_3$

Methylthioethane (Methylethyl sulfide)

CH_3SCH_3

Methylthiomethane (Dimethyl Sulfide)

Amines

Amines are organic derivatives of ammonia. Replacement of one, two or three hydrogens of ammonia gives primary secondary or tertiary amines.

CH_3NH_2

Methanamine
(Methyl amine)

A primary amine

$(CH_3)_2NH$

N-Methylmethanamine
(Dimethyl amine)

A Secondary amine

$(CH_3)_3N$

N,N-Dimethylmethanamine
(Trimethylamine)

A Tertiary amine

1. According to IUPAC nomenclature for primary amines 'e' of parent hydrocarbon is dropped and suffix -amine is added. The nitrogen atom is neither counted nor numbered to establish the name of compound.

2. Secondary and Tertiary amines are named as substituents on parent primary amine, the letter 'N' shows the bonding of substituent alkyl group to the nitrogen atom.

3. Another acceptable nomenclature for amines is naming them as hydrocarbon derivatives of ammonia. A hydrocarbon root name with suffix -ylamine is the basis of naming amines like CH_3NH_2 is named as Methylamine.

$$\underset{\substack{| \\ CH_3}}{CH_3CHNH_2}$$

2-Propanamine
(isopropylamine)

$$(CH_3CH_2)_2NH$$

N-Ethylethanamine
(Diethylamine)

$$CH_3-\underset{\substack{| \\ CH_3}}{\overset{\substack{CH_3 \\ |}}{C}}-CH_2CH_2N(CH_3)_2$$

3,3,N,N-Tetramethylbutanamine
(3,3,N,N-Tetramethylbutylamine)

$$CH_3-\underset{\substack{| \\ CH_3}}{\overset{\substack{CH_3 \\ |}}{C}}-NH_2$$

2-Methyl-2-propanamine
(tert butylamine)

Aldehydes and Ketones

1. Aldehydes are named by replacing 'e' of parent hydrocarbon by 'al'.

$$\underset{\substack{| \\ CH_3}}{CH_3}-\underset{}{CHCH_2}\overset{\substack{H \\ |}}{C}=O$$

3-Methylbutanal

$$CH_3CH_2\overset{\substack{H \\ |}}{C}=O$$

Propanal

2. If aldehyde group (-CHO) is present on a carboxylic acid as an substituent it is named as methanoyl or formyl as -COOH group takes priority over -CHO group.

$$H-\overset{\substack{O \\ ||}}{C}CH_2 CH_2\overset{\substack{O \\ ||}}{C}-OH \quad \textbf{3-Formylpropanoic acid}$$

3. If two aldehyde groups are present in a compound they are termed as -dial.

$$CHO(CH_2)_6CHO \quad \textbf{Octanedial}$$

4. Ketones are named by replacing 'e' of parent compound by '-one'. The numbering indicates the position of carbonyl group along the parent chain.

$$CH_3CH_2\overset{\substack{O \\ ||}}{C}CH_3$$

2-Butanone
(Ehtylmethyl ketone)

$$CH_3CH_2\overset{\substack{O \\ ||}}{C}CH(CH_3)_2$$

2-Methyl-3-pentanone
(Ethylisopropyl ketone)

$$CH_3\overset{\substack{O \\ ||}}{C}CH_2\overset{\substack{O \\ ||}}{C}CH_3$$

2,4-Pentanedione

5. When another functional group having priority over ketone, is present in the compound, the oxygen atom of carbonyl group is regarded as a substituent and designated by prefix -oxo, for example, in the following compound aldehyde group takes priority over ketone:

$$CH_3\overset{\substack{O \\ ||}}{C}CH_2CH_2CH_2\overset{\substack{H \\ |}}{C}=O \quad \textbf{5-Oxohexanal}$$

Carboxylic Acid

1. When carboxylic group is present at terminal position, number '1' is not included while naming the compound and suffix -oic acid is employed to root hydrocarbon.

CH₃   O
CH₃CHCH₂—C—OH
 4   3   2   1
3-Methylbutanoic acid

CH₃ O   O
CH₃—CH—C—CH₂—C—OH
 5   4   3   2   1
4-Methyl-3-oxopentanoic acid

2. Alternatively, the carboxyl group may be regarded as substituent and is denoted by suffix carboxylic acid to the parent hydrocarbon chain, for example,

CH₃COOH

Methanecarboxylic acid

CH₃
CH₃CHCH₂CH₂COOH
 4   3   2   1

3-Methylbutanecarboxylic acid

3. When –COOH functional group is attached to a cyclic structure, the suffix –carboxylic acid is added to the root word of the cyclic compound, for example,

COOH

Cyclohexanecarboxylic acid

COOH

Benzenecarboxylic acid (Benzoic acid)

4. When the -COOH functional group is present as substitutent, it is named as -carboxy.

5. When two or three carboxylic functional groups are present in a molecule, it is named by adding suffix -dioic and -trioic acid respectively.

COOHCH₂COOH
Propanedioic acid

CH₂COOHCHCOOHCH₂COOH
1,2,3-Propanetrioic acid

Carboxylic Acid Derivatives

Acyl halides: The suffix –oyl is added to parent hydrocarbon by deleting 'e' of alkane followed by a separate word for the halogen atom present there,

O
‖
CH₃CH₂CCl
Propanoyl chloride

CH₃ O
     ‖
CH₃CH₂CHCH₂C–Br
3-Methylpentanoyl Bromide

Acid Anhydrides

Theoretically acid anhydrides are formed by removal of a water molecule from two molecule of carboxylic acid, thus they are named according to the name of carboxylic acid.

1. If both groups attached to oxygen are same, prefix 'di' is not used, for example,

(CH₃CH₂CO)₂O
Propanoic anhydride

(CH₃CO)₂O
Ethanoic anhydride

2. If two different groups are attached to oxygen , their carboxylic acid names are employed as separate words, for example,

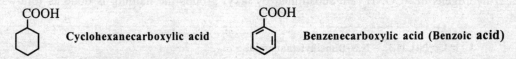

O   O
‖   ‖
CH₃CH₂C—O—CCH₂CH₂CH₃   **Butanoic propanoic anhydride**

Acid Esters

Since they are formed by replacement of 'H' of –COOH group by an alkyl group, they are named as alkyl salts of parent acids, for example, if 'H' of ethanoic acid is replaced by methyl group it is named as methylethanoate, suffix -oate replaces -oic acid from the parent carboxylic acid.

$$CH_3\overset{\overset{O}{\|}}{C}-OCH_3 \quad \textbf{Methylethanoate}$$

Amides

1. Acid amides are the compounds formed by replacement of -OH group of carboxylic acid by an -NH$_2$ group. According to IUPAC rules the suffix 'e' of parent hydrocarbon is replaced by 'amide' to name these compounds.

$$CH_3\overset{\overset{O}{\|}}{C}-NH_2 \qquad\qquad CH_3CH=CH\overset{\overset{O}{\|}}{C}-NH_2$$

Ethanamide (acetamide)   **2-Butenamide**

2. If hydrogens of –CONH$_2$ are substituted by alkyl groups the naming is done as follows:

$$CH_3\overset{\overset{O}{\|}}{C}-N(CH_3)_2 \quad \textbf{N,N-Dimethylethanamide}$$

Alkyl Nitriles

1. The alkyl nitriles with general formula R–C≡N are named by adding suffix -nitrile to the parent hydrocarbon. They may also be named from the common names of the acids to which they are hydrolysed. The ending -ic acid is dropped and suffix onitrile is added. For example, CH$_3$CN is named as ethanenitrile or acetonitrile (CH$_3$CN forms acetic acid on hydrolysis).

$$CH_3CH_2C≡N \qquad\qquad \overset{Br}{\underset{}{CH_3\overset{|}{C}HC≡N}}$$

Propanenitrile (Propionitrile)   **2-Bromopropanenitrile**
                                          (2-Bromopropionitrile)

2. If –C≡N group is present as substituent, it is named by using prefix 'cyano'.

$$\overset{C≡N}{\underset{}{CH_3\overset{|}{C}HCH_2CH_2COOH}} \quad \textbf{4-cyanopentanoic acid}$$

3.4   PRIORITIES OF FUNCTIONAL GROUPS

When a compound has two or more functional groups, a choice must be made among them to decide the basis for parent name. We have taken various examples earlier also to this effect like the following compound is named as 5-Hydroxy-2-hexanone as a ketone group takes priority over hydroxy group.

$$\overset{OH}{\underset{}{CH_3\overset{|}{C}HCH_2CH_2\overset{\overset{O}{\|}}{C}-CH_3}} \quad \textbf{5-Hydroxy-2-hexanone (not 5-Oxo-2-hexanol)}$$

The decreasing order of priority of functional group established for IUPAC nomenclature is given below:

Acid > Acid derivatives (except nitriles) > Aldehydes > Nitriles > Ketones > alcohols > Mercaptans > Amines > Ethers > Alkenes > Alkynes > Alkyl halides > Nitro compounds > Alkanes

Below is given various groups with their prefixes and suffixes in Table 3.5.

Table 3.5:   Prefixes and Suffixes of Various Groups

Sl.No	Functional groups	Prefix name	Suffix name
1	$-COOH$	Carboxy	-oic acid
2	$-COOR$	Alkoxy carbonyl	–
3	$-SO_3H$	Sulpho	Sulphonic acid
4	$-CONH_2$	Carbamoyl	-amide
5	$-CHO$	Formyl	-al
6	$-CN$	Cyano	-nitrile
7	$>C=O$	Oxo	-one
8	$-OH$	Hydroxy	-ol
9	$-SH$	Mercapto	-thiol
10	$-NH_2$	Amino	-amine
11	$>C=C<$	–	-ene
12	$-C\equiv C-$	–	-yne

The nomenclature of various classes of organic compounds is also discussed in details in individual chapters.

3.5   PROBLEMS

1. Explain the following:
    (a) Homologous series
    (b) Alicyclic and aromatic compounds
    (c) Carbocyclic and Heterocyclic compounds
    (d) Trivial names
2. Provide IUPAC names of the following:
    (a) $CH_3CH_2CH(CH_3)C\equiv CH$
    (b) $CH_3CH(CH_3)COCH_3$
    (c) $CH_3CH(CH_3)CH_2CH_2CH(Cl)CH_2CH_3$
    (d) $CH_2=CHCH(OH)CH_2CH_2OH$
    (e) $CH_3CH_2CH(NO_2)CH_2CH(CH_3)_2$
    (f) $CH_3CH(NH_2)CH=CHCH_2OH$
    (g) $CH\equiv CCH_2HC=CHOH$
    (h) $(CH_3)_2C=CHCH_2CH_2CH_2OH$
    (i) $CH_3C\equiv CCH(OH)CH=CH_2$
    (j) $CH_2OHCHOHCHOHCH_2OH$
    (k) $(CH_3)_2CHCH(CH_3)CH_2OH$
    (l) $CH_3CH_2CH_2OC_2H_5$
    (m) $(CH_3)_2CHCH_2COOC_2H_5$
    (n) $OHCCH_2CH_2COOH$
    (o) $CH_2ClCHOHCH_2CH_2CH_3$
    (p) $CH_3CH(CH_3)CONH_2$
3. Draw structural formula for the following:
    (a) 4-ethyl-3,4-dimethylcyclohexanone
    (b) 4-chloro-4-ethylheptanoic acid
    (c) 5-oxanonane
    (d) Methyl isopropyl sulphide
    (e) 3-methyl-2-heptanol
    (f) (1-chloropropyl)cyclopentane
    (g) Trimethyl ammonium chloride
    (h) 2-ethyl-3-phenylpropanoate
    (i) 2-isopropyl-3-phenylcyclohexanone
4. Write the structural formulas and provide IUPAC names for the following:
    (a) Ketones with molecular formula $C_5H_{10}O$
    (b) Esters with molecular formula $C_5H_{10}O_2$
    (c) Acids with molecular formula $C_5H_{10}O_2$
    (d) Alcohols with molecular formula $C_5H_{10}O$
    (e) Ethers with molecular formula $C_5H_{10}O$
    (f) Aldehydes with molecular formula $C_5H_{10}O$
    (g) Alkanes with molecular formula C_5H_{12}
    (h) Alkenes or Cycloalkanes with molecular formula C_5H_{10}
    (i) Alkynes with molecular formula C_5H_8

5. Provide IUPAC names for the following:

(a)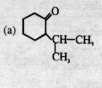

(b)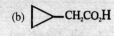

(c)

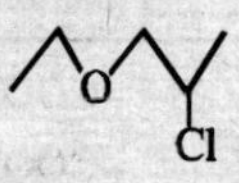

(d)

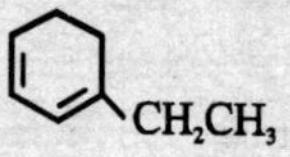

(e)

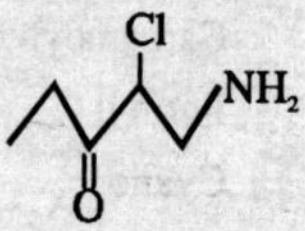

(f)

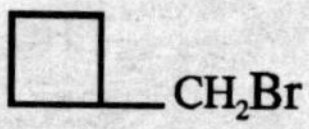

(g)

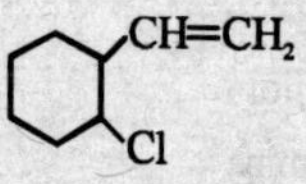

(h)

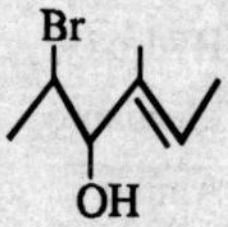

(i)

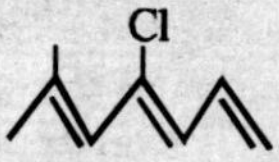

4

Organic Reactions and Mechanism

4.1 INTRODUCTION

A chemical reaction can be represented in the form of following equation.

$$Substrate + Reagent \rightarrow Product$$

The molecules undergoing attack by a *reagent* are called *substrate* and are converted into *products* but, the above representation of a reaction does not indicates the step wise progress of the reaction. Thus to understand a reaction completely, one must know its *mechanism i.e., the actual pathway, the substrate, the attacking species, the kinetics and the thermodynamics of the reaction.* The knowledge of the mechanism is important in devising the synthesis of new molecules which finds use in every sphere of life.

4.2 FACTORS AFFECTING REACTIVITY

As mentioned earlier, a chemical reaction commences in presence of an attacking reagent, thus, the progress of reaction is dependent upon the reactivity of the substrate. We shall now discuss various factors which influence the reactivity of substrate molecules.

4.2.1 Inductive Effect

As has been discussed earlier that a covalent bond develops polarity in the molecule if the two atoms forming the bond are unequal in electronegativity. For example, if carbon is attached to chlorine, the shared pair of σ (sigma) electrons will drift towards chlorine atom developing slightly positive (δ^+) charge on carbon and slightly negative (δ^-) charge on chlorine.

$$-\overset{|}{\underset{|}{C}}\overset{\delta^+}{-}\overset{\delta^-}{Cl} \qquad or \qquad -\overset{|}{\underset{|}{C}}\overset{\delta^+}{\rightarrow}\overset{\delta^-}{Cl}$$

Inductive effect (I effect), a permanent effect, develops in the molecule due to polarity produced by the higher electronegativity of one atom compared to other in a covalent bond. The carbon hydrogen bond is assumed to have zero polarity (due to marginal difference in electronegativity). The atoms or groups which release electrons towards a carbon atom are called +I groups and those which withdraws electrons away from carbon are called –I groups. Some common groups which have +I effect (electron releasing) are as follows:

$$(CH_3)_3C-; (CH_3)_2 CH-; CH_3CH_2-; CH_3-$$

Groups having –I effect (electron withdrawing) are as follows:

NO_2, F, Cl, Br, I, OH, C_6H_5

The inductive effect is not confined to the polarisation of one bond, but it transmits along the chain although it diminished beyond second carbon atom.

$$C_4 - C_3 - C_2 \rightarrow \overset{\delta^+}{C_1} \overset{\delta^-}{\ggg} \overset{\delta^-}{Cl}$$

Table 4.1 Shows the pK_a values of isomeric monochlorobutanoic acid which clearly indicates that acidity decreases as the distance between carboxyl group and chloro group increases. The influence of inductive effect of chloro group is greatest on the strength of acid when it is on the carbon atom next to the carboxyl group.

Table 4.1: pK_a Value of Monochlorobutanoic Acid

Name	Structure	pK_a
Butanoic acid	$CH_3CH_2CH_2\overset{O}{\overset{\|}{C}}\text{-OH}$	4.82
2-Chlorobutanoic acid	$CH_3CH_2CHCl\overset{O}{\overset{\|}{C}}\text{-OH}$	2.86
3-Chlorobutanoic acid	$CH_3CHClCH_2\overset{O}{\overset{\|}{C}}\text{-OH}$	4.05
4-Chlorobutanoic acid	$ClCH_2CH_2CH_2\overset{O}{\overset{\|}{C}}\text{-OH}$	4.52

4.2.2 Electromeric Effect

It involves π electrons. When a double or triple bond is exposed to attack by an attacking reagent, the two π electrons are shifted to either of the atoms (forming the π bond) depending upon the electronegativity of the atoms forming bond as shown below:

$$A=B \xrightarrow{E^+} A^+ - B\colon^- \text{ or: } A^+ - B^-$$

$$CH_2=CH_2 \xrightarrow{E^+} \overset{+}{C}H_2 - \overset{-}{C}H_2$$

$$\overset{H}{\underset{|}{H-C}}=O \xrightarrow{E^+} \overset{+}{C}H_2 - O^-$$

Thus one of the atoms acquires a positive charge whereas the other which receives the electron pair acquires the negative charge or in other words we can say that *electromeric effect (E–effect) refers to the polarity induced in a multiple bond system when it is approached by a reagent.* This is a temporary effect and takes place only in the presence of an attacking reagent.

4.2.3 Resonance

When a molecule can be given several structures but none of them represent it completely, the molecule is said to be a resonance hybrid of those structures and the phenomenon is called *resonance.* The structures representing the molecule are called contributing structures. For example the acetate anion formed by the dissociation of acetic acid is represented by structures I and II.

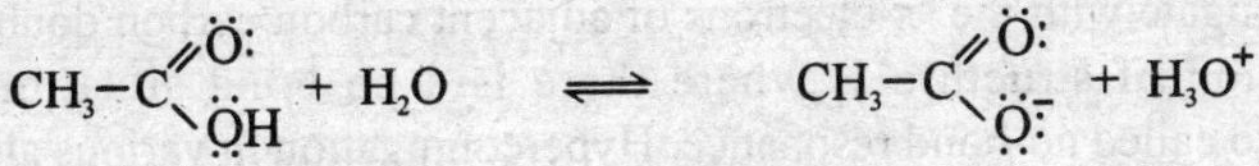

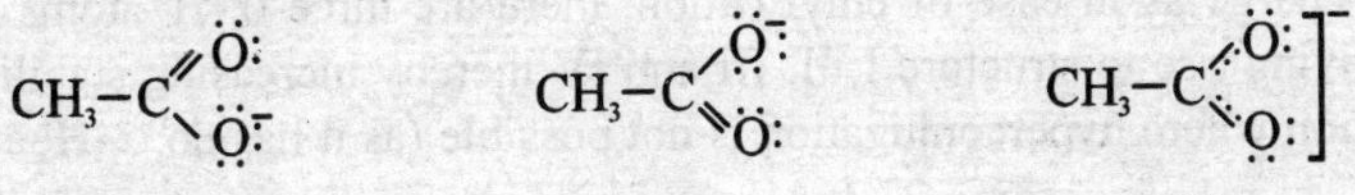

$$CH_3-C{\overset{\ddot{O}:}{\underset{\ddot{O}H}{}}} + H_2O \rightleftharpoons CH_3-C{\overset{\ddot{O}:}{\underset{\ddot{O}:^-}{}}} + H_3O^+$$

$$CH_3-C{\overset{\ddot{O}:}{\underset{\ddot{O}:^-}{}}} \qquad CH_3-C{\overset{\ddot{O}:^-}{\underset{\ddot{O}:}{}}} \qquad \left[CH_3-C{\overset{\ddot{O}:}{\underset{\ddot{O}:}{}}} \right]$$

[I] **[II]** **[III]**

The true structure of acetate anion is neither represented by structure I nor by structure II, but by a resonance hybrid III. Structure I and II are called contributing structures and in both the structures, one of the oxygen is attached to carbon by a double bond and the other by a single bond. On the contrary acetate anion has no carbon oxygen single bond and carbon oxygen double bond, but both oxygens are attached to carbon by a bond order of 1½ or in other words carbon oxygen bond length in acetate anion is 1.26 Å, which is between the bond length of C=O double bond (1.20 Å) and C−O single bond (1.43 Å). Thus structure III represents the resonance hybrid of acetate anion. The phenomenon of resonance is shown by a number of organic compounds. The main conditions for resonance to occur are:

(i) When non bonding electrons and electrons from multiple bond change positions (delocalise) from one atom to another and no electron from covalent bond is involved. For example,

$$CH_2=CH-\overset{+}{C}H_2 \longleftrightarrow \overset{+}{C}H_2-CH=CH_2$$
(Allyl Cation)

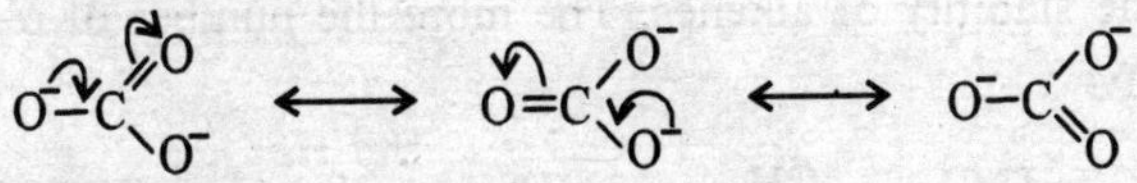

(Carbonate anion)

(ii) When all the resonance structures have same number of paired and unpaired electrons.

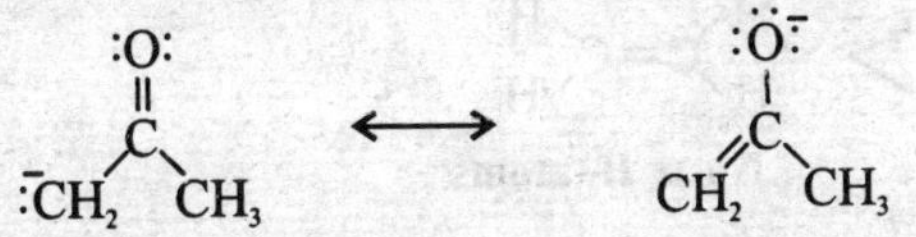

It is important to note at this juncture that resonance is important when the contributing structures are of same energy making the resonance hybrid more stable (low in energy content) than any of the contributing structures. Resonance energy of a molecule is a measure of its stability, larger the resonance energy stabler the molecule.

4.2.4 Hyperconjugation
Hyperconjugation involves the conjugation of sigma electrons with π electrons, p electrons and empty orbitals present in adjacent position.

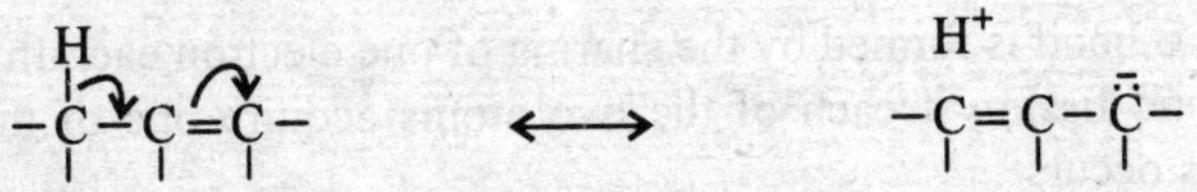

$$\overset{H}{\underset{|}{-C}}-C=C- \longleftrightarrow \overset{H^+}{} \quad -C=C-\overset{..}{C}-$$

I **II**

The σ electrons of C–H bond conjugate with the π electrons of adjacent carbon carbon double bond of structure I resulting in the formation of structure II where there is a no bond situation between carbon hydrogen bond, thus, this is also called no-bond resonance. Hyperconjugation in various alkyl cations shows delocalisation of positive charge as in case of ethyl cation, there are three α–H–atoms which can participate in hyperconjugation giving rise to structure I, II, III and IV thereby increasing stability of ethyl carbocation over methyl carbocation where hyperconjugation is not possible (as it has no α–H–atoms).

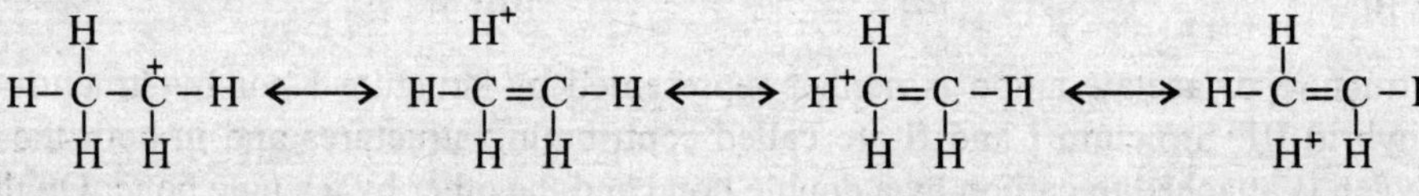

$$
\underset{[I]}{H-\overset{\overset{\displaystyle H}{|}}{\underset{\underset{\displaystyle H}{|}}{C}}-\overset{+}{\underset{\underset{\displaystyle H}{|}}{C}}-H} \longleftrightarrow
\underset{[II]}{\overset{\displaystyle H^+}{H-C}=\overset{\overset{\displaystyle H}{|}}{\underset{\underset{\displaystyle H}{|}}{C}}-H} \longleftrightarrow
\underset{[III]}{\overset{\overset{\displaystyle H}{|}}{H^+C}=\overset{\overset{\displaystyle H}{|}}{\underset{\underset{\displaystyle H}{|}}{C}}-H} \longleftrightarrow
\underset{[IV]}{H-\overset{\overset{\displaystyle H}{|}}{C}=\overset{\overset{\displaystyle H}{|}}{\underset{\underset{\displaystyle H}{|}}{C}}-H}
$$

Similarly, the secondary carbocation (isopropyl cation) has six α–H–atoms which can participate in hyperconjugation and tertiary carbocation (t-butyl cation) has nine α–H–atoms which can participate in hyperconjugation. More the delocalisation of positive charge, greater will be the stability thus the order of stability of various carbocations is as follows:

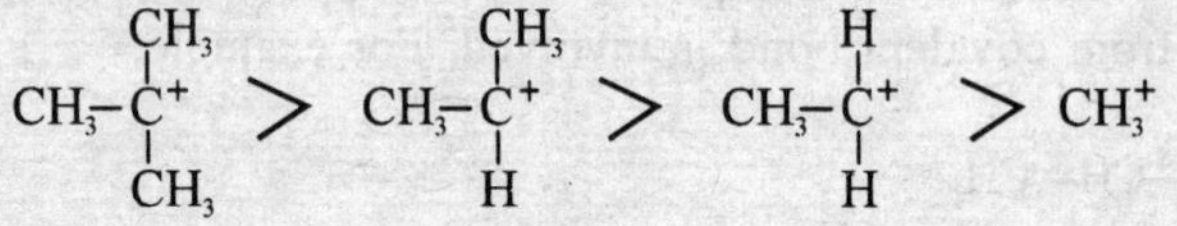

$$
\underset{\underset{\displaystyle CH_3}{|}}{\overset{\overset{\displaystyle CH_3}{|}}{CH_3-\overset{+}{C}}} > \underset{\underset{\displaystyle H}{|}}{\overset{\overset{\displaystyle CH_3}{|}}{CH_3-\overset{+}{C}}} > \underset{\underset{\displaystyle H}{|}}{\overset{\overset{\displaystyle H}{|}}{CH_3-\overset{+}{C}}} > CH_3^+
$$

The same explanation is extended to the stability of alkenes. The more the number of α–H–atoms, stabler will be the alkene as shown below:

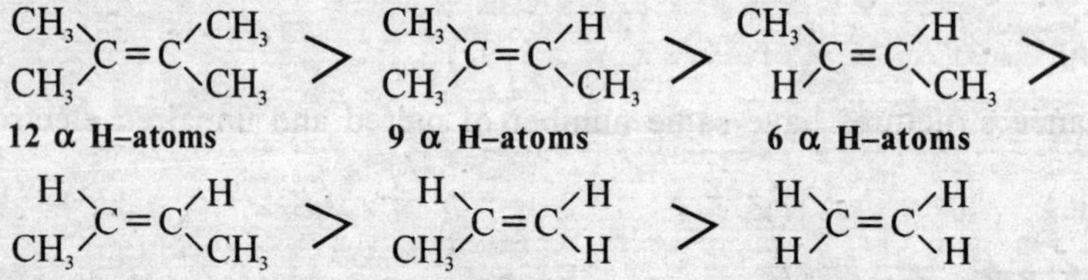

$$
\underset{\text{12 } \alpha \text{ H–atoms}}{\overset{CH_3}{\underset{CH_3}{>}}C=C\overset{CH_3}{\underset{CH_3}{}}} > \underset{\text{9 } \alpha \text{ H–atoms}}{\overset{CH_3}{\underset{CH_3}{>}}C=C\overset{H}{\underset{CH_3}{}}} > \underset{\text{6 } \alpha \text{ H–atoms}}{\overset{CH_3}{\underset{H}{>}}C=C\overset{H}{\underset{CH_3}{}}} >
$$

$$
\underset{\text{6 } \alpha \text{ H–atoms}}{\overset{H}{\underset{CH_3}{>}}C=C\overset{H}{\underset{CH_3}{}}} > \underset{\text{3 } \alpha \text{ H–atoms}}{\overset{H}{\underset{CH_3}{>}}C=C\overset{H}{\underset{H}{}}} > \underset{\text{No } \alpha \text{ H–atoms}}{\overset{H}{\underset{H}{>}}C=C\overset{H}{\underset{H}{}}}
$$

Hyperconjugation is used to explain many unconnected phenomenon although it itself is controversial as it involves the formation of a weak π bond at the expense of a strong σ bond.

4.3   HOMOLYTIC AND HETEROLYTIC BOND FISSION

We have already seen that two atomic orbitals of comparable energy overlap to give a covalent bond. During the bond formation energy is released or we can say the bonding molecular orbital is at a lower energy than the two atomic orbitals. So during the breaking of a covalent bond the same procedure will be reversed and energy will be required to break the bond. A covalent bond can undergo fission in two different ways.

(i)   *Homolytic Fission or Homolysis:* σ bond is formed by the sharing of one electron each from the two atoms forming the bond. During the bond fission if each of the two atoms acquire one of the bonding electrons, homolytic fission or homolysis occurs.

$$\text{A:B} \longrightarrow \text{A}^{\cdot} + \text{B}^{\cdot}$$

A˙ and B˙ are called *free radicals*. They are electrically neutral and have one odd electron. They are extremely reactive since they have the tendency to pair up the odd electron. Homolysis commonly occurs in presence of heat, light or organic peroxides or in vapor phase.

(ii)   *Heterolytic Fission or Heterolysis:* In this type of bond fission one atom or group takes away the bonding pair of electron and becomes negatively charged and the other gets positive charge. When B is more electronegative than A, Heterolysis occurs as follows:

$$A\text{–}B \longrightarrow A^+ + B^-$$

If atom A is more electronegative heterolysis occurs as follows:

$$A\text{–}B \longrightarrow A^- + B^+$$

Heterolytic fission furnishes carbocation (positive ions) and carbanions (negative ions) and it occurs readily with polar compounds in polar solvents.

4.4   REACTION INTERMEDIATES
Homolytic and heterolytic bond fission results in the formation of short lived, reactive species called *reaction intermediates*.

4.4.1   Carbocations
Carbocation is the charged ion having a positive charge on carbon atom and is formed by the heterolytic fission of a σ bond.

$$-\overset{|}{\underset{|}{C}}-X \xrightarrow{\text{Heterolysis}} -\overset{|}{\underset{|}{C}}{}^+ + X^-$$

The carbocations thus formed are called primary, secondary or tertiary carbocations after the name of parent alkyl group. For example,

Methyl Carbocation (Primary) $\overset{+}{C}H_3$

Ethyl Carbocation (Primary) $CH_3\overset{+}{C}H_2$

Isopropyl Carbocation (Secondary) $CH_3\overset{+}{C}H\ CH_3$

tert Butyl Carbocation (Tertiary) $(CH_3)_3\overset{+}{C}$

The geometry of the carbon having positive charge is trigonal with three sp^2 hybrid orbitals forming sigma bond with three other atoms or groups and one empty unhybridised p-orbital which lies above and below the plane of the other three sp^2 hybrid orbitals. This empty p orbital makes the carbon an electron deficient reactive intermediate.

Stability of carbocations is influenced by resonance, Hyperconjugation and Inductive effect. Benzyl and allyl carbocations, for example, are more stable than propyl carbocation as the former two are stabilised by resonance. Resonance structures of allyl carbocation is given below:

$$CH_2=CH-\overset{+}{C}H_2 \longleftrightarrow \overset{+}{C}H_2-HC=CH_2$$

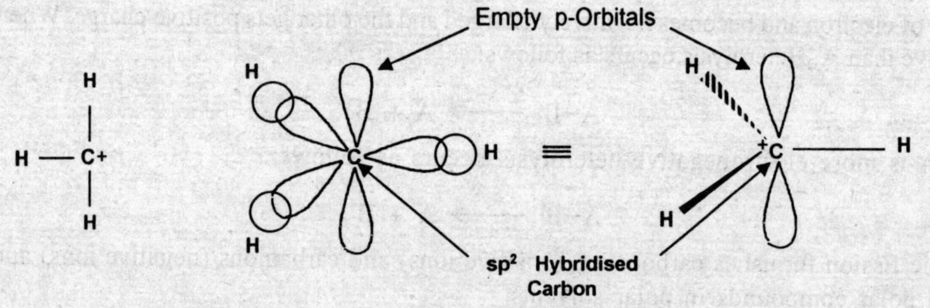

Fig. 4.1:   Structure of Methyl Carbocation

Resonance structures of benzyl carbocation:

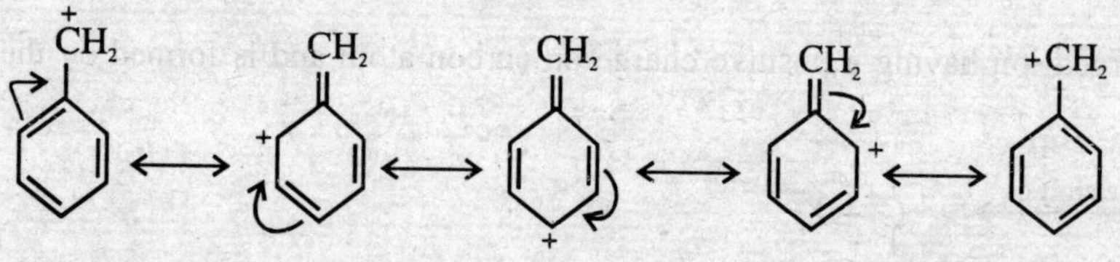

Electron releasing groups (+I groups) stablises the carbocations by partial neutralisation of positive charge. Thus, tertiary butyl carbocation having three methyl groups (+I groups) is more stable than isopropyl carbocation having two methyl groups which in turn is more stable than ethyl carbocation and methyl carbocation having one and no methyl group respectively as shown below:

In other words, the stability of carbocations follows the order given below:

tert-carbocation > sec-carbocation > pri-carbocation

The electron withdrawing (–I groups) groups like –NO$_2$, –Br reduces the stability of carbocation.

Hyperconjugation increases the stability of carbocations by the delocalisation of positive charge with sigma electrons of α–H–atoms. More the number of α–H–atoms, stabler will be the carbocation. Thus due to hyperconjugation the order of stability of carbocations follows the same order as in case of +I effect i.e.,

tert-carbocation > sec-carbocation > pri-carbocation

Reactions of Carbocations: Carbocations involve following reactions.

(i) Loss of a proton from alkyl carbocations forms an alkene. For example, ethyl carbocation on elimination of proton furnishes ethylene.

$$CH_3-\overset{+}{C}H_2 \longrightarrow CH_2=CH_2 + H^+$$

(ii) It may combine with a nucleophile or an electron rich molecule.

$$CH_3-\overset{+}{C}H-CH_3 + Cl^- \longrightarrow CH_3-\overset{\overset{\displaystyle Cl}{|}}{C}H-CH_3$$

(iii) It combines with an alkene to form a bigger carbocation like isobutylene dimerisation.

$$CH_3-\overset{\overset{\displaystyle CH_3}{|}}{C}=CH_2 + H^+ \longrightarrow CH_3-\overset{\overset{\displaystyle CH_3}{|}}{\underset{+}{C}}-CH_3$$

$$CH_3-\overset{\overset{\displaystyle CH_3}{|}}{C}=CH_2 + \overset{\overset{\displaystyle CH_3}{|}}{\underset{\underset{\displaystyle CH_3}{|}}{_+C}}-CH_3 \longrightarrow CH_3-\overset{\overset{\displaystyle CH_3}{|}}{\underset{+}{C}}-CH_2-\overset{\overset{\displaystyle CH_3}{|}}{\underset{\underset{\displaystyle CH_3}{|}}{C}}-CH_3$$

The resultant carbocation on proton elimination produces dimerised isobutylene.

(iv) Carbocations undergo rearrangement to form a more stable carbocations.

(a) n-Butyl carbocation rearranges to more stable sec butyl carbocation.

$$CH_3CH_2CH_2\overset{+}{C}H_2 \longrightarrow CH_3CH_2\overset{+}{C}HCH_3$$

(b) Primary carbocation rearranges to tertiary carbocation by hydride shift.

$$CH_3\,CH_2\overset{\overset{\displaystyle CH_3}{|}}{C}H\,\overset{+}{C}H_2 \longrightarrow CH_3CH_2\overset{\overset{\displaystyle CH_3}{|}}{\underset{+}{C}}CH_3$$

(c) Secondary carbocation rearranges to tertiary carbocation by methyl shift.

$$CH_3-\overset{\overset{\displaystyle CH_3}{|}}{\underset{\underset{\displaystyle CH_3}{|}}{C}}-\overset{+}{C}H-CH_3 \longrightarrow CH_3-\overset{\overset{\displaystyle CH_3}{|}}{\underset{+}{C}}-\overset{\overset{\displaystyle CH_3}{|}}{C}H-CH_3$$

The driving force for these rearrangements is obviously the gain in stability. During these rearrangements either hydrogen or methyl group migrates with its pair of electrons to the adjacent electron deficient carbon atom. Such type of migrations are called hydride shift or methyl shift and are referred to as 1,2-shift.

4.4.2 Carbanions

Carbanions are negatively charged ions having a negative charge on the carbon atom. They are formed by heterolytic bond fission. For example,

$$-\overset{|}{\underset{|}{C}}:M \xrightarrow{\text{Heterolysis}} -\overset{|}{\underset{|}{C}}:^- + M^+$$

where electronegativity of M is less than carbon. Carbanions are named as primary, secondary or tertiary depending upon the nature of alkyl group from which they are formed.

$CH_3\bar{C}H_2$ Ethyl carbanion (pri)

$CH_3\bar{C}HCH_3$ Isopropyl carbanion (sec)

$(CH_3)_3-\bar{C}$ tert Butyl carbanion (tert)

Geometry of carbon having negative charge is tetrahedral with sp³ hybridisation in aliphatic compounds whereas it is sp² in aromatic compounds. In case of methyl carbanion out of four sp³ hybrid orbitals three half filled orbitals form sigma bond by the overlap of s orbital of hydrogen whereas the fourth sp³ hybrid orbital is completely filled giving a negative charge to carbon atom.

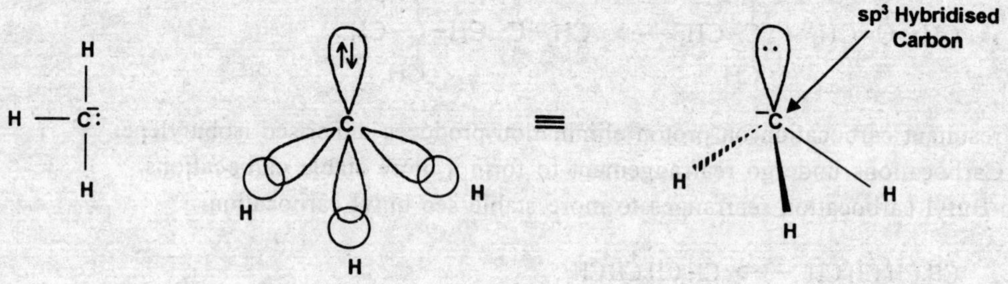

Fig. 4.2:   Structure of Methyl Carbanion

In case of cyclopentadienyl anion, the carbon having negative charge is sp² hybridised and has a trigonal geometry. The pair of electron is present in its unhybridised p orbital as shown below:

The stability of carbanions is also influenced by resonance effects. Any effect which stabilises the negative charge on carbon atom will give stability to carbanion. The benzyl carbanion is much more stable than propyl carbanion because the former one is stablised by resonance whereas no resonance is possible in case of propyl carbanion. Below are given the resonating structures of benzyl anion.

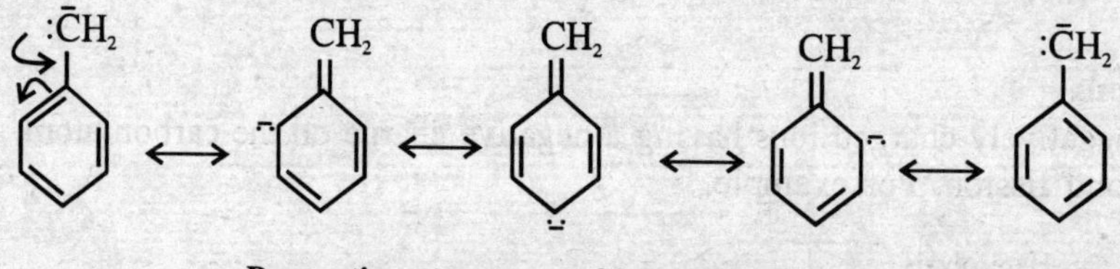

Resonating structures of benzyl carbanion

Carbanions are stabilised by inductive effect. Electron releasing (+I) groups increase the electron density on carbon making it less stable thus the order of stability of carbanions follows the following order.

$$CH_3-\overset{\overset{\displaystyle CH_3}{|}}{\underset{\underset{\displaystyle CH_3}{|}}{C}}:^- \;<\; CH_3-\overset{\overset{\displaystyle CH_3}{|}}{\underset{\underset{\displaystyle H}{|}}{C}}:^- \;<\; CH_3-\overset{\overset{\displaystyle H}{|}}{\underset{\underset{\displaystyle H}{|}}{C}}:^- \;<\; CH_3:^-$$

Electron withdrawing (–I) groups like $-NO_2$, –Br stabilise carbanions by partial dispersion of negative charge on the carbon.

Reactions of Carbanions: Carbanions are nucleophilic i.e. they are electron rich and can add to the electron deficient carbon of carbonyl group.

$$H-\overset{\overset{\displaystyle O}{||}}{C}-\bar{C}H_2 \;+\; \overset{\delta+}{C}=\overset{\delta-}{O} \longrightarrow H-\overset{\overset{\displaystyle O}{||}}{C}-CH_2-\overset{|}{\underset{|}{C}}-O^-$$

Carbanions can undergo substitution reactions at an electron deficient carbon atom as sodium salt of malonic ester reacts with methyl iodide.

$$Na^+\overset{\overset{\displaystyle COOC_2H_5}{|}}{\underset{\underset{\displaystyle COOC_2H_5}{|}}{C}}H^- \;+\; CH_3-I \longrightarrow CH_3-\overset{\overset{\displaystyle COOC_2H_5}{|}}{\underset{\underset{\displaystyle COOC_2H_5}{|}}{C}}H$$

4.4.3 Free Radicals

As mentioned earlier free radicals are formed when a sigma bond undergoes homolytic bond fission and each atom gets an unpaired electron. They are formed by:

(i) *Photochemical Fission*: Bond fission occurs due to the absorption of electromagnetic radiation. Chlorine, bromine, acetone and peroxides undergo photochemical fission as follows:

$$Cl-Cl \xrightarrow{h\nu} Cl^{\cdot} + Cl^{\cdot}$$

$$CH_3-\overset{\overset{\displaystyle O}{||}}{C}-CH_3 \xrightarrow{h\nu} {}^{\cdot}CH_3 + {}^{\cdot}COCH_3 \longrightarrow CO + {}^{\cdot}CH_3$$

(ii) *Thermal Fission*: Pyrolysis also is capable of forming free radicals from a variety of compounds like lead tetraethyl, azo compounds, organic peroxides etc.

$$Pb\,(C_2H_5)_4 \xrightarrow{\Delta} Pb^{\cdot} + 4\,C_2H_5^{\cdot}$$

$$CH_3-N=N-CH_3 \xrightarrow{300^\circ} 2\,{}^{\cdot}CH_3 + N_2$$

$$C_6H_5-\overset{\overset{\displaystyle O}{||}}{C}-O-O-\overset{\overset{\displaystyle O}{||}}{C}-C_6H_5 \longrightarrow 2C_6H_5-\overset{\overset{\displaystyle O}{||}}{C}-O^{\cdot} \longrightarrow 2C_6H_5^{\cdot} + 2CO_2$$

(iii) Oxidation reduction reactions involving inorganic ions, metals or electrolysis too furnishes free radicals.

$$H_2O_2 + Fe^{2+} \longrightarrow HO^{\cdot} + OH^- + Fe^{3+}$$

Geometry of free radical may be:

(i) *Planer* i.e., sp^2 hybridisation of carbon with unpaired electron in p-orbital (structure I).

(ii)   *Pyramidal* i.e., sp³ hybridisation of carbon with unpaired electron in one of the four sp³ hybrid orbital (Structure II).

It is suggested that the shape of a free radical is some where between structure I and structure II.

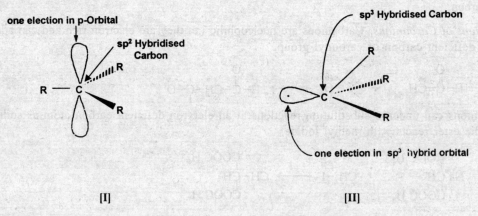

[I]   [II]

Fig. 4.3:   Structure of Alkyl Free Radical

Stability of free radicals is influenced by the bond dissociation energy which is required to break the bond homolytically. Smaller the bond dissociation energy stabler will be the free radical. On the basis of homolytic bond dissociation energy, stabilities of free radicals follows the following order:

$$(CH_3)_3\dot{C} > (CH_3)_2\dot{C}H > CH_3\dot{C}H_2 > \dot{C}H_3$$

Free radicals are also stabilised by resonance.

Reactions of Free Radicals: Free radicals undergo

(i) Combination with other free radicals

$$Cl^· + Cl^· \longrightarrow Cl_2$$
$$Cl^· + ^·CH_3 \longrightarrow CH_3Cl$$

(ii) Disproportionation of ethyl radical gives ethane and ethylene

$$2CH_3{-}\dot{C}H_2 \longrightarrow CH_3{-}CH_3 + CH_2{=}CH_2$$

(iii) Abstraction of an atom to form another free radical

$$^·CH_3 + H{-}C_2H_5 \longrightarrow CH_4 + CH_3\dot{C}H_2$$

4.4.4   Carbenes

Carbenes are neutral species having a bivalent carbon. They are highly reactive and act as strong electrophiles as the valency shell of the carbon in carbene intermediate is short of two electrons. It may be generated with in the reaction by following reactions.

(i) By reaction of chloroform in presence of strong alkali

$$CHCl_3 + R\bar{O} \longrightarrow ROH + :\bar{C}Cl_3$$

$$Cl_3\bar{C}: \longrightarrow Cl_2C: + Cl^-$$

(ii) By the decomposition of diazomethane or ketene in UV light

$$CH_2=\overset{+}{N}=\bar{N} \xrightarrow{UV} :\bar{C}H_2 + N_2$$

$$CH_2=C=O \xrightarrow{UV} :\bar{C}H_2 + CO$$

Geometry: It has been observed that carbenes exist in two different forms:

(i) *Singlet carbene* (methylene) in which carbon is sp^2 hybridised. Each of the three sp^2 hybrid orbitals contain two electrons and one empty p orbital which lie above and below the plane of hybrid orbitals.

(ii) *Triplet carbene* (methylene) where carbon is in sp hybrid state with two electrons in each of the sp hybrid orbitals and one electron each in the two unhybridised p-orbitals. Triplet state is more stable than singlet state.

Carbenes are stabilised when substituted with groups having unshared pairs of electrons, for example, chloro group gives a stabilising effect in dichloro carbene.

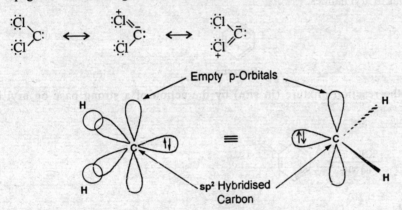

Fig. 4.4: **Structure of Singlet Carbene**

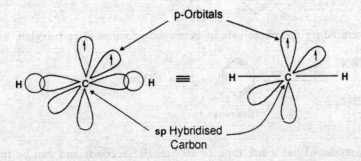

Fig. 4.5: **Structure of Triplet Carbene**

Phenyl groups again can stabilise carbene through resonance as in case of diphenyl carbene a number of resonating structures are possible and a few have been given below:

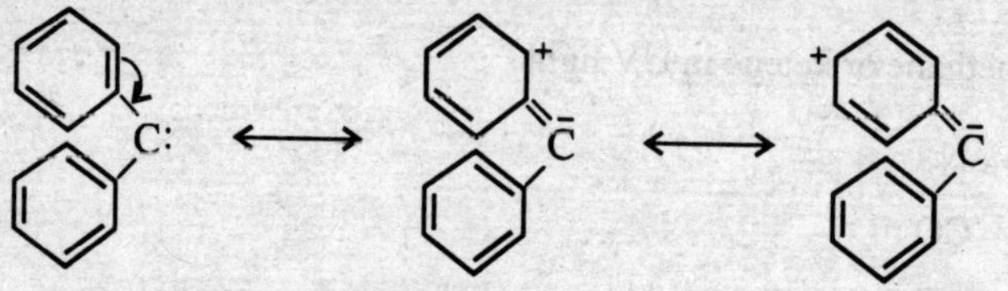

An important reaction of carbene is its addition to a carbon carbon double bond to form a cyclopropane ring. For example, propene on reaction with diazomethane gives methyl cyclopropane.

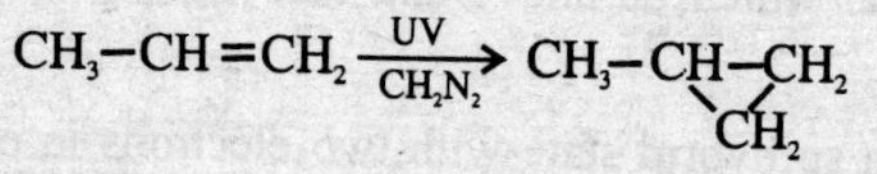

$$CH_3-CH=CH_2 \xrightarrow[CH_2N_2]{UV} CH_3-CH-CH_2$$

Propene   Methylcyclopropane

4.4.5 Benzynes

Benzynes or Arynes or 1,2-dehydrobenzenes are highly reactive reaction intermediates formed during nucleophilic substitutions of aryl halides.

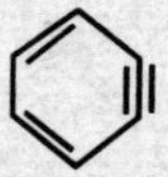

Benzyne

It is *generated* in the reaction mixture (in situ) by the action of a strong base on aryl halide as follows.

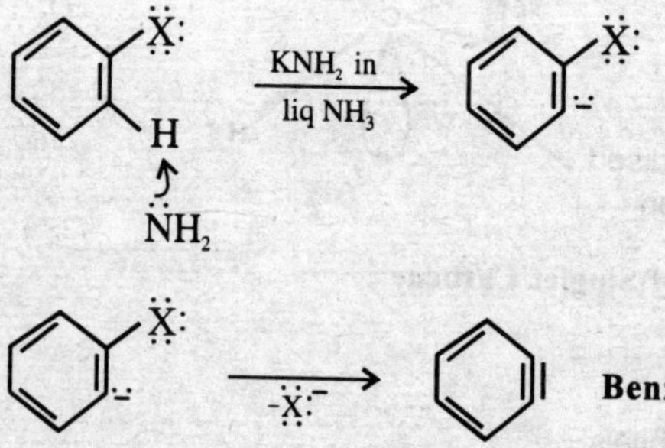

It can also be generated by the photolysis of benzenediazonium-O-carboxylate as follows:

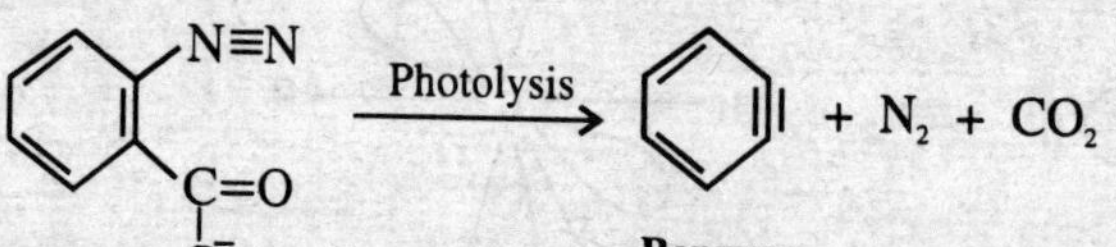

The benzyne thus produced has a life time of 10^{-5} to 10^{-4} seconds and can be trapped in solvents like furan or anthracene. In absence of suitable trapping agent they dimerise as they are highly reactive.

Geometry of benzyne intermediate involves the side ways overlap of sp² hybrid orbitals of two carbon atoms, one originally holding the halogen and the other hydrogen to form an additional π bond which is out of the plane of the aromatic π electron cloud. The new orbital lies along the side of the ring and has minor interactions with the π electron cloud lying above and below the ring.

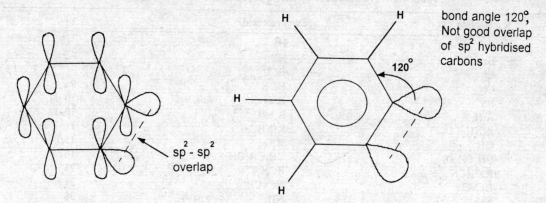

Fig. 4.6: Structure of Benzyne

The structure of the intermediate accounts for its instability and high reactivity.

Reactions of Benzyne: Benzyne intermediate undergoes a variety of reactions like addition, nucleophilic substitution, dimerisation etc. Nucleophilic substitution to benzyne intermediate is shown below:

In the absence of an appropriate trapping reagent it dimerises to give diphenylene.

Diphenylene

4.5 NUCLEOPHILES AND ELECTROPHILES

It has been mentioned earlier that the substrate and reagent reacts together to give the products in a chemical reaction. To understand a chemical reaction we must have knowledge of attacking reagents which can be classified on the basis of their electronic structure.

Nucleophiles: These are electron rich reagents having at least one nonbonded pair of electron and attack an electron deficient site in the substrate. They are nucleus loving and tend to donate an electron pair. Nucleophiles are negatively charged or neutral molecules having at least one lone pair. For example, Nucleophiles with negative charge

Cl^-, Br^-, I^-, CN^-, OH^-, RO^-, HS^-

Neutral molecules $H_2\ddot{O}$, $\ddot{N}H_3$, $R\ddot{N}H_2$, $R\ddot{O}H$

Nucleophiles can donate electron pair i.e. according to Lewis concept they are bases and the strength of nucleophiles can be estimated in terms of their basicity. The basicity of two nucleophiles can be compared in terms of the pK_a values of their respective conjugate acids. A higher value of pK_a indicates a weak conjugate acid or a strong base i.e. a strong nucleophile. Table 4.2 shows the pK_a values of conjugate acids of some common nucleophiles.

Table 4.2: Nucleophiles—Their Conjugate Acids with pK$_a$ Values

Nucleophile	Conjugate acid	pK$_a$
I^-	HI	−10
Br^-	HBr	−9
Cl^-	HCl	−7
CN^-	HCN	9.2
RS^-	RSH	10–11
R_3N	R_3NH^+	10–11
R_2NH	$R_2NH_2^+$	11
OH^-	H_2O	15.7
$CH_3CH_2O^-$	CH_3CH_2OH	16
$R\text{-}CHCN$	RCH_2CN	25
$HC{\equiv}C^-$	$HC{\equiv}CH$	25
$\bar{N}H_2$	NH_3	38
$CH_2{=}CH^-$	$CH_2{=}CH_2$	44
$\bar{C}H_3$	CH_4	48

The basicity of a molecule is influenced by the following factors:

Inductive effect: +I Groups increase basicity of amines

$$NH_3 < CH_3NH_2 < (CH_3)_2NH$$

Resonance: This also affects the base strength as it is related to the availability of lone pair of electrons. For example, cyclohexylamine is a stronger base than aniline as the availability of lone pair of electron on nitrogen in aniline is reduced due to the delocalisation of electrons through resonance.

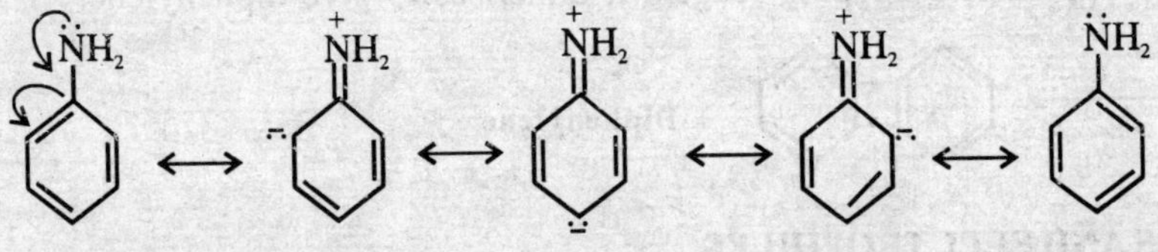

Electrophiles: Electrophiles are electron deficient reagents which can accept electron pair. These reagents attack the electron rich site in the molecule during chemical reactions. Since they can accept electron pair they are acids according to Lewis concept. They are positively charged ions or neutral molecules as listed in Table 4.3.

Table 4.3:   List of Electrophiles

Positively charged	Neutral
H^+	$I{-}Cl$, CCl_2
M^+, MX^+	
Br^+, Cl^+	$R{-}\overset{O}{\overset{\|}{C}}{-}Cl$
NO_2^+, NO^+, NH_4^+	$R{-}\overset{O}{\overset{\|}{C}}{-}O{-}\overset{O}{\overset{\|}{C}}{-}R$
H_3O^+	CO_2, SO_3
R_3C^+, $Ar{-}N^+{\equiv}N$, $R{-}C^+{=}O$	BF_3, $ZnCl_2$, $AlCl_3$, $FeCl_3$

4.6 TYPES OF ORGANIC REACTIONS

The organic compounds show a wide variety of reactions and they can be catagorised as follows.

Substitution Reaction: When an atom or a group replaces an atom or a group from the substrate, the reaction is called substitution reaction.

$$CH_4 + Br_2 \longrightarrow CH_3Br + HBr \qquad \text{...... (I)}$$

$$CH_3Br + OH^- \longrightarrow CH_3OH + Br^- \qquad \text{...... (II)}$$

In reaction number I, Bromine atom has replaced the hydrogen from methane and in reaction number II, hydroxide ion has replaced Bromide ion from methyl bromide. (For details see Section 7.10.)

Addition Reaction: Addition Reaction involves addition of atoms or molecules in unsaturated compounds without the elimination of any atom or group to give a saturated compound. For example,

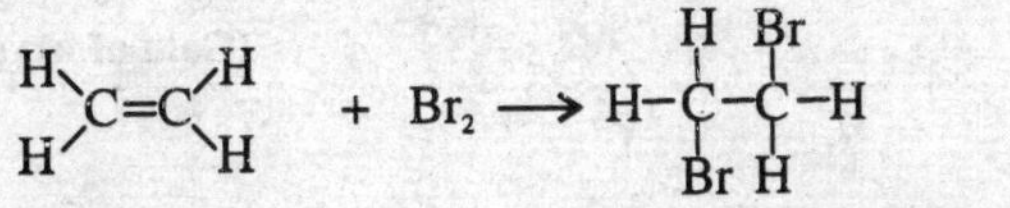

Ethylene   **1,2-Dibromoethane**

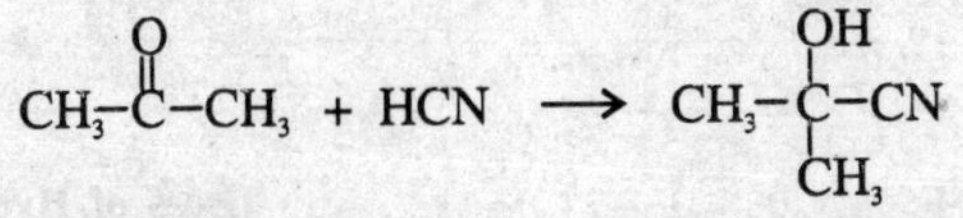

Acetone   **Acetonecyanohydrin**

In the first reaction, two bromine atoms have been added to carbon carbon double bond and in the second reaction addition has taken place at carbon oxygen double bond. (For details see Section 9.9.)

Elimination Reactions: Elimination reactions involve the removal of atoms or groups from two adjacent atoms of the substrate molecule and introduce unsaturation in an originally saturated compound. For example, hydrogen bromide is removed from adjacent carbon atoms of alkyl bromide in presence of alcoholic potassium hydroxide and a double bond is introduced in the molecule. (For details see Section 9.6.)

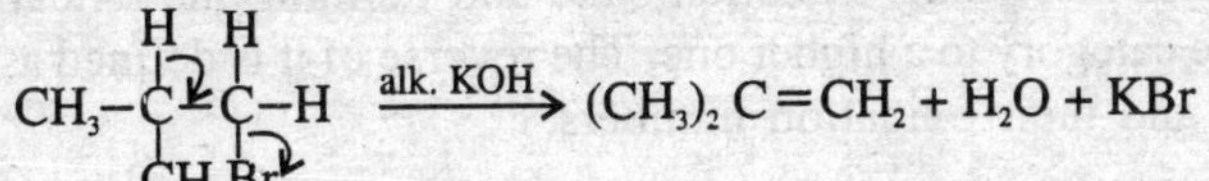

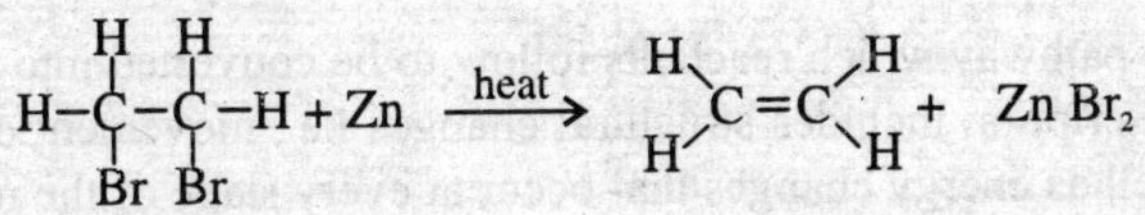

Rearrangement: During the reaction an atom or a group migrates from one position to another within the molecule. Such type of reactions are called rearrangements. For example,

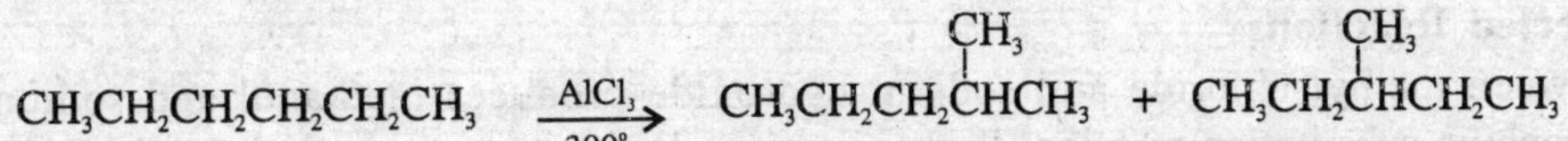

n-Hexane   **2-Methylpentane**   **3-Methylpentane**

n-Butyl bromide   **2-Bromobutane**

Polymerisation: Polymerisation is the union of two or more molecules of a compound to give a compound of same empirical formula but of higher molecular weight. For example, ethylene polymerises to polyethylene in presence of catalyst.

$$nCH_2 = CH_2 \longrightarrow (CH_2 - CH_2)_n$$
Ethylene   **Polyethylene**

Oxidation and Reduction Reactions: In *organic reactions* it is difficult to define oxidation or reduction. In *inorganic chemistry,* oxidation is either termed as loss of electron or increase in oxidation number and reduction as gain of electron or decrease in oxidation number. In organic molecules, however, to determine whether the molecule has undergone oxidation or reduction is found out by *Thumps rule* which states:

(i) If a molecule gains electronegative element like nitrogen, halogen or oxygen or loses electropositive element like hydrogen, it is oxidised.

$$CH_3CH_2OH \xrightarrow[\text{Agent}]{\text{Oxidising}} CH_3COOH$$ **(Gain of oxygen)**

(Gain of Br)

$$CH_3-\overset{\overset{\displaystyle OH}{|}}{C}H-CH_3 \xrightarrow[\text{Agent}]{\text{Oxidising}} CH_3-\overset{\overset{\displaystyle O}{||}}{C}-CH_3$$ **(Loss of Hydrogen)**

(ii) If a molecule loses an electronegative element like nitrogen, halogen or oxygen or gains electropositive element like hydrogen, it is reduced.

$$CH_2=CH_2 \xrightarrow[\text{Agent}]{\text{Reducing}} CH_3-CH_3$$ **(Gain of Hydrogen)**

$$CH_3COOH \xrightarrow[\text{Agent}]{\text{Reducing}} CH_3CH_2OH$$ **(Loss of Oxygen, Gain of Hydrogen)**

(iii) *Another way to determine* that the molecule has undergone oxidation or reduction is by arranging various functional groups in the order of increasing oxidation state and defining the oxidation as the change in the functional group from one category to a higher one. The reverse of it is defined as reduction. Table 4.4. Shows functional groups and their oxidation numbers.

4.7   REACTION MECHANISM

The detailed step by step description of the pathway which reactants follow to be converted into products is known as the *reaction mechanism.* The description includes structural changes i.e., movement of electrons leading to bond breaking and making as well as energy changes that occur at every stage of the reaction. So far we have discussed the structural factors that control a chemical reaction. Now we shall discuss some factors related to reaction energetics.

4.7.1   Concerted Reactions

Reaction between methyl chloride and sodium hydroxide produces methanol and sodium chloride through nucleophilic substitution reaction. In the reaction electron pair from nucleophile (hydroxide ion) forms a new bond between oxygen and carbon whereas chlorine atom departs with the electron pair as chloride ion.

Table 4.4: Oxidation Numbers of Various Functional Groups

Approximate Oxidation number	−4	−2	0	+2	+4
		→ OXIDATION →			
F U N C T I O N A L G R O U P	RX	$>C=C<$	$R-C\equiv C-R$	RCO_2H	CO_2
		$R-OH$	$R-\overset{\overset{O}{\parallel}}{C}-R$	$R-\overset{\overset{O}{\parallel}}{C}-NH_2$	CCl_4
		$R-Cl$	$\overset{H}{\underset{H}{}}CCl_2$		
		$R-NH_2$	$Cl-\overset{R}{\underset{H}{C}}-\overset{R}{\underset{H}{C}}-Cl$	$-CCl_3$	
			$R-\overset{OH}{\underset{H}{C}}-\overset{OH}{\underset{H}{C}}-R$		
		← REDUCTION ←			

$$HO^- + CH_3-Cl \longrightarrow CH_3OH + Cl^-$$

The synchronised reaction as shown above where bond making and bond breaking takes place at the same time is known as *concerted reaction*.

4.7.2 The Transition State

The reaction of chloromethane and sodium hydroxide is a rapid reaction under normal condition but if a dilute solution of chloromethane is made to react in aqueous base, the reaction takes weeks to reach equilibrium state or in other words we can say a rapid reaction can not be guaranteed as it is dependent upon a number of factors such as:

(i) The reactants must have appropriate energy to collide.

(ii) The reactants must come together in favorable orientation

(iii) The system must have sufficient energy to break the bond undergoing change.

The configuration of reactants in which all the requirements for an effective collision are met is known as *Transition state or activated complex,* a hypothetical description of the reactant molecules at the point of highest energy along the path of reaction. Transition state can not be isolated or detected whereas the reaction intermediates can be detected and in some cases even isolation is also possible.

4.7.3 Activation Energy

The energy-structure relationship of a reaction can be represented by an energy profile diagram. Energy is plotted along the vertical axis and reaction co-ordinates are plotted horizontally. Reaction co-ordinates represents the change in the geometry of reactants as reaction proceeds. Fig. 4.7. shows the energy profile diagram for the formation of methanol from chloromethane.

According to energy profile diagram, energy must be provided to the reactants in order to attain the transition state. This energy, the difference between average free energy of the reactants and the transition state is known as free *Energy of Activation (Ea).* It is this energy of activation which governs the rate of a reaction. Organic reactions have energy of activation of the order of 10-50 kcal/mol.

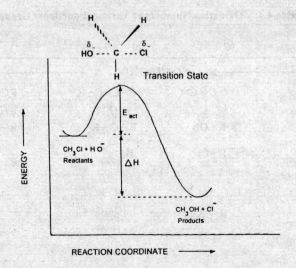

Fig. 4.7:   Energy Profile Diagram for Concerted Reaction

The standard free energy of the reaction which is also called *Heat of reaction* (ΔH) is the difference in energy between reactants and products.

4.7.4   Multistep Reactions

The reaction of chloromethane with sodium hydroxide is a one step, concerted reaction having one energy maximum (transition state) as shown in energy profile diagram (Fig. 4.7), but some reactions involves more than one step. For example, the reaction of sodium hydroxide with t-butylchloride is a two step reaction. The first step is the heterolytic cleavage of carbon chlorine bond to give a high energy, unstable tert-butyl carbocation intermediate which reacts with water in the rapid second step to produce protonated tert-butanol.

$$(CH_3)_3C\text{–}Cl \rightleftharpoons (CH_3)_3\overset{+}{C} + Cl^-$$

$$(CH_3)_3\overset{+}{C} + \overset{\cdot\cdot}{O}H_2 \rightleftharpoons (CH_3)_3C\text{–}\overset{+}{O}H_2$$

The energy profile diagram for such reaction shows (Fig. 4.8) two transition states associated with the formation of carbocation intermediate and the product alcohol. The unstable carbocation intermediate corresponds to the energy minimum between two transition states.

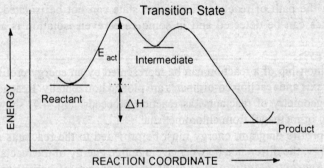

Fig. 4.8:   Energy Profile Diagram for a Two-Step Reaction

4.7.5   The Rate-Controlling Step
The rate of a reaction is dependent upon various factors as discussed earlier in this chapter. On an energy profile diagram, the rate controlling step is the one which involves the highest energy transition state. In a multistep reaction as shown in Fig. 4.8, the first step is the rate controlling step which involves the heterolysis of carbon chlorine bond as the first transition state is of higher energy.

4.7.6   The Reaction Kinetics
To understand reaction kinetics, the experiments are done by controlling the temperature and concentration of reactants. As the reaction proceeds, the reactant concentration decreases (which can be measured by various analytical methods) or product concentration increases in relation to the time passed. These concentrations are incorporated into a mathematical expression known as rate equation as follows

$$\text{Rate} = k[A][B]$$

[A] and [B] represents concentrations of reactants A and B. The reaction becomes a *first order reaction* if the rate varies in relation to the concentration of only one reactant. The reaction becomes *second order,* when the rate is proportional to the concentration of two reactants or the square of one reactant. The commonly encountered kinetic situations have been shown in Table 4.5.

Table 4.5:   Kinetic Observations and Common Mechanistic Deductions

Kinetic Observation		Mechanistic Deduction	
Rate Expression	Order	Molecularity	Probable Reaction
k[A]	First	1	A → T.S.
k[A][B]	Second	2	A+B → T.S.
k[A][A] or k[A]^2	Second	2	2A → T.S.

T.S. ≡ transition state

With the above observations we can infer that in first order reactions only one molecule goes to transition state where as in second order reaction two molecules go to the transition states. *The rate of a chemical reaction can be altered easily by changing the temperature conditions.* An increase in temperature increases the kinetic energy of the molecules thus more molecules have sufficient energy to reach the transition state as compared to the molecules at a lower temperature as shown in Fig. 4.9. (A catalyst can also enhance the rate of a chemical reaction by decreasing its energy of activation.)

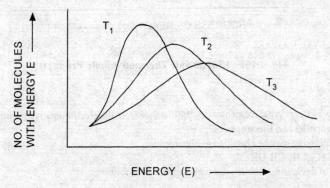

Fig. 4.9:   Relationship of Average Energy to the Temperature

The temperatures are increasing in the order $T_1 < T_2 < T_3$ and the area under each curve represents the total number of molecules with sufficient energy to react, thus indicating clearly that rate of a chemical reaction can be enhanced by the increase of temperature.

4.7.7   Kinetic Verses Thermodynamic Product

Let us now consider following reactions

$$A \rightarrow B \qquad \qquad ......(i)$$

$$A \rightarrow C \qquad \qquad ......(ii)$$

Energy of activation for reaction number (i) is lower than for reaction number (ii) so product B is formed faster than product C or in other words, B is termed as *Kinetic Product.*

Energy of activation for reaction (ii) is higher but product C is more stable and is associated with higher heat of reaction. Thus if the reaction is allowed to proceed for a long time product B reverts back to reactant A and than A is converted to more stable product C or A, B and C will remain in equilibrium and since product C is more stable it is predominantly termed as *thermodynamic product.*

We can conclude from the above observations that low temperature and short reaction time favors the formation of kinetic product whereas thermodynamic product is favoured by higher temperature and longer reaction time.

When more stable product has lower energy of activation, the kinetic and thermodynamic products are similar.

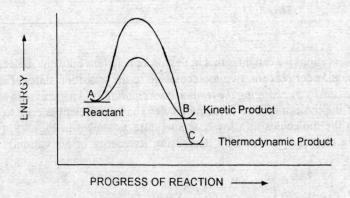

Fig. 4.10:   Kinetic and Thermodynamic Products

4.9   PROBLEMS

1. What is Inductive effect? What happens when the distance from the electronegative group or atom increases and why?
2. What are Nucleophiles and Electrophiles?
3. Which of the following may be classified as electrophile
   NH_3, $AlCl_3$, $(CH_3)_2CH$, CH_3OH
4. Why the order of decreasing stability for carbocations is as follows:

   $(CH_3)_3\overset{+}{C} > (CH_3)_2\overset{+}{CH} > \overset{+}{CH}_3$

5. What are concerted reactions?
6. Classify the following reactions:
   (a) $CH_3CH=CH_2 + Cl_2 \rightarrow CH_2ClCH=CH_2$ (c) $CH_3CH_2OH + HI \rightarrow CH_3CH_2I + H_2O$
   (b) $CH_2=CH_2 + Cl_2 \rightarrow CH_2Cl-CH_2Cl$ (d) $CH_3CHCl \rightarrow CH_3CH=CH_2$
7. Which of the following may be classified as Nucleophile:
   HO^-, Br^-, BF_3, CN^-, H_2O, C_2H_5OH
8. Draw an enthalpy diagram for a one step exothermic reaction.
9. Why tert butyl carbocation is more stable than sec. butyl carbocation?
10. Which is a better nucleophile among phenoxide and ethoxide?
11. Write note on:
    (a) Energy of activation (b) Transition State (c) Benzyne (d) Hyperconjugation (e) Resonance

5

Stereochemistry

5.1 INTRODUCTION

Stereochemistry is the chemistry of compounds in three dimensions, the term has been derived from Greek word 'stereos' meaning 'solid'. Stereochemistry is an important branch of chemistry as it is not only concerned with the geometry of molecules but it is also helpful in understanding the pathway of a chemical reaction and chemical equilibrium. In this chapter, we shall confine our discussion mainly to the stereoisomerism beginning with the concept of isomerism in general.

The phenomenon of existence of two or more compounds possessing same molecular formula but different physical and chemical properties is known as isomerism and these compounds are individually referred to as isomers. The following flow chart clearly indicates different types of isomerism.

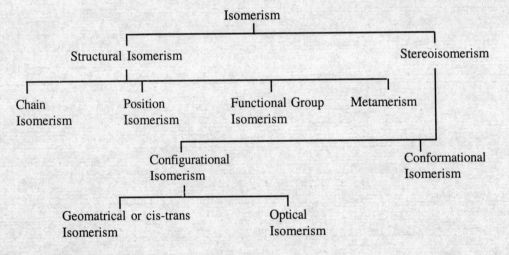

5.2 STRUCTURAL ISOMERISM

Compounds having structural differences are classified as structural isomers. This type of isomerism arises due to structural differences and can further be divided as follows:

1. *Chain Isomerism:* when two or more compounds have similar molecular formula but different carbon skeleton, they are referred to as chain isomers and the phenomenon is termed as chain isomerism.

For example, molecular formula C_5H_{12} can represent three compounds viz. n-pentane, 2-methylbutane and 2,2-dimethylpropane

$$CH_3CH_2CH_2CH_2CH_3$$

Pentane
(n-Pentane)

$$CH_3 \atop \overset{|}{CH_3CHCH_2CH_3}$$

2-Methylbutane
(Isopentane)

$$CH_3 \atop CH_3-\overset{|}{\underset{|}{C}}-CH_3 \atop CH_3$$

2,2-Dimethylpropane
(Neopentane)

This type of isomerism is shown by hydrocarbons as well as other classes of compounds having a long hydrocarbon chain, viz, C_4H_9OH can be represented as n-butanol and isobutanol.

$$CH_3CH_2CH_2CH_2OH$$

1-Butanol
(n-Butyl alcohol)

$$CH_3 \atop \overset{|}{CH_3CHCH_2OH}$$

2-Methyl-1-propanol
(Isobutyl alcohol)

2. *Position Isomerism*: When two or more compounds differ only in the position of substituent atom or group on the carbon skeleton, they are called *position isomers* and this phenomenon is termed as position isomerism.

Position Isomerism is shown by all the classes of organic compounds except for alkanes (which show only chain isomerism). Molecular formula C_3H_8O represents 1-propanol and 2-propanol.

$$CH_3CH_2CH_2OH$$

1-Propanol
(n-Propanol)

$$OH \atop CH_3-\overset{|}{CH}-CH_3$$

2-Propanol
(Isopropanol)

C_5H_{10} represents 1-pentene and 2-pentene

$$CH_3CH_2CH_2CH=CH_2$$
1-Pentene

$$CH_3CH_2CH=CHCH_3$$
2-Pentene

3. *Functional Group Isomerism:* Two or more compounds having similar molecular formula but different functional groups are called *functional isomers* and this phenomenon is termed as *functional group isomerism.* for example molecular formula C_3H_6O represents two compounds, acetone having –C=O functional group and propanal having a –CH=O functional group.

$$O \atop \overset{||}{CH_3CCH_3}$$

Acetone

$$H \atop \overset{|}{CH_3CH_2C=O}$$

Propanal

Molecular formula C_3H_8O represents methyl ethyl ether (function group –O–) and propanol (functional group –OH)

$$C_2H_5OCH_3$$
Methoxyethane
(Ethyl methyl ether)

$$CH_3CH_2CH_2OH$$
Propanol
(n-Propyl alcohol)

4. *Metamerism:* Metamerism arises due to unequal distribution of alkyl groups on either side of functional group in the molecule. Metamers are the members of same classes. This isomerism is shown by classes having central functional groups flanking between two alkyl groups such as ethers, ketones and amines. For example, $C_4 H_{10} O$ represents

$CH_3OC_3H_7$
Methoxypropane

$C_2H_5OC_2H_5$
Ethoxyethane

and, $C_5H_{10}O$ represents

$$\underset{\text{3-Pentanone}}{C_2H_5\overset{\overset{\displaystyle O}{\|}}{C}\,C_2H_5}$$

$$\underset{\text{2-Pentanone}}{CH_3\overset{\overset{\displaystyle O}{\|}}{C}C_3H_7}$$

and, C_2H_7N represents

$CH_3CH_2NH_2$
Ethanamine

CH_3NHCH_3
N-Methylmethanamine

5.3   REPRESENTATION OF MOLECULES IN SPACE

1. *Molecular Models:* To understand chemical reactions in proper perspective, molecular models are used. They help in visualising the interaction between atoms within a molecule. Molecular models are of three types.

(i)   Frame work model

(ii) Space-filling model

(iii) Ball and stick model

Frame work models are not very expensive and clearly show the position of atoms and bonds.

Space filling models do not show the skeleton but give some idea about steric strain in the molecule.

Ball and stick models can represent a molecule better as they clearly show the attachment of bonds and atoms as well as give some idea about non-bonded interactions due to the size of substituted atoms or groups in a molecule.

Molecular models, although, are helpful in visualising molecules, do not furnish all the information about it like molecular strains (Fig. 5.1).

2. *Representation of Molecules in Two Dimensions:* It is necessary to represent three dimensional molecules in two dimensions because our most common method of communication is either paper or black board or a TV screen and all of them are two dimensional surfaces. Following are illustrated various methods of representation of three dimensional molecules in two dimensions.

(i)   *Wedge Formula:* It is the most common method of representation of molecule in which a solid wedge represents the bond that comes forward or towards the reader, a broken wedge represents the bond that goes backward or away from the reader and the solid line represents the bond on the plane of paper (Fig. 5.2).

(ii) *Sawhorse Formula:* In this representation the molecule is viewed from slightly above and side of one carbon bond. The C-C bond is on the plane of paper and the remaining six bonds attached to each carbon atoms are represented as straight lines. Fig. 5.3 represents Sawhorse formula of ethane.

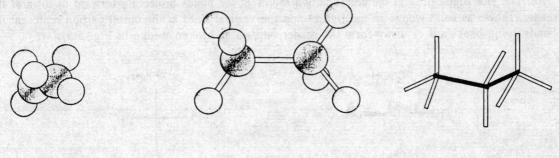

Space Filling Model   Ball and Stick Model   Frame Work Model

Fig. 5.1: Molecular Models

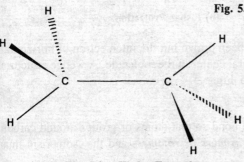

Fig. 5.2: Wedge Formula

(iii) *Newman Projection Formula:* Newman Projections are used to draw conformational isomers. The front carbon represents its three covalent bonds in the shape of alphabet 'Y'. The back carbon is represented by a circle with three bonds again in the form of an incomplete Y. Here carbon carbon bond lies along the line of vision. The Fig. 5.4 shows the Newman projection formula of ethane.

(iv) *Fischer Projection Formula:* Fischer projection is a standard method of projection of compounds having one or more chiral centers. It can be drawn easily and quickly and in addition it gives the comparison of stereoisomers which may otherwise be complex. We shall discuss the illustration of the simplest optically active compound, the lactic acid in the subsequent lines.

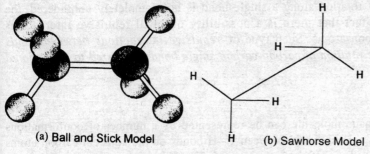

(a) Ball and Stick Model   (b) Sawhorse Model

Fig. 5.3: Sawhorse Formula

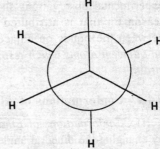

Fig 5.4: Newman Projection Formula

(a) The molecule is visualized with its main carbon chain vertical and carbon number 1 at the top. Thus in our molecule carbon of COOH group takes the top position.

(b) The second step is to mentally flatten the structure at each chiral carbon into a two dimensional projection.

(c) The Horizontal lines at the chiral carbon represent the bonds projecting forward or towards the reader (shown as solid wedge in the figure) and the vertical lines at the chiral carbon represent the bonds going backward or away form the reader (shown as broken wedge in Fig. 5.5).

(a)   (b)

Fig. 5.5:   Lactic Acid (a) Wedge Formula   (b) Fisher Projection

It is important to note here that the chiral carbon has not been drawn but the intersection of horizontal and vertical lines represent the position of asymmetric or chiral center in the molecule. A wedge structure can be easily translated into fisher projection as shown in Fig. 5.5.

5.4   CONFORMATIONAL ISOMERISM

Isomers arising due to rotation about a carbon-carbon single bond so that atoms or groups around carbon get different spatial arrangement, are called *conformational isomers or rotamers* and the isomerism thus arising is conformational isomerism.

Rotation about a single bond is relatively rapid as the sigma bond is formed by the overlap of sp^3 hybrid orbitals of two sp^3 hybridised carbon atoms. As we have discussed earlier (Section 2.2) the sigma bond arising due to overlap of sp^3 hybrid orbitals has cylindrical symmetry of electron density around carbon-carbon bond axis, there is no possibility of hindered rotation. But various experimental observations show that rotation along a single bond is not completely unhindered, the reason for that is attributed to the fact that there is a possibility of small repulsive interactions between the bonds on adjacent carbon atoms. Such type of *repulsive interactions between bonds on adjacent atoms which restricts the rotation of carbon-carbon single bond is referred to as torsional strain.*

5.4.1   Conformations of Alkanes

(i)   *Conformations of Ethane:* Ethane molecule can be represented by a number of conformations by changing the dihedral angle (The angle observed between C–H bonds on the adjacent carbon atoms as we look along C–C bond) which have been shown in the Fig. 5.6. The figure shows energy relationship for ethane conformations.

When dihedral angle is 0° or 120° the hydrogen on both carbon atoms are parallel and each C–H bond is at a minimum distance form the C–H bond of adjacent carbon atom as a result the electrons forming these bonds repel each other making this conformation unstable and it is represented at an energy maximum in the above figure and is known as *eclipsed conformation.*

The other conformation called staggered conformation is possible when dihedral angle is 60°. It is a situation when all the C–H bonds on adjacent carbon atoms are at a maximum distance from each other making it stable by 2.9 kcal/mol (12.13 kJ/mol) compared to eclipsed conformation of ethane molecule. It takes its position at an energy minimum as shown in Fig. 5.6.

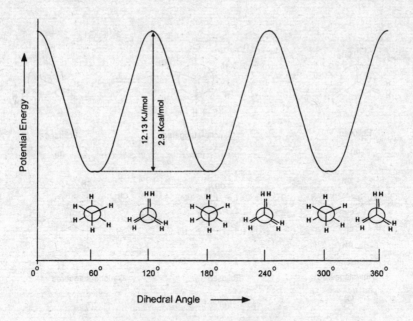

Fig. 5.6: Conformations of Ethane

The experimental observations now clearly indicate that rotation around carbon carbon single bond of ethane is not completely free and the answer to this restricted rotation is *torsional strain* which arises due to repulsive interaction between bonds (or electron pairs involved in bond formation) on adjacent atoms. The magnitude of energy associated with torsional strain in simple molecules is of the order of 5.0 kcal/mol (21.0kJ/mol). At room temperature thermal energy can easily promote processes with energy barrier of 15-20 kcal/mol (60-80kJ/mol) or we can say thermal energy at room temperature is sufficient enough to overcome torsional strain associated with eclipsed conformation of ethane making the interconversion of two isomers (staggered and eclipsed) a rapid process (about 10^{11} times per second). It is also observed that, for most of the time ethane molecule exists in staggered conformation passing only transiently through its eclipsed conformation.

(ii) *Conformations of Butane:* Rotation along the central carbon-carbon bond of butane gives rise to various conformational isomers associated with torsional strain. As one side of the molecule rotates through 360° relative to the other side, three eclipsed and three staggered rotamers are encountered as shown in Fig. 5.7. When dihedral angle along C_2 and C_3 of butane is zero, the conformation is called eclipsed conformation (Fig. 5.7(a)) in which methyl groups on both the carbons are at a minimum possible distance making the conformation energetically least favorable and this conformation is considered as transitional orientation between staggered forms.

Rotation along C_2–C_3 bond by 60° gives gauche or skew conformation (Fig. 5.7(b)) which is at a lower energy than the preceding eclipsed form. Another 60° rotation gives eclipsed conformations (Fig. 5.7(c)) where CH_3 and H are eclipsing whereas in the earlier case CH_3 and CH_3 were eclipsing so that it takes a lower energy position in energy relationship diagram than the rotamer shown in Fig. 5.7(a).

Antistaggered conformation (Fig. 5.7(d)) is obtained by further rotation of 60°. This conformation is associated with lowest energy (or is most stable) since the bulky methyl groups on central carbon atoms are farthest apart from each other. Another eclipsed and gauche staggered conformations (Fig. 5.7(e) and (f)) are obtained by 60° rotation along C_2–C_3 bond of butane.

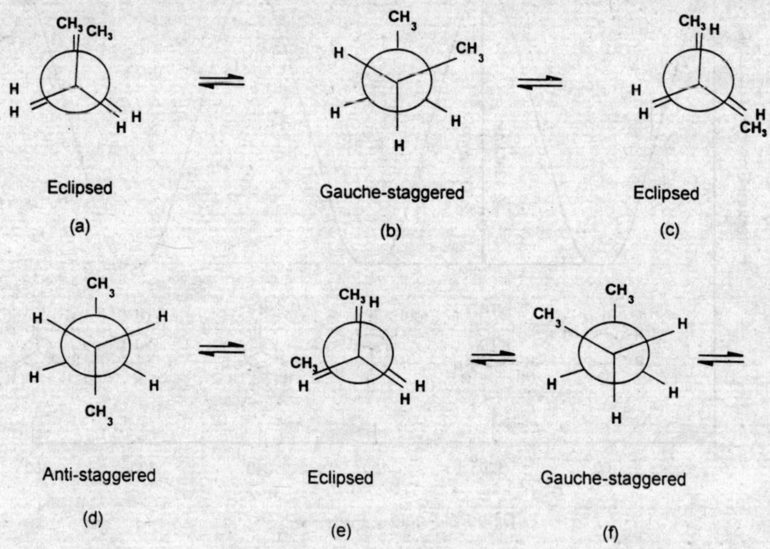

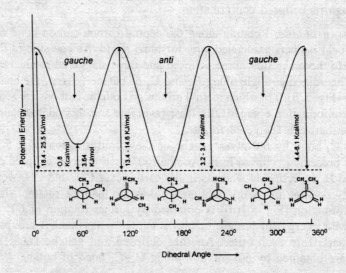

Fig. 5.7:   Conformations of Butane

In the following Fig. 5.8 is given the variation in potential energy for various conformations of butane.

As has been mentioned earlier, the energy barrier about most single bonds is so small that essentially free rotation occurs at room temperature but difference in average rotamer population occurs and it has been found that at room temperature butane is a mixture of 70% 'anti' and 15% each of the 'gauche' conformations. The interconversion of these rotational isomers is so rapid that they can not be separated at room temperatures whereas at low temperature (43k) separation can be accomplished as the interconversion become slow.

Fig. 5.8:   Energy Diagram for Conformations of Butane

5.4.2 Conformations of Cyclic Compounds

In nineteenth century cycloalkanes were supposed to be planer molecules and the internal bond angles for cycloalkanes, according to the German chemist Baeyer, were the same as those of corresponding regular polygon i.e., cyclopentane corresponds to regular pentagon having an internal angle of 108° as shown in Fig. 5.9.

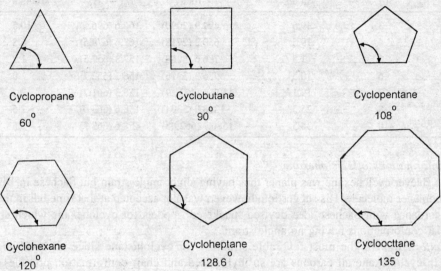

Fig. 5.9: Angle Strain in Cycloalkane Rings

In the year 1885, Baeyer proposed a theory of angle strain for cycloalkanes in which he used the difference of tetrahedral angle (109.5°) and internal angle of corresponding polygon as a measure of molecular stability. Taking this in consideration we can state very clearly that the order of stability for the following cycloalkane rings will be as follows.

cyclopropane < cyclobutane < cyclopentane

Cyclopentane, according to Baeyer's prediction, is the most stable molecule whereas cyclohexane and other higher cycloalkanes are less stable as the angle of larger polygons deviate more from the regular tetrahedral angle as given in Table 5.1.

Baeyer's Theory is not consistent with the experimental measurements as small cycloalkanes do show some angle strain but large cycloalkane rings are associated with only small strain.

The best method to measure angle strain related to cycloalkane rings is by comparison of their *Heats of Combustion* (HOC) with that of cyclohexane. It is assumed that heat released on combustion of hydrocarbons to give carbon dioxide and water is a measure of the bond energies and energy associated with molecular strain

$$-(CH_2)_n- + (3/2)\, nO_2 \longrightarrow nCO_2 + nH_2O$$

Cyclohexane has been taken as standard strain free molecule. HOC for cyclohexane being 936.9 kcal/mol (3916 kJ/mol) or 156.1 kcal/mol (652.7 kJ/mol) per-CH_2-Group is used to calculate the strain in other cycloalkane rings as shown in the Table 5.1

The conclusion we draw from the data given in the Table 5.1 is that small rings i.e., C_3 and C_4 have large angle strain, medium sized rings (C_5 to C_{12}) show moderate angle strain and large rings show very small strain.

Table 5.1: Heats of Combustion and Relative Strain Energies for Cycloalkanes

Cycloalkane	Ring Size	Baeyer Angle Strain	Heats of Combustion kcal/mol (kJ/mol)		Strain per $-CH_2-^*$	
			Total	Per $-CH_2-$	kcal/mol	(kJ/mol)
Cyclopropane	3	49.5	499.9 (2090)	166.6 (696.5)	10.5	(43.8)
Cyclobutane	4	19.5	650.2 (2718)	162.6 (679.5)	6.5	(26.8)
Cyclopentane	5	1.5	786.6 (3288)	157.3 (657.5)	1.2	(4.9)
Cyclohexane	6	10.5	936.9 (3916)	156.1 (652.7)	0	(0)
Cycloheptane	7	19.1	1108.1 (4631.9)	158.3 (661.7)	2.2	(9)
Cyclooctane	8	25.5	1268.9 (5304)	158.6 (663.0)	2.5	(10.3)
Cyclodecane	10	34.5	1586.1 (6629.8)	158.6 (662.9)	2.5	(10.2)

5.4.2.1 Conformations of Cyclohexane

According to Baeyer cyclohexane was planer thus having slight angle strain but Sachese in 1895 proposed that it is a nonplaner molecule. His suggestion, however, was not accepted at that time but in the year 1918, Mohr, in accordance with Sachese idea devised 'puckered' models for cyclohexane which are known as chair and boat conformations having no angle strain.

Chair Conformation is the most favorable structure for cyclohexane since all bond angles remain tetrahedral. In cyclohexane all carbons are sp^3 hybridised and chair conformation provides appropriate positions for all the atoms and bonds as it is free from angle strain. It is also free from torsional strain as all the groups are staggered as shown in the Fig. 5.10 through Newman projection. Ball and stick model also gives some idea about the staggering of groups.

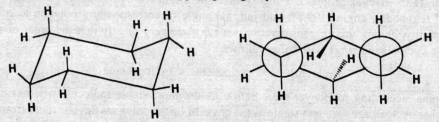

Fig 5.10:   Chair Conformations of Cyclohexane

Boat Conformation of cyclohexane is another nonplaner model having no angle strain but there is significant torsional strain in the molecule due to eclipsing of C–H bonds along C_2–C_3 and C_5–C_6 bonds which is very clear in Newman projection of the isomer (Fig. 5.11).

Fig. 5.11:   Boat Conformation of Cyclohexane

In addition to eclipsing torsional strain, there is nonbonded interaction between two H-atoms across the ring (C_1 and C_4 often called flagpole hydrogens). The flagpole hydrogens are 1.83Å apart, a distance less than the sum of their Van der Waal radii (2.40Å) resulting in repulsion between them and making the conformation unstable.The boat conformation of cyclohexane is about 6.5 kcal/mol (27 kJ/mol) higher in energy than chair conformation at 25°. This is the reason that over 99.9 percent of cyclohexane exists in chair form at room temperature.

Another conformational isomer of cyclohexane is *twist boat or skew boat* (Fig. 5.12) which arises due to the twisting of bond along C_2–C_3 and bond along C_5–C_6 so that flagpole hydrogen move away from each other releasing the non-bonded interaction between them. Torsional strain is also reduced in this conformation making it more stable than boat conformation by about 1.5 kcal/mol (6.0 kJ/mol).

Fig. 5.12:   Twist Boat Conformations of Cyclohexane

Cyclohexane can also be represented in the form of Half chair (Fig. 5.13) which is a transition state conformation associated with angle strain and torsional strain. This form is less stable than chair form by 11.0 kcal/mol (45.0 kJ/mol).

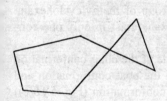

Fig. 5.13:   Conformation of Half Chair

Among the conformations discussed above for cyclohexane the equilibrium lies between chair form and twist boat form in a ratio of 10,000 to 1 at room temperature as the chair form is the most stable conformation. Fig. 5.16 shows various conformation of cyclohexane with respect to their potential energy.

5.4.2.2   Axial and Equatorial Bonds

Cyclohexane has 12 hydrogen atoms which occupy different positions. In chair conformation of cyclohexane, six hydrogen atoms are perpendicular to the average plane of the molecule and six are directed outward from the ring, slightly above or below the molecular plane (Fig. 5.14). Bonds which are perpendicular to the ring (directing towards the axis) are called *axial bonds* (Fig. 5.14a), and those which extend outward from the ring (directing towards equator) are called *equatorial bonds* (Fig. 5.14b).

Fig. 5.14   Axial and Equatorial Bonds

Out of six axial bonds, three direct upward originate from alternate C–atoms. In the same way out of six equatorial bonds, three direct slightly upwards originate from alternate carbon atoms. Hydrogen atoms attached to all these six bond are above molecular plane as shown in Fig. 5.14. Similarly rest of the six bonds (three axial and three equatorial) take positions below molecular plane. The molecular plane in chair conformation is defined as an imaginary plane running through the molecule so that all the carbon atoms are equidistant from that plane.

5.4.2.3 Interconversion of Chair Conformations

Cyclohexane is interconverting between two chair conformations at room temperature as shown in Fig. 5.15. These interconversions are possible due to small rotations along all C–C single bonds. When one chair form is converted into another as a result of ring flip, all the equatorial bonds are converted into axial and all the axial bonds are converted into equatorial. This interconversion is so rapid at room temperature that all hydrogen on cyclohexane molecule are considered equivalent.

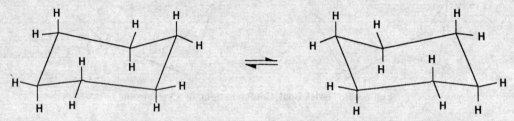

Fig. 5.15: Interconversion of Chair Conformations

In monosubstituted cyclohexane (for example methylcyclohexane) methyl group rapidly interconverts between equatorial and axial positions but it is energetically more favorable in equatorial position. It has been further supported by the fact that at room temperature, methyl group of methylcyclohexane exists in 95% equatorial and 5% axial position or monosubstituted cyclohexane exist in only one conformation.

The inversion of cyclohexane ring takes place by the movement of one side of the chair conformation Fig. 5.16(a) to produce a twist boat Fig. 5.16(b) which is more stable than the boat conformation and less stable than chair conformation. Twist boat than is converted into boat conformation (Fig. 5.16(c)) which on further twisting of bonds is converted in to another chair form Fig. 5.16(e) via a twist boat form Fig. 5.16(d).

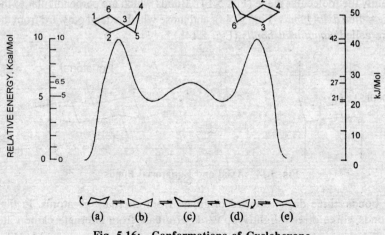

Fig. 5.16:   Conformations of Cyclohexane

A transition state conformation or half chair conformation of cyclohexane with angle strain and torsional strain lies between chair and twist boat conformations.

It is important to note that ring flipping requires 11.0 kcal/mol (45.0 kJ/mol) of energy and even at room temperature it is very fast.

5.4.2.4 Conformations of Monosubstituted Cyclohexane Derivatives

In methylcyclohexane (monosubstituted), the methyl group can occupy either axial or equatorial position as shown below in Fig. 5.17. Ring flipping interconverts the two conformations with the methyl group changing from axial to equatorial and vice-versa. The two conformational isomers are not enantiomers but they are diastereomers (Section 5.6.3) thus, have different stabilities and energies. When methyl group occupies axial position (Fig. 5.18) there is nonbonded interaction between CH_3 group and H-atoms at position 1 and 3 commonly known as *1,3-diaxial interaction* as the distance between H of CH_3 and H at C_3 & C_5 of cyclohexane is less than the sum of Van der Waal's radii for two hydrogens. This repulsion destabilises axial conformation of methyl cyclohexane whereas no such interaction is possible with equatorial conformation of methyl cyclohexane making it a more stable conformation.

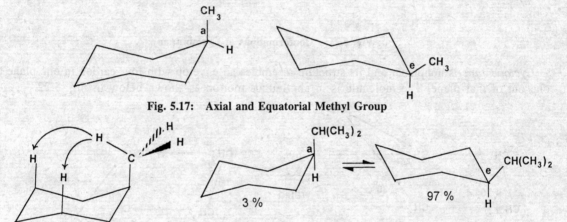

Fig. 5.17: Axial and Equatorial Methyl Group

Fig. 5.18: 1,3-Diaxial Interactions   **Fig. 5.19: Percentage of Equatorial and Axial Forms**

With the variation in the size of the substituent, the equilibrium between equatorial and axial form may shift as shown in Fig. 5.19.

5.4.2.5 Other Cycloalkanes

Cyclopropane: Cyclopropane is triangular with flat structure and bond angle of 60°. It is a significant deviation from the normal tetrahedral angle of 109.5°. The ring is associated with severe angle strain due to non linear overlap of sp^3 hybrid orbitals of carbon. The highest electron density of C–C bonds lie principally outside the triangular internuclear line forming *bent bonds* (Fig. 5.20).

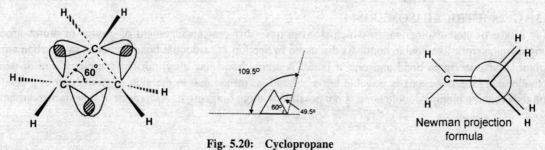

Newman projection formula

Fig. 5.20: Cyclopropane

In addition to angle strain, cyclopropane ring is associated with torsional strain which arise due to eclipsing of C–H bonds as shown in the Newman projection formula.

Cyclobutane has less angle strain than cyclopropane. It assumes a slightly puckered structure to reduce torsional strain which otherwise arise in planer (square) structure due to the eclipsing of C–H bonds. Although there is a slight increase in angle strain as the C–C–C angle changes from 90° to 88.5° increasing the difference from normal tetrahedral angle, the puckered structure of cyclobutane is not rigid and it flips over rapidly from one form to another (Fig. 5.21).

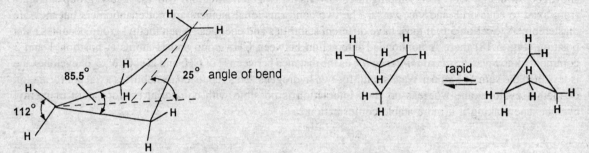

Fig. 5.21:   Conformations of Cyclobutane

Cyclopentane is nonplaner and its structure resembles an envelop with four carbon in one plane and one out of that plane. The molecule is in continuous motion as shown below in Fig. 5.22.

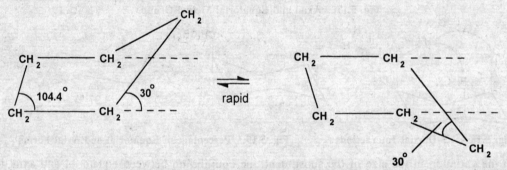

Fig. 5.22:   Conformations of Cyclopentane

As you can see that even though the normal pentagon angle (108°) can furnish an almost strain free cyclopentane, the structure is nonplaner with angle 104.4°. The puckering essentially reduces torsional strain due to eclipsing of C–H bonds in a planer arrangement of atoms.

The medium sized rings (7-12 ring carbon) are relatively free of angle strain but they are associated with partial eclipsing of hydrogen atoms and nonbonded repulsions between hydrogens.

5.5 GEOMETRICAL ISOMERISM

It is a type of stereoisomerism in which isomers have different arrangement of groups or atoms around a rigid frame work like double bond. As discussed in Section 2.2 a double bond between two carbon atoms comprise of one sigma bond and one π bond. A sigma bond has a cylindrical symmetry due to which rotation along the bond axis is possible but a π bond is formed due to the side ways overlap of p-orbitals and thus rotation along a double bond is not possible without breaking it. As a result of this hindered rotation the positions of various atoms or groups is fixed in space. For example, 2-butene can be represented by two structures (I and II) as shown below:

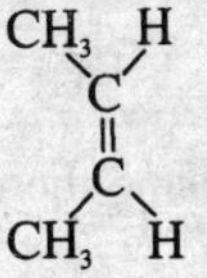

$$CH_3 \quad H$$
$$\underset{\|}{C}$$
$$\underset{C}{}$$
$$CH_3 \quad H$$

I: cis-2-Butene

$$CH_3 \quad H$$
$$\underset{\|}{C}$$
$$\underset{C}{}$$
$$H \quad CH_3$$

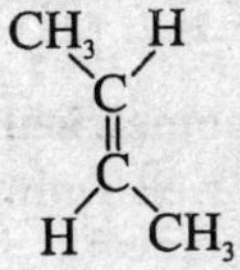

II: trans-2-Butene

The isomer I in which similar groups are on same side is called *cis isomers (Latin cis-same side)* and isomer II, having similar groups on opposite sides is called the *trans isomer (Latin trans-across)*. The isomers arising due to the difference in spatial arrangement of atoms or groups about a carbon-carbon double bond are called *Geometrical isomers* and the phenomenon is known as *Geometrical isomerism or cis-trans isomerism*. Below are given cis and trans isomers of but-2-ene-1,4-dioic acid:

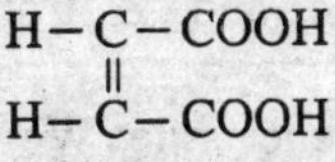

$$H-C-COOH$$
$$\|$$
$$H-C-COOH$$

Maleic Acid (cis form)

$$H-C-COOH$$
$$\|$$
$$COOH-C-H$$

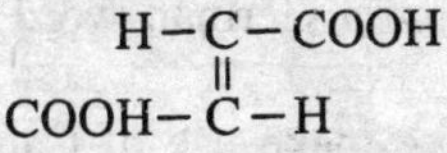

Fumaric Acid (trans form)

Hindered rotation is also possible with carbon-nitrogen double bond, thus oximes of aldehydes also show cis trans isomerism.

$$C_6H_5-C-H$$
$$\|$$
$$:N-OH$$

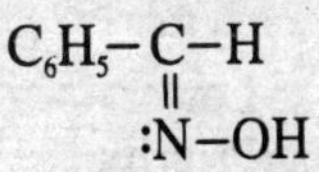

α–Benzaldoxime
(cis or syn form)

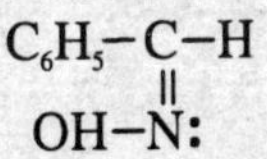

$$C_6H_5-C-H$$
$$\|$$
$$OH-N:$$

β–Benzaldoxime
(trans or anti form)

There is hindered rotation about all carbon-carbon double bonds but geometrical isomerism is possible only in certain special cases, for example, propene, 1-butene or isobutylene have double bonds but do not show geometrical isomerism

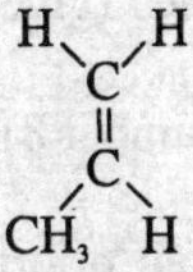

$$H \quad H$$
$$\underset{\|}{C}$$
$$\underset{C}{}$$
$$CH_3 \quad H$$

Propene

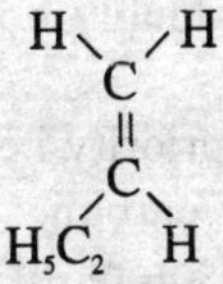

$$H \quad H$$
$$\underset{\|}{C}$$
$$\underset{C}{}$$
$$H_5C_2 \quad H$$

1-Butene

$$H \quad H$$
$$\underset{\|}{C}$$
$$\underset{C}{}$$
$$CH_3 \quad CH_3$$

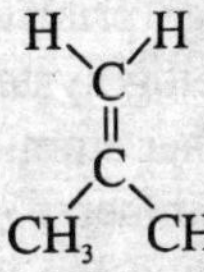

Isobutylene

In the same way 1,1-dichloroethene does not show cis-trans isomerism but 1,2-dichloroethene does.

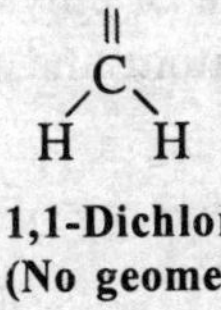

$$Cl \quad Cl$$
$$\underset{\|}{C}$$
$$\underset{C}{}$$
$$H \quad H$$

1,1-Dichloroethene
(No geometrical Isomerism)

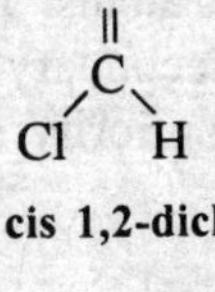

$$Cl \quad H$$
$$\underset{\|}{C}$$
$$\underset{C}{}$$
$$Cl \quad H$$

cis 1,2-dichloroethene

$$Cl \quad H$$
$$\underset{\|}{C}$$
$$\underset{C}{}$$
$$H \quad Cl$$

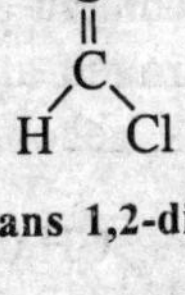

trans 1,2-dichloroethene

With the above discussion we can conclude that the phenomenon of geometrical isomerism is possible with baC=Cab or baC=Ccd type of molecules whereas it is not possible with aaC=Cbb or aaC=Cab type of molecules.

5.5.1   E & Z Notation of Geometrical Isomers

The prefix cis and trans works well with disubstituted ethenes but it is difficult to name the compounds of the type given below.

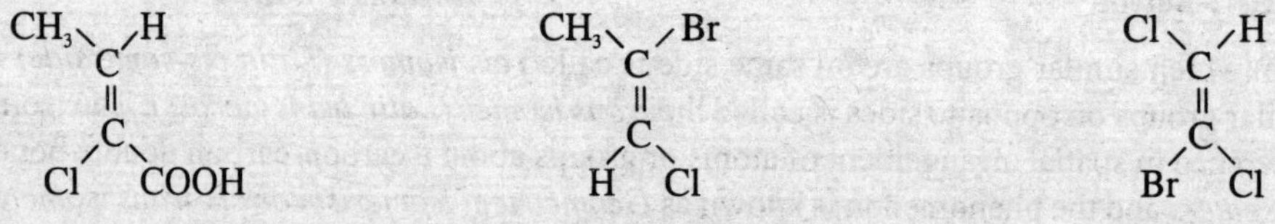

$$CH_3 \diagdown \diagup H \qquad\qquad CH_3 \diagdown \diagup Br \qquad\qquad Cl \diagdown \diagup H$$
$$C \qquad\qquad\qquad C \qquad\qquad\qquad C$$
$$\| \qquad\qquad\qquad \| \qquad\qquad\qquad \|$$
$$C \qquad\qquad\qquad C \qquad\qquad\qquad C$$
$$\diagup \diagdown \qquad\qquad \diagup \diagdown \qquad\qquad \diagup \diagdown$$
$$Cl \quad COOH \qquad\qquad H \quad Cl \qquad\qquad Br \quad Cl$$

Therefore, a new system of nomenclature has been developed which is known as E & Z notation where 'E' has been derived from a German word *Entgagen* (means opposite) and 'Z' from *Zusamen* (means together).

To designate such isomers we make use of Cahn-Ingold and Prelog sequence rule. We first assign priorities to the groups attached to two doubly bonded carbons. If the groups of higher priority are on opposite sides of the double bond, the isomer is designated as 'E' isomer and when the groups of higher priority are on the same side of the double bond, the isomer is designated as Z isomer.

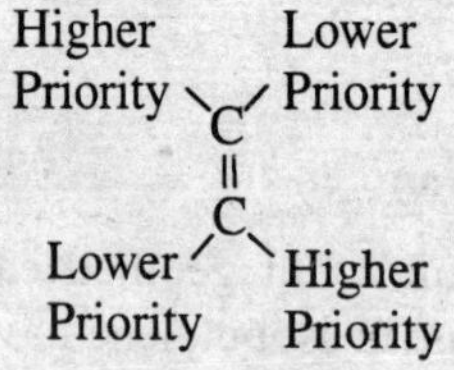

Higher   Lower
Priority $\diagdown \diagup$ Priority
C
$\|$
C
Lower $\diagup \diagdown$ Higher
Priority   Priority

E-isomer

Higher   Lower
Priority $\diagdown \diagup$ Priority
C
$\|$
C
Higher $\diagup \diagdown$ Lower
Priority   Priority

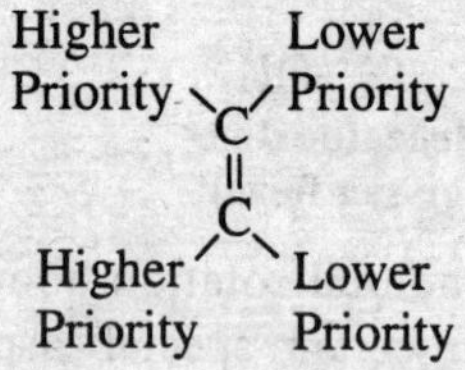

Z-isomer

To assign priority to various groups about doubly bonded carbons following rules (given by Cahn-Ingold-Prelog system) are observed.

(1) Higher atomic number atoms get higher priority i.e., among oxygen (atomic number 8) and carbon (atomic number 6) oxygen gets priority over carbon.

(2) Among the isotopes of same element, isotope of higher atomic mass is given priority over the one with lower atomic mass. Among tritium ($_1H^3$) and hydrogen ($_1H^1$), tritium gets priority.

(3) When the doubly bonded carbon is attached to two groups which are same at the point of attachment as shown below, the priorities are assigned on the basis of first point of difference. Since doubly bonded carbon is attached to a ethyl and a secondary butyl group, there is no difference at the point of attachment as both of them are carbon atoms. The second carbon, in case of ethyl is attached to three hydrogen atoms whereas in case of secondary butyl to one hydrogen atom and two carbon atoms. Thus, secondary butyl group will get priority over ethyl group.

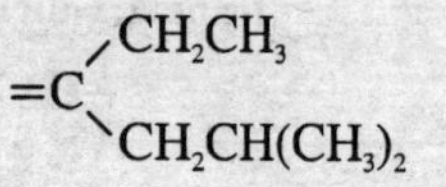

$$=C \diagup^{CH_2CH_3}_{\diagdown CH_2CH(CH_3)_2}$$

(4) When double or triple bonds are present, they are considered to be duplicated or triplicated. Thus,

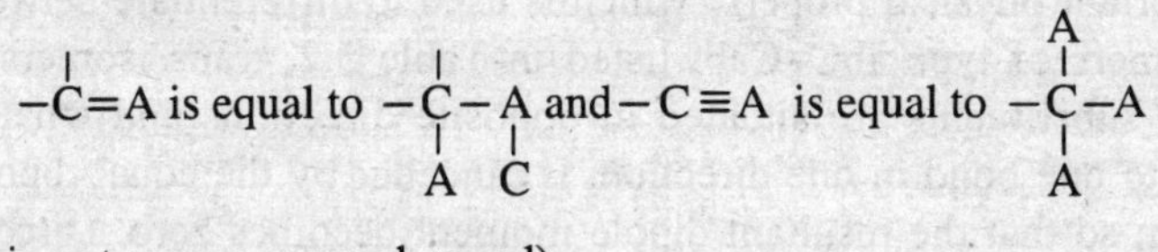

$$-\overset{|}{C}=A \text{ is equal to } -\overset{|}{\underset{A}{\overset{|}{C}}}-\overset{|}{\underset{C}{A}} \text{ and} -C \equiv A \text{ is equal to } -\overset{A}{\underset{A}{\overset{|}{C}}}-A$$

(This rule is not very commonly used)

If one of the positions is occupied by lone pair (as in case of oximes of aldehydes), it is given priority over the bonded group as shown below.

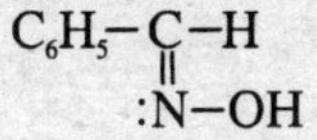

$$C_6H_5-\overset{||}{\underset{:N-OH}{C}}-H \qquad\qquad C_6H_5-\overset{||}{\underset{HO-N:}{C}}-H$$

α–Benzaldoxime   β–Benzaldoxime
(cis or syn form)   (trans or anti form)

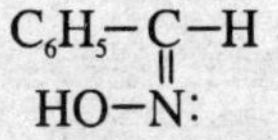

Now we shall use the above priority rules to assign E & Z notation to various geometrical isomers.

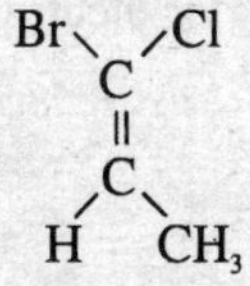

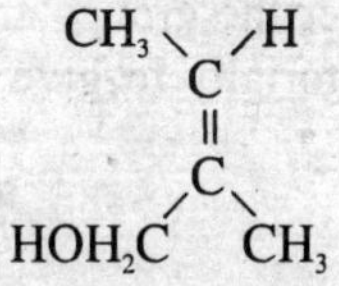

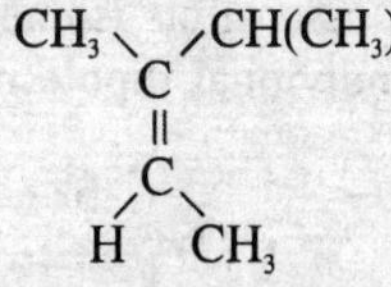

Br > Cl CH_3 > H $CH(CH_3)_2$ > CH_3
CH_3 > H CH_2OH > CH_3 CH_3 > H
E-isomer   Z-isomer   Z-isomer

5.5.2 Characterisation of Geometrical Isomers

Geometrical isomers can be characterised by their physical properties and in some cases by their chemical properties too.

Geometrical isomers have different physical properties like different melting point, boiling point, refractive index, solubility, density and dipole moments. Some of the physical properties of a few geometrical isomers have been summaries in the Table 5.2.

Table 5.2: Physical Properties of Some Geometrical Isomers

Compound	Melting point(K)	Boiling point(K)	Dipole Moment in Debye unit
cis-2-butane	134	277	0.33
trans-2-butane	167	274	0
cis-1,2-dichloroethene	193	333	1.85
trans-1,2-dichloroethene	223	321	0
cis-1,2-dibromoethene	220	383	1.35
trans-1-2-dibromoethene	267	381	0
cis-1,2-diiodoethene	259	345	0.75
trans-1,2-diiodoethene	461	465	0

It is clear from the above data that trans isomers have higher melting point than the corresponding cis isomer as the former is more symmetrical and fits well into the crystal lattice.

Dipole moment is another important physical property which is used to differentiate between cis and trans isomers. In geometrical isomers of type abC=Cab, listed in Table 5.2, trans isomers have zero dipole moment because the same substituents are situated in opposite directions and whatever is the magnitude of dipole moment due to one bond in one direction, is cancelled by the equal dipole moment operating in the opposite direction; so that the resultant dipole moment becomes zero. Direction of the dipole moment due to individual bonds in an isomer depends on the nature of substituents, However, weather the substituents are electron releasing or electron withdrawing, the resultant dipole moment of trans isomer is zero.

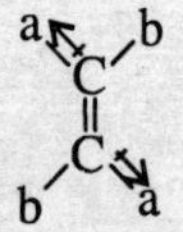

a = Cl, Br or I μ = 0D

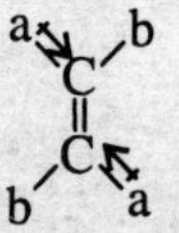

a = alkyl or CH₃ μ = 0D

In case of cis isomers, However, wheather the groups are electron withdrawing or electron donating, the individual dipole moment is added vectorially to give dipole moment.

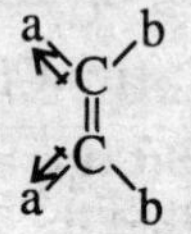

a = Cl
μ = 1.85 D

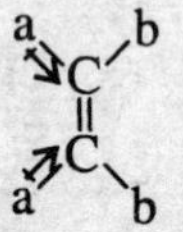

a = CH₃
μ = 0.33 D

Let us now consider a situation when both electron withdrawing and electron releasing groups are present together, both cis and trans isomers have dipole moment which is the vectorial addition of individual bond moments.

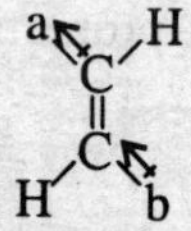

a = Cl, b= CH₃
trans isomer, μ = 1.97 D

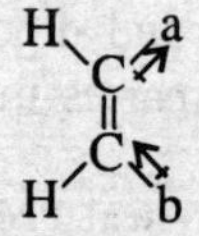

a = Cl, b = CH₃
cis isomers, μ = 1.70 D

Geometrical isomers can also be differentiated by chemical means as they contain certain functional groups which react with same reagents but at a different rate or sometimes a reaction is possible with a certain isomer only. For example, maleic acid which is a cis but-2-ene-1,4-dioic acid is converted into its anhydride by elimination of a water molecule whereas the same reaction is not possible with fumaric acid which is a trans isomer.

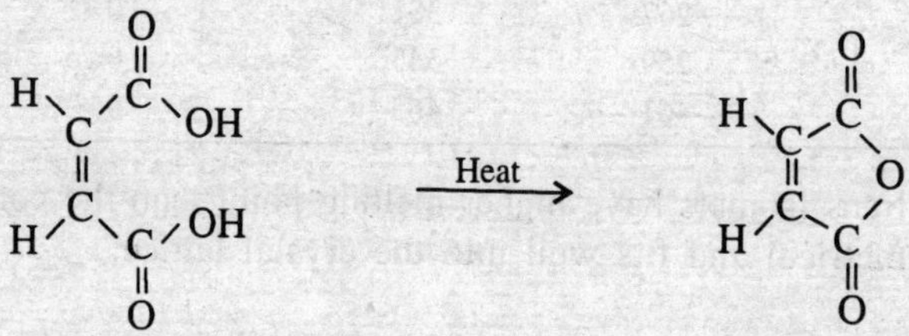

Maleic acid
(cis isomer)

Maleic anhydride

**Fumaric Acid
(trans isomer)**

5.5.3 Inter Conversion of Geometrical Isomers

Geometrical isomers are quite stable at room temperature but one can be converted into another or an equilibrium mixture of both can be obtained catalytically or by heating or exposing to UV radiations. For example cis 2-butene can be converted into trans 2-butene by providing enough heat energy so that breaking of bond followed by rotation and formation of bond is possible.

cis-2-Butene (47%)   **trans-2-Butene (53%)**

5.6 OPTICAL ISOMERISM

Optical isomerism arises due to difference in rotation of plane polarised light. So far we have discussed conformational and geometrical isomerism in which isomers have different physical and in few cases chemical properties but optical isomers have similar chemical and physical properties differing only in rotation of plane of plane polarised light. To understand this isomerism we shall first discuss the nature, origin & application of plane polarised light.

5.6.1 Plane Polarised Light and Optical Activity

You may be aware of the fact that light may be regarded as an electromagnetic radiation having oscillating electric and magnetic field associated with it and the vectors describing these fields are at right angles to each other. Ordinary light, having vibrations in all directions (Fig. 5.23(a)) can be converted into plane polarised light by passing it through a nicol prism or polaroid lens. *Plane polarised light* is the light having single wave length and vibrating only in one direction *A monochromatic light* vibrating in different planes and having a single wavelength ($\lambda = 589$ nm) is called *Sodium D line.* It is obtained from sodium lamp and is used in the experiments. It is passed through a polariser and converted into plane polarised light (Fig. 5.23(b)).

It is observed that substances like quartz crystal and organic compounds like camphor, tartaric acid and lactic acid etc. rotate the plane of plane polarised light (Fig. 5.23(c)). Such compounds are called *optically active* and the instruments used to determine the extent of optical activity associated with these compounds is called *polarimeter* (Fig. 5.24).

The extent of rotation (α) for an optically active compound depends upon the thickness of sample (given by the length of the polarimeter tube 'l') concentration of sample, nature of solvent, temperature and wave length of the light used. When 'l' is taken in decimeters (1 decimeter=10 centimeters) and 'c' is taken in kgdm^{-3}, the rotation [α] in degrees is termed as specific rotation.

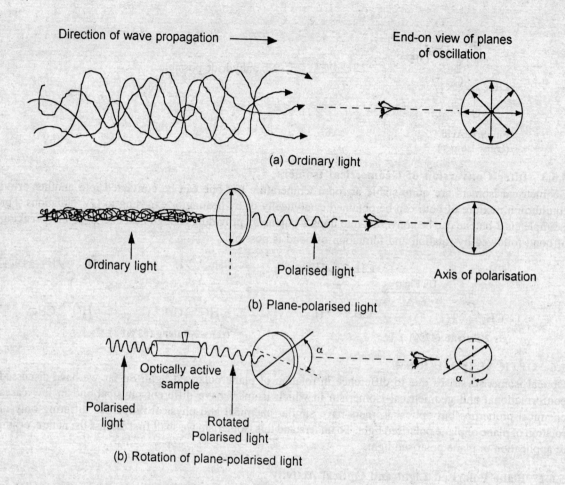

Fig. 5.23:   **Plane Polarised Light and Optical Activity**

Thus *Specific rotation*, a physical constant characteristic of an optically active substance, is *defined as the number of degrees of rotation of plane polarised light when the light passed through 1 decimeter of its solution having a concentration of 1 kgdm⁻³*. The specific rotation can be calculated using the following expression:

$$[\alpha]_\lambda^t = \frac{\alpha}{l \times c}$$

The temperature 't' and wave length of light (λ) have been specified as superscript and subscript respectively. Thus $[\alpha]_D^{20}$ denotes the specific rotation at 20° when the measurement is done by using D line of sodium with a wavelength of 589 nm.

The terms dextrorotatory and laevorotatory specify the direction of rotation of plane polarised light. When a compound rotates the of plane polarised light in clockwise direction, it is called dextrorotatory,

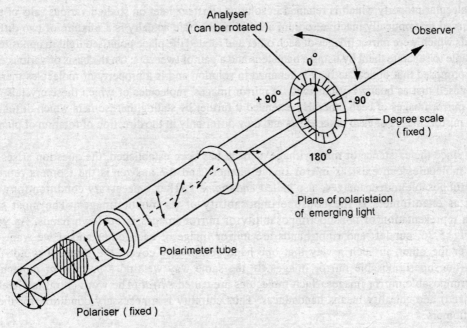

Analyser
(can be rotated)

Observer

0°

+ 90°

- 90°

180°

Degree scale
(fixed)

Plane of polaristaion
of emerging light

Polarimeter tube

Polariser (fixed)

Fig. 5.24: Polarimeter

Fig. 5.25(a) and is denoted by a positive (+) sign prefix to the name of the compound whereas if the compound rotates the plane in anti-clockwise direction, it is called laevorotatory and a negative(−) sign is prefixed to the name of the compound, Fig. 5.25(b).

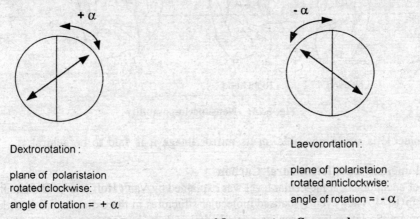

+ α

- α

Dextrorotation :

plane of polaristaion
rotated clockwise:

angle of rotation = + α

Laevorortation :

plane of polaristaion
rotated anticlockwise:

angle of rotation = - α

Fig. 5.25: Dextrorotatory and Laevorotatory Compounds

5.6.2 Origin of Optical Activity

Optical activity was discovered in the year 1815 at the college of de France by the physicist *Jean-Baptiste Biot* who discovered the existence of two types of quartz crystal which rotated the plane polarised light into opposite directions. He later observed that this property was not only associated with crystalline structure but certain compounds in solution also exhibit the same property. This optical activity in solution is due to some molecular property which is retained in solution. *Pasteur* later on studied various salts of tartaric acid and observed that optically inactive sodium ammonium tartrate actually is a mixture of two different kinds of crystals which were mirror images of each other and rotated the plane polarised light in opposite directions. He was able to separate them by using a hand lens and a pair of tweezers. On the basis of various experiments *Pasteur* proposed that optical activity is retained in solution and is a property of molecules themselves. He also proposed that as quartz crystals exist as mirror images, molecules of which these crystals are formed, are also mirror images of each other. He extended it further by stating that isomers which differ structurally in being mirror images of each other exists and they differ only in the direction of rotation of plane polarised light.

Now, since the existence of mirror image isomers has been established, the question arises as to what kind of molecules can exist as mirror image isomers and the answer is the isomers represented by nonsuperimposable mirror images, also called *enantiomers*. Thus a necessary condition for a compound to exist as enantiomers is the nonsuperimposability of its mirror images. The most simple and common representation of nonsuperimposability of mirror images is our own hands. As you can see in the Fig. 5.26, our left and right hands are mirror image of each other and if we want to put one hand over the other, in such a way that one hand completely covers the other, it is not possible as they are nonsuperimposable mirror images. In the same way we find a number of molecules having nonsuperimposable mirror images. Such molecules are called *chiral* (The word chiral in Greek language means hand) and chirality means handedness. Thus chirality is a necessary condition for the existence of enantiomers.

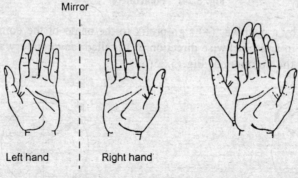

Fig. 5.26: Nonsuperimposability

When a molecule is superimposable on its mirror image it is said to be achiral.

5.6.3   Enantiomerism and Tetrahedral Carbon

The existence of enantiomers at molecular level was explained by Van't Hoff and Le Bel simultaneously and independently in 1874. They even visualised molecular structures in three dimensions in order to solve the problem of isomers. So let us first visualise a tetrahedral carbon having four different group w, x, y, and z attached to it. The mirror image isomers arising from it have been shown below in ball and stick model (Fig. 5.27) as well as in wedge formulas.

If we look at them carefully we find that the two mirror image isomers are not superimposable on each other. We may twist them or turn them as we please but only two of the four groups attached to the carbon coincide or in other words we can say that the two isomers represent enantiomers as they have nonsuperimposable mirror images. Following are given a few compounds and their mirror images.

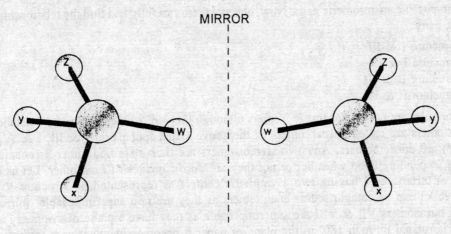

MIRROR

Fig. 5.27: Nonsuperimposable Mirror Images

| COOH | COOH | C₂H₅ | C₂H₅ |

$$H - \overset{\displaystyle COOH}{\underset{\displaystyle CH_3}{C}} - OH \qquad HO - \overset{\displaystyle COOH}{\underset{\displaystyle CH_3}{C}} - H \qquad HOH_2C - \overset{\displaystyle C_2H_5}{\underset{\displaystyle CH_3}{C}} - H \qquad H - \overset{\displaystyle C_2H_5}{\underset{\displaystyle CH_3}{C}} - CH_2OH$$

Lactic acid   **2-Methyl-1-butanol**

All the pairs shown above are nonsuperimposable and therefore represent enantiomers. It is interesting to note here that even two isotopes of the same element like protium and deuterium are different enough to permit detectable isomerism by showing difference in the direction of rotation of plane polarised light. Thus,

$$D - \overset{\displaystyle CH_3}{\underset{\displaystyle C_6H_5}{C}} - H \qquad\qquad H - \overset{\displaystyle CH_3}{\underset{\displaystyle C_6H_5}{C}} - D$$

α-Deuterioethylbenzene

With the discussion we have had so far, we may say that carbon which is attached to four different groups or atoms is a chiral center or asymmetric center.

So far, we have discussed the molecules having only one chiral center. We shall now *discuss optical isomerism in compounds having more than one chiral center.*

When an organic compound contains two chiral centers, following four structures (I to IV) are possibly given, out of which structure I & II and structure III & IV are mirror image isomers as they

$$
\begin{array}{cccc}
\overset{\displaystyle x}{|} & \overset{\displaystyle x}{|} & \overset{\displaystyle x}{|} & \overset{\displaystyle x}{|} \\
a - C - b & b - C - a & a - C - b & b - C - a \\
| & | & | & | \\
a - C - b & b - C - a & b - C - a & a - C - b \\
| & | & | & | \\
y & y & y & y \\
\\
I & II & III & IV
\end{array}
$$

are nonsuperimposable enantiomeric pairs. Now, observe them carefully and find the relationship between the following pairs:

Structure I & III
Structure I & IV
Structure II & III
Structure II & IV

All the four pairs are not mirror image isomers although they are isomers. Thus stereoisomers which are not enantiomers are called diastereomers or the above stated four pairs (I & III, I & IV, II & III and II & IV) are diastereomers. Any two stereoisomers are thus classified either as enantiomers or diastereomers, depending upon wheather or not they are mirror images of each other. Let us now take the example of tartaric acid having two asymmetric centers as represented by structure V to VIII.

Isomer V & VI can be catagorised as enantiomers as they are non superimposable mirror images of each other but isomers VII & VIII are superimposable as they have a plane of symmetry and when formula VIII is rotated through 180° in the plane of paper it becomes identical with formula VII thus structure VII & VIII are same isomer. In general for a compound having 'n' chiral centers number of possible stereoisomers is given by 2^n. isomer V & VII and VI & VII are diastereomers as they are not mirror images.

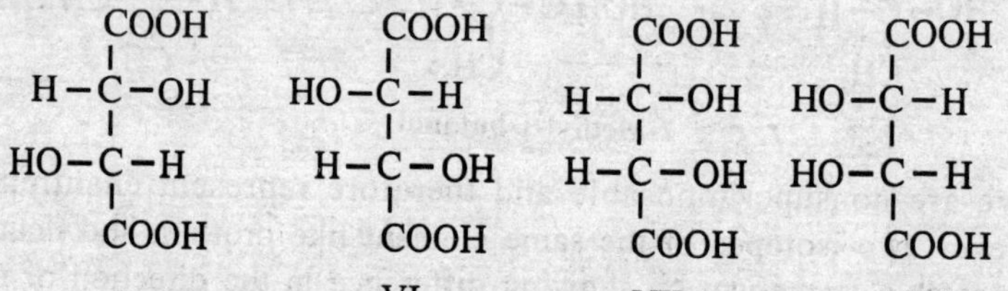

V VI VII VIII

Tartaric acid, thus, exists as a pair of enantiomers each of which is diastereomeric with optically inactive meso-tartaric acid as shown below:

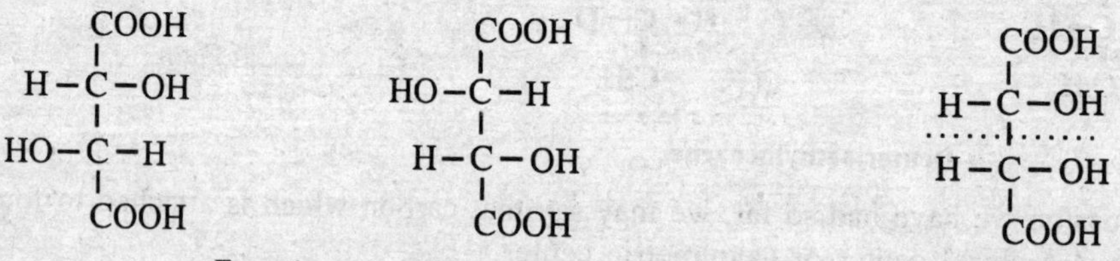

Enantiomeric pair **meso-Tartaric Acid**

± Tartaric acid or dl tartaric acid is the mixture of equal amounts of (+) and (–) tartaric acids. The physical properties of all of them have been summarised in Table 5.3.

Table 5.3: Physical Constants of Tartaric Acid

Name	MP(°C)	Density	[α]
(+) Tartaric acid	179	1.760	+12
(–) Tartaric acid	170	1.760	–12
(±) Tartaric acid	206	1.697	0
meso-Tartaric acid	140	1.666	0

External and Internal Compensation : If equimolar amounts of (+) and (–) isomers are mixed in a solvent, the resultant solution becomes optically inactive. The rotation of each isomer is balanced by the equal and opposite rotation of the other. Optical inactivity of this origin is known to exist due to external compensation. Such mixtures of dextrorotatory and levorotatory isomers are called *Recemic Mixtures* and they can be separated into their active principals. For example, ± Tartaric acid may be separated into (+) tartaric acid and (−) tartaric acid.

In case of meso tartaric acid, inactivity is due to the internal arrangement of atoms at the chiral center. The force of rotation due to upper half of the molecule is balanced by the equal and opposite force of rotation due to lower half of the molecule. The optical inactivity, thus, rendered is due to internal compensation. It occurs when the compound has two or more chiral centers with plane or point of symmetry. They can not be separated into optically active components.

5.6.4 Elements of Symmetry and Chirality

Chirality in a molecule can be recognised by the presence of chiral center. A general test for chirality is nonsuperimposability of the molecule and its mirror image, however, there is one more method to test for chirality in a molecule and it is the absence of elements of symmetry. In other words we can say that achiral molecules have one or more elements of symmetry. There are three main elements of symmetry which have been discussed below.

1. *Plane of Symmetry (r):* A plane of symmetry is defined as an imaginary plane which divides the molecule into two halves which are mirror image of each other. A few examples have been shown in Fig. 5.28 below.

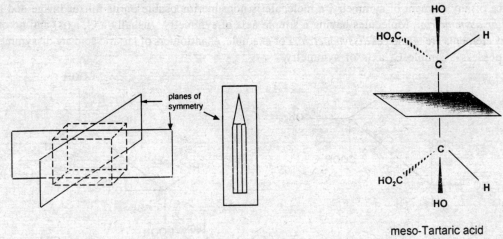

meso-Tartaric acid

Fig. 5.28: Elements of Symmetry—Plane of Symmetry

2. *Center of Symmetry (i):* Center of symmetry is a point in the center of the molecule from which any line drawn through the molecule finds similar environment in opposite directions. This has been illustrated with the help of examples shown in Fig. 5.29.

Fig. 5.29: Center of Symmetry

3.   *Alternating Axis of Symmetry (C_s):* Axis of symmetry (C_n) is defined as a line which passes through the molecule so that a rotation of 360°/n about this axis leads to a three dimensional structure which is similar to the original one. All molecules possess a one fold axis of symmetry (n = 1) since rotation of 360° about an axis leads to the identical structure. The water molecule has a two fold axis of symmetry (C_2) as rotation of the molecule by 180° gives away the original arrangement of atoms [Fig. 5.30(a)]. Similarly, ammonia molecule has a three fold axis of symmetry (C_3).

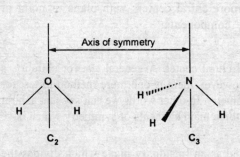

Fig. 5.30(a):   Axis of Symmetry

An alternating or improper or rotation reflection axis of symmetry of the order n, is an axis (n-fold), rotation of 360°/n around it followed by reflection in a plane perpendicular to the said axis shows a structure similar to the original one. For example, the conformation of meso-tartaric acid (I) in Fig. 5.30(b) when reflected in a plane placed at the center of C–C axis and at right angle to it gives an orientation (II) which on rotation by 180° along the axis becomes superimposable with structure I. Meso-tartaric acid, therefore, contains a S_2 axis.

A molecule having plane of symmetry or center of symmetry or an alternating axis of symmetry is superimposable on its mirror image and hence is *achiral.* On the contrary in absence of any element of symmetry, a molecule is nonsuperimposable on its mirror image and thus is *chiral or asymmetric.* Molecules having a simple axis of symmetry (usually a C_2 axis) and no other symmetry elements are termed as *dissymmetric.* For example, enantiomers of tartaric acid are dissymmetric as they possess a simple C_2 axis of symmetry.

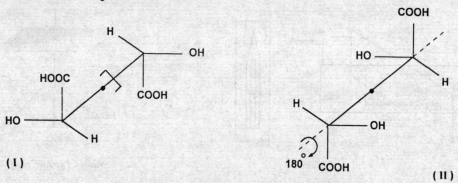

Fig. 5.30(b):   Alternating Axis of Symmetry

5.6.5   Chiral Molecules Without Chiral Center

In the following few examples you will find that even though there is no chiral center in the molecule, the compound shows optical activity.

Biphenyls substituted at ortho positions by large groups show enantiomerism and can be resolved into optical isomers, as the free rotation along the central bond is restricted due to steric hindrance of bulky groups at ortho positions. As shown in Fig. 5.31, the plane of two rings of biphenyls are perpendicular to each other and their mirror images are nonsuperimposable.

Another class of compounds which show optical activity in absence of chiral center is allenes having cumulative double bonds. The molecular dissymmetry has been clearly indicated in Fig. 5.32 showing the two end of allene molecule perpendicular.

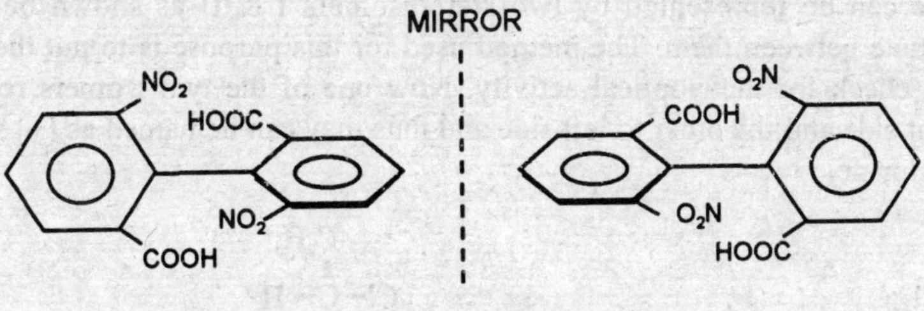

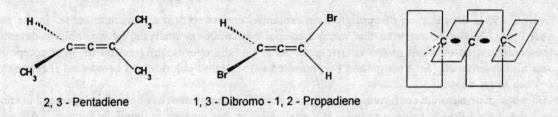

Fig. 5.31: Optical Activity in Biphenyls

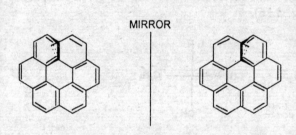

2, 3 - Pentadiene   1, 3 - Dibromo - 1, 2 - Propadiene

Fig. 5.32: Optical Activity in Allenes

An interesting example of existence of molecular asymmetry without chiral center is shown by hexahelicene which is a twisted or spiral molecule (Fig. 5.33). It can be resolved into stable enantiomers corresponding to right and left helices.

The spirane has structural similarity with allenes if both the double bonds in allene are replaced by rings and shows optical activity (Fig. 5.34).

Compounds with exocyclic double bonds are chiral without having an asymmetric carbon as they do not have any plane or center of symmetry (Fig. 5.35).

Fig. 5.33: Optical Activity in Hexahelicene

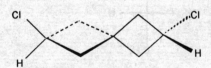

Fig. 5.34: Optical Activity in Spirane

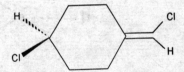

Fig. 5.35: Optical Activity in Compounds having Exocyclic Double Bond

5.6.6 Configuration and Fischer Projection Formula

Configuration is the arrangement of atoms that characterises a particular stereoisomer.

sec Butyl chloride can be represented by two stereoisomers I & II as shown below but how are we going to differentiate between them. The method used for this purpose is to put the two compounds in a polarimeter and check for their optical activity. Now one of the two isomers rotate the plane of polarised light to right side and the other to left side and thus they can be named as (+) sec butyl chloride and (–) sec butyl chloride.

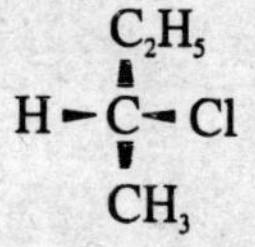

C_2H_5

$H-C-Cl$

CH_3

I, (–) sec Butyl chloride

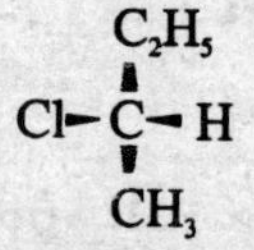

C_2H_5

$Cl-C-H$

CH_3

II, (+) sec Butyl chloride

Until 1951, the question of configuration remained unanswered in an absolute sense. In the same year Prof. J.M.Bijvoet reported that using a special type of X-ray analysis, he was able to determine the configuration of a salt of (+) tartaric acid. After this discovery configuration of other compounds was also determined. Thus compound I represented (–) sec butyl chloride and compound II represented (+) sec butyl chloride.

To designate absolute configuration to various steroisomers a system has been developed and is known as cahn-Ingold-Prelog priority system. We shall discuss this system in detail in Section 5.6.7.

We have already discussed the representation of Fischer projection formula in the Section 5.3.2. Now we shall learn to interconvert the Fischer projection formula into perspective formula without changing the configuration. This is important because several orientations of one molecule are possible depending upon which two substituents are chosen to face the observer or in other words we can say that several Fischer projections can be drawn for the same molecule. Following are given Perspective drawing and Fischer projections of 2-butanol (Fig. 5.36).

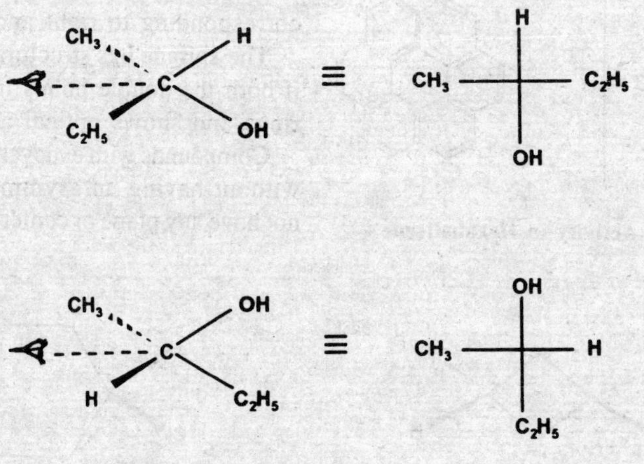

Fig. 5.36: Perspective View and Fischer Projection of 2-butanol

Therefore, it is necessary to know how to write various Fischer projections for the same molecule. Certain rules have been drawn to do this which are given below:

1. Rotation of the given Fischer projection formula by 180° in the plane of paper gives another projection formula of the same configuration, for example,

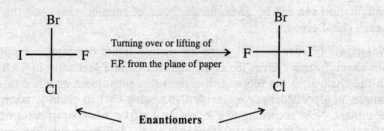

Or we can say that rotation of a Fischer projection by 180° in the plane of paper does not alter the original configuration.

2. Rotation of a Fischer projection formula of a compound by 90° in the plane of paper gives projection formula of its enantiomer or we can say that rotation by 90° alters the original configuration.

$$
\text{H} \overset{\displaystyle \text{Cl}}{\underset{\displaystyle \text{F}}{\rule{0pt}{1pt}\!\!-\!\!\!\!-\!\!\!\!-\!\!}} \text{Br}
\quad \xrightarrow[\text{in the plane of paper}]{\text{Rotation by 90°}} \quad
\text{F} \overset{\displaystyle \text{H}}{\underset{\displaystyle \text{Br}}{\rule{0pt}{1pt}\!\!-\!\!\!\!-\!\!\!\!-\!\!}} \text{Cl}
$$

←   **Enantiomers**   →

3. A Fischer projection formula should not be lifted out of the plane of the paper or turned over because this turning over leads to the change in original configuration.

$$
\text{I} \overset{\displaystyle \text{Br}}{\underset{\displaystyle \text{Cl}}{\rule{0pt}{1pt}\!\!-\!\!\!\!-\!\!\!\!-\!\!}} \text{F}
\quad \xrightarrow[\text{F.P. from the plane of paper}]{\text{Turning over or lifting of}} \quad
\text{F} \overset{\displaystyle \text{Br}}{\underset{\displaystyle \text{Cl}}{\rule{0pt}{1pt}\!\!-\!\!\!\!-\!\!\!\!-\!\!}} \text{I}
$$

←   **Enantiomers**   →

4. Inter change of two pairs of substituents does not alter the configuration. This can be illustrated with the help of following example.

$$
\text{I} \overset{\displaystyle \text{CH}_3}{\underset{\displaystyle \text{C}_3\text{H}_7}{\rule{0pt}{1pt}\!\!-\!\!\!\!-\!\!\!\!-\!\!}} \text{OH}
$$

Interchange of one pair of substituents (i.e. CH_3 and C_3H_7) leads to Fischer projection as follows

$$
\text{H} \overset{\displaystyle \text{C}_3\text{H}_7}{\underset{\displaystyle \text{CH}_3}{\rule{0pt}{1pt}\!\!-\!\!\!\!-\!\!\!\!-\!\!}} \text{OH}
$$

Second interchange between substituents (H and OH) gives Fischer formula as:

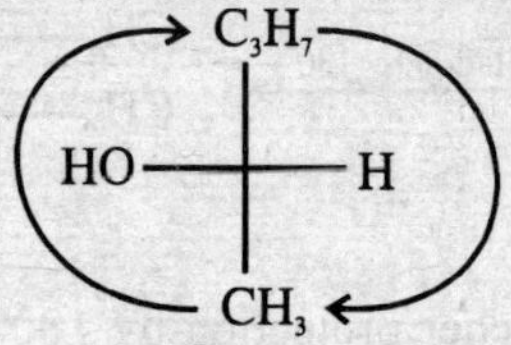

$$C_3H_7$$
$$HO \underline{\quad\quad\quad} H$$
$$CH_3$$

Rotation of the above Fischer projection yields the Fischer projection which is nothing but the isomer we had started with or we can say that interchange of two pairs of substituents gives another Fischer projection of the same isomer.

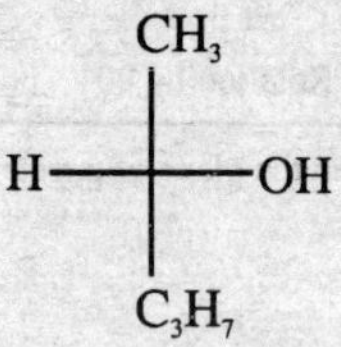

$$CH_3$$
$$H \underline{\quad\quad\quad} OH$$
$$C_3H_7$$

5.6.7 Nomenclature of Configurational Isomers

Configurational isomers differ from each other in the rotation of plane polarised light thus prefix 'D' or 'L' was used to designate dextrorotatory and laevorotatory isomers respectively but it was later realised that absolute configuration can not be given on the basis of rotation only, and the configuration must be specified at each chiral center.

D & L Notations: Fischer in 1891 gave the nomenclature of chiral compounds as D & L on the basis of main chain. In this system the compound is oriented vertically in such a way that the carbon number 1 in the chain is at the top and then the main substituent attached to chiral center is looked for. For example, in glyceraldehyde molecule CHO being C−1 in chain is taken at the top and OH as the main substituent. Now according to Fischer's rule if the main substituent is on right, the molecule has D configuration and if the main substituent is on the left hand side, the molecule is designated as L isomer.

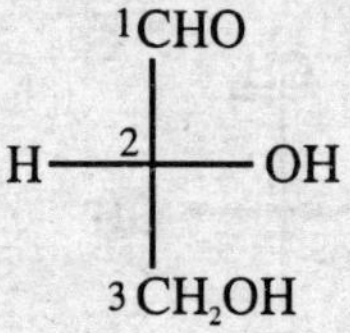

$$^1CHO$$
$$H \underline{\overset{2}{\quad\quad}} OH$$
$$3\ CH_2OH$$

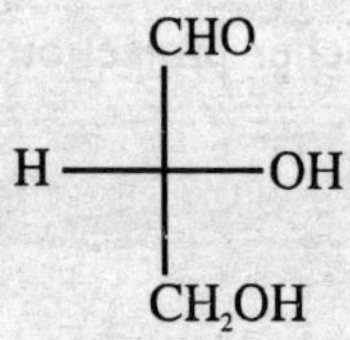

$$CHO$$ $$H \underline{\quad\quad} OH$$ $$CH_2OH$$	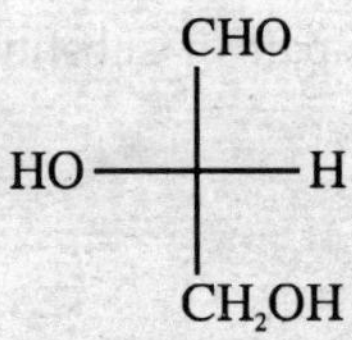 $$CHO$$ $$HO \underline{\quad\quad} H$$ $$CH_2OH$$
D(+) Glyceraldehyde	**L(−) Glyceraldehyde**

All the chiral molecules which can be chemically related to D glyceraldehyde were designated as D-isomers and the molecules related to L glyceraldehyde were designated as L-isomers. In polyhydroxy compounds having more than one chiral center, the system can be extended. A sugar belongs to D series if it has stereochemistry similar to D glyceraldehyde at the stereocenter farthest from the carbonyl group. The sugars having opposite configuration at that center belongs to L series.

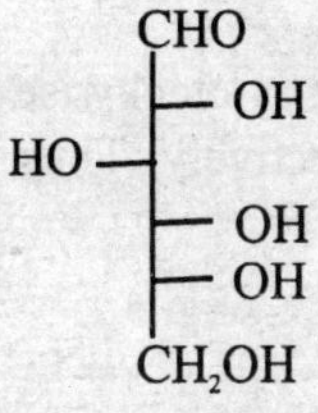

CHO — OH
HO — OH — OH
CH$_2$OH

D Glucose

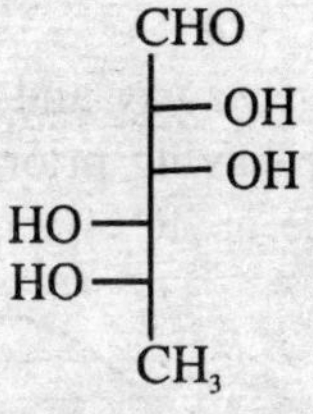

CHO — OH — OH
HO — HO —
CH$_3$

L Rhamnose

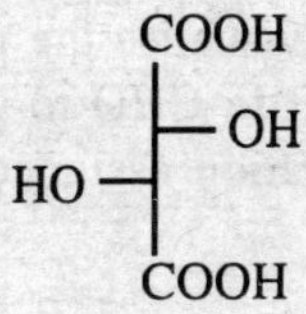

COOH — OH
HO —
COOH

L Tartaric Acid

D and L notation can be assigned to compounds having structural similarity to glyceraldehyde or in other words we can say that it does not indicate more than the relative configuration. At the same time it is not possible to assign configuration to molecules of the type ClFBrCH, so another systematic method of nomenclature has been developed.

R & S Notations: R and S system of notation is based on a set of arbitrary but consistent sequence rules which assign priority to the substituent on the chiral center. The center is then viewed with the substituent with lowest priority pointing away from the viewer. Now if the path for remaining three substituents going from highest priority to the lower priority is *clockwise,* the configuration R and if the path is anti-clockwise, the configuration S is assigned at that chiral center.

The sequence is assigned to various substituents around a chiral center on the basis of cahn-ingold-Prelog priority rules which we have already discussed in Section 5.5.1 of this chapter. Now we shall take up an example of D glyceraldehyde to assign the R or S notation. In D glyceraldehyde priority numbers 1,2,3, and 4 have been given to the groups as shown below.

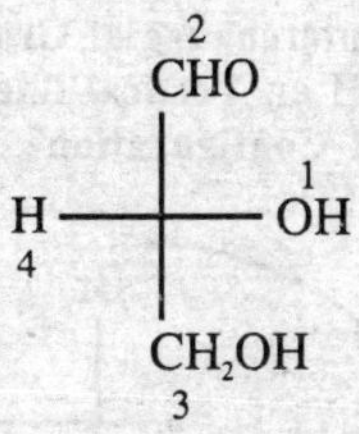

2
CHO
1
H —|— OH
4
CH$_2$OH
3

Now the molecule is viewed in such a way that the group of lowest priority i.e. 'H' points away from the viewer. So the transformed Fischer projection will be as follows (two pairs of substituents interchanged)

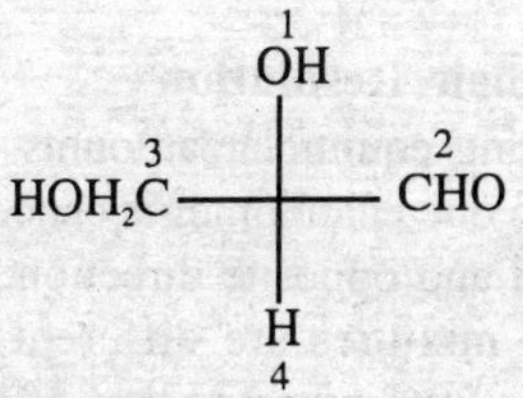

1
OH
3   2
HOH$_2$C —|— CHO

H
4

The above Fischer projection can be converted into perspective drawing as follows.

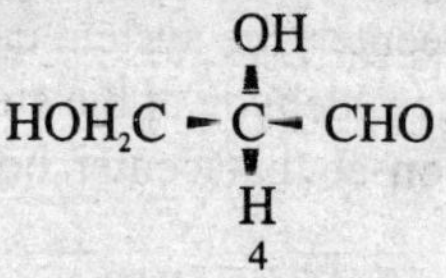

$$HOH_2C - \overset{\overset{\displaystyle OH}{\mid}}{\underset{\underset{\displaystyle 4}{\mid}}{C}} - CHO$$

Now visualise the molecule with hydrogen away and the pathway we get is clockwise while going from OH to CHO to CH$_2$OH i.e. the group having priority 1, 2 and 3 respectively. Thus the absolute configuration assigned to D glyceraldehyde is 'R'.

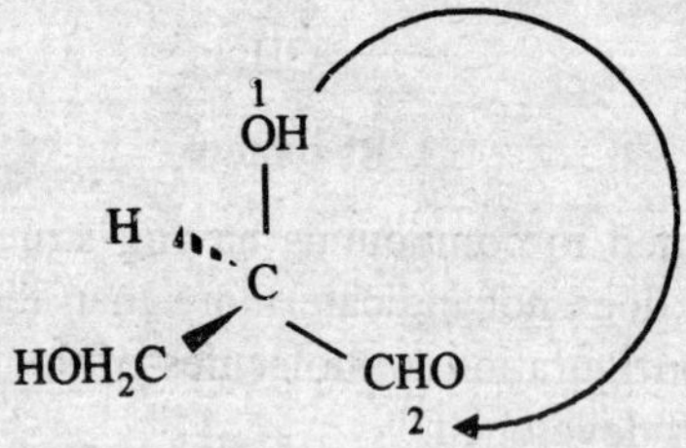

A simpler way of assigning absolute configurate using Fischer projection is by writing the Fischer projection in such a way that the group of lowest priority come on the lower end of vertical line, followed by neglecting this substituent and tracing the pathway from priority 1 to 2 to 3 as stated earlier.

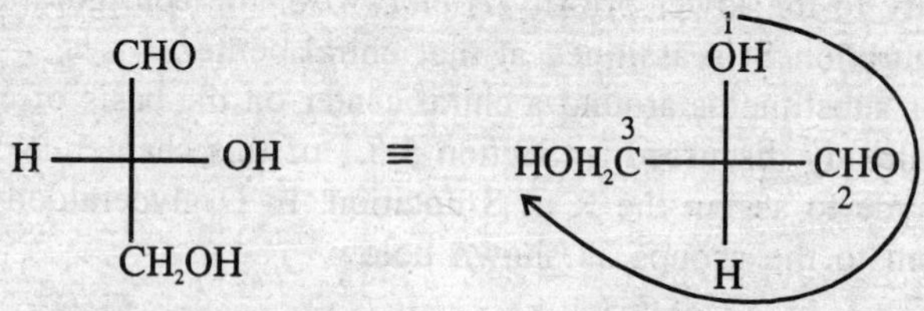

D Glyceraldehyde
(H on horizontal line)

Interchange of Groups
(H on Vertical line)
R Configuration

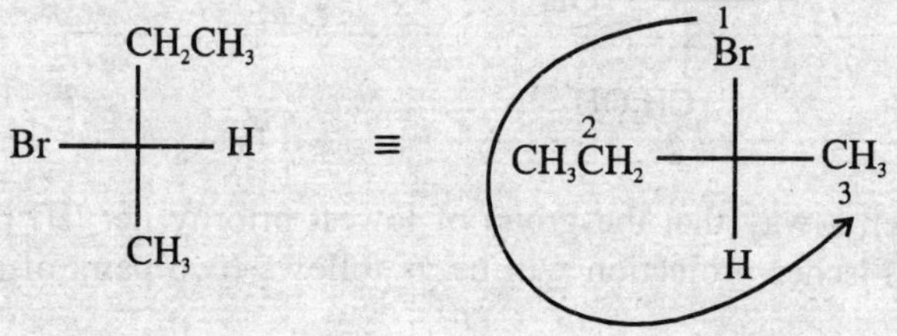

2-Bromobutane

S Configuration

5.6.8 Recemic Modifications and their Resolution

Recemic modification is a mixture having equimolar amounts of two enantiomers. It does not rotate the plane of polarised light as rotation due to one enantiomer is canceled by the rotation due to other enantiomer which rotate the polarised light in equal and opposite direction. It is often indicated by (±) sign.

The physical properties of recemic mixtures are different from those of enantiomers, for example melting point of recemic tartaric acid is 206° whereas that of its enantiomers is 170° as shown in Table

5.3. They differ from their enantiomers as the recemic mixtures are optically inactive where as the enantiomers are optically active.

The recemic mixtures can be obtained from pure enantiomers by recemisation, simple mixing of two enantiomers in equimolar amount or by chemical reactions. (you will learn about nucleophilic substitution reaction which give recemic modifications in Section 12.3.7).

The separation of recemic mixture into enantiomers is termed as resolution. The resolution of recemic mixture is done by the methods mentioned below. (1) Mechanical separation (2) Biochemical methods (3) Chemical methods (4) Selective absorption.

Mechanical Separation: This method of resolution of recemic mixture applies only to solids which have well defined crystals. The crystals of recemic modifications consist of two types of crystals which are mirror images of each other and they can be separated by the use of a magnifying lens and tweezer. The method is too tedious to be used for practical purposes and is of only historic interest. Pasteur in 1848, separated the crystals of sodium ammonium tartarate by the above method. The Fig. 5.37 shows the dextrorotatory and laevorotatory sodium ammonium tartarate along with its recemic modification which crystallises at a higher temperature.

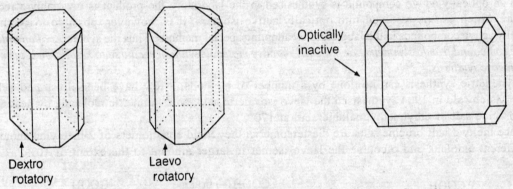

Fig. 5.37: Optically Active and Inactive Ammonium Tartarate Crystals

Biochemical Method : A biochemical process utilising one of the enantiomers leaving behind the other is the basis of resolultion by this method. When certain micro-organism such as yeast, mould or bacteria are allowed to grow in a solution of recemic mixtures, they utilise one form selectively leaving the other in solution. The separation by this method is almost quantitative as the form left in the solution can be isolated by fractional crystallisation.

A solution of (±) tartaric acid is converted into a laevorotatory tartaric acid when an ordinary mould *Penicillium glaucum* is added to the solution. The mould selectively utilised dextrorotatory tartaric acid due to the presence of optically active enzymes in the micro-organisms.

This method was developed by Pasteur in the year 1958, and is no doubt better than the earlier one but it too suffers a few disadvantages:

(i) One enantiomer which may or may not be useful is sacrificed.

(ii) Toxic racemic mixtures can not be separated by this method as they will kill the micro-organism and desired reaction will not commence.

Chemical Methods: In this method the recemic mixture is made to react with another suitable optically active compound and the products thus formed are not mirror image isomers and are separated by crystallisation. For example, a solution of recemic lactic acid is treated with optically active (–) Brucine which is a base so that the products formed are as follows:

(i)   (+) Acid.(–) base

(ii)   (–) Acid.(–) base

The above two salts do not show mirror image isomerism and are diastereomers and can be separated by crystallisation. Once the two salts are separated they can be further separated into optically active acids by treatment with sulphuric acid.

The same scheme maybe used for the separation of recemic mixtures of a base by using an isomer of optically active tartaric acid. Optically active acids are also used for the separation of recemic mixture of alcohols as the two react to give an ester.

Selective Adsorption: Resolution of recemic modifications can also be achieved by passing the solution of recemic mixture over a column of finely powdered optically active adsorbent such as sugar, Quartz or starch which selectively adsorbs one enantiomer leaving behind the other one in the reaction mixture thus the solution becomes rich in one kind of enantiomer. This method again is of limited significance as only partial resolution of recemic mixture is obtained.

5.6.9   Asymmetric Synthesis

When an optically active compound is synthesised in the laboratory the product is invariably a recemic mixture which can be resolved into optically active isomers. It is, however, possible to synthesise optically active compounds in the laboratory by adopting special methods. Thus the *syntheses of an optically active compound from a symmetric or optically inactive molecule without recourse to resolution is termed as Asymmetric synthesis.*

Asymmetric synthesis can be done by a number of methods, a few have been illustrated below.

1. Marchwald in 1904 synthesised the laevo isovaleric acid, an asymmetric molecule by heating the half brucine salt of ethylmethylmalonic acid at 170°C.

Since the two half brucine salts are diastereomeric, they yield enantiomers of 2-methylbutanoic acid in different amounts and ofcourse the laevo isomer in larger amount to the extent of 10%.

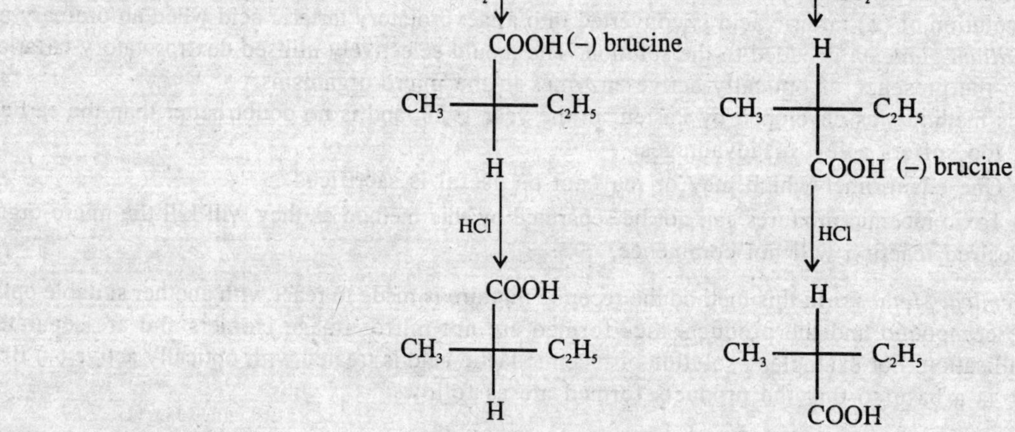

Enantiomers of 2-Methylbutanoic acid or Isovaleric acid

2. The reduction of pyruvic acid in laboratory yields a recemic mixture of lactic acid. However, if it is reduced by yeast, it produces laevorotatory lactic acid as reported by Mackenzie in the year 1905.

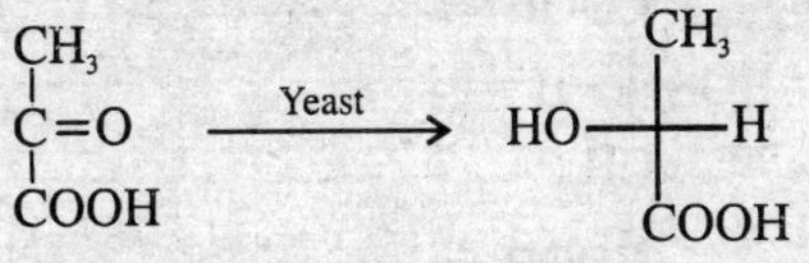

$$CH_3 - C(=O) - COOH \xrightarrow{\text{Yeast}} HO \underset{COOH}{\overset{CH_3}{\vert}} H$$

Pyruvic Acid   **(–) Lactic Acid**

3. Laevorotatory mandelic acid may be synthesised from benzaldehyde and hydrogen cyanide in presence of an enzyme emulsin followed by hydrolysis.

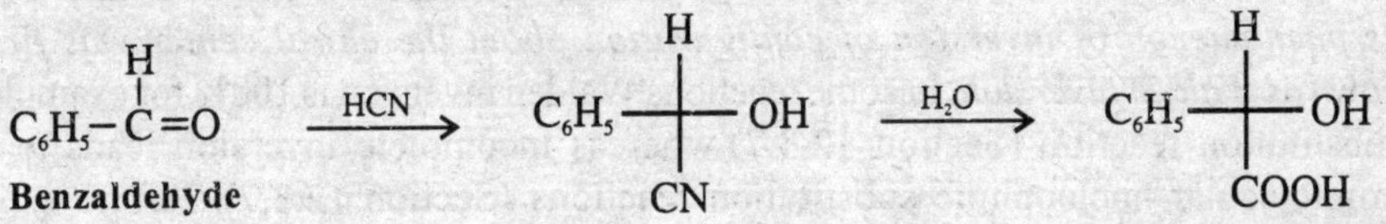

$$C_6H_5 - \overset{H}{\underset{}{C}} = O \xrightarrow{HCN} C_6H_5 \underset{CN}{\overset{H}{\vert}} OH \xrightarrow{H_2O} C_6H_5 \underset{COOH}{\overset{H}{\vert}} OH$$

Benzaldehyde   **(–) Mandelic Acid**

The laevorotatory mandelic acid is obtained as a major product.

4. The epoxidation of allyl alcohols with t-butyl-hydroperoxide and titanium tetraisopropoxide is commonly known as *sharpless asymmetric epoxidation*. The reaction is highly steroselective when pure (+) or (–) tartrate esters of substituted allyl alcohol are used. (+) tartrate ester produces (+) epoxide whereas (–) tartrate ester, (–) epoxide.

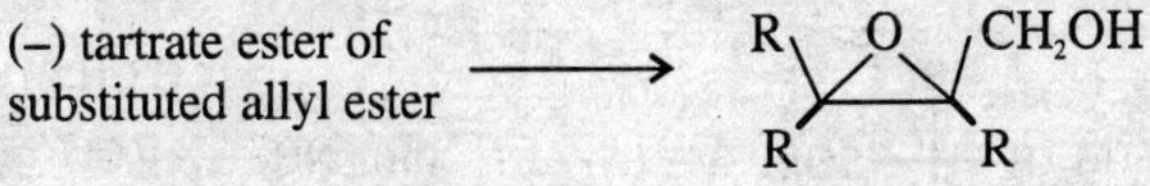

(–) tartrate ester of substituted allyl ester $\longrightarrow$

(–) Epoxide

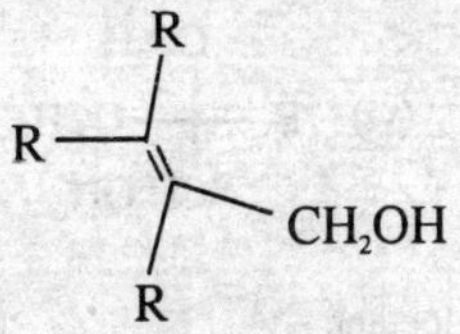

Substituted allyl alcohol

(+) tartarate ester of substituted allyl alcohol $\longrightarrow$

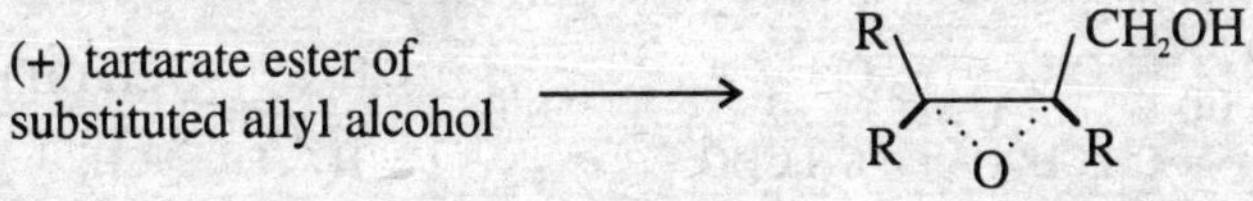

(+) Epoxide

5.6.10 Walden Inversion

Walden in 1893 observed the following conversions associated with enantiomers of malic acid

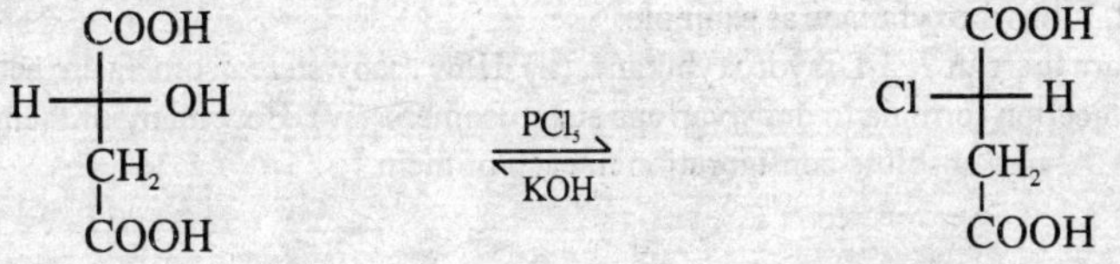

$$H \underset{\underset{COOH}{\overset{CH_2}{\vert}}}{\overset{COOH}{\vert}} OH \underset{\text{KOH}}{\overset{PCl_5}{\rightleftharpoons}} Cl \underset{\underset{COOH}{\overset{CH_2}{\vert}}}{\overset{COOH}{\vert}} H$$

D (+) Malic Acid   **L (–) Chlorosuccinic Acid**

$$\uparrow AgOH$$

```
        COOH
H ——|—— Cl
        CH₂
        COOH
```

$$\xrightleftharpoons[\text{KOH}]{\text{PCl}_5}$$

$$\downarrow AgOH$$

```
        COOH
HO ——|—— H
        CH₂
        COOH
```

D (+) Chlorosuccinic Acid   **L (–) Malic Acid**

The above transformations indicate the inversion of configuration at the asymmetric carbon atom converting the D(+) malic acid into its L(–) isomer and L(–) chlorosuccinic acid into D(+) isomer in two steps. *This phenomenon of inversion of configuration about the chiral center was first studied by Walden thus known as Walden inversion.* In some reactions Walden inversion is 100% for example bimolecular nucleophilic substitution reaction (Section 12.3.7) whereas incomplete inversion leads to recemisation for example unimolecular nucleophilic substitution reactions (Section 12.3.7).

5.7   PROBLEMS

1. Explain the term conformation and configuration.
2. Explain with the help of examples:
   (i) D and L notations   (ii) E and Z notations
3. Write possible stereoisomers of the following.
   (i) $CH_3CHBrCBr_2CH_3$ (iii) $CH_3CHOH\ CHOH\ CH_3$ (ii) $CH_3CHCl\ CHCl\ CH_3$ (iv) $CH_3CH(Cl)CH(CH_3)COOH$
4. Assign R and S notations to the following.

(i)
```
        Cl
H ——|—— Br
        I
```

(ii)
```
        Br
H ——|—— SO₃H
        Cl
```

(iii)
```
        CH₂CH₃
NH₂ ——|—— Cl
        C₆H₅
```

(iv)
```
        CH(CH₃)₂
CH₂=CH ——|—— H
        C≡CH
```

(v)
```
        CO₂H
NH₂ ——|—— CH₃
        Br
```

(vi)
```
        CO₂H
H ——|—— OCH₃
        CH₂OH
```

5. Draw Fischer projection formulae for the following.
   (i) R and S 2-Bromopentane   (iii) R and S 3-Chloro-3-methyloctane
   (ii) R and S 3-Chloro-1-pentene   (iv) R and S 2-pentanol
6. Assign E and Z notation to the following.

(i)
$$\begin{matrix} Cl \\ Br \end{matrix}\!\!>\!\!C=C\!\!<\!\!\begin{matrix} CH_3 \\ H \end{matrix}$$

(ii)
$$\begin{matrix} CH_3 \\ CH_3CH_2 \end{matrix}\!\!>\!\!C=C\!\!<\!\!\begin{matrix} Cl \\ CH_2CHO \end{matrix}$$

(iii)
$$\begin{matrix} C_5H_{11} \\ H \end{matrix}\!\!>\!\!C=C\!\!<\!\!\begin{matrix} CH(CH_3)_2 \\ CH_3 \end{matrix}$$

7. Explain why
   (i) The heat of combustion for methylene in cyclopropane is a higher value than in cyclohexane.
   (ii) Conformation of cyclopentane is puckered.
   (iii) It is difficult to synthesize large cycloalkane (C_{12}-C_{16}) rings.
   (iv) Why chair form of cyclohexane is a strain free conformation.
8. Answer the following taking 2,3-dihydroxybutane as example:
   (i) How many chiral carbons are there in 2, 3-Dihydroxybutane. (ii) How many stereoisomers are actually observed. (iii) Use Fischer projection formula to draw various stereoisomers. (iv) How many of them are optically active and why ?. (v) Assign absolute configuration to each of them.

9. Write various structural isomers and enantiomers if any, for the following.
   (i) Monochloropentane (ii) 3-methyl-3-hexanol (iii) 3-Phenyl-3-chloropropanol (iv) 3-bromo-3-pentanol
   Assign R and S notation also.
10. Explain, How the differences in solubilities of diastereomers be used to resolve a recemic mixture into individual enantiomers.
11. Write short note on the following.
    (i) Walden inversion.   (iv) Optical activity.
    (ii) Asymmetric synthesis.   (v) Elements of symmetry.
    (iii) Recemisation.   (vi) Resolution of recemic mixture.

6
Spectroscopy

6.1 INTRODUCTION

The usual methods of structure elucidation of organic compounds involved elemental analysis, determination of empirical formula, molecular formula, determination of functional groups and other chemical degradations. These methods, though dependable, consume lot of time and material. Compared to the chemical analysis, the spectroscopic methods are faster, accurate and require much less sample.

In general, spectroscopy is referred to as an area of study that gives information about the molecular structure with electromagnetic radiations. The instruments used are called spectrometers. Though different type of spectroscopic methods are available, the present discussion is restricted to the ultraviolet, infrared and nuclear magnetic resonance spectroscopy.

6.2 ULTRAVIOLET SPECTROSCOPY

Molecules absorb ultraviolet/visible light and undergo electronic transitions, thus UV spectroscopy is alternatively referred to as electronic spectroscopy. It gives the information about the extent of multiple bond and conjugation in an organic compound.

The UV region of primary interest to an organic chemist is from 200-380 nm. This is known as near or Quartz ultraviolet. Most of the unsaturated organic compounds absorb in this region. The saturated compounds absorb below 200 nm (Vacuum UV). The region extending from 380-800 nm is known as visible range. All colored compounds absorb in this region. The ultraviolet portion of the electromagnetic spectrum is indicated in Fig. 6.1.

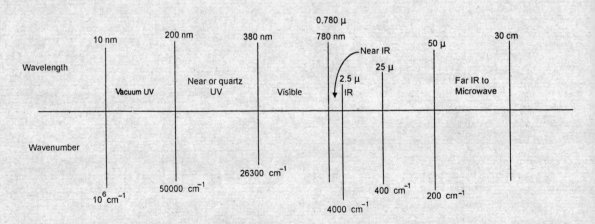

Fig. 6.1: UV portion of Electromagnetic Spectrum

6.2.1 Theory of Ultraviolet Spectroscopy

On subjecting a compound with multiple bonds to electromagnetic radiations in the ultraviolet/ visible region, a portion of the radiation is absorbed by the compound. The amount of absorption depends on the structure of the compound and the wave length of radiation. The ultraviolet spectrum is a plot of wavelength (or frequency) of absorption vs the absorption intensity (absorbance or transmittance). Thus, the ultraviolet spectrum records the wave length of an absorption maximum which is represented by λ_{max} in nanometers, nm (1 nm = 10^{-9} m). The intensity of absorption is given by the Beer-Lambert Law.

$$\log (I_o/I) = \varepsilon \, c.l$$

$$\varepsilon = A/c.l$$

or where I_o is the intensity of incident light; I is the intensity of transmitted radiation; A is absorbance (optical density); ε is molar absorptivity (molar extinction coefficient); c is the concentration of the solution (g.mol/l) and l is the length of the cell containing the solution (in cm).

The plot of curve of molar absorptivity (ε) or logarithm of molar absorptivity (log ε) versus wave length (λ) constitutes the absorption band (Fig. 6.2).

The molar absorptivity ε is constant for an organic compound at a given wavelength, and is expressed as ε_{max}, i.e., molar absorptivity at an absorption maximum.

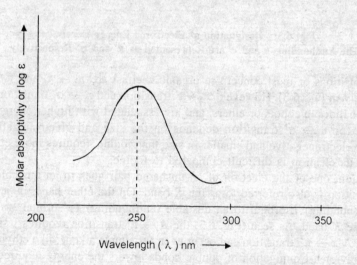

Fig. 6.2: **Representation of a UV Absorption Band**

It will be of interest to explain some terms that we frequently use in the discussion of ultraviolet spectroscopy. A *chromophore* is a covalently bonded unsaturated group and is responsible for electronic absorption like for example C=C, C=O and NO_2 groups. On the other hand, an *auxochrome* is a saturated group (e.g. –OH, –NH_2, –Cl) with nonbonded electrons, which when attached to a chromophore increases the intensity of color.

A *bathochromic shift (red shift)* is a shift of absorption to a longer wave length which may be due to substitution or solvent effect. On the other hand, a *hypsochromic shift (blue shift)* is the shift of absorption to a shorter wave length.

Hyperchromic effect is the increase in absorption intensity whereas, *hypochromic effect* is the decrease in absorption intensity.

It has already been mentioned that when a compound is subjected to electromagnetic radiations in the ultraviolet and visible region, a portion of the radiation is absorbed. This produces changes in the electronic state of the molecule due to transitions of valence electrons in the molecule. In these transitions the electrons from an occupied molecular orbital are excited to the next higher unoccupied orbital, i.e., the antibonding orbital. The relative energies of important orbitals are illustrated in Fig. 6.3.

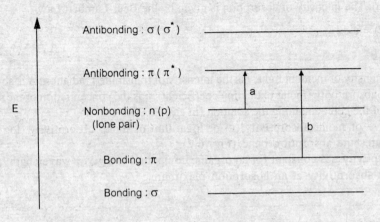

Fig. 6.3: Designation of Electronic Energy Levels.
The Antibonding π and σ are Represented as π* and σ* Respectively

The electronic transitions of most concern to organic chemist are $n \rightarrow \pi^*$ and $\pi \rightarrow \pi^*$ (represented as a and b respectively in Fig. 6.3). However, $\sigma \rightarrow \pi^*$ transition and $n \rightarrow \sigma^*$ transition involve saturated hydrocarbons, alkyl halides, alcohols or ethers, and are associated with higher energy. In alkyl halides, the energy required for $n \rightarrow \sigma^*$ transition depends on the size and electronegativity of halogens. Chlorine being more electronegative and smaller in size than iodine, requires more energy for transition as the excitation of its electron is difficult compared to iodine.

In $n \rightarrow \pi^*$ transition, one of the electron of an unshared pair goes to an unstable (antibonding) π^* orbital. $n \rightarrow \pi^*$ transition is also referred to as the R-band. On the other hand, in $\pi \rightarrow \pi^*$ an electron goes from a stable (bonding) π orbital to an unstable (antibonding) π^* orbital. $\pi \rightarrow \pi^*$ transition is also referred to as the K-band. As seen (Fig. 6.3) the $n \rightarrow \pi^*$ transition requires less energy compared to $\pi \rightarrow \pi^*$ transition. A $\pi \rightarrow \pi^*$ transition can occur even for simple alkene, like ethylene, but absorption occurs in the far ultraviolet. Conjugation of double bonds lowers the energy required for transition and absorption moves to longer wave length. In case there are enough double bonds in conjugation, absorption will move into the visible region and the compound will be colored. For example, β-carotene is a yellow pigment found in carrots and green leaves and is a precursor of vitamin A. It contains eleven carbon carbon double bonds in conjugation and owes its color to absorption at the violet end of the visible spectrum (λ_{max} 451nm).

B-Bands (benzenoid bands) are characteristic of the spectra of aromatic and heteroaromatic nucleus. The E-bands like the B-bands are characteristic of aromatic structures.

A transition may be an allowed or a forbidden one on the basis of extinction coefficient (ε_{max}) which is related to transition probability by following relation:

$$\varepsilon_{max} = 0.87 \times 10^{20} \times P \times a$$

where, P is the transition probability and a is the target area of absorption.

The target area of absorption in a molecule is a chromophore. A chromophore with a low transition probability has ε_{max} less than 1000 or in other words, absorption intensity is directly related to the nature of chromophore. Thus, on the basis of symmetry of molecule and ε_{max}, the transition may be an allowed or a forbidden one. Generally, the transitions having ε_{max} value more than 10^4 are allowed whereas those having a lesser values are forbidden transitions. Symmetrical molecules (for example benzene) have more restrictions on transition and are associated with forbidden transitions compared to less symmetrical molecules (having allowed transitions).

6.2.2   The Ultraviolet Spectrometer

It consists of a radiation source (a hydrogen or deuterium discharge lamp is normally used) giving radiation in the region 180-400 nm, a monochromator (which resolves the radiation from radiation source into narrow bands; silica prisms of different types or diffraction gratings are used as monochromators) and a sample cell (made of quartz or fused silica). The spectra are recorded in dilute solutions. Common solvents used are hexane, ethanol and cyclohexane as these are cheap and are transparent down to about 210 nm. The solution of the substance in a suitable solvent is taken in a cell and placed after the monochromator. A similar cell containing only the pure solvent is also placed along with the sample cell and both cells are mounted in a spectral holder so that two equal beams of ultraviolet light are passed - one through the sample cell and the other through the solvent cell. The transmitted beam is passed through a detector (phototubes or photovoltaic cells); this detects the photons in the transmitted beam of ultraviolet radiation.

6.2.3   Interpretation of UV Spectra: Conjugated Dienes, Trienes, Polyenes and Eneynes

The ultraviolet spectrum of conjugated dienes, trienes and tetraenes is shown in Fig. 6.4 which clearly indicates the increase in wave length of absorption with the increase in conjugation. This progression continues to a limit of 450 nm as the number of conjugated double bonds reaches to 11 or 12. The orange/red color of tomatoes and carrots is due to such conjugation.

A $\pi \rightarrow \pi^*$ transition occurs in simple alkenes and in conjugated dienes and trienes, but the difference in λ_{max} is due to the difference in the energy levels of highest occupied molecular orbital (HOMO) and lowest unoccupied molecular orbital (LUMO). The difference in HOMO and LUMO is less in case of

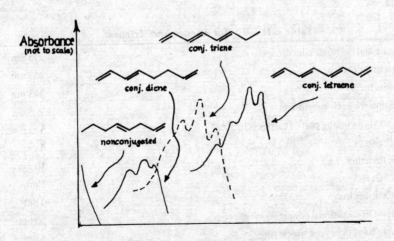

Fig. 6.4:   UV Spectra of Dienes, Trienes and Tetraenes

butadiene as shown in Fig. 6.5, thereby requiring less energy for $\pi \rightarrow \pi^*$ transition and absorbing the UV radiation of longer wavelength.

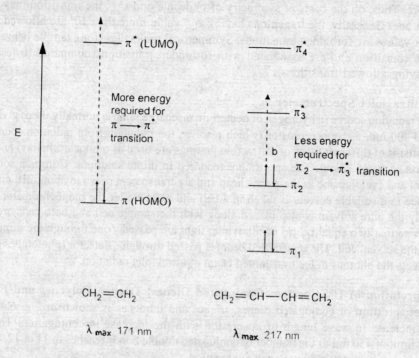

Fig. 6.5: Energy Levels of π MOs in Ethylene and Butadiene

The presence of most intense band may be correlated with the help of values given in Table 6.1 for λ_{max} commonly known as *Woodword Fieser rules or simply Woodword rules*. The λ_{max} for a particular compound is calculated by making a correct choice of base value and adding the contribution due to various substituents.

Table 6.1: λ_{max} for Dienes and Trienes

acyclic and heteroannular dienes	215 nm
homoannular dienes	253 nm
acyclic triene	243 nm
addition for each substituent	
–R alkyl (including part of carbocyclic ring)	5 nm
–OR alkoxy	6 nm
–SR thioether	30 nm
–Cl, -Br	5 nm
–OCOR acyloxy	0 nm
–CH=CH- additional conjugation	30 nm
double bond exocyclic to one ring	5 nm
double bond exocyclic to two rings	10 nm

The calculation of λ_{max} with the help of Woodward and Fieser rule is shown in the following examples.

Base value	253 nm	Base value	215 nm
4 ring residue	20 nm	3 ring residue	15 nm
2 exocyclic double bond	10 nm	exocyclic double bond	5 nm
		1 methyl residue	5 nm
Calculated λ_{max}	283 nm		240 nm
Observed λ_{max}	285 nm		239 nm

Base value	215 nm	Base value	243 nm
2 methyl residue	10 nm	3 methyl residue	15 nm
Calculated λ_{max}	225 nm		258 nm
Observed λ_{max}	226 nm		257 nm

In case of poly-ynes and eneynes, the intensities and values of λ_{max} increase with the increase in conjugation and substitution.

Woodword and Fieser rule holds good for the compounds whose structures are close to models.

6.2.4 Interpretation of UV Spectra: α, β–Unsaturated Carbonyl Compounds

α,β-unsaturated carbonyl compounds in which carbonyl double bond is in conjugation with olefinic double bond show weak $n \rightarrow \pi^*$ (R-band) and strong $\pi \rightarrow \pi^*$ (K-band) transitions. Fig. 6.6 shows the difference between a conjugated and non-conjugated carbonyl compound's absorption.

The position of $\pi \rightarrow \pi^*$ transitions associated with α,β-unsaturated carbonyl compounds may be calculated by use of data given in Table 6.2. The Woodword rule does not provide data for weak $n \rightarrow \pi^*$ transitions.

Polarity of solvent also plays an important role in electronic transitions as the carbonyl group is polar. The R-band or $n \rightarrow \pi^*$ transition due to polar solvent is associated with hypsochromic or blue shift whereas K-band or $\pi \rightarrow \pi^*$ transition with bathochromic or red shift (see Table 6.2).

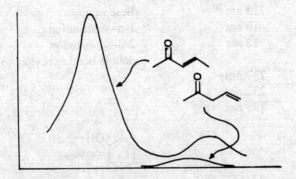

Fig. 6.6: **UV Spectra of Conjugated and Non-conjugated Carbonyl Compounds**

Table 6.2: λ_{max} for α , β -Unsaturated Carbonyl Compounds

ketones –C=C–CO– acyclic or six ring cyclic			215 nm	
five ring cyclic			202 nm	
aldehydes –C=C–CHO			207 nm	
acids and esters –C=C–CO$_2$H(R)			197 nm	
extended conjugation –C=C–C=C–CO–			30 nm	
homoannular second double bond			39 nm	

Addition for each substituents:

	α	β	τ	δ
–R alkyl including part of a carbocyclic ring	10 nm	12 nm	17 nm	17 nm
–OR alkoxy	35 nm	30 nm	17 nm	31 nm
–OH hydroxy	35 nm	30 nm	30 nm	50 nm
–SR thioether	–	80 nm	–	–
–Cl chloro	15 nm	12 nm	12 nm	12 nm
–Br bromo	25 nm	30 nm	25 nm	25 nm
–OCOR acyloxy	6 nm	6 nm	6 nm	6 nm
–NH$_2$, -NHR, –NR$_2$	–	95 nm	–	–
one exocyclic double bond one ring		5 nm		
exocyclic double bond to two rings together		10 nm		

Solvent Shifts:

methanol	0
chloroform	– 1 nm
dioxan	– 5 nm
diethylether	– 7 nm
hexane, cyclohexane	–11 nm
water	+ 8 nm

Following examples illustrate the method of calculation of λ_{max} for various compounds.

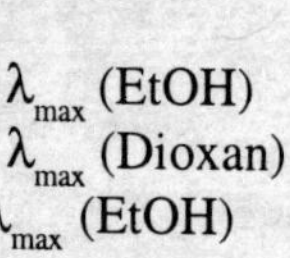

Base value	215 nm	Base value		215 nm
α–substitution	10 nm	1 α–substitution		10 nm
β–substitution	12 nm	2 β–substitution		24 nm
		double bond exocyclic to 2 rings		10 nm
Calculated λ$_{max}$ (EtOH)	237 nm			259 nm
Calculated λ$_{max}$ (Dioxan)	232 nm			254 nm
Observed λ$_{max}$ (EtOH)	235 nm			257 nm

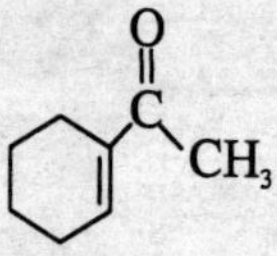

Base value	215 nm	Base value	215 nm

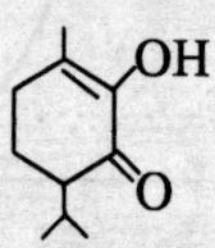

α–substitution	10 nm	α–hydroxy	35 nm
β–substitution	12 nm	2 β–substitution	24 nm
Calculated λ_{max} (EtOH)	237 nm		274 nm
Observed λ_{max} (EtOH)	232 nm		270 nm

Base value	215 nm	Base value	202 nm
α–substitution	10 nm	α–hydroxy	35 nm
2 β–substitution	24 nm	β–substitution	12 nm
Calculated λ_{max} (EtOH)	249 nm		249 nm
Observed λ_{max} (EtOH)	243 nm		247 nm

Base value	197 nm
2 β–substitution	24 nm
Calculated λ_{max} (EtOH)	221 nm
Observed λ_{max} (EtOH)	220 nm

6.2.5 Interpretation of UV Spectra: Benzene and its Derivatives

Benzene shows three absorption bands at 184 nm (ε_{max} 60000), 204 nm (ε_{max} 7900) and at 254 nm (ε_{max} 200). These bands originate as a result of $\pi \rightarrow \pi^*$ transitions. The intense transition (ε_{max} 60000) at 184 nm is due to allowed transition while weak bands around 200 nm and 250 nm are due to forbidden transitions of highly symmetrical benzene molecule. Also, benzene and its derivatives show fine B-band (near 200 nm) in either vapor phase or in nonpolar solvents. In polar solvent, however, interaction between solute and solvent reduces the fine structure.

The B-band of benzene shows a bathochromic shift when substituted with methyl group due to hyperconjugation in which σ electrons of C–H bond conjugates with ring electrons. In case of two alkyl groups, the effect is maximum when the alkyl groups are at para position (steric effect reduces hyperconjugation and ε_{max} in case of ortho disubstituted benzene derivatives).

Table 6.3 gives the λ_{max} for various bands of benzene derivatives in ethanol.

Table 6.3: λ_{max} for Benzene Derivatives

Parent chromophore, Ar=C_6H_5		
G = alkyl or ring residue (ArCOR)		246 nm
G = H (ArCHO)		250 nm
G = OH, OR (ArCOOH, ArCOOR)		230 nm
Addition for each substituent on Ar		
	o,m	p
alkyl or ring residue	3 nm	10 nm
–OH, –OCH$_3$, –OR	7 nm	25 nm
–Cl	0	10 nm
–Br	2 nm	15 nm
–NH$_2$	13 nm	58 nm
–NHCOCH$_3$	20 nm	45 nm
–NHCH$_3$	–	73 nm
–N(CH$_3$)$_2$	20 nm	85 nm

Following examples will illustrate the method of calculation of λ_{max} by Woodward and Fieser rule.

Base value	246 nm
o-ring residue	3 nm
p-hydroxy	25 nm
Calculated λ_{max} (EtOH)	274 nm
Observed λ_{max} (EtOH)	276 nm

Base value	246 nm
o-ring residue	3 nm
p-methoxy	25 nm
m-methoxy	7 nm
	281 nm
	279 nm

6.2.6 Interpretation of UV Spectra: Polynuclear Hydrocarbons

The electronic spectra of aromatic compounds becomes more and more complicated with the fusion of benzene rings as shown in Fig. 6.7 which incorporates the spectra of benzene (180 nm), naphthalene (314 nm) and anthracene. This provides useful fingerprints for identification. Thus phenanthrene (Fig. 6.7b) can be identified from anthracene (Fig. 6.7a) with the help of UV spectra.

6.2.7 Application of UV Spectroscopy

As mentioned earlier, the ultraviolet spectroscopy is mainly a diagnostic device for conjugated systems. The organic chemist can make its use to find out the extent of conjugation and to distinguish between conjugated and unconjugated systems. The electronic spectra also gives an account of strain in the organic molecule due to the presence of bulky groups and the geometrical isomerism and tautomerism associated with them. The UV spectra may also be used to study the structural changes in a compound with the change of solvent.

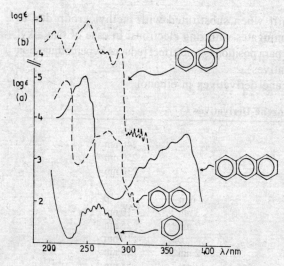

Fig. 6.7: UV Spectra of Polynuclear Hydrocarbons

6.3 INFRARED SPECTROSCOPY

Infrared spectroscopy is an important technique, for the identification of functional groups in organic compounds. It is equally important in providing information about the functional groups absent. Infrared spectroscopy does not provide complete information about the structure of the compound. However, it is used for establishing the identity (superimposable IR) or non-identity (non-superimposable IR) of the compounds.

Organic compounds absorb some frequencies of infrared radiations, while transmit others. The plot of percent absorbance/percent transmittance against frequency is nothing but the IR spectra.

6.3.1 Molecular Vibrations and the IR Spectrum

The infrared radiations do not cause excitation of electrons (as in case of UV spectroscopy) due to less energy associated with them. The radiations, however, cause vibration of atoms within the molecules which are quantised and absorb infrared radiations of a particular energy.

Atoms in a molecule vibrate in a variety of ways. Fig. 6.8 shows the various stretching and bending vibrations in an AX_2 system (the ball like structure represents the atoms in a molecule).

As shown in the Fig. 6.8, stretching alters the distance between the two atoms without affecting the bond axis. So these vibrations require high energy and occur at a higher frequency. In the bending vibrations or deformations, however, the distance between the two atoms remains constant but the bond axis may change as a result of twisting, rocking, scissoring or wagging etc. Therefore, they require less energy and occur at a lower frequency or a higher wave length. For a nonlinear molecule having n atoms, the number of possible vibrations are *3n-6*. Thus, a molecule of ethane has theoretically *18* and butane has *36* possible infrared absorption bands.

Only those vibrational changes that result in a change in the dipole moment are observed in the IR spectrum and are said to be IR active. The mass of the atoms and the strength of the bond involved in vibrations is also an important factor which influences the vibrational frequency. The two C=O bonds present in the molecule of carbon dioxide can show symmetric and unsymmetric stretching shown below, but the symmetric stretching does not result in the change in the dipole moment and hence is IR inactive. The asymmetric stretch, on the other hand, is IR active. Similarly the C=C stretch in p-xylene is IR inactive.

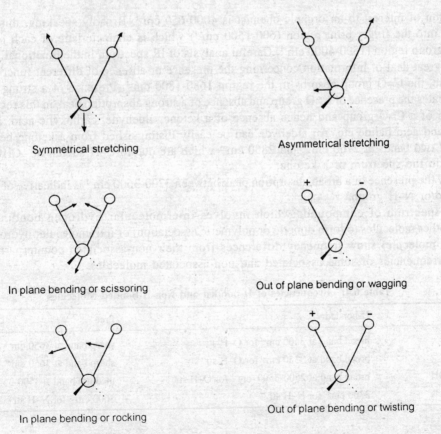

Symmetrical stretching   Asymmetrical stretching

In plane bending or scissoring   Out of plane bending or wagging

In plane bending or rocking   Out of plane bending or twisting

Fig. 6.8: **Stretching and Bending Vibrations**

An isolated C–H bond has only one stretching vibration, whereas C–H bonds in CH_2 or CH_3 have more than one stretching vibrations ($3n$-6) which can combine together to give coupled vibrations of same or different frequency. This is referred to as *vibrational coupling*. Vibrational coupling arises as a result of fundamental vibrations of two or more bonds (present at a reasonable distance with respect to each other) with a similar frequency. Coupling of two fundamental vibrations produces two new modes of vibrations with frequencies higher and lower than that observed when there is no interaction. Interaction, however, can occur between fundamental and overtone vibrations. When fundamental vibrations couples with the overtone of some other vibration, the phenomenon is referred to as *Fermi resonance*, first described by Enrico Fermi. For example, carbon dioxide exhibits Fermi resonance and gives two bands near 1286 cm^{-1} and 1388 cm^{-1}. The splitting is due to the coupling of fundamental C=O stretching vibration and overtone of bending vibrations.

The energy required for causing changes in these modes of vibrations is provided by the IR radiations. The exact amount of energy required depends on the strength of the bonds and the atoms attached through the bonds. The plot of % energy absorption (or % transmittance) vs the wavelength (in microns, $1\mu=10^4 Å$) or wave number (in cm^{-1}) of radiations is known as the infrared spectrum. The IR spectra of an organic compound shows a wide variety of absorptions because the bonds present in them are associated with various types of vibrational changes. Thus, the straightforward interpretation of IR spectra is difficult. But, same functional groups present in different compounds appear at a similar frequency.

The region of interest to an organic chemist is 4000-650 cm^{-1}. Broadly speaking, this region may be divided into the finger print region (600-1500 cm^{-1}) which is characteristic of each molecule and functional group region (1500-4000 cm^{-1}). Careful analysis of IR spectrum in the functional group region provides a great deal of information concerning the presence or absence of different functional groups. For example, the C=O group absorbs in the region 1680-1800 cm^{-1}. Presence of a strong peak in this region indicates the presence of C=O group and absence of a strong absorption peak in this region indicates the absence of a C=O group and hence absence of a ketone, aldehyde, carboxylic acid, amide, ester, anhydride and acid halide etc. An aldehyde can be easily distinguished from a ketone because of the presence of two bands at 2760 cm^{-1} and 2850 cm^{-1} which are due to C–H stretch of –CHO group and are absent in the spectrum of a ketone.

Similarly the presence of a broad absorption peak between 3200-3600 cm^{-1} is indicative of the presence of O–H and/or N–H group.

The IR spectrum of compounds which involves intermolecular hydrogen bonding (with the solvent or other molecules to form dimeric or polymeric association) or intramolecular hydrogen bonding (within the molecule) show frequency differences from their non-associated counterparts. Table 6.4 shows the frequencies of some associated and non-associated molecules.

Table 6.4:   Frequencies of H-bonded and Non-H-bonded Molecules

Group	H-bonded	Free
R-OH	broad band at 3500 cm^{-1} for O–H str	sharp band at 3650 cm^{-1}
Ar-OH	broad band at 3350 cm^{-1} for O–H str	sharp band at 3650 cm^{-1}
R-COOH	broad band at 2500-3500 cm^{-1} for O–H str	medium band at 3500-3550 cm^{-1}
R-NH$_2$	3300 cm^{-1} for N–H str	3600 cm^{-1} for N–H str

The extent of inter and intramolecular hydrogen bonding depends on temperature and concentration. Intermolecular hydrogen bonding diminishes at a very low concentration in nonpolar solvent. Intramolecular hydrogen bonding, on the other hand, is an internal phenomenon and exists even at a very low concentration.

Alcohols, phenols, carboxylic acids and amines show a broad band due to O–H or N–H stretch when their IR spectrum is recorded in polar solvents or even in dilute solutions along with the usual sharp band. The reason for the appearance of broad band is the H–bonding associated with these molecules. In inert solvents (or vapor phase) and dilute solutions, however, proportion of free molecules increases, thus the spectra gives absorption bands for free molecules.

The broadening of the band in associated molecules may be shown by taking the example of alcohols which may be represented as following structures:

$$\underset{\textbf{I}}{R-O-H \quad \overset{\overset{\displaystyle H}{|}}{O}-R} \longleftrightarrow \underset{\textbf{II}}{R-\bar{O} \quad \overset{+}{\underset{}{H}}-\overset{\overset{\displaystyle H}{|}}{\underset{+}{O}}-R} \longleftrightarrow \underset{\textbf{III}}{R-\overset{\delta^-}{O}\cdots H\cdots\overset{\overset{\displaystyle H}{|}}{\underset{\delta+}{O}}-R}$$

The O–H bond in structure III shows lengthening due to hydrogen bonding, which weakens the bond. This results in lower frequency of stretching vibration.

6.3.2   The IR Spectrometer

A simplified diagram of the various parts of the IR spectrophotometer is shown in Fig. 6.9.

A continuous beam of IR radiation produced by a heated filament is separated into radiation of different wave lengths by passing through a prism (made from sodium chloride). Rotation of the prism permits the different wave lengths to pass through the focussing slit to produce IR radiation of a single wave length. The beam then splits into two beams of equal intensity, one passing through the sample and other through the reference. The beams then strike an IR detector. The detector converts the transmittance into an electric signal that is amplified and sent to the recorder that traces the IR spectrum.

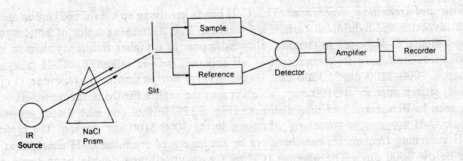

Fig. 6.9:   The IR Spectrometer

IR spectra can be determined for solids, liquids as well as gases. Glass containers can not be used for taking the sample or its solution as glass absorbs very strongly in the IR region. For same reasons quartz, common organic solvents (except $CHCl_3$, CCl_4 and CS_2) cannot be used. The sample should be free from moisture also.

IR spectra of a solid sample may be determined in one of the following ways:

(i) *KBr Pellets*: Solid compound (50 mg) is finely powdered with spectral grade potassium bromide (0.5 g). It is then converted into a pellet by applying high pressure to the mixture. The pellet is placed in the spectrometer and IR recorded. Though this method is generally used for solids, special precautions should be taken to use dry potassium bromide and to exclude moisture during pellet formation. The presence of moisture interferes in detection of those functional groups which absorb in this region, e.g., O–H, N–H, S–H etc.

(ii) *Solution method:* The dry solid compound is dissolved in anhydrous CCl_4, $CHCl_3$ or CS_2 (20% solution) and the solution is placed in the sample cell. Pure solvent is taken in the solvent cell and the IR is recorded in a double beam spectrometer which automatically cancels the absorption due to solvent. Small absorption peaks are always observed because of unequal concentration of solvent in the two cells, these peaks can however be ignored.

Note: CS_2 absorbs at 1500 cm^{-1} and absorption due to C-Cl occurs between 600-800 cm^{-1}.

(iii) *Mull method:* A dispersion of the finely powdered organic compound in Nujol (A mixture of saturated hydrocarbon containing 20-24 carbon atoms) or perfluorokerosene (fluorinated kerosene containing C_{12}-C_{18} linear hydrocarbons) can be used instead of the solution of the organic compound and the spectrum recorded.

IR spectrum of pure (neat) liquid compound can be recorded using NaCl windows. A small amount of the liquid is placed on a moisture free NaCl plate (window), it is covered with another NaCl plate and the spectrum recorded. If the compound is a low boiling liquid, some of it tends to evaporate during the time taken to record the spectrum and a uniform concentration of the compound is not maintained throughout. Also a uniform thickness of the sample is not possible.

6.3.3 Characterisation of Organic Compounds by IR Spectroscopy

The infrared spectra of organic compounds is used to find out the presence or absence of functional groups, hydrogen bonding, Fermi resonance etc. Fig. 6.10 gives a correlation of absorption bands of some important groups.

(i) *Aliphatic and Aromatic hydrocarbons:* The C–H bonds involving sp hybridised carbon are strongest and those involving sp^3 hybridised carbon are weakest. The decreasing order of bond strength is sp > sp^2 > sp^3. Hence C–H bonds in terminal alkynes appear at a higher frequency than in terminal alkenes or alkanes. Stretching frequencies of –C–H (alkane), =C–H (alkene), ≡C–H (alkynes) are 2800-3000 cm^{-1}, 3000-3100 cm^{-1}, 3300 cm^{-1} respectively. For example, in 1-octene (Fig. 6.11) =C–H (olefinic) stretch appears at 3100 cm^{-1}, which is also the region for the aromatic =C–H (olefinic) stretching as seen in IR spectrum of propylbenzene (Fig. 6.12). In fact one may detect several bands due to aromatic C–H asymmetric stretching vibrations in the 3000-3100 cm^{-1} region. The olefinic and aromatic C–H stretching frequencies can therefore be distinguished from the C–H absorptions, as the latter occurs below 3000 cm^{-1}. The alkenes (CH_2=C) with vinylidene double bond show olefinic =C–H bend at 900 and 980 cm^{-1}. Alkene structure introduces a new mode of vibration, C=C stretching at 1667-1640 cm^{-1} (moderate to weak absorption for unconjugated alkene) as shown in Fig. 6.11. Conjugation always lowers the absorption frequency and being predominant in aromatic compounds, absorption of C=C stretching within the ring occur at 1600-1585 and 1500-1400 cm^{-1} (Fig. 6.12). Absorption frequencies for alkanes, alkenes and alkynes are given in Table 6.5.

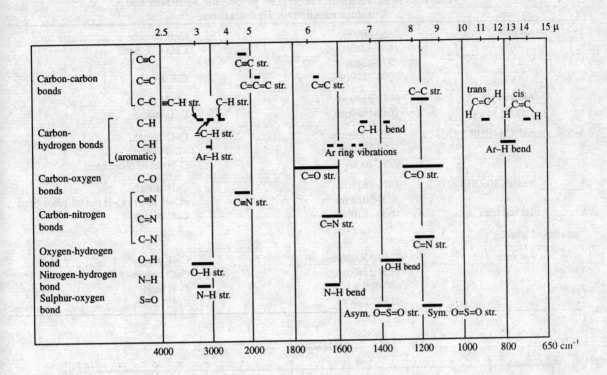

Fig. 6.10:   Correlation of Absorption Bands of Some Important Bondings

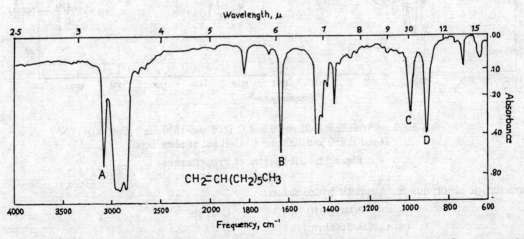

$$CH_2=CH(CH_2)_5CH_3$$

A 3100 cm^{-1} (olefinic C–H str); B 1645 cm^{-1} (C=C str) C and D 980 and 900 cm^{-1} (=C–H out of plane bend)

Fig. 6.11:   IR Spectra of 1-Octene

**Table 6.5:   Characteristic Frequencies in cm⁻¹ for Saturated and
Unsaturated Aliphatic Hydrocarbons**

Alkanes	2850-3000(s)	C–H stretch
	1480-1440(s)	CH_2, CH_3 bend
	1370-1380(s)	
	2900-2880(w)	CH stretch
Alkenes	3080-3140(m)	=C–H stretch
	1640-1650(m)	C=C stretch
cisRCH=CHR	1655-1660(m)	C=C stretch
	3020-3010(m)	=C–H stretch
	675-720(m)	C–H out of plane bend
transRCH=CHR	1670-1675(m)	C–H stretch
	3020-3010(m)	=C–H stretch C–H out of plane bend
Tri and tetra	1660-1670(w)	C=C stretch
substituted alkenes		
Alkynes	3310-3300(s)	C–H stretch
	2100-2140(m)	C≡C stretch
	660-700(w)	C–H bend
	2190-2260(w)	C≡C stretch

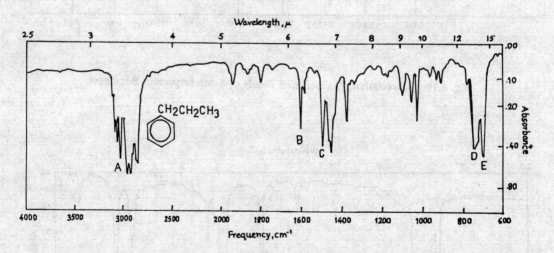

A 3050 cm⁻¹ (olefinic C–H str; B and C 1600 and 1500 cm⁻¹ (C=C str)
D and E 750 and 700 cm⁻¹ (=C–H out of plane bend)

Fig. 6.12:   IR Spectra of Propylbenzene

Characteristic absorption for Aromatic hydrocarbon

3040-3010 (m)	C–H str
~ 1600 (m)	
~ 1580 (m)	C=C ring stretch
~ 1500 (m)	
below 900	=C–H bend out of plane

(ii) *Alcohols and Phenols:* Alcohols and phenols may be recognised from their IR spectra in the region 3200-3600 cm⁻¹ and C–O stretch in 1000–1200 cm⁻¹. The latter region is used to distinguish between primary, secondary and tertiary alcohols. Order of absorption of stretching frequency is 3º > 2º >1º alcohols. Like alcohols, phenols also show both O–H and C–O stretching. The C–O stretching of phenols appear at a higher frequency than in alcohols (compare Fig. 6.13 and Fig. 6.14 for benzyl alcohol and phenol).

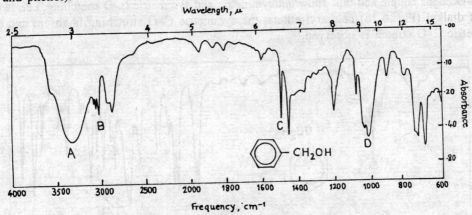

A 3300 cm⁻¹ (O–H str); B 3100 cm⁻¹ C 1500 cm⁻¹ (C=C ring str); D 1017 cm⁻¹ (C–O str)

Fig. 6.13: IR Spectra of Benzyl Alcohol

Characteristic absorptions for

RCH₂OH	3600-3200(s)	O–H Stretch
	1350-1250(s)	O–H Bend
	1080-1010	C–O Stretch
R₂CHOH	3600-3200(s)	O–H Stretch
	1410-1500(s)	O–H Bend
	1120-1030	C–O Stretch
R₃COH	3600-3200(s)	O–H Stretch
	1400-1500(s)	O–H Bend
	1170-1100	C–O Stretch

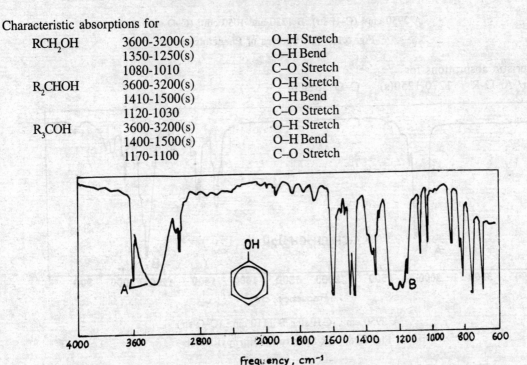

A 3350 and 3600 cm⁻¹ (O–H str); B 1200 cm⁻¹ (C–O str)

Fig. 6.14: IR Spectra of Phenol

Characteristic absorptions for
Intermolecular hydrogen bonded phenols
              3500-3450(s) O–H Stretch

(iii)   *Ethers:* Ethers may be recognised by their strong C–O stretching bands at 1060-1275 cm^{-1} which falls in the finger print region. In case of highly unsymmetrical ethers e.g., alkyl aryl ethers (Fig. 6.15) the two C–O bonds couple and thus show antisymmetric and symmetric C–O stretching, i.e., two bands. In case of dialkyl (Fig. 6.16) or diaryl ethers, the symmetric C–O stretching is absent and only the antisymmetric C–O stretch is observed.

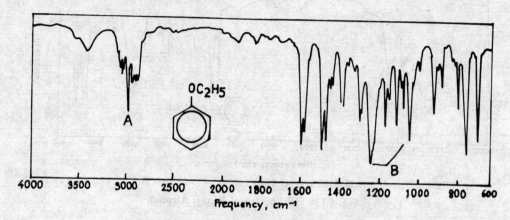

A 2950 cm^{-1} (C–H str); B 1230 and 1050 cm^{-1} (C–O str)

Fig. 6.15: IR Spectra of Phenetole

Characteristic absorptions for
Ar-O-Ar/ Ar-O-R   1270-1250(s)   C–O Stretch

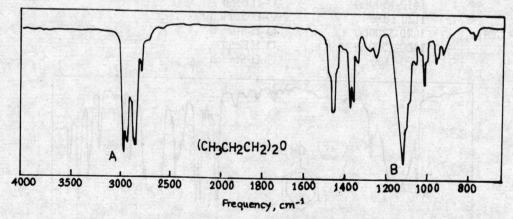

A 2950 cm^{-1} (C–H str); B 1110 cm^{-1} (C–O str)

Fig. 6.16:   IR Spectra of Dipropyl Ether

Characteristic absorptions for
Saturated acyclic ether –CH$_2$–O–CH$_2$ 1150-1060(s) C–O Stretch

(iv)   *Aldehydes and Ketones:* A strong absorption in the region 1640-1760 cm⁻¹ denotes C=O stretching which may be due to presence of ketones, aldehydes, carboxylic acids, esters, amides and other acid derivatives. Aliphatic aldehydes generally absorb 10-15 cm⁻¹ above the corresponding ketones as acetaldehyde absorbs at 1730 cm⁻¹. Conjugation lowers the C=O stretching frequency by about 30 cm⁻¹ as seen in IR spectra of benzaldehyde (Fig. 6.17). Aldehydes and ketones can be distinguished easily by observing two characteristic C–H stretching for aldehydes only between 2830-2695 cm⁻¹ (no such peak is seen in the spectra of ketone, Fig. 6.18).

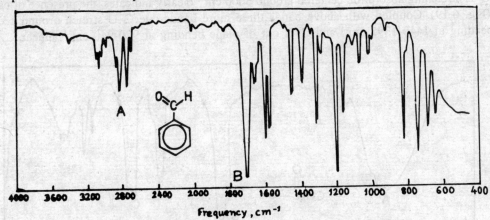

A 2800 and 2780 cm⁻¹ (C–H Str of aldehyde); B 1700 cm⁻¹(C=O Str)

Fig. 6.17:   IR Spectra of Benzaldehyde

Characteristic absorptions for

Aliphatic aldehyde	1740-1720 (s)	C=O Stretch
RCHO	2870-2810	C–H Stretch
	975-780 (w)	C–H Bend
Aromatic aldehyde		
ArCHO	1715-1695 (s)	C=O Stretch

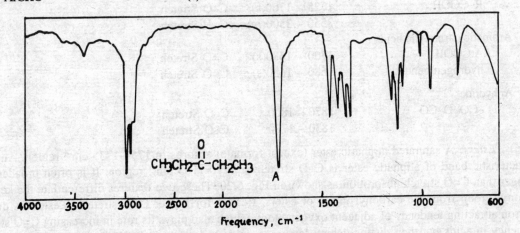

A 1715 cm⁻¹ (C = O Str)

Fig. 6.18:   IR Spectra of 3-Pentanone

Characteristic absorptions for

Aliphatic acyclic ketones, R–CO–R   1720-1700 (s)   C=O Stretch

Aryl alkyl ketones, (Ar–CO–R)   1700-1680 (s)   C=O Stretch

Diaryl ketones, (Ar–CO–Ar)   1670-1660 (s)   C=O Stretch

(v)   *Carboxylic Acids:* The C=O stretching bands in carboxylic acids are considerably more intense than the ketonic C=O stretching bands. Copresence of C=O stretching at 1700-1725 cm^{-1} with very broad O–H stretching absorption centered around 3000 cm^{-1} clearly indicates the presence of carboxylic acids (Fig. 6.19). Coupled with above bands three more bands due C–O stretch around 1320-1210, O–H bending at 1440-1395 cm^{-1} and O–H out of plane bending at 920-940 cm^{-1} appear.

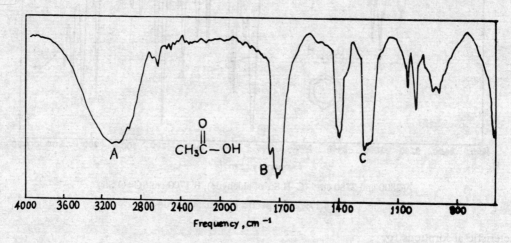

A 3100 cm^{-1} (O–H str); B 1700 cm^{-1} (C=O str); C 1290 cm^{-1} (C–O str)

Fig. 6.19:   IR Spectra of Acetic Acid

Characteristic absorptions for

Aliphatic carboxylic acids

R–COOH	1725 – 1700(s)	C–O Stretch
	3550 – 3500(m)	O–H Stretch
Aromatic carboxylic acids		
ArCOOH	1700 – 1680(s)	C=O Stretch
hydrogen bonded	1680 – 1650(s)	C=O Stretch
Anhydride		
–CO–O–CO–	1870 – 1800	C–O Stretch
	1830 – 1720	C–O Stretch

(vi)   *Esters:* A saturated aliphatic ester (except formate) absorbs in 1750-1735 cm^{-1} region. Another characteristic band of aliphatic ester is C–O stretch in 1210-1163 cm^{-1} region. It is often broader and stronger than C=O stretch absorption as shown in Fig. 6.20. These two features differentiate the ketonic carbonyl group from the carbonyl group of esters. Reason for higher C=O stretch in esters is due to electron attracting tendency of adjacent oxygen atom, which also plays its role in increasing C=O stretch frequency in acids but then distinguishing feature in acids and esters is O–H absorption (Fig. 6.19 and Fig. 6.20).

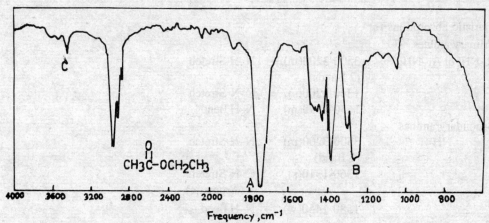

A 1740 cm⁻¹ (C=O str); B 1250 cm⁻¹.(C–O str); C 3478 cm⁻¹
(overtone of C=O str normally twice that of C=O str)

Fig. 6.20: IR Spectra of Ethyl Acetate

Characteristic absorptions for

Esters	1750-1735(s)	C=O Stretch
(saturated, acyclic)	1280-1050	C–O Stretch
	2 bends)	

(vii) *Amines:* Stretching frequencies of N–H bonds of amines and amides give absorption in about the same region where O–H absorbs and thus interpretation may be difficult. However, due to oxygen being more electronegative than nitrogen, O–H stretching band is stronger than N–H stretching (Fig. 6.21). Primary amines display two bands in region 3200-3500 cm⁻¹ of N–H stretching. Secondary amines show a single weak band in 3350-3310 cm⁻¹ region. The other bands in the spectrum of amines are N–H bending vibrations in the region 1580-1650 cm⁻¹, medium to strong and C–N stretching vibrations in 1020-1360 cm⁻¹. Aromatic amines show strong C–N stretching absorption near 1280-1300 cm⁻¹. The higher frequency of aromatic amines is due to the resonance of lone pair of electron on nitrogen with benzene ring.

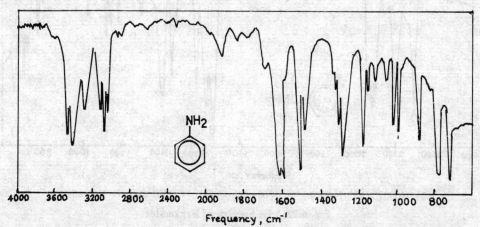

A 3440 and 3350 cm⁻¹ (N–H str); B 1280 cm⁻¹ (C–N str); C 1620 cm⁻¹ (N–H bend)

Fig. 6.21: IR Spectra of Aniline

Characteristic absorptions for

Primary amines		
R–NH$_2$ or Ar–NH$_2$	3500-3200(m) (2 Bands)	N–H Stretch
	1340-1200(s)	C–N Stretch
	1650-1590(m)	N–H Bend
Secondary amines		
–NH–	3500-3200(m) (1 Band)	N–H Stretch
	1360-1310(s)	C–N Stretch (Aromatic)
	1580-1450(w)	N–H Bend
Tertiary amines		
| –N–	1360-1310(s)	C–N Stretch (Aromatic)
Imines		
=NH	3400-3000(m)	N–H Stretch

(viii)   *Amides:* Another nitrogen containing functional group is amide which shows a carbonyl absorption band around 1630-1700 cm^{-1} known as amide I band. Resonance of lone pair of nitrogen with carbonyl group in amide, shifts the carbonyl absorption to lower frequency region from normal carbonyl absorption (1720-1680 cm^{-1}). The N–H bending in the region of 1650-1515 cm^{-1}, known as amide II bands, differentiate primary and secondary amides from tertiary amides (Fig. 6.22). Secondary amides absorb at a lower frequency region (1570-1651 cm^{-1}) as compared to their primary counterparts (1650-1600 cm^{-1}). Like amines, primary and secondary amides also exhibit N–H stretching frequency of medium intensity in the region 3140-3520 cm^{-1}; two bands for primary amides and one band for secondary amides.

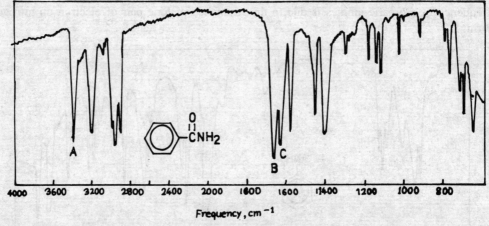

A 3370 cm^{-1} (N–H str);   B 1660 cm^{-1} (C=O str) C 1631 cm^{-1} (N–H Bend)

Fig. 6.22:   IR Spectra of Benzamide

Characteristic absorptions for Amides

Free –$CONH_2$	1630-1700(s)	C=O Stretch (Amide I)
	3400-3500(m)	N–H Stretch
	1650-1550	N–H Bend (Amide II)
Bonded –$CONH_2$	3350-3180	N–H Stretch
Free –CONH–	1670-1700(s)	C=O Stretch
	3430(m)	N–H Stretch
	1550-1510	N–H Bend
Bonded –CONH–	3320-3140(m)	N–H Stretch

tert Amides

–CO$\overset{\mid}{N}$–	1670-1630(s)	C=O Stretch

(ix)   *Nitro and Nitrile Group:* For nitro group two bands appear in the region 1370-1570cm⁻¹ due to symmetrical and antisymmetrical N–O stretching (Fig. 6.23). Aromatic nitro compounds again absorb at a slightly lower frequency region than aliphatic compounds.

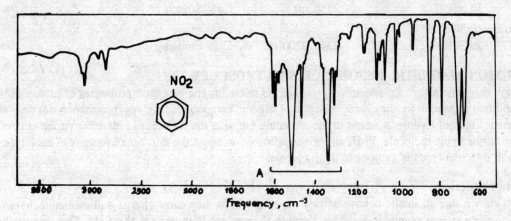

A 1520 cm⁻¹ and 1345 cm⁻¹ (N=O str)

Fig. 6.23:   IR Spectra of Nitrobenzene

Characteristic absorptions for
    Aliphatic and Aromatic nitro compounds

-NO_2	1570-1550(s)	N=O Stretch
	1380-1370(s)	N=O Stretch

Spectrum of nitriles depicts clearly C≡N stretch for nitriles at 2200-2250 cm⁻¹. For example, benzonitrile absorbs at 2210 cm⁻¹ (Fig. 6.24).

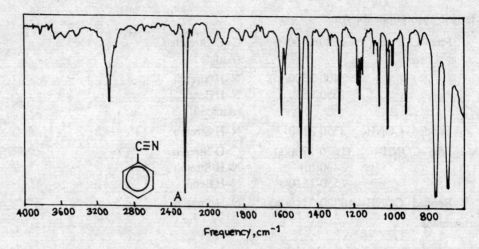

A 2210 cm^{-1} (C≡N str)

Fig. 6.24: IR Spectra of Benzonitrile

Characteristic absorptions for

Aliphatic Nitriles		
RC≡N	2250(m)	C≡N Stretch
Aromatic Nitriles		
ArC≡N	2240-2200(m)	C≡N Stretch

6.4 PROTON MAGNETIC RESONANCE SPECTROSCOPY

The proton magnetic resonance spectroscopy (PMR) is related to the magnetic properties of proton. It is the most important technique for the characterisation of organic compounds. It gives information not only about the different kinds of hydrogen atoms in the molecule but also the different kinds of environments of the hydrogen atoms in the molecule. PMR also gives information about the number of protons of each type and the ratio of different types of protons in the molecule.

6.4.1 Introduction

It is well known that all nuclei behave as tiny bar magnets as they carry charge and mechanical spin and respond to the external magnetic field by aligning themselves with/against the field. The orientation of nuclei aligned with the external magnetic field is in a lower energy state compared to the orientation aligned opposed to the field.

The nuclei or proton (in case of PMR) in particular, under the influence of external magnetic field, under-goes *precessional motion* also as it behaves like a spinning magnet (shown in Fig. 6.25). The precessional frequency is influenced by the external field and is represented as:

$$v \propto H_o$$

where, v is the frequency of electromagnetic radiation and H_o the external magnetic field.

In other words, we can change the precessional frequency of a nuclei by altering field strength. For example a proton precesses about 60 million times per second when the external magnetic field is 1.4 τ (14,092 gauss) and its precessional frequency is equal to 60 MHz. The value of v is ≈ 100 MHz for field strength 2.3 τ and ≈ 220 MHz for 5.1 τ.

The precessing proton in aligned or parallel orientation may absorb energy from an energy source and pass into opposed or antiparallel orientation. It may later come back to its original low energy state by releasing this extra energy. If such a nuclei is exposed to a radiofrequency energy of appropriate frequency, the low energy (aligned) orientation may absorb energy and move to the high energy (opposed) orientation. The precessing nuclei will absorb the energy only when its frequency will be equal to the frequency of radiofrequency beam. When the two frequencies are equal, the nuclei and the radiofrequency source are said to be in resonance and this is reason, the technique is termed as nuclear magnetic resonance.

6.4.2   Theory of PMR Spectroscopy

It is well known that all nuclei carry charge. The nuclei of some elements (or their isotopes) 'spins' along the nuclear axis, and this spinning of nuclear charge generates a magnetic dipole. The angular momentum of the spinning charge is described in terms of spin quantum number I. The spin quantum number is related to atomic mass and atomic number of the nuclei.

Atomic mass	Atomic number	Spin quantum number
odd	odd or even	1/2, 3/2, 5/2 ...
even	even	0
even	odd	1, 2, 3

The nuclei like ^{12}C and ^{16}O having $I = 0$, are non-magnetic; whereas, ^2H (deuterium) and ^{14}N have $I = 1$ and involve spin-spin splitting. The important magnetic nuclei that have been studied by NMR are ^1H, ^{11}B, ^{13}C, ^{19}F and ^{31}P. The intrinsic magnitude of the generated dipole is expressed in terms of nuclear magnetic moment (μ). Thus the spinning of the charged nuclei makes them behave like tiny bar magnets.

The spinning nucleus of an ordinary hydrogen atom (^1H or proton) is the simplest and is commonly encountered in organic chemistry and the present discussion is restricted to PMR spectroscopy. Under the influence of external magnetic field, a nucleus can take up $(2I + 1)$ orientations with respect to

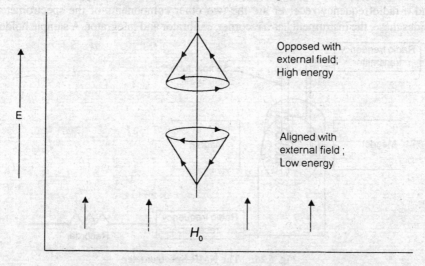

Fig. 6.25:   The External Magnetic Field and the Precessing Nuclei

that field. The hydrogen nucleus has a magnetic moment, $\mu = 2.79268$ and its spin number is 1/2. Hence in an applied magnetic field, its magnetic moment may have two possible orientations. The orientation in which the magnetic moment is aligned with the applied magnetic field is more stable (lower energy) than in which the magnetic moment is aligned opposed or antiparallel to the applied field (higher energy). The energy required for flipping the proton from its lower energy alignment to the higher energy alignment depends upon the difference in energy (ΔE) between the two states and equal to $h\nu$ ($\Delta E = h\nu$), where h is the Plank's constant. Just how much energy is needed to flip the proton over depends on the strength of external field; the stronger the field, the greater the tendency to remain lined up with it, and the higher the frequency of the radiation to undergo the spin flip;

i.e.,
$$\nu = \frac{\tau H_o}{2\pi}$$

where, ν is the frequency in Hz, H_o is the strength of the magnetic field in Gauss, τ is a nuclear constant; the magnetogyric ratio, 26750 for the proton.

As mentioned earlier, in a field of 14092 Gauss, for example, the energy required corresponds to electromagnetic radiation of frequency 60 MHz. For higher frequencies of operation, proportionately higher field has to be used.

In principle, the substance could be placed in a magnetic field of constant strength and then the spectrum can be obtained in the same way as in case of an infrared or an ultraviolet spectrum by passing radiation of steadily changing frequency through the substance and observing the frequency at which radiation is absorbed. In practice, however, it is found more convenient to keep the radiation frequency constant and to vary the strength of the magnetic field. At some value of the field strength, the energy required to flip the proton matches the energy of the radiation (or the proton comes to resonance with the energy source), absorption occurs and a signal is obtained. Such a spectrum is called a proton magnetic resonance (PMR) spectrum.

6.4.3   The NMR Spectrometer

It consists of a strong powerful magnet with a homogeneous field that can be continuously and precisely varied over a relatively narrow range; this is accomplished by a sweep generator. A radiofrequency transmitter and a radiofrequency receiver are the two other components of the spectrometer as shown in Fig. 6.26. Besides these, the instrument has a recorder, calibrator and integrator. A sample holder positions the

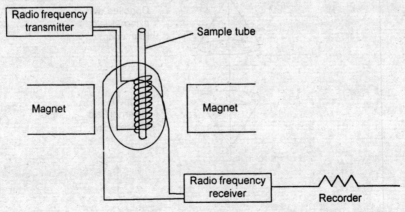

Fig. 6.26:   The NMR Spectrometer

sample between the magnetic poles; the holder also spins the sample by air-driven turbine to average the magnetic field over the sample. The solution of the sample is normally prepared in a solvent containing no protons. The most common solvents are deuterated chloroform ($CDCl_3$) and carbon tetrachloride (CCl_4). A small amount of (1%) tetramethylsilane (TMS) is added as an internal standard; this is helpful for the standardization of the NMR spectrum.

6.4.4 Chemical Shift

All the protons present in an organic compound do not precess with the same frequency under the influence of an external magnetic field. Actually, it depends on the environment, the proton is having. For example, the PMR spectra of toluene shows two precessional frequencies (see Fig. 6.27) as two different environments are there in toluene molecule for protons. This is termed as chemical shift.

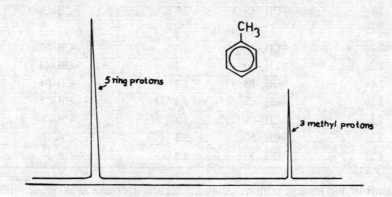

Fig. 6.27: PMR Spectrum of Toluene

The measurement of precessional frequencies of protons in different environment in frequency units is very difficult thus a difference in frequencies is measured with respect to some reference compound. The most commonly used reference compound is tetramethylsilane (TMS) as it gives an intense sharp signal even at a very low concentration. Also it is chemically inert and has a very low boiling point and can be removed from the sample easily by simple evaporation.

The chemical shifts are expressed in a form (scale) that makes it independent of the instrumental frequency and the intrinsic strength of the magnetic field that the NMR spectrometer employs. The most commonly used scale is the (delta) scale. The position of tetramethylsilane signal is taken as zero. Most of the chemical shifts have values between 0 and 10 ppm. Another scale, i.e., (tau) scale is also used some times in which the TMS signal is taken as 10 ppm. Most values lie between 0 and 10 ppm. The two scales are related as $\tau = 10 - \delta$.

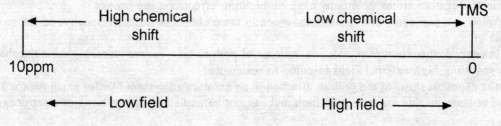

6.4.5   Factors Influencing Chemical Shift

Shielding and Deshielding: Chemical shift of a proton is influenced by the electronegativities of the atoms attached to it as is shown in Table 6.6.

Table 6.6:   Chemical Shift Values for Methyl Protons
Methylene Protons and Methyne Protons

Methyl protons	δ	Methylene protons	δ	Methyne protons	δ
TMS	0.0				
CH_3–C–	0.9	–CH_2–C–	1.4	$\overset{\mid}{C}H$–C–	1.5
CH_3–C=C	1.6	–CH_2–C=C	2.3	–CH–C=C	2.6
CH_3–C=O	2.1	–CH_2–C=O	2.4	–CH–C=O	2.5
CH_3–I	2.15				
CH_3–NR_2	2.2	–CH_2–NR_2	2.5	–CH–NR_2	2.9
CH_3–Ar	2.3	–CH_2–Ar	2.7	–CH–Ar	3.0
CH_3Br	2.84	–CH_2–Br	3.3	–CH–Br	4.1
CH_3Cl	3.10	–CH_2–Cl	3.4	–CH–Cl	4.1
CH_3–O–	3.3	–CH_2–O–	3.4	–CH–O–	3.7
CH_3–N–	3.3	–CH_2–N–	2.5	–CH–N–	2.8
CH_3F	4.25	–CH_2–F	4.5	–CH–F	5.2

The chemical shift of the methyl protons shows a regular increase in δ value with the increase in the electronegativity of atoms attached to it.

The hydrogen nuclei are surrounded by electrons which to some extant shields them from the influence of applied field. In applied magnetic field, the electrons surrounding the hydrogen nuclei undergo induced circulation resulting in the development of another magnetic field opposed to the applied field. This results in shielding of the proton which in turn require more energy (to overcome the shielding effect) to come to resonance with the applied field. The most electronegative fluorine attracts the electron density (in C–F bond) towards itself thereby causing a deshielding effect. In other words, lower applied field will bring the methyl protons of CH_3F to resonance resulting in a high chemical shift compared to CH_3Cl or CH_3Br.

Similar pattern of chemical shift may be seen in protons next to quaternary ammonium ions (R_4N^+) as they are highly deshielded and require lower applied field to come to resonance with.

In addition to methyl protons, Table 6.6 gives the chemical shift values of methylene and methyne protons also and Table 6.7 chemical shifts for some common functional groups.

The chemical shifts of these protons help us to infer that:

(i) Electronegative atoms or groups have deshielding effects on the proton.

(ii) Proton in a particular environment show nearly same chemical shifts irrespective of the molecule it is part of.

(iii) In equivalent environment, more the number of protons on a carbon atom, higher is the shielding thereby requiring high external field to come to resonance.

(iv) The chemical shifts of the protons attached to an unsaturated system like for example, the alkene, carbonyl or benzene ring are unusually high and can not be explained on the basis of electronegativity

Table 6.7: Chemical Shift Values in δ for Common Functional Groups

RCHO	9.4-9.5	HC=C(isolated)	5.1-5.8
ArCHO	9.7-10.0	HC=C(conjugated)	5.8-6.6
R−NH$_2$	5.0-8.0	H$_2$C=C(isolated)	4.5-5.0
Ar−NH$_2$	3.4-6.0	H$_2$C=C(conjugated)	5.3-5.8
R−OH	0.5-4.0	HC≡C(isolated)	2.4-2.7
Ar−OH	4.5	HC≡C(conjugated)	2.7-3.1
R−SH	1.0-2.0	Ar-H	6.5-8.0
Ar−SH	3.0-4.0		
R−COOH	10.5-12.0		
R−C−OCH$_2$R'	3.7-4.1		
R−O−C−H	3.3-4.0		

only. It is actually the π electrons, which induce secondary magnetic field. This induced magnetic field is diamagnetic around the carbon atoms, whereas paramagnetic in the region of protons as shown in the Fig. 6.28, Fig. 6.29 and Fig. 6.30 for alkene, carbonyl compounds and benzene respectively.

When an alkene, carbonyl compound or benzene is placed in an applied magnetic field with the direction of the later perpendicular with respect to former, a secondary magnetic field is induced. The circulation of π electrons generates a field which augments the applied field in the region of electrons. Thus, these proton require less energy to come to resonance (the direction of external field and induced field is same) and are absorbed relatively downfield. These effects are pronounced when π electrons are involved as they are comparatively loosely bonded.

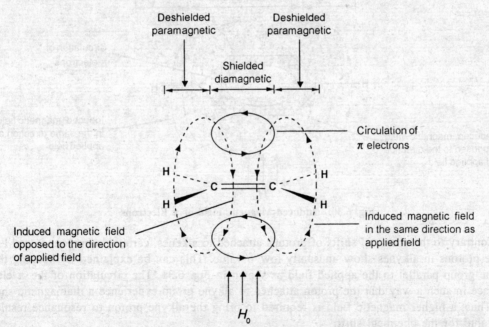

Fig. 6.28: Induced Magnetic Field Around C=C

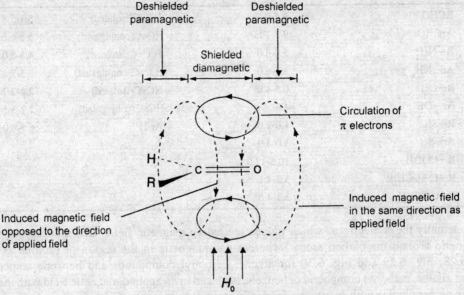

Fig. 6.29: Induced Magnetic Field Around C=O

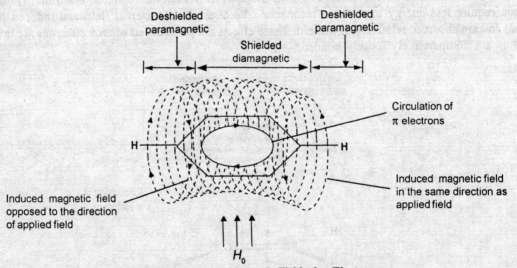

Fig. 6.30: Induced Magnetic Field of π Electrons

(v) Contrary to the chemical shifts of protons attached to alkenes, carbonyl compounds and benzene ring, the protons in alkynes show unusually low δ value. This can be explained by keeping the axis of alkyne group parallel to the applied field as shown in Fig. 6.31. The circulation of the π electrons takes place in such a way that the proton attached to alkyne group experience a diamagnetic shielding effect. Thus, a higher magnetic field is required to bring the alkyne proton to resonance resulting in low δ value for the chemical shift.

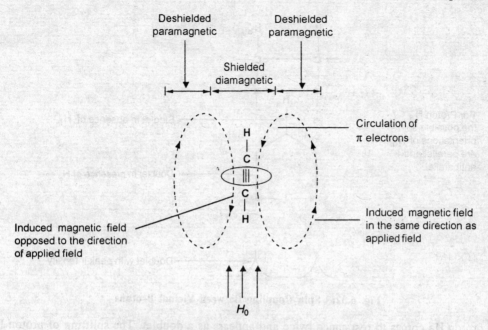

Fig. 6.31: Induced Magnetic Field Around C≡C

6.4.6 Spin Coupling/Spin-Spin Splitting

The intensities of the signals indicates the number of different types of protons in the molecule. The splitting of a signal into several peaks indicates the environment of a proton with respect to other nearby protons. The understanding of the splitting of a signal is the most important step for the structure determination of an organic compound. Consider the NMR spectra of 1,1,2-tribromoethane, 1,1-dibromoethane and ethyl bromide having only two kinds of protons but instead of two peaks the spectra show five, six and seven peaks respectively.

$$\begin{array}{c} H \\ | \\ Br-C-CH_2Br \\ | \\ Br \end{array}$$

1,1,2-Tribromoethane
(Five peaks)

$$\begin{array}{c} H \\ | \\ Br-C-CH_3 \\ | \\ Br \end{array}$$

1,1-Dibromoethane
(Six peaks)

$$\begin{array}{c} H \\ | \\ Br-C-CH_3 \\ | \\ H \end{array}$$

Ethyl bromide
(Seven peaks)

This arises due to splitting of the NMR signals caused by spin-spin coupling. The signals we expect from each set of equivalent protons appear as a group of peaks (and not a single peak). Thus splitting reflects the environment of the absorbing protons with respect to nearby protons.

Let us first consider the case of vicinal protons Ha and Hb which do not give two singlets. On the contrary, two doublets are observed for them as they have different magnetic environment and come to resonance at different positions as shown in Fig. 6.32. The theory of spin-spin splitting can be understood by considering the mutual magnetic influence between proton Ha and Hb. For proton Ha, the possible spin orientations of proton Hb are aligned (parallel) or opposed (antiparallel), which in turn correspond to two different magnetic fields.

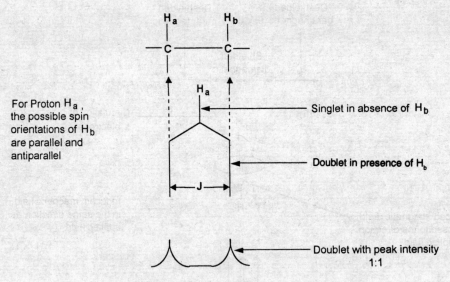

For Proton H_a, the possible spin orientations of H_b are parallel and antiparallel

Singlet in absence of H_b

Doublet in presence of H_b

Doublet with peak intensity 1:1

Fig. 6.32: Spin Coupling Between Vicinal Protons

Thus, proton Ha comes to resonance twice and appears as a doublet. The splitting of proton Hb may be explained similarly. The PMR spectra of trans-cinnamic acid shows splitting of vicinal protons into two doublets as shown in Fig. 6.33.

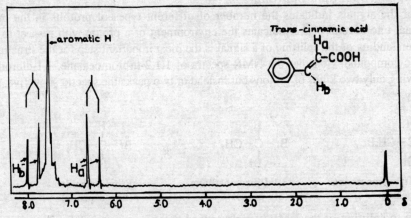

Fig. 6.33:   PMR Spectra of Trans-cinnamic Acid

Let us now consider the case of adjacent carbon atoms carrying a pair of secondary protons and a tertiary proton respectively.

First consider the absorption by one of the secondary protons (Hb). The magnetic field that a secondary proton feels at a particular instant is slightly increased or decreased by the spin of the neighboring tertiary proton (Ha); increased if the tertiary proton happens at that instant to be aligned with the applied field, and decreased if the tertiary proton happens to be opposed to the applied field. Thus, for half the molecules, absorption by a secondary proton is shifted slightly downfield and for the other half of the molecules the absorption is shifted slightly upfield. The signal is split into two peaks; a doublet with equal peak intensities.

The absorption by a tertiary proton (Ha) is affected by the spins of the two neighboring secondary protons (Hb); its signal splits into a triplet with relative peak intensities 1:2:1 as shown in Fig. 6.34.

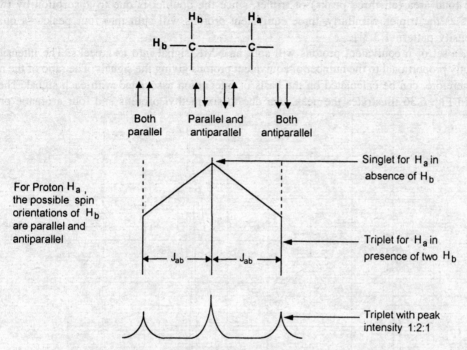

Fig. 6.34: Spin Coupling Between a Tert Proton and Two Sec Protons

The PMR spectra of 1,1,2-trichloroethane shows the splitting of –CH– into a triplet and –CH$_2$– into a doublet) see Fig. 6.35).

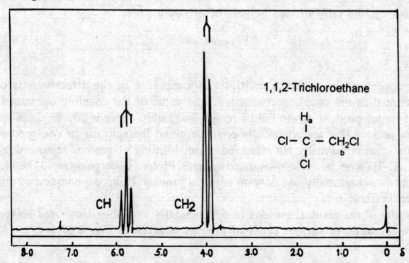

Fig. 6.35: PMR Spectra of 1,1,2-Trichloroethane

Thus, in the NMR spectrum of a compound containing the group $-CH-CH_2-$, we see a 1:1 doublet (from the $-CH_2$) and a 1:2:1 triplet (from $-CH-$). The total area (both peaks) under the doublet is twice as big as the total area (all three peaks) of triplet, since the doublet is due to absorption by twice as many protons as the triplet. Similarly, three equivalent protons will split into four peaks—a quartet—with the intensity pattern 1:3:3:1.

In general, a set of n equivalent protons will split an NMR signal into n+1 peaks. The intensity of a signal is directly proportional to the number of equivalent protons giving the signals. The ratio of the number of protons, therefore, can be calculated on the basis of integration associated with each signal. The PMR of p-xylene in Fig. 6.36 illustrates the peak area due to six methyl protons and four aromatic protons.

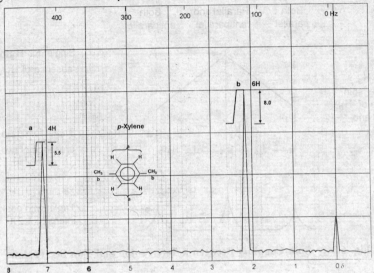

Fig. 6.36:   PMR Spectra of p-xylene

For Proton Counting, the ratio of step heights of proton a : b is

$$5.5 : 8.0$$
$$\text{or} \quad 1 : 1.5$$
$$\text{or} \quad 4 : 6$$

The distance between peaks within a multiplet is a measure of the effectiveness of spin-spin coupling, and is termed as the coupling constant, J. The value of the coupling constant is expressed in cps or Hertz; it is independent of the field strength and rarely exceeds 20 cps. Spin-spin coupling is not observed for protons that are chemically equivalent and the splitting of one proton by another on the same carbon atom is normally not observed. Also, splitting of protons separated by more than two atoms (H–C–C–C–H) is not possible in saturated systems. Protons undergoing rapid chemical exchange like that of alcohols do not normally couple with adjacent protons. Coupling constant of different types of protons is given in Table 6.8.

The chemical shifts of the residual protons in commercially available deuterated solvents and also the range of chemical shifts of various types of protons is given in Fig. 6.37.

Table 6.8: Spin-spin Coupling Constants (J_{ab} in Hz)

Hb—C—C—C—C—Ha		0
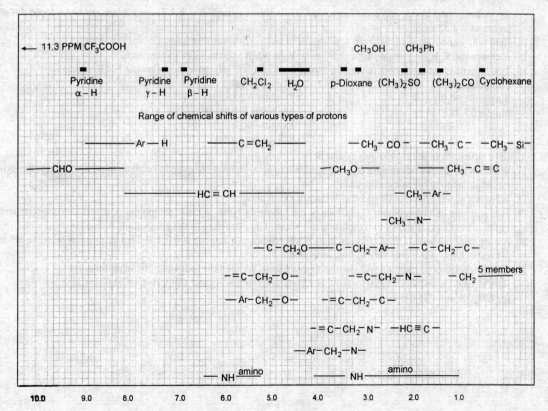 Hb, Ha on C	(Gem)	10-15
C Ha—C Hb	(Vic)	6-8
C=C (Ha, Hb)	(Gem)	0-3
Ha C=C Hb	(Cis)	5-12
Ha C=C Hb	(trans)	12-19
ortho		7-10
meta		2-3
para		0-1

Range of chemical shifts of various types of protons

Fig. 6.37: Chemical Shift of Residual Protons in Commercially Available Deutrated Solvents

6.5 PROBLEMS

1. What is spectroscopy? What information can be obtained from IR, UV and NMR spectroscopy?
2. Using IR spectroscopy how will you distinguish between:
   (a) CH_3COCH_3 and CH_3COOCH_3 (c) $CH_3CH_2N(CH_3)_2$ and $CH_3CH_2NH_2$
   (b) CH_3COOH and CH_3COOCH_3 (d) CH_3CH_2OH and CH_3COCH_3
3. What special features to look forward in the IR spectra of C_6H_5CN, C_6H_5CHO, $CH_3COC_2H_5$, CH_3COOH and $C_6H_5NH_2$?
4. Discuss the basic principles involved in UV spectroscopy.
5. Explain n, π^*, π and σ bonding in UV spectroscopy. Explain the different types of transitions you come across in UV spectroscopy.
6. State Woodward Fieser rules for calculating the λ_{max} of conjugated dienes. Using this rule, calculate the λ_{max} of $(CH_3)_2C=CHCOCH_3$.
7. Explain the following terms:
   (a) Chromophore (b) Auxochrome (c) Bathochromic shift (d) Hyperchromic effect (e) Hypsochromic effect.
8. Discuss the principle involved in NMR spectroscopy.
9. What is an internal standard in NMR spectroscopy? Why is TMS a good standard?
10. In case TMS is not available how will you determine the chemical shifts?
11. How many signals do you expect in the NMR spectrum of
    (a) $CH_3CH_2CH_3$ (b) CH_3OCH_3 (c) CH_3COOCH_3 (d) $C_6H_5CH_2CH_3$
12. Using NMR spectroscopy how will you differentiate between:
    (a) Methyl alcohol and ethyl alcohol (b) Water and deuterium oxide (c) Acetone and acetaldehyde.
13. Give a structure consistent with the following NMR data
    (a) A compound having molecular formula CH_4O shows a singlet of one proton at δ 1.43 and another singlet of three protons at δ 3.47.
    (b) A compound having molecular formula C_3H_7Br shows a doublet at δ 1.71 for six protons and a septet for one proton at δ 4.32.
14. How will you differentiate between 1,5-hexadiene and 1,3,5-hexatriene using UV spectroscopy?
15. A compound C_4H_9Br gives a single sharp peak in its NMR spectra. Give its structure.
16. Two organic compounds both having the molecular formula $C_{10}H_{16}$ gave the following NMR signal.

Compound A	Compound B
Singlet δ 1.30,9H	Doublet δ 0.88,6H
Singlet δ 7.28,5H	Multiplet δ 0.86,1H
	Doublet δ 2.45,2H
	Singlet δ 7.12,5H

    Give structures to the compounds A and B
17. What are the special advantages in using spectroscopic techniques compared to conventional methods for the structure elucidation of organic compounds?
18. What is a superimposable IR spectra? What are its advantage?
19. Two isomeric chloropropanes, C_3H_7Cl show the following NMR signals: compound A shows two doublets and a septet. Compound B shows two triplets and a sextet. Give structure to the compounds A and B.
20. An organic compound $C_3H_5Cl_3$ shows in its NMR spectra two singlets at δ 2.2 and δ 4.02 corresponding to 3H and 2H respectively. Give structure to the compound.
21. Which spectroscopic method will you use to distinguish between the following giving reasons:
    (a) CH_3COCD_3 and CH_3COCH_3 (b) $CH_3CH=CH-CH=CHCH_3$ and $CH_3CH=CHCH_2CH=CH_2$
    (c) CH_3OH and CH_3CH_2OH

7

Alkanes

7.1 INTRODUCTION

Carbon forms a large number of compounds with hydrogen commonly known as *Hydrocarbons.* These are subdivided into two main categories—Aliphatic and Aromatic. The aliphatic hydrocarbons are further divided into saturated (known as alkanes), unsaturated (alkenes and alkynes) and alicyclic hydrocarbons (the cyclic analogues are also called cycloalkanes).

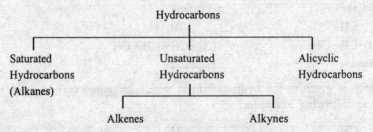

Alkanes are also known as *Paraffins,* the name is derived from two Latin words, 'paraum' and 'affins', which mean 'little affinity'. In accordance with the name, the paraffins, under ordinary conditions are inert towards reagents like acids, alkalis, oxidising and reducing agents. However, under suitable conditions, alkanes do undergo different types of reactions like halogenation, nitration, pyrolysis etc.

Alkanes are also known as *Saturated Hydrocarbons.* These form a homologous series with the general formula C_nH_{2n+2}, where n is the number of carbon atoms in the molecule. Thus, the first member of the series has the molecular formula CH_4 (n = 1) and is known as methane and the second member of the series has the molecular formula C_2H_6 (n = 2) and is called ethane.

7.2 NOMENCLATURE

There are three systems in use for naming alkanes.

(a) *The Common System or the Trivial System:* The first four members of the series are called by their trivial or common names, viz. methane, ethane, propane and butane. Alkanes having five, six, seven and eight carbons are named by prefixing pent, hex, hept and oct (for 5, 6, 7 and 8 carbons) to the terminal -ane.

(1)	CH_4 Methane	(2)	C_2H_6 Ethane	(3)	C_3H_8 Propane
(4)	C_4H_{10} Butane	(5)	C_5H_{12} Pentane	(6)	C_6H_{14} Hexane
(7)	C_7H_{16} Heptane	(8)	C_8H_{18} Octane	(9)	C_9H_{20} Nonane
(10)	$C_{10}H_{22}$ Decane	(11)	$C_{12}H_{26}$ Dodecane	(12)	$C_{20}H_{42}$ Eicosane
(13)	$C_{30}H_{62}$ Triacontane	(14)	$C_{50}H_{102}$ Pentacontane	(15)	$C_{70}H_{142}$ Heptacontane
(16)	$C_{80}H_{162}$ Octacontane				

The hydrocarbon side chain formed by the removal of one hydrogen atom of the paraffin is called *Alkyl group*. Following is given structure of common alkyl groups, their name and representation:

Alkyl group	Name	Representation
CH_3-	Methyl	Me
CH_3CH_2-	Ethyl	Et
$CH_3CH_2CH_2-$	Propyl	n–Pr
$(CH_3)_2CH-$	Isopropyl	iso–Pr
$CH_3CH_2CH_2CH_2-$	n-Butyl	n–Bu
$CH_3-\overset{\overset{CH_3}{|}}{CH}-CH_2-$	Isobutyl	iso–Bu

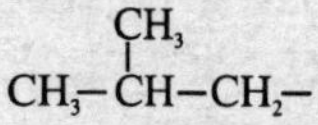

In the trivial or common system of nomenclature, the straight-chain compounds are termed normal (designated as n) and those with a branched chain are referred to as iso hydrocarbons (designated as iso). Following examples explain the nomenclature of normal and the branched chain hydrocarbons.

$CH_3CH_2CH_2CH_3$
n-Butane

$CH_3CH_2CH_2CH_2CH_3$
n-Pentane

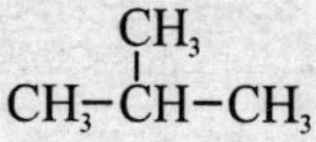

$CH_3-\overset{\overset{CH_3}{|}}{CH}-CH_3$
Isobutane

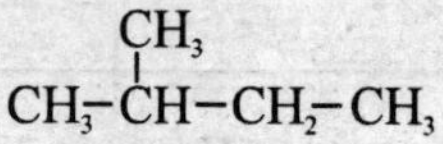

$CH_3-\overset{\overset{CH_3}{|}}{CH}-CH_2-CH_3$
Isopentane

A suffix 'neo' is given to the hydrocarbon in which branched carbon atom is quaternary in nature as shown in the following examples.

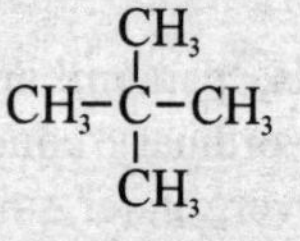

$CH_3-\overset{\overset{CH_3}{|}}{\underset{\underset{CH_3}{|}}{C}}-CH_3$
Neopentane

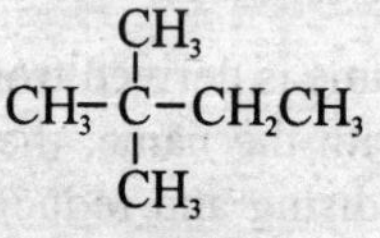

$CH_3-\overset{\overset{CH_3}{|}}{\underset{\underset{CH_3}{|}}{C}}-CH_2CH_3$
Neohexane

Normally, we come across four types of carbon atoms. These are primary (1°), secondary (2°), tertiary (3°) and quaternary (4°) carbons bonded to one, two, three and four carbon atoms respectively. The following structure depicts all types of carbon atoms.

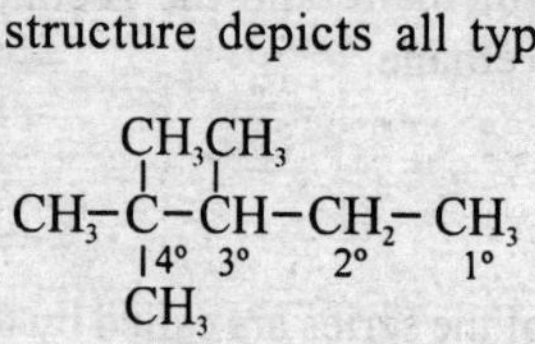

$CH_3-\overset{\overset{CH_3}{|}}{\underset{\underset{CH_3}{|}}{C}}-\overset{\overset{CH_3}{|}}{CH}-CH_2-CH_3$
$\quad 4° \quad 3° \quad 2° \quad 1°$

(b) As Derivatives of Methane

In this system, the branched chain paraffins are named as substituted methanes. The most highly branched carbon atom is termed the methane parent and the alkyl groups attached to this carbon atom are named in alphabetical order.

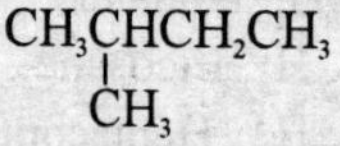

$CH_3CHCH_2CH_3$
$\quad\;\,\overset{|}{CH_3}$ **Dimethylethylmethane**

$$CH_3-\underset{\underset{CH_3}{|}}{\overset{\overset{CH_3}{|}}{C}}-CH_2CH_3 \qquad \textbf{Trimethylethylmethane}$$

(c) The IUPAC System

In IUPAC system of nomenclature the common names of first ten (C_1 to C_{10}) alkanes are retained. The longest carbon chain present in the molecule is selected and the compound is designated as the derivative of the parent alkane containing this chain. The position of the substituents attached to the parent chain are determined by numbering the chain from the end which puts them on carbons having the lowest numbers. Finally, the IUPAC name is given by naming the substituents with position numbers in alphabetical order. This is illustrated by the following examples.

$$\underset{1}{CH_3}-\underset{2}{\overset{\overset{CH_3}{|}}{CH}}-\underset{3}{CH_2}-\underset{4}{CH_2}-\underset{5}{CH_3}$$

2-Methylpentane

$$\underset{1}{CH_3}-\underset{2}{\overset{\overset{CH_3}{|}}{CH}}-\underset{3}{\overset{\overset{CH_2CH_3}{|}}{CH}}-\underset{4}{CH_2}-\underset{5}{CH_3}$$

3-Ethyl-2-methylpentane

$$\underset{1}{CH_3}-\underset{2}{\overset{\overset{CH_3}{|}}{CH}}-\underset{3}{\overset{\overset{CH_3}{|}}{CH}}-\underset{4}{CH_2}-\underset{5}{CH_3}$$

2,3-Dimethylpentane

$$\underset{1}{CH_3}-\underset{2}{CH_2}-\underset{3}{\overset{\overset{CH_2CH_3}{|}}{\underset{\underset{CH_3}{|}}{C}}}-\underset{4}{CH_2}-\underset{5}{CH_2}-\underset{6}{CH_3}$$

3-Ethyl-3-methylhexane

7.3 ORBITAL STRUCTURE

The carbon atom in all the alkanes is attached to four different atoms or groups as they are saturated molecules and uses its sp³ hybrid orbitals for the formation of bonds. It, therefore, has a tetrahedral geometry with a bond angle of 109.5°. The orbital structure of methane and ethane has been given below (see Section 2.3 also):

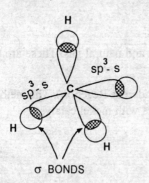

Fig. 7.1: Orbital Structure of Methane

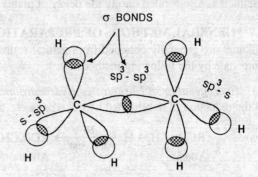

Fig. 7.2: Orbital Structure of Ethane

7.4 HOMOLOGOUS SERIES

On examination of the formula of various alkanes, we find that the formula of each individual differs from that of its neighbor by CH_2, for example, CH_4, C_2H_6, C_3H_8, C_4H_{10} etc. A set of compounds, such as alkanes, in which the members differ in composition from one another by CH_2, is known as homologous series and the individual members are known as homologues. Similar types of homologous series are encountered in other compounds, like, alkyl halides, carboxylic acids, alcohols, aldehydes and ketones.

7.5   ISOMERISM IN ALKANES

As discussed earlier organic compounds possessing the same molecular formula but different physical or chemical properties are called isomers and the phenomenon is referred to as isomerism. The alkanes offer an example of chain isomerism. The first three members of the alkane series do not exhibit isomerism but the next in series i.e., butane (C_4H_{10}) can be represented in two forms:

$$CH_3-CH_2-CH_2-CH_3 \quad \text{and} \quad CH_3-\overset{\underset{\displaystyle CH_3}{|}}{CH}-CH_3$$

   n-Butane   Isobutane

Both these hydrocarbons are known and have boiling points 1° and –10.2° respectively.
The next hydrocarbon, pentane (C_5H_{12}) can be represented by three structures.

$$CH_3-CH_2-CH_2-CH_2-CH_3 \qquad \text{n-Pentane}$$

$$CH_3-\overset{\underset{\displaystyle CH_3}{|}}{CH}-CH_2-CH_3 \qquad \text{Isopentane or 2-Methylbutane}$$

$$CH_3-\overset{\overset{\displaystyle CH_3}{|}}{\underset{\underset{\displaystyle CH_3}{|}}{C}}-CH_3 \qquad \text{Neopentane or 2,2-Dimethylpropane}$$

Similarly, hexane (C_6H_{14}) can be represented by five structures, heptane (C_7H_{16}) by nine structures and decane by 75 structures.

7.6   OCCURRENCE

Alkanes occur widely in nature. The natural gas (obtained from petroleum region) contains methane (80%), ethane (10%) and other hydrocarbons (10%). The crude petroleum is the chief source of alkanes. The well known fuel gas obtained from coal, for example coal gas, contains these hydrocarbons in small amounts. Methane is also formed during the decay of plants and animal tissues.

7.7   GENERAL METHODS OF PREPARATION

Alkanes are generally obtained from natural sources, viz. petroleum and natural gas. These are obtained in pure state by the following methods.

(i)   *Hydrogenation of Unsaturated Hydrocarbons.* (Sabatier and Senderens Reaction). The method consists in the reduction of unsaturated hydrocarbons viz. alkenes and alkynes with nickel catalyst.

$$RCH=CH_2 + H_2 \xrightarrow[200-300°]{Ni} RCH_2CH_3$$
   Alkene   Alkane

$$CH_2=CH_2 + H_2 \xrightarrow[200-300°]{Ni} CH_3-CH_3$$
   Ethene   Ethane

$$CH_3CH=CH_2 + H_2 \xrightarrow[200-300°]{Ni} CH_3CH_2CH_3$$
   Propene   Propane

$$R-C\equiv C-R + 2H_2 \xrightarrow[200-300°]{Ni} RCH_2CH_2R$$
   Alkyne   Alkane

$$HC\equiv CH + 2H_2 \xrightarrow[200-300^\circ]{Ni} CH_3-CH_3$$

Acetylene   Ethane

The hydrogenation can also be carried out with platinum or palladium although they are more expensive than nickel.

This method is used for the industrial preparation, since the starting materials, viz. alkenes and alkynes are obtained in large quantities by the cracking of petroleum.

(ii) *From Alkyl Halide*

(a) *Wurtz Reaction*: In this reaction an alkyl halide is treated with sodium in presence of dry ether to form alkane.

$$2RX + 2Na \xrightarrow{Ether} R-R + 2NaX$$

$$2CH_3Br + 2Na \xrightarrow{Ether} CH_3-CH_3 + 2NaBr$$

Methyl   Ethane
bromide

This method is useful only with one alkyl halide. If two different alkyl halides are used, a mixture of three different alkanes is obtained, which is difficult to separate. For example, a mixture of methyl bromide and ethyl bromide reacts to give ethane, propane and butane.

$$CH_3Br + C_2H_5Br + Na \Bigg\langle \begin{array}{l} \longrightarrow CH_3-CH_3 \\ \longrightarrow CH_3-C_2H_5 + NaBr \\ \longrightarrow C_2H_5-C_2H_5 \end{array}$$

Thus, Wurtz reaction is useful for the preparation of only those alkanes which contain an even number of carbon atoms.

Mechanism: A free radical mechanism has been suggested for the Wurtz reaction which may be given as follows:

$$CH_3-I + Na \longrightarrow CH_3^\cdot + NaI$$

                        Free
                        radical

$$CH_3^\cdot + CH_3^\cdot \longrightarrow CH_3-CH_3$$

                        Ethane

The free radicals can also undergo disproportionation, i.e., one radical can gain one hydrogen at the expense of the other which loses hydrogen.

$$CH_3^\cdot + CH_3\dot{C}H_2 \longrightarrow CH_4 + CH_2=CH_2$$

                           Methane   **Ethylene**

Ionic Mechanism has also been suggested for Wurtz reaction which involves the bimolecular nucleophilic substitution as well as bimolecular elimination.

$$C_2H_5Br + 2Na \longrightarrow C_2H_5^-Na^+ + NaBr$$

$$C_2H_5^-Na^+ + C_2H_5Br \longrightarrow C_2H_5-C_2H_5 + NaBr \text{ (S\textsc{n}2)}$$

$$Na^+CH_3CH_2^- + H-CH_2-CH_2-Br \longrightarrow CH_3CH_3 + CH_2=CH_2 + NaBr \text{ (E2)}$$

A versatile method for coupling together two different alkyl groups is the *Corey and House* method. In this method an alkyl halide (primary, secondary or tertiary) is first converted into lithium dialkyl copper (R_2CuLi), and then reacted with another alkyl halide (preferably primary alkyl halide) to give the corresponding alkane in good yield.

$$R-X \xrightarrow{\text{Li}} R-Li \xrightarrow{\text{CuI}} R_2-Cu-Li \xrightarrow{R'-X} R-R' + R-Cu + LiX$$

Alkyl halide   Alkyl lithium   Lithium dialkyl copper   Alkane

The following reactions represent the preparation of n-nonane from n-octyl iodide and methyl bromide.

$$CH_3Br \xrightarrow{\text{Li}} CH_3-Li \xrightarrow{\text{CuI}} CH_3-\overset{\overset{\displaystyle CH_3}{|}}{CuLi}$$

Methyl bromide   Methyl lithium   Lithium dimethyl copper

$$(CH_3)_2CuLi + CH_3(CH_2)_6CH_2I \longrightarrow CH_3CH_2(CH_2)_6CH_3 + CH_3Cu + LiI$$

            n-Octyl iodide   n-Nonane

(b) *Reduction of Alkyl Halides:* Alkanes may be obtained from alkyl halides by reduction with zinc and acetic acid or zinc-copper couple in hydrochloric acid.

$$RX \xrightarrow[\text{Zn-Cu/HCl}]{\text{Zn-HOAc or}} RH$$

Zinc-copper couple is granulated zinc (small pieces) coated with copper and is obtained by immersing zinc pieces in copper sulphate solution.

Reduction can also be carried out with lithium aluminum hydride ($LiAlH_4$) which is an excellent reducing agent. Though it reduces many unsaturated functional groups like C=O, C≡N, it does not reduce isolated double or triple bonds.

$$CH_2=CH-CH_2Cl \xrightarrow[\text{Ether}]{\text{LiAlH}_4} CH_2=CH-CH_3 + LiCl + AlCl_3$$

3-Chloropropene   **Propene**

(c) *Grignard Reagent:* Alkyl halides react in anhydrous ether with magnesium to form alkyl magnesium halide, known as grignard reagent. These decompose on treatment with water or dilute acid to give alkanes.

$$Rx + Mg \xrightarrow{\text{Ether}} RMgX \xrightarrow{\text{H}^+/\text{H}_2\text{O}} RH + Mg(OH)X$$

                      Grignard   **Alkane**
                      reagent

$$CH_3Br + Mg \xrightarrow{\text{Ether}} CH_3MgBr \xrightarrow{\text{H}^+/\text{H}_2\text{O}} CH_4 + Mg(OH)Br$$

                                            Methane

(iii) *From Carboxylic Acids*

(a) *Decarboxylation:* The sodium salts of carboxylic acids on heating with soda lime undergo decarboxylation to give alkanes. Soda lime is a mixture of sodium hydroxide and calcium oxide, the active ingredient of which is sodium hydroxide whereas lime keeps the reaction mixture porous.

$$RCOONa \quad + \quad NaOH \xrightarrow{\text{CaO}} RH \quad + \quad Na_2CO_3$$

Sodium salt   **Alkane**
of carboxylic acid

The process of eliminating carbon dioxide from a carboxylic acid is known as decarboxylation. The alkane thus produced contain one carbon atom less than the original acid. Although methane is obtained from ethanoic acid in good yield, other acids give only 10-20% yield of the corresponding hydrocarbon.

Decarboxylation can also be effected by heating the carboxylic acid with an organic base, such as pyridine using copper chromite ($CuO.Cr_2O_3$) as catalyst.

$$CH_3CH_2COOH \quad + \quad C_5H_5N \xrightarrow[\text{heat}]{\text{CuO.Cr}_2\text{O}_3} CH_3CH_3 \quad + \quad CO_2$$

Propanoic acid   **Pyridine**   **Ethane**

(b) *Kolbe's Electrolytic Method:* When a concentrated solution of sodium or potassium salt of a carboxylic acid is electrolysed, a higher alkane is formed.

$$RCOOK + RCOOK \xrightarrow{\text{Electrolysis}} R-R + 2CO_2 + H_2 + 2KOH$$

$$CH_3COOK + CH_3COOK \xrightarrow{\text{Electrolysis}} CH_3CH_3 + 2CO_2 + H_2 + 2KOH$$

Mechanism: When potassium salt of a carboxylic acid is electrolysed,acetate ions migrates towards the anode, gives up one electron to produce acetate free radical ($CH_3COO^\bullet$), which decomposes to give a methyl free radical ($CH_3^\bullet$) and carbon dioxide. Two such methyl radicals combine to give ethane.

$$CH_3COOK \xrightarrow{\text{Electrolysis}} CH_3COO^- + K^+$$

At Anode

$$CH_3COO^- \xrightarrow{-e} CH_3COO^\bullet \longrightarrow CH_3^\bullet + CO_2$$

$$CH_3^\bullet + CH_3^\bullet \longrightarrow CH_3 - CH_3$$
                          Ethane

$$CH_3 : \overset{\overset{\displaystyle O}{\|}}{C} - \ddot{O}: \longrightarrow CH_3^\bullet + CO_2$$
                  Methyl radical

$$CH_3^\bullet + CH_3^\bullet \longrightarrow CH_3 - CH_3$$
                          Ethane

In case a mixture of two carboxylic acids is electrolysed, mixture of alkanes is obtained.

$$RCOOK + R'COOK \xrightarrow{\text{Electrolysis}} R-R + R'-R + R'-R'$$

This reaction has limited synthetic applications as it forms a number of side products, however, this method is useful for the synthesis of symmetrical alkanes.

(iv)   *Reduction of Alcohols, Aldehydes and Ketones:* Alkanes can be obtained by the reduction of alcohols, aldehydes, ketones and carboxylic acids with hot concentrated hydriodic acid in combination with red phosphorus.

$$R-OH + 2HI \xrightarrow[150°]{P} R-H + I_2 + H_2O$$
Alcohol   **Alkane**

$$\underset{\text{Aldehyde}}{R-\overset{\displaystyle O}{\overset{\|}{C}}-H} + 4HI \xrightarrow[150°]{P} \underset{\text{Alkane}}{RCH_3 + 2I_2 + H_2O}$$

$$\underset{\text{Ketone}}{R-\overset{\displaystyle O}{\overset{\|}{C}}-R'} + 4HI \xrightarrow[150°]{P} \underset{\text{Alkane}}{RCH_2R' + 2I_2 + H_2O}$$

$$\underset{\substack{\text{Carboxylic} \\ \text{acid}}}{R-\overset{\displaystyle O}{\overset{\|}{C}}-OH} + 6HI \xrightarrow[150°]{P} \underset{\text{Alkane}}{RCH_3 + 3I_2 + 2H_2O}$$

In the above reductions, hydriodic acid is an effective reducing agent. Red phosphorus reacts with iodine in the reactions to regenerate HI, which is used again.

$$2P + 3I_2 + 6H_2O \xrightarrow[150°]{P} 6HI + 2H_3PO_3$$

This method is of special interest for the preparation of higher alkanes from carboxylic acids obtained by hydrolysis of fats.

It is appropriate to mention here that the ketone can also be reduced with zinc amalgam and concentrated hydrochloric acid. This reaction is called *Clemmensen reduction.* Alternatively, reduction can also be effected with hydrazine and sodium ethoxide *(Wolff Kishner reduction).*

$$RCOR \xrightarrow[\text{Conc. HCl}]{\text{Zn-amalgam}} RCH_2R$$

$$\underset{\textbf{Ketone}}{RCOR} \xrightarrow[\text{NaOEt}]{NH_2NH_2} \underset{\textbf{Alkane}}{RCH_2R}$$

7.8 PHYSICAL PROPERTIES

The alkanes from C_1 to C_4 (methane to butane) are colorless gases and the alkanes from C_5 to C_{17} are colorless liquids. The higher members from C_{18} onwards are waxy solids. The alkane molecules are nonpolar or very weakly polar and so are insoluble in water. They are, however, soluble in non-polar solvents like carbon tetrachloride, benzene etc.

The boiling point of n-alkanes increases with molecular weight. It is increased by 20° to 30° for each CH_2 unit that is added to the chain. In case of alkanes which exist in isomeric forms, the branched chain isomer has a lower boiling point than the corresponding normal chain isomer. Thus, n-butane has a higher boiling point than isobutane. The boiling point is still lower if the branching is more. Thus, neopentane with two branching boils at lower temperature than isopentane with one branching only. This is because the branching in a chain reduces the surface area and therefore decreases intermolecular attractive forces resulting in the decrease in the boiling point. The boiling points of some alkanes are given below.

Name	Formula	B.P. (°C)	M.P. (°C)	Relative density (at 20 °C)
Methane	CH_4	−162	−183	
Ethane	CH_3CH_3	−88.5	−172	
Propane	$CH_3CH_2CH_3$	−42	−187	
n-Butane	$CH_3CH_2CH_2CH_3$	0	−138	
Isobutane	CH_3CHCH_3 $\quad\vert$ $\quad CH_3$	−12	−159	
n-Pentane	$CH_3CH_2CH_2CH_2CH_3$	36	−130	0.626
Isopentane	$CH_3CHCH_2CH_3$ $\quad\vert$ $\quad CH_3$	28	−160	0.620
Neopentane	$\quad\ CH_3$ $\quad\ \vert$ $CH_3{-}C{-}CH_3$ $\quad\ \vert$ $\quad\ CH_3$	9.5	−17	
n-Hexane	$CH_3CH_2CH_2CH_2CH_2CH_3$	69	−95	0.659
3-Methylpentane	$CH_3CH_2CHCH_2CH_3$ $\qquad\quad \vert$ $\qquad\quad CH_3$	63	−118	0.676
2,2-Dimethylbutane	$\quad\ CH_3$ $\quad\ \vert$ $CH_3{-}C{-}CH_2CH_3$ $\quad\ \vert$ $\quad\ CH_3$	50	−98	0.649
2,3-Dimethylbutane	$CH_3CH{-}CHCH_3$ $\quad\ \vert \quad\ \vert$ $\quad\ CH_3\ \ CH_3$	58	−129	

The melting points of alkanes also increase with the increase in molecular weight but the increase is not so régular. This is because the intramolecular forces in a crystal depend not only on the size of the molecule but also upon how they fit within a crystal lattice.

The relative densities increase with size of the alkanes but they level off at about 0.8. Thus all alkanes are lighter than water.

7.9 SPECTRAL PROPERTIES

Ultraviolet (UV) spectroscopy is not of much help in the characterisation of alkanes, since they do not show any absorption band above 200 nm.

In the infrared (IR) spectra of alkanes, the absorption due to C−H stretching depends on the nature of the hydrogen atom i.e. whether it is attached to a primary, secondary or tertiary carbon atom. The C−H stretching absorption frequencies for various carbons are given in Sec. 6.3.3.

The NMR spectra of alkanes give characteristic signals at δ 0.9(CH_3), δ 1.4($-\overset{\vert}{C}H_2$) and δ 1.5($-\overset{\vert}{\underset{\vert}{C}}H$).

7.10 CHEMICAL PROPERTIES

It has already been stated that the paraffins or alkanes under ordinary conditions are inert towards acids, alkalis, oxidising and reducing agents. However, under suitable conditions, alkanes do undergo different types of reactions like halogenation, nitration, sulphonation and some thermal and catalytic reactions.

The non reactivity of alkanes under normal conditions may be explained on the basis of the nonpolarity of the bonds forming them. The electronegativities of carbon (2.6) and of hydrogen (2.1) do not differ appreciably and the bonding electrons between C–H and C–C are equally shared between them making them almost nonpolar. In view of this, the ionic reagents such as acids and alkalis find no reaction sites in the alkane molecules to which they could be attracted. Alkanes, however, undergo substitution reactions by free radical chain mechanism. Some important reactions of alkanes are given below.

Halogenation

Alkanes react with halogens at 250°-400° or in presence of UV light to form halogen substituted alkanes and hydrogen halide. The reaction is called *Substitution reaction*. Methane forms methyl chloride, which on further substitution forms methylene chloride or dichloromethane.

$$CH_4 + Cl_2 \longrightarrow CH_3Cl$$
$$\text{Chloromethane}$$

$$CH_3Cl + Cl_2 \longrightarrow CH_2Cl_2$$
$$\text{Dichloromethane}$$

Chlorination continues further to form chloroform or trichloromethane, and carbon tetrachloride or tetrachloromethane.

$$CH_2Cl_2 + Cl_2 \longrightarrow CHCl_3 \longrightarrow CCl_4$$
$$\text{Trichloromethane} \quad \text{Tetrachloromethane}$$

The extent to which each product is formed depends on the initial chlorine to methane ratio. Thus, excess of methane gives chloromethane as the major product and excess of chlorine gives tetrachloromethane as the major product. The mixture of products is separated by fractionation.

Methane reacts with bromine under similar conditions but less vigorously to form methyl bromide, methylene bromide, tribromomethane and tetrabromomethane.

$$CH_4 \xrightarrow[-HBr]{Br_2} CH_3Br \xrightarrow[-HBr]{Br_2} CH_2Br_2 \xrightarrow[-HBr]{Br_2} CHBr_3 \xrightarrow[-HBr]{Br_2} CBr_4$$
$$\text{Methane} \quad \text{Methyl} \quad \text{Methylene} \quad \text{Tribromo-} \quad \text{Tetrabromo-}$$
$$\text{bromide} \quad \text{bromide} \quad \text{methane} \quad \text{methane}$$

The reaction of methane with iodine, however, is reversible. The reaction can be made to proceed in one direction only if it is carried out in presence of an oxidising agent which consumes hydriodic acid formed during the reaction.

$$CH_4 + I_2 \rightleftharpoons CH_3I + HI$$
$$5HI + HIO_3 \longrightarrow 3I_2 + 3H_2O$$

The Iodo compound can however be conveniently prepared by treating alkyl chlorides and bromides with sodium iodide in acetone or methanol.

$$CH_3Cl + NaI \xrightarrow{\text{Acetone}} CH_3I + NaCl$$

The reaction is possible because of the solubility of sodium iodide in acetone whereas sodium chloride and bromide are insoluble. The reaction is known as *Finkelstein reaction*.

The reaction of methane with fluorine is explosive and even in dark the reaction is controlled by mixing the reactants in inert solvent at low pressure.

The order of reactivity of halogens in substitution is $F_2 > Cl_2 > Br_2 > I_2$.

The reaction has only limited synthetic utility, since a mixture of products is obtained.

Mechanism: The mechanism of halogenation is believed to involve the following steps.

$$Cl_2 \longrightarrow 2Cl\cdot \tag{i}$$

$$Cl\cdot + CH_4 \longrightarrow CH_3^\cdot + HCl \tag{ii}$$

$$CH_3^\cdot + Cl_2 \longrightarrow CH_3Cl + Cl\cdot \tag{iii}$$

The first step involves the homolysis of chlorine molecule to form free radicals (initiation step). In the second step the chlorine free radical attacks the C—H bond of methane molecule to generate a methyl radical along with the formation of a molecule of hydrogen chloride (propagation step). The third step involves the reaction of methyl free radical with chlorine molecule to give methyl chloride and chlorine free radical. The second and the third steps are repeated again and again till the chain is exhausted.

The reaction is terminated by the combination of two reactive intermediates like two chlorine radicals, two methyl radicals or one methyl and one chlorine radical as shown below.

$$Cl\cdot + Cl\cdot \longrightarrow Cl_2$$

$$CH_3^\cdot + CH_3^\cdot \longrightarrow CH_3CH_3$$

$$CH_3^\cdot + Cl\cdot \longrightarrow CH_3Cl$$

The methyl chloride formed combines with a chlorine free radical to produce chloromethyl radical and hydrogen chloride. This radical in turn participates in further chain reaction to form dichloromethane and chlorine free radical.

$$Cl\cdot + CH_3Cl \longrightarrow HCl + CH_2Cl\cdot$$

$$CH_2Cl\cdot + Cl-Cl \longrightarrow CH_2Cl_2 + Cl\cdot$$

Similarly, trichloromethane and tetrachloromethane are obtained by further chain reaction.

The free radical mechanism finds support by the following experiment. If the halogenation is carried out in presence of a catalytic amount of dibenzoyl peroxide the reaction takes place in the dark at room temperature.

$$C_6H_5CO-O-O-COC_6H_5 \longrightarrow 2C_6H_5COO\cdot$$

Dibenzoyl peroxide   **Benzoyl free radical**

$$C_6H_5COO\cdot \longrightarrow CO_2 + C_6H_5^\cdot$$

The phenyl free radical reacts with chlorine molecule to produce a chlorine free radical.

$$C_6H_5^\cdot + Cl-Cl \longrightarrow C_6H_5Cl + Cl\cdot$$

Once the chlorine free radicals are produced, the reaction proceeds in a manner described earlier.

Enthalpy Changes and Energy of Activation: Let us discuss the change in enthalpy and energy of activation during the chlorination of methane.

The initiation step requires 58 Kcal of heat per mol and, therefore, is a highly endothermic reaction.

$$Cl-Cl \longrightarrow 2Cl^\cdot \qquad \Delta H = +58 \ Kcal/mol$$

The enthalpy changes during propagation step may be given as follows.

$$CH_3-H + Cl^{\bullet} \longrightarrow HCl + CH_3^{\bullet} \qquad \Delta H = +1 \text{ Kcal/mol}$$
$$CH_3^{\bullet} + Cl-Cl \longrightarrow CH_3Cl + Cl^{\bullet} \qquad \Delta H = -25.5 \text{ Kcal/mol}$$

Here, one bond is breaking and another bond is forming, therefore, the ΔH for this step is the energy difference of the two processes. In the first reaction C–H bond breaking requires 104 Kcal/mol of heat, whereas, 103 Kcal/mol of heat is released during the formation of H–Cl bond. The second step, however, is an exothermic step since the energy released during C–Cl bond formation (83.5 Kcal/mol) is higher than the energy required for the homolysis of Cl–Cl bond (58 Kcal/mol).

The termination step involves only bond formation, thus is a highly exothermic reaction.

$$CH_3^{\bullet} + Cl^{\bullet} \longrightarrow CH_3Cl \qquad \Delta H = -83.5 \text{ Kcal/mol}$$
$$CH_3^{\bullet} + CH_3^{\bullet} \longrightarrow CH_3-CH_3 \qquad \Delta H = -88.0 \text{ Kcal/mol}$$
$$Cl^{\bullet} + Cl^{\bullet} \longrightarrow Cl-Cl \qquad \Delta H = -58.0 \text{ Kcal/mol}$$

Energy of Activation: It is the minimum amount of energy that must be provided by a collision for the reaction to occur.

In the initiation step, since it involves bond breaking, the energy of activation is equal to the heat of reaction as shown in Fig. 7.3.

$$Cl-Cl \longrightarrow 2 Cl^{\bullet} \qquad \Delta H = 58.0 \text{ Kcal/mol}$$

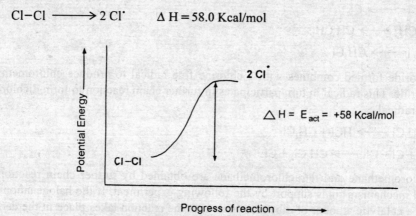

Fig. 7.3:   Energy Profile Diagram for Initiation Step

The propagation step involves the attack of chlorine radical on C–H bond of methane molecule to form hydrogen chloride and methyl radical.

$$Cl^{\bullet} + CH_3-H \longrightarrow H-Cl + CH_3^{\bullet} \qquad E_{act} = 4 \text{ Kcal/mol}$$

The energy of activation for this step is 4 Kcal/mol, whereas heat of reaction is only 1 Kcal/mol. Since the bond breaking and bond forming are not perfectly synchronised and the energy liberated by one process is not completely available for other, the reaction requires more energy (4 Kcal/mol) for activation. The source of this energy is the kinetic energy of moving particles. The collision of sufficient energy and proper orientation gives the product.

The formation of methyl chloride is exothermic and requires very small energy for its activation as shown in Fig. 7.4.

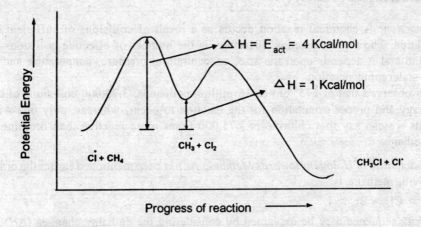

Fig. 7.4:   **Energy Profile Diagram for Propagation Step**

The heat of reaction and energy of activation for the similar step involving the attack of bromide radical instead of chlorine are 16 Kcal/mol and 18 Kcal/mol as shown in the energy profile diagram in Fig. 7.5.

$$Br^{•} + CH_3 - H \longrightarrow H - Br + CH_3^{•} \qquad E_{act} = 18 \text{ Kcal/mol}$$

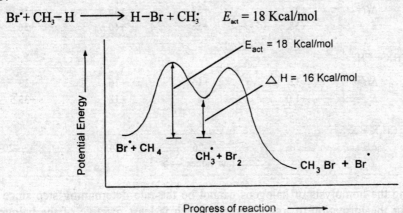

Fig. 7.5:   **Energy Profile Diagram of Bromination of Methane**

The termination step of halogenation involves bond formation only, therefore, is an exothermic reaction and does not require any energy for its activation as shown in Fig. 7.6.

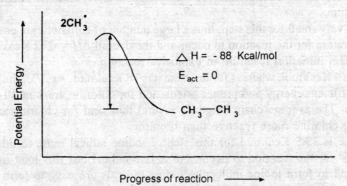

Fig. 7.6:   **Energy Profile Diagram of Termination Step**

Rate of Reaction: A chemical reaction occurs as a result of collisions of sufficient energy and of proper orientation. The rate of reaction, therefore, is the number of effective collisions that occur per liter per second and it depends upon the size, concentration, pressure, temperature and orientation of the particles undergoing reaction.

It has been observed that at 275° out of 15 million collisions, 3,75,000 collisions of chlorine are of sufficient energy and proper orientation for the reaction to occur, whereas, only one with bromine or in other words we can say that chlorine is 3,75,000 times more reactive than bromine when reacted upon with methane.

Relative Reactivities of Halogens towards Methane: As has been mentioned earlier the order of reactivity of halogens with methane is:

$$F_2 > Cl_2 > Br_2 > I_2$$

The reactivity sequence may be explained by considering the enthalpy changes (ΔH) and energy of activation (E_{act}) of each step as given below in Kcal/mol:

		X=	F	Cl	Br	I
Reaction Steps:						
1.	$X_2 \longrightarrow 2X\cdot$	$\Delta H=$	+38	+58	+46	+36
		$E_{act}=$	+38	+58	+46	+36
2.	$X\cdot + CH_4 \longrightarrow HX + CH_3\cdot$					
		$\Delta H=$	−32	+1	+16	+33
		$E_{act}=$	+1.0	+4	+18	+33.5
3.	$CH_3\cdot + X_2 \longrightarrow CH_3X + X\cdot$					
		$\Delta H=$	−70	−26	−24	−20
		$E_{act}=$		$\longleftarrow$ low values $\longrightarrow$		

First step involving the homolysis of halogens cannot be the rate determining step since minimum energy is required for the formation of iodine radical which is least reactive of the halogens.

Third step also cannot be the rate determining step as the reaction of methyl radical with halogens is exothermic and is a fast reaction.

Second step involves the abstraction of hydrogen from methane by halide radical and is exothermic for fluorine, slightly endothermic for chlorine, moderately endothermic for bromine and highly endothermic for iodine.

E_{act} for fluorine is very small for this step, thus a large number of collisions are energetically favourable even at room temperature for the reaction to occur and the overall $\Delta H= -102$ Kcal/mol, which increases the frequency of chain initiation resulting in explosive reaction.

E_{act} for chlorine is 4 Kcal/mol, whereas for bromine it is 18 Kcal/mol. At 275°, the fraction of successful collisions having sufficient energy and proper orientation for chlorine are 1 in 40 whereas for bromine it is 1 in 15 million. The average chain length is several thousand for chlorine and less than hundred for bromine making chlorine more reactive than bromine.

The E_{act} for iodine is 33.5 Kcal/mol for this step or iodine radical must collide at least ten million million times at 275° for the reaction to occur. No iodine radical last this long instead combines with another iodine radical to form iodine molecule. Iodine radicals are easy to form but can not abstract the hydrogen from methane (step 2), therefore prevent the iodination to occur.

Halogenation of Higher Alkanes: Under the influence of ultraviolet light or at 250-400°, alkanes form a number of isomeric monohalogenated products. For example, reaction with chlorine gives following products.

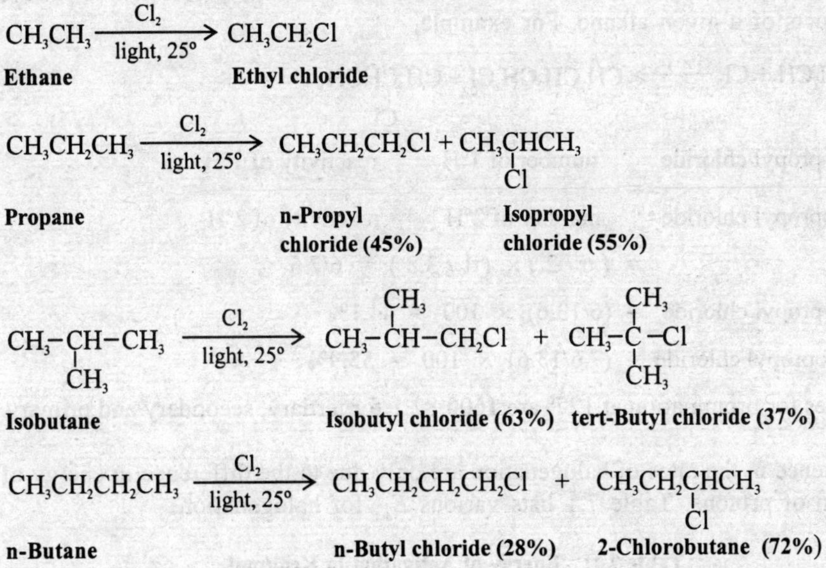

$$CH_3CH_3 \xrightarrow[\text{light, 25°}]{Cl_2} CH_3CH_2Cl$$

Ethane **Ethyl chloride**

$$CH_3CH_2CH_3 \xrightarrow[\text{light, 25°}]{Cl_2} CH_3CH_2CH_2Cl + CH_3\underset{\underset{Cl}{|}}{C}HCH_3$$

Propane **n-Propyl chloride (45%)** **Isopropyl chloride (55%)**

$$CH_3-\underset{\underset{CH_3}{|}}{C}H-CH_3 \xrightarrow[\text{light, 25°}]{Cl_2} CH_3-CH-CH_2Cl + CH_3-\underset{\underset{CH_3}{|}}{\overset{\overset{CH_3}{|}}{C}}-Cl$$

Isobutane **Isobutyl chloride (63%)** **tert-Butyl chloride (37%)**

$$CH_3CH_2CH_2CH_3 \xrightarrow[\text{light, 25°}]{Cl_2} CH_3CH_2CH_2CH_2Cl + CH_3CH_2\underset{\underset{Cl}{|}}{C}HCH_3$$

n-Butane **n-Butyl chloride (28%)** **2-Chlorobutane (72%)**

Bromination, similarly, gives corresponding bromides but in different proportions as mentioned below.

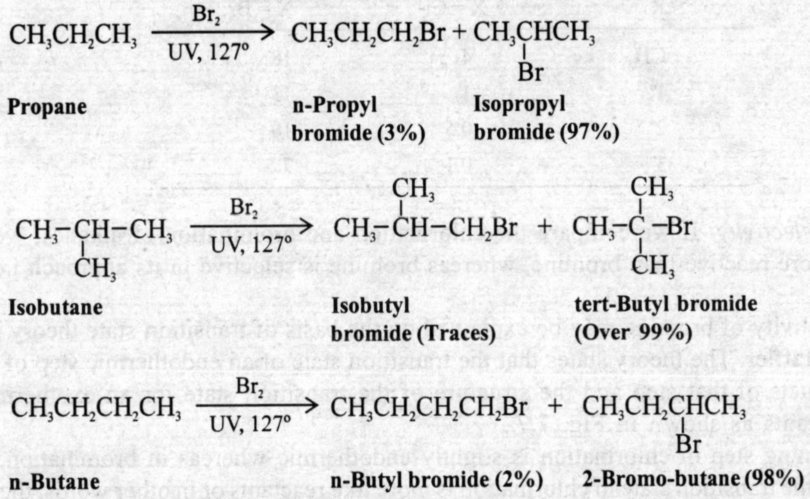

$$CH_3CH_2CH_3 \xrightarrow[\text{UV, 127°}]{Br_2} CH_3CH_2CH_2Br + CH_3\underset{\underset{Br}{|}}{C}HCH_3$$

Propane **n-Propyl bromide (3%)** **Isopropyl bromide (97%)**

$$CH_3-\underset{\underset{CH_3}{|}}{C}H-CH_3 \xrightarrow[\text{UV, 127°}]{Br_2} CH_3-CH-CH_2Br + CH_3-\underset{\underset{CH_3}{|}}{\overset{\overset{CH_3}{|}}{C}}-Br$$

Isobutane **Isobutyl bromide (Traces)** **tert-Butyl bromide (Over 99%)**

$$CH_3CH_2CH_2CH_3 \xrightarrow[\text{UV, 127°}]{Br_2} CH_3CH_2CH_2CH_2Br + CH_3CH_2\underset{\underset{Br}{|}}{C}HCH_3$$

n-Butane **n-Butyl bromide (2%)** **2-Bromo-butane (98%)**

Alkanes on chlorination and bromination form mixtures of isomers. Chlorination gives mixtures where no isomer predominates whereas bromination selectively forms only one product.

Let us now discuss the orientation in various alkanes. Propane, for example, forms n-propyl chloride and isopropyl chloride in 45% and 55% yield respectively which is dependent upon the relative rates of formation of n-propyl and isopropyl radical. In other words, we can say that the collisions with secondary protons are more successful than with primary protons or the E_{act} for the abstraction of secondary protons is less than that for primary protons.

Similarly, the E_{act} for the abstraction of tertiary protons is even less than for secondary protons or the reactivity sequence of hydrogen atoms is 3° > 2° > 1°.

The relative rates per hydrogen atom at room temperature are 5.0:3.8:1.0 for tertiary, secondary and primary hydrogens respectively and can be used successfully for the prediction of ratio of isomeric chlorination products of a given alkane. For example,

$$CH_3CH_2CH_2 + Cl_2 \longrightarrow CH_3CH_2CH_2Cl + CH_3\underset{\underset{Cl}{|}}{C}HCH_3$$

$$\frac{\text{n-propyl chloride}}{\text{isopropyl chloride}} = \frac{\text{number of 1°H}}{\text{number of 2°H}} \times \frac{\text{reactivity of 1°H}}{\text{reactivity of 2°H}}$$

$$= (6/2) \times (1/3.8) = 6/7.6$$

% of n-propyl chloride $= (6/13.6) \times 100 = 44.1\%$

% of isopropyl chloride $= (7.6/13.6) \times 100 = 55.9\%$

The relative rates for bromination at 127° are 1600:82:1 for tertiary, secondary and primary hydrogens respectively.

Thus, the difference in the rates of halogenation are only due to the difference in energy of activation for the abstraction of protons. Table 7.1 lists various E_{act} for halogenation.

Table 7.1: Energy of Activation in Kcal/mol

$$R-H + X^{\cdot} \longrightarrow R^{\cdot} + HX$$

R	X = Cl	X = Br
CH_3	4	18
1°	1	13
2°	0.5	10
3°	0.1	7.5

Reactivity and Selectivity: If we compare the chlorination and bromination of alkanes, we find that chlorine is much more reactive than bromine, whereas bromine is selective in its approach i.e., it gives only one product.

The greater selectivity of bromine may be explained on the basis of transition state theory postulated by Hammond and Haffler. The theory states that the transition state of an endothermic step of a reaction resembles the products of that step and the structure of the transition state for an exothermic step is more like the reactants as shown in Fig. 7.7.

The rate determining step in chlorination is slightly endothermic whereas in bromination, is highly endothermic. Thus, the transition state in chlorination is more like reactants or in other words, the transition state is reached early and the alkyl group gains very little radical character and can be represented as follows.

$$R-H + Cl^{\cdot} \longrightarrow [R^{\cdot} \ldots\ldots H \ldots\ldots\ldots Cl^{\cdot}] \longrightarrow R^{\cdot} + HCl$$

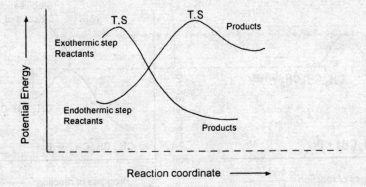

Fig. 7.7:   Transition States of Exothermic and Endothermic Reactions

In case of bromination, on the other hand, the transition state is more like the products or the transition state is reached late and the alkyl group gains considerable radical character (as the reaction is endothermic).

$$R-H + Br^{\cdot} \longrightarrow [\,\overset{\cdot}{R}..........H......\overset{\cdot}{Br}\,] \longrightarrow R^{\cdot} + HBr$$

Selectivity of a reaction is directly related to the rate of formation of various classes of free radicals. A more stable radical is formed faster because of the factor that stabilises it (i.e., the delocalisation of odd electron) and also stabilises the incipient radical in transition state. If this is so, then the higher the radical character in transition state, more will be the stabilisation of the transition state and faster will be the formation of radical.

The stability sequence of the free radicals may be given as follows which is dependent upon the energy content of the free radical. The less the energy associated with the radical the more stable it will be. Below are given the bond dissociation energies for the formation of various free radicals.

$$3° > 2° > 1° > CH_3$$

$$CH_3-H \longrightarrow CH_3^{\cdot} + H^{\cdot} \qquad\qquad \Delta H = 104 \text{ Kcal/mol}$$

$$CH_3CH_2-H \longrightarrow CH_3CH_2^{\cdot} + H^{\cdot} \qquad \Delta H = 98 \text{ Kcal/mol}$$

$$CH_3CH_2CH_2-H \longrightarrow CH_3CH_2CH_2^{\cdot} + H^{\cdot} \qquad \Delta H = 98 \text{ Kcal/mol}$$

$$CH_3-CH-CH_3 \longrightarrow CH_3CH^{\cdot}CH_3 + H^{\cdot} \qquad \Delta H = 95 \text{ Kcal/mol}$$
$$\underset{H}{|}$$

$$CH_3-\overset{\overset{\displaystyle CH_3}{|}}{\underset{\underset{\displaystyle H}{|}}{C}}-CH_3 \longrightarrow CH_3-\overset{\overset{\displaystyle CH_3}{|}}{\underset{\cdot}{C}}-CH_3 + H^{\cdot} \quad \Delta H = 92 \text{ Kcal/mol}$$

The isopropyl radical is 3 Kcal/mol more stable than the n-propyl radical, thus, we can presume that the difference in E_{act} during bromination will also be the same (see Table 7.1) resulting in higher percentage of isopropyl radical compared to n-propyl radical making it more selective as shown in Fig. 7.8.

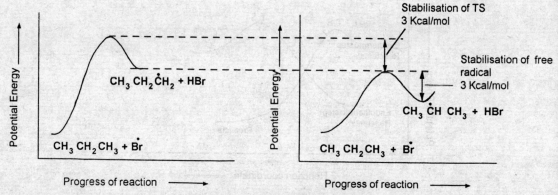

Fig. 7.8: Selectivity of Bromine

In chlorination, on the contrary, the difference in E_{act} of n-propyl and isopropyl radical is only 0.5 Kcal/mol (see Table 7.1) or the difference in the percentage of the two radicals (n-propyl and isopropyl) will not be large making it less selective than bromine as shown in Fig. 7.9.

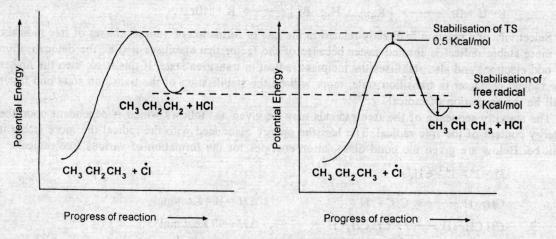

Fig.7.9: Nonselectivity of Chlorine

Nitration

Alkanes undergo vapor phase nitration under drastic conditions (i.e., at 400-500°) to give nitroalkanes. Nitration of methane gives nitromethane.

$$CH_4 + HNO_3 \xrightarrow{400°} CH_3 NO_2$$

Nitromethane

Like halogenation, nitration also follows free radical mechanism.

$$HO-NO_2 \longrightarrow HO^{\cdot} + NO_2^{\cdot}$$

$$RCH_2-H + HO^{\cdot} \longrightarrow RCH_2^{\cdot} + H_2O$$

Since $NO_2^{\cdot}$ exists as $O=N^{\cdot} \longrightarrow O$ or $O=N-O^{\cdot}$

$$RCH_2^* + \overset{\cdot}{N}O_2 \longrightarrow RCH_2-NO_2$$

$$RCH_2^* + \overset{\cdot}{O}-N=O \longrightarrow RCH_2-\overset{\cdot}{O} + \overset{\cdot}{N}=O$$

$$RCH_2-O^\cdot \longrightarrow R^\cdot + CH_2=O$$

$$R^\cdot + NO_2^* \longrightarrow R-NO_2$$

The nitration of propane gives a mixture of four nitroalkanes which may be separated by fractionation.

$$CH_3CH_2CH_3 + HNO_3 \xrightarrow{400°} CH_3CH_2CH_2NO_2 + CH_3\overset{NO_2}{\underset{|}{C}HCH_3} + CH_3CH_2NO_2 + CH_3NO_2$$

Like halogenation, nitration takes place easily at tertiary positions. Nitroalkanes generally find use as fuels, solvents and in organic synthesis.

Sulphonation

When alkanes are treated with fuming sulphuric acid (oleum), a hydrogen atom is replaced by sulphonic acid group, $-SO_3H$. This is known as sulphonation and the products obtained are alkylsulphonic acids.

$$R-H + HOSO_3H \longrightarrow R-SO_3H + H_2O$$

Sulphuric acid   **Alkyl sulphonic acid**

$$CH_3\overset{CH_3}{\underset{|}{C}H}CH_3 + HOSO_3H \longrightarrow CH_3-\overset{CH_3}{\underset{\underset{CH_3}{|}}{C}}-SO_3H$$

Isobutane

2-methylpropane-2-sulphonic acid

The ease of replacement of hydrogen atoms is tertiary > secondary > primary.
The mechanism suggested for sulphonation is a free radical mechanism.

$$HO-SO_3H \longrightarrow HO^\cdot + {}^\cdot SO_3H$$

$$R-H + HO^\cdot \longrightarrow R^\cdot + H_2O$$

$$R^\cdot + {}^\cdot SO_3H \longrightarrow R-SO_3H$$

Alkanes undergo chlorosulphonation on treatment with a mixture of sulphur dioxide and chlorine in presence of ultraviolet light to form alkanesulphonyl chlorides.

$$CH_3CH_2CH_3 + 2SO_2 + 2Cl_2 \xrightarrow[50°]{UV\ light} CH_3CH_2CH_2SO_2Cl + CH_3\overset{|}{\underset{\underset{SO_2Cl}{|}}{C}HCH_3} + 2HCl$$

A free radical mechanism has been suggested.

$$Cl-Cl \longrightarrow 2\,Cl^\cdot$$

$$R-H + Cl^\cdot \longrightarrow R^\cdot + HCl$$

$$R^\cdot + SO_2 \longrightarrow RSO_2^\cdot$$

$$RSO_2^\cdot + Cl-Cl \longrightarrow RSO_2Cl + Cl^\cdot$$

Chlorosulphonation of $C_{14}-C_{16}$ fraction of petroleum followed by saponification forms sodium alkylsulphonates which are used as detergents.

Oxidation

Normally alkanes are inert towards usual oxidising agents. However, potassium permanganate oxidises a tertiary hydrogen to a hydroxyl group. Thus isobutane on oxidation gives tert-butyl alcohol.

$$(CH_3)_3CH + [O] \xrightarrow{KMnO_4} (CH_3)_3COH$$

Isobutane   **tert-Butyl alcohol**

Alkanes can be oxidised under catalytic conditions. Thus if a mixture of methane and oxygen is passed over a copper tube under high temperature and pressure, it gives methyl alcohol. This method is used for industrial preparation of methyl alcohol.

$$CH_4 + O_2 \xrightarrow[100 \text{ atm, } 200°]{\text{Catalyst}} 2CH_3OH$$

Methane   **Methyl alcohol**

However, controlled oxidation of methane yields formaldehyde.

$$CH_4 + O_2 \xrightarrow[MoO, 400°/200 \text{ atms}]{\text{Controlled Oxidn.}} HCHO + H_2O$$

Methane   **Formaldehyde**

On ignition in presence of excess of oxygen, alkanes burn to form carbon dioxide and water. The reaction is accompanied by evolution of large quantities of heat. The hydrocarbons are used as fuels due to their high heat of combustion which increases with molecular weight.

$$CH_4 + 2O_2 \longrightarrow CO_2 + 2H_2O + 211 \text{ Kcal}$$

$$2CH_3(CH_2)_6CH_3 + 25O_2 \longrightarrow 16CO_2 + 18H_2O + 1223 \text{ Kcal}$$

The slow combustion of lower alkanes at low temperature and high pressure undergo a series of changes finally forming carbon dioxide and water.

$$2CH_4 \xrightarrow{[O]} CH_3OH \xrightarrow{[O]} \left[CH_2 \begin{array}{c} OH \\ OH \end{array} \right]$$

$$\downarrow -H_2O$$

$$CO_2 + H_2O \xleftarrow{[O]} HCOOH \xleftarrow{[O]} CH{=}O$$

                    Formic acid   **Formaldehyde**

Isomerisation

The straight chain alkanes are converted into branched chain alkanes in presence of aluminum chloride and hydrogen chloride.

$$CH_3CH_2CH_2CH_3 \xrightarrow[25°]{AlCl_3, -HCl} CH_3{-}\overset{\overset{\displaystyle CH_3}{|}}{CH}{-}CH_3$$

n-Butane   **Isobutane**

The molecular rearrangement of one compound to another compound or compounds is called isomerisation. This technique is used to increase the branching in lower alkanes, since branched chain alkanes are more valuable than n-alkanes in motor spirit.

Pyrolysis or Cracking

The alkanes or other organic compounds on heating to high temperature in the absence of oxygen decompose into smaller fragments. This is known as *Pyrolysis*. This word is taken from Greek words, pyro (fire) and lysis (disintegration). The process when applied to alkanes is known as *Cracking*. Propane on cracking gives a number of products.

$$CH_3CH_2CH_3 \xrightarrow[\text{or Catalyst } (350°)]{\text{Cracking } 450-500°} CH_3CH=CH_2 + CH_2=CH_2 + CH_4 + H_2$$

Propane   **Propene**   **Ethene**

The mechanism of cracking is believed to be a free radical mechanism.

$$CH_3CH_2CH_3 \longrightarrow CH_3^{\bullet} + CH_3CH_2^{\bullet}$$

$$CH_3^{\bullet} + CH_3CH_2CH_3 \longrightarrow CH_4 + CH_3CH_2CH_2^{\bullet}$$

$$CH_3CH_2CH_2^{\bullet} \longrightarrow CH_3CH=CH_2 + H^{\bullet}$$

$$CH_3CH_2^{\bullet} \longrightarrow CH_2=CH_2 + H^{\bullet}$$

$$H^{\bullet} + H^{\bullet} \longrightarrow H_2$$

Pyrolysis in presence of catalyst (oxides of molybdanum, chromium or vanadium) is used to manufacture alkenes.

Aromatisation

The process of converting aliphatic or alicyclic compounds into aromatic hydrocarbons is known as aromatisation. Thus, alkanes with six or more carbon atoms on heating strongly under pressure in presence of a catalyst give aromatic hydrocarbons. Aromatisation involves cyclisation, isomerisation and dehydrogenation. Thus, n-hexane when passed over chromium oxide supported on alumina at 500° gives benzene as the main product.

$$CH_3 (CH_2)_4CH_3 \xrightarrow[500°]{Cr_2O_3/Al_2O_3} \bigcirc + 4H_2$$

Aromatisation of gasoline increases their octane number, since unsaturated hydrocarbons are better fuels.

7.11 INDIVIDUAL MEMBERS

7.11.1 Methane [CH_4]

The first member of the alkane series is methane. It is found in marshy places where it is produced by the bacterial decomposition of vegetable matter. So it is also called *marsh gas*. It forms explosive mixture with air. The best source of methane is natural gas obtained from petroleum wells, which contains 80% methane. It also occurs in coal gas up to 30%.

Preparation

In the laboratory pure methane is obtained by the following methods.

  (i)   Decarboxylation of sodium acetate with sodalime forms methane.

$$CH_3COONa + NaOH \xrightarrow[\text{heat}]{CaO} CH_4 + Na_2CO_3$$

Sod. acetate   **Methane**

 (ii)   Reduction of methyl iodide with zinc and acetic acid or zinc-copper couple forms methane.

$$CH_3I + 2[H] \xrightarrow[\text{Zn-Cu couple}]{\text{Zn-HOAc} \atop \text{or}} CH_4 + HI$$

Methyl iodide   **Methane**

(iii) Decomposition of methyl magnesium bromide with water produces methane in good yield.

$$CH_3Br + Mg \xrightarrow{\text{dry ether}} CH_3MgBr \xrightarrow{H_2O} CH_4 + Mg(OH)Br$$

Methyl   **Methyl**   **Methane**

bromide   **magnesium bromide**

Industrial Method

Methane is obtained commercially by the following methods.

(i) By fractional distillation of wet natural gas, which contains 80% methane. It is also obtained by the cracking of petroleum.

(ii) From aluminum carbide by the action of water.

$$Al_4C_3 + 12H_2O \longrightarrow 3CH_4 + 4Al(OH)_3$$

Aluminum   **Methane**

carbide

(iii) Direct combination of carbon and hydrogen in the electric arc.

$$C + 2H_2 \xrightarrow{1200°} CH_4$$

Methane

Since methane is obtained from its elements, viz. carbon and hydrogen, this method constitutes the 'Total synthesis' of methane.

(iv) By passing a mixture of carbon monoxide and hydrogen over heated nickel.

$$CO + 3H_2 \xrightarrow[250-500°]{Ni} CH_4 + H_2O$$

Methane

Properties

Methane is colorless, odorless and non-poisonous gas. Its boiling point is $-162°/760$ mm and melting point $-183°$. It is sparingly soluble in water but is quite soluble in ethanol and ether. It burns with nonluminous flame in air or oxygen forming carbon dioxide and water.

$$CH_4 + 2O_2 \longrightarrow CO_2 + 2H_2O + 211.0 \text{ Kcal}$$

It explodes with violence on being mixed with air or oxygen and ignited. This is believed to be the cause. of explosions in coal mines and where methane is known as fire-damp. It has the properties of alkanes discussed earlier. Some of its characteristic chemical properties are given below.

(i) *Pyrolysis:* On pyrolysis at 1000° methane gives carbon and hydrogen. The carbon produced is known as 'Carbon black' and finds use in paints, printing ink and in the production of automobile tyres.

$$CH_4 \xrightarrow[\text{Pyrolysis}]{1000°} C + 2H_2$$

Methane   **Carbon black**

(ii) *Reaction with Steam in presence of Nickel catalyst:* Methane reacts with steam at 1000° in presence of nickel forming carbon monoxide and hydrogen. This method is used for the industrial preparation of hydrogen.

$$CH_4 + H_2O \xrightarrow[Ni]{1000°} CO + 3H_2$$
Carbon monoxide

(iii) On *controlled oxidation* in presence of molybdanum oxide and air methane is converted into formaldehyde.

$$CH_4 + O_2 \xrightarrow{MoO, \Delta} HCHO + H_2O$$
Formaldehyde

(iv) On *catalytic oxidation* by passing a mixture of methane and oxygen through a copper tube under high pressure and temperature, methyl alcohol is obtained. This method is used for industrial preparation of methyl alcohol.

$$2CH_4 + O_2 \xrightarrow[100 \text{ atms } 300–350°]{Catalyst} 2CH_3OH$$
Methanol

(v) Methane is converted into hydrogen cyanide by reaction with nitrogen or ammonia. This is an industrial method for the preparation of hydrogen cyanide.

$$2CH_4 + N_2 \xrightarrow{Electric \ arc} 2HCN + 3H_2$$

$$CH_4 + NH_3 \xrightarrow[Al_2O_3]{Electric \ arc} HCN + 3H_2$$
Hydregon cyanide

Uses
(i) As illuminant and domestic fuel.
(ii) For the preparation of methyl chloride, methylene chloride, chloroform and carbon tetrachloride.
(iii) For the industrial preparation of methyl alcohol, formaldehyde and hydrogen cyanide.
(iv) For the preparation of 'Carbon black' used in paints, printing inks and automobile tyres.

7.11.2 Ethane [C_2H_6]
The second member of the alkane series is ethane. It occurs with methane in natural gas.

Preparation
It is prepared by all the general methods of preparation of alkanes discussed earlier. In the laboratory it is prepared by the following methods.

(i) Electrolysis of concentrated solution of sodium or potassium acetate (Kolbe's method).

$$2CH_3COONa + H_2O \xrightarrow{Electrolysis} CH_3CH_3 + 2CO_2 + 2KOH + H_2$$
Sod. acetate   **Ethane**

(ii) Decarboxylation of sodium propionate with sodalime.

$$CH_3CH_2COONa + NaOH \xrightarrow[Heat]{CaO} C_2H_6 + Na_2CO_3$$
Sod. propionate   **Ethane**

(iii) Reduction of ethyl iodide with zinc-copper couple and ethyl alcohol.

$$C_2H_5I + 2H \xrightarrow[\text{C}_2\text{H}_5\text{OH}]{\text{Zn–Cu Couple}} C_2H_6 + HI$$

Industrial Method: Ethane is obtained commercially by the fractional distillation of wet natural gas. It is also obtained by cracking of long-chain alkanes.

Properties

Ethane is a colorless gas, b.p. $-89°$. It is sparingly soluble in water but readily soluble in ethanol. It gives all the general reactions of alkanes.

(i) On burning in air or oxygen it forms carbon dioxide and water.

$$2C_2H_6 + 7O_2 \longrightarrow 4CO_2 + 6H_2O + 736.8 \text{ Kcal}$$

(ii) Ethane undergoes halogenation faster than methane. It forms two dichloro-ethanes, viz. geminal and vicinal dichloroethane.

CH_3CHCl_2	$CH_2Cl–CH_2Cl$
1,1-Dichloroethane	**1,2-Dichloroethane**
(geminal dichloroethane)	**(vicinal dichloroethane)**

(iii) Ethane dehydrogenates on heating in the absence of air and in presence of hydrogen at $450°$ to form ethylene.

$$CH_3–CH_3 \xrightarrow[\text{Catalyst}]{400°-450°} CH_2{=}CH_2 + H_2$$

Ethylene

This is a convenient method for the industrial preparation of ethylene from natural gas.

Uses

Ethane is mainly used as a fuel. It is also used for the industrial preparation of ethylene (by catalytic dehydrogenation), which is a starting material for a number of important products.

7.12   PROBLEMS

1. Fill in the blanks.
   - (a) Alkane molecules are very weakly polar or nonpolar because there is very little _____ difference between the carbon and hydrogen atoms.
   - (b) Boiling points of alkanes _____ with the increase in the length of carbon chain.
   - (c) Branching of carbon chain _____ the boiling point.
   - (d) Alkanes are soluble in _____ solvents.
2. How is butane obtained from 1-Chlorobutane.
3. Complete the following reactions.

   (a) $CH_3CH_3 + Cl_2 \xrightarrow{\text{hv}}$

   (b) $CH_3CH_3 + HNO_3 \longrightarrow$

   (c) $CH_3(CH_2)_5CH_3 \xrightarrow{\text{Pt/400°}}$

   (d) $CH_3CH_2Br + CH_3CH_2Br \xrightarrow{\text{Na}}$

   (e) $CH_3CH_2CH_2Br \xrightarrow{\text{LiAlH}_4}$

   (f) $CH_3CH{=}CH–CH_2Cl \xrightarrow{\text{H}_2/\text{Pt}}$

   (g) $\underset{\underset{CH_3}{|}}{CH_3CH–CH_2CH_3} \xrightarrow{\text{Cl}_2}$

4. Starting with $CH_2=CHCH_2Br$, how will you prepare $CH_3(CH_2)_4CH_3$

5. Starting with $(CH_3)_3CCl$, how will you prepare $(CH_3)_3CH$ and $(CH_3)_3CC(CH_3)_2CH_2Cl$.

6. Write equation to show Wurtz reaction.

7. Give the structural formula of
   (a) 2,2,3,3 -Tetramethylpentane
   (b) 2,3-Dimethylbutane
   (c) 2,5-Dimethylhexane
   (d) 3-Chloro-2-methylpentane
   (e) 3-Methyl-2-ethylpentane

8. What are alkanes? Why they are called saturated hydrocarbons ?

9. Give IUPAC names of the following alkanes.

   (a) $CH_3-CH_2-CH_2-CH_2-CH_3$

   (b) $CH_3-CH_2-CH-CH_2-CH_3$
               |
               CH_3

   (c) $CH_3-\overset{\overset{\displaystyle CH_3}{|}}{\underset{\underset{\displaystyle CH_3}{|}}{C}}-\overset{}{\underset{\underset{\displaystyle CH_3}{|}}{CH}}-CH_3$

10. Give the structures of all hydrocarbons of the molecular formula C_5H_{12}. Give IUPAC names for all and identify primary, secondary and tertiary hydrogen atoms in these structures.

11. How will you prepare the following.
    (a) n-Butane from ethane   (b) Benzene from n-Hexane   (c) 2,3-Dimethylbutane from isopropyl bromide.

12. Complete the following equations:

    (a) $CH_3CH_2CH_2OH + HI \xrightarrow[150°]{P}$

    (b) $CH_3CH_2Br + Mg \xrightarrow{Ether}$

    (c) $CH_4 + Cl_2 \xrightarrow{UV\ light}$

13. Discuss the free radical mechanism for the chlorination of methane.

14. Write notes on the following (give their mechanisms too):
    (a) Wurtz reaction
    (b) Sabatier and Senderens reaction
    (c) Corey and House method
    (d) Kolbe's electrolytic method.

15. What products are obtained in the following:

    (a) $CH_3CH_2CH_3 \xrightarrow[500–600°]{Cracking}$

    (b) $CH_3(CH_2)_4CH_3 \xrightarrow[500°]{Cr_2O_3\ /\ Al_2O_3}$

16. Discuss the free radical mechanism for the chlorination of methane. How the mechanism actually proved ?

8

Cycloalkanes

8.1 INTRODUCTION

Cycloalkanes are cyclic aliphatic compounds also known as *alicyclic* compounds. Like alkanes, the unsubstituted cycloalkanes form a homologous series having the general molecular formula C_nH_{2n}, where n is the number of carbon atoms in the molecule. The first member of the series (when n=3) is cyclopropane, C_3H_6 and the second member of the series (when n=4) is cyclobutane, C_4H_8. Thus, cycloalkanes have two hydrogen atoms less than the corresponding alkanes and are the structural isomers of alkenes.

8.2 NOMENCLATURE

The cycloalkanes are named by prefixing *Cyclo* to the name of the corresponding open-chain hydro-carbons.

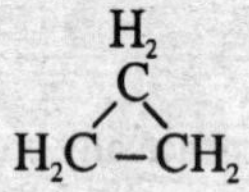

Cyclopropane 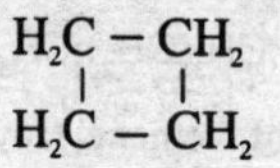   **Cyclobutane** 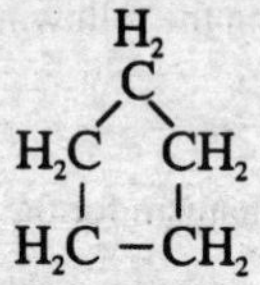   **Cyclopentane** 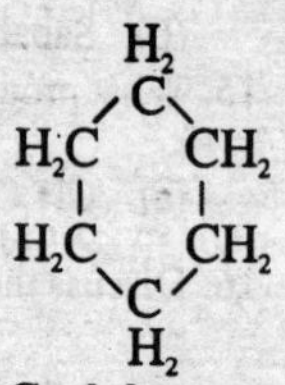   **Cyclohexane**

The substituents on the ring are named and their positions indicated by the numbers. The rings are numbered so that the carbon having the substituent have the lowest possible numbers.

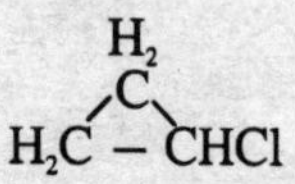

1-Chlorocyclopropane 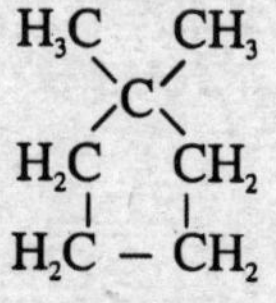   **1,1-Dimethylcyclopentane** 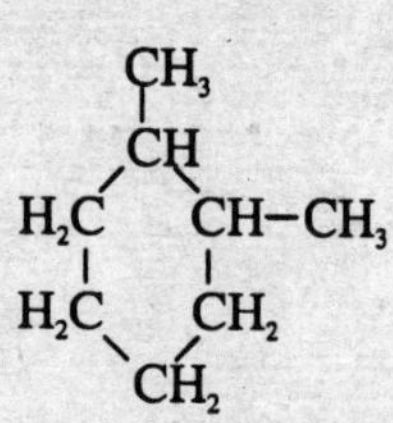   **1,2-Dimethylcyclohexane**

In practice, for the sake of convenience, the rings are represented by simple geometric figures. For example, a triangle for cyclopropane, a square for cyclobutane, a pentagon for cyclopentane and a hexagon for cyclohexane etc. It is implied that a CH_2 group is at each corner of the figure unless there is some other group indicated.

Cyclopropane   Cyclobutane    Cyclopentane 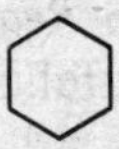   Cyclohexane

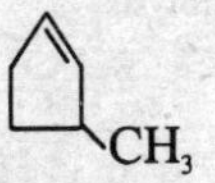

3-Methylcyclopentene 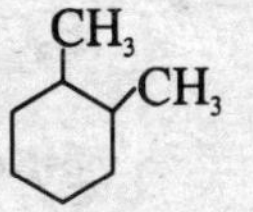   1,2-Dimethylcyclohexane 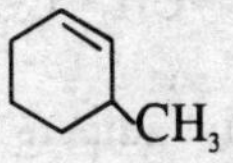   3-Methylcyclohexene

8.3 PREPARATION

Cycloalkanes, particularly cyclohexane, methylcyclohexane, methylcyclopentane etc. are obtained from petroleum as well as from natural gas. These may also be obtained in pure state by the following methods.

(i) *Cyclisation of Dihalogen Compounds:* The method consists in the treatment of suitable dihalogen compounds with sodium or zinc to give the corresponding cycloalkanes.

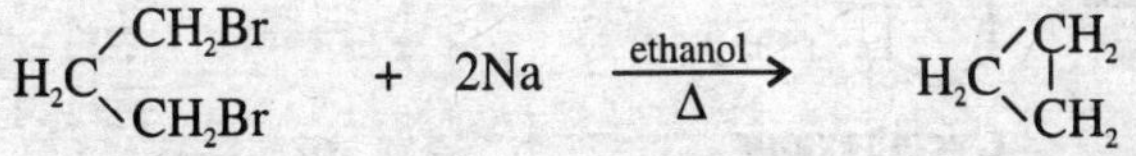

1, 3-Dibromopropane   Cyclopropane

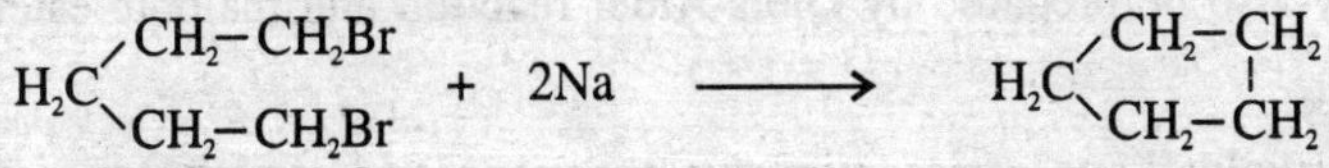

1, 5-Dibromopentane   Cyclopentane

This reaction may be regarded as an extension of Wurtz reaction and holds good for the preparation of rings upto six carbons. For the preparation of higher rings ($>C_6$) intermolecular Wurtz reaction is employed.

$$Br(CH_2)_nBr + Br(CH_2)_nBr \longrightarrow Br(CH_2)_n-(CH_2)_n-Br$$

(ii) *From Other Aliphatic Compounds:* Preparation of cycloalkanes from other aliphatic compounds generally involves two stages. The first stage is the conversion of an open-chain compound into a cyclic compound (known as cyclisation) and the second step is the conversion of the cyclic compound thus obtained into the kind of compound that is required.

(a) *From Calcium or Barium Salts of Dicarboxylic Acids:* When calcium or barium salt of a dicarboxylic acid is heated, a cyclic ketone is obtained which on clemmensons reduction gives the cycloalkane. For example barium adipate gives cyclopentanone which on treatment with zinc and hydrochloric acid forms cyclopentane.

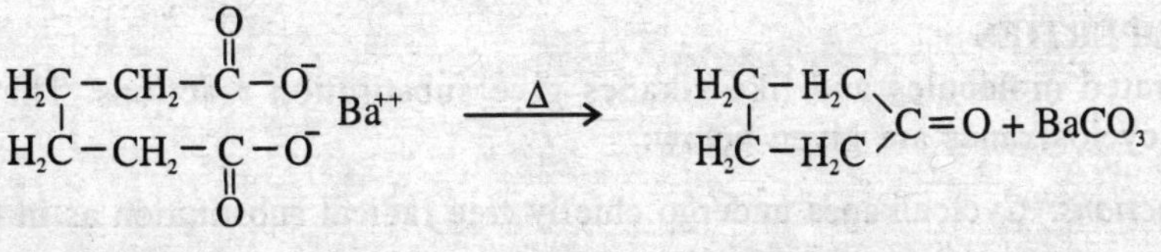

Barium adipate   Cyclopentanone

(b) *From Esters of Dicarboxylic Acids:* The diester of adipic acid on treatment with sodium undergoes cyclisation (intramolecular condensation) to give a β-keto ester. This cyclisation is known as

Dieckmann Reaction. The β-keto ester on hydrolysis gives the cyclic β-keto acid, which on decarboxylation followed by reduction gives the corresponding cycloalkane.

$$H_2C-CH_2-COOC_2H_5$$
$$H_2C-CH_2-COOC_2H_5 \xrightarrow[Na]{C_2H_5ONa}$$

Diethyl adipate

$$\begin{array}{c} H_2C-H_2C \\ | \\ H_2C-HC \end{array} C=O \\ \backslash COOC_2H_5 \xrightarrow[Hydrolysis]{aq.HCl}$$

β-Keto ester

$$\begin{array}{c} H_2C-H_2C \\ | \\ H_2C-HC \end{array} C=O \\ \backslash COOH \xrightarrow[-CO_2]{\Delta} \quad \square =O \xrightarrow[HCl]{Zn/Hg} \square$$

β-Keto acid   **Cyclopentanone**   **Cyclopentane**

(iii) *From Aromatic Compounds:* The catalytic hydrogenation of benzene and its homologues yields pure and substituted cyclohexane

$$\hexagon + 3H_2 \xrightarrow[25\ atms]{Ni,\ 150\text{-}200°} \hexagon$$

Benzene   **Cyclohexane**

In addition, cycloalkanes may also be prepared by Diels-Alder reaction and malonic ester synthesis.

8.4 PHYSICAL PROPERTIES

The first two members of the cycloalkane series, viz. cyclopropane and cyclobutane are gases at ordinary temperatures. The other members are liquids. The melting and boiling points show a gradual rise with increase in molecular weight. The density of cycloalkanes is less than one and are thus lighter than water. They are insoluble in water but soluble in solvents like ethers and alcohols. The physical constant of some important members are given below:

Cycloalkanes	M.P. (°C)	B.P. (°C)	Relative Density (at 20°C)
Cyclopropane	−127	−33	–
Cyclobutane	−80	13	–
Cyclopentane	−94	49	0.746
Cyclohexane	6.5	81	0.778
Cycloheptane	−12	118	0.810
Cyclooctane	14	149	0.830
Cyclononane	11	178	0.850
Cyclodecane	10	202	0.858

8.5 CHEMICAL PROPERTIES

Cycloalkanes are saturated molecules and like alkanes give substitution reactions. The important chemical properties of cycloalkanes are given below:

(i) *Substitution Reactions:* Cycloalkanes undergo chiefly free radical substitution as in the case of alkanes.

(ii) *Special Reactions:* The cyclopropane and cyclobutane are not as stable as their higher homologues. Therefore, the bonds in cyclopropane and cyclobutane are attacked by certain reagents, and open chain compounds are obtained. Thus, cyclopropane on hydrogenation, treatment with chlorine or bromine in dark and with concentrated hydrogen bromide and hydrogen iodide give the corresponding open chain compounds.

$$
\begin{array}{ll}
\xrightarrow[\text{Ni, H}_2,\ 80°]{\text{hydrogenation}} & \underset{\ \ \ \ H\ \ \ \ \ \ \ \ \ \ \ \ H}{CH_2-CH_2-CH_2} \qquad \textbf{Propane} \\[2mm]
\xrightarrow[\text{dark}]{\text{Cl}_2,\ \text{CCl}_4} & \underset{\ \ \ \ Cl\ \ \ \ \ \ \ \ \ \ \ Cl}{CH_2-CH_2-CH_2} \qquad \textbf{1,3-Dichloropropane} \\[2mm]
\xrightarrow{\text{Br}_2} & \underset{\ \ \ \ Br\ \ \ \ \ \ \ \ \ \ Br}{CH_2-CH_2-CH_2} \qquad \textbf{1,3-Dibromopropane} \\[2mm]
\xrightarrow{\text{HI}} & \underset{\ \ \ \ H\ \ \ \ \ \ \ \ \ \ \ \ I}{CH_2-CH_2-CH_2} \qquad \textbf{1-Iodopropane} \\[2mm]
\xrightarrow{\text{conc. H}_2\text{SO}_4} & \underset{\ \ \ \ H\ \ \ \ \ \ \ \ \ \ \ OH}{CH_2-CH_2-CH_2} \qquad \textbf{n-Propyl alcohol}
\end{array}
$$

Cyclopropane

In each of these reactions a carbon-carbon bond is broken to give a saturated open chain propane derivative.

$$
\underset{Y}{\overset{X}{|}} + \underset{H_2C-CH_2}{CH_2} \longrightarrow \underset{X \qquad\qquad Y}{CH_2-CH_2-CH_2}
$$

However, cyclobutane does not undergo most of the ring opening reactions of cyclopropane. It is hydrogenated only under more vigorous conditions.

$$
\begin{array}{c}
H_2C-CH_2 \\
| \quad\ \ \ | \\
H_2C-CH_2
\end{array}
+ H_2 \xrightarrow[200°]{\text{Ni}} CH_3CH_2CH_2CH_3
$$

Cyclobutane   **n-Butane**

Cyclopentane and higher members do not give this reaction.

8.6 RELATIVE STABILITY OF CYCLOALKANES—BAEYER STRAIN THEORY

The relative stability of cycloalkanes is explained by the Baeyer's strain theory. The strain theory is based on the classical hypothesis of Le Bel and Vant Hoff that the four valencies of a carbon atom are directed towards the four corners of a regular tetrahedron. Thus, the normal angle between carbon atoms is 109.5°. According to Baeyer any deviation of bond angle from the normal value (109.5°) causes a strain on the ring. Thus, cyclopropane ring having an angles of 60° and cyclobutane having an angle of 90° are strained rings and hence are unstable compared to molecules in which the bond angles are tetrahedral. Cyclopropane and cyclobutane undergo ring opening reactions since these relieve the strain and yield more stable open chain compounds.

Since the deviation of the bond angle in cyclopropane ($109.5°-60° = 49.5°$) is greater than in cyclobutane ($109.5° - 90° = 19.5°$), cyclopropane is more strained and more unstable and more prone to ring opening reactions than cyclobutane.

The angles of a regular pentagon ($108°$) is very close to the tetrahedral angle ($109.5°$) and hence cyclopentane is virtually free of angle strain. As the ring size increases the deviation from the normal tetrahedral angle also increases causing increased difficulty in the synthesis of higher rings.

Drawbacks of Baeyer's Strain Theory and Modern View: The Baeyer's strain theory holds good for lower rings but is not applicable to rings larger than four carbon atoms. The theory is based on the assumption that all the rings are planer but it is not so except for cyclopropane. The theory has no experimental support also. The strain in various cycloalkanes may be shown by taking into consideration the relative heats of combustion as discussed in Section 5.4.2.   The data clearly indicates the instability of cyclopropane (HOC 166.6 Kcal/ mol/ CH_2) compared to cyclohexane (HOC 157.4 Kcal/ mol/ CH_2). The rings having four or more carbons show puckering so as to relieve the angle strain.

The difficulty in the preparation of larger rings is not due to their instability (as postulated by Baeyer) but it is difficult to bring the two ends of a large chain compound close enough to undergo condensation without undergoing any side reaction. (For details see section 5.4.2.)

8.7   INDIVIDUAL MEMBERS

8.7.1   Cyclopropane [C_3H_6]

The first member of the cycloalkane series is cyclopropane also called trimethylene. It is colorless, pleasant smelling gas and is sparingly soluble in water but freely soluble in alcohol. It is used as a general anesthetic. A mixture of oxygen and cyclopropane (20-30%) produces insensibility to pain but not unconsciousness. However, the disadvantage is that the mixture is explosive.

Cyclopropane is obtained in pure form by treating 1,3- dibromopropane with metallic sodium in boiling xylene.

Cyclopropane undergoes substitution reactions and some other special reactions (see Section 8.5).

The orbital picture of angle strain associated with the cyclopropane ring has been discussed in Section 5.4.2.5.

8.7.2   Cyclobutane [C_4H_8]

The second member of the cycloalkane series is cyclobutane also called tetramethylene. It is a colorless, pleasant-smelling gas (b.p $13°$). It is prepared in pure state by the reaction of 1,4-dibromobutane with sodium.

$$\begin{array}{c} H_2C-CH_2-Br \\ | \\ H_2C-CH_2-Br \end{array} + 2Na \longrightarrow \begin{array}{c} H_2C-CH_2 \\ | \quad\quad | \\ H_2C-CH_2 \end{array}$$

1,4-dibromobutane   **Cyclobutane**

Cyclobutane is comparatively more stable than cyclopropane and is unaffected by bromine, hydrogen iodide etc. However, it can be catalytically hydrogenated to give butane.

The structure of cyclobutane has been discussed in Section 5.4.2.5.

8.8 PROBLEMS

1. Write the equations stating synthesis of the following
   (a) Cyclopropane from 1,3-dibromopropane   (b) Cyclohexane from cyclohexene

2. Complete the following reaction

   (a) $CH_3-CH_2-CH-CH_2 + Br_2 \longrightarrow$
       $\underset{CH_2}{|}$

   (c) $\underset{H_2C-CH_2}{\overset{H_2C-CH_2}{|\quad|}} + H_2 \longrightarrow$

   (b) $CH_3-CH_2-CH-CH_2 + conc.H_2SO_4 \longrightarrow$
       $\underset{CH_2}{|}$

   (d) $\underset{H_2C-CH_2}{\overset{H_2C-CH_2}{|\quad|}} + Br_2 \longrightarrow$

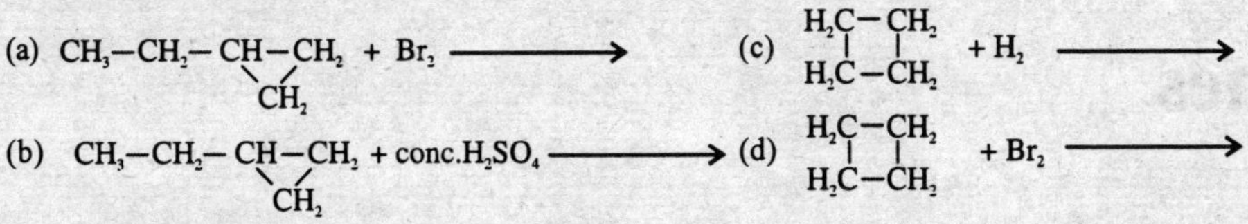

3. Give nomenclature to the following:

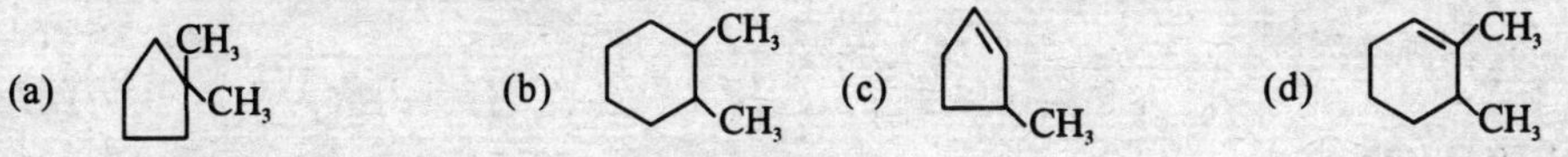

   (a)   (b)   (c)   (d)

4. Complete the following equations

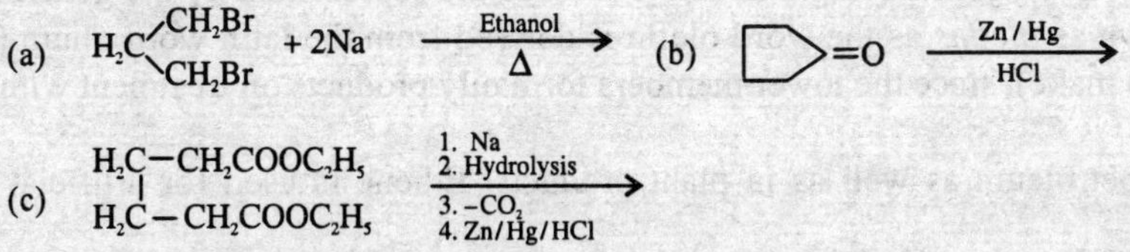

   (a) $H_2C\overset{CH_2Br}{\underset{CH_2Br}{<}} + 2Na \xrightarrow[\Delta]{Ethanol}$

   (b) $\rhd = O \xrightarrow[HCl]{Zn/Hg}$

   (c) $\underset{H_2C-CH_2COOC_2H_5}{\overset{H_2C-CH_2COOC_2H_5}{|}} \xrightarrow[\substack{3.\ -CO_2 \\ 4.\ Zn/Hg/HCl}]{\substack{1.\ Na \\ 2.\ Hydrolysis}}$

5. What happens when cyclopropane is subjected to the following reactions.

   (a) $\xrightarrow{H_2,\ Ni,\ 80°}$

   (d) $\xrightarrow{HI}$

   (b) $\xrightarrow{Cl,\ CCl_4,\ Dark}$

   (e) $\xrightarrow{conc.\ H_2SO_4}$

   (c) $\xrightarrow{Br_2}$

9
Alkenes

9.1 INTRODUCTION

Alkenes are unsaturated hydrocarbons having a double bond and are represented by the general formula C_nH_{2n}. They are also known as *olefins* as the word olefin is derived from the latin word olium (meaning oil) and ficare (meaning to make), since the lower members form oily products on treatment with chlorine or bromine.

Alkenes are found in petroleum as well as in plant products. Ethene is used for artificial ripening of fruits.

9.2 TYPES OF ALKENES

Alkenes having one double bond are known as monoenes and are represented by the general formula C_nH_{2n}.

Alkenes containing two double bonds are called *diolefins* or *dienes* and have the general formula C_nH_{2n-2}. These are isomeric with alkynes. Similarly alkenes having three or four double bonds are known as *trienes or tetraenes*. The term *Polyene* is used for hydrocarbons containing more than four double bonds.

The dienes are of three types depending on the relative positions of the two double bonds in the molecule. These are isolated or nonconjugated, conjugated and cumulated dienes. In conjugated dienes the two double bonds are separated by a single bond. However, in isolated dienes, the two double bonds are separated by two or more single bonds. On the other hand, in cumulated dienes, the two double bonds are around the same carbon atom. Such a system is known but uncommon.

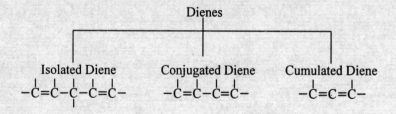

9.3 NOMENCLATURE

The *common system* of naming alkenes is through the corresponding alkane by changing the ending 'ane' of the latter by 'ylene'. The first member, C_2H_4, is therefore named as ethylene and the second member, C_3H_6, propylene. For n-butylenes greek letters (α or β) are used to distinguish between isomers having double bond at the first or the second carbon atom.

According to the *IUPAC system,* alkenes are named by adding suffix 'ene' to the root word given for a particular carbon chain. Thus, the first member C_2H_4 is named as ethene and the second member C_3H_6 propene. The IUPAC names of higher alkenes are obtained by applying the following rules:

(a) The name of the alkene is based on the name of the parent alkene having the longest carbon chain of which double bond is a part.

(b) The chain is numbered in such a way so as to provide lowest number to the doubly bonded carbon.

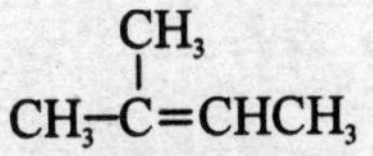

$$CH_3\text{--}\underset{\underset{CH_3}{|}}{C}\text{=}CHCH_3$$

2-Methyl-2-butene

$$CH_3\text{--}CH_2\text{--}\underset{\underset{CH_2CH_3}{||}}{C}\text{--}\underset{}{CH}\text{--}CH_3$$

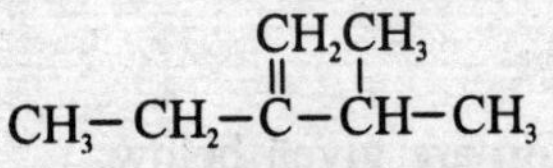

2-Ethyl-3-methyl-1-butene

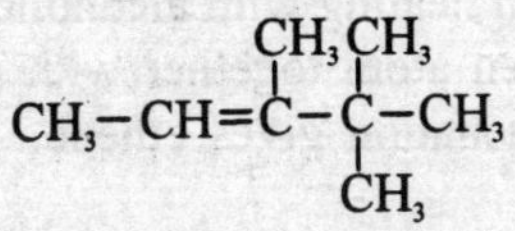

$$CH_3\text{--}CH\text{=}C\text{--}\underset{\underset{CH_3}{|}}{\overset{\overset{CH_3}{|}}{C}}\text{--}CH_3$$

3,4,4-Trimethyl-2-pentene

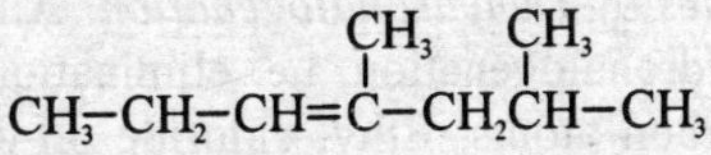

$$CH_3\text{--}CH_2\text{--}CH\text{=}\underset{\underset{CH_3}{|}}{C}\text{--}CH_2\underset{\underset{CH_3}{|}}{CH}\text{--}CH_3$$

4,6-Dimethyl-3-heptene

If one hydrogen atom from an alkene is removed, the monovalent group thus obtained is called the *alkenyl group.* The name of these groups are derived by replacing the 'e' of the parent alkene by 'yl'.

$$CH_2\text{=}CH\text{--}$$
Ethenyl
(Vinyl)

$$CH_2\text{=}CHCH_2\text{--}$$
2-Propenyl
(Allyl)

9.4 ORBITAL STRUCTURE

The orbital structure of alkenes with special reference to ethene has been discussed in Section 2.3 of this book.

9.5 ISOMERISM

Alkenes normally show following types of isomerism:

(i) *Position Isomerism:* The first two members of the alkene series do not show isomerism. However, butene exhibits position isomerism as 1-butene and 2-butene differ in the position of the double bond.

$$CH_3CH_2CH\text{=}CH_2$$
1-Butane

$$CH_3CH\text{=}CHCH_3$$
2-Butane

(ii) *Chain Isomerism:* Butene also shows chain isomerism as isobutene is a branched chain structure whereas n-butene is a straight chain structure.

$$CH_3CH_2CH\text{=}CH_2$$
n-Butene

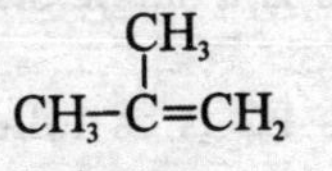

$$CH_3\text{--}\underset{\underset{CH_3}{|}}{C}\text{=}CH_2$$
Isobutene

(iii) *Geometrical Isomerism:* Alkenes exhibit geometrical or cis-trans isomerism depending upon substitution around the double bond.When the groups of higher priority are on one side of the double bond, the isomer is designated as *cis or 'Z'* and when they are on the opposite sides, as *trans or 'E'.* 2-Butene exists as cis and trans isomers as shown below:

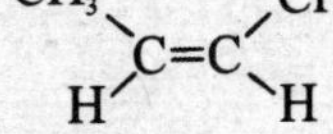

$$\underset{H}{\overset{CH_3}{\diagdown}}C\text{=}C\underset{H}{\overset{CH_3}{\diagup}}$$

cis-2-Butene

$$\underset{CH_3}{\overset{H}{\diagdown}}C\text{=}C\underset{H}{\overset{CH_3}{\diagup}}$$

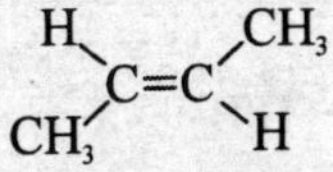

trans-2-Butene

Geometrical isomerism has been discussed in detail in Section 5.5 of this book.

9.6   GENERAL METHODS OF PREPARATION

Alkenes are generally obtained by introducing a double bond in a saturated system. The methods involve elimination of atoms or groups from two adjacent carbon atoms.

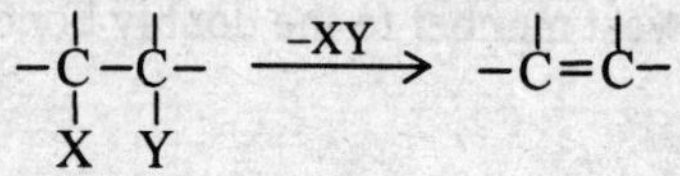

$$-\overset{|}{\underset{X}{C}}-\overset{|}{\underset{Y}{C}}- \xrightarrow{\;-XY\;} -\overset{|}{C}=\overset{|}{C}-$$

Some important methods of preparation are given below:

(i)   *From Alkyl Halides by Dehydrohalogenation:* Alkyl halides on heating with alcoholic potassium hydroxide undergo dehydrohalogenation, i.e. elimination of a halogen atom together with a hydrogen atom from   adjacent carbon atoms. Ethyl chloride on dehydrohalogenation gives ethene.

$$CH_3CH_2Br \xrightarrow{KOH,\ C_2H_5OH} CH_2{=}CH_2 + HBr + H_2O$$

Ethyl bromide   **Ethene**

The reaction forms only one product in some cases whereas more than one in others depending upon the structure of alkyl halide.

$$CH_3CH_2CH_2CH_2Cl \xrightarrow{alc.\ KOH} CH_3CH_2CH{=}CH_2$$

1-Chlorobutane   **1 Butene (only Product)**

$$\underset{\textbf{2-Chlorobutane}}{CH_3CH_2\overset{\overset{\textstyle Cl}{|}}{C}HCH_3} \xrightarrow{alc.\ KOH} CH_3CH_2CH{=}CH_2 + CH_3CH{=}CHCH_3$$

                              1-Butene (20% yield) 2-Butene (80% yield)

The Russian chemist Alexander Saytzeff gave a rule for the dehydrohalogenation associated with more than one product which states that *the preffered product in dehydrohalogenation is the alkene having greater number of alkyl groups attached to doubly bonded carbons.*

Mechanism: The dehydrohalogenation may proceed via unimolecular or bimolecular elimination.

The E2 Mechanism: When a concentrated solution of strong base is used the rate of elimination depends on the concentration of both alkyl halide and base.

Rate of alkene formation = k[RX] [B]

The reaction follows second order kinetics and is a single step reaction as proposed by Hughes and Ingold.

The reactivity order for alkyl halides towards E2 is as follows.

RF < RCl < RBr < RI
tert RX > sec RX > pri RX

The E2 mechanism proposed for the elimination reaction involves the abstraction of proton from one carbon by base and removal of halide ion from the adjacent carbon resulting in the formation of a double bond.

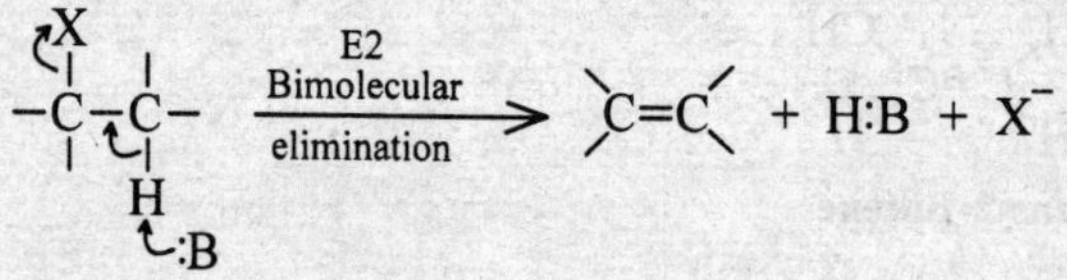

$$-\overset{\curvearrowleft X}{\underset{\underset{\underset{\curvearrowleft :B}{H}}{|}}{C}}\!-\!\overset{|}{C}- \xrightarrow[\text{elimination}]{\overset{E2}{\text{Bimolecular}}} \;\; {>}C{=}C{<} + H{:}B + X^-$$

The transition state involved in E2, as shown below has alkene character, which stabilises it and lowers the energy of activation. The net result is a more substituted alkene.

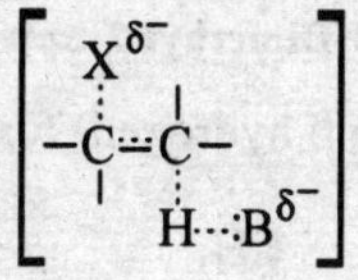

$$
\left[
\begin{array}{c}
X^{\delta-} \\
\vdots \\
-C \cdots C- \\
 H \cdots :B^{\delta-}
\end{array}
\right]
$$

The E1 Mechanism: The rate of formation of alkene depends only on the concentration of alkyl halide when it is treated with a weak base. The reaction is first order and is known as *unimolecular elimination,* E1.

Rate = k[RX]

The reactivity order of alkyl halides towards E1 is as follows.

tertiary RX > secondary RX > primary RX

The mechanism involves two steps. The first step is the heterolytic cleavage of C–X bond to form a carbocation and the second step involves the removal of proton.

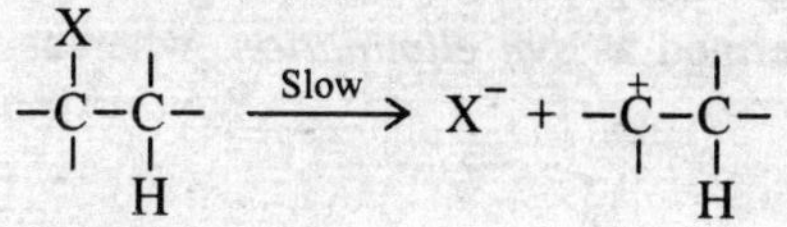

$$
-\overset{X}{\underset{H}{\underset{|}{C}}}-\overset{|}{\underset{|}{C}}- \xrightarrow{\text{Slow}} X^- + -\overset{+}{\underset{H}{\underset{|}{C}}}-\overset{|}{\underset{|}{C}}- \qquad \textbf{(Step I)}
$$

$$
-\overset{+}{\underset{\underset{\underset{:B}{\large\curvearrowright}}{H}}{C}}\!-\!\overset{|}{\underset{|}{C}}- \xrightarrow{\text{Fast}} \diagup C=C\diagdown + H:B \qquad \textbf{(Step II)}
$$

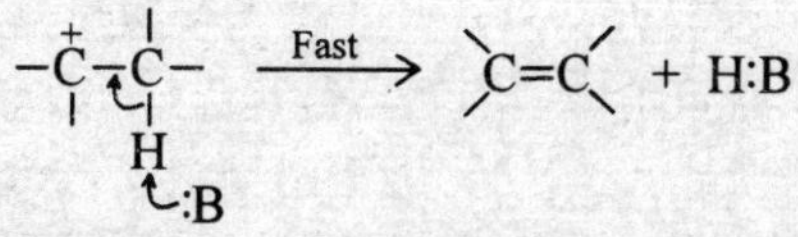

The first step (involving the substrate only), being the slow step, determines the rate of reaction. The orientation of alkene formation is governed by the second step in which transition state develops a double bond character as shown below:

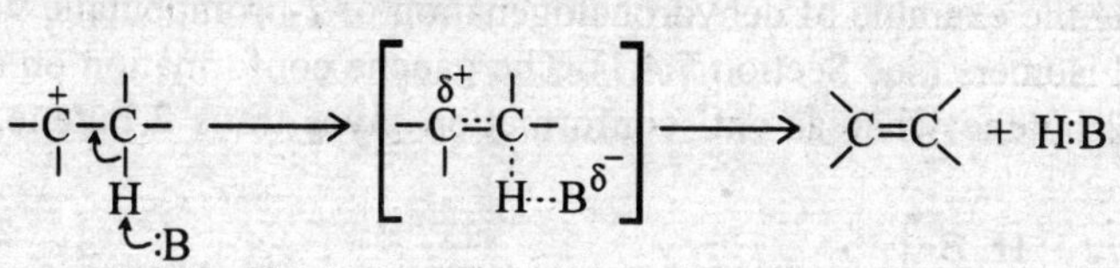

$$
-\overset{+}{\underset{\underset{:B}{H}}{C}}\!-\!\overset{|}{\underset{|}{C}}- \longrightarrow \left[-\overset{\delta+}{\underset{H\cdots B^{\delta}}{C}}\cdots\overset{|}{\underset{|}{C}}- \right] \longrightarrow \diagup C=C\diagdown + H:B
$$

Double bond character in transition state

The factors which stabilise the alkene also stabilise the transition state forming an alkene. The energy of activation is, thus, lowered and alkene is formed faster.

The E1 reaction involves carbocation, thus associated with rearrangements. 3,3-Dimethyl-2-bromobutane forms 2,3-dimethyl-2-butene and 2,3-dimethyl-1-butene which is possible only via rearrangement of carbocation as shown below:

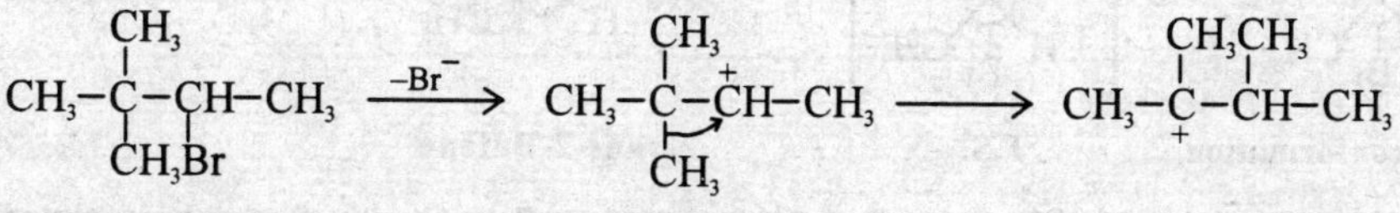

$$
\underset{\text{CH}_3\text{Br}}{\overset{\text{CH}_3}{\text{CH}_3-\text{C}-\text{CH}-\text{CH}_3}} \xrightarrow{-\text{Br}^-} \underset{\text{CH}_3}{\overset{\text{CH}_3}{\text{CH}_3-\text{C}-\overset{+}{\text{C}}\text{H}-\text{CH}_3}} \longrightarrow \underset{+}{\overset{\text{CH}_3\text{CH}_3}{\text{CH}_3-\text{C}-\text{CH}-\text{CH}_3}}
$$

3,3-Dimethyl-2-bromobutane   **a 2° cation**   **a 3° cation**

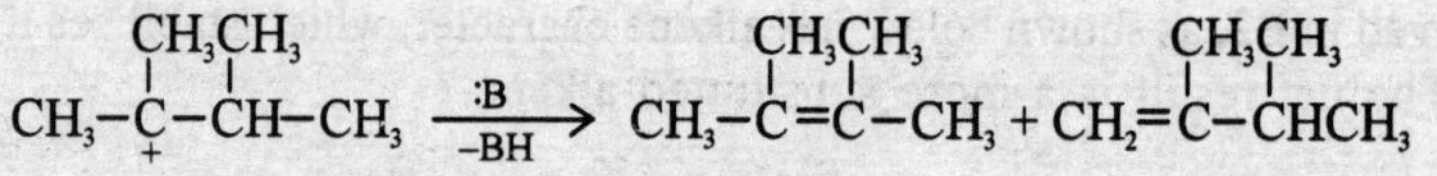

$$CH_3-\overset{\overset{\displaystyle CH_2CH_3}{|}}{\underset{+}{C}}-\overset{\displaystyle CH-CH_3}{|} \quad \xrightarrow[-BH]{:B} \quad CH_3-\overset{\overset{\displaystyle CH_2CH_3}{|}}{C}=C-CH_3 + CH_2=\overset{\overset{\displaystyle CH_2CH_3}{|}}{C}-CHCH_3$$

2,3-Dimethyl-2-butene   2,3-Dimethyl-1-butene

Similarly neopentyl bromide (having no β hydrogen) forms 2-methyl-2-butene and 2-methyl-1-butene.

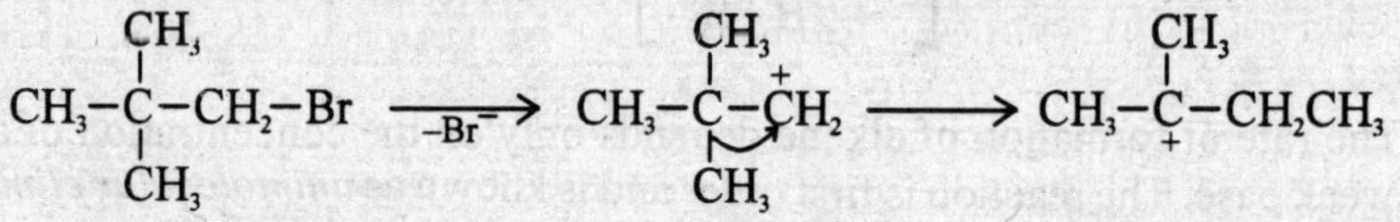

$$CH_3-\overset{\overset{\displaystyle CH_3}{|}}{\underset{\underset{\displaystyle CH_3}{|}}{C}}-CH_2-Br \xrightarrow{-Br^-} CH_3-\overset{\overset{\displaystyle CH_3}{|}}{\underset{\underset{\displaystyle CH_3}{|}}{C}}-\overset{+}{C}H_2 \longrightarrow CH_3-\overset{\overset{\displaystyle CH_3}{|}}{\underset{+}{C}}-CH_2CH_3$$

Neopentylbromide   a 1° cation   a 3° cation

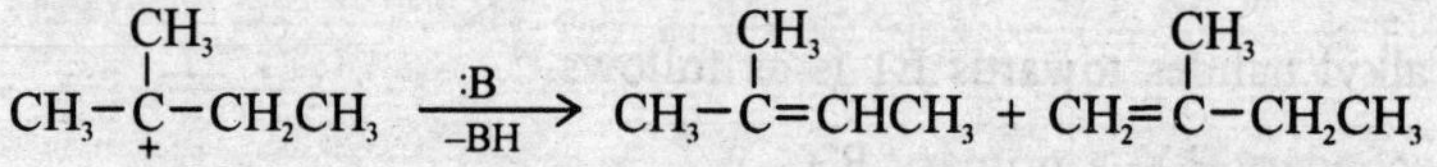

$$CH_3-\overset{\overset{\displaystyle CH_3}{|}}{\underset{+}{C}}-CH_2CH_3 \xrightarrow[-BH]{:B} CH_3-\overset{\overset{\displaystyle CH_3}{|}}{C}=CHCH_3 + CH_2=\overset{\overset{\displaystyle CH_3}{|}}{C}-CH_2CH_3$$

2-Methyl-2-butene   2-Methyl-1-butene

Streochemistry of Elimination: The alkene molecule has a planer geometry along the doubly bonded carbons due to sp^2 hybridisation of carbons. If in dehydrohalogenation, the two groups are removed from the same side of the molecule the elimination is termed as *syn elimination*, whereas when they are removed from opposite sides it is *anti elimination*.

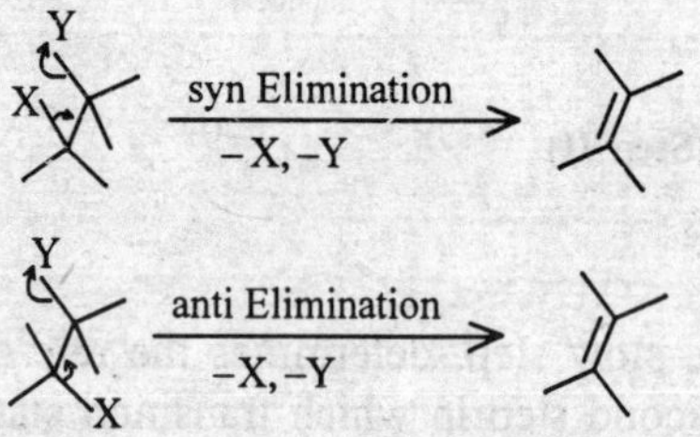

This may be explained by taking the example of dehydrohalogenation of 2-bromobutane which exists as anti and gauche conformational isomers (see Section 5.4.1). The gauche conformation on elimination of hydrogen bromide forms cis 2-butene whereas anti conformation gives trans 2-butene.

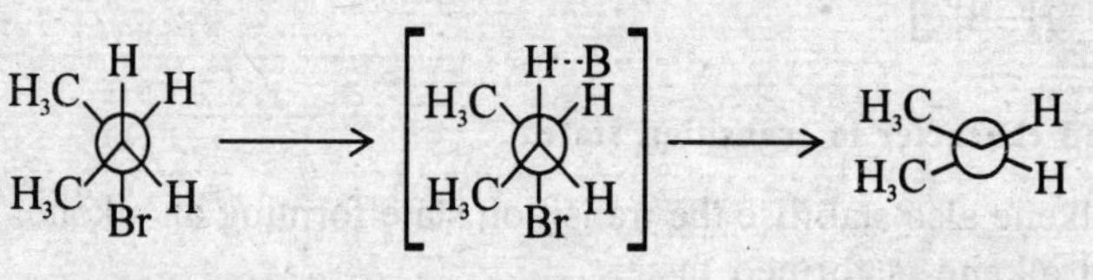

Gauche conformation   T.S.   cis-2-Butene

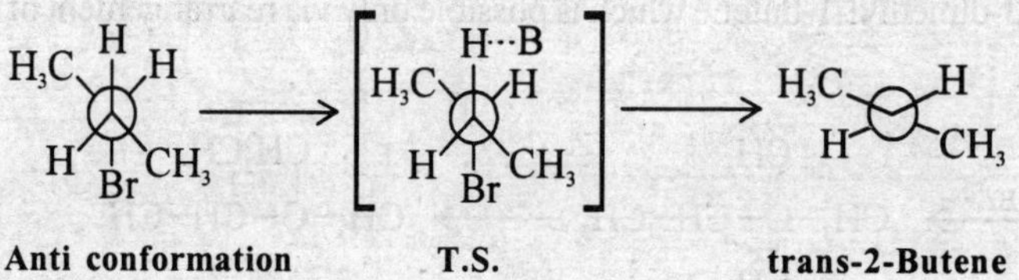

Anti conformation   T.S.   trans-2-Butene

Since the anti form is more stable than the syn form, trans-2-butene is formed six times more than the cis 2-butene. It is important to note that E2 reaction is always an anti elimination i.e. in transition state the hydrogen and the leaving group are located in anti relationship.

We shall now consider the dehydrohalogenation of 1-bromo-1,2-diphenylpropane having two chiral centres under E2 conditions.

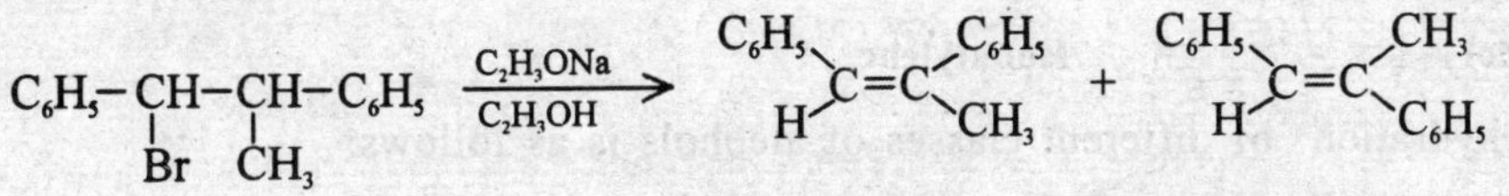

$$C_6H_5-\underset{\underset{Br}{|}}{CH}-\underset{\underset{CH_3}{|}}{CH}-C_6H_5 \xrightarrow[C_2H_5OH]{C_2H_5ONa} \underset{H}{\overset{C_6H_5}{>}}C=C\underset{CH_3}{\overset{C_6H_5}{<}} + \underset{H}{\overset{C_6H_5}{>}}C=C\underset{C_6H_5}{\overset{CH_3}{<}}$$

Z-isomer   E-isomer

1,2-Diphenyl-1-propene

The 1-bromo-1,2-diphenylpropane exists as erythro(I) and threo(II) form. Erythro form on elimination forms 'Z' alkene whereas threo form gives 'E' alkene exclusively. The reaction, therefore, is termed as stereoselective and stereospecific. *Stereoselective reactions* are the reactions that yield only one streoisomer or a pair of enantiomer of the several possible diastereomers. *Streospecific reactions* on the other hand are the reactions in which stereochemically different molecules react differently. Below are given the E2 reactions with erythro and threo isomers.

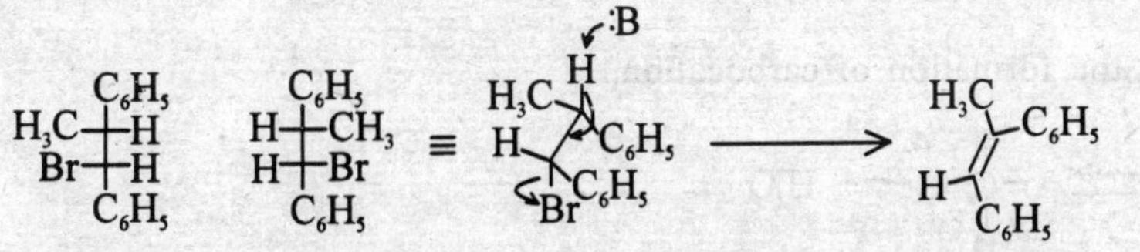

Erythro, I
1-Bromo-1,2-diphenylpropane   **(Z)-1,2-Diphenyl-1-propene**

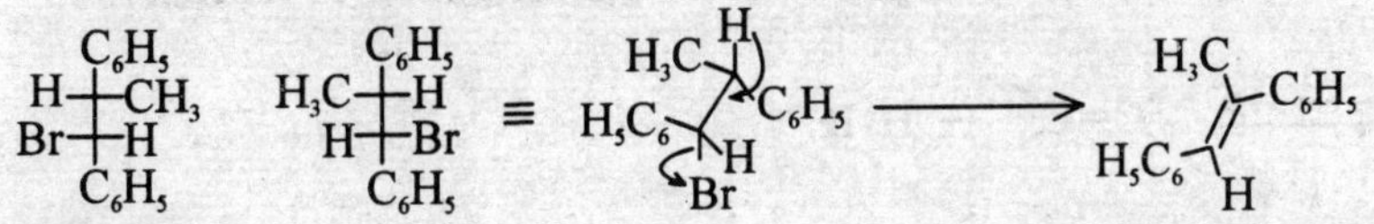

Threo, II
1-Bromo-1,2-diphenylpropane   **(E)-1,2-Diphenyl-1-propene**

(ii) *From Alcohols by Dehydration:* The alcohols on heating with conc. sulphuric acid or phosphoric acid result in elimination of a water molecule with the formation of an alkene.

$$CH_3CH_2OH \xrightarrow[170°-180°]{95\% \ H_2SO_4} CH_2=CH_2 + H_2O$$

Ethyl alcohol   **Ethene**

$$CH_3CH_2CH_2OH \xrightarrow[170°-180°]{95\% \ H_2SO_4} CH_3CH=CH_2 + H_2O$$

1-Propanol   **Propene**

With secondary or tertiary alcohols, where β hydrogen is available, a mixture of two alkenes is obtained out of them one is the major product.

$$CH_3CH_2CH_2CH_2OH \xrightarrow[140°]{75\% \ H_2SO_4} CH_3CH_2CH=CH_2 + CH_3CH=CHCH_3$$

n-Butyl alcohol   **1-Butene (minor)**   **2-Butene (major)**

$$CH_3CH_2\underset{\underset{OH}{|}}{CH}CH_3 \xrightarrow[100°]{60\% \ H_2SO_4} CH_3CH_2CH=CH_2 + CH_3CH=CHCH_3$$

see-Butyl alcohol   **1-Butene (minor)**   **2-Butene (major)**

$$CH_3-\underset{\underset{OH}{|}}{\overset{\overset{CH_3}{|}}{C}}-CH_3 \xrightarrow[80-90°]{20\% \ H_2SO_4} CH_3-\overset{\overset{CH_3}{|}}{C}=CH_2$$

t-Butanol   Isobutylene

The rate of dehydration of different classes of alcohols is as follows:

tertiary > secondary > primary

The 1,2-elimination in case of alkyl halides is base promoted as discussed earlier whereas dehydration of alcohols is an acid catalysed reaction. The mechanism for secondary and tertiary alcohols may be summarised as follows.

Step-1 involves protonation and is fast.

$$-\underset{\underset{H}{|}}{C}-\underset{\underset{OH}{|}}{C}- + H:B \rightleftharpoons -\underset{\underset{H}{|}}{C}-\underset{\underset{+OH_2}{|}}{C}- + B^-$$

Step-2 involves the formation of carbocation.

$$-\underset{\underset{H \ +OH_2}{|}}{C}-\underset{|}{C}- \rightleftharpoons -\underset{\underset{H}{|}}{C}-\underset{\underset{+}{|}}{C}- + H_2O$$

Step-3 involves the loss of β hydrogen to base and formation of alkene.

$$-\underset{\underset{B:}{\overset{H}{\nearrow}}}{C}-\underset{\underset{+}{|}}{C}- \rightleftharpoons \underset{}{C}=\underset{}{C} + H:B$$

The dehydration is reversible and alkene being volatile, is driven off the reaction mixture, shifting the equilibrium of Step-3 to the right.

The dehydration of alcohols involve rearrangement as a carbocation is formed during the reaction. 3,3-Dimethyl-2-butanol on dehydration gives 2,3-dimethyl-2-butene and 2,3-dimethyl-1-butene.

$$CH_3-\underset{\underset{CH_3}{|}}{\overset{\overset{CH_3}{|}}{C}}-\underset{\underset{OH}{|}}{C}H-CH_3 \longrightarrow CH_3-\underset{\underset{CH_3}{|}}{C}=\overset{\overset{CH_3}{|}}{C}-CH_3 + CH_2=\overset{\overset{CH_3}{|}}{C}-\underset{\underset{CH_3}{|}}{C}H-CH_3$$

3-3-Dimethyl-2-butanol   2,3-Dimethyl-2-butene   2,3-Dimethyl-1-butene
                               (major)   (minor)

The dehydration always follows Saytzeff rule i.e. where more than one product can be formed, the preffered product is a more substituted alkene. 2-Methyl-1-butanol forms 2-methyl-2-butene as the major product.

$$CH_3CH_2\overset{\overset{CH_3}{|}}{C}HCH_2OH \longrightarrow CH_3CH=\overset{\overset{CH_3}{|}}{C}-CH_3 + CH_3CH_2\overset{\overset{CH_3}{|}}{C}=CH_2$$

2-Methyl-1-butanol   2-Methyl-2-butene   2-Methyl-1-butene
                             (major)   (minor)

The dehydration can also be brought about by passing the alcohol Vapors over heated alumina (350°-400°). Other dehydrating agents which may be used are phosphorus pentoxide (P_2O_5) and phosphoric acid (H_3PO_4).

(iii) *From Vicinal Dihalides by Dehalogenation:* Vicinal dihalides (having halogen atoms on adjacent carbon atoms) on heating with zinc dust and alcohol form alkene.

$$CH_3\text{-}CH\text{-}CH_2Br + Zn \xrightarrow[\text{alcohol}]{\Delta} CH_3\text{-}CH=CH_2 + ZnBr_2$$
$$\quad\quad |$$
$$\quad\quad Br$$

1,2-Dibromopropane   **Propene**

(iv) *From Alkynes by controlled Hydrogenation:* Alkynes on controlled hydrogenation in the presence of Lindlar's catalyst (palladium poisoned with calcium carbonate and quinoline) give alkenes.

$$CH_3\text{-}C\equiv CH + H_2 \xrightarrow[\text{Quinoline}]{\text{Pd-CaCO}_3} CH_3CH=CH_2$$

Propyne   **Propene**

(v) *From Alkanes by Cracking or Pyrolysis:* Alkanes on heating at 500°–600° or in presence of a catalyst at 225° yield alkenes.

$$CH_3CH_2CH_3 \longrightarrow CH_3CH=CH_2 + CH_2=CH_2 + CH_4 + H_2$$

Propane   **Propene**   **Ethene**

(vi) *From Carbonyl Compounds by Wittig Reaction:* Alkenes are obtained from carbonyl compounds (ketones or aldehydes) by Wittig reaction. This reaction leads to the replacement of a carbonyl group by the group =CRR′ (where R and R′ are hydrogen or alkyl group).

$$-\overset{|}{C}=O \xrightarrow[\text{reaction}]{\text{Wittig}} -\overset{|}{C}=\overset{|}{C}-$$

Carbonyl compound   Alkene

The first step in Wittig reaction is the reaction of triphenyl phosphine with primary or secondary alkyl halide to give the phosphonium salt. The phosphonium salt reacts with a strong base, which abstracts a weakly acidic α hydrogen to give alkylidenephenyl phosphorane (the phosphorus ylide) commonly known as the Wittig reagent.

$$Ph_3\text{-}P + R\text{-}\overset{\overset{R'}{|}}{C}H\text{-}X \longrightarrow Ph_3\text{-}P^+\text{-}\overset{\overset{R'}{|}}{C}H\text{-}RX^-$$

Triphenyl   Alkyl halide   Phosphonium salt
phosphine

$$Ph_3\text{-}P^+\text{-}CH\text{-}RX^- + C_6H_5Li \longrightarrow Ph_3\text{-}P=\overset{\overset{R'}{|}}{C}\text{-}R + C_6H_6 + LiX$$

Phosphonium salt   Phenyl lithium   Ylide
(Strong base)

The phosphorus ylide has a hybrid structure, and it is the negative charge on carbon that is responsible for the characteristic reaction.

$$Ph_3\text{-}P^+\text{-}\overset{|}{\underset{R'}{\bar{C}}}\text{-}R \longleftrightarrow Ph_3\text{-}P=\overset{|}{\underset{R'}{C}}\text{-}R$$

The second step is the attack of phosphorus ylide onto the carbonyl carbon to form an intermediate *betanine,* which spontaneously undergoes elimination to yield alkene (Betanine is a molecule having non-adjacent opposite charges). The mechanism of the reaction may be summerised as follows:

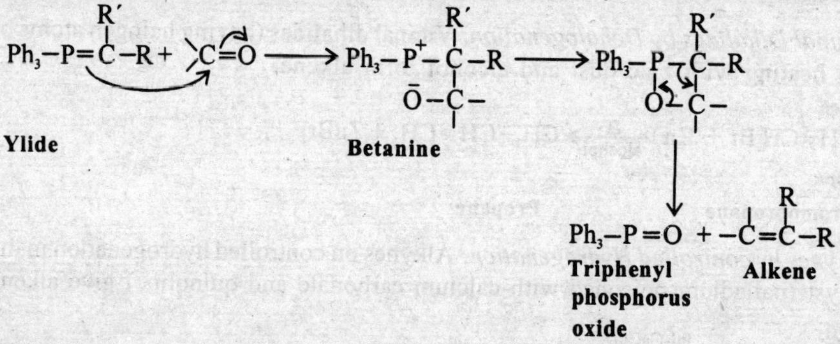

$$Ph_3-P=\overset{R'}{\underset{}{C}}-R + \overset{}{\underset{}{>}}C=O \longrightarrow Ph_3-\overset{+}{P}-\overset{R'}{\underset{}{C}}-R \longrightarrow Ph_3-P-\overset{R'}{\underset{}{C}}-R$$

| Ylide | | Betanine | |

$$\downarrow$$

$$Ph_3-P=O + -\overset{R}{\underset{}{C}}=\overset{}{\underset{}{C}}-R$$

Triphenyl   Alkene
phosphorus
oxide

The Wittig reaction was discovered by George Wittig (1965). Because of the numerous synthetic applications of this reaction he was awarded Noble Prize.

(vii) *Kolbe's Electrolytic Method:* Electrolysis of a concentrated solution of dialkali salt of dicarboxylic acid produces alkene.

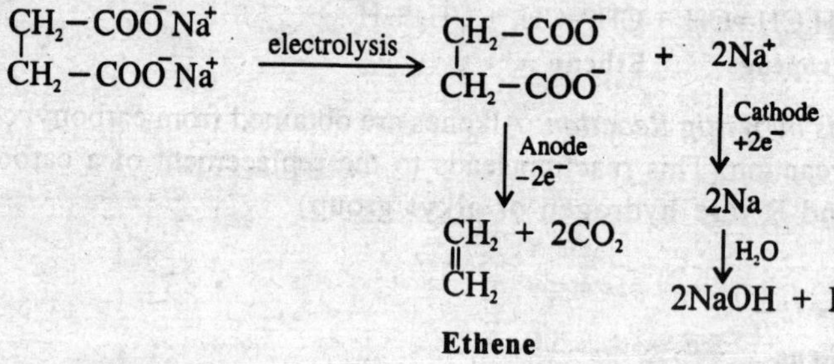

$$\begin{array}{l} CH_2-COO^-Na^+ \\ | \\ CH_2-COO^-Na^+ \end{array} \xrightarrow{\text{electrolysis}} \begin{array}{l} CH_2-COO^- \\ | \\ CH_2-COO^- \end{array} + 2Na^+$$

Anode −2e⁻

$$\begin{array}{l} CH_2 + 2CO_2 \\ || \\ CH_2 \end{array}$$

Ethene

Cathode +2e⁻

2Na

H₂O

2NaOH + H₂

The reaction is not very popular as the yields are poor.

(viii) *From Grignard Reagent:* Lower alkenyl halides when reacted with grignard reagents form higher alkenes.

$$RMgX + CH_2=CHX \longrightarrow R-CH=CH_2 + MgX_2$$

Alkenyl halide   Alkene

$$CH_3MgI + CH_2=CHCl \longrightarrow CH_3-CH=CH_2 + Mg(Cl)I$$

Vinyl chloride   Propene

(ix) *Hofmann Elimination Reaction:* When tetraalkylammonium hydroxide is heated strongly, it forms alkene along with tertiary amine. The reaction involves the elimination of β hydrogen and has been discussed in detail in Section 25.9.2.

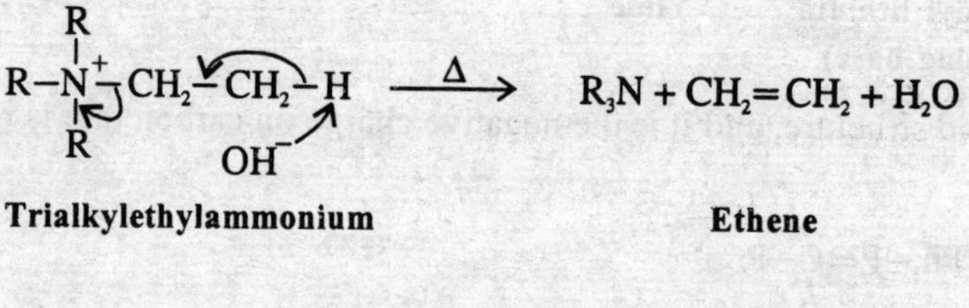

$$R-\overset{R}{\underset{R}{\overset{+}{N}}}-CH_2-CH_2-H \xrightarrow{\Delta} R_3N + CH_2=CH_2 + H_2O$$

$$OH^-$$

Trialkylethylammonium   Ethene

$$(CH_3)_3\overset{+}{N}CH_2-\underset{\underset{H}{|}}{CH}-CH_3 \; OH^- \quad \xrightarrow{\;OH^-\;} \quad (CH_3)_3N \; + \; CH_3-CH=CH_2 \; + \; H_2O$$

Propene

9.7 PHYSICAL PROPERTIES

The alkenes resemble alkanes in most of their physical properties, like the boiling points of alkenes increases with the increases in carbon atoms. As in alkanes, branching in an alkene also lowers the boiling point. Alkenes containing two to four carbon atoms are gases, those containing five to fifteen carbon atoms are liquids and higher alkenes are solids. They are insoluble in water but soluble in organic solvents. They are lighter than water.

9.8 SPECTRAL PROPERTIES

Ultravoilet spectroscopy is not of much help in the characterisation of alkenes, since they do not show any absorption band above 200 nm. Though the ethylenic chromophore shows absorption band below 200 nm, it is not of much importance, since the UV spectrum in this region is influenced by the absorptions of air and solvent molecules.

The Infrared spectrum of alkenes is quite useful in structure determination. The unsymmetrical alkenes being polar absorb in the region 1600-1700 cm^{-1}. However, the symmetrical alkenes being nonpolar do not absorb in this region. Also depending upon the substituent, one or more bands of medium intensity appear in the region 3000-3100 cm^{-1} for =C–H stretching. The cis and trans isomers of the type RCH=CHR can be ditinguished by C-H deformation frequencies. The cis isomer absorbs at 675-730 cm^{-1} and the trans isomer at 960-975 cm^{-1}.

In the NMR spectra, the chemical shifts of olefinic protons are shifted towards lower field than those of alkane protons. The exact absorption depends on the location of the double bond in the chain. Generally speaking the proton on terminal alkenyl carbon absorbs near δ 4.7 while the protons on nonterminal carbon absorb further downfield at δ 5.3. The protons α to a double bond (–CH$_2$CH=CHCH$_2$–) absorb at δ 2.06.

In conjugated dienes, the olefinic protons are more deshielded compared to the unconjugated alkenes.

9.9 CHEMICAL PROPERTIES

The chemical properties of alkenes are essentially due to the breaking of weak π bond.

(i) *Addition of Halogens:* The alkenes on treatment with halogens in inert solvent give 1,2-dihalogenated products.

$$\overset{}{\underset{}{C}}=\overset{}{\underset{}{C} } + X_2 \quad \longrightarrow \quad -\overset{|}{\underset{\underset{X}{|}}{C}}-\overset{|}{\underset{\underset{X}{|}}{C}}-$$

Alkene   Vicinal dihalide

Bromine and chlorine are effective electrophilic reagents. Fluorine is too reactive and the reaction is difficult to control. Iodine does not react with alkenes. This is the best method for the preparation of vicinal dihalides.

$$CH_3-CH=CH_2 + Br_2 \quad \xrightarrow{\;CCl_4\;} \quad CH_3-\underset{\underset{Br}{|}}{CH}-\underset{\underset{Br}{|}}{CH_2}$$

Propene   1,2-Dibromopropane

The *mechanism* of addition of halogens is *electrophilic addition* and involves two steps.

$$\text{C=C} \quad Br\!-\!Br \quad \longrightarrow \quad Br^- + \overset{+}{\underset{|}{C}}\text{Br}\overset{}{\underset{|}{C}}$$

Bromonium ion

$$\overset{+}{\underset{|}{Br}}\ -C-C- \ +\ Br^- \quad \longrightarrow \quad \underset{Br}{\underset{|}{-C}}-\underset{|}{C}-$$

Step-1 involves the polarisation of bromine by the π electron cloud of alkene molecule and formation of a cyclic bromonium ion in which electrophilic bromine is attached to both the carbons having double bond. Step-2 involves the reaction of nucleophilic bromide ion with bromonium ion to form the dibromide.

Alkenes having electron releasing groups undergo addition faster than those having electron withdrawing groups attached to it.

The evidence in favor of this mechanism may be given by the reaction of ethylene and bromine in presence of sodium chloride which gives 1,2-dibromoethane and 2-bromo-1-chloroethane. This is possible only when bromonium ion is formed and is capable of attack by the nucleophile i.e. bromide ion or chloride ion as the aqueous salt solution alone is inert towards alkene.

$$CH_2=CH_2 \xrightarrow[NaCl]{Br_2} \overset{+}{\underset{}{CH_2\!-\!CH_2}} \longrightarrow \begin{cases} \xrightarrow{Br^-} CH_2Br-CH_2Br \\ \textbf{1,2-Dibromoethane} \\ \\ \xrightarrow{Cl^-} CH_2Br-CH_2Cl \\ \textbf{2-Bromo-1-chloroethane} \end{cases}$$

Ethene

The addition of halogens to alkenes is strictly stereoselective and stereospecific. This is shown by taking the example of bromination of 2-butene which exists as cis and trans isomers. cis 2-butene on bromination gives exclusively racemic 2,3-dibromobutane whereas trans 2-butene gives meso 2,3-dibromobutane.

$$\underset{H}{\overset{CH_3}{>}}C=C\underset{H}{\overset{CH_3}{<}} \xrightarrow{\text{Bromination}} CH_3-\underset{Br}{\underset{|}{CH}}-\underset{Br}{\underset{|}{CH}}-CH_3$$

cis-2-Butene   **rac-2,3-Dibromobutane**

$$\underset{H}{\overset{CH_3}{>}}C=C\underset{CH_3}{\overset{H}{<}} \xrightarrow{\text{Bromination}} CH_3-\underset{Br}{\underset{|}{CH}}-\underset{Br}{\underset{|}{CH}}-CH_3$$

trans-2-Butene   **meso-2,3-Dibromobutane**

This is possible only when a cyclic bromonium ion is formed as intermediate. A noncyclic intermediate can form mixture of products (racemic and meso) due to rotation along the single bond as shown in cation I.

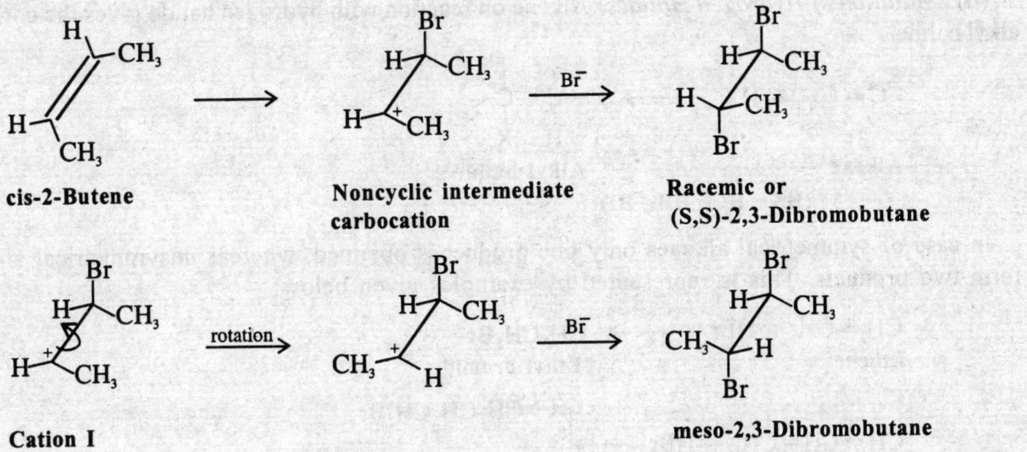

cis-2-Butene

Noncyclic intermediate carbocation

Racemic or (S,S)-2,3-Dibromobutane

Cation I

meso-2,3-Dibromobutane

The addition of halogens is an anti addition i.e. addition takes place on the opposite sides of the double bond.

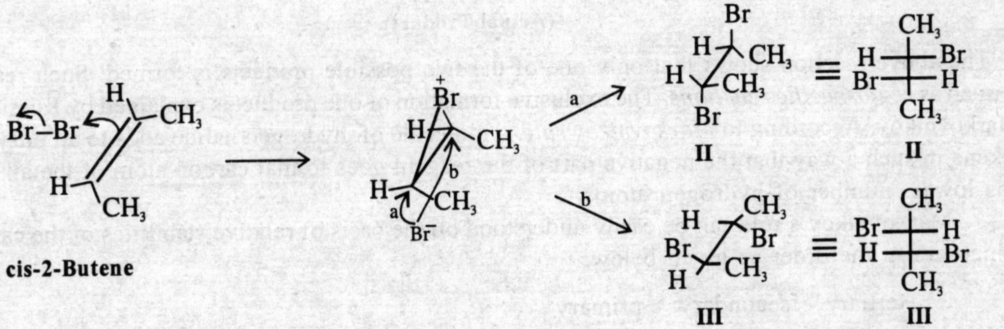

cis-2-Butene

II

II

III

III

II, III-enantiomers
rac-2,3-Dibromobutane

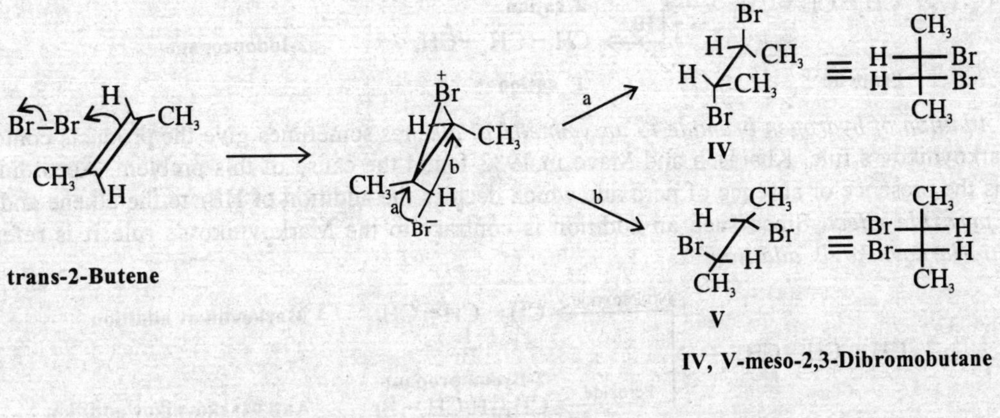

trans-2-Butene

IV

V

IV, V-meso-2,3-Dibromobutane

(ii)   *Addition of Hydrogen Halides:* Alkene on reaction with hydrogen halide gives the corresponding alkyl halide.

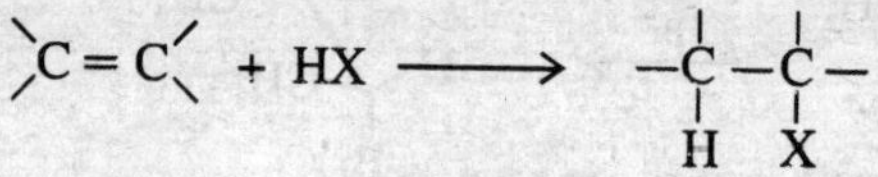

$$\underset{\text{Alkene}}{>C=C<} \;+\; HX \;\longrightarrow\; \underset{\underset{\text{Alkyl halide}}{H \quad X}}{-\overset{|}{\underset{|}{C}}-\overset{|}{\underset{|}{C}}-}$$

(HX = HCl, HBr, HI)

In case of symmetrical alkenes only one product is obtained, whereas unsymmetrical alkenes may form two products. This is represented by examples given below:

$$\underset{\text{Ethene}}{CH_2=CH_2} \;+\; HBr \;\longrightarrow\; \underset{\text{Ethyl bromide}}{CH_3CH_2Br}$$

$$\underset{\text{Propene}}{CH_3-CH=CH_2} \;+\; HBr \;\longrightarrow\;
\begin{cases}
\xcancel{\longrightarrow} CH_3CH_2CH_2Br \\[4pt]
\longrightarrow \underset{\underset{\underset{\text{(Actual Product)}}{\text{2-Bromopropane}}}{Br}}{CH_3\overset{|}{CH}CH_3}
\end{cases}$$

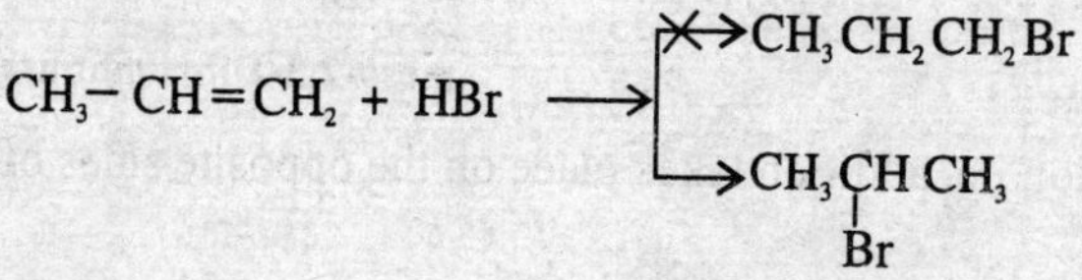

The above reaction shows that only one of the two possible products is formed. Such reactions are termed as *regiospecific reactions*. The exclusive formation of one product is explained by Russian chemist Markovnikov. According to *Markovnikov rule,* a molecule of hydrogen halide adds to an unsymmetrical alkene in such a way that the negative part of the reagent goes to that carbon atom of the alkene which has lower number of hydrogen atoms.

Markovnikov's rule can be easily understood on the basis of relative stabilities of the carbocations which are in the order as given below:

tertiary > secondary > primary

In case of propene the addition of HI gives exclusively 2-iodopropane via the more stable intermediates.

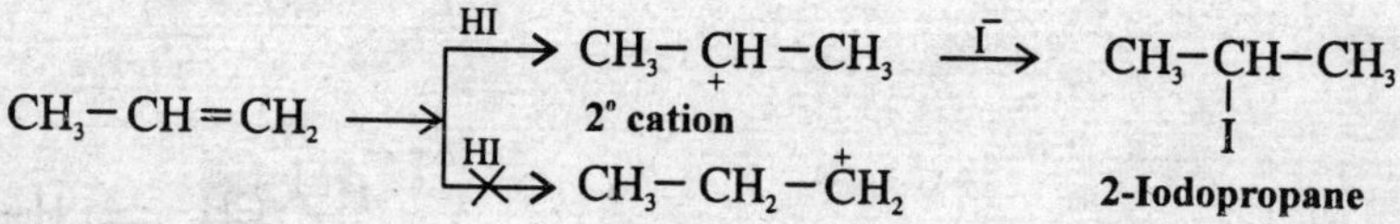

$$\underset{\text{Propene}}{CH_3-CH=CH_2} \;\longrightarrow\;
\begin{cases}
\overset{HI}{\longrightarrow} \underset{\text{2° cation}}{CH_3-\overset{+}{CH}-CH_3} \overset{\bar{I}}{\longrightarrow} \underset{\underset{\text{2-Iodopropane}}{I}}{CH_3-\overset{|}{CH}-CH_3} \\[8pt]
\overset{HI}{\xcancel{\longrightarrow}} \underset{\text{1° cation}}{CH_3-CH_2-\overset{+}{CH_2}}
\end{cases}$$

Addition of hydrogen bromide to unsymmetrical alkenes sometimes give the products contrary to the Markovnikov's rule. Kharasch and Mayo in 1933 found the cause of this problem. According to them it is the presence or absence of peroxide which decides the addition of HBr to the alkene and is known as *peroxide effect.* Since such an addition is contrary to the Markovnikov's rule it is referred to as *anti-Markovnikov's additions.*

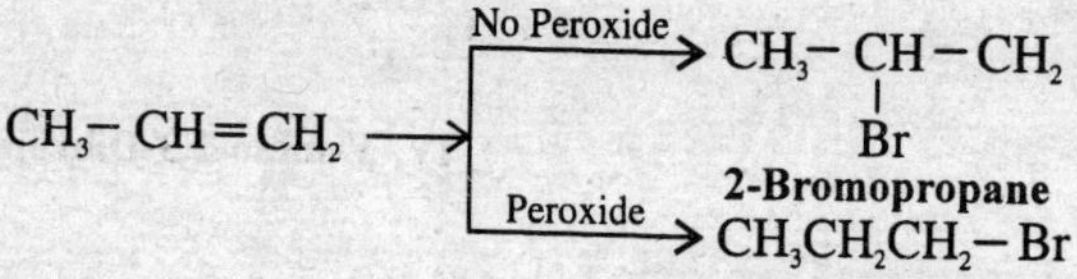

$$\underset{\text{Propene}}{CH_3-CH=CH_2} \;\longrightarrow\;
\begin{cases}
\overset{\text{No Peroxide}}{\longrightarrow} \underset{\underset{\underset{\text{2-Bromopropane}}{Br}}{}}{CH_3-\overset{|}{CH}-CH_2} \qquad \textbf{Markovnikov addition} \\[10pt]
\overset{\text{Peroxide}}{\longrightarrow} \underset{\text{n-Propyl bromide}}{CH_3CH_2CH_2-Br} \qquad \textbf{Anti Markovnikov addition}
\end{cases}$$

The formation of anti Markovnikov's addition product is explained on the basis of free radical mechanism. It is believed that the peroxide dissociates to give two alkoxy free radicals, which attack hydrogen bromide to form a bromine free radical. The bromine free radical than attacks the alkene giving two possible bromoalkyl free radicals.

$$Peroxide \longrightarrow Rad^{\cdot}$$

$$Rad^{\cdot} + H–Br \longrightarrow Rad–H + B\dot{r}$$

$$CH_3– \overset{\cdot}{C}H–CH_2Br \xrightarrow[-B\dot{r}]{HBr} CH_3– CH_2– CH_2Br$$

2° Free radical   n-Propyl bromide

$$CH_3– CH = CH_2$$

$$CH_3– \underset{Br}{\overset{|}{C}H}– \overset{\cdot}{C}H_2$$

1° Free radical

The order of stability of free radicals is 3°>2°>1°. Therefore in presence of peroxide the secondary free radical in the above step is predominently formed and reacts with HBr to form anti-Markovnikov's product.

It is found that HCl and HI give normal Markovnikov's addition products even in presence of peroxides. With HCl, the homolytic cleavage of H–Cl bond to give chloride free radical is unfavorable as H–Cl bond is stronger than H–Br bond. In case of HI, on the other hand, iodine free radical is formed easily (since the H–I bond is weak) but the iodine radicals combine together to form Iodine molecule instead of the formation of relatively unstable carbon iodine bond.

(iii) *Addition of Sulphuric Acid:* Alkenes react with cold concentrated sulphuric acid to form alkyl hydrogen sulphate. The addition is strictly Markovnikov's and the alkyl hydrogen sulphates on heating with water form alcohols.

$$CH_3CH = CH_2 \xrightarrow{80\% H_2SO_4} CH_3– \underset{OSO_3H}{\overset{|}{C}H}–CH_3 \xrightarrow[\Delta]{H_2O} CH_3– \underset{OH}{\overset{|}{C}H}–CH_3$$

Propylene   Isopropyl hydrogen sulphate   Isopropanol

$$CH_3– \overset{\overset{\displaystyle CH_3}{|}}{C}=CH_2 \xrightarrow{60\% H_2SO_4} CH_3–\underset{OSO_3H}{\overset{\overset{\displaystyle CH_3}{|}}{\underset{|}{C}}}–CH_3 \xrightarrow[\Delta]{H_2O} CH_3–\underset{OH}{\overset{\overset{\displaystyle CH_3}{|}}{\underset{|}{C}}}–CH_3$$

Isobutylene   tert-Butyl hydrogen sulphate   tert-Butyl alcohol

Since the alkenes are soluble in cold concentrated acid the method is used to separate them from compounds insoluble in acid like alkanes and alkyl halides.

(iv) *Addition of Hypohalous Acid:* Ethene forms ethylene halohydrin on reaction with halogens in presence of water.

$$CH_2 = CH_2 + Cl_2 + H_2O \longrightarrow Cl\,CH_2–CH_2OH$$

Ethylene   Ethylene chlorohydrin

With higher homologues, the addition takes place according to Markovnikov's rule.

(v) *Addition of Water (Hydration):* Water adds to alkenes in presence of acid catalyst (H_2SO_4) to form alcohol.

$$\underset{\text{Alkene}}{\overset{\diagdown}{\diagup}C = C\overset{\diagup}{\diagdown}} + H_2O \xrightarrow{\;H^+\;} \underset{\underset{H\quad OH}{|\quad|}}{-C-C-} \underset{\text{Alcohol}}{}$$

Addition of water to unsymmetrical alkenes follows Markovnikov's rule.

$$\underset{\text{Propene}}{CH_3CH = CH_2} + H_2O \xrightarrow{\;H^+\;} \underset{\text{2-Propanol}}{CH_3 - \overset{\overset{\displaystyle OH}{|}}{CH} - CH_3}$$

$$\underset{\text{1-Butene}}{CH_3CH_2CH = CH_2} + H_2O \xrightarrow{\;H^+\;} \underset{\text{2-Butanol}}{CH_3 - CH_2 - \overset{\overset{\displaystyle OH}{|}}{CH} - CH_3}$$

The Markovnikov's hydration of alkenes is also accomplished by oxymercuration demercuration reaction which involves the reaction with mercuric acetate in presence of water followed by reduction with sodium borohydride. This procedure gives better yields of alcohols than by the addition of water in presence of sulphuric acid.

$$CH_3CH_2CH = CH_2 + Hg(OCOCH_3)_2 \xrightarrow[\text{Oxymercuration}]{\;H^+\;} \underset{\text{Hydroxymercurial compound}}{CH_3 - CH_2 - \overset{\overset{\displaystyle OH}{|}}{CH} - CH_2 - HgO.COCH_3}$$

$$\Big\downarrow NaBH_4 \text{(Demercuration)}$$

$$\underset{\text{2-Butanol}}{CH_3 - CH_2 - \overset{\overset{\displaystyle OH}{|}}{CH} - CH_3}$$

The mechanism involved is *electrophilic addition* like the addition of halogens (discussed earlier) and may be summarised as follows:

$$\overset{\diagdown}{\diagup}C = C\overset{\diagup}{\diagdown} + HZ \xrightarrow{\;\text{Slow}\;} \underset{\underset{H}{|}}{-C} - \overset{+}{C} - \qquad \textbf{Step 1}$$

$$\underset{\underset{H}{|}}{-C} - \overset{+}{C} - \; + :Z^- \xrightarrow{\;\text{Fast}\;} \underset{\underset{H\quad Z}{|\quad|}}{-C - C -} \qquad \textbf{Step 2}$$

$$HZ = HCl,\ HBr,\ HI,\ H_2SO_4,\ H_2O$$

Step-1 involves the transfer of hydrogen from HZ to form a cation and Step-2 is the combination of cation and the base. Following is given the addition of hydrogen chloride to propene.

$$CH_3 - CH = CH_2 + HCl \longrightarrow CH_3 - \overset{+}{CH} - CH_3 + Cl^-$$

$$CH_3 - \overset{+}{CH} - CH_3 + Cl^- \longrightarrow \underset{\text{2-Chloropropane}}{CH_3 - \underset{\underset{Cl}{|}}{CH} - CH_3}$$

When an unsymmetrical alkene is protonated it forms two carbocations out of them the more stable one is the preffered intermediate as the order of formation and stability of cabocation is as follows.

$$3^\circ > 2^\circ > 1^\circ > CH_3^+$$

Also in the transition state the positive charge developed is stabilised by electron releasing groups. Stabilisation of transition state brings down the E_{act} permitting a faster reaction.

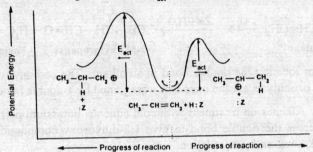

Fig. 9.1:   Stability of Transition State and Carbocation

The electrophilic addition reactions are associated with rearrangements as they involve carbocation intermediate. 3,3-Dimethyl-1-butene on addition to hydrogen iodide forms 2-iodo-3,3-dimethylbutane and 2-iodo-2,3-dimethylbutane.

$$CH_3-\underset{\underset{CH_3}{|}}{\overset{\overset{CH_3}{|}}{C}}-CH=CH_2 \xrightarrow{HI} CH_3-\underset{\underset{CH_3}{|}}{\overset{\overset{CH_3}{|}}{C}}-\overset{+}{CH}-CH_3 \xrightarrow{-CH_3Shift} CH_3-\overset{+}{\underset{\underset{CH_3}{|}}{C}}-CH-CH_3$$

3,3-Dimethyl-1-butene

$$\downarrow I^- \qquad\qquad\qquad\qquad \downarrow I^-$$

$$CH_3-\underset{\underset{CH_3\,I}{|\;\;|}}{\overset{\overset{CH_3}{|}}{C}}-CH-CH_3 \qquad\qquad CH_3-\underset{\underset{I\;\;CH_3}{|\;\;|}}{\overset{\overset{CH_3}{|}}{C}}-CH-CH_3$$

2-Iodo-3,3-dimethylbutane   **2-Iodo-2,3-dimethylbutane**

(vi) *Ozonolysis:* When ozone is passed into a solution of alkene in an inert solvent like carbon tetrachloride or chloroform at low temperature a cyclic product called ozonide is formed which on treatment with a reducing agent like zinc and water or hydrogen in presence of palladium gives aldehydes and/or ketones depending upon the structure of alkene.

$$\underset{/}{\overset{\backslash}{C}}=\underset{\backslash}{\overset{/}{C}} \xrightarrow{O_3} \quad \text{Ozonide} \xrightarrow{Zn + H_2O} \quad \overset{O}{\overset{||}{C}} + \overset{O}{\overset{||}{C}} + H_2O$$

Ozonide   **Carbonyl compounds**

The process of preparing an ozonide and then decomposing it to get carbonyl compounds is called ozonolysis.

$$CH_2=CH_2 \xrightarrow{O_3} \xrightarrow{Zn + H_2O} CH_2O + CH_2O$$

Ethene   Aldehyde

$$\underset{\text{Isobutene}}{CH_3-\overset{\displaystyle CH_3}{\underset{|}{C}}=CH_2} \xrightarrow{O_3} \xrightarrow{Zn+H_2O} \underset{\text{Acetone}}{CH_3-\overset{\displaystyle CH_3}{\underset{|}{C}}=O} + \underset{\text{Formaldehyde}}{H-\overset{\displaystyle H}{\underset{|}{C}}=O}$$

$$\underset{\text{3-Methyl-1-butene}}{CH_3-\overset{\displaystyle CH_3}{\underset{|}{CH}}-CH=CH_2} \xrightarrow{O_3} \xrightarrow{Zn+H_2O} \underset{\text{2-Methylpropanal}}{CH_3-\overset{\displaystyle CH_3}{\underset{|}{CH}}-CH=O} + \underset{\text{Formaldehyde}}{H_2C=O}$$

Ozonolysis is used for locating the double bond in an alkene. The carbonyl carbons of the carbonyl compounds obtained by ozonolysis are the ones that were joined by a double bond in the original alkene.

(vii) *Hydroxylation:* Alkenes on treatment with cold aqueous potassium permanganate solution add two hydroxy groups across the double bond to give 1,2-dihydroxy compounds known as glycols or diols. This reaction is known as hydroxylation.

$$\underset{\text{Ethene}}{CH_2=CH_2} + H_2O \xrightarrow{KMnO_4} \underset{\text{1,2-Ethanediol}}{\underset{\displaystyle \overset{|}{OH} \quad \overset{|}{OH}}{CH_2-CH_2}}$$

$$\underset{\text{Propene}}{CH_3-CH=CH_2} + H_2O \xrightarrow{KMnO_4} \underset{\text{1,2-Propanediol}}{\underset{\displaystyle \overset{|}{OH} \quad \overset{|}{OH}}{CH_3-CH-CH_2}}$$

This reaction is also the basis of an analytical test for showing the presence of unsaturation in the molecule and is known as *Baeyer's test*. A dilute solution of potassium permanganate when added to alkene decolorises immediately because of the oxidation of double bond.

Osmium tetraoxide, reacts with alkenes via cyclic osmic ester which on hydrolysis in presence of sodium sulphite yields glycol.

$$\underset{\text{Alkene}}{\overset{\displaystyle R-CH}{\underset{\displaystyle R-CH}{\|}}} + \underset{\text{Osmium tetraoxide}}{\overset{O}{\underset{O}{\overset{\diagdown}{\underset{\diagup}{Os}}}}\overset{\diagup}{\underset{\diagdown}{O}}} \longrightarrow \overset{\displaystyle R-CH-O}{\underset{\displaystyle R-CH-O}{}}\overset{O}{\underset{O}{\overset{\diagdown}{\underset{\diagup}{Os}}}}\overset{\diagup}{\underset{\diagdown}{O}}$$

$$\downarrow \text{Hydrolysis}$$

$$\underset{\text{1,2-Diol}}{\overset{\displaystyle R-CH-OH}{\underset{\displaystyle R-CH-OH}{|}}}$$

(viii) *Epoxidation:* Alkenes on treatment with peracids, like perbenzoic acid (C_6H_5COOOH), mono perphthalic acid ($HOOCC_6H_4COOOH$) and p-nitroperbenzoic acid ($p-NO_2-C_6H_4COOOH$) give the corresponding epoxides. This reaction of alkene giving epoxides is known as epoxidation.

$$\underset{\text{Ethene}}{CH_2=CH_2} + \underset{\text{Perbenzoic acid}}{C_6H_5-\overset{\displaystyle O}{\overset{\|}{C}}-O-OH} \longrightarrow \underset{\substack{\text{Ethylene}\\\text{oxide}}}{\overset{O}{\overset{\diagup\diagdown}{CH_2-CH_2}}} + C_6H_5COOH$$

$$CH_3-CH=CH_2 + C_6H_5-\overset{\overset{O}{\|}}{C}-O-OH \longrightarrow CH_3-\overset{O}{\overset{\diagdown}{CH}}-\overset{\diagup}{CH_2} + C_6H_5COOH$$

Propene   **Epoxypropane**
                                                                         (Methyl oxirane)

(ix) *Hydroboration-oxidation:* Alkenes react with diborane to give alkylboranes which on oxidation form alcohols.

$$(BH_3)_2 \xrightarrow{RCH=CH_2} RCH_2-CH_2-BH_2 \xrightarrow{RCH=CH_2} (RCH_2CH_2)_2BH \xrightarrow{RCH=CH_2} (RCH_2CH_2)_3B$$

Diborane   **Trialkylborane**

$$(RCH_2CH_2)_3B + 3H_2O_2 \xrightarrow{OH^-} 3\,RCH_2CH_2OH + B(OH)_3$$

                        Alcohol   **Boric acid**

Diborane is the dimer of hypothetical borane (BH_3) which reacts with alkene to form monoalkyl, dialkyl and trialkylborane. The trialkylborane is then oxidised with hydrogen peroxide in presence of hydroxide to form alcohol. The net addition in the reaction is of a water molecule but it follows anti-Markovnikov's rule. Also the addition here is syn addition as a cyclic transition state is involved in which hydrogen transfer from borane and attachment of boron takes place at same side of the molecule.

$$\overset{\diagdown}{C}=\overset{\diagup}{C} + H-B\overset{\diagup H}{\diagdown H} \longrightarrow \overset{\diagdown}{\underset{H-\overset{\delta}{\underset{|}{B}}-H}{\overset{\delta^+}{C}\cdots\overset{\diagup}{C}}} \longrightarrow$$
$$\underset{H}{}$$

$$\left[\begin{array}{c} \overset{\diagdown}{\underset{\delta^+}{C}}\overset{\overset{H}{|}}{\underset{\cdots}{\overset{\cdots}{B}}}{}^{\delta^-}\!-H \\ \underset{C}{\cdots}\overset{\diagup}{\underset{H}{}} \end{array}\right] \longrightarrow \begin{array}{c} \overset{H}{\underset{|}{}} \\ -\overset{|}{C}-B-H \\ -\overset{|}{C}-H \\ | \end{array} \xrightarrow{H_2O_2,OH^-} \begin{array}{c} -\overset{|}{C}-OH \\ -\overset{|}{C}-H \\ | \end{array}$$

Transition state   **Alcohol**

The hydroboration-oxidation reaction is not associated with rearrangements and 3,3-dimethyl-1-butene forms 3,3-dimethyl-1-butanol as the only product.

$$CH_3-\overset{\overset{CH_3}{|}}{\underset{\underset{CH_3}{|}}{C}}-CH=CH_2 \xrightarrow{(BH_3)_2} \xrightarrow{H_2O_2,OH^-} CH_3-\overset{\overset{CH_3}{|}}{\underset{\underset{CH_3}{|}}{C}}-CH_2-CH_2OH$$

3,3-Dimethyl-1-butene   **3,3-Dimethyl-1-butanol**

(x) *Alkylation:* Alkenes add on to alkanes in presence of acid catalyst (H_2SO_4, HF) to give addition products. This reaction is called alkylation of alkenes and has been used for large scale preparation of 2,2,4-trimethylpentane commonly known as isooctane, a principal ingredient of premium gasoline.

$$CH_3-\overset{\overset{CH_3}{|}}{C}=CH_2 + CH_3-\overset{\overset{CH_3}{|}}{C}HCH_3 \xrightarrow{H_2SO_4,0^\circ} CH_3-\overset{\overset{CH_3}{|}}{\underset{\underset{CH_3}{|}}{C}}-CH_2-\overset{\overset{CH_3}{|}}{\underset{\underset{H}{|}}{C}}-CH_3$$

Isobutene   **Isobutane**   **2,2,4-Trimethylbutane**

(xi) *Dimerisation:* Alkenes dimerise in presence of acid catalyst. The dimerisation of isobutylene forms 2,2,4-trimethyl-2-pentene which upon hydrogenation form isooctane.

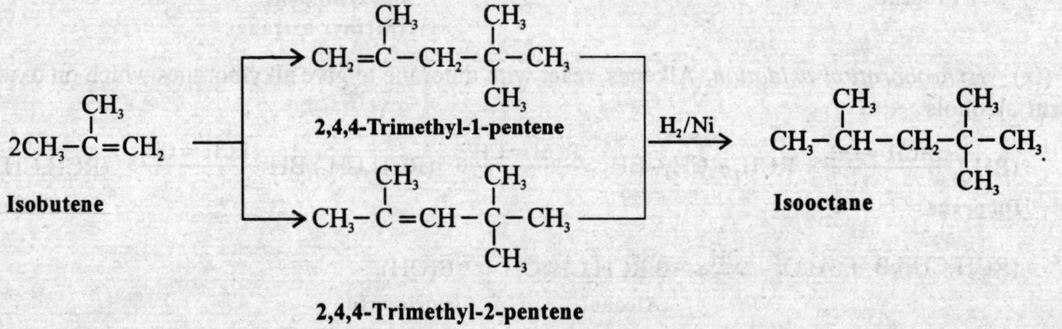

$$2CH_3-\underset{\underset{CH_3}{|}}{C}=CH_2 \quad \xrightarrow{\qquad}$$

Isobutene

$$CH_2=\underset{\underset{CH_3}{|}}{C}-CH_2-\underset{\overset{|}{CH_3}}{\underset{|}{C}}-CH_3$$

2,4,4-Trimethyl-1-pentene

$$CH_3-\underset{\underset{CH_3}{|}}{C}=CH-\underset{\overset{|}{CH_3}}{\underset{|}{C}}-CH_3$$

2,4,4-Trimethyl-2-pentene

$$\xrightarrow{H_2/Ni}$$

$$CH_3-\underset{\overset{|}{CH_3}}{\underset{|}{CH}}-CH_2-\underset{\overset{|}{CH_3}}{\underset{|}{C}}-CH_3$$

Isooctane

(xii) *Free radical Halogenation of Alkenes:* Propylene reacts with chlorine in gas phase at 500-600° to form chiefly substitution product i.e. 3-chloro-1-propene or allyl chloride. Whereas it reacts with chlorine in carbon tetrachloride at 20° to give 1,2-dichloropropane, an addition prodcut (as discussed earlier).

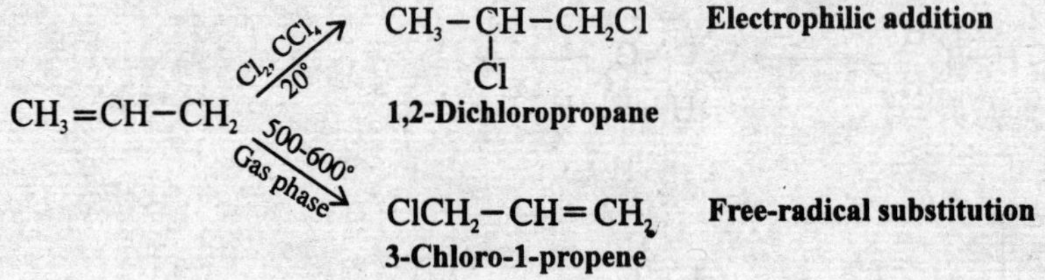

$$CH_3=CH-CH_2 \quad \underset{\underset{Gas\,phase}{500\text{-}600°}}{\overset{\overset{Cl_2, CCl_4}{20°}}{\rlap{\Big\langle}}}$$

$$CH_3-\underset{\underset{Cl}{|}}{CH}-CH_2Cl \qquad \textbf{Electrophilic addition}$$
1,2-Dichloropropane

$$ClCH_2-CH=CH_2 \qquad \textbf{Free-radical substitution}$$
3-Chloro-1-propene

The gas phase reaction follows the free radical mechanism and undergoes substitution and not addition because the allyl radical (I) is more stable than the isopropyl radical (III) formed during the reaction.

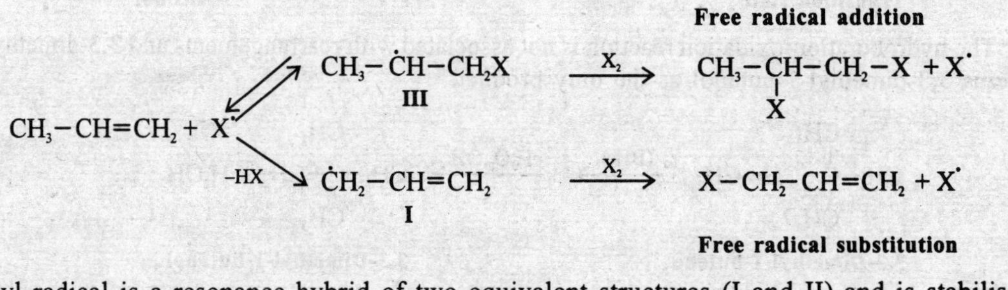

Free radical addition

$$CH_3-CH=CH_2 + X^{\cdot} \quad \nearrow \quad \underset{\textbf{III}}{CH_3-\dot{C}H-CH_2X} \quad \xrightarrow{X_2} \quad \underset{\overset{|}{X}}{CH_3-CH-CH_2-X} + X^{\cdot}$$

$$\underset{-HX}{\searrow} \quad \underset{\textbf{I}}{\dot{C}H_2-CH=CH_2} \quad \xrightarrow{X_2} \quad X-CH_2-CH=CH_2 + X^{\cdot}$$

Free radical substitution

The allyl radical is a resonance hybrid of two equivalent structures (I and II) and is stabilised due to delocalisation of electrons which are equally distributed over terminal carbons as shown in Fig. 9.2.

$$\underset{\textbf{(I)}}{CH_2=CH-\dot{C}H_2} \quad \longleftrightarrow \quad \underset{\textbf{(II)}}{\dot{C}H_2-CH=CH_2} \quad or \quad CH_2\text{--}CH\text{--}CH_2$$

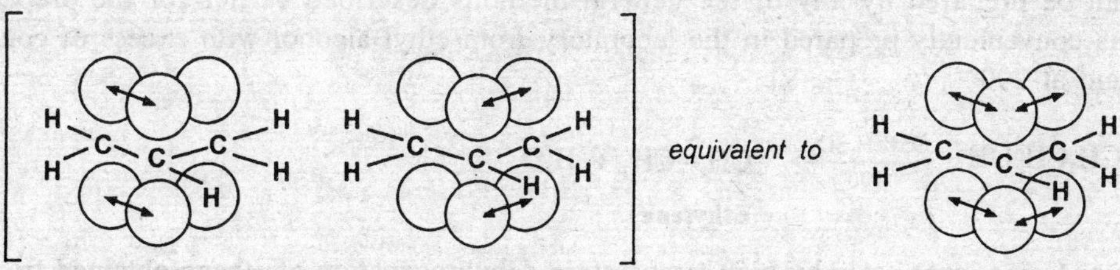

Fig. 9.2: Orbital Picture of Allyl Radical

The free radical substitution in alkenes is associated with rearrangement and is known as *allylic rearrangement*. When 1-octene reacts with N-bromosuccinimide, 3-bromo-1-octene and 1-bromo-2-octene is obtained.

$$CH_3(CH_2)_4CH_2-CH=CH_2 \xrightarrow{\text{N-Bromosuccinimide}} CH_3(CH_2)_4\underset{Br}{CH}-CH=CH_2 + CH_3(CH_2)_4CH=CH-CH_2Br$$

1-Octene 　　　　　　　　　　　　　　**3-Bromo-1-octene** 　　　　**1-Bromo-2-octene**

This is possible because of the existence of allyl radical as a resonance hybrid as shown below:

$$Br_2 + CH_3(CH_2)_4-\underset{\displaystyle \downarrow a}{CH_2}\cdots CH\cdots \underset{\displaystyle \downarrow b}{CH_2} \underset{b}{\overset{a}{\nearrow}} \begin{array}{l} CH_3(CH_2)_4\underset{Br}{CH}-CH=CH_2 \\[2mm] CH_3(CH_2)_4-CH=CH-CH_2-Br \end{array}$$

(xiii) *Polymerisation:* Alkenes polymerise in presence of catalysts and/or suitable conditions.

$$nCH_2=CH_2 \xrightarrow[\text{Pressure}]{\text{High temperature}} -(CH_2-CH_2)_n$$

Ethylene 　　　　　　　　　　　　　**Polyethylene**

$$nCH_3-CH=CH_2 \xrightarrow{\text{High temp., Pressure}} -(\underset{\displaystyle CH_3}{CH}-CH_2)_n$$

Propylene 　　　　　　　　　　　　**Polypropylene**

The process of formation of a polymer by joining together of many simple compounds is *termed polymerisation*.

9.10 INDIVIDUAL MEMBERS

9.10.1 Ethylene/Ethene [$CH_2=CH_2$]

It is the most important alkene and is highly reactive. It occurs in coal gas to the extent of 6% and is also present in natural gas. The gaseous mixture produced by cracking of high boiling fraction of petroleum contain a high percentage of ethylene.

Preparation

Ethylene can be prepared by any of the general methods described earlier for the preparation of alkenes. It is conveniently prepared in the laboratory from ethyl alcohol with excess of concentrated sulphuric acid at 200°

$$CH_3CH_2OH \xrightarrow[200°]{95\% \; H_2SO_4} CH_2=CH_2 + H_2O$$

Ethylene

It is obtained on a large scale by high temperature dehydrogenation of ethane obtained from natural gas.

$$CH_3CH_3 \xrightarrow{\Delta} CH_2=CH_2 + H_2$$

It is also obtained by the partial reduction of acetylene in presence of Lindlar catalyst.

$$CH \equiv CH + H_2 \xrightarrow{Lindiar's \; Catalyst} CH_2=CH_2$$

Ethene

Properties: Ethylene is a colourless gas with sweet odor. It is slightly soluble in water but is more soluble in organic solvents like ethanol and ether. When inhaled, it produces general anaesthesia.

It gives all the general reactions of alkenes. In addition it adds to sulphur monochloride to produce *Mustard gas.* It was used as poisonous gas in World War-I (1914).

Uses: Ethylene is used for artificial ripening of green fruits, as a general anesthetic and as a starting material for the manufacture of ethylene glycol (anti-freeze), ethanol, diethyl ether and polyethylene (used for making plastic goods).

9.10.2   Propylene/Propene [$CH_3-CH=CH_2$]

Propene can be obtained by any of the general methods of preparation of alkenes. It is also obtained by the cracking of high boiling fractions of petroleum.

Propene is a colourless gas. It undergoes all reactions described under alkenes. It is used for the commercial preparation of isopropyl alcohol, acetone and allyl chloride. It polymerises to polypropylene which is an important plastic material.

9.11   DIENES

It has already been stated (Section 9.2) that dienes are of three types, viz. conjugated, non conjugated and cumulated. The present discussion is restricted only to conjugated dienes, viz 1,3-butadiene.

9.11.1   1,3-Butadiene [$CH_2=CH-CH=CH_2$]

Heat of Hydrogenation and Stability of Alkenes and Dienes: Heats of hydrogenation provides concrete information about the stability of unsaturated compounds. Lower the heat of hydrogenation higher is the stability of alkene. For example cis 2-butene has a heat of hydrogenation of 28.6 kcal/mol and trans 2-butene has 27.6 kcal/mol showing the higher stability of trans isomer by 1 kcal/mol. On an average one double bond requires 30.0 kcal/mol of energy for hydrogenation. A lower value than this is due to stability of alkene. Table 9.1 shows the heats of hydrogenation values for various alkenes.

Table 9.1: Heats of Hydrogenation of Mono and Dienes

Monoenes/ Dienes	Heat of Hydrogenation [ΔH] kcal/mol
1-Butene	30.3
cis 2-Butene	28.6
trans 2-Butene	27.6
1-Pentene	30.1
cis 2-Pentene	28.6
trans 2-Pentene	27.6
1,4-Pentadiene	60.8
1,5-Hexadiene	60.5
1,3-Butadiene	57.0
2,3-dimethyl-1,3-butadiene	54.0

For nonconjugated dienes the calculations for heat of hydrogenation (30 × 2 = 60 kcal/mol) holds good but for conjugated dienes these values are lower than expected. For example 1,3-butadiene is expected to have a ΔH value of 60.0 kcal/mol but the actual value is only 57.0 kcal/mol. This is due to the stability of butadiene.

Resonance in Butadiene: Butadiene can be represented by two resonating structures (I) and (II).

$$-\overset{|}{\underset{|}{C}}=\overset{|}{\underset{|}{C}}-\overset{|}{\underset{|}{C}}=\overset{|}{\underset{|}{C}}- \qquad -\overset{|}{\underset{|}{C}}-\overset{|}{\underset{|}{C}}=\overset{|}{\underset{|}{C}}-\overset{|}{\underset{|}{C}}-$$

$$\text{I} \qquad\qquad\qquad \text{II}$$

Since each carbon has r electron, the overlap of r orbitals to form a conjugated system is possible. This delocalisation of π electrons provides stability to the conjugated double bond system as shown in Fig. 9.3.

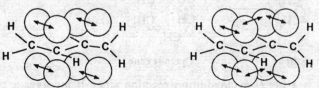

Fig. 9.3: Orbital Structure of Butadiene

Preparation

(i) *Dehydration of Butane-1,4-diol:* 1,4-Dihydroxybutane on dehydration in presence of acid form 1,3-butadiene.

$$\underset{\overset{|}{OH}}{CH_2}-CH_2-CH_2-\underset{\overset{|}{OH}}{CH_2} \xrightarrow[-2H_2O]{H_2SO_4} CH_2=CH-CH=CH_2$$

1,4-Dihydroxybutane   **1,3-Butadiene**

(ii)   *Cracking Process:* It may also be prepared from butane by cracking using Cr_2O_3 as catalyst.

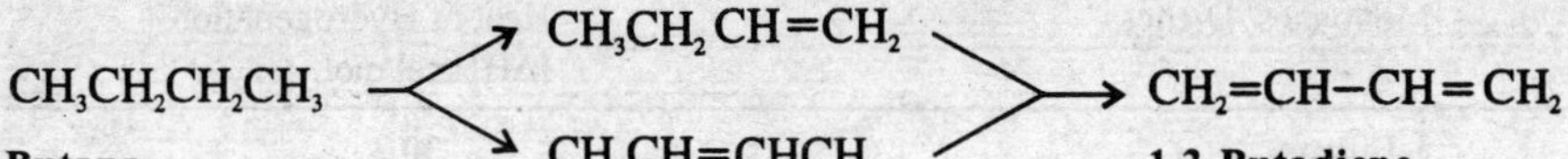

$$CH_3CH_2CH_2CH_3 \nearrow\searrow \begin{matrix} CH_3CH_2\,CH{=}CH_2 \\ CH_3CH{=}CHCH_3 \end{matrix} \searrow\nearrow CH_2{=}CH{-}CH{=}CH_2$$

Butane   1,3-Butadiene

(iii)   *Retro Diels-Alder Reaction:* 1,3-Butadiene is prepared by passing vapors of cyclohexene over heated nichrome (Ni-Cr-Fe) alloy.

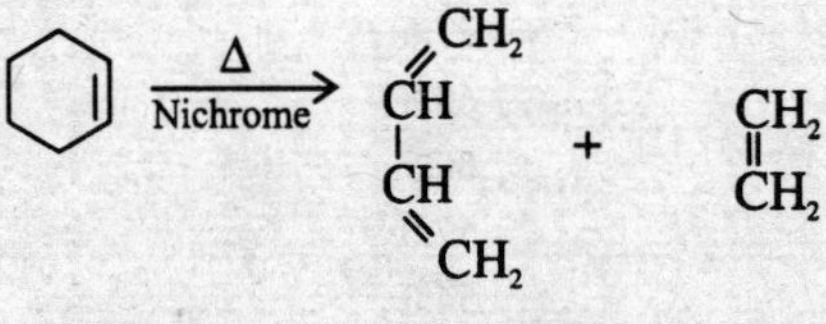

Cyclohexene   1,3-Butadiene   Ethylene

This reaction is the reverse of Diels-Alder reaction and is known as retro Diels-Alder reaction.

Properties

Butadiene is a colorless gas. It is a conjugated diene and undergoes abnormal addition reactions and readily gets polymerised. some typical reactions of conjugated dienes are given below.

(i)   *Diels-Alder Reaction:* In this reaction, a conjugated diene is treated with an unsaturated compound called the dienophile to yield a cyclic system. The reaction is known as Diel-Alder reaction after the names of the two German Chemists, Diel and Alder. Both Diel and Alder Jointly received Noble Prize for their work in 1950. This is a very useful reaction for synthesis of cyclic systems. The simplest Diels-Alder reaction is of 1,3- butadiene with ethylene to yield cyclohexene.

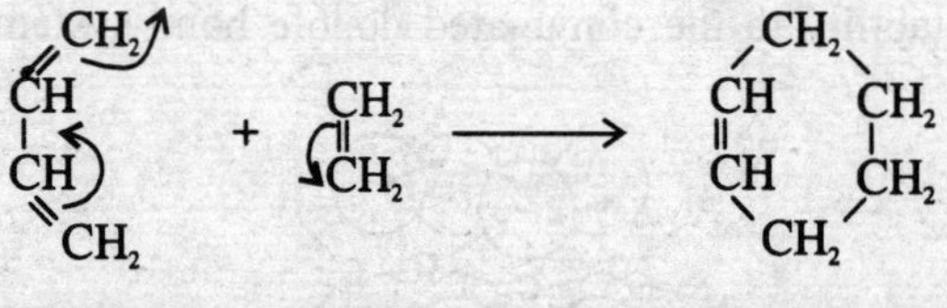

1,3-Butadiene Ethene   Cyclohexene

Diels-Alder reaction is a [4+2] cycloaddition reaction since it involves a system of **4π electrons** (1,3-butadiene) and a system of **2π electrons** (ethylene). The reactivity of the diene is enhanced by electron releasing groups, while the dienophile reactivity is enhanced by electron withdrawing groups.

The reaction is useful not only because it forms cyclic system but also because it takes place with a variety of reactants readily. 1,3-Butadiene and acrolein forms 1,2,3,6-tetrahydrobenzaldehyde.

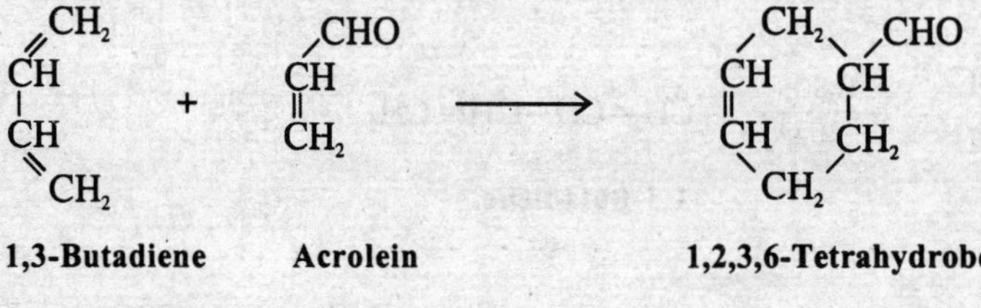

1,3-Butadiene   Acrolein   1,2,3,6-Tetrahydrobenzaldehyde

(ii)   *Addition Reaction:* Conjugated dienes undergo abnormal addition reactions. For example, 1,3-butadiene on treatment with bromine water forms two dibromo derivatives, viz. 3,4-dibromo-1-butene and 1,4-dibromo-2-butene. The 1,4-addition product is obtained in major amount. However, with excess of bromine, the 1,4 as well as the 1,2-addition products give the same 1,2,3,4-tetrabromobutane.

$$
\begin{array}{ccc}
& & \overset{\text{Br}}{\underset{}{|}}\;\;\overset{\text{Br}}{\underset{}{|}} \\
& \nearrow & \text{CH}_2-\text{CH}-\text{CH}=\text{CH}_2 \\
& & \text{3,4-Dibromo-} \\
\text{CH}_2=\text{CH}-\text{CH}=\text{CH}_2 & \longrightarrow & \text{1-butene} \\
& & \overset{\text{Br}}{\underset{}{|}}\qquad\qquad\overset{\text{Br}}{\underset{}{|}} \\
& \searrow & \text{CH}_2-\text{CH}=\text{CH}-\text{CH}_2
\end{array}
\xrightarrow{\text{Br}_2}
\begin{array}{c}
\overset{\text{Br}}{\underset{}{|}}\;\overset{\text{Br}}{\underset{}{|}}\;\overset{\text{Br}}{\underset{}{|}}\;\overset{\text{Br}}{\underset{}{|}} \\
\text{CH}_2-\text{CH}-\text{CH}-\text{CH}_2
\end{array}
$$

1,3-Butadiene   **1,4-Dibromo-2-butene**   **1,2,3,4-Tetrabromobutane**

Similar addition pattern is shown when hydrogen halides and hydrogen is used.

$$
\text{CH}_2=\text{CH}-\text{CH}=\text{CH}_2
\begin{cases}
\xrightarrow{\text{HX}} & \overset{\text{H}}{\underset{}{|}}\;\overset{\text{X}}{\underset{}{|}} & \overset{\text{H}}{\underset{}{|}}\qquad\quad\overset{\text{X}}{\underset{}{|}} \\
& \text{CH}_2-\text{CH}-\text{CH}=\text{CH}_2 + & \text{CH}_2-\text{CH}=\text{CH}-\text{CH}_2 \\
\xrightarrow{\text{H}_2} & \overset{\text{H}}{\underset{}{|}}\;\overset{\text{H}}{\underset{}{|}} & \overset{\text{H}}{\underset{}{|}}\qquad\quad\overset{\text{H}}{\underset{}{|}} \\
& \text{CH}_2-\text{CH}-\text{CH}=\text{CH}_2 + & \text{CH}_2-\text{CH}=\text{CH}-\text{CH}_2
\end{cases}
$$

1,2-addition   **1,4-addition**

It has been found that the addition reaction at low temperature (– 80°) yields a mixture containing 20% 1,4-product and 80% 1,2-product whereas, at higher temperature (40°) 80% 1,4-product and 20% 1,2-product is obtained. To account for the facts we shall look at the mechanism of addition which involves the attack of electrophile in Step-1 to form an allylic cation only which is stabilised due to delocalisation and has two resonating structures (I and II) equivalent to structure III. Structure III in the next step combines with the anion to form 1,2 and 1,4-addition product.

$$
\text{CH}_2=\text{CH}-\text{CH}=\text{CH}_2 + \text{H}^+
\begin{cases}
\longrightarrow & \text{CH}_3-\overset{+}{\text{C}}\text{H}-\text{CH}=\text{CH}_2 & \text{allyl cation} \\
\xcancel{\longrightarrow} & \text{CH}_2-\text{CH}_2-\text{CH}=\text{CH}_2 \;\; \underset{+}{} & \text{1° cation}
\end{cases}
$$

$$
\text{CH}_3-\overset{+}{\text{C}}\text{H}-\text{CH}=\text{CH}_2 \quad\longleftrightarrow\quad \text{CH}_3-\text{CH}=\text{CH}-\overset{+}{\text{C}}\text{H}_2 \equiv \text{CH}_3-\text{CH}\overset{\cdots}{=}\text{CH}\overset{\cdots}{=}\text{CH}_2
$$

$$
\quad\quad\quad \text{I} \qquad\qquad\qquad\qquad\qquad \text{II} \qquad\qquad\qquad\qquad \overset{+}{\text{III}}
$$

$$
\overset{a}{\underset{}{\swarrow}}\quad\overset{b}{\underset{}{\searrow}}
$$

$$
\text{CH}_3-\text{CH}\overset{\cdots}{=}\text{CH}\overset{\cdots}{=}\underset{+}{\text{CH}_2} \;\;\xrightarrow{\text{X}^-}
\begin{cases}
\overset{a}{\nearrow} & \overset{\text{X}}{\underset{}{|}} \\
& \text{CH}_3-\text{CH}-\text{CH}=\text{CH}_2 & \text{1,2-addition product} \\
\overset{b}{\searrow} & \text{CH}_3-\text{CH}=\text{CH}-\text{CH}_2-\text{X} & \text{1,4-addition product}
\end{cases}
$$

The proportions of the products at low (– 80°) temperatures are determined by rates of addition as the 1,2-product is formed faster (low E_{act}) than 1,4-product (high E_{act}). High temperature addition, on the other hand is determined by the equilibrium between the two isomers. The 1,4-product is more stable than the 1,2-product as shown in Fig. 9.4.

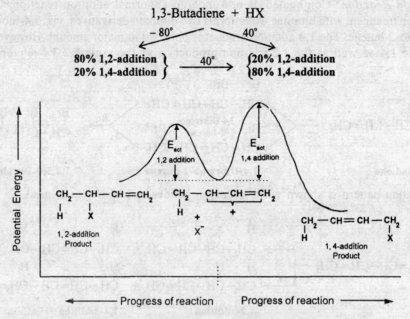

Fig. 9.4:   1,2-Addition vs 1,4-Addition in Butadiene

(iii) *Polymerisation:* 1,3-Butadiene polymerises in presence of peroxides to form a rubber like product called polybutadiene.

$$CH_2=CH-CH=CH_2 \xrightarrow{\text{Peroxide}} \text{+}CH_2-CH=CH-CH_2\text{+}_n$$

1,3-Butadiene   **Polybutadiene**

The structure of polybutadiene suggest that 1,4-addition takes place predominately.

9.13   PROBLEMS

1. Which of following is isolated, conjugated or cumulated dienes

   (a) $-\overset{|}{C}=\overset{|}{C}-\overset{|}{C}-\overset{|}{C}=\overset{|}{C}-$

   (b) $-\overset{|}{C}=\overset{|}{C}-\overset{|}{C}-\overset{|}{C}-\overset{|}{C}=\overset{|}{C}-$

   (c) $-\overset{|}{C}=\overset{|}{C}-\overset{|}{C}=\overset{|}{C}-$

   (d) $-\overset{|}{C}=C=\overset{|}{C}-$

2. Draw the orbital structure of ethylene and butadiene.

3. What happens when

   (a) $CH_3CH_2Br \xrightarrow[C_2H_5OH]{KOH}$

   (b) $CH_3CH_2CH_2Br \xrightarrow[C_2H_5OH]{KOH}$

   (c) $CH_3CH_2\underset{\underset{Cl}{|}}{C}HCH_3 \xrightarrow[\text{Alcoholic}]{KOH}$

   (d) $CH_3CH_2CH_2OH \xrightarrow[475\,K]{H_2SO_4}$

4. Discuss the ease of dehydration of primary, secondary and tertiary alcohols.

5. Discuss the ease of dehydrohalogenation of primary, secondary and tertiary alkyl halides.

6. Give the conditions necessary to effect the following conversions.

   (a) $CH_3-C\equiv C-H + H_2 \longrightarrow CH_3-CH=CH_2$

   (b) $CH_3CH_2CH_3 \longrightarrow CH_3-CH=CH_2$

(c) $CH_3-CO-CH_3 \longrightarrow CH_3-\overset{\overset{\displaystyle CH_2}{\|}}{C}-CH_3$

7. How will you distinguish between cis and trans forms by the spectroscopy.

8. Explain:
   (a) Markovnikov's rule   (b) Saytzeff rule
   (c) Wittig reaction   (d) Baeyer's test
   (e) Retro Diels-Alder reaction.

9. Explain the fact that addition of hydrogen to alkane gives trans addition product.

10. Write a note on hydroboration, ozonolysis. What are their uses?

11. Discuss Diel's Alder reaction.

12. Explain what product or products are obtained by the reaction of 1,3 -butadiene and bromine. Discuss the mechanism involved.

13. Fill in the Blanks.
   (a) Polymers contain more than .... double bonds.
   (b) Dienes are isomeric with ....
   (c) Double bonds that alternate with single bonds are ...... dienes.
   (d) In allene the central atom is..........hybridised and the terminal atoms are..........hybridised.

14. Give equations for the preparation of alkenes from the following starting material.

   (a) $CH_3CH_2\underset{\underset{\displaystyle Br}{|}}{CH}-CH_3$

   (b) $CH_3-\overset{\overset{\displaystyle CH_3}{|}}{\underset{\underset{\displaystyle CH_3OH}{|}}{C}}-CH-CH_3$

   (c) $\overset{\displaystyle CH_3}{\underset{\displaystyle CH_3}{}}\!\!>\!C=O$

15. Complete the following reactions.

   (a) $CH_3\underset{\underset{\displaystyle CH_3}{|}}{CH}CH=CH_2 + HBr \xrightarrow{\text{Peroxide}}$

   (b) $CH_3-\underset{\underset{\displaystyle CH_3}{|}}{CH}-CH=CH_2 + HCl \longrightarrow$

   (c)  ⬡ $+ Br_2 \longrightarrow$

16. Predict the products in the following reactions.

   (a) $CH_3-\underset{\underset{\displaystyle CH_3}{|}}{C}=CH_2 \xrightarrow[\text{(2) Zn-H}_2\text{O}]{\text{(1) O}_3}$

   (b) $CH_3CH_2CH=\overset{\overset{\displaystyle CH_3}{|}}{C}-CH_3 \xrightarrow[\text{(2) Zn / H}_2\text{O}]{\text{(1) O}_3}$

10
Alkynes

10.1 INTRODUCTION

Unsaturated hydrocarbons which contain a carbon-carbon triple bond are called alkynes or acetylenes. They have the general molecular formula C_nH_{2n-2} and contain two hydrogen atoms less than the corresponding alkenes. Alkynes may be terminal or internal depending upon the position of triple bond in the alkyne molecule as shown below:

$CH_3C \equiv CH$
1-Propyne
(terminal alkyne)

$CH_3CH_2C \equiv CCH_2CH_3$
3-Hexyne
(internal alkyne)

$CH_3C \equiv C-C \equiv CCH_3$
2,4-Hexadiyne
(internal alkyne)

10.2 NOMENCLATURE

In the common system of nomenclature, the first member is named as acetylene and the higher members as alkyl derivatives of acetylene.

In the *IUPAC System*, the *alkyne* is named by adding the suffix 'yne' to the root word given to a carbon chain. The carbon containing the triple bond is assigned the lowest number possible. Examples in the following table make the nomenclature easily understood.

Alkyne	Common name	IUPAC name
$H-C \equiv C-H$	Acetylene	Ethyne (acetylene)
$CH_3-C \equiv C-H$	Methylacetylene	Propyne
$CH_3-C \equiv C-CH_3$	Dimethylacetylene	2-Butyne
$CH_3CH_2-C \equiv C-H$	Ethylacetylene	1-Butyne
$CH_3-\overset{\displaystyle |}{\underset{\displaystyle CH_3}{CH}}-C \equiv CH$	Isopropylacetylene	3-Methyl-1-butyne

If one hydrogen atom of the alkyne molecule is removed, the monovalent group thus obtained is called an alkynyl group. These are named by replacing 'e' of the corresponding alkyne with 'yl'.

$HC \equiv C-$ **Ethynyl**
$H_3C-C \equiv C-$ **Propynyl**
$HC \equiv C-CH_2-$ **2-Propynyl**

10.3 ORBITAL STRUCTURE

The orbital structure of alkynes have already been discussed in Section 2.3 taking the example of acetylene. The carbon atoms having the triple bond in acetylene are sp hybridized and have a linear geometry. They are attached to each other by a σ bond and two π bonds. Fig. 10.1 shows the orbital structure of acetylene molecule.

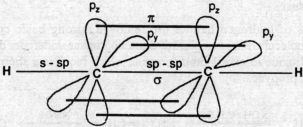

Fig. 10.1: Orbital Structure of Acetylene

Thus the triple bond of alkyne is made up of one strong σ bond and two weak π bonds. A carbon-carbon triple bond (199.0 kcal/mol) is stronger than a carbon-carbon double bond (115.8 kcal/mol) or a carbon-carbon single bond (82.0 kcal/mol). Electron diffraction, X-ray diffraction and spectroscopic analysis reveal that acetylene is a linear molecule.

10.4 ISOMERISM

Unlike alkenes, the alkynes do not exhibit cis-trans isomerism since the molecule is linear. We normally encounter three types of isomerism in alkynes.

(i) *Chain Isomerism:* The alkynes exhibit chain isomerism as illustrated by the following example.

$$CH_3CH_2CH_2C\equiv CH \qquad\qquad CH_3\overset{\overset{\displaystyle CH_3}{|}}{C}HC\equiv CH$$

1-Pentyne   **3-Methyl-1-butyne**

(ii) *Position Isomerism:* This arises due to the difference in the position of the triple bond (e.g., terminal and nonterminal alkynes).

$$CH_3CH_2C\equiv CH \qquad\qquad\qquad CH_3C\equiv CCH_3$$

1-Butyne (Terminal)   **2-Butyne (Non-terminal)**

(iii) *Functional Isomerism:* Alkynes exhibit functional isomerism with alkadienes (diolefins).

$$CH_3C\equiv CCH_3 \qquad\qquad\qquad\qquad CH_2{=}CH{-}CH{=}CH_2$$

2-Butyne   **1,3-Butadiene**

10.5 GENERAL METHODS OF PREPARATION

There are two approaches for the synthesis of alkynes. In the first approach, a triple bond is formed by elimination reactions. In the second approach alkyl groups are added to compounds containing a triple bond. Some important methods of preparation of alkynes are given below:

(i) *Dehydrohalogenation of Dihalides:* Alkyne can be prepared by elimination of two molecules of hydrogen halides from a dihalide. The dihalide may be geminal or vicinal in nature.

1,2-Dihalides (vicinal dihalides) on treatment with a base eliminates two molecules of hydrogen halide from adjacent carbon atoms to give an alkyne.

$$\underset{\substack{\text{1,2-Dihalides} \\ \text{(Vicinal dihalide)}}}{R-\overset{\overset{\displaystyle H}{|}}{\underset{\underset{\displaystyle X}{|}}{C}}-\overset{\overset{\displaystyle X}{|}}{\underset{\underset{\displaystyle H}{|}}{C}}-R} \xrightarrow[-2HX]{\text{Base}} \underset{\text{Alkyne}}{R-C\equiv C-R}$$

Since two molecules of hydrogen halides are removed, strong basic conditions are needed to remove the second HX molecule. For example, 1,2-dibromopropane undergoes dehydrohalogenation with alcoholic potassium hydroxide to give 1-bromopropene which requires a strong base like sodamide for further reaction to give propyne.

$$\underset{\substack{\text{1,2-Dibromo-} \\ \text{propane}}}{CH_3-\overset{\overset{\displaystyle Br}{|}}{CH}-\overset{\overset{\displaystyle Br}{|}}{CH_2}} \xrightarrow[-HBr]{KOH/C_2H_5OH} \underset{\text{1-Bromo-1-propene}}{CH_3CH=CHBr} \xrightarrow[-HBr]{NaNH_2} \underset{\text{Propyne}}{CH_3C\equiv CH}$$

The reaction can be carried out in one step by using sodamide. Since the starting dihalides can be prepared by addition of halogen to an alkene, the sequence of halogenation dehydrohalogenation provides an excellent route for the preparation of an alkyne from an alkene.

$$\underset{\text{Propene}}{CH_3CH=CH_2} \xrightarrow{Br_2/CCl_4} \underset{\text{1,2-Dibromopropane}}{CH_3-\overset{\overset{\displaystyle Br}{|}}{CH}-\overset{\overset{\displaystyle Br}{|}}{CH_2}} \xrightarrow[-2HBr]{\text{Base}} \underset{\text{Propyne}}{CH_3C\equiv CH}$$

1,1-Dihalides (geminal dihalides) on treatment with base bring about double dehydrohalogenation to give alkynes.

$$\underset{\substack{\text{Germinal} \\ \text{Dihalide}}}{RCH_2CHX_2} \xrightarrow[-HX]{\text{Alc. KOH}} \underset{\text{Alkenyl halide}}{RCH=CHX} \xrightarrow[-HX]{NaNH_2} \underset{\text{Alkyne}}{RC\equiv CH}$$

$$\underset{\text{1,1-Dibromopropane}}{CH_3CH_2CHBr_2} \xrightarrow[-HBr]{\text{Alc. KOH}} CH_3CH=CHBr \xrightarrow[-HBr]{NaNH_2} \underset{\text{Propyne}}{CH_3C\equiv CH}$$

The 1,1-dihalides may be prepared by the action of phosphorus halides (PCl_5 or PBr_5) on aldehydes and ketones, therefore the above method is useful for conversion of an aldehyde or a ketone to the alkynes.

Alkynes may also be prepared by the treatment of trihalogen derivatives with silver powder.

$$CHX_3 + 6Ag + X_3CH \longrightarrow CH\equiv CH + 6AgX$$

(ii) *Dehalogenation of Tetrahalides:* Alkynes may also be prepared by dehalogenation of tetrahalides. 1,1,2,2-tetrabromopropane on treatment with zinc and alcohol yield propyne.

$$\underset{\substack{\text{1,1,2,2,-Tetrabromo-} \\ \text{propane}}}{CH_3CBr_2CHBr_2} \xrightarrow[-ZnBr_2]{Zn/alcohol} \underset{\text{Propyne}}{CH_3C\equiv CH}$$

Since the tetrahalides are obtained from alkynes, this method is not of any synthetic importance. However, this method can be used for the separation and purification of alkynes.

(iii) *Alkylation of Terminal Alkynes:* The reaction of terminal alkynes with sodium in liquid ammonia gives the corresponding sodium salt (sodium alkylide), which reacts with alkyl halides to give substituted alkynes.

$$RC \equiv CH + Na \xrightarrow{\text{liq } NH_3} RC \equiv \overset{-}{C}\overset{+}{Na}$$
$$\textbf{Sodium alkylide}$$

$$RC \equiv \overset{-}{C}\overset{+}{Na} + CH_3I \longrightarrow RC \equiv C-CH_3 + NaI$$
$$\quad\quad\quad\quad\quad \textbf{Methyl} \quad\quad\quad\quad\quad \textbf{Alkyne}$$
$$\quad\quad\quad\quad\quad \textbf{iodide}$$

Using this procedure a lower alkyne can be converted into a higher alkyne.

$$CH_3C \equiv CH \xrightarrow{\text{Na / liq } NH_3} CH_3C \equiv \overset{-}{C}\overset{+}{Na} \xrightarrow[-NaI]{CH_3I} CH_3C \equiv CCH_3$$
$$\textbf{Propyne} \quad\quad\quad\quad\quad\quad\quad\quad\quad\quad\quad\quad\quad\quad\quad\quad \textbf{2-Butyne}$$

The alkynes may also be alkylated by treatment with an appropriate Grignard reagent followed by reaction with an alkyl halide.

$$HC \equiv CH + CH_3CH_2MgI \longrightarrow CH \equiv \overset{-}{C}\overset{+}{Mg}I + CH_3CH_3$$
$$\textbf{Ethyne} \quad\quad\quad\quad\quad\quad\quad\quad\quad\quad\quad\quad\quad \textbf{Ethane}$$

$$HC \equiv \overset{-}{C}\overset{+}{Mg}I + CH_3CH_2I \longrightarrow CH \equiv CCH_2CH_3 + MgI_2$$
$$\quad\quad\quad\quad\quad\quad\quad\quad\quad\quad \textbf{1-Butyne}$$

This procedure is also used to estimate the active hydrogens in an organic molecule. From the amount of methane liberated when methyl iodide is used, the number of active hydrogens can be determined.

10.6 PHYSICAL PROPERTIES

Alkynes resemble alkenes in their physical properties. They are all colorless and odorless with the exception of ethyne which has a characteristic odor. The first three members (ethyne, propyne and butyne) are gases at room temperature. The next eight members are liquids and the higher members are solids. The melting points, boiling points and densities increase gradually with the increase in molecular weight. Alkynes have slightly higher boiling points compared to the corresponding alkenes and alkanes. They are non polar and dissolve readily in solvents like ether, benzene and carbon tetrachloride.

10.7 SPECTRAL PROPERTIES

Ultraviolet spectroscopy is not of much help in the characterisation of alkynes, since the alkynyl chromophore absorbs below 200 nm (π-π^* transition), which is difficult to detect. However, conjugation with a multiple bond results in a bathochromic shift.

The infrared spectra of alkynes is of diagnostic value. In terminal alkynes, RC≡CH, there is a absorption band in the region $3300-3100$ cm^{-1} due to ≡C–H stretching and another in the region $2140-2100$ cm^{-1} due to –C≡C– stretching. In the internal alkynes (RC≡CR), however, the absorption in the region $2260-2190$ cm^{-1} corresponds to the –C≡C– stretching.

In the NMR spectra, the terminal alkynes (RC≡CH) give an absorption signal between δ 2–3, characteristic of alkynyl proton. Such an absorption is absent in internal alkynes (RC≡CR).

10.8   CHEMICAL PROPERTIES

Alkynes having two π bonds give electrophilic addition, reduction and oxidation reactions:

(I)   *Reduction of Alkynes:* Alkynes add up two molecules of hydrogen in the presence of catalysts like finely divided platinum, palladium or Raney nickel to form first an alkene and finally the alkane.

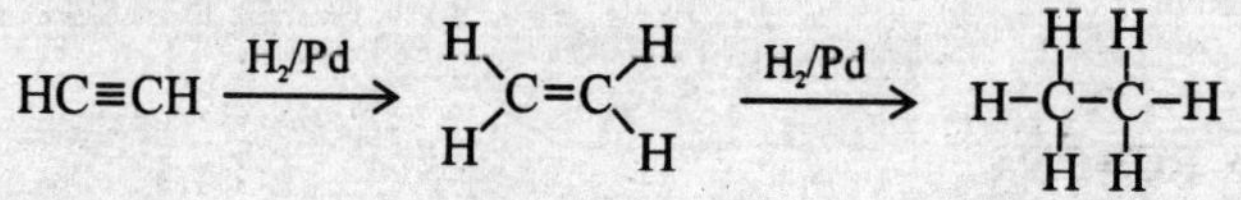

$$HC{\equiv}CH \xrightarrow{H_2/Pd} \underset{\text{Ethene}}{\overset{H}{\underset{H}{>}}C{=}C\overset{H}{\underset{H}{<}}} \xrightarrow{H_2/Pd} \underset{\text{Ethane}}{H-\overset{H}{\underset{H}{C}}-\overset{H}{\underset{H}{C}}-H}$$

Ethyne   Ethene   Ethane

The reaction can, however, be stopped at the alkene stage by using Lindlar's catalyst (Pb/BaSO$_4$ poisoned with quinoline).

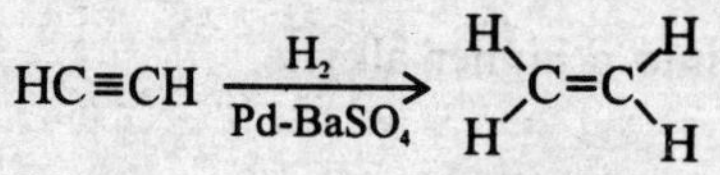

$$HC{\equiv}CH \xrightarrow[\text{Pd-BaSO}_4]{H_2} \overset{H}{\underset{H}{>}}C{=}C\overset{H}{\underset{H}{<}}$$

Ethyne   Ethene

Disubstituted alkynes, RC≡CR, on catalytic hydrogenation with Lindlar's catalysts give exclusively the cis alkene. If, however, the reduction is carried out with sodium or lithium in liquid ammonia, trans alkene is the predominant product.

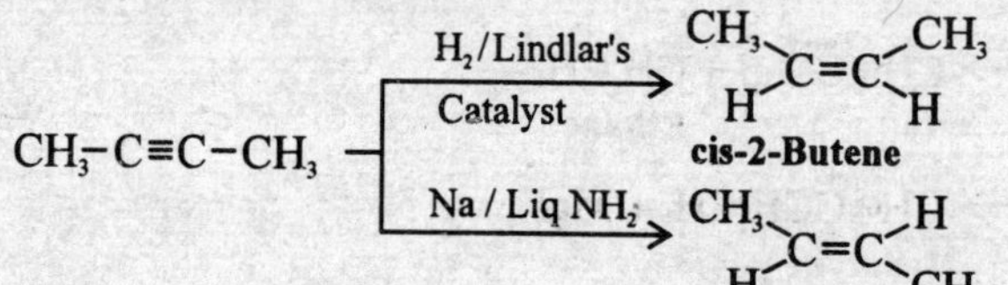

$$CH_3-C{\equiv}C-CH_3$$

H$_2$/Lindlar's Catalyst → cis-2-Butene

Na / Liq NH$_2$ → trans-2-Butene

Such reactions which predominantly produce only one stereoisomer are called *'stereoselective reactions'*.

(II)   *Electrophilic Addition:* Alkynes add on two molecules of a reagent to give a saturated compound.

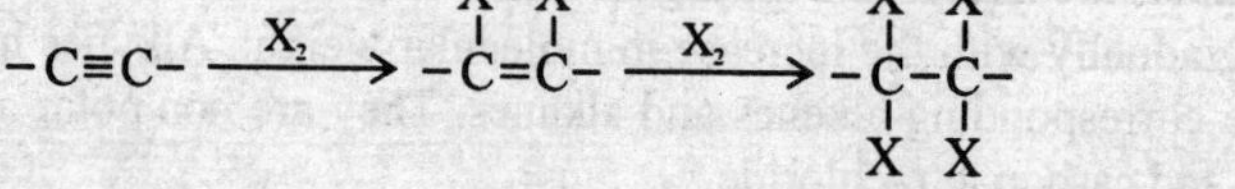

$$-C{\equiv}C- \xrightarrow{X_2} -\overset{X}{C}{=}\overset{X}{C}- \xrightarrow{X_2} -\overset{X}{\underset{X}{C}}-\overset{X}{\underset{X}{C}}-$$

Alkyne   Alkene   Saturated product

(a)   *Addition of Halogens:* Alkynes add on two molecules of halogen forming first a dihaloalkene and then a tetrahaloalkane.

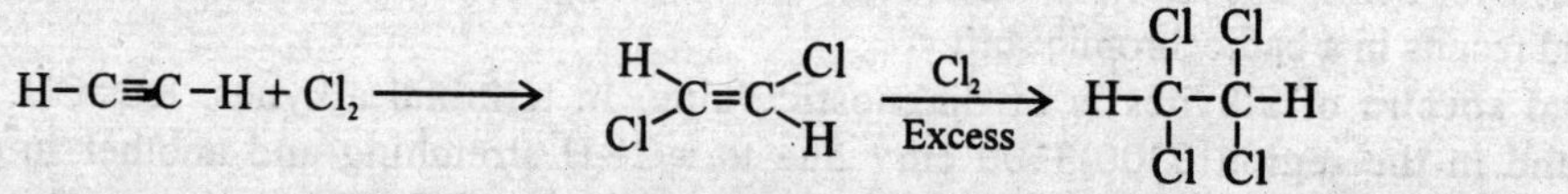

$$H-C{\equiv}C-H + Cl_2 \longrightarrow \underset{\text{trans-1,2-Dichloroethene}}{\overset{H}{\underset{Cl}{>}}C{=}C\overset{Cl}{\underset{H}{<}}} \xrightarrow[\text{Excess}]{Cl_2} \underset{\text{1,1,2,2-Tetrachloroethane}}{H-\overset{Cl}{\underset{Cl}{C}}-\overset{Cl}{\underset{Cl}{C}}-H}$$

Acetylene   trans-1,2-Dichloroethene   1,1,2,2-Tetrachloroethane

The alkynes are considerably less reactive compared to alkenes towards addition of halogens. The reaction involves the formation of a cyclic halonium ion intermediate which is formed with difficulty in case of alkynes (see Section 9.9).

(b) *Addition of Hydrogen Halides:* Alkynes add halogen acids in accordance with Markovnikov's rule. Acetylene combines with hydrogen bromide to form first bromoethene and then 1,1-dibromoethane.

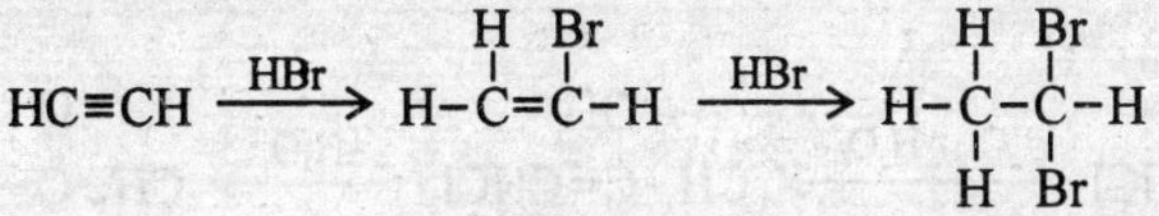

$$HC \equiv CH \xrightarrow{HBr} \underset{\substack{H \quad Br}}{H-C=C-H} \xrightarrow{HBr} \underset{\substack{H \quad Br}}{H-C-C-H}$$

Acetylene   **Vinyl bromide**   **1,1-Dibromoethane**

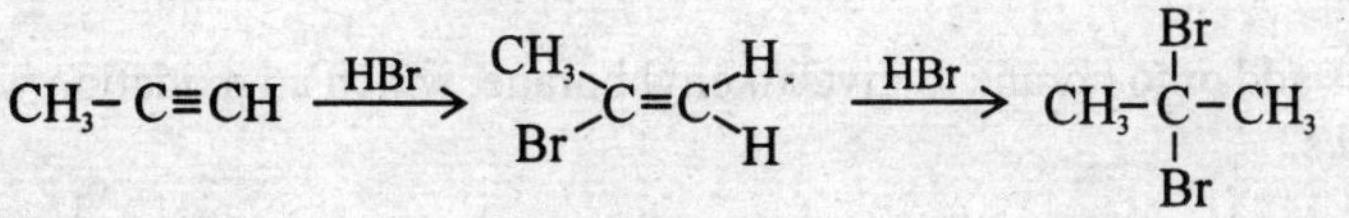

$$CH_3-C \equiv CH \xrightarrow{HBr} \underset{Br}{\overset{CH_3}{>}}C=C\overset{H}{\underset{H}{<}} \xrightarrow{HBr} \underset{Br}{\overset{Br}{CH_3-C-CH_3}}$$

Propyne   **2-Bromopropene**   **2,2-Dibromopropane**

The mechanism involved is electrophilic addition and is similar to that discussed in alkenes. The only difference here is the formation of a *vinylic cation* as shown below:

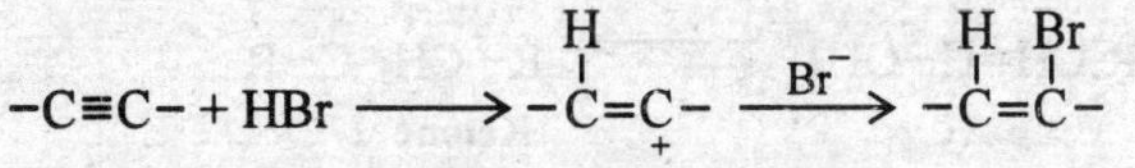

$$-C \equiv C- + HBr \longrightarrow \underset{+}{\overset{H}{-C=C-}} \xrightarrow{Br^-} \underset{\substack{H \quad Br}}{-C=C-}$$

a vinylic cation

The vinylic cation formed above is less stable than the saturated carbocation formed in case of alkenes, even then the addition of HX to alkynes takes place at much the same rate as that of alkenes. This is due to almost equal energy difference of alkyne to a vinyl cation and an alkene to a saturated carbocation.

In the presence of peroxides, anti Markovnikov's addition of HBr takes place with alkynes.

$$CH_3CH_2C \equiv CH \xrightarrow[\text{Peroxide}]{HBr} CH_3CH_2CH=CHBr$$

1-Butyne   **1-Bromo-1-butene**

(c) *Addition of Water:* Alkynes on hydration in presence of mercuric sulphate dissolved in sulphuric acid at 75° add a molecule of water to form vinyl alcohol, an unstable enol. The enol has the hydroxy group attached to a doubly bonded carbon atom and isomerises (or tautomerises) to an aldehyde or a ketone. This process is called *Keto-enol tautomerism* and the keto and enol forms are in equilibrium with each other which lies in favor of more stable keto form.

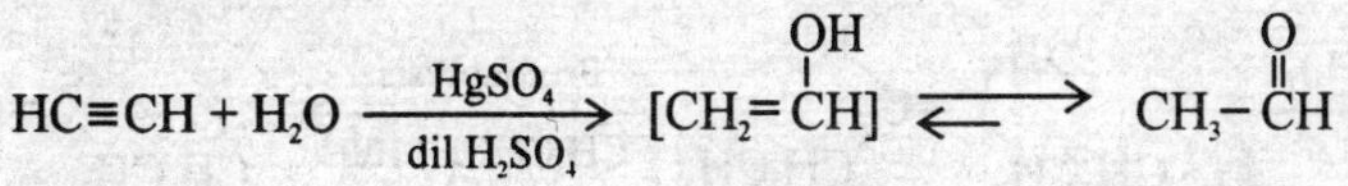

$$HC \equiv CH + H_2O \xrightarrow[\text{dil } H_2SO_4]{HgSO_4} [CH_2 \overset{OH}{=}CH] \rightleftharpoons CH_3-\overset{O}{\overset{\|}{C}}H$$

Ethyne   **Vinyl alcohol**   **Acetaldehyde**

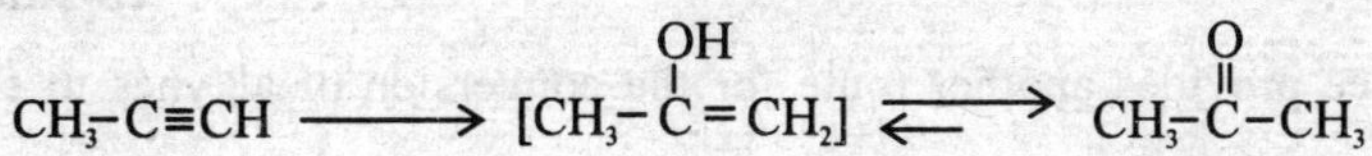

$$CH_3-C \equiv CH \longrightarrow [CH_3-\overset{OH}{\overset{|}{C}}=CH_2] \rightleftharpoons CH_3-\overset{O}{\overset{\|}{C}}-CH_3$$

Propyne   **Acetone**

$$CH_3-C \equiv C-CH_3 \xrightarrow{HgSO_4, H_2SO_4} [CH_3-\overset{OH}{\overset{|}{C}}=\overset{H}{\overset{|}{C}}-CH_3] \rightleftharpoons CH_3-\overset{O}{\overset{\|}{C}}-CH_2-CH_3$$

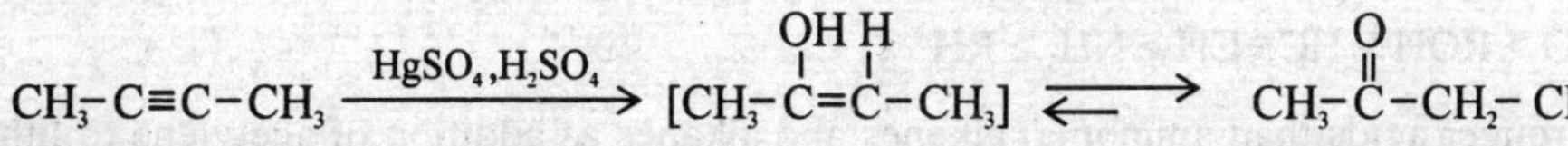

2-Butyne   **2-Butanone**

(d) *Addition of Hypohalous Acid:* Alkynes form haloaldehyde or haloketone when chlorine or bromine is passed through its aqueous solution. The addition is in accordance with Markovnikov's rule.

$$HC{\equiv}CH \xrightarrow{Cl_2/H_2O} HO-HC{=}CH-Cl \xrightarrow{Cl_2/H_2O} (HO)_2CH-CHCl_2 \xrightarrow{-H_2O} O{=}CHCHCl_2$$

Ethyne   **Dichloroacetaldehyde**

$$CH_3C{\equiv}CH \xrightarrow{Cl_2/H_2O} CH_3\overset{\overset{\displaystyle HO}{|}}{C}{=}CHCl \xrightarrow{Cl_2/H_2O} CH_3\overset{\overset{\displaystyle OH}{|}}{\underset{\underset{\displaystyle OH}{|}}{C}}{=}CHCl_2 \xrightarrow{-H_2O} CH_3\overset{\overset{\displaystyle O}{\|}}{C}-CHCl_2$$

Propyne   **1,1-Dichloroacetone**

(e) *Hydroboration:* Alkynes add onto borane to give alkenyl borane, which an oxidation with alkaline hydrogen peroxide yield ketones.

$$R-C{\equiv}C-R' \xrightarrow{BH_3} R-\overset{\overset{\displaystyle B{<}}{|}}{C}{=}CHR' \xrightarrow{RC{\equiv}CR} (R'CH{=}C)_2{-}\overset{\overset{\displaystyle R}{|}}{BH} \xrightarrow{RC{\equiv}CR'} (R'CH{=}\overset{\overset{\displaystyle R}{|}}{C})_3{-}B$$

Alkenyl borane

$$(R'CH{=}\overset{\overset{\displaystyle R}{|}}{C})_3{-}B \xrightarrow{H_2O_2/OH^-} [R'CH{=}\overset{\overset{\displaystyle R}{|}}{C}{-}OH] \rightleftharpoons R{-}CH_2{-}\overset{\overset{\displaystyle O}{\|}}{C}{-}R$$

Ketone

The symmetrical internal alkynes give a single product. However, unsymmetrical alkynes give mixtures. For example, 3-hexyne gives 3-hexanone only while 2-hexyne gives a mixture of 2-hexanone and 3-hexanone.

$$CH_3CH_2C{\equiv}CCH_2CH_3 \xrightarrow[\text{2.H}_2\text{O}_2/\text{OH}^-]{\text{1.BH}_3} CH_3CH_2\overset{\overset{\displaystyle O}{\|}}{C}CH_2CH_2CH_3$$

3-Hexyne   **3-Hexanone**

$$CH_3C{\equiv}CCH_2CH_2CH_3 \xrightarrow[\text{2.H}_2\text{O}_2/\text{OH}^-]{\text{1. BH}_3} CH_3\overset{\overset{\displaystyle O}{\|}}{C}CH_2CH_2CH_2CH_3 + CH_3CH_2\overset{\overset{\displaystyle O}{\|}}{C}CH_2CH_2CH_3$$

2-Hexyne   **2-Hexanone**   **3-Hexanone**

The terminal alkynes on hydroboration give aldehydes.

The organo boranes may be subjected to *protonolysis* on treatment with acetic acid to yield cis alkenes.

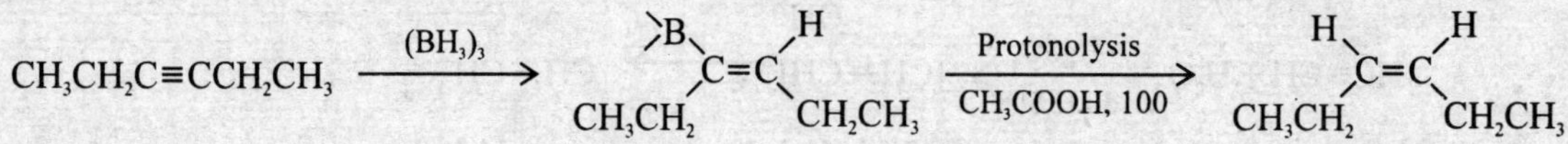

$$CH_3CH_2C{\equiv}CCH_2CH_3 \xrightarrow{(BH_3)_3} \overset{\displaystyle CH_3CH_2}{\underset{}{}}\overset{\displaystyle >B}{C}{=}C\overset{\displaystyle H}{\underset{\displaystyle CH_2CH_3}{}} \xrightarrow[\text{CH}_3\text{COOH, 100}]{\text{Protonolysis}} \overset{\displaystyle H}{\underset{\displaystyle CH_3CH_2}{}}C{=}C\overset{\displaystyle H}{\underset{\displaystyle CH_2CH_3}{}}$$

3-Hexyne   **cis-3-Hexene**

The above reaction sequence provides another route for the conversion of alkynes to cis alkenes.

(III) *Acidity of Alkynes:* Alkynes are very weak acids as given in the acidity order below:

$$H_2O > ROH > HC{\equiv}CH > NH_3 > RH$$

They are stronger acids than ammonia, alkenes and alkanes as addition of acetylene to lithium amide or ethylmagnesium bromide in ether forms ammonia and ethane respectively.

$$HC{\equiv}CH + Li^+NH_2^- \rightleftharpoons H{-}NH_2 + HC{\equiv}\bar{C}Li^+$$

$$HC\equiv CH + C_2H_5MgBr \rightleftharpoons H-C_2H_5 + HC\equiv \overset{-}{C}\overset{+}{M}gBr$$

Stronger   **Stronger**   **Weaker**   **Weaker**
acid   **base**   **acid**   **base**

However, acetylene is a weaker acid than water or alcohol as they liberate acetylene from sodium acetylide.

$$HOH + HC\equiv C^-Na^+ \rightleftharpoons HC\equiv CH + Na\overset{+}{\overset{}{O}}H$$

$$C_2H_5OH + HC\equiv C^-Na^+ \rightleftharpoons HC\equiv CH + Na\overset{+}{\overset{}{O}}C_2H_5$$

Stronger   **Stronger**   **Weaker**   **Weaker**
acid   **base**   **acid**   **base**

Acetylene when reacts with active metals like lithium or sodium, forms metal acetylide and hydrogen gas is liberated.

$$HC\equiv CH + Li \longrightarrow HC\equiv \overset{-}{C}\overset{+}{L}i + \tfrac{1}{2}H_2$$

Lithium acetylide

The acidity in alkynes is due to the triply bonded carbon attached to hydrogen in terminal alkynes. Let us now consider the electronic configuration of acetylide anion and ethyl anion to compare the acidity of alkynes and alkanes.

$$HC\equiv CH \rightleftharpoons \overset{+}{H} + HC\equiv \overset{-}{C}$$

Stronger acid   **Acetylide anion**
                            Weaker base

$$CH_3CH_2-H \longleftarrow \overset{+}{H} + CH_3CH_2^-$$

Weaker acid   **Ethyl anion**
                           Stronger base

In acetylide anion, the unshared pair of electron occupies sp orbital while in ethyl anion it is in sp^3 orbital. Since the sp^3 orbital is elongated (more p character), the distance of electrons from nucleus is more and they are more available for sharing with acids. On the other hand electrons in alkylide anion are closer to the nucleus (sp orbital has 50% s character), held tightly and are less available for sharing with acids. The acetylide anion, thus, is a weaker base than ethyl anion or acetylene is a stronger acid than ethane.

(IV) *Formation of Metal Salts:* It has already been stated that alkynes on reaction with sodium in liquid ammonia as solvent or with sodamide give the ionic compounds, viz. sodium alkylides. These salts are stable when dry but are readily hydrolysed by water regenerating the alkynes.

$$HC\equiv CH \xrightarrow{Na/liq\ NH_3} HC\equiv \overset{-}{C}\overset{+}{N}a \xrightarrow[-NaOH]{H_2O} HC\equiv CH$$

Acetylene   **Acetylene**

Alkynes react with ammonical solution of copper sulphate and silver nitrate to give the corresponding copper and silver alkylides which are red and white in color. Unlike sodium alkylides, these are stable in water but can be decomposed with dilute hydrochloric acid to regenerate the parent alkyne. These are sensitive to shock and explode in dry state.

$$HC\equiv CH + 2Cu(NH_3)_2Cl \longrightarrow Cu^+C^-\equiv C^-Cu^+ + 2NH_3$$

Copper acetylide (red)

$$HC \equiv CH + 2Ag(NH_3)_2NO_2 \longrightarrow Ag^+\bar{C} \equiv \bar{C}Ag^+ + 2NH_3$$

Silver acetylide (white)

$$R - C \equiv CH + Ag(NH_3)_2NO_3 \longrightarrow R - C \equiv C^-Ag^+ + NH_4NO_3 + NH_3$$

Silver Alkylide (white)

$$R - C \equiv C - R + Ag(NH_3)_2NO_3 \longrightarrow \text{No reaction}$$

The reaction with ammonical silver nitrate is used to distinguish between terminal and nonterminal alkynes and used as a diagnostic test for $-C \equiv CH$ group.

(V)   *Oxidation:* The alkynes on oxidation with potassium permanganate give carboxylic acid by cleavage of the molecule at the site of the triple bond (compare with alkenes, which give diols).

$$R - C \equiv C - H + 4[O] \xrightarrow{KMnO_4} RCOOH + CO_2$$
$$R - C \equiv C - R' + 4[O] \longrightarrow RCOOH + R'COOH$$

This degradation is used for structure elucidation of alkynes.

However, on treatment with ozone, the alkynes give the ozonides (the structure is different with the ozonides obtained from alkenes), which on treatment with water yield carboxylic acids.

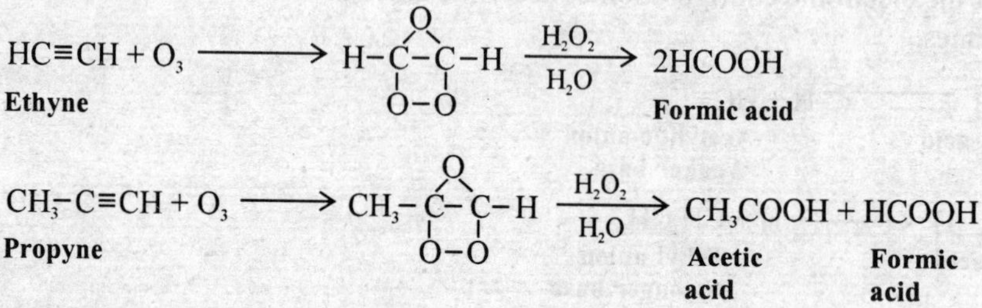

Ethyne   **Formic acid**

Propyne   **Acetic acid**   **Formic acid**

The ozonolysis of alkynes is used for structure elucidation of alkynes.

(VI)   *Polymerisation:* Alkynes undergo polymerisation to form linear or cyclic products. On passing through red hot iron tube acetylene polymerises to give benzene whereas in presence of cuprous chloride and ammonium chloride it forms a linear polymer.

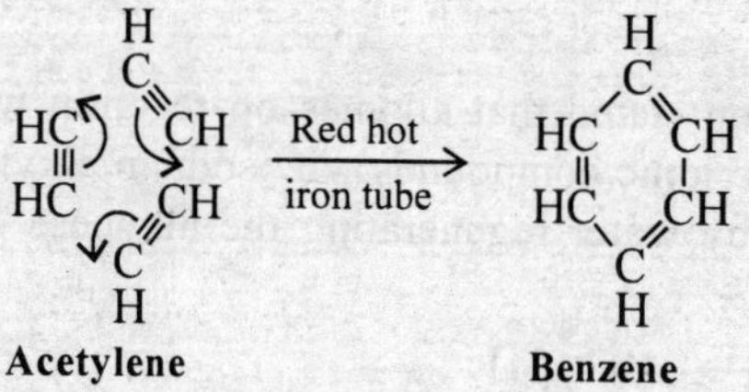

Acetylene   **Benzene**

$$CH \equiv CH \xrightarrow{Cu_2Cl_2, NH_4Cl} CH_2 = CH - C \equiv CH \xrightarrow{HC \equiv CH} CH_2 = CH - C \equiv C - CH = CH_2$$

Acetylene   **Vinyl acetylene**   **Divinyl acetylene**

(VII)   *Oxidative Coupling:* It has already been mentioned that terminal alkynes on treatment with ammonical solution of copper sulphate give the corresponding copper acetylides. These on heating in acetic acid in presence of air undergo oxidative coupling to form 1,3-dialkyne.

$$R-CH{\equiv}CH + Cu(NH_3)_2Cl \longrightarrow R-C{\equiv}C^-Cu^+ \xrightarrow[\text{AcOH}]{O_2} R-C{\equiv}C-C{\equiv}C-R$$

Alkyne   **1,3-Dialkyne**

$$CH_3-C{\equiv}CH \xrightarrow[\text{2. O}_2/\text{AcOH}]{\text{1. Cu(NH}_3)_2\text{Cl}} CH_3-C{\equiv}C-C{\equiv}C-CH_3$$

Propyne   **2,4-Hexadiyne**

(VIII)   *Isomerisation:* 1-Butyne may be isomerised to a more stable 2-butyne in presence of alcoholic potassium hydroxide and the reverse of this reaction may occur with sodamide.

$$CH_3-CH_2-C{\equiv}CH \underset{\text{NaNH}_2}{\overset{\text{Alk. KOH}}{\rightleftharpoons}} CH_3-C{\equiv}C-CH_3$$

1-Butyne   **2-Butyne**

10.9   INDIVIDUAL MEMBERS

10.9.1   Acetylene/Ethyne [HC≡CH]

Acetylene, the first member of the alkyne series, is the only alkyne which is of significant industrial importance.

Preparation

(i)   *From Calcium Carbide:* Acetylene is prepared conveniently in the laboratory by the action of water on calcium carbide.

$$CaC_2 + 2H_2O \longrightarrow H-C{\equiv}C-H + Ca(OH)_2$$

Calcium   **Acetylene**
carbide

This method is also used for the industrial preparation of acetylene. For this purpose calcium carbide is obtained by heating a mixture of coke and lime stone in an electric furnace.

$$CaCO_3 \xrightarrow{\Delta} CaO + CO_2$$

Lime

$$CaO + 3C \xrightarrow{2000\text{-}3000^\circ} CaC_2 + CO_2$$

(ii)   *From Methane:* Acetylene is prepared now a days on industrial scale by the partial combustion of methane, which is obtained from natural gas.

$$6CH_4 + O_2 \xrightarrow{1500^\circ} 2HC{\equiv}CH + 2CO + 10H_2$$

Acetylene

(iii)   *By Dehydrohalogenation of Dihalides:* Ethylene dibromide on treatment with alcoholic potassium hydroxide forms first the vinyl bromide, which on treatment with a strong base like sodamide gives acetylene.

$$CH_3CHBr_2 \text{ or } BrCH_2-CH_2Br \xrightarrow{\text{Alk. KOH}} CH_2{=}CHBr \xrightarrow{\text{NaNH}_2} HC{\equiv}CH$$

Ethylidene   **Ethylene**   **Vinyl bromide**   **Acetylene**
bromide   **dibromide**

(iv) *By Kolbe's Reaction:* Electrolysis of a concentrated solution of sodium or potassium salt of maleic or fumaric acid gives acetylene.

$$\begin{matrix} CH-COO^-K^+ \\ \| \\ CH-COO^-K^+ \end{matrix} \xrightarrow{2H_2O} \begin{matrix} CH \\ \||| \\ CH \end{matrix} + 2CO_2 + 2KOH + H_2$$

Potassium　　　　　　　**Acetylene**
fumarate

(v) *From Iodoform:* Iodoform on heating with silver powder gives acetylene.

$$2CHI_3 + 6Ag \longrightarrow HC\equiv CH + 6AgI$$

Properties

Acetylene is a colorless gas with sweet smell (when pure). It is lighter than air, sparingly soluble in water but soluble in alcohol and acetone. It can be liquefied at $-84°$ under ordinary pressure. However, liquefied acetylene is very explosive and so its storage is prohibited. For transportation acetylene is dissolved in acetone soaked on porous material under pressure in steel cylinders.

It gives all the general reactions of alkynes. However, some special reactions are given below.

(i) *Reaction with Alcohols and Carboxylic Acids:* Acetylene adds to alcohols in presence of alkali to give vinyl ether.

$$HC\equiv CH + CH_3OH \xrightarrow[120-180°]{KOH} CH_3OCH=CH_2$$

Methyl vinyl ether

Similarly, carboxylic acid and acetylene in presence of mercuric salts yield the vinyl ester.

$$HC\equiv CH + CH_3COOH \xrightarrow[80°]{Hg^{++}} CH_3-\overset{\overset{\displaystyle O}{\|}}{C}-O-CH=CH_2$$

Vinyl acetate

(ii) *Reaction with Hydrogen Cyanide:* Acetylene adds on hydrogen cyanide in presence of copper salt as catalyst to give acrylonitrile, which is a starting material for the manufacture of synthetic rubber 'Orlon'.

$$HC\equiv CH + HCN \xrightarrow[500°]{CuCl} CH_2=CHCN$$

Vinyl cyanide
(Acrylonitrile)

(iii) *Reaction with Arsenic Trichloride:* Acetylene reacts with arsenic trichloride in presence of anhydrous aluminum chloride to form β-chlorovinyl dichloroarsine, which was used as a poison gas under the name Lewisite in World War II.

$$HC\equiv CH + AsCl_3 \xrightarrow{anhydr.\ AlCl_3} \overset{\overset{\displaystyle Cl\ \ H}{|\ \ \ |}}{H-C=C}-As\overset{\diagup Cl}{\diagdown Cl}$$

β-Chlorovinyldi-
chloroarsine (Lewisite)

(iv) *Reaction with Formaldehyde:* Acetylene reacts with formaldehyde in presence of copper acetylide to form 2-butyne-1,4-diol. Small quantity of propargyl alcohol is also obtained in the above reaction.

$$HC\equiv CH + H-\overset{\overset{O}{\|}}{C}-H \xrightarrow{Cu_2C_2} HC\equiv C-CH_2OH$$

$$HC\equiv CH + 2H-\overset{\overset{O}{\|}}{C}-H \xrightarrow{Cu_2C_2} HOCH_2-C\equiv C-CH_2OH$$

2-Butyne-1,4-diol

This reaction is known as *ethynylation.*

(v) *Combustion:* On burning in oxygen, acetylene gives very high temperature (2500-3000°).

$$2HC\equiv CH + 5O_2 \xrightarrow{\Delta} 4CO_2 + 2H_2O + heat$$

Uses: Acetylene is used for illumination purposes, in oxy-acetylene welding and artificial ripening of fruits. It is used for the manufacture of acetaldehyde, acetic acid, ethyl alcohol and also for the preparation of plastics, synthetic rubbers and synthetic fiber 'orlon'.

10.10   PROBLEMS

1. Discuss the orbital structure of acetylene.
2. Starting with ethyne, how will you obtain 2-heptyne and 3-heptyne.
3. Give the products obtained in the following.

$$CH_2=\overset{\overset{Br}{|}}{C}-CH_2Br \xrightarrow{KOH/EtOH}$$

$$CH\equiv CH \xrightarrow[\text{(ii) CH}_3I]{\text{(i) Na/liq.NH}_3}$$

$$RC\equiv CH \xrightarrow[\text{(ii) CH}_3I]{\text{(i) CH}_3MgI}$$

4. Starting with 2-hexyne how will you obtain cis-2-hexene, trans-2-hexene and hexane.
5. How will you carry out the following transformation:

   (a) $CH_3CH_2C\equiv CH \xrightarrow{?} CH_3CH_2COCH_3$

   (b) $CH_3CH_2C\equiv CH \xrightarrow{?} CH_3CH_2CH_2CHO$

   (c) $CH_3CH_2C\equiv CCH_3 \xrightarrow{?} CH_3CH_2COOH + CH_3COOH$

   (d) $CH_3CH_2CH_2CH=CH_2 \dashrightarrow^{?} CH_3CH_2CH_2C\equiv CH$

6. Predict the products obtained in the following reactions

$$CH_3CH_2CH_2C\equiv CH \begin{array}{l} \xrightarrow{2HBr} \\ \xrightarrow{H_2/\text{Lindlar's Catalyst}} \\ \xrightarrow[\text{2. CH}_3CH_2I]{\text{1. NaNH}_2/NH_3} \end{array}$$

7. How will you obtain:
   (a) Propyne from 2-bromopropane   (b) 2-Hexyne from 1-bromopropane
8. What products are obtained if acetylene is treated with:
   (a) Methanol in presence of potassium hydroxide at 120-180°.
   (b) Acetic acid in presence of Hg^{2+} salts at 80°.
   (c) Hydrogen cyanide in presence of copper salt.
   (d) Arsenic trichloride in presence of anhydrous aluminum chloride.
   (e) Formaldehyde (excess) in presence of copper acetylide.
9. Explain the acidity of acetylene.

11
Petroleum

11.1 INTRODUCTION

Petroleum is an oily, thick and usually dark colored liquid. It is found deep below earth's crust trapped within rock structure. The origin of the word petroleum is from the Latin words petra (meaning rock) and oleum (meaning oil). The biggest oil-producing country of the world is USA. The other major oil producing countries are Russia, Venezuela, Iran, Gulf countries, Rumania, Mexico, Pakistan and India.

In India, Petroleum industry has made tremendous head way only after independence. In the last 30-35 years, petroleum production has increased almost 20 times. Even with this increased production, only about two third of our needs are met.

Large quantities of gas is associated with liquid petroleum. This gas is referred to as *natural gas*. Its composition varies with the source, and consists chiefly the first six alkanes. Other gases such as water vapor, hydrogen, nitrogen, carbon dioxide and hydrogen sulphide may also be present in amounts that vary with the locality of occurence.

A mixture of propane and butane obtained from natural gas is liquefied at room temperature under a pressure of 3-5 atmospheres. This is known as *LPG* (liquefied petroleum gas) and is stored in 40-100 liter capacity steel cylinders and is used in gas stoves for heating purposes. Though the LPG is odorless, but to warn against leakage a very small amount of foul smelling substance is added. Natural gas also serves as starting material for making useful petrochemicals.

11.2 COMPOSITION

The composition of crude petroleum varies with occurrence. However, they all contain alkanes (Straight and branched-chain from C_1 to C_{40}), cycloalkanes, naphthenes and aromatic hydrocarbons. The low boiling fractions of petroleum are composed of alkanes. It is the composition of the higher boiling fractions which differ according to the source of petroleum. Besides hydrocarbons, petroleum also contains compounds containing oxygen, nitrogen and sulphur.

11.3 ORIGIN OF PETROLEUM IN NATURE

A number of theories have been proposed to explain the formation of petroleum in nature.

(i) *The Carbide Theory:* According to this theory petroleum is regarded as of inorganic origin. This theory was originally put forth by Mendeleef and subsequently supported by Moissan, Sabatier and Senderens. Petroleum is believed to be formed by the action of steam or water on metallic carbides in the interior portion of earth's crust. The acetylene thus formed is reduced to ethylene and ethane with hydrogen in presence of metallic catalyst at high temperature (the hydrogen needed is obtained by the action of steam on hot metals). Also, the acetylene formed polymers in presence of hot metals and gave aromatic hydrocarbons, cycloalkanes and other open chain hydrocarbons. The carbides are believed to be formed by the reaction of metals (Al and Ca) with coal in the hot interior of the earth.

$$4Al + 3C \longrightarrow Al_4C_3 \xrightarrow{H_2O} CH_4$$

$$\text{Aluminum} \qquad\qquad \text{Methane}$$
$$\text{carbide}$$

$$Ca + 2C \longrightarrow CaC_2 \xrightarrow{H_2O} HC \equiv CH \longrightarrow C_6H_6$$

$$\text{Calcium} \qquad \text{Acetylene} \qquad \text{Benzene}$$
$$\text{carbide}$$

$$C_6H_{12} \xleftarrow{\text{Polymerisation}} CH_2 = CH_2 \xrightarrow{\text{Polymerisation}} CH_3CH=CHCH_3 \xrightarrow{\text{Redn}} CH_3CH_2CH_2CH_3$$

$$\text{Cyclohexane} \qquad \text{Ethylene} \qquad\qquad \text{2-butene} \qquad\qquad \text{Butane}$$

Though the carbide theory explains the formation of petroleum, the theory does not explain the presence of sulphur and nitrogen compounds and also the presence of optically active substances in petroleum.

(ii) *The Engler's Theory:* According to this theory, petroleum is regarded as of organic origin. Petroleum is believed to be formed by the decomposition of marine animals under high pressure and temperature. This is supported by the fact that fish oil and certain other fats on heating under pressure yield a product resembling petroleum. The marine animals are a rich source of nitrogen and sulphur compounds and so the theory proposed by Engler also explains the presence of nitrogen and sulphur compounds and also the optical activity of petroleum. The association of petroleum with brine suggests its origin from sea.

(iii) *Present Views:* Many oils contain compounds that could be derived from chlorophyll which is present in plants thus pointing to vegetable origin of petroleum. The high resin contents of Burma oil can also be accounted for by the vegetable origin of the oil.

It is now believed that Petroleum is originated by partial decomposition of marine animals (Englers theory) as well as vegetable matter. It is quite likely that changes, in earth's crust must have brought forests under the earth crust, where by the action of heat and water, the plants are converted into petroleum products. The association of petroleum with brine (sea-water) and the presence of coal deposits (from prehistoric forests) in the neighbor of oil deposits indicate that the origin of petroleum is a combination of organic as well as vegetable origin.

11.4 LOCATION OF OIL FIELDS

In early times the location of oil bearing deposits depended on certain visible signs on the surface of earth. For example, the explorer by smelling the soil could locate the presence of oil below the surface of earth. Alternatively the 'natural fire' resulting due to the seepage of petroleum products and gases from the ground was also used as a good indication for the presence of oil underneath the earth. These days the oil fields are located by certain scientific methods. Some of the methods used are given below:

(i) *Gravity Survey:* A geologist by 'gravity survey' can reasonably guess the location of oil fields. The principle of this method depends on the fact that the pull of gravity in any area varies with the density of the rocks beneath. The sedimentary rocks below the oil reserves are less dense compared to rocks elsewhere and show a relatively lower gravitational attraction. Therefore gravity-meters show a fall in the value of gravity near the petroleum fields.

(ii) *Determination of Magnetic Field:* Generally speaking, the oil deposits are found in the vicinity of salt reserves within the earth's crust. These salt reserves are diamagnetic and thus weaken the intensity of magnetic field of the earth. By determination of the magnetic field or the magnetic gradient with the help of magnetometers at various places, the presence of petroleum rich rocks can be located.

(iii)   *Seismic Method:* In this method a miniature earthquake is created by exploding a dynamite under the earth surface. This sends shock waves through the ground. These waves are reflected from the petroleum surface and the dense underlying strata are returned to the surface. These vibrations or shock waves are recorded by the geophones (special types of microphones) connected to the seismometer, an automatic recording device. The time noted between the dynamite blast and the reception of vibrations on the surface reveals the nature of the rocks beneath and also the existence and depth of the oil reserves. This method is the most reliable and effective for locating oil fields.

On the basis of the data collected by the above methods tentative sites for the presence of oil reserves are chosen. The final proof, however, is to bore down into the rock and find out the actual amount of petroleum present.

11.5   DRILLING OF PETROLEUM

Petroleum is obtained by drilling holes in the earth's crust and sinking pipes until the drill penetrates the oil bearing sands. Due to the pressure exerted by the natural gas, the oil rushes out through the pipes. Once the pressure of gas diminishes, two coaxial pipes are lowered into the bore and compressed air is forced through the pipe.The result is the oil coming out in the form of froth through the other pipe. From the oil-fields the crude oil is taken to the refineries through pipelines. The refineries are generally located at a distance of few kilometers.

11.6   FRACTIONATION OF PETROLEUM

The first step in the refining of Petroleum is the separation by fractional distillation. The crude oil is heated in a furnace at 475° and hot liquid is passed through a fractionating column. The column is fitted with horizontal stainless steel trays. Each tray is provided with chimneys covered with a loose cap called ball cap as shown in Fig 11.1. As the vapors ascend, they gradually cool and condense at different heights.

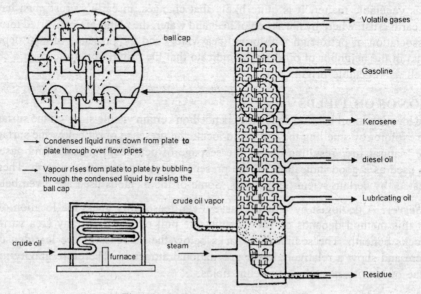

ball cap

→ Condensed liquid runs down from plate **to** plate through over flow pipes

→ Vapour rises from plate to plate by bubbling through the condensed liquid by raising the ball cap

crude oil vapor

crude oil   steam

furnace

→ Volatile gases

→ Gasoline

→ Kerosene

→ diesel oil

→ Lubricating oil

→ Residue

Fig. 11.1:   Fractionation of Petroleum

The high boiling fractions condense in the lower portion of the column. By this process the crude oil is separated into a number of fractions. Each fraction is a mixture of different hydrocarbons and so is purified before use.

The important petroleum fractions along with their boiling ranges and important uses are given below.

11.7 PETROL OR GASOLINE

Due to the use of petrol or gasoline as motor fuel it is the most important fraction of crude oil. However, the yield of this fraction is quite low. Due to increased demand of this product, the yield of this fraction is increased considerably by *Cracking*.

Fraction	b.p.(°C)	Approx. composition	Uses
Natural Gas	–	C_1-C_4	Fuel
Light Petrol	20-100	C_5-C_7	Solvent
Benzene	70-90	C_6-C_7	Dry Cleaning
Ligroin	80-120	C_6-C_8	Solvent
Petrol(Gasoline)	70-200	C_6-C_{11}	Motor Fuel
Kerosine	200-300	C_{12}-C_{16}	Lighting, Fuel
Heavy Oil (gas oil)	> 300	C_{13}-C_{18}	Fuel Oil
Lubricating Oil	> 300	C_{16}-C_{20}	Lubricants
Greases, Vaseline, Petroleum	> 300	C_{18}-C_{22}	Pharmaceutical preparations
Paraffin Wax	> 300	C_{20}-C_{30}	Candles, wax, Paper
Residue (asphaltic bitumen)	–	C_{30}-C_{40}	Asphalt tar, Petroleum coke.

Cracking is the process of getting lower boiling hydrocarbons from higher boiling hydrocarbon by application of heat. There are two main types of cracking. The first is the *Liquid Phase Cracking*, in which the heavy oil is cracked by heating at a suitable temperature (475-525°) and under 10-70 atmosphere of pressure. High pressure maintains the cracked material in liquid condition. The yield is 60-65%. The second is the *Vapor phase cracking*. In this process kerosine or any other similar oil is heated to 600° under 3-10 atmospheres of pressure. Heavy oils cannot be cracked by this process since they cannot be completely vaporized under the conditions. Below are given the products of cracking of various hydrocarbons like butane, octane and dodecane.

$$C_4H_{10} \longrightarrow CH_3CH{=}CHCH_3 + CH_3CH_2CH{=}CH_2 + CH_3CH{=}CH_2 + CH_2{=}CH_2 + CH_4 + H_2$$
Butane

$$C_8H_{18} \longrightarrow C_8H_6 + C_6H_{14} + C_4H_{10} + C_4H_8 + C_2H_4 + CH_4 + H_2$$
Octane

$$C_{12}H_{26} \longrightarrow C_7H_{16} + C_7H_{14} + C_6H_{14} + C_6H_{12} + C_5H_{12} + C_5H_{10}$$
Dodecane

The mechanism involved in cracking is free radical and may be summerised as follows:

$$CH_3CH_2CH_3 \longrightarrow CH_3CH_2^{\cdot} + CH_3^{\cdot} \longrightarrow CH_2{=}CH_2 + CH_4$$
$$CH_3CH_2CH_2CH_3 \longrightarrow CH_4 + CH_3CH^{\cdot}CH_3 \longrightarrow CH_3CH{=}CH_2 + H^{\cdot}$$
$$H^{\cdot} + H^{\cdot} \longrightarrow H_2$$

The chemistry of cracking is very complicated and involves not only molecular disintegration but also polymerisation, alkylation, cyclisation, aromatisation etc. A few of these processes have been given below.

Polymerisation: n-Butane, n-butene, iso-butene etc. produced during the cracking undergo polymerisation. Isobutene on dimerisation gives isooctene which on hydrogenation gives isooctane having high octane number.

$$\underset{\text{Isobutene}}{\underset{CH_3}{\overset{CH_3}{>}}C=CH_2} + \underset{CH_3}{\overset{CH_3}{>}}C=CH_2 \xrightarrow{H^+} \underset{\text{Isooctane}}{\underset{CH_3}{\overset{CH_3}{}}CH_3-\underset{CH_3}{\overset{}{C}}-CH=C\overset{CH_3}{<}_{CH_3}} + CH_3-\underset{CH_3}{\overset{}{C}}-CH_2-C\overset{CH_2}{<}_{CH_3}$$

Alkylation: Highly branched hydrocarbons which serve as gasoline are formed during alkylation involving the addition of saturated hydrocarbons to alkenes. Addition of isobutane to isobutene forms isooctane.

$$\underset{\text{Isobutene}}{\underset{CH_3}{\overset{CH_3}{>}}C=CH_2} + \underset{\text{Isobutane}}{\underset{CH_3}{\overset{CH_3}{>}}CH-CH_3} \longrightarrow \underset{\text{Isooctane}}{\underset{CH_3}{\overset{CH_3}{}}CH_3-\underset{CH_3}{\overset{}{C}}-CH_2-\underset{CH_3}{\overset{}{C}H}\overset{CH_3}{<}}$$

Reforming is used to raise the octane number. In this process, the gasoline is heated at a temperature of 600° under pressure (28-50 atm). The pressure is maintained in such a way that the molecule does not crack but is reformed. The chemical changes that occur in reforming are isomerisation, dehydration, cyclisation and aromatisation.

$$\underset{\text{n-Heptane}}{CH_3CH_2CH_2CH_2CH_2CH_2CH_3} \longrightarrow \underset{\text{2-Methylhexane}}{CH_3\overset{\overset{CH_3}{|}}{C}HCH_2CH_2CH_2CH_3}$$

$$\underset{\text{n-Heptane}}{n\text{-}C_7H_{16}} \longrightarrow \underset{\text{Methylcyclohexane}}{\bigcirc\!\!-CH_3} \qquad \underset{\text{Methylcyclohexane}}{\bigcirc\!\!-CH_3} \longrightarrow \underset{\text{Toluene}}{\bigcirc\!\!-CH_3} + 3H_2$$

The effect of reforming is the increase in octane number of the fuel. The reforming carried out in presence of hydrogen is known as *hydro-reforming*. Reforming can also be carried out in presence of catalyst like hondary catalyst (four parts silica, one part alumina and 1% MnO_2) or oxides of chromium, vanadium, molybdenum or aluminum at 150-300° or lead sulphate at 480-550°.

11.8 SYNTHETIC PETROL

The demand for petrol has increased considerably due to its use in automobiles and airplane industry. Therefore successful attempts have been made for developing other synthetic methods. Two methods are used for this purpose.

Bergius Process: In this process, finely powdered coal is hydrogenated in presence of catalyst such as tin and lead to give a mixture of hydrocarbons. During this process the carbon rings in coal undergo fission to give smaller fragments which on hydrogenation give open chain and cyclic hydrocarbons. The mixture of the products obtained is fractionally distilled and give gasoline (b.p. up to 200°) and kerosine (b.p. upto 300°).

Fishcher-Tropsch Process: This method was developed by two German chemists, Franz Fishcher and Hans Tropsch in 1923. Steam is blown over red hot coke to get water gas, a mixture of $CO+H_2$. This water gas is mixed with half its volume of hydrogen and then passed over catalyst at 150-600° under 1-10 atmospheric pressure to give a crude oil which is refined by fractional distillation as in the case of petroleum. The catalyst used in this process is a mixture of cobalt (100 parts), thoria (5 Parts), magnisia (8 Parts) and Kieselguhr (200 parts).

$$X\ CO + Y\ H_2 \longrightarrow \text{Mixture of hydrocarbons} + H_2O$$

11.9 CRITERIA OF A GOOD GASOLINE

The most commonly used fuel for automobile is gasoline. In an internal combustion engine the mixture of gasoline and air burn within the cylinder. The mixture is fired by means of a spark from the spark plug. For best results, the fuel mixture is highly compressed before firing in the cylinder. Some times, the total fuel does not burn smoothly and a sound like that of explosion is observed giving rise to metallic rattle called *Knocking*. This knocking is caused by incomplete combustion of fuel and thus causes loss of power. It has been found that the tendency to knock falls off in the following order with the change in the nature of fuel as follows:

Straight-chain paraffins > Branched chain paraffins > Olefins > Naphthenes > Aromatics

The antiknock quality of gasoline is described by *Octane number*. The 2,2,4-trimethylpentane is considered to be a good fuel and is given an octane number 100. On the other hand, n-heptane, a very poor fuel is given an octane number of zero. Mixture of these two compounds are used to define octane numbers between 0 and 100. Octane number is the percentage of 2,2,4-trimethylpentane present in a mixture of 2,2,4-trimethylpentane and n-heptane which has similar ignition properties as the fuel under examination. For example, a fuel that performs as well as 1:1 mixture of 2,2,4-trimethylpentane and n-heptane has an octane number 50. Commercial gasoline has octane number 81, 74 and 65 for premium, regular and third grade gasoline. Normally a good motor fuel used in modern automobiles has octane number in the range 87-95.It has been found that the branching of the hydrocarbon chain, presence of more unsaturated hydrocarbons compared to saturated hydrocarbons and presence of cycloalkanes increase the octane number. However octane number decreases as the chain length increases.

Certain additives, such as tetraethyl lead [$(C_2H_5)_4Pb$] and tertiary-butyl methyl ether [$(CH_3)_3\ COCH_3$)] are added to gasoline to boost the octane number of gasoline. The use of tetraethyl lead is being curtailed due to environmental reasons. Other methods for raising the octane number of a poor gasoline are 'alkylation', 'reforming', 'hydroforming'; the discussion of these methods is beyond the scope of this book. In diesel engines (whose working is different than that of gasoline engine) fuels having a lower octane number are much more useful than those having a higher octane number. In other words, the straight chain hydrocarbons constitutes a superior fuel than the branched chain hydrocarbons. The quality of diesel fuel is expressed in terms of *Cetane number*.

Cetane (hexadecane, $C_{16}H_{34}$) is considered to be a good fuel and is given a cetane number 100. On the other hand, α-methylnaphthalene is a poor fuel and is given cetane number zero. Cetane number is defined as the percentage of cetane in a mixture of cetane and α-methylnaphthalene which has similar ignition properties as the fuel under examination. Good diesel for modern diesel engine have cetane number greater than 45.

11.10 CRITERIA OF A GOOD KEROSENE OIL

For illuminating purposes, Kerosene oil is used in lamps. The oil used should not be sufficiently volatile at normal temperatures, otherwise its vapors will form an explosive mixture with air. Thus the explosive

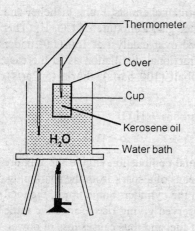

Fig. 11.2:   Abel's Apparatus

nature of kerosene oil is determined by its volatility. The term '*Flash Point*' is used to describe the explosive nature of an oil. It is defined as the minimum temperature at which an oil gives off enough vapors to form an explosive mixture with air.

An oil used in tropical climates like India must be high boiling and the one used in cold climates like England must necessarily be low boiling. In other words, the flash point fixed for hot countries is high (India: 44°C) and that fixed for cold countries is low (England: 22.8°C).

The flash point of an oil is determined by Abel's apparatus. The oil is placed in a small cup and heated slowly by means of a surrounding water bath. From time to time, the cover of the metal cup is opened and a spark introduced momentarily by mean of an automatic device. The temperature at which the vapor ignites with a minute explosion is noted and gives the Flash-point (Fig.11.2).

11.11   PETROCHEMICALS

Numerous organic chemicals are derived from petroleum sources. These are obtained either from natural gas or any other chemical obtained during the refining of crude oil. All such chemicals that are derived from petroleum sources are designated as *petrochemicals*. Petroleum is now a source of vary large number of organic compounds. Approximately one million organic compounds could be synthesised from hydrocarbons derived from natural gas and petroleum fractions. It is not possible to record all the chemicals synthesised. Only a few important ones are given below which may be obtained from petroleum. For example from natural gas by catalytic cracking, pyrolysis, catalytic aromatisation etc and also some of the important products obtained from these hydrocarbons.

Methane, a component of natural gas, refining gases and cracked gas can be chlorinated to give methyl chloride, methylene chloride, chloroform and carbon tetrachloride. Also oxidation of methane under various conditions gives methyl alcohol, formaldehyde and formic acid. Pyrolysis of methane gives acetylene and carbon black. With steam, methane gives a synthetic gas ($CO + H_2$) which is used for hydrogenation.

Ethane is used for the manufacture of ethyl chloride, ethyl alcohol, ethylene and ethylene oxide.

Propane and Butanes are used to manufacture acetone, acetic acid, propylene, butadiene and other alkylated products.

Ethylene is used in the manufacture of ethyl alcohol, ethylene oxide, dichloroethane, vinyl chloride, acetaldehyde.

Propylene and *Butenes* are used for the manufacture of isopropyl alcohol, isopropyl benzene (from propylene) and butadiene and butanol (from butenes).

Acetylene is used for the manufacture of vinyl chloride, vinyl acetate, acetaldehyde and acrylonitrile.

Cyclohexane is derived from gasoline fraction and is used mainly for the preparation of adipic acid by oxidation.

Benzene and Toluene are very useful starting materials for a large number of aromatic compounds. Styrene is used in rubber industry and is obtained from benzene.

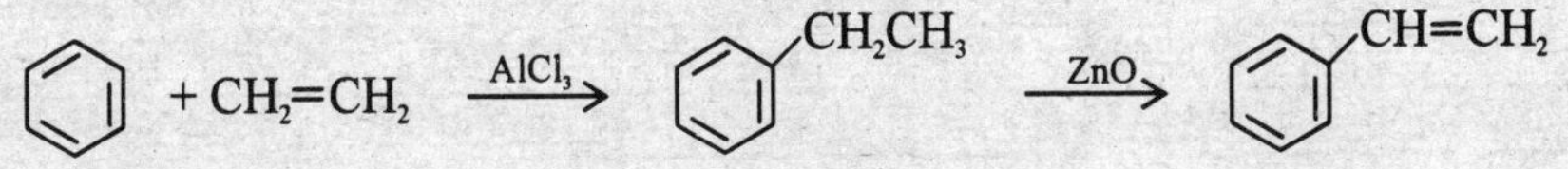

Benzene   Ethene   Ethyl benzene   Styrene

Toluene is used for the preparation of benzyl chloride, benzal chloride, benzotrichloride, benzyl alcohol, benzaldehyde, benzoic acid etc.

o-, and p-Xylenes are used for the manufacture of phthalic acid and terephthalic acid.

A number of chemicals synthesised from petroleum products are used for the manufacture of polymers, plastics, synthetic fibers, synthetic rubbers, synthetic detergents and fertilizers.

11.12   PROBLEMS

1. State which compound has best octane rating in each of the following pairs:
   (a) $CH_3(CH_2)_4CH_3$ and $CH_3CH_2CH_2CH_3$
   (b) $CH_2=CHCH_2CH_3$ and $CH_3CH_2CH_2CH_3$
   (c) $CH_3-\underset{\underset{CH_3}{|}}{CH}-\underset{\underset{CH_3}{|}}{CH}-CH_3$ and $CH_3(CH_2)_4CH_3$
   (d) $CH_3-\underset{\underset{CH_3}{|}}{C}=CH-\underset{\underset{CH_3}{|}}{CH}-CH_3$ and $CH_3-\underset{\underset{CH_3}{|}}{CH}-CH_2-\underset{\underset{CH_3}{|}}{CH}-CH_3$

2. What is petroleum ? Why is it called liquid gold ?
3. What is natural gas? Discuss its applications.
4. Discuss briefly the methods used to find petroleum deposits.
5. How petroleum is formed in nature ?
6. How is petrol obtained synthetically ?
7. What are petrochemicals ? What important chemicals are obtained from various hydrocarbons derived from petroleum ?
8. Write notes on:
   (a) Octane number   (b) Cetane number   (c) LPG   (d) Cracking of petroleum.

12

Halogen Derivatives

12.1 INTRODUCTION

Halogen derivatives are considered to be derived from hydrocarbons by the replacement of one or more hydrogen atom(s) with halogens. These are not found in nature and are purely synthetic compounds. They are used as solvents and intermediates for the synthesis of other organic compounds.

12.2 CLASSIFICATION

The halogen derivatives may be divided into three categories depending upon the nature of carbon residues, to which the halogens are attached.

The Alkyl halides in which the halogen atom is attached to a saturated hydrocarbon chain, are represented as RX, where R is an alkyl group and X is halogen. Some examples are given below:

$$CH_3-Cl$$
Chloromethane (Methyl chloride)

$$CH_3CH_2-Cl$$
Chloroethane (Ethyl chloride)

The Aryl halides are the compounds in which the halogen atom is attached to aromatic ring. For example, C_6H_5Cl is named as chlorobenzene. These will be discussed separately.

The Alkenyl halides are the compounds in which the halogen atom is attached to an unsaturated hydrocarbon chain. Some examples are:

$$CH_2=CHCl$$
Chloroethene (Vinyl Chloride)

$$CH_3-CH=\overset{\overset{\displaystyle Br}{|}}{C}-CH_3$$
2-bromo-2-butene

Depending upon the number of halogens present in the molecule, halogen derivatives may be classified as mono-, di-, tri- and tetrahalogen derivatives as given below:

CH_3Cl	CH_2Cl_2	$CHCl_3$	CCl_4
Methyl chloride	Methylene chloride	Chloroform	Carbon tetrachloride
Monohalogen derivative	Dihalogen derivative	Trihalogen derivative	Tetrahalogen derivative

The halogen derivatives are excellent solvents.

The *monohalogen derivatives or alkyl halides* are further subdivided into primary (1°), RCH_2-X; secondary (2°), R_2CH-X; and tertiary (3°), R_3C-X depending upon the nature of the alkyl group or the position of the halogen atom in the molecule. An example of each type is given below:

CH_3CH_2Cl

Ethyl chloride
primary (1°)

$$\underset{\displaystyle CH_3\underset{\displaystyle \overset{\displaystyle |}{Cl}}{C}HCH_3}{}$$

2-Chloropropane
secondary (2°)

$$CH_3-\underset{\underset{CH_3}{|}}{\overset{\overset{CH_3}{|}}{C}}-Cl$$

2-Chloro-2-methylpropane
tertiary(3°)

12.3   ALKYL HALIDES

The alkyl halides or monohalogen derivatives of paraffins have the general molecular formula $C_nH_{2n+1}X$ or RX, where X may be F, Cl, Br or I.

12.3.1   Nomenclature

The alkyl halides are named depending on the nature of the alkyl group and the halogen present.

CH_3I

Methyl iodide

$$CH_3-\underset{\overset{|}{Cl}}{\overset{\overset{Cl}{|}}{C}H}-CH_3$$

Isopropyl chloride

$$CH_3-\underset{\underset{CH_3}{|}}{\overset{\overset{CH_3}{|}}{C}}-Cl$$

tert-Butyl chloride

   According to IUPAC nomenclature the names of the alkyl halides are obtained by prefixing the halogen (e.g., fluoro, chloro, bromo or iodo) onto the name of the hydrocarbon. In higher members the carbon chain is numbered.

C_2H_5Cl

Chloroethane

$CH_3CH_2CH_2Br$

1-Bromopropane

$$\underset{4\quad\ 3\quad\ 2\quad\ 1}{CH_3-CH_2-\underset{\overset{|}{CH_3}}{\overset{\overset{CH_3}{|}}{C}H}-CH_2-Br}$$

1-Bromo-2-methylbutane

12.3.2   Isomerism

The alkyl halides exhibit chain and position isomerism. For example, butane has four isomeric mono halogen derivatives out of which two illustrate position isomerism and rest the chain isomerism.

$CH_3-CH_2-CH_2-CH_2-Cl$
1-Chlorobutane

$$CH_3-CH_2-\underset{\overset{|}{Cl}}{\overset{\overset{Cl}{|}}{C}H}-CH_3$$
2-Chlorobutane

$\left.\begin{array}{c}\\ \\ \end{array}\right\}$ Position isomers

$$CH_3-\underset{\overset{|}{CH_3}}{\overset{\overset{CH_3}{|}}{C}H}-CH_2Cl$$
1-Chloro-2-methylpropane

$$CH_3-\underset{\underset{CH_3}{|}}{\overset{\overset{CH_3}{|}}{C}}-Cl$$
2-Chloro-2-methylpropane

$\left.\begin{array}{c}\\ \\ \end{array}\right\}$ Chain isomers

12.3.3   Orbital Structure

The orbital structure of alkyl halide is discussed taking methyl chloride as an example. In methyl chloride, the carbon atom is sp^3 hybridised and forms a σ bond by the overlap of sp^3 hybrid orbital of carbon and half filled 3p orbital of chlorine. The C–Cl bond thus formed has a bond length of 1.77Å. The orbital structure of methyl chloride is shown in Fig. 12.1. The rest of the structure (of methyl group) is the same as discussed earlier in Sec. 2.3

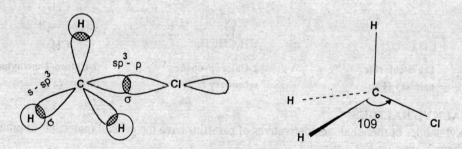

Fig. 12.1: Orbital Structure of Methyl Chloride

The bonds between C–F, C–Br, C–I also are formed by the overlap of sp³ orbital of carbon and 2p orbital of fluorine, 4p orbital of bromine and 5p orbital of iodine respectively. The bond energies and bond lengths of these bonds (Table 12.1) indicates that the C–F bond is the strongest and C–I bond weakest in nature.

Table 12.1:   Nature of Carbon Halogen Bonds

Bond	Bond energy kcal/mol	Bond length(Å)
C–F	105.4	1.42
C–Cl	78.5	1.77
C–Br	65.5	1.90
C–I	57.5	2.10

12.3.4   General Methods of Preparation

(i)   *From Alcohols*

(a)   *Action of Hydrogen Halides:* Alkyl chlorides are obtained from alcohols by the action of concentrated hydrochloric acid in presence of anhydrous zinc chloride as catalyst.

$$R-OH + HCl \xrightarrow[\text{anhydrous}]{ZnCl_2} R-Cl + H_2O$$

Alkyl
Chloride

$$CH_3CH_2OH + HCl \xrightarrow[\text{anhydrous}]{ZnCl_2} CH_3CH_2Cl + H_2O$$

Ethyl chloride

The order of reactivity of different alcohols in this reaction is tertiary > secondary > primary. Thus, tertiary alcohols give good yields of tertiary alkyl chlorides in cold and in the absence of zinc chloride.

$$CH_3-\underset{\underset{CH_3}{|}}{\overset{\overset{CH_3}{|}}{C}}-OH + HCl \xrightarrow{\text{Cold}} CH_3-\underset{\underset{CH_3}{|}}{\overset{\overset{CH_3}{|}}{C}}-Cl + H_2O$$

tert-Butanol   **tert-Butyl chloride**

Alkyl bromides and iodides are prepared by the action of hydrogen bromide or iodide respectively on alcohol.

$$R-OH + HBr\ (HI) \longrightarrow R-Br\ (RI) + H_2O$$

    Alcohol   **Alkylhalide**

(b) *Action of Phosphorus Halides :* Alkyl halides are obtained from alcohols by the action of phosphorus halides.

$$CH_3CH_2OH + PCl_5 \longrightarrow CH_3CH_2Cl + POCl_3 + H_2O$$

$$(CH_3)_2CHOH + PCl_5 \longrightarrow (CH_3)_2CHCl + POCl_3 + H_2O$$

Phosphorus tribromide or phosphorus triiodide required for the preparation of alkyl bromides or iodides are obtained in situ by the action of red phosphorus and bromine or iodine.

$$\overset{\displaystyle OH}{\underset{\displaystyle}{3CH_3CHCH_3}} + PBr_3\,(or\ P + Br_2) \longrightarrow \overset{\displaystyle Br}{\underset{\displaystyle}{3CH_3CHCH_3}} + H_3PO_3$$

Isopropyl alcohol   **Isopropyl bromide**

$$3CH_3CH_2OH + PI_3\,(or\ P + I_2) \longrightarrow 3CH_3CH_2I + H_3PO_3$$

Ethyl alcohol   **Ethyl iodide**

(c) *Action of Thionyl Chloride:* Alcohols react with thionyl chloride in presence of pyridine (mild base) to form alkyl chlorides.

$$R-OH + SOCl_2 \xrightarrow{\text{Pyridine}} R-Cl + SO_2 + HCl$$

          Thionyl   **Alkyl**

          chloride   **chloride**

$$CH_3CH_2CH_2OH + SOCl_2 \xrightarrow{\text{Pyridine}} CH_3CH_2CH_2Cl + SO_2 + HCl$$

n-Propanol   **n-Propyl chloride**

(ii) *From Alkenes and Alkynes:* Alkenes and alkynes on treatment with hydrogen halides in dark at room temperature form alkyl halides.

$$RCH=CH_2 + HX \longrightarrow \overset{\displaystyle X}{\underset{\displaystyle}{R-CH-CH_3}}$$

Alkene   **Alkyl halide**

$$CH_3CH_2CH=CH_2 + HBr \longrightarrow \overset{\displaystyle Br}{\underset{\displaystyle}{CH_3-CH_2-CH-CH_3}}$$

1-Butene   **2-Bromobutane**

The addition takes place as per Markovnikov's rule. However in presence of peroxides, anti-Markovnikov addition takes place.

$$\overset{\displaystyle CH_3}{\underset{\displaystyle}{CH_3-C=CH_2}} \xrightarrow[\text{Peroxide}]{\text{HBr}} \overset{\displaystyle CH_3}{\underset{\displaystyle}{CH_3-CH-CH_2Br}}$$

Isobutene   **Isobutyl bromide**

Alkynes on reaction with hydrogen halides form vinyl halides and then geminal dihalides.

$$H-C\equiv C-H + HBr \longrightarrow CH_2=CHBr \xrightarrow{\text{HBr}} CH_3-CHBr_2$$

Acetylene   **Vinyl bromide**   **1,1-dibromoethane**

For detailed discussion and reaction mechanism see Section 9.9.

(iii) *From Carboxylic Acids:* The method involves the treatment of silver salt of carboxylic acid with bromine in carbon tetrachloride. It is known as *Hunsdiecker Reaction* and is useful for the preparation of long chain alkyl bromides from fatty acids.

$$R-CH_2-COOAg + Br_2 \xrightarrow[\Delta]{CCl_4} R-CH_2-Br + CO_2 + AgBr$$

Silver salt of   Alkyl bromide
Carboxylic acid

The reaction proceeds by a free radical mechanism.

$$RCOO^-Ag^+ \xrightarrow[-AgBr]{Br_2} RCOOBr \longrightarrow RCOO^{\cdot} + Br^{\cdot}$$

                                                    Carboxylate   Bromine
                                                    radical   radical

$$RCOO^{\cdot} \longrightarrow R^{\cdot} + CO_2$$
$$R^{\cdot} + RCOOBr \longrightarrow R-Br + RCOO^{\cdot}$$

(iv) *Halide Exchange:* This is a good method for preparing alkyl iodides and fluorides, since these can not be prepared by direct iodination or fluorination of alkanes. This method involves the reaction of the alkyl chlorides or bromides with sodium iodide in acetone (or methanol) and is known as *Finkelstein reaction.*

$$RCH_2Cl + NaI \xrightarrow{Acetone} RCH_2I + NaCl$$

Similarly alkyl fluorides may be obtained by treating alkyl chlorides or bromides with inorganic fluorides.

$$2CH_3Cl + Hg_2F_2 \longrightarrow 2CH_3F + Hg_2Cl_2$$

                   Mercurous   Methyl   Mercurous
                   fluoride   fluoride   chloride

This method finds application in the preparation of Freon-12.

$$3CCl_4 + 2SbF_3 \longrightarrow 3CCl_2F_2 + 2SbCl_3$$

Carbon   Antimony   Dichlorodifl-
tetrachloride   fluoride   uoromethane
                                           (Freon-12)

It will be appropriate to mention that the chlorofluorocarbons (CFC) also known as Freons are inert, nontoxic gases used as refrigerant in air conditioners and refrigerators. Freon-12 is the most commonly used refrigerant. The freons have a damaging effect on environment. They catalyse the decomposition of ozone and destroy the protective layer that surrounds the earth. For this reason most of the countries in the world have banned the use of Freons.

(v) *From Alkanes by Direct Halogenation:* This method has been discussed in detail in Section 7.10.

12.3.5 Physical Properties

The lower alkyl halides are gases, whereas alkyl halides upto C_{18} are liquids. The alkyl halides are colorless when pure and have pleasant sweetish odor. Alkyl halides beyond C_{18} are colorless solids. They are generally insoluble in water but are soluble in alcohol, ether and benzene.

Alkyl chlorides and fluorides are lighter than water. However, alkyl iodides and bromides are heavier than water. The boiling points and densities of alkyl halides show a regular pattern in the order iodide > bromide > chloride > fluoride.

12.3.6 Spectral Properties

The ultraviolet - visible spectra of alkyl halides show a weak absorption band between 170-258 nm due to the presence of loosely held non bonding electrons of halogen. These electrons undergo n-π* transition. [C–Cl = 173 nm; C–Br = 240 nm and C–I = 258 nm]

The infrared spectra of alkyl halides show C–X stretching frequency depending on the nature of the halogen present. Thus, for C–F bond, 1100 -1000 cm^{-1}; C–Cl bond 750-700 cm^{-1}; C–Br bond 600-500 cm^{-1} and C–I bond 500 cm^{-1} are the characteristic absorptions.

The nuclear magnetic resonance spectra of alkyl halides exhibit a downfield chemical shift of the α–protons (deshielding) due to the marked electron attracting nature of the halogens. The order of deshielding, resulting in lowering of the chemical shift, is in the order of electronagativity, i.e., I < Br < Cl < F. Thus the δ value of the methyl protons in CH_3I is 2.17; CH_3Br is 2.65; CH_3Cl is 3.2; and CH_3F is 4.3.

12.3.7 Chemical Properties

Alkyl halides are reactive compounds and give following reactions.

(i) *Substitution Reactions:* The alkyl halides undergo *Nucleophilic aliphatic substitution* as the carbon halogen bond in alkyl halides is polar.The weakly basic halide ion, being a good leaving group is replaced by nucleophiles easily. Methyl bromide on reaction with sodium hydroxide forms methanol.

$$CH_3Br + HO^- \longrightarrow CH_3OH + Br^-$$
Methyl   Methanol
bromide

The order of reactivity of alkyl halides increases from fluoride to iodide. Thus, R–F< R–Cl< R–Br< R–I towards substitutions.

The S_N2 reaction: A reaction between ethyl bromide and the hydroxide ion to yield ethanol follows second order kinetics i.e. the rate of the reaction is dependent upon the concentration of substrate and nucleophile both and is called bimolecular nucleophilic substitution.

$$CH_3CH_2-Br + HO^- \longrightarrow CH_3CH_2-OH + Br^-$$
Rate = k[CH_3CH_2Br] [OH^-]

The *Mechanism* proposed is a one step concerted mechanism in which the attack of nucleophile and the displacement of the halide ion takes place simultaneously. The reaction passes through a transition state in which the C–OH bond is half formed and the C–X bond is half broken or in the transition state the carbon is attached to five odd groups as represented below:

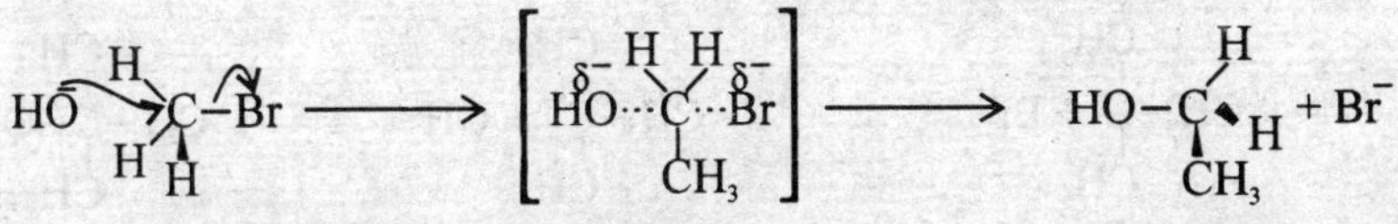

Transition state

Primary alkyl halides are more reactive towards bimolecular nucleophilic substitution than secondary and tertiary alkyl halides. The reason is that, the larger groups block the approach of the nucleophile or the position of attack is sterically hindered. The order of reactivity amongst alkyl halides is:

$$CH_3-Br > CH_3CH_2-Br > (CH_3)_2CH-Br > (CH_3)_3C-Br$$

As seen, in S_N2 reaction, the attack of the nucleophile takes place from the opposite side to that of the leaving group, thus the reaction proceeds via *complete inversion of configuration*. When laevorotatory 2-bromooctane reacts with sodium hydroxide via S_N2 it forms dextrorotatory 2-octanol.

$$H-\underset{CH_3}{\overset{C_6H_{13}}{C}}-Br \xrightarrow[S_N2]{NaOH} HO-\underset{CH_3}{\overset{C_6H_{13}}{C}}-H$$

(–)-2-Bromooctane   (+)-2-Octanol
$[\alpha]= -39.6°$ $[\alpha]= +10.3°$

The S_N2 reactions are favored by a concentrated solution of strong base and in a nonpolar solvent. The transition state here has dispersed charges thus bonding with the solvent is weak compared to the reactants. The polar solvent, thus, stabilises the reactants rather than the transition state and this is the reason that S_N2 reaction takes place faster in aprotic solvents like dimethylsulphoxide(DMSO), Dimethylformamide(DMF) etc.

$$HO^- + R-X \longrightarrow [\overset{\delta-}{HO}\cdots R\cdots \overset{\delta-}{X}] \longrightarrow HO-R + X^-$$

Reactants   **Transition state**   **Product**
Concentrated charge   **(Dispersed charge)**
Stabilised by solvent

The S_N1 Reaction: A typical S_N1 reaction is the hydrolysis of tertiary butyl bromide with aqueous sodium hydroxide. The reaction is found to be of first order meaning thereby that the rate of this reaction is proportional to the concentration of only one reacting species, i.e., tertiary butyl bromide and is independent of the concentration of the nucleophile or hydroxide ion.

$$CH_3-\underset{CH_3}{\overset{CH_3}{C}}-Br + HO^- \longrightarrow CH_3-\underset{CH_3}{\overset{CH_3}{C}}-OH + Br^- \qquad \text{Rate = k [tert Butyl bromide]}$$

tert-Butyl bromide   **tert-Butanol**

The Mechanism: A two step mechanism has been proposed in which the first step (rate determining step) is the ionisation of an alkyl halide by heterolytic bond cleavage to form a carbocation intermediate. In the second step, the nucleophile may approach the central carbon atom (having the positive charge) from either side with equal probability. The carbocation has a planer geometry, thus the nucleophile may engage the empty p orbital from either side of the molecule as represented below:

$$CH_3-\underset{CH_3}{\overset{CH_3}{C}}-Br \xrightarrow{slow} CH_3-\underset{CH_3}{\overset{CH_3}{C^+}} + Br^- \qquad CH_3-\underset{CH_3}{\overset{CH_3}{C^+}} + OH^- \xrightarrow{fast} CH_3-\underset{CH_3}{\overset{CH_3}{C}}-OH$$

Step I   **Step II**

Thus, the reaction of an optically active alkyl halide with sodium hydroxide according to S_N1 gives racemic substitution products along with some inverted product. *The S_N1 reaction is associated with recemisation and inversion.* Laevorotatory 2-bromooctane on hydrolysis via S_N1 forms 90% recemic 2-octanol and rest dextrorotatory 2-octanol.

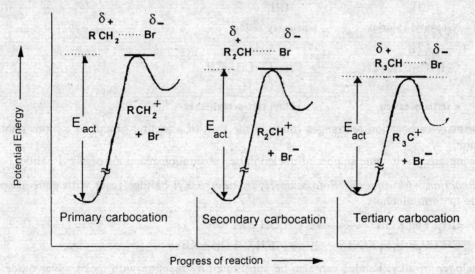

In S_N1 reaction the relative order of reactivities of alkyl halides follows the following order.

$3° > 2° > 1°$

The reactivity is dependent upon the stability of carbocations. The stable the carbocation, easily it is formed (low E_{act}) and faster it reacts with the nucleophile to form the product. Fig. 12.2 shows the stability order of primary secondary and tertiary carbocations.

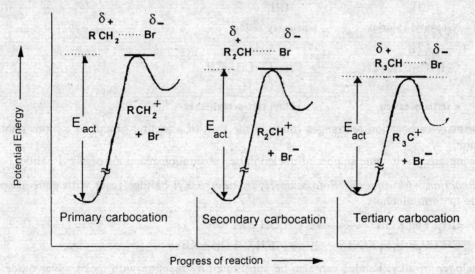

Fig. 12.2: **Stability of 1°, 2° and 3° Carbocations**

However, the order of reactivity of alkyl halides increases from fluoride to iodide as in the case of S_N2 reaction.

The S_N1 reaction is favored by a dilute solution of weak base as the rate determining step is independent of the nature of base. The reaction is faster in polar solvents as the transition state involves ions which are stabilised by ion dipole interaction or solvolysis more than the reactants as shown below:

$$R-X \longrightarrow [R^+\cdots X^-] \longrightarrow R^+ + X^-$$

Reactant   **Transition state**   **Product**

More polar than reactant

Stabilised by solvolysis

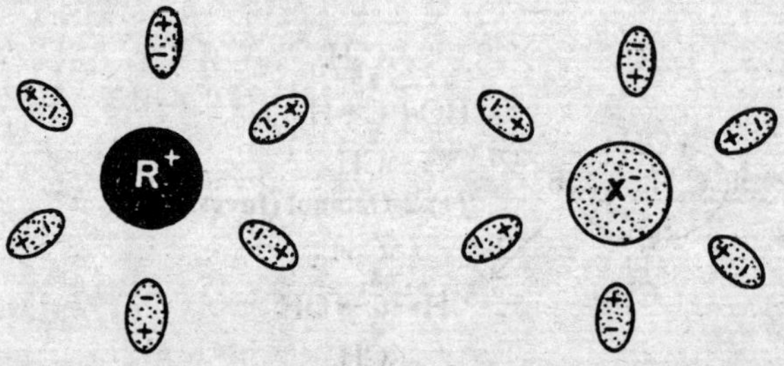

Fig. 12.3 Ion-dipole Interaction : Solvated Carbocation and Anion

As discussed in Section 4.4.1, *S$_N$1 reaction is associated with rearrangements* as the intermediate carbocation may rearrange to a more stable carbocation. Neopentyl chloride on treatment with ethanol forms ethyl tert-pentyl ether.

$$\underset{\overset{|}{CH_3}}{\overset{\overset{CH_3}{|}}{CH_3-C-CH_2-Cl}} \longrightarrow \underset{\overset{|}{CH_3}}{\overset{\overset{CH_3}{|}}{CH_3-\overset{+}{C}-CH_2}} \xrightarrow[\text{1,2-Methyl shift}]{\text{Rearrangement}}$$

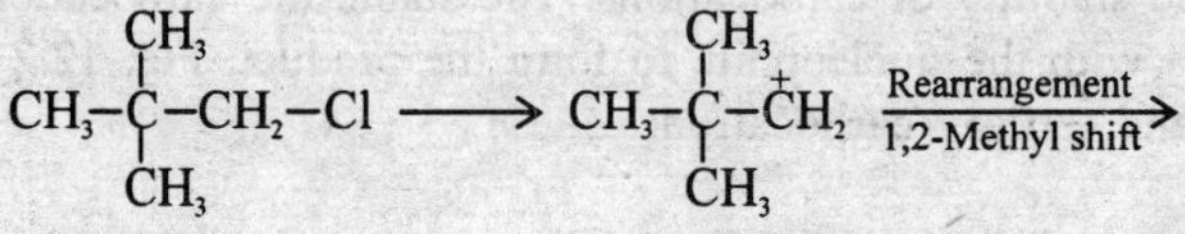

Neopentyl chloride   **a primary cation**

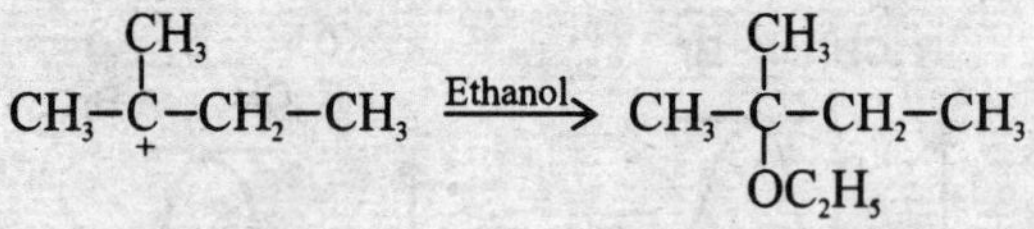

$$\underset{\overset{|}{CH_3}}{\overset{\overset{CH_3}{|}}{CH_3-\underset{+}{C}-CH_2-CH_3}} \xrightarrow{\text{Ethanol}} \underset{\overset{|}{OC_2H_5}}{\overset{\overset{CH_3}{|}}{CH_3-C-CH_2-CH_3}}$$

a tertiary cation   **Ethyl tert-pentyl ether**

The primary carbocation rearranges (through the shift of a methyl group) to a more stable tertiary carbocation.

Following are given some important nucleophilic substitution reactions of alkyl halides.

(a) *Reaction with Aqueous Potassium Hydroxide:* Alkyl halides react with aqueous potassium hydroxide to form alcohols.

$$CH_3I + aq.KOH \longrightarrow CH_3OH + KI$$

$$CH_3CH_2Br + aq.KOH \longrightarrow CH_3CH_2OH + KBr$$

The halogen of alkyl halides can also be substituted by reaction with moist silver oxide.

$$Ag_2O + H_2O \longrightarrow 2AgOH$$

$$CH_3CH_2Br + AgOH \longrightarrow CH_3CH_2OH + AgBr$$

Ethyl bromide   **Ethyl alcohol**

(b) *Reaction with Sodium Alkoxides:* Alkyl halides on treatment with metal alkoxides produce ethers.

$$CH_3CH_2\overset{-}{O}\overset{+}{Na} + CH_3CH_2I \xrightarrow{S_N2} CH_3CH_2OCH_2CH_3 + NaI$$

Sodium ethoxide Ethyl iodide   **Diethyl ether**

This is called *Williamson Synthesis*
Ethers can also be obtained by heating alkyl halides with dry silver oxide.

$$2CH_3I + Ag_2O \longrightarrow CH_3OCH_3 + 2AgI$$

(c) *Reaction with Sodium Cyanide:* Alkyl halides react with sodium cyanide in a suitable solvent to give alkyl cyanides or alkyl nitriles.

$$NaCN \longrightarrow Na^+ + CN^-$$

$$\overset{\frown}{CN} + CH_3CH_2\overset{\frown}{-}Br \xrightarrow{S_N2} CH_3CH_2CN + Br^-$$
Ethyl cyanide

Alkyl cyanides are useful starting materials for the synthesis of acids, amides and amines.

$$R-C\equiv N + 2H_2O \xrightarrow{dil\ acid} R-COOH + NH_4Cl$$
Acid

$$R-C\equiv N + H_2O \xrightarrow{alk.\ H_2O_2} R-CONH_2$$
Amide

$$R-C\equiv N + 4H \xrightarrow{LiAlH_4} RCH_2NH_2$$
Primary amine

However, alkyl halides react with silver cyanide in aqueous ethanolic solution to give alkyl isocyanide as the major product.

$$CH_3I + AgCN \longrightarrow CH_3NC + AgI$$
| Methyl | Silver | Methyl isocyanide |
| iodide | cyanide | (Major product) |

(d) *Reaction with Potassium Hydrogen Sulphide:* Alkyl halides react with potassium hydrogen sulphide in alcoholic solution to give thioalcohols.

$$K^+SH^- + CH_3CH_2-I \xrightarrow{S_N1} CH_3CH_2SH + KI$$
Potassium   **Ethanethiol**
hydrogen
sulphide

However, alkyl halides react with potassium sulphide or a hot alcoholic solution of a mercaptide to give thioethers.

$$2C_2H_5I + K_2S \longrightarrow C_2H_5-S-C_2H_5 + 2KI$$
Dielhyl thioether

$$C_2H_5-\bar{S}-Na^+ + CH_3CH_2-Br \longrightarrow C_2H_5-S-CH_2CH_3 + NaBr$$
Dielhyl thioether

(e) *Reaction with Silver Salt of Carboxylic Acid:* Alkyl halides on heating with an alcoholic solution of the salt of a carboxylic acid give an ester.

$$CH_3-\overset{\overset{\displaystyle O}{\|}}{C}-\bar{O}Ag^+ + CH_3CH_2-Br \xrightarrow[\Delta]{ethanol} CH_3CH_2O-\overset{\overset{\displaystyle O}{\|}}{C}-CH_3 + AgBr$$
Silver acetate   **Ethyl acetate**

(f) *Reaction with Ammonia:* Alkyl halides on heating with alcoholic solution of ammonia under pressure (in a sealed tube) give primary amines. If excess of alkyl halide is used the reaction continues to form secondary amines, tertiary amines and finally quaternary ammonium salts.

$$\ddot{N}H_3 + CH_3CH_2-Br \xrightarrow[\Delta]{\text{ethanol}} CH_3CH_2\ddot{N}H_2 + HBr$$
Ethyl amine

$$CH_3CH_2 \ddot{N}H_2 + CH_3CH_2Br \longrightarrow (CH_3CH_2)_2\ddot{N}H + HBr$$
Diethyl amine

$$(CH_3CH_2)_2 \ddot{N}H + CH_3CH_2Br \longrightarrow (CH_3CH_2)_3\ddot{N} + HBr$$
Triethyl amine

$$(CH_3CH_2)_3 \ddot{N} + CH_3CH_2Br \longrightarrow (CH_3CH_2)_4\overset{+}{N}\overline{Br}$$
Tetraethyl ammonium bromide
(Quaternary ammonium salt)

The reaction is also known as *Hofmann Ammonolysis.*

(g) *Reaction with Silver Nitrite:* Alkyl halides react with silver nitrite to form the nitroparaffins.

$$CH_3CH_2I + AgNO_2 \longrightarrow CH_3CH_2NO_2 + AgI$$
Ethyl iodide   **Nitroethane**

However, the reaction of alkyl halides with sodium or potassium nitrite gave the corresponding alkyl nitrites as the attack on alkyl halide takes place through oxygen atom of nitrite group. This is due to the difference in the electronegativities of sodium and silver.

$$NaNO_2 \equiv \overset{+}{N}a[:\overset{\cdot\cdot}{\underset{\cdot\cdot}{O}}-\ddot{N}=\underset{\cdot\cdot}{O}:] \longleftrightarrow \overset{+}{N}a[\overset{\cdot\cdot}{\underset{\cdot\cdot}{O}}=\ddot{N}-\overset{\cdot\cdot}{\underset{\cdot\cdot}{O}}:]$$

$$Na^+:\overset{\cdot\cdot}{\underset{\cdot\cdot}{O}}-\ddot{N}=\ddot{O} + CH_3-CH_2-I \longrightarrow CH_3CH_2-\ddot{O}-N=\ddot{O}$$
Ethyl nitrite

$$A\overset{+}{g}-:\overset{\cdot\cdot}{\underset{\cdot\cdot}{O}}-\ddot{N}=\ddot{O} + CH_3-CH_2-I \longrightarrow CH_3CH_2-\overset{+}{N}\overset{\ddot{O}:}{\underset{\overset{-}{\underset{\cdot\cdot}{O}}:}{}}$$
Nitroethane

(h) *Reaction with sodium alkylides:* Alkyl halides react with sodium alkylides to form higher alkynes.

$$R-C\equiv\overset{-}{C} \overset{+}{Na} + CH_3-Br \longrightarrow R-C\equiv C-CH_3 + NaBr$$
Sodium alkylide

$$H-C\equiv\overset{-}{C} \overset{+}{Na} + CH_3-Br \longrightarrow HC\equiv C-CH_3 + NaBr$$
Sodium acetylide   **Propyne**

(ii) *Elimination Reaction (Dehydrohalogenation):* We have seen that alkyl halides on treatment with aqueous potassium hydroxide form alcohols and on treatment with sodium ethoxide form ethers. However, under appropriate conditions such as use of a strong base and high temperature, dehydrohalogenation occurs to form alkenes as the main product.

The 1,2-elimination reactions of alkyl halides like the nucleophilic substitution can proceed either by a first order or a second order mechanism as discussed in Section 9.6.

The tertiary and to some extent secondary alkyl halides undergo E1 reaction in a dilute solution of weak base to form an alkene.

$$CH_3-\underset{\underset{CH_3}{|}}{\overset{\overset{CH_3}{|}}{C}}-Br \xrightarrow[\Delta]{\overset{E1}{CH_3CH_2OH}} CH_3-\underset{}{\overset{\overset{CH_3}{|}}{C}}=CH_2$$

tert-Butyl bromide   **Isobutene**

The E2 reaction is generally carried out by heating the alkyl halide with potassium hydroxide or sodium ethoxide in ethanol.

$$CH_3-\underset{\underset{}{}}{\overset{\overset{Br}{|}}{CH}}-CH_3 + CH_3CH_2\overset{-}{O}\overset{+}{N}a \xrightarrow[\Delta]{\overset{E2}{CH_3CH_2OH}} CH_3CH=CH_2 + CH_3CH_2OH + Br^-$$

Isopropyl bromide   **Propene**   **Ethanol**

It has already been stated that elimination and substitution reactions are competitive, i.e., one reaction occurs at the cost of another. The course of the reaction depends on the structure of the alkyl halide and the nature of the base.

It is found that branching in alkyl halides increases the rate of E2 reaction and decreases the rate of S_N2 reaction. Thus the rate of reaction follows the order given below:

$$(CH_3)_3C-X > (CH_3)_2CH-X > CH_3CH_2-X > CH_3-X \text{ for E2}$$
$$CH_3-X > CH_3CH_2-X > (CH_3)_2CH-X > (CH_3)_3C-X \text{ for } S_N2$$

In case of unsymmetrical alkyl halide, for example, in secondary butyl chloride, the course of the elimination is determined by *Saytzeff rule*, i.e., hydrogen is eliminated preferentially from the carbon atom which has less number of hydrogen atoms and thus the highly substituted alkene is the major product.

$$CH_3CH_2-\underset{\underset{}{}}{\overset{\overset{Cl}{|}}{CH}}-CH_3 \xrightarrow{\text{alk. KOH}} CH_3CH=CHCH_3 + CH_3CH_2CH=CH_2$$

2-Chlorobutane   **2-Butene**   **1-Butene**
                                 80% yield   **20% yield**

(iii) *Reduction of Alkyl Halides:* Alkyl halides on reduction with Zn-Cu couple and alcohol, zinc and hydrochloric acid or with lithium aluminium hydride give alkanes.

$$CH_3CH_2-Br \xrightarrow[\substack{\text{or} \\ \text{Zn/HCl or} \\ \text{Li AlH}_4}]{Zn-Cu/C_2H_5OH} CH_3CH_3 + HBr$$

Ethyl   **Ethane**
bromide

However, alkyl iodides can be reduced with red phosphorus and hydriodic acid.

$$R-I + HI \xrightarrow[150°]{\text{red P}} R-H + I_2$$

Alkyl iodide   **Alkane**

Alkanes can be obtained from alkyl halides by catalytic hydrogenation. This is a good method for obtaining alkanes.

$$R-X + H_2 \xrightarrow[\text{Pd/alk-solution}]{\text{Hydrogenation}} R-H + HX$$

Alkane

(iv) *Reaction with Metals:* Alkyl halides react with metals like magnesium, lithium and sodium to form metal alkyls.

(a) *With Magnesium:* Alkyl halides form Grignard reagents called alkylmagnesium halides on reaction with magnesium in ether solution.

$$CH_3I + Mg \xrightarrow{\text{Ether}} CH_3MgI$$

Methylmagnesium iodide

$$\underset{\substack{|\\ Br}}{CH_3CH_2-CH-CH_3} \xrightarrow[\text{Mg}]{\text{Ether}} \underset{\substack{|\\ CH_3}}{CH_3CH_2-CHMgBr}$$

2-Bromobutane   **sec-Butylmagnesium iodide**

For a detailed discussion see Section 13.2.

(b) *With Lithium:* Lithium alkyls are formed. These are more reactive than Grignard reagents (see Section 13.3).

$$\underset{\substack{|\\ Br}}{CH_3-CH-CH_3} + 2Li \longrightarrow \underset{\substack{|\\ Li}}{CH_3-CH-CH_3} + LiBr$$

Isopropyl lithium

(c) *With Sodium:* A coupling reaction, known as Wurtz reaction takes place to produce a symmetrical alkane.

$$2CH_3CH_2-Cl + 2Na \longrightarrow CH_3CH_2CH_2CH_3 + 2NaCl$$

Ethyl chloride   **n-Butane**

12.3.8   Individual Members

12.3.8.1   Methyl Iodide/Iodomethane [CH₃I]

Methyl iodide is a sweet smelling liquid (b.p. 42.5°, specific gravity 2.27). It is mostly used as a methylating agent in the laboratory for organic synthesis. It is obtained on a commercial scale as well as in the laboratory by the following two methods.

1. From methyl alcohol by heating with red phosphorus and iodine.

$$6CH_3OH + 2P + 3I_2 \longrightarrow 6CH_3I + 2H_3PO_3$$

Methyl   **Methyl**
alcohol   **iodide**

2. From dimethyl sulphate by reaction with potassium iodide solution in presence of calcium carbonate.

$$(CH_3)_2SO_4 + 2KI \xrightarrow{CaCO_3} 2CH_3I + K_2SO_4$$

Dimethyl
sulphate

12.3.8.2   Ethyl Chloride/Chloroethane [C₂H₅Cl]

Ethyl chloride is a pleasant smelling liquid (b.p. 12.5°). It is used as a local anesthetic, as a refrigerant and in the preparation of ethyl cellulose, tetraethyl lead and as a ethylating agent. It is obtained as follows.

1. Industrial Preparation: By chlorination of ethane.

$$C_2H_6 + Cl_2 \xrightarrow{400°} C_2H_5Cl + HCl$$

Ethane   Ethyl
chloride

2. From ethylene by addition of hydrogen chloride in presence of a catalyst.

$$CH_2=CH_2 + HCl \xrightarrow{CaCO_3} CH_3CH_2Cl$$

Ethylene   Ethyl chloride

12.3.8.3 Ethyl Iodide/Iodoethane [C_2H_5I]

Ethyl iodide is a colorless, sweet smelling liquid and is an important reagent for organic synthesis. It is obtained from ethyl alcohol by the action of red phosphorus and iodine.

$$P_4 + 6I_2 \longrightarrow 4PI_3$$
$$3C_2H_5OH + PI_3 \longrightarrow 3C_2H_5I + H_3PO_3$$

12.4 DIHALOGEN DERIVATIVES

12.4.1 Introduction

The dihalogen derivatives of paraffins have the general molecular formula $C_nH_{2n}X_2$ and are obtained by replacing two hydrogen atoms of a hydrocarbon with halogens. The common dihalogen derivatives which can be obtained from methane are dichloromethane (methylene chloride), CH_2Cl_2; dibromomethane (methylene bromide), CH_2Br_2 and diiodomethane (methylene iodide), CH_2I_2.

In higher members, the two halogen atoms may be linked to the different or same carbon atoms giving rise to vicinal or geminal dihalides. Thus, in case of ethane the dihalogen derivatives are:

ClCH$_2$CH$_2$Cl   CH$_3$CHCl$_2$
1,2- or vic-Dichloroethane   **1,1- or gem-Dichloroethane**

The other dihalogen derivatives are named depending on the location of the halogen in the carbon chain.

Br–CH$_2$CH$_2$CH$_2$Br   BrCH$_2$CH$_2$CH$_2$CH$_2$Br
1,3-Dibromopropane   **1,4-Dibromobutane**

12.4.2 General Methods of Preparation

(i) *From Aldehydes and Ketones:* Geminal dichlorides or dibromides are obtained by the action of phosphorus pentahalides on aldehydes or ketones.

$$CH_3CHO + PCl_5 \longrightarrow CH_3CHCl_2 + POCl_3$$
$$\text{1,1–Dichloroethane}$$
$$CH_3COCH_3 + PCl_5 \longrightarrow CH_3CCl_2CH_3 + POCl_3$$
$$\text{2,2-Dichloropropane}$$

(ii) *From Alkenes and Alkynes:* Alkenes on treatment with halogens give vicinal dihalo compounds.

$$CH_2=CH_2 + Br_2 \longrightarrow BrCH_2-CH_2Br$$
Ethene   **1,2–Dibromoethane**
$$CH_3CH=CH_2 + Br_2 \longrightarrow CH_3CHBrCH_2Br$$
Propene   **1,2–Dibromopropane**

Alkynes, on the other hand, react with halogen acids to give geminal dihalo compounds.

$$HC \equiv CH \xrightarrow{\text{HBr}} CH_2=CHBr \xrightarrow{\text{HBr}} CH_3CHBr_2$$

Acetylene   Vinyl bromide   Ethylidene bromide

(iii) *From Monohalogen and Trihalogen Derivatives:* The monohalogen derivatives, for example, methyl chloride on chlorination give methylene chloride.

$$CH_3Cl + Cl_2 \longrightarrow CH_2Cl_2 + HCl$$

Methyl   Methylene
chloride   chloride

The trihalogen derivatives on reduction give the dihalogen derivatives.

$$CHCl_3 + 2H \longrightarrow CH_2Cl_2 + HCl$$

Dichloromethane

(iv) *From Glycols:* The glycols on reaction with hydrogen halides or phosphorus halides yield 1,2-dihalides.

$$\begin{array}{l} CH_2OH \\ | \\ CH_2OH \end{array} + HCl \longrightarrow \begin{array}{l} CH_2Cl \\ | \\ CH_2Cl \end{array}$$

Ethylene   Ethylene chloride
glycol

12.4.3   Properties

The dihalogen compounds are colorless and sweet smelling. Their boiling points are comparatively higher than the mono halogen derivatives. The higher members are solids. They are insoluble in water but soluble in organic solvents. They are generally heavier than water.

The dihalides give the usual reactions of monohalides. The reaction takes place with both the halogen atoms. Besides the usual reaction, they give the following typical reactions.

(i) *Hydrolysis with Aqueous Alkali:* On hydrolysis with aqueous alkali the 1,2-dihalides give the corresponding glycols.

$$\begin{array}{l} CH_2Cl \\ | \\ CH_2Cl \end{array} + 2KOH \longrightarrow \begin{array}{l} CH_2OH \\ | \\ CH_2OH \end{array} + 2KCl$$

Ethylene chloride   Ethylene glycol

However, the gem dihalides give the corresponding dihydroxy compounds, which are unstable and eliminate a molecule of water to give an aldehyde or a ketone.

$$CH_3CHCl_2 + 2KOH \longrightarrow 2KCl + \left[CH_3CH{\overset{\displaystyle OH}{\underset{\displaystyle OH}{}}} \right] \xrightarrow{-H_2O} CH_3CHO$$

Ethylidene   . Unstable   Acetaldehyde
chloride

$$\begin{array}{l} CH_3 \\ C \\ CH_3 \end{array}{\overset{Cl}{\underset{Cl}{}}} + 2KOH \longrightarrow 2KCl + \left[\begin{array}{l} CH_3 \\ C \\ CH_3 \end{array}{\overset{OH}{\underset{OH}{}}} \right] \xrightarrow{-H_2O} \begin{array}{l} CH_3 \\ C=O \\ CH_3 \end{array}$$

Isopropylidene   Unstable   Acetone
chloride

(ii) *Dehydrohalogenation:* The 1,1- and 1,2-dihalides on treatment with *alcoholic potassium hydroxide* undergo dehydrohalogenation and give *alkynes.*

$$CH_3CHCl_2 + 2KOH \longrightarrow CH\equiv CH + 2KCl + H_2O$$
Ethylidene   **Acetylene**
chloride

$$CH_3CHBrCH_2Br + 2KOH \longrightarrow CH_3C\equiv CH + 2KCl + H_2O$$
1, 2-Dibromopropane   **Propyne**

(iii) *Dehalogenation:* The 1,2–dihalides on treatment with zinc dust and alcohol undergo dehalogenation to give alkenes.

$$CH_3CHBrCH_2Br + Zn \xrightarrow{\text{ethanol}} CH_3CH=CH_2 + ZnBr_2$$
1,2–Dibromopropane   **Propene**

Other dihalides in which the two halogens are separated by 3 to 6 carbon atoms give cyclic compounds.

$$BrCH_2CH_2CH_2Br + Zn \xrightarrow{\text{ethanol}} H_2C\overset{\displaystyle CH_2}{-}CH_2 + ZnBr_2$$
1, 3–Dibromopropane   **Cyclopropane**

12.4.4 Methylene Chloride [CH_2Cl_2]

It is the most important member of the dihalogen compounds. It is obtained by direct halogenation of methane. It is used as an industrial solvent.

12.5 TRIHALOGEN DERIVATIVES

12.5.1 Introduction

These are prepared by replacement of three hydrogen atoms of a hydrocarbon by halogens . The most important trihalogen derivatives are chloroform (trichloromethane), $CHCl_3$; bromoform (tribromomethane), $CHBr_3$ and iodoform (triiodomethane), CHI_3.

12.5.2 Chloroform/Trichloromethane [$CHCl_3$]

Preparation

(i) *From Alcohol or Acetone:* Chloroform is prepared by heating ethyl alcohol with bleaching powder. It is first oxidised to acetaldehyde with the chlorine available from bleaching powder followed by the reaction of acetaldehyde with chlorine to form trichloroacetaldehyde (chloral). Finally, chloral on hydrolysis with lime present in bleaching powder forms chloroform and calcium formate.

$$CH_3CH_2OH + Cl_2 \xrightarrow{[O]} CH_3CHO + 2HCl$$
                         Acetaldehyde

$$CH_3CHO + 3Cl_2 \xrightarrow{\text{chlorination}} CCl_3CHO + 3HCl$$
                        Chloral

$$CCl_3CHO + H_2O \xrightarrow{\text{Hydrolysis}} CHCl_3 + HCOOH$$
                     Chloroform Formic acid

$$2HCOOH + Ca(OH)_2 \longrightarrow (HCOO)_2Ca + 2H_2O$$
                        Calcium formate

In case, acetone is used as starting material, the reaction takes place in two steps as shown below

$$CH_3COCH_3 + 3Cl_2 \longrightarrow CCl_3COCH_3 + 3HCl$$
Acetone   Trichloro acetone

$$CCl_3COCH_3 + HOH \longrightarrow CHCl_3 + CH_3COOH$$
                       Chloroform   Acetic acid

(ii) *From Carbon Tetrachloride:* It forms chloroform when reduced by iron and water.

$$CCl_4 + H_2 \xrightarrow{Fe/H_2O} CHCl_3 + HCl$$
Carbon   Chloroform
tetrachloride

This method is used for industrial preparation of chloroform.

Properties: Chloroform is colorless liquid (b.p. 61°) having a sweet sickly odor and taste. It is insoluble in water but is soluble in organic solvents. Chloroform vapors when inhaled cause temporary unconsciousness and hence is used as anesthetic.

(i) *Oxidation:* In presence of light, chloroform is oxidised slowly by the oxygen of air to form phosgene.

$$CHCl_3 + \tfrac{1}{2}O_2 \longrightarrow Cl-\overset{\overset{\displaystyle O}{\|}}{C}-Cl + HCl$$
Chloroform   Phosgene

It is stored in dark bottles to prevent the formation of phosgene, a highly poisonous substance.

(ii) *Reduction:* Chloroform can be reduced by nascent hydrogen generated by the action of zinc on ethanolic hydrogen chloride to give dichloromethane (methylene chloride).

$$CHCl_3 + 2H \xrightarrow{Zn-alcohol} CH_2Cl_2 + HCl$$
Chloroform   Methylene
                             chloride

(iii) *Hydrolysis:* On treatment with hot sodium hydroxide solution, chloroform undergoes hydrolysis to form sodium chloride and sodium formate.

$$CHCl_3 + 4NaOH \longrightarrow HCOONa + 3NaCl + 2H_2O$$
Chloroform   Sodium formate

The mechanism proposed for the hydrolysis of chloroform involves the formation of dichlorocarbene

$$CHCl_3 + HO^- \rightleftharpoons :\bar{C}Cl_3 + H_2O$$
                (Carbanion)

$$:\bar{C}Cl_3 \longrightarrow :CCl_2 + Cl^-$$
               Dichloro-
               carbene

$$:CCl_2 + H-O-H \longrightarrow H-O-CHCl_2 \xrightarrow{-HCl} H-\overset{\overset{\displaystyle O}{\|}}{C}-Cl$$

$$\xrightarrow{H_2O} H-\underset{OH}{\overset{H-O}{\underset{|}{C}}}-Cl \xrightarrow{-Cl^-,-H} H-\overset{\overset{\displaystyle O}{\|}}{C}-OH \xrightarrow[-H_2O]{OH} H-\overset{\overset{\displaystyle O}{\|}}{C}-O^-$$
                                                            Formate anion

(iv) *Chlorination:* Chloroform on chlorination yields carbon tetrachloride.

$$CHCl_3 + Cl_2 \longrightarrow CCl_4 + HCl$$

Chloroform   Carbon
                  tetrachloride

(v) *Nitration:* On reaction with concentrated nitric acid, chloroform undergoes nitration to form chloropicrin or mononitro chloroform, which is used as an insecticide and also as a war gas.

$$CHCl_3 + HONO_2 \longrightarrow CCl_3NO_2 + H_2O$$

Chloroform   Nitric acid   Chloropicrin

(vi) *Carbylamine Reaction:* Chloroform on warming with primary amine and ethanolic potassium hydroxide solution forms isonitrile (carbylamine) having characteristic unpleasant smell. This is a very sensitive test for primary amine.

$$CHCl_3 + 3KOH + R-NH_2 \longrightarrow R-\overset{+}{N}=\bar{C}: + 3 KCl + 3 H_2O$$

Chloroform   Primary amine   Isonitrile

The *Mechanism* of the reaction involves the formation of dichlorocarbene.

$$CHCl_3 + HO^- \xrightarrow[-H_2O]{} :\bar{C}Cl_3 \longrightarrow :CCl_2 + Cl^-$$

                                        Dichlorocarbene

$$R-NH_2 + :CCl_2 \longrightarrow R-NH-CH-Cl_2$$

$$R-NH-CH-Cl_2 \xrightarrow{OH^-} R-\overset{+}{N}=\bar{C} + 2Cl^- + H_2O$$

                           Alkyl
                         isonitrile

(vii) *Dehalogenation:* Chloroform on warming with silver powder gives acetylene.

$$HCCl_3 + 6Ag + Cl_3CH \longrightarrow H-C\equiv C-H + 6AgCl$$

Chloroform   Acetylene

(viii) *Condensation with Acetone:* Chloroform on heating with acetone in presence of alkali undergoes condensation to give chloretone, which is used as a hypnotic.

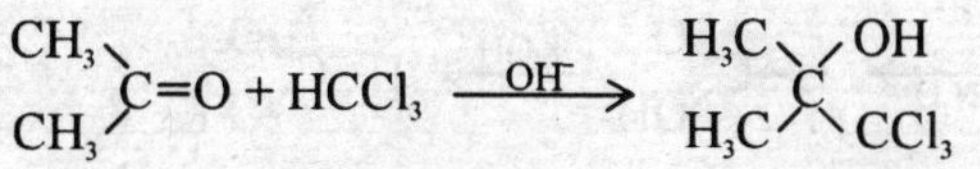

Acetone   Chloroform   Chloretone

(ix) *Reimer Tiemann Reaction:* On heating the concentrated solution of chloroform and phenol, salicylaldehyde is obtained.

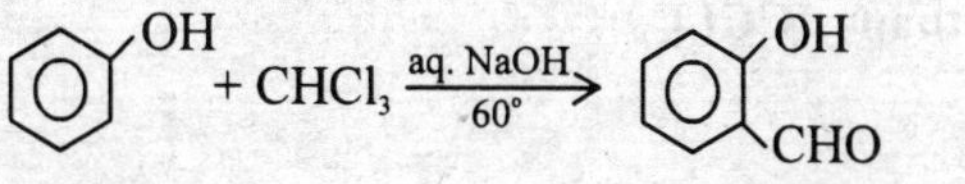

Phenol   Salicylaldehyde

This reaction has been discussed in Section 34.5.

12.5.3 Iodoform/Triiodomethane [CHI$_3$]

Preparation

(i) *From Ethanol or Acetone:* Ethanol or acetone by the action of iodine and alkali form iodoform.

The mechanism is similar to that described under chloroform.

$$CH_3CH_2OH + I_2 \xrightarrow{[O]} CH_3CHO + 2HI$$

Ethanol   Acetaldehyde

$$CH_3CHO + 3I_2 \longrightarrow CI_3CHO + 3HI$$

Acetaldehyde   Triiodoacetaldehyde

$$CI_3CHO + KOH \longrightarrow CHI_3 + HCOOK$$

Triiodoacetaldehyde   Iodoform

In place of potassium hydroxide, sodium carbonate can also be used.

$$CH_3CH_2OH + 4I_2 + 3Na_2CO_3 \longrightarrow CHI_3 + HCOONa + 5NaI + 2H_2O + CO_2$$

Iodoform

With acetone, the reaction takes place in the following way

$$2CH_3COCH_3 + 3I_2 \longrightarrow 2CI_3COCH_3 + 6HI$$

Acetone   Triiodoacetone

$$2CI_3COCH_3 + 2Na_2CO_3 \longrightarrow CHI_3 + CH_3COONa + 3NaI + 2CO_2 + H_2O$$

Triiodoacetone   Iodoform

(ii) *Electrolysis:* Electrolysis of an aqueous solution of alcohol or acetone containing sodium carbonate and sodium or potassium iodide produces iodoform. The solution temperature is maintained at 60-70° and a current of carbon dioxide is passed to neutralise the sodium hydroxide produced as a result of electrolysis. Iodine and sodium carbonate obtained during electrolysis react with ethanol or acetone to give iodoform (as given above). The method is used for industrial preparation of iodoform. Ethyl alcohol gives better yields.

Properties: Iodoform is a yellow crystalline solid, m.p. 119°. It has a characteristic unpleasant smell. It is insoluble in water but soluble in alcohol, chloroform and ether. It has a marked antiseptic action.

It resembles chloroform in its chemical properties. It is, however, less stable than chloroform. On heating with alkali, iodoform is hydrolysed to form alkali formate.

$$CHI_3 + 3NaOH \longrightarrow \left[HC{\overset{\displaystyle OH}{\underset{\displaystyle OH}{-OH}}} \right] \xrightarrow{-H_2O} H-C{\overset{\displaystyle O}{\underset{\displaystyle OH}{\diagup\diagdown}}} \xrightarrow{KOH} H-C{\overset{\displaystyle O}{\underset{\displaystyle O^-K^+}{\diagup\diagdown}}}$$

Iodoform   Formic acid   Potassium formate

12.6   TETRAHALOGEN DERIVATIVES

12.6.1   Carbon Tetrachloride/Tetrachloromethane [CCl₄]

Preparation

(i) *By Chlorination of Methane:* It is prepared on industrial scale by the chlorination of methane

$$CH_4 + 4Cl_2 \longrightarrow CCl_4 + 4HCl$$

(ii) *By the action of chlorine* on carbon disulphide in presence of anhydrous aluminum chloride as catalyst.

$$CS_2 + 3Cl_2 \xrightarrow{AlCl_3} CCl_4 + S_2Cl_2$$
$$\text{Sulphur monochloride}$$

Carbon tetrachloride is separated by fractional distillation from the above reaction mixture.

Properties: Carbon tetrachloride is a colorless liquid, b.p. 77°. It is insoluble in water but soluble in all organic solvents. It is a good solvent for fatty substances and is used as a fire extinguisher. It is a relatively inert substance. It gives the following reactions.

(i) *Action of Steam:* Carbon tetrachloride reacts with steam at high temperature to give phosgene gas.

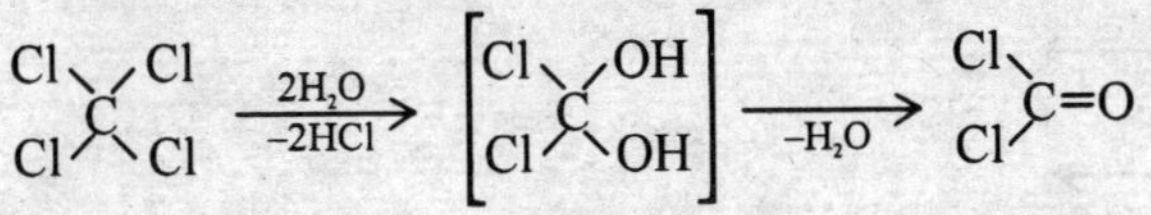

Carbon tetrachloride **Unstable** **Phosgene**

(ii) *Hydrolysis:* On heating with alcoholic potassium hydroxide, it gives potassium carbonate.

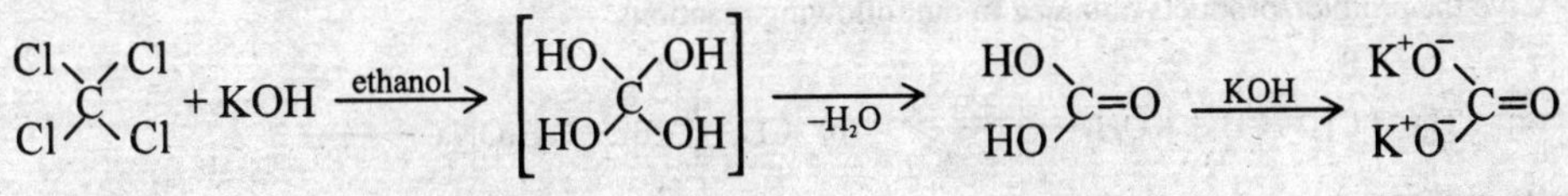

Carbonic acid **Potassium carbonate**

(iii) *Reduction:* Carbon tetrachloride can be reduced by moist iron filings to chloroform. This is an industrial method for the preparation of chloroform.

$$CCl_4 + H_2 \longrightarrow CHCl_3 + HCl$$

(iv) *Reaction with Hydrogen fluoride:* When hydrogen fluoride is passed into carbon tetrachloride in presence of antimony pentachloride, dichlorodifluoromethane (freon) is obtained.

$$CCl_4 + 2HF \xrightarrow{SbCl_5} CCl_2F_2 + 2HCl$$
$$\text{Freon}$$

Freon has been widely used for refrigeration purposes.

12.7 PROBLEMS

1. What are alkyl halides ? Give IUPAC names of various monochlorobutanes.
2. Describe the mechanism of S_N1 and S_N2 reactions in the case of alkyl halides.
3. What are Elimination reactions ? Discuss different types of elimination reactions taking one example of each.
4. How is chloroform obtained ? Discuss the mechanism involved in its preparation from acetone and bleaching powder.
5. Write notes on: (a) Finkelstein Reaction; (b) Hunsdiecker Reaction and (c) Saytzeff Rule
6. Classify each of the following alkyl halides as 1°, 2° or 3°.

(a) CH_3CHCH_2Br (b) 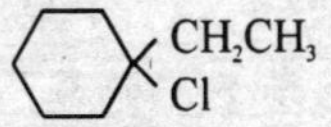 (c)

7. Give the preparation of:
   (a) $C_6H_5CHBrCH_3$ from $C_6H_5CH_2CH_3$ (c) 1-Bromopropane from propene
   (b) $CH_3CHBrCH_3$ from $CH_3CHOHCH_3$ (d) 2-Chloropropane from 2-propanol
8. Which of the members in following pairs will undergo faster S_N2 reaction.

(a) 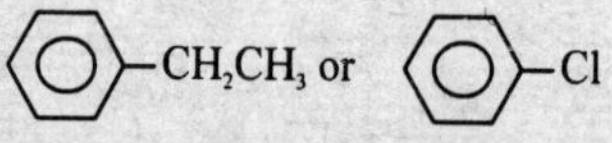 $\langle\bigcirc\rangle-CH_2CH_3$ or $\langle\bigcirc\rangle-Cl$ (b) $CH_3-\overset{\overset{\displaystyle CH_3}{|}}{\underset{\underset{\displaystyle CH_3}{|}}{C}}-Cl$ or $(CH_3)_2 \, CHCH_2Cl$

(c) $CH_2=CHCH_2Cl$ or $CH_3CH_2CH_2Cl$

9. What product / products will be obtained in the following reaction ?

$$CH_3-CH_2-\overset{\overset{\displaystyle Br}{|}}{\underset{\underset{\displaystyle Br}{|}}{C}}-CH_3 \ \xrightarrow[C_2H_5OH]{C_2H_5ONa}$$

10. Complete the following reactions:

   (a) $CH_3CH_2CH_2Br + NaOH$ (aq) $\longrightarrow$

   (b) $C_6H_5CH_2Cl + H_2O \longrightarrow$

   (c) $CH_3CH_2CH_2CH_2Br + NaSH \longrightarrow$

11. Complete the equations given below:

   (a) $CH_3CH_2CH_2CH_2Br + NaOC_2H_5 \longrightarrow$

   (b) $C_6H_5CH_2Cl + AgNO_3 + H_2O \longrightarrow$

   (c) $C_6H_5Cl + AgNO_3 \longrightarrow$

12. Give the product/products obtained in the following reactions.

   (a) $CH_3CH_2\overset{\overset{\displaystyle Br}{|}}{C}H-CH_3 + KOH \xrightarrow{\ alcohol\ }$ (b) $CH_3-\overset{\overset{\displaystyle CH_3}{|}}{\underset{\underset{\displaystyle CH_3}{|}}{C}}-Br + C_2H_5ONa \longrightarrow$

13. What happens when

   (a) Ethyl bromide is treated with alcoholic KOH ?   (b) Chloroform is treated with KOH ?

14. Discuss the orbital structure of methyl chloride.

15. Starting with alkyl halides, how will you obtain thioalcohols, esters and amines ?

16. How is methyl iodide obtained ?

17. How will you prepare iodoform ?

13

Organometallic Compounds

13.1 INTRODUCTION

An organic compound in which a metal is directly linked to a carbon is known as organometallic compound

$$-\overset{|}{\underset{|}{C}}-Metal \qquad or \qquad -\overset{|}{\underset{|}{C}}-M$$

where M = Mg, Li, Pb, Zn, Na etc.

Names of the organometallic compounds are derived by suffixing the name of the metal to that of the organic group attached to the metal.

CH$_3$MgI	CH$_3$CH$_2$MgI	Zn(C$_2$H$_5$)$_2$	Pb(C$_2$H$_5$)$_4$	CH$_3$Li
Methyl	**Ethyl**	**Diethyl zinc**	**Lead tetraethyl**	**Methyl**
magnesium iodide	**magnesium iodide**			**lithium**

The organometallic compounds may be covalent or ionic:

$$-\overset{|}{\underset{|}{C}}-M \qquad\qquad -\overset{|}{\underset{|}{C}}:^- M^+$$

Covalent **Ionic**

The ionic character of carbon metal bond depends on the nature of the metal (or the difference between the electronegativity of carbon and metal). The ionic character of carbon metal bond decreases in the order:

$$Na > Li > Mg > Al > Zn > Cd > Hg.$$

Discussion on the organometallic compounds is restricted to Grignard reagents and alkyl lithium.

13.2 GRIGNARD REAGENTS

The organomagnesium halides are known as *Grignard Reagents*. These were discovered by Victor Grignard, hence named Grignard reagents. Victor Grignard demonstrated tremendous synthetic potential of Grignard reagents and was awarded Nobel Prize (1912) for his discovery.

The Grignard reagents are represented as R—Mg—X, where R=Alkyl, alkenyl, alkynyl or aryl group and X=Cl, Br or I.

13.2.1 Preparation

Grignard reagents are prepared in the laboratory by the action of alkyl halide on magnesium metal in dry ether.

$$R-X + Mg \xrightarrow[\text{ether}]{\text{reflux}} R-Mg-X$$

Alkyl halide　　　　**Alkylmagnesium halide**

$$CH_3-I + Mg \xrightarrow[\text{ether}]{\text{reflux}} CH_3-Mg-I$$

Methylmagnesium iodide

$$C_2H_5-I + Mg \xrightarrow[\text{ether}]{\text{reflux}} C_2H_5-Mg-I$$

Ethylmagnesium iodide

For an alkyl halide, the ease of formation of Grignard reagent is in the order $RI > RBr > RCl$. However, for a given halogen the reactivity order of alkyl group is $CH_3 > C_2H_5 > C_3H_7$ or in other words with the increase in carbon atoms Grignard reagent formation becomes difficult. The reaction is performed with absolutely dry reagents and under anhydrous conditions. Use of a catalyst (e.g. iodine) starts the initial sluggish reaction. When the reaction is complete as indicated by a clear solution, a solution of Grignard reagent in ether is obtained. This is then treated in situ with various substances to get the required product.

Mechanism: The reaction most likely occurs via free radical mechanism.

$$2 R-X + Mg \longrightarrow R-R + MgX_2$$

$$Mg + MgX_2 \longrightarrow 2 \overset{\bullet}{M}gX$$

$$R-X + \overset{\bullet}{M}gX \longrightarrow \overset{\bullet}{R} + MgX_2$$

$$\overset{\bullet}{R} + \overset{\bullet}{M}gX \longrightarrow R-Mg-X$$

This mechanism finds support by the fact that addition of catalytic amount of iodine enhances considerably the formation of Grignard reagent. The iodine reacts with magnesium to give magnesium iodide.

$$Mg + I_2 \longrightarrow MgI_2$$

13.2.2　Structure

The formula of Grignard reagent was suggested as the loose molecular complex $R_2Mg.MgX_2$ (Jolibois, 1912). The following equilibrium (Schlenk, 1929) has been suggested for Grignard reagent.

$$2RMgX \rightleftharpoons R_2Mg + MgX_2$$

It has been found that an equimolar mixture of dialkylmagnesium and magnesium halide behaved exactly as a Grignard reagent. This can be explained by the dissolution of the Joliboi's complex $R_2Mg.MgX_2$.

$$R_2Mg.MgX_2 \rightleftharpoons R_2Mg + MgX_2$$

It is suggested that RMgX is dimerised and solvated in ether and the following equilibria are possible (Ubbelohde, 1955).

$$R_2Mg.MgX_2 \rightleftharpoons R_2Mg + MgX_2 \rightleftharpoons 2RMgX$$

Grignard reagent in ether can exist in either of the two forms given below.

$$R-Mg-X \xrightarrow[\text{ether}]{\text{Solvolysis}}$$

Monomer　　　　**Dimer**

In subsequent discussions the formula RMgX will be used to designate a Grignard reagent.

13.2.3 Properties

Grignard reagents are colorless solids. They are generally not isolated in the free state since they are explosive in nature. For synthetic purposes they are used in ether solution.

Grignard reagents are very reactive compounds since carbon-magnesium bond is strongly polarised thereby making the carbon atom nucleophilic. The most characteristic reactions of Grignard reagents are the nucleophilic substitution and nucleophilic addition reactions with a variety of organic compounds.

13.2.4 Synthetic Applications

Grignard reagents are used for the synthesis of a variety of organic compounds.

(i) *Synthesis of Hydrocarbons:* Grignard reagents form alkanes when reacted with compounds containing active hydrogen, for example, water, alcohol, ammonia and amines.

$$CH_3MgI + H-OH \longrightarrow CH_4 + Mg^+(OH)I^-$$

Methyl-
magnesium-
iodide   **Water**

$$CH_3MgI + R-OH \longrightarrow CH_4 + Mg^+(OR)I^-$$
                Alcohol

$$CH_3MgI + NH_3 \longrightarrow CH_4 + Mg^+(NH_2)I^-$$
                Ammonia

$$CH_3MgI + R-NH_2 \longrightarrow CH_4 + Mg^+(NHR)I^-$$
                Amine

This reaction also forms the basis for the *Zerevitinoff determination* of active hydrogens. Thus by measuring the amount of methane liberated from a known weight of the compound containing active hydrogen, the number of active hydrogens present can be estimated.

Alkanes are also obtained by the reaction of Grignard reagents with saturated alkyl halides.

$$\overset{\delta^-}{C}H_3-\overset{\delta^+}{Mg}I + CH_3\overset{\delta^+}{C}H_2-\overset{\delta^-}{Br} \longrightarrow CH_3CH_2CH_3 + Mg\begin{smallmatrix}Br\\I\end{smallmatrix}$$

Alkenes are also obtained by the reaction of Grignard reagents with unsaturated alkyl halides.

$$\overset{\delta^-}{C}H_3-\overset{\delta^+}{Mg}I + CH_2=CH\overset{\delta^+}{C}H_2-\overset{\delta^-}{Br} \longrightarrow CH_2=CHCH_2CH_3 + MgI(Br)$$
        Allyl bromide   **1-Butene**

Higher Alkynes are obtained by the treatment of terminal alkynes (viz. propyne, $CH_3C\equiv C-H$) with the Grignard reagent. The resulting Grignard reagent ($CH_3C\equiv CMgI$) on treatment with alkyl halide forms higher alkynes.

$$CH_3-C\equiv C-H + CH_3MgI \longrightarrow CH_3-C\equiv C-MgI \xrightarrow{CH_3I} CH_3-C\equiv C-CH_3 + MgI_2$$
Propyne   **Propynylmagnesium**   **2-Butyne**
                               iodide

(ii) *Synthesis of Alcohols: Primary Alcohols* are obtained by the reaction of Grignard reagents with methanal (formaldehyde) followed by hydrolysis with dilute acids.

$$HCHO + CH_3MgI \longrightarrow \underset{\text{Adduct}}{\overset{H}{\underset{H}{>}}C\overset{CH_3}{\underset{OMgI}{<}}} \xrightarrow[H^+]{H_2O} \underset{\text{Ethanol}}{CH_3CH_2OH} + Mg(I)OH$$

Methanal

Primary alcohols are also obtained by the reaction of Grignard reagents with *epoxides* followed by acid hydrolysis of the formed adduct.

$$\overset{\delta^-}{CH_3}-\overset{\delta^+}{MgI} + \underset{CH_2}{\overset{CH_2}{|}}O \longrightarrow \begin{bmatrix} H_2C-\overset{-}{O}Mg^+I \\ | \\ H_2C-CH_3 \end{bmatrix} \longrightarrow \underset{H_2C-CH_3}{\overset{H_2C-OH}{|}} + Mg(OH)I$$

Ethylene oxide   **Adduct**   **Primary alcohol**
(epoxide)   **(n-propyl alcohol)**

Alternatively, treatment of Grignard reagent with dry *oxygen* followed by subsequent acid hydrolysis also gives primary alcohols.

$$CH_3MgI + O_2 \longrightarrow CH_3O_2MgI \xrightarrow{CH_3MgI} 2CH_3\overset{-}{O}Mg^+I \xrightarrow{H^+} 2CH_3OH + 2Mg(OH)I$$

Secondary alcohols are obtained by treatment of the Grignard reagent with an *aldehyde* (other than methanal, which gives primary alcohols) followed by decomposition of the formed adduct with dilute acid.

$$CH_3CHO + \overset{\delta^-}{CH_3}\overset{\delta^+}{MgI} \longrightarrow \begin{bmatrix} H \\ | \\ H_3C-C-\overset{-}{O}Mg^+I \\ | \\ CH_3 \end{bmatrix} \xrightarrow[H^+]{H_2O} \begin{matrix} H \\ | \\ H_3C-C-OH \\ | \\ CH_3 \end{matrix} + Mg(OH)I$$

Acetaldehyde   **Isopropyl alcohol**

Tertiary alcohols are obtained on treatment of the Grignard reagent with a *ketone* and subsequent hydrolysis of the adduct.

$$\underset{H_3C}{\overset{H_3C}{>}}C=O + H_3\overset{\delta^-}{C}\overset{\delta^+}{MgI} \longrightarrow \begin{bmatrix} H_3C & \overset{-}{O}Mg^+I \\ & C \\ H_3C & CH_3 \end{bmatrix} \xrightarrow{H_2O} \underset{H_3C}{\overset{H_3C}{>}}C\overset{CH_3}{\underset{OH}{<}} + Mg(OH)I$$

Acetone   **tert-Butyl alcohol**

(iii) *Synthesis of Carboxylic Acids:* Carboxylic acids are obtained by the reaction of a Grignard reagent with carbon dioxide followed by hydrolysis of the adduct with dilute acid.

$$H_3CMgI + O=C=O \longrightarrow \begin{bmatrix} O=C-\overset{-}{O}Mg^+I \\ | \\ CH_3 \end{bmatrix} \xrightarrow[HCl]{H^+} H_3CCOOH + Mg(I)Cl$$

Acetic acid

(iv) *Synthesis of Aldehydes:* Grignard reagents on reaction with ethylformate, ethyl orthoformate and hydrogen cyanide followed by hydrolysis of the adduct give aldehydes.

$$H_3CMgI + HC \overset{O}{\underset{OC_2H_5}{\Big<}} \longrightarrow \left[\begin{matrix} H \\ H_3C \end{matrix} C \overset{OMg^+I}{\underset{OC_2H_5}{\Big<}} \right] \xrightarrow{HOH} CH_3-C \overset{O}{\underset{H}{\Big<}} + Mg(OC_2H_5)Cl$$

Ethyl formate   **Acetaldehyde**

In case excess of Grignard reagent is present, alcohol may be obtained by subsequent reaction.

$$H_3C-MgI + H-C \overset{OC_2H_5}{\underset{OC_2H_5}{\overset{|}{-}OC_2H_5}} \longrightarrow H_3CCH(OC_2H_5)_2 + Mg(OC_2H_5)I$$

Ethyl orthoformate   **Acetaldehyde acetal**

$$H_3CCH(OC_2H_5)_2 \xrightarrow{H^+} H_3CCHO + 2C_2H_5OH$$

Acetaldehyde

This is a convenient method of making aldehydes as in this case the formation of secondary alcohol is prevented by the formation of an acetal. Aldehydes are also obtained by the reaction of Grignard reagent with hydrogen cyanide.

$$H_3CMgI + H-C \equiv N \longrightarrow H-\overset{CH_3}{\underset{}{\overset{|}{C}}}=\bar{N}\overset{+}{M}gI \xrightarrow[-Mg(OH)I]{H^+/H_2O} \left[H-\overset{CH_3}{\underset{}{\overset{|}{C}}}=NH \right] \xrightarrow[-NH_3]{H^+/H_2O} H-\overset{CH_3}{\underset{}{\overset{|}{C}}}=O$$

Acetaldehyde

(v)   *Synthesis of Ketones:* Grignard reagents on reaction with alkyl cyanides followed by decomposition of the adduct with dilute acid give ketones.

$$H_3C-C \equiv N + H_3CMgI \longrightarrow \overset{H_3C}{\underset{H_3C}{\Big>}}C=NMgI \xrightarrow{H^+/H_2O} \overset{H_3C}{\underset{H_3C}{\Big>}}C=O + NH_3$$

Methyl cyanide   **Adduct**   **Acetone**

Alternatively the action of acid chlorides on Grignard reagents also give ketones.

$$H_3C-C \overset{O}{\underset{Cl}{\Big<}} + H_3CMgI \longrightarrow \left[\overset{H_3C}{\underset{H_3C}{\Big>}}C \overset{O-MgI}{\underset{Cl}{\Big<}} \right] \xrightarrow{HOH} \overset{H_3C}{\underset{H_3C}{\Big>}}C=O + Mg(Cl)I$$

Acetone

(vi)   *Synthesis of Ethers:* Grignard reagents react with lower halogenated ethers to produce higher ethers.

$$H_3C-H_2C-MgI + CH_3OCH_2-Cl \longrightarrow H_3C-H_2C-H_2COCH_3$$

Methyl propyl ether

(vii)   *Synthesis of Alkyl Cyanides:* The reaction of Grignard reagents with cyanogen or cyanogen chloride yields alkyl cyanides.

$$H_3CMgI + NC-CN \longrightarrow H_3CC \equiv N \quad + Mg(CN)I$$

Methyl cyanide

$$\overset{\delta^- \;\; \delta^+}{H_3CMgI} + NC-Cl \longrightarrow H_3CC\equiv N \quad + Mg(Cl)I$$

**Cyanogen
chloride**   **Methyl cyanide**

(viii) *Synthesis of Primary Amines:* Grignard reagents react with chloramines to give primary amines.

$$\overset{\delta^- \;\; \delta^+}{H_3CMgI} + H_2N-Cl \longrightarrow H_3CNH_2 \;\; + Mg(Cl)I$$

Chloramine   **Methylamine**

This method is very useful for the preparation of primary amines containing a branched hydrocarbon chain. For example,

$$\begin{matrix} H_3C \\ \diagdown \\ \diagup \\ H_3C \end{matrix} CH-CH_2-MgBr + H_2N-Cl \longrightarrow \begin{matrix} H_3C \\ \diagdown \\ \diagup \\ H_3C \end{matrix} CH-CH_2-NH_2 + Mg(Cl)Br$$

**2-Methylpropyl   Chloramine   2-Methylpropylamine
magnesium bromide**

(ix) *Synthesis of Thioalcohols:* Thioalcohols are obtained by the action of sulphur on Grignard reagents followed by hydrolysis of the adduct.

$$RMgX + S \longrightarrow Mg\overset{SR}{\underset{X}{\diagup}} \xrightarrow{H_2O} RSH \quad + Mg(OH)X$$

Thioalcohol

(x) *Synthesis of Dithionic Acids:* The reaction of Grignard reagents with carbon disulphide followed by hydrolysis of the adduct give dithionic acids (compare the formation of carboxylic acids with carbon dioxide).

$$\overset{\delta^- \;\; \delta^+}{H_3CMgI} + S=C=S \longrightarrow S=\overset{\overset{CH_3}{\|}}{C}-S^-Mg^+I \xrightarrow{H_2O} H_3C-\overset{\overset{S}{\|}}{C}-SH + Mg(OH)I$$

Dithioacetic acid

(xi) *Synthesis of Organometallic Compounds:* The reaction of inorganic halides with Grignard reagents give organometallic compounds.

$$4C_2H_5MgBr + \quad 2PbCl_2 \longrightarrow (C_2H_5)_4Pb + 4Mg(Br)Cl$$

**Ethylmagnesium   Lead   Tetraethyl lead
bromide   chloride**

$$2C_2H_5MgI + HgCl_2 \longrightarrow (C_2H_5)_2Hg + 2MgI(Cl)$$

**Mercuric   Diethyl mercury
chloride**

$$4CH_3MgI \quad + SiCl_4 \longrightarrow (CH_3)_4Si \quad + 4MgI(Cl)$$

**Silicon   Tetramethyl
tetrachloride   silane**

$$3CH_3MgBr + PCl_3 \longrightarrow (CH_3)_3P + 3Mg(Cl)Br$$

**Phosphorus   Trimethyl
trichloride   phosphine**

(xii) *Synthesis of Alkyl Halides:* The reaction of Grignard reagent with iodine or bromine gives corresponding alkyl halide.

$$CH_3MgCl + I_2 \longrightarrow CH_3I + Mg(I)Cl$$

Limitations of Grignard Reagents

(i) The Grignard reagent is difficult to handle as it reacts with even moisture, air and carbon dioxide.

(ii) Due to its reactivity Grignard reagent can not be prepared from alkyl halides having $-COOH$, $-OH$, $-NH_2$, $-C=O$, $-CN$, $-COOR$, $-SO_3H$, $-NO_2$ groups.

(iii) When reacted with polyfunctional compounds the Grignard reagents gives a mixture of products.

13.3   ORGANOLITHIUM COMPOUNDS

13.3.1   Introduction

The organolithium compounds have a bond between carbon and lithium.

CH_3-Li	CH_3CH_2-Li
Methyllithium	**Ethyllithium**

They behave in the same way as Grignard reagents but have the additional advantage of exhibiting enhanced reactivity. Therefore, they are extensively used in organic synthesis.

13.3.2   Preparation

Organolithium compounds are obtained by the reaction of alkyl halides on metallic lithium in an inert solvent.

$$CH_3Br \quad + 2Li \xrightarrow[-10°, N_2]{\text{dry ether}} CH_3-Li \quad + LiBr$$

Methyl   **Methyllithium**
bromide

$$CH_3CH_2CH_2CH_2Br +2Li \xrightarrow[-10°, N_2]{\text{dry ether}} CH_3CH_2CH_2CH_2-Li + LiBr$$

n-Butyl bromide   **n-Butyllithium**

$$C_6H_5Br + 2Li \xrightarrow{\text{dry ether}} C_6H_5Li + LiBr$$

Bromo-   **Phenyl-**
benzene   **lithium**

Hydrocarbons serve as better solvent for the preparation of alkyllithium compounds as the use of ether normally gives a Wurtz reaction product and also lithium alkyls attack ethers as follows.

$$CH_3CH_2OCH_2CH_3 +Bu-Li \longrightarrow BuH + CH_3CH-LiOCH_2CH_3 \longrightarrow CH_2=CH_2 + CH_3CH_2OLi$$

The desired organolithium compounds may also be prepared by halogen metal exchange reaction.

$$Bu-Li + R-X \longrightarrow R-Li + Bu-X$$

13.3.3   Properties

The $C-Li$ bond has greater ionic character (43%) in comparison to $C-Mg$ bond (35%) in Grignard reagents. This accounts for the increased reactivity of organolithium compounds over Grignard reagents. However, organolithium compounds are particularly sensitive towards air and moisture.

Highly sterically hindered tertiary alcohols can be prepared from lithiumalkyls whereas Grignard reagent does not form tert alcohols in good yield.

Lithium compounds add to an ethylenic double bond whereas Grignard reagents do not.

Organolithium compounds react with compounds containing active hydrogens, aldehydes, ketones, carbon dioxide, ethylene oxide etc in the same way as Grignard reagents as discussed in section 13.2.4.

13.4   PROBLEMS

1. What are Grignard reagents ? How they are prepared ?
2. Discuss the structure of Grignard reagents. What is the role of ether in its structure ?
3. Give a synthesis of ethyl magnesium bromide. How will you prepare ethane, propionic acid, isopropyl alcohol and ethyl methyl ketone from it.
4. Give with help of one example the mechanism of nucleophilic substitution and nucleophilic addition reactions of alkyl magnesium halides.
5. What products are obtained by the reaction of methylmagnesium iodide with formaldehyde, acetaldehyde, ethyl formate, carbon dioxide and carbon disulphide.
6. Explain why organolithium compounds are more reactive than the organomagnesium compounds.
7. How will you prepare primary, secondary and tertiary alcohols starting with a Grignard reagent. Give one reaction for each case.
8. Complete the following:
   (a) $CH_3MgI + NH_2Cl \longrightarrow$ (d) $C_2H_5MgI + PbCl_2 \longrightarrow$
   (b) $CH_3MgI + S \longrightarrow$ (e) $CH_3MgI + SiCl_4 \longrightarrow$
   (c) $CH_3MgI + CS_2 \longrightarrow$ (f) $C_6H_5Cl + Li \longrightarrow$
9. Using Grignard reagents, how will you estimate the number of active hydrogens in an organic compound.

14

Monohydric Alcohols

14.1 INTRODUCTION

Alcohols are considered to be derived from hydrocarbons by the replacement of one hydrogen atom with a hydroxy group, and have the general formula $C_nH_{2n+1}OH$.

14.2 CLASSIFICATION

Alcohols are classified according to the number of hydroxy groups present in the molecule. Thus, alcohols containing one hydroxy group are called *monohydric alcohols* whereas alcohols with two, three or more hydroxy groups are called *dihydric, trihydric* or *polyhydric alcohols* respectively. Chapter 15 discusses polyhydric alcohols.

CH_3CH_2OH $\begin{array}{c} CH_2OH \\ | \\ CH_2OH \end{array}$ $\begin{array}{c} CH_2OH \\ | \\ CHOH \\ | \\ CH_2OH \end{array}$

Ethanol (monohydric)   **Glycol (dihydric)** 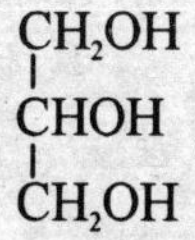   **Glycerol (trihydric)**

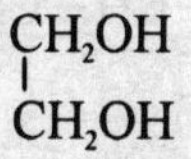

The monohydric alcohols are further subdivided into a primary (1°), secondary (2°) and tertiary (3°) alcohols on the basis of carbon to which hydroxy group is attached.

$CH_3CH_2CH_2CH_2OH$ $\begin{array}{c} OH \\ | \\ CH_3-CH_2-CH-CH_3 \end{array}$ $\begin{array}{c} CH_3 \\ | \\ CH_3-C-OH \\ | \\ CH_3 \end{array}$

1-Butanol   **2-Butanol**   **2-Methyl-2-propanol**
n-Butyl alcohol   **sec Butyl alcohol**   **tert Butyl alcohol**
(Primary alcohol)   **(Secondary alcohol)**   **(Tertiary alcohol)**

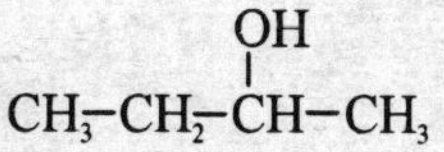

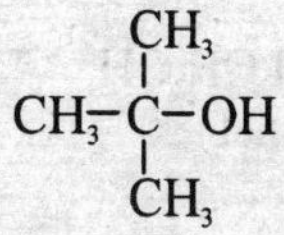

14.3 NOMENCLATURE

Three systems are in use for the nomenclature of monohydric alcohols.

(i) *Common system:* According to this system the names of the alcohols are derived by suffixing the word 'alcohol' to the 'alkyl' group.

CH_3OH CH_3CH_2OH $\begin{array}{c} CH_3 \\ | \\ CH_3-CH-OH \end{array}$ $CH_3CH_2CH_2OH$

Methyl alcohol   **Ethyl alcohol**   **Isopropyl alcohol**   **n-Propyl alcohol**

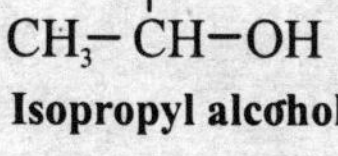

(ii) *Carbinol System:* According to this system the names of the alcohols are derived from the methyl alcohol or carbinol. The carbinol designates the carbinol carbon and the attached hydroxy group. The alkyl groups attached to the carbinol carbon are named in alphabetical order and then the suffix carbinol is added.

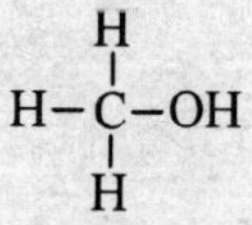

$$\underset{\textbf{Carbinol}}{H-\overset{\displaystyle H}{\underset{\displaystyle H}{C}}-OH} \qquad \underset{\textbf{Methyl carbinol}}{CH_3-\overset{\displaystyle H}{\underset{\displaystyle H}{C}}-OH} \qquad \underset{\textbf{Ethylmethyl carbinol}}{CH_3CH_2-\overset{\displaystyle CH_3}{\underset{\displaystyle H}{C}}-OH} \qquad \underset{\textbf{tert Butyl carbinol}}{CH_3-\overset{\displaystyle CH_3}{\underset{\displaystyle CH_3}{C}}-\overset{\displaystyle H}{\underset{\displaystyle H}{C}}-OH}$$

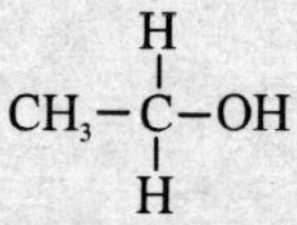

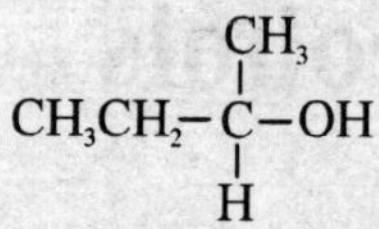

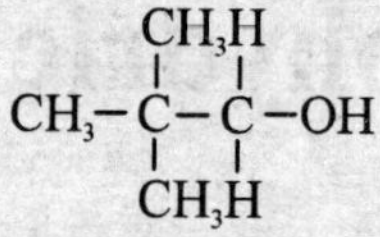

(iii) *IUPAC System:* According to this system the alcohols are designated as *alkanols*. The name of an alcohol is derived from the name of the corresponding alkane by replacing 'e' with 'ol'. Thus, methyl alcohol, a derivative of methane is named methanol and ethyl alcohol as ethanol.

For higher alcohols, the longest continuous carbon chains containing the $-OH$ group is selected as the basic hydrocarbon chain. The position of the $-OH$ group, other group (if any) and the side chain are indicated by the numbers and the lowest possible number is given to the carbon atom to which the $-OH$ group is attached.

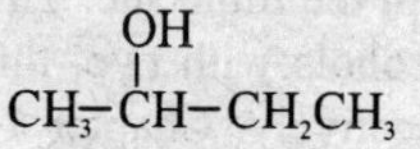

$$\underset{\textbf{2-Butanol}}{CH_3-\overset{\displaystyle OH}{CH}-CH_2CH_3} \qquad\qquad \underset{\textbf{2-Methyl-1-butanol}}{CH_3-CH_2-\overset{\displaystyle CH_3}{CH}-CH_2OH}$$

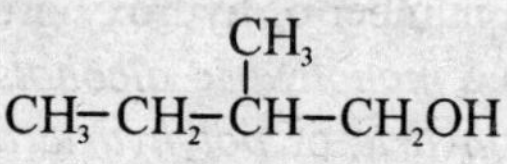

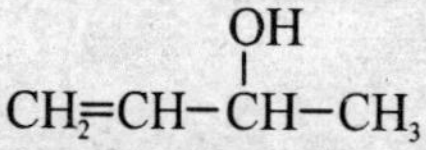

$$\underset{\textbf{3-Buten-2-ol}}{CH_2=CH-\overset{\displaystyle OH}{CH}-CH_3} \qquad\qquad \underset{\textbf{3-Pentene-1-ol}}{CH_3-CH=CH-CH_2-CH_2OH}$$

14.4   ISOMERISM

The alcohols exhibit position and chain isomerism. The former arises due to different position of $-OH$ group in the hydrocarbon chain and the later is due to different structures of the carbon chain. Isomeric butyl alcohols, C_4H_9OH, show position and chain isomerism as shown below.

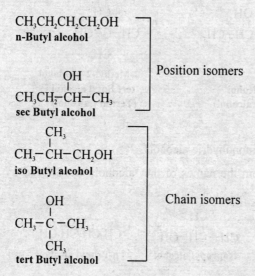

CH$_3$CH$_2$CH$_2$CH$_2$OH
n-Butyl alcohol

$$\underset{\textbf{sec Butyl alcohol}}{CH_3CH_2-\overset{\displaystyle OH}{CH}-CH_3}$$

Position isomers

$$\underset{\textbf{iso Butyl alcohol}}{CH_3-\overset{\displaystyle CH_3}{CH}-CH_2OH}$$

$$\underset{\textbf{tert Butyl alcohol}}{CH_3-\overset{\displaystyle OH}{\underset{\displaystyle CH_3}{C}}-CH_3}$$

Chain isomers

Alcohols also exhibit functional isomerism with ethers. Thus, C_2H_6O represents ethanol and dimethyl ether having different functional groups.

$$CH_3CH_2OH \qquad\qquad CH_3-O-CH_3$$

Ethyl alcohol   **Dimethyl ether**

14.5   ORBITAL STRUCTURE

As has been discussed in Section 2.3, the orbital structure of alcohols involves sp^3 hybridized carbon and oxygen atoms. Below is given the orbital picture of methanol having a C–O–H bond angle of 105° (less than the normal tetrahedral angle of 109.5°). The deviation from the normal angle is due to greater repulsion by completely filled sp^3 orbitals (having lone pairs) of oxygen than the bond orbitals (Fig. 14.1).

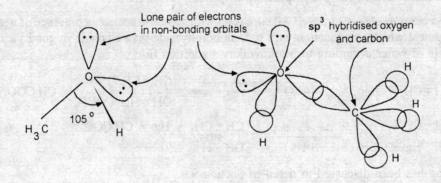

Fig. 14.1:   Orbital Structure of Methyl Alcohol

14.6   GENERAL METHODS OF PREPARATION

(i)   *From Alkenes*

(a)   *acid catalysed hydration of alkenes:* As has already been mentioned (Section 9.9) water adds to alkenes in presence of acid catalyst to form alcohols.

Addition of water to an unsymmetrical alkenes follows Markovnikov's rule.

$$CH_3CH{=}CH_2 + H_2O \xrightarrow[\text{catalyst}]{H_2SO_4} \overset{\displaystyle OH}{\underset{\displaystyle |}{CH_3{-}CH{-}CH_3}}$$

Propene   **2-Propanol**
                              (isopropyl alcohol)

This method is suitable for the preparation of secondary and tertiary alcohols only with the exception of ethyl alcohol, a primary alcohol which is obtained by hydration of ethylene. These reactions are also useful for industrial preparation of alcohols.

(b) *Hydroboration-oxidation of alkenes:* As has been discussed in Section 9.9, the reaction leads to antimarkovnikov's addition of water to alkene. The alkene is reacted with diborane in an inert solvent (for example ether) to give an intermediate alkylborane, which on oxidation with alkaline hydrogen peroxide yields an alcohol.

$$CH_3-\overset{\overset{\displaystyle CH_3}{|}}{C}=CH_2 + BH_3 \longrightarrow CH_3-\overset{\overset{\displaystyle CH_3}{|}}{CH}-CH_2BH_2 \xrightarrow{CH_3-\overset{\overset{\displaystyle CH_3}{|}}{C}=CH_2} (CH_3-\overset{\overset{\displaystyle CH_3}{|}}{CH}CH_2)_2BH$$

Isobutene

$$\xrightarrow{CH_3-\overset{\overset{\displaystyle CH_3}{|}}{C}=CH_2} (CH_3-\overset{\overset{\displaystyle CH_3}{|}}{CH}-CH_2)_3B$$

$$(CH_3-\overset{\overset{\displaystyle CH_3}{|}}{CH}CH_2)_3B \xrightarrow{H_2O_2 / OH^-} CH_3-\overset{\overset{\displaystyle CH_3}{|}}{CH}-CH_2OH + H_3BO_3$$

Triisobutyl borane   **Isobutanol**

It is an excellent method for the preparation of primary alcohols.

(c) *Oxymercuration-demercuration:* Alkenes react with mercuric acetate in presence of water to form mercury compound which on reduction with sodium borohydride give alcohols in good yield. Alkenes add a molecule of water according to Markovnikov's addition rule.

$$CH_2{=}CH_2 + Hg(OAc)_2 + H_2O \xrightarrow{Oxymercuration} \underset{\underset{\displaystyle OH \quad HgOCOCH_3}{|\qquad\quad|}}{CH_2-CH_2} + CH_3COOH$$

$$\underset{\underset{\displaystyle OH \quad HgOCOCH_3}{|\qquad\quad|}}{CH_2-CH_2} \xrightarrow[NaBH_4]{Demercuration} \underset{\underset{\displaystyle OH \quad H}{|\qquad|}}{CH_2-CH_2} + Hg + CH_3COO^-$$

The reaction has been discussed in detail in Section 9.9.

(ii) *From Carbonyl Compounds*

(a) *Reduction:* The carbonyl compounds, viz. aldehydes and ketones on reduction with sodium and alcohol or hydrogen in presence of nickel or lithium aluminum hydride give the corresponding alcohols. Aldehydes give primary alcohols whereas ketones give secondary alcohols.

$$R-\overset{\overset{\displaystyle O}{||}}{C}-H \xrightarrow[\substack{or\ Ni\ /\ H_2 \\ or\ LiAlH_4}]{Na-C_2H_5OH} R-CH_2-OH \qquad\qquad R-\overset{\overset{\displaystyle O}{||}}{C}-R \longrightarrow R-\overset{\overset{\displaystyle R}{|}}{CH}-OH$$

Aldehyde   **Primary alcohol**   **Ketone**   **Secondary alcohol**

(b) *Grignard synthesis:* As has been discussed in Section 13.2.4 , the method is used for the preparation of primary, secondary and tertiary alcohols using suitable carbonyl compounds and Grignard reagents in dry ether followed by hydrolysis.

$$\underset{}{\diagup}C{=}O + RMgX \xrightarrow[2.\ H^+/\ H_2O]{1.\ Dry\ ether} -\overset{\overset{\displaystyle |}{}}{\underset{\underset{\displaystyle R}{|}}{C}}-OH$$

(iii) *From Alkyl Halides:* Treatment of alkyl halides with dilute aqueous sodium or potassium hydroxide solution gives alcohols (S$_N$1 and S$_N$2).

$$R-X \xrightarrow{\text{dil NaOH solution}} R-OH + NaX$$
Alkyl halide   **Alcohol**

The method suffers from the drawback that dehydrohalogenation may take place to produce simultaneously an alkene. This shortcoming is overcome by using either aqueous potassium carbonate or moist silver oxide for hydrolysis. The reaction is discussed in detail in section 12.3.7.

(iv) *From Carboxylic Acids, Acid Chlorides, Acid Anhydrides and Esters by Reduction:* Carboxylic acids and their derivatives on reduction with conventional reducing agents form primary alcohols.

$$RCOCl + H_2 \xrightarrow{\text{Ni}} RCH_2OH + HCl$$
Acid chloride

$$RCOOR' + H_2 \xrightarrow{\text{Ni}} RCH_2OH + R'OH$$
Ester

$$RCOOH \xrightarrow{\text{LiAlH}_4} RCH_2OH$$
Carboxylic   **Alcohol**
acid

(v) *From Ethers:* The ethers on heating with dilute sulphuric acid under pressure yield alcohols.

$$C_2H_5OC_2H_5 + H_2O \xrightarrow{H_2SO_4} 2C_2H_5OH$$
Diethyl ether   **Ethanol**

This method is discussed in Section 16.8.

(vi) *From Primary Amines:* The primary amines on treatment with nitrous acid give alcohols.

$$C_2H_5NH_2 + HONO \xrightarrow{0-5°} C_2H_5OH + N_2 + H_2O$$
Ethylamine   **Ethanol**

14.7 PHYSICAL PROPERTIES

The alcohols are generally colorless volatile liquids having characteristic odor and strange burning taste. As the molecular weight increases the smell and the taste becomes less pronounced.

The alcohols especially the lower members are completely miscible in water, the solubility decreases as the number of carbon atoms increases in the molecule. The solubility of lower alcohols can be explained by their ability to form hydrogen bonds (shown in dotted line) with water.

$$H-\underset{R}{\overset{..}{O}}\cdots H-\underset{H}{\overset{..}{O}}\cdots H-\underset{R}{\overset{..}{O}}\cdots H-\underset{H}{\overset{..}{O}}$$

The boiling points of alcohols increase in a regular fashion (about 20° from one member to the next higher member). However the alcohols have much higher boiling points compared to the corresponding alkanes or most other compounds of approximately the same molecular weight. This is explained on the basis of the structure of alcohols and that of water. The oxygen in alcohol (analogous to oxygen in water) is in the sp^3 hybridised state and has two unshared pairs of electrons. Also the hydroxy group in water and in alcohol is polar.

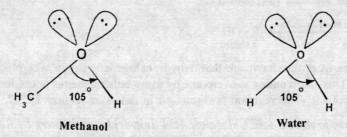

Thus, as expected, alcohols like water are strongly hydrogen bonded. The formation of hydrogen bonds results in association of large number of alcohol molecules. Before boiling occurs, these molecular associations have to be broken.

$$H-\overset{..}{O}:\cdots H-\overset{..}{O}:\cdots H-\overset{..}{O}:\cdots H-\overset{..}{O}:\cdots$$
$$\quad\ \ |\qquad\quad\ |\qquad\quad\ |\qquad\quad\ |$$
$$\quad\ \ R\qquad\quad H\qquad\quad R\qquad\quad H$$

Hydrogen bonding in alcohol molecules

In case of isomeric alcohols, the primary alcohols have the highest boiling points and the tertiary alcohols the lowest. The order of boiling point is primary alcohol > secondary alcohol > tertiary alcohol. This is because in straight chain compounds, the Van der Waal's attractive forces are relatively large due to large surface area. However, in branched chain structure, the molecule has comparatively less surface area since the molecule tends to become spherical.

14.8   SPECTRAL PROPERTIES

The ultra violet (UV) spectra of alcohols is not of much use since they show absorption only in the region 180-188 nm which is normally not covered in UV spectroscopy.

In the infrared spectra, the alcohols show O–H stretching in the region 3650-3590 cm^{-1}. This O–H stretching is shifted to 3520-3200 cm^{-1} (broad band) due to intermolecular hydrogen bonding and to 3590-3420 cm^{-1} due to intramolecular hydrogen bonding. The C–O stretching for primary, secondary and tertiary alcohols are observed in the region 1050, 1110 and 1150 cm^{-1} respectively.

In the NMR spectrum also the chemical shift of the hydroxy group is influenced by the hydrogen bonding. Generally the hydroxy proton in an alcohols absorbs in the δ 1-5.

14.9   CHEMICAL PROPERTIES

Alcohols show following chemical properties:
  (i)   Reactions due to O–H bond breaking
  (ii) Reactions due to C–O bond breaking
  (iii) Oxidation of alcohols

Alcohols as Acids and Bases: Alcohols behave both as an acid and a base (amphoteric), though they are neutral to litmus. In acidic solution, alcohols are protonated and the acid base equilibrium with alcohol acting as a base is established. The reaction is similar to that occurs between water and an acid.

$$R-O-H + HCl \rightleftharpoons R-OH_2^+ + Cl^-$$

Alcohol   **Protonated**
                                  alcohol

Alternatively, alcohols can also lose a proton to a strong base yielding an alkoxide ion or liberate hydrogen gas when reacted with an active metal showing its acidic nature. The acidic nature of alcohols is due to the polarity of oxygen hydrogen bond which facilitates the removal of proton.

$$ROH + NaH \longrightarrow RO^- Na^+ + H_2$$
Alcohol   Sodium   Sodium
          hydride   alkoxide

$$ROH + Na \longrightarrow RO^- Na^+ + \frac{1}{2} H_2$$

Alcohols have approximately same pk_a values as water in dilute aqueous solutions. Thus, the pk_a of methanol in water is 15.5 and that of pure water is 15.74. Therefore an alcohol is as acidic as water. The order of decreasing acid strength amongst alcohols is as follows.

$$CH_3OH > H_2O > pri\ alcohol > sec\ alcohol > tert\ alcohol$$

(i) *Reactions Due to O–H Bond Breaking*

(a) *Reaction with active metals:* Active metals like K, Na, Mg, Al and Zn which are strongly electropositive liberate hydrogen from alcohols and form alkoxides.

$$2C_2H_5OH + 2Na \longrightarrow 2C_2H_5O^- Na^+ + H_2$$
Ethanol   Sodium ethoxide

(b) *Esterification:* Alcohols react with carboxylic acids in presence of an acid catalyst to form esters. This reaction, known as esterification, involves the breaking of O-H bond in alcohols. If the reaction is carried out in dry hydrogen chloride, it is called *Fischer Spier* method.

$$R-OH + R'COOH \xrightarrow{H^+} R'COOR + H_2O$$
Alcohol   Carboxylic acid   Ester

This reaction will be discussed in detail in Section 19.7.
    The alcohols form esters when react with acid chlorides and anhydrides.

$$C_2H_5OH + CH_3COCl \longrightarrow CH_3COOC_2H_5 + HCl$$
Ethanol   Acetyl chloride   Ethyl acetate
$$C_2H_5OH + (CH_3CO)_2O \longrightarrow CH_3COOC_2H_5 + CH_3COOH$$
Ethanol   Acetic anhydride   Ethyl acetate

(c) *Grignard Reaction:* Alcohols (being more acidic than alkanes) react with Grignard reagents to form alkanes.

$$ROH + R'MgX \xrightarrow{ether} R'H + Mg(OR)X$$
$$C_2H_5OH + CH_3MgI \xrightarrow{ether} CH_4 + Mg(OC_2H_5)I$$
Ethanol   Methyl   Methane
               magnesium
               iodide

(ii) *Reactions Due to C–O Bond Breaking*

(a) *Reaction with hydrogen halides and phosphorus halides:* The alcohols react with hydrogen halides (HX), phosphorus halides (PX_3 and PX_5) and with thionyl chloride ($SOCl_2$) to give alkyl halides.

$$R-OH + HX \longrightarrow R-X + H_2O$$
Alcohol   Hydrogen   Alkyl
             halide   halide

$$R-OH + PX_3 \longrightarrow R-X + H_3PO_3$$
Alcohol   Phosphorus   Alkyl
             trihalide   halide

$$R-OH + PCl_5 \longrightarrow R-Cl + POCl_3 + HCl$$
Alcohol   Phosphorus   Alkyl
             pentachloride   chloride

$$R-OH + SOCl_2 \longrightarrow R-Cl + SO_2 + HCl$$
Alcohol   Thionyl   Alkyl
             chloride   chloride

We will consider the reaction of alcohols with hydrogen halides in detail. The alcohols undergo substitution reactions with hydrogen halides under acidic conditions or in presence of Lewis acids like anhydrous zinc chloride.

$$CH_3CH_2CH_2OH + HBr \xrightarrow[\text{Heat}]{ZnCl_2} CH_3CH_2CH_2Br + H_2O$$
n-Propyl alcohol   1-Bromopropane

$$(CH_3)_3C-OH + HCl \xrightarrow[\text{Heat}]{ZnCl_2} (CH_3)_3C-Cl + H_2O$$
tert Butyl alcohol   tert Butyl chloride

As seen the alcohols require catalytic conditions like acid or $ZnCl_2$ for the reaction to occur. They do not undergo substitution reactions under neutral or alkaline conditions as in the case of alkyl halides (see Section 12.3.7). We have seen that the halides viz. Cl^-, Br^- and I^- are good leaving groups as they are weak bases. However, OH^- is a strong base and so is a poor leaving group. In acidic solution the hydroxy group gets protonated and is converted into a good leaving group, water, which is a weak base. Now even a weak nucleophile, such as halide ion can displace the water molecule to yield an alkyl halide.

$$R-\ddot{O}-H \rightleftharpoons R-\overset{+}{\underset{\underset{H}{|}}{\ddot{O}}}-H \xrightarrow{X^-} R-X + H_2O$$
Alcohol   Oxonium ion   Alkyl
             (protonated   halide
             alcohol)

Similarly, anhydrous zinc chloride is a powerful Lewis acid with empty orbitals that can accept electrons from the oxygen of alcohol. The formation of a complex of $ZnCl_2$ with the oxygen of alcohol weakens the C–O bond and thus enhances the leaving ability of the OH group.

$$R-\ddot{O}H + ZnCl_2 \rightleftharpoons R-\overset{+}{\underset{\underset{H}{|}}{\ddot{O}}}-ZnCl_2 \xrightarrow{Cl^-} R-Cl + [Zn(OH)Cl_2]^-$$

$$[Zn(OH)Cl_2]^- + H^+ \rightleftharpoons ZnCl_2 + H_2O$$

In the substitution reactions of alcohols, the reactivity order observed for hydrogen halides and alcohols is as follows

HF < HCl < HBr < HI

Methanol < pri < sec < tert

This order of reactivity forms the basis for the *Lucas test* for differentiating primary, secondary and tertiary alcohols (see Section 14.10).

Mechanism: Primary alcohols undergo substitution reaction by S$_N$2 mechanism, tertiary alcohols by S$_N$1 mechanism and secondary alcohols by either S$_N$1 or S$_N$2 mechanism (as in the case of alkyl halides). This is because a tertiary alcohol (R$_3$C–OH) very easily forms a carbocation (R$_3$C$^+$) and reacts by S$_N$1 mechanism. On the other hand, it is very difficult for a primary alcohol to form a carbocation (less stable), but the primary carbocation structure is open to backside attack and so S$_N$2 reaction is possible. A secondary alcohol may react by either S$_N$1 or S$_N$2 mechanism depending upon the reaction conditions.

The primary alcohols involve S$_N$2 mechanism when reacted upon with HX as given below as the transition state is unhindered.

$$R-\overset{..}{\underset{..}{O}}-H \overset{H^+}{\rightleftharpoons} R-\overset{+.}{\underset{\underset{H}{|}}{O}}-H \xrightarrow[S_N2]{X^-} [\overset{\delta-}{X}\cdots R\cdots \overset{\delta+}{\overset{..}{O}H_2}] \longrightarrow R-X \quad + H_2O$$

Primary alcohol   **Protonated alcohol**   **Alkyl halide**

The tertiary alcohols, on the contrary, involve S$_N$1 mechanism when reacted with HX as follows.

$$R_3C-\overset{..}{\underset{..}{O}}H \overset{H^+}{\rightleftharpoons} R_3C-\overset{+.}{\underset{\underset{H}{|}}{O}}-H \xrightarrow[S_N1]{-H_2O} [R_3C^+] \xrightarrow{X^-} R_3C-X$$

Tertiary alcohol   **Protonated alcohol**   **Alkyl halide**

The secondary alcohols tend to undergo rearrangements during the reaction.

3,3-Dimethyl-2-butanol   **Protonated alcohol**

sec. Carbocation   **ter.Carbocation**   **2-Bromo-2,3-dimethylbutane**

As seen, the protonated alcohol eliminates a water molecule to give a secondary carbocation, which rearranges to a more stable tertiary carbocation. It is this more stable tertiary carbocation which reacts with Nucleophile to give the substitution product.

(b) *Reaction with nitric acid and sulphuric acid:* Alcohols react with nitric acid and sulphuric acid to form alkyl nitrates and alkyl hydrogen sulphates respectively.

$$CH_3CH_2OH + HO-NO_2 \longrightarrow CH_3CH_2-ONO_2 + H_2O$$
Ethyl alcohol   **Ethyl nitrate**

$$CH_3CH_2OH + HO-SO_3H \longrightarrow CH_3CH_2-OSO_3H + H_2O$$
Ethyl alcohol   **Ethyl hydrogen sulphate**

$$CH_3CH_2OSO_3H + CH_3CH_2OH \longrightarrow CH_3CH_2O-SO_2-OCH_2CH_3 + H_2O$$
Ethyl hydrogen   **Diethyl sulphate**
sulphate

(c) *Dehydration of Alcohols:* The dehydration of alcohols involve cleavage of the C–O bond along with the loss of a proton from the ß-position. It is generally effected by heating alcohols at 400-700° or heating at a lower temperature in presence of a catalyst such as alumina or a mineral acid like sulphuric acid. The product of dehydration is an alkene or a mixture of alkenes. The order of dehydration of alcohols is tert > sec > pri.

Primary alcohols on dehydration give only one product whereas secondary or tertiary alcohols form mixture of alkenes.

$$CH_3CH_2CH_2CH_2OH \xrightarrow[160-170°]{H_2SO_4} CH_3CH_2CH=CH_2$$
Primary alcohol   **1-Butene**

$$CH_3CH_2CHOHCH_3 \xrightarrow[140-150°]{H_2SO_4} CH_3CH=CHCH_3 + CH_3CH_2CH=CH_2$$
Secondary alcohol   **2-Butene (major)   1-Butene (minor)**

The reaction has been discussed in detail in section 9.6.

(d) *Reaction with Ammonia:* Alcohols form mixture of amines when reacted with ammonia in presence of a catalyst like alumina or copper chromite.

$$R-OH + HNH_2 \longrightarrow R-NH_2 + H_2O$$
               Ammonia

$$R-OH + RNH_2 \longrightarrow R_2NH + H_2O$$
               Primary
               amine

$$R-OH + R_2NH \longrightarrow R_3N + H_2O$$
               Secondary
               amine

(iii) *Oxidation of Alcohols:* The common oxidising agents used for the oxidation of alcohols are acidic potassium dichromate, acidic or alkaline potassium permanganate, concentrated nitric acid, chromic acid or chromium trioxide complex with pyridine. The nature of the oxidation product depends on the nature of the alcohol i.e., whether the alcohol is primary, secondary or tertiary.

Primary alcohols on oxidation give an aldehyde, which on further oxidation gives a carboxylic acid. Both oxidation products i.e., the aldehyde and the carboxylic acid have the same number of carbon atoms as the alcohol. In case aldehyde is required as the major product, oxidation is carried out with chromium trioxide pyridine complex. Yield of aldehyde with other reagents is poor since the oxidation continues until the carboxylic acid is formed.

$$CH_3CH_2OH \xrightarrow{[O]} CH_3CHO \xrightarrow{[O]} CH_3COOH$$
Ethyl alcohol   **Acetaldehyde**   **Acetic acid**

Secondary alcohols on oxidation give ketone with the same number of carbon atoms as the alcohol. The ketone on further oxidation under drastic conditions give a mixture of carboxylic acids containing fewer carbon atoms than the alcohols.

$$CH_3CHOHCH_2CH_3 \xrightarrow{[O]} CH_3COCH_2CH_3$$

2-Butanol   **Ethylmethyl ketone**

Tertiary alcohols are not easily oxidised under acidic or alkaline conditions. The acidic oxidising agents convert a tertiary alcohol into an alkene, which is then oxidised to give a mixture of ketone and carboxylic acid.

$$CH_3-\underset{\underset{CH_3}{|}}{\overset{\overset{CH_3}{|}}{C}}-OH \xrightarrow{[O]} CH_3-\underset{}{\overset{\overset{CH_3}{|}}{C}}=CH_2 \xrightarrow{[O]} CH_3-\overset{\overset{CH_3}{|}}{C}=O + HCOOH$$

tert. Butanol   **Isobutene**   **Acetone**   **Formic acid**

Oxidation of alcohols can also be achieved by *catalytic dehydrogenation* in presence of heated metal catalyst (Ag or Cu). In this method vapors of alcohol is passed over heated reduced copper. Primary, secondary and tertiary alcohol exhibit different behaviour.

$$CH_3CH_2OH \xrightarrow{Cu, 300°} CH_3CHO \quad + H_2$$

Ethanol   **Acetaldehyde**

$$CH_3CHOHCH_3 \xrightarrow{Cu, 300°} CH_3\overset{\overset{O}{\|}}{C}CH_3 \quad + H_2$$

Isopropanol   **Acetone**

The *tertiary alcohol* as has already been mentioned are resistant to oxidation. When vapors of tertiary alcohols are passed over heated copper, they undergo dehydration to give alkenes.

$$CH_3-\underset{\underset{CH_3}{|}}{\overset{\overset{CH_3}{|}}{C}}-OH \xrightarrow{Cu, 300°} CH_3-\overset{\overset{CH_3}{|}}{C}=CH_2 + H_2O$$

tert. Butyl alcohol   **2-Methylpropene**

The dehydrogenation is used for the industrial preparation of aldehydes and ketones.

14.10   TESTS FOR AN ALCOHOLIC HYDROXY GROUP

The organic compound (dissolved in an inert solvent) is tested for hydroxy group as follows:

(i) Add a small piece of clean sodium to a portion of the above solution. If bubbles of hydrogen gas are given off, the compound contains a −OH group.

$$2R-OH + 2Na \longrightarrow 2R-ONa + H_2$$

(ii) Add phosphorus pentachloride to a portion of the above solution. If the mixture becomes warm with the evolution of HCl gas, the given compound contains −OH group.

$$R-OH + PCl_5 \longrightarrow R-Cl + POCl_3 + HCl$$

Distinction Between Primary, Secondary and Tertiary Alcohols

Following tests are used to distinguish between primary secondary and tertiary alcohols.

(i)   *Lucas Test:* Alcohols are treated with hydrochloric acid and zinc chloride (Lucas reagent) to form alkyl halide.

$$R-OH + HCl \xrightarrow{ZnCl_2} R-Cl + H_2O$$

Alcohol   Alkyl
                            halide

Tertiary alcohols react with Lucas reagent rapidly. Secondary alcohols react slowly while primary alcohols react with Lucas reagent only on heating. The test is performed by mixing alcohol at room temperature with concentrated hydrochloric acid and zinc chloride. The formed alkyl chloride is insoluble in the medium. It causes the solution to become cloudy before its separation as a distinct layer. With tertiary alcohols cloudiness appears immediately while with secondary alcohols cloudiness appears with in 5 minutes. The solution remains clear with Primary alcohols since they do not react with Lucas reagent at room temperature. Higher temperatures are required for the reaction to procceed.

(ii)   *Oxidation:* The alcohol is treated with sodium dichromate in sulphuric acid at room temperature. Identification of the oxidation products give information regarding the type of alcohol.
Primary alcohols give a carboxylic acid containing the same number of carbon atoms.
Secondary alcohols give a ketone containing the same number of carbons. However on further oxidations carboxylic acids with less number of carbon atoms are obtained.
In the above two cases the orange color of sodium dichromate in concentrated sulphuric acid changes to green.
Tertiary alcohols do not react under these conditions.

(iii)   *Victor Meyer Test:* It is carried out by the reaction of alcohol with cold hydrogen iodide or red phosphorus and iodine to form alkyl iodide. The iodide thus formed is treated with silver nitrite to give the corresponding nitroalkane. To the nitroalkane is added nitrous acid ($NaNO_2$ + dil H_2SO_4) and the solution is rendered alkaline and the color observed. Primary alcohols give blood red color, secondary alcohols give blue color and tertiary alcohols do not give any color. The reactions involved are given below:

$$CH_3CH_2OH \xrightarrow{HI} CH_3CH_2I \xrightarrow{AgNO_2} CH_3CH_2NO_2 \xrightarrow[2.\ OH]{1.\ HNO_2} CH_3-C\begin{smallmatrix}NO_2\\N-OH\end{smallmatrix}$$

Primary alcohol   Nitrolic acid
                                                                             (Blood red color)

$$\begin{smallmatrix}CH_3\\CH_3\end{smallmatrix}CHOH \xrightarrow{HI} \begin{smallmatrix}CH_3\\CH_3\end{smallmatrix}CHI \xrightarrow{AgNO_2} \begin{smallmatrix}CH_3\\CH_3\end{smallmatrix}CHNO_2 \xrightarrow[2.\ OH]{1.\ HNO_2} \begin{smallmatrix}CH_3\\CH_3\end{smallmatrix}C\begin{smallmatrix}NO_2\\NO\end{smallmatrix}$$

Secondary alcohol   Pseudo nitrol
                                                                             (Blue color)

$$CH_3-\underset{\underset{CH_3}{|}}{\overset{\overset{CH_3}{|}}{C}}-OH \xrightarrow{HI} CH_3-\underset{\underset{CH_3}{|}}{\overset{\overset{CH_3}{|}}{C}}-I \xrightarrow{AgNO_2} CH_3-\underset{\underset{CH_3}{|}}{\overset{\overset{CH_3}{|}}{C}}-NO_2 \xrightarrow{HNO_2} \text{No reaction}$$

Tertiary alcohol

4.11 INDIVIDUAL MEMBERS

4.11.1   Methyl Alcohol/Methanol/Carbinol [CH₃OH]

This is the first member of alkanol series. It is also called wood spirit (or wood naphtha) as it was earlier prepared from wood by destructive distillation.

Manufacture: Methyl alcohol can be prepared by any of the general methods of preparation of alcohols. It is manufactured by the following methods.

(i)   *From Water Gas:* Water gas is first obtained by passing steam over red hot coke and then is mixed with half its volume of hydrogen, compressed to about 300 atmospheres and passed over heated catalyst ($ZnO.CrO_3$).

$$C + H_2O \longrightarrow CO + H_2$$
Red   Steam   Water gas
hot coke

$$CO + 2H_2 \xrightarrow[300°]{ZnO.CrO_3} CH_3OH$$
Methyl alcohol

(ii)   *From Natural Gas:* Methane obtained from natural gas is used for the manufacture of methyl alcohol. A mixture of methane and oxygen (9:1) is passed over a catalyst at 260° and 100 atmospheric pressure.

$$CH_4 + \tfrac{1}{2}O_2 \xrightarrow[\text{100 atms;260°}]{\text{Cu catalyst}} CH_3OH$$
Methyl alcohol

Properties: Methyl alcohol is a colorless liquid, b.p. 65°, soluble in water. It burns with a faintly luminous flame. With air or oxygen, its vapors form an explosive mixture. Methanol is poisonous in nature and if taken internally it can cause blindness and death. Its poisonous action is due to its oxidation to formaldehyde in the human body which is poisonous and causes complications.

Uses: Methanol is used
  (a) as a solvent for paints and varnishes
  (b) for making antifreeze mixtures for automobiles
  (c) for denaturing alcohol, i.e., rendering it unfit for drinking purposes
  (d) for the manufacture of formaldehyde
  (e) as a motor fuel (mixed with petrol up to 20%).

Tests
  (a) On warming with a few drops of concentrated sulphuric acid and salicylic acid methanol gives a characteristic fragrance of methyl salicylate (known as oil of winter green).
  (b) Methyl alcohol on coming in contact with a red hot copper spiral gives a pungent smell of formaldehyde.

14.11.2   Ethyl Alcohol/Ethanol/Methyl Carbinol [C₂H₅OH]

Ethyl alcohol is the most important member of the alcohol series and is simply known as *alcohol*. It is an important constituent of all wines and is also called *'Spirit of Wine'*. Since it is manufactured from starchy grain it is also known as *grain alcohol*.

Manufacture: It can be obtained by any of the general methods of preparation of primary alcohols. It is manufactured by any of the following methods:

(i)   *Fermentation of Sugars:* This is the oldest method of alcohol manufacture and is still used. The alcohol is produced by the fermentation of sugars in the presence of yeast (see Section 14.13).

$$C_{12}H_{22}O_{11} + H_2O \xrightarrow{\text{Invertase}} C_6H_{12}O_6 + C_6H_{12}O_6$$
Sucrose   Glucose   Fructose

$$C_6H_{12}O_6 \xrightarrow{\text{Invertase}} 2C_2H_5OH + 2CO_2$$
Ethyl alcohol

The cheapest source of glucose, fructose and cane sugar is molasses, the byproduct obtained after obtaining sucrose from cane juice. It is also produced from cheap starchy materials like potatoes, maize and barley. The starchy materials are first converted into sugar and then fermented.

(ii)   *From Ethylene by Hydration:* Ethylene (from cracked petroleum) is passed into concentrated sulphuric acid (98%) to give ethyl hydrogen sulphate which on hydrolysis by boiling water forms ethanol.

$$H_2C=CH_2 + H_2SO_4 \xrightarrow{100°} CH_3CH_2HSO_4$$
Ethylene   Ethyl hydrogen sulphate

$$CH_3CH_2HSO_4 + H_2O \xrightarrow{\Delta} CH_3CH_2OH + H_2SO_4$$
Ethanol

The alcohol produced is recovered by distillation.

The direct hydration of ethylene is carried out by passing a mixture of ethylene and water vapor over phosphoric acid.

$$CH_2=CH_2 + H_2O \xrightarrow[300°]{H_3PO_4,\ 60atm} CH_3CH_2OH$$
Ethanol

The reaction products are cooled and passed through water. The aqueous solution of alcohol thus obtained is distilled to get pure alcohol.

(iii)   *From Acetylene:* Acetylene is first converted into acetaldehyde by passing into dilute sulphuric acid in presence of a catalyst (mercurous sulphate) , which on catalytic hydrogenation over reduced copper or nickel gives ethyl alcohol.

$$HC \equiv CH + H_2O \xrightarrow[100°]{Hg_2SO_4} CH_3CHO$$
Acetaldehyde

$$CH_3CHO + H_2 \xrightarrow[\text{Catalyst}]{\text{Ni or Cu}} CH_3CH_2OH$$
Ethyl alcohol

Properties: Ethyl alcohol is a colorless liquid (b.p. 78.5°) with a pleasant odor and a burning taste. It is miscible with water in all proportions. When taken orally in small amount, it is a good stimulant and produces heat due to oxidation in blood. In large amounts it acts as a poison.

It gives all the general reactions of alcohol as described earlier. In addition, ethanol gives haloform reaction on heating with iodine and alkali.

$$CH_3CH_2OH + 4I_2 + 6KOH \longrightarrow CHI_3 + HCOOK + 5KI + 5H_2O$$
Ethyl alcohol   Iodoform

Like methyl alcohol, it cannot be dried over anhydrous calcium chloride with which it gives an adduct $CaCl_2.4C_2H_5OH$.

Uses: Ethyl alcohol finds use in the preparation of beverages. It is also used as:
   (a) raw material for the preparation of organic compounds like ethylene, ether, acetic acid, iodoform, chloroform etc.
   (b) solvent for drugs, tinctures, oils, perfumes, inks, dyes, varnishes etc.
   (c) substitute for petrol in internal combustion engines.
   (d) preservative for biological specimens.
   (e) an antifreeze for automobile radiators.

Tests: Ethanol gives following tests:
   (a) *Iodoform test:* Ethyl alcohol gives a yellow precipitate of iodoform on warming with iodine and alkali.
   (b) *Fruity odor test:* On heating gently with acetic acid and few drops of concentrated sulphuric acid, a characteristic fruity odor of ethyl acetate is obtained.

Ethanol can be distinguished from methanol by the fact that it gives iodoform test while methanol does not. Also ethanol on oxidation gives acetic acid while methanol gives formic acid.

14.12 INTERCONVERSION OF DIFFERENT TYPES OF ALCOHOLS

(i) *Conversion of Ethyl Alcohol to Methyl Alcohol*

$$CH_3CH_2OH \xrightarrow{[O]} CH_3CHO \xrightarrow{[O]} CH_3COOH \xrightarrow{NaOH} CH_3COONa$$
Ethyl alcohol

$$\xrightarrow[\Delta]{Sodalime} CH_4 \xrightarrow{Cl_2} CH_3Cl \xrightarrow{AgOH} CH_3OH$$
Methyl alcohol

(ii) *Conversion of Methyl Alcohol to Ethyl Alcohol*

$$CH_3OH \xrightarrow{HI} CH_3I \xrightarrow{KCN} CH_3CN \xrightarrow{4H}$$
Methyl alcohol

$$CH_3CH_2NH_2 \xrightarrow{HNO_2} CH_3CH_2OH$$
Ethyl alcohol

(iii) *Conversion of Primary Alcohol to Secondary Alcohol*

$$CH_3CH_2CH_2OH \xrightarrow{Conc.H_2SO_4} CH_3CH=CH_2 \xrightarrow{HI}$$
n-Propyl alcohol

$$CH_3CHICH_3 \xrightarrow{aqKOH} CH_3CHOHCH_3$$
Isopropyl alcohol

(iv) *Conversion of Secondary Alcohol to Tertiary Alcohol*

This is achieved by the oxidation of secondary alcohol to a ketone, which on reaction with a Grignard reagent gives tertiary alcohol.

$$\begin{matrix} CH_3 \\ CH_3 \end{matrix}\!\!\diagdown\!CHOH \xrightarrow{[O]} \begin{matrix} CH_3 \\ CH_3 \end{matrix}\!\!\diagdown\!C=O \xrightarrow{CH_3MgI} CH_3-\overset{\overset{\displaystyle CH_3}{|}}{\underset{\underset{\displaystyle CH_3}{|}}{C}}-OH + Mg(I)OH$$

Isopropyl alcohol   tert. Butyl alcohol

(v)　Conversion of Primary Alcohol to Tertiary Alcohol

A primary alcohols is first converted into secondary alcohol (as in iii above) and then converted into a tertiary alcohol (as in iv above). Another method of conversion is given below:

$$CH_3\!\!\diagdown\!\!CHCH_2OH \xrightarrow{\text{Conc.H}_2\text{SO}_4} CH_3\!\!\diagup\!\!\diagdown\!\!C=CH_2 \xrightarrow{\text{HI}} CH_3-\underset{\underset{CH_3}{|}}{\overset{\overset{CH_3}{|}}{C}}-I \xrightarrow{\text{aq.KOH}} CH_3-\underset{\underset{CH_3}{|}}{\overset{\overset{CH_3}{|}}{C}}-OH$$

Isobutyl alcohol　　　　　　　　　　　　　　　　　　　　　　　　　**tert Butyl alcohol**

(vi)　Conversion of a Secondary Alcohol to Primary Alcohol

The secondary alcohol is first converted into the alkene, which on hydroboration gives primary alcohol.

$$CH_3\!\!\diagdown\!\!CHOH \xrightarrow{\text{Al}_2\text{O}_3} CH_3-CH=CH_2 \xrightarrow[\text{2. H}_2\text{O}_2,\text{HO}]{\text{1. B}_2\text{H}_6} CH_3CH_2CH_2OH$$

Isopropyl alcohol　　　　　　　　　　　　　　　　　**n-Propyl alcohol**

(vii)　Conversion of a Tertiary Alcohol to a primary Alcohol

The tertiary alcohol like secondary alcohol is first converted into an alkene, which on hydroboration gives primary alcohol.

$$CH_3-\underset{\underset{CH_3}{|}}{\overset{\overset{CH_3}{|}}{C}}-OH \xrightarrow{\text{H}_2\text{SO}_4} CH_3-\overset{\overset{CH_3}{|}}{C}=CH_2 \xrightarrow[\text{2. H}_2\text{O}_2,\text{OH}]{\text{1. B}_2\text{H}_6} CH_3-\overset{\overset{CH_3}{|}}{CH}CH_2OH$$

tert Butyl alcohol　　　　　　　　　　　　　　　　**Isobutyl alcohol**

14.13　FERMENTATION

A solution of glucose in distilled water on treatment with yeast begins to froth and gives the impression that the solution is boiling due to the evolution of carbon dioxide during the reaction. The product obtained is ethyl alcohol.

$$\underset{\textbf{Glucose}}{C_6H_{12}O_6} \xrightarrow{\text{Yeast}} \underset{\textbf{Ethyl alcohol}}{2C_2H_5OH} + 2CO_2$$

This process of conversion of sugars into ethyl alcohol in presence of yeast is termed *Fermentation*.

Yeast is a single celled plant. It grows and multiplies rapidly under suitable conditions. It was earlier believed (Pasteur) that fermentation of sugar to alcohol was brought about by the physiological action of yeast cells. It is now believed that fermentation is brought about by some non-living complex nitrogenous substances (called *enzymes*) present in yeast cells.

The enzymes act like a catalyst during fermentation. Also, the enzymes have selective and specific action. i.e. one particular enzyme can bring about only one chemical change. For example, enzyme penicilium attacks D-tartaric acid only and not the L-tartaric acid. The activity of an enzyme may be enhanced by the presence of another substance called *'co-enzyme'*. Following are given some common enzymes, their origin and utility.

Enzyme	Origin	Utility
Diastage	Malt	Changes malt sugar into dextrin
Maltase	yeast	Changes malt sugar into glucose
Zymase	yeast	Converts glucose into alcohol and CO_2
Invertase	yeast	Converts canesugar into glucose and fructose

The best results in all the above enzymes are obtained at about 27°.

It is now known that a large number of decomposition reactions are brought about by complex organic compounds obtained from yeast or other living organisms. So the term fermentation covers all such reactions which involve slow decomposition of big molecules into simpler ones under the catalytic influence of non-living complex substances. These non-living complex substances are called *Fermants*. The yeast has a number of fermants, which are called *Enzymes*.

The conversion of sugar into ethyl alcohol by yeast is now called *Alcoholic Fermentation*. This distinguishes it from other fermentation reactions.

The scope of the present discussions is limited only to the manufacture of ethyl alcohol.

14.13.1 Manufacture of Ethyl Alcohol

Two types of raw materials are used for the manufacture of ethyl alcohol. The first type are substances containing sugars which can be fermented, e.g. cane juice, beets, molasses, fruit juices etc. The second type are those substances which can not be fermented but can be converted into materials which in turn can be fermented, for example potatoes, rice, barly and maize.

(I) *Ethanol from Molasses:* Molasses are the mother liquor left after the crystallisation of cane sugar from concentrated cane juice. It is a dark colored syrupy liquid containing 50-60% fermentable sugars (glucose, fructose and sucrose). Molasses are excellent and cheap raw material for the manufacture of ethyl alcohol.

In the *first step* molasses having 50-60% fermentable sugars are diluted with water so as to obtain 8-10% sugar in the solution. The resulting solution is acidified with a little dilute sulphuric acid (acid favours growth of yeast but retard bacterial growth). Excess of acid is avoided as the yeast may get destroyed.

Molasses normally contain enough nitrogenous matter to act as food for yeast during fermentation. In case the nitrogenous content of molasses is low, some ammonium sulphate or ammonium phosphate is added.

The *second step* involves the addition of yeast and the temperature is maintained at about 30° for a few days. During this period fermentation sets in and the enzyme invertase (present in yeast) converts sucrose into glucose and fructose. These are further converted into ethyl alcohol by zymase (another enzyme present in yeast).

$$C_{12}H_{22}O_{11} + H_2O \xrightarrow{\text{Invertase}} C_6H_{12}O_6 + C_6H_{12}O_6$$
Sucrose   **Glucose**   **Fructose**

$$C_6H_{12}O_6 \xrightarrow{\text{Zymase}} 2C_2H_5OH + 2CO_2$$
Ethyl alcohol

The fermentation is carried out in fermentation tanks. During fermentation carbon dioxide is evolved which is recovered and used for various purposes.

The *third step* involves distillation. The fermented liquid, known as *wash*, contains 10-15% ethyl alcohol along with water and other impurities. It is subjected to fractional distillation in specially designed continuous distillation apparatus. The fractional distillation gives the following four fractions:

(i) First fraction contains acetaldehyde, an important by-product.

(ii) The second contains 90-95% ethyl alcohol and is used as *Rectified Spirit.*

(iii) Third fraction (120-140°) is called *Fusel Oil*. It mainly contains amyl alcohol and is used as a solvent for paints, varnishes and liquors.

(iv) The fourth is obtained as residue from the still. It is known as *spent wash* and is rich in nitrogen and is used as manure or cattle fodder.

(II) *Ethanol from Starch:* Raw materials containing starch are potatoes, rice, maize, barley and other cereals.

First step is the treatment of raw materials to get starch in solution. For this, potatoes (which are commonly employed) are cut into small pieces and crushed. The resultant mass is steamed at 140-150° under pressure. The starch in cell walls is thus brought into solution called *mash.*

In *Second step,* starch is hydrolysed to maltose with the help of diastase which is obtained by the germination of moist barley in dark at about 15°. Germination is stopped by heating to 50-60°. Dried and germinated barley are called *malt* and contains the enzyme diastase.

To the 'mash' obtained in first step is added 'malt' and the temperature raised to 50°. Within 25-30 minutes the diastase present in malt converts starch into maltose.

$$2(C_6H_{10}O_5)_n + nH_2O \xrightarrow[\text{in malt}]{\text{Diastase}} n\ C_{12}H_{22}O_{11}$$
Starch   **Maltose**

Third step is the fermentation of maltose into alcohol. Yeast is added to Maltose solution (obtained from second step) which forms ethyl alcohol (as discussed earlier).

$$C_{12}H_{22}O_{11} + H_2O \xrightarrow[\text{in yeast}]{\text{Maltase}} 2C_6H_{12}O_6$$
Maltose   **Glucose**

$$C_6H_{12}O_6 + H_2O \xrightarrow[\text{in yeast}]{\text{Zymase}} 2C_2H_5OH + 2CO_2$$
Glucose   **Ethyl alcohol**

The *fourth step* involves distillation and is carried out as the earlier method (ethanol from molasses).

According to a modification of the above method, the starch may be hydrolysed directly to glucose by heating with dilute sulphuric acid. The excess of acid is then neutralised with lime and the separated calcium sulphate is filtered. The solution is then fermented as described above.

By-products: During the manufacture of ethyl alcohol by fermentation following by-products are obtained.

(i)   *Carbon dioxide:* Considerable quantities of carbon dioxide gas is evolved during fermentation. It is stored in iron cylinders under pressure and used for making aerated water. It is also used as dry ice.

(ii) *Acetaldehyde:* It is obtained in the more volatile fraction during the alcohol distillation.

(iii) *Fusel oil:* It is obtained in the fraction 120-140° during the distillation and consists of a mixture of isoamyl alcohol (major content), isobutyl alcohol and n-propyl alcohol.

$$CH_3CH_2CHCH_2OH$$
$$|$$
$$CH_3$$

Isoamyl alcohol

$$CH_3CHCH_2OH$$
$$|$$
$$CH_3$$

Isobutyl alcohol

$$CH_3CH_2CH_2OH$$

n-Propyl alcohol

It is interesting to note that these alcohols are not produced by the fermentation of sugars but are obtained from proteins (present in the raw materials) by the action of yeast. The fusel oil is used for the preparation of amyl acetate, which is a valuable solvent for varnishes.

(iv) *Spent wash:* The residual liquor from which ethyl alcohol has been removed by distillation is called 'spent wash'. It is rich in nitrogen content and is used as manure or cattle food.

(v) *Potassium hydrogen tartarate:* It is obtained during the fermentation of grape juice. Being insoluble in alcohol, it separates during fermentation. It is used for the manufacture of tartaric acid and Rochelle salt.

14.13.2   Absolute Alcohol

The alcohol obtained by the fermentation process is 95% alcohol and is called rectified spirit. It is not possible to get an alcohol of higher concentration by fractional distillation of rectified spirit. this is because a mixture of 95.6% alcohol and water boil at a lower temperature (78.1°) than the boiling point of pure alcohol (78.5°). Pure or absolute alcohol is obtained by refluxing rectified spirit for 6-8 hours over quick lime and then distilling. The fore run is discarded and the main distillate is 100% alcohol or absolute alcohol.

Absolute alcohol can also be obtained by the azeotropic distillation of rectified spirit with benzene. A small quantity of benzene is added to rectified spirit and the distillation is carried out. First a ternary mixture of water, alcohol and benzene distills at 65° till all the water is removed. The boiling point then rises and the remaining benzene comes over as binary mixture with alcohol at 68°. Finally absolute alcohol distills at 78.5°.

14.13.3   Denatured Alcohol

Since ethyl alcohol is used for the preparation of alcoholic beverages, its manufacture and sale is controlled by the government. Heavy excise duty is levied on the sale of alcoholic beverages. However, for industrial use, the alcohol has to be cheap and duty free. Therefore the industrial alcohol is denatured by addition of some poisonous substances like methyl alcohol, acetone or pyridine. This renders alcohol unfit for human consumption. The alcohol obtained after adding poisonous substances is called 'denatured alcohol' and is used under the name of 'methylated spirit'. Normally methylated spirit contains 4-5% methyl alcohol together with traces of pyridine and some coloring matter.

14.13.4   Power Alcohol

The demand of petroleum is increasing at an alarming rate while the petroleum resources are limited and are likely to be exhausted. This has made scientists to discover alternate fuels. It has been found that alcohol can be mixed with petrol and used in internal combustion engines ,though rectified spirit is insoluble in petrol (due to presence of 5% water), it is rendered soluble by mixing with some benzene. Alcohol, thus, used for generation of power is called *power alcohol.*

14.13.5   Alcoholic Beverages

Alcoholic beverages are liquors used for drinking purposes and contain alcohol as the main intoxicating agent. These contain different percentages of alcohols and are obtained from different materials. alcoholic beverages are of two types the undistilled and distilled ones.

The undistilled beverages are obtained by fermentation of juices or grains. Following are some common undistilled beverages along with their source and alcohol content:

Name	Source	% Alcohol
Beer	Barley	2-6
Cidar	Apples	3-6
Wine	Grapes	8-10
Port and Sherry	Grapes	15-20 (fortified)

In case the alcoholic content of any beverage is low, it is fortified by addition of extra ethyl alcohol to achieve the desired concentration of alcohol.

Distilled beverages are obtained by distillation of the fermented liquids. By distillation, most of the alcohol along with other volatile products like flavours, essential oils, esters etc are passed over as distillate. The distilled beverages have higher alcohol content (45-50%). Some of the common distilled beverages along with their source and percentage alcohol are:

Name	Source	% Alcohol
Whiskey	Barley	40-50
Brandy	Peeches, Apples	40-50
Gin	Maize	35-40
Rum	Molasses	35-40

14.13.6   Alcoholometry (Strength of Alcohol)

The determination of the percentage of alcohol is termed alcoholometry. A simple method commonly used for the determination of percentage of alcohol is by determining the specific gravity of the sample by means of the hydrometer and then finding the exact percentage of alcohol by refering to reference tables.

The percentage of alcohol in pharmaceutical preparation is determined by taking 100 ml of the preparation, diluting with water (100 ml) and then distilling it. 100 ml of the distillate is collected in a measuring flask and its specific gravity determenined and from this the % alcohol found out by refering to the standard table.

For excise purposes the percentage of alcohol is expressed in terms of proof spirit. Proof spirit is defined as an alcohol water mixture having specific gravity 0.91976 at 15° and contain 57.1% alcohol by weight. A sample is called over proof (OP) when it is stronger than proof spirit and under proof (UP) when it is weaker than the proof spirit. The strength of any liquor is expressed as °OP or °UP.

14.14   PROBLEMS

1. Explain why the lower alcohols are completely miscible in water and higher alcohols are not.
2. (a) Explain why alcohols have much higher boiling points compared to other molecules of the same molecular weight.
   (b) Explain how alcohols behaves both as an acid and a base.
   (c) The alcohol on treatment with thionyl chloride and phosphorus trichloride give alkyl halides. Give their mechanism of formation.
3. Give the products obtained in the following reaction giving mechanism:

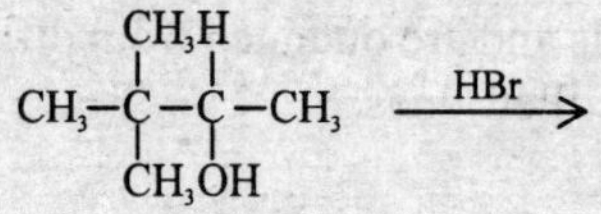

$$CH_3-\overset{\overset{\displaystyle CH_3}{|}}{C}-\overset{\overset{\displaystyle H}{|}}{\underset{\underset{\displaystyle CH_3OH}{|}}{C}}-CH_3 \xrightarrow{HBr}$$

4. How is 2-butanol obtained by the following methods:
   (a) Hydration of an alkene.   (c) Use of a Grignard reagent
   (b) Hydrolysis of an alkyl halide   (d) Reduction of Ketone
5. Give structural formula for the following alcohols and mention whether it is primary, secondary or tertiary:
   (a) 3-Pentanol   (d) 3-methyl-2-pentanol
   (b) 2,3-dimethyl-1-propanol   (e) 1-methylcyclopentanol
   (c) 2-methyl 1-butanol
6. Discuss the orbital structure of methyl alcohol.
7. Give the products obtained in:

$$CH_3-CH=CH_2 \xrightarrow[\text{(ii) } H_2O_2, OH^-]{\text{(i) } B_2H_6}$$

$$C_2H_5NH_2 \xrightarrow{HNO_2}$$

$$CH_3COOH + C_2H_5OH \xrightarrow{H^1}$$

$$CH_3-\underset{\underset{CH_3}{|}}{\overset{\overset{CH_3}{|}}{C}}-OH \xrightarrow{[O] \atop HNO_3/H_2O}$$

8. How will you differentiate between a primary, secondary and a tertiary alcohol.
9. How will you obtain ethyl alcohol from ethylene and acetone.
10. How will you carry out the following conversions..
    (a) $CH_3CH_2CH_2OH \longrightarrow CH_3OH$ (c) $CH_3CHOHCH_3 \longrightarrow (CH_3)_3C-OH$
    (b) $CH_3CH_2CH_2OH \longrightarrow CH_3CHOHCH_3$ (d) $CH_3CHOHCH_3 \longrightarrow CH_3CH_2CH_2OH$
    (e) $(CH_3)_3COH \longrightarrow CH_3\underset{\overset{|}{CH_3}}{CH}-CH_2OH$

11. Write a note on alcoholic fermentation.
12. What are enzymes. Give their uses.
13. What is absolute alcohol, denatured alcohol and power alcohol.
14. How will you determine the concentration of ethyl alcohol in alcoholic beverages.

15

Polyhydric Alcohols and Lipids

Polyhydric alcohols, as the name indicates contain more than one hydroxy group per molecule. They are further subdivided into dihydric alcohols (containing two hydroxy groups) or trihydric alcohols (containing three hydroxy groups).

15.1 DIHYDRIC ALCOHOLS OR GLYCOLS

Dihydric alcohols are the compounds containing two hydroxy groups. They are classified α-, β- or γ-glycols depending upon the relative positions of the two hydroxy groups like 1,2-diols are α-, 1,3-diols are β- and 1,4-diols are γ-diols. However, diols having two hydroxy groups on same carbon atom are unstable. The α-glycols are most common amongst the various classes.

15.2 NOMENCLATURE

The common names to various diols are assigned from the corresponding alkanes or polymethylenes from which they can be prepared by direct hydroxylation.

IUPAC system of nomenclature adds suffix 'diol' to the parent alkane and the numbers are used to indicate the position of hydroxy groups and the side chains if any. Below is given a few glycols and their common and IUPAC names.

Glycols	Formula	Common Name	IUPAC Name
α	$HOCH_2CH_2OH$	Ethylene glycol	1,2-Ethanediol
α	$HOCH_2CH(OH)CH_3$	Propylene glycol	1,2-Propanediol
α	$(CH_3)_2C(OH)CH_2OH$	Isobutylene glycol	2-Methyl-1,2-propanediol
β	$HOCH_2CH_2CH_2OH$	Trimethylene glycol	1,3-Propanediol
γ	$HOCH_2CH_2CH_2CH_2OH$	Tetramethylene glycol	1,4-Butanediol

15.3 EHTYLENE GLYCOL/GLYCOL/1,2-ETHANEDIOL

It is the simplest and most common member of the class.

15.3.1 Preparation

(i) *Hydroxylation of Ethylene:* Ethylene glycol may be prepared by passing ethylene into cold dilute alkaline potassium permanganate solution.

$$CH_2=CH_2 + H_2O + [O] \xrightarrow{KMnO_4} HOCH_2CH_2OH$$
Ethylene $\qquad\qquad\qquad\qquad$ Ethylene glycol

(ii) Ethylene glycol may be prepared by passing ethylene into hypochlorous acid followed by hydrolysis of ethylene chlorohydrin by boiling with aqueous sodium bicarbonate.

$$CH_2=CH_2 + HOCl \longrightarrow HOCH_2CH_2Cl \longrightarrow HOCH_2CH_2OH$$
Ethylene $\qquad\qquad\qquad$ Ethylene $\qquad\qquad$ Ethylene glycol
$\qquad\qquad\qquad\qquad\quad$ chlorohydrin

(iii) *Hydrolysis of ethylene dibromide* with hot aqueous sodium carbonate is a laboratory method of preparation of ethylene glycol.

$$BrCH_2CH_2Br + Na_2CO_3 + H_2O \longrightarrow HOCH_2CH_2OH + 2\,NaBr + CO_2$$
Ethylene dibromide $\qquad\qquad\qquad\qquad$ Ethylene glycol

The yield in the above reaction is only 50% due to the conversion of some of the ethylene dibromide into vinyl bromide.

$$BrCH_2CH_2Br + Na_2CO_3 \longrightarrow CH_2=CHBr + NaBr + NaHCO_3$$
Ethylene dibromide $\qquad\qquad\qquad$ **Vinyl bromide**

Better yields are obtained by heating ethylene dibromide with potassium acetate in glacial acetic acid followed by hydrolysis of diacetate with hydrogen chloride in methanolic solution.

$$BrCH_2CH_2Br + 2CH_3COO^-K^+ \xrightarrow{AcOH} \begin{matrix} CH_2OCOCH_3 \\ | \\ CH_2OCOCH_3 \end{matrix} + 2KBr$$
Ethylene dibromide $\qquad\qquad\qquad\qquad\qquad\qquad$ Ethylene diacetate

$$\begin{matrix} CH_2OCOCH_3 \\ | \\ CH_2OCOCH_3 \end{matrix} + H_2O \xrightarrow[CH_3OH]{HCl} \begin{matrix} CH_2OH \\ | \\ CH_2OH \end{matrix} + 2CH_3COOH$$
$\qquad\qquad\qquad\qquad\qquad\qquad\qquad$ Ethylene glycol

(iv) *Hydrolysis of ethylene oxide* with warm dilute hydrochloric acid produces ethylene glycol.

$$H_2C\overset{\displaystyle O}{\overset{\diagup\diagdown}{-}}CH_2 + H_2O \xrightarrow[65°]{HCl} HOCH_2CH_2OH$$
Ethylene oxide $\qquad\qquad\qquad\qquad$ Ethylene glycol

This method is used for the industrial production of ethylene glycol.

15.3.2 Physical Properties

Ethylene glycol is a colorless viscous liquid b.p. 197°, m.p. −11.5° and has a sweet taste. It is miscible with water in all proportions but immiscible in ether. The high boiling point and high solubility in water is due to hydrogen bonding involving both hydroxy groups.

15.3.3 Chemical Properties

Ethylene glycol undergoes reactions of primary alcohols as it contains two primary alcoholic groups but more vigrous conditions are sometimes required for the reaction at the second hydroxy group.

(1) Reaction of ethylene glycol with sodium yields mono and dialkoxides under different conditions.

$$HOCH_2CH_2OH + Na \xrightarrow{50°} HOCH_2CH_2O^-Na^+ + ½H_2$$
Ethylene glycol $\qquad\qquad\qquad$ **Monosodium glycolate**

$$HOCH_2CH_2O^-Na^+ + Na \xrightarrow{160°} Na^+O-CH_2CH_2O^-Na^+ + \tfrac{1}{2}H_2$$

Disodium glycolate

(2)   When reacted upon with hydrogen chloride, ethylene chlorohydrin and ethylene dichloride are formed.

$$HOCH_2CH_2OH + HCl \xrightarrow{160°} ClCH_2CH_2OH + H_2O$$

Ethylene glycol   **Ethylene chlorohydrin**

$$ClCH_2CH_2OH + HCl \xrightarrow{200°} ClCH_2CH_2Cl + H_2O$$

Ethylene chlorohydrin   **Ethylene dichloride**

(3)   Phosphorus trichloride and phosphorus tribromide react with glycol to give corresponding dichloride and dibromide.

$$HOCH_2CH_2OH + PBr_3 \longrightarrow 3BrCH_2CH_2Br + 2H_3PO_3$$

Ethylene glycol   **Ethylene dibromide**

Phosphorus triiodide produces unstable ethylene diiodide which decomposes into ethylene and iodine.

$$HOCH_2CH_2OH + PI_3 \longrightarrow [CH_2I-CH_2I] \longrightarrow CH_2=CH_2 + I_2$$

Ethylene glycol   **Ethylene diiodide**   **Ethylene**

(4)   Reaction of ethylene glycol with carboxylic acids produces mono and diesters.

$$HOCH_2CH_2OH \xrightarrow{CH_3COOH/H_2SO_4} HOCH_2CH_2OCOCH_3 \longrightarrow CH_3OCOCH_2CH_2OCOCH_3$$

Glycol   **Glycol monoacetate**   **Glycol diacetate**

Dibasic acids form polymers with ethylene glycol. The well known synthetic fibre 'terylene' is a polymerisation product of terephthalic acid and glycol.

$$n\ HOOC-\langle\bigcirc\rangle-COOH + n\ HOCH_2CH_2OH \longrightarrow \{CO-\langle\bigcirc\rangle-COOCH_2CH_2O\}_n$$

(5)   Aldehydes and ketones give cyclic acetals and ketals respectively.

$$HOCH_2CH_2OH + O=C\langle^H_R \longrightarrow \begin{matrix} CH_2-O \\ | \\ CH_2-O \end{matrix}\rangle C\langle^H_R$$

Ethylene glycol   **Aldehyde**   **Cyclic acetal**

$$HOCH_2CH_2OH + O=C\langle^R_R \longrightarrow \begin{matrix} CH_2-O \\ | \\ CH_2-O \end{matrix}\rangle C\langle^R_R$$

Ethylene glycol   **Ketone**   **Cyclic ketal**

This method is used for the protection of carbonyl group. The carbonyl compound may be regenerated by the action of periodic acid.

$$\begin{matrix} CH_2-O \\ | \\ CH_2-O \end{matrix}\rangle C\langle^{CH_3}_{CH_3} \xrightarrow{HIO_4} 2CH_2O + \begin{matrix} CH_3 \\ CH_3 \end{matrix}\rangle C=O$$

Formal-   **Acetone**
dehyde

(6) Ethylene glycol yields a number of oxidation products. When heated with nitric acid as one or both the alcoholic groups get oxidised as follows.

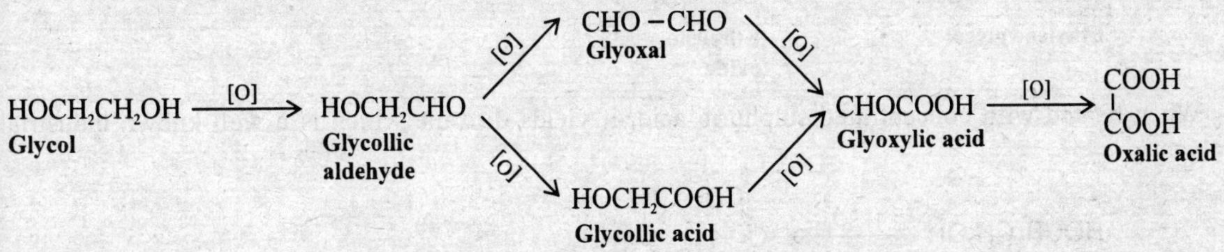

$$HOCH_2CH_2OH \xrightarrow{[O]} HOCH_2CHO$$

Glycol (under $HOCH_2CH_2OH$)

Glycollic aldehyde (under $HOCH_2CHO$)

$$\begin{array}{c} CHO-CHO \\ \text{Glyoxal} \end{array} \qquad \begin{array}{c} HOCH_2COOH \\ \text{Glycollic acid} \end{array}$$

$$CHOCOOH$$
Glyoxylic acid

$$\xrightarrow{[O]} \begin{array}{c} COOH \\ | \\ COOH \end{array}$$
Oxalic acid

When oxidised with lead tetraacetate or periodic acid, glycol yields formaldehyde.

$$HOCH_2CH_2OH \xrightarrow[\text{or HIO}_4]{Pb(OCOCH_3)_4} \overset{O}{\underset{}{\overset{\|}{H-C-H}}}$$

Ethylene glycol   Formaldehyde

The mechanism involved with lead tetraacetate is as follows.

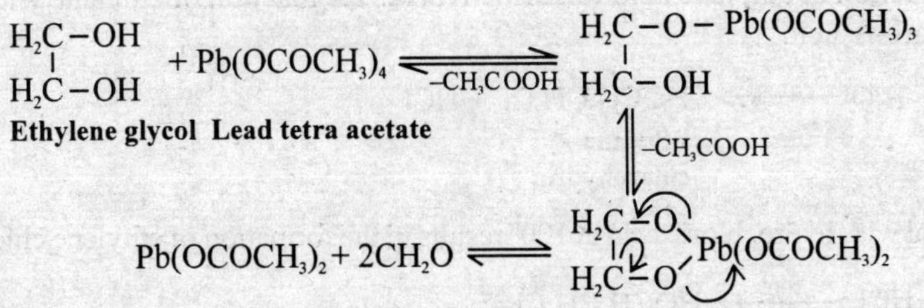

$$\begin{array}{c} H_2C-OH \\ | \\ H_2C-OH \end{array} + Pb(OCOCH_3)_4 \underset{-CH_3COOH}{\overset{}{\rightleftharpoons}} \begin{array}{c} H_2C-O-Pb(OCOCH_3)_3 \\ | \\ H_2C-OH \end{array}$$

Ethylene glycol Lead tetra acetate

$$\Big\updownarrow -CH_3COOH$$

$$Pb(OCOCH_3)_2 + 2CH_2O \rightleftharpoons \begin{array}{c} H_2C-O \\ | \quad \rangle Pb(OCOCH_3)_2 \\ H_2C-O \end{array}$$

With periodic acid, it involves the following steps.

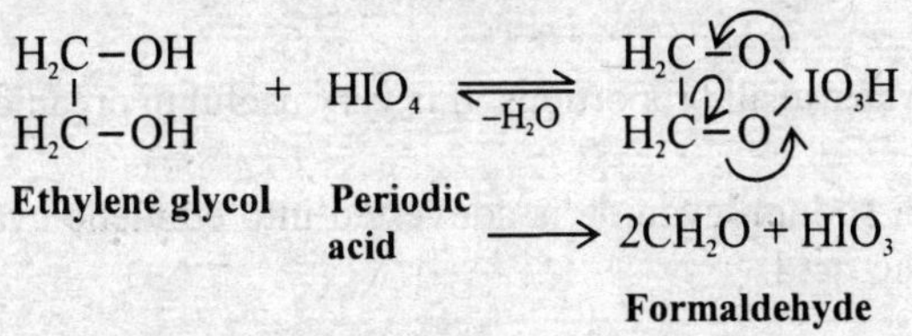

$$\begin{array}{c} H_2C-OH \\ | \\ H_2C-OH \end{array} + HIO_4 \underset{-H_2O}{\rightleftharpoons} \begin{array}{c} H_2C-O \\ | \quad \rangle IO_3H \\ H_2C-O \end{array}$$

Ethylene glycol   Periodic acid

$$\longrightarrow 2CH_2O + HIO_3$$
Formaldehyde

Ethylene glycol may be oxidised to formic acid on treatment with potassium permanganate or potassium dichromate in acidic medium.

$$HOCH_2CH_2OH \xrightarrow[K_2Cr_2O_7/H^+]{KMnO_4/H^+} \overset{O}{\underset{}{\overset{\|}{H-C-OH}}}$$

Ethylene glycol   Formic acid

(7) Action of heat: Ethylene glycol is dehydrated when heated with zinc chloride.

$$HOCH_2CH_2OH \xrightarrow{ZnCl_2} CH_3CHO + H_2O$$

Ethylene glycol   Acetaldehyde

When heated alone at 500°, it gives ethylene oxide.

$$HOCH_2CH_2OH \xrightarrow{500°} \overset{O}{\overset{\diagup\diagdown}{CH_2-CH_2}} + H_2O$$

Ethylene glycol   **Ethylene oxide**

When heated with concentrated sulphuric acid, it yields dioxane which is a well known industrial solvent.

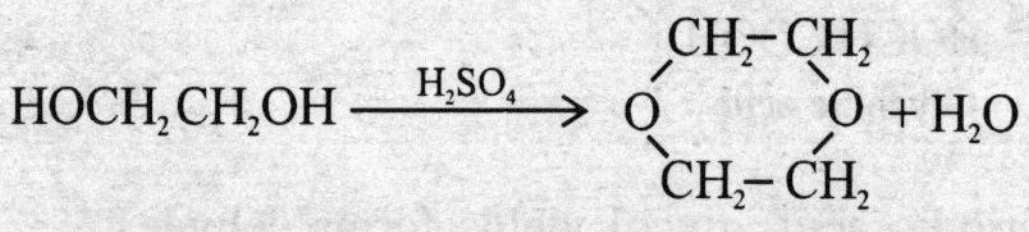

$$HOCH_2CH_2OH \xrightarrow{H_2SO_4} \begin{matrix} CH_2-CH_2 \\ O \qquad\qquad O \\ CH_2-CH_2 \end{matrix} + H_2O$$

Ethylene glycol   **Dioxane**

Amongst the various uses of ethylene glycol, the uses as antifreeze agent and as a solvent are most important.

15.4   ETHYLENE CHLOROHYDRIN/2-CHLOROETHANOL

15.4.1   Preparation

(1)   The large scale production of ethylene chlorohydrin involves the reaction of ethylene with chlorine water or aqueous hypochlorous acid.

$$CH_2=CH_2 + Cl_2 + H_2O \longrightarrow HOCH_2CH_2Cl + HCl$$

Ethylene   **Ethylene chlorohydrin**

(2)   The reaction of glycol with hydrochloric acid at 100° results in the formation of ethylene chlorohydrin.

$$HOCH_2CH_2OH + HCl \xrightarrow{100°} HOCH_2CH_2Cl$$

Ethylene glycol   **Ethylene chlorohydrin**

15.4.2   Properties

It is a colorless liquid, b.p.128.8° and soluble in water in all proportions. It is very useful in organic synthesis as it contains two different reactive groups.

(1)   Ethylene chlorohydrin on warming with sodium cyanide is converted into ethylene cyanohydrin which on hydrolysis yields β-hydroxypropanoic acid.

$$HOCH_2CH_2Cl + NaCN \longrightarrow HOCH_2CH_2CN + NaCl$$

Ethylene chlorohydrin   **Ethylene cyanohydrin**

$$HOCH_2CH_2CN \xrightarrow{HCl/H_2O} HOCH_2CH_2COONH_4 \xrightarrow{-NH_4Cl} HOCH_2CH_2COOH$$

Ethylene cyanohydrin   **β-Hydroxy-propanoic acid**

(2)   When distilled with concentrated alkali ethylene chlorohydrin forms ethylene oxide.

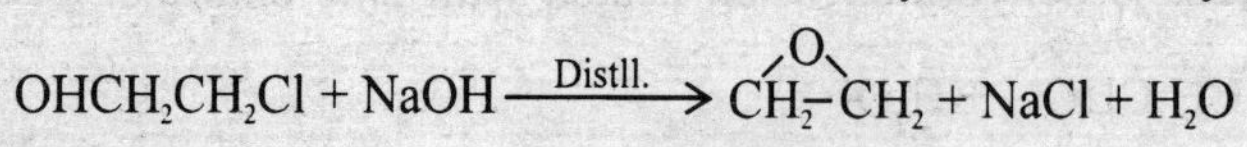

$$OHCH_2CH_2Cl + NaOH \xrightarrow{Distll.} \overset{O}{\overset{\diagup\diagdown}{CH_2-CH_2}} + NaCl + H_2O$$

Ethylene chlorohydrin   **Ethylene oxide**

15.5   ETHYLENE OXIDE/EPOXYETHANE/OXIRANE

Since the oxygen atom is linked to two carbon atoms in the molecule, ethylene oxide is called epoxyethane. It is also known as oxirane as it contains oxirane ring (three membered ring having an oxygen atom) or may also be referred to as cyclic ether.

15.5.1   Preparation

(1)   It is manufactured by passing a mixture of ethylene and oxygen (air) over silver catalyst under pressure.

$$CH_2=CH_2 + \tfrac{1}{2}O_2 \xrightarrow{\;Ag\;} \overset{O}{\overbrace{CH_2\!-\!CH_2}}$$

Ethylene   Ethylene oxide

(2)   It may also be prepared by epoxydation of ethylene with perbenzoic acid.

$$CH_2=CH_2 + C_6H_5COOOH \longrightarrow \overset{O}{\overbrace{CH_2\!-\!CH_2}}$$

Ethylene   Perbenzoic acid   Ethylene oxide

(3)   Distillation of ethylene chlorohydrin in concentrated potassium hydroxide yields ethylene oxide.

$$HOCH_2CH_2Cl + KOH \longrightarrow \overset{O}{\overbrace{CH_2\!-\!CH_2}} + KCl + H_2O$$

Ethylene   Ethylene
chlorohydrin   oxide

15.5.2   Properties

It is a gas at ordinary temperature and pressure, b.p. 10.7°, soluble in water, ethanol and ether.

(1)   Ethylene oxide undergoes rearrangement on heating to acetaldehyde.

$$\overset{O}{\overbrace{CH_2\!-\!CH}}\!\!\!\overset{|}{\underset{H}{}} \xrightarrow{\;\text{Rearrangement}\;} CH_3CHO$$

                                            Acetaldehyde
Ethylene oxide

(2)   Ethylene oxide undergoes ring opening in acid medium to form addition products as follows:

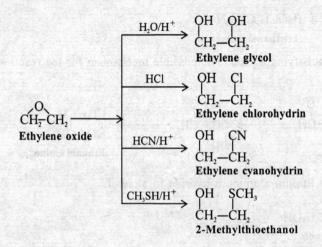

$\xrightarrow{H_2O/H^+}$
OH   OH
| |
CH$_2$—CH$_2$
Ethylene glycol

$\xrightarrow{HCl}$
OH   Cl
| |
CH$_2$—CH$_2$
Ethylene chlorohydrin

$\overset{O}{\overbrace{CH_2\!-\!CH_2}}$ ⟶
Ethylene oxide

$\xrightarrow{HCN/H^+}$
OH   CN
| |
CH$_2$—CH$_2$
Ethylene cyanohydrin

$\xrightarrow{CH_3SH/H^+}$
OH   SCH$_3$
| |
CH$_2$—CH$_2$
2-Methylthioethanol

In all the reactions stated above ethylene oxide under the influence of acid forms oxonium ion which in turn is attacked by nucleophile as shown below.

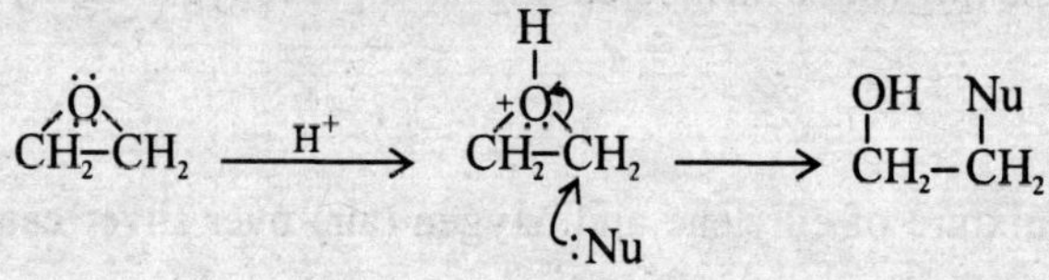

$$\underset{CH_2-CH_2}{\overset{\ddot{O}}{\diagup}} \xrightarrow{H^+} \underset{\underset{\ddot{:}Nu}{\overset{H}{\underset{|}{+O}}}}{CH_2-CH_2} \longrightarrow \underset{CH_2-CH_2}{\overset{OH \quad Nu}{\overset{|}{}\overset{|}{}}}$$

[:Nu = HO–, Cl–, CN–, CH$_3$S–]

(3)   Ethylene oxide undergoes ring opening in basic medium also to form addition products.

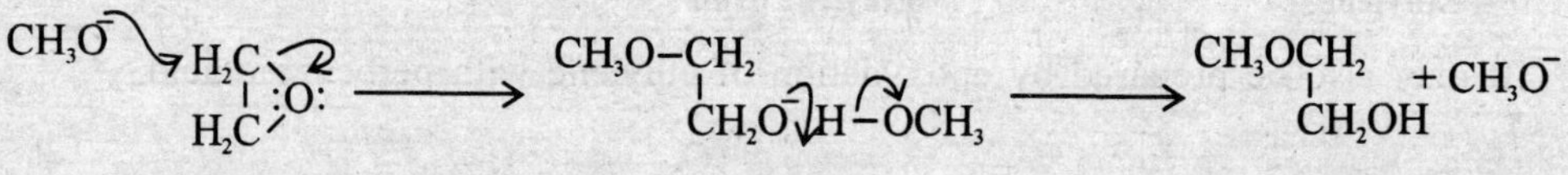

$$CH_3O \searrow \underset{H_2C}{\overset{H_2C}{\diagdown}} \overset{\frown}{\underset{:O:}{|}} \longrightarrow \underset{CH_2O\overset{\frown}{\rfloor}H\overset{\frown}{-}OCH_3}{CH_3O-CH_2} \longrightarrow \underset{CH_2OH}{\overset{CH_3OCH_2}{|}} + CH_3O^-$$

When ethylene oxide is heated with methanol under pressure monomethyl ether of glycol is produced which is known as methyl cellosolve.

$$\underset{\textbf{Ethylene oxide}}{\overset{\ddot{O}}{\overset{\diagup}{CH_2-CH_2}}} + CH_3OH \longrightarrow \underset{\textbf{Methyl cellosolve}}{HOCH_2-CH_2OCH_3}$$

The corresponding ethyl ether is called ethyl cellosolve. Cellosolves are useful as solvents. Cellosolves on further reaction with ethylene oxide produces carbitols.

$$\underset{\textbf{Methyl Cellosolve}}{HOCH_2CH_2OCH_3} + \underset{\textbf{Ethylene oxide}}{\overset{\ddot{O}}{\overset{\diagup}{CH_2-CH_2}}} \longrightarrow \underset{\textbf{Carbitol}}{HOCH_2CH_2OCH_2CH_2OCH_3}$$

(4)   Ethylene oxide reacts with ammonia to form a mixture of three amino alcohols commonly known as ethanolamines.

$$\underset{CH_2-CH_2}{\overset{\ddot{O}}{\diagup}} + NH_3 \longrightarrow \underset{\textbf{Monoethanol amine}}{HOCH_2CH_2NH_2} \xrightarrow{\overset{\ddot{O}}{\overset{\diagup}{CH_2-CH_2}}}$$

$$\underset{\textbf{Diethanol amine}}{(HOCH_2CH_2)_2NH} \xrightarrow{\overset{\ddot{O}}{\overset{\diagup}{CH_2-CH_2}}} \underset{\textbf{Triethanol amine}}{(HOCH_2CH_2)_3N}$$

The ethanol amines find use as emulsifying agents. The possible mechanism for the reaction can be given as below:

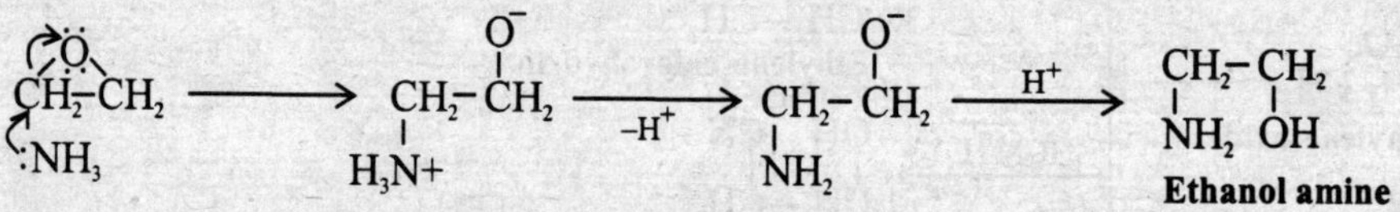

$$\underset{\underset{:NH_3}{CH_2-CH_2}}{\overset{\overset{\frown}{\ddot{O}}}{\diagup}} \longrightarrow \underset{\underset{H_3N+}{CH_2-CH_2}}{\overset{O^-}{\overset{|}{}}} \xrightarrow{-H^+} \underset{\underset{NH_2}{CH_2-CH_2}}{\overset{O^-}{\overset{|}{}}} \xrightarrow{H^+} \underset{\underset{NH_2 \quad OH}{CH_2-CH_2}}{\overset{CH_2-CH_2}{}}$$

Ethanol amine

(5)   Ethylene oxide is reduced by lithium aluminum hydride to alcohol.

$$\underset{\textbf{Ethylene oxide}}{\overset{O}{\overset{\diagup}{CH_2-CH_2}}} \xrightarrow{LiAlH_4} \underset{\textbf{Ethanol}}{CH_3CH_2OH}$$

(6) Ethylene oxide reacts with Grignard reagent to give primary alcohols.

$$\underset{\text{Ethylene oxide}}{CH_2\text{-}CH_2} + RMgI \xrightarrow{\text{Ether}} RCH_2CH_2OMgI \xrightarrow{H_2O/H^+} \underset{\text{Primary alcohol}}{RCH_2CH_2OH}$$

Ethylene oxide is used as insecticide for grains, dry fruits and tobacco. It is also used as effective fumigant. It also serves as a starting material for the synthesis of various chemicals like cellosolves, ethanolamines, diglyme and carbitols.

15.6 PINACOL/TETRAMETHYLENE GLYCOL/2,3-DIMETHYLBUTANE-2,3-DIOL

15.6.1 Preparation

The most common method for preparation of pinacol is by reduction of acetone with magnesium amalgam.

$$2CH_3COCH_3 \xrightarrow[\text{(ii) } H_2O]{\text{(i) Mg–Hg}} \underset{\text{Pinacol}}{CH_3\text{-}\underset{OH}{\underset{|}{\overset{CH_3}{\overset{|}{C}}}}\text{-}\underset{OH}{\underset{|}{\overset{CH_3}{\overset{|}{C}}}}\text{-}CH_3}$$

Acetone   Pinacol

The mechanism of this reaction is believed to proceed as

$$2\ \underset{\text{Acetone}}{CH_3COCH_3} + 2e^- \xrightarrow{\text{Mg}} [(CH_3)_2\ C\text{-}O^-]_2\ Mg^{2+}$$

$$\longrightarrow \left[CH_3\text{-}\underset{O^-}{\underset{|}{\overset{CH_3}{\overset{|}{C}}}}\text{-}\underset{O^-}{\underset{|}{\overset{CH_3}{\overset{|}{C}}}}\text{-}CH_3 \right] Mg^{2+} \xrightarrow{\text{(i) } H^+/H_2O} \underset{\text{Pinacol}}{CH_3\text{-}\underset{OH}{\underset{|}{\overset{CH_3}{\overset{|}{C}}}}\text{-}\underset{OH}{\underset{|}{\overset{CH_3}{\overset{|}{C}}}}\text{-}CH_3}$$

Pinacol undergoes rearrangement in presence of dilute sulphuric acid to form pinacolone.

$$\underset{\text{Pinacol}}{(CH_3)_2\text{-}\underset{OH}{\underset{|}{C}}\text{-}\underset{OH}{\underset{|}{C}}\text{-}(CH_3)_2} \xrightarrow{H_2SO_4} \underset{\text{Pinacolone}}{CH_3\text{-}\underset{O}{\overset{||}{C}}\text{-}C\,(CH_3)_3}$$

This reaction is the general reaction for ditertiary alcohols (pinacols) and is known as pinacol-pinacolone rearrangement. Its mechanism is believed to involve initial formation of a carbocation followed by migration of an alkyl or aryl group or a hydride ion.

$$CH_3\text{-}\underset{OH}{\underset{|}{\overset{CH_3}{\overset{|}{C}}}}\text{-}\underset{OH}{\underset{|}{\overset{CH_3}{\overset{|}{C}}}}\text{-}CH_3 \underset{}{\overset{H^+}{\rightleftharpoons}} CH_3\text{-}\underset{HO}{\underset{|}{\overset{CH_3}{\overset{|}{C}}}}\text{-}\underset{^+OH_2}{\underset{|}{\overset{CH_3}{\overset{|}{C}}}}\text{-}CH_3 \xrightarrow[-H_2O]{}$$

$$CH_3\text{-}\underset{OH}{\underset{|}{C}}\overset{CH_3}{\underset{+}{\overset{|}{=}}}C\text{-}CH_3 \xrightarrow[\text{shift}]{\text{1,2-Methyl}} CH_3\text{-}\underset{+}{\overset{:\ddot{O}H}{\overset{|}{C}}}\text{-}\underset{CH_3}{\underset{|}{\overset{CH_3}{\overset{|}{C}}}}\text{-}CH_3 \longleftrightarrow$$

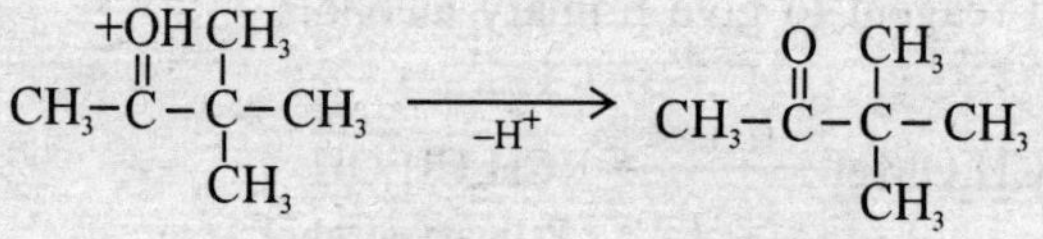

$$CH_3-\overset{\overset{+OH}{\|}}{C}-\overset{\overset{CH_3}{|}}{\underset{\underset{CH_3}{|}}{C}}-CH_3 \xrightarrow{-H^+} CH_3-\overset{\overset{O}{\|}}{C}-\overset{\overset{CH_3}{|}}{\underset{\underset{CH_3}{|}}{C}}-CH_3$$

Pinacolone

Similarly, benzpinacolone is obtained from benzpinacol.

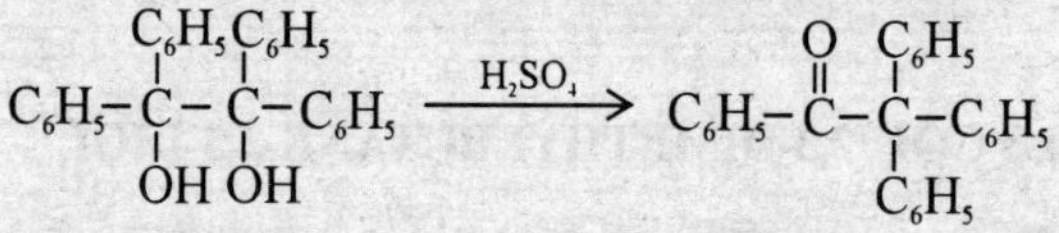

$$C_6H_5-\overset{\overset{C_6H_5C_6H_5}{|}}{\underset{\underset{OH}{|}}{C}}-\overset{|}{\underset{\underset{OH}{|}}{C}}-C_6H_5 \xrightarrow{H_2SO_4} C_6H_5-\overset{\overset{O}{\|}}{C}-\overset{\overset{C_6H_5}{|}}{\underset{\underset{C_6H_5}{|}}{C}}-C_6H_5$$

Benzpinacol   **Benzpinacolone**

15.7   TRIHYDRIC ALCOHOLS/GLYCEROL/GLYCERINE/1,2,3-PROPANETRIOL

Glycerol is the only important trihydric alcohol, b.p. 290°. It occurs in almost all animal and vegetable oils and fats as the glyceryl ester of palmitic, stearic and oleic acids.

15.7.1   Preparation

(1)   Glycerol is prepared in large quantities as a by-product of soap manufacture or stearic acid manufacture required in candle industry.

Alkaline hydrolysis of natural fats and oils (which are triester of glycerol and higher carboxylic acids), produces glycerol and sodium salts of higher fatty acids or soaps.

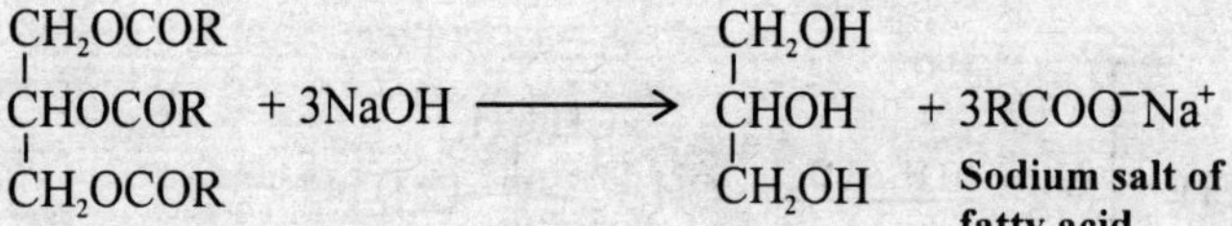

$$\begin{array}{l} CH_2OCOR \\ |\\ CHOCOR \\ |\\ CH_2OCOR \end{array} + 3NaOH \longrightarrow \begin{array}{l} CH_2OH \\ |\\ CHOH \\ |\\ CH_2OH \\ \textbf{Glycerol} \end{array} + 3RCOO^-Na^+ \\ \quad\quad\quad\quad\quad\quad\quad\quad\quad\quad\quad\quad\quad\quad \textbf{Sodium salt of} \\ \quad\quad\quad\quad\quad\quad\quad\quad\quad\quad\quad\quad\quad\quad \textbf{fatty acid}$$

Glycerol tripalmitate ($R=C_{15}H_{31}$) forms sodium palmitate, whereas glycerol tristearate ($R=C_{17}H_{35}$) forms sodium salt of stearic acid and glycerol trioleate ($R=C_{17}H_{33}$) forms sodium salt of oleic acid.

In the above reaction soaps precipitates out leaving behind glycerol in solution known as spent lye. Glycerol is separated as by-product by various purification steps.

(2)   Various synthetic methods of preparation of glycerol have been developed starting from propene which is a petroleum product.

$$CH_3CH=CH_2 \xrightarrow[450-500°]{Cl_2} ClCH_2CH=CH_2 \xrightarrow[150°,12\ atm]{aq\ Na_2CO_3} HOCH_2CH=CH_2$$

Propene   **Allyl chloride**   **Allyl alcohol**

$$\xrightarrow{HOCl} HOCH_2CHClCH_2OH \xrightarrow{NaOH} HOCH_2CHOHCH_2OH$$

Glycerol-β-monochlorohydrin   **Glycerol**

An alternative route via epichlorohydrin is as follows:

$$ClCH_2CH=CH_2 \xrightarrow{HOCl} \left.\begin{array}{l} ClCH_2CHOHCH_2Cl \\ ClCH_2CHClCH_2OH \end{array}\right\} \xrightarrow{CaO}$$

Allyl chloride

$$\overset{\overset{O}{\diagup\,\diagdown}}{CH_2-CH}-CH_2Cl \xrightarrow{NaOH} HOCH_2CHOHCH_2OH$$

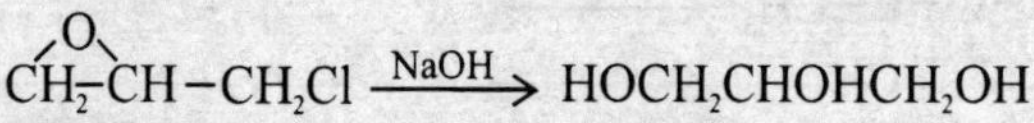

Epichlorohydrin   **Glycerol**

Another synthetic route involves the oxidation of propylene to acrolein with oxygen in presence of CuO catalyst at 350°.

$$CH_3CH=CH_2 \xrightarrow[350°]{CuO} CH_2=CHCHO \xrightarrow[MgO+ZnO,350-400°]{(CH_3)_2\,CHOH}$$

Propylene   Acrolein

$$CH_2=CHCH_2OH \xrightarrow[WO_3]{H_2O_2} HOCH_2CHOHCH_2OH$$

Allyl alcohol   Glycerol

15.7.2 Properties

It is a colorless, odorless, thick syrupy liquid having sweet taste. It has one secondary and two primary alcoholic groups and undergoes common reactions of these classes. The three carbon atoms in glycerol are indicated as α, β and α' or γ ($\overset{\alpha}{C}H_2OH-\overset{\beta}{C}HOH-\overset{\alpha'\text{ or }\gamma}{C}H_2OH$)

(1) When glycerol reacts with sodium, one α-hydroxy groups is readily attacked, other α-hydroxy group less readily and β-hydroxy group is not attacked at all.

$$HOCH_2CHOHCH_2OH \xrightarrow{Na} NaOCH_2CHOHCH_2OH \xrightarrow{Na} NaOCH_2CHOHCH_2ONa$$

(2) When hydrogen chloride is passed into glycerol at 110°, both α- and β-glycerolmonochlorohydrin are formed with α-chlorohydrin as the predominant product (66% yield).

Prolonged action of hydrogen chloride produces glycerol α, α'-dichlorohydrin as the major product (57%) and glycerol α, β-dichlorohydrin as the minor one.

$$HOCH_2CHOHCH_2OH \xrightarrow[110°]{HCl} ClCH_2CHOHCH_2OH + HOCH_2CHClCH_2OH$$

$$\xrightarrow{HCl} ClCH_2CHOHCH_2Cl + ClCH_2CHClCH_2OH$$

Glycerol-α, α'-   Glycerol-α, β-
dichlorohydrin   dichlorohydrin

(3) When glycerol or either of the dichlorohydrins are heated with phosphorus pentachloride, glyceryl trichloride is obtained.

$$HOCH_2CHOHCH_2OH + PCl_5 \longrightarrow ClCH_2CHClCH_2Cl$$

Glyceryl trichloride

(4) Glycerol reacts with hydrogen bromide or phosphorus bromide to give glycerol tribromide but analogous iodine compounds behave differently depending upon the amount of reagent used. When glycerol is heated with small amount of hydrogen iodide or phosphorus triiodide, allyl iodide is the main product.

$$HOCH_2CHOHCH_2OH \xrightarrow{PI_3} [ICH_2CHICH_2I] \xrightarrow{-I_2} CH_2=CHCH_2I$$

Glycerol   Allyl iodide

With excess phosphorus iodide, however, isopropyl iodide is the main product.

$$CH_2=CHCH_2I \xrightarrow{HI} [CH_3CHICH_2I] \xrightarrow{-I_2} CH_2=CHCH_3 \xrightarrow{+HI} CH_3CHICH_3$$

Isopropyl iodide

(5) When glycerol is treated with mono carboxylic acid, mono, di or triesters are obtained depending upon the amount of carboxylic acid used and the temperature at which the reaction is performed.

$$\text{HOCH}_2\text{CHOHCH}_2\text{OH} \xrightarrow{\text{CH}_3\text{COOH}} \text{CH}_3\text{COOCH}_2\text{CHOHCH}_2\text{OH}$$

Glycerol

$$\xrightarrow{\text{CH}_3\text{COOH}} \text{CH}_3\text{COOCH}_2\text{CHOHCH}_2\text{OCOCH}_3$$

$$\xrightarrow{\text{CH}_3\text{COOH}} \text{CH}_3\text{COOCH}_2 - \overset{\overset{\displaystyle \text{OCOCH}_3}{|}}{\text{CH}} - \text{CH}_2\text{OCOCH}_3$$

Glyceryl triacetate

(6) Glycerol reacts with nitric acid in presence of sulphuric acid at 25° to form glyceryl trinitrate commonly known as nitroglycerine.

$$\text{HOCH}_2\text{CHOHCH}_2\text{OH} \xrightarrow[25°]{\text{HNO}_3/\text{H}_2\text{SO}_4} \text{O}_2\text{NOCH}_2 - \overset{\overset{\displaystyle \text{ONO}_2}{|}}{\text{CH}}\text{CH}_2\text{ONO}_2$$

Glyceryl trinitrate

(7) When heated with oxalic acid at 110°, glycerol forms its monoformate which upon hydrolysis gives back glycerol and formic acid. This method is used for the preparation of formic acid.

$$\text{HOCH}_2\text{CHOHCH}_2\text{OH} + \overset{\overset{\displaystyle \text{COOH}}{|}}{\underset{\displaystyle \text{COOH}}{}} \xrightarrow[-\text{CO}_2]{-\text{H}_2\text{O}} \text{HOOC}-\text{OCH}_2\text{CHOHCH}_2\text{OH}$$

Glycerol **Oxalic acid** **Glyceryl monoformate**

$$\xrightarrow{\text{Hydrolysis}} \text{HOCH}_2\text{CHOHCH}_2\text{OH} + \text{H}-\overset{\overset{\displaystyle \text{O}}{||}}{\text{C}}-\text{OH}$$

Glycerol **Formic acid**

However, at 230° reaction with oxalic acid yields allyl alcohol.

$$\text{HOCH}_2\text{CHOHCH}_2\text{OH} + \overset{\overset{\displaystyle \text{COOH}}{|}}{\underset{\displaystyle \text{COOH}}{}} \longrightarrow \left[\begin{array}{c} \text{CO}-\text{CO} \\ | \quad\quad | \\ \text{O} \quad\quad \text{O} \\ | \quad\quad | \\ \text{CH}_2 - \text{CHCH}_2\text{OH} \end{array} \right] \longrightarrow \text{CH}_2{=}\text{CHCH}_2\text{OH} + 2\text{CO}_2$$

Glycerol **Allyl alcohol**

(8) A variety of oxidation products are obtained when glycerol is oxidised depending upon the oxidising agent used.

$$\text{HOCH}_2\text{CHOHCH}_2\text{OH} \xrightarrow{[\text{O}]} \text{OHCCHOHCH}_2\text{OH} \xrightarrow{[\text{O}]} \text{HOOCCHOHCH}_2\text{OH}$$

Glycerol **Glyceraldehyde** **Glyceric acid**

$$\downarrow [\text{O}] \qquad\qquad\qquad\qquad\qquad\qquad\qquad\qquad\qquad \downarrow [\text{O}]$$

$$\text{HOCH}_2\text{COCH}_2\text{OH} \xrightarrow{[\text{O}]} \text{HOOCCOCOOH} \xleftarrow{[\text{O}]} \text{COOHCHOHCOOH}$$

Dihydroxy acetone **Mesoxalic acid** **Tartonic acid**

Glycerol undergoes oxidative cleavage when treated with periodic acid

$$\text{HOCH}_2\text{CHOHCH}_2\text{OH} \xrightarrow{\text{HIO}_4} 2\text{CH}_2\text{O} + \text{HCOOH} + \text{HIO}_3 + \text{H}_2\text{O}$$

It burns with a blue flame when dropped on solid potassium permanganate due to exothermic reaction.

(9) Glycerol undergoes dehydration to acrolein when heated with potassium hydrogen sulphate.

$$\text{HOCH}_2\text{CHOHCH}_2\text{OH} \xrightarrow{\text{KHSO}_4} \text{CH}_2\!=\!\text{CHCHO} + 2\text{H}_2\text{O}$$

Glycerol   Acrolein

(10) Glycerol forms glyptal resin when reacted upon with phthalic anhydride which is a thermo-setting plastic used for making synthetic fibre. Here all three hydroxy groups of glycerol form an ester linkage with the anhydride.

Uses

(a) as a sweetening agent
(b) in cosmetics manufacture
(c) in the preservation of tobacco
(d) as antifreeze in automobile radiators
(e) in the preparation of nitroglycerin
(f) it can be fermented to produce propionic acid, succinic acid, acetic acid, n-butanol, lactic acid etc.

15.8 NITROGLYCERINE/GLYCERYL TRINITRATE

The correct name for nitroglycerine is glyceryl trinitrate since it is an ester and not a nitrocompound.

It is manufactured by addition of glycerol (1 part) in a thin stream to a cold mixture of concentrated sulphuric acid (5 parts) and fuming nitric acid (3 parts).

$$\begin{array}{l} \text{CH}_2\text{OH} \\ \text{CHOH} \\ \text{CH}_2\text{OH} \end{array} + \text{HNO}_3 \xrightarrow{\text{H}_2\text{SO}_4} \begin{array}{l} \text{CH}_2-\text{ONO}_2 \\ \text{CH}-\text{ONO}_2 \\ \text{CH}_2-\text{ONO}_2 \end{array}$$

Glycerol   Nitroglycerine

It is a poisonous colorless, oily liquid, insoluble in water. It usually burns quietly when ignited but when heated, rapidly struck or detonated, explodes violently due to the formation of gases which occupy 10,900 times the volume occupied by nitroglycerine.

$$4\text{C}_3\text{H}_5(\text{ONO}_2)_3 \longrightarrow 12\text{CO}_2 + 10\text{H}_2\text{O} + 6\text{N}_2 + \text{O}_2$$

It is used in the manufacture of explosives like dynamite, blasting gelatin or gelatin and cordite. It is also used as a medicine for the treatment of angina pectoris.

15.9 LIPIDS

The term lipid (Greek, Lipos meaning fat) is used for a variety of naturally occurring compounds having solubility in organic solvents but are insoluble or sparingly soluble in water. They also yield saturated and unsaturated mono carboxylic acids on hydrolysis. They have been subdivided into following classes: (1) oils and fats, (2) waxes and (3) phospholipids.

15.9.1 Oils and Fats

This class of lipids is often termed as simple lipids. They are glyceryl esters or glycerides of higher fatty acids and on alkaline hydrolysis or saponification form glycerol and salts of higher fatty acids.

$$\text{Oils and fats} \xrightarrow{\text{Saponification}} \text{Glycerol + Mixture of salt of fatty acids}$$

Oils which are liquid at ordinary temperature contain a large proportion of unsaturated acids like oleic acid and linoleic acid whereas fats are solids at ordinary temperature and contain large proportions of saturated acids such as lauric acid, myristic acid, palmitic acid and stearic acid. The acids present in glycerides are almost always straight chain acids and invariably contain even number of carbon atoms.

Natural oils and fats are usually complex mixtures of triesters of glycerol, However, some of them contain simple glycerides. Like for example, coconut oil is a mixed glyceride of myristic, palmitic, stearic, oleic and linolcic acid in variable proportions.

15.9.2   Origin

Oils and fats originate from natural sources and according to the source, they have been classified as:

Animal fats are present in adipose tissue cells. In human body 12% of the total body weight is fat which forms heat insulating subcutaneous layer, intermuscular connecting tissues etc.

Vegetable oils are present in seeds and nuts of various plants. Plants like ground nut, soybean, mustard, coconut etc. are important sources of edible oils, whereas nonedible oils may be extracted from cotton seeds, linseed and caster seeds.

Marine oils are extracted from marine animals like sardines, salmons, whales, dolphins, seals etc.

Essential oils are the oils having essence like lemon oil, turpentine oil, clove oil. Their origin may be plant or animal sources and they are used in perfume industry.

15.9.3   Extraction

The methods employed for the extraction of fats and oils are as follows.

Heating or Melting: Animal fat is extracted by heating the chopped animal tissue under pressure so that the melted fat separates out as an oily layer.

Crushing or Pressing from oil seeds by expeller which is a continuous screw press. The oil separates on pressing, leaving behind pressed oil cake which still contains about 5% oil and is used as cattle feed. Cotton seed oil, groundnut oil, mustard oil are extracted by this method from their seeds.

Solvent Extraction: Oils and fats may be extracted from a mixture by the careful selection of organic solvents such as petroleum ether. This method is effective as only 1-2% of the oil is left in the crushed seeds which then can be used as cattle feed.

15.9.4   Properties

15.9.4.1   Physical Properties

Oils and fats are liquids or non-crystalline solids at ordinary temperature. They are colorless, odorless and tasteless when pure. They are immiscible in water but miscible in organic solvents like petroleum ether, benzene, carbon tetrachloride etc. They are lighter than water and form oily upper layer when mixed in water and undergo emulsification when agitated with water in presence of emulsifiers like soap, gelatin etc.

15.9.4.2   Chemical Properties

1. *Hydrolysis*: Oils and fats being triesters of glycerol undergo *acid hydrolysis* with enzyme lipase in the biological system to give glycerol and fatty acids which play an important role in the metabolism within the body.

They undergo *alkaline hydrolysis or saponification* by sodium or potassium hydroxide to give glycerol and sodium or potassium salts of fatty acids which are used as soaps (See Section 15.7.1).

2. *Hydrogenation*: Oils are glycerides of unsaturated fatty acids and they can be converted into fats by passing hydrogen gas through them under pressure in presence of activated nickel catalyst. This

process is also called *hardening of oil* since it increases the melting point of the oil. Glyceryl trioleate, an oil having melting point in the range of –17° after hydrogenation produces glyceryl tristearate, a fat having m.p. 70°.

$$CH_2O\overset{O}{\overset{\|}{C}}(CH_2)_7CH=CH(CH_2)_7CH_3$$
$$|$$
$$CHO\overset{O}{\overset{\|}{C}}(CH_2)_7CH=CH(CH_2)_7CH_3 \xrightarrow[\text{Heat}]{H_2/Ni}$$
$$|$$
$$CH_2O\overset{O}{\overset{\|}{C}}(CH_2)_7CH=CH(CH_2)_7CH_3$$

Glyceryl trioleate m.p. -17°

$$CH_2O\overset{O}{\overset{\|}{C}}(CH_2)_{16}CH_3$$
$$|$$
$$CHO\overset{O}{\overset{\|}{C}}(CH_2)_{16}CH_3$$
$$|$$
$$CH_2O\overset{O}{\overset{\|}{C}}(CH_2)_{16}CH_3$$

Glyceryl tristearate m.p 70°

The edible oil to be hydrogenated is first freed from free acids by treatment with dilute sodium hydroxide followed by washing with water. The oil is then bleached and deodorised with animal charcoal followed by hydrogenation of oil in presence of nickel. The fats thus produced are marketed under difference names such as Rath, Dalda etc.

Non-edible oils after hydrogenation are used for making soaps and cosmetics.

Hydrogenation is important since it converts vegetable oils having offensive color and odor into color-less, odorless fats which have various applications.

3. *Hydrogenolysis*: Oils upon hydrogenation at high temperature and pressure in presence of copper chromite split up into glycerol and long chain saturated alcohols.

$$CH_2OCO(CH_2)_{16}CH_3$$
$$|$$
$$CHOCO(CH_2)_{16}CH_3 \xrightarrow[\text{250°,Pressure}]{H_2/CuCr_2O_4}$$
$$|$$
$$CH_2OCO(CH_2)_{16}CH_3$$

Glyceryl tristearate

$$CH_2OH$$
$$|$$
$$CHOH + 3C_{17}H_{35}CH_2OH$$
$$|$$
$$CH_2OH$$

Glycerol   n-Octadecanol

The long chain saturated alcohols thus produced are used in the manufacture of synthetic detergents.

4. *Halogenation*: Oils having unsaturated glycerides add halogens at the site of double bond just as alkenes do. Iodine adds at the double bond to give diiodide and the amount of iodine consumed by a glyceride is a measure of the extent of unsaturation in a given oil or fat.

5. *Rancidity*: When fats and oils are left exposed to warm, moist air for a long time they develop some disagreeable odor due to the hydrolysis of ester linkage and oxidation of double bonds. The fats and oils are then called rancid as they develop offensive odor due to the formation of volatile, low molecular weight acids by hydrolysis and oxidation.

Hydrolytic rancidity: Butter glycerides on hydrolysis in presence of enzyme lipase present in air gives glycerol, butyric acid, caperic acid and caprylic acid having offensive odor.

Oxidative rancidity: Triglyceride of unsaturated fatty acids like glyceryl trioleate first undergo hydrolysis to give glycerol and oleic acid which on oxidation cleaves at the double bond to give pelargonic acid and azelaic acid having offensive odor.

15.9.5   Analysis of Oils and Fats

Oils and fats are natural products having variable composition. Their composition may vary during storage due to hydrolysis and/or oxidation and a specified application requires raw material of proper specifications thus analysis of these natural products is important and various methods (physical and chemical) have been developed to evaluate a given oil or fat. They are characterised by physical constants such as melting point, specific gravity and refractive index. Modern techniques such as X-ray analysis, absorption, mass and NMR spectroscopy have also been used for their structure elucidation. Various methods of analysis of oils and fats have been discussed below.

15.9.5.1   Saponification Value

As has been discussed earlier, saponification is the term used for the hydrolysis of fats and oils in alkaline medium and is defined as the *number of milligrams of potassium hydroxide required to saponify one gram of fat or oil*. The saponification value gives an indication of the length of carbon chain of the fatty acid component of triglyceride. The higher the saponification value of an oil or fat, the higher the percentage of low molecular weight glycerides it contains. Values of common fat and oils have been given in Table 15.1.

The saponification value of pure glycerides may be calculated as follows:

$$(C_{15}H_{31}COO)_3C_3H_5 + 3KOH \longrightarrow 3C_{15}H_{31}COO^- K^+ + C_3H_5(OH)_3$$

Glyceryl tripalmitate   **Potassium salt of**
Mol. wt = 836   **Palmitic acid**

Here $3 \times 56 \times 1000$ milligrams KOH is required to saponify 836 grams of glyceryl tripalmitate. Thus,

$$\text{Saponification value of glyceryl tripalmitate} = \frac{3 \times 56 \times 1000}{836} = 208$$

15.9.5.2   Iodine Value

The extent of unsaturation is expressed in term of iodine number in fats and oils. It is defined as number of grams of iodine which combines with 100 grams of fat or oil.

Iodine value for a saturated fat is zero. It can be calculated by dissolving a weighed amount of oil in chloroform or carbon tetrachloride. The oil is then reacted upon with iodine chloride in acetic acid (the Wij's solution) which adds at the double bond to form diiodide. The excess iodine is then titrated with standard sodium thiosulphate solution using starch indicator.

Iodine values of some common oils have been given in Table 15.1.

Table 15.1:   Analytical Data for Some Common Oils and Fats

Oils and Fats	Saponification value	Iodine value
Butter	210-230	26-30
Coconut oil	250	—
Cotton seed oil	190-196	105-112
Soyabean oil	190-195	127-135
Linseed oil	185-195	170-185
Sunflower oil	185-195	140-155

15.9.5.3   Reichert Meissl Value

It is the number of milligrams of 0.1 N potassium hydroxide solution required to neutralise the distillate of 5 grams of hydrolysed fat. It is a measure of steam volatile fatty acids present in oils and fats.

Five grams of the fat is saponified with potassium hydroxide and the mixture is acidified with sulphuric acid followed by steam distillation. The distillate is cooled, filtered and titrated against standard alkali.

15.9.6 Soaps or Surfactants

Soaps are sodium or potassium salts of higher fatty acids such as palmetic, stearic or lauric acid. They are used in household and in industries for their cleansing action and are also called surfactants. Soaps derived from potassium salts of fatty acids are comparatively soft and have better solubility in water and are termed as toilet soaps. Methods for manufacture of soap are discussed in the following sections.

15.9.6.1 The Hot Process

The hot process includes following steps in the manufacture of soap.

1. *Saponification*: Saponification of an oil or a fat is done by heating it with 10% sodium hydroxide solution and steam in a giant iron pan called soap kettle. The reaction takes several hours to complete and a frothy mixture of soap and glycerol is obtained.

$$\text{Oil or Fat} \xrightarrow{\text{NaOH}} \text{Glycerol} + \text{Soap}$$

2. *Salting Out*: On completion of saponification which is indicated by excess of alkali in reaction mixture, the soap is precipitated by addition of common salt or brine. Heating is continued till a thick mass of soap separates in the upper layer and the lower layer contains mainly glycerol along with unreacted alkali which is sent to glycerol recovery unit.

3. *Finishing*: The soap obtained from salting out process may contain some unreacted oil or fat which is again boiled with caustic soda to completely saponify it. The solid soap is then boiled with water to dissolve out any excess alkali and then allowed to settle. The impure soap called nigre forms the lower layer whereas the upper layer of pure soap is sent to a steam jacketed tank called crutcher through a swing pipe.

In crutcher, soap is finally mixed with glycerol, color, perfumes etc., dried to the requisite moisture content and then run into moulds or frames and after solidification cut into various sizes and stamped.

15.9.6.2 The Cold Process

In the cold process of manufacture of soap, with each part of oil or fat, alkali (1 part), water (7 parts) and starch (1 part) is mixed in an iron pan and stirred thoroughly till soap separates and sets whereas glycerine remains in the soap. At this stage the soap is run into frames.

The cold process is used for the manufacture of soft coconut oil soap or potassium soaps which cannot be prepared by earlier process as they are highly soluble in water and fail to salt out.

15.9.6.3 The Modern Process

This process hydrolyses the oil or fat efficiently within 15 minutes with hot water under pressure in presence of catalyst like lime or zinc oxide (Ittner process), or dilute sulphuric acid and aryl sulphonic acids (Twitchel process). The free acid obtained on hydrolysis, is neutralised with caustic soda or soda ash. This method of soap manufacture is simple and economical.

For large scale continuous hydrolysis of oils and fats, the apparatus consists of a tower of about 65 feet height. Fat or oil is introduced from the lower side and hot water at about 250° enters at the top. Fats or oils get hydrolysed in presence of catalyst. The fatty acids taken out from the top are neutralised to get soaps whereas glycerol leaves from the bottom of the tower and sent to glycerol recovery unit.

Soaps thus prepared can be finished into various varieties by the use of different additives.

(1) Laundry soaps are obtained from cheap quality oils and fats. They have certain filling material and perfumes to conceal the unpleasant smell of raw soap.

(2) Toilet soaps are prepared from good quality oils and fats. They may either be named as medicated soaps (by addition of neem) or transparent soaps (by addition of glycerol and ethanol). They also contain a variety of perfumes like lcmon, lavender, jasmine etc. Shaving soaps have potassium sodium stearate with glycerine and gum so that they can provide lasting lather.

15.9.6.4 Cleansing Action of Soaps

A soap molecule consists of mainly two parts: one the nonpolar hydrocarbon chain which is oil soluble or lipophilic and the other is the polar carboxylate ion which is water soluble or hydrophilic as shown below:

$$C_{17}H_{35}-\overset{\overset{\text{O}}{\|}}{C}-\overset{-}{O}\,Na^{+}$$

**Non polar,   Polar, water
oil soluble,   soluble,
lipophilic   hydrophilic**

When soap is added into water, its molecules spread uniformly on the surface of water with their hydrophilic ends embedded in water and lyophilic ends standing out of water as shown in Fig. 15.1.

Now, when the dirty piece of cloth is immersed in this soap solution, the hydrocarbon part of soap dissolves the dirt and grease present therein by forming spherical clusters called micelles (Fig. 15.2). The increased attractive forces between soap and grease particles lift the later off the water surface in the form of small globules which can then be washed away. The negatively charged groups repel the globules so that they do not coagulate or redeposit on the fabric.

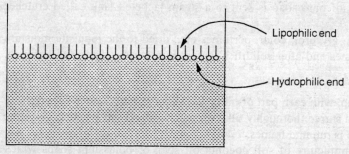

Lipophilic end

Hydrophilic end

Fig. 15.1:   Distribution of Soap on Water

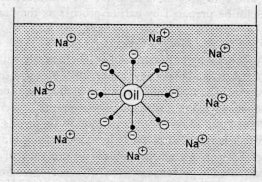

Fig. 15.2:   A Soap Micelle

According to another version of mechanism of action of soap, it is believed that soap when dissolved in water concentrate on its surface reducing its surface tension which causes foaming. This helps soap penetrate the fabric. Soap also has the ability to emulsify fats and oils by micelle formation which is then washed away by water.

Soaps are used for their cleansing action but if the water available for washing is hard, it limits their usefulness. Sodium soaps when come in contact with Ca^{++} and Mg^{++} ions present in hard water form calcium and magnesium salts of fatty acids which appear on the surface of fabric as sticky white scum. This not only wastes the sodium soap but also discolors and hardens the fabric.

15.9.7 Detergents

Scientifically the term 'Detergent' covers both soaps and synthetic detergents or 'syndets' but it is widely used to indicate synthetic cleansing compounds, as distinguished from soaps. Detergents differ from soaps in their action in hard water as they form soluble calcium and magnesium salts in hard water. They can also be used in acidic solutions.

Detergents are salts of sulphonic acids or alkyl hydrogen sulphates whereas soaps are salts of carboxylic acids. They can be classified into three categories.

15.9.7.1 Anionic Detergents

Anionic detergents comprise the largest group of synthetic detergents. The detergents having lyophilic hydrocarbon chain and hydrophilic anionic group such as sulphate or sulphonate are catagorised as anionic detergents.

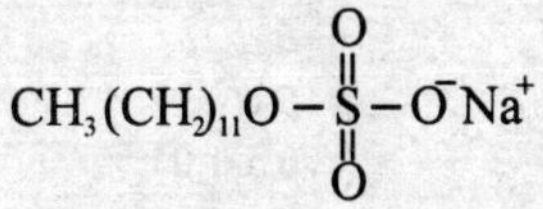

$$CH_3(CH_2)_{11}O - \overset{\overset{O}{\|}}{\underset{\underset{O}{\|}}{S}} - O^- Na^+ \qquad\qquad CH_3(CH_2)_{11} - \bigcirc - \overset{\overset{O}{\|}}{\underset{\underset{O}{\|}}{S}} - O^- Na^+$$

Sodium lauryl sulphate 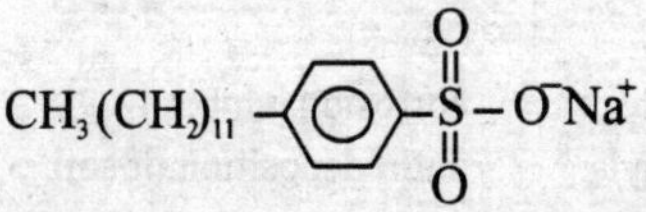   **Sodium p-n-dodecylbenzene sulphonate**

15.9.7.2 Cationic Detergents

In cationic or invert detergents lyophilic chain remains the same but the water soluble part comprise of cations. Quaternary ammonium compound and dialkyl dimethyl quaternary ammonium compounds forms cationic detergents as shown below.

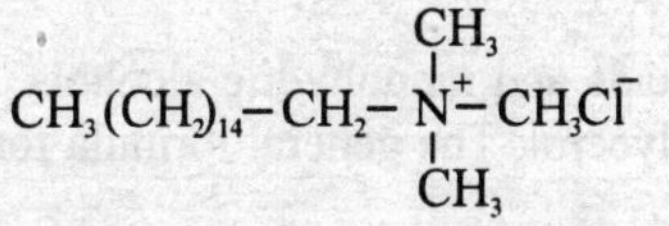

$$CH_3(CH_2)_{14} - CH_2 - \overset{\overset{CH_3}{\underset{|}{}}}{\underset{\underset{|}{CH_3}}{N^+}} - CH_3 Cl^-$$

n-Hexadecyltrimethyammonium chloride

15.9.7.3 Nonionic Detergents

Nonionic detergents have hydrophilic groups at the end of long hydrocarbon chain which provide them solubility in water. They include mono esters of polyhydric alcohols or polyethers derived from ethylene oxide.

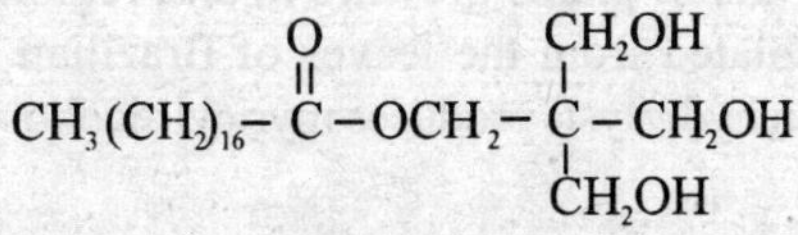

$$CH_3(CH_2)_{16} - \overset{\overset{O}{\|}}{C} - OCH_2 - \overset{\overset{CH_2OH}{\underset{|}{}}}{\underset{\underset{|}{CH_2OH}}{C}} - CH_2OH \qquad\qquad H_{19}C_9 - \bigcirc - (OCH_2CH_2)_6 OH$$

Monoester of polyhydric alcohol 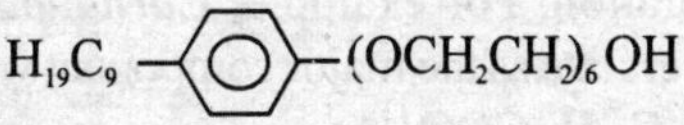   **Polyether of Nonylphenol**

15.9.7.4   Composition of a Common Detergent

The basic components of a detergent are given in Table 15.2 along with their function.

The cleansing action of detergents is similar to that of soaps. The soil removal is accomplished by wetting, emulsification and dispersion. The detergent molecules come together in water in the form of spherical clusters called micelles (Fig. 15.2). Oil soluble and water insoluble compounds like dyes are often dissolved in the centre of micelles due to the attraction of hydrocarbon groups.

During the years 1960 to 1970, detergents underwent rapid changes due to environmental pollution caused as a result of their non biodegradability. This problem was solved to some extent by using linear, long chain alkyl sulphonates (LAS detergents) which are biodegradable. The high phosphorus content in detergents also causes eutrophication of lakes.

Table 15.2:   Composition of Detergent

Ingredient	Function	Ingredient in % by weight
1. Surfactants Sulphonates	removal of oil, soil and cleansing	18-20
2. Builders		
Sodium tri-polyphosphate	removal of inorganic soil	50
Sodium sulphate	filler with some building action in soft water	3-5
Soda ash	filler with building action	3-5
3. Additives		
Sodium silicate	corrosion inhibitor	5-6
Carboxymethyl-cellulose	antiredeposition of soil	0.5-1.0
Fluorescent dye	optical brightner	0.05-0.1
Perfume, dye or pigment	aesthetic, improved product characteristics	0.1
4. Water	filler and binder	10-15

15.9.8   Waxes

Waxes are lipids which on hydrolysis furnish long chain fatty acids and monohydric alcohols. They differ from oils and fats as they are not esters of trihydric alcohol i.e. glycerol. The general formula for waxes can be:

$$R - \overset{\overset{\displaystyle O}{\displaystyle \|}}{C} - OR' \quad \text{where R and R' are long, even numbered carbon chains}$$

Natural waxes are variable mixture of several esters and in addition they also contain small quantities of free fatty acids, alcohols and some long chain alkanes. They are widely distributed in nature and are found in plants as well as in animals.

Plant Waxes: Waxes are found on leaves, stems and fruits of plants growing in arid regions to protect them from dehydration. For example, *Carnauba wax,* isolated from the leaves of Brazilian palm trees, contains myricyl cerolate as the major component which upon hydrolysis produces myricyl alcohol ($C_{31}H_{63}OH$) and cerotic acid ($C_{25}H_{51}COOH$).

Animal Waxes: Animal waxes are found in hairs, feathers or skin of the animals to protect them from the environment. Various varieties of waxes can be obtained from whales and bees also.

Beeswax is obtained from honeycomb of bees and its main component is myricyl palmitate($C_{15}H_{31}COOC_{30}H_{61}$).

Supermaceti is obtained from the oil exuded from the head cavity of sperm whale. It consists mainly of cetyl palmitate ($C_{15}H_{31}COOC_{16}H_{33}$).

Waxes are used in the manufacture of shoe polishes, varnishes, candles, ointments and cosmetics etc.

15.9.9 Phospholipids

We have seen that simple lipids are esters of glycerol with various organic acids, i.e., they are glyceryl esters or glycerides. However, the phospholipids differ from simple lipids in that they contain phosphorus and nitrogen in addition. Thus, they are glycerides, in which one organic acid residue (α and β) is replaced by a group containing phosphoric acid and a base.

$$\begin{array}{l} \text{CH}_2\text{O} \dotplus \text{COR'] From fatty acid}\\ \text{From fatty acid [R''CO} \dotplus \text{OCH} \quad \text{OH}\\ \text{CH}_2\text{O} \dotplus \text{P–OH] From phosphoric acid}\\ \qquad\qquad\qquad \text{O} \end{array}$$

From glycerol

The phosphate group in phospholipids is bonded through another phosphate ester linkage to one of the following nitrogen containing compound.

$$\text{HOCH}_2\text{CH}_2\text{NH}_2 \qquad \text{HOCH}_2\overset{+}{\text{CH}}\text{N(CH}_3)_3\text{O}\overline{\text{H}} \qquad \text{HOCH}_2-\overset{\overset{\displaystyle +NH_3}{|}}{\underset{\underset{\displaystyle H}{|}}{\text{C}}}-\text{COO}^-$$

Ethanolamine   **Choline**   **L-Serine**
(2-Aminoethanol)

The important classes of phospholipids and their structure has been given below.

Kephalines/Cephalines: In phospholipids, if the base is $CH_2OHCH_2NH_2$, it is called kelphalins ethanolamine or cephalins. Thus, an α-kephalin structure is

$$\begin{array}{l} \text{CH}_2\text{OCOR'}\\ \text{R''COOCH} \qquad \text{O}^-\\ \text{CH}_2\text{O}-\text{P}-\text{OCH}_2\text{CH}_2\overset{+}{\text{NH}}_3\\ \qquad\qquad\text{O} \end{array}$$

α-kephalin

The acidic components in α-kephalin can be stearic, oleic, linoleic or achidonic acids.

Lecithins: If the base is choline [$CH_2OHCH_2\overset{+}{N}(CH_3)_3$] $\overline{O}H$, the phospholipid is called a lecithin and is represented as

$$\begin{array}{l} \text{CH}_2\text{OCOR'}\\ \text{R''COOCH} \qquad \text{O}^-\\ \text{CH}_2\text{O}-\text{P}-\text{OCH}_2\text{CH}_2\overset{+}{\text{N}}(\text{CH}_3)_3\\ \qquad\qquad\text{O} \end{array}$$

α-Lecithins

The acidic components are similar to that present in kephalins.

Phospholipids having L-serine are called *phosphatidyl serine*. Phospholipids resemble soaps and detergents as they have both polar and nonpolar groups and dissolve in aqueous medium by forming micelles. The hydrophilic and lipophilic part of phospholipids are best suited for their function in biological system as they create an interface between an organic and aqueous medium. They are found in biological system in association with proteins and glycolipids.

The above two phospholipids, viz. kephalins and lecithins are called *phosphoglycerides*. Another type of phospholipids are the plasmalogens.

Plasmalogens/Acetal Phospholipids: The structure of plasmalogens is more or less similar to the general structure of kephalins or lecithins and differs only in that one of the fatty acids is replaced by aldehydrogenic group.

$$CH_2OCH=CHR'$$
$$R''COOCH \quad O^-$$
$$CH_2O-\overset{\displaystyle O}{\underset{\displaystyle \|}{P}}-OCH_2CH_2\overset{+}{N}H_3$$

Plasmalogens

The phosphoglycerides occur in brain and in the spinal cord.

Sphingolipids : These phospholipids are derived from sphingosine and on hydrolysis yield fatty acid, phosphoric acid, choline and sphingosine (having trans configuration about the double bond). The sphingolipids are also known as sphingomyelins and have the structure

$$CH_3(CH_2)_{12}C-H$$
$$H-\underset{\displaystyle OH}{\overset{\displaystyle }{C}}-\underset{\displaystyle NH}{\overset{\displaystyle }{CH}}-CH-CH_2-O-\overset{\displaystyle O}{\underset{\displaystyle \|}{P}}-OCH_2CH_2\overset{+}{N}(CH_3)_3$$
$$H_{47}C_{23}-C=O$$

Cerebrosides/Glycolipids: These, like spingolipids, contain spingosine but instead of the phosphocholine have carbohydrate as a polar group:

$$CH_3(CH_2)_{12}-C-H$$
$$H-\underset{\displaystyle OH}{\overset{\displaystyle }{C}}-\underset{\displaystyle NH}{\overset{\displaystyle }{CH}}-CH-CH_2OCH(CHOH)_3CHCH_2OH$$
$$\qquad \overset{\displaystyle }{\underset{\displaystyle \qquad O \qquad}{\lfloor \qquad \qquad \rfloor}}$$
$$H_{47}C_{23}-C=O$$

Only two sugar, viz. glucose and galactose are found to be present in sphingolipids. The sphingolipids together with proteins and polysaccharides form *myelin* which forms a protective coating that encloses nerve fibers.

15.10   PROBLEMS

1. How is ethylene glycol prepared ? How does it react with acetone, zinc chloride and acetic acid ?
2. What are various methods of preparation of glycerol ?
3. What are oils, fats and waxes ? How will you distinguish between vegetable, mineral and essential oil ?
4. How is vegetable fat manufactured from oil seeds ?
5. What are soaps ? Describe the cleansing action of soap.

6. What are alkyl benzene sulphonates ? What is their function in detergents.
7. What is the composition of a normal detergent formulation ? What is the function of fluorescent dye and soda ash in detergent formulation ?
8. Discuss the mechanism of action of detergents. How they differ from soaps ?
9. How is soap manufactured ? What are various additives used in the process ?
10. How is oils and fats analysed ?
11. What are lipids ?

16

Ethers

16.1 INTRODUCTION

The general molecular formula of ethers is $C_nH_{2n+2}O$, which is the same as that for monohydric alcohols. They have the general structure R–O–R and may be regarded as alkyl oxides or the anhydrides of alcohol.

Like alcohols, which are considered as monoalkyl derivatives of water, ethers may be considered as dialkyl derivatives of water. They may also be considered as derivatives of alcohols in which the hydroxy hydrogen is replaced by an alkyl group.

$$\underset{\textbf{Water}}{\text{H–O–H}} \qquad \underset{\textbf{Alcohol}}{\text{R–O–H}} \qquad \underset{\textbf{Ether}}{\text{R–O–R}}$$

The ethers can be simple or mixed depending upon whether the two alkyl group attached to oxygen are the same or different.

$$\underset{\textbf{Dimethyl ether}}{CH_3-O-CH_3} \qquad \underset{\textbf{Ethylmethyl ether}}{CH_3-O-C_2H_5}$$

Ethers can be open chain or cyclic. When the ring size is five or more, the chemistry of cyclic ethers is similar to that of an open chain ether. The three membered cyclic ethers, which are often known as epoxides are also called oxiranes. These are more reactive than ethers having large rings due to Baeyer strain associated with small rings. Some examples of cyclic ethers are given below:

$$\underset{\substack{\textbf{Oxirane (Ethylene}\\\textbf{oxide)}}}{\overset{\overset{\displaystyle O}{\diagdown}}{H_2C-CH_2}} \qquad \underset{\textbf{Tetrahydrofuran}\\\textbf{(THF)}}{} \qquad \underset{\textbf{1,4-Dioxane}}{} \qquad \underset{\textbf{18-Crown-6 (Crown ether)}}{}$$

Ethers having large rings with repeating $-OCH_2CH_2-$ units are called *Crown ethers*. 18-Crown-6 is a typical example of a crown ethers (in which 18 represents the ring size and 6 represents the number of oxygen atoms present). The unique feature of crown ethers is that they can chelate metal ions and give metal complexes, which are soluble in non polar organic solvents. For example, the complex of 18-crown-6 with potassium permanganate is soluble in benzene and is called purple benzene.

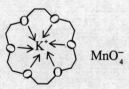

MnO_4^-

18-Crown-6 complex

16.2 NOMENCLATURE

Ethers are generally given the common names by putting the names of the two alkyl groups (in alphabetic order) with the word ether.

$$CH_3OCH_3 \qquad\qquad C_2H_5OC_2H_5 \qquad\qquad CH_3OC_2H_5$$
Dimethyl ether   **Diethyl ether**   **Ethyl methyl ether**

According to the IUPAC system, ethers are named as alkoxyalkanes. The larger alkyl group is considered to be the alkane. The name of the alkoxy group is prefixed to the alkane and the position of attachment is also given.

$$CH_3OCH_3 \qquad CH_3OCH_2CH_3 \qquad \overset{\textstyle CH_3}{\underset{}{CH_3OCH-CH_3}} \qquad \overset{\textstyle OCH_2CH_3}{\underset{}{CH_3-CH-CH_2-CH_3}}$$
Methoxymethane   **Methoxyethane**   **2-Methoxypropane**   **2-Ethoxybutane**

16.3 ORBITAL STRUCTURE

The electronic structure of ethers is represented as given below.

$$R\ \overset{..}{\underset{..}{O}}\ R' \qquad or \qquad R-\overset{..}{\underset{..}{O}}-R' \qquad or \qquad R-O-R'$$

In this representation, the oxygen atom in ethers form a bridge between two alkyl groups by forming two single bonds.

In the orbital representation of ethers, the oxygen atom and the carbon of the alkyl group is sp^3 hybridised. The two σ bonds are formed by the overlap between sp^3 hybridised orbitals of oxygen and carbon of alkyl groups with an R–O–R angle equal to 110° (slightly greater than the normal tetrahedral angle). It may be mentioned that the H–O–H bond angle in water is 105° and this divergence from the normal tetrahedral angle of 109.5° may be attributed to the repulsion of lone pairs in nonbonding orbitals. In case of ethers, the repulsion between the bulky alkyl groups is more than the repulsion of the lone pairs in nonbonding orbitals and the tetrahedral angle is almost retained. Fig. 16.1 shows the orbital representation of an ether molecule.

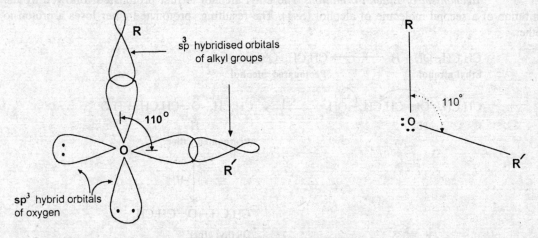

Fig. 16.1: Orbital Structure of an Ether

16.4 ISOMERISM

Ethers exhibit two types of isomerism.

Functional group isomerism is due to the difference in the nature of functional groups, for example C_2H_6O represents dimethyl ether (CH_3-O-CH_3) and ethyl alcohol (CH_3CH_2OH). Similarly molecular formula $C_4H_{10}O$ represents diethyl ether ($C_2H_5-O-C_2H_5$), n-butyl alcohol ($CH_3CH_2CH_2CH_2OH$) and t-butyl alcohol ($(CH_3)_3COH$)

Metamerism, the second type of isomerism is exhibited between simple and mixed ethers. Thus, molecular formula $C_4H_{10}O$ represents a simple ether, $C_2H_5OC_2H_5$ (diethyl ether) and mixed ether, $CH_3-O-CH_2CH_2CH_3$ (methy-n-propyl ether) or $CH_3-O-CH(CH_3)_2$ (isopropyl methyl ether).

16.5 GENERAL METHODS OF PREPARATION

(1) *Dehydration of Alcohols:* Primary alcohols on heating with concentrated sulphuric acid at 140° form ethers.

$$2\,ROH \xrightarrow[140°]{Conc.\ H_2SO_4} R-O-R + H_2O$$
$$\textbf{Ether}$$

This method is useful for preparing symmetrical ethers only.

$$2CH_3CH_2OH \xrightarrow[140°]{Conc.\ H_2SO_4} CH_3CH_2OCH_2CH_3 + H_2O$$
$$\textbf{Ethanol} \qquad\qquad \textbf{Diethyl ether}$$

In this reaction the conditions have to be carefully controlled. Thus, excess alcohol and keeping the temperature at 140° gives maximum yield of ether. However, at 150° sufficient amount of ethylene is obtained.

$$C_2H_5OH \xrightarrow[150°]{Conc.\ H_2SO_4} CH_2{=}CH_2 + H_2O$$

Mechanism of Ether Formation: The ethyl alcohol is first protonated followed by the substitution of a second molecule of alcohol (S_N1). The resulting protonated ether loses a proton to give ether.

$$CH_3CH_2-\overset{..}{\underset{..}{O}}H + H^+ \longrightarrow CH_3CH_2-\overset{+}{O}H_2$$
$$\textbf{Ethyl alcohol} \qquad\qquad \textbf{Protonated alcohol}$$

$$CH_3CH_2-\underset{\underset{H}{|}}{\overset{..}{O}}: + CH_3CH_2-\overset{+}{O}H_2 \xrightarrow{S_N1} CH_3CH_2-\underset{\underset{H}{|}}{\overset{+}{O}}-CH_2CH_3 + H_2O$$
$$\textbf{Protonated Ether}$$

$$\Big\downarrow {-H^+}$$

$$CH_3CH_2-\overset{..}{\underset{..}{O}}-CH_2CH_3$$
$$\textbf{Diethyl ether}$$

(2) *Williamson Ether Synthesis:* The reaction of alkyl halide with sodium or potassium alkoxide (obtained by the action of an alcohol with metallic sodium or potassium) forms ethers.

$$2C_2H_5OH + 2Na \longrightarrow 2C_2H_5ONa \quad + H_2$$

Ethyl alcohol   **Sodium ethoxide**

$$2C_2H_5ONa + C_2H_5Br \xrightarrow{\Delta} C_2H_5OC_2H_5 + NaBr$$

Diethyl ether

By taking different alkyl halides, mixed ethers may be obtained.

$$C_2H_5ONa + CH_3I \longrightarrow C_2H_5OCH_3 + NaI$$

Methyl	**Methoxy**
iodide	**ethane**

The Williamson's synthesis involves S_N2 mechanism.

$$R-\ddot{O}: + R-X \longrightarrow R-\ddot{O}-R + X^-$$

The tendency of the alkyl halides to undergo this reaction is in the following order.

Primary > Secondary > Tertiary

(3) *From Alcohols:* The lower members of primary alcohols can be dehydrated by passing vapors of alcohol over heated alumina at 240-260°.

$$2C_2H_5OH \xrightarrow[240-260°]{Al_2O_3} C_2H_5OC_2H_5 + H_2O$$

Ethyl alcohol   **Diethyl ether**

This method is unsatisfactory with secondary or tertiary alcohols as they form alkenes under these conditions.

(4) *From Alkyl Halides:* Simple ethers can be obtained by heating alkyl halides with dry silver oxide.

$$2C_2H_5I + Ag_2O \xrightarrow{\Delta} C_2H_5OC_2H_5 + 2AgI$$

Diethyl ether

16.6 PHYSICAL PROPERTIES

With the exception of dimethyl ether and ethyl methyl ether which are gases, all other ethers are colorless liquids with pleasant odors. Ethers have lower boiling points as compared to those of alcohols containing the same number of carbon atoms. This is due to the fact that ethers do not have hydrogen attached to oxygen (unlike alcohols) and so cannot form hydrogen bonds. However, ethers can form hydrogen bond with water, alcohols and phenols. Due to hydrogen bonding with water, ethers are soluble in water to a certain extent.

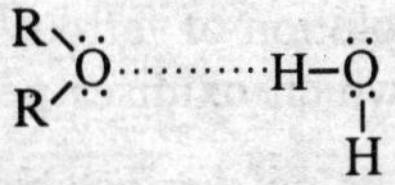

$$\begin{array}{c} R \\ \\ R \end{array} \ddot{O} \cdots\cdots H-\underset{\underset{H}{|}}{\ddot{O}}$$

16.7 SPECTRAL PROPERTIES

In infra red spectrum, ethers like alcohols show C–O stretching vibration at 1060-1300 cm^{-1}. However, the O–H bond characteristic of alcohol is absent.

In the NMR spectra, as in the case of alcohols, the oxygen atoms cause deshielding of the protons adjacent to carbon (α protons) and the absorption is shifted downfield. The NMR absorptions of different protons in case of dipropyl ether is shown below:

$$\overset{\gamma}{CH_3}\overset{\beta}{CH_2}\overset{\alpha}{CH_2}O\overset{\alpha}{CH_2}\overset{\beta}{CH_2}\overset{\gamma}{CH_3}$$

δ 0.9 CH_3 at γ-carbons; δ 1.5 CH_2 at β-carbons; δ 3.4 CH_2 at α-carbons

16.8 CHEMICAL PROPERTIES

The bond between carbon and oxygen in an ether is called the *ether linkage*. The ethers are quite unreactive and behave more like alkanes. The chemical properties of ethers can be discussed under the following heads.

(i) *Inert nature*: Ethers are comparatively inert substances. The ether linkage is not affected by bases, oxidising agents and reducing agents.

(ii) *Reactions at the site of oxygen atom.*

(a) *Formation of Oxonium Salts:* Ethers on treatment with cold concentrated sulphuric acid or hydrochloric acid form protonated ethers called oxonium salts, which are soluble in cold acid solution.

$$C_2H_5\ddot{O}-C_2H_5 + H_2SO_4 \xrightarrow{\text{Cold}} [C_2H_5-\overset{+}{\underset{|}{\overset{H}{O}}}-C_2H_5]HSO_4^-$$

Diethyl ether   **Oxonium salt of diethyl ether**

The oxonium salts on treatment with water regenerate the ether. This reaction is used to distinguish between ethers and alkanes (the alkanes do not react with cold concentrated sulphuric acid).
Similarly ethers react with Lewis acids to give oxonium salts. Diethyl ether reacts with boron trifluoride to give a stable complex viz. boron trifluoride etherate.

$$C_2H_5-\underset{\underset{C_2H_5}{|}}{\ddot{O}:} + BF_3 \longrightarrow H_5C_2-\underset{\underset{C_2H_5}{|}}{\ddot{O}:} BF_3$$

Diethyl ether   **Boron trifluoride etherate**

(b) *Formation of Peroxides (Autooxidation):* In air, most aliphatic ethers are slowly converted into unstable peroxides. These peroxides although present in low concentrations, but are very dangerous and may cause violent explosion during distillations.

$$C_2H_5-O-C_2H_5 + [O] \xrightarrow{\text{air}} (C_2H_5)_2\ddot{O}: \rightarrow O$$

Diethyl ether   **Peroxide**

The presence of peroxide is indicated by the formation of a red color when ether is shaken with aqueous solution of ferrous ammonium sulphate and potassium thiocyanate. The peroxide oxidises ferrous ion to ferric ion, which reacts with thiocyanate ion to give the characteristic blood-red color of the complex. The peroxides can be removed either by washing ether with a solution of ferrous ion (which reduces the peroxides) or distillation with concentrated sulphuric acid (which oxidises peroxides).

(iii) *Reactions involving C−O bonds:* The carbon to oxygen linkage is not very stable compared to carbon to carbon linkage. The linkage is ruptured in presence of a number of reagents giving rise to a variety of products.

(a) *Hydrolysis:* On heating ethers with dilute sulphuric acid under pressure alcohols are formed.

$$C_2H_5-O-C_2H_5 \xrightarrow[\text{Pressure}]{\text{dil } H_2SO_4 \, \Delta} 2C_2H_5OH$$

Diethyl ether   **Ethyl alcohol**

(b) *Reaction with Phosphorus Pentachloride*: Ethers on heating with phosphorus pentachloride give alkyl chloride.

$$CH_3OCH_3 + PCl_5 \xrightarrow{\Delta} 2CH_3Cl + POCl_3$$

$$CH_3OCH_2CH_3 + PCl_5 \xrightarrow{\Delta} CH_3Cl + CH_3CH_2Cl + POCl_3$$

(c) *Reaction with Hydrogen Iodide*: Ethers cleave by hydriodic acid to form a variety of products depending upon the reaction conditions. In cold, ether gives alkyl halide and alcohol.

$$C_2H_5-O-C_2H_5 + HI \xrightarrow{Cold} C_2H_5I + C_2H_5OH$$

However, in hot, both alkyl groups give alkyliodides.

$$C_2H_5-O-C_2H_5 + HI \xrightarrow{Hot} 2C_2H_5I + H_2O$$

In mixed ethers, the iodide goes to the smaller alkyl group.

$$CH_3-O-C_2H_5 + HI \xrightarrow{Cold} CH_3I + C_2H_5OH$$

Cleavage of ethers may also be accomplished by the use of concentrated hydrobromic acid (48%) when alkyl bromides are obtained.

$$\overset{\overset{\displaystyle CH_3}{|}}{CH_3-CH}-O-\overset{\overset{\displaystyle CH_3}{|}}{CH}-CH_3 \xrightarrow[\Delta]{48\% \ HBr} 2CH_3-\overset{\overset{\displaystyle CH_3}{|}}{CH}-Br$$

Di-isopropyl ether   **Isopropyl bromide**

The cleavage of ether with hydriodic acid or hydrobromic acid proceeds by the same pattern as the reaction of alcohol with halogen acids, i.e. protonation of ether, followed by S$_N$1 or S$_N$2 reaction.

$$R-\overset{..}{\underset{..}{O}}-R \underset{}{\overset{H^+}{\rightleftharpoons}} R-\overset{\overset{\displaystyle H}{|}}{\underset{+}{O}}-R + I^- \xrightarrow{S_N2} [\underset{\delta}{I}\cdots R \cdots \underset{\delta}{\overset{\overset{\displaystyle H}{|}}{O}}-R]$$

Protonated ether   **Transition state**

$$\downarrow$$

$$R-I + ROH$$

(iv) *Reactions involving Alkyl groups (Substitution reactions): Halogenation*: On treatment with chlorine or bromine in dark, substitution products are obtained. The extent of substitution depends on the reaction conditions. α-Positions are preferentially halogenated.

$$CH_3CH_2OCH_2CH_3 \xrightarrow[dark]{Cl_2} CH_3CHClOCH_2CH_3 \xrightarrow[dark]{Cl_2} CH_3CHClOCHClCH_3$$

Diethyl ether   **α, α′-Dichlorodiethyl ether**

However, in presence of light, completely substituted product is obtained.

$$(C_2H_5)_2O \xrightarrow{Cl_2}{light} (C_2Cl_5)_2O$$

Diethyl ether   **Pentachlorodiethyl ether**

Uses: Ethers are extensively used as industrial solvents. Lower members are used as general anesthetics. They produce intense local cooling when sprayed on skin. Ethers are also used as local anesthetics for minor surgical operations.

16.9   INDIVIDUAL MEMBERS

16.9.1   Dimethyl Ether [CH_3–O–CH_3]

It is the first member of the series and can be obtained by using any of the general methods of preparation discussed earlier. On a large scale it is obtained by passing methyl alcohol vapors over heated aluminum phosphate catalyst under pressure.

$$2\ CH_3OH \xrightarrow[350\text{-}400°]{Al_2(PO_4)_3,\ 15\ atm} CH_3OCH_3 + H_2O$$

Methyl alcohol   Dimethyl ether

It is a gas (b.p. – 24°), soluble in water and gives all characteristic reactions of ethers. It is used as a refrigerant and as a low temperature solvent.

16.9.2   Diethyl Ether/Ether [C_2H_5–O–C_2H_5]

This is the most important member of this class and is commonly known as ether. It is also called ether sulphuric, since it is obtained by heating alcohol with sulphuric acid.

It is obtained by any of the general methods of preparations. On a commercial scale, ether is obtained by the reaction of ethylene with concentrated sulphuric acid.

$$CH_2{=}CH_2 + H_2SO_4 \xrightarrow{140°} CH_3CH_2OSO_3H$$

Ethyl hydrogen sulphate

$$CH_3CH_2OSO_3H + H_2O \longrightarrow CH_3CH_2OH + H_2SO_4$$

Ethyl alcohol

$$CH_3CH_2OSO_3H + CH_3CH_2OH \xrightarrow{conc.H_2SO_4} CH_3CH_2OCH_2CH_3 + H_2SO_4$$

Diethyl ether

Properties: Ether is a colorless, volatile liquid (b.p. 34.6°) and has a pleasant odor with sweet burning taste. It is lighter than water (specific gravity 0.720 at 15°) and is sparingly soluble in water (approximately 3-4% at 20°). It gives all the chemical reactions discussed in section 16.8.

Uses: It is an excellent solvent for organic reactions. It is one of the best known anesthetics and is superior to chloroform in that it causes loss of consciousness without interfering much with the function of heart and lungs. It is also used as a refrigerant. A mixture of ether and dry ice gives a temperature as low as –77°.

16.10   PROBLEMS

1. How will you explain the solubility of ether in water ?
2. How are ethers obtained from alcohol ? Discuss the mechanism of their formation.
3. What happen when:
   (a) $C_2H_5I + Ag_2O \xrightarrow{\Delta}$ (b) $C_2H_5OH \xrightarrow[240\text{-}260°]{Al_2O_3}$ (c) $C_2H_5ONa + C_2H_5Br \xrightarrow{\Delta}$
4. What are oxonium salts? What is its application ?
5. Discuss the reaction of ether with:
   (a) Dilute H_2SO_4 /heat/ pressure   (c) Hydrogen iodide
   (b) Phosphorus pentachloride   (d) Chlorine in dark and sunlight
6. How will you remove peroxides from ether ?
7. Give tests to differentiate between ethers and alcohols.
8. Write structural formula of the isomeric ethers of molecular formula $C_4H_{10}O$ and give names according to IUPAC system.
9. Discuss the orbital structure of ether.
10. What are crown ethers ? Give their important applications.

17
Thioalcohols and Thioethers

17.1 INTRODUCTION

Sulphur forms organic compounds similar to those containing oxygen such as alcohols and ethers. These sulphur containing organic compounds are known as organic sulphur compounds. The sulphur analogue of an alcohol is called a thioalcohol, mercaptan , an alkanethiol or simply a thiol. Whereas, the sulphur analogue of a phenol is called thiophenol. The sulphur analogue of an ether is called a sulphide or thioether.

R$-$OH	R$-$SH	C_6H_5-OH	C_6H_5-SH
Alcohol	**Thioalcohol**	**Phenol**	**Thiophenol**
R$-$O$-$R	R$-$S$-$R		
Ether	**Thioether**		

17.2 THIOALCOHOLS

The thioalcohols are commonly known as alkyl mercaptans. These names are derived by suffixing 'mercaptan' to the names of the alkyl groups. Thus, CH_3SH is called methyl mercaptan. This system is not used commonly.

According to the IUPAC system of nomenclature, thioalcohols are named as alkanethiols. These names are derived by suffixing 'thiol' to the name of parent hydrocarbon.

		CH_3CH-CH_2SH
CH_3SH	C_2H_5SH	$\overset{\textstyle }{\underset{\textstyle CH_3}{\vert}}$
Methanethiol	**Ethanethiol**	**2-Methylpropanethiol**

17.2.1 Preparation

Following methods are used for the preparation of thioalcohols.

1. *From Alkyl Halides*

(i) *Reaction of ethyl bromide with excess of sodium hydrosulphide* gives ethanethiol.

$$CH_3CH_2-Br + Na^+SH^- \longrightarrow CH_3CH_2-SH + Na^+Br^-$$

Ethyl bromide   **Ethanethiol**

(ii) *Reaction with Thiourea:* The alkyl halides on treatment with thiourea give S-alkylisothiouronium salt, which on hydrolysis with aqueous alkali give thioalcohol.

$$CH_3CH_2\,Br \; + S=C\overset{\textstyle NH_2}{\underset{\textstyle NH_2}{\diagup}} \longrightarrow Br^-CH_3CH_2-\overset{+}{S}=C\overset{\textstyle NH_2}{\underset{\textstyle NH_2}{\diagup}}$$

Ethyl bromide   **Thiourea**   **S-Ethylisothiouronium salt**

$$\text{Br}^-\text{CH}_3\text{CH}_2-\overset{+}{\text{S}}=\text{C}\overset{\nearrow\text{NH}_2}{\searrow\text{NH}_2} + \text{NaOH} \longrightarrow \text{CH}_3\text{CH}_2\text{SH} + \text{O}=\text{C}\overset{\nearrow\text{NH}_2}{\searrow\text{NH}_2}$$

Ethanethiol   Urea

(iii) *The Grignard reagent* obtained from alkyl halide is reacted with sulphur to give the intermediate sulphur compound which on acid hydrolysis gives thioalcohol.

$$(\text{CH}_3)_3\text{CBr} + \text{Mg} \xrightarrow{\text{Ether}} (\text{CH}_3)_3\text{CMgBr} \xrightarrow{\text{S}} (\text{CH}_3)_3\text{CSMgBr}$$

tert Butyl   **Grignard reagent**
bromide

$$\downarrow \text{HCl}$$

$$(\text{CH}_3)_3\text{C}-\text{SH} \quad + \quad \text{MgBrCl}$$

2-Methylpropa-
nethiol

2. *From Alcohols*

(i) By heating with phosphorus pentasulphide.

$$5\text{C}_2\text{H}_5\text{OH} + \text{P}_2\text{S}_5 \longrightarrow 5\text{C}_2\text{H}_5\text{SH} + \text{P}_2\text{O}_5$$

Ethanethiol

(ii) By passing a mixture of alcohol vapor and hydrogen sulphide over thorium catalyst at 400°. This method is used for commercial preparation of thiols.

$$\text{C}_2\text{H}_5\text{OH} + \text{H}-\text{SH} \xrightarrow[400°]{\text{ThO}_2} \text{C}_2\text{H}_5-\text{SH} + \text{H}_2\text{O}$$

Ethanethiol

3. *From Alkenes:* Alkenes on reaction with hydrogen sulphide in presence of acid catalyst form thiols. The addition takes place as per Markovnikov's rule. This method is useful for the preparation of tertiary thiols.

$$\underset{\text{2-Methylpropene}}{\text{CH}_3-\overset{\overset{\text{CH}_3}{|}}{\text{C}}=\text{CH}_2} + \text{H}-\text{SH} \xrightarrow{\text{H}^+} \underset{\text{2-Methylpropanethiol}}{\text{CH}_3-\overset{\overset{\text{CH}_3}{|}}{\underset{\underset{\text{CH}_3}{|}}{\text{C}}}-\text{SH}}$$

4. *From Disulphides by Reduction with Zinc*

$$\text{C}_2\text{H}_5.\text{S}.\text{S}.\text{C}_2\text{H}_5 + 2\text{H} \xrightarrow{\text{Zn/H}_2\text{SO}_4} 2\text{C}_2\text{H}_5\text{SH}$$

Ethyl disulphide   **Ethanethiol**

17.2.2   Properties

17.2.2.1   Physical Properties

The first member of the series i.e. methanethiol is a gas. The other members are colorless volatile liquids. The boiling points of thiols are much lower than the corresponding alcohols due to weak intermolecular hydrogen bonding in thiols than in alcohols. Unlike alcohols, thiols are insoluble in water. This is due to the fact that thiol molecules do not form hydrogen bonds to an appreciable extent with water molecules. However, thiols are soluble in organic solvents. The lower members have extremely disagreeable odors. Their presence can be detected by smell at levels of 0.02 parts thiol per billion of air. For this reason methyl sulphide is added to natural or liquefied petroleum gas as an odor for safety precaution.

Thiols are weak acids. However, compared to alcohols, they are fairly strong acids. This is due to relatively weaker hydrogen sulphur bond, unlike alcohols which can be deprotonated only by strong base such as sodium metal or $NaNH_2$. Thiols can be ionised by the hydroxide ion.

$$CH_3CH_2SH + HO^- \longrightarrow CH_3CH_2S^- + H_2O$$

17.2.2.2   Chemical Properties

Thiols undergo reactions similar to those of alcohols. However, they show considerable difference from alcohols in respect of acidity and oxidation reaction.

1.   *Reaction with sodium:* Like alcohols, thiols react with metallic sodium with the evolution of hydrogen.

$$2C_2H_5SH + 2Na \longrightarrow 2C_2H_5SNa + H_2$$
Sodium ethyl mercaptide

2.   *Reaction with carboxylic acids and acid chlorides:* Thiols resemble alcohols in their reaction with carboxylic acid (in presence of acid catalysts) and acid chlorides to form thioesters.

$$\underset{\textbf{Acetic acid}}{CH_3-\overset{\overset{O}{\|}}{C}-OH} + HS-C_2H_5 \xrightarrow{H^+} \underset{\textbf{Ethyl thioacetate}}{CH_3-\overset{\overset{O}{\|}}{C}-SC_2H_5}$$

$$\underset{\textbf{Acetylchloride}}{CH_3-\overset{\overset{O}{\|}}{C}-Cl} + HS-C_2H_5 \xrightarrow{H^+} CH_3-\overset{\overset{O}{\|}}{C}-SC_2H_5$$

3.   *Reaction with aldehydes and ketones:* Like alcohols, thioalcohols react with aldehydes and ketones in presence of acids to form thioacetals (mercaptals) or thioketals (mercaptols) respectively.

$$\underset{\textbf{Acetaldehyde}\quad\textbf{Ethanethiol}}{\overset{CH_3}{\underset{H}{>}}C=O + 2C_2H_5SH} \xrightarrow{H^+} \underset{\substack{\textbf{Diethyl methyl mercaptal}\\ \textbf{(Acetaldehyde diethyl thioacetal)}}}{\overset{CH_3}{\underset{H}{>}}C\overset{SC_2H_5}{\underset{SC_2H_5}{<}}} + H_2O$$

$$\underset{\textbf{Acetone}\quad\textbf{Ethanethiol}}{\overset{CH_3}{\underset{CH_3}{>}}C=O + 2C_2H_5SH} \xrightarrow{H^+} \underset{\substack{\textbf{Diethyl dimethyl mercaptol}\\ \textbf{(Acetone diethyl thioketal)}}}{\overset{CH_3}{\underset{CH_3}{>}}C\overset{SC_2H_5}{\underset{SC_2H_5}{<}}}$$

4.   *Acidic nature of thiols:* One of the main differences between the thiols and alcohols is the greater acidity of thiols. The acidity of thiols enable them to react with aqueous alkali to give water soluble salts known as mercaptides.

$$\underset{\textbf{Ethane thiol}}{CH_3CH_2SH} + NaOH \longrightarrow \underset{\textbf{Sodium ethyl mercaptide}}{CH_3CH_2SNa} + H_2O$$

5.   *Oxidation reactions:* A significant difference in the chemical behavior of alcohols and thiols is the ease with which thiols are oxidised. Thiols on oxidation with mild oxidising agents like iodine, hypochlorous

acid or oxygen (in presence of iron or copper catalyst) or hydrogen peroxide yields disulphides.

$$2CH_3CH_2-SH + [O] \longrightarrow CH_3CH_2-S-S-CH_2CH_3$$

Ethanethiol   **Diethyldisulphide**

However, on vigorous oxidation with nitric acid, sulphonic acids are obtained.

$$2CH_3CH_2-SH + 3[O] \longrightarrow CH_3CH_2-\underset{\underset{O}{\|}}{\overset{\overset{O}{\|}}{S}}-OH$$

Ethanethiol   **Ethylsulphonic acid**

Uses: The lower thiols are used to detect minor leakages in natural and coal gas pipes due to their extremely unpleasant smell. They are also used in preparation of sulphonal and other hypnotics.

17.2.3 Sulphonal

It represents an important class of sleep producing or hypnotic drugs. It is obtained by oxidation of acetone mercaptol or diethyl dimethyl mercaptol with potassium permangnate.

$$\underset{CH_3}{\overset{CH_3}{>}}C=O + 2HSC_2H_5 \xrightarrow{H^+} \underset{CH_3}{\overset{CH_3}{>}}C\underset{SC_2H_5}{\overset{SC_2H_5}{<}} \xrightarrow[kMnO_4]{4[O]} \underset{CH_3}{\overset{CH_3}{>}}C\underset{SO_2C_2H_5}{\overset{SO_2C_2H_5}{<}}$$

Acetone   **Ethyl-**   **Diethyldimethyl-**   **Sulphonal**
            mercaptan   **mercaptol**

17.3 THIOETHERS

Thioethers are sulphur analogous of ethers and may be considered as dialkyl derivatives of hydrogen sulphide.

$$R-O-R \xrightarrow[+S]{-O} R-S-R \xleftarrow[+2R]{-2H} H-S-H$$

Ether   **Thioether**   **Hydrogen sulphide**

The common names of thioethers are obtained by prefixing thio before the name of ether. As per IUPAC system, they are known as dialkyl sulphide. Their name is derived by prefixing the names of the two alkyl group (bonded with S) in alphabetical order to the word sulphide.

Formula	Common Name	IUPAC Name
CH_3-S-CH_3	Dimethylthioether	Dimethylsulphide
$CH_3-S-C_2H_5$	Ethylmethylthioether	Ethylmethylsulphide
$C_2H_5-S-C_2H_5$	Diethylthioether	Diethylsulphide

17.3.1 Preparation

1. *From Alkyl Halides*

(i) By heating with potassium sulphide.

$$2C_2H_5I + K_2S \xrightarrow{\Delta} C_2H_5SC_2H_5 + 2KI$$

Ethyliodide   **Diethylsulphide**

(ii) By heating sodium alkyl mercaptide with alkyl halide. This method is similar to Williamson synthesis of ethers.

$$C_2H_5Cl + NaSC_2H_5 \longrightarrow C_2H_5SC_2H_5 + NaCl$$

Ethyl   Sodiumethyl-   Diethylsulphide
chloride   mercaptide

2. *From Ethers:* By heating with phosphorus pentasulphide.

$$5C_2H_5OC_2H_5 + P_2S_5 \longrightarrow 5C_2H_5SC_2H_5 + P_2O_5$$

Diethyl ether   Diethylsulphide

3. *From Thioalcohols*

(i) By passing vapors of thioalcohols over heated catalyst.

$$2C_2H_5SH \xrightarrow[300°]{Al_2O_3/ZnS} C_2H_5SC_2H_5 + H_2S$$

(ii) By addition to an olefin in presence of peroxides.

$$CH_2=CH_2 + C_2H_5SH \xrightarrow{Benzoyl\ peroxide} C_2H_5SC_2H_5$$

Ethene   Diethylsulphide

$$CH_3CH=CH_2 + C_2H_5SH \xrightarrow{Benzoyl\ peroxide} CH_3CH_2CH_2SC_2H_5$$

Propene   Ethylpropylsulphide

17.3.2 Properties

Physical Properties

Thioethers are colorless, unpleasant smelling volatile liquids. They are insoluble in water but soluble in organic solvents. Their boiling points are higher than those of the corresponding ethers.

Chemical Properties

Thioethers have no ionisable hydrogen and so are less reactive than thiols. They resemble ethers in their chemical properties. Thioethers undergo addition reactions much more readily than ethers since sulphur has a greater tendency to act as an electron pair doner.

1. *Addition Reactions*

(i) *Addition of Halogen*: Addition products are obtained with halogens.

$$\begin{array}{c} C_2H_5 \\ \diagdown \\ S \\ \diagup \\ C_2H_5 \end{array} + Br_2 \longrightarrow \begin{array}{c} C_2H_5 \quad Br \\ \diagdown \diagup \\ S \\ \diagup \diagdown \\ C_2H_5 \quad Br \end{array}$$

Diethylsulphide   Diethylsulphide dibromide

Ethers do not undergo this reactions.

(ii) *Addition of Alkyl Halides*: Thioethers on treatment with alkyl halide give sulphonium salts, which are similar to oxonium salts formed by ethers.

$$C_2H_5SC_2H_5 + C_2H_5I \longrightarrow (C_2H_5)_3S^+I^-$$

Diethylsulphide   Triethylsulphonium iodide

The sulphonium salts on treatment with moist silver oxide form sulphonium hydroxides, which on heating decompose to form thioethers and alkenes.

$$(C_2H_5)_3S^+I^- + AgOH \longrightarrow (C_2H_5)_3S^+OH^- + AgI$$

Triethylsulphonium iodide   **Triethylsulphonium hydroxide**

$$(C_2H_5)_3S^+OH^- \xrightarrow{\Delta} (C_2H_5)_2S + CH_2=CH_2 + H_2O$$

Diethyl-   Ethene
sulphide

2. *Hydrolysis:* On boiling with alkali, thioethers hydrolyse to alcohols.

$$C_2H_5SC_2H_5 + H_2O \xrightarrow{NaOH} 2C_2H_5OH + H_2S$$

Diethylsulphide   Ethyl alcohol

3. *Oxidation:* Thioethers on mild oxidation with hydrogen peroxide, or chlorine water give sulphoxides which on further oxidation give sulphones.

$$C_2H_5SC_2H_5 \xrightarrow[H_2O_2]{[O]} (C_2H_5)_2S \rightarrow O \xrightarrow{[O]} (C_2H_5)_2S \overset{O}{\underset{O}{\lessgtr}}$$

Diethylsulphide   Diethylsulphoxide   Diethylsulphone

conc. HNO_3 or $KMnO_4$

However with strong oxidising agents like concentrated nitric acid or potassium permanganate they are directly oxidised to sulphones.

Uses: Thioethers are used for the preparation of sulphoxides and sulphones. Sulphoxides, especially dimethylsulphoxide (DMSO) is a versatile solvent (b.p. 86° / 25 mm). They are also used in the preparation of highly poisonous mustard gas.

17.3.3   Mustard Gas/β,β-Dichlorodiethylsulphide [($ClCH_2CH_2)_2S$]

It was used as a highly poisonous gas in world war-I, and may be obtained from ethylene as follows:

$$\underset{\text{Ethylene}}{\overset{CH_2}{\underset{CH_2}{\|}}} \xrightarrow{HOCl} \underset{\substack{\text{Ethylene} \\ \text{chlorohydrin}}}{\overset{CH_2OH}{\underset{CH_2Cl}{|}}} \xrightarrow[-2NaOH]{Na_2S} \underset{\text{Mustard gas}}{\overset{CH_2Cl \quad CH_2Cl}{\underset{CH_2-S-CH_2}{|\qquad\quad|}}}$$

It can also be obtained by passing ethylene into sulphur monochloride.

$$\underset{\text{Ethylene}}{\overset{CH_2}{\underset{CH_2}{\|}}} + \underset{\substack{\text{Sulphur} \\ \text{monochloride}}}{\overset{Cl \;\; Cl}{\underset{S-S}{|\;\;|}}} + \underset{\text{Ethylene}}{\overset{CH_2}{\underset{CH_2}{\|}}} \longrightarrow \underset{\text{Mustard gas}}{\overset{CH_2Cl \quad CH_2Cl}{\underset{CH_2-S-CH_2}{|\qquad\quad|}}} + S$$

Properties

It is not a gas but an oily liquid, b.p. 215°. However, it vapourises when sprinkled by means of burning shells. It produces painful blisters and affects lungs and causes death in 4-5 days. The vapors of mustard gas can penetrate rubber. Its vapors are absorbed by active charcoal and oxidised to harmless sulphoxide by chlorine or bleaching powder.

1. *Hydrolysis:* It can be hydrolysed by the action of water.

$$(ClCH_2CH_2)_2S + H_2O \longrightarrow (HOCH_2CH_2)_2S + 2HCl$$

**2,2'-dihydroxy-
diethylthioether**

2. *Oxidation:* On oxidation with hydrogen peroxide, it forms sulphoxide, which on further oxidation gives sulphone.

$$(ClCH_2CH_2)_2S \xrightarrow[\text{H}_2\text{O}_2/\text{CH}_3\text{COOH}]{[O]} (ClCH_2CH_2)_2S \rightarrow O$$

**2,2'-Dichlorodi-
ethylsulphoxide**

$$\Big\downarrow [O].H_2O_2$$

$$(ClCH_2CH_2)_2S \underset{O}{\overset{O}{\diagup}}$$

2,2'-Dichlorodiethylsulphone

3. *Reaction with Chlorine:* It reacts with chlorine or bleaching powder to form addition products.

$$(ClCH_2CH_2)_2S + Cl_2 \longrightarrow (ClCH_2CH_2)_2SCl_2$$

**2,2'-Dichlorodiethyl-
thioetherdichloride**

17.4 PROBLEMS

1. Complete the following:

(a) $CH_3CH_2SH + NaOH \longrightarrow$

(b) $CH_3CH_2SH \xrightarrow[[O]]{HOCl}$

(c) $C_2H_5SH + CH_3COCH_3 \xrightarrow{H^+}$

(d) $C_2H_5SH + CH_3COOH \xrightarrow{H^+}$

(e) $C_2H_5SH + CH_3CHO \xrightarrow{H^+}$

(f) $C_2H_5-S-S-C_2H_5 \xrightarrow{Zn/H_2SO_4}$

2. How is dimethyl sulphide obtained from ethyliodide ?

3. Complete the following:

(a) $C_2H_5SH \xrightarrow{Al_2O_3/ZnS}$

(b) $C_2H_5SC_2H_5 \xrightarrow{Br_2}$

(c) $C_2H_5SC_2H_5 + C_2H_5I \longrightarrow A \longrightarrow B$

(d) $C_2H_5SC_2H_5 + H_2O \xrightarrow[\Delta]{NaOH}$

(e) $C_2H_5SC_2H_5 + H_2O \xrightarrow[H_2O_2]{[O]}$

4. How is mustard gas obtained. Why it is called a gas when its boiling point is 215° ?

5. How is mustard gas destroyed ?

6. Explain why the boiling point of thiols are lower than the corresponding alcohols.

7. Explain the insolubility of thiols in water compared to alcohols.

18
Aldehydes and Ketones

18.1 INTRODUCTION

Aldehydes and ketones belong to an important class of organic compounds having a carbonyl group (>C=O). In aldehydes, the two valencies of carbonyl carbon are satisfied either by hydrogen (formaldehyde) or by hydrogen and an alkyl group (acetaldehyde, propanaldehyde etc.). In ketones, on the contrary, both the valencies of carbonyl carbon are satisfied by same or different alkyl groups. The ketones may be classified as symmetrical or unsymmetrical ketones depending on whether the two alkyl groups are same or different.

$$\underset{\text{Aldehyde}}{\overset{\displaystyle H}{R-C=O}} \qquad \underset{\text{Formaldehyde}}{\overset{\displaystyle H}{H-C=O}} \qquad \underset{\text{Acetaldehyde}}{\overset{\displaystyle H}{CH_3-C=O}}$$

$$\underset{\text{Ketone}}{\overset{\displaystyle R}{R-C=O}} \qquad \underset{\substack{\text{Acetone}\\\text{Symmetrical ketone}}}{\overset{\displaystyle CH_3}{CH_3-C=O}} \qquad \underset{\substack{\text{Ethyl methyl ketone}\\\text{Unsymmetrical ketone}}}{\overset{\displaystyle CH_3}{C_2H_3-C=O}}$$

Thus an aldehydic group may be represented as $-\overset{\displaystyle H}{C}=O$ and a ketonic group as $-\overset{\displaystyle |}{C}=O$.

18.2 ORBITAL STRUCTURE

Aldehydes and ketones have a σ bond and a π bond along the carbonyl carbon and oxygen as shown in Fig.18.1. The carbon and oxygen atoms involved in bond formation utilise their sp² hybridised orbitals for the formation of sigma bonds. The π bond is formed by the sideways overlap of the unhybridised p orbitals of carbon and oxygen.

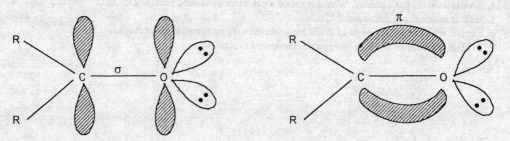

Fig. 18.1: Orbital Structure of Aldehydes and Ketones

As seen only one sp^2 hybridised orbital of oxygen is involved in σ bond formation and the other two are occupied by electron pairs. Thus, the double bond between carbon and oxygen is pulled towards the more electronegative oxygen atom and has about 40-45% ionic character. This explains the high dipole moment and higher boiling points of these compounds compared to the corresponding hydrocarbons.

On the basis of electron diffraction and spectroscopic studies, it is found that carbon, oxygen and the other two groups attached to carbonyl carbon lie in the same plane and have bond angles of roughly 120°.

The carbonyl double bond differs from the carbon-carbon double bond of alkene which is non polar.

18.3 NOMENCLATURE

The common names of *aldehydes* are derived from corresponding carboxylic acids which are obtained on their oxidation. Thus, the aldehyde which on oxidation gives formic acid is named as formaldehyde and the aldehyde which on oxidation gives acetic acid as acetaldehyde.

According to the IUPAC system, the names of aldehydes are derived from that of corresponding hydrocarbons by replacing the terminal *'e'* with *'al'*. Thus, in case of CH_3CHO, since the corresponding hydrocarbon is ethane, the IUPAC name 'ethanal' is given. The common and IUPAC names of some aldehydes are given in Table 18.1.

Table 18.1

Aldehyde (Formula)	Corresponding Acid	Common Name	IUPAC Name
HCHO	HCOOH (Formic acid)	Formaldehyde	Methanal
CH_3CHO	CH_3COOH (Acetic acid)	Acetaldehyde	Ethanal
CH_3CH_2CHO	CH_3CH_2COOH (Propionic acid)	Propionaldehyde	Propanal
$CH_3CH_2CH_2CHO$	$CH_3CH_2CH_2COOH$ (Butyric acid)	Butyraldehyde	Butanal
$CH_3CH_2CH_2CH_2CHO$	$CH_3CH_2CH_2CH_2COOH$ Valeric acid	Valeraldehyde	Pentanal

For substituted aldehydes, the position of the substituent or the side chain is indicated by numbering the carbon atoms of the parent chain by numerals or Greek letters giving the carbonyl carbon number and letter α to the carbon next to carbonyl carbon as shown below.

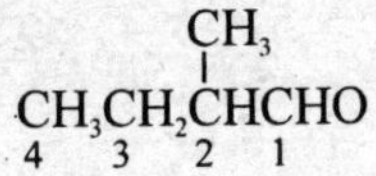

$$\underset{4}{CH_3}\underset{3}{CH_2}\underset{2}{\overset{\overset{\displaystyle CH_3}{|}}{CH}}\underset{1}{CHO} \qquad \textbf{2-Methylbutanal or α-Methylbutanal}$$

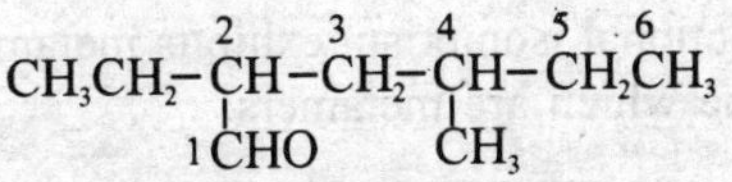

$$\underset{2}{CH_3CH_2-}\underset{}{\overset{}{CH}}\underset{1 CHO}{|}-\underset{3}{CH_2-}\underset{}{\overset{}{CH}}\underset{CH_3}{|}-\underset{5}{CH_2}\underset{6}{CH_3} \qquad \textbf{2-Ethyl-4-methylhexanal}$$

The common names of *ketones* are derived by naming the alkyl groups separately in alphabetical order and adding ketone to them. Thus, CH_3COCH_3 is commonly known as acetone or dimethyl ketone and $CH_3COCH_2CH_3$ as ethyl methyl ketone.

In the IUPAC system the ketones are termed alkanones. The name of an individual member is derived by dropping the ending *'e'* of the parent hydrocarbon and adding the suffix *'one'*, thus CH_3COCH_3 is named as propanone and $CH_3COCH_2CH_3$ butanone.

For higher ketones, the position to the carbonyl group and other substituents if any are assigned by numerals with the lowest possible number given to carbonyl carbon.

$$\underset{\textbf{2-Pentanone}}{CH_3\overset{\overset{\displaystyle O}{\|}}{C}-CH_2-CH_2-CH_3}$$

$$\underset{\underset{\textbf{3-Methyl-2-hexanone}}{}}{CH_3\overset{\overset{\displaystyle O}{\|}}{C}-\underset{\underset{CH_3}{|}}{CH}-CH_2-CH_2-CH_3}$$

$$\underset{\textbf{3-Penten-2-one}}{CH_3\overset{\overset{\displaystyle O}{\|}}{C}-CH=CH-CH_3}$$

Greek letters can also be used to specify the position of substituents. Carbon atoms adjacent to ketonic group are termed α and β on one side and α', β' on the other side. Thus, $CH_3-CHCl-CO-CH_2CH_2Cl$ is named α, β'-dichlorodiethylketone.

18.4   ISOMERISM

Aldehydes and ketones show chain isomerism amongst themselves and functional group isomerism with each other and with compounds like cyclic ethers and unsaturated alcohols.

(i)   *Chain Isomerism:* C_4H_8O represents two chain isomers.

$$\underset{\textbf{n-Butyraldehyde}}{CH_3CH_2CH_2CHO} \qquad\qquad \underset{\textbf{Isobutyraldehyde}}{(CH_3)_2CHCHO}$$

Methyl n-propyl ketone and isopropyl methyl ketone are chain isomers (both have molecular formula $C_5H_{10}O$).

$$\underset{\textbf{Methyl n-propylketone}}{CH_3COCH_2CH_2CH_3} \qquad\qquad \underset{\textbf{Isopropyl methyl ketone}}{(CH_3)_2CH-CO-CH_3}$$

(ii)   *Functional Group Isomerism*: C_3H_6O can be an aldehyde, a ketone, an oxide or an unsaturated alcohol.

$$\underset{\textbf{Propanal}}{CH_3CH_2CHO} \qquad \underset{\textbf{Acetone}}{CH_3COCH_3} \qquad \underset{\textbf{Allyl alcohol}}{CH_2=CHCH_2OH}$$

$$\underset{\underset{\textbf{α, γ,-propylene oxide}}{}}{\overset{CH_2-CH_2}{\underset{CH_2-O}{|\quad\ |}}} \qquad\qquad \underset{\underset{\textbf{α, β-propylene oxide}}{}}{CH_3-\overset{}{\underset{\diagdown O \diagup}{CH-CH_2}}}$$

(iii)   *Metamerism:* Ketones, in addition to chain and functional isomerism exhibits metamerism also. $C_5H_{10}O$ may be represented as 2-pentanone and 3-pentanone which are metamers.

$$\underset{\textbf{2-Pentanone}}{CH_3\overset{\overset{\displaystyle O}{\|}}{C}CH_2CH_2CH_3} \qquad\qquad \underset{\textbf{3-Pentanone}}{CH_3CH_2\overset{\overset{\displaystyle O}{\|}}{C}CH_2CH_3}$$

18.5 GENERAL METHODS OF PREPARATION OF ALDEHYDES AND KETONES

1. From Alcohols

(i) *By Oxidation:* Primary alcohols on oxidation with acidified potassium permanganate or dichromate or manganese dioxide or chromic anhydride in glacial acetic acid yield aldehydes.

$$R-CH_2OH + [O] \longrightarrow RCHO + H_2O$$
Primary alcohol   Aldehyde

$$CH_3CH_2OH + O \longrightarrow CH_3CHO + H_2O$$
Ethanol   Acetaldehyde
                 (Ethanal)

The aldehydes formed above are readily oxidised to carboxylic acid, therefore, they must be removed from the reaction mixture. This is easily done by carrying out the oxidation slightly above the boiling point of the aldehyde so that it is removed by distillation. The aldehydes have lower boiling points than the corresponding alcohols.

Secondary alcohols under similar conditions give ketones.

$$\begin{matrix} R \\ R' \end{matrix}\!\!\diagdown\!\!CHOH + [O] \longrightarrow \begin{matrix} R \\ R' \end{matrix}\!\!\diagdown\!\!C=O + H_2O$$
Secondary alcohol   Ketone

$$\begin{matrix} CH_3 \\ CH_3 \end{matrix}\!\!\diagdown\!\!CHOH + [O] \longrightarrow \begin{matrix} CH_3 \\ CH_3 \end{matrix}\!\!\diagdown\!\!C=O + H_2O$$
Isopropyl alcohol   Acetone

Oxidation of primary and secondary alcohols can also be carried out by passing their vapors mixed with air over silver catalyst at 250°. This method gives good yield.

Secondary alcohols can be oxidised in good yields to ketones using aluminum tert. butoxide in excess of acetone. This oxidation is known as *Oppenauer's oxidation.*

$$\begin{matrix} R \\ R' \end{matrix}\!\!\diagdown\!\!CHOH + CH_3COCH_3 \xrightarrow{[(CH_3)_3CO]_3Al} \begin{matrix} R \\ R' \end{matrix}\!\!\diagdown\!\!C=O + \begin{matrix} CH_3 \\ CH_3 \end{matrix}\!\!\diagdown\!\!CHOH$$
Secondary   Acetone   Ketone   Isopropanol
alcohol

In this reaction secondary alcohol is oxidised at the cost of acetone, which is reduced.

(ii) *By Dehydrogenation:* Both primary and secondary alcohols on passing over heated catalyst (copper) at 300° give aldehydes and ketones respectively.

$$CH_3CH_2OH \xrightarrow{Cu}{300°} CH_3CHO + H_2$$
Ethyl alcohol   Acetaldehyde

$$CH_3-\underset{\underset{OH}{|}}{CH}-CH_3 \xrightarrow{Cu}{300°} CH_3-\underset{\underset{O}{\|}}{C}-CH_3 + H_2$$
Isopropyl alcohol   Acetone

2.   From Carboxylic acids and their derivatives

(i)   *From Calcium Salts of Carboxylic Acids*: When calcium formate is heated alone, formaldehyde is obtained.

$$(HCOO)_2Ca \xrightarrow[\text{Pyrolysis}]{\Delta} HCHO + CaCO_3$$

Calcium formate   Formaldehyde

Other aldehydes may be obtained by heating calcium formate and calcium salt of a fatty acid together.

$$(HCOO)_2Ca + (CH_3COO)_2Ca \xrightarrow{\Delta} 2CH_3CHO + 2CaCO_3$$

Calcium formate   Calcium acetate   Acetaldehyde

The yield in this reaction is poor due to side reactions. Simple ketones may also be obtained by the pyrolysis of calcium salt of a fatty acid (other than formic acid).

$$(CH_3COO)_2Ca \xrightarrow{\Delta} CH_3-\overset{\overset{\displaystyle O}{\|}}{C}-CH_3 + CaCO_3$$

Calcium acetate   Acetone

A mixed ketone, can be obtained by heating together a mixture of calcium salts of different fatty acids.

$$(CH_3COO)_2Ca + (C_2H_5COO)_2Ca \xrightarrow{\Delta} CH_3\overset{\overset{\displaystyle O}{\|}}{C}C_2H_5 + CaCO_3$$

Calcium acetate   Calcium propionate   Ethyl methyl ketone

The yield of mixed ketones is poor due to side reactions.

(ii)   *Catalytic Decomposition of Carboxylic Acids:* Vapors of fatty acids mixed with formic acid when passed over heated thoria (350°), alumina (400°) or manganese dioxide gives aldehydes.

$$CH_3COOH + HCOOH \xrightarrow[400°]{Al_2O_3} CH_3CHO + CO_2 + H_2O$$

Acetic acid   Formic acid   Acetaldehyde

If formic acid alone is used, formaldehyde is obtained.

$$2\,HCOOH \xrightarrow[400°]{Al_2O_3} HCHO + CO_2 + H_2O$$

Formic acid   Formaldehyde

However, vapors of higher acid alone under similar conditions give ketones.

$$RCOOH \xrightarrow[500°]{MnO_2} RCOR + CO_2 + H_2O$$

Fatty acid   Ketone

Use of two different acids (other than formic acid) give mixed ketones in low yields.

(iii)   *Rosenmund's Reduction of Acid Chlorides:* Aldehydes may be obtained in good yields when acid chlorides are reduced in presence of palladium catalyst supported on barium sulphate.

$$RCOCl + H_2 \xrightarrow{Pd/BaSO_4} RCHO + HCl$$

$$CH_3CH_2COCl + H_2 \xrightarrow{BaSO_4-Pd} CH_3CH_2CHO + HCl$$

Propanoyl   Propanal
chloride

(iv) *From Nitriles:* Aldehydes can be obtained by the reduction of a nitrile dissolved in ether with stannous chloride and hydrochloric acid and hydrolysing the aldimine hydrochloride thus obtained by warm water *(Stephens method)*.

$$CH_3-C\equiv N \xrightarrow[\text{Ether}]{\text{SnCl}_2 + \text{HCl}} CH_3-\overset{\overset{\displaystyle H}{|}}{C}=\overset{+}{N}H_2\bar{C}l \xrightarrow{2H_2O} CH_3CHO + 2NH_4Cl$$

| Ethanenitrile | Aldimine hydrochloride | | Acetaldehyde |

An aldehyde is also obtained when an alkyl cyanide is reduced with lithium aluminum hydride and the resultant product hydrolysed.

$$2R-C\equiv N + LiAlH_4 \longrightarrow (R-CH=N\overset{}{\rightarrow}_3Al \xrightarrow{2H_2O} RCHO$$

Alkyl cyanide   **Aldehyde**

3. *From Olefins by Ozonolysis:* Aldehydes and ketones may be obtained by ozonolysis of substituted alkenes followed by reduction of ozonide with zinc and acetic acid.

$$R-CH=CH-R' \xrightarrow{O_3} R-\underset{\underset{\displaystyle O-O}{}}{\overset{\overset{\displaystyle O}{\diagdown\diagup}}{CH\quad CH}}-R' \xrightarrow[\text{Catalyst}]{H_2} RCHO + R'CHO$$

Olefin   **Ozonide**   **Aldehyde**

$$\underset{R}{\overset{R}{\diagdown}}C=C\underset{R'}{\overset{R'}{\diagup}} \xrightarrow{O_3} \underset{R}{\overset{R}{\diagdown}}\underset{}{C}\underset{\underset{\displaystyle O-O}{}}{\overset{\overset{\displaystyle O}{\diagdown\diagup}}{}}\underset{R'}{\overset{R'}{\diagup}}C \xrightarrow[\text{Catalyst}]{H_2} \underset{R}{\overset{R}{\diagdown}}C=O + \underset{R'}{\overset{R'}{\diagdown}}C=O$$

Olefin   **Ozonide**   **Ketone**

The procedure is used to find the position of the double bond in the olefin molecule.

4. *From Alkynes by Hydration:* Acetylene on hydration in presence of a solution of mercuric sulphate and dilute sulphuric acid gives acetaldehyde.

$$HC\equiv CH + H_2O \xrightarrow[\text{H}_2\text{SO}_4]{Hg^{++}} H-\underset{}{\overset{\overset{\displaystyle H}{|}}{C}}=\underset{}{\overset{\overset{\displaystyle H}{|}}{C}}-OH \xrightarrow{\text{isomerises}} CH_3-\overset{\overset{\displaystyle H}{|}}{C}=O$$

                                                       Acetaldehyde

Homologues of acetylene on similar hydration yield methyl ketones.

$$CH_3-C\equiv C-CH_3 + H_2O \xrightarrow[\text{H}_2\text{SO}_4]{Hg^{++}} CH_3-\overset{\overset{\displaystyle O}{\|}}{C}-CH_2CH_3$$

2-Butyne   **2-Butanone**

5. *From gem-Dihalides by Hydrolysis:* gem-Dihalides in which two halogen atoms are attached to the terminal carbon atom on alkaline hydrolysis form aldehydes.

$$R-CH\underset{Cl}{\overset{Cl}{\diagup}} \xrightarrow{OH} \left[R-CH\underset{OH}{\overset{OH}{\diagup}}\right] \xrightarrow{-H_2O} R-\overset{\overset{\displaystyle H}{|}}{C}=O$$

gem-Dihalide   **Aldehyde**

$$CH_3-CH\begin{array}{c}Cl\\Cl\end{array} \xrightarrow{OH^-} \left[CH_3-CH\begin{array}{c}OH\\OH\end{array}\right] \xrightarrow{-H_2O} CH_3-\overset{H}{\underset{}{C}}=O$$

1, 1-Dichloroethane   **Acetaldehyde**

However, gem-dihalides in which two halogen atoms are attached to any carbon atom other than the terminal carbon on alkaline hydrolysis give ketones.

$$R-\overset{Cl}{\underset{Cl}{C}}-R \xrightarrow{HO^-} \left[R-\overset{OH}{\underset{OH}{C}}-R\right] \xrightarrow{-H_2O} R-\overset{O}{\overset{\|}{C}}-R$$

gem-Dihalide   **Ketone**

$$CH_3-\overset{Br}{\underset{Br}{C}}-CH_3 \xrightarrow{HO^-} \left[CH_3-\overset{OH}{\underset{OH}{C}}-CH_3\right] \xrightarrow{-H_2O} CH_3-\overset{O}{\overset{\|}{C}}-CH_3$$

2, 2-Dibromopropane   **Acetone**

6. *From Grignard reagent:* An aldehyde may be prepared by treatment of Grignard reagent with excess of formic ester.

$$H-\overset{OC_2H_5}{\overset{|}{C}}=O + C_2H_5MgBr \longrightarrow \overset{H}{\underset{H_3C_2}{C}}\overset{\overset{+}{O}MgBr}{\underset{OC_2H_5}{}} \xrightarrow{H_2O/H^+} \overset{H}{\underset{H_3C_2}{C}}=O + Mg\overset{Br}{\underset{OC_2H_5}{}}$$

Ethyl   **Ethyl**   **Propanal**
formate   **magnesium**
              bromide

Ketones can be prepared from organo cadmium compounds by reaction with acid halides. This method is preferable over Grignard reaction, since they do not react further with ketone obtained to form alcohol.

$$R-MgCl + CdCl_2 \longrightarrow RCdCl + MgCl_2$$
$$R'COCl + RCdCl \longrightarrow R'COR + CdCl_2$$
$$\text{\textbf{Ketone}}$$

7. *From Acetoacetic esters:* Alkyl derivatives of acetoacetic ester on hydrolysis with dilute alkali give the corresponding ketones.

$$CH_3COCH_2COOC_2H_5 \xrightarrow[2.\ CH_3I]{1.\ C_2H_5ONa} CH_3CO\overset{}{\underset{CH_3}{C}}HCOOC_2H_5 \xrightarrow[2.\ H^+]{1.\ dil\ KOH} CH_3COCH_2CH_3 + CO_2 + C_2H_5OH$$

Ethyl acetoacetate   **2-Butanone**

18.6 PHYSICAL PROPERTIES

With the exception of formaldehyde (which is a gas at ordinary temperature), the lower members of aldehydes and ketones (upto C_{10}) are colorless, volatile liquids. The higher members of both aldehydes and ketones are solids.

The lower aldehydes have unpleasant odor but the higher aldehydes have a fruity odor. The ketones possess pleasant odor.

As already mentioned aldehydes and ketones are polar compounds and therefore possess intermolecular dipole-dipole interaction. Due to these interactions both aldehydes and ketones have higher boiling points than the non polar alkanes of similar molecular weights. The boiling points of aldehydes and ketones are however, much lower than the boiling points of the corresponding alcohols. This is because the alcohol molecules are held together by strong hydrogen bonds, whereas the molecules of aldehydes and ketones are held together by much weaker electrostatic interactions between the dipoles.

The lower members of the aldehydes and ketones are soluble in water but the solubility decreases with increase in molecular weight. The higher members with more than five carbon atoms are practically insoluble in water. However, they are soluble in organic solvents like alcohol and ether. The partial solubility in water can be explained by the formation of hydrogen bonds between carbonyl compounds and water molecule.

18.7 SPECTRAL PROPERTIES

The ultraviolet spectra of aldehydes and ketones shows two absorption bands for the carbonyl group; in aldehydes at 180nm and 295nm for $\pi \rightarrow \pi^*$ and $n \rightarrow \pi^*$ transitions and in ketones at 190nm and 270-280nm for $\pi \rightarrow \pi^*$ and $n \rightarrow \pi^*$ transitions respectively.

The infrared spectrum is the most diagnostic for the carbonyl group. The C=O stretching for aldehyde is 1700-1740 cm^{-1} and for ketones is 1660-1750 cm^{-1}. In addition, the aldehydes show C$-$H stretching of $-$CHO at 2700-2900 cm^{-1}.

18.8 CHEMICAL PROPERTIES

The reactions of aldehydes and ketones can be grouped in the following categories.

1. *Reactions of the Carbonyl Group*: It has already been stated that the carbon oxygen double bond in carbonyl group is polar as oxygen being more electronegative attracts the bonding electron pair towards itself and attains a negative charge compared to carbon which becomes positive as represented below.

$$\text{>C=}\ddot{\text{O}}\text{:} \longleftrightarrow \text{>}\overset{+}{\text{C}}-\ddot{\bar{\text{O}}}\text{:}$$

The carbonyl compounds, therefore, are good substrates for nucleophilic and electrophilic attack at carbon and oxygen of carbonyl group respectively.

$$\text{>C=}\ddot{\text{O}}\text{:} \longleftrightarrow \text{>}\overset{+}{\text{C}}-\ddot{\bar{\text{O}}}\text{:}$$

Position for   **Position for**
Nucleophilic attack   **Electrophilic attack**

The nucleophilic addition to the carbonyl carbon is catalysed by acids, since proton adds to the carbonyl oxygen.

$$\text{>C=}\ddot{\text{O}}\text{:}+\text{H}^+ \longleftrightarrow \text{>C=}\overset{+}{\text{O}}\text{H} \longleftrightarrow \text{>}\overset{+}{\text{C}}-\ddot{\text{O}}\text{H}$$

In addition reactions, the relative reactivity of aldehydes and ketones is attributed to the extent of polarisation on the carbonyl carbon. The more polarised the carbonyl group the greater is the positive charge on the carbonyl carbon and the higher the reactivity. Ketones are less reactive than aldehydes because of the presence of two alkyl group in the former. The alkyl groups are electron releasing groups and they make the carbon of the carbonyl group less electron deficient in ketones than in aldehydes (having only one alkyl group). Further methanal with no alkyl group attached the carbonyl carbon is more reactive than ethanal and other aldehydes. The increasing reactivity of the carbonyl group towards nucleophilic addition in aldehydes and ketones can be depicted as follows.

$$Ketones < Aldehydes < Formaldehyde$$

The reactions of the carbonyl group have been discussed below:

(i) *Nucleophilic Addition Reactions:* The mechanism of nucleophilic addition involves the following steps:

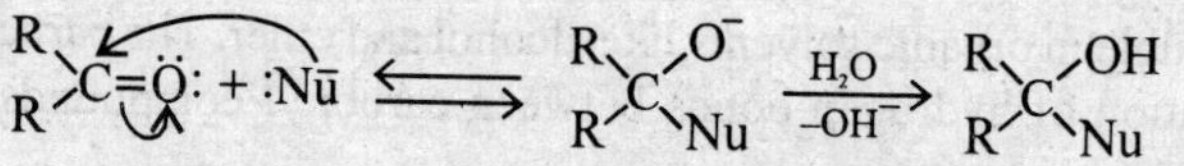

where, nucleophile may be HO^-, RO^-, CN^-, NH_3, H_2O, ROH, $C_6H_5NHNH_2$, $R–NH_2$, $RMgX$ etc.

(a) *Addition of Hydrogen Cyanide:* Hydrogen cyanide adds onto aldehydes and ketones to yield *cyanohydrins.*

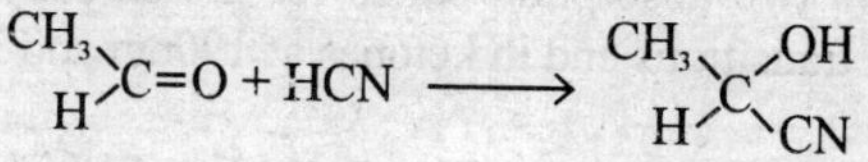

Acetaldehyde   Acetaldehyde cyanohydrin

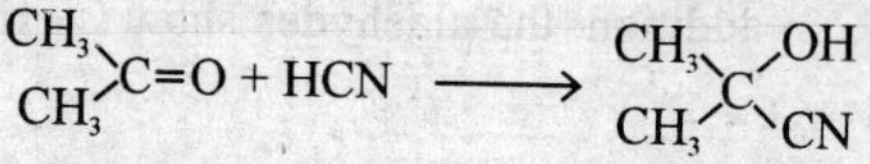

Acetone   Acetone cyanohydrin

Hydrogen cyanide is extremely poisonous. It is produced in the reaction mixture by the action of sodium or potassium cyanide with dilute sulphuric acid.

The cyanohydrins are important synthetic reagents and can be converted to hydroxy and amino acids (*Strecker's synthesis*) by treatment with water and ammonia.

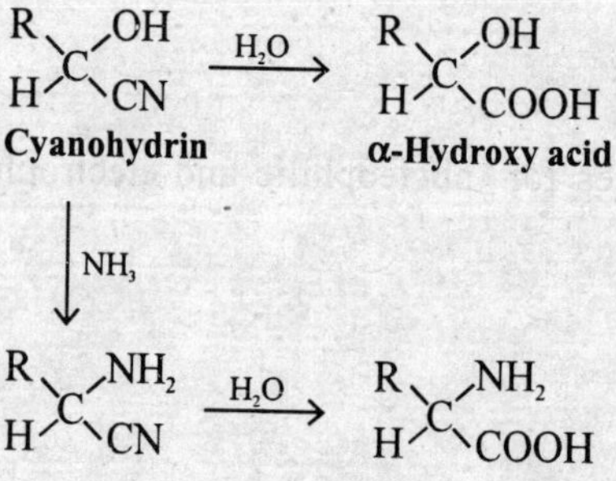

Cyanohydrin   α-Hydroxy acid

α-Amino acid

(b) *Addition of Sodium Bisulphite:* The reaction of carbonyl compounds with sodium bisulphite forms the bisulphite addition products.

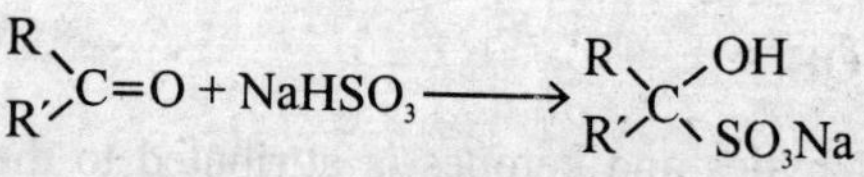

Bisulphite addition product

The addition products are crystalline solids and on heating with dilute sulphuric acid or aqueous sodium carbonate regenerate the carbonyl compound. This method is used for separation and purification of aldehydes and ketones.

(c) *Addition of Ammonia:* Aldehydes react with ammonia to form aldehyde-ammonia. These addition products are unstable and lose a water molecule to form dehydrated product, which polymerises to cyclic trimers.

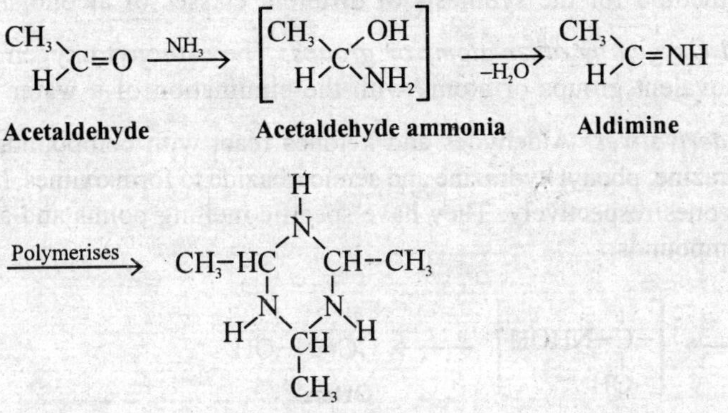

In practice, only the final product is obtained. However, formaldehyde reacts differently to give hexamethylene tetramine (see Section 18.9.1)

Ketones react with ammonia to give ketone-ammonia which cannot be isolated. Acetone, however, reacts slowly with ammonia at low temperature to give acetone-ammonia and at higher temperature diacetone-amine.

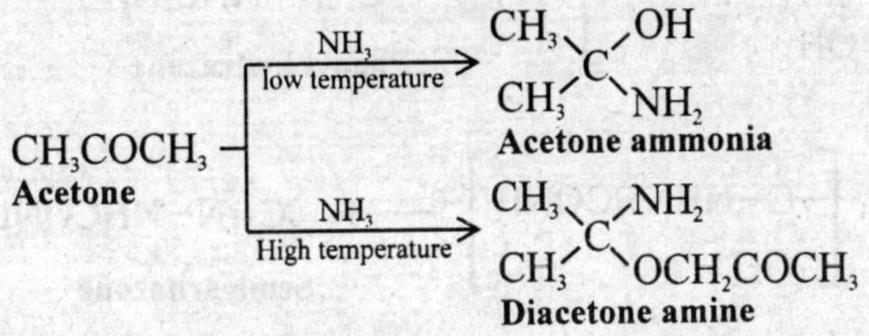

(d) *Addition to Grignard Reagents:* All aldehydes and ketones react with Grignard reagents to form adducts which on hydrolysis give secondary alcohols (in case of aldehydes except formaldehyde) and tertiary alcohols (in case of ketones).

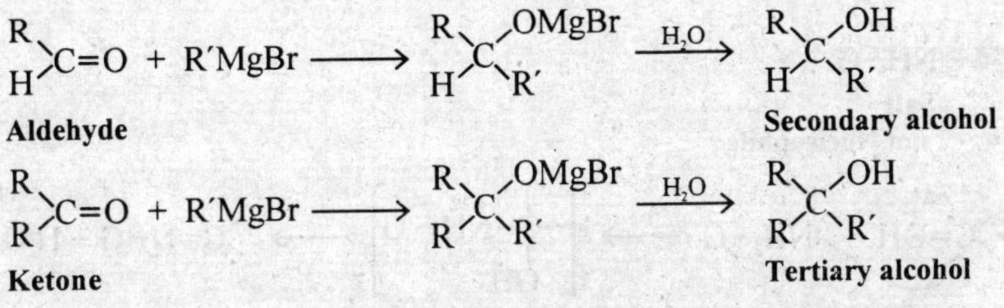

Formaldehyde, however, gives a primary alcohol.

$$\underset{\text{Formaldehyde}}{\overset{\text{H}}{\underset{\text{H}}{>}}C=O} \; + \; \underset{\substack{\text{Methyl}\\\text{magnesium}\\\text{bromide}}}{CH_3MgBr} \; \longrightarrow \; \underset{}{\overset{\text{H}}{\underset{\text{H}}{>}}C\overset{\text{OMgBr}}{\underset{\text{CH}_3}{<}}} \; \overset{H_2O}{\longrightarrow} \; \underset{\text{Ethanol}}{CH_3CH_2OH}$$

This reaction is an excellent method for the synthesis of different classes of alcohols.

(ii) *Replacement of Carbonyl Oxygen by other atoms or groups:* The carbonyl oxygen is replaced by a bivalent group or two monovalent groups or atoms with the elimination of a water molecule.

(a) *Reaction with Ammonia derivatives:* Aldehydes and ketones react with compounds related to ammonia like hydroxylamine, hydrazine, phenyl hydrazine and semicarbazide to form oximes, hydrazones, phenylhydrazones and semicarbazones respectively. They have specific melting points and are used for the identification of carbonyl compounds.

$$>C=O \; + \; \underset{\substack{\text{Hydroxyl}\\\text{amine}}}{NH_2OH} \; \longrightarrow \; \left[-\overset{|}{\underset{\underset{OH}{|}}{C}}-NHOH\right] \; \longrightarrow \; \underset{\text{Oxime}}{>C=N-OH}$$

$$>C=O \; + \; \underset{\text{Hydrazine}}{NH_2NH_2} \; \longrightarrow \; \left[-\overset{|}{\underset{\underset{OH}{|}}{C}}-NHNH_2\right] \; \longrightarrow \; \underset{\text{Hydrazone}}{>C=N-NH_2}$$

$$>C=O \; + \; \underset{\text{Phenyl hydrazine}}{NH_2NHC_6H_5} \; \longrightarrow \; \left[-\overset{|}{\underset{\underset{OH}{|}}{C}}-NHNHC_6H_5\right] \; \longrightarrow \; \underset{\text{Phenyl hydrazone}}{>C=N-NHC_6H_5}$$

$$\underset{\substack{\text{Carbonyl}\\\text{compound}}}{>C=O} \; + \; \underset{\text{Semicarbazide}}{NH_2NHCONH_2} \; \longrightarrow \; \left[-\overset{|}{\underset{\underset{OH}{|}}{C}}-NHNHCONH_2\right] \; \longrightarrow \; \underset{\text{Semicarbazone}}{>C=N-NHCONH_2}$$

The reaction takes place in slightly acidic conditions as strongly acidic conditions convert nucleophilic ammonia derivative into its salt which is not nucleophilic in nature and does not undergo addition. The mechanism of the reaction is given below.

$$\underset{\substack{\text{Free base}\\\text{Nucleophile}}}{\overset{..}{N}H_2{-}G} \; \underset{}{\overset{H^+}{\rightleftharpoons}} \; \underset{\substack{\text{Salt}\\\text{not Nucleophile}}}{\overset{+}{N}H_3{-}G}$$

$$>C=O \; \overset{H^+}{\longrightarrow} \; >C\overset{+}{=}OH \; + \; :NH_2{-}G \; \longrightarrow \; \left[-\overset{|}{\underset{\underset{OH}{|}}{C}}-\overset{+}{N}H_2{-}G\right] \; \longrightarrow \; >C=N-G + H_2O + H^+$$

$$G = -OH, -NH_2, -NHC_6H_5, -NHCONH_2$$

(b) *Reaction with Alcohols:* Aldehydes react with alcohols in presence anhydrous acids to form acetals.

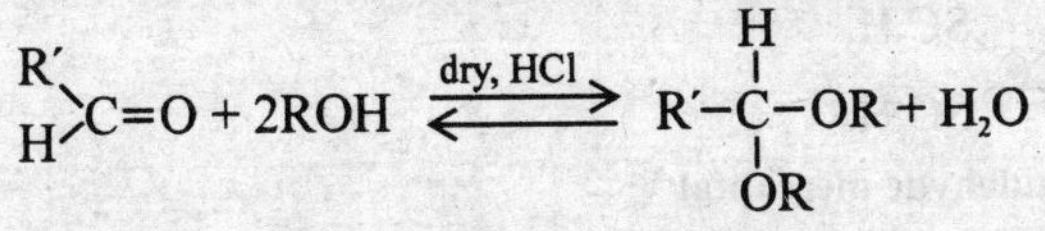

$$\underset{H}{\overset{R'}{>}}C=O + 2ROH \underset{}{\overset{dry,\ HCl}{\rightleftharpoons}} \underset{\underset{OR}{|}}{\overset{\overset{H}{|}}{R'-C-OR}} + H_2O$$

Aldehyde   Alcohol   Acetal

The aldehydes in alcoholic solutions exists in equilibrium with a compound called hemiacetal having both hydroxy and ether groups. Hemiacetal in acidic solution reacts with the alcohol to form acetal.

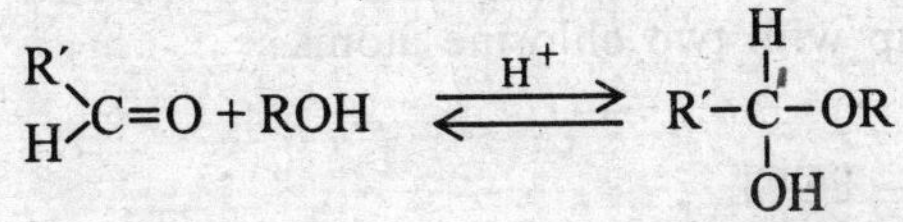

$$\underset{H}{\overset{R'}{>}}C=O + ROH \overset{H^+}{\rightleftharpoons} \underset{\underset{OH}{|}}{\overset{\overset{H}{|}}{R'-C-OR}}$$

Aldehyde   Hemiacetal

$$\underset{\underset{OH}{|}}{\overset{\overset{H}{|}}{R'-C-OR}} + ROH \overset{H^+}{\rightleftharpoons} \underset{\underset{OR}{|}}{\overset{\overset{H}{|}}{R'-C-OR}}$$

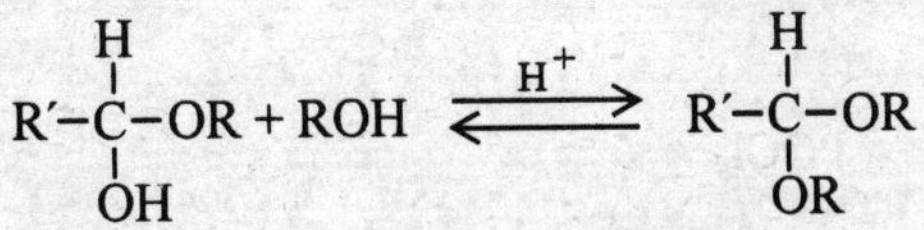

Hemiacetal   Acetal

similarly, ketones form ketals.

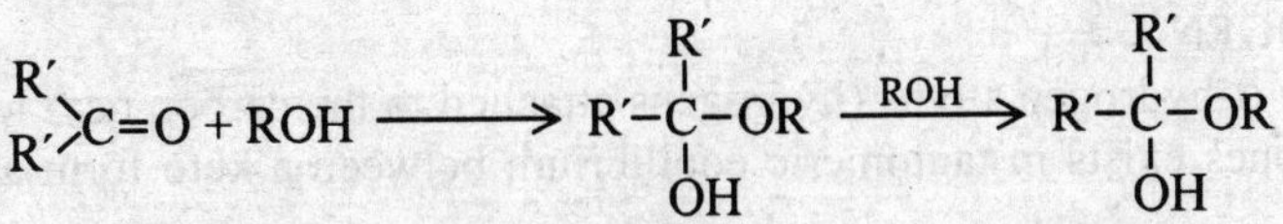

$$\underset{R'}{\overset{R'}{>}}C=O + ROH \longrightarrow \underset{\underset{OH}{|}}{\overset{\overset{R'}{|}}{R'-C-OR}} \overset{ROH}{\longrightarrow} \underset{\underset{OH}{|}}{\overset{\overset{R'}{|}}{R'-C-OR}}$$

Ketone   Hemiketal   Ketal

The mechanism involved is S$_N$1 and involves following steps.

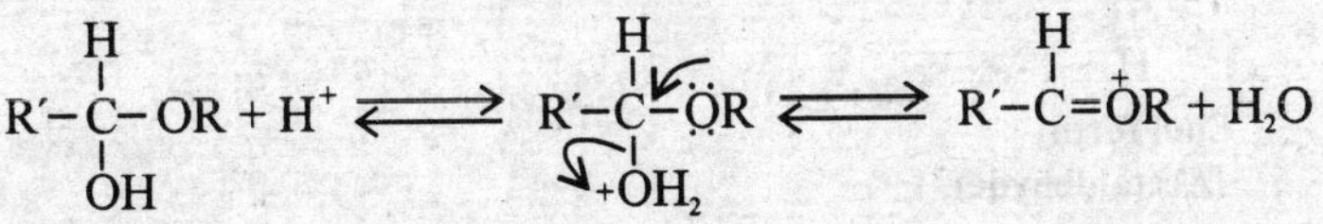

$$\underset{\underset{OH}{|}}{\overset{\overset{H}{|}}{R'-C-OR}} + H^+ \rightleftharpoons \underset{\underset{+OH_2}{|}}{\overset{\overset{H}{|}}{R'-C-OR}} \rightleftharpoons R'-\overset{\overset{H}{|}}{C}=\overset{+}{O}R + H_2O$$

Hemiacetal

$$R'-\overset{\overset{H}{|}}{C}=\overset{+}{O}R + R\overset{..}{O}H \rightleftharpoons \underset{\underset{\underset{H}{|}}{+O-R}}{\overset{\overset{H}{|}}{R'-C-OR}} \rightleftharpoons \underset{\underset{OR}{|}}{\overset{\overset{H}{|}}{R'-C-OR}}$$

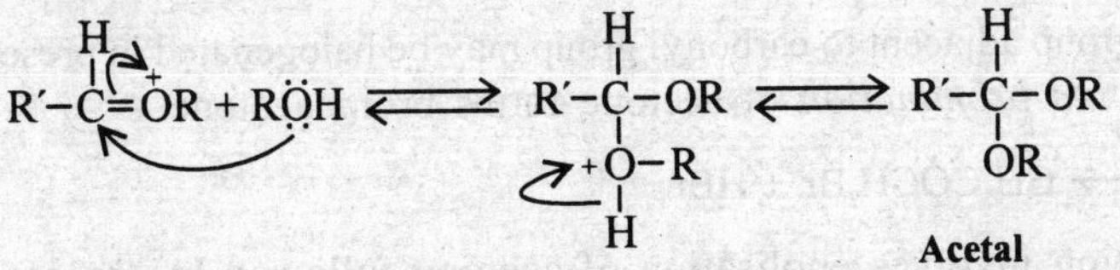

Acetal

Acetals and ketals regenerate aldehydes and ketones via acid hydrolysis, whereas they are stable towards bases. This method is used for protecting the carbonyl group of aldehydes and ketones.

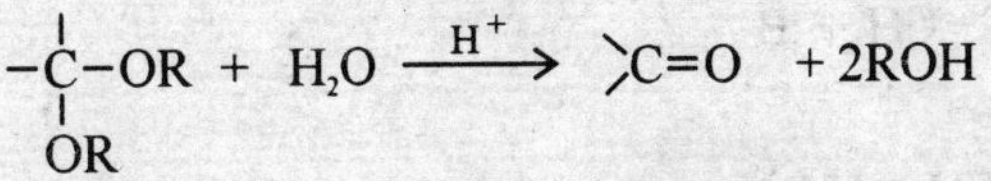

$$\underset{\underset{OR}{|}}{\overset{\overset{|}{|}}{-C-OR}} + H_2O \overset{H^+}{\longrightarrow} {>}C=O + 2ROH$$

Acetal or   Aldehyde   Alcohol
ketal   or ketone

(c) *Reaction with Thioalcohols:* Aldehydes and ketones react with thioalcohols (mercaptans) to form thioacetals (mercaptals) and thioketals (mercaptols) respectively.

$$CH_3 \atop H \Large{>}\normalsize C=O + C_2H_5SH \xrightarrow[-H_2O]{HCl} \; {CH_3 \atop H}\Large{>}\normalsize C{SC_2H_5 \atop SC_2H_5}$$

Acetaldehyde   Acetaldehyde mercaptal

$$CH_3 \atop CH_3 \Large{>}\normalsize C=O + C_2H_5SH \xrightarrow[-H_2O]{HCl} \; {CH_3 \atop CH_3}\Large{>}\normalsize C{SC_2H_5 \atop SC_2H_5}$$

Acetone   Acetone mercaptol

(d) *Reaction with Phosphorus Pentachloride:* It reacts with both aldehydes and ketones resulting in the replacement of oxygen of the carbonyl group with two chlorine atoms.

$$CH_3 \atop H \Large{>}\normalsize C=O + PCl_5 \longrightarrow \; {CH_3 \atop H}\Large{>}\normalsize C{Cl \atop Cl} + POCl_3$$

Acetaldehyde   Ethylidene
                                chloride

$$CH_3 \atop CH_3 \Large{>}\normalsize C=O + PCl_5 \longrightarrow \; {CH_3 \atop CH_3}\Large{>}\normalsize C{Cl \atop Cl} + POCl_3$$

Acetone   Isopropylidene
                                chloride

2. REACTIONS OF α-HYDROGEN

Carbonyl compounds have acidic α-hydrogen atoms (hydrogens attached to the carbon next to carbonyl) due to which aldehydes and ketones exists in tautomeric equilibrium between a keto form and an enol form.

$$H{-}\underset{\underset{H}{|}}{\overset{\overset{H}{|}}{C}}{-}\underset{\underset{H}{|}}{C}{=}O \rightleftharpoons H{-}\underset{\underset{H}{|}}{C}{=}\overset{\overset{H}{|}}{C}{-}OH$$

Keto form   Enol form
(Acetaldehyde)   (Acetaldehyde)

Enolisation is catalysed by both acids and bases.

(i) *Halogenation:* The alkyl group adjacent to carbonyl group may be halogenated in presence of acid as well as base. The *acid catalysed bromination* of acetone forms bromoacetone.

$$CH_3COCH_3 + Br_2 \longrightarrow CH_3COCH_2Br + HBr$$

The acid catalysed halogenation involves enolisation of acetone followed by the reaction with halogen as shown in the following steps.

$$CH_3{-}\overset{\overset{O}{\|}}{C}{-}CH_3 + H{:}B^+ \underset{fast}{\rightleftharpoons} CH_3{-}\overset{\overset{+OH}{\|}}{C}{-}CH_3 + {:}B$$

$$CH_3{-}\overset{\overset{+OH}{\|}}{C}{-}CH_3 + {:}B \underset{slow}{\rightleftharpoons} CH_3{-}\overset{\overset{OH}{|}}{C}{=}CH_2 + H{:}B^+$$

$$CH_3-\overset{\overset{\displaystyle OH}{|}}{C}=CH_2 + X-X \xrightarrow{\text{fast}} CH_3-\overset{\overset{\displaystyle +OH}{\|}}{C}-CH_2X + X^-$$

$$CH_3-\overset{\overset{\displaystyle +OH}{\|}}{C}-CH_2X + :B \underset{}{\overset{\text{fast}}{\rightleftharpoons}} CH_3-\overset{\overset{\displaystyle O}{\|}}{C}-CH_2X + H:B^+$$

Base promoted halogenation, on the contrary utilises the base (which may be hydroxide or acetate ion) during the rate determining step.

$$CH_3-\overset{\overset{\displaystyle O}{\|}}{C}-CH_3 + :\bar{B} \underset{}{\overset{\text{slow}}{\rightleftharpoons}} H:B + CH_3-\overset{\overset{\displaystyle O}{\|}}{C}-\bar{C}H_2$$

$$CH_3-\overset{\overset{\displaystyle O}{\|}}{C}-\bar{C}H_2 \longleftrightarrow CH_3-\overset{\overset{\displaystyle O^-}{|}}{C}=CH_2 \equiv CH_3-\overset{\overset{\displaystyle O}{|}}{C}-CH_2$$

$$CH_3-\overset{\overset{\displaystyle O}{}}{C}-CH_2 + X-X \xrightarrow{\text{fast}} CH_3-\overset{\overset{\displaystyle O}{\|}}{C}-CH_2X$$

(ii) *Aldol Condensation:* As already seen, the removal of proton leaves an electron pair on carbon and gives a carbanion or enolate ion which is resonance stablised. The carbanion thus produced is a good nucleophile and can attack carbonyl group of another molecule. This process of formation of the carbanion followed by its addition to a carbonyl group, is involved in aldol and aldol type condensations of aldehydes and ketones.

Two molecules of an aldehyde or a ketone having α-hydrogen condense together in presence of base or dilute acid to form β-hydroxyaldehyde or β-hydroxyketone and the reaction is referred to as aldol condensation.

(a) When two molecules of acetaldehyde condense in presence of dilute alkali, β-hydroxybutyraldehyde is formed.

$$2CH_3CHO \xrightarrow{\text{dil alkali}} CH_3-\overset{\overset{\displaystyle OH}{|}}{\underset{\underset{\displaystyle H}{|}}{C}}-CH_2CHO$$

Acetaldehyde   **β-Hydroxybutyraldehyde**

On heating, aldol loses a molecule of water to form an α,β-unsaturated aldehyde.

$$CH_3-\overset{\overset{\displaystyle OH}{|}}{\underset{\underset{\displaystyle H}{|}}{C}}-CH_2CHO \xrightarrow[-H_2O]{\Delta} CH_3CH=CHCHO$$

Aldol   **α,β-unsaturated aldehyde (crotonaldehyde)**

The base catalysed mechanism of aldol condensation involves the formation of carbanion by the action of base.

$$HO^- + H-CH_2CHO \rightleftharpoons H_2O + \bar{C}H_2CHO$$

$$CH_3-\overset{\overset{\displaystyle O}{\|}}{\underset{\underset{\displaystyle H}{|}}{C}} + \bar{C}H_2CHO \rightleftharpoons CH_3-\overset{\overset{\displaystyle \bar{O}}{|}}{\underset{\underset{\displaystyle H}{|}}{C}}-CH_2CHO \xrightarrow{H_2O} CH_3-\overset{\overset{\displaystyle OH}{|}}{\underset{\underset{\displaystyle H}{|}}{C}}-CH_2CHO + HO^-$$

(b)   Condensation of two molecules of acetone give diacetone alcohol, which on heating loses a molecule of water forming mesityl oxide (α, β-unsaturated ketone).

$$2CH_3\overset{O}{\overset{||}{C}}-CH_3 \xrightarrow{Ba(OH)_2} CH_3-\underset{\underset{CH_3}{|}}{\overset{\overset{OH}{|}}{C}}-CH_2COCH_3 \longrightarrow CH_3-\underset{\underset{CH_3}{|}}{\overset{}{C}}=CHCOCH_3$$

     Acetone   **Diacetone alcohol**   **Mesityl oxide**

Acetone undergoes aldol condensation in presence of dry hydrogen chloride gas to give mesityl oxide and phorone. However, in presence of concentrated sulphuric acid, mesitylene is obtained.

$$2CH_3\overset{O}{\overset{||}{C}}CH_3 \xrightarrow{HCl} CH_3-\underset{\underset{CH_3}{|}}{\overset{}{C}}=CHCOCH_3 \underset{}{\overset{CH_3COCH_3}{\rightleftharpoons}} CH_3-\underset{\underset{CH_3}{|}}{\overset{}{C}}=CHCOCH=\underset{\underset{CH_3}{|}}{\overset{}{C}}-CH_3$$

     Acetone   **Mesityl oxide**   **Phorone**

$$2CH_3COCH_3 \xrightarrow[3H_2O]{Conc. H_2SO_4}$$

                                  Mesitylene

The mechanism of base catalysed condensation of acetone is analogous to that of acetaldehyde.

$$HO^- + H-CH_2-\overset{O}{\overset{||}{C}}-CH_3 \rightleftharpoons H_2O + \bar{C}H_2-\overset{O}{\overset{||}{C}}-CH_3$$

$$CH_3-\overset{O}{\overset{||}{C}} + \bar{C}H_2COCH_3 \rightleftharpoons CH_3-\underset{\underset{CH_3}{|}}{\overset{\overset{O^-}{|}}{C}}-CH_2COCH_3 \underset{}{\overset{H_2O}{\rightleftharpoons}} CH_3-\underset{\underset{CH_3}{|}}{\overset{\overset{OH}{|}}{C}}-CH_2COCH_3$$

The mechanism of acid catalysed aldol condensation, however, proceeds by the reaction between the conjugate acid and the enol form of the carbonyl compound.

$$CH_3-\overset{O}{\overset{||}{C}}-CH_3 \underset{}{\overset{H^+}{\rightleftharpoons}} CH_3-\overset{\overset{+OH}{||}}{C}-CH_3 \longleftrightarrow CH_3-\overset{\overset{OH}{|}}{\underset{+}{C}}-CH_3$$

$$CH_3-\underset{\underset{CH_3}{|}}{\overset{\overset{OH}{|}}{C^+}} CH_2=\underset{\underset{CH_3}{|}}{\overset{}{C}}-\overset{}{O}-H \rightleftharpoons CH_3-\underset{\underset{CH_3}{|}}{\overset{\overset{OH}{|}}{C}}-CH_2-\underset{\underset{CH_3}{|}}{\overset{}{C}}=O + H^+ \rightleftharpoons$$

$$CH_3-\underset{\underset{CH_3}{|}}{\overset{\overset{+OH_2}{|}}{C}}-CH_2-\underset{\underset{CH_3}{|}}{\overset{}{C}}=O \underset{-H_2O}{\rightleftharpoons} CH_3-\underset{\underset{CH_3}{|}}{\overset{+}{C}}-CH_2-\underset{\underset{CH_3}{|}}{\overset{}{C}}=O \underset{+H^+}{\overset{-H^+}{\rightleftharpoons}} CH_3-\underset{\underset{CH_3}{|}}{\overset{}{C}}=CH-\underset{\underset{CH_3}{|}}{\overset{}{C}}=O$$

(c)   Aldol condensation can occur between two aldehydes, two ketones or an aldehyde and a ketone having α-hydrogen atom. With two different aldehydes four products are obtained.

$$CH_3CHO + CH_3CH_2CHO \underset{-H_2O}{\rightleftharpoons} CH_3CH_2\overset{\overset{OH}{|}}{C}HCH_2CHO + CH_3\overset{\overset{OH}{|}}{C}H-\overset{\overset{CH_3}{|}}{C}HCHO + CH_3\overset{\overset{OH}{|}}{C}HCH_2CHO$$

$$+ CH_3CH_2\overset{\overset{OH}{|}}{C}H-\overset{\overset{CH_3}{|}}{C}HCHO$$

However, by the use of different catalyst, only one product may be obtained as the major product.

$$CH_3CHO + CH_3CH_2CHO \underset{}{\overset{HCl}{\rightleftharpoons}} \underset{OH \quad CH_3}{CH_3\overset{OH}{\underset{|}{C}}H-\overset{CH_3}{\underset{|}{C}}HCHO}$$

$$CH_3CHO + CH_3CH_2CHO \underset{}{\overset{NaOH}{\rightleftharpoons}} CH_3CH_2\overset{OH}{\underset{|}{C}}HCH_2CHO$$

When condensation between an aldehyde and a ketone takes place, α-hydrogen of ketone is involved. Reaction of acetaldehyde and acetone forms 4-hydroxypentan-2-one along with aldol and diacetone alcohol.

$$\underset{\substack{\text{Acetaldehyde} \quad \text{Acetone}}}{CH_3CHO \ + \ CH_3COCH_3} \underset{}{\overset{NaOH}{\rightleftharpoons}} \underset{\substack{\text{4-Hydroxypentan-2-one}}}{CH_3\overset{OH}{\underset{|}{C}}HCH_2COCH_3}$$

(d)   Aldehydes lacking α hydrogens enter into a mixed aldol condensation with aldehydes having an α-hydrogen. For example, benzaldehyde reacts with acetaldehyde to form cinnamaldehyde, an α,β-unsaturated aromatic aldehyde. This is also referred to as *Claisen-Schmidt reaction or Claisen reaction.*

Benzaldehyde   Acetaldehyde   Cinnamaldehyde

3.   *Oxidation:* Aldehydes can be easily oxidised under mild conditions to give carboxylic acids containing same number of carbon atoms. Ketones, on the other hand are difficult to oxidise. These can be oxidised by heating with strong oxidising reagents resulting in the rupture of carbon carbon bonds to produce acids containing lesser number of carbon atoms compared to the starting ketones.

$$\underset{\substack{\text{Acetaldehyde}}}{CH_3CHO} \xrightarrow[\text{Mild Oxidation}]{[O]} \underset{\substack{\text{Acetic acid}}}{CH_3COOH}$$

$$\underset{\substack{\text{2-Butanone}}}{CH_3CH_2\overset{O}{\overset{||}{C}}-CH_3} \xrightarrow[\text{KMnO}_4,\text{H}^+,\Delta]{[O]} \underset{\substack{\text{Acetic acid}}}{CH_3COOH}$$

The methyl group of aldehydes and ketones is oxidised to aldehydic group by selenium dioxide.

$$\underset{\substack{\text{Acetaldehyde}}}{CH_3CHO + SeO_2} \longrightarrow \underset{\substack{\text{Glyoxal}}}{OHCCHO} + Se + H_2O$$

$$\underset{\substack{\text{Acetone} \qquad \text{Selenium} \\ \text{dioxide}}}{CH_3CO\ CH_3 + SeO_2} \longrightarrow \underset{\substack{\text{Methyl glyoxal}}}{CH_3COCHO} + Se + H_2O$$

The extreme cases of oxidation of aldehydes provide a simple method for distinguishing between aldehydes and ketones. Mild oxidising reagents like Tollen's reagent and Fehling's solution can be used for this purpose.

Tollen's Reagent: It is ammonical silver nitrate solution.

$$AgNO_3 + 2NH_3 + NaOH \longrightarrow Ag(NH_3)_2OH$$

$$RCHO + 2Ag(NH_3)_2OH \xrightarrow{\Delta} RCOO\overset{+}{N}H_4 + 2Ag + 2NH_3 + H_2O$$

The silver ions are reduced by the aldehyde to the metallic silver and form a mirror of silver if the reaction is carried out in a clean test tube. The reaction is not given by ketones.

Fehling's Solution: It is prepared by mixing equal volumes of Fehling's solution A(which contains copper sulphate) and Fehling's solution B (which contains sodium hydroxide and Rochelle's salt i.e. sodium potassium tartarate). During the oxidation of aldehydes to acids, the cupric ions are reduced to cuprous ions which are precipitated as red cuprous oxide.

Haloform Reaction: Aldehydes and ketones having CH_3CO group undergo haloform reaction when dissolved in dioxan followed by the addition of dilute sodium hydroxide and excess of potassium iodide, warming and finally diluting with water. The bright yellow solid iodoform is precipitated during the reaction.

$$CH_3CHO + I_2 \xrightarrow{KI} CI_3CHO + 3HI$$

$$CI_3CHO + HO^- \longrightarrow CHI_3 + HCOO^-$$
Iodoform

This reaction is given by most compounds having acetyl group attached to carbon or hydrogen or by the compounds oxidised to a derivative having acetyl group under the conditions of reaction.

4. REDUCTION

Aldehydes and ketones on reduction form various products depending upon the nature of reducing agent. Generally speaking aldehydes form primary alcohols and ketones secondary alcohols on reduction.

(i) *Catalytic Hydrogenation* or reduction with sodium and alcohol or metallic hydrides (lithium aluminum hydride or sodium borohydride) gives alcohols.

$$\underset{\text{Aldehyde}}{R-\overset{\overset{H}{|}}{C}=O} + H_2 \xrightarrow{\text{Pt or Ni}} \underset{\text{Primary alcohol}}{RCH_2OH}$$

$$\underset{\text{Ketone}}{\overset{R}{\underset{R'}{}}C=O} \xrightarrow[\text{2. Ether}]{\text{1. LiAlH}_4} (R-\overset{R}{\underset{H}{|}}{C}-O)_4 AlLi \xrightarrow{H^+/H_2O} \underset{\text{sec. Alcohol}}{R-\overset{R}{\underset{H}{|}}{C}-OH}$$
1. NaBH$_4$; 2. H$^+$/H$_2$O

(ii) *Clemmensons Reduction:* Aldehydes and ketones may be reduced with zinc amalgam and hydrochloric acid to give alkanes; the carbonyl group is reduced to CH_2. This reaction is known as *Clemmensons reduction.*

$$\underset{\substack{\text{A carbonyl}\\\text{compound}}}{>C=O} \xrightarrow{\text{Zn–Hg, HCl}} \underset{\text{Alkane}}{>CH_2 + H_2O}$$

$$\underset{\text{2-Butanone}}{CH_3CH_2\overset{\overset{O}{\|}}{C}CH_3} \xrightarrow{\text{Zn–Hg, HCl}} \underset{\text{n-Butane}}{CH_3CH_2CH_2CH_3}$$

(iii) *Wolff Kishner Reduction:* In this method the hydrazones of aldehydes or ketones on heating with potassium hydroxide or sodium ethoxide give the corresponding alkanes. Like Clemmensons reduction, in this case also the carbonyl group is reduced to methylene group.

$$\begin{array}{c} R \\ \diagdown \\ R \diagup \end{array} C=NNH_2 \xrightarrow[\text{Ethylene glycol}]{\text{KOH, }\Delta} RCH_2R + N_2$$

Hydrazone   **Alkane**

In this case, the semicarbazones can also be used.

(iv) *Meerwein-Ponndorf-Verley Reduction:* The carbonyl compounds when heated with aluminum isopropoxide in isopropanol solution are reduced to corresponding alcohols. The isopropoxide is oxidised to acetone which is continuously removed from the reaction mixture by distillation.

$$R_2C=O + [(CH_3)_2CHO]_3 Al \xrightarrow{\text{dil }H_2SO_4} R_2CHOH + CH_3COCH_3$$

Carbonyl   **Alcohol**
compound

This reducing agent specifically reduces the carbonyl group without affecting the double bond, nitro or other reducible groups present in the compound.

The reduction involves hydride transfer from aluminum isopropoxide to carbonyl compound via a cyclic transition state.

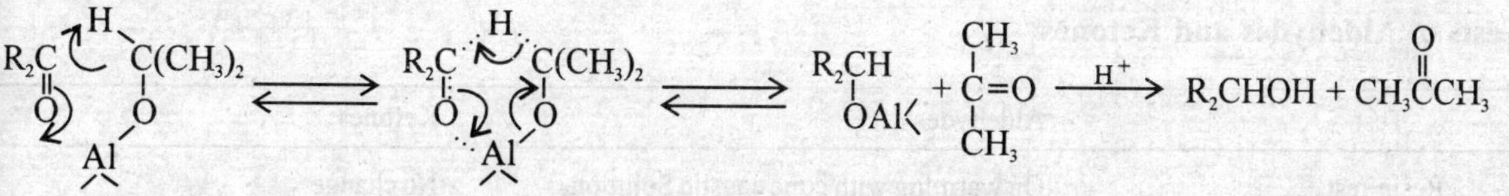

5. MISCELLANEOUS REACTIONS

(i) *Wittig Reaction:* It is a very important and useful method of conversion of carbonyl compounds into alkenes. It involves the reaction between an aldehyde or a ketone and a phosphorus ylide to form a betaine which involves elimination of base to form alkene.

$$\diagup C=O + (C_6H_5)_3 - \overset{+}{P} - \overset{-}{C}H_2 \longrightarrow \diagup C=CH_2 + (C_6H_5)_3 - PO$$

Aldehyde **Triphenylpho-**   **Alkene** **Triphenyl phosphine**
or ketone **sphonium ylide**   **oxide**

In this reaction, the oxygen of the carbonyl groups is substituted by a methylene group. The reaction involves the nucleophilic attack of ylide on electron deficient carbon. The reaction has been discussed in detail in Section 9.6.

(ii) *Reaction with Alkali:* Aldehydes having α-hydrogens give brown resinous mass when treated with strong solution of caustic alkali. Ketones do not give this reaction. The formation of resin is explained through a series of condensation steps as shown below in case of acetaldehyde.

$$2CH_3CHO \longrightarrow CH_3CH(OH)CH_2CHO \xrightarrow{-H_2O}$$

$$CH_3CH=CHCHO \xrightarrow{CH_3CHO} CH_3CH=CHCH(OH)CH_2CHO$$

$$\xrightarrow{-H_2O} CH_3CH=CHCH=CHCHO \longrightarrow$$

Formaldehyde, which does not contain α-hydrogen undergoes self oxidation reduction when treated with 50% aqueous or alcoholic alkali at room temperature. In this case one molecule of formaldehyde is oxidised at the expense of other which is reduced to methanol. This reaction is known as *Cannizaro's reaction.*

$$2HCHO + NaOH \longrightarrow HCOONa + CH_3OH$$

Formaldehyde   **Sodium**   **Methanol**
                          formate

The mechanism of Cannizaro reaction involves following steps.

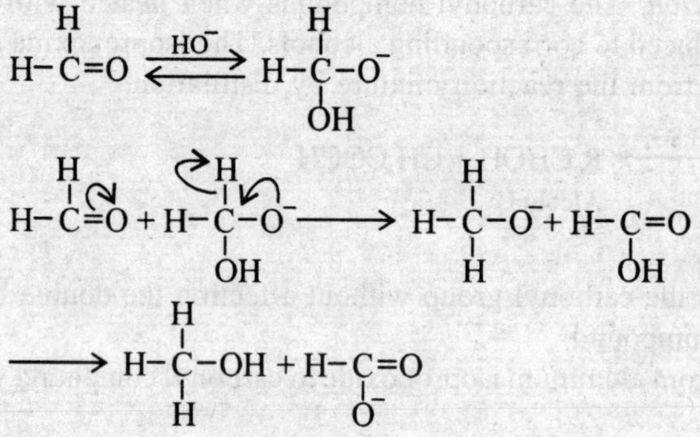

Tests of Aldehydes and Ketones

	Test	Aldehydes	Ketones
1.	Resin test	On warming with conc caustic Solution, dark colored resins are obtained. However, formaldehyde do not give this test.	No change
2.	Schiff's Reagent	Intense red color.	No change
3.	Silver mirror test	On warming with ammonical $AgNO_3$ solution silver mirror is obtained.	No change
4.	Fehling's test	On warming with an equal volume of Fehling's A and Fehling's B solutions red cuprous oxide is obtained.	No change
5.	Sodium bisulphite solution	On shaking with cold saturated solution of sodium bisulphite a crystalline bisulphite compound separates.	Crystalline bisulphite compound separates
6.	2,4-Dinitrophenylhydrazone test	An orange, yellow or red 2,4-DNP is obtained.	An orange 2,4-DNP is obtained
7.	Hydroxyl amine or phenyl hydrazine test	Crystalline oximes or phenyl hydrazone are formed	Same as in case of aldehydes

18.9   INDIVIDUAL MEMBERS

18.9.1   Formaldehyde/Methanal [HCHO]
It is the first member of the aldehyde series.

Preparation: It is prepared by the general methods of preparation of aldehydes discussed in Section 18.5. It is obtained in laboratory by the oxidation of methyl alcohol in presence of platinum.

$$CH_3OH + \tfrac{1}{2}O_2 \longrightarrow HCHO + H_2O$$

Manufacture

(i)   Controlled oxidation of methane or natural gas under pressure in presence of molybdenum oxides yields formaldehyde.

$$CH_4 + O_2 \xrightarrow{\text{Molybdenum oxide}} \underset{\textbf{Formaldehyde}}{HCHO} + H_2O$$

(ii)   Water gas (a mixture of carbon monoxide and hydrogen) on passing at low pressure through an electric discharge of low intensity gives formaldehyde.

$$\underset{\textbf{Water Gas}}{CO + H_2} \xrightarrow{\text{Electric discharge}} \underset{\textbf{Formaldehyde}}{HCHO}$$

Formaldehyde is a gas and its 40 percent aqueus solution is called *Formalin.*

Properties: It is a colorless, pungent smelling gas and is extremely soluble in water. It can be easily condensed to a liquid (b.p. −21°). It causes irritation to skin, eyes, nose and throat. In gaseous state or in solution it is a powerful disinfectant. In dilute aqueous solution it is present as hydrated form to form methylene glycol.

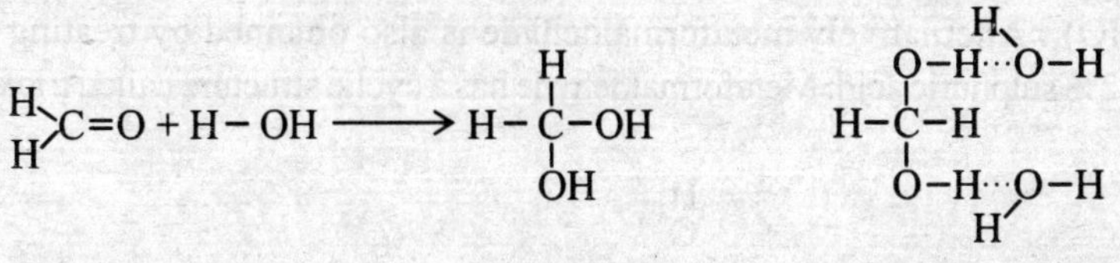

$$\underset{\textbf{Formaldehyde}}{\overset{H}{\underset{H}{>}}C=O + H-OH} \longrightarrow \underset{\textbf{Hydrated form}}{H-\overset{H}{\underset{OH}{C}}-OH}$$

Hydrogen bonding between water and hydrate

The solubility of formaldehyde in water is due to hydrogen bonding between water molecules and formaldehyde hydrate.

   Structurally, formaldehyde differs from other aldehydes in that it has a hydrogen linked to the −CHO group instead of the alkyl group. So it does not give all general reactions of higher aldehydes. Some typical reactions of formaldehyde are given below:

(i)   *Reaction with Ammonia:* Formaldehyde, unlike other aldehydes does not form aldehyde-ammonia instead gives hexamethylene tetramine, a white crystalline compound having cyclic structure.

$$6CH_2O + 4NH_3 \longrightarrow (CH_2)_6N_4 + 6H_2O$$

Hexamethylene tetramine is used in medicine as a urinary antiseptic under the name *Urotropine.* Oxidation of urotropine with nitric acid gives a well known explosive, cyclonite, commonly known as RDX.

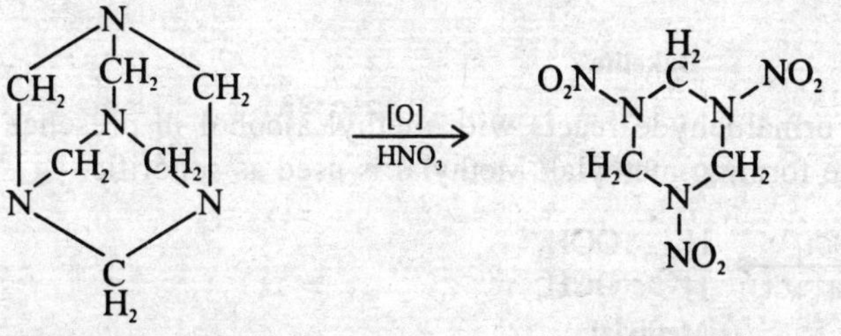

**Cyclic structure of
Hexamethylene tetramine**

Cyclonite (RDX)

(ii) *Reaction with Sodium Hydroxide:* Unlike other aldehydes, formaldehyde undergoes self oxidation reduction when reacted with 50% aqueous or alcoholic alkali at room temperature. In this case one molecule is oxidised to formic acid at the expense of the other molecule which is reduced to methanol. This reaction is called *Cannizzaro reaction.*

$$2HCHO + NaOH \longrightarrow HCOONa + CH_3OH$$

Formaldehyde   Sod. formate   Methanol

(iii) *Reaction with Amines:* Methanal, being a methylating agent, on heating with ethylamine in presence of formic acid forms N-methylethylamine.

$$C_2H_5NH_2 + HCHO + HCOOH \longrightarrow C_2H_5NHCH_3 + CO_2 + H_2O$$

Ethylamine   N-Methylethylamine

(iv) *Polymerisation:* Formaldehyde readily undergoes polymerisation to form various products depending upon the reaction conditions.

(a) *Paraformaldehyde:* When formalin solution in water is evaporated to dryness, a white crystalline solid with fishy odor is obtained. The paraformaldehyde thus obtained is a long chain polymer with formula $(CH_2O)_n.H_2O$, where, *n* may be from 6 to 50.

(b) *Metaformaldehyde:* Formalin when allowed to stand at room temperature, slowly polymerises to metaformaldehyde $(HCHO)_3$. Alternatively metaformaldehyde is also obtained by treating methanol's aqueous solution (60%) with 2% sulphuric acid. Metaformaldehyde has a cyclic structure called trioxymethylene or trioxane.

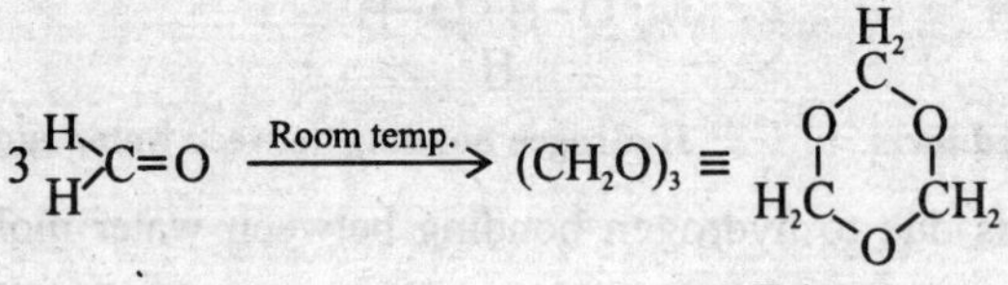

Formaldehyde   Trioxane

(c) *Polyoxymethylenes:* It is obtained by treating formaldehyde with concentrated sulphuric acid. It is a white, water insoluble polymer having the formula $(CH_2O)_n.H_2O$, where, n is more than 100.

(v) *Condensation with Phenol:* When formalin is heated with phenol in presence of a catalyst (ammonia or dilute sodium hydroxide solution) a resinous mass is obtained. It is used under the name *Bakelite* for making electric equipments.

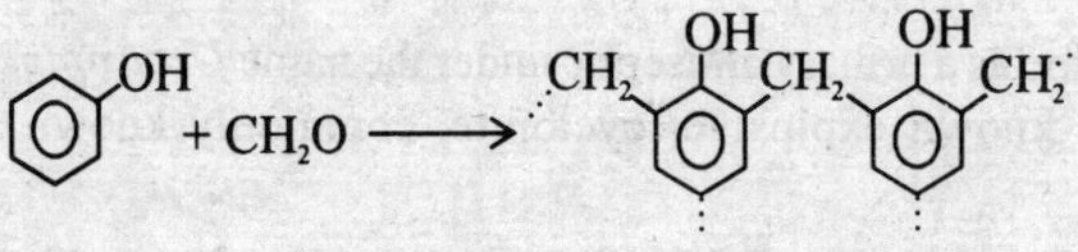

Bakelite

(vi) *Reaction with Alcohol:* Formaldehyde reacts with methyl alcohol in presence of hydrogen chloride and fused calcium chloride forming methylal. Methylal is used as soporific.

Formaldehyde   Methylal

Uses : Formaldehyde is used in the form of 40% aqueous solution under the name formalin for most purposes.

It is used as disinfectant and for sterilising surgical instruments. It is also used for preserving biological and anatomical specimens. It is used for making Bakelite. It is also used in tanning industry and in the manufacture of dyestuffs.

18.9.2 Acetaldehyde/Ethanal [CH_3CHO]

Preparation: Acetaldehyde may be prepared by any of the general methods of preparation already discussed. In the laboratory, it in obtained by the oxidation of ethyl alcohol with sodium dichromate and sulphuric acid. As soon as it is formed, it is distilled off to avoid its oxidation to acetic acid. The oxidation product is purified by conversion into acetaldehyde-ammonia by treatment with ammonia. The pure acetaldehyde-ammonia is distilled with dilute sulphuric acid and the regenerated acetaldehyde is collected in ice cold receiver.

$$CH_3CH_2OH \xrightarrow[Na_2CrO_7/dil\ H_2SO_4]{[O]} CH_3CHO + H_2O$$

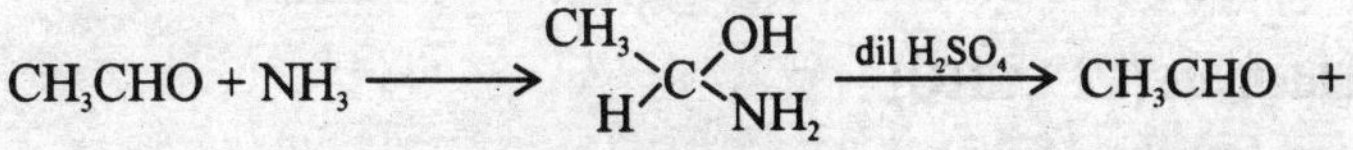

Ethanol · · · · · · · · · Acetaldehyde

$$CH_3CHO + NH_3 \longrightarrow \underset{\substack{\text{Acetaldehyde} \\ \text{ammonia}}}{\overset{CH_3}{\underset{H}{>}}C\overset{OH}{\underset{NH_2}{<}}} \xrightarrow{dil\ H_2SO_4} CH_3CHO + NH_3$$

Acetaldehyde

Manufacture

(i) It is obtained commercially either by catalytic oxidation of ethyl alcohol with air in presence of silver at 250° or by passing vapors of ethyl alcohol over copper at 300°.

$$CH_3CH_2OH \xrightarrow[Cu/300°]{Ag/Air/250°} CH_3CHO$$

Ethanol · · · · · · · Acetaldehyde

(ii) By hydration of acetylene in presence of sulphuric acid and mercury salt.

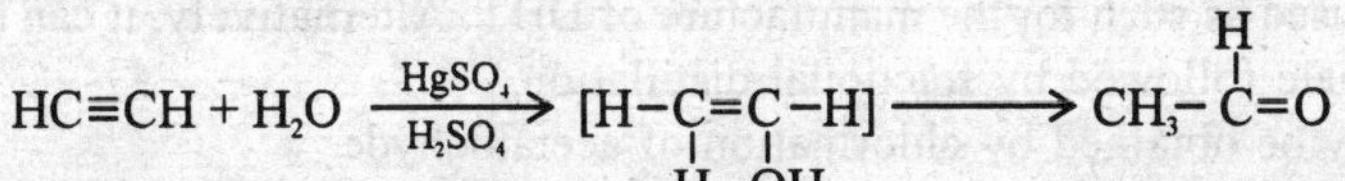

$$HC\equiv CH + H_2O \xrightarrow[H_2SO_4]{HgSO_4} [H-\underset{H}{C}=\underset{OH}{C}-H] \longrightarrow CH_3-\overset{H}{\underset{}{C}}=O$$

Acetylene · · · · · · · · · · · · · · · · Acetaldehyde

Properties: It is a colorless volatile liquid (b.p. 21°) and has a characteristic pungent odor. It is soluble in water, alcohol, and ether. It undergoes all the general reactions of aldehydes. Some typical reactions of acetaldehyde are given below.

Polymerisation: On treatment with a small amount of concentrated sulphuric acid, acetaldehyde polymerises to give a trimer having a cyclic structure commonly known as paraldehyde.

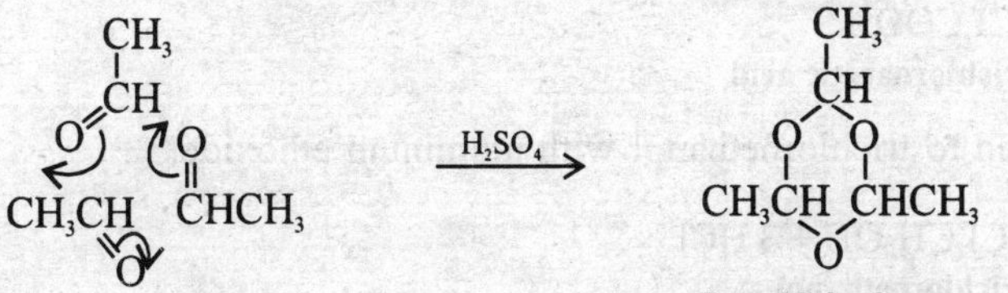

Paraldehyde

Paraldehyde is a sweet smelling liquid (b.p. 128°) and is used in medicines as hypnotic. It does not possess reducing properties.

However, polymerisation of acetaldehyde with few drops of concentrated sulphuric acid at 0° gives tetramer, a cyclic structure, $(CH_3CHO)_4$, known as metaldehyde.

$$4CH_3CHO \xrightarrow[O]{H_2SO_4}$$

$$
\begin{array}{c}
O \\
H_3CCH \quad CHCH_3 \\
O \qquad\qquad O \\
H_3CCH \quad CHCH_3 \\
O
\end{array}
$$

Metaldehyde

Both the polymers viz. paraldehyde and metaldehyde regenerate acetaldehyde on distilling with dilute sulphuric acid.

Uses: Acetaldehyde is used in the preparation of acetic acid, ethyl acetate and n-butyl alcohol. Paraldehyde is used in medicine as a hypnotic and also for the silvering of mirrors. It is also used for the preparation of chloral.

18.9.3   Chloral/Trichloroacetaldehyde [CCl₃CHO]

Preparation: It is prepared on a commercial scale by chlorination of ethyl alcohol. Chlorine is passed through cold ethyl alcohol and then at 60° till it is saturated. The chloral alcoholate is obtained as a white crystalline solid. It is separated and distilled with concentrated sulphuric acid to give chloral.

$$CH_3CH_2OH \xrightarrow[[O]]{Cl_2} CH_3CHO \xrightarrow{Cl_2} CCl_3CHO$$

Ethanol   **Acetaldehyde**   **Chloral**

$$\xrightarrow{C_2H_5OH} CCl_3\overset{\overset{\displaystyle OH}{|}}{CH}-OC_2H_5 \xrightarrow{Conc.\ H_2SO_4} CCl_3CHO$$

Chloral alcoholate   **Chloral**

Chloral thus obtained can be used as such for the manufacture of DDT. Alternatively, it can be purified by shaking with calcium carbonate followed by fractional distillation.

In the laboratory, chloral may be obtained by chlorination of acetaldehyde.

Properties: Chloral is a colorless oily liquid (b.p. 98°) with pungent odor and sweetish taste. It is soluble in water.

Chloral gives most of the reactions of aldehydes (already discussed) like nucleophilic addition reactions and condensation reactions. It is more reactive than other aldehydes. It also reduces Fehling's and Tollen's reagents. Some typical reactions of chloral are given below.

(i)   *Oxidation:* On treatment with dilute nitric acid, chloral is oxidised to trichloroacetic acid.

$$CCl_3CHO \xrightarrow{dil\ HNO_3} CCl_3COOH$$

Chloral   **Trichloroacetic acid**

(ii)   *Reduction:* Chloral is reduced to trichloroethanol with aluminum ethoxide.

$$CCl_3CHO \xrightarrow{(C_2H_5O)_3Al} CCl_3CH_2OH + 3\ HCl$$

Chloral   **Trichloroethanol**

However, on reduction with zinc and hydrochloric acid it gives acetaldehyde.

$$CCl_3CHO \xrightarrow{\text{Zn / HCl}} CH_3CHO \quad + 3\ HCl$$

$\qquad$ **Chloral** $\qquad\qquad\qquad$ **Acetaldehyde**

(iii) *Reaction with Sodium Hydroxide:* On warming with sodium hydroxide solution, chloral undergoes *hydrolysis* to form chloroform.

$$CCl_3CHO + NaOH \xrightarrow{\Delta} CHCl_3 \quad + \quad HCOONa$$

$\qquad$ **Chloral** $\qquad\qquad\qquad\qquad$ **Chloroform** $\quad$ **Sodium formate**

(iv) *Hydration:* On treatment with a small amount of water chloral yields chloral hydrate, $CCl_3CH(OH)_2$, a stable crystalline compound (m.p. 57°). Similarly, on treatment with ethyl alcohol, chloral alcoholate (m.p. 46°) is obtained.

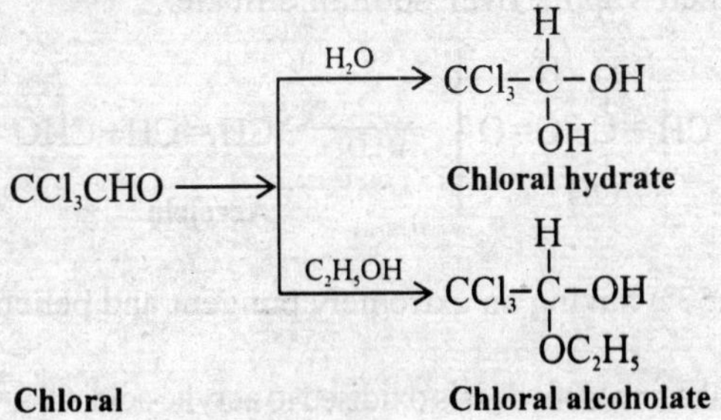

Chloral hydrate is a stable compound inspite of the fact that the two hydroxy groups are attached to the same carbon atom. The reason for its stability is the presence of strong electron withdrawing chlorine atoms to the adjacent carbon atom. This makes the aldehydic carbon electron deficient resulting in firm attachment to the hydroxy group.

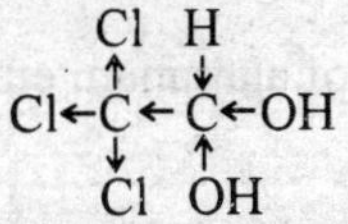

Also, Davies (1940) from infrared studies showed the formation of intramolecular hydrogen bonding rendering addition stability to chloral hydrate.

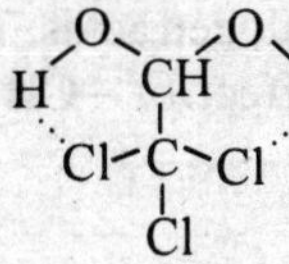

Uses: Chloral hydrate is used as a hypnotic and stimulant. Chloral is used for the manufacture of DDT a well known insecticide.

18.9.4 Acrolein/Acraldehyde [CH₂=CHCHO]

Acrolein is an α,β-unsaturated aldehyde and is named as 2-propenal.

Preparation:

(i) By dehydration of glycerol with potassium hydrogen sulphate.

$$\begin{array}{c} CH_2OH \\ | \\ CHOH \\ | \\ CH_2OH \end{array} \xrightarrow[-H_2O]{KHSO, \ \Delta} \left[\begin{array}{c} CH_2 \\ || \\ C \\ || \\ CHOH \end{array}\right] \xrightarrow{\text{Isomerise}} \left[\begin{array}{c} CH_2 \\ || \\ CH \\ | \\ CHO \end{array}\right]$$

Glycerol   Unstable   Acrolein

(ii) By aerial oxidation of propene in presence of copper oxide.

$$CH_3CH=CH_2 \xrightarrow[CuO]{[O]} CH_2=CHCHO + H_2O$$

Propene   Acrolein

(iii) On a commercial scale, acroline is prepared by the reaction of acetaldehyde and formaldehyde in presence of a base or by passing the combined vapors over sodium silicate.

$$\underset{\text{Acetaldehyde}}{CH_3-\overset{\overset{H}{|}}{C}=O} + \underset{\text{Formaldehyde}}{H-\overset{\overset{H}{|}}{C}=O} \xrightarrow[\Delta]{\text{sod. Silicate}} \left[CH_3-\overset{\overset{H}{|}}{\underset{\underset{OH}{|}}{C}}-\overset{\overset{H}{|}}{C}=O\right] \xrightarrow{-H_2O} \underset{\text{Acrolein}}{CH_2=CH-CHO}$$

Properties: Acrolein is a colorless liquid (b.p. 53°) having an extremely pungent and penetrating odor. It causes watering of eyes and nose.

Acrolein behaves as an aldehyde and as an olefin. Like an aldehyde it is oxidised to acrylic acid, $CH_2=CHCOOH$. However, reduction with metal and acid gives a mixture of allyl alcohol, propanal and propanol. It does not undergo aldol condensation with alkalis as the carbon chain breaks at the double bond to give a mixture of methanal and ethanal.

$$CH_2=CHCHO \xrightarrow{NaOC_2H_5} HCHO \ + CH_3CHO$$

Acrolein   Methanal   Ethanal

A typical reaction of acrolein is the formation of allylacrylate in presence of aluminum ethoxide. This is known as *Tischenko's reaction*.

$$2CH_2=CHCHO \xrightarrow{(C_2H_5O)_3Al} CH_2=CHCOOCH_2CH=CH_2$$

Allylacrylate

As an aldehyde, it gives an oxime, phenyl hydrazone and cyanohydrin.

As an olefinic compound, acrolein gives addition products with halogens and halogen acids. The addition of halogen acid occurs contrary to Markovnikov's rule due to −I and −R effect of −CHO group.

$$CH_2=CHCHO + Br_2 \longrightarrow CH_2BrCHBrCHO$$
$$CH_2=CHCHO + HCl \longrightarrow CH_2ClCH_2CHO$$

On reduction with Raney nickel and hydrogen at low temperature and high pressure, only the double bond is reduced to give propanal (CH_3CH_2CHO).

Reductive ozonolysis of acrolein gives a mixture of formaldehyde and glyoxal.

$$CH_2=CH-CHO \xrightarrow[2. \ H_2]{1. \ \text{Ozonolysin}} HCHO \ + \ OHC-CHO$$

Acrolein   Formal-   Glyoxal
                       dehyde

An important reaction of acrolein is that it undergoes Diels-Alder reaction with butadiene to form 1,2,3,6-tetrahydrobenzaldehyde.

$$
\begin{array}{c}
\underset{\substack{\text{Butadiene}}}{
\begin{array}{c}
CH_2 \\
\parallel \\
CH \\
\mid \\
CH \\
\diagdown CH_2
\end{array}}
+
\underset{\substack{\text{Acrolein}}}{
\begin{array}{c}
CH_2 \\
\parallel \\
CHCHO
\end{array}}
\xrightarrow[\text{Diels Alder reaction}]{\text{anlyd. AlCl}_3, \Delta}
\underset{\substack{\text{1,2,3,6-Tetrahydrobenzaldehyde}}}{
\begin{array}{c}
CH_2 \\
\diagup \diagdown \\
CH \quad CH_2 \\
\parallel \qquad \mid \\
CH \quad CHCHO \\
\diagdown CH_2 \diagup
\end{array}}
\end{array}
$$

Another typical reaction of acrolein is that it undergoes Michael reaction on treatment with diethyl malonate in presence of sodium ethoxide.

$$
\underset{\substack{\text{Acrolein}}}{CH_2=CH-CHO}
+
\underset{\substack{\text{Diethyl malonate}}}{
\begin{array}{c}
COOC_2H_5 \\
\diagup \\
CH_2 \\
\diagdown \\
COOC_2H_5
\end{array}}
\xrightarrow{\text{NaOC}_2H_5}
\underset{\substack{\text{4,4-Dicarboethoxybutanal}}}{
\begin{array}{c}
CH_2-CH_2-CHO \\
\mid \\
CH(COOC_2H_5)_2
\end{array}}
$$

Uses: It is used in the preparation of insecticides and acrolein-urea-formaldehyde resin. It is also used as tear gas and for detection of any leakage in refrigeration with methyl chloride.

18.9.5   Crotonaldehyde/2-Butenal [CH$_3$CH=CHCHO]

It is prepared by the aldol condensation of acetaldehyde followed by dehydration of the formed aldol.

$$
\underset{\substack{\text{Acetaldehyde}}}{CH_3CHO}
\xrightarrow[\text{Condensation}]{\text{Aldol}}
\begin{array}{c}
OH \\
\mid \\
CH_3-CH-CH_2CHO
\end{array}
\xrightarrow[\text{ZnCl}_2]{\Delta}
\underset{\substack{\text{Crotonaldehyde}}}{CH_3CH=CHCHO} + H_2O
$$

It is a colorless liquid (b.p. 104°) and resembles acrolein in chemical properties. It exhibits geometrical isomerism.

$$
\underset{\substack{\text{Cis-isomer}}}{
\begin{array}{c}
H-C-CH_3 \\
\parallel \\
H-C-CHO
\end{array}}
\qquad\qquad
\underset{\substack{\text{Trans-isomer}}}{
\begin{array}{c}
H-C-CH_3 \\
\parallel \\
OHC-C-H
\end{array}}
$$

It is used as an insecticide and for the preparation of crotonic acid and butyraldehyde.

18.9.6   Dimethyl Ketone/Acetone/Propanone [CH$_3$COCH$_3$]

Preparation: It is prepared in the laboratory by dry distillation of calcium acetate.

$$
\underset{\substack{\text{Calcium acetate}}}{(CH_3COO)_2Ca}
\xrightarrow{\text{Dry distillation}}
\underset{\substack{\text{Acetone}}}{CH_3COCH_3} + CaCO_3
$$

On commercial scale it is prepared by the following methods.

(i)   By passing vapors of isopropyl alcohol over heated copper.

$$
\underset{\substack{\text{Isopropanol}}}{
\begin{array}{c}
OH \\
\mid \\
CH_3CHCH_3
\end{array}}
\xrightarrow[300-400°]{\text{Cu}}
\underset{\substack{\text{Acetone}}}{
\begin{array}{c}
O \\
\parallel \\
CH_3-C-CH_3
\end{array}} + H_2
$$

(ii)   *From Acetylene*: By passing a mixture of acetylene and steam over heated catalyst (magnesium and zinc vanadates) at 420°.

$$CH \equiv CH + H_2O \xrightarrow[420°]{Catalyst} CH_3COCH_3 + CO_2 + 2H_2$$

Acetylene   Acetone

(iii)   *From Propene*: Propene (obtained as a byproduct in the cracking of heavy oil) is first absorbed in concentrated sulphuric acid. The formed isopropyl hydrogen sulphate on boiling with water gives isopropyl alcohol, which on dehydrogenation in presence of copper gives acetone.

$$CH_3CH = CH_2 + H_2SO_4 \longrightarrow \underset{\underset{\text{hydrogen sulphate}}{\text{Isopropyl}}}{CH_3\overset{OSO_3H}{\underset{|}{C}}HCH_3} \xrightarrow[\Delta]{H_2O} \underset{\text{Isopropanol}}{CH_3\overset{OH}{\underset{|}{C}}HCH_3} \xrightarrow[300°]{Cu} \underset{\text{Acetone}}{CH_3 - \overset{O}{\overset{\|}{C}} - CH_3}$$

Propene

(iv)   *By Fermentation*: Fermentation of starch or molasses by bacterium Clostridium acetobutylicum (Weizmann process) at 30-35° gives sugar, which undergo further decomposition to butyl alcohol, acetone (30%). From the mixture, acetone is obtained by fractional distillation.

$$3C_6H_{12}O_6 \longrightarrow 2C_4H_9OH + CH_3COCH_3 + 7CO_2 + 4H_2 + H_2O$$

Glucose   n-Butanol   Acetone

Properties: Acetone is a colorless, inflammable liquid (b.p. 56°) and has a characteristic pleasant smell. It is miscible with water, alcohol and ether.

It gives all the general reactions of ketones. Some other reactions are given below.

(i)   *Reaction with Nitrous Acid:* Acetone on treatment with nitrous acid (mixture of sodium nitrite and hydrochloric acid) gives oximino acetone (isonitroso acetone).

$$CH_3COCH_3 + HNO_2 \longrightarrow CH_3COCH = NOH$$

Acetone   Oximino acetone

(ii)   *Reduction with Magnesium-Amalgam and Water:* Acetone on reduction with magnesium-amalgam and water gives pinacol.

$$2CH_3COCH_3 \xrightarrow[H_2O]{Mg-Hg} (CH_3)_2\overset{OH}{\underset{|}{C}} - \overset{OH}{\underset{|}{C}}(CH_3)_2$$

Acetone   Pinacol

(iii)   *Reaction with Peracids:* The reaction of ketones with peracids gives esters. This reaction is known as *Baeyer-Villiger oxidation*.

$$CH_3COCH_3 + R'COOOH \longrightarrow CH_3COOCH_3 + R'COOH$$

Acetone   Peracid   Methyl acetate   Acid

(iv)   *Haloform Reaction:* Like ethyl alcohol and acetaldehyde, acetone also undergoes haloform reaction. Thus, on heating with a solution of iodine in potassium iodide and sodium hydroxide it forms yellow shining crystals of iodoform.

$$CH_3COCH_3 + 3I_2 \longrightarrow CI_3COCH_3 + 3HI$$

Triiodoacetone

$$CI_3COCH_3 + NaOH \longrightarrow CHI_3 + CH_3COONa$$

Iodoform

Uses: It is used as a solvent for rayon, celluloid, varnishes etc. It is also used in the preparation of chloroform (anesthetic), iodoform (antiseptic), sulphonal (soporific) and chloretone (hypnotic and sedative). It is also used for preparing acetylene and in the manufacture of Cordite-a smokeless powder. It finds use as a nail polish remover.

18.9.7 Ethyl methyl ketone/2-Butanone [CH₃COCH₂CH₃]

Preparation: It may be prepared by any of the general methods already discussed. It may be manufactured by following methods.

(i) The oxidation of 2-butanol with air using silver catalyst at 250-300° yields 2-butanone.

$$CH_3CH_2\overset{\underset{|}{OH}}{C}HCH_3 + O_2 \xrightarrow[250-300°]{Ag} CH_3CH_2\overset{\underset{\|}{O}}{C}CH_3 + 2H_2O$$

2-Butanol   **2-Butanone**

(ii) The dehydrogenation of 2-butanol in presence of copper at 300° yields 2-butanone.

$$CH_3CH_2\overset{\underset{|}{OH}}{C}HCH_3 \xrightarrow{Cu/300°} CH_3CH_2\overset{\underset{\|}{O}}{C}CH_3 + H_2$$

2-Butanol   **2-Butanone**

Properties: It is a colorless liquid (b.p. 80°) insoluble in water. It gives all the characteristic reactions of ketones. It is used as a solvent for vinyl resins, synthetic rubber etc.

18.9.8 Methyl isopropyl ketone/3-Methylbutan-2-one [CH₃COCH(CH₃)₂]

It is prepared from tertiary amyl alcohol by bromination followed by hydrolysis of the formed dibromo derivative.

$$(CH_3)_2\overset{\underset{|}{OH}}{C}CH_2CH_3 \xrightarrow{Br_2} (CH_3)_2\overset{}{C}-\overset{\underset{|}{Br}}{C}HCH_3 \xrightarrow{KOH} (CH_3)_2\overset{\underset{|}{OH}}{C}-\overset{\underset{|}{OH}}{C}HCH_3$$

tert-Amyl alcohol   **Dibromo derivative**

$$\xrightarrow[\text{Rearrangement}]{H^+} (CH_3)_2CH-\overset{\underset{\|}{O}}{C}-CH_3$$

Methyl isopropyl ketone

18.9.9 Tertiary butyl methyl ketone/Pinacolone/Pinacone [CH₃COC(CH₃)₃]

It is prepared by distilling pinacol or its hydrate with sulphuric acid.

$$CH_3-\overset{\underset{|}{OH}}{\underset{}{C}}\overset{CH_3}{\underset{}{}}-\overset{\underset{|}{OH}}{\underset{}{C}}\overset{CH_3}{\underset{}{}}-CH_3 \xrightarrow{H_2SO_4} CH_3-\overset{CH_3}{\underset{\underset{CH_3}{|}}{C}}-\overset{\underset{\|}{O}}{C}-CH_3$$

Pinacol   **Pinacolone**

It is also prepared in 94 % yield by heating a solution of pincaol dissolved in dioxane over heated catalyst (silica gel impregnated with phosphoric acid). The conversion of pinacol to pinacone is called pinacol-pinacolone rearrangement and takes place in presence of acid (see Section 15.6.1 for detailed mechanism).

Pinacolone is a colorless liquid (b.p. 119°) with camphor like odor. It can be readily oxidised to trimethyl acetic acid with alkaline sodium hypobromite.

$$(CH_3)_3C-\overset{\overset{\displaystyle O}{\|}}{C}-CH_3 \xrightarrow{Br/NaOH} (CH_3)_2C-\overset{\overset{\displaystyle O}{\|}}{C}-OH$$

Pinacolone   **Trimethylacetic acid**

18.10   PROBLEMS

1. Discuss general methods of preparation of aldehydes and ketones.
2. Describe common properties of aldehydes and ketone.
3. How will you explain the difference in properties of formaldehyde as compared to other aldehydes ?
4. Explain the difference in properties of aldehydes and ketones.
5. Using infra-red spectroscopy how will you differentiate between an aldehyde and a ketone ?
6. Comment on the statement that carbonyl double bond differs from carbon-carbon double bonds of alkenes.
7. Give IUPAC names to the following:

   (a) $CH_3CH_2CH_2CHO$ (b) $CH_3COCH\overset{\overset{\displaystyle CH_3}{|}}{CH_2}CH_3$,   (c) $CH_3COCH=CHCH_3$

8. Give the structure of the following:
   (a) α,β-Dichlorodiethylketone
   (b) β-Methylbutanone
   (c) 2-Ethyl-4-methylhexanal
9. Discuss various types of isomerism that are possible in case of aldehydes and ketones.
10. Complete the following equations:

   $CH_2CH_2COCl + H_2 \xrightarrow{Pd/BaSO_4}$

   $\overset{\displaystyle R}{\underset{\displaystyle R}{\diagdown}}CHOH + CH_3COCH_3 + [(CH_3)_3CO]_3Al \longrightarrow$

   $CH_3C\equiv N \xrightarrow{LiAlH_4}$

   $CH_3C\equiv N \xrightarrow{SnCl_2, HCl}$

   $HC\equiv CH + H_2O \xrightarrow[HgSO_4]{H^+}$

11. Explain why aldehydes and ketones have higher boiling points than alkanes of similar molecular weight ?
12. Explain the reaction of α-hydrogen in aldehydes.
13. Give the mechanism of nucleophilic addition to aldehydes and ketones.
14. Explain
   (a) Aldol condensation   (b) Strecker's synthesis
   (c) Stephens method   (d) Rosenmund's reduction
   (e) Oppenauer's oxidation   (f) Haloform reaction
15. Which of the following compounds will not give haloform reaction:
   CH_3CHO; $HCHO$; CH_3COCH_3; $CH_3COCH_2CH_3$;
   $CH_3COC(CH_3)_3$; C_6H_5CHO
16. Discuss the products obtained by reduction of aldehydes and ketones under different conditions.
17. Explain Wittig reaction for the synthesis of alkenes from an aldehyde or a ketone.
18. How is chloral obtained? what product is obtained in presence of water and alcohol ?
19. Explain the product obtained in the following
   (a) butadiene + acrolein $\longrightarrow$
   (b) diethylmalonate + acrolein $\longrightarrow$
20. Discuss Pinacol-Pinacolone rearrangement.

19
Monocarboxylic Acids

19.1 INTRODUCTION

Compounds having carboxyl functional group (–COOH) are called carboxylic acids. These may also be considered as the carboxyl derivatives of hydrocarbons in which one or more hydrogen atoms are replaced by carboxyl group.

Acids may be mono-, di- or tri-carboxylic acids depending on whether they have one, two or three carboxyl groups. The monocarboxylic acids are called *Fatty Acids,* since many higher members like stearic acid ($C_{17}H_{35}COOH$), palmitic acid ($C_{15}H_{31}COOH$) and oleic acid ($C_{17}H_{33}COOH$) occur in fats as glycerides.

Monocarboxylic acids, are saturated or unsaturated depending upon whether the hydrocarbon group to which the carboxyl is attached is saturated or not.

CH_3CH_2COOH	$CH_2=CHCOOH$	$CH_3CH=CHCOOH$
Propionic acid	**Acrylic acid**	**Crotonic acid**
(Saturated)	**(Unsaturated)**	**(Unsaturated)**

19.2 NOMENCLATURE

Lower members are commonly known by their *Trival Names,* which are derived from the source of the individual acids. For example, formic acid was first obtained from ants (Latin: formica, ant). Similarly, the name acetic acid is derived from the Latin word *acetum* for vinegar of which it is the chief constituent. The names butyric acid (C_3H_7COOH) and valeric acid (C_4H_9COOH) are similarly derived from the name of the source (butyrum in case of butyric acid and roots of valerian plant in case of valeric acid).

For the nomenclature of substituted acids, the longest carbon chain including the carboxyl group is taken as the main chain. The position of the substituents are indicated by numerals or by Greek letters α, β, γ etc. (α carbon is next to the COOH group).

$$\begin{array}{c} \qquad CH_3 \\ \qquad | \\ CH_3CHCH_2COOH \\ \gamma \quad \beta \quad \alpha \end{array}$$

β-Methylbutyric acid or
3-Methylbutanoic acid

IUPAC System: In this system, the carboxylic acids are named by replacing the ending *'e'* of alkane by *'oic acid'*. Following are given the IUPAC and trivial names of a few carboxylic acids.

Formula	IUPAC Name	Trivial Name
HCOOH	Methanoic acid	Formic acid
CH_3COOH	Ethanoic acid	Acetic acid
CH_3CH_2COOH	Propanoic acid	Propionic acid
$CH_3CH_2CH_2COOH$	Butanoic acid	Butyric acid
$CH_3CH_2CH_2CH_2COOH$	Pentanoic acid	Valeric acid
$CH_3CH_2CH_2CH_2CH_2COOH$	Hexanoic acid	Caproic acid

Alternatively, the carboxyl group is taken as a substituent and is denoted by adding the suffix 'carboxylic acid' to the name of the corresponding alkane.

$$\overset{\overset{\displaystyle CH_3}{|}}{CH_3CH_2CHCOOH}$$

1-Methylpropane-1-carboxylic acid

$$\overset{\overset{\displaystyle CH_3}{|}}{CH_3CH_2CHCH_2COOH}$$

2-Methylbutane-1-carboxylic acid

19.3   ISOMERISM

Monocarboxylic acids exhibit chain, position and functional isomerism.

The chain and position isomerism is explained on the basis of possible structures of the acid having molecular formula $C_6H_{12}O_2$.

(i) $CH_3CH_2CH_2CH_2CH_2COOH$ **Hexanoic acid**

(ii) $CH_3CH_2CH_2CH(COOH)CH_3$ **2-Methylpentanoic acid**

(iii) $CH_3CH_2CH(COOH)CH_2CH_3$ **2-Ethylbutanoic acid**

(iv) $CH_3CH(CH_3)CH_2CH_2COOH$ **4-Methylpentanoic acid**

(v) $CH_3C(CH_3)_2CH_2COOH$ **3,3-Dimethylbutanoic acid**

(vi) $CH_3CH_2C(CH_3)_2COOH$ **2,2-Dimethylbutanoic acid**

The carboxylic acid, $C_6H_{12}O_2$, can be represented by six possible structures having chain isomerism as shown above. However, the structure (i), (ii) and (iii) exhibit position isomerism.

Monocarboxylic acids exhibit functional isomerism with ester containing the same number of carbon atoms. For example, molecular formula $C_3H_6O_2$ represents propanoic acid (CH_3CH_2COOH) and methyl acetate (CH_3COOCH_3).

19.4   GENERAL METHODS OF PREPARATION

1. *Oxidation of Alcohols:* Primary alcohols on oxidation with acidified potassium permanganate or potassium dichromate give carboxylic acids containing the same number of carbon atoms as the starting alcohol. In this case aldehydes are obtained as intermediates.

$$RCH_2OH \xrightarrow{[O]} RCHO \xrightarrow{[O]} R-COOH$$

Primary   **Aldehyde**   **Carboxylic**
alcohol   **acid**

However, secondary alcohols on similar oxidation give a mixture of carboxylic acid each having less number of carbon atoms than the starting alcohol. In this case ketones are obtained as intermediates.

$$\overset{R}{\underset{R'}{>}}CHOH \xrightarrow{[O]} \overset{R}{\underset{R'}{>}}C=O \xrightarrow{[O]} RCOOH + R'COOH$$

Secondary   **Ketone**   **Carboxylic**
alcohol   **acid**

2. *Oxidation of Methyl Ketones (Haloform reaction):* Methyl ketones on oxidation with halogen nd sodium hydroxide give carboxylic acids containing one carbon atom less than the starting ketone. Ialoform is obtained as a by-product.

$$R-\overset{\displaystyle O}{\overset{\|}{C}}-CH_3 \xrightarrow{X_2,NaOH} R-\overset{\displaystyle O}{\overset{\|}{C}}-CX_3 \xrightarrow{NaOH} R-\overset{\displaystyle O}{\overset{\|}{C}}-ONa + CHX_3$$

Methyl ketone   **Sodium salt of**
                                     carboxylic acid   Haloform

3. *Hydrolysis of Nitriles:* The alkyl cyanides on hydrolysis with acid or base give the corresponding arboxylic acid.

$$R-C\equiv N \xrightarrow{H_2O} R-\overset{\displaystyle O}{\overset{\|}{C}}-NH_2 \xrightarrow{H_2O} R-\overset{\displaystyle O}{\overset{\|}{C}}-OH$$

Alkyl cyanide   **Amide**   **Carboxylic acid**

In case acid is used for hydrolysis, ammonium salt of carboxylic acid is obtained, which on treatment with a mineral acid gives the carboxylic acid. The use of alkali for hydrolysis, however, gives the sodium salt which on acidification gives the carboxylic acid.

4. *Hydrolysis of Esters:* The esters on hydrolysis with mineral acids or alkalis yield the corresponding arboxylic acid and alcohol.

$$R-\overset{\displaystyle O}{\overset{\|}{C}}-OR' + H_2O \xrightarrow{Hydrolysis} R-\overset{\displaystyle O}{\overset{\|}{C}}-OH + R'-OH$$

Ester   **Carboxylic   Alcohol**
                             acid

This method finds application in the hydrolysis of oils and fats, which are triglycerides of higher carboxylic acids. Thus, hydrolysis of tripalmitin with alkali yields glycerol and sodium or potassium salt of palmitic acid. These salts of higher acids are soaps. The process is known as saponification. For details see section 15.7.1.

$$\begin{array}{l}
CH_2-O-\overset{\displaystyle O}{\overset{\|}{C}}(CH_2)_{14}CH_3 \\
CH-O-\overset{\displaystyle O}{\overset{\|}{C}}(CH_2)_{14}CH_3 \\
CH_2-O-\overset{\displaystyle O}{\overset{\|}{C}}(CH_2)_{14}CH_3
\end{array} \xrightarrow{NaOH} \begin{array}{l} CH_2OH \\ CHOH \\ CH_2OH \end{array} + 3CH_3(CH_2)_{14}COO^-Na^+$$

Tripalmitin   **Glycerol**   **Sodium palmitate**

$$CH_3(CH_2)_{14}COO^-Na^+ \xrightarrow{H^+} CH_3(CH_2)_{14}COOH$$

Sodium palmitate   **Palmitic acid**

5. *Carbonation of Grignard Reagent:* Grignard reagents react with carbon dioxide to give salts of carboxylic acids which on treatment with mineral acids give carboxylic acids having one carbon more than the starting alkyl halide.

$$R-MgBr + O=C=O \longrightarrow R-\overset{\displaystyle O}{\overset{\|}{C}}-O^-{}^+MgBr \xrightarrow{H^+/H_2O} R-\overset{\displaystyle O}{\overset{\|}{C}}-OH + Mg\overset{OH}{\underset{Br}{\big<}}$$

Alkyl   **Carboxylic acid**
magnesium bromide

Grignard reagents, in turn, are prepared by the reaction of an alkyl halide with magnesium metal : dry ether. For example, n-butyl chloride forms n-pentanoic acid.

$$CH_3CH_2CH_2CH_2Cl \xrightarrow[Ether]{Mg} CH_3CH_2CH_2CH_2MgCl$$

n-Butyl chloride

$$\xrightarrow{CO_2} CH_3CH_2CH_2CH_2\overset{O}{\overset{\|}{C}}-\bar{O}\overset{+}{Mg}Cl \xrightarrow{H^+/H_2O} CH_3CH_2CH_2CH_2\overset{O}{\overset{\|}{C}}-OH$$

n-Pentanoic acid

6. *From Dicarboxylic Acids:* Dicarboxylic acids having two carboxyl group attached to the sam carbon atom on heating undergo decarboxylation of one carboxyl group to give monocarboxylic acid

$$CH_2\overset{\diagup COOH}{\diagdown COOH} \xrightarrow{Heat} CH_3COOH + CO_2$$

Malonic acid   **Acetic acid**

7. *From Active Methylene Compounds:* Fatty acids may be easily obtained by hydrolysis of diethy malonate or ethyl acetoacetate or their alkyl substituted derivatives.

$$CH_3COCH_2C_2H_5 \xrightarrow[2.\ C_2H_5I]{1.\ C_2H_5ONa} CH_3CO\overset{C_2H_5}{\overset{|}{CH}}CO_2C_2H_5 \xrightarrow[2.\ C_2H_5I]{1.\ C_2H_5ONa}$$

Ethyl acetoacetate

$$CH_3CO\overset{C_2H_5}{\underset{C_2H_5}{\overset{|}{\underset{|}{C}}}}CO_2C_2H_5 \xrightarrow[2.\ H^+]{1.\ KOH} CH_3COOH + CH_3CH_2\overset{C_2H_5}{\underset{H}{\overset{|}{\underset{|}{C}}}}COOH + C_2H_5OH$$

                                    Acetic acid

2-Ethylbutanoic acid

$$CH_2(COOC_2H_5)_2 \xrightarrow[2.\ CH_3I]{1.\ C_2H_5ONa} CH_3CH(COOC_2H_5)_2 \xrightarrow[2.\ CH_3I]{1.\ C_2H_5ONa}$$

Diethyl malonate

$$\overset{CH_3}{\underset{CH_3}{>}}C(COOC_2H_5)_2 \xrightarrow[\Delta]{KOH} \overset{CH_3}{\underset{CH_3}{>}}C(COOK)_2 \xrightarrow{HCl} \overset{CH_3}{\underset{CH_3}{>}}C\overset{\diagup COOH}{\diagdown COOH}$$

$$\xrightarrow[\Delta]{150-200°} CH_3-\overset{CH_3}{\overset{|}{CH}}-COOH$$

Dimethylacetic acid

19.5 PHYSICAL PROPERTIES

The first three members are colorless, pungent smelling liquids. The next member, butyric acid has the odor of rancid butter. The next five members (C_5 to C_9) have goat like odor and those above C_{10} are colorless and odorless waxy solids.

Carboxylic acids are polar in nature. They form hydrogen bonds in the solid as well as in the liquid state. This accounts for higher boiling points of carboxylic acids as compared to those of alcohols o similar molecular weight. Thus, propanoic acid (molecular weight 72) boils at 141°, while n-butanol (molecular weight 74) boils at 118°. This indicates greater strength of the hydrogen bonds in acids than in alcohols.

The molecular weights for lower members of the series are found to be twice the molecular weight expected due to the existence of hydrogen bonding between two molecules.

$$CH_3-C\underset{O-H\cdots O}{\overset{O\cdots H-O}{\lessgtr}}C-CH_3$$

Dimer of Acetic Acid in Vapor Phase

The presence of eight membered ring has been confirmed by electron diffraction studies and also by molecular weight determination.

A comparison of the melting points of carboxylic acids shows that the acids having an even number of carbon atoms have higher melting points as compared to the carboxylic acid having an odd number of carbon atoms immediately above and below in the series. Thus, it follows a 'saw-tooth' pattern. As an example, the melting point of acetic acid (17°) having even number of carbon atoms is higher than the melting point of formic acid (8.4°) and propanoic acid (–22°) having odd number of carbon atoms.

Due to hydrogen bonding the carboxylic acids show appreciable solubility in water. The first four members are miscible in water in all proportions. As the chain length increases, the solubility in water decreases.

Acidity of Carboxylic Acids: Most of the carboxylic acids are only slightly ionised in water and so they are fairly weak acids.

$$R-\overset{O}{\overset{\|}{C}}-\ddot{O}-H + H-\ddot{O}-H \rightleftharpoons R-\overset{O}{\overset{\|}{C}}-\ddot{O}: + H-\overset{+}{\underset{}{\overset{}{\ddot{O}}}}\overset{H}{\underset{H}{\diagdown}}$$

Carboxylic acid   **Carboxylate anion**

The acid character decreases with increase in molecular weight. Thus, formic acid, being the lowest in molecular weight, is the strongest of all fatty acids.

Applying the law of Mass action, the dissociation or acidity constant K_a for the following equilibrium is expressed as follows.

$$R-\overset{O}{\overset{\|}{C}}-OH \rightleftharpoons R-\overset{O}{\overset{\|}{C}}-O^- + H^+$$

$$K_a = \frac{[RCOO^-][H^+]}{[RCOOH]}$$

The value of K_a is directly proportional to the concentration of hydrogen ions and is thus a measure of the acidity of the acid under consideration.

The carboxylate anion is a resonance hybrid of the following two structures (I) and (II). Both these forms are of equal stability and contribute equally. Thus the negative charge on the carboxylate anion is evenly distributed over both oxygen atoms as shown in the resonance hybrid (III).

$$R-C\underset{\bar{O}}{\overset{O}{\diagup}} \longleftrightarrow R-C\underset{O}{\overset{\bar{O}}{\diagup}} \quad \text{equivalent to} \quad R-C\underset{O}{\overset{O}{\lessgtr}}\Big\}^-$$

$$\textbf{I} \qquad\qquad \textbf{II} \qquad\qquad\qquad \textbf{III}$$

The resonance in carboxylate anion has been confirmed by X-Ray and electron diffraction studies.

It has been shown that the two carbon oxygen bond length in formic acid are different while in sodium formate they are equal.

Formic acid   **Sodium formate**

Furthermore, the C–O bond lengths are intermediate between those of normal double and single carbon-oxygen bond.

We have seen that the acidity of carboxylic acids is due to their ability to release proton. Thus, any factor that stabilises the carboxylate anion would facilitate the release of protons and increase the acidity. Thus, electron withdrawing substituents (like Cl, NO_2, CN etc.) in a carboxylic acid would disperse the negative charge of the anion, stabilise it and thus enhance its acid strength. In other words the following generalisations can be made.

(i) Acidity decreases with larger alkyl group. Thus, the acidity is in the following order:

$$HCOOH > CH_3COOH > CH_3CH_2COOH > CH_3CH_2CH_2COOH$$

(ii) Acidity increases with increasing number of electron withdrawing substituents on the α-carbon.

$$CCl_3COOH > CHCl_2COOH > CH_2ClCOOH > CH_3COOH$$

(iii)   Acidity increases with increasing electronegativity of substituents.

$$FCH_2COOH > ClCH_2COOH > BrCH_2COOH > ICH_2COOH$$

(iv)   The acidity declines with increasing distance between electron withdrawing group and carboxyl group. This is represented as follows.

$$CH_3CH_2XCHCOOH > CH_3XCHCH_2COOH > XCH_2CH_2CH_2COOH$$

19.6   SPECTRAL PROPERTIES

(1)   *IR Spectra:* In carboxylic acids, the carboxyl group consists of a carbonyl group and a hydroxy group. The carbonyl group absorbs in the region 1725-1700 cm^{-1}, while the free Hydroxy group stretching give a strong broad band at about 1250 cm^{-1} and O–H bending band near 1400 cm^{-1} and 920 cm^{-1}. Strong hydrogen bonding (in case of acetic acid) is observed as a broad and intense band at about 3000 cm^{-1}.

(2)   *UV Spectra:* Saturated carboxylic acids do not show any band. However, conjugation of carboxyl group with a double bond produces an intense band. For example, crotonic acid, $CH_3CH=CHCOOH$, has λ_{max} at 208 nm.

(3)   *NMR Spectra:* NMR spectra shows a characteristic absorption of hydroxy proton of the carboxyl group which is observed down field between δ 9-13 compared to the hydroxy proton in alcohols (δ 2-3). This is due to deshielding effect of carbonyl group present adjacent to the hydroxy group. The signal of this proton disappears by adding D_2O to the sample, as deuterated water converts –COOH into –COOD.

19.7 CHEMICAL PROPERTIES

A carboxylic acid molecule is made up of a carbonyl group, a hydroxy group and an alkyl group and shows various reactions due to them.

1. Reactions Involving Hydroxy Proton

It has already been stated that the carboxylic acids undergo ionisation into proton and the carboxylate anion. However, the degree of ionisation is small as compared to that of inorganic acids. The carboxylic acids react with strong electropositive metals, carbonates, alkalis, ammonia etc to form salts.

$$2RCOOH + Zn \longrightarrow (RCOO)_2Zn + H_2$$
$$2RCOOH + Na_2CO_3 \longrightarrow 2RCOONa + H_2O + CO_2$$
$$RCOOH + NaOH \longrightarrow RCOONa + H_2O$$
$$RCOOH + NH_3 \longrightarrow RCOONH_4$$

2. Reactions Involving the Hydroxy Group

(i) *Esterification:* Carboxylic acids on reaction with an alcohol in presence of a dehydrating agent (viz. concentrated sulphuric acid, anhydrous zinc chloride, hydrogen chloride or p-toluenesulphonic acid) form esters.

$$RCOOH + R'OH \underset{}{\overset{H^+}{\rightleftharpoons}} RCOOR' + H_2O$$

Carboxylic **Alcohol** **Ester**
acid

$$CH_3COOH + C_2H_5OH \underset{}{\overset{H^+}{\rightleftharpoons}} CH_3COOC_2H_5 + H_2O$$

Acetic acid **Ethanol** **Ethyl acetate**

In this esterification reaction the water is formed by the elimination of −OH from carboxylic acid and a proton from alcohol and not vice-versa (i.e., −OH from alcohol and proton from carboxylic acid). This has been proved by using isotopically labeled alcohol. It has been found that the oxygen of the water formed, comes exclusively from the carboxylic acid.

$$\underset{}{R-\overset{\overset{O}{\|}}{C}-OH} + HO^{18}CH_3 \overset{H^+}{\rightleftharpoons} R-\overset{\overset{O}{\|}}{C}-O^{18}CH_3 + H_2O$$

$$R-\overset{\overset{O}{\|}}{C}-OH + HO^{18}CH_3 \overset{H^+}{\underset{}{\rightleftharpoons}}\!\!\!\!\!/\ \ R-\overset{\overset{O}{\|}}{C}-OCH_3 + H_2O^{18}$$

Mechanism of esterification involves three steps.

Step 1: involves protonation of the hydroxy group of carboxylic acid by a strong acid used as catalyst which makes it more electrophilic and enables it to react with the alcohol which is a weak nucleophile.

$$R-\overset{\overset{:O:}{\|}}{C}-\ddot{O}H \underset{-H^+}{\overset{H^+}{\rightleftharpoons}} \left[\begin{array}{ccc} \overset{\overset{+}{:}\ddot{O}-H}{\|} & :\ddot{O}-H & :\ddot{O}-H \\ R-C-\ddot{O}-H & \longleftrightarrow & R-\overset{+}{C}-\ddot{O}-H & \longleftrightarrow & R-C=\overset{+}{\ddot{O}}-H \end{array} \right]$$

Carboxylic acid **Protonated carboxylic acid**

Step 2: involves the attack by alcohol, the nucleophiles.

$$R-\overset{\overset{\displaystyle +\ddot{O}-H}{\|}}{\underset{\overset{\displaystyle \ddot{O}:}{H}}{C}} + CH_3\ddot{O}H \rightleftharpoons R-\overset{\overset{\displaystyle \ddot{O}-H}{\|}}{\underset{\overset{\displaystyle \ddot{O}:}{H}}{C}}-\overset{\displaystyle \ddot{O}{\overset{CH_3}{\underset{+}{\diagup}}}}{\underset{}{}}H \overset{-H^+}{\rightleftharpoons} R-\overset{\overset{\displaystyle \ddot{O}-H}{\|}}{\underset{\overset{\displaystyle \ddot{O}:}{H}}{C}}-\ddot{O}-CH_3$$

Intermediate

Step 3: involves the elimination of water from the above intermediate followed by deprotonation.

$$R-\overset{\overset{\displaystyle \ddot{O}-H}{\|}}{\underset{\overset{\displaystyle \ddot{O}-H}{}}{C}}-\ddot{O}-CH_3 \overset{H^+}{\rightleftharpoons} R-\overset{\overset{\displaystyle \ddot{O}-H}{\|}}{\underset{\overset{\displaystyle H-\overset{+}{\underset{}{O}}-H}{}}{C}}-\ddot{O}-CH_3 \overset{-H_2O}{\rightleftharpoons} R-\overset{\overset{\displaystyle \ddot{O}-H}{\|}}{\underset{}{C}}-\ddot{O}-CH_3$$

Intermediate

$$R-\overset{\overset{\displaystyle :O:}{\|}}{\underset{}{C}}-\ddot{O}-CH_3 \overset{-H^+}{\rightleftharpoons} R-\overset{\overset{\displaystyle \ddot{O}-H}{\|}}{\underset{+}{C}}-\ddot{O}-CH_3$$

Ester

Tertiary alcohols can not be esterified by acid catalysed esterification as the reaction is very slow due to steric factors and they undergo elimination instead of esterification.

(ii) *Conversion into Acid Chlorides*: Thionyl chloride, phosphorus trichloride and phosphorus pentachloride react with carboxylic acid to give acid chlorides.

$$RCOOH + SOCl_2 \longrightarrow RCOCl + HCl + SO_2$$
Thionyl
chloride

$$3RCOOH + PCl_3 \longrightarrow 3RCOCl + H_3PO_3$$
Phosphorus
trichloride

$$RCOOH + PCl_5 \longrightarrow RCOCl + POCl_3 + HCl$$
Carboxylic   Phosphorus   Acid
acid   pentachloride   chloride

(iii) *Dehydraton:* Carboxylic acids undergo dehydration in presence of a dehydrating agent like phosphorus pentoxide to form acid anhydrides.

$$2R\,COOH \xrightarrow[-H_2O]{P_2O_5} R-\overset{\overset{\displaystyle O}{\|}}{C}-O-\overset{\overset{\displaystyle O}{\|}}{C}-R$$
Carboxylic acid   acid anhydride

$$2CH_3COOH \xrightarrow[-H_2O]{P_2O_5} CH_3-\overset{\overset{\displaystyle O}{\|}}{C}-O-\overset{\overset{\displaystyle O}{\|}}{C}-CH_3$$
Acetic acid   Acetic anhydride

(iv) *Conversion into amides:* Carboxylic acids on reaction with ammonia give ammonium salts, which on heating give amides.

$$RCOOH + NH_3 \longrightarrow R-\overset{\overset{\displaystyle O}{\|}}{C}-\bar{O}\overset{+}{N}H_4 \xrightarrow[-H_2O]{\Delta} R-\overset{\overset{\displaystyle O}{\|}}{C}-NH_2$$
Carboxylic   Ammonium salt   Amide
acid

3. Reactions Involving Carboxyl Group

(i) *Reaction with Diazomethane:* Monocarboxylic acids react with diazomethane in ether to form the corresponding methyl ester.

$$R-\overset{\overset{\displaystyle O}{\|}}{C}-OH \ + \ CH_2N_2 \ \longrightarrow \ RCOOCH_3 + N_2$$

Diazo-   Ester
methane

Mechanism

$$R-\overset{\overset{\displaystyle O}{\|}}{C}-\overset{..}{\underset{..}{O}}-H + :\bar{C}H_2-\overset{+}{N}\equiv N: \longrightarrow R-\overset{\overset{\displaystyle O}{\|}}{C}-\overset{..}{\underset{..}{O}}:^- + CH_3-\overset{+}{N}\equiv N:$$

$$R-\overset{\overset{\displaystyle O}{\|}}{C}-\overset{..}{\underset{..}{O}}: + CH_3-\overset{+}{N}\equiv N \longrightarrow R-\overset{\overset{\displaystyle O}{\|}}{C}-OCH_3 + :N\equiv N:$$

Ester

Note: Diazomethane is a toxic gas and is explosive in nature. It also produces allergy so this method is used only for small scale preparation of esters. The advantage of the reaction is that the reaction is clean and gives pure products since nitrogen is the only by product.

(ii) *Decarboxylation:* Fatty acids undergo decarboxylation when their sodium salts are heated with soda lime to produce alkanes.

$$RCOONa + NaOH \xrightarrow{\ \Delta\ } RH + Na_2CO_3$$

Sodium   **Alkane**
salt of
carboxylic acid

The mechanism of this reaction is uncertain.

$$R-C\overset{\overset{\displaystyle O-H}{}}{\underset{\displaystyle O}{}} \quad OH \longrightarrow CO_2 + H_2O + R^-$$

$$R^- \xrightarrow{\ H^+\ } RH$$

Decarboxylation is favoured by electron withdrawing groups.

Alternatively, carboxylic acid can also be decarboxylated by heating in quinoline at 170-180° in presence of copper bronze.

$$R-C\overset{\overset{\displaystyle O}{}}{\underset{\displaystyle O-H}{}} \xrightarrow[\text{Cu/quinoline}]{170\text{-}180^\circ} H^+ + R-C\overset{\overset{\displaystyle O}{}}{\underset{\displaystyle O}{}} \longrightarrow RH + CO_2$$

Carboxylic acid   **Alkane**

(iii) *Kolbe's Reaction:* Electrolysis of concentrated aqueous solution of sodium or potassium salt of a fatty acid yields an alkane.

$$R-COONa \longrightarrow RCOO^- + Na^+$$

$$2RCOO^- \longrightarrow R-R + CO_2 + 2e^- \text{ [at anode]}$$

Alkane

$$2Na^+ + 2e^- \longrightarrow 2Na \text{ [at cathode]}$$

$$2Na + H_2O \longrightarrow 2NaOH + H_2 \text{ [at cathode]}$$

(iv) *Formation of Aldehydes and Ketones:* Calcium salt of a fatty acid (other than formic acid) on heating strongly gives a ketone. On the other hand, if calcium salt is heated with calcium formate an aldehyde is obtained.

$$(CH_3COO)_2Ca \longrightarrow CH_3COCH_3 + CaCO_3$$
Calcium acetate   **Acetone**

$$(CH_3COO)_2Ca + (HCOO)_2Ca \longrightarrow 2CH_3CHO + 2CaCO_3$$
Calcium acetate   **Calcium formate**   **Acetaldehyde**

(v) *Formation of Alkyl Halides (Hunsdiecker's reaction):* Silver salts of fatty acids on heating with a halogen give alkyl halides.

$$RCOOAg + X_2 \longrightarrow R-X + CO_2 + AgX$$
Silver   **Alkyl**
salt of fatty   **halide**
acid

(vi) *Formation of Primary Amines (Schmidt's Reaction):* Fatty acids when heated with hydrazoic acid at 50-55° in presence of concentrated sulphuric acid, a vigorous reaction sets in with the evolution of nitrogen. The resulting solution on warming with water give amines with the evolution of carbon dioxide.

$$RCOOH + HN_3 \xrightarrow[\Delta]{H_2SO_4} RNH_2 + N_2 + CO_2$$
           Hydrazoic   **Primary**
           acid   **amine**

The mechanism of the reaction involves following steps:

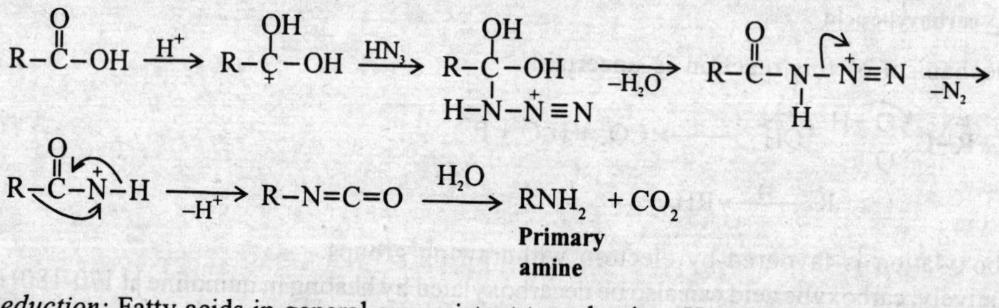

(vii) *Reduction:* Fatty acids in general are resistant to reduction. However, under drastic conditions, i.e. heating with hydriodic acid in presence of small amount of red phosphorus produce paraffins.

$$RCOOH + HI \xrightarrow{red\ P} RCH_3$$

Reduction can easily be effected by treating with hydrogen at high temperature under pressure in presence of nickel catalyst.

$$RCOOH + 3H_2 \xrightarrow{Ni} RCH_3 + 2H_2O$$

Carboxylic acids may be easily reduced to the corresponding alcohols by lithium aluminum hydride.

$$RCOOH + LiAlH_4 \longrightarrow RCH_2OH$$
Carboxylic   **Primary alcohol**
acid

4. Reactions Involving the α-Carbon

Aliphatic carboxylic acids on reaction with bromine or chlorine in presence of small amount of red phosphorus give α-halo acids. This reaction is known as *Hell-Volhard Zelinsky* reaction.

$$CH_3CH_2CH_2COOH \xrightarrow{Br_2 + red\ P} CH_3CH_2-\underset{\underset{Br}{|}}{C}H-COOH + HBr$$

Butyric acid   **α-Bromobutyric acid**

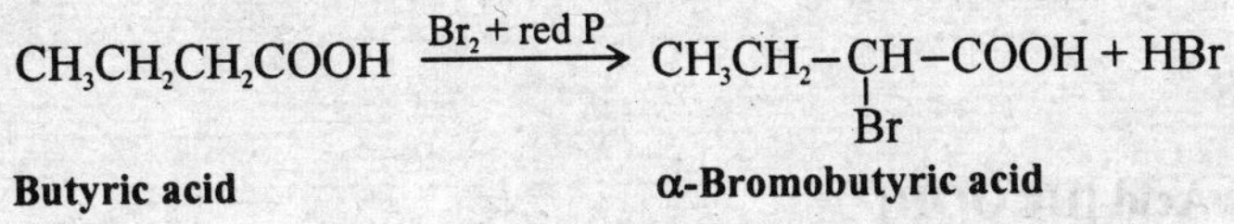

It is important to note here that phosphorus reacts with bromine to give phosphorus tribromide. If more than one equivalent of halogen is used in the above reaction, then dihalo or trihalo acids are obtained.

Mechanism: Phosphorus tribromide formed by the reaction of red phosphorus with bromine converts the carboxylic acid into acid bromide (I), which upon enolisation gives (II). It undergone bromination at α-carbon followed by the reaction of the intermediate (III) with a second molecule of carboxylic acid by interchange of bromo group and hydroxy group to form the α-halogenated acid.

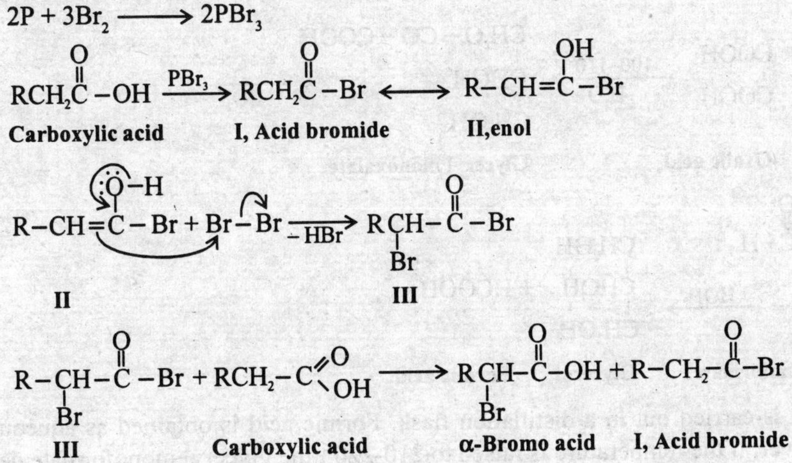

$$2P + 3Br_2 \longrightarrow 2PBr_3$$

The acid halide (I) generated in the final step is reused in the second step.

The Hell-Volhard-Zelinsky reaction is of special interest since the α-halo acids obtained can be used for the preparation of α-hydroxy, α-amino and α-cyano carboxylic acids.

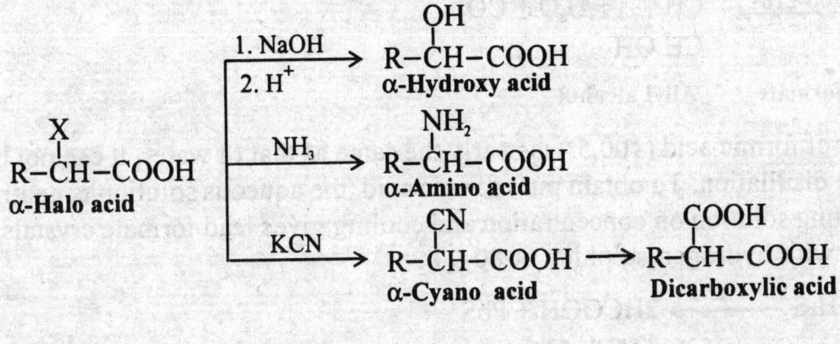

Tests of Carboxyl group

A carboxylic acid gives effervescence with sodium bicarbonate solution due to evolution of carbon dioxide. In aqueus solution it turns blue litmus red. On warming with ethanol and concentrated sulphuric acid the carboxylic acid forms an ester which is detected by its fruity odor.

19.8   INDIVIDUAL MEMBERS

19.8.1   Formic Acid/Methanoic Acid [HCOOH]

The first member of the acid series is formic acid. It was first obtained by distillation of red ants. It also occurs in stringing nettles, pine nettles and fruits.

Preparation: It can be obtained by the general methods of preparation of carboxylic acids.

Laboratory Preparation: It is conveniently prepared by heating glycerol with oxalic acid at 100-110° to first give glyceryl mono-oxalate which decomposes into glyceryl monoformate and carbon dioxide. After the evolution of carbon dioxides has ceased, more of crystalline oxalic acid is added. The glyceryl monoformate gets hydrolysed to formic acid and glycerol is regenerated, which reacts with fresh oxalic acid.

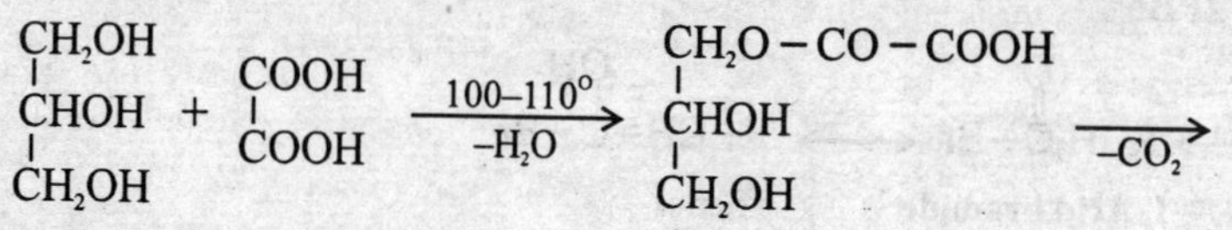

$$\begin{array}{l} CH_2OH \\ | \\ CHOH \\ | \\ CH_2OH \end{array} + \begin{array}{l} COOH \\ | \\ COOH \end{array} \xrightarrow[-H_2O]{100-110^\circ} \begin{array}{l} CH_2O-CO-COOH \\ | \\ CHOH \\ | \\ CH_2OH \end{array} \xrightarrow{-CO_2}$$

Glycerol   **Oxalic acid**   **Glyceryl monoxalate**

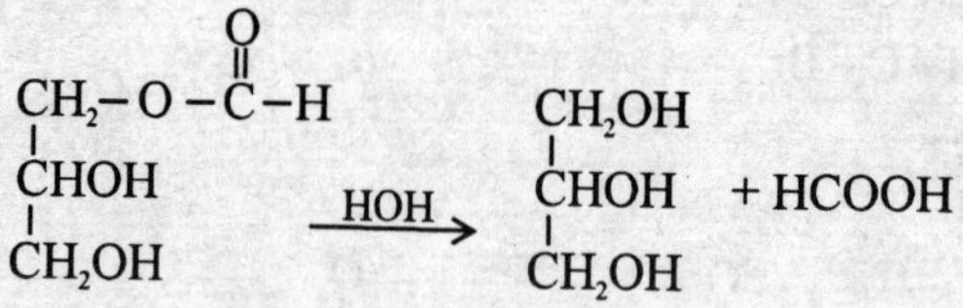

$$\begin{array}{l} CH_2-O-\overset{\overset{O}{\|}}{C}-H \\ | \\ CHOH \\ | \\ CH_2OH \end{array} \xrightarrow{HOH} \begin{array}{l} CH_2OH \\ | \\ CHOH \\ | \\ CH_2OH \end{array} + HCOOH$$

Glyceryl monoformate   **Glycerol**   **Formic acid**

The above reaction is carried out in a distillation flask. Formic acid is obtained as aqueous solution in the distillate.However, if the temperature is raised to 210-220°, the glyceryl monoformate decomposes to give allyl alcohol.

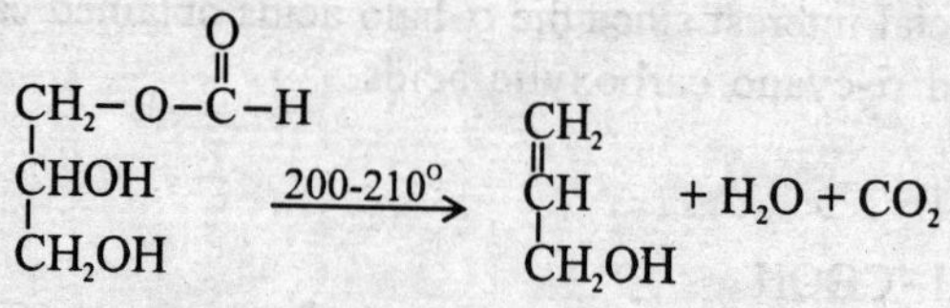

$$\begin{array}{l} CH_2-O-\overset{\overset{O}{\|}}{C}-H \\ | \\ CHOH \\ | \\ CH_2OH \end{array} \xrightarrow{200-210^\circ} \begin{array}{l} CH_2 \\ \| \\ CH \\ | \\ CH_2OH \end{array} + H_2O + CO_2$$

Glyceryl monoformate   **Allyl alcohol**

Since the boiling point of formic acid (100.5°) is nearly the same as that of water, it can not be obtained in pure form by fractional distillation. To obtain pure formic acid, the aqueous solution is neutralised with lead carbonate. The resulting solution on concentration and cooling gives lead formate crystals. These are filtered and decomposed with hydrogen sulphide at 80-90°.

$$(HCOO)_2Pb + H_2S \longrightarrow 2HCOOH + PbS$$
Lead formate   **Formic acid**

The pure formic acid is obtained by distillation.

Industrial Preparation: Formic acid is manufactured by heating powdered sodium hydroxide and carbon monoxide at 150° under pressure.

$$NaOH + CO \xrightarrow[\text{8 atm}]{150°} HCO\bar{O}\overset{+}{Na}$$
$$\text{Sodium formate}$$

The sodium formate is distilled with dilute sulphuric acid when an aqueous solution of formic acid is obtained. It is purified in the same manner as described above.

Properties: Formic acid is a colorless, pungent smelling liquid (b.p. 100.5°). It is soluble in water, ether and alcohol. It is a corrosive liquid and causes blisters on the skin.

As already stated, formic acid is the strongest of all other members of the homologous series. The special behaviour of formic acid is because, besides containing a carboxyl group it has an aldehyde group also. So it behaves both as an aldehyde and an acid.

As carboxylic acid, formic acid gives all reactions of fatty acids. Thus, it turns blue litmus red, neutralises alkalis, and decomposes carbonates.

As an aldehyde, formic acid behaves as a reducing agent. It reduces ammonical silver nitrate, Fehling solution and decolorises acidified potassium permanganate solution. During these reductions, formic acid is oxidised to carbonic acid.

$$\underset{\text{Formic acid}}{H-\overset{\overset{O}{\|}}{C}-OH} \xrightarrow{[O]} \underset{\text{Carbonic acid}}{\left[HO-\overset{\overset{O}{\|}}{C}-OH\right]} \longrightarrow CO_2 + H_2O$$

Some other reactions of formic acid are given below:

(1) *Dehydration:* On warming with concentrated sulphuric acid, formic acid is dehydrated to give carbon monoxide and water.

$$\underset{\text{Formic acid}}{HCOOH} \xrightarrow[\Delta]{H_2SO_4} CO + H_2O$$

This reaction is used as a laboratory preparation of carbon monoxide.

(2) *Decomposition:* On heating under pressure, formic acid decomposes to give carbondioxide and hydrogen.

$$HCOOH \xrightarrow[\text{pressure}]{\Delta, 160°} CO_2 + H_2$$

(3) *Action of heat on Sodium formate:* On heating sodium formate, sodium oxalate is obtained. This reaction has been used for the manufacture of oxalic acid.

$$\underset{\text{Sodium formate}}{2HCOONa} \xrightarrow[-H_2]{350°} \underset{\text{Sodium Oxalate}}{\overset{\displaystyle COONa}{\underset{\displaystyle COONa}{|}}} \xrightarrow{HCl} \underset{\text{Oxalic acid}}{\overset{\displaystyle COOH}{\underset{\displaystyle COOH}{|}}}$$

Uses: Formic acid is used in tanning for removing lime from leather. It is also used as an antiseptic and for the preservation of fruits and coagulating rubber. Nickel formate is used as a catalyst in the hydrogenation of oils.

Tests: It reduces Tollen's solution and Fehling's solution (difference from other fatty acids). With neutral ferric chloride solution, a neutral solution of formic acid gives red color.

19.8.2 Acetic Acid/Ethanoic acid/[CH₃COOH]

Acetic acid is known since very early times in the form of vinegar. It is the chief constituent of vinegar and hence its name (Latin: acetum-Vinegar). It is found in fruit juices which have become sour by fermentation.

Preparation: Acetic acid is prepared by the general methods of preparation described earlier.

1. *From Acetylene:* Acetylene is bubbled through 40% sulphuric acid at 60° in presence of 1% mercuric sulphate catalyst when acetaldehyde is obtained. The acetaldehyde is further oxidised to acetic acid by passing a mixture of acetaldehyde and air at 70° under pressure over manganese acetate.

$$HC \equiv CH + H_2O \xrightarrow[HgSO_4]{H_2SO_4} CH_3CHO \xrightarrow{[O]} CH_3COOH$$
$$\text{Acetylene} \qquad\qquad \text{Acetaldehyde} \qquad \text{Acetic acid}$$

The yield is excellent (80%) and the acid is pure (97%). Most of the acetic acid is now a days manufactured by this method.

2. *From Pyroligneous Acid:* The pyroligneous acid is obtained by the destructive distillation of wood. It contains 10% acetic acid, 2-4% methyl alcohol, 0.5% acetone and water. Its vapors are passed through hot milk of lime when calcium acetate is obtained. It is decomposed with calculated amount of concentrated sulphuric acid when a dilute solution (10-15%) of acetic acid is obtained. The dilute acetic acid is neutralised with sodium carbonate and is converted into sodium acetate, which crystallised out. It is distilled with calculated amount of concentrated sulphuric acid when pure acetic acid called glacial acetic acid is obtained.

3. *From Vinegar:* Vinegar contains 5-7% acetic acid and is obtained by fermentation of a dilute solution of alcohol (10-15%), cane juice or malt. The process is known as 'quick vinegar process'. In this process, a dilute solution of alcohol (10-15%) containing phosphate and inorganic salts (needed for fermentation) is fermented by bacterium *acetic* in presence of air. The fermentation is carried out in wooden vats containing loosely packed beech wood shavings impregnated with vinegar (as a source of bacterium *acetic*). A dilute solution of alcohol (10-15%) is trickled from the top and air enters the system through the holes at the bottom. The temperature is maintained at 30-35°. By this process alcohol is oxidised to acetic acid. By repeating the process vinegar containing 8-10% acetic acid is obtained.

$$C_2H_5OH + O_2 \longrightarrow CH_3COOH + H_2O$$
$$\text{Ethanol} \qquad\qquad \text{Acetic acid}$$

Vinegar is treated with lime to form calcium acetate, which is processed as discussed above (method 2) to give pure acetic acid.

The fermentation process gives good results if the concentration of alcohol is not more than 15% (at higher concentrations the bacteria becomes inactive) and the supply of air is regulated (with less air the oxidation proceeds upto acetaldehyde while with excess oxygen the acetic acid is further oxidised to carbon dioxide and water).

Properties: Acetic acid is a colorless, corrosive liquid (b.p. 118°) with a sharp vinegar odor and sour taste. Below 16° it solidifies to an icy mass and so is called glacial acetic acid. It is miscible in water, ether and alcohol. Acetic acid is a good solvent for sulphur, iodine and many organic compounds. As already mentioned, it is dimeric due to association.

Acetic acid gives all the general reactions of monocarboxylic acids described earlier.

The halogenation of acetic acid is worth mentioning. When chlorine is passed into hot acetic acid in presence of UV light one, two or all the three hydrogens attached to α-carbon atom can be successively replaced by chlorine.

$$CH_3COOH + Cl_2 \xrightarrow{\ UV\ } CH_2ClCOOH + HCl$$
Chloroacetic acid

$$CH_2ClCOOH + Cl_2 \xrightarrow{\ UV\ } CHCl_2COOH + HCl$$
Dichloroacetic acid

$$CHCl_2COOH + Cl_2 \xrightarrow{\ UV\ } CCl_3COOH + HCl$$
Trichloroacetic acid

This method is used for the preparation of trichloroacetic acid.

Uses: It is used in the manufacture of various dye stuffs, perfumes, rayons, rubber from latex and casein from milk. Its salts are used in medicine. Acetates of aluminum and chromium are used as mordants. Its dilute solution (5-6%) in water is used as vinegar.

Tests: Its neutral solution with neutral ferric chloride solution gives a red color. On heating with ethyl alcohol and concentrated sulphuric acid a fruity smell of ethyl acetate is obtained.

Comparison of Formic Acid and Acetic Acid: Both formic acid and acetic acid contain the same functional group, thus, show similar properties with respect to the reaction of this group. However, formic acid differs from acetic acid since formic acid has only one hydrogen atom attached to carboxyl group while in acetic acid there is a methyl group. The methyl group in acetic acid shows usual substitution reactions but the hydrogen atom of formic acid does not. Formic acid shows the reaction of an aldehyde which are not shown by acetic acid.

Resemblance: Both formic and acetic acid give salts with alkalis and carbonates. On heating with alcohol and sulphuric acid both give esters (formates and acetates).Both give acid chlorides with phosphorus pentachloride of which acetyl chloride is stable whereas formyl chloride is unstable. Finally, ammonium salts of both acids on heating give amides (formamide and acetamide).

Difference:
1. Action of heat: Formic acid on heating gives carbon dioxide and hydrogen. Acetic acid on the other hand is stable.
2. Action of sulphuric acid: Formic acid gives carbon monoxide and water. Acetic acid does not evolve CO.
3. Action of halogen: Acetic acid gives substitution products where as formic acid does not react.
4. Reducing properties: Formic acid reduces ammonical silver nitrate and Fehlings solution. Acetic acid has no action.
5. Action of heat on calcium salt: Calcium formate on heating gives formaldehyde. However calcium acetate on heating gives acetone.
6. Action of heat on sodium salt: Sodium formate on heating gives sodium oxalate and hydrogen. Sodium acetate has no action.
7. Electrolysis: Electrolysis of salt of formic acid gives hydrogen. On the other hand electrolysis of sodium salt of acetic acid gives ethane.

19.8.3   Propionic Acid/Propanoic Acid [CH_3CH_2COOH]

Preparation: Propionic acid is prepared by any of the general methods of preparation of acids already discussed. On a large scale, it is obtained by the oxidation of n-propyl alcohol with potassium dichromate and sulphuric acid.

$$CH_3CH_2CH_2OH \xrightarrow[K_2Cr_2O_7 / H_2SO_4]{[O]} CH_3CH_2COOH$$

n-Propyl alcohol   **Propionic acid**

It is also manufactured by the interaction of ethylene, carbon monoxide and water at high temperature in presence of catalyst.

$$CH_2{=}CH_2 + CO + H_2O \xrightarrow[High\ Temp\ \&\ Pressure]{H_3PO_4} CH_3CH_2COOH$$

Ethylene   **Propionic acid**

Properties: It is a colorless oily liquid (b.p. 141°) having an acid odor. It is miscible with water, ether and alcohol. It gives all the general reactions of monocarboxylic acids.

19.8.4   Butyric Acids

There are two isomeric butyric acids, viz n-butyric acid $CH_3CH_2CH_2COOH$ and isobutyric acid $(CH_3)_2CHCOOH$.

19.8.4.1   *n-Butyric Acid/Butanoic Acid* [$CH_3CH_2CH_2COOH$]

It occurs in butter as its glyceryl ester. The characteristic smell of rancid butter is due to the liberation of butyric acid due to hydrolysis.

It is prepared on a large scale by oxidation of n-butyl alcohol.

$$CH_3CH_2CH_2CH_2OH \xrightarrow{[O]} CH_3CH_2CH_2COOH$$

n-Butyl alcohol   **n-Butyric acid**

It can also be manufactured by the fermentation of carbohydrate (glucose or starch) solution with *Bacillus butyrious* (present in sour milk).

n-Butyric acid is rancid-smelling viscous liquid (b.p. 162°). It is soluble in water, ether and alcohol. It shows all the reaction of carboxylic acids.

19.8.4.2   *Isobutyric Acid/2-Methylpropanoic Acid/*[$(CH_3)_2CHCOOH$]

It occurs in free state as well as in some plants. On a large scale, it is obtained by the oxidation of iso-butyl alcohol.

$$(CH_3)_2CHCH_2OH \xrightarrow{[O]} (CH_3)_2CHCOOH$$

Isobutyl alcohol   **Isobutyric acid**

Isobutyric acid is a colorless oily liquid (b.p. 154°). It is sparingly soluble in water.

It is used for the preparation of esters, which are used in flavoring essences and in perfumery.

19.8.5   Higher Fatty Acids

A number of higher fatty acids occur as esters of trihydric alcohol (glycerol) i.e. as glycerides in fats and oils. The most important are palmitic acid, $C_{15}H_{31}COOH$ and stearic acid, $C_{17}H_{35}COOH$ obtained by hydrolysis of tallow. Lauric acid, $C_{11}H_{23}COOH$ is also obtained from coconut oil. All these acids are obtained as byproduct in soap industry.

19.9   UNSATURATED MONOCARBOXYLIC ACIDS

Unsaturated monocarboxylic acid having one double bond in the molecule have the general formula

$C_nH_{2n-1}COOH$. These are derivatives of alkenes and are obtained by replacement of a hydrogen atom by a carboxyl group.

19.10 NOMENCLATURE

Most of the unsaturated monocarboxylic acids are known by their common names. Their IUPAC names are derived by considering the longest carbon chain containing the carboxyl group and naming the acid as *Alkenoic acid*. The position of the double bond with respect to carboxyl group is indicated by a number. A few important members of the series are:

Formula	Common Name	IUPAC Name
$CH_2{=}CH{-}COOH$	Acrylic acid	Prop-2-enoic acid
$CH_3CH{=}CHCOOH$	Crotonic acid	But-2-enoic acid
$\underset{\displaystyle CH_3{=}\overset{\textstyle CH_3}{C}{-}COOH}{}$	Methacrylic acid	2-Methylprop-2-enoic acid
$\underset{H}{\overset{CH_3}{}}C{=}C\underset{COOH}{\overset{CH_3}{}}$	Tiglic acid	E-2-Methylbut-2-enoic acid
$\underset{H}{\overset{CH_3}{}}C{=}C\underset{CH_3}{\overset{COOH}{}}$	Angelic acid	Z-2-Methylbut-2-enoic acid
$\underset{H}{\overset{CH_3(CH_2)_7}{}}C{=}C\underset{H}{\overset{(CH_2)_7COOH}{}}$	Oleic acid	Z-octadec-9-enoic acid

As seen, the unsaturated monocarboxylic acids exhibit geometrical isomerism.

In general, the acidity of unsaturated monocarboxylic acids in which the double bond and the carboxyl group are far apart is of the same order as that in the corresponding saturated analogue. However, if the double bond is adjacent to the carboxyl group (for example, α, β-unsaturated acids) the acidity is greater than the corresponding saturated acid. This is because in α, β-unsaturated acids carboxyl group is attached to sp^2 hybridised carbon while in saturated acids to a sp^3 hybridised carbon.

19.11 INDIVIDUAL MEMBERS

19.11.1 Acrylic Acid/Prop-2-enoic Acid [$CH_2{=}CH{-}COOH$]

It is obtained from acrolein ($CH_2{=}CHCHO$), hence called acrylic acid.

Preparation:

(i) Oxidation of acrolein with ammonical silver nitrate produces acrylic acid.

$$CH_2{=}CH{-}CHO \xrightarrow[\text{ammo. AgNO}_3]{[O]} CH_2{=}CH{-}COOH$$

Acrolein   **Acrylic acid**

(ii) *Oxidation of allyl alcohol*: In this case the double bond is first protected by bromination followed by oxidation and removal of bromine.

$$\underset{\text{Allyl alcohol}}{\overset{\displaystyle CH_2}{\underset{\displaystyle CH_2OH}{\overset{\displaystyle \|}{\underset{\displaystyle |}{CH}}}}} \xrightarrow{Br_2} \underset{}{\overset{\displaystyle CH_2Br}{\underset{\displaystyle CH_2OH}{\overset{\displaystyle |}{\underset{\displaystyle |}{CHBr}}}}} \xrightarrow{[O]} \underset{}{\overset{\displaystyle CH_2Br}{\underset{\displaystyle COOH}{\overset{\displaystyle |}{\underset{\displaystyle |}{CHBr}}}}} \xrightarrow[-ZnBr_2]{Zn,CH_3OH} \underset{\text{Acrylic acid}}{\overset{\displaystyle CH_2}{\underset{\displaystyle COOH}{\overset{\displaystyle \|}{\underset{\displaystyle |}{CH}}}}}$$

(iii)   From propionic acid by bromination in the α-position followed by heating with alcoholic potassium hydroxide.

$$\underset{\text{Propionic acid}}{CH_3CH_2COOH} \xrightarrow[\text{red P}]{Br_2} \underset{\text{Bromopropionic acid}}{CH_3-\overset{\overset{\displaystyle Br}{\displaystyle |}}{CH}-COOH} \xrightarrow{\text{alk.KOH}} \underset{\text{Acrylic acid}}{CH_2=CH-COOH}$$

(iv)   From α, β-dibromopropionic acid by refluxing it with zinc and alcohol.

$$\underset{\text{α, β-Dibromopropionic acid}}{BrCH_2-\overset{\overset{\displaystyle Br}{\displaystyle |}}{CH}-COOH} \xrightarrow{Zn/C_2H_5OH} \underset{\text{Acrylic acid}}{CH_2=CH-COOH} + ZnBr_2$$

(v)   From β-hydroxypropionic acid by distilling with zinc chloride. The required β-hydroxy propionic acid is obtained from ethylene.

$$\underset{\text{Ethylene}}{CH_2=CH_2} \xrightarrow{HOCl} \underset{\text{Ethylene chlorohydrin}}{HOCH_2-CH_2Cl} \xrightarrow{KCN} \underset{\text{Ethylene cyanohydrin}}{HOCH_2CH_2CN}$$

$$\xrightarrow{H^+} \underset{\text{β-Hydroxypropionic acid}}{HOCH_2-CH_2-COOH} \xrightarrow[-H_2O]{Zn\,Cl_2} \underset{\text{Acrylic acid}}{CH_2=CH-COOH}$$

(vi)   Acrylic acid is manufactured from hydrocarbonylation of acetylene in presence of nickel carbonyl.

$$\underset{\text{Acetylene}}{CH \equiv CH} + CO + H_2O \xrightarrow[HCl]{Ni(CO)_4} \underset{\text{Acrylic acid}}{CH_2=CH-COOH}$$

(vii)   From acrylonitrile by hydrolysis. Acrylonitrile is obtained either from ethylene oxide or acetylene as follows.

$$\underset{\substack{\text{Ethylene}\\ \text{oxide}}}{H_2\overset{\displaystyle O}{\overset{\displaystyle \diagup\!\!\diagdown}{C}}-CH_2} \xrightarrow{HCN} \underset{}{\overset{\displaystyle CH_2OH}{\underset{\displaystyle CH_2CN}{\overset{\displaystyle |}{\underset{\displaystyle |}{}}}}} \xrightarrow{\text{Conc. }H_2SO_4} \underset{\text{Acrylonitrile}}{CH_2=CH-CN} \xrightarrow{H_2O/H^+} \underset{\text{Acrylic acid}}{CH_2=CH-COOH}$$

$$\underset{\text{Acetylene}}{HC \equiv CH} + HCN \nearrow$$

This method is also used for industrial preparation of acrylic acid.

Properties: It is a colorless pungent smelling liquid (b.p. 142°) and is miscible with water in all proportions.

Acrylic acid gives the reactions of a carboxylic acid as well as an alkene.

As a carboxylic acid, it forms salts with metal carbonates and alkalis, reacts with phosphorus halides to form acryl chloride, which on treatment with ammonia gives the amide. Further it can be esterified to form ethyl acrylate.

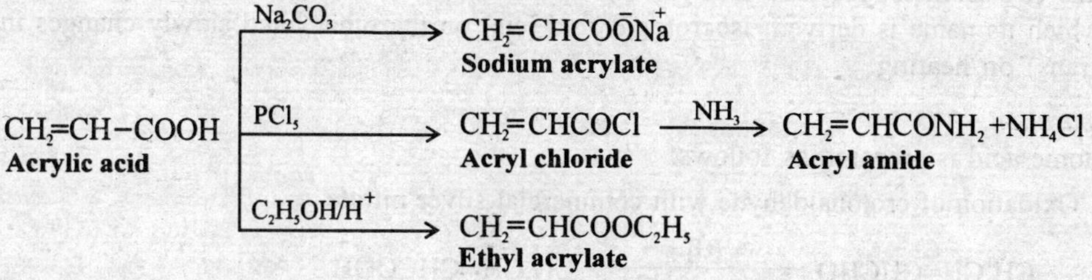

$$CH_2=CH-COOH \xrightarrow{Na_2CO_3} CH_2=CHCOO\overset{-}{O}\overset{+}{Na}$$
Sodium acrylate

$$\xrightarrow{PCl_5} CH_2=CHCOCl \xrightarrow{NH_3} CH_2=CHCONH_2 + NH_4Cl$$
Acryl chloride   **Acryl amide**

$$\xrightarrow{C_2H_5OH/H^+} CH_2=CHCOOC_2H_5$$
Ethyl acrylate

Acrylic acid (left side, $CH_2=CH-COOH$)

As an alkene, it can be hydrogenated and can form addition products with halogen and halogen acids. The addition of halogen acid to acrylic acid is according to Antimarkovnikov rule due to the inductive effect of carboxyl group.

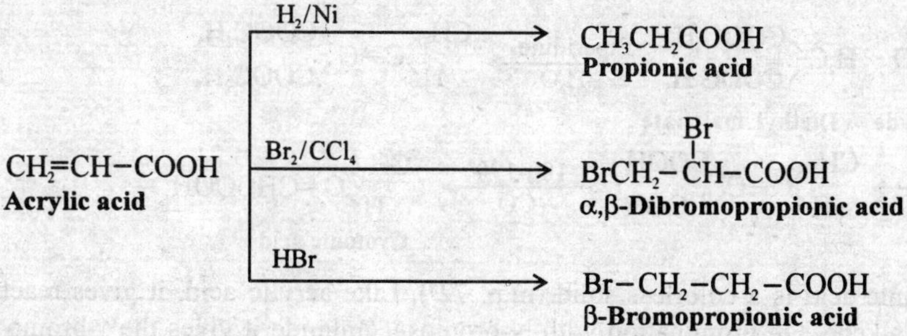

$$CH_2=CH-COOH \xrightarrow{H_2/Ni} CH_3CH_2COOH$$
Acrylic acid   **Propionic acid**

$$\xrightarrow{Br_2/CCl_4} BrCH_2-\underset{\underset{Br}{|}}{CH}-COOH$$
α,β-Dibromopropionic acid

$$\xrightarrow{HBr} Br-CH_2-CH_2-COOH$$
β-Bromopropionic acid

The double bond of acrylic acid may be hydroxylated to form a diol which on further oxidation gives oxalic acid.

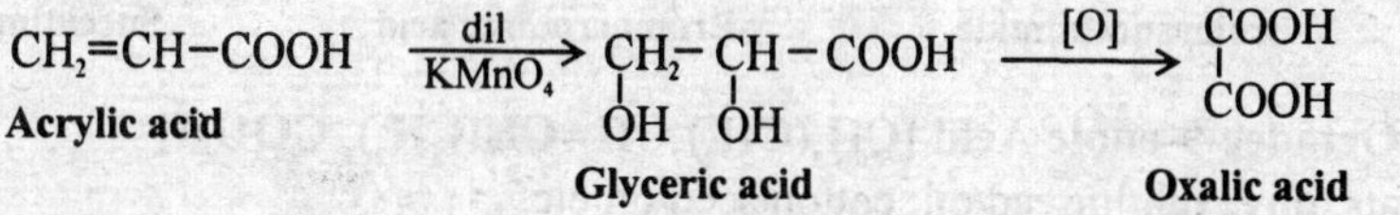

$$CH_2=CH-COOH \xrightarrow[KMnO_4]{dil} \underset{\underset{OH}{|}}{CH_2}-\underset{\underset{OH}{|}}{CH}-COOH \xrightarrow{[O]} \underset{\underset{COOH}{|}}{COOH}$$
Acrylic acid   **Glyceric acid**   **Oxalic acid**

On hydration in presence of $HgSO_4$ and H_2SO_4, acrylic acid forms β-hydroxypropionic acid (Antimarkovnikov addition).

$$CH_2=CH-COOH + H_2O \xrightarrow[HgSO_4]{H_2SO_4} HOCH_2-CH_2-COOH$$
Acrylic acid   **β-Hydroxypropionic acid**

On standing, acrylic acid polymerises to a solid.

Uses: It is used to make esters, which are important source of plastics and other valuable polymers. Acrylonitrile, a derivative of acrylic acid is used in the manufacture of acrilon, orlon and other acrylic fibers.

19.11.2 Crotonic Acid/But-2-enoic Acid [$CH_3CH=CH-COOH$]
Crotonic acid is the trans but-2-enoic acid. The cis isomer is known as *isocrotonic acid*

$$\underset{\underset{HOOC-H}{||}}{H-C-CH_3}$$
Crotonic Acid (trans)

$$\underset{\underset{H-C-COOH}{||}}{H-C-CH_3}$$
Isocrotonic Acid (cis)

Of the two isomers, crotonic acid (trans) is more stable and occurs as glyceryl ester in croton oil from which its name is derived. Isocrotonic acid (cis) on the other hand slowly changes into crotonic acid (trans) on heating.

Preparation:
Crotonic acid is prepared as follows.

(i)   Oxidation of crotonaldehyde with commercial silver nitrate.

$$CH_3CH=CHCHO \xrightarrow[\text{ammo. AgNO}_3]{[O]} CH_3CH=CHCOOH$$

   Crotonaldehyde   **Crotonic Acid**

(ii)   Reaction of acetaldehyde with diethyl malonate in presence of piperidine or diethylamine (*knoevenagel reaction*) forms crotonic acid.

$$CH_3-\overset{\overset{H}{|}}{C}=O + H_2C\diagdown^{COOC_2H_5}_{COOC_2H_5} \xrightarrow[-H_2O]{\text{Piperidine}} \overset{CH_3}{\diagdown}C=C\diagup^{COOC_2H_5}_{COOC_2H_5}$$

   Acetaldehyde   Diethyl malonate

$$\xrightarrow[\Delta]{\text{aqKOH}} \overset{CH_3}{\diagdown}C=C\diagup^{COOH}_{COOH} \xrightarrow[-CO_2]{180\text{-}190°} \overset{CH_3}{\diagdown}C=CHCOOH$$

                                                                          Crotonic acid

Properties: Crotonic acid is a colorless solid (m.p. 72°). Like acrylic acid, it gives reaction of both carboxylic acid and alkene. On bromination with N-bromosuccinimide it gives the γ-bromo compound.

$$CH_3CH=CHCOOH + \overset{CH_2-CO}{\underset{CH_2-CO}{\diagup}}N-Br \longrightarrow BrCH_2-CH=CH-COOH + \overset{CH_2-CO}{\underset{CH_2-CO}{\diagup}}N-H$$

   Crotonic acid   **N-Bromosuccinimide**   **γ-Bromocrotonic acid**   **Succinimide**

19.11.3   Oleic Acid/cis-Octadec-9-enoic Acid [$CH_3(CH_2)_7CH=CH(CH_2)_7$ COOH]

It occurs as glycerides in fats, olive oil, linseed oil, cotton seed oil etc.

Oleic acid is prepared by the hydrolysis of olive oil with sodium hydroxide, when a mixture of sodium salts of palmitic, stearic and oleic acid is obtained.

$$\begin{matrix} CH_3OCOC_{15}H_{31} \\ | \\ CHOCOC_{17}H_{35} \\ | \\ CH_2OCOC_{17}H_{33} \end{matrix} + 3NaOH \longrightarrow \begin{matrix} CH_2OH \\ | \\ CHOH \\ | \\ CH_2OH \end{matrix} + C_{15}H_{31}COONa + C_{17}H_{35}COONa + C_{17}H_{33}COONa$$

   Olive Oil   **Glycerol**   **Sodium palmitate   Sodium sterate   Sodium oleate**

The sodium salts of palmitic, stearic and oleic acids are water soluble. The aqueous solution is heated with aqueous lead acetate when the insoluble lead salts of the three acids are precipitated. The lead salts are filtered, dried and treated with ether when only the lead oleate dissolves. The etheral solution is filtered, evaporated and the residue heated with mineral acid to liberate oleic acid as oily layer. This is separated, dried and distilled under reduced pressure.

Oleic acid is colorless oily liquid (b.p. 286°, m.p. 16°). It is insoluble in water but soluble in ether and alcohol. It gives general reactions of carboxylic acids and also of alkenes. Thus, as alkenes, it can be hydrogenated to give stearic acid, and brominated to give dibromostearic acid. On oxidation with dilute potassium permanganate, it gives 9,10-dihydroxysteric acid, $CH_3(CH_2)_7-CHOH-CHOH(CH_2)_7COOH$. However, oxidation with acidified potassium permanganate solution gives a mixture of nonanoic and azelaic acids.

$$CH_3(CH_2)_7 CH=CH(CH_2)_7 COOH \xrightarrow[KMnO_4]{[O]} CH_3(CH_2)_7 COOH + HOOC (CH_2)_7 COOH$$

Oleic acid   Nonanoic acid   Azelaic acid

Ozonolysis of oleic acid followed by decomposition of the formed ozonide with zinc and water gives pelargonic aldehyde and azelaic acid half aldehyde.

$$CH_3(CH_2)_7 CH=CH(CH_2)_7 COOH \xrightarrow{O_3} CH_3(CH_2)_7-\overset{\overset{O}{\|}}{C}\underset{O-O}{}\overset{O}{C}-(CH_2)_7COOH$$

$$\longrightarrow CH_3(CH_2)_7CHO + OHC (CH_2)_7 COOH$$

Pelargonic aldehyde   **Azelaic acid half aldehyde**

Oleic acid exhibits geometrical isomerism.

$$\begin{array}{l} H-\underset{\|}{C}-(CH_2)_7CH_3 \\ H-C-(CH_2)_7COOH \end{array} \qquad\qquad \begin{array}{l} CH_3(CH_2)_7-\underset{\|}{C}-H \\ H-C-(CH_2)_7COOH \end{array}$$

Oleic acid (cis)   **Elaidic acid (trans)**
(m.p. 16°)   **(m.p. 51°)**

19.12 PROBLEMS

1. Give general methods of preparation of monocarboxylic acids.
2. Explain why formic acid is more reactive than acetic acid.
3. How will you synthesise n-hexanoic acid using Grignard reaction?
4. Using Haloform reaction how will you obtain propionic acid?
5. Explain the:
   (a) higher boiling points of carboxylic acids as compared to those of alcohols of similar molecular weight.
   (b) solubility of carboxylic acids in water.
6. Discuss the acidity of carboxylic acids. Arrange the following pair of acids in ascending order:
   (a) Formic acid, propionic acid, n-butyric acid, acetic acid.
   (b) Acetic acid, mono-, di- and trichloroacetic acids.
   (c) α, β and γ-chloroacetic acids.
   (d) Propionic acid, acrylic acid.
7. Discuss the mechanism of esterification. Justify the statement, that in the esterification reaction, the water is formed by the elimination of OH of carboxylic acid and H of alcohol and not vice-versa.
8. Give the products obtained in:
   (a) $CH_3COOAg + Cl_2 \longrightarrow$
   (b) $CH_3CH_2COOH + NH_3 \xrightarrow{\Delta}$
   (c) $CH_3COOH \xrightarrow{LiAlH_4}$
9. Discuss the mechanism of Kolbe reaction, and Hell-Volhard-Zelinsky reaction.
10. Starting with propionic acid, how will you obtain the corresponding α-hydroxy, α-amino and α-cyano acids.
11. Give the industrial method of preparation of formic and acetic acid. How are they obtained in 100% pure form.
12. Give the structures of:
    (a) 2-Methylprop-2-enoic acid
    (b) cis-2-Methylbut-2-enoic acid
    (c) cis-Octadec-9-enoic acid

20

Dicarboxylic Acids

20.1 INTRODUCTION

The dicarboxylic acids, as the name implies, contain two carboxyl group. They may be saturated or unsaturated. The saturated dicarboxylic acids have the general formula $C_nH_{2n}(COOH)_2$ (for oxalic acid, $n = 0$).

20.2 NOMENCLATURE

Most of the saturated dicarboxylic acids are known by their common names which indicate the source from which they have been isolated. For example, oxalic acid is obtained from plants of oxalis group. According to IUPAC system, the suffix '*dioic acid*' is added after the name of the parent alkane. For example $(COOH)_2$ is called ethanedioic acid. The names of some important saturated dicarboxylic acids are:

Formula	Common Name	IUPAC Name
$(COOH)_2$	Oxalic acid	Ethanedioic acid
$CH_2(COOH)_2$	Malonic acid	Propanedioic acid
$(CH_2)_2(COOH)_2$	Succinic acid	Butanedioic acid
$(CH_2)_3(COOH)_2$	Glutaric acid	Pentanedioic acid
$(CH_2)_4(COOH)_2$	Adipic acid	Hexanedioic acid
$(CH_2)_5(COOH)_2$	Pimelic acid	Heptanedioic acid
$(CH_2)_6(COOH)_2$	Suberic acid	Octanedioic acid
$(CH_2)_7(COOH)_2$	Azelaic acid	Nonanedioic acid
$(CH_2)_8(COOH)_2$	Sebacic acid	Decanedioic acid

Alternatively, in the IUPAC system, the position of the two carboxyl groups is indicated by arabic numerals if necessary. However, in the trivial system, the position of the substituents is indicated by the Greek letters.

Formula	Common Name	IUPAC Name
$HOOCCH_2CH_2CH_2COOH$	Glutaric acid	1,3-Propanedicarboxylic acid or Propane-1,3-dicarboxylic acid or Pentanedioic acid
$HOOCCH_2CH_2CH-COOH$ Br	α-Bromoglutaric acid	2-Bromopentanedioic acid
$HOOC-CH-CH_2-CH_2-CH-COOH$ CH_3 Cl	α-chloro-α′-methyladipic acid	2-Chloro-5-methylhexanedioic acid

20.3   GENERAL METHODS OF PREPARATION

(1) *Oxidation of Glycols*: The glycols or dihydric alcohols on oxidation give dicarboxylic acids.

$$\begin{array}{c} CH_2OH \\ | \\ CH_2OH \end{array} \xrightarrow[KMnO_4]{[O]} \begin{array}{c} COOH \\ | \\ COOH \end{array}$$

Ethylene glycol   **Oxalic acid**

This method is of limited use since higher glycols are difficult to obtain.

(2) *Hydrolysis of Dinitriles:* The dinitriles or cyano-monocarboxylic acids with boiling dilute hydrochloric acid give dicarboxylic acids.

$$(CH_2)_n \begin{array}{c} \nearrow CN \\ \searrow CN \end{array} \xrightarrow[Hydrolysis]{HCl,\ \Delta} (CH_2)_n \begin{array}{c} \nearrow COOH \\ \searrow COOH \end{array} + NH_4Cl$$

Dinitrile   **Dicarboxylic acid**

$$\begin{array}{c} CH_2CN \\ | \\ CH_2CN \end{array} \xrightarrow{Hydrolysis} \begin{array}{c} CH_2COOH \\ | \\ CH_2COOH \end{array} + NH_4Cl$$

Ethylene cyanide   **Succinic acid**

$$(CH_2)_n \begin{array}{c} \nearrow CN \\ \searrow COOH \end{array} \xrightarrow{Hydrolysis} (CH_2)_n \begin{array}{c} \nearrow COOH \\ \searrow COOH \end{array} + NH_4Cl$$

Cyanocarboxylic acid   **Dicarboxylic acid**

$$CH_2 \begin{array}{c} \nearrow CN \\ \searrow COOH \end{array} \xrightarrow{Hydrolysis} CH_2 \begin{array}{c} \nearrow COOH \\ \searrow COOH \end{array} + NH_4Cl$$

Cyanoacetic acid   **Malonic acid**

(3) *By the Action of Silver or Zinc on Halogenated Monocarboxylic Esters:* A reaction similar to Wurtz reaction takes place to form dicarboxylic acid.

$$2Ag + \begin{array}{c} ICH_2CH_2COOC_2H_5 \\ ICH_2CH_2COOC_2H_5 \end{array} \xrightarrow{-2AgI} \begin{array}{c} CH_2-CH_2-COOC_2H_5 \\ | \\ CH_2-CH_2-COOC_2H_5 \end{array} \xrightarrow{H_2O/H^+} \begin{array}{c} CH_2-CH_2-COOH \\ | \\ CH_2-CH_2-COOH \end{array}$$

Ethyl-β-iodopropionate

(4) *Electrolysis of Potassium Salt of Carboxylic Acids*: The electrolysis of aqueous solution of potassium salt of lower acids to produce dicarboxylic acids is known as Crum-Brown and Walker method.

$$\begin{array}{c} KOOC-CH_2-COOC_2H_5 \\ KOOC-CH_2-COOC_2H_5 \end{array} + 2H_2O \xrightarrow{Electrolysis} \begin{array}{c} CH_2-COOC_2H_5 \\ | \\ CH_2-COOC_2H_5 \end{array} + 2CO_2 + 2KOH + H_2$$

Pot. ethyl malonate   **Diethyl succinate**

$$\begin{array}{c} CH_2-COOC_2H_5 \\ | \\ CH_2-COOC_2H_5 \end{array} \xrightarrow{Hydrolysis} \begin{array}{c} CH_2COOH \\ | \\ CH_2COOH \end{array}$$

Diethyl succinate   **Succinic acid**

(5) *From Dihaloalkanes:* The Grignard reagent obtained from dihaloalkanes on treatment with carbon dioxide followed by hydrolysis give the corresponding dicarboxylic acids.

$$\begin{array}{c} CH_2-Br \\ | \\ CH_2-Br \end{array} \xrightarrow[\text{dry ether}]{Mg} \begin{array}{c} CH_2-MgBr \\ | \\ CH_2-MgBr \end{array} \xrightarrow{CO_2} \begin{array}{c} CH_2-COOMgBr \\ | \\ CH_2-COOMgBr \end{array}$$

Ethylene   **Grignard reagent**
bromide

$$\xrightarrow{H_2O/H^+} \begin{array}{c} CH_2-COOH \\ | \\ CH_2-COOH \end{array} + 2Mg(Br)OH$$

Succinic acid

(6) *Oxidation of Cyclic Ketones:* The cyclic ketones undergo ring opening on oxidation with nitric acid to form dicarboxylic acids.

$$\text{Cyclohexanone} \xrightarrow[HNO_3]{[O]} HOOC(CH_2)_4COOH$$

Cyclohexanone   **Adipic acid**

(7) *Oxidation of Unsaturated Acids:* Unsaturated acid like oleic acid on oxidation with nitric acid or permanganate give pelargonic acid and azelaic acid.

$$CH_3(CH_2)_7CH=CH(CH_2)_7COOH \xrightarrow[KMnO_4]{[O]} CH_3(CH_2)_7\,COOH + HOOC\,(CH_2)_7\,COOH$$

Oleic acid   **Pelargonic acid**   **Azelaic acid**
                        (Nonanoic acid)   **(Nonanedioic acid)**

(8) *From Diethyl Malonate or Acetoacetic Esters:* Diethyl malonate on treatment with sodium ethoxide gives monosodium salt, which on treatment with dibromoalkanes followed by hydrolysis give dicarboxylic acid.

$$CH_2(COOC_2H_5)_2 \xrightarrow{NaOEt} NaCH(COOC_2H_5)_2 \xrightarrow{BrCH_2-CH_2Br}$$

Diethyl malonate   **Monosodium salt**

$$\begin{array}{c} CH_2-CH(COOC_2H_5)_2 \\ | \\ CH_2-CH(COOC_2H_5)_2 \end{array} \xrightarrow{\text{Hydrolysis}} \begin{array}{c} CH_2-CH(COOH)_2 \\ | \\ CH_2-CH(COOH)_2 \end{array} \xrightarrow[-2CO_2]{\Delta} \begin{array}{c} CH_2-CH_2COOH \\ | \\ CH_2-CH_2COOH \end{array}$$

Adipic acid

The mono sodium salt of ethyl acetoacetate is treated with ethylchloroacetate and the α-substituted ethyl acetoacetate on hydrolysis gives dicarboxylic acid.

$$CH_3COCH_2COOC_2H_5 \xrightarrow{C_2H_5ONa} CH_2-CO-\underset{\underset{Na}{|}}{CH}-COOC_2H_5 \xrightarrow{ClCH_2COOC_2H_5}$$

Ethyl acetoacetate

Mono sodium salt

$$CH_3-CO-\underset{\underset{CH_2-COOC_2H_5}{|}}{CH}-COOC_2H_5 \xrightarrow{\text{Acid hydrolysis}} CH_3COOH + \begin{array}{c} CH_2COOH \\ | \\ CH_2COOH \end{array} + 2C_2H_5OH$$

Acetic acid   **Succinic acid**

(9) *Reduction of Unsaturated Dicarboxylic Acids:* Maleic acid on reduction forms succinic acid.

$$\begin{array}{c} H-C-COOH \\ \parallel \\ H-C-COOH \end{array} \xrightarrow{H_2/\text{Catalyst}} \begin{array}{c} CH_2-COOH \\ | \\ CH_2-COOH \end{array}$$

Maleic acid   **Succinic acid**

20.4 PHYSICAL PROPERTIES

The dicarboxylic acids are crystalline solids. The lower members of the series are sufficiently soluble in water but are slightly soluble in organic solvents. The solubility in water keeps on decreasing with the increase in molecular weight.

The melting points of acids with even number of carbon atoms are higher than those containing odd number of carbon atoms. This is attributed to zig-zag arrangements of carbon atoms with the two carboxyl groups on the same side in odd number acids but on opposite sides in even number acids.

$$HOOC \overset{109.5°}{\diagup} COOH \qquad \overset{109.5°}{\diagup} COOH \qquad HOOC \overset{109.5°}{\diagup} COOH$$

The *acid strength* of dicarboxylic acids decreases with increase in molecular weight. They dissociate in two steps, the dissociation constant (K_1) of the first carboxyl group being greater than that of the second (K_2). The higher value of K_1 is due to -I effect of one carboxyl group on the other. This effect decreases as the carbon chain increases. The lower value of K_2 is due to +I effect of carboxylate anion.

The following table gives the physical constants of dicarboxylic acids, $HOOC(CH_2)_nCOOH$.

Table 20.1: Physical Constants of Dicarboxylic Acid

Name	n	Solubility g/100 g H_2O at 20°	m.p. °C	Dissociation constants (25°)	
				$K_1 \times 10^5$	$K_2 \times 10^5$
Oxalic acid	0	9	189	5400	5.20
Malonic acid	1	74	136	140	0.20
Succinic acid	2	6	185	6.4	0.23
Glutaric acid	3	64	98	4.5	0.38
Adipic acid	4	2	151	3.7	0.39
Pimelic acid	5	5	105	3.1	0.37
Suberic acid	6	0.2	144	3.0	0.39
Azelaic acid	7	0.3	106	2.2	0.39
Sebacic acid	8	0.1	134	2.6	0.40

20.5 CHEMICAL PROPERTIES

1. *Reactions of the Carboxyl Group*: Due to the presence of two carboxyl groups these form two series of salts, esters, amides and acid derivatives.

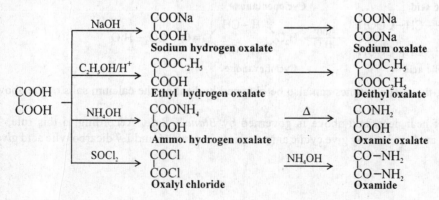

The two carboxyl groups can also be made to react successively with different reagents to give mixed derivatives. These are useful intermediates for organic synthesis.

$$\begin{array}{c} COOH \\ | \\ COOH \end{array} \xrightarrow[H^+]{C_2H_5OH} \begin{array}{c} COOC_2H_5 \\ | \\ COOH \end{array} \xrightarrow{SOCl_2} \begin{array}{c} COOC_2H_5 \\ | \\ COCl \end{array}$$

Oxalic acid   Ethyl hydrogen
                     oxalate

2.   *Action of Heat*: The nature of the product formed depends on the number of carbon atoms between the two carboxyl groups.

(i)   Dicarboxylic acids in which the two carboxyl groups are linked directly to each other or to the same carbon atom on heating loose a molecule of CO_2 to form monocarboxylic acids.

$$\begin{array}{c} COOH \\ | \\ COOH \end{array} \xrightarrow[-CO_2]{190°} HCOOH$$

Oxalic acid   Formic acid

$$CH_2 \begin{array}{c} {}^{\nearrow COOH} \\ {}_{\searrow COOH} \end{array} \xrightarrow[-CO_2]{140°} CH_3COOH$$

Malonic acid   Acetic acid

(ii)   Dicarboxylic acids, in which the two carboxyl groups are separated by 2 or 3 carbon atoms on heating loose a molecule of water to form anhydrides. The anhydride formation is facilitated if the heating is done with acetic anhydride.

$$\begin{array}{c} CH_2COOH \\ | \\ CH_2COOH \end{array} \xrightarrow[-H_2O]{100-150°} \begin{array}{c} CH_2CO \\ | \quad\quad {}^{\searrow}O \\ CH_2CO {}^{\nearrow} \end{array}$$

Succinic acid   Succinic anhydride

$$CH_2 \begin{array}{c} {}^{\nearrow CH_2COOH} \\ {}_{\searrow CH_2COOH} \end{array} \xrightarrow[-H_2O]{100-150°} H_2C \begin{array}{c} {}^{\nearrow CH_2 - CO} \\ \quad\quad\quad\quad {}^{\searrow}O \\ {}_{\searrow CH_2 - CO} \end{array}$$

Glutaric acid   Gultaric anhydride

(iii)   Dicarboxylic acid, in which the two carboxyl groups are separated by 4 or 5 carbon atoms on heating alone or distilling with acetic anhydride give cyclic ketones.

$$\begin{array}{c} CH_2 - CH_2COOH \\ | \\ CH_2 - CH_2COOH \end{array} \xrightarrow[-H_2O]{300°} \begin{array}{c} CH_2 - CH_2 \\ | \quad\quad\quad\quad {}^{\searrow}C=O + CO_2 + H_2O \\ CH_2 - CH_2 {}^{\nearrow} \end{array}$$

Adipic acid   Cyclopentanone

$$CH_2 \begin{array}{c} {}^{\nearrow CH_2 - CH_2 - COOH} \\ {}_{\searrow CH_2 - CH_2 - COOH} \end{array} \xrightarrow[-H_2O]{300°} H_2C \begin{array}{c} {}^{\nearrow CH_2 - CH_2} \\ \quad\quad\quad\quad\quad {}^{\searrow}C=O + CO_2 + H_2O \\ {}_{\searrow CH_2 - CH_2} \end{array}$$

Pimelic acid   Cyclohexanone

Alternatively, the cyclic ketones can also be obtained by heating the calcium salts of the above acids at 300-320°.

The action of heat described above is governed by *Blanc's Rule*. According to this rule, 1,4 and 1,5 dicarboxylic acids on heating give cyclic anhydrides, where as 1,6 and 1,7 dicarboxylic acid give cyclic ketones.

3. *Oxidation:* In general, the dicarboxylic acids are stable to oxidising agents. Oxalic acid, however, is an exception. It can be oxidised to give carbon dioxide and water.

$$\begin{matrix} COOH \\ | \\ COOH \end{matrix} \xrightarrow[KMnO_4/H^+]{[O]} 2CO_2 + H_2O$$

Oxalic acid

4. *Acyloin Condensation*: Higher dicarboxylic acids on heating with sodium in xylene followed by hydrolysis give acyloins.

$$(CH_2)_n \begin{matrix} COOH \\ \\ COOH \end{matrix} \xrightarrow[\Delta]{Na/Xylene} (CH_2)_R \begin{matrix} C-O^- Na^+ \\ \| \\ C-O^- Na^+ \end{matrix} \xrightarrow[\Delta]{H^+} (CH_2)_R \begin{matrix} C-OH \\ \| \\ C-OH \end{matrix} \rightleftharpoons (CH_2)_R \begin{matrix} C=O \\ | \\ CHOH \end{matrix}$$

Acyloin

20.6 INDIVIDUAL MEMBERS

20.6.1 Oxalic Acid/Ethanedioic Acid $(COOH)_2$

Oxalic acid is one of the earliest dicarboxylic acid known. It occurs as potassium hydrogen oxalate in the wood sorrel (*Oxalis acetozella*) and as calcium oxalate in some plants such as spinach, cabbage, tomatoes etc. In urine oxalic acid is present as ammonium oxalate.

Preparation:

(i) In the laboratory, oxalic acid is prepared by oxidation of cane sugar with nitric acid in presence of a catalyst.

$$C_{12}H_{22}O_{11} + 18[O] \xrightarrow{HNO_3} 6 \begin{matrix} COOH \\ | \\ COOH \end{matrix} + 5H_2O$$

Sucrose   **Oxalic acid**

(ii) *From Sodium Formate*: On a commercial scale, oxalic acid is obtained by heating sodium formate at 370-400°. The formed sodium oxalate is converted into calcium oxalate by treatment with lime from which oxalic acid is obtained by reaction with sulphuric acid. The sodium formate required is obtained by passing carbon monoxide over powdered sodium hydroxide.

$$CO + NaOH \xrightarrow[8-10\ atm]{200°} HCOONa$$

Sod. formate

$$2HCOONa \xrightarrow{370-400°} \begin{matrix} COONa \\ | \\ COONa \end{matrix} + H_2$$

Sod. formate   **Sod. oxalate**

$$\begin{matrix} COONa \\ | \\ COONa \end{matrix} \xrightarrow{Ca(OH)_2} \begin{matrix} COO \\ | \\ COO \end{matrix} Ca + 2NaOH$$

Sod. oxalate   **Calcium oxalate**

$$\begin{matrix} COO \\ | \\ COO \end{matrix} Ca + H_2SO_4 \longrightarrow \begin{matrix} COOH \\ | \\ COOH \end{matrix} + CaSO_4$$

Calcium oxalate   **Oxalic acid**

(iii)   *From carbon dioxide and sodium*: The direct combination of carbon dioxide and metallic sodium forms sodium oxalate.

$$2CO_2 + 2Na \xrightarrow{360°} \begin{matrix} COONa \\ COONa \end{matrix}$$
Sodium oxalate

Oxalic acid is obtained from sodium oxalate as described above.

Physical Properties: Oxalic acid is a colorless crystalline solid containing two molecules of water of crystallisation.The anhydrous acid (m.p. 189.5°) is obtained by heating carefully the hydrated acid at 150°. It is soluble in water and alcohol, but is practically insoluble in anhydrous ether. It is poisonous.

Chemical Properties: Oxalic acid is the strongest of all the dicarboxylic acids. It is even stronger than phosphoric acid.

1.   *Formation of Salts:* As already stated oxalic acid forms two series of salts, esters, amides and acid halides (see general chemical properties of dicarboxylic acids).

2.   *Action of Heat:* It has already been mentioned that oxalic acid on heating at 190° loses a molecule of carbon dioxide to form formic acid.

3.   *Action of Concentrated Sulphuric Acid:* On heating with concentrated sulphuric acid, oxalic acid decomposes to give carbon monoxide and carbon dioxide.

$$(COOH)_2 \xrightarrow{\text{cons } H_2SO_4} H_2O + CO_2 + CO$$

This method is used to prepare carbon monoxide in lab.

4.   *Reaction with Glycerol:* Oxalic acid on heating with glycerol gives formic acid or allyl alcohol depending on the reaction conditions (see Sec. 19.8.1, preparation of formic acid).

5.   *Oxidation:* Oxalic acid can be readily oxidised by heating with acidic potassium permanganate to give carbon dioxide and water.

$$(COOH)_2 \xrightarrow[KMnO_4, H_2SO_4]{[O]} 2CO_2 + H_2$$

6.   *Reduction:* Reduction of oxalic acid with zinc and dilute sulphuric acid gives glycollic acid. However, electrolytic reduction using lead cathode forms glycollic and glyoxylic acids.

$$\begin{matrix} COOH \\ | \\ COOH \end{matrix} \left[\begin{matrix} \xrightarrow{Zn + H_2SO_4} & \begin{matrix} CH_2OH \\ | \\ COOH \end{matrix} + H_2O \\ & \textbf{Glycollic acid} \\ \xrightarrow[\text{reduction}]{\text{Electrolytic}} & \begin{matrix} CH_2OH \\ | \\ COOH \end{matrix} + \begin{matrix} CHO \\ | \\ COOH \end{matrix} \end{matrix} \right.$$

Oxalic acid   **Glycollic   Giyoxylic**
                                             acid   acid

7.   *Reaction with Ethylene Glycol:* Oxalic acid on heating with ethylene glycol gives a cyclic compound, ethylene oxalate.

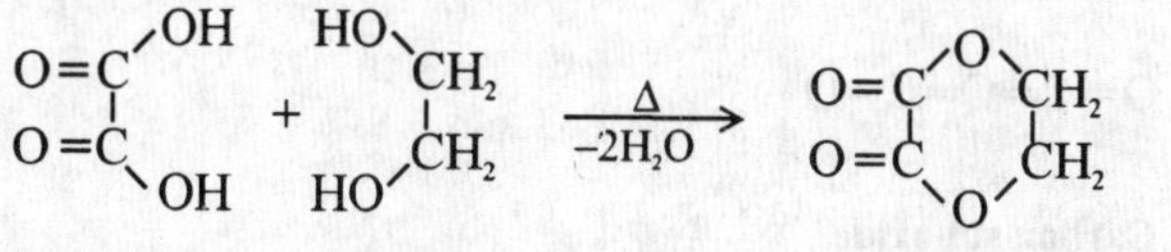

Oxalic acid   Ethylene glycol   Ethylene oxalate

Uses: Oxalic acid is used as a mordant in dyeing and calico printing and in the manufacture of dyes, inks, formic acid and allyl alcohol. It is used as a reagent and also as a standard in volumetric analysis.

Tests:

(1) It decolorises acidified potassium permanganate solution on warming.

(2) On heating with concentrated sulphuric acid it gives a mixture of carbon monoxide and carbon dioxide.

(3) A neutral solution of oxalic acid on treatment with calcium chloride solution yields a white precipitate of calcium oxalate, which is insoluble in acetic acid.

20.6.2   Malonic Acid/Propanedioic Acid [$CH_2(COOH)_2$]

It was first obtained from malic acid (hydroxy succinic acid found in apples) and thus the name is derived.

Preparation:

(1)   It is usually prepared by heating potassium chloroacetate with potassium cyanide followed by hydrolysis of the formed potassium cyanoacetate with hydrochloric acid.

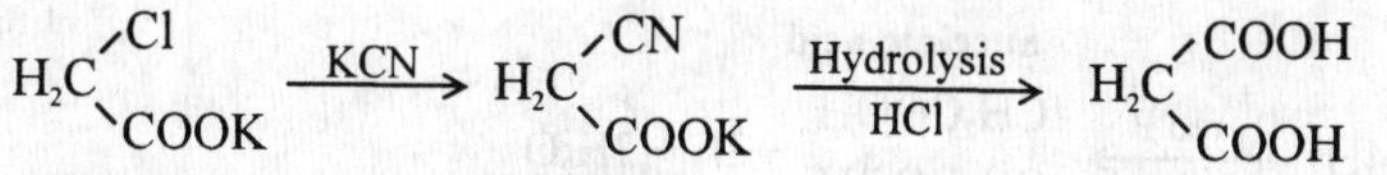

Potassium chloroacetate   Potassium cyanoacetate   Malonic acid

Since malonic acid is soluble in water it is obtained by ether extraction. Potassium chloroacetate required for the preparation is obtained by the action of potassium bicarbonate on chloroacetic acid.

(2)   Oxidation of malic acid with acidic dichromate produces malonic acid.

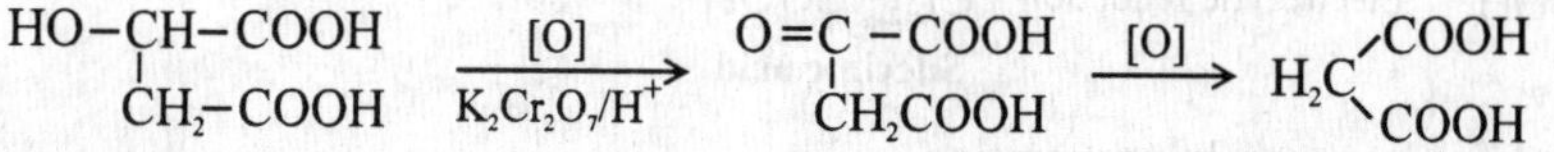

Malic acid   Oxalacetic acid   Malonic acid

Properties

Malonic acid is the colorless, crystalline solid (m.p. 136°). It is readily soluble in water and alcohol. It gives all the usual reactions of a dicarboxylic acids.

Chemical Properties

(i)   *Action of Heat*: On heating at 140° malonic acid decomposes into carbon dioxide and acetic acid.

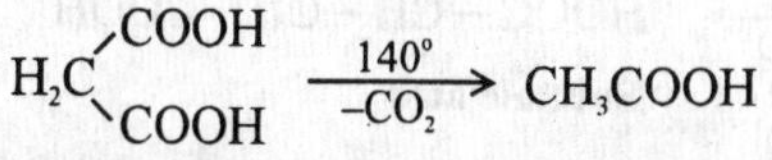

Malonic acid   Acetic acid

(ii)   *Action of Phosphorus Pentoxide*: On heating with phosphorus pentoxide malonic acid gives a small amount of carbon suboxide.

$$O=\overset{\underset{|}{OH}}{\underset{}{C}}-\overset{\underset{|}{H}}{\overset{|}{C}}-\overset{\underset{|}{H}}{\underset{}{C}}=O \xrightarrow[-2H_2O]{P_2O_5,\Delta} O=C=C=C=O$$

Malonic acid   **Carbon suboxide**

Uses: Diethyl malonate commonly known as malonic ester finds great synthetic applications.

20.6.3   Succinic Acid/Butanedioic Acid [HOOCCH$_2$CH$_2$COOH]

It was first obtained by distillation of amber (known as succinum in Latin) and hence its name. It is also formed during fermentation of sugars .

Preparation

(1)   From ethylene bromide by treatment with potassium cyanide followed by hydrolysis of the formed ethylene cyanide with hydrochloric acid.

$$\begin{matrix} CH_2Br \\ | \\ CH_2Br \end{matrix} \xrightarrow{KCN} \begin{matrix} CH_2CN \\ | \\ CH_2CN \end{matrix} \xrightarrow{H_2O/H^+} \begin{matrix} CH_2COOH \\ | \\ CH_2COOH \end{matrix}$$

Ethylene   **Ethylene**   **Succinic acid**
bromide   **cyanide**

(2)   From malic or tartaric acid by heating in a sealed tube with red phosphorus and hydriodic acid.

$$\begin{matrix} CH(OH)COOH \\ | \\ CH_2COOH \end{matrix} + 2HI \xrightarrow[\Delta]{red\ P} \begin{matrix} CH_2COOH \\ | \\ CH_2COOH \end{matrix} + I_2 + H_2O$$

Malic acid   **Succinic acid**

$$\begin{matrix} CH(OH)COOH \\ | \\ CH(OH)COOH \end{matrix} + 4HI \xrightarrow[\Delta]{red\ P} \begin{matrix} CH_2COOH \\ | \\ CH_2COOH \end{matrix} + 2I_2 + 2H_2O$$

Tartaric acid   **Succinic acid**

(3)   From maleic acid by catalytic reduction or electrolytic reduction.

$$\begin{matrix} H-C-COOH \\ \parallel \\ H-C-COOH \end{matrix} \xrightarrow[\text{Electrolytic reduction}]{\text{Catalytic reduction}} \begin{matrix} CH_2-COOH \\ | \\ CH_2-COOH \end{matrix}$$

Maleic acid   **Succinic acid**

This method is used for large scale preparation.

(4)   From malonic ester by treatment with sodium ethoxide and ethyl chloroacetate followed by hydrolysis.

$$H_2C \overset{COOC_2H_5}{\underset{COOC_2H_5}{\diagdown}} \xrightarrow[\text{2. ClCH}_2\text{COOC}_2\text{H}_5]{\text{1. C}_2\text{H}_5\text{ONa}} C_2H_5OOC-CH_2-CH \overset{COOC_2H_5}{\underset{COOC_2H_5}{\diagdown}}$$

Diethyl malonate

$$\xrightarrow{\text{Hydrolysis}} HOOC-CH_2-CH \overset{COOH}{\underset{COOH}{\diagdown}} \xrightarrow[-CO_2]{\Delta} HOOC-CH_2-CH_2-COOH$$

                                                                                  Succinic acid

Properties: Succinic acid is a white crystalline solid (m.p. 185°), moderately soluble in water and alcohol, but sparingly soluble in ether.

It gives the general reactions of dicarboxylic acids mentioned earlier.

1. *Action of Heat*: On heating, succinic acid sublimes. However, a small amount of anhydride is also obtained on strong heating. Heating with acetyl chloride gives excellent yield of the anhydride.

$$CH_2-COOH \quad \xrightarrow{-H_2O} \quad \begin{array}{l} CH_2-CO \\ CH_2-CO \end{array} \! \! O$$

Succinic acid   **Succinic anhydride**

Succinic anhydride can be easily converted into tetramethylene glycol by its reduction (sodium and alcohol) followed by reduction of the formed γ-butyrolactone.

$$\begin{array}{l} CH_2-CO \\ CH_2-CO \end{array} \! \! O \quad \xrightarrow[4H;-H_2O]{Na-C_2H_5OH} \quad \begin{array}{l} CH_2-CO \\ CH_2-CH_2 \end{array} \! \! O \quad \xrightarrow{4H} \quad \begin{array}{l} CH_2-CH_2OH \\ CH_2-CH_2OH \end{array}$$

Succinic anhydride   **γ-Butyrolactone**   **1,4-Butanediol**

2. *Action of Ammonia*: On heating with ammonia, succinic acid (or its anhydride or ester) is converted into succinimide.

$$CH_2COOH \quad \xrightarrow{\Delta} \quad \begin{array}{l} CH_2-CO \\ CH_2-CO \end{array} \! \! O \quad \xrightarrow[-H_2O]{NH_3, \Delta} \quad \begin{array}{l} CH_2-CO \\ CH_2-CO \end{array} \! \! NH$$

Succinic acid   **Succinic anhydride**   **Succinimide**

Succinimide on reaction with bromine and alkali yields N-bromosuccinimide, an important brominating agent.

$$\begin{array}{l} CH_2-CO \\ CH_2-CO \end{array} \! \! NH + Br_2 + NaOH \quad \longrightarrow \quad \begin{array}{l} CH_2-CO \\ CH_2-CO \end{array} \! \! NBr + NaBr + H_2O$$

Succinimide   **N-Bromosuccinimide**

Uses: It is used in medicine and in the manufacture of dyes, perfumes and polyester resins.

20.6.4 Glutaric Acid/Pentanedioic Acid [HOOC(CH₂)₃COOH]

Preparation

(1) Hydrolysis of trimethylene cyanide forms glutaric acid.

$$CH_2 \! \! \begin{array}{l} CH_2-CN \\ \\ CH_2-CN \end{array} \quad \xrightarrow{Hydrolysis} \quad CH_2 \! \! \begin{array}{l} CH_2-COOH \\ \\ CH_2-COOH \end{array}$$

Trimethylene cyanide   **Glutaric acid**

(2) Oxidation of cyclopentanone with nitric acid forms glutaric acid.

$$\begin{array}{l} CH_2-CH_2 \\ CH_2-CH_2 \end{array} \! \! C=O \quad \xrightarrow[HNO_3]{[O]} \quad CH_2 \! \! \begin{array}{l} CH_2-COOH \\ \\ CH_2-COOH \end{array}$$

Cyclopentanone   **Glutaric acid**

(3) Diethyl malonate produces glutaric acid by following sequence of reactions.

$$CH_2 \! \! \begin{array}{l} COOC_2H_5 \\ \\ COOC_2H_5 \end{array} \xrightarrow[2.\ CH_2=CHCOOC_2H_5]{1.\ C_2H_5ONa} \begin{array}{l} CH_2-CH_2-COOC_2H_5 \\ CH(COOC_2H_5)_2 \end{array} \xrightarrow[2.\ H^+]{1.\ KOH/H_2O} CH_2 \! \! \begin{array}{l} CH_2COOH \\ \\ CH_2COOH \end{array}$$

Diethyl malonate   **Glutaric acid**

Properties: Glutaric acid is a crystalline solid (m.p. 98°). It gives the general reactions of dicarboxyli acids. On heating alone or with acetic anhydride or thionyl chloride it gives glutaric anhydride (see propertie of dicarboxylic acids).

20.6.5 Adipic Acid/Hexanedioic Acid [HOOCCH$_2$CH$_2$CH$_2$CH$_2$COOH]

It was first obtained by oxidation of fats (Latin-adeps meaning fat) and hence its name.

Preparation

(1) *Oxidation of Cyclohexanol:* It is prepared on a commercial scale by oxidation of cyclohexano which is obtained by catalytic hydrogenation of phenol.

Phenol	**Cyclohexanol**	**Adipic acid**

(2) *From Tetrahydrofuran*: Adipic acid is also manufactured from tetrahydrofuran by reaction wit carbon monoxide and water.

Properties: Adipic acid is a crystalline solid (m.p. 150°). It gives all the reactions of dicarboxylic acids. O heating (280-300°) in presence of barium hydroxide, it gives cyclopentanone (see general properties o dicarboxylic acids).

Uses: Its most important use is in the preparation of polyesters and manufacture of nylon.

20.7 UNSATURATED DICARBOXYLIC ACIDS

Maleic acid and fumaric acids are the two simplest and most important unsaturated dicarboxylic acids. Thes are represented as:

$$H-C-COOH$$
$$\| $$
$$H-C-COOH$$
Maleic acid (cis)

$$H-C-COOH$$
$$\|$$
$$HOOC-C-H$$
Fumaric acid (trans)

These two acids are the classical examples for explaining geometrical isomerism. Maleic acid is th cis-isomer (i.e., both COOH groups on the same side of double bond). Thus, the other isomer (in whicl the two carboxyl groups are on the opposite sides of the double bond) is fumaric acid. The anhydrid of fumaric acid is unknown. However, on heating for longer periods fumaric acid is converted int maleic acid, which in turn forms maleic anhydride. This transformation occurs by rotation about the doubl bond followed by elimination of a molecule of water.

Looking at the structures of maleic and fumaric acid it is believed that both the acids may be of equa strength since they have identical structure. But it is found that maleic acid is a much stronger acid tha fumaric acid. This is explained by the fact that the maleate anion left after the removal of a proton is stabilise by H-bonding due to the close proximity of the second carboxyl group; on the other hand, this is not possibl with fumarate anion.

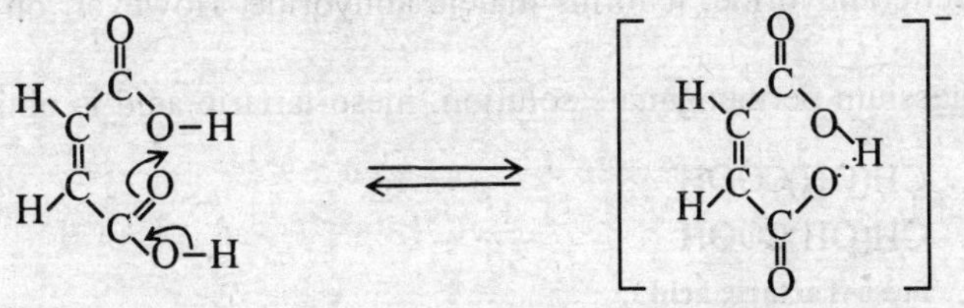

Maleic acid   Maleate ion (H-bonding)

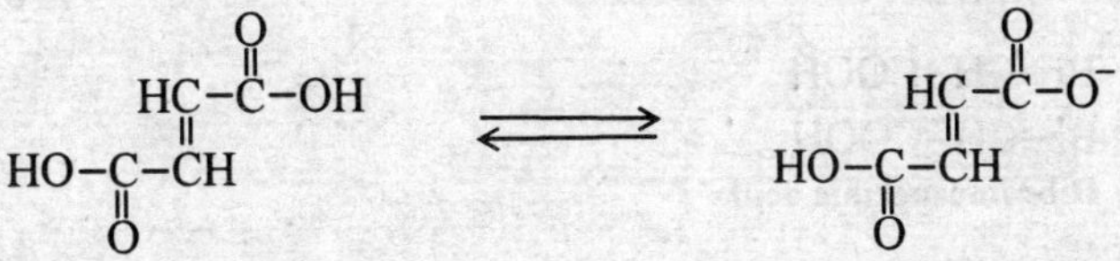

Fumaric acid   Fumarate ion (No H-bonding)

20.7.1   Maleic Acid/cis-Ethylene-1,2-dicarboxylic Acid

It is synthesised as follows:

1. *From Malic acid*: On heating at 250-260° malic acid forms maleic anhydride, which on hydrolysis gives maleic acid.

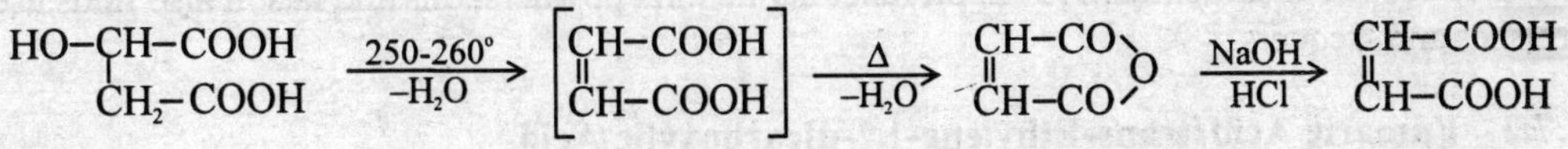

Malic acid   Maleic acid   Maleic anhydride   Maleic acid

2. *From Benzene*: Maleic acid is obtained commercially by the oxidation of benzene in presence of vanadium pentoxide at 450°. The formed anhydride is hydrolysed with water.

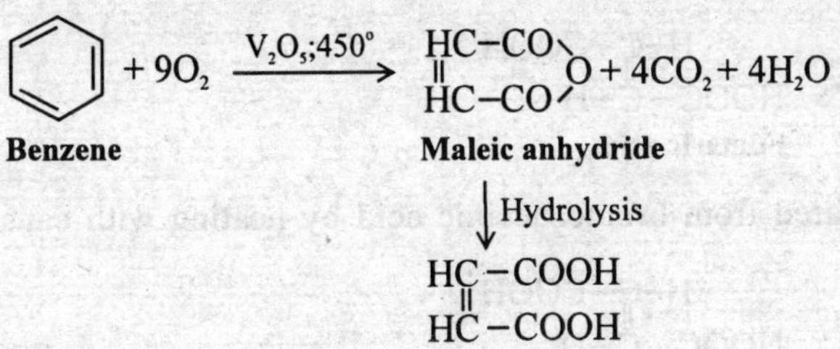

Benzene   Maleic anhydride

Maleic acid

3. *From 2-Butene*: 2-Butene (obtained from cracked petroleum) on catalytic oxidation gives maleic anhydride, which on hydrolysis gives maleic acid. This is also a method for large scale preparation.

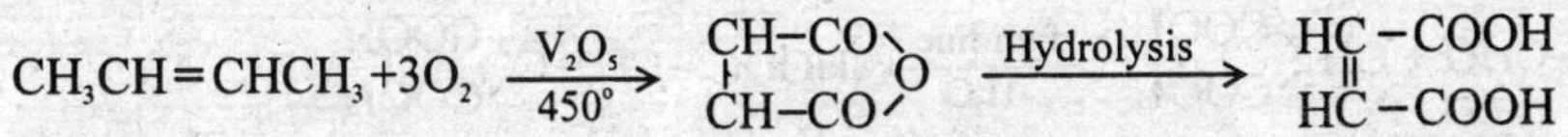

2-Butene   Maleic anhydride   Maleic acid

Properties: It is a colorless, crystalline solid (m.p. 135°) and is soluble in water.

It gives reactions of the two carboxyl group as well as of alkenes.

(1)   It forms two series of salts, esters and amides.

(2)   On heating alone or with acetic anhydride, it forms maleic anhydride. However, on prolonged heating it gives fumaric acid.

(3)   On oxidation with dilute potassium permanganate solution, meso-tartaric acid is obtained.

$$
\begin{array}{c}
\text{H-C-COOH} \\
\text{||} \\
\text{H-C-COOH} \\
\text{\textbf{Maleic acid}}
\end{array}
\xrightarrow[\text{KMnO}_4]{\text{[O]}}
\begin{array}{c}
\text{CH(OH)COOH} \\
\text{CH(OH)COOH} \\
\text{\textbf{meso-Tartaric acid}}
\end{array}
$$

(4)   It forms addition products with halogens and halogen acids.

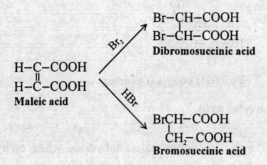

$$
\begin{array}{c}
\text{H-C-COOH} \\
\text{||} \\
\text{H-C-COOH} \\
\text{\textbf{Maleic acid}}
\end{array}
\nearrow^{Br_2}
\begin{array}{c}
\text{Br-CH-COOH} \\
\text{|} \\
\text{Br-CH-COOH} \\
\text{\textbf{Dibromosuccinic acid}}
\end{array}
$$

$$
\searrow_{HBr}
\begin{array}{c}
\text{BrCH-COOH} \\
\text{|} \\
\text{CH}_2\text{-COOH} \\
\text{\textbf{Bromosuccinic acid}}
\end{array}
$$

Uses: It is used as an antioxidant to inhibit rancidity in milk powders, oils and fats. It also finds use in making varnishes and lacquers.

20.7.2   Fumaric Acid/trans-Ethylene-1,2-dicarboxylic Acid

It is present in the sap of the plant Fumaria officianilis and hence its name is derived. It also occurs in many moulds.

Preparation

(1)   By heating malic acid at 150-170° for a long time, fumaric acid is produced.

$$
\begin{array}{c}
\text{HO-CH-COOH} \\
\text{|} \\
\text{CH}_2\text{-COOH} \\
\text{\textbf{Malic acid}}
\end{array}
\xrightarrow[\text{-H}_2\text{O}]{150\text{-}170°}
\begin{array}{c}
\text{H-C-COOH} \\
\text{||} \\
\text{HOOC-C-H} \\
\text{\textbf{Fumaric acid}}
\end{array}
$$

(2)   Fumaric acid may also be prepared from bromosuccinic acid by heating with caustic potash.

$$
\begin{array}{c}
\text{Br-CH-COOH} \\
\text{|} \\
\text{CH}_2\text{COOH} \\
\text{\textbf{Bromosuccinic acid}}
\end{array}
\xrightarrow{\text{KOH},\Delta}
\begin{array}{c}
\text{H-C-COOH} \\
\text{||} \\
\text{HOOC-C-H} \\
\text{\textbf{Fumaric acid}}
\end{array}
$$

(3)   By the condensation of malonic acid with glyoxylic acid in presence of a base like pyridine (Knoevenagel reaction) followed by heating at 130-140°.

$$
\underset{\text{\textbf{Glyoxylic acid}}}{\text{HOOC-CHO}} + \underset{\text{\textbf{Malonic acid}}}{\text{CH}_2{\Large\langle}_{\text{COOH}}^{\text{COOH}}}
\xrightarrow[\text{-H}_2\text{O}]{\text{Pyridine}}
\text{HOOC-CH=C}{\Large\langle}_{\text{COOH}}^{\text{COOH}}
$$

$$
\xrightarrow[\text{-CO}_2]{\Delta}
\begin{array}{c}
\text{HOOC-C-H} \\
\text{||} \\
\text{H-C-COOH} \\
\text{\textbf{Fumaric acid}}
\end{array}
$$

(4)   On a commercial scale it is prepared by heating maleic acid with hydrochloric acid or sodium hydroxide followed by acidification.

Properties: It is a colorless solid (m.p. 287°) and is sparingly soluble in water. Its chemical properties are similar to that of maleic acid.

20.8   PROBLEMS

1. Give IUPAC names to the following:

   (a) HOOC−CH$_2$−CH$_2$−CH−COOH
                     |
                     Br

   (b) HOOC−CH−CH$_2$−CH$_2$−CH−COOH
             | |
             CH$_3$ Br

2. (a) How is adipic acid obtained from diethyl malonate ?
   (b) Starting from ethylene bromide how will you obtain succinic acid via Grignard reagent ?
   (c) How will you distinguish between adipic acid and glutaric acid?

3. Arrange the following in the increasing order of their acidity: oxalic acid, adipic acid, malonic acid, glutaric acid.

4. How is oxalic acid obtained on a commercial scale ?

5. What happens when:

   (a) Oxalic acid $\xrightarrow{\text{electrolytic redn}}$

   (b) Oxalic acid + Ethylene glycol $\xrightarrow{\Delta}$

   (c) Oxalic acid + Glycerol $\xrightarrow{\Delta}$

6. How is malonic acid obtained ?

7. Starting with tetrahydrofuran, how will you obtain adipic acid ?

8. Which is more acidic, maleic acid or fumaric acid ? Give reasons.

9. How will you prove that in fumaric acid both the carboxyl group are on opposite side of the double bond ?

10. How is maleic acid obtained on commercial scale ?

21

Substituted Carboxylic Acids

21.1 INTRODUCTION

Substituted carboxylic acids are the derivatives of monocarboxylic acids having substitution in their hydrocarbon chain. Thus, they become bifunctional compounds and are called hydroxy acids, halo acids, amino acids and nitro acids according to the substituent in their hydrocarbon chain.

Structure	Name	Example
$HO(CH_2)_nCOOH$	Hydroxy acid	Glycollic acid, Lactic acid
$X(CH_2)_nCOOH$	Halo acid	Chloroacetic acid, Bromopropionic acid
$NH_2(CH_2)_nCOOH$	Amino acid	Glycine
$NO_2(CH_2)_nCOOH$	Nitro acid	Nitroacetic acid

21.2 HYDROXY ACIDS

Nomenclature: The usual method to name hydroxy acids is as derivatives of parent acid, the position of hydroxy group is indicated by Greek letters.

$$CH_3CH_2\underset{\underset{OH}{|}}{CH}CO_2H \qquad \textbf{α-Hydroxybutyric acid}$$

In IUPAC nomenclature position of hydroxy group is indicated by numbers.

$$CH_3CH_2\underset{\underset{OH}{|}}{CH}CO_2H \qquad \textbf{2-Hydroxybutanoic acid}$$

Some other examples are:

Formula	Common Name	IUPAC Name
$OHCH_2CO_2H$	Hydroxyacetic acid or Glycollic acid	Hydroxyethanoic acid
$CH_3CH(OH)CO_2H$	α-Hydroxypropionic acid or lactic acid	2-Hydroxypropanoic acid
$\begin{array}{l}CH(OH)CO_2H\\ |\\ CH_2CO_2H\end{array}$	Monohydroxysuccinic acid or Malic acid	2-Hydroxybutanedioic acid
$\begin{array}{l}CH(OH)CO_2H\\ |\\ CH(OH)CO_2H\end{array}$	Dihydroxysuccinic acid or Tartaric acid	2,3-Dihydroxybutanedioic acid
$\begin{array}{l}CH_2CO_2H\\ |\\ C(OH)CO_2H\\ |\\ CH_2CO_2H\end{array}$	Citric acid or β-Hydroxytricarballylic acid	2-Hydroxy-1,2,3-propane-panetrioic acid

21.3 GENERAL METHODS OF PREPARATION

(1) α-Hydroxy acids are prepared by the hydrolysis of corresponding haloacid with aqueous silver oxide or sodium carbonate solution.

$$\underset{\textbf{2-Bromopropanoic acid}}{CH_2-\overset{\overset{\displaystyle Br}{|}}{CH}-COOH} + AgOH \longrightarrow \underset{\textbf{Lactic acid}}{CH_3-\overset{\overset{\displaystyle OH}{|}}{CH}-COOH} + AgBr$$

(2) α-Hydroxy acids may also be prepared by cyanohydrin synthesis as follows:

$$\underset{}{R-\overset{\overset{\displaystyle O}{\|}}{C}-H} + HCN \longrightarrow R-\overset{\overset{\displaystyle OH}{|}}{CH}-CN \xrightarrow{H_2O/H^+} R-\overset{\overset{\displaystyle OH}{|}}{CH}-COOH$$

(3) Controlled oxidation of glycols with dilute nitric acid is another useful method for the preparation of α-hydroxy acids.

$$OHCH_2CH_2OH + 2[O] \xrightarrow{HNO_3} \underset{\textbf{Glycollic acid}}{OHCH_2COOH} + H_2O$$

$$CH_3CHOHCH_2OH + 2[O] \xrightarrow{HNO_3} \underset{\textbf{Lactic acid}}{CH_3CH(OH)COOH} + H_2O$$

(4) Naturally occurring α-amino acids may be treated with nitrous acid to give α-hydroxy acids.

$$\underset{\textbf{Alanine}}{CH_3-\overset{\overset{\displaystyle NH_2}{|}}{CH}-COOH} + NaNO_2 + HCl \longrightarrow \underset{\textbf{Lactic acid}}{CH_3-\overset{\overset{\displaystyle OH}{|}}{CH}-CO_2H} + N_2 + H_2O$$

(5) β-Hydroxy acids can be prepared by Reformatsky reaction.

$$\underset{}{CH_3-\overset{\overset{\displaystyle O}{\|}}{C}-CH_3} + \underset{\textbf{Ethylbromoacetate}}{BrCH_2CO_2C_2H_5} \xrightarrow{Zn} CH_3-\overset{\overset{\displaystyle OZnBr}{|}}{\underset{\underset{\displaystyle CH_3}{|}}{C}}-CH_2CO_2C_2H_5$$

$$\downarrow H_2O/H^+$$

$$\underset{\textbf{3-Hydroxy-3-methylbutanoic acid}}{(CH_3)_2-\overset{\overset{\displaystyle OH}{|}}{C}-CH_2CO_2H}$$

γ-Hydroxy acids can be prepared similarly starting from β-bromoesters.

$$CH_3-\overset{\overset{\displaystyle O}{\|}}{C}-CH_3 + \underset{\textbf{β-Bromoester}}{CH_3-\overset{\overset{\displaystyle}{\underset{\underset{\displaystyle Br}{|}}{CH}}}-CH_2CO_2C_2H_5} \xrightarrow[H_2O/H^+]{Zn} \underset{\textbf{γ-Hydroxy acid}}{(CH_3)_2-\overset{\overset{\displaystyle OH}{|}}{C}-\underset{\underset{\displaystyle CH_3}{|}}{CH}CH_2CO_2H}$$

(6) α, β, γ or δ-Hydroxy acids may be synthesised by catalytic reduction of appropriate keto esters.

$$CH_3COCH_2CO_2C_2H_5 \xrightarrow[150°]{H_2/Ni} CH_3CHOHCH_2CO_2C_2H_5 \xrightarrow{KOH/H^+} CH_3CHOHCH_2CO_2H$$

β-Hydroxy acid

21.4   PROPERTIES

Physical Properties: Hydroxy acids are colorless, crystalline solids or syrupy liquids. Their melting points and boiling points are much higher than those of corresponding monocarboxylic acid. They are much more soluble in water than the corresponding carboxylic acids mainly due to the presence of –OH and –COOH groups which can form hydrogen bonds with water. They exhibit additional acidity due to the presence of strongly electronegative hydroxy group which withdraws electrons from the carbon of carboxyl group facilitating ionisation of proton.

$$HO \leftarrow \overset{\overset{H}{|}}{\underset{\underset{H}{|}}{C}} \leftarrow \overset{\overset{O}{\parallel}}{C} \leftarrow OH \longrightarrow HO - \overset{\overset{H}{|}}{\underset{\underset{H}{|}}{C}} - \overset{\overset{O}{\parallel}}{C} - O^- + H^+$$

The inductive effect of hydroxy group diminishes with the increase of distance between –OH and –COOH group. The order of decreasing acidity in following hydroxy acids is as indicated:

$$HOCH_2CO_2H > HOCH_2CH_2CO_2H > HOCH_2CH_2CH_2CO_2H$$

Chemical Properties: Since they have hydroxy and carboxylic groups, they undergo reactions of both the groups. When the two groups are widely separated they do not interfere with each other in their reactions. Hydroxy acids give a wide variety of reactions

$$\overset{OH}{\underset{|}{R-CH(CH_2)_nCO_2H}} \Bigg\lbrace$$

$$+ Na \longrightarrow R - \overset{\overset{\bar{O} \; Na^+}{|}}{CH} - (CH_2)_nCO_2H$$

$$+ NaOH \xrightarrow{H_2O} R - \overset{\overset{OH}{|}}{CH} - (CH_2)_nCO\bar{O}Na^+$$

$$+ C_2H_5OH \xrightarrow{H^+} R - \overset{\overset{OH}{|}}{CH} - (CH_2)_nCOOC_2H_5$$

$$+ NH_3 \longrightarrow R - \overset{\overset{OH}{|}}{CH} - (CH_2)_nCOO\overset{+}{N}H_4 \xrightarrow{P_2O_5} R - \overset{\overset{OH}{|}}{CH}(CH_2)_nCONH_2$$

$$+ SOCl_2 \longrightarrow R - \overset{\overset{Cl}{|}}{CH}(CH_2)_nCOCl \xrightarrow{H_2O/H^+} R - \overset{\overset{Cl}{|}}{CH}(CH_2)_nCO_2H$$

$$+ 2HI \longrightarrow RCH_2(CH_2)_nCOOH$$

$$+ [O] \longrightarrow R - \overset{\overset{O}{\parallel}}{C} - (CH_2)_nCOOH$$

$$+ CH_3COCl \longrightarrow R - \overset{\overset{OCOCH_3}{|}}{CH} - (CH_2)_n - CO_2H$$

In addition to the reactions shown above, hydroxy acids form ester by the reaction of -OH and -COOH groups within the molecule. This type of reaction depend upon the nature of hydroxy acid involved.

α-Hydroxy acids undergo intermolecular esterification on heating to form six membered ring compounds called *lactides*

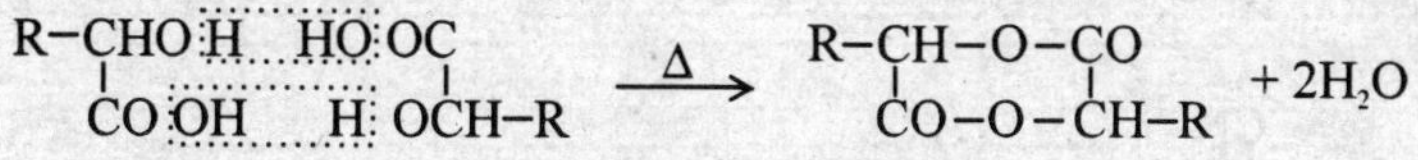

$$R-CHO\underset{CO\underset{}{\overset{}{OH}}}{\overset{}{H}}\ \underset{H}{\overset{HO}{}}\underset{OCH-R}{\overset{OC}{}} \xrightarrow{\Delta} \underset{CO-O-CH-R}{\overset{R-CH-O-CO}{}} + 2H_2O$$

Lactide

lactides undergo alkaline hydrolysis readily to yield acids.

β-Hydroxy acids on heating eliminate a water molecule to form α-β-unsaturated acid and a small amount of β, γ-unsaturated acid.

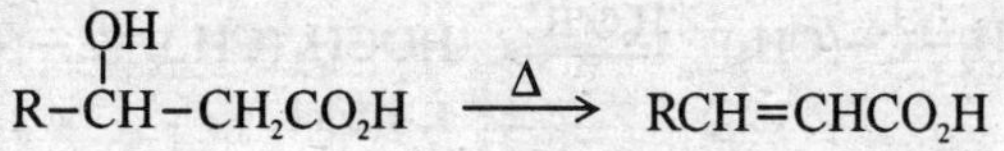

$$\overset{\overset{OH}{|}}{R-CH-CH_2CO_2H} \xrightarrow{\Delta} RCH=CHCO_2H$$

α, β-Unsaturated acid

γ and δ-Hydroxy acids undergo intramolecular esterification to form cyclic esters called *lactones*.

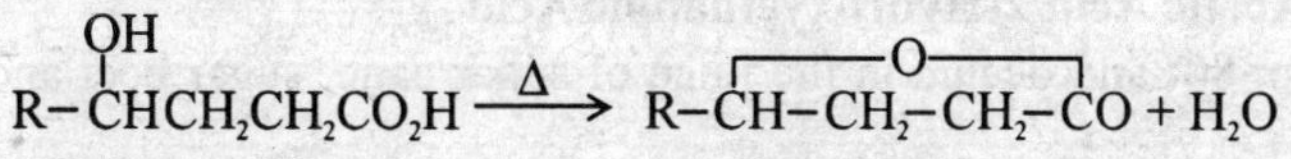

$$\overset{\overset{OH}{|}}{R-CHCH_2CH_2CO_2H} \xrightarrow{\Delta} R-CH-CH_2-CH_2-CO + H_2O$$

γ-Hydroxy acid　　　　　　**γ-Butyrolactone**

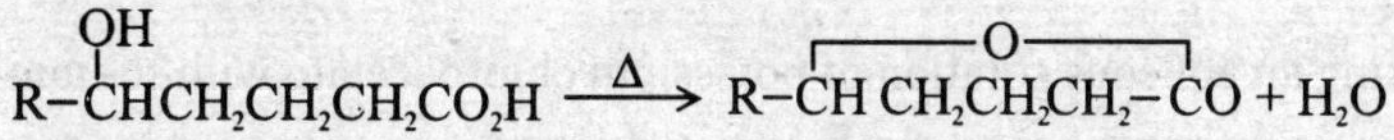

$$\overset{\overset{OH}{|}}{R-CHCH_2CH_2CH_2CO_2H} \xrightarrow{\Delta} R-CH\ CH_2CH_2CH_2-CO + H_2O$$

δ-Hydroxy acid　　　　　　**δ-Valerolactone**

ε-Hydroxy acids have little tendency to form lactones instead they undergo intermolecular esterification to form linear polyesters.

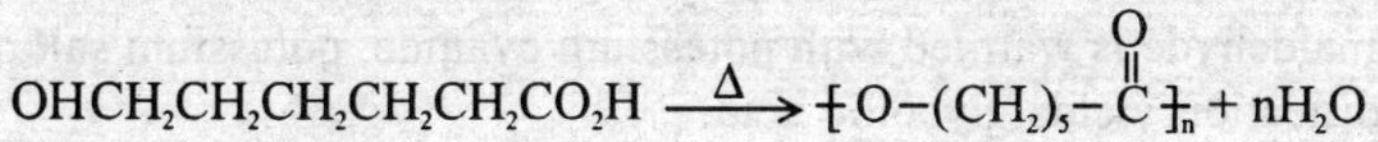

$$OHCH_2CH_2CH_2CH_2CH_2CO_2H \xrightarrow{\Delta} \left\{ O-(CH_2)_5 - \overset{\overset{O}{\|}}{C} \right\}_n + nH_2O$$

ε-Hydroxy acid　　　　　　**a linear Polyester**

Reactions of Lactones: lactones are important intermediate for a number of products.

(1)   They are converted into salts of corresponding hydroxy acid when refluxed with excess of alkali.

$$R\overset{\frown}{CH\ CH_2CH_2CO} + NaOH \longrightarrow RCHOHCH_2CH_2CO_2Na$$

(2)   When treated with sodium amalgam in acid solution lactones form corresponding acids.

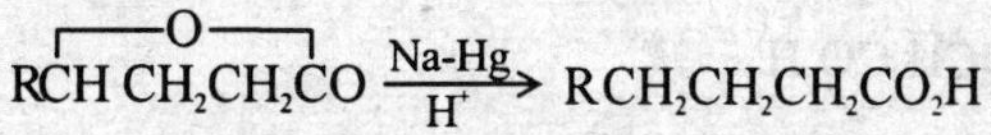

$$R\overset{\frown}{CH\ CH_2CH_2CO} \xrightarrow[H^+]{Na-Hg} RCH_2CH_2CH_2CO_2H$$

(3)   With concentrated hydrogen halide they give corresponding halogen substituted acid.

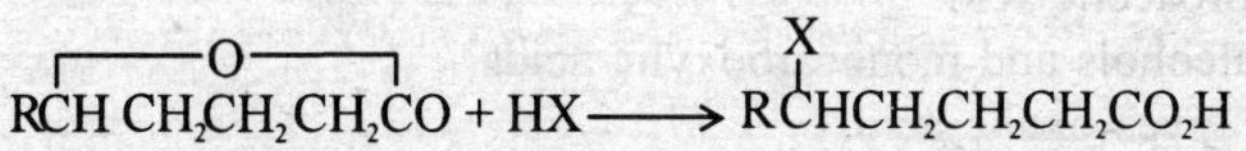

$$R\overset{\frown}{CH\ CH_2CH_2\ CH_2CO} + HX \longrightarrow R\overset{\overset{X}{|}}{C}HCH_2CH_2CH_2CO_2H$$

(4)   When reacted with concentrated ammonia solution, hydroxyamide or in some cased γ-lactam is formed.

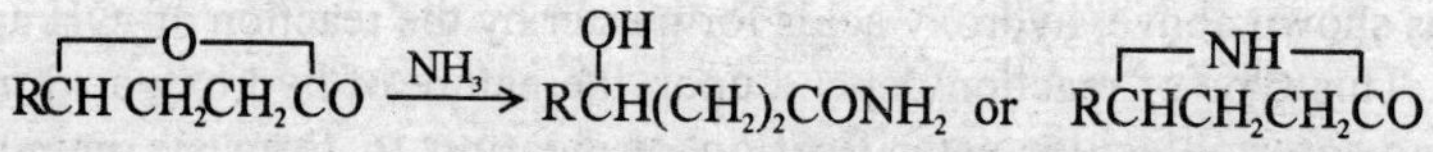

$$\overset{\boxed{}\,O\,\boxed{}}{RCH\ CH_2CH_2CO} \xrightarrow{NH_3} RCH\overset{OH}{(CH_2)_2}CONH_2 \text{ or } \overset{\boxed{}\,NH\,\boxed{}}{RCHCH_2CH_2CO}$$

γ-Lactam

(5)   Lactones react with potassium cyanide to give cyano acids which on hydrolysis yield dicarboxylic acids.

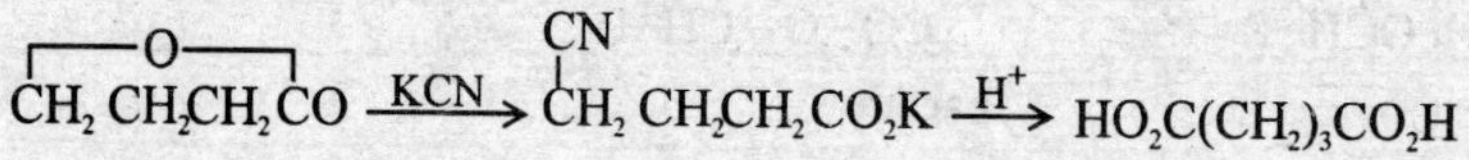

$$\overset{\boxed{}\,O\,\boxed{}}{CH_2\ CH_2CH_2CO} \xrightarrow{KCN} \overset{CN}{CH_2\ CH_2CH_2CO_2K} \xrightarrow{H^+} HO_2C(CH_2)_3CO_2H$$

Propane dicarboxylic acid

(6)   With Grignard reagent, lactones react to form diols.

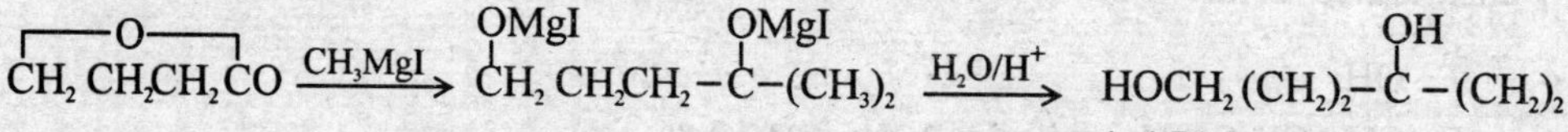

$$\overset{\boxed{}\,O\,\boxed{}}{CH_2\ CH_2CH_2CO} \xrightarrow{CH_3MgI} \overset{OMgI}{CH_2\ CH_2CH_2}-\overset{OMgI}{C}-(CH_3)_2 \xrightarrow{H_2O/H^+} HOCH_2(CH_2)_2-\overset{OH}{C}-(CH_2)_2$$

1, 4-Diol

21.5   INDIVIDUAL MEMBERS

21.5.1   Glycollic Acid/Hydroxy Acetic Acid/2-Hydroxyethanoic Acid

It is the simplest hydroxy acid, m.p. 80° and occurs in the juice of sugar cane, sugar beet and in unripe grapes.

Preparation:

(1)   It is prepared by refluxing an aqueous solution of potassium chloroacetate with sodium carbonate.

$$ClCH_2CO_2K + H_2O \xrightarrow[\text{2. HCl}]{\text{1. Na}_2CO_3} HOCH_2CO_2H + KCl$$

Potassium
chloroacetate

Glycollic acid

(2)   When a solution of formaldehyde is warmed with potassium cyanide, potassium salt of glycollic acid is produced which on acidification yields glycollic acid.

$$HCHO + KCN + 2H_2O \longrightarrow HOCH_2CO_2K \xrightarrow{HCl} HOCH_2CO_2H$$

(3)   On industrial scale, it is prepared by electrolytic reduction of oxalic acid.

$$(COOH)_2 + 4[H] \longrightarrow HOCH_2CO_2H$$

It can also be prepared by heating a mixture of formaldehyde, carbon monoxide and water in acidic medium at 160-170° under pressure.

$$HCHO + CO + H_2O \xrightarrow[\text{H+, Pressure}]{160-170°} HOCH_2CO_2H$$

Properties: It is a white crystalline solid, readily soluble in water, ethanol and ether. It is a stronger acid than acetic acid and weaker than chloroacetic acid.

It gives the reactions of primary alcohols and monocarboxylic acids.

When oxidised with nitric acid it gives oxalic acid.

Its lactide is called glycollide.

Use: It is used in printing of fabrics.

21.5.2 Lactic Acid/α-Hydroxypropionic Acid/2-Hydroxypropanoic Acid [CH₃CHOHCOOH]

It is the main ingredient of sour milk hence name lactic acid (Latin, Lac, milk). It is found in blood and muscles where it is formed due to the metabolism of glycogen, and so is also called sarcolactic acid.

Preparation : It can be prepared by any of the general methods used for the preparation of α-hydroxy acids. On industrial scale, lactic acid is prepared by fermentation of lactose by Bacillus acidi lactiti.

$$\underset{\textbf{Lactose}}{C_{12}H_{22}O_{11}} + H_2O \xrightarrow{\text{Fermentation}} \underset{\textbf{Lactic acid}}{4CH_3CHOHCO_2H}$$

Fermentation of sucrose by Rhizopus oryzae also yields lactic acid.

Since lactic acid contains one asymmetric carbon atom, theoretically, it can exhibit optical activity (see Section 5.6.3). Three forms of lactic acid are known.

(a) L (+) Lactic acid, m.p. 26°

(b) D (−) Lactic acid, m.p. 26°

(c) DL or (±) Lactic acid, m.p. 18°

The lactic acid prepared by above methods is recemic (±) having equal amounts of (+) and (−) lactic acids. Lactic acid extracted from meat is L(+) and is also known as sarcolactic acid (Greek, sarkos, flesh). D (−) Lactic acid may be obtained by fermentation of sucrose by Bacillus acidi laevolactiti.

Properties: It is a colorless, syrupy liquid, hygroscopic in nature, and soluble in water in all proportions. It has sour smell and taste.

It has a secondary alcoholic group and a carboxylic group and gives all the reactions of α-hydroxy acids mentioned in general reactions.

(1) When oxidised with hydrogen peroxide, it gives pyruvic acid, an α-ketoacid.

$$CH_3CHOHCO_2H \xrightarrow{H_2O_2/Fe^{++}} \underset{\textbf{Pyruvic acid}}{CH_3COCO_2H}$$

Oxidation with potassium permanganate gives acetic acid.

$$CH_3CHOHCO_2H \xrightarrow{KMnO_4} CH_3COOH + CO_2 + H_2O$$

(2) When treated with hydriodic acid at 130°, it is reduced to propanoic acid.

$$CH_3CHOHCO_2H + HI \longrightarrow CH_3CH(I)CO_2H + H_2O$$

$$CH_3CH(I)CO_2H + HI \longrightarrow CH_3CH_2CO_2H + I_2$$

(3) It gives Iodoform when treated with iodine and alkali.

$$CH_3CHOHCO_2H + I_2 \longrightarrow CH_3COCO_2H + 2HI$$

$$CH_3COCO_2H + NaOH + I_2 \longrightarrow \underset{\textbf{Triiodopyruvic acid}}{CI_3COCOOH + H_2O}$$

$$I_3CCOCOOH + NaOH \longrightarrow \underset{\textbf{Iodoform}}{CHI_3} + (COONa)_2 + H_2O$$

Uses
1. It is used in dying and tanning industries.
2. Ethyl lactate is used as a solvent for cellulose nitrate.
3. Calcium and iron lactate are used in medicines.

21.5.3   Malic Acid/Monohydroxysuccinic Acid/2-Hydroxy-butane-1, 4-dioic Acid [$HO_2CCH_2CH(OH)CO_2H$]

It occurs in sour apples (Latin, malum, apple), fruits and berries.

Preparation

(1)   It is synthesised by heating maleic acid in dilute sulphuric acid under pressure.

$$\begin{array}{l} CHCOOH \\ \| \\ CHCOOH \end{array} + H_2O \xrightarrow[\text{Pressure}]{H_2SO_4,\ \Delta} \begin{array}{l} CH(OH)CO_2H \\ CH_2CO_2H \end{array}$$
Malic acid

(2)   In the laboratory, it is prepared by heating bromosuccinic acid with silver oxide.

$$\begin{array}{l} CHBrCO_2H \\ CH_2CO_2H \end{array} + AgOH \xrightarrow{\Delta} \begin{array}{l} CH(OH)CO_2H \\ CH_2CO_2H \end{array} + AgBr$$

Malic acid contains one asymmetric carbon atom thus exists in following three forms, the ± or recemic mixture is a mixture of (+) and (−) malic acid in equal amounts.

(a) L(−) malic acid m.p. 100° obtained from natural sources.

(b) D(+) malic acid m.p. 100° may be obtained by careful reduction of D(+) tartaric acid with concentrated hydriodic acid.

$$\begin{array}{c} COOH \\ | \\ H-C-OH \\ | \\ OH-C-H \\ | \\ COOH \end{array} + 2HI \longrightarrow \begin{array}{c} COOH \\ | \\ H-C-OH \\ | \\ CH_2 \\ | \\ COOH \end{array} + I_2 + H_2O$$

D(+) Tartaric acid   **D(+) Malic acid**

(c) DL or ± malic acid m.p. 130° may be prepared by any of the synthetic methods discussed above.

Properties: The reactions undergone by malic acid are due to secondary hydroxy group and dibasic acid.

(1)   It undergoes dehydration, characteristic of β-hydroxy acids when heated to yield maleic anhydride and fumaric acid.

$$\begin{array}{l} CH-CO \\ \| \qquad\ \ \diagdown O \\ CH-CO \diagup \end{array} \xleftarrow{\Delta} \begin{array}{l} CH(OH)CO_2H \\ CH_2CO_2H \end{array} \xrightarrow{\Delta} \begin{array}{l} H-C-COOH \\ \| \\ HO_2C-C-H \end{array}$$
Maleic anhydride   **Malic acid**   **Fumaric acid**

(2)   When reduced with hydriodic acid, it forms succinic acid.

$$\begin{array}{l} CH(OH)CO_2H \\ CH_2CO_2H \end{array} + 2HI \longrightarrow \begin{array}{l} CH_2CO_2H \\ CH_2CO_2H \end{array} + I_2 + H_2O$$
Malic acid   **Succinic acid**

(3) It forms oxalacetic acid on careful oxidation which exhibits keto-enol tautomerism

$$\begin{matrix} CH(OH)CO_2H \\ | \\ CH_2CO_2H \end{matrix} \xrightarrow{[O]} \begin{matrix} COCO_2H \\ | \\ CH_2CO_2H \end{matrix} \rightleftharpoons \begin{matrix} C(OH)CO_2H \\ || \\ CHCO_2H \end{matrix}$$

Malic acid **Oxalacetic acid** (keto form) **Oxalacetic acid** (enol form)

Uses
1. It is used in medicine formulations.
2. It is also used in beverages.

21.5.4 Tartaric Acid/α-α'-Dihydroxysuccinic Acid/ 2,3-Dihydroxybutanedioic Acid [HO$_2$CCHOHCHOHCO$_2$H]

It occurs in free state and as potassium hydrogen tartarate in grapes, tamarind and other fruits. During the fermentation of grapes, the potassium salt of acid separates as a reddish brown crystalline mass which is known as argol.

Tartaric acid contains two structurally identical asymmetric carbons and therefore can exist in following forms (see Section 5.6.3):

(1) (+) Tartaric acid, m.p. 179°, specific rotation +12°
(2) (−) Tartaric acid, m.p. 170° specific rotation −12°
(3) (±) Tartaric acid, m.p. 260°, specific rotation, 0
(4) meso-Tartaric acid, m.p. 140°, specific rotation, 0

D(−)Tartaric acid   **L(+)Tartaric acid**   **meso-Tartaric acid**

Preparation:

(1) Both (±) and meso tartaric acid are formed when α,α'-dibromosuccinic acid is boiled with silver oxide in water.

$$\begin{matrix} CHBrCO_2H \\ | \\ CHBrCO_2H \end{matrix} + 2AgOH \xrightarrow{\Delta} \begin{matrix} CHOHCO_2H \\ | \\ CHOHCO_2H \end{matrix} + 2AgBr$$

Tartaric acid

(2) They are also formed by hydrolysis of glyoxalcyanohydrin.

$$\begin{matrix} CHO \\ | \\ CHO \end{matrix} + 2\ HCN \xrightarrow{\Delta} \begin{matrix} CHOHCN \\ | \\ CHOHCN \end{matrix} \xrightarrow{H_2O} \begin{matrix} CHOHCO_2H \\ | \\ CHOHCO_2H \end{matrix}$$

(3) Kiliani-Fischer synthesis also produces tartaric acid from glyceraldehyde.

$$\begin{matrix} CHO \\ | \\ CHOH \\ | \\ CH_2OH \end{matrix} \xrightarrow{HCN} \begin{matrix} CN \\ | \\ CHOH \\ | \\ CHOH \\ | \\ CH_2OH \end{matrix} \xrightarrow{H_2O/H^+} \begin{matrix} CO_2H \\ | \\ CHOH \\ | \\ CHOH \\ | \\ CH_2OH \end{matrix} \xrightarrow[Oxid^n]{HNO_3} \begin{matrix} CO_2H \\ | \\ CHOH \\ | \\ CHOH \\ | \\ CO_2H \end{matrix}$$

Glyceraldehyde   **Tartaric acid**

(4)   (±) Tartaric acid is formed when fumaric acid is treated with dilute alkaline potassium permanganate, whereas the same reaction with maleic acid produces meso tartaric acid.

$$\begin{array}{c} H-C-COOH \\ \| \\ HOOC-C-H \end{array} + [O] \xrightarrow{\text{alk. KMnO}_4} \begin{array}{c} CHOHCOOH \\ | \\ CHOHCOOH \end{array}$$

(5)   (+)-Tartaric acid occurs in nature and is isolated from argol, the brown crystalline solid from the fermentation of grapes. Argol crystallises into cream of tartar, a white crystalline solid from which (+) tartaric acid is obtained in batches.

$$\begin{array}{c} CHOHCOOK^+ \\ | \\ CHOHCOOH \end{array} + Ca(OH)_2 \longrightarrow \begin{array}{c} CHOHCOO^-K^+ \\ | \\ CHOHCOO^-K^+ \end{array} + \begin{array}{c} CHOHCOO^- \\ | \\ CHOHCOO^- \end{array} Ca^{++} + 2H_2O$$

Pot. Hydrogen tartarate   **Potassium tartaratae**

$$\begin{array}{c} CHOHCOO^-K^+ \\ | \\ CHOHCOO^-K^+ \end{array} + CaCl_2 \longrightarrow \begin{array}{c} CHOHCOO^- \\ | \\ CHOHCOO^- \end{array} Ca^{++} + 2KCl$$

Calcium tartarate

The insoluble calcium tartarate is separated by filtration and decomposed with calculated quantity of dilute sulphuric acid.

$$\begin{array}{c} CHOHCOO^- \\ | \\ CHOHCOO^- \end{array} Ca^{++} + H_2SO_4 \longrightarrow \begin{array}{c} CHOHCOOH \\ | \\ CHOHCOOH \end{array} + CaSO_4$$

The precipitated calcium sulphate is separated and the filtrate on evaporation gives tartaric acid.

(6)   (–) Tartaric acid does not occur in nature and is obtained by resolution of (±) tartaric acid.

Properties

(+) and (–) Tartaric acids are colorless crystalline solids without any water of crystallisation. (±)-Tartaric acid or recemic tartaric acid is a crystalline hemihydrate having chemical formula $(C_4H_6O_6)_2 \cdot H_2O$. It contains equal amounts of (+) and (–) tartaric acids. meso-Tartaric acid crystallises as monohydrate. It is optically inactive due to internal compensation.

(1)   Since it contains two secondary hydroxy groups and two carboxyl groups, it forms two series of salts when treated with alkalis.

$$\begin{array}{c} CHOHCO_2K \\ | \\ CHOHCO_2H \end{array} \qquad\qquad \begin{array}{c} CHOHCO_2K \\ | \\ CHOHCO_2K \end{array}$$

pot. hydrogen tartarate   **Potassium tartarate**
or
Acid potassium tartarate

(2)   On prolonged heating at high temperature (about 200°), it forms anhydride.

$$\begin{array}{c} CHOHCO_2H \\ | \\ CHOHCO_2H \end{array} \xrightarrow{\Delta} \begin{array}{c} CHOHCO \\ | \\ CHOHCO \end{array}\!\!\!\! \raisebox{-0.5ex}{\Large$\rangle$} O + H_2O$$

Tartaric acid   **Tartaric anhydride**

On strong heating, however, pyruvic acid is formed with a smell of burnt sugar

$$\begin{array}{c} CHOHCO_2H \\ | \\ CHOHCO_2H \end{array} \longrightarrow CH_3COCO_2H + CO_2 + H_2O$$

Tartaric acid   **Pyruvic acid**

(3) It is reduced to malic acid and succinic acid with concentrated hydriodic acid.

$$\underset{\overset{|}{\text{CHOHCO}_2\text{H}}}{\text{CHOHCO}_2\text{H}} \xrightarrow{\text{HI}} \underset{\overset{|}{\text{CH}_2\text{CO}_2\text{H}}}{\text{CHOHCO}_2\text{H}} \xrightarrow{\text{HI}} \underset{\overset{|}{\text{CH}_2\text{CO}_2\text{H}}}{\text{CH}_2\text{CO}_2\text{H}}$$

Tartaric acid **Malic acid** **Succinic acid**

(4) On reaction with hydrobromic acid, it forms dibromosuccinic acid.

$$\underset{\overset{|}{\text{CHOHCO}_2\text{H}}}{\text{CHOHCO}_2\text{H}} + 2\text{HBr} \longrightarrow \underset{\overset{|}{\text{CHBrCO}_2\text{H}}}{\text{CHBrCO}_2\text{H}} + 2\text{H}_2\text{O}$$

Dibromosuccinic acid

(5) Oxidation of tartaric acid with mild oxidising agents yields tartonic acid. With strong oxidising agents, however, it forms oxalic acid.

$$\underset{\overset{|}{\text{CHOHCO}_2\text{H}}}{\text{CHOHCO}_2\text{H}} \xrightarrow{[\text{O}]} \underset{\overset{|}{\text{CO}_2\text{H}}}{\text{CHOHCO}_2\text{H}} \xrightarrow{[\text{O}]} \underset{\overset{|}{\text{CO}_2\text{H}}}{\text{CO}_2\text{H}}$$

Tartonic acid **Oxalic acid**

With Fenton's reagent (Fe^{++} and H_2O_2) it is oxidised to dihydroxymaleic acid.

$$\underset{\overset{|}{\text{CHOHCO}_2\text{H}}}{\text{CHOHCO}_2\text{H}} \xrightarrow{[\text{O}]} \underset{\overset{||}{\text{COHCO}_2\text{H}}}{\text{COHCO}_2\text{H}}$$

Dihydroxymaleic acid

(6) In Fehlings solution ($CuSO_4$ + Rochelle salt+ NaOH) insoluble copper hydroxide is first formed which dissolves later on due to the formation of complex which may be represented as follows:

$$\left[\begin{array}{c} \text{O}^- - \overset{\overset{\text{O}}{||}}{\text{C}} - \text{CHO} \\ \text{O}^- - \underset{\underset{\text{O}}{||}}{\text{C}} - \text{CHO} \end{array} \right. \overset{2-}{\underset{\text{Cu}}{\diagdown}} \left. \begin{array}{c} \text{OCH} - \overset{\overset{\text{O}}{||}}{\text{C}} - \text{O}^- \\ \text{OCH} - \underset{\underset{\text{O}}{||}}{\text{C}} - \text{O}^- \end{array} \right] 6\,\text{Na}^+$$

Uses
1. (+) Tartaric acid is used in effervescent drinks.
2. The acid and tartar, both are used in dyeing and printing.
3. Sodium potassium tartarate (rochelle salt) is used in the preparation of Fehlings solution.
4. Potassium antimonyl tartarate (tartar emetic) is used in medicines and in dyeing as mordent.

21.5.5 Citric Acid/β-Hydroxytricarballylic Acid/2-Hydroxy-1,2,3-Propanetricarboxylic Acid

Citric acid occurs in many fruits especially in unripe fruits of citrus family. Lemon juice contains about 6-10% citric acid.

Preparation
(1) It is extracted from the juice of lemon, galgal or oranges. The juice is boiled to coagulate the albumin and filtered. The filtrate is neutralised with calcium carbonate followed by filtration of calcium citrate which is deposited while neutralisation. Calcium citrate is treated with calculated amount of dilute sulphuric acid to separate calcium sulphate as solid and citric acid is collected from the filtrate by evaporation.

(2)   Citric acid may be manufactured from a dilute solution of molasses by fermentation. Fermentation requires a proper supply of inorganic salts as nutrients for the micro organisms such as *Citromyces pfefferianus* or *Aspergillus wentii*.

(3)   It may be synthesised from glycerol by the following sequence of reactions.

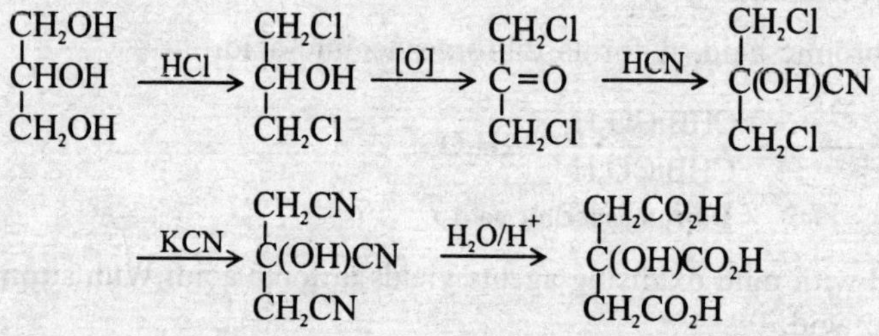

$$\underset{CH_2OH}{\overset{CH_2OH}{CHOH}} \xrightarrow{HCl} \underset{CH_2Cl}{\overset{CH_2Cl}{CHOH}} \xrightarrow{[O]} \underset{CH_2Cl}{\overset{CH_2Cl}{C=O}} \xrightarrow{HCN} \underset{CH_2Cl}{\overset{CH_2Cl}{C(OH)CN}}$$

$$\xrightarrow{KCN} \underset{CH_2CN}{\overset{CH_2CN}{C(OH)CN}} \xrightarrow{H_2O/H^+} \underset{\underset{\text{Citric acid}}{CH_2CO_2H}}{\overset{CH_2CO_2H}{C(OH)CO_2H}}$$

Properties: It is a crystalline solid having one molecule of water of crystallisation, m.p. 153°. When heated to 130°, it loses its water of crystallisation. It behaves as an alcohol and a tribasic acid. It gives reactions of both α and β hydroxy acids since –OH group is α to one carboxyl group and β to the other two.

(1)   It forms three series of salts as follows:

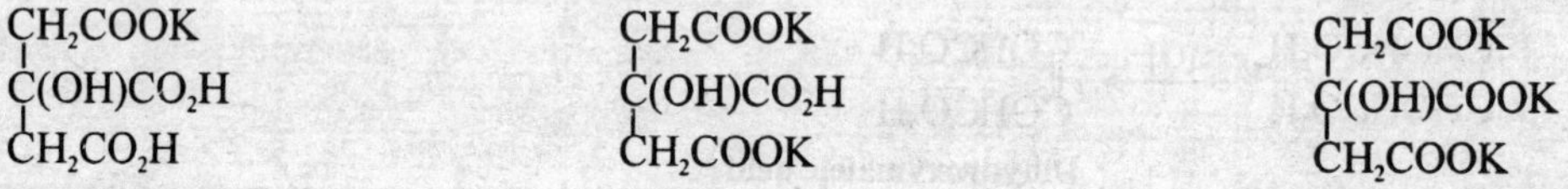

$$\underset{\underset{\text{Monopotassium citrate}}{CH_2CO_2H}}{\overset{CH_2COOK}{C(OH)CO_2H}} \qquad \underset{\underset{\text{Dipotassium citrate}}{CH_2COOK}}{\overset{CH_2COOK}{C(OH)CO_2H}} \qquad \underset{\underset{\text{Tripotassium citrate}}{CH_2COOK}}{\overset{CH_2COOK}{C(OH)COOK}}$$

(2)   It forms monoacetyl citric acid when treated with acetic anhydride or acetyl chloride.

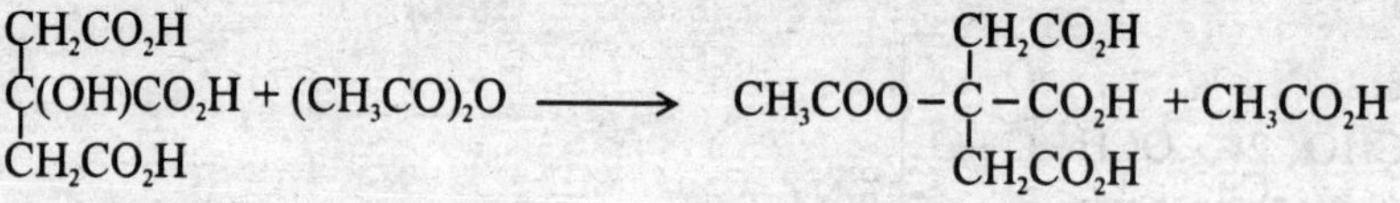

$$\underset{CH_2CO_2H}{\overset{CH_2CO_2H}{C(OH)CO_2H}} + (CH_3CO)_2O \longrightarrow CH_3COO-\underset{CH_2CO_2H}{\overset{CH_2CO_2H}{C}}-CO_2H + CH_3CO_2H$$

Monoacetylcitric acid

(3)   When reduced with hydriodic acid, citric acid forms tricarballylic or 1,2,3-tripropanoic acid.

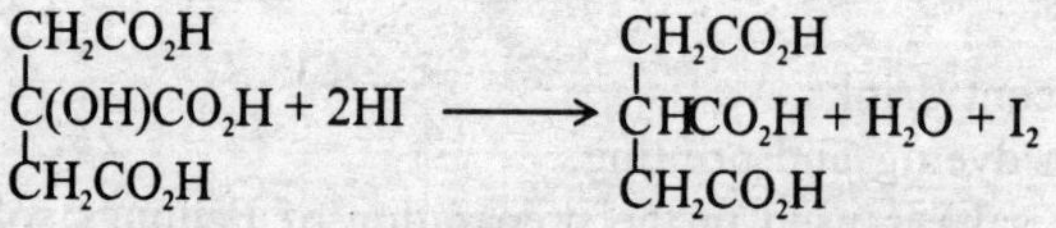

$$\underset{CH_2CO_2H}{\overset{CH_2CO_2H}{C(OH)CO_2H}} + 2HI \longrightarrow \underset{CH_2CO_2H}{\overset{CH_2CO_2H}{CHCO_2H}} + H_2O + I_2$$

Tricarballylic acid

(4)   Citric acid when heated to 150° undergoes dehydration (similar to β-hydroxy acids)

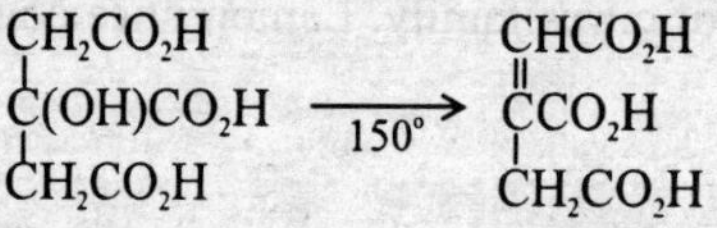

$$\underset{CH_2CO_2H}{\overset{CH_2CO_2H}{C(OH)CO_2H}} \xrightarrow{150°} \underset{CH_2CO_2H}{\overset{CHCO_2H}{CCO_2H}}$$

Aconitic acid

On heating with fuming sulphuric acid it gives acetone dicarboxylic acid.

$$\begin{array}{c} CH_2CO_2H \\ | \\ C(OH)CO_2H \\ | \\ CH_2CO_2H \end{array} \quad \xrightarrow[150°]{\text{Fuming } H_2SO_4} \quad \begin{array}{c} CH_2CO_2H \\ | \\ C=O \\ | \\ CH_2CO_2H \end{array}$$

Acetone dicarboxylic acid

(5)   Citric acid forms soluble complex salts with Benedict solution which contains copper sulphate, sodium carbonate and sodium citrate.

$$\left[\begin{array}{c} ^-OOC\,CH_2 \\ \\ ^-OOC\,CH_2 \end{array} \!\!\! C \!\!\! \begin{array}{c} O \\ \diagdown \\ CO-O \end{array} \!\!\! Cu \!\!\! \begin{array}{c} O-CO \\ \diagup \\ O \end{array} \!\!\! C \!\!\! \begin{array}{c} CH_2COO^- \\ \diagdown \\ CH_2COO^- \end{array} \right]^{2-} 6\,Na^+$$

Sodium Cupricitrate

Uses
1. It is used in beverages compositions.
2. It is also used as a mordant in dyeing and printing.
3. Its ferric, magnesium, potassium or sodium salts are used in medicines.
4. Esters of citric acid (tributyl citrate) are used as plasticizers for varnishes and lacquers.

21.6   HALOGEN SUBSTITUTED ACIDS

Nomenclature: These acids are derived from carboxylic acids by replacing one or more hydrogen of alkyl groups by halogens. They are classified as α, β or γ-halogeno acids, depending upon the position of halogen with respect to carboxyl group.

Formula	Common Name	IUPAC Name
$ClCH_2CO_2H$	Chloroacetic acid	Chloroethanoic acid
$CH_3CHClCO_2H$	α-Chloropropionic acid	2-Chloropropanoic acid
$CH_3CHBrCH_2CO_2H$	β-Bromobutyric acid	3-Bromobutanoic acid
$CH_2ClCH_2CH_2CO_2H$	γ-Chlorobutyric acid	4-Chlorobutanoic acid

21.7   GENERAL METHODS OF PREPARATION
α-Halogeno Acids
(1)   α-Chloro and α-Bromo acids may be prepared by Hell-Volhard Zelinsky (HVZ) reaction which involves the reaction of acid with chlorine or bromine in presence of small amount of red phosphorus.

$$CH_3CO_2H + Cl_2 \xrightarrow{\text{red P}} \underset{\textbf{Chloroacetic acid}}{ClCH_2CO_2H} + HCl$$

In excess of chlorine second and third hydrogens of alkyl group may also be replaced giving dichloro and trichloroacetic acid.

$$\underset{\textbf{Chloroacetic acid}}{ClCH_2CO_2H} \xrightarrow{Cl_2 + P} \underset{\textbf{Dichloroacetic acid}}{Cl_2CHCO_2H} \xrightarrow{Cl_2 + P} \underset{\textbf{Trichloroacetic acid}}{Cl_3CCO_2H}$$

The mechanism of HVZ reaction involves enol intermediate as discussed in Sec. 19.7.

The reaction does not stop at monohalogenation and second or third α-hydrogens may also be halogenated in excess of halogen. Bromine replaces only α-hydrogens selectively whereas chlorination may proceed at other positions also.

(2)   α-Iodoacids are prepared by treatment of α-chloro or β-bromoacids with potassium iodide.

$$CH_3CHBrCO_2H + KI \longrightarrow CH_3CHICO_2H + KBr$$

(3)   They can also be prepared by halogenation of alkyl malonic acid followed by decarboxylation

$$CH_3CH \underset{CO_2H}{\overset{CO_2H}{\diagdown}} \xrightarrow[solvent]{Br_2} CH_3C \underset{CO_2H}{\overset{CO_2H}{-Br}} \xrightarrow[140-150°]{\Delta} CH_3CHBrCO_2H + CO_2$$

(4)   α-hydroxy acids may be converted to α-halogeno acids when treated with hydrogen halide or phosphorus halide.

$$CH_3CHOHCO_2H + HBr \longrightarrow CH_3CHBrCO_2H + H_2O$$

Lactic acid   **α-Bromopropanoic acid**

β-Halogeno acids

(1)   β-Halogeno acids may be prepared by treating α,β-unsaturated acids with hydrogen halides.

$$CH_2=CH-\overset{\overset{O}{||}}{C}-OH \xrightarrow{HBr} Br-CH_2-CH=\overset{\overset{OH}{|}}{C}-OH \rightleftharpoons BrCH_2-CH_2-\overset{\overset{O}{||}}{C}-OH$$

(2)   α,β-Unsatured aldehydes on reaction with hydrogen halides followed by oxidation furnish β-halogeno acids.

$$CH_2=CHCHO + HCl \longrightarrow ClCH_2CH_2CHO \xrightarrow{[O]} ClCH_2CH_2CO_2H$$

The addition of hydrogen halide to α,β- and β,γ- unsaturated carbonyl compounds is antimarkovnikov's due to inductive effect of carbonyl group.

(3)   It may also be prepared by refluxing ethylene cyanohydrin with halogen acids.

$$OHCH_2CH_2CN + 2HBr + H_2O \longrightarrow BrCH_2CH_2CO_2H + NH_4Br$$

Ethylene cyanohydrin   **β-Bromopropionic acid**

(4)   Oxidation of various cholrohydrins with concentrated nitric acid furnish halogen acids.

$$ClCH_2CH_2CH_2OH \xrightarrow{[O]} ClCH_2CH_2CO_2H$$

Trimethylene chlorohydrin   **3-Chloropropanoic acid**

γ-Halogeno Acids

(1)   γ-Halogeno acids may be prepared by addition of hydrogen halides to β, γ-unsaturated acids (antimarkovikov's addition).

$$CH_3CH=CHCH_2CO_2H + HBr \longrightarrow CH_3CHBrCH_2CH_2CO_2H$$

β, γ-Unsaturated acid   **γ-Halogeno acid**

(2)   γ-Halogeno acids may also be prepared from γ-hydroxy acids when treated with concentrated halogen acids or phosphorus halide or sulphonyl halide.

$$HOCH_2CH_2CH_2CO_2H + HCl \longrightarrow ClCH_2CH_2CH_2CO_2H + H_2O$$

γ-Halogeno acid

The methods for the preparation of a particular type of halogen acid have already been described. Below is given a method which is common to all halogen acid preparation and is known as *Hunsdiecker reaction* (sec Section 12.3.7) starting from diethyl adipate. The following sequence of reactions produces δ-bromoacid.

$$\underset{\text{Ethyl hydrogen adipate}}{\overset{\displaystyle CH_2CH_2CO_2C_2H_5}{CH_2CH_2CO_2H}} \xrightarrow[\text{KOH}]{\text{AgNO}_3} \overset{\displaystyle CH_2CH_2CO_2C_2H_5}{CH_2CH_2CO_2Ag} \xrightarrow[\text{CCl}_4]{\text{Br}_2} \overset{\displaystyle CH_2CH_2CO_2C_2H_5}{CH_2CH_2Br} \xrightarrow[\Delta]{\text{HO}^-} \underset{\text{δ-Bromo acid}}{Br(CH_2)_4CO_2H}$$

21.8 PROPERTIES

The halogen acids are colorless solids or syrupy liquids which dissolve in water to give acidic solution. They are much stronger acids than the parent carboxylic acids due to the presence of electron withdrawing halogen atoms which exert – I effect favoring the release of proton from carboxyl group.

$$X \leftarrow CH_2 \overset{\displaystyle O}{\underset{\displaystyle \|}{\leftarrow C}} \leftarrow O \leftarrow H \qquad \overset{\displaystyle X}{\underset{\displaystyle X}{>}} CH \overset{\displaystyle O}{\underset{\displaystyle \|}{\lll C}} \lll O \lll H \qquad \overset{\displaystyle X}{\underset{\displaystyle X}{\overset{\displaystyle X}{\Subset}} C} \lll \overset{\displaystyle O}{\underset{\displaystyle \|}{C}} \lll O \lll H$$

The more the number of halogen atoms attached to α-carbon, the stronger is the inductive effect and stronger is the acid. There is, however, a decrease in acid strength if the distance of halogen atom increases from the carboxyl group due to diminishing – I effect of halogens. Thus, the order of decreasing acid strength is as follows:

α-halogeno acids > β-halogeno acids > γ-halogeno acids.

Halogen acids give reactions of both alkyl halides and carboxylic acid.

(1) As has already been discussed earlier, halogen atom of α-halogeno acids are more reactive (due to electron withdrawl by the carboxyl group) and they undergo a variety of bimolecular nucleophilic substitutions (S$_N$2) when treated with appropriate reagents.

$$ClCH_2CO_2H \longrightarrow \begin{cases} +AgOH \xrightarrow{H_2O} HOCH_2CO_2H + AgCl \\[2mm] +KI \xrightarrow{CH_3COCH_3} ICH_2CO_2H + KCl \\[2mm] +KCN \xrightarrow{C_2H_5OH} NCCH_2CO_2H + KCl \\[2mm] +2NH_3 \xrightarrow{C_2H_5OH} NH_2CH_2CO_2NH_4 + HCl \\[2mm] +2NH_3 \xrightarrow{H_2O/H^+} NH_2CH_2CO_2H + NH_4Cl \end{cases}$$

β- and γ-halogeno acids undergo similar reactions but they are slow since β and γ-halogeno acids are less reactive than α-halogeno acids (less electron withdrawl by carboxyl group as the distance is increased).

(2) Halogeno acids react with aqueous alkali to give a variety of products depending upon the position of halogen with respect to carboxyl group.

α-Halogeno acids give α-hydroxy acids.

$$CH_3CHClCO_2H + NaOH \longrightarrow CH_3CHOHCH_2H$$

β-Halogeno acids are converted into α,β-unsaturated acids.

$$CH_3CHClCH_2CO_2H + NaOH \longrightarrow CH_3CH=CHCO_2H$$
β-Halogeno acid   **α, β-Unsaturated acid**

γ and δ-halogenoacids are converted into γ and δ lactones respectively.

$$ClCH_2CH_2CH_2CO_2H + NaOH \longrightarrow \underset{\text{γ-Lactone}}{\overset{\displaystyle CH_2-O}{\underset{\displaystyle CH_2-CO}{CH_2}}} + NaCl$$

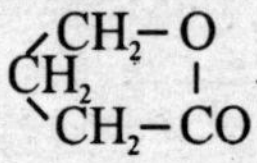

γ-Halogeno acid

$$ClCH_2CH_2CH_2CH_2CO_2H + NaOH \longrightarrow \underset{\text{δ-Lactone}}{CH_2\overset{\displaystyle CH_2-O}{\underset{\displaystyle CH_2-CH_2}{}}CO} + NaCl$$

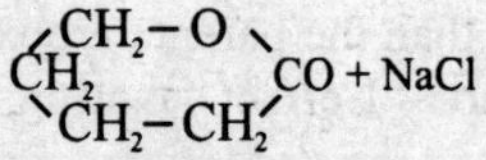

δ-Halogeno acid

This is an example of neighboring group participation.

ε-Halogenoacids form ε-hydroxy acids

$$ClCH_2(CH_2)_4CO_2H + NaOH \longrightarrow HOCH_2(CH_2)_4CO_2H + NaCl$$

21.9   INDIVIDUAL MEMBERS

21.9.1   Chloroacetic Acid/Chloroethanoic Acid [ClCH₂CO₂H]

Preparation

(1)   It can be prepared by HVZ reaction by passing chlorine in acetic acid in presence of red phosphorus.

(2)   On commercial scale chloroacetic acid is prepared by heating trichloroethylene with 75% sulphuric acid.

$$CHCl=CCl_2 + 2H_2O \xrightarrow{\underset{140°}{H_2SO_4}} ClCH_2CO_2H + 2HCl$$
Trichloroethene   **Chloroacetic acid**

(3)   It can also be prepared from acetic acid by its chlorination in acetic anhydride-sulfuric acid mixture.

Properties

It is a deliquescent solid, m.p. 61°, soluble in water and ethanol. It gives all the reaction of α-halogeno acids discussed earlier.

It is readily esterified in presence of sulfuric acid.

$$ClCH_2CO_2H + C_2H_5OH \xrightarrow{H_2SO_4} ClCH_2CO_2C_2H_5$$
Chloroacetic acid   **Ethyl chloroacetate**

Ethyl chloroacetate forms chloroacetamide on reaction with ammonia which undergoes dehydration to yield chloroacetonitrile.

$$CLCH_2CO_2C_2H_5 \xrightarrow[-C_2H_5OH]{NH_3} CLCH_2CONH_2 \xrightarrow[-H_2O]{P_2O_5} CLCH_2C\equiv N$$

Ethyl chloroacetate   **Chloroacetamide**   **Chloroacetonitrile**

21.9.2   Dichloroacetic Acid [Cl$_2$CHCO$_2$H]

Preparation

It is prepared in laboratory as well as in industries by adding calcium carbonate to a warm aqueous solution of chloral hydrate followed by addition of aqueous sodium cyanide.

$$2CCl_3CH(OH)_2 + 2CaCO_3 \xrightarrow{NaCN} (CHCl_2CO_2)_2Ca + 2CO_2 + CaCl_2 + 2H_2O$$

$$(CHCl_2CO_2)_2Ca + 2HCl \longrightarrow 2CHCl_2CO_2H + CaCl_2$$

Dichloroacetic acid

The role of sodium cyanide is not well understood. Most probably sodium cyanide helps in the elimination of a molecule of hydrochloric acid from chloral hydrate yielding unstable enol which readily rearranges to dichloroacetic acid.

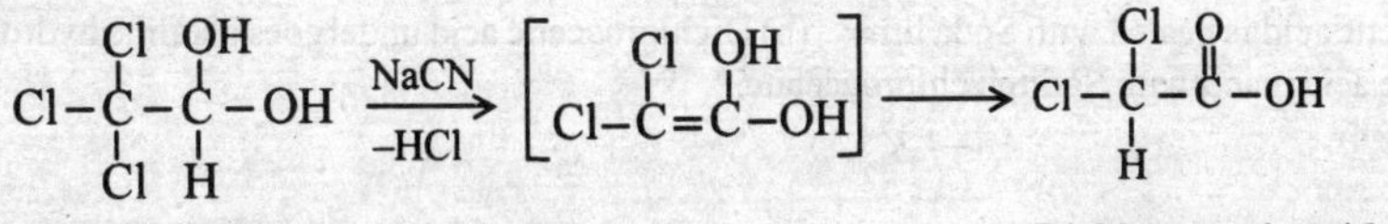

Chloral hydrate   **Diol**   **Dichloroacetic acid**

Properties

It is a colorless liquid, b.p. 194°, soluble in water. When hydrolysed with dilute alkali, it yields glyoxylic acid.

$$Cl_2CHCO_2H + H_2O \xrightarrow[2.\ H^+]{1.\ NaOH} \begin{matrix} CHO \\ | \\ CO_2H \end{matrix} + 2HCl$$

Dichloroacetic acid   **Glyoxylic acid**

When dichloroacetic acid is hydrolysed with strong alkali it gives sodium oxalate and sodium glycollate as glyoxylic acid first produced undergo Cannizaro reaction.

$$2\begin{matrix} CHO \\ | \\ CO_2H \end{matrix} + H_2O \xrightarrow{NaOH} HOCH_2CO_2Na + (COONa)_2$$

Sodium glycollate   Sodium oxalate

21.9.3   Trichloroacetic Acid [Cl$_3$CCO$_2$H]

Preparation

Oxidation of chloral hydrate with concentrated nitric acid yields trichlroacetic acid.

$$CCl_3CH(OH)_2 + [O] \xrightarrow{HNO_3} CCl_3CO_2H + H_2O$$

Properties

It is a deliquescent solid, m.p 58° and is one of the strongest organic acids.

The presence of three chlorine atoms on a carbon next to carboxylic group breaks the C–C bond very easily. Thus, when boiled with dilute alkali or even water, chloroform is obtained.

$$CCl_3CO_2H \longrightarrow CHCl_3 + CO_2$$

21.10   PROBLEMS

1. Discuss the methods of preparation of α-hydroxycaboxylic acids.
2. What happens when lactic acid reacts with:
   (a) H_2O_2 (b) Hot, dilute H_2SO_4 (c) Dilute acidified potassium permanganate.
3. Discuss the isomerism in lactic acid.
4. How is tartaric acid synthesised from acetylene? What happens when it reacts with:
   (a) Hydriodic acid   (b) Concentrated sulphuric acid   (c) PCl_5
5. How is citric acid obtained from 1,3-dichloro-2-propanone?
6. How will you synthesize:
   (a) Tartaric acid from glyceraldehyde   (b) Citric acid from glycerol.
   (c) Citric acid from acetic acid   (d) Malonic acid from acetic acid.
7. What is the action of alkali on:
   (a) α, β and γ-chlorobutyric acids   (b) mono, di and trichloroacetic acids.
8. What is the effect of chlorination of methyl group of acetic acid on its acid strength.
9. Explain that Chloroacetic acid is a better proton doner compared to acetic acid.
10. Arrange the following in the order of increasing acid strength.
    (a) CH_3COOH, $ClCH_2COOH$, Cl_3CCOOH and $Cl_2CHCOOH$.
    (b) $ClCH_2COOH$, $BrCH_2COOH$, CH_3COOH and ICH_2COOH.
11. What happens when:
    (a) Trichloroacetic acid is heated with Soda lime   (b) Dichloroacetic acid undergoes alkaline hydrolysis.
    (c) Chloroacetic acid reacts with Sodium chloroacetate.

22

Functional Derivatives of Carboxylic Acids

22.1 INTRODUCTION

The functional derivatives of carboxylic acids are obtained from the acids by repacement of −OH of the carboxyl group by other groups or atoms.

Acid halides	RCOX(X=Cl,Br,I)	Obtained by the replacement of OH by halogen
Acid amide	RCONH$_2$	Obtained by the replacement of OH by amino group
Acid anhydride	(RCO)$_2$O	Obtained by the replacement of OH by acyloxy group
Esters	RCOOR′	Obtained by the replacement of OH by alkoxy group

The above functional derivatives of carboxylic acids have been discussed in this chapter.

22.2 ACID HALIDES

Acid halides, RCOX, are commonly known as acyl halides.

22.2.1 Nomenclature

The common names of acid chlorides are derived by replacing the suffix 'ic acid' of the parent acid with 'yl chloride'. In the IUPAC system, the names are derived by replacing ending 'e' of the parent alkane with 'oyl halide'. The common and IUPAC names of some acid chlorides are given below:

Formula	Common Name	IUPAC Name
HCOCl	Formyl chloride	Methanoyl chloride
CH$_3$COCl	Acetyl chloride	Ethanoyl chloride
CH$_3$CH$_2$COCl	Propionyl chloride	Propanoyl chloride
CH$_3$CH$_2$CH$_2$COCl	Butyryl chloride	Butanoyl chloride
CH$_3$CH(CH$_3$)COCl	iso-Butyryl chloride	2-Methylpropanoyl chloride

22.2.2 General Methods of Preparation

1. *From Carboxylic Acids:* The carboxylic acids on heating with Phosphorus trichloride, phosphorus pentachloride or thionyl chloride give the corresponding acid chloride.

$$3RCOOH + PCl_3 \longrightarrow 3RCOCl + H_3PO_3$$

Acid **Phosphoric**
chloride **acid**

$$CH_3COOH + PCl_5 \longrightarrow CH_3COCl + POCl_3 + HCl$$
$$CH_3COOH + SOCl_2 \longrightarrow CH_3COCl + SO_2 + HCl$$
Acetyl chloride

2. *From Salts of Carboxylic Acids:* The salts of carboxylic acids when distilled with phosphorus trichloride, phosphorus oxychloride or sulphuryl chloride form acid chloride.

$$3CH_3COONa + PCl_3 \longrightarrow 3CH_3COCl + Na_3PO_3$$
Sodium acetate

$$2CH_3COONa + POCl_3 \longrightarrow 2CH_3COCl + NaCl + NaPO_3$$

$$(CH_3COO)_2Ca + SO_2Cl_2 \longrightarrow 2CH_3COCl + CaSO_4$$
Calcium acetate   **Acetyl chloride**

The acid bromides, iodides or fluorides are prepared from carboxylic acid chlorides by the reaction with HBr, HI or HF.

$$RCOCl + HX \longrightarrow R-\overset{\overset{\text{O}}{\|}}{C}-X + HCl$$
(HX = HBr, HI or HF)

22.2.3   Physical Properties

Acid chlorides have pungent, irritating odors. The lower members are colorless liquids while the higher members are solids. Their boiling points are lower than those of the corresponding acids. This is due to the absence of intermolecular hydrogen bonding.

22.2.4   Spectral Properties

In IR spectra, the acid chlorides show carbonyl group (stretching) absorption in the region 1815-1785 cm^{-1}. Besides they exhibit C−Cl bond absorption in the region 800-600 cm^{-1}.

22.2.5   Chemical Properties

1. *Nucleophilic Acyl Substitution:* All acid derivatives (having carbonyl group) undergo nucleophilic substitution in a similar way to form various products(where the carbonyl group is retained) by replacement of Cl$^-$, RCOO$^-$, NH$_2^-$ and RO$^-$ by some other basic group.

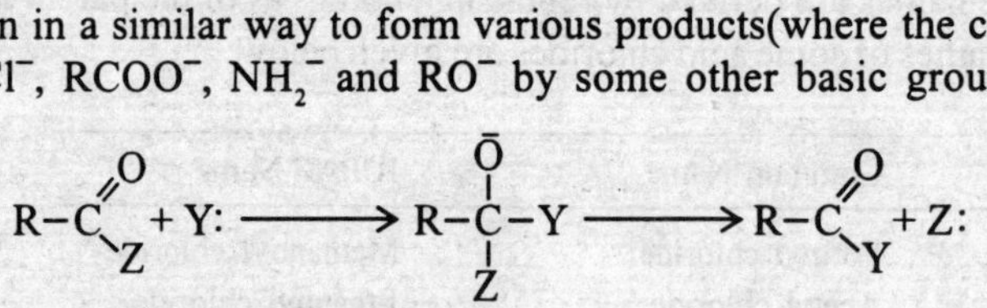

$$R-\overset{\overset{\text{O}}{\diagup}}{\underset{Z}{C}} + Y: \longrightarrow R-\overset{\overset{\bar{\text{O}}}{|}}{\underset{Z}{C}}-Y \longrightarrow R-\overset{\overset{\text{O}}{\diagup}}{C}_{Y} + Z:$$

Z: = −Cl, −OCOR, −NH$_2$, −OR
Y: = HO$^-$, RO$^-$

Functional derivatives of acids differ from aldehydes and ketones which undergo nucleophilic addition as shown below.

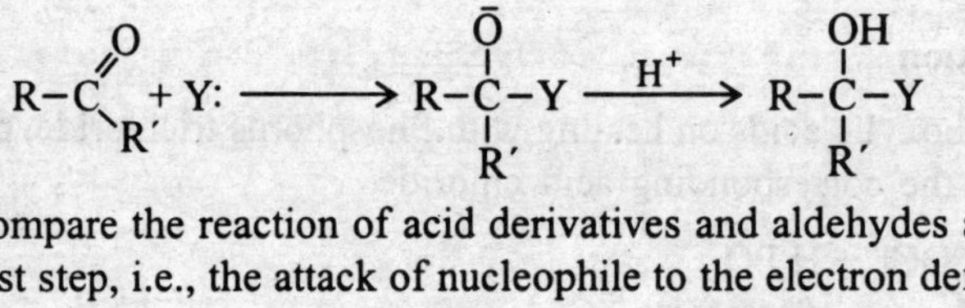

$$R-\overset{\overset{\text{O}}{\diagup}}{\underset{R'}{C}} + Y: \longrightarrow R-\overset{\overset{\bar{\text{O}}}{|}}{\underset{R'}{C}}-Y \xrightarrow{H^+} R-\overset{\overset{\text{OH}}{|}}{\underset{R'}{C}}-Y$$

If we compare the reaction of acid derivatives and aldehydes and ketones with nucleophiles, we find that the first step, i.e., the attack of nucleophile to the electron deficient carbon is same but in the second step acid derivatives undergo substitution whereas aldehydes and ketones undergo addition. This is because the removal of weakly basic Cl^- in acid chloride, moderately basic $RCOO^-$ in anhydride and strongly basic NH_2^- in amide and RO^- in esters is easier compared to the removal of H^- (hydride ion) in aldehydes and R^- (alkide ion) in ketone which are highly basic in nature.

The nucleophilic acyl substitution in presence of acid involves following steps.

$$\underset{Z}{\overset{R}{>}}C=O + H^+ \longrightarrow \underset{Z}{\overset{R}{>}}C \overset{+}{=} OH \rightleftharpoons \ddot{Y}: \longrightarrow R-\overset{\overset{Y}{|}}{\underset{\underset{Z}{|}}{C}}-OH \longrightarrow \underset{Y}{\overset{R}{>}}C=O + HZ$$

The mechanism of nucleophilic acyl substitution involves the attack of nucleophile on carbonyl carbon followed by removal of leaving group.

$$\underset{Z}{\overset{R}{>}}C=O + \bar{\ddot{Y}}: \longrightarrow R-\overset{\overset{Y}{|}}{\underset{\underset{Z}{\wedge}\bar{O}}{C}} \longrightarrow \underset{Y}{\overset{R}{>}}C=O + Z:$$

Leaving group

The substitution reactions are favoured by electron withdrawing groups which stabilise the negative charge on the oxygen in the intermediate. On the other hand bulky groups slows down the reaction as they crowd the transition state.

Acid chlorides undergo nucleophilic substitution more readily than alkyl halides due to the carbonyl group.

In alkyl halides, nucleophilic attack (S_N2) on tetrahedral carbon involves a crowded transition state having pentavalent carbon.

$$CH_3Br + Y: \xrightarrow{S_N2} [Y \cdots \overset{\overset{H \diagdown \diagup H}{|}}{\underset{H}{C}} \cdots Br] \longrightarrow Y-CH_3 + B\bar{r}$$

Whereas the nucleophilic attack on sp^2 hybridised carbonyl carbon involves a relatively unhindered transition state leading to a tetrahedral intermediate which on removal of halide forms the substitution product.

$$R-\overset{\overset{O}{||}}{C}-Cl + :Y-H \longrightarrow R-\overset{\overset{\bar{O}}{|}}{\underset{\underset{+Y-H}{|}}{C}}Cl \xrightarrow{-C\bar{l}}$$

$$R-\overset{\overset{O}{||}}{\underset{\underset{+Y-H}{|}}{C}} \xrightarrow{-H^+} R-\overset{\overset{O}{||}}{C}-Y$$

Substituted product

Some nucleophilic substitution reactions of acetyl chloride are given below.

(i) *Action of Water (Hydrolysis):* On reaction with water, acid chlorides are hydrolysed to form parent acid.

$$\underset{\text{Acetyl chloride}}{CH_3-\overset{\overset{O}{||}}{C}-Cl} + H_2O \longrightarrow \underset{\text{Acetic acid}}{CH_3-\overset{\overset{O}{||}}{C}-OH} + HCl$$

(ii) *Action of Alcohol (Esterification)*: On reaction with alcohols, acid chlorides form esters.

$$CH_3-\overset{\overset{O}{\|}}{C}-Cl + C_2H_5\overset{..}{O}H \longrightarrow CH_3-\overset{\overset{O}{\|}}{C}-OC_2H_5 + HCl$$
Acetyl chloride   **Ethyl acetate**

(iii) *Action of Ammonia (Ammonolysis)*: Acid chlorides react with ammonia to form amides. Similar reaction also takes place with primary and secondary amines.

$$CH_3-\overset{\overset{O}{\|}}{C}-Cl + NH_3 \longrightarrow CH_3-\overset{\overset{O}{\|}}{C}-NH_2 + HCl$$
Ammonia   **Acetamide**

$$CH_3-\overset{\overset{O}{\|}}{C}-Cl + H_2NC_2H_5 \longrightarrow CH_3-\overset{\overset{O}{\|}}{C}-NHC_2H_5 + HCl$$
Ethanamine   **N-Ethylacetamide**

$$CH_3-\overset{\overset{O}{\|}}{C}-Cl + HN(C_2H_5)_2 \longrightarrow CH_3-\overset{\overset{O}{\|}}{C}-N(C_2H_5)_2 + HCl$$
N-Ethylethanamine   N,N-Diethylacetamide

(iv) *Action of Hydrazine:* Acid chlorides react with hydrazine to form hydrazides.

$$RCOCl + H_2NNH_2 \longrightarrow RCONHNH_2 + HCl$$
                      Hydrazide

(v) *Action of Hydroxylamine:* Hydroxylamine reacts with acid chlorides to form hydroxamic acid.

$$RCOCl + NH_2OH \longrightarrow RCONHOH + HCl$$
                      Hydroxamic acid

2. *Reduction (Rosenmund's reduction)*: Reduction of acid chlorides with hydrogen in presence o palladised barium sulphate gives aldehydes (−COCl is converted into −CHO).

$$RCOCl + H_2 \xrightarrow{Pd/BaSO_4} RCHO + HCl$$
Acid chloride   **Aldehyde**

In the above reduction barium sulphate acts as 'poison' and reduces the acitivity of the catalyst to the extent that the reduction stops at the aldehyde stage.

However, reduction of acid chlorides with lithium aluminum hydride gives alcohols (-COCl is converted into CH$_2$OH).

$$RCOCl \xrightarrow{LiAlH_4} R-CH_2OH + HCl$$
Acid chloride   **Primary alcohol**

3. *Halogenation:* Acid chlorides undergo halogenation, the halogen entering the more active α-position. Due to electronegative chlorine attached to the carbonyl carbon, the α-hydrogen atom becomes more active.

$$\overset{\beta}{CH_3}\overset{\alpha}{CH_2}COCl + Cl_2 \longrightarrow CH_3CHClCOCl + HCl$$
Propanoyl chloride   **α-chloropropanoyl chloride**

It has already been stated (properties of monocarboxylic acids) that the acids can be chlorinated in presence of small amount of red phosphorus and chlorine (Hell Volhard-Zelinskey reaction).

4. *Friedel Crafts Reaction*: Acid chlorides react with aromatic hydrocarbons in presence of anhydrous aluminum chloride to form aromatic ketones.

$$C_6H_6 + CH_3COCl \xrightarrow{AlCl_3} C_6H_5COCH_3$$
$$\textbf{Acetophenone}$$

This reaction is known as Friedel Crafts acylation and will be discussed in detail in aromatic hydrocarbons.

5. *Action of Potassium Cyanides*: Acid chlorides on reaction with potassium cyanide form acyl cyanide, which on hydrolysis give keto acids.

$$CH_3COCl + KCN \xrightarrow[-KCl]{} CH_3COCN \xrightarrow[H^+]{H_2O} CH_3COCOOH$$

Acetyl chloride   **Acetyl cyanide**   **Pyruvic acid**

6. *Reaction with Grignard Reagents*: Acid chlorides react with Grignard reagent to form tertiary alcohol.

$$CH_3COCl + CH_3MgCl \longrightarrow CH_3COCH_3 + MgCl_2$$
$$\textbf{Acetone}$$

$$CH_3COCH_3 + CH_3MgCl \longrightarrow \begin{bmatrix} \overset{O^-M^+gCl}{\underset{CH_3}{CH_3-C-CH_3}} \end{bmatrix} \xrightarrow{H_2O} \underset{CH_3}{CH_3-C-CH_3} + Mg\overset{OH}{\underset{Cl}{\diagdown}}$$

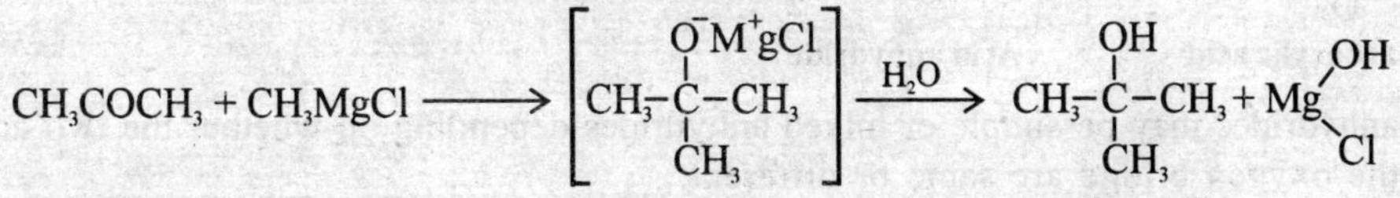

Acetone   **tert-Butanol**

7. *Reaction with Sodium Salt of Fatty Acid*: Acid chlorides on heating with sodium salt of a fatty acid give acid anhydrides.

$$CH_3COCl + CH_3COONa \longrightarrow CH_3.CO.O.CO.CH_3 + NaCl$$

Acetyl   **Sodium**   **Acetic anhydride**
chloride   **acetate**

22.2.6 Individual Members

22.2.6.1 *Formyl Chloride/Methanoyl Chloride* [HCOCl]

Formyl chloride is an unstable compound and does not exist at ordinary temperature. However, there is evidence of its existance at low temperature (–80°). A mixture of carbon monoxide and hydrogen chloride behaves as formyl chlorides.

$$CO + HCl \longrightarrow H-\overset{O}{\overset{\|}{C}}-Cl$$
$$\textbf{Formyl chloride}$$

This mixture is used for formylation of aromatic compounds in presence of anhydrous aluminum chloride.

$$\text{Benzene} + CO + HCl \xrightarrow{anhyd.\ AlCl_3} \text{Benzaldehyde (CHO)}$$

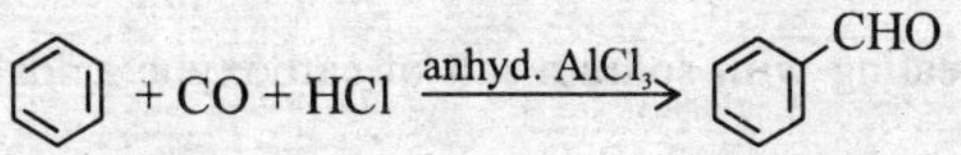

Benzene   **Benzaldehyde**

This reaction is known as *Gatterman koch aldehyde synthesis.*

22.2.6.2 *Acetyl Chloride/Ethanoyl Chloride* [CH_3COCl]

Acetyl chloride is the most important acid chloride and can be prepared by the general methods described earlier. It is prepared in the laboratory by the reaction of acetic acid with phosphorus trichloride.

$$3CH_3COOH + PCl_3 \longrightarrow 3CH_3COCl + H_3PO_3$$

Acetic acid   Acetyl chloride

It is a colorless, pungent smelling, volatile liquid (b.p. 52°). It fumes in moist air and is soluble in ether, acetone and acetic acid. It is stored in bottles having tight stoppers.

Acetyl chloride gives all general reactions of acid chlorides described earlier.

It is used as an acetylating agent for the preparation of acetic anhydride, acetamide and acetanilide. It is redily used in organic synthesis. It is also used to detect and estimate the number of hydroxy groups in organic compounds.

22.3 ACID ANHYDRIDES

The acid anhydrides are regarded as being derived from two molecules of acid by elimination of a molecule of water.

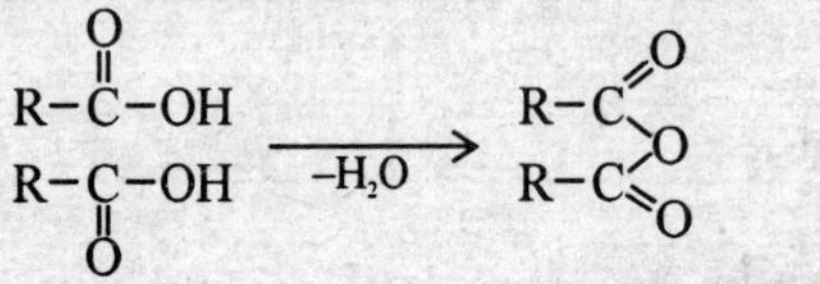

Carboxylic acid   Acid anhydride

The acid anhydrides may be simple or mixed anhydrides depending on whether the two acyl groups attached to the oxygen bridge are same or different.

Except formic acid, all monocarboxylic acids form the corresponding anhydrides. However, mixed anhydrides of formic acid and other carboxylic acids are known.

Formic acid   Acetic acid   Acetic formic anhydride

22.3.1 Nomenclature

The names of the acid anhydrides are derived by dropping the word *'acid'* from the name of the parent acid and adding the word *'anhydride'*. This method is used in both common and IUPAC systems. For mixed anhydrides, the names of the two parent acids are written in alphabetical order. Some examples given as follows:

Formula	Common Name	IUPAC Name
$CH_3CO-O-COCH_3$	Acetic anhydride	Ethanoic anhydride
$CH_3CH_2CO-O-COCH_2CH_3$	Propionic anhydride	Propanoic anhydride
$HCO-O-COCH_3$	Acetic formic anhydride	Ethanoic methanoic anhydride

22.3.2 General Methods of Preparation

1. *From Acid Chlorides*: Acid chlorides on heating with sodium salt of carboxylic acids give acid anhydrides.

$$CH_3-\overset{O}{\overset{\|}{C}}-Cl + CH_3COONa \longrightarrow CH_3CO-O-COCH_3 + NaCl$$

Acetyl chloride Sodium acetate Acetic anhydride

2. *From Carboxylic Acids*: Carboxylic acids on heating with a suitable dehydrating agent (for example P_2O_5) give anhydrides.

$$2CH_3COOH \xrightarrow[-H_2O]{P_2O_5,\ \Delta} \begin{matrix} CH_3-CO \\ CH_3-CO \end{matrix}\!\!\!\diagdown O$$

Acetic acid   Acetic anhydride

Another method to convert carboxylic acids into anhydrides is by heating with acetic anhydride.

$$2R-\overset{O}{\overset{\|}{C}}-OH + CH_3CO-O-COCH_3 \longrightarrow R-\overset{O}{\overset{\|}{C}}-O-\overset{O}{\overset{\|}{C}}-R + CH_3COOH$$

Carboxylic acid   Acetic anhydride   Acid anhydride

The second method is useful for the preparation of high molecular weight acid anhydrides by distilling the corresponding carboxylic acid with acetic anhydrides.

22.3.3 Physical Properties

The lower member of the series (except formic anhydride which is unstable) are colorless liquids with sharp pungent odor. The higher members are solids. Their boiling points increase with increase in the molecular weight. They are only slightly soluble in water but soluble in alcohol and ether.

22.3.4 Spectral Properties

The infrared absorption of the carbonyl group (stretching) in acid anhydrides is in the region 1840-1800 cm^{-1} and 1780-1740 cm^{-1}. The C−O (stretching) of the group 'C−O−C' is noticed in the region 1180-1030 cm^{-1}.

22.3.5 Chemical Properties

1. *Nucleophilic Acyl Substitution*: The nucleophilic acyl substitution reactions of acid anhydrides are analogues to those of acid halides discussed earlier. The difference is that in case of acid anhydride the leaving group is a carboxylate anion instead of the halide anion in the case of acid halides. Some simple nucleophilic substitution reactions of acid anhydrides are given below.

(i) *Action of Water (Hydrolysis)*: The acid anhydrides on reaction with water get hydrolysed to form carboxylic acid.

$$R-\overset{O}{\overset{\|}{C}}-O-\overset{O}{\overset{\|}{C}}-R + H_2O \longrightarrow 2R-\overset{O}{\overset{\|}{C}}-OH$$

Acid anhydride   Carboxylic acid

(ii) *Reaction with Alcohols*: On reaction with alcohols, the acid anhydrides form esters.

$$R-\overset{\overset{\text{O}}{\|}}{C}-O-\overset{\overset{\text{O}}{\|}}{C}-R + R'OH \longrightarrow R-\overset{\overset{\text{O}}{\|}}{C}-O-R' \quad 2R-\overset{\overset{\text{O}}{\|}}{C}-OH$$

Acid anhydride   Ester   Carboxylic acid

(iii) *Reaction with Ammonia and Amines*: The acid anhydrides react with ammonia to form an amide and a carboxylic acid, which further reacts with ammonia to form ammonium salt.

$$R-\overset{\overset{\text{O}}{\|}}{C}-O-\overset{\overset{\text{O}}{\|}}{C}-R + 2NH_3 \longrightarrow R-\overset{\overset{\text{O}}{\|}}{C}-NH_2 + R-\overset{\overset{\text{O}}{\|}}{C}-\bar{O}\overset{+}{N}H_4$$

Acid anhydride   Ammonia   Amide   Ammonium salt of acid

Amines react in a similar way.

$$R-\overset{\overset{\text{O}}{\|}}{C}-O-\overset{\overset{\text{O}}{\|}}{C}-R + 2R'_2NH \longrightarrow R-\overset{\overset{\text{O}}{\|}}{C}-NR'_2 + R-\overset{\overset{\text{O}}{\|}}{C}-\bar{O}\overset{+}{N}H_2R'_2$$

Acid anhydride   sec. amine   Amide   Ammonium carboxylate salt

2. *Friedel Crafts Reaction*: Acid anhydrides serve as a source of alkanoyl cation and can be used in Friedel-crafts alkanoylations.

$$\bigcirc + CH_3-\overset{\overset{\text{O}}{\|}}{C}-O-\overset{\overset{\text{O}}{\|}}{C}-CH_3 \xrightarrow[\Delta]{AlCl_3} \bigcirc^{COCH_3} + CH_3COOH$$

Acetophenone

22.3.6 Individual Members

22.3.6.1 *Acetic Anhydride/Ethanoic Anhydride* $[CH_3\overset{\overset{\text{O}}{\|}}{C}-O-\overset{\overset{\text{O}}{\|}}{C}CH_3]$

Preparation

(1) In the laboratory it is prepared by heating acetyl chloride with anhydrous sodium acetate.

$$CH_3COCl + CH_3COONa \longrightarrow CH_3CO-O-COCH_3 + NaCl$$

Acetyl chloride   Sodium acetate   Acetic anhydride

(2) It is obtained on a commercial scale by passing acetylene into glacial acetic acid containing some mercuric sulphate (catalyst) at 60-85°. The formed ethylidene acetate is distilled to give acetic anhydride.

$$HC\equiv CH + 2CH_3COOH \xrightarrow[60-85°]{Hg^{2+}} CH_3CH(OCOCH_3)_2 \xrightarrow{distn.} (CH_3CO)_2O + CH_3CHO$$

Acetylene   Ethylidene acetate

(3) On a commercial scale it is also prepared by passing ketene into glacial acetic acid. The ketene required is obtained by pyrolysis of acetone.

$$CH_3COCH_3 \xrightarrow{Pyrolysis} CH_2=C=O + CH_4$$

      Acetone   Ketene

$$CH_2=C=O + CH_3COOH \longrightarrow (CH_3CO)_2O$$

      Ketene   Acetic anhydride

Properties: Acetic anhydride is a colorless liquid (b.p. 139.5°) with a pungent irritating smell. It is sparingly soluble in water, but readily soluble in ether and benzene.

It gives reactions similar to those of acetyl chloride, but is much less reactive. Some of its reactions are given below.

$$CH_3CO.O.COCH_3 + H_2O \longrightarrow 2CH_3COOH$$
Acetic anhydride   **Acetic acid**

$$CH_3CO.O.COCH_3 + 2C_2H_5OH \longrightarrow 2CH_3COOC_2H_5$$
Ethyl acetate

$$CH_3CO.O.COCH_3 + NH_3 \longrightarrow CH_3CONH_2 + CH_3COOH$$
Acetamide

$$CH_3CO.O.COCH_3 + HCl \longrightarrow CH_3COCl + CH_3COOH$$
Acetyl chloride

$$CH_3CO.O.COCH_3 + N_2O_5 \longrightarrow 2CH_3COONO_2$$
Acetyl nitrate

$$CH_3CO.O.COCH_3 + C_6H_6 \xrightarrow{AlCl_3} C_6H_5COCH_3 + CH_3COOH$$
Acetophenone

$$CH_3CO.O.COCH_3 + CH_3CHO \longrightarrow CH_3CH(OCOCH_3)_2$$
Ethylidene acetate

$$CH_3CO.O.COCH_3 + 8[H] \xrightarrow{LiAlH_4} 2CH_3CH_2OH + H_2O$$
Ethyl alcohol

$$CH_3CO.O.COCH_3 + [O] \longrightarrow CH_3 - CO - O - O - CO - CH_3$$
Acetyl peroxide

Uses: Acetic anhydride is used as an acetylating agent. Due to its stability it is preferred over acetyl chloride. It is used in the preparation of medicines like aspirin and phenacetin and in the manufacture of plastics and polyvinyl acetate.

22.4 ACID AMIDES

Acid amides are obtained by the replacement of hydroxy group of COOH by an amino group.

$$\underset{\textbf{Acid}}{R-\overset{\overset{\textstyle O}{\|}}{C}-OH} \longrightarrow \underset{\textbf{Acid amide}}{R-\overset{\overset{\textstyle O}{\|}}{C}-NH_2}$$

Alternatively, they may be regarded as acyl derivatives of ammonia obtained by the replacement of hydrogen atom of ammonia by an acyl group.

$$\underset{\textbf{Ammonia}}{NH_3} \longrightarrow \underset{\textbf{Acid Amide}}{RCONH_2}$$

22.4.1 Nomenclature

The common names are derived from the common name of the parent carboxylic acid by dropping the suffix 'ic acid' and adding 'amide'. In the IUPAC system the ending 'e' of the parent hydrocarbon is replaced by 'amide'. A few examples are given below.

Formula	Common Name	IUPAC Name
$HCONH_2$	Formamide	Methanamide
CH_3CONH_2	Acetamide	Ethanamide
$CH_3CH_2CONH_2$	Propionamide	Propanamide
$CH_3CH_2CH_2CONH_2$	Butyramide	Butanamide
$CH_3CONHCH_3$	N-methylacetamide	N-Methylethanamide

22.4.2   General Methods of Preparation

(1) *From Ammonium Salts of Carboxylic Acids*: Simple amides are obtained by heating the ammonium salts of the corresponding carboxylic acids.

$$\underset{\text{Ammonium carboxylate}}{R-\overset{\overset{\displaystyle O}{\|}}{C}-\bar{O}\overset{+}{N}H_4} \xrightarrow{180-240°} \underset{\text{Amide}}{R-\overset{\overset{\displaystyle O}{\|}}{C}-NH_2} + H_2O$$

(2) *From Alkyl Nitriles*: Controlled hydrolysis of alkyl nitriles gives acid amides.

$$\underset{\text{Alkyl nitrile}}{R-C\equiv N} + H_2O \xrightarrow[\text{Room temp.}]{\text{Conc. HCl}} \underset{\text{Acid amide}}{R-\overset{\overset{\displaystyle O}{\|}}{C}-NH_2}$$

(3) *From Acid Chlorides, Anhydrides or Esters*: Amides are obtained by the ammonolysis of acid chlorides, anhydrides or esters with concentrated ammonia.

$$\underset{\text{Acid chloride}}{RCOCl} + NH_3 \longrightarrow \underset{\text{Amide}}{RCONH_2} + HCl$$

$$\underset{\text{Acid anhydride}}{(RCO)_2O} + NH_3 \longrightarrow RCONH_2 + RCOOH$$

$$\underset{\text{Ester}}{RCOOR'} + NH_3 \longrightarrow RCONH_2 + R'OH$$

22.4.3 Physical Properties

The amides (with the exception of formamide) are colorless crystalline solids. They are odorless when pure. The unpleasant smell associated with simple amides is due to the presence of impurities. They have high melting points due to hydrogen bonding.

$$\cdots O=\underset{R}{\overset{|}{C}}-\underset{H}{\overset{|}{N}}-H\cdots O=\underset{R}{\overset{|}{C}}-\underset{H}{\overset{|}{N}}-H\cdots$$

[Intermolecular hydrogen bonding]

The lower member (up to 5 carbon atoms) are soluble in water. This is due to association with water through hydrogen bonding.

$$\cdots O=\underset{R}{\overset{|}{C}}-\underset{H}{\overset{|}{N}}-H\cdots \underset{H}{\overset{|}{O}}-H\cdots O=\underset{R}{\overset{|}{C}}-\underset{H}{\overset{|}{N}}-H\cdots \underset{H}{\overset{|}{O}}-H\cdots$$

[Hydrogen bonding with water]

The water solubility decreases with increase in molecular weight.

22.4.4   Spectral Properties

The carbonyl group (stretching) show Infrared absorption in the region 1699 and 1650 cm^{-1}. Primary and secondary amides show the N−H stretching absorption in the region 3200-3400 cm^{-1} and a strong N−H bending absorption near 1640 cm^{-1}.

22.4.5   Chemical Properties

(1) *Basic and Acidic Nature of Amides*: Amides behave both as a weak base and weak acid as they are resonance hybrid of the following two structures.

$$R-\overset{\overset{\text{O}}{\|}}{C}-\overset{\overset{\cdot\cdot}{N}}{\underset{H}{}}-H \longleftrightarrow R-\overset{\overset{\bar{O}}{|}}{C}=\overset{+}{\underset{H}{N}}-H \equiv R-\overset{\overset{\delta-}{O}}{C}\overset{\delta+}{\underset{H}{=N}}-H$$

Resonance hybrid

As seen there is delocalization of the non-bonding electron pair of nitrogen between O−C−N. So the electron pair is not available for protonation. Hence amides are neutral or weakly basic in character compared to amines in which the electron pair of nitrogen is available for protonation. On the other hand, due to the delocalisation of the non-bonding electron pair of nitrogen, oxygen acquires a partial negative charge and nitrogen a partial positive charge (see the hybrid structure). Thus the amides (as in the case of carboxylic acids) are expected to exhibit acid character. Due to weakening of the N−H bond as a result of delocalization of the electron pair, the release of the proton (H$^+$) is facilitated. Thus acid amides act as very weak acids.

(i) As weak bases, amides form salts with strong mineral acids. These salts are stable in aqueous solution.

$$R-\overset{\overset{\text{O}}{\|}}{C}-NH_2 + HCl \longrightarrow \left[R-C\overset{\overset{+}{O}-H}{\underset{\cdot\cdot NH_2}{}} \longleftrightarrow R-C\overset{O-H}{\underset{+NH_2}{}} \right] Cl^-$$

Amide

(ii) As weak acids, amides react with sodamide in ether solution to form salts. In this case the proton attached to nitrogen atom is released.

$$R-\overset{\overset{\text{O}}{\|}}{C}-NH_2 + Na\overset{+}{N}H_2 \longrightarrow \left[R-\overset{\overset{\text{O}}{\|}}{C}-\bar{N}H \right] Na^+ + NH_3$$

Amide   **Sodamide**   **Sodium salt of amide**

(2) *Hydrolysis*: On heating with dilute acids or alkalis, the amides are hydrolysed to give the parent acid and ammonia.

$$R-\overset{\overset{\text{O}}{\|}}{C}-NH_2 + H_2O \xrightarrow{\overset{+}{H}Cl^-} R-\overset{\overset{\text{O}}{\|}}{C}-OH + NH_4Cl$$

Amide   **Carboxylic acid**

$$R-\overset{\overset{\text{O}}{\|}}{C}-NH_2 + H_2O \xrightarrow{\overset{+}{Na}OH^-} R-\overset{\overset{\text{O}}{\|}}{C}-\bar{O}\overset{+}{Na} + NH_4OH$$

The *Mechanism of Acid Hydrolysis* of amides involves the steps given below.

$$\text{R-}\overset{\overset{\text{O}}{\|}}{\text{C}}\text{-NH}_2 \underset{}{\overset{\text{H}^+}{\rightleftharpoons}} \text{R-}\overset{\overset{+\text{OH}}{\|}}{\text{C}}\text{-}\overset{..}{\text{N}}\text{H}_2 + \text{H}_2\overset{..}{\text{O}} \rightleftharpoons \text{R-}\overset{\overset{\text{OH}}{|}}{\underset{\underset{}{\text{H-O-H}}}{\text{C}}}\text{-}\overset{..}{\text{N}}\text{H}_2$$

$$\overset{-\text{H}^+}{\rightleftharpoons} \text{R-}\overset{\overset{\text{OH}}{|}}{\underset{\underset{}{\text{OH}}}{\text{C}}}\text{-}\overset{..}{\text{N}}\text{H}_2 \overset{\text{H}^+}{\rightleftharpoons} \text{R-}\overset{\overset{\text{OH}}{|}}{\underset{\underset{}{\text{O-H}}}{\text{C}}}\text{-}\overset{+}{\text{N}}\text{H}_3 \longrightarrow \underset{\textbf{Carboxylic acid}}{\text{R-C=O} + \text{NH}_4^+}$$

Basic hydrolysis involves the attack of base followed by removal of ammonia and formation of sal of carboxylic acid.

$$\text{R-}\overset{\overset{\text{O}}{\|}}{\text{C}}\text{-NH}_2 + \text{HO}^- \rightleftharpoons \text{R-}\overset{\overset{\text{O}^-}{|}}{\underset{\underset{}{\text{O-H}}}{\text{C}}}\text{-NH}_2 \longrightarrow \text{R-}\overset{\overset{\text{O}}{\|}}{\text{C}}\text{-}\overset{-}{\text{O}} + \text{NH}_3$$

(3) *Reaction with Nitrous Acid*: The amides on treatment with nitrous acid (sodium nitrite and hy drochloric acid) give acid and nitrogen.

$$\underset{\textbf{Amide}}{\text{RCONH}_2} + \text{HO-NO} \longrightarrow \underset{\textbf{Carboxylic acid}}{\text{RCOOH}} + \text{N}_2 + \text{H}_2\text{O}$$

(4) *Dehydration*: On heating with phosphorus pentoxide, the amide loses a molecule of water to form cyanides.

$$\underset{\textbf{Amide}}{\text{RCONH}_2} \underset{-\text{H}_2\text{O}}{\overset{\text{P}_2\text{O}_5, \Delta}{\longrightarrow}} \underset{\textbf{Alkyl cyanides}}{\text{RCN}}$$

(5) *Reduction*: The amides on reduction with sodium and alcohol or lithium aluminum hydride yield primary amines.

$$\underset{\textbf{Amide}}{\text{RCONH}_2} \underset{\text{or Na-Ethanol}}{\overset{\text{LiAlH}_4}{\longrightarrow}} \underset{\textbf{Primary amine}}{\text{RCH}_2\text{NH}_2} + \text{RCH}_2$$

(6) *Hofmann reaction*: In this reaction a mixture of amide, bromine and alkali is heated. The product is a primary amine containing one carbon atom less than the amide.

$$\text{CH}_3\text{CONH}_2 + \text{Br}_2 + \text{KOH} \longrightarrow \underset{\textbf{Acetobromamide}}{\text{CH}_3\text{CONHBr}} + \text{KBr} + \text{H}_2\text{O}$$

$$\text{CH}_3\text{CONHBr} + \text{KOH} \longrightarrow \underset{\textbf{Methyl isocyanate}}{\text{CH}_3\text{NCO}} + \text{KBr} + \text{H}_2\text{O}$$

$$\text{CH}_3\text{NCO} + 2\text{KOH} \longrightarrow \underset{\textbf{Methanamine}}{\text{CH}_3\text{NH}_2} + \text{K}_2\text{CO}_3$$

$$\overline{\text{CH}_3\text{CONH}_2 + \text{Br}_2 + 4\text{KOH} \longrightarrow \text{CH}_3\text{NH}_2 + 2\text{KBr} + 2\text{H}_2\text{O} + \text{K}_2\text{CO}_3}$$

Mechanism

$$\text{R-}\overset{\overset{\text{O}}{\|}}{\text{C}}\text{-NH}_2 + \text{Br}_2 + \text{NaOH} \longrightarrow \text{R-}\overset{\overset{\text{O}}{\|}}{\text{C}}\text{-}\underset{\underset{}{\text{H}}}{\text{N}}\text{-Br} + \text{NaBr} + \text{H}_2\text{O}$$

$$R-\overset{\overset{O}{\|}}{C}-\overset{H}{\underset{|}{N}}-Br + :\overset{..}{O}H \longrightarrow R-\overset{\overset{O}{\|}}{C}-\overset{..}{N}: + \overset{-}{Br} + H_2O$$
$$\text{(Unstable)}$$

$$\overset{\overset{O}{\|}}{\textcircled{R}-C-\overset{..}{N}:} \longrightarrow R-\overset{..}{N}=C=O$$
$$\textbf{Isocyanate}$$

$$R-\overset{..}{N}=C=O + 2H\overset{-}{O} \xrightarrow{H_2O} RNH_2 + CO_3^{--}$$
$$\textbf{Amine}$$

The reaction is also known as Hofmann rearrangement. Here alkyl group with its bonding pair of electrons migrates to electron deficient nitrogen (an example of 1,2-shift). The reaction is associated with complete retention of configuration in the migrating group which is a common feature of all 1,2-shifts.

22.4.6   Individual Members

22.4.6.1   Formamide/Methanamide [HCONH₂]

It is prepared by heating ammonium formate in an atmosphere of ammonia.

$$\underset{\textbf{Amm. formate}}{HCOONH_4} \xrightarrow{\Delta} \underset{\textbf{Formamide}}{HCONH_2 + H_2O}$$

It is manufactured by the reaction of carbon monoxide and ammonia under high pressure.

$$CO + NH_3 \longrightarrow \underset{\textbf{Formamide}}{HCONH_2}$$

Formamide is a hygroscopic liquid (b.p. 193°) and is soluble in water and alcohol. It is used as a solvent and plasticiser.

22.4.6.2   Acetamide/Ethanamide [CH₃CONH₂]

It is prepared by the general methods of preparatios of acid amides already discussed.

Acetamine is a colorless crystalline compound (m.p. 82°, b.p. 222°) having a mice like odor but after crystallisation with acetone it becomes odorless. It is soluble in water and alcohol.

Acetamide undergoes all general reactions typical of amides.

$$\underset{\textbf{Acetamide}}{CH_3CONH_2 + NaOH} \xrightarrow{\Delta} \underset{\textbf{Sodium acetate}}{CH_3COONa + NH_3}$$

$$CH_3CONH_2 + HCl \longrightarrow \underset{\textbf{Acetic acid}}{CH_3COOH + NH_4Cl}$$

$$CH_3CONH_2 + P_2O_5 \xrightarrow[-H_2O_2]{\Delta} \underset{\textbf{Acetonitrile}}{CH_3-C\equiv N}$$

$$CH_3CONH_2 + Br_2 + NaOH \longrightarrow \underset{\textbf{Methyl amine}}{CH_3NH_2}$$

$$\underset{\textbf{Acetamide}}{CH_3CONH_2 + HNO_2} \longrightarrow \underset{\textbf{Acetic acid}}{CH_3COOH + N_2 + H_2O}$$

It is used in organic synthesis and in leather tanning. It is also used as a plasticiser in cloth, films etc.

22.5 ESTERS

22.5.1 Introduction

Esters are described as derivatives of carboxylic acids in which the $-OH$ of the carboxyl group is replaced by alkoxy group $(-OR)$. The esters have the general formula $RCOOR'$.

Esters are the most important derivatives of carboxylic acids and are widely distributed in nature. Esters are often responsible for the pleasant flavour and odor of most of the fruits and flowers. For example, amyl acetate is present in bananas, octyl acetate in oranges, amyl propionate in apricots, ethyl butyrate in peaches, n-butyl butyrate in pineapples and isoamyl valerate in apples. We have already seen that fats and oils are esters derived from glycerol and higher fatty acids.

22.5.2 Nomenclature

The esters are named by first naming the alkyl group followed by the name of the acid and changing 'ic acid' by 'oate'. Following are given the common and IUPAC names of some esters.

Formula	Common Name	IUPAC Name
$HCOOCH_3$	Methyl formate	Methylmethanoate
$CH_3COOC_2H_5$	Ethyl acetate	Ethylethanoate
$CH_3CH_2COOCH_2CH_3$	Ethyl propionate	Ethylpropanoate

22.5.3 Isomerism

Esters exhibit following three types of isomerism.

(i) Esters show chain isomerism. For example, $C_5H_{10}O_2$ represent two chain isomers.

$$CH_3CH_2CH_2COOCH_3 \qquad CH_3CHCOOCH_3$$
$$\qquad\qquad\qquad\qquad\qquad |$$
$$\qquad\qquad\qquad\qquad\qquad CH_3$$

Methyl butyrate **Methyl isobutyrate**

(ii) Esters are isomeric with acids and show functional isomerism. Thus, $C_3H_6O_2$ represents acid as well as ester.

$$HCOOC_2H_5 \qquad CH_3CH_2COOH$$
Ethyl formate **Propionic acid**

(iii) Esters show metamerism. For example $C_3H_6O_2$ represents ethyl formate and methyl acetate.

$$HCOOC_2H_5 \qquad CH_3COOCH_3$$
Ethyl formate **Methyl acetate**

22.5.4 General Methods of Preparation

(1) *Esterification*: Esters are generally prepared by heating a mixture of a carboxylic acid and an alcohol in presence of mineral acid as catalyst. This reaction is described as Fischer-Speir esterification.

$$\underset{\text{Acid}}{R-\overset{\overset{O}{\|}}{C}-OH} + \underset{\text{Alcohol}}{R'-OH} \xrightarrow{\;\;H^+\;\;} \underset{\text{Ester}}{R-\overset{\overset{O}{\|}}{C}-OR'} + H_2O$$

$$\underset{\text{Acetic acid}}{CH_3-\overset{\overset{O}{\|}}{C}-OH} + \underset{\text{Ethanol}}{CH_3CH_2OH} \xrightarrow{\;\;H^+\;\;} \underset{\text{Ethyl acetate}}{CH_3-\overset{\overset{O}{\|}}{C}-OCH_2CH_3} + H_2O$$

The mechanism of esterification has already been discussed in Section 19.7. The ease of estification of alcohols is as follows.

$$\text{Primary} > \text{Secondary} > \text{Tertiary}$$

Also estification with normal chain acids is faster than with branched chain acids.
Esters can also be obtained by passing vapours of acid and alcohol over heated thorium oxide.

(2) *From Acid Chlorides and Anhydrides (Alcoholysis)*: Esters are prepared conveniently by the action of alcohols on acid chlorides or acid anhydrides.

$$\underset{\substack{\text{Acid} \\ \text{chloride}}}{RCOCl} + \underset{\text{Alcohol}}{R'OH} \longrightarrow \underset{\text{Ester}}{RCOOR'} + HCl$$

$$\underset{\substack{\text{Acid} \\ \text{anhydride}}}{(RCO)_2O} + \underset{\text{Alcohol}}{R'OH} \longrightarrow \underset{\text{Ester}}{RCOOR'} + RCOOH$$

(3) *From Silver Salt of Carboxylic Acids:* An alcoholic solution of an alkyl halide on treatment with silver salt of a carboxylic acid gives esters.

$$\underset{\substack{\text{Silver salt} \\ \text{of carboxylic} \\ \text{acid}}}{RCOO\overset{+}{A}g} + R'X \longrightarrow \underset{\text{Ester}}{RCOOR'} + AgX \downarrow$$

(4) *From Carboxylic Acids*: Methyl esters are obtained by the action of etheral solution of diazo methane on carboxylic acids.

$$\underset{\text{Acetic acid}}{CH_3COOH} + \underset{\substack{\text{Diazo-} \\ \text{methane}}}{CH_2N_2} \longrightarrow \underset{\text{Methyl acetate}}{CH_3COOCH_3} + N_2$$

(5) *From Ketenes*: Esters are obtained by the reaction of alcohols with ketenes.

$$\underset{\text{Ketene}}{CH_2=C=O} + \underset{\text{Alcohol}}{ROH} \longrightarrow \left[\underset{}{CH_2=\overset{\overset{\displaystyle OH}{|}}{C}=OR} \right] \longrightarrow \underset{\text{Ester}}{CH_3-\overset{\overset{\displaystyle O}{\|}}{C}-OR}$$

(6) *From Alkenes*: Treatment of a carboxylic acid with an alkene in presence of boron trifluoride as catalyst yields an ester.

$$\underset{\substack{\text{Carboxylic} \\ \text{acid}}}{RCOOH} + \underset{\text{Ethylene}}{CH_2=CH_2} \xrightarrow{BF_3} \underset{\text{Ester}}{RCOOCH_2CH_3}$$

(7) *Tischenko Reaction*: The condensation of aldehydes in presence of aluminum ethoxide yield the esters.

$$\underset{\text{Acetaldehyde}}{CH_3-\overset{\overset{\displaystyle O}{\|}}{C}-H} + \underset{\text{Acetaldehyde}}{CH_3CHO} \xrightarrow{Al(OC_2H_5)_3} \underset{\text{Ethyl acetate}}{CH_3COOCH_2CH_3}$$

22.5.5 Physical Properties

Esters are liquids (lower members) or solids with fruity or flowery odor. Esters with lower molecular weight are soluble in water, the solubility decreases with increase in molecular weight. However, they are soluble in most organic solvents. In esters there is no association due to hydrogen bonding and so they have low melting and boiling points as compared to those of corresponding carboxylic acids.

22.5.6 Spectral Properties

In the Infrared absorption, the carbonyl group (stretching) in saturated aliphatic esters is at 1750-1735 cm⁻¹. This absorption is higher than the carbonyl absorption in ketones and acids. So the esters can be easily identified.

22.5.7 Chemical Properties

(1) *Nucleophilic Acyl Substitution*: Esters undergo nucleophilic acyl substitution less readily than acid chlorides and acid anhydrides. This is because esters are resonance hybrids of the two forms:

$$R-\overset{\overset{\textstyle O}{\|}}{C}-\overset{..}{\underset{..}{O}}-R' \longleftrightarrow R-\overset{\overset{\textstyle O^-}{|}}{C}=\overset{+}{\underset{..}{O}}-R'$$

As a result of this, the carbonyl carbon is not as electron deficient (or positive) as in acid halides and anhydries. Also, the alkoxide leaving group in esters is more basic than halide ion in acid halides (as discussed in Section 22.2.5). Therefore nucleophilic substitution takes place less readily. However, esters undergo a number of nucleophilic substitution reactions given below.

(i) *Hydrolysis*: Esters can be hydrolysed by heating with mineral acids or alkalis to corresponding carboxylic acid and alcohol.

(a) *Acid Hydrolysis*: Esters on heating with mineral acids (H_2SO_4 or HCl) undergo hydrolysis to give the parent carboxylic acid and alcohol.

$$\underset{\text{Ester}}{R-\overset{\overset{\textstyle O}{\|}}{C}-OR'} + H_2O \underset{}{\overset{H^+}{\rightleftharpoons}} \underset{\text{Carboxylic acid}}{R-\overset{\overset{\textstyle O}{\|}}{C}-OH} + \underset{\text{Alcohol}}{R'OH}$$

$$\underset{\text{Ethyl acetate}}{CH_3-\overset{\overset{\textstyle O}{\|}}{C}-OC_2H_5} + H_2O \overset{H^+}{\rightleftharpoons} \underset{\text{Acetic acid}}{CH_3-\overset{\overset{\textstyle O}{\|}}{C}-OH} + \underset{\text{Ethanol}}{C_2H_5OH}$$

The acid hydrolysis is a reversible reaction. To hydrolyse most of the ester, the equilibrium must be pushed towards right side by using excess of one of the reagents (water). Due to this reason acid hydrolysis is not convenient.

(b) *Alkaline Hydrolysis (Saponification):* Esters on heating with alkali undergo hydrolysis to give the parent alcohol and sodium or potassium salt of carboxylic acid.

$$\underset{\text{Ester}}{R-\overset{\overset{\textstyle O}{\|}}{C}-OR'} + NaOH \overset{H_2O}{\underset{\Delta}{\longrightarrow}} \underset{\substack{\text{Sodium salt of}\\\text{carboxylic acid}}}{R-\overset{\overset{\textstyle O}{\|}}{C}-\overset{+}{O}Na} + \underset{\text{Alcohol}}{R'OH}$$

Alkaline hydrolysis was first used for the preparation of soaps from esters of higher fatty acids (oils and fats). So this process is known as saponification.

The alkaline hydrolysis unlike acid hydrolysis is a irreversible process and procceeds to completion. It is, therfore more convenient than acid hydrolysis.

Mechanism

The hydrolysis of an ester can proceed either by the cleavage of the ester group at the C−O bond (designated as A) or at O−R bond (designated as B). That the hydrolysis takes place by the cleavage of the ester group at the C−O bond (A) has been confirmed by using water containing the isotope of oxygen, O^{18} instead of the usual O^{16}. The acid produced contains O^{18}.

$$
\begin{array}{cc}
\underset{\displaystyle \overset{|}{\underset{HO^{18}}{}} \ H}{R-\overset{\displaystyle O}{\overset{\|}{C}}\!\!+\!\!O-R'} &
\underset{H \ \ O^{18}H}{R-\overset{\displaystyle O}{\overset{\|}{C}}-O\!\!+\!\!R'} \\
\textbf{(A)} & \textbf{(B)}
\end{array}
$$

The mechanism of *acid hydrolysis* is exactly the reverse of the acid catalysed esterification of carboxylic acids.

Step (i): Protonation of carbonyl oxygen to give cation I, which is mesomeric with cation II.

$$
R-\overset{\overset{\displaystyle :O:}{\|}}{C}-O-R' + H^+ \rightleftharpoons R-\overset{\overset{\displaystyle :\overset{+}{O}H}{\|}}{C}-O-R' \rightleftharpoons R-\overset{\overset{\displaystyle :\overset{..}{O}H}{|}}{\underset{+}{C}}-O-R'
$$

$$\textbf{I} \qquad\qquad \textbf{II}$$

Step (ii): Attack by water molecule at electron deficient carbonyl carbon and proton transfer.

$$
\underset{\overset{|}{\underset{H}{\overset{+}{\underset{}{}}}\overset{..}{O}-H}}{R-\overset{\overset{\displaystyle :\overset{..}{O}-H}{|}}{C}-\overset{..}{O}-R'} \rightleftharpoons
\underset{\overset{|}{\underset{H \ \ H}{\overset{+}{\underset{}{}}:\overset{..}{O}:}}}{R-\overset{\overset{\displaystyle :\overset{..}{O}-H}{|}}{C}-\overset{..}{O}-R'} \rightleftharpoons
\underset{\overset{|}{\underset{H}{}}\overset{..}{:O:} \ H}{R-\overset{\overset{\displaystyle :\overset{..}{O}-H}{|}}{C}-\overset{..}{\underset{+}{O}}-R'}
$$

$$\textbf{III}$$

Step (iii): Elimination of R′OH to form acid.

$$
\underset{\overset{|}{\underset{H}{}}:\overset{..}{O}: \ H}{R-\overset{\overset{\displaystyle :\overset{..}{O}-H}{|}}{\underset{+}{C}}-\overset{..}{O}-R'} \xrightarrow[\rightleftharpoons]{-R'OH}
\underset{:\overset{..}{O}H}{R-\overset{\overset{\displaystyle +\overset{..}{O}-H}{\|}}{C}} \xrightarrow[\rightleftharpoons]{-H^+}
R-\overset{\overset{\displaystyle :O:}{\|}}{C}-\overset{..}{O}H
$$

$$\textbf{III}$$

Alkaline hydrolysis involves the following steps.

Step (i): Attack of nucleophile on carbonyl carbon.

$$
R-\overset{\overset{\displaystyle :\overset{..}{O}:}{\|}}{C}-\overset{..}{O}-R' + \overset{-}{O}H \rightleftharpoons R-\overset{\overset{\displaystyle :\overset{..}{O}:^{-}}{|}}{\underset{\overset{|}{OH}}{C}}-\overset{..}{O}-R'
$$

$$\textbf{I}$$

Step (ii): Formation of acid (II) and alkoxy ion.

$$R-\overset{\overset{\displaystyle :O:}{\|}}{\underset{\underset{\displaystyle OH}{|}}{C}}-\ddot{O}-R' \rightleftharpoons R-\overset{\overset{\displaystyle :O:}{\|}}{C}-\ddot{O}-H +:\overset{\displaystyle -}{\ddot{O}}-R'$$

$$\qquad\quad \textbf{I} \qquad\qquad\qquad\qquad \textbf{II}$$

Step (iii): The reversible step forming acid anion and alcohol.

$$R-\overset{\overset{\displaystyle :O:}{\|}}{C}-\ddot{O}-H +:\overset{\displaystyle -}{\ddot{O}}-R' \rightleftharpoons R-\overset{\overset{\displaystyle :O:}{\|}}{C}-\overset{\displaystyle -}{\ddot{O}}:+ R'-\ddot{O}-H$$

$$\qquad\qquad\qquad\qquad\qquad\qquad \textbf{Acid anion} \quad \textbf{Alcohol}$$

(2) *Alcoholysis (Transesterification):* In transesterification the alcoholic part (OR') of an ester, RCOOR' is replaced by the alcoholic part OR" of another alcohol R"OH.

$$\underset{\textbf{Ester}}{R-\overset{\overset{\displaystyle O}{\|}}{C}-OR'} + \underset{\textbf{Alcohol}}{R''-OH} \xrightarrow{R''\bar{O}Na^+} \underset{\textbf{New ester}}{R-\overset{\overset{\displaystyle O}{\|}}{C}-OR''} + \underset{\textbf{Alcohol}}{R'-OH}$$

$$\underset{\textbf{Ethyl acetate}}{CH_3-\overset{\overset{\displaystyle O}{\|}}{C}-OC_2H_5} + \underset{\textbf{n-Butanol}}{C_4H_9OH} \xrightarrow{C_2H_5ONa} \underset{\textbf{n-Butyl acetate}}{CH_3-\overset{\overset{\displaystyle O}{\|}}{C}-OC_4H_9} + \underset{\textbf{Ethanol}}{C_2H_5OH}$$

The reaction is catalysed by base or acid. Basic catalyst is preferred since the reaction goes to complition. The basic catalyst corresponding to the reacting alcohol (R"OH) is most effective. Since, transesterification is a reversible process, it can be made to go to completion by using excess of alcohol (R"OH). This method is mostly used for preparing esters of a higher alcohol from that of a lower alcohol, since separation by distillation is easier.

Mechanism

The mechanism is similar to that of hydrolysis of ester. The reacting alcohol serves as nucleophile in place of water. Thus, the base catalysed reaction is represented as follows.

$$R-\overset{\overset{\displaystyle O}{\|}}{C}-O-R' +\overset{\displaystyle -}{\bar{O}}R'' \rightleftharpoons R-\overset{\overset{\displaystyle \bar{O}}{|}}{\underset{\underset{\displaystyle OR''}{|}}{C}}-OR' \rightleftharpoons R-\overset{\overset{\displaystyle O}{\|}}{C}-OR'' + R'\bar{O}$$

(3) *Reaction with Ammonia (Ammonolysis):* Esters on reaction with concentrated ammonia form acid amides.

$$\underset{\textbf{Ester}}{R-\overset{\overset{\displaystyle O}{\|}}{C}-OR'} + NH_3 \longrightarrow \underset{\textbf{Amide}}{R-\overset{\overset{\displaystyle O}{\|}}{C}-NH_2} + R'OH$$

(4) *Reduction:* Esters on reduction form alcohols. Thus, reduction with lithium aluminum hydride is of special use, since the double bonds if present are not reduced by this reagent.

$$\text{RC}\overset{\text{O}}{\underset{\|}{-}}\text{OR}' \begin{cases} \xrightarrow{\text{LiAlH}_4} \text{RCH}_2\text{OH} + \text{R}'\text{OH} \\ \\ \xrightarrow[\text{Bouveault Blanck reduction}]{\text{Na/C}_2\text{H}_5\text{OH}} \text{RCH}_2\text{OH} + \text{R}'\text{OH} \\ \\ \xrightarrow[\text{Adkin's method}]{\text{Copper chromite; 250-300°; 300 Atm}} \text{RCH}_2\text{OH} + \text{R}'\text{OH} \end{cases}$$

Ester

(5) *Reaction with Grignard Reagents*: Esters on reaction with Grignard reagents form addition products which decompose to form ketones. The formed ketone reacts with excess of the reagent to form tertiary alcohol.

$$\underset{\text{Ester}}{\text{R}-\overset{\text{O}}{\underset{\|}{\text{C}}}-\text{OR}'} + \underset{\text{Grignard reagent}}{\text{R}''-\text{MgX}} \longrightarrow \text{R}-\overset{\overset{+}{\text{O}}\text{Mg}X}{\underset{\underset{\text{R}''}{|}}{\overset{|}{\text{C}}}}-\text{OR}' \xrightarrow{\text{H}_2\text{O/H}^+} \underset{\text{Ketone}}{\text{R}-\overset{\text{O}}{\underset{\|}{\text{C}}}-\text{R}''} + \text{R}'\text{OMgX}$$

$$\underset{\text{Ketone}}{\text{R}-\overset{\text{O}}{\underset{\|}{\text{C}}}-\text{R}''} + \text{R}''-\text{MgX} \longrightarrow \left[\text{R}-\overset{\overset{+}{\text{O}}\text{Mg}X}{\underset{\underset{\text{R}''}{|}}{\overset{|}{\text{C}}}}-\text{R}''\right] \xrightarrow{\text{H}_2\text{O/H}^+} \underset{\text{tert-Alcohol}}{\text{R}-\overset{\text{OH}}{\underset{\underset{\text{R}''}{|}}{\overset{|}{\text{C}}}}-\text{R}''}$$

(6) *Reaction with Hydrazine*: Esters on reaction with hydrazine form the corresponding hydrazides.

$$\underset{\text{Ester}}{\text{R}-\overset{\text{O}}{\underset{\|}{\text{C}}}-\text{OR}'} + \underset{\text{Hydrazine}}{\text{H}_2\text{N}-\text{NH}_2} \longrightarrow \underset{\text{Hydrazide}}{\text{R}-\overset{\text{O}}{\underset{\|}{\text{C}}}-\text{NHNH}_2} + \underset{\text{Alcohol}}{\text{R}'-\text{OH}}$$

Similar reaction with *hydroxylamine* give the corresponding hydroxamic acids.

$$\underset{\text{Ester}}{\text{R}-\overset{\text{O}}{\underset{\|}{\text{C}}}-\text{OR}'} + \underset{\substack{\text{Hydroxyl-}\\\text{amine}}}{\text{NH}_2\text{OH}} \longrightarrow \underset{\substack{\text{Hydroxamic}\\\text{acid}}}{\text{R}-\overset{\text{O}}{\underset{\|}{\text{C}}}-\text{NHOH}} + \underset{\text{Alcohol}}{\text{R}'\text{OH}}$$

(7) *Hell-Volhard-Zelinsky Reaction*: The esters on halogenation with halogen in presence of red phosphorus give α-halogenated esters.

$$\underset{\text{Ethyl acetate}}{\text{CH}_3-\overset{\text{O}}{\underset{\|}{\text{C}}}-\text{OC}_2\text{H}_5} + \text{Br}_2 \xrightarrow{\text{red P}} \underset{\text{α-Bromoethyl acetate}}{\text{BrCH}_2-\overset{\text{O}}{\underset{\|}{\text{C}}}-\text{OC}_2\text{H}_5} + \text{HBr}$$

(8) *Claison Condensation*: Esters undergo self condensation in presence of sodium ethoxide to give β-keto esters. (For details see properties of aldehydes and ketones.)

$$\underset{\text{Ethyl acetate}}{2\text{CH}_3-\overset{\text{O}}{\underset{\|}{\text{C}}}-\text{OC}_2\text{H}_5} \xrightarrow{\text{C}_2\text{H}_5\text{O}^-\text{Na}^+} \underset{\text{Ethyl acetoacetate}}{\text{CH}_3\text{COCH}_2\text{COOC}_2\text{H}_5}$$

Uses: Esters are used as solvents for oils, fats, cellulose, paints, varnishes etc. An important use of esters is for making artificial flavours and essences.

22.5.8 Individual Members

22.5.8.1 Ethyl Acetate/Ethyl Ethanoate [$CH_3COOC_2H_5$]

It is prepared in the laboratry as well as on a large scale by refluxing glacial acetic acid and absolute ethyl alcohol in presence of concentrated sulphuric acid or hydrogen chloride gas.

$$CH_3COOH + C_2H_5OH \underset{}{\overset{H^+}{\rightleftharpoons}} CH_3COOC_2H_5 + H_2O$$

| Acetic acid | Ethanol | Ethyl acetate |

On a commercial scale it is also prepared by passing a mixture of glacial acetic acid and absolute ethyl alcohol over a heated catalyst (thorium oxide or titanium dioxide at 200-300°).

Ethyl acetate is a colorless, sweet smelling liquid (b.p. 77.5°). It is sparingly soluble in water but more soluble in alcohol and ether. It gives all general reactions of esters as described earlier.

Ethyl acetate is mostly used as a solvent and for the preparation of ethyl acetoacetate (Claisen condensation).

22.6 PROBLEMS

1. How are acid bromides, iodides and fluorides obtained?
2. Compare the spectral properties of acid halides, amides, anhydrides and esters. How IR spectra is helpful for differentiation between these?
3. Why are esters less reactive than acid chlorides in nucleophilic substitution reactions? Give reasons.
4. How is acetic anhydride and ethyl acetate obtained on large scale?
5. Amides behave both as very weak bases and acids. Justify your answer.
6. Give the mechanism of the hydrolysis of esters under basic and acidic conditions.
7. What type of hydrolysis of esters (acidic or basic) is preferred? Give reasons.
8. What do you understand by 'Transesterification'?
9. Write notes on:
   (a) Rosenmund reduction   (d) Hofmann's reaction
   (b) Friedel Crafts reaction.   (e) Bouveault Blanck reduction
   (c) Gatterman Koch Aldehyde synthesis (f) Adkins reduction
10. Complete the following:

    (a) $CH_3\overset{\overset{O}{\|}}{C} - OC_2H_5 + CH_3MgBr \longrightarrow$

    (b) $CH_3.CO.O.CO.CH_3 + CH_3CH_2\overset{\overset{O}{\|}}{C}OH \longrightarrow$

    (c) $HC \equiv CH + CH_3COOH \xrightarrow{Hg^{2+}}$

    (d) $CH_2 = C = O + CH_3COOH \longrightarrow$

23
Derivatives of Carbonic Acid

23.1 INTRODUCTION

Carbonic acid, $HO-\overset{\overset{O}{\|}}{C}-OH$, containing two hydroxy groups attached to the same carbon atom is unstable and has not been isolated. An aqueous solution of carbon dioxide is believed to contain carbonic acid.

$$O=C=O + H_2O \longrightarrow HO-\overset{\overset{O}{\|}}{C}-OH$$
Carbonic Acid (Unstable)

Though unstable, carbonic acid behaves as a dibasic acid, since it contains two hydroxy groups bonded to the same carbonyl group. Each hydroxy group behaves like the hydroxy group of a carboxylic acid. Carbonic acid forms monofunctional and bifunctional derivatives. The monofunctional derivatives, viz. chlorocarbonic acid ($Cl-\overset{\overset{O}{\|}}{C}-OH$), carbamic acid ($HO-\overset{\overset{O}{\|}}{C}-NH_2$), ethyl hydrogen carbonate ($HO-\overset{\overset{O}{\|}}{C}-OC_2H_5$) are very unstable but their salts are known. The bifunctional derivatives, on the other hand, are stable. Some example are phosgene ($Cl-\overset{\overset{O}{\|}}{C}-Cl$, carbonyl chloride), urea ($NH_2-\overset{\overset{O}{\|}}{C}-NH_2$, carbamide), ethyl chloroformate ($Cl-\overset{\overset{O}{\|}}{C}-OC_2H_5$), carbamyl chloride ($Cl-\overset{\overset{O}{\|}}{C}-NH_5$), diethyl carbonate ($C_2H_5O-\overset{\overset{O}{\|}}{C}-OC_2H_5$) and ethyl carbamate ($NH_2-\overset{\overset{O}{\|}}{C}-OC_2H_5$, urethane).

23.2 PHOSGENE/CARBONYL CHLORIDE [$Cl-\overset{\overset{O}{\|}}{C}-Cl$]

It is the diacid chloride of carbonic acid.

Preparation

1. *From Chloroform*: The oxidation of chloroform by oxygen or air in presence of sunlight or with acidified potassium dichromate forms phosgene.

$$2CHCl_3 + O_2 \xrightarrow{\text{Sun light}} 2Cl-\overset{\overset{O}{\|}}{C}-Cl + 2HCl$$
Phosgene

2. *From Carbon Tetrachloride*: By the action of fuming sulphuric acid (oleum) on carbon tetrachloride at 78-80°.

$$CCl_4 + H_2SO_4SO_3 \longrightarrow Cl-\overset{\overset{\displaystyle O}{\|}}{C}-Cl + 2ClSO_2H$$

    Oleum   Phosgene   Chlorosulphonic acid

3. *From Carbon Monoxide*: A mixture of carbon monoxide and chlorine is passed over activated charcoal at 220-250° to form phosgene (commercial method).

$$CO_2 + Cl_2 \xrightarrow[200-250°]{Charcoal} Cl-\overset{\overset{\displaystyle O}{\|}}{C}-Cl$$

                                         Phosgene

Properties: Phosgene is a colorless gas (b.p. 8°) with suffocating odor. When inhaled death occurs in a few minutes. It was used as a chemical warfare agent in World War-I. It is soluble in benzene and toluene, the solution, therefore, is used for keeping and transporting it.

It behaves like acid chloride. Some of the reactions are given below:

1. *Hydrolysis*: Phosgene on treatment with water gets hydrolysed to give hydrogen chloride and carbon dioxide.

$$O=C\overset{\displaystyle Cl}{\underset{\displaystyle Cl}{\big\langle}} + 2H_2O \longrightarrow 2HCl + \left[O=C\overset{\displaystyle OH}{\underset{\displaystyle OH}{\big\langle}} \right] \longrightarrow CO_2 + H_2O$$

    Phosgene   Carbonic acid

2. *Reaction with Alcohols*: On reaction with alcohol (one mole), in an inert solvent at low temperature phosgene yields corresponding alkyl chloroformate. With ethyl alcohol ethyl chloroformate is obtained.

$$Cl-\overset{\overset{\displaystyle O}{\|}}{C}-Cl + H-O-C_2H_5 \xrightarrow[0°]{C_2H_6} Cl-\overset{\overset{\displaystyle O}{\|}}{C}-OC_2H_5 + HCl$$

    Phosgene   Ethyl chloroformate

However, with excess ethyl alcohol in presence of pyridine, diethyl carbonate is obtained.

$$O=C\overset{\displaystyle Cl}{\underset{\displaystyle Cl}{\big\langle}} + 2C_2H_5OH \xrightarrow{Pyridine} O=C\overset{\displaystyle OC_2H_5}{\underset{\displaystyle OC_2H_5}{\big\langle}} + 2HCl$$

    Phosgene   Diethyl Carbonate

3. *Reaction with Amines*: Phosgene reacts with excess of primary amine to form substituted urea.

$$O=C\overset{\displaystyle Cl}{\underset{\displaystyle Cl}{\big\langle}} + 2C_2H_5NH_2 \longrightarrow O=C\overset{\displaystyle NHC_2H_5}{\underset{\displaystyle NHC_2H_5}{\big\langle}} + 2HCl$$

    Phosgene   N,N-Diethylurea

Similar reaction with ammonia yields urea.

$$O=C\overset{\displaystyle Cl}{\underset{\displaystyle Cl}{\big\langle}} + 2NH_3 \longrightarrow O=C\overset{\displaystyle NH_2}{\underset{\displaystyle NH_2}{\big\langle}} + 2HCl$$

    Phosgene   Urea

4. *Friedel-Crafts Reaction*: Phosgene on reaction with benzene in presence of anhydrous aluminum chloride yields benzophenone.

$$O=C{\displaystyle {{Cl \atop Cl}}} + C_6H_6 \xrightarrow{AlCl_3} O=C{\displaystyle {{C_6H_5 \atop C_6H_5}}}$$

Phosgene   **Benzophenone**

5. *Reaction with N,N-dimethylaniline*: On reaction with N,N-dimethylaniline in presence of anhydrous zinc chloride Michler's ketone is obtained.

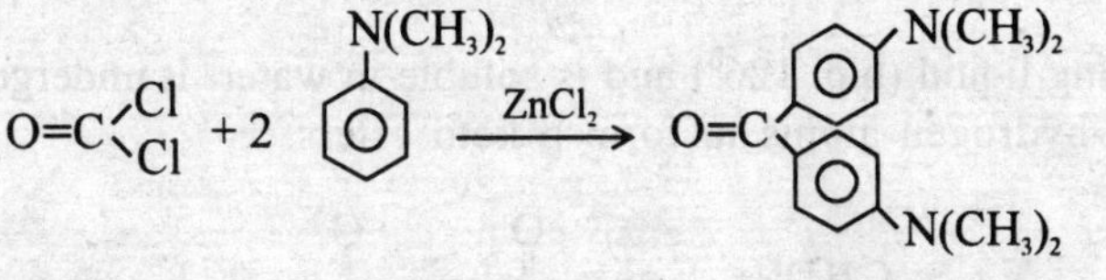

Phosgene   **N,N-Diethyl aniline**   **Michler's ketone**

6. *Reaction with Benzyl Alcohol*: On reaction with benzyl alcohol phosgene forms carbobenzoxy chloride (benzyl chloroformate).

$$\underset{\textbf{Phosgene}}{Cl-\overset{O}{\overset{\|}{C}}-Cl} + \underset{\textbf{Benzyl alcohol}}{C_6H_5CH_2OH} \longrightarrow \underset{\textbf{Carbobenzoxy chloride}}{C_6H_5CH_2O-\overset{O}{\overset{\|}{C}}-Cl} + HCl$$

Uses: Though highly poisonous, phosgene is an important industrial chemical. It is used in the manufacture of Michler's ketone for dye industry, carbobenzoxy chloride for peptide synthesis and toluenediisocyanate for foam rubber industry.

23.3   CHLOROFORMIC ACID/CHLOROCARBONIC ACID [$Cl-\overset{O}{\overset{\|}{C}}-OH$]

It is the monochloride of carbonic acid. It is unstable but two of its esters, viz. ethyl chloroformate and ethyl carbonate are stable and discussed in Sec. 23.4.

23.4   ETHYL CHLOROFORMATE [$Cl-\overset{O}{\overset{\|}{C}}-OC_2H_5$]

It is prepared by the reaction of phosgene with one mole of ethyl alcohol in benzene at low temperature (see properties of phosgene).

Ethyl chloroformate is a colorless liquid (b.p. 94°) having suffocating odor. It behaves like an acid chloride and an ester. It reacts with compounds containing active hydrogen atoms. Thus, it reacts with alcohol, ammonia and amines.

$$\underset{\substack{\textbf{Ethyl} \\ \textbf{alcohol}}}{C_2H_5OH} + \underset{\substack{\textbf{Ethyl} \\ \textbf{chloroformate}}}{Cl-\overset{O}{\overset{\|}{C}}-OC_2H_5} \longrightarrow \underset{\textbf{Ethyl carbonate}}{C_2H_5O-\overset{O}{\overset{\|}{C}}-OC_2H_5} + HCl$$

$$\underset{\textbf{Ammonia}}{NH_3} + Cl-\overset{O}{\overset{\|}{C}}-OC_2H_5 \longrightarrow \underset{\textbf{Ethyl carbamate}}{H_2N-\overset{O}{\overset{\|}{C}}-OC_2H_5} + HCl$$

$$CH_3NH_2 + Cl-\overset{\overset{\displaystyle O}{\|}}{C}-OC_2H_5 \longrightarrow CH_3NH-\overset{\overset{\displaystyle O}{\|}}{C}-OC_2H_5 + HCl$$

Methyl amine   **N-Methylurethane**

Ethyl chloroformate is used in organic synthesis.

23.5 ETHYL CARBONATE [$C_2H_5O-\overset{\overset{\displaystyle O}{\|}}{C}-OC_2H_5$]

It is prepared by heating phosgene with excess of ethyl alcohol in presence of pyridine. (see properties of phosgene).

Ethyl carbonate is a pleasant smelling liquid (b.p. 126°) and is soluble in water. It undergoes Claisen condensation with ketones having α-hydrogen atoms to form β-keto esters.

$$C_6H_5-\overset{\overset{\displaystyle O}{\|}}{C}-CH_3 + C_2H_5O-\overset{\overset{\displaystyle O}{\|}}{C}-OC_2H_5 \xrightarrow{C_2H_5ONa} C_6H_5-\overset{\overset{\displaystyle O}{\|}}{C}-CH_2-\overset{\overset{\displaystyle O}{\|}}{C}-OC_2H_5 + C_2H_5OH$$

Acetophenone   **Ethyl carbonate**   **Ethyl benzoylacetate**

23.6 CARBAMIC ACID [$HO-\overset{\overset{\displaystyle O}{\|}}{C}-NH_2$]

Carbamic acid, the monoamide of carbonic acid is unstable. However its salts and esters are useful compounds.

23.7 AMMONIUM CARBAMATE [$H_2N-\overset{\overset{\displaystyle O}{\|}}{C}-ONH_4$]

It is obtained by the reaction of carbon dioxide with ammonia.

$$O=C=O + 2NH_3 \longrightarrow H_2N-\overset{\overset{\displaystyle O}{\|}}{C}-\overset{+}{\overset{-}{O}}NH_4$$

Ammonium carbamate

Ammonium carbamate is a crystalline solid and is soluble in water. Its aqueous solution on heating at 60° gets hydrolysed to give ammonium carbonate. However on heating at 140-150° in vacuum, ammonium carbamate gives urea.

23.8 ETHYL CARBAMATE/URETHANE [$H_2N-\overset{\overset{\displaystyle O}{\|}}{C}-OC_2H_5$]

It is prepared as follows:

1. *From Ethyl Chloroformate*: by the action of ammonia (see properties of ethyl chloroformate)

2. *From Carbamyl Chloride*: by heating with alcohol.

$$H_2N-\overset{\overset{\displaystyle O}{\|}}{C}-Cl + C_2H_5OH \longrightarrow H_2N-\overset{\overset{\displaystyle O}{\|}}{C}-OC_2H_5 + HCl$$

Carbamyl chloride   **Ethyl carbamate**

3. *From Urea*: by heating with ethyl alcohol. The reaction proceeds via the formation of isocyanic acid.

$$H_2N-\overset{\overset{\displaystyle O}{\|}}{C}-NH_2 \longrightarrow [\,H-N=C=O \rightleftharpoons HO-C\equiv N\,] + NH_3$$

Urea   Isocyanic acid   Cyanic acid

$$HN=C=O + C_2H_5OH \longrightarrow H_2N-\overset{\overset{\displaystyle O}{\|}}{C}-OC_2H_5$$

Isocyanic acid   Urethane

Ethyl carbamate (urethane) is a crystalline solid (m.p.50°) soluble in water, alcohol and ether. On treatment with ammonia, urea is obtained.

$$H_2N-\overset{\overset{\displaystyle O}{\|}}{C}-OC_2H_5 + NH_3 \longrightarrow H_2N-\overset{\overset{\displaystyle O}{\|}}{C}-NH_2 + C_2H_5OH$$

Ethyl carbamate   **Urea**   **Ethyl alcohol**

On heating with aqueous sodium hydroxide, it is decomposed:

$$H_2N-\overset{\overset{\displaystyle O}{\|}}{C}-OC_2H_5 + 2NaOH \longrightarrow C_2H_5OH + NH_3 + Na_2CO_3$$

Ethyl carbamate   **Ethyl alcohol**

Ethyl carbamate has a mild hypnotic action and is used in the treatment of lukemia.

23.9   UREA/CARBAMIDE [NH$_2$CONH$_2$]

Urea was first isolation from urine and is the main product of protein metabolism in man and mammals. A normal adult excretes about 30g urea daily.

Preparation

1.   *From Urine*: Urine is concentrated and treated with concentrated nitric acid to give urea nitrate, which crystallyses out. The crystals of urea nitrate are filtered and decomposed with barium carbonate. Urea is extracted from the reaction mixture with alcohol. The alcoholic extract on evaporation gives urea.

$$NH_2CONH_2HNO_3 + BaCO_3 \longrightarrow 2NH_2CONH_2 + Ba(NO_3)_2 + H_2O + CO_2$$

Urea nitrate   Urea

2.   *From Ammonium Cyanate (Wohler's Synthesis)*: By heating aqueous solution of ammonium cyanate.

$$NH_4NCO \longrightarrow H_2N-\overset{\overset{\displaystyle O}{\|}}{C}-NH_2$$

Ammonium   **Urea**
cyanate

3.   *From Phosgene*: By the action of ammonia. (See properties of phosgene.)

4.   *Commercial Method*: On a commercial scale urea is obtained by heating ammonia and carbon dioxide under pressure.

$$CO_2 + 2NH_3 \xrightarrow[\text{Pressure}]{140\text{-}150°} H_2NCONH_2$$

Urea

Properties: Urea is a white crystalline solid (m.p. 132°) and is readily soluble in water and alcohol.

Urea is considered to be a resonance hybrid of the following structures.

$$H_2N-\overset{\overset{O}{\|}}{C}-NH_2 \longleftrightarrow H_2\overset{+}{N}=\overset{\overset{O^-}{|}}{C}-NH_2 \longleftrightarrow H_2N-\overset{\overset{O^-}{|}}{C}=\overset{+}{N}H_2$$

From X-ray diffraction studies, it is found that both C–N bonds in urea are of the same length (1.33 Å) which is shorter than a C–N single bond in amine (1.48 Å) and longer than the normal C–N double bond (1.28 Å). This suggests that both C–N bonds in urea are identical and posses significant double bond character. This is the reason for the hybrid nature of the urea molecule.

Urea is a weak base. However it is a stronger base than ordinary amides. This is attributed to the resonance stabilisation of the cation.

$$H_2N-\overset{\overset{O}{\|}}{C}-NH_2+H^+ \rightleftharpoons \left[H_2N-\underset{+OH}{\overset{\overset{O}{\|}}{C}}-NH_2 \longleftrightarrow H_2\overset{+}{N}=\underset{OH}{\overset{|}{C}}-NH_2 \longleftrightarrow H_2N-\underset{OH}{\overset{|}{C}}=\overset{+}{N}H_2 \right]$$

Urea gives all characteristic reactions of an amide and also of amines. some of the important reactions are given below:

1. *Basic Nature*: As already mentioned urea is a weak base and behaves like a monoacid base. It forms crystalline salts with acids like nitric and oxalic acids.

$$\underset{\textbf{Urea}}{H_2N-CO-NH_2} + HNO_3 \longrightarrow \underset{\textbf{Urea nitrate}}{H_2N-CO-NH_2.HNO_3}$$

Urea nitrate on reaction with cold concentrated sulphuric acid forms nitrourea.

$$NH_2CONH_2HNO_3 \xrightarrow{H_2SO_4} NH_2CONHNO_2 + H_2O$$

2. *Hydrolysis*: Urea gets hydrolysed on heating with dilute mineral acid or alkali.

$$H_2N-CO-NH_2 + H_2O \xrightarrow[\text{or Alkali}]{\text{Acid}} CO_2 + 2NH_3$$

The exzyme urease (occurs in soyabean) brings about the same change.

3. *Reaction with Thionyl Chloride*: Urea on treatment with thionyl chloride undergoes dehydration yielding cyanamide.

$$H_2N-\overset{\overset{O}{\|}}{C}-NH_2 + SOCl_2 \longrightarrow \underset{\textbf{Cyanamide}}{H_2N-C \equiv N} + SO_2 + 2HCl$$

4. *Reaction with Nitrous Acid*: Urea reacts with nitrous acid to give nitrogen.

$$NH_2CONH_2 + NaNO_2 + HCl \longrightarrow 2H_2O + N_2 + \left[O=C\overset{OH}{\underset{OH}{\diagdown}} \right] \longrightarrow CO_2 + H_2O$$

5. *Reaction with Hydrazine*: Urea reacts with hydrazine to give semicarbazide.

$$\underset{\textbf{Urea}}{H_2N-\overset{\overset{O}{\|}}{C}-NH_2} + \underset{\textbf{Hydrazine}}{H_2N-NH_2} \xrightarrow{-NH_3} \underset{\textbf{Semicarbazide}}{H_2N-\overset{\overset{O}{\|}}{C}-NHNH_2}$$

6.   *Action of Heat*: Urea on heating above its melting point (132°) evolves ammonia and a crystalline product known as biuret is formed.

$$2H_2N-\overset{O}{\overset{\|}{C}}-NH_2 \xrightarrow[-NH_3]{132°} H_2N-\overset{O}{\overset{\|}{C}}-NH-\overset{O}{\overset{\|}{C}}-NH_2$$
Urea   Biuret

The alkaline solution of biuret on treatment with a drop of copper sulphate solution gives a violet color. This is known as *Biuret test* for urea and also for proteins, peptides and other compounds containing –CONH– linkage.

The mechanism of the reaction is summerised in following steps.

$$H-NH-\overset{O}{\overset{\|}{C}}-NH_2 \underset{-NH_3}{\rightleftarrows} [H-N=C=O \longleftrightarrow HO-C\equiv N]$$
                                        Isocyanic acid   Cyanic acid

$$H_2N-\overset{O}{\overset{\|}{C}}-\overset{H}{\underset{..}{N}}H + \overset{O}{\overset{\|}{C}}=N-H \rightleftharpoons H_2N-\overset{O}{\overset{\|}{C}}-\overset{H}{\underset{+}{N}}H-\overset{O}{\overset{\|}{C}}-\overset{\bar{}}{N}-H$$

$$\rightleftharpoons H_2N-\overset{O}{\overset{\|}{C}}-NH-\overset{O}{\overset{\|}{C}}-NH_2$$
Biuret

On strong heating, however, isocyanic acid is obtained which polymerises to cyanuric acid.

7.   *Reaction with Alkaline Solution of Hypohalites*: On treatment with an alkaline solution of hypobromite (Br_2 + NaOH) it gives hydrazine, which is further oxidised by sodium hypobromite to nitrogen.

$$H_2N-\overset{O}{\overset{\|}{C}}-NH_2 + NaOBr + 2NaOH \longrightarrow H_2N-NH_2 + NaBr + Na_2CO_3 + H_2O$$
Urea   Hydrazine

$$NH_2 NH_2 + 2NaOBr \longrightarrow 2NaBr + N_2 + 2H_2O$$

This reaction is used for the estimation of urea in various biological fluids, e.g. urine by measuring the volume of nitrogen evolved when a test sample is mixed with alkaline hypobromite.

8.   *Reaction with Fuming Sulphuric Acid*: Urea reacts with fuming sulphuric acid (oleum) to yield sulphamic acid.

$$H_2N-\overset{O}{\overset{\|}{C}}-NH_2 + H_2SO_4SO_3 \longrightarrow 2NH_2 SO_3H + CO_2$$
Urea   Oleum   Sulphamic acid

Two of the derivatives of sulphamic acid, i.e. ammonium sulphamate and sodium cyclohexylsulphamate find industrial applications.

Sodium cyclohexylsulphamate is used as an artificial sweetener under the name cyclamate or sucaryl.

9.   *Reaction with Malonic Ester and Malonic Acid*: On heating with malonic acid in presence of phosphorus oxychloride, urea yields malonyl urea (a tautomeric mixture of barbituric acid)). It can also be obtained by heating urea with malonic ester in presence of sodium ethoxide.

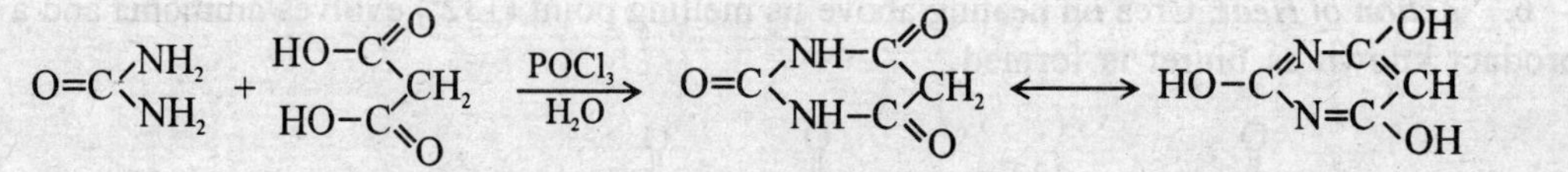

Urea   Malonic acid   Barbituric acid (Keto form)   Barbituric acid (Enol form)

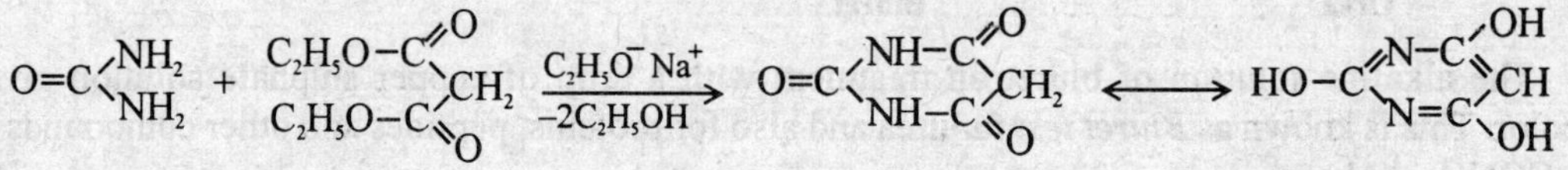

Urea   Malonic ester   Keto and Enol forms of Barbituric acid

10.   *Reaction with Formaldehyde (Urea-formaldehyde resins)*: In presence of base or acid, urea reacts with formaldehyde to form methylolurea and dimethylolurea.

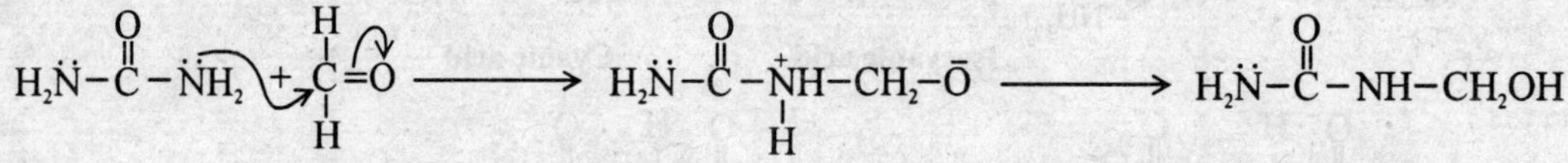

Urea   Formaldehyde   Methylol urea

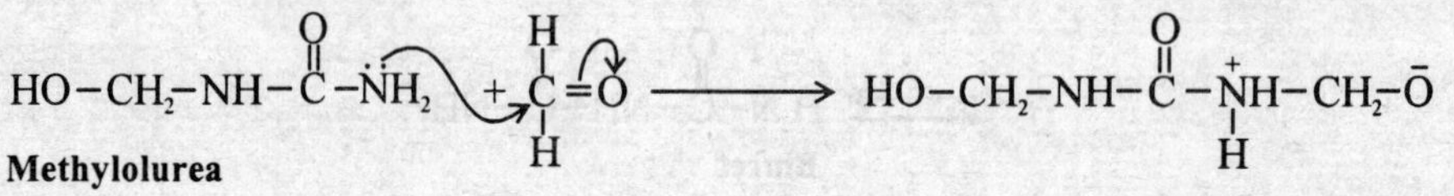

Methylolurea

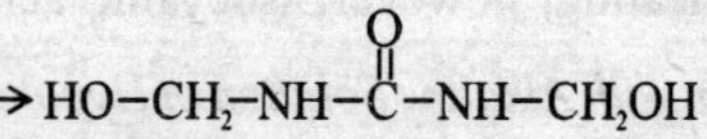

Dimethylolurea

Methylolurea and dimethylolurea can further condense with more urea molecules to form linear and cross linked polymers. The linear and cross linked polymers are known as urea-formaldehyde resin. By changing the reaction conditions, resins of different types are obtained. These resins find industrial applications.

Uses: Urea is largely used as a fertiliser. It is used in the manufacture of barbiturates (drugs for inducing sleep), urea formaldehyde resins and in the manufacture of non-ionic detergents.

23.10   PROBLEMS

1. Why is carbonic acid unstable?
2. How is phosgene prepared? How it is converted into ethyl chloroformate, ethyl carbonate, urea, N,N-diethyl urea, Michler's ketone and carbobenzoxy chloride?
3. How is ethyl chloroformate obtained?
4. What happens if acetophenone is condensed with ethyl carbonate in presence of sodium ethoxide?
5. How is urea obtained on a commercial scale? Explain why urea behaves as a weak base?
6. Starting with urea, how will you obtain cyanamide, semicarbazide, sulphamic acid, barbituric acid?
7. Write notes on   (a) Biuret test and   (b) Urea formaldehyde resins.

24

Active Methylene Compounds

24.1 INTRODUCTION

When a methylene group is situated between two electron withdrawing groups, the hydrogens bonded to carbon become highly acidic and reactive and the compounds having such methylene groups are termed as active methylene compounds.

$$C_2H_5OOCCH_2COOC_2H_5 \qquad CH_3COCH_2COOC_2H_5 \qquad N{\equiv}CCH_2COOC_2H_5$$

Diethyl malonate   **Ethyl acetoacetate**   **Ethyl cyanoacetate**

The enhanced activity of methylene group is due to the adjacent electron withdrawing -I groups which weakens the carbon-hydrogen bond of methylene group, making the hydrogen acidic and labile. The active methylene compounds when treated with base like sodium ethoxide, form a carbanion which stabilises itself by resonance.

$$CH_3CO\overset{\overset{\displaystyle H}{|}}{\underset{\underset{\displaystyle H}{|}}{C}}COOC_2H_5 + Na\bar{O}C_2H_5 \longrightarrow CH_3CO-\overset{\overset{\displaystyle H}{|}}{\underset{\underset{\displaystyle Na^+}{|}}{\bar{C}}}-COOC_2H_5 + C_2H_5OH$$

The carbanion, thus formed, acts as a strong nucleophile and undergoes various substitution reactions. The active methylene compounds are useful in the synthesis of a wide variety of organic compounds.

24.2 ETHYL ACETOACETATE/ACETOACETIC ESTER [CH$_3$COCH$_2$COOC$_2$H$_5$]

It is the ethyl ester of acetoacetic acid (β-keto butyric acid), thus, named as ethyl acetoacetate. The name acetoacetic ester has also been given since it can be regarded as acetyl derivative of acetic ester or ethyl acetate.

24.2.1 Preparation

It is prepared by refluxing ethyl acetate in excess of sodium ethoxide.

$$2CH_3CO_2C_2H_5 + C_2H_5\bar{O}\,Na^+ \longrightarrow CH_3\overset{\overset{\displaystyle O}{\|}}{C}-CH_2-\overset{\overset{\displaystyle O}{\|}}{C}-OC_2H_5 + C_2H_5OH$$

Ethyl acetoacetate

This is an example of claisen condensation in which a keto ester is formed by reaction of two ester molecules having α-hydrogen atoms. The most widely accepted mechanism for claisen condensation involves the formation of carbanion which attacks the electropositive carbonyl carbon to give ethyl acetoacetate. Since all the reactions are reversible, a large excess of sodium ethoxide or sodamide is required which results in the formation of sodium salt of acetoacetic ester from which ester is generated by acidification.

$$C_2H_5O + H-CH_2COOC_2H_5 \rightleftharpoons C_2H_5OH + \bar{C}H_2COOC_2H_5$$

$$CH_3-\overset{O}{\underset{OC_2H_5}{\overset{\|}{C}}} + \bar{C}H_2COOC_2H_5 \rightleftharpoons CH_3-\overset{O}{\underset{OC_2H_5}{\overset{\|}{C}}}-CH_2COOC_2H_5 \rightleftharpoons$$

$$CH_3-\overset{O}{\overset{\|}{C}}\overset{H}{\underset{}{CHCOOC_2H_5}} + C_2H_5O^- \rightleftharpoons CH_3-\overset{O^-}{\overset{}{C}}=CHCOOC_2H_5 + C_2H_5OH$$

$$CH_3-\overset{O^-}{\overset{}{C}}=CHCOOC_2H_5 \xrightarrow{CH_3COOH} CH_3-\overset{OH}{\overset{}{C}}=CHCOOC_2H_5 \rightleftharpoons CH_3-\overset{O}{\overset{\|}{C}}-CH_2-COOC_2H_5$$

Industrial Preparation

Acetoacetic ester is produced on industrial scale by dimerisation of ketene in cold acetone. The diketene thus formed reacts with ethanol to form the ester.

$$\begin{matrix} CH_2=C=O \\ CH_2=C=O \end{matrix} \longrightarrow \begin{matrix} CH_2=C-O \\ H_2C-C=O \end{matrix} \xrightarrow{EtOH} \begin{matrix} CH_2=C-O \\ H_2C-C=O \end{matrix}\overset{+}{\underset{OEt}{O}}\overset{H}{} \longrightarrow$$

Ketene

$$\left[\begin{matrix} \bar{O}\cdots H - \overset{+}{O}-Et \\ CH_2=C-CH_2-C=O \end{matrix} \longrightarrow \begin{matrix} O-H\cdots O \\ CH_2=C-CH_2-C-OEt \end{matrix} \right] \longrightarrow CH_3-\overset{O}{\overset{\|}{C}}-CH_2-\overset{O}{\overset{\|}{C}}-OEt$$

The ketene required is prepared as follows.

$$CH_3-\overset{CH_3}{\overset{|}{C}}=O \xrightarrow[\text{Pyrolysis}]{700-750°} 2CH_2=C=O + 2CH_4$$

Acetone   **Ketene**

24.2.2 Tautomerism in Ethyl Acetoacetate

Ethyl acetoacetate exists in two isomeric forms, I and II, which exist in dynamic equilibrium

$$CH_3-\overset{O}{\overset{\|}{C}}-CH_2-\overset{O}{\overset{\|}{C}}-OC_2H_2 \rightleftharpoons CH_3-\overset{OH}{\overset{}{C}}=CH-\overset{O}{\overset{\|}{C}}-OC_2H_5$$

         I   **II**

Form I is referred to as keto form since it contains a keto groups and form II as enol form since it contains a double bond and hydroxy group. This particular type of isomerism which involves a dynamic equilibrium between keto form and enol form of a compound is termed as *keto-enol tautomerism*. The two forms are called *tautomers* or *tautomerides*. When one tautomer is more stable than the other under ordinary conditions, the one which is lower in energy is termed as stable form whereas the other is called 'labile form'. In practice, However, a slight change in conditions such as temperature or solvent shifts the equilibrium from one form to another. Tautomerism is rare in solid state but the two forms exist in an equilibrium in solution or liquid state and in gaseous state.

In ethyl acetoacetate, enol form is more volatile and the change from enol to keto form is extremely sensitive to catalyst. The keto form (I) was first suggested by Frankland and Duppa in 1863 whereas as the enol form (II) was independently proposed by Geuther in 1865, both the structures were supported by various evidences.

Evidence in Favour of Keto Form

(1) It forms a bisulphite compound with sodium hydrogen sulphite.

(2) It forms cyanohydrin with hydrogen cyanide.

(3) It reacts with hydroxyl amine and phenyl hydrazine to form oxime and phenyl hydrazone respectively.

(4) When reduced with sodium amalgam, it yields a secondary alcohol.

Evidence in Favour of Enol Form

(1) The presence of hydroxy group is indicated by the reaction of ester with sodium to give its sodium salt and hydrogen gas.

(2) When reacted with phosphorus pentachloride, it gives a chloro derivative indicating the presence of hydroxy group.

(3) When it is treated with ethanolic solution of bromine, the color of bromine immediately discharged showing the presence of double bond or unsaturation.

(4) It gives a reddish-violet color with ferric chloride indicating the presence of $-\overset{\displaystyle OH}{\underset{\displaystyle |}{C}}H=\overset{\displaystyle |}{C}-$ group.

The controversy of ethyl acetoacetate having two structures continued till 1910 when chemists came to the conclusion that both the structures are correct and the two compound exist together in equilibrium in solution.

Composition of Ethyl Acetoacetate

keto and enol forms of ethyl acetoacetate were separated by Ludwig Knorr in 1911, The composition of tautomeric mixture may be estimated by physical and chemical methods.

The physical methods includes refractive index measurements, optical rotation measurments and proton magnetic resonance spectroscopy. The PMR of ethyl acetoacetate has been shown in Fig. 24.1, which shows the olefinic protons at δ 5.1, the enolic protons at δ12.0 and methyl singlet peak at δ 1.9. The quartet and the triplet peaks from ethyl group of both enol and keto form appear at δ 4.2 and δ 1.2 respectively. The $-CH_2$ and $-CH_3$ groups of keto form appear as singlet at δ 3.5 and δ 2.1.

The PMR spectra of ethyl acetoacetate shows major peaks at usual positions due to keto form and only minor peaks due to enol form (marked by arrows) indicating the presence of enol form in very small proportions compared to keto form.The ratio of the keto and enol form can be determined by the intensity of the enolic and CH_2 proton.

The chemical method utilises direct and indirect methods introduced by Meyer to find out the composition of tautomers. *The direct method* involves the rapid titration of ethanolic solution of the mixture of keto and enol form with etanolic bromine solution at $0°$ (to slow down the interconvertion of tautomers). The reaction is not dependable as with time the keto form is converted into enol form giving higher percentage of enol form.

Indirect method, however, is reliable in which excess of bromine is added in ethanolic solution of ethyl acetoacetate followed by addition of β-naphthol in ethanol (which removes the excess of bromine immmediately) to give the bromoester. The bromoester then reacts with acidified potassium iodide to

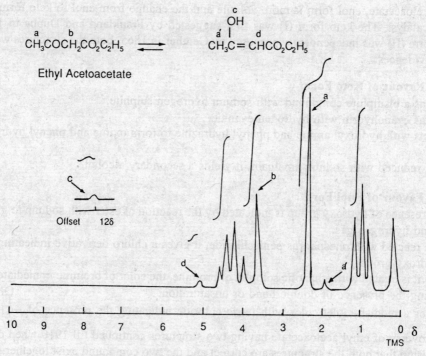

Fig. 24.1: PMR Spectra of Ethyl Acetoacetate

give iodine which is titrated with standard sodium thiosulphate. The liberated iodine is a measure of enol content of the isomeric mixture. At room temperature ethyl acetoacetate is 7% enol form and 93% keto form.

$$CH_3COCH_2COOC_2H_5 + Br_2 \longrightarrow CH_3COCHBrCOOC_2H_5$$
$$CH_3COCHBrCOOC_2H_5 + KI \longrightarrow CH_3COCHICOOC_2H_5 + KBr$$
$$CH_3COCHICOOC_2H_5 + HI \longrightarrow CH_3COCH_2COOC_2H_5 + I_2$$

24.2.3 Properties of Ethyl Acetoacetate

Physical Properties: Ethyl acetoacetate is a colorless, pleasant smelling liquid, b.p. 181° (with slight decomposition). It is sparingly soluble in water but freely soluble in organic solvents.

Chemical Properties: Since ethyl acetoacetate is a mixture of keto form and enol form, it shows the reactions due to keto form, enol form and active methylene group.

Reactions of Enol Form

(1) Enol form of ethyl acetoacetate, being slightly acidic, forms sodium salt with sodium hydroxide.

$$\overset{OH}{\underset{|}{CH_3-C}}=CHCO_2C_2H_5 + NaOH \rightleftharpoons \overset{\bar{O}\ Na^+}{\underset{|}{CH_3-C}}=CHCO_2C_2H_5$$

(2) It forms a green colored copper compound with copper acetate which is soluble in organic solvents indicating that it is not ionic but a chelate compound.

$$\begin{array}{c} \overset{\displaystyle OH}{\underset{\displaystyle |}{}} \\ CH_3\text{-}C\text{=}CHCO_2C_2H_5 \\ + \\ Cu(OCOCH_3)_2 \end{array} \longrightarrow \begin{array}{c} CH_3\text{-}C\text{-}O \qquad O\text{=}C\text{-}OC_2H_5 \\ HC \qquad Cu \qquad CH \\ C_2H_5O\text{-}C\text{=}O \qquad O\text{-}C\text{-}CH_3 \end{array}$$

(3)   When treated with acetyl chloride in pyridine as solvent, it forms O-acetyl derivative of ethyl acetoacetate.

$$\overset{\displaystyle OH}{\underset{\displaystyle |}{CH_3\text{-}C}}\text{=}CHCO_2C_2H_5 + CH_3COCl \xrightarrow{\text{Pyridine}} \overset{\displaystyle OCOCH_3}{\underset{\displaystyle |}{CH_3\text{-}C}}\text{=}CHCO_2C_2H_5$$

Ethyl O-acetylacetoacetate

(4)   With ethanolic solution of bromine it forms ethyl α-bromoacetoacetate.

$$Br\text{-}Br + \overset{\displaystyle :\ddot{O}H}{\underset{\displaystyle |}{CH\text{=}C}}\text{-}CH_3 \longrightarrow Br^- + \overset{\displaystyle \overset{+}{O}H}{\underset{\displaystyle |}{BrCH\text{-}C}}\text{-}CH_3 \xrightarrow{-H^+} \overset{\displaystyle O}{\underset{\displaystyle ||}{CH_3\text{-}C}}\text{-}CHBr\,CO_2\,C_2H_5$$

Ethyl α-bromoacetoacetate

(5)   It reacts with Grignard reagent to form alkane.

$$\overset{\displaystyle OH}{\underset{\displaystyle |}{CH_3\text{-}C}}\text{=}CHCO_2C_2H_5 + C_2H_5MgI \longrightarrow C_2H_6 + \overset{\displaystyle OMgI}{\underset{\displaystyle |}{CH_3\text{-}C}}\text{=}CHCO_2C_2H_5$$

Ethane

Reactions of Keto Form

(6)   Keto form of ethyl acetoacetate undergoes reduction with sodium amalgam to form ethyl β-hydroxy-butyrate.

$$CH_3COCH_2CO_2C_2H_5 \xrightarrow{Na\text{-}Hg/H_2O} CH_3CHOHCH_2CO_2C_2H_5$$

Ethyl-β-hydroxybutyrate

The same product is obtained on reduction with lithium aluminum hydride in pyridine whereas in absence of pyridine 1,3-butanediol is obtained.

$$CH_3COCH_2CO_2C_2H_5 \xrightarrow{LiAlH_4} CH_3CHOHCH_2CH_2OH + C_2H_5OH$$

1, 3-Butanediol

(7)   It reacts with sodium hydrogen sulphite and hydrogen cyanide to give bisulphite compound and cyanohydrin, respectively.

$$CH_3COCH_2CO_2C_2H_5 \left\{ \begin{array}{l} \xrightarrow{NaHSO_3} \overset{\displaystyle HO\quad SO_3Na}{\underset{\displaystyle |\qquad |}{CH_3\text{-}C\text{-}CH_2CO_2C_2H_5}} \\ \textbf{Bisulphite compound} \\[2mm] \xrightarrow{HCN} \overset{\displaystyle HO\quad CN}{\underset{\displaystyle |\qquad |}{CH_3\text{-}C\text{-}CH_2CO_2C_2H_5}} \\ \textbf{Cyanohydrin} \end{array} \right.$$

(8)   *Ketonic Hydrolysis*: When ethyl acetoacetate cleaves into acetone, as a result of its reaction with dilute aqueous or ethanolic alkali followed by acidification, the reaction is known as *ketonic hydrolysis*.

$$CH_3COCH_2CO_2C_2H_5 \xrightarrow[\text{2. H}^+/\text{H}_2\text{O}]{\text{1. KOH}} CH_3COCH_3 + CO_2 + C_2H_5OH$$

Acetone

The first step in the hydrolysis is the nucleophilic attack at electron deficient carboxyl carbon followed by decarboxylation.

$$CH_3-\overset{O}{\overset{\|}{C}}-CH_2-\overset{O}{\overset{\|}{C}}-OC_2H_5 + HO^- \rightleftharpoons CH_3-\overset{O}{\overset{\|}{C}}-CH_2-\overset{O^-}{\overset{|}{\underset{OH}{C}}}-OC_2H_5 \xrightarrow{NaOH}$$

$$CH_3-\overset{O}{\overset{\|}{C}}-CH_2-\overset{O}{\overset{\|}{C}}-\bar{O}Na^+ \xrightarrow{H^+} CH_3-\overset{O}{\overset{\|}{C}}-CH_2-\overset{O}{\overset{\|}{C}}-OH \xrightarrow{\Delta} CH_3-\overset{O}{\overset{\|}{C}}-CH_3 + CO_2$$

The electron attracting carbonyl group helps in the elimination of carbondioxide.

(9)   Acid Hydrolysis: When ethyl acetoacetate is hydrolysed with concentrated aqueous or ethanolic potassium hydroxide followed by acidification, it yields acetic acid as the main product.

$$CH_3COCH_2CO_2C_2H_5 \xrightarrow{\text{conc. KOH/HCl}} 2CH_3CO_2H + C_2H_5OH$$

Acetic acid

The key step in acid hydrolysis is the reversal of claisen condensation.

$$HO^- + CH_3-\overset{O}{\overset{\|}{C}}-CH_2-\overset{O}{\overset{\|}{C}}-OC_2H_5 \rightleftharpoons CH_3-\overset{O^-}{\overset{|}{C}}-CH_2-\overset{O}{\overset{\|}{C}}-OC_2H_5 \rightleftharpoons$$

$$[CH_3-\overset{O}{\overset{\|}{C}}-OH + \bar{C}H_2-\overset{O}{\overset{\|}{C}}-OC_2H_5] \rightleftharpoons [CH_3-\overset{O}{\overset{\|}{C}}-O^- + CH_3-\overset{O}{\overset{\|}{C}}-OC_2H_5]$$

$$\xrightarrow{HO^-} 2CH_3-\overset{O}{\overset{\|}{C}}-OH + C_2H_5OH$$

Reactions of Active Methylene Group

(10)   The methylene group in ethyl acetoacetate is reactive due to the presence of two electron withdrawing carbonyl groups adjacent to it. It reacts with sodium ethoxide to form a carbanion which undergoes nucleophilic substitution reaction with alkyl halides to form monoalkyl derivative of ethyl acetoacetate.

$$CH_3COCH_2CO_2C_2H_5 \xrightarrow{C_2H_5\bar{O}Na^+} CH_3CO\bar{C}HCO_2C_2H_5Na^+$$

$$CH_3CO\bar{C}HCO_2C_2H_5Na^+ + RBr \longrightarrow CH_3COCHRCO_2C_2H_5$$

Ethyl monoalkylacetoacetate

Monoalkyl ethyl acetoacetate still having an active hydrogen can form dialkyl derivatives.

$$CH_3COCHRCO_2C_2H_5 \xrightarrow{C_2H_5\bar{O}Na^+} CH_3CO\bar{C}RCO_2C_2H_5Na^+$$

$$CH_3CO\bar{C}RCO_2C_2H_5Na^+ \xrightarrow{R-Br} CH_3COCR_2CO_2C_2H_5$$

Ethyl dialkylacetoacetate

(11)   When ethyl acetoacetate is treated with acetyl chloride in benzene in presence of magnesium metal, C-acetyl derivative of ethyl acetoacetate is produced.

$$CH_3COCH_2CO_2C_2H_5 + CH_3COCl \xrightarrow{Mg} (CH_3CO)_2CH-CO_2C_2H_5 + HCl$$
Ethyl C-acetylacetoacetate

(12)   With nitrous acid, ethyl oximinoacetoacetate is obtained.

$$CH_3COCH_2CO_2C_2H_5 + HNO_2 \longrightarrow CH_3-\overset{\overset{O}{\|}}{C}-\overset{\overset{N-OH}{\|}}{C}-CO_2C_2H_5 + H_2O$$
Ethyl oximinoacetoacetate

(13)   It undergoes Knoevenagel condensation when reacted upon with aldehydes in pyridine. With acetaldehyde it gives α-ethylidene ethyl acetoacetate.

$$CH_3COCH_2CO_2C_2H_5 + CH_3CHO \xrightarrow{Pyridine} CH_3CO\overset{\overset{CHCH_3}{\|}}{C}CO_2C_2H_5 + H_2O$$
Ethyl α-ethylideneacetoacetate

24.2.4   Ethyl Acetoacetate in Organic Synthesis

(1)   *Synthesis of Alkyl Acetoacetic Esters*: As discussed earlier ethyl acetoacetate forms monoalkyl and dialkyl derivatives when treated with strong base and alkyl halide.

$$CH_3-\overset{\overset{O}{\|}}{C}-CH_2CO_2C_2H_5 \xrightarrow[-C_2H_5OH]{C_2H_5\bar{O}N\overset{+}{a}} CH_3CO\bar{C}HCO_2C_2H_5N\overset{+}{a} \xrightarrow{RX}$$

$$CH_3-\overset{\overset{O}{\|}}{C}-\overset{\overset{R}{|}}{C}H-CO_2C_2H_5 \xrightarrow[-C_2H_5OH]{C_2H_5\bar{O}N\overset{+}{a}} CH_3CO\bar{C}RCO_2C_2H_5N\overset{+}{a} \xrightarrow{R'X}$$
Monoalkyl derivative

$$CH_3-\overset{\overset{O}{\|}}{C}-\overset{\overset{R}{|}}{\underset{\underset{R'}{|}}{C}}-CO_2C_2H_5$$
Dialkyl derivative

(2)   *Synthesis of Fatty Acids*: Fatty acids and monosubstituted fatty acids are prepared by acidic hydrolysis of mono and dialkyl derivatives of ethyl acetoacetate respectively.

$$CH_3\overset{\overset{O}{\|}}{C}-CH_2CO_2C_2H_5 \xrightarrow[2.\ n\text{-}C_4H_9Br]{1.\ C_2H_5\bar{O}N\overset{+}{a}} CH_3CO\overset{\overset{}{\underset{\underset{CH_2CH_2CH_2CH_3}{|}}{C}}}HCO_2C_2H_5 \xrightarrow{Acid\ hydrolysis}$$
Ethyl acetoacetate

$$CH_3COOH + CH_3CH_2CH_2CH_2CH_2COOH + C_2H_5OH$$
Hexanoic acid

$$CH_3CO\overset{\overset{}{\underset{\underset{CH_2CH_2CH_2CH_3}{|}}{C}}}H\overset{\overset{O}{\|}}{C}OC_2H_5 \xrightarrow[2.\ CH_3Br]{1.\ C_2H_5\bar{O}N\overset{+}{a}} CH_3CO\ \overset{\overset{CH_3}{|}}{\underset{\underset{CH_2CH_2CH_2CH_3}{|}}{C}}-\overset{\overset{O}{\|}}{C}-OC_2H_5$$

$$\xrightarrow{\text{Acid hydrolysis}} CH_3COOH + CH_3(CH_2)_3\overset{\overset{\displaystyle CH_3}{|}}{C}HCOOH + C_2H_5OH$$
2-Methylhexanoic acid

(3) *Synthesis of Dicarboxylic Acids*: Sodium salt of ethyl acetoacetate is treated upon with appropriate bromoethyl ester followed by the acid hydrolysis of the product.

$$CH_3COCH_2CO_2C_2H_5 \xrightarrow[\text{2. Br(CH}_2)_2CO_2C_2H_5]{\text{1. C}_2H_5\overset{-}{O}\overset{+}{Na}} CH_3CO\overset{\overset{\displaystyle}{|}}{C}HCO_2C_2H_5 \xrightarrow[\text{Hydrolysis}]{\text{Acid}}$$
$$\qquad\qquad CH_2CH_2CO_2C_2H_5$$

$$CH_3COOH + HO_2CCH_2CH_2CH_2CO_2H + 2C_2H_5OH$$
Glutaric acid

Similarly other dicarboxylic acids can be prepared by altering the number of carbon atoms in halo-ester.

(4) *Synthesis of α,β-Unsaturated Acids*: Unsaturated acids are obtained by the reactions of EAA with aldehydes in presence of base followed by acid hydrolysis.

$$CH_3COCH_2CO_2C_2H_2 + CH_3CHO \xrightarrow[-H_2O]{(C_2H_5)_2NH} CH_3CO\overset{\overset{\displaystyle}{\underset{\displaystyle CHCH_3}{||}}}{C}-CO_2C_2H_2$$

$$\xrightarrow{\text{Acid hydrolysis}} CH_3COOH + CH_3CH=CHCO_2H + C_2H_5OH$$
Crotonic Acid

(5) *Synthesis of Keto Acids*: Sodium derivative of EAA reacts with α-haloester to give acetosuccinic ester which on ketonic hydrolysis forms γ-keto acids.

$$CH_3CO\overset{-}{C}HCO_2C_2H_5\overset{+}{Na} + ClCH_2CO_2C_2H_5 \longrightarrow CH_3CO\overset{\overset{\displaystyle}{|}}{C}HCO_2C_2H_5$$
$$\qquad\qquad\qquad\qquad\qquad\qquad CH_2CO_2C_2H_5$$

$$\xrightarrow[\text{Hydrolysis}]{\text{Ketonic}} CH_3COCH_2CH_2CO_2H + CO_2 + 2C_2H_5OH$$
γ-keto acid

(6) *Synthesis of Ketones*: Higher ketones are obtained on ketonic hydrolysis of mono and dialkyl derivatives of EAA.

$$CH_3CO\overset{\overset{\displaystyle}{\underset{\displaystyle CH_3}{|}}}{C}HCO_2C_2H_2 \xrightarrow[\text{Hydrolysis}]{\text{Ketonic}} CH_3COCH_2CH_3 + CO_2 + C_2H_2OH$$

2-Butanone

$$CH_3CO\overset{\overset{\displaystyle CH_2CH_3}{|}}{\underset{\underset{\displaystyle CH_2CH_3}{|}}{C}}CO_2C_2H_2 \xrightarrow[\text{Hydrolysis}]{\text{Ketonic}} CH_3CO\overset{\overset{\displaystyle CH_2CH_3}{|}}{C}HCH_2CH_3 + CO_2 + C_2H_5OH$$
3-Ethyl-2-pentanone

(7) *Synthesis of 1,3-Diketones*: Sodium salt of EAA reacts with acetyl chloride followed by ketonic hydrolysis of product to give acetyl acetone.

$$CH_3CO\overset{-}{C}HCO_2C_2H_5\overset{+}{Na} \xrightarrow{CH_3COCl} CH_3CO\overset{\overset{\displaystyle}{\underset{\displaystyle COCH_3}{|}}}{C}HCO_2C_2H_5 \xrightarrow[\text{Hydrolysis}]{\text{Ketonic}}$$
$$CH_3COCH_2COCH_3 + CO_2 + C_2H_5OH$$
Acetyl acetone

(8) *Synthesis of 1,4-Diketones*: 1,4-diketones are obtained when excess of sodium salt of EAA is treated with iodine followed by ketonic hydrolysis of the product.

$$2CH_3CO\bar{C}HCO_2C_2H_5Na^+ + I_2 \longrightarrow \begin{array}{c} CH_3COCH-CO_2C_2H_5 \\ | \\ CH_3COCH-CO_2C_2H_5 \end{array} \xrightarrow[\text{Hydrolysis}]{\text{Ketonic}}$$

$$CH_3COCH_2CH_2COCH_3 + 2CO_2 + 2C_2H_5OH$$

(9) *Synthesis of Hydrocarbons:* EAA and its alkyl derivatives form paraffins on catalytic reduction.

$$CH_3COCH_2CO_2C_2H_5 + 10\,[H] \longrightarrow CH_3CH_2CH_2CH_3 + 2H_2O + C_2H_5OH$$
$$\textbf{n-Butane}$$

$$\begin{array}{c} CH_3COCHCO_2C_2H_5 + 10\,[H] \longrightarrow CH_3CH_2CH(CH_3)CH_2CH_2CH_3 + 2H_2O + C_2H_5OH \\ | \\ CH_2CH_2CH_3 \end{array}$$
$$\textbf{3-Methylhexane}$$

(10) *Synthesis of Alicyclic Compounds*: Sodium salt of EAA reacts with dihalogen compounds to form alicyclic compounds of ring having five to seven carbons.

$$\begin{array}{c} CH_2CH_2Br \\ | \\ CH_2CH_2Br \end{array} + Na\bar{C}H\!\!<\!\!\begin{array}{c} COCH_3 \\ CO_2C_2H_5 \end{array} \xrightarrow[-NaBr]{\Delta} \begin{array}{c} CH_2-CH_2-CH\!\!<\!\!\begin{array}{c} COCH_3 \\ CO_2C_2H_5 \end{array} \\ | \\ CH_2-CH_2Br \end{array} \longrightarrow$$

$$\textbf{1,4-Dibromobutane}$$

$$\xrightarrow{C_2H_5\bar{O}Na^+} \begin{array}{c} CH_2-CH_2\!\!\!\overset{Na^+}{\overset{|}{C}}\!\!<\!\!\begin{array}{c} COCH_3 \\ CO_2C_2H_5 \end{array} \\ | \\ CH_2-CH_2-Br \end{array} \xrightarrow{-NaBr}$$

$$\begin{array}{c} CH_2-CH_2 \\ | \qquad \quad C\!\!<\!\!\begin{array}{c} COCH_3 \\ CO_2C_2H_5 \end{array} \\ CH_2-CH_2 \end{array} \xrightarrow[\text{Hydrolysis}]{\text{Ketonic}} \begin{array}{c} CH_2-CH_2 \\ | \qquad \quad CHCOCH_3 + CO_2 + C_2H_5OH \\ CH_2-CH_2 \end{array}$$

$$\textbf{Acetylcyclopentane}$$

Similarly other alicyclic compounds can be prepared by changing the alkyl chain of dihalogen compounds.

(11) *Synthesis of Heterocyclic Compounds*: EAA on reaction with urea, hydroxylamine and phenyl hydrazine forms a variety of heterocyclic compounds.

$$\begin{array}{c} NH_2 \\ | \\ C=O \\ | \\ NH_2 \end{array} + \begin{array}{c} C_2H_5O-CO \\ | \\ CH \\ || \\ HO-C-CH_3 \end{array} \xrightarrow{POCl_3} \begin{array}{c} NH-CO \\ | \\ C=OCH \\ || \\ NH-C-CH_3 \end{array} + H_2O + C_2H_5OH$$

$$\textbf{Urea} \qquad\qquad\quad \textbf{EAA} \qquad\qquad\quad \textbf{4-Methyluracil}$$

$$CH_3COCH_2CO_2C_2H_5 + NH_2OH \xrightarrow{-H_2O}$$
$$\textbf{EAA} \qquad\qquad \textbf{Hydroxyl amine}$$

$$\begin{array}{c} CH_3-C-CH_2 \\ || \qquad\quad | \\ N \quad C=O \\ | \quad\; | \\ OH\; OC_2H_5 \end{array}$$

$$\Big\downarrow -C_2H_5OH$$

$$\begin{array}{c} CH_3-C-CH_2 \\ || \qquad\quad | \\ N \quad C=O \\ \backslash \;\; / \\ O \end{array}$$

$$\textbf{Methyl isooxazolone}$$

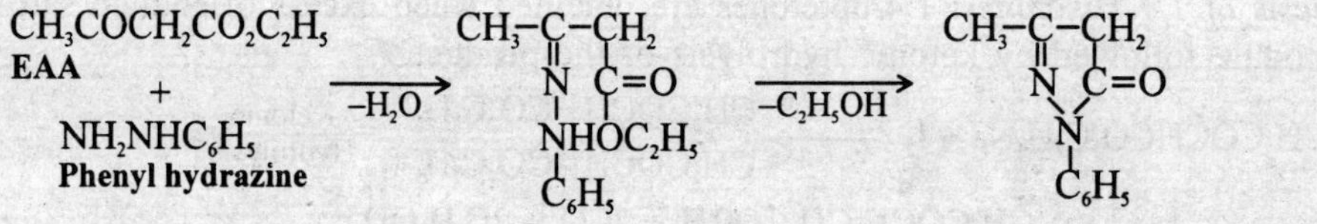

$$CH_3COCH_2CO_2C_2H_5$$
EAA
+
$$NH_2NHC_6H_5$$
Phenyl hydrazine

$\xrightarrow{-H_2O}$

(structure: pyrazolone intermediate with NHOC$_2$H$_5$ and C$_6$H$_5$)

$\xrightarrow{-C_2H_5OH}$

(structure: 3-Methyl-1-phenylpyrazolone)

3-Methyl-1-phenylpyrazolone

24.3   DIETHYL MALONATE/MALONIC ESTER [CH$_2$(COOC$_2$H$_5$)$_2$]

24.3.1   Preparation

(1)   Diethyl malonate is readily prepared by heating sodium chloroacetate and sodium cyanide. The sodium cyanoacetate thus formed is heated with absolute ethanol and concentrated hydrochloric acid.

$$ClCH_2COONa + NaCN \longrightarrow NC-CH_2COONa + NaCl$$
Sodium chloroacetate

$$NC-CH_2COONa + C_2H_5OH + HCl \longrightarrow H_5C_2OOC-CH_2-COOC_2H_5 + 2NH_4Cl$$
Diethyl malonate

In the above reaction, hydrolysis of cyano group and esterification takes place together.

(2)   Malonic ester may also be prepared by hydrolysis and esterification of methylene cyanide which in turn is prepared from the reaction of methylene chloride with sodium cyanide.

$$Cl-CH_2-Cl + NaCN \xrightarrow[-NaCl]{\Delta} CN-CH_2-CN \xrightarrow[C_2H_5OH]{H^+/H_2O}$$
Methylene chloride

$$C_2H_5OOC-CH_2-COOC_2H_5 + 2NH_4Cl$$
Diethyl malonate

24.3.2   Properties

Physical: It is a colorless liquid having pleasant odor (b.p. 198°). It is sparingly soluble in water but soluble in ethanol, benzene and chloroform.

Chemical: Diethyl malonate contains a methylene group flanked between two carbonyl groups as the case was with ethyl acetoacetate. Since the carbonyl group is electron withdrawing, the hydrogen atoms of methylene group becomes quite acidic and mobile. The C–H bond of methylene is weakened due to electron withdrawing effect of carbonyl group.

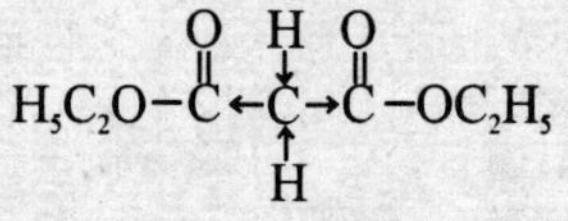

$$H_5C_2O-\overset{\overset{O}{\|}}{C}\leftarrow\overset{\overset{H}{\downarrow}}{\underset{\overset{\uparrow}{H}}{C}}\rightarrow\overset{\overset{O}{\|}}{C}-OC_2H_5$$

The acidity of the hydrogens is also attributed to the fact that diethyl malonate anion formed by the removal of proton is stabilised by resonance. Below are given the resonating structures of diethyl malonate anion.

$$C_2H_5O-\overset{\overset{O}{\|}}{C}-\bar{C}H-\overset{\overset{O}{\|}}{C}-OC_2H_5 \longleftrightarrow C_2H_5O-\overset{\overset{\bar{O}}{|}}{C}=CH-\overset{\overset{O}{\|}}{C}-OC_2H_5$$

$$\longleftrightarrow C_2H_5O-\overset{\overset{O}{\|}}{C}-CH=\overset{\overset{\bar{O}}{|}}{C}-OC_2H_5$$

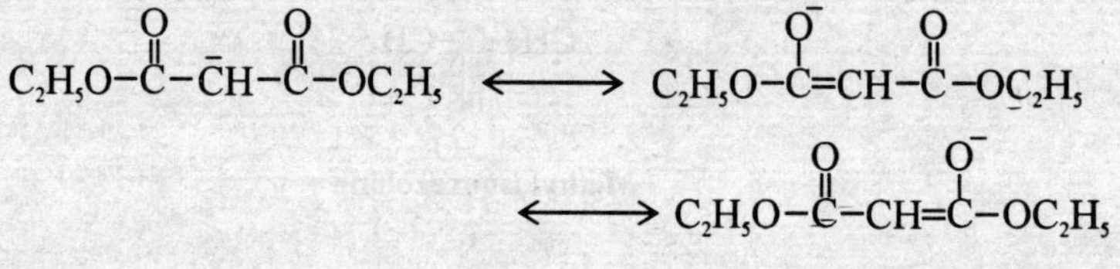

Due to resonance stabilisation of anion diethyl malonate is a strong acid. In the equilibrium mixture, however, diethyl malonate exists in ketoform and only a minute fraction represents enol form.

Reactions Due to Active Methylene Group

(1) Diethyl malonate reacts with sodium ethoxide to form its sodium salt.

$$CH_2(COOC_2H_5)_2 + C_2H_5\overset{+}{O}\overset{-}{N}a \longrightarrow \overset{+}{N}a\ \overset{-}{C}H\ (COOC_2H_5)_2$$

Sodium diethyl malonate

(2) Sodium diethyl malonate serves as a strong nucleophile and replaces halogen from alkyl halides forming mono and dialkyl malonic ester.

$$\overset{+}{N}a(COOC_2H_5)_2\ \overset{\frown}{\overset{-}{C}H} + R\overset{\frown}{-}\overset{-}{Br} \xrightarrow[-NaBr]{} R-CH(COOC_2H_5)_2$$

Monoalkylmalonic ester

$$\xrightarrow[-C_2H_5OH]{C_2H_5\overset{-}{O}\ \overset{+}{N}a} R-\overset{-}{C}(COOC_2H_5)_2\overset{+}{N}a + R'-Br \longrightarrow \overset{R}{\underset{R}{>}}C-(COOC_2H_5)_2$$

Dialkylmalonic ester

(3) The active methylene group of malonic ester reacts with nitrous acid followed by hydrolysis give ketomalonic acid.

$$\begin{matrix} COOC_2H_5 \\ CH_2 \\ COOC_2H_5 \end{matrix} + O=N-OH \longrightarrow \begin{matrix} COOC_2H_5 \\ C=N-OH \\ COOC_2H_5 \end{matrix} \xrightarrow{H_2O/H^+} \begin{matrix} COOH \\ C=O \\ COOH \end{matrix}$$

Diethyl malonate   **Ketomalonic Acid**

(4) With bromine it forms C-bromodiethyl malonate.

$$CH_2(COOC_2H_5)_2 + Br_2 \longrightarrow BrCH(COOC_2H_5)_2$$

C-bromodiethyl malonate

Reactions of Ester Group

(5) Diethyl malonate and its mono and dialkyl derivatives on treatment with aqueous potassium hydroxide followed by acidification give geminal dicarboxylic acids (having the two -COOH groups on the same carbon) which on heating above their melting points undergo decarboxylation to yield substituted acetic acids.

$$CH_2(COOC_2H_5)_2 \xrightarrow[2.\ H^+/H_2O]{1.\ KOH/H_2O/\Delta} H_2C(COOH)_2 \xrightarrow[-CO_2]{\Delta} CH_3COOH$$

Acetic acid

$$RCH(COOC_2H_5)_2 \xrightarrow[2.\ H^+/H_2O]{1.\ KOH/H_2O/\Delta} RCH(COOH)_2 \xrightarrow[-CO_2]{\Delta} RCH_2COOH$$

Substituted acetic acid

$$\overset{R}{\underset{R'}{>}}C(COOC_2H_5)_2 \xrightarrow[2.\ H^+/H_2O]{1.\ KOH/H_2O/\Delta} \overset{R}{\underset{R'}{>}}C(COOH)_2 \xrightarrow[-CO_2]{\Delta} \overset{R}{\underset{R'}{>}}CHCOOH$$

Dialkylacetic acid

24.3.3   Diethyl Malonate in Organic Synthesis

(1)   *Synthesis of Fatty Acids*: Malonic ester as stated earlier, forms malonic acid on hydrolysis which when heated above its melting point undergoes decarboxylation. Higher fatty acids are synthesised by the reaction of sodium salt of the ester with appropriate alkyl halide followed by hydrolysis and decarboxylation.

$$CH_2(COOC_2H_5)_2 \xrightarrow{C_2H_5O^-Na^+} Na^+\bar{C}H(COOC_2H_5)_2 \xrightarrow{n-C_4H_9Br}$$

$$nC_4H_9CH(COOC_2H_5)_2 \xrightarrow{1.\ KOH/H_2O/\Delta} nC_4H_9CH(COOH)_2$$

$$\xrightarrow[-CO_2]{\Delta} nC_4H_9CH_2COOH$$

$$\text{Caproic acid}$$

Similarly substituted fatty acids can be prepared from monoalkyl and dialkyl malonic ester.

$$CH_2(COOC_2H_5)_2 \xrightarrow[2.\ CH_3Br]{1.\ C_2H_5O^-Na^+} CH_3CH(COOC_2H_5)_2 \xrightarrow[2.\ CH_3Br]{1.\ C_2H_5O^-Na^+}$$

Malonic ester   **Diethyl methylmalonate**

$$\begin{matrix} CH_3 \\ CH_3 \end{matrix}\!\!\!>\!\!C(COOC_2H_5)_2 \xrightarrow{KOH/H_2O/\Delta} \begin{matrix} CH_3 \\ CH_3 \end{matrix}\!\!\!>\!\!C(COOH)_2 \xrightarrow[150°-200°]{\Delta} \begin{matrix} CH_3 \\ CH_3 \end{matrix}\!\!\!>\!\!CHCOOH$$

Diethyl dimethyl-   **Dimethylacetic acid**
malonate

(2)   *Synthesis of Dicarboxylic Acids*: Treatment of sodium salt of diethyl malonate with ethylene dibromide followed by hydrolysis and decarboxylation furnishes adipic acid.

$$\begin{matrix} CH_2Br \\ | \\ CH_2Br \end{matrix} + \begin{matrix} Na^+\bar{C}H(COOC_2H_5)_2 \\ Na^+\bar{C}H(COOC_2H_5)_2 \end{matrix} \longrightarrow \begin{matrix} CH_2-CH(COOC_2H_5)_2 \\ | \\ CH_2-CH(COOC_2H_5)_2 \end{matrix}$$

Ethylene
Bromide

$$\xrightarrow[2.\ 150°-200°]{1.\ KOH/H_2O/\Delta} \begin{matrix} CH_2-CH_2COOH \\ | \\ CH_2-CH_2COOH \end{matrix}$$

Similarly, other higher dicarboxylic acids can be prepared by using dihalides.

$$2Na^+\bar{C}H(COOC_2H_5)_2 + Br(CH_2)_nBr \longrightarrow (C_2H_5OOC)_2CH(CH_2)_nCH(COOC_2H_5)_2$$

$$\xrightarrow{Hydrolysis} (HOOC)_2CH(CH_2)_nCH(COOH)_2 \xrightarrow[-2CO_2]{Decarboxylation} HOOCCH_2(CH_2)_nCH_2COOH$$

Dicarboxylic acid

Dicarboxylic acids may also be prepared by the treatment of malonic ester with appropriate chloro ester. In the subsequent synthesis chloro ethylacetate is taken.

$$Na^+\bar{C}H(COOC_2H_5)_2 + Cl-CH_2COOC_2H_5 \longrightarrow$$

$$C_2H_5OOCCH_2CH(COOC_2H_5)_2 \xrightarrow[2.\ Decarboxylation]{1.\ Hydrolysis} HOOCCH_2CH_2COOH$$

Succinic acid

(3)   *Synthesis of α,β-Unsaturated Acids*: Diethyl malonate condenses with aldehydes in presence of

bases such as pyridine or diethylamine (Knoevenagel condensation) to give alkylidene diethyl malonate which upon hydrolysis and decarboxylation produces α, β-unsaturated acids.

$$\underset{H}{CH_3-\overset{\overset{\textstyle H}{|}}{C}=O} + CH_2(COOC_2H_5)_2 \xrightarrow{\text{Pyridine}} CH_3CH=C(COOC_2H_5)_2$$

$$\xrightarrow[\text{2. Decarboxylation}]{\text{1. Hydrolysis}} CH_3CH=CHCOOH$$
$$\textbf{Crotonic acid}$$

Similarly other α, β-unsaturated acids may be prepared by using higher aliphatic and aromatic aldehydes. Benzaldehyde gives cinnamic acid. Formaldehyde, on the contrary, does not give α, β-unsaturated acids but glutaric acid.

$$\underset{H}{H-\overset{\overset{\textstyle H}{|}}{C}=O} + 2CH_2(COOC_2H_5)_2 \cdot \xrightarrow{(C_2H_5)_2NH} H_2C{\overset{\textstyle CH(COOC_2H_5)_2}{\underset{\textstyle CH(COOC_2H_5)_2}{\big\langle}}}$$

$$\xrightarrow[\text{2. Decarboxylation}]{\text{1. Hydrolysis}} H_2C{\overset{\textstyle CH_2COOH}{\underset{\textstyle CH_2COOH}{\big\langle}}}$$
$$\textbf{Glutaric acid}$$

(4) *Synthesis of Keto Acids*: Treatment of malonic ester with acid chloride and subsequent hydrolysis followed by decarboxylation produces β-keto acids.

$$CH_3COCl + Na^+ \, \bar{C}H(COOC_2H_5)_2 \longrightarrow CH_3COCH(COOC_2H_5)_2 \xrightarrow[\text{2. Decarboxylation}]{\text{1. Hydrolysis}} CH_3COCH_2COOH$$
$$\textbf{β-Ketobutyric acid}$$

(5) *Synthesis of Amino Acids*: Treatment of nitrous acid with malonic ester followed by reduction produces amino malonic ester which upon usual hydrolysis and decarboxylation produces amino acids.

$$HO-N=O + \underset{\textbf{Malonic ester}}{CH_2(COOC_2H_5)_2} \longrightarrow \underset{\textbf{Diethyl oximinomalonate}}{HO-N=C(COOC_2H_5)_2}$$

$$\xrightarrow{Zn/CH_3COOH} \underset{\textbf{Diethyl aminomalonate}}{NH_2-CH(COOC_2H_5)_2} \xrightarrow{CH_3COCl} \underset{\textbf{Diethyl N-acetylaminomalonate}}{CH_3CONHCH(COOC_2H_5)_2}$$

$$\xrightarrow[\text{2. Decarboxylation}]{\text{1. Hydrolysis}} \underset{\textbf{Glycine}}{NH_2CH_2COOH}$$

(6) *Synthesis of Ketones*: Treatment of appropriate acid chloride with diethyl malonate followed by usual reactions forms ketones.

$$\underset{\textbf{Propanoyl chloride}}{CH_3CH_2COCl} + CH_2(COOC_2H_5)_2 \longrightarrow CH_3CH_2COCH(COOC_2H_5)_2$$

$$\xrightarrow[\text{2. Decarboxylation}]{\text{1. Hydrolysis}} CH_3CH_2COCH_2COOH \xrightarrow[\text{Decarboxylation}]{\Delta} \underset{\textbf{2-Butanone}}{CH_3CH_2COCH_3}$$

(7) *Synthesis of Alicyclic Compounds*: Treatment of dihalides with one mole of sodium diethyl malonate produces alicyclic compounds as shown below:

$$CH_2Br \atop CH_2Br + Na\overset{+}{\overset{}{C}}\overset{-}{H}(COOC_2H_5)_2 \longrightarrow \begin{matrix} CH_2-CH(COOC_2H_5)_2 \\ CH_2Br \end{matrix} \xrightarrow[-C_2H_5OH]{C_2H_5\overset{-}{O}\ Na^+}$$

$$\begin{matrix} CH_2-C(COOC_2H_5)_2Na^+ \\ CH_2-Br \end{matrix} \longrightarrow \begin{matrix} CH_2 \\ | \\ CH_2 \end{matrix}\!\!>\!\!C(COOC_2H_5)_2 \xrightarrow[\text{2. Decarboxylation}]{\text{1. Hydrolysis}} \begin{matrix} CH_2 \\ | \\ CH_2 \end{matrix}\!\!>\!\!CHCOOH$$

Cyclopropane-carboxylic acid

(8) *Synthesis of Heterocyclic Compounds*: Malonic ester condenses with urea in presence of phosphorus oxychloride to give barbiturates which have medicinal use as hypnotics, sedatives and anticonvulsants. Veronal (barbitone) may be prepared from diethyl derivative of diethyl malonate.

$$\begin{matrix} (C_2H_5)_2C(COOC_2H_5)_2 \\ + \\ NH_2CONH_2 \end{matrix} \longrightarrow \begin{matrix} C_2H_5 \\ C_2H_5 \end{matrix}\!\!>\!\!C\!\!<\!\!\begin{matrix} CO-NH \\ CO-NH \end{matrix}\!\!>\!\!CO$$

Barbitone

(9) *Synthesis of Glutaric Acid*: Malonic ester undergoes Michael condensation with α,β-unsaturated esters in presence of base followed by usual hydrolysis and decarboxylation to produces glutaric acid.

$$CH_2(COOC_2H_5)_2 + CH_2=CHCOOC_2H_5 \longrightarrow \begin{matrix} CH_2CH_2COOC_2H_5 \\ CH(COOC_2H_5)_2 \end{matrix}$$

$$\xrightarrow[\text{2. Decarboxylation}]{\text{1. Hydrolysis}} HOOCCH_2CH_2CH_2COOH$$

Glutaric acid

Michael condensation of N-acetylaminomalonic ester (obtained as given above in the synthesis of amino acid) with ethyl acrylate followed by hydrolysis and decarboxylation yields racemic glutamic acid.

$$CH_2=CHCOOC_2H_5 + CH_3CONH\overset{\overset{\displaystyle COOC_2H_5}{|}}{\underset{\underset{\displaystyle COOC_2H_5}{|}}{CH}} \longrightarrow$$

Diethyl N-acetylaminomalonate

$$CH_3CONH-\overset{\overset{\displaystyle COOC_2H_5}{|}}{\underset{\underset{\displaystyle COOC_2H_5}{|}}{C}}-CH_2CH_2COOC_2H_5 \xrightarrow[\text{2. Decarboxylation}]{\text{1. Hydrolysis}} \overset{\overset{\displaystyle COOH}{|}}{H_2N-CHCH_2CH_2COOH}$$

(±) Glutamic acid

24.4 PROBLEMS

1. What is tautomerism? How will you differentiate between tautomerism and resonance?
2. Why the enol form of a dicarbonyl compound more stable than that of a monocarbonyl compound?
3. What is meant by active methylene group?
4. Why keto form of Acetoacetic ester is more stable than its enolic form?
5. Discuss the isomerism exhibited by malonic ester.

6. How is a pinacolone synthesised from ethyl acetoacetate?

7. The sodium salt of malonic ester when condensed with 1,4-dibromobutane forms a cyclic monocarboxylic acid (A), and an open chain dicarboxylic acid (B). Write down the structure of A & B.

8. Identify A and B and C in the following:

$$EAA + Ethyl\ acrylate \xrightarrow[EtOH]{NaOEt} A \xrightarrow[H_2O]{Heat} B$$

$$EAA + (CH_3)_2C=CHCOCH_3 \xrightarrow[EtOH]{NaOEt} A \xrightarrow{NaOEt} B \xrightarrow{H_3O^+} C$$

9. (a) To what class does the reaction between sodium salt of EAA and an alkyl halide belong?

   (b) What will be the relative yields with primary, secondary and tertiary alkyl halides?

   (c) Will there be any reaction between EAA and aryl halide?

10. How will you prepare the following compounds starting from diethylmalonate.

   (a) Succinic acid   (b) Adipic acid

   (c) Barbituric acid   (d) Crotonic acid.

11. How will you prepare the following compounds starting from ethyl acetoacetate.

   (a) Butane-1,4-dioic acid   (b) Methyl-isopropyl ketone

   (c) 4-Methyluracil   (d) Crotonic acid

25
Aliphatic Amines

25.1 INTRODUCTION

The amines are regarded as derivatives of ammonia, in which one or more hydrogen atoms have been replaced by alkyl groups. They may be classified as primary, secondary or tertiary amines depending upon whether one, two or three hydrogen atoms of ammonia are replaced by alkyl groups.

$$
\underset{\textbf{Ammonia}}{H-\overset{\overset{H}{|}}{N}-H}
\qquad
\underset{\textbf{Primary amines}}{R-\overset{\overset{H}{|}}{N}-H}
\qquad
\underset{\textbf{Secondary amines}}{R-\overset{\overset{H}{|}}{N}-R}
\qquad
\underset{\textbf{Tertiary amines}}{R-\overset{\overset{R}{|}}{N}-R}
$$

The primary, secondary and tertiary amines are also represented as 1°, 2° and 3° amines, respectively. The secondary and tertiary amines are further classified as simple or mixed depending upon whether the alkyl groups are identical or different. The functional groups in primary, secondary and tertiary amines are $-NH_2$, $>NH$, $->N$ respectively. The quaternary ammonium salts are the tetra-alkyl derivatives of ammonium salts. For example $(CH_3)_4N^+Cl^-$ is called tetramethyl ammonium chloride.

25.2 NOMENCLATURE

In *Common System,* amines are named after the alkyl groups attached to nitrogen atom and adding suffix *'amine'*.

In IUPAC system, amines are named as amino alkanes, alkylamino alkanes and dialkylamino alkanes, respectively depending upon whether the amine is primary, secondary or tertiary. Amines may also be named by replacing *'e'* of the alkyl group and adding suffix *'amine'* to the name of the parent chain to which amino group is attached. Following are some examples of amines with their common and IUPAC names.

Amine	Common name	IUPAC name
CH_3NH_2	Methylamine	Methylamine or Methanamine
$\overset{CH_3}{\underset{CH_3}{>}}NH$	Dimethylamine	Dimethylamine or Methylaminomethane or N-Methylmethanamine
$\overset{CH_3}{\underset{CH_3}{CH_3->}}N$	Trimethylamine	Trimethylamine or Dimethylaminomethane or N,N-Dimethylmethanamine
$CH_3CH_2NH_2$	Ethylamine	Ethylamine or Ethanamine
$\overset{CH_3}{\underset{C_2H_5}{>}}NH$	Ethylmethylamine	Ethylmethylamine or Methylaminoethane or N-Methylethanamine
$\overset{CH_3}{\underset{C_3H_7}{C_2H_5->}}N$	Ethylmethylpropylamine	Ethylmethylpropylamine amine or N-Ethyl-N-methylpropanamine

5.3 ISOMERISM

ollowing types of isomerism are exhibited by amines.

1. *Chain isomerism* is due to the difference in the nature of the carbon chain of alkyl groups attached
→ the amino group. $C_4H_{11}N$ represents chain isomerism.

$$CH_3CH_2CH_2CH_2NH_2$$
n-Butylamine

$$\underset{\displaystyle CH_3-CH-CH_2-NH_2}{\overset{\displaystyle CH_3}{|}}$$
Isobutylamine

2. *Position Isomerism* is due to the difference in the position of the amino group in the carbon chain.C_3H_9N
epresents n-propylamine and isopropylamine.

$$CH_3CH_2CH_2NH_2$$
n-Propylamine

$$\underset{\displaystyle CH_3-CH-CH_3}{\overset{\displaystyle NH_2}{|}}$$
Isopropylamine

3. *Functional isomerism* is due the difference in the nature of the amino group. C_3H_9N represents
the following isomers.

$$CH_3CH_2CH_2NH_2$$
n-Propylamine
(Primary amine)

$$\underset{\displaystyle CH_3-N-CH_2CH_3}{\overset{\displaystyle H}{|}}$$
N-Methylethylamine
(Secondary amine)

$$\underset{\displaystyle CH_3-N-CH_3}{\overset{\displaystyle CH_3}{|}}$$
Trimethylamine
(Tertiary amine)

4. *Metamerism* is due to the nature of the alkyl group attached to the same functional group. $C_4H_{11}N$
epresents the following isomers.

$$\underset{\displaystyle C_2H_5-N-C_2H_5}{\overset{\displaystyle H}{|}}$$
Diethylamine
(secondary)

$$\underset{\displaystyle CH_3-N-CH_2CH_2CH_3}{\overset{\displaystyle H}{|}}$$
N-Methylpropylamine
(secondary)

25.4 STRUCTURE

Aliphatic amines have a pyramidal shape or if we regard the lone pair of electrons as a group, an approximately
tetrahedral shape. The three valencies of the tetrahedron are occupied by three substituent groups and the
fourth is occupied by the lone pair.

If we consider the orbital structure of carbon and nitrogen (Fig. 25.1), it is observed that in contrast
to carbon, one of the sp^3 hybrid orbital in nitrogen is occupied by a pair of electron. The three half
filled orbitals form σ bonds by overlap with s orbital of hydrogen atom or sp^3 orbitals of carbon of
alkyl group. The orbital structure of 1°, 2° and 3° amines has been represented in Fig. 25.2. The structural
representation of three types of amines has been shown in Fig. 25.3.

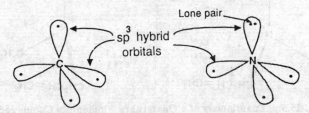

Fig. 25.1: Orbital Structure of Carbon and Nitrogen

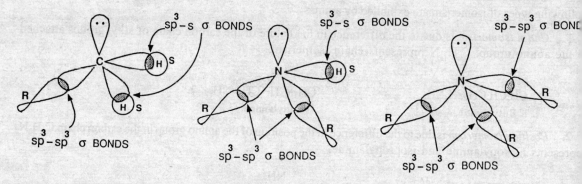

Fig. 25.2:   Orbital Structure of 1°, 2° and 3° Amines

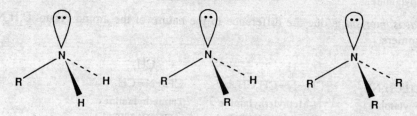

Fig. 25.3:   Structural Representation of 1°, 2° and 3° Amines

If the three substituents on nitrogen atom in amine are different, the nitrogen being chiral leads to the possibility of existence of enantiomers. However, in the absence of steric factor, amines undergo a rapid inversion at nitrogen via a planer transition state (Fig. 25.4). So it is not possible to isolate the enantiomers. Such an inversion, however, is not possible in quaternary ammonium compounds and so they can be separated into enantiomers. The enantiomers of such a quaternary ammonium ion is shown in Fig. 25.5.

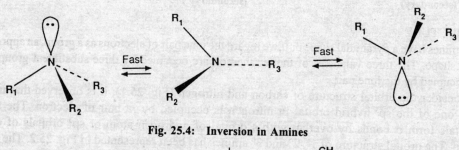

Fig. 25.4:   Inversion in Amines

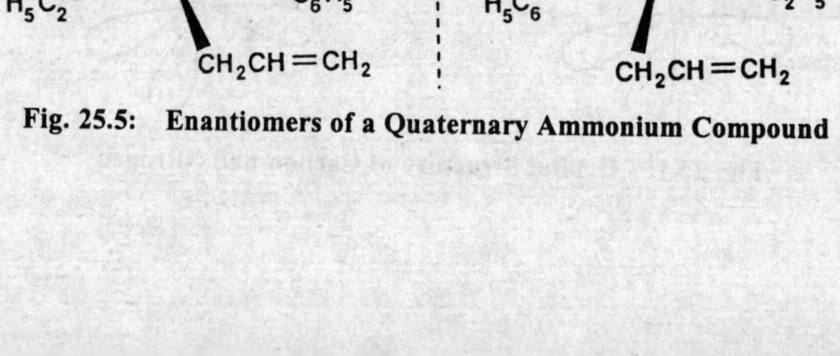

Fig. 25.5:   Enantiomers of a Quaternary Ammonium Compound

25.5 GENERAL METHODS OF PREPARATION

25.5.1 Primary Amines

(1) *From Nitroparaffins*: Nitroparaffins on reduction with tin and hydrochloric acid or by lithium aluminum hydride form primary amines.

$$R-NO_2 + 6H \longrightarrow R-NH_2 + 2H_2O$$

(2) *From Alkyl Cyanides*: Catalytic reduction of alkyl cyanide with sodium and alcohol or with lithium aluminum hydride yields primary amines.

$$\underset{\text{Methyl cyanide}}{CH_3C \equiv N} + 4H \longrightarrow \underset{\text{Ethylamine}}{CH_3CH_2NH_2}$$

This reaction is known as *Mendius Reaction*

(3) *From Oximes:* Reduction of oximes with sodium and alcohol forms primary amines.

$$\underset{\text{Acetaldoxime}}{CH_3CH=NOH} + 4H \longrightarrow \underset{\text{Ethylamine}}{CH_3CH_2NH_2} + H_2O$$

This method is used to convert aldehydes and ketones into amines.

(4) *From Acid Amides*: Acid amides by reduction with sodium and alcohol or by lithium aluminum hydride form 1° amines.

$$\underset{\text{Acetamide}}{CH_3CONH_2} + 4H \longrightarrow \underset{\text{Ethylamine}}{CH_3CH_2NH_2} + H_2O$$

This method is used to convert acids into amines.

$$\underset{\text{Acid}}{RCOOH} \xrightarrow{SOCl_2} RCOCl \xrightarrow{NH_3} RCONH_2 \xrightarrow{\text{reduction}} \underset{\text{Amines}}{RCH_2NH_2}$$

(5) *Gabriel's Phthalimide Synthesis*: Phthalimide when reacted with caustic potash forms potassium phthalimide which on treatment with an alkyl halide gives N-alkylphthalimide. Its hydrolysis with hydrochloric acid gives primary amines.

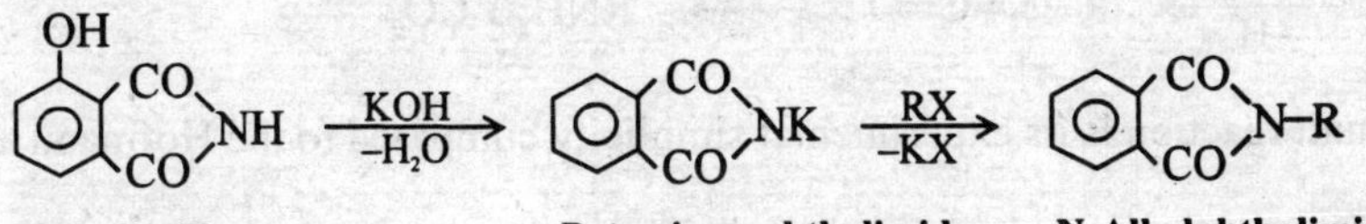

Phthalimide   Potassium phthalimide   N-Alkylphthalimide

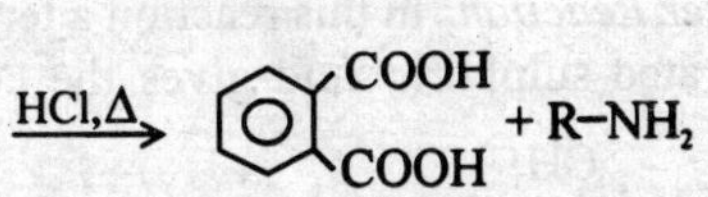

Phthalic acid   Primary amine

(6) *Hoffmann's Bromide Method:* Acid amides (upto six carbon atoms) form primary amines having one carbon atom less than the starting amide when treated with bromine and caustic potash.

$$RCONH_2 + Br_2 + 4KOH \longrightarrow RNH_2 + 2KBr + K_2CO_3 + 2H_2O$$

This is a convenient method for the preparation of primary amines. Details of the reaction have been discussed in Section 22.4.5.

(7) *Curtius Degradation:* In this method an acyl azide (obtained from the reaction of carboxylic acid chloride and sodium azide) in acidic or alkaline medium forms primary amine. The reaction involves an intermediate isocyanate which is hydrolysed to give primary amines.

$$R-COOH \xrightarrow{SOCl_2} RCOCl \xrightarrow{NaN_3} RCON_3$$

Acid   Acid chloride   Acyl azide

$$RCON_3 \xrightarrow{-N_2} R-N=C=O \xrightarrow[H^+]{H_2O} R-NH_2 + CO_2$$

Acyl azide   Alkyl isocyanate   1° Amine

The mechanism of curtius rearrangement resembles hofmann reaction and may be represented as follows.

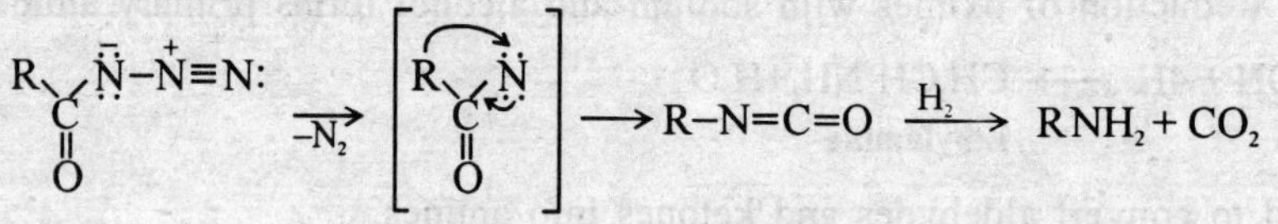

A modification of the curtius degradation is also used which is known as *Schmidt reaction.* In this reaction, a carboxylic acid is warmed with sodium azide in concentrated sulphuric acid to give primary amine having one carbon atom less than the starting acid.

$$R-COOH \xrightarrow[\text{conc. } H_2SO_4, \text{ warm}]{NaN_3} R-NH_2 + N_2 + CO_2$$

Acid   1° Amine

The mechanism of Schmidt reaction involves following steps.

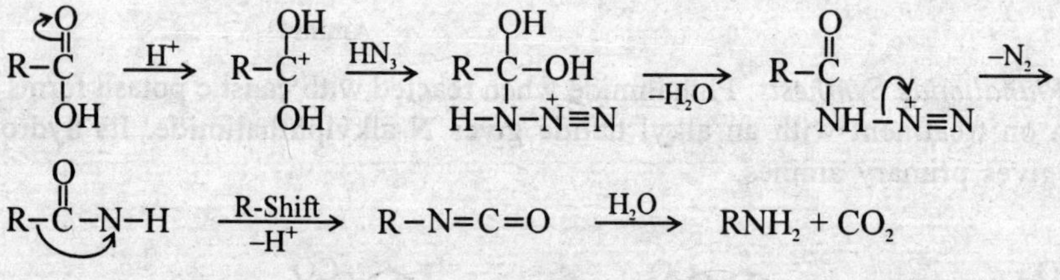

The advantage of Schmidt reaction is its experimental simplicity compared to the Hofmann and Curtius degradations.

(8) *Ritter Reaction:* In this reaction a tertiary alcohol on treatment with hydrogen cyanide in presence of concentrated sulphuric acid gives the corresponding primary amines.

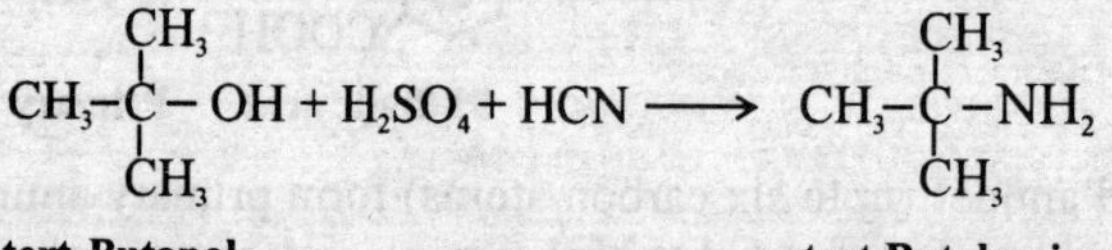

tert Butanol   tert Butylamine

The reaction is believed to proceed as follows.

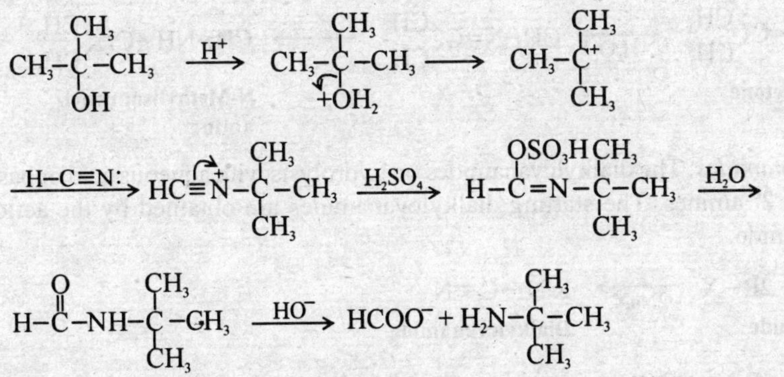

$$CH_3-\underset{\underset{OH}{|}}{\overset{\overset{CH_3}{|}}{C}}-CH_3 \xrightarrow{H^+} CH_3-\underset{\underset{+OH_2}{|}}{\overset{\overset{CH_3}{|}}{C}}-CH_3 \longrightarrow CH_3-\underset{\underset{CH_3}{|}}{\overset{\overset{CH_3}{|}}{C^+}}$$

$$\xrightarrow{H-C\equiv N:} HC\overset{+}{=}\overset{\curvearrowright}{N}-\underset{\underset{CH_3}{|}}{\overset{\overset{CH_3}{|}}{C}}-CH_3 \xrightarrow{H_2SO_4} H-\underset{}{C}=N-\underset{\underset{CH_3}{|}}{\overset{\overset{OSO_3HCH_3}{|}}{C}}-CH_3 \xrightarrow{H_2O}$$

$$H-\overset{\overset{O}{\|}}{C}-NH-\underset{\underset{CH_3}{|}}{\overset{\overset{CH_3}{|}}{C}}-GH_3 \xrightarrow{HO^-} HCOO^- + H_2N-\underset{\underset{CH_3}{|}}{\overset{\overset{CH_3}{|}}{C}}-CH_3$$

25.5.2 Secondary Amines

(1) *From Primary Amines*: Secondary amines may be prepared by heating primary amines with calculated amount of alkyl halides.

$$R-NH_2 + R'-I \xrightarrow{\Delta} R-NH-R'.HI$$
1° Amine

$$C_2H_5NH_2 + C_2H_5I \xrightarrow{\Delta} (C_2H_5)_2NH.HI$$
Ethylamine   Diethylamine hydriodide

(2) *From Alkyl Isocyanide*: Alkyl isocyanides on reduction with sodium and alcohol furnish secondary amines.

$$R-NC + 4H \xrightarrow{Na/C_2H_5OH} RNHCH_3$$
Alkyl isocyanide   2° Amine

(3) *From p-Nitrosodialkylanilines*: p-Nitrosodialkylanilines on boiling with strong alkali form secondary amines. The required p-nitroso derivatives are obtained from N,N-dialkylaniline.

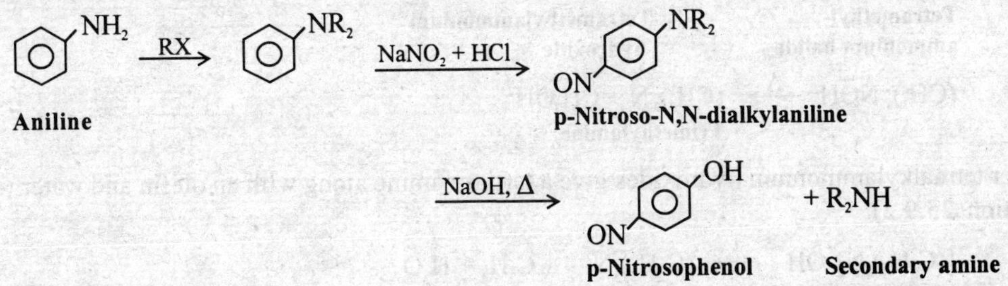

Aniline   p-Nitroso-N,N-dialkylaniline

p-Nitrosophenol   Secondary amine

(4) *From Aldehydes and Ketones*: In this method a primary amine is reacted with an aldehyde or a ketone to form imines, which on reduction with hydrogen and Raney nickel give secondary amine.

$$R-NH_2 + O=C\overset{R}{\underset{R}{<}} \xrightarrow[-H_2O]{} R-N=C\overset{R}{\underset{R}{<}} \xrightarrow{H_2/Raney\ Ni} R-NH-CH\overset{R}{\underset{R}{<}}$$
Primary   Ketone   Secondary amine
amine

$$CH_3NH_2 + O=C\underset{CH_3}{\overset{CH_3}{<}} \xrightarrow{-H_2O} CH_3-N=C\underset{CH_3}{\overset{CH_3}{<}} \longrightarrow CH_3-NH-CH\underset{CH_3}{\overset{CH_3}{<}}$$

Methyl-   **Acetone**   **N-Methylisopropyl-**
amine   **amine**

(5) *From Dialkylcyanamides*: The dialkylcyanamides on hydrolysis with aqueous acid or base followed
by decarboxylation form 2° amines. The starting dialkylcyanamides are obtained by the action of alkyl
halide on sodium cyanamide.

$$Na_2N-C\equiv N + 2R-X \xrightarrow{-2NaX} R_2N-C\equiv N$$

Sodium cyanamide   **Dialkylcyanamide**

$$\xrightarrow[-NH_3]{H^+ \text{ or } OH^-} R_2N-COOH \xrightarrow[-CO_2]{\Delta} R_2-NH$$

Dialkylcarbamic acid   **Secondary amine**

25.5.3   Tertiary Amines

(1) *From Secondary Amines*: A secondary amine on heating with an appropriate alkyl halide in ethanolic
solution gives the corresponding trialkylammonium salt, which on treatment with sodium hydroxide
liberates tertiary amine.

$$R-NH-R + R-X \xrightarrow{alcohol, \Delta} R-\overset{\overset{\displaystyle R}{|}}{\underset{\underset{\displaystyle R}{|}}{N}}-HX^- \xrightarrow{NaOH} R-\overset{\overset{\displaystyle R}{|}}{N}-R + NaX + H_2O$$

Secondary   **Alkyl halide**   **Trialkylammonium**   **Tertiary amine**
amine   **halide**

(2) *From Hofmann Elimination*: The tetramethylammonium hydroxides (obtained by the treatment
of tetramethylammonium halides with moist silver oxide) on heating decomposes to give tertiary amine
and methanol.

$$(CH_3)_4\overset{+}{N}X^- + AgOH \longrightarrow (CH_3)_4\overset{+}{N}\overset{-}{O}H$$

Tetramethyl-   **Tetramethylammonium**
ammonium halide   **hydroxide**

$$(CH_3)_4\overset{+}{N}\overset{-}{O}H \xrightarrow{\Delta} (CH_3)_3N + CH_3OH$$

Trimethylamine

Other tetraalkylammonium hydroxides give a tertiary amine along with an olefin and water (see details
in Section 25.9.2).

$$[(C_2H_5)_4N]^+OH^- \longrightarrow (C_2H_5)_3N + C_2H_4 + H_2O$$

Tetraethyl-   **Triethylamine**   **Ethylene**
ammonium hydroxide

(3) *Reductive Alkylation*: Tertiary amines may be prepared by methylation of a primary or a secondary amine by a mixture of formic acid and formaldehyde.

$$R-NH_2 + 2CH_2O + 2HCOOH \longrightarrow R-N\underset{CH_3}{\overset{CH_3}{<}} + 2H_2O + 2CO_2$$

1° Amine   **3° Amine**

25.5.4   Methods Yielding Mixture of Amines

(1) *From Alkyl Halides (Hofmann's method)*: In this method an aqueous or alcoholic solution of ammonia is heated with an alkyl halide in a sealed tube at 100°. A mixture of all the three types of amines along with quaternary ammonium salt is formed.

$$C_2H_5I + HNH_2 \longrightarrow C_2H_5NH_2 + HI$$
Ethylamine

$$C_2H_5NH_2 + C_2H_5I \longrightarrow (C_2H_5)_2NH + HI$$
Diethylamine

$$(C_2H_5)_2NH + C_2H_5I \longrightarrow (C_2H_5)_3N + HI$$
Triethylamine

$$(C_2H_5)_3N + C_2H_5I \longrightarrow (C_2H_5)_4\overset{+}{N}\overset{-}{I}$$
Tetraethylammonium iodide

The order of reactivity of alkyl halides is RI > RBr > RCl. In this method use of excess ammonia gives primary amine as the main product. On the other hand, use of excess of alkyl halide gives tertiary amine as the main product.

(2) *From Alcohols:* When vapors of alcohol and ammonia are passed over heated alumina at 360°, a mixture of all the three types of amines and quaternary ammonium salt is obtained.

$$C_2H_5OH + HNH_2 \longrightarrow C_2H_5NH_2 + H_2O$$
Ethylamine

$$C_2H_5OH + C_2H_5NH_2 \longrightarrow (C_2H_5)_2NH + H_2O$$
Diethylamine

$$C_2H_5OH + (C_2H_5)_2NH \longrightarrow (C_2H_5)_3N + H_2O$$
Triethylamine

$$C_2H_5OH + (C_2H_5)_3N \longrightarrow (C_2H_5)_4\overset{+}{N}\overset{-}{O}H$$
Tetraethylammonium hydroxide

25.6   SEPARATION OF A MIXTURE OF AMINES

The mixture of primary, secondary and tertiary amines along with quaternary ammonium salt is distilled with potassium hydroxide to get 1°, 2° and 3° amines as distillate and the non-volatile quaternary salt is left behind in the distillation flask. The mixture of amines thus obtained is separated by the following methods.

(1) *Fractional Distillation*: Since the boiling points of primary, secondary and tertiary amines are fairly apart, the separation can be done by fractional distillation. This method is most efficient and is employed in industries.

(2) *Hofmann's Method*: The mixture of amines is treated with diethyl oxalate. Primary amines from a solid oxamide whereas secondary amine gives a liquid oxamic ester. The tertiary amines do not react with diethyl oxalate.

$$\begin{matrix}COOC_2H_5 \\ | \\ COOC_2H_5\end{matrix} + 2RNH_2 \overset{\Delta}{\longrightarrow} \begin{matrix}CO-NHR \\ | \\ CO-NHR\end{matrix} + 2C_2H_5OH$$

Diethyl Oxalate   **Primary amine**   **Dialkyl oxamide (solid)**

$$\begin{matrix}COOC_2H_5 \\ | \\ COOC_2H_5\end{matrix} + R_2NH \xrightarrow{\Delta} \begin{matrix}CO-NR_2 \\ | \\ COOC_2H_5\end{matrix} + C_2H_5OH$$

Diethyl	Secondary	Dialkyl oxamic ester
Oxalate	amine	(liquid)

The reaction mixture is distilled when the tertiary amine distills over and is collected. The residual mixture contains a solid (dialkyl oxamide) and a liquid (dialkyl oxamic ester). These are easily separated by simple filtration and treated separately with strong alkalis when the corresponding amines are regenerated which are purified by distillation.

$$\begin{matrix}CONHR \\ | \\ CONHR\end{matrix} + KOH \longrightarrow \begin{matrix}COOK \\ | \\ COOK\end{matrix} + 2RNH_2$$

Dialkyl oxamide   Primary amine
(solid)

$$\begin{matrix}CONR_2 \\ | \\ COOC_2H_5\end{matrix} + KOH \longrightarrow \begin{matrix}COOK \\ | \\ COOK\end{matrix} + C_2H_5OH + R_2NH$$

Dialkyl oxamic ester	Potassium	Secondary
(liquid)	oxalate	amine

(3)   *Hinsberg's Method:* In this method the mixture of amines is treated with benzenesulphonyl chloride and shaken with potassium hydroxide solution (5%). The primary amine form N-alkylbenzenesulphonamide, which is soluble in base. The secondary amine on the other hand forms N,N-dialkyl benzenesulphonamide, which is insoluble in potassium hydroxide solution. Tertiary amines do not react with benzenesulphonyl chloride.

$$C_6H_5SO_2Cl \quad + \quad RNH_2 \quad \longrightarrow \quad C_6H_5SO_2NHR + HCl$$

Benzenesulphonyl	Primary	N-Alkylbenzenesulphonamide
chloride	amine	(soluble in aq KOH)

$$C_6H_5SO_2Cl \quad + \quad R_2NH \quad \longrightarrow \quad C_6H_5SO_2NR_2 + HCl$$

Benzenesulphonyl	Secondary	N, N-Dialkylbenzenesulphonamide
chloride	amine	(insoluble in aq KOH)

The reaction mixture obtained after treatment with benzenesulphonyl chloride is distilled when the tertiary amine distills over. The remaining mixture is filtered and the filtrate containing N-alkylbenzenesulphonamide on acidification gives the sulphonamide of the primary amine. The solid residue is the sulphonamide of the secondary amine. The two sulphonamides thus separated are hydrolysed with acid to regenerate the individual primary and secondary amines which are fractionated for purification.

$$C_6H_5SO_2NHR \quad + H_2O \xrightarrow{H^+} C_6H_5SO_3H \quad + RNH_2$$

N-Alkylbenzene-	Benzenesulphonic	Primary
sulphonamide	acid	amine

$$C_6H_5SO_2NR_2 \quad + H_2O \xrightarrow{H^+} C_6H_5SO_3H \quad + R_2NH$$

N-N-Dialkylbenzene-	Benzenesulphonic	Secondary
sulphonamide	acid	amine

The benzenesulphonyl chloride can be profitably replaced by p-toluenesulphonyl chloride. In the later case the substituted sulphonamides are stable and can be easily crystallised in case of primary amines.

25.7   PROPERTIES

25.7.1   Physical Properties

The lower members are gases, higher amines (C_4 to C_{11}) are volatile liquids, while still higher members are solids. The lower members, such as methylamine and dimethylamine have fishy ammonical odor. Also, the lower members are soluble in water, but the solubility decreases with increase in molecular weight. The boiling points also increase with increase in molecular weight.

Amines are polar compounds, primary amines have two polar $N-H$ bonds, secondary amines have one polar $N-H$ bond, but tertiary amines have none. In view of this the primary and secondary amines are capable of forming intermolecular hydrogen bonds represented below.

$$
\begin{array}{ccc}
CH_3 & H & CH_3 \\
| & | & | \\
\cdots N-H\cdots N-H\cdots N-H\cdots \\
| & | & | \\
H & CH_3 & H
\end{array}
\qquad\qquad
\begin{array}{ccc}
CH_3 & CH_3 & CH_3 \\
| & | & | \\
\cdots N-H\cdots N-H\cdots N-H\cdots \\
| & | & | \\
CH_3 & CH_3 & CH_3
\end{array}
$$

Intermolecular H-bonding in methylamine   **Intermolecular H-bonding in dimethylamine**

The intermolecular H-bonding in primary and secondary amine is not so strong as in the case of alcohols and carboxylic acids. So the amines have lower boiling points than those of alcohols or acids of nearly the same molecular weight.

25.7.2   Spectral Properties

In *UV Spectra,* the absorptions due to n-σ* transition of saturated amine occur at short wavelengths (220 nm) and, therefore, are not of much use for identification purposes.

The *Infrared Spectra* of primary and secondary amines show a characteristic broad band due to $N-H$ stretching absorption in the region 3300cm^{-1}. The primary amines show two bands in this region whereas secondary amines show only one band. In case of primary amines the $N-H$ bending absorption is observed near 1600 cm^{-1}.

In *NMR Spectra,* the $H-C-N$ protons of alkylamines show absorption in the range δ2.5-3.0. The $N-H$ proton also undergoes deuterium exchange as in case of protons of OH of alcohols.

25.7.3   Chemical Properties

Amines are basic in nature due to the presence of unshared electron pair on nitrogen which can accept proton. The electron density on nitrogen increases due to positive inductive effect (+I) of the alkyl groups making the unshared pair of electron more available for protonation.

$$
\begin{array}{ccc}
H & R & R \\
| & \downarrow & \downarrow \\
R\rightarrow N: & R\rightarrow N: & R\rightarrow N: \\
| & | & \uparrow \\
H & H & R
\end{array}
$$

It is thus expected that the basicity of amines should be in the order tertiary > secondary > primary. The amines, like ammonia, are basic as they dissolve in water to abstract a proton from water molecule to form alkyl substituted ammonium ions and hydroxide ions.

$$
\begin{array}{c}
H \\
| \\
H-\overset{..}{N}: + H-\overset{\frown}{\overset{..}{O}}-H \\
| \\
H
\end{array}
\longrightarrow
\left[\begin{array}{c}
H \\
| \\
H-N-H \\
| \\
H
\end{array}\right]^{+}
+ :\overset{..}{O}H
$$

Ammonia Water

Similarly the primary, secondary and tertiary amines react with water to form $R-NH_3^+$, $R_2-NH_2^+$ and R_3NH^+ and the hydroxide ions, respectively. Because of the formation of hydroxide ions, the aqueous solutions of amines are alkaline in nature.

The basicity of the amines is also dependent on the size of alkyl group/groups present.

Alkyl group	Relative basicity
CH_3-	Sec > Pri > Tert > NH_3
CH_3CH_2-	Sec > Pri > NH_3 > Tert
$(CH_3)_2CH-$	Pri > NH_3 = Sec > Tert
$(CH_3)_3C-$	NH_3 > Pri > Sec > Tert

The reactions of primary, secondary and tertiary amines have been discussed separately.

25.7.3.1   Reactions Given by Primary Amines Only

(i) *Carbylamine Reaction*: Primary amines on heating with chloroform and alcoholic solution of potassium hydroxide form isocyanides or carbylamine having extremely disagreeable odor. This is a very sensitive test for primary amines.

$$C_2H_5NH_2 + CHCl_3 + 3KOH \xrightarrow{\Delta} C_2H_5N{\equiv}C + 3\,KCl + 3H_2O$$

Ethylamine   **Ethylisocyanide**

The reaction proceeds via the formation of dichlorocarbene, which adds onto the primary amine and subsequent elimination of hydrogen chloride yields isocyanides.

$$CHCl_3 + \bar{O}H \longrightarrow H_2O + \bar{C}Cl_3 \xrightarrow{Slow} :CCl_2 \quad + C\bar{l}$$

Dichlorocarbene

$$\underset{\underset{H}{|}}{\overset{\overset{H}{|}}{R-N:}} + CCl_2 \longrightarrow \underset{\underset{H}{|}}{\overset{\overset{H}{|}}{R-\overset{+}{N}-\bar{C}Cl_2}} \xrightarrow{-H^+} R-\dot{N}H-\bar{C}Cl_2$$

$$\xrightarrow{-HCl} R-\dot{N}=\bar{C}Cl \xrightarrow[-Cl]{} R-\overset{+}{\dot{N}}{\equiv}\bar{C} \longleftrightarrow R-\dot{N}=C:$$

Alkylisocyanide

(ii) *Reaction with Carbon Disulphide*: Primary amines on heating with carbon disulphide and mercuric chloride form isothiocyanate having pungent smell similar to that of mustard oil. This reaction is called *Hofmann mustard oil reaction* and is used as a test for primary amines.

$$R-NH_2 + S{=}C{=}S \xrightarrow{\Delta} \left[\underset{\underset{H}{|}}{R-N-\overset{\overset{S}{\|}}{C}-SH} \right] \xrightarrow{HgCl_2} R-N{=}C{=}S + HgS + 2HCl$$

Dithiocarbamic acid   **Alkylisothiocyanate**

Though secondary amines also react with carbon disulphide but in this case the formed dithiocarbamic acid is not decomposed with mercuric chloride.

$$\underset{\underset{H}{|}}{\overset{\overset{R}{|}}{R-NH}} + S{=}C{=}S \xrightarrow{\Delta} \underset{\underset{\overset{\|}{S}}{}}{\overset{\overset{R}{|}}{R-N-C-SH}} \xrightarrow{HgCl_2} \text{No reaction}$$

Dithiocarbamic acid

25.7.3.2   Reactions Given by Primary and Secondary Amines

(i)   *Acylation*: Primary and secondary amines react with acid chloride or anhydride to form acyl derivatives.

$$R-NH_2 + R'COCl \longrightarrow R-NHCO-R' + HCl$$
1° Amine   Acyl chloride   N-Alkylamide

$$CH_3CH_2NH_2 + CH_3COCl \longrightarrow CH_3CH_2NHCOCH_3 + HCl$$
Ethylamine   Acetyl chloride   N-Ethylacetamide

$$R-NH-R + R'COCl \longrightarrow \frac{R}{R}N-COR' + HCl$$
2° Amine   N, N-Dialkylamide

$$C_2H_5NHC_2H_5 + CH_3COCl \longrightarrow \frac{C_2H_5}{C_2H_5}N-CO-CH_3 + HCl$$
Diethylamine   N,N-Diethylacetamide

Acid anhydrides react in a similar way.

$$R-NH_2 + (CH_3CO)_2O \longrightarrow RNH-COCH_3 + CH_3COOH$$
1° Amine   N-Alkylacetamide

$$\frac{R}{R}NH + (CH_3CO)_2O \longrightarrow \frac{R}{R}N-COCH_3 + CH_3COOH$$
2° Amine   N,N-Dialkylacetamide

In the above acylation reaction, a tertiary amine (pyridine) is used as an 'acid scavenger' to remove hydrogen chloride or acetic acid.

(ii)   *Reaction with Benzenesulphonyl Chloride*: As mentioned earlier (Section 25.6), the primary and secondary amines react with benzenesulphonyl chloride to form N-alkylbenzenesulphonamide (soluble in alkali solution) and N,N-dialkylbenzenesulphonamide (insoluble in alkali solution) respectively.

$$R-NH_2 + C_6H_5SO_2Cl \longrightarrow C_6H_5SO_2NHR + HCl$$
Primary   Benzenesulphonyl   N-Alkylbenzene-
amine   chloride   sulphonamide

$$\frac{R}{R}NH + C_6H_5SO_2Cl \longrightarrow C_6H_5SO_2N\frac{R}{R} + HCl$$
Secondary   Benzenesulphonyl   N,N-Dialkylbenzene-
amine   chloride   sulphonamide

Most of the sulphonamide derivatives are crystalline compounds and are used for the identification of primary and secondary amines.

(iii)   *Reaction with Aldehydes and Ketones*: Primary amines readily add to aldehydes and ketones to form α-hydroxyamines (also called carbinol amines), which spontaneously lose a molecule of water to form imines or Schiffs bases. The reaction is catalysed by acids.

$$R-NH_2 + R-\overset{H}{\underset{}{C}}=O \xrightarrow{H^+} R-NH-\overset{H}{\underset{OH}{C}}-R \xrightarrow{-H_2O} R-N=CHR$$

1° amine   Aldehyde   α-Hydroxyamine   Aldimine (Schiff's base)

$$R-NH_2 + R'-\overset{\overset{R'}{|}}{C}=O \xrightarrow{H^+} R-NH-\overset{\overset{R'}{|}}{\underset{\underset{OH}{|}}{C}}-R' \xrightarrow{-H_2O} R-N=C\overset{R'}{\underset{R'}{<}}$$

1° amine   Ketone   Ketimine

However, the reaction of secondary amines with aldehydes and ketones (having α-hydrogen atoms) gives the intermediate α-hydroxy amine, which instantaneously lose a molecule of water to form compound known as enamine.

$$\overset{R}{\underset{R'}{>}}N-H + O=\overset{\overset{R'}{|}}{\underset{\underset{H}{|}}{C}}-\overset{\overset{R'}{|}}{C}-R' \longrightarrow \overset{R}{\underset{R'}{>}}N-\overset{\overset{R'}{|}}{\underset{\underset{OH}{|}}{C}}-\overset{\overset{R'}{|}}{\underset{\underset{H}{|}}{C}}-R' \xrightarrow{-H_2O} \overset{R}{\underset{R'}{>}}N-\overset{\overset{R'}{|}}{C}=C\overset{R'}{\underset{R'}{<}}$$

Secondary   Ketone   Enamine
amine

In enamines, the basic nitrogen is directly attached to an olefinic carbon. Enamines are useful synthetic intermediates. Enamines derived from aldehydes are less stable than those from ketones.

(iv) *Halogenation*: Primary and secondary amines react with halogens in presence of alkali to give halogenated amines.

$$R-NH_2 \xrightarrow{Cl_2/NaOH} R-NH-Cl \xrightarrow{Cl_2/NaOH} R-N\overset{Cl}{\underset{Cl}{<}}$$

Primary amine   **N-Chloroalkylamine**   **N,N-Dichloroalkylamine**

$$\overset{R}{\underset{R'}{>}}N-H \xrightarrow{Cl_2/NaOH} \overset{R}{\underset{R'}{>}}N-Cl$$

Secondary amine   **N-Chlorodialkylamine**

(v) *Reaction with Grignard Reagents*: Primary and secondary amines react with Grignard reagents to form alkanes.

$$RNH_2 + R'MgX \xrightarrow{ether} R'H + RNHMgX$$
$$R_2NH + R'MgX \xrightarrow{ether} R'H + R_2NMgX$$
                                    Alkane

(vi) *Reaction with Phenylisocyanate*: Both primary and secondary amines react with phenylisocyanate to form crystalline substituted ureas which are helpful for the identification of amines.

$$RNH_2 + C_6H_5NCO \longrightarrow RNHCONHC_6H_5$$
$$R_2NH + C_6H_5NCO \longrightarrow R_2NCONHC_6H_5$$
                            Substituted urea

(vii) *Reaction with Phenyl Lithium*: On heating with phenyl lithium, primary and secondary amines form lithium derivatives.

$$\overset{R}{\underset{R'}{>}}N-H + C_6H_5-Li \xrightarrow{ether} \overset{R}{\underset{R'}{>}}\bar{N}-\overset{+}{Li} + C_6H_6$$

Phenyl lithium   **Benzene**

25.7.3.3 Reactions given by Primary, Secondary and Tertiary Amines

(i) *Reaction with Water*: Amines combine with water to form alkylammonium hydroxides which dissociates into alkyl substituted ammonium ions and hydroxide ions.

$$R-NH_2 + H_2O \longrightarrow R-\overset{+}{N}H_3\overset{-}{O}H \rightleftharpoons R\overset{+}{N}H_3 + OH^-$$

$$R_2-NH + H_2O \longrightarrow R_2\overset{+}{N}H_2\overset{-}{O}H \rightleftharpoons R_2\overset{+}{N}H_2 + OH^-$$

$$R_3-N + H_2O \longrightarrow R_3\overset{+}{N}H\overset{-}{O}H \rightleftharpoons R_3\overset{+}{N}H + OH^-$$

As already stated, this accounts for the basic nature of amines.

The aqueous solution of amines on treatment with ferric chloride gives brown precipitate of ferric hydroxide.

$$3R\overset{+}{N}H_3\overset{-}{O}H + FeCl_3 \longrightarrow 3R\overset{+}{N}H_3\overset{-}{Cl} + Fe(OH)_3$$

(ii) *Salt Formation*

(a) Amines being basic combine with acids to form salts.

$$RNH_2 + HCl \longrightarrow R\overset{+}{N}H_3\overset{-}{Cl} \ (or \ RNH_2.HCl)$$
1° Amine   **Amine hydrochloride**

$$R_2NH + H_2SO_4 \longrightarrow (R_2NH)_2.H_2SO_4$$
2° Amine   **Amine sulphate**

$$R_3N + HCl \longrightarrow R_3N.HCl$$
3° Amine   **Amine hydrochloride**

(b) Amines react with auric and platinic chlorides to form double salts.

$$CH_3NH_2 + AuCl_3 \xrightarrow{HCl} [CH_3NH_3]AuCl_4$$
Primary amine   **Methylammonium chloroaurate**

$$(CH_3)_2NH + PtCl_4 \xrightarrow{HCl} [(CH_3)_2NH_2]_2PtCl_6$$
Secondary amine   **Dimethylammonium Chloroplatinate**

$$(CH_3)_3N + AuCl_3 \xrightarrow{HCl} [(CH_3)_3NH]AuCl_4$$
Tertiary amine   **Trimethylammonium Chloroaurate**

These double salts on ignition decompose and pure metal is left. This reaction is used to determine the molecular weight of amines.

(iii) *Alkylation:* Primary, secondary and tertiary amines react with alkyl halides to form quaternary ammonium salts as the final product.

$$CH_3NH_2 \xrightarrow[-HI]{CH_3I} (CH_3)_2NH \xrightarrow[-HI]{CH_3I} (CH_3)_3N \xrightarrow{CH_3I} (CH_3)_4N^+ I^-$$
Methyl-   **Dimethyl-**   **Trimethyl-**   **Tetramethyl-**
amine   **amine**   **amine**   **ammonium iodide**

(iv) *Reaction with Nitrous Acid*: Primary amines react with nitrous acid to form alcohol with the evolution of nitrogen.

$$RNH_2 + HONO \longrightarrow ROH + N_2 + H_2O$$

The reaction is not as simple as represented above. In actual practice a number of products are obtained which include alcohols, alkenes and halides (if halide ion is present) with the evolution of nitrogen. The formation of these products may be explained on the basis of carbocation formed during the reaction. n-Propylamine reacts with nitrous acid as given below.

$$CH_3CH_2CH_2NH_2 \xrightarrow{NaNO_2/HCl} CH_3CH_2\overset{+}{C}H_2 \quad + N_2 + Cl^-$$
n-Propylamine   **n-Propylcarbocation**

$$CH_3CH=CH_2$$
$$\uparrow -H^+$$
$$CH_3CH_2CH_2Cl \xleftarrow{Cl^-} CH_3CH_2\overset{+}{C}H_2 \xrightarrow{H_2O} CH_3CH_2CH_2\overset{+}{O}H_2 \xrightarrow{-H^+} CH_3CH_2CH_2OH$$
$$\downarrow H^- \text{ Shift}$$

$$\underset{Cl}{CH_3CHCH_3} \xleftarrow{Cl^-} CH_3\overset{+}{C}HCH_3 \xrightarrow{H_2O} CH_3-\underset{\overset{|}{+OH_2}}{CH}-CH_3 \xrightarrow{-H^+} CH_3-\underset{\overset{|}{OH}}{CH}-CH_3$$

The secondary amines react with nitrous acid to give a neutral oil called nitrosamines without the evolution of nitrogen. The nitrosamines on warming with a little phenol and a few drops of concentrated sulphuric acid produce a green solution, which turns deep blue or violet on addition of sodium hydroxide. This reaction is known as *Liebermann's nitroso reaction* and is used as a test for secondary amines.

$$HO-NO + H^+ \rightleftarrows H_2O^+-NO \rightleftarrows H_2O + NO^+$$
Nitrous acid

$$R_2-NH + NO^+ \rightleftarrows R_2\overset{+}{N}HNO \longrightarrow R_2N-N=O$$
Secondary amine   **Nitrosamine**

The process involving the replacement of the available hydrogen by the nitroso group ($-N=O$) is called *Nitrosation*.

The tertiary amines on treatment with nitrous acid give a soluble trialkylammonium salt.

$$\underset{\underset{R}{|}}{\overset{\overset{R}{|}}{R-N}}: + HO-N=O \longrightarrow \left[\underset{\underset{R}{|}}{\overset{\overset{R}{|}}{R-N-H}}\right]^+ N\bar{O}_2$$

Tertiary amine   **Trialkylammonium nitrite**

(v) *Oxidation*: The amines on oxidation give a variety of products depending upon the nature of the oxidising agent, the type of amine and the nature of the alkyl group.

(a) Primary amines on oxidation with acidified potassium permanganate give aldehydes, ketones, nitroalkanes and ammonia depending upon the nature of alkyl chain attached to amino group.

$$R-CH_2NH_2 \xrightarrow{KMnO_4/H^+} R-CH=NH \xrightarrow{H_2O/H^+} R-\overset{\overset{H}{|}}{C}=O + NH_3$$
                                        Aldehyde

$$\underset{R}{\overset{R}{>}}CHNH_2 \xrightarrow{KMnO_4/H^+} \underset{R}{\overset{R}{>}}C=NH \xrightarrow{H_2O/H^+} \underset{R}{\overset{R}{>}}C=O + NH_3$$
                                        Ketone

$$R-\underset{\underset{R}{|}}{\overset{\overset{R}{|}}{C}}-NH_2 \xrightarrow{KMnO_4/H^+} R-\underset{\underset{R}{|}}{\overset{\overset{R}{|}}{C}}-NO_2$$

Primary amine   **Nitroalkane**

With Caro's acid (H_2SO_5), however, primary amines give following oxidation products.

$$RCH_2NH_2 \xrightarrow{H_2SO_5} RCH_2NHOH + RCH=NOH + R\overset{\overset{OH}{|}}{C}=NOH$$

N-Alkylhydroxyl-   **Aldoxime**   **Hydroxamic acid**
amine

$$R_2CHNH_2 \xrightarrow{H_2SO_5} R_2C=NOH$$

Ketoxime

$$R_3CNH_2 \xrightarrow{H_2SO_5} R_3CNO$$

Nitrosoalkane

(b) Secondary amines on oxidation with potassium permanganate give tetraalkylhydrazine. However oxidation with hydrogen peroxide or per-acid yields dialkylhydroxylamine.

$$\underset{R}{\overset{R}{>}}N-H \xrightarrow{KMnO_4/H^+} \underset{R}{\overset{R}{>}}N-N\underset{R}{\overset{R}{<}} + H_2O$$

Secondary amine   **Tetraalkylhydrazine**

$$\underset{R}{\overset{R}{>}}N-H \xrightarrow{H_2O_2 \text{ or per-acid}} \underset{R}{\overset{R}{>}}N-OH$$

Secondary amine   **Dialkylhydroxylamine**

(c) Tertiary amines on oxidation with peracids or hydrogen peroxide form amine oxides.

$$R_3N + [O] \xrightarrow{H_2O_2 \text{ or per-acid}} R_3\overset{+}{N}-O^-$$

Tertiary amine   **Amine oxide**

25.8 INDIVIDUAL MEMBERS

25.8.1 Methylamine/Methanamine [CH_3NH_2]

Preparation: It may be prepared by any of the general methods of preparations of primary amines. It is most conveniently prepared in the laboratory by the Hofmann's bromide method.

$$CH_3CONH_2 + Br_2 + 4KOH \longrightarrow CH_3NH_2 + 2KBr + K_2CO_3 + H_2O$$

Acetamide   **Methylamine**

Industrially, it is obtained by heating ammonium chloride with formaldehyde (formalin).

$$2HCHO + NH_4Cl \longrightarrow CH_3\overset{+}{N}H_3C\bar{l} + HCOOH$$

                         Methylammonium
                         chloride

Free amine is liberated by treating the salt with a base.

Properties: Methylamine is a colorless gas with characteristic ammonia like fishy odor. It is soluble in water giving alkaline solution. It is a typical primary amine and gives all general reactions of primary amines. It is used as a refrigerant.

25.8.2 Ethylamine/Ethanamine [$CH_3CH_2NH_2$]

Preparation: It is prepared in the laboratory by the Hofmann's bromide reaction.

$$C_2H_5CONH_2 + Br_2 + 4KOH \longrightarrow C_2H_5NH_2 + 2KBr + K_2CO_3 + 2H_2O$$

    Propanamide   **Ethylamine**

On a commercial scale it is obtained by passing a mixture of ethylene and ammonia over cobalt catalyst at 450° and under 20 atmospheric pressure.

$$CH_2{=}CH_2 + H{-}NH_2 \xrightarrow[\text{450°, 20 atm}]{\text{Co Catalyst}} CH_3CH_2NH_2$$

Properties: Ethylamine is a colorless liquid (b.p. 19°) having ammonical odor and is soluble in water. It gives all the characteristic reactions of primary amines.

25.8.3 Dimethylamine/Methylaminomethane [$(CH_3)_2NH$]

Preparation: It is prepared by the hydrolysis of p-nitrosodimethyl aniline with alkali.

p-Nitrosodimethylaniline   Dimethyl-   Sod. salt of p-Nitrosophenol
                              amine

Properties: It is a colorless gas with strong ammonical odor. On cooling it condenses to a liquid (b.p. 7°). It is soluble in water and gives all general reactions of secondary amines.

It is used as an accelerator in vulcanisation of rubber and for dehairing of hides.

25.8.4 Trimethylamine/Dimethylaminomethane [$(CH_3)_3N$]

It is the simplest tertiary amine and occurs in the excreta of fish and some plants. Industrially it is obtained by the distillation of molasses (beat sugar residue), which contains betaine.

$$2\,(CH_3)_3N^+CH_2COO^- \longrightarrow 2\,(CH_3)_3N + CH_2{=}CH_2 + 2CO_2$$

                                    Trimethylamine

It is also obtained by the thermal decomposition of tetramethylammonium hydroxide.

$$(CH_3)_4N^+OH^- \xrightarrow{\Delta} (CH_3)_3N + CH_3OH$$

                      Trimethyl-
                      amine

Trimethylamine is a gas (b.p. 3.5°). It gives all chemical reactions of tertiary amines.

On heating with hydrochloric acid under pressure it decomposes to give methyl chloride. Thus, trimethyl-amine is used as a source for industrial preparation of methyl chloride.

$$(CH_3)_3N + 4HCl \xrightarrow[\text{pressure}]{\Delta} 3CH_3Cl + NH_4Cl$$

Trimethyl-   **Methyl-**
amine   **chloride**

It is used in the manufacture of disinfectants, resins and quaternary ammonium salts.

25.9   QUATERNARY AMMONIUM SALTS

These are tetraalkyl derivatives of ammonium salts obtained by replacement of all the four hydrogen atoms by alkyl groups. Some of these are used in industry as detergents and compounds of medicinal interest.

25.9.1   Tetraalkylammonium Halides [$R_4 \overset{+}{N} X$]

These are obtained (as already mentioned) by heating ammonia with excess of alkyl halide.

$$NH_3 + 4RX \longrightarrow R_4 \overset{+}{N} \overset{-}{X} + 3HX$$

$$NH_3 + 4CH_3Cl \longrightarrow (CH_3)_4 \overset{+}{N} \overset{-}{Cl} + 3HCl$$
$$\text{Tetramethylammonium}$$
$$\text{chloride}$$

They can also be obtained by heating primary, secondary or tertiary amines with excess of alkyl halides.

25.9.2   Tetraalkylammonium Hydroxides [$R_4 \overset{+}{N} \overset{-}{OH}$]

These are obtained by the treatment of quaternary ammonium halides with moist silver oxide.

$$R_4 \overset{+}{N} - \overset{-}{Cl} + AgOH \longrightarrow R_4 \overset{+}{N} - \overset{-}{OH} + AgCl$$

$$\underset{\substack{| \\ CH_3}}{\overset{\substack{CH_3 \\ |}}{CH_3 - \overset{+}{N} - CH_2CH_2CH_3 \overset{-}{Cl}}} \xrightarrow{AgOH} \underset{\substack{| \\ CH_3}}{\overset{\substack{CH_3 \\ |}}{CH_3 - \overset{+}{N} - CH_2CH_2CH_3 \overset{-}{OH}}}$$

Trimethyl-n-propylammonium   Trimethyl-n-propylammonium
chloride   hydroxide

On thermal decomposition, the tetraalkylammonium hydroxides yield tertiary amines along with an alkene. The reaction is commonly known as *Hofmann elimination*.

$$\underset{\substack{| \\ CH_3}}{\overset{\substack{CH_3 \\ |}}{CH_3 - \overset{+}{N} - CH_2CH_2CH_3 \overset{-}{OH}}} \xrightarrow{Heat} \underset{\substack{| \\ CH_3 \ \textbf{Propene}}}{\overset{\substack{CH_3 \\ |}}{CH_3 - N:}} + CH_3CH=CH_2 + H_2O$$

Trimethyl-n-propylammonium   Trimethyl-
hydroxide   amine

The mechanism of the reaction is most commonly E2 (E1 is also known) and is initiated by the abstraction of β hydrogen by base followed by the breaking of C−N bond to form tertiary amine and alkene. Base other than hydroxide ion may also be taken.

$$\underset{\substack{| \\ CH_3}}{\overset{\substack{CH_3 \\ |}}{CH_3 - \overset{+}{N} - CH_2 - CH - CH_3}} \overset{-}{OH} \longrightarrow \underset{\substack{| \\ CH_3}}{\overset{\substack{CH_3 \\ |}}{CH_3 - N:}} + CH_2 = CH - CH_3 + H_2O$$

If two β hydrogen atoms are present in the quaternary ammonium hydroxide, elimination can produce mixture of alkenes. The orientation of elimination is in accordance with Hofmann's rule (and not in accordance with Saytzeff's rule as discussed in Section 9.6) i.e. *the least branched alkene is the preffered product* . Such an orientation is known as *Hofmann orientation*. 2-Pentyltrimethylammonium hydroxide when treated with sodium ethoxide in ethanol forms 1-pentene and 2-pentene in 96% and 4% yields respectively along with trimethyl amine.

$$
\underset{\substack{\text{2-Pentyltrimethyl-}\\\text{ammonium hydroxide}}}{CH_3CH_2CH_2CH-\overset{+}{N}(CH_3)_3OH^-} \xrightarrow[C_2H_5OH]{C_2H_5ONa} \underset{\text{1-Pentene (96\%)}}{CH_3CH_2CH_2CH=CH_2} + \underset{\text{2-Pentene (4\%)}}{CH_3CH_2CH=CHCH_3}
$$

To account for the Hofmann orientation we shall observe the two β protons carefully in the above 2-pentyltrimethylammonium hydroxide and the nature of transition state arising as a result of attack of base (compare the transition state in case of alkenes in Section 9.6).

$$
\underset{\substack{\underset{+}{N}(CH_3)_3OH^-}}{CH_3CH_2-\underset{\beta}{CH}-CH-\underset{\beta}{CH_2}}
$$

2-Pentyltrimethylammonium hydroxide

The transition state here has a considerable carbanion character due to strong electron withdrawal by positively charged nitrogen and the β proton is abstracted from the carbon that can best accomodate the negative charge. In this case the abstraction is preffered from primary carbon compared to secondary one.

$$
\left[\underset{+N(CH_3)_3}{CH_3CH_2-CH_2-CH-\overset{\overset{H\cdots\overset{\delta}{O}H}{\delta:}}{CH_2}} \right]
$$

Transition state; carbanion character

Steric factor is also a governing factor in Hofmann orientation as R_3N, a large leaving group causes crowding in the transition state and the proton from less substituted carbon is abstracted by the base.

25.10   TESTS TO DISTINGUISH PRIMARY, SECONDARY AND TERTIARY AMINES

Following tests are useful to distinguish between the three types of amines.

Primary amines	Secondary amines	Tertiary amines
(a) *Carbylamine reaction:* Carbylamine, RNC having obnoxious odor	No Reaction	No Reaction
(b) *Reaction with nitrous acid:* Gives alcohol (along with other products) with evolution of nitrogen.	gives oily nitrosamine, which on heating with phenol and conc H_2SO_4 followed by addition of base gives deep blue color.	Forms salt which is soluble in water.
(c) *Reaction with benzenesulphonyl chloride:* Forms N-alkylbenzenesulphon-amide, soluble in alkali.	Forms N,N-dialkylbenzenesulphona-mide, insoluble in alkali.	No Reaction

25.11 ASCENT AND DESCENT OF HOMOLOGOUS SERIES

The reactions of amines studied are useful for the ascent and descent of homologues series.

1. Conversion of ethylamine to methylamine.

$$C_2H_5NH_2 \xrightarrow{\text{HONO}} C_2H_5OH \xrightarrow{[O]} CH_3CHO \xrightarrow{[O]}$$

Ethylamine   Ethanol   Acetaldehyde

$$CH_3COOH \xrightarrow{NH_3} CH_3COO\bar{\overset{+}{N}}H_4 \xrightarrow{\Delta} CH_3CONH_2 \xrightarrow{Br_2/KOH} CH_3NH_2$$

Amm₄ acetate   Acetamide   Methylamine

2. Conversion of methylamine to ethylamine.

$$CH_3NH_2 \xrightarrow{\text{HONO}} CH_3OH \xrightarrow{P + I_2} CH_3I$$

Methylamine   Methanol   Methyl iodide

$$\xrightarrow{KCN} CH_3CN \xrightarrow{Na-C_2H_5OH} CH_3CH_2NH_2$$

Methylcyanide   Ethylamine

25.12 PROBLEMS

1. What are amines? What are the functional groups in the three types of amines?
2. Give the structural formula of the following amines and indicate whether they are primary, secondary or tertiary:
   (a) Methylethylamine   (b) Methylethylpropylamine   (c) Propylamine
3. In methylethylpropylamine is it possible to separate the two enantiomers? Give reasons.
4. How are primary, secondary and tertiary amines prepared? Give two methods each for their preparation.
5. How will you separate a mixture of the three amines and quaternary ammonium salt?
6. Explain why the boiling points of amines are lower than those of alcohols or acids of nearly the same molecular weight?
7. Arrange the following in order of their increasing basicity:
   (a) Primary, secondary and tertiary amines   (b) Ethylamine, diethylamine, triethylamine and ammonia
8. Give the reaction of ethylamine with the following reagents:
   Nitrous acid, Acetyl chloride, Benzenesulphonyl chloride and Carbon disulphide in presence of Mercuric chloride.
9. How will you distinguish between different types of amines?
10. Complete the equations:

   (a) $CH_3CONH_2 + Br_2 + KOH \longrightarrow$

   (b) $CH_2{=}CH_2 + NH_3 \xrightarrow[\text{450°/20 atm pressure}]{\text{Co Catalyst}}$

   (c) $(CH_3)_4 N^+OH^- \xrightarrow{\Delta}$

   (d) $CH_3{-}\overset{\overset{\displaystyle CH_3}{|}}{\underset{\underset{\displaystyle CH_3}{|}}{N^+}}{-}CH_2CH_2OH^- \xrightarrow{\Delta}$

11. How will you convert ethylamine into methylamine and vice versa?
12. Discuss the following reactions:
   (a) Hofmann elimination   (b) Gabriel's phthalimide synthesis
   (c) Hofmann bromide reaction   (d) Curtius degradation.

26
Cyano Compounds

26.1 INTRODUCTION
The cyano compounds contain cyano or cyanide group and are considered as alkyl derivatives of hydrogen cyanide. Hydrogen cyanide exists as a tautomeric equilibrium mixture of cyanide form (I) and the isocyanide form (II) of which the cyanide form predominates.

$$H-N\equiv C \qquad\qquad H-C\equiv N$$
Cyanide form (I)   **Isocyanide form (II)**

The organic compounds derived from cyanide form are called alkyl cyanides or nitriles, whereas those derived from isocyanide form are called alkyl isocyanides or isonitriles.

26.2 ALKYL CYANIDES OR NITRILES
The alkyl cyanides or nitriles have the general formula $R-C\equiv N$.

26.2.1 Nomenclature
These are called alkyl cyanides or nitriles of the acid they produce on hydrolysis. In the IUPAC system they take the name of the parent hydrocarbon and the word nitrile is suffixed. Some examples are:

Formula	as Cyanides	as Nitriles	IUPAC Name
CH_3CN	Methyl Cyanide	Acetonitrile	Ethanenitrile
CH_3CH_2CN	Methyl Cyanide	Propionitrile	Propanenitrile
$CH_3CH_2CH_2CN$	Propyl Cyanide	Butyronitrile	Butanenitrile
$(CH_3)_2CHCN$	Isopropyl Cyanide	Isobutyronitrile	2-Methylpropanenitrile

26.2.2 General Methods of Preparation

1. *From Alkyl Halides*: Alkyl halides on heating with aqueous ethanolic solution of sodium or potassium cyanide yield alkyl cyanides.

$$RX + KCN \xrightarrow{100^\circ} RCN + KX$$

A small amount of isocyanide is also obtained in the above reaction.

2. *From Amides*: Amides on dehydration by heating with a suitable dehydrating agent like phosphorus pentoxide yield nitriles.

$$\underset{\text{Amides}}{R-\overset{\overset{\displaystyle O}{\|}}{C}-NH_2} \xrightarrow[-H_2O]{P_2O_5;\,\Delta} \underset{\text{Nitrile}}{R-C\equiv N}$$

$$\underset{\text{Acetamide}}{CH_3-\overset{\overset{\displaystyle O}{\|}}{C}-NH_2} \xrightarrow[-H_2O]{P_2O_5;\,\Delta} \underset{\text{Acetonitrile}}{CH_3-C\equiv N}$$

3. *From Oximes*: The oximes of aldehydes on heating with acetic anhydride yield the cyanides.

$$\underset{\text{Aldoxime}}{RCH=NOH} \xrightarrow[-H_2O,\,\Delta]{\text{Acetic anhydride}} \underset{\text{Nitrile}}{RC\equiv N}$$

$$\underset{\text{Acetaldoxime}}{CH_3CH=NOH} \xrightarrow[-H_2O]{(CH_3CO)_2O,\,\Delta} \underset{\text{Acetonitrile}}{CH_3C\equiv N}$$

4. *From Grignard Reagent*: The reaction of a Grignard reagent with cyanogen chloride yields alkyl cyanides.

$$RMgX + ClCN \longrightarrow RCN + Mg{\overset{\displaystyle X}{\underset{\displaystyle Cl}{<}}}$$

5. *From Carboxylic Acids* (*Industrial Method*): A mixture of carboxylic acid and ammonia are passed over heated alumina at 500°.

$$RCOOH + NH_3 \longrightarrow RCOONH_4 \xrightarrow[-H_2O]{Al_2O_3,\,250°} \underset{\text{Amide}}{RCONH_2} \xrightarrow[-H_2O]{Al_2O_3,\,500°} \underset{\text{Alkylnitrite}}{RCN}$$

26.2.3 Physical Properties

The nitriles are pleasant smelling substances. The lower members (up to C-15) are liquids, while the higher members are solids. They are sparingly soluble in water, but the solubility decreases with increase in molecular weight. The nitriles are dipolar compounds. Due to dipolar association between the molecules, their boiling points are higher than alkanes of approximately same molecular weight.

26.2.4 Spectral Properties

The nitriles absorb in their infrared spectra in the region 2200 to 2500 cm^{-1} due to the stretching vibrations of $C\equiv N$ group.

26.2.5 Chemical Properties

The nitriles are resonance hybrid of the following two structures:

$$R-C\equiv\ddot{N}: \longleftrightarrow R-\overset{+}{C}=\ddot{\ddot{N}}: \equiv R-\overset{\delta+}{C}=\overset{\delta-}{\ddot{N}}$$

Thus, nitriles can be attacked by electrophiles on nitrogen and by nucleophiles on carbon. Further, the $C\equiv N$ group being electron attracting activates the α-hydrogen which can be easily removed by a strong base. In other words cyano group may be compared to the carbonyl group in its chemical behaviour.

1. *Basic Nature of Cyanides*: Inspite of the presence of unshared pair of electrons on the nitrogen atom, the nitriles are not basic enough to form salts with aqueous acids. Thus, they are less basic than amines and ammonia. They, however, form addition compounds with strong acids in the absence of water.

2. *Hydrolysis*: Alkyl cyanides on heating with mineral acid or aqueous alkali get hydrolysed to yield carboxylic acid through the intermediate formation of amide (not isolated).

$$R-C\equiv N \xrightarrow[\text{H}^+\text{or OH}^-]{\text{H}_2\text{O}} \underset{\textbf{Amide}}{R-\overset{\overset{\displaystyle O}{\|}}{C}-NH_2} \xrightarrow[\text{H}^+\text{or OH}^-]{\text{H}_2\text{O}} \underset{\textbf{Acid}}{R-\overset{\overset{\displaystyle O}{\|}}{C}-OH}$$

Alkyl cyanide

Use of mineral acid directly gives carboxylic acid. However, use of dilute sodium hydroxide gives the sodium salt of the carboxylic acid.

(a) *Acid Hydrolysis*: In presence of acid the alkyl cyanide is first protonated to give the protonated cyanide (I), which is in resonance with the form (II). Cation II undergoes nucleophilic addition of water followed by the loss of proton to give the amide, which in turn gets hydrolysed to carboxylic acid.

$$R-C\equiv N:+\overset{+}{H} \longrightarrow \underset{I}{R-C\equiv \overset{+}{N}-H} \longleftrightarrow \underset{II}{R-\overset{+}{C}=\ddot{N}-H}$$

$$\underset{H}{R-\overset{+}{C}=\ddot{N}-H+:\overset{\cdot\cdot}{O}-H} \longrightarrow R-\overset{\overset{\displaystyle H-\overset{+}{O}-H}{|}}{C}=\ddot{N}-H \xrightarrow{-H^+} R-\overset{\overset{\displaystyle O-H}{|}}{C}=\ddot{N}-H$$

$$\longrightarrow \underset{\textbf{amide}}{R-\overset{\overset{\displaystyle O}{\|}}{C}-\overset{\overset{\displaystyle H}{|}}{N}-H} \underset{H^+}{\rightleftharpoons} R-\overset{\overset{\displaystyle +OH}{|}}{C}-NH_2, +:\overset{\cdot\cdot}{O}-H \rightleftharpoons$$

$$\underset{H-\overset{+}{O}-H}{R-\overset{\overset{\displaystyle OH}{|}}{C}-\ddot{N}H_2} \xrightarrow{-H^+} \underset{\overset{\displaystyle |}{OH}}{R-\overset{\overset{\displaystyle OH}{|}}{C}-\ddot{N}H_2} \xrightarrow{H^+} \underset{\overset{\displaystyle |}{O-H}}{R-\overset{\overset{\displaystyle OH}{|}}{C}-\overset{+}{N}H_3} \longrightarrow R-\overset{\overset{\displaystyle O}{\|}}{C}-OH + \overset{+}{N}H_4$$

Carboxylic acid

(b) *Alkaline Hydrolysis*: Hydrolysis in presence of base involves the nucleophilic attack at the carbon atom to give the species (III), which takes a proton from water regenerating hydroxide ions. The intermediate (IV) rearranges to give the amide, which gets hydrolysed to give sodium salt of carboxylic acid.

$$H\overset{-}{O} +R-\overset{-}{C}\equiv \ddot{N}: \longrightarrow \underset{III}{R-\overset{\overset{\displaystyle O-H}{|}}{C}=\ddot{N}:+H-\overset{-}{O}} \xrightarrow{-H\overset{-}{O}} \underset{IV}{R-\overset{\overset{\displaystyle O-H}{|}}{C}=\ddot{N}-H} \rightleftharpoons R-\overset{\overset{\displaystyle O}{\|}}{C}-NH_2,$$

$$R-\overset{\overset{\displaystyle O}{\|}}{C}-NH_2+\overset{-}{/HO} \rightleftharpoons R-\overset{\overset{\displaystyle O^-}{|}}{\underset{O-H}{C}}\overset{}{N}H_2 \longrightarrow R-\overset{\overset{\displaystyle O}{\|}}{C}-O^-+NH_3$$

3. *Reduction:* Nitriles on reduction with lithium aluminum hydride or sodium and alcohol gives primary amines.

$$R-C \equiv N + [H] \xrightarrow[\text{or Na} - C_2H_5OH]{\text{LiAlH}_4/\text{ether}} R-CH_2NH_2$$

Alkyl cyanide **Primary amine**

However, reduction with stannous chloride and hydrochloric acid gives an aldehyde *(Stephens reaction).*

$$R-C \equiv N \xrightarrow{\text{SnCl}_2/\text{HCl}} [R-CH=NH.HCl] \xrightarrow{H_2O} RCHO + NH_4Cl$$

Alkyl cyanide **Imine hydrochloride** **Aldehyde**

4. *Alkylation:* Nitriles having α-hydrogen atoms on alkylation with an alkyl halide in presence of sodium amide gives the alkylated products.

$$R-CH_2C \equiv N + R'-X \xrightarrow{\text{NaNH}_2} R-\overset{\overset{\displaystyle R'}{\displaystyle |}}{C}H-C \equiv N$$

Alkyl cyanide **Alkyl halide** **Alkylated product**

$$R-\overset{\overset{\displaystyle H}{\displaystyle |}}{\underset{\underset{\displaystyle H}{\displaystyle |}}{C}}-C \equiv N + :\ddot{N}H_2 \xrightarrow{-NH_3} R-\overset{\overset{\displaystyle H}{\displaystyle |}}{\underset{}{\ddot{C}}}-C \equiv N + R'-X \xrightarrow{-X} R-\overset{\overset{\displaystyle H}{\displaystyle |}}{\underset{\underset{\displaystyle R'}{\displaystyle |}}{C}}-C \equiv N$$

5. *Reaction with Grignard Reagent:* Nitriles on reaction with Grignard reagents form ketimines, which on hydrolysis form ketones.

$$R-C \equiv N + R'MgX \longrightarrow R-\overset{\overset{\displaystyle R'}{\displaystyle |}}{C}=NMgX \xrightarrow{H_2O/H^+} R-\overset{\overset{\displaystyle R'}{\displaystyle |}}{C}= O + Mg(OH)X + NH_3$$

Alkyl **Grignard**
cyanide **regent**

6. *Reaction with Alcohol (Alcoholysis):* Alkyl cyanides on refluxing with absolute ethyl alcohol in presence of dry hydrogen chloride yield an imido ester, which on hydrolysis forms ester.

$$R-C \equiv N + C_2H_5OH + HCl \longrightarrow [R-\overset{\overset{\displaystyle +NH_2}{\displaystyle ||}}{C} - OC_2H_5] \; C\bar{l} \xrightarrow{H_2O} R-\overset{\overset{\displaystyle O}{\displaystyle ||}}{C}-OC_2H_5 + NH_4Cl$$

Alkyl **Imido ester** **Ester**
cyanide

Uses : Alkyl cyanides are important intermediates for organic synthesis. Adiponitrile $[NC(CH_2)_4CN]$ is used as an intermediate for the synthesis of Nylon-66.

26.3 ALKYL ISOCYANIDES/ISONITRILES [R—N≡C]

The alkyl isocyanides or isonitriles are also known as carbylamines. They are represented as a resonance hybrid of the following two forms.

$$R-\overset{+}{N} \equiv \bar{C}: \longleftrightarrow R-\dot{N}=C:$$

In contrast to alkyl cyanides, in which the alkyl group is attached to carbon, in alkyl isocyanides the alkyl group is attached to nitrogen.

26.3.1 Nomenclature

The common names are derived by adding the suffix 'isocyanide' to the name of the alkyl group. In the IUPAC system, the name 'alkyl carbylamine' is used. It is derived by adding the suffix 'carbylamine' to the name of the alkyl group. Some examples are:

Formula	Common Name	IUPAC Name
CH_3NC	Methyl isocyanide	Methylcarbylamine
CH_3CH_2NC	Ethyl isocyanide	Ethylcarbylamine

26.3.2 General Methods of Preparation

1. *From Primary Amines (Hoffman Carbylamine Reaction):* Primary amines on heating with chloroform and alcoholic potassium hydroxide yield isocyanides.

$$R-NH_2 + CHCl_3 + 3KOH \xrightarrow{\Delta} RNC + 3KCl + 3H_2O$$

Primary   **Alkyl**
amine   **isocyanide**

Mechanism of the reaction is discussed in Section 32.2.2.

2. *From Alkyl Halides:* Alkyl iodides on heating with an aqueous alcoholic solution of silver cyanide yield isocyanides along with small amount of alkyl cyanides.

$$R-I + AgCN \longrightarrow RNC + AgI$$

3. *From N-Alkylformamide:* By heating with phosphorus oxychloride in presence of pyridine.

$$R-\overset{\overset{H}{|}}{N}-\overset{\overset{O}{||}}{C}-H + POCl_3 \xrightarrow[\Delta]{Pyridine} RN{\cong}C + H_2O$$

N-Alkylformamide   **Alkyl isocyanide**

26.3.3 Physical Properties

The alkyl isocyanides are colorless, unpleasant smelling liquids and are insoluble in water. They have low boiling points compared to alkyl cyanides.

26.3.4 Spectral Properties

The $-N{\cong}C$ stretching frequency of alkyl isocyanides in IR is in the region 2180-2144 cm^{-1} which is lower then that for $-C{\equiv}N$ group of alkyl cyanides.

26.3.5 Chemical Properties

1. *Hydrolysis:* Alkyl isocyanides on heating with dilute mineral acids get hydrolysed to form primary amines. These cannot be hydrolysed with alkalis.

$$R-N{\cong}C + 2H_2O \xrightarrow{H^+} RNH_2 + HCOOH$$

Mechanism: The acidic hydrolysis involves the attack of nucleophile on electron deficient carbon of alkyl isocyanide.

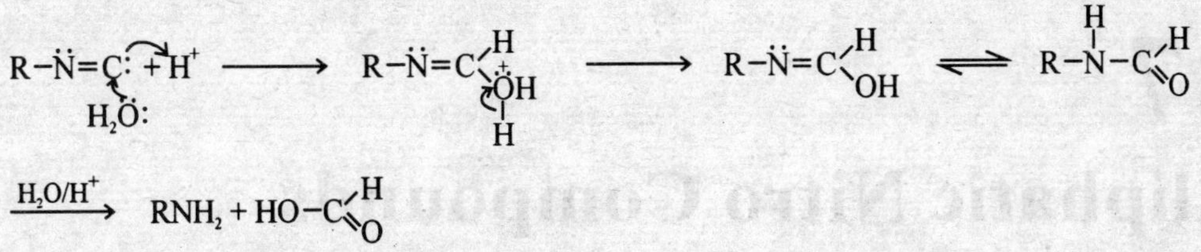

$$R-\overset{..}{N}=\underset{H_2\overset{..}{O}:}{C}:\overset{\frown}{+}H^+ \longrightarrow R-\overset{..}{N}=C\overset{H}{\underset{\underset{H}{\overset{+}{O}H}}{\big\langle}} \longrightarrow R-\overset{..}{N}=C\overset{H}{\underset{OH}{\big\langle}} \rightleftharpoons R-\overset{\overset{H}{|}}{N}-C\overset{H}{\underset{O}{\big\langle}}$$

$$\xrightarrow{H_2O/H^+} RNH_2 + HO-C\overset{H}{\underset{O}{\big\langle}}$$

2. *Reduction*: On reduction with hydrogen in presence of nickel, secondary amines are obtained.

$$R-N{\equiv}C + 2H_2 \xrightarrow[H_2/Ni]{Na-C_2H_5OH \text{ or}} RNHCH_3$$

Alkyl isocyanide   **Secondary amine**

3. *Action of Heat*: Isocyanides on heating with potassium or sodium cyanide rearrange to more stable cyanides. Here the strong nucleophile cyanide ion displaces the isocyanide ion.

$$R-N{\equiv}C \xrightarrow[100^\circ]{KCN} R-C{\equiv}N$$

Alkyl isocyanide   **Alkyl cyanide**

4. *Addition Reactions:* Due to the presence of unshared pair of electrons on carbon, isocyanides are very reactive. Thus, they add on halogen and sulphur to form addition compounds.

$$R-\overset{+}{N}{\equiv}\bar{C}: + Cl_2 \longrightarrow R-N=C\overset{Cl}{\underset{Cl}{\big\langle}}$$

Alkyliminocarbonyl
chloride

$$R-\overset{+}{N}{\equiv}\bar{C}: + S \longrightarrow R-N=C=S$$

Alkyl isocyanide   **Alkyl isothiocyanate**

Uses: These are used in organic synthesis for the preparation of complex nitrogen compounds.

26.4 PROBLEMS

1. Give IUPAC names of the following:
    (a) $CH_3CH_2CH_2CN$
    (b) $(CH_3)_2CHCN$
    (c) CH_3CH_2NC
2. How will you differentiate between alkyl cyanides and alkyl isocyanides.
3. Discuss the basic nature of cyanides. Give mechanism of hydrolysis of alkyl cyanides under acidic and basic conditions.
4. Complete the following:
    (a) $CH_3CONH_2 \xrightarrow{\Delta, P_2O_5}$

    (b) $CH_3CH_2CH=NOH \xrightarrow[\Delta]{(CH_3CO)_2O}$

    (c) $CH_3COOH + NH_3 \xrightarrow{\Delta}$

    (d) $CH_3CH_2CN \xrightarrow[2.\ H_2O]{1.\ SnCl_2/HCl}$

    (e) $CH_3NHCHO \xrightarrow[POCl_3]{Pyridine}$

27

Aliphatic Nitro Compounds

27.1 INTRODUCTION

Nitroalkanes are considered as derivatives of alkanes in which a hydrogen atom is replaced by a nitro group. The general molecular formula of nitroalkanes is RNO_2. They are isomeric with alkyl nitrites, $R-O-N=O$.

Both nitroalkanes and alkyl nitrites can be considered to be derived from the two tautomeric forms of nitrous acid (represented as I and II).

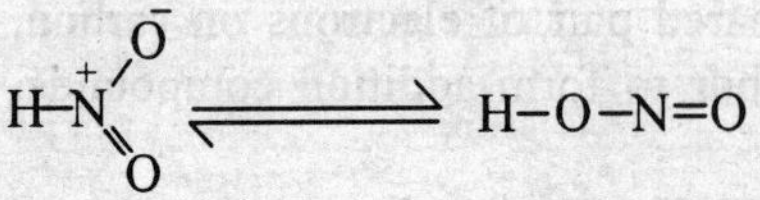

Thus, the alkyl derivatives derived from I are known as aliphatic nitro compounds and those derived from II are called alkyl nitrites; the later are the esters of nitrous acid. As seen, in nitroalkane, the nitrogen is directly attached to the alkyl group (C−N) and in alkyl nitrite it is attached to carbon through oxygen (C−O−N).

The nitroalakanes may be primary, secondary and tertiary depending on whether the nitro group is attached to a primary, secondary or a tertiary carbon atom.

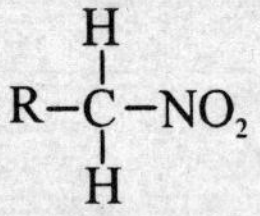

Primary nitroalkane 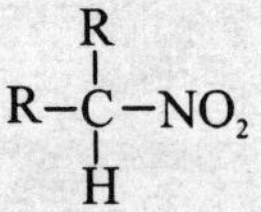   **Secondary nitroalkane** 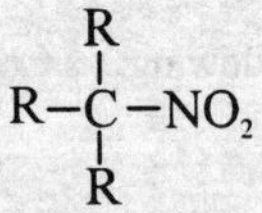   **Tertiary nitroalkane**

27.2 NOMENCLATURE

Nitroalkanes are named by prefixing 'nitro' to the name of the parent hydrocarbon. The position of the nitro group is indicated by a number (IUPAC System). Some examples are:

CH_3NO_2	Nitromethane
$CH_3CH_2NO_2$	Nitroethane
$CH_3CH_2CH_2NO_2$	Nitropropane
$CH_3CH(NO_2)CH_3$	2-Nitropropane
$(CH_3)_2CHCH(NO_2)CH_3$	2-Nitro-3-methylbutane

27.3 ISOMERISM

The nitroalkanes exhibit chain, position and functional isomerism. Examples of each type are given below:

Chain isomerism $CH_3CH_2CH_2CH_2NO_2$ $CH_3-CH-CH_2NO_2$
$\qquad\qquad\qquad\qquad\qquad\qquad\qquad\qquad\qquad\qquad\qquad |$
$\qquad\qquad\qquad\qquad\qquad\qquad\qquad\qquad\qquad\qquad\quad CH_3$

                     Nitrobutane   **2-Methylnitropropane**

Position isomerism $CH_3CH_2CH_2NO_2$ CH_3CHCH_3
$\qquad\qquad\qquad\qquad\qquad\qquad\qquad\qquad\qquad\qquad\qquad\qquad\quad |$
$\qquad\qquad\qquad\qquad\qquad\qquad\qquad\qquad\qquad\qquad\qquad\quad\;\; NO_2$

                     1-Nitropropane   **2-Nitropropane**

Functional isomerism $CH_3CH_2-\overset{+}{N}\!\!\begin{smallmatrix}\nearrow\bar{O}\\\searrow O\end{smallmatrix}$ $CH_3CH_2-O-N=O$

                     Nitroethane   **Ethylnitrite**

27.4 STRUCTURE

The nitro group is a resonance hybrid of the two equivalent contributing structures A and B.

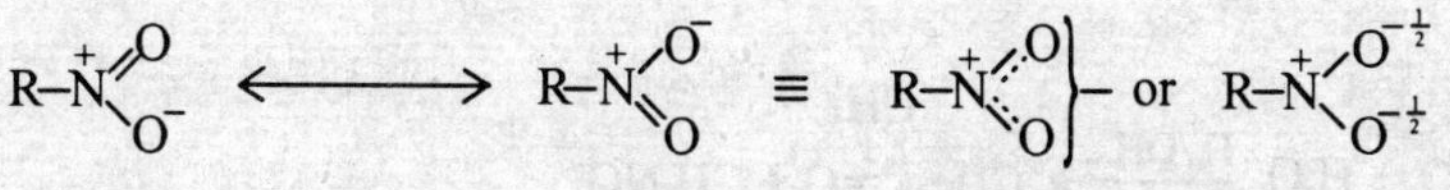

$$R-\overset{+}{N}\!\!\begin{smallmatrix}\nearrow O\\\searrow O^-\end{smallmatrix} \longleftrightarrow R-\overset{+}{N}\!\!\begin{smallmatrix}\nearrow O^-\\\searrow O\end{smallmatrix} \equiv R-\overset{+}{N}\!\!\begin{smallmatrix}\nearrow O\\\searrow O\end{smallmatrix}\!\! \Big] - \text{ or } R-\overset{+}{N}\!\!\begin{smallmatrix}\nearrow O^{-\frac{1}{2}}\\\searrow O^{-\frac{1}{2}}\end{smallmatrix}$$

        (A)   **(B)**   **Hybrid structure**

The hybrid structure suggests that the two $N-O$ bonds are equivalent. On the basis of spectroscopic and diffraction studies the bond distance for each of the $N-O$ bond is found to be 1.22 Å which is less than $N-O$ single bond (1.36 Å) and more than $N=O$ double bond (1.15 Å).

27.5 GENERAL METHODS OF PREPARATION

1. *From Alkanes by Vapor Phase Nitration:* A mixture of alkane and nitric acid is passed through a metal tube at 400° in gaseous state.

$$CH_4 + HNO_3 \xrightarrow{350\text{-}400°} CH_3NO_2 + H_2O$$

Methane   **Nitromethane**

With higher alkanes, a mixture of nitro compounds is obtained, which can be separated by fractional distillation.

$$CH_3-CH_3 + HNO_3 \xrightarrow{350\text{-}400°} CH_3CH_2NO_2 + CH_3NO_2$$

Ethane   **Nitroethane**   **Nitormethane (27%)**
                             (73%)

$$CH_3CH_2CH_2 + HNO_3 \xrightarrow{350\text{-}400°} CH_3CH_2CH_2NO_2 + CH_3\overset{\overset{\textstyle NO_2}{|}}{C}HCH_3$$

Propane   **1-Nitropropane**   **2-Nitropropane (40%)**
                             (25%)

$$+ CH_3CH_2NO_2 \quad + \quad CH_3NO_2$$

                      Nitroethane (10%)   **Nitromethane (25%)**

2. *From Alkyl Halides*: The alkyl halides on treatment with sodium nitrite using N,N-dimethylformamide as solvent give nitro compounds.

$$R-Br + NaNO_2 \xrightarrow{\text{DMF}} RNO_2 + RO-N=O + NaBr$$

Alkyl halide (1° or 2°)   **(55-60%)**

In this reaction a small amount of alkyl nitrite ($R-O-N=O$) is also obtained.

Better yields are obtained by using silver nitrite in place of sodium nitrite.

Tertiary halides give mainly the alkyl nitrites and alkenes.

3. *From α-Halogeno Carboxylic Acid*: The lower members of the series are conveniently prepared by heating an aqueous solution of sodium nitrite with sodium salt of α-halogenocarboxylic acid.

$$ClCH_2\overset{O}{\overset{\|}{C}}-ONa + NaNO_2 \xrightarrow[\text{aq. soln}]{\Delta} O_2NCH_2\overset{O}{\overset{\|}{C}}-ONa \xrightarrow[\Delta]{H_2O} CH_3NO_2 + NaHCO_3$$

Sodium chloroacetate   **Nitromethane**

This method is particularly useful for the preparation of nitromethane.

4. *From α-Nitroalkenes*: By hydrolysis of α-nitroalkenes with water, acids or alkalis, nitroalkanes are formed.

$$CH_3-\overset{CH_3}{\overset{|}{C}}=CH-NO_2 + H_2O \xrightarrow{H^+/OH^-} CH_3-\overset{CH_3}{\overset{|}{C}}=O + CH_3NO_2$$

2-Methyl-1-nitropropene   **Acetone**   **Nitromethane**

5. *Oxidation of tert Alkylamines:* Oxidation of tertiary alkylamines with aqueous potassium permanganate yield tertiary nitroalkanes.

$$R-\overset{R}{\underset{R}{\overset{|}{\underset{|}{C}}}}-NH_2 + 3 [O] \xrightarrow{KMnO_4} R-\overset{R}{\underset{R}{\overset{|}{\underset{|}{C}}}}-NO_2 + H_2O$$

3° Alkylamines   **3° Nitroalkanes**

27.6   PHYSICAL PROPERTIES

Nitroalkanes are colorless and pleasant smelling liquids. Nitromethane is sparingly soluble in water but the higher members are insoluble. However, they are soluble in organic solvents. They have high boiling point and high dipole moment. This is due to the fact that nitroalkanes are highly polar compounds.

27.7   SPECTRAL PROPERTIES

Primary and secondary nitroalkane show IR absorption in the region 1560-550 cm^{-1} and 1380-1360 cm^{-1}. The tertiary nitroalkanes absorb in the range 1540-1530 cm^{-1} and 1360-1345 cm^{-1}.

27.8   CHEMICAL PROPERTIES

1. *Acidic Character*: The α-hydrogen atoms attached to the carbon atom carrying the nitro group are acidic in nature as in the case of aldehydes and ketones. The nitro compounds dissolve in sodium hydroxide to form salts. The acidic behaviour is due to strongly electron withdrawing effect of nitro group and resonance stabilisation of the resulting anion.

$$CH_3-\overset{+}{N}\overset{O}{\underset{O^-}{\diagup}} \xrightarrow{HO^-} \left[:\overset{..}{C}H_2-\overset{+}{N}\overset{O}{\underset{O^-}{\diagup}} \longleftrightarrow :\overset{..}{C}H_2-\overset{+}{N}\overset{O^-}{\underset{O}{\diagup}} \longleftrightarrow CH_2=\overset{+}{N}\overset{O}{\underset{O^-}{\diagup}} \right]$$

The solution of the nitro compound in alkali on acidification yields the acidic isomer of nitromethane known as the *aci form,* which slowly changes to the more stable nitromethane.

$$CH_2=\overset{+}{N}\overset{O}{\underset{O^-}{\diagup}} \rightleftharpoons \overset{H^+}{\quad} CH_2=\overset{+}{N}\overset{OH}{\underset{O^-}{\diagup}} \xrightarrow{Slow} CH_3NO_2$$

$$\textbf{Aci form} \qquad\qquad \textbf{Nitromethane}$$

The aci form can behave both as nucleophile as well as electrophile.

as Nucleophile

$$\diagdown C=\overset{+}{N}\overset{OH}{\underset{O}{\diagdown}}$$

and as electrophile

$$Nu: \frown \diagdown C=\overset{+}{N}\overset{OH}{\underset{O^-}{\diagdown}}$$

The pK_a of nitromethane, nitroethane and nitropropane is 10.2, 8.5 and 7.8, respectively. A typical nucleophilic reaction of nitroalkanes is given below.

2. *Reaction with Aldehydes*: Nitroalkanes having α-hydrogen can undergo nucleophilic addition reaction with aldehydes similar to aldol type addition reaction.

$$NO_2CH_3 \xrightarrow{\bar{O}H} NO_2\bar{C}H_2 + R-\overset{O}{\overset{\|}{C}}-H \longrightarrow R-\overset{O^-}{\underset{H}{\overset{|}{C}}}-CH_2NO_2 \longrightarrow R-\overset{OH}{\underset{H}{\overset{|}{C}}}-CH_2NO_2$$

Nitromethane

The addition product obtained has α-hydrogen to the nitro group and hence can react further.

3. *Reduction*: On catalytic hydrogenation or reduction with iron and hydrochloric acid, the nitro alkanes give the corresponding amines.

$$RNO_2 + 6[H] \xrightarrow[\text{or Fe}-HCl]{\text{cat. redn}} RNH_2 + H_2O$$

However, neutral reducing agents like zinc dust and ammonium chloride give N-alkylhydroxyl amines which reduce ammonical silver nitrate.

$$RNO_2 + 4[H] \xrightarrow{Zn/NH_4Cl} RNHOH + H_2O$$

4. *Hydrolysis*: Primary nitroalkanes on heating with hydrochloric acid or 80% sulphuric acid give hydroxyl amine and the corresponding carboxylic acid.

$$RCH_2NO_2 + H_2O \xrightarrow[H_2SO_4]{HCl \text{ or}} RCOOH + NH_2OH$$

Primary nitroalkane   **Hydroxyl amine**

This reaction is used for the manufacture of hydroxylamine. However, secondary nitroalkanes on hydrolysis yield ketones and nitrous oxide.

$$2R_2CHNO_2 \xrightarrow[H_2SO_4]{HCl \ or} 2R_2CO + N_2O + H_2O$$

Secondary   **Ketone**
nitroalkane

Tertiary nitroalkanes are unaffected by acids.

5. *Reaction with Nitrous Acid*: Primary nitroalkanes on reaction with nitrous acid give nitrolic acids which give red color with alkali.

$$CH_3CH_2NO_2 + O=N-O-H \xrightarrow[-H_2O]{} CH_3-\overset{\displaystyle NOH}{\underset{}{C}}-NO_2 \xrightarrow{NaOH} CH_3-\overset{\displaystyle NONa}{\underset{}{C}}-NO_2$$

Nitroethane   Nitrous acid   Nitrolic acid   Sodium nitrolate (red)

Secondary nitroalkane produce pseudonitrols with nitrous acid which give blue color with alkali.

$$R_2CHNO_2 + H-O-N=O \longrightarrow R_2\overset{\displaystyle N=O}{\underset{}{C}}-NO_2 + H_2O$$

Secondary   Nitrous acid   Pseudonitrol
nitroalkane

Tertiary nitroalkane do not react, since they lack α-hydrogen atom. This test is known as *victor Meyer test* and is used for the differentiation of primary, secondary and tertiary nitroalkanes.

6. *Halogenation*: Primary and secondary nitroalkanes may be halogenated with chlorine or bromine in alkaline solution; the α-hydrogens are replaced by halogen.

$$RCH_2NO_2 \xrightarrow[NaOH]{Br_2} R-\overset{\displaystyle Br}{\underset{\displaystyle H}{C}}-NO_2 \xrightarrow[NaOH]{Br_2} R-\overset{\displaystyle Br}{\underset{\displaystyle Br}{C}}-NO_2 + NaBr$$

Primary nitroalkane   **Monobromo derivative   Dibromo derivative**

$$R_2CHNO_2 + Br_2 \xrightarrow{NaOH} R_2-\overset{\displaystyle Br}{\underset{}{C}}-NO_2 + NaBr$$

Secondary   Monobromo derivative
nitroalkane

$$R_3CNO_2 + Br_2 \xrightarrow{NaOH} \text{No Reaction}$$

Tertiary nitroalkane

7. *Action of Heat*: On heating nitroalkanes decompose to form alkenes.

$$R-CH_2-CH_2NO_2 \xrightarrow[300°]{\Delta} RCH=CH_2 + HNO_2$$

Alkene

Uses: Nitroalkanes are used as solvents for oils, fats, shellac and cellulose derivatives. They are used for the manufacture of primary amines, carboxylic acids and their derivatives.

27.9 TAUTOMERISM

Primary and secondary nitro compounds which have α- hydrogen atoms exhibit tautomerism as shown below:

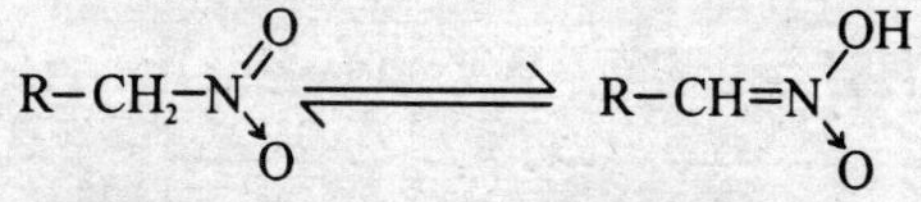

$$R-CH_2-N\begin{matrix}O\\\\O\end{matrix} \rightleftharpoons R-CH=N\begin{matrix}OH\\\\O\end{matrix}$$

Nitro form   **Aci form**
(Pseudo acid form)   **(Nitronic acid)**

The nitronic acid dissolve in aqueous alkali but not in aqueous sodium carbonate, which disturbs the nitro-aci equilibrium.

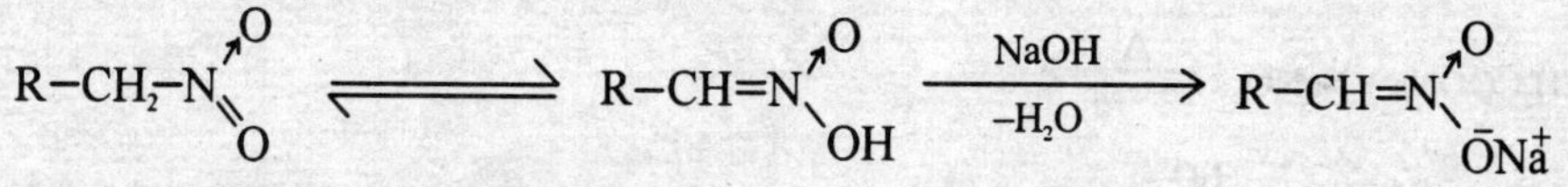

$$R-CH_2-N\begin{matrix}O\\\\O\end{matrix} \rightleftharpoons R-CH=N\begin{matrix}O\\\\OH\end{matrix} \xrightarrow[-H_2O]{NaOH} R-CH=N\begin{matrix}O\\\\\bar{O}Na^+\end{matrix}$$

The primary and secondary nitroalkanes behave as acids in presence of strong alkalis but not in their absence. Such compound are called *pseudo acids*.

27.10 NITROALKANES vs ALKYL NITRITES

As already stated nitroalkanes and alkyl nitrites are functional isomers.

$$R-\overset{+}{N}\begin{matrix}O\\\\O^-\end{matrix} \qquad R-O-N=O$$

(Nitroalkanes)   **(Alkyl Nitrites)**

They can be distinguished by their specific chemical behaviour.

1. *Reduction*: Nitroalkanes on catalytic reduction give primary amine.

$$CH_3CH_2NO_2 \xrightarrow{[H]} CH_3CH_2NH_2$$

Nitroethane   **Ethanamine**

The formation of primary amine shows the presence of C−N bond in the original nitralkane. The reduction of isomeric alkyl nitrite, R−O−N=O does not form amine. In fact, alkyl nitrites produce alcohol on reduction showing that nitrogen is bonded to carbon through oxygen (C−O−N).

$$CH_3CH_2-O-N=O \xrightarrow{[H]} CH_3CH_2OH + NH_3 + H_2O$$

Ethyl nitrite   **Ethanol**

2. *Hydrolysis*: The alkyl nitrites (as esters of nitrous acid) on heating with sodium hydroxide solution yield parent alcohol.

$$R-O-N=O + NaOH \xrightarrow{\Delta} R-O-H + NaNO_2$$

Alkyl nitrite   **Alcohol**

Nitroalkanes are not affected by sodium hydroxide.

3. *Action of Nitrous Acid*: Alkyl nitrites have no action on nitrous acid. The nitroalkanes (as already stated) especially the primary and secondary react with nitrous acid. (See properties of nitroalkanes).

27.11   PROBLEMS

1. Discuss the structures of nitroalkanes.
2. Discuss the tautomerism exhibited by nitroalkanes.
3. How are nitroalkanes prepared?
4. What happens when:

$$CH_3-\underset{\underset{CH_3}{|}}{\overset{\overset{CH_3}{|}}{C}}-NH_2 + [O] \xrightarrow{KMnO_4}$$

$$CH_3-\underset{\underset{CH_3}{|}}{C}=CH-NO_2 + H_2O \xrightarrow[\Delta]{H^+}$$

$$ClCH_2COONa + NaNO_2 \xrightarrow[\text{aq. soln}]{\Delta}$$

$$CH_3NO_2 + CH_3CHO \xrightarrow{HO^-}$$

5. Discuss the acidic character of nitroalkanes.
6. What are pseudo acids?
7. How will you differentiate between primary, secondary and tertiary nitroalkanes?
8. How will you prove that in nitroalkanes there is a C—N bond while alkyl nitrite has the C—O—N bond?

28
Aromatic Hydrocarbons

AROMATIC HYDROCARBONS

Earlier chapters have discussed the chemistry of acyclic and cyclic aliphatic hydrocarbons. Now we shall discuss another class of hydrocarbons called Aromatic Hydrocarbons. The term *Aromatic* (Greek: *Aroma*, fragrant smell) was used for such compounds because they had a pleasant odor although the structure of these compounds was not known at that time. Now the term *Aromatic* is used to represent benzene and related compounds.

28.1 NOMENCLATURE

Aromatic hydrocarbons are generally classified as arenes. Most arenes are related to six membered conjugated benzene. Thus aromatic compounds are named as derivatives of benzene or related structure. For example,

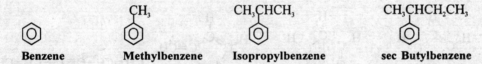

Benzene   **Methylbenzene**   **Isopropylbenzene**   **sec Butylbenzene**

When two substitutents are present in benzene ring, their position is indicated by numbering or more commonly by the prefix ortho(means substituents are in 1,2-relation) meta (means substituents are in 1,3-relation) or para(means substituent are in 1,4-relation).

1,2-Dimethylbenzene   **1-Isopropyl-3-methylbenzene**   **1-Ethyl-4-methylbenzene**
or o-Dimethylbenzene   **or m-Isopropylmethylbenzene**   **or p-Ethylmethylbenzene**
or o-Xylene

When three or more substituents are present in benzene ring, IUPAC nomenclature is preferred. The groups are numbered along the ring in clock wise or anticlock wise direction in such a way so as give lowest number to the substituents. The subsituents are arranged in alphabetical order.

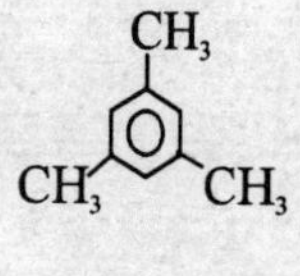

1,3,5-Trimethylbenzene
(Not Mesitylene)

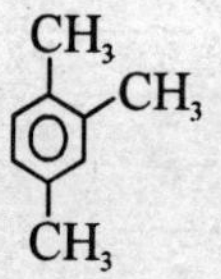

1,2,4-Trimethylbenzene
(Not 1,3,4-Trimethylbenzene)

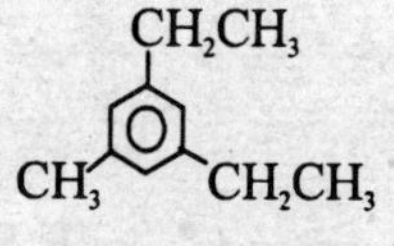

1,3-Diethyl-5-methylbenzene

The substituent group derived from benzene by removing one H-atom is named as phenyl (C_6H_5-) whereas as the groups derived from toluene are ($C_6H_5CH_2$-) benzyl.

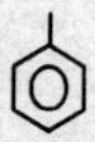

Phenyl 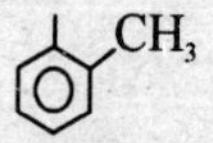   **o-Tolyl** 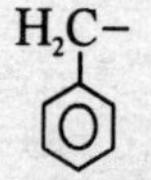   **Benzyl**    **Benzylidene** 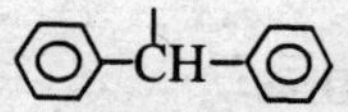   **Benzhydryl**

28.2 STRUCTURE OF BENZENE

Benzene has been known since 1825 when it was first isolated by Michael Faraday. The physical and chemical properties of benzene were better known than its structure until about 1931. We shall now discuss some facts upon which the structure of benzene was built.

(a) *The Kekule Structure:* From elemental analysis and molecular weight determinations, the molecular formula of benzene comprise of six carbon atoms and six hydrogen atoms. In 1865, August kekule, proposed a ring structure for benzene(I). Other structures consistent with formula C_6H_6 were also proposed (II–V; given below) of which structure I was found to be the most satisfactory representation.

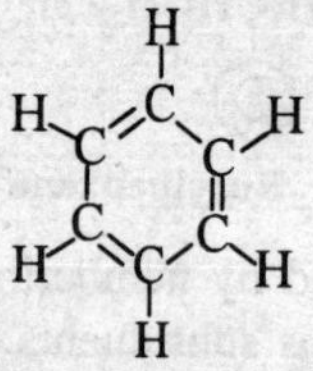

'Kekule' (I) 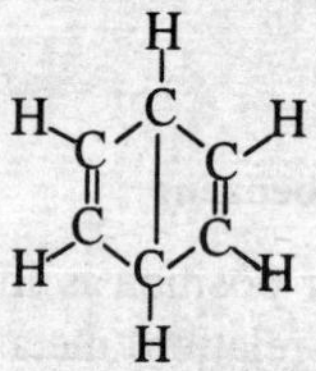   **'Dewar' (II)** 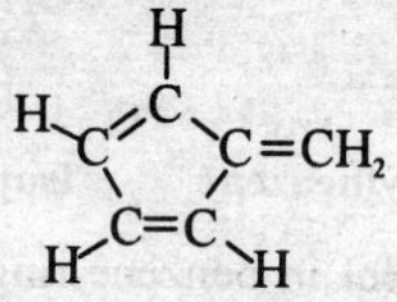   **(III)**   **(IV)**

$$CH_3-C\equiv C-C\equiv C-CH_3$$
(V)

(b) *Benzene Yields only one Mono Substituted Product, C_6H_5X:* When benzene is treated with chlorine in presence of aluminum chloride, it undergoes substitution producing only C_6H_5Cl. Similarly only one bromobenzene or nitrobenzene is obtained when reacted with bromine or nitric acid. Thus, the reaction clearly indicates that all the hydrogen atoms in benzene ring are equivalent, rejecting formula II, III and IV which have more that one type of H-atoms.

(c) *Benzene forms Three Isomeric Disubstituted Products, Dibromobenzene $C_6H_4Br_2$ or Dichlorobenzene $C_6H_4Cl_2$:* This evidence further limits our choice of structures rejecting structure V. Structure I, on the contrary can give three isomeric dibromo derivatives.

| 1,2-Dibromobenzene | 1,3-Dibromobenzene | 1,4-Dibromobenzene |

(d) *Stability of Benzene Ring:* The benzene ring is unusually stable even though it contains three double bonds as shown in kekule structure.

(I) *Benzene Undergoes Substitution Rather than Addition*

When we compare the Kekule structure of benzene (cyclohexatriene) from cyclohexene or cyclohexadiene, we expect it to readily undergo reaction of alkenes. Table 28.1. clearly shows that benzene is either inert or reacts very slowly towards the addition reactions undergone by alkenes rapidly.

Table 28.1: Benzene vs Cyclohexene

Reagent	Benzene	Cyclohexene
1. KMnO$_4$ Cold, dilute, aqueous	no reaction	oxidation reaction
2. Br$_2$/CCl$_4$	no reaction	addition reaction
3. HI	no reaction	addition reaction
4. H$_2$+Ni	slow hydrogenation at 100-200° and 1500 lb/cm³	fast hydrogenation at 25° and 20 lb/cm³

Benzene, on the contrary, undergoes substitution reactions readily.

$$C_6H_6 + HNO_3 \xrightarrow{H_2SO_4} C_6H_5NO_2$$
Nitrobenzene

$$C_6H_6 + Br_2 \xrightarrow{Fe^{+++}} C_6H_5Br$$
Bromobenzene

(II) *Heats of Hydrogenation and Combustion of Benzene are Lower than Expected*

Heat of hydrogenation is the amount of energy evolved when one mole of an unsaturated compound is hydrogenated and in most cases this value is about 28-30 Kcal/mole for each double bond present in the compound. The heat evoluled on hydrogenation of cyclohexene and cyclohexadiene are 28.6 Kcal/mole and 55.4 kcal/mole respectively. On the basis of these values we can expect the heat of hydrogenation for cyclohexatriene or benzene to be 28.6x3=85.8 Kcal/mole, whereas the actual value for benzene is only 49.8 Kcal/mole which is 36 Kcal/mole less than the expected value.

The fact that benzene evolves 36 Kcal/mole less energy when hydrogenated means it is stable by 36 Kcal/mole than we had expected cyclohexatriene to be.

The heat of combustion of benzene, similarly, is lower than expected, by similar amount.

(e) *Carbon-Carbon Bond Length in Benzene*: As expected from Kekule structure, there are three single and three double bonds in benzene between the carbon atoms but it has been found out by X-ray diffraction studies that all the carbon-carbon bond lengths in benzene are equal and are 1.39 Å which is intermediate between C–C single bond (1.54 Å) and C=C=(1.34 Å)

(f) *Resonance in Benzene:* The concept of resonance exists in benzene as the Kekule's structure can be represented by structure I and II, the contributing structures, which differ only in the arrangement of

π electrons. Since structure I and II are equivalent, the stabilisation due to resonance is large and to the extent of 36 Kcal/mole.

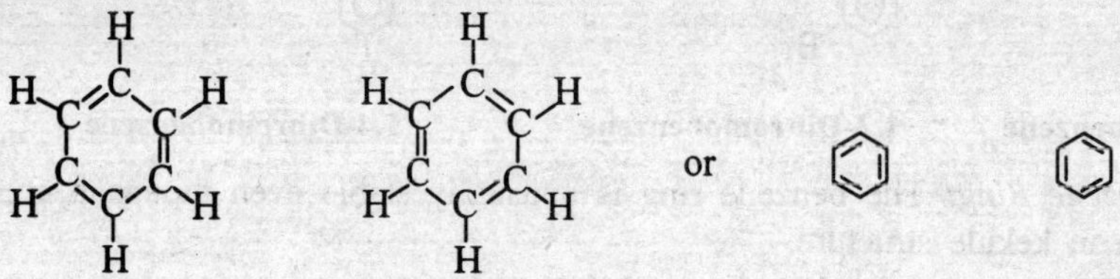

(I)   **(II)**   **(I)**   **(II)**

Thus benzene is best represented as a resonance hybrid of structure I and II.

(g) *Orbital Picture of Benzene:* According of molecular orbital theory, the molecule of benzene is flat and planer having the shape of a regular hexagon. The C−C−C bond angle is 120°. Each carbon atom is sp² hybridised having three sp² hybrid orbitals, of which two orbitals overlap with the sp² hybrid orbital of adjacent carbons whereas remaining sp² hybrid orbital overlaps with the s orbital of hydrogen forming σ bond. The σ bond skeleton of benzene has been shown in Fig. 28.1. Each carbon still contains one electron in its p-orbital lying perpendicular to the plane of the ring. Since all the p-orbitals are equivalent they can overlap on either sides equally well, thus resulting in delocalisation of π electrons in the form of cloud above and below the plane of the ring.

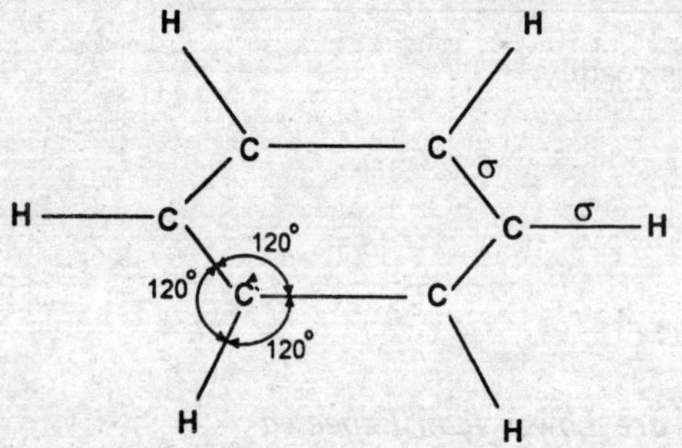

σ **Bond** 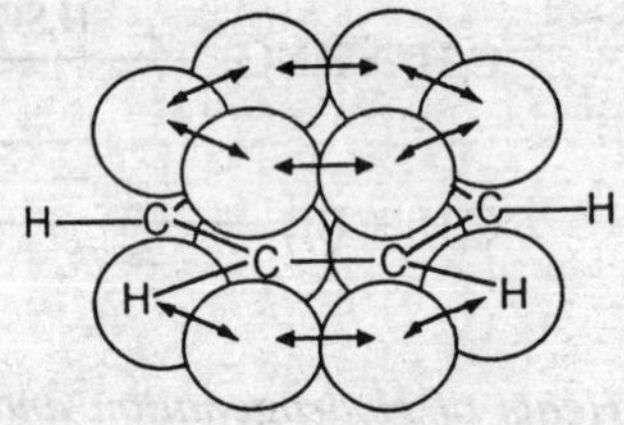   **p-Orbitals overlap**

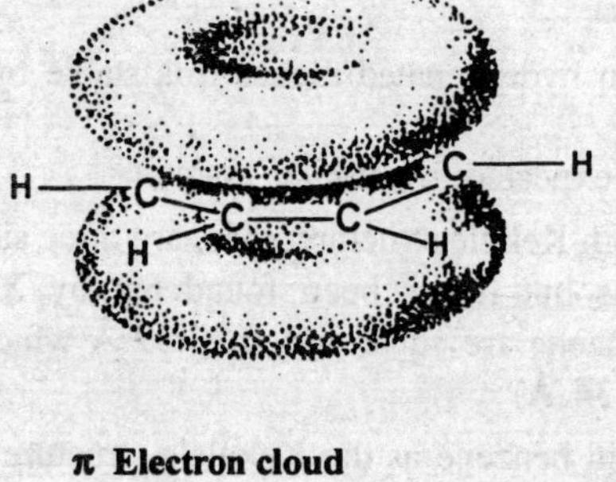

π **Electron cloud** 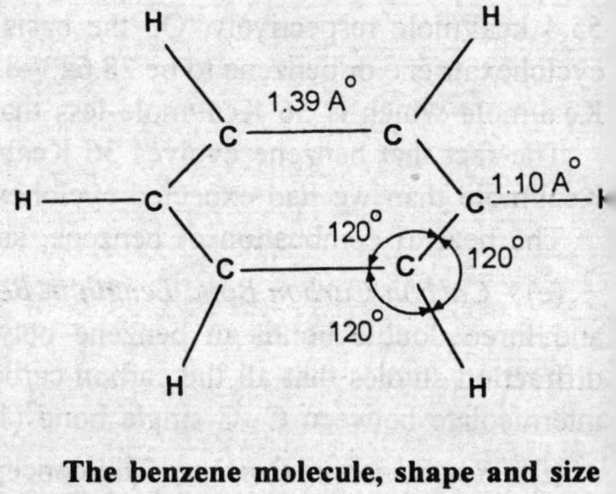   **The benzene molecule, shape and size**

Fig. 28.1: Orbital Picture of Benzene

The orbital approach reveals the significance of planer structure of benzene ring. The ring is flat because all carbon atoms have trigonal geometry which fit in the angles of a regular hexagon. The flatness of the molecule allows the overlap of p orbitals in both the directions resulting in delocalisation of π electrons and stabilisation.

8.3 AROMATICITY: THE HÜCKEL 4n+2 RULE

A compound may be termed as aromatic, in general opinion, if it has benzene like structure but the question arises that just what properties of benzene should a compound have to become aromatic. There are a number of compounds which exhibit aromatic character even though they do not resemble benzene.

On the basis of experimental data discussed in the earlier section a compound is aromatic when it shows high degree of unsaturation (even then it undergoes substitution and not addition), has low heats of hydrogenation and combustion and possess high stability.

Theoretically, as advanced by E. Hückel in 1931, a compound is aromatic when it contains a cyclic cloud of delocalised π electrons above and below the plane of the molecule and the π electron cloud must contain a total of $(4n+2)$ π electrons. According to this rule only delocalisation of electrons is not sufficient but the number of π electrons is also important. The compound should be cyclic and planer too. The requirement of π electrons according to Hückel 4n+2 rule is based upon quantum mechanics and related to the filling of orbitals that make up the π cloud. The rule is supported by facts which have been discussed below.

(a) Benzene, naphthalene, anthracene are aromatic since they have 6, 10 and 14 π electrons respectively which are Hückel numbers corresponding to n = 1, 2 and 3.

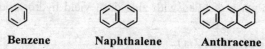

Benzene   **Naphthalene**   **Anthracene**

(b) Cyclobutadiene, cyclopentadiene and cyclooctatetraene are not aromatic as the π electrons 4, 4 and respectively are not hückel numbers.

Cyclobutadiene   **Cyclopentadiene**   **Cyclooctatetraene**

(c) Nonbenzenoid heterocyclic compounds like furan, thiophene and pyrrole are aromatic as the heteratom contributes the nonbonded electrons to complete the sextet.

Furan   **Thiophene**   **Pyrrole**

(d) Cyclopentadienyl anion shows unusually high stability even though cyclopentadienyl cation and cyclopentadienyl radical can be represented by five contributing structures due to the rotation of charge or odd electron on each carbon.

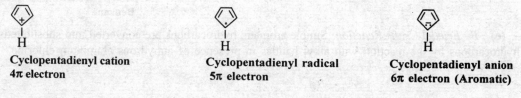

Cyclopentadienyl cation   **Cyclopentadienyl radical**   **Cyclopentadienyl anion**
4π electron   **5π electron**   **6π electron (Aromatic)**

(e) Tropylium ion (cycloheptatrienyl cation) is highly stable compared to cycloheptatrienyl radical and cycloheptatrienyl anion even though the later ones can also be represented by seven contributing structures due to the rotation of odd electron or negative charge on each carbon.

**Cycloheptatriethyl cation
(Tropylium ion)
6π electron (Aromatic)**

**Cycloheptatrienyl radical
7π electron**

**Cycloheptatrienyl anion
8π electron**

28.4 GENERAL METHODS OF PREPARATION

The general methods of preparation of aromatic hydrocarbons are as follows.

(a) *By Decarboxylation*: When aromatic acids or their sodium or potassium salts are heated with soda lime, carbon dioxide is eliminated to form the hydrocarbon.

$$\text{C}_6\text{H}_5-\text{COONa} + \text{NaOH} \xrightarrow{\text{CaO}} \text{C}_6\text{H}_6 + \text{Na}_2\text{CO}_3$$

Benzene

$$\text{CH}_3-\text{C}_6\text{H}_4-\text{COONa} + \text{NaOH} \xrightarrow{\text{CaO}} \text{C}_6\text{H}_5-\text{CH}_3 + \text{Na}_2\text{CO}_3$$

Toluene

(b) *By Deoxygenation of Phenols*: Phenols when distilled with zinc dust, yield hydrocarbon.

$$\text{C}_6\text{H}_4(\text{OH})(\text{CH}_2\text{CH}_3) + \text{Zn} \xrightarrow{\text{Distill}} \text{C}_6\text{H}_5-\text{CH}_2\text{CH}_3 + \text{ZnO}$$

Ethylbenzene

(c) *By Hydrolysis of Aryl Sulphonic Acids*: Hydrolysis of aryl sulphonic acids with super heated steam or by heating with dilute hydrochloric acid at 150° under pressure forms aromatic hydrocarbons due to the removal of -SO$_3$H group.

$$(\text{CH}_3)_2\text{C}_6\text{H}_3-\text{SO}_3\text{H} \xrightarrow[\text{dil HCl at 150°}]{\text{Steam or}} \text{CH}_3-\text{C}_6\text{H}_4-\text{CH}_3$$

m-Xylene

(d) *By the Reduction of Aryl Diazonium Salt:* Diazonium salts formed by the reaction of aromatic amine and nitrous acid, when heated with hypophosphorus acid undergo removal of diazo group forming aromatic hydrocarbons.

$$\text{C}_6\text{H}_5-\text{NH}_2 \xrightarrow[0-5°]{\text{NaNO}_2/\text{HCl}} \text{C}_6\text{H}_5-\overset{+}{\text{N}}\equiv\text{NCl}^- \xrightarrow{\text{H}_3\text{PO}_4} \text{C}_6\text{H}_6 + \text{N}_2 + \text{HCl}$$

Benzene

(e) *By Friedel Crafts Reaction*: Simple aromatic hydrocarbons are converted into substituted aromatic hydrocarbons by the reaction with alkyl halides in presence of anhydrous aluminum chloride.

$$\text{C}_6\text{H}_6 + \text{CH}_3\text{CHBrCH}_3 \xrightarrow{\text{AlCl}_3} \text{C}_6\text{H}_5-\overset{\text{CH}_3}{\underset{|}{\text{CHCH}_3}}$$

Isopropylbenzene

$$\text{C}_6\text{H}_6 + \text{CH}_3\text{Cl} \xrightarrow{\text{AlCl}_3} \text{CH}_3-\overset{\text{CH}_3}{\underset{\text{CH}_3}{\text{C}_6\text{H}_2}}-\text{CH}_3$$

Benzene   **Durene**

(f) *By Wurtz Fittig Reaction*: Aromatic hydrocarbons can be obtained by the action of sodium metal on a mixture of alkyl halide and aryl halide in ether solution.

$$\text{CH}_3-\text{C}_6\text{H}_4-\text{Br} + \text{Na} + \text{BrCH}_3 \xrightarrow{\text{Ether}} \text{CH}_3-\text{C}_6\text{H}_4-\text{CH}_3 + 2\text{NaBr}$$

p-Bromotoluene   **p-Xylene**

The mechanism of Wurtz Fittig reaction is uncertain and it is supposed to involve free radicals. Thus, byproducts are also formed along with the main hydrocarbons which may be separated by fractionation since there is a considerable difference in their boiling points. The mechanism involved resembles that of Wurtz reaction

$$\text{Na} + \text{C}_6\text{H}_5\text{Br} \rightarrow \text{C}_6\text{H}_5^{\bullet} + \text{NaBr}$$
$$\text{Na} + \text{CH}_3\text{Br} \rightarrow \text{CH}_3^{\bullet} + \text{NaBr}$$
$$\text{C}_6\text{H}_5^{\bullet} + \text{CH}_3^{\bullet} \rightarrow \text{C}_6\text{H}_5\text{CH}_3$$
$$2\text{C}_6\text{H}_5^{\bullet} \rightarrow \text{C}_6\text{H}_5-\text{C}_6\text{H}_5 \quad \text{(biphenyl)}$$
$$2\text{CH}_3^{\bullet} \rightarrow \text{CH}_3-\text{CH}_3 \quad \text{(ethane)}$$

(g) *From Grignard Reagent*: Benzene homologues may be prepared by the action of phenyl magnesium bromide with alkyl halides in ether solution.

$$\text{C}_6\text{H}_5-\text{MgBr} + \text{CH}_3\text{CH}_2\text{Br} \longrightarrow \text{C}_6\text{H}_5-\text{CH}_2\text{CH}_3 + \text{MgBr}_2$$

Ethyl Benzene

(h) *By Cyclisation of Long Chain Alkanes:* Long chain alkanes containing 6 to 9 carbon atoms obtained from petroleum are passed over a metal catalyst generally platinum supported over alumina at 500° to get aromatic hydrocarbons. The reaction involves cyclisation followed by aromatisation.

$$\text{CH}_3(\text{CH}_2)_5\text{CH}_3 \xrightarrow[500°]{\text{Pt}-\text{Al}_2\text{O}_3} \text{C}_6\text{H}_{11}-\text{CH}_3 \xrightarrow{\text{Pt}-\text{Al}_2\text{O}_3} \text{C}_6\text{H}_5-\text{CH}_3$$

n-Heptane   **Toluene**

The method is now a days used for the preparation of aromatic hydrocarbons on commercial scale.

(i) *By Polymerisation of Alkynes*: Alkynes on polymerisation at high temperature or in presence of catalyst yield benzene and its homologues.

$$3\text{CH}\equiv\text{CH} \longrightarrow \text{C}_6\text{H}_6$$

Acetylene   **Benzene**

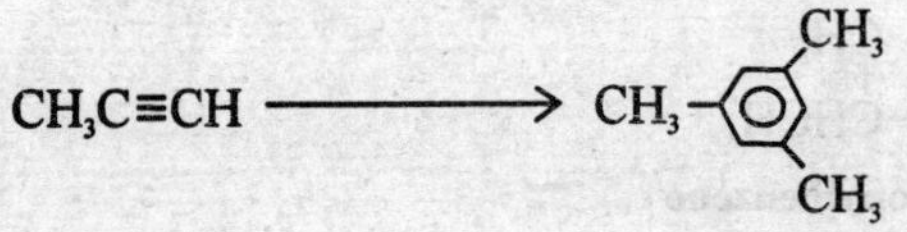

$CH_3C \equiv CH \longrightarrow$ CH₃-(ring with CH₃ groups) — Mesitylene

Propyne   **Mesitylene**

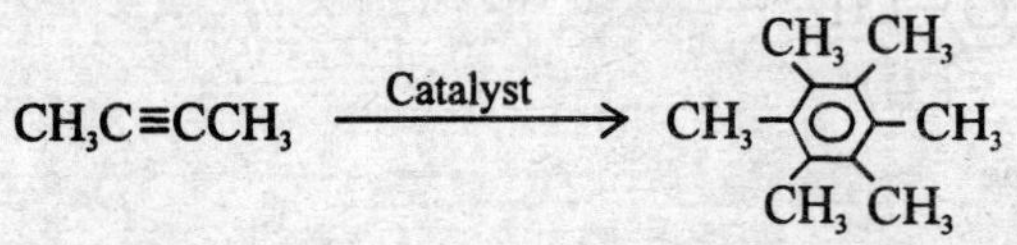

$CH_3C \equiv CCH_3 \xrightarrow{\text{Catalyst}}$ CH₃-(ring with CH₃ groups)

2-Butyne   **Hexamethylbenzene**

(j) *By Clemmensons Reduction:* Homologues of benzene are formed when aromatic ketones are reduced with amalgamated zinc and hydrochloric acid.

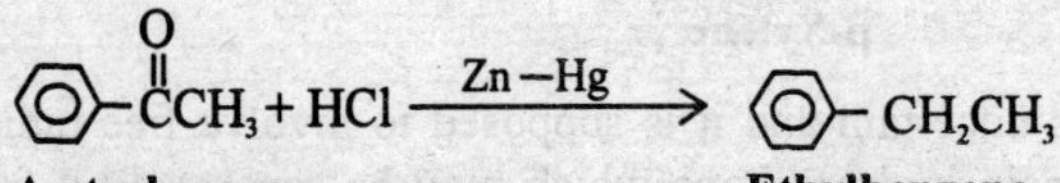

(structure) $-\overset{\overset{O}{\|}}{C}CH_3 + HCl \xrightarrow{Zn-Hg}$ (structure) $-CH_2CH_3$

Acetophenone   **Ethylbenzene**

28.5 PHYSICAL PROPERTIES

Aromatic hydrocarbons are usually colorless, refractive liquids, insoluble in water but soluble in organic solvents in all proportions. They are inflammable, burn with sooty flame and have a characteristic odor. They are toxic and carcinogenic in nature. Their boiling points increases with the increase in the molecular mass as shown in Table 28.2 but their melting points do not show any specific pattern.

Table 28.2: Physical Constants of Some Aromatic Hydrocarbons

Name	Formula	b.p. °C	m.p. °C
Benzene	C_6H_6	80	5.5
Toluene	$C_6H_5CH_3$	111	–95
Ethylbenzene	$C_6H_5CH_2CH_3$	136	–94
n-Propylbenzene	$C_6H_5CH_2CH_2CH_3$	159	–99
Isopropylbenzene	$C_6H_5CH(CH_3)_2$	152	–96
o-Xylene	$C_6H_4(CH_3)_2$-1,2	144	–25
m-Xylene	$C_6H_4(CH_3)_2$-1,3	139	–47
p-Xylene	$C_6H_4(CH_3)_2$-1,4	138	–13
Mesitylene	$C_6H_3(CH_3)_3$-1,3,5	165	–50
Durene	$C_6H_2(CH_3)_4$-1,2,4,5	191	–80

28.6 SPECTRAL PROPERTIES

The ultra violet spectrum can detect aromatic ring in a compound fairly well. They show a series of strong absorptions around 205 nm and less intense absorptions in the range of 255-275 nm. Increased conjugation increases the absorption.

The IR spectrum of aromatic hydrocarbons shows weak absorption bands near 3030 cm⁻¹ for aryl C–H stretching vibrations and a series of bands are observed in the region 1450 to 1600 cm⁻¹ due C=C stretching in the benzene nucleus. Other bands in the region of 970-1225 cm⁻¹ are not very useful for the identification due to overlapping.

The NMR spectrum is also very useful for the determination of structure of benzene and related compounds. For benzene, the NMR spectrum gives a singlet at δ 7.27, since all the six hydrogen atoms are equivalent. Electron withdrawing substituents shifts the absorption of adjacent protons downfield whereas electron releasing groups shift it upfield for adjacent protons.

The mass spectrum of benzene gives prominent molecular ion peak (M^+), M+1 and M+2 peaks due to ^{13}C and ^2H. Benzene and almost all benzene derivatives show prominent peaks at m/z $78(C_6H_6)^+$, m/z $77(C_6H_5)^+$, m/z $53(C_4H_5)^+$, m/z $51(C_4H_3)^+$, m/z 50 $(C_4H_2)^+$ and m/z $39(C_3H_3)^+$.

28.7   CHEMICAL PROPERTIES

Benzene and related compounds undergo substitution reactions under normal conditions. They undergo addition reactions and oxidation reactions too.

(a) *Aromatic Substitution Reactions:* Aromatic hydrocarbons undergo electrophilic substitution reaction when treated with appropriate reagents which furnishes electrophiles. (Detailed discussion is in section 28.8)

Nitration: Reaction of aromatic hydrocarbons (ArH) with nitric acid in presence of sulphuric acid forms aromatic nitro compound.

$$ArH + HO\text{-}NO_2 \longrightarrow \underset{\textbf{Nitro compound}}{Ar\text{-}NO_2} + H_2O$$

Sulphonation: Reaction with concentrated sulphuric acid yields aryl sulfonic acid.

$$ArH + HOSO_3H \longrightarrow \underset{\textbf{Sulfonic acid}}{ArSO_3H} + H_2O$$

Halogenation: Reaction with halogens in presence of iron produces an aryl halide.

$$ArH + Cl_2 \xrightarrow{\text{Fe}} \underset{\textbf{Aryl chloride}}{ArCl} + HCl$$

Friedel Crafts Alkylation: Reaction with alkyl halides in presence of anhydrous aluminum chloride yields alkylbenzenes.

$$ArH + R\text{-}Cl \longrightarrow \underset{\textbf{Alkylbenzene}}{Ar\text{-}R} + HCl$$

Friedel Crafts Acylation: Reaction with acyl chloride in presence of anhydrous aluminum chloride yields aromatic ketone.

$$ArH + RCOCl \longrightarrow \underset{\textbf{Ketone}}{ArCOR} + HCl$$

(b) *Thallation:* Reaction of aromatic hydrocarbons with thallium trifluoroacetate in trifluoroacetic acid yields arylthallium-di-trifluoroacetates which are stable crystalline compounds.

$$\text{Thallium trifluoracetate} \qquad \text{Phenylthallium-di-trifluoroacetate}$$

The reaction proceeds via electrophilic attack of acidic thallium on π electron cloud of benzene ring.

(c) *Mercuration*: Aromatic hydrocarbons react with mercuric acetate to form aryl mercury acetates.

Aryl mercurial compounds thus produced find use in pharmaceutical industry.

$$\text{C}_6\text{H}_6 + \text{Hg (OOCCH}_3)_2 \longrightarrow \text{C}_6\text{H}_5\text{—HgOOCCH}_3$$

Mercuric acetate   **Acetoxymercurybenzene**

This reaction also proceed via electrophilic attack of mercury acetate on benzene ring.

(d) *Catalytic Hydrogenation*: Aromatic hydrocarbons add hydrogen in presence of finely divided platinum metal to form cyclic aliphatic compounds. Toluene yields methylcyclohexane on hydrogenation.

$$\text{C}_6\text{H}_5\text{—CH}_3 + 3\text{H}_2 \xrightarrow{\text{Pt catalyst}} \text{C}_6\text{H}_{11}\text{—CH}_3$$

Methylcyclohexane

Partial reduction with sodium in liquor ammonia alcohol mixture yields partially reduced compounds (Birch reaction).

$$\overset{R}{\bigcirc} \xrightarrow[\text{C}_2\text{H}_5\text{OH}]{\text{Na/liq NH}_3} \overset{R}{\bigcirc}$$

(e) *Addition of Halogens*: Aromatic hydrocarbons add three molecules of chlorine or bromine under the influence of UV light to form hexahalides. Benzene on chlorination forms benzene hexachloride.

$$\bigcirc + \text{Cl}_2 \xrightarrow{\text{uv light}} \text{C}_6\text{H}_6\text{Cl}_6$$

Benzene hexachloride

The γ-isomer of benzene hexachloride, called lindane, has insecticidal properties.

(f) *Ozonolysis*: Aromatic hydrocarbons form unstable ozonide when treated with ozone which on hydrolysis yield dialdehydes. Benzene forms three molecules of glyoxal.

$$\underset{\textbf{Benzene}}{\bigcirc} \xrightarrow[\text{2. H}_2\text{O/Zn}]{\text{1. O}_3} 3 \underset{\textbf{Glyoxal}}{\overset{\displaystyle \text{H—C=O}}{\underset{\displaystyle \text{H—C=O}}{}}}$$

(g) *Oxidation:* Benzene is highly stable to oxidation and is not affected by hot alkaline potassium permanganate or acidic dichromate. Vapor phase oxidation in presence of vanadium pentoxide catalyst, at 500°, however, ruptures the ring to form maleic anhydride.

$$\underset{}{\bigcirc} \xrightarrow[\text{500°}]{\text{O}_2/\text{V}_2\text{O}_5} \overset{\displaystyle \text{CH—CO}}{\underset{\displaystyle \text{CH—CO}}{}}\rangle\text{O}$$

Maleic anhydride

The reaction is used for the commercial preparation of maleic anhydride.

Side chain of the aromatic hydrocarbons, on the contrary, undergoes oxidation with conventional oxidising agents.

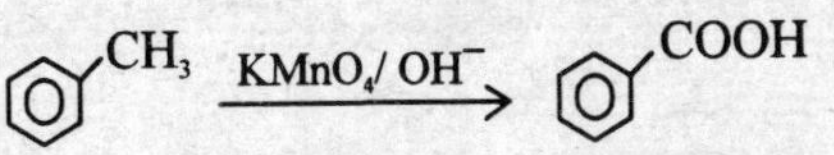

Benzoic Acid

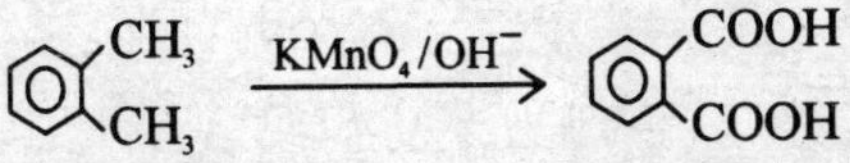

o -Xylene   **Phthalic acid**

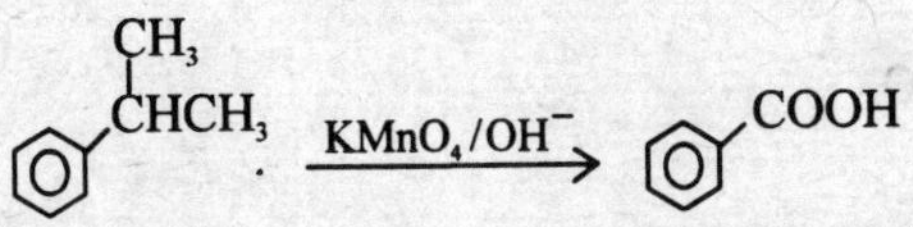

Isopropylbenzene   **Benzoic acid**

Mild oxidising agents such as chromyl chloride yields aromatic aldehydes when reacted with aromatic hydrocarbons having side chain (Etard reaction).

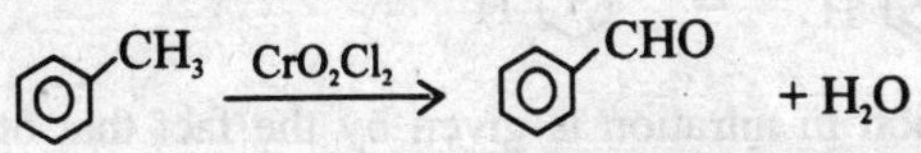

Benzaldehyde

The oxidation reactions are used to elucidate the structure of hydrocarbons as the position of carboxyl group indicates the attachment of side chain to the parent hydrocarbon.

(h) *Combustion:*Aromatic hydrocarbons burn with sooty flame giving carbon dioxide & water vapors when ignited.

$$2C_6H_6 + 15O_2 \longrightarrow 12CO_2 + 6H_2O$$

The combustion is also used as a test for aromatic compounds.

28.8 ELECTROPHILIC SUBSTITUTION REACTIONS

Electrophilic substitution reaction includes a wide variety of reactions such as nitration, halogenation sulfonation, Friedel Crafts alkylation and acylation. Almost all aromatic hydrocarbons undergo this reaction.

Nitration: Substitution of a hydrogen by the nitro group in an aromatic ring is called nitration. Nitration of benzene is carried out with concentrated nitric acid and concentrated sulphuric acid.

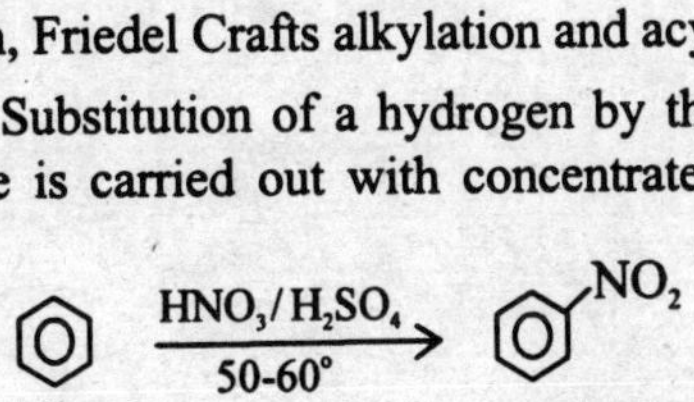

Nitrobenzene

The mechanism of nitration involves the interaction of nitronium ion (I), the electrophile generated by the reaction of nitric acid, and sulphuric acid, with π electron cloud of benzene ring forming the carbocation (II). The attack of base (HSO_4^-) on carbocation (II) then forms nitrobenzene (III).

$$HONO_2 + H_2SO_4 \rightleftharpoons H-\overset{H}{\underset{+}{O}}-NO_2 + HSO_4^-$$

$$H-\overset{H}{\underset{+}{O}}-NO_2 \;\rightleftharpoons\; H_2O + NO_2^+$$

(I)

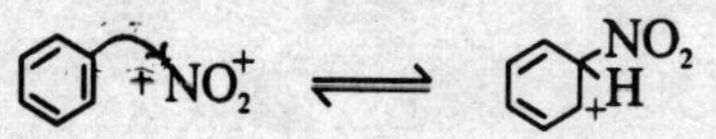

(II)

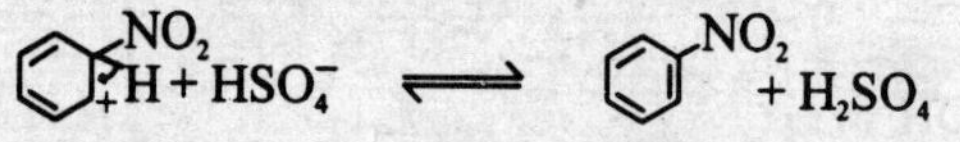

(III)

The carbocation (II) thus formed stabilises itself by resonance. Following are shown the resonating structures of carbocation II and its resonance hybrid.

The evidence for the participation of nitronium ion in nitration is given by the fact that other species furnishing nitronium ion such as $NO_2^+BF_4^-$, $NO_2^+NO_3^-$ and $NO_2^+ClO_4^-$ also form nitrated aromatic hydrocarbons.

Halogenation: Aromatic hydrocarbons react with halogens in presence of ferric chloride or ferric bromide as catalyst to yield halogen substituted products commonly termed as aryl halides.

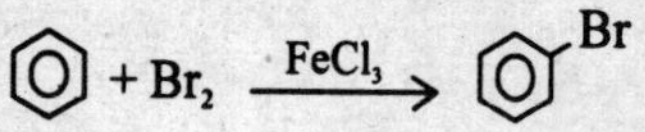

Bromobenzene

Fluorine is too reactive under similar conditions and give very poor yield whereas chlorine and bromine react smoothly giving excellent yields. Iodine is unreactive but iodination of benzene is carried out in presence of oxidising agents such as hydrogen peroxide or cupric chloride which convert molecular iodine into an electrophile, I^+.

$$I_2 + 2Cu^{++} \longrightarrow 2I^+ + 2Cu^+$$

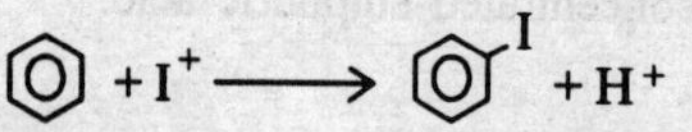

Iodobenzene

Halogenation can also be done with hypochlorous or hypobromous acid in presence of concentrated mineral acids.

$$H-O-Cl + H^+ \longrightarrow H-\overset{H}{\underset{+}{O}}-Cl$$

Mechanism of halogenation involves the formation of electrophilic halogen atom by partial or complete polarisation of halogen-halogen bond.

$$Cl_2 + FeCl_3 \rightleftharpoons Cl_3Fe-Cl^+-Cl \rightleftharpoons FeCl_4^-Cl^+$$

$$FeCl_4^-Cl^+ + \;\bigcirc\; \rightleftharpoons \underset{+}{\bigcirc}\!\!\overset{Cl}{\underset{H}{}} + FeCl_4^-$$

$$\underset{+}{\bigcirc}\!\!\overset{Cl}{\underset{H}{}} + FeCl_4^- \rightleftharpoons \bigcirc\!\!\overset{Cl}{} + HCl + FeCl_3$$

The electrophilic halogen forms the carbocation when attacked by π electron cloud of the benzene ring which stabilise itself by resonance. The carbocation then loses its proton to $FeCl_4^-$ forming chloro substituted product.

Sulphonation: Replacement of hydrogen from an aromatic ring by sulphonic group is called sulphonation and the compound thus produced is called aryl sulphonic acid. It is usually carried out by heating aromatic hydrocarbons with concentrated sulphuric acid or fuming sulphuric acid containing varying proportions of sulphur trioxide.

$$\bigcirc + H_2SO_4 + SO_3 \xrightarrow{40\text{-}50^\circ} \bigcirc\!\!\overset{SO_3H}{}$$

Benzene Sulphonic acid

The reactive electrophile here is neutral sulphur trioxide produced by the reaction of sulphuric acid as is evident from the structures given below.

$$2H_2SO_4 \rightleftharpoons SO_3 + HSO_4^- + H_3O^+$$

$$\underset{+}{\bigcirc}\,SO_3 \rightleftharpoons \underset{+}{\bigcirc}\!\!\overset{SO_3^-}{\underset{H}{}}$$

$$\underset{+}{\bigcirc}\!\!\overset{SO_3^-}{\underset{H}{}} + HSO_4^- \rightleftharpoons \bigcirc\!\!\overset{SO_3^-}{} + H_2SO_4$$

$$\bigcirc\!\!\overset{SO_3^-}{} + H_3O^+ \rightleftharpoons \bigcirc\!\!\overset{SO_3H}{} + H_2O$$

Unlike other electrophilic substitution reactions, the sulfonation is a highly reversible reaction. Concentrated or fuming sulphuric acid favors sulphonation whereas hot and dilute sulphuric acid favors desulphonation.

Friedel-Crafts Alkylation: Replacement of a hydrogen atom in aromatic ring by an alkyl group is termed as Friedel Crafts alkylation. This reaction help introduce various alkyl groups to readily available hydrocarbons like benzene or toluene to prepare complex organic hydrocarbons. The reaction takes place in presence of aluminum chloride.

$$\bigcirc + CH_3CH_2Cl \xrightarrow{AlCl_3} \bigcirc\!\!\overset{CH_2CH_3}{}$$

Ethylbenzene

The electrophile 'R$^+$' is generated by the reaction of alkyl halide with Lewis acid such as AlCl$_3$, BF$_3$,FeCl$_3$,AlBr$_3$ etc. Tertiary alkyl halide, furnish a tertiary carbocation as electrophile whereas with primary and secondary alkyl halides, instead of formation of a carbocation the electrophilic species is a alkyl halide-Lewis acid complex having positively polarised carbon.

$$(CH_3)_3C{-}Cl + AlCl_3 \longrightarrow (CH_3)_3\overset{+}{C} + AlCl_4^-$$
tert-Butyl chloride

+ $(CH_3)_3C^+$ ⟶ $C(CH_3)_3$

tert-Butylbenzene

$$CH_3CH_2Cl + AlCl_3 \longrightarrow CH_3CH_2Cl.....AlCl_3 \qquad CH_3\overset{+}{CH_2}.....AlCl_4^-$$
Alkyl halide Lewis acid complex

+ $CH_3\overset{+}{CH_2}.....AlCl_4^-$ ⟶ CH_2CH_3 + $AlCl_4^-$

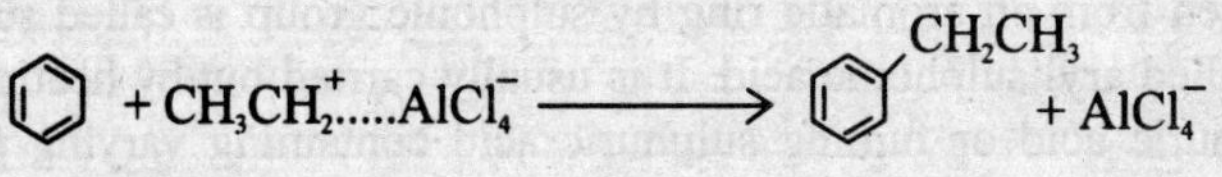

Limitations: Though the reaction is widely used for the preparation of aromatic hydrocarbons it has some limitations given below.

1. Alkylation of benzene forms alkylbenzene and the alkyl group activates the ring for further substitution.

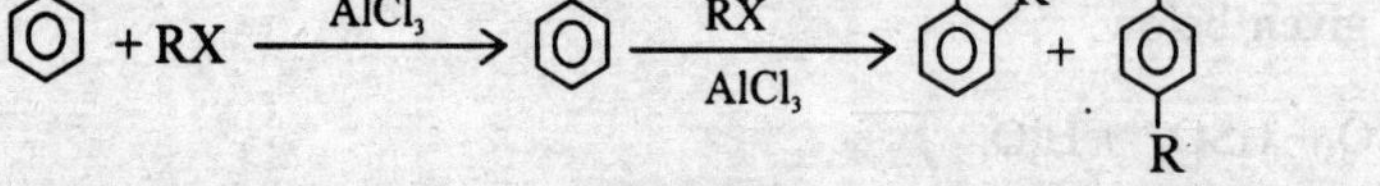

Benzene　　　　　**Alkylbenzene**　　　**Disubstituted benzene**

2. Friedel crafts alkylation is limited to alkyl halides only. Aryl halides and alkenyl halides do not react since the aryl and alkenyl carbocations are too unstable to be formed.

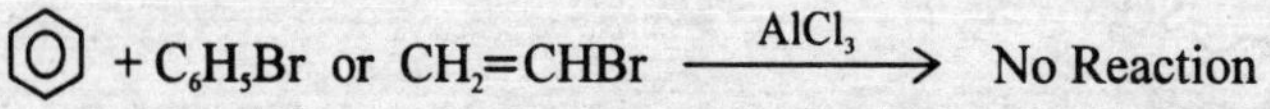

+ C_6H_5Br or $CH_2{=}CHBr$ $\xrightarrow{AlCl_3}$ No Reaction

3. In presence of electron withdrawing groups, the benzene ring does not undergo Friedel Crafts alkylation reaction as the electron withdrawing group deactivate the ring.

Aromatic amines too fail to undergo this reaction largely due to the same reason.

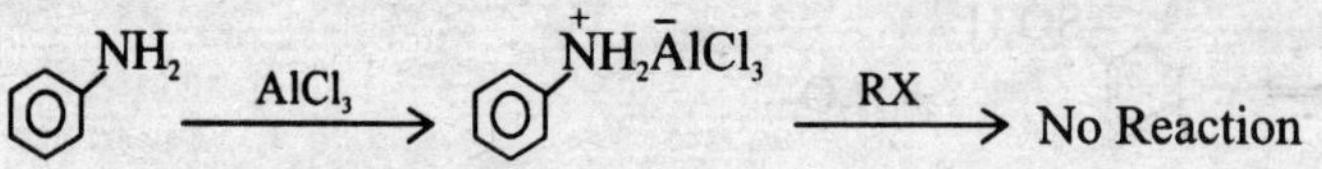

The amino group-Lewis acid complex deactivates the aromatic ring for electrophilic substitution.

4. When long chain alkyl halides are used, rearrangement in alkyl side chain may occur. n-Propyl chloride when reacts with benzene in presence of aluminum chloride forms n-propylbenzene as a minor product whereas isopropylbenzene is a major product.

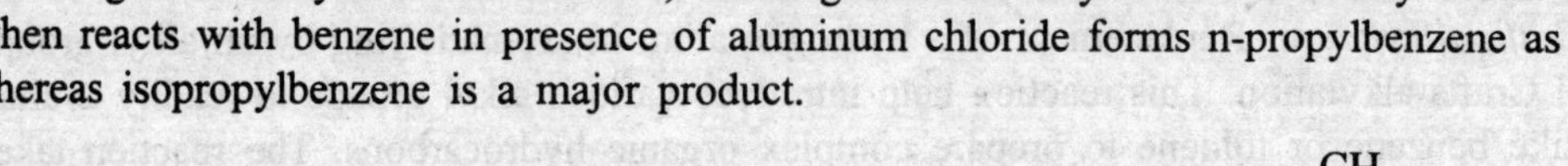

n-Propylbenzene　　　　　**Isopropylbenzene**
(minor)　　　　　　　　　**(major)**

It is due to the rearrangement of n-propyl carbocation into more stable isopropyl carbocation.

Friedel Crafts Acylation: The substitution of an acyl group into the aromatic ring in presence of Lewis acid as catalyst is called Friedel Crafts acylation. The reaction of benzene with acetyl chloride forms acetophenone.

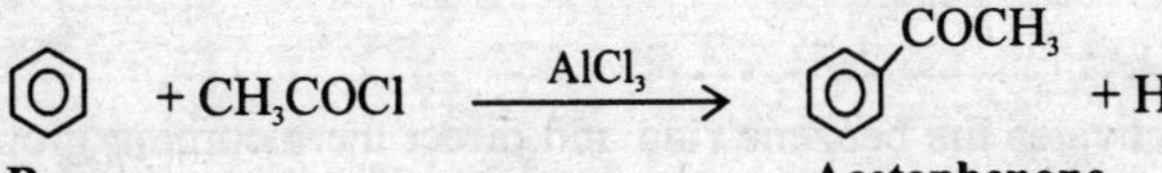

The mechanism of Friedel Crafts acylation involves the attack of resonance stablised acylium cation which is formed by reaction of acid chloride and Lewis acid.

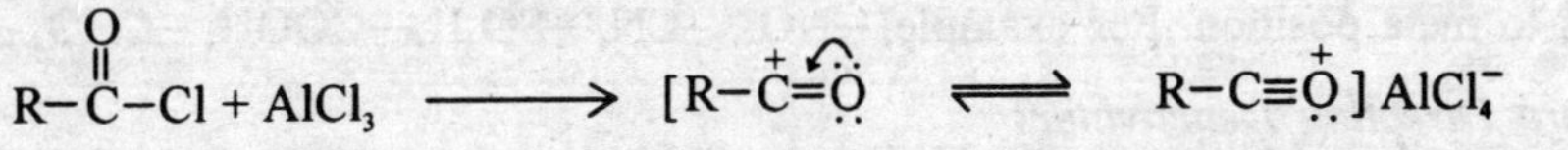

$$R-\overset{\overset{\displaystyle O}{||}}{C}-Cl + AlCl_3 \longrightarrow [R-\overset{+}{C}=\overset{..}{\overset{..}{O}} \rightleftharpoons R-C\equiv\overset{+}{\overset{..}{O}}]\ AlCl_4^-$$

Friedel Crafts acylation reactions are neither accompanied by rearrangement within the acyl group nor do they undergo polysubstitution as the ring gets deactivated after the introduction of an acyl group.

Mechanism of Aromatic Substitution: The mechanism of electrophilic substitution reaction involves the attack of π electron cloud of benzene on the electrophile, forming a σ bond and the +ve charge of electrophile is transferred to the adjacent carbon of the ring. This is a slow step, and is, therefore a rate determining step.

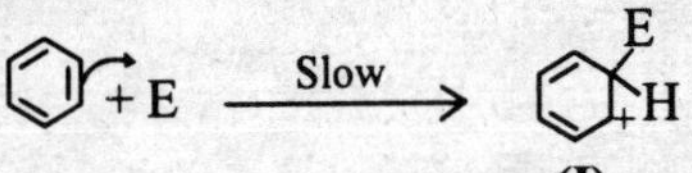

The carbocation is stablised by resonance (structure I to III) and the resonance hybrid may be represented as IV.

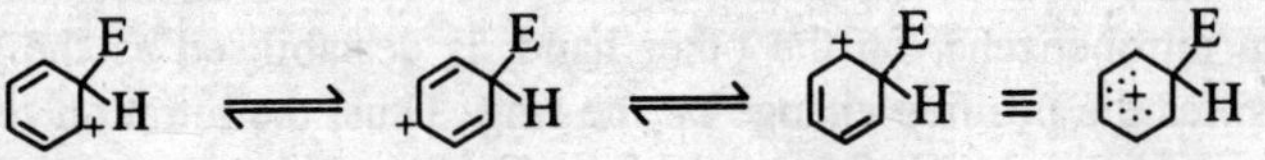

The next step is the attack of base or nucleophile to abstract a ring proton to yield substituted aromatic product. Addition of nucleophile (as was the case with aliphatic electrophilic addition) does not take place here as it would destroy the resonence stabilisation of benzene ring.

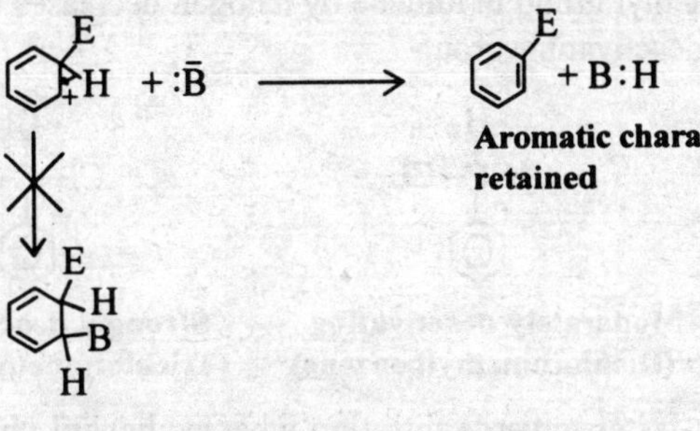

Aromatic character retained

Aromatic character destroyed

28.9 ORIENTATION IN AROMATIC DISUBSTITUTION

We have seen that benzene forms only one monosubstituted product. Now let us see what happens when a substitution reaction is carried out on a substituted benzene. It has been observed that the substituents affect the reactivity and orientation in electrophilic substitution reactions and on this basis they have been divided into three groups

(I) *Ortho and Para Directing Activators*
The substituents that release electrons, activates the benzene ring and direct the incoming groups to ortho and para positions. For example, $-NH_2$, $-OH$, $-OCH_3$, $-NHCOCH_3$, $-C_6H_5$, $-CH_3$ etc.

(II)   *Meta Directing Deactivators*
The substituents that withdraw electrons, deactivate the benzene ring for further substitution and direct the incoming groups to meta position. For example, $-NO_2$, $-CN$, $-SO_3H$, $-COOH$, $-CHO$, $-COR$.

(III)   *Ortho and Para Directing Deactivators*
The substituents that withdraw the electrons, deactivate the ring and direct the incoming group to ortho and para position. For example, $-Cl$, $-Br$ etc.

Effect of Substituents on Reactivity: To study the effect of substitution on reactivity let us compare the reactivity of benzene, methylbenzene (toluene) and nitrobenzene towards nitration. The structure of carbocations formed from the three compounds are as follows:

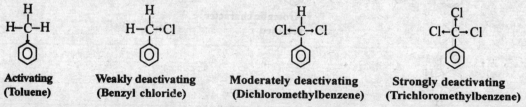

The carbocation, II, formed from toluene tends to stabilise the positive charge by electron release thus activates the ring for reaction faster than benzene (I).

The carbocation III, formed from nitrobenzene, on the other hand, is destabilised as the nitro group is electron withdrawing and it intensifies the positive change on the ring. Thus, the nitration reaction with nitrobenzene is even slower than it is in benzene and the order of reactivity towards electrophilic aromatic substitution is as follows.

Toluene > Benzene > Nitrobenzene.

Similar effects are seen in the nitration of toluene, benzyl chloride, dichloromethylbenzene and trichloromethylbenzene. The replacement of hydrogen of methyl group in toluene by halogen decreases the electron releasing tendency of methyl group making it a deactivating group.

Activating **(Toluene)**	**Weakly deactivating** **(Benzyl chloride)**	**Moderately deactivating** **(Dichloromethylbenzene)**	**Strongly deactivating** **(Trichloromethylbenzene)**

Compared to benzene, toluene reacts 25 times faster towards nitration whereas benzyl chloride only one third as reactive as benzene.

Effect of Substituents on Orientation: Nitration of toluene gives o- and p-nitrotoluene as the major product. m-Nitro toluene, however, is formed in minor amounts.

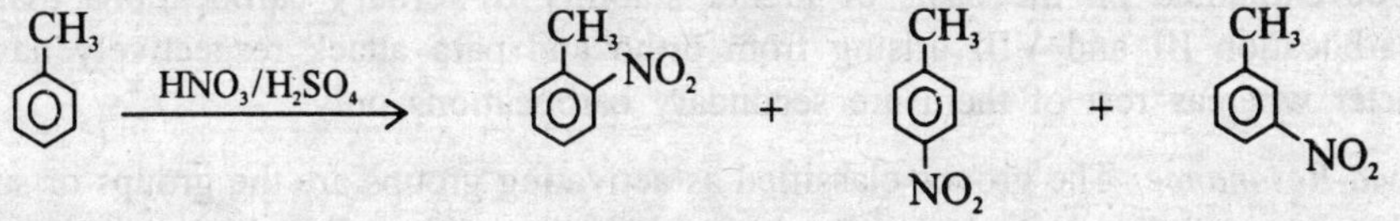

o-Nitrotoluene   p-Nitrotoluene   m-Nitrotoluene

On the other hand nitration of nitrobenzene forms m-dinitrobenzene as a major product along with o- and p- products in traces.

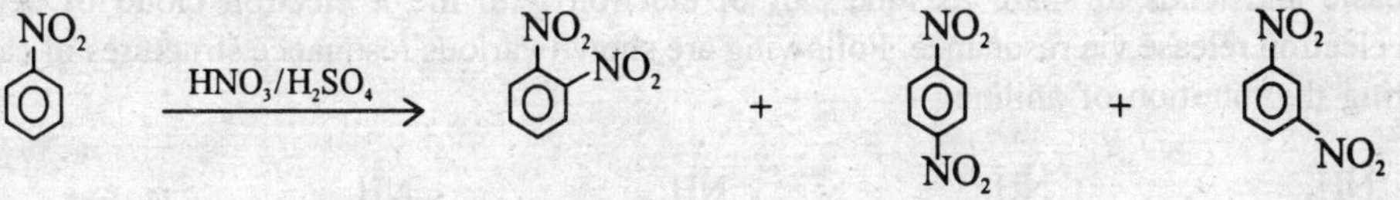

o-Dinitro-   p-Dinitro-   m-Dinitrobenzene, 7%   benzene, 2%   benzene, 91%

It is important to note here that activating groups activate all the positions of the ring. They are o- and p- directing because they activate ortho and para positions more than the meta positions. Similarly deactivating groups deactivate ortho and para positions much more than the meta position making them meta directors. Thus both ortho, para orientation and meta orientation arise in the same way i.e. the effect of the substituent group (whether activating or deactivating), is strongest at the ortho and para positions.

Ortho and Para Directing Activators: We shall understand the effect of substituents on orientation by taking the example of nitration of toluene. Let us inspect the possible resonance structures of carbocations formed during the reaction.

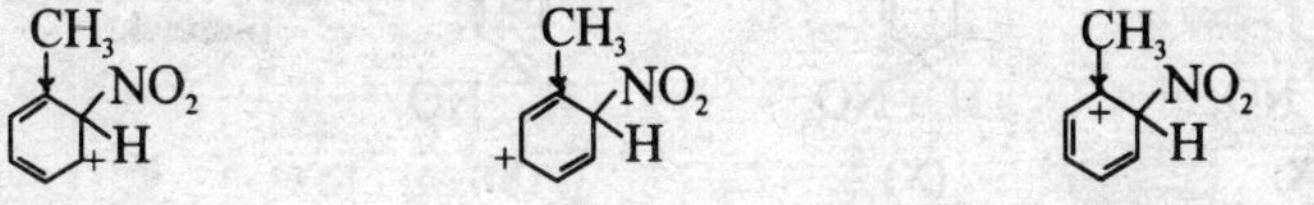

(I)   (II)   (III)   o-attack

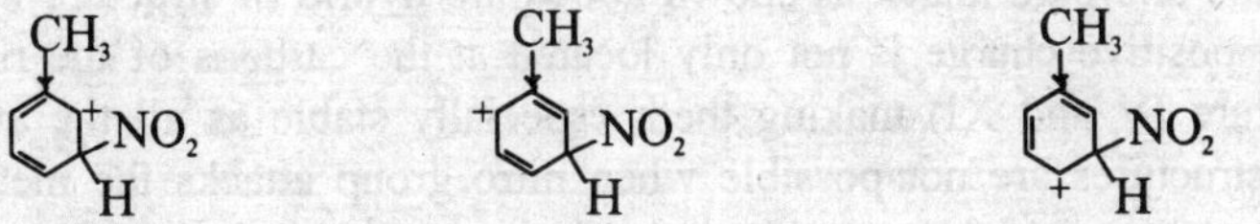

(IV)   (V)   (VI)   m- attack

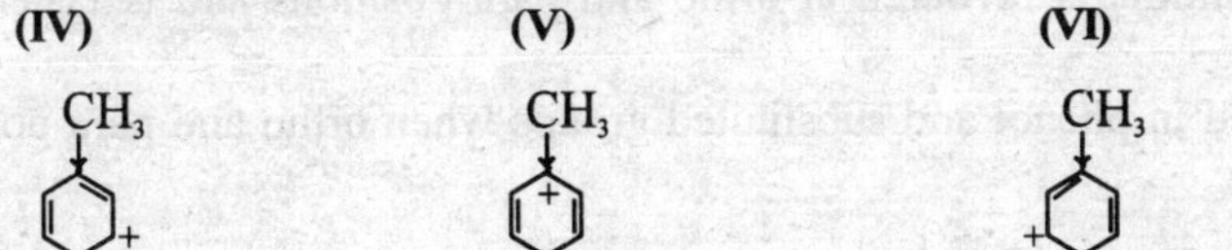

(VII)   (VIII)   (IX)   p-attack

In structure III and VIII, resulting from ortho and para attack respectively, the positive charge is located at the carbon having electron releasing methyl group making them especially stable by inductive effect. As the contribution from structure III and VIII stabilises the hybrid carbocation arising from ortho and para attack more than the carbocations arising from meta attack, the ortho and para substitution occur faster than meta substitution.

This can also be explained on the basis of higher stability of tertiary carbocations than that of secondary. The carbocation III and VIII arising from ortho and para attack respectively have tertiary carbocation character whereas rest of them are secondary carbocations only.

Electron Release via Resonance: The groups classified as activating groups are the groups or substituents in benzene ring that increase the electron density due to inductive effect. But this is not the case with $-NH_2$, $-NHR$, $-OH$, $-OR$ groups as although, they exert electron withdrawing inductive effect due to the attachment of the ring with electronegative oxygen and nitrogen atoms, they strongly activate the ring for ortho and para substitution. This is due to resonance and although electronegative, the nitrogen of $-NH_2$ groups is basic and tends to share its lone pair of electron with the π electron cloud of benzene ring resulting in electron release via resonance. Following are shown various resonance structures of carbocations formed during the nitration of aniline.

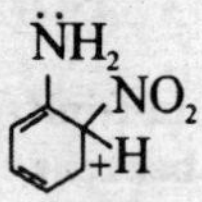
(I)

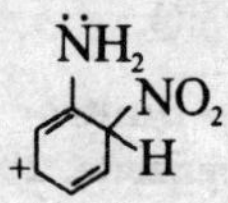
(II)

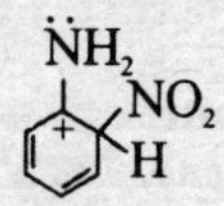
(III)

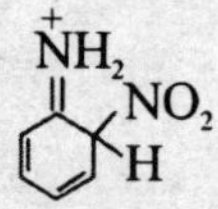
(IV)

o-attack

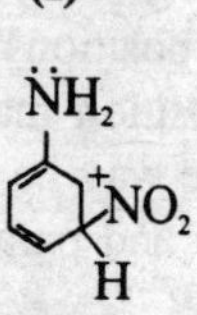
(V)

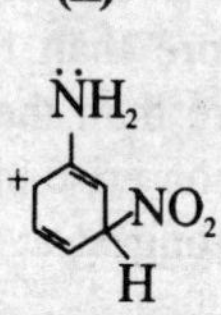
(VI)

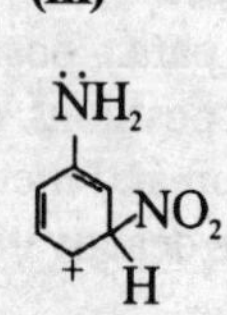
(VII)

m-attack

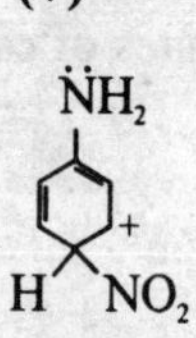
(VIII)

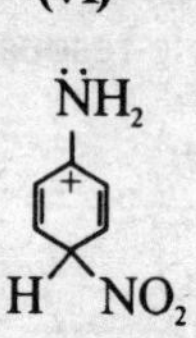
(IX)

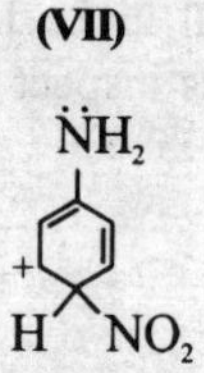
(X)

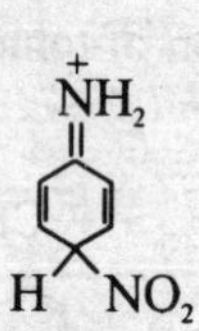
(XI)

p-attack

The carbocations formed by ortho and para attack as shown above are hybrid of structure I to IV and VIII to XI respectively where the positive charge is not only located at the carbons of the ring but on the nitrogen of amine also (structure IV and XI) making them especially stable as all the atoms have complete octet of electron. Such structures are not possible when nitro group attacks the meta position of the aniline. Thus substitution in aniline is favoured at ortho and para positions and is faster than that in benzene or toluene.

Similar electron release is observed in phenol and substituted groups when ortho and para positions are attacked.

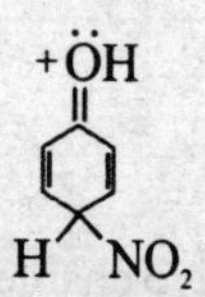
Para attack in –OH

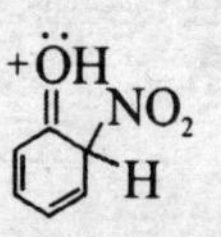
Ortho attack in –OH

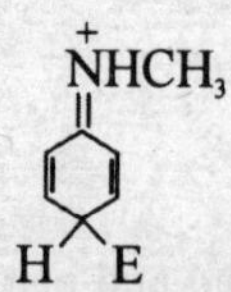
Para attack in $-NHCH_3$

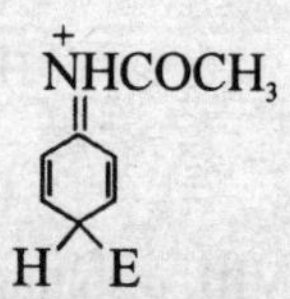
Para attack in $-NHCOCH_3$

Meta Directing Deactivators: We shall now examine the orientation when a deactivating group or electron withdrawing group like $-NO_2$, $-CHO$, $-COOH$ etc. is attached to benzene ring. Following are shown various resonance structures of carbocations formed during the nitration of nitrobenzene.

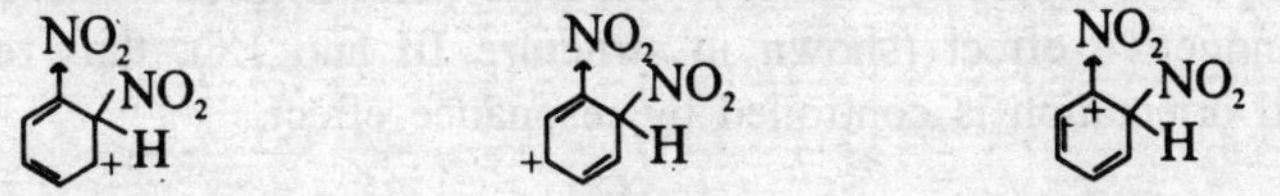

(I)   (II)   (III)   o-attack

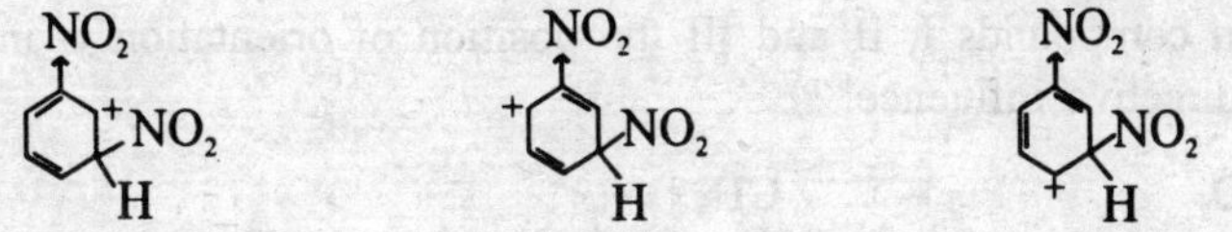

(IV)   (V)   (VI)   m-attack

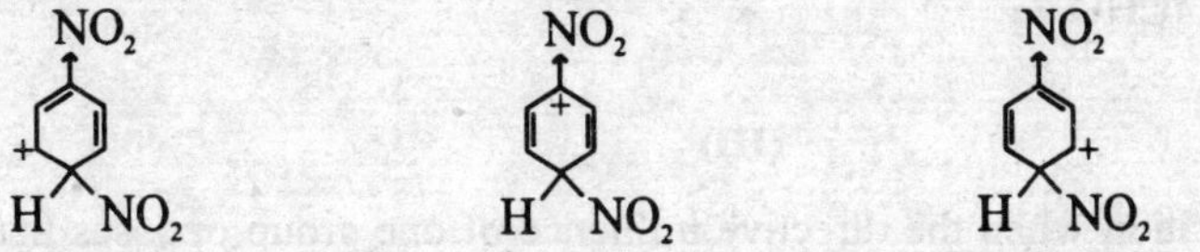

(VII)   (VIII)   (IX)   p-attack

Here the number of resonating structure due to ortho, meta and para attack are equal i.e. three but structure III and VIII are unfavorable as the positive charge is present on the carbon having electron withdrawing group whereas no such structure is possible with meta attack making it a more favorable position for attack by an electrophile compared to ortho and para.

Ortho and Para Directing Deactivators: Halogens show unusual deactivation towards electrophilic substi-tution. They direct the incoming group at ortho and para position even though they exert electron withdrawing -I effect. To understand their action we shall inspect various resonance structures of cabocations formed during the nitration of chlorobenzene.

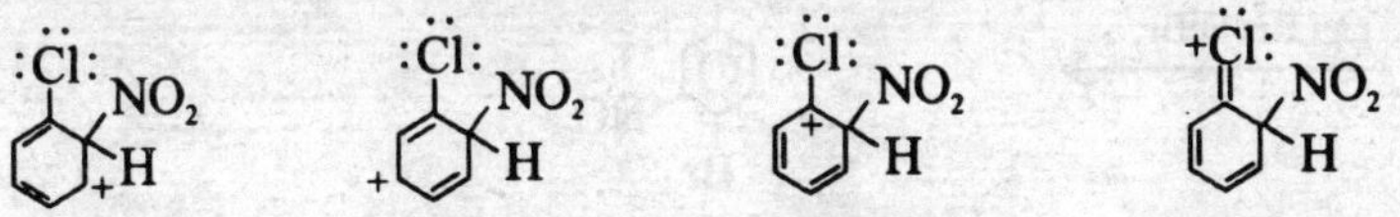

(I)   (II)   (III)   (IV)   o-attack

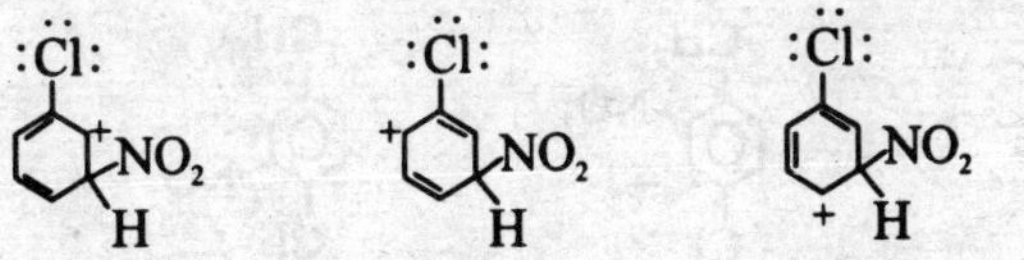

(V)   (VI)   (VII)   m-attack

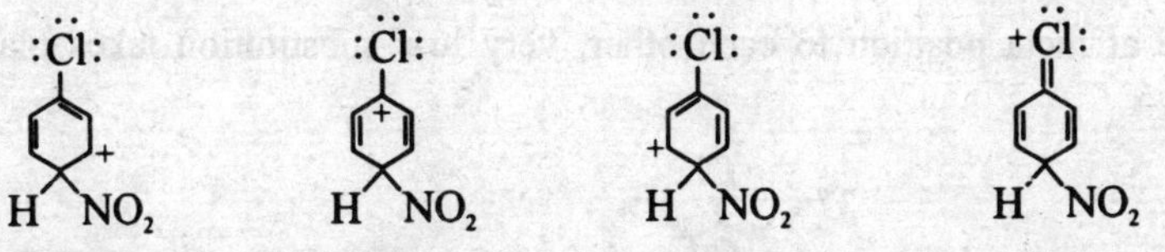

(VIII)   (IX)   (X)   (XI)   p-attack

Structure IV and XI arising from ortho and para attack are more stable as every atom has its octet of electrons complete making chlorobenzene ortho and para directing and no such stabilisation is possible in meta attack. Structure III and IX are unfavorable because chlorine withdraws electrons through inductive effect making it a deactivating group. The resonance effect of chlorine (shown in structure IV and XI) outweighs the instability due to inductive effect (shown in structure III and IX), thus reactivity is controlled by   inductive effect and orientation is controlled by resonance effect.

28.10   ORIENTATION IN DISUBSTITUTED BENZENES

In presence of two subtituents in the ring, the third group orients itself under the directive influence of the present groups. For example, in compounds I, II and III the position of orientation is indicated by arrows which is due to collective directive influence.

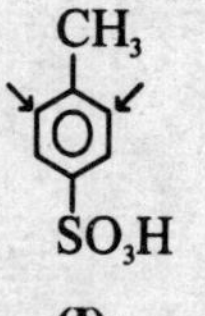

(I) 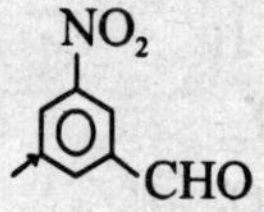   (II) 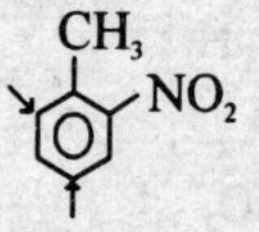   (III)

It is difficult to predict the orientation when the directive influence of one group opposes that of other. Under such circumstances, mixture of products are obtained. When a weakly activating or deactivating group is present with a highly activating group, the incoming group orients itself under the influence of the later.

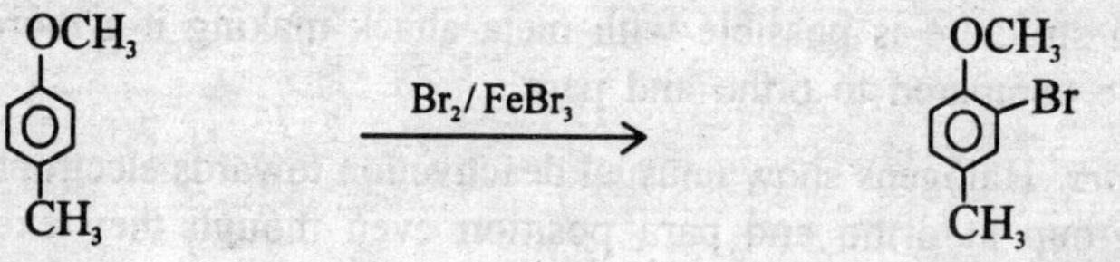

Major product

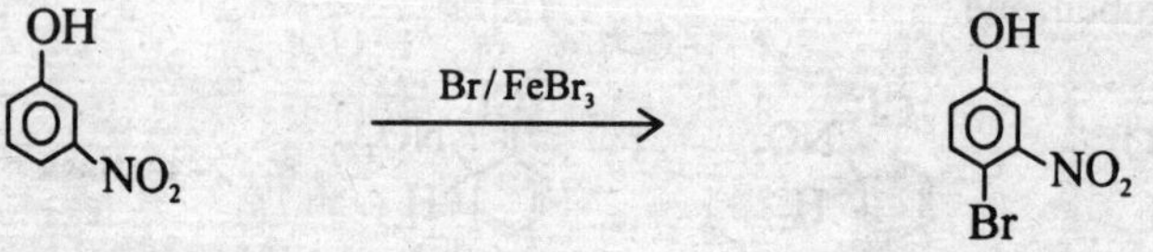

Major product

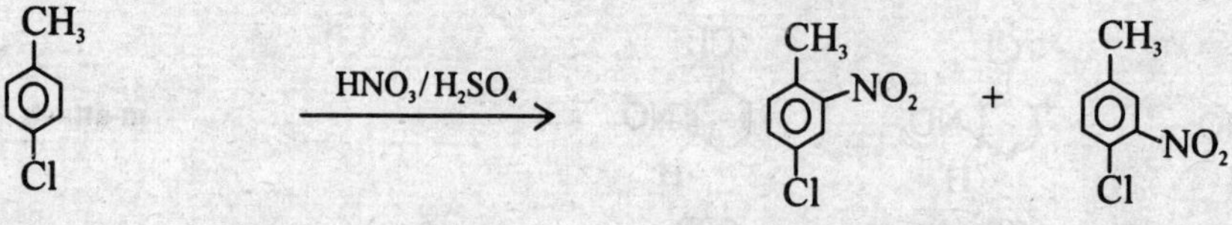

58%   42%

When the two groups are at meta position to each other, very low substitution takes place between them.

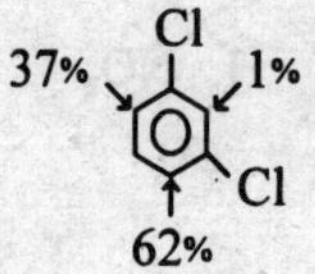

37%   1%

62%

28.11 INDIVIDUAL MEMBERS

28.11.1 Benzene

Benzene was first isolated by Faraday in 1825 from oil condensed in cylinders containing compressed illuminating gas produced by destructive distillation of vegetable oils. Later in 1849, Hofmann isolated benzene from coal tar.

Manufacture:

(1) *Platforming:* It is the process of cyclisation and aromatisation of n-hexane in presence of platinum catalyst supported over alumina. Large scale production of benzene is done from gasoline and naphtha fraction of petroleum. Over 90% of the commercial benzene is prepared by this method.

(2) *Hydrodealkylation of Toluene:* Toluene, obtained in large scale from petroleum, is converted into benzene by treatment with hydrogen at 600° in presence of platinum catalyst supported over alumina.

(3) *Distillation of Coal Tar:* Benzene produced by this source accounts for less than 10% of the total production of commercial benzene. The method has been discussed in detail in Section 11.6.

Physical Properties: Benzene is a colorless, highly refractive liquid having a characteristic aromatic odor m.p. 5.5°, b.p. 80° and specific gravity 0.8790 at 20°. It is insoluble in water and soluble in organic solvents such as ethanol, ether etc. It is used as a solvent in various reactions. However, the use of benzene has been banned in some countries due to its carcinogenic nature.

Chemical Properties: Chemical properties of benzene have already been discussed in Section 28.7.

Uses: Benzene is used
1. As a solvent for fats, oils and resins.
2. For dry cleaning .
3. As a motor fuel under the name of benzol.
4. For the manufecture of styrene (used in the synthesis of plastics and rubber), phenol (used in dyes and plastic manufecture), maleic anhydride and many other aromatic compounds.

28.11.2 Toluene/Methylbenzene

Toluene was first obtained from dry distillation of Tolu balsum. Its industrial name is toluol.

Preparation

It can be prepared from any of the general methods for the preparation of aromatic hydrocarbons in Section 28.4.

It is prepared on large scale from coal and petroleum by the methods discussed for benzene.

Physical Properties: It is a colorless liquid having a characteristic odor b.p. 111°, specific gravity 0.866 at 20°. It is insoluble in water but soluble in organic solvents. It is a good solvent for many organic reactions.

Chemical Properties: Toluene undergoes all the reactions discussed in Section 28.7. They take place, however, at a faster rate since methyl group of toluene activates the ring for ortho and para orientation.

Toluene when reacts with excess of nitric acid and sulphuric acid for a long time yields 2,4,6-trinitrotoluene, an explosive commonly known as TNT.

$$\underset{\text{Toluene}}{\overset{CH_3}{\underset{}{\bigcirc}}} + HNO_3 \xrightarrow{H_2SO_4} \underset{\text{2,4,6-Trinitrotoluene (TNT)}}{\overset{NO_2 \quad CH_3 \quad NO_2}{\underset{NO_2}{\bigcirc}}}$$

Reaction of Side Chain: Toluene and other alkyl benzenes offer two sites for the attack of halogens—the ring and the side chain. The choice of the proper reaction conditions controls the position of attack.

If chlorine is bubbled into hot toluene in presence of UV light, substitution occurs exclusively in the side chain whereas in absence of light and in presence of ferric ions, ring substitution results.

$$\underset{\text{Toluene}}{\overset{CH_3}{\bigcirc}} + Cl_2 \xrightarrow{hv} \underset{\text{Benzyl chloride}}{\overset{CH_2Cl}{\bigcirc}}$$

$$\underset{\text{Toluene}}{\overset{CH_3}{\bigcirc}} + Cl_2 \xrightarrow{Fe^{+++}} \underset{\text{o-Chlorotoluene}}{\overset{CH_3}{\underset{Cl}{\bigcirc}}} + \underset{\text{p-Chlorotoluene}}{\overset{CH_3}{\underset{Cl}{\bigcirc}}}$$

Chlorination of side chain in presence of light proceed via free radical chain mechanism (analogous to alkanes) whereas ring substitution takes place according to electrophilic aromatic substitution (dicussed in Section 28.8.)

The side chain oxidation of methyl benzene furnishes aromatic aldehyde or carboxylic acid by the proper choice of reagents.

$$\underset{\text{Toluene}}{\overset{CH_3}{\bigcirc}} \xrightarrow[\text{2. H}_2\text{O}]{\text{1. Cr}_2\text{O}_3, \text{ acetic anhydride}} \underset{\text{Benzaldehyde}}{\overset{CHO}{\bigcirc}}$$

$$\underset{\text{Toluene}}{\overset{CH_3}{\bigcirc}} \xrightarrow[\text{2. K}_2\text{Cr}_2\text{O}_7/\text{H}^+]{\text{1. KMnO}_4/\text{ HO}^- \text{ or}} \underset{\text{Benzoic acid}}{\overset{COOH}{\bigcirc}}$$

Uses: Toluene is used
1. For the preparation of various aromatic halogen compounds which on hydrolysis give benzyl alcohol, benzaldehyde and benzoic acid having diverse uses.
2. As a solvent for resins and adhesives.
3. For the preparation of TNT

28.11.3 Cumene/Isopropylbenzene

Preparation: It is synthesised by the reaction of propylene and benzene in presence of sulphuric acid or aluminum chloride.

$$\underset{\text{Propylene}}{CH_3-CH=CH_2} + \underset{\text{Benzene}}{\bigcirc} \xrightarrow[\text{AlCl}_3]{\text{H}_2\text{SO}_4 \text{ or}} \underset{\text{Cumene}}{\overset{CH_3}{\underset{CHCH_3}{\bigcirc}}}$$

The reaction proceeds via electrophilic attack of isopropyl cation on benzene.

Properties: It is a colorless liquid with a b.p. 153°. It is converted into commercial phenol and acetone via following sequence of reactions.

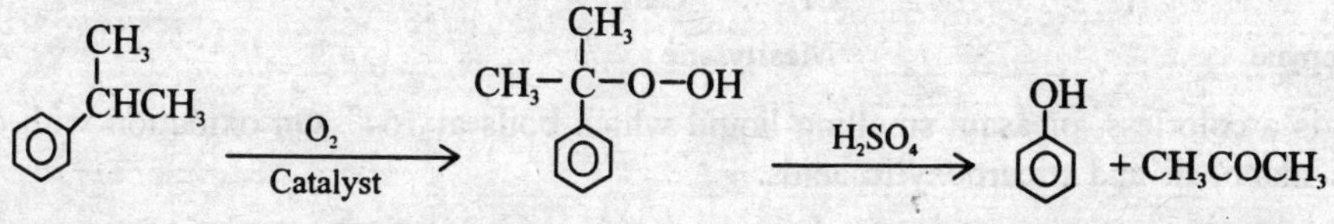
CH3
CHCH3
O2
Catalyst
CH3
CH3–C–O–OH
H2SO4
OH
+ CH3COCH3

Cumene **Phenol Acetone**

28.11.4 Xylenes/Dimethylbenzenes

Xylenes exhibit postion isomerism and are isomeric with ethylbenzene as shown below.

CH3
CH3
CH3
CH3
CH3
CH3
CH2CH3

o-Xylene, b.p. 144° **m-Xylene, b.p. 139°** **p-Xylene, b.p. 138°** **Ethylbenzene, b.p. 136°**

Preparation: The three isomeric xylenes are commercially produced these days by reforming C_6 to C_8 fraction of light naphtha at 400-500° over a platinum- alumina catalyst taken over an inert support at 25 atmosphere, followed by cyclisation and aromatisation.

Properties: Xylenes are colorless liquids used as solvents in the manufacture of paints and lacquers. Xylenes undergo electrophilic substitution much in the same way as other hydrocarbons.

They oxidise to give dicarboxylic acids. O-xylene on air oxidation in presence of vanadium pentoxide at high temperature gives phthalic anhydride whereas p-xylene on oxidation gives terephthalic acid which is used in polyester preparation.

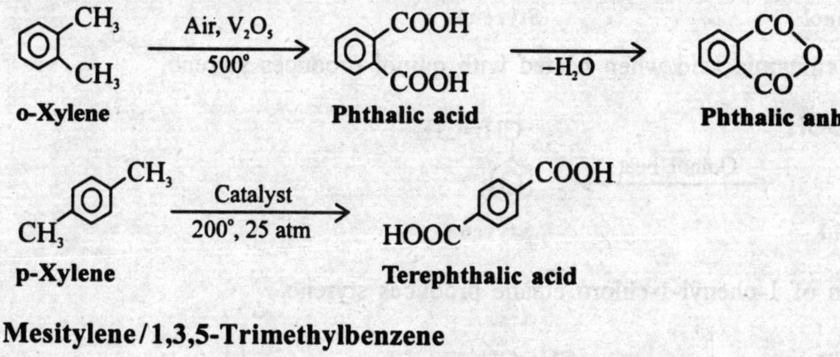
CH3
CH3
Air, V2O5
500°
COOH
COOH
–H2O
CO
O
CO
CH3
CH3
Catalyst
200°, 25 atm
COOH
HOOC

o-Xylene **Phthalic acid** **Phthalic anhydride**

p-Xylene **Terephthalic acid**

28.11.5 Mesitylene/1,3,5-Trimethylbenzene

It occurs in coal tar but is prepared conveniently by distilling acetone with sulfuric acid.

3CH3COCH3
Conc. H2SO4
CH3
CH3
CH3

Acetone **Mesitylene**

Similarly, polymerisation of propyne also produces mesitylene.

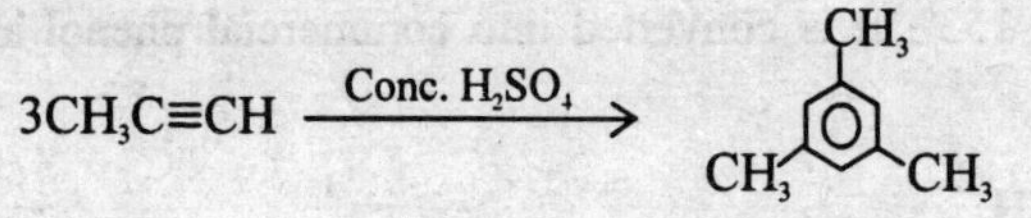

3CH₃C≡CH
Conc. H₂SO₄
CH₃
CH₃
CH₃

Propyne **Mesitylene**

Properties: It is a colorless, pleasant smelling liquid which boils at 164°. On oxidation with dilute nitric acid, it forms mono, di and tricarboxylic acids.

CH₃
CH₃
COOH
COOH
CH₃
COOH
COOH
HOOC
COOH

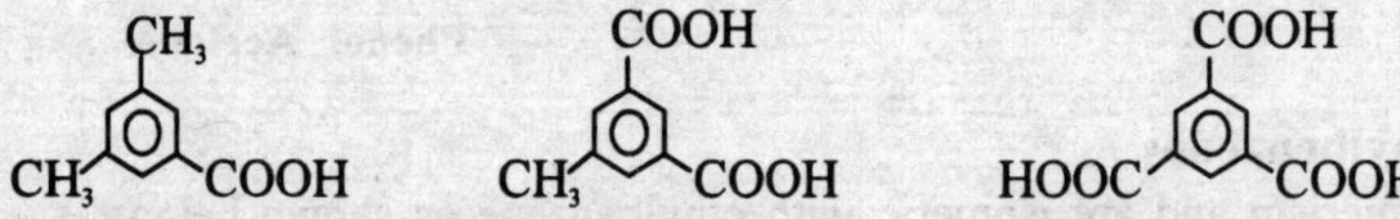

Mesitylenic acid **Uvitic acid** **Trimeric acid**

28.11.6 Styrene/Phenylethene/Vinylbenzene

It occurs in coaltar and in plant storex (hence named styrene), a balsam.

Preparation: Styrene is manufactured by the Friedel Craft reaction of benzene with ethylene to form ethylbenzene which on catalytic dehydrogenation produces styrene.

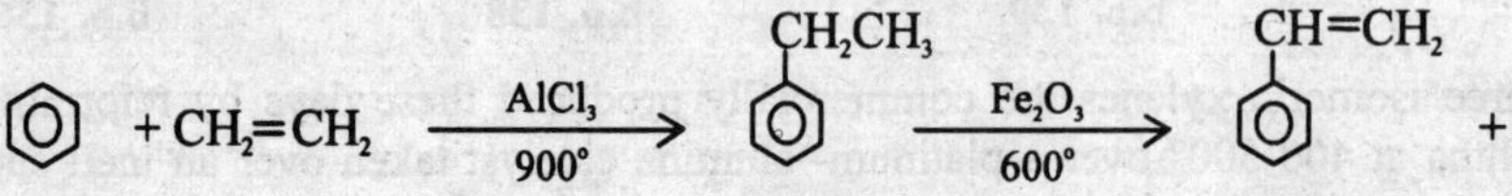

+ CH₂=CH₂
AlCl₃
900°
CH₂CH₃
Fe₂O₃
600°
CH=CH₂
+ H₂

Benzene **Ethylbenzene** **Styrene**

It is the industrial method of preparation of styrene. Styrene can also be prepared by dehydration of 1-phenylethanol.

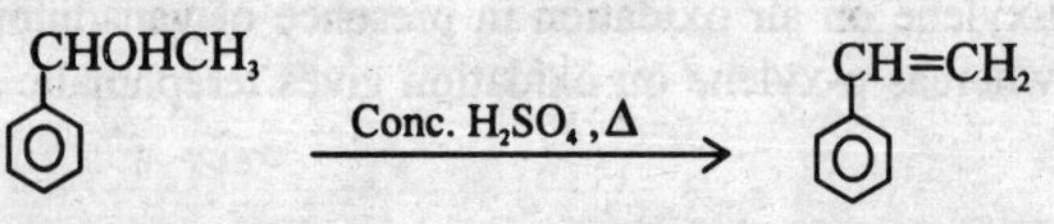

CHOHCH₃
Conc. H₂SO₄ ,Δ
CH=CH₂

1-phenylethanol **Styrene**

Decarboxylation of cinnamic acid when heated with quinol produces styrene.

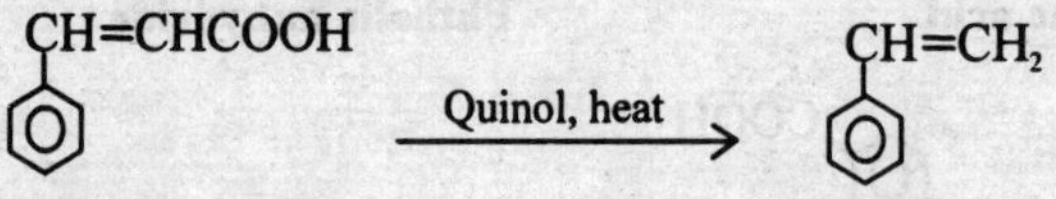

CH=CHCOOH
Quinol, heat
CH=CH₂

Cinnamic acid **Styrene**

Dehydrohalogenation of 1-phenyl-1-chloro ethane produces styrene.

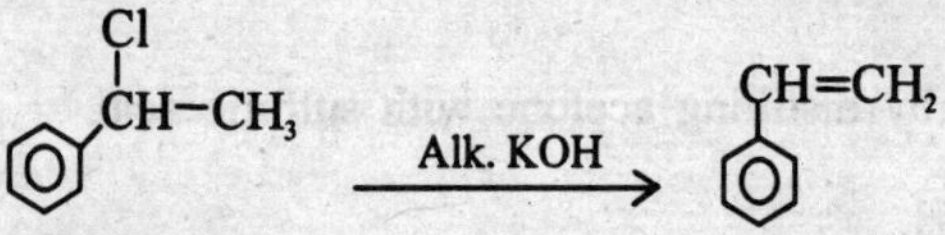

Cl
CH−CH₃
Alk. KOH
CH=CH₂

1-Phenyl-1-chloroethane **Styrene**

Properties: It is a colorless liquid (b.p. 145°) having a benzene ring with alkene side chain. It undergoes two sets of reactions: substitution in the ring and addition to the double bond in side chain.

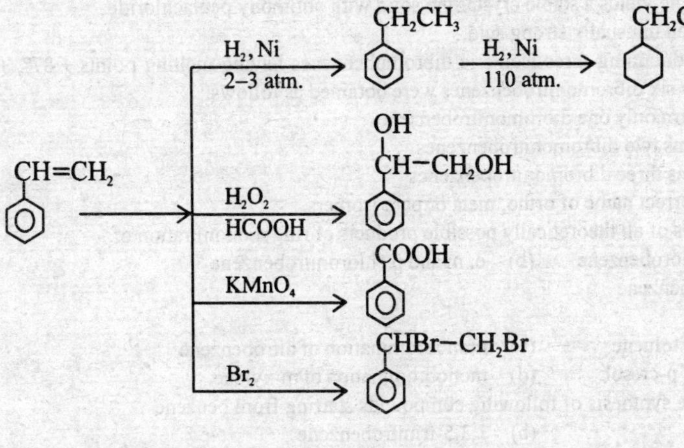

$$
\underset{\text{styrene}}{C_6H_5\text{—CH=CH}_2}
\begin{cases}
\xrightarrow[\text{2–3 atm.}]{H_2, Ni} \ \underset{\text{(CH}_2\text{CH}_3\text{)}}{C_6H_5\text{—CH}_2\text{CH}_3} \xrightarrow[\text{110 atm.}]{H_2, Ni} \ \text{(CH}_2\text{CH}_3) \\[2mm]
\xrightarrow[\text{HCOOH}]{H_2O_2} \ \underset{\substack{\text{OH}\\ \text{CH—CH}_2\text{OH}}}{} \\[2mm]
\xrightarrow{KMnO_4} \ \text{COOH} \\[2mm]
\xrightarrow{Br_2} \ \text{CHBr—CH}_2\text{Br}
\end{cases}
$$

Similarly, styrene undergoes other reactions characteristics of carbon-carbon double bond.

In styrene the phenyl ring is comparatively stable and independent in its reaction as the p-orbitals of alkene side chain are in a different plane than that of the p-orbitals of benzene ring and stability due to delocalisation is not possible. In general, the double bond shows higher reactivity towards electrophilic reagents than the ring.

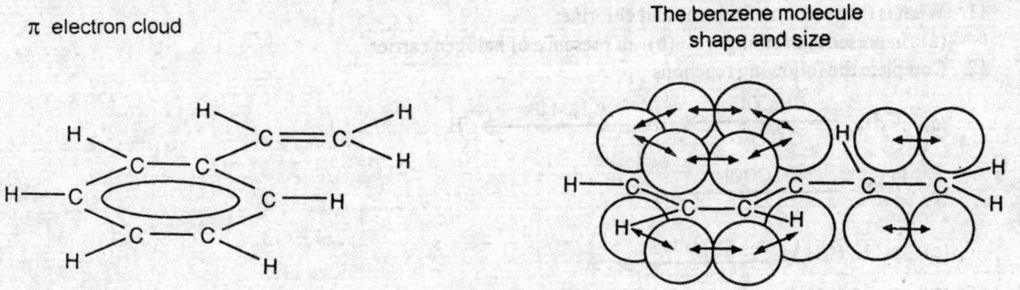

π electron cloud

The benzene molecule shape and size

Fig. 28.2: Orbital Picture of Styrene

Styrene undergoes rapid polymerisation when exposed to sunlight to form meta styrene, $(C_8H_8)_n$. However, in presence of dibenzoyl peroxide it gives polystyrene.

$$
n\ C_6H_5CH=CH_2 \longrightarrow \underset{\substack{C_6H_5 \qquad\ C_6H_5}}{-CH_2\text{—}(CH\text{—}CH_2)_n CH\text{—}}
$$

Styrene in its polymerised form has diverse uses in plastic and rubber industry. Polystryene is also used in making light weight packaging materials.

28.12 PROBLEMS

1. What is meant by Aromaticity? Why benzene having three double bonds undergo substitution and not addition?
2. Discuss the stability of benzene by taking into account its heat of hydrogenation and heat of combustion.
3. The homolytic bond dissociation energy for C—H bond in benzene is considerably larger than for cyclohexane. Discuss with the help of orbital pictures of the two.

4. Give reasons:
   (a) 3-Chlorocyclopropene yields a stable crystalline solid with antimony pentachloride.
   (b) Cyclopentadiene is an unusually strong acid.
5. There are three bottles containing three isomeric dibromobenzenes having melting points $+ 87°$, $+ 6°$ and $- 7°$ respectively. On nitration six dibromonitrobenzenes were obtained as follows.
   isomer A (m.p. $+ 87$ °)form only one dibromonitrobenzene.
   isomer B (m.p. $+ 6°$)forms two dibromonitrobenzenes
   isomer C (m.p. $- 7°$) forms three dibromonitrobenzencs
   Label A, B, C with the correct name of ortho, meta or para isomers.
6. Give structures and names of all theoretically possible products of ring mononitration of
   (a) o, m and p bromochlorobenzene   (b) o, m and p chloronitrobenzene
   (c) o, m and p dibromobenzene
7. Predict the product of
   (a) monobromination of toluene   (b) monobromination of nitrobenzene
   (c) monobromination of p-cresol   (d) monobromination of m-xylene.
8. Outline all the steps in the synthesis of following compounds starting from benzene
   (a) p-nitrotoluene   (b) 1,3,5-trinitrobenzene
   (c) 3,5-dinitrobenzoic acid   (d) m-bromobenzoic acid
   (e) p-bromonitrobenzene   (f) p-chloronitrobenzene
9. Explain why towards electrophilic substitution
   (a) Aniline reacts faster than nitrobenzene   (b) Nitrobenzene reacts slower than benzene
   (c) Toluene reacts faster than benzene   (d) Phenol reacts faster than toluene
10. Starting with styrene how will you prepare
   (a) Benzoic acid   (b) polystyrene   (c) Ethylcyclohexane   (d) p-chlorostyrene
11. What is the reaction of toluene and chlorine:
   (a) in presence of sunlight   (b) in presence of halogen carrier
12. Complete the following reactions

   (a) C_6H_6 $\xrightarrow[C_2H_5OH]{CH_3COCl}$ A $\xrightarrow{C_2H_5MgBr}$ B

   (b) C_6H_6 $\xrightarrow[AlCl_3]{CH_3COCl}$ A

   (c) $pC_6H_4 (CH_3)_2$ $\xrightarrow{\text{Oxidation}}$ A

14. Why Friedel Crafts alkylation of benzene form polysubstituted products.

29

Aryl Halides

Aryl halides or aromatic halohydrocarbons are the compounds having halogen atom (or atoms) attached to aromatic nucleus. If more than one halogen is attached to the ring, the prefix di,tri or tetra is used prior to the name of compounds. The position of halogen is also indicated by numbers.

Chlorobenzene **1,2-Dibromobenzene** **2,4-Dichlorotoluene** **1,3,5-Trichlorobenzene**

29.1 GENERAL METHODS OF PREPARATION

Aryl halides are prepared from aromatic hydrocarbons and their derivatives.

(a) *Direct Halogenation of Aromatic Hydrocarbons:* Aryl chlorides and aryl bromides are prepared by the reaction of chlorine or bromine with aromatic hydrocarbons in presence of Lewis acid (e.g. $AlCl_3$, or $FeCl_3$).

Chlorobenzene

o-Bromotoluene **p-Bromotoluene**

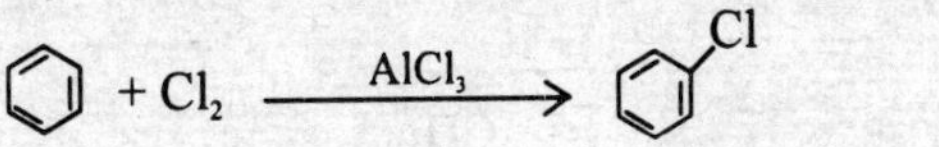

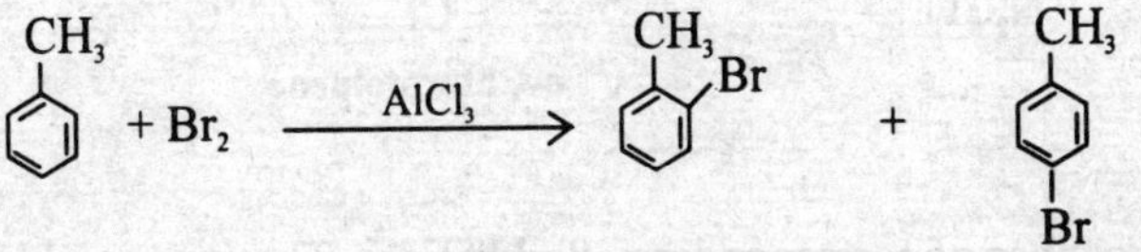

Benzene on chlorination gives chlorobenzene whereas toluene on bromination yields a mixture of o- and p-bromotoluene of which the later predominates. Aryl iodides can not be prepared by the same method as the reaction is reversible with iodine. They are obtained in good yield by carrying out the reaction in presence of oxidising agents such as HIO_3, HNO_3 or HgO.

$$5HI + HIO_3 \longrightarrow 3I_2 + 3H_2O$$

Iodobenzene

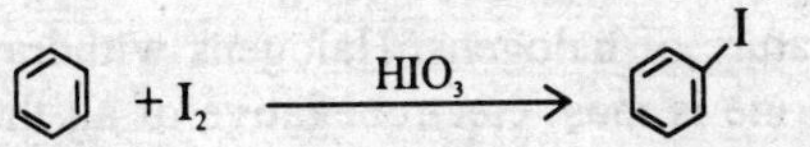

(b) *From Aromatic Amines:* Aryl halides are prepared in good yields from diazonium salts which in turn are prepared from aromatic amines.

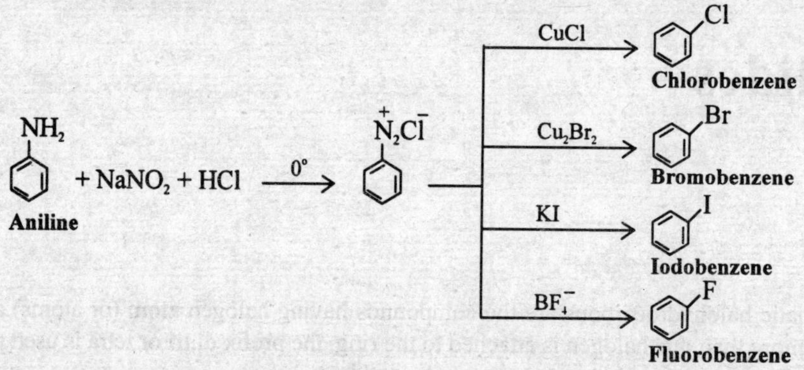

The above reaction is known as Sandmeyer reaction.

(c) *Rasching Process* is the process for commercial preparation of chlorobenzene in which vapors of benzene, air and hydrogen chloride are passed over copper chloride.

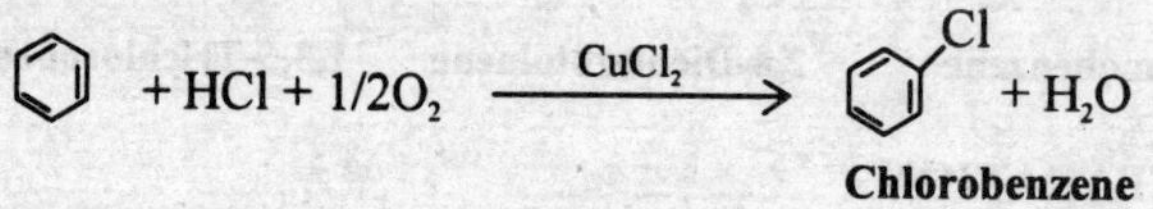

The preparation of aryl halides from aryl diazonium salts is more important than the other methods as aryl fluorides and iodide can be prepared easily by this method. Direct halogenation usually furnishes mixture of ortho and para halogenated compounds while in this method pure ortho or para isomers are prepared depending upon the starting amino compounds. o-Toluidine exclusively gives o-chlorotoluene by sandmeyer reaction.

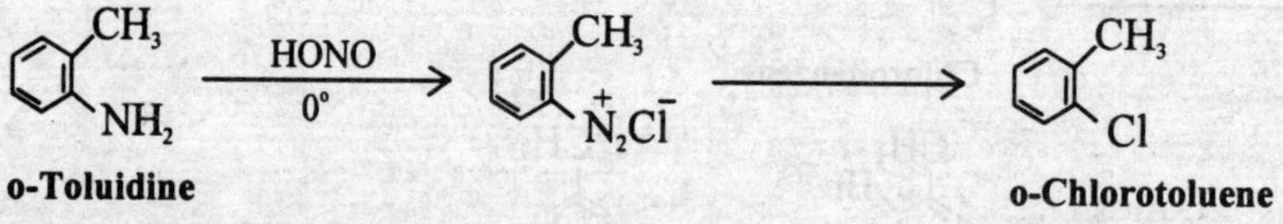

29.2 PHYSICAL PROPERTIES

Aryl halides show similar properties like those of corresponding alkyl halides. They are soluble in organic solvents and insoluble in water like alkyl halides. Chloro and bromobenzene have almost similar boiling points as those of n-hexylchloride and n-hexylbromide.

29.3 SPECTRAL PROPERTIES

The UV-visible spectra of aryl halides show a weak absorption between 170 and 258 nm due to the presence of nonbonded electrons of halogen. Absorption due to aromatic hydrocarbon is also exhibited.

The IR spectra of aryl halides show C−X stretching bond near 1100 cm^{-1}.

The NMR spectra of aryl halides depend upon the nature of halogens. Halogens withdraw electron density from the ring in the order F > Cl > Br > I. Fluorine is most electronegative of all the halogens and causes maximum deshielding and lowering in chemical shift.

The mass spectra show strong molecular ion peak and further fragmentation pattern is usual.

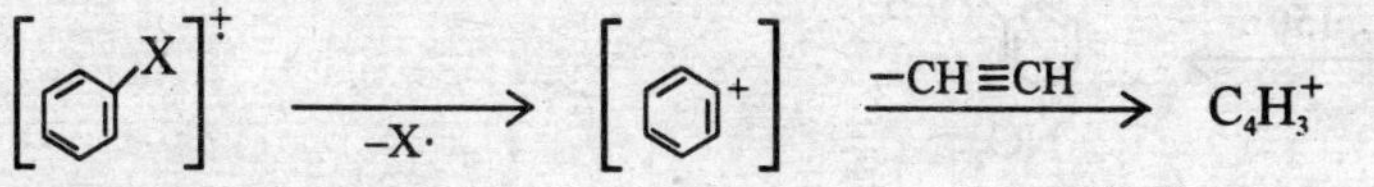

$$\left[\bigcirc\!\!-\!X\right]^{\ddagger} \xrightarrow[-X\cdot]{} \left[\bigcirc\right]^{+} \xrightarrow{-CH\equiv CH} C_4H_3^{+}$$

m/z 77   m/z 51

29.4 CHEMICAL PROPERTIES

The reaction shown by aryl halides are due to the aromatic ring and the halogen atom.

(a) *Formation of Grignard Reagent:* Aryl halides form Grignard reagent when react with magnesium in dry ether or tetrahydrofuran (THF).

$$Ar-Br + Mg \xrightarrow{dry\ ether} ArMgBr$$

$$Ar-Cl + Mg \xrightarrow{THF} ArMgCl$$

Aryl halides having carboxyl, hydroxy or amino group do not undergo the above reaction as these groups are acidic enough to decompose Grignard reagent. The nitro group oxidises the Grignard reagent.

Aryl halides form aryllithium compound when react with lithium metal in ether. Aryllithium compounds have similar synthetic utility.

$$Ar-Br + 2Li \xrightarrow{dry\ ether} Ar-Li + LiBr$$

$$Ar-Li + CO_2 \xrightarrow{} Ar-CO\bar{O}\ Li^{+} \xrightarrow{H^{+}} ArCOOH$$

(b) *Substitution in the Ring:* Aryl halides undergo electrophilic substitution in the ring as discussed in Section 28.8 and orient the incoming groups at ortho and para positions.

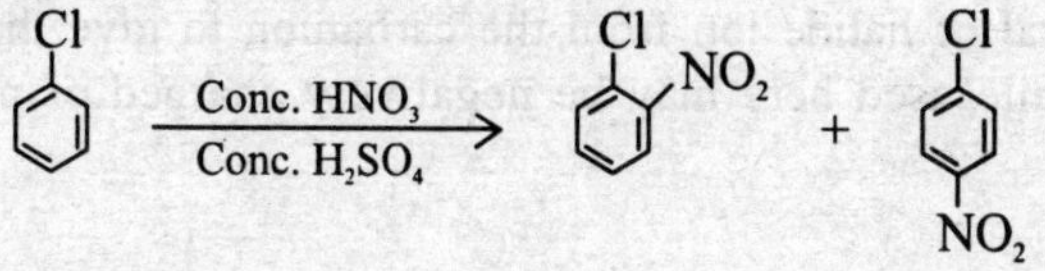

Halogen atom deactivates the ring for electrophilic substitution. (Detailed discussion in Section 28.8.)

(c) *Nucleophilic Aromatic Substitution:* Aryl halides show low reactivity towards nucleophiles like OH−,RO−,CN− etc., but, the presence of electron withdrawing groups at certain position in the ring enhances the reactivity. Chlorobenzene having nitro group at ortho and/or para positions undergoes bimolecular nucleophilic substitution under mild conditions as stated below.

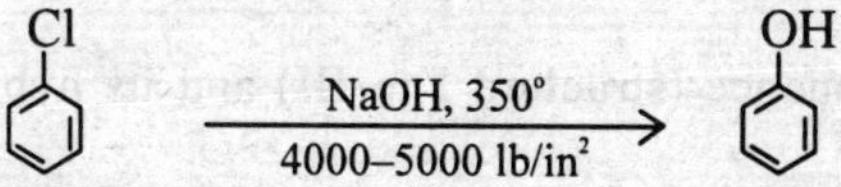

Chlorobenzene   Phenol

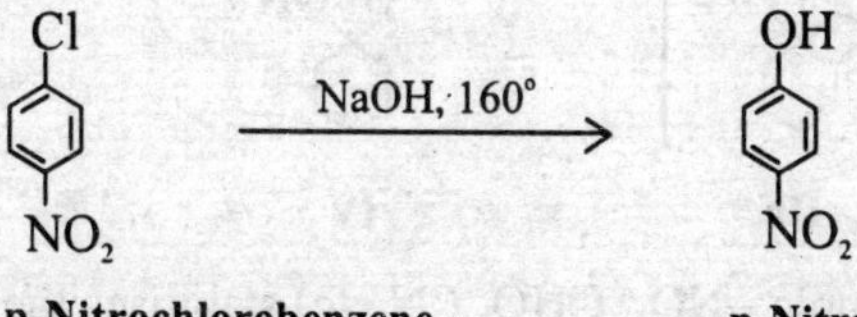

p-Nitrochlorobenzene   p-Nitrophenol

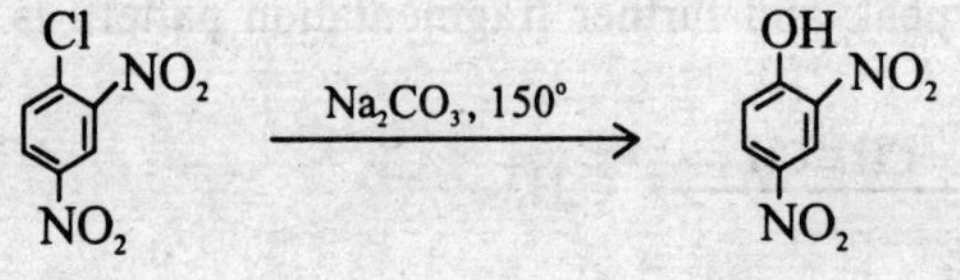

2,4-Dinitrochlorobenzene   **2,4-Dinitrophenol**

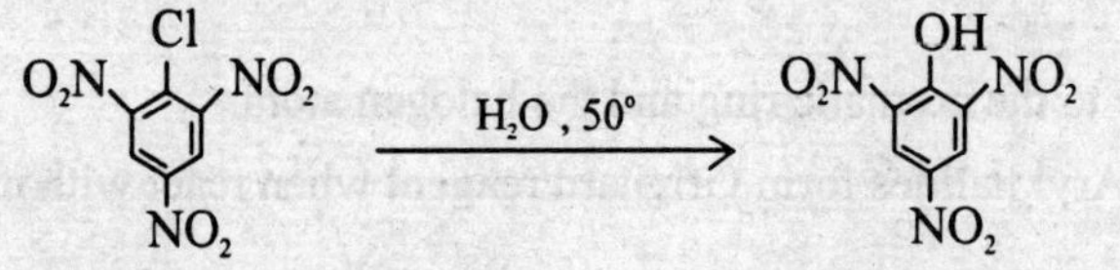

2,4,6-Trinitrochlorobenzene   **2,4,6-Trinitrophenol**

Similarly other nucleophiles undergo reactions with aryl halides. Chlorobenzene reacts with ammonia as shown below.

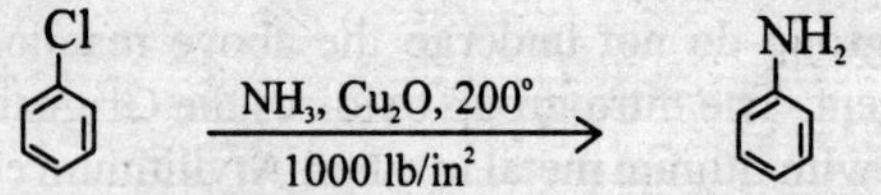

Chlorobenzene   **Aniline**

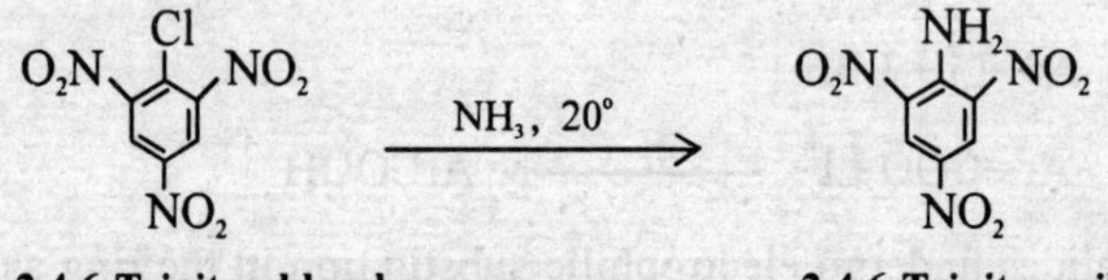

2,4,6-Trinitrochlorobenzene   **2,4,6-Trinitroaniline**

The mechanism of bimolecular nucleophilic substitution involves two steps: attack of nucleophile upon the ring to form carbanion (I) and the removal of halide ion from the carbanion to give the product. The first step is the slow step. The nucleophile used here may be negatively charged or neutral but its attack on the ring creates a carbanion.

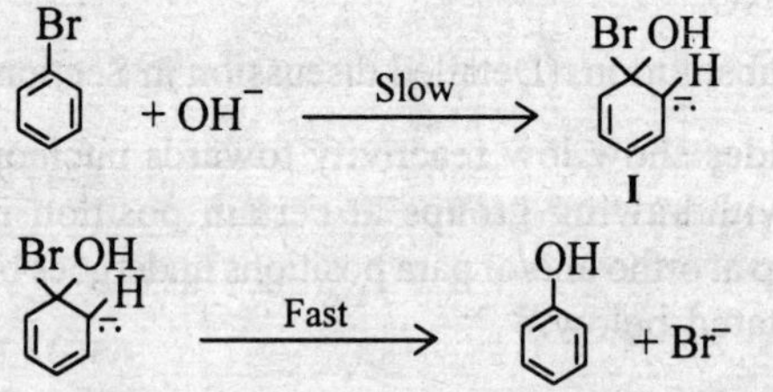

The intermediate carbanion, I, is stabilised by resonance (structure I to III) and its hybrid is represented as IV.

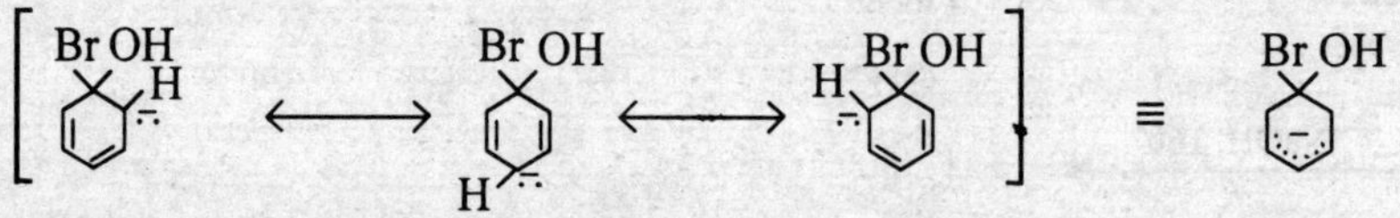

    I   **II**   **III**   **IV**

Now any group which withdraws electrons (for example $-NO_2$, CHO, CN etc) stabilises the carbanion and activates the ring for nucleophilic substitution, whereas electron releasing groups (for example NH_2,

H, R) destabilises the carbanion and deactivates the ring for substitution reaction. Below are given esonating structures of carbanion of p-nitrochlorobenzene formed during the hydrolysis of p-nitrochlorobenzene.

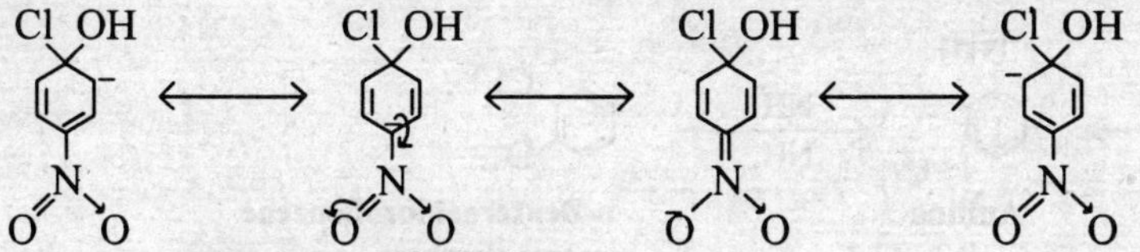

Especially stabile, electron withdrawal via resonance

As shown the intermediate carbanion formed by nucleophilic attack on p-nitrochlorobenzene is a hybrid of structures having negative charge on carbon as well as on oxygen of nitro group. Oxygen being electronegative readily accommodates the negative charge extending additional stability to the structure. Thus nitro group activates the aryl halides towards nucleophilic substitution.

(d) *Nucleophilic Aromatic Substitution - Benzyne Mechanism:* Nucleophilic substitution occurs even in absence of electron withdrawing groups by the treatment of aryl halides with very strong bases. When chlorobenzene is treated with very strongly basic amide ions in liquid ammonia, it yields aniline.

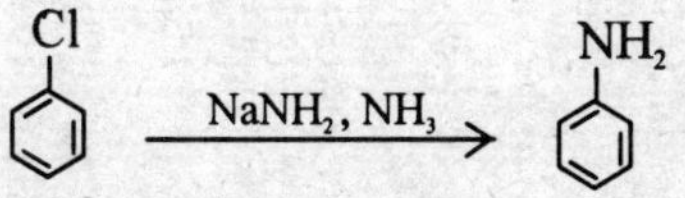

The reaction is not a simple displacement reaction but involves two steps, the elimination and the addition. The mechanism suggested for the above reaction involves the formation of a highly unstable reaction intermediate called benzyne.

Elimination

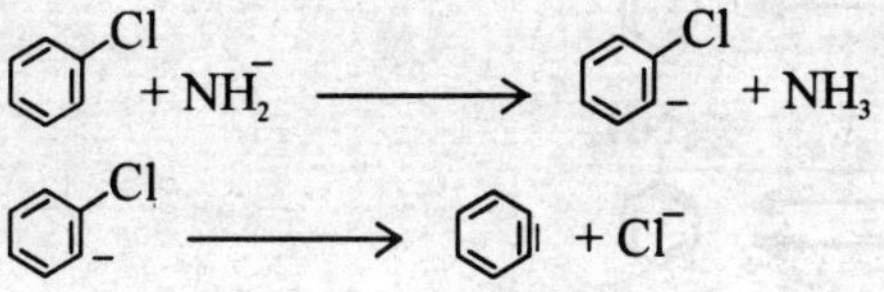

Benzyne

Addition

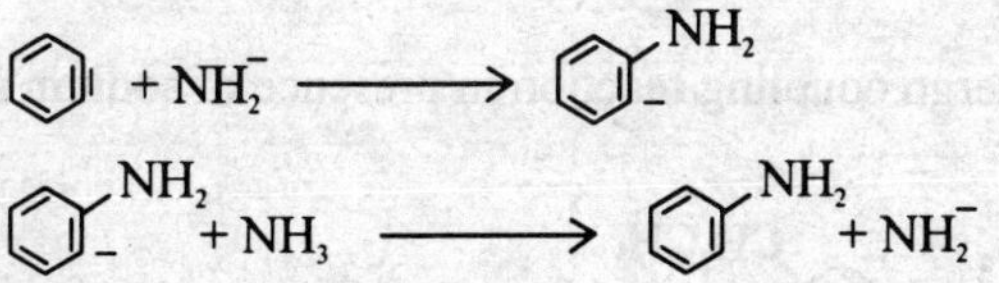

The mechanism of this reaction is supported by the following facts.

(I) 2-Chloro-3-ethylanisol does not react with sodamide because with ortho positions occupied, benzyne intermediate can not be formed.

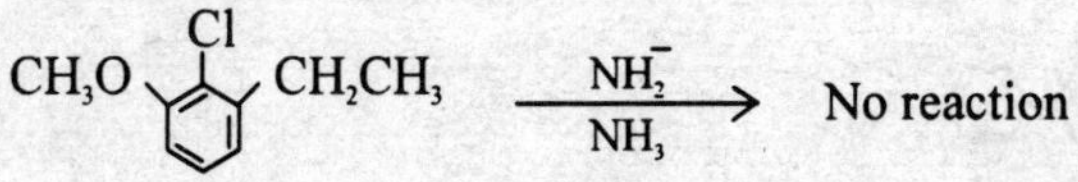

No reaction

(II) When a mixture of chlorobenzene and o-deuterochlorobenzene in equal amounts is reacted wit a limited amount of sodamide in liquid ammonia, the recovered unreacted material is mostly o-deuterate chlorobenzene.

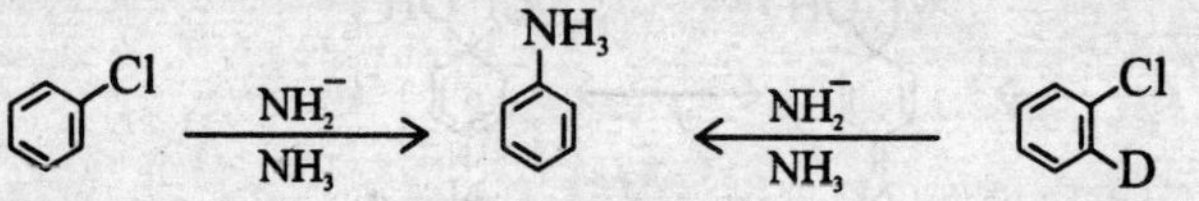

Chlorobenzene   **Aniline**   **o-Deuterochlorobenzene**
reacts faster   **reacts slowly**

The reason is slow elimination of deuterium in the first step which slows down further reaction compare to the elimination of hydrogen in case of chlorobenzene.

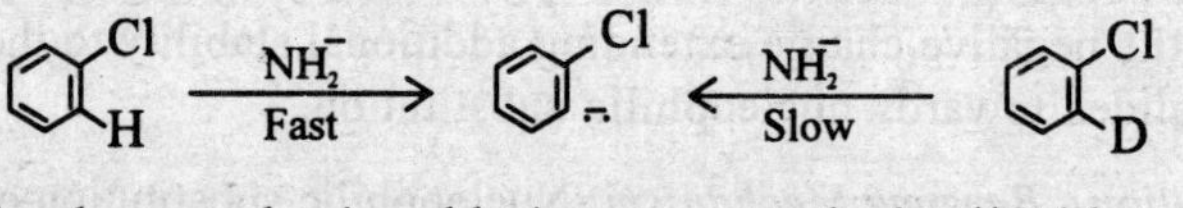

(III) Chlorobenzene having chlorine atom attached to ^{14}C (shown by an asteric mark on the ring carbon on treatment with amide gives aniline, half of which has amino group attached to labeled carbon an half having amino group next to labeled carbon.

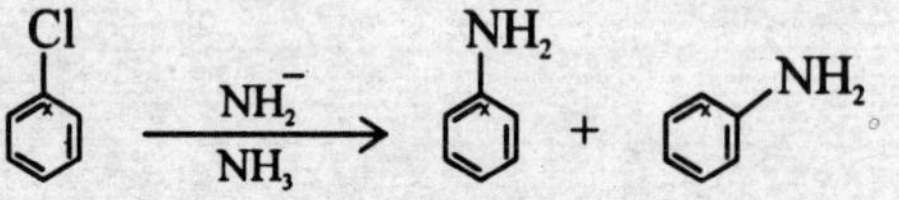

The Benzyne intermediate formed as a result of abstraction of adjacent protons is equivalent and form aniline by random addition of amide except for a small isotopic effect.

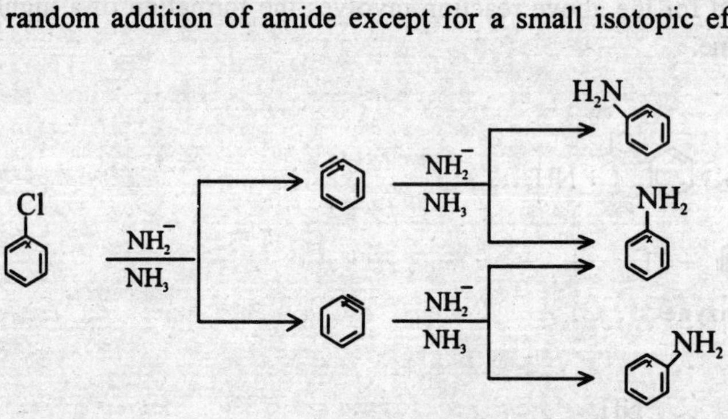

(e) *Wurtz-Fittig Reaction:* Aryl halides undergo coupling reaction in presence of sodium in dry ether to form alkylbenzenes.

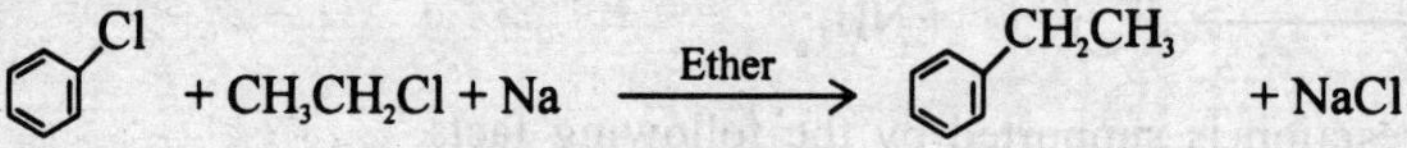

(f) *Ullmann Reaction:* The reaction developed by ullmann is another example of coupling of aryl halides where two molecules of aryl halides couple together when heated with copper powder.

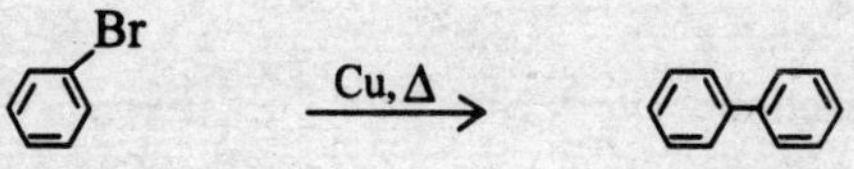

Bromobenzene   **Biphenyl**

The reaction is activated by using electron withdrawing groups at ortho and/or para positions.

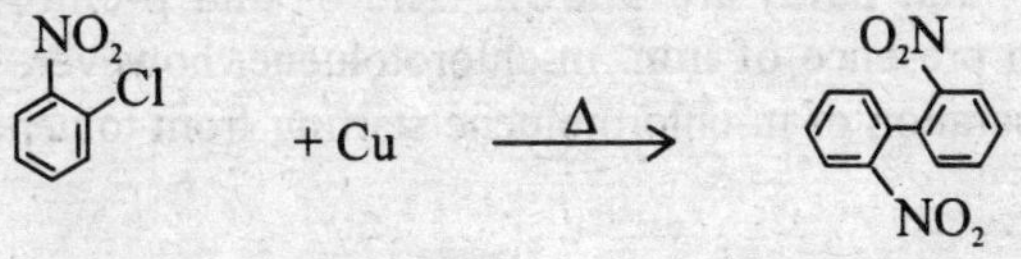

2-Nitrochlorobenzene   2,2'-Dinitrobiphenyl

9.5 INDIVIDUAL MEMBERS

9.5.1 Chlorobenzene

Preparation: Chlorobenzene may be prepared by any of the general methods discussed in Sec. 29.1. Commercially it is prepared by passing a mixture of benzene vapors, air and hydrogen chloride over copper chloride (Raschig method) whereas laboratory method involves reaction of benzene diazonium chloride with cuprous chloride.

Properties: Chlorobenzene is a colorless, pleasant smelling, steam volatile liquid (b.p. 132°) insoluble in water but soluble in organic solvents. It gives all the reactions characteristic of aryl halides discussed in Sec. 29.4.

When heated with chloral in presence of concentrated sulphuric acid, chlorobenzene forms DDT (m.p. 109-110°), a powerful insecticide.

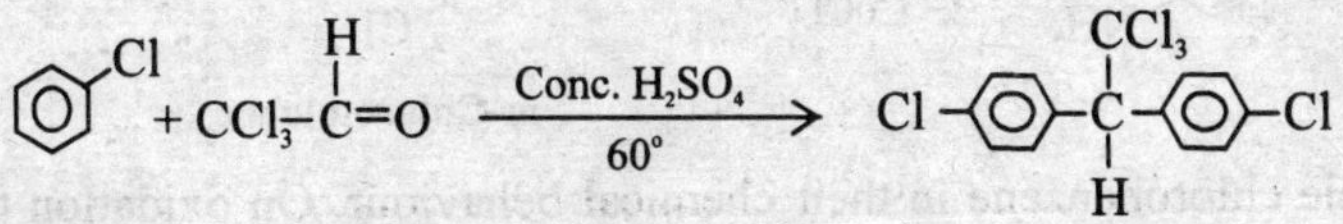

29.5.2 Bromobenzene

Preparation: It may be prepared by any of the general methods of preparation of aryl halides but a convenient laboratory preparation involves bromination of benzene in presence of iron.

Properties: It is a colorless, heavy liquid (b.p. 155°) and undergoes all the reactions characteristic of aryl halides.

29.5.3 Iodobenzene

Preparation: It is prepared by the reaction of potassium iodide on benzene diazonium chloride.

Properties: Iodobenzene is a colorless liquid which turns yellow on exposure to light. It is steam volatile, insoluble in water having b.p. 188°.

Amongst the chemical reactions, iodobenzene undergoes all the reaction undergone by chloro or bromobenzene. The only difference is its reactivity, as iodobenzene is more reactive than the other two aryl halides.

It reacts with chlorine to form iodobenzene dichloride which gives iodosobenzene on treatment with dilute sodium hydroxide. Iodosobenzene when boiled with water forms iodoxybenzene. Both iodoso and iodoxybenzenes explode without melting on heating.

$$C_6H_5I \xrightarrow{Cl_2} C_6H_5ICl_2 \xrightarrow{NaOH} C_6H_5IO \xrightarrow{H_2O, 100°} C_6H_5IO_2$$

       Iodobenzene   Iodoso-   Iodoxybenzene
       dichloride   benzene

29.5.4 Chlorotoluenes

Three isomeric chlorotoluenes (ortho, meta and para) are known. The o- and p-chlorotoluenes are prepared by direct chlorination of toluene in presence of iron. m-chlorotoluene, however, cannot be prepared by direct chlorination. The indirect preparation of m-chlorotoluene starting from toluene involves following steps:

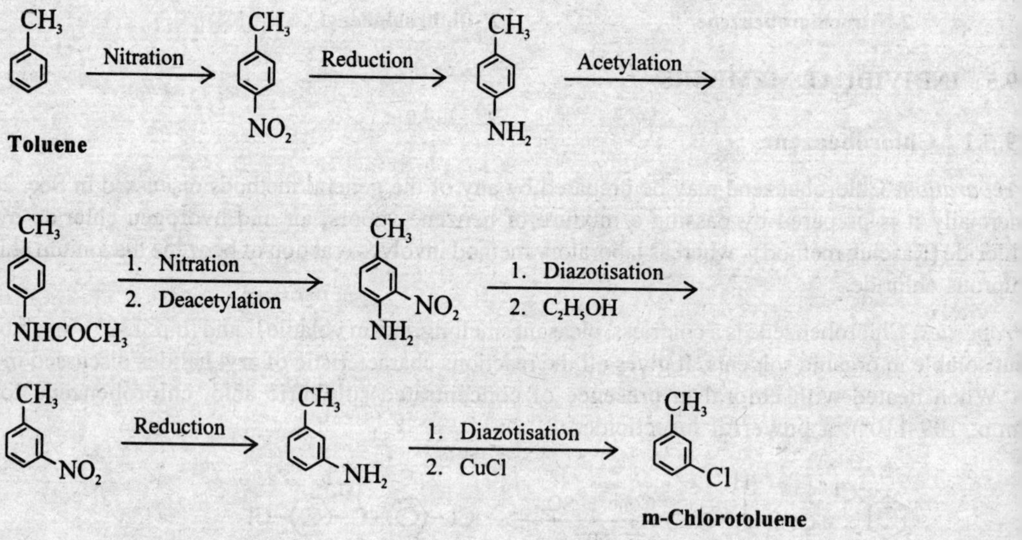

m-Chlorotoluene

Chlorotoluenes closely resemble chlorobenzene in their chemical behaviour. On oxidation they form chlorobenzoic acids. o-chlorotoluene forms o-chlorobenzoic acid.

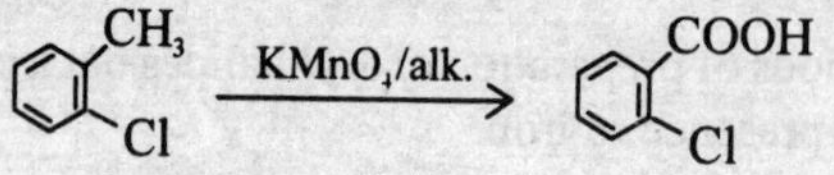

o-Chlorotoluene   **o-Chlorobenzoic acid**

29.6 ARYL ALKYL HALIDES

Aromatic compounds having halogen atoms in the side chain are termed as aryl alkyl halides.

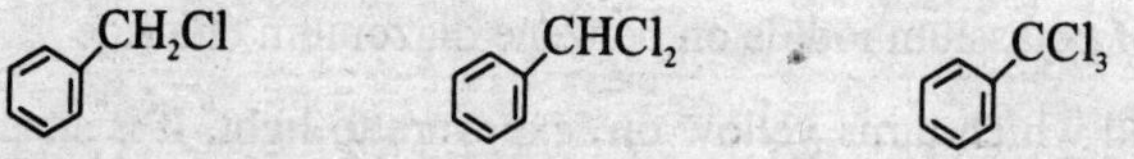

Benzyl chloride   **Benzal chloride**   **Benzo trichloride**

29.6.1 Benzyl Chloride

It is isomeric with o- m- and p-chlorotoluenes but differs from them in various reactions.

Preparation:

(a) *Chlorination of Toluene:* Benzyl chloride is prepared by refluxing toluene and chlorine in presence of light and in absence of halogen carrier (iron). The reaction involves free radical substitution on the methyl side chain of toluene.

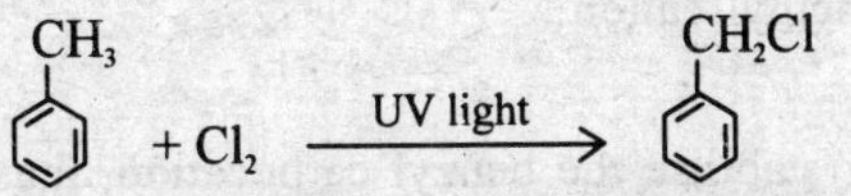

Benzyl chloride

Under controlled conditions, benzyl chloride undergoes further chlorination to form benzal chloride and benzo trichloride.

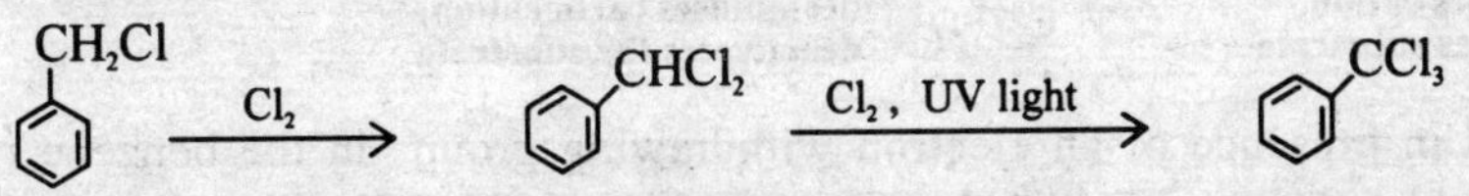

Benzyl chloride **Benzal chloride** **Benzo trichloride**

(b) *By the Action of Sulphonyl Chloride:* Sulphonyl chloride reacts with toluene to give benzyl chloride.

$$C_6H_5CH_3 + SO_2Cl_2 \longrightarrow C_6H_5CH_2Cl + HCl + SO_2$$

(c) *By the Action of N-Bromosuccinimide*

$$C_6H_5CH_3 + \begin{matrix} CH_2-CO \\ | \\ CH_2-CO \end{matrix}\!\!>NBr \longrightarrow C_6H_5CH_2Br$$

Toluene **Benzyl Bromide**

(d) *From Benzyl Alcohol:* Benzyl alcohol undergoes substitution when heated with phosphorus pentachloride.

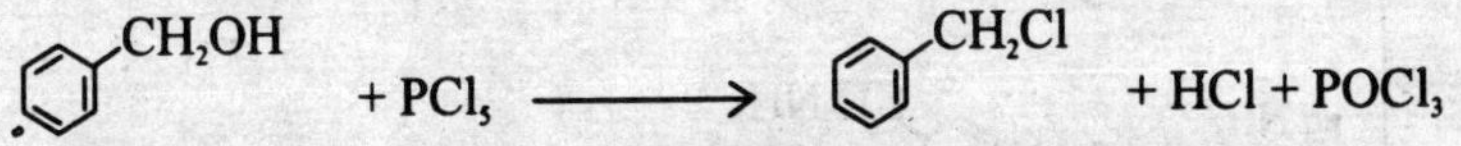

(e) *By Chloromethylation:* Chloromethylation is the introduction of chloromethyl group directly in the nucleus and is done by passing hydrogen chloride in a mixture of benzene and formalin (or paraformaldehyde) in presence of zinc or aluminum chloride.

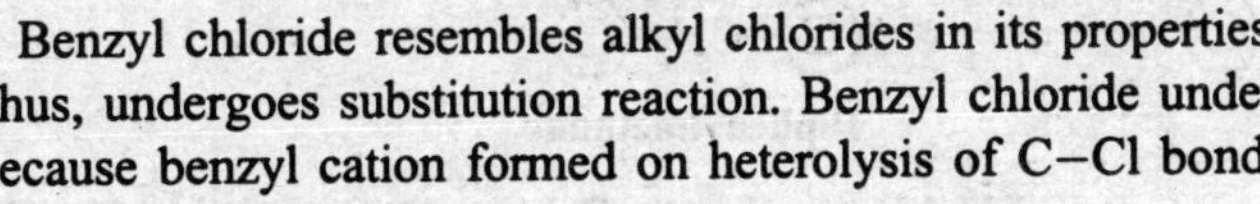

The reaction involves substitution of electrophile to the benzene ring.

Properties: Benzyl chloride is a colorless unpleasant smelling liquid (b.p. 179°), insoluble in water but soluble in organic solvents. Its vapors have an irritating action on eyes and nose causing tears.

Chemical: Benzyl chloride resembles alkyl chlorides in its properties as it is a phenyl substituted methyl chloride, thus, undergoes substitution reaction. Benzyl chloride undergoes substitution faster than methyl chloride because benzyl cation formed on heterolysis of C–Cl bond is stablised by resonance as shown below.

$$C_6H_5CH_2Cl \longrightarrow C_6H_5\overset{+}{C}H_2 + Cl^-$$

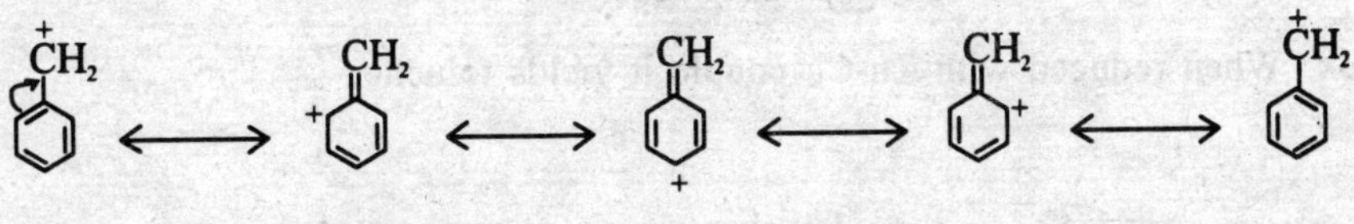

No such resonance stabilisation is possible with methyl cation.

Effect of Substituents on Nucleophilic Substitution

Electron releasing groups present in the benzene ring stabilise the benzyl carbocation, thus, activates the substrate and increase the rate of nucleophilic substitution.

G release electrons,
stabilises cation,
activates substrate

G withdraws electrons,
destabilises carbocation,
deactivates the substrate

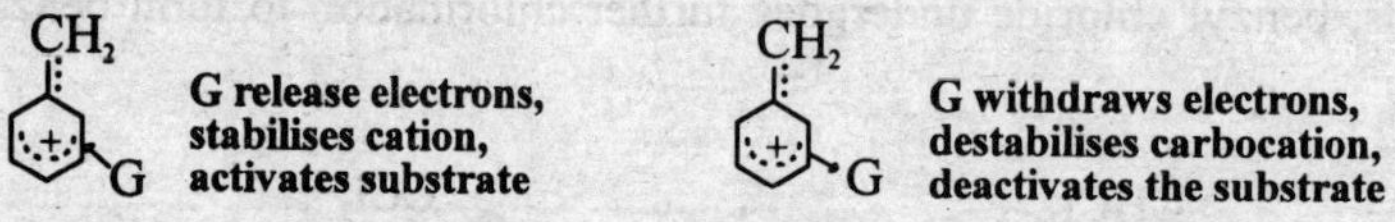

The effect is reversed in presence of an electron withdrawing group in the benzene ring. Thus, amongst p-methylbenzyl chloride, benzyl chloride and p-nitrobenzyl chloride, the order of reactivity towards nucleophilic substitution is as follows.

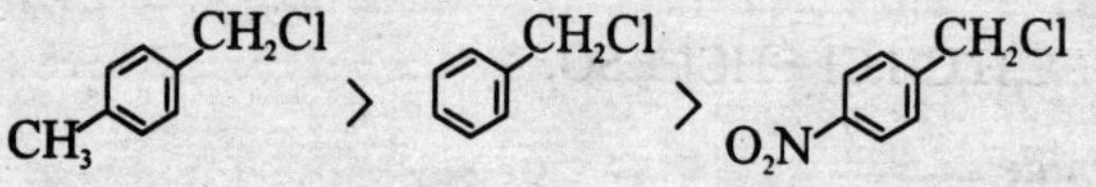

(a) *Displacement Reaction:* Benzyl chloride undergoes following displacement reactions.

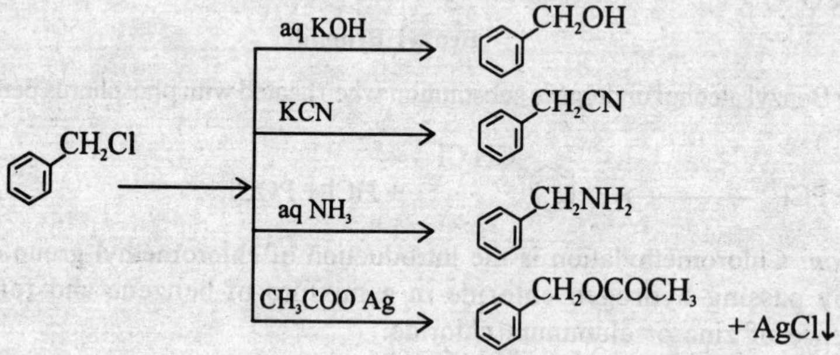

aq KOH → CH_2OH

KCN → CH_2CN

aq NH_3 → CH_2NH_2

$CH_3COO\,Ag$ → CH_2OCOCH_3 + AgCl↓

(b) *Fittig or Wurtz-Fittig Reaction:* Benzyl chloride undergoes reaction with alkyl halides (Wurtz) or aryl halides (Wurtz Fittig) in presence of metallic sodium in dry ether.

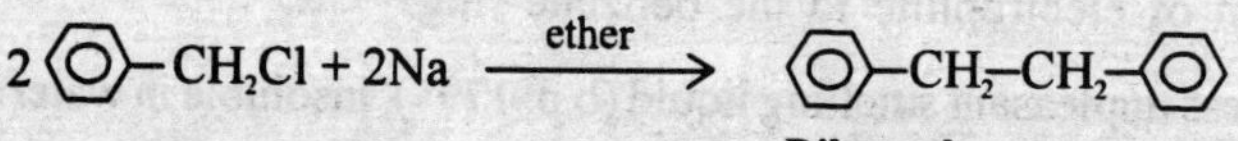

$2\;\bigcirc-CH_2Cl + 2Na \xrightarrow{ether} \bigcirc-CH_2-CH_2-\bigcirc$

Dibenzyl

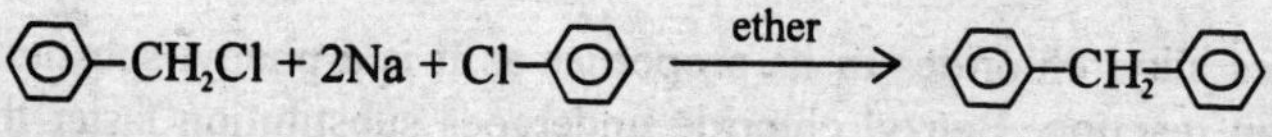

$\bigcirc-CH_2Cl + 2Na + Cl-\bigcirc \xrightarrow{ether} \bigcirc-CH_2-\bigcirc$

Diphenylmethane

(c) *Formation of Grignard Reagent:* Benzyl chloride forms Grignard reagent when refluxed with magnesium turning in dry ether.

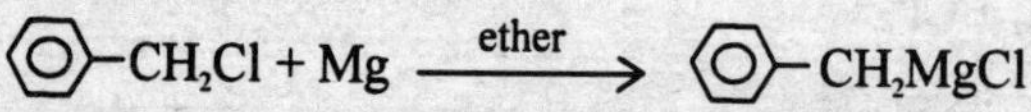

$\bigcirc-CH_2Cl + Mg \xrightarrow{ether} \bigcirc-CH_2MgCl$

(d) *Reduction:* When reduced with Zn-Cu couple it yields toluene.

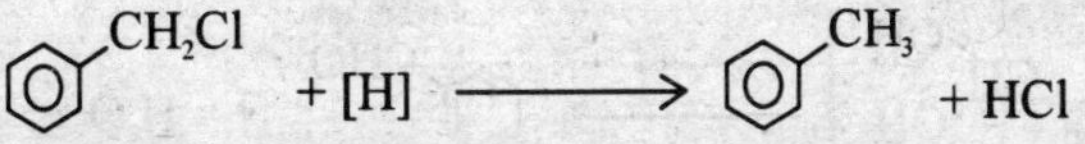

$$\text{C}_6\text{H}_5\text{CH}_2\text{Cl} + [\text{H}] \longrightarrow \text{C}_6\text{H}_5\text{CH}_3 + \text{HCl}$$

(e) *Oxidation:* When oxidised with nitric acid or alkaline potassium permanganate the $-\text{CH}_2\text{Cl}$ group oxidises to $-\text{COOH}$ to form benzoic acid (a reaction used to distinguish it from isomeric ortho, meta and para chlorotoluenes).

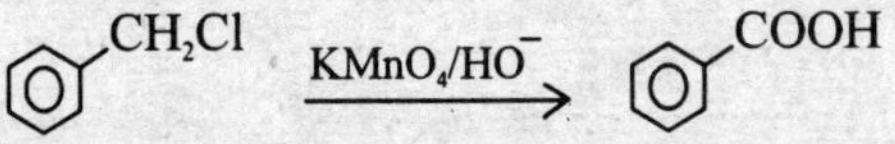

$$\text{C}_6\text{H}_5\text{CH}_2\text{Cl} \xrightarrow{\text{KMnO}_4/\text{HO}^-} \text{C}_6\text{H}_5\text{COOH}$$

Mild oxidising agents like cupric nitrate converts it into benzaldehyde.

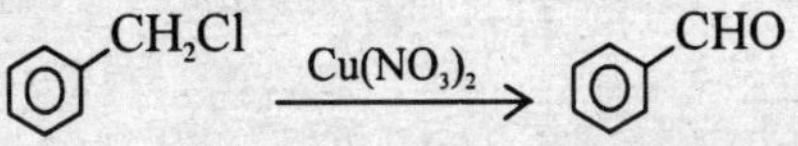

$$\text{C}_6\text{H}_5\text{CH}_2\text{Cl} \xrightarrow{\text{Cu(NO}_3)_2} \text{C}_6\text{H}_5\text{CHO}$$

Benzaldehyde

(f) *Ring Substitution:* Benzyl chloride undergoes electrophilic substitution directing the incoming groups at ortho and para positions. Nitration of benzyl chloride gives o- and p-nitrobenzyl chloride.

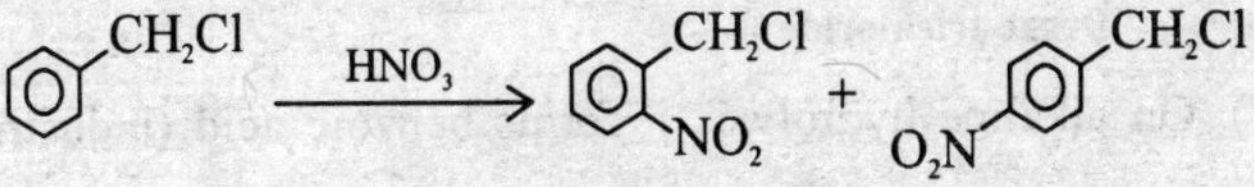

$$\text{CH}_2\text{Cl} \xrightarrow{\text{HNO}_3} \overset{\text{CH}_2\text{Cl}}{\underset{\text{NO}_2}{\bigcirc}} + \overset{\text{CH}_2\text{Cl}}{\underset{\text{O}_2\text{N}}{\bigcirc}}$$

Similarly, it undergoes sulphonation, to yield o- and p-substituted benzyl chloride. *Friedel Crafts alkylation* on the other hand, yields higher hydrocarbons instead of ring substitution when reacted with aryl halides.

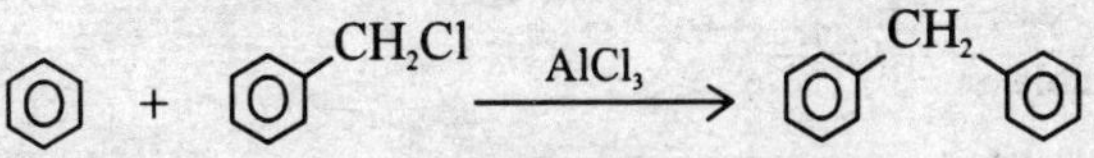

$$\bigcirc + \overset{\text{CH}_2\text{Cl}}{\bigcirc} \xrightarrow{\text{AlCl}_3} \bigcirc^{\text{CH}_2}\bigcirc$$

29.6.2 Benzal Chloride/Benzylidene Chloride

Preparation:

(a) Benzal chloride may be prepared by passing calculated amount of chlorine into boiling toluene in presence of light.

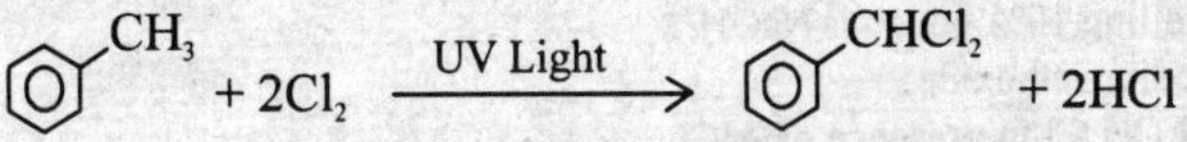

$$\text{C}_6\text{H}_5\text{CH}_3 + 2\text{Cl}_2 \xrightarrow{\text{UV Light}} \text{C}_6\text{H}_5\text{CHCl}_2 + 2\text{HCl}$$

Benzal chloride

(b) It may also be prepared by the reaction of phosphorus pentachloride on benzaldehyde.

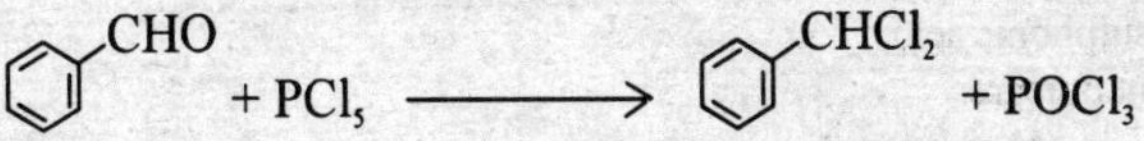

$$\text{C}_6\text{H}_5\text{CHO} + \text{PCl}_5 \longrightarrow \text{C}_6\text{H}_5\text{CHCl}_2 + \text{POCl}_3$$

Benzal chloride

Properties: It is a colorless liquid (b.p. 207°). It undergoes hydrolysis to form benzaldehyde, an industrial method of preparation of benzaldehyde.

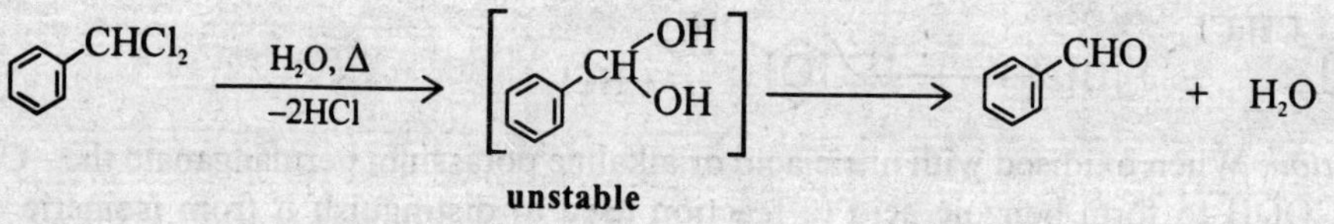

$$\underset{}{\text{C}_6\text{H}_5\text{CHCl}_2} \xrightarrow[-2\text{HCl}]{\text{H}_2\text{O}, \Delta} \left[\underset{}{\text{C}_6\text{H}_5\text{CH}} \begin{matrix} \text{OH} \\ \text{OH} \end{matrix} \right] \longrightarrow \underset{}{\text{C}_6\text{H}_5\text{CHO}} + \text{H}_2\text{O}$$

unstable

On oxidation with conventional oxidising agents it forms benzoic acid.

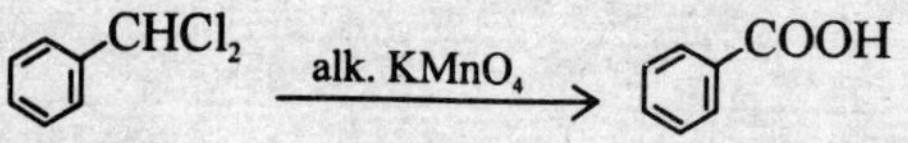

$$\underset{}{\text{C}_6\text{H}_5\text{CHCl}_2} \xrightarrow{\text{alk. KMnO}_4} \underset{}{\text{C}_6\text{H}_5\text{COOH}}$$

It undergoes typical electrophilic substitutions like nitration, sulphonation and halogenation to form ortho, meta and para substituted benzal chloride.

29.6.3   Benzo Trichloride/Phenylchloroform

It is prepared by passing chlorine in boiling toluene.

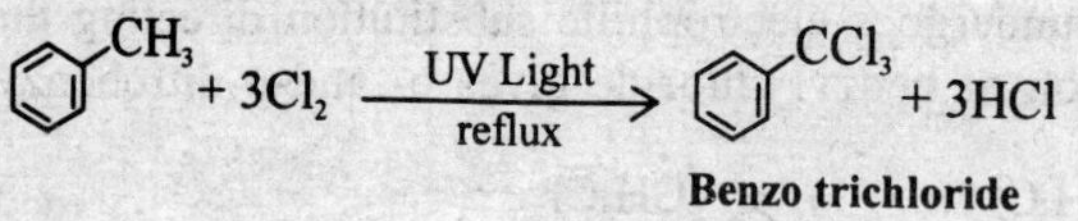

$$\underset{}{\text{C}_6\text{H}_5\text{CH}_3} + 3\text{Cl}_2 \xrightarrow[\text{reflux}]{\text{UV Light}} \underset{}{\text{C}_6\text{H}_5\text{CCl}_3} + 3\text{HCl}$$

Benzo trichloride

It is a colorless liquid (b.p. 214°). On alkaline hydrolysis it forms benzoic acid (industrial method of preparation of benzoic acid)

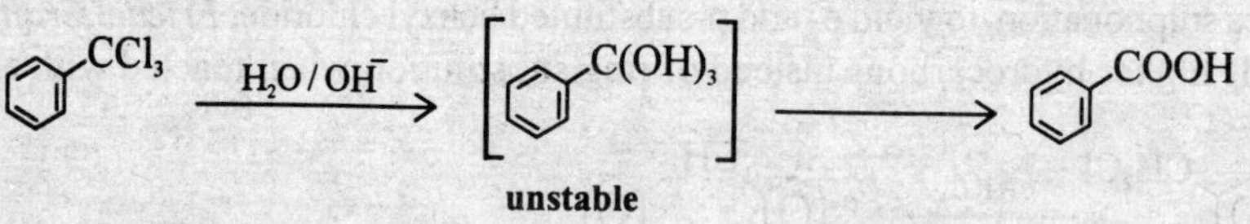

$$\underset{}{\text{C}_6\text{H}_5\text{CCl}_3} \xrightarrow{\text{H}_2\text{O}/\text{OH}^-} \left[\underset{}{\text{C}_6\text{H}_5\text{C(OH)}_3} \right] \longrightarrow \underset{}{\text{C}_6\text{H}_5\text{COOH}}$$

unstable

Benzoic acid is also formed by oxidation of benzo trichloride. It undergoes ring substitution to form mainly meta substituted benzo trichloride.

29.7   PROBLEMS

1. Give structure and name of the product of following reactions:
    (a) Bromobenzene with Cl_2 and Fe.
    (b) Bromobenzene with NH_3 at 100°.
    (c) 2,4-dinitrobromobenzene with boiling 10% aqueous NaOH.
    (d) 2,4-dinitrobromobenzene with sodium ethoxide.
    (e) Chlorobenzene with $CH_3CH_2CH_2CH_2Cl$ in presence of $AlCl_3$
2. How will you convert:
    (a) bromobenzene into aniline   (b) bromobenzene to 1,2,4-tribromobenzene
    (c) chlorobenzene to benzyl alcohol   (d) chlorobenzene to 2,4-dinitrophenol
    (e) chlorobenzene to p-chlorobenzenesulphonic acid
3. From benzene and toluene how will you prepare:
    (a) m-bromobenzoic acid   (b) p-bromobenzoic acid
    (c) 3,4-dibromonitrobenzene   (d) p-chlorobenzal chloride
    (e) p-bromostyrene   (f) p-bromobenzenesulphonic acid
4. What is the order of reactivity of following compounds towards NaOH:
    (a) chlorobenzene   (b) o-chloronitrobenzene
    (c) m-chloronitrobenzene   (d) 2,4-dinitrochlorobenezene
    (e) 2,4,6-trinitrochlorobenzene

5. What is the order of reactivity of following compounds towards nitrating mixture and why.
   (a) Benzene   (b) Toluene
   (c) Chlorobenzene   (d) Nitrobenzene
6. Which of following will undergo nucleophilic substitution fastest and why.
   (a) Chlorobenzene   (b) Benzyl chloride   (c) Ethyl chloride
7. Why chlorine atom in chlorobenzene is less reactive than in benzyl chloride?
8. Chlorobenzene does not give a precipitate with ammonical silver nitrate whereas tertiary butyl chloride does. why?
9. Write down the conditions required for following conversion.
   (a) Bromocyclohexane $\longrightarrow$ Cyclohexene.
   (b) p-Nitrotoluene $\longrightarrow$ p-Toluidine.
   (c) Phenyldiazonium chloride $\longrightarrow$ Fluorobenzene.
10. Two isomeric compounds A & B are represented by formula C_7H_7Cl. Oxidation of A produces carboxylic acid without chlorine whereas B produces a halogen substituted acid on oxidation. Deduce the structure of A and B.
11. Which chlorine atom of 1,2,4 -trichlorobenzene is most reactive and why?
12. Why 2-bromoanisole reacts with sodamide in ammonia but not 2-bromo-3-methylanisole.
13. Why ortho and meta-bromoanisole yield the same product i.e. m-anisidine, when treated with sodamide in liquid ammonia.
14. o-Chlorotoluene on reaction with sodamide in liquid ammonia produces a mixture of o-toluidine and m-toluidine. Explain.
15. What is the reaction of 2-pentene with N-bromosuccinimide in CCl_4.
16. How is delocalisation of odd electron in benzyl radical responsible for its stability?
17. What happens when toluene reacts with:
    (a) Cl_2 in presence of halogen carrier and   (b) Cl_2 in sunlight.

30

Aromatic Sulphonic Acids

The derivatives of aromatic hydrocarbons, in which one or more hydrogen atom of the aromatic ring are replaced by $-SO_3H$ function, are called aromatic sulphonic acids or arylsulphonic acids. They can also be referred to as derivatives of sulphuric acid in which one hydroxy group is replaced by an aryl group.

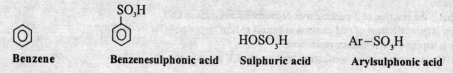

| Benzene | Benzenesulphonic acid | Sulphuric acid | Arylsulphonic acid |

$HOSO_3H$ — Sulphuric acid

$Ar-SO_3H$ — Arylsulphonic acid

30.1 NOMENCLATURE

The aromatic sulphonic acids are generally named by adding suffix *sulphonic acid* to the name of parent hydrocarbon.

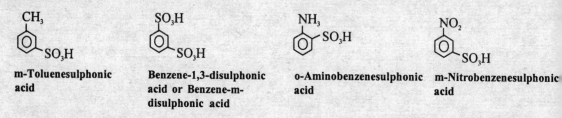

m-Toluenesulphonic acid

Benzene-1,3-disulphonic acid or Benzene-m-disulphonic acid

o-Aminobenzenesulphonic acid

m-Nitrobenzenesulphonic acid

30.2 GENERAL METHODS OF PREPARATION

1. *Sulphonation:* Direct sulphonation is one of the most convenient method of preparation of arylsulphonic acids. Treatment of aromatic hydrocarbons with concentrated sulphuric acid forms arylsulphonic acid, the number of sulphonic acid groups entering the nucleus depend upon temperature, concentration of acid and nature of substrate.

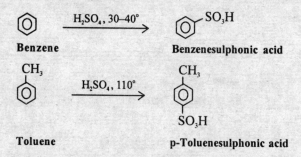

Benzene $\xrightarrow{H_2SO_4, 30-40°}$ Benzenesulphonic acid

Toluene $\xrightarrow{H_2SO_4, 110°}$ p-Toluenesulphonic acid

Sulphonation by concentrated sulphuric acid generally forms mixture of ortho and para monosulphonated products. With oleum, however, di- and trisulphonated products may be obtained.

In general, high temperature favors the formation of p-sulphonated product and low temperature favors o-sulphonated product.

Activating groups and halogens present in the ring orient sulphonic acid group at o-and p-positions whereas deactivating groups at m-position.

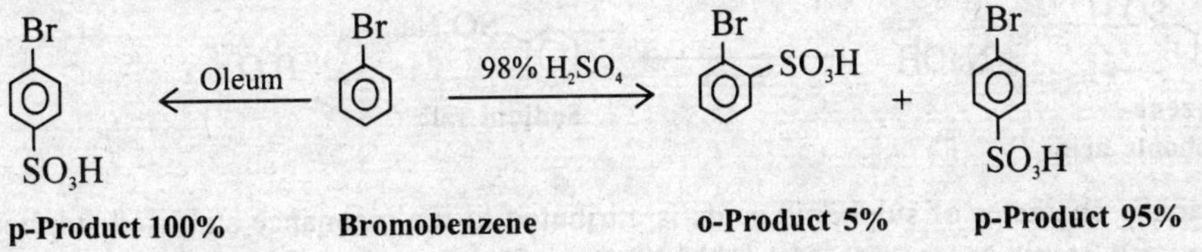

p-Product 100%   **Bromobenzene**   **o-Product 5%**   **p-Product 95%**

Di- and tri-benzenesulphonic acids are prepared using fuming sulphuric acid at high temperature.

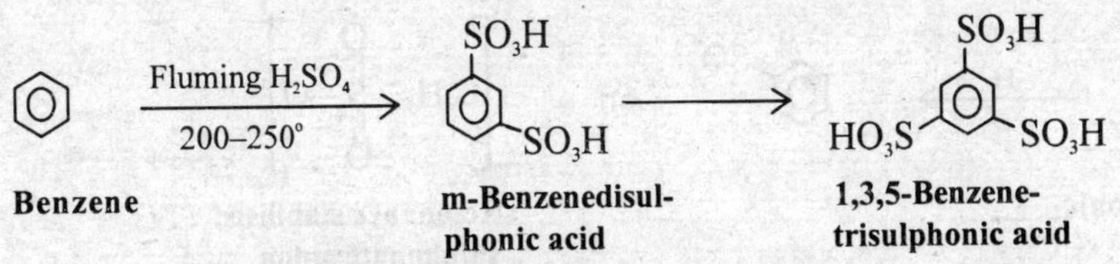

Benzene   **m-Benzenedisul-**   **1,3,5-Benzene-**
                phonic acid   **trisulphonic acid**

Sulphonation proceeds via electrophilic substitution discussed in Sec. 28.7.

2. *Action with Chlorosulphonic Acid:* Aromatic hydrocarbons when treated with chlorosulphonic acid at room temperature forms arylsulphonylchloride which on hydrolysis gives arylsulphonic acid.

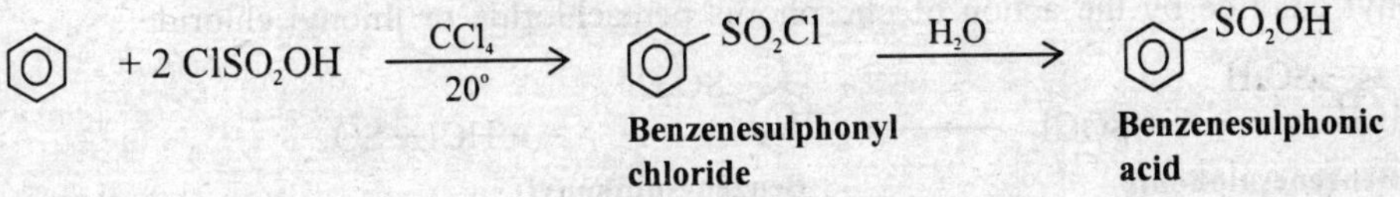

Benzenesulphonyl   **Benzenesulphonic**
chloride   **acid**

3. *Oxidation of Thiophenols:* The oxidation of thiophenols with alkaline potassium permanganate yields arylsulphonic acids.

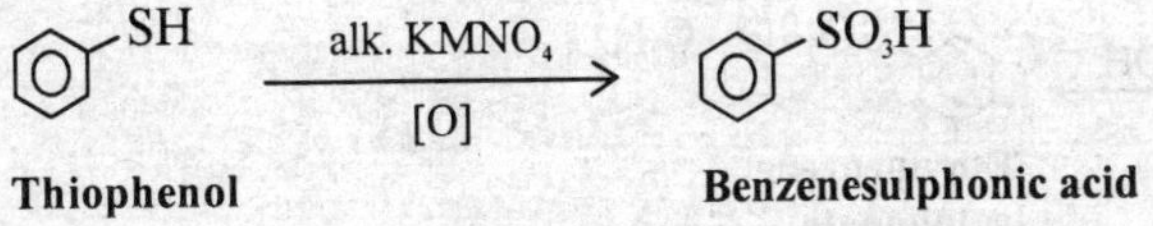

Thiophenol   **Benzenesulphonic acid**

Isolation of free sulphonic acids from reaction mixture is very difficult due to their high solubility in water and nonvolatility. To overcome this problem following methods are employed for their isolation.

(i) Salting out method in which the reaction mixture is run into a saturated solution of sodium chloride resulting in the precipitation of arylsulphonic acid as its sodium salt.

(ii) In another method, arylsulphonic acid is separated from the reaction mixture by neutralising the acid with barium hydroxide or lead carbonate. The precipitated sulphates are filtered off and the filtrate is treated with calculated amount of sulphuric acid and the resulting mixture evaporated to dryness under reduced pressure to separate out arylsulphonic acid.

30.3 PROPERTIES

Physical Properties: Aromatic sulphonic acids are generally colorless, crystalline, hygroscopic substance having sour taste. They are freely soluble in water and are strongly acidic in nature. They are nonvolatile and   have lower melting points than those of the corresponding carboxylic acids.

Chemical Properties: Arylsulphonic acids undergo reactions due to $-SO_3H$ group and benzene nucleus.

(i)   *Acidic Character:* Arylsulphonic acids are strong acids with acid strength about the same as that of sulphuric acid. They form stable water soluble salts called sulphonates, with inorganic bases.

Benzene-
sulphonic acid

Sodium salt

The strong acidic character of sulphonic acids is attributed to the resonance stablised sulphonate anion having three oxygen atoms to accommodate the negative charge.

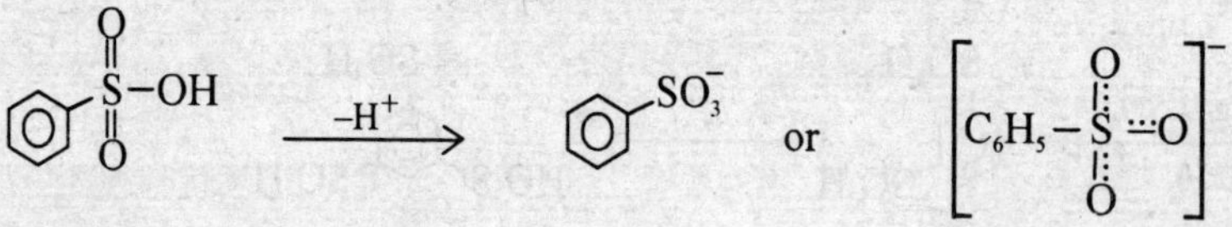

Benzenesulphonic
acid

Resonance stabilised
sulphonate anion

Electronegative groups increase the acidic character of sulphonic acids particularly when they are present at ortho and para positions.

(ii)   *Formation of Sulphonyl Chloride, Ester and Anhydride:* Arylsulphonic acids are converted into aryl sulphonyl chloride by the action of phosphorus pentachloride or thionyl chloride.

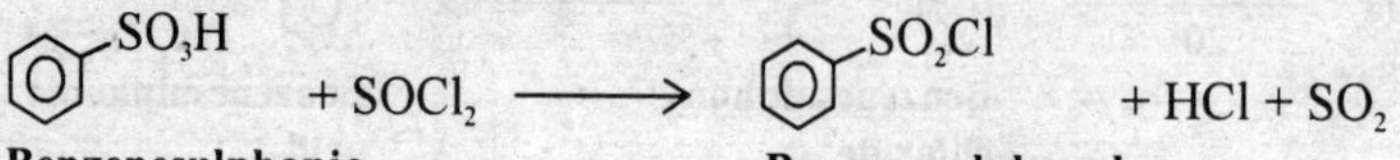

Benzenesulphonic
acid

Benzenesulphonyl
chloride

Arylsulphonyl chloride forms esters when treated with alcohols in presence of alkali.

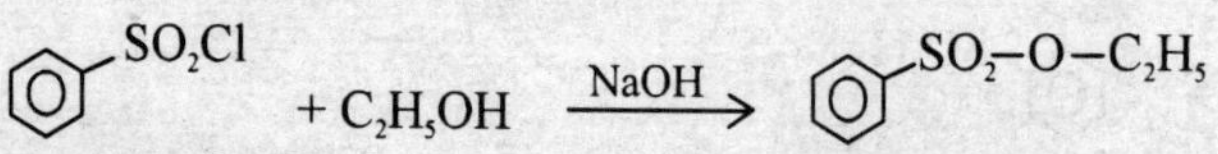

Benzenesulphonyl
chloride

Ethylbenzene
sulphonate

Arylsulphonyl chlorides reacts readily with ammonia to form sulphonamide.

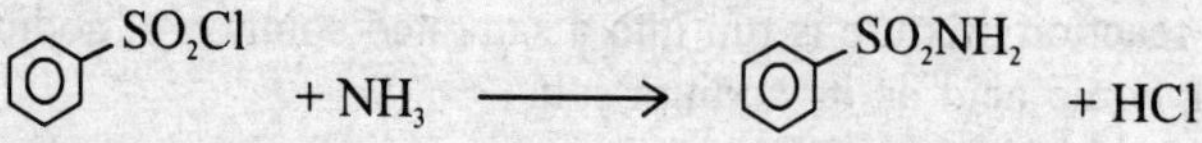

Benzenesulphonyl
chloride

Phenyl
sulphonamide

Arylsulphonic acids when heated with excess of phosphorus pentachloride undergo dehydration forming anhydride.

$$ArSO_3H + PCl_5 \xrightarrow{150°} (ArSO_2)_2O + H_2O$$

Arylsulphonic
acid

Arylsulphonic
anhydride

(iii) *Formation of Sulphones:* Arylsulphonic acids when react with hydrocarbons form neutral compounds called sulphones.

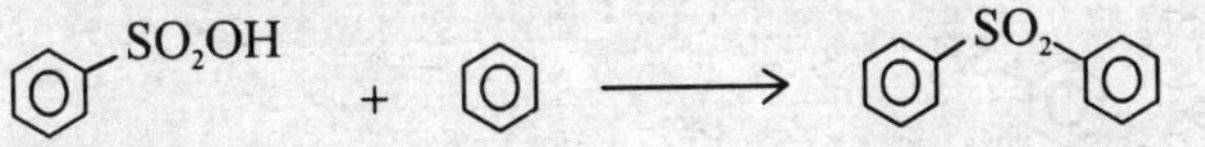

Benzenesulphonic acid **Benzene** **Diphenyl sulphone**

(iv) *Desulphonation:* Arylsulphonic acids when heated to 100-175° with aqueous acid, are converted into sulphuric acid and an aromatic hydrocarbon. This reaction is known as desulphonation.

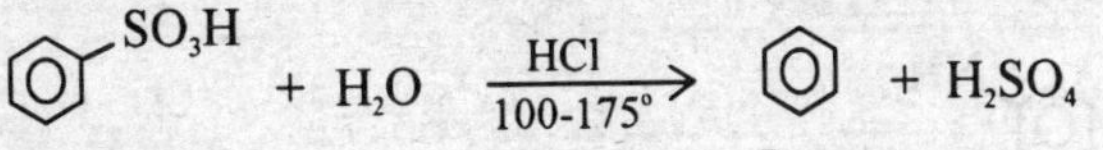

Benzene-sulphonic acid **Benzene**

The reaction is reversible and the mechanism of desulphonation is the reverse of that of sulphonation. Here, the electrophile is the proton, H^+, thus it is also referred to as protodesulphonation. The reaction is favoured by high concentration of water, high temperature and removal of relatively nonvolatile hydrocarbon by steam distillation.

$$C_6H_5SO_3^- + H^+ \rightleftharpoons C_6H_5^+(H)(SO_3^-)$$

$$C_6H_5^+(H)(SO_3^-) \rightleftharpoons C_6H_6 + SO_3$$

(v) *Formation of Phenols:* When sodium salt of sulphonic acids are heated with fused sodium hydroxide, they form phenols.

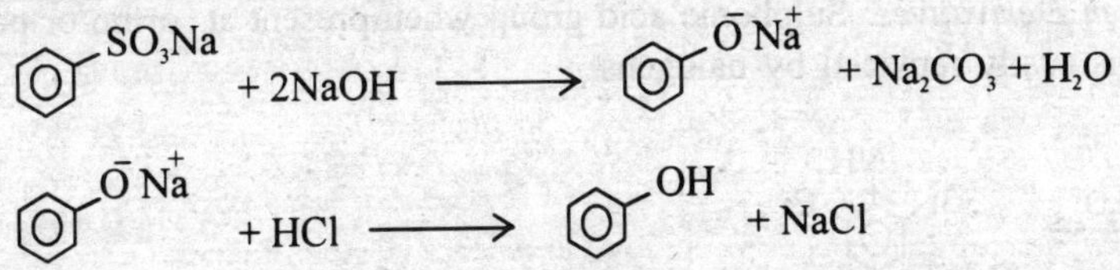

Phenol

(vi) *Formation of Cyanides:* When treated with cyanides sodium sulphonates form aryl cyanides.

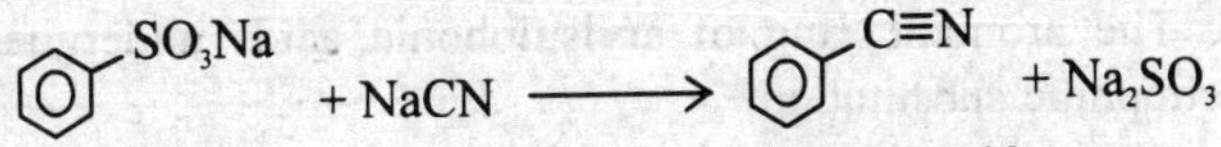

Phenyl cyanide

(vii) *Formation of Carboxylic Acids:* When treated with sodium formate, sodium benzenesulphonate forms sodium salt of benzoic acid which on acidification forms free benzoic acid.

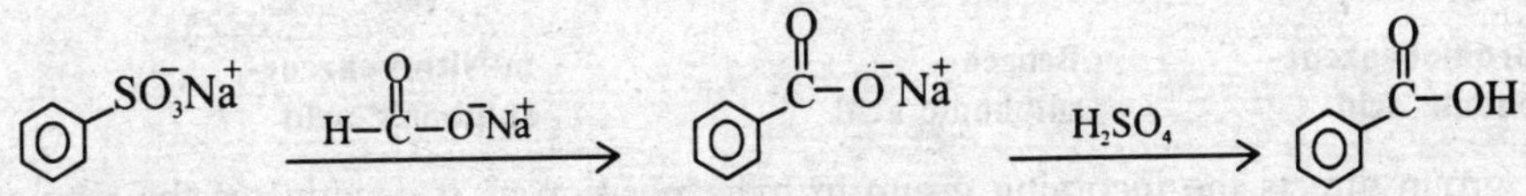

Sodium benzene-sulphonate **Sodium benzoate** **Benzoic acid**

(viii) *Formation of Thiophenols:* When heated with potassium hydrogen sulphide sodium benzene-sulphonate forms thiophenols.

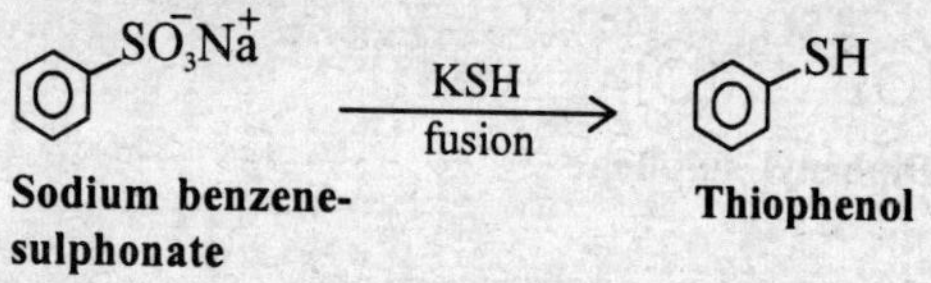

Sodium benzene-sulphonate   Thiophenol

(ix) *Formation of Aromatic Amines:* Aromatic amines are formed when sodium sulphonates are treated with sodamide.

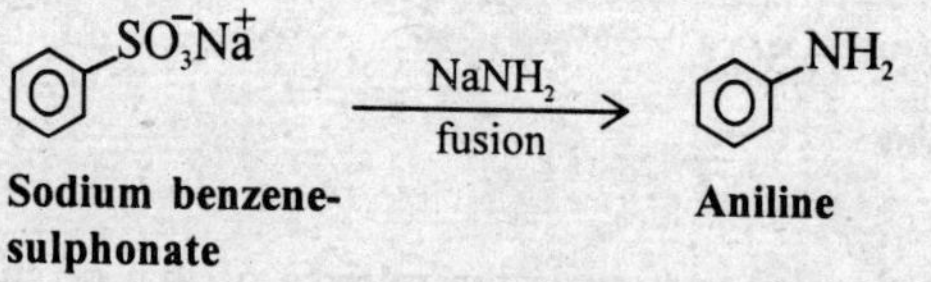

Sodium benzene-sulphonate   Aniline

(x) *Formation of Nitro Derivatives:* Sulphonic acid group is replaced by nitro group if it is present at ortho or para position to phenolic or amino group by treatment with nitric acid.

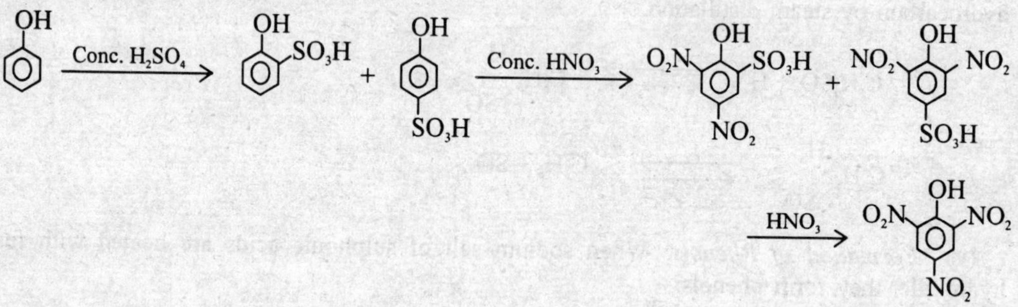

(xi) *Formation of Halogen Derivatives:* Sulphonic acid group when present at ortho or para position to phenolic or amino group is easily replaced by halogens.

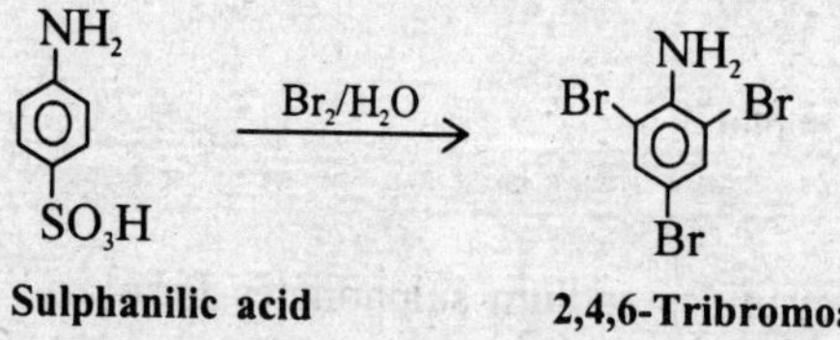

Sulphanilic acid   2,4,6-Tribromoaniline

(xii) *Ring Substitution Reaction:* The aromatic ring of arylsulphonic acids undergoes nitration, sulphonation and halogenation via electrophilic substitution.

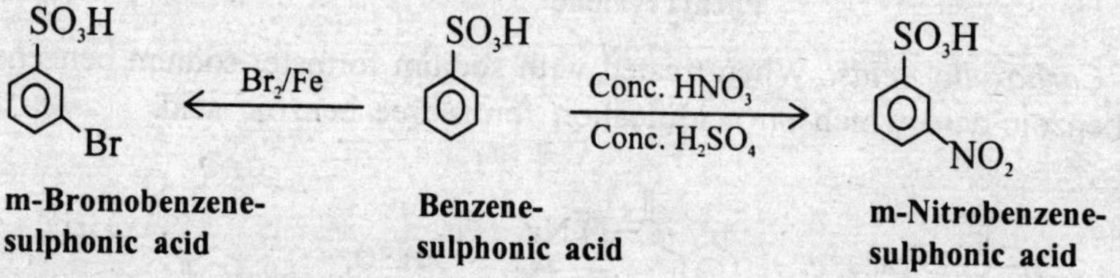

m-Bromobenzene-sulphonic acid   Benzene-sulphonic acid   m-Nitrobenzene-sulphonic acid

The $-SO_3H$ group directs the incoming group to meta position as it deactivates the ring at ortho and para position. It withdraws electrons from the ring via inductive and mesomeric effect.

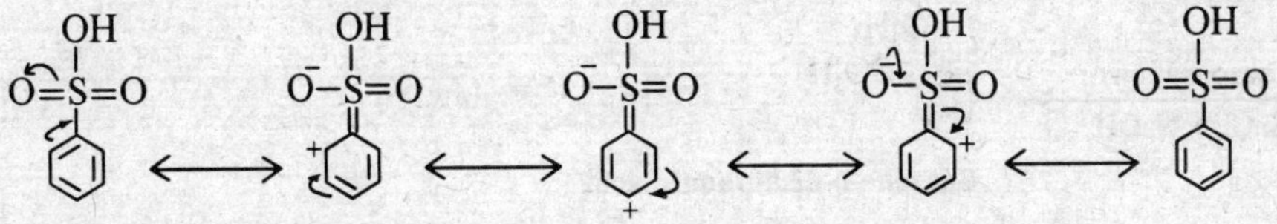

Uses

1. Arylsulphonic acids are used as valuable synthetic reagents as –SO₃H group can be replaced by other groups easily.
2. Sodium p-dodecylbenzenesulphonate is used as synthetic detergent.
3. Their high molecular weight barium or calcium salts are used as additive in lubricating oils for internal combustion engines.
4. They are used as acid catalyst in a number of reactions due to their strong acidic and mild oxidising properties.

30.4   INDIVIDUAL MEMBERS AND THEIR DERIVATIVES

30.4.1   Benzenesulphonic Acid

Preparation: It is most conveniently prepared by heating benzene with conc. sulphuric acid or oleum at 80°

Benzenesulphonic acid

Properties: Benzenesulphonic acid forms colorless crystals with $1\frac{1}{2}$ molecules of water, m.p. 44°, whereas the anhydrous products melt at 51°. It is deliquescent and highly soluble in water. The aqueous solution is highly acidic as that of sulphuric acid. It exhibits all the characteristic properties of arylsulphonic acids discussed in Sec. 30.3.

30.4.2   Benzenedisulphonic Acids

Three isomeric benzenedisulphonic acids (o, m and p) are possible.

Preparation: Benzene-m-disulphonic acid may be prepared by heating benzene with excess of fuming sulphuric acid at 180-200°.

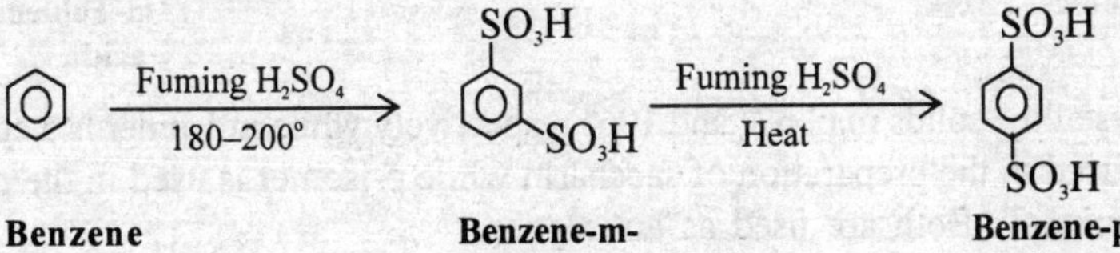

Benzene   **Benzene-m-disulphonic acid**   **Benzene-p-disulphonic acid**

The m-isomer on prolonged heating with fuming sulphuric acid forms p-isomer. The o-isomer, however, is prepared by indirect method.

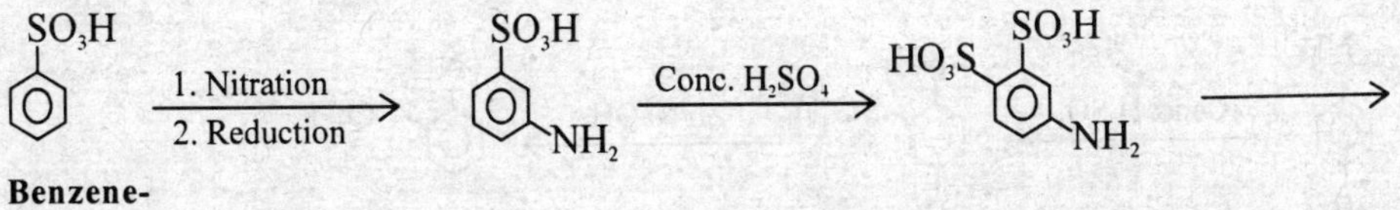

Benzene-sulphonic acid

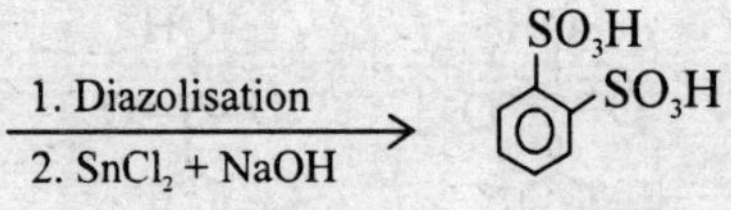

Benzene-o-disulphonic acid

Reactions: When fused with caustic potash both meta and para benzenedisulphonic acids form resorcinol (m-dihydroxybenzene) while the o-isomer forms catachol (o-dihydroxybenzene).

30.4.3   Toluenesulphonic Acids

Toluene on sulphonation forms o-and p-toluenesulphonic acids in almost equal amounts. If the reaction is carried out above 100°, p-isomer predominates whereas at room temperature o-isomer is the main product.

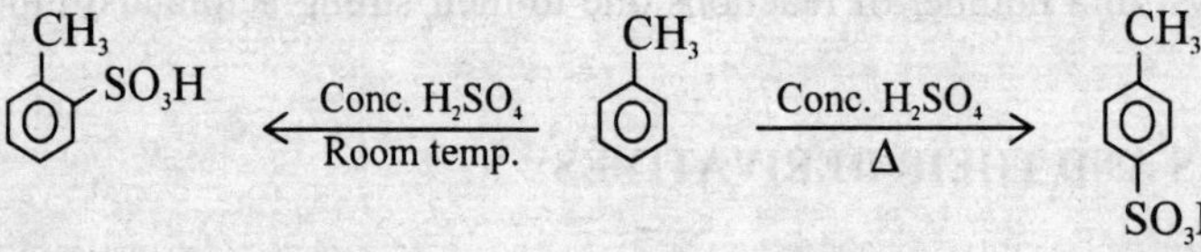

o-Toluenesulphonic acid   **Toluene**   **p-Toluenesulphonic acid**

o-Toluenesulphonic acid may also be prepared in good yield by sulphonating toluene with chlorosulphonic acid at room temperature followed by hydrolysis.

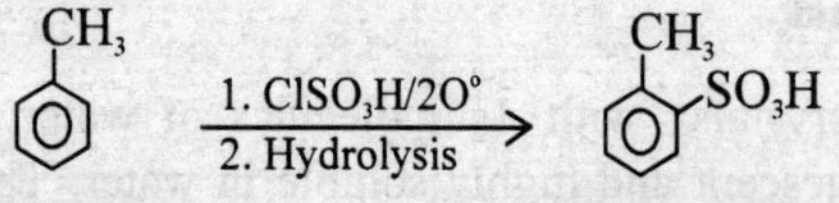

Toluene   **o-Toluenesulphonic acid (85%)**

m-Toluenesulphonic acid may be prepared by indirect methods.

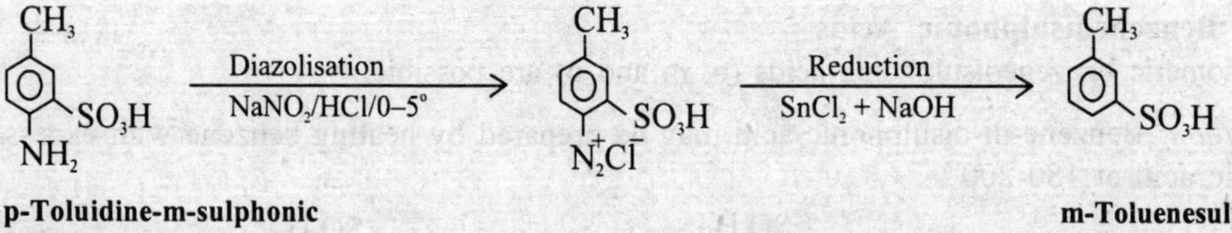

p-Toluidine-m-sulphonic acid   **m-Toluenesulphonic acid**

Both o-and p-isomers are crystalline solids m.p. 67° and 106° respectively while m-isomer is a colorless oil.
o-Toluenesulphonic acid is used in the preparation of saccharin while p-isomer is used in the preparation of chloramine-T and dichloramine-T. Both are used as antiseptics.

30.4.4   Aminosulphonic Acids

Orthanilic or *o-Aminobenzenesulphonic* acid is prepared by sulphonation of p-bromoaniline followed by removal of bromo group with zinc dust and alkali.

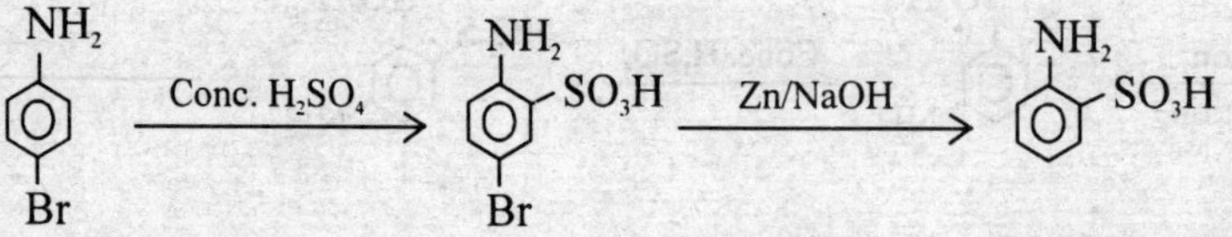

p-Bromoaniline   **Orthanilic acid**

It is a crystalline solid which isomerises to its p- isomer i.e. sulphanilic acid on heating with concentrated sulphuric acid.

Metanilic or *m-Aminobenzenesulphonic* acid may be prepared from benzenesulphonic acid by following sequence of reactions.

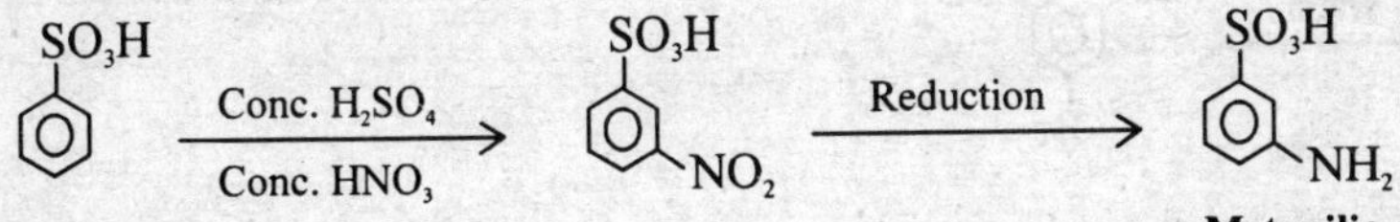

SO₃H structure:

Benzene-
sulphonic
acid

$\xrightarrow[\text{Conc. HNO}_3]{\text{Conc. H}_2\text{SO}_4}$

SO₃H / NO₂ structure

$\xrightarrow{\text{Reduction}}$

SO₃H / NH₂ structure

Metanilic acid

It is a crystalline solid used in the manufacture of azodyes.

Sulphanilic or *p-Aminobenzenesulphonic* acid is prepared on commercial scale by heating aniline and sulphuric acid at 200°.

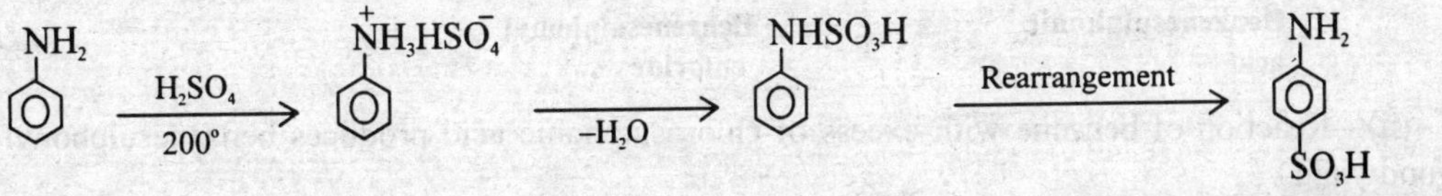

NH₂ structure

Aniline

$\xrightarrow[200°]{\text{H}_2\text{SO}_4}$

$\overset{+}{\text{N}}\text{H}_3\text{HS}\bar{\text{O}}_4$ structure

$\xrightarrow{-\text{H}_2\text{O}}$

NHSO₃H structure

$\xrightarrow{\text{Rearrangement}}$

NH₂ / SO₃H structure

Sulphanilic
acid

It is a white crystalline solid, insoluble in cold water but soluble in small proportions in hot water. It melts at 288° with decomposition.

The sulphanilic acid has an acid group (–SO₃H) and a basic group (–NH₂). The interaction of the two groups present in the molecule forms internal salt or dipolar ion called the zwitterion.

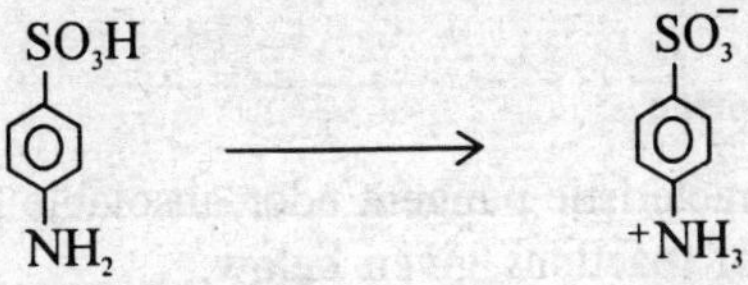

SO₃H / NH₂ structure

Sulphanilic acid

$\longrightarrow$

SO₃⁻ / ⁺NH₃ structure

Zwitterion

Sulphanilic acid dissolves in aqueous sodium hydroxide forming water soluble salt.

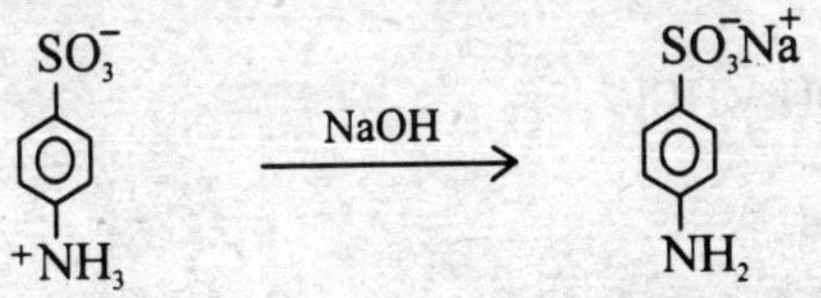

SO₃⁻ / ⁺NH₃ structure

$\xrightarrow{\text{NaOH}}$

SO₃⁻Na⁺ / NH₂ structure

Sodium-p-aminobenzene
sulphonate

Sulphanilic acid when heated with concentrated nitric acid forms p-nitroaniline (nitro group replaces –SO₃H).

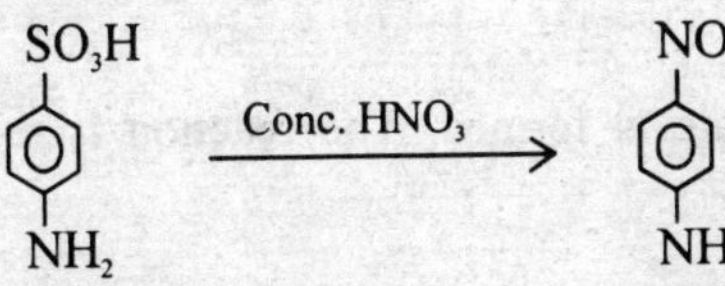

SO₃H / NH₂ structure

$\xrightarrow{\text{Conc. HNO}_3}$

NO₂ / NH₂ structure

Sulphanilic acid   **p-Nitroaniline**

The sulphonic acid group of sulphanilic acid is replaced by bromo group when it is treated with bromine water to give 2,4,6-tribromoaniline.

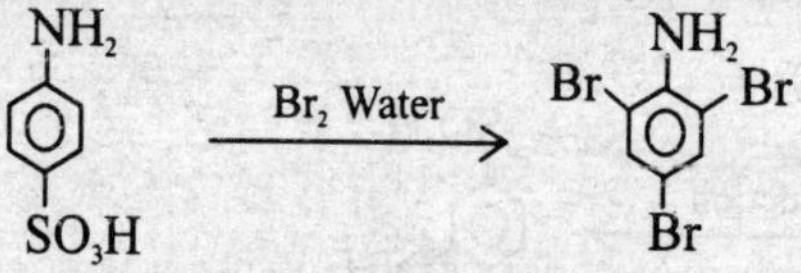

It is used as an important dye intermediate.

30.4.5   Benzenesulphonyl Chloride

Preparation:

(i)   It may be prepared by the reaction of benzenesulphonic acid with phosphorus pentachloride.

$$C_6H_5SO_2OH + PCl_5 \longrightarrow C_6H_5SO_2Cl + POCl_3 + HCl$$

Benzenesulphonic   **Benzenesulphonyl**
acid   **chloride**

(ii)   Reaction of benzene with excess of chlorosulphonic acid produces benzenesulphonyl chloride in good yield.

$$C_6H_6 + 2ClSO_3H \longrightarrow C_6H_5SO_2Cl + HCl + H_2SO_4$$

Benzene   Chlorosul-   **Benzenesul-**
phonic acid   **phonyl chloride**

(iii)   It may also be prepared by the reaction of benzene and sulphuryl chloride in presence of AlCl₃.

$$C_6H_6 + Cl-SO_2-Cl \xrightarrow{AlCl_3} C_6H_5SO_2Cl + HCl$$

Benzenesul-
phonyl chloride

Properties: It is a colorless liquid, b.p. 250°, having a characteristic pungent odor, insoluble in water but soluble in most organic solvents. It undergoes a number of reactions given below.

(i)   *Hydrolysis:* It decomposes slowly in cold water but the reaction rate is enhanced by the addition of alkali at elevated temperature.

$$C_6H_5SO_2Cl + H_2O \xrightarrow{NaOH} C_6H_5SO_2OH + HCl$$

Benzenesulphonyl   **Benzenesul-**
chloride   **phonic acid**

(ii)   *Reaction with Alcohols:* With alcohols, benzenesulphonyl chloride forms sulphonic esters in presence of pyridine.

$$C_6H_5SO_2Cl + ROH \xrightarrow{Pyridine} C_6H_5SO_2OR + HCl$$

Benzenesulphonyl   **Alkyl benzene-**
chloride   **sulphonate**

With p-toluenesulphonyl chloride (tosyl chloride), tosylate is formed. The reaction is referred to as tosylation.

(iii)   *Reaction with Ammonia and Amines:* Benzenesulphonyl chloride when reacts with ammonia forms benzenesulphonamide.

$$C_6H_5SO_2Cl + NH_3 \xrightarrow{150°} C_6H_5SO_2NH_2 + HCl$$

Benzenesulphonyl
chloride

Benzene-
sulphonamide

Similarly reaction with primary and secondary amines yield N-alkylsulphonamide and N,N-dialkylsulphonamide respectively.

$$C_6H_5SO_2Cl + RNH_2 \longrightarrow C_6H_5SO_2{-}NHR + HCl$$

Benzene-   pri. Amine
sulphonyl chloride

N-Alkylsulphonamide

$$C_6H_5SO_2Cl + R_2NH \longrightarrow C_6H_5SO_2{-}NR_2 + HCl$$

Benzene-   sec. Amine
sulphonyl chloride

N-N-Dialkylsulphonamide

The sulphonamides are crystalline compounds having sharp melting points. This reaction, therefore, forms the basis of Hinsberg's method for the separation and identification of amines.

(iv)   *Reduction:* Mild reducing agents convert benzenesulphonyl chloride into benzenesulphinic acid whereas strong reducing agents such as lithium aluminum hydride forms thiophenol.

$$HCl + C_6H_5SO_2H \xleftarrow{Zn + HCl} C_6H_5SO_2Cl \xrightarrow{LiAlH_4} C_6H_5SH + HCl + H_2O$$

Benzenesul-
phinic acid

Benzene-
sulphonyl chloride

Thiophenol

(v)   *Friedel Crafts Reaction:* Aromatic hydrocarbons react with benzenesulphonyl chloride in presence of Lewis acid to form sulphones.

$$C_6H_6 + C_6H_5SO_2Cl \xrightarrow{AlCl_3} C_6H_5{-}SO_2{-}C_6H_5$$

Benzenesulphonyl
chloride

Diphenylsulphone

30.4.6   Saccharin/o-Sulphobenzoicimide

It is the cyclic imide of o-sulphobenzoic acid. It is a colorless crystalline solid, m.p. 224°, insoluble in water but its sodium salt is freely soluble in water. It is about 500-600 times sweeter than sugar but has no calorific value hence given to diabetic patients.

Preparation of saccharin involve following steps.

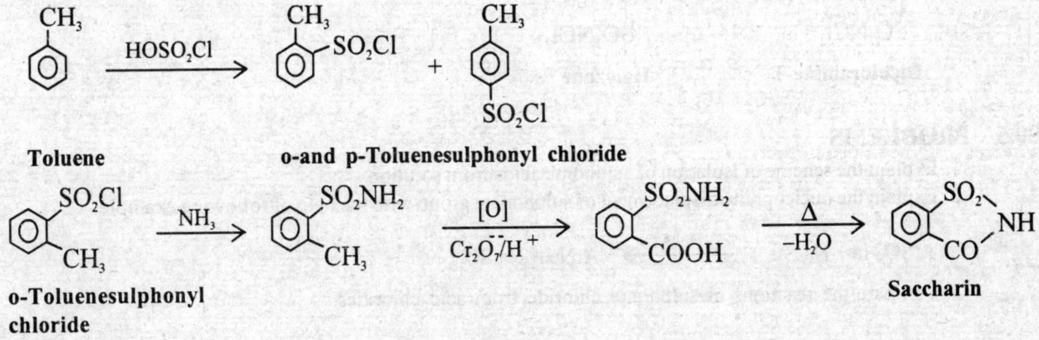

Toluene

o-and p-Toluenesulphonyl chloride

o-Toluenesulphonyl
chloride

Saccharin

Another commercial method makes use of anthranilic acid.

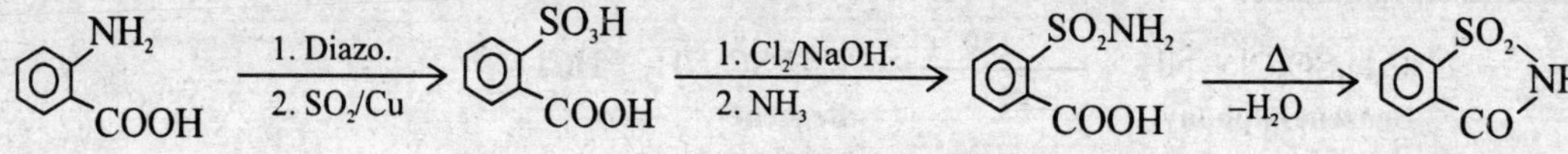

Anthranilic acid   **Saccharin**

It is very sweet in dilute solutions but concentrated solutions of saccharin are bitter in taste. It is used as sweetening agents.

30.4.7   Chloramine-T and Dichloramine-T

Chloramine-T is sodium salt of N-chloro-p-toluenesulphonamide whereas dichloramine-T is N,N-dichloro-p-toluenesulphonamide. Both of them are prepared from p-toluenesulphonamide which in turn is produced from toluene as follows.

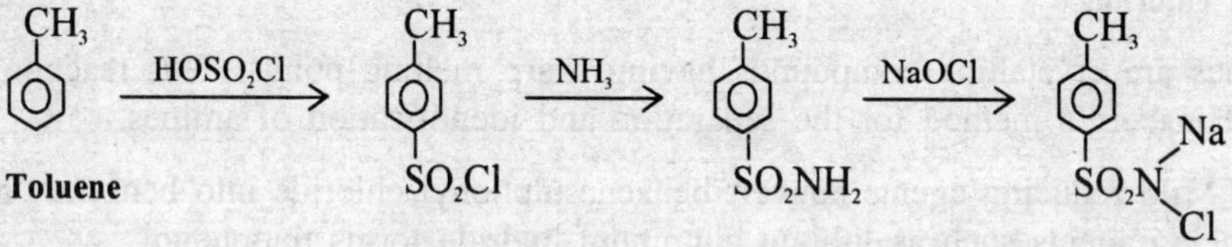

Toluene   **p-Toluene-**   **Chloramine**
             sulphonyl chloride

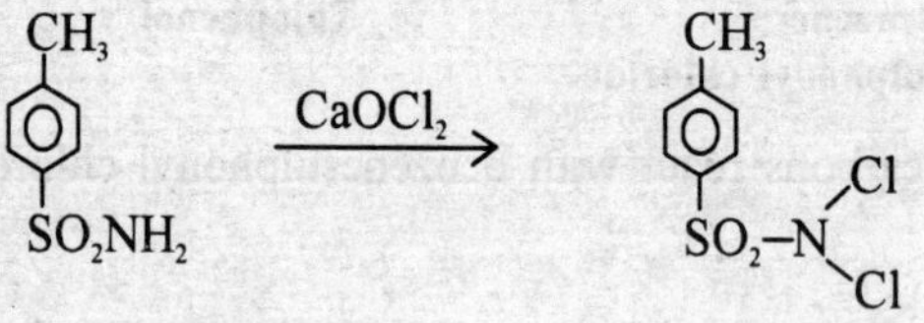

p-Toluenesulphonamide   **Dichloramine**

Chloramine-T is white solid, soluble in water containing 12.5% active chlorine. Its dilute aqueous solution is used in mouth wash and for washing wounds. It is a good disinfectant.

Dichloroamine-T is a yellow solid, sparingly soluble in water containing 28-30% active chlorine and is used as effective antiseptic. It forms halozone on oxidation which is used for the disinfection of drinking water.

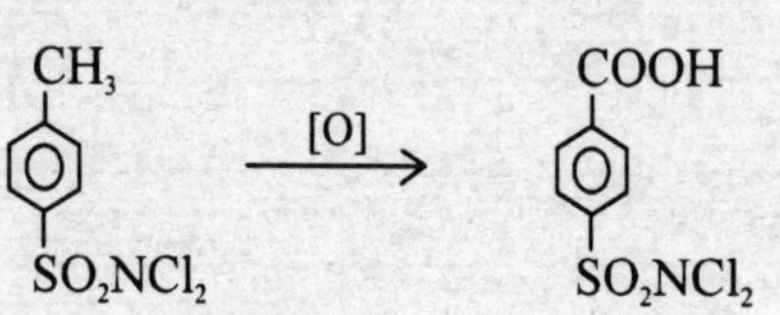

Dichloramine-T   **Halozone**

30.5   PROBLEMS

1. Explain the scheme of isolation of sulphonic acids from solution.
2. Explain the nucleophilic displacement of sulphonate group with the help of following example:

$$ArSO_3Na + \overset{+}{Na}\overset{-}{Nu} \xrightarrow{\text{Heat}} ArNu + Na_2SO_3$$

3. Compare the reactions of sulphonyl chloride from acid chlorides.

4. Design the synthesis of o-bromotoluene from toluene making use of sulphonation and desulphonation.

5. Identity the following:

Compound A $C_9H_{11}O_2SCl$ $\xrightarrow[\text{H}_3\text{O}^+]{\text{NaOH}}$ Compound B water soluble $\xrightarrow{\text{H}_3\overset{+}{\text{O}}/\Delta}$ Compound C C_9H_{12}

Compound A $\xrightarrow{\text{aq. AgNO}_3}$ AgCl ↓

Compound C $\xrightarrow{\text{Br}_2/\text{Fe}}$ Compound D a monobromo derivative

6. How will you prepare
   (a) p-Methylthiophenol from benzene?   (c) sym-Tribromoaniline from sulphanilic acid?
   (b) o-Methylanisol from toluene?   (d) Picric acid from phenol?

7. How will you prepare chloramine-T and Saccharin?

8. Discuss the reversible reaction between benzene and sulphuric acid.

9. What is the effect of temperature and concentration of acid on the orientation of sulphonic acid group.

10. Discuss the use of benzenesulphonyl chloride or p- toluenesulphonyl chloride in the separation of primary, secondary and tertiary amines.

11. What happens when
    (a) Benzenesulphonic acid reacts with bromine in presence of iron bromide?
    (b) Benzenesulphonyl chloride reacts with methanamine?
    (c) Benzenesulphonic acid reacts with PCl_5?
    (d) Benzenesulphonic acid reacts with dilute HCl under pressure at 200°?

12. Explain why benzenesulphonic acid is less acidic than sulphuric acid but more acidic than carboxylic acids?

31

Aromatic Nitro Compounds

Aromatic nitro compounds are the derivatives of nitric acid having one hydroxy group replaced by an aromatic ring. They may either be nuclear or side chain nitro compounds.

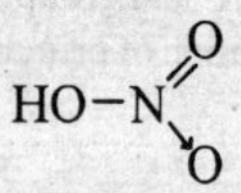

Nitric acid

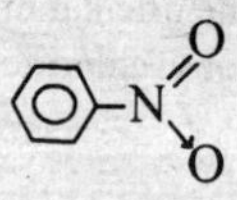

Nitrobenzene
Nuclear nitro compound

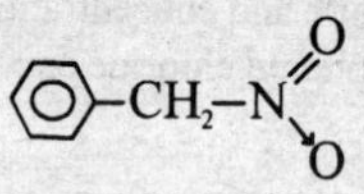

Phenyl nitromethane
Side chain nitro compound

The nuclear nitro compounds have synthetic importance and are used for the synthesis of various pharmaceuticals, vitamins explosives etc, whereas, side chain nitro compounds resemble aliphatic nitro compounds discussed in Chapter 27.

31.1 NOMENCLATURE

Aromatic nitro compounds are nitro substitution on the parent hydrocarbon therefore named as nitro hydrocarbons. The position of nitro group is indicated by prefix o, m, p or by numerals.

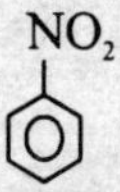

Nitrobenzene

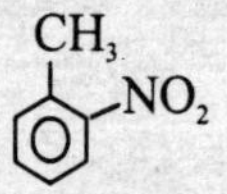

o-Nitrotoluene
or 2-Nitrotoluene

p-Nitrotoluene
or 4-Nitrotoluene

When two or more nitro groups are present in a compound the position as well as the number is prefixed. For example,

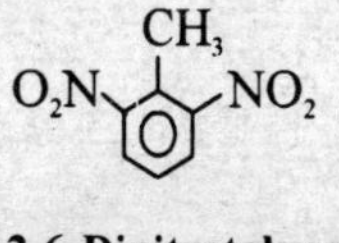

2,6-Dinitrotoluene

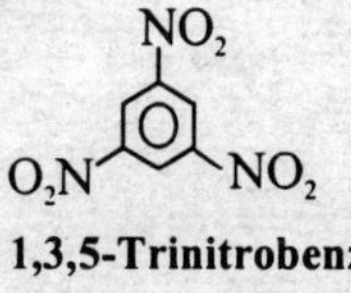

1,3,5-Trinitrobenzene

31.2 GENERAL METHODS OF PREPARATIONS

1. *Direct Nitration:* It is the most commonly used method for the introduction of nitro group into

n aromatic ring. The reaction involves refluxing of hydrocarbon with nitric acid and sulphuric acid.

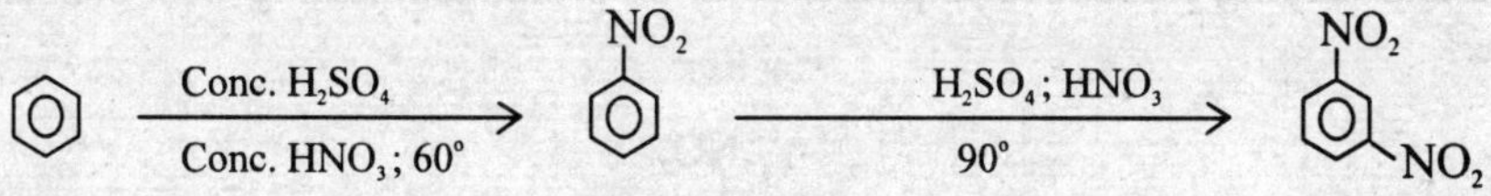

Benzene   **Nitrobenzene**   **m-Dinitrobenzene**

The nitration may also be done by using other reagents like pyridinium nitrate in presence of pyridine or acetyl nitrate. The later one, although explosive, has the advantage of introducing nitro group at ortho position to the group already present in the hydrocarbon. Toluene produces o-nitrotoluene exclusively with acetyl nitrate.

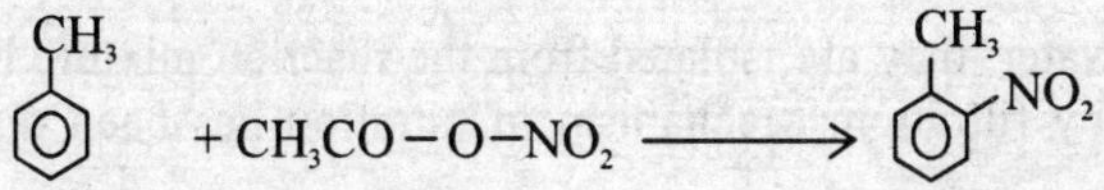

Toluene   **Acetyl nitrate**   **o-Nitrotoluene**

Nitration may also be done by the reaction of appropriate hydrocarbon with liquid or gaseous nitrogen peroxide in presence of anhydrous aluminum chloride.

Boron trifluoride catalyses the nitration process for the yields are very high and compound prepared is quite pure.

Another very useful nitrating agent for the nitration is nitronium tetrafluoroborate.

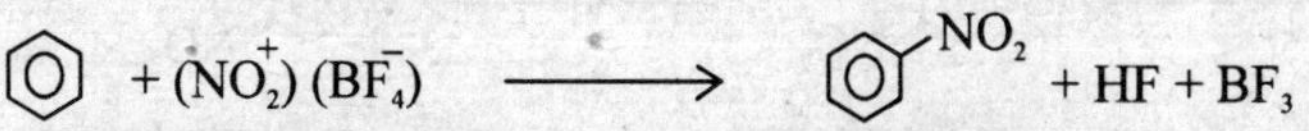

Nitronium   **Nitrobenzene**
tetrafluoroborate

The method gives a direct preparative proof of nitration involving electrophilic substitution (nitronium ion in nitronium tetrafluoroborate has electrophilic character).

The extent of nitration varies with temperature, nature of nitrating agent and the nature of the substrate. Low temperature favors mononitration whereas high temperature favors dinitro derivative formation.

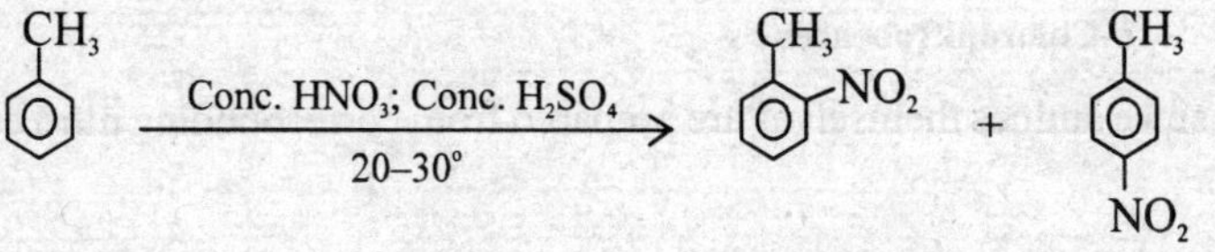

Toluene   **o-Nitrotoluene**   **p-Nitrotoluene**

Toluene at low temperature produces mononitrotoluene whereas by raising the temperature of the reaction mixture, 2,6- and 2,4-dinitrotoluene is produced.

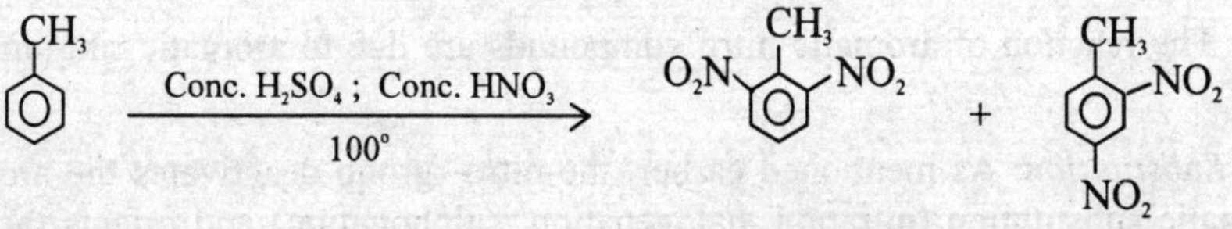

Toluene   **2,6-Dinitrotoluene**   **2,4-Dinitrotoluene**

Concentrated nitric acid produces mono and dinitro derivatives whereas fuming nitric acid and nitrating mixtures (nitric acid and varying amount of sulphuric acid) on prolonged heating with benzene produce sym-trinitrobenzene.

Benzene   sym-Trinitrobenzene

Nature of substrate governs the orientation of nitro group and has been discussed in Sec. 28.8. Electron releasing groups direct the nitro group to ortho and/or para positions whereas electron withdrawing groups to meta position.

Since nitro compounds are insoluble in water, they are isolated from the reaction mixture by pouring in water and separating the solid product by filtration. Mechanism of nitration has been discussed in Sec. 28.7.

2. *From Aromatic Amines via Diazotisation:* Reaction of amines with nitrous acid produces diazonium salt which is converted into the nitro derivative on treatment with sodium nitrite and copper powder. m-Dinitrobenzene may be prepared from m-nitroaniline.

m-Nitroaniline   m-Nitrobenzene   m-Dinitrobenzene
                            diazonium sulphate

3. *Oxidation of Aromatic Amines:* Aromatic amines undergo oxidation to aromatic nitro compounds with persulphuric acid ($H_2S_2O_8$) or peroxytrifluoroacetic acid (CF_3CO_3H).

o-Chloroaniline   o-Chloronitrobenzene

The methods is not very useful since amines themselves are prepared from corresponding nitro compounds.

31.3 PROPERTIES

Physical Properties: Most of the nitro compounds are yellow crystalline solids, except for a few like nitrobenzene and nitrotoluenes which are yellow liquids. Many of them are steam volatile and except for a few mononitro derivatives, can not be distilled under atmospheric pressure as they decompose with explosive violence on strong heating. They are insoluble in water and soluble in organic solvents.

Chemical Properties: The reaction of aromatic nitro compounds are due to aromatic ring and the nitro group.

(1) *Electrophilic Substitution*: As mentioned earlier, the nitro group deactivates the aromatic ring for electrophilic aromatic substitution (nitration, halogenation, sulphonation) and orients the incoming group to meta position. Nitration of nitrobenzene requires drastic conditions i.e., high temperature and nitrating mixture.

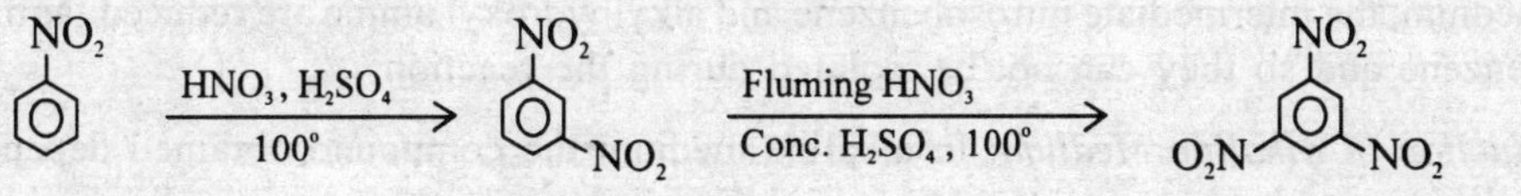

Nitrobenzene ──HNO₃, H₂SO₄ / 100°──→ m-Dinitrobenzene ──Fluming HNO₃ / Conc. H₂SO₄ , 100°──→ sym-Trinitrobenzene

The introduction of third nitro group to prepare sym- trinitrobenzene requires extremely drastic conditions i.e., heating with fuming nitric acid and concentrated sulphuric acid.

(2) *Nucleophilic Substitution*: Nitro group undergoes nucleophilic substitution when heated with aqueous potassium hydroxide.

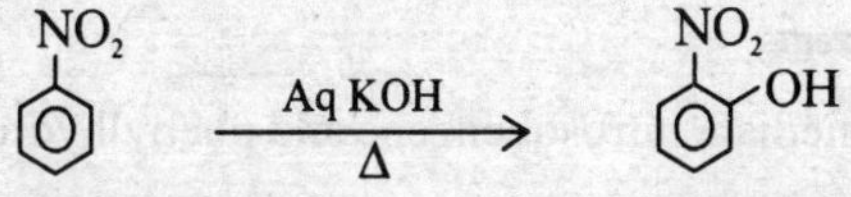

Nitrobenzene ──Aq KOH / Δ──→ o-Nitrophenol

When two nitro groups are present at ortho or para position to each other, one of them undergoes nucleophilic displacement.

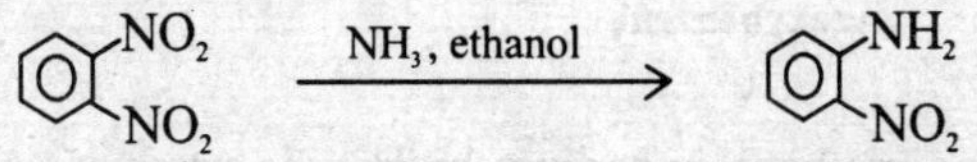

o-Dinitrobenzene ──NH₃, ethanol──→ o-Nitroaniline

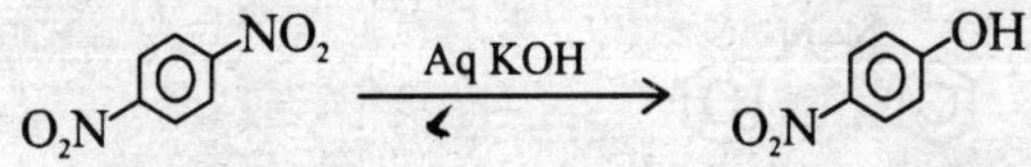

p-Dinitrobenzene ──Aq KOH──→ p-Nitrophenol

(3) *Reduction of Nitro Group*: Aromatic nitro compound is reduced to give a variety of products (I to III) depending upon the nature of reducing agent.

$$Ar{-}NO_2 \longrightarrow Ar{-}NO \longrightarrow Ar{-}NHOH \longrightarrow Ar{-}NH_2$$

Aromatic nitro compound	Aromatic nitroso compound	Aromatic hydroxyl amine	Aromatic amines
	I	II	III

(i) *Reduction in Acidic Medium*: The reduction of aromatic nitro compounds with metals like iron, zinc, titanium, tin and hydrochloric acid produces aromatic amines.

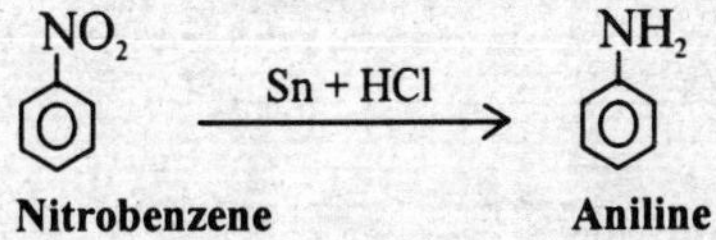

Nitrobenzene ──Sn + HCl──→ Aniline

Acetic acid is used with metals to reduce o-nitroacetanilide to o-aminoacetanilide as the former one undergoes hydrolysis also with stronger acid.

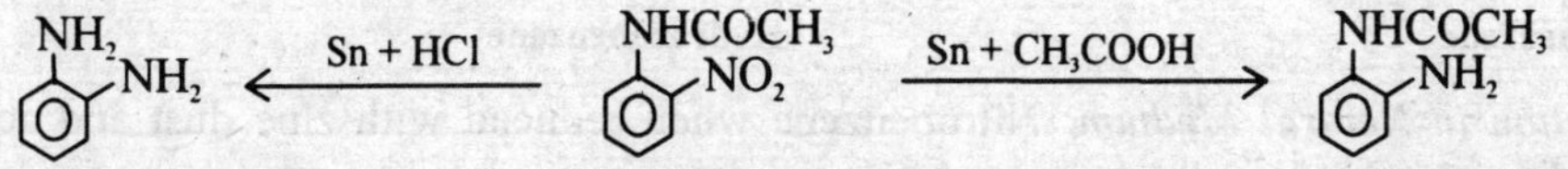

o-Aminoaniline ←──Sn + HCl── o-Nitroacetanilide ──Sn + CH₃COOH──→ o-Aminoacetanilide

In acid medium, the intermediate nitrosobenzene and alkylhydroxyl amine are reduced far more rapidly than nitrobenzene and so they can not be isolated during the reaction.

(ii) *Reduction in Alkaline Medium*: In alkaline medium the compound obtained depend upon the nature of reducing agent.

(a)   Sodium arsenite in aqueous sodium hydroxide reduces nitrobenzene to azoxybenzene.

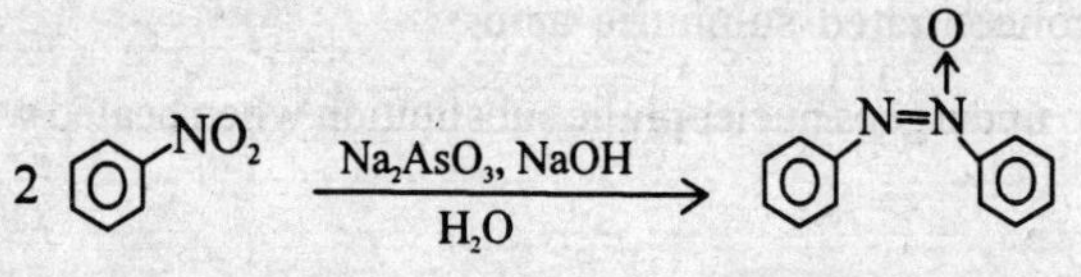

Nitrobenzene   Azoxybenzene

This is believed to be formed by the reaction of intermediate nitrosobenzene and phenylhydroxyl amine.

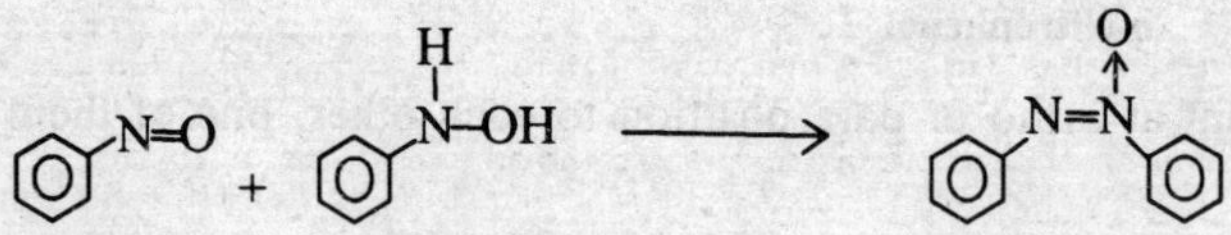

Nitroso-   N-Phenylhydroxyl   Azoxybenzene
benzene   amine

(b)   Nitrobenzene when reduced with zinc and methanolic sodium hydroxide produces azobenzene.

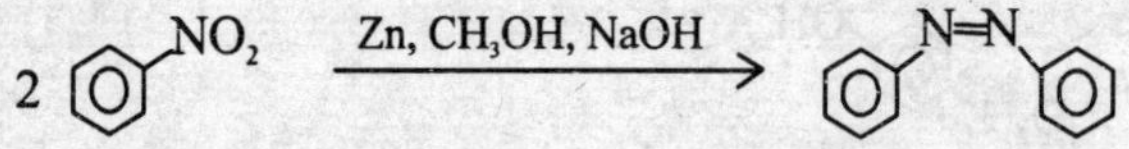

Azobenzene

This is believed to be formed by the reaction between two molecules of phenylhydroxylamine.

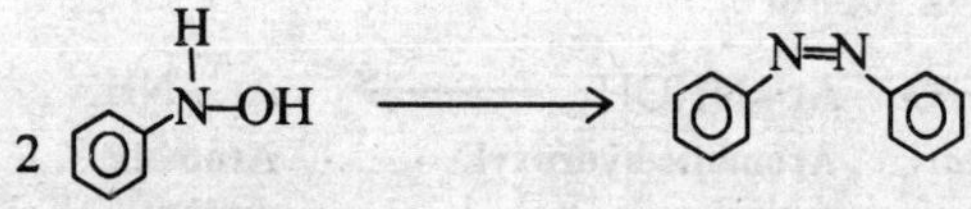

Phenylhydroxyl amine   Azobenzene

Azobenzene is also formed by reduction of azoxybenzene with iron.

(c)   Zinc in sodium hydroxide converts nitrobenzene directly to hydrazobenzene.

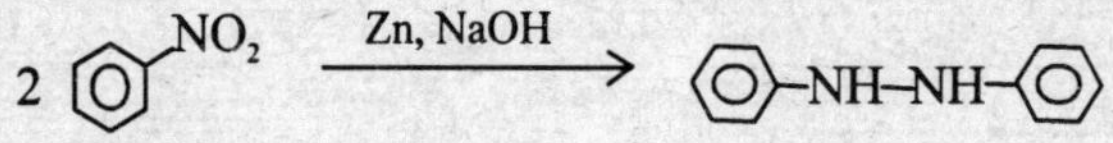

Nitrobenzene   Hydrazobenzene

This is believed to be formed by the reduction of azobenzene.

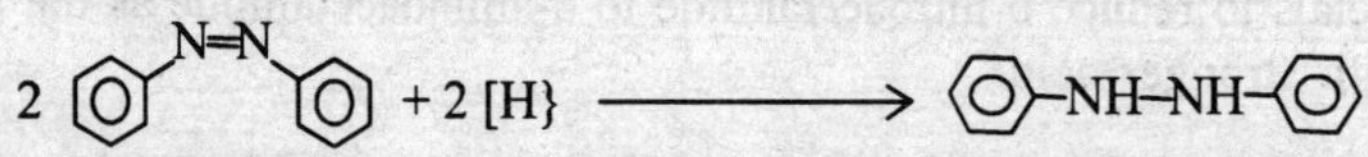

Azobenzene   Hydrazobenzene

(iii) *Reduction in Neutral Medium:* Nitrobenzene when reduced with zinc dust and aqueous ammonium chloride yields N-phenylhydroxylamine.

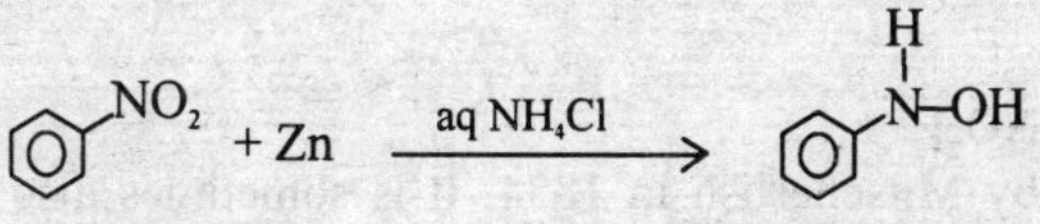

Nitrobenzene + Zn $\xrightarrow{\text{aq NH}_4\text{Cl}}$ N-Phenylhydroxyl amine

(iv) *Electrolytic Reduction*: Electrolytic reduction of nitrobenzene in acid or alkaline medium takes place through the following stages.

$$C_6H_5NO_2 \longrightarrow C_6H_5NO \longrightarrow C_6H_5NHOH \longrightarrow C_6H_5NH_2$$

Nitrobenzene   Aniline

In weakly acidic solution, the main product is aniline but in strongly acidic conditions p-aminophenol is formed by the rearrangement of phenylhydroxylamine.

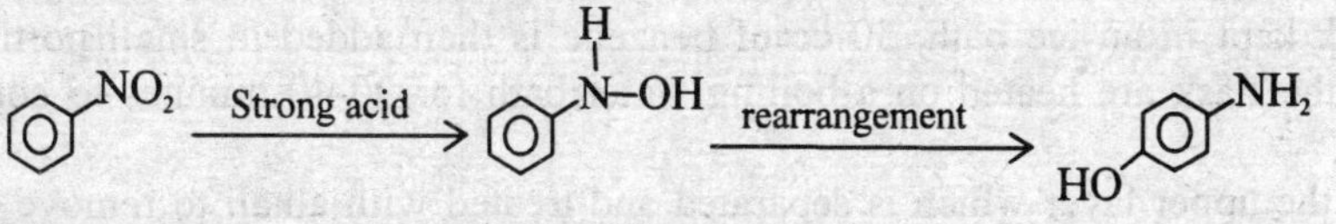

Nitrobenzene $\xrightarrow{\text{Strong acid}}$ N-Phenylhydroxyl amine $\xrightarrow{\text{rearrangement}}$ p-Aminophenol

In alkaline medium, the product obtained are as follows:

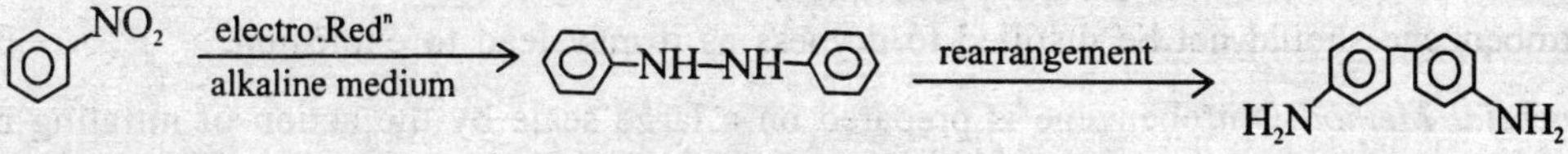

Nitrobenzene $\xrightarrow[\text{alkaline medium}]{\text{electro.Red}^n}$ NH–NH $\xrightarrow{\text{rearrangement}}$ Benzidine

(v) *Catalytic Reduction:* Reduction of nitrobenzene with hydrogen in presence of metal catalyst (platinum or nickel) produces aniline.

Lithium aluminum hydride, on the other hand, reduces nitrobenzene to azobenzene.

(vi) *Selective Reduction*: It is possible to reduce selectively only one nitrogroup in a compound having two or more nitro groups by sodium or ammonium sulphide.

m-dinitrobenzene yields m-nitroaniline when reduced with ammonium sulphide.

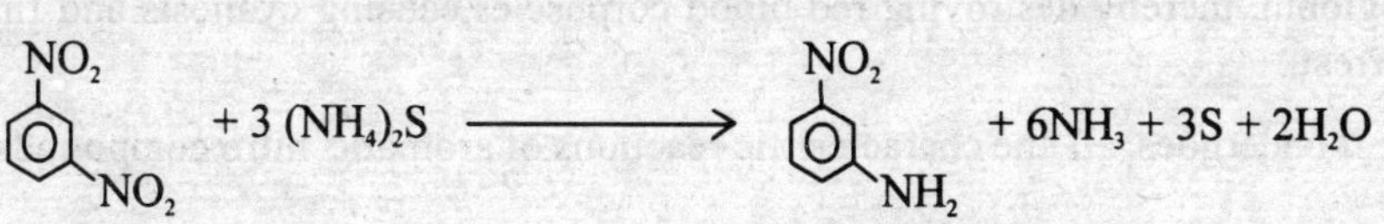

m-Dinitrobenzene + 3 $(NH_4)_2S \longrightarrow$ m-Nitroaniline + $6NH_3$ + 3S + $2H_2O$

The nitro group in o-nitrostyrene is reduced to amino group without affecting the unsaturated side chain by alkaline ferrous sulphate.

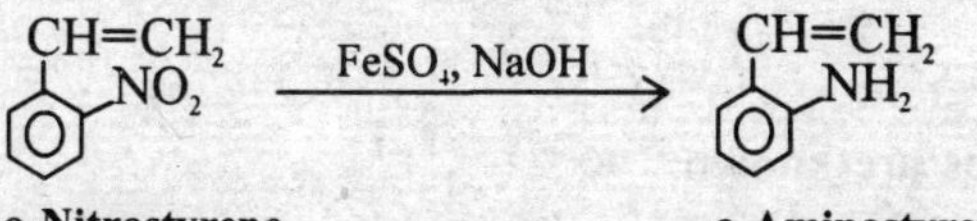

o-Nitrostyrene $\xrightarrow{\text{FeSO}_4,\ \text{NaOH}}$ o-Aminostyrene

The nitro group of o-nitrobenzaldehyde is also reduced similarly to amino group without affecting the aldehydic side chain.

31.4   INDIVIDUAL MEMBERS

31.4.1   Nitrobenzene/Oil of Mirbane [$C_6H_5NO_2$]

Nitrobenzene was first prepared and named by Mitscherlish in 1834. It is sometimes also known as artificial oil of bitter almond (or oil of mirbane) because of its odor like that of bitter almond.

Preparation

Laboratory Method: In laboratory it is prepared by heating a mixture of nitric acid and sulphuric acid with benzene below 70°.

$$C_6H_6 + HONO_2 \xrightarrow[70°]{\text{Conc. } H_2SO_4,} C_6H_5NO_2 + H_2O$$

Benzene   **Nitrobenzene**

Concentrated nitric acid (60 cc) and concentrated sulphuric acid (80 cc) are mixed together slowly in a round bottom flask kept in an ice bath. 50 cc of benzene is then added in small portions in the flask. The contents of the flask are heated on a boiling water bath for 30-40 minutes to complete the reaction.

Nitrobenzene forms the upper layer which is separated and treated with alkali to remove the excess of acid. The lower nitrobenzene layer is then separated, washed with water, dried over anhydrous calcium chloride and distilled using air condenser. The fraction distilling between 207-211° is nitrobenzene which may be redistilled for further purification.

Nitrobenzene should not be distilled to dryness as it may lead to explosion.

Commercial Method: Nitrobenzene is prepared on a large scale by the action of nitrating mixture on benzene. The method of preparation is the same as the laboratory method but differs only in the nature of reactor. Instead of a round bottom flask, nitration is carried out here in large cast iron pans fitted with stirrers which are externally cooled with cold water.

Properties

Physical Properties: It is a pale yellow oily liquid, b.p. 211°, with the odor of bitter almond. It is heavier than water and is insoluble in it but dissolves in organic solvents. It is volatile in steam and can dissolve aluminum chloride into it, thus, used as a solvent in Friedel-Craft reaction. It is highly toxic, forms complex with heamoglobin, thereby destroying red blood corpuscles, causing cyanosis and finally death due to respiratory arrest.

Chemical Properties: It undergoes all the characteristic reactions of aromatic nitro compounds discussed in Section 31.3.

Uses

1. It is mainly used in the manufacture of aniline and aniline dyes.
2. It is also used as high boiling industrial solvent.
3. As an oxidising agent in organic reactions.

31.4.2   Dinitrobenzenes [$C_6H_4(NO_2)_2$]

The three isomeric (o, m and p) dinitrobenzenes are known.

o-Dinitrobenzene

Preparation: It may be isolated from the mother liquor after removing m-dinitrobenzene when nitrobenzene is nitrated.

Nitration of acetanilide provides a mixture o-and p- nitroacetanilides which may be separated by steam distillation.

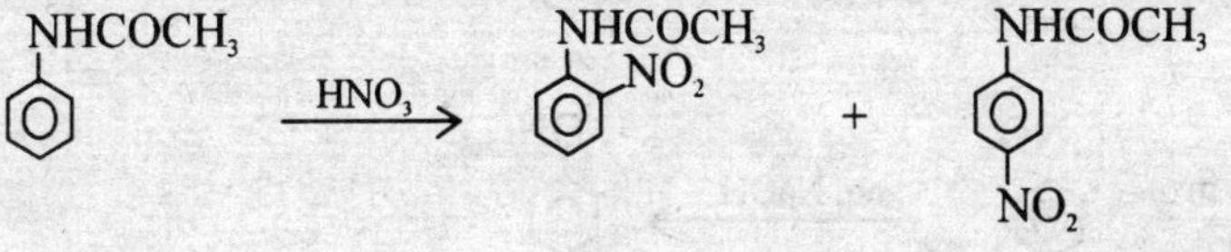

| Acetanilide | o-Nitroacetanilide | p-Nitroacetanilide |

o-Nitroacetanilide by following sequence of reactions produces o-dinitrobenzene.

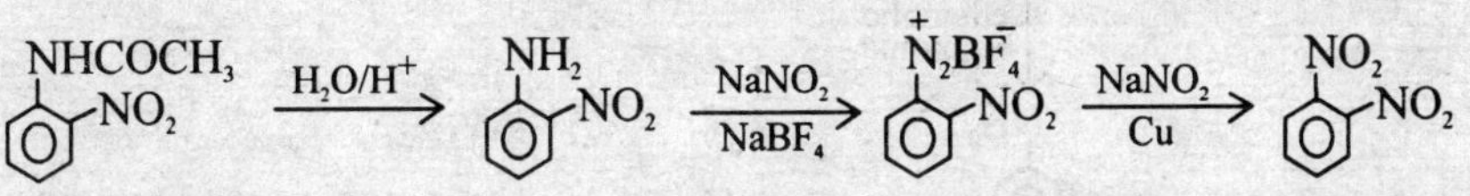

o-Nitroacetanilide   o-Dinitrobenzene

o-Dinitrobenzene can also be prepared by the oxidation of o-nitroaniline with caro's acid to give o-nitronitrosobenzene followed by oxidation with nitric acid and hydrogen peroxide mixture.

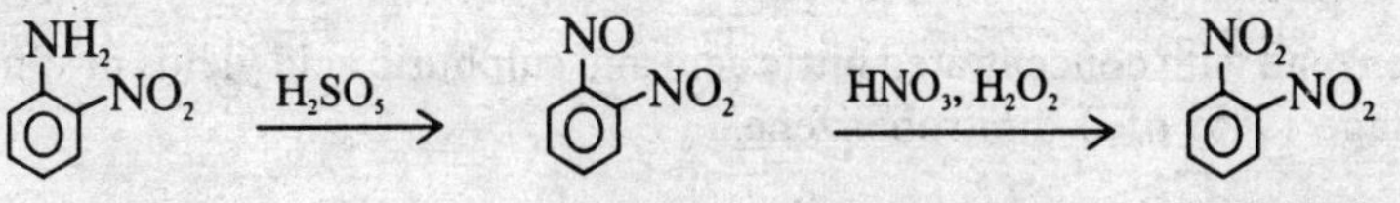

o-Nitroaniline   o-Nitrosonitrobenzene   o-Dinitrobenzene

o-Nitroaniline on direct oxidation with peroxytrifluoroacetic acid produces o-dinitrobenzene.

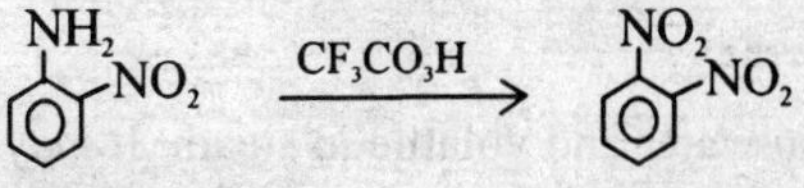

o-Nitroaniline   o-Dinitrobenzene

Properties: It is a colorless solid, m.p. 118°, resembles other nitro compounds in its physical and chemical properties.

It differs from its m-isomer in that, one of the nitro group is easily replaces by nucleophiles like –OH, –NH$_2$, –OCH$_3$.

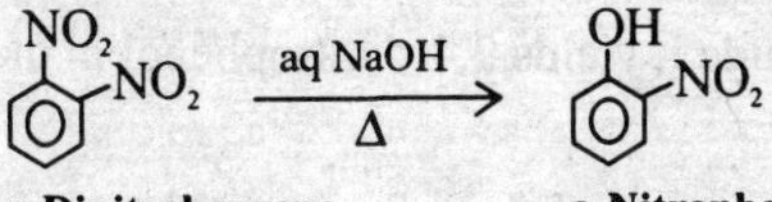

o-Dinitrobenzene   o-Nitrophenol

p-Dinitrobenzene

p-Dinitrobenzene may be prepared by the methods given for the preparation of o-dinitrobenzene starting from p-nitroaniline (diazotisation and oxidation)

It can also be prepared by the oxidation of p-benzoquinonedioxime with nitric acid.

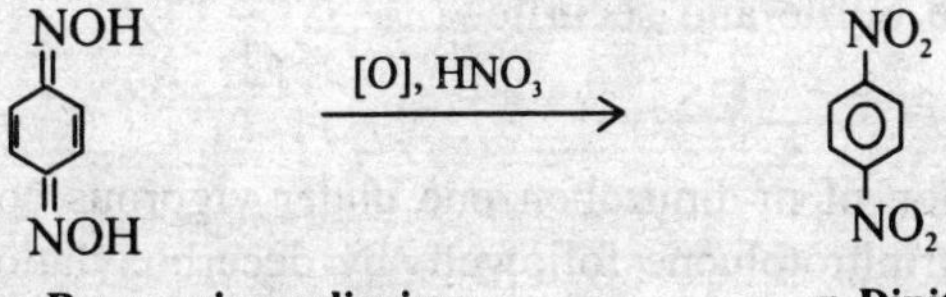

p-Benzoquinonedioxime   p-Dinitrobenzene

Properties: It is a colorless solid, m.p. 173°. It resembles m-isomer in its properties but differs from that in its reaction with aqueous sodium hydroxide, ethanolic ammonia and methanolic sodium methoxide. One of the nitro group undergoes nucleophlilic displacement by –OH, –NH₂ or –OCH₃ groups as a follows:

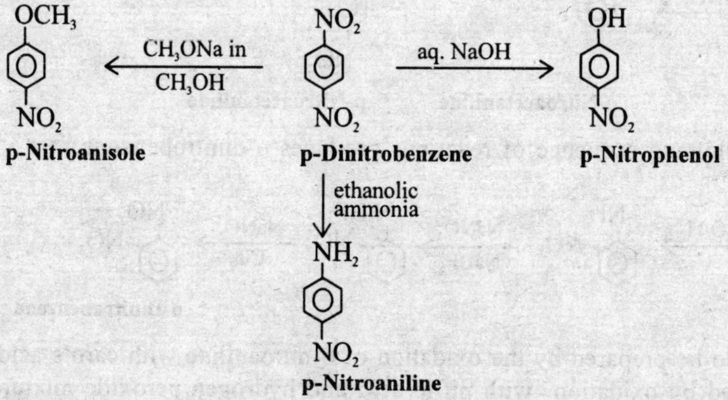

p-Nitroanisole   p-Dinitrobenzene   p-Nitrophenol

p-Nitroaniline

m-Dinitrobenzene

Vigorous nitration of nitrobenzene with concentrated nitric acid and sulphuric acid yields m-dinitrobenzene along with a small percentage (7%) of o-dinitrobenzene.

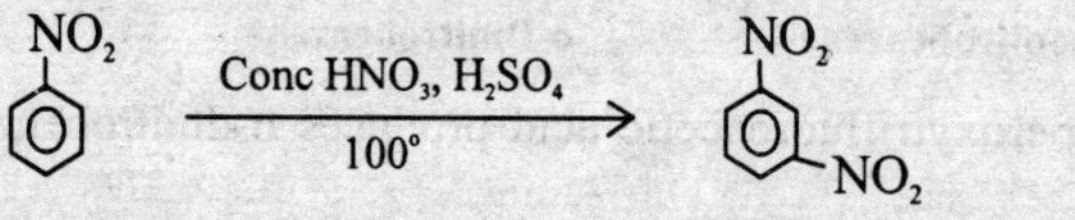

Nitrobenzene   m-Dinitrobenzene

Properties: It is a pale yellow solid, m.p. 90°, insoluble in water and volatile in steam. It may be reduced stepwise by sodium or ammonium sulphide to m-nitroaniline and than m-phenylenediamine.

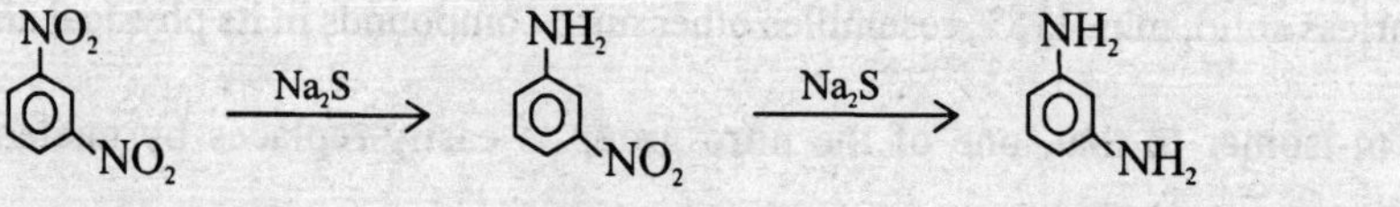

m-Dinitrobenzene   m-Nitroaniline   m-Phenylenediamine

On oxidation with aqueous alkaline potassium ferricyanide, it yields 2,4-dinitrophenol in major amount along with 2,6-dinitrophenol (as a minor product).

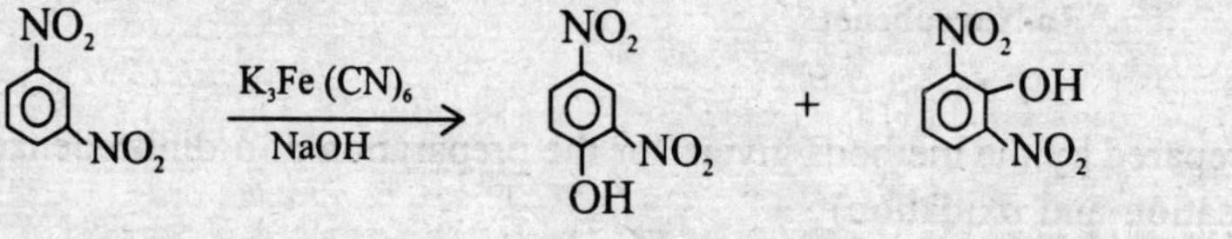

m-Dinitrobenzene   2,4-Dinitrophenol   2,6-Dinitrophenol

It is used in the production of explosives like burite and securite.

31.4.3 sym-Trinitrobenzene [$C_6H_3(NO_2)_3$]

1,3,5-Trinitrobenzene may be prepared by nitration of m-dinitrobenzene under vigorous conditions. It may also be prepared by the oxidation of 2,4,6-trinitrotoluene followed by decarboxylation of 2,4,6-trinitrobenzoic acid in about 45% yields.

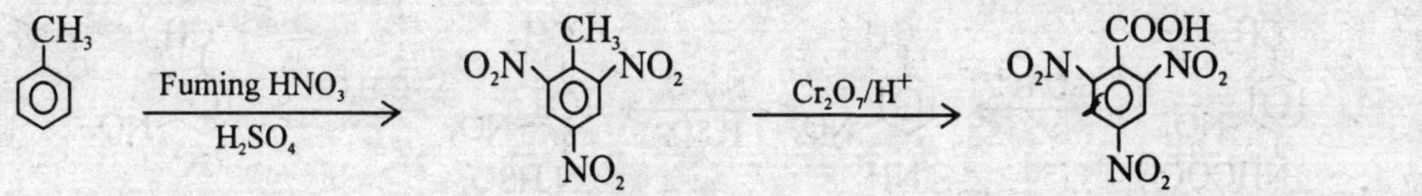

Toluene → 2,4,6-Trinitrotoluene → 2,4,6-Trinitrobenzoic acid → 1,3,5-Trinitrobenzene

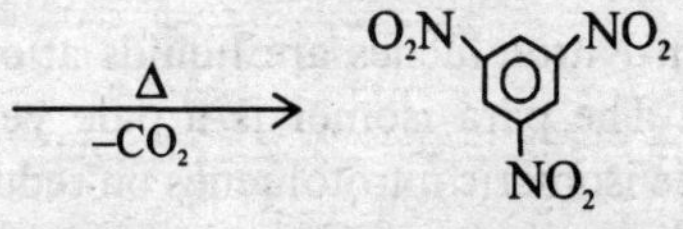

Another method of preparation with better yields (70%) involves reaction of picryl chloride with hydridic acid in acetone solution.

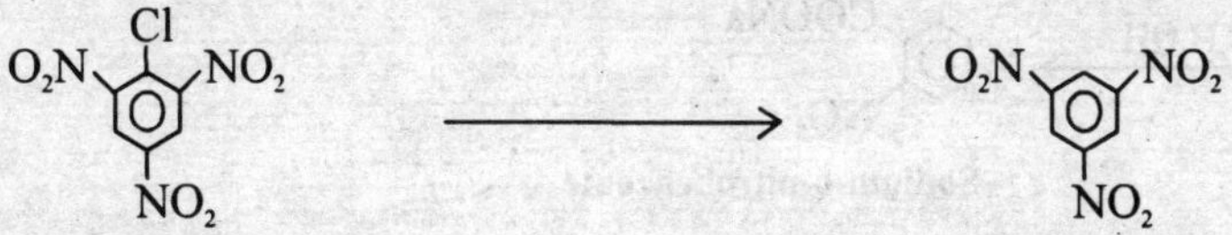

Picryl chloride → 1,3,5-Trinitrobenzene

sym-Trinitrobenzene is a colorless solid, m.p. 122°, forms well defined addition compounds with hydrocarbons and phenols. One of the nitro group may be replaced by −OH, −NH$_2$ or −OCH$_3$ group easily when it is treated with appropriate reagent.

It is a more powerful explosive than TNT, but is not used much as an explosive due to the difficulty in its preparation.

1.4.4 Nitrotoluenes [CH$_3$C$_6$H$_4$NO$_2$]

Toluene on nitration with conventional methods gives a mixture of o-and p-nitrotoluene which may be separated by fractionation under reduced pressure.

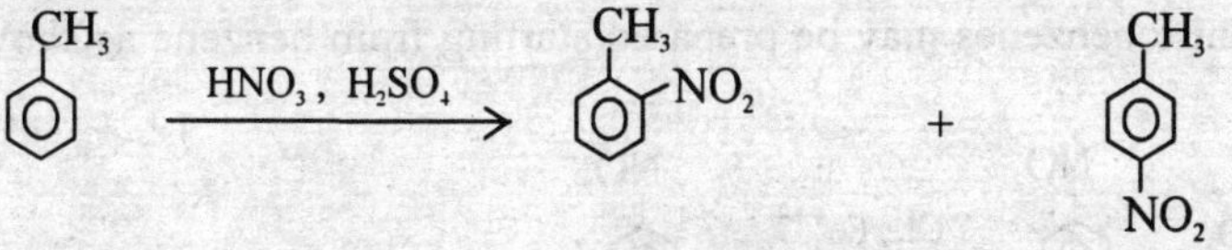

Toluene → o-Nitrotoluene 70 % + p-Nitrotoluene 30 %

An even higher yields (90%) of o-nitrotoluene is prepared when toluene is nitrated with acetyl nitrate.

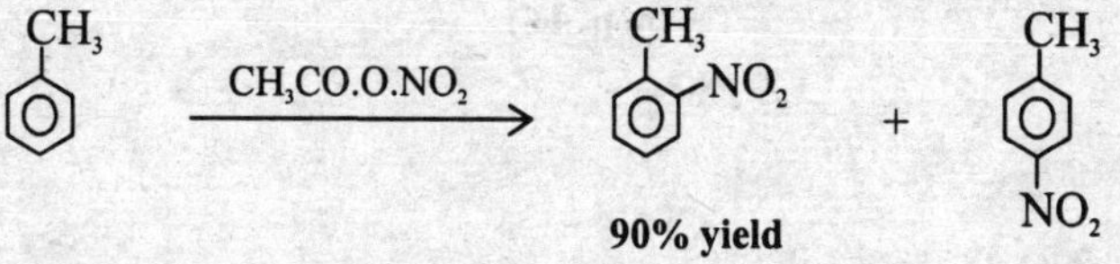

90% yield

m-Nitrotoluene may be prepared by following sequence of reactions starting from p-nitrotoluene.

p-Nitrotoluene

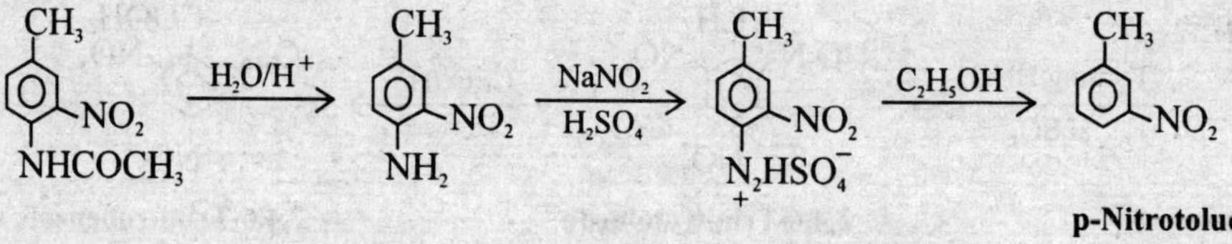

p-Nitrotoluene

o- and m-Nitrotoluenes are liquids at ordinary temperature (ortho m.p. -4°, b.p. 222°; meta m.p. 16°, b.p. 227°). The para isomer is a pale yellow solid (m.p. 54°, b.p. 238°).

The three isomeric nitrotoluenes on reduction give corresponding toluidines whereas on oxidation form corresponding nitrobenzoic acid.

o-nitrotoluene, when heated with ethanolic sodium hydroxide forms sodium salt of o-nitrobenzoic acid as a result of oxidation.

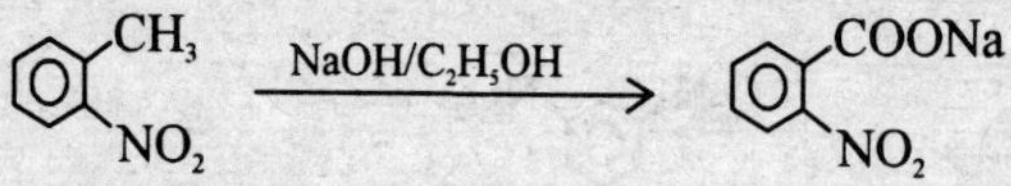

o-Nitrotoluene   Sodium-o-nitrobenzoate

m- and p-Isomers do not give this reaction.
Nitrotoluenes are used in the manufacture of explosives and dyes.

31.4.5   2,4,6-Trinitrotoluene/TNT [$CH_3C_6H_2(NO_2)_3$]

This may be prepared by nitration of toluene with fuming nitric acid and fuming sulphuric acid. The reaction takes place far more readily than with benzene because of the activation of the ring by methyl group.

It is a yellow crystalline solid, m.p. 81°. It is used as an explosive in shells, bombs and torpedoes under the name of 'trotyl'. It forms 'amotol' when mixed with 80% ammonium nitrate.

31.4.6   Chloronitrobenzene [$ClC_6H_4NO_2$]

All the three (o, m and p) chloronitrobenzenes may be prepared starting from benzene as shown below:

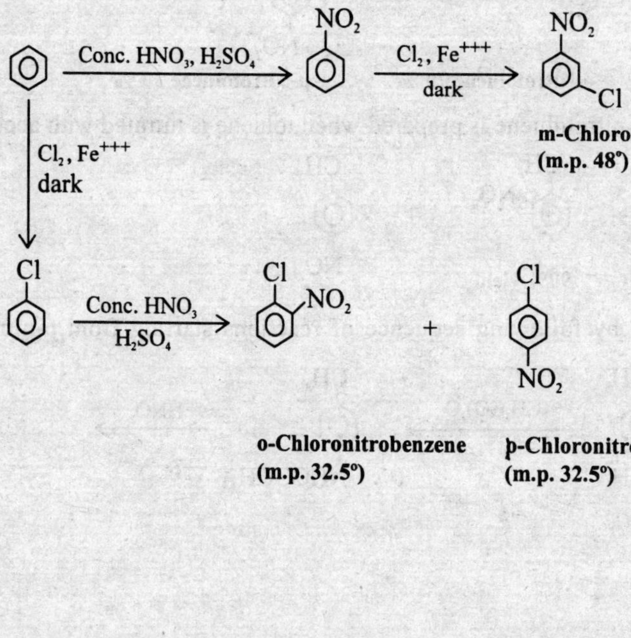

m-Chloronitrobenzene
(m.p. 48°)

o-Chloronitrobenzene   p-Chloronitrobenzene
(m.p. 32.5°)   (m.p. 32.5°)

31.4.7 Phenylnitromethane [$C_6H_5CH_2NO_2$]

Phenylnitromethane may be prepared by heating toluene with dilute nitric acid in a sealed tube at 100° or by heating benzyl chloride with aqueous ethanolic silver nitrate solution.

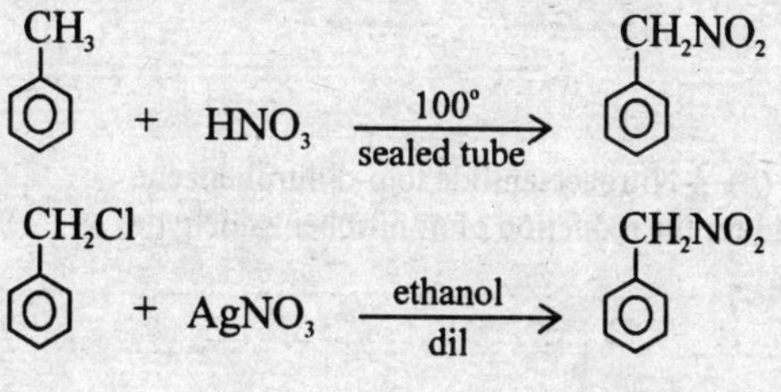

Phenylnitromethane

Another method of preparation, starting from benzyl cyanide gives better yields of phenylnitromethane as shown below:

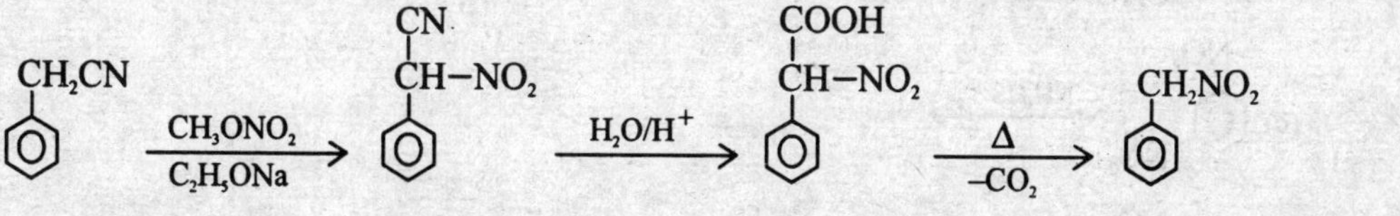

Benzyl cyanide   **Phenylnitromethane**

It is a yellow oil, b.p. 226°. It resembles aliphatic nitrocompounds and exists in two forms nitro and acinitro of which nitro form is more stable.

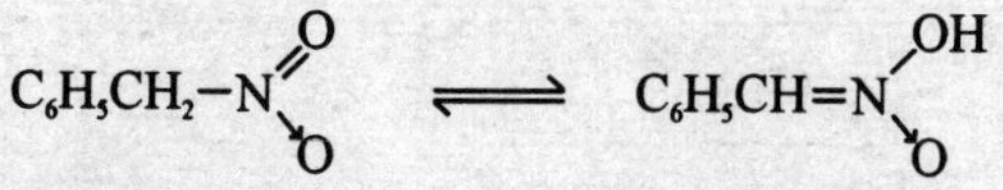

Nitro form   **Acinitro form**
yellow oil, b.p. 226°   solid, m.p. 84°

It is isomeric with nitrotoluenes and can be distinguished from later by oxidation. Phenylnitromethane on oxidation give benzoic acid whereas nitrotoluenes give nitrobenzoic acids.

31.5 PROBLEMS

1. Write the mechanism of nitration:
   (a) with mixed acid   (b) with concentrated nitric acid.
2. Discuss the effect of a nitro group on the reactivity of other nuclear substituents.
   (a) a methyl group   (b) hydroxy group and   (c) a chloro group.
3. Give an account of the reduction products of nitrobenzene.
4. Discuss the difference in o, m and p-dinitrobenzene towards various chemical reactions.
5. How will you prepare the following starting from benzene.
   (a) $C_6H_5NO_2$ (b) o, m and p-$C_6H_4(NO_2)_2$ (c) sym-$C_6H_3(NO_2)_3$
   (d) TNT   (e) o, m and p-$CH_3C_6H_4NO_2$ (f) $C_6H_5CH_2NO_2$
6. Answer question number 5 starting from toluene.
7. Explain why:
   (a) It is difficult to introduce more than one nitro group in the benzene ring.
   (b) Nitro groups increase the reactivity of nuclear halogen atom towards nucleophilic displacement.
   (c) p-Nitrobenzoic acid is a stronger acid than m- nitrobenzoic acid.
   (d) p-Nitrobenzoic is a weak base.

8. Complete the following reactions:

    (a) $C_6H_5NO_2$ $\xrightarrow{\text{Zn + dil HCl}}$

    (b) $C_6H_5NO_2$ $\xrightarrow{\text{Zn + NH}_4\text{Cl}}$

    (c) $C_6H_5NO_2$ $\xrightarrow{\text{LiAlH}_4}$

9. How are following conversions brought about?
    (a) p-Nitrotoluene to m-nitrotoluene   (b) p-Nitroacetanilide to p-dinitrobenzene
10. Why alkaline $FeSO_4$ is used instead of metal and acid for the reduction of m-nitrobenzaldehyde.
11. Complete the following reactions

    (a) 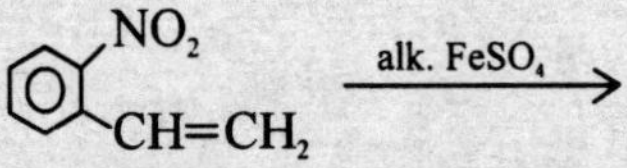 NO$_2$ / CH=CH$_2$ $\xrightarrow{\text{alk. FeSO}_4}$

    (b) 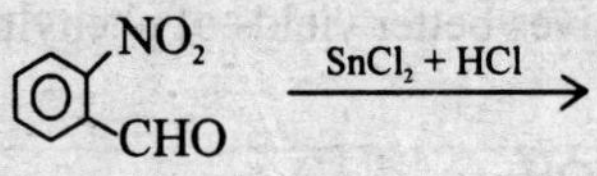 NO$_2$ / CHO $\xrightarrow{\text{SnCl}_2 + \text{HCl}}$

    (c) 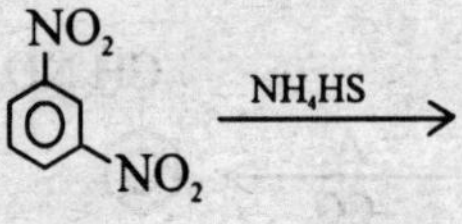 NO$_2$ / NO$_2$ $\xrightarrow{\text{NH}_4\text{HS}}$

    (d) 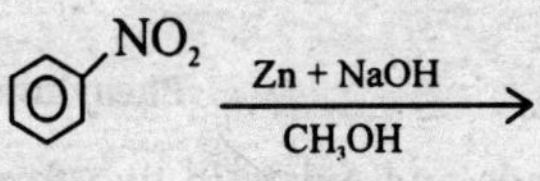 NO$_2$ $\xrightarrow[\text{CH}_3\text{OH}]{\text{Zn + NaOH}}$

    (e) 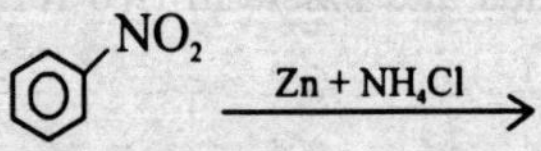 NO$_2$ $\xrightarrow{\text{Zn + NH}_4\text{Cl}}$

    (f) 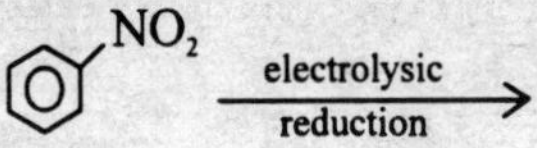 NO$_2$ $\xrightarrow[\text{reduction}]{\text{electrolysic}}$

32

Aromatic Amines

Aromatic amines are of two types, one in which the amino group is attached to the ring directly (I) and the other in which the amino group is in the side chain (II).

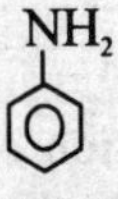

NH$_2$

I, Aniline

CH$_2$CH$_2$NH$_2$

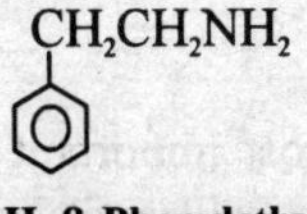

II, ß-Phenylethylamine

Amines of the type II resemble aliphatic amines very much and are regarded as aryl substituted alkyl amines or aryl alkyl amines whereas amines of the type I are termed as aromatic amines.

Aromatic amines (like aliphatic amines) are divided into three classes, viz, primary amines, ArNH$_2$; secondary amines, Ar$_2$NH or ArNHR and tertiary amines, Ar$_3$N or Ar$_2$NR or ArNR$_2$.

Purely aromatic quarternary compounds of the type Ar$_4$N$^+$X$^-$, have not yet been prepared but compounds like Ar$_3$N$^+$RX$^-$ are known.

32.1 NOMENCLATURE

Aromatic amines may be named by adding suffix amine to the parent aromatic nucleus. The number of aromatic rings are indicated by prefix di- or tri- as shown below.

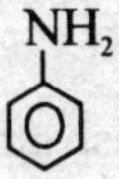

NH$_2$

Aniline

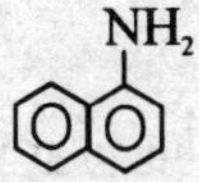

NH$_2$

Naphthylamine

H
N

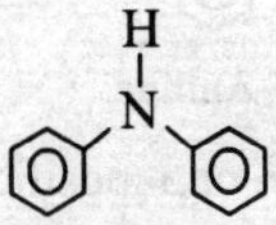

Diphenylamine

When two or more amino groups are present in the ring, the number and position of the amino groups are prefixed as di-, tri- and o, m, p (or 1, 2, 3) respectively as shown below:

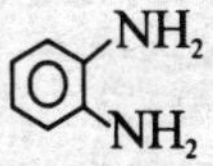

NH$_2$
NH$_2$

**o-Phenylenediamine
(or 1,2-Benzenediamine)**

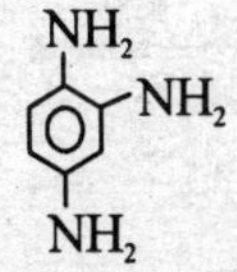

NH$_2$
NH$_2$
NH$_2$

1,2,4-Benzenetriamine

The common names are frequently used for various aromatic amines:

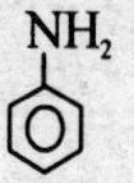

Aniline
(Phenylamine)

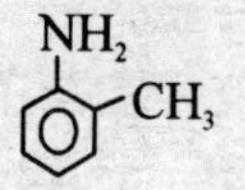

o-Toluidine
(2-Aminotoluene)

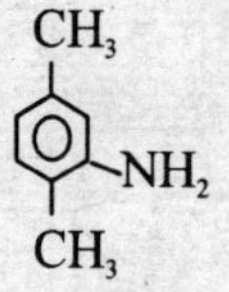

p-Xylidene
(Aminoxylene)

Secondary and tertiary aromatic amines are N-alkyl or N-aryl derivatives of the parent hydrocarbons

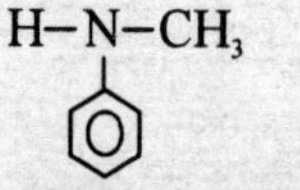

N-Methylaniline

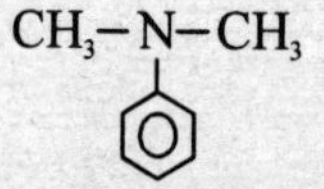

N,N-Dimethylaniline

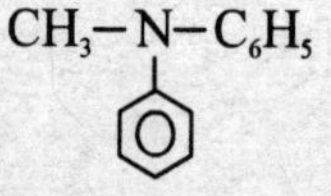

N-Methyl-N-phenylaniline

32.2   AROMATIC PRIMARY AMINE/MONOAMINES

32.2.1   General Methods of Preparation

1. *Reduction of Nitro Compounds:* This is the most important method of preparation of aromatic amines. The reduction is brought about chemically, catalytically and electrolytically as discussed in Sec. 31.3. Nitrobenzene when reduced with metals like tin, zinc or iron and mineral acid like hydrochloric or sulphuric acid, forms aniline.

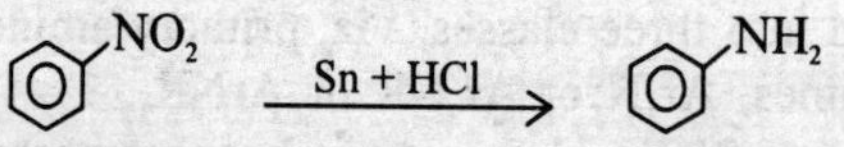

Nitrobenzene

Aniline

2. *Ammonolysis of Aryl Halides:* The reaction of aryl halides with ammonia under reduced pressure and at high temperature in presence of copper salts (catalyst) forms amines in 80-90% yields.

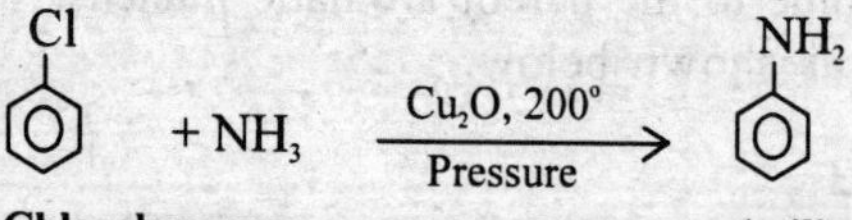

Chlorobenzene

Aniline

Electron withdrawing groups if present at o- or p- position to the halogen group, accomplish ammonolysis with aqueous ammonia.

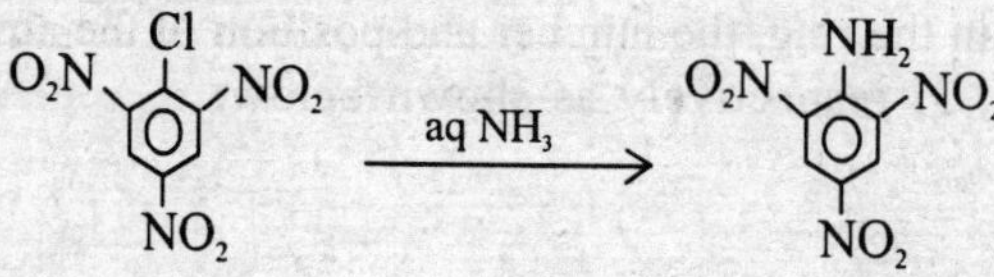

2,4,6-Trinitro-
chlorobenzene

2,4,6-Trinitro-
aniline

The details of the reaction are discussed in Sec. 29.4.

3. *Ammonolysis of Phenols:* Phenols react with ammonia at 300° in presence of zinc chloride to form aniline.

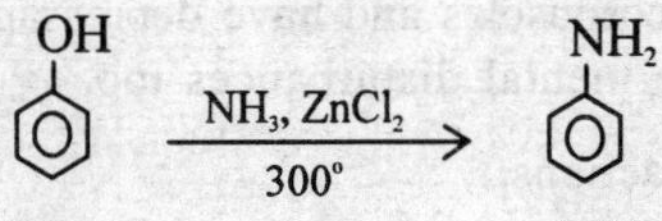

OH → NH₂

Phenol —NH₃, ZnCl₂ / 300°→ Aniline

4. *Hofmann Rearrangement:* Aromatic amides on reaction with hot chlorine or bromine in aqueous sodium hydroxide undergo degradation to yield aromatic amines (analogous to aliphatic amides). Benzamide gives aniline under these conditions.

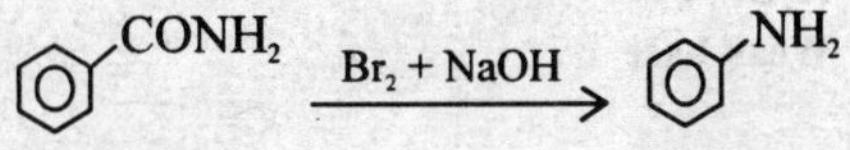

Benzamide (C₆H₅CONH₂) —Br₂ + NaOH→ Aniline (C₆H₅NH₂)

5. *Hofmann Martius Rearrangement:* Secondary and tertiary N-alkylarylamine hydrochloride undergo rearrangement when heated at 300° in a sealed tube to give homologues of aniline.

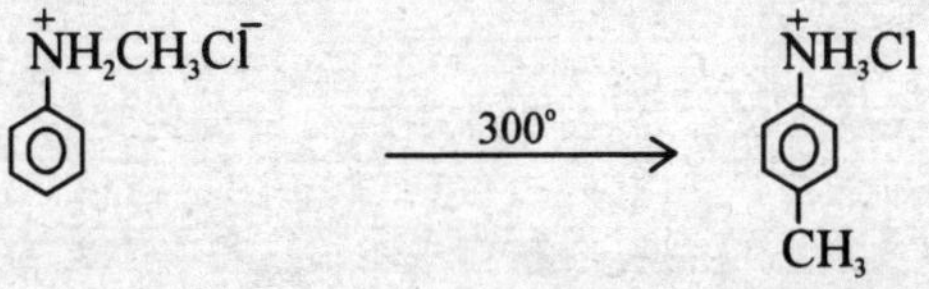

$\overset{+}{N}H_2CH_3\overset{-}{Cl}$ —300°→ $\overset{+}{N}H_3Cl$ (with CH₃)

N-Methylanilinium chloride → **4-Methylanilinium chloride**

N(CH₃)₂HCl —300°→ NH₂ (with 2 CH₃)

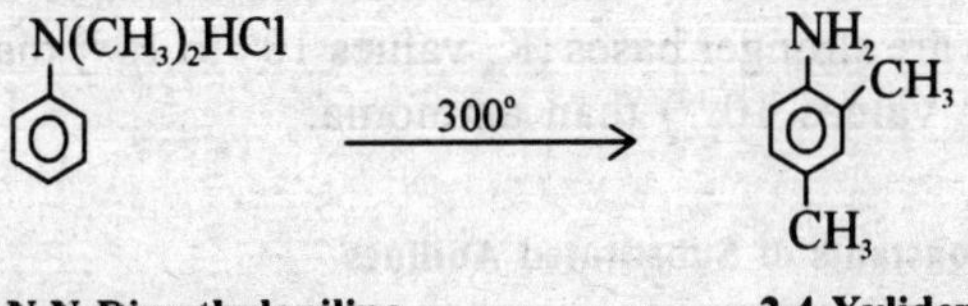

N,N-Dimethylaniline hydrochloride → **2,4-Xylidene**

6. *Schmidt Reaction:* The reaction of aromatic carboxylic acid with hydrazoic acid in chloroform and mineral acid furnishes primary aromatic amines.

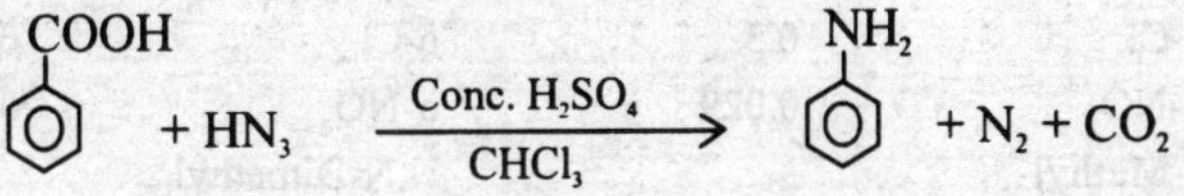

COOH + HN_3 —Conc. H_2SO_4 / $CHCl_3$→ NH₂ + N_2 + CO_2

7. *Action of Hydroxylamine on Hydrocarbons:* Aromatic hydrocarbons react with hydroxylamine in presence of catalyst like iron or aluminum chloride to give amines.

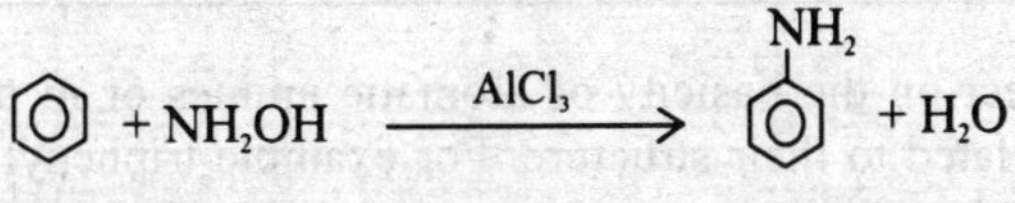

C₆H₆ + NH_2OH —AlCl₃→ NH₂ (aniline) + H_2O

32.2.2 General Properties of Monoamines

Physical Properties: Aromatic monoamines are generally colorless liquids or solids which may turn brown on exposure to light due to atmospheric oxidation. They are sparingly soluble in water but soluble in conventional organic solvents. They are volatile in steam.

They are highly toxic in nature as they destroy red blood corpuscles and have depressant action on heart muscles. Prolonged exposure of arylamines may cause mental disturbances too.

Chemical Properties: Aromatic amines undergo following reactions.

(i)   *Basicity of Amines:* Like ammonia, amines react with aqueous mineral acids to form salts and are liberated from their salts by the action of aqueous hydroxides. Thus, amines are more basic than water and less basic than hydroxide ion.

$$RNH_2 + H_3\overset{+}{O} \longrightarrow R\overset{+}{N}H_3 + H_2O$$

Strong base   **Weak base**

$$R\overset{+}{N}H_3 + OH^- \longrightarrow RNH_2 + H_2O$$

Strong base   **Weak base**

To compare the basicities of various amines, their basicity constant (K_b) is found out which is a measure of the extent to which amines accept hydrogen ion from water.

$$RNH_2 + H_2O \rightleftharpoons R\overset{+}{N}H_3 + OH^-$$

$$K_b = \frac{[R\overset{+}{N}H_3][OH^-]}{[RNH_2]}$$

Below are given basicity constants, K_b, for various aromatic amines (Table 32.1). Higher the value of K_b stronger is the base. If we compare the basicities of aromatic amines and aliphatic amines with that of ammonia we find that the aliphatic amines are stronger bases (K_b values 10^{-3} to 10^{-4}) than ammonia whereas aromatic amines are weaker bases (K_b values 10^{-10}) than ammonia.

Table 32.1:   Basicity Constants of Substituted Anilines

Compound	$K_b(10^{-10})$	Compound	$K_b(10^{-10})$	Compound	$K_b(10^{-10})$
p-NH$_2$	140	m-NH$_2$	10	o-NH$_2$	3
p-OCH$_3$	20	m-OCH$_3$	2	o-OCH$_3$	3
p-CH$_3$	12	m-CH$_3$	5	o-CH$_3$	2.6
p-Cl	1	m-Cl	0.3	o-Cl	0.05
p-NO$_2$	0.001	m-NO$_2$	0.029	o-NO$_2$	0.00006
Aniline	4.2	N-Methyl-aniline	7.1	N,N-Dimethyl aniline	11
		Diphenyl-amine	6.3×10^{-4}	Triphenyl-amine	1×10^{-10}

Substituents in the ring have a marked effect on the basicity of aromatic amines or in other words we can say that the basicities of amines are related to their structure. For example triphenyl amine and diphenyl amines are much less basic compared to aniline.

To account for the weak basic nature of aromatic amines, let us compare the structure of aniline and anilinium cation with that of ammonia and ammonium ion. Below are given structures of ammonia (I), ammonium ion (II), aniline (III-VII) and anilinium ion (VIII and IX).

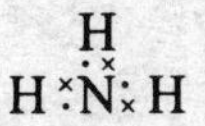

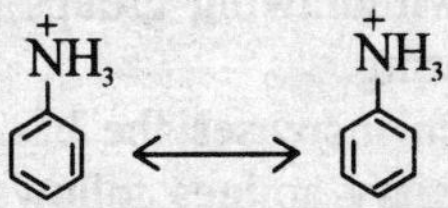

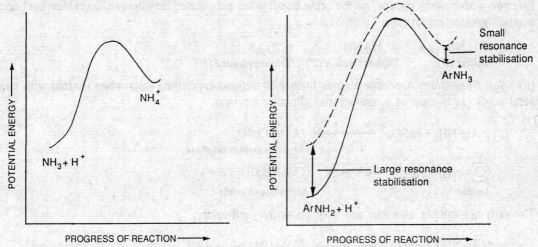

Aniline is represented by five resonating structures (III to VII) whereas only two resonating structures are possible for anilinium ion (VIII, IX). Thus, low basicity of aromatic amines is due to the fact that amine is stabilised by resonance to a greater extent than its ion.

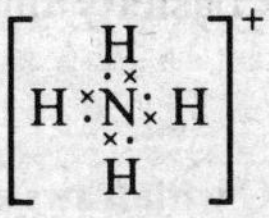

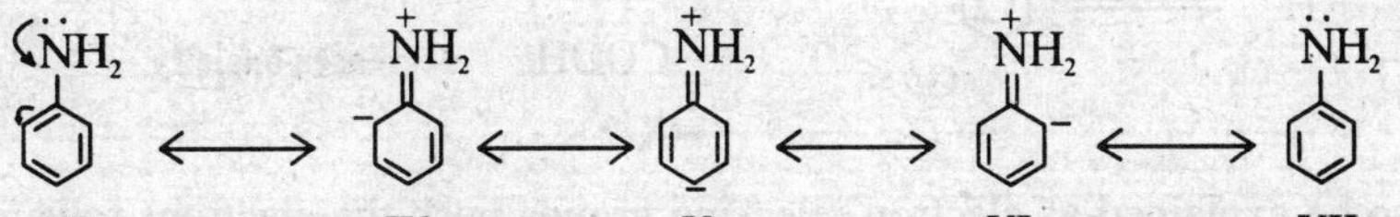

According to another view point, aniline is a weaker base than ammonia because the pair of electron on nitrogen is partially shared with the ring and is less available for sharing with a proton. At the same time the tendency of amines to share the lone pair of electron with the ring through resonance makes the ring more reactive towards electrophilic attack.

Basicity of aromatic amines is affected by the substitution in the ring (as shown in Table 32.1). Electron releasing groups like $-CH_3$, $-OCH_3$ and $-NH_2$ increases the basicity whereas electron withdrawing groups like $-NO_2$, $-CHO$, $-X$ decreases the basicity. Electron releasing groups tend to disperse the positive charge on nitrogen and stabilise the anilinium ion relative to amine as shown below.

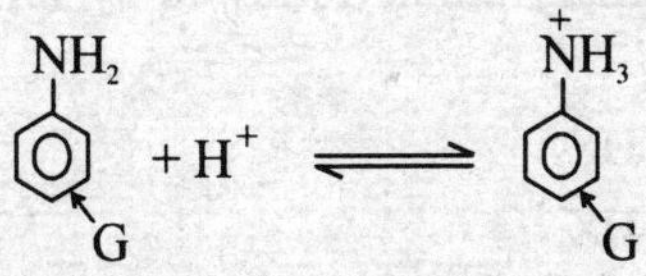

Electron withdrawing groups, on the other hand, decreases basicity due to the reverse effect shown below.

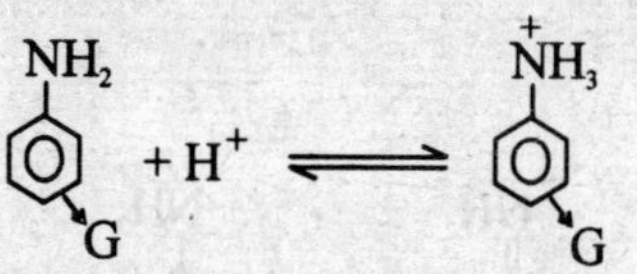

$$NH_2 \quad \overset{+}{N}H_3 \qquad G = -\overset{+}{N}H_3 \qquad \text{G withdraws electrons,}$$

$$\underset{G}{\bigcirc} + H^+ \rightleftharpoons \underset{G}{\bigcirc} \qquad \begin{array}{l} -NO_2 \\ -COOH \\ -X \end{array} \qquad \begin{array}{l} \text{destabilises cation,} \\ \text{decreases basicity} \end{array}$$

This may also be explained as electron releasing groups push the electrons towards nitrogen making the lone pair available for protonation whereas electron withdrawing groups pulls the electrons away from nitrogen making the lone pair less available for sharing.

The electron releasing groups if substituted on nitrogen increases the basicity to a greater extent. Thus the order of basicity for primary, secondary and tertiary amines follows the order given below.

$$\underset{\text{N,N-Dimethylaniline (3°)}}{C_6H_5-N-(CH_3)_2} \quad > \quad \underset{\text{N-Methylaniline (2°)}}{C_6H_5-NHCH_3} \quad > \quad \underset{\text{Aniline (1°)}}{C_6H_5NH_2}$$

Electron withdrawing groups, on the other hand, when substituted on nitrogen decreases the basicity to a much greater extent (see Table 32.1)

$$\underset{\text{Aniline (1°)}}{C_6H_5NH_2} \quad > \quad \underset{\text{Diphenylamine (2°)}}{(C_6H_5)_2NH} \quad > \quad \underset{\text{Triphenylamine (3°)}}{(C_6H_5)_3N}$$

(ii) *Salt Formation:* Aromatic amines form well defined crystalline salts when reacted with strong mineral acids (as in case of ammonia and aliphatic amines).

$$2\,Ar-NH_2 + H_2SO_4 \longrightarrow \underset{\text{Arylammonium sulphate}}{(Ar\,\overset{+}{N}H_3)_2SO_4^-}$$

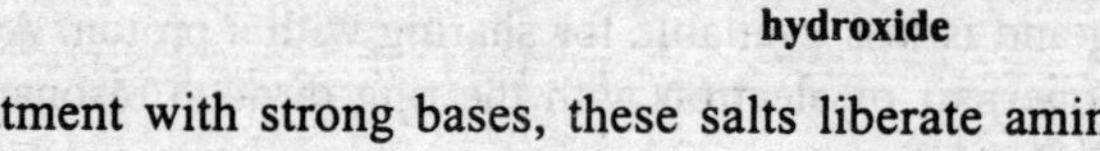

$$\underset{\text{Aniline}}{\bigcirc-NH_2} + HCl \longrightarrow \underset{\text{Anilinium chloride}}{\bigcirc-\overset{+}{N}H_3Cl^-}$$

The salts are soluble in water and may be hydrolysed easily.

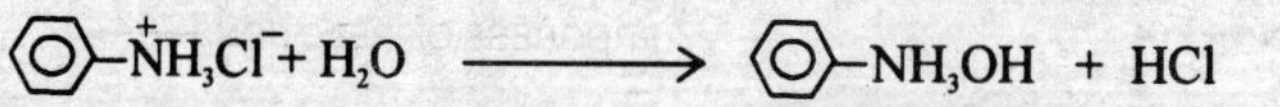

$$\bigcirc-\overset{+}{N}H_3Cl^- + H_2O \longrightarrow \underset{\substack{\text{Phenylammonium} \\ \text{hydroxide}}}{\bigcirc-NH_3OH} + HCl$$

On treatment with strong bases, these salts liberate amine.

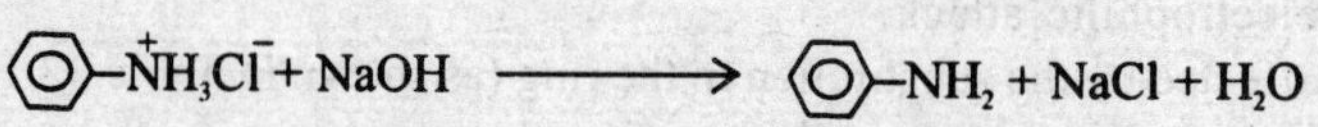

$$\bigcirc-\overset{+}{N}H_3Cl^- + NaOH \longrightarrow \bigcirc-NH_2 + NaCl + H_2O$$

(iii) *Reaction with Alkyl Halides:* Alkyl halides undergo nucleophilic substitution with basic amines serving as nucleophilic reagent. A primary amine reacts with alkyl halide to form secondary amine which in turn reacts with another molecule of alkyl halide to form tertiary amine followed by formation of quaternary ammonium salt.

$$ArNH_2 \;+\; RX \longrightarrow ArNHR \;+\; HX$$
$$ArNHR \;+\; RX \longrightarrow ArNR_2 \;+\; HX$$
$$ArNR_2 \;+\; RX \longrightarrow Ar\overset{+}{N}R_3X^-$$

The reaction of aniline with methyl iodide is shown as follows.

$$C_6H_5NH_2 \xrightarrow[-HI]{CH_3I} C_6H_5NHCH_3 \xrightarrow[-HI]{CH_3I}$$
N-Methylaniline

$$C_6H_5N(CH_3)_2 \xrightarrow{CH_3I} C_6H_5\overset{+}{N}(CH_3)_3\bar{I}$$
N,N-Dimethylaniline   **Trimethylphenyl-ammonium iodide**

When an aqueous solution of quaternary ammonium halide is treated with silver oxide, silver halide precipitates leaving behind quaternary ammonium hydroxide in solution which is almost as basic as sodium or potassium hydroxide.

$$Ar\overset{+}{N}R_3X^- + Ag_2O \longrightarrow Ar\overset{+}{N}R_3OH^- + AgX \downarrow$$

When quaternary ammonium hydroxide is heated strongly, it decomposes to yield a tertiary amine, an alkene and water. Triethylphenylammonium hydroxide on heating produces N,N-diethylaniline, ethene and water.

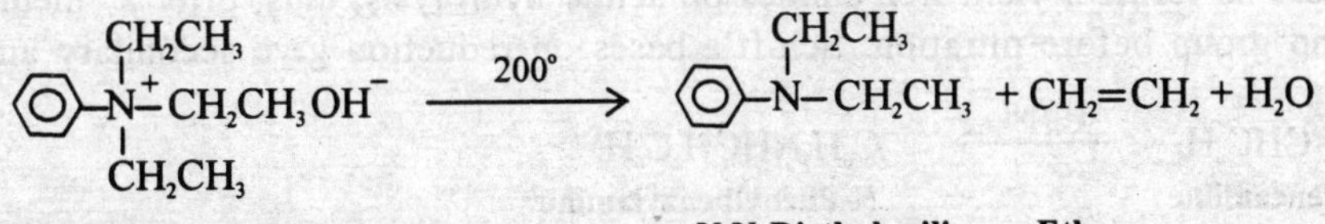

Triethylphenylammonium hydroxide   **N,N-Diethylaniline**   **Ethene**

This reaction is called hofmann elimination reaction and is often used for the determination of structure of complicated amines.

(iv) *Acylation:* Aromatic amines react with acid chloride or acid anhydride in presence of base to yield anilides.

$$C_6H_5NH_2 + CH_3COCl \xrightarrow{Pyridine} C_6H_5NHCOCH_3 + HCl$$
       Acetyl chloride   **Acetanilide**

$$C_6H_5NH_2 + C_6H_5COCl \xrightarrow{OH^-} C_6H_5NHCOC_6H_5 + HCl$$
       Benzoyl chloride   **Benzanilide**

The reaction with benzoyl chloride is known as *Schotten Baumann reaction*.

(v) *Carbylamine Reaction:* Primary aromatic amines when treated with chloroform and alcoholic caustic potash form isocyanide having offensive odor (a test for primary amines)

$$ArNH_2 + CHCl_3 + 3KOH \longrightarrow ArNC + 3KCl + 3 H_2O$$
                                             Arylisocyanide

$$C_6H_5NH_2 + CHCl_3 + 3KOH \longrightarrow C_6H_5NC + 3KCl + 3 H_2O$$
Aniline   **Phenylisocyanide**

The mechanism involves electrophilic attack of dichlorocarbene on electron rich nitrogen of primary amine.

$$CHCl_3 + OH^- \longrightarrow :CCl_3^- + H_2O$$

$$:CCl_3^- \xrightarrow[-Cl^-]{} :CCl_2$$

Dichlorocarbene

$$C_6H_5\overset{\cdot\cdot}{N}H_2 + :CCl_2 \longrightarrow C_6H_5-\overset{+}{N}H_2-\bar{C}Cl_2 \xrightarrow[-H^+]{}$$

$$C_6H_5NH\bar{C}Cl_2 \xrightarrow[-HCl]{} C_6H_5N=\bar{C}Cl \xrightarrow[-Cl^-]{} C_6H_5N\cong C$$

Phenylisocyanide

(vi) *Reaction with Aldehydes:* Primary aromatic amines condense with aldehydes to give *Schiff's base or Anils*.

$$ArNH_2 + ArCHO \longrightarrow ArN=CHAr + H_2O$$

$$C_6H_5NH_2 + C_6H_5\overset{H}{\underset{|}{C}}=O \longrightarrow C_6H_5-N=\overset{H}{\underset{|}{C}}-C_6H_5 + H_2O$$

Aniline   Benzaldehyde   Benzylideneaniline

The Schiff's bases so formed, yield free amines on acidic hydrolysis, thus, offer a method for the protection of amino group before nitration. Schiff's bases on reduction give secondary amines.

$$C_6H_5N=CHC_6H_5 \xrightarrow{H_2/Ni} C_6H_5NHCH_2C_6H_5$$

Benzylideneaniline   N-Phenylbenzylamine

(vii) *Reaction with Phosgene:* With phosgene, primary amines form phenyl isocyanates which are important intermediates for the synthesis of various pesticides.

$$C_6H_5NH_2 + COCl_2 \longrightarrow C_6H_5N=C=O + 2HCl$$

Phenylisocyanate

(viii) *Reaction with Grignard Reagent:* Primary aromatic amines form alkanes with alkyl magnesium halides.

$$ArNH_2 + RMgX \longrightarrow RH + ArNHMgX$$

$$C_6H_5NH_2 + CH_3CH_2CH_2MgBr \longrightarrow CH_3CH_2CH_3 + C_6H_5NHMgBr$$

n-Propyl magnesium   n-Propane
bromide

(ix) *Reaction with Carbon Disulphide:* Primary aromatic amines react with carbon disulphide in ethanolic potassium hydroxide to give diarylthiourea.

$$2C_6H_5NH_2 + CS_2 + 2KOH \xrightarrow{80°} \begin{array}{c} C_6H_5NH \\ \diagdown \\ \diagup \\ C_6H_5NH \end{array} C=S + K_2S + 2H_2O$$

Diphenyl thiourea

(x) *Reaction with Nitrous Acid (Diazotisation):* Primary, secondary and tertiary amines react with nitrous acid to give different products.

Primary amines reacts with nitrous acid (a mixture of $NaNO_2$ and HCl) in acidic solution at low temperature to form water soluble diazonium salt. The reaction is known a diazotisation. Aniline reacts with nitrous acid to yield phenyldiazonium chloride.

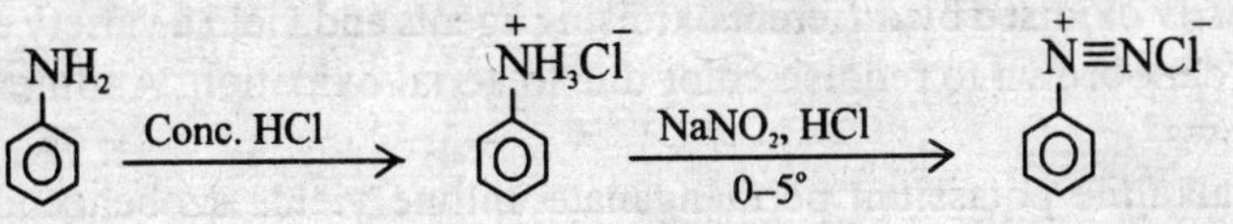

The diagram shows the conversion:

NH_2 (on benzene ring) $\xrightarrow{\text{Conc. HCl}}$ $\overset{+}{N}H_3\bar{Cl}$ (on benzene ring) $\xrightarrow[0-5°]{\text{NaNO}_2, \text{HCl}}$ $\overset{+}{N}\equiv N\bar{Cl}$ (on benzene ring)

Phenyldiazonium chloride

The diazonium salt decomposes at higher temperature. Thus, aniline at higher temperature forms phenol and nitrogen.

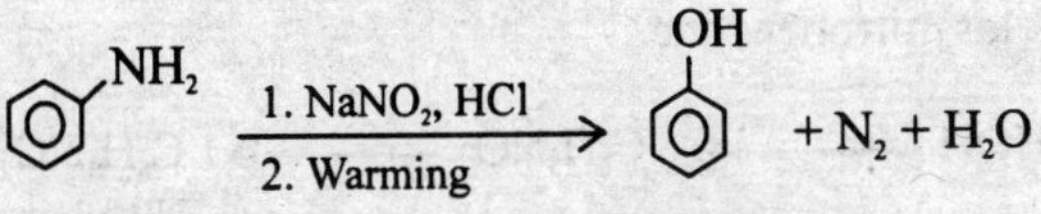

NH_2 (on benzene ring) $\xrightarrow[\text{2. Warming}]{\text{1. NaNO}_2, \text{HCl}}$ OH (on benzene ring) $+ N_2 + H_2O$

Phenol

Aliphatic primary amines do not form stable diazonium salt under similar conditions, on the contrary they yield alcohol and nitrogen. Thus, this reaction is used as a distinguishing test between aliphatic and aromatic primary amines.

Secondary aromatic amines react with nitrous acid (like secondary aliphatic amine) to yield nitrosamine. N-Ethylaniline forms N-ethyl-N-nitrosamine.

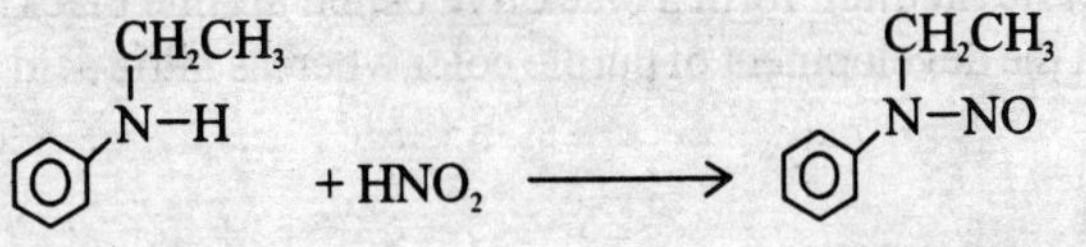

CH_2CH_3, N–H (on benzene ring) $+ HNO_2 \longrightarrow$ CH_2CH_3, N–NO (on benzene ring)

N-Ethylamine   **N-Ethyl-N-nitrosamine**

The nitrosamines, thus formed are yellow colored stable substances. They undergo Fischer-Hepp rearrangement when treated with concentrated hydrochloric acid in ethanol. The rearrangement involves the migration of nitroso group to para position of the ring.

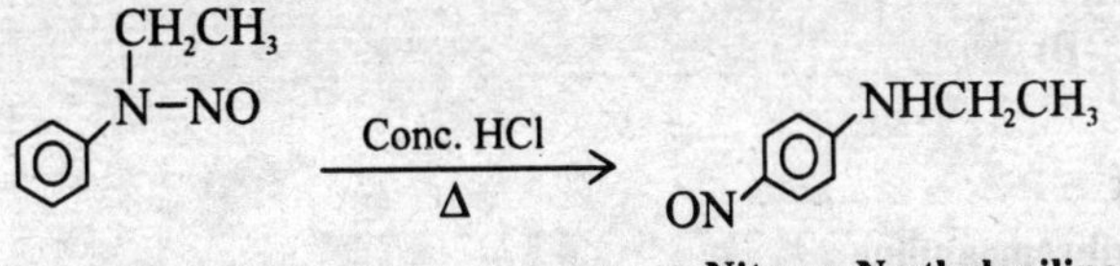

CH_2CH_3, N–NO (on benzene ring) $\xrightarrow[\Delta]{\text{Conc. HCl}}$ ON—(benzene ring)—$NHCH_2CH_3$

p-Nitroso-N-ethylaniline

Tertiary amines form p-nitroso-N,N-dialkyl amines when reacted upon with nitrous acid which on boiling with sodium hydroxide yields dialkyl amine (an aliphatic secondary amine) along with p-nitrosophenol.

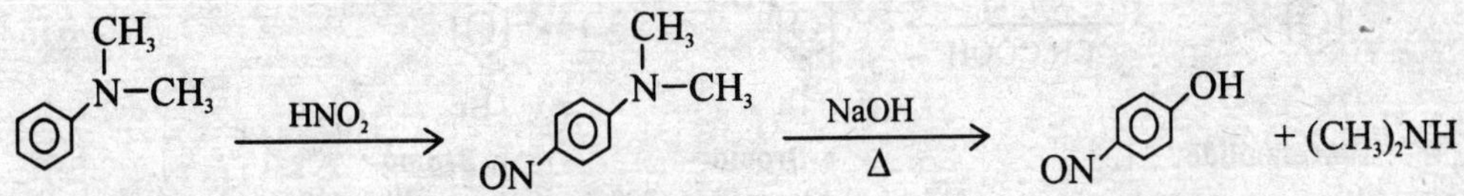

CH_3, N–CH$_3$ (on benzene ring) $\xrightarrow{\text{HNO}_2}$ ON—(benzene ring)—N–CH$_3$ with CH$_3$ $\xrightarrow[\Delta]{\text{NaOH}}$ ON—(benzene ring)—OH $+ (CH_3)_2NH$

N,N-Dimethylaniline   **p-Nitroso-N,**   **p-Nitrosophenol**
                              N-dimetylaniline

Nitrosation takes place due to the strong activation of p-position by dimethylamino group. If p-position is occupied, nitrosation takes place at o-position.

(xi) *Oxidation:* Amines are easily oxidised by different oxidising agents and yield a variety of products. Pure, colorless aniline acquires a dark brown to reddish color due to aerial oxidation. Aniline reacts with various oxidising agents as follows.

1.   When reacted upon with alkaline potassium permanganate aniline yields azobenzene.

$$C_6H_5NH_2 \xrightarrow{\text{alk. KMnO}_4} C_6H_5NHOH \longrightarrow C_6H_5NO \longrightarrow C_6H_5N=NC_6H_5$$

Aniline   Phenylhydroxyl   Nitrosobenzene   Azobenzene
                                        amine

2.   When oxidised with caro's acid, it yields nitrobenzene.

$$C_6H_5NH_2 \xrightarrow{H_2SO_5} C_6H_5NHOH \longrightarrow C_6H_5NO \longrightarrow C_6H_5NO_2$$

Aniline   Nitrobenzene

3.   Oxidation with chromic acid produces p-benzoquinone

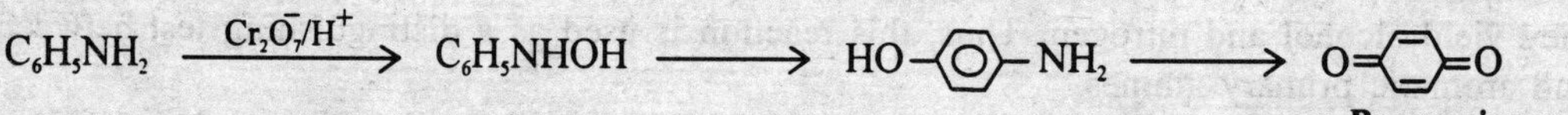

$$C_6H_5NH_2 \xrightarrow{Cr_2O_7/H^+} C_6H_5NHOH \longrightarrow HO-\!\!\!\langle O \rangle\!\!\!-NH_2 \longrightarrow O=\!\!\langle = \rangle\!\!=O$$

p-Benzoquinone

With neutral potassium permanganate, azobenzene and nitrobenzene are obtained. Potassium permanganate and potassium dichromate in dilute acidic medium form a black dye called aniline black. Bleaching powder oxidises aniline to p-aminophenol with the development of purple color whereas nitric acid completely decomposes aniline.

Reaction of the Benzene Nucleus

(xii) *Halogenation:* As discussed earlier (Sec. 28.8) amino group is a highly activating group and directs the substitutents at ortho and para position. Aniline when treated with bromine water or chlorine in chloroform forms tribromo or trichloroaniline.

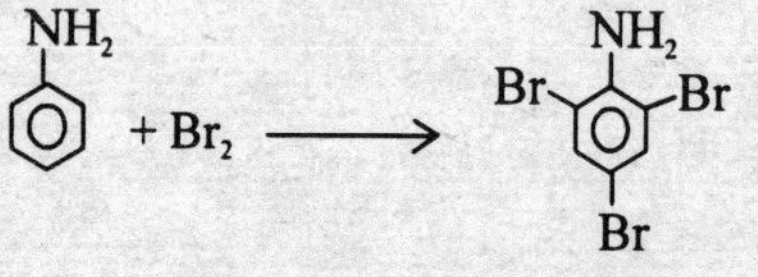

2,4,6-Tribromoaniline

Monochloro or monobromo aniline may be prepared by chlorination or bromination of acetanilide.

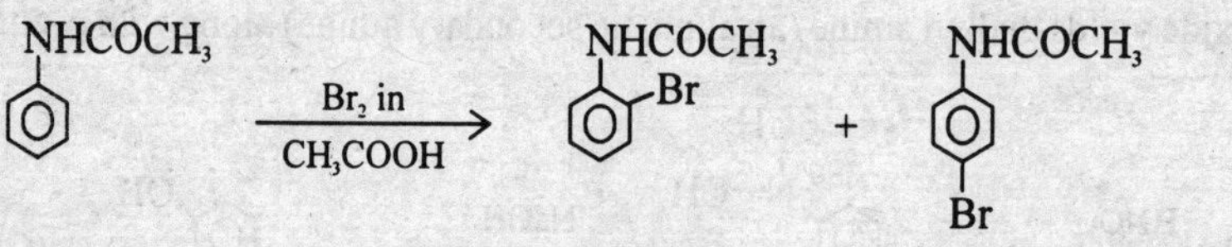

Acetanilide   o-Bromo-   p-Bromo-
                                 acetanilide (20%)   acetanilide (80%)

The ortho and para bromoacetanilide may be separated by crystallisation followed by alkaline hydrolysis to produce mono bromoaniline.

Acetanilide deactivates the ring for aromatic electrophilic substitution because the lone pair of electron on nitrogen enters into resonance with the carbonyl group (as shown below) and not available completely for resonance with the benzene ring.

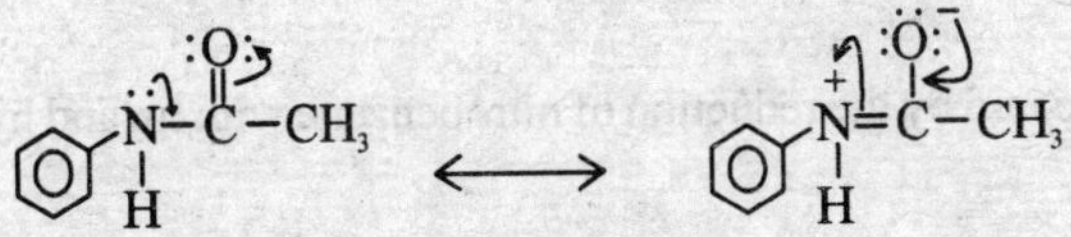

(xiii) *Nitration:* Direct nitration of aniline is not possible as it is associated with oxidation of aniline. Aniline is first acetylated (protected) and then nitration by usual reagents produce a mixture of ortho and para nitro aniline of which p-nitroaniline is the major product. The two isomers are separated and hydrolysed back to corresponding nitroanilines.

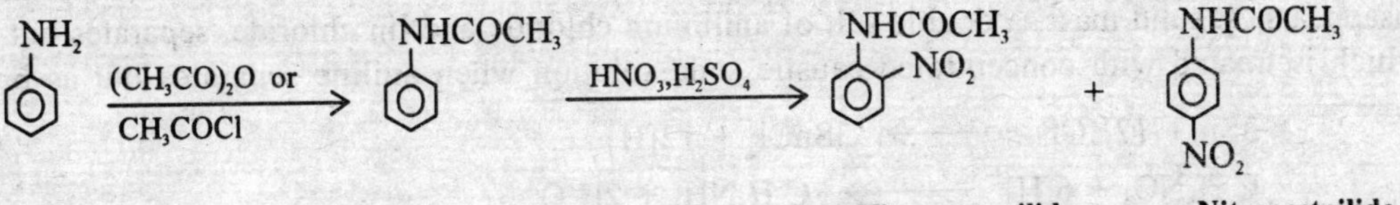

o-Nitroacetanilide   **p-Nitroacetnilide**

Direct nitration of aniline produces m-nitroaniline. The first step here is the protonation of amino group to form anilinium ion which deactivates the benzene ring for ortho and para substitution.

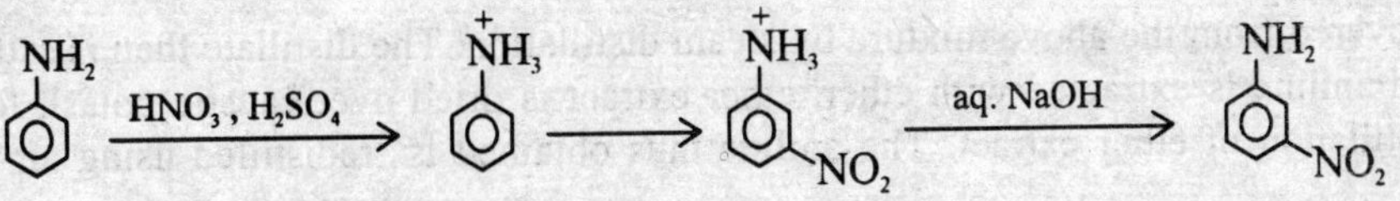

m-Nitroaniline

(xiv) *Sulphonation:* Direct sulphonation of aniline produces p-aminobenzenesulphonic acid or sulphanilic acid.

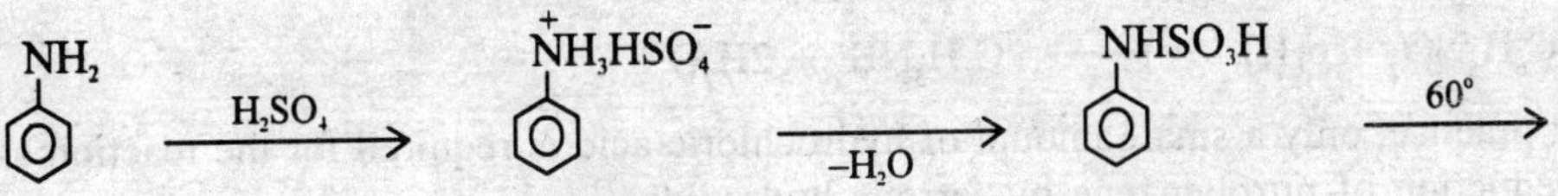

Anilinium sulphate   **Phenylsulphamic acid**

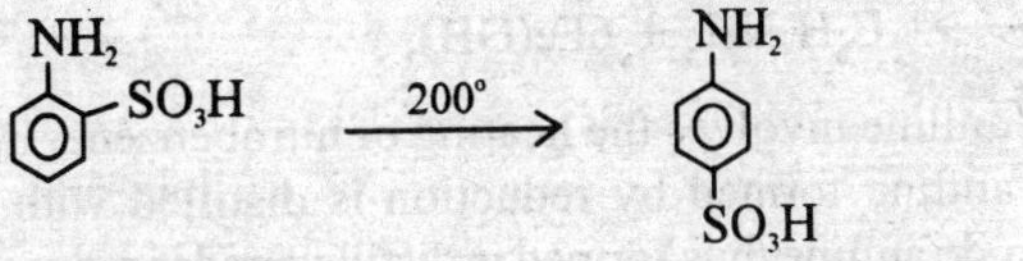

o-Aminobenzene-sulphonic acid   **p-Aminobenzene-sulphonic acid**

Direct sulphonation is possible in case of aniline because sulphuric acid is a weak oxidising agent compared to nitric acid. p-Aminobenzenesulphonic acid is produces because it is the most stable isomer.

32.3 INDIVIDUAL MEMBERS

32.3.1 Aniline [$C_6H_5NH_2$]

Unverdorben obtained aniline first in 1826 by dry distillation of indigo and the name aniline was given by Fritzsche who obtained it from anil (Portuguese, anil, indigo).

Preparation

Laboratory Method: In laboratory, aniline is prepared by the reduction of nitrobenzene with tin and hydrochloric acid.

$$C_6H_5NO_2 + 6\,[H] \xrightarrow{\;Sn + HCl\;} C_6H_5NH_2$$
Aniline

Nitrobenzene (10g) and granulated tin (20g) are taken in round bottom flask fitted with reflux condenser. To this is added concentrated hydrochloric acid (50cc) in small amounts with constant shaking. The temperature is kept below 100°. The contents of the flask are refluxed till the smell of nitrobenzene disappears. A solid mass, a double salt of anilinium chloride and tin chloride, separates out on cooling which is treated with concentrated caustic soda solution when aniline separates out as brown oil.

$$3Sn + 12HCl \longrightarrow 3SnCl_4 + 12[H]$$
$$C_6H_5NO_2 + 6[H] \longrightarrow C_6H_5NH_2 + 2H_2O$$
$$2C_6H_5NH_2 + 2HCl + SnCl_4 \longrightarrow (C_6H_5NH_2.HCl)_2\,SnCl_4$$
Double salt
$$(C_6H_5NH_2.HCl)_2\,SnCl_4 + 8NaOH \longrightarrow 2C_6H_5NH_2 + Na_2SnO_2 + 6NaCl + 5H_2O$$

Aniline is recovered from the above mixture by steam distillation. The distillate then is saturated with common salt and aniline is extracted with ether. ether extract is dried over caustic potash followed by filtration and distillation of ether extract. The aniline thus obtained is redistilled using air condenser.

Industrial Method

(1) Aniline is prepared on large scale by the reduction of nitrobenzene with scrap iron and concentrated hydrochloric acid.

$$3Fe + 6HCl \longrightarrow 3FeCl_2 + 6[H]$$
$$C_6H_5NO_2 + 6[H] \longrightarrow C_6H_5NH_2 + 2H_2O$$

In actual practice, only a small amount of hydrochloric acid is required for the reaction as evidences show the reduction of nitrobenzene by ferrous hydroxide.

$$FeCl_2 + 2H_2O \longrightarrow Fe(OH)_2 + 2HCl$$
$$C_6H_5NO_2 + 6Fe(OH)_2 + 4H_2O \longrightarrow C_6H_5NH_2 + 6Fe(OH)_3$$

The process for the continuous production of aniline involves the heating of nitrobenzene, iron turning, hydrochloric acid and water with steam. The aniline formed by reduction is distilled with steam and separated as aniline forms the lower layer. The crude aniline thus formed is distilled under reduced pressure.

(2) Catalytic reduction of nitrobenzene with nickel and concentrated hydrochloric acid is another method for the commercial production of aniline.

(3) Aniline may also be manufactured by heating chlorobenzene with excess of aqueous ammonia in presence of cuprous oxide at 200° under pressure. The cuprous oxide renders the reaction irreversible as it decomposes ammonium chloride formed during the reaction.

$$2C_6H_5Cl + 2NH_3 + Cu_2O \longrightarrow 2C_6H_5NH_2 + 2CuCl + H_2O$$
Chlorobenzene   **Aniline**

Properties

Physical: Freshly prepared pure aniline is colorless liquid, b.p. 184° having an unpleasant odor. When exposed to air, it quickly darkens due to oxidation. It is insoluble in water but is steam volatile and is readily soluble in organic solvents.

Chemical Properties: It gives all the reactions characteristics of primary aromatic amines (discussed in Sec. 32.2.2). In addition, aniline give following reactions also.

(I)   Aniline reacts with active metals like sodium or potassium on heating to form metal anilides with the evolution of hydrogen.

$$C_6H_5NH_2 + K \longrightarrow [C_6H_5NH]^- K^+ + 1/2\ H_2$$

(II)   When heated with aniline hydrochloride in a sealed tube at 200°, aniline yields diphenylamine, a secondary amine.

$$C_6H_5NH_2 + C_6H_5\overset{+}{N}H_3\overset{-}{Cl} \longrightarrow C_6H_5NHC_6H_5 + NH_4Cl$$
$$\textbf{Diphenylamine}$$

(III)   Hydrogen atoms of amino group may be replaced by halogens when aniline is treated with hypohalous acid.

$$C_6H_5NH_2 + HOCl \longrightarrow C_6H_5NCl_2 + 2H_2O$$

(IV)   It undergoes mercuration at ortho position when heated with mercuric acetate.

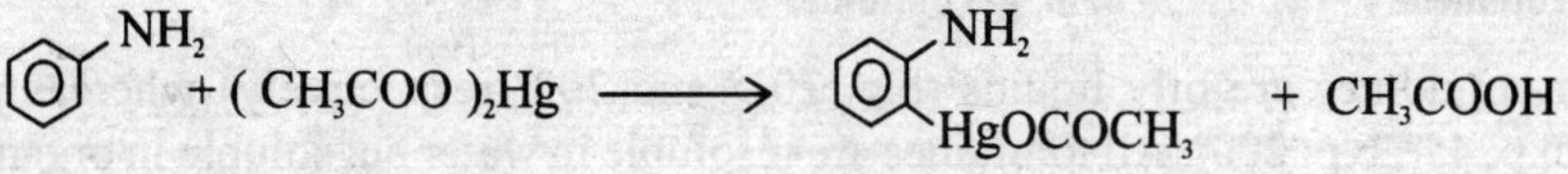

Mercuric acetate   **o-Acetoxymercury aniline**

(V)   Aniline couples with benzene diazonium salt to give p-aminoazobenzene (aniline yellow) in acidic medium.

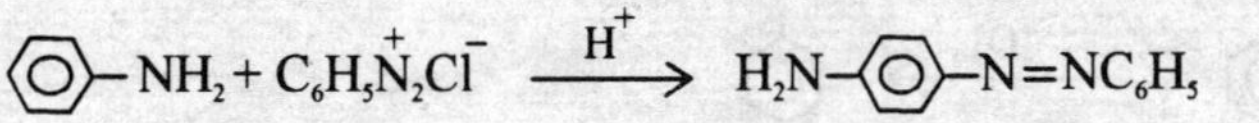

p-aminoazobenzene

Uses
1. It is used in the preparation of diazonium compounds which are used in dye industry in the preparation of azodyes.
2. Schiff's bases are used as antioxidants in rubber industry.
3. It is used in the preparation of synthetic drugs.

32.3.2   Acetanilide/N-Phenylacetamide [$C_6H_5NHCOCH_3$]

It is prepared by acetylation of aniline using acetyl chloride, acetic anhydride or acetic acid. In laboratory it is prepared by heating aniline, acetic acid and anhydrous zinc chloride in a round bottom flask fitted with reflux condenser. The contents of the flask are poured into ice cold water after cooling. The precipitated acetanilide is filtered and purified by recrystallisation.

$$C_6H_5NH_2 + CH_3COOH \xrightarrow[\text{ZnCl}_2]{\text{anhydrous}} C_6H_5NHCOCH_3 + H_2O$$
$$\textbf{Aniline} \qquad\qquad\qquad\qquad\qquad \textbf{Acetanilide}$$

It is a white crystalline solid, m.p. 114°, dissolves in hot water but practically insoluble in cold water.

It hydrolyses back to aniline in presence of dilute acid or base, thus used as a method of protection of amino group.

32.3.3   Toluidines/Aminotoluenes [$CH_3C_6H_4NH_2$]

The ortho, meta and para aminotoluenes are known as o, m and p-toluidines respectively. They are isomeric with benzylamine ($C_6H_5CH_2NH_2$), an arylalkyl amine.

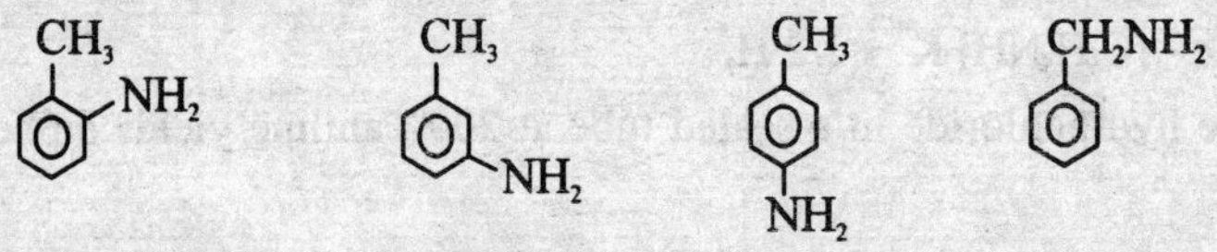

o-Toluidine   m-Toluidine   p-Toluidine   Benzyl amine
(2-Aminotoluene)   (3-Aminotoluene)   4-Aminotoluene)

All the toluidines are prepared by the reduction of corresponding nitrotoluenes with conventional reducing agents.

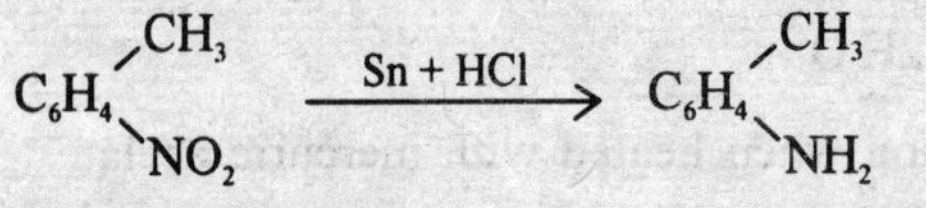

$$C_6H_4 \underset{NO_2}{\overset{CH_3}{\diagup}} \xrightarrow{Sn + HCl} C_6H_4 \underset{NH_2}{\overset{CH_3}{\diagup}}$$

o, m, p-Nitrotoluene   o, m, p-Toluidine

Properties: o- and m-toluidines are oily liquids (b.p. 201° and 200° respectively) whereas p-toluidine is a crystalline solid m.p. 45°, b.p. 200°. All toluidines are insoluble in water but soluble in organic solvents and are toxic in nature. They exhibit all the reactions of primary aromatic amines. They are oxidised to corresponding amino benzoic acids. p-toluidine forms p-aminobenzoic acid by following sequence of reactions.

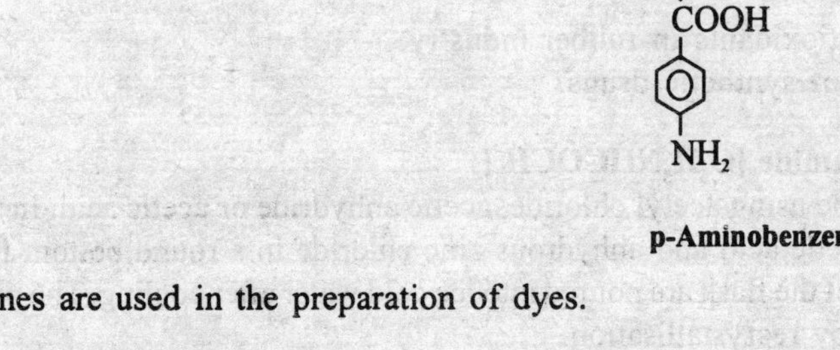

p-Aminobenzene acid

Toluidines are used in the preparation of dyes.

32.3.4   Nitroanilines [$NO_2C_6H_4NH_2$]

Three isomeric nitroanilines (o-, m- and p-) are known. o- and p-nitroanilines are prepared by nitration of acetanilide followed by separation of o- and p-product and hydrolysis.

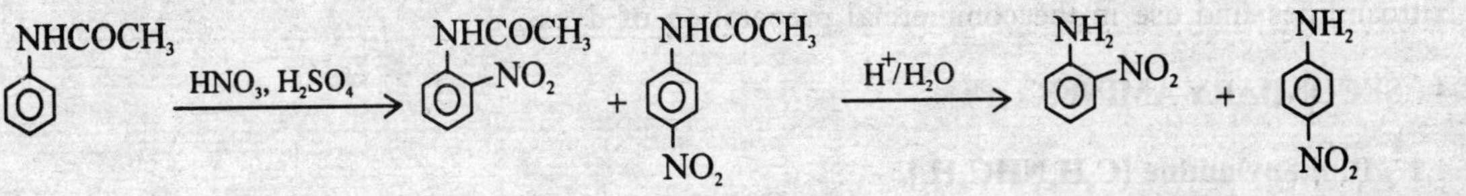

NHCOCH₃ → (HNO₃, H₂SO₄) → NHCOCH₃–NO₂ + NHCOCH₃ (NO₂) → (H⁺/H₂O) → NH₂–NO₂ + NH₂ (NO₂)

o-Nitroaniline   p-Nitroaniline

o-Nitroacetanilide, being solid, is separated from p-nitroacetanilide (liquid) by simple filtration.

o- and p-Nitroaniline may also be prepared by heating corresponding chloronitrobenzenes with ammonia. The o-nitroaniline may also be prepared in good yield by following sequence of reactions.

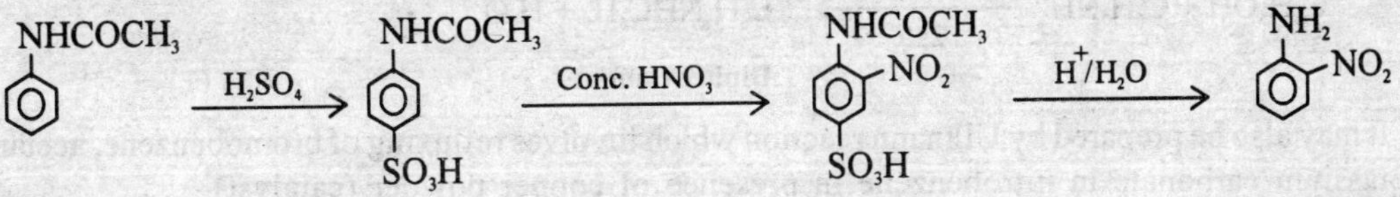

NHCOCH₃ → (H₂SO₄) → NHCOCH₃ (SO₃H) → (Conc. HNO₃) → NHCOCH₃–NO₂ (SO₃H) → (H⁺/H₂O) → NH₂–NO₂

Acetanilide   **o-Nitroaniline**

m-Nitroaniline is obtained by treatment of m-dinitrobenzene with ammonium hydrogen sulphide or sodium sulphide.

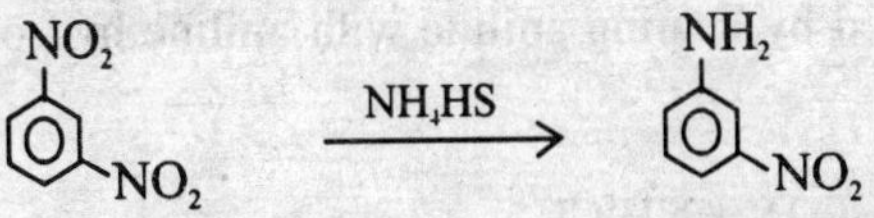

NO₂ (NO₂) → (NH₄HS) → NH₂ (NO₂)

m-Dinitrobenzene   **m-Nitroaniline**

Properties: All the three isomeric nitroanilines are solids (o-, m.p. 61°; m-, m.p. 114°; p-, m.p. 148°) having low solubility in water but are freely soluble in alcohol.

They are weak bases compared to aniline due to the electron withdrawing deactivating effect of nitro group. They give corresponding diamines on reduction. o-Nitroaniline on reduction gives o-phenylenediamine.

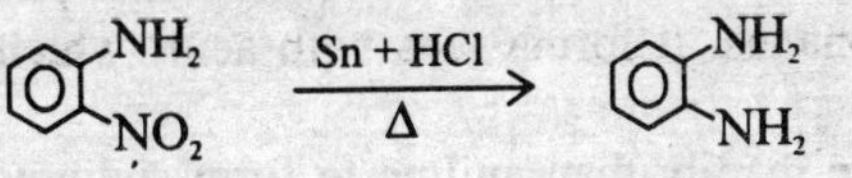

NH₂ (NO₂) → (Sn + HCl, Δ) → NH₂ (NH₂)

o-Nitroaniline   **o-Phenylenediamine**

o- and p-Nitroanilines (not m-nitroaniline) react with hot aqueous caustic soda to from corresponding nitrophenols.

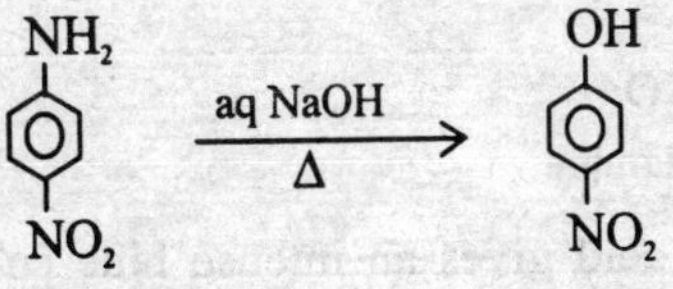

NH₂ (NO₂) → (aq NaOH, Δ) → OH (NO₂)

p-Nitroaniline   **p-Nitrophenol**

m-Nitroaniline on nitration forms 2,3,4,6-tetranitroaniline (TNA), an explosive.

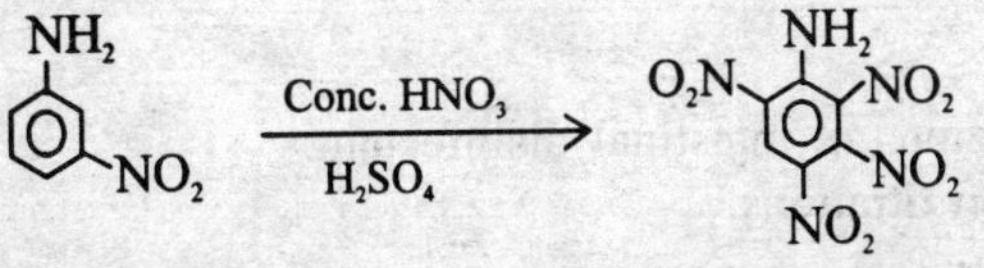

NH₂ (NO₂) → (Conc. HNO₃, H₂SO₄) → O₂N–NH₂–NO₂ (NO₂) (NO₂)

m-Nitroaniline   **2,3,4,6-Tetranitroaniline**

Nitroanilines find use in the commercial preparation of dyes.

32.4   SECONDARY AMINES

32.4.1   Diphenylamine [$C_6H_5NHC_6H_5$]

Preparation

1.   Diphenylamine may be prepared by heating phenol and aniline in presence of anhydrous zinc chloride at 250°.

$$C_6H_5OH + C_6H_5NH_2 \xrightarrow[250°]{ZnCl_2} C_6H_5NHC_6H_5 + H_2O$$
Diphenylamine

2.   It may also be prepared by Ullmann reaction which involves refluxing of bromobenzene, acetanilide and potassium carbonate in nitrobenzene in presence of copper powder (catalyst).

$$C_6H_5NHCOCH_3 + C_6H_5Br + K_2CO_3 \xrightarrow{Cu} C_6H_5NHC_6H_5 + CH_3COOK + KBr + CO_2$$
Diphenylamine

3.   On commercial scale diphenylamine is prepared by heating aniline with aniline hydrochloride in a sealed tube at 200°.

$$C_6H_5NH_2 + C_6H_5\overset{+}{N}H_3\bar{C}l \longrightarrow C_6H_5NHC_6H_5 + NH_4Cl$$
Diphenylamine

Properties: It is a white, pleasant smelling crystalline solid, m.p 54°, insoluble in water but soluble in ethanol and ether. It shows the usual reactions shown by aliphatic secondary amines. In addition, it gives the following reactions also.

(I)   Diphenylamine is a weaker base than aniline (see Table 32.1) since it contains two phenyl groups sharing the lone pair of electron of nitrogen for resonance. It forms salts with acids which hydrolyse completely in aqueous solution.

(II)   It reacts with sodium metal or sodamide more readily than aniline to form sodium derivative.

$$(C_6H_5)_2NH + Na \longrightarrow (C_6H_5)_2NNa + 1/2\ H_2$$

(III)   With nitrous acid, N-nitrosodiphenylamine is produce

$$(C_6H_5)_2NH + HNO_2 \xrightarrow{0-5°} (C_6H_5)_2NNO + H_2O$$
N-Nitrosodiphenylamine

(IV)   In presence of nitrates, diphenylamine in sulphuric acid gives an intense blue color (a confirmatory test for nitrates)

Uses

1. It is used in the preparation of dyes.
2. As a reagent for detection of nitrates.
3. It is used in the preparation of phenothiazine, an intestinal disinfectant.
4. As internal indicator in oxidation reduction titrations.
5. As a stabiliser for explosives.

32.4.2 N-Methylaniline/N-Methylphenylamine [$C_6H_5NHCH_3$]

Preparation

(I) N-methylaniline may be prepared by heating equimolar amounts of aniline and methyl iodide.

$$C_6H_5NH_2 + CH_3I \longrightarrow C_6H_5NHCH_3 + HI$$
N-Methylaniline

(II) On large scale, it is prepared by heating aniline with excess of methanol in presence of concentrated sulphuric acid at 250° under pressure.

$$C_6H_5NH_2 + CH_3OH \xrightarrow[250°]{H_2SO_4} C_6H_5NHCH_3 + H_2O$$
N-Methylaniline

Properties: It is a colorless liquid, b.p. 195°. It is a stronger base than aniline (see Table 32.1) due to electron releasing +I effect of methyl group. It resembles (in some of its reactions) aliphatic secondary amines but differs from them in electrophilic ring substitution. It directs the substituents at ortho and para positions of the ring.

(I) With nitrous acid, it forms N-nitrosomethylaniline.

$$C_6H_5NHCH_3 + HNO_2 \longrightarrow \begin{array}{c} C_6H_5 \\ {}^{\diagdown}NNO + H_2O \\ CH_3{}^{\diagup} \end{array}$$
N-Nitrosomethylaniline

The hydrolysis of N-nitroso derivative gives back N-methylaniline.

(II) When heated strongly, it forms p-toluidine (involving the migration of methyl group.)

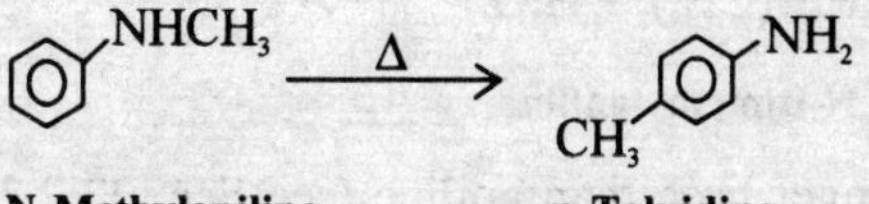

N-Methylaniline   p-Toluidine

It is used in the preparation of dyes.

32.5 TERTIARY AMINES

32.5.1 Triphenylamine [$(C_6H_5)_3N$]

1. Triphenylamine may be prepared by heating sodium salt of diphenylamine with chlorobenzene or bromobenzene.

$$(C_6H_5)_2NNa + C_6H_5Br \longrightarrow (C_6H_5)_3N + NaBr$$
Triphenylamine

2. It may also be prepared by heating diphenylamine and iodobenzene in nitrobenzene with potassium carbonate in presence of copper powder (Ullmann reaction).

$$2\,(C_6H_5)_2NH + 2\,C_6H_5I + K_2CO_3 \xrightarrow{Cu} 2\,(C_6H_5)_3N + CO_2 + 2\,KI + H_2O$$
Triphenylamine

Properties: It is a colorless, crystalline solid, m.p 127°. It does not form salt with acids as it is an extremely weak base due to the sharing of lone pair of electron of nitrogen with the three phenyl groups. It, however gives a blue color when dissolved in concentrated sulphuric acid.

32.5.2   N,N-Dimethylaniline [$C_6H_5N(CH_3)_2$]

1. N,N-Dimethylaniline may be prepared by alkylation of aniline.

$$C_6H_5NH_2 \xrightarrow[-HI]{CH_3I} C_6H_5NHCH_3 \xrightarrow[-HI]{CH_3I} C_6H_5N(CH_3)_2$$

Aniline   N-Methylaniline   N,N-Dimethylaniline

2.   It may also be prepared by heating aniline and methanol (in excess) under pressure in presence of sulphuric acid.

$$C_6H_5NH_2 + 2CH_3OH \xrightarrow[250°]{H_2SO_4} C_6H_5N(CH_3)_2 \quad + 2H_2O$$

N,N-Dimethylaniline

The yield of N,N-dimethylaniline in the above reaction is over 95%.

3.   Another method involves catalytic hydrogenation of a mixture of aniline and formaldehyde in aqueous alcoholic mineral acid.

$$C_6H_5NH_2 + 2CH_2O + 2H_2 \xrightarrow[H^+/H_2O]{Pt} C_6H_5N(CH_3)_2 + 2H_2O$$

4.   When a mixture of aniline and dimethyl ether vapors are passed over heated alumina N,N-Dimethylaniline is formed.

$$C_6H_5NH_2 + CH_3OCH_3 \xrightarrow[200-300°]{Al_2O_3} C_6H_5N(CH_3)_2 + H_2O$$

N,N-Dimethylaniline

Properties: It is an oily liquid, b.p 193°. It is a stronger base than aniline (see Sec. 32.2.2) but forms salts with mineral acids. The methyl groups on nitrogen activates the ring for o- and p- substitution.

1.   It forms p-nitrosodimethylaniline when treated with nitrous acid.

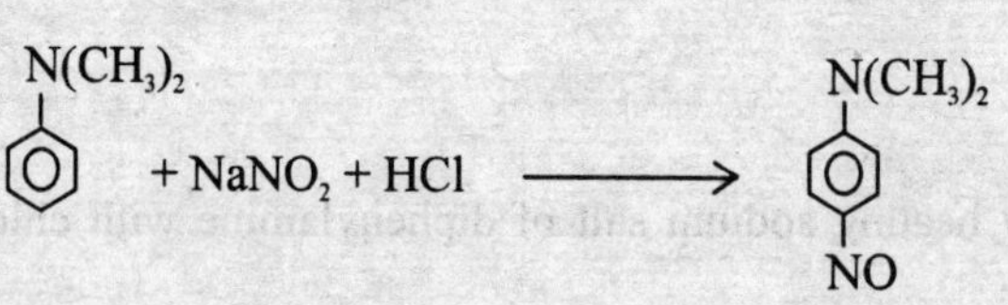

p-Nitrosodimethylaniline

The nitroso substituted dimethylaniline when boiled with alkali splits to form p-nitrosophenol.

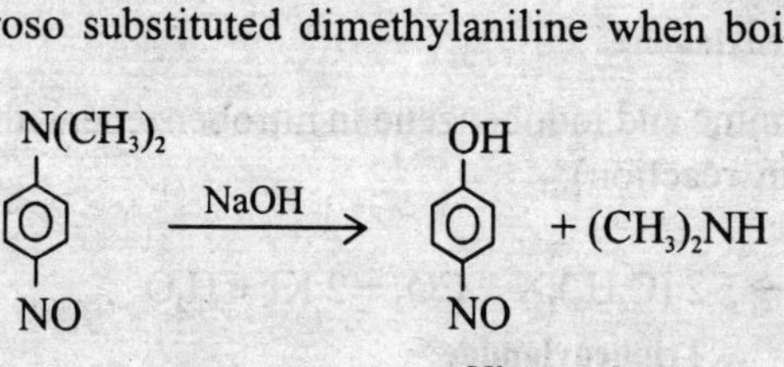

p-Nitrosophenol

2. N,N-Dimethylaniline condenses with formaldehyde and carbonyl chloride to form diphenylmethane and benzophenone derivatives respectively.

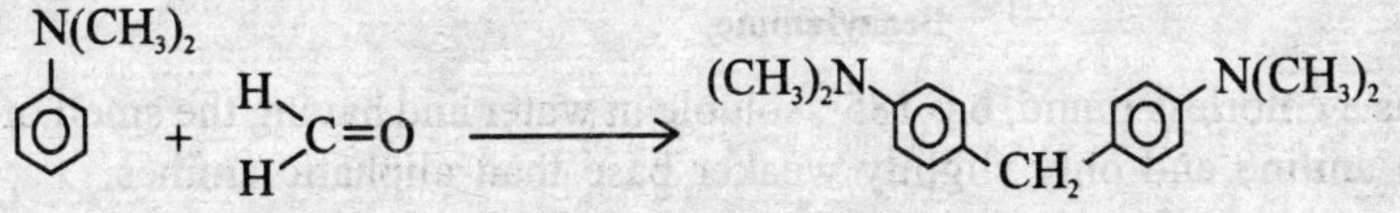

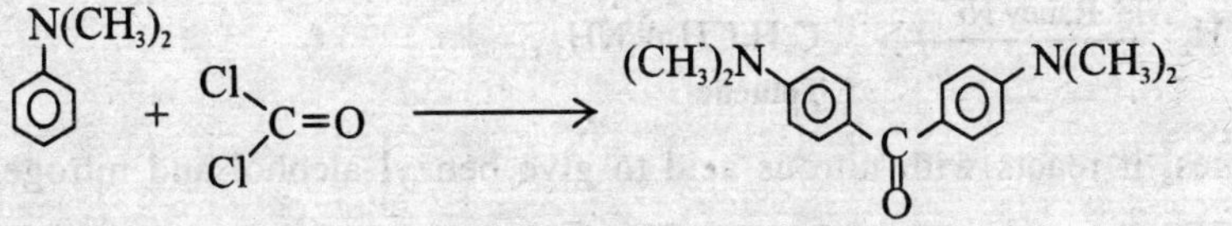

3. With aromatic aldehyde, triphenylmethane derivatives are formed.

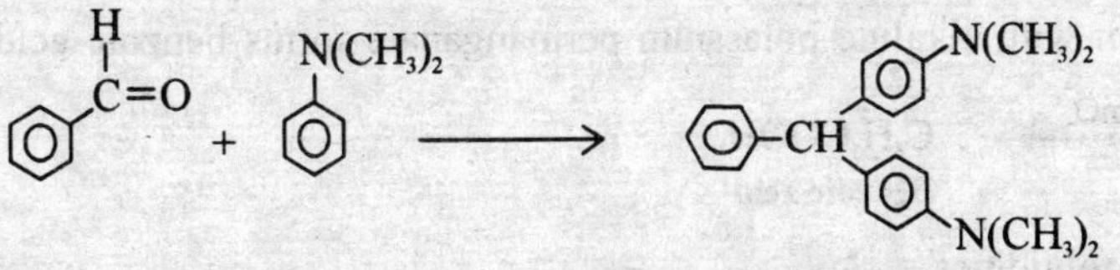

4. Dimethylanilinium chloride when heated strongly under pressure forms 2,4-dimethylaniline due to migration of methyl groups from nitrogen to the ring.

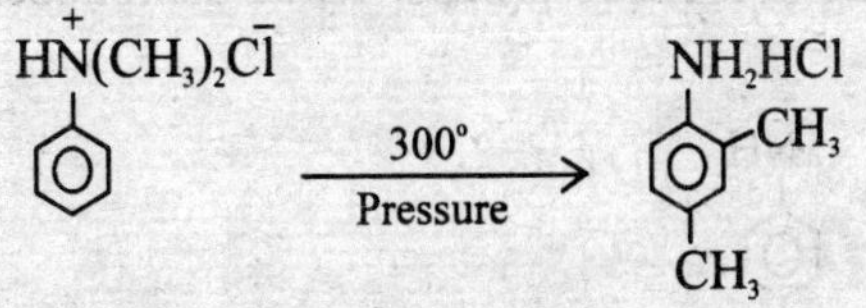

This reaction is known as Hofmann martius rearrangement and is employed to prepare alkyl substituted anilines.

It is used in various dye preparations. It is also used for the preparation of dimethylamine.

32.6 ARYLALKYLAMINES

32.6.1 Benzylamine [$C_6H_5CH_2NH_2$]

1. It is prepared by the reduction of phenyl cyanide or benzaldoxime.

$$C_6H_5CN \xrightarrow{\text{Na, } C_2H_5OH} C_6H_5CH_2NH_2$$

Phenyl cyanide   **Benzylamine**

$$C_6H_5CH=NOH \xrightarrow{\text{Na, } C_2H_5OH} C_6H_5CH_2NH_2$$

Benzaldoxime   **Benzylamine**

2. The reaction of benzyl chloride with ammonia, substitution nucleophilic, yields benzylamine.

$$C_6H_5CH_2Cl + NH_3 \longrightarrow C_6H_5CH_2NH_2 + HCl$$

                                       Benzylamine

3.   Phenylacetamide undergoes Hofmann degradation to from benzylamine.

$$C_6H_5CH_2CONH_2 \xrightarrow{\text{Br}_2,\ \text{NaOH}} C_6H_5CH_2NH_2$$
Benzylamine

Properties: Benzylamine is a colorless liquid, b.p 185°, soluble in water and having the smell of ammonia. It is a stronger base than aniline and only slightly weaker base than aliphatic amines.

1.   It resembles aliphatic amines in the reactions of amino group. It forms toluene on catalytic hydrogenation.

$$C_6H_5CH_2NH_2 + H_2 \xrightarrow{\text{Raney Ni}} C_6H_5CH_3 + NH_3$$
Toluene

2.   Like aliphatic amines, it reacts with nitrous acid to give benzyl alcohol and nitrogen.

$$C_6H_5CH_2NH_2 + HNO_2 \longrightarrow C_6H_5CH_2OH + N_2 + H_2O$$
Benzyl alcohol

3.   Benzylamine on oxidation with alkaline potassium permanganate forms benzoic acid.

$$C_6H_5CH_2NH_2 \xrightarrow{\text{KMnO}_4} C_6H_5COOH$$
Benzoic acid

It is isomeric with the three toluidines.

32.7   DIAMINES

The diamines comprise the aromatic compounds having two amino groups. Three diaminobenzenes are known.

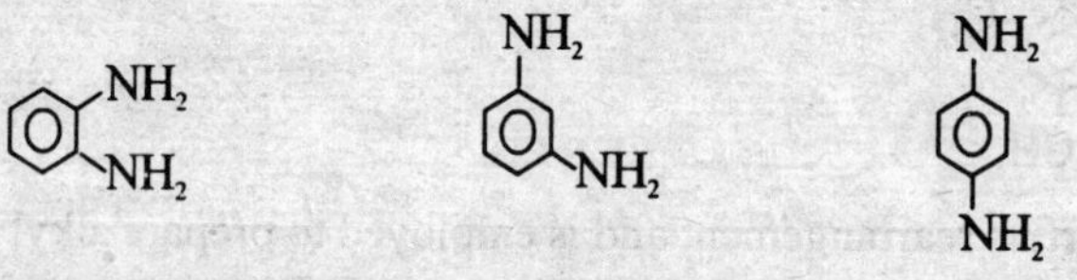

o-Phenylenediamine
(1,2-Benzenediamine)   **m-Phenylenediamine**
(1,3-Benzenediamine)   **p-Phenylenediamine**
(1,4-Benzenediamine)

32.7.1   o-Phenylenediamine/1,2-Benzenediamine

It is prepared by the reduction of o-dinitrobenzene or o-nitroaniline with zinc dust and sodium hydroxide.

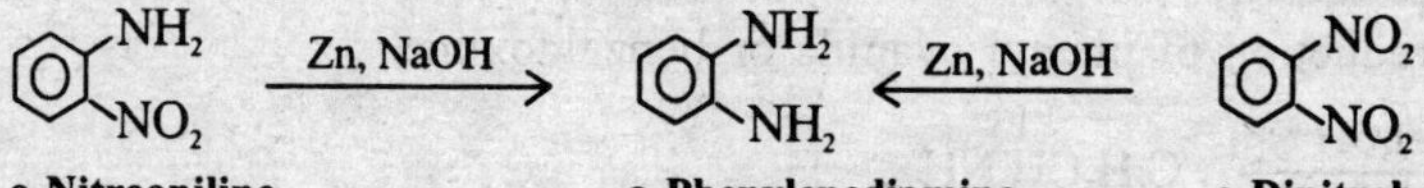

o-Nitroaniline   o-Phenylenediamine   o-Dinitrobenzene

Properties: It is a white crystalline solid, m.p. 101°, readily turns brown due to aerial oxidation. It gives usual reactions of amino group. It forms a variety of heterocyclic compounds.

1.   With acetic acid, it forms 2-methylbenzimidazole.

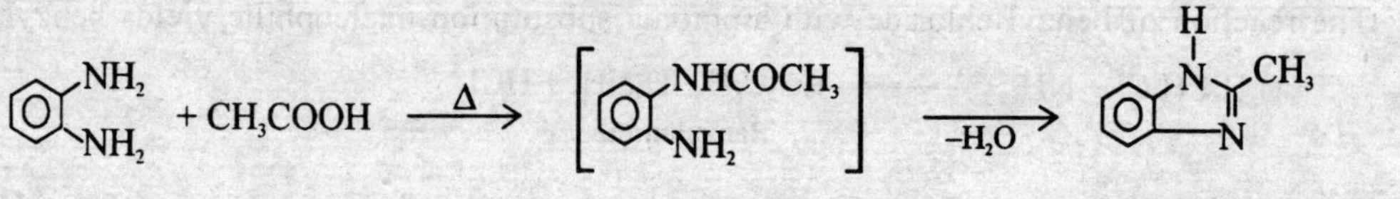

2-Methylbenzimidazole

2. With nitrous acid, benztriazoles are formed.

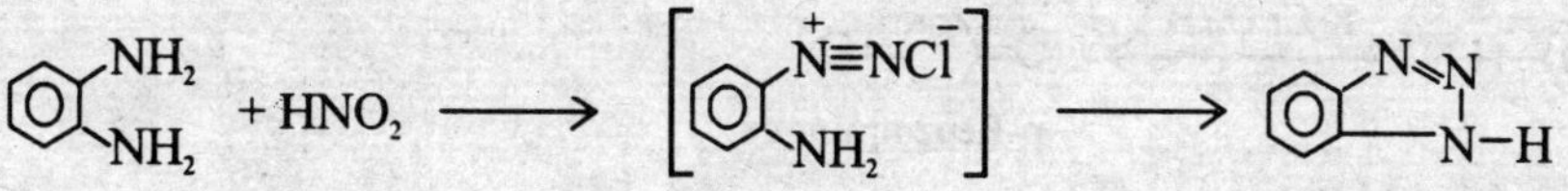

Benztriazole

3. With glyoxal, it forms quinoxaline.

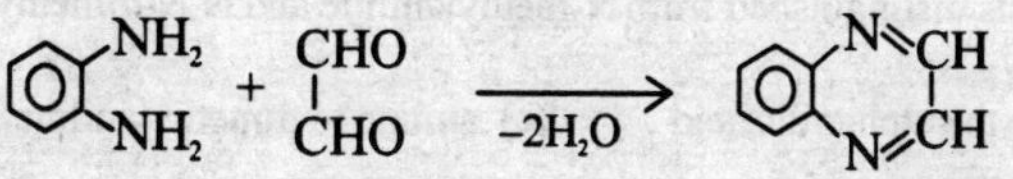

Quinoxaline

4. It gives dark red color with ferric chloride due to the formation of 2,3-diaminophenozine.

2.7.2 m-Phenylenediamine/1,3-Benzenediamine

t is best prepared by the reduction of m-dinitrobenzene.

It is a white crystalline solid, m.p. 63° which turns brown on standing in air. It forms bismark brown a brown dye) when treated with nitrous acid. Both monoazo and bis-diazo compounds are formed.

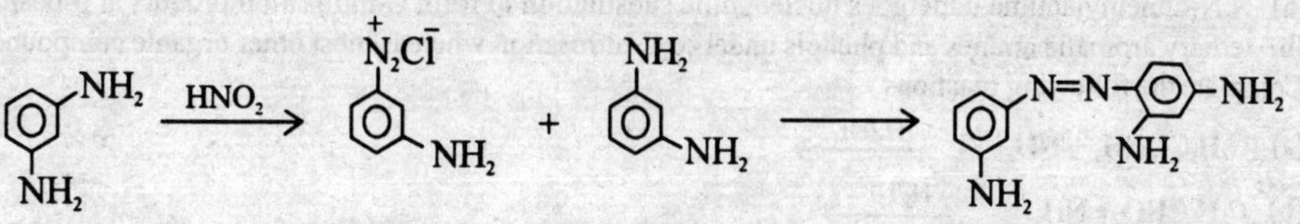

Monoazodye (Brown)

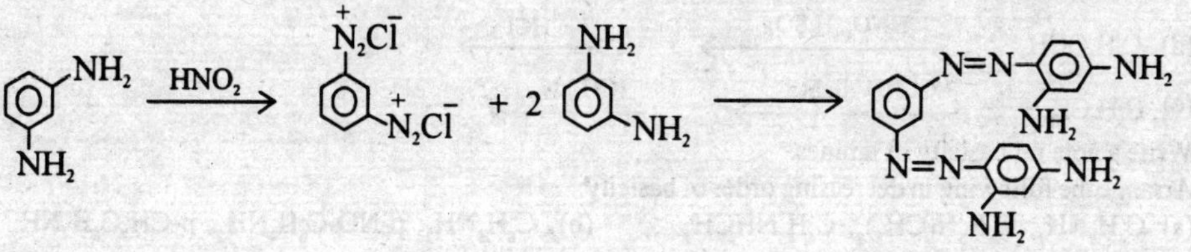

Bismark Brown

This reaction occurs in very dilute solutions, thus used in colorimetry for the detection and estimation f nitrites in drinking water

Bismark brown dye is used in boot polish preparations.

2.7.3 p-Phenylenediamine/1,4-Benzenediamine

t may be prepared by the reduction of p-nitroaniline with usual reducing agents.

Another convenient method involves reduction of p-aminoazobenzene with sodium thiosulphate.

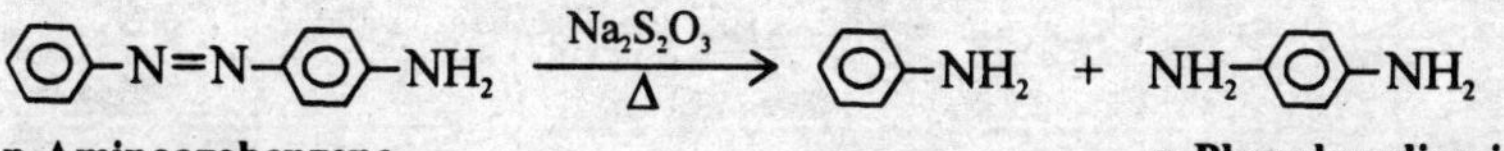

p-Aminoazobenzene   p-Phenylenediamine

It is a white crystalline solid, m.p. 147°. It reacts with nitrous acid to give bis-diazonium salt which ouples with the diamine to form azodyes.

On vigrous oxidation with acidic potassium dichromate, it forms p-benzoquinone.

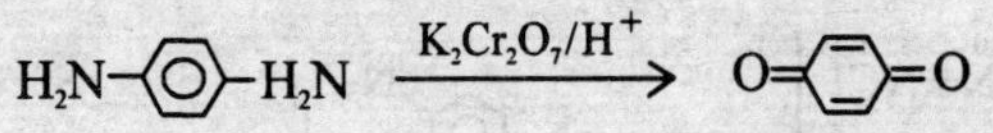

$$H_2N-\langle O \rangle-H_2N \xrightarrow{K_2Cr_2O_7/H^+} O=\langle \rangle=O$$

p-Benzoquinone

It is used as an excellent photographic developer. It is also used as hair dye.

32.8 PROBLEMS

1. How is aniline prepared in laboratory? How it is distinguished from N-methylaniline and N,N-dimethylaniline?

2. How will you convert:
   (a) aniline to p-bromoaniline   (b) aniline to sulphanilic acid   (c) aniline to dimethylaniline

3. What happens when aniline reacts with:
   (a) Methyl iodide   (b) Acetyl chloride
   (c) Benzenesulphonyl chloride (d) Nitrous acid
   (e) Chloroform + KOH   (f) Carbon disulphide
   (g) Potassium dichromate in acid (h) Bromine water
   (i) Conc. sulphuric acid   (j) Conc. sulphuric and nitric acid.

4. Write a note on
   (a) Hofmann degradation   (b) Hofmann Martius rearrangement

5. Give reasons
   (a) N,N-dimethylaniline undergoes nucleophilic substitution to form C-nitrosation product at p-position.
   (b) tertiary aromatic amines and phenols undergo C-nitrosation whereas most other organic compounds do not.

6. Complete the following reactions.
   (a) $C_6H_5COCH_3 + NH_3 \xrightarrow{H_2/Ni}$

   (b) $C_6H_5CHO + NH_3 \xrightarrow{H_2/Ni}$

   (c) $C_6H_5COOH \xrightarrow{PCl_5} \xrightarrow{C_2H_5NH_2} \xrightarrow{H_2/Ni}$

   (d) $C_6H_5OCH_3 \xrightarrow{HNO_3, H_2SO_4} \xrightarrow{Fe, HCl}$

   (e) $C_6H_5CH_3 \xrightarrow{Cl_2/UV\ light} \xrightarrow{(CH_3)_3N}$

7. Write a note on basicity of amines.

8. Arrange the following in decreasing order of basicity:
   (a) $C_6H_5NH_2$; $C_6H_5N(CH_3)_2$; $C_6H_5NHCH_3$ (b) $C_6H_5NH_2$; $p-NO-C_6H_4NH_2$; $p-CH_3C_6H_4NH_2$
   (c) $C_6H_5NH_2$; $(C_6H_5)_3N$; $(C_6H_5)_2NH$

9. How will you synthesise following compounds from benzene or toluene.
   (a) 4-amino-2-chlorotoluene   (b) 4-amino-3-chlorotoluene
   (c) 4-aminoacetanilide   (d) 4-aminobenzylamine
   (e) 3-bromoaniline   (f) phenylethylamine
   (g) N-ethylaniline   (h) 2,4-dinitroaniline.

10. How will you convert p-toluenediazonium salt into:
    (a) toluene   (b) 4-hydroxy-4-methylazobenzene.

11. Discuss various methods of distinguishing primary, secondary and tertiary amines.

33

Diazonium Salt and Related Compounds

33.1 INTRODUCTION

Primary aromatic amines when treated with nitrous acid in cold form an unstable compound known as diazonium salt.

$$ArNH_2 + HNO_2 + HCl \longrightarrow ArN_2Cl + H_2O$$

Aromatic amine   **Aryldiazonium chloride**

The same reaction, however, with primary aliphatic amine gives alcohol and a quantitative amount of nitrogen.

$$RNH_2 + HNO_2 \longrightarrow ROH + N_2 + H_2O$$

Aliphatic amines may also yield diazo compounds provided there is an electron withdrawing group (e.g. cyano, acyl or ester) attached to the carbon next to amino group.

$$CH_2(NH_2)CO_2C_2H_5 \xrightarrow{\text{NaNO}_2,\ \text{HCl}} N_2CHCO_2C_2H_5$$

Ethyl diazoacetate

Thus, the reaction of primary aromatic amine and nitrous acid in ice cold solution produces diazonium salt and is known as diazotisation.

33.2 NOMENCLATURE

The diazonium salts are named by adding suffix diazonium chloride or diazonium sulphate to the parent hydrocarbon. for example,

Benzenediazonium sulphate   p-Toluenediazonium chloride   m-Bromobenzene-diazonium fluoroborate   β-Naphthalene-diazonium bromide

33.3 PREPARATION

Benzenediazonium salt, in particular, is prepared by cooling one mole of aniline and 3 moles of hydrochloric acid to 0-5° in an ice bath. A cold, aqueous sodium nitrite is then added slowly with constant stirring. The excess of sodium nitrite (tested with starch iodide paper) is destroyed by addition of urea.

$$NH_2CONH_2 + HNO_2 \longrightarrow CO_2 + 2N_2 + 3H_2O$$

The diazonium chloride thus produced is soluble in water and decomposes slowly even at 0-5°, therefore, used immediately for further reactions.

$$C_6H_5NH_2 + NaNO_2 + HCl \xrightarrow{0-5°} C_6H_5N_2Cl + NaCl + 2H_2O$$

Aromatic amines substituted with nitro or chloro groups require more concentrated acids (than for unsubstituted amines) due to the weak basic character of amines. Di- or trisubstituted amines require concentrated sulphuric acid to dissolve the amine and sodium nitrite too is taken in concentrated sulphuric acid.

The mechanism of diazotisation is not confirmed but the rate determining step involves free base. A probable mechanism has been outlined below.

$$Ar-\overset{..}{N}H_2 + HO-N{=}O \rightleftharpoons Ar-\overset{+}{N}H-H \rightleftharpoons Ar-\overset{..}{N}-H$$
$$HO-N-O \qquad\qquad N{=}O$$

$$H_2O + [\ Ar-\overset{+}{N}{\equiv}N\ ]\bar{C}l \rightleftharpoons Ar-\overset{..}{N}$$
$$N{-}OH$$

Diazonium chloride   **Diazonium hydroxide**

33.4 PROPERTIES

Physical Properties: Aryl diazonium compounds are salts of strong base diazonium hydroxide (ArN_2OH) which has not been isolated so far, but known in aqueous solutions. Aryl diazonium salts are colorless, crystalline solids which turn brown when come in contact with air. They are unstable and explode when dry. They are extremely soluble in water but only sparingly soluble in ethanol and insoluble in ether. Their double salts with metals like zinc $[(ArN_2)_2^{2+}ZnCl_4^{2-}]$ are stable in solution, hence offer a method of stabilisation of diazonium salts.

Chemical Properties: The diazonium salts have synthetic importance as they are used for the synthesis of various organic compounds, dyes and drugs. They involve mainly two types of reactions; one replacement reactions, in which the diazo group is replaced by some other group and the other, in which the two nitrogen atoms of the diazo group are retained.

Replacement Reactions

1. *Replacement by Hydroxy group:* Steam distillation of diazonium salt solution replaces the diazo group by hydroxy group. The solution should be acidic enough to avoid coupling which leads to tar formation.

$$C_6H_5N_2HSO_4 + H_2O \longrightarrow C_6H_5OH + N_2 + H_2SO_4$$

2. *Replacement by Hydrogen:* Hypophosphorus acid replaces the diazo group by hydrogen specially in aliphatic aromatic diamines.

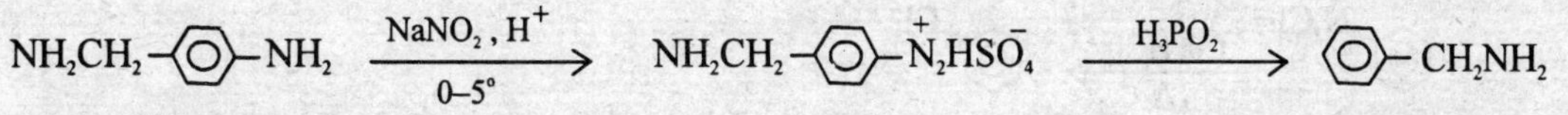

$$NH_2CH_2-\langle\bigcirc\rangle-NH_2 \xrightarrow[0-5°]{NaNO_2,\,H^+} NH_2CH_2-\langle\bigcirc\rangle-\overset{+}{N_2}HS\bar{O}_4 \xrightarrow{H_3PO_2} \langle\bigcirc\rangle-CH_2NH_2$$

p-Aminobenzylamine   **Benzylamine**

Another common method of replacing diazo group by hydrogen is by warming the salt in ethanol in presence of copper catalyst.

$$C_6H_5\overset{+}{N_2}HS\bar{O}_4 + C_2H_5OH \xrightarrow{Cu} C_6H_6 + CH_3CHO + N_2 + H_2SO_4$$
                                                            Benzene

The replacement here takes place via free radical mechanism.

$$C_6H_5N=NCl \longrightarrow C_6\dot{H}_5 + N_2 + C\dot{l}$$
$$C_6\dot{H}_5 + CH_3CH_2OH \longrightarrow C_6H_6 + CH_3\dot{C}HOH$$
$$CH_3\dot{C}HOH + C\dot{l} \longrightarrow CH_3CHO + HCl$$

A better yield is obtained when aryl diazonium fluoroborate and ethanol are warmed in presence of zinc dust.

$$ArN_2BF_4 + C_2H_5OH \xrightarrow{Zn} ArH$$

3. Replacement by Halogen

(a) *Sandmaeyer Reaction:* Diazonium salts when react with cuprous halide dissolved in corresponding halogen acid, replace the azo group by a halogen atom.

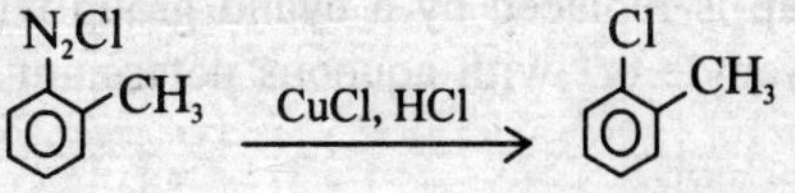

o-Chlorotoluene

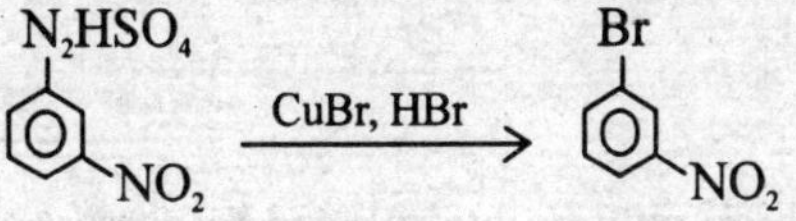

m-Nitrobromobenzene

In the above reaction it is the halogen atom joined to copper, that enters the benzene ring. The mechanism is uncertain but initiated most probably with the reaction of cuprous chloride with a halide ion to form reactive $CuCl_2^-$ ion, which decomposes the diazonium ion into aryl free radical.

$$HCl + CuCl \longrightarrow CuC\bar{l}_2 + H^+$$
$$Ar\overset{+}{N_2} + CuC\bar{l}_2 \xrightarrow{Slow} A\dot{r} + CuCl_2 + N_2$$
$$A\dot{r} + CuCl_2 \longrightarrow Ar-Cl + CuCl$$

(b) *Gattermann Reaction:* It is a modification of Sandmeyer reaction and involves the warming of diazonium salt in halogen acid in presence of copper powder to yield corresponding halo compound.

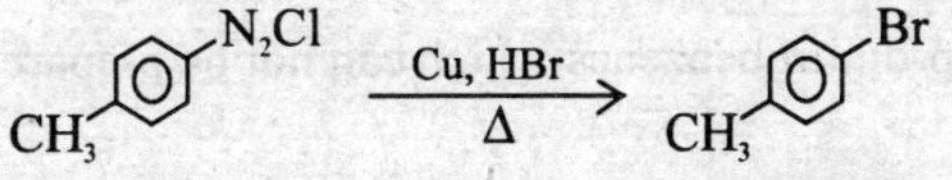

p-Bromotoluene

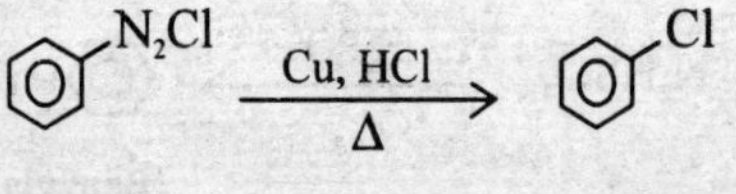

Chlorobenzene

(c) *Replacement by Iodine:* Diazonium salt reacts readily with sodium or potassium iodide to give iodo compounds.

$$ArN_2Cl + KI \longrightarrow Ar-I + N_2 + KCl$$

The mechanism involves nucleophilic substitution of iodide in place of diazo group.

(d) *Replacement by Fluorine* (Balz-Schiemann reaction): Fluoroboric acid forms stable diazonium fluoroborate which on slow heating in absence of solvent decomposes to give aryl fluoride.

$$Ar\overset{+}{N_2}\overset{-}{Cl} + HBF_4 \longrightarrow Ar\overset{+}{N_2}\overset{-}{BF_4} + HCl$$

$$Ar\overset{+}{N_2}\overset{-}{BF_4} \xrightarrow{\Delta} Ar-F + BF_3 + N_2$$

This is the best method of introduction of fluorine in the benzene ring.

Fluoro compounds may also be prepared by adding solid sodium nitrite in a well cooled solution of amine in anhydrous hydrofluoric acid and evaporating the excess of hydrofluoric acid.

$$ArNH_2 \xrightarrow{NaNO_2, HF} ArN_2F \xrightarrow{\Delta} ArF + N_2$$

4. *Replacement by a Cyano Groups:* The diazo group is replaced by a cyano group when treated with cuprous cyanide dissolved in aqueous potassium cyanide or with aqueous potassium cyanide in presence of copper powder.

$$ArN_2Cl \xrightarrow{CuCN, KCN} ArCN + N_2$$

$$ArN_2Cl \xrightarrow[KCN]{Cu} ArCN + N_2$$

It is a useful method of preparation of aryl cyanide which may be hydrolysed to corresponding acid easily.

5. *Replacemnt by a Nitro Group:* Diazonium salt when treated with nitrous acid in presence of cuprous oxide forms nitro compound.

$$C_6H_5N_2Cl + HNO_2 \xrightarrow{Cu_2O} C_6H_5NO_2 + HCl + N_2$$

Nitrobenzene

A better method, however, is the reaction of diazonium borofluoride with aqueous sodium nitrite in presence of copper powder.

$$Ar-\overset{+}{N_2}\overset{-}{BF_4} \xrightarrow[Cu]{NaNO_2} Ar-NO_2 + NaBF_4 + N_2$$

The reaction is used to synthesise o-dinitro and p-dinitrobenzenes which can not be prepared by direct nitration.

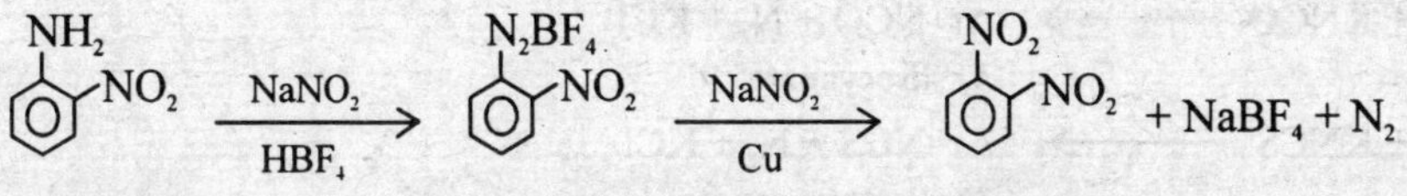

o-Nitroaniline   o-Dinitrobenzene

6. *Replacement by Arsonic Acid Group* (Bart's reaction): The decomposition of diazonium salt with sodium arsenite in presence of copper sulphate forms arylarsonic acid.

$$Ar-N_2Cl + Na_3AsO_3 \xrightarrow{CuSO_4} Ar-AsO_3Na_2 + NaCl + N_2$$

$$Ar-AsO_3Na_2 + 2HCl \longrightarrow Ar-AsO_3H_2 + 2\ NaCl$$

Arylarsonic acid

It is an important reaction for preparing other aromatic nuclear arsenic compounds.

7. *Replacement by an Aryl Group:* Treatment of diazonium salt with ethanol and copper powder produces a diaryl.

$$2\ C_6H_5N_2Cl \xrightarrow{C_2H_5OH}_{Cu} C_6H_5-C_6H_5 + N_2 + 2HCl$$

Diphenyl

The reaction actually is a special case of Gattermann reaction. Alternatively, the diazo group is replaced by an aryl group by adding an aromatic compound into the alkaline solution of diazonium salt and the reaction is known as Gomberg reaction. 2-Bromobiphenyl is produced by the reaction of o-bromobenzene diazonium chloride with benzene in sodium hydroxide or sodium acetate.

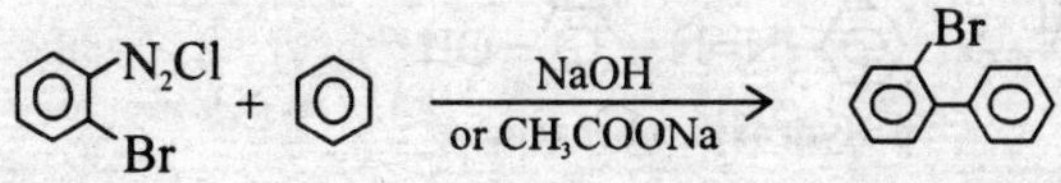

o-Bromodiphenyl

The mechanism of Gomberg reaction was suggested by Hey et al. as a free radical substitution. In the following example p-nitrodiphenyl results due to the reaction of benzene diazonium chloride and nitrobenzene.

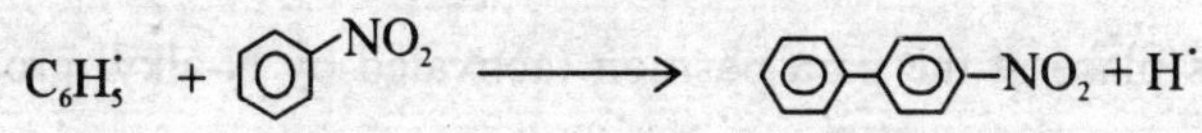

p-Nitrodiphenyl

$$H^{\cdot} + Cl^{\cdot} \longrightarrow HCl$$

8. *Replacement by Thio Group:* Diazonium salt decomposes to thiophenols when reacted upon with hydrogen sulphide.

$$ArN_2Cl + H_2S \longrightarrow ArSH + N_2 + HCl$$

$$2\ ArN_2Cl + H_2S \longrightarrow Ar-S-Ar + 2\ HCl + N_2$$

Diaryl sulphide

9. *Replacement by Isocyanides and Isothiocyanides:* Diazonium salt decompose to form arylisocyanates and arylisothiocyanates when treated with potassium isocyanide or isothiocyanide respectively.

$$ArN_2Cl + KNCO \longrightarrow Ar–NCO + N_2 + KCl$$
$$\textbf{Arylisocyanate}$$

$$ArN_2Cl + KNCS \longrightarrow Ar–NCS + N_2 + KCl$$
$$\textbf{Arylisothiocyanate}$$

10.   *Replacement by Chloromercury:* The double salt of Diazonium chloride and mercuric chloride when heated in ethanol or acetone in presence of copper powder forms chloromercurated aromatic compounds.

Cl—⟨O⟩—$N_2Cl.HgCl_2$ + 2 Cu $\longrightarrow$ Cl—⟨O⟩—$HgCl$ + N_2 + 2 CuCl

Reactions Proceeding without the loss of Nitrogen

1.   *Reduction:* Diazonium salts when reduced with stannous chloride and hydrochloric acid form aryl hydrazines.

$$ArN_2Cl + 4[H] \xrightarrow{SnCl_2/HCl} ArNHNH_2.HCl$$

In presence of strong reducing agent such as zinc and hydrochloric acid, aromatic amines are produced

$$ArN_2Cl \xrightarrow{Zn, HCl} ArNHNH_2 \xrightarrow{Zn, HCl} ArNH_2 + NH_3$$

2.   *Coupling Reactions:* Diazonium salts undergo coupling readily to form azo dyes. The compounds that couple with diazonium salt are phenols, naphthols, primary, secondary and tertiary aromatic amines. With phenol, p-hydroxyazobenzene, an orange dye, is produced.

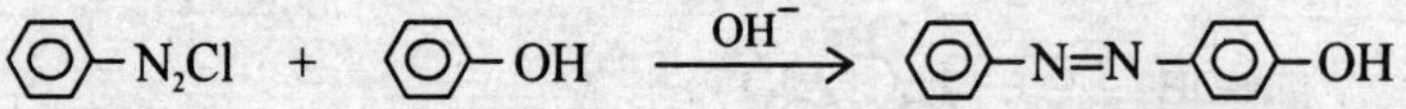

⟨O⟩—N_2Cl + ⟨O⟩—OH $\xrightarrow{OH^-}$ ⟨O⟩—N=N—⟨O⟩—OH

$$\textbf{p-Hydroxyazobenzene}$$

When primary or secondary amine couples with diazonium salt the product may be an N-azo compound (diazoamino compound) or a C-azo compound (azoamino compound) depending upon the nature of compound and the pH of solution.

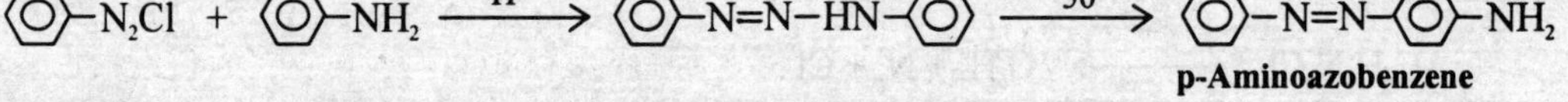

⟨O⟩—N_2Cl + ⟨O⟩—NH_2 $\xrightarrow{H^+}$ ⟨O⟩—N=N—HN—⟨O⟩ $\xrightarrow{50°}$ ⟨O⟩—N=N—⟨O⟩—NH_2

$$\textbf{p-Aminoazobenzene}$$

With tertiary amines, coupling takes place at the para position (activated by N-alkyl groups)

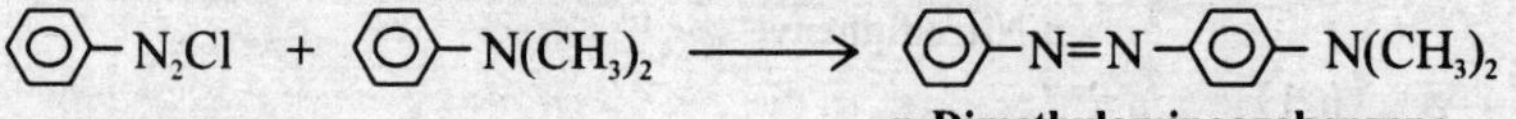

⟨O⟩—N_2Cl + ⟨O⟩—$N(CH_3)_2$ $\longrightarrow$ ⟨O⟩—N=N—⟨O⟩—$N(CH_3)_2$

$$\textbf{p-Dimethylaminoazobenzene}$$

Phenols couple in alkaline medium with the introduction of azo group mainly at para position with respect to phenolic group. In case, para position is occupied, coupling may take place at ortho position too but in no case does the coupling take place at meta position.

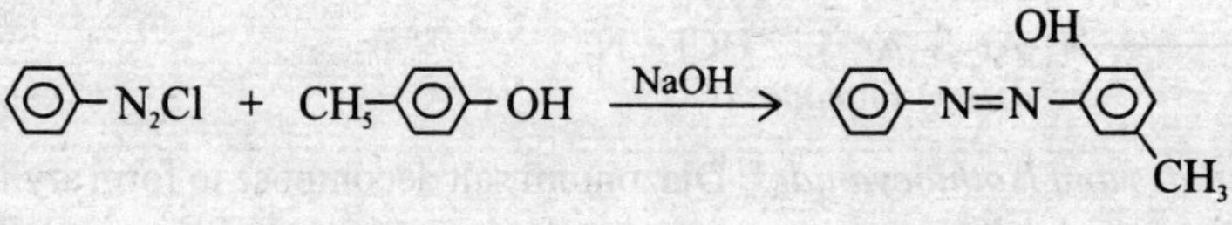

⟨O⟩—N_2Cl + CH_3—⟨O⟩—OH $\xrightarrow{NaOH}$ ⟨O⟩—N=N—⟨O⟩ (OH, CH_3)

When excess of diazonium salt is used, bisazo- and trisazo- compounds are produced.

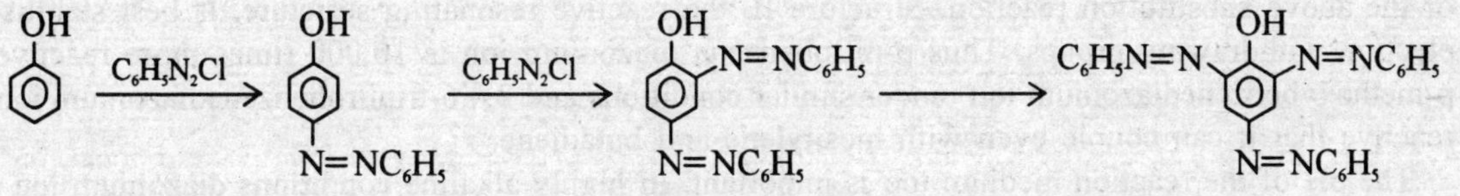

Bisazo compound **Trisazo compound**

Formation of bisazo and trisazo compounds is facilitated by the presence of alkyl group at para position to the hydroxyl group and by two hydroxy groups at meta positions. Resorcinol readily forms trisazo-derivative.

resorcinol **Bisazo compound** **Trisazo compound**

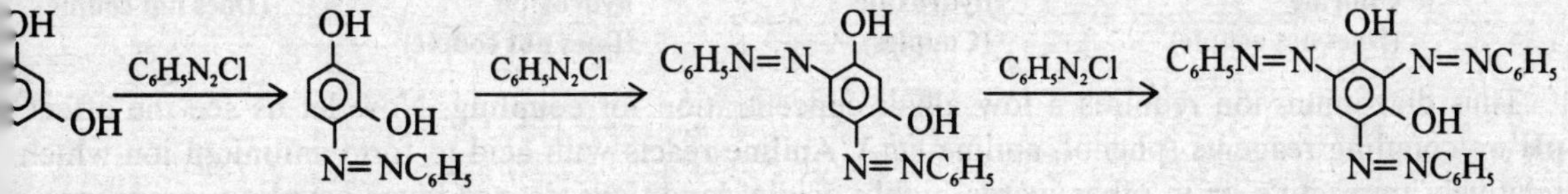

When p-aminophenol couples with diazonium salt in acidic medium, ortho position with respect to amino group is occupied whereas in alkaline medium diazo group couples at ortho position to hydroxy group.

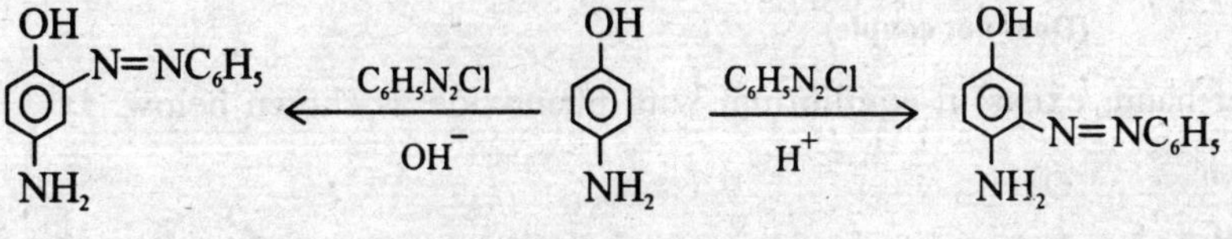

When carboxyl or sulphonic acid groups are present at para position to hydroxy group, they are replaced by azo group.

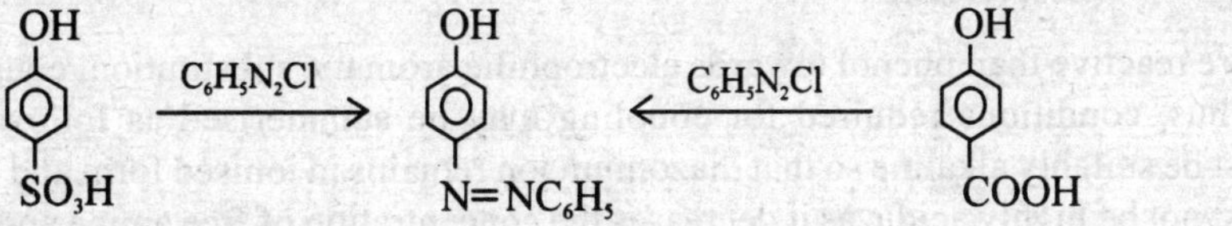

p-Hydroxybenzene sulphonic acid **p-Hydroxy-azobenzene** **p-Hydroxybenzoic acid**

Mechanism of coupling involves the attack of electrophilic aryldiazonium ion. Thus, it is an electrophilic substitution reaction. Aryldiazonium ion is a very weak electrophile and requires compounds having activating groups such as $-OH$, $-NH_2$, $-NHR$ or $-NR_2$.

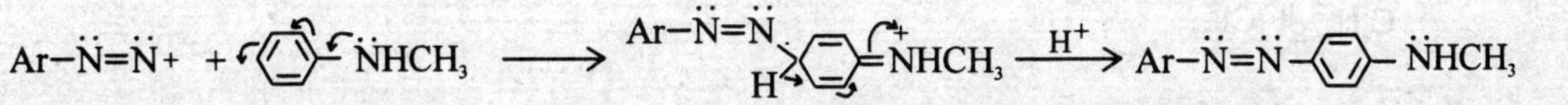

Aryldiazonium ion, a resonance hybrid of structure I and II, plays an important role in the reactivity of the above substitution reaction. Structure II, the reactive resonating structure, is best stabilised [by] electron withdrawing groups. Thus p-nitrobenzene diazonium ion is 10,000 times more reactive th[an] p-methoxybenzenediazonium ion under similar conditions and 2,4,6-trinitrobenzenediazonium ion is [so] reactive that it can couple even with mesitylene and butadiene.

The pH of the reaction medium too is important. In highly alkaline conditions diazonium ion exi[sts] as hydroxide and its sodium salt which do not couple.

$$Ar-\overset{+}{N}\equiv N\bar{C}l \xrightarrow{\text{NaOH}} Ar-\overset{+}{N}\equiv \overset{..}{N}OH^- \underset{H^+}{\overset{OH^-}{\rightleftharpoons}} Ar-N=N-OH \underset{H^+}{\overset{OH^-}{\rightleftharpoons}} Ar-N=N-\bar{O}\overset{+}{N}a$$

Aryl diazonium chloride (Does not couple)	Aryl diazonium hydroxide (Couples)	Aryl diazonium hydroxide (Does not couple)	Sodium aryl diazoa[te] (Does not couple)

Thus diazonium ion requires a low alkali concentration for coupling. Now let us see the effect [of] pH on coupling reagents (phenol, aniline etc.). Aniline reacts with acid to form anilinium ion which [is] relatively unreactive or in other words highly acidic conditions do not favor coupling in aniline.

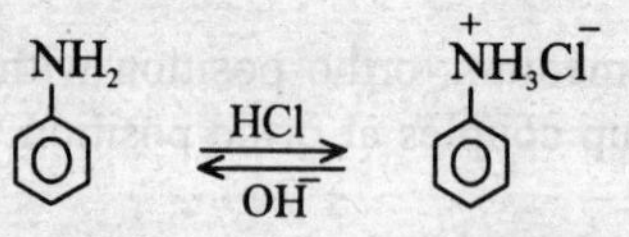

NH$_2$ $\overset{+}{N}H_3\bar{C}l$

$$\underset{OH^-}{\overset{HCl}{\rightleftharpoons}}$$

Aniline   **Anilinium chloride**
(Couples)   **(Does not couple)**

Phenol on the other hand, exists in equilibrium with phenoxide as shown below.

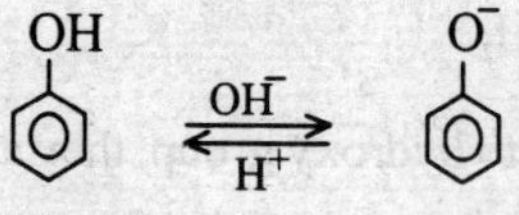

OH   O$^-$

$$\underset{H^+}{\overset{OH^-}{\rightleftharpoons}}$$

Phenol   **Phenoxide ion**
(Couples slow)   **(Couples fast)**

Phenoxide being more reactive than phenol towards electrophilic aromatic substitution, couples readi[ly] in alkaline medium. Thus, conditions required for coupling may be summerised as follows.

1. The solution must be suitably alkaline so that diazonium ion remains in ionised form and can coupl[e]
2. The solution must not be highly acidic as it decreases the concentration of free amine and phenoxi[de] ion (as shown above). In other words amines undergo coupling fastest in mild acidic conditions wherea[s] phenols in mild alkaline conditions.

33.5 STRUCTURE OF DIAZONIUM ION

Various structures were proposed to diazonium ion which have been discussed below.

1. *Griess in 1864,* proposed a structure for benzenediazonium chloride in which each nitrogen wa[s] attached to the benzene ring.

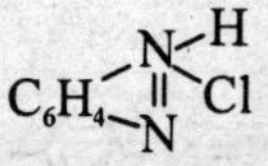

$$C_6H_4 \overset{N-H}{\underset{N}{\underset{\|}{\diagdown}}} Cl$$

2. *Kekule in 1866,* forwarded a structure in which only one nitrogen atom was attached to the ring.

$$C_6H_5-N=N-Cl$$

The Kekule's formula was supported by the fact that:

(a) In various reactions of diazonium salt, the N_2Cl group is replaced by univalent radicals.

(b) Benzene diazonium chloride forms hydrazine on reduction.

(c) It forms azo-compounds on coupling.

3. *Blomstrand in 1869,* suggested a formula where nitrogen is in pentavalent state (as in ammonium chloride). The reason for this belief was that both diazonium chloride and ammonium chloride resemble closely in basicity and solubility in water. Both have pentavalent nitrogen and formation of diazonium salt from primary amine can easily be shown as follows.

$$C_6H_5-\overset{H}{\underset{Cl}{\overset{|}{N}}}\overset{H}{\underset{H}{-}} \ + \ \overset{O}{\underset{HO}{\diagdown}}N \longrightarrow C_6H_5-\underset{Cl}{\overset{|}{N}}\equiv N + 2H_2O$$

The diazonium chloride, now a days, is represented by structure $[C_6H_5-\overset{+}{N}\equiv N]\ Cl^-$ which is further supported by the following facts.

(a) *Goldschmidt in 1890,* showed through electrical conductivity measurements that a diazonium salt dissociates into two ions.

(b) *Strecker in 1871,* supported the above structure on the basis of difference in the properties of diazonium salt with that of azo-compounds. The former is unstable and explode in dry state whereas the later ones are very stable.

(c) *Le Fevre et al. in 1955,* supported the blomstrand structure as he found the presence of a triple bond in the infrared spectrum of diazonium salt.

4. *Von Pechmann in 1892,* proposed the nitrosamine structure for benzene diazonium hydroxide (C_6H_5NHNO). He also explained the existence of structural isomerism by diazoates i.e. the n-diazo and isodiazohydroxides.

$$C_6H_5N=N.OH \longrightarrow C_6H_5N=N.ONa$$
n-Diazohydroxide   **Sodium-n-diazoate**

$$C_6H_5NHNO \longrightarrow C_6H_5NNa.NO$$
Isodiazohydroxide   **Sodium isodiazoate**

5. *Hatzsch in 1895,* however, proposed the existence of geometrical isomerism in diazoates. He adopted kekule structure for diazoates since blomstrand structure having $-N\equiv N-$ can not show geometrical isomerism. The two forms (syn and anti) for sodium diazoate are as follows.

$$\begin{array}{cc} C_6H_5-N & C_6H_5-N \\ \| & \| \\ NaO-N & N-O.Na \end{array}$$
Sodium-n-diazoate   **Sodium isodiazoate**
Syn form   **Anti form**

He further explained that benzene diazonium chloride exists in blomstrand structure in acidic medium and in kekule structure in alkaline medium as shown below.

$$[C_6H_5N\equiv N:]^+Cl^- \xrightarrow{\ NaOH\ } [C_6H_5N\equiv N:]^+OH^-$$
Benzenediazonium hydroxide

Benzenediazonium chloride is a salt of the base benzenediazonium hydroxide which has not been isolated so far. In presence of alkali, it is converted into diazonium hydroxide which rearranges into benzenediazohydroxide. The diazoates are produced by the reaction with alkali as shown below.

$$C_6H_5\overset{+}{N}\equiv N: \longleftrightarrow C_6H_5-\overset{+}{N}=\overset{..}{N}: \overset{OH^-}{\rightleftharpoons} C_6H_5\overset{..}{N}=\overset{..}{N}-OH \xrightarrow{NaOH} C_6H_5\overset{..}{N}=\overset{..}{N}-\overset{-}{O}Na^+$$

Benzenediazo-   **Diazoate**
hydroxide

33.6   PHENYLHYDRAZINE [$C_6H_5NHNH_2$]

Phenylhydrazine is prepared by the reduction of benzenediazonium chloride.

$$C_6H_5N_2Cl + 4\,[H] \xrightarrow{SnCl_2/HCl} C_6H_5NHNH_2.HCl$$

The reduction of benzenediazonium chloride may also be accomplished with sodium bisulphite.

Properties: Phenylhydrazine is a colorless liquid, b.p. 240°, slightly soluble in water but fairly soluble in organic solvents.

1. It is highly basic in nature and forms well defined crystalline salts when treated with acids.

$$C_6H_5NHNH_2 + HCl \longrightarrow C_6H_5NH\overset{+}{N}H_3\overset{-}{Cl}$$

Phenylhydrazine hydrochloride

2. It is a strong reducing agent and reduces fehlings solution in cold.

$$C_6H_5NHNH_2 + 2\,CuO \longrightarrow C_6H_6 + Cu_2O + N_2 + H_2O$$

3. It is reduced to aniline and ammonia when treated with strong reducing agents.

$$C_6H_5NHNH_2 + 2\,[H] \xrightarrow{Zn/HCl} C_6H_5NH_2 + NH_3$$

4. It reacts with aldehydes and ketones to form crystalline solids known as hydrazones.

$$\overset{\diagup}{\underset{\diagdown}{}}C=O + C_6H_5NHNH_2 \longrightarrow \overset{\diagup}{\underset{\diagdown}{}}C=NNHC_6H_5 + H_2O$$

Phenylhydrazone

With sugars, phenylhydrazine forms osazone. Both hydrazone and osazone are used for the identification of carbonyl compounds and sugars respectively.

It is used as an important laboratory reagent for the identification of carbonyl compounds and sugars and also in various pharmaceutical preparation.

33.7   DIAZOAMINOBENZENE [$C_6H_5N=NNHC_6H_5$]

Benzenediazonium chloride reacts with aniline in weakly acidic medium in presence of sodium acetate to yield diazoaminobenzene.

$$C_6H_5N_2Cl + C_6H_5NH_2 \xrightarrow{CH_3COONa} C_6H_5N=NNHC_6H_5 + HCl$$

Diazoaminobenzene

It may also be prepared by diazotising aniline hydrochloride with calculated amount of sodium nitrite which is just sufficient to diazotise half of aniline, then adding sodium acetate.

$$C_6H_5NH_2.HCl + NaNO_2 \longrightarrow C_6H_5N_2Cl + H_2O + NaOH$$
$$C_6H_5N_2Cl + C_6H_5NH_2 \longrightarrow C_6H_5N=NNHC_6H_5$$

Properties: It exists in two forms, the golden yellow prisms, m.p. 98° and yellow prisms, m.p. 80°. It is insoluble in water but soluble in ethanol and ether. It explodes on heating. It is weakly basic in nature and does not form stable salts with acids.

1. When boiled with dilute sulphuric acid, it forms aniline and phenol.

$$C_6H_5N=NNHC_6H_5 + H_2O \xrightarrow{H^+} C_6H_5OH + C_6H_5NH_2 + N_2$$

2. When boiled with concentrated hydrobromic acid, it forms bromobenzene and aniline.

$$C_6H_5N=NNHC_6H_5 + HBr \longrightarrow C_6H_5Br + C_6H_5NH_2 + N_2$$

3. When reacted upon with nitrous acid, it forms benzenediazonium chloride.

$$C_6H_5N=NNHC_6H_5 + NaNO_2 + HCl \longrightarrow 2\ C_6H_5N_2Cl + 2H_2O$$

4. It undergoes rearrangement to form p-aminoazobenzene on heating with a small amount of aniline.

$$C_6H_5N=NNH-\langle\bigcirc\rangle \xrightarrow[\text{aniline}]{50°} C_6H_5N=N-\langle\bigcirc\rangle-NH_2$$
p-Aminoazobenzene

33.8 p-AMINOAZOBENZENE/ANILINE YELLOW [$C_6H_5N=NC_6H_4NH_2$-p]

It is prepared by warming diazoaminobenzene with a small amount of aniline (see Sec. 33.7).

It may also be prepared by coupling benzenediazonium chloride and aniline in fairly acidic medium.

$$C_6H_5N_2Cl + \langle\bigcirc\rangle-NH_2 \xrightarrow{H^+} C_6H_5N=N-\langle\bigcirc\rangle-NH_2$$
p-Aminoazobenzene

Properties: It is a yellow dye, m.p. 126°. It is insoluble in water and forms two series of salts with acids.

$$C_6H_5N=N-\langle\bigcirc\rangle-NH_2 \xrightarrow{HCl} C_6H_5N=N-\langle\bigcirc\rangle-\overset{+}{N}H_3Cl^- \xrightarrow{HCl} C_6H_5NH-N=\langle\bigcirc\rangle=\overset{+}{N}H_2Cl^-$$
Aniline yellow　　　　　　　　　**Unstable yellow salt**　　　　　　　　　**Stable violet salt**

On reduction with tin and hydrochloric acid, it forms aniline and p-phenylenediamine.

$$\langle\bigcirc\rangle-N=N-\langle\bigcirc\rangle-NH_2 \xrightarrow{Sn/HCl} \langle\bigcirc\rangle-NH_2 + NH_2-\langle\bigcirc\rangle-NH_2$$

On oxidation it forms p-benzoquinone. It is used in the preparation of bis-azodyes.

33.9 AZOBENZENE [$C_6H_5N=NC_6H_5$]

Azobenzene is prepared by the reaction of nitrosobenzene and aniline.

$$C_6H_5NO + C_6H_5NH_2 \longrightarrow C_6H_5N=NC_6H_5 + H_2O$$

It may also be produced by alkaline reduction of nitrobenzene with zinc dust and methanolic sodium hydroxide or lithium aluminum hydride.

$$2\ C_6H_5NO_2 + 8\ [H] \xrightarrow[\substack{CH_3OH\ or \\ LiAlH_4}]{Zn + NaOH} C_6H_5N=NC_6H_5 + 4\ H_2O$$

Reduction of azoxybenzene with iron yields azobenzene.

$$C_6H_5-\overset{\overset{O}{\uparrow}}{N}=N-C_6H_5 \xrightarrow[\text{Fe}]{\text{Red}^n} C_6H_5N=NC_6H_5 + H_2O$$

Azoxybenzene   **Azobenzene**

Properties: It is a crystalline red solid, m.p. 68°, insoluble in water but soluble in organic solvents. It shows geometrical isomerism with the ordinary azobenzene being the anti form.

$$\begin{array}{c} C_6H_5-N \\ \parallel \\ C_6H_5-N \end{array} \qquad\qquad \begin{array}{c} C_6H_5-N \\ \parallel \\ N-C_6H_5 \end{array}$$

Syn form, m.p. 71.5°   Anti form, m.p. 68°

It forms bisulphite addition product when treated with sodium bisulphite.

$$C_6H_5N=NC_6H_5 + NaHSO_3 \longrightarrow C_6H_5NHNC_6H_5 \overset{\overset{SO_3Na}{|}}{}$$

When reduced with mild reducing agents, it forms hydrazobenzene which on further reduction with stronger reducing agent yields aniline.

$$C_6H_5N=NC_6H_5 \xrightarrow[\text{NaOH}]{\text{Zn}} C_6H_5NHNHC_6H_5 \xrightarrow[\text{HCl}]{\text{SnCl}_2} 2\,C_6H_5NH_2$$

Azobenzene   **Hydrazobenzene**   **Aniline**

It is oxidised to azoxybenzene when reacted with hydrogen peroxide in acetic acid.

$$C_6H_5N=NC_6H_5 \xrightarrow{[O]} C_6H_5N=\overset{\overset{O}{\uparrow}}{N}C_6H_5$$

Azoxybenzene

33.10   AZOXYBENZENE [$C_6H_5NO=NC_6H_5$]

Reduction of nitrobenzene with sodium methoxide in methanol furnishes azoxybenzene.

$$4\,C_6H_5NO_2 + 3\,CH_3ONa \xrightarrow{CH_3OH} 2\,C_6H_5N=\overset{\overset{O}{\uparrow}}{N}C_6H_5 + 3HCOONa + 3H_2O$$

Azoxybenzene

Other reducing agents such as glucose in alkaline solution or metallic thallium in alcohol or alkaline sodium arsenite may also be used for the above reaction.

$$4\,C_6H_5NO_2 + 3\,As_2O_3 + 18\,NaOH \longrightarrow 2\,C_6H_5N=\overset{\overset{O}{\uparrow}}{N}C_6H_5 + 6\,Na_3AsO_4 + 9H_2O$$

Oxidation of aniline with per acetic acid also yields azoxylbenzene.

$$C_6H_5NH_2 + 2\,CH_3CO_3H \longrightarrow C_6H_5N=\overset{\overset{O}{\uparrow}}{N}C_6H_5 + 2\,CH_3COOH + H_2O$$

Azobenzene on oxidation with hydrogen peroxide forms azoxybenzene.

Properties: It is a yellow crystalline solid, m.p. 36°, insoluble in water but soluble in organic solvents.

Like azobenzene, it also exhibits geometrical isomerism, the ordinary azoxybenzene being the anti or trans isomer.

It undergoes reduction to form azobenzene, hydrazobenzene and finally aniline by using various reducing agents.

$$
\begin{array}{c}
\underset{\text{Azoxybenzene}}{\text{C}_6\text{H}_5\text{N}=\overset{\overset{\displaystyle O}{\uparrow}}{\text{N}}\,\text{C}_6\text{H}_5} \xrightarrow{\text{Iron fillings}} \underset{\text{Azobenzene}}{\text{C}_6\text{H}_5\text{N}=\text{N}\,\text{C}_6\text{H}_5}
\end{array}
$$

$$
\xrightarrow{(\text{NH}_4)_2\text{S}} \underset{\text{Hydrazobenzene}}{\text{C}_6\text{H}_5\text{NHNHC}_6\text{H}_5} \xrightarrow{\text{Sn} + \text{HCl}} \underset{\text{Aniline}}{2\,\text{C}_6\text{H}_5\text{NH}_2}
$$

When treated with warm concentrated sulphuric acid, it rearranges to hydroxyazobenzene.

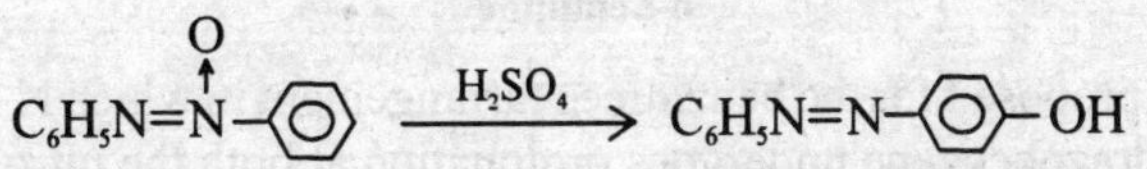

$$
\underset{}{\text{C}_6\text{H}_5\text{N}=\overset{\overset{\displaystyle O}{\uparrow}}{\text{N}}-\!\!\bigcirc} \xrightarrow{\text{H}_2\text{SO}_4} \underset{\text{p-Hydroxyazobenzene}}{\text{C}_6\text{H}_5\text{N}=\text{N}-\!\!\bigcirc\!\!-\text{OH}}
$$

The above rearrangement is known as *Wallach rearrangement* and an intramolecular mechanism has been proposed for it.

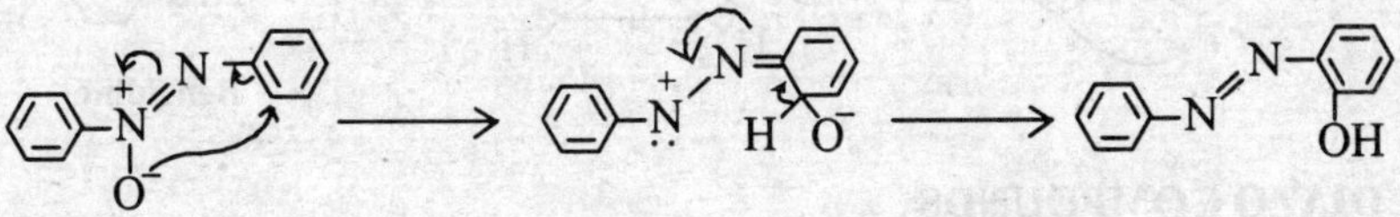

33.11 HYDRAZOBENZENE/sym DIPHENYL HYDRAZINE [C₆H₅NHNHC₆H₅]

Reduction of nitrobenzene or azobenzene with zinc dust and aqueous sodium hydroxide produces hydrazobenzene.

$$
\underset{\text{Nitrobenzene}}{\text{C}_6\text{H}_5\text{NO}_2} \xrightarrow{\text{Zn/NaOH}} \underset{\text{Hydrazobenzene}}{\text{C}_6\text{H}_5\text{NHNHC}_6\text{H}_5} \xleftarrow{\text{Zn/NaOH}} \underset{\text{Azobenzene}}{\text{C}_6\text{H}_5\text{N}=\text{NC}_6\text{H}_5}
$$

It may also be prepared by electrolytic reduction of azoxybenzene.

Properties: It is a colorless crystalline solid, m.p. 126°, insoluble in water. It is oxidised to azobenzene on exposure to atmosphere and turns orange. On reduction it produces aniline.

It undergoes rearrangement when warmed with concentrated hydrochloric acid. The product formed is 4,4′-diaminodiphenyl also known as benzidine along with a small amount of 2,4′-diaminodiphenyl (diphenylene).

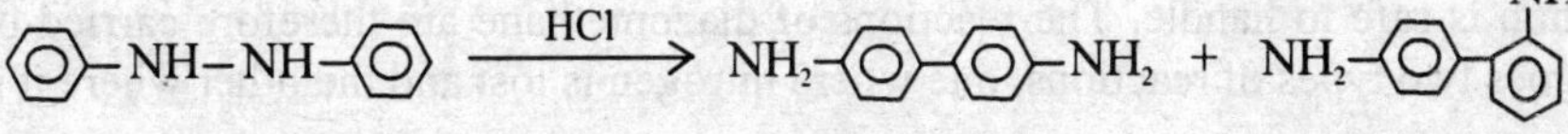

The above reaction is known as *Benzidine rearrangement*. When para position of one of the ring of hydrazobenzene is occupied by strong electron releasing group, the major product is o-semidine together with a small amount of p-semidine and the rearrangement is known as *Semidine rearrangement*.

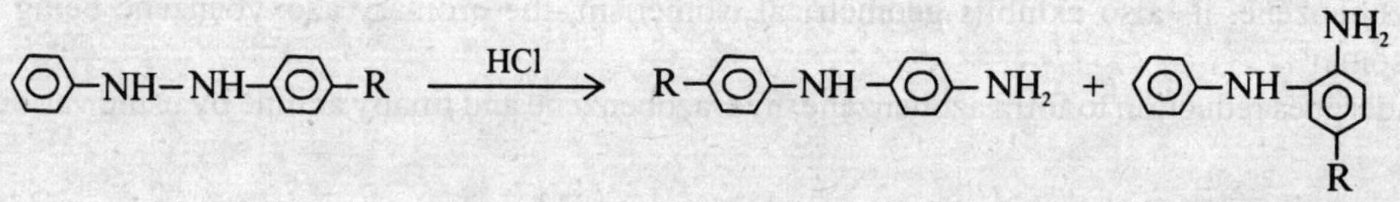

p-Semidine   o-Semidine

When para position of both the rings are occupied, the main product is o-semidine.

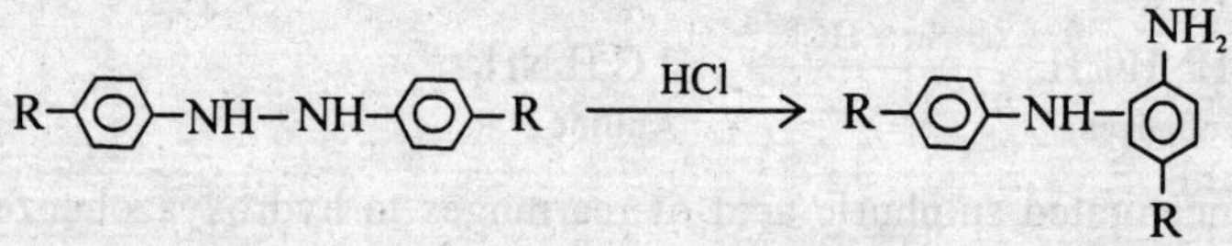

o-Semidine

An intramolecular mechanism has been proposed for the benzidine rearrangement in which N−N linkage has rearranged to a C−C linkage. The hydrazobenzene undergoes protonation at both the nitrogen atoms.

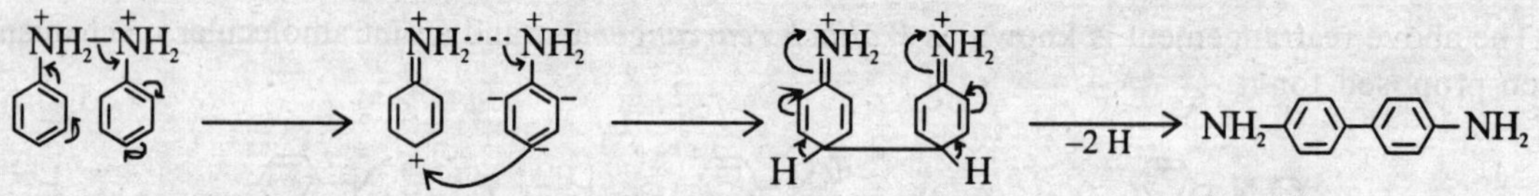

Benzidine

33.12 ALIPHATIC DIAZO COMPOUNDS

The aliphatic diazo compounds are characterised by the presence of CN_2 moiety. The most important of these is diazomethane which is used as an important methylating agent.

33.12.1 Diazomethane [CH_2N_2]

Diazomethane may be prepared by the following sequence of reactions starting with methanamine.

$$CH_3NH_2 + ClCO_2C_2H_5 \xrightarrow[-HCl]{} CH_3NHCO_2C_2H_5 \xrightarrow{HNO_2}$$

Methanamine   Chloroethyl   N-Methylurethane
               formate

$$CH_3N(NO)CO_2C_2H_5 \xrightarrow{KOH/CH_3OH} CH_2N_2 + CO_2 + C_2H_5OH$$

N-Methyl-N-nitrosourethane   Diazomethane

Aliphatic amines can not be diazotised like aromatic amines to give diazo compounds.

Properties: It is a poisonous gas, soluble in ether. Liquid diazomethane, b.p. -24°, is explosive in nature but its ether solution is safe to handle. The reactions of diazomethane are therefore carried out in ether solution. It undergoes two types of reactions, one where nitrogen is lost and the other where it is retained.

[A] *Reactions where Nitrogen is Lost:* Diazomethane, as nucleophile, attacks compounds having active hydrogen to convert −CH_2 into CH_3 and N_2 evolves as nitrogen gas.

1. Diazomethane reacts with hydrogen bromide to give methyl bromide.

$$CH_2N_2 + HBr \longrightarrow CH_3Br + N_2 \uparrow$$

2. It methylates hydroxy group readily in carboxylic acids, sulphonic acids, phenols etc.

$$CH_2N_2 + RCO_2H \longrightarrow RCO_2CH_3 + N_2\uparrow$$

3. Alcohols are also methylated by diazomethane in presence of aluminum alkoxide as catalyst.

$$CH_2N_2 + ROH \xrightarrow{Al(OR)_3} ROCH_3 + N_2\uparrow$$

4. Diazomethane reacts with aldehydes to yield methyl ketones and in some cases ethylene oxide derivatives are also formed.

$$RCHO + CH_2N_2 \xrightarrow{-N_2} RCOCH_3 + R-\underset{\underset{O}{\diagdown\diagup}}{CH-CH_2}$$

5. Ketones react similarly with diazomethane.

$$CH_3COCH_3 + CH_2N_2 \xrightarrow{-N_2} CH_3COCH_2CH_3 + (CH_3)_2-\underset{\underset{O}{\diagdown\diagup}}{C-CH_2}$$

The mechanism of reaction with carbonyl compounds involve nucleophilic addition.

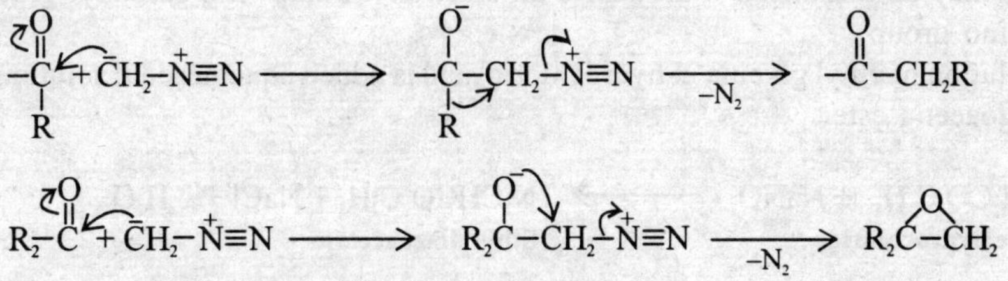

$$\underset{R}{\overset{O}{-C}} + \overset{+}{CH_2 - N} \equiv N \longrightarrow \underset{R}{-\overset{O}{\underset{|}{C}} - CH_2 - \overset{+}{N} \equiv N} \xrightarrow{-N_2} \overset{O}{-C} - CH_2R$$

$$R_2\overset{O}{C} + \overset{+}{CH_2 - \overset{..}{N}} \equiv N \longrightarrow R_2C - CH_2 - \overset{+}{N} \equiv N \xrightarrow{-N_2} R_2\overset{O}{C} - CH_2$$

6. It methylates primary amines to secondary amines.

$$RNH_2 + CH_2N_2 \longrightarrow RNHCH_3 + N_2\uparrow$$

7. With amides, it forms N-methylamide.

$$RCONH_2 + CH_2N_2 \dashrightarrow RCONHCH_3 + N_2\uparrow$$

[B] *Reactions where Nitrogen is Retained*
1. Diazomethane on reduction with sodium amalgam produces methyl hydrazine.

$$CH_2N_2 + 4 [H] \xrightarrow{Na-Hg/H_2O} CH_3NHNH_2$$

2. When reacted with alkenes and alkynes, diazomethane forms cyclic addition products. Alkenes form pyrazolins and alkynes form pyrazole.

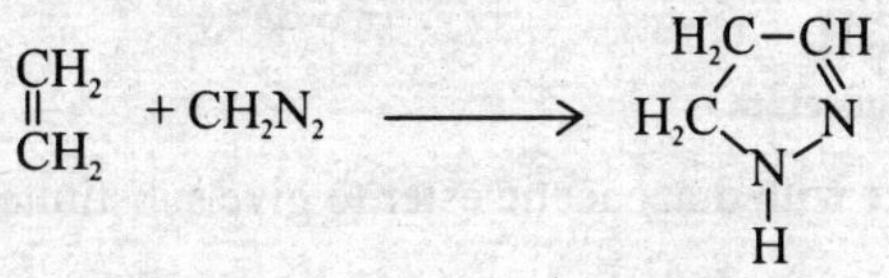

$$\overset{CH_2}{\underset{CH_2}{\|}} + CH_2N_2 \longrightarrow \begin{array}{c} H_2C-CH \\ H_2C \diagup \qquad \diagdown N \\ \diagdown N \diagup \\ | \\ H \end{array}$$

Pyrazolin

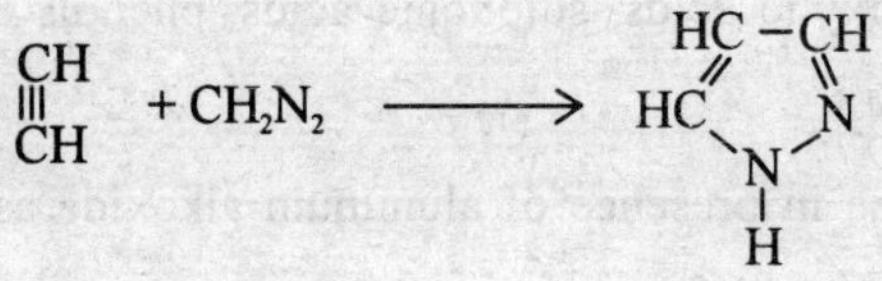

$$\begin{array}{c} CH \\ \|\| \\ CH \end{array} + CH_2N_2 \longrightarrow$$

(Pyrazole structure)

Pyrazole

It is best represented as a resonance hybrid of linear structures I, II, III and IV.

$$CH_2\overset{+}{=}\overset{..}{N}=\overset{..}{N}: \longleftrightarrow \overset{..}{C}H_2-\overset{+}{N}\equiv N: \longleftrightarrow \overset{+}{C}H_2-\overset{..}{N}=\overset{..}{N}: \longleftrightarrow \overset{..}{C}H_2-\overset{..}{N}\equiv\overset{+}{N}:$$

$$\quad\quad\quad I \quad\quad\quad\quad\quad\quad II \quad\quad\quad\quad\quad\quad III \quad\quad\quad\quad\quad\quad IV$$

It has a low dipole moment value which may be accounted only if diazomethane is a resonance hybrid of the four linear structures (I to IV) shown above. Under suitable condition, diazomethane can behave as an electrophile (III), or as a nucleophile (II), or as a 1,3-dipole (IV).

33.12.2   Diazoacetic Ester/Ethyl Diazoacetate [$CHN_2CO_2C_2H_5$]

Diazoacetic ester is prepared by diazotisation of ethyl glycine hydrochloride (as discussed in Section 33.1, aliphatic amines may be diazotised if they have an electron withdrawing carbonyl group attached to carbon having amino group).

To a well cooled solution of ethyl glycine in hydrochloric acid is added aqueous cold solution of sodium nitrite to prepare diazoacetic ester.

$$\underset{\textbf{Ethyl glycine hydrochloride}}{Cl^-\overset{+}{N}H_3CH_2CO_2C_2H_5} + NaNO_2 \longrightarrow \underset{\textbf{Ethyl diazoacetate}}{N_2CHCO_2C_2H_5} + NaCl + 2 H_2O$$

Properties: Ethyldiazoacetate is a yellow oil, b.p. 139°, insoluble in water but soluble in ether and alcohol. It undergoes mainly two types of reactions.

[A]   *Reactions where Nitrogen is Lost*

1. When heated with dilute hydrochloric acid, diazoacetic ester gives glycollic ester.

$$N_2CHCO_2C_2H_5 + H_2O \xrightarrow[\Delta]{HCl} \underset{\textbf{Ethylglycollate}}{CH_2OHCO_2C_2H_5} + N_2$$

With concentrate halogen acid, however ethyl halogenoacetate is produced.

$$N_2CHCO_2C_2H_5 + HBr \longrightarrow \underset{\textbf{Ethylbromoacetate}}{CH_2BrCO_2C_2H_5} + N_2$$

2. When reacted upon with iodine, it forms ethyl diiodoacetate.

$$N_2CHCO_2C_2H_5 + I_2 \longrightarrow \underset{\textbf{Ethyldiiodoacetate}}{CHI_2CO_2C_2H_5} + N_2$$

3. Compounds having active hydrogen atoms react with diazoacetic ester to give substituted glycollic esters.

$$N_2CHCO_2C_2H_5 + CH_3COOH \longrightarrow \underset{\textbf{Acetylglycollic ester}}{CH_3COOCH_2CO_2C_2H_5} + N_2$$

$$N_2CHCO_2C_2H_5 + C_2H_5OH \longrightarrow C_2H_5OCH_2CO_2C_2H_5 + N_2$$
Ethyl ether of glycollic ester

[B] *Reactions where Nitrogen is Retained:* Like diazomethane, diazoacetic ester reacts with ethylene and acetylene to form heterocyclic compounds.

$$\begin{matrix} CH_2 \\ \| \\ CH_2 \end{matrix} + N_2CHCO_2C_2H_5 \longrightarrow$$

$$\begin{matrix} H_2C-C-COOC_2H_5 \\ H_2C \quad N \\ N \\ | \\ H \end{matrix}$$

Pyrazolin-3-carboxylic ester

$$\begin{matrix} CH \\ \| \| \| \\ CH \end{matrix} + N_2CHCO_2C_2H_5 \longrightarrow$$

$$\begin{matrix} HC-C-COOC_2H_5 \\ HC \quad N \\ N \\ | \\ H \end{matrix}$$

Pyrazole-3-carboxylic ester

When reduced with zinc and acetic acid, it gives ethyl glycine ester and ammonia.

$$N_2CHCO_2C_2H_5 + 4\,[H] \xrightarrow[CH_3COOH]{Zn} NH_2CH_2CO_2C_2H_5 + NH_3$$
Glycine ester

33.13   PROBLEMS

1. What is diazotisation? How is benzenediazonium chloride prepared? Discuss its mechanism.
2. Starting with benzene, toluene and chlorobenzene how will you prepare the following:
   (a) m-Nitrotoluene   (b) 2,4-Diaminophenol   (c) p-Chlorotoluene   (d) 2,4-Dinitrophenol
   (e) m-Chlorotoluene   (f) p-Dinitrobenzene   (g) p-Dichlorobenzene   (h) m-Nitroaniline
3. What are azodyes? How are they produced starting from benzene?
4. Give reasons for the following:
   (a) Diazonium salts require alkaline conditions for coupling with phenols.
   (b) Diazonium salts require acidic conditions for coupling with amines.
   (c) Excess of mineral acid is used in diazotisation.
   (d) Benzenediazonium chloride couples with phenol but not with anisole.
   (e) 2,4-Dinitrobenzenediazonium chloride couples with methoxybenzene.
   (f) 2,4,6-Trinitrobenzenediazonium chloride couples with mesitylene.
5. What is the difference between diazo compounds and azo compounds? Explain with the help of their preparation and stability.
6. What happens when benzenediazonium chloride is treated with the following:
   (a) Cuprous chloride and hydrochloric acid   (b) Potassium iodide
   (c) Potassium cyanide in presence of copper   (d) Tin chloride and hydrochloric acid
   (e) Phenol   (f) p-Hydroxy-N-methylaniline
7. How is diazomethane prepared? How is it useful in organic synthesis?
8. Complete the following
   (a) $C_2H_2 + CH_2N_2 \longrightarrow$
   (b) $CHN_2CO_2C_2H_5 \xrightarrow{\text{dil HCl}}$
   (c) $CHN_2CO_2C_2H_5 \xrightarrow{\text{Conc. HCl}}$

(d) $CH_2NH_2CO_2C_2H_5$ $\xrightarrow{\text{NaNO}_2,\ \text{HCl},\ 0-5°}$

(e) $CH_3COCH_3 + CH_2N_2$ $\xrightarrow{\hspace{2cm}}$

9. Give preparation and uses of
    (a) Phenyl hydrazine   (b) Hydrazobenzene   (c) Azobenzene   (d) Azoxybenzene

10. Discuss the following reactions and their mechanism.
    (a) Sandmeyer reactions   (b) Gomberg reactions   (c) Diazotisation reaction   (d) Coupling reaction

11. How will you synthesise the following starting from benzenediazonium chloride.
    (a) Diazoaminobenzene   (b) Aminoazobenzene   (c) Phenylhydrazine   (d) Fluorobenzene
    (e) p-Cresol   (f) o-Dichlorobenzene   (g) m-Dinitrobenzene   (h) 1,3,5-Trichlorobenzene
    (i) Diphenyl mercury

34

Phenols

Aromatic compounds having hydroxy group directly attached to the benzene ring are termed as *Phenols*. They are classified as mono, di and trihydric phenols on the basis of the number of hydroxy groups attached to the aromatic ring. The name phenol is derived from the Greek word '*phene*' means benzene and '*ol*' means OH. Another class of aromatic compounds having hydroxy group in side chain of aromatic nucleus is termed as *Aromatic alcohols*.

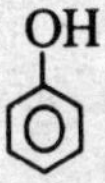

OH

Phenol

CH₂OH

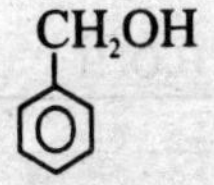

**Benzyl alcohol,
an Aromatic alcohol**

34.1 NOMENCLATURE

Phenols are named as a derivative of their simplest member, hydroxybenzene, commonly known as phenol.

OH CH₃

**2-Methylphenol
(o-Cresol)**

OH Cl

**3-Chlorophenol
(m-Chlorophenol)**

OH OH

**1,2-Dihydroxybenzene
(Catechol)**

If functional groups such as carboxyl or carbonyl are present in phenols they take precedence in assigning number over hydroxyl group.

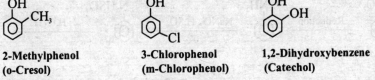

COOH OH

**2-Hydroxybenzoic Acid
(not 2-Carboxylphenol)**

CHO OH

**3-Hydroxybenzaldehyde
(not 3-Formylphenol)**

Dihydric phenols are the phenols having two hydroxy group in the aromatic nucleus.

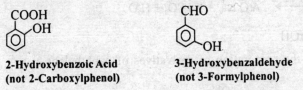

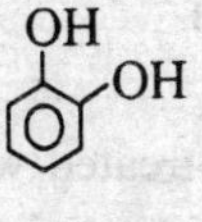

OH OH

**1,2-Dihydroxybenzene
(Catechol)**

OH OH

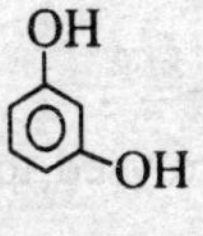

**1,3-Dihydroxybenzene
(Resorcinol)**

OH OH

**1,4-Dihydroxybenzene
(Hydroquinone)**

Trihydric phenols are the phenols having three hydroxy group in the aromatic nucleus.

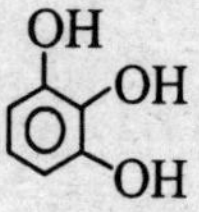

1,2,3-Trihydroxybenzene   **1,2,4-Trihydroxybenzene**   **1,3,5-Trihydroxybenzene**
(Pyrogallol)   **(Hydroxyquinol)**   **(Phloroglucinol)**

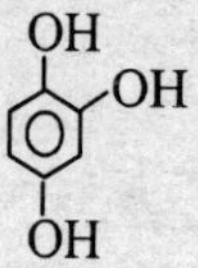

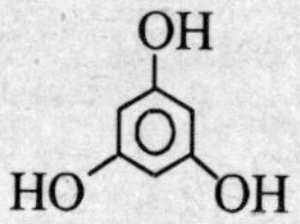

34.2   GENERAL METHODS OF PREPARATION

Phenols are synthesised in laboratory by following methods.

1.   *Hydrolysis of Diazonium Salts:* Phenols are prepared by heating aqueous solutions of diazonium salts.

$$Ar-\overset{+}{N_2}C\bar{l} + H_2O \longrightarrow Ar-OH + N_2 + HCl$$

Diazonium salts are prepared from corresponding aromatic amines. m-Chloroaniline, for example, yields m-chlorophenol.

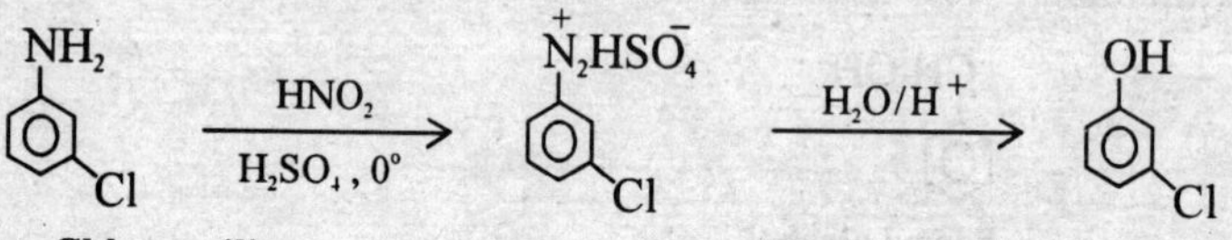

m-Chloroaniline   **m-Chlorophenol**

This method is a versatile method for the preparation of phenols and is the final step in the synthetic route beginning from hydrocarbons.

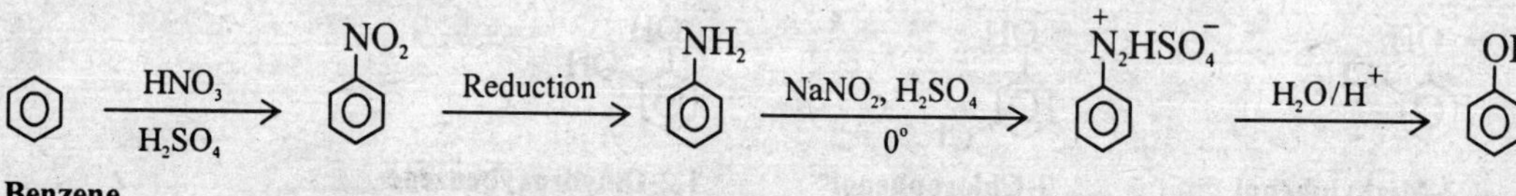

Benzene   **Phenol**

2.   *Alkali Fusion of Sulphonates:* The fusion of sodium salt of arylsulphonic acids with sodium hydroxide at 300° yields phenols.

$$Ar\overset{+}{SO_3}N\overset{+}{a} + 2\,NaOH \longrightarrow Ar\bar{O}\,N\overset{+}{a} + Na_2SO_3 + H_2O$$

$$Ar\bar{O}\,Na^+ \xrightarrow{dil\ H_2SO_4} ArOH$$

This method is generally used for the preparation of derivatives of naphthalene.

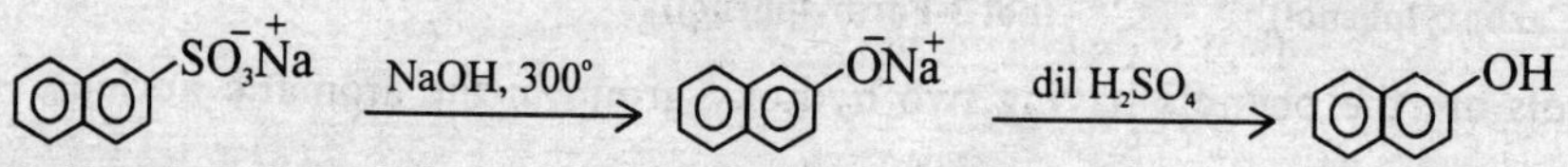

Sodium-2-naphthalene   **Sodium-2-**   **2-Naphthol**
sulphonate   **naphthoxide**

3.   *Hydrolysis of Aryl Halides:* Aryl halides are hydrolysed to phenols when treated with sodium hydroxide at high temperature and high pressure.

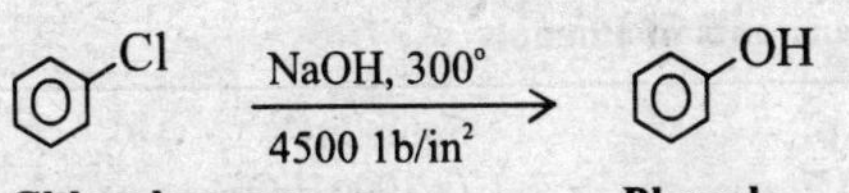

Chlorobenzene $\xrightarrow[\text{4500 1b/in}^2]{\text{NaOH, 300}^\circ}$ **Phenol**

The reaction is facilitated by electron withdrawing groups present at the ortho and para positions to halogen (see Section 28.1). 2,4-Dinitrophenol and 2,4,6-Trinitrophenol are produced on a large scale by this method.

4. *Decarboxylation of Phenolic Acids:* When sodium salt of phenolic acids are distilled with sodalime, they undergo decarboxylation to form sodium phenoxide.

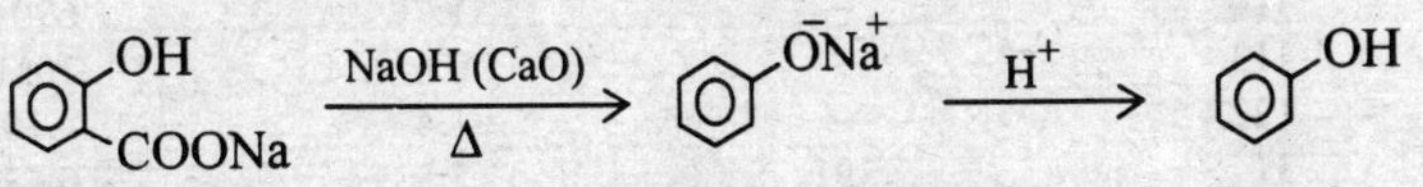

Sodium salicylate $\xrightarrow[\Delta]{\text{NaOH (CaO)}}$ **Sodium phenoxide** $\xrightarrow{\text{H}^+}$ **Phenol**

5. *From Arylmagnesium Halides (Grignard Reagent):* Arylmagnesium halide adds to molecular oxygen to give phenols by following sequence of reactions.

$$ArMgBr + O_2 \longrightarrow Ar-O-O-MgBr \xrightarrow{ArMgBr}$$

$$2\ Ar-O-MgBr \xrightarrow{\text{H}_2\text{O/H}^+} 2\ ArOH + 2\ Mg\ (OH)\ Br$$

6. *From Natural Sources:* Phenol is prepared commercially from the natural sources like coal tar and heavy oils. They provide considerable amounts of phenol, cresols, dihydroxy and trihydroxyphenols.

7. *From Aromatic Hydrocarbons:* Catalytic oxidation of toluene by air in presence of cupric salt yields phenol on a large scale.

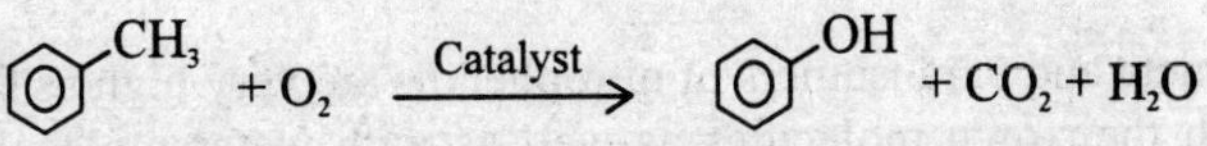

$+ O_2 \xrightarrow{\text{Catalyst}}$ $+ CO_2 + H_2O$

8. *The Cumene Process:* Almost all of the total production of phenol is achieved by this method now a days. Cumene (isopropyl benzene) on air oxidation followed by hydrolysis yields phenol and a good deal of acetone as a by-product.

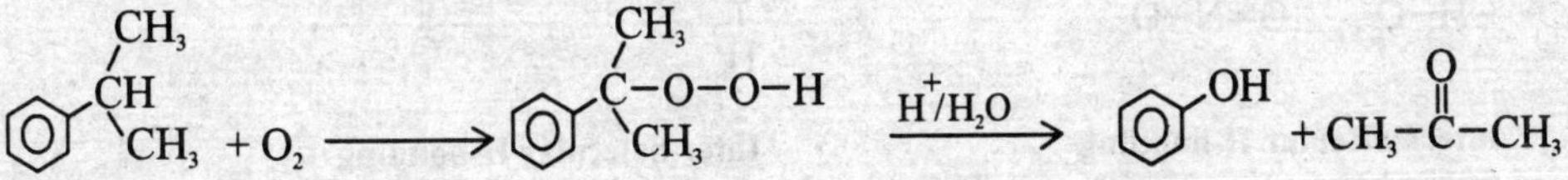

Cumene $+ O_2 \longrightarrow$ $\xrightarrow{\text{H}^+/\text{H}_2\text{O}}$ **Phenol** $+ CH_3-\overset{O}{\overset{\|}{C}}-CH_3$ **Acetone**

The last three methods are used for the commercial preparation of phenol.

34.3 PROPERTIES

Physical Properties: The simplest phenols are liquids or low melting solids. Like their alkyl counter parts (alcohols), they have high boiling points and moderate solubility in water due to hydrogen bonding. They undergo autooxidation on exposure to air and light.Physical constants for some phenols have been summerised in Table 34.1.

Table 34.1: Physical Constants of Phenols

Name	m.p. (°C)	b.p. (°C)	Solubility (g/100g H_2O at 25°)	K_a
Phenol	41	182	9.3	1.1×10^{-10}
Catachol	104	246	45.0	1.0×10^{-10}
Resorcinol	110	281	123.0	3.0×10^{-10}
Hydroquinone	45	286	8.0	2.0×10^{-10}
o-Nitrophenol	45	217	0.2	600×10^{-10}
m-Nitrophenol	96	–	1.4	50×10^{-10}
p-Nitrophenol	114	–	1.7	690×10^{-10}
2,4-Dinitrophenol	113	–	0.6	10×10^{-6}
2,4,6-Trinitrophenol	122	–	1.4	$> 10 \times 10^{-6}$
o-Methylphenol	31	191	2.5	0.63×10^{-10}
m-Methylphenol	11	201	2.6	0.98×10^{-10}
p-Methylphenol	35	202	2.3	0.67×10^{-10}

The physical constants shown in the Table 34.1 are peculiar for isomeric nitrophenols and needs discussion. The melting point of o-nitrophenol is much lower than that of its meta and para isomer due to intramolecular hydrogen bonding. It does not associate with other molecules or with water and hence has low melting point and solubility and is volatile in steam.

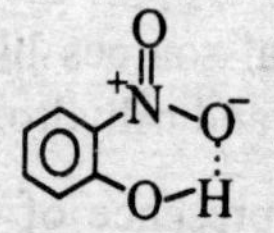

**Intramolecular H-bonding
(o-Nitrophenol)**

The melting point and solubility of meta and para isomers of nitrophenols are very high as they exhibit intermolecular hydrogen bonding with their own molecules as well as with water molecules.

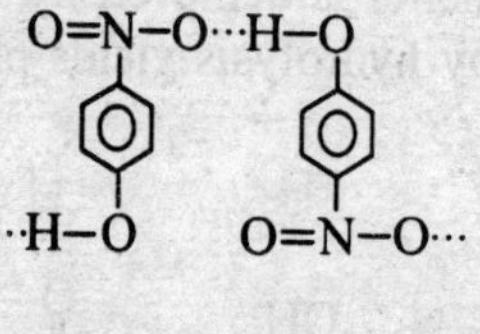

**Intermolecular H-bonding
(p-Nitrophenol)**

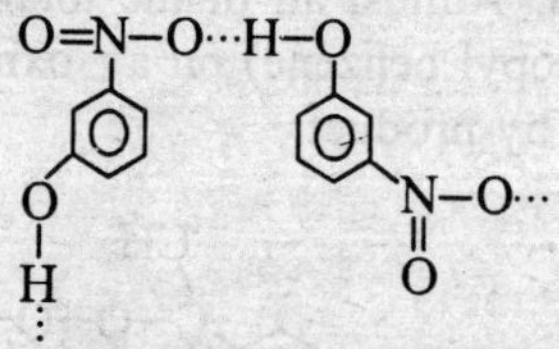

**Intermolecular H-bonding
(m-Nitrophenol)**

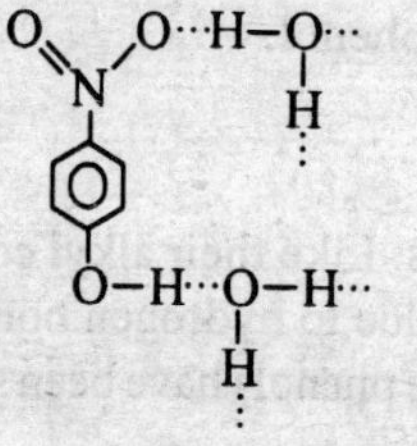

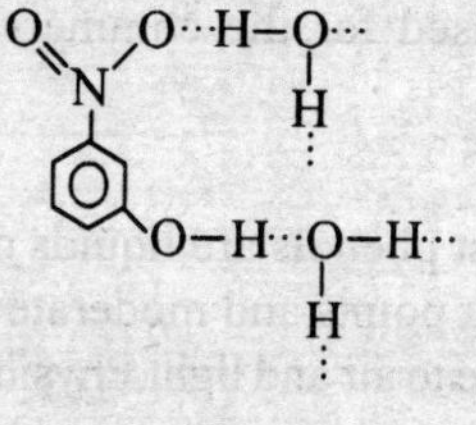

**Association with water molecules
(p-Nitrophenol)**

(m-Nitrophenol)

34.4   SPECTRAL PROPERTIES

The IR spectra of phenols show a strong broad band in the region 3600-3200 cm^{-1} due to O–H stretching and another strong broad band due to C–O stretching which is observed in the region 1200-1250 cm^{-1}.

In NMR, the phenolic hydroxy proton appears at much lower field (δ 4-12) than those of alcohols (δ 3-4.5) and aromatic protons appear around δ 7.

The mass spectra of simple phenol show very intense molecular ion peaks. The common fragmentation pattern follows loss of CO (M–28) and CHO (M–29) which can be represented as shown below:

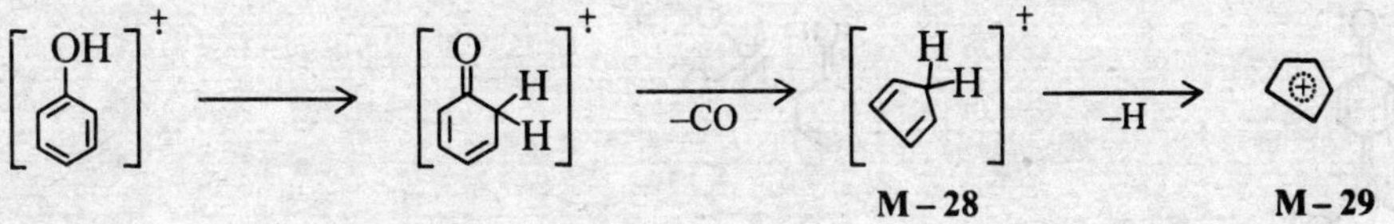

M–28   **M–29**

Phenols with alkyl side chain also undergo benzylic fission.

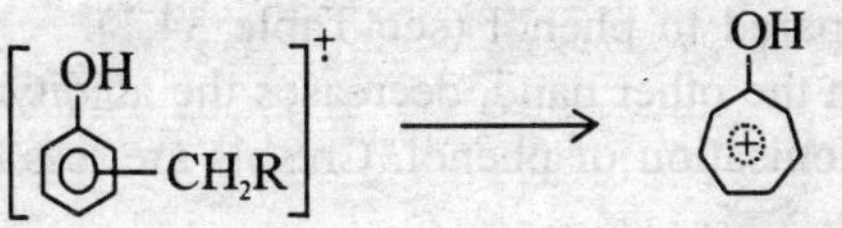

Hydroxy tropylium ion, m/z = 107

34.5   CHEMICAL PROPERTIES

Phenols exhibit reactions due to its phenolic hydroxy group and the aromatic ring.

1. *Acidity of Phenols:* Phenols are acidic and form salts with aqueous hydroxide but not with aqueous bicarbonates. The salts are converted into free phenol by aqueous mineral acids.

$$ArOH + OH^- \longrightarrow ArO^- + H_2O$$
$$ArO^- + H_2SO_4 \longrightarrow ArOH + HSO_4^-$$

Therefore, phenol must be a stronger acid than water but considerably weaker acid than carboxylic acids (K_a 10^{-5}). Acidity constants for some phenols have been given in Table 34.1.

Phenols are much more acidic than alcohols (having k_a values in the range of 10^{-16} to 10^{-18}). This can be explained on the basis of relative stabilities of anion and unionised alcohols and phenols in question.

$$\underset{\text{Alcohol}}{R-O-H} \rightleftharpoons \underset{\substack{\text{Alkoxide} \\ \text{ion}}}{R\bar{O}} + H^+$$

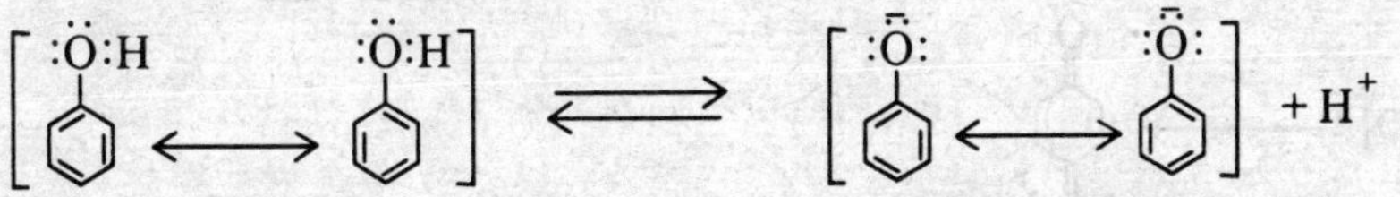

Phenols and phenoxide ion are stabilised by additional structures given below as oxygen can share an electron pair with the ring.

The structures shown for phenol have higher energy (as energy must be supplied to separate opposite charges) and hence are less stable than structures of phenoxide ion. The net result is greater stabilisation of phenoxide ion thus the equilibrium shifts to towards ionisation making phenols more acidic.

No such effect is possible for the stabilisation of alcohol and alkoxide ion.

Effect of Substituents on Acidity of Phenol: Any group which attracts the electrons like halogens or nitro group will increase the acidity of phenols as it stablises the phenoxide ion.

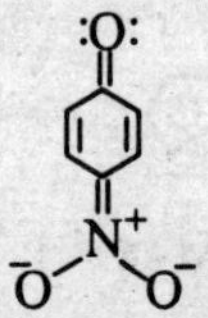

p-Nitrophenoxide ion 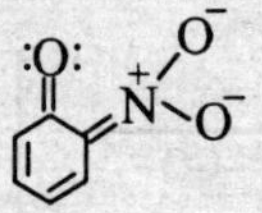   o-Nitrophenoxide ion

ortho and para nitrophenols are highly acidic compared to phenol (see Table 34.1).

Electron releasing substituents like methyl group, on the other hand, decreases the acidity of phenols as they destabilise the phenoxide ion and reduce the ionisation of phenol. Cresols are less acidic than phenols (see Table 34.1).

2.   *Salt Formation:* Like alcohols, phenol reacts with alkali metals, Grignard reagent and lithium aluminum hydride to form metal alkoxides.

$$C_6H_5-OH + Na \longrightarrow C_6H_5-ONa + 1/2H_2$$

Phenol   **Sodium**
                                              phenoxide

$$C_6H_5-OH + CH_3MgBr \longrightarrow C_6H_5-OMgBr + CH_4$$

Phenol   **Phenoxy magnesium**
                                              bromide

$$C_6H_5-OH + LiAlH_4 \longrightarrow (C_6H_5-O)_4AlLi + 4H_2$$

Phenol   **Lithium aluminum**
                                              phenoxide

3.   *Oxidation of Phenols:* Phenols are easily oxidised with mild oxidising agents or even with air (autooxidation). This property of phenols make them a useful additive in gasoline, rubber etc. as antioxidants. Phenols readily react with oxygen shielding the other organic chemicals from oxidation. Phenol on oxidation give 1,4-benzoquinone.

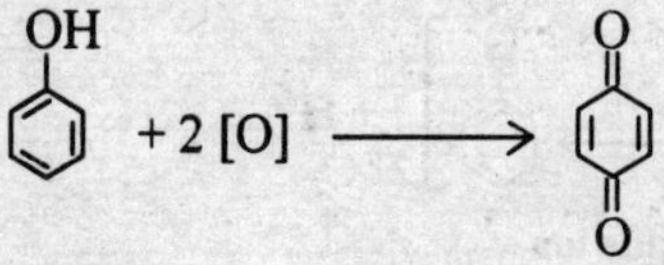

Phenol   **1,4-Benzoquinone**

Similarly catachol and hydroquinone are also oxidised to 1,2-benzoquinone and 1,4-benzoquinone respectively when treated with mild oxidising agents.

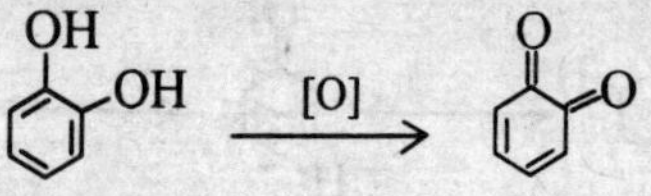

Catechol   **1,2-Quinone**

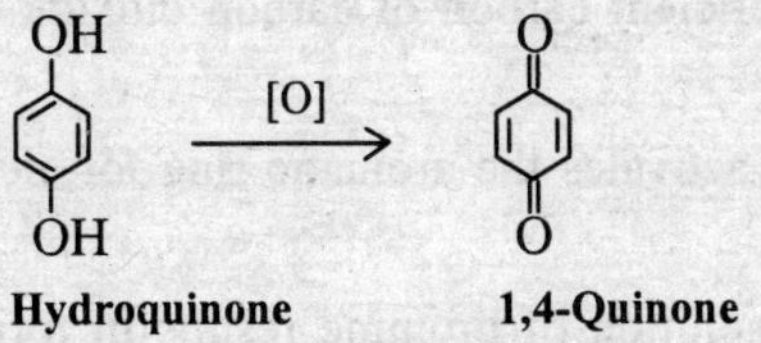

Hydroquinone   1,4-Quinone

4. *Reimer-Tiemann Reaction:* Phenols when reacted with chloroform in aqueous sodium hydroxide solution followed by acidification form 2-hydroxybenzaldehyde (salicylaldehyde).

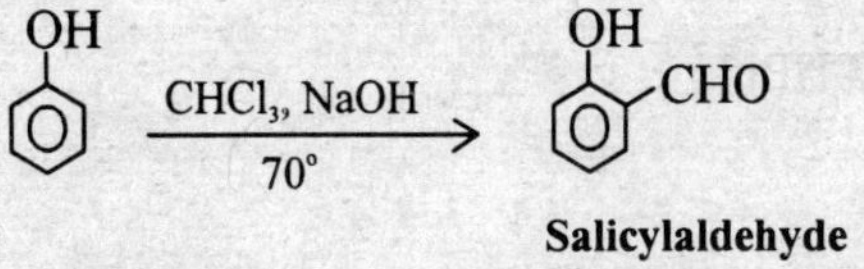

Salicylaldehyde

The reaction is known as Reimer-Tiemann reaction and is employed to prepare aromatic aldehydes having a hydroxy group at ortho position. The reaction involves electrophilic substitution of dichlorocarbene generated from chloroform to the highly reactive phenoxide ring.

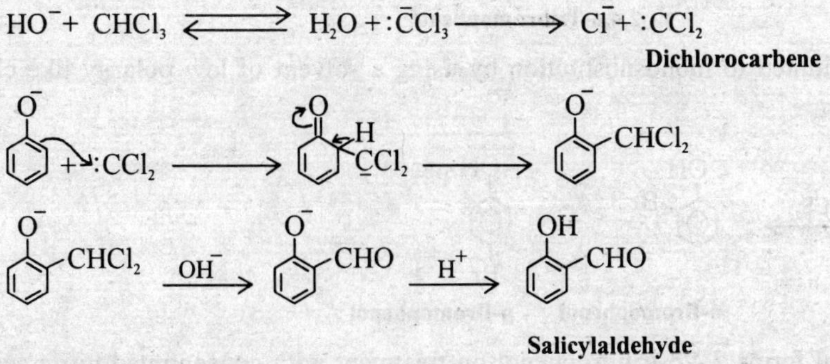

$$HO^- + CHCl_3 \rightleftharpoons H_2O + :\bar{C}Cl_3 \longrightarrow C\bar{l} + :CCl_2$$

                                                            Dichlorocarbene

Salicylaldehyde

5. *Kolbe Reaction:* Reaction of salt of phenol with carbon dioxide brings about substitution of a carboxyl group ortho to hydroxy.

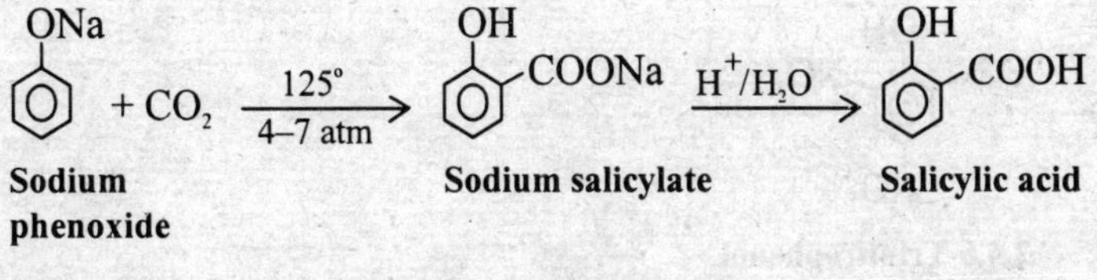

Sodium   Sodium salicylate   Salicylic acid
phenoxide

If the reaction is carried out at high temperature p-hydroxybenzoic acid is produced.

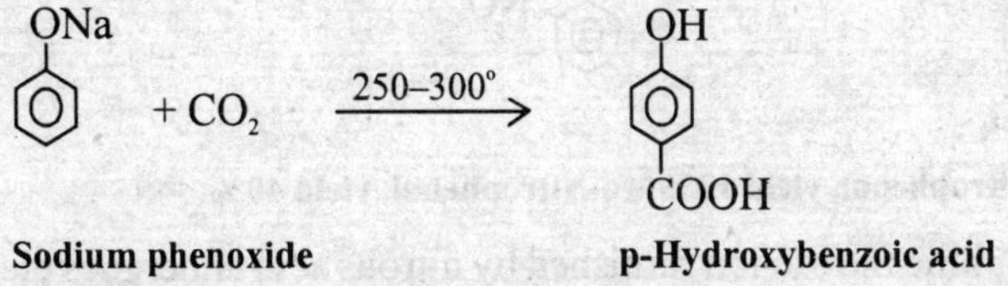

Sodium phenoxide   p-Hydroxybenzoic acid

The mixture of ortho and para hydroxybenzoic acid is separated readily by steam distillation as ortho isomer is more volatile.

The reaction involves electrophilic attack of electron deficient carbon of carbon dioxide on highly reactive phenoxide ion.

6. *Electrophilic Aromatic Substitution:* Phenolic group activates the aromatic ring for electrophilic substitution at ortho and para positions (see Sec. 28.8).

(I) *Halogenation:* Treatment of phenol with aqueous solution of bromine results in formation of 2,4,6-tribromophenol. If deactivating groups like SO_3H is present, it is replaced by bromo group.

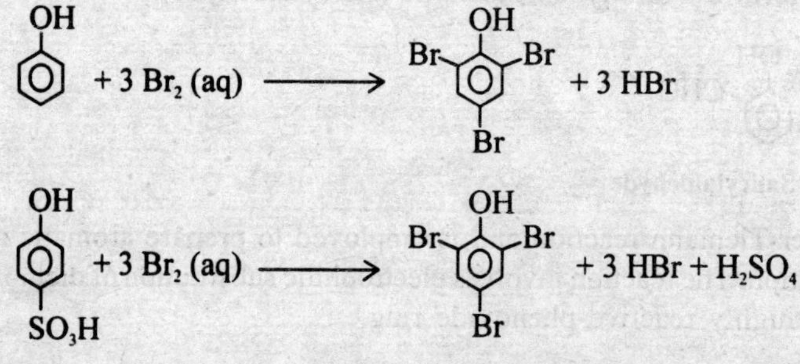

2,4,6-Tribromophenol

The reaction can be limited to monosubstitution by using a solvent of low polarity like chloroform or carbon disulphide.

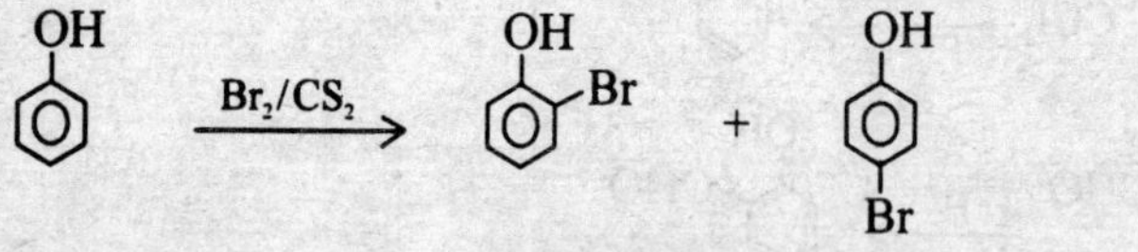

Phenol   **o-Bromophenol** **p-Bromophenol**

(II) *Nitration:* Phenol forms 2,4,6-trinitrophenol on treatment with concentrated nitric acid but the reaction is accompanied by considerable oxidation. For this reason nitration is performed with dil nitric acid at low temperature. The yield in the reaction is low.

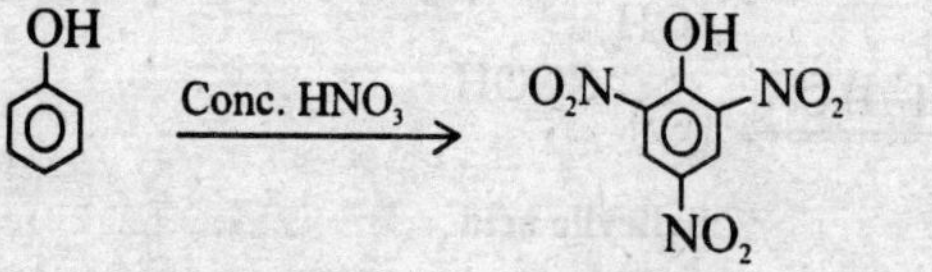

2,4,6-Trinitrophenol

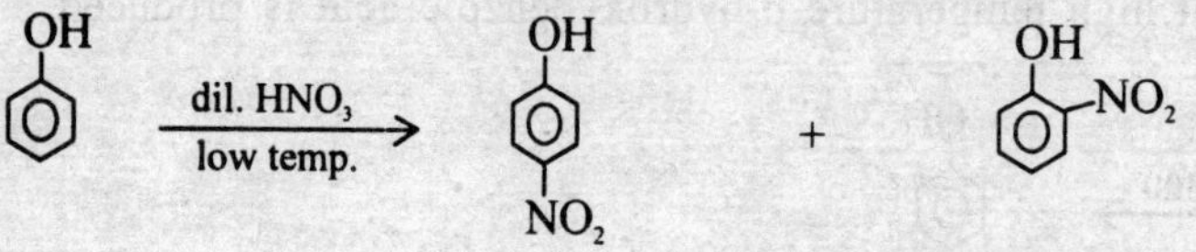

p-Nitrophenol, yield 13%   **o-Nitrophenol, yield 40%**

(III) *Nitrosation:* The weakly electrophilic nitroso ion furnished by nitrous acid undergoes electrophilic substitution in the ring to form p-nitrosophenol (80% yield) and o-nitrosophenol (20% yield).

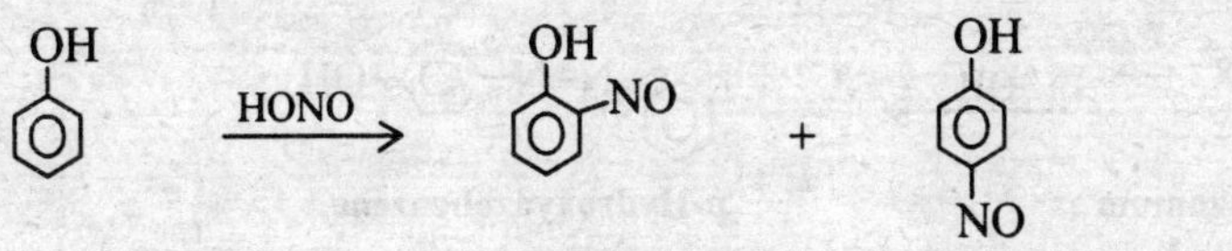

Phenol   **o-Nitrosophenol**   **p-Nitrosophenol**
                (20% yield)   **(80% yield)**

(IV) *Sulphonation:* Phenol on treatment with concentrated sulphuric acid at low temperature forms mainly ortho substitution product whereas at high temperature para product.

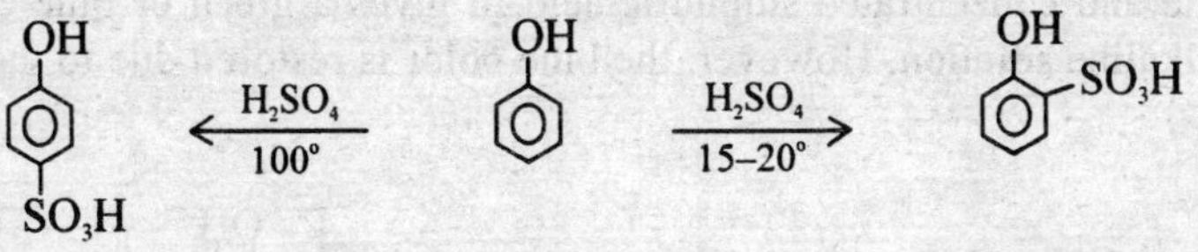

p-Phenolsulphonic   **Phenol**   **o-Phenolsulphonic**
acid (80% yield)   **acid (80% yield)**

(V) *Friedel Crafts Alkylation and Acylation:* Treatment with alkyl halide in presence of Lewis acid, results in the formation of ortho and para alkylphenols which may be separated through steam distillation.

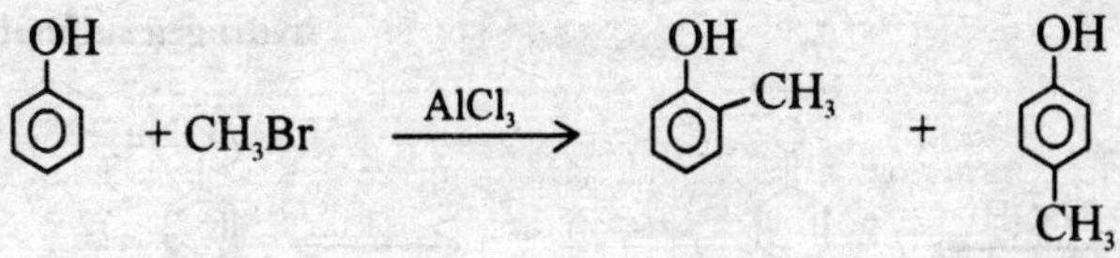

                         o-Cresol   **p-Cresol**

Similarly, o-acyl and p-acylphenols are obtained by treatment of phenol with acyl chloride in presence of aluminum chloride.

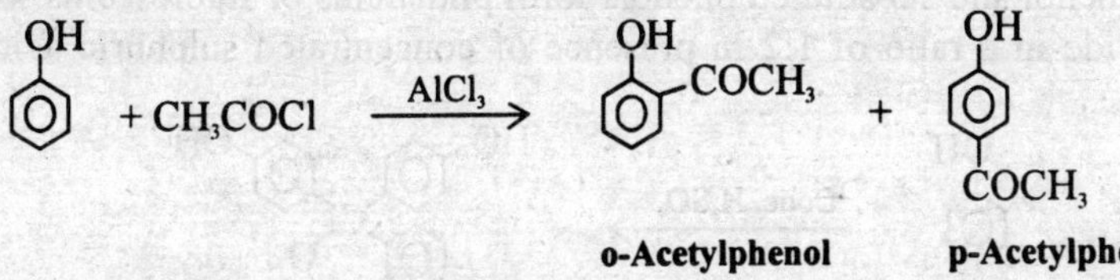

                       o-Acetylphenol   **p-Acetylphenol**

The yields are generally poor in these reactions, thus, phenolic ketones are prepared by *Fries rearrangement.* (see next page)

(VI) *Mercuration:* Phenol undergoes electrophilic substitution when refluxed with mercuric acetate.

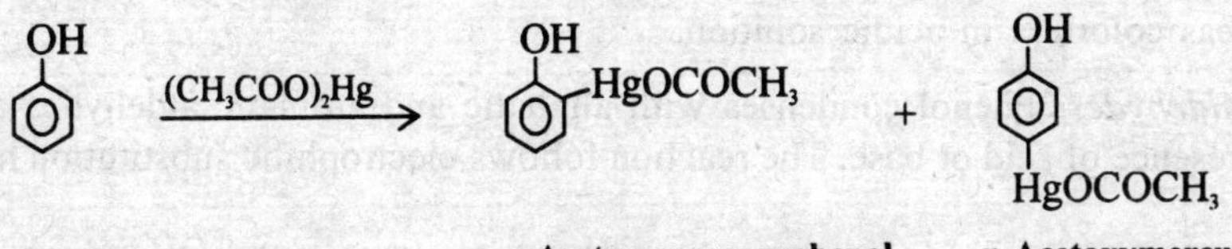

         o-Acetoxymercuryphenol   **p-Acetoxymercuryphenol**

(VII) *Coupling with Diazonium Salts:* Phenols couple with diazonium salts in alkaline solution to form p-hydroxyazocompounds.

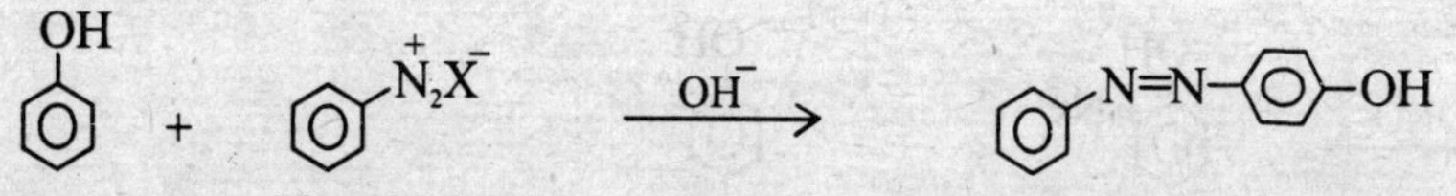

Phenyldiazonium halide   p-Hydroxyazobenzene

An ortho coupling product is formed when para position of phenol is occupied. If both ortho and para positions are occupied, either the group at para is knocked off or no coupling takes place.

7. *Liebermann's Nitroso Reaction:* The reaction is used as a test for identification of phenols. When reacted upon with sodium nitrite and concentrated sulphuric acid, it gives a green or blue color which changes to red on dilution. In alkaline solution, However, the blue color is restored due to the following sequence of reaction.

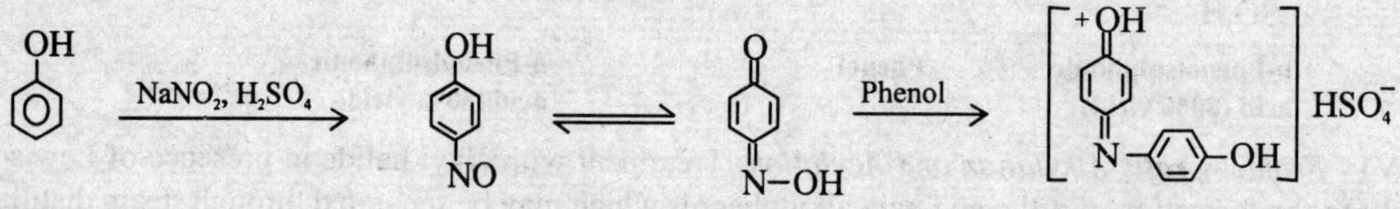

Phenol   p-Nitrosophenol   Quinone monoxime   Indophenol hydrogen sulphate (Blue)

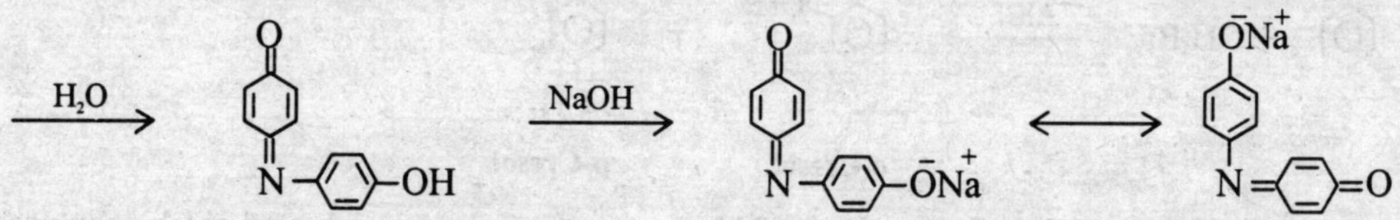

Indophenol (red)   Sodium salt (blue)

8. *Phthalein Reaction:* Phenol and substituted phenols form phthaleins or fluoresceins when treated with phthalic acid or anhydride in a ratio of 1:2 in presence of concentrated sulphuric acid or $ZnCl_2$.

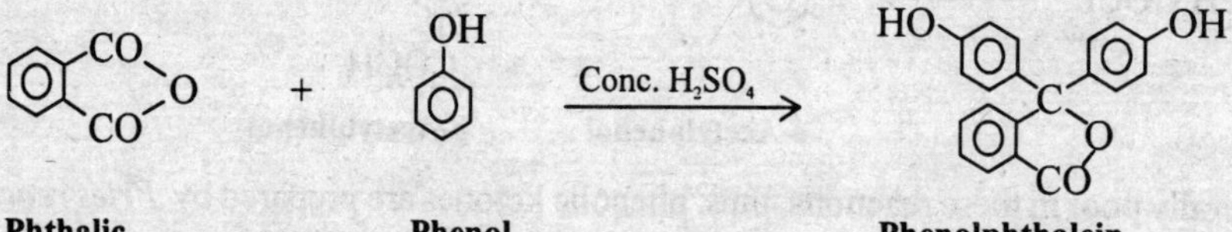

Phthalic anhydride   Phenol   Phenolphthalein

Phenol forms phenolphthalein which is used as an indicator in acid base titrations as it is pink in alkaline solution whereas colorless in acidic solution.

9. *Reaction with Aldehydes:* Phenol condenses with aliphatic and aromatic aldehydes at its ortho and para positions in presence of acid or base. The reaction follows electrophilic substitution mechanism.

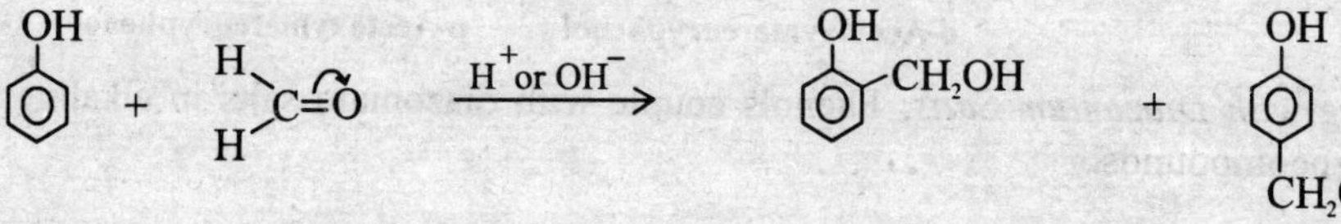

o-Hydroxybenzyl alcohol   p-Hydroxybenzyl alcohol

The reaction does not stop at this stage. On the contrary, a chain reaction starts which results in the formation of cross linked polymer known as phenol formaldehyde resin (or Bakelite).

10. *Esterification or Acylation:* Phenol reacts with carboxylic acids in presence of polyphosphoric acid, p-toluenesulphonic acid or sulphuric acid to form esters.

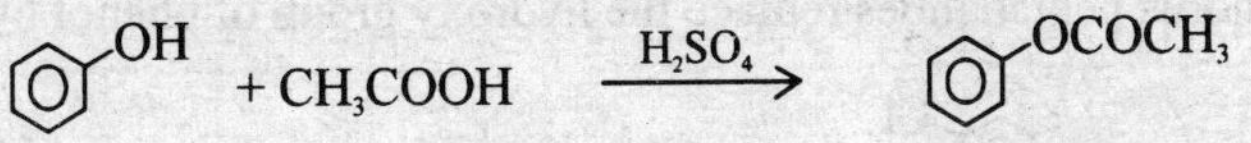

Phenyl acetate

Better yields are obtained if acid chloride or anhydride replaces the acid in presence of base.

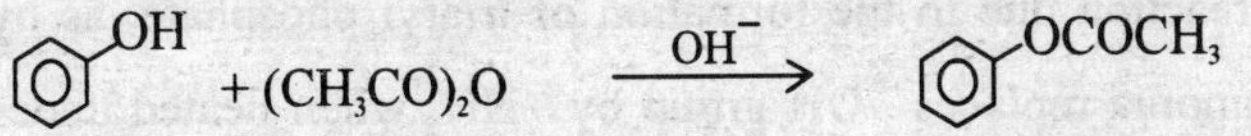

Acetic anhydride   **Phenyl acetate**

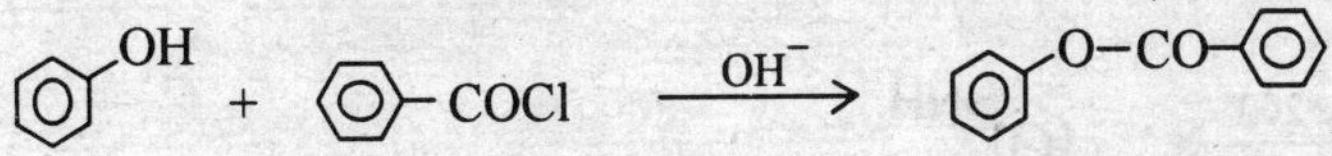

Benzoyl chloride   **Phenyl benzoate**

The reaction of phenol with benzoyl chloride to form phenyl benzoate is also known as *Schotten Baumann reaction.*

The esters of phenols undergo *Fries rearrangement* in presence of anhydrous aluminum chloride to form o- and p-hydroxyacetophenones.

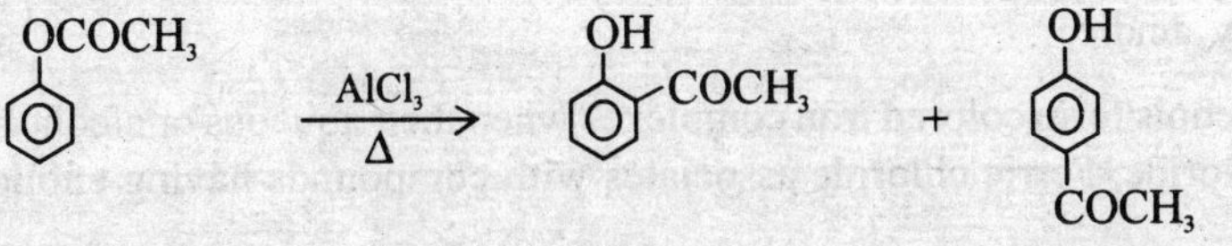

Phenyl acetate   **o-Hydroacetophenone**   **p-Hydroxyacetophenone**

o-Isomer is steam volatile, hence, can be easily separated from p-isomer. Fries rearrangement is therefore used for the preparation of phenolic ketones. The mechanism is believed to involve the attack of electrophilic acylium cation separated from the ester (see Friedel Crafts acylation in Section 28.8).

Esterification reactions are used to protect the phenolic group as they easily hydrolyse back to phenols in presence of acid or base.

11. *Ether Formation or Alkylation:* Phenols react with alkyl halides or alkyl sulphates in alkaline solution to form phenolic ethers.

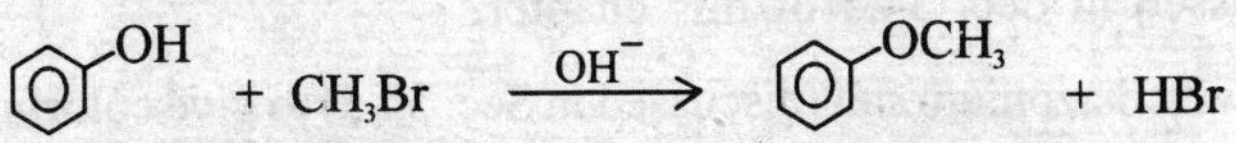

Methyl   **Methoxybenzene**
bromide   **or Anisole**

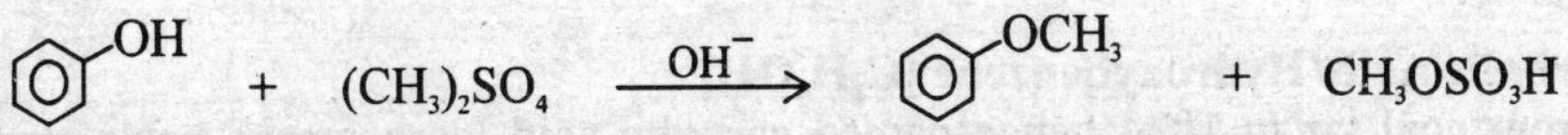

Phenol   **Dimethyl sulphate**   **Phenyl methyl ether**

In alkaline solution, phenoxide ion, the nucleophile replaces halogen atom of alkyl halide resulting in nucleophilic substitution.

$$C_6H_5\overset{\frown}{O} + R\overset{\frown}{-}X \longrightarrow [C_6H_5\overset{\delta^-}{O}\cdots R\cdots\overset{\delta^-}{X}] \longrightarrow C_6H_5OR + X^-$$

Aryl halides do not undergo this reaction due to their inert nature.

12. *Reaction with Phosphorus Halides:* It is difficult to replace the –OH group of phenol by simple HX or PCl_3 but highly reactive phosphorus pentahalides replace the hydroxy group of phenol by halogen.

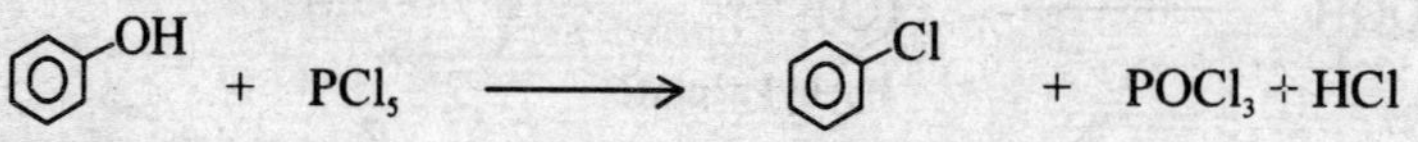

Phenol   + PCl_5 $\longrightarrow$ **Chlorobenzene**   + $POCl_3 + HCl$

The yields are poor in the above reaction due to the formation of triaryl phosphate, as by-product.

13. *Reaction with Ammonia:* Ammonia replaces –OH group by $-NH_2$ when heated under pressure in presence of anhydrous zinc chloride.

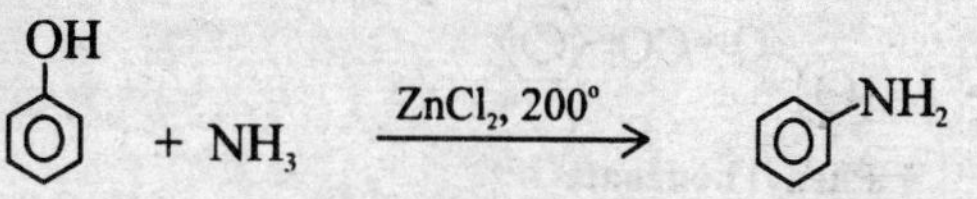

+ NH_3 $\xrightarrow{ZnCl_2, 200°}$ **Aniline**

34.6   TESTS OF PHENOLS

Following tests may be used to identify the phenols.

1. *Solubility Test:* Phenols are soluble in sodium hydroxide but insoluble in sodium carbonate and bicarbonate as they are weakly acidic.

2. *Ferric Chloride Test:* Phenols form colored iron complexes when their aqueous or alcoholic solution is treated with neutral ferric chloride. Ferric chloride associates with compounds having enolic structure as shown below in phenol.

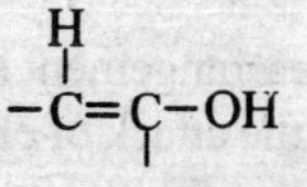

Enolic structure 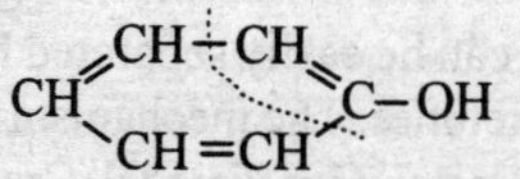   **Phenol having enolic structure**

3. *Bromine Water Test:* Aqueous solution of phenol forms tribromophenol when treated with bromine water.

4. *Libermann's Nitroso Test:* Phenols when warmed with concentrated sulphuric acid and sodium nitrite give a unique color test discussed in Sec. 34.5 of this chapter.

5. *Coupling Test:* Phenol couples with diazonium salt (discussed in Sec. 34.5) to give colorful azodyes.

34.7   INDIVIDUAL MONOHYDRIC PHENOLS

34.7.1   Phenol/Carbolic Acid/Hydroxybenzene [C_6H_5OH]

It was discovered from coal tar in 1834 hence named carbolic acid (carb means coal; oleum means oil) It may be prepared by any of the general methods of preparation of phenols (see Sec. 34.2). A few other commercial methods of preparation of phenol have been summarised below.

Dows Process: Chlorobenzene is quantitatively converted into phenol by exhaustive alkaline hydrolysis. An emulsion of chlorobenzene in 10% sodium hydroxide is heated to 350° under 200 atmospheric pressure in presence of copper catalyst.

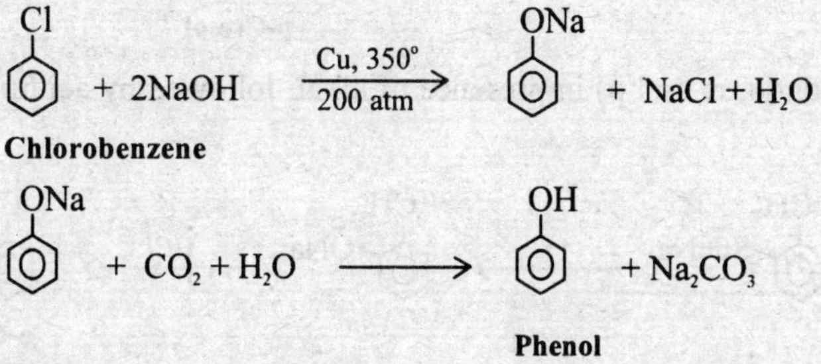

Phenol is liberated from the formed sodium salt by passing carbon dioxide in aqueous solution.

Raschig Process: Hydrolysis of chlorobenzene by superheated steam forms phenol. Chlorobenzene, in turn, is produced by the action of hydrogen chloride and oxygen on benzene in presence of catalyst.

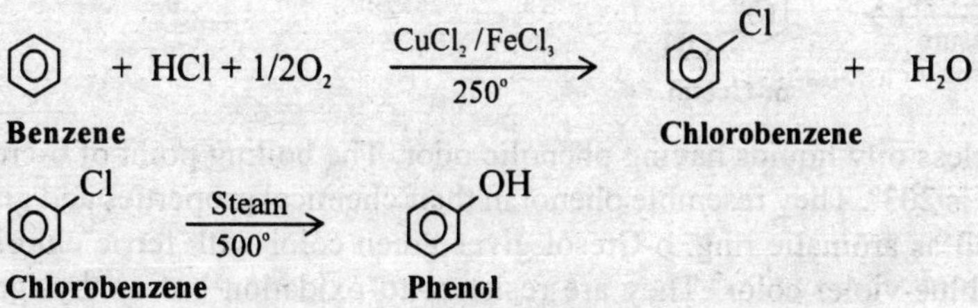

Properties: Phenol (m.p. 43°, b.p. 182°) is a colorless, crystalline, deliquescent solid turns pink on exposure to light or air. It has a peculiar phenolic odor having corrosive action on skin. It is sparingly soluble in water but readily soluble in organic solvents.

It exhibits all the reactions characteristic of phenols discussed in Sec. 34.5.

Uses: Phenol is used in various industries. Some of the important uses have been given below:
1. As an antiseptic. 2,4-Dichloro-3,5-dimethyl phenol is used in the preparation of dettol.
2. In the preparation of phenol formaldehyde resin (bakelite)
3. In the preparation of cyclohexanol, salicylic acid, picric acid, phenolphthalein etc. which have diverse uses.
4. It is also used in the preparation of fungicides, herbicides and bactericides.
5. It is used as an ink preservative.

34.7.2 Cresols

Cresols are ortho, meta or para methyl substituted phenols constituting the major fraction of creosote oil available from tar fraction of coal, wood or petroleum. They may be separated from this fraction by fractionation.

Preparation

(I) ortho, meta or para cresols may be synthesised by diazotisation of corresponding toluidines and warming the aqueous solutions.

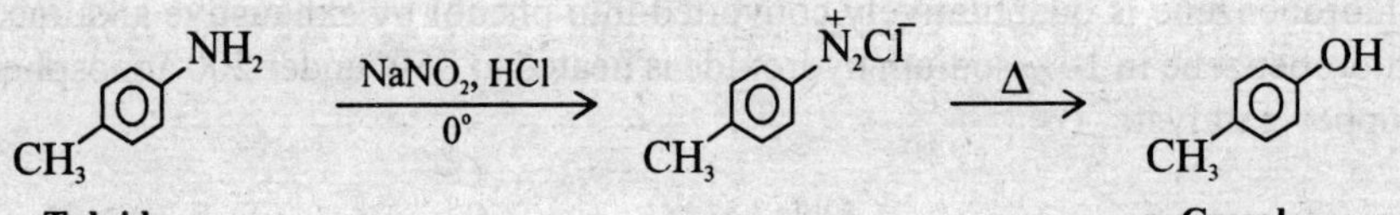

p-Toluidene → → p-Cresol

(II)   Hydrolysis of sulphonic acids (o, m and p) in presence of alkali followed by acidification form cresols.

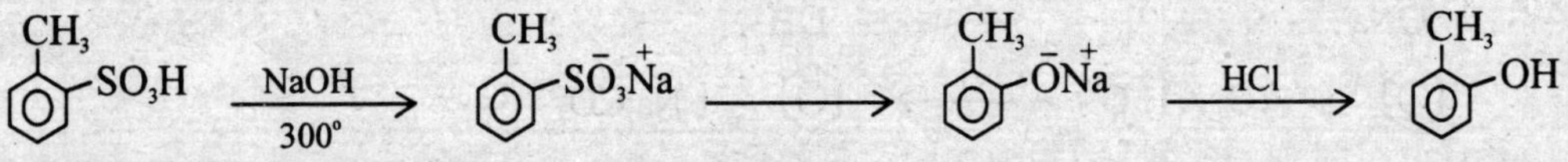

o-Toluene-
sulphonic acid   o-Cresol

(III)   Hydrolysis of corresponding chlorotoluenes also yields cresols.

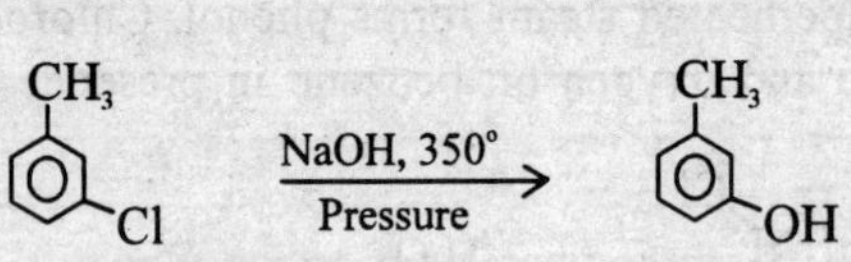

m-Chlorotoluene   m-Cresol

Properties: Cresols are colorless oily liquids having phenolic odor. The boiling point of o-cresol is 190° m-cresol is 201° and p-cresol is 203°. They resemble phenol in their chemical properties and give reaction due to phenolic group as well as aromatic ring. o-Cresol gives green color with ferric chloride wherea meta and para cresols give blue-violet color. They are resistant to oxidation but methyl group can be oxidised to carboxylic acid easily with chromic acid if the –OH group is protected by alkylation or acyla tion.

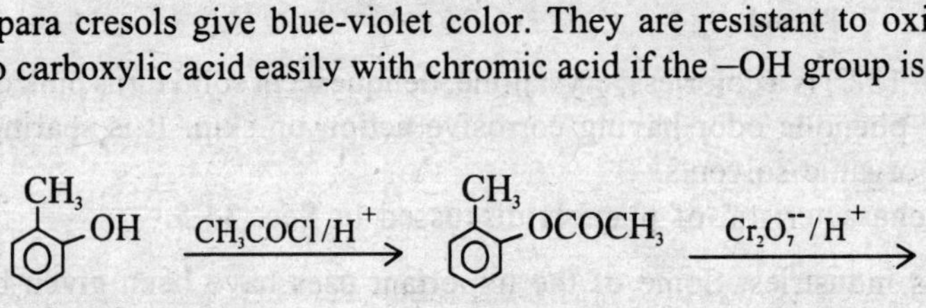

o-Cresol

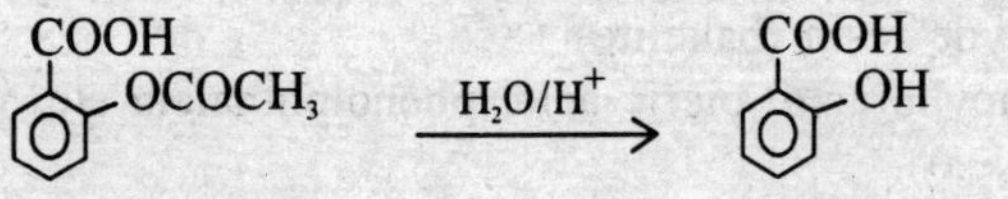

Acetyl salicylic   Salicylic acid
acid (Aspirin)

p-Cresol forms p-hydroxybenzoic acid by similar sequence of reactions.

Uses
1. Mixture of o-, m- and p-cresol, called creosote, is used in wood preservation.
2. Mixture of cresols is also used as a disinfectant commonly known as lysol in soaps.
3. It is also used for the manufacture of synthetic resins, dyes and explosives.
4. They find use in rubber industry for the preparation of antioxidants.

34.7.3   Nitrophenols

Preparation: Ortho and para nitrophenol may be prepared by direct nitration of phenol with dilute nitric cid at room temperature followed by separation via steam distillation. (o-nitrophenol is steam volatile). ֺ-Nitrophenol, however, is obtained by indirect method starting from m-nitroaniline.

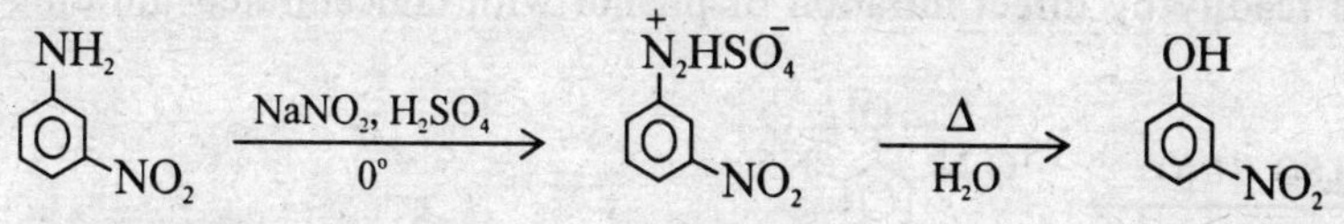

m-Nitroaniline   **m-Nitrophenol**

Properties: o-Nitrophenol is a yellow crystalline solid, m.p. 45° with a strong odor whereas m- and p-ֺsomers are colorless solids having melting points 97° and 114° respectively. They are sparingly soluble ֺn water. The intramolecular hydrogen bonding is possible only in o-isomer, thus, prevents it from association ֺith many molecules making it steam volatile whereas m- and p-isomer having intermolecular hydrogen ֺonding have high m.p. and are low in volatility (see also Sec. 34.3).

They exhibit general reactions of phenols discussed earlier (Sec. 34.5). Some additional reactions ֺhown by nitrophenols are given below.

(i) o- or p-Nitrophenol on treatment with bromine water forms 2,4,6-tribromophenol.

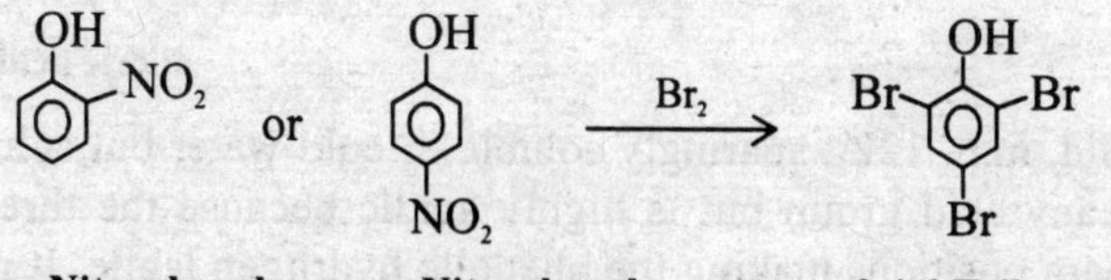

o-Nitrophenol   **p-Nitrophenol**   **2,4,6-Tribromophenol**

Nitro group is knocked off by bromine. m-Isomer does not undergo this reaction.

(ii) o-and p-Nitrophenols give two series of ethers n-form or benzenoid form which is colorless and ֺhe other aci-form or quinonoid form which is colored.

Colorless n-ether   **Colored aci-ether**
(Benzenoid form)   **(Quinonoid form)**

This indicates that nitrophenols may exist in tautomeric forms but they have not been isolated as ֺet.

Enolic form   **Keto form**
                    p-Nitrophenol

The colorless n-ethers of o- and p-nitrophenols are stable and undergo hydrolysis slowly whereas ֺhe colored aci-ethers are unstable and readily hydrolyse back.

m-Nitrophenol forms only colorless ethers indicating that it does not show taumerism as it can not ֺxist in quinonoid form.

(iii) All the cresols give yellow red color in alkaline solution.

34.7.4   Picric Acid/ 2,4,6-Trinitrophenol

Preparation

(I)   It can be prepared readily by direct nitration of phenol with concentrated nitrating mixture.

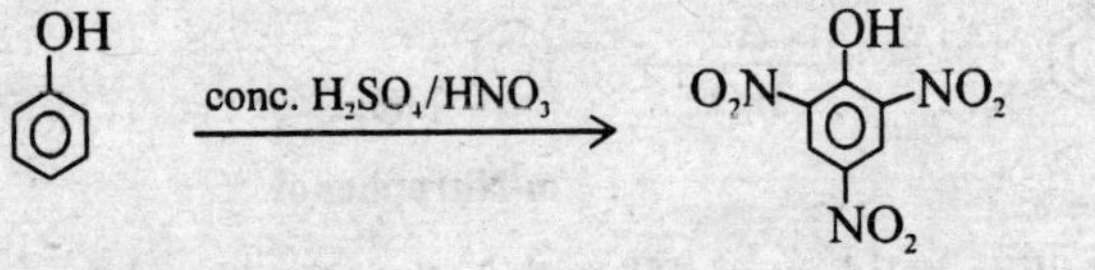

Phenol   **2,4,6-Trinitrophenol**

(II)   There is considerable yield loss due to oxidation of phenol by nitric acid, so picric acid now a days is prepared from chlorobenzene.

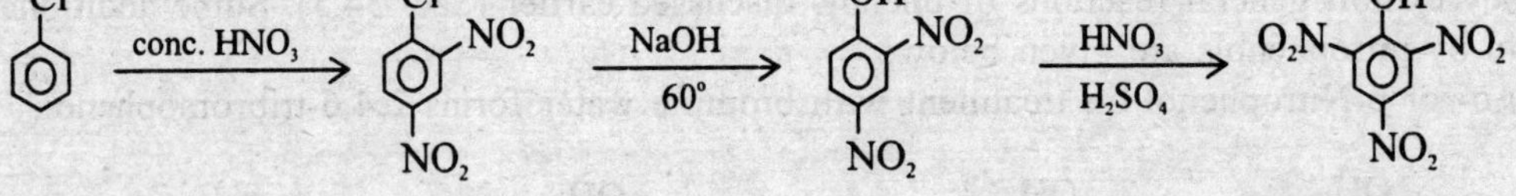

Chlorobenzene   **Picric acid**

Properties: It is a yellow crystalline solid, m.p. 122°, sparingly soluble in cold water but soluble in hot water and alcohol. It does not contain any acid group but is highly acidic because the three electron withdrawing nitro groups at ortho and para positions making the phenolic hydrogen labile. It reacts with sodium carbonate and bicarbonate with effervesence (a characteristic of carboxylic acids).

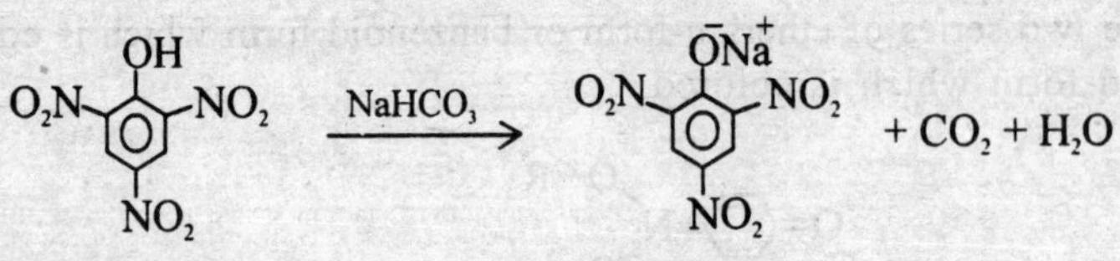

Picric acid   **Sodium picrate**

When reacted with phosphorus pentachloride it forms picryl chloride which forms picramide on reaction with ammonia.

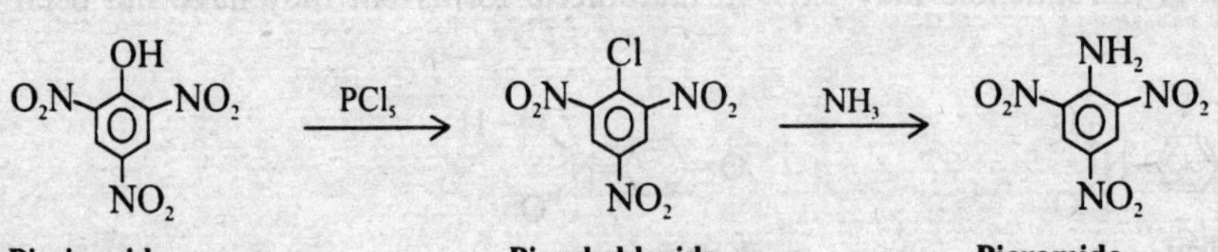

Picric acid   **Picryl chloride**   **Picramide**

When reduced with sodium sulphide, it forms picramic acid.

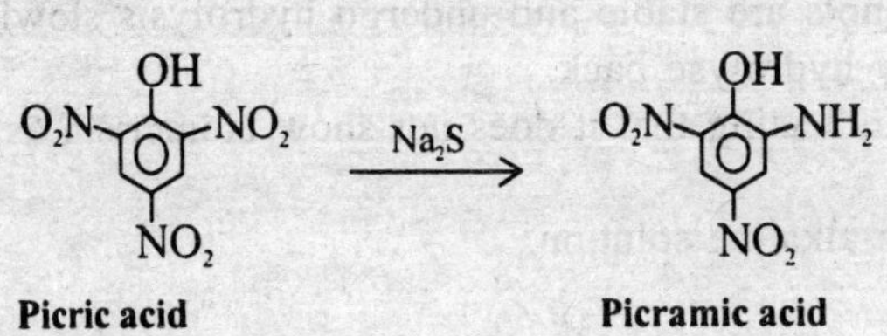

Picric acid   **Picramic acid**

It forms well defined crystalline charge transfer complexes or addition compounds (called picrates) having characteristic melting points with aromatic hydrocarbons, phenols and amines. Picrates are, therefore, used to identify these organic compounds.

Uses

1. As a laboratory reagent for the isolation and identification of organic compounds.
2. In manufacture of explosives *lyddite and melinite.*
3. As an antiseptic and analgesic in the treatment of burns.
4. For dying silk and wool.

4.7.5 Aminophenols

Preparation: o-, m- and p-Aminophenols are prepared by the reduction of corresponding nitrophenols with tin and hydrochloric acid.

m-Aminophenol is prepared from resorcinol on commercial scale by treatment with ammonia in aqueous ammonium chloride under pressure.

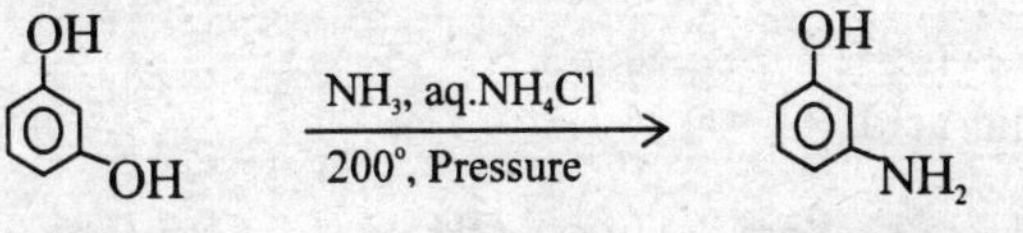

Resorcinol → m-Aminophenol

$$\xrightarrow[\text{200°, Pressure}]{\text{NH}_3, \text{ aq.NH}_4\text{Cl}}$$

p-Aminophenol may be prepared by the reduction of p-hydroxyazobenzene with sodium bisulphite.

$$\bigcirc-N=N-\bigcirc-OH \longrightarrow H_2N-\bigcirc-OH$$

p-Hydroxyazobenzene **p-Aminophenol**

Rearrangement of phenyl hydroxylamine in presence of acid also produces p-aminophenol.

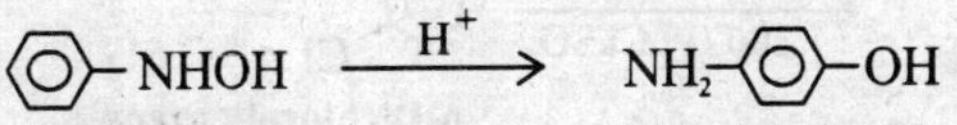

$$\bigcirc-NHOH \xrightarrow{H^+} NH_2-\bigcirc-OH$$

Phenyl hydroxyl amine **p-Aminophenol**

Properties: Aminophenols are colorless solids which turn yellow to brown on exposure to air. The melting points of o,m and p isomers are 137°, 123° and 185° respectively.

o- and p-Aminophenols are less acidic than phenol due to the presence of electron releasing amino group making the phenoxide ion unstable. They do not form phenoxide ion with alkalis. They, however, form salt with acids due to the presence of $-NH_2$ group.

The o- and p-isomer undergo oxidation to form quinones whereas m-isomer does not form such products.

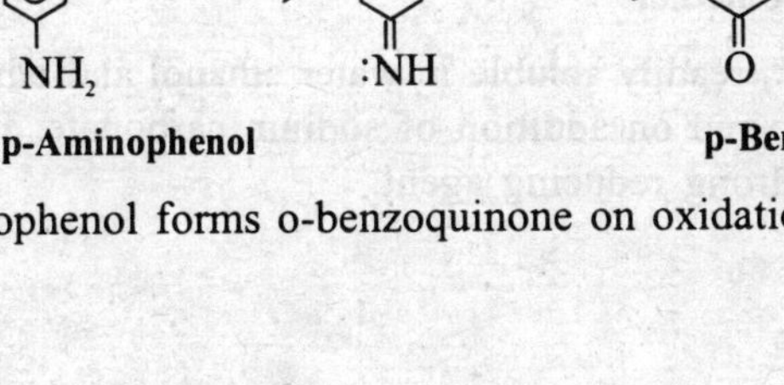

p-Aminophenol **p-Benzoquinone**

o-Aminophenol forms o-benzoquinone on oxidation.

On acetylation with conventional acetylating agents, N-acetylation takes place readily than phenol hydroxy group.

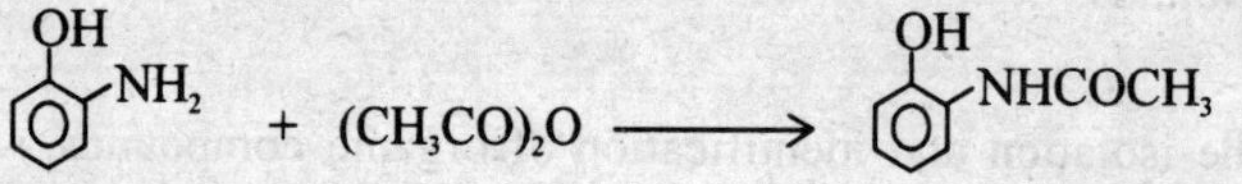
OH
NH2
+ (CH3CO)2O
OH
NHCOCH3

o-Aminophenol **o-Acetamidophenol**

Uses

1. p-Aminophenol is an important photographic developer and used as a dye intermediate.
2. m-Aminophenol is also used in dye manufacture.

34.8 DIHYDRIC PHENOLS

34.8.1 Catechol/Pyrocatechol/o-Dihydroxybenzene

It occurs in *gum catechu*, a secretion from *acacia catechu* tree, and was first isolated from this source hence named catechol.

Preparation

(I) By alkaline fusion of o-phenolsulphonic acid.

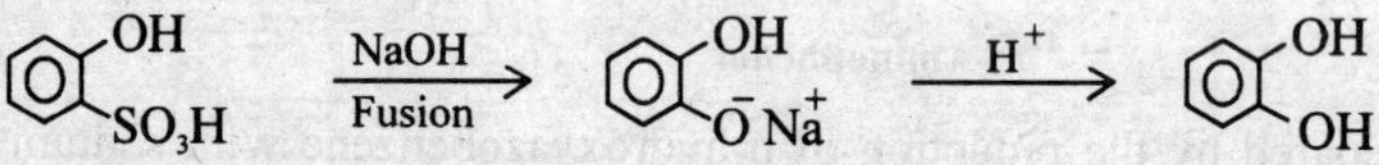
OH
SO3H
NaOH
Fusion
OH
O⁻ Na⁺
H⁺
OH
OH

o-Phenolsulphonic acid **Catechol**

(II) o-Chlorophenol or o-dichlorobenzene on hydrolysis with 30% NaOH at 200° under pressure and in presence of copper sulphate catalyst produces catachol.

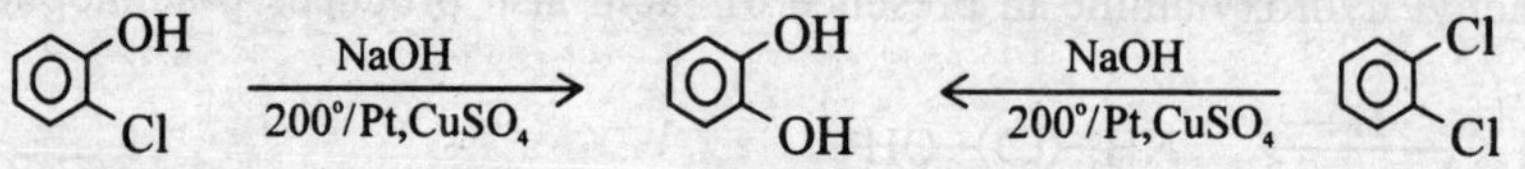
OH
Cl
NaOH
200°/Pt,CuSO4
OH
OH
NaOH
200°/Pt,CuSO4
Cl
Cl

o-Chlorophenol **Catechol** **o-Dichlorobenzene**

(III) Guaiacol, monomethylether of catechol, obtained from wood tar yields catechol on acid hydrolysis.

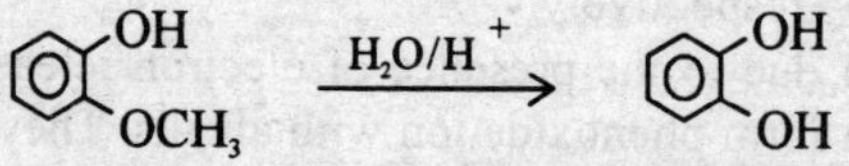
OH
OCH3
H2O/H⁺
OH
OH

Guaiacol **Catechol**

(IV) Oxidation of salicylaldehyde with alkaline hydrogen peroxide produces catechol.

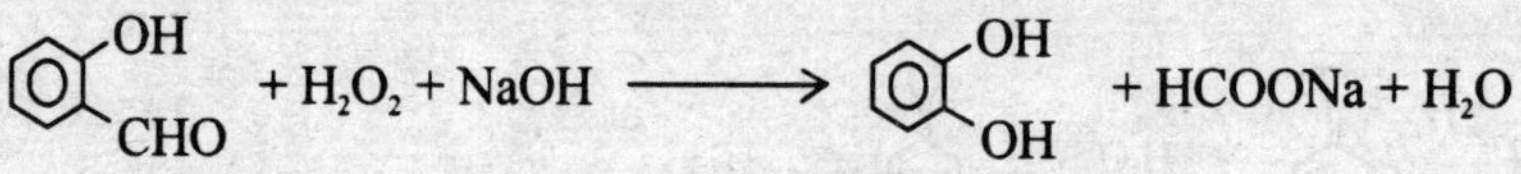
OH
CHO
+ H2O2 + NaOH
OH
OH
+ HCOONa + H2O

Salicylaldehyde **Catechol**

Properties: It is a while crystalline solid, m.p.104°, readily soluble in water ethanol and ether. It gives green color with ferric chloride which changes to red on addition of sodium carbonate. It is readily oxidised to o-benzoquinone and hence act as a strong reducing agent.

It reacts with phthalic anhydride to form alizarin, a dye.

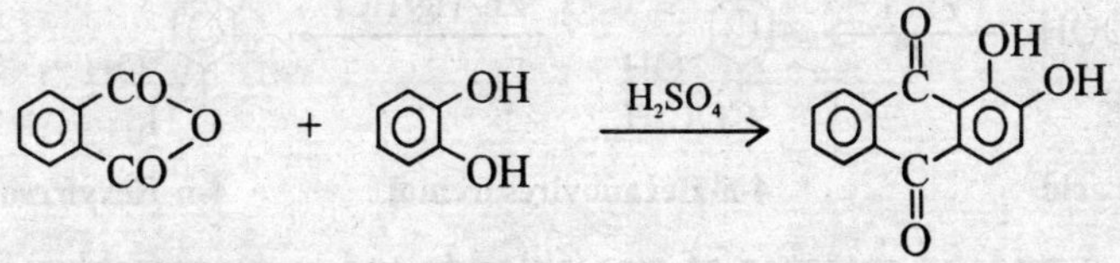

Phthalic anhydride   **Catechol**   **Alizarin**

ses
1. As a photographic developer.
2. In the manufacture of alizarin dye.
3. Guaiacol, the monomethyl ether of catechol is used as an antioxidant in gasoline.

4.8.2   Resorcinol/m-Dihydroxybenzene

reparation: Resorcinol is prepared by fusion of sodium m-benzenedisulphonate with sodium hydroxide ollowed by acidification.

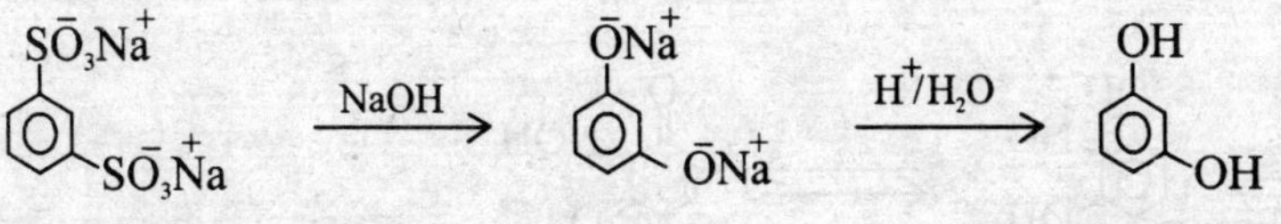

Sodium-m-benzene-   **Resorcinol**
disulphonate

roperties: It is a colorless, crystalline solid m.p.110°, soluble in water, alcohol and ether, darkens on xposure to air. Its aqueous solution gives violet color with ferric chloride. Like phenol, bromination nd nitration of resorcinol gives 2,4,6-trisubstituted product. It couples with diazonium salts to form .yes and undergo phthalein reaction with phthalic anhydride to form fluorescein.

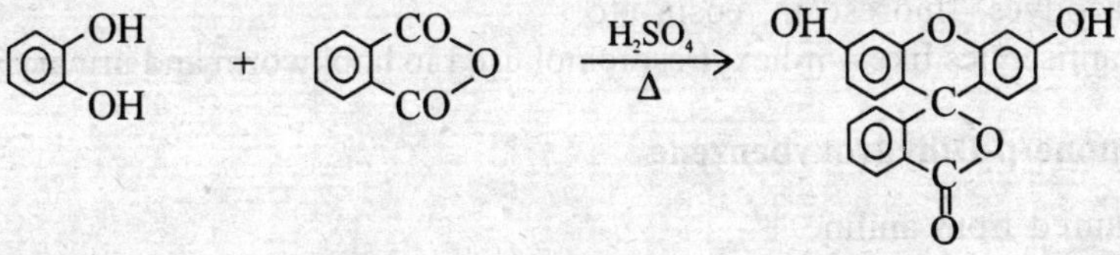

Resorcinol   **Phthalic anhydride**   **Fluorescein**

Resorcinol exhibits keto-enol tautomerism, as it forms dioxime and bisulphite compounds.

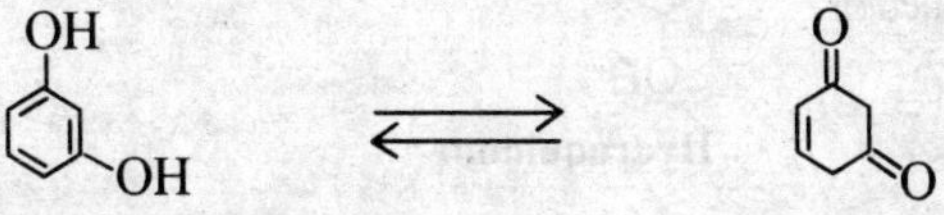

Enol form (forms bisulphite)   **Keto form (forms dioxime)**

When heated with n-caproic acid and zinc chloride, it forms 4-n-hexanoylresorcinol which on Clemmensen eduction yields 4-n-hexylresorcinol which is a powerful antiseptic.

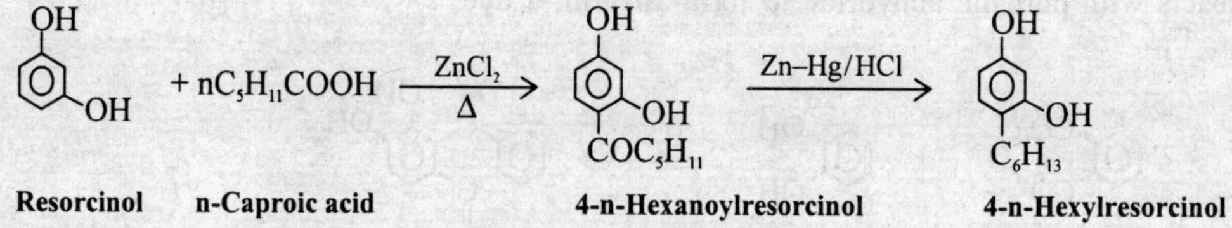

Resorcinol   n-Caproic acid   4-n-Hexanoylresorcinol   4-n-Hexylresorcinol

A similar reaction with alkyl cyanide in presence of zinc chloride and hydrogen chloride produce phenolic ketones.

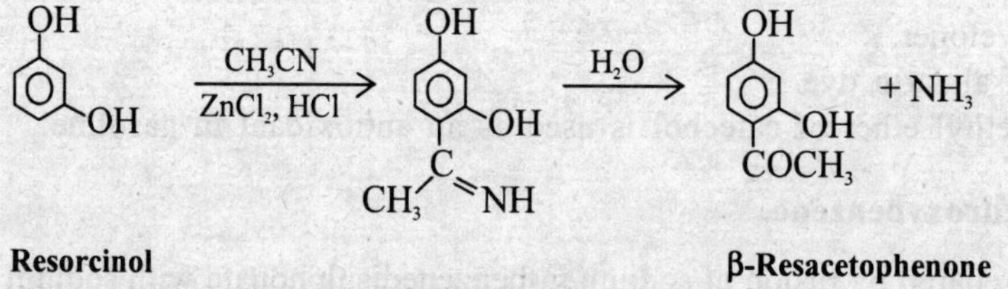

Resorcinol   β-Resacetophenone

The reaction is a characterisation of phenols having two hydroxy groups at m-position and is named as *Houben Hoesch reaction*. When treated with nitrous acid, resorcinol forms dinitrosoresorcinol.

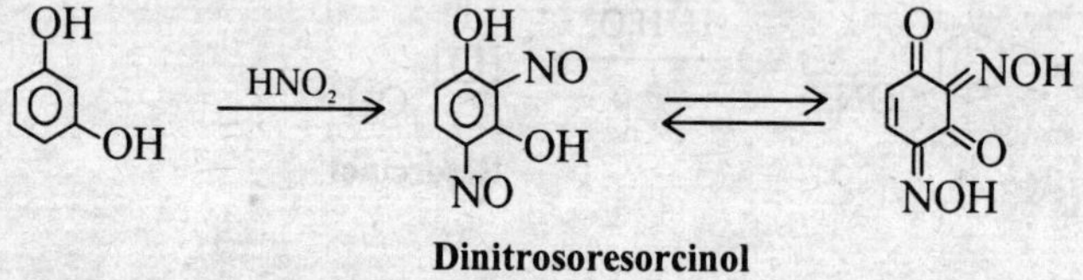

Dinitrosoresorcinol

Uses
1. As an antiseptic for the treatment of eczema.
2. In the preparation of styphnic acid (2,4,6-trinitroresorcinol) used in the identification of certain organic compounds.
3. In the synthesis of azodyes, fluorescein, eosin etc.
4. In the preparation of antiseptics like 4-n-hexylresorcinol used in hookworm and urinary infections

34.8.3   Quinol/Hydroquinone/p-Dihydroxybenzene

Preparation: It is manufactured from aniline.

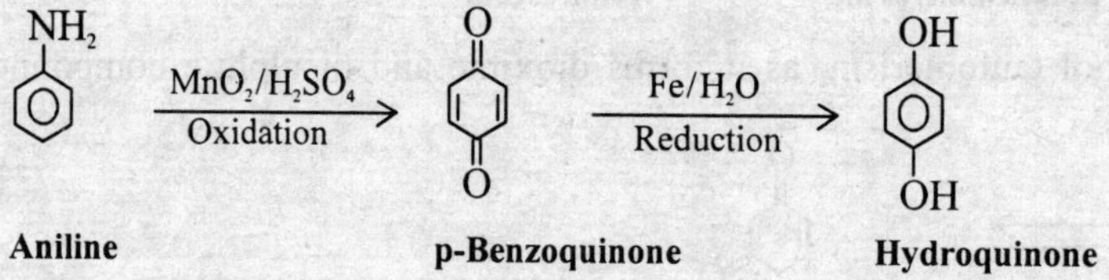

Aniline   p-Benzoquinone   Hydroquinone

Properties: It is a colorless, crystalline solid, m.p. 170°, soluble in water, alcohol and ether. Its aqueous solution turns brown on exposure to air. It is strongly reducing in nature, reduces fehlings solution and ammonical silver nitrate. With mild oxidising agents like ferric chloride, it forms p-benzoquinone (a green solid).

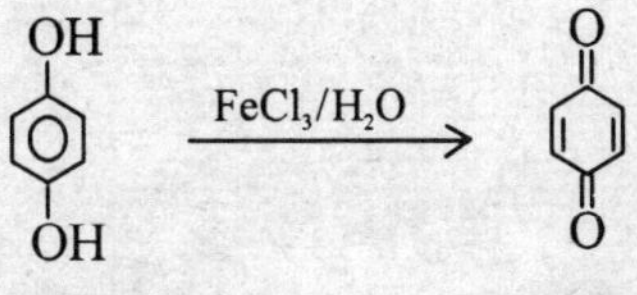

OH
FeCl₃/H₂O →
(structure)

Hydroquinone   **p-Benzoquinone**

Like resorcinol, it exhibits keto enol tautomerism and forms oxime with hydroxylamine and a bisulphite addition compound with sodium bisulphite.

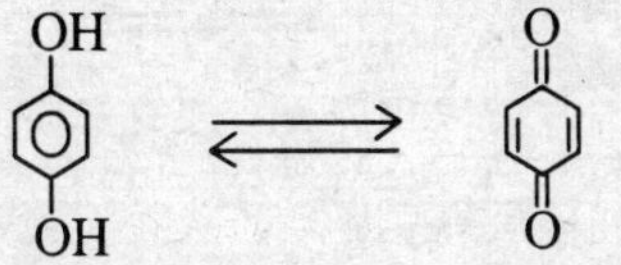

OH
(structure) ⇌ (structure)
OH

Enol form   **Keto form**
(form Bisulphite)   **(forms Dioxime)**

Ammonolysis of hydroquinone with primary amine in presence of zinc chloride yields N-alkyl substituted p-phenylenediamine which is an important antioxidant for preserving gasoline and rubber.

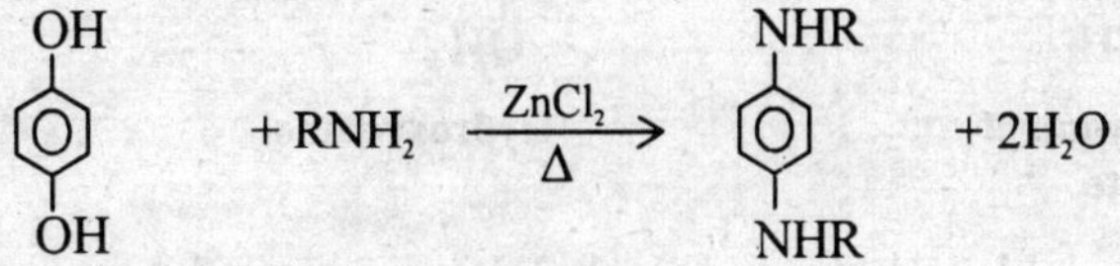

OH   NHR
(structure) + RNH₂ $\xrightarrow{\underset{\Delta}{ZnCl_2}}$ (structure) + 2H₂O
OH   NHR

Uses
1. As a photographic developer.
2. As an antioxidant or inhibitor of autooxidation and polymerisation reactions.
3. In quinone-hydroquinone electrodes.
4. As an antiseptic in the treatment of conjunctivitis.

14.9 TRIHYDRIC PHENOLS

14.9.1 Pyrogallol/Pyrogallic Acid/1,2,3-Trihydroxybenzene

Preparation: It is prepared by decarboxylation of gallic acid by heating under pressure.

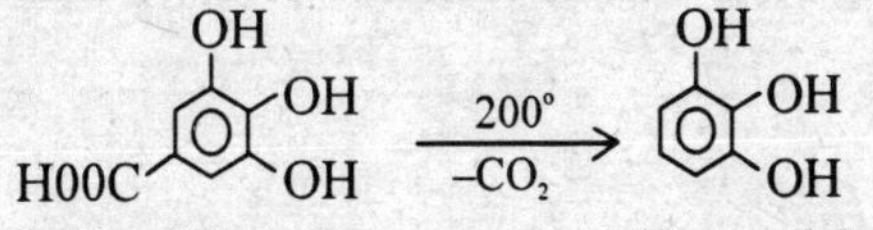

OH   OH
HOOC (structure) OH $\xrightarrow[-CO_2]{200°}$ (structure) OH
OH   OH

Gallic acid   **Pyrogallol**

Properties: It is a white crystalline solid, m.p. 133°, soluble in water, ethanol and ether, forms red color with ferric chloride solution.

Alkaline solution of pyrogallol absorbs oxygen rapidly and turns brown. This property is used for the absorption of oxygen in gas analysis.

It is the most powerful of all the phenols and reduces silver nitrate and fehlings solution in cold. It can also reduce the salts of other metals like gold, platinum and mercury to the respective metals.

Uses

1. Used as an excellent photographic developer.
2. As antiseptic and antioxidant.
3. In gas analysis as oxygen absorber.
4. As a hair dye.

34.9.2 Hydroxyquinol/1,2,4-Trihydroxybenzene

It is prepared by alkali fusion of quinol in air.

Quinol **Hydroxyquinol**

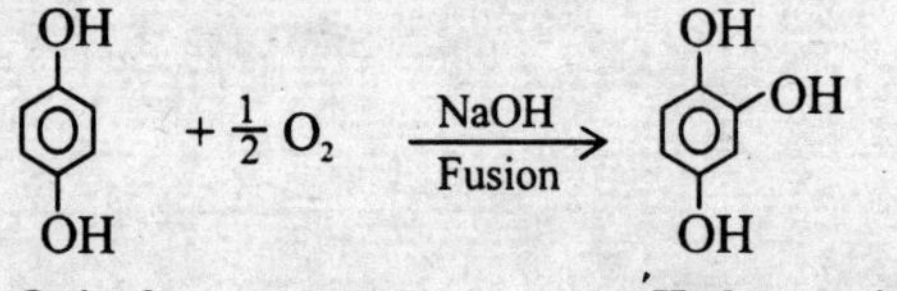

It can also be prepared from p-benzoquinone as follows.

p-Benzoquinone **Hydroxyquinol-triacetate** **Hydroxyquinol**

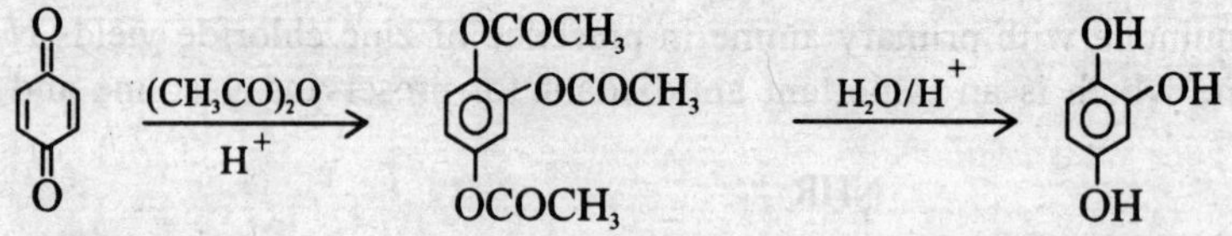

It is a white crystalline solid, m.p.140°, soluble in water and alcohol. Its aqueous solution gives greenis brown color with $FeCl_3$. It is a powerful reducing agent.

34.9.3 Phloroglucinol/1,3,5-Trihydroxybenzene

Preparation: It is prepared by fusion of resorcinol with alkali in air.

Resorcinol **Phloroglucinol**

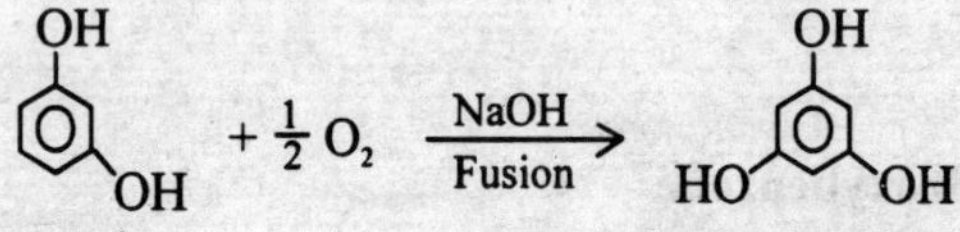

It is conveniently prepared in laboratory by heating 1,3,5-triaminobenzene or 2,4,6-triaminobenzoi acid with acids.

1,3,5-Triaminobenzene **Phloroglucinol**

2,4,6-Triaminobenzoic acid **Phloroglucinol**

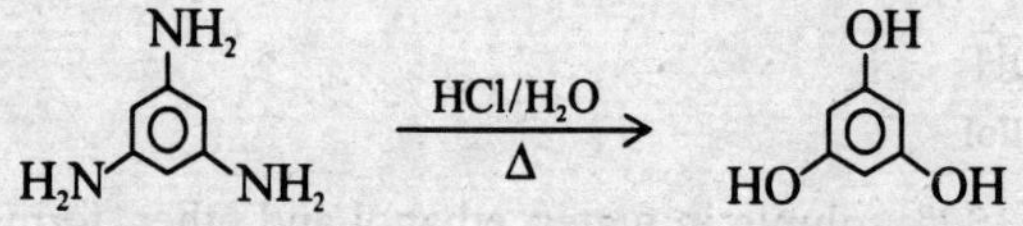

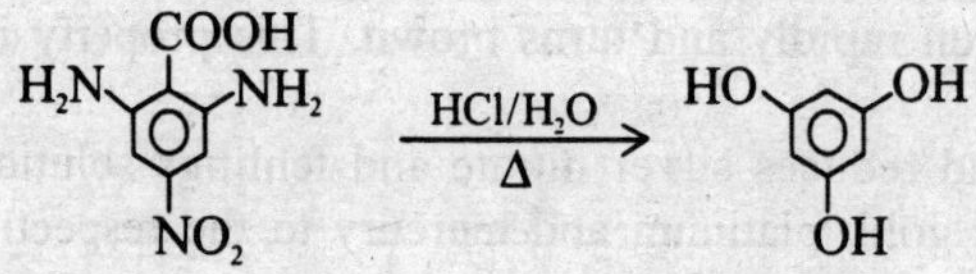

operties: It is a colorless crystalline solid, m.p. 218°, fairly soluble in water. Its aqueous solution tastes ~veet and give bluish violet color with ferric chloride. Alkaline solution darkens rapidly on exposure air.

It forms triacetate when treated with acetic anhydride and trioxime when treated with hydroxyl amine dicating that it exhibits keto-enol tautomerism.

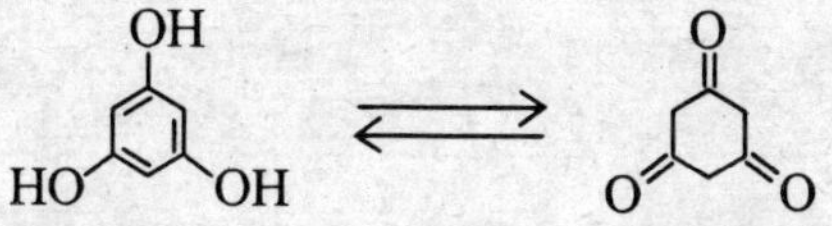

Enol form
(forms triacetate)

Keto form
(forms trioxime)

ses
1. It forms red color with carbohydrates in presence of sulphuric acid thus used for identification of carbohydrates.
2. In the preparation of dyes.
3. In estimation of pentoses (carbohydrates).

4.10   AROMATIC ALCOHOLS

4.10.1   Benzyl Alcohol/Phenyl Carbinol

is the simplest member of the class of aromatic alcohols and is isomeric with cresols. It occurs as ee alcohol in Peru and Tolu balsams and as its ester in Jasmine oil and other essential oils in small nounts.

reparation
(I) It is prepared by alkaline hydrolysis of benzyl chloride.

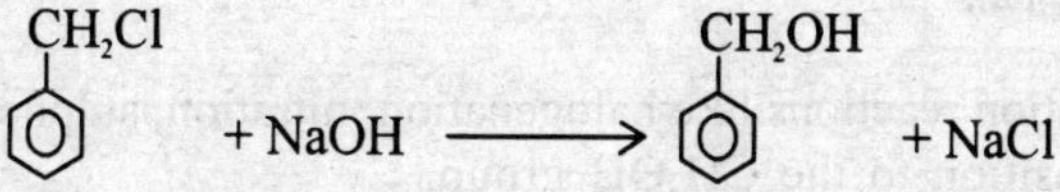

Benzyl chloride   **Benzyl alcohol**

(II) Reduction of benzaldehyde by conventional method produces benzyl alcohol.

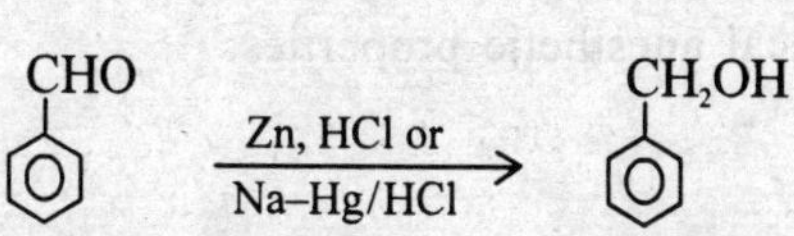

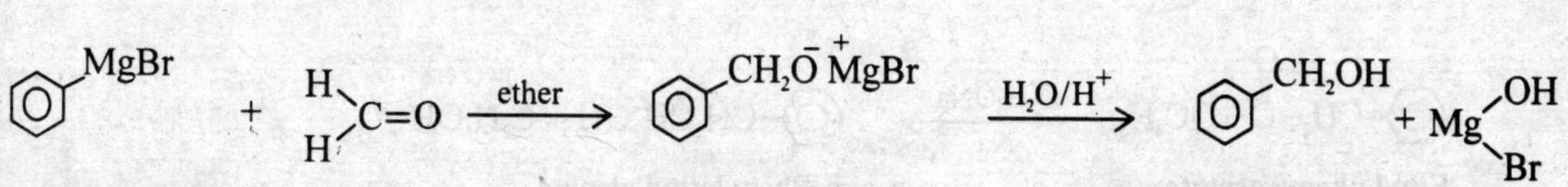

Benzaldehyde   **Benzyl alcohol**

(III) Grignard synthesis: Phenylmagnesium bromide on reaction with formaldehyde followed by hydrolysis :oduces benzyl alcohol.

Phenylmagnesium
bromide

Benzyl alcohol

(IV) Benzyl alcohol may also be prepared from benzaldehyde and formaldehyde in alkaline solutic (Canizzaro reaction).

Properties: It is a colorless, pungent smelling liquid, b.p. 206°, sparingly soluble in water but readi soluble in most organic solvents.

It undergo reactions of aliphatic alcohols as shown below.

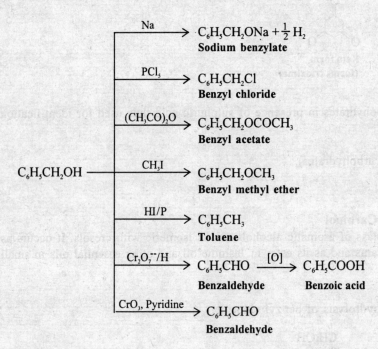

$C_6H_5CH_2OH$

Na $\rightarrow$ $C_6H_5CH_2ONa + \frac{1}{2}H_2$
Sodium benzylate

PCl_5 $\rightarrow$ $C_6H_5CH_2Cl$
Benzyl chloride

$(CH_3CO)_2O$ $\rightarrow$ $C_6H_5CH_2OCOCH_3$
Benzyl acetate

CH_3I $\rightarrow$ $C_6H_5CH_2OCH_3$
Benzyl methyl ether

HI/P $\rightarrow$ $C_6H_5CH_3$
Toluene

$Cr_2O_7^{--}/H$ $\rightarrow$ C_6H_5CHO $\xrightarrow{[O]}$ C_6H_5COOH
Benzaldehyde **Benzoic acid**

CrO_3, Pyridine $\rightarrow$ C_6H_5CHO
Benzaldehyde

It undergoes electrophilic aromatic substitution reactions like halogenation, nitration, sulphonation etc and directs the groups at ortho and para position to the CH_2OH group.

Uses
1. Benzyl chloride and its acetates and benzoates are used in perfume industry.
2. It is used in ointments as it has antiseptic and local anesthetic properties.
3. It is also used in asthma and cough.

34.10.2 β-Phenylethyl Alcohol/Phenylethanol
It occurs in rose oil and used in perfume industry.

Preparation
(I) By reducing ethyl phenyl acetate with sodium and ethanol.

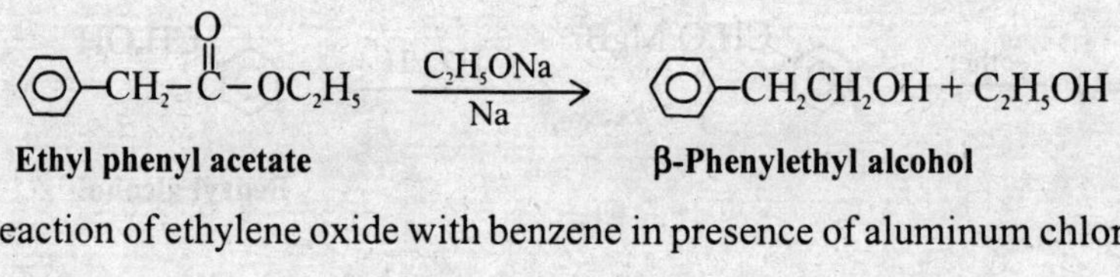

$$\text{Ethyl phenyl acetate} \quad \bigcirc\!\!-CH_2-\overset{\overset{O}{\|}}{C}-OC_2H_5 \xrightarrow[Na]{C_2H_5ONa} \bigcirc\!\!-CH_2CH_2OH + C_2H_5OH$$
Ethyl phenyl acetate **β-Phenylethyl alcohol**

(II) Reaction of ethylene oxide with benzene in presence of aluminum chloride produces β-phenylethy alcohol.

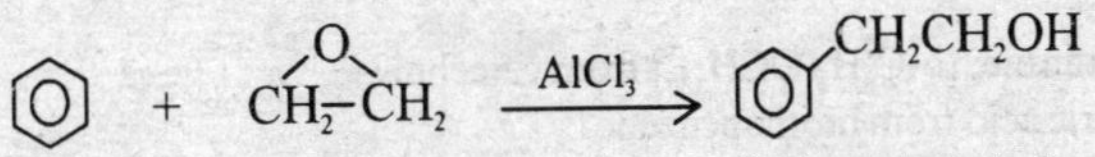

Benzene + Ethylene oxide → β-Phenylethyl alcohol

$$\text{Benzene} + \underset{\text{Ethylene oxide}}{CH_2-CH_2 (O)} \xrightarrow{AlCl_3} \underset{\beta\text{-Phenylethyl alcohol}}{C_6H_5CH_2CH_2OH}$$

Properties: It is a colorless liquid having strong rose odor, b.p. 221°, slightly soluble in water but freely soluble in ethanol and ether. On heating with alkali it forms styrene.

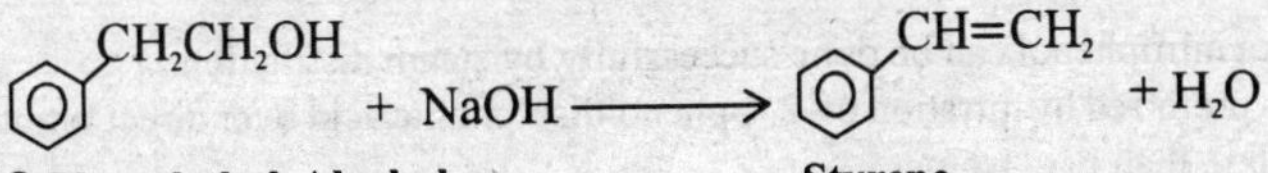

$$\underset{\beta\text{-Phenylethyl Alcohol}}{CH_2CH_2OH} + NaOH \longrightarrow \underset{\text{Styrene}}{CH=CH_2} + H_2O$$

34.10.3 Diphenylcarbinol/Benzhydrol

It is a secondary alcohol having phenyl ring as substituent. It is prepared by reduction of benzophenone.

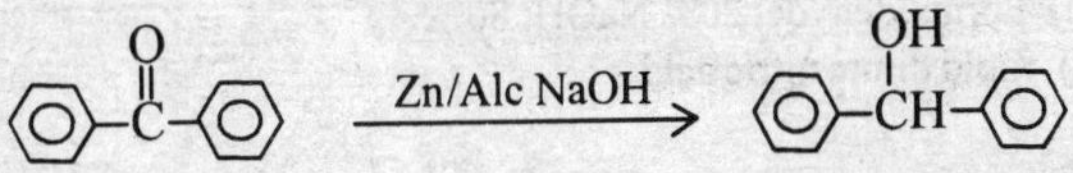

$$\underset{\text{Benzophenone}}{C(=O)} \xrightarrow{\text{Zn/Alc NaOH}} \underset{\text{Diphenylcarbinol}}{CH(OH)}$$

It is a colorless solid, m.p. 68°, sparingly soluble in water but freely soluble in alcohol and ether.

It gives characteristic reactions of secondary alcohols and electrophilic substitution in benzene ring.

It is used for the preparation of antihistaminic drugs. Benadryl, an antihistaminic drug, is used for the treatment of allergic conditions.

34.10.4 Triphenyl Carbinol

It is a tertiary alcohol having phenyl substituents and prepared by the reaction of phenylmagnesium bromide with ethyl benzoate or benzophenone.

$$\underset{\text{Ethyl benzoate}}{C_6H_5-C(=O)OC_2H_5} \xrightarrow[H_2O/H^+]{C_6H_5 MgBr} \underset{\text{Benzophenone}}{C_6H_5COC_6H_5} \xrightarrow[H_2O/H^+]{C_6H_5 MgBr} \underset{\text{Triphenyl carbinol}}{(C_6H_5)C-OH}$$

It is a colorless solid, m.p. 165°, soluble in organic solvents. It gives the characteristic reactions of tertiary alcohols.

It reacts with HCl to forms triphenylmethyl chloride or trityl chloride which reacts with primary alcohols in pyridine to form trityl ethers.

$$\underset{\text{Trityl chloride}}{RCH_2OH + (C_6H_5)_3C-Cl} \xrightarrow{\text{Pyridine}} \underset{\text{Trityl ether}}{RCH_2OC(C_6H_5)_3 + HCl}$$

This reaction is also known as *Tritylation.*

Triphenyl carbinol also exhibits electrophilic substitution reactions in the aromatic rings.

34.11 PROBLEMS

1. Write structural formulae for
   (a) m-cresol
   (b) Salicylaldehyde
   (c) Resorcinol
   (d) Anisole
   (e) 4-n-Propylcatechol
   (f) Methyl salicylate

2. How will you prepare
   (a) Catechol from phenol?   (b) Veratrole, [o-C$_6$H$_4$(OCH$_3$)$_2$] from catechol?
   (c) Resorcinol from benzene? (d) Picric acid from bromobenzene?
3. Arrange the following sets in the order of increasing acidity (give reasons for your answer).
   (a) Benzenesulphonic acid, Benzyl alcohol, Phenol, Benzoic acid
   (b) m-Nitrophenol, m-Bromophenol, m-Cresol and Phenol
   (c) 4-Chlorophenol, 2,4-Dichlorophenol, 2,4,6-Trichlorophenol
4. Give reasons:
   (a) Separation of ortho and para nitrophenol can be done successfully by steam distillation.
   (b) Preparation of picric acid is preferred by nitration of 2,4-phenoldisulphonic acid over direct nitration of phenol
   (c) o-Nitrophenol has a lower b.p. than para isomer.
   (d) Phenol is acidic while ethanol is neutral.
   (e) Boiling points of phenols are much higher than that of the corresponding hydrocarbons.
5. What is the product of reaction of phenol with:
   (a) Bromine in CS$_2$ (b) Carbon dioxide, NaOH, 125 °, 5 atm.
   (c) Chloroform, aqueous NaOH, 75°   (d) Formaldehyde, 20% NaOH, 80°
   (e) Bromine water   (f) Cold dilute nitric acid
   (g) Methyl sulphate, aqueous NaOH
6. Answer problem 5 starting with o-Cresol.
7. What happens when
   (a) Resorcinol is heated with phthalic anhydride and the product is treated with bromine.
   (b) Catachol is heated with phthalic anhydride.
8. How will you differentiate between cresols and benzyl alcohol?
9. Explain why unlike most phenols, 2,4-dinitrophenol and 2,4,6- trinitrophenol are soluble in aqueous sodium bicarbonate?
10. Cyclohexanol when refluxed with HBr yields bromocyclohexane whereas phenol does not. Explain your answer with the help of mechanism.
11. What is the use of conc. sulphuric acid in nitration of o-cresol?
12. Why phenol couples readily in slightly basic than in acidic solution?
13. How will you separate a mixture of phenol, benzyl alcohol and benzoic acid?
14. Identify the following.

   o–Nitrophenol $\xrightarrow[\text{NaOH}]{\text{(CH}_3)_2\text{SO}_4}$ A $\xrightarrow{\text{Zn, HCl}}$ B $\xrightarrow[\text{HCl, 0°}]{\text{NaNO}_2}$ C $\xrightarrow{\text{C}_2\text{H}_5\text{OH}}$ D

15. How will you prepare
    (a) p-Benzoquinone from phenol?   (b) p-Benzoquinone dioxime?
16. Show the possible products in the following:

    (a) p-Cresol + CHCl$_3$ + NaOH $\xrightarrow{70°}$

    (b) C$_6$H$_5$CH$_2$OH $\xrightarrow{\text{Cr}_2\text{O}_7^{--}/\text{H}^+}$

    (c) p-C$_2$H$_5$C$_6$H$_4$OCOC$_2$H$_5$ $\xrightarrow{\text{AlCl}_3}$

    (d) p-ClC$_6$H$_4$OCOCH$_3$ $\xrightarrow{\text{AlCl}_3}$

17. How will you synthesise aspirin from methyl salicylate and phenol?

35

Aromatic Aldehydes and Ketones

35.1 AROMATIC ALDEHYDES

Aromatic compounds having an aldehyde group directly attached to the benzene ring are termed as aromatic aldehydes (I), whereas those in which aldehyde group is attached to a side chain are called aryl substituted aliphatic aldehydes (II).

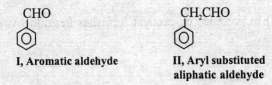

I, Aromatic aldehyde

II, Aryl substituted aliphatic aldehyde

35.1.1 Nomenclature

The aromatic aldehydes are named by replacing 'ene' of the parent hydrocarbon by 'al'.

Benzal

4-Methylbenzal

2-Nitrobenzal

The common names of aromatic aldehydes are derived from corresponding carboxylic acid by replacing 'oic acid' or 'ic acid' by 'aldehyde'. For example, salicylaldehyde is derived from salicylic acid by replacing 'ic acid' with 'aldehyde'.

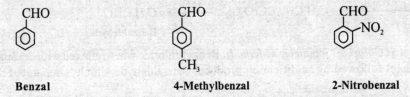

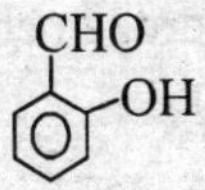

Salicylaldehyde

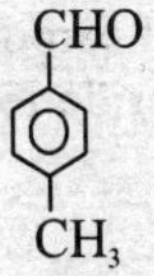

p-Tolualdehyde

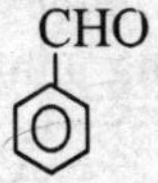

Benzaldehyde

35.1.2 Benzaldehyde/Benzal [C$_6$H$_5$CHO]

Benzaldehyde is found in nature as the glycoside amygdalin in bitter almonds. When amygdalin is boiled with dilute acid or hydrolysed with enzyme emulsin, benzaldehyde is produced along with glucose and hydrogen cyanide.

$$C_{20}H_{27}O_{11}N + 2H_2O \longrightarrow C_6H_5CHO + 2C_6H_{12}O_6 + HCN$$

Amygdalin   Benzaldehyde   Glucose

Amygdalin is poisonous as it forms hydrogen cyanide on hydrolysis. It is also found in reactor gland of millipedes.

Preparation: Benzaldehyde and substituted aromatic aldehydes may be prepared by following methods

(1)   *Oxidation of Toluene*
(i)   Oxidation of toluene with chromic anhydride and acetic anhydride yields benzaldehyde.

$$C_6H_5CH_3 \xrightarrow[\text{(CH}_3\text{CO)}_2\text{O}]{\text{CrO}_3} C_6H_5CH(OCOCH_3)_2 \xrightarrow{H_2O} C_6H_5CHO + 2CH_3COOH$$

Toluene   Benzylidene acetate   Benzaldehyde

Formation of benzylidene acetate prevents further oxidation to benzoic acid.
(ii)   Oxidation of toluene in liquid phase with manganese dioxide and sulphuric acid yields benzaldehyde.

$$C_6H_5CH_3 + 2[O] \xrightarrow[200°]{MnO_2} C_6H_5CHO + H_2O$$

Toluene   Benzaldehyde

(iii)   Oxidation of toluene (in vapor phase) in presence of catalyst furnishes benzaldehyde.

$$C_6H_5CH_3 + O_2 \xrightarrow[500°]{MnO_2 / Mo_2O_3 / ZrO_2} C_6H_5CHO$$

Toluene   Benzaldehyde

(iv)   Oxidation of toluene (Etard's reaction) with chromyl chloride in carbon tetrachloride forms a complex which upon decomposition produces benzaldehyde.

$$C_6H_5CH_3 \xrightarrow{CrO_2Cl_2} C_6H_5CH_3 . 2CrO_2Cl_2 \xrightarrow{H_2O} C_6H_5CHO$$

Toluene   Benzaldehyde

(2)   *Oxidation of Benzyl Chloride (Sommelet's Method):* Benzyl chloride when refluxed with hexamethylene tetramine in dilute ethanol followed by acidification produces benzaldehyde which is separated by steam distillation.

$$C_6H_5CH_2Cl \xrightarrow[H_2O/H^+]{(CH_2)_6N_4/ C_2H_5OH} C_6H_5CHO$$

Benzyl chloride   Benzaldehyde

(3)   *Gattermann-Koch Reaction:* Equimolar amounts of carbon monoxide and hydrogen chloride are refluxed in benzene containing anhydrous aluminum chloride to form benzaldehyde.

$$CO + HCl \longrightarrow H-\overset{O}{\underset{}{C}}-Cl$$

$$C_6H_6 + Cl-\overset{O}{\underset{}{C}}-H \longrightarrow C_6H_5-\overset{O}{\underset{}{C}}-H + HCl$$

Benzene   Benzaldehyde

This is a convenient method for the preparation of homologues of benzaldehyde.

(4) *Gattermann Cyanide Synthesis:* Benzene is treated with a mixture of hydrogen cyanide and hydrogen chloride in presence of anhydrous aluminum chloride. The addition product, formimino chloride formed reacts with benzene to give imine which upon decomposition produces benzaldehyde.

$$HCl + HCN \longrightarrow \underset{\textbf{Formimino chloride}}{HN=\overset{\overset{\displaystyle H}{|}}{C}-Cl}$$

$$\underset{\textbf{Benzene}}{C_6H_6} + Cl-\overset{\overset{\displaystyle H}{|}}{C}=NH \xrightarrow[-HCl]{AlCl_3} \underset{\textbf{Imine}}{C_6H_5-\overset{\overset{\displaystyle H}{|}}{C}=NH} \xrightarrow{H_2O} \underset{\textbf{Benzaldehyde}}{C_6H_5-\overset{\overset{\displaystyle H}{|}}{C}=O} + NH_3$$

This reaction is chiefly used on phenol or activated substrates and it is an electrophilic substitution reaction.

(5) *Rosenmund Reaction:* Benzoyl chloride when reduced in presence of palladium poisoned with barium sulphate yields benzaldehyde.

$$\underset{\textbf{Benzoyl chloride}}{C_6H_5COCl} + H_2 \xrightarrow{Pd-BaSO_4} \underset{\textbf{Benzaldehyde}}{C_6H_5CHO} + HCl$$

(6) *Stephen's Reaction:* Phenyl cyanide on reduction with stannous chloride and hydrogen chloride forms a complex, the aldimine, which upon hydrolysis produces benzaldehyde.

$$C_6H_5CN + SnCl_2 + 4HCl \longrightarrow \left[C_6H_5-\overset{\overset{\displaystyle H}{|}}{C}=NH_2\right]_2^+ SnCl_6^{2-}$$

$$\xrightarrow{H_2O} \underset{\textbf{Benzaldehyde}}{C_6H_5CHO} + 2NH_4Cl + SnCl_4$$

(7) *Hydrolysis of Benzal Chloride* yields benzaldehyde.

$$\underset{\textbf{Benzal chloride}}{C_6H_5CH_3} + 2Cl_2 \longrightarrow C_6H_5CHCl_2 \xrightarrow{H_2O} \underset{\textbf{Unstable}}{\left[C_6H_5-CH\overset{\displaystyle OH}{\underset{\displaystyle OH}{\Big\langle}}\right]} \xrightarrow{-H_2O} \underset{\textbf{Benzaldehyde}}{C_6H_5CHO}$$

(8) *Grignard Reaction:* Benzaldehyde may also be prepared by the reaction of phenylmagnesium bromide with ethyl formate. Phenylmagnesium bromide in turn is produced by the reaction of bromobenzene and magnesium in dry ether.

$$\underset{\textbf{Bromobenzene}}{C_6H_5Br} + Mg \xrightarrow{Ether} \underset{\textbf{Phenylmagnesium bromide}}{C_6H_5MgBr}$$

$$C_6H_5MgBr + \underset{\textbf{Ethyl formate}}{H-\overset{\overset{\displaystyle O}{\|}}{C}-OC_2H_5} \longrightarrow \underset{\textbf{Benzaldehyde}}{C_6H_5CHO} + Mg\overset{\displaystyle OC_2H_5}{\underset{\displaystyle Br}{\Big\langle}}$$

(9)   *Dry Distillation* of a mixture of calcium formate and calcium benzoate yields benzaldehyde

$$(C_6H_5COO)_2Ca + (HCOO)_2Ca \longrightarrow 2C_6H_5CHO + 2CaCO_3$$

Calcium   Calcium   Benzaldehyde
benzoate   formate

Physical Properties: It is a colorless oily liquid, b.p. 179°, having the odor of bitter almonds. It is insolubl
in water but soluble in organic solvents. It is steam volatile.

Chemical Properties: Benzaldehyde (and aromatic aldehydes in general) resemble aliphatic aldehydes i
the following reactions.

(1)   It gives pink color with Schiff's reagent (the reagent is obtained by disolving rosaniline hydrochloride in water and bubbling sulphur dioxide gas till a colorless solution results).

(2)   It is a strong reducing agent and is readily oxidised. It reduces ammonical silver nitrate (Tollen'
reagent)to silver and is oxidised to benzoic acid. Similar oxidation is possible in air also. According t
Baeyer and Villiger, autooxidation in air takes places via perbenzoic acid formation.

$$C_6H_5\overset{O}{\overset{\|}{C}}-H + O_2 \longrightarrow C_6H_5\overset{O}{\overset{\|}{C}}-O-OH \xrightarrow{C_6H_5CHO} C_6H_5\overset{O}{\overset{\|}{C}}-OH$$

Benzaldehyde   Perbenzoic acid   Benzoic acid

(3)   It forms benzyl alcohol and hydrobenzoin when reduced with zinc and hydrochloric acid o
with sodium amalgam and water(aliphatic aldehydes give primary alcohols).

$$C_6H_5\overset{O}{\overset{\|}{C}}-H + 2[H] \longrightarrow C_6H_5CH_2OH$$

Benzaldehyde   Benzyl alcohol

$$C_6H_5\overset{O}{\overset{\|}{C}}-H + 2[H] \longrightarrow C_6H_5\overset{OH}{\overset{|}{C}H}-\overset{OH}{\overset{|}{C}H}-C_6H_5$$

Benzaldehyde   Hydrobenzoin

(4)   It undergoes nucleophilic addition reaction with sodium bisulphite and hydrogen cyanide.

$$C_6H_5\overset{O}{\overset{\|}{C}}-H + NaHSO_3 \longrightarrow C_6H_5\overset{OH}{\overset{|}{C}H}-SO_3Na$$

Bisulphite compound

$$C_6H_5\overset{O}{\overset{\|}{C}}-H + HCN \longrightarrow C_6H_5\overset{OH}{\overset{|}{C}H}-CH$$

Benzaldehyde   Cyanohydrin

The first reaction is often used to purify aldehydes.

(5)   Benzaldehyde and related compounds undergoes addition elimination reactions when treated with
derivatives of ammonia.

$$C_6H_5\overset{O}{\overset{\|}{C}}-H + NH_2OH \xrightarrow{-H_2O} C_6H_5\overset{H}{\overset{|}{C}}=NOH$$

Benzaldehyde   Hydroxylamine   Benzaldoxime

$$C_6H_5-\overset{\overset{O}{\|}}{C}-H \ + C_6H_5NHNH_2 \ \xrightarrow{-H_2O} \ C_6H_5-\overset{\overset{H}{|}}{C}=NNHC_6H_5$$

Benzaldehyde   **Phenyl hydrazine**   **Phenyl hydrazone of benzaldehyde**

The benzaldoxime exists in two geometrical isomeric forms.

$$C_6H_5-\overset{\|}{C}-H \\ \quad N-OH$$

$$C_6H_5-\overset{\|}{C}-H \\ \quad HO-N$$

Syn-form   **Anti-form**

The above reaction first involves addition of ammonia derivative followed by elimination of water.

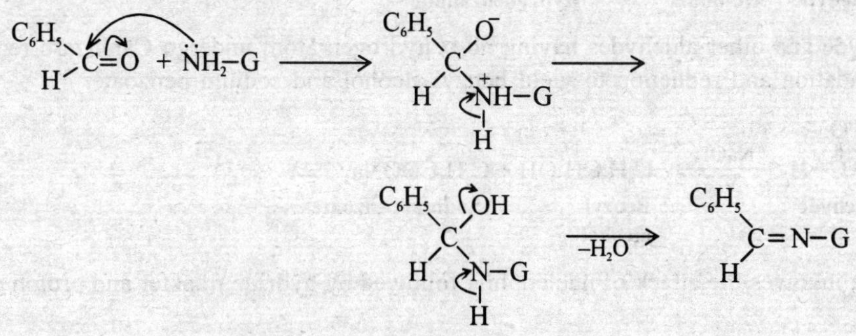

$$\underset{H}{\overset{C_6H_5}{>}}C=O \ + \ \ddot{N}H_2-G \ \longrightarrow \ \underset{H}{\overset{C_6H_5}{}}\overset{\overset{O^-}{|}}{\underset{\overset{+}{N}H-G}{C}} \ \longrightarrow$$

$$\underset{H}{\overset{C_6H_5}{}}\overset{\overset{OH}{|}}{\underset{\overset{N-G}{|}}{C}} \ \xrightarrow{-H_2O} \ \underset{H}{\overset{C_6H_5}{>}}C=N-G$$

$$\mathbf{G=HO-, \ C_6H_5NH-}$$

Benzaldehyde reacts with hydrazine to form benzylideneazine

$$2 \ C_6H_5-\overset{\overset{O}{\|}}{C}-H + NH_2NH_2 \ \longrightarrow \ C_6H_5CH=N-N=CHC_6H_5 + 2H_2O$$

Benzaldehyde   **Benzylideneazine**

(6)   Benzaldehyde reacts with phosphorus pentachloride to form benzylidene chloride.

$$C_6H_5-\overset{\overset{O}{\|}}{C}-H + PCl_5 \ \longrightarrow \ C_6H_5CHCl_2 \qquad + POCl_3$$

Benzaldehyde   **Benzylidene chloride**

(7)   Benzaldehyde undergoes intermolecular oxidation and reduction with aluminum alkoxide in presence of anhydrous aluminum chloride or zinc chloride to form acid and alcohol (similar to Cannizaro reaction) but the acid and alcohol react together to form esters. benzaldehyde when reacted with aluminum isopropoxide forms benzyl benzoate

$$C_6H_5-\overset{\overset{O}{\|}}{C}-H \ \xrightarrow{Al \ [OCH(CH_3)_2]_3} \ C_6H_5CH_2COOC_6H_5$$

Benzaldehyde   **Benzylbenzoate**

The reaction is known as *Tischenko reaction*.

(8)   Benzaldehyde undergoes wittig reaction when reacted upon with phosphorus ylide to give unsaturated hydrocarbons.

$$C_6H_5-\overset{\overset{\displaystyle O}{\|}}{C}-H + (C_6H_5)_3P=CH_2 \longrightarrow C_6H_5CH=CH_2 + (C_6H_5)_3P=O$$

Benzaldehyde   Methylene triphenyl   Styrene   Triphenylphosphine oxide
                phosphorane

Benzaldehyde (and other Aromatic Aldehydes) Differ from Aliphatic Aldehydes in the following Reactions
1. Benzaldehyde does not reduce Fehling's solution.
2. It gives a complex product, hydrobenzamide, when treated with ammonia.

$$\overset{\qquad\qquad\qquad\qquad\quad C_6H_5}{3C_6H_5CHO + 2NH_3 \longrightarrow C_6H_5CH=N-\overset{|}{C}H-N=CHC_6H_5}$$

Benzaldehyde   Ammonia   Hydrobenzamide

3. Benzaldehyde and other aldehydes having no α hydrogen atom undergo Cannizaro reaction, the intermolecular oxidation and reduction to yield benzyl alcohol and sodium benzoate.

$$2C_6H_5-\overset{\overset{\displaystyle O}{\|}}{C}-H \xrightarrow{NaOH} C_6H_5CH_2OH + C_6H_5COO\overset{+}{Na}$$

Benzaldehyde   Benzyl   Sodium benzoate
                                    alcohol

The mechanism involves the attack of nucleophile followed by hydride transfer and proton exchange.

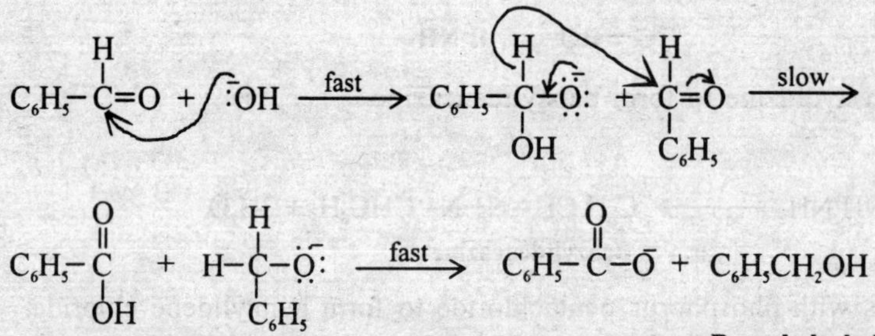

$$C_6H_5-\overset{\overset{\displaystyle H}{|}}{C}=O + \text{:OH} \xrightarrow{fast} C_6H_5-\overset{\overset{\displaystyle H}{|}}{\underset{\underset{\displaystyle OH}{|}}{C}}-\overset{..}{O}: + \overset{\overset{\displaystyle H}{|}}{\underset{\underset{\displaystyle C_6H_5}{|}}{C}}=O \xrightarrow{slow}$$

$$C_6H_5-\overset{\overset{\displaystyle O}{\|}}{\underset{\underset{\displaystyle OH}{|}}{C}} + H-\overset{\overset{\displaystyle H}{|}}{\underset{\underset{\displaystyle C_6H_5}{|}}{C}}-\overset{..}{O}: \xrightarrow{fast} C_6H_5-\overset{\overset{\displaystyle O}{\|}}{C}-\bar{O} + C_6H_5CH_2OH$$

Benzyl alcohol

When benzaldehyde reacts with formaldehyde (cross Cannizaro reaction) benzyl alcohol and sodium formate are formed as formaldehyde is oxidised more readily than benzaldehyde.

$$C_6H_5-\overset{\overset{\displaystyle O}{\|}}{C}-H + H-\overset{\overset{\displaystyle O}{\|}}{C}-H \xrightarrow{NaOH} C_6H_5CH_2OH + H-\overset{\overset{\displaystyle O}{\|}}{C}-O\overset{+}{Na}$$

Benzaldehyde   Formaldehyde   Benzyl alcohol   Sodium formate

4. Benzaldehyde and other aromatic aldehydes undergo *Perkins reaction*. When heated with the anhydride of an aliphatic acid in presence of sodium salt of the same acid, they yield α,β-unsaturated acids. Benzaldehyde reacts with acetic anhydride in presence of sodium acetate at 200° to yield cinnamic acid.

$$C_6H_5-\overset{\overset{\displaystyle O}{\|}}{C}-H + (CH_3CO)_2O \xrightarrow{CH_3COO\overset{+}{Na}} C_6H_5CH=CHCOOH + CH_3COOH$$

Benzaldehyde   Acetic anhydride   Cinnamic acid

When reacted with propionic anhydride and sodium propionate, it gives α-methyl cinnamic acid.

$$C_6H_5-\overset{\overset{O}{\|}}{C}-H + (CH_3CH_2CO)_2O \xrightarrow{CH_3CH_2COO^-\overset{+}{Na}} C_6H_5CH=\overset{\overset{CH_3}{|}}{C}-COOH + CH_3CH_2COOH$$

Benzaldehyde **Propanoic anhydride** **α-Methylcinnamic acid**

The mechanism of Perkins reaction involves the formation of carbanion by the attack of base, which then attacks the carbonyl carbon of benzaldehyde as given below:

$$CH_3COO^- + CH_3CO-O-COCH_3 \longrightarrow CH_3COOH + \bar{C}H_2CO-O-COCH_3$$

$$C_6H_5-\overset{\overset{O}{\diagup}}{\underset{H}{C}} + \bar{C}H_2CO-O-COCH_3 \longrightarrow C_6H_5-\overset{\overset{O^-}{|}}{\underset{H}{C}}-CH_2CO-O-COCH_3$$

$$\xrightarrow{H^+} C_6H_5-\overset{\overset{OH}{|}}{\underset{H}{C}}-CH_2CO-O-COCH_3 \xrightarrow{-H_2O} C_6H_5CH=CH-CO-O-COCH_3$$

$$\xrightarrow{H_2O/H^+} C_6H_5CH=CHCOOH + CH_3COOH$$

Cinnamic acid **Acetic acid**

5. Benzaldehyde undergoes *Benzoin condensation*, when heated with aqueous ethanolic sodium or potassium cyanide. It dimerises to benzoin, the α-hydroxy ketone in presence of cyanide ions as catalyst.

$$2\ C_6H_5CHO \xrightarrow[C_2H_5OH]{NaCN} C_6H_5-\overset{}{\underset{OH}{CH}}-\overset{\overset{O}{\|}}{C}-C_6H_5$$

Benzaldehyde **Benzoin**

The mechanism involves nucleophilic addition of cyanide ion to the substrate molecule. The various steps involved in the addition are summerised below:

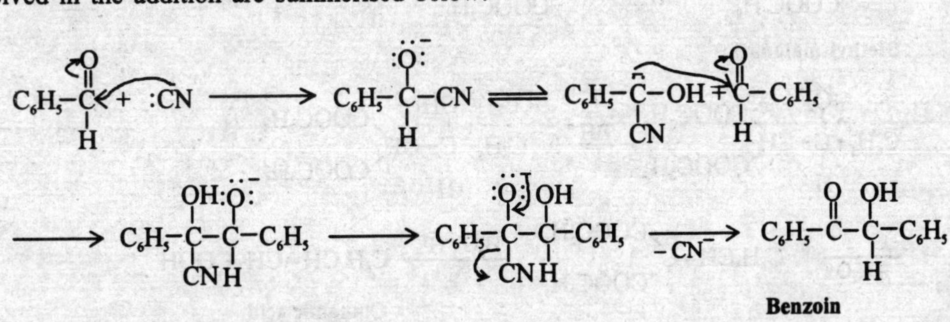

$$C_6H_5-\overset{\overset{O}{\diagup}}{\underset{H}{C}} + :CN^- \longrightarrow C_6H_5-\overset{\overset{:\overset{..}{O}:^-}{|}}{\underset{H}{C}}-CN \rightleftharpoons C_6H_5-\overset{\overset{}{|}}{\underset{CN}{C}}-OH + \overset{\overset{O}{\diagup}}{\underset{H}{C}}-C_6H_5$$

$$\longrightarrow C_6H_5-\overset{\overset{OH}{|}}{\underset{CN}{C}}-\overset{\overset{:\overset{..}{O}:^-}{}}{\underset{H}{C}}-C_6H_5 \longrightarrow C_6H_5-\overset{\overset{:\overset{..}{O}:^-}{}}{\underset{CN}{C}}-\overset{\overset{OH}{|}}{\underset{H}{C}}-C_6H_5 \xrightarrow{-CN^-} C_6H_5-\overset{\overset{O}{\|}}{C}-\overset{\overset{OH}{|}}{\underset{H}{C}}-C_6H_5$$

Benzoin

Benzoin is oxidised readily to a diketone, benzil and reduced to a number of products as given below:

$$\underset{\textbf{Benzoin}}{C_6H_5\overset{\overset{\displaystyle O}{\|}}{C}-\overset{\overset{\displaystyle OH}{|}}{CH}-C_6H_5}$$

$$\xrightarrow[\text{CuSO}_4\text{ , Pyridine}]{[O]} \underset{\textbf{Benzil}}{C_6H_5CO-COC_6H_5 + H_2O}$$

$$\xrightarrow[\text{Reduction}]{\text{Sn + HCl}} \underset{\textbf{Deoxybenzoin}}{C_6H_5COCH_2C_6H_5 + H_2O}$$

$$\xrightarrow[\text{Reduction}]{\text{Na–Hg / ethanol}} \underset{\textbf{Hydrobenzoin}}{C_6H_5CHOHCHOHC_6H_5}$$

$$\xrightarrow[\text{Reduction}]{\text{Zn–Hg / HCl}} \underset{\textbf{Stilbene}}{C_6H_5CH= CHC_6H_5 + 2H_2O}$$

$$\xrightarrow[\text{Reduction}]{\text{H}_2\text{ / Ni}} \underset{\textbf{Bibenzyl}}{C_6H_5CH_2- CH_2C_6H_5 + 2H_2O}$$

6.   Benzaldehyde undergoes *Knoevenagel Reaction* to yield cinnamic acid when treated with compounds having active methylene group such as diethyl malonate, ethyl acetoacetate etc.

$$\underset{\textbf{Benzaldehyde \ Diethyl malonate}}{C_6H_5\overset{\overset{\displaystyle O}{\|}}{C}-H + CH_2(COOC_2H_5)_2} \xrightarrow[\text{Piperidine}]{\text{Pyridine}} C_6H_5\overset{\overset{\displaystyle H}{|}}{C}=C\overset{\diagup COOC_2H_5}{\diagdown COOC_2H_5} \xrightarrow{\text{Hydrolysis}}$$

$$C_6H_5\overset{\overset{\displaystyle H}{|}}{C}=C\overset{\diagup COOH}{\diagdown COOH} \xrightarrow[-CO_2]{\Delta} \underset{\textbf{Cinnamic acid}}{C_6H_5CH=CH-COOH}$$

The mechanism of the reaction involves the formation of carbanion due to proton abstraction by the base.

$$\underset{\textbf{Diethyl malonate}}{CH_2\overset{\diagup COOC_2H_5}{\diagdown COOC_2H_5}} \xrightarrow[-BH^+]{\text{Base}} \overset{-}{CH}\overset{\diagup COOC_2H_5}{\diagdown COOC_2H_5} + \overset{\overset{\displaystyle O}{\|}}{\underset{H}{C}}-C_6H_5 \longrightarrow$$

$$C_6H_5-\overset{\overset{\displaystyle :\ddot{O}:^-}{}}{\underset{H}{C}}-CH\overset{\diagup COOC_2H_5}{\diagdown COOC_2H_5} \xrightarrow{BH^+} C_6H_5-\overset{\overset{\displaystyle OH}{|}}{\underset{H}{C}}-CH\overset{\diagup COOC_2H_5}{\diagdown COOC_2H_5}$$

$$\xrightarrow{-H_2O} C_6H_5CH=C\overset{\diagup COOC_2H_5}{\diagdown COOC_2H_5} \xrightarrow[\Delta]{H_2O / H^+} \underset{\textbf{Cinnamic acid}}{C_6H_5CH=CH-COOH}$$

7.   Benzaldehyde undergoes *Claisen or Claisen Schmidt reaction* when condensed with aliphatic aldehydes or ketones containing α hydrogen atoms in presence of alkali to give α,β-unsaturated aldehyde.

$$C_6H_5-\overset{\overset{O}{\|}}{C}-H + CH_3CHO \xrightarrow{\text{NaOH}} C_6H_5CH=CHCHO + H_2O$$

$\quad$ Benzaldehyde $\quad$ Acetaldehyde $\qquad\qquad$ **Cinnamaldehyde**

When reacted with acetone, it gives benzylideneacetone used in perfume industry.

$$C_6H_5-\overset{\overset{O}{\|}}{C}-H + CH_3COCH_3 \longrightarrow C_6H_5CH=CHCOCH_3$$

$\quad$ Benzaldehyde $\quad$ Acetone $\qquad\qquad$ **Benzylideneacetone**

When reacted is presence of aqueous alcoholic sodium hydroxide dibenzylideneacetone is produced.

$$C_6H_5-\overset{\overset{O}{\|}}{C}-H + CH_3COCH_3 \xrightarrow[\text{alc alkali}]{\text{aqueous}} C_6H_5CH=CH-\overset{\overset{O}{\|}}{C}-CH=CHC_6H_5$$

$\quad$ Benzaldehyde $\quad$ Acetone $\qquad\qquad$ **Dibenzylideneacetone**

Benzaldehyde condenses with acetophenone to form phenyl styryl ketone or chalcone.

$$C_6H_5-\overset{\overset{O}{\|}}{C}-H + CH_3COC_6H_5 \xrightarrow{\text{NaOH}} C_6H_5CH=CHCOC_6H_5 + H_2O$$

$\qquad\qquad\qquad$ **Acetophenone** $\qquad\qquad$ **Benzylideneacetophenone**

Benzylideneacetophenone may be catalytically reduced to benzylacetophenone. This reaction is used in the detection of $-CH_2CO-$ group in carbonyl compounds.

The *mechanism of Claisen reaction* involves following steps.

$$HO^- + CH_3CHO \longrightarrow H_2O + \overset{..}{C}H_2CHO$$

$$C_6H_5-\overset{\overset{O}{\|}}{\underset{H}{C}}+\overset{-}{C}H_2CHO \longrightarrow C_6H_5-\overset{\overset{:\overset{..}{O}^-}{|}}{\underset{H}{C}}-CH_2CHO \xrightarrow{H^+}$$

$$C_6H_5-\overset{\overset{OH}{|}}{\underset{H}{C}}-CH_2CHO \xrightarrow[-H_2O]{HO^-} C_6H_5-\overset{\overset{OH}{|}}{\underset{H}{C}}\overset{\frown}{C}HCHO \longrightarrow C_6H_5CH=CH-CHO + HO^-$$

8. Benzaldehyde reacts with primary aliphatic and aromatic amines to form Schiff's bases.

$$C_6H_5-\overset{\overset{O}{\|}}{C}-H + CH_3NH_2 \longrightarrow C_6H_5-\overset{\overset{H}{|}}{C}=N-CH_3 + H_2O$$

$\quad$ Benzaldehyde $\quad$ Methylamine $\qquad$ **N-Methylbenzyl-**
$\qquad\qquad\qquad\qquad\qquad\qquad\qquad$ **ideneimine**

$$C_6H_5-\overset{\overset{O}{\|}}{C}-H + C_6H_5NH_2 \longrightarrow C_6H_5-\overset{\overset{H}{|}}{C}=N-C_6H_5$$

$\quad$ Benzaldehyde $\quad$ Aniline $\qquad\qquad$ **Benzylideneaniline**

The products with aromatic amines are also called anils. Schiff's bases are reduced easily to secondary aromatic amines.

$$C_6H_5CH=N-C_6H_5 + H_2 \xrightarrow{\text{Raney Ni}} C_6H_5CH_2NHC_6H_5$$

Benzylideneaniline   **N-Benzylaniline**

9.   Benzaldehyde undergoes *Reformatsky reaction* when reacted with α-halogenated esters in presence of zinc to yield β-hydroxy esters.

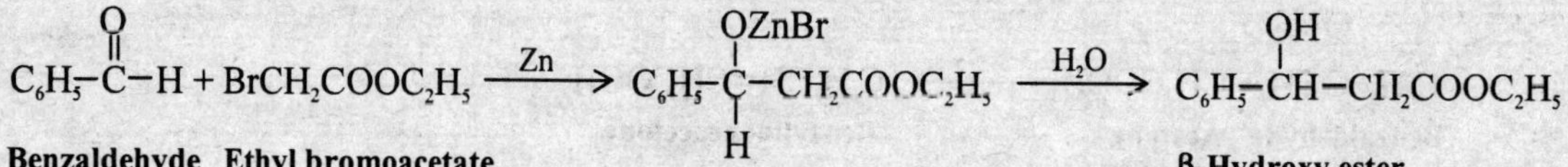

$$C_6H_5-\overset{\overset{O}{\parallel}}{C}-H + BrCH_2COOC_2H_5 \xrightarrow{\text{Zn}} C_6H_5-\overset{\overset{OZnBr}{|}}{\underset{H}{C}}-CH_2COOC_2H_5 \xrightarrow{\text{H}_2O} C_6H_5-\overset{\overset{OH}{|}}{CH}-CH_2COOC_2H_5$$

Benzaldehyde   Ethyl bromoacetate   **β-Hydroxy ester**

10.   When condensed with tertiary aromatic amines, benzaldehyde forms dyes of triphenylmethane group.

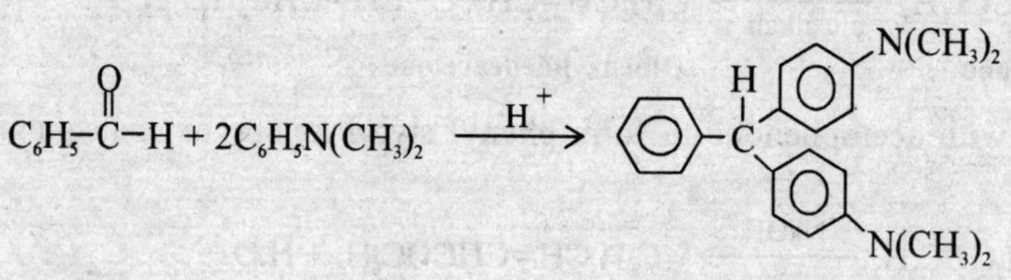

$$C_6H_5-\overset{\overset{O}{\parallel}}{C}-H + 2C_6H_5N(CH_3)_2 \xrightarrow{\text{H}^+}$$

Benzaldehyde   Dimethylaniline   Triphenylmethane derivative

11.   It forms benzoyl chloride when reacted with chlorine alone.

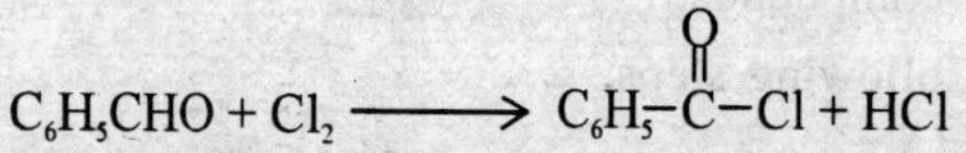

$$C_6H_5CHO + Cl_2 \longrightarrow C_6H_5-\overset{\overset{O}{\parallel}}{C}-Cl + HCl$$

Benzaldehyde   **Benzoyl chloride**

Benzaldehyde (and other Aldehydes) Undergo Electrophilic Substitution Reactions in the Benzene Ring:
The –CHO group in aldehydes, being electron withdrawing, withdraws the electron cloud of benzene ring towards itself making ortho and para positions of benzene ring especially electron deficient and thus, orient the second substituent in benzaldehyde at meta positions.

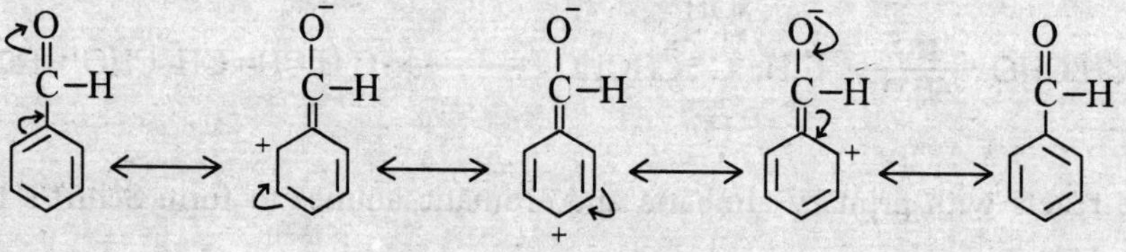

Benzaldehyde undergoes nitration, sulphonation and halogenation under usual conditions to give m-substituted benzaldehyde. When reacted with concentrated nitric acid and concentrated sulphuric acid, benzaldehyde yields m-nitrobenzaldehyde as a major product.

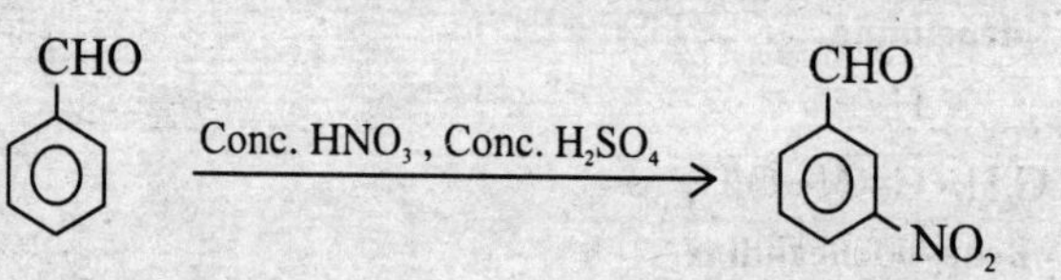

Benzaldehyde   **m-Nitrobenzaldehyde**

Uses: Benzaldehyde is used
1. As a flavoring agent
2. In perfume industry
3. As a synthetic intermediate
4. In the manufacture of certain dyes.

35.1.3   Toluic Aldehyde [$CH_3C_6H_4CHO$]

The three isomeric toluic aldehydes are prepared by Rosenmund reduction of corresponding toluic acid chlorides. Thus o-toluic acid chloride on reduction furnishes o-toluic aldehyde.

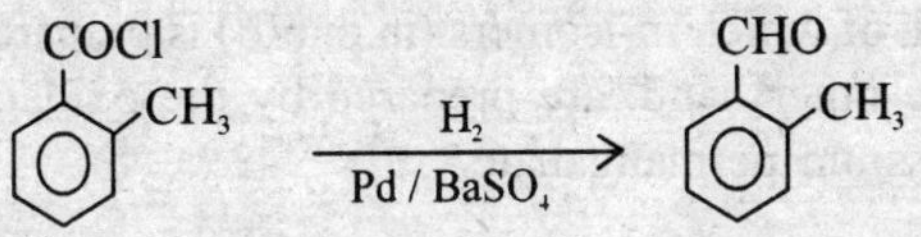

o-Toluic acid chloride   o-Toluic aldehyde

They resemble benzaldehyde in their physical and chemical properties. They are oxidised by mild oxidising agents to corresponding toluic acids and on vigrous oxidation, forms corresponding dicarboxylic acids.

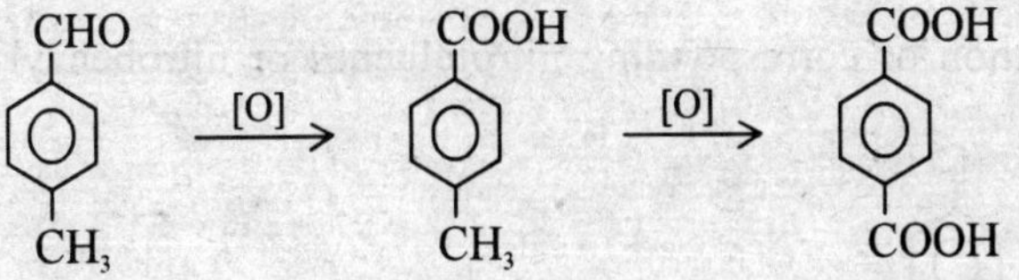

p-Toluic   p-Toluic acid   Terephthalic acid
aldehyde

35.1.4   Halo-Benzaldehydes [ClC_6H_4CHO]

All the three isomeric halogen derivative of benzaldehydes are known of which o- and p-isomers may be prepared from corresponding chlorotoluenes.

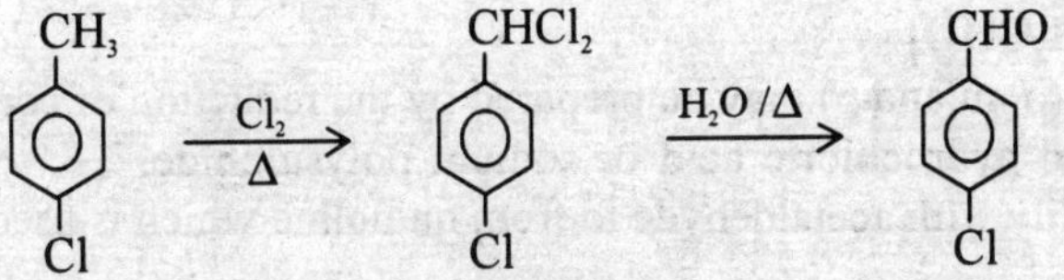

p-Chlorotuluene   p-Chlorobenzal   p-Chlorobenzaldehyde
                         chloride

The side chain of chlorotoluic acids may also be oxidised by chromyl chloride (Etard's reaction)

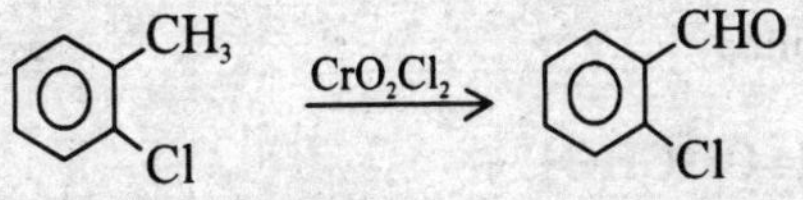

o-Chlorotuluene   o-Chlorobenzaldehyde

The m-chlorobenzaldehyde is prepared from benzaldehyde by either direct chlorination or via nitration.

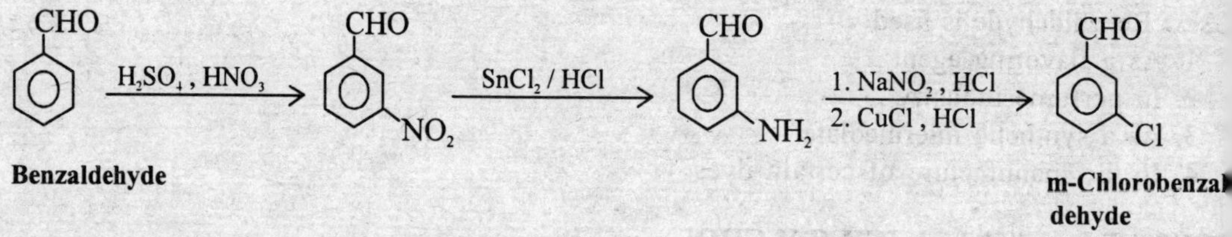

Benzaldehyde   **m-Chlorobenzadehyde**

They show all the reaction of benzaldehyde discussed earlier and also of nuclear halogen.

35.1.5   Nitrobenzaldehydes [NO₂C₆H₄CHO]

All the three isomeric nitrobenzaldehydes are known of which m-isomers (m.p. 50°) is prepared by direc
nitration of benzaldehyde.o- and p-isomers, on the other hand, are prepared by the oxidation of cor
responding nitrocinnamic acids with alkaline potassium permanganate.

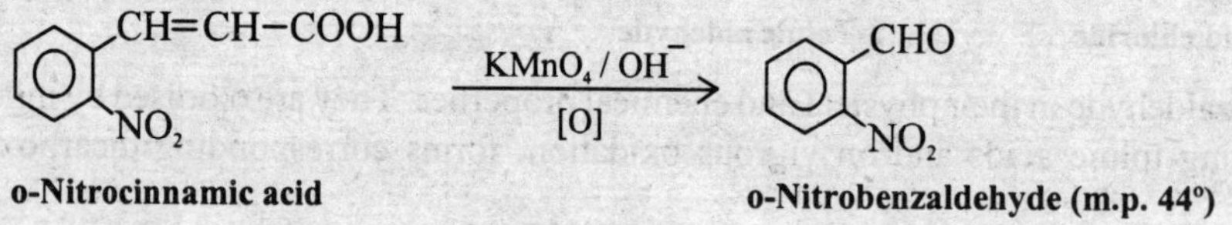

o-Nitrocinnamic acid   **o-Nitrobenzaldehyde (m.p. 44°)**

They may also be prepared by oxidation of corresponding nitrotoluenes or nitrobenzyl chloride.

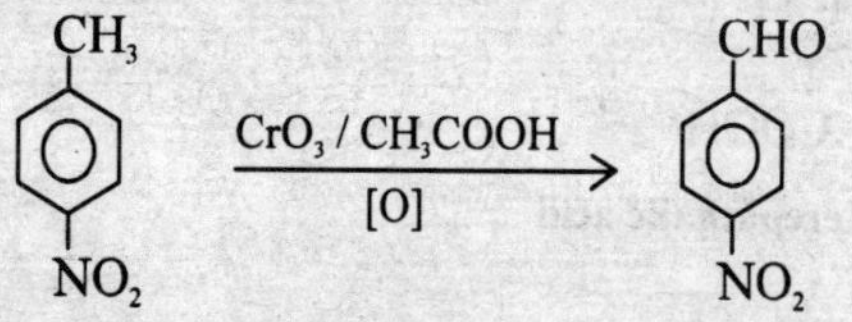

All the three isomeric nitrobenzaldehydes are reduced to corresponding aminobenzaldehydes by conventional
reducing agents. Sodium borohydride, lithium aluminum hydride-aluminum chloride and aluminum isopropoxide,
however, reduce nitrobenzaldehydes to corresponding nitrobenzyl alcohols.

35.1.6   Aminobenzaldehydes [NH₂C₆H₄CHO]

The three isomeric aminobenzaldehydes (o, m and p) may be prepared by the reduction of corresponding
nitrobenzaldehydes with tin chloride and hydrochloric acid or sodium polysulphide.

o-Aminobenzaldehyde condenses readily with acetaldehyde to from quinoline which is used in various
dye preparations.

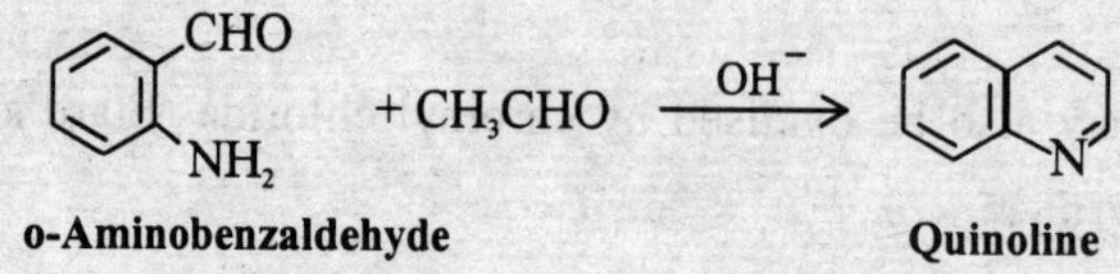

o-Aminobenzaldehyde   **Quinoline**

35.1.7   Cinnamaldehyde/ 3-Phenylpropenal [C₆H₅CH=CHCHO]

It is the chief constituent of cinnamon oil obtained from the bark of Cinnamonum ceylonicum and oil
of cassia obtained from the bark of Cinnamonum cassia. It may be isolated from these oils by forming
sodium bisulphite addition compound.

It is synthesised by Claisen Schmidt condensation involving the reaction of benzaldehyde and acetaldehyde
in presence of base.

$$C_6H_5CHO + CH_3CHO \xrightarrow{\text{NaOH}} C_6H_5CH=CH-CHO + H_2O$$

It is a pleasant smelling, colorless oil, b.p. 246°, insoluble in water but soluble in organic solvents.

It gives the general reactions characteristic of aromatic aldehydes, since -CHO group in cinnamaldehyde s in conjugation with aromatic ring through $-C=C-$. In addition, it gives the reactions of carbon carbon double bonds also. It forms normal bisulphite compound with sodium bisulphite (characteristic of aldehyde group) but on prolonged heating sodium salt of disulphonic acid (1,4-addition) is formed.

$$C_6H_5CH=CH-CHO + NaHSO_3 \rightleftharpoons C_6H_5CH=CH-\overset{\overset{\displaystyle OH}{|}}{CH}-SO_3Na$$

Cinnamaldehyde   **Bisulphite compound**

$$C_6H_5\underset{\underset{\displaystyle SO_3Na}{|}}{CH}-CH_2CHO \rightleftharpoons C_6H_5-\underset{\underset{\displaystyle SO_3Na}{|}}{CH}-CH_2-\overset{\overset{\displaystyle OH}{|}}{CH}-SO_3Na$$

Sod. salt of Disulphonic acid

The normal addition is reversible whereas the 1,4- addition is irreversible, thus cinnamaldehyde is converted into disulphonic acid gradually.

With bromine cinnamaldehyde forms dibromide.

$$C_6H_5CH=CH-CHO + Br_2 \longrightarrow C_6H_5CHBr-CHBrCHO$$
Cinnamaldehyde   **Cinnamaldehyde dibromide**

It is oxidised to cinnamic acid by air or by Tollen's reagent. With strong oxidising agents, however, it is converted into benzoic acid.

$$C_6H_5CH=CH-CHO \xrightarrow[\text{Air}]{[O]} C_6H_5CH=CH-COOH$$
Cinnamaldehyde   **Cinnamic acid**

$$C_6H_5CH=CH-CHO \xrightarrow[\text{KMnO}_4/H^+]{[O]} C_6H_5COOH$$
Cinnamaldehyde   **Benzoic acid**

Cinnamaldehyde is reduced to cinnamyl alcohol when treated with aluminum isopropoxide (Meerwein Pondorf Verley reduction.)

$$C_6H_5CH=CHCHO \xrightarrow{[(CH_3)_2CHO]_3Al} C_6H_5CH=CH-CH_2OH$$
Cinnamaldehyde   **Cinnamyl alcohol**

Reduction may also be done by lithium aluminum hydride or sodium borohydride.

35.1.8 Phenolic Aldehydes/Hydroxybenzaldehydes

Compounds having one or more hydroxy group along with an aldehyde group directly attached to benzene ring are termed phenolic aldehydes. They exhibit properties of both phenols and aromatic aldehydes.

35.1.8.1 *Salicylaldehyde/ o-Hydroxybenzaldehyde* [OHC_6H_4CHO]

It occurs in the oil of Spirea ulmaria. It may be prepared by Gattermann aldehyde synthesis starting with phenol.

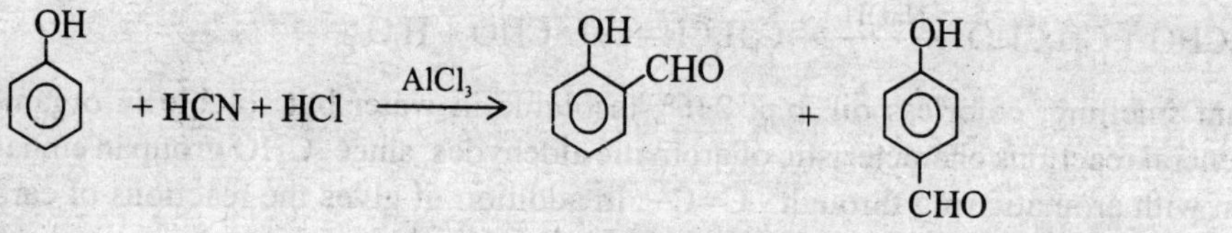

Phenol   Salicylaldehyde   p-Hydroxybenzaldehyde

The o-isomer is separated from the p-isomer by steam distillation. Salicylaldehyde may also be prepared in good yield by Reimer Tiemann reaction.

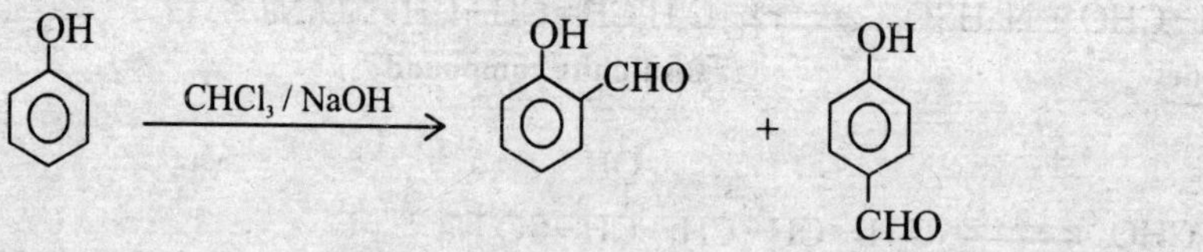

Phenol   Salicylaldehyde   p-Hydroxybenzaldehyde

It may also be prepared by Duff's reaction which involves heating of phenol with hexamethylene tetramine in glycerol and boric acid. The product is separated from the reaction mixture by steam distillation.

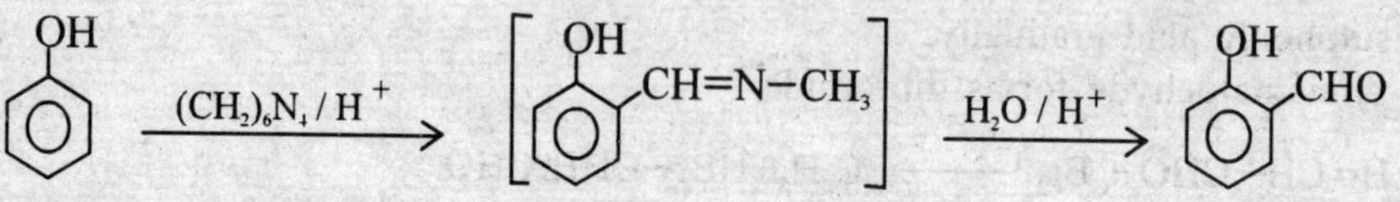

Phenol   Salicylaldehyde

On commercial scale, it is prepared by the oxidation of o-hydroxybenzyl alcohol (Saligenin) with nitrobenzene or phenylhydroxyl amine in presence of copper catalyst.

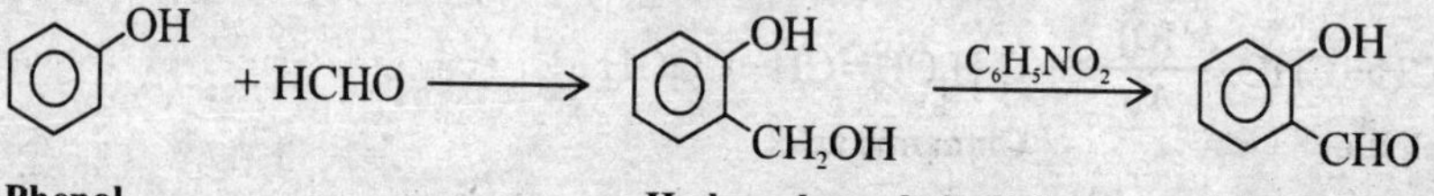

Phenol   o-Hydroxybenzyl alcohol   Salicylaldehyde

Salicylaldehyde is a pleasant smelling oil, b.p. 197°. It is sparingly soluble in water but soluble in alkali. It resembles phenol as well as benzaldehyde in its chemical properties. It gives violet color with ferric chloride. It is easily oxidised to salicylic acid and reduced to form saligenin or o-hydroxybenzyl alcohol.

The phenolic hydroxy group of salicylaldehyde is less reactive than corresponding m- and p-isomers due to the existence of intramolecular H-bonding which is also the reason for its steam volatility (m- and p-isomers are not steam volatile)

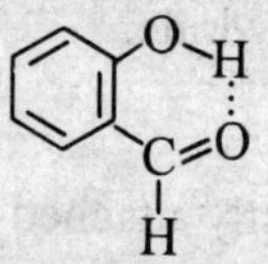

Intramolucular H-bonding in Salicylaldehyde

35.1.8.2   Vanillin/ 4-Hydroxy-3-methoxybenzaldehyde [OHC$_6$H$_3$(OCH$_3$)CHO]
It is the active odoriferous constituent of Vanilla plantifolia and is extracted by ethanol.

It may be synthesised from eugenol, available from plant sources by following sequence of reactions.

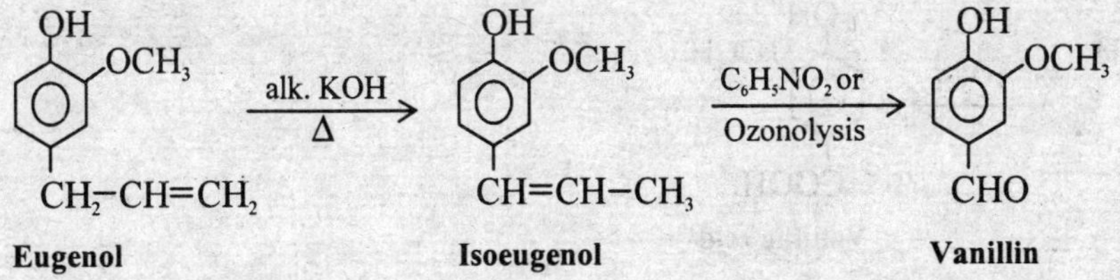

| Eugenol | Isoeugenol | Vanillin |

It may also be synthesised from guaiacol by Reimer Tiemann reaction or Gattermann reaction.

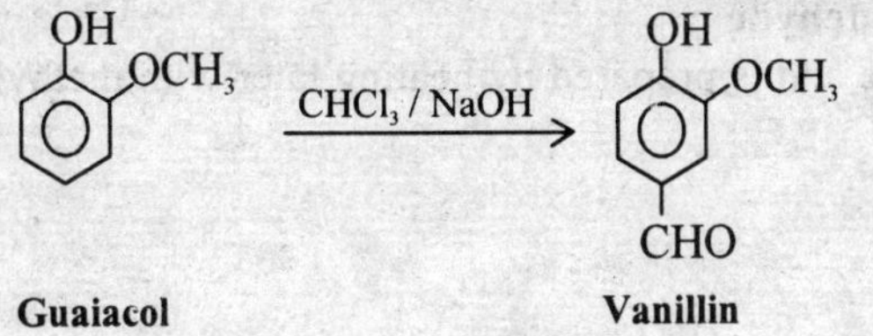

| Guaiacol | Vanillin |

Guaiacol produces vanillin via Lederer-Manasse reaction as follows.

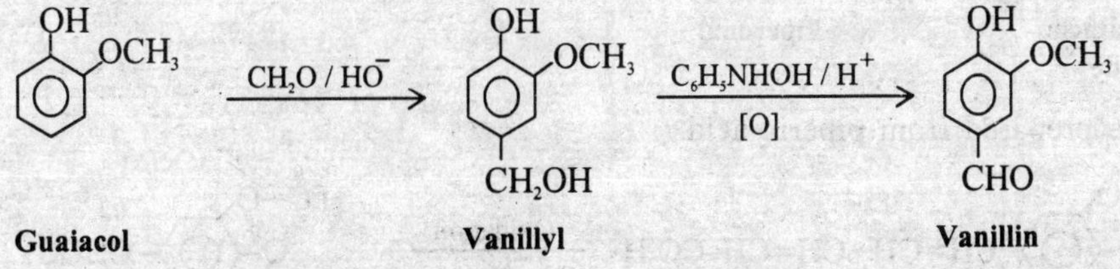

| Guaiacol | Vanillyl | Vanillin |

Vanillin may also be produced by the hydrolysis of lignins, a by-product of paper industry.

Vanillin is a pleasant smelling white crystalline solid, m.p. 81°. It exhibits the general reactions of phenol as well as aldehydes. it gives blue color with ferric chloride. It forms protocatachuic aldehyde when hydrolysed with dilute acids.

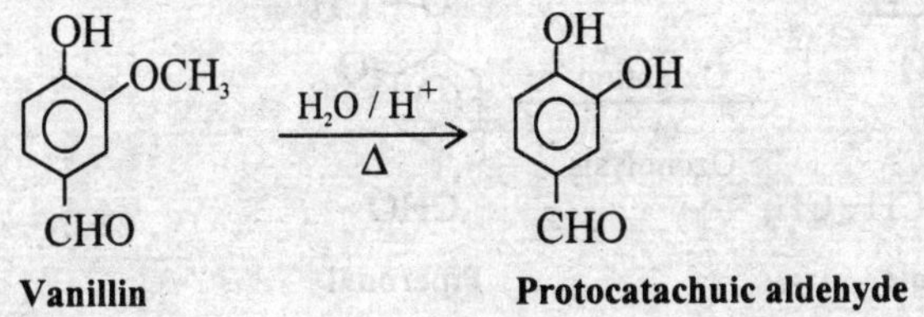

| Vanillin | Protocatachuic aldehyde |

It forms veratraldehyde on methylation and ethavan on ethylation with dimethylsulphate and diethyl-sulphate respectively.

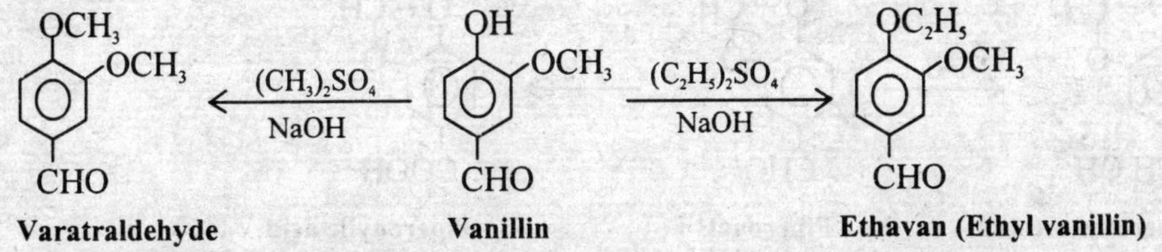

| Varatraldehyde | Vanillin | Ethavan (Ethyl vanillin) |

Ethavan has about 3.5 times more flavoring power than that of vanillin.

It is not oxidised in air but oxidises to vanillic acid when treated with alkaline silver oxide solution

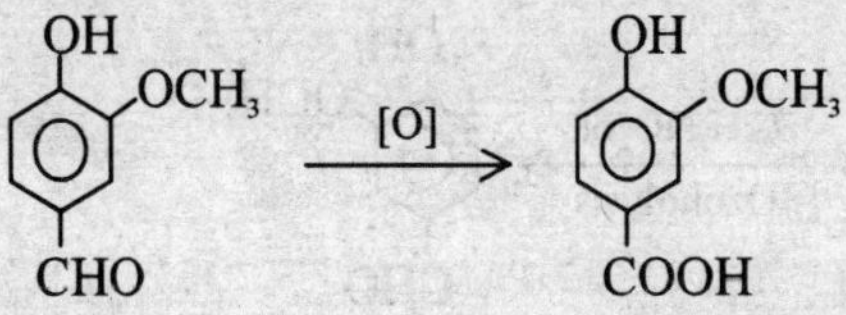

Vanillin   Vanillic acid

Vanillin is widely used as flavoring agent and also to mask various undesirable odors.

35.1.9   Piperonal/ 3,4-Methylenedioxybenzaldehyde

It is the methylene ether of protocatechualdehyde and is prepared by heating later with methylene iodide and sodium hydroxide.

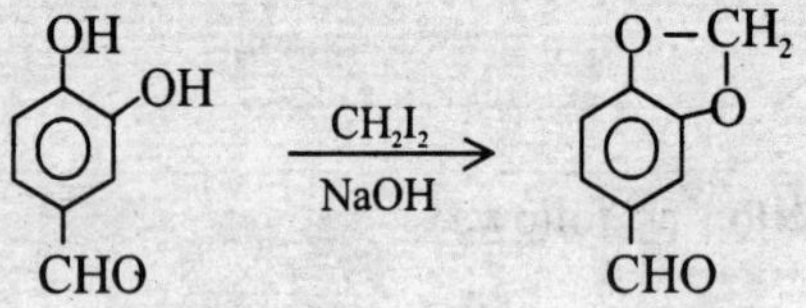

Protocatachu-   Piperonal
aldehyde

It may also be prepared from piperic acid.

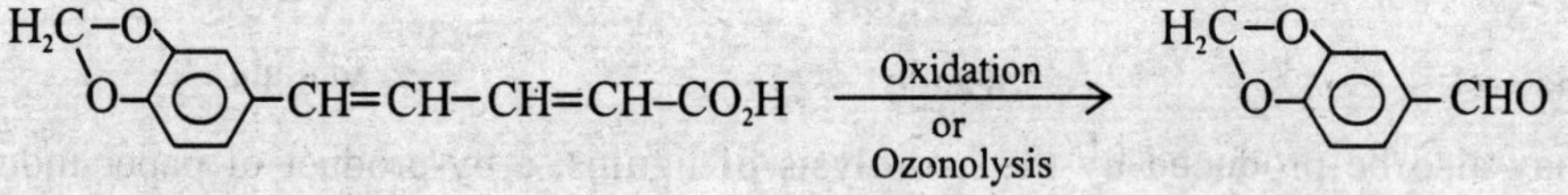

Piperic acid   Piperonal

It is prepared on large scale from safrole, main constituent of oil of Sassafras.

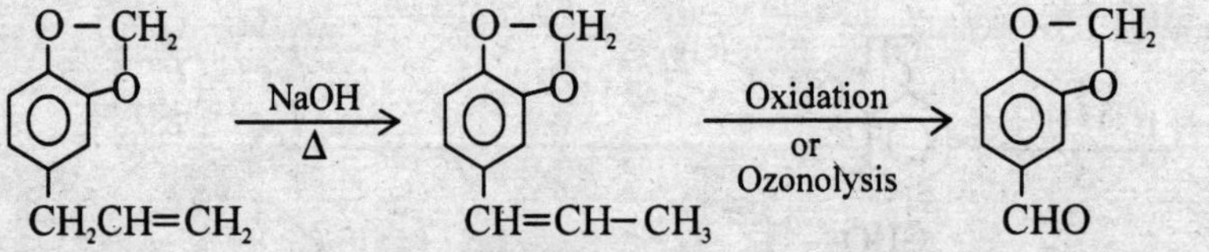

Safrole   Isosafrole   Piperonal

It is a white solid, m.p. 37°, with a characteristic odor of helitrope (hence also named helitropin). It is oxidised to piperonylic acid and reduced to piperonyl alcohol.

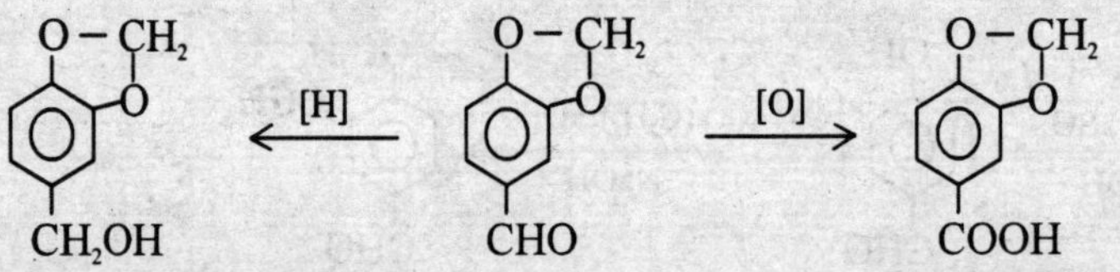

Piperonyl alcohol   Piperonal   Piperonylic acid

When treated with dilute hydrochloric acid, it decomposes back to protocatachualdehyde.

35.2 AROMATIC KETONES

Aromatic ketones may be classified as *aryl alkyl ketones* when one of the hydrocarbon group attached to carbonyl group is aryl and other alkyl (I). They are, however, known as *diaryl ketones* when both the hydrocarbon groups attached to carbonyl group are aromatic in nature (II).

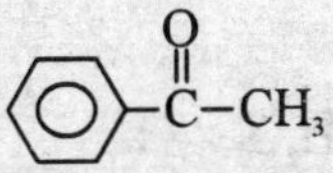

I, Acetophenone
(Methyl phenyl ketone)

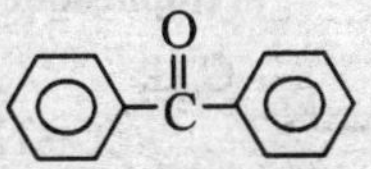

II, Benzophenone
(Diphenyl ketone)

Aromatic ketones, in general, may be prepared by any of the general methods used for the preparation of aliphatic ketones but the most commonly used method is the Friedel Crafts acylation reaction.

Aromatic ketones resemble aliphatic ketones in their reactions except for the reaction with sodium bisulphite and reactions of aromatic ring.

35.2.1 Acetophenone/Methyl Phenyl Ketone/Acetylbenzene [$C_6H_5COCH_3$]

Acetophenone may be prepared by distillation of a mixture of calcium benzoate and calcium acetate.

$$C_6H_5COO \diagdown Ca + Ca \diagup OCOCH_3 \xrightarrow{\text{Distillation}} 2\,C_6H_5COCH_3 + 2\,CaCO_3$$
$$C_6H_5COO \diagup \qquad \diagdown OCOCH_3$$

Calcium benzoate   Calcium acetate   **Acetophenone**

It may also be prepared by Friedel Crafts acylation involving the reaction of benzene with acetyl chloride in presence of anhydrous aluminum chloride.

$$C_6H_6 + CH_3COCl \xrightarrow{AlCl_3} C_6H_5COCH_3 + HCl$$

Benzene   Acetyl chloride   **Acetophenone**

Acetophenone may be prepared on commercial scale by catalytic oxidation of ethylbenzene in presence of manganese acetate at 130° under pressure.

$$C_6H_5CH_2CH_3 + O_2 \xrightarrow[130°,\,\text{Pressure}]{\text{Catalyst}} C_6H_5COCH_3 + H_2O$$

Ethylbenzene   **Acetophenone**

Properties: It is a colorless liquid, b.p. 202°, sparingly soluble in water but dissolves readily in ethanol and ether. It is steam volatile.

It resembles aliphatic ketones in its chemical properties but does not form bisulphite compounds with sodium bisulphite like aliphatic ketones. It also shows usual electrophilic substitution at meta position. Other important reactions of acetophenone are given below:

(i) *Condensation Reactions:* Acetophenone condenses with derivatives of ammonia like hydroxylamine, phenylhydrazine, semicarbazide to give oxime, phenylhydrazone and semicarbazone respectively.

$$\begin{array}{c} C_6H_5 \\ \diagdown \\ CH_3 \end{array} C=O + NH_2OH \longrightarrow \begin{array}{c} C_6H_5 \\ \diagdown \\ CH_3 \end{array} C=N-OH + H_2O$$

Acetophenone Hydroxyl amine   **Oxime**

$$\underset{\substack{\text{Acetophenone}}}{\overset{\displaystyle C_6H_5}{\underset{\displaystyle CH_3}{\Large{>}}} C=O} + \underset{\text{Phenylhydrazine}}{C_6H_5NHNH_2} \longrightarrow \underset{\substack{\text{Phenylhydrazone of}\\ \text{acetophenone}}}{\overset{\displaystyle C_6H_5}{\underset{\displaystyle CH_3}{\Large{>}}} C=N-NHC_6H_5} + H_2O$$

$$\underset{}{\overset{\displaystyle C_6H_5}{\underset{\displaystyle CH_3}{\Large{>}}} C=O} + \underset{\text{Semicarbazide}}{NH_2NHCSNH_2} \longrightarrow \overset{\displaystyle C_6H_5}{\underset{\displaystyle CH_3}{\Large{>}}} C=N-NHCSNH_2 + H_2O$$

In the presence of aluminum tertiary butoxide, acetophenone condenses to form dypnone.

$$2\,\underset{\text{Acetophenone}}{C_6H_5COCH_3} \xrightarrow[\text{tert-butoxide}]{\text{Aluminum}} \underset{\text{Dypnone}}{C_6H_5-\overset{\overset{\displaystyle CH_3}{|}}{C}=CH-COC_6H_5}$$

It condenses with acetic anhydride in presence of boron trifluoride to give benzoylacetone.

$$\underset{\text{Acetophenone}}{C_6H_5COCH_3} + \underset{\text{Acetic anhydride}}{(CH_3CO)_2O} \xrightarrow{BF_3} \underset{\text{Benzoylacetone}}{C_6H_5COCH_2COCH_3} + CH_3COOH$$

Same product is produced with ethyl acetate.

$$\underset{\text{Acetophenone}}{C_6H_5COCH_3} + \underset{\text{Ethyl acetate}}{CH_3COOC_2H_5} \xrightarrow{C_2H_5ONa} \underset{\text{Benzoylacetone}}{C_6H_5COCH_2COCH_3} + C_2H_5OH$$

With ethyl benzoate, acetophenone forms dibenzoylmethane.

$$\underset{\text{Acetophenone}}{C_6H_5COCH_3} + \underset{\text{Ethyl benzoate}}{C_6H_5COOC_2H_5} \xrightarrow{C_2H_5ONa} \underset{\text{Dibenzoylmethane}}{C_6H_5COCH_2COC_6H_5} + C_2H_5OH$$

The mechanism of condensation involves formation of carbanion from acetophenone by the attack of base (analogous to aldol type of condensation).

$$C_6H_5COCH_3 \xrightarrow[-ROH]{RO^-} C_6H_5CO\overset{..}{C}H_2 + \underset{\displaystyle OR}{\overset{\overset{\displaystyle R}{|}}{C}=O} \longrightarrow C_6H_5COCH_2-\underset{\displaystyle OR}{\overset{\overset{\displaystyle R}{|}}{C}-O} \xrightarrow{-RO^-} C_6H_5COCH_2COR$$

(ii) *Mannich Reaction:* Acetophenone condenses with formaldehyde and ammonia or an amine (preferably secondary amine) to form β-aminoketones.

$$\underset{\text{Acetophenone}}{C_6H_5COCH_3} + \underset{\substack{\text{Formal-}\\ \text{dehyde}}}{HCHO} + \underset{\substack{\text{Secondary}\\ \text{amine}}}{R_2NH} \xrightarrow[-H_2O]{H^+} \underset{\text{β-Aminoketone}}{C_6H_5COCH_2CH_2NR_2}$$

The products are often referred to as Mannich bases.

(iii) *Reaction with Chlorine:* Acetophenone reacts with chlorine in absence of catalyst to give phenacyl chloride.

$$C_6H_5COCH_3 + Cl_2 \xrightarrow{\text{Catalyst}} C_6H_5COCH_2Cl + HCl$$

Acetophenone   **Phenacyl chloride**

Phenacyl chloride, also known as 'mob gas', is a powerful lachrymator and is used to disperse the mob. Phenacyl bromide, on the other hand, is used for the identification and characterisation of acids which form crystalline phenacyl esters.

$$RCOONa + C_6H_5COCH_2Br \longrightarrow RCOOCH_2COC_6H_5 + NaBr$$

Sodium salt   **Phenacyl bromide**   **Phenacyl ester**
of acid

(iv) *Oxidation:* Acetophenone on oxidation with potassium permanganate gives phenylglyoxylic acid, which on further oxidation forms benzoic acid.

$$C_6H_5COCH_3 \xrightarrow[\text{[O]}]{\text{KMnO}_4} C_6H_5COCOOH \xrightarrow[\text{[O]}]{\text{KMnO}_4} C_6H_5COOH$$

Acetophenone   **Phenylglyoxylic acid**   **Benzoic acid**

When oxidised with selenium oxide, it gives phenylglyoxal.

$$C_6H_5COCH_3 + SeO_2 \xrightarrow{\text{Oxidation}} C_6H_5COCHO$$

Acetophenone   **Phenylglyoxal**

(v) *Reduction:* Acetophenone is reduced to secondary alcohols (like aliphatic ketones) when reduced with sodium metal and ethanol.

$$C_6H_5COCH_3 \xrightarrow[\text{Reduction}]{\text{Na} + C_2H_5OH} C_6H_5CHOHCH_3$$

Acetophenone   **Phenylmethylcarbinol**

With strong reducing agent, however, ethylbenzene is produced.

$$C_6H_5COCH_3 \xrightarrow[\text{Clemmenson reduction}]{\text{Zn—Hg / HCl}} C_6H_5CH_2CH_3$$

Acetophenone   Clemmenson   **Ethylbenzene**
          reduction

(vi) *Haloform Reaction:* Acetophenone (like other methyl ketones) undergo haloform reaction.

Uses: Acetophenone is used in the preparation of pharmaceuticals and in perfume industry.

35.2.2 Benzophenone/Diphenylketone [$C_6H_5COC_6H_5$]

Benzophenone, a diarylketone, may be prepared by distilling calcium benzoate.

$$(C_6H_5COO)_2Ca \xrightarrow[\Delta]{\text{Distillation}} C_6H_5COC_6H_5 + CaCO_3$$

Calcium benzoate   **Benzophenone**

It may also be prepared by the Friedel Crafts acylation reaction between benzene and benzoyl chloride in presence of Lewis acid.

$$C_6H_6 + C_6H_5COCl \xrightarrow{\text{AlCl}_3} C_6H_5COC_6H_5 + HCl$$

Benzene   **Benzoyl chloride**   **Benzophenone**

It is prepared on commercial scale by Friedel Crafts reaction between benzene and carbon tetrachloride followed by the hydrolysis of geminal dihalides.

$$C_6H_6 + CCl_4 \xrightarrow[-2HCl]{AlCl_3} C_6H_5CCl_2C_6H_5 \xrightarrow{H_2O} C_6H_5-\overset{\overset{\displaystyle O}{\|}}{C}-C_6H_5$$

Benzene   Benzophenone

Benzophenone is a colorless pleasant smelling solid which exists in two forms. The unstable, monoclinic, crystalline form melts at 27° and stable, rhombic crystalline form at 49°. It gives the general reactions of aliphatic and aromatic ketones, but does not form bisulphite compound. It does not undergo condensation reactions like acetophenone since it lacks α hydrogen atom.

It is reduced to benzhydrol or diphenylcarbinol when reduced with zinc and ethanolic potassium hydroxide.

$$C_6H_5COC_6H_5 \xrightarrow[C_2H_5OH]{Zn / KOH} C_6H_5CHOHC_6H_5$$

Benzophenone   Diphenylcarbinol

Benzophenone is reduced to benzopinacol with zinc and acetic acid.

$$C_6H_5COC_6H_5 \xrightarrow{Zn / CH_3COOH} C_6H_5-\overset{\overset{\displaystyle H_5C_6}{|}}{\underset{\underset{\displaystyle HO}{|}}{C}}-\overset{\overset{\displaystyle C_6H_5}{|}}{\underset{\underset{\displaystyle OH}{|}}{C}}-C_6H_5$$

Benzophenone   Benzopinacol

Benzopinacol undergoes *pinacol-pinacolone rearrangement* when heated in presence of acid and iodine as catalyst to yield benzopinacolone.

$$C_6H_5-\overset{\overset{\displaystyle H_5C_6}{|}}{\underset{\underset{\displaystyle HO}{|}}{C}}-\overset{\overset{\displaystyle C_6H_5}{|}}{\underset{\underset{\displaystyle OH}{|}}{C}}-C_6H_5 \xrightarrow{H^+/I_2} C_6H_5-\overset{\overset{\displaystyle }{}}{\underset{\underset{\displaystyle O}{\|}}{C}}-\overset{\overset{\displaystyle C_6H_5}{|}}{\underset{\underset{\displaystyle C_6H_5}{|}}{C}}-C_6H_5$$

Benzopinacol   Benzopinacolone

Benzophenone is oxidised to give benzoic acid, carbon dioxide and water.

$$C_6H_5COC_6H_5 \xrightarrow{[O]} C_6H_5COOH + 6 CO_2 + 2H_2O$$

Benzophenone   Benzoic acid

When fused with potassium hydroxide, it splits to form benzene and potassium benzoate.

$$C_6H_5COC_6H_5 + KOH \longrightarrow C_6H_6 + C_6H_5CO\overset{-}{O}\overset{+}{K}$$

Benzophenone   Benzene   Potassium benzoate

Benzophenone forms crystalline deep violet thiobenzophenone when reacted upon with dry hydrogen chloride and hydrogen sulphide in ethanol.

It is used in perfume industry and in organic synthesis.

35.3 QUINONES

Benzoquinones are unsaturated conjugated, cyclic diketones. They are not truly aromatic in nature but are discussed here sine they can be easily converted into aromatic compounds. Two benzoquinones are known, the p-benzoquinone (I) and the o-benzoquinone (II).

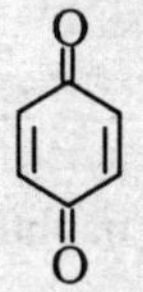

I, p-Benzoquinone 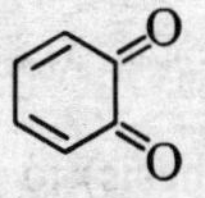   **II, o-Benzoquinone**

m-Benzoquinone is not known.

35.3.1 p-Benzoquinone/p-Quinone

p-Benzoquinone may be prepared by the oxidation of quinol with acidic dichromate, lead tetraacetate, manganese dioxide, ferric chloride etc. in presence of V_2O_5 catalyst.

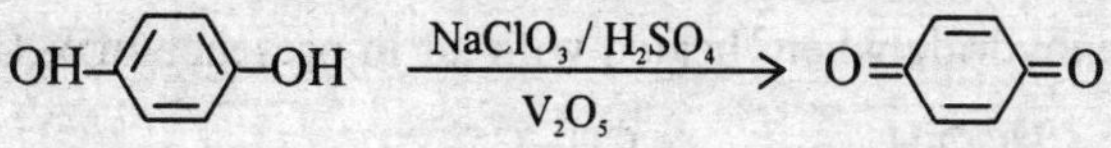

          Quinol   **p-Benzoquinone**

In laboratory, it may be prepared by the oxidation of aniline with acidic potassium dichromate.

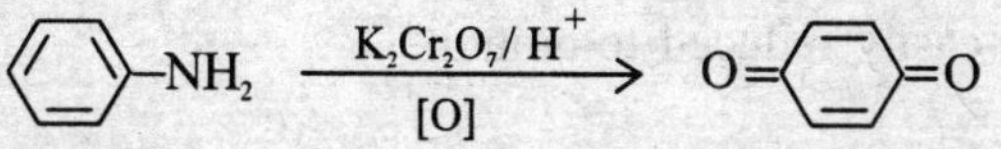

          Aniline   **p-Benzoquinone**

It is a yellow, crystalline solid m.p. 116°, having a pleasant odor. It is steam volatile, sublimes and turns brown when exposed to light.

It shows the reactions of α,β-unsaturated ketones rather than those of aromatic compounds. It undergoes addition of bromine to form dibromide and tetrabromide of benzoquinone.

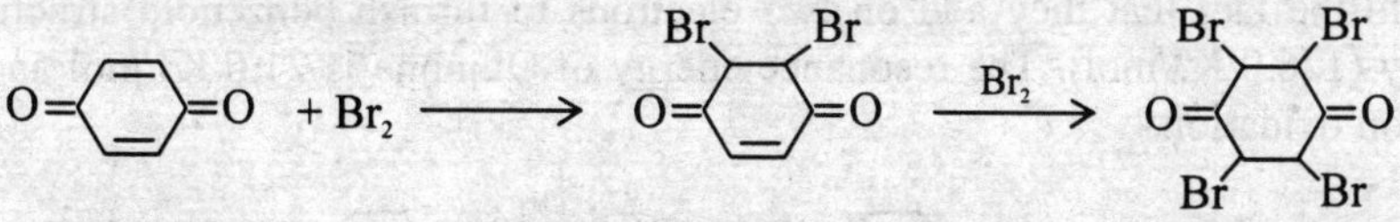

          p-Benzoquinone   **Dibrome-p-benzoquinone**   **Tetrabromo-p-benzoquinone**

When hydrochloric acid is added to p-benzoquinone, it forms chloroquinol by 1,4-addition followed by enolisation.

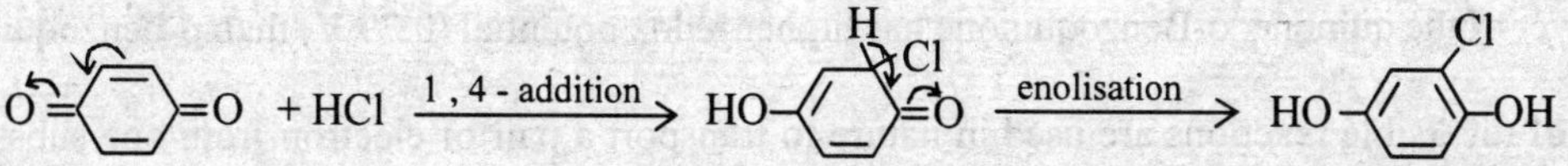

          2-Chloro-1,4-dihydroxybenzene
          (2-chloroquinol)

2-Chloro-1,4-dihydroxybenzene (chloroquinol) on oxidation forms chloro-p-benzoquinone which on reaction with hydrochloric acid and subsequent oxidation gives chloranil (tetrachloro-p-benzoquinone), which is used as a fungicide.

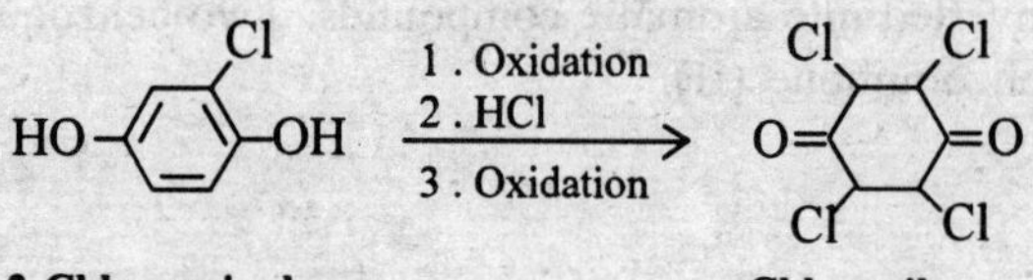

2-Chloroquinol → **Chloranil**
(1. Oxidation, 2. HCl, 3. Oxidation)

p-Benzoquinone adds to dienes (Diels-Alder diene synthesis) to yield hydrogenated naphthaquinone derivative.

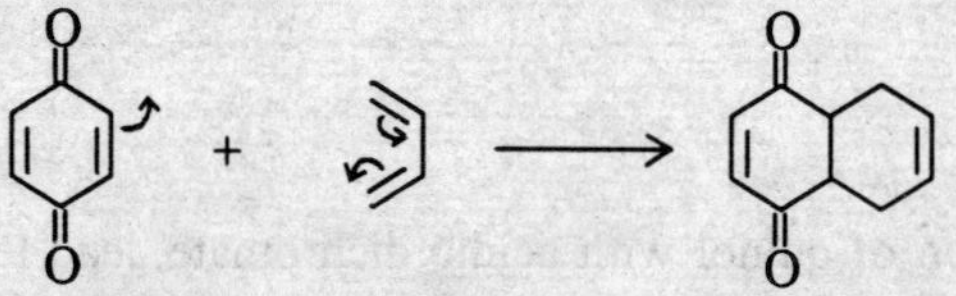

p-Benzoquinone   Butadiene

It is oxidised to maleic acid and maleic anhydride when heated with air in presence of V_2O_5 catalyst.

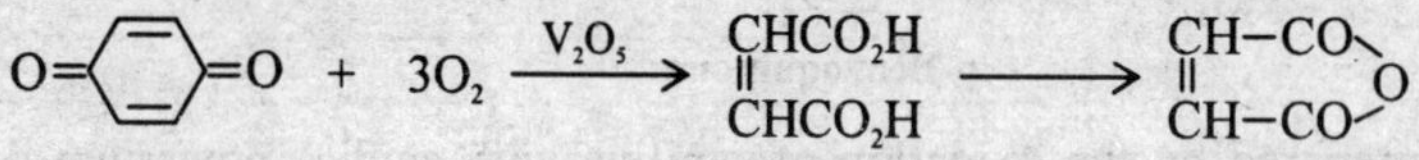

p-Benzoquinone   **Maleic acid**   **Maleic anhydride**

It is a very good oxidising agent since it is easily reduced to quinol.

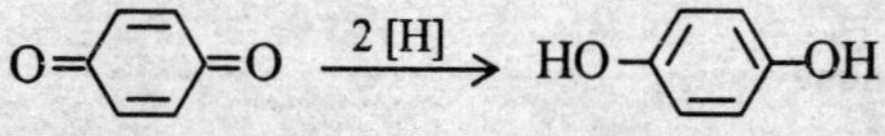

p-Benzoquinone   **p-Benzoquinol**

Chloranil is generally employed for the oxidation purpose in the reaction where inorganic oxidising agents are avoided. It liberates iodine from acidified potassium iodide solution. The oxidising properties of Quinones are due to the fact that they add on two electrons to furnish benzenoid structure having large resonance energy (126.0 KJ/mol). The resonance energy of Quinone is 21.0 KJ/mol and it attains additional stability upon oxidation.

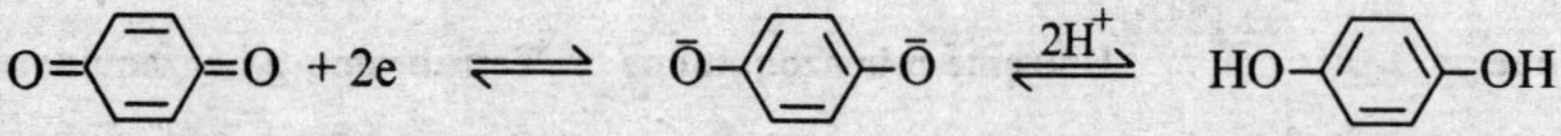

The reversible oxidation-reduction is the characteristic of all quinones and a measure of the oxidising power of the system is given by its redox potential. Higher the positive redox potential, higher is the oxidising power of the quinone. o-Benzoquinone has higher redox potential (0.79 V) than p-benzoquinone (0.70 V).

This kind of reversible reactions are used in nature to transport a pair of electron from one substance to another. Ubiquinone also known as coenzyme Q is found in biological systems. Vitamin K_1, an important dietary factor contains 1,4-naphthaquinone moiety.

It reacts with phosphorus pentachloride to give p-dichlorobenzene.

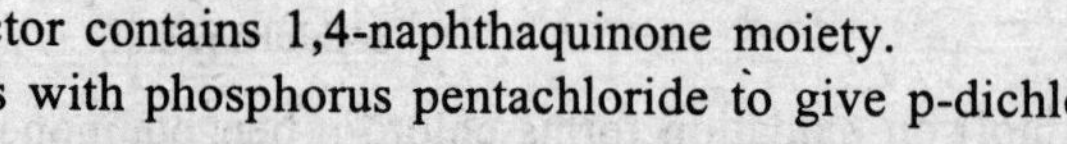

p-Benzoquinone   **p-Dichlorobenzene**

Quinone reacts with hydroxylamine to form oxime and dioxime showing the presence of two carbonyl groups.

p-Benzoquinone Benzoquinone oxime Benzoquinone dioxime

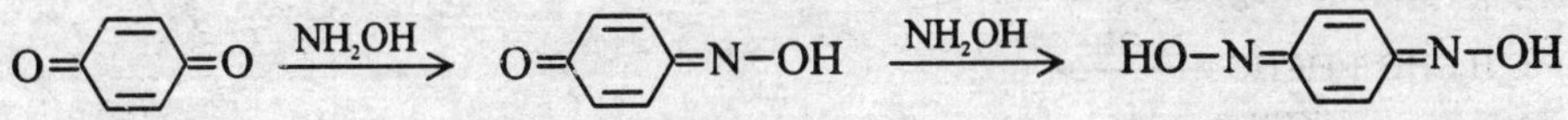

The monooxime is a tautomer of p-nitrosophenol.

Benzoquinone oxime p-Nitrosophenol

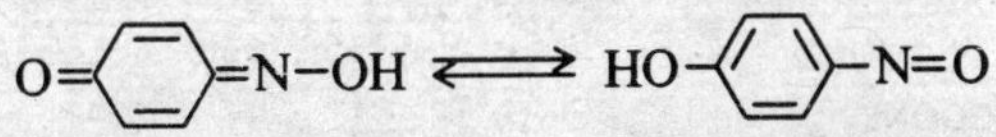

p-Benzoquinone reacts with acetic anhydride in presence of concentrated sulphuric acid to give hydroxyquinol triacetate. This type of acetylation is known as Thiele acetylation.

p-Benzoquinone

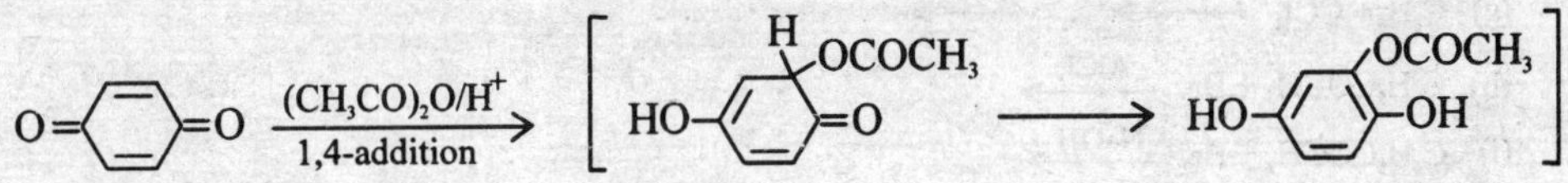

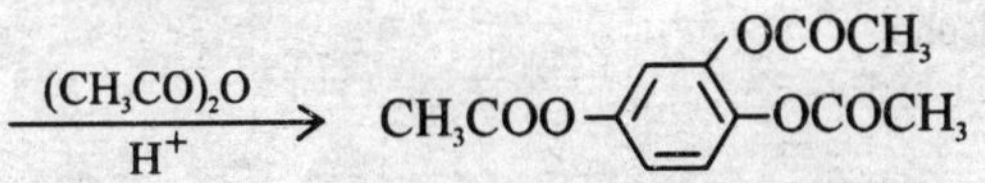

Hydroxyquinol triacetate

35.3.2 o-Benzoquinone/o-Quinone

It may be prepared by the oxidation of catechol with silver oxide in ether solution in presence of anhydrous sodium sulphate.

Catechol o-Benzoquinone

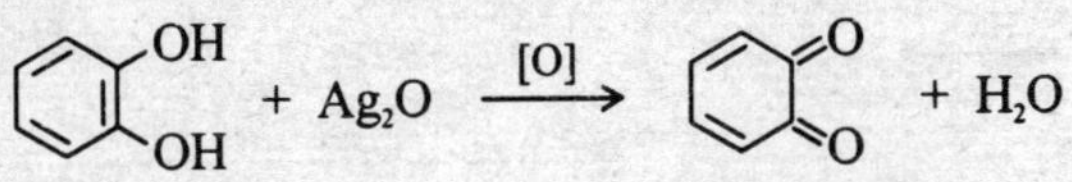

It decomposes in excess of silver oxide in presence of water, thus, anhydrous conditions favour its preparation from catechol.

It exists in two forms, the unstable, green, needle shaped crystals and stable, red, crystalline plates.

It is a strong oxidising agent since it easily reduce back to catechol by the action of sulphurous acid. It liberates iodine from acidic potassium iodide solution.

35.4 PROBLEMS

1. What are aromatic aldehydes? compare them with aryl alkyl aldehydes and aliphatic aldehydes.
2. How is benzaldehyde prepared from toluene? Give two important reactions of benzaldehyde.
3. Discuss the mechanism of
   (a) Perkins reaction (b) Benzoin condensation
   (c) Cannizarro reaction (d) Pinacol pinacolone rearrangement
   (e) Reimer-Tiemann reaction (f) Knoevenagel reaction
4. Give reasons for the following:
   (a) Benzyl alcohol is not acidic in nature whereas o- hydroxybenzaldehyde is acidic.
   (b) Salicylaldehyde has much lower melting point compared to p-hydroxybenzaldehyde.

(c) In presence of base acetaldehyde undergoes aldol condensation whereas benzaldehyde undergoes Cannizaro reaction.

5. Complete the following reactions:

(a) $C_6H_5CHO + HCHO \xrightarrow{NaOH}$

(b) $C_6H_5CHO \xrightarrow{KCN}$

(c) $C_6H_5CHO + (CH_3CO)_2O \xrightarrow{CH_3COONa}$

(d) $C_6H_5CHO \xrightarrow{NH_2OH}$, $\xrightarrow{PCl_5}$

(e) 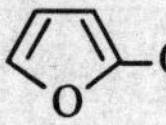 $\underset{O}{\bigcirc}$—$CHO + (CH_3CH_2CO)_2O \xrightarrow[\Delta]{CH_3CH_2COONa}$

(f) $(C_6H_5COO)_2Ca \xrightarrow{\Delta}$

(g) $C_6H_6 + CCl_4 \xrightarrow{AlCl_3}$, $\xrightarrow{H_2O}$

(h) $C_6H_5COCH_3 + Br_2 \xrightarrow{AlCl_3}$

(i) $C_6H_5COCH_3 + Br_2 \xrightarrow{NaOH}$

6. Starting from benzaldehyde how will you prepare:
    (a) Cinnamic acid   (b) Benzyl alcohol
    (c) Benzoin   (d) Cinnamaldehyde
    (e) Malachite green

7. Starting from Benzophenone how will you prepare
    (a) Phenylglyoxal   (b) Dypnone
    (c) Phenacyl bromide   (d) Benzoyl acetone

8. Complete the following reactions.

(a) $C_6H_5CHO + CH_3CHO \xrightarrow{NaOH}$

(b) $C_6H_5CH{=}CHCHO + NaHSO_3 \longrightarrow$

(c) $C_6H_5OH + HCN + HCl \xrightarrow{AlCl_3}$, $\xrightarrow{H_2O}$

(d) $C_6H_5OH \xrightarrow[\text{2. } H^+]{\text{1. } (CH_2)_6N_4}$

9. Give the product in following reactions.
    (a) Guaiacol undergoes Reimer Tiemann reaction.   (b) Isoeuginol is oxidised.
    (c) Vanillin is ethylated.   (d) Protocatachualdehyde reacts with CH_2I_2 and NaOH.
    (e) Aniline is oxidised with acidic $K_2Cr_2O_7$.   (f) Bromine is added to p-quinone.
    (g) p-Quinone reacted with hydroxylamine.   (h) Catachol is oxidised.

10. Write note on the following:
    (a) Fries rearrangement   (b) Gattermann koch aldehyde synthesis
    (c) Etard reaction   (d) Stephen's reaction

36

Aromatic Carboxylic Acid and Related Compounds

36.1 INTRODUCTION

Aromatic carboxylic acids have one or more carboxyl group attached directly to the benzene ring or to the side chain.

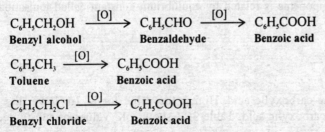

Benzoic acid   o-Toluic acid   Salicylic acid (o-Hydroxybenzoic acid)   Phthalic acid (Benzene-o-dicarboxylic acid)

When carboxyl group is present in the side chain, the acid is known as aryl substituted aliphatic acid. The aryl substituted aliphatic carboxylic acids resemble aliphatic acids in their reaction.

Phenylacetic acid   Cinnamic acid

36.2 GENERAL METHODS OF PREPARATION

Aromatic carboxylic may be prepared by

(i) *Oxidation* of a variety of substrates like alcohols, aldehydes, hydrocarbons etc.

$$C_6H_5CH_2OH \xrightarrow{[O]} C_6H_5CHO \xrightarrow{[O]} C_6H_5COOH$$
Benzyl alcohol   Benzaldehyde   Benzoic acid

$$C_6H_5CH_3 \xrightarrow{[O]} C_6H_5COOH$$
Toluene   Benzoic acid

$$C_6H_5CH_2Cl \xrightarrow{[O]} C_6H_5COOH$$
Benzyl chloride   Benzoic acid

$$C_6H_5CH{=}CH_2 \xrightarrow{[O]} C_6H_5COOH$$
Styrene   Benzoic acid

(ii) *Hydrolysis of Aryl Nitriles*   or cyanides in presence of mineral acid.

$$C_6H_5CN \xrightarrow{H^+/H_2O} C_6H_5COOH$$
Phenyl   Benzoic acid
cyanide

(iii) *From Grignard Reagent:* Aryl magnesium bromide when treated with carbon dioxide.

$$C_6H_5MgBr \ +CO_2 \longrightarrow C_6H_5COOMgBr \xrightarrow{H_2O/H^+} C_6H_5COOH$$
Phenyl magnesium   Benzoic acid
bromide

(iv) *Hydrolysis of Trihalogen Side Chain:* Toluene forms trichlorotoluene by side chain halogenation which on hydrolysis forms benzoic acid.

$$C_6H_5CH_3 \xrightarrow[UV]{Cl_2} C_6H_5CCl_3 \xrightarrow{NaOH} \left[C_6H_5C{-}OH \begin{matrix} OH \\ \\ OH \end{matrix} \right] \xrightarrow{-H_2O} C_6H_5COOH$$
Toluene   Benzoic acid

(v) *Friedel Crafts Reaction:* Benzene and carbonyl chloride in presence of aluminum chloride forms benzoyl chloride which upon hydrolysis yields benzoic acid.

$$C_6H_6 \ + \ COCl_2 \xrightarrow{AlCl_3} C_6H_5COCl \xrightarrow{H_2O} C_6H_5COOH$$
Benzene   Carbonyl   Benzoyl   Benzoic
          chloride   chloride   acid

36.3 PROPERTIES

Aromatic acids are generally crystalline solids having low water solublity compared to aliphatic carboxylic acids. They are, however, freely soluble in hot water and ethanol.

(i) *Acidity of Aromatic Acids:* Aromatic acids are slightly stronger acids than aliphatic acids. They ionise to give carboxylate anion and hydronium ion.

$$Ar\,COOH + H_2O \rightleftharpoons Ar\,COO^- + H_3O^+$$

The concentration of various components is related by equilibrium constant called ionisation constant K_a, i.e.

$$K_a = \frac{[ArCOO^-]\,[H_3O^+]}{[ArCOOH]}$$

K_a is a measure of acidity of the carboxylic acid. Higher the value for K_a (or lower the value for pK_a), greater will be the acidity of carboxylic acid. Table 36.1 shows pK_a values of substituted benzoic acids.

Table 36.1: Acidity Constants of Substituted Benzoic Acids

Name	pK$_a$
Benzoic acid	4.2
o-Toluic acid	3.9
m-Toluic acid	4.3
p-Toluic acid	4.4
o-Hydroxybenzoic acid	3.0
m-Hydroxybenzoic acid	4.1
p-Hydroxybenzoic acid	4.5
o-Nitrobenzoic acid	2.2
m-Nitrobenzoic acid	3.5
p-Nitrobenzoic acid	3.4
o-Chlorobenzoic acid	2.9
m-Chlorobenzoic acid	3.8
p-Chlorobenzoic acid	4.0

Acidity of benzoic acid is affected by the substitutents as shown in Table 36.1. Methyl and hydroxy groups make benzoic acid weaker whereas nitro and chloro groups have acid strengthening affect on benzoic acid. For example both p-nitrobenzoic and p-chlorobenzoic acids are stronger than benzoic acid. The order or descending acid strength is.

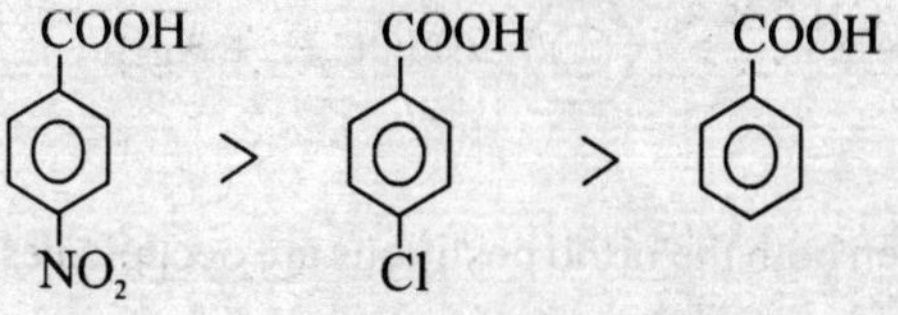

p-Nitrobenzoic is stronger acid than p-chlorobenzoic as it exerts electron withdrawing inductive and electron withdrawing resonance effects which stabilises the corresponding p-nitrobenzoate anion more than p-chlorobenzoate anion furnished by p-chlorobenzoic acid.

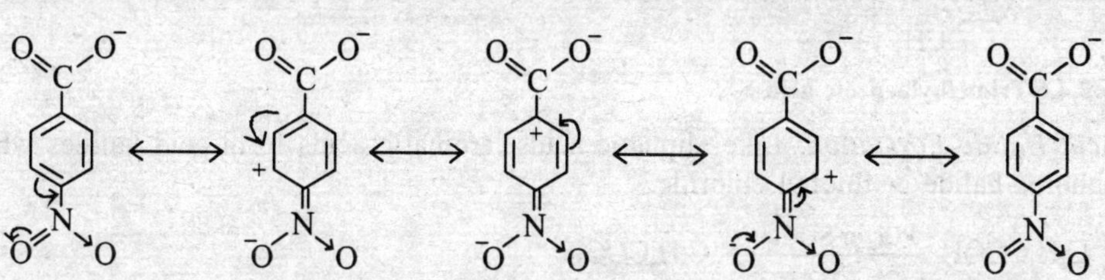

Electron withdrawing groups decrease electron density at ortho and para positions thereby delocalising negative charge on carboxylate anion making it stable. Such delocalisation is not possible with electron releasing groups like −OH and −CH$_3$.

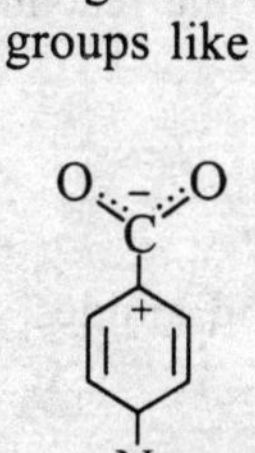

p-Nitrobenzoate anion
More acidic
−NO$_2$ withdraws electrons
Delocalises charge
Stabilises anion

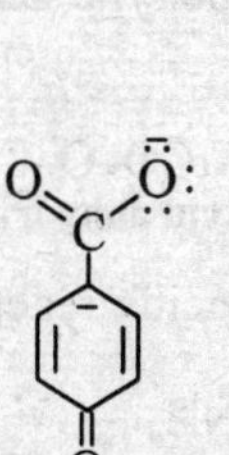

p-Hydroxybenzoate anion
Less acidic
−OH releases electrons
Unable to delocalise charge
Does not stabilise anion

o-Nitrobenzoic acid is a stronger acid than p-nitrobenzoic acid due to its strong inductive effect (being closer to anion than its p-isomer) which causes additional delocalisation of negative charge.

If we compare pK_a values of varoius o-substituted benzoic acids (Table 36.1), the ortho substituted acids do not fit into the pattern set by their meta and para isomers. All *ortho substituents* (electron realeasing or withdrawing) show acid strengthening effect. This *ortho effect* responsible for increase in acid strength was first explained on the basis of steric hinderence and later as *proximity effect* which has something to do with the nearness of the groups involved.

(ii)   *Salt Formation:* Aromatic carboxylic acids form salts with bases.

$$ArCOOH + NaHCO_3 \longrightarrow ArCOO^{-} Na^{+} + H_2O + CO_2$$

This method is used for identification and separation of carboxylic acids.

(iii)   *Esterification:* Aromatic carboxylic acids react with alcohols in presence of mineral acid to yield esters.

$$ArCOOH + C_2H_5OH \xrightarrow{H^+} ArCOOC_2H_5 + H_2O$$

The rate of esterification of carboxylic acids is greatly affected by substituents at ortho posititons principally because of steric hinderance. The o-substituted carboxylic acids are esterified with great difficulty and in low yields by the reaction of their silver salts with alkyl halides.

o-Toluic acid

o-Substituted ester cannot be hydrolysed easily. When both the ortho postitions are occupied esterification does not occur at all.

2,4,6-Trimethylbenzoic acid

(iv)   *Acid Halide Formation:* Like aliphatic acids, aromatic acids form acid halides when treated with phosphorus halide or thionyl chloride.

$$C_6H_5COOH \xrightarrow[\Delta]{PCl_5 \text{ or } SOCl_2} C_6H_5COCl$$

Benzoic acid   **Benzoyl chloride**

(v)   *Anhydride Formation:* Aromatic carboxylic acids form acid anhydrides by the reaction of acid chloride and the sodium salt of the acid.

$$ArCOCl + ArCOO^{-}Na^{+} \longrightarrow ArCO-O-COAr + NaCl$$

Acid chloride   Acid anhydride

(vi)   *Acid Amide Formation:* Aromatic acids react with ammonia to form acid amides.

$$ArCOOH \xrightarrow[NH_3]{NaOH} ArCOONH_4 \xrightarrow[-H_2O]{\Delta} ArCONH_2$$

Carboxylic acid   **Ammonium salt**   **Acid amide**

Acid amides dehydrate in presence of phosphorus pentoxide to nitriles.

$$ArCONH_2 \xrightarrow[-H_2O]{P_2O_5} Ar-C\equiv N$$

Acid amide   **Aryl nitriles**

(vii)   *Decarboxylation:* When sodium salt of aromatic acids are heated with soda lime, they form hydrocarbons and a molecule of carbon dioxide is removed from the acid.

$$ArCOONa + NaOH \longrightarrow ArH + Na_2CO_3$$

(viii)   *Reduction:* Carboxylic acids when reduced with lithium aluminum hydride form corresponding alcohols.

$$ArCOOH \xrightarrow{LiAlH_4} ArCH_2OH$$

(ix)   *Ring Substitution Reactions:* Aromatic carboxylic acids undergo electrophilic substitution in the ring orienting the substituent at m-postition with respect to carboxyl group since it is a deactivating group. The meta directing effect of carboxyl group operates through inductive and resonance effect which decreases the electron density at o- and p-positions making m-positions susceptible for electrophilic attack.

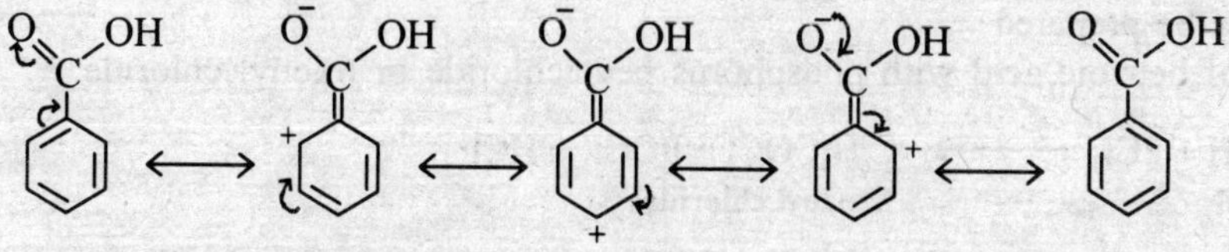

Aromatic acids undergo nitration, sulphonation and halogenation in usual conditions.

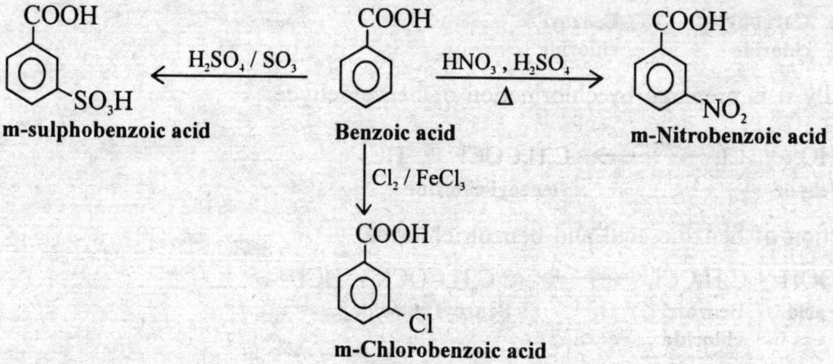

36.4   INDIVIDUAL MEMBERS

36.4.1   Benzoic Acid [C$_6$H$_5$COOH]

It is the simplest aromatic acid and may be prepared by any of the methods discussed in Sec. 36.2. It may also be prepared by the oxidation of naphthalene or xylene.

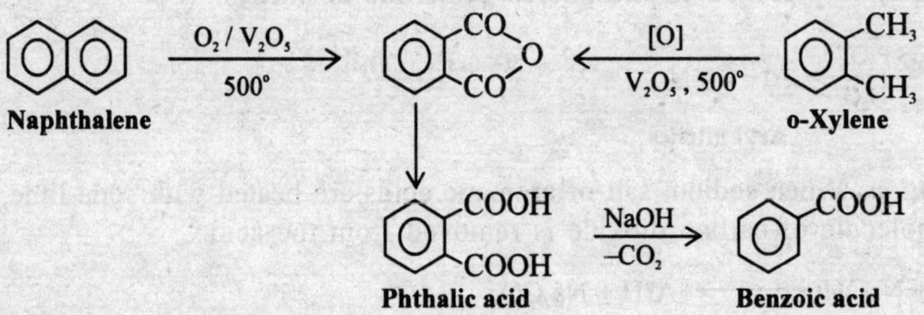

Commercially it is prepared by oxidation of toluene and by hydrolysis of benzotrichloride.

Benzoic acid is a white crystalline solid, m.p. 121°, sparingly soluble in water but freely soluble in hot water, alocohol and ether. It is steam volatile and may be sublimed.

It gives all the characteristic reactions of aromatic carboxylic acids discussed in Sec. 36.3.

Uses: Benzoic acid is used:
1. as urinary antiseptic and as disinfectant of bronchial tubes.
2. in dyes industry for the preparation of aniline blue.
2. as preservative.

36.4.2   Benzoyl Chloride [C_6H_5COCl]
Benzoyl chloride may be prepared
1. by distillation of benzoic acid with phosphorus pentachloride or thionyl chloride.

$$C_6H_5COOH + PCl_5 \longrightarrow C_6H_5COCl + POCl_3 + HCl$$
Benzoic acid   **Benzoyl chloride**

2. by Friedel Crafts reaction involving benzene and carbonyl chloride in presence of aluminum chloride.

$$C_6H_6 + COCl_2 \xrightarrow[\text{anhyd.}]{AlCl_3} C_6H_5COCl + HCl$$
Benzene   Carbonyl   **Benzoyl**
chloride   **chloride**

3. Commercially it is prepared by chlorination of benzaldehyde.

$$C_6H_5CHO + Cl_2 \xrightarrow{150\text{-}160°} C_6H_5COCl + HCl$$
Benzaldehyde   **Benzoyl chloride**

4. by the reaction of benzoic acid and benzotrichloride.

$$C_6H_5COOH + C_6H_5CCl_3 \longrightarrow 2\,C_6H_5COCl + HCl$$
Benzoic acid   Benzotri   **Benzoyl chloride**
chloride

Benzoyl chloride is a pungent smelling colorless liquid b.p. 197°. It has itching and lacrymatory action and affects the repiratory track also.

It is hydrolysed slowly by water or by dilute alkali to benzoic acid.

$$C_6H_5COCl + H_2O \longrightarrow C_6H_5COOH + HCl$$
Benzoyl chloride   **Benzoic acid**

Benzoyl chloride reacts with compounds having active hydrogen like phenols and amines to give benzoates and anilides.

$$C_6H_5COCl + C_6H_5OH + NaOH \longrightarrow C_6H_5COOC_6H_5 + NaCl + H_2O$$

Benzoyl   Phenol   **Phenyl benzoate**
chloride

$$C_6H_5COCl + C_6H_5NH_2 + NaOH \longrightarrow C_6H_5CONHC_6H_5 + NaCl + H_2O$$

Benzoyl   Aniline   **Benzanilide**
chloride

$$C_6H_5COCl + C_2H_5OH \longrightarrow C_6H_5COOC_2H_5$$

Benzoyl   Ethanol   **Ethyl benzoate**
chloride

$$C_6H_5COCl + 2\ NH_3 \longrightarrow C_6H_5CONH_2 + NH_4Cl$$

Benzoyl   Ammonia   **Benzamide**
chloride

The benzoylation reactions mentioned above are commonly known as *Schotten Baumann reaction*. The reaction can also be performed with sodium carbonate or pyridine. The benzoyl derivatives are crystalline compounds and are used to identify phenols and anilines. They hydrolyse back to phenols and anilines easily, thus used to protect these groups.

Benzoyl chloride undergoes Friedel-Crafts reaction when reacted with benzene to give benzophenone.

$$C_6H_5COCl + C_6H_6 \xrightarrow{AlCl_3} C_6H_5COC_6H_5 + HCl$$

Benzoyl   **Benzene**   **Benzophenone**
chloride

It is reduced to benzaldehyde when reacted with hydrogen in presence of paladium poisoned with barium sulphate or quinoline.

$$C_6H_5COCl + H_2 \xrightarrow[Quinoline]{Pd-BaSO_4} C_6H_5CHO + HCl$$

Benzoyl chloride   **Benzaldehyde**

The above reaction is known as *Rosenmund reaction*.

It reacts with sodium benzoate to form benzoic anhydride.

$$C_6H_5COCl + C_6H_5COONa \xrightarrow{\Delta} C_6H_5CO-O-COC_6H_5 + HCl$$

Benzoyl   **Sodium**   **Benzoic anhydride**
chloride   **benzoate**

36.4.3 Benzamide [$C_6H_5CONH_2$]

It may be prepared by ammonolysis of benzoyl chloride.

$$C_6H_5COCl + NH_3 \longrightarrow C_6H_5CONH_2 + HCl$$

Benzoyl   **Benzamide**
chloride

It may also be prepared by partial hydrolysis of phenyl cyanide with alkaline hydrogen peroxide.

$$C_6H_5C \equiv N + H_2O \xrightarrow{H_2O_2 / OH^-} C_6H_5CONH_2$$

Phenyl cyanide   **Benzamide**

Benzamide is a white crystalline solid, m.p. 130°, insoluble in cold water but soluble in hot water. It resembles aliphatic amides in its properties.

It undergoes acidic as well as alkaline hydrolyisis to give benzoic acid with the liberation of ammonia.

$$C_6H_5CONH_2 + H_2O \xrightarrow{H^+} C_6H_5COOH + NH_4^+$$

Benzamide   Benzoic acid

$$C_6H_5CONH_2 + NaOH \longrightarrow C_6H_5COO^-Na^+ + NH_3$$

Benzamide   Sodium benzoate

It, being a weak acid, reacts with sodium or sodamide in ether solution to give salt.

$$C_6H_5CONH_2 \xrightarrow[\text{ether}]{\text{Na or NaNH}_2} C_6H_5CONH^-Na^+$$

Benzamide   Sodium salt of benzamide

The salts of benzamide exist in two tautomeric forms (I and II) which form two types of ethers.

$$\underset{\textbf{I}}{\overset{\displaystyle O}{C_6H_5-\overset{\|}{C}-\overset{-}{N}H\overset{+}{N}a}} \longleftrightarrow \underset{\textbf{II}}{\overset{\displaystyle O^-}{C_6H_5-\overset{|}{C}=NH\overset{+}{N}a}}$$

The sodium salt reacts with ethyl iodide to form N-ethylbenzamide which on hydrolysis gives benzoic acid and ethyl amine.

$$C_6H_5CONH^-Na^+ \xrightarrow{C_2H_5I} C_6H_5CONHC_2H_5 \xrightarrow{H_2O/H^+} C_6H_5COOH + C_2H_5NH_2$$

                    N-Ethylbenzamide   Benzoic acid   Ethyl amine

Silver salts, on the other hand, under similar conditions give O-ethylbenzamide which upon hydrolysis yields benzoic acid, ethanol and ammonia.

$$\overset{\displaystyle O^-}{C_6H_5-\overset{|}{C}-NH\overset{+}{A}g} \xrightarrow{C_2H_5I} \overset{\displaystyle OC_2H_5}{C_6H_5-\overset{|}{C}=NH} \xrightarrow{H_2O/H^+} C_6H_5COOH + C_2H_5OH + NH_3$$

Silver salt of benzamide   O-Ethylbenzamide   Benzoic acid   Ethanol

Benzamide dehydrates to benzonitrile when heated with phosphorus pentoxide.

$$C_6H_5CONH_2 \xrightarrow[-H_2O]{P_2O_5} C_6H_5C\equiv N$$

Benzamide   Benzonitrile

36.4.4 Benzonitrile/Phenyl Cyanide [C₆H₅C≡N:]

Benzonitrile may be prepared by dehydration of benzamide. It may also be prepared by the fusion of sodium salt of benzenesulphonic acid with sodium cyanide.

$$C_6H_5CONH_2 \xrightarrow[-H_2O]{P_2O_5} C_6H_5C\equiv N$$

Benzamide   Benzonitrile

$$C_6H_5SO_3Na + NaCN \longrightarrow C_6H_5C\equiv N + Na_2SO_3$$

Sod. salt of Benzene-   Benzonitrile
sulphonic acid

It is prepared from benzenediazonium chloride by Sandmeyer reaction.

$$C_6H_5N_2Cl + CuCN \xrightarrow{\Delta} C_6H_5C\equiv N + CuCl + N_2$$

Benzenedia-
zonium chloride   **Benzonitrile**

Benzonitrile is a colorless oily liquid, b.p. 191°, having smell of bitter almonds. It resembles aliphatic nitriles in its rections. It forms benzoic acid on hydrolysis and benzyl amine on reduction.

$$C_6H_5C\equiv N \xrightarrow{H_2O/H^+} C_6H_5CONH_2 \xrightarrow{H_2O/H^+} C_6H_5COOH + NH_3$$

Benzonitrile   **Benzamide**   **Benzoic acid**

$$\downarrow LiAlH_4$$

$$C_6H_5CH_2NH_2$$

Benzyl amine

Phenyl cyanide undergoes electrophilic substitution in the benzene ring like halogenation, nitration and sulphonation but not Friedel-Crafts reaction. The –CN group is electron withdrawing deactivating group and directs the incoming substituent at the m-positions.

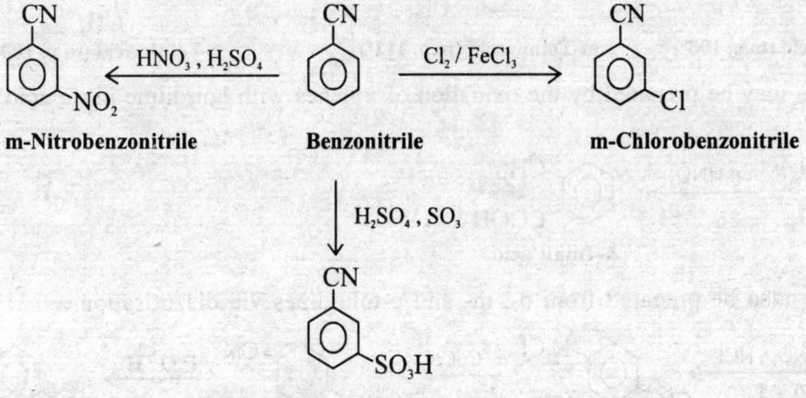

m-Nitrobenzonitrile   **Benzonitrile**   **m-Chlorobenzonitrile**

m-Cyanobenzenesulphonic acid

36.4.5 Benzoic Anhydride [C₆H₅CO)₂O]

It is prepared by heating sodium benzoate with benzoyl chloride.

$$C_6H_5COONa + C_6H_5COCl \longrightarrow (C_6H_5CO)_2O + NaCl$$

Sodium   **Benzoyl**   **Benzoic**
benzoate   **chloride**   **anhydride**

It may also be prepared by distilling a mixture of benzoic acid and acetic anhydride.

$$2C_6H_5COOH + (CH_3CO)_2O \longrightarrow (C_6H_5CO)_2O + 2CH_3COOH$$

Benzoic acid   **Acetic anhydride**   **Benzoic anhydride**   **Acetic acid**

Benzoic anhydride is a white crystalline solid, m.p. 42°. It resembles aliphatic anhydrides in its reactions. Unlike aectic anhydride, it hydrolyses slowly to benzoic acid.

It can replace benzoyl chloride in the benzoylation reactions involving phenols and amines.

36.4.6 Benzoyl Peroxide [$C_6H_5CO-O-O-COC_6H_5$]

It may be prepared by the reaction of benzoyl chloride with sodium peroxide.

$$2C_6H_5COCl + Na_2O_2 \xrightarrow{-NaCl} C_6H_5CO-O-O-COC_6H_5$$

Benzoyl chloride   Sodium   Benzoyl peroxide
                        peroxide

It may also be prepared by the reaction of benzoyl chloride and hydrogen peroxide in alkaline medium.

$$2C_6H_5COCl + H_2O_2 + 2NaOH \longrightarrow C_6H_5CO-O-O-COC_6H_5 + 2NaCl + 2H_2O$$

Benzoyl chloride   Benzoyl peroxide

Benzoyl peroxide is a colorless solid, m.p. 104°. It is a harmless bleaching agent, hence used for the bleaching of edible oils and fats. It is often used as initiator in varius polymerisation reactions.

36.4.7 Toluic Acids [$CH_3C_6H_4COOH$]

The three isomeric toluic acids are:

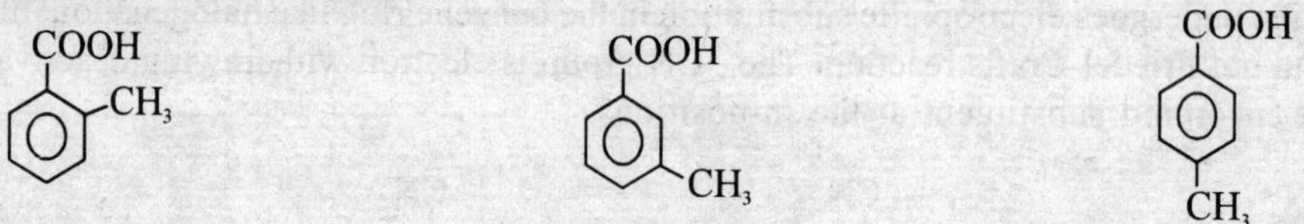

o-Toluic acid (m.p. 105°)   m-Toluic acid (m.p. 111°)   p-Toluic acid (m.p. 180°)

These toluic acids may be prepared by the oxidation of xylenes with hot dilute nitric acid. o-Xylene gives o-toluic acid.

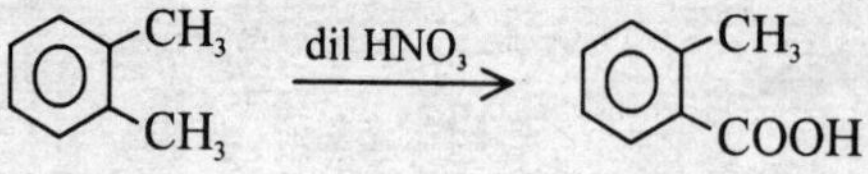

o-Xylene   o-Toluic acid

Toluic acids may also be prepared from o-, m- and p-toluidines via diazotisation.

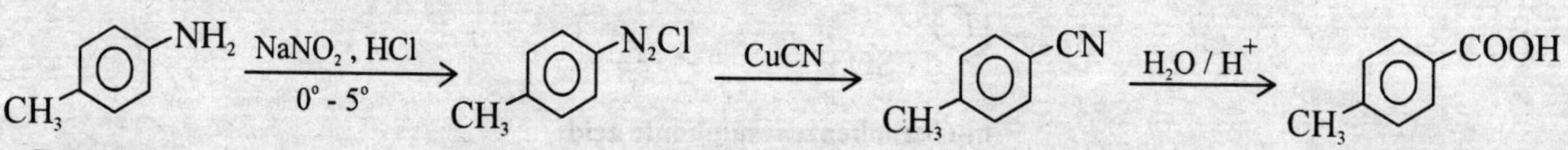

p-Toluidine   p-Toluic acid

Toluic acids resemble benzoic acid in their chemical behaviour. Toluic acids are slightly less acidic than benzoic acid (see Table 36.1) due to the inductive effect of methyl group.

36.4.8 Aminobenzoic Acids [$NH_2C_6H_4COOH$]

Aminobenzoic acids may be prepared by the reduction of corresponding nitrobenzoic acids.

Anthranilic acid/ o-Aminobenzoic Acid

Anthanilic acid is prepared commercially by the oxidation of phthalimide with sodium hypochlorite and sodium hydroxide (Hofmann reaction).

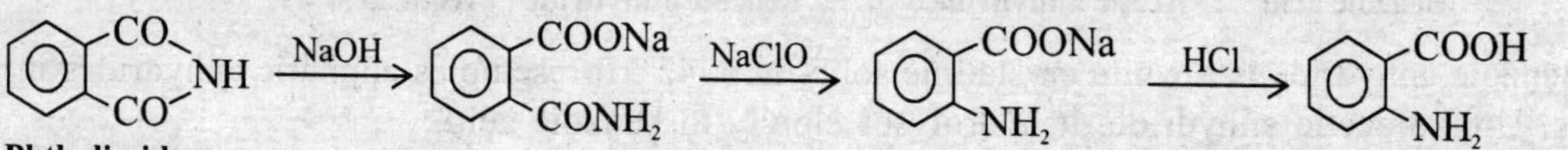

Phthalimide   Anthranilic acid

It is a white crystalline solid, m.p. 145°, soluble in water, alcohol and ether. It gives usual reactions of amines as well as acids. It form aniline when heated with the evolution of carbon dioxide.

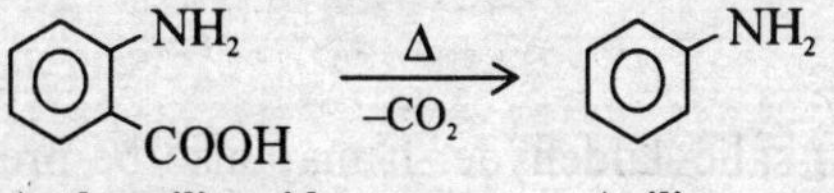

Anthranilic acid   Aniline

Uses: Anthranilic acid is used:
1. In the manufacture of indigo, a dye.
2. In perfume industry methyl anthranilate is used to imitate jasmine and orange.
3. In the manufacture of saccharin.

p-Aminobenzoic Acid (PABA)

It is obtained by the reduction of p-nitrobenzoic acid which, in turn, is produced by the oxidation of p-nitrotoluene.

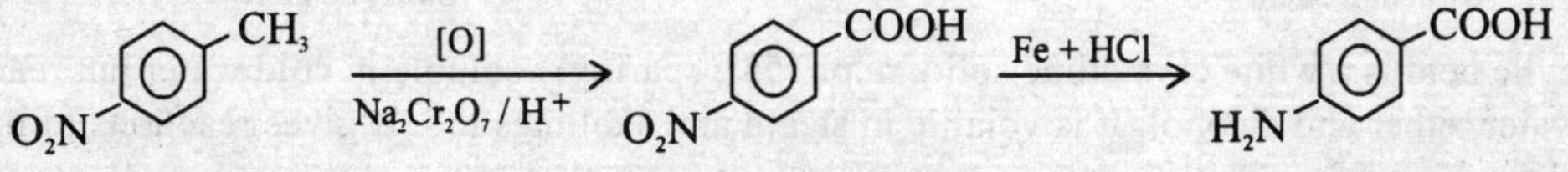

p-Nitrotoluene   p-Nitrobenzoic acid   p-Aminobenzoic acid

It is a white crystalline solid, m.p. 186°, soluble in water. It resembles anthranilic acid in its chemical properties and gives reactions of both amino and carboxyl groups.

It is used in the synthesis of local anaesthetics like benzocaine and novocaine. It is also a component of vitamin B-complex and is an essential growth factor, antigrey hair factor and used to cure rocky mountain spotty fever.

36.4.9   Phenolic Acids: Salicylic Acid/o-Hydroxybenzoic acid

It occurs in oil of winter green in the form of its methyl ester. It may be prepared by following methods.

1. *By Kolbe's Reaction:* Sodium phenoxide on heating with carbon dioxide under pressure forms salicylic acid.

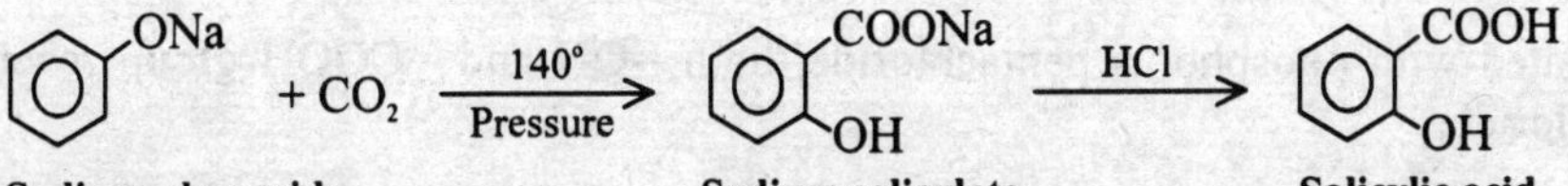

Sodium phenoxide   Sodium salicylate   Salicylic acid

A small amount of p-hydroxybenzoic acid is also formed during the reaction. The amount of p-isomer increases with the increase in temperature. Potassium phenoxide yields p-isomer as the major product. The mechanism of Kolbe's reaction involves weak electrophile carbon dioxide.

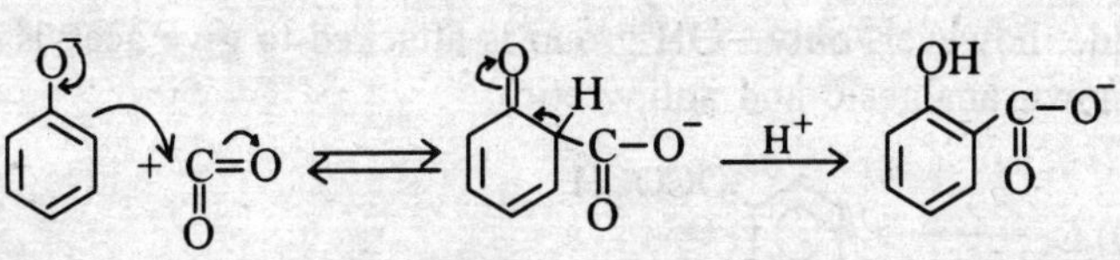

2. *By Riemer Tiemann Reaction:* Phenol reacts with carbon tetrachloride in alkaline solution to from salicylic acid along with some p-isomer. The two isomers are separated by steam distillation since salicylic acid is steam volatile and the p-isomer is not.

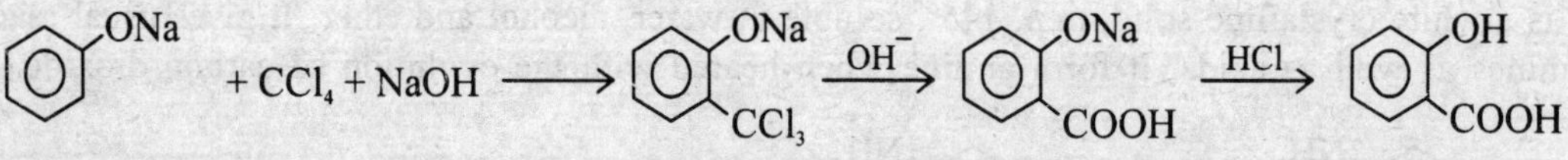

[Reaction scheme: Sodium phenoxide + CCl$_4$ + NaOH → intermediate with ONa and CCl$_3$ →$^{OH^-}$ intermediate with ONa and COOH →^{HCl} Salicylic acid with OH and COOH]

Sodium phenoxide   **Salicylic acid**

3. Salicylic acid may be prepared by oxidation of salicylaldehyde. It may also be prepared from anthranilic acid.

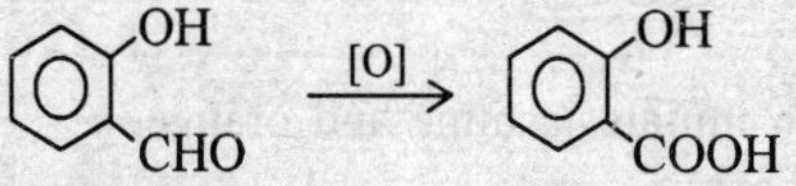

[Reaction scheme: Salicylaldehyde (OH, CHO) →$^{[O]}$ Salicylic acid (OH, COOH)]

Salicylaldehyde   **Salicylic acid**

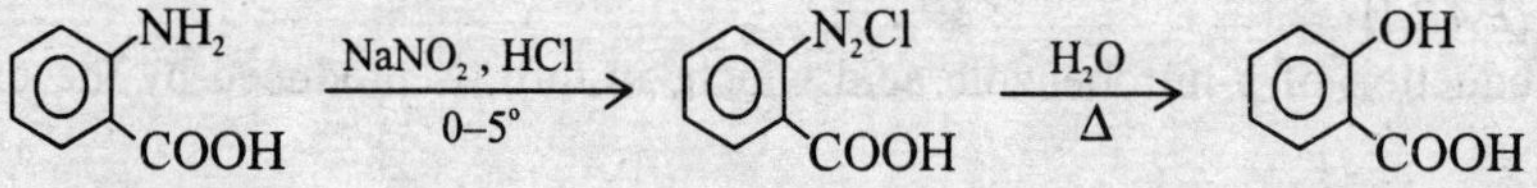

[Reaction scheme: Anthranilic acid (NH$_2$, COOH) →$^{NaNO_2, HCl}_{0-5°}$ intermediate (N$_2$Cl, COOH) →$^{H_2O}_{\Delta}$ Salicylic acid (OH, COOH)]

Anthranlic acid   **Salicylic acid**

Salicylic acid is a white crystalline solid, m.p. 155°, sparingly soluble in cold water but readily soluble in hot water, ether and alcohol. It is volatile in steam and sublimes too. It gives reactions of both phenols and carboxylic acids.

(a) Like phenols it gives violet color with ferric chloride.

(b) It reacts with sodium carbonate and bicarbonate to form its sodium salt along with the evolution of carbon dioxide. With sodium hydroxide, however, a disodium salt is obtained.

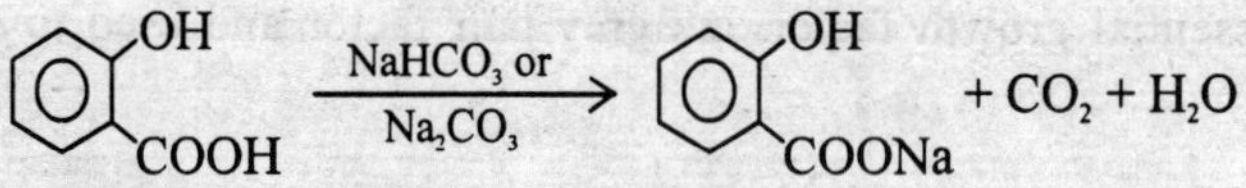

[Reaction scheme: Salicylic acid (OH, COOH) →$^{NaHCO_3 \text{ or}}_{Na_2CO_3}$ Sodium salicylate (OH, COONa) + CO$_2$ + H$_2$O]

Sàlicylic acid   **Sodium salicylate**

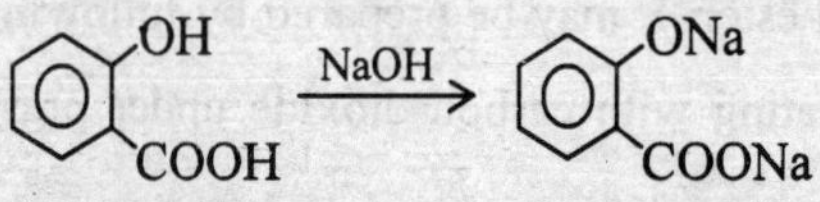

[Reaction scheme: Salicylic acid (OH, COOH) →^{NaOH} Disodium salt (ONa, COONa)]

Salicylic acid   **Disodium salt**

(c) When treated with phosphorus pentachloride, both −OH and −COOH group react to give o-chlorobenzoyl chloride.

[Reaction scheme: Salicylic acid (OH, COOH) →^{PCl_5} o-Chlorobenzoyl chloride (Cl, COCl)]

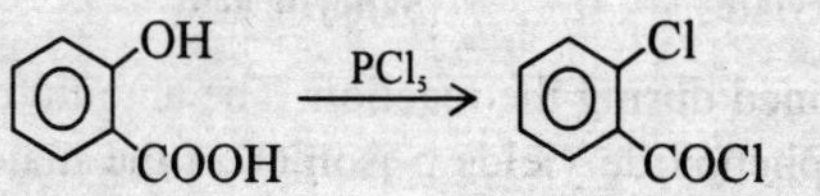

Salicylic acid   **o-Chlorobenzoyl chloride**

With acid chloride and anhydride, however, only −OH group is attacked to give acetylsalicylic acid (also known as aspirin; a well known analgesic and antipyretic)

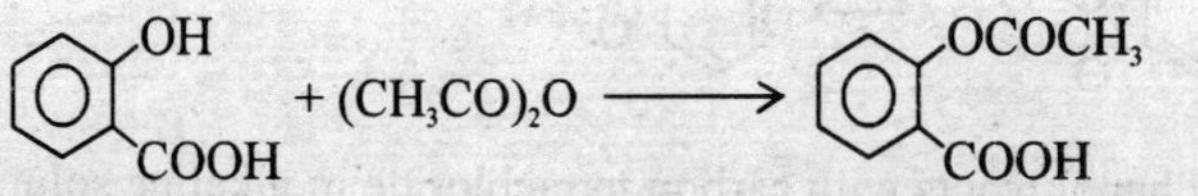

[Reaction scheme: Salicylic acid (OH, COOH) + (CH$_3$CO)$_2$O → Acetylsalicylic acid (OCOCH$_3$, COOH)]

Salicylic acid   **Acetylsalicylic acid**

(d)   With alcohols, in presence of acid salicylic acid forms esters.

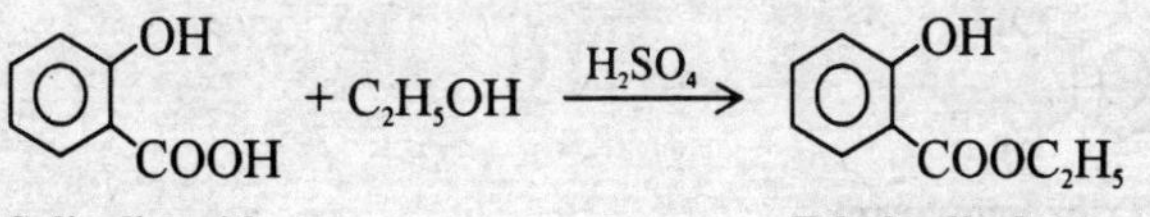

| **Salicylic acid** | | **Ethyl salicylate** |

(e)   Sodium salicylate undergoes decarboxylation when heated with soda lime.

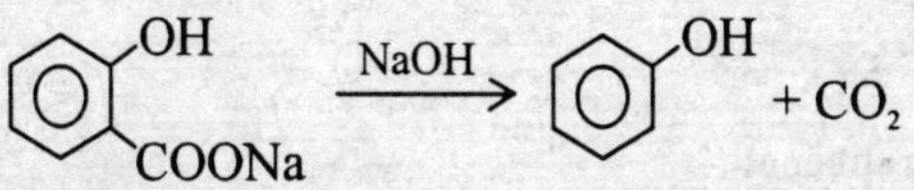

Sodium salicylate   **Phenol**

When heated alone above 200°, it forms phenyl salicylate.

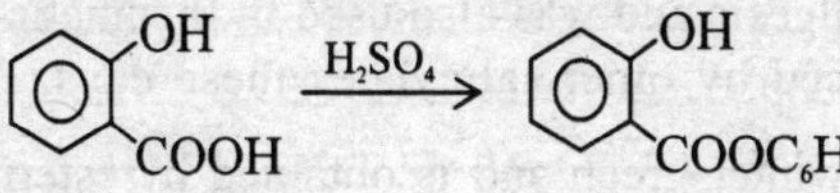

Salicylic acid   **Phenyl salicylate**

When potassium salicylate is heated above 200°, it rearranges to yield p-hydroxybenzoic acid.

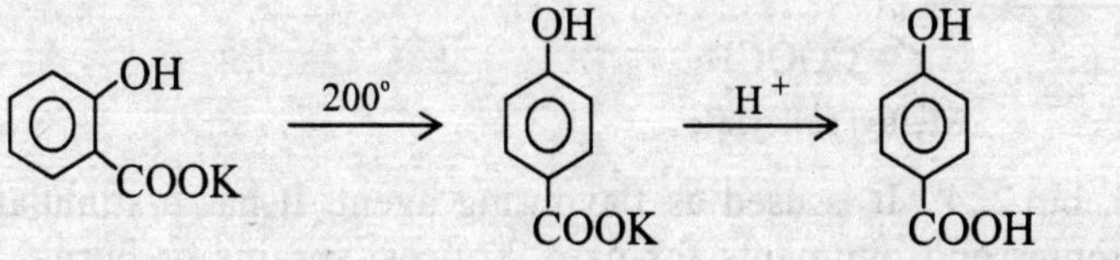

Potassium salicylate   **p-Hydroxybenzoic acid**

(f)   Like phenols, it couples with diazonium salts to form azodyes.

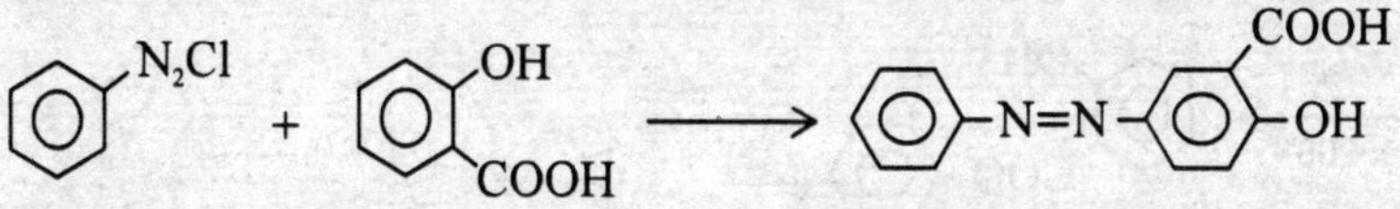

Benzene-   Salicylate acid   3-Carboxy-4-hydroxyazobenzene
diazonium chloride

(g)   When reduced with sodium and isoamyl alcohol, it is hydrogenated to pimelic acid.

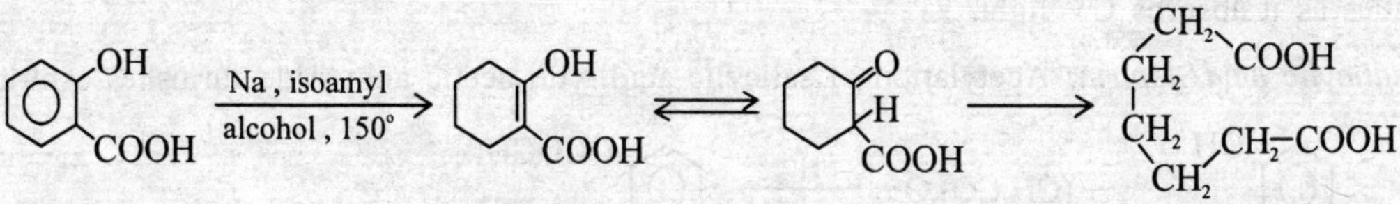

Salicylic acid   **Pimelic acid**

(h)   Salicylic acid undergoes electrophilic substitution reactions to form o- and p-substituted phenols, the carboxyl group is replaced by the incoming groups. It form 2,4,6- tribromo and 2,4,6-trinitrophenol on bromination and nitration respectively.

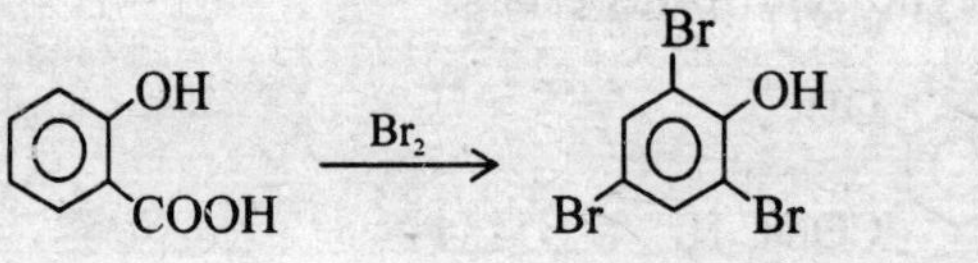

2,4,6-Tribromophenol

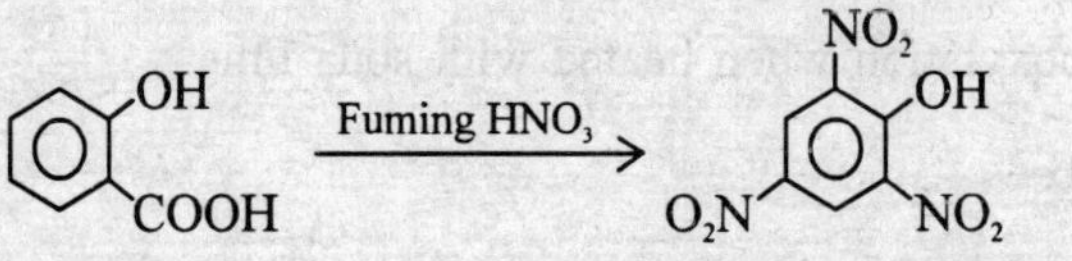

Salicylic acid   **2,4,6-Trinitrophenol**

This reaction is used to estimate salicylic acid quantitatively. Salicylic acid and its derivatives are used in the preparation of azodyes, antiseptics, analgesics, antipyretics and also in perfume industry.

Sodium salicylate: It is an effective antipyretic and analgesic and was also used in rheumatic fever. But due to its stomach irritating effect, it has been replaced by other salicylates these days.

Methyl salicylate: It is the chief component of oil of winter green and is obtained by esterification of salicylic acid.

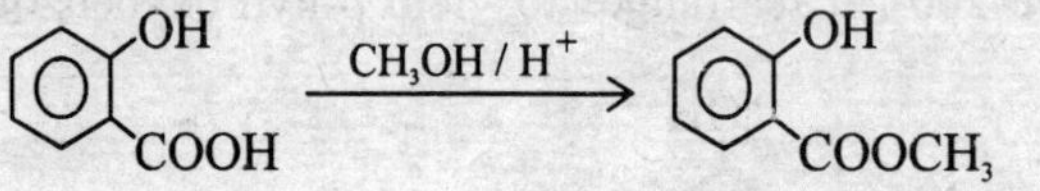

Salicylic acid   **Methyl salicylate**

It is a pleasant smelling liquid, b.p.224°. It is used as flavouring agent. It has a stimulating action on skin, therefore, used in hair tonics and ointments for pain, bruises, sprains or burns.

Phenyl salicylate/Salol: Phenyl salicylate is produced by heating salicylic acid in a sealed tube above 200°. The reaction possibly produces phenol initially which reacts with salicylic acid to give phenyl salicylate.

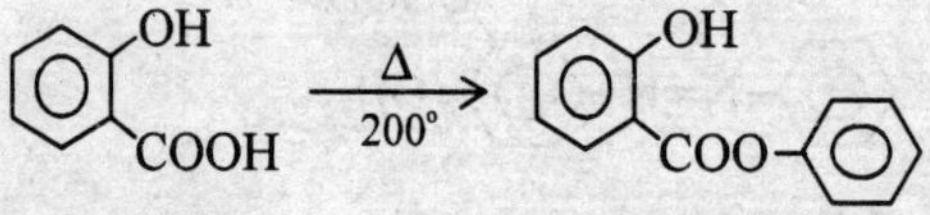

Salicylic acid   **Phenyl salicylate**

It is a white solid, m.p. 42°. It is used as an internal antiseptic under the name of salol as it passes through the acidic medium of stomach unchanged. It also finds use as an sun screening agent and stabiliser of plastics as it absorbs UV light.

Acetylsalicylic acid/Aspirin: Acetylation of salicylic acid with acetic anhydride furnishes aspirin.

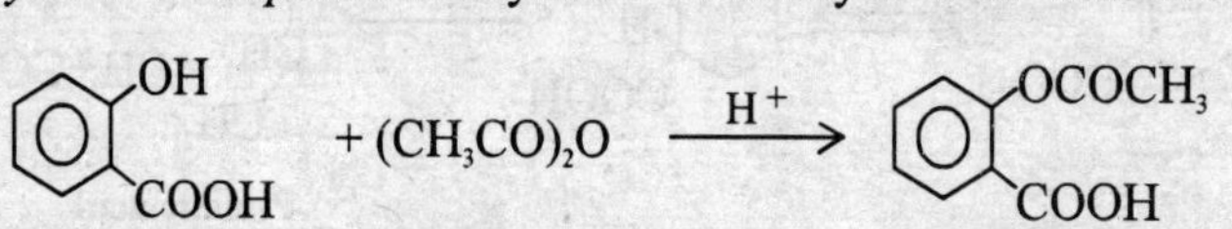

Salicylic acid   **Acetic anhydride**   **Acetylsalicylic acid**

It is a white, crystalline solid, m.p. 134°. It is an important ingradient of varius pain relieving formulations. Like salol, it passes through the acidic medium of somach unchanged and is hydrolysed in alkaline medium of intestine.

36.4.10   Gallic Acid/3,4,5-Trihydroxybenzoic Acid

It occurs in free state in tea, oak bark and many other plants. It is prepared by the acidic hydrolysis of tannin present in nut-galls. Tannin or tanic acid is a mixture of gallic acid esters of glucose.

It may also be prepared from 4-bromo-3,5-dihydroxybenzoic acid or 3-bromo-4,5-dihydroxy benzoic acid by alkali fusion.

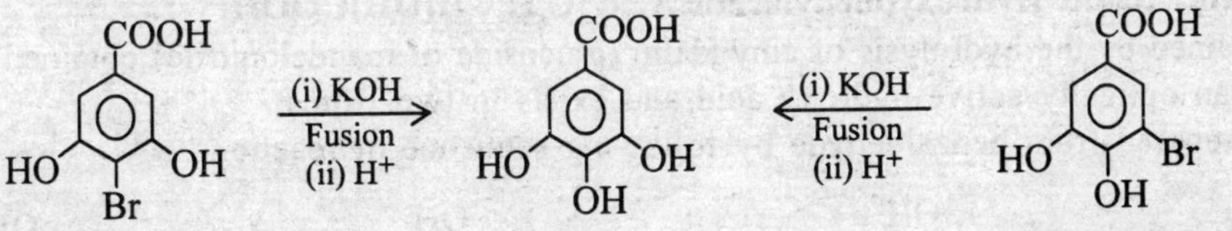

| 4-Bromo-3,5-dihydroxybenzoic acid | Gallic acid | 3-Bromo-4,5-dihydroxybenzoic acid |

It is a white solid, m.p. 252°, readily soluble in water and alcohol. it exhibits reactions of both phenol and carboxylic acid. On heating it forms pyrogallol with the evolution of carbon dioxide.

It is mainly used for the preparation of blue-black ink which is a mixture of gallic acid, ferrous sulphate and a blue dye in water. Small amounts of additives such a gum (protective agent) sulphuric acid(to slow down oxidation of ferrous sulphate), phenol (preservative)are also added in the above preparation.

It is also used in the manufecture of dyes, pyrogallol and as a photographic developer.

36.5   MONOBASIC ACIDS WITH CARBOXYL GROUPS IN SIDE CHAIN

36.5.1   Phenylacetic Acid [$C_6H_5CH_2COOH$]

It is ismoeric with toluic acids. It may be prepared by heating benzyl chloride and alcoholic potassium cyanide followed by the hydrolysis of benzyl cyanide.

$$C_6H_5CH_2Cl \xrightarrow[\Delta]{KCN} C_6H_5CH_2CN \xrightarrow{H_2O/H^+} C_6H_5CH_2COOH$$

| Benzyl chloride | Benzyl cyanide | Phenylacetic acid |

It may also be produced by reducing mandelonitrile with phosphorus and hydriodic acid.

$$C_6H_5CHO \xrightarrow{HCN} C_6H_5\overset{OH}{\underset{}{CH}}-CN \xrightarrow{P/HI} C_6H_5CH_2COOH$$

| Benzaldehyde | Mandelonitrile | Phenylacetic acid |

It is a white crystalline solid, m.p. 86°, soluble in water. It is a stronger acid than acetic acid due to electron withdrawing effect of phenyl group. It exhibits the reactions of carboxylic acids as well as the side chain. the product of halogenation in presence and absence of halogen carrier are as follows:

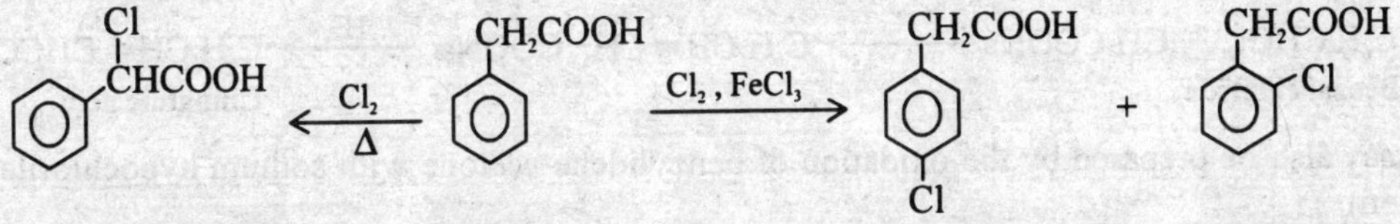

| α-Chlorophenylacetic acid | Phenylacetic acid | p-Chlorophenyl-acetic acid | o-Chlorophenylacetic acid |

It is oxidised to benzoic acid when treated with chromic acid.

$$C_6H_5CH_2COOH \xrightarrow{Cr_2O_7^-/H^+} C_6H_5COOH$$

Phenylacetic acid   **Benzoic acid**

It is used in organic synthesis.

36.5.2 Mandelic Acid/α-Hydroxyphenylacetic Acid [C₆H₅CH(OH)COOH]

It was first obtained by the hydrolysis of amygdalin (glucoside of mandelonitrile) obtained from bitter almonds. It is an optically active hydroxy acid and exists in two froms.

It may be prepared from benzaldehyde by following sequence of reactions.

$$C_6H_5CHO \xrightarrow{\cdot NaHSO_3} C_6H_5-\underset{\underset{OH}{|}}{CH}-SO_3Na \xrightarrow{NaCN} C_6H_5-\underset{\underset{OH}{|}}{CH}-CN \xrightarrow{H_2O/H^+} C_6H_5-\underset{\underset{OH}{|}}{CH}-COOH$$

Benzaldehyde   **Mandelic acid**

It is a white crystalline solid, soluble in water. The acid obtained from amygdalin is laevorotatory, m.p. 133° wheareas the acid obtained from synthetic methods is a recemic mixture, m.p. 118°. It exhitbits the reactions of α-hydroxy acids. When reduced with hydriodic acid and phosphorus, it forms phenyl-acetic acid.

$$C_6H_5-\underset{\underset{OH}{|}}{CH}-COOH \xrightarrow{HI+P} C_6H_5-CH_2-COOH$$

Mandellic acid   **Phenylacetic acid**

It is used as an internal antiseptic in the treatment of urinary infections. p-Bromomandelic acid is used to determine the ratio of hafnium to zirconium in their mixture.

36.5.3 Cinnamic Acid/β-Phenylacrylic Acid

β-Phenylacrylic acid exhibits geometical isomerism due to restricted rotation along the double bond. The two isomeric forms are as follows:

$$\begin{array}{c} C_6H_5-C-H \\ \| \\ H-C-COOH \end{array} \qquad\qquad \begin{array}{c} C_6H_5-C-H \\ \| \\ HOOC-C-H \end{array}$$

trans form, Cinnamic acid   **cis form, Allocinnamic acid**

Cinnamic acid, the trans-β-phenylacrylic acid occurs free or in the form of its ester in oil of Cinnamon, resins and balsams.

Cinnamic acid may be prepared by Perkin's reaction, Claisen reaction, Knovenagel reaction starting with benzaldehyde as discussed in Sec. 35.1.2.

It is manufectured by heating benzal chloride with sodium acetate.

$$C_6H_5CHCl_2 + CH_3COONa \longrightarrow C_6H_5CH=CH-COONa \xrightarrow{H^+} C_6H_5CH=CHCOOH$$

Benzal chloride   **Cinnamic acid**

It may also be prepared by the oxidation of benzylidene acetone with sodium hypochlorite (haloform reaction).

$$C_6H_5CHO + CH_3COCH_3 \longrightarrow C_6H_5CH=CHCOCH_3 \xrightarrow{NaOCl} C_6H_5CH=CHCOOH$$

Benzaldehyde Acetone   **Benzylideneacetone**   **Cinnamic acid**

It is a white crystalline solid, m.p. 133°, sparingly soluble in water but readily soluble in alcohol. It exhibits the reactions of both carboxyl group as well as the double bond.

1. *Reactions of Double Bond:* It adds hydrogen and bromine at the site of the double bond.

$$C_6H_5CHBrCHBrCOOH \xleftarrow{\text{Br}_2} C_6H_5CH=CHCOOH \xrightarrow{\text{H}_2/\text{Ni}} C_6H_5CH_2CH_2COOH$$

β-Phenyl-α,β-dibormo-   **Cinnamic acid**   **β-Phenylpropanoic acid**
propanoic acid

It forms benzaldehyde on oxidation with chromic acid. Strong oxidising agents convert it into benzoic acid.

$$C_6H_5CH=CHCOOH \xrightarrow[\text{[O]}]{\text{CrO}_3} C_6H_5CHO \xrightarrow[\text{[O]}]{\text{alk.KMnO}_4} C_6H_5COOH$$

Cinnamic acid   **Benzaldehyde**   **Benzoic acid**

On ozonolysis, it yields benzaldehyde and glyoxalic acid.

$$C_6H_5CH=CHCOOH \xrightarrow{O_3} \left[\begin{array}{c} O-O \\ \mid \quad \mid \\ C_6H_5CH \quad CHCOOH \\ \diagdown O \diagup \end{array} \right] \longrightarrow C_6H_5CHO + \begin{array}{c} CHO \\ \mid \\ COOH \end{array}$$

Cinnamic acid   **Ozonide**   **Benzaldehyde Glyoxalic acid**

With osmium tetraoxide, it gives 1,2-dihydroxy derivative which on further oxidation gives benzaldehyde and glyoxalic acid.

$$C_6H_5CH=CHCOOH \xrightarrow{\text{OsO}_4} \underset{\underset{\text{OH OH}}{\mid \quad \mid}}{C_6H_5CH-CH-COOH} \xrightarrow{\text{NaIO}_4} C_6H_5CHO + \begin{array}{c} CHO \\ \mid \\ COOH \end{array}$$

Cinnamic acid   **Benzaldehye Glyoxalic acid**

When exposed to direct sunlight, it undergoes dimerisation to form truxinic acid and truxillic acid.

$$C_6H_5CH=CHCOOH \xrightarrow{\text{hv}} \begin{array}{c} C_6H_5-CH-CH-COOH \\ \mid \quad \mid \\ C_6H_5-CH-CH-COOH \end{array} + \begin{array}{c} C_6H_5-CH-CH-COOH \\ \mid \quad \mid \\ HOOC-CH-CH-C_6H_5 \end{array}$$

Cinnamic acid   **Truxinic acid**   **Truxillic acid**

2. *Reaction of Carboxyl Group:* Cinnamic acid undergoes decarboxylation when heated alone to form styrene.

$$C_6H_5CH=CHCOOH \xrightarrow{\Delta} C_6H_5CH=CH_2 + CO_2$$

Cinnamic acid   **Styrene**

It is reduced to cinnamyl alcohol when treated with lithium aluminum hydride at low tempperature in ether solution.

$$C_6H_5CH=CHCOOH \xrightarrow{\text{LiAlH}_4} C_6H_5CH=CHCH_2OH$$

Cinnamic acid   **Cinnamyl alcohol**

At higher temperature, however, γ-phenylpropanol is the major product.

3. *Reactions of Benzene Ring:* Cinnamic acid undergoes electrophilic substitution directing the incoming groups at o- and p-positions with respect to —CH=CHCOOH group. The rate of reaction, however,

is slow due to electron withdrawing effect of carboxyl group. It undergoes nitration, halogenation and sulphonation to from o- and p-substituted cinnamic acid of which p-isomer is the major product.

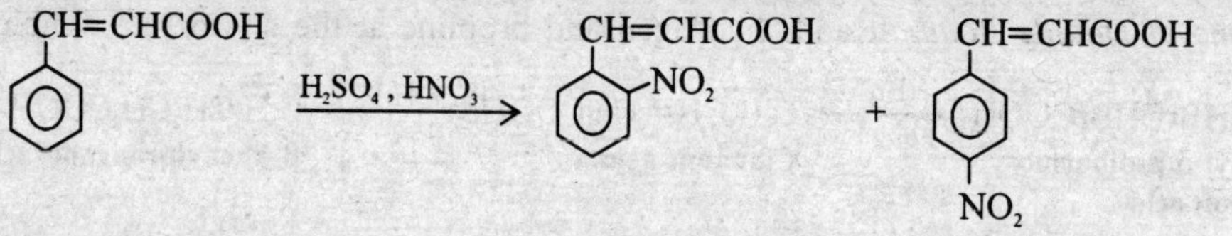

Cinnamic acid	**o-Nitrocinnamic acid** **(minor)**	**p-Nitrocinnamic acid** **(major)**

36.5.4 Coumarin

It may be prepared from salicylaldehyde by Perkin's reaction.

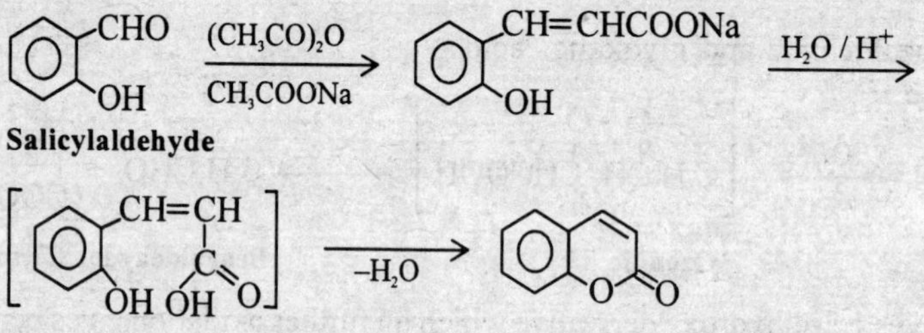

Salicylaldehyde

Coumarin (γ-lactone of coumarinic acid)

It is a white crystalline solid, m.p. 67°. It is used as perfume and as flavouring agent.

36.6 DICARBOXYLIC ACIDS

The three isomeric benzene dicarboxylic acids are:

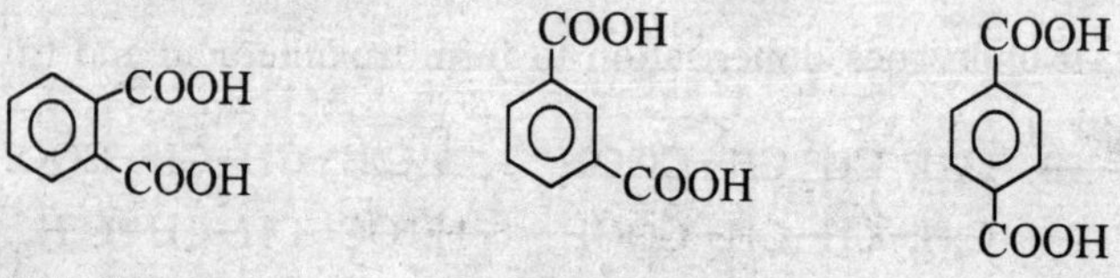

Phthalic acid	**Isophthalic acid**	**Terephthalic acid**

They may be prepared by the oxidation of corresponding xylenes or toluic acids.

36.6.1 Phthalic Acid/Benzene-1,2-dicarboxylic Acid

It is manufactured by catalytic oxidation of naphthalene or o-xylene.

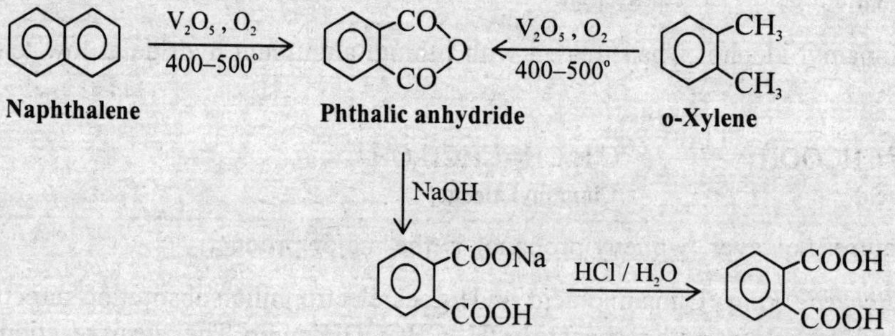

Naphthalene	**Phthalic anhydride**	**o-Xylene**

Phthalic acid

It is a white crystalline solid, m.p. 231°, insoluble in cold water but soluble in hot water. It gives typical reactions of carboxylic acids. When heated alone, it readily eliminates a water molecule to form phthalic anhydride. When heated with sodalime, however, it forms benzoic acid and then benzene by decarboxylation.

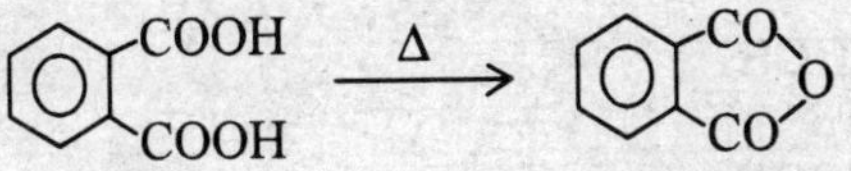

Phthalic acid   Phthalic anhydride

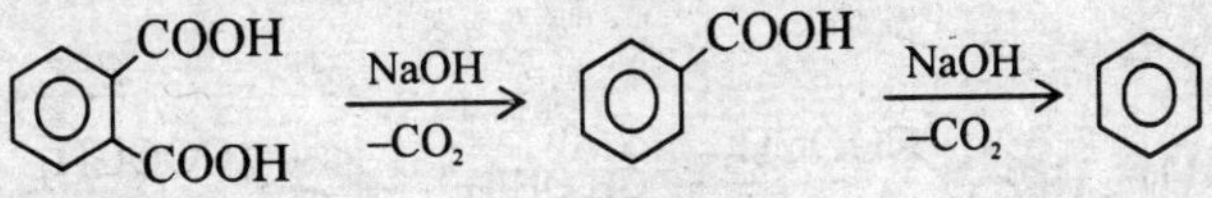

Phthalic acid   Benzoic acid   Benzene

It forms monosodium and disodium phthalate when treated with alkalis.

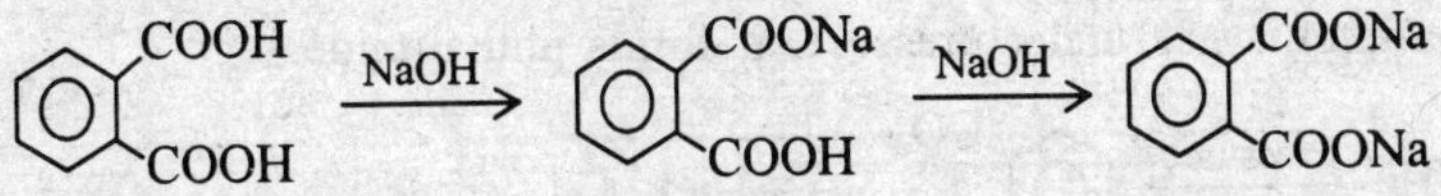

Phthalic acid   Monosodium phthalate   Disodium phthalate

When treated with ethanol in presence of mineral acid, it froms monoethyl phthalate and diethyl phthalate.

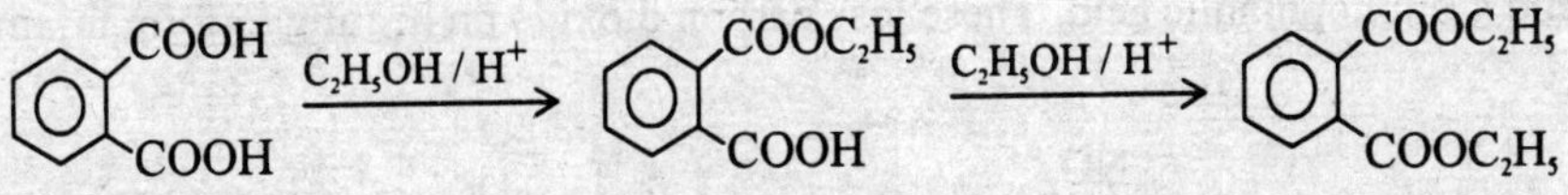

Phthalic acid   Monoethyl phthalate   Diethyl phthalate

When heated with ammonia, it forms phthalimide.

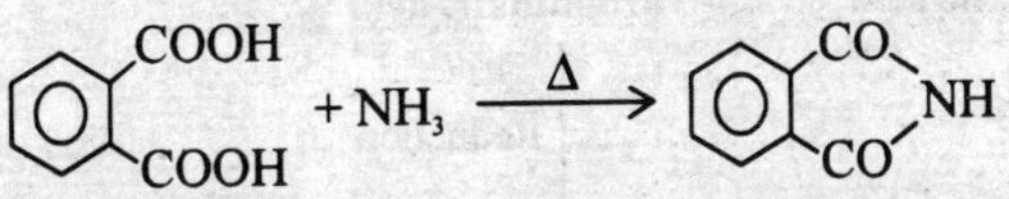

Phthalic acid   Phthalimide

Phthalic acid is used in the manufacture of plastics. It is also used in various dye preparations.

36.6.2 Isophthalic Acid/Benzene-1,3-dicarboxylic Acid

It may be prepared by oxidation of m-xylene or m-toluic acid with alkaline potassium permanganate.

It is a crystalline solid, m.p. 346°. It resembles phthalic acid in its reactions except that it does not form anhydride.

It is used in the manufecture of alkyd resins.

36.6.3 Terephthalic Acid/Benzene-1,4-dicarboxylic Acid

It is obtained by the oxidation of p-xylene or p-toluic acid by air in liquid phase in presence of cobalt or manganese salts as catalyst. It is generally isolated as its dimethyl ester.

It sublimes without melting. It is extensively used in the manufecture of polyester as it polymerises with ethylene glycol to give polyethylene terphthalate (PET).

36.6.4   Phthalic Anhydride

It is manufactured by thermal dehydration of phthalic acid obtained by the oxidation of naphthalene.

It is a white solid, m.p. 128°, insoluble in water. It resembles anhydrides of aliphatic dicarboxylic acids in its reactions. It hydrolyses slowly with hot water but rapidly with acid or alkali to phthalic acid.

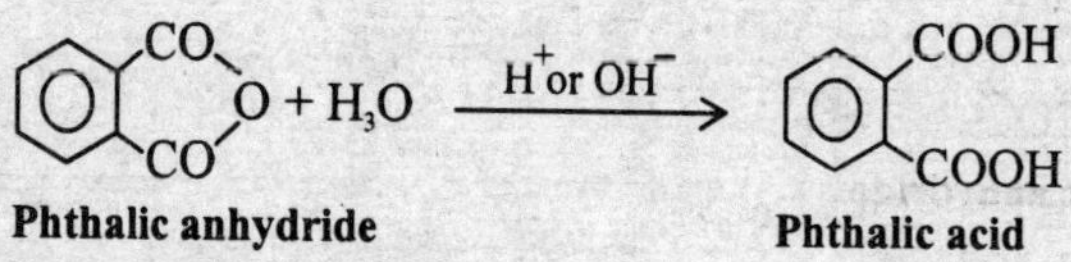

Phthalic anhydride   Phthalic acid

It forms esters with alcohols

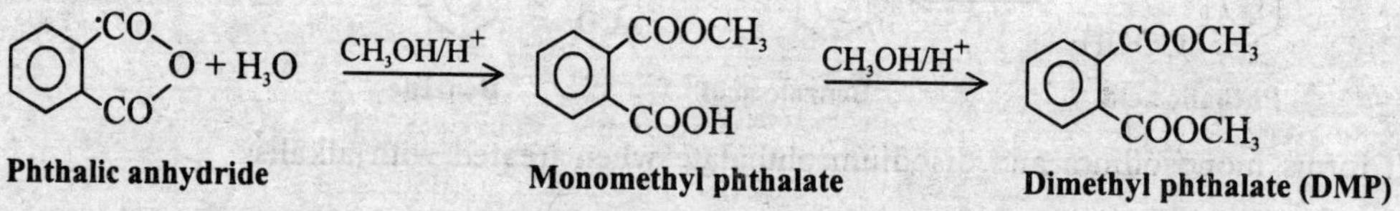

Phthalic anhydride   Monomethyl phthalate   Dimethyl phthalate (DMP)

When heated with dry ammonia gas under pressure, it forms phthalimide.

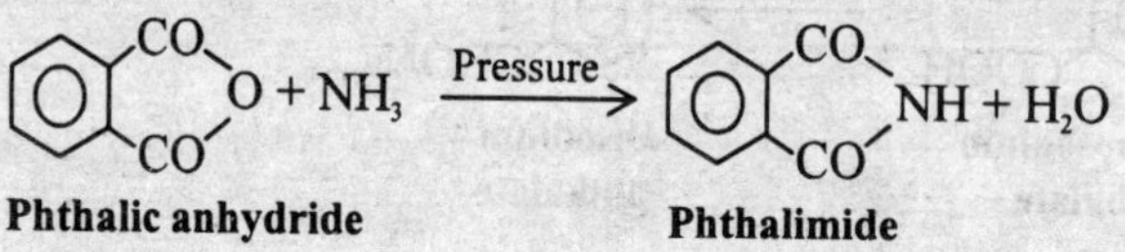

Phthalic anhydride   Phthalimide

Phthalic anhydride undergoes nitration to form 3-nitro- and 4-nitrophthalic acid which on reduction forms 3-amino- and 4-aminophthalic acid. These lose carbon dioxide on heating to form m-aminobenzoic acid.

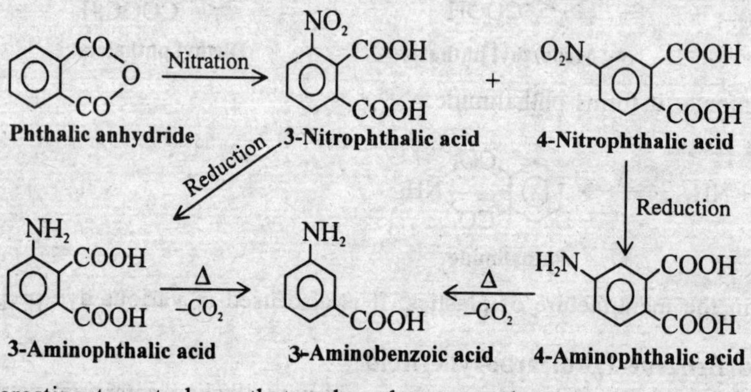

Phthalic anhydride   3-Nitrophthalic acid   4-Nitrophthalic acid

3-Aminophthalic acid   3-Aminobenzoic acid   4-Aminophthalic acid

It is interesting to note here, that carboxyl group only ortho or para to amino group is eliminated during the reaction.

Phthalic anhydride condenses with benzene in presence of aluminum chloride to give o-benzoylbenzoic acid which on cyclisation forms anthraquinone.

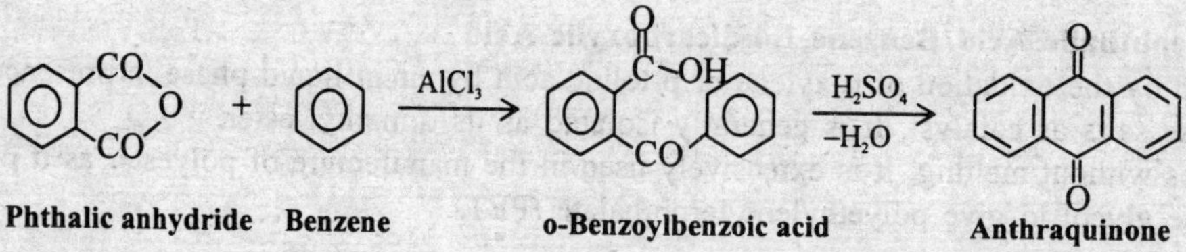

Phthalic anhydride   Benzene   o-Benzoylbenzoic acid   Anthraquinone

It condenses with phenol in presence of acid to give phenolphthalein and with resorcenol to give fluorescein.

It reacts with glycerol to form glyptal type resin, which further reacts to form three dimentional cross linked alkyd resin used in drying enamels and wall paints.

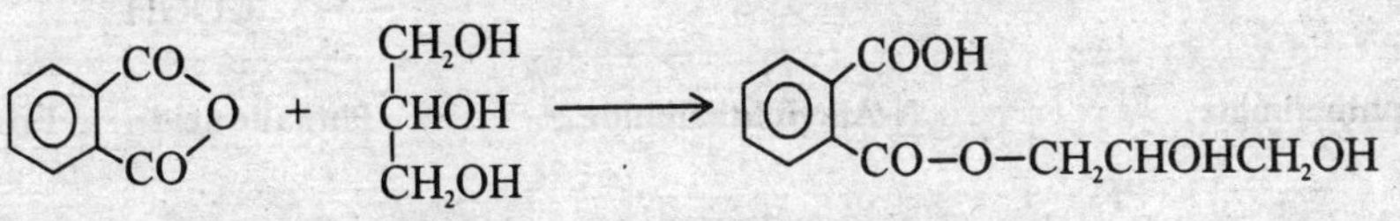

Phthalic anhydride **Glycerol**   **Glyptal type resin**

It reacts with phosphorus pentachloride to form phthalyl chloride

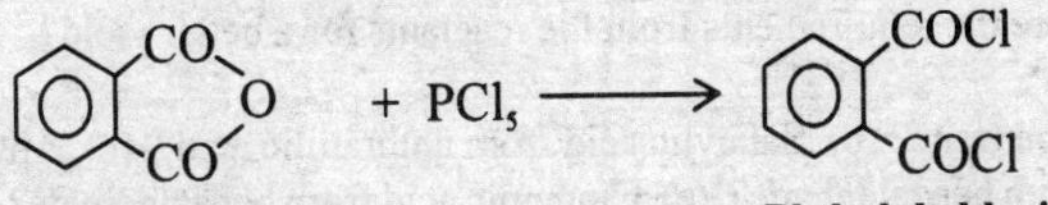

Phthalic anhydride   **Phthalyl chloride**

It forms phthalide when reduced with zinc dust and alkali.

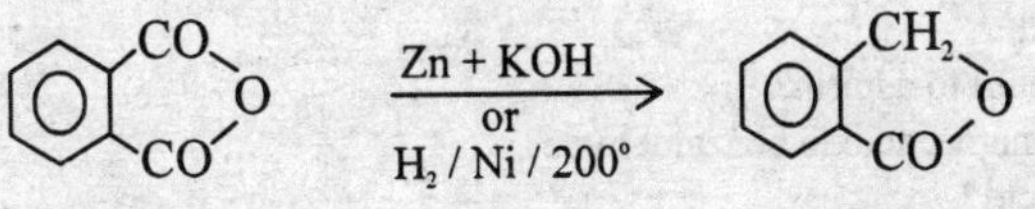

Phthalic anhydride   **Phthalide**

Uses: Phthalic anhydride is used in the :
1. manufacture of esters which in turn are used as plasticisers and insect repellants.
2. manufacture of polyester resins (glyptal).
3. preparation of anthraquinone which is used for various dye preparations.
4. organic synthesis.

36.6.5 Phthalimide

It is prepared by heating phthalic acid or phthalic anhydride with ammonia at $200°$ under pressure.

It is a white solid, m.p. $238°$, insoluble in water. It is weakly acidic in nature and forms salt with potassium hydroxide.

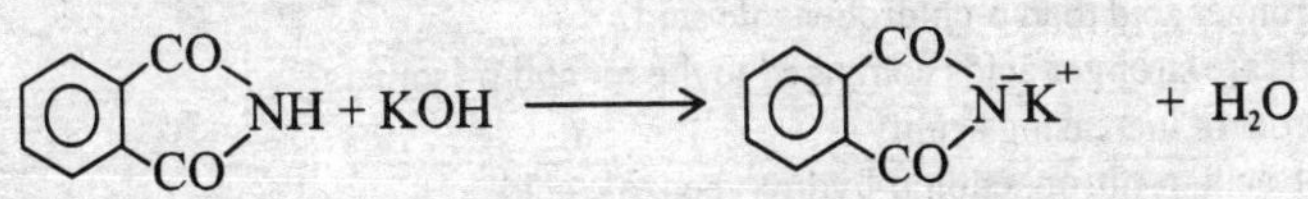

Phthalimide   **Potassium phthalimide**

It forms phthalic acid on alkaline hydrolysis. When oxidised with alkaline sodium hypochlorite, it gives anthranilic acid.

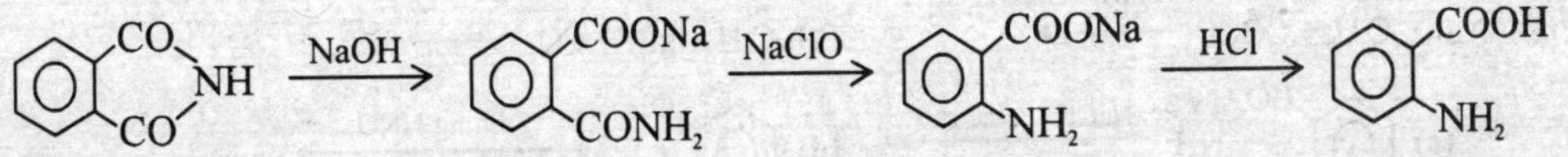

Phthalimide   **Anthranilic acid**

It is used in the preparation of primary amines (Gabriel's phthalimide synthesis) and phthalic acid.

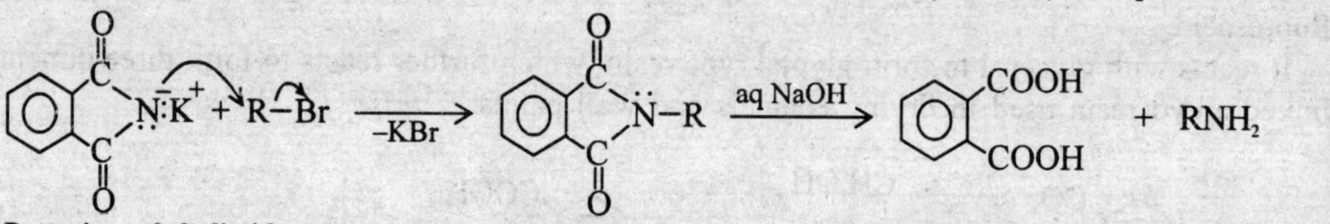

| Potassium phthalimide | N-Alkylphthalimide | Phthalic acid | Primary amine |

36.7 PROBLEMS

1. Discuss various methods of preparation of aromatic carboxylic acids. Starting from benzoic acid how will you prepare:
   (a) Benzoic anhydride   (b) Benzonitrile   (c) Benzamide   (d) Methyl benzoate
2. What is Schotten Bauman reaction ? What are the requirements from the reactants for a better yield?
3. How will you prepare:
   (a) Benzoic acid from phosgene and bromobenzene? (b) Salicylic acid from anthranilic acid? (c) Anthranilic acid from o-xylene? (d) Benzoyl chloride from benzaldehyde? (e) Cinnamic acid from benzaldehyde?
4. How will you distinguish between:
   (a) Phthalic acid and terephthalic acid?   (b) o-Chlorobenzoic acid and benzoyl chloride?
   (c) Benzoic acid and benzamide?   (d) Toluic acid and phenylacetic acid?
5. How are following conversions accomplished.
   (a) Salicylic acid to benzene?   (b) Phenol to aspirin?
   (c) Salicylaldehyde to coumarin?   (d) Cinnamic acid to benzaldehyde?
   (e) Phthalic anhydride to m-aminobenzoic acid?
6. Discuss synthesis and uses of the following:
   (a) Aspirin (b) Methyl salicylate (c) Glyptal (d) Salol (e) DMT
7. Give reasons:
   (a) Salicylic acid forms 2,4,6-trinitrophenol on nitration.
   (b) Carboxylic acid is meta directing whereas carboxylate anion is ortho and para directing.
   (c) Benzoic acid forms methyl benzoate with methanol and hydrogen chloride but under similar conditions 2,4,6-trimethylbenzoic acid does not form its ester.
   (d) Isophthalic acid does not form anhydride.
8. Give reasons why:
   (a) aromatic carboxylic acids are stronger acids than aliphatic acids?
   (b) monochloroacetic acid is stronger acid than acetic acid?
   (c) 2,4,6-trinitrophenol is called picric acid?
   (d) electron withdrawing substituents like $-Br$ or $-NO_2$ increase the acidity of benzoic acid?
   (e) p-nitrobenzoic acid is a stronger acid than p-chlorobenzoic acid?
   (f) o-Substituted benzoic acids are stronger acids compared to the m- and p-isomers?
9. Arrange the following in the order of increasing acidity.
   (a) Benzoic acid, nitrobenzoic acid, p-nitrophenol, p-hydroxybenzoic acid
   (c) Benzoic acid, o-nitrobenzoic acid, m-nitrobenzoic acid, p-nitrobenzoic acid.
   (d) Benzoic acid, benzamide, phthalimide.
10. Write a note on acidity of aromatic carboxylic acid. How is it affected by substituents at ortho positions?
11. Complete the following reactions and give their mechanism.

   (a) $C_6H_6 + COCl_2 \xrightarrow{AlCl_3}$ (b) $C_6H_5MgBr + CO_2 \longrightarrow$

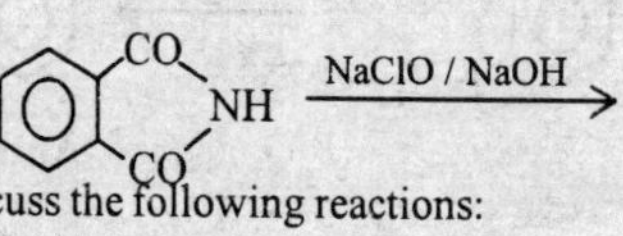

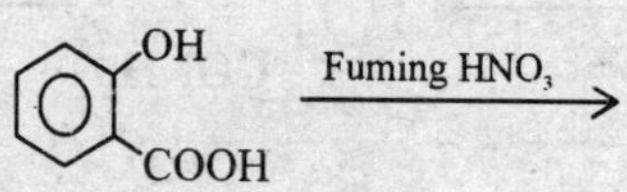

12. Discuss the following reactions:
   (a) Perkin's reaction   (b) Kolbe's reaction
   (c) Gabriel's phthalimide synthesis (d) Reimer Tiemann reaction.

37

Polynuclear Hydrocarbons

37.1 INTRODUCTION

Polynuclear hydrocarbons are divided into two groups: (i) in which the rings are isolated, e.g., biphenyl, diphenylmethane etc. and (ii) in which two or more rings are fused together and are known as condensed polynuclear hydrocarbons. In these the fusion of the rings is due to sharing of two or more carbon atoms by two or more rings. Examples of this type include naphthalene, anthracene and phenanthrene etc.

The present discussion is restricted to condensed polynuclear hydrocarbons.

37.2 NAPHTHALENE [$C_{10}H_8$]

It is the simplest polynuclear hydrocarbon and is the single largest constituent of coal tar (9%). It is obtained by cooling the middle and heavy fractions of coal tar distillate, when naphthalene crystallises out. The crude product obtained is treated with concentrated sulphuric acid (to remove basic impurities), washed with water and then with sodium hydroxide solution (to remove acidic impurities). Finally pure naphthalene is obtained by distillation of the crude product.

Naphthalene may also be synthesised from petroleum. The petroleum fractions are passed over heated catalyst (copper at 680-700°) at atmospheric pressure when naphthalene and higher aromatics (methyl-naphthalenes) are obtained. Naphthalene is isolated from the mixture by usual methods. Alternatively the mixture containing methylnaphthalene is hydrodealkylated to naphthalene.

37.2.1 Nomenclature of Naphthalene Derivatives

The naphthalene derivatives are named by indicating the position occupied by the substituents. The numbering is done in the following manner:

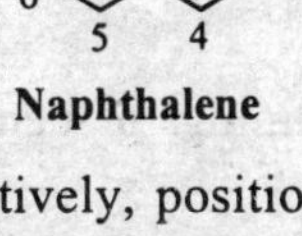

Naphthalene 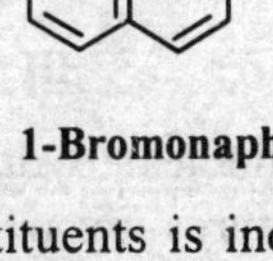   **1-Bromonaphthalene** 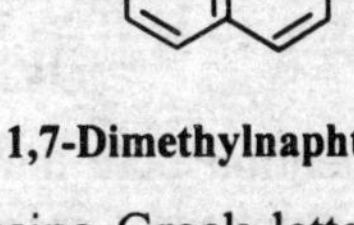   **1,7-Dimethylnaphthalene**

Alternatively, position of the substituents is indicated by using Greek letters α, β.

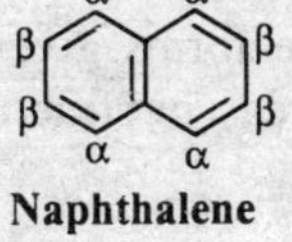

Naphthalene 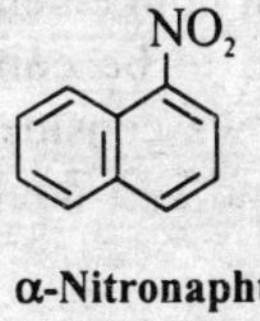   **α-Nitronaphthalene** 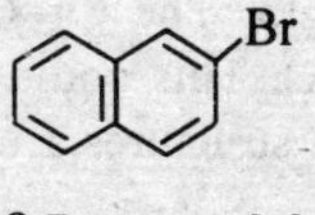   **β-Bromonaphthalene**

37.2.2 Structure of Naphthalene

The molecular formula of naphthalene as determined by elemental analysis was found to be $C_{10}H_8$. It was shown (Erlenmeyer 1866; Graebe 1869) that naphthalene consists of two benzene rings fused together in o-position. Its structure is based mainly on oxidation studies. The oxidation of naphthalene gave phthalic acid proving the presence of following structure, i.e., it has a benzene ring having two side chains in the o-position.

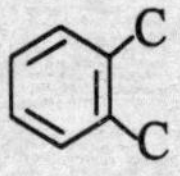

Further, nitration of naphthalene gave nitronaphthalene, which on oxidation gave o-nitrophthalic acid. This proves that the nitro group is in the benzene ring and that it is the side chain that is oxidised. However, aminonaphthalene, (obtained by reduction of nitronaphthalene) on oxidation gave phthalic acid. We have seen, an amino group attached to the nucleus renders the latter extremely sensitive to oxidation and usually results in ring rupture. Thus, it is clear that during the oxidation of aminonaphthalene, the ring containing the amino group is oxidised. On the basis of the above it is concluded that naphthalene is represented by the structure (I). The above oxidation reactions can be explained as follows:

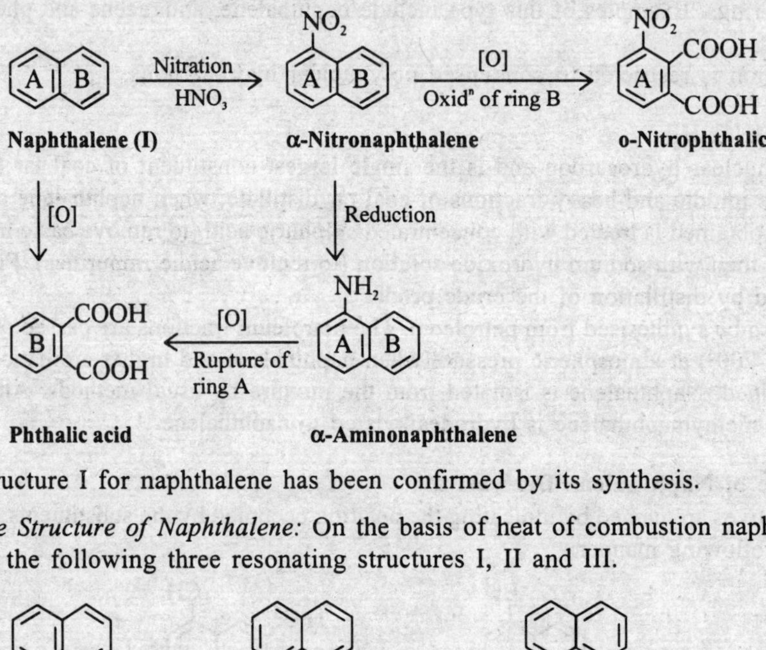

Naphthalene (I) **α-Nitronaphthalene** **o-Nitrophthalic acid**

Phthalic acid **α-Aminonaphthalene**

The structure I for naphthalene has been confirmed by its synthesis.

Resonance Structure of Naphthalene: On the basis of heat of combustion naphthalene is a resonance hybrid of the following three resonating structures I, II and III.

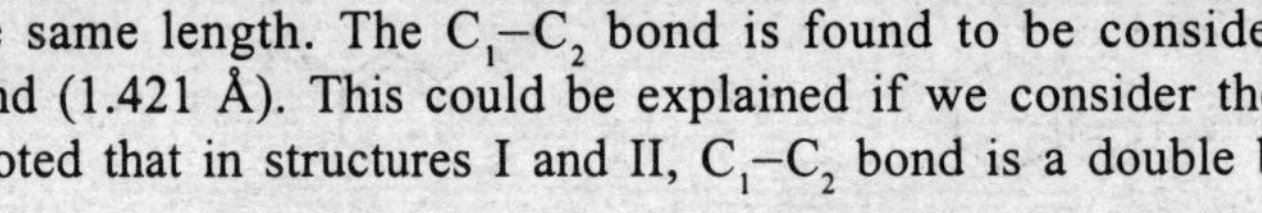

I **II** **III**

It is appropriate to mention here that for a polynuclear hydrocarbon containing 'n' benzene rings in a linear manner, the number of resonating structures are n+1.

X-Ray diffraction studies of naphthalene show that unlike benzene all C–C bonds in naphthalene are not of the same length. The C_1–C_2 bond is found to be considerably shorter (1.361 Å) than the C_2–C_3 bond (1.421 Å). This could be explained if we consider the resonance forms given above. It is to be noted that in structures I and II, C_1–C_2 bond is a double bond, whereas in structure III it is a single bond. Thus it is expected that C_1-C_2 bond will have more double bond character (shorter bond length) and the C_2-C_3 bond will have more single bond character. The bond distances and the bond character in naphthalene is shown in the structures IV and V respectively.

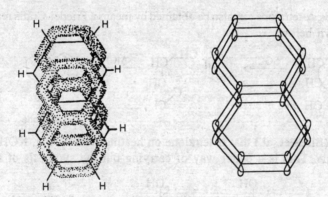

The resonance energy of naphthalene is approximately 61 kcal/mol which is less than twice the amount of benzene (36 kcal/mole). Consequently, naphthalene is somewhat less aromatic (more reactive) than benzene.

The Molecular Orbital approach: All carbon atoms in naphthalene are sp^2 hybridised and lie at the corners of two fused hexagons. The sp^2 hybrid orbitals overlap with each other and with s orbitals of the eight hydrogen atoms forming C–C and C–H σ bonds. All carbon and hydrogen atoms in naphthalene lie in one plane as sp^2 hybridisation has trigonal geometry. This has also been confirmed by X-ray diffraction studies.

Further, each carbon atom in naphthalene possess an unhybridised p orbital containing one electron. The p orbitals are perpendicular to the plane containing the σ bonds. Thus lateral overlap of the p orbitals produces π molecular orbital containing ten electrons. One half of this π molecular orbital lies above the plane of the σ bonds and the other half lies below the plane of the σ bonds thus imparting aromatic character to naphthalene. Since it contains ten π electrons, a number required for exhibiting aromatic character (Huckel's rule), naphthalene shows aromatic properties. Fig. 37.1 shows the molecular orbital picture of the naphthalene molecule.

Fig. 37.1: Molecular Orbital Picture of Naphthalene

37.2.3 Synthesis of Naphthalene

(i) When 4-phenyl-1-butene is passed over red hot calcium oxide, naphthalene is formed.

4-Phenyl-1-butene $\xrightarrow[\Delta]{\text{CaO}}$ Naphthalene

(ii)   4-Phenylbut-3-enoic acid on heating with concentrated sulphuric acid gives 1-naphthol, which on distillation with zinc dust gives naphthalene.

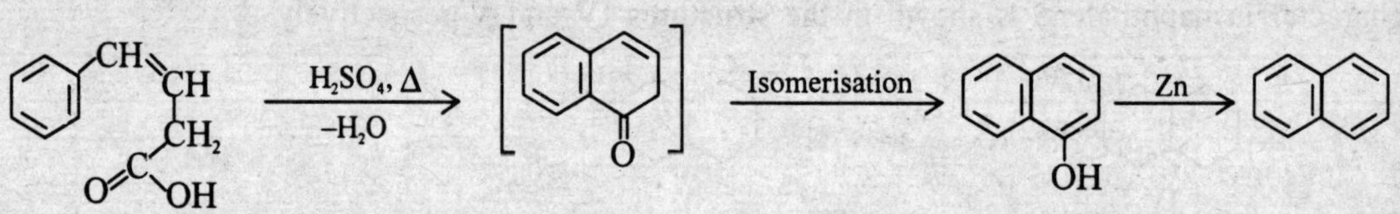

4-Phenylbut-3-enoic acid   **1-Naphthol**   **Naphthalene**

(iii)   Haworth Synthesis: The reaction of benzene with succinic anhydride in presence of aluminum chloride (Friedel-Crafts acylation) gave β-benzoylpropanoic acid, which on Clemmensen's reduction and subsequent ring closure gave α-tetralone. It was reduced (Clemmensen method) to give tetrahydronaphthalene, which on dehydrogenation with selenium or palladised charcoal gave naphthalene.

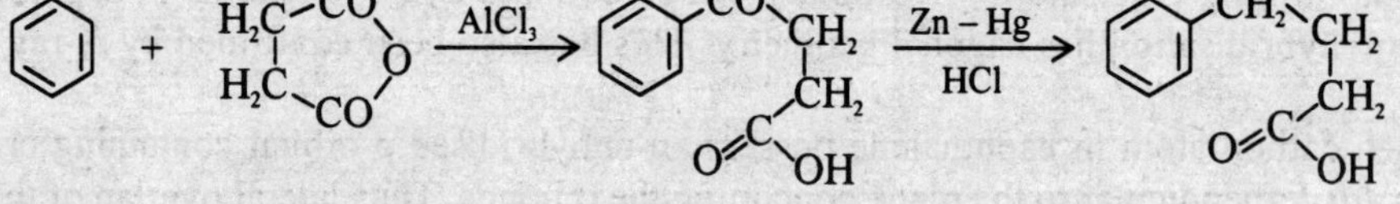

Benzene   Succinic anhydride   β-Benzoylpropanoic acid

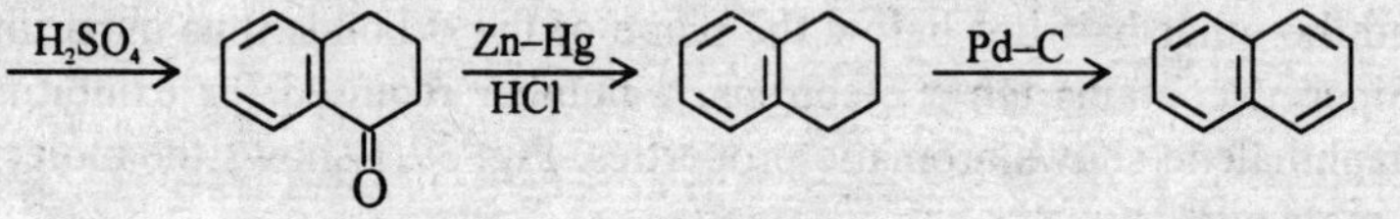

α-Tetralone   **Naphthalene**

In the above synthesis, α-tetralone can also be obtained by means a Friedel-Crafts reaction of corresponding acid chloride as shown below.

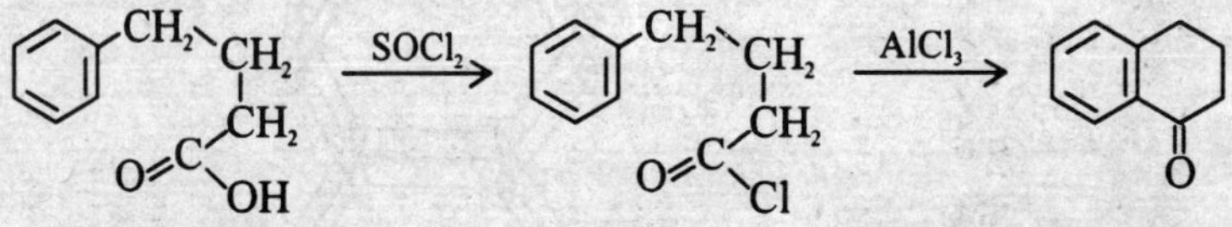

α-Tetralone

It has been shown (Birch et. al.) that α-tetralone on heating with NaOH, KOH mixture (1:1) at 220° gives 58% naphthalene. This is a better way of carrying out the synthesis of naphthalene.

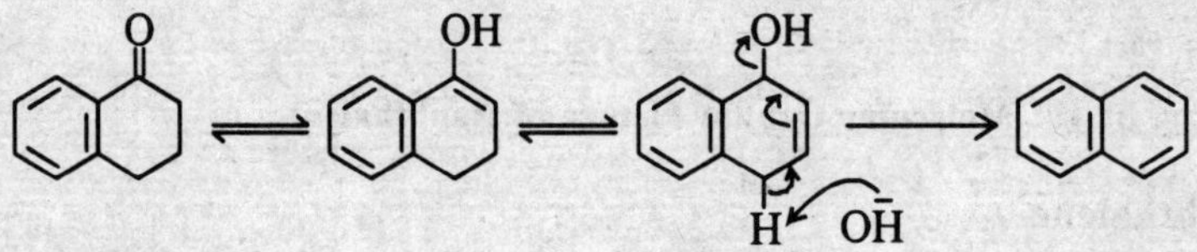

α-Tetralone   **Naphthalene**

37.2.4   Physical Properties

Naphthalene, m.p. 80° has a characteristic odor and is very volatile. It is used as an insecticide, moth repellent and in the preparation of phthalic anhydride and dyes. It is insoluble in water but soluble in alcohol, ether and benzene.

Chemical Properties: It has already been shown that naphthalene is less aromatic than benzene and consequently is more reactive than benzene.

Naphthalene undergoes oxidation or reduction more readily as compared to benzene, but only to a stage where substituted benzene is formed, further oxidation or reduction requires more vigrous conditions. Considering logically, i.e., by resonance energy concept, naphthalene is resonance stabilised by 51 kcal/mol and benzene by 36 kcal/mol. In reactions where aromatic character of one ring of naphthalene is destroyed, only 25 kcal/mol (61 − 36 = 25) of resonance energy is utilized and in the next stage, still 36 kcal/mol of energy has to be sacrificed. This raises the energy of the system, and hence reaction stops at substituted benzene formation.

(I) *Oxidation Reactions:* Naphthalene when oxidised in the presence of vanadium pentoxide yields phthalic anhydride, whereas chromium trioxide in presence of acetic acid at 25° forms naphthaquinone.

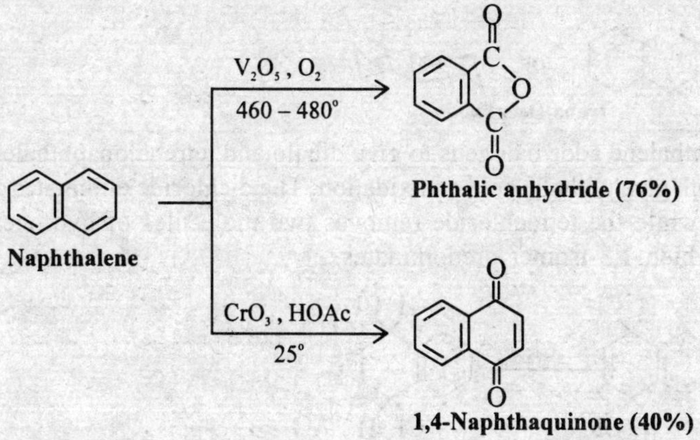

Naphthalene

V$_2$O$_5$, O$_2$
460 – 480°

Phthalic anhydride (76%)

CrO$_3$, HOAc
25°

1,4-Naphthaquinone (40%)

(II) *Addition Reactions:*

(a) *Addition of hydrogen:* A number of reduction products of naphthalene are obtained depending upon the reducing agent used.

On reduction with sodium and alcohol, naphthalene forms 1,4-dihydronaphthalene (1,4-dialin), which readily isomerises to 1,2-dialin. However, on reduction with sodium and isoamyl alcohol, naphthalene forms 1,2,3,4-tetrahydronaphthalene (tetralin). Complete reduction of naphthalene may be effected by reduction with hydrogen in presence of nickel catalyst to give decahydronaphthalene (decalin).

Na, C$_2$H$_5$OH
75–80°

1,4-Dihydrona-
phthalene
(1,4-dialin)

1,2-Dialin

Na, Isoamyl alc.
130–140°

(1,2,3,4-Tetrahydronaphthalene)
Tetralin

H$_2$ / Ni

Decahydronaphthalene
(Decalin)

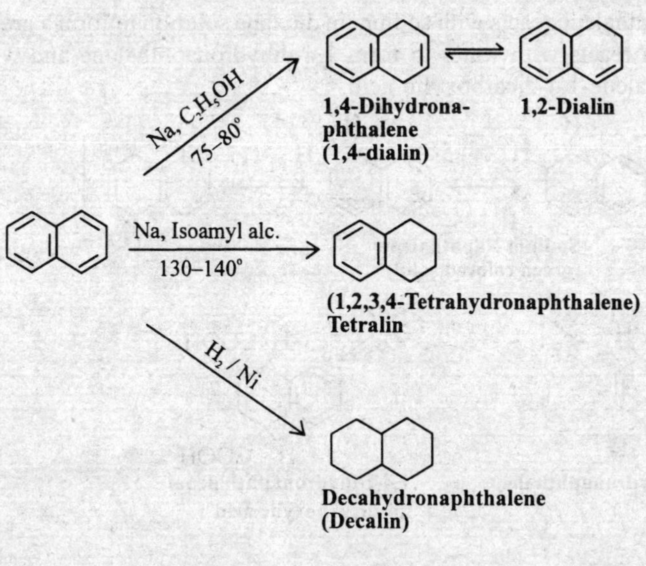

Both tetralin and decalin are used as solvents for varnishes and lacquers. Decalin exists in two geometrical isomeric forms, cis and trans decalins. When nickel is used for hydrogenation of nephthalene, the main product is trans isomer whereas, with platinum catalyst the cis isomer is formed. The configuration of cis and trans forms is shown below:

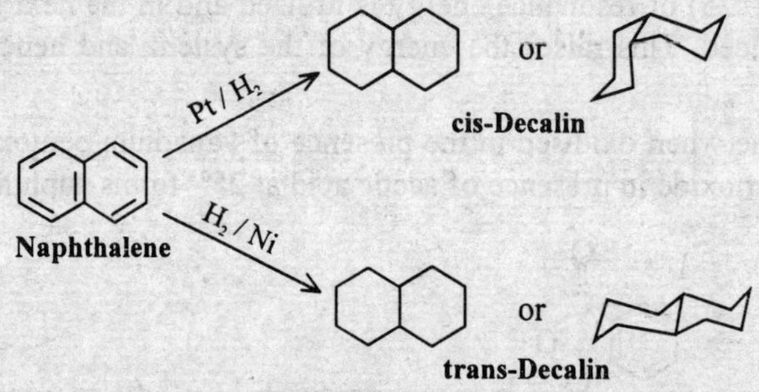

(b) *Addition of halogen:* Naphthalene adds halogens to give dihalo and tetrahalonaphthalene, having the halogens in the same ring as both give phthalic acid on oxidation. The dichloride eliminates a molecule of hydrogen chloride on heating while the tetrachloride removes two molecules of hydrogen chloride to give dichloronaphthalene of which 1,3-isomer predominates.

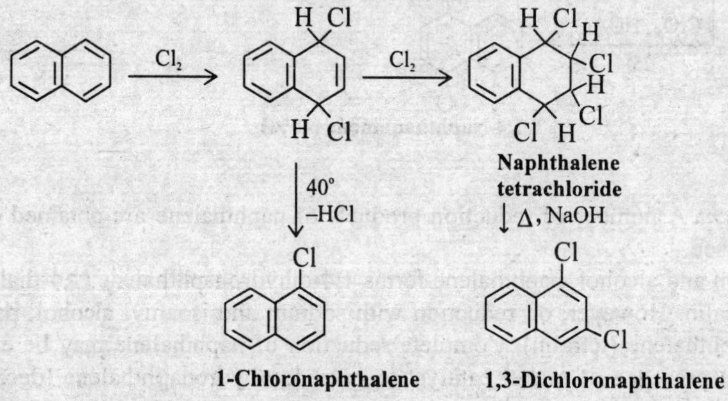

(c) *Reaction with sodium:* Naphthalene reacts with sodium in dioxane solution to form a green colored solid (sodium naphthalene), which reacts with water to form 1,4-dihydronaphthalene and with carbon dioxide to give 1,4-dihydronaphthalene-1,4-dicarboxylic acid.

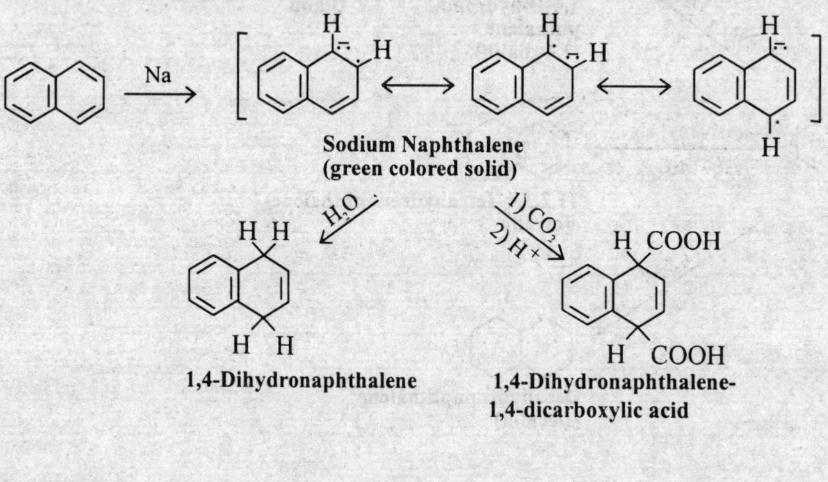

(III) *Substitution Reactions:* Naphthalene (as benzene) undergoes electrophilic substitution reactions. Substitution at C-1 predominates, However substitution at C-2 occurs only when the reaction is carried out at high temperatures. The mechanism of electrophilic substitution is given below:

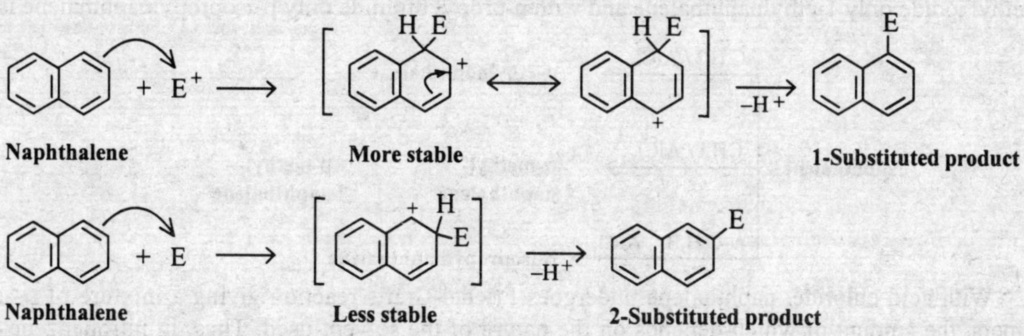

Naphthalene　　　　　　**More stable**　　　　　　**1-Substituted product**

Naphthalene　　　　　　**Less stable**　　　　　　**2-Substituted product**

As seen above two resonance forms of intermediate carbocation are obtained when substitution takes place at C-1, whereas only one with C-2, stabilising the intermediate more in former, C-1 substituted product, therefore, predominates.

(a) *Nitration:* Naphthalene undergoes nitration with concentrated nitric acid in presence of concentrated sulphuric acid at 60° and forms 1-nitronaphthalene as the main product.

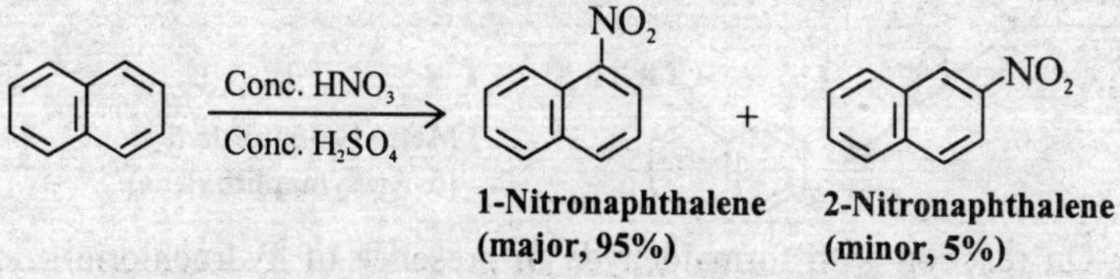

1-Nitronaphthalene　　**2-Nitronaphthalene**
(major, 95%)　　　　　　**(minor, 5%)**

However nitration at higher temperature gives a mixture of 1,5- and 1,8-dinitronaphthalenes.

2-Nitronaphthalene can also be prepared by heating 2-naphthalenediazonium fluoroborate with sodium nitrite and copper powder. The behaviour of nitronaphthalene is similar to that of nitrobenzene.

(b) *Sulphonation:* Naphthalene undergoes sulphonation and the products obtained depend on the reaction conditions. Sulphonation with concentrated sulphuric acid (98%) at 40-80° gives α-naphthalenesulphonic acid as the major product (96%). However, at 150° β-naphthalenesulphonic acid is obtained as major product (85%). The former (α) can also be converted into latter (β) by heating at 150°.

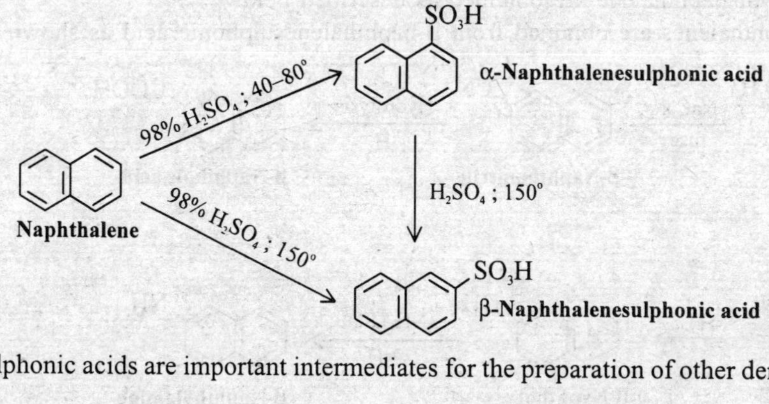

α-Naphthalenesulphonic acid

β-Naphthalenesulphonic acid

The sulphonic acids are important intermediates for the preparation of other derivatives of naphthalene.

(c) *Friedel-Crafts reaction:* The reaction is carried out successfully under mild conditions, since under vigorous conditions binaphthyls and other compounds are obtained in which one of the rings of naphthalene is opened. Thus, with methyl iodide a mixture of α- and β-methylnaphthalene is formed. With ethyl iodide only 1-ethylnaphthalene and with n-propyl bromide only β-isopropylnaphthalene is obtained.

Naphthalene $\xrightarrow{C_2H_5I / AlCl_3}$ **α-ethylnaphthalene**

Naphthalene $\xrightarrow{CH_3I / AlCl_3}$ **α-methylnaphthalene** + **β-methylnaphthalene**

Naphthalene $\xrightarrow{n\text{–}C_3H_7Br,\ AlCl_3}$ **β-isopropylnaphthalene**

With acid chloride, naphthalene undergoes Friedel-Crafts reaction giving a mixture of α- and β-isomers, the amount of which depends on the nature of the solvent used. Thus, in nitrobenzene as solvent β-isomer predominates, and in carbon disulphide α-isomer is the main product.

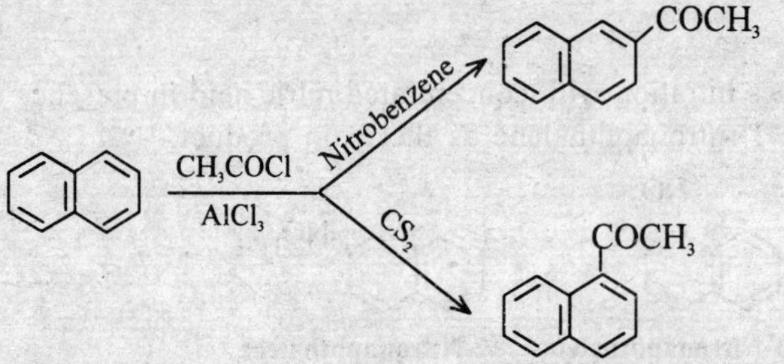
COCH3
Nitrobenzene
CH3COCl
AlCl3
CS2
COCH3

Methyl-β-naphthylketone (β-Acetylnaphthalene)

Methyl-α-naphthylketone (α-Acetylnaphthalene)

(d) *Chloromethylation:* On reaction with formaldehyde in presence of hydrochloric acid and zinc chloride naphthalene gives 1-chloromethylnaphthalene.

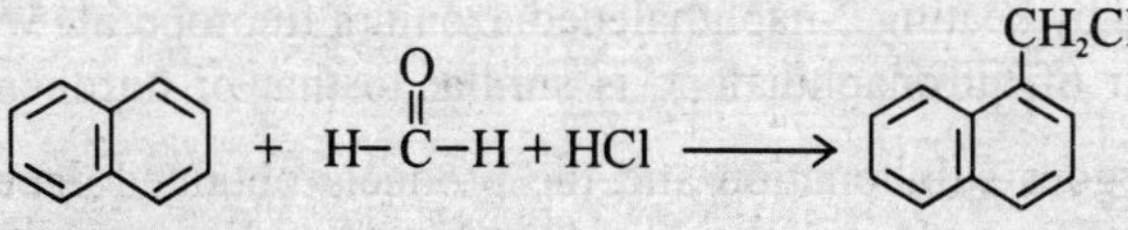
+ H–C–H + HCl
O
CH2Cl

Naphthalene **1-Chloromethylnaphthalene**

37.2.5 Derivatives of Naphthalene

Some of the derivatives of naphthalene are obtained as described below.

The β-substituted naphthalenes are obtained from β-naphthalenesulphonic acid as shown below:

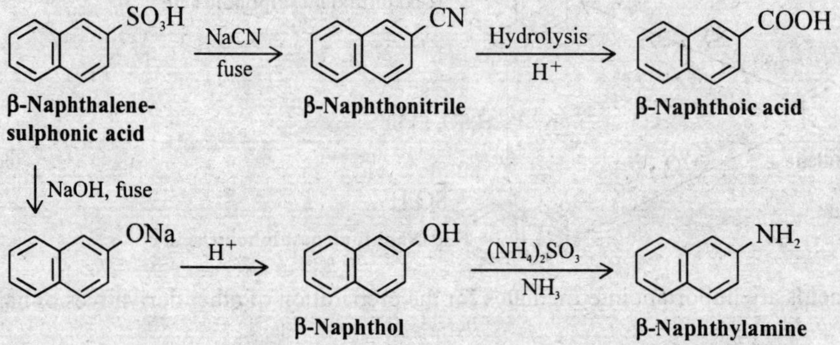
SO3H
NaCN
fuse
CN
Hydrolysis
H+
COOH
β-Naphthalene-
sulphonic acid
β-Naphthonitrile
β-Naphthoic acid
NaOH, fuse
ONa
H+
OH
(NH4)2SO3
NH3
NH2
β-Naphthol
β-Naphthylamine

β-Substituted halides, nitro derivatives etc. can be obtained from β-naphthylamine via the diazonium salt. The α-substituted derivatives of naphthalene are obtained from the corresponding nitro or chloro naphthalenes as follows.

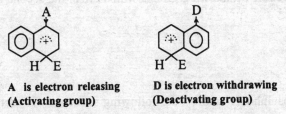

α-Nitronaphthalene   α-Naphthylamine   α-Diazonium salt

Sn + HCl →   HNO₂ →   α–Substituted halides, nitro or hydroxy compounds

α-Chloronaphthalene   α-Naphthyl magnesium chloride

Mg / THF →   Various α–substituted compounds via usual reactions of Grignard reagent

37.2.6 Orientation in Naphthalene Disubstitution

It has already been discussed that naphthalene undergoes nitration and halogenation chiefly at α-position, and sulfonation and Friedel-Crafts acylation at either α or β-position, depending upon the conditions. Now the next step is the orientation of the second substituent, its attachment etc. under the influence of group already present in naphthalene.

Orientation of second substitution in the naphthalene is more complicated than in benzene since there are seven different positions open to attack in contrast to only three positions in monosubstituted benzene. Also entering group may attach itself to the ring that already carries the first substituent or to the other ring.

The major products of further substitution in monosubstituted naphthalene can usually be predicted by the following rules.

(i) An activating group (electron releasing group) tends to direct further substitution into the same ring. An activating group in position 1 or α directs further substitution to position 4 (and to a lesser extent to position 2). An activating group in position 2 or β, on the other hand directs further substitution to position 1.

A is electron releasing
(Activating group)

D is electron withdrawing
(Deactivating group)

(ii) A deactivating group (electron withdrawing) tends to direct further substitution into the other ring. When NO₂ or SO₃H groups are present at position 1 or 2, heteronuclear substitution occurs at position 5 or 8, whereas NH₂ and X present at position 2 directs the second substituent at position 5 or 8.

Mechanism of Orientation

Orientation in naphthalene derivatives can be accounted for by the formation of more stable intermediate carbocation and this can only happen if aromatic sextet is preserved the most. The structures preserving an aromatic sextet are those in which the positive charge is carried by ring under attack. It is in this ring, therefore, that the charge chiefly develops. As a result, attack occurs most readily in the ring tha can best accommodate the positive charge i.e. the ring that carries an electron releasing group or the ring that does not carry an electron withdrawing group.

An electron releasing group located at position 1 or α can best accommodate the positive charge if attack occurs at position 4 (or position 2) through contributing structures I and II.

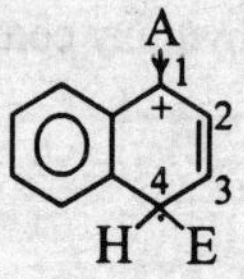

(I) 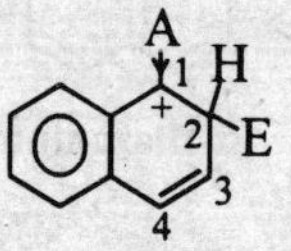   (II)

This is true for the groups which release electrons via inductive effect or resonance effect as shown below:

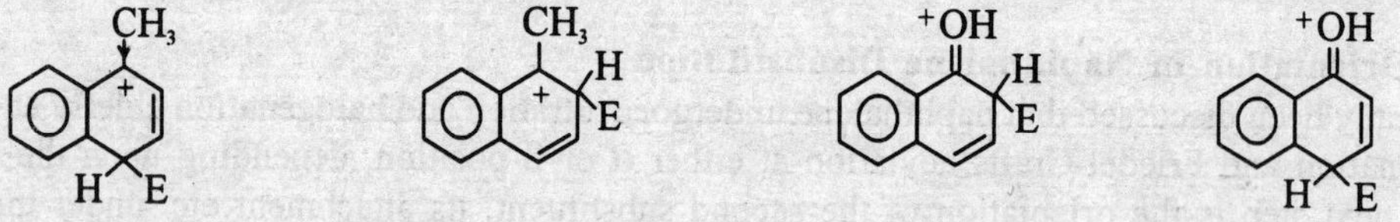

An electron releasing group present at position 2 or β can help accommodate the positive charges if attack occurs at position 1 (structure III) or if attack occurred at position 3 (structure IV).

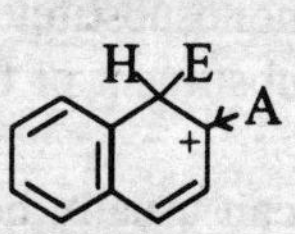

III More stable   **IV Less stable**
(Aromatic sextet preserved)   **(Aromatic sextet destroyed)**

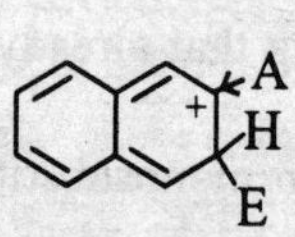

Since structure III preserves the aromatic sextet, it is much more stable than structure IV and hence substitution occurs entirely at position 1.

Similar reasoning may be given for electron withdrawing groups which direct the second substituent to the other ring.

37.2.7   Individual Members

37.2.7.1   1-Naphthol/α-Naphthol

Preparation:

(i) It is obtained from 1-naphthalenesulphonic acid by the following sequence of reactions:

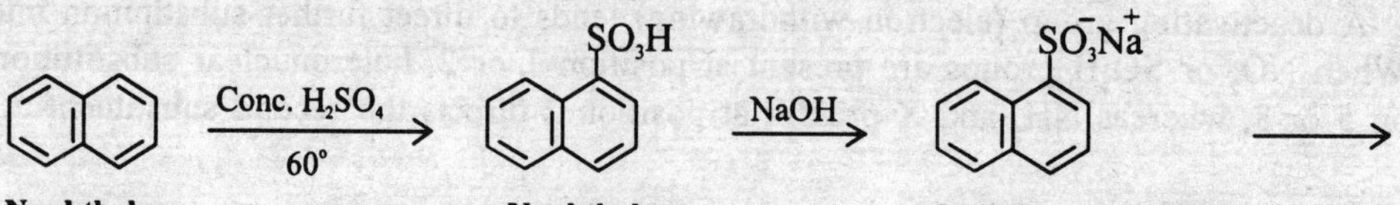

Naphthalene   **α-Naphthalene-**   **Sodium-α-naphthalene**
                         sulphonic acid   **sulphonate**

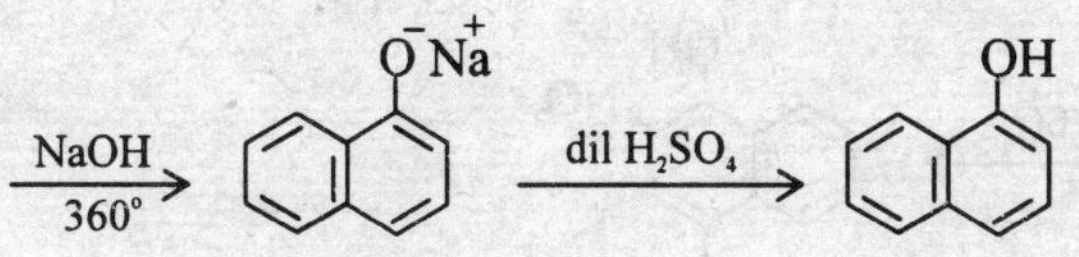

Sodium-1-
naphthoxide

α-Naphthol

(ii) Pure α-naphthol is prepared industrially in 95% yield by heating α-naphthylamine with dilute sulphuric acid under pressure.

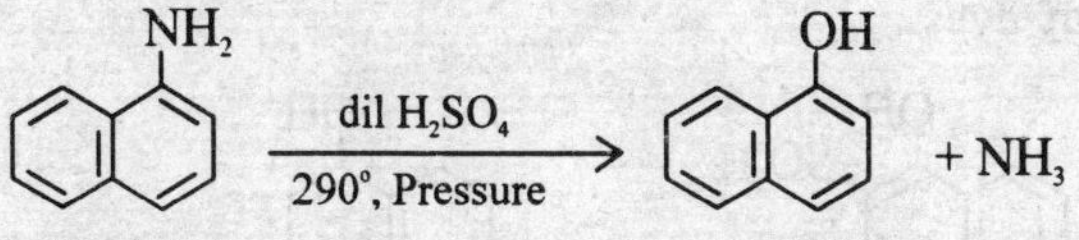

α-Naphthyl-
amine

α-Naphthol

Properties: α-Naphthol is colorless solid, m.p. 94°. It is slightly soluble in water but readily soluble in alcohol and ether. It resembles phenol in its chemical properties.

(i) *Acidic Character:* α-Naphthol is a weak acid. It dissolves in sodium carbonate or sodium hydroxide solution to form sodium-1-naphthoxide. α-Naphthol is a stronger acid than phenol due to greater stability of naphthoxide ion compared to phenoxide ion.

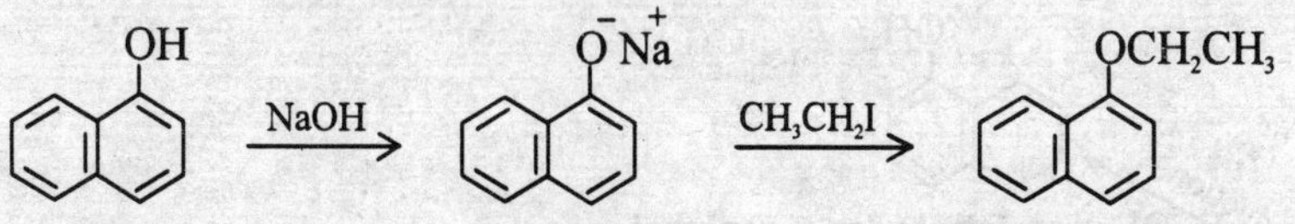

α-Naphthol

Sodium-1-
Naphthoxide

Ethyl-1-naphthyl ether
(1-Ethoxynaphthalene)

The sodium salt on heating with alkyl halides give the corresponding ether (Williamson's synthesis).

(ii) *Bucherer Reaction:* α-Naphthol on heating with ammonia and ammonium sulphite under pressure at 150° yields α-Naphthylamine.

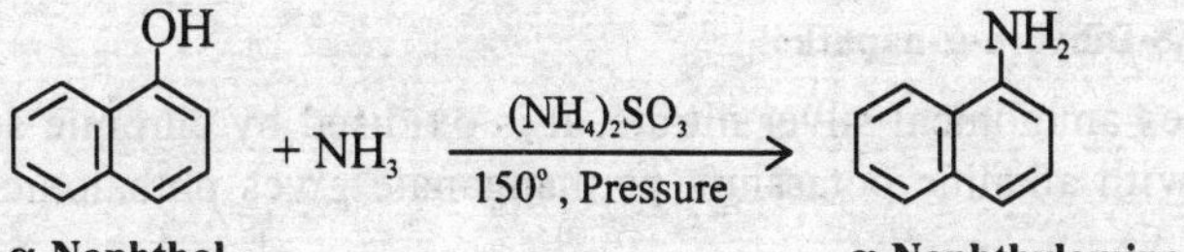

α-Naphthol

α-Naphthylamine

It has already been stated that the reverse of this reaction is used for the industrial preparation of α-naphthol.

(iii) *Nitration:* α-Naphthol on nitration with concentrated nitric acid and concentrated sulphuric acid at 20° gives 2,4-dinitro-α-naphthol. Its sodium salt is used as a dye for wool and silk under the name Martius yellow.

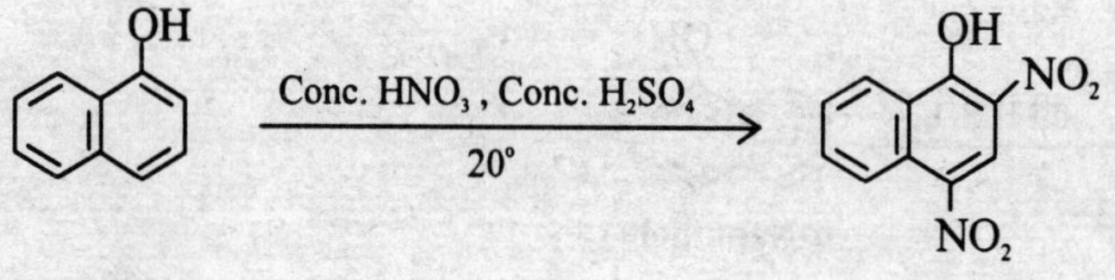

α-Naphthol

Conc. HNO₃, Conc. H₂SO₄ / 20°

2,4-Dinitro-α-naphthol

(iv) *Sulphonation:* α-Naphthol on reaction with concentrated sulphuric acid yields corresponding 2-sulphonic acid and 4-sulphonic acid. Both of these are used under the name Schaeffer's acid and Nevile-Winther's acid in the manufacture of dyes.

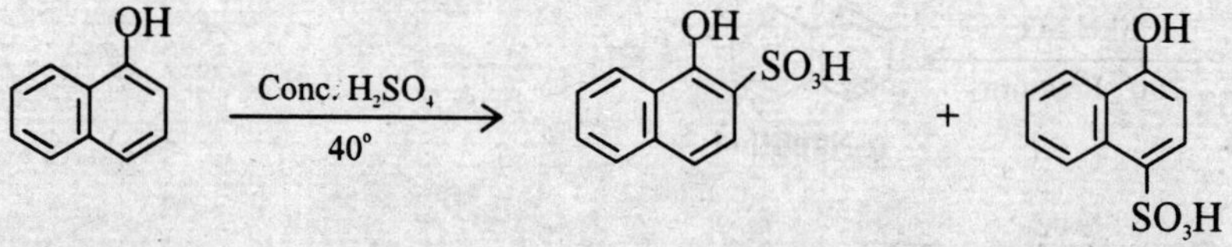

α-Naphthol

Conc. H₂SO₄ / 40°

α-Naphthol-2-sulphonic acid (Schaeffer's acid)

+

α-Naphthol-4-sulphonic acid (Nevile-Winther's acid)

(v) *Reduction:* α-Naphthol on reduction with sodium and isopentanol gives ar-tetrahydro-α-naphthol ('ar'indicates aromatic, i.e., the benzene ring containing OH group is not reduced). However, on reduction with lithium metal in liquid ammonia, 5,8-dihydro-α-naphthol is obtained (Birch reduction).

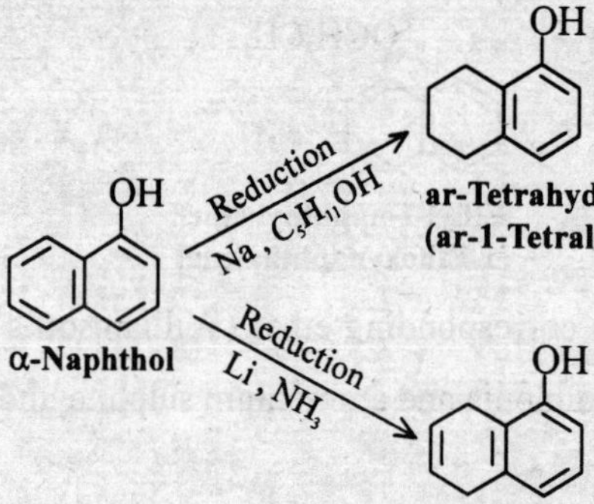

α-Naphthol

Reduction / Na, C₅H₁₁OH

ar-Tetrahydro-α-naphthol (ar-1-Tetralol)

Reduction / Li, NH₃

5,8-Dihydro-α-naphthol

(vi) *Oxidation:* α-Naphthol reduces ammonical silver nitrate. It is oxidised by chromic acid to 1,4-naphthaquinone. However oxidation with alkaline potassium permanganate gives phthalonic acid.

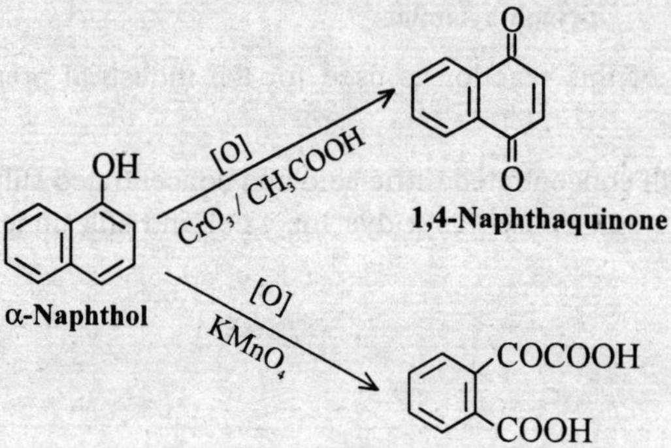

α-Naphthol

[O] / CrO₃, / CH₃COOH

1,4-Naphthaquinone

[O] / KMnO₄

Phthalonic acid

(vii) *Azocoupling:* α-Naphthol couples with benzenediazonium chloride to give 4-phenylazo-1-naphthol.

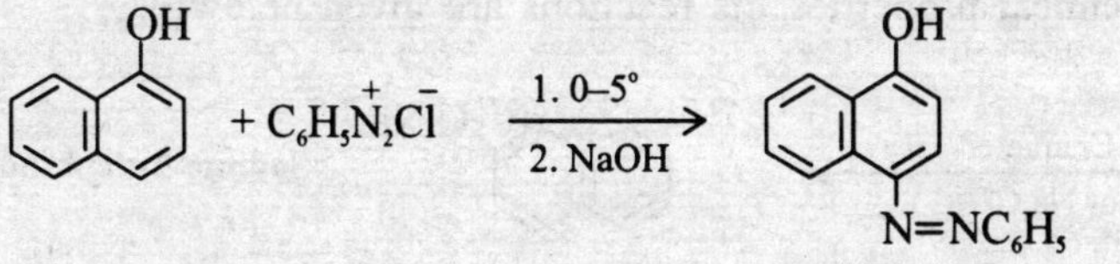

α-Naphthol   4-Phenylazo-1-naphthol

(viii) *Reaction with Nitrous Acid:* Treatment of α-naphthol with nitrous acid gives 2-oxime of 1,2-naphthaquinone as the major product, and a small amount of 4-oxime of 1,4-naphthaquinone.

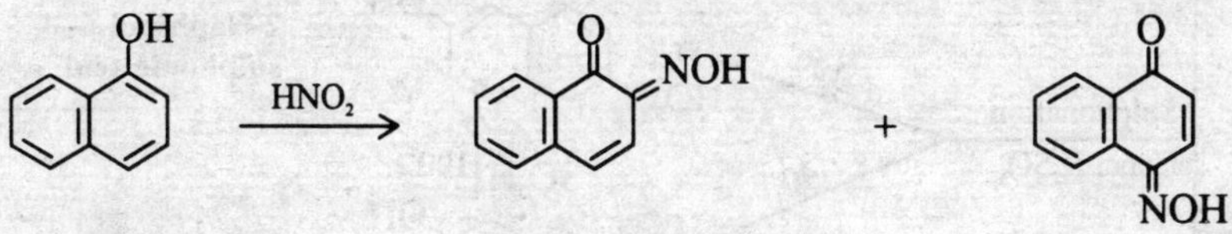

α-Naphthol   2-Oxime of 1,2-naphtha-   4-Oxime of 1,4-naphthaquinone
                    quinone

(ix) *Oxidative Coupling:* α-Naphthol on reaction with ferric chloride undergoes oxidative coupling to give a blue-violet precipitate of 4,4′-bis-1-naphthol or α-binaphthol.

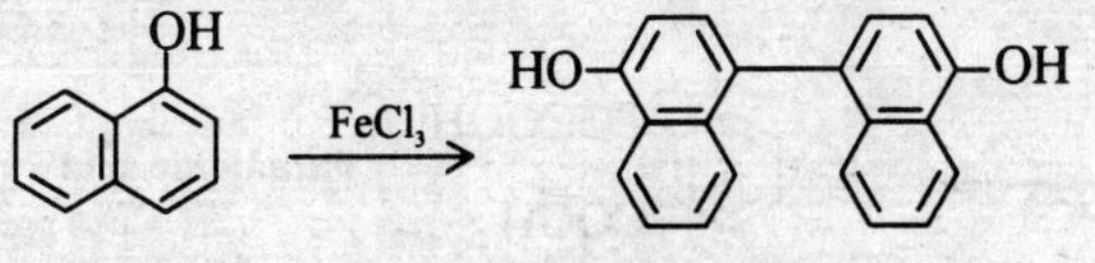

α-Naphthol   α-Binaphthol

Uses

α-Naphthol is used in the manufacture of dyes and insecticides.

37.2.7.2 β-Naphthol/2-Naphthol

Preparation: β-Naphthol is prepared from sodium 2-naphthalene sulphonate by fusing with sodium hydroxide followed by treatment of the product with dilute sulphuric acid.

Naphthalene   2-Naphthalene-   Sodium-2-naphthalene
                     sulphonic acid   sulphonate

Sodium-2-   β-Naphthol
naphthoxide

Properties: β-Naphthol is a colorless solid, m.p. 123°. It is sparingly soluble in water but readily soluble in ether and alcohol.

It resembles α-naphthol in chemical properties. Its reactions are given below:

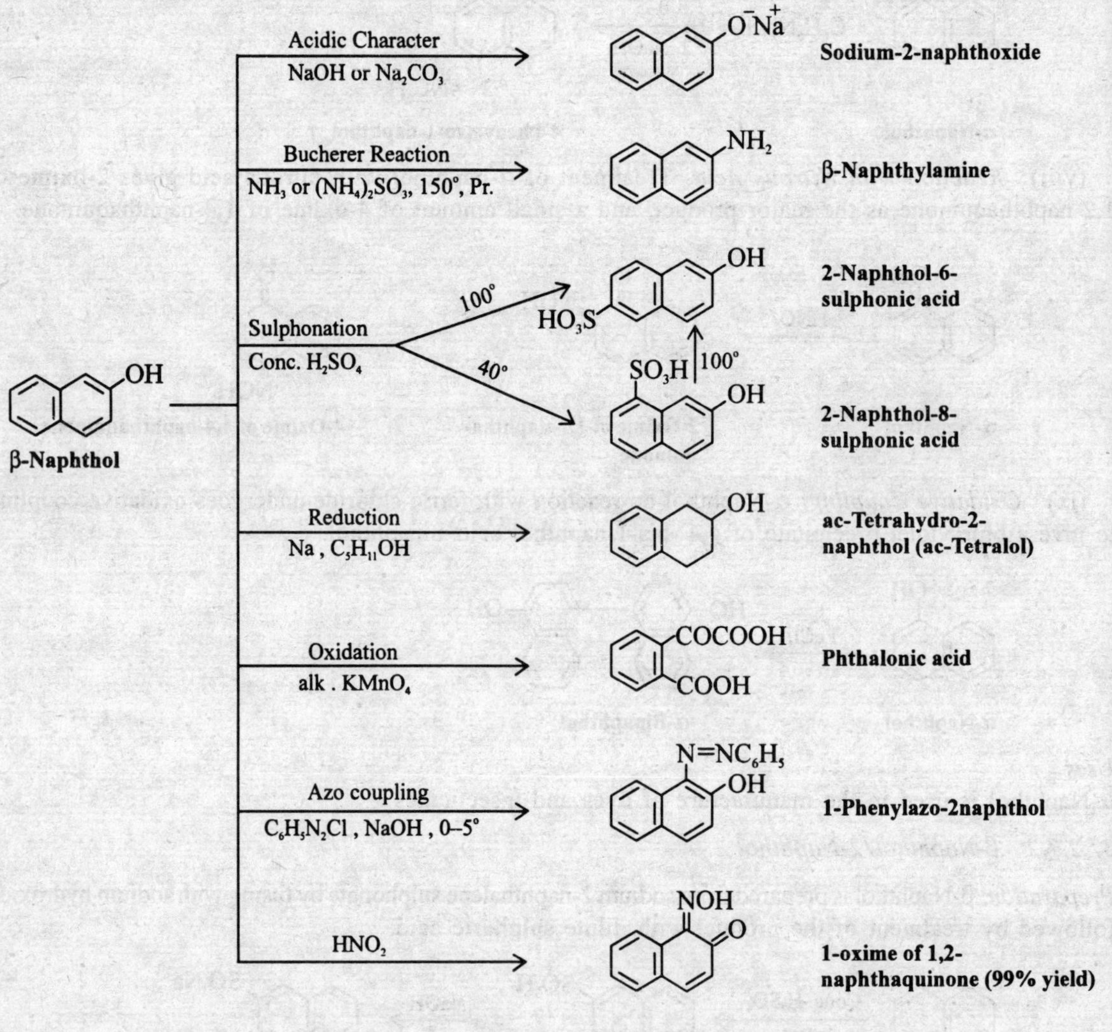

Uses

β-Naphthol is used in the manufacture of dyes, as an antioxidant in the manufacture of synthetic rubber. Its methyl ether is used in perfumery under the name nerolin.

37.2.7.3   1-Naphthylamine/α-Naphthylamine

Preparation: It is prepared by the reduction of 1-nitronaphthalene with iron and hydrochloric acid (commercial method). It is also obtained by heating α-naphthol with ammonia and ammonium sulphite at 150° under pressure (Bucherer reaction).

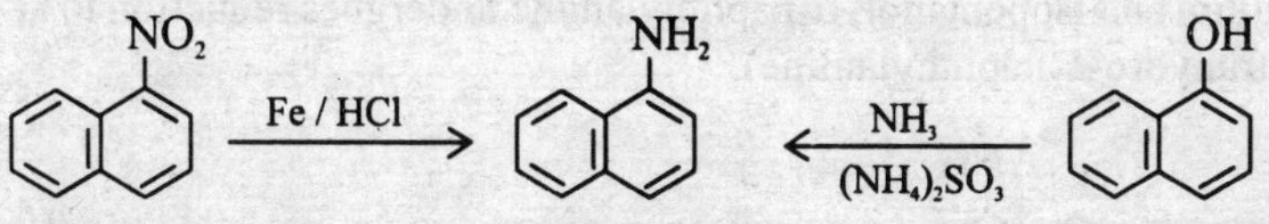

1-Nitronaphthalene — Fe / HCl → **1-Naphthylamine** ← NH₃ (NH₄)₂SO₃ — **1-Naphthol**

Properties: It is a colorless, crystalline solid, m.p.50°, turns brown on exposure to air. It is insoluble in water but soluble in ether and alcohol. It has an unpleasant odor and gives a blue precipitate with ferric chloride solution.

α-Naphthylamine resembles aniline in chemical properties.

(i) *Basic Nature:* Like aniline and other amines, it is basic in nature and dissolve in dilute mineral acids forming salts.

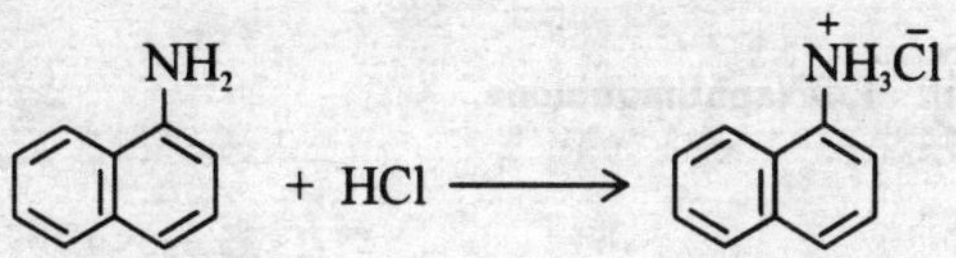

α-Naphthyl-amine + HCl → α-Naphthylamine hydrochloride

It is slightly weaker base than aniline due to extensive delocalisation of lone pair of electron on nitrogen in case of 1-naphthylamine.

(ii) *Diazotisation:* α-Naphthylamine on diazotisation with sodium nitrite and hydrochloric acid gives 1-naphthalenediazonium chloride, which couples with phenols and aromatic amines. It is also used for the preparation of α-substituted naphthalene derivatives.

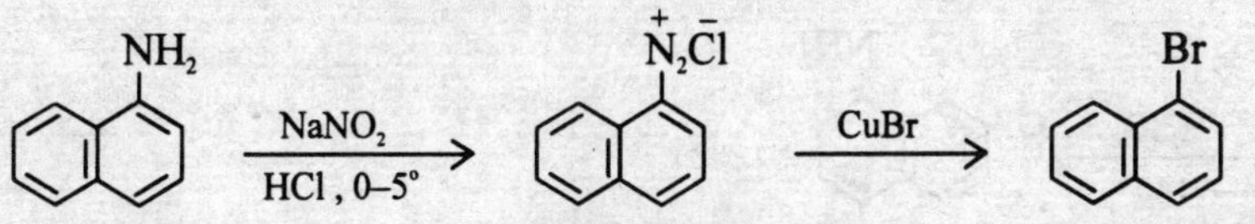

α-Naphthyl-amine — NaNO₂ / HCl, 0–5° → 1-Naphthalene-diazonium chloride — CuBr → 1-Bromonaphthalene

— Mg / ether → 1-Naphthyl magnesium bromide → Other Derivation

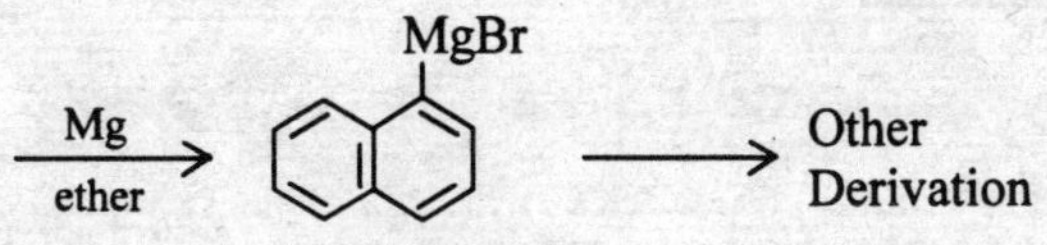

(iii) *Azocoupling:* α-Naphthylamine on reaction with benzenediazonium chloride in alkaline solution at 0-5° give 4-phenylazo-1-naphthylamine.

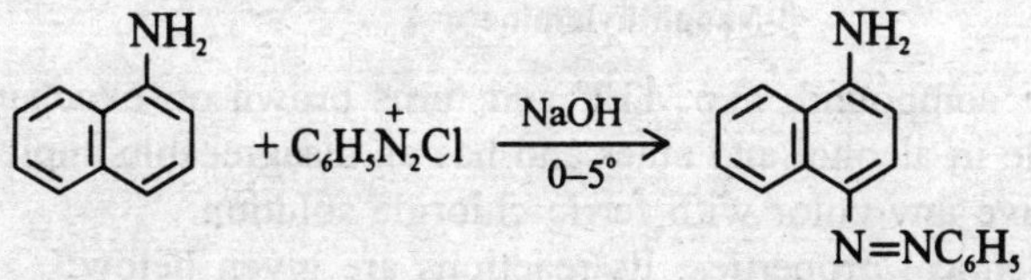

1-Naphthylamine + C₆H₅N₂Cl — NaOH / 0–5° → **4-Phenylazo-1-naphthylamine**

(iv)   *Reduction:* With sodium and isopentanol, α-naphthylamine undergoes reduction to ar-tetrahydro 1-naphthylamine (5,6,7,8-tetrahydro-1-naphthylamine).

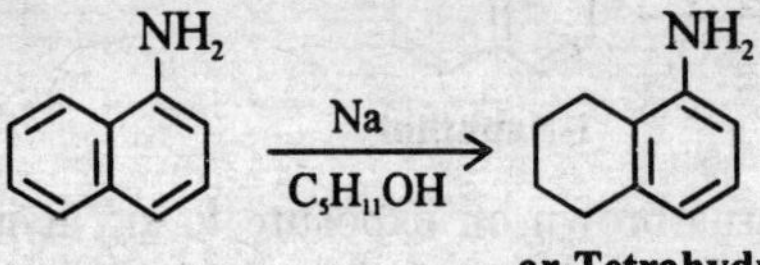

ar-Tetrahydro-1-naphthylamine

(v)   *Oxidation:* It reduces ammonical silver nitrate and is oxidised by chromium trioxide in acetic acid to give 1,4-naphthaquinone. However, oxidation with alkaline potassium permanganate, forms phthalic acid.

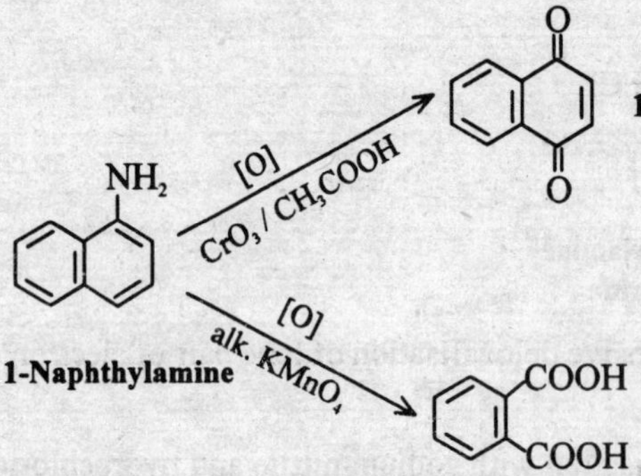

1-Naphthylamine

1,4-Naphthaquinone

Phthalic acid

(vi)   *Sulphonation:* α-Naphthylamine reacts with concentrated sulphuric acid at 140° to give 1-naphthyl-amine-4-sulphonic acid or naphthionic acid.

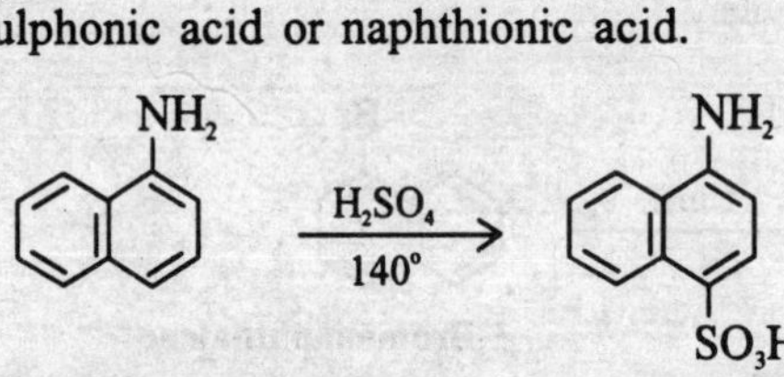

1-Naphthylamine   **Naphthionic acid**

It is used in the manufacture of dyes.

37.2.7.4   *2-Naphthylamine/β-Naphthylamine*

Preparation: It is obtained by the reaction of β-naphthol with ammonia and ammonium sulphite at 150° under pressure.

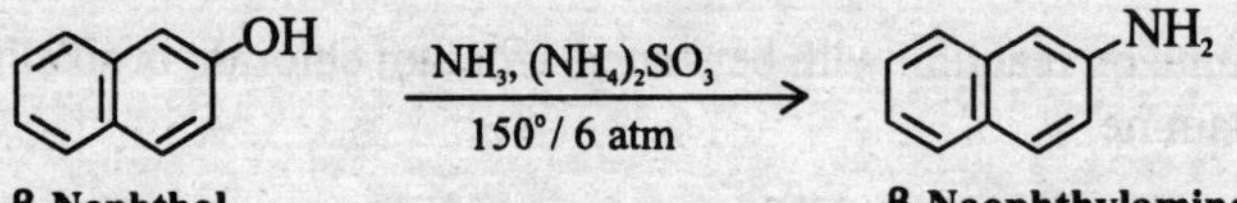

β-Naphthol   **β-Naophthylamine**

Properties: It is a colorless crystalline compound, m.p. 112° and turns brown on exposure to air or light. It is insoluble in water but soluble in alcohol and ether and has an disagreeable unpleasant odor. Unlike 1-naphthylamine, it does not give any color with ferric chloride solution.

It resembles α-naphthylamine in chemical properties. Its reactions are given below:

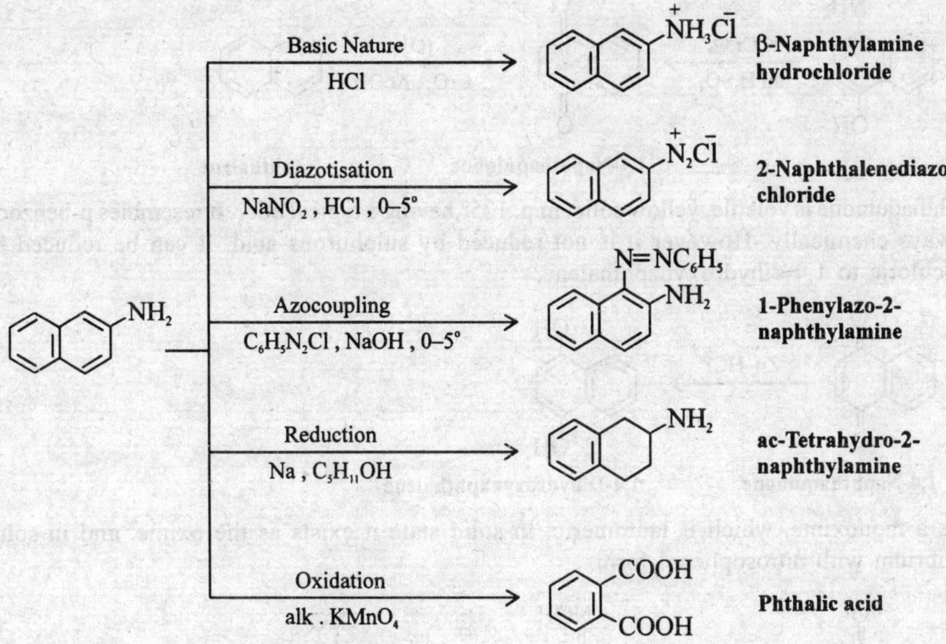

It is used in the manufacture of dyes.

37.2.7.5 Naphthalenecarboxylic Acids

Both α and β naphthoic acids are obtained by the hydrolysis of corresponding cyanides or naphthylmagnesium bromide or acetyl compounds. Below is given the preparation of 1-naphthoic acid or α-naphthoic acid.

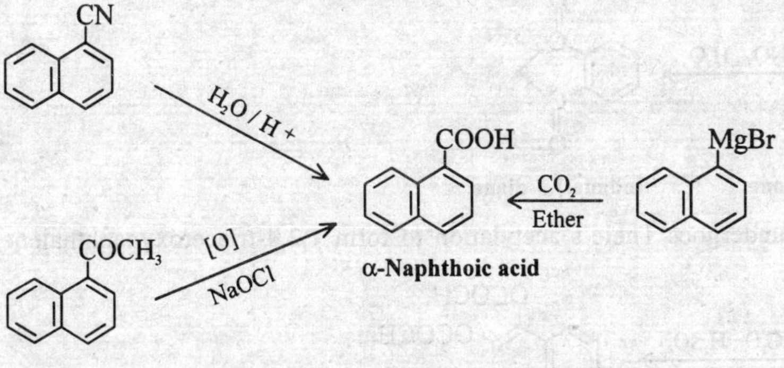

Both naphthoic acids undergo usual reactions of carboxylic acids.

37.2.7.6 Naphthaquinones

Six naphthaquinones are theoretically possible. Out of these only three, i.e., 1,2-, 1,4- and 2,6- are known.

1,4-Naphthaquinone: It is prepared by the oxidation of 1,4- diamino-, dihydroxy- or aminohydroxynaphthalene with potassium dichromate and sulphuric acid.

It is also prepared by the direct oxidation of naphthalene with chromic acid in acetic acid.

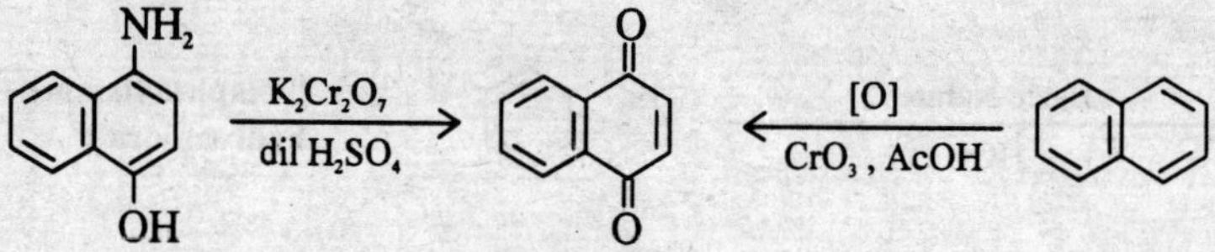

1,4-Naphthaquinone   **Naphthalene**

1,4-naphthaquinone is volatile, yellow solid, m.p. 125°, having a typical odor. It resembles p-benzoquinone in many ways chemically. However it is not reduced by sulphurous acid. It can be reduced by zinc and hydrochloric to 1,4-dihydroxynaphthalene.

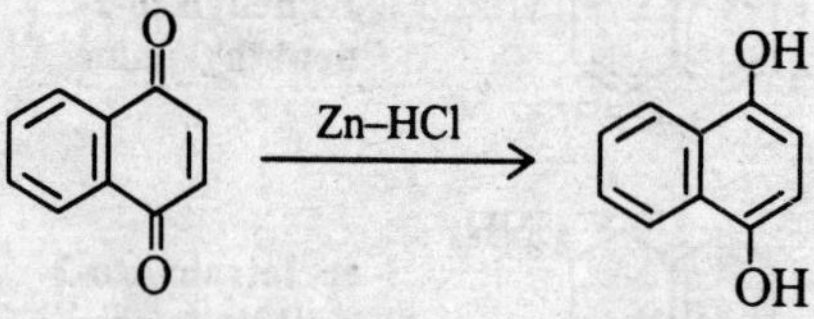

1,4-Naphthaquinone   **1,4-Dihydroxynaphthalene**

It forms a monoxime, which is tautomeric. In solid state it exists as the oxime, and in solution it is in equilibrium with nitrosophenol form.

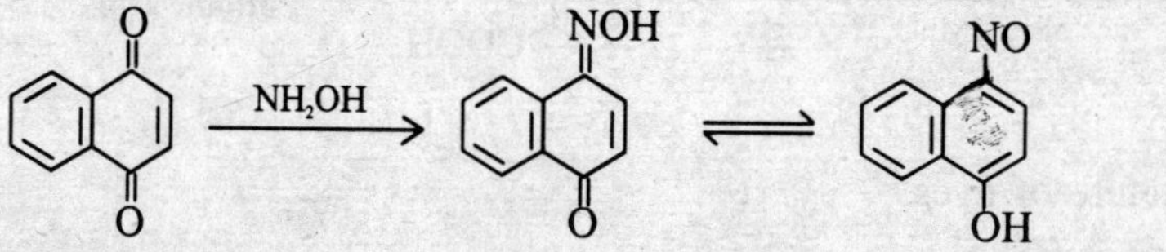

1,4-Naphthaquinone   **Monoxime**   **Nitrosophenol form**

It is converted into indane-1,3-dione on treatment with nitrous acid.

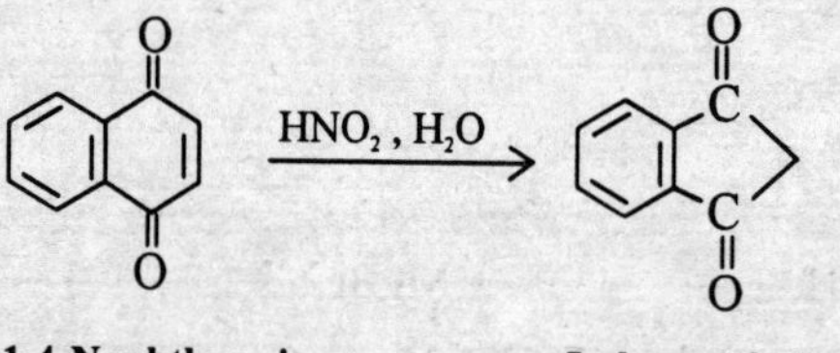

1,4-Naphthaquinone   **Indane-1,3-dione**

Like hydroquinone, it undergoes Thele's acetylation to form 1,2,4-triacetoxynaphthalene.

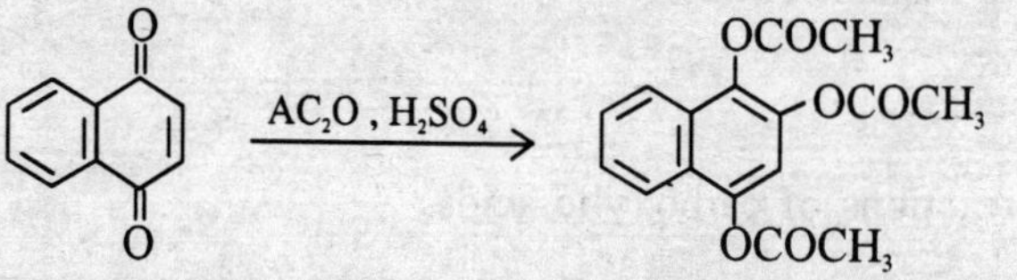

1,4-Naphthaquinone   **1,2,4-Triacetoxynaphthalene**

1,2-Naphthaquinone: It is prepared by the oxidation of 1-amino-2-naphthol hydrochloride with ferric chloride in hydrochloric acid.

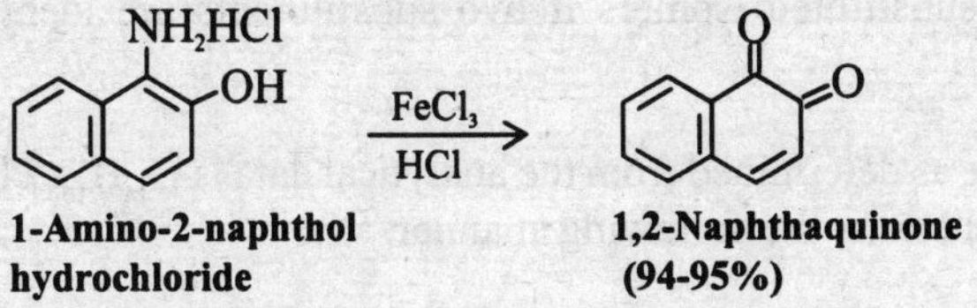

1-Amino-2-naphthol
hydrochloride

$\xrightarrow[\text{HCl}]{\text{FeCl}_3}$

1,2-Naphthaquinone
(94-95%)

It is a nonvolatile, colorless solid, m.p. 115-120°.

1,4-Naphthaquinone nucleus is present in vitamin K_1 and K_2 (antihaemorrhigic factor).

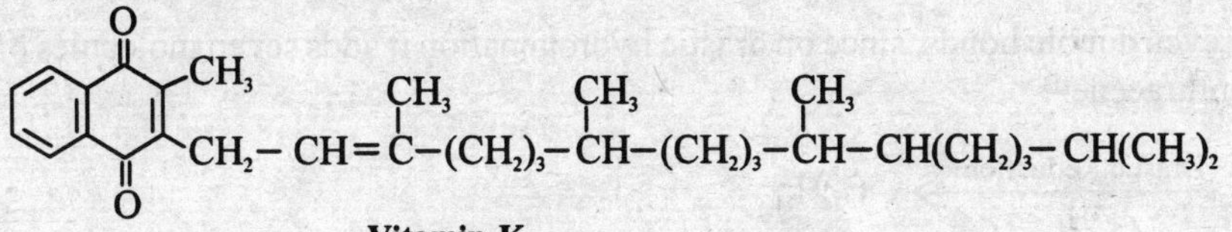

Vitamin K_1

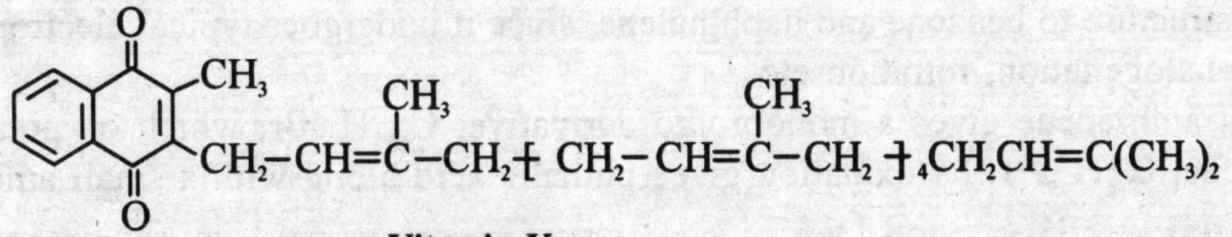

Vitamin K_2

Another vitamin K, known as menadione is 2-methyl-1,4- naphthaquinone and is obtained by the oxidation of 2-methtylnaphthalene.

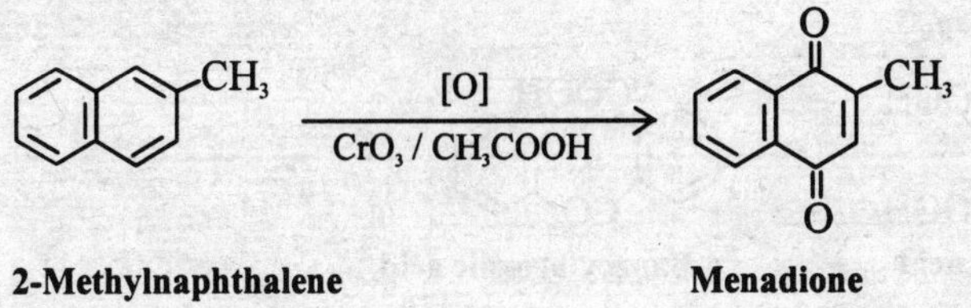

2-Methylnaphthalene

$\xrightarrow[\text{CrO}_3 / \text{CH}_3\text{COOH}]{\text{[O]}}$

Menadione

37.3 ANTHRACENE [$C_{14}H_{10}$]

It is obtained from the anthracene oil fraction of coal tar. This fraction is cooled to give crude anthracene containing phenanthrene and carbazole. The crude product is powdered and washed with solvent naphta, which dissolves phenanthrene. The remaining solid is then washed with pyridine which dissolves carbazole. Finally, anthracene is purified by sublimation.

37.3.1 Nomenclature

The position of substituents is indicated in anthracene by numbers or Greek letters as shown below:

37.3.2 Isomerism of Anthracene Derivatives

As seen above in anthracene three monosubstitution products are possible, viz., α (or 1), β (or 2) and

γ (or 9). Further, there are 15 possible disubstituted isomers if two substitutents are identical.

37.3.3   Structure of Anthracene

(1)   The molecular formula of anthracene as determined from the analytical data is $C_{14}H_{10}$. The formula suggests that it may be related to naphthalene in the following manner.

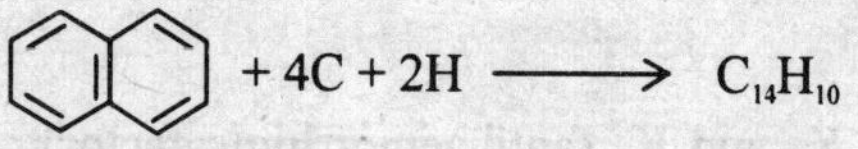

Naphthalene   **Anthracene**

(2)   It contains seven double bonds, since on drastic hydrogenation it adds seven molecules of hydrogen to form perhydroanthracene.

$$C_{14}H_{10} \xrightarrow[+ 7H_2]{\text{Drastic Reduction}} C_{14}H_{24}$$

Anthracene   **Perhydroanthracene**

(3)   It is similar in structure to benzene and naphthalene, since it undergoes typical electrophilic substitution reactions like halogenation, nitration etc.

(4)   Bromination of anthracene gives a monobromo derivative, $C_{14}H_9Br$, which on potash fusion forms hydroxyanthracene, $C_{14}H_9OH$. Its oxidation gives phthalic acid along with a small amount of o-benzoylbenzoic acid.

$$C_{14}H_{10} \xrightarrow{Br_2} C_{10}H_9Br \xrightarrow[\text{Fusion}]{KOH} C_{10}H_9OH$$

Anthracene   **Monobromo-**   **Hydroxyanthracene**
                        anthracene

$$C_{14}H_{10} \xrightarrow{[O]}$$

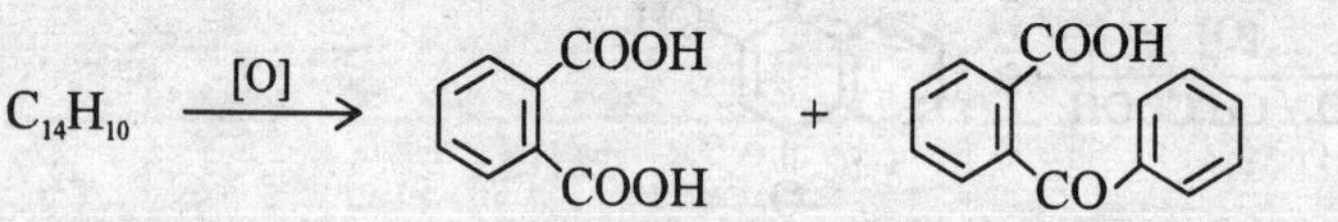

Anthracene   **Phthalic acid**   **o-Banzoylbenzoic acid**

(5)   Oxidation of anthracene with chromic acid gives anthraquinone, which on potash fusion gives two molecules of benzoic acid.

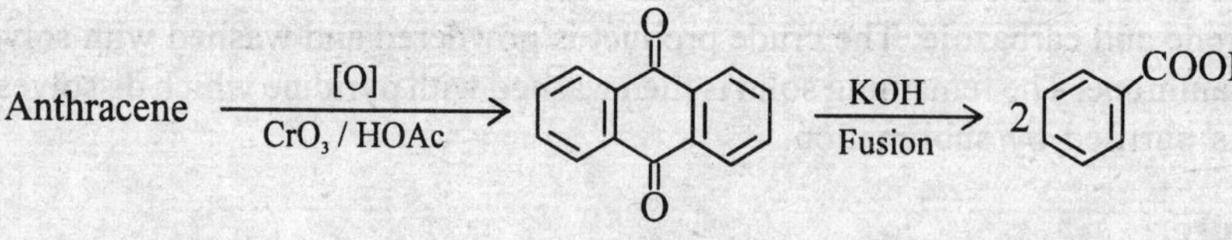

Anthracene $\xrightarrow[CrO_3 / HOAc]{[O]}$ ... $\xrightarrow[\text{Fusion}]{KOH}$ 2 ...

Anthraquinone   **Benzoic acid**

On the basis of the above discussion anthracene is given the structure I.

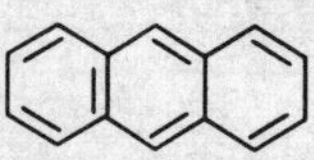

I, Anthracene

(6)   The structure I for anthracene is finally confirmed by its synthesis which involves Friedel-Crafts reaction between two molecules of benzyl chloride to give dihydroanthracene which readily loses two hydrogen atoms under the reaction conditions to give anthracene.

Benzyl chloride   **Dihydroanthracene**

(7) According to the resonance theory, anthracene is considered to be a hybrid of the following four resonating forms.

      (I)   **(II)**   **(III)**   **(IV)**

(8) On the basis of X-ray diffraction studies, it is found that all carbon and hydrogen atoms in anthracene lie in the same plane and that like naphthalene all C−C bonds in anthracene are not of the same length. In fact the C_1−C_2 bond is shorter (1.37 Å) than the C_2−C_3 bond (1.42 Å). This difference in bond length is understood on examining the four resonating structures (I to IV). The C_1−C_2 bond is a double bond in three structures (I, II and III) and a single bond in structure IV. The C_2−C_3 bond, on the other hand, is a single bond in three structures (I, II and III) and a double bond only in one (IV). Thus, it is expected that the C_1−C_2 bond has more double bond character (shorter bond length) and the C_2−C_3 bond more single bond character (longer bond length).

Further the resonance energy of anthrancene is 84 kcal/mol. This means that the resonance energy per ring averages to 28 kcal/mol, which is substantially lower than that in benzene (36 kcal/mol). Also the resonance energy of naphthalene (61 kcal/mol) averages to 30.5 kcal/mol per ring. As a result, anthracene is much less aromatic than benzene or even naphthalene.

(9) Molecular orbital approach to the structure of anthracene shows a planer structure for it as shown in Fig. 37.2. All carbon atoms in anthracene are sp^2 hybridized. The sp^2 hybrid orbitals overlap with each other and with the s orbitals of the ten hydrogen atoms forming C−C and C−H σ bonds. The bonds resulted from the overlap of trigonal sp^2 orbitals. Therefore all carbon and hydrogen atoms in anthracene lie in the same place. This has been confirmed by X-ray diffraction studies.

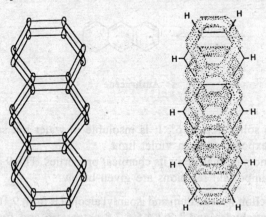

Fig. 37.2: Molecular Orbital Picture of Anthracene

Further each carbon atom in anthracene possess an unhybridised p orbital containing one electron.

These p orbitals are perpendicular to the plane of the σ bonds. Thus, the lateral overlap of these orbitals produces a π molecular orbital containing fourteen electrons half of these π molecular orbital lie above and the other half lie below the place of the σ bonds.

37.3.4   Synthesis of Anthracene

(i) *Diels Alder Reaction:* The Diels Alder reaction of 1,4- naphthaquinone with 1,3-butadiene gives an adduct, which on oxidation with chromium trioxide in glacial acetic acid forms 9,10-anthraquinone. Its distillation with zinc dust yields anthracene.

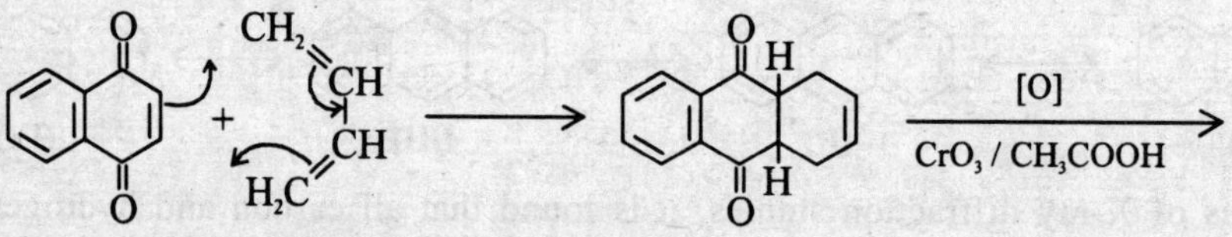

1,4-Naphthaquinone

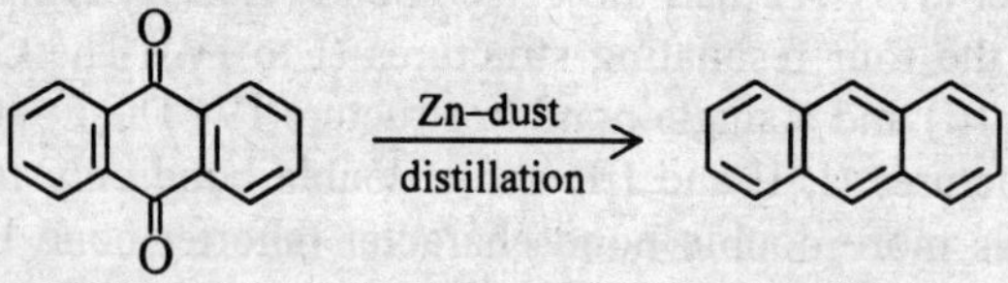

9,10-Anthraquinone   **Anthracene**

(ii) *Haworth's Synthesis:* The reaction of benzene with phthalic anhydride in presence of anhydrous aluminum chloride forms o-benzoylbenzoic acid, which on heating with concentrated sulphuric acid undergoes cyclisation to give 9,10-anthraquinone, distillation of which with zinc dust forms anthracene.

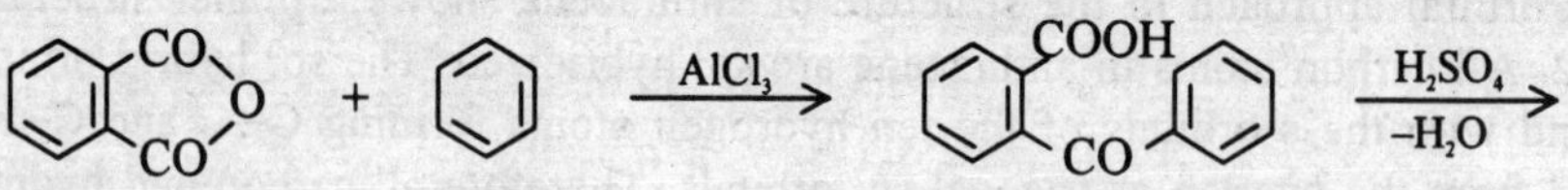

Phthalic anhydride   Benzene   **o-Benzoylbenzoic acid**

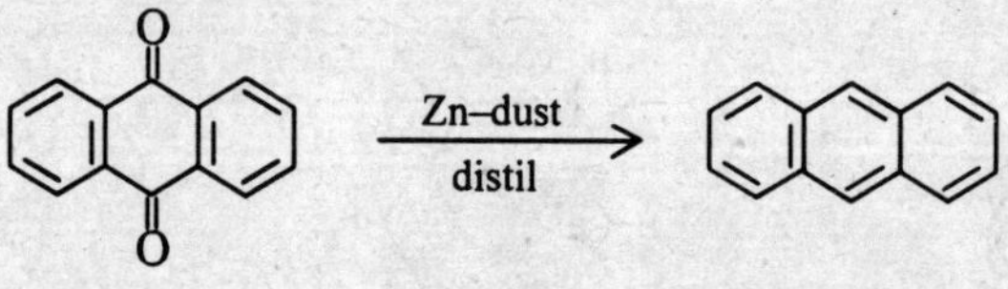

Anthraquinone   **Anthracene**

37.3.5   Properties

Anthracene is a colorless solid, m.p. 216°. It is insoluble in water but soluble in benzene. It shows a blue fluorescence when exposed to ultra violet light.

It resembles benzene and naphthalene in its chemical properties. The 9 and 10 positions of anthracene are very reactive. Some important reactions are given below:

(i) *Reduction:* On reduction with sodium and isoamyl alcohol it form 9,10-dihydroanthracene. Catalytic reduction gives 1,2,3,4-tetrahydro and 1,2,3,4,5,6,7,8-octahydro and finally perhydroanthracene.

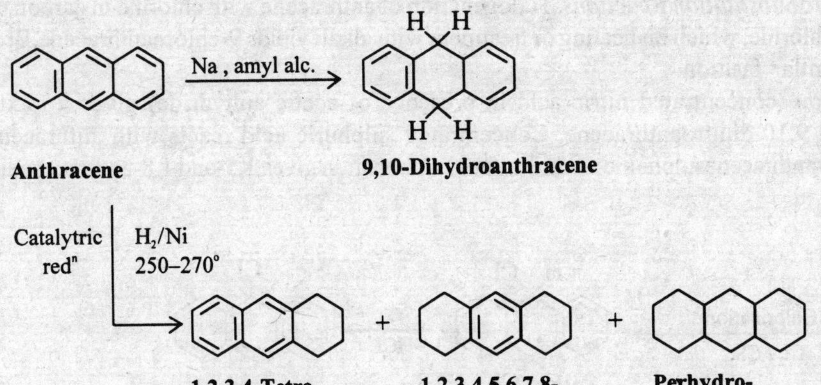

Anthracene   **9,10-Dihydroanthracene**

Catalytric red" | H$_2$/Ni 250–270°

1,2,3,4-Tetra-hydroanthracene   **1,2,3,4,5,6,7,8-octahydroanthracene**   **Perhydro-anthracene**

(ii) *Oxidation:* Oxidation of anthracene, with sodiom dichromate, sulphuric acid or chromium trioxide, acetic acid or air in presence of V$_2$O$_5$ gives 9,10-anthraquinone.

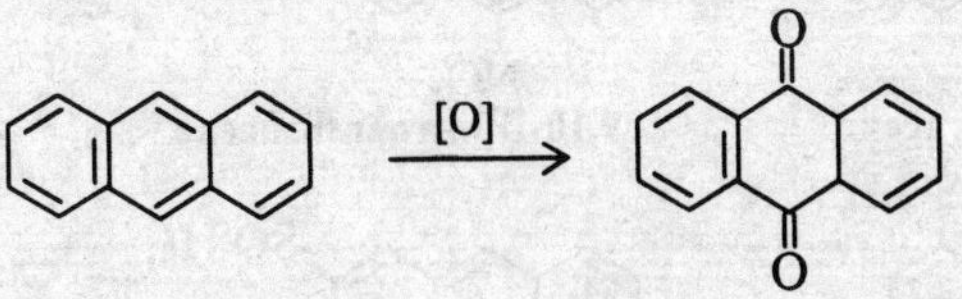

Anthracene   **9,10-Anthraquinone**

(iii) *Diel's Alder Reaction:* Anthracene undergoes Diels Alder reaction with maleic anhydride.

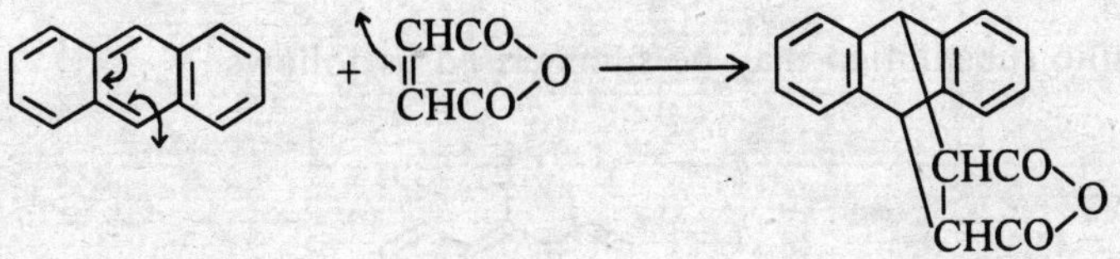

Anthracene   **Maleic anhydride**   **Diel's Alder adduct**

(iv) *Reaction with Oxygen:* Anthracene adds on a molecule of oxygen in presence of light to form a peroxide.

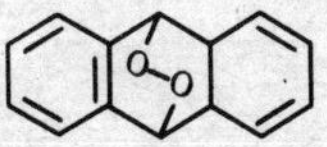

Anthracene peroxide

(v) *Dimerisation:* Anthracene dimerises on exposing its saturated solution in xylene to sunlight.

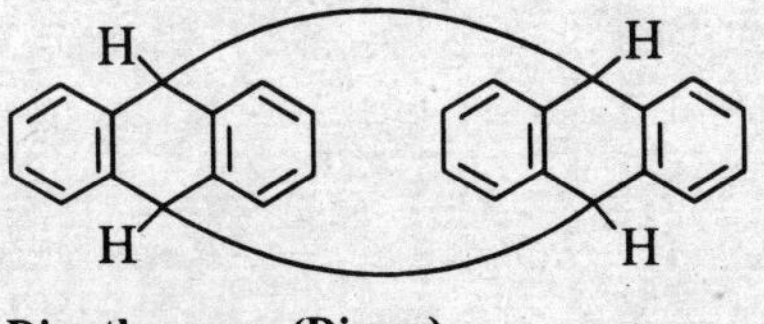

Dianthracene (Dimer)

(vi) *Electrophilic Substitution Reactions:* Halogenation of anthracene with chlorine in carbon disulphide gives anthracene dichloride, which on heating or treatment with alkali yields 9-chloroanthracene. Bromination takes place in a similar fashion.

Similarly, nitration (concentrated nitric acid in presence of acetic anhydride) gives a mixture of 9-nitroanthracene and 9,10-dinitroanthracene. Concentrated sulphuric acid reacts with anthracene to give a mixture of 1- and 2-anthracenesulphonic acids. Further sulphonation gives 1,5-and 1,8-anthracenedisulphonic acids.

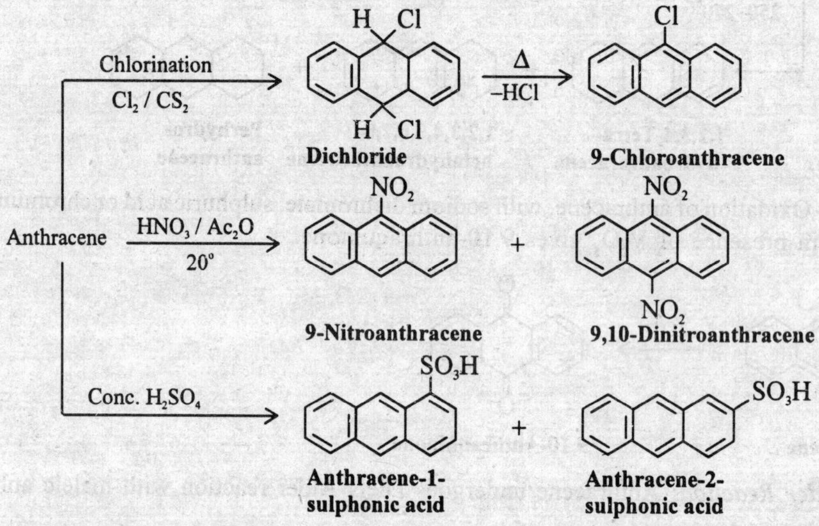

The *Mechanism* of electrophilic substitution may be summerised as follows.

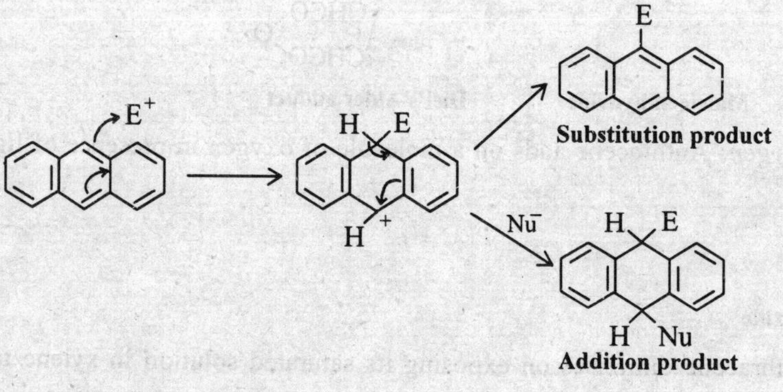

37.3.6   Derivatives of Anthracene

37.3.6.1   Anthraquinone

Preparation:

(i) *From Anthracene:* Oxidation of anthracene with sodium dichromate and sulphuric acid gives anthraquinone.

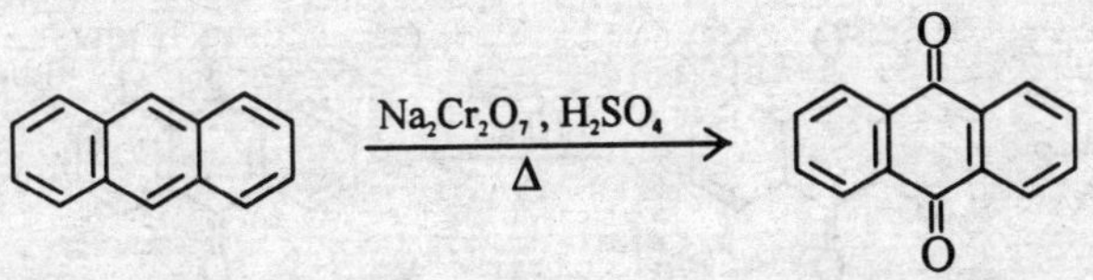

Anthracene   **9,10-Anthraquinone**

On a commercial scale it is manufactured by the vapor phase air oxidation in presence of V_2O_5 at 300-500° of crude anthracene obtained from anthracene oil fraction of coal tar. Any carbazole present in the crude anthracene is oxidised to carbon dioxide and water.

(ii) *Friedel-Crafts Reaction:* The condensation of phthalic anhydride with benzene in presence of anhydrous aluminum chloride gives o-benzoylbenzoic acid, which on cyclisation with concentrated sulphuric acid gives anthraquinone.

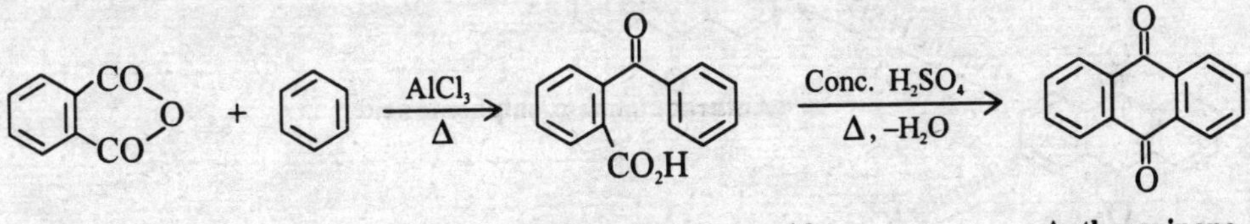

Phthalic anhydride   **o-Benzoylbenzoic acid**   **Anthraquinone**

Properties: It is pale yellow solid, m.p. 286°. It is insoluble in water but soluble in benzene or acetic acid. It is comparatively stable and shows little resemblance to p-benzoquinone. Some important chemical properties are given below.

(i) *Reduction:* It undergoes reduction on treatment with zinc and alkali to form 9,10-dihydroxyanthracene which is isomeric with oxanthrol. However, reduction with tin and hydrochloric acid gives anthrone, which tautomerisés to anthrol (9-hydroxyanthracene) with hot alkali. When reduced with zinc and hydrochloric acid in acetic acid the main product is dianthryl.

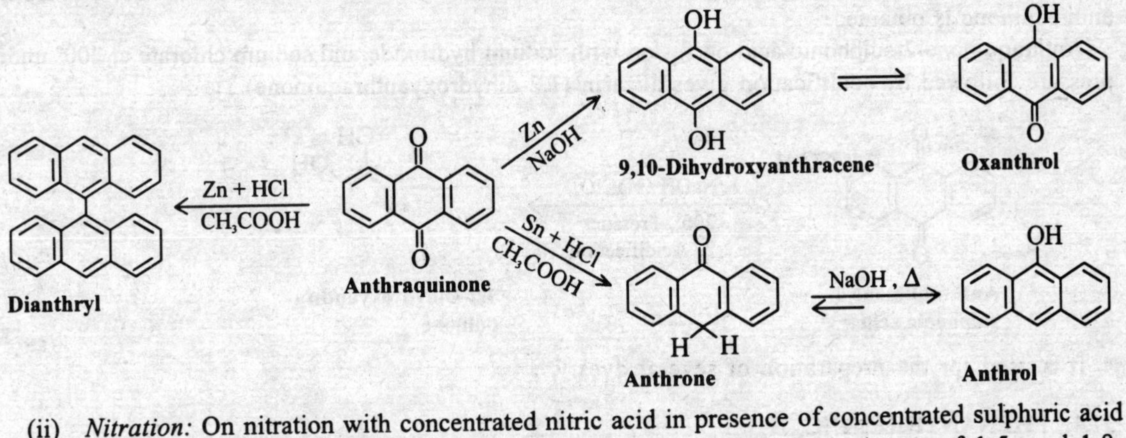

Dianthryl   **Anthraquinone**   **9,10-Dihydroxyanthracene**   **Oxanthrol**

Anthrone   **Anthrol**

(ii) *Nitration:* On nitration with concentrated nitric acid in presence of concentrated sulphuric acid 1-nitroanthraquinone is obtained. However, under drastic conditions, a mixture of 1,5- and 1,8-dinitroanthraquinones is obtained. The nitro group in α- or 1-position is very reactive and on heating with ammonia forms α-aminoanthraquinone.

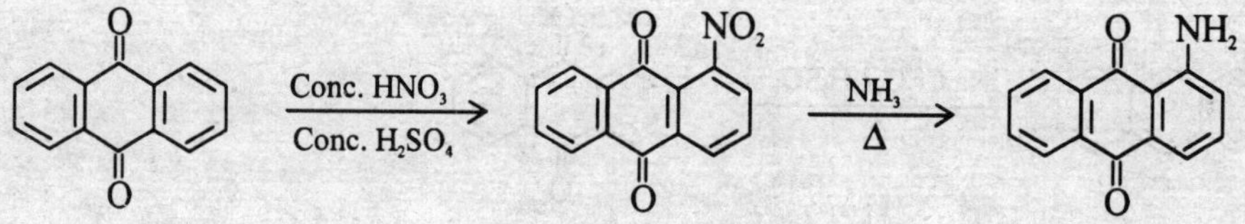

Anthraquinone   1-Nitroanthraquinone   1-Aminoanthraquinone

(iii) *Sulphonation:* Anthraquinone undergoes sulphonation with fuming sulphuric acid at 160° to give β-anthraquinonesulphonic acid. However, sulphonation in presence of mercuric sulphate catalyst at 135° gives α-sulphonic acid.

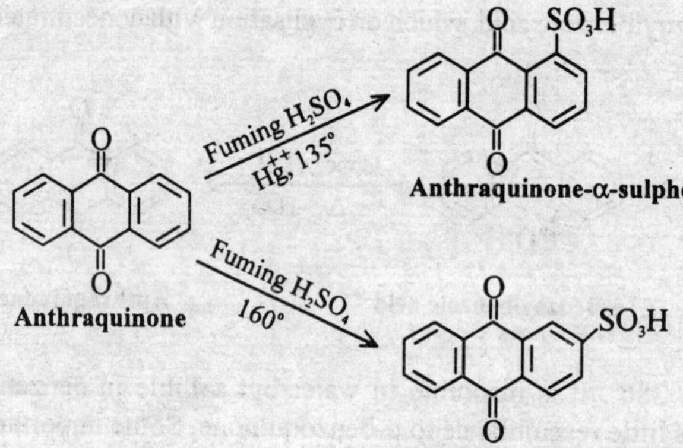

Anthraquinone-α-sulphonic acid

Anthraquinone

Anthraquinone-β-sulphonic acid

Further sulphonation of α-isomer gives a mixture of 1,5- and 1,8-disulphonic acids and further sulphonation of the β-isomer gives a mixture of 2,6- and 2,7-disulphonic acids.

The sulphonic acid group in α- and β-position is very reactive and can easily be replaced by other group. For example with chlorine chloroanthraquinone and with calcium hydroxide at 185°, hydroxyanthraquinone is obtained.

Anthraquinone-2-sulphonic acid on fusion with sodium hydroxide and sodium chlorate at 200° under pressure followed by acidification gives alizarin (1,2-dihydroxyanthraquinone).

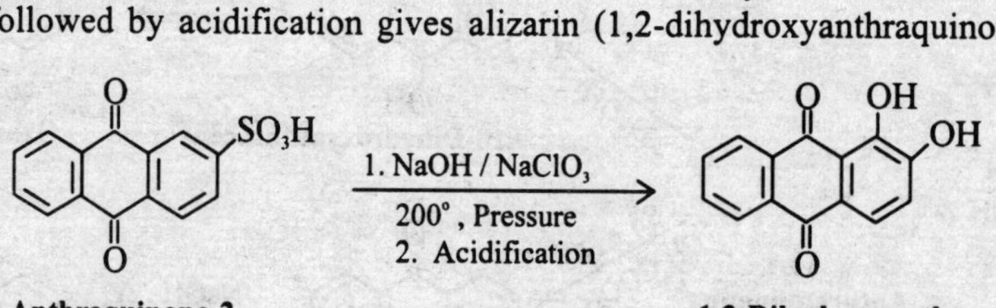

Anthraquinone-2-
sulphonic acid

1. NaOH / NaClO$_3$
   200°, Pressure
2. Acidification

1,2-Dihydroxyanthra-
quinone

It is used for the preparation of several dyes.

37.4 PHENANTHRENE [C$_{14}$H$_{10}$]

It is isomeric with anthracene and is an example of angular polynuclear hydrocarbon. It occurs in anthracene oil fraction of coal tar and is separated from anthracene by means of solvent naphtha. The numbering of various carbon atoms and the position of substituents in phenanthrene is indicated as shown below:

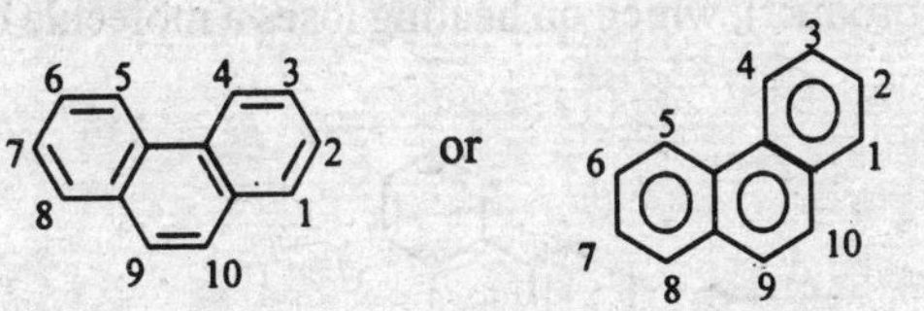

37.4.1 Synthesis of Phenanthrene

It is obtained by the Haworth method from naphthalene and succinic anhydride.

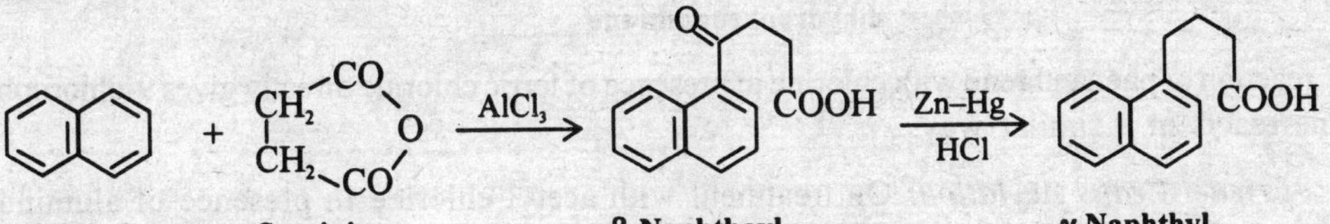

| Naphthalene | Succinic anhydride | β-Naphthoyl-propionic acid | γ-Naphthyl-butyric acid |

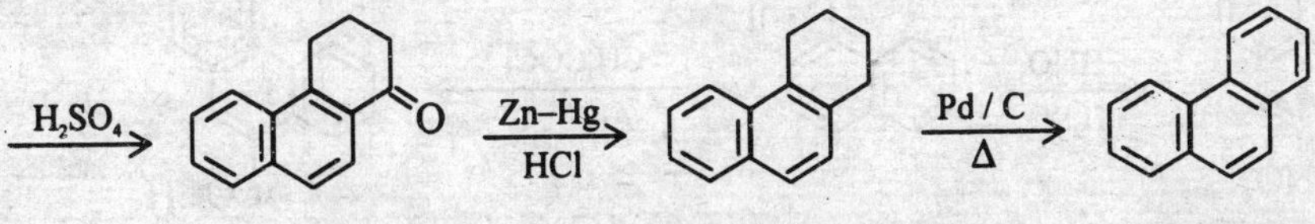

1,2,3,4-Tetrahydro-phenanthrene

Phenanthrene

37.4.2 Properties

Phenanthrene is a colorless solid, m.p. 100°. It is insoluble in water but readily soluble in ethanol, benzene and ether. It shows blue fluorescence in benzene solution.

Some important reactions of phenanthrene are given below.

(i) *Reduction:* With sodium and isopentanol, phenanthrene gives 9,10-dihydrophenanthrene.

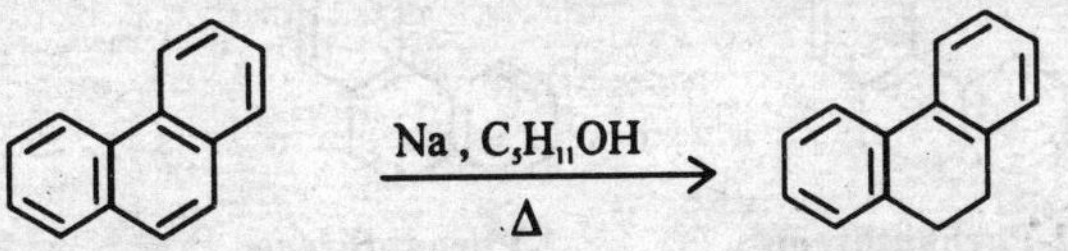

Phenanthrene   9,10-Dihydrophenanthrene

(ii) *Oxidation:* It can be oxidised with potassium dichromate and sulphuric acid or chromium trioxide in acetic acid to form 9,10-phenanthraquinone. However, on further oxidation with hydrogen peroxide in acetic acid diphinic acid is formed.

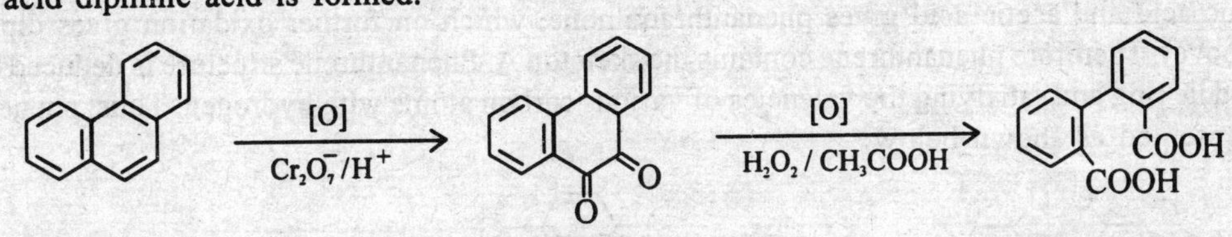

Phenanthrene   9,10-Phenanthra-   Diphinic acid
                      quinone

(iii) *Reaction with Halogens:* Phenanthrene on reaction with chlorine in carbon tetrachloride gives

9,10-dichloro-9,10-dihydrophenanthrene (addition product), which on heating loses a molecule of hydrogen chloride to yield 9-chlorophenanthrene.

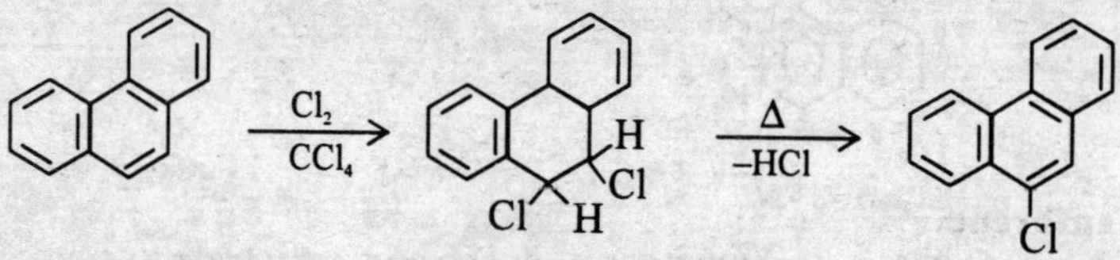

| Phenanthrene | 9,10-Dichloro-9,10-dihydrophenanthrene | 9-Chlorophenanthrene |

The reaction of phenanthrene with chlorine in presence of ferric chloride directly gives 9-chlorophenanthrene. Bromine reacts in a similar way.

(iv) *Friedel-Crafts Acylation:* On treatment with acetyl chloride in presence of aluminum chloride at low temperature, 9-acetylphenanthrene is obtained.

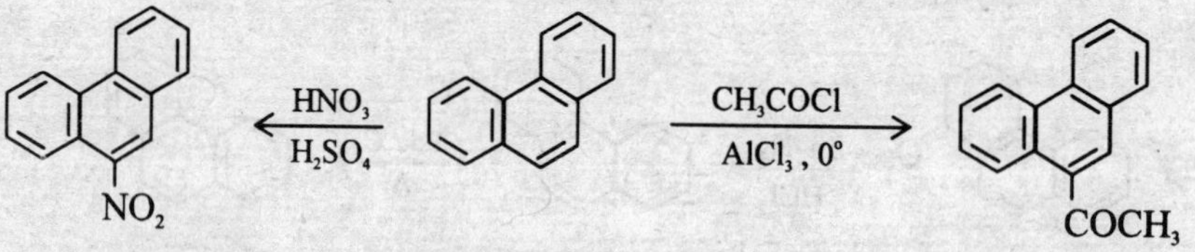

| 9-Nitrophenanthrene | Phenanthrene | 9-Acetylphenanthrene |

(v) *Nitration:* Nitration of phenanthrene with concentrated nitric acid and concentrated sulphuric acid gives 9-nitrophenanthrene.

(vi) *Sulphonation:* Phenanthrene reacts with concentrated sulphuric acid at 120° to give a mixture of 2-phenanthrenesulphonic acid and 3-phenanthrenesulphonic acid.

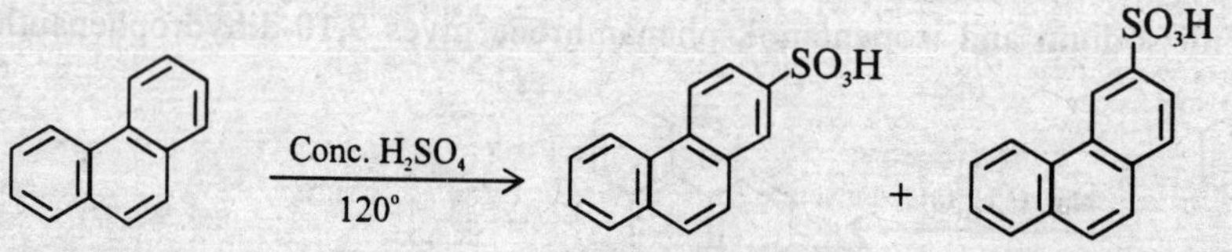

| Phenanthrene | 2-Phenanthrene-sulphonic acid | 3-Phenanthrene-sulphonic acid |

37.4.3 Structure of Phenanthrene

Phenanthrene, $C_{14}H_{10}$ has the same molecular formula as anthracene. Oxidation of phenanthrene with chromic acid and acetic acid gives phenanthraquinone, which on further oxidation gives diphinic acid (see above). Therefore phenanthrene contains the skeleton A. Phenanthrene structure is deduced by closing the middle ring and satisfying the valencies of various carbon atoms with hydrogen. Thus, phenenanthrene is represented as shown below:

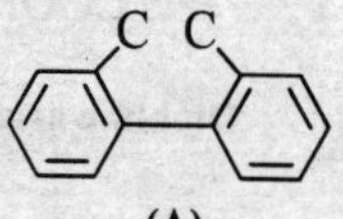

(A) 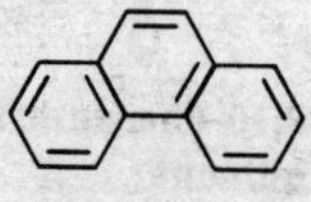   Phenanthrene

The structure of phenanthene is further confirmed by its synthesis as discussed earlier.

Phenanthrene is considered to be a hybrid of the following five resonating structures.

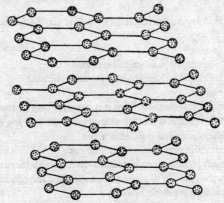

37.5 MISCELLANEOUS

Graphite might be considered under the fused ring aromatic systems. X-rays analysis shows that the carbon atoms are arranged in layers. Each layer is a continuous network of planar, hexagonal rings. The lubricating properties of graphite may be due to slipping of layers over one another. Fig. 37.3 shows the structure of graphite.

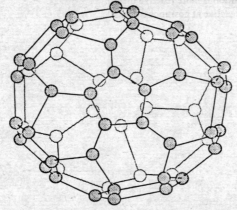

Fig. 37.3:   Structure of Graphite

Fig. 37.4:   Structure of C_{60} Buckminsterfullerene

A newly discovered (1985) allotrope of carbon is buckminsterfullerene. Unlike diamond and graphite, which have indefinite number of carbons, the fullerene has a definite formula i.e. C_{60} which has been constructed using sixty carbon units and utilising sp² hybridised carbons. Here the fused ring aromatic system bends around and closes to form a soccer ball shaped molecule with twenty, six membered rings and twelve, five membered rings as shown in Fig. 37.4. It is called the most symmetrical molecule possible. By the year 1991 fullerene had been reported to possess remarkable properties as superconductors.

37.6   PROBLEMS

1. What are polynuclear hydrocarbons? How they are classified? Give an example of each type.
2. How will you establish the structure of naphthalene? Also give molecular orbital approach for the structure of naphthalene.
3. Give reason why is naphthalene somewhat less aromatic or more reactive than benzene?
4. Give one synthesis of naphthalene. How is it obtained from petroleum fractions?
5. Discuss the reduction products of naphthalene.
6. Give the mechanism of electrophilic substitution reactions of naphthalene. Why substitution goes to position 1 in contrast to position 2?
7. How is α-naphthol obtained? Why is α-naphthol a stronger acid than Phenol? How it is converted into α-naphthylamine?
8. How is β-Naphthol obtained? Compare the reactions of both α and β-naphthols. In what respects these differ from phenol?
9. How will you obtaine the following:
   (a) 1-Bromonaphthalene.   (b) 1,4-Naphthaquinone.
   (c) 1-Phenylazo-2-naphthylamine. (d) Naphthalene-1-carboxylic acid.
   (e) 1,4-Dihydroxynaphthalene.
10. How is the structure of anthracene established? How it is obtained on a commercial scale?
11. Give reason to support that anthracene is much less aromatic than benzene or naphthalene.
12. Give different reduction products obtained from anthracene.
13. Give mechanism of the electrophilic substitution in anthracene.
14. How is anthraquinone obtained? How it is converted to 9- hydroxyanthracene and 1,2-dihydroxyanthraquinone?
15. Give the product obtained by the reaction of anthracene and maleic anhydride.
16. How is the structure of phenanthrene established?
17. Explain:
   (a) In naphthalene α position is more reactive than β.
   (b) Anthracene is more reactive at positions 9 and 10.

38

Heterocyclic Compounds

38.1 INTRODUCTION

Cyclic compounds in which one or more carbon atoms are replaced with hetero atoms, like oxygen, nitrogen and sulphur are known as heterocyclic compounds.

A number of compounds (ethylene oxide, lactones, acid anhydrides and acid amides) containing a heterocyclic ring are not considered to be heterocyclic compounds, as the rings may be opened up easily and they do not possess aromatic properties. Thus, heterocyclic compounds are the compounds having five or six membered ring and at least one hetero atom in the ring. They are comparatively stable and exhibit aromatic character. The heterocyclic compounds are also designated as *Heteroaromatics*.

38.2 CLASSIFICATION

The heterocyclic compounds are classified on the basis of ring size, number and nature of hetero atoms and the presence of fused ring system. The three classes of heterocyclic compounds are as follows.

(1) *Five Membered Heterocyclic Compounds:* These are considered to be derived from five membered rings by replacement of one or more carbon atom with hetero atom. They are further divided into two types.

(a) *Compounds with One Hetero Atom:* Simple examples of this type are furan, thiophene and pyrrole containing oxygen, sulphur and nitrogen respectively as the hetero atom. These are also referred to as oxole, thiole and azole respectively.

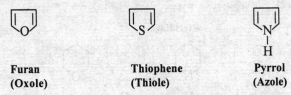

Furan	Thiophene	Pyrrol
(Oxole)	(Thiole)	(Azole)

(b) *Compounds with more than One Hetero Atom:* In this case, the compounds contain two or more hetero atoms which may be same or different. Some examples are given below:

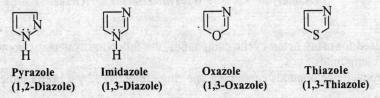

Pyrazole	Imidazole	Oxazole	Thiazole
(1,2-Diazole)	(1,3-Diazole)	(1,3-Oxazole)	(1,3-Thiazole)

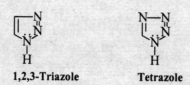

1,2,3-Triazole   **Tetrazole**

(2)   *Six Membered Heterocyclic Compounds:* These are considered to be derived from six membered rings by replacement of one or more carbon atoms with hetero atoms.

(a)   *Compounds with One Hetero Atom*, for example pyridine, pyran and thiopyran etc.

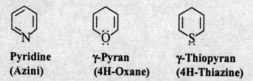

Pyridine   **γ-Pyran**   **γ-Thiopyran**
(Azini)   **(4H-Oxane)**   **(4H-Thiazine)**

(b)   *Compounds with two Hetero Atoms*, for example pyrimidine, pyrazine and pyridazine, etc.

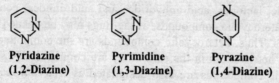

Pyridazine   **Pyrimidine**   **Pyrazine**
(1,2-Diazine)   **(1,3-Diazine)**   **(1,4-Diazine)**

(3)   *Condensed Heterocyclic Compounds:* These contain two or more rings fused together. The rings can be carbocyclic or heterocyclic. Some examples are given below:

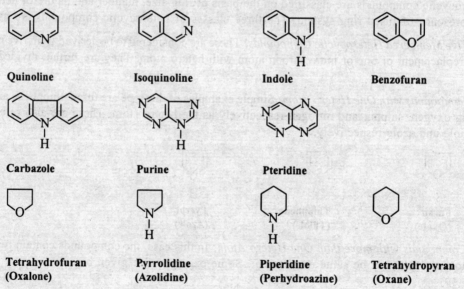

Quinoline   **Isoquinoline**   **Indole**   **Benzofuran**

Carbazole   **Purine**   **Pteridine**

Tetrahydrofuran   **Pyrrolidine**   **Piperidine**   **Tetrahydropyran**
(Oxalone)   **(Azolidine)**   **(Perhydroazine)**   **(Oxane)**

38.3   NOMENCLATURE

Though trival names are used for many heterocyclic compounds, the following systematic nomenclature is used for heterocyclic compounds.

The names of the heterocyclic compounds containing one ring are derived by a prefix indicating the nature of the heteroatom present. For example, *oxa* for oxygen, *thia* for sulphur and *aza* for nitrogen. If two or more same hetero atoms are present, the prefix *di* or *tri* is used. In case the two hetero atoms are different, priority is given to the hetero atom present in the higher group in the periodic table and in a group the atom of lower atomic number gets priority. Thus, the order of priority of hetero atoms is O, S, N, P, Si. For example, in thiazole containing hetero atoms sulphur and nitrogen, the former takes priority over the later.

In monocyclic compounds containing only one hetero atom, numbering starts from the hetero atom. The ring is numbered to give substitutents or other hetero atoms the lowest numbers possible. If the hetero atoms are different, the numbering starts from the atom of higher priority and the proceeds round the ring in order of preference. Further, the state of hydrogenation is indicated by the prefixes dihydro, tetrahydro etc.

Some heterocyclic compounds and their nomenclature are given in preceeding Sec. 38.2.

38.4   FIVE MEMBERED HETEROCYCLIC COMPOUNDS

Furan, pyrrole and thiophene are the important members of this group with one hetero atom. The ring atoms are numbered by giving numerals with the lowest number to the heteroatom. Alternatively, carbon atoms next to the hetero atom are often designated as α while those next to them as β carbon atoms.

(X = O, N or S)

Since two positions (α and β) are available for substitution, these heterocycles can form two monosubstitution products. The number of isomers becomes four when the two substituents are alike and six if they are different.

38.4.1   Preparation

The Paal-Knorr Synthesis: This is the most commonly used method of synthesising five membered heterocyclic rings and their derivatives. It consists in the reaction of a 1,4- dicarbonyl compounds with an appropriate source of hetero atom, viz. P_2O_5, $(NH_4)_2CO_3$ or P_2S_5.

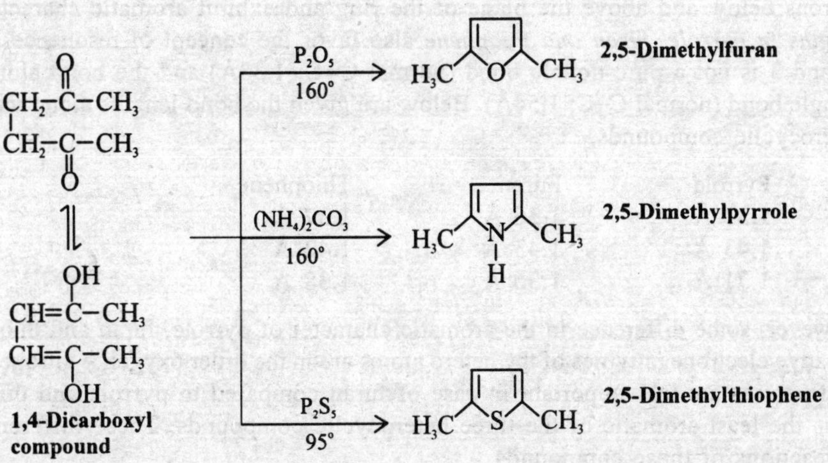

The mechanism has been illustrated in the case of a thiophene derivative.

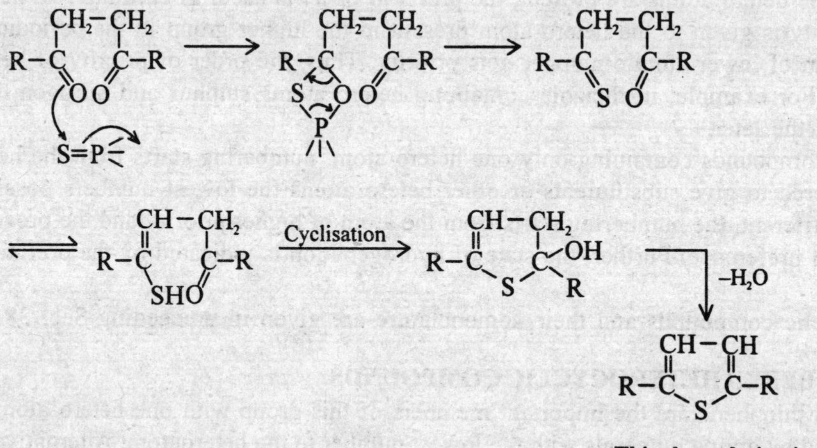

Thiophene derivative

38.4.2   General Properties

These are colorless liquids having aromatic character like benzene and are resonance stabilised. Their resonance energies are much lower than that of benzene and are in the order thiophene > pyrrole > furan. The resonance contributing structures of these compounds are given below.

(Resonance stabilized heterocyclic compounds)
Z = O, S or NH

All the resonating structures are not equivalent but have dipolar ionic character.

These heterocyclic compounds have a planar pentagonal structure as the four carbon atoms and the hetero atom (N, O or S) are in sp^2 hybridised state. They have a total of six π electrons (in two π and one p orbital) available for delocalisation. Thus, like benzene, these heterocyclic compounds form a cloud of electrons below and above the plane of the ring and exhibit aromatic character.

The bond lengths in pyrrole, furan and thiophene also favor the concept of resonance as the bond along carbon 2 and 3 is not a pure double bond (normal C=C; 1.33Å) and the bond along carbon 3 and 4 is not a single bond (normal C–C; 1.54Å). Below are given the bond lengths along various bonds in the three heterocyclic compounds.

	Pyrrole	Furan	Thiophene
C_2–C_3	1.37 Å	1.36 Å	1.37 Å
C_3–C_4	1.43 Å	1.43 Å	1.42 Å
C–X	1.71 Å	1.36 Å	1.38 Å

There are, however, some difference in the aromatic character of pyrrole, furan and thiophene. This is because the relative electronegativities of the hetero atoms are in the order oxygen > nitrogen > sulphur. The resonance structures are less important in case of furan compared to pyrrole and thiophene and therefore furan is the least aromatic of the three heterocyclic compounds. These differences will be clear from the reactions of these compounds.

Spectroscopic Properties: In UV region they exhibit absorption maxima as shown below:

Pyrrole   :   172, 183, 211, 304 nm
Furan   :   191, 205 nm
Thiophene   :   215, 231 nm

In the IR spectra all the three show C−H stretching bands near 3000-3100 cm^{-1} along with bands at 1590, 1490 and 1400 cm^{-1} characteristic of the aromatic system. However, pyrrole exhibits in addition to the above, a band in the region 3400-3500 cm^{-1} due to N−H stretching.

Chemical Reactivity: The reactivity of these compounds is so much that they even undergo reactions like Reimer-Tiemann, formylation and coupling with diazonium salts.

The electrophilic substitution occurs either at α or β position, but the α substitution is preferred. This is attributed to the greater resonance stabilisation of the intermediate carbocation formed during α substitution (three contributing structures) compared to the intermediate carbocation formed in the β substitution (two contributing structures). The α and β substitution is represented below:

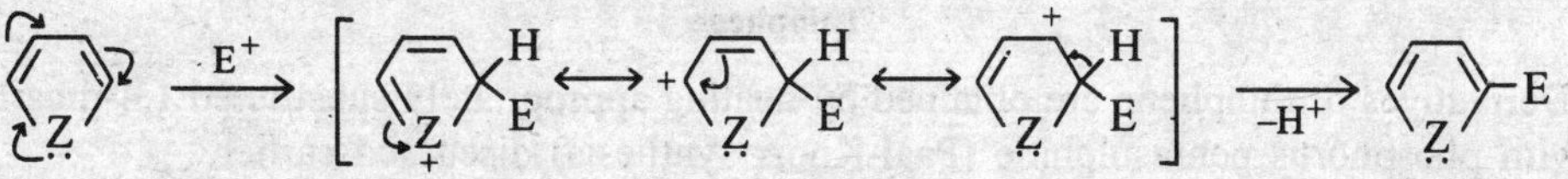

α-Substitution (2 or 5 substitution)

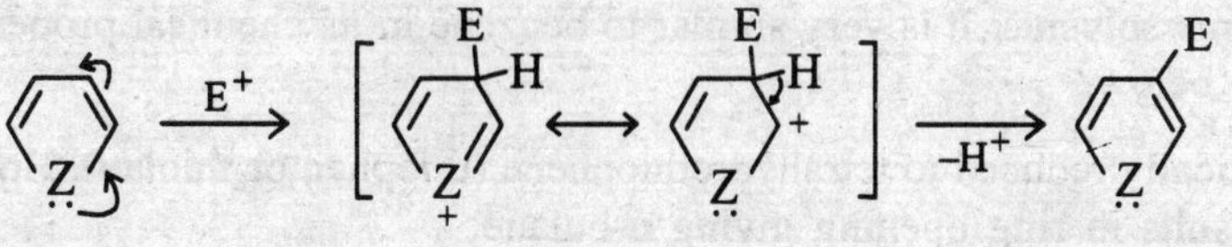

β-Substitution (3 or 4 substitution)

For a detailed discussion on chemical reactivities see individual members.

38.4.3   Individual Members

38.4.3.1   Thiophene [C$_4$H$_4$S]

It occurs in coal tar and shale oils. Its b.p. 84° is close to the boiling point of benzene and so it is difficult to separate from the benzene fraction obtained from coal-tar. The separation is effected by shaking benzene with concentrated sulphuric acid in cold when thiophene forms water soluble sulphonic acid. Benzene remains unaffected under these conditions. Thiophene is obtained by steam distillation of the thiophenesulphonic acid.

In an alternative method, benzene is refluxed with aqueous mercuric acetates when thiophene forms acetoxymercury derivative, which is insoluble in benzene. Thiophene is recovered from its mercurated derivative by distilling with hydrochloric acid.

The presence of thiophene is detected by the indophenine reaction which involves the development of a blue color on treatment of thiophene with isatin and sulphuric acid.

Preparation

(i) Thiophene is obtained commercially by passing a mixture of acetylene and hydrogen sulphide through a tube containing alumina at 400°.

$$2CH{\equiv}CH + H_2S \xrightarrow[-H_2O]{Al_2O_3 / 400°} \langle\!\!\langle \ \rangle\!\!\rangle_S$$

Acetyline **Hydrogen sulphide** **Thiophene**

(ii) It is also manufactured by heating sodium succinate with phosphorus trisulphide.

$$\begin{array}{c} CH_2COONa \\ | \\ CH_2COONa \end{array} + P_2S_3 \xrightarrow{\Delta} \langle\!\!\langle \ \rangle\!\!\rangle_S$$

Sodium succinate **Thiophene**

(iii) The reaction of n-butane and sulphur in vapor state at 650-700° gives thiophene (industrial method).

$$\begin{array}{c} CH_2{-}CH_2 \\ \diagup \qquad \diagdown \\ CH_3 \qquad CH_3 \end{array} + 4S \xrightarrow{650-700°} \langle\!\!\langle \ \rangle\!\!\rangle_S + 3H_2S$$

Thiophene

(iv) Derivatives of thiophene are obtained by heating appropriately substituted 1,4-dicarbonyl compounds with phosphorus pentasulphide (Paal-Knorr synthesis) discussed earlier.

Properties: Thiophene is a colorless liquid, b.p. 84° with a characteristic disagreeable odor. It is insoluble in water but is soluble in organic solvents. It is very similar to benzene in its chemical properties. Some important reactions are given below.

(i) *Reduction:* It is catalytically reduced to tetrahydrothiophene (thiophan or thiolan). However, use of raney nickel as catalyst results in ring opening giving n-butane.

$$\langle\!\!\langle \ \rangle\!\!\rangle_S \xrightarrow{Reduction} \begin{cases} \xrightarrow{H_2/ Pd} \langle\!\!\langle \ \rangle\!\!\rangle_S & \textbf{Tetrahydrothiophene} \\ \\ \xrightarrow{H_2/ Ni} CH_3CH_2CH_2CH_3 + NiS & \textbf{n-Butane} \end{cases}$$

(ii) *Reaction with Butyllithium:* Thiophene reacts with butyllithium to give 2-thienyllithium, which shows the usual reactions of organo lithium compounds. For example, it reacts with carbon dioxide to give thiophene-2-carboxylic acid.

$$\langle\!\!\langle \ \rangle\!\!\rangle_S + C_4H_9Li \longrightarrow \langle\!\!\langle \ \rangle\!\!\rangle_S{-}Li \xrightarrow[2.\ H^+]{1.\ CO_2} \langle\!\!\langle \ \rangle\!\!\rangle_S{-}COOH$$

Thiophene **2-Thienyllithium** **Thiophene-2-carboxylic acid**

(iii) *Electrophilic Substitution Reactions:* Thiophene undergoes electrophilic substitution reactions under mild conditions. As already stated, substitution taken place in the α position. Some of the electrophilic substitution reactions are given below.

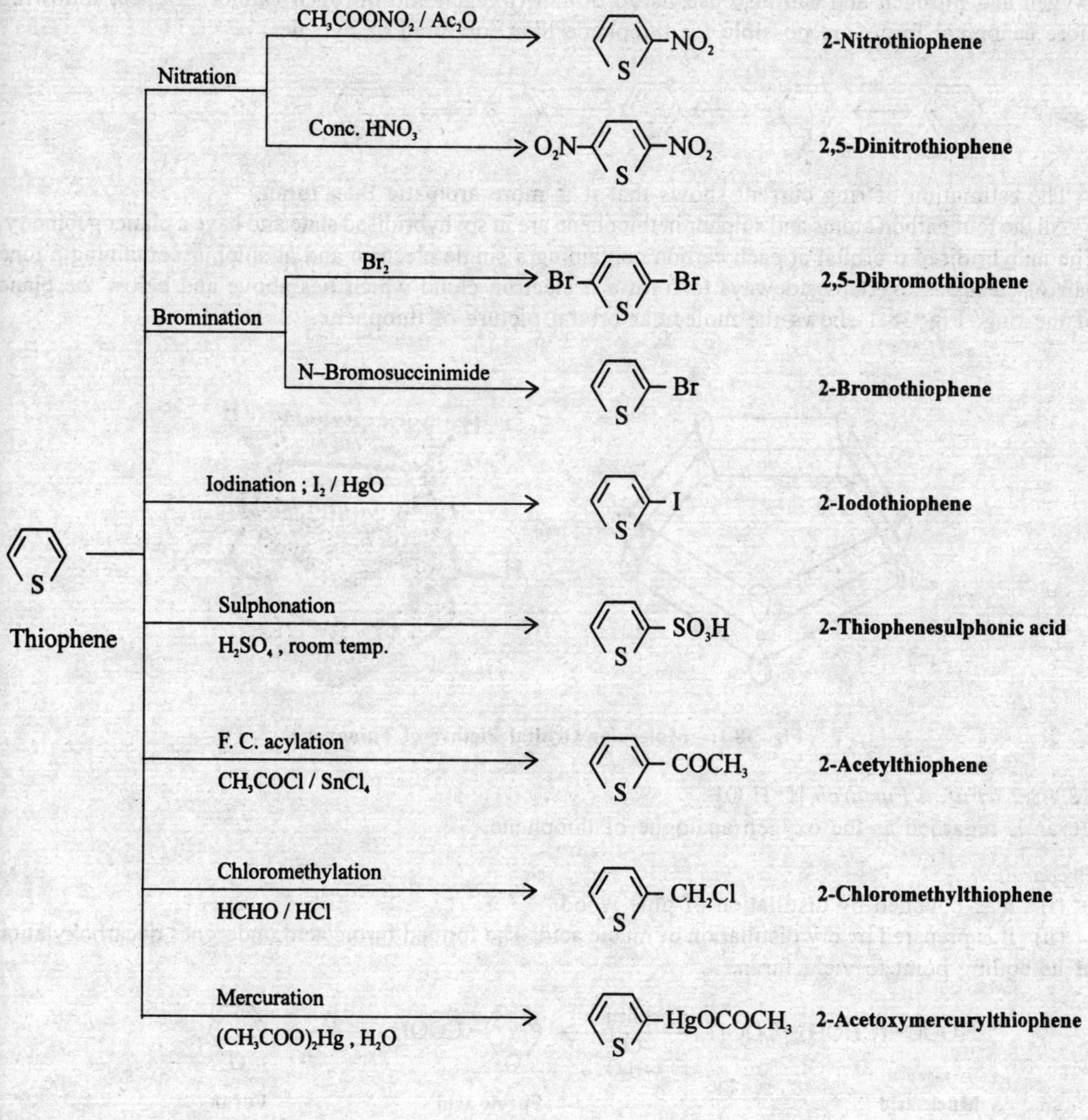

Reaction	Reagent	Product	Name

Nitration — CH₃COONO₂ / Ac₂O → (thiophene)–NO₂ **2-Nitrothiophene**

Nitration — Conc. HNO₃ → O₂N–(thiophene)–NO₂ **2,5-Dinitrothiophene**

Bromination — Br₂ → Br–(thiophene)–Br **2,5-Dibromothiophene**

Bromination — N–Bromosuccinimide → (thiophene)–Br **2-Bromothiophene**

Iodination ; I₂ / HgO → (thiophene)–I **2-Iodothiophene**

Sulphonation
H₂SO₄ , room temp. → (thiophene)–SO₃H **2-Thiophenesulphonic acid**

F. C. acylation
CH₃COCl / SnCl₄ → (thiophene)–COCH₃ **2-Acetylthiophene**

Chloromethylation
HCHO / HCl → (thiophene)–CH₂Cl **2-Chloromethylthiophene**

Mercuration
(CH₃COO)₂Hg , H₂O → (thiophene)–HgOCOCH₃ **2-Acetoxymercurylthiophene**

(Thiophene structure at left)

Structure of Thiophene

Thiophene is a resonance hybrid of the following structures.

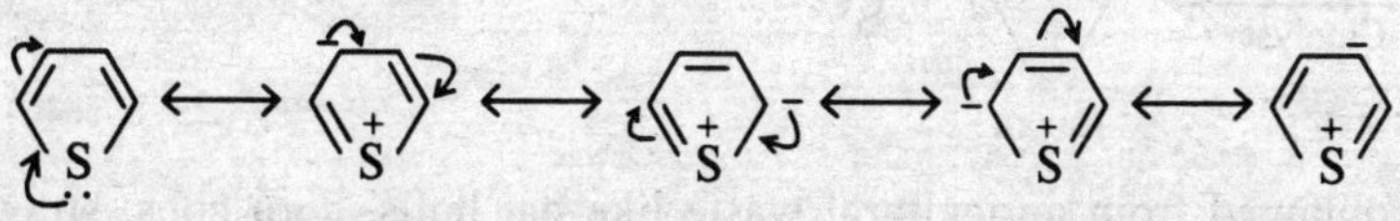

Its resonance energy is 28-31 kcal/mol. The sulphur atom contributes two electrons to form a $4\pi+2$ electron system which provides stability to the molecule. However, sulphur is less electronegative than oxygen and nitrogen and can also use its 3d orbital (oxygen and nitrogen cannot). Hence, following more canonical forms are possible for thiophene than for furan or pyrrole.

$$\langle\!\!\!\langle \overset{.}{\underset{S}{}} \rangle\!\!\!\rangle \longleftrightarrow \langle\!\!\!\langle \overset{.}{\underset{\bar{S}}{}} \rangle\!\!\!\rangle^+ \longleftrightarrow \langle\!\!\!\langle \overset{+}{\underset{\bar{S}}{}} \rangle\!\!\!\rangle \longleftrightarrow \langle\!\!\!\langle \overset{+}{\underset{\bar{S}}{}} \rangle\!\!\!\rangle \longleftrightarrow {}^+\langle\!\!\!\langle \overset{.}{\underset{S}{}} \rangle\!\!\!\rangle$$

The estimation of ring current shows that it is more aromatic than furan.

All the four carbon atoms and sulphur in thiophene are in sp^2 hybridised state and have a planer geometry. The unhybridised p orbital at each carbon containing a single electron and at sulphur containing a lone pair of electrons overlaps sideways to form a π electron cloud which lies above and below the plane of the ring. Fig. 38.1 shows the molecular orbital picture of thiophene.

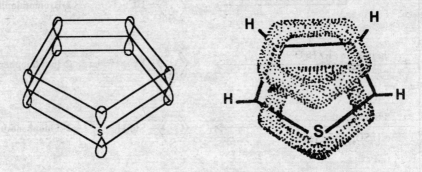

Fig. 38.1: Molecular Orbital Picture of Thiophene

38.4.3.2 Furan/Furfuran [C_4H_4O]
Furan is regarded as the oxygen analogue of thiophene.

Preparation
(i)   It is obtained by distillation of pine wood.

(ii) It is prepared by dry distillation of mucic acid. The formed furoic acid undergoes decarboxylation at its boiling point to yield furan.

$$HOOC-(CHOH)_4-COOH \xrightarrow[\substack{-CO_2 \\ -H_2O}]{\text{Distillation}} \langle\!\!\!\langle \underset{O}{} \rangle\!\!\!\rangle-COOH \xrightarrow[-CO_2]{\Delta} \langle\!\!\!\langle \underset{O}{} \rangle\!\!\!\rangle$$

Mucic acid   **Furoic acid**   **Furan**

(iii)   It is manufactured by catalytic decomposition of furfural in steam in presence of an oxide catalyst.

$$\langle\!\!\!\langle \underset{O}{} \rangle\!\!\!\rangle-CHO \xrightarrow[\text{Catalyst}]{\text{Steam / 400}^\circ} \langle\!\!\!\langle \underset{O}{} \rangle\!\!\!\rangle + CO$$

Furfural   **Furan**

The furfural required is obtained from agricultural waste like oat hulls, corn cobs, straw, rice hull etc containing pentosans with hot dilute hydrochloric acid.

$$(C_5H_8O_4)_n \xrightarrow{\text{H}^+/\text{H}_2\text{O}} \underset{\text{CH}_2\text{OH}}{\overset{\text{CHO}}{(\text{CHOH})_3}} \xrightarrow[-3\text{H}_2\text{O}]{\text{H}^+} \left\langle\!\!\left\langle\underset{O}{}\right\rangle\!\!\right\rangle\!\!-\!\text{CHO}$$

Pentosan   **Pentose**   **Furfural**

(iv)   Derivatives of furan are obtained by heating appropriately substituted 1,4-dicarbonyl compounds with phosphorus pentoxide (Paal-Knorr synthesis) already discussed.

(v)   Furan derivatives are also obtained by Feist-Benary synthesis which involves the condensation of α-chloroketones with a β-ketoester in presence of pyridine.

$$\underset{\text{CH}_3-\text{CO}}{\overset{\text{Et O}_2\text{C}-\text{CH}_2}{|}} + \underset{\text{CH}_2\text{Cl}}{\overset{\text{CO}-\text{CH}_3}{|}} \xrightarrow{\text{C}_5\text{H}_5\text{N}} \underset{\text{H}_3\text{C}}{\overset{\text{H}_5\text{C}_2\text{OOC}}{}}\left.\!\!\!\left\langle\!\!\left\langle\underset{O}{}\right\rangle\!\!\right\rangle\right.\!\!\!\overset{\text{CH}_3}{}$$

β-Ketoester   **α-chloroketone**   **Ethyl 2,4-dimethylfuran-3-carboxylate**

(vi)   Another method for the preparation of furan derivatives is from ethyl acetoacetate. The steps involved in the synthesis are as follows:

$$2\text{CH}_3\text{CO}\overset{-}{\text{CH}}\overset{+}{\text{Na}}\text{COOC}_2\text{H}_5 + \text{I}_2 \longrightarrow 2\text{NaI} + \underset{\text{CH}_3\text{COCH}-\text{COOC}_2\text{H}_5}{\overset{\text{CH}_3\text{COCH}-\text{COOC}_2\text{H}_5}{|}}$$

Sod. salt of EAA   **Diacetosuccinic ester**

$$\xrightarrow{\text{H}_2\text{SO}_4} \underset{\text{H}_3\text{C}}{\overset{\text{HOOC}}{}}\left.\!\!\!\left\langle\!\!\left\langle\underset{O}{}\right\rangle\!\!\right\rangle\right.\!\!\!\overset{\text{COOH}}{\underset{\text{CH}_3}{}}$$

2,5-Dimethylfuran-3,4-dicarboxylic acid

Properties: Furan is a colorless liquid, b.p. 31°. It is practically insoluble in water but is soluble in alcohol and ether. It resembles thiophene in its chemical reactions, but is more reactive. Some important reactions of furan are given below:

(i)   *Reduction:* It may be reduced catalytically to form tetrahydrofuran (THF), which is commonly used as solvent.

$$\left\langle\!\!\left\langle\underset{O}{}\right\rangle\!\!\right\rangle + 2\text{H}_2 \xrightarrow[\text{Raney Ni}]{\overset{\text{Pd / PdO}}{\text{or}}} \left\langle\!\!\left\langle\underset{O}{}\right\rangle\!\!\right\rangle$$

Furan   **Tetrahydrofuran (THF)**

(ii)   *Reaction with n-Butyllithium:* Like thiophene, it reacts with n-butyllithium to give 2-furanlithium, which exhibits usual reactions of organo lithium compounds. For example, it reacts with carbon dioxide to give furoic acid.

$$\left\langle\!\!\left\langle\underset{O}{}\right\rangle\!\!\right\rangle + \text{C}_4\text{H}_9\text{Li} \longrightarrow \left\langle\!\!\left\langle\underset{O}{}\right\rangle\!\!\right\rangle\!\!-\!\text{Li} \xrightarrow{\text{CO}_2} \left\langle\!\!\left\langle\underset{O}{}\right\rangle\!\!\right\rangle\!\!-\!\text{COOH}$$

Furan   **2-Furanlithium**   **Furoic acid**

(iii)   *Diels Alder Reaction:* Furan exhibits considerable diene character. It undergoes Diels-Alder reaction with maleic anhydride. Thiophene and pyrrole do not give this reaction.

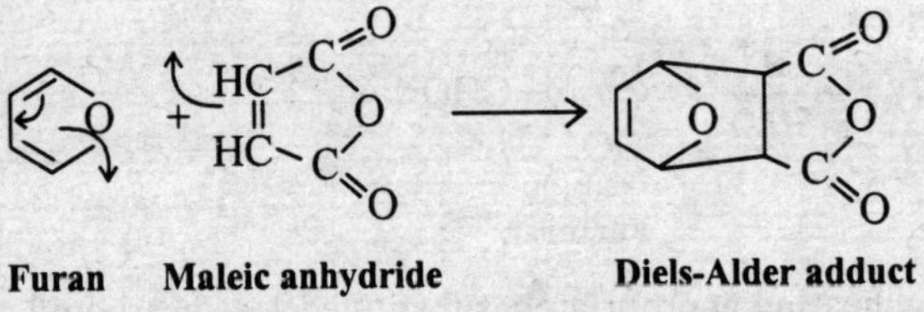

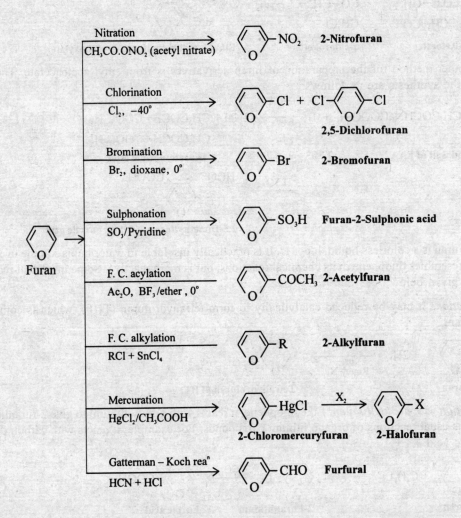

Furan   Maleic anhydride   Diels-Alder adduct

(iv)   *Electrophilic Substitution Reactions:* Furan is more reactive than benzene and thiophene in electrophilic substitution reactions. It gives disubstituted products and undergoes 2,5- additions in preference to substitution reactions. some of the important reactions are given below:

Structure of Furan
Furan is a resonance hybrid of the following structures.

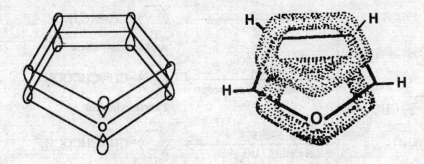

Its resonance energy (16 kcal/mol) is considerably less than that of benzene. Therefore, furan is expected to behave both as an aromatic compound and a diene. The X-ray analysis has shown that the C–O bond (1.37 Å) in furan is shorter than the normal C–O bond (1.43Å) which also confirms the hybrid character of furan molecule.

Fig. 38.2 shows the molecular orbital picture of furan which is similar to that of thiophene except that the hetero atom sulphur is replaced by oxygen.

Fig. 38.2:   Molecular Orbital Picture of Furan

38.4.3.3   Furfural

It is an important derivative of furan and is prepared by distilling bran (containing pentoses) with sulphuric acid.

$$\text{HOCH}_2\,(\text{CHOH})_3\text{CHO} \xrightarrow[\text{dil H}_2\text{SO}_4]{\text{Distillation}} \underset{O}{\langle \rangle}\text{–CHO} + 3\text{H}_2\text{O}$$

Pentoses   **Furfural**

The method is used for the industrial preparation of furfural.

Furfural is a colorless liquid, b.p. 162° having a pleasant characteristic odor. It is similar to benzaldehyde in its chemical reactions. Some important reactions are given below:

(i)   *Cannizzaro Reaction:* With aqueous sodium hydroxide furfural forms furfuryl alcohol and furoic acid.

$$\underset{O}{\langle \rangle}\text{–CHO} + \text{NaOH} \longrightarrow \underset{O}{\langle \rangle}\text{–CH}_2\text{OH} + \underset{O}{\langle \rangle}\text{–COOH}$$

Furfural   **Furfuryl alcohol**   **Furoic acid**

(ii)   *Benzoin Condensation:* With ethanolic potassium cyanide it forms furoin, which upon oxidation gives furil.

$$\underset{O}{\langle \rangle}\text{–CHO} + \text{KCN} \xrightarrow{\text{C}_2\text{H}_5\text{OH}} \underset{O}{\langle \rangle}\text{–CHOHCO–}\underset{O}{\langle \rangle}$$

Furfural   **Furoin**

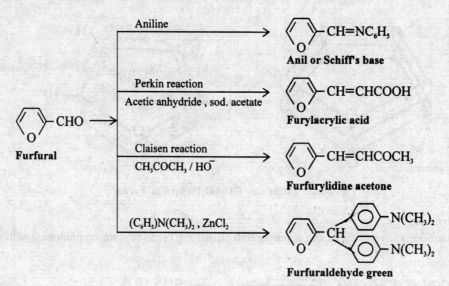

Furil   **Furilic acid (Potassium salt)**

(iii)   *Benzilic Acid Rearrangement:* Furil formed by benzoin condensation [as given in (ii) above] undergoes benzilic acid rearrangement on heating with aqueous potassium hydroxide to give furilic acid.

(iv)   *Miscellaneous Reactions:* Furfural reacts with aniline to form anil, undergoes Perkins reaction and the Claisen reaction. It condenses with dimethylaniline in presence of zinc chloride to form furfuraldehyde green (analogous to malachite green, a dye).

Tests
(i) It gives red color with aniline and hydrochloric acid.
(ii) It turns green a pine splint moistened with hydrochloric acid.

Uses: It is used for the preparation of dyes, plastics and furmaric acid. It is also used as a solvent in the manufacture of synthetic rubber.

Furfuryl Alcohol
It is prepared by reducing furfural by cannizzaro reaction. On a commercial scale, it is also obtained by the catalytic reduction of furfural with copper chromite.

Furoic Acid/Furan-2-carboxylic Acid
As mentioned earlier, it is obtained by direct distillation of mucic acid. Also the oxidation of furfural with cuprous oxide-silver oxide in aqueous sodium hydroxide gives 86-90% furoic acid.

Furoic acid behaves like an unsaturated aliphatic acid and not like benzoic acid. It adds on four bromine atoms when reacted with bromine vapor. On oxidation with bromine water it gives fumaric acid.

Tetrahydrofuran/THF
It is obtained commercially from acetylene as follows.

$$\underset{\text{Acetylene}}{\overset{\text{CH}}{\underset{\text{CH}}{\mathbin{\vert\vert\vert}}}} + \text{Ag} \longrightarrow \overset{\text{CAg}}{\underset{\text{CAg}}{\mathbin{\vert\vert\vert}}} \xrightarrow{2\text{HCHO}} \underset{\text{2-Butyne-1,4-diol}}{\overset{\displaystyle \text{C}\!\equiv\!\text{C}}{\underset{\text{OH} \quad \text{OH}}{\text{CH}_2 \qquad \text{CH}_2}}}$$

$$\xrightarrow{\text{H}_2/\text{Ni}} \underset{\text{Butane-1,4-diol}}{\overset{\text{CH}_2-\text{CH}_2}{\underset{\text{OH} \quad \text{OH}}{\text{CH}_2 \qquad \text{CH}_2}}} \xrightarrow[-\text{H}_2\text{O}]{\Delta} \underset{\text{THF}}{\overset{\text{H}_2\text{C}-\text{CH}_2}{\underset{\text{O}}{\text{H}_2\text{C} \qquad \text{CH}_2}}}$$

It is a colorless liquid with a characteristic odor and is soluble in water. It behaves like a cyclic ether and does not show aromatic properties. The ring opening takes place with acids. This reaction is used for the preparation of adipic acid, hexamethylene diamine and tetramethylene glycol.

$$\underset{\text{THF}}{\langle\text{O}\rangle} \xrightarrow{\text{HCl}} \underset{\substack{\text{Tetramethylene}\\\text{chlorohydrin}}}{\langle\text{HO Cl}\rangle} \xrightarrow[\text{ZnCl}_2]{\text{HCl}} \underset{\text{Cl Cl}}{\langle\;\rangle} \xrightarrow{\text{KCN}} \underset{\text{NC CN}}{\langle\;\rangle} \xrightarrow{\text{H}_2/\text{Ni}} \underset{\substack{\text{Hexamethylene diamine}}}{\text{H}_2\text{N}(\text{CH}_2)_6\text{NH}_2}$$

$$\downarrow \text{H}_2\text{O}/\text{H}^+ \qquad\qquad \downarrow \text{H}_2\text{O}/\text{H}^+$$

$$\underset{\substack{\text{HO OH}\\\text{Tetramethylene}\\\text{glycol}}}{\langle\;\rangle} \qquad \underset{\text{Adipic acid}}{\text{HOOC}(\text{CH}_2)_4\text{COOH}}$$

Uses: THF is used as a nonaqueous solvent in organic reactions. It is also used for the preparation of adipic acid and hexamethylene diamine (for manufacture of nylon) and tetramethylene glycol (for manufacture of polyurethanes).

38.4.3.3 Pyrrole

The pyrrole nucleus is present in many naturally occurring compounds like alkaloids, chlorophyll etc. It may be considered as the nitrogen analogue of furan and thiophene and occurs in coal tar and bone oil.

Preparation

(i) *From Bone Oil:* Bone oil is first washed with dilute alkali (to remove acidic constituents) and then with acid to remove basic constituents like pyridine bases. Pyrrole is finally separate by fractional distillation. The fraction boiling between 110-140° is collected and pyrrole obtained by fusion with potassium hydroxide when solid potassium salt of pyrrole is obtained. This on distillation yields pyrrole.

(ii) *From Ammonium Mucate:* Ammonium mucate on distillation with glycerol at 200° forms pyrrole.

$$\underset{\text{Ammonium mucate}}{\text{H}_4\text{NOOC}-(\text{CHOH})_2-\text{COONH}_4} \xrightarrow[\text{glycerol}]{\text{Distill, 200}^\circ} \underset{\text{Pyrrole}}{\overset{\displaystyle \langle\;\rangle}{\underset{\text{H}}{\text{N}}}} + 2\text{CO}_2 + \text{NH}_3 + 4\text{H}_2\text{O}$$

(iii) *From Succinimide:* Succinimide when distilled with zinc dust forms pyrrole.

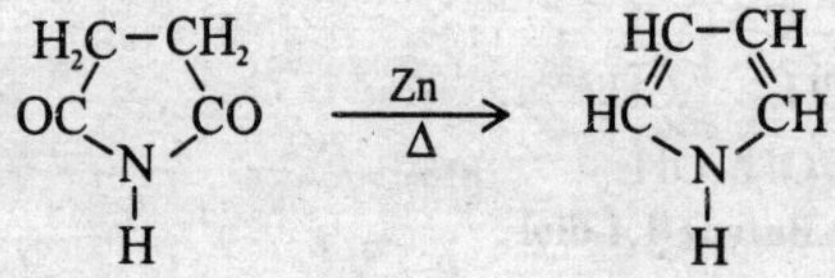

Succinimide   Pyrrole

(iv) *From Furan:* Pyrrole is manufactured by passing a mixture of furan, ammonia and steam over alumina catalyst.

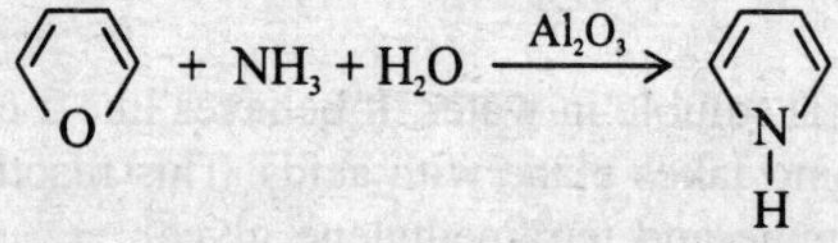

Furan   Pyrrole

(v) *Paal-Knorr Synthesis:* 1,4-Diketones on heating with ammonia or primary amines give pyrrole and its derivatives.

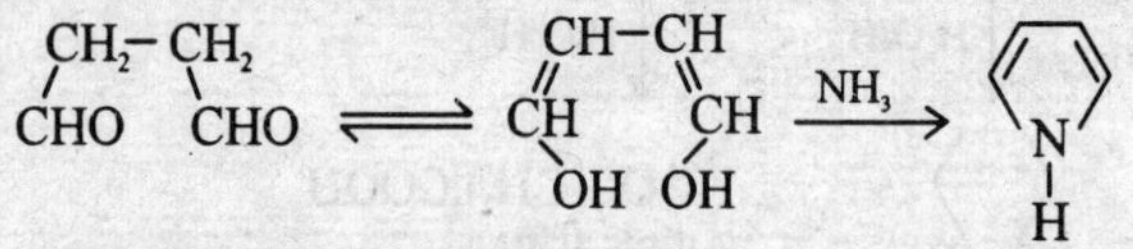

Succinaldehyde   Pyrrole

(vi) *Knorr Pyrrole Synthesis:* Condensation of a β-ketoester or a β-diketone with an α-aminoketone gives pyrrole derivatives.

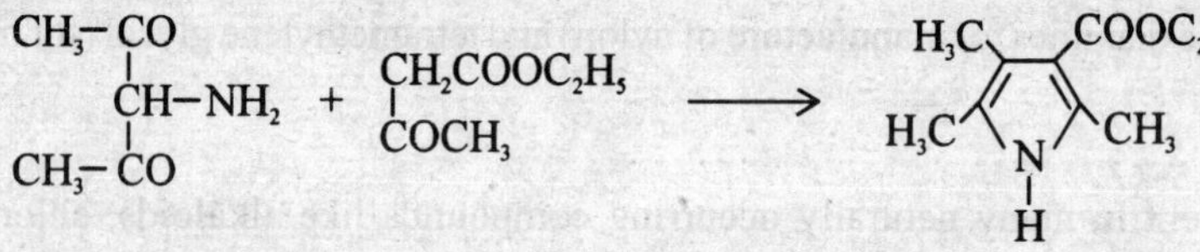

α-Aminoketone   Pyrrole derivative

Properties: Pyrrole is a colorless liquid, b.p. 131°, sparingly soluble in water and freely soluble in ethanol and ether. It darkens on exposure to air.

In chemical properties, pyrrole differs markedly from thiophene and furan. However, like thiophene and furan, it undergoes electrophilic substitution reactions but is much more reactive. Some reactions of pyrrole are given below.

(i) *Amphoteric Nature:* Pyrrole behaves as a very weak base. This is because the lone pair of electrons on nitrogen is involved in aromatic sextet formation and is not available for protonation with acids.

It is weakly acidic in nature as indicated by its reaction with metallic potassium or potassium hydroxide to form a potassium salt, which on treatment with water gives back pyrrole. This property is used in the preparation of pure pyrrole from bone-oil.

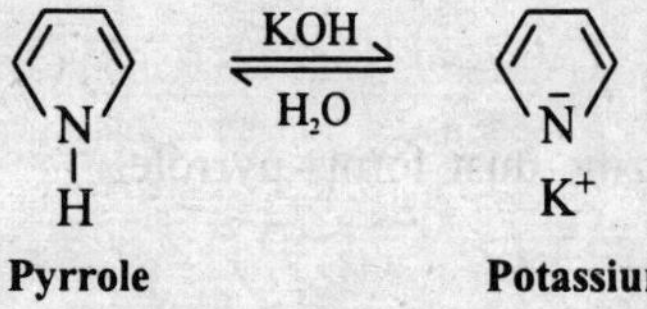

Pyrrole   Potassium pyrrole

The acidic character of pyrrole is attributed to the greater stability of the pyrrole anion as compared to pyrrole. As will be seen later, pyrrole is a resonance hybrid of various contributing forms carrying a positive charge at nitrogen which helps in release of proton. The acidic character is also exhibited in the formation of pyrrolemagnesium bromide (on treatment with methylmagnesium bromide). The structure of the salt is shown as a resonance hybrid of the following structure.

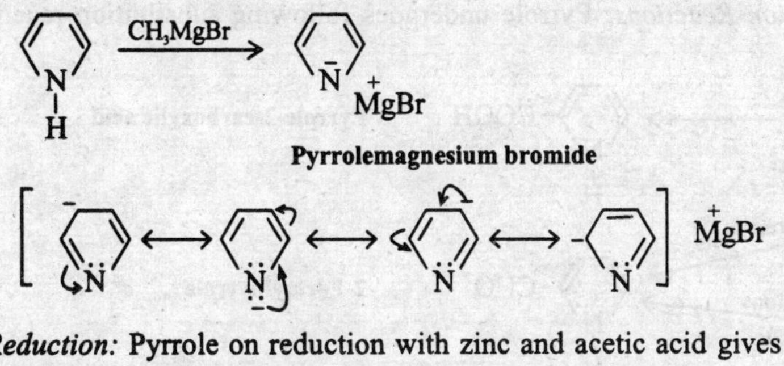

Pyrrolemagnesium bromide

(ii) *Reduction:* Pyrrole on reduction with zinc and acetic acid gives 2,5-dihydropyrrole (pyrroline), b.p. 91°, which on heating with hydriodic acid and red phosphorus forms tetrahydropyrrole (pyrrolidine),b.p. 88°.

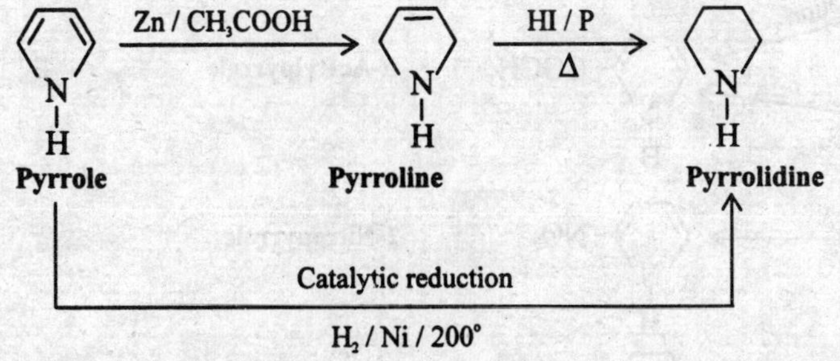

Pyrrolidine is prepared in one step by catalytic reduction of pyrrole in presence of nickel at 200°. Both pyrroline and pyrrolidine are strong bases.

(iii) *Oxidation:* Pyrrole, on oxidation with chromium trioxide forms maleinimide.

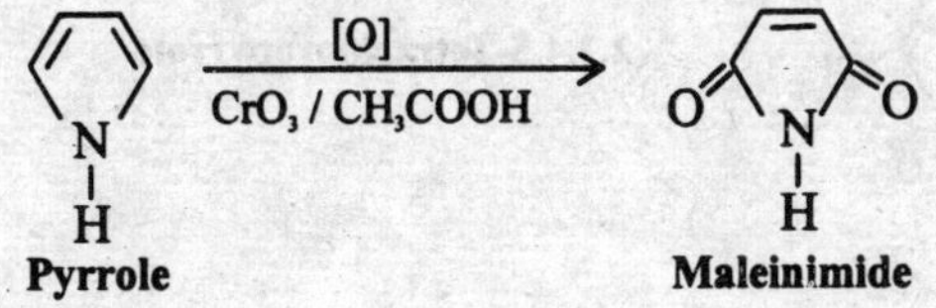

(iv) *Reaction with Dichlorocarbene:* Dichlorocarbene generated in situ by the reaction of chloroform and potassium ethoxide adds onto the potassium salt of pyrrole. The intermediate formed rearranges to give β-chloropyridine.

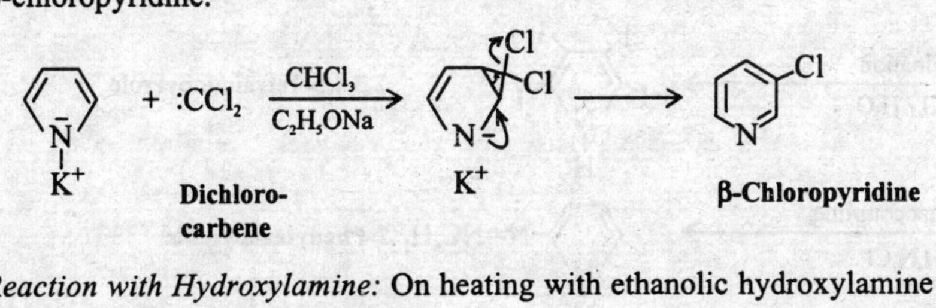

(v) *Reaction with Hydroxylamine:* On heating with ethanolic hydroxylamine pyrrole undergoes ring opening forming succinaldehyde dioxime.

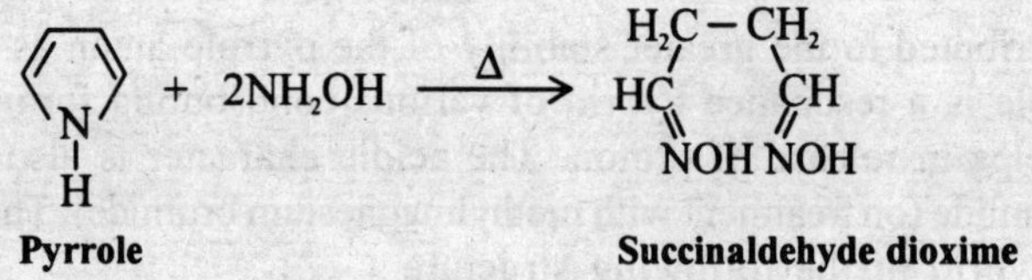

Pyrrole + 2NH$_2$OH $\xrightarrow{\Delta}$ Succinaldehyde dioxime

Pyrrole   **Succinaldehyde dioxime**

(vi)   *Electrophilic Substitution Reactions:* Pyrrole undergoes following substitution reactions.

Kolbe reaction
CO$_2$, Δ → — COOH   **Pyrrole-2-carboxylic acid**

Reimer Tiemann reactions
CHCl$_3$ + ROH → — CHO   **2-Formylpyrrole**

Gatterman reactions
HCN / HCl

Houber–Hoesch reaction
CH$_3$CN , HCl → — COCH$_3$ **α-Acetylpyrrole**

F. C. acylation
(CH$_3$CO)$_2$O / SnCl$_4$

Nitration
HNO$_3$, Ac$_2$O , 0° → — NO$_2$ **2-Nitropyrrole**

Sulphonation
SO$_3$ / Pyridine / 100° → — SO$_3$H   **Pyrrole-2-sulphonic acid**

Chlorination
SO$_2$Cl$_2$, ether , 0° → Cl, Cl, Cl, Cl   **2,3,4,5-Tetrachloropyrrole**

Bromination
Br$_2$ / CH$_3$COOH / 0° → Br, Br, Br, Br   **2,3,4,5-Tetrabromopyrrole**

Iodination
I$_2$ / KI / H$_2$O → I, I, I, I   **2,3,4,5-Tetraiodopyrrole**

Diazocoupling
C$_6$H$_5$N$_2$Cl → — N=NC$_6$H$_5$ **2-Phenylazopyrrole**

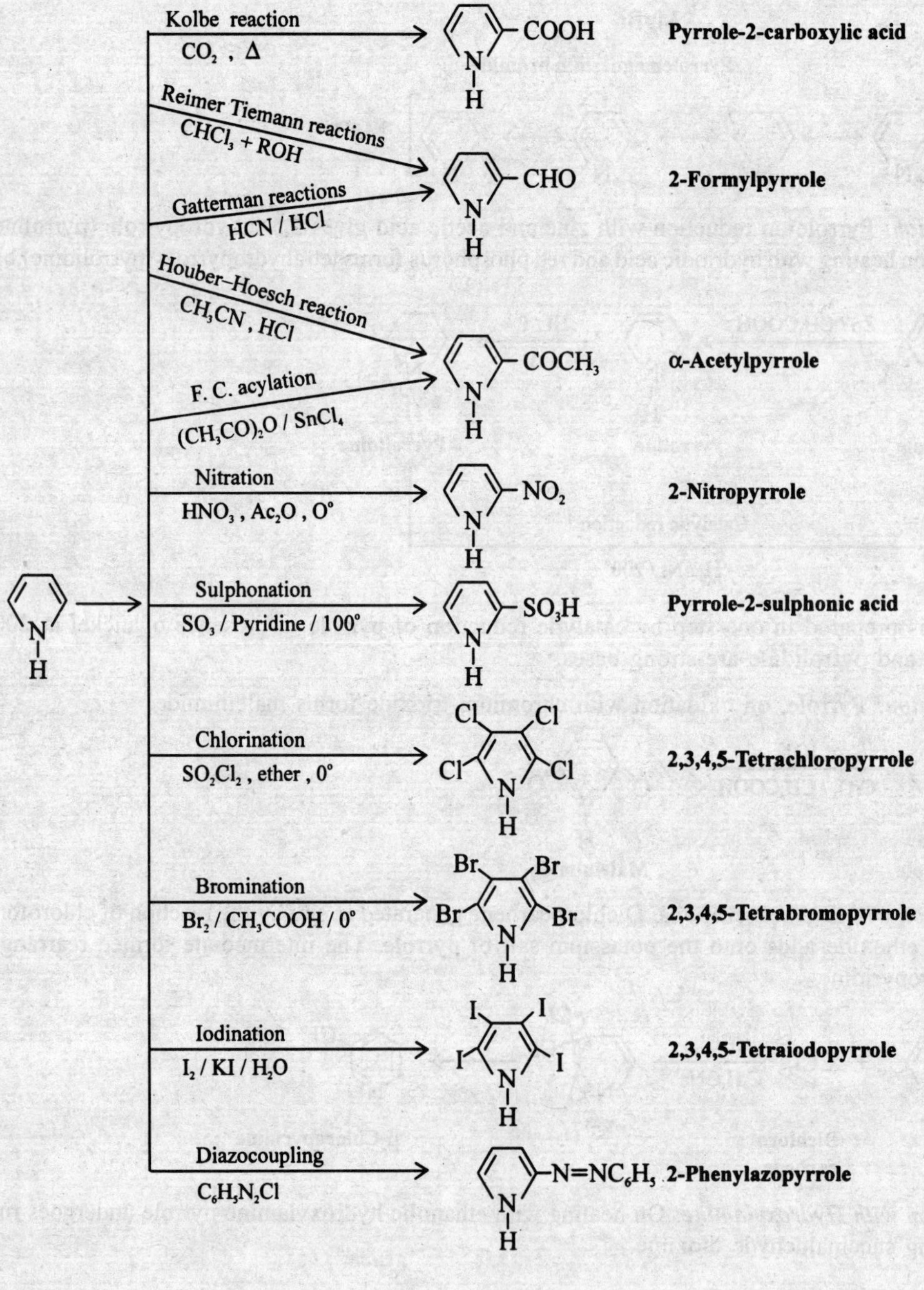

In the above reactions pyrrole resembles phenols.

(vii)   *Reaction with Alkyl Halides and Acyl Halides:* Pyrrole resembles aromatic amines in its rections. It reacts with acetyl chloride or methyl iodide to give N-acyl or N-alkyl derivatives, which rearrange
n heating to give the α-substituted derivatives.

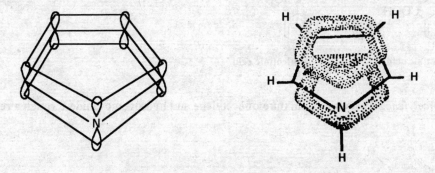

N-Acetylpyrrole   **2-Acetylpyrrole**

N-Methylpyrrole   **2-Methylpyrrole**

tructure of Pyrrole

yrrole resembles furan and thiophene in its structure and is a resonance hybrid of the following structures.

Its resonance energy is 21 kcal/mol.

Like thiophene and furan, the carbon and nitrogen atoms in pyrrole are sp^2 hybridised and have a
laner geometry. Each carbon atom has one electron in its unhybridised p orbital and nitrogen atom
as a pair of electron in its unhybridised p orbital which overlaps sideways to form a π electron cloud
vhich lies above and below the plane of the ring. The π molecular orbital in pyrrole contains six electrons
a Hückel number), thus show aromatic properties. The molecular orbital picture of pyrrole is shown
n Fig. 38.3.

Fig. 38.3:   Molecular Orbital Picture of Pyrrole

38.4.4   Six Membered Heterocyclic Compounds

38.4.4.1   Pyridine [C₅H₅N]

It is the most important member of six membered heterocyclic compounds containing one hetero atom. The pyridine nucleus is present in a number of natural products like alkaloids (nicotine, piperidine); vitamins (nicotinic acid, vitamin B₆) etc.

Pyridine occurs in light oil fraction of coal tar and in bone oil. It is obtained from light oil by treatment with dilute sulphuric acid, which dissolves pyridine and other basic substances. The acid solution is neutralised with sodium hydroxide and the basic fraction thus obtained is repeatedly fractionally distilled. Pyridine is obtained commercially by this process.

Nomenclature

As per IUPAC system of nomenclature, pyridine is called azine, but this name is rarely used. The numbering in pyridine is done by the numerals or Greek letters as shown below.

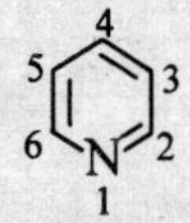

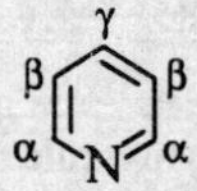

In case of pyridine derivatives, common names are used. Thus, methyl substituted pyridines are called picolines, dimethylpyridines as lutidines and trimethylpyridines as collidines. The pyridine carboxylic acid are named as shown below:

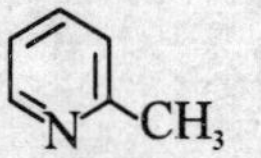

α-Methylpyridine
(2-Picoline)

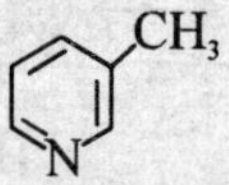

β-Methylpyridine
(3-Picoline)

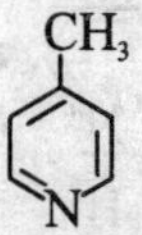

γ-Methylpyridine
(4-Picoline)

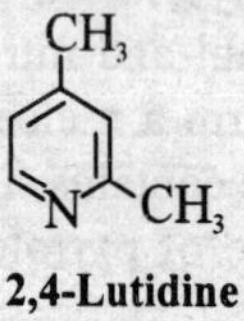

2,4-Lutidine

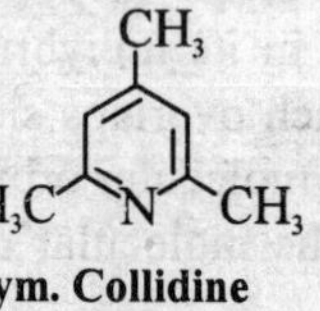

Sym. Collidine
(2,4,6-Trimethylpyridine)

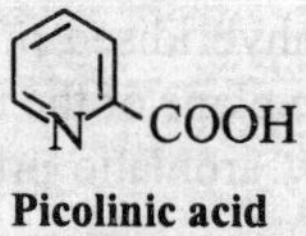

Picolinic acid

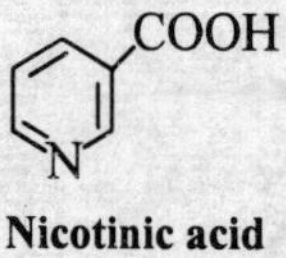

Nicotinic acid

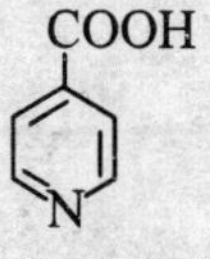

Isonicotinic acid

Preparation

(i)   *From Acetylene:* By passing a mixture of acetylene and hydrogen cyanide through a red hot tube.

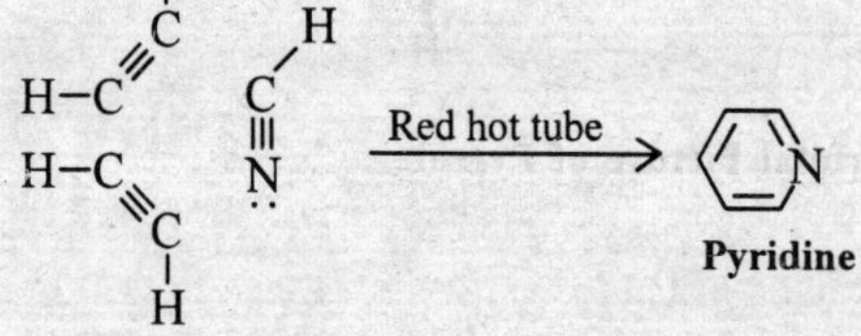

Red hot tube

Pyridine

(ii) *From Pentamethylenediamine Dihydrochloride:* By heating pentamethylenediamine dihydrochloride, piperidine is formed which when dehydrogenated catalytically or heated with concentrated sulphuric acid at 300° forms pyridine.

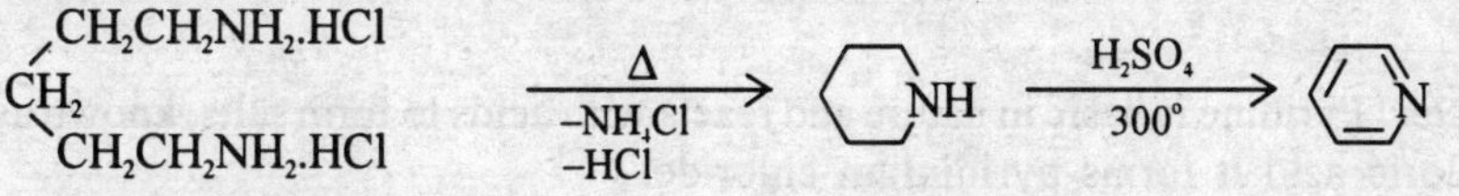

Pentamethylenediamine dihydrochloride   **Piperidine**   **Pyridine**

(iii) *Hantzsch Synthesis:* It is useful for the preparation of pyridine derivatives. The condensation of two moles of a suitable β-carbonyl compound (e.g., ethyl acetoacetate) with one mole each of an aldehyde and ammonia yields the dihydropyridine derivative, which on oxidation with nitric acid gives the pyridine derivative.

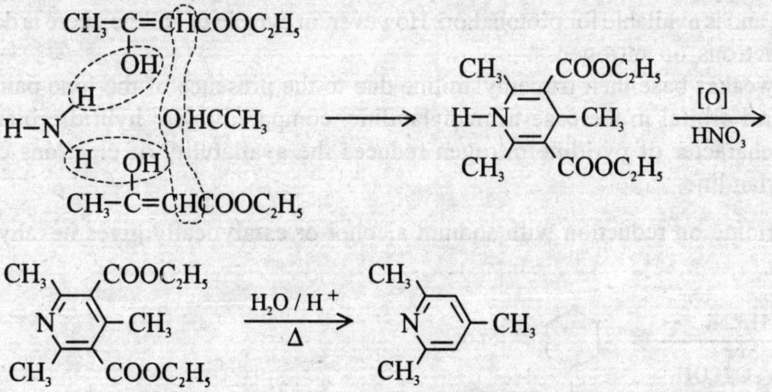

2,4,6-Trimethylpyridine (sym. Collidine)

(iv) *From Pyrrole:* The reaction of pyrrole with dichlorocarbene gives 2-chloropyridine (see Sec. 38.4.3.3).

(v) *From Tetrahydrofurfuryl Alcohol:* The reaction of tetrahydrofurfuryl alcohol with ammonia at 500° gives pyridine (commercial method).

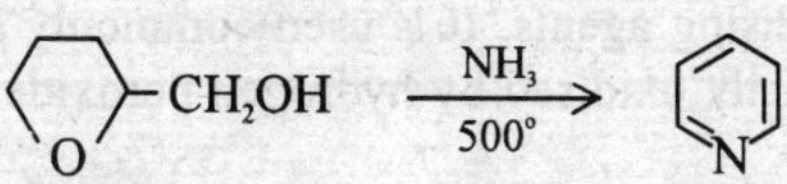

Tetrahydrofurfuryl alcohol   **Pyridine**

The required tetrahydrofurfuryl alcohol is obtained by catalytic reduction of furfuryl alcohol.

(vi) *Pure Pyridine* required for the synthesis of medicines is obtained commercially by passing a mixture of acetylene, formaldehyde acetal and ammonia over aluminum oxide at 500°.

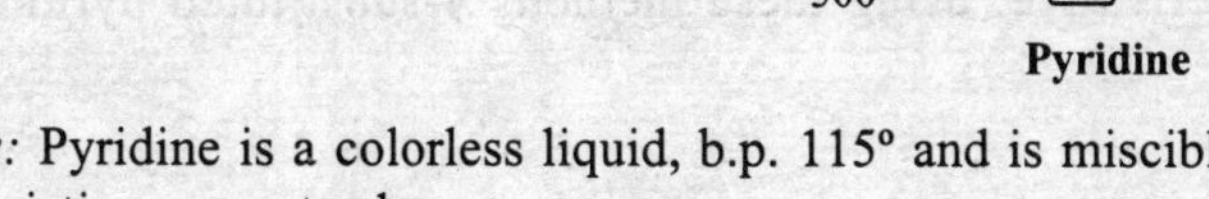

Pyridine

Properties: Pyridine is a colorless liquid, b.p. 115° and is miscible with water in all proportions. It has a characteristic pungent odor.

The IR spectrum of pyridine shows C–H stretching band in the region 3000-3100 cm⁻¹ in addition to bands characteristic of aromatic system (1605, 1575, 1480 and 1430 cm⁻¹).

In UV region, pyridine shows absorption maxima at 178 nm, 251nm and 270nm.

Chemical Properties

(i)   *Basic Character:* Pyridine is basic in nature and reacts with acids to form salts, known as pyridinium salts. With hydrochloric acid it forms pyridinium chloride.

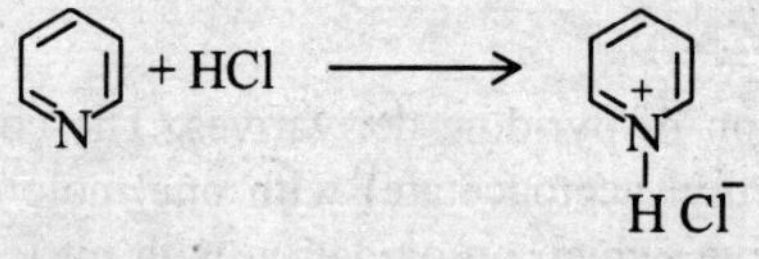

Pyridine   **Pyridinium chloride**

The basic character of pyridine is because of the presence of a lone pair of electron on the nitrogen which is not delocalised and is available for protonation. However, in pyrrole or aniline, there is delocalisation of the lone pair of electrons on nitrogen.

Pyridine is a much weaker base then trimethylaniline due to the presence of the lone pair of electron on nitrogen in sp^3 hybrid orbital in the case trimethylaniline, compared to sp^2 hybrid orbital in case of pyridine. The more s character of pyridine nitrogen reduces the availability of electrons compared to sp^3 hybridised trimethylaniline.

(ii)   *Reduction:* Pyridine on reduction with sodium alcohol or catalytically gives hexahydropyridine known as piperidine.

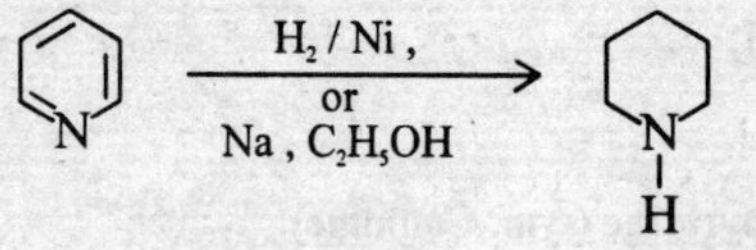

Pyridine   **Piperidine**

Piperidine is a stronger base compared to pyridine. The piperidine nucleus is present in a number of natural products like alkaloids. The piperidine ring can be easily opened up. This method is used for structure elucidation of alkaloids

(iii)   *Oxidation:* Pyridine is unaffected by usual oxidising agents. It is used commonly as a solvent for chromium trioxide oxidations. However, it can be easily oxidised by hydrogen peroxide or peracids to N-oxide.

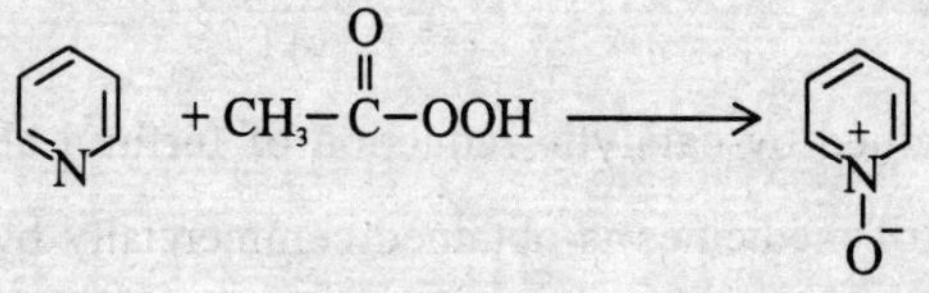

Pyridine   Peracetic acid   **Pyridine-N-oxide**

Pyridine-N-oxide can be reduced to pyridine under catalytic conditions or with iron and acid. Further, it can be nitrated to give 4-nitro derivative. using these methods 4-substituted pyridines can be obtained.

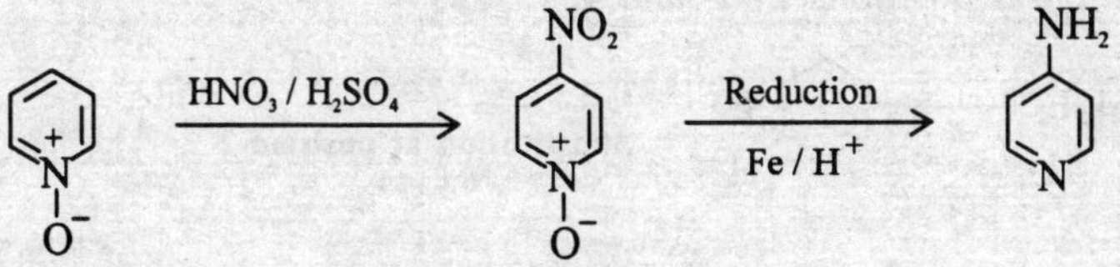

Pyridine-N-oxide   4-Nitro-   4-Aminopyridine
pyridine-N-oxide

(iv)   *Electrophilic Substitution:* Pyridine is considered to be a resonance hybrid of the following structures.

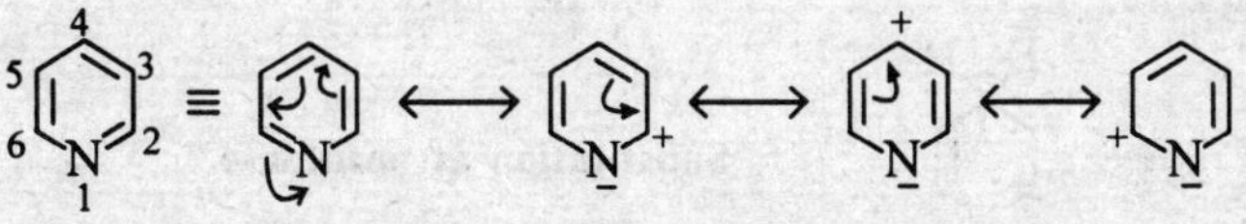

On the basis of the above resonating structures, the position 3 and 5 are the sites for electrophilic attack and the positions 2, 4 and 6 for nucleophilic attack. It is also noted that the ring is deactivated towards electrophilic reagents (rather it resembles the benzene ring in nitrobenzene). This is attributed to the withdrawal of electrons from the ring carbon atoms towards the nitrogen atom. Further, in strongly acidic medium, pyridine is protonated and the positively charged nitrogen atom deactivates the ring considerably. Therefore, drastic conditions are necessary for electrophilic substitution reactions some of which are shown below.

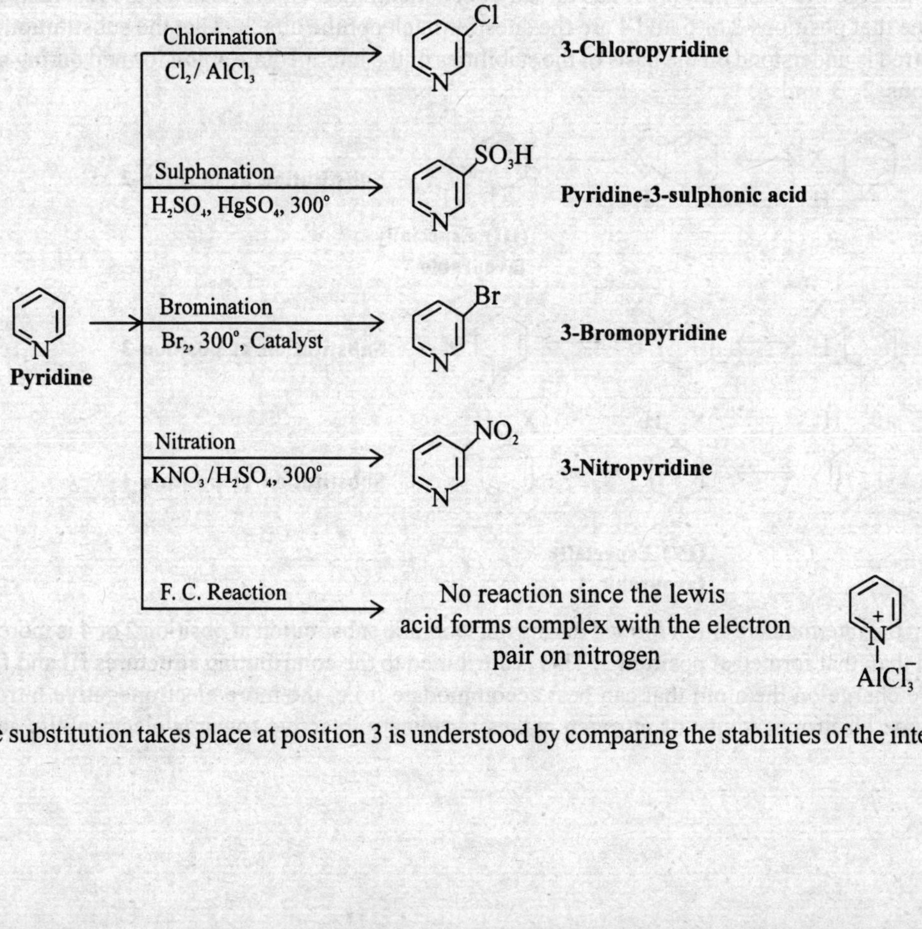

That the substitution takes place at position 3 is understood by comparing the stabilities of the intermediate cations formed during substitution at positions 2-,3- and 4.

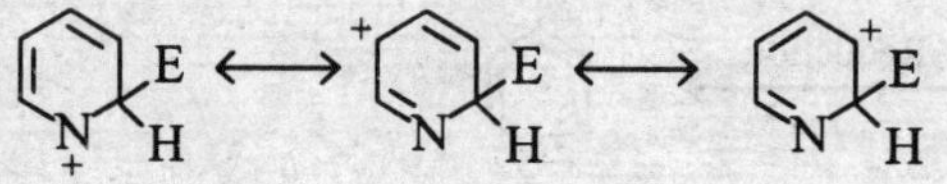

Substitution at position-2

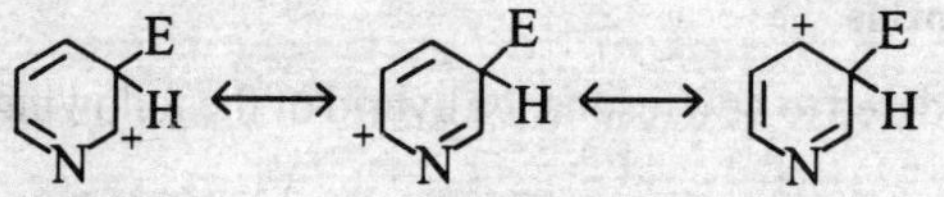

(I) Especially
unfavourable

Substitution at position-3

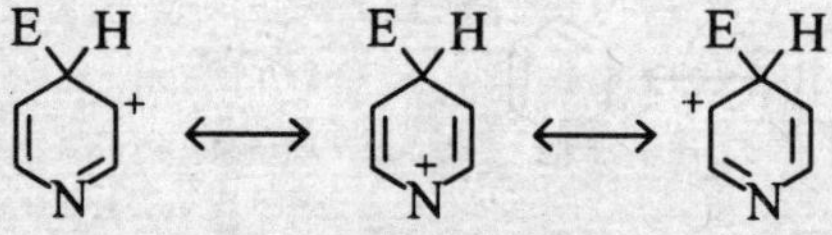

Substitution at position-4

(II) Especially
unfavourable

It is seen that the cation I and II resulting from attack at C-2 and C-4 positions respectively are especially unstable, as electronegative nitrogen atom has only six electrons. As a result attack at carbon 2 and 4 is unfavourable energetically, hence slow and is ignored. Therefore, the product with substitution at C-3 predominates and is favourable. However, if the C-3 position is already blocked substitution takes place at the C-5 position.

(v) *Nucleophilic Substitution:* It has already been mentioned on the basis of the resonance structures of pyridine that positions 2 or 6 and 4 are the sites for nucleophilic attack. That the substitution at position 2 is preferred is understood on the basis of the stabilities of the intermediate anion formed during substitution at positions 2, 3 and 4.

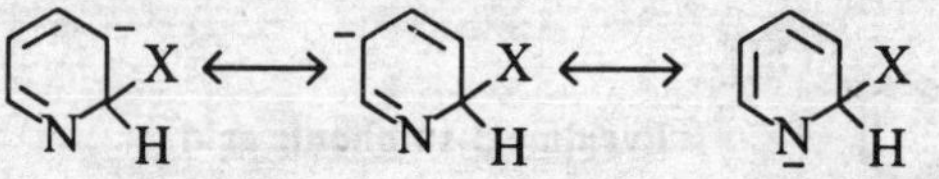

Substitution at position-2

(III) Especially
favourable

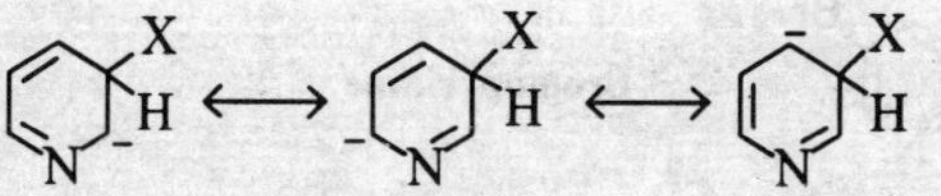

Substitution at position-3

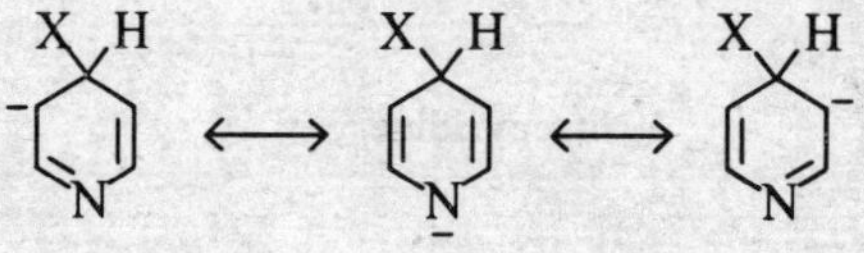

Substitution at position-4

(IV) Especially
favourable

As seen the intermediate anion formed during nucleophilic substitution at position 2 or 4 is more resonance stabilised than that formed at position 3. This is attributed to the contributing structures III and IV carrying a negative charge on the atom that can best accommodate it i.e. the more electronegative nitrogen atom. The same electronegativity of nitrogen makes pyridine unreactive towards electrophilic substitution.

Some important reactions are given below.

(a) *Reaction with Sodamide (Chichibabian Reaction):* Pyridine on heating with sodamide in toluene gives 2-aminopyridine. However, use of excess sodamide gives 2,6-diaminopyridine.

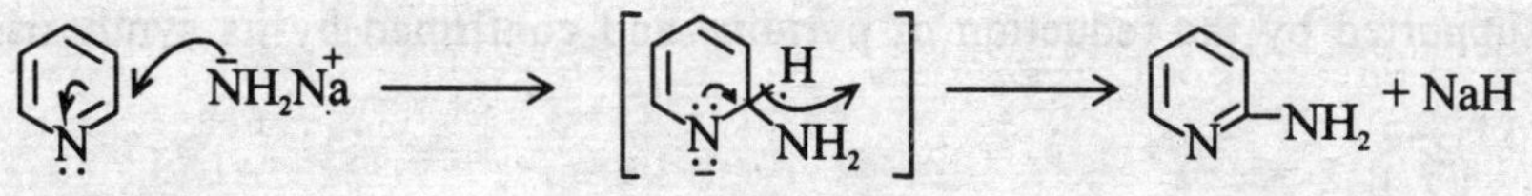

Pyridine   2-Aminopyridine

$$(NaH + NH_3 (liq) \longrightarrow Na^+ \bar{N}H_2 + H_2)$$

(b) *Alkylation or Arylation:* Alkylation or arylation is achieved at position 2 by the reaction of pyridine with n-butyllithium or phenyllithium.

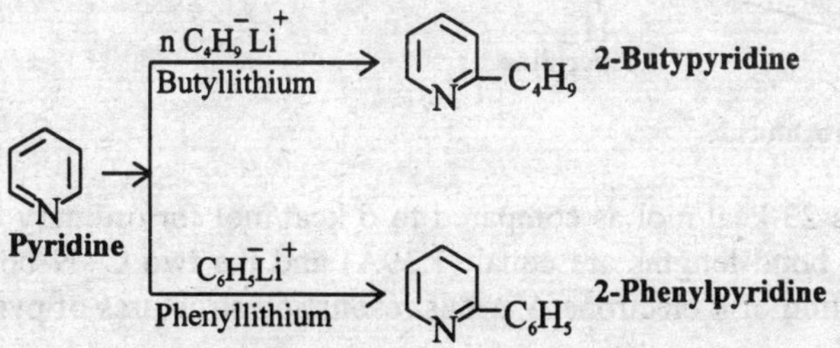

2-Butypyridine

Pyridine

2-Phenylpyridine

(c) *Reaction with Sodium Hydroxide:* Pyridine on reaction with sodium hydroxide in presence of potassium ferricyanide gives α-pyridone.

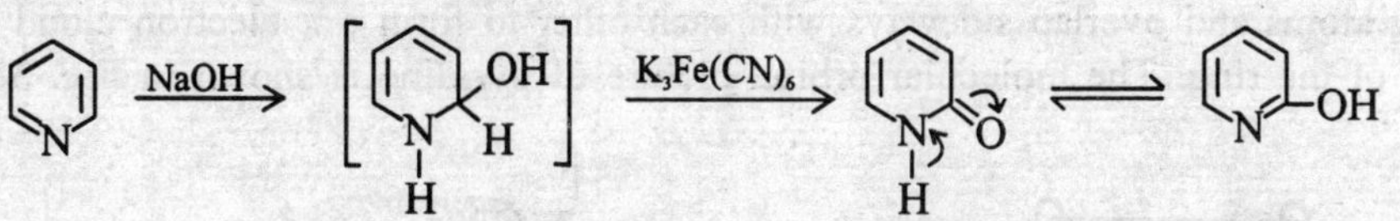

Pyridine   α-Pyridone   2-Hydroxypyridine

Structure of Pyridine
1. Its molecular formula as determined by analytical methods is C_5H_5N.
2. Nitrogen in pyridine is tertiary in nature, since it does not give reactions of primary and secondary amines but forms a quaternary ammonium salt with methyl iodide.

$$C_5H_5N + CH_3I \longrightarrow C_5H_5\overset{+}{N} CH_3\bar{I}$$

3. It is resistant to common oxidising agents and usual addition reactions. It does not decolorise bromine water.
4. It undergoes electrophilic as well as nucleophilic substitution reactions (see properties of pyridine).

On the basis of the above, it is inferred that pyridine resembles benzene and exhibits aromatic character.
5. The alkyl derivatives (viz. picolines) on oxidation give corresponding carboxylic acids.

$$CH_3C_5H_4N \xrightarrow{[O]} HOOCC_5H_4N$$

α,β or γ-   α, β or γ Pyridine-
Picoline   carboxylic acids

6. Like amines, the amino derivatives of pyridine can be diazotised.

7.   In halogen substituted pyridines, halogen atoms can be displaced (nucleophilic displacement) by nucleophiles.

On the basis of the above and by analogy with benzene, pyridine is assigned the following structure given by Korner.

The structure is supported by the reduction of pyridine and confirmed by its synthesis.

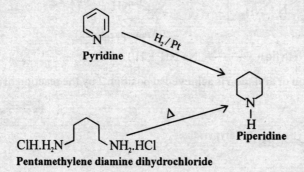

The resonance energy of pyridine is 23 kcal/mol as compared to 6 kcal/mol for ordinary conjugated diene. Also in pyridine the four C−C bond lengths are equal (1.39Å) and the two C−N bond lengths too (1.36Å), which is due to delocalisation of π electrons. Various resonating structures of pyridine have already been discussed in Sec. 38.4.1.

In pyridine, the nitrogen and each of the five carbon atoms are in sp^2 hybridised state having trigonal geometry. The unhybridised p orbitals at each of the carbon and nitrogen atoms are perpendicular to plane of the ring atoms and overlap sideways with each other to form a π electron cloud above and below the plane of the ring. The molecular orbital picture of pyridine is shown in Fig. 38.4.

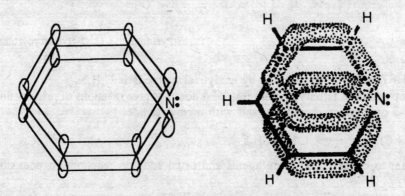

Fig. 38.4:   Molecular Orbital Picture of Pyridine

38.4.5   Condensed Heterocyclic Compounds

The important members of this class are quinoline and isoquinoline.

38.4.5.1   Quinoline [C_9H_7N]

Quinoline nucleus has a benzene ring fused to pyridine at the α and β positions and is designated as α,β-benzopyridine.

It occurs in coal tar and in bone oil. It was obtained by the distillation of quinine alkaloid with alkali and so the name quinoline given to it. The quinoline nucleus is present in a number of alkaloids (quinine, cinchonine etc), antimalarials (chloroquine) and analgesics (uricophen).

Preparation

(i)   *From Heavy oil Fraction of Coal-tar:* The procedure is similar to that used for pyridine.

(ii)   *Skarup Synthesis:* When a mixture of aniline and glycerol is heated in presence of concentrated sulphuric acid and nitrobenzene (which acts as a mild oxidising agent), quinoline is produced. Since the reaction becomes violent, ferrous sulphate is generally added to make the reaction go smoothly.

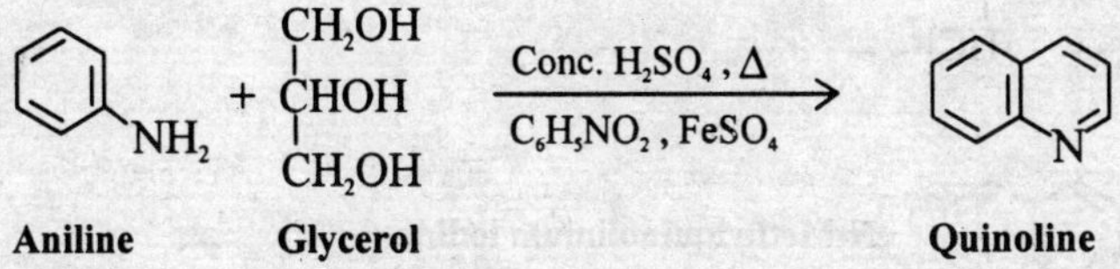

Aniline	**Glycerol**		**Quinoline**

The mechanism of the reaction is uncertain. It is believed that in the first step glycerol is converted into acrolein, which then undergoes 1,4-addition.Various steps involved are given below:

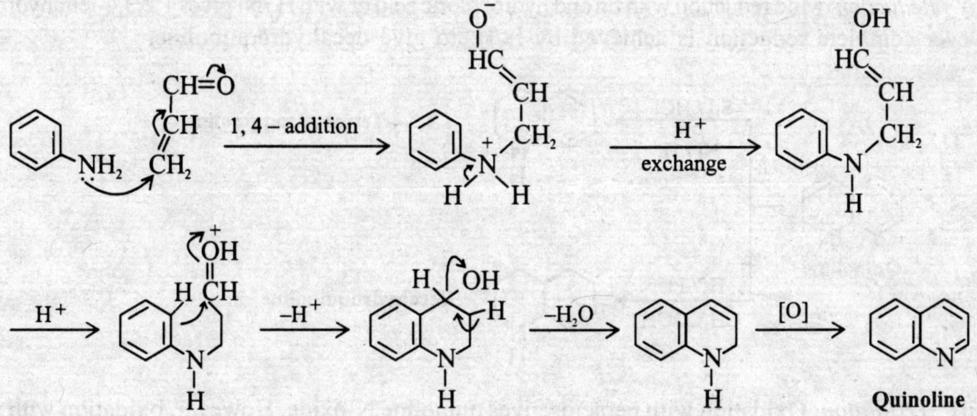

In Skarup synthesis, acraldehyde itself is not used since it polymerises under the reaction conditions. However, this is a general method of preparation of quinoline derivatives starting with appropriately substituted anilines.

(iii)   *Friedlander's Synthesis:* It involves the reaction of o-aminobenzaldehyde with acetaldehyde in presence of dilute alkali.

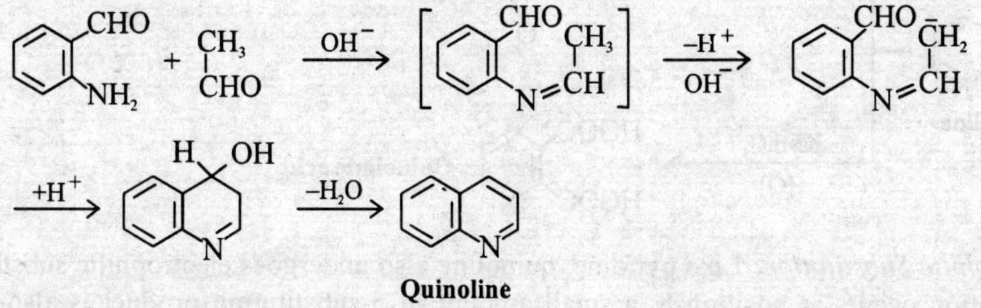

Quinoline

This method is also used for the synthesis of quinoline derivatives.

Properties: Quinoline is a colorless liquid, b.p. 237°. It is sparingly soluble in water but miscible with ether and alcohol. It has an unpleasant odor and is steam volatile. It exhibits reactions of both pyridine and benzene. Some important chemical properties are given below.

(i)   *Basic Characters:* Quinoline is a tertiary base and forms salts with inorganic acids. With methyl iodide it forms N-methylquinolinium iodide.

Quinoline hydrochloride

N-Methylquinolinium iodide

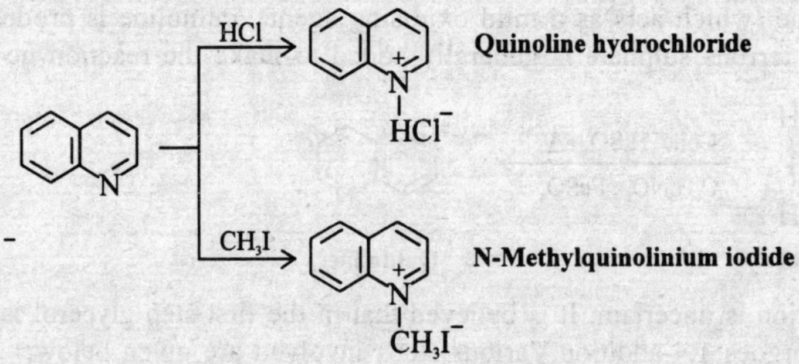

(ii)   *Reduction:* Mild reduction with tin and hydrochloric acid or with H_2/Ni gives 1,2,3,4-tetrahydroquinoline. However complete reduction is achieved by Pt/H_2 to give decahydroquinoline.

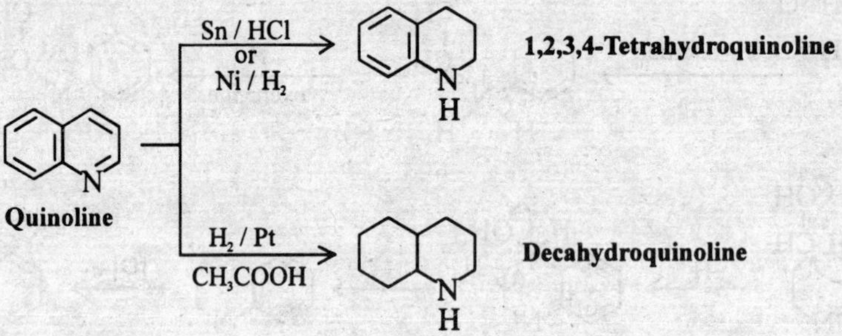

1,2,3,4-Tetrahydroquinoline

Quinoline

Decahydroquinoline

(iii)   *Oxidation:* Oxidation with peracids gives quinoline N-oxide. However, oxidation with potassium permanganate yields pyridine-2,3-dicarboxylic acid (quinolinic acid). The later reaction shows the presence of pyridine ring in quinoline.

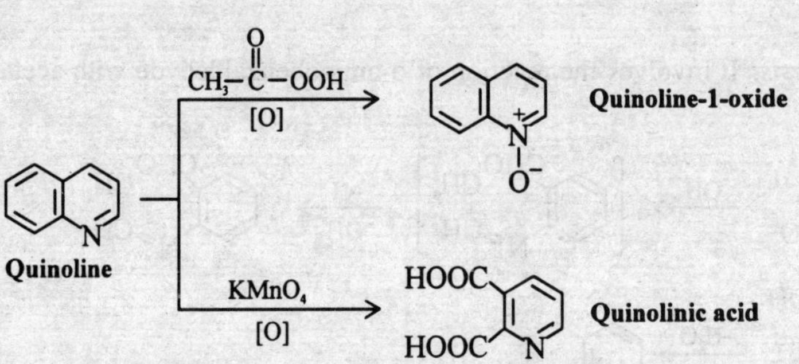

Quinoline-1-oxide

Quinoline

Quinolinic acid

(iv)   *Electrophilic Substitution:* Like pyridine, quinoline also undergoes electrophilic substitution reactions. Substitution occurs at position 8, a small amount of 5-substitution product is also obtained.

Some of the electrophilic substitution reactions are given below.

(a) *Bromination:* Under acidic conditions, bromination gives a mixture of 8-bromo and 5-bromoquinoline. However, vapor phase bromination gives 3-bromoquinoline and at 500° the product is 2-bromoquinoline.

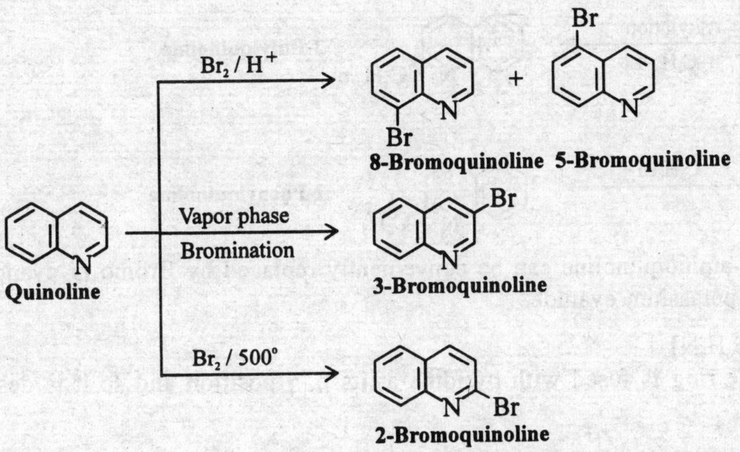

(b) *Nitration:* Nitration of quinoline with usual nitrating agent gives a mixture of 8-nitro and 5-nitroquinolines. However, nitration with nitric acid and acetic anhydride gives 3-nitroquinoline.

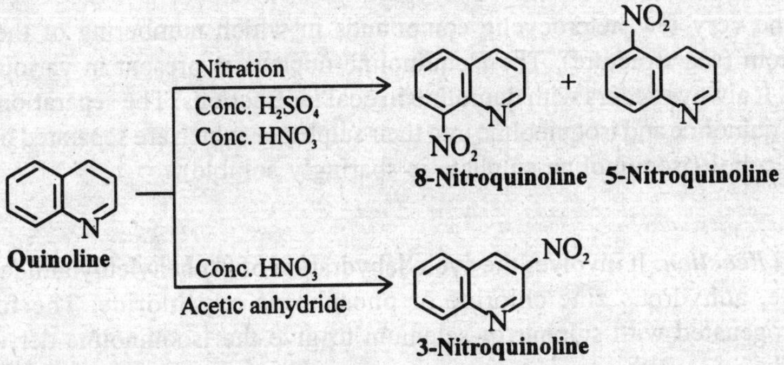

(c) *Sulphonation:* Quinoline on treatment with fuming sulphuric acid at 220° gives quinoline-8-sulphonic acid as the main product, together with a small amount of 5-substitution. The former on heating to 300° rearranges to quinoline-6-sulphonic acid.

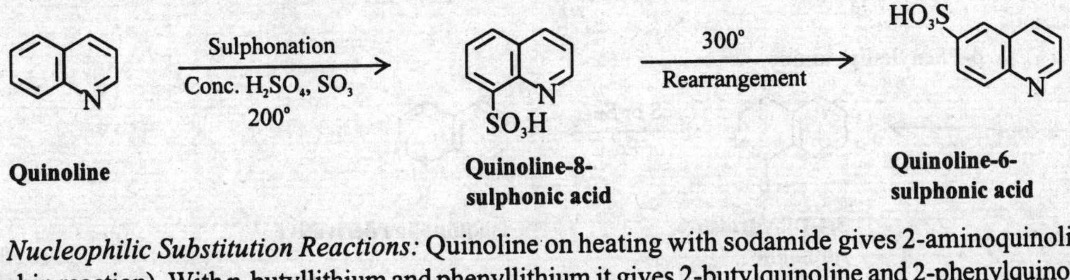

(v) *Nucleophilic Substitution Reactions:* Quinoline on heating with sodamide gives 2-aminoquinolines (chichibabin reaction). With n-butyllithium and phenyllithium it gives 2-butylquinoline and 2-phenylquinoline respectively.

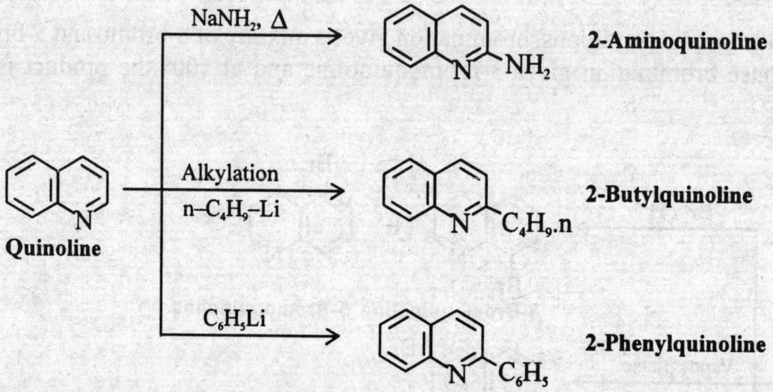

The amino group in 2-aminoquinoline can be conveniently replaced by Bromo or cyano group on heating with bromine or potassium cyanide.

38.4.5.2   Isoquinoline [C_9H_7N]

In isoquinoline a benzene ring is fused with pyridine at its β, γ position and so it is designated as β, γ-benzopyridine.

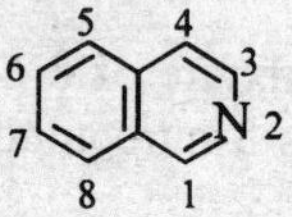

Isoquinoline is one of the very few heterocyclic compounds in which numbering of the ring does not start with the hetero atom (see structure). The isoquinoline nucleus is present in various alkaloids (papaverine, narcotine etc.). It always occurs with quinoline in coal tar fractions. The separation is effected by converting the mixture of quinoline and isoquinoline into their sulphates, which are separated by fractional crystallisation with ethyl alcohol (isoquinoline sulphate is sparingly soluble).

Preparation

(i)   *Bischler-Napieralski Reaction:* It involves the cyclodehydration of β-phenylethylamide on heating with phosphorus pentoxide, anhydrous zinc chloride or phosphorus oxychloride. The formed 3,4-dihydroquinoline is dehydrogenated with sulphur or selenium to give the isoquinoline derivative.

β-Phenylethyl amide

3,4-Dihydroiso-
quinoline

Isoquinoline derivative

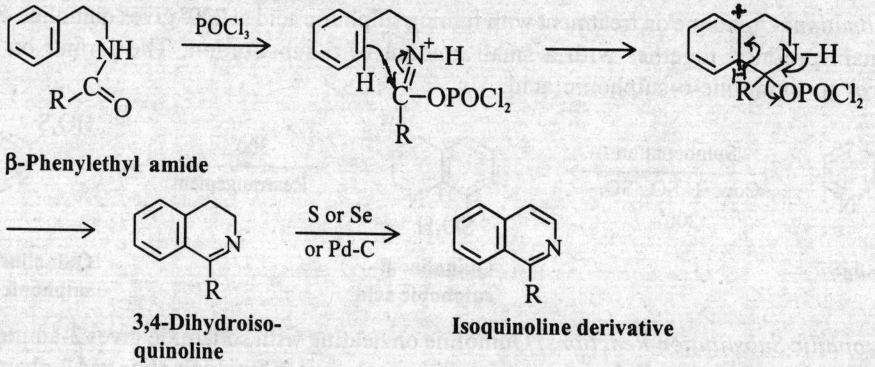

(ii) *Pictet-Spengler Reaction:* The condensation of β-arylethylamine with an aldehyde in presence of hydrochloric acid at 100° forms 1,2,3,4-tetrahydroisoquinoline which upon dehydrogenation with palladium carbon gives isoquinoline.

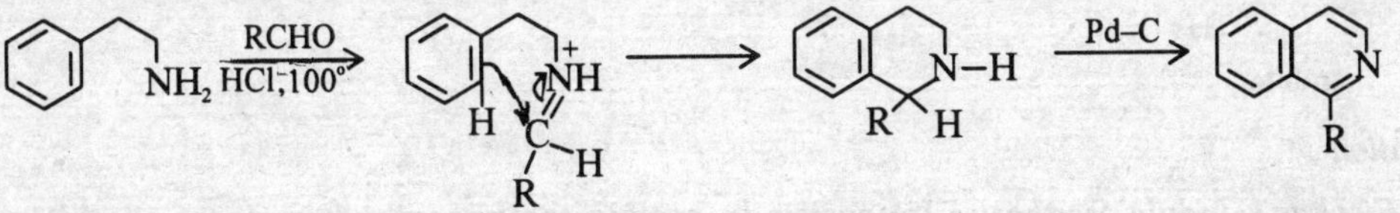

β-Arylethylamine   **Isoquinoline**

(iii) *From Cinnamaldehyde Oxime:* The oxime of cinnamaldehyde on heating with phosphorus pentoxide gives isoquinoline.

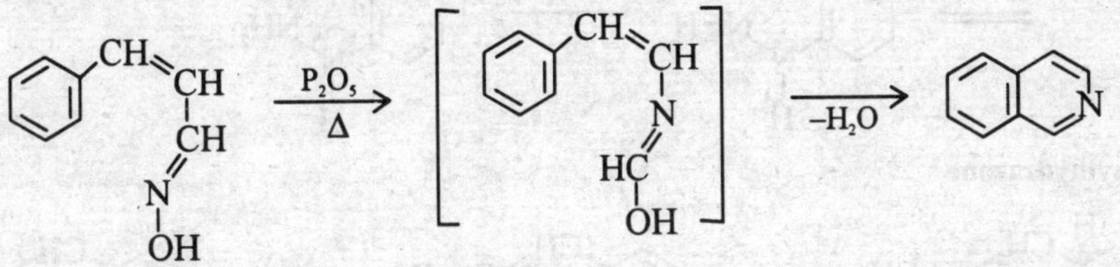

Cinnamaldehyde oxime   **Isoquinoline**

In the above reaction the expected product is quinoline and not isoquinoline. The formation of iso-quinoline is explained by assuming that the oxime first undergoes Beckmann rearrangement followed by ring closure.

Properties: It is a colorless solid, m.p. 23°. It gives chemical reactions similar to quinoline.

It is reduced by sodium in liquid ammonia to give 1,2-dihydro compound. However, with tin and hydrochloric acid 1,2,3,4-tetrahydroisoquinoline and under catalytic conditions octahydroisoquinoline is obtained.

Electrophilic attack occurs mainly at position 5 though a small amount of 8-substituted product is also obtained. Thus, nitration and sulphonation gives predominantly 5-substituted product. However, mercuration (with mercuric acetate) and bromination give the 4-substituted product.

Nucleophilic substitutions occurs at position-1. On heating with sodamide, 1-aminoisoquinoline is formed. It also forms quaternary salts and on oxidation with peracids gives N-oxide.

Oxidation of isoquinoline with potassium permanganate gives a mixture of phthalic and cinchomeronic acid (pyridine-3,4-dicarboxylic acid).

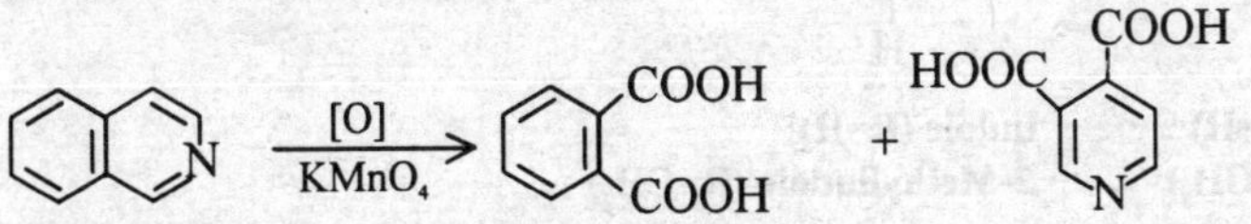

Isoquinoline   **Phthalic acid**   **Cinchomeronic acid**

38.4.5.3 *Indole*

In indole a benzene ring is fused to the α, β position of a pyrrole ring and it is called benzopyrrole. It occurs in coal tar and in the oils of Jasmine and orange blossoms. Indole nucleus is present in a number of alkaloids and amino acids.

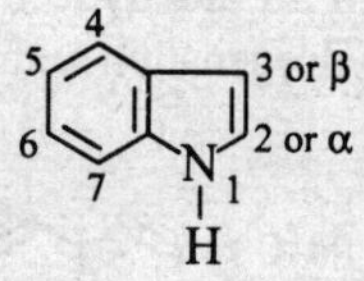

Indole

Preparation

(i) *Fischer's Indole Synthesis:* It consists in heating the phenylhydrazone of an aldehyde, ketone or a ketonic acid with zinc chloride or polyphosphoric acid. A possible mechanism is given below.

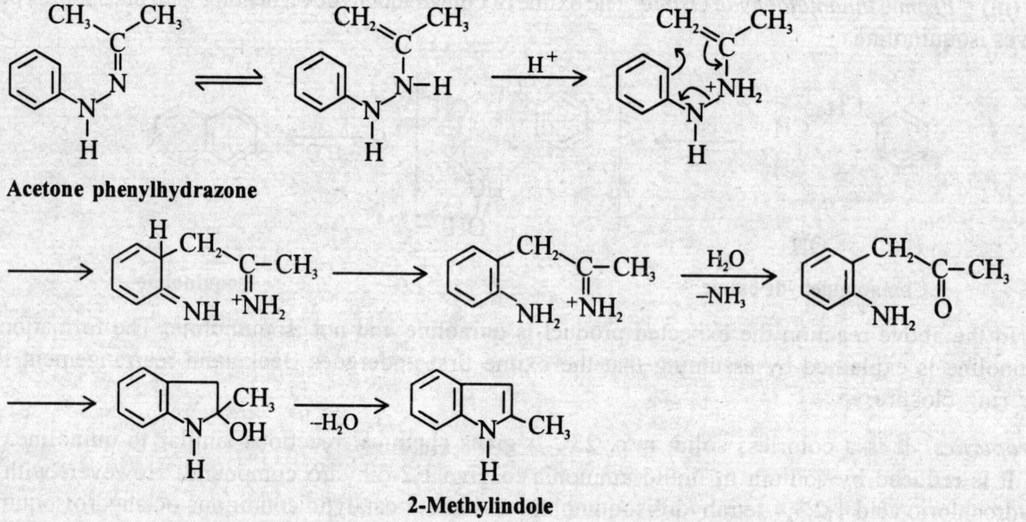

Acetone phenylhydrazone

2-Methylindole

This is the most important method of preparation of indole derivatives.

(ii) *Madelung Synthesis:* It involves the cyclisation of o-acylaminotoluene by strong base like potassium tert-butoxide, sodamide etc. Thus, indole or 2-methylindole is obtained from o-formamidotoluene or o-acetamidotoluene.

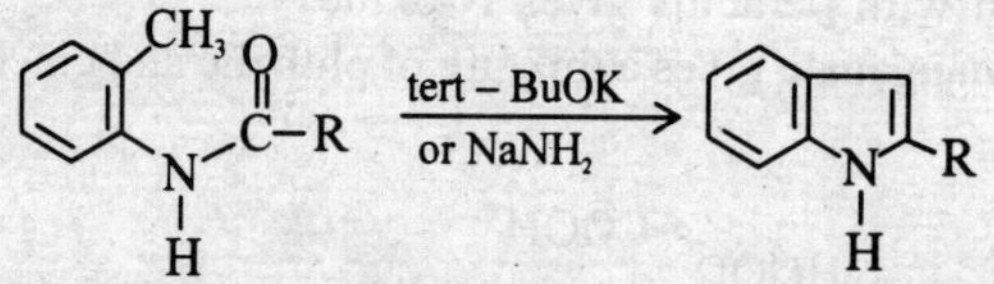

o-Formamidotuluene (R=H)   Indole (R=H)
o-Acetamidotuluene (R=CH₃)   2-Methylindole (R=CH₃)

The mechanism of this reaction is uncertain. One possible pathway is given below.

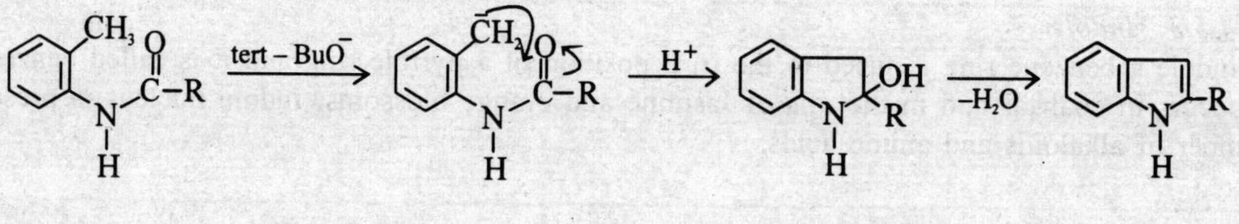

(iii) *Reissert Synthesis:* In this method o-nitrotoluene (or its substituted derivative) is condensed with ethyl oxalate. Various steps are shown below:

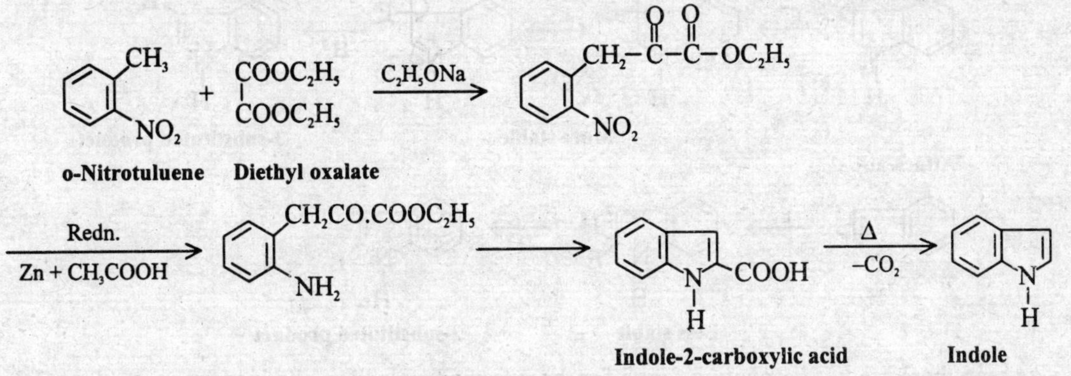

o-Nitrotoluene **Diethyl oxalate**

Indole-2-carboxylic acid   **Indole**

Properties: It is a colorless, volatile solid, m.p. 52°. It has a powerful pleasant odor and is used as a perfume. Some chemical properties of indole are given below.

(i) *Amphoteric Nature:* Like pyrrole, indole is a weak acid and a weak base.

(ii) *Oxidation:* Indole is oxidised with peroxy acids or ozone. The heterocyclic ring opens up to give 2-formamidobenzaldehyde.

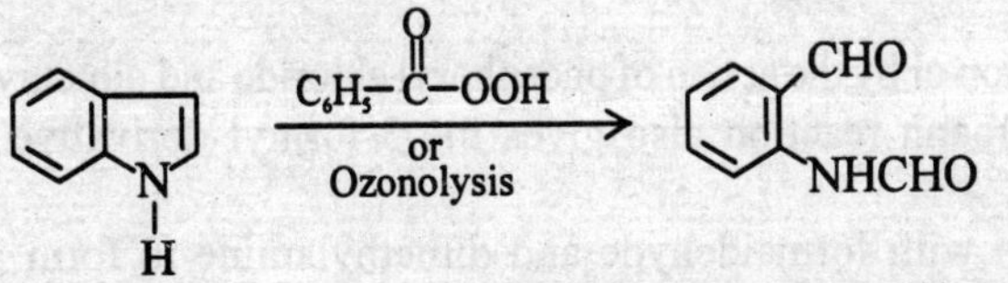

Indole   **2-Formamidobenzaldehyde**

(iii) *Reduction:* Zinc and hydrochloric acid reduces indole to 2,3-dihydroindole. However, catalytic hydrogenation gives octahydroindole.

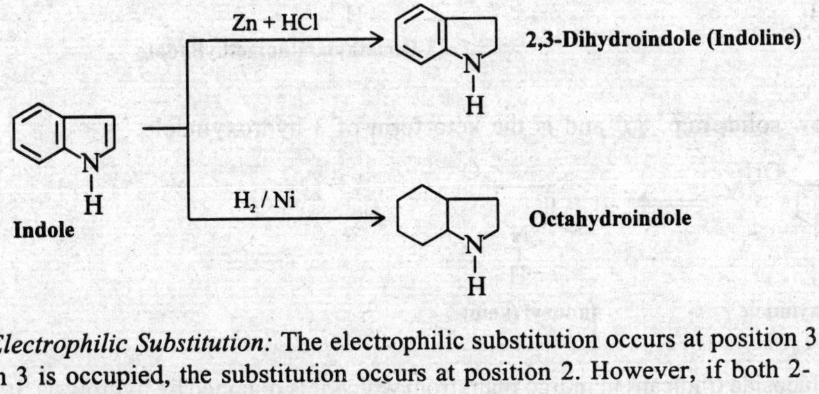

Indole

(iv) *Electrophilic Substitution:* The electrophilic substitution occurs at position 3 (compare pyrrole). If position 3 is occupied, the substitution occurs at position 2. However, if both 2- and 3- position are occupied, substitution occurs at position 6.

The formation of 3-substituted derivative is explained by considering the carbocations formed by the attack of electrophile at C-3 and C-2.

Attack at C-3

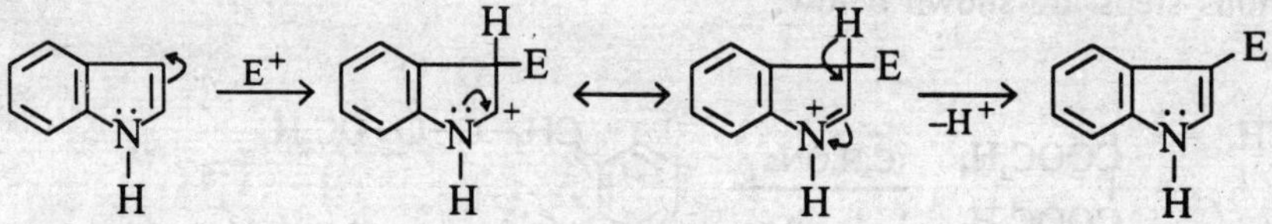

More stable   3-substituted product

Attack at C-2

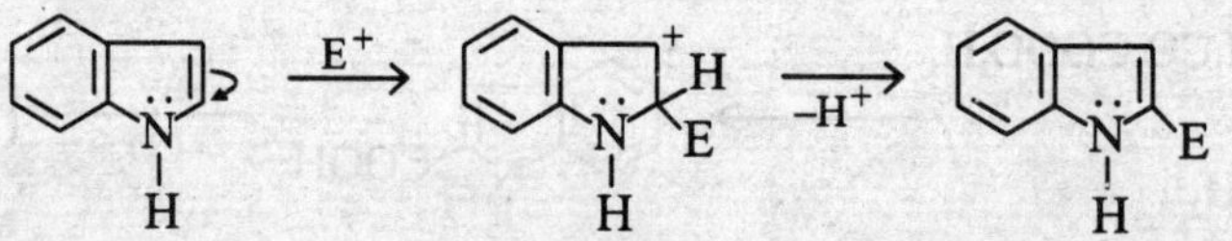

Less stable   2-substituted product

As seen there are two resonating forms for the intermediate cation obtained from attack at C-3 where as only one such form is possible for substitution at C-2. Therefore substitution occurs preferentially at position-3.

Thus, bromination or iodination in indole gives the 3-halogeno derivatives. The 3-chloroindole is obtained by the action of sulphuryl chloride on indole.

Nitration of indole with ethyl nitrate and sodium ethoxide gives 3-nitroindole.

An usual product, indole-2-sulphonic acid is obtained by sulphonation with sulphur trioxide in pyridine at 100-120°.

Formylation of indole by Gattermann reaction or by the action of phosphoryl chloride and dimethylformamide gives the 3-formyl derivative. Reimer-Tiemann reaction also gives the 3-formyl derivative along with 3-chloroquinoline (compare pyrrole).

Indole also undergoes Mannich reaction with formaldehyde and dimethylamine to form 3-dimethyl-aminomethylindole.

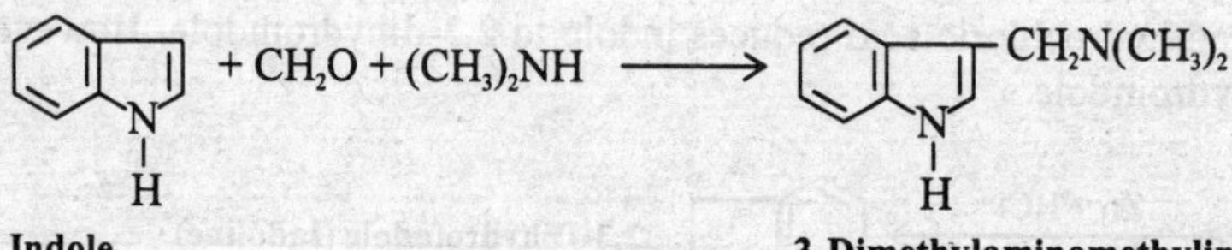

Indole   3-Dimethylaminomethylindole

38.4.5.4 Indoxyl

It is a bright yellow solid, m.p. 85° and is the keto form of 3-hydroxyindole.

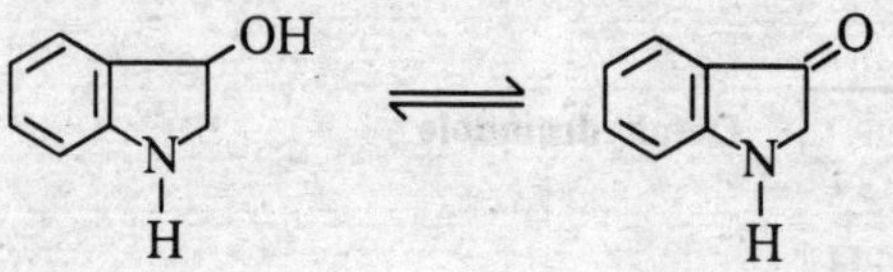

3-Hydroxyindole   Indoxyl (keto)
(enol)

It occurs as its glucoside (indican) in indigo plant from which it is obtained by hydrolysis. It is prepared from aniline as shown below.

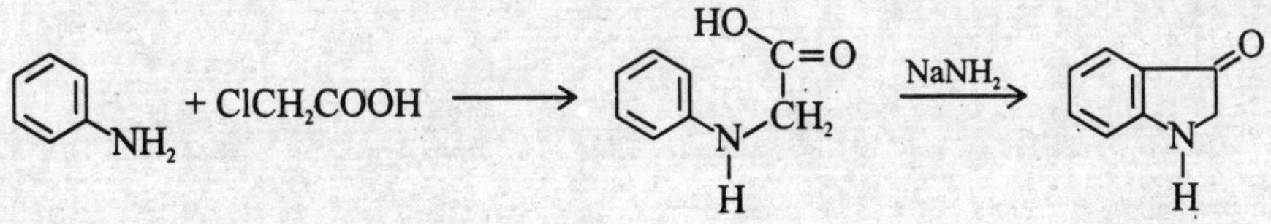

Aniline	**Chloroacetic acid**	**Indoxyl**

On oxidation indoxyl forms indigotin.

38.5 PROBLEMS

1. In electrophilic substitution reactions of thiophene, furan or pyrrole the substitution takes place predominantly at position 2. Give reasons.
2. Discuss the aromatic character of thiophene. What is the difference in the aromatic character of pyrrole, furan and thiophene?
3. How is thiophene obtained from coal tar? Give its molecular orbital picture.
4. Give Paal-Knorr synthesis of pyrrole, furan and thiophene derivatives.
5. Explain Diel Alder reaction of furan with maleic anhydride. What products are obtained in case of thiophene and pyrrole?
6. How is furfural, furoic acid and tetrahydrofuran obtained?
7. How does pyrrole react with (give products obtained):
   (a) Chloroform and sodium ethoxide   (b) Chloroform and potassium hydroxide
   (c) Hydroxyl amine   (d) Chromium trioxide in acetic acid
8. How is pyridine obtained from coal tar fractions? Discuss Hantzsch synthesis for pyrrole.
9. How is the structure of pyridine arrived at?
10. The electrophilic substitution of pyridine takes place at position 3. Explain why?
11. The nucleophilic substitution of pyridine occurs mainly at position 2. Explain why?
12. What products are obtained in the following reactions of pyridine?
    (a) Sodium hydroxide in presence of potassium ferricyanide.
    (b) Sodamide (heat)   (c) Phenyllithium   (d) Friedel-Crafts reaction
14. Discuss Skarup synthesis of quinoline. Give the mechanism involved.
15. Give the products obtained by oxidation and reduction of pyridine under different conditions.
16. How is the structure of quinoline and isoquinoline deduced?
17. Cinnamaldehyde oxime on heating with phosphorus pentoxide gives isoquinoline and not quinoline. Give reasons?
18. Give Fischer's indole synthesis and Madelung synthesis.
19. Explain why indole gives 3-substituted product in preference to 2-substituted products?

39

Carbohydrates

39.1 INTRODUCTION

Carbohydrates are naturally occurring compounds. They are composed of carbon, hydrogen and oxygen only, the later two elements viz H and O are in the ratio of 2:1 as in water and are referred to as hydrates of carbon. Carbohydrates have the general formula $C_x(H_2O)_y$. However, it should be noted that all compounds having the formula $C_x(H_2O)_y$ are not necessarily carbohydrates, e.g., formaldehyde (CH_2O), acetic acid (CH_3COOH or $C_2H_4O_2$) etc. Also, there are some carbohydrates, which do not contain the usual proportions of hydrogen and oxygen, e.g., rhamnose ($C_6H_{12}O_5$). It is to be noted that there are some carbohydrates which contain nitrogen or sulphur in addition to carbon, hydrogen and oxygen.

All carbohydrates can be described as polyhydroxy aldehydes or ketones or substances that are converted to these on hydrolysis.

In plants, simple compounds like carbon dioxide and water combine to form sugar (+)-glucose. This process is known as *photosynthesis*. This process is catalysed by chlorophyll and the required energy is supplied by sun in the form of light.

$$6CO_2 + 6H_2O + \text{Solar energy} \xrightarrow{\text{Chlorophyll}} C_6(H_2O)_6 + 6O_2 \uparrow$$

The glucose units combine to form large molecule of cellulose. These can also combine to form starch molecule, which is stored in the seed to serve as food for new growing plants.

On the other hand, when animals eat plants the starch (and in certain cases also the cellulose) is broken into the original (+)-glucose units. These are carried by the blood stream to the liver to form glycogen, or animal starch. Whenever necessary, the glycogen can be broken down once again into (+)-glucose, which is carried by the blood stream to the tissues, where it is oxidised to carbon dioxide and water with the release of energy originally supplied as sunlight.

39.2 CLASSIFICATION

The carbohydrates are divided into three major classes depending upon whether they could be hydrolysed or not and if they can be hydrolysed, the number of products formed.

(i) *Monosaccharides:* These are polyhydroxy aldehydes or polyhydroxy ketones which cannot be hydrolysed to simpler compounds. Examples of this type are glucose, fructose. A monosaccharide is known as an *aldose* (e.g. glucose) if it contains an aldehyde group or a *ketose* (e.g. fructose) if it contains a keto group. Depending upon the number of carbon atoms it contains, a monosaccharide is known as a triose, tetrose, pentose or hexose. Thus an aldohexose is a six carbon monosaccharide containing an aldehyde group (e.g., glucose) and a ketohexose is a six membered monosaccharide containing a keto group (e.g. fructose).

(ii)   *Oligosaccharides:* The oligosaccharides are carbohydrates which yield 2-10 molecules of monosaccharides on hydrolysis. This group includes,

(a)   *Disaccharides:* These on hydrolysis give two molecules of monosaccharides. Examples are sucrose and maltose.

$$\underset{C_{12}H_{22}O_{11}}{Sucrose} + H_2O \xrightarrow{H^+} \underset{C_6H_{12}O_6}{Glucose} + \underset{C_6H_{12}O_6}{Fructose}$$

$$\underset{C_{12}H_{22}O_{11}}{Maltose} + H_2O \xrightarrow{H^+} \underset{C_6H_{12}O_6}{2\ Glucose}$$

(b)   *Trisaccharides:* These on hydrolysis yield three molecules of monosaccharides, e.g., raffinose ($C_{18}H_{32}O_{16}$) on hydrolysis give glucose ($C_6H_{12}O_6$), fructose ($C_6H_{12}O_6$) and galactose ($C_6H_{12}O_6$).

(c)   *Tetrasaccharides:* These an hydrolysis yield four molecules of monosaccharides.

(iii)   *Polysaccharides:* These are carbohydrates, which on hydrolysis yield a number of monosaccharides. Thus, starch and cellulose both having the molecules formula, $(C_6H_{10}O_5)_n$ on hydrolysis give glucose.

$$\underset{\substack{\textbf{Starch or}\\\textbf{cellulose}}}{(C_6H_{10}O_5)_n} + nH_2O \xrightarrow{H^+} \underset{\textbf{Glucose}}{nC_6H_{12}O_6}$$

The carbohydrates can also be classified as sugars and non-sugars. The sugars include monosaccharides and oligosaccharides. The non-sugars, on the other hand, include polysaccharides.

Another method of classification of carbohydrates depends on their reducing properties. Thus, they can be classified as reducing or non-reducing sugars. The reducing sugars reduce Fehling's solution and Tollen's reagent. All monosaccharides (whether aldose or ketose) are reducing sugars. Most disaccharides are reducing sugars with the exception of sucrose which is a non-reducing sugar.

39.3   THE MONOSACCHARIDES

The monosaccharides are polyhydroxy aldehydes (aldose) or polyhydroxy ketones (ketose). All carbohydrates are either monosaccharides or are converted into monosaccharides on hydrolysis and are therefore the basis of carbohydrate chemistry. As already stated the monosaccharides may be a biose, triose, tetrose, pentose or a hexose depending on the number of carbon atoms it contains. A brief account of these is given below:

(i)   *Bioses:* The simplest monosaccharides is glycolaldehyde, $HOCH_2CHO$, which is a biose. It is optically inactive, since it does not contain an asymmetric carbon atom. It is excluded from the group of carbohydrates, since all naturally occurring sugars are optically active. It is more appropriate to define sugars as optically active polyhydric aldehydes or ketones.

(ii)   *Trioses:* Glyceraldehyde, $HOCH_2\ CHOH\ CHO$, an aldotriose, is the simplest of all the carbohydrates that fit into the definition given. The enantiomers of glyceraldehyde have been chosen as arbitary standard for the D and L series in sugar chemistry. Following points are of interest in understanding the *D- and L-Terminology.*

(a) Using a ball and stick model, the D- and L- forms of glyceraldehyde are represented as shown in Fig. 39.1(a). This shows the spacial relationship between the groups attached to the asymmetric carbon in each case. As seen, the two forms are mirror images that are not superimposable thus are *enantiomers*

(b) The D- and L-glyceraldehydes are best represented using a two dimensional projection—the *Fischer Projection*. In this projection, glyceraldehyde has the aldehyde group at the top. The horizontal bonds are coming out towards the viewer and the vertical bonds are away from the viewer [Fig. 39.1(b)]

(c) A still simpler way of drawing Fischer Projection is to use horizontal lines to show the position of the hydroxy group on the asymmetric carbon atom. In this projection, the hydrogen atoms are not shown. Thus D-glyceraldehyde is represented as shown in Fig. 39.1(c).

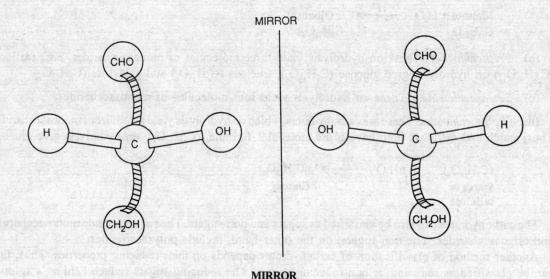

MIRROR

Fig. 39.1(a):   Representation of D- and L-Glyceraldehyde Using Ball and Stick Model

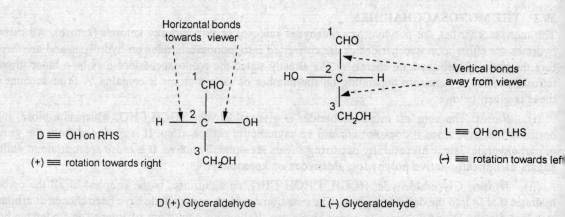

D ≡ OH on RHS

(+) ≡ rotation towards right

L ≡ OH on LHS

(–) ≡ rotation towards left

D (+) Glyceraldehyde   L (–) Glyceraldehyde

Fig. 39.1(b):   Fischer Projection Representation of D- and L-Glyceraldehyde

```
                    CHO                    CHO

        H                 OH      ≡               OH

                   CH₂OH                   CH₂OH
```

Fig. 39.1(c): Simple Ways of Drawing Fischer Projection

(d) The D- and L-concept: If the hydroxy group on the asymmetric carbon atom farthest from aldehyde or ketone group projects to the right, the compound belongs to the D-series. On the other hand, if the hydroxy group on the asymmetric carbon atom farthest from aldehyde or ketonic group projects to the left, the compound belongs to the L-series (see the representation of D- and L-glyceraldehyde in Fig. 39.1(b).

(e) The (+)- and (–)-Concept: We have already seen (Chapter 5) that optically active compound rotates the plane of the polarised light to right or left. It the compound rotates the plane of the polarised light to the right, it is said to be *dextrorotatory* and is represented by the sign (+). On the other hand, if the plane of the polarised light is rotated towards left, the compound is said to be *laevorotatory* and is represented by the sign (–) (see the representation of D (+)-glyceraldehyde and L(–)-glyceraldehyde in Fig. 39.1(b).

(f) The relationship of D- and L- and (+) and (–) concepts: It should be understood that the letters D- and L- refer to the absolute configuration around the asymmetric carbon atom. The signs (+) and (–) refer to the directions of rotation of the polarised light. The two are not necessarily related. A compound of the D- group may rotate the polarisd light to the right or to the left.

(g) Number of Optical Isomers: The maximum number of optical isomers of a carbohydrate is related to the number of asymmetric carbon atoms in the molecule.

Maximum number of optical isomers = 2^n

where, n is the number of asymmetric carbon atoms.

Dihydroxyacetone, $HOCH_2CO.CH_2OH$, may be considered a ketotriose but it is not optically active and so by definition it is not considered to be a carbohydrate.

(iii) *Tetroses:* The aldotetrose, $HOCH_2.\overset{*}{C}HOH.\overset{*}{C}HOH.CHO$ has two asymmetric carbons (marked *) and four optical isomers (2^n), all prepared synthetically and represented as the following two pairs (in simple Fischer Projection, Fig. 39.2):

```
   CHO            CHO            CHO            CHO




   CH₂OH          CH₂OH          CH₂OH          CH₂OH

 D(–)-Erythrose  L(+)-Erythrose  D(–)-Threose   L(+)-Threose
```

Fig. 39.2:   Mirror Images, Recemic Pair

The D(–)-erythrose and L(+)-erythrose are mirror images i.e., they are enantiomers. They have the same degree of rotation but in opposite direction. Equal amounts of two isomers constitutes a *racemic mixture* (which does not rotate the plane of the polarised light) but can be separated into the dextro and laevorotatory isomers. The same is true for the recemic pair of D(–) threose and L(+) threose.

It will be appropriate to introduce the term *diastereomers* which refers to optical isomers that are not mirror images. The degree of rotation of each may differ. For example D(–)-Erythrose and L(+) threose are not mirror images of each other; they are optical isomers called diastereomers.

From above it is seen that D- and L-sugars confirm the definition given earlier and that no connection exists between D-sugars and dextrorotatory, and L-sugar and laevorotatory i.e. there is no relation between D- and L-concept and (+) and (–)-concept.

The ketotetrose, $HOCH_2-CO-CHOH-CH_2OH$, contains one asymmetric carbon and two optical isomers called D- and L-erythrulose (Fig. 39.3):

CH₂OH

CH₂OH
C=O

CH₂OH
C=O

CH₂OH

CH₂OH

D- Erythrulose

L-Erythrulose

Fig. 39.3

(iv) *Pentoses:* The *aldopentoses* are important group of monosaccharides represented as CHO.ĊHOH.ĊHOH.ĊHOH.CH₂OH.

These contain three different asymmetric carbon atoms (carbon with * mark) and can exist in eight optically active forms (2^3). They correspond to D- and L-forms of arabinose, xylose, ribose and lyxose. Each of these exist as racemic pair.

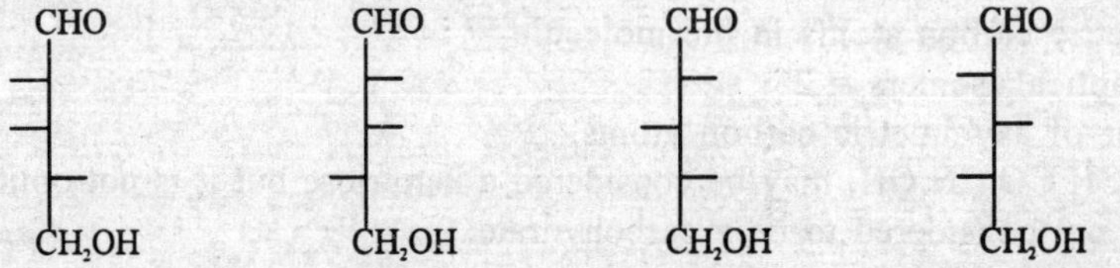

CHO

CH₂OH

CHO

CH₂OH

CHO

CH₂OH

CHO

CH₂OH

D (–)-Lyxose   **L(+)-Lyxose**   **D(–)-Xylose**   **L(+)- Xylose**

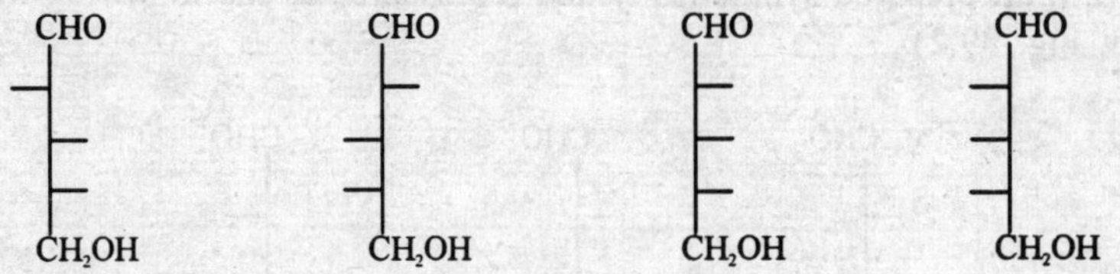

CHO

CH₂OH

CHO

CH₂OH

CHO

CH₂OH

CHO

CH₂OH

D(–)- Arabinose   **L(+)- Arabinose**   **D(–)- Ribose**   **L(+)- Ribose**

Of the eight isomers, D-arabinose, D-ribose, D-xylose and L-arabinose occur in nature. The others are synthetic.

An important aldopentose is 2-deoxyribose, $OHC-CH_2-CHOH-CHOH-CH_2OH$, which occurs in nucleic acids. The prefix 'deoxy' indicates the replacement of a hydroxy group by hydrogen at C-2.

The *Ketopentoses,* HOH$_2$C.CO.ĊHOH.ĊHOH.CH$_2$OH have two structurally different asymmetric carbon atoms and can therefore exist in four optically active forms. All are known and correspond to D- and L-forms of ribulose and xylulose.

(v) *Hexoses:* The *aldohexoses* are the most important group of monosaccharides represented as OCH.ĊHOH.ĊHOH.ĊHOH.ĊHOH.CH$_2$OH.

As seen, these contain four structurally different asymmetric carbon atoms and so exist in sixteen optically active forms. All are known and correspond to the D- and L-forms of glucose, mannose, galactose, allose, altrose, gulose, iodose and talose. Each of these exists as racemic pairs. These are represented as:

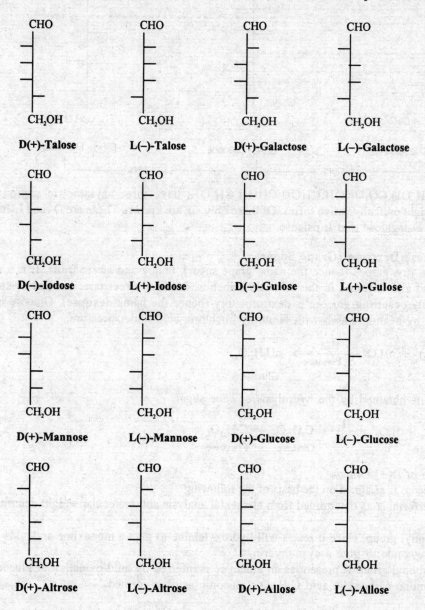

D(+)-Talose	L(–)-Talose	D(+)-Galactose	L(–)-Galactose
D(–)-Iodose	L(+)-Iodose	D(–)-Gulose	L(+)-Gulose
D(+)-Mannose	L(–)-Mannose	D(+)-Glucose	L(–)-Glucose
D(+)-Altrose	L(–)-Altrose	D(+)-Allose	L(–)-Allose

Of the sixteen aldohexoses (which have been prepared synthetically), only three, viz. D-glucose, D-mannose and D-galactose are found in nature. all the three are diastereomers since any of these optical isomers is not the mirror image of any of the others.

It may be appropriate to introduce the term *epimers*. A pair of diastereomers that differ only in the configuration about a single carbon atom are designated as epimers. Thus, out ot the above sixteen optical isomers, D(+)-Mannose and D(+)-galactose are epimers of D(+)-glucose. This is represented as:

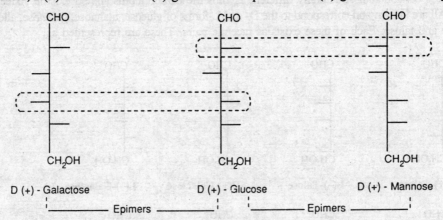

Ketohexoses, $CH_2OH.CO.\overset{*}{C}HOH.\overset{*}{C}HOH.\overset{*}{C}HOH.CH_2OH$, have three asymmetric carbon atoms and hence exist in eight optically active forms. Of these only six are known. These are D- and L-fructose, D- and L-sorbose, D-tagalose and L-psicose.

39.3.1   D(+)-Glucose/Dextrose (Grape Sugar)

Glucose is found in ripe grapes (hence the name grape sugar), honey and sweet fruits. It is a normal constituent of blood and is found in the urine of diabetics. It has a sweet taste but not as sweet as cane sugar. Naturally occurring glucose is dextrorotatory (hence the name dextrose). Glucose is commercially prepared by heating starch with dilute hydrochloric acid under pressure:

$$(C_6H_{10}O_5)_n + nH_2O \xrightarrow[\Delta, Pressure]{HCl} nC_6H_{12}O_6$$

Starch   **Glucose**

In laboratory, it is obtained by the hydrolysis of cane sugar.

$$C_{12}H_{22}O_{11} + H_2O \xrightarrow{H^+} C_6H_{12}O_6 + C_6H_{12}O_6$$

Cane sugar   **Glucose**   **Fructose**

39.3.1.1   Structure of D(+)-Glucose

The structure of glucose is assigned on the basis of the following.

(i) Its molecular formula as determined from elemental analysis and molecular weight determination is $C_6H_{12}O_6$.

(ii) It has a carbonyl group, since it reacts will hydroxylamine to give a monoxime and adds on one mole of hydrogen cyanide to give a cyanohydrin.

(iii) That the carbonyl group is present as an aldehyde is inferred by mild oxidation of glucose with bromine water -a mono carboxylic acid $C_6H_{12}O_7$ (gluconic acid) is formed.

$$-CHO \xrightarrow{\text{Br}_2 \text{ Water}} -COOH$$

Glucose   **Gluconic acid**

$$C_5H_{11}O_5-CHO \xrightarrow{[O]} C_5H_{11}O_5-COOH$$

Glucose   **Gluconic acid**

(iv) The presence of aldehyde group is also confirmed by the fact that glucose reduces ammonical silver nitrate solution (Tollen's reagent) and also Fehling's solution. It is also confirmed by its reduction with sodium amalgam when sorbitol (CHO → CH$_2$OH) is obtained.

$$C_5H_{11}O_5-CHO \xrightarrow[\text{Redn.}]{\text{Sod. amalgam}} C_5H_{11}O_5-CH_2OH$$

Gluconic acid   **Sorbitol**

(v) Glucose contains a primary alcoholic group. This is proved by further oxidation of gluconic acid with concentrated nitric acid to give a dicarboxylic acid, $C_6H_{10}O_8$, glucaric acid. The CH$_2$OH is oxidised to COOH. Thus,

$$HOCH_2-C_4H_8O_4-COOH \xrightarrow[\text{Conc. HNO}_3]{[O]} HOOC-C_4H_8O_4-COOH$$

Gluconic   **Glucaric acid**
acid

(vi) It contains five hydroxy groups, since on acetylation with acetic anhydride a pentaacetate is formed. Thus,

$$\begin{matrix} HOCH_2-C_4H_8O_4-CHO \\ \text{or} \\ HOCH_2-[CHOH]_4-CHO \end{matrix} \xrightarrow[\text{Acetylation}]{\text{Ac}_2\text{O}} AcOCH_2-[CHOAc]_4-CHO$$

Glucose (Partial structure)   **Glucose pentaacetate**

(vii) All the hydroxy groups are present on separate carbon atoms. In case of two hydroxy groups on the same carbon atom, it will loose a molecule of water to form a carbonyl group, which is not the case.

$$\underset{OH}{\overset{OH}{C}} \xrightarrow{-H_2O} C=O$$

(viii) On the basis of the above, it is inferred that in glucose molecule each of the five hydroxy groups are attached to a different carbon atom.

(ix) In glucose all the six carbon atoms constitute a straight chain. This is supported by the observation that glucose on prolonged heating with hydriodic acid forms n-hexane. This is also proved by the fact that the cyanohydrin of glucose on hydrolysis gives the corresponding carboxylic acid (CN → COOH), which on reduction with hydriodic acid gives n-heptanoic acid

$$\text{Glucose} \xrightarrow[\Delta]{\text{HI}} CH_3CH_2CH_2CH_2CH_2CH_3$$

n-Hexane

$$C_5H_{11}O_5-CH(OH)CN \xrightarrow{H^+} C_5H_{11}O_5-CH(OH)COOH$$

$$\xrightarrow[\Delta]{\text{HI}} CH_3CH_2CH_2CH_2CH_2CH_2COOH$$

n-Heptanoic acid

On the basis of the above, it is concluded that glucose has −CHO and −CH$_2$OH group and that all carbon atoms are present in a straight chain. The proposed open chain structure.

HOH$_2$C.CHOH.CHOH.CHOH.CHOH.CHO. was assigned by Bayer (1870) keeping in view the tetracovalency of carbon and all its evidences are summarised in Table 39.1.

Table 39.1:   Evidence in Favour of Open Chain Structure of Glucose

Experiment		Observation
	Elemental analysis and molecular wt. determination	Molecular formula C$_6$H$_{12}$O$_6$
NH$_2$OH	CH=NOH | (CHOH)$_4$ | CH$_2$OH Oxime	Presence of ⟩C=O group
HCN	CH⟨OH,CN | (CHOH)$_4$ | CH$_2$OH Cyanohydrin	Presence of ⟩C=O group
[O] Br$_2$ water	COOH | (CHOH)$_4$ | CH$_2$OH Gluconic acid	Presence of CHO group
[H] Na–Hg	CH$_2$OH | (CHOH)$_4$ | CH$_2$OH Sorbitol	Presence of CHO group
Fehling's Solution	Reduction takes place Red cuprous oxide is formed	Presence of CHO group
Tollen's reagent	Reduction takes place Metallic silver is formed	Presence of CHO group
[O] Br$_2$ water	COOH | (CHOH)$_4$ | CH$_2$OH Gluconic acid →[O] Conc. HNO$_3$ → COOH | (CHOH)$_4$ | COOH Glucaric acid	Presence of CH$_2$OH group
Ac$_2$O C$_5$H$_5$N	CHO | (CHOAc)$_4$ | CH$_2$OAc Glucose pentaacetate	Presence of five hydroxy groups

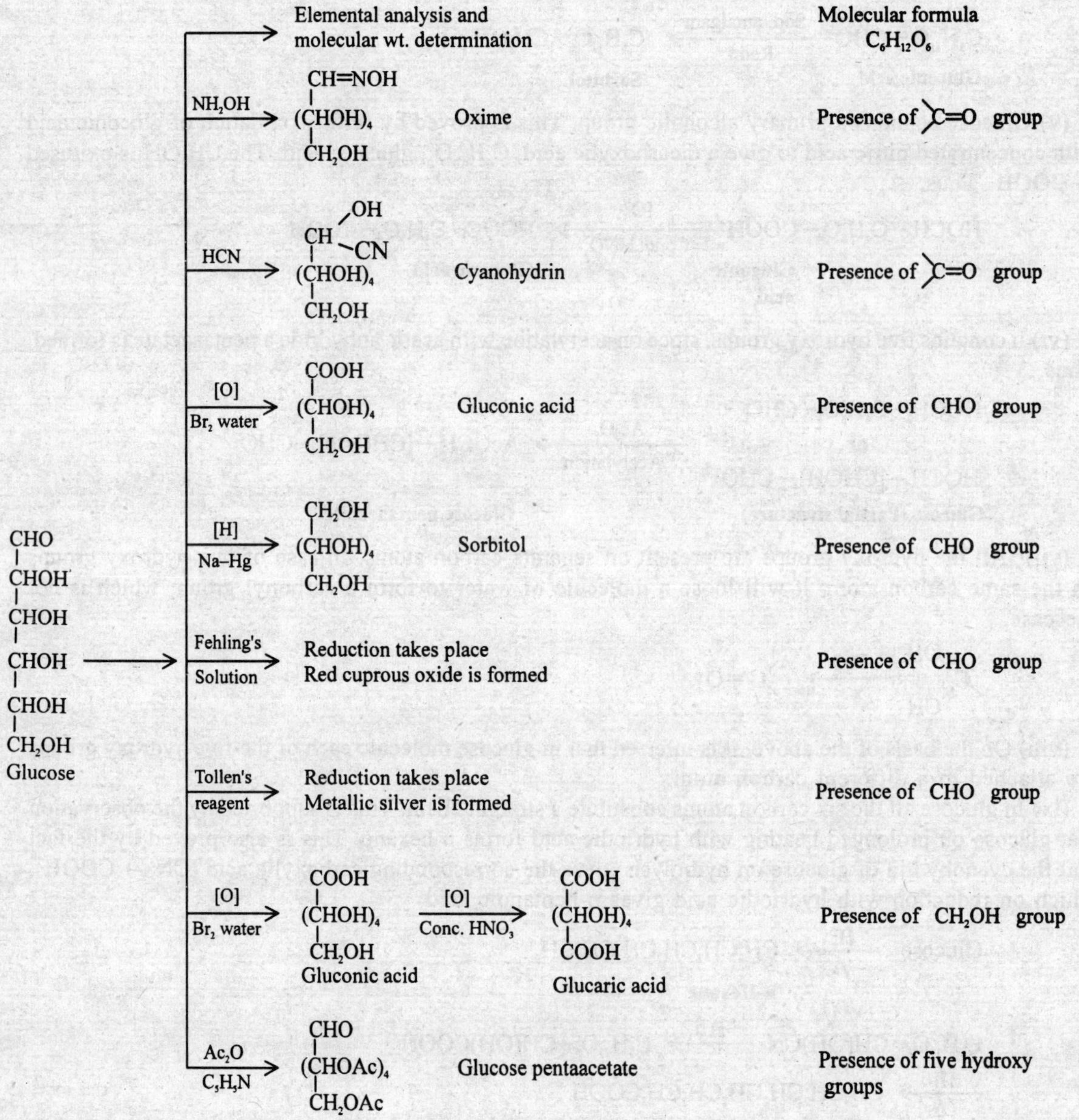

Glucose structure (left side):

CHO
|
CHOH
|
CHOH
|
CHOH
|
CHOH
|
CH$_2$OH
Glucose

(Contd.)

Experiment	Observation

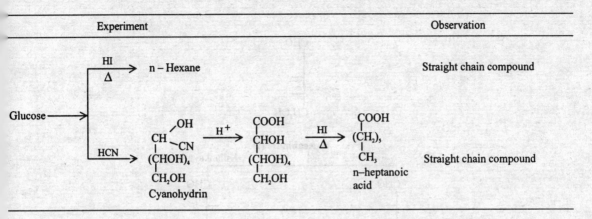

Glucose →	Straight chain compound
Cyanohydrin → COOH/CHOH/(CHOH)₄/CH₂OH → n–heptanoic acid	Straight chain compound

39.3.1.2 Configuration of Hydroxy Groups in Glucose (Stereochemistry of Glucose)

The open chain structure of glucose contains four asymmetric carbon atoms (numbers 2,3,4 and 5) and sixteen stereoisomers out of which only one configuration represents glucose. All are known and their configurations established.

The configuration of glucose defines the arrangement of H and OH groups on carbon atoms 2,3,4 and 5 of glucose. It is determined on the basis of the known configuration of D-arabinose (I) (see page 714). It is now known that D-arabinose on Kiliani's synthesis gives a mixture of glucose (II) and mannose (III). Thus, glucose and mannose differ in configurations only at C-2; the configurations at C-3, C-4 and C-5 in glucose and mannose is the same as in D-arabinose. The configuration at C-2 in glucose and mannose is determined by subjecting each of these to Kiliani's syntheses; both of these give two heptoses each. The heptoses (IV and V) obtained from glucose on oxidation give two dibasic acids (VIII and IX) only one of which (IX) is optically active and the other (VIII) is meso or optically inactive. However, heptoses (VI and VII) obtained from mannose yield two dibasic acids, (X and XI) both of which are optically active.

In the above sequence of reactions, the dibasic acid (VIII) is meso or inactive since it possesses a plane of symmetry. It (VIII) could only be derived by the oxidation of heptose (IV), which is obtained only from the hexose (II) by Kiliani's syntheses. Thus, glucose has the configuration corresponding to (II); the other configuration (III) represents mannose.

39.3.1.3 Cyclic Structure (Pyranose Structure) (Determination of the Ring size)

The open chain structure of glucose proposed by Baeyer accounted for most of its properties. However, it failed to explains the following:

(i) Although glucose has an aldehydic group, it does not give the Schiff's test for aldehydes.

(ii) It does not give other characteristic reactions of aldehydes. For example, it does not react with sodium bisulphite and ammonia.

(iii) Glucose pentaacetate does not react with hydroxylamine which indicates the absence of free CHO group in this derivative.

(iv) Isolation of two anomers of glucose, which are called α-glucose and β-glucose. The former, m.p. 146°C, is obtained by crystallisation of glucose from a concentrated aqueous solution at 30°C. On the other hand the β-form m.p. 150°C is obtained by crystallisation of glucose from a hot saturated aqueous solution. The α- and β-forms have different specific rotation ($[\alpha]_D = 111°$ and 19.2° respectively).

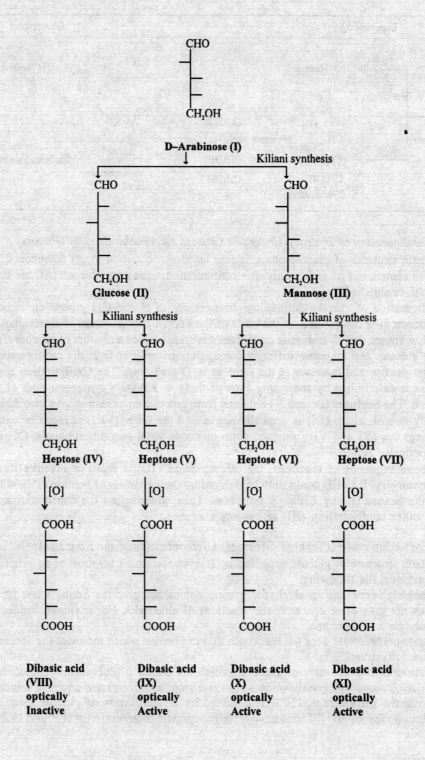

CHO

CH₂OH

D–Arabinose (I)

Kiliani synthesis

CHO

CH₂OH
Glucose (II)

CHO

CH₂OH
Mannose (III)

Kiliani synthesis

Kiliani synthesis

CHO

CH₂OH
Heptose (IV)

CHO

CH₂OH
Heptose (V)

CHO

CH₂OH
Heptose (VI)

CHO

CH₂OH
Heptose (VII)

[O]

[O]

[O]

[O]

COOH

COOH

COOH

COOH

COOH

COOH

COOH

COOH

**Dibasic acid
(VIII)
optically
Inactive**

**Dibasic acid
(IX)
optically
Active**

**Dibasic acid
(X)
optically
Active**

**Dibasic acid
(XI)
optically
Active**

If either of the two forms is dissolved in water and allowed to stand, the specific rotation of the solutions changes gradually until a final value +52.5° in obtained. The equilibrium is attained faster in presence of acid or base catalyst.

α-D(+) Glucose $\rightleftharpoons$ Equilibrium $\rightleftharpoons$ β-D(+) Glucose
                                          Mixture
Specific rotation : +111°   +52.5°   +19.2°

A change in the specific rotation of an optically active compound is called *Mutarotation*. On the basis of the above, it is suggested that glucose exists in two forms, viz α- and β-. The open chain structure of glucose does not account for this behavior.

(v) Glucose on treatment with methanol in presence of dilute hydrochloric acid does not give the acetal (as is expected of aldehydes but instead two isomeric monomethyl derivatives (methyl glucosides) are obtained. The formation of acetal in case of aldehydes is explained in the following way:

$$R-CHO \xrightarrow[H^+]{CH_3OH} \underset{\underset{OCH_3}{|}}{\overset{\overset{OH}{|}}{R-C-H}} \xrightarrow{CH_3OH} \underset{\underset{OCH_3}{|}}{\overset{\overset{OCH_3}{|}}{R-C-H}}$$

Hemiacetal   **Acetal**

However, in the case of glucose two methyl glucosides are obtamied. One is called methyl α-D-glucoside (m.p. 165°C, sp.rotation +158°) and the other methyl β-D-glucoside (m.p. 107°C, sp.rotation −33°). These glucosides do not reduce Fehling's solution and also do not react with hydroxylamine or hydrogen cyanide indicating thereby that glucosides do not have a free −CHO group. Both the glucosides are not hydrolysed by water but require dilute acid for their hydrolysis.

In case of glucose, only one molecule of methyl alcohol is used—the other hydrogen atom for elimination of a water molecule comes from some alcoholic group in glucose resulting in the intramolecular hemiacetal formation. The formation of the two methyl glucosides can be represnted as shown below.

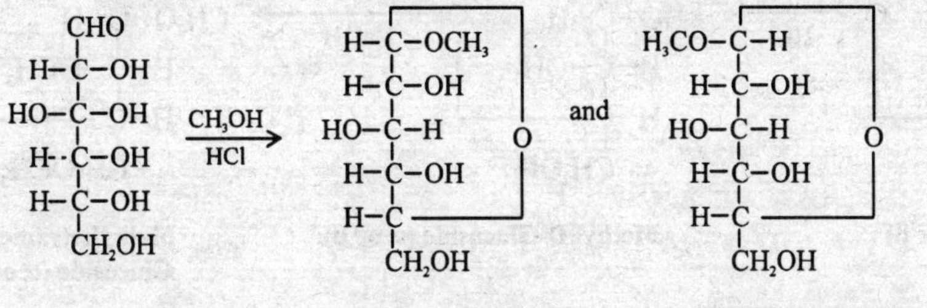

```
      CHO                    H-C-OCH₃              H₃CO-C-H
      |                      |                     |
   H-C-OH                 H-C-OH               H-C-OH
      |                      |                     |
  HO-C-OH    CH₃OH        HO-C-H        O   and  HO-C-H        O
      |      ----→           |                     |
  H··C-OH     HCl         H-C-OH               H-C-OH
      |                      |                     |
   H-C-OH                 H-C                  H-C
      |                      |                     |
    CH₂OH                  CH₂OH                 CH₂OH
```

Methyl-α-D-glucoside   **Methyl-β-D-glucoside**

On the basis of Baeyer's strain theory a six or a five membered ring is more stable than the 3, 4 or 7 membered. The six membered ring system has less bond angle strain than the 5-membered ring and so it is possible that the hydroxy at C-5 is involved in ring formation.

By analogy of the formation of two methyl glucosides, it was believed that glucose exists in two forms, α-D-glucose and β-D-glucose. These two forms change into one another through an open chain structure showing mutarotation.

```
   H–C–OH ┐               CHO                 HO–C–H ┐
   H–C–OH │             H–C–OH                H–C–OH │
  HO–C–H  │ O  ⇌      HO–C–H       ⇌       HO–C–H  │ O
   H–C–OH │             H–C–OH                H–C–OH │
   H–C ───┘             H–C–OH                H–C ───┘
    CH₂OH                CH₂OH                 CH₂OH
```

α-D(+) Glucose   Open Chain form β-D(+) Glucose

As a result of cyclisation, the C-1 of glucose becomes asymmetric. This implies that when hydrogen of OH group at C_5 adds to the oxygen of the aldehyde group, the OH group formed may move either to the left or right resulting in the formation of two isomers. The isomer having OH group to the right side of C-1 is designated as α-D-glucose and the one having OH group to the left of C-1 is designated as β-D-glucose. Thus, the α- and β-D-glucose differ only in configurations at C-1; they are not enantiomers (i.e. are not mirror images of each other). The configuration of all other carbon atoms (C-2 to C-5) is the same. Such pairs of stereoisomeric ring forms of any sugar are known as *anomers*.

The presence of six membered ring in glucose (as stated above) was established by the work of Howarth et al. as follows. Glucose on treatment with methyl alcohol in presence of hydrogen chloride gas gives methyl-D-glucoside, which on methylation with dimethyl sulphate gives methyltetramethyl-D-glucoside. Its hydrolysis with hydrochloric acid to tetramethyl D-glucose followed by oxidation with nitric acid affords xylotrimethoxyglutaric acid whose structure is known. The above sequence of reactions are represented below.

```
   CHOH ┐                          CHOCH₃ ┐                       H–C–OCH₃ ┐
   H–C–OH │                        H–C–OH  │                      H–C–OCH₃ │
  HO–C–H  │ O   CH₃OH            HO–C–H   │ O   (CH₃)₂SO₄        CH₃O–C–H   │ O
   H–C–OH │   ───────→            H–C–OH  │   ───────→            H–C–OCH₃ │
   H–C ───┘     HCl               H–C ────┘      NaOH             H–C ──────┘
    CH₂OH                          CH₂OH                           CH₂OCH₃
```

D-Glucose (α or β)   Methyl D-Glucoside (α or β)   Methyltetramethyl-D-Glucoside (α or β)

```
                    CH–OH ┐                            COOH
                  H–C–OCH₃ │                         H–C–OCH₃
   HCl          CH₃O–C–H   │ O    [O]             CH₃O–C–H
 ───────→         H–C–OCH₃ │    ───────→            H–C–OCH₃
                  H–C ─────┘       HNO₃              COOH
                   CH₂OCH₃
```

Tetramethyl D-Glucose (α or β)   Xylotrimethyoxyglutaric acid

The formation of xylotrimethoxyglutaric acid indicates the presence of a six-membered ring in glucose. If a five membered ring would have been present, the above sequence of reactions would have produced dimethoxysuccinic acid.

The cyclic form of glucose is represented by the Howarth projection as shown below. The lower thickened edge of the ring is assumed to be nearest to the observer. The groups projected to the right, in the Fischer projection go below the plane of the ring, while those to the left side go above the ring.

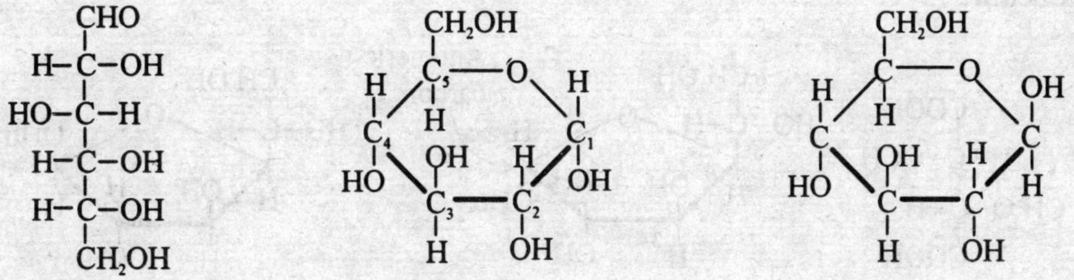

D-Glucose
Fischer Projection

α-D-Glucose

β-D-Glucose

Howarth Projections

The simplest way of drawing Howarth Projections is to omit the ring carbons. Thus α- and β-D-glucose are represented as shown below:

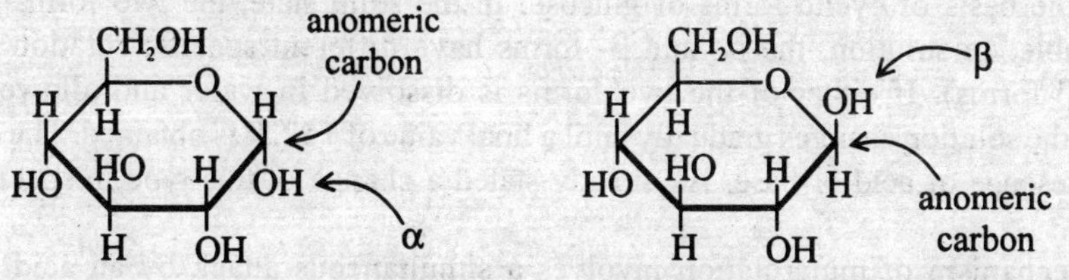

α-D-Glucose

β-D-Glucose

It has been shown on the basis of X-ray studies that the cyclic forms of glucose preferentially exists in the non-planer chair conformation like those of cyclohexanes.

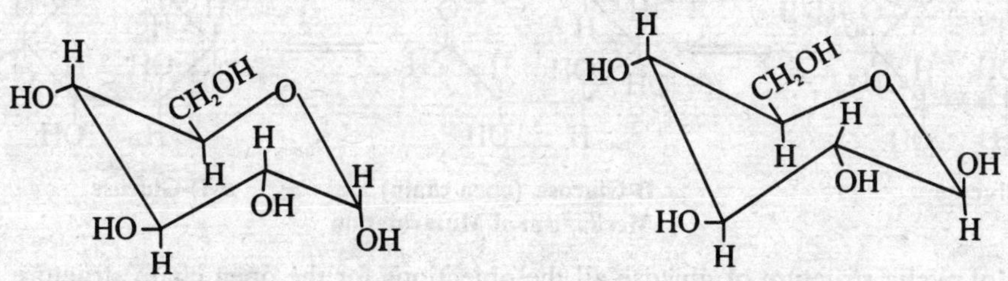

α-D-Glucose

β-D-Glucose

Chair Conformation of Glucose

In the α-D-glucose, the glycosidic hydroxy is axial and in the β-D-glucose it is equatorial. Since a more stable isomer is the one with large number of equatorial substituents, β-form may be expected to be more stable.

39.3.1.4   The Furanose Structure

By the methylation of glucose with methanol and hydrochloric acid and keeping the mixture at O°, Fische obtained a methyl glucoside in small amount, which was different from the methyl glucosides, obtaine earlier. This new methyl glucoside was called methyl γ-glucoside and was found to contain a five membere ring. The size of the ring as five membered was determined by Haworth's method, viz. methylation wit dimethylsulphate, hydrolysis by acid and final nitric acid oxidation when dimethoxysuccinic acid was ob tained. On this basis Howarth proposed a furanose structure for γ-sugars. The structures are represente as shown below:

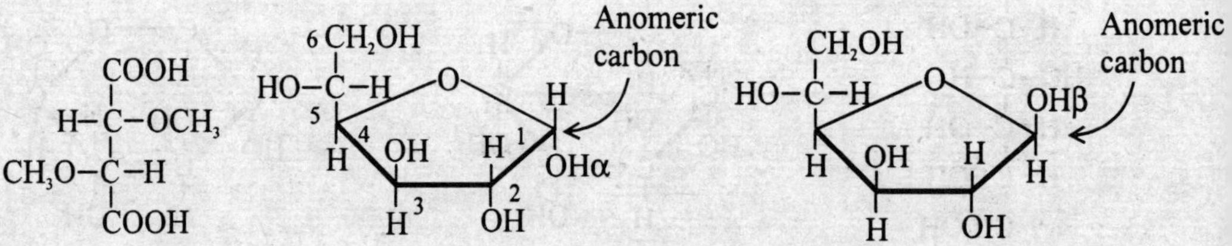

Dimethoxysuccinic acid   **α-D-Glucofuranose**   **β-D-Glucofuranose**
(Furanose Structure of D-glucose)

39.3.1.5   Mutarotation

The phenomenon of mutarotation, which could not be explained on the basis of open chain structure can be explained on the basis of cyclic forms of glucose. In the solid state, the two forms of glucose viz. α- and β- are stable. In solution, the α- and β- forms have different specific rotation ($[\alpha]_D$=111' and 19.2° for α-and β-forms). If either of the two forms is dissolved in water and allowed to stand, the specific rotation of the solution changes gradually until a final value of +52.5° is obtained. The equilibrium is attained faster in presence of acid or base. As already stated a change of this type in specific rotation is called mutarotation.

The most likely mechanism of mutarotation involves a simultaneous attack by an acid and a base to yield an open chain aldehyde form, which then recyclises to give the other form. Thus,

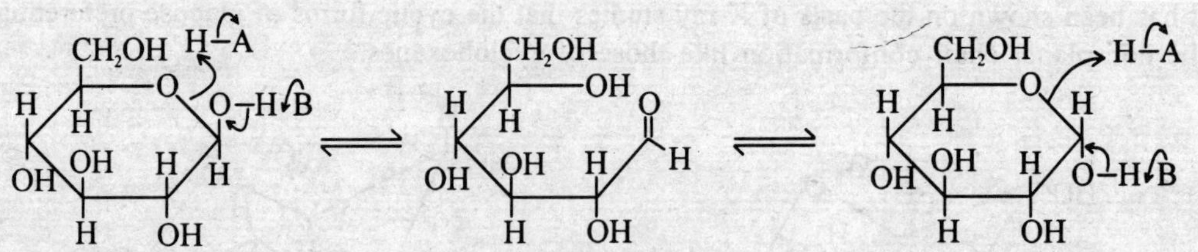

β-D-Glucose   **D-Glucose (open chain)**   **α-D-Glucose**
Mechanism of Mutarotation

On the basis of cyclic structure of glucose all the objections for the open chain structure (mentioned above) can be easily explained.

Glucoside Formation

Formation of two glucosides can be explained with the help of cyclic form or hemiacetal structure of glucose. D-glucose on reaction with alcohol forms acetal just as simple hemiacetals react with alcohols to form acetal derivative. The acetal of glucose are known as glucosides. The hydroxy group produced at the oxo group by ring formaion is known as glycosidic hydroxy group. The glucoside derivatives are decomposed readily by various reagents.

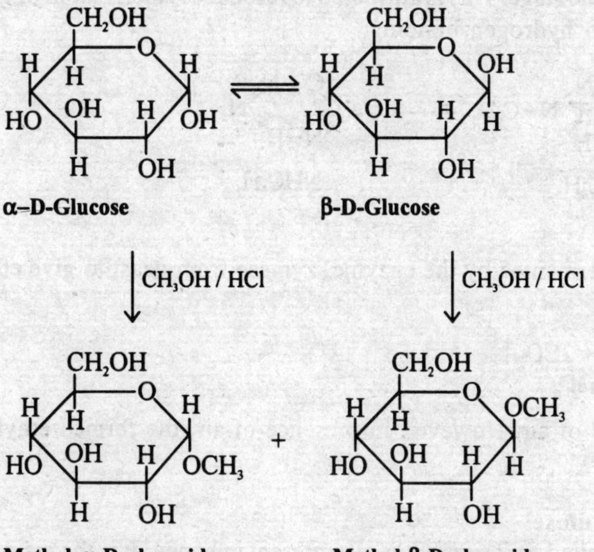

α–D-Glucose   β-D-Glucose

CH_3OH / HCl CH_3OH / HCl

Methyl α-D-glucoside   +   **Methyl β-D-glucoside**
$[\alpha] = +159°$ [m.p. 166°C] $[\alpha] = -34°$ [m.p. 107°C]

Some typical reactions of glucose are given below:

Osazone Formation
Like hydroxylamine, D-glucose reacts with phenylhydrazine to give D-glucose phenylhydrazone (soluble). If excess of phenylhydrazine is used, a dihydrazone called osazone is formed.

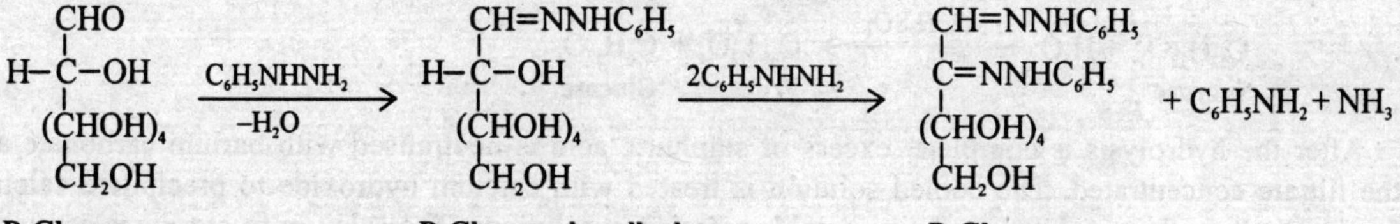

D-Glucose   D-Glucose phenylhydrazone   D-Glucosazone

The above reaction was discovered by Emil Fischer (1887). The generally accepted mechanism was proposed by Weygand and Semykain (1965).

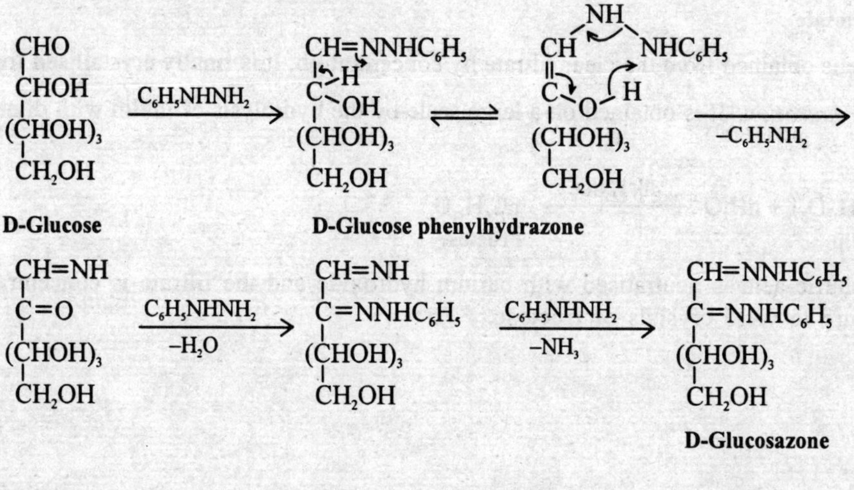

D-Glucose   D-Glucose phenylhydrazone

D-Glucosazone

The above reaction stops at the osazone stage. This is attributed to resonance stabilisation of glucosazone in the form of cyclic structure due to hydrogen bonding.

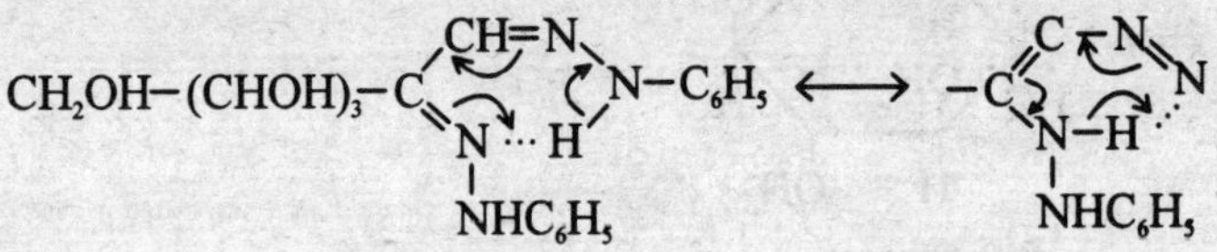

$$CH_2OH-(CHOH)_3-C\underset{\underset{NHC_6H_5}{|}}{\overset{\overset{CH=N}{|}}{\diagdown}}\underset{N\cdots H}{\overset{}{\diagup}}N-C_6H_5 \longleftrightarrow -C\underset{\underset{NHC_6H_5}{|}}{\overset{\overset{C-N}{|}}{\diagdown}}\underset{N-H}{\overset{}{\diagup}}N$$

Fermentation

An aqueous solution of D-glucose is fermented by the enzyme, zymase from yeast to give ethyl alcohol and carbon dioxide.

$$C_6H_{12}O_6 \longrightarrow 2C_2H_5OH + 2CO_2 \uparrow$$

D-Glucose   Ethyl alcohol

The reaction takes place in absence of air. However, in presence of air, the formed ethyl alcohol is oxidised to acetic acid (vinegar).

39.3.2   Fructose/Fruit Sugar/Laevulose

Fructose is present in fruits and hence called fruit sugar. It is also present in honey. It is the main constituent of inulin, a polysaccharide found in artichokes and dahlias. Fructose is laevorotatory and so is called laevulose.

39.3.2.1   Preparation

Laboratory Preparation: It is prepared in the laboratory by the hydrolysis of cane sugar by boiling with dilute sulphuric acid.

$$C_{12}H_{22}O_{11} + H_2O \xrightarrow{\text{dil. H}_2\text{SO}_4} C_6H_{12}O_6 + C_6H_{12}O_6$$

Sucrose   Fructose   Glucose

After the hydrolysis is complete, excess of sulphuric acid is neutralised with barium carbonate and the filtrate concentrated. The cooled solution is treated with calcium hydroxide to precipitate calcium fructosate, the calcium gluconate being soluble remains in solutions. The calcium fructosate is converted into fructose by passing carbon dioxide through its suspension in water.

$$C_6H_{11}O_6\,CaOH + CO_2 \longrightarrow C_6H_{12}O_6 + CaCO_3$$

Calcium D-   Fructose
fructosate

The fructose is obtained from the clear filtrate by concentration. It is finally crystallised from alcohol.

Commercial Preparation: It is obtained on a large scale by the hydrolysis of inulin with dilute sulphuric acid

$$(C_6H_{10}O_5)_n + nH_2O \xrightarrow{\text{dil. H}_2\text{SO}_4} nC_6H_{12}O_6$$

Inulin   Fructose

Excess sulphuric acid is neutralised with barium hydroxide and the filtrate is concentrated   under reduced pressure to yield crystals of fructose.

39.3.2.2 Structure

The structure of fructose is based on the basis of following evidences.

(i) The molecular formula of fructose $C_6H_{12}O_6$ is assigned on the basis of elemental analysis and molecular weight determination.

(ii) It has a carbonyl group - since it forms a monoxime and adds on one molecule of hydrogen cyanide to form a cyanohydrin.

(iii) That the carbonyl group is present as keto group is inferred by its oxidation to glycollic acid and tartaric acid; the oxidation occurs by rupture of the carbon chain.

$$
\begin{array}{c}
CH_2OH \\
| \\
C=O \\
..|.. \\
(CHOH)_3 \\
| \\
CH_2OH
\end{array}
\xrightarrow[HNO_3]{[O]}
\begin{array}{c}
CH_2OH \\
| \\
COOH
\end{array}
+
\begin{array}{c}
COOH \\
| \\
(CHOH)_2 \\
| \\
COOH
\end{array}
$$

Fructose   **Glycollic acid**   **Tartaric acid**

Oxidation with nitric acid also proves the presence of CH_2OH group.

(iv) It contains five hydroxy groups, since on acetylation with acetic anhydride a pentaacetate is formed. Thus,

$$HOCH_2COC_3H_6O_3CH_2OH$$

or

$$HOCH_2CO[CHOH]_3.CH_2OH \xrightarrow{Ac_2O} AcOCH_2-CO-[CHOAc]_3-CH_2OAc$$

Fructose   **Fructose Pentaacetate**

(v) All the hydroxy groups are present on separate carbon atoms. In case two hydroxy groups are present on the same carbon atom, then it will loose a molecule of water, which is not the case.

(vi) In fructose (as in the case of glucose) all the six carbon atoms constitute a straight chain. This is supported by the fact that fructose on heating with hydriodic acid forms n-hexane.

$$Fructose \xrightarrow{HI} CH_3CH_2CH_2CH_2CH_2CH_3$$

n-Hexane

On the basis of the above it is concluded that fructose has a keto and two primary hydroxy groups and that all carbon atoms are present in a straight chain. The following open chain structures has been proposed.

$$HOCH_2.CO.CHOH.CHOH.CHOH.CH_2OH$$

(vii) The above open chain structure is also confirmed by the following observation:

(a) On partial reduction with sodium amalgam and water, a mixture of two epimeric alcohols, sorbitol and mannitol is obtained. This is due to creation of a new asymmetric carbon atom at C-2. This confirms the presence of a keto group

$$
\begin{array}{c}
CH_2OH \\
| \\
C=O \\
| \\
(CHOH)_3 \\
| \\
CH_2OH
\end{array}
+ 2[H]
\xrightarrow[H_2O]{Na-Hg}
\begin{array}{c}
CH_2OH \\
| \\
H-C-OH \\
| \\
(CHOH)_3 \\
| \\
CH_2OH
\end{array}
+
\begin{array}{c}
CH_2OH \\
| \\
HO-C-OH \\
| \\
(CHOH)_3 \\
| \\
CH_2OH
\end{array}
$$

Fructose   **Sorbitol**   **Mannitol**

(b) Fructose cyanohydrin on hydrolysis and subsequent reduction with hydriodic acid and red phosphorus gives 2-methylhexanoic acid. This proves that the ketonic group is adjacent to the terminal carbon atom.

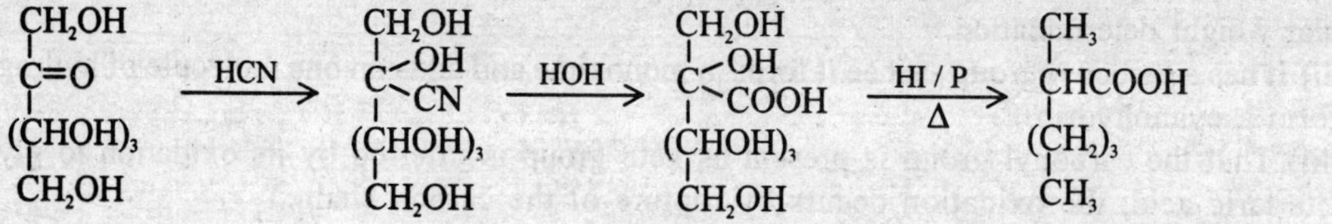

$$
\begin{array}{ccccc}
CH_2OH & & CH_2OH & & CH_2OH & & CH_3 \\
| & & |\,^-OH & & |\,^-OH & & | \\
C=O & \xrightarrow{HCN} & C\,^-CN & \xrightarrow{HOH} & C\,^-COOH & \xrightarrow[\Delta]{HI\,/\,P} & CHCOOH \\
| & & | & & | & & | \\
(CHOH)_3 & & (CHOH)_3 & & (CHOH)_3 & & (CH_2)_3 \\
| & & | & & | & & | \\
CH_2OH & & CH_2OH & & CH_2OH & & CH_3
\end{array}
$$

2-Methylhexanoic acid

As seen, the structure of fructose has three asymmetric carbon atoms. The configuration of D-fructose is determined as follows: Both glucose and fructose on treatment with excess phenylhydrazine give identical osazone.

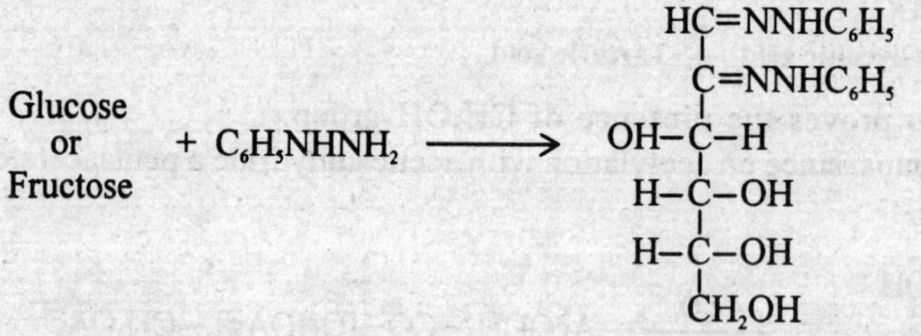

$$
\begin{array}{l}
\text{Glucose} \\
\text{or} \quad + \; C_6H_5NHNH_2 \longrightarrow \\
\text{Fructose}
\end{array}
\qquad
\begin{array}{c}
HC=NNHC_6H_5 \\
| \\
C=NNHC_6H_5 \\
| \\
OH-C-H \\
| \\
H-C-OH \\
| \\
H-C-OH \\
| \\
CH_2OH
\end{array}
$$

Glucosazone or Fructosazone, m.p. 206°

The osazone formation involves only the carbonyl group at C-2 and a hydroxy group at C-1. Therefore both glucose and fructose have the same configuration at the remaining carbon atoms.

39.3.2.3 Cyclic Structure

The above open chain structure of fructose dose not account for the following

(i) Although fructose has a carbonyl group it does not react with sodium bisulphite.

(ii) The phenomenon of mutarotation is exhibited by fructose as in the case of glucose. A freshly prepared solution of the two forms of fructose has a specific rotation of −133° and −21° and this attains an equilibrium value of −92° on standing.

(iii) The existence of methyl fructoside in two forms, viz. α-D-fructoside and β-D-fructoside.

In order to explain the above it is believed (by analogy to glucose) that fructose exists in α- and β- forms as shown below:

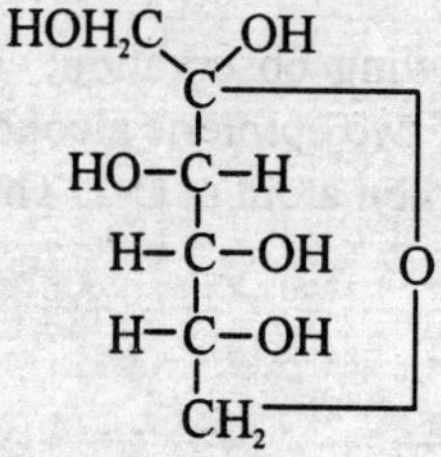

$$
\begin{array}{cc}
\begin{array}{c}
HOH_2C \quad\; OH \\
\diagdown C \diagup \\
| \\
HO-C-H \\
| \\
H-C-OH \\
| \\
H-C-OH \\
| \\
CH_2
\end{array}
&
\begin{array}{c}
HO \quad\; CH_2OH \\
\diagdown C \diagup \\
| \\
HO-C-H \\
| \\
H-C-OH \\
| \\
H-C-OH \\
| \\
CH_2
\end{array}
\end{array}
$$

α-D-Fructose
Sp. rotation −21°

β-D-Fructose 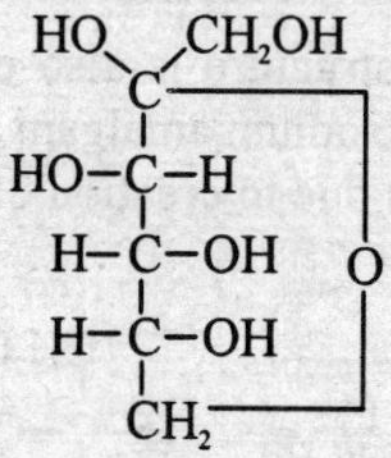
Sp. rotation −133°

The size of the ring (whether 5 or 6 membered) is determined by Haworth's method (as in the case of glucose). The final product obtained is D-arabinotrimethoxyglutaric acid, which proves the presence of a six membered ring.

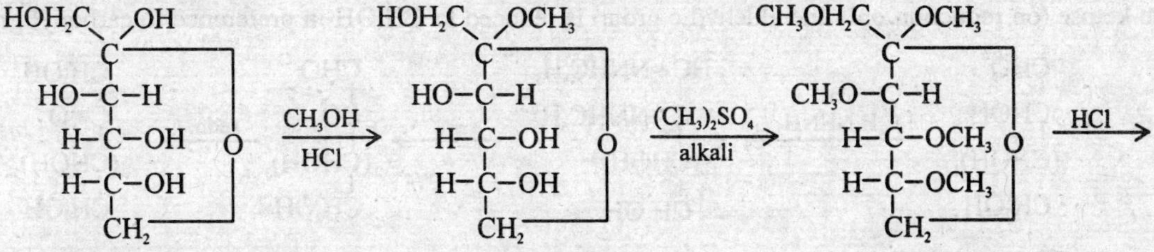

D-Fructose $\xrightarrow[\text{HCl}]{\text{CH}_3\text{OH}}$ Methyl D-fructoside $\xrightarrow[\text{alkali}]{(\text{CH}_3)_2\text{SO}_4}$ Methyl tetra-O-methylfructoside $\xrightarrow{\text{HCl}}$

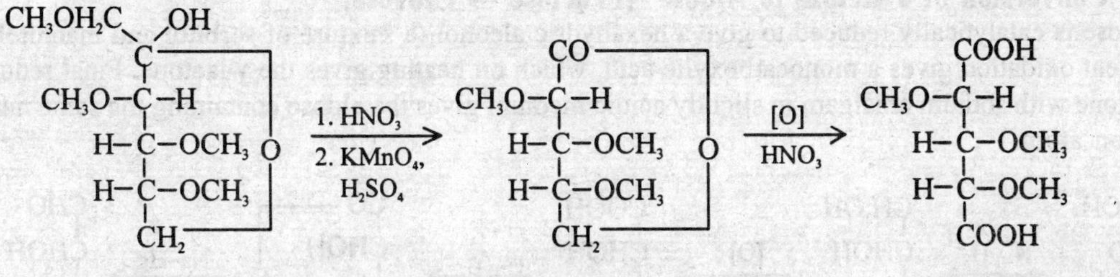

Tetra-O-methylfructose $\xrightarrow[\substack{2.\ \text{KMnO}_4, \\ \text{H}_2\text{SO}_4}]{1.\ \text{HNO}_3}$ Tri-O-methyl D-arabinolactone $\xrightarrow[\text{HNO}_3]{[\text{O}]}$ Arabinotrimethoxy-glutaric acid

On the basis of the above, it is established that free fructose exists as a six membered (δ-oxide) ring. However, in sucrose or other sugars having a fructose unit, it is also known that fructose exists in a five membered (γ-oxide) ring form. It has been suggested that hydrolysis of sucrose yields fructose with a γ-oxide ring, which then readily changes to the δ-oxide ring. Following are given Haworth pyranose (six mambered ring) and furanose (five membered ring) structures for fructose.

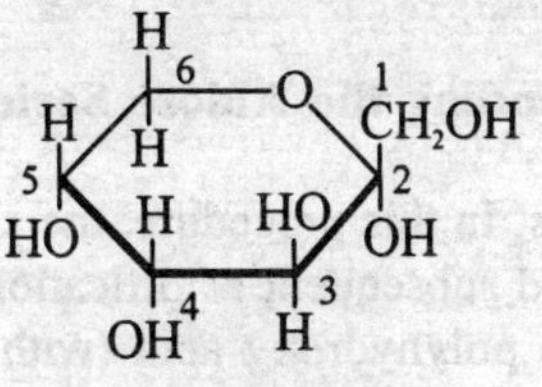

α-D-Fructopyranose 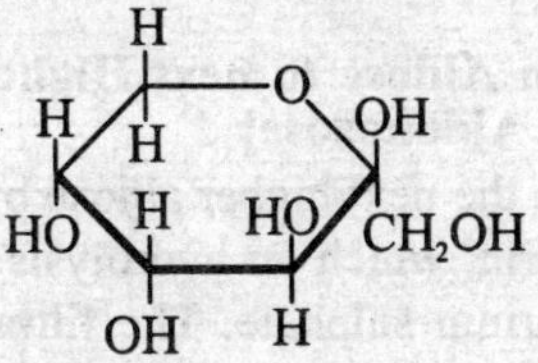   β-D-Fructopyranose

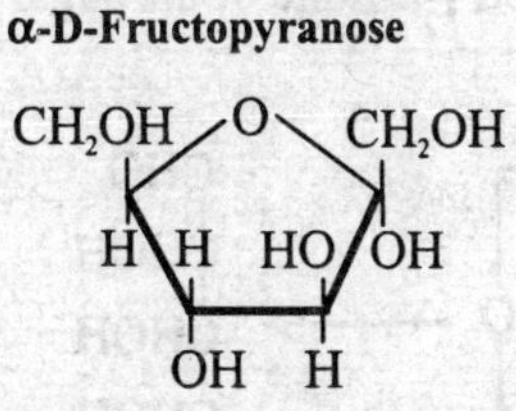

α-D-Fructofuranose 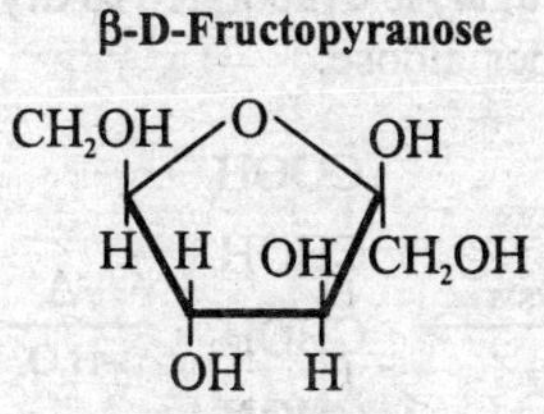   β-D-Fructofuranose

39.4   SOME TYPICAL CONVERSIONS IN MONOSACCHARIDES

39.4.1   Conversion of an Aldose to Ketose [Glucose → Fructose]

The aldose is converted into its osazone, which is hydrolysed to an osone. The osone on reduction gives the ketose (on reduction only the aldehydic group is reduced to CH_2OH in preference to carbonyl group).

$$
\begin{array}{c}
CHO \\
CHOH \\
(CHOH)_3 \\
CH_2OH
\end{array}
\xrightarrow{3\ C_6H_5NHNH_2}
\begin{array}{c}
HC=NNHC_6H_5 \\
C=NNHC_6H_5 \\
(CHOH)_3 \\
CH_2OH
\end{array}
\longrightarrow
\begin{array}{c}
CHO \\
CO \\
(CHOH)_3 \\
CH_2OH
\end{array}
\xrightarrow{redn.}
\begin{array}{c}
CH_2OH \\
C=O \\
(CHOH)_3 \\
CH_2OH
\end{array}
$$

Glucose (Aldose)   Glucosazone   Glucosone   Fructose (Ketose)

39.4.2   Conversion of a Ketose to Aldose   [Fructose → Glucose]

The ketose is catalytically reduced to give a hexahydric alcohol (a mixture of sorbitol and mannitol). Its subsequent oxidation gives a monocarboxylic acid, which on heating gives the γ-lactone. Final reduction of γ-lactone with sodium amalgam in slightly acidic medium gives the aldose containing the same number of carbon atom.

$$
\begin{array}{c}
CH_2OH \\
C=O \\
(CHOH)_3 \\
CH_2OH
\end{array}
\xrightarrow{Ni,\ H_2}
\begin{array}{c}
CH_2OH \\
CHOH \\
(CHOH)_3 \\
CH_2OH
\end{array}
\xrightarrow[Br_2/H_2O]{[O]}
\begin{array}{c}
COOH \\
CHOH \\
(CHOH)_3 \\
CH_2OH
\end{array}
\xrightarrow{\Delta}
\begin{array}{c}
CO \\
CHOH \\
CHOH \\
HC \\
CHOH \\
CH_2OH
\end{array}O
\xrightarrow[H^+]{Na-Hg}
\begin{array}{c}
CHO \\
CHOH \\
CHOH \\
CHOH \\
CHOH \\
CH_2OH
\end{array}
$$

Fructose   Hexahydric alcohol   Monocarboxylic   γ-Lactone   Glucose
(Ketose)   (Mixture of Sorbitol   acid   (Aldose)
                  and mannitol)

39.4.3   Conversion of an Aldose to next Higher Aldose   (Ascending the Aldose Series) [Aldopentose → Aldohexose]

An aldose is converted into the next higher aldose by *Kiliani synthesis*. In this procedure, the aldopentose is converted into cyanohydrin, which on hydrolysis with $Ba(OH)_2$ and subsequent acidification with sulphuric acid precipitates barium sulphate. The filtrate containing the polyhydroxy acid (with one more carbon atom than the original sugar) is evaporated to dryness. The formed γ-lactone of the sugar is reduced (Na-Hg) to give the next higher aldose.

$$
\begin{array}{c}
CHO \\
CHOH \\
CHOH \\
CHOH \\
CH_2OH
\end{array}
\xrightarrow[\substack{Ba(OH)_2 \\ 3.\ H^+}]{\substack{1.\ HCN \\ 2.\ Hydrolysis}}
\begin{array}{c}
COOH \\
CHOH \\
CHOH \\
CHOH \\
CHOH \\
CH_2OH
\end{array}
\xrightarrow[-H_2O]{\Delta}
\begin{array}{c}
CO \\
CHOH \\
CHOH \\
CH \\
CHOH \\
CH_2OH
\end{array}O
\longrightarrow
\begin{array}{c}
CHO \\
CHOH \\
CHOH \\
CHOH \\
CHOH \\
CH_2OH
\end{array}
$$

Aldopentose   Polyhydroxy acid   γ-lactone   Aldohexose

In the above method, only one cyanohydrin is obtained in major amount though theoretically two are possible. So the final product is mainly one type of hexose with very little of its epimer.

39.4.4. Conversion of an Aldose to next Lower Aldose (Descending the aldose series) [Aldohexose → Aldopentose]

(a) *Wohl's Method:* In this procedure an aldose is converted into its oxime, which on heating with acetic anhydride gives the nitrile (with the hydroxy groups acetylated). Final treatment with ammonical silver nitrate (silver hydroxide) gives the aldopentose.

$$\underset{\textbf{aldohexose}}{\begin{matrix}CHO\\|\\CHOH\\|\\(CHOH)_3\\|\\CH_2OH\end{matrix}} \xrightarrow{NH_2OH} \underset{\textbf{Oxime}}{\begin{matrix}CH=NOH\\|\\CHOH\\|\\(CHOH)_3\\|\\CH_2OH\end{matrix}} \xrightarrow[-H_2O]{(CH_3CO)_2O,\ \Delta} \underset{\textbf{Acetylated Nitrile}}{\begin{matrix}CN\\|\\CHOCOCH_3\\|\\(CHOCOCH_3)_3\\|\\CH_2OCOCH_3\end{matrix}} \xrightarrow[-AgCN]{AgOH} \underset{\textbf{Aldopentose}}{\begin{matrix}CHO\\|\\(CHOH)_3\\|\\CH_2OH\end{matrix}}$$

(b) *Ruffs Method:* In this method the aldose is oxidised to aldonic acid. Its calcium salt on treatment with Fenton's reagent gives an aldose containing one carbon atom less than the starting aldose

$$\underset{\textbf{Aldohexose}}{\begin{matrix}CHO\\|\\CHOH\\|\\(CHOH)_3\\|\\CH_2OH\end{matrix}} \xrightarrow[Br_2,\ H_2O]{[O]} \underset{\textbf{Aldonic acid}}{\begin{matrix}COOH\\|\\CHOH\\|\\(CHOH)_3\\|\\CH_2OH\end{matrix}} \xrightarrow{CaCO_3} \underset{\textbf{Calcium salt}}{\left(\begin{matrix}COO^-\\|\\CHOH\\|\\(CHOH)_3\\|\\CH_2OH\end{matrix}\right)_2 Ca^{++}} \xrightarrow{H_2O_2\ /\ Fe^{++}} \underset{\textbf{Aldopentose}}{\begin{matrix}CHO\\|\\(CHOH)_3\\|\\CH_2OH\end{matrix} + CO_2 + H_2O}$$

39.5 THE DISACCHARIDES
The disaccharides are made up of two molecules of monosaccharides. As such, they on hydrolysis yield two molecules of monosaccharides. The monosaccharides obtained may be same or different.

$$Sucrose \xrightarrow{Hydrolysis} Glucose + Fructose$$

$$Lactose \xrightarrow{Hydrolysis} Glucose + Galactose$$

$$Maltose \xrightarrow{Hydrolysis} Glucose + Glucose$$

39.5.1 Sucrose/Cane Sugar [$C_{12}H_{22}O_{11}$]
It is the most common disaccharide and is widely distributed in plants especially sugar cane and sugar-beet. Sugar obtained from sugar-beet is known as beet sugar. It is manufactured from both these sources in approximately equal amounts. Sugar is also present in fruits like pineapple, ripe banana and ripe mangoes.

39.5.1.1 Manufacture
Sugar cane is the main raw material for the manufacture of sugar in India. However in temperate climates, it is obtained from sugar beat. Following steps are involved in its manufacturing.

(i) *Extraction of Juice:* The sugar cane is cleaned, cut into short lengths and then passed through a set of roller crushers. After most of the juice has been extracted in the first two crushers (expellers), water is sprinkled over the bagasse, where the remaining juice is extracted by further milling. The cellulose left over is called bagasse and is used as a fuel under boilers and also for the manufacture of insulating material known as celotex.

In case sugar beet is used as staring material, it is cut into small pieces (slices) and put in hot water when the sugar present is dissolved in water. a number of tanks are used in succession. The solution obtained is treated like cane-juice.

(ii) *Purification of Juice:* The juice obtained above is heated by steam and treated with lime. By this process the organic acids and phosphates are separated as insoluble calcium salts. The proteins and the colloidal colouring matters separate in the form of scum on the surface. The precipitated calcium salts and the scum are removed by filtration through canvas. The whole process is known as *defecation.*

The defecated juice (obtained above) contains excess of lime and soluble calcium sucrate. Carbon dioxide gas is passed through the juice. By this process excess of lime is removed as $CaCO_3$ and the calcium sucrate is decomposed back to sugar and $CaCO_3$, which is removed by filtrations. This process is known as *carbonation*

$$C_{12}H_{21}O_{11}CaOH + CO_2 \longrightarrow C_{12}H_{22}O_{11} + CaCO_3$$
Calcium sucrate   **Sucrose**

The juice obtained after carbonation is subjected to *sulphitation.* In this process a current of sulphur dioxide is passed through the juice. The excess lime if present, is decomposed and any calcium sucrate is also decomposed. Also, the colour of juice is bleached. The juice is filtered to remove precipitated calcium sulphate.

(iii) *Concentration of Juice:* The purified and clarified juice is concentrated by boiling under reduced pressure in multiple effect evaporators. Three evaporators are used in succession. The steam is passed in steam coils through the first evaporator. The juice gets concentrated and the steam generated is passed through the second evaporator at lower pressure and from second to the third evaporator at still lower pressure. In the last evaporator crystallisation starts. The syrup known as massicuite is sent to crystallisation tanks.

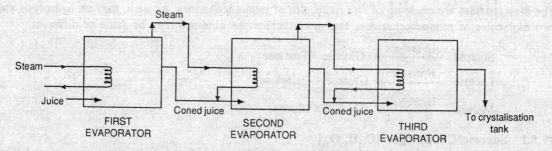

Fig. 39.4:   Concentration of Juice in Multiple Effect Evaporators

(iv) *Crystallisation of Sugar:* In the crystallisation tank the sugar crystals grow. These are separated from the mother liquor (molasses) by centrifugal machine. Impurities sticking on to the surface of crystals are removed by washing with a spray of water in the centrifugal machine. The sugar crystals are dried by hot air on counter current principle.

The molasses obtained above still contain appreciable amounts of sugar. It can be further concentrated to get a fresh crop of crystals. The remaining molasses, which contain glucose and fructose are fermented to obtain ethyl alcohol.

39.5.1.2 Properties

Cane sugar is a colorless, crystalline substance and is sweet in taste. It is soluble in water and its solution is dextrorotatory, $[\alpha]_D = +66.5°$. Following are given some typical chemical properties of cane sugar.

(i) *Action of Heat:* On heating above its melting point, it forms a brown substance known as caramel, which has a characteristic odor and is used as a flavouring material in foods and candies.

(ii) *Hydrolysis:* On heating with dilute acids, cane sugar undergoes hydrolysis to form equimolecular mixture of D(+)-glucose and D(−)-fructose.

$$C_{12}H_{22}O_{11} + H_2O \xrightarrow{HCl} C_6H_{12}O_6 + C_6H_{12}O_6$$

Cane Sugar	Glucose	Fructose
$[\alpha]_D = +66.5°$	$[\alpha]_D = +52.5°$	$[\alpha]_D = -92.4°$

The hydrolysis can also be brought about by enzyme invertase.

Sucrose is dextrorotatory and on hydrolysis gives dextrorotatory glucose and laevorotatory fructose. The mixture is laevorotatory due to greater leavorotation of fructose. Thus, hydrolysis of cane sugar brings about a change (inversion) of rotation of the reaction mixture from dextro to laevo towards polarised light, the phenomenon is called *inversion* and the mixture is known as *invert sugar*.

(iii) *Oxidation:* Sucrose on oxidation with nitric acid under different conditions gives either oxalic acid (80%), tartaric acid (35%-40%) or glucaric acid (30%-55%).

$$C_{12}H_{22}O_{11} \xrightarrow[HNO_3]{[O]} \begin{cases} \text{Oxalic acid} \\ \text{Tartaric acid} \\ \text{Glucaric acid} \end{cases}$$

Sucrose

However, oxidation of sucrose in alkaline solution with air gives D-arabonic acid.

(iv) *Hydrogenation:* Sucrose on hydrogenation under controlled conditions gives a mixture of sorbitol and mannitol, which can be separated by fractional crystallisation

$$C_{12}H_{22}O_{11} \xrightarrow{H_2} C_6H_{14}O_6 + C_6H_{14}O_6$$

Sucrose	Sorbitol	Mannitol

(v) *Action of Sulphuric Acid:* Sucrose on heating with concentrated sulphuric acid gives CO_2 and SO_2 with charring.

$$C_{12}H_{22}O_{11} + H_2SO_4 \xrightarrow{\Delta} 12C + 11H_2O + H_2SO_4$$

Sucrose

$$C + 2H_2SO_4 \longrightarrow CO_2 + 2SO_2 + 2H_2O$$

(vi) *Fermentation:* Sucrose solution is fermented by yeast when the enzyme invertase hydrolyses sucrose to glucose and fructose and zymase converts them to ethyl alcohol

$$C_{12}H_{22}O_{11} + H_2O \xrightarrow{Invertase} C_6H_{12}O_6 + C_6H_{12}O_6$$

Sucrose   Glucose   Fructose

$$C_6H_{12}O_6 \xrightarrow{Zymase} 2C_2H_5OH + 2CO_2$$

Glucose or Fructose

Uses: It is used as a sweetening agent for various food preparations, jams, syrups, sweets and also as food preservative. Sucrose octaacetate is used in the manufacture of transparent papers and as the ingredient of non aqueous adhesives.

39.5.1.3 *Structure*

(i) Its molecular formula is $C_{12}H_{22}O_{11}$.

(ii) It has eight hydroxy group, since it gives an octaacetate and an octamethyl derivative.

(iii) It does not reduce Fehlings solution or Tollens reagent. It also does not react with hydroxylamine, phenylhydrazine. Also it does not show mutarotation.

On the basis of these negative tests it is concluded that sucrose does not have a free aldehyde or ketonic group.

(iv) On hydrolysis sucrose gives an equimolecular mixture of D(+) glucose and D(–) fructose. Further as seen above, it does not have a free aldehydic or ketonic group. It is therefore obvious that the glucose and fructose must have been linked through C-1 carbon of glucose (carrying CHO group) and C-2 carbon of fructose (carrying a ketonic group).

(v) The octamethyl derivative of sucrose on hydrolysis results in the formation of 2,3,4,6-tetra-O-methyl derivative of glucose and 1,3,4,6-tetra-O-methyl derivative of fructose. The formation of the former viz. 2,3,4,6-tetra-O-methyl derivative shows the presence of ring between C-1 and C-5 of glucose unit and formation of 1,3,4,6-tetra-O-methyl derivative of fructose indicates a ring between C-2 and C-5 of fructose unit. Hence glucose has pyranoside and fructose has furanoside structure.

(vi) Sucrose can be hydrolysed by maltase, which hydrolyses only α-glucosides. It is also hydrolysed by invertase, that hydrolyses β- but not α-fructosides. Therefore on the basis of enzymic hydrolyses, it is concluded that sucrose is both an α-glucoside and a β-fructoside.

On the basis of the above, Haworth (1927) suggested the following formula for sucrose.

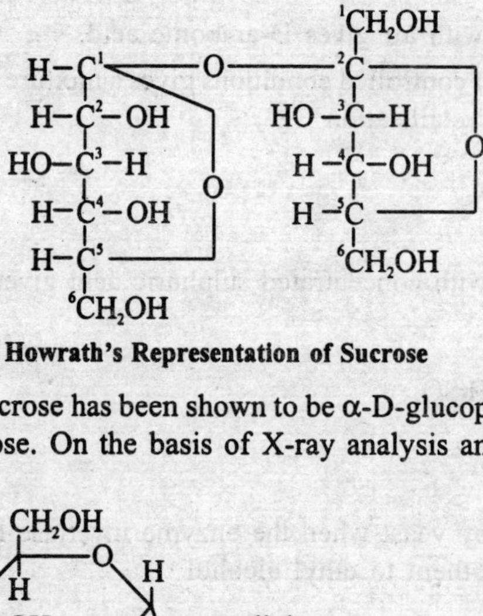

Howrath's Representation of Sucrose

Thus sucrose has been shown to be α-D-glucopyranosyl-β-D-fructofuranoside, i.e., α-glucose is linked to β-fructose. On the basis of X-ray analysis and total synthesis sucrose has been assigned following structure.

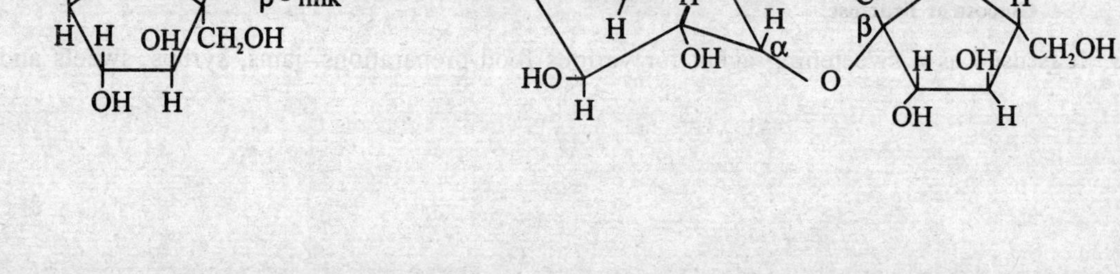

39.5.2   Lactose/Milk-Sugar ($C_{12}H_{22}O_{11}$)

Lactose occurs in milk of all animals. For example, cows milk contains 5-6% and human milk contains 6-8 % lactose. It is prepared commercially from cows milk after removal of cheese and casein and the remaining mother liquor (called whey) is concentrated, decolourised using animal charcoal and crystallised.

39.5.2.1   Properties

It is a white crystalline solid, soluble in water but insoluble in solvents like alcohol, ether, benzene etc. Lactose is dextrorotatory and exhibits mutarotation and is present in two forms α- and β-. The two forms are obtained by crystallisation from water under different conditions.

α-form	equilibrium	β-form
$[\alpha]_D^{20}$=+89.5° m.p. 203°	$[\alpha]_D^{20}$=+55.5°	$[\alpha]_D^{20}$=+35° m.p. 252°

(i) On hydrolysis with dilute acid and emulsin, it gives glucose and galactose.

$$C_{12}H_{22}O_{11} + H_2O \longrightarrow C_6H_{12}O_6 + C_6H_{12}O_6$$

(+)Lactose   (+)Glucose (+)Galactose

Since it is hydrolysed by emulsin (β-glycosidic splitting enzyme) Lactose is β-glucoside.

(ii) Lactose reduces Fehling's solutions and Tollens reagent. It also reacts with hydrogen cyanide and forms an osazone. The reactions indicate that one free hemiacetal group must be present and this is in equilibrium with free aldehydic form.

(iii) Oxidation of lactose with bromine water gives lactonic acid, which on hydrolysis yields a mixture of D-galactose and D-gluconic acid. It is thus concluded that it is the glucose unit that contains free hemiacetal-aldehyde group.

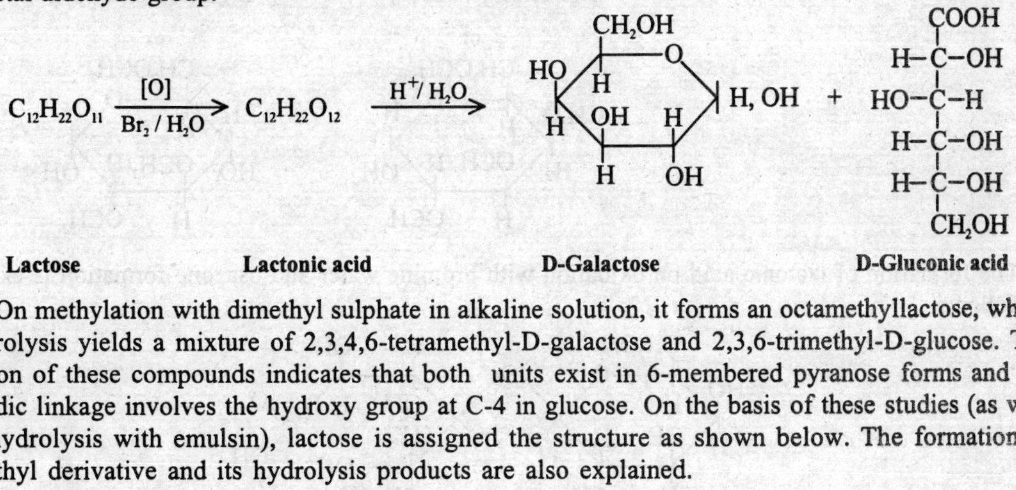

Lactose	Lactonic acid	D-Galactose	D-Gluconic acid

(iv) On methylation with dimethyl sulphate in alkaline solution, it forms an octamethyllactose, which on hydrolysis yields a mixture of 2,3,4,6-tetramethyl-D-galactose and 2,3,6-trimethyl-D-glucose. The formation of these compounds indicates that both units exist in 6-membered pyranose forms and the glucosidic linkage involves the hydroxy group at C-4 in glucose. On the basis of these studies (as well as by hydrolysis with emulsin), lactose is assigned the structure as shown below. The formation of octamethyl derivative and its hydrolysis products are also explained.

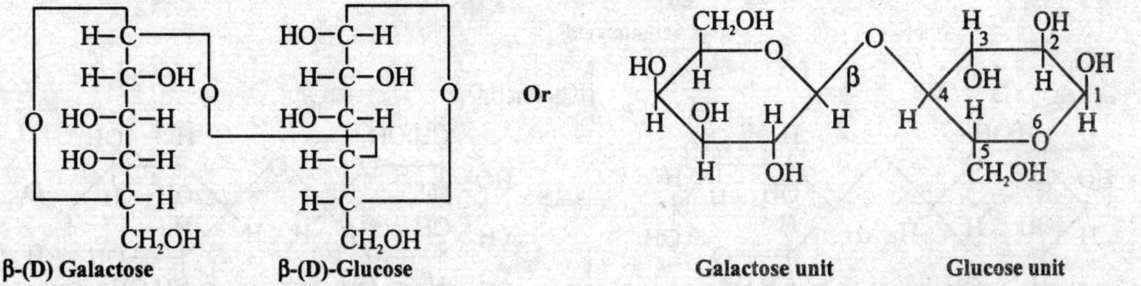

β-(D) Galactose	β-(D)-Glucose		Galactose unit	Glucose unit

Lactose		Or	Lactose	

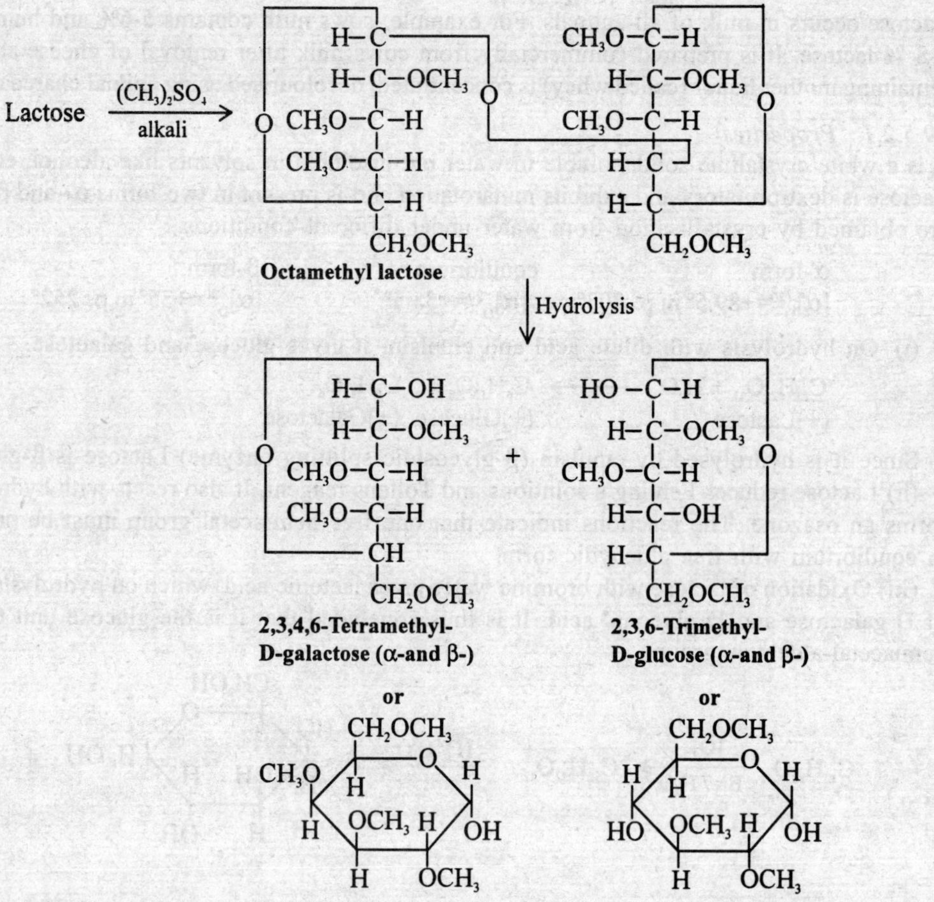

Lactose $\xrightarrow{\underset{\text{alkali}}{(CH_3)_2SO_4}}$

Octamethyl lactose

$\downarrow$ Hydrolysis

2,3,4,6-Tetramethyl-
D-galactose (α-and β-)

+

2,3,6-Trimethyl-
D-glucose (α-and β-)

or

or

The formation of lactonic acid on oxidation with bromine water and osazone formation is explained as follows:

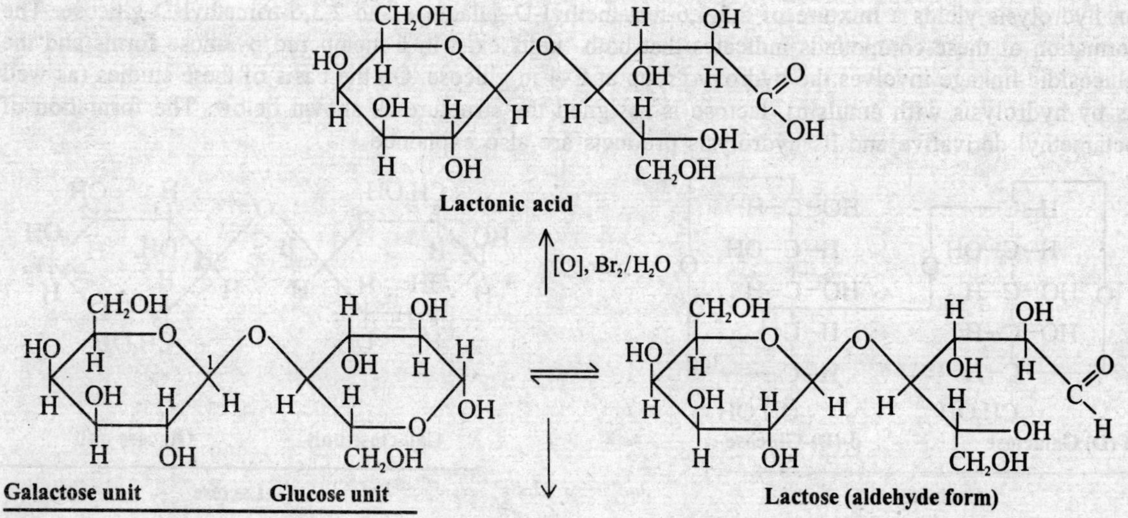

Lactonic acid

$\uparrow$ [O], Br$_2$/H$_2$O

Galactose unit   **Glucose unit**

Lactose (β-form)

$\rightleftharpoons$

Lactose (aldehyde form)

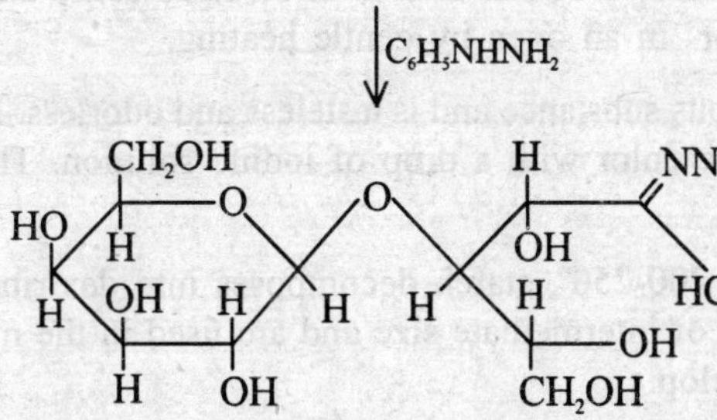

$C_6H_5NHNH_2$

CH_2OH

HO H

H OH H H H H OH

H OH

Lactosazone

NNHC_6H_5

HC=NNHC_6H_5

OH

CH_2OH

on the basis of the above evidence, Lactose is designated as 4-O-(β-D-galactopyranosyl)-D-glucopyranose.

39.5.3   Maltose/Malt sugar ($C_{12}H_{22}O_{11}$)

It is obtained by the partial hydrolysis of starch by diastase [an enzyme present in malt (sprouted barley seeds)].

$$(C_6H_{10}O_5)_n + H_2O \xrightarrow{\text{Diastase}} C_{12}H_{22}O_{11}$$

Starch   **Maltose**

Like lactose, it exists in α- and β- forms. It reduces Tollen's reagent and Fehling's solution and hence is a reducing sugar. Also it yields an osazone.

On hyrolysis maltose gives only D-glucose. This proves that maltose is made up of two glucose units, the two glucose units are joined by an α-glycosidic linkage between C-1 of one unit and C-4 of other. This is established by oxidation of maltose with bromine water followed by methylation of the formed maltobionic acid and its subsequent methylation and hydrolysis.

On the basis of the above studies, maltose is given the structure, 4-O-(α-D-glucopyranosyl)-D-glucopyranose.

CHOH

CHOH

CHOH

HC

CH

CH_2OH

CH

CHOH

CHOH

CHOH

HC

CH_2OH

≡

CH_2OH

H

H

HO OH H

H OH

CH_2OH

H

1 O 4

H OH

H

H

OH H OH

H OH

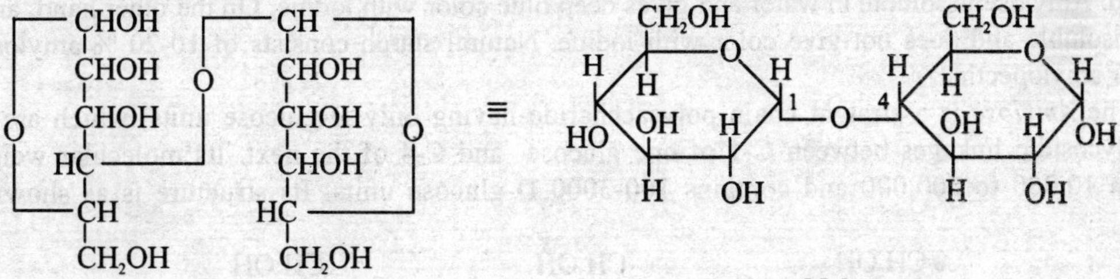

α-Maltose

39.6   POLYSACCHARIDES

These are carbohydrates in which hundreds of monosaccharide units are joined together by glycosidic linkages. The most important polysaccarides are starch, glycogen and cellulose.

39.6.1   Starch ($C_6H_{10}O_5)_n$

It occurs in all green plants, the chief sources are maize, wheat, barley, rice, potatoes and sorghum.

39.6.1.1   Manufacture

The grains (wheat, maize or rice) are soaked in water (8-16 hours) and are crushed to break the cell walls. The pulp obtained is mixed with water and filtered using a fine sieve. The cell tissues are left behind (used as fodder) and a milky suspension is obtained. It is allowed to settle and starch is separated by decantation and finally dried in air or in an oven by gentle heating.

Properties: Starch is a white, amorphous substance and is tasteless and odorless. It has no definite melting point. Its solution in water gives blue color with a drop of iodine solution. The blue color disappears on heating and reappears on cooling.

(i) *Action of Heat:* on heating to 200-250°, starch decomposes into dextrins and other compounds. Dextrins are glucose polysaccharides of intermediate size and are used in the manufacture of adhesives for use on postage stamps and envelop.

(ii) *Hydrolysis:* Hydrolysis of starch with hot dilute acids or by enzymes gives dextrins of different complexity, maltose and finally D-glucose. In these formulas both n and n1 are not known but n is greater than n1.

$$(C_6H_{10}O_5)_n \longrightarrow (C_6H_{10}O_5)_{n1} \longrightarrow C_{12}H_{22}O_{11} \longrightarrow C_6H_{12}O_6$$

Starch   **Dextrin**   **Maltose**   **Glucose**

Diastase

(iii) Starch does not reduce Fehling's solution and Tollen's reagent and does not form an osazone. It cannot be fermented by yeast.

Uses: starch is used as food in the form of potatoes, bread, cakes, rice. It is used in laundering and in the manufacture of dextrins, glucose and ethyl alcohol. It is also used for the manufacture of nitro starch, which is used as an explosive.

39.6.1.2 Structure

Starch is obtained as a mixture of two polysaccharides, *amylose* and *amylopectin;* these two can be separated. Amylose is soluble in water and gives deep blue color with iodine. On the other hand, amylopectin is insoluble and does not give color with iodine. Natural starch consists of 10-20 % amylose and 80-90% amylopectin.

The *Amylose* is a straight chain polysaccharide having only D-glucose units, which are joined by α-glycosidic linkages between C-1 of one glucose and C-4 of the next. Its molecular weight ranges from 10,000 to 500,000 and contains 100-3000 D-glucose units. Its structure is as shown below:

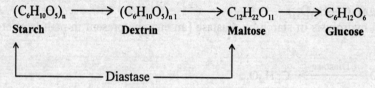

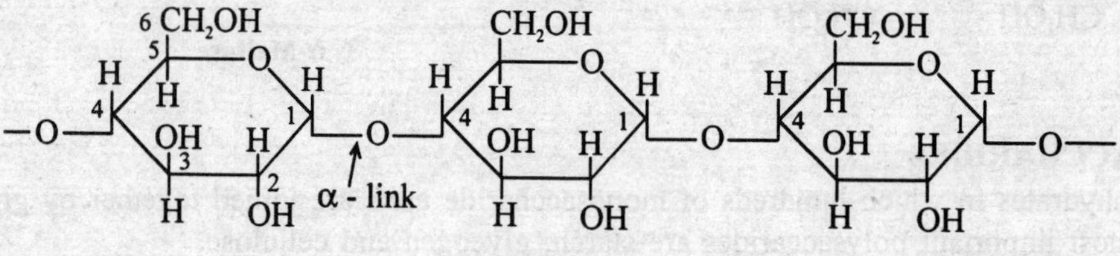

Amylose

The *Amylopectin* is a branched chain polysaccharide and is composed of chains of 25-30 D-glucose units joined by α-glycosidic linkages between C-1 of one glucose unit and C-4 of the next glucose unit (as in the case of amylose). These chains are connected to each other by 1,6-linkages. The molecular weight of amylopectin is 50,000-1,00,00,000 consisting of 300-6000 D-glucose units. Its structure is

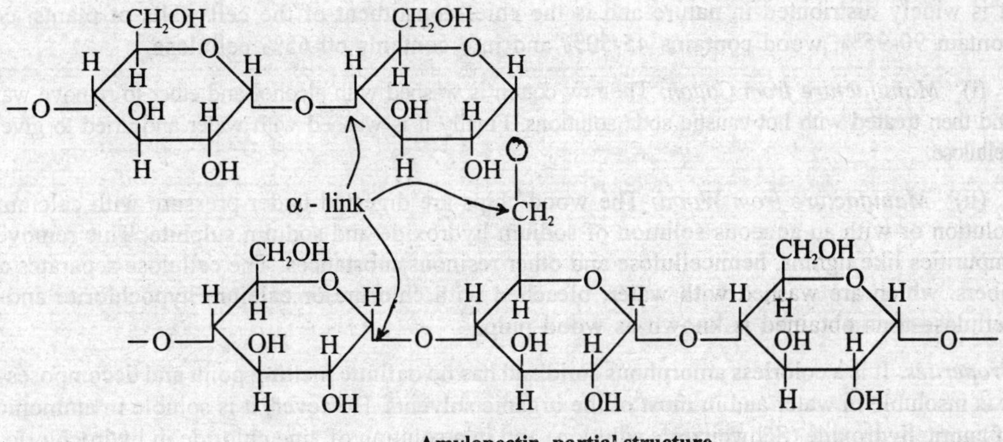

Amylopectin, partial structure

39.6.1.3 Glycogen/Animal Starch $(C_6H_{10}O_5)_n$

It occurs in liver and muscles of animals. It also occurs in yeast and mushrooms. It is a colourless, amorphous powder, soluble in water and gives a reddish brown color with iodine solution. On hydrolysis with hot dilute acid only D-glucose is obtained. It neither reduces Fehlings solutions nor forms an osazone and nor is fermented by yeast.

The structure of glycogen is similar to that of amylopectin in that it has 1,6- as well as 1,4-glycosidic linkages. Its molecular weight (1,000,000 to 14,000,000) is higher than amylopectin.

39.6.1.4 Inulin $(C_6H_{10}O_5)_n.H_2O$

Inulin, a special variety of starch occurs in dahlia tubers and jerusalem artichokes. It is a straight chain polysaccharide containing about 30 D-fructofuranose units, which are joined by β-glucosidic linkages between C-1 of one fructose unit and C-2 of the next fructose unit. Its structure is as

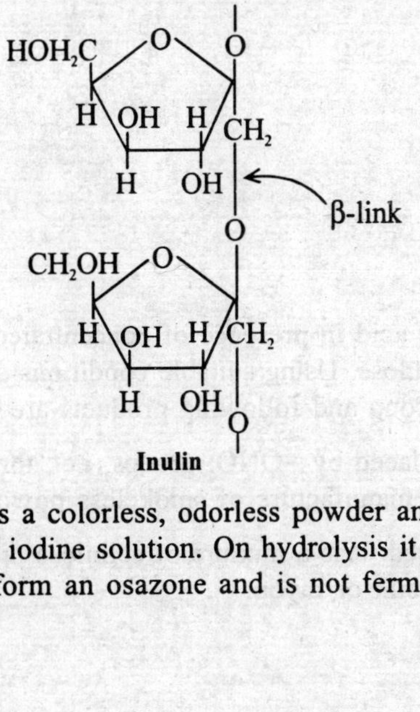

Inulin

Inulin is a colorless, odorless powder and forms a colloidal suspension in water. It does not give color with iodine solution. On hydrolysis it give D-fructose. It does not reduce Fehling's solution and does not form an osazone and is not fermented by yeast.

39.6.2 Cellulose $(C_6H_{10}O_5)_n$

It is widely distributed in nature and is the chief constituent of the cell walls of plants. cotton seeds contain 90-95%, wood contains 45-50% and jute contains 60-65% cellulose.

(i) *Manufacture from Cotton:* The raw cotton is washed with alcohol and ether to remove waxes and fats and then treated with hot caustic soda solutions. Finally it is washed with water and dried to give amorphous cellulose.

(ii) *Manufacture from Wood:* The wood chips are digested under pressure with calcium bisulphite solution or with an aqueous solution of sodium hydroxide and sodium sulphite. This removes the chief impurities like lignins, hemicellulose and other resinous substances. The cellulose separates as insoluble fibers, which are washed with water, bleached with chlorine or calcium hypochlorite and dried. The cellulose thus obtained is known as wood pulp.

Properties: It is a colorless amorphous solid and has no definite melting point and decomposes on heating. It is insoluble in water and in most of the organic solvents. However, it is soluble in ammonical solution of cupric hydroxide (Schweitzer's solution) and in a solution of zinc chloride in hydrochloric acid (cross and Bevans reagent). It is also soluble in a mixture of sodium hydroxide and carbon disulphide.

It is not hydrolysed as easily as starch. However, on heating with dilute sulphuric acid under pressure only D-glucose is obtained. Cellulose does not reduce Fehling's solution or Tollen's reagent and does not form an osazone and is not fermented by yeast.

Structure: Cellulose is a straight chain polysaccharide and is composed of D-glucose units, which are joined by β-glycosidic linkage between C-1 of one glucose unit and C-4 of the next glucose unit. Its molecular weight is 50,000 -500,000 and contains 300-2500 D-glucose units. Its structure is

(Cellulose)

Chemical Properties

(i) *Nitration:* Cellulose reacts with concentrated nitric acid in presence of concentrated sulphuric acid to form esters known as cellulose nitrates or nitrocellulose. Using suitable conditions some or all the OH groups in cellulose can be replaced by $-ONO_2$ group and following products are obtained:

Gun Cotton: In this almost all the $-OH$ groups are replaced by $-ONO_2$ groups, i.e., three $-ONO_2$ group per glucose unit. It is used as an explosive in the manufacture of smokeless powders.

Proxylin: In this only some of the OH groups are replaced. It is a mixture of cellulose dinitrate and mononitrate and is used for the manufacture of artificial silk or rayon.

(ii) *Acetylation.* Cellulose on reaction with acetic anhydride and catalytic amount of concentrated sulphuric acid gives cellulose acetate, the three OH groups are replaced by $-OCOCH_3$ per glucose unit.

Cellulose acetate is used in molding plastics for objects like automobile steering etc. It is also used in acetone solution in varnishes and lacquers. It finds applications in the manufacture of shatter-proof glass.

(iii) *Ether Formation:* Cellulose on reaction with ethyl iodide in presence of sodium hydroxide gives ethers known as ethyl cellulose, which is used as an emulsifying and thickening agent for creams, shampoos and toothpastes. Ethyl cellulose is also used in the manufacture of paints, lacquers, varnishes etc.

(iv) *Action of Alkali:* When cotton is treated with strong solution of sodium hydroxide and then washed free of alkali—the product obtained is known as mercerised cotton, which has a better luster and has a greater absorption capacity for dyes.

Artificial Silk or Rayons

All synthetic fibers made from cellulose are known as Rayons. Following process are used for its manufacture. The basic principle of all the methods is same, i.e., a solution of cellulose is first made and then forced through tiny jets (spironerts), when cellulose is precipitated from the stream of solution. Alternatively the solution is evaporated in hot air to give thin films, which have characteristics of silk.

(i) *Cuprammonium process:* In this process, a solution of cotton in ammonical solution of cupric hydroxide (Schweitzer's reagent) is forced through tiny jets into a bath of sulphuric acid when cellulose gets precipitated as threads, which are dried, spun together and woven. The final product is called *cupra silk* or *Cuprammonium Rayon.*

(ii) *Acetate process:* In this process, cellulose acetate (obtained by treatment of cellulose with acetic anhydride in glacial acetic acid) is dissolved in acetone. The viscous solution is forced through tiny jets into hot air. The solvent evaporates leaving behind threads of cellulose acetate, which are spun and woven. The silk obtained is called *Acetate Silk* or *Cleaness Silk.*

(iii) *Viscose process:* In this process, the cellulose is first treated with sodium hydroxide solution to give cellulose alkoxide, which on treatment with carbon disulphide give cellulose xanthate. A viscous solution of cellulose xanthate in water is known as *viscose.* It is carefully forced through tiny jets into a bath of dilute sulphuric acid, which hardens the gum-like threads into rayon fibers. The finished product is called *Viscose rayon.*

Manufacture of Paper

The most important use of cellulose is for the manufacture of paper. The raw materials used are wood, cotton rags, waste paper and cereal straws etc. The process involves conversion of wood into pulp and its final conversion into paper.

Wood Pulp: It is obtained by one of the following processes:

(i) *Mechanical process:* In this process, the short soft wood (like pine) logs are subjected to mechanical grinding in presence of water to get the pulp. The mechanical pulp is difficult to bleach and is used for the manufacture of cheap, inferior paper, e.g., news print, wall paper etc.

(ii) *Chemical process:* In this process the raw material is cut into small chips and digested under pressure either with a solution of calcium bisulphite (*sulphite Process*) or with a caustic soda solution (*soda process*). To the pulpy material, thus obtained, is added chlorine water or bleach-liquor. The bleached pulp is beaten with water in a pulpbeater. During beating some fillers (like clay-calcium carbonate, calcium sulphate, talc, titanium dioxide, zinc sulphide etc.—these make better printing surface and increase the opacity), colouring material (like ultramarines to give the desired shade) and sizers (like rosin, starch, glue or casein) are added.

Paper: The wood pulp obtained above is fed on to a wire screen on an endless belt. Most of the water is removed by this process and a consistent mat is obtained. It is conveyed by means of belts made of felts through a series of rollers where it is made compact. Finally, it is passed through driers (steam heated cast iron cylinders). The paper is lastly finished or polished by pressing between a number of horizontal, highly polished, chilled cast iron rollers. It is then rolled on reels.

If no sizing material is used, the product obtained is porous and resembles filter paper. Such a paper on immersing in 75% sulphuric acid for short duration and washing gives *parchment paper.*

39.7   PROBLEMS

1. What are carbohydrates. How they are classified?
2. Give evidences leading to the cyclic structure of glucose.
3. How is glucose manufactured? What is the action of bromine water, nitric acid and conc. HI+ Red P, HCN and $C_6H_5NHNH_2$ on glucose.
4. Explain why both glucose and fructose give the same osazone.
5. How will you prove the presence of furanose ring in glucose.
6. How is glucose converted into fructose and vice-versa.
7. What is mutarotation? How it is explained?
8. Discuss the configuration of glucose and fructose.
9. How does fructose occur in nature. How it is manufactured? discuss its important reactions.
10. Describe the method for
    (i) Descending and ascending the sugar series. (ii) Converse an aldose into a ketose and vice versa.
11. Discuss the structure of sucrose. How it is obtained on a commercial scale.
12. Describe the manufacture of starch. Give the industrial applications of starch and cellulose.
13. What is the basic difference in the structure of starch and cellulose.
14. Write notes on:
    (i) Artificial silk   (ii) Manufacture of paper   (iii) Killiani synthesis
15. Write the pyranose structure for α- and β-D-glucose.
16. Glucose is subjected to the following processes successively:
    (i) CH_3OH/HCl (ii) $(CH_3)_2SO_4/NaOH$ (iii) HCl   (iv) [O] HNO_3
    The final product obtained has the structure xylotrimethylglutanic acid. Based on the above reactions, explain the various products obtained in the above steps. What structure will you assign to glucose.
17. Discuss the mechanism of the osazone formation of glucose.
18. Fructose is subjected to the following reactions:

    Fructose $\xrightarrow{\text{HCN}}$? $\xrightarrow{\text{HOH}}$? $\xrightarrow{\text{HI/P}}$?

    What products are obtained in each step? What will you infer from the above sequence of reactions?
19. Give the Howorth's pyranose and Furanose structure for fructose.
20. What is invert sugar? How it is obtained?
21. What happens when sucrose is subjected to the following reactions:
    (i)   HCl/Heat $\longrightarrow$
    (ii) [O]/HNO₃ $\longrightarrow$
    (iii) H_2 $\longrightarrow$
    (iv) Conc. H_2SO_4/ $\longrightarrow$
    (v)   Fermentation $\longrightarrow$
22. Give the structures of cane sugar, maltose and lactose.

40
Amino Acids, Peptides and Proteins

40.1 INTRODUCTION

The proteins are the most complex compounds produced in nature. The proteins have very high molecular weights and on hydrolysis they give amino acid via peptides.

$$\text{Proteins} \xrightarrow{\text{Hydrolysis}} \text{Peptides} \longrightarrow \text{Amino acids}$$

40.2 AMINO ACIDS

The amino acids are carboxylic acids having an amino group. Simple amino acids having carboxyl group and amino group are classified as α, β or γ-amino acids depending on the position of the amino group with respect to carboxyl group (IUPAC system of nomenclature). Some examples are given below.

α-Amino acids	NH₂CH₂COOH	Aminoacetic acid
	CH₃CH—COOH | NH₂	α-Aminopropionic acid (Alanine)
	C₆H₅CH₂CH—COOH | NH₂	α-Amino-β-phenylpropionic acid (Phenylalanine)
β-Amino acids	H₂NCH₂CH₂COOH	β-Aminopropionic acid
γ-Amino acids	H₂NCH₂CH₂CH₂COOH	γ-Aminobutyric acid

Out of the above amino acids, the α-amino acids are the most important since they are the final products of hydrolysis of proteins.

A convenient way of classification is to classify amino acids as neutral, acidic or basic amino acids.

Neutral Amino Acids: They have equal number of carboxyl and amino groups.

H₂N—CH₂—COOH	Aminoacetic acid (glycine)
HOOC—CH—CH₂—S—S—CH₂—CH—COOH |NH₂ |NH₂	Bis-(2-amino-2-carboxyethyl) disulphide

Acidic Amino Acids: These have more number of carboxyl group than the amino groups, for example,

HOOC—CH₂—CH—COOH | NH₂	Aspartic acid (aminosuccinic acid)
HOOC—CH₂—CH₂—CH—COOH | NH₂	Glutamic acid (α-aminoglutaric acid)

Basic Amino Acids: These have more number of amino groups than the carboxyl groups, for example,

$$H_2NCH_2CH_2CH_2CH_2\underset{\underset{NH_2}{|}}{C}HCOOH$$ Lysine (α,ε-Diaminocaproic acid)

$$\underset{\underset{NH}{||}}{H_2N}\,C\,NH-CH_2CH_2CH_2\underset{\underset{NH_2}{|}}{C}HCOOH$$ Arginine (α-Amino-δ-guanidinovaleric acid)

About 25 amino acids have been isolated from the hydrolysis of proteins. Most of these are α-amino acids. Amongst these the body can synthesis some but not all. The amino acids that cannot be synthesised in the body have to be supplied in the diet and are called *essential amino acids.* On the other hand, amino acids that can be synthesised from other compounds by the tissues of the body are called *non-essential amino acids.* Both essential and nonessential amino acids are needed for our growth and good health. A deficiency of any of the essential amino acid prevents the growth in young animals and can cause even death.

Except glycine, almost all α-amino acids obtained from natural sources are optically active due to asymmetric α-carbon atom. It is noteworthy that all amino acids obtained from animal or plant sources have L-configuration at the 2- position. Table 40.1 lists the essential and nonessential amino acids (the letter 'e' indicates the essential amino acid).

Table 40.1

Name	Systematic Name	Formula	Symbol
Neutral Amino Acids:			
Glycine	Aminoacetic acid	H_2NCH_2COOH	Gly
Alanine	α-Aminopropionic acid	$CH_3\underset{\underset{NH_2}{|}}{C}HCOOH$	Ala
Valine (e)	α-Aminoisovaleric acid	$CH_3-\underset{\underset{CH_3}{|}}{C}H-\underset{\underset{NH_2}{|}}{C}H-COOH$	Val
Leucine(e)	α-Aminocaproic acid	$CH_3-\underset{\underset{CH_3}{|}}{C}H-CH_2-\underset{\underset{NH_2}{|}}{C}H-COOH$	Leu
Isoleucine(e)	α-Amino-β-methyl-n-valeric acid	$CH_3CH_2\underset{\underset{CH_3}{|}}{C}H-\underset{\underset{NH_2}{|}}{C}H-COOH$	Ile
Phenylalanine(e)	α-Amino-β-phenylpropionic acid	$C_6H_5-CH_2-\underset{\underset{NH_2}{|}}{C}H-COOH$	Phe
Tyrosine	α-Amino-β-(p-hydroxyphenyl)propionic acid	$HO-\langle O \rangle-CH_2-\underset{\underset{NH_2}{|}}{C}H-COOH$	Tyr
Serine	α-Amino-β-hydroxypropionic acid	$HOCH_2-\underset{\underset{NH_2}{|}}{C}H-COOH$	Ser
Cysteine	α-Amino-β-mercaptopropionic acid	$HS-CH_2-\underset{\underset{NH_2}{|}}{C}H-COOH$	Cys

(Contd.)

Name	Systematic Name	Formula	Symbol
Cystine	Bis-(α-aminopropionic-acid)-β-disulphide	$\left[-SCH_2-\underset{\underset{NH_2}{\vert}}{CH}-COOH\right]_2$	Cysscy
Threonine(e)	α-Amino-β-hydroxy-n-butyric acid	$CH_3-\underset{\underset{OH}{\vert}}{CH}-\underset{\underset{NH_2}{\vert}}{CH}-COOH$	Thr
Methionine(e)	α-Amino-γ-methylthio-n-butyric acid	$CH_3SCH_2CH_2\underset{\underset{NH_2}{\vert}}{CH}-COOH$	Met
Iodogorgic acid	3,5-Diiodotyrosine	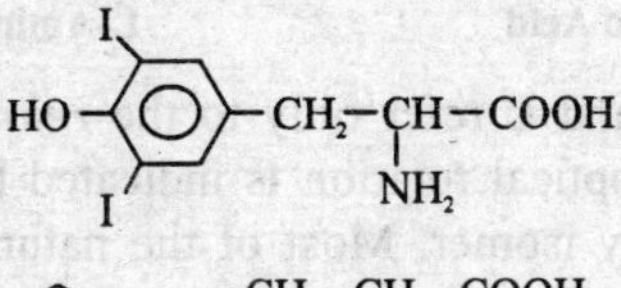 $HO-\underset{I}{\overset{I}{\bigcirc}}-CH_2-\underset{\underset{NH_2}{\vert}}{CH}-COOH$	-
Tryptophan(e)	α-Amino-β-indolepropionic acid	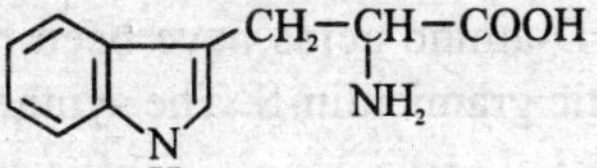 $CH_2-\underset{\underset{NH_2}{\vert}}{CH}-COOH$	Try
Proline	Pyrrolidine-α-carboxylic acid	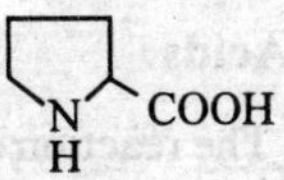 (pyrrolidine ring) $\underset{H}{N}$ COOH	Pro
Hydroxyproline	γ-Hydroxypyrrolidine-α-carboxylic acid	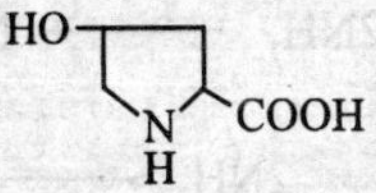 HO— (pyrrolidine ring) $\underset{H}{N}$ COOH	Hyp
Acidic Amino Acids:			
Aspartic acid	α-Aminosuccinic acid	$HOOC-CH_2-\underset{\underset{NH_2}{\vert}}{CH}-COOH$	Asp
Asparagnine	α-Aminosuccinamic acid	$H_2NCOCH_2\underset{\underset{NH_2}{\vert}}{CH}-COOH$	-
Glutamic acid	α-Aminoglutaric acid	$HOOC-CH_2CH_2\underset{\underset{NH_2}{\vert}}{CH}COOH$	Glu
Glutamine	α-Aminoglutaramic acid	$H_2NCOCH_2CH_2\underset{\underset{NH_2}{\vert}}{CH}-COOH$	-
Basic Amino Acids:			
Ornithine	α, δ-Diamino-n-valeric acid	$H_2NCH_2CH_2CH_2\underset{\underset{NH_2}{\vert}}{CH}-COOH$	-
Arginine(e)	α-Amino-δ-guanido-n-valeric acid	$HN=\underset{\underset{NH_2}{\vert}}{C}-NHCH_2CH_2CH_2\underset{\underset{NH_2}{\vert}}{CH}COOH$	Arg
Lysine(e)	α,ε-Diaminocaproic acid	$H_2NCH_2CH_2CH_2CH_2\underset{\underset{NH_2}{\vert}}{CH}COOH$	Lys
Histidine(e)	α-Amino-β-imidazolepropionic acid	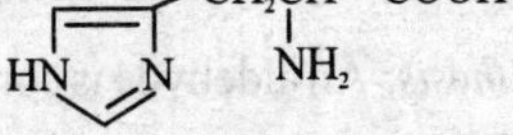 (imidazole ring) $CH_2\underset{\underset{NH_2}{\vert}}{CH}-COOH$	His

40.2.1   Optical Isomers of Amino Acids

As already stated, except glycine, all amino acids have an asymmetric carbon atom at the α-position and are therefore, optically active. These exist in D and L forms, which are non superimposable mirror images. As a matter of convention, the carboxyl group is written on the top. The D form refers to the isomer with -NH$_2$ group on the right side and the L-form refers to the isomer with -NH$_2$ group on the left side. The reference compound for this assignment is glyceraldehyde.

$$\begin{array}{c} COOH \\ | \\ H-C-NH_2 \\ | \\ R \end{array} \qquad\qquad \begin{array}{c} COOH \\ | \\ H_2N-C-H \\ | \\ R \end{array}$$

D-Amino Acid   **L-Amino Acid**

The letters D and L refer only to the *relative configuration* around the asymmetric carbon atom. The direction of optical rotation is indicated by a (+) sign for a dextrorotatory isomer and a (−) sign for a laevorotatory isomer. Most of the naturally occurring amino acids have the L-configuration.

However, a few D-amino acids have been found in few sources, e.g., D-phenylalanine occurs in the polypeptide antibiotic gramicidin-S. The synthetic amino acids, on the other hand, are generally optically inactive.

40.2.2   Synthesis of α-Amino Acids

(i)   *From α-halogenated Acids:* The reaction of α-haloacids with liquor ammonia gives the α-amino acids.

$$ClCH_2COOH + 2NH_3 \longrightarrow NH_2CH_2COOH$$
Chloroacetic acid   **Glycine**

$$\begin{array}{c} CH_3-CH-COOH \\ | \\ Br \end{array} + 2NH_3 \longrightarrow \begin{array}{c} CH_3-CH-COOH \\ | \\ NH_2 \end{array}$$
α-Bromopropionic acid   **Alanine**

The required α-halogenated acids are obtained by the Hell-Volhard-Zelinsky halogenation of the corresponding unsubstituted carboxylic acids (see Section 19.7).

(ii)   *Gabriel's Phthalimide Synthesis:* It consists in the treatment of an ester of α-halo acid with potassium phthalimide. The substituted phthalimide, thus formed, is hydrolysed to give phthalic acid and an amino acid.

Potassium   **Ethyl chloroacetate**
phalimide

This procedure gives better yields compared to the first method.

(iii)   *Strecker Synthesis:* An aldehyde is converted into an amino acid by this method. Various steps involved are given as follows:

$$CH_3CHO + HCN \longrightarrow CH_3CH \begin{smallmatrix} OH \\ CN \end{smallmatrix} \xrightarrow{NH_3}$$

Acetaldehyde   **Acetaldehyde cyanohydrin**

$$CH_3CH \begin{smallmatrix} NH_2 \\ CN \end{smallmatrix} \xrightarrow{H_2O} CH_3-\underset{\underset{NH_2}{|}}{CH}-COOH$$

Aminonitrile   **Alanine**

It is more convenient to obtain the aminonitrile directly from the aldehyde in one step by treatment with an equimolar mixture of ammonium chloride and potassium cyanide.

This method is used for the preparation of glycine, alanine, serine, valine, methionine, glutamic acid, leucine, isoleucine and phenylalanine.

(iv) *From Diethyl Malonate:* This method is basically an extension of the first two method since it offers a means of preparing α-halo acids. Various steps involved are given below.

$$CH_2(COOC_2H_5)_2 \xrightarrow[RX]{C_2H_5ONa} RCH(CO_2C_2H_5)_2 \xrightarrow[(2) HCl]{(1) KOH} RCH(COOH)_2$$

Diethyl malonate

$$\xrightarrow{Br_2} RCBr(COOH)_2 \longrightarrow RCHBrCOOH \xrightarrow{NH_3} \underset{\underset{NH_2}{|}}{RCHCOOH}$$

α-Amino acid

α-Amino acids, which can be conveniently prepared by this method (using easily available starting materials) are phenylalanine, proline, leucine, isoleucine and methionine.

The malonic ester synthesis can be combined with the Gabriel phthalimide synthesis to obtain phenylalanine, tyrosine, proline, cystine, serine, aspartic acid, methionine and lysine. This is known as *Sovensen synthesis* The synthesis of some typical amino acids is given below:

Phenylalanine

$$CH_2(COOC_2H_5)_2 \longrightarrow BrCH(COOC_2H_5)_2$$

Diethyl malonate   **Monobromomalonic ester**

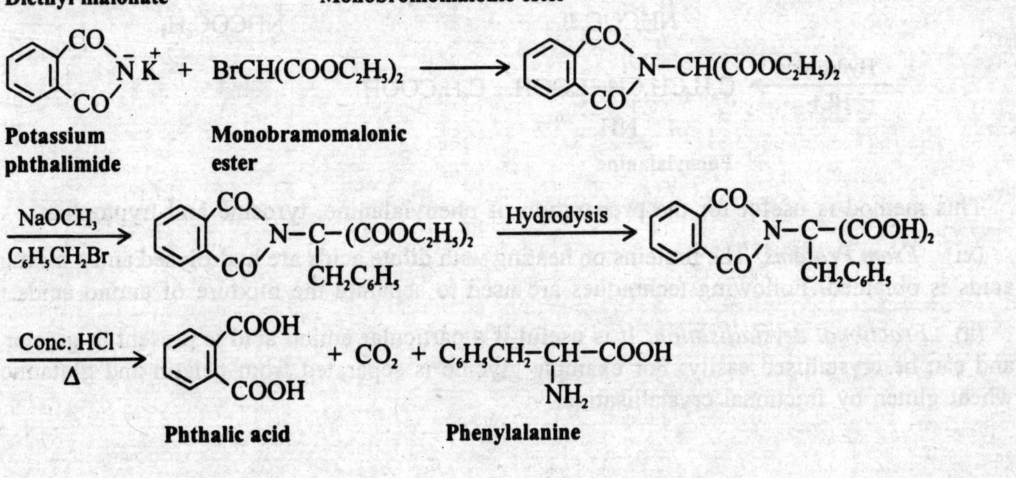

Potassium   **Monobramomalonic**
phthalimide   **ester**

Phthalic acid   **Phenylalanine**

Cystine

$$C_6H_5CH_2SH + HCHO + HCl \longrightarrow C_6H_5CH_2SCH_2Cl$$

Benzyl thiol   **Benzyl thiomethyl chloride**

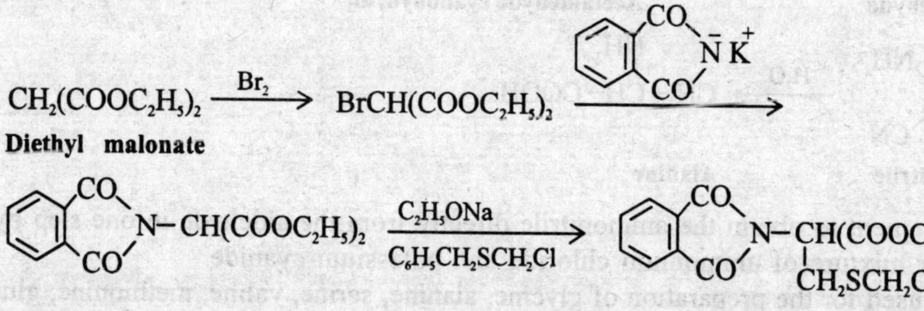

$$CH_2(COOC_2H_5)_2 \xrightarrow{Br_2} BrCH(COOC_2H_5)_2 \longrightarrow$$

Diethyl malonate

$$\underset{CO}{\overset{CO}{\big\rangle}}N-CH(COOC_2H_5)_2 \xrightarrow[C_6H_5CH_2SCH_2Cl]{C_2H_5ONa} \underset{CO}{\overset{CO}{\big\rangle}}N-\underset{CH_2SCH_2C_6H_5}{\overset{|}{CH(COOC_2H_5)_2}}$$

$$\xrightarrow[\text{2. HCl}]{\text{1. NaOH}} H_2N-\underset{COOH}{\overset{|}{CH}}-CH_2-S-CH_2C_6H_5 \xrightarrow[NH_3]{Na/liq} H_2N-\underset{COOH}{\overset{|}{CH}}-CH_2SH$$

(±)-Cysteine

$$\xrightarrow{air} H_2N-\underset{COOH}{\overset{|}{CH}}-CH_2-S-S-CH_2-\underset{COOH}{\overset{|}{CH}}-NH_2$$

(±)-Cystine

(v)   *The Azlactone Synthesis* (*Erlenmeyer Azlactone Synthesis*): It consists in heating an aromatic aldehyde with hippuric acid (benzoyl glycine), acetic anhydride and fused sodium acetate. The formed azlactone (4-benzylidene-2- phenyloxazol-5-one) is hydrolysed and subsequently reduced and hydrolysed to give α-amino acid.

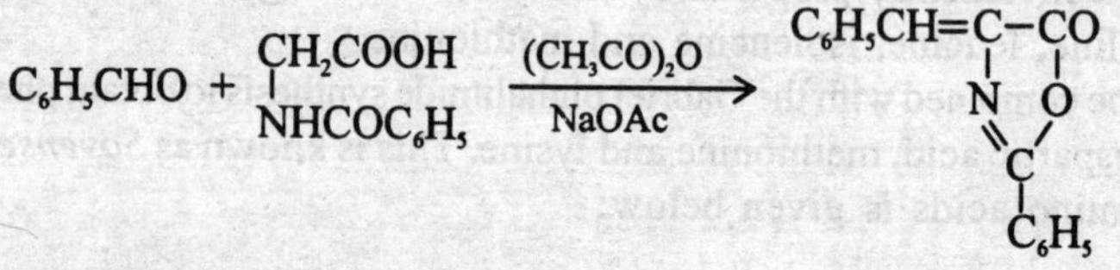

$$C_6H_5CHO + \underset{NHCOC_6H_5}{\overset{CH_2COOH}{|}} \xrightarrow[NaOAc]{(CH_3CO)_2O}$$

Benzaldehyde   Benzoyl glycine   **Azlactone**

$$\xrightarrow{NaOH} C_6H_5CH=\underset{NHCOC_6H_5}{\overset{|}{C}}-COOH \xrightarrow{Reduction} C_6H_5CH_2\underset{NHCOC_6H_5}{\overset{|}{CH}}-COOH$$

$$\xrightarrow[HCl]{Hydrolysis} C_6H_5CH_2\underset{NH_2}{\overset{|}{CH}}-COOH + C_6H_5COOH$$

Phenylalanine

This method is useful for the preparation of phenylalanine, tyrosine and tryptophan.

(vi)   *From Proteins:* The proteins on heating with dilute acids are hydrolysed and a mixture of amino acids is obtained. Following techniques are used to separate the mixture of amino acids.

(a)   *Fractional crystallisation:* It is useful if a particular amino acid is present in greater proportion and can be crystallised easily. For example glycine is separated from gelatin and glutamic acid from wheat gluten by fractional crystallisation.

(b) *Fischer's method:* In this technique the mixture of amino acids is converted into esters, which are then separated by fractional distillation. The pure esters thus obtained are hydrolysed to give the amino acids. This method is useful for the separation of neutral amino acids.

(c) *Dakins method:* The amino acids are extracted with butanol saturated with water. They are then separated either by fractional distillation or fractional crystallisation. After extraction with butanol saturated with water, the remaining residue is treated with phosphotungstic acid when the basic amino acids are precipitated as salts.

(d) *Chromatographic methods:* The separation can be effected by column, paper and gas chromatography.

(e) *Electrophoresis:* This technique is widely used for the separation of amino acids (for details see Section 40.2.3.4).

40.2.3 Physical Properties

The amino acids are colorless, crystalline compounds. They are soluble in water, acids and alkalis but are sparingly soluble in organic solvents. The amino acids may be regarded as salts since a basic group ($-NH_2$) and an acidic group ($-COOH$) is present in the same molecule. This accounts for the higher melting points (or melting with decomposition) of the amino acids.

40.2.3.1 Zwitterionic Nature

As already stated, amino acids contain a carboxyl and an amino group. The acidic carboxyl group can lose a proton and the basic amino group can gain a proton.

Zwitter ion

As seen a salt like structure is formed. This is known as a *Zwitter ion or dipolar ion.* Although, it is neutral, it contains both positive and negative charge. The zwitterionic nature of amino acids accounts for their solubility in water and insolubility in organic solvents and also their higher melting points.The zwitterionic nature is confirmed by spectroscopic studies. Thus, in IR spectrum an absorption peak at 1600 and 1400 cm^{-1} correspond to the stretching vibrations of the carboxylate ion. Also a broad absorption band at 3000-2500 cm^{-1} corresponds to the presence of the ammonium ion.

Due to the existence of dipolar ion, amino acids are amphoteric substances. Therefore, they react with both acids and alkalis. Base converts the ammonium ion($-NH_3^+$) to an amino group and an acid converts the carboxylate ion to a carboxyl group. Thus, in acidic solution, the amino acids exists as positive ions (cations). On the other hand, in basic solution they exist as negative ions (anion).

Zwitter ion

Thus, in acidic solution, the amino acid behaves as a positive ion and migrates towards the cathode when placed in an electric field. However, in alkaline solution, the amino acids behave as a negative ion and migrates towards the anode as represented in Fig. 40.1.

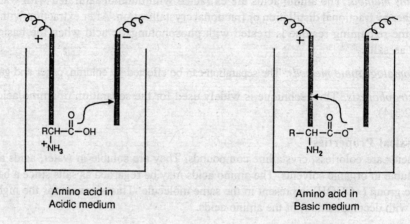

Amino acid in
Acidic medium

Amino acid in
Basic medium

Fig. 40.1: Behavior of Amino Acid in Electrical Field

40.2.3.2 The Isoelectric Point

In the context to what has been said above (behavior of acidic and alkaline solutions of amino acid in electrical field) at a constant pH, the concentration of anions and cations becomes equal and thus net movement towards any electrode is zero. This pH is known as the Isoelectric point of that amino acid. The isoelectric point is not the same for all amino acids and depends on the nature of other functional groups present. The neutral, acidic and basic amino acids have isoelectric points in the pH range of 5.5-6.3, 3, 10 respectively. The Table 40.2 gives the isoelectric points of some amino acids.

Table 40.2:   Isoelectric Points of Amino Acids

Amino acid	Isoelectric point	$[\alpha]_D^{25}$ (H$_2$O)
Neutral Amino Acid		
Threonine	5.7	−28.3
Valine	6.0	+ 6.4
Alanine	6.1	+ 2.7
Acidic Amino Acid		
Aspartic acid	2.8	+ 4.7
Glutamic acid	3.2	+11.5
Basic Amino Acid		
Lysine	9.7	+14.6
Arginine	10.8	+12.6

At isoelectric points, the amino acids have minimum solubility in water. This has been utilised in the separation of α-amino acids from protein hydrolysis.

40.2.3.3 pK Values

The amino acids behave both as an acid and a base. Thus, monoamino monocarboxylic acids have two pK values; one as an acid (on titration with a base) and the other as a base (on titration with an acid). By convention, pK$_1$ is the one that corresponds to the group titrated at the most acid region, i.e., the carboxyl group. However, in presence of salts, the ions of the salts may combine with the dipolar ion. The pH for maximum concentration of dipolar ion varies. Also, the rotations of the dipolar ion, conjugate acid and conjugate base are different. Therefore, the specific rotation of a given amino acid depends on the pH of the solution and also on the presence or absence of salts. Table 40.2 gives the specific rotations of the acids.

The pK$_2$ value of an amino acid is determined by titration with alkali. In order to titrate the carboxyl group with alkali, the amino group must be 'masked'. This method of titrating amino acids with alkali is known as the *Sorensen formal titration* (the NH$_2$ group is masked by formalin).

40.2.3.4 Electrophoresis

This technique is useful for the separation of a mixture of amino acids from protein hydrolysis. It is based on the fact that different amino acids migrate towards anode or cathode at pH values other than the isoelectric point. Thus, on passing electric current, the positively charged diaminomonocarboxylic acids (viz. lysine, arginine etc.) go to the cathode. On the other hand, the negatively charged monoaminodicarboxylic acids (viz. aspartic, glutamic acid etc.) go to the anode. The monoaminocarboxylic acids (neutral amino acids, e.g., leucine) remain at or near the starting point. Thus, the separation of a mixture of amino acids containing lysine, leucine and glutamic acid can be represented as shown below.

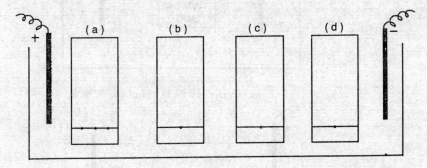

Spotting in (a) Mixture of amino acids (b) Lysine (c) Leucine and (d) Glutamic acid

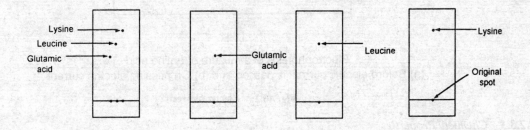

Spots after the development of electrophotogram

The migratory aptitude or mobility of the an ion depends on the experimental conditions. It is more convenient to carry out paper electrophoresis in which the mixture of amino acids viz. lysine, leucine and glutamic acid is spotted on a strip of Whatmann filter paper near the bottom of the strip. Similarly, three more strips are prepared and each is spotted with a solution containing only one of the reference amino acid. (lysine, leucine and glutamic acid respectively). All the strips are dried and saturated with a buffer solution and placed horizontally on a glass plate. The two ends of each strip are bent and immersed in tank containing the buffer solution and a platinum electrode.

A potential difference is applied across them. The positively charged ions (in this case lysine) move towards the cathode and the negatively charged ion (in this case glutamic acid) moves towards the anode; the neutral amino acid remains more or less at the same point. After 2-3 hours, the strips are dried and sprayed with ninhydrin reagent and again dried. By comparing the distances traveled by the components in the mixture with those of the reference compounds, the identification of the mixture can be done as shown in Fig. 40.2.

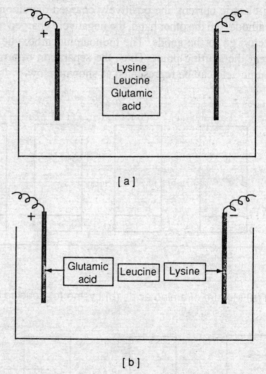

Electrophoresis of a mixture of amino acid
(a) Before electric current is passed and (b) On passing electric current

Fig. 40.2: Electrophoresis

40.2.3.5 Chemical Properties
The amino acids show properties of both the amino and the carboxyl group. Some important properties are given below.

(I) *Reactions Due to the Amino Group*

(i) *Salt formation:* The amino acids form salts with strong acids.

$$\underset{\text{Glycine}}{H_2NCH_2COOH} + HCl \longrightarrow \underset{\text{Glycine hydrochloride}}{Cl\overset{-}{H_3}\overset{+}{N}CH_2COOH}$$

$$\underset{\underset{NH_2}{|}}{\underset{\text{Alanine}}{CH_3- CH-COOH}} + HCl \longrightarrow \underset{\underset{+NH_3Cl^-}{|}}{\underset{\text{Alanine hydrochloride}}{CH_3- CH-COOH}}$$

(ii) *Acylation:* With acetic anhydride or acetyl chloride, the amino acids are acetylated.

$$\underset{\text{Glycine}}{H_2NCH_2COOH} + CH_3COCl \longrightarrow \underset{\text{Acetyl glycine}}{CH_3CONHCH_2COOH}$$

Similarly, benzoylation may be done by benzoyl chloride.

$$\underset{\text{Glycine}}{H_2NCH_2COOH} + \underset{\text{Benzoyl chloride}}{C_6H_5COCl} \longrightarrow \underset{\text{Benzoyl glycine}}{C_6H_5CONHCH_2COOH}$$

(iii) *Reaction with nitrous acid:* Amino acids react with nitrous acid like primary amines to give the corresponding hydroxy acids.

$$\underset{\text{Glycine}}{H_2NCH_2COOH} + HNO_2 \longrightarrow HOCH_2COOH + N_2\uparrow + H_2O$$

Mechanism:

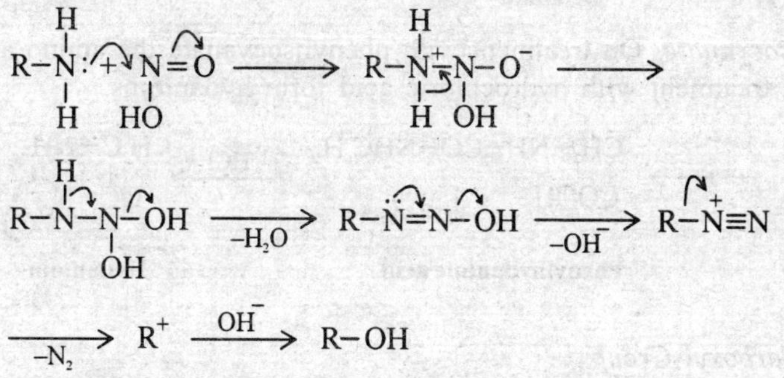

In the above reaction, nitrogen is quantitatively evolved and measured. From this, the amount of amino acid can be estimated. As seen, one half molecule of nitrogen is eliminated from each free amino group. This is the basis of *Van Slyke method* for determination of free amino groups in proteins or for the estimation of amino acids.

(iv) *Carbylamine reaction:* Like primary amines, the amino acids on heating with chloroform and alcoholic potash give carbylamine (isocyanides).

$$H_2NCH_2COOH + CHCl_3 + 3KOH \xrightarrow{\Delta} :\overset{-}{C}=\overset{+}{N}CH_2COOH$$

In this case the isocyanides are not volatile due to salt formation with alkali and so sometimes the odor is not detected.

(v) *Reaction with formaldehyde:* Amino acids react with formaldehyde to give the corresponding N-methyleneamino acids.

$$CH_2O + H_2NCH_2COOH \xrightarrow[-H_2O]{} H_2C=NCH_2COOH$$

Formal-   Glycine   N-Methyleneglycine
dehyde

In the above reaction, the amino group is blocked and the resulting acid can be titrated with alkali in the usual manner. This reaction is the basis of *Sorenson Formal method* for the determination of neutralisation equivalents or estimation of amino acids.

(vi) *Deamination:* On heating with hydriodic acid, the amino group is knocked out giving rise to a carboxylic acid.

$$NH_2CH_2COOH + 3HI \longrightarrow CH_3COOH + NH_4I + I_2$$

Glycine   Acetic acid

(vii) *Reaction with alkyl halides:* The amino acids on heating with methyl iodide in methanolic solution give the trialkyl derivatives known as betanines. These exist as dipolar ions.

$$H_3\overset{+}{N}CH_2CO\overset{-}{O} + 3\,CH_3I \longrightarrow (CH_3)_3\overset{+}{N}CH_2CO\overset{-}{O} + 3HI$$

Glycine   Betanine

Betanine is also prepared by warming an aqueous solution of chloroacetic acid with trimethyl amine.

$$(CH_3)_3N + ClCH_2COOH \longrightarrow (CH_3)_3\overset{+}{N}CH_2CO\overset{-}{O} + HCl$$

**   Chloroacetic acid   Betanine**

(viii) *Reaction with phenylisocyanate:* On treatment with phenylisocyanate the amino acids form phenylhydantoic acid, which on treatment with hydrochloric acid form hydantoins.

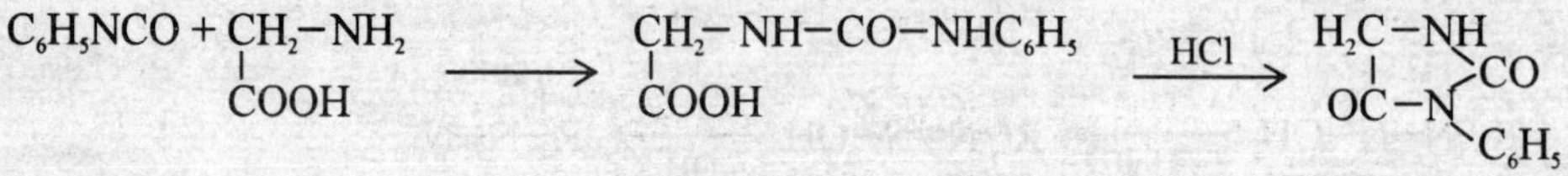

$$C_6H_5NCO + \underset{\underset{COOH}{|}}{CH_2-NH_2} \longrightarrow \underset{\underset{COOH}{|}}{CH_2-NH-CO-NHC_6H_5} \xrightarrow{HCl} \begin{array}{c} H_2C-NH \\[-2pt] | \qquad \diagdown CO \\[-2pt] OC-N \diagup \\[-2pt] \qquad \diagdown C_6H_5 \end{array}$$

Phenyl   Glycine   Phenylhydantoic acid   Hydantoin
isocyanate

(II) *Reactions Due to the Carboxyl Group*

(i) *Salt formation:* With alkalis the amino acids form salts.

$$H_2NCH_2COOH \xrightarrow{NaOH} H_2NCH_2COONa + H_2O$$

Glycine   Sodium salt of glycine

(ii) *Esterification:* The amino acids can be esterified by refluxing with absolute alcohol in presence of anhydrous hydrogen chloride. The formed hydrochloride of the ester on treatment with silver hydroxide (moist Ag_2O) gives the esters.

$$H_2N-CH_2-COOH + HCl \longrightarrow \overset{-}{Cl}H_3\overset{+}{N}-CH_2COOH \xrightarrow[-H_2O]{C_2H_5OH}$$

$$\overset{-}{Cl}H_3\overset{+}{N}CH_2COOC_2H_5 \xrightarrow{AgOH} H_2NCH_2COOC_2H_5 + AgCl + H_2O$$

Ethyl-α-aminoacetate

(iii) *Decarboxylation:* On heating with barium hydroxide, the amino acids undergo decarboxylation.

$$H_2NCH_2COOH + Ba(OH)_2 \xrightarrow{\Delta} CH_3NH_2 + BaCO_3 + H_2O$$

Glycine   Methyl amine

(iv) *Reduction:* Lithium aluminum hydride reduces the amino acids to the corresponding alcohols.

$$H_2NCH_2COOH \longrightarrow H_2NCH_2CH_2OH$$

Glycine   Amino alcohol

(III) *Reactions Involving both the Amino and Carboxyl Group*

(i) *Action of heat:* In case of α-amino acids (e.g. glycine), two molecules combine to form cyclic diamides known as diketopiperazines.

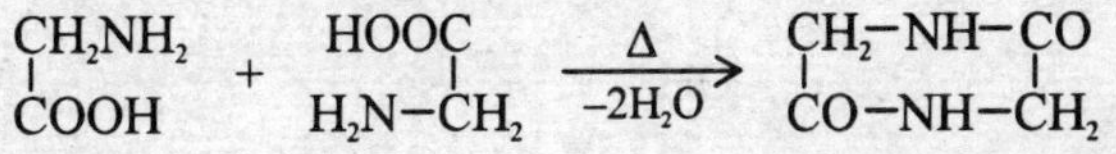

Glycine   2,5-Diketopiperazine

On the other hand, the β-keto acids on heating lose a molecule of ammonia to give α,β-unsaturated acid (as in the case of β-hydroxy acids).

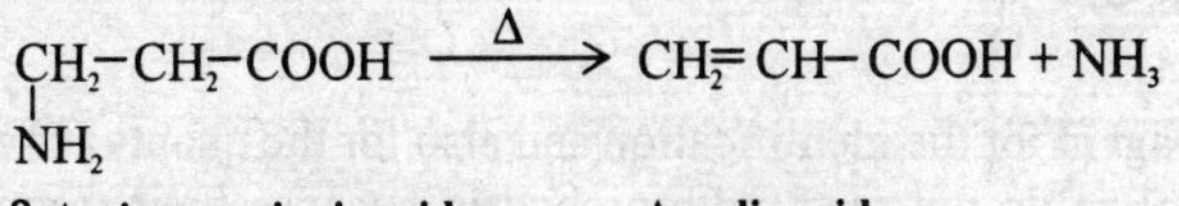

β-Aminopropionic acid   Acrylic acid

However, γ- and δ-amino acids on heating give cyclic amides called lactams.

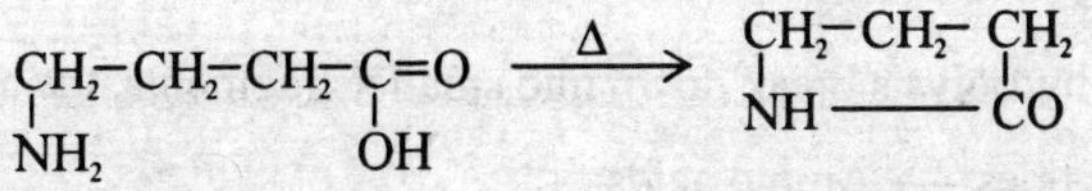

γ-Aminobutyric acid   γ-Butyrolactam

(ii) *Reaction with cupric oxide:* Amino acids on treatment with cupric oxide in water give deep blue complex.

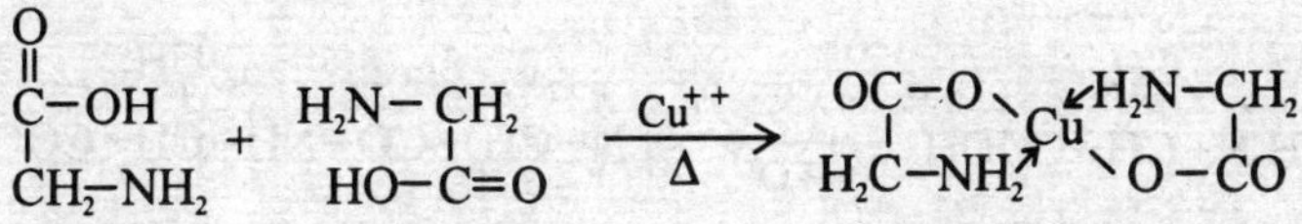

Glycine   Cupric Glycinate

(iii) *Reaction with ninhydrin:* Amino acids react with ninhydrin (indane-1,2,3-trione hydrate) to form purple colored complex.

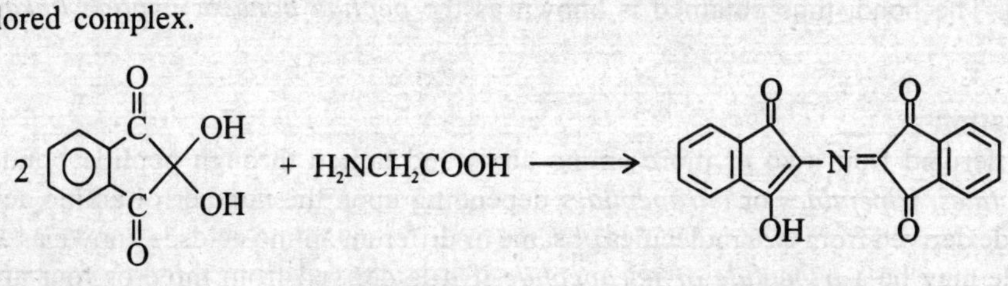

Ninhydrin   Glycine   Purple complex

Mechanism:

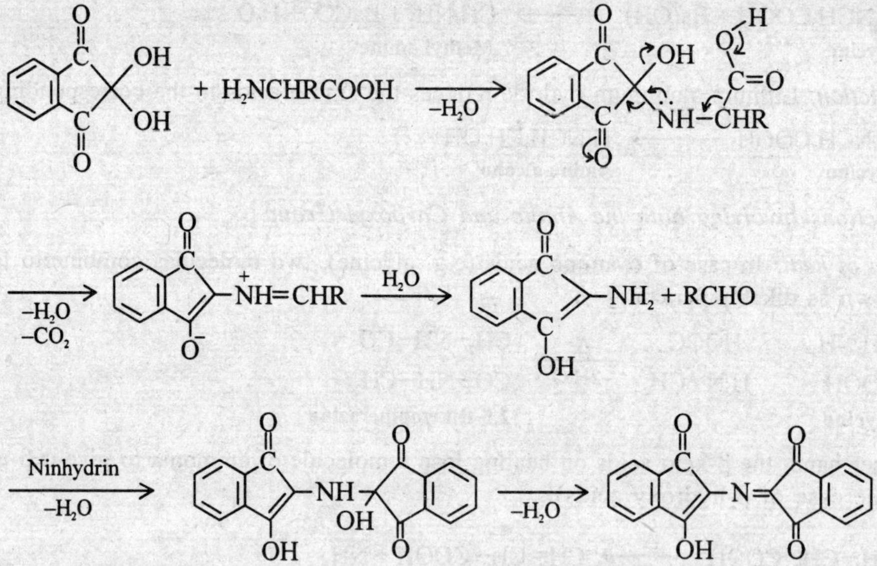

The ninhydrin is used as a spraying reagent for the identification and also for the quantitative estimation of amino acids. Proline and hydroxyproline do not react with ninhydrin because their α-amino groups are part of a five membered ring.

40.3 PEPTIDES

We have seen (Section 40.1) that proteins on hydrolysis break down into smaller fragments, the amino acids.

Proteins $\longrightarrow$ Polypeptides $\longrightarrow$ Peptides $\longrightarrow$ Amino acids

The Peptide Linkage: An amino acid molecule has both the carboxyl and amino groups. When two amino acids combine via an intermolecular reaction, the carboxyl group of one amino acid and the amino group of another amino acid combine with the elimination of a molecule of water.

$$\underset{\text{Glycine}}{H_2N-CH_2-COOH} + \underset{\text{Alanine}}{H_2N-\overset{\overset{\displaystyle CH_3}{|}}{CH}-COOH} \xrightarrow{-H_2O} \underset{\text{Glycylalanine}}{H_2N-CH_2-CO-NH-\overset{\overset{\displaystyle CH_3}{|}}{CH}-COOH}$$

The formula of the resulting product show that the two amino acids are linked by means of a –CO–NH– group. The bond, thus obtained is known as the *peptide bond* or *peptide linkage* and the product *peptide.*

40.3.1 Classification

The peptides are derived from two or more amino acids and united through peptide bond. They are classified as *dipeptides, tripeptides* or *tetrapeptides* depending upon the number of amino acid residues in it. Thus, a peptide derived from two molecules of same or different amino acids is known as a *dipeptide.* Similarly a peptide may be a *tripeptide or tetrapeptide* if it is derived from three or four amino acids. A peptide derived from a large number of amino acids is known as *polypeptide.*

40.3.2   Representation of Polypeptides

The two amino acids (e.g. glycine and alanine) may condense in two different ways to give two dipeptides as represented below.

$$H_2NCH_2COOH + H_2N-\overset{\overset{\displaystyle CH_3}{|}}{C}H-COOH \xrightarrow[-H_2O]{} H_2NCH_2CONH\overset{\overset{\displaystyle CH_3}{|}}{C}HCOOH$$

Glycine   Alanine   Glycylalanine

or

$$H_2N\overset{\overset{\displaystyle CH_3}{|}}{C}HCOOH + H_2NCH_2COOH \xrightarrow[-H_2O]{} H_2N\overset{\overset{\displaystyle CH_3}{|}}{C}HCONHCH_2COOH$$

Alanine   Glycine   Alanylglycine

In a similar manner either of the two dipeptides may react with an amino acid (say glycine) at either side to form two different tripeptides.

$$H_2NCH_2CONH\underset{\underset{\displaystyle CH_3}{|}}{C}HCOOH + H_2NCH_2COOH \longrightarrow H_2NCH_2CONH\overset{\overset{\displaystyle CH_3}{|}}{C}HCONHCH_2COOH$$

Glycylalanine   Glycine   Glycylalanylglycine (a tripeptide)

or

$$NH_2CH_2COOH + H_2NCH_2CONH\underset{\underset{\displaystyle CH_3}{|}}{C}HCOOH \longrightarrow H_2NCH_2CONHCH_2CONH\overset{\overset{\displaystyle CH_3}{|}}{C}HCOOH$$

Glycine   Glycylalanine   Glycylglycylalanine (a tripeptide)

In practice it is not possible to synthesise polypeptides in this fashion due to the formation of a mixture of products. (For the actual method of synthesis see Section 40.3.4). In a polypeptide, the amino acid that contains the free amino group is called the *N-terminal residue* and the amino acid that contains the free carboxyl group is called the *C-terminal residue*. As a matter of convention N-terminal residue is written on the left hand side of the polypeptide chain and the C-terminal residue is written on the right hand side of the chain. Thus, a tripeptide (glycylalanylphenylalanine) from glycine, alanine and phenylalanine is represented as shown below:

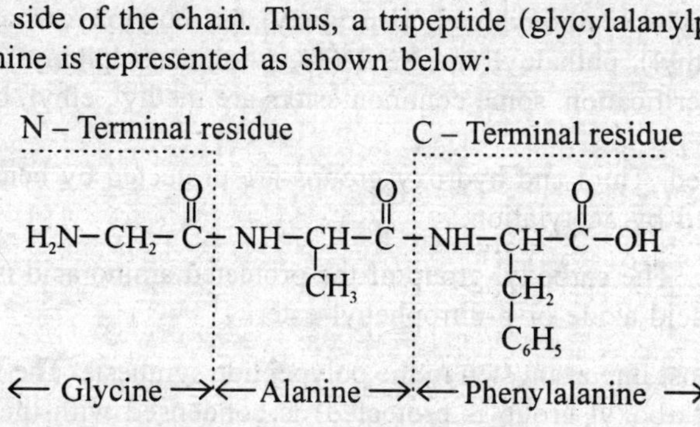

N – Terminal residue   C – Terminal residue

$$H_2N-CH_2-\overset{\overset{\displaystyle O}{||}}{C} \vdots NH-\underset{\underset{\displaystyle CH_3}{|}}{C}H-\overset{\overset{\displaystyle O}{||}}{C} \vdots NH-\underset{\underset{\underset{\displaystyle C_6H_5}{|}}{CH_2}}{C}H-\overset{\overset{\displaystyle O}{||}}{C}-OH$$

$\longleftarrow$ Glycine $\longrightarrow \longleftarrow$ Alanine $\longrightarrow \longleftarrow$ Phenylalanine $\longrightarrow$

40.3.3   Nomenclature

The peptides are named (as indicated above) by listing the amino acid in the order they are linked. The amino acid suffix *'ine'* is replaced by the suffix *'yl'* (as glycine to glycyl) for all amino acids except the C-terminal acids. A few illustrations are given below:

$$H_2N-CH_2-\overset{\overset{\displaystyle O}{\displaystyle \|}}{C}-NH-\underset{\underset{\displaystyle CH_3}{|}}{CH}-COOH$$

Glycylalanine (A dipeptide from glycine and alanine)

$$H_2N-CH_2-\overset{\overset{\displaystyle O}{\displaystyle \|}}{C}-NH-\underset{\underset{\displaystyle CH_3}{|}}{C}-\overset{\overset{\displaystyle O}{\displaystyle \|}}{C}-NH-\underset{\underset{\underset{\displaystyle C_6H_5}{|}}{\underset{\displaystyle CH_2}{|}}}{CH}-COOH$$

Glycylalanylphenylglycine
(a tripeptide from glycine, alanine and phenylalanine)

The above nomenclature is not frequently used. Instead, a three letter abbreviation (the first three alphabets of the amino acid) is used (see Table 40.1). For example, glycylalanylphenylalanine is represented as Gly-Ala-Phe.

40.3.4 Synthesis of Polypeptides

Polypeptides are polyamides formed by the step by step condensation of amino group of one amino acid with carboxyl group of another amino acid. The process of synthesis of polypeptides is not as simple as it appears due to the presence of both amino and carboxyl group in the same amino acid. There is a possibility of self condensation of amino acid that result in the formation of undesirable products. This problem is overcome by protecting the amino group of one amino acid so that it does not react in the subsequent reactions and then activating the carboxyl group of the same amino acid followed by condensation with another amino acid. Another way of synthesis is to protect the amino group of one amino acid and carboxyl group of another amino acid and condensing them together in presence of dehydrating agent. The general principle of polypeptide synthesis constitute the following four steps.

(i) *Protection of the Functional Groups:* The protection of various groups is essential to avoid the interaction of two molecules of the same acid and also formation of mixture of products. It is important that the protecting groups should be easily introduced and also easily removable under mild conditions so that the peptide bond is not hydrolysed and that no racemisation or rearrangements occur.

Some useful protecting groups are : benzyloxycarbonyl (carbobenzyloxy), tert-butyloxycarbonyl (BOC; carbo-tert-butyloxy), trityl (triphenyl methyl), phthaloyl and tosyl (Ts, p-toluenesulphonyl).

The carboxyl group is protected by esterification. some common esters are methyl, ethyl, benzyl and tert-butyl.

The reactive side chain is also protected. Thiol and hydroxy groups are protected by benzyl group. The hydroxy group can also be protected by acetylation.

(ii) *Activation of the Carboxyl Group:* The carboxyl group of the protected amino acid is activated by converting it into the acid chloride, acid azide or p-nitrophenyl ester.

(iii) *The Coupling Step:* This is the most important step in the polypeptide synthesis. The free amino group of the amino acid (in which the carboxyl group is protected) is condensed with the protected and activated amino acid (in which the NH$_2$ group has been protected and the COOH group has been activated).

$$P-NH-CH_2COCl \; + \; \overset{\overset{\displaystyle R}{\displaystyle |}}{H_2NCHCOOCH_3}$$

Protected and   **Amino acid in which COOH**
activated amino acid   **group is protected**

$$\longrightarrow \; P-NHCH_2CONH-\overset{\overset{\displaystyle R}{\displaystyle |}}{CH}-COOCH_3$$

Dipeptide ester
(P is the protecting group)

The dipeptide ester is hydrolysed and the formed acid group is further activated and reacted with an amino acid to form a tripeptide and so on.

The coupling step is done under suitable conditions so that there is no racemisation or side reactions and yields are good. The most important condensing agent is dicyclohexylcarbodiimide (DCC) in organic solvent (methylene dichloride, tetrahydrofuran etc.)

(iv) *Removal of the Protecting Group:* After the polypeptide chain of the required length has been obtained, the final step is the removal of the protecting group. The carbobenzyloxy group is removed by catalytic reduction (if the amino acid contains sulphur, then catalytic reduction cannot be used, since the sulphur poisons the catalyst; in such cases, the carbobenzyloxy group is removed either by sodium metal in liquid ammonia at 40° or by triethylsilane and palladium chloride). The N-phthalyl group is removed easily by means of hydrazine. The tosyl group is removed by hydriodic acid at 50-60°, the trityl group is removed by heating with acetic acid and the benzyl group is removed by sodium and liquid ammonia.

Following methods have been used for the synthesis peptides and polypeptides.

[A] *Methods Involving Protection of Amino Group of One of the Amino Acids*

(i) *Fischer's method (1903):* In this method the amino group of amino acid ester is protected by carbethoxy group. The method is illustrated by the synthesis of glycylalanine.

$$ClCOOC_2H_5 + H_2NCH_2COOC_2H_5 \; \xrightarrow[-C_2H_5OH]{} \; C_2H_5OOC-NH-CH_2-COOC_2H_5$$

Ethyl chloro-   **Glycine ethyl ester**   **Carbethroxy glycine ethyl ester**
formate

$$\xrightarrow{\overset{\overset{\displaystyle CH_3}{\displaystyle |}}{H_2N-CH-COOC_2H_5}} \; C_2H_5OOC-NH-CH_2-CO-NH-\overset{\overset{\displaystyle CH_3}{\displaystyle |}}{CH}-COOC_2H_5 \; \xrightarrow{Hydrolysis}$$

a dipeptide

$$H_2NCH_2CONH-\overset{\overset{\displaystyle CH_3}{\displaystyle |}}{CH}-COOH$$

Glycylalanine (a dipeptide)

The dipeptide obtained above can be converted into tripeptide by condensation with another molecule of amino acid ester. In this way a polypeptide may be prepared.

A serious drawback of the above method is the removal of the protecting group without effecting the peptide bond and so the yield is low.

A modification of the above method involves the use of tosyl group as the protecting group (Fischer, 1915).

$$p-CH_3C_6H_4SO_2Cl + H_2NCH_2COOH \xrightarrow[\text{2. }CH_3COOH]{\text{1. NaOH}}$$

p-Toluenesulphonyl   Glycine
chloride

$$p-CH_3C_6H_4SO_2NHCH_2COOH \xrightarrow[\text{2. } H_2N-\underset{\underset{CH_3}{|}}{CH}-COOH]{\text{1. }SOCl_2}$$

protected glycine

$$p-CH_3C_6H_4SO_2NHCH_2CONH\underset{\underset{CH_3}{|}}{CH}-COOH \xrightarrow{Na / liq NH_3}$$

$$H_2N-CH_2-CONH-\underset{\underset{CH_3}{|}}{CH}-COOH + p-CH_3C_6H_4SH$$

Glycylalanine

No racemisation occurs in this modified procedure.

(ii) *Bergmann method (1932)*: This is a modification of the Fischer's method. In this method the amino group is protected by the carbobenzyloxy group ($-OCOCH_2C_6H_5$). This group is introduced by using benzylchloroformate, which is obtained readily by the action of carbonyl chloride on benzyl alcohol in toluene solution. The protecting group is removed at the desired stage by catalytic hydrogenation.

$$C_6H_5CH_2OH + COCl_2 \longrightarrow C_6H_5CH_2OCOCl$$

Benzyl alcohol   Carbonyl   **Benzyl chloroformate**
chloride

$$C_6H_5CH_2OCOCl + H_2NCH_2COOH \longrightarrow C_6H_5CH_2OCONHCH_2COOH$$
Glycine

$$\xrightarrow[\text{2. } H_2N-\underset{\underset{CH_3}{|}}{CH}-COOH]{\text{1. }PCl_5} C_6H_5CH_2OCONHCH_2CONH-\underset{\underset{CH_3}{|}}{CH}-COOH$$

$$\xrightarrow{H_2 / Pd} H_2NCH_2CONH\underset{\underset{CH_3}{|}}{CH}COOH + C_6H_5CH_3 + CO_2$$

Glycylalanine (a dipeptide)

The process can be repeated to give polypeptide.

In the above method, if the amino acid contains sulphur, the catalyst is poisoned. In such a situation, the protecting group is removed by sodium and liquid ammonia or preferably by triethylsilane and palladium chloride.

(iii) *Sheehan's method (1949)*: In this method, the amino group is protected by phthaloyl group. The protecting group is removed at any stage by treatment with hydrazine. This method is analogous to the Gabriel's synthesis.

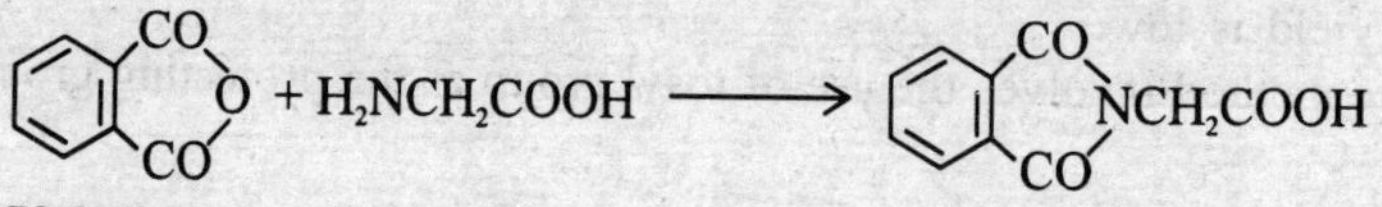

Phthalic   Glycine   **Phthaloyl glycine**
anhydride

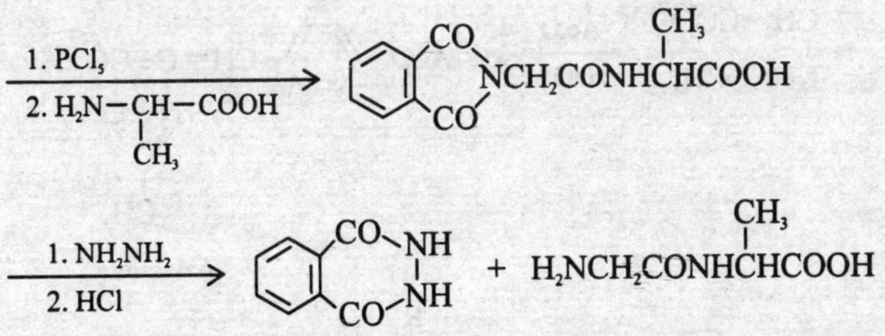

In the above synthesis no racemisation takes place provided the temperature is kept below 150°. Optically pure phthaloyl derivatives are obtained in good yields by phthaloylation with N-carbethoxyphthalimide (obtained from potassium phthalimide and ethyl chloroformate) in aqueous sodium bicarbonate.

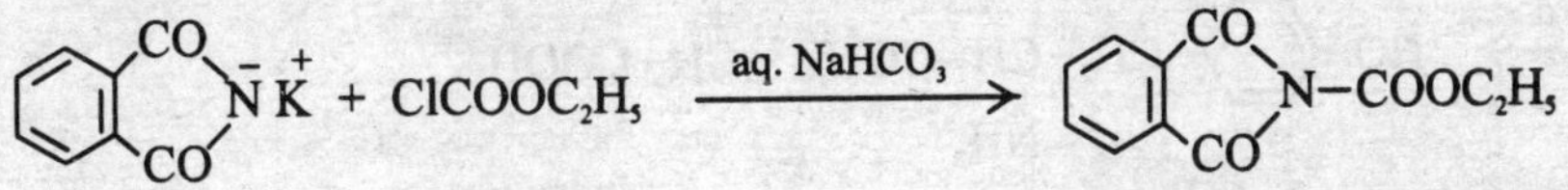

Potassium phthalimide **Ethyl chloroformate** **N-Carbethoxyphthalimide**

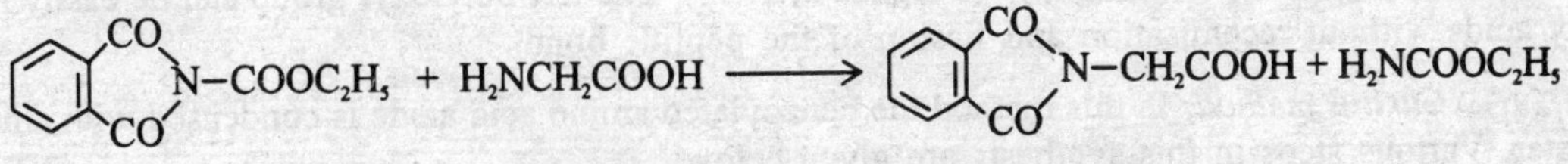

Glycine **N-Phthaloiglycine**

The phthaloyl derivatives have also been prepared (Weygand, 1961) without racemisation by heating the amino acids with diethyl phthalate and triethylamine in phenol.

(iv) *Halpern's method:* The amino group is protected by reaction with dimedone. The protected group from the peptide is removed by bromination (Halpern, 1964).

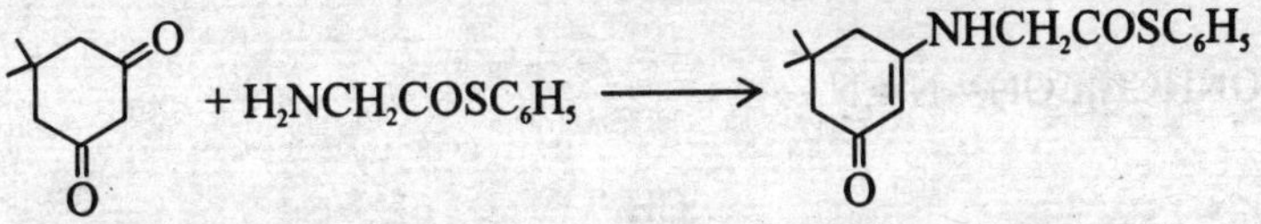

Dimedone **Thiophenyl ester of glycine** **Enamine derivative**

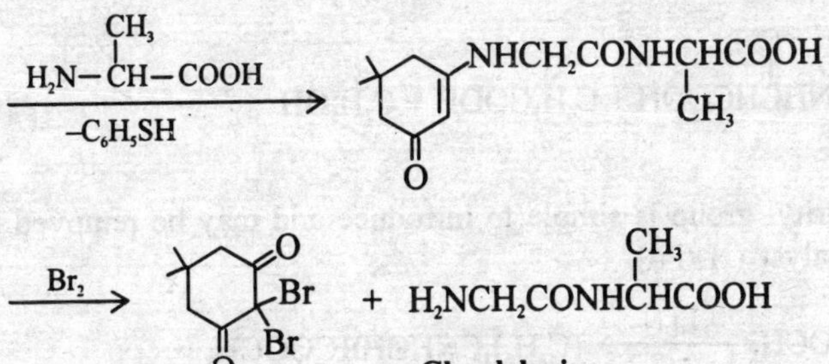

Glycylalanine

(v) *Azlactone synthesis:* This method is useful for the synthesis of peptides containing both hydroxy and amino groups (which can not be synthesised by other methods discussed so far).

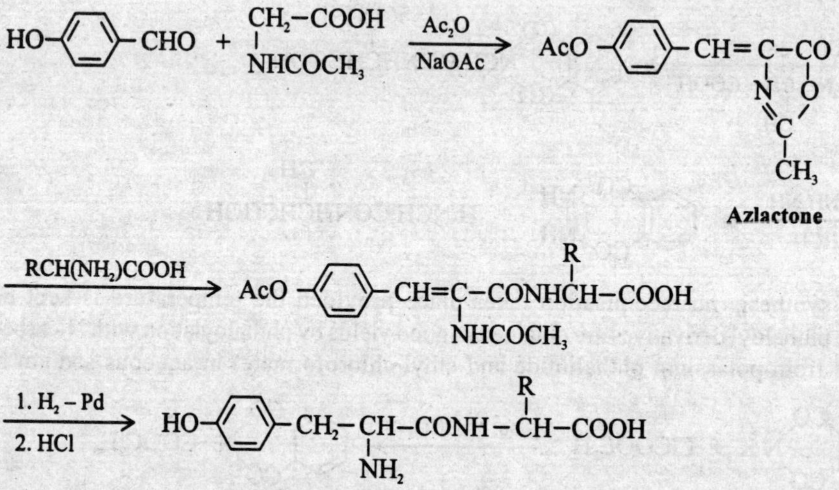

$$HO-\langle\text{ring}\rangle-CHO + \underset{NHCOCH_3}{\overset{CH_2-COOH}{|}} \xrightarrow[NaOAc]{Ac_2O} AcO-\langle\text{ring}\rangle-CH=\underset{\underset{CH_3}{\overset{|}{C}}}{\overset{O}{\underset{N}{\overset{\diagup}{C}}}}-CO$$

Azlactone

$$\xrightarrow{RCH(NH_2)COOH} AcO-\langle\text{ring}\rangle-CH=\underset{NHCOCH_3}{\overset{\overset{R}{|}}{C}}-CONHCH-COOH$$

$$\xrightarrow[2.\ HCl]{1.\ H_2-Pd} HO-\langle\text{ring}\rangle-CH_2-\underset{NH_2}{\overset{|}{C}H}-CONH-\overset{\overset{R}{|}}{C}H-COOH$$

The hydroxy group can also be protected by introducing tertiary butyloxy group (by addition of isobutene to the acid) in place of acetyl group (Bergmann, 1961). The tert-butyloxyl group can be easily removed by acids without racemisation and fission of the peptide bond.

(vi) *Curtius method:* In this method the benzoylated amino acid azide is condensed with amino acid ester. Various steps in this synthesis are given below.

$$H_2NCH_2COOH \xrightarrow[NaOH]{C_6H_5COCl} C_6H_5CONHCH_2COOH \xrightarrow{C_2H_5OH}$$

Glycine **Benzoylglycine**

$$C_6H_5CONHCH_2COOC_2H_5 \xrightarrow{NH_2NH_2} C_6H_5CONHCH_2CONHNH_2$$

$$\xrightarrow[0°]{HONO} C_6H_5CONHCH_2CON=\overset{+}{N}=\overset{-}{N}$$

$$\xrightarrow{\underset{NH_2-CH-COOC_2H_5}{\overset{CH_3}{|}}} C_6H_5CONHCH_2CONH\overset{\overset{CH_3}{|}}{C}H-COOC_2H_5$$

$$\xrightarrow{Hydrolysis} H_2NCH_2CONH\overset{\overset{CH_3}{|}}{C}HCOOH + C_6H_5COOH + C_2H_5OH$$

Glycylalanine

(vii) *Use of trityl reagent:* The trityl group is simple to introduce and may be removed by heating in acetic acid or in presence of catalyst (H$_2$/Pd).

$$(C_6H_5)_3CCl + H_2NCHRCOOCH_3 \xrightarrow{Et_3N} (C_6H_5)_3CNHCHRCOOCH_3$$

$$\xrightarrow[2.\ CH_3COOH]{1.\ NaOH} (C_6H_5)_3CNHCHRCOOH \longrightarrow$$

$$\xrightarrow[\text{2. NH}_2\text{CHR'COOH}]{\text{1. SOCl}_2} \quad (C_6H_5)_3\text{CNHCHRCONHCHR'COOH}$$

$$\xrightarrow{\text{CH}_3\text{COOH}} \quad (C_6H_5)_3\text{COOCH}_3 + \text{H}_2\text{NCHRCONHCHR'COOH}$$

[B] *Anhydride Method:* This is a novel approach for the synthesis of polypeptides. The N-carboxyanhydride (NCA) required for the synthesis may be prepared by a number of methods. A simple method utilises N-benzyloxycarbonyl derivative of an amino acid as follows:

$$\text{PhCH}_2\text{OCONHCHRCOOH} \xrightarrow{\text{PCl}_5} \text{PhCH}_2\text{OCONHCHRCOCl}$$

$$\xrightarrow[\text{Vacuo}]{\text{Heat in}} \quad \begin{array}{c} \text{R—CH—CO} \\ | \qquad\quad \rangle\text{O} \\ \text{NH—CO} \end{array} + \text{PhCH}_2\text{Cl}$$

$$\textbf{NCA}$$

The NCA derivative is used to build up a peptide chain by reacting with an amino acid in alkaline solution (pH 10) followed by acidification.

$$\begin{array}{c} \text{R—CH—CO} \\ | \qquad\quad \rangle\text{O} \\ \text{NH—CO} \end{array} + \text{H}_2\text{NCHR'COO}^- \xrightarrow{\text{OH}^-} \bar{\text{O}}\text{OCNHCHRCONHCHR'COO}^-$$

$$\begin{array}{c}\textbf{NCA} \qquad\qquad\qquad \textbf{Amino acid}\end{array}$$

$$\xrightarrow[\text{—CO}_2]{\text{H}^+} \text{H}_2\text{NCHRCONHCHR'COOH}$$

$$\textbf{a Dipeptide}$$

The dipeptide couples with another NCA derivative and the process is repeated.

[C] *Intermolecular Dehydration using DCC:* The N-carbobenzyloxy protected amino acid and an amino acid ester are made to condense in presence of N,N-dicyclohexylcarbodiimide (DCC) to give the carbobenzyloxy derivative of a dipeptide ester.

$$\overset{\text{R}}{\underset{|}{\text{C}_6\text{H}_5\text{CH}_2\text{OCONHCHCOOH}}} + \text{H}_2\text{N}-\overset{\text{R}}{\underset{|}{\text{CH}}}-\text{COOC}_2\text{H}_5$$

N-Carbobenzyloxy protected   **Amino acid ester**
amino cid

$$\downarrow \quad \bigcirc\!\!-\!\text{N=C=N}\!-\!\bigcirc$$
$$\textbf{DCC}$$

$$\overset{\text{R}}{\underset{|}{\text{C}_6\text{H}_5\text{CH}_2\text{OCONHCHCONHCHCOOC}_2\text{H}_5}} + \bigcirc\!\!-\!\text{NH—CO—NH}\!-\!\bigcirc$$

Carbobenzyloxy deriv. of a dipeptide ester   **N,N′-Dicyclohexyl urea**

The free dipeptide is obtained by catalytic hydrogenolysis followed by hydrolysis.

[D] *Solid phase Peptide Synthesis:* This versatile procedure introduced by Merrifield (1964) consists in binding chemically an amino acid or a peptide to an insoluble synthetic resin and then the chain is built up with one amino acid residue at a time to the free end. When the desired peptide has been synthesised, it is liberated from the solid support. This method has been automated, i.e., each addition of the appropriate amino acid is carried out automatically at the predetermined time. Some special advantages of this method are:

(i) Since the solid support is insoluble, purification of the products is not necessary. Excess of reagents are removed by simple washing with suitable solvents.

(ii) The time has been considerably reduced for the synthesis of peptides and proteins.

(iii) Excellent yields.

The method utilises a solid support or resin of copolymer of styrene and divinyl benzene. It is chloromethylated resulting in the attachment of the $-CH_2Cl$ group. The BOC (tert-butyloxycarbonyl) protected amino acid is reacted with $-CH_2Cl$ so that this first amino acid is attached as its benzyl ester. This first amino acid is the C-terminal of the peptide. The BOC protecting group is selectively removed by HCl-AcOH and the formed dihydrochloride is converted into the free amino group by the addition of triethyl amine. The benzyl ester of the first amino acid residue is coupled with N-tert-butyloxycarbonyl derivative of second amino acid by means of DCC. The process is then repeated with the N-protected third acid and so on. When the desired peptide has been obtained, the ester bond linking it to the resin is split by dry hydrogen bromide in trifluoroacetic acid. Various steps involved are given below.

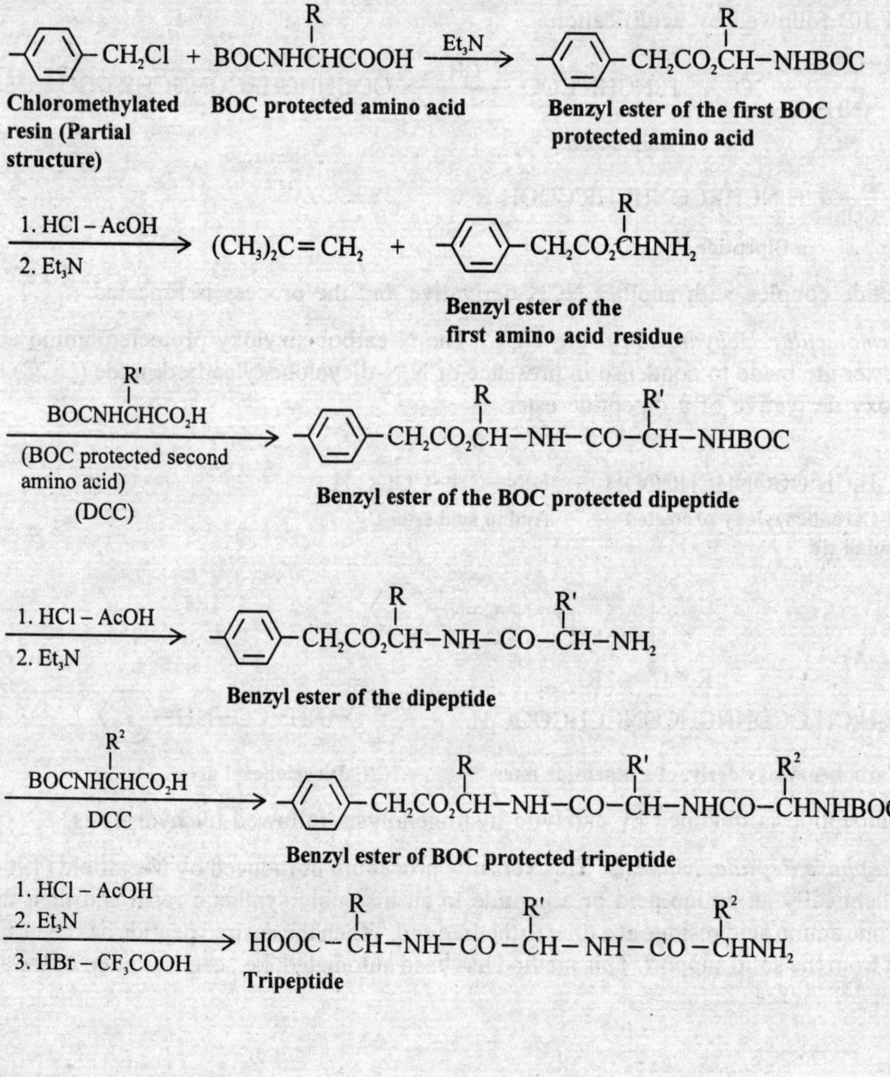

Chloromethylated resin (Partial structure)   **BOC protected amino acid**   **Benzyl ester of the first BOC protected amino acid**

Benzyl ester of the first amino acid residue

Benzyl ester of the BOC protected dipeptide

Benzyl ester of the dipeptide

Benzyl ester of BOC protected tripeptide

Tripeptide

40.4 PROTEINS

Proteins are nitrogeneous substances which occur in the protoplasm of all animal and plant cells. These are essential for the growth and maintenance of life and constitute approximately 75% of the dry material of most living systems. Proteins are synthesised by plants using carbon dioxide, water, nitrates and ammonium salts in presence of energy obtained from the sun(photosynthesis). Human beings and animals take plant in their food; the proteins of the plants are hydrolysed to amino acids by enzymes in the human or animal system. The proteins are resynthesised in the human and animal system from these amino acids. Following table gives the sources and percentage of proteins present in them.

Table 40.2: Sources and Percentage of Proteins

Source	% Proteins
Soyabean	43-44
Pea	28-30
Milk Powder	26-27
Ground Nut	25-27
Fish	20-22
Cashewnut	20-22
Mutton	18-20
Almonds	20-21
Egg Yolk	13-14
Oatmeal	13-14
Wheat	10-12

The composition of proteins vary with the source. The following table gives the approximate composition:

Table 40.3: Composition of Proteins

Carbon	45-55%	Nitrogen	10-30%
Hydrogen	6-8%	Sulphur	0.2-0.3%
Oxygen	12-20%	Phosphorus	0.1-1%

As already stated proteins are build up from α-amino acid residues and may be hydrolysed to amino acids.

$$\text{Proteins} \longrightarrow \text{polypeptides} \longrightarrow \text{peptides} \longrightarrow \text{Amino acids}$$

There is no sharp dividing line between peptides, polypeptides and proteins. Arbitrarily, proteins are the molecules with molecular weight more than 10,000 and peptides or polypeptides are the molecules with molecular weight less than 10,000.

40.4.1 Classification

Proteins may be classified in a number of ways. Following are given two classifications.

1. *Classification based on Molecular Structure and Behaviour:* The proteins are divided into two types on the basis of their structure and behaviour.

(i)   *Fibrous proteins:* These are insoluble in common organic solvents but are soluble in concentrated acids and alkalis. They are mostly the linear condensation products of neutral amino acids; in some cases the molecules are held together by intramolecular hydrogen bonds. Examples of this type are:

Collagen-   the protein connecting tissues
Elastin-   the protein of elastic connective tissues.
Keratin-   the proteins of hair, nails, skin, wool and feathers etc.

(ii)   *Globular proteins:* These are soluble in water, dilute acids, dilute alkalis and salts. Unlike fibrous proteins, the globular proteins are mostly branched and cross linked condensation products of acidic or basic amino acids. The globular proteins are folded to three dimensional shape and the peptide chain is stabilised by intramolecular hydrogen bonds. Examples of this type are enzyme and some hormones like insulin and thyroglobin, casein of milk, globin of haemoglobin and albumin in eggs.

2.   *Classification on the Basis of Physical Properties:* According to this method the proteins are classified on the basis of increasing complexity in their structures and physical properties.

(i)   *Simple proteins:* These give only α-amino acids on hydrolysis and are further subdivided on the basis of their solubility into the following classes.

(a)   *Albumins:* These are soluble in water, acids and alkalis and are coagulated by heat and are precipitated by saturating their solution with diammonium sulphate. They are mostly devoid of glycine. Examples of this type are egg albumin, serum albumin and lactalbumin.

(b)   *Globulins:* These are insoluble in water, but are soluble in dilute solutions of strong alkalis and acids. They are coagulated by heat and precipitated by saturating their solutions with ammonium sulphate. They contain glycine. Some examples are serum globulin, tissue globulin and vegetable globulin.

(c)   *Prolamins:* These are insoluble in water and salt solution but are soluble in dilute alkalis or acids and in dilute ethanol (70-90%). They contain comparatively large amounts of proline and are deficient in lysine. Examples of this type are zein (from maize), hordein (from barley) and gliadin (from wheat).

(d)   *Glutelins:* These are insoluble in water or salt solution, but are soluble in dilute alkalis or acids. They are coagulated by heat and are comparatively rich in arginine, proline and glutamic acid. Some examples are glutenin (from wheat) and oyrzenin (from rice).

(e)   *Scleroproteins (albuminoids):* These are insoluble in water or salt solution but are soluble in concentrated alkalis or acids and are not attacked by enzymes. Some examples are keratin (from hair), and fibroin (from silk).

The Scleroproteins are of two types:

Collagens: These are found in skin, tendons and bones and on boiling with water form gelatin, a water soluble protein. These are attacked by pepsin or trypsin.

Elastins: These are found in arteries, tendon and other elastic tissues and are slowly attacked by trypsin. They are not converted into gelatin.

(f)   *Basic proteins:* The basic proteins are of two types:

Histones: Histones are soluble in water and dilute acids but are insoluble in dilute ammonium hydroxide solution. They contain comparatively large amounts of histidine and argenine but do not contain tryptophan. They can be hydrolysed by pepsin and trypsin and are the proteins of the nucleic acids, haemoglobin etc.

Protamins: These are comparatively more basic than histones and are soluble in water, dilute acids or dilute ammonium hydroxide solution. They are not coagulated by heat and are precipitated from solution by ethanol. They occur in various nucleic acids and contain comparatively large amounts of arginine and can be hydrolysed by enzymes like trypsin, papain etc.

(ii) *Conjugated proteins:* These are simple protein molecules which contain non-protein groups. The non-protein group is known as the *prosthetic group* and can be separated from the protein part by careful hydrolysis. Further classification of conjugated proteins depends on the nature of the prosthetic group.

(a) *Nucleoproteins:* These contain nucleic acid as the prosthetic group.

(b) *Chromoproteins:* These contain a colored prosthetic group. The color is due to the presence of a metal in their structure. The most common examples of this class are chlorophyll and haemoglobin. In haemoglobin, globin is the protein and haem is the prosthetic group.

(c) *Phosphoproteins:* They contain phosphoric acid as the prosthetic group in some form other than in nucleic acids or in lipoproteins. For example, casein of milk.

(d) *Glycoproteins:* These contain a carbohydrate or its derivative as the prosthetic group like mucin of saliva. These are also known as *mucoproteins*.

(e) *Lipoproteins:* These contain lecithin, kephalin etc. as the prosthetic group. For example lipoproteins of serum.

(iii) *Derived proteins:* These are obtained by the hydrolysis of proteins by acids, alkalis or enzymes and may be regarded as intermediate hydrolysis products. These are classified on the basis of the progressive cleavage into the following:

Denatured proteins: These are insoluble proteins formed by the action of heat on proteins.

Primary proteoses (metaproteins): These are insoluble in water or dilute salt solution but are soluble in alkalis or acids. These are precipitated by half saturation with ammonium sulphate.

Secondary proteoses: They are soluble in water and are not coagulated by heat. They are precipitated by saturation with ammonium sulphate.

Peptones, polypeptides and simple peptides: These are soluble in water and are neither coagulated by heat nor precipitated by saturation with ammonium sulphate.

Amino acids: The final degradation product of derived proteins. Soluble in water but sparingly soluble in organic solvents.

40.4.2 Characteristics of Proteins

1. Most of the proteins are amorphous, colorless, tasteless and odorless. However, some are crystalline when pure. They have no definite melting points. All proteins are optically active.

2. *Solubility:* Most of the proteins are insoluble in water and alcohol but are soluble in dilute alkalis and acids. However, some proteins are completely insoluble.

3. *Amphoteric Nature:* Proteins are amphoteric. Their behaviour as a cation or an anion depends on the pH of the solution. At some definite pH, the positive and the negative charges are exactly balanced i.e., there is no effective charge on the protein molecule. This means that the molecule will not migrate in an electric field. In this state, the protein is said to be at its isoelectric point. The isoelectric point is the characteristic of each protein and at this pH the protein has least solubility and is readily precipitated.

The viscosity as well as the osmotic pressure of protein solution are minimum at the isoelectric point. The isoelectric points of some proteins are given below:

Caesin	4.6
Gelatin	4.80-4.85
Insulin	4.30
Haemoglobin	6.80-6.83

4.   *Denaturation:* Proteins can be coagulated and precipitated from aqueous solution by heat, addition of alkalis, acids or salts. The proteins in this precipitated state are said to be *denatured*. The process is known as *denaturation*. It occurs most readily at the isoelectric point. Denaturation is believed to result in the change of conformation or unfolding of the protein molecule. Though denaturation is generally irreversible but some examples are known where the process has been reversed and is called *renaturation* or *refolding*. In case denaturation is effected by heat, renaturation may be carried out by very slow cooling. The biological activity of proteins (if present) is lost during denaturation. This process of renaturation is referred to as *annealing*.

5.   *Molecular Weight:* Proteins have high molecular weight. The approximate molecular weight of some proteins are given below:

Insulin	12,000
Egg albumin	45,000
casein	180,000
Haemoglobin	70,000

The molecular weights of proteins cannot be determined by the usual methods due to their high molecular weight, non-volatile nature and insolubility in common solvents. In case of proteins the molecular weights are determined by means of ultracentrifugal sedimentation, osmotic pressure measurements, X-ray diffraction, light scattering method, gel filtration and by chemical analysis.

6.   *Color Reactions:* Proteins on treatment with some specific reagents give characteristic color. The tests are specific for certain group present in protein and should not be taken as general tests for all proteins.

(i)   *Biuret test:* Treatment of an alkaline solution of a protein with a very dilute solution of copper sulphate gives a red or violet color. This reaction is due to the presence of the –CO–NH–CHR–CO–NH– group. Except dipeptides, this test is given by all proteins, peptones and peptides.

(ii)   *Millon's reaction:* Treatment of protein solution with Millon's reagent (mercuric nitrate in nitric acid containing traces of nitrous acid) gives a white precipitate, which turns red on heating. This reaction is characteristic of phenols and is given by proteins containing tyrosine (Tyrosine is the only phenolic amino acid that occurs in proteins).

(iii)   *Xanthoproteic reaction:* Proteins on warming with concentrated nitric acid give a yellow color which becomes orange when the solution is made alkaline. This reaction is due to the nitration of benzene ring in phenylalanine, tyrosine and tryptophan.

(iv)   *Ninhydrin reaction:* Peptides and proteins give this test (see Section 40.2.3.5). The colors are, however, different from that of amino acids.

Uses of Proteins

1. The most important use of proteins is as food.

2. A number of amino acids are obtained from proteins.

3. Industrial uses: Casein obtained from milk is used in the production of plastics. Gelatin is used to relieve fatigue and to increase energy.

4. Proteins are finding increasing applications in medicines. Blood plasma (obtained after removal of blood cells by centrifugation) is used for the treatment of shock produced by serious injury.

5. Biological functions: Proteins regulate various metabolic process. For example, insulin is used for maintaining blood sugar levels. Besides, some proteins functions as catalysts for biological reactions. Proteins like haemoglobin and other peptide hormones are necessary for the working of the human system.

40.4.3 Structure of Proteins

The proteins are considered to be polymers containing large number of amino acids joined to each other by peptide bonds and have definite three dimensional structure. A number of factors are responsible for the determination of the exact shape of protein molecules. It is most appropriate to consider the structure of protein at four different levels, viz. primary, secondary, tertiary and quaternary structure. Thus, the spacial arrangement of the polypeptide in a protein molecule is determined by the *primary structure* (the amino acid sequence) of the protein. The conformation that the polypeptide chain assumes is called the *secondary structure.* The way, the molecule folds to produce a specific shape is called the *tertiary structure* of the protein. The *quaternary structure* describes the arrangement and ways in which the subunits of proteins are held together. Following is given a brief account of the various types of structure of proteins.

40.4.3.1 Primary Structure

It refers to the number, nature and sequence of amino acids present in a protein molecule. The primary structure of protein is determined as follows.

(i) *Purification:* The protein must be isolated in a pure state for this purpose. The crude sample of protein is precipitated out near its isoelectric point. The sample is freed from salts by dialysis or gel filtration and then fractionated by chromatographic methods. Finally, the purified sample is isolated by freeze drying. The purity of the protein is assessed by electrophoresis and counter current distribution. In case the protein is insoluble, its structure is determined by X-ray studies.

(ii) *Determination of the nature of peptide chain present:* It is necessary to ascertain whether the protein molecule consists of a single peptide chain or is composed of a number of subunits. In case there are a number of subunits, these are separated and examined separately. The nature of the peptide chain is determined by partial hydrolysis by different types of enzymes.

(iii) *Determination of amino acid composition:* The proteins under investigation is hydrolysed by means of acid (6N HCl), alkali (2N Ba(OH)$_2$) at 100° or enzymes to its constituent amino acids. These are separated and identified by means of ion exchange chromatography and reaction with ninydrin. The color produced with ninhydrin is also used for the quantitative estimation of a particular amino acid.

The amount of each amino acid can also be determined by isotopic dilution method. In this procedure, a known amount of a ^{14}C labeled amino acid, (whose analysis is desired) is added to the mixture. The separation procedure gives a mixture of labeled amino acids along with the corresponding non labeled acid. Determination of the new level of radioactivity in this mixture will give the amount of the non labeled amino acid in the protein.

(iv) *Molecular weight determination:* From the amino acid percentage composition the minimum molecular weight is determined. The molecular weight of proteins is also determined by means of ultracentrifugal sedimentation, osmotic pressure measurement, X-ray diffraction, light scattering effects and molecular sieves (gel filtration). From the data (ii and iii above), the molecular formula is determined.

(v) *Sequence determination:* The sequence in which the amino acid residues are arranged along the peptide chain is determined by following methods:

A. *End Group Analysis:* By this procedure the nature of the N- and C-terminal groups (see also Section 40.3.2) is determined.

(a) *Determination of N-terminal amino acid:* As already stated in a polypeptide (and so in a protein), the amino acids that contain the free amino group is called the N- terminal residue. It is determined as given below:

(aa) *Sanger's method:* The method consist in treating the protein or the polypeptide with 1-fluoro-2,4-dinitrobenzene (FDNB) in presence of sodium bicarbonate solution at room temperature. The amino acid of the polypeptide (or the protein) reacts with FDNB to form a 2,4-dinitrophenyl derivative. The product is hydrolysed with acid (the cleavage of the peptide bond connecting the N-terminal amino acid to the rest of the polypeptide molecule takes place) to form the dinitrophenyl derivative of the N-terminal amino acid and the remaining polypeptide molecule or amino acid residues.

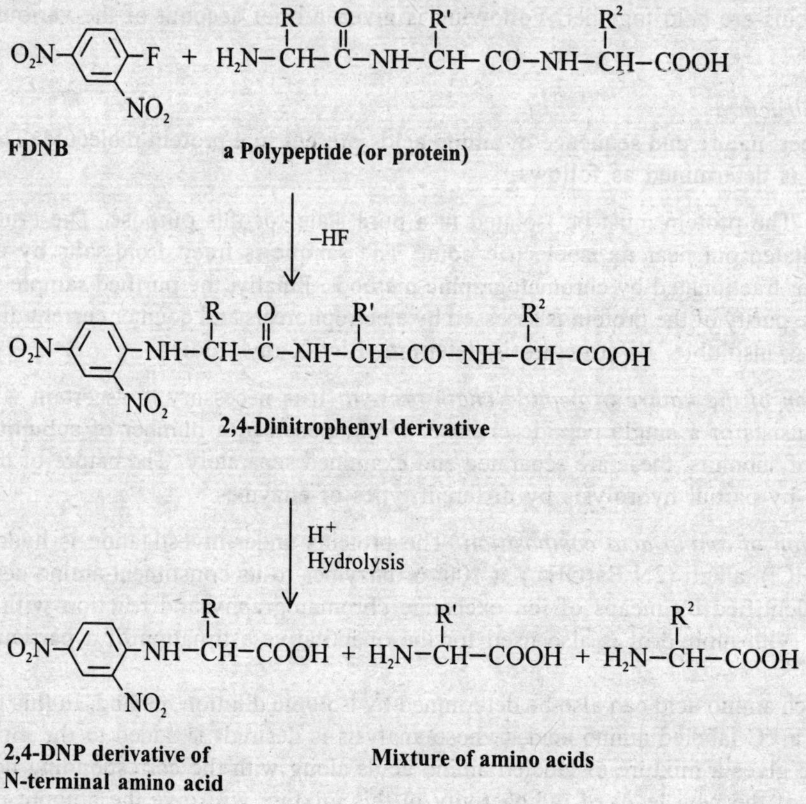

$$O_2N-\langle\!\!\!\bigcirc\!\!\!\rangle-F \quad + \quad H_2N-\overset{R}{\underset{|}{C}H}-\overset{O}{\underset{\|}{C}}-NH-\overset{R'}{\underset{|}{C}H}-CO-NH-\overset{R^2}{\underset{|}{C}H}-COOH$$
$$\underset{NO_2}{}$$

FDNB   **a Polypeptide (or protein)**

$$\downarrow -HF$$

$$O_2N-\langle\!\!\!\bigcirc\!\!\!\rangle-NH-\overset{R}{\underset{|}{C}H}-\overset{O}{\underset{\|}{C}}-NH-\overset{R'}{\underset{|}{C}H}-CO-NH-\overset{R^2}{\underset{|}{C}H}-COOH$$
$$\underset{NO_2}{}$$

2,4-Dinitrophenyl derivative

$$\downarrow \begin{array}{c} H^+ \\ \text{Hydrolysis} \end{array}$$

$$O_2N-\langle\!\!\!\bigcirc\!\!\!\rangle-NH-\overset{R}{\underset{|}{C}H}-COOH \quad + \quad H_2N-\overset{R'}{\underset{|}{C}H}-COOH \quad + \quad H_2N-\overset{R^2}{\underset{|}{C}H}-COOH$$
$$\underset{NO_2}{}$$

2,4-DNP derivative of   **Mixture of amino acids**
N-terminal amino acid

The 2,4-dinitrophenyl derivative of all amino acids are known and can be identified either by TLC separation or by determination of its melting point.

As seen the DNP derivatives are formed with any free amino group, the diaminiocarboxylic acids like lysine also reacts even if it is not an N-terminal acid. Also the hydroxy group of tyrosine, the thiol group of cysteine and the imidazole nucleus of histidine also form the DNP derivative (though very slowly). So by this method a number of DNP derivatives are formed. These are isolated and identified by TLC.

(ab)   *Edman method:* It consist in the treatment of the protein (or the peptide) with phenyl isothiocyanate in presence of dilute alkali to form the phenylthiocarbamyl (PTC) -peptide (or protein), which on treatment with dilute acid (hydrochloric acid or trichloroacetic acid) is converted into a phenylthiohydantoin (PTH) and the remaining protein (or peptide) is left intact. The PTH is separated and identified by paper chromatography.

$$C_6H_5NCS + \overset{\displaystyle R}{\underset{\displaystyle |}{H_2N-CH}}-\overset{\displaystyle O}{\overset{\displaystyle ||}{C}}-NH-\overset{\displaystyle R'}{\underset{\displaystyle |}{CH}}-CO-NH-\overset{\displaystyle R^2}{\underset{\displaystyle |}{CH}}-COOH$$

Phenyl   **a Protein or a Polypeptide**
isothiocyanate

$$\downarrow OH^-$$

$$C_6H_5NHCSNH\overset{\displaystyle R}{\underset{\displaystyle |}{CH}}-\overset{\displaystyle O}{\overset{\displaystyle ||}{C}}-NH-\overset{\displaystyle R'}{\underset{\displaystyle |}{CH}}-CO-NH-\overset{\displaystyle R^2}{\underset{\displaystyle |}{CH}}-COOH$$

PTC-peptide (or protein)

$$\downarrow H^+$$

$$\begin{matrix} NH-CHR \\ | \quad\quad | \\ SC \quad\quad CO \\ \backslash\;\; N\;/ \\ | \\ C_6H_5 \end{matrix} \quad + \quad \overset{\displaystyle R'}{\underset{\displaystyle |}{H_2N-CH}}-CO-NH-\overset{\displaystyle R^2}{\underset{\displaystyle |}{CH}}COOH$$

Phenyl thio-   **Remaining protein or polypeptide**
hydantoin (PTH)

The process is repeated on the remaining part of the protein (or polypeptide) to identify the second amino acid and so on. This method has been automated for determination of the amino acid sequence in protein (or polypeptides).

(ac)   *The dansyl method:* This is a modification of the Sanger's method. It consists in the reaction of the protein (or the polypeptide) with 5-dimethylaminonaphthalene-1-sulphonyl chloride (dansyl chloride; DNS-Cl) in place of FDNB.

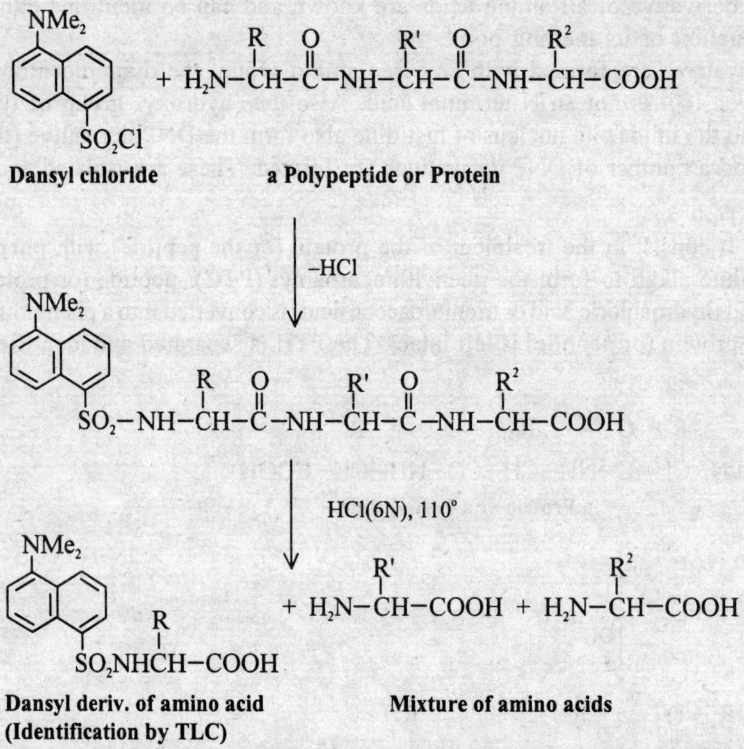

Dansyl chloride   a Polypeptide or Protein

$-HCl$

Dansyl deriv. of amino acid   Mixture of amino acids
(Identification by TLC)

The dansyl method is widely used since the dansyl group is highly fluorescent; the dansyl derivative of amino acid may be detected and estimated in minute amounts by fluorimetric methods.

(ad) *Enzymic method:* The enzyme leucine aminopeptidase attacks proteins (or peptides) only on the end which contains the free amino group. The method proceeds to liberate in succession each new terminal amino acid. After a fixed time of hydrolysis, the amount of free amino acid is estimated. This method gives the sequence of amino acids.

(b) *Determination of C-terminal amino acid:* It has already been stated that a polypeptide (or a protein) that contain the free carboxyl group is called the C-terminal residue. It is determined as given below.

(ba) *Hydrazinolysis (akabori et.al., 1956):* This is the most commonly used method. It involves the heating of protein (or polypeptide) with anhydrous hydrazine, which converts all amino acids except the C-terminal ones into amino acid hydrazides.

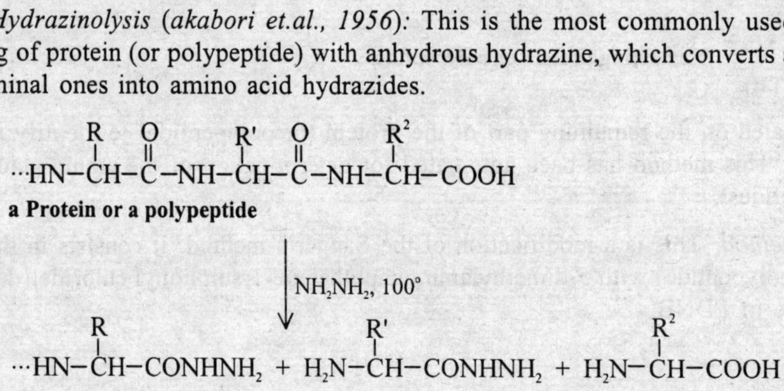

a Protein or a polypeptide

$NH_2NH_2, 100°$

The mixture of products obtained above is separated by column chromatography on a strong cation-exchange resin followed by elution. The basic hydrazides are retained and the free amino acid is eluted and identified.

(bb) *Reduction:* The protein (or the polypeptide) is reduced with lithium aluminum hydride or lithium borohydride. By this process the free terminal carboxyl group is converted into a primary alcoholic group. Subsequent hydrolysis gives a mixture of amino acids and an amino alcohol. The amino alcohol is separated and identified (paper chromatography).

(bc) *Enzymic method:* The enzyme carboxypeptidase attacks proteins (or peptides) only at the end which contains the free α-carboxyl group. Thus, the terminal amino acid is liberated after a fixed time. The enzymic hydrolysis is continued and the new terminal free carboxyl group is attacked by the enzyme.

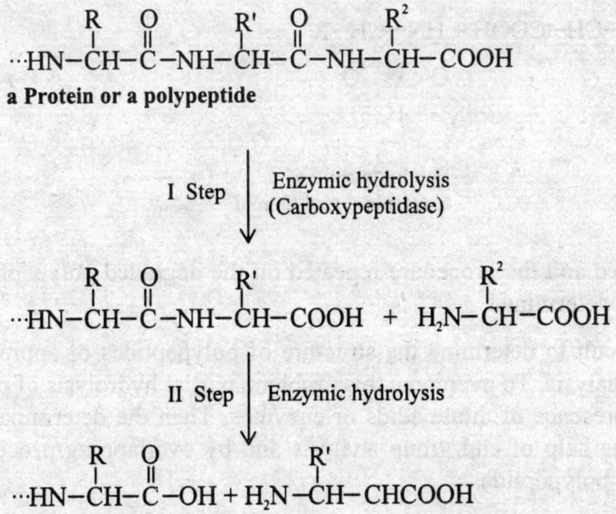

$$\cdots HN-\underset{\underset{R}{|}}{CH}-\underset{\underset{O}{\|}}{C}-NH-\underset{\underset{R'}{|}}{CH}-\underset{\underset{O}{\|}}{C}-NH-\underset{\underset{R^2}{|}}{CH}-COOH$$

a Protein or a polypeptide

I Step | Enzymic hydrolysis (Carboxypeptidase)

$$\cdots HN-\underset{\underset{R}{|}}{CH}-\underset{\underset{O}{\|}}{C}-NH-\underset{\underset{R'}{|}}{CH}-COOH \ + \ H_2N-\underset{\underset{R^2}{|}}{CH}-COOH$$

II Step | Enzymic hydrolysis

$$\cdots HN-\underset{\underset{R}{|}}{CH}-\underset{\underset{O}{\|}}{C}-OH + H_2N-\underset{\underset{R'}{|}}{CH}-CHCOOH$$

This method is useful for the identification, quantitative estimation and determination of the sequence of amino acids present in a protein (or a polypeptide).

(bd) *Schlack and Kumpf method:* The amino group is first protected by benzoylation and then C-terminal amino acid is converted into thiohydantoin, which is hydrolysed to the amino acid as in the Edman method.

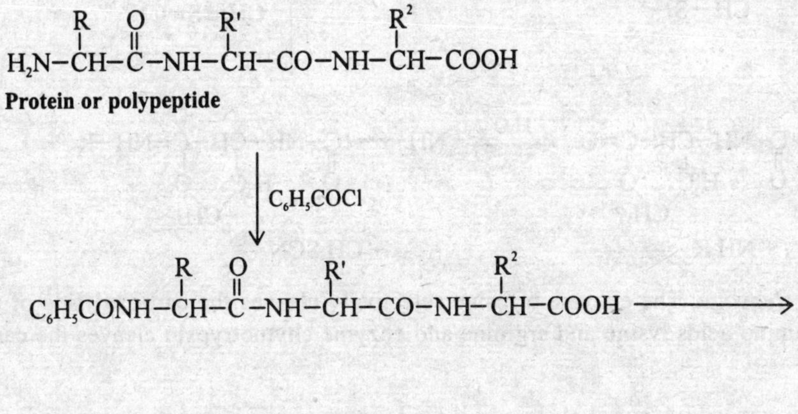

$$H_2N-\underset{\underset{R}{|}}{CH}-\underset{\underset{O}{\|}}{C}-NH-\underset{\underset{R'}{|}}{CH}-CO-NH-\underset{\underset{R^2}{|}}{CH}-COOH$$

Protein or polypeptide

C_6H_5COCl

$$C_6H_5CONH-\underset{\underset{R}{|}}{CH}-\underset{\underset{O}{\|}}{C}-NH-\underset{\underset{R'}{|}}{CH}-CO-NH-\underset{\underset{R^2}{|}}{CH}-COOH \longrightarrow$$

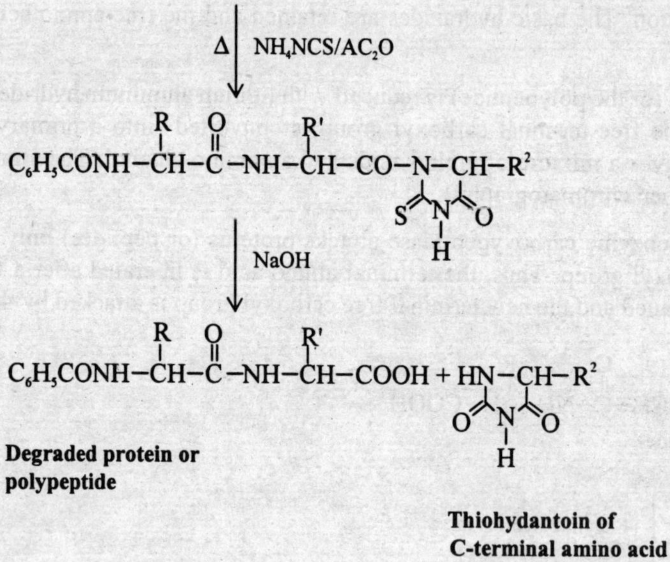

$$\Delta \Big\downarrow \quad NH_4NCS/AC_2O$$

$$\underset{\substack{| \\ \text{R} \quad \text{O} \quad \text{R}'}}{C_6H_5CONH-CH-\overset{\text{O}}{\overset{\|}{C}}-NH-CH-CO-N-CH-R^2}$$

(with thiohydantoin ring drawn below, labelled S, N, O, H)

$$\Big\downarrow \quad NaOH$$

$$\underset{\substack{\text{R} \quad \text{O} \quad \text{R}'}}{C_6H_5CONH-CH-\overset{\text{O}}{\overset{\|}{C}}-NH-CH-COOH} + HN-CH-R^2$$

(with ring O, N, O, H drawn below)

**Degraded protein or
polypeptide**

**Thiohydantoin of
C-terminal amino acid**

The formed amino acid is identified and the procedure repeated on the degraded polypeptide. In this way the sequence of amino acid is determined.

B.   *Partial Hydrolysis:* It is difficult to determine the structure of polypeptides of appreciable size by the above methods of sequential analysis. To overcome this problem, partial hydrolysis of polypeptide to smaller peptides can be done in presence of dilute acids or enzymes. Then the determination of the structure of smaller peptides with the help of end group analysis and by overlapping process is done to find out the structure of original polypeptide.

Specific chemical cleavage: The peptide bond formed from the carboxyl group of methionine is cleaved by making use of cyanogen bromide according to the following reaction.

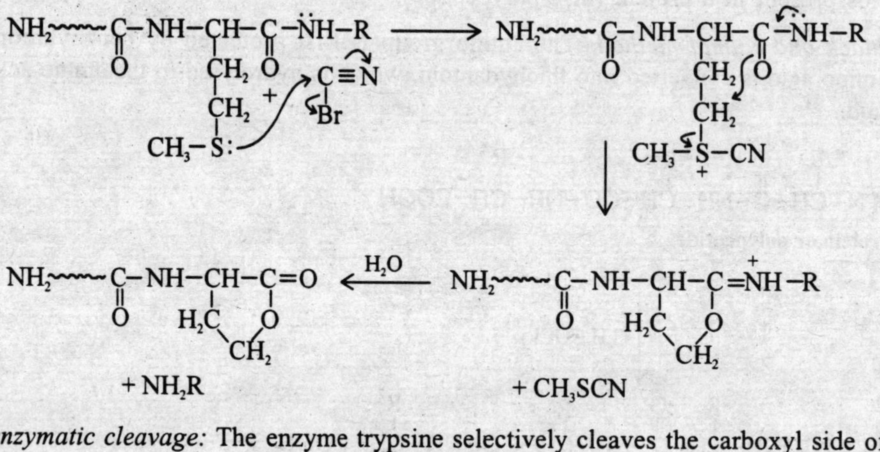

$$NH_2\text{\textasciitilde}\text{\textasciitilde}\underset{\substack{O}}{C}-NH-CH-\underset{\substack{| \\ CH_2 \\ | \\ CH_2 \\ | \\ CH_3-S:}}{\overset{O}{C}}-\overset{..}{N}H-R \quad \xrightarrow{\quad C\equiv N \quad +\quad Br \quad} \quad NH_2\text{\textasciitilde}\text{\textasciitilde}\underset{\substack{O}}{C}-NH-CH-\underset{\substack{CH_2 \\ CH_2}}{\overset{O}{C}}-NH-R$$

$$\Big\downarrow \quad CH_3-\overset{+}{S}-CN$$

$$NH_2\text{\textasciitilde}\text{\textasciitilde}\underset{\substack{O}}{C}-NH-CH-C=O \quad \xleftarrow{\;H_2O\;} \quad NH_2\text{\textasciitilde}\text{\textasciitilde}\underset{\substack{O}}{C}-NH-CH-\overset{+}{C}=NH-R$$

$$+ NH_2R \qquad\qquad + CH_3SCN$$

Selective enzymatic cleavage: The enzyme trypsine selectively cleaves the carboxyl side of the peptide bonds of the basic amino acids lysine and arginine and enzyme chymotrypsin cleaves the carboxyl side of the peptide bonds of the amino acids containing an aromatic ring viz. phenylalanine, tryptophan and tyrosine.

(vi) *Determination of amino acid sequence:* The amino acid sequence is determined by Edman automated N-terminal method wherever possible. In case the protein is relatively small, it is hydrolysed under controlled conditions to give a number of simple peptides; these are isolated and amino acid sequence determined. Finally the amino acid sequence in the proteins is deduced by the overlapping procedure.

In case a disulphide bond is present in the protein molecule, these are cleaved before determination of the sequence. The disulphide bond may be cleaved either by oxidation or by reduction. In the oxidation procedure the protein is oxidised with performic acid to give cysteic acid (the sulphonic acid). However, in the reduction procedure, the protein is treated with sodium borohydride to give the thiol, which is treated with iodoacetic acid.

$$
\begin{array}{ccc}
& \underset{|}{NH} & \underset{|}{NH} \\
& CH-CH_2-S-S-CH_2-CH \\
& \underset{|}{CO} & \underset{|}{CO}
\end{array}
$$

Protein having disulphide bond

$$[O] \diagup HCO_3H \qquad \underset{NaBH_4}{\searrow}$$

$$
\begin{array}{cccc}
& & \underset{|}{NH} & \\
2\underset{|}{CH}-CH_2SH & \xrightarrow{ICH_2COOH} & 2\ \underset{|}{CH}-CH_2-S\ CH_2COOH \\
\underset{|}{CO} & & \underset{|}{CO}
\end{array}
$$

$$
\begin{array}{c}
\underset{|}{NH} \\
2\ CH-CH_2SO_3H \\
\underset{|}{CO}
\end{array}
$$

Cysteic acid (sulphonic acid)

The primary structure of the oxidation and reduction products is determined (as described earlier) and the position of the disulphide linkage deduced.

(vii) *Mass spectrometry:* This technique has also been used to determine the primary structure of proteins or peptides. In this procedure the protein or polypeptide being non volatile, is first converted into volatile product. One procedure is to reduce the protein with lithium aluminum hydride or sodium borohydride to a more volatile protein in which the peptide bond is reduced and the terminal carboxyl group gives a polyamino alcohol. The volatile product thus obtained is subjected to electron bombardment. The fission can take place at the bonds a, b and c.

$$
\begin{array}{ccc}
\underset{|}{R} & \underset{|}{R'} & \underset{|}{R_2} \\
H_2NCH-CO-NH-CH-CO-NH-CH-COOH
\end{array}
$$

$$\downarrow LiAlH_4$$

$$
\begin{array}{ccc}
d\underset{\vdots}{\overset{R}{|}} & e\underset{\vdots}{\overset{R'}{|}} & f\underset{\vdots}{\overset{R_2}{|}} \\
H_2N-CH\vdots CH_2-NH-CH\vdots CH_2-NH-CH\vdots CH_2OH \\
\underset{a}{\downarrow} & \underset{b}{\downarrow} & c
\end{array}
$$

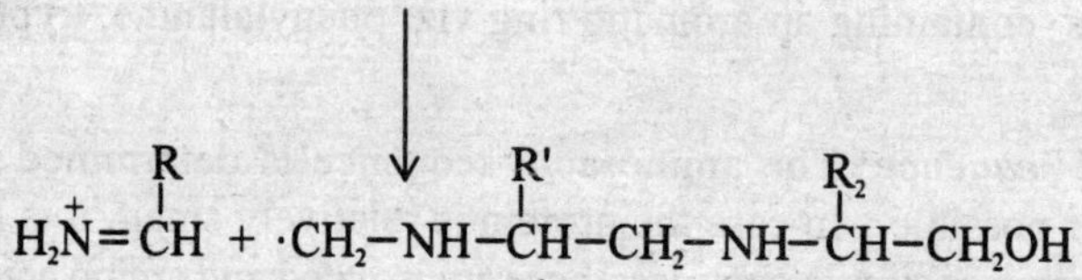

$$\text{H}_2\overset{+}{\text{N}}=\text{CH} + \cdot\text{CH}_2-\text{NH}-\overset{\text{R'}}{\underset{|}{\text{CH}}}-\text{CH}_2-\text{NH}-\overset{\text{R}_2}{\underset{|}{\text{CH}}}-\text{CH}_2\text{OH}$$

with R above the first CH.

The peak for the amine ion gives the size of R and from this, the size of (R_1+R_2) can be deduced. Similarly fission at b will give the size of R_2 and of $(R+R_1)$.

With more complicated peptides, peaks due to additional fissions at d, e and f are also taken into account.

Alternatively the methyl, ethyl and higher esters of peptides are acetylated at the free amino group and then subjected to mass spectrometry.

40.4.3.2   Secondary Structure

The secondary structure of a protein deals with the shape in which the long amino acid chain exists. In other words, it deals with the conformation of the peptide linkage present in the protein molecule. Two different conformation, viz, α-helix and β-conformation are found to exist.

The α-helix model was postulated by Pauling et.al., in 1951 on theoretical consideration and was subsequently verified experimentally. It arose due to the resonance of the peptide linkage and the hydrogen bonding between −NH− and −C=O group in the protein chain.

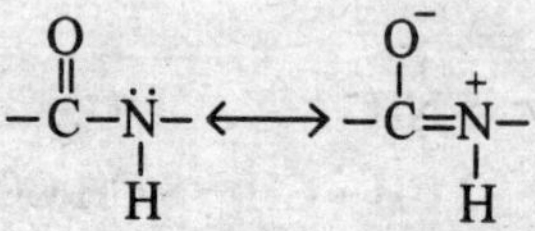

Resonance of the peptide linkage

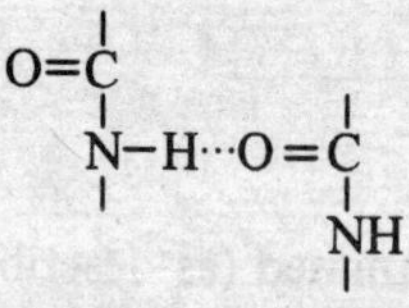

Hydrogen bonding between -NH and -C=O group

In the α-helix structure, each turn of the helix has approximately 3.7 amino acids and it is at a distance of 5.4Å from the other. It is represented as shown below. Each hydrogen bond is formed between the carbonyl group of one amino acid residue and the −NH− group of the fourth amino acid residue in the chain. This hydrogen bonding prevents free rotation and so the helix is rigid. Also, three adjacent hydrogen bonds must be broken before free rotation can occur in a segment of the helix. The α-helix may be left handed or right handed.

Moffitt (1956) on the basis of theoretical considerations deduced that the right handed helix (for L-amino acids) is more stable than the left handed helix. So the right handed helix is expected to be present is protein. This is supported by the fact that except glycine, the common amino acids are optically active and all have the L-configuration.

The β-conformation was also proposed by Pauling et al (1951).

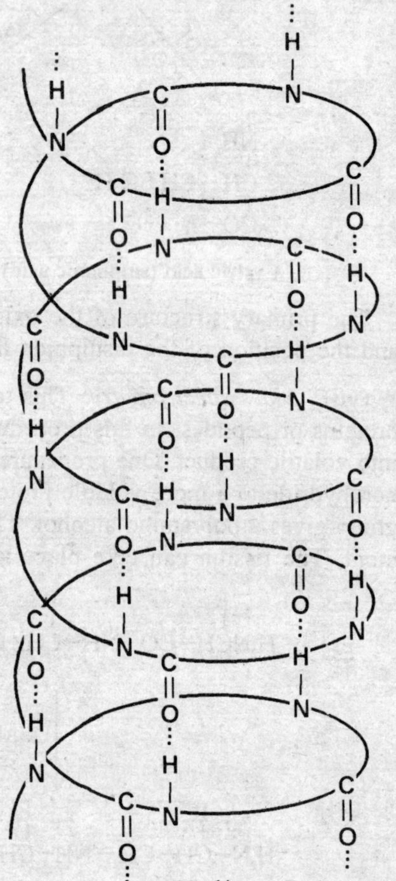

a-Helix structure

In the β-conformation or the pleated sheet, the polypeptide chain is extended and the chains are held together by intermolecular hydrogen bonds. The pleated sheets can be parallel or antiparallel. In the parallel pleated sheet, all the polypeptides run in the same direction (e.g., as in keratin). However, in anti-parallel pleated sheets the chains run in opposite direction (e.g. as in fibroin).

Parallel β-conformation   **Antiparallel β-conformation**

X-Ray Analysis: On the basis of X-ray analysis, the α-helical structure in proteins in the solid state has been established. Further, the X-ray data show two types of repeat units. One is the distance between two successive units which is 5.0-5.5Å. The second is the distance, in the direction of the helical axis, between two like atoms in the chain, which is 1.5Å. The diameter of the helix has been estimated to be about 10Å. It has been found that all polypeptide chains are not capable of forming the α-helix. The stability of this helix depends on the nature and sequence of the side chain (R groups). The amount of helical content varies from 0 to 100 % and can be estimated by means of optical rotations, optical rotatory dispersion, IR and NMR studies.

Also, the X-ray analysis has established the existence of the pleated sheet structure in solid proteins. The calculated bond distance between two CHR groups in the same side of the chain is 7.2Å. However, the experimental value is 7.0Å. This difference is attributed to the crowding caused by the side chain (R), which prevents the chain from being fully extended.

40.4.3.3 Tertiary Structure

It refers to the three dimensional structure of proteins. In this the coiling of the long peptide chain takes place with or without a helical region into the final structure. The three dimensional structure describes the overall spatial arrangement of the polypeptide chain and gives an exact account of the molecular shape of most of the small and medium sized proteins. Two major molecular shapes are found, viz., fibrous and globular. The fibrous proteins have a large helical content and are essentially rigid molecules with rod-like shape. The globular proteins, on the other hand, have a polypeptide chain consisting partly of helical sections and folded about the random cut to give a spherical shape.

The tertiary structures of proteins have been best elucidated by X-ray analysis, viscosity measurements, diffusion, light-scattering, ultracentrifuge method, and electron microscopy.

40.4.3.4   Quaternary Structure

It is concerned with those proteins which contain subunits and is a description of the arrangement and ways in which the subunits are held together. Both the fibrous and globular proteins consist of only one polypeptide chain. If several chains are present, the globular protein is said to be oligomeric. The individual chains are known as protomers or subunits, which may or may not be identical. These subunits are held together by hydrogen bonds and can be separated by dissolving in water containing urea. Thus, mycoglobin consists of a single polypeptide chain, which contains about eight straight subunits (α-helices) and are folded in an irregular manner at the random-coil section. On the other hand, haemoglobin contains four subunits, two identical α-chains and two identical β-chains. Each of these subunits have a tertiary structure similar to that of mycoglobin.

The fibrous proteins are made up of bundles of fibrils which are packed together in a parallel arrangement.

40.5   PROBLEMS

1. What are amino acids? Explain neutral, acidic and basic amino acids with one example each.
2. Explain the terms essential and non-essential amino acids. Which type of amino acids (α, β or γ) are obtained from the proteins?
3. Explain the term D and L amino acids. What do you understand by relative configuration?
4. Discuss Gabriel's Phthalimide, Strecker and malonic ester synthesis of amino acids.
5. How is phenylalanine and tyrosine obtained by the azlactone synthesis?
6. How are amino acids obtained from proteins?
7. Explain the terms: Zwitterionic nature of amino acids, Isoelectric point and Electrophoresis.
8. What products are obtained by the reaction of glycine with nitrous acid, formaldehyde, hydriodic acid and phenylisocyanate?
9. Discuss Van slyke method for determination of free amino groups in proteins and amino acids.
10. What are peptides? How are they classified? Explain by giving examples how are they named?
11. Discuss the general principles of the synthesis of polypeptides.
12. Give Fischers, Bergmann and Sheehan's method for the Synthesis of peptides.
13. How are peptides obtained by the anhydride method?
14. Discuss solid-phase peptide synthesis. What are the especial advantages of this method?
15. What are proteins? How are they classified?
16. Discuss the amphoteric nature of proteins. What is denaturation and renaturation?
17. Give color reactions of proteins. What are the uses of proteins?
18. Discuss the primary, secondary, tertiary and quaternary structure of protein.
19. How is the N-terminal amino acid and C-terminal amino acid determined in proteins? Discuss the chemical and enzymic methods.
20. How is amino acid sequence determined in proteins?
21. Discuss the α-helix structure of proteins.
22. Write a note on β-conformation of proteins.

41

Ureides, Pyrimidines and Purines

41.1 UREIDES

The acyl derivatives of urea are known as *ureides*. The ureides are classified as:

(i) *Simple Ureides or Open Chain Ureides:* The ureides obtained from the acetylation of urea are classified as simple or open chain ureides. For examples acetylurea, diacetylurea etc.

(ii) *Cyclic Ureides:* These are derived from dicarboxylic acids. Oxalic acid reacts with urea to give oxalurea (parabanic acid) and malonic acid reacts with urea to give malonylurea (barbituric acid).

Urea **Oxalic acid** **Oxalurea (Parabanic acid)**

Urea **Malonic acid** **Malonyl urea (Barbituric acid)**

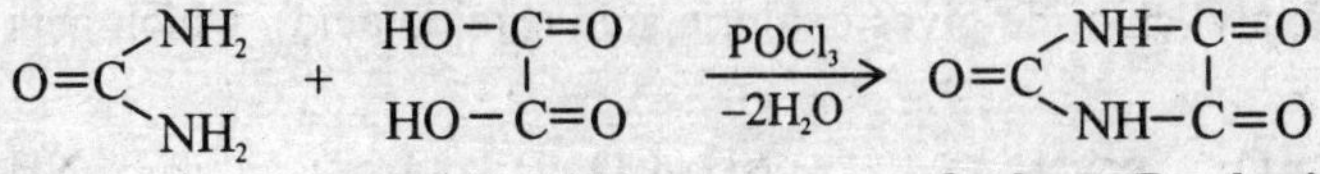

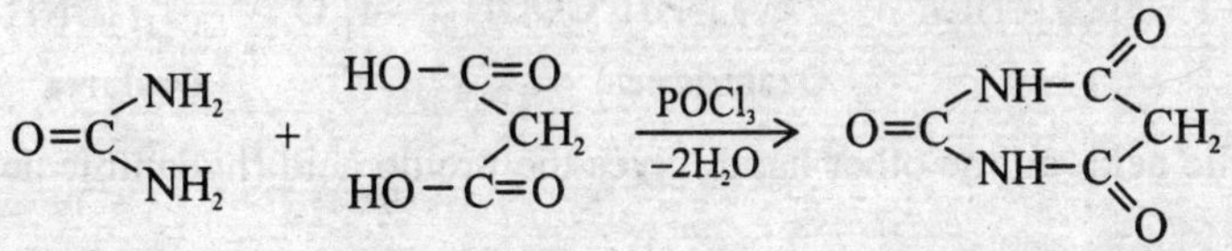

As seen, the cyclic ureides can be five membered (oxalurea) or six membered (malonyl urea).

41.1.1 Simple Ureides or Open Chain Ureides

Preparation

These are prepared by the action of acyl halide, ester or acid anhydride of monocarboxylic acid on urea. Urea reacts with acetyl chloride to give acetylurea.

$$H_2NCONH_2 + CH_3COCl \longrightarrow \underset{\textbf{Acetylurea}}{H_2NCONHCOCH_3} + HCl$$

Acetylurea further reacts with acetyl chloride to give diacetylurea.

$$\underset{\textbf{Acetylurea}}{H_2NCONHCOCH_3} + CH_3COCl \longrightarrow \underset{\textbf{Diacetylurea}}{CH_3CONHCONHCOCH_3}$$

The diacetylurea is also obtained by the action of carbonyl chloride on acid amides.

Carbonyl chloride **Acetamide** **Diacetylurea**

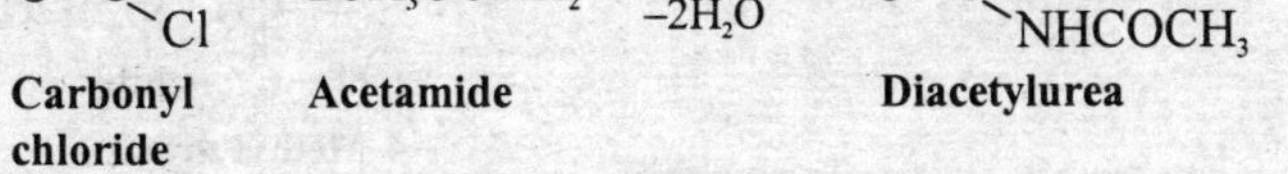

Properties

The simple or open chain ureides contain the amide group and are similar to amides in properties. For example, on treatment with dilute alkali they undergo hydrolysis to give sod. salt of acid.

$$CH_3CONHCONH_2 + NaOH \longrightarrow CH_3COONa + H_2NCONH_2$$

Acetylurea   **Sod. acetate**   **Urea**

Under the reaction conditions, urea further gets hydrolysed into ammonia and carbon dioxide.

$$H_2NCONH_2 + H_2O \longrightarrow 2NH_3 + CO_2$$

Urea

41.1.2 Cyclic Ureides

Preparation

(i) Condensation of dibasic acid chlorides with urea forms cyclic ureides. Oxalyl chloride reacts with urea to give oxalurea.

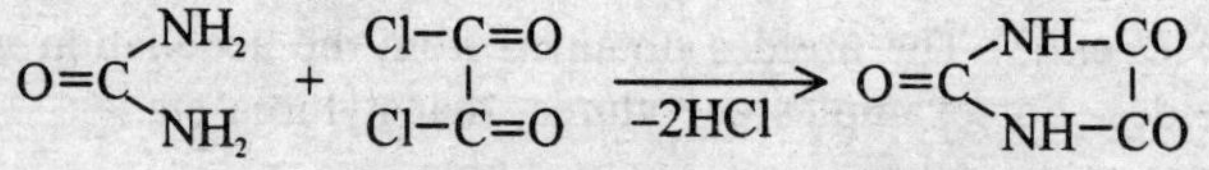

Urea   **Oxalyl chloride**   **Oxalurea**
(**Parabanic acid**)

Use of oxalic acid in place of oxalyl chloride gives oxaluric acid (ureido acid), which gets converted into oxalurea.

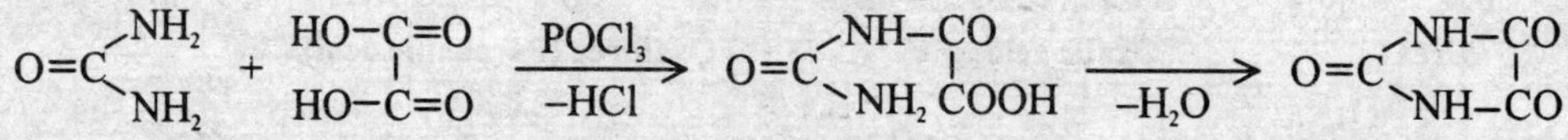

                                   Oxaluric acid   **Oxalurea**

(ii) Condensation of glycollic acid, on the other hand, gives the ureido acid (hydantoic acid), which gives the ureide (hydantoin).

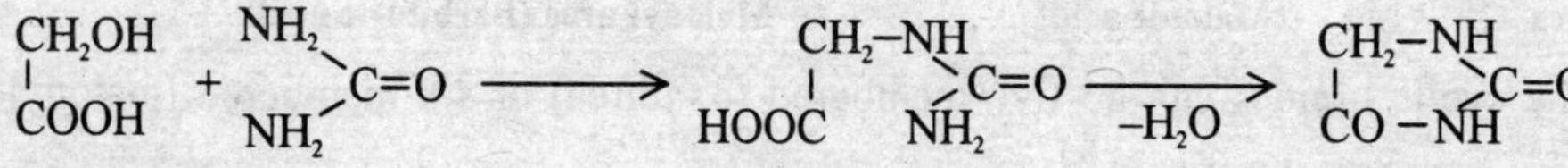

Glycollic   **Urea**   **Hydantoic acid**   **Hydantoin**
acid   **(an ureide)**

(iii) Condensation of a β-ketoester (ethylacetoacetate) with urea gives the cyclic ureides (uracil).

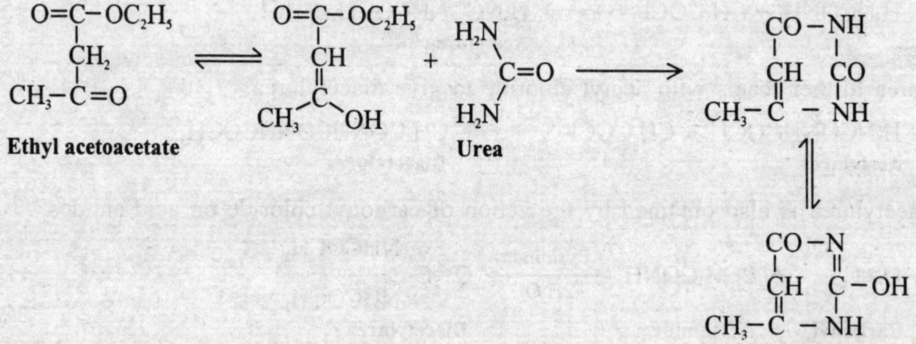

Ethyl acetoacetate   **Urea**

4-Methyl uracil

Properties

The cyclic ureides contain an imide linkage, $-\overset{\overset{\displaystyle OH}{|}}{C}=N-CO$. As such the H of $-NH$ is very reactive and can be easily replaced by metals and subsequently by alkyl group. Due to the presence of imide group, the ureides are soluble in alkalis.

Besides the imide linkage, the cyclic ureides also contain the $-CO-NH-CO-$ linkage. This renders the cyclic ureides labile due to the conversion of $-CO-NH-CO-$ into $-\overset{\overset{\displaystyle OH}{|}}{C}=N-CO-$; In this form it reacts with phosphorus pentachloride to give a chloro derivative.

Further, the presence of $-NH-CO-NH-$ group renders the cyclic ureides easily hydrolysable into acid and urea.

41.1.3 Individual Members

41.1.3.1 Allophanic Acid [H₂NCONHCOOH]

This is a simple open chain ureide and is derived from urea by reaction with formyl chloride. Allophanic acid is not stable in free state but is known as its ester.

Preparation

(i) By the condensation of urea with ethyl chloroformate.

$$H_2NCONH_2 + ClCOOC_2H_5 \xrightarrow{-HCl} H_2NCONHCOOC_2H_5$$

Urea   **Ethyl chloro-**   **Ethyl allophanate**
          formate

(ii) From cyanic acid by reaction with ethyl carbamate (urethane) or with alcohol.

$$HNCO + H_2NCOOC_2H_5 \longrightarrow H_2NCONHCOOC_2H_5$$

Cyanic acid   Ethyl carbamate   **Ethyl allophanate**

$$C_2H_5OH + HNCO \longrightarrow H_2NCOOC_2H_5 \xrightarrow{HNCO} H_2NCONHCOOC_2H_5$$

Ethyl alcohol   Cyanic acid   **Ethyl carbamate**   **Ethyl allophanate**

41.1.3.2 Parabanic Acid/Oxalurea

It is a five membered cyclic ureide and is obtained by the reaction of urea with oxalic acid or oxalyl chloride (see Section 41.1.2).

Properties

(i) *Acidic Nature:* Ureides are acidic in nature due to the presence of the imide linkage. The hydrogen of the imide group can be replaced by metals. Oxalurea forms disilver salt on treatment with alcoholic solution of silver nitrate and ammonia. The acidic nature is also explained due to the equilibrium shown below.

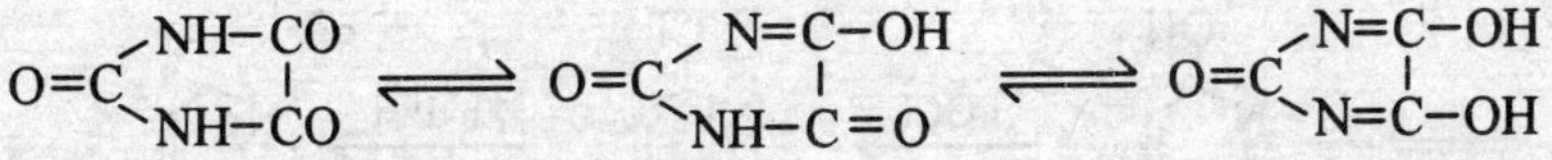

$$O=C\overset{\displaystyle \diagup NH-CO}{\underset{\displaystyle \diagdown NH-CO}{}} \rightleftharpoons O=C\overset{\displaystyle \diagup N=C-OH}{\underset{\displaystyle \diagdown NH-C=O}{}} \rightleftharpoons O=C\overset{\displaystyle \diagup N=C-OH}{\underset{\displaystyle \diagdown N=C-OH}{}}$$

Oxalurea

(ii) *Hydrolysis:* Due to the presence of the $-NH-CO-NH$ group, it is easily hydrolysed into urea and oxalic acid.

Oxalurea **Oxaluric acid** **Urea** **Oxalic acid**

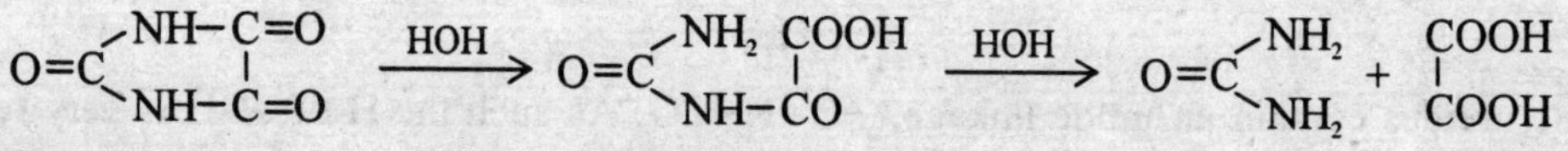

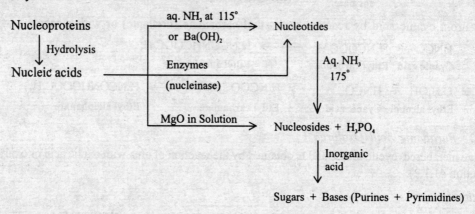

(iii) *Reduction:* On electrolytic reduction, one of the carbonyl group is reduced to give hydantoin.

Oxalurea **Hydantoin**

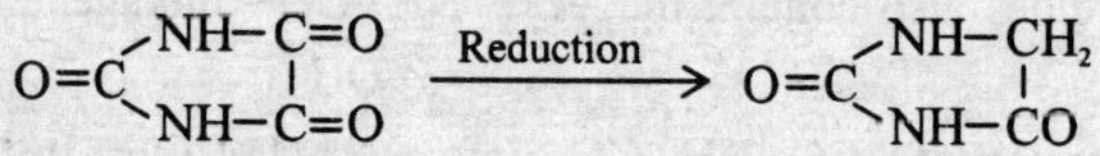

Hydantoin can be converted into allantoin by bromination followed by reaction with urea.

Hydantoin **Bromo deriv.** **Allantoin**

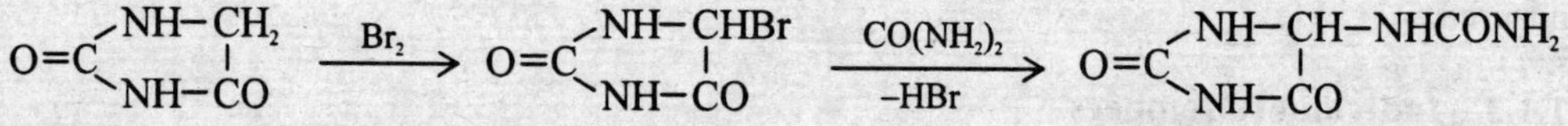

41.2 PYRIMIDINES

Pyrimidines and purines are the bases obtained by hydrolysis of nucleic acids. The nucleic acids are colorless solids having C, H, O, N and P and are made up of three units: bases (purines and pyrimidines), sugar and phosphoric acid. The nucleic acids are obtained by the hydrolysis of nucleoproteins which is a class of conjugated proteins. The nucleic acid part is the prosthetic group and the protein part consists of protamins and histones. In other words protamins and histones are basic compounds which form salt like compounds (such as nucleoproteins) with nucleic acid. The following chart shows the nature of products obtained by the hydrolysis of nucleic acids under different conditions.

An elementary treatment of nucleic acids is given in Section 41.4.

41.2.1 Pyrimidine

Pyrimidine is obtained from barbituric acid as shown below:

Barbituric acid **Pyrimidine**

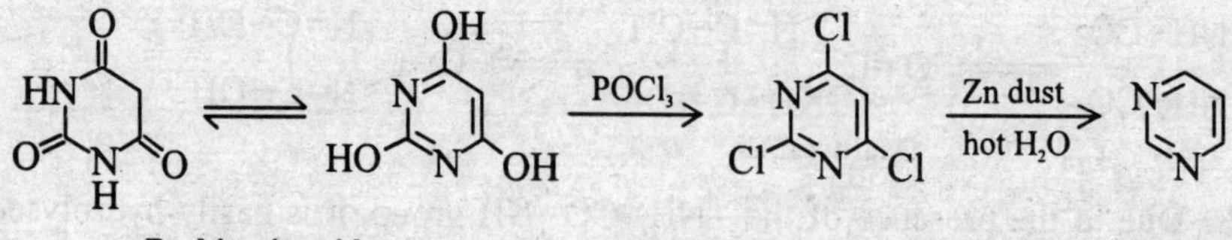

It can also be prepared from 1,1,3,3-tetraethoxypropane and formamide.

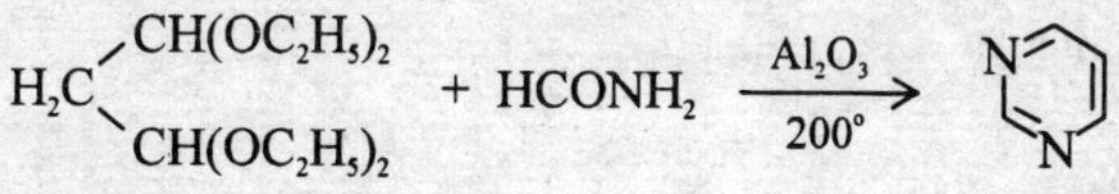

1, 1, 3, 3-Tetraethoxy-propane   **Pyrimidine**

The numbering in pyrimidine is done as per IUPAC system.

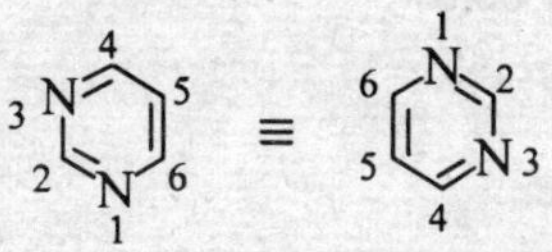

Pyrimidine is a resonance hybrid of the following structures.

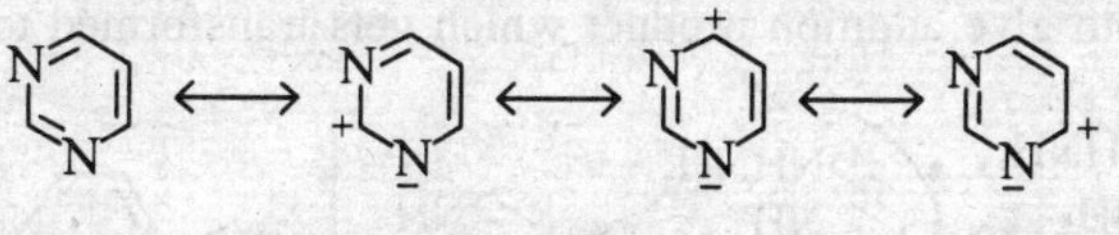

As seen, the pyrimidine ring is deactivated and position 5 has greatest electron density. Therefore it is expected that the attack by electrophilic reagents is difficult. However, the attack by nucleophile at positions 2, 4 and 6 is facilitated. The chlorine atoms at positions 2, 4 or 6 may be readily replaced by hydroxy or amino groups. Further, an amino group at position 2 or 4 can be conveniently replaced by hydroxy group on boiling with water.

The introduction of $-OH$ or $-NH_2$ group at positions 2, 4 or 6 diminishes the aromatic character (see barbituric acid, uracil).

Reactions of Pyrimidines

1. *Basic Character:* Pyrimidine is much weaker base (pK_{a1} 1.30) than pyridine (pK_a 8.8). The reason is that, the cation formed by protonation at one nitrogen gets destabilised due to election withdrawing inductive effect of second nitrogen. The pK_{a2} value for pyrimidine is -6.9 showing that second protonation in the molecule is very difficult.

In substituted pyrimidine, electron releasing groups like methyl and methoxy increases the basic character (for example, pK_a value for 4-methoxy-6-methylpyrimidine is 3.65 and for 4-methylpyrimidine is only 1.98).

2. *Electrophilic Substitution Reactions:* Pyrimidine do not undergo nitration and sulfonation but is halogenated at position 5 when treated with bromine in nitrobenzene.

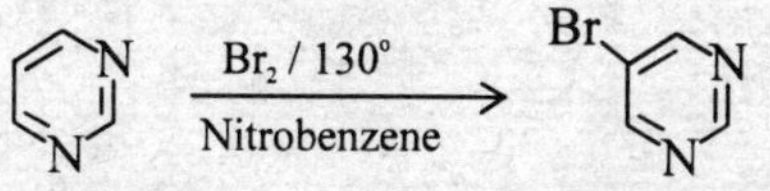

Pyrimidine   **5-Bromopyrimidine**

Pyrimidines form monoquaternary salts on reaction with alkyl halides at room temperature. Diquaternary salts, however, are obtained when treated with more reactive reagents like triethyloxonium borofluoride.

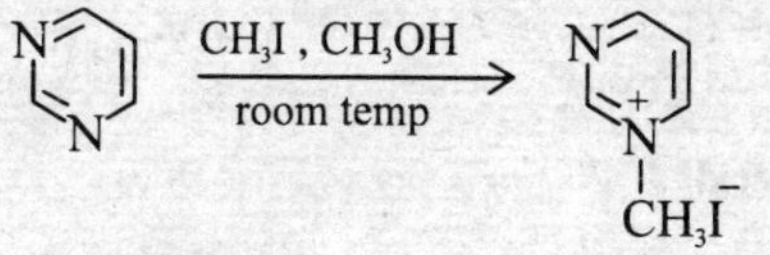

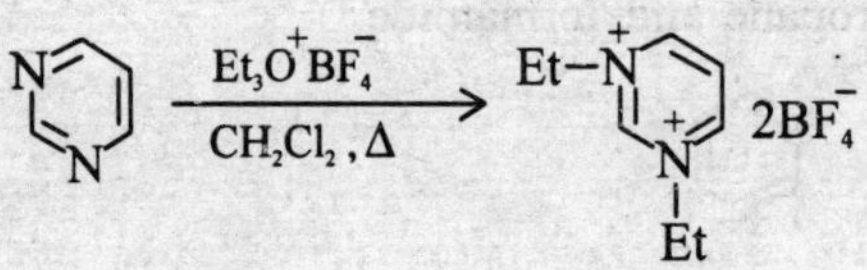

3.   *Reaction with Nucleophilic Reagents:* The pyrimidines easily undergo nucleophilic addition reactions.

(i)   Aqueous sodium hydroxide adds onto pyrimidine as shown below.

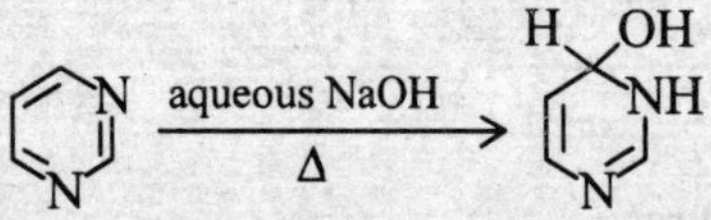

(ii)   Pyrimidine reacts with hydrazine to give addition product which gets transformed to pyrazole.

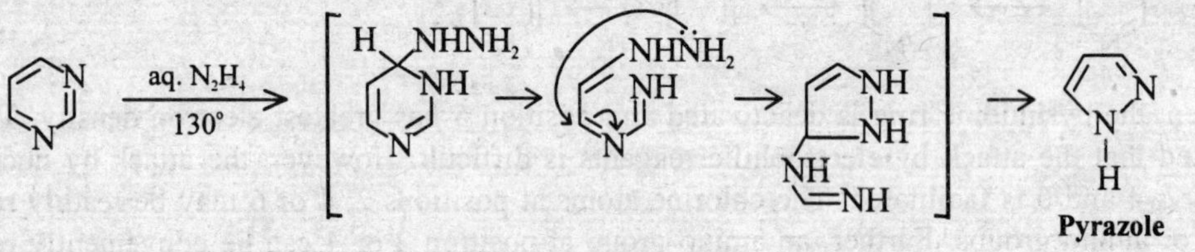

Pyrazole

(iii)   It reacts with Grignard reagent and alkyl or aryllithium compounds to give addition products at position 4. The aromatization of addition product is done by oxidation.

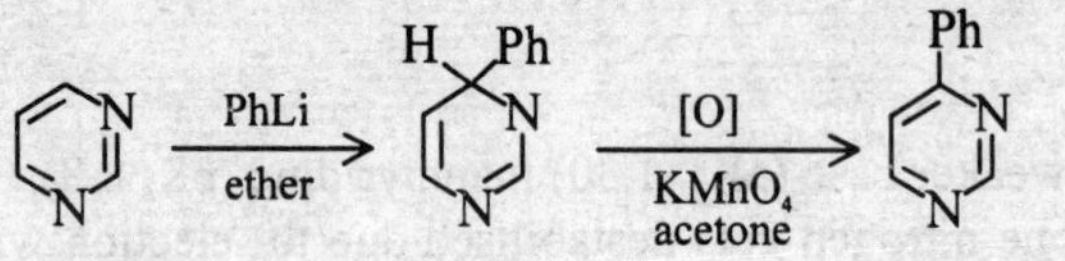

4-Phenylpyrimidine

(iv)   Pyrimidine reacts with sodamide in liquid ammonia to give the sodium salt, the aromatization of which is not possible.

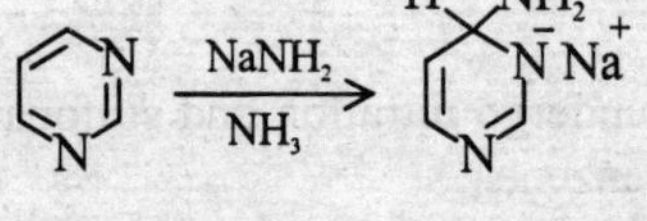

41.2.2   Pyrimidine Derivative: Barbituric Acid

Preparation

(I)   It is prepared by the condensation of urea with malonic acid in presence of phosphoryl chloride.

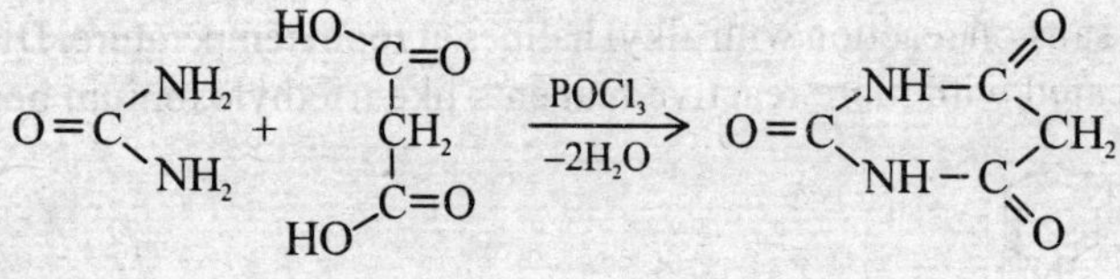

Urea   Malonic acid   Barbituric acid

However, a more convenient synthesis utilises diethylmalonate and urea in presence of sodium ethoxide.

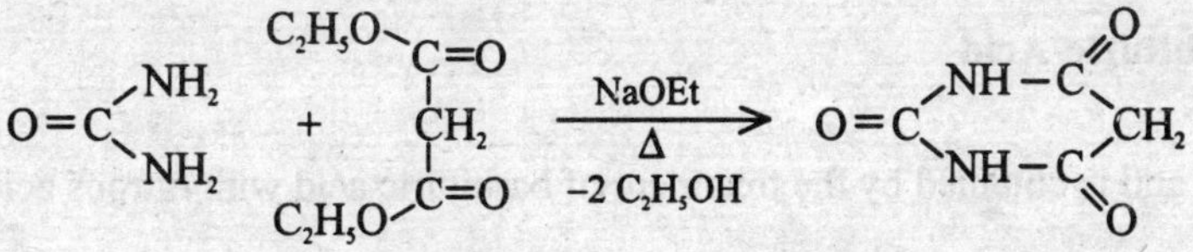

| Urea | Diethyl malonate | Barbituric acid |

(II) Alkyl or dialkyl derivatives of barbituric acid are prepared by using mono or disubstituted diethylmalonate.

Properties

Barbituric acid is a crystalline solid, m.p. 253° and is sparingly soluble in water. Due to enolisation it exhibits lactam-lactins tautomerism. It is regarded as 2,4,6-trihydroxypyrimidine.

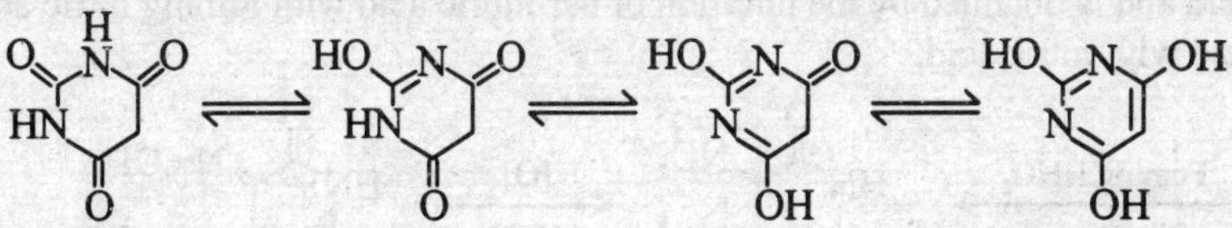

I   II   III   IV, 2,4,6-trihydroxy-pyrimidine

It is difficult to acylate hydroxypyrimidines (II, III or IV). This suggests that structure I is more probable. Further barbituric acid forms oximino derivative with nitrous acid indicating the presence of an active methylene group. Also, methylation of hydroxypyrimidine with methyl iodide in presence of alkali gives the N-methyl derivative indicating the presence of imido group. X-ray analysis indicates that structure I is the predominant form (in the solid state) though the molecule is planer.

Barbituric acid lacks aromatic character and can be hydrolysed to urea and malonic acid.

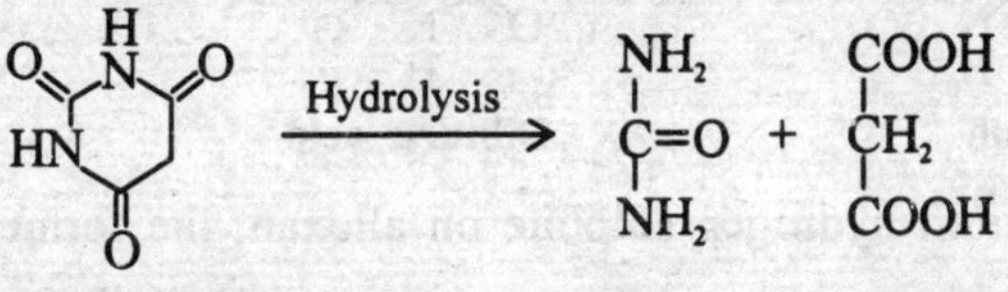

Barbituric acid   Urea   Malonic acid

Barbituric acid can be easily alkylated at position 5. Some of these derivatives are medicinally important compounds. Thus, 5,5-diethylbarbituric acid, commonly known as veronal is a hypnotic and is obtained as follows.

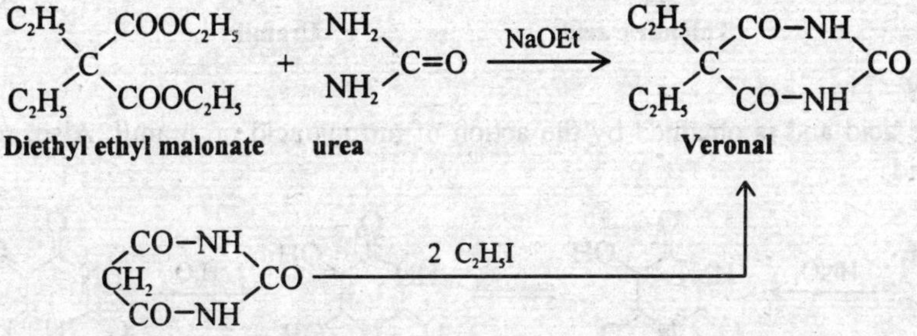

Diethyl ethyl malonate   urea   Veronal

Barbituric acid

Another useful alkyl substituted barbituric acid is Luminal, 5-ethyl-5-phenylbarbituric acid.

41.2.3   Derivatives of Barbituric Acid

41.2.3.1   Violuric Acid

It is 5-oximinobarbituric acid and is obtained by the treatment of barbituric acid with nitrous acid.

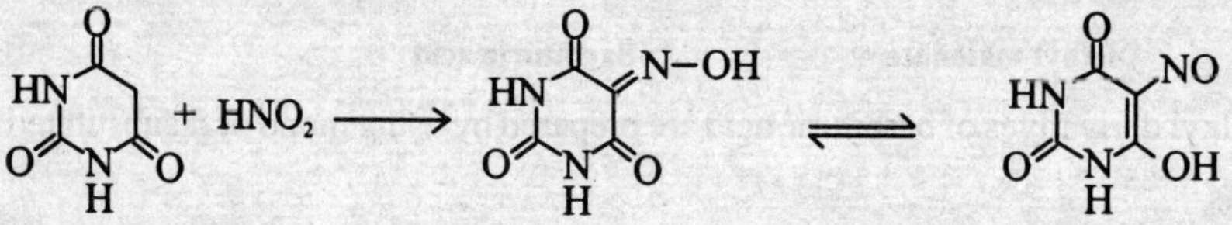

Barbituric acid   **Violuric acid**

41.2.3.2   Diluturic Acid

It is 5-nitrobarbituric acid and is obtained by the nitration of barbituric acid with fuming nitric acid or by the oxidation of violuric acid with nitric acid.

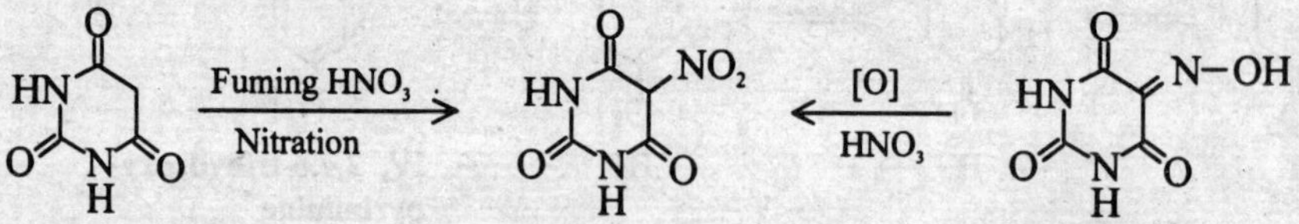

Barbituric acid   **Diluturic acid**   **Violuric acid**

41.2.3.3   Uramil

It is 5-aminobarbituric acid and is obtained by reduction of diluturic acid or violuric acid.

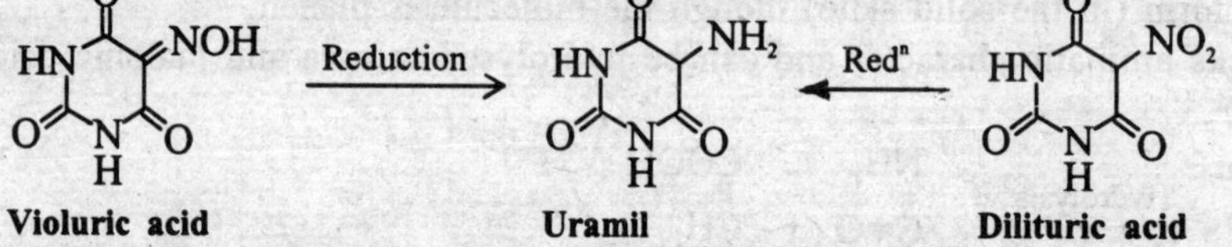

Violuric acid   **Uramil**   **Diluturic acid**

It is also obtained by the action of ammonium hydrogen sulphite on alloxan; the formed thionuric acid is boiled with water to give uramil.

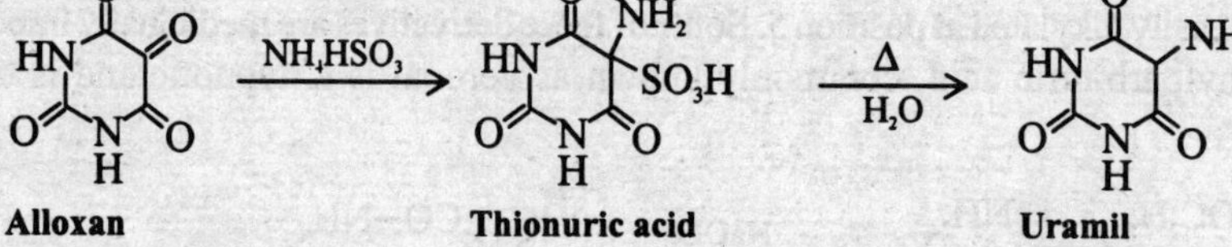

Alloxan   **Thionuric acid**   **Uramil**

41.2.3.4   Dialuric Acid

It is 5-hydroxybarbituric acid and is obtained by the action of nitrous acid on uramil. Also, reduction of alloxan gives dialuric acid.

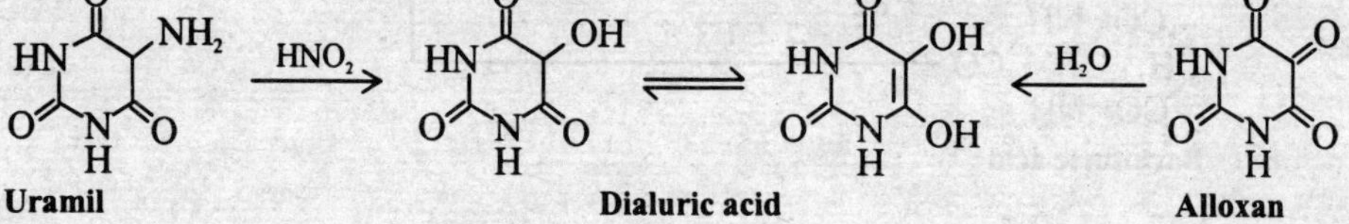

Uramil   **Dialuric acid**   **Alloxan**

41.2.3.5   Alloxan/Mesoxalyl Urea

It is obtained from barbituric acid as given below.

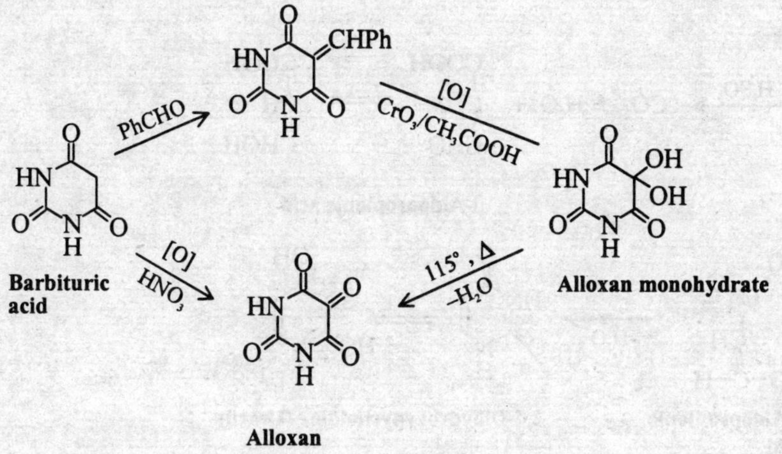

Alloxan

Alloxan is acidic in nature and forms salts. The central carbonyl group is very reactive and behaves like a ketone. Thus, it forms a bisulphite derivative and an oxime (violuric acid) on treatment with sodium bisulphite and hydroxyl amine respectively. Further, on reduction it gives dialuric acid and alloxantin under different conditions.

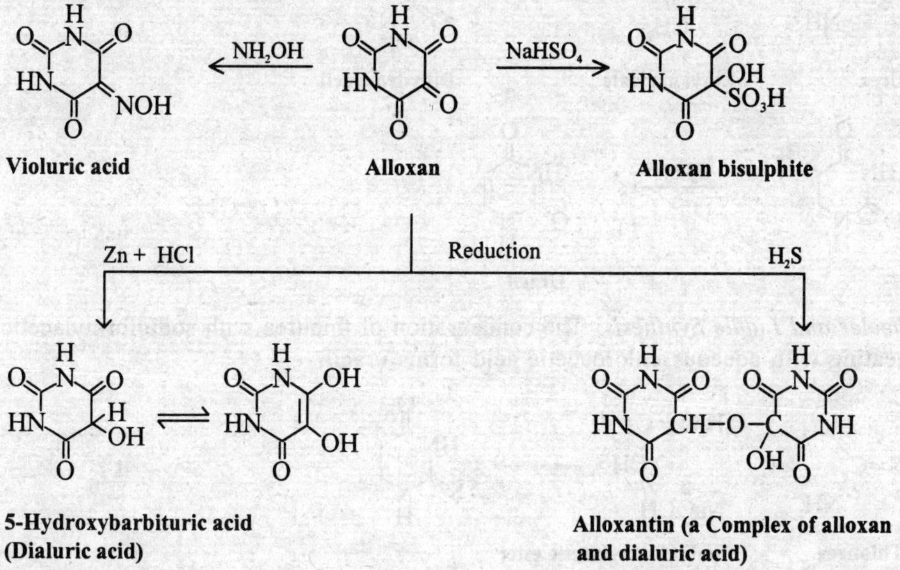

Violuric acid   **Alloxan**   **Alloxan bisulphite**

5-Hydroxybarbituric acid   **Alloxantin (a Complex of alloxan**
(Dialuric acid)   **and dialuric acid)**

Alloxan can also be converted into uramil (see Section 41.2.3.3). The structure elucidation of alloxan has been discussed in Section 41.3.3.1.

41.2.3.6   Uracil/2,4-Dihydroxypyrimidine

Uracil was earlier designated as 2,6-dihydroxypyrimidine. However as per IUPAC system it should be 2,4-dihydroxypyrimidine.

It is a hydrolytic product of nucleic acids. However, it is prepared by the following methods.

(i) *Davidsons Synthesis:* It consists in treating a mixture of urea and malic acid with fuming sulphuric acid.

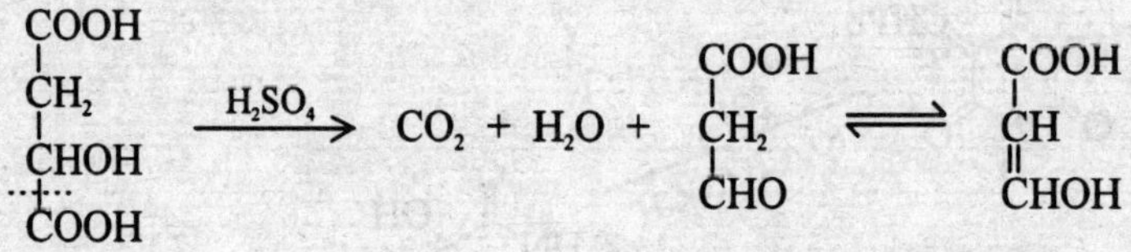

Malic acid   β-Aldopropionic acid

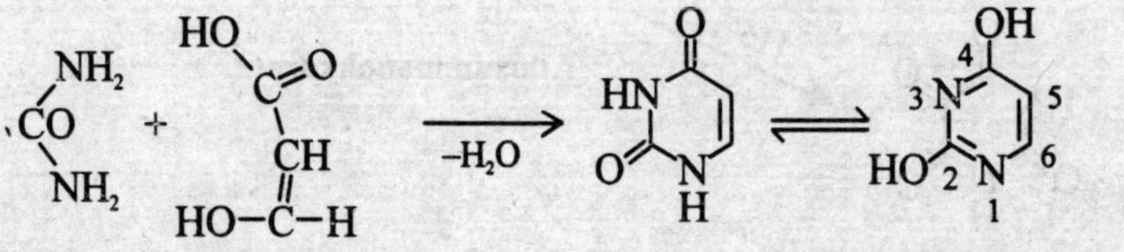

Urea   β-Aldopropionic   2,4-Dihydroxypyrimidine (Uracil)
          acid

(ii) *Fischer and Roeder's Synthesis:* It consists in the condensation of urea and ethyl acrylate. The formed dihydrouracil is brominated and the bromo product heated in pyridine to give uracil.

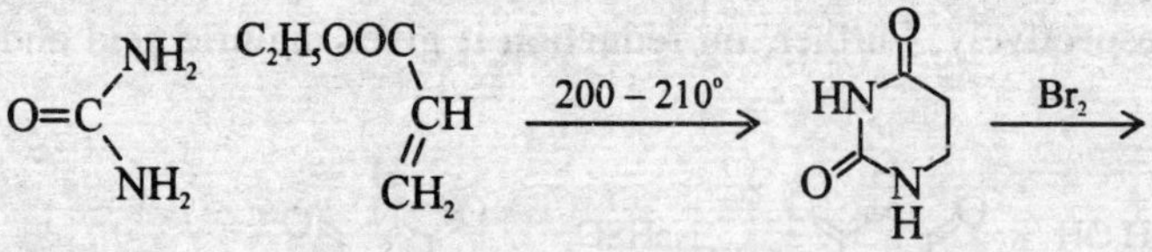

Urea   Ethyl acrylate   Dihydrouracil

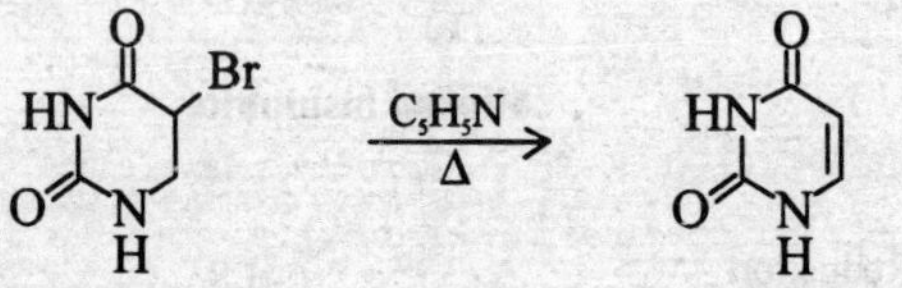

Uracil

(iii) *Wheeler and Liddle Synthesis:* The condensation of thiourea with sodioformylacetic ester followed by heating with aqueous chloroacetic acid forms uracil.

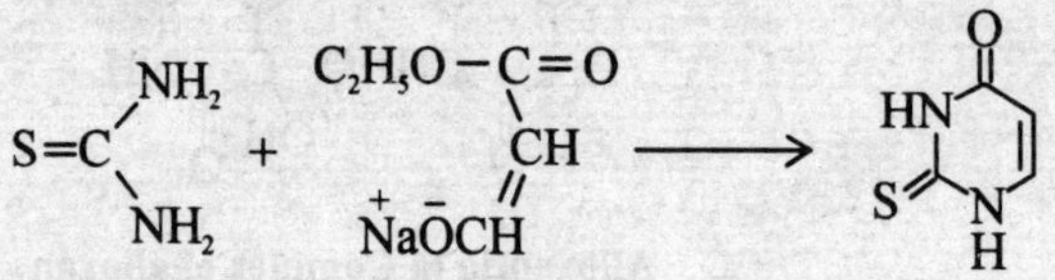

Thiourea   Sodio formylacetic ester

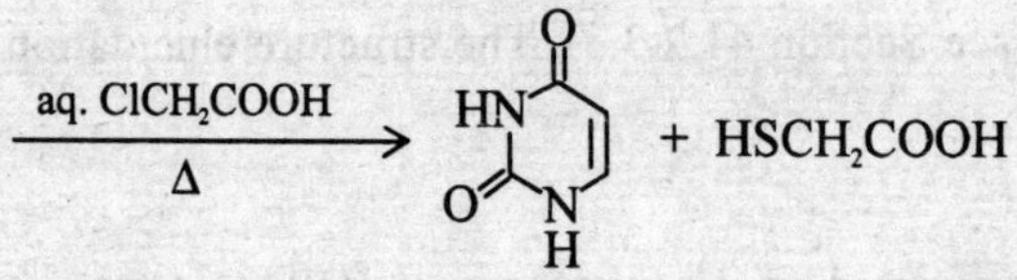

Uracil

41.2.3.7 Thymine

Thymine is 5-methyluracil or 2,4-dihydroxy-5- methylpyrimidine. Like uracil, thymine is also a hydrolytic product of nucleic acids. It is obtained by methods similar to those for the synthesis of uracil.

(i) *Fischer and Roeder Synthesis:* It consists in the condensation of urea wih ethyl methacrylate.

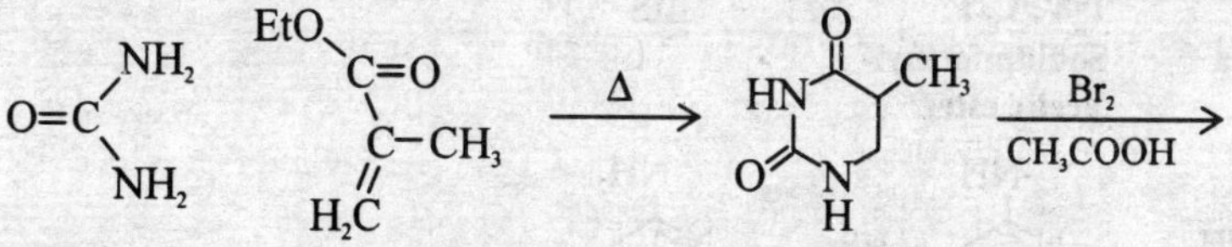

Urea Ethyl methacrylate 5-Methyldihydrouracil

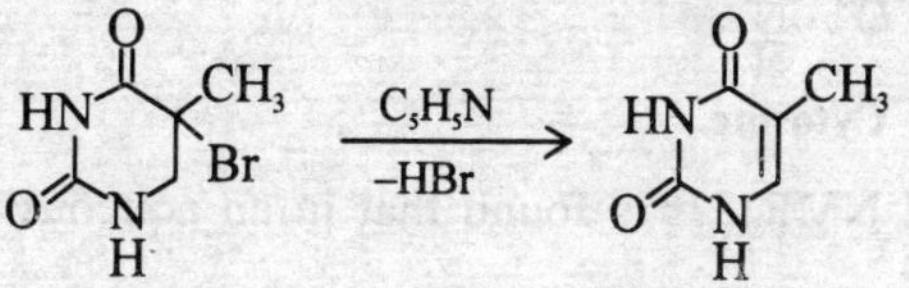

Thymine

(ii) *Wheeler and Liddle Synthesis:* In this synthesis, thiourea is condensed with sodiumformylpropionic ester.

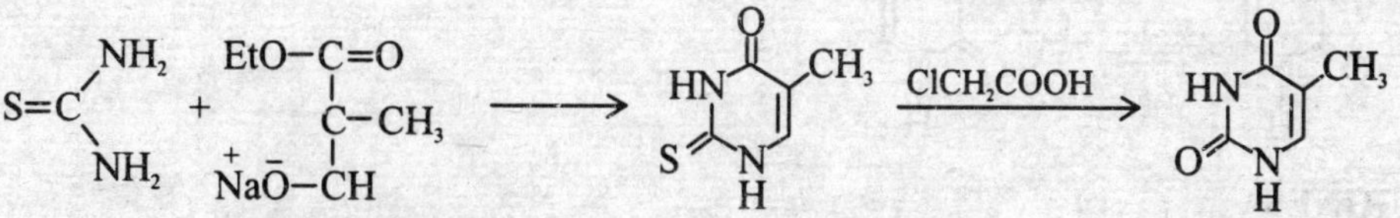

Thiourea Sodiumformyl-propionic ester Thymine

(iii) *Bergmann Synthesis*

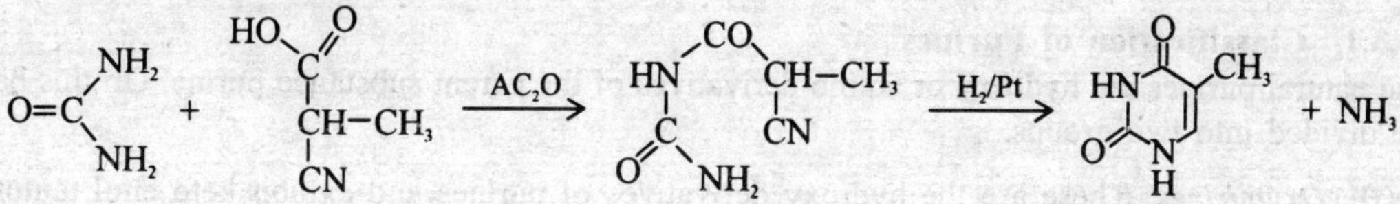

Urea Thymine

41.2.3.8 Cytosine

It is 4-amino-2-hydroxypyrimidine and like uracil and thymine, is also a hydrolytic product of nucleic acids.

Cytosine is obtained from malondialdehyde acetal by reaction with hydroxylamine. The formed β-ethoxyacrylonitrile is reacted with urea in presence of a base to give cytosine (Tarsio et.al.).

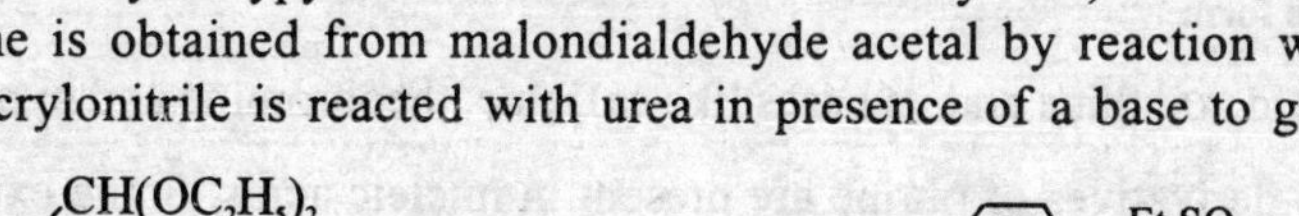

Malondialdehyde acetal Hydroxylamine hydrochloride Isoxazole

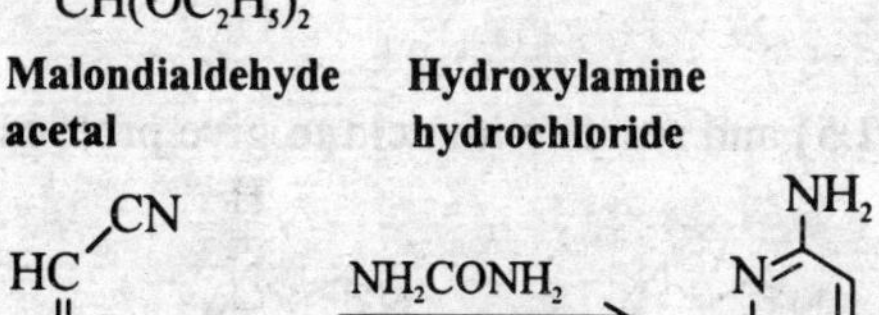

β-Ethoxyacrylonitrile Cytosine

It is also obtained from S-ethylisothiourea and sodiumformylacetic ester (Wheeler and Johnson).

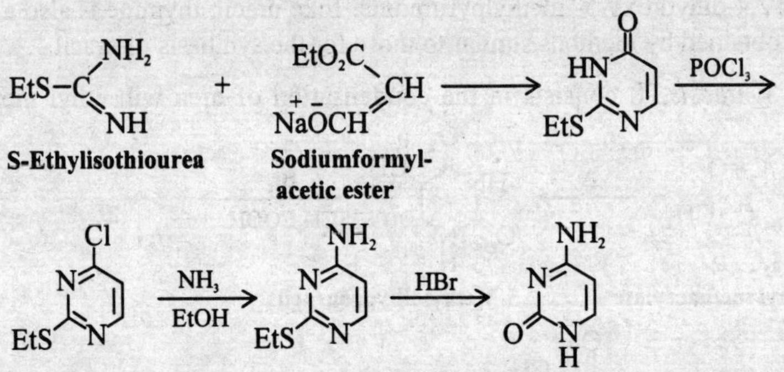

S-Ethylisothiourea   **Sodiumformyl-**
                            acetic ester

Cytosine

On the basis of spectroscopic studies (UV and NMR). It is found that in an aqueous solution of cytosine the following two species are present.

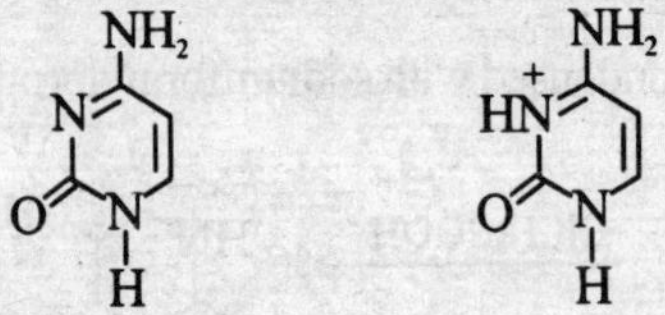

41.3   PURINES

Purines are derived from parent substance 'Purine' and are, therefore, called Purines. The structure contains a pyrimidine ring fused to an imidazole ring. These are considered to be built up from two molecules of urea and one molecule of dicarboxylic acid.

41.3.1   Classification of Purines

The natural purines are hydroxy or amino derivatives of the parent substance purine. On this basis they are divided into two groups.

(i)   *Oxypurines:* These are the hydroxy derivatives of purines and exhibit keto enol tautomerism.

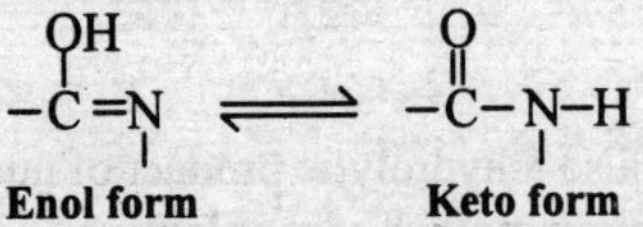

Enol form   **Keto form**

Examples of this type are uric acid, xanthine and its bases like caffeine, theobromine and hypoxanthine.

(ii)   *Aminopurines:* The amino derivatives of purine are present in nucleic acids. Some examples are adenine, guanine etc.

Reactions of Purines and Alkylpurines

(I)   *Basic Character:* Purine is a weak base (pK_a 2.5) and reacts with acid to give protonated purine.

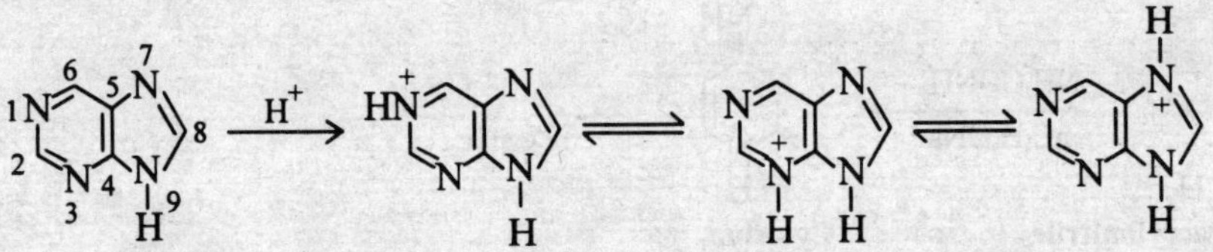

(II)   *Acidic Character:* Purine is a stronger acid (pK$_a$ 8.9) than phenol (pK$_a$ 9.98), imidazole (pK$_a$ 14.2) and benzimidazole (pK$_a$ 12.3). The acidic character is due to the stabilisation of anion by delocalisation.

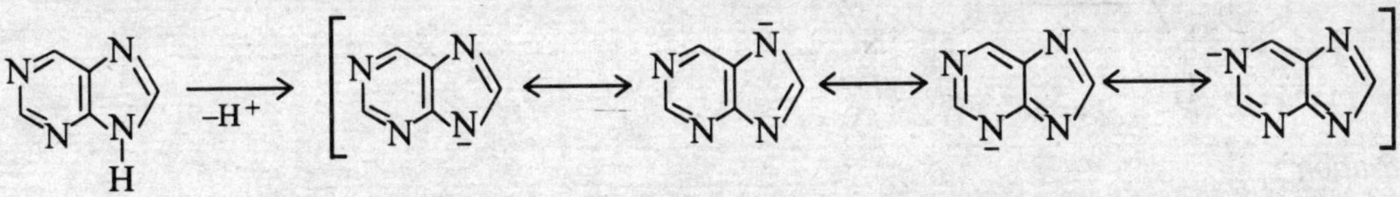

(III)   *Alkylation:* Alkylation of purine with dimethyl sulphate or methyl iodide produces 9-alkylpurines. The reaction proceeds via the formation of intermediate purine anion.

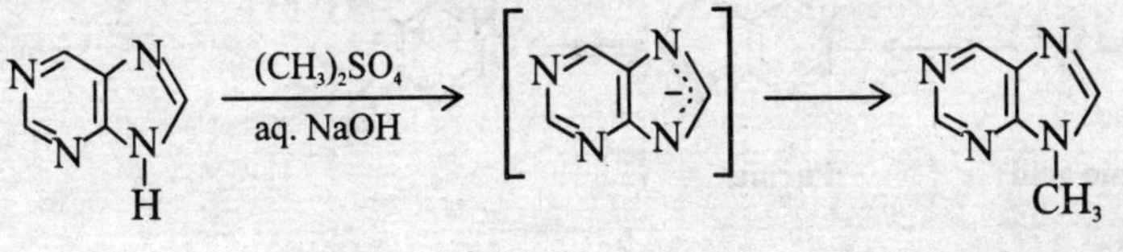

9-Methylpurine

(IV)   *Reactions with Nucleophiles:* The nucleophilic addition of hydroxide ion takes place at C-8 in 9-methylpurine and finally produces 5-amino-4-methylaminopyrimidine.

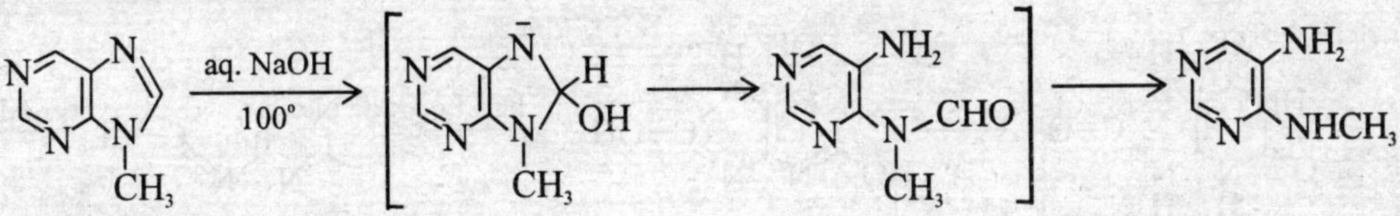

9-Methylpurine   **5-Amino-4-methylamino-pyrimidine**

Purine itself is stable in hot aqueous sodium hydroxide due to the formation of stable purine anion.

(V)   *Oxidation:* Oxidation with peracetic acid produces N-oxide at N-1.

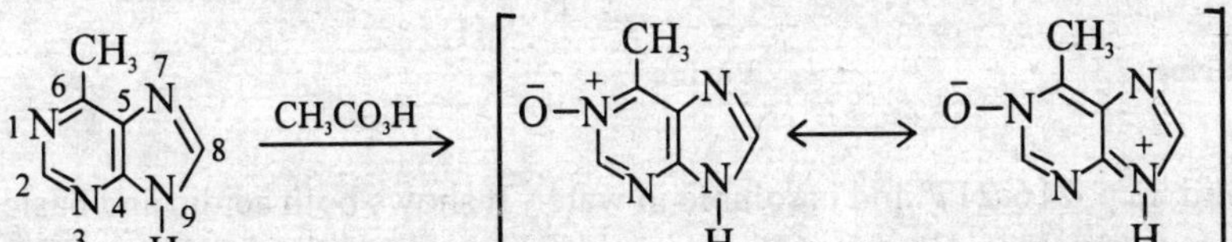

Purines do not undergo electrophilic substitution reactions like nitration, sulphonation or halogenation. It is also resistant to hydrogenation in presence of palladium and charcoal.

41.3.2   Purine

Purine, the parent substance of the group of compounds 'Purines' was earlier represented as:

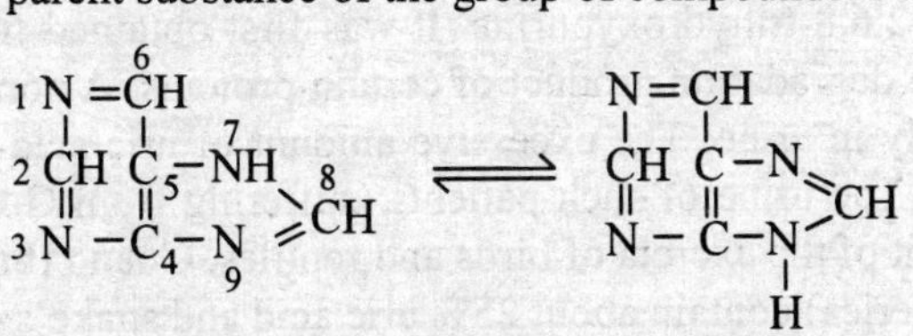

Purine

However, purine is now represented as:

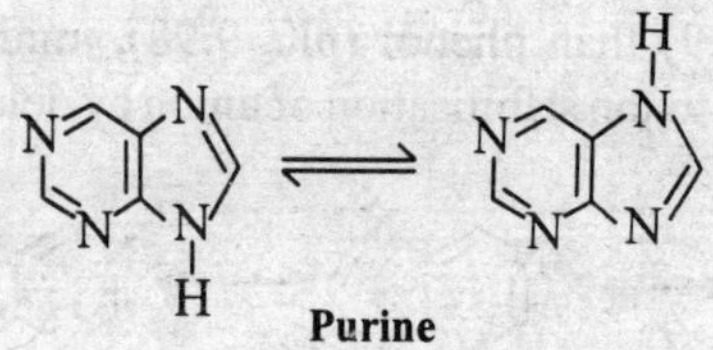

Purine

Preparation

(I) Condensation of 4,5-diaminopyrimidine with formic acid yields purine (Albert and Broun).

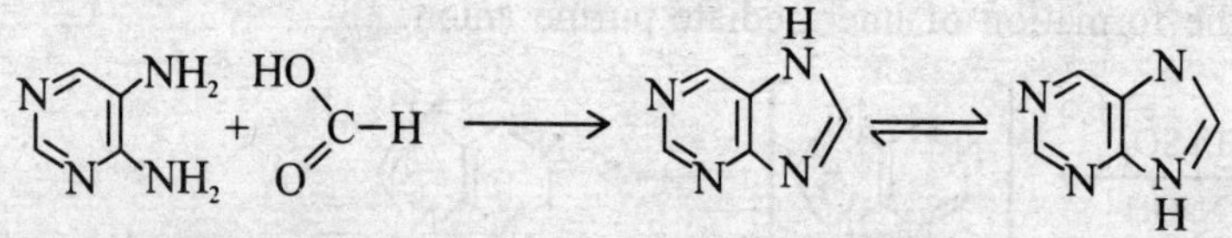

4, 5-Diamino-   Formic acid   Purine
pyrimidine

(II) Uric acid on treatment with phosphorus oxychloride gives 2,6,8-trichloropurine, which on treatment with hydriodic acid affords 2,6-diiodopurine. The iodine is replaced with hydrogen when treated with zinc dust to give purine.

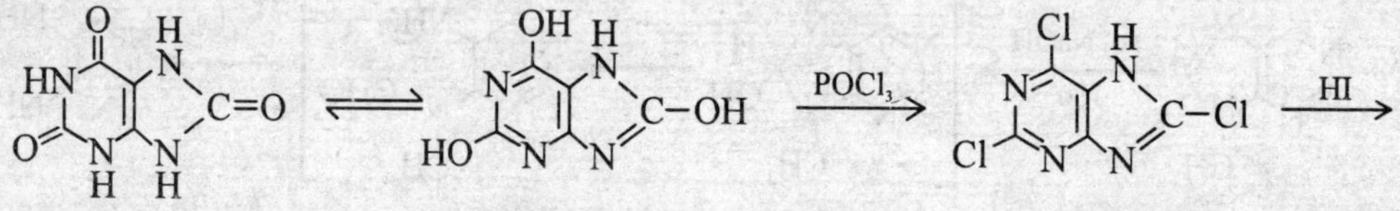

Uric acid   (enolic form)   2, 6, 8-Trichloropurine

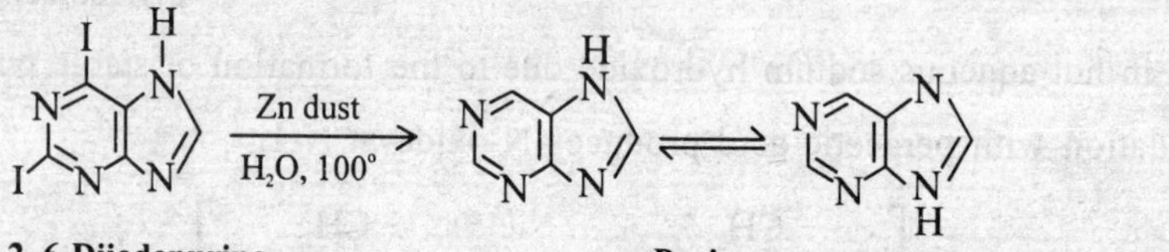

2, 6-Diiodopurine   Purine

Properties

Purine is a colorless solid, m.p. 216-217° and is soluble in water. It shows both acidic and basic properties. Purine itself does not occur in nature. However, its derivatives like uric acid, xanthine, caffeine, adenine, guanine etc are of natural origin.

41.3.3 Individual Members

41.3.3.1 Uric Acid

Uric acid is the most important purine and is 2,6,8-trihydroxypurine. It was first obtained from human urine (Scheele and Bergmann, 1776) and is the degradation product of certain proteins. A normal human being excretes about 200-400mg uric acid daily in urine. The excessive amount of uric acid sometimes gets accumulated in bladder, kidney or joints. The urine of such patients (suffering from Gout) contain more uric acid. Uric acid is the chief constituent of the excreta of birds and reptiles. Guano (birds excreta from island near the western coast of south America) contain about 25% uric acid and snake's excretment contain 90% in the form of ammonium ureate.

(i) *From Guano:* Guano is the birds excretment from islands in south America and contains 25% uric acid in the form of ammonium salt. The dry excretement is powdered and boiled with sodium hydroxide solution until the evolution of ammonia ceases. The clear hot solution contains sodium ureate, which is acidified and the solution cooled to give uric acid crystals.

(ii) *From Human Urine:* The human urine is concentrated and treated with hydrochloric acid. The uric acid crystallises out on cooling.

Structure

1. The molecular formula of uric acid as determined by analytical methods is $C_5H_4N_4O_3$.

2. *Oxidation with Nitric Acid:* Uric acid on oxidation with dilute nitric acid gives one molecule of alloxan and one molecule of urea.

$$C_5H_4N_4O_3 \xrightarrow[\text{HNO}_3]{[O]} C_4H_2N_2O_4 + H_2NCONH_2$$

   Uric acid   **Alloxan**   **Urea**

3. *Structure of Alloxan:* The structure of uric acid depends on the structure of alloxan, which is determined as follows. The hydrolysis of alloxan with sodium hydroxide gives urea and mesoxalic acid in equimolar proportion.

$$C_4H_2N_2O_4 + 2H_2O \xrightarrow{\text{NaOH}} H_2NCONH_2 + HOOC-CO-COOH$$

   Alloxan   **Urea**   **Mesoxalic acid**

Since alloxan does not contain a free carboxyl group or amino group, it must be mesoxalurea. This is confirmed by synthesis of alloxan from urea and mesoxalic acid (Liebig and Wohler, 1839).

   Urea   **Mesoxalic acid**   **Alloxan**

Alloxan has also been prepared from barbituric acid (see Section 41.2.3.5).

It has already been stated that alloxan gives 5- oximine (violuric acid) on treatment with hydroxylamine. Further on reduction (Zn + HCl) it give 5-hydroxybarbituric acid known as dialuric acid. However, on reduction with hydrogen sulphide, alloxantin is obtained (see Section 41.2.3.5).

4. *Oxidation with Lead Dioxide:* On oxidation with aqueous suspension of lead dioxide (Liebig and Wohler,1838), the uric acid gives allantoin and carbon dioxide. These products are obtained in quantitative yields by oxidation with alkaline potassium permanganate (Behrend,1904).

$$C_5H_4N_4O_3 + H_2O + [O] \xrightarrow[\text{or}]{\text{PbO}_2} C_4H_6N_4O_3 + CO_2$$

   Uric acid   Alk. KMnO₄   **Allantoin**

5. *Structure of Allantoin:* The structure of uric acid also depends on the structure of allantoin. Its structure is assigned as follows:

(i) Hydrolysis of allantoin with alkali gives two molecules of urea and one molecule of glyoxalic acid.

$$C_4H_6N_4O_3 + 2H_2O \xrightarrow{\text{Hyrolysis}} 2\,NH_2CONH_2 + OHC-COOH$$

   Allantoin   **Urea**   **Glyoxalic acid**

On the basis of the results of hydrolysis it is believed that allantoin is diureide of glyoxalic acid.

(ii)   Oxidation of allantoin with nitric acid gives urea and parabanic acid (1:1).

$$C_4H_6N_4O_3 + [O] \xrightarrow{HNO_3} H_2NCONH_2 + C_3H_2N_2O_3$$

Allantoin   **Urea**   **Parabanic acid**

Parabanic acid does not contain a free carboxyl or amino group. Further, on hydrolysis it gives urea and oxalic acid. On this basis, parabonic acid is oxalurea. This is confirmed by its synthesis.

$$O=C\begin{array}{c} NH_2 \\ NH_2 \end{array} + \begin{array}{c} Cl \\ | \\ C=O \\ | \\ C=O \\ | \\ Cl \end{array} \xrightarrow[-2H_2O]{POCl_3} O=C\begin{array}{c} NH-CO \\ | \\ NH-CO \end{array}$$

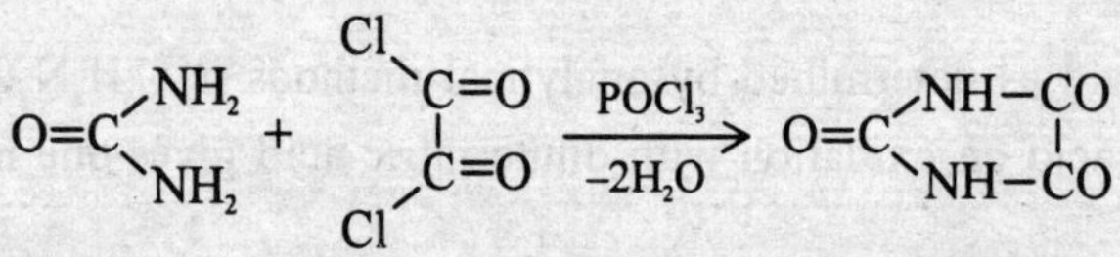

Urea   **Oxalyl chloride**   **Oxalurea (Parabanic acid)**

On the basis of the above facts it is clear that in allantoin, parabanic acid is joined to a molecule of urea.

(iii)   *Reduction* of allantoin yields urea and hydantoin.

$$C_4H_6N_4O_3 + 2H \xrightarrow{HI} H_2NCONH_2 + C_3H_4N_2O_2$$

Allantoin   **Urea**   **Hydantoin**

Further, hydantoin on controlled hydrolysis yields hydantoic acid (ureidoacetic acid), which on further hydrolysis gives glycine, ammonia and carbon dioxide. On the basis of these results, it is believed that hydantoin is glycollylurea (ureide of glycollic acid).

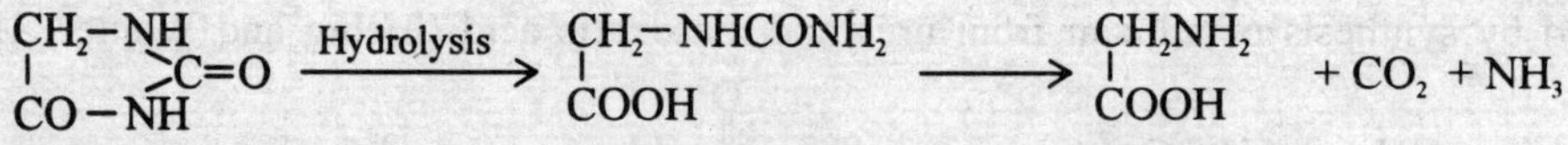

$$\begin{array}{c} CH_2-NH \\ | \quad\quad >C=O \\ CO-NH \end{array} \xrightarrow{Hydrolysis} \begin{array}{c} CH_2-NHCONH_2 \\ | \\ COOH \end{array} \longrightarrow \begin{array}{c} CH_2NH_2 \\ | \\ COOH \end{array} + CO_2 + NH_3$$

Hydantoin   **Hydantoic acid**   **Glycine**

The above structure of hydantoin has been confirmed by its synthesis.

$$\begin{array}{c} CH_2NH_2 \\ | \\ COOH \end{array} \xrightarrow{\begin{array}{c} KCNO \\ CH_3COOH \end{array}} \begin{array}{c} CH_2NHCONH_2 \\ | \\ COOH \end{array} \xrightarrow[HCl]{\Delta} \begin{array}{c} CH_2-NH \\ | \quad\quad >C=O \\ CO-NH \end{array}$$

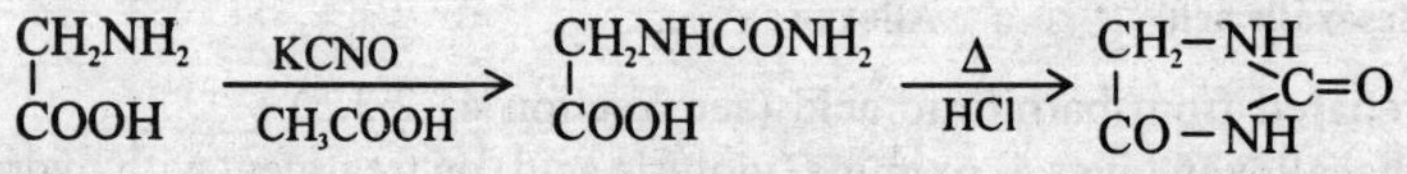

Glycine   **Hydantoic acid**   **Hydantoin**

On the basis of the above evidences, allantoin is given the structure as shown below. This structure explains the products obtained by oxidation, reduction and hydrolysis of allantoin.

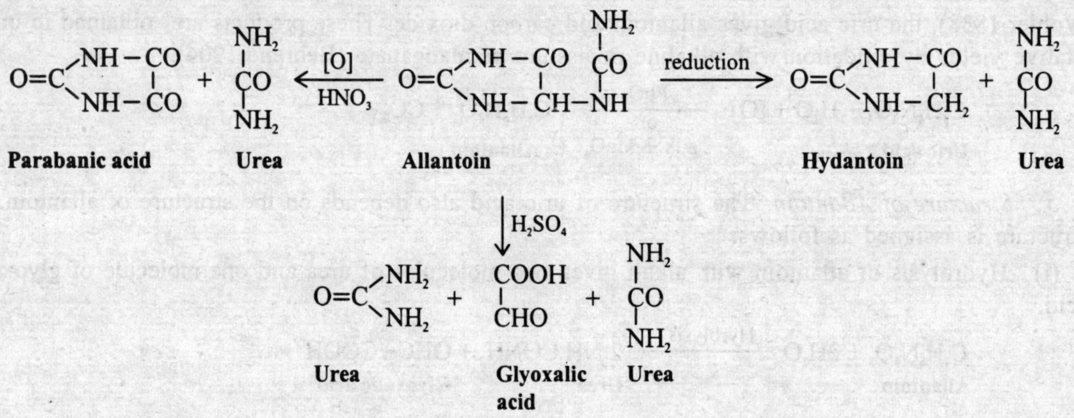

$$O=C\begin{array}{c} NH-CO \\ | \\ NH-CO \end{array} + \begin{array}{c} NH_2 \\ | \\ CO \\ | \\ NH_2 \end{array} \xleftarrow[HNO_3]{[O]} O=C\begin{array}{c} NH-CO \quad CO \\ | \quad\quad\quad\quad | \\ NH-CH-NH \end{array}^{NH_2} \xrightarrow{reduction} O=C\begin{array}{c} NH-CO \\ | \\ NH-CH_2 \end{array} + \begin{array}{c} NH_2 \\ | \\ CO \\ | \\ NH_2 \end{array}$$

Parabanic acid   **Urea**   **Allantoin**   **Hydantoin**   **Urea**

$$\downarrow H_2SO_4$$

$$O=C\begin{array}{c} NH_2 \\ NH_2 \end{array} + \begin{array}{c} COOH \\ | \\ CHO \end{array} + \begin{array}{c} NH_2 \\ | \\ CO \\ | \\ NH_2 \end{array}$$

Urea   **Glyoxalic**   **Urea**
                                    acid

The above structure of allantoin has finally been confirmed by its synthesis from glyoxalic acid and urea (Grimacix, 1876).

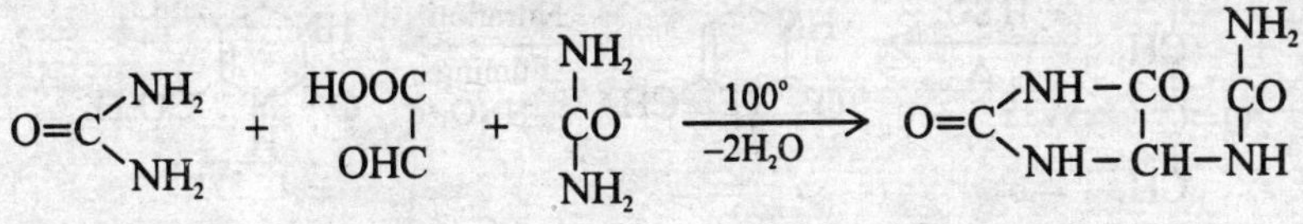

| Urea | Glyoxalic acid | Urea | | Allantoin |

As seen, the above structure of allantoin contains a chiral centre. Therefore two optically active forms are possible. Both forms have been isolated and are found to racemise rapidly in solution; the racemisation occurs via enolisation.

As seen above, oxidation of uric acid with lead dioxide gives allantoin and a molecule of carbon dioxide. This implies that uric acid contains allantoin with one additional cabron atom. Following two structures were proposed for uric acid. Both accounted for the fact known at the time the structures were proposed.

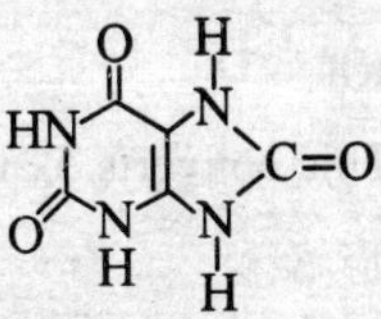

Medicus formula
(1875)
(Unsymmetrical)

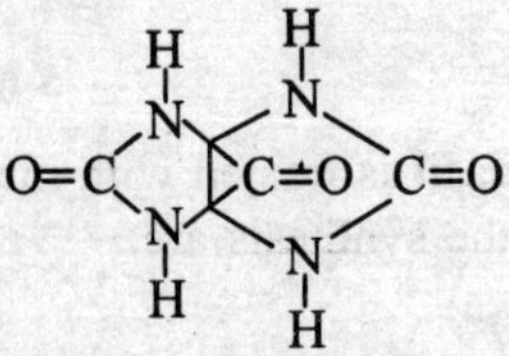

Fittig Formula
(1878)
(Symmetrical)

Uric acid was assigned the Medicus formula on the basis of the following evidences. Two isomeric monomethyluric acids were prepared by Fisher (1884). One of these gave methylalloxan and urea on oxidation with nitric acid and the other gave alloxan and methylurea. Fittig's formula is symmetrical and can give one monomethyluric acid and hence this structure is untenable. However, Medicus formula may give two isomeric monomethyl derivatives; having methyl group at position 1 or 3 in the pyrimidine nucleus and will give methylalloxan and urea. A methyl group at position 7 or 9 in the imidazole nucleus will produce alloxan and urea. It was subsequently shown (Fischer) that the two monomethyluric acids contained the methyl group at the 3 and 9 positions. In fact Fischer prepared all the possible four monomethyl, six dimethyl and four trimethyl derivatives of uric acid. This provided strong support to the Medicus formula.

Synthesis
Final confirmation of the structure is made on the basis of its synthesis.

(i)   The first synthesis was carried out by Behrand and Roosen (1888) starting from urea and ethyl acetoacetate. The various steps involved are given as follows.

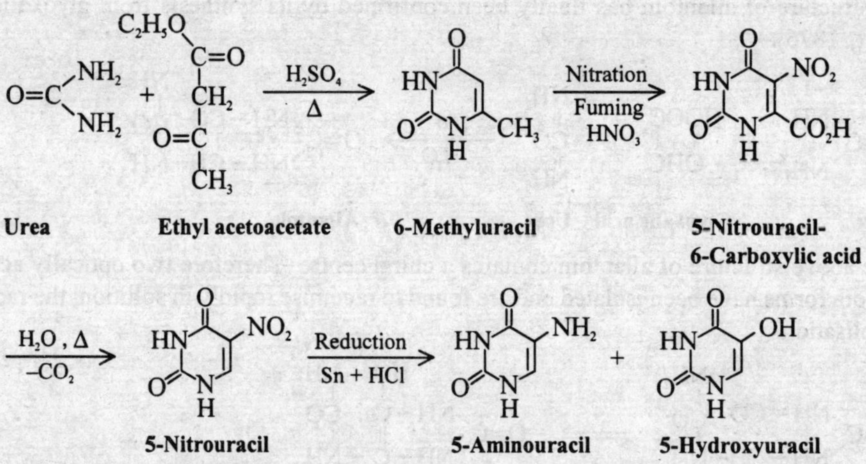

Urea   Ethyl acetoacetate   6-Methyluracil   5-Nitrouracil-
                                                                6-Carboxylic acid

5-Nitrouracil   5-Aminouracil   5-Hydroxyuracil

The formation of 5-hydroxyuracil is explained as follows:

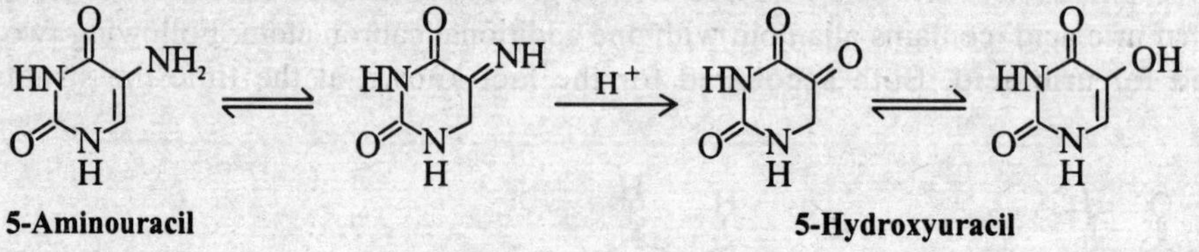

5-Aminouracil   5-Hydroxyuracil

The mixture of 5-amino and 5-hydroxyuracil is treated with nitrouacid. This converts 5-aminouracil into 5-hydroxyuracil. Subsequent steps in the synthesis are:

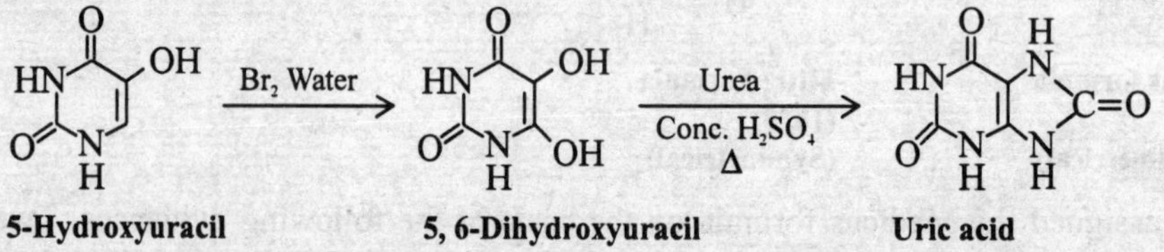

5-Hydroxyuracil   5, 6-Dihydroxyuracil   Uric acid

(ii) *Baeyer-Fischer Synthesis:* Baeyer (1863) was successful only in obtaining pseudouric acid. He could not achieve the final step of dehydration by the usual dehydrating agents. However, Fischer carried out the final dehydration by heating with hydrochloric acid (20%).

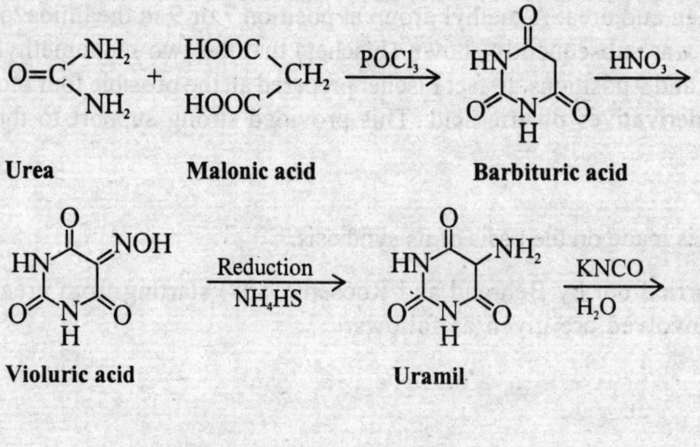

Urea   Malonic acid   Barbituric acid

Violuric acid   Uramil

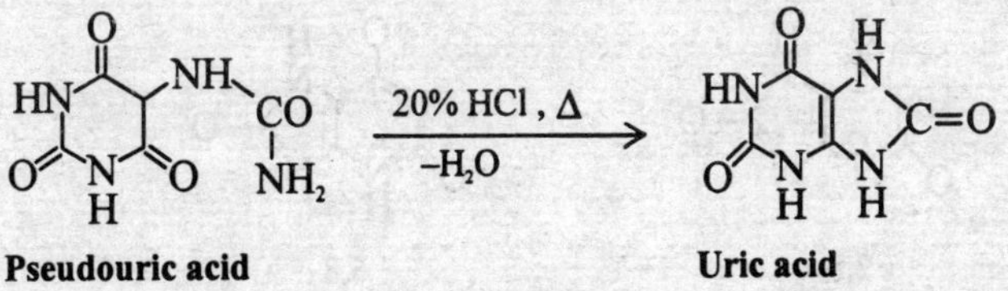

Pseudouric acid → 20% HCl, Δ, –H_2O → **Uric acid**

(iii) *Traube's Synthesis:* This is used for the preparation of purines. However, uric acid is obtained from urea and ethylcyano acetate.

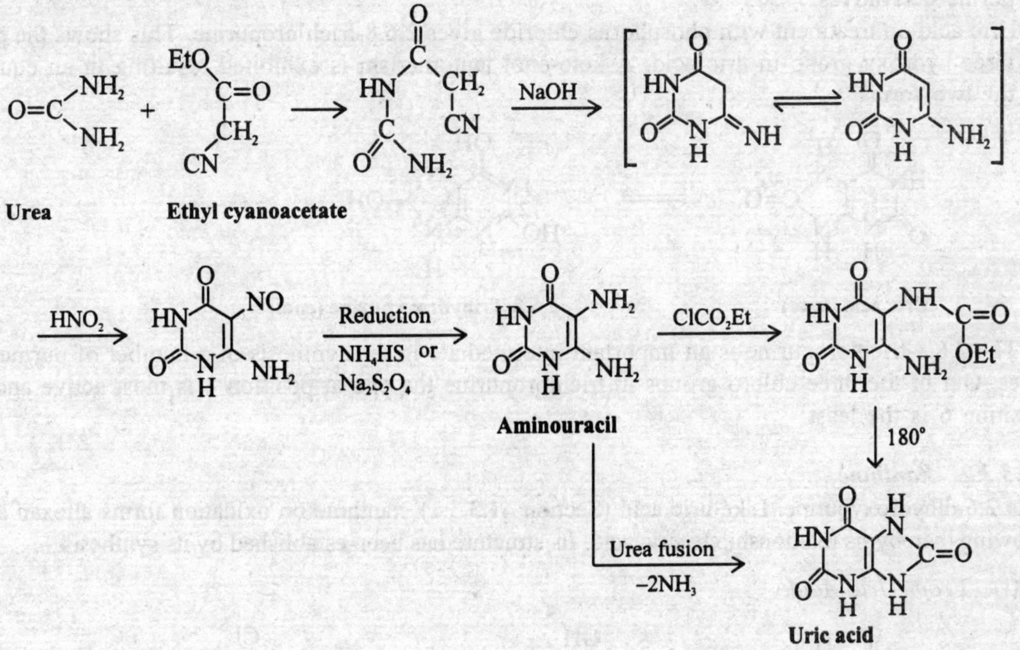

Urea   **Ethyl cyanoacetate**

Aminouracil

Uric acid

It has been shown (Clusius et. al., 1953) using labeled studies that in the formation of uric acid from 5,6-diaminouracil by fusion with labeled urea, the nitrogen atoms of diaminouracil are retained.

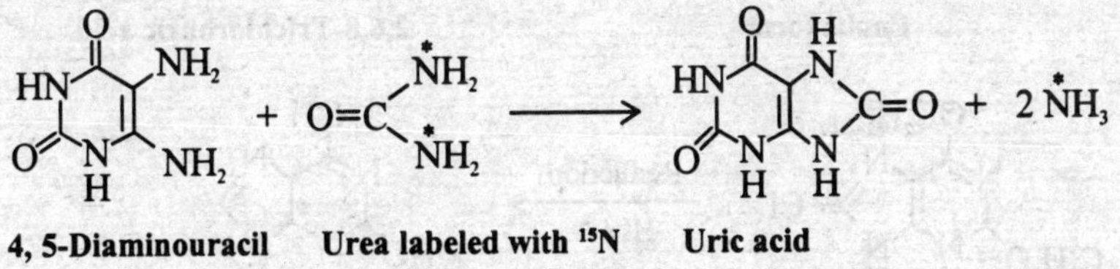

4, 5-Diaminouracil   Urea labeled with ^{15}N Uric acid

Properties
Uric acid is a white, crystalline solid, sparingly soluble in water and is insoluble in alcohol. Being acidic in nature, it is soluble in alkali. It behaves like a weak dibasic acid and gives two series of salts viz. mono- and disodiumurate. The disodium salt may be 2,6-, 2,8- or 6,8- of which the correct representation is not certain.

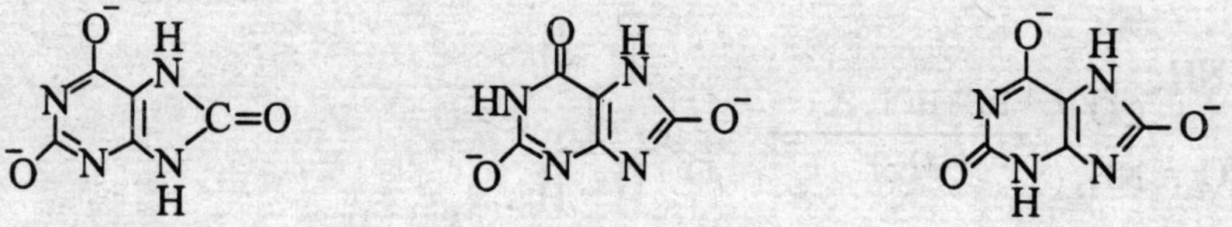

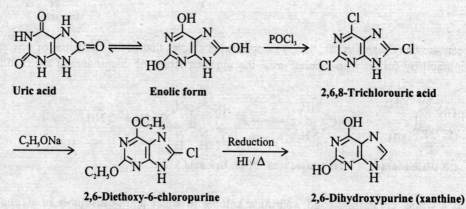

| 2,6- | 2,8- | 6,8 |

It was believed by Fischer that 2,6-disodium salt is the correct representation, since in this arrangement the pyrimidine ring is 'aromatic' and resonance stabilised. This is also supported by the ultraviolet spectra of purine derivatives.

Uric acid on treatment with phosphorus chloride gives 2,6,8-trichloropurine. This shows the presence of three hydroxy group in uric acid. A keto-enol tautomerism is exhibited resulting in an equilibrium of the two forms.

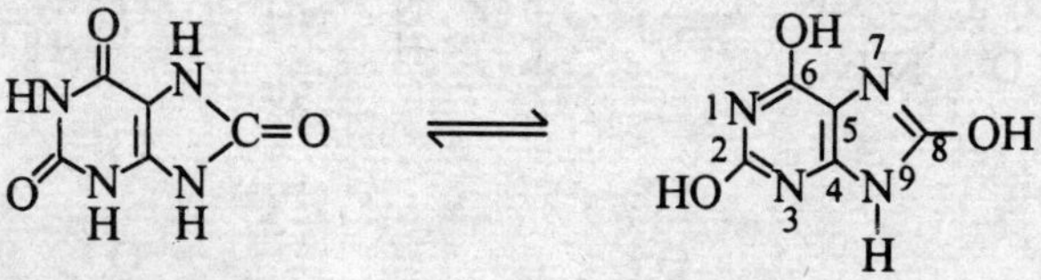

Uric acid (keto)   2,6,8-Trihydroxypurine (enol)

The 2,6,8-trichloropurine is an important intermediate in the synthesis of a number of purine derivatives. Out of the three chloro groups in trichloropurine the one at position 8 is most active and that at position 6 is the least.

41.3.3.2 Xanthine

It is 2,6-dihydroxypurine. Like uric acid (Section 41.3.3.1), xanthine on oxidation forms alloxan and urea showing thereby its relationship to uric acid. Its structure has been established by its synthesis.

(i) *From Uric Acid:*

Uric acid   Enolic form   2,6,8-Trichlorouric acid

C_2H_5ONa

2,6-Diethoxy-6-chloropurine   2,6-Dihydroxypurine (xanthine)

(ii) *Traube's Synthesis:* 4,5-Diaminouracil is first obtained from ethyl cyanoacetate and urea (see Traube's synthesis of uric acid). This is then condensed with formic acid followed by heating the sodium salt of the formed product.

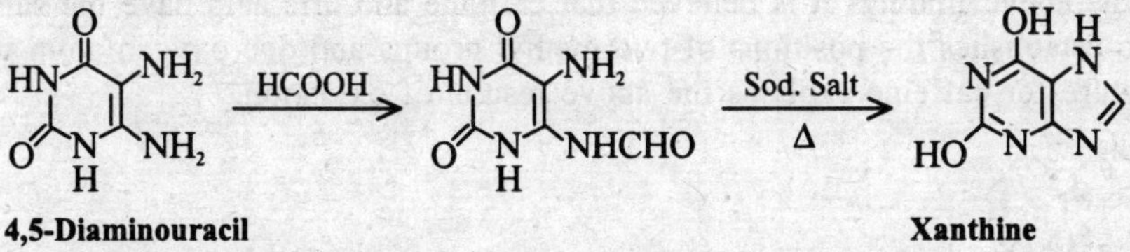

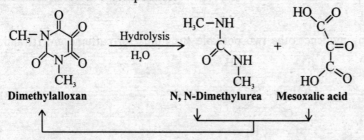

4,5-Diaminouracil → **Xanthine**

(reactions: HCOOH; Sod. Salt Δ)

Xanthine occurs in animal tissues and tea extract. It is a crystalline powder, sparingly soluble in water and forms salts with acids as well as with bases. It exists in tautomeric equilibrium.

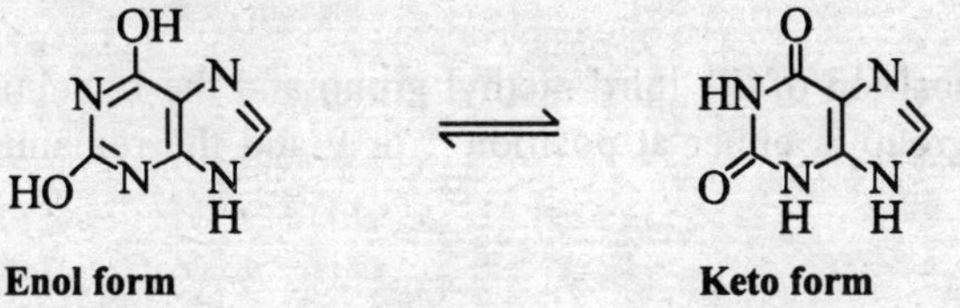

Enol form ⇌ **Keto form**

Its methyl derivatives, viz, caffeine (1,3,7-trimethylxanthine), theophylline (1,3-dimethylxanthine) and theobromine (3,7-dimethylxanthine) are referred to as xanthine bases.

41.3.3.3 Caffeine

It is 1,3,7-trimethylxanthine and occurs in coffee beans, tea leaves and coca nuts.

Isolation from Tea Leaves: Tea contains about 5% caffeine. The leaves are boiled with water and filtered. The filtrate is treated with lead acetate. Albumin (protein) and tannin get precipitated. The filtrate obtained is treated with dilute sulphuric acid and the precipitated lead sulphate is filtered. The filtrate is concentrated, decolorised with animal charcoal and cooled. Caffeine separates on cooling.

Structure

(i)   Its molecular formula is $C_8H_{10}N_4O_2$.

(ii)   *Oxidation:* On oxidation with potassium chlorate in hydrochloric acid, caffeine gives dimethylalloxan and methylurea. It may be recalled that uric acid on oxidation (Section 41.3.3.1) gives alloxan and urea. This shows similarity in the structures of caffeine and uric acid.

$$C_8H_{10}N_4O_2 \xrightarrow[\text{KClO}_4 / \text{HCl}]{[O]} C_6H_6N_2O_4 + CH_3NHCONH_2$$

Caffeine   **Dimethylalloxan Methylurea**

On the basis of the analogy of the structure of alloxan, dimethylalloxan is represented as shown below. This structure is established by its conversion into N,N′-dimethylurea and mesoxalic acid and confirmed by synthesis from these two compounds.

Dimethylalloxan → (Hydrolysis H₂O) → **N, N-Dimethylurea** + **Mesoxalic acid**

On the basis of the above findings it is believed that caffeine and uric acid have the same skeletal frame work. It also establishes the positions of two methyl groups and one oxygen atom in caffeine. The following structure for caffeine explains the above results of oxidation.

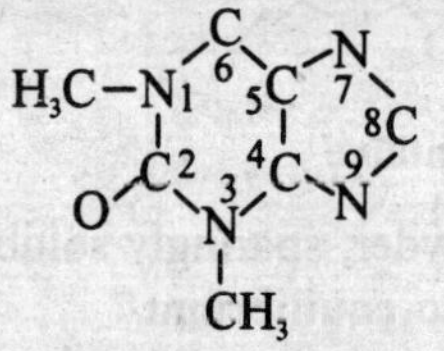

Skeletal Framework of Caffeine

The only question to be decided is the position of the third methyl group and the remaining oxygen atom. It can be imagined that the methyl group is either at position 7 or 9 and the remaining oxygen atom is at position 6 or 8.

Position of the Methyl Group
Besides the two oxidation products viz. dimethylalloxan and methylurea, Fischer isolated a third oxidation product, which on hydrolysis gave N-methylglycine, carbon dioxide and ammonia. On the basis of this findings, the third oxidation product must be N-methylhydantoin.

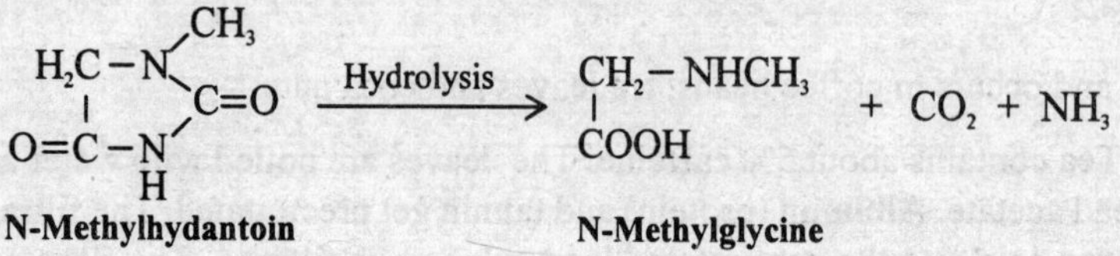

N-Methylhydantoin   **N-Methylglycine**

On the basis of the above results, it is established that caffeine contains two rings, one ring of dimethylalloxan and another ring of methylhydantoin. The following two structures are possible, since both will give the required oxidation products.

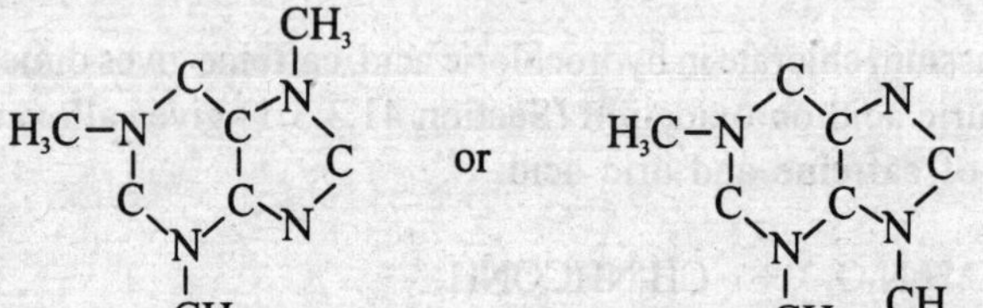

The differentiation between I and II was made by Fischer, who isolated a fourth oxidation product, N,N′-dimethyloxamide ($CH_3NHCOCONHCH_3$). Since only structure I can give this oxamide, caffeine has skeletal frame work of I.

Position of Oxygen Atom
On the basis of the foregoing evidences the two possible structures for caffeine are III and IV.

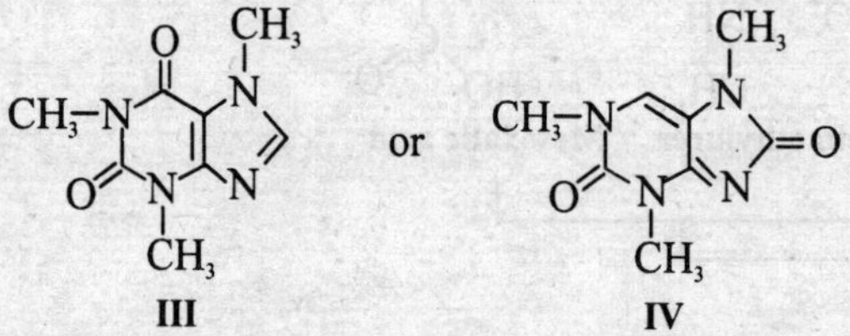

III   IV

Structure III is the more likely structure by analogy with uric acid. This structure was assigned by Fischer on the basis of the following evidences.

$$\text{Caffeine} \xrightarrow{Cl_2} \text{Chlorocaffeine} \xrightarrow[\text{NaOH}]{CH_3OH} \text{Methoxycaffeine}$$

$$C_8H_{10}N_4O_2 \qquad\qquad C_8H_9ClN_4O_2 \qquad\qquad C_8H_9N_4O_2\text{--}OCH_3$$

$$\xrightarrow[\text{HCl}]{\text{dil}} \text{Oxycaffeine} \xrightarrow[CH_3I + NaOH]{\text{Methylation}} \text{Tetramethyluric acid}$$

$$C_8H_{10}N_4O_3 + CH_3Cl$$

The oxycaffeine obtained above was found to be identical with trimethyluric acid, since on methylation with methyl iodide and sodium hydroxide solution oxycaffeine was converted into tetramethyluric acid. On the basis of these studies it was concluded that methoxycaffeine is either V or VI and oxycaffeine is VII or VIII.

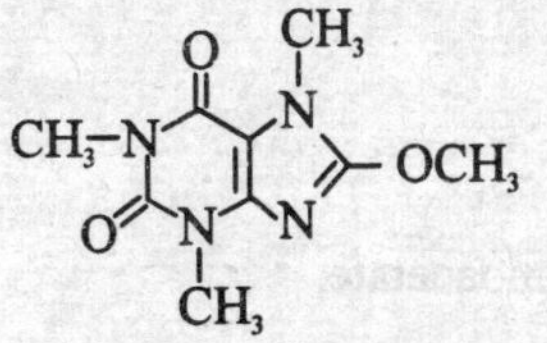

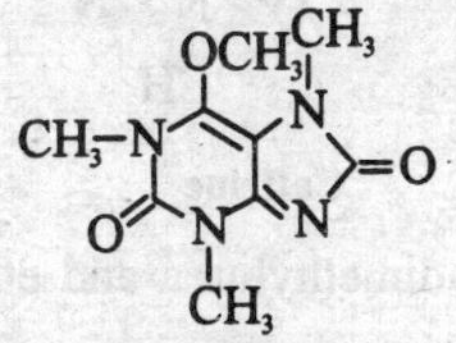

(V) **Methoxycaffeine** (VI)

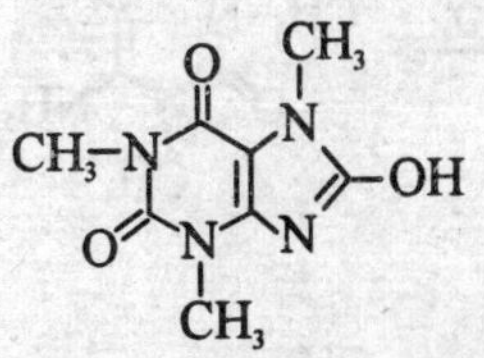

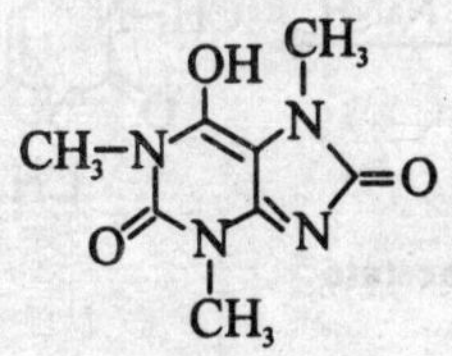

(VII) **Oxycaffeine** (VIII)

The silver salt of oxycaffine on heating with methyl iodide gives a mixture of tetramethyluric acid (containing four N–CH$_3$ groups) and methoxycaffeine (containing three N–CH$_3$ groups and one OCH$_3$ group). This suggests that oxycaffeine is tautomeric with amido-imidol triad system.

$$-NH-\overset{|}{C}=O \rightleftharpoons -N=\overset{|}{C}-OH$$

It is clear that out of pyrimidine and imidazole rings in oxycaffeine, the triad system may be present only in the imidazole nucleus. A close examination of the structures VII and VIII for oxycaffeine shows that only the structure VII can give the above tautomeric system. Thus the methoxy group in methoxycaffeine is in the imidazole nucleus and so the chlorine atom in chlorocaffeine is also in this nucleus. Thus, caffeine is IX and chlorocaffeine is X.

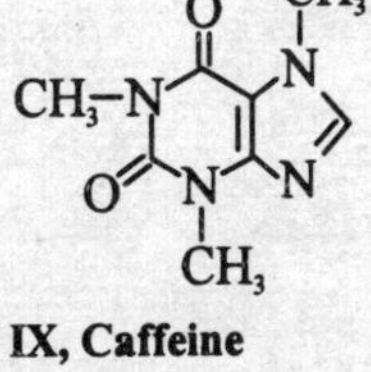

IX, Caffeine 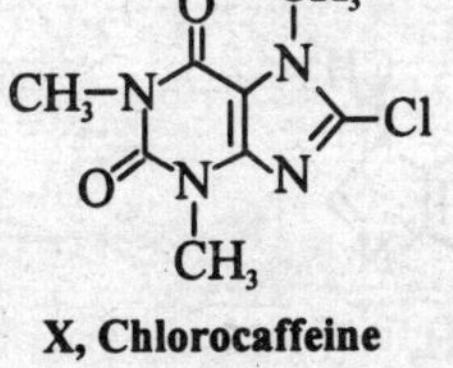 X, Chlorocaffeine

The structure of caffeine has finally been confirmed by its synthesis.

Synthesis

(i) *From Uric Acid (Fischer, 1899)*

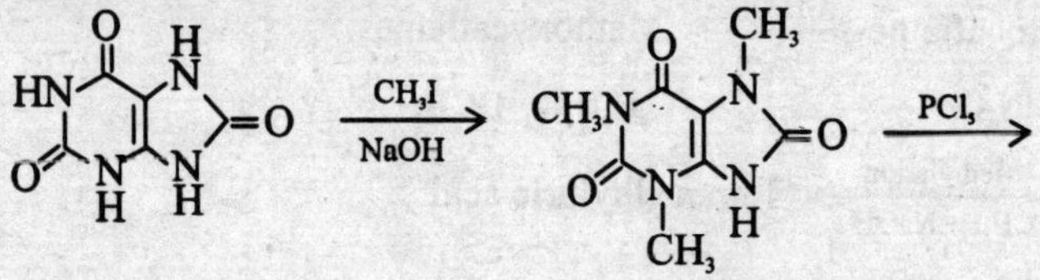

Uric acid   1, 3, 7-Trimethyluric acid

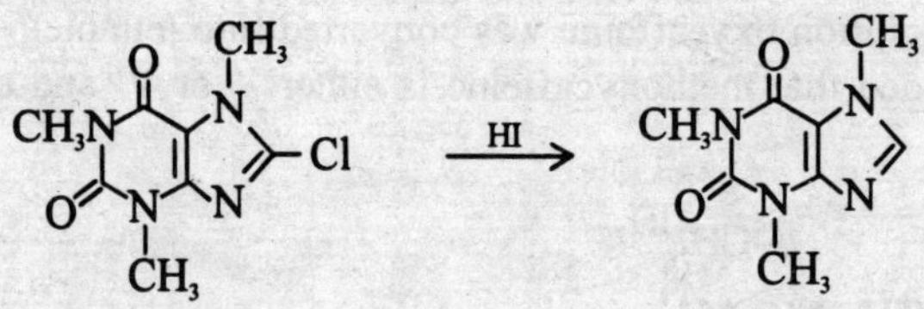

Chlorocaffeine   Caffeine

(ii) *Traube's Synthesis:* From N,N-dimethylurea and ethyl cyanoacetate.

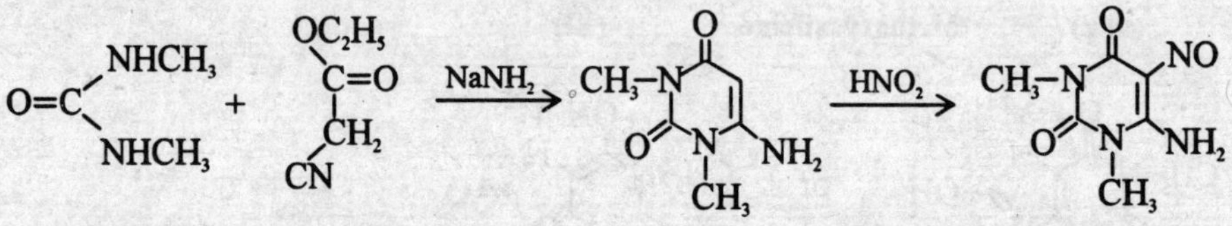

N, N-Dimethylurea   Ethyl cyanoacetate

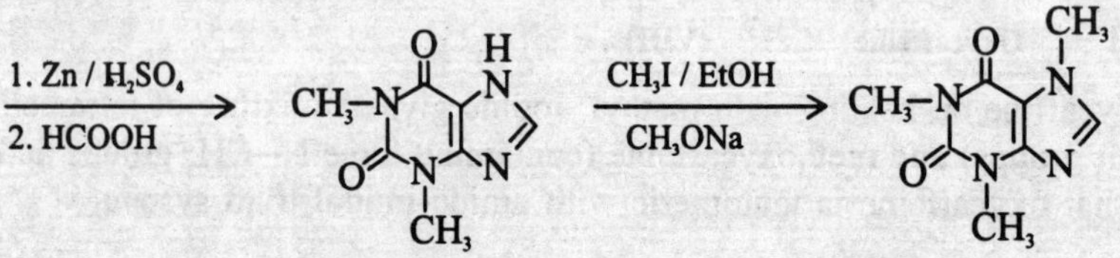

Theophylline   Caffeine

This procedure is used for commercial production of caffeine.

41.3.3.4 Theobromine

It is 3,7-dimethylxanthine and occurs in coca beans and tea. Its structure is assigned on the basis of its oxidation with potassium chlorate and hydrochloric acid to give methylalloxan and methylurea. Also, theobromine can be converted into caffeine by heating the silver salt of theobromine with methyl iodide.

On the basis of the above studies, theobromine is either I or II.

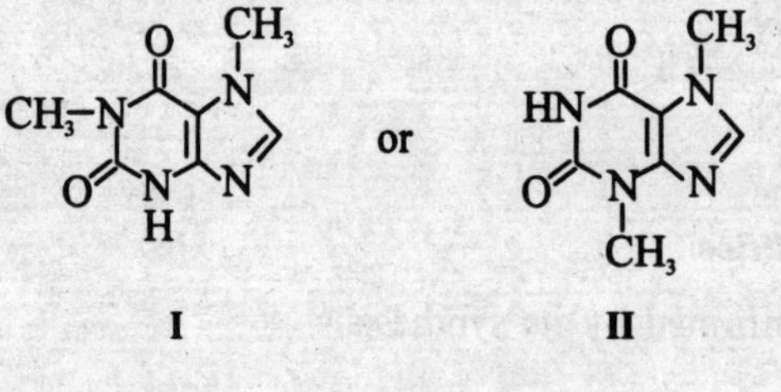

I   or   II

The structure II for theobromine was confirmed by its synthesis using Traube's method.

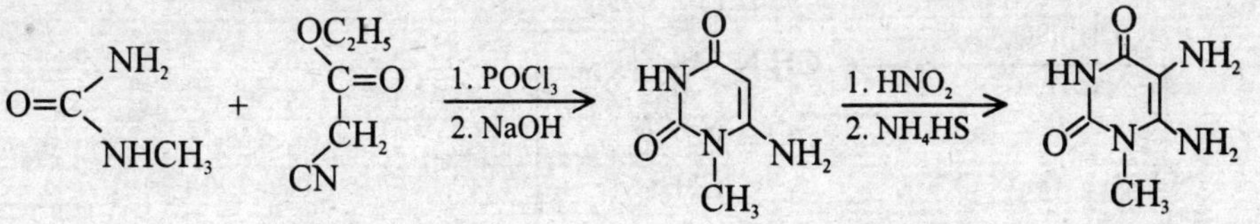

Methylurea   **Ethyl cyanoacetate**

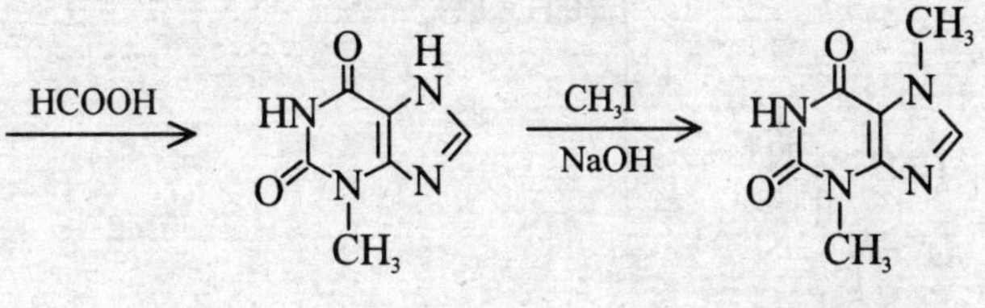

Theobromine

Theobromine is also obtained from uric acid (Fischer, 1899).

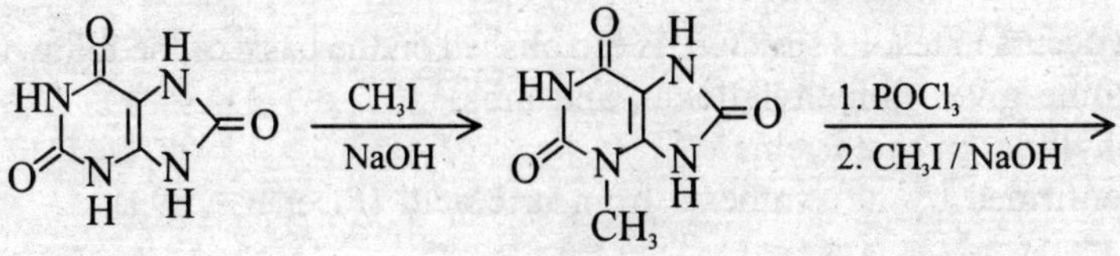

Uric acid   **3-Methyluric acid**

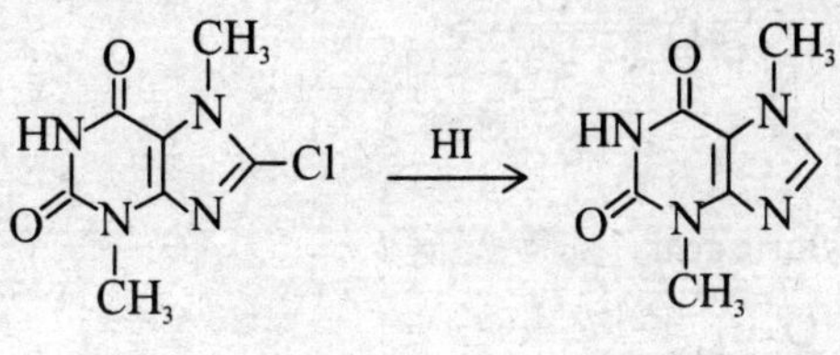

Theobromine

Uric acid can be very conveniently converted into xanthine, caffeine and theobromine (Bredereck, 1950, 1959).

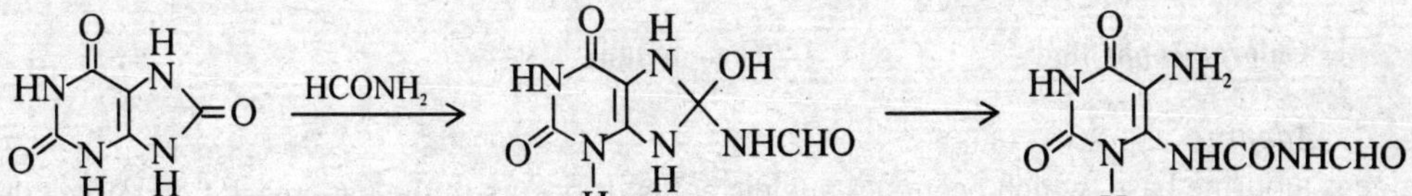

Uric acid

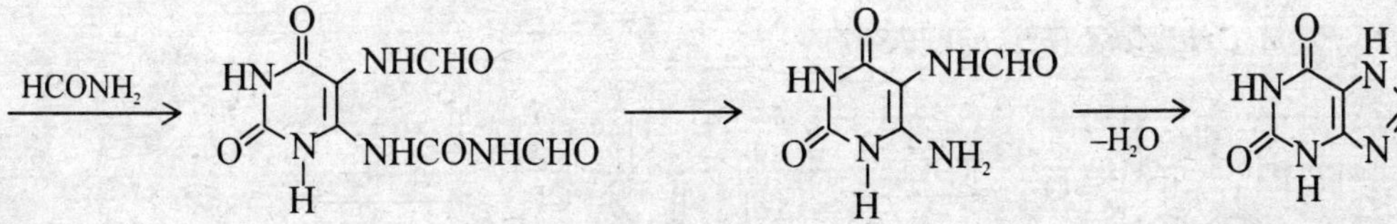

Xanthine

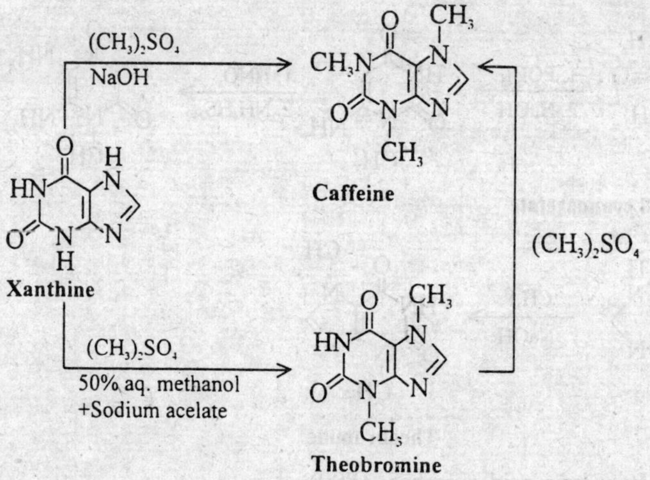

41.3.3.5   Theophylline

It is 1,3-dimethylxanthine and occurs in tea. Its structure is established on the basis of the following findings.
   (i) On oxidation theophylline gives dimethylalloxan and urea.
   (ii) On methylation theophylline gives caffeine.
The structure is finally confirmed by its synthesis from uric acid (Fischer, 1899).

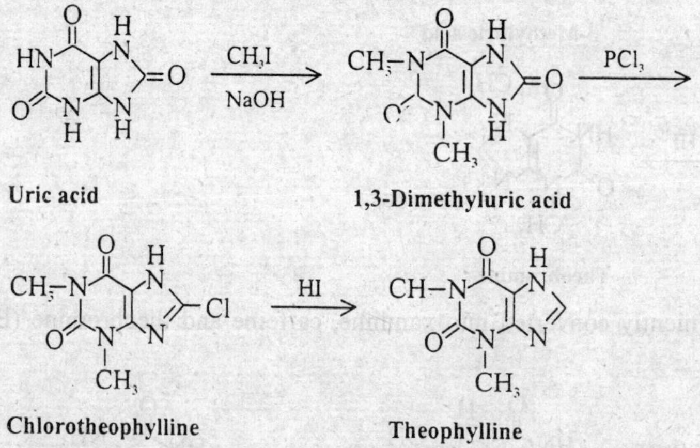

41.3.3.6   Adenine

Adenine is a purine base, which occurs in nucleic acids. It occurs in the pancreas of certain cattles and in tea extract. It is 6-aminopurine and its structure has been established on the basis of the fact that it resembles purine in its general reactions. The structure is confirmed by its synthesis.

   (i) *From Uric Acid (Fischer, 1897)*

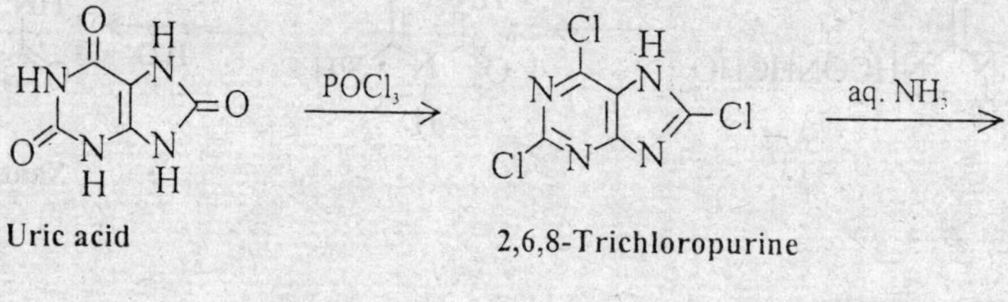

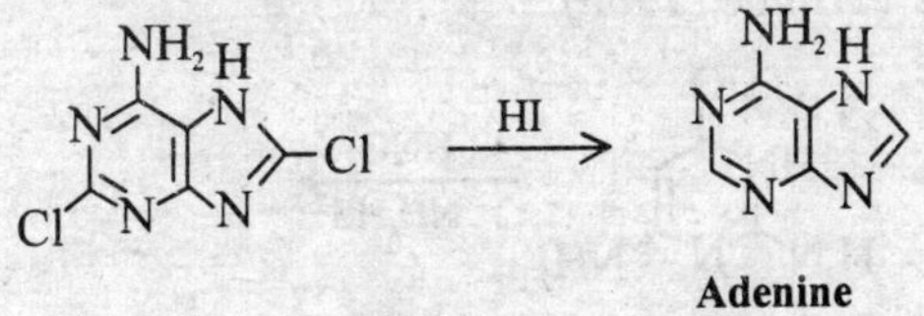

Adenine

(ii) *From Thiourea and Malanonitrile (Traube, 1904)*

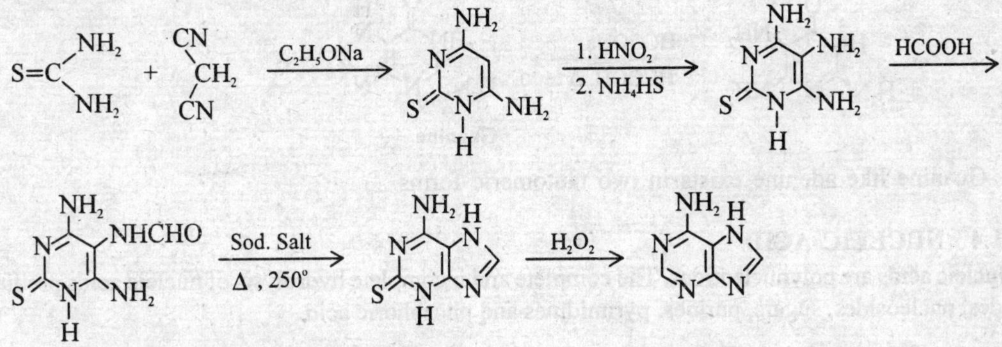

Adenine

Adenine is a tautomeric mixture of the two forms represented below.

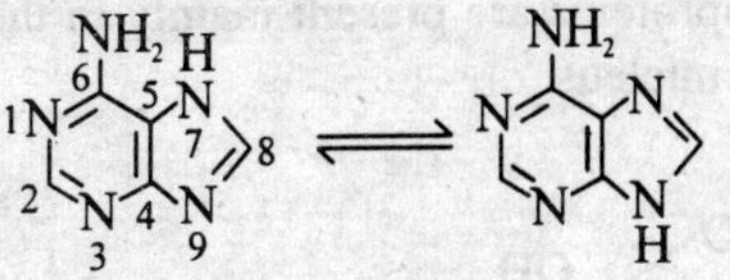

Adenine

41.3.3.7 *Guanine*

Guanine is a pyrimidine base, which occurs in nucleic acids. It occurs in the pancreas of guano and in certain fish scales. It is 2-amino-6-hydroxypurine. Its structure is established on the basis of the following.

(a) Reaction with nitrous acid gives xanthine.

(b) On heating with 25% hydrochloric acid xanthine is obtained.

Final confirmation is provided by its synthesis.

(i) *From Uric Acid (Fischer, 1897)*

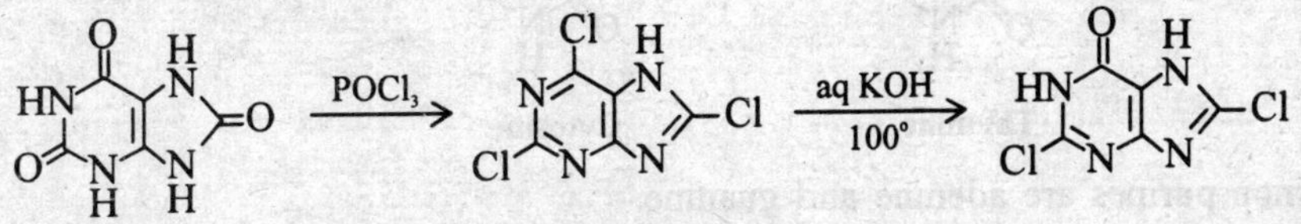

Uric acid · **2,6,8-Trichloropurine**

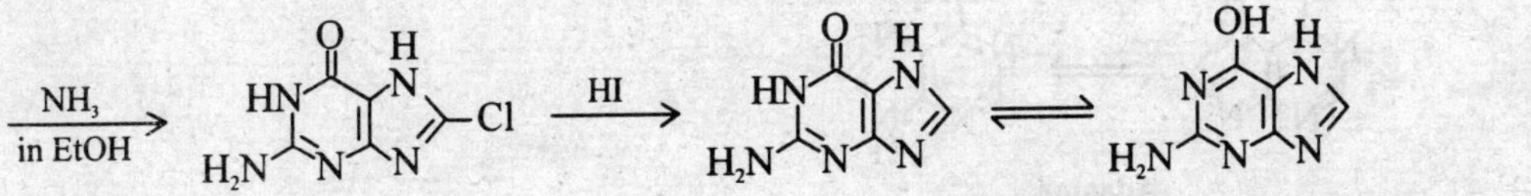

Guanine

(ii)   *From Guanidine and Ethyl Cyanoacetate (Traube, 1900)*

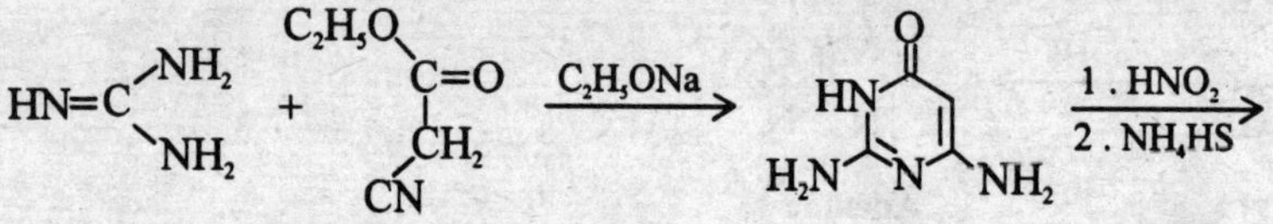

| Guanidine | Ethyl cyanoacetate |

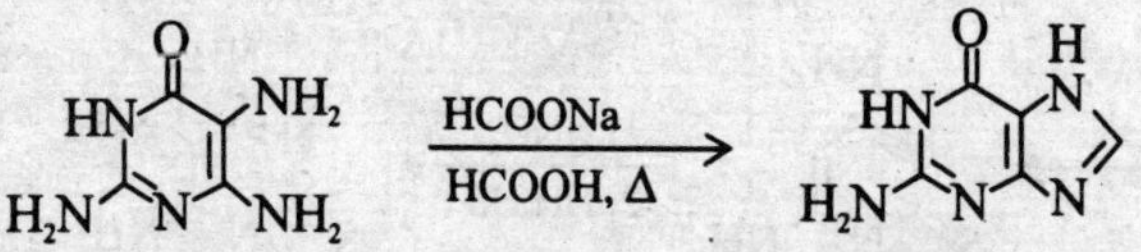

Guanine

Guanine like adenine exists in two tautomeric forms.

41.4   NUCLEIC ACID

Nucleic acids are polynucleotides. The complete and incomplete hydrolysis of nucleic acids produces nucleotides, nucleosides, sugars, purines, pyrimidines and phosphoric acid.

Sugars: The complete hydrolysis of nucleic acid produces two sugars, one is D(−)-ribose and other is 2-deoxy-D-(−)-ribose.

Depending upon the nature of sugar present, nucleic acids are divided into two types, the ribonucleic acid (RNA) and deoxyribonucleic acid (DNA). Ribonucleoproteins are present mainly in the cytoplasm of the cell and deoxy-ribonucleoproteins are found in all nucleus.

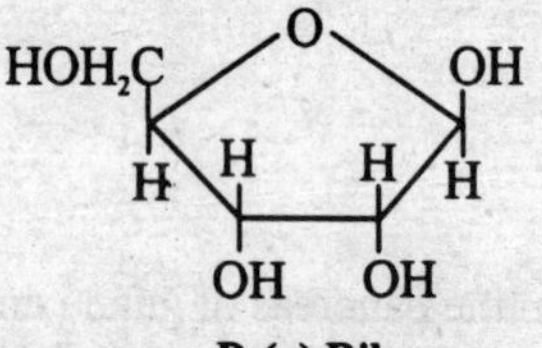

D-(−) Ribose 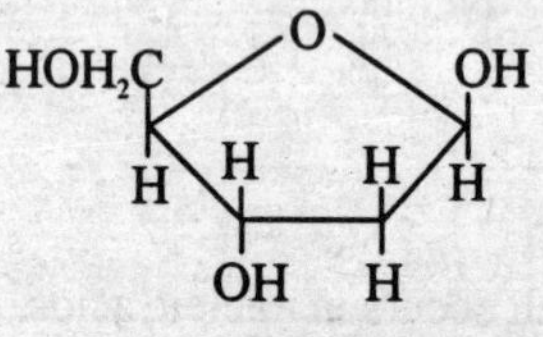   **2-Deoxy-D-(−)-ribose**

Bases: In nucleic acids, two types of bases are present, the pyrimidines and the purines.
The most common pyrimidines are uracil, thymine and cytosine.

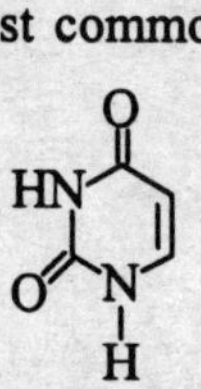

Uracil 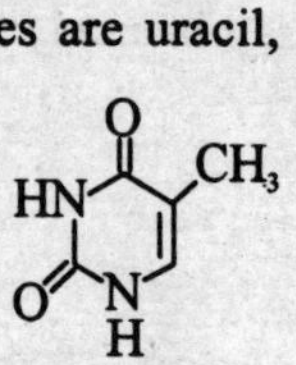   **Thymine** 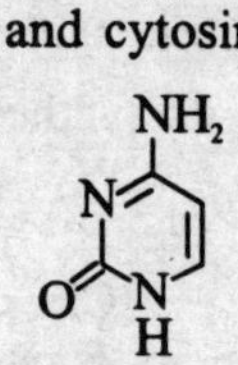   **Cytosine**

The most common purines are adenine and guanine.

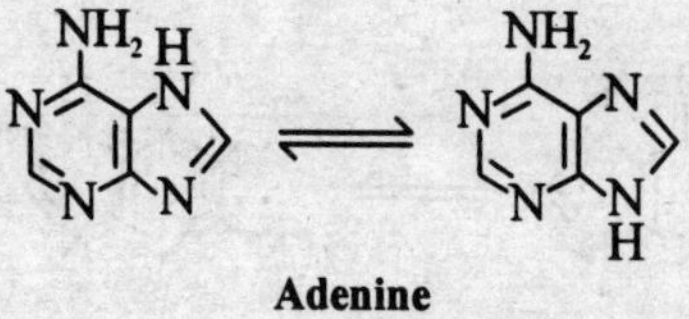

Adenine

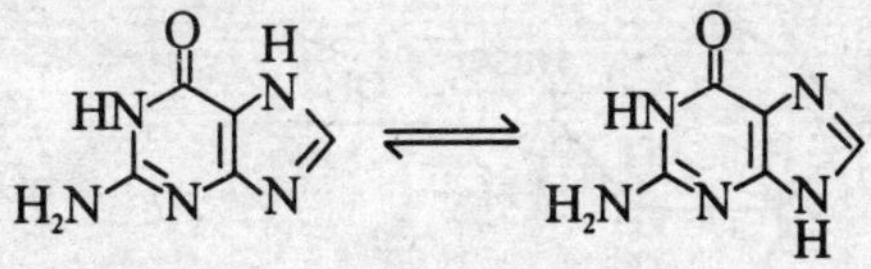

Guanine

Nucleosides: The combination of purine or pyrimidine base with a sugar produces nucleosides, for example adenine and ribose combine to form adenosine. Other nucleosides are guanosine (guanine + ribose), cytidine (cytosine + ribose), uridine (uracil + ribose), thymidine(thymine + deoxyribose) etc.

Purine bases are attached at N-9 position to the sugar moiety while pyrimidine bases are attached at N-1 position to sugar. The linkage is O-glycosidic linkage.

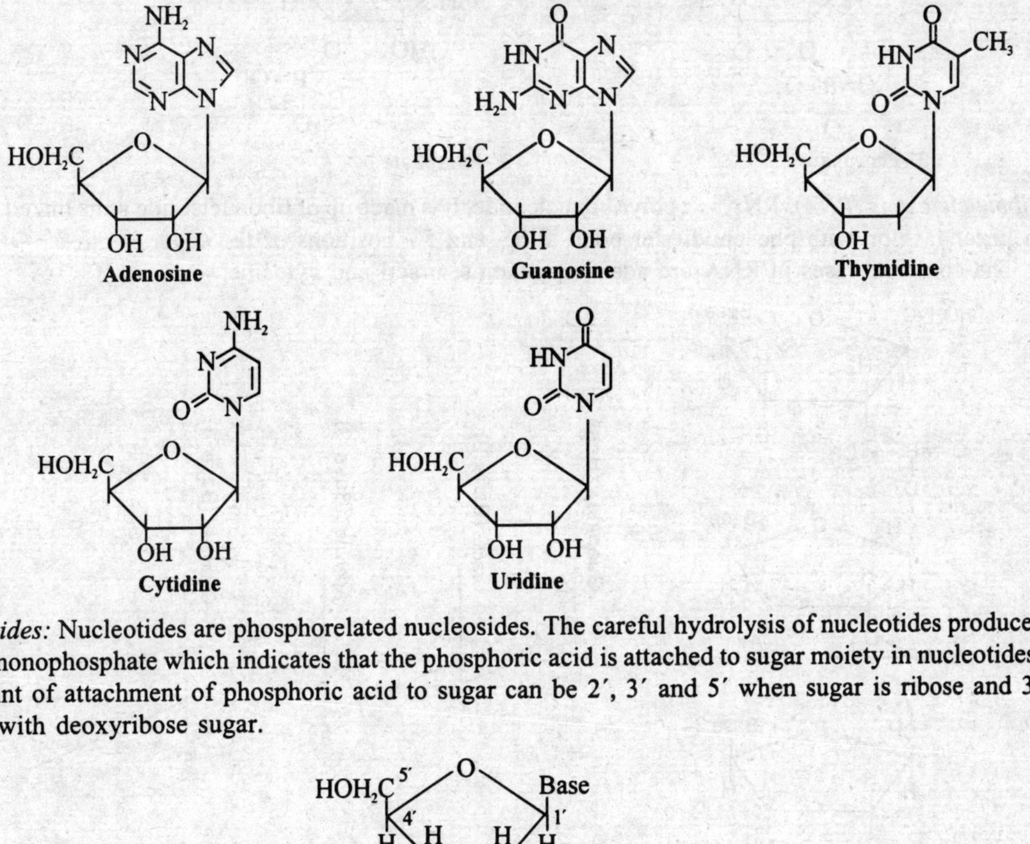

Adenosine   **Guanosine**   **Thymidine**

Cytidine   **Uridine**

Nucleotides: Nucleotides are phosphorelated nucleosides. The careful hydrolysis of nucleotides produces ribose monophosphate which indicates that the phosphoric acid is attached to sugar moiety in nucleotides. The point of attachment of phosphoric acid to sugar can be 2′, 3′ and 5′ when sugar is ribose and 3′ and 5′ with deoxyribose sugar.

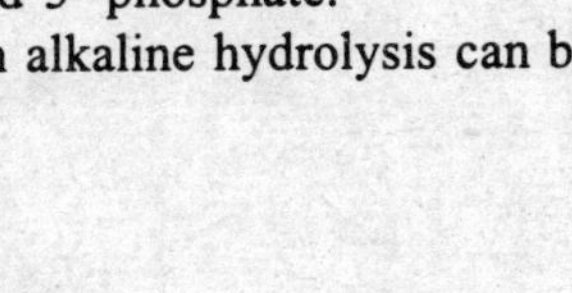

The enzymatic hydrolysis of nucleic acid produces 3′- or 5′-phosphates and alkaline hydrolysis gives a mixture of 2′- (in RNA only) and 3′-phosphate.

The formation of 2′-phosphate in alkaline hydrolysis can be explained on the basis of the following proposed mechanism.

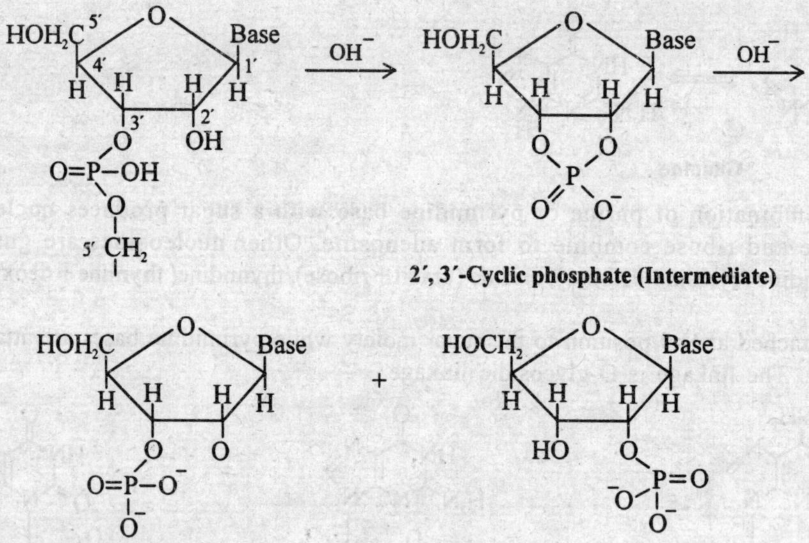

2′, 3′-Cyclic phosphate (Intermediate)

3′-Phosphate   **2′-Phosphate**

Ribonucleic acid (RNA): RNA is a polyribonucleotide. It is made up of ribonucleotide units linked together in linear fashion with phosphodiester bond at 3′- and 5′- positions of the sugar.

The common bases in RNA are adenine, guanine, uracil and cytosine.

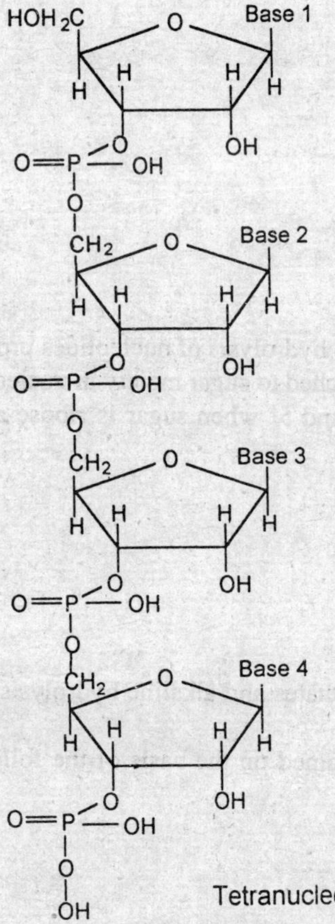

Tetranucleotide

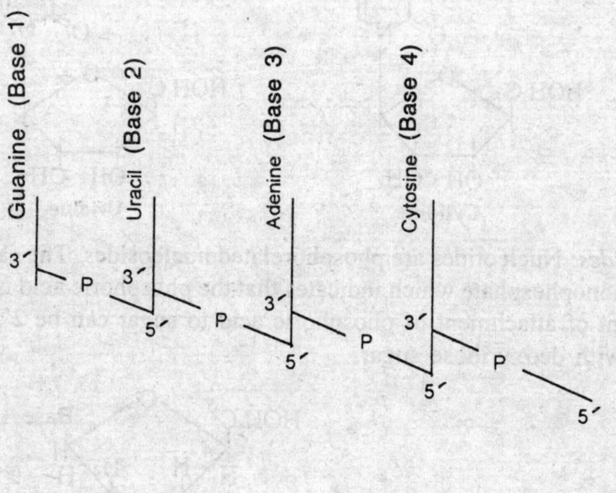

Primary structure of RNA

Deoxyribonucleic Acid (DNA): DNA is also a polynucleotide. The common bases in DNA are adenine (A), guanine(G), thymine(T) and cytosine(C). Watson and Click (1953) proposed the helix structure of DNA, in which the two linear deoxyribonucleotide chains wound as right handed helices round a common axis. The two nucleotide chains are antiparallel as shown in the structure.

Polynucleotide chains are so oriented in space that adenine (purine base) is always located opposite to thymine (pyrimidine base) and guanine (purine base) is opposite to cytosine (pyrimidine base). There exists hydrogen bonding between pairs of bases that holds the helices together. Adenine (A)-Thymine(T) pair is held together by two hydrogen bonds and guanine (G) -cytosine (C) pair by three hydrogen bonds.

The backbone of DNA consists of deoxyriboses linked by 3′,5′ phosphodiester bridges.

The X-ray studies have shown that the pairs of bases (A−T and G−C) are planar and hydrogen bonds (2.8-2.9Å) are almost colinear. Each turn of the helix contains ten nucleotide pairs and diameter of the helix is about 20Å. The spacing between adjacent pairs is 3.4 Å.

The two helices or the two nucleotide chains must be complementary to each other i.e. a chain with a given sequence of bases can pair only with another chain which has the complementary sequence of bases.

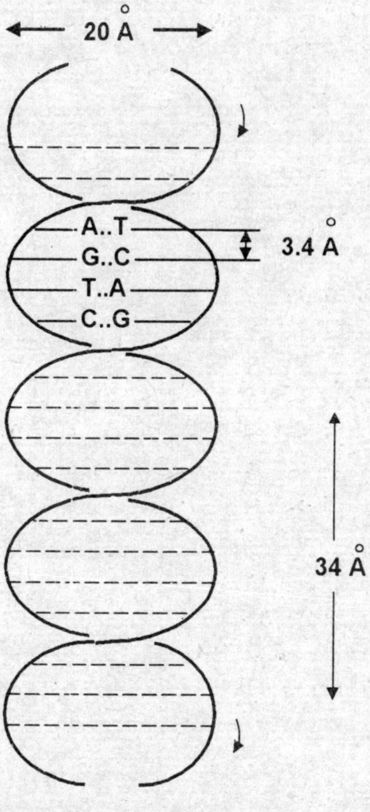

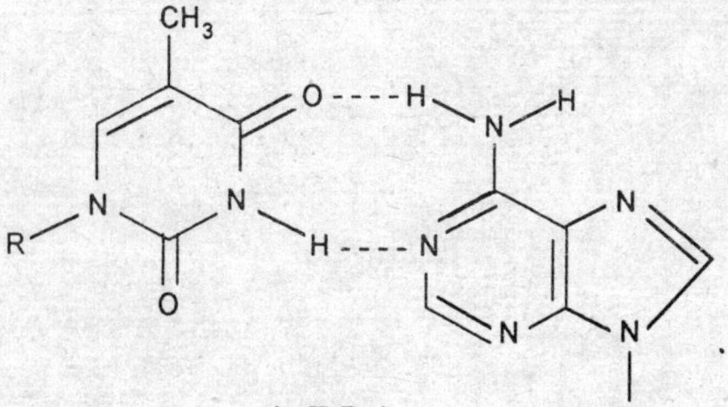

A -T Pair

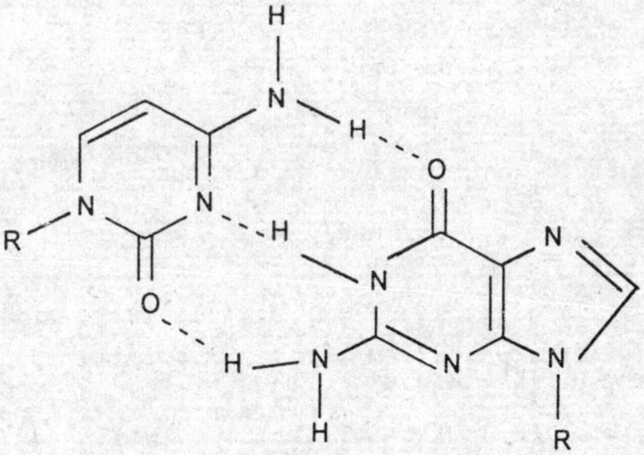

G -C Pair

41.5 PROBLEMS

1. What are ureides? How they are classified? Give the preparation of allophanic acid and parabanic acid.
2. How is pyrimidine obtained? Discuss its reactivity.
3. How is the structure of barbituric acid established? How it is converted into violuric acid, uramil and dialuric acid?

4. Give the IUPAC name for uracil. Give one of its Synthesis.
5. Give synthesis of thiamine and cytosine.
6. What are purines? How are they classified? How is simple purine obtained?
7. How is the structure of uric acid assigned?
8. Give oxidation products of uric acid using different oxidising agents.
9. In the following reaction:

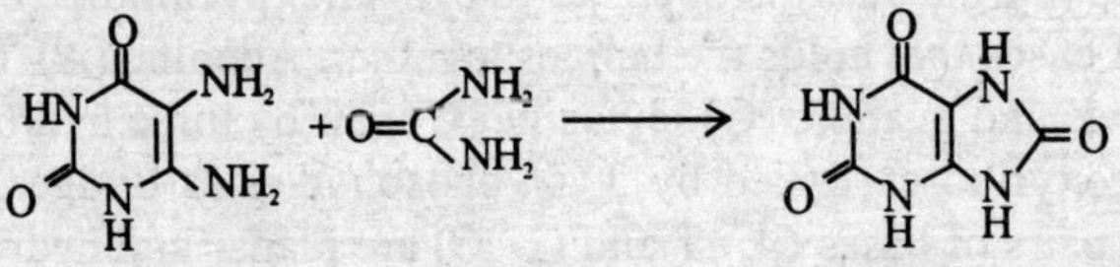

4,5-Diaminouracil   urea

The N atoms of the imidazoline ring are derived from the NH_2 groups of pyrimidine or urea. Explain giving evidences.

10. Explain the utility of 2,6,8-trichloropurine in the synthesis of other derivatives of purine.
11. How is the structure of caffeine arrived at? How it is obtained on a commercial scale?
12. How will you obtain xanthine, caffeine, theobromine, theophylline, adenine and guanine from uric acid?

42
Alkaloids

42.1 INTRODUCTION
The term alkaloid (means alkali like) was used earlier for all organic bases isolated from plants. This definition covers a wide variety of compounds. Alkaloids are now defined as basic nitrogenous compounds of vegetable origin generally having complex molecular structure and significant pharmacological activity.

42.2 OCCURRENCE
More than two thousand five hundred alkaloids are known to occur in nature. They are rarely found in cryptogamia, gymnosperms or monocotyledons. However, there are some exceptions, for example, ergot alkaloids. They occur widely in certain dicotyledons and have been isolated from the roots, seeds, leaves or bark of the plants. They are generally found to occur as salts of various plant acids like acetic acid, oxalic acid, citric acid, malic acid, tartaric acid etc.

42.3 ISOLATION OF ALKALOIDS FROM PLANTS
A general method of isolation of alkaloids consist in extraction of the dried and powdered plant material by refluxing with a suitable solvent, the most commonly used solvent is methanol. The solvent is removed by distillation and the residue is extracted with petroleum ether to remove the plant oils. The remaining residual fraction is treated with inorganic acid and the basic alkaloids are isolated as their insoluble salts. Addition of sodium carbonate to the salts liberates the free bases (alkaloids), which are isolated by extraction with a suitable solvent like ether or chloroform. The mixture of alkaloids thus obtained are separated by various methods into individual compounds. Now a days chromatographic methods are also used for the separation and purification of the alkaloids.

42.4 GENERAL PROPERTIES
Most of the alkaloids are colorless solids, a few are colored, for example, berberine is yellow. They are generally insoluble in water but are soluble in solvents like ether or chloroform. Some alkaloids like coniine and nicotine are liquids and are soluble in water. Normally, the alkaloids have bitter taste and are optically active (laevorotatory) and are used for the resolution of racemic acids. They normally have one or two tertiary nitrogen in a ring system. Besides nitrogen, most of the alkaloids contain oxygen also.

The alkaloids form insoluble precipitates with solutions of picric acid, potassium mercuric iodide, phosphomolybdic acid and phosphotungstic acid (these are referred to as alkaloid reagents). Some of the precipitates have crystalline shapes and are helpful in the identification of alkaloids.

42.5 DETERMINATION OF STRUCTURE
The compound is first obtained in pure state and its molecular formula determined by elemental analysis (C,

H, N and O) and molecular weight determination (especially by mass spectrometry). In case the alkaloid is optically active, its specific rotation is measured. Its structure is then determined systematically as follows:

1. *Functions of Oxygen:* If the alkaloid contains oxygen, its nature is determined by the following tests:

(a) *Carboxyl group:* Its presence is determined by the solubility of the compound in sodium bicarbonate or by the formation of esters.

(b) *Hydroxy group:* Its presence is indicated by acetylation (using acetic anhydride or acetyl chloride and benzoylation (using benzoyl chloride). This reaction must be used in conjunction with the nature of nitrogen. It is well known that amino compounds also react with acetic anhydride, acetyl chloride and benzoyl chloride. The number of hydroxy groups is then estimated. The number of hydroxy group is generally determined by acetylation followed by hydrolysis of a known weight of acetate with a known volume of standard sodium hydroxide. The excess or unused alkali is estimated by back titration with standard acid and from the volume of normal(1N) NaOH used. The number of acetyl groups or hydroxy groups is then calculated. The next step is to find whether the hydroxy group is phenolic or alcoholic. The presence of phenolic hydroxy group is indicated by a color with ferric chloride and by dissolving in sodium hydroxide solution followed by reprecipitation by carbon dioxide. In case the phenolic nature is not there, the hydroxy function may be alcoholic. This is confirmed by treatment with a suitable dehydrating agent (phosphorus pentoxide or concentrated sulphuric acid). The presence of alcoholic hydroxy group is also indicated by the behaviour of the compound towards oxidising agents.

The nature of the alcoholic hydroxy group (viz. primary, secondary or tertiary) is determined by the behaviour of the compound towards oxidising agents. Thus, primary alcohols on oxidation give an aldehyde and then a carboxylic acid, both containing the same number of carbon atoms as the original compound. Secondary alcohols, on the other hand give a ketone (having the same number of carbon atoms) and then an acid having lesser number of carbon atoms. The tertiary alcohols on oxidation give ketones and then carboxylic acids, both containing lesser number of carbon atoms than the original compounds.

(c) *Carbonyl group:* Its presence is detected by the formation an oxime, semicarbazone and phenylhydrazone.

(d) *Ester, lactone, amide groups:* The hydrolysis of the alkaloid with dilute acid or alkali and identification of the formed products indicates the presence of an ester, lactone, amide, lactam or a betaine.

(e) *Methylenedioxy group* ($-OCH_2O-$): Its presence is indicated by the formation of formaldehyde when the alkaloid is heated with hydrochloric or sulphuric acid.

(f) *Methoxy group:* The presence of methoxy groups and their number is determined by Zeisel's method. In Zeisel's method the alkaloid is heated with concentrated hydriodic acid at 128°, the methoxy group is converted into methyl iodide. The formed methyl iodide is absorbed in ethanolic silver nitrate solution and the silver iodide obtained is weighed.

It is appropriate to mention that the phenolic ethers can also be decomposed by heating with pyridine hydro chloride at 170-240°. This method is particularly useful for ethers that are sensitive to other demethylating agents.

(g) *The number of active hydrogens:* These are determined by Zerewitinoff active hydrogen determination method. In alkaloids, an active hydrogen atom is the one which is joined to oxygen (like hydroxy group) and nitrogen (amino and imino groups). In this method methyl magnesium iodide is made to react with the alkaloid, the liberated methane is measured by volume, one molecule of methane being equivalent to one active hydrogen atom.

$$ROH + CH_3MgI \longrightarrow CH_4 + Mg(OR)I$$

However, in case of primary amines, only one active hydrogen reacts at room temperature.

$$RNH_2 + CH_3MgI \longrightarrow CH_4 + RNHMgI$$
$$R_2NH + CH_3MgI \longrightarrow CH_4 + R_2NMgI$$

At reasonably high temperature, the active hydrogen atom in the magnesium derivative of primary amine reacts further with a molecule of methyl magnesium iodide.

$$RNHMgI + CH_3MgI \longrightarrow CH_4 + RN(MgI)_2$$

Thus, it is possible to estimate the number of amino and imino groups in compounds containing both. Generally, it is not possible to get enough high temperature for a second reaction with ether. A satisfactory solvent for complete Zerewitinoff determination is pyridine.

Lithium aluminum hydride also reacts with compounds containing active hydrogen, and so can be used for the determination of their active hydrogen.

$$4ROH + LiAlH_4 \longrightarrow 4H_2 + (RO)_4LiAl$$
$$4RNH_2 + LiAlH_4 \longrightarrow 4H_2 + (RNH)_4LiAl \xrightarrow{LiAlH_4} 4H_2 + 2(RN)_2LiAl$$

2. Functions of Nitrogen

(a) *Nature of nitrogen:* Most of the alkaloids (with few exceptions like alkaloids of phenylalkylamino group) contains nitrogen in the heterocyclic ring. The general reactions of alkaloids with acetic anhydride, methyl iodide and nitrous acid normally show the nature of nitrogen. Nitrogen may be present as secondary function (its presence detected by acetylation or Liebermann's nitroso reaction) or as tertiary function. If all the reactions are negative, than the nitrogen is most likely tertiary. The chief difficulty is that alkaloid may undergo ring fission and the product may be an N-acetyl derivative. The tertiary nature of nitrogen is indicated by the formation of amine oxide with 30% hydrogen peroxide.

(b) *The number and nature of alkyl group attached to nitrogen* is inferred by distillation of an alkaloid with aqueous potassium hydroxide. The identification of volatile products of hydrolysis as methyl amine, dimethyl amine or trimethyl amine indicates the attachment of one, two or three methyl groups to the nitrogen atom respectively. However, evolution of ammonia on hydrolysis shows the presence of an amide group ($-CONH_2$). The presence of higher N-alkyl groups have not been reported in alkaloids. The only exception is of aconite which contains an N-ethyl group.

(c) *The number and the presence of N-methyl groups* is determined by Herzig-Meyer method. The method consists in heating the alkaloid with hydriodic acid at 150° when the N- methyl groups are converted into methyl iodide.

$$ArNHCH_3 + HI \xrightarrow{150°} ArNH_2 + CH_3I$$

The formed methyl iodide is passed into ethanolic silver nitrate solution and the silver iodide obtained is weighed. (Compare this method with Zeisel's method for the estimation of methoxy groups, in which the methoxy compound is heated at 100° with hydriodic acid). Then by a combination of Zeisel's method and Herzig Meyer method both the methoxy and methylamino groups can be estimated separately when both are present in the same compound.

(d) *Amide, lactam or betaine functions:* The presence of an amide, lactam or betaine is shown by the hydrolysis.

. (e) *Degradation of alkaloids:* A number of degradative methods are used with a view to find a rec-ognisable moiety.

(i) *Hofmann's exhaustive methylation method:* This method has been extensively used in structure elucidation of alkaloids and consists in opening the heterocyclic ring with the elimination of nitrogen to give a carbon fraction, which is easily recognisable. In this method the alkaloid is first hydrogenated (in case it is unsaturated) and than converted into the quaternary methylammonium hydroxide, which on heating loses a molecule of water. The hydroxy group is eliminated from the tetramethylammonium hydroxide and the hydrogen atom from the β-position with respect to the nitrogen atom resulting in ring opening at the nitrogen atom on the same side from which the β-hydrogen was eliminated. The process is repeated on the formed product till the nitrogen is eliminated and an unsaturated hydrocarbon is left, which isomerises to a conjugated diene. Thus, starting with pyridine, piperylene is obtained as shown below:

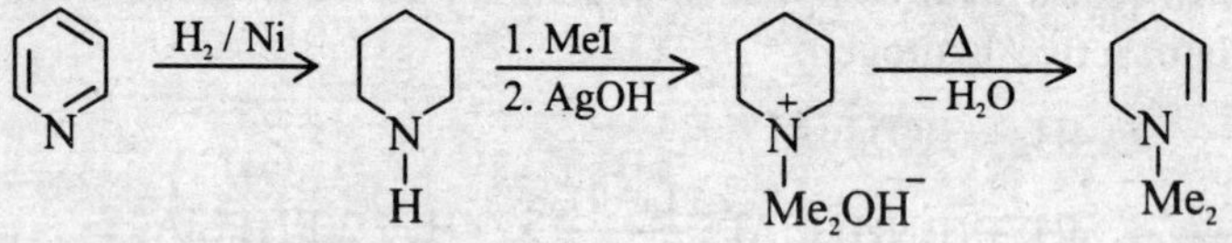

Pyridine   **Piperidine**

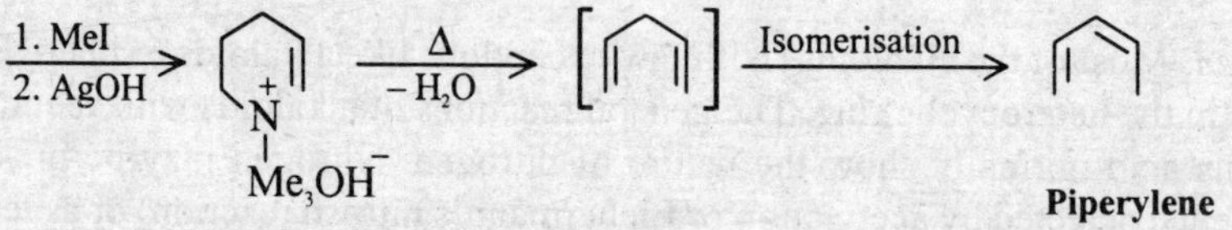

Piperylene

The Hofmann's exhaustive methylation fails in cases where the alkaloids do not contain a β-hydrogen atom. In such cases Emde's degradation is used.

(ii) *Emde's Degradation:* In this method, the quaternary ammonium halide is reduced with sodium amalgam in aqueous ethanol or with sodium in liquid ammonia or it is catalytically hydrogenated.

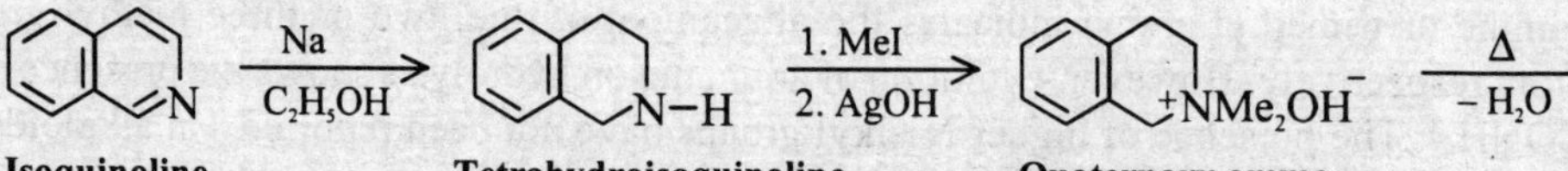

Isoquinoline   **Tetrahydroisoquinoline**   **Quaternary ammo. hydroxide**

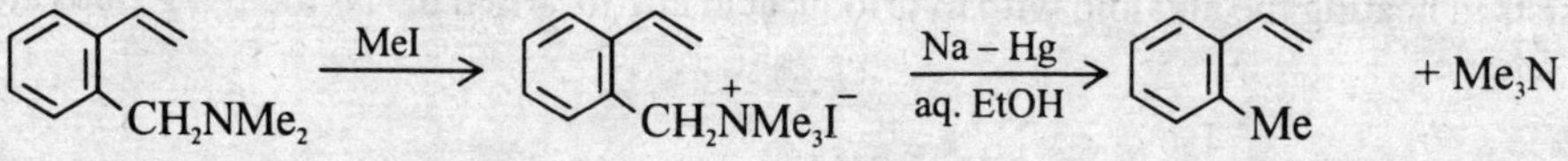

(I)   **o-Methylstyrene**

As seen, the intermediate (I) cannot be degraded by Hofmann's method as it has no β-hydrogen but it can be degraded by Emde's method.

It is appropriate to mention that in case of tetrahydroquinoline, though it contains a β-hydrogen atom, the Hofmann's degradation fails. The ring, however, may be opened up by Emde's degradation.

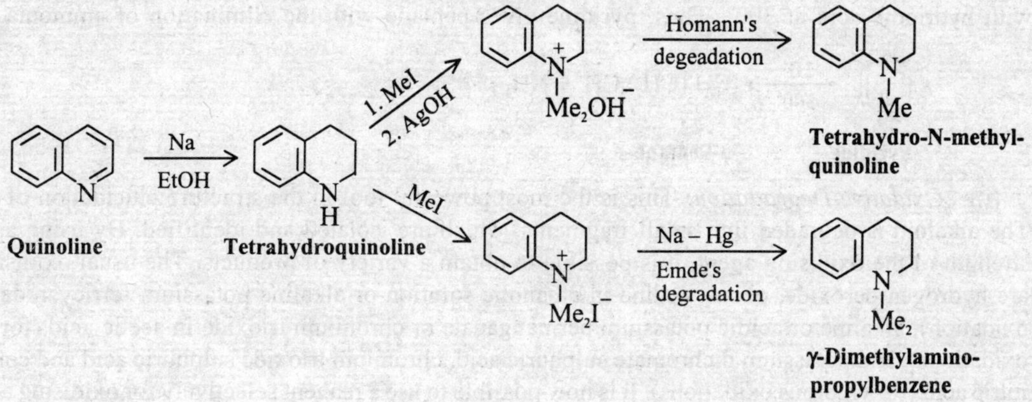

Quinoline → Tetrahydroquinoline → Tetrahydro-N-methyl-quinoline / γ-Dimethylamino-propylbenzene

(iii) *Von Braun Method:* This method is divided into two categories depending on whether the alkaloid is a secondary cyclic amine or a tertiary cyclic amine.

(a) *In the case of a secondary cyclic amine,* it is first converted into its benzoyl derivative, which on treatment with phosphorus halide followed by distillation under reduced pressure forms α-ω-dihalo compounds, the nitrogen being eliminated as benzonitrile.

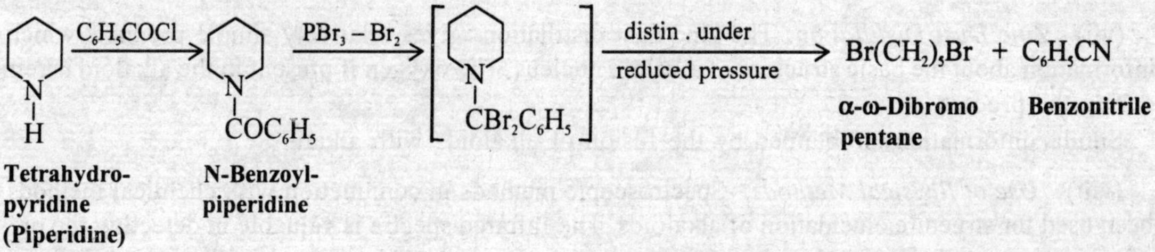

Tetrahydro-pyridine (Piperidine) → N-Benzoyl-piperidine → α-ω-Dibromo pentane + Benzonitrile

(b) *In the case of a tertiary cyclic amines,* it is reacted with cyanogen bromide. The formed product on heating with hydrogen bromide gives the bromo compound.

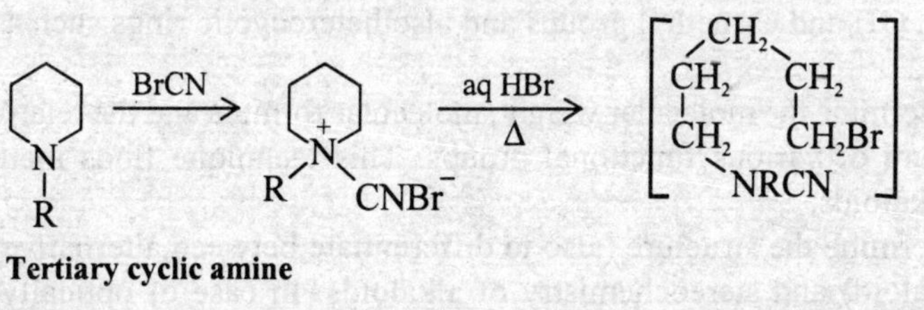

Tertiary cyclic amine

A special point about the cyanogen bromide method is that it is generally successful with compounds that fail with Hofmann's method. However, in cases where both methods can be used, ring opening occurs at different positions.

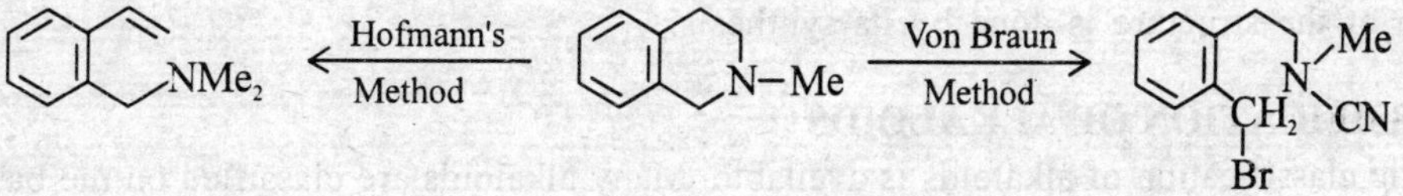

Hofmann's Method ← → Von Braun Method

(iv) *Ring Opening by Hydriodic Acid:* In some cases the alkaloid ring may be opened by heating with hydriodic acid at 300°. Thus, pyridine gives pentane with the elimination of ammonia.

$$\text{Pyridine} \quad \xrightarrow[300°]{HI} \quad CH_3(CH_2)_3CH_3 + NH_3$$

Pyridine   n-Pentane

(v) *Oxidative Degradation:* This is the most powerful tool in the structure elucidation of alkaloids. The alkaloid is degraded into small fragments, which are isolated and identified. By using appropriate strength of the oxidising agent, it is possible to obtain a variety of products. The usual oxidising agents are hydrogen peroxide, ozone, iodine in ethanolic solution or alkaline potassium ferricyanide (for mild oxidation), alkaline or acidic potassium permanganate or chromium trioxide in acetic acid (for moderate oxidations), and potassium dichromate sulphuric acid, chromium trioxide sulphuric acid and concentrated nitric acid (for vigorous oxidations). It is now possible to use a reagent selectively for oxidising a particular group in a molecule.

(vi) *Reductive Degradation:* Reduction by means of sodium amalgam, sodium and ethyl alcohol, tin and hydrochloric acid, hydriodic acid and catalytic hydrogenation etc. are used to show the presence of unsaturation. The presence of unsaturation is also demonstrated by the addition of bromine, halogen acid or dilute potassium permanganate solution. The reduction degrades the alkaloids in recognisable fragments and sodium in liquid ammonia gives the Emde's type of degradations.

(vii) *Zinc Dust Distillation:* The zinc dust distillation gives relatively simple products which give information about the basic structure of alkaloid nucleus. The oxygen if present in the alkaloid is removed during the process.

Similar information is obtained by the fusion of alkaloids with alkali.

(viii) *Use of Physical Methods:* Spectroscopic methods in conjunction with chemical methods have been used for structure elucidation of alkaloids. The Infrared spectra is valuable in detecting the presence of many functional groups (like hydroxy group gives a characteristic band in the region $3400 cm^{-1}$, an oxo group in 1700-$1750 cm^{-1}$ and an α-β-unsaturation in 1660-$1700 cm^{-1}$). The UV spectroscopy is also used to indicate the structure. The NMR spectroscopy is used for detection of many functional groups, for example, olefinic protons, NH, OH and C-methyl groups and also heterocyclic rings such as pyridine, pyrrole etc.

Mass spectrometry is used to determine the molecular weight, molecular formula and the relative position of the double bonds and the nature of various functional groups. This technique finds increasing use in the structure elucidation of alkaloids.

X-Ray analysis is useful to determine the structure (also to differentiate between alternative structures that fit equally well with the alkaloid) and stereochemistry of alkaloids. In case of optically active alkaloids, the stereochemical assignments have also been made by means of optical rotatory dispersion and circular dichroism.

(ix) *Synthesis:* The use of above methods normally give a tentative structure for the alkaloid. Final confirmation of the structure is done by its synthesis.

42.6 CLASSIFICATION OF ALKALOIDS

No systematic classification of alkaloids is available. Many alkaloids are classified on the basis of their sources. With the structures of over 2500 alkaloids known, the classification of the alkaloids is still arbitrary.

Still, the most satisfactory method of classification is on the basis of nucleus present in the molecule. The following groups of alkaloids have been discussed.

I. Phenylethylamine group
II. Pyrrolidine group
III. Pyridine Piperidine group
IV. Pyrrolidine Pyridine group
V. Quinoline group
VI. Isoquinoline group
VII. Phenanthrene group.
VIII. Indole group
IX. Colchinine.

Some typical examples of different group of alkaloids are given below:

I. Phenylethylamine group

$C_6H_5CH_2CH_2NH_2$

β-Phenylethylamine

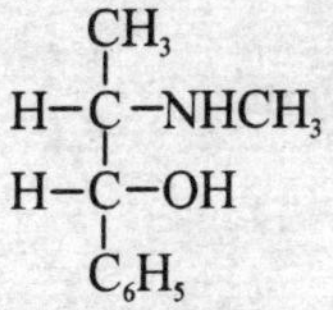

$$\begin{array}{c} CH_3 \\ | \\ H-C-NHCH_3 \\ | \\ H-C-OH \\ | \\ C_6H_5 \end{array}$$

D(–)-Ephedrine

$C_6H_5CH_2CH(CH_3)NH_2$

Benzedrine (Amphetamine)

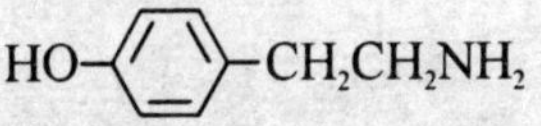

HO—⟨⟩—$CH_2CH_2NH_2$

β-p-Hydroxyphenylethylamine (Tyramine)

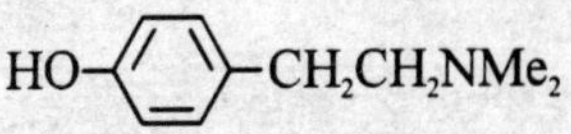

HO—⟨⟩—$CH_2CH_2NMe_2$

Hordenine (Anhaline)
(β-p-hydroxyphenyldimethylamine)

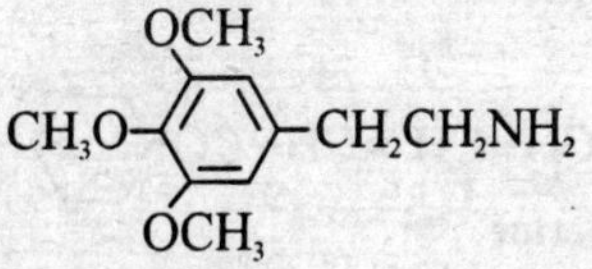

CH_3O—⟨⟩—$CH_2CH_2NH_2$
with OCH₃ above and OCH₃ below the ring

$$\begin{array}{c} OCH_3 \\ CH_3O-\langle\ \rangle-CH_2CH_2NH_2 \\ OCH_3 \end{array}$$

Mescaline

HO—⟨⟩—$CHOHCH_2NHCH_3$
HO

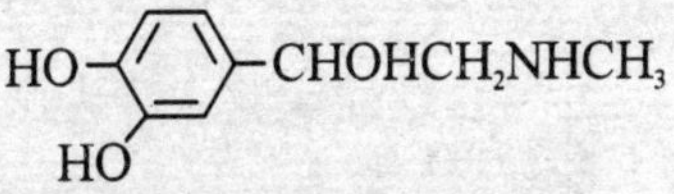

(–) Adrenaline (Epinephrine)

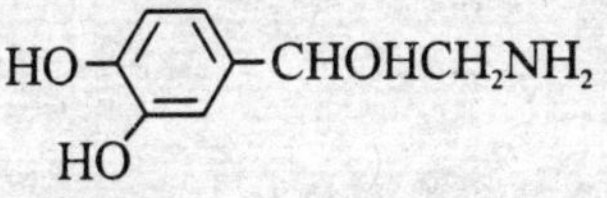

HO—⟨⟩—$CHOHCH_2NH_2$
HO

(–) Noradrenaline (Norepinephrine)

II. Pyrrolidine group

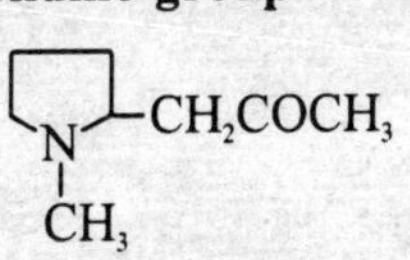

⟨pyrrolidine ring⟩—CH_2COCH_3
N
CH_3

(–) Hygrine

⟨pyrrolidine⟩—CH_2COCH_2—⟨pyrrolidine⟩
N   N
CH_3 CH_3

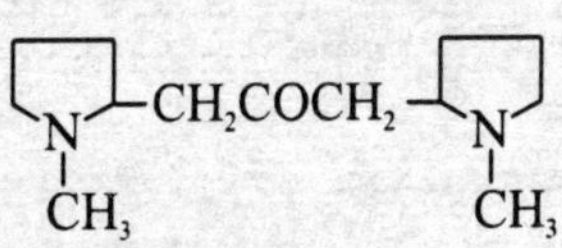

Cuscohygrine

III. Pyridine and Piperidine group

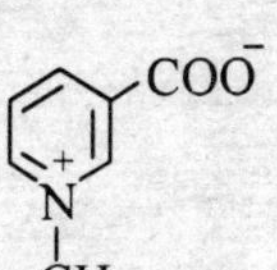

⟨pyridinium ring⟩—COO^-
N^+
CH_3

Trigonelline

$$\begin{array}{c} OCH_3 \\ CN \\ \langle ring \rangle O \\ N \\ CH_3 \end{array}$$

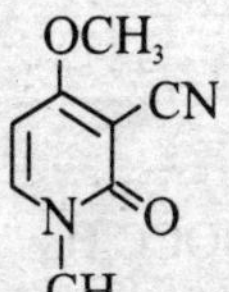

Ricinine

Areca (or Betal) Nut Alkaloids

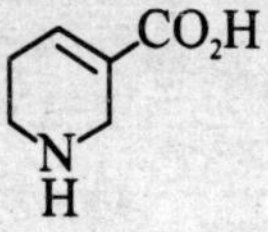

Guvacine

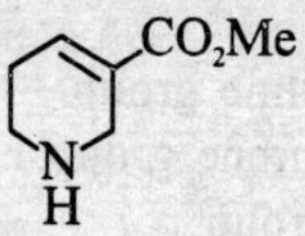

Guvacoline

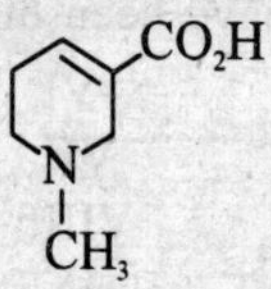

Arecaidine

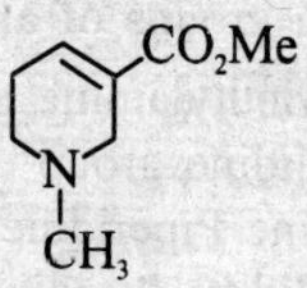

Arecoline

Hemlock Alkaloids

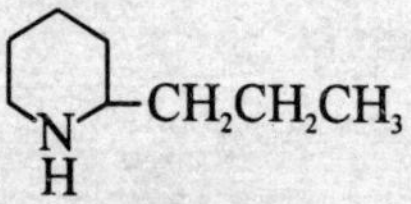

(+) Coniine $-CH_2CH_2CH_3$

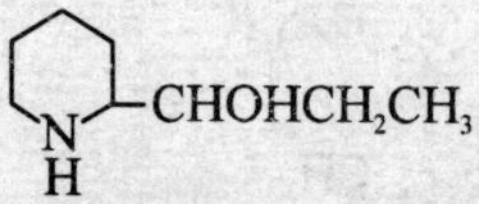

Conhydrine $-CHOHCH_2CH_3$

γ-Coniceine $-CH_2CH_2CH_3$

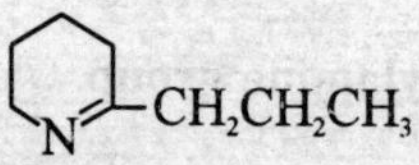

Pomegranate Alkaloids

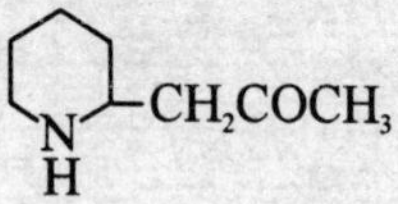

Isopelletierine $-CH_2COCH_3$

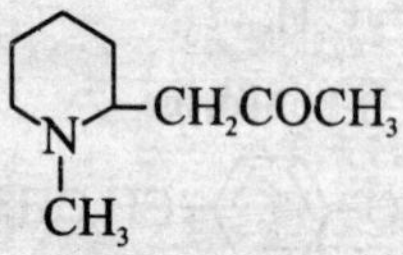

N-Methylisopelletierine $-CH_2COCH_3$

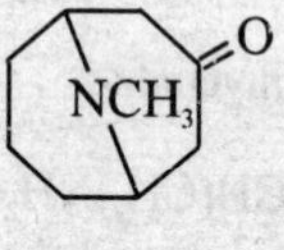

Pseudopelletierine

Piperine Alkaloids

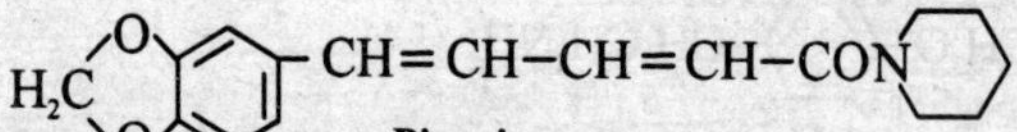

H_2C —— $CH=CH-CH=CH-CON$

Piperine

IV   Pyrrolidine–Pyridine group

Tobacco Alkaloids

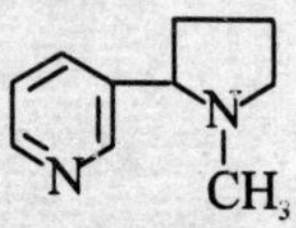

(–) Nicotine

Solanaceous Alkaloids

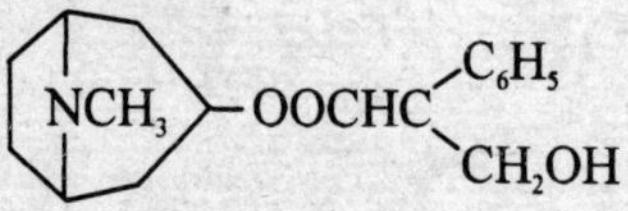

NCH_3 — $OOCHC$ $\genfrac{}{}{0pt}{}{C_6H_5}{CH_2OH}$

Atropine (±) Hyoscyamine

Coca Alkaloids

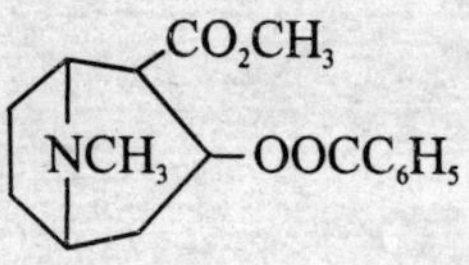

NCH_3 — $\genfrac{}{}{0pt}{}{CO_2CH_3}{OOCC_6H_5}$

(–) Cocaine

NCH_3 — $OOCC_6H_5$

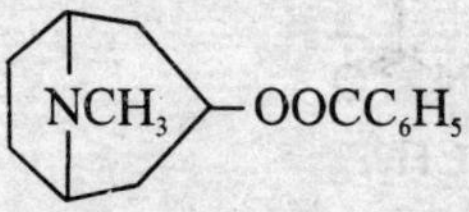

Tropacocaine

V   Quinoline group

Angostura Alkaloids

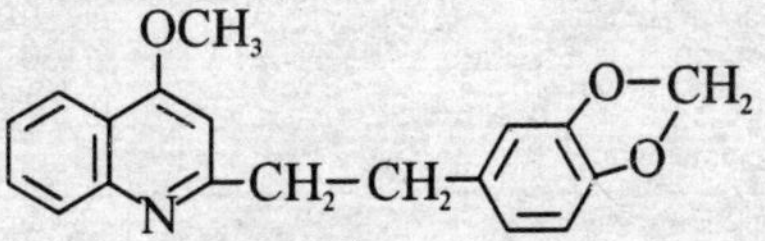

Cusparine

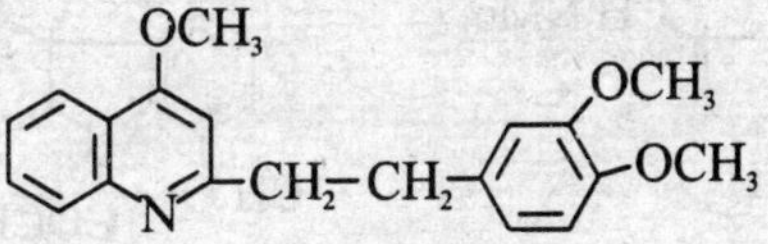

Galipine

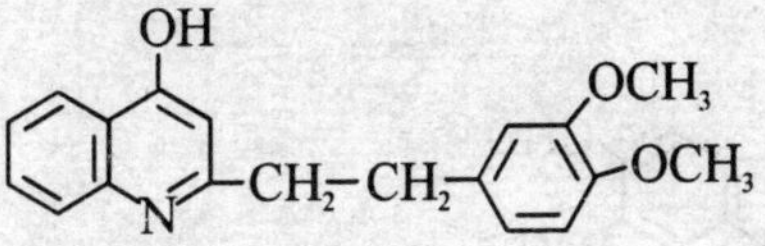

Galipoline

Cinchona Alkaloids

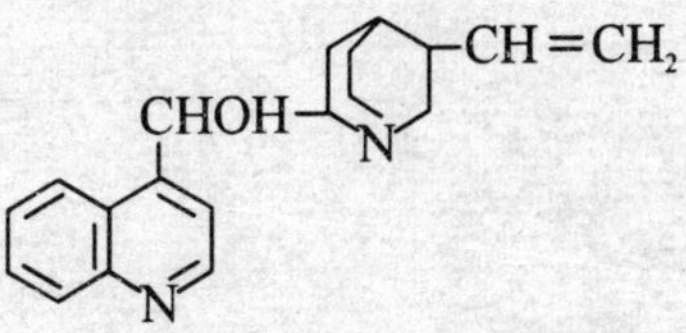

(+) Cinchonine

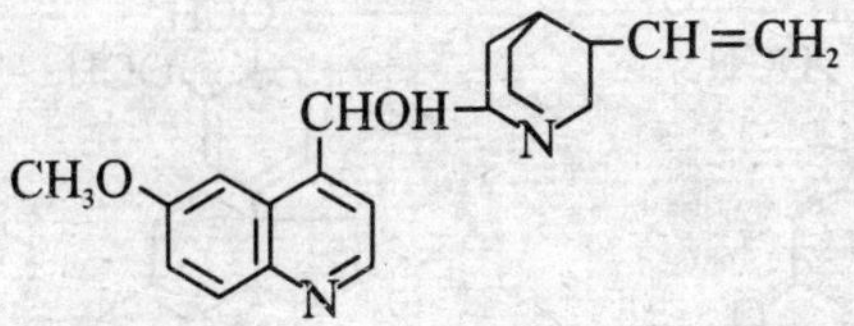

(−) Quinine

VI   Isoquinoline group

Opium Alkaloids

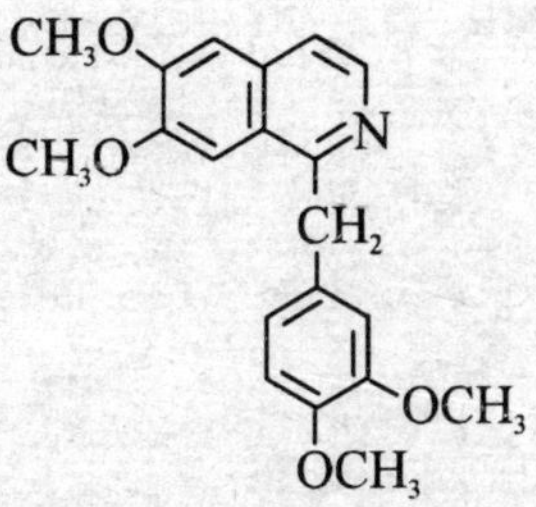

Papaverine

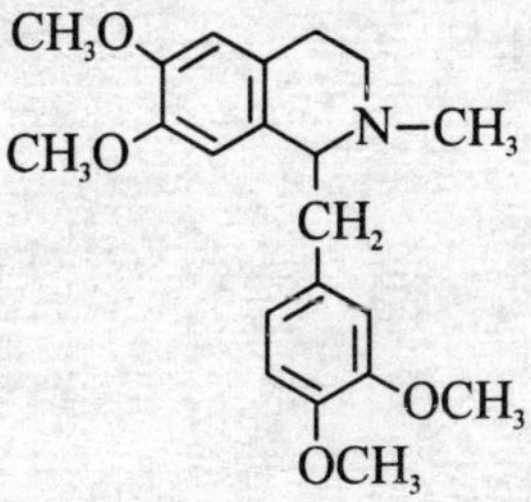

Laudanosine

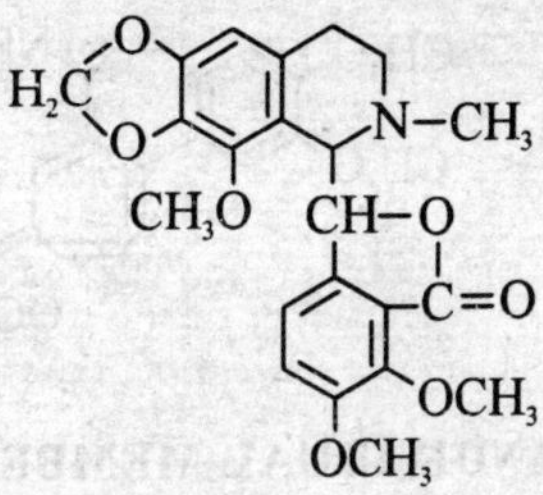

Narcotine

VII   Phenanthrene group

Opium Alkaloids

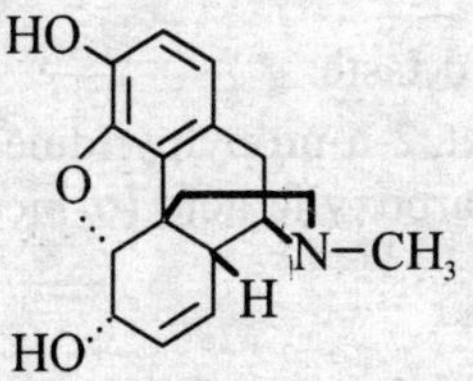

(−) Morphine

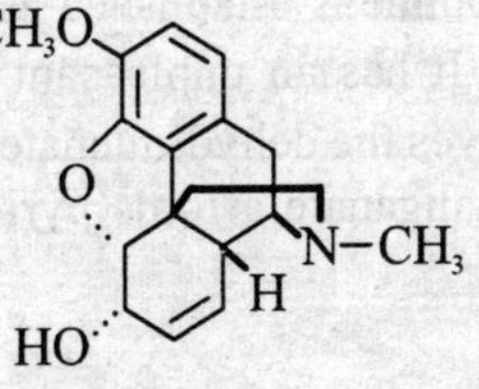

(−) Codeine

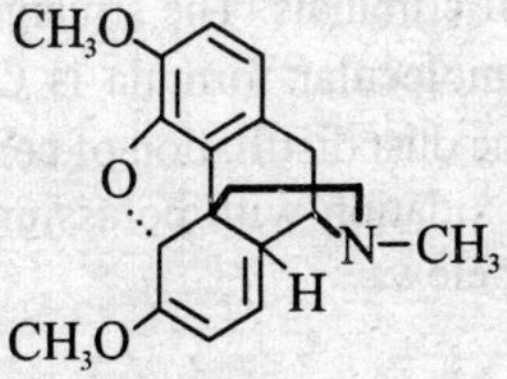

(−) Thebaine

VIII   Indole group

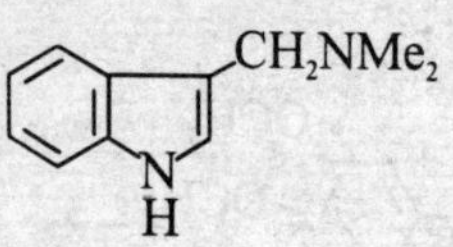

Gramini

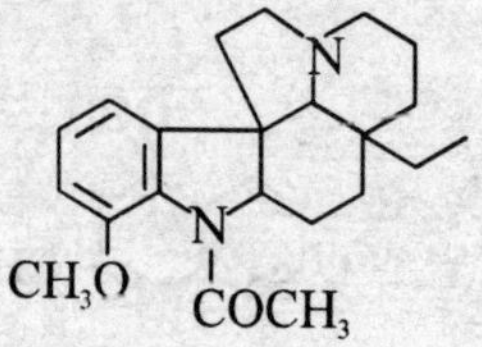

Aspidospermine

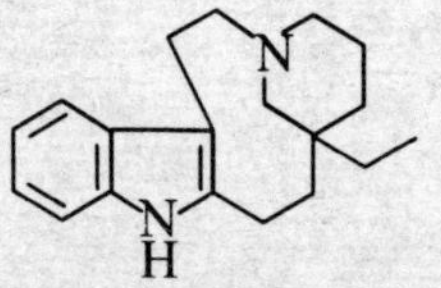

Quebrachamine

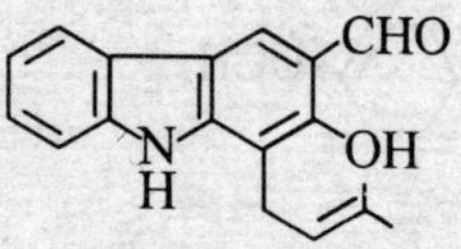

Heptaphylline

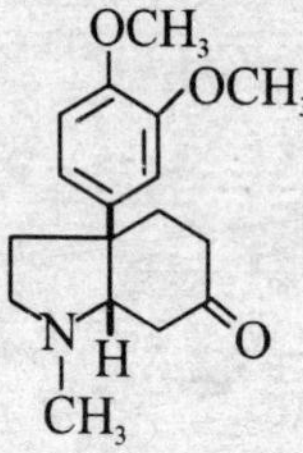

Mesembrine

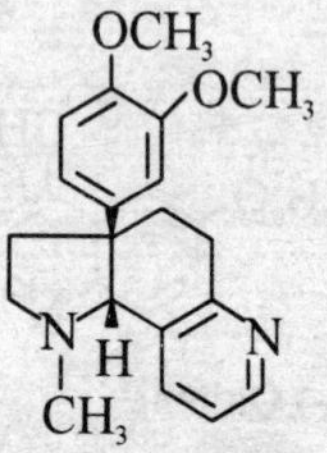

Sceletcum alkaloid A$_4$

IX   Colchinine

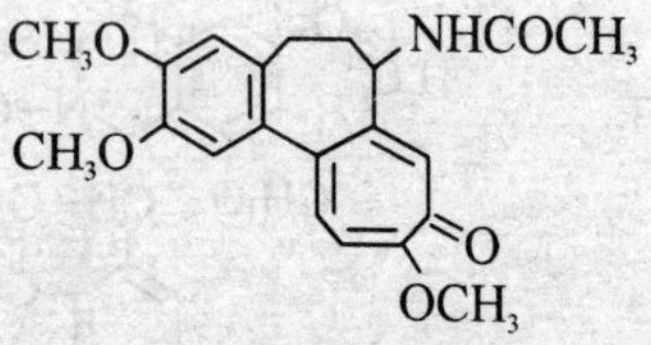

42.7   INDIVIDUAL MEMBERS

42.7.1   (+)-Coniine

Coniine belongs to the pyridine piperidine group of alkaloids. It is one of the most important member of the Hemlock alkaloids (the other members are γ-coniceine, conhydrine, pseudoconhydrine and N-methylconiine). The Hemlock extract (containing a mixture of the above alkaloids) has been used by the Greeks for the execution of criminals. The structure of coniine is established as follows:

(i)   Its molecular formula is $C_8H_{17}N$. It has an unpleasant odor and taste.

(ii)   Zinc dust distillation of coniine gives the dehydrogenated product, 2-n-propylpyridine (conyrine), which on oxidation with potassium permanganate affords pyridine-2-carboxylic acid (α-picolinic acid) as shown below:

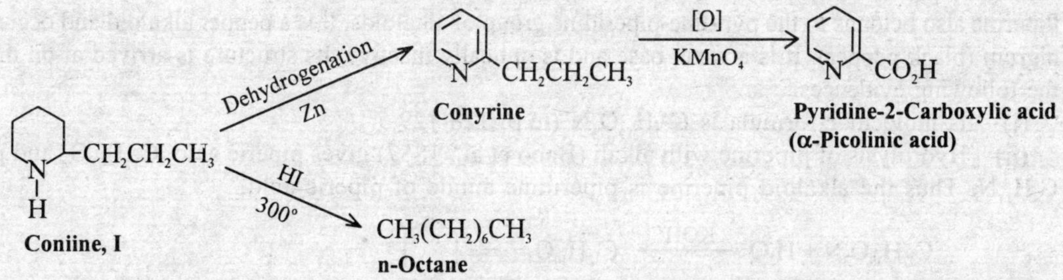

Coniine, I

Dehydrogenation / Zn → **Conyrine** (with CH$_2$CH$_2$CH$_3$ side chain)

[O] / KMnO$_4$ → **Pyridine-2-Carboxylic acid** (α-Picolinic acid) (CO$_2$H)

HI / 300° → CH$_3$(CH$_2$)$_6$CH$_3$ **n-Octane**

(iii)   On the basis of the above, coniine is probably a piperidine derivative with a three carbon atom side chain in 2-position. The side chain may either be n-propyl or isopropyl group. It has been shown that the side chain is n-propyl, since coniine on heating with hydriodic acid at 300° gives n-octane (an isopropyl side chain is likely to give iso-octane). Thus coniine is α-n-propylpiperidine (I).

(iv)   The structure (I) for coniine has finally been confirmed by its synthesis.

(a)   *Ladenburg's Synthesis (1886).*

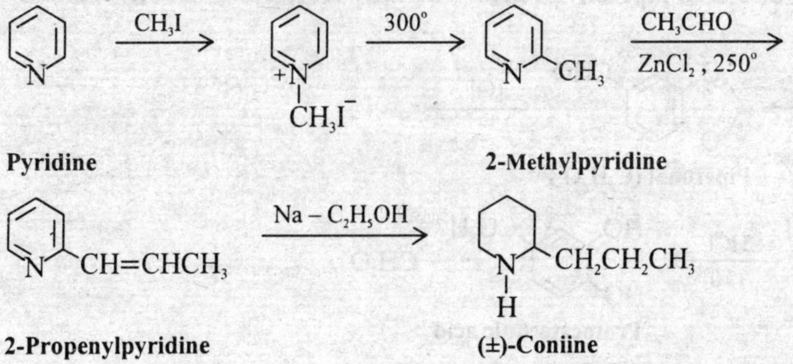

Pyridine   CH$_3$I →   (CH$_3$I$^-$) → 300° → **2-Methylpyridine** (CH$_3$)   CH$_3$CHO / ZnCl$_2$, 250° →

2-Propenylpyridine (CH=CHCH$_3$)   Na – C$_2$H$_5$OH →   **(±)-Coniine** (CH$_2$CH$_2$CH$_3$)

The (±)-coniine is resolved by (+)-tartaric acid and the (+)-coniine thus obtained is identical with the natural compound.

(b)   *Bergmann's Synthesis (1932)*

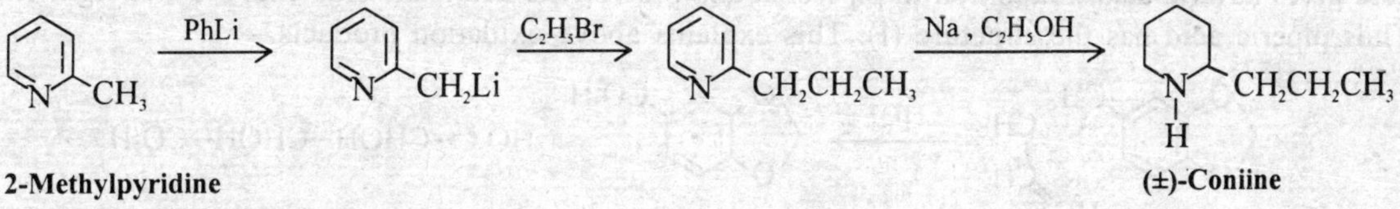

2-Methylpyridine (CH$_3$)   PhLi →   (CH$_2$Li)   C$_2$H$_5$Br →   (CH$_2$CH$_2$CH$_3$)   Na, C$_2$H$_5$OH →   **(±)-Coniine** (CH$_2$CH$_2$CH$_3$)

Other Hemlock alkaloids are:

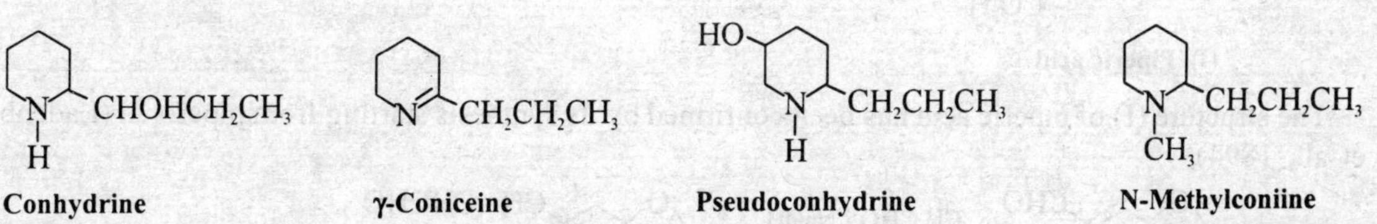

Conhydrine (CHOHCH$_2$CH$_3$)   **γ-Coniceine** (CH$_2$CH$_2$CH$_3$)   **Pseudoconhydrine** (HO-, CH$_2$CH$_2$CH$_3$)   **N-Methylconiine** (CH$_2$CH$_2$CH$_3$, CH$_3$)

42.7.2   Piperine

Piperine also belongs to the pyridine-piperidine group of alkaloids. It is a pepper alkaloid and occurs in Piper nigrum (black pepper). It is a weak base and is optically inactive. Its structure is arrived at on the basis of the following evidences.

(i)   Its molecular formula is $C_{17}H_{19}O_3N$ (m.p. 128-129.5°).

(ii)   Hydrolysis of piperine with alkali (Babo et al., 1857) gives piperic acid, $C_{12}H_{10}O_4$ and piperidine $C_5H_{11}N$. Thus the alkaloid piperine is piperidine amide of piperic acid.

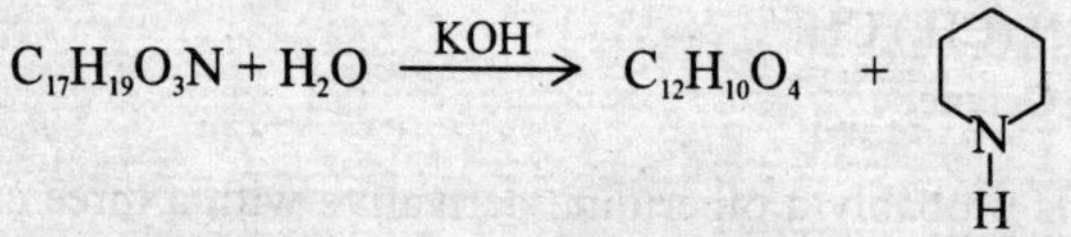

$$C_{17}H_{19}O_3N + H_2O \xrightarrow{KOH} C_{12}H_{10}O_4 \ +$$

Piperine   **Piperic acid**   **Piperidine**

The structure of piperidine as hexahydropyridine is well known. Thus, the structure of piperine depends on the structure of piperic acid.

(iii)   The structure of piperic acid, $C_{12}H_{10}O_4$ is determined as follows. Usual tests show that piperic acid has one carboxylic group and two double bonds. Oxidation of piperic acid with potassium permanganate gives first piperonal and then piperonylic acid. The later on heating with hydrochloric acid gives protocatechuic acid.

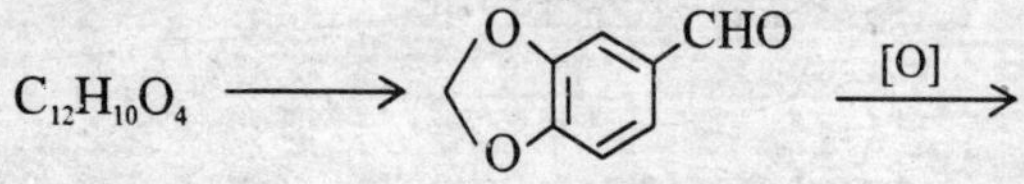

$$C_{12}H_{10}O_4 \longrightarrow \quad \xrightarrow{[O]}$$

Piperic acid   **Piperonal ($C_8H_6O_3$)**

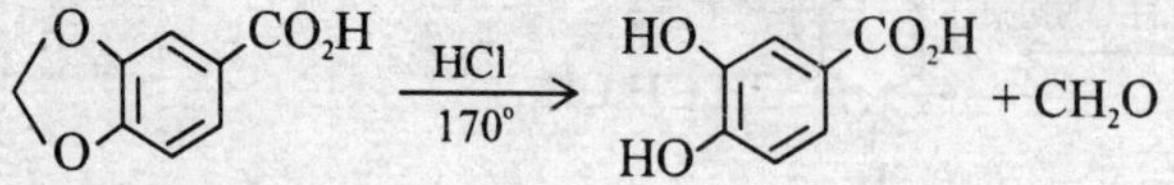

$$\xrightarrow[170°]{HCl} \quad + CH_2O$$

Piperonylic acid   **Protocatechuic acid**
($C_8H_6O_4$)

The structure of piperonal (section 35.1.9), piperonylic acid and protocatechuic acid is well known.

On the basis of the above oxidative degradation, piperic acid is a benzene derivative having one side chain which contains two double bonds and a carboxylic group. Further, since careful oxidation of piperic acid gives tartaric acid in addition to piperonal and piperonylic acid, the side chain is a straight chain. Thus piperic acid has the structure (I). This explains above oxidation products.

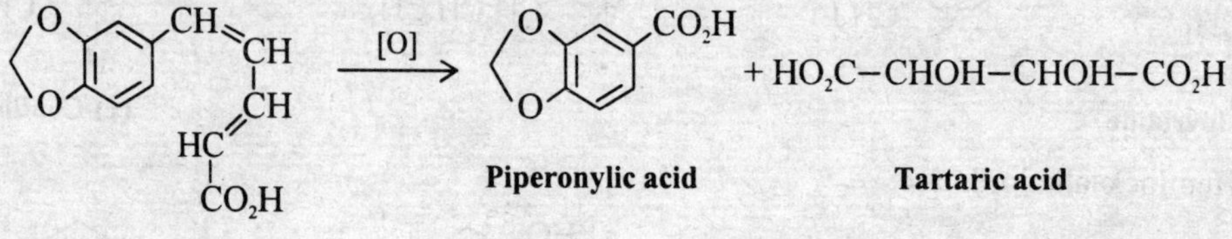

$$\xrightarrow{[O]} \quad + HO_2C-CHOH-CHOH-CO_2H$$

(I), Piperic acid   **Piperonylic acid**   **Tartaric acid**

The structure (I) of piperic acid has been confirmed by its synthesis starting from piperonal (Ladenburg et al., 1894).

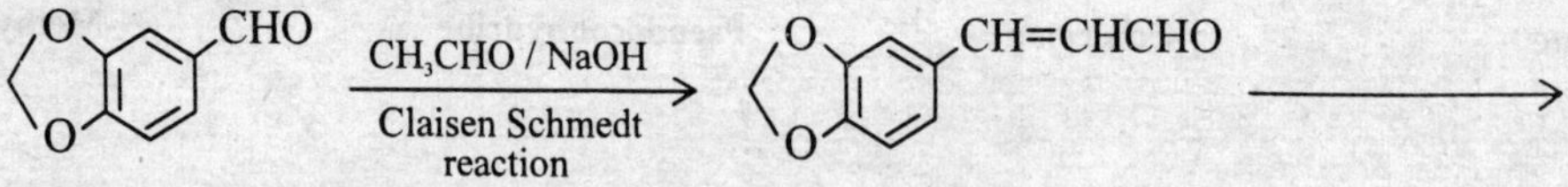

$$\xrightarrow[\text{reaction}]{CH_3CHO \, / \, NaOH \atop \text{Claisen Schmedt}} \longrightarrow$$

Piperonal

$$\xrightarrow[\substack{\text{NaOAC} \\ \text{Perkin reaction}}]{(CH_3CO)_2O}$$

[structure] CH=CHCH=CHCO$_2$H

Piperic acid

On the basis of the above, Piperine has the structure (II), which is confirmed by its synthesis.

Piperic acid, (I) $\xrightarrow[\text{2. HN}]{\text{1. SOCl}_2}$ [structure] CH=CHCH=CHCON

Piperine (II)

The stereochemistry of piperine is shown to be trans-trans about the double bonds. The cis-cis stereoisomer is chavicine, which also occurs in pepper along with piperine.

42.7.3 Nicotine

Nicotine belongs to the pyrrolidine -pyridine group of alkaloids and is a well known tobacco alkaloid. It is the chief alkaloid isolated from the leaves of tobacco plant of Nicotiana species. The other minor alkaloidal constituents of tobacco plant are nor-nicotine, nicotimine etc. Nicotine is poisonous in nature and is used as an insecticide. The structure of nicotine is established as:

(i)   Its molecular formula is $C_{10}H_{14}N_2$ (b.p. 247°) and occurs as the (–)-form, $[\alpha]_D -169°$.

(ii)   Nicotine on oxidation with dichromate, sulphuric acid or potassium permanganate gives nicotinic acid, pyridine-3-carboxylic acid (1a). Thus, nicotine has a pyridine nucleus with a side chain at position-3:

[structure] $C_5H_{10}N$

(iii)   The nature of the side chain is

$(C_{10}H_{14}N_2)$ Nicotine $\xrightarrow[\text{Na–Amyl alcohol}]{\text{Reduction}}$

[structure]

$\begin{array}{cc} & CH_2 \\ CH_2 & CH-C_5H_{10}N \\ CH_2 & CH_2 \\ & N \\ & H \end{array}$ $(C_{10}H_{20}N_2)$

Hexahydro derivative of Nicotine

(a) Reduction of nicotine with sodium and amyl alcohol gives a hexahydro derivative with the addition of six hydrogen atoms which suggests that the side chain of nicotine is saturated.

(b) Nicotine forms a crystalline addition compound with zinc chloride which on heating with lime gives pyridine, pyrrole and methylamine. This indicates that the side chain $C_5H_{10}N$ is a pyrrole derivative.

(c) Further, nicotine on heating with concentrated hydriodic acid (Herzig-Meyer method) gives methyl iodide.

Thus, the side chain has a N-methyl group which can be N-methylpyrrolidine and its point of attachment is decided by a series of reactions.

(d) The conversion of nicotine into hygrinic acid, (1d) [nicotine on treatment with methyl iodide gives nicotine isomethiodide (1b), which on oxidation with potassium ferricyanide yields nicotone (1c). Its further oxidation with chromium trioxide produces hygrinic acid (1d)] proves conclusively that the side chain is saturated N-methylpyrrolidine and attached to the C-3 of pyridine nucleus.

Nicotine is given structure (I) on the basis various reactions explained.

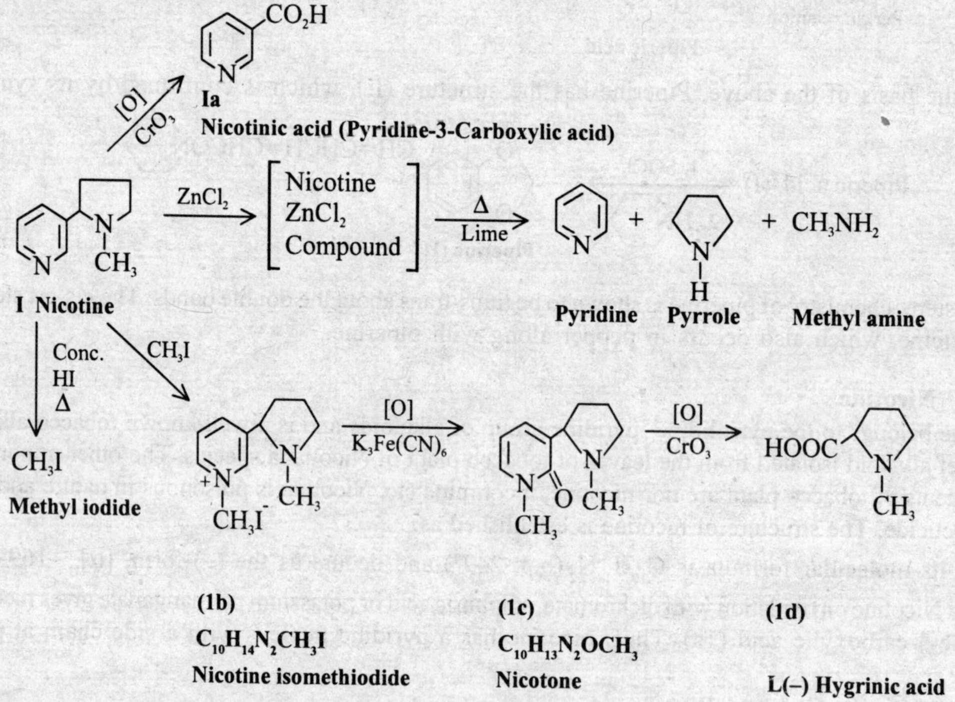

Ia
Nicotinic acid (Pyridine-3-Carboxylic acid)

I Nicotine

Nicotine ZnCl₂ Compound → Δ/Lime → Pyridine + Pyrrole + CH_3NH_2 Methyl amine

Methyl iodide

(1b)
$C_{10}H_{14}N_2CH_3I$
Nicotine isomethiodide

(1c)
$C_{10}H_{13}N_2OCH_3$
Nicotone

(1d)
L(–) Hygrinic acid

(iv)   Structure (I) for nicotine also explains following two reactions:

(a) Nicotine on treatment with bromine in acetic acid gives the hydrobromide perbromide (as the main product), which on treatment with aqueous sulphurous acid affords dibromocotinine, $C_{10}H_{10}Br_2N_2O$. Its subsequent heating with a mixture of sulphurous acid and sulphuric acid (130-140°) gives 3-acetylpyridine, oxalic acid and methyl amine.

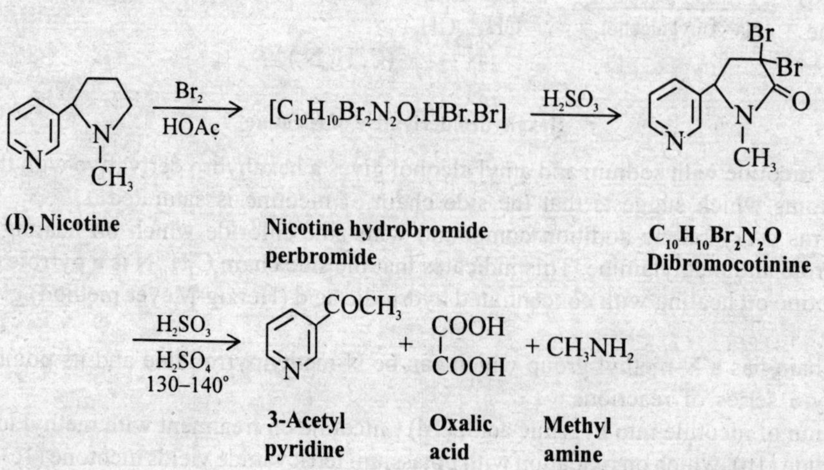

(I), Nicotine

$\xrightarrow[\text{HOAc}]{Br_2}$ $[C_{10}H_{10}Br_2N_2O.HBr.Br]$ **Nicotine hydrobromide perbromide**

$\xrightarrow{H_2SO_3}$ $C_{10}H_{10}Br_2N_2O$ **Dibromocotinine**

$\xrightarrow[\substack{H_2SO_4 \\ 130-140°}]{H_2SO_3}$ **3-Acetyl pyridine** + **Oxalic acid** + CH_3NH_2 **Methyl amine**

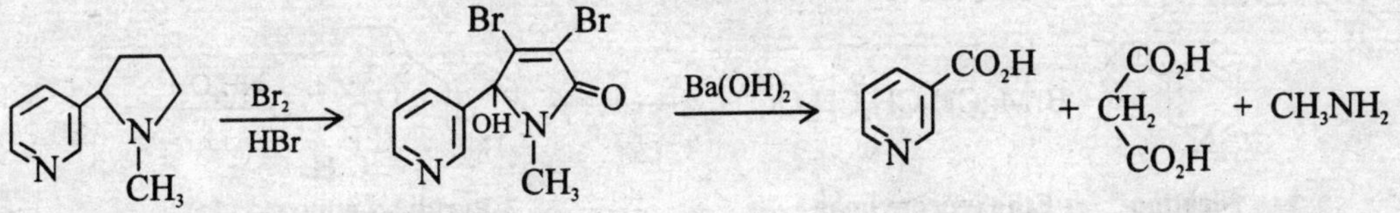

(I), Nicotine   →[Br$_2$ / HBr]   C$_{10}$H$_8$Br$_2$N$_2$O$_2$ Dibromoticonine   →[Ba(OH)$_2$]   Nicotinic acid + Malonic acid + Methyl amine

(b) Nicotine on treatment with bromine in presence of hydrobromic acid gives dibromoticonine, C$_{10}$H$_8$Br$_2$N$_2$O$_2$, which on heating with barium hydroxide yields nicotinic acid, malonic acid and methyl amine as shown above.

(v)   The structure of nicotine has finally been confirmed by its synthesis.

(a)   *Spath and Bretschncider (1928)*

First Step

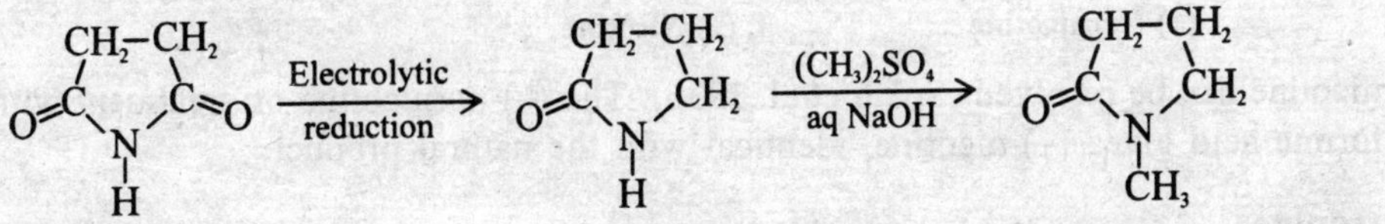

Succinimide   →[Electrolytic reduction]   2-Pyrrolidone   →[(CH$_3$)$_2$SO$_4$ / aq NaOH]   N-Methyl-2-Pyrrolidone

Second Step

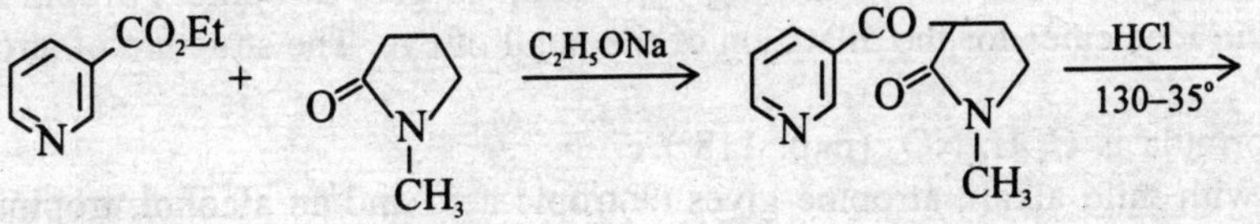

Ethyl nicotinate + N-Methyl-2-pyrrolidone   →[C$_2$H$_5$ONa]   (intermediate)   →[HCl 130–35°]

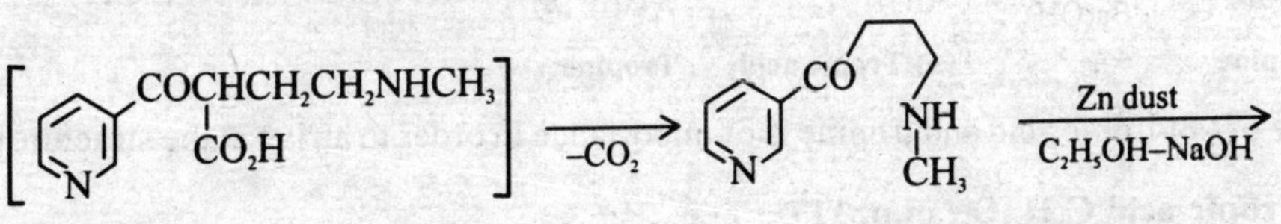

[β-ketonic acid]   →[–CO$_2$]   →[Zn dust C$_2$H$_5$OH–NaOH]

→[HI 100°]   →[NaOH]   (±) – Nicotine

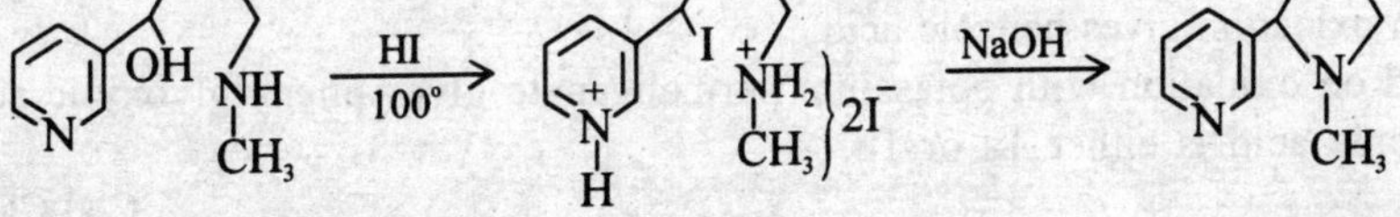

The (±)-nicotine is resolved with (+)-tartaric acid. The synthetic (–)-nicotine is identical with the natural compound.

(b) *Craig (1933)*

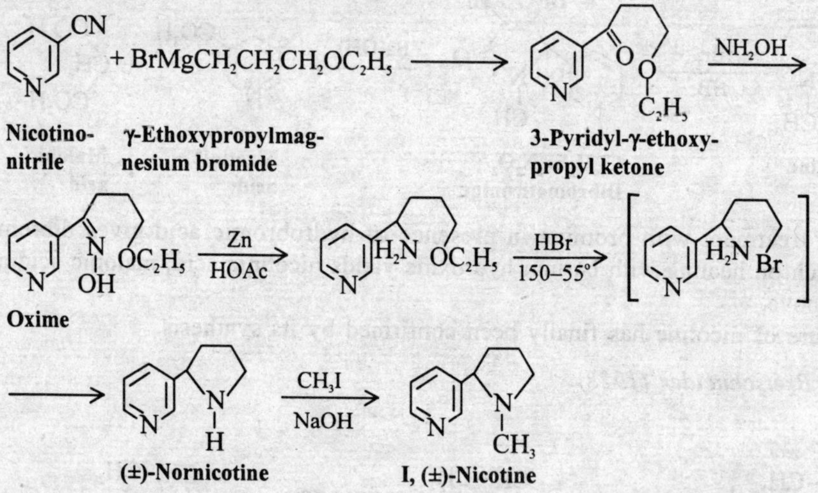

Nicotino-   γ-Ethoxypropylmag-   3-Pyridyl-γ-ethoxy-
nitrile   nesium bromide   propyl ketone

Oxime

(±)-Nornicotine   I, (±)-Nicotine

(±)-Nornicotine can be resolved (Spath et al.,1936). The (–)-nornicotine on treatment with formalde-hyde and formic acid gives (–)-nicotine, identical with the natural product.

42.7.4 Atropine

Atropine also belongs to the pyrrolidine-pyridine group of alkaloids and is known as a solanaceous alkaloid. It is present in Atropa belladonna (deadly nightshade) together with hyoscyamine. Hyoscyamine is optically active, $[\alpha]_D - 22°$ and readily recemises on warming with alkali to give atropine. Atropin is thus (±)-hyoscyamine. It is used in medicines for the dilatation of the pupil of eye. The structure of atropine is es-tablished as follows:

  (i)   Its molecular formula is $C_{17}H_{23}NO_3$ (m.p. 118°).
  (ii) On hydrolysis with mild alkali, atropine gives (±)tropic acid and an alcohol, tropine. Thus, at-ropine is tropine ester of tropic acid, i.e. tropine tropate.

$$C_{17}H_{23}NO_3 \xrightarrow{Ba(OH)_2} C_9H_{10}O_3 \ + \ C_8H_{15}NO$$

Atropine   **(±)-Tropic acid**   **Tropine**

The structure of (±)-tropic acid and tropine is of importance in order to arrive at the structure of atropine.

Structure of Tropic acid $C_9H_{10}O_3$, m.p. 117°

Usual tests show that tropic acid contains one carboxyl group and one alcoholic group and is a saturated compound (as shown by bromine test). On heating, tropic acid loses a molecule of water and gives atropic acid, $C_9H_8O_2$, which on oxidation gives benzoic acid.

Further, atropic acid on oxidation with potassium permanganate gives phenylglyoxylic acid. On the basis of the above, tropic acid is either 1a or 1b.

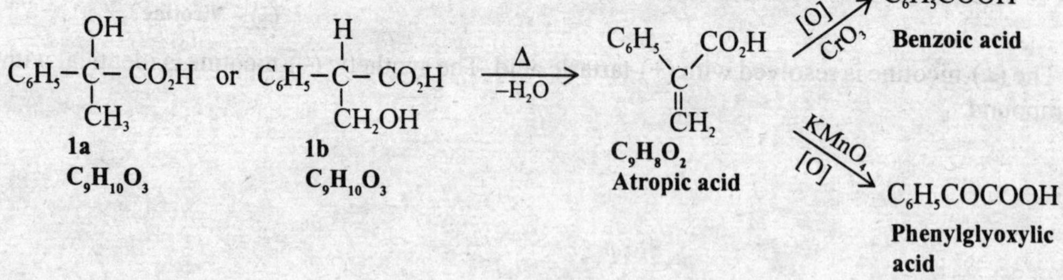

$$C_6H_5-\overset{\overset{\displaystyle OH}{|}}{\underset{\underset{\displaystyle CH_3}{|}}{C}}-CO_2H \quad or \quad C_6H_5-\overset{\overset{\displaystyle H}{|}}{\underset{\underset{\displaystyle CH_2OH}{|}}{C}}-CO_2H$$

1a   1b
$C_9H_{10}O_3$ $C_9H_{10}O_3$

$C_9H_8O_2$
Atropic acid

C_6H_5COOH
Benzoic acid

$C_6H_5COCOOH$
Phenylglyoxylic acid

The structure Ib for tropic acid has been confirmed by its synthesis (Mackenzie and Wood, 1919).

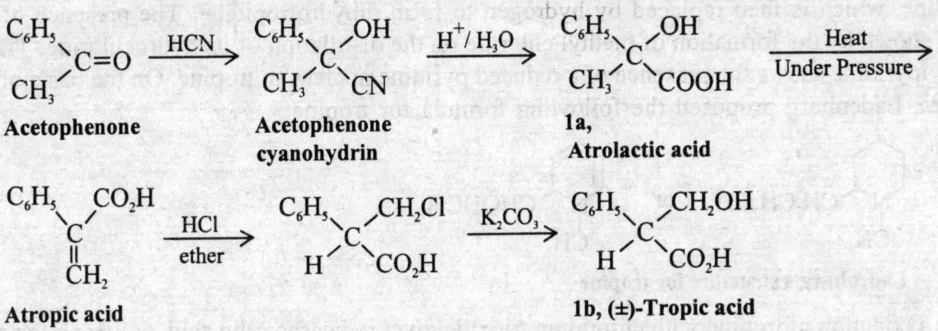

$$C_6H_5\!-\!\underset{CH_3}{\overset{}{C}}\!=\!O \xrightarrow{HCN} \underset{CH_3}{\overset{C_6H_5}{C}}\!\!\!\!\!\!\underset{CN}{\overset{OH}{}} \xrightarrow{H^+/H_2O} \underset{CH_3}{\overset{C_6H_5}{C}}\!\!\!\!\!\!\underset{COOH}{\overset{OH}{}} \xrightarrow[\text{Under Pressure}]{\text{Heat}}$$

| Acetophenone | Acetophenone cyanohydrin | 1a, Atrolactic acid | |

$$\underset{\overset{\|}{CH_2}}{\overset{C_6H_5}{\underset{}{C}}}\!\!\!\!\!\!\overset{CO_2H}{} \xrightarrow[\text{ether}]{HCl} \underset{H}{\overset{C_6H_5}{C}}\!\!\!\!\!\!\underset{CO_2H}{\overset{CH_2Cl}{}} \xrightarrow{K_2CO_3} \underset{H}{\overset{C_6H_5}{C}}\!\!\!\!\!\!\underset{CO_2H}{\overset{CH_2OH}{}}$$

Atropic acid   **1b, (±)-Tropic acid**

As seen in the above synthesis the dehydration of atrolactic acid (1a) to atropic acid, confirms the structure of atropic acid. It is of interest to note that addition of hydrogen chloride to atropic acid takes place contrary to Markovnikov's rule. In case, the addition has been in accordance to Markovnikov's rule, the atrolactic acid (1a) would have been obtained instead of tropic acid (1b). The (±)-tropic acid (1b) obtained above can be resolved by quinine.

(±)-Tropic acid has also been synthesised (Blicke et al. 1952) as follows.

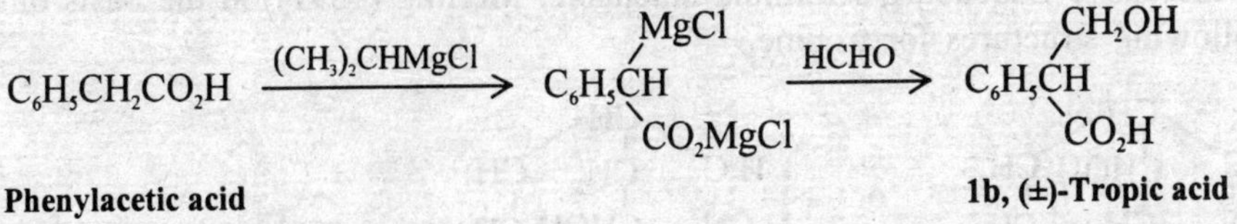

$$C_6H_5CH_2CO_2H \xrightarrow{(CH_3)_2CHMgCl} \underset{CO_2MgCl}{\overset{MgCl}{C_6H_5CH}} \xrightarrow{HCHO} \underset{CO_2H}{\overset{CH_2OH}{C_6H_5CH}}$$

Phenylacetic acid   **1b, (±)-Tropic acid**

The absolute configuration of (–)-tropic acid has been established (Fodor et al, 1961) by its correlation with (–)-alanine.

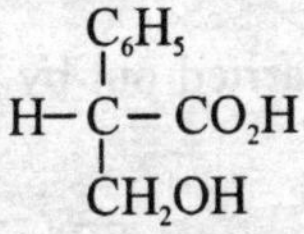

$$H\!-\!\underset{CH_2OH}{\overset{C_6H_5}{C}}\!\!\!\!\!\!-CO_2H$$

(–)-Tropic acid

Structure of Tropine $C_8H_{15}NO$, m.p. 63°C

(i)   It is shown to have a secondary alcoholic group, as on oxidation tropine gives a ketone, tropinone, $C_8H_{13}NO$.

(ii)   It contains a tertiary nitrogen, as on treatment with methyl iodide, it gives the methiodide, $C_8H_{15}NO.CH_3I$.

(iii)   Tropine contaíns a reduced pyridine nucleus. This was shown by Laderburg (1883, 1887) by the sequence of reactions shown below.

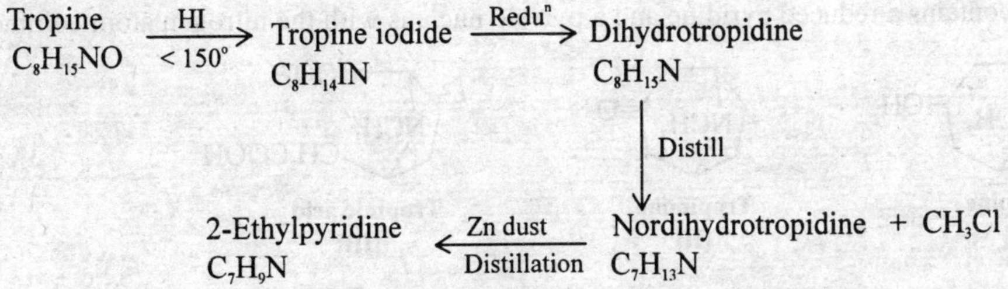

$$\text{Tropine} \atop C_8H_{15}NO \xrightarrow[<150°]{HI} {\text{Tropine iodide} \atop C_8H_{14}IN} \xrightarrow{\text{Redu}^n} {\text{Dihydrotropidine} \atop C_8H_{15}N}$$

$$\downarrow \text{Distill}$$

$${\text{2-Ethylpyridine} \atop C_7H_9N} \xleftarrow[\text{Distillation}]{\text{Zn dust}} {\text{Nordihydrotropidine} \atop C_7H_{13}N} + CH_3Cl$$

In the above reactions tropine iodide is obtained by the replacement of alcoholic group of tropine with iodine, which is then replaced by hydrogen to form dihydrotropidine. The presence of N-methyl group is shown by the formation of methyl chloride on the distillation of its hydrochloride. The isolation of 2-ethylpyridine shows the presence of a reduced pyridine nucleus in tropine. On the basis of the above evidences, Ladenburg proposed the following formula for tropine:

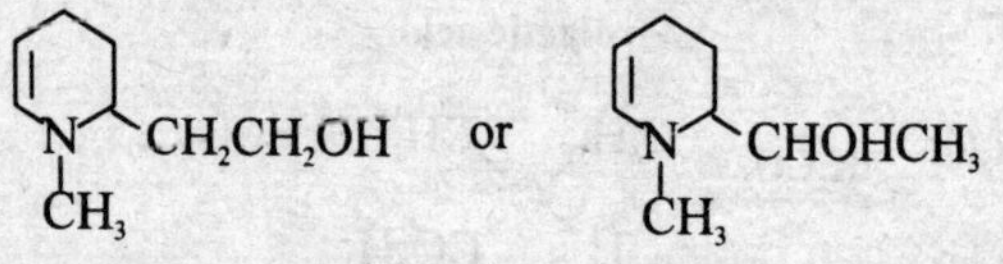

Ladenburg's structure for tropine

(iv)   Oxidation of tropine with chromium trioxide gives a dicarboxylic acid, (±)-tropinic acid, having the same number of carbon atoms as tropine.

$$C_8H_{15}NO \xrightarrow[CrO_3]{[O]} C_8H_{13}NO_4$$

Tropine   **(±) Tropinic acid**

On the basis of the above, it is concluded that the hydroxy group in tropine must be in a ring system, thereby making the above Ladenburg's formula untenable. Merling (1891) on the basis of the above, proposed the following structures for tropine.

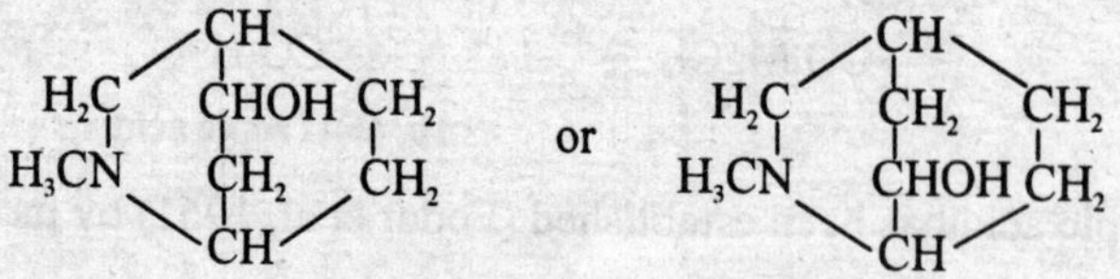

Merling's structure for tropine

(v)   The systematic examination of the oxidation products of tropine was carried out by Willstatter (1895-1901).

$$\text{Tropine} \atop C_8H_{15}NO \quad \xrightarrow[CrO_3]{[O]} \quad {\text{Tropinone} \atop C_8H_{13}NO} \quad \xrightarrow[CrO_3]{[O]} \quad {\text{(±)-Tropinic acid} \atop C_8H_{13}NO_4} \quad \xrightarrow[CrO_3]{[O]}$$

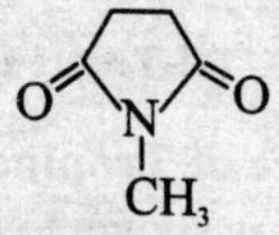

N-Methylsuccinimide

Tropinone obtained upon oxidation behaved like a ketone indicating the presence of secondary alcoholic group in tropine (see also Merling's formula). Further, tropinone formed a dibenzylidene derivative with benzaldehyde and a dioximino derivative with amyl nitrite and hydrochloric acid. Thus, tropinone contains the group, $-CH_2COCH_2-$, and so Merling's formula is untenable. Out of the three structures proposed by Willstatter, the following structure (II) explains all the above observations (the other two eliminated). This structure contains a reduced pyridine and a pyrrole nucleus with the nitrogen atom common to both.

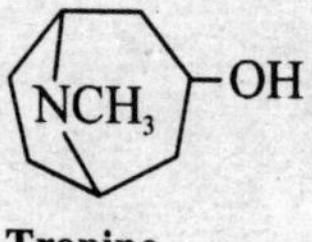

Tropine   **Tropinone**   **Tropinic acid**
[I]   **[II]**   **[III]**

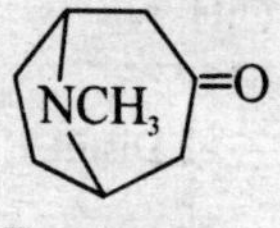

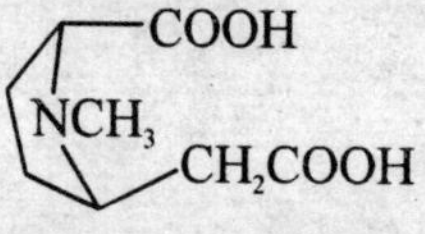

On this basis the structures for tropine, tropinone and tropinic acid are given as I, II and III respectively. The Willstatter's formula (I) for tropine accounts for all the reactions of tropine.

(a) Formation of 2-ethylpyridine from tropine (I).

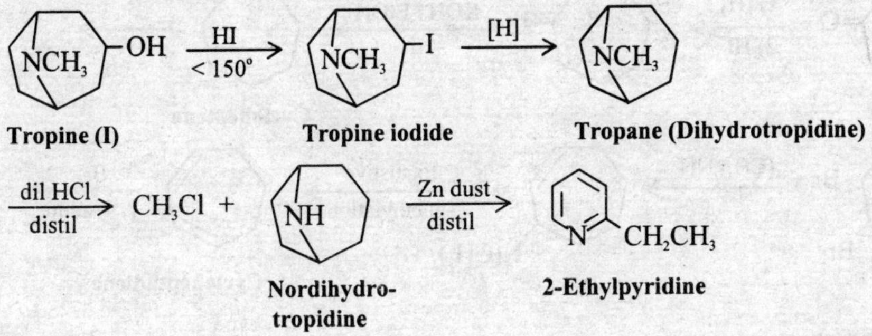

Tropine (I)   Tropine iodide   Tropane (Dihydrotropidine)

Nordihydro-   2-Ethylpyridine
tropidine

(b) Formation of tropinone, tropinic acid and N-methylsuccinimide from tropine (I).

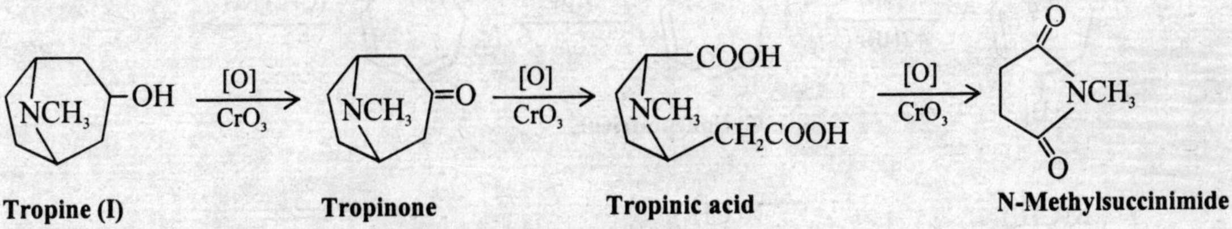

Tropine (I)   Tropinone   Tropinic acid   N-Methylsuccinimide

(c) Exhaustive methylation of tropine gives tropilidene (Cycloheptatriene).

Tropine (I)

Tropilidene (Cycloheptatriene)

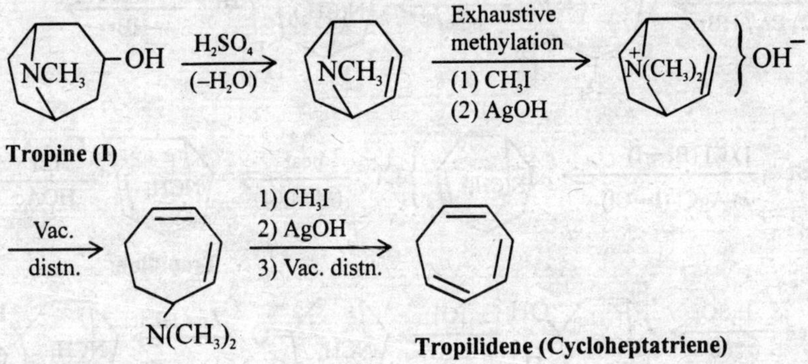

(d) Exhaustive methylation of tropinic acid gives an unsaturated dicarboxylic acid, which on reduction forms pimelic acid.

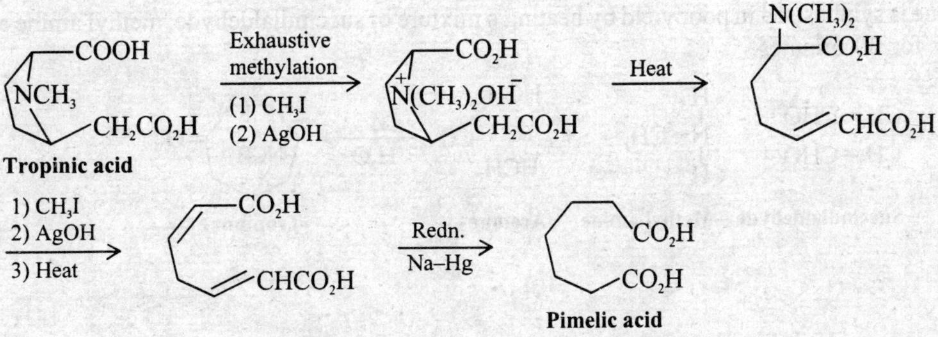

Tropinic acid

Pimelic acid

(vi)   The structure (I) for tropine has finally been confirmed by its synthesis.

Willstatter's Synthesis

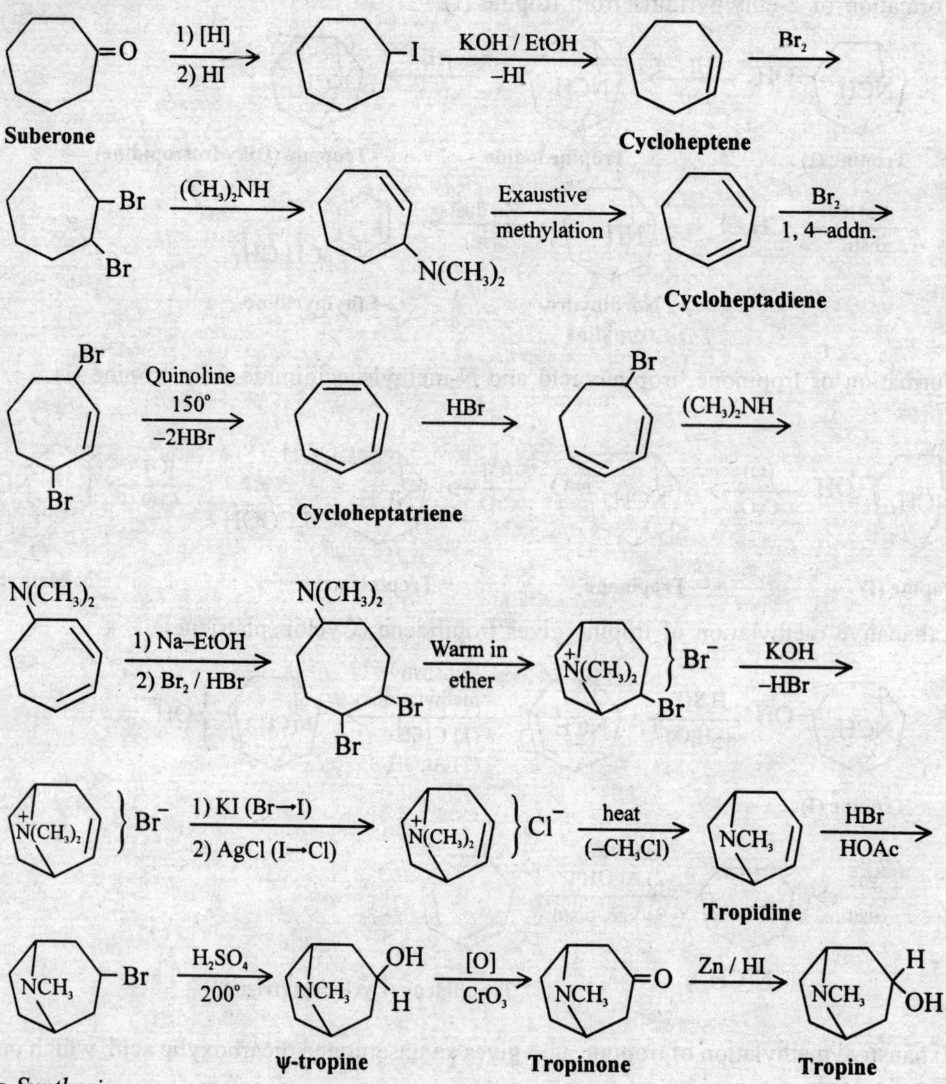

Suberone $\xrightarrow[\text{2) HI}]{\text{1) [H]}}$ (cycloheptyl iodide) $\xrightarrow[\text{-HI}]{\text{KOH / EtOH}}$ Cycloheptene $\xrightarrow{\text{Br}_2}$

(dibromocycloheptane) $\xrightarrow{\text{(CH}_3)_2\text{NH}}$ (N(CH$_3$)$_2$ cycloheptene) $\xrightarrow[\text{methylation}]{\text{Exaustive}}$ Cycloheptadiene $\xrightarrow[\text{1, 4–addn.}]{\text{Br}_2}$

(dibromocycloheptadiene) $\xrightarrow[\text{-2HBr}]{\substack{\text{Quinoline} \\ \text{150°}}}$ Cycloheptatriene $\xrightarrow{\text{HBr}}$ (Br cycloheptadiene) $\xrightarrow{\text{(CH}_3)_2\text{NH}}$

(N(CH$_3$)$_2$ cycloheptadiene) $\xrightarrow[\text{2) Br}_2\text{ / HBr}]{\text{1) Na–EtOH}}$ (N(CH$_3$)$_2$ dibromocycloheptane) $\xrightarrow[\text{ether}]{\text{Warm in}}$ $\left\{ \overset{+}{\text{N(CH}_3)_2} \right\}$ Br$^-$ $\xrightarrow[\text{–HBr}]{\text{KOH}}$

$\left\{ \overset{+}{\text{N(CH}_3)_2} \right\}$ Br$^-$ $\xrightarrow[\text{2) AgCl (I→Cl)}]{\text{1) KI (Br→I)}}$ $\left\{ \overset{+}{\text{N(CH}_3)_2} \right\}$ Cl$^-$ $\xrightarrow[\text{(–CH}_3\text{Cl)}]{\text{heat}}$ (NCH$_3$) **Tropidine** $\xrightarrow[\text{HOAc}]{\text{HBr}}$

(NCH$_3$ Br) $\xrightarrow[\text{200°}]{\text{H}_2\text{SO}_4}$ (NCH$_3$ OH H) **ψ-tropine** $\xrightarrow[\text{CrO}_3]{\text{[O]}}$ (NCH$_3$ =O) **Tropinone** $\xrightarrow{\text{Zn / HI}}$ (NCH$_3$ H OH) **Tropine**

Robinson's Synthesis

Tropinone is synthesised in poor yield by heating a mixture of succindialdehyde, methyl amine and acetone in water for 30 minutes.

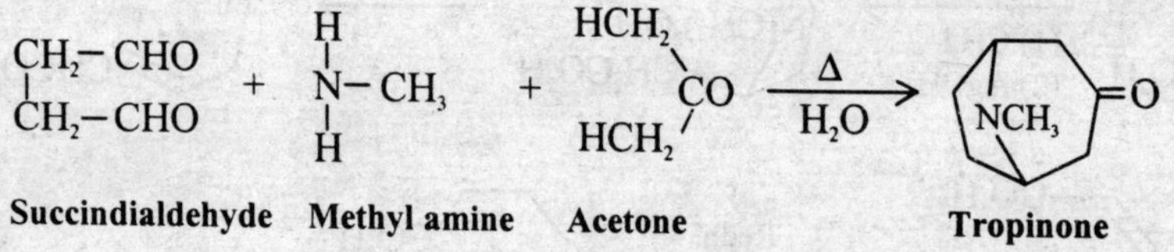

$$\begin{matrix} \text{CH}_2\text{—CHO} \\ | \\ \text{CH}_2\text{—CHO} \end{matrix} \quad + \quad \begin{matrix} \text{H} \\ | \\ \text{N—CH}_3 \\ | \\ \text{H} \end{matrix} \quad + \quad \begin{matrix} \text{HCH}_2 \\ \diagdown \\ \quad\quad \text{CO} \\ \diagup \\ \text{HCH}_2 \end{matrix} \quad \xrightarrow[\text{H}_2\text{O}]{\Delta} \quad \text{(NCH}_3 \text{ =O)}$$

Succindialdehyde   Methyl amine   Acetone   Tropinone

A better yields (40%) of tropinone is obtained by using calcium salt of acetone dicarboxylate or ethyl acetone dicarboxylate instead of acetone.

$$CH_2-CHO \atop CH_2-CHO \quad + \quad {H \atop N-CH_3 \atop H} \quad + \quad {Ca^+\bar{O}OCCH_2 \atop CO \atop Ca^+\bar{O}OCCH_2} \longrightarrow$$

Succindial-dehyde	Methyl-amine	Ca salt of acetone dicarboxylate

$$\xrightarrow{\text{Warm} \atop \text{HCl}}$$

Tropinone

Elming's Method: Tropinone is synthesised in 80% yield by condensing methyl amine, acetone dicarboxylic acid and succindialdehyde (generated in situ by the action of acid on 2,5-dimethoxy tetrahydrofuran).

$$H_3CO \diagup O \diagdown OCH_3 \xrightarrow[H_2O]{H^+} {CH_2-CHO \atop CH_2-CHO} \xrightarrow{CH_3NH_2.HCl \atop OC(CH_2CO_2H)}$$

2, 5-Dimethoxy-tetrahydrofuran		Succindi-aldehyde	Tropinone

Finally, the structure of atropine is established by its synthesis by condensing tropic acid and tropine in presence of hydrogen chloride.

$$NCH_3 - OH \quad + \quad HO_2C-\underset{CH_2OH}{\overset{C_6H_5}{\underset{|}{\overset{|}{C}}}}-H \xrightarrow{HCl} NCH_3 -OOC-\underset{CH_2OH}{\overset{C_6H_5}{\underset{|}{\overset{|}{C}}}}-H$$

Tropine	(+) Tropic acid	(+) Atropine

(–) Tropic acid forms (–) atropine.

42.7.5 Cocaine

Like nicotine and atropine, cocaine also belongs to the pyrrolidine pyridine group of alkaloids. It is also known as coca alkaloid.

It is the chief alkaloid constituent of the coca leaves (Erythrxylon coca) and is used as a local anaesthetic. It is a colorless, crystalline solid, m.p. 98° and sparingly soluble in water. It is a strong base and forms salts with acids. Its structure is arrived at as follows.

(i)   Its molecular formula is $C_{17}H_{21}NO_4$.

(ii)   Hydrolysis of cocaine gives methanol and benzoylecgonine, $C_{16}N_{19}NO_4$. Thus, cocaine contains a carbomethoxy group and benzoylecgonine has a carboxyl group.

(iii)   Benzoylecgonine on further hydrolysis by heating with barium hydroxide gives benzoic acid and ecgonine, $C_9H_{15}NO_3$. Ecgonine shows the reactions of an alcohol and so benzoylecgonine is the benzoyl derivative of an hydroxycarboxylic acid.

(iv)   The structure of ecgonine is deduced from the nature of its oxidation products. Thus ecgonine on oxidation with chromic acid gives tropinone ($C_8H_{13}NO$), which on subsequent oxidation gives tropinic acid ($C_8H_{13}NO_4$) and ecgoninic acid ($C_7H_{11}NO_3$). On the basis of the above results, it is concluded that ecgonine has the tropane structure. Further, the formation of tropinone shows that ecgonine has the alcoholic group in the same position as in tropine.

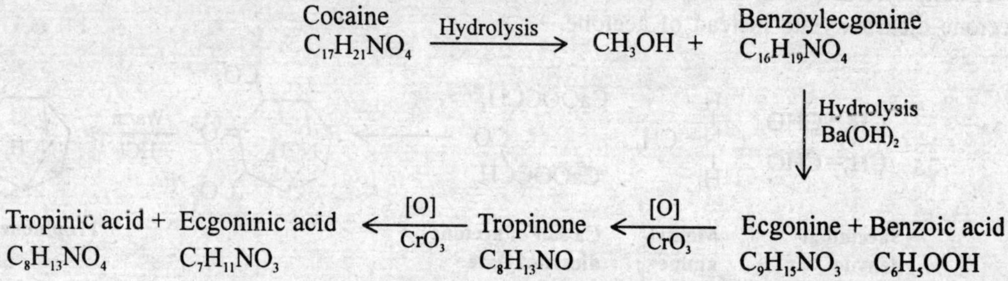

$$\text{Cocaine} \atop C_{17}H_{21}NO_4 \xrightarrow{\text{Hydrolysis}} CH_3OH + {\text{Benzoylecgonine} \atop C_{16}H_{19}NO_4}$$

$$\Big\downarrow {\text{Hydrolysis} \atop Ba(OH)_2}$$

$${\text{Tropinic acid} + \text{Ecgoninic acid} \atop C_8H_{13}NO_4 \qquad C_7H_{11}NO_3} \xleftarrow[CrO_3]{[O]} {\text{Tropinone} \atop C_8H_{13}NO} \xleftarrow[CrO_3]{[O]} {\text{Ecgonine} + \text{Benzoic acid} \atop C_9H_{15}NO_3 \quad C_6H_5OOH}$$

It has been shown that ecgonine contains a carboxyl group which is lost during the formation of tropinone. Thus, the carboxyl group in ecgonine is in such a position that the oxidation of the secondary alcoholic group to a keto group is accompanied by elimination of the carboxyl group. This type of elimination is characteristic of β-keto acids. It has actually been shown by Willstatter that oxidation of ecgonine takes place through an intermediate keto acid, which loses carbon dioxide with great ease to form tropinone. On the basis of the above ecgonine has following structure.

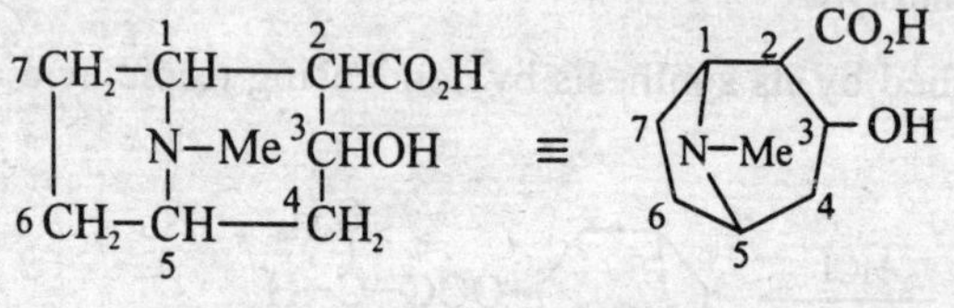

Ecgonine

Thus, cocaine has the structure I (the carboxyl group of ecgonine is converted into the methyl ester and the secondary alcoholic group into the benzoyl derivative). All the above transformations of cocaine (I) can be explained as follows:

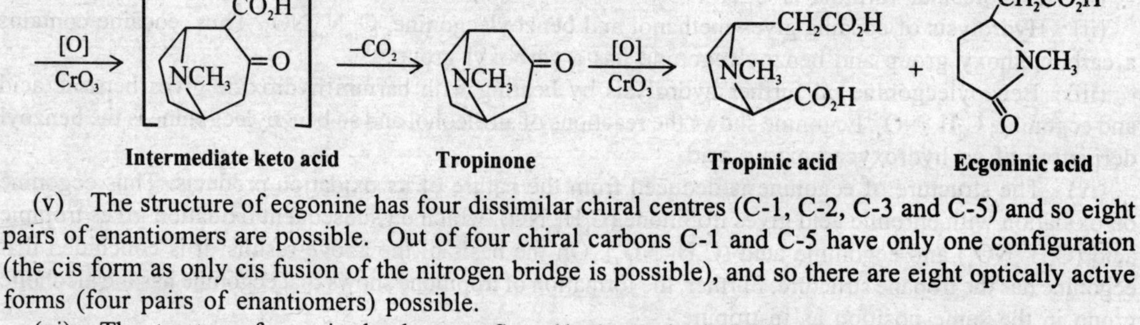

Cocaine (I)   **Benzoylecgonine**   **Ecgonine**   **Benzoic acid**

Intermediate keto acid   **Tropinone**   **Tropinic acid**   **Ecgoninic acid**

(v)   The structure of ecgonine has four dissimilar chiral centres (C-1, C-2, C-3 and C-5) and so eight pairs of enantiomers are possible. Out of four chiral carbons C-1 and C-5 have only one configuration (the cis form as only cis fusion of the nitrogen bridge is possible), and so there are eight optically active forms (four pairs of enantiomers) possible.

(vi)   The structure of ecgonine has been confirmed by its synthesis (Willstatter, 1921) using the Robinson's method.

$$\begin{matrix} CH_2-CHO \\ | \\ CH_2-CHO \end{matrix} \quad + \quad \begin{matrix} H \\ | \\ N-CH_3 \\ | \\ H \end{matrix} \quad + \quad \begin{matrix} CH_2CO_2Et \\ | \\ C=O \\ | \\ CH_2CO_2H \end{matrix} \quad \xrightarrow{KOH} \quad [\text{bicyclic structure with } NCH_3, CO_2Et, =O, CO_2H] \quad \xrightarrow[-CO_2]{heat}$$

Succindialdehyde   **Methyl-amine**   **Monoethyl ester of acetone dicar-boxylic acid**

[bicyclic structure with NCH_3, CO_2Et, $=O$] $\xrightarrow[\text{2) hydrolysis}]{\text{1) Redn. Na–Hg}}$ [bicyclic structure with NCH_3, CO_2H, OH]

(±)-Ecgonine

(vi) Finally, the structure of cocaine as (I) has been established by its synthesis. The (±)-ecgonine obtained above is resolved and the (−)- form is esterified with methanol and then benzoylated to give (−)-cocaine, identical with the natural product.

[bicyclic structure with NCH_3, CO_2H, OH] $\xrightarrow[\text{2) Benzoylation} \atop (C_6H_5COCl)]{\text{1) Esterification} \atop (CH_3OH / HCl)}$ [bicyclic structure with NCH_3, CO_2CH_3, $OOCC_6H_5$]

(−)-Ecgonine   **(−)-Cocaine, I**

The conformations of ecgonine and cocaine have been established by Fodor et al. (1953, 1954) and Findlay (1953, 1954).

[conformational structure with H_3C, N, CO_2H, H, OH, H]   [conformational structure with H_3C, N, CO_2CH_3, H, $OOCC_6H_5$, H]

Ecgonine   **Cocaine, I**

42.7.6 Quinine

Quinine belongs to the quinoline group of alkaloids and is known as a cinchona alkaloid. It has long been used medicinally as an antimalarial. Its structure is established as

(i)   Its molecular formula is $C_{20}H_{24}O_2N_2$ (m.p.177°)

(ii)   Both the nitrogens are tertiary, since quinine adds on two molecules of methyl iodide to form a diquaternary salt, $C_{20}H_{24}O_2N_2.2CH_3I$.

(iii)   It has one hydroxy group, since it forms monoacetate and monobenzoate. Quinine on oxidation with chromium trioxide gives a ketone, quinone $C_{20}H_{22}N_2O_2$, so the hydroxy group is secondary. It also contains one methoxy group.

(iv)   It has one ethylenic double bond, since quinine adds on one molecule of hydrogen, bromine or halogen acid. Further, the ethylenic double bond is present as vinyl group, since quinine on oxidation, gives a monocarboxylic acid and formic acid.

$$C_{18}H_{21}O_2N_2[-CH=CH_2 \xrightarrow[KMnO_4]{[O]} C_{18}H_{21}O_2N_2[-COOH + HCOOH$$

Quinine   Monocarboxylic acid   Formic acid

(v) Vigorous oxidation of quinine with chromic acid gives quininic acid $C_{11}H_9NO_3$ and a compound, designated as the 'second half', and called meroquinene, $C_9H_{15}NO_2$.

$$C_{20}H_{24}O_2N_2 \xrightarrow[CrO_3]{[O]} C_{11}H_9NO_3 + C_9H_{15}NO_2$$

Quinine   Quininic acid   Meroquinene

Thus the structure of quinine depends on the structure of quininic acid and meroquinene.

(vi) Structure of quininic acid $C_{11}H_9NO_3$.

(a) Quininic acid on heating with soda lime undergoes decarboxylation to a methoxyquinoline, identified as 6-methoxyquinoline. Thus, quininic acid has a quinoline nucleus.

(b) Oxidation of quininic acid with chromic acid gives pyridine-2,3,4-tricarboxylic acid. This shows the presence of methoxy group in benzene ring (of quinoline) and carboxyl group at position-4.

(c) Quininic acid on heating with hydrochloric acid undergoes demethylation and decarboxylation to give 6-hydroxyquinoline, a known product. Thus quininic acid is 6-methoxycinchoninic acid (I).

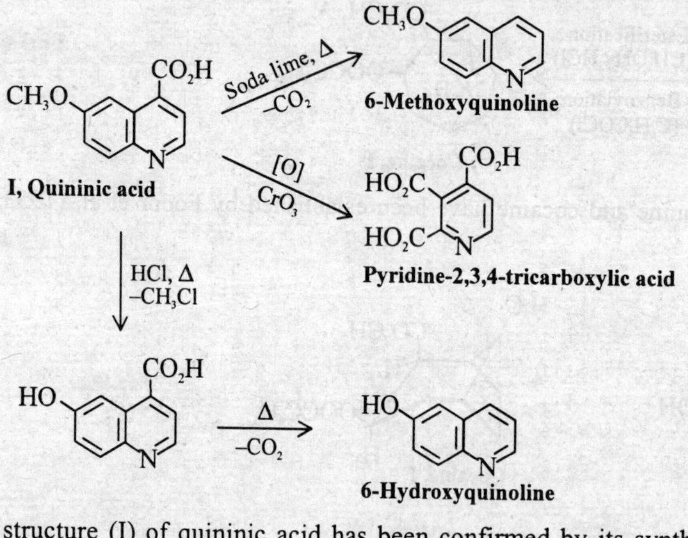

6-Methoxyquinoline

I, Quininic acid

Pyridine-2,3,4-tricarboxylic acid

6-Hydroxyquinoline

(d) The structure (I) of quininic acid has been confirmed by its synthesis (Rabe et al., 1931).

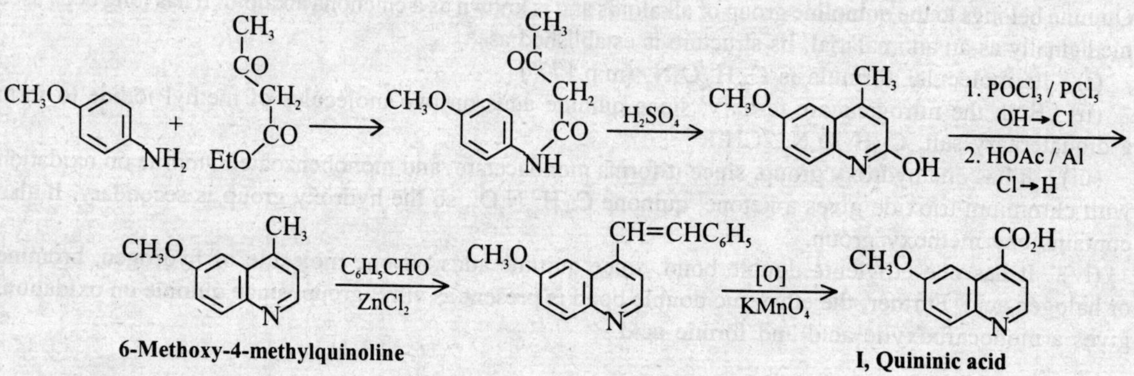

6-Methoxy-4-methylquinoline

I, Quininic acid

In the above synthesis the direct oxidation of methyl group of 6-methoxy-4-methylquinoline to quininic acid (I) is extremely difficult (direct oxidation of methyl group is accompanied by the oxidation of the benzene ring to give pyridine-2,3,4-tricarboxylic acid).

On the basis of above, quinine may be represented by its partial structure as

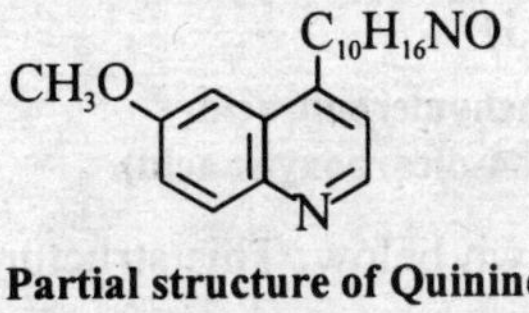

Partial structure of Quinine

The main problem is to find the structure of the 'second half', i.e. meroquinene.

(vii) Structure of meroquinene, $C_9H_{15}NO_2$.

(a) Meroquinene contains one carboxyl group and one double bond as shown by routine tests.

(b) Oxidation of meroquinene with cold acidic potassium permanganate gives formic acid, and a dicarboxylic acid, $C_8H_{13}NO_4$, Cincholoiponic acid. The formation of formic acid indicates the presence of vinyl side chain in meroquinene. The presence of this group is also demonstrated by ozonolysis of meroquinene, which gives formaldehyde. Also meroquinene on heating with hydrochloric acid at 240° gave 3-ethyl-4-methylpyridine.

(c) Cincholoiponic acid on further oxidation with cold acidic permanganate results in the formation of loiponic acid, $C_7H_{11}NO_4$ (also a dicarboxylic acid) which exist in two isomeric forms (cis and trans). However, on heating with potassium hydroxide, it isomerises to more stable form, hexahydrocinchomeronic acid (piperidene-3,4-dicarboxylic acid).

Loiponic acid or its isomerised product contains one methylene less than its precursor, cincholoiponic acid. This suggests that the later contains a side chain $-CH_2CO_2H$.

(d) Furthermore, cincholoiponic acid on treatment with concentrated sulphuric acid gives γ-picoline. This suggests that the additional $-CH_2$ group is present at position 4 in cincholoiponic acid.

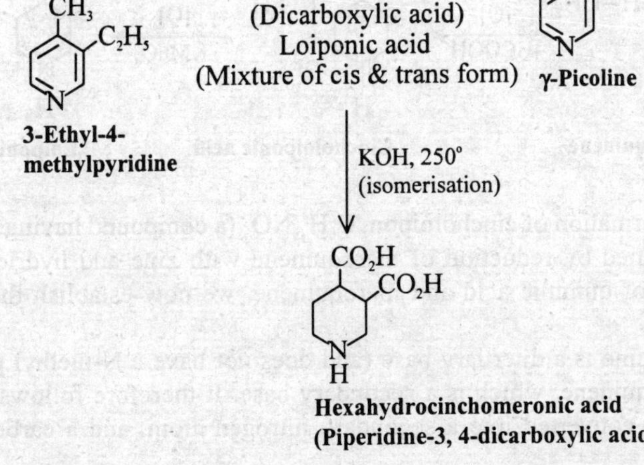

The structure of hexahydrocinchomeronic acid as piperidene-3,4-dicarboxylic acid has been confirmed by its synthesis as

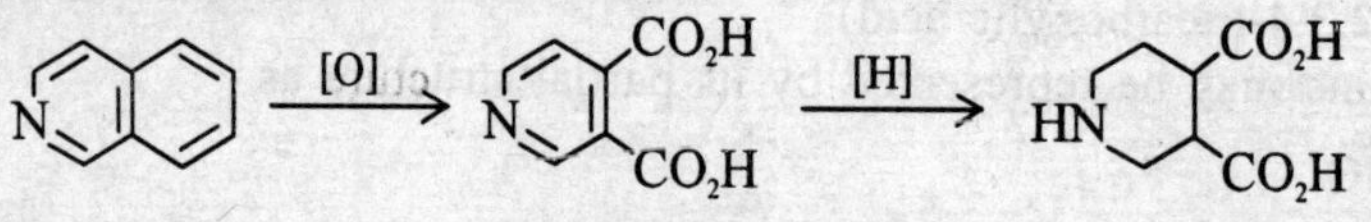

Isoquinoline	**Cinchomeronic acid**	**Hexahydrocinchomeronic acid**
		(Piperidine-3, 4-dicarboxylic acid)

On the basis of the above cincholoiponic acid is represented as shown below. This structure has been confirmed by its synthesis (Wohl and Losanitsch; 1907).

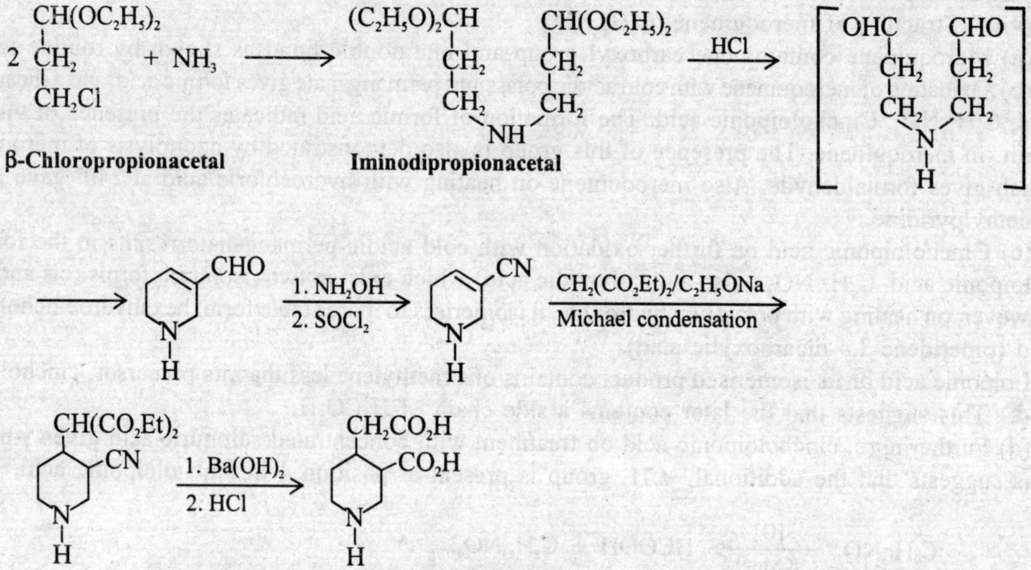

β-Chloropropionacetal   **Iminodipropionacetal**

(±)-Cincholoiponic acid

A careful consideration of the above results show that the structure of meroquinene is (II) as shown below. It also explains all the reactions of meroquinene.

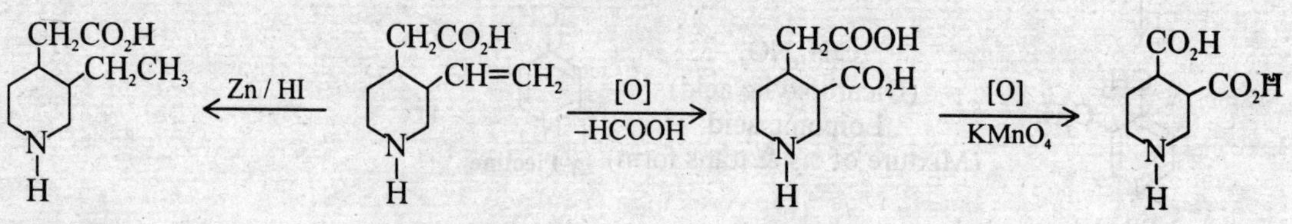

$C_9H_{17}NO_2$	**II, Meroquinene**	**Cincholoiponic acid**	**Loiponic acid**
Cincholoipon			

The structure II also explains the formation of cincholoipon, $C_9H_{17}NO_2$ (a compound having a carboxyl group and an ethyl group) and obtained by reduction of meroquinene with zinc and hydriodic acid.

Having established the structures of quininic acid and meroquinene, we now establish the structure of quinine.

It has already been stated that quinine is a ditertiary base (and does not have a N-methyl group) and on oxidative degradation gives meroquinene, which is a secondary base. It therefore follows that in its formation a tertiary nitrogen atom is converted into a secondary nitrogen atom, and a carboxyl group is also produced at the same time. A reasonable explanation for this observation is that the tertiary nitrogen atom is the part of a bridged ring, one C−N bond is broken when quinine is oxidised. A junction of this type is shown in a synthetic compound, 3-vinylquinuclidine.

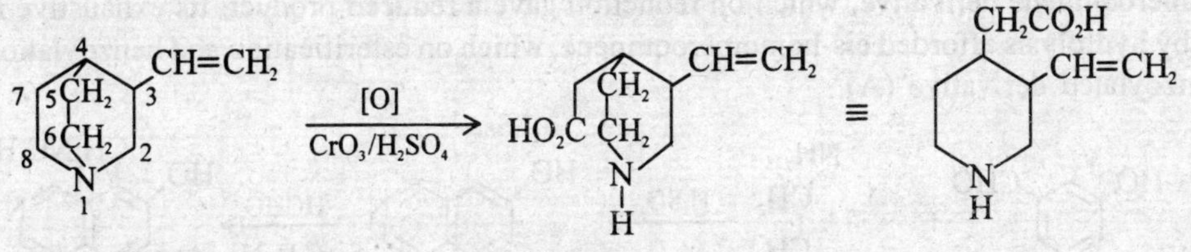

3-Vinylquinuclidine   Meroquinene

The point of linkage of the quinuclidine group to the rest of the molecule is settled by Rabe (1908), Who converted quinine into a ketone, quininone by mild oxidation with chromic acid. Both the nitrogen atoms in this ketone are still tertiary and on treatment with amyl nitrite and hydrogen chloride gave quininic acid and an oxime. Formation of an oxime and an acid indicates the presence of a methine group adjacent to a carbonyl group, viz. −COCH. Structure of the oxime as 8-oximino-3-vinylquinuclidine is assigned by its hydrolysis to hydroxylamine and meroquinene. From above, quinine is given structure (III) which explains the above facts.

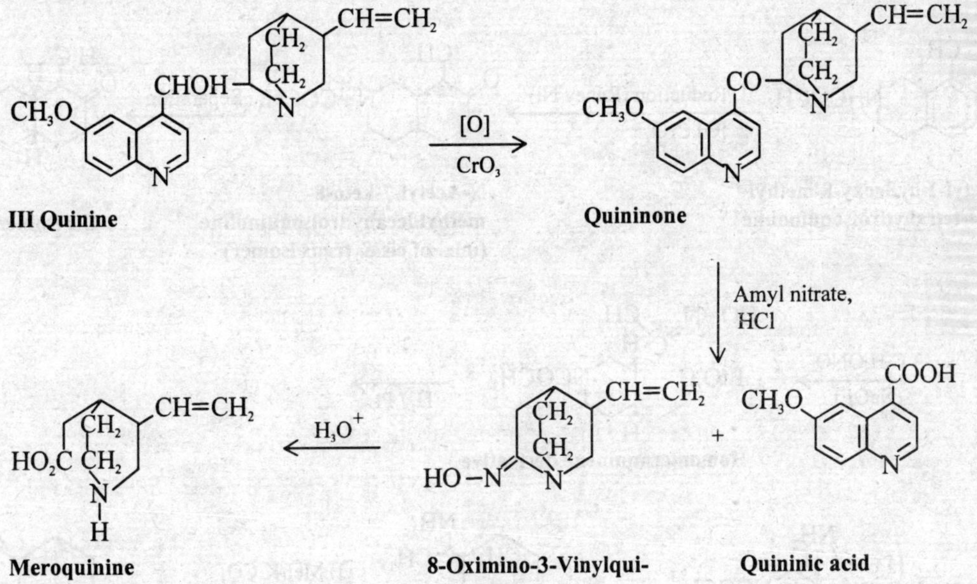

III Quinine   Quininone

Amyl nitrate, HCl

Meroquinine   8-Oximino-3-Vinylqui-
                                 nuclidine   Quininic acid

The structure of quinine as (III) has finally been confirmed by its synthesis. A partial synthesis (Rabe et al,1918) starts from quinotoxine, which was obtained by heating quinine in acetic acid. Quinotoxine was synthesised by Woodward and Doering (1944). Thus a total synthesis of quinine has been achieved.

The special features of the synthesis are described below. Condensation of m-hydroxybenzaldehyde with aminoacetal dehyde acetal and cyclisation of the product with sulphuric acid gave 7-hydroxyisoquinoline. Its treatment with formaldehyde in methanol containing piperidine afforded the complex, which on heating with sodium methoxide gave 7-hydroxy-8-methylisoquinoline which on catalytic reduction followed by acetylation afforded N-acyl-7-hydroxy-8-methyl-1,2,3,4-tetrahydroisoquinoline. Its further reduction with

Raney nickel under pressure followed by oxidation with chromic acid gave N-acetyl-7-keto-8-methyldecahydroisoquinoline, which was a mixture of cis and trans forms and was separated via their crystalline hydrates. The cis isomer, on treatment with ethyl nitrite in presence of sodium ethoxide gave the homomeroquinene derivative, which on reduction gave a reduced product. Its exhaustive methylatin followed by hydrolysis afforded cis-homomeroquinene, which on esterification and benzoylation afforded the N-benzoylated derivative (A).

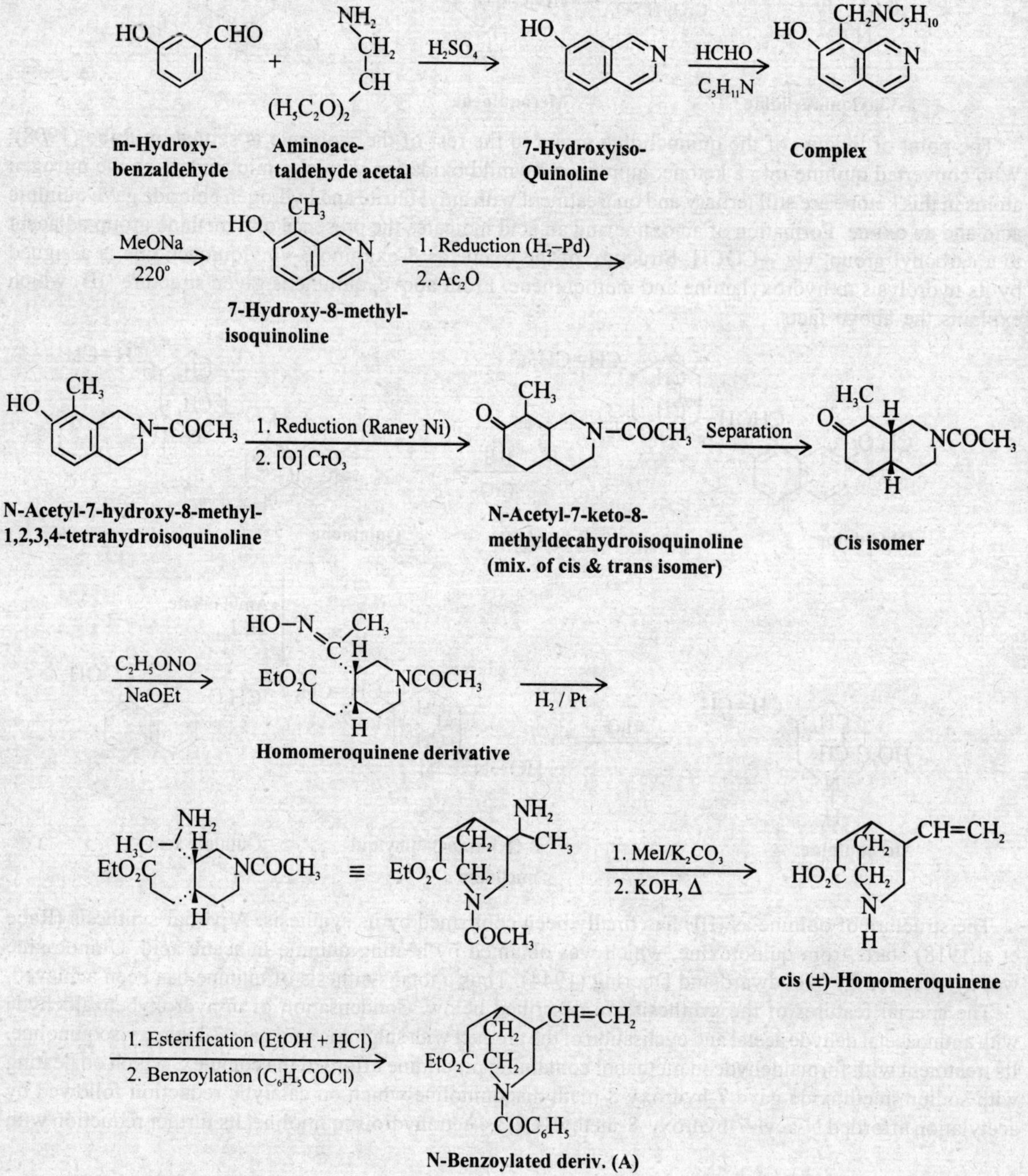

m-Hydroxy-benzaldehyde   Aminoace-taldehyde acetal   7-Hydroxyiso-Quinoline   Complex

7-Hydroxy-8-methyl-isoquinoline

N-Acetyl-7-hydroxy-8-methyl-1,2,3,4-tetrahydroisoquinoline   N-Acetyl-7-keto-8-methyldecahydroisoquinoline (mix. of cis & trans isomer)   Cis isomer

Homomeroquinene derivative

cis (±)-Homomeroquinene

N-Benzoylated deriv. (A)

Claisen condensation of the N-benzoylated derivative(A) with ethyl quinate gave a β-keto ester, which ·n heating with hydrochloric acid afforded quinotoxine. It was resolved via the dibenzoyl tartarate and ·he (+)-isomer reacted with sodium hypobromide followed by alkali treatment to give (±)-quininone. This ·n reduction afforded (±)-quinine (III). It could be resolved to give (–)-quinine, identical with the natural ·roduct.

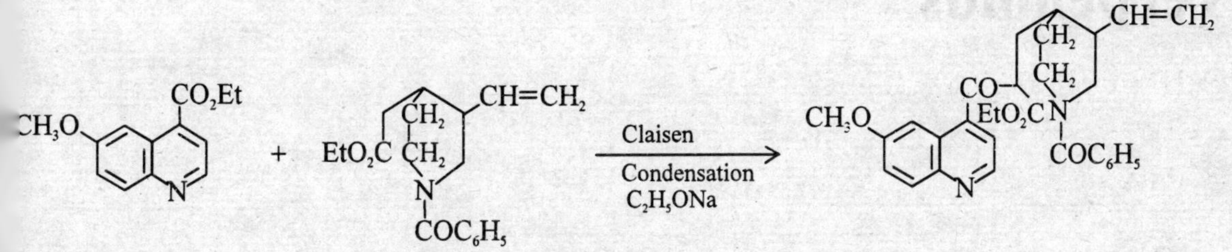

Ethyl Quinate   **N-Benzoylated derivative (A)**   **β-ketoester**

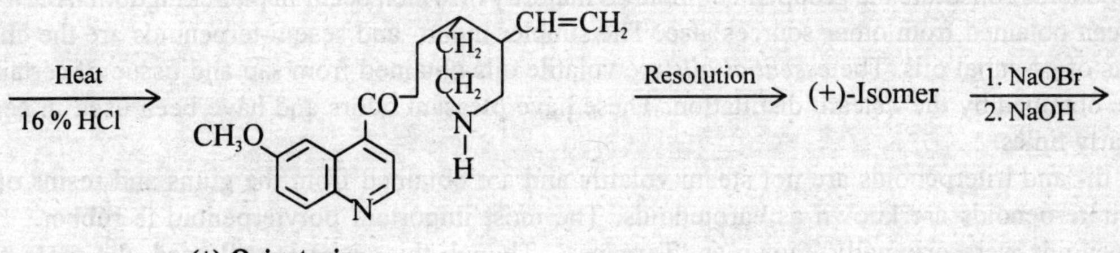

(±)-Quinotoxine

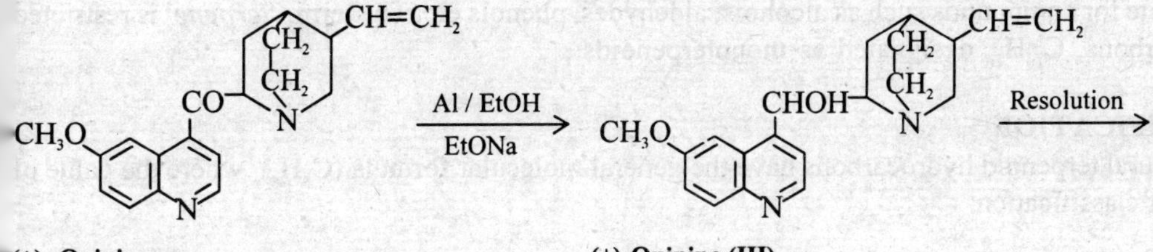

(+) -Quininone   **(±)-Quinine (III)**

42.8 PROBLEMS

1. What are alkaloids? How are they isolated from plants?
2. Discuss general methods employed for determination of the structure of alkaloids.
3. How will you estimate the number of N-methyl groups in alkaloids?
4. Write notes on:
   (i) Hofmann's exhaustive methylation (ii) Emde's degradation
5. How will you establish the structure of (–)-coniine.
6. Give one synthesis of nicotine.
7. How will you establish the presence of pyridine ring in nicotine?
8. Give one example of each of the following group of alkaloids (give their structures).
   (a) Phenylethylamine   (b) Pyridine piperidine   (c) Isoquinoline
   (d) Quinoline   (e) Indole   (f) Cinchronine
9. How is the structure of quinine established?
10. How will you test if an organic compound is an alkaloid or not?

43

Terpenoids

43.1 INTRODUCTION

The terpenoids constitute the group of compounds majority of which occur in plant kingdom. A few of them have been obtained from other sources also. The simpler mono- and sesqui-terpenoids are the chief constituents of essential oils. The *essential oils* are volatile oils obtained from sap and tissue of certain plants and are obtained by the steam distillation. These have pleasant odors and have been used in perfumery from early times.

The di- and triterpenoids are not steam volatile and are obtained from the gums and resins of plants. The tetraterpenoids are known as carotenoids. The most important polyterpenoid is rubber.

Terpenoids were originally known as 'Terpenes'. Though this name is still used, the more common name 'Terpenoids' is preferred. The suffix *'ene'* in terpene indicates unsaturation. Thus, the name *'terpene'* is not appropriate for compounds such as alcohols, aldehydes, phenols etc. The term, *'terpene'* is restricted to the hydrocarbons, $C_{10}H_{16}$ designated as monoterpenoids.

43.2 CLASSIFICATION

Most of the natural terpenoid hydrocarbons have the general molecular formula $(C_5H_8)_n$ where the value of n is the basis of classification.

	Number of carbon atoms	Molecular formula	Class of compound
(i)	10	$C_{10}H_{16}$	Monoterpenoids
(ii)	15	$C_{15}H_{24}$	Sesquiterpenoids
(iii)	20	$C_{20}H_{32}$	Diterpenoids
(iv)	25	$C_{25}H_{40}$	Sesterterpenoids
(v)	30	$C_{30}H_{48}$	Triterpenoids
(vi)	40	$C_{40}H_{64}$	Tetraterpenoids (Carotenoids)
(vii)	>40	$(C_5H_8)_n$	Polyterpenoids

Following are given typical examples of each class of terpenoids.

(I) *Monoterpenoids:* Monoterpenoids having the general formula $C_{10}H_{16}$ are subdivided into three classes.

a) *Acyclic Monoterpenoids:*

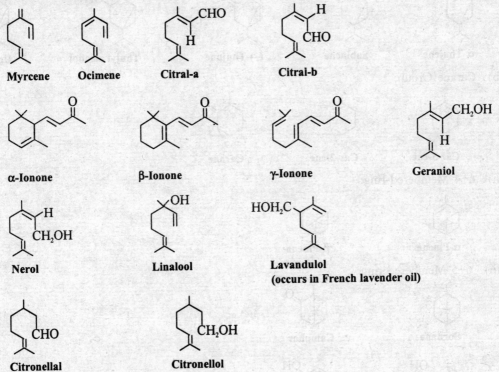

Myrcene   Ocimene   Citral-a   Citral-b

α-Ionone   β-Ionone   γ-Ionone   Geraniol

Nerol   Linalool   Lavandulol
(occurs in French lavender oil)

Citronellal   Citronellol

(b) *Monocyclic Monoterpenoids:*

(±)α-Terpineol   β-Terpineol   γ-Terpineol

Carvone   Diosphenol   (+) or (–)-Limonene

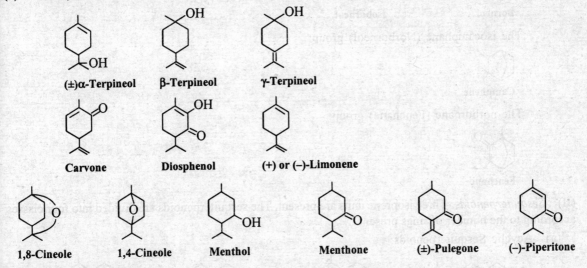

1,8-Cineole   1,4-Cineole   Menthol   Menthone   (±)-Pulegone   (–)-Piperitone

(c) *Bicyclic Monoterpenoids:*

Divided into three classes according to the size of the second ring.

(i) 6+3 Membered ring:

(a) Thujane Group:

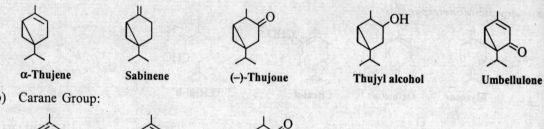

α-Thujene   Sabinene   (−)-Thujone   Thujyl alcohol   Umbellulone

(b) Carane Group:

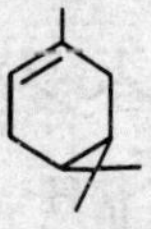

Car-3-ene 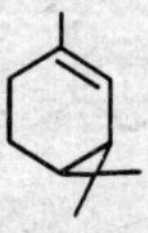   Car-2-ene 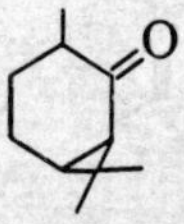   Carone

(ii) 6+4 Membered ring:

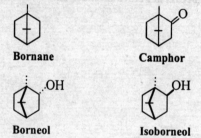

α-Pinene

β-Pinene

(iii) 6+5 Membered ring:

Bornane   Camphor

Borneol   Isoborneol

The isocamphane (Norborneol) group:

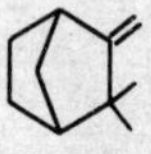

Camphene

The norbornane (Fenchane) group:

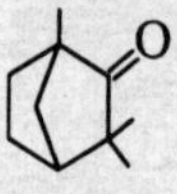

Fenchone

(II) *Sesquiterpenoids:* Three isoprene units are present. The sesquiterpenoids are divided into four classes according to the number of rings present.

(a) Acyclic Sesquiterpenoids:

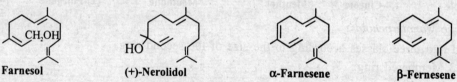

Farnesol   (+)-Nerolidol   α-Farnesene   β-Fernesene

(b) Monocyclic Sesquiterpenoids:

Bisbolane group:

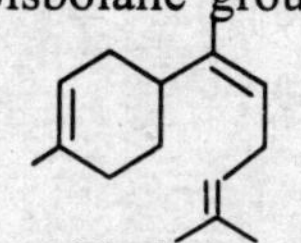

α-Bisabolene 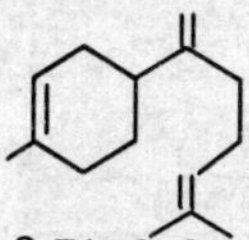   **β-Bisabolene** 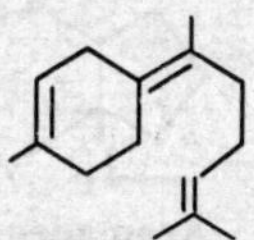   **γ-Bisabolene**

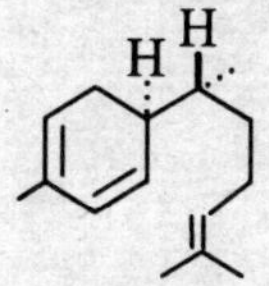

(–)-Zingiberene 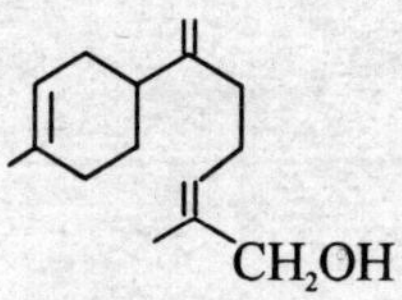   **Lanceol** 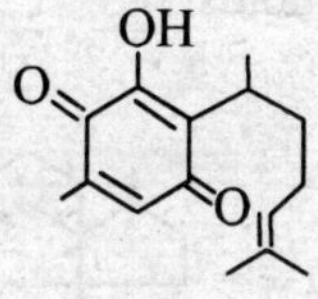   **Perezone**

Elemane Group:

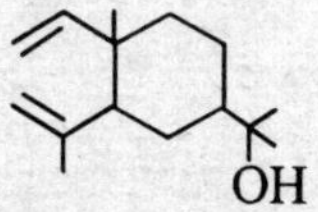

Elemol

Humulane and Germacrane Group:

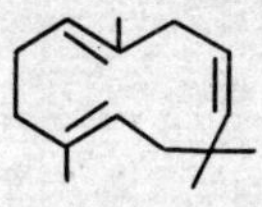

Humulene 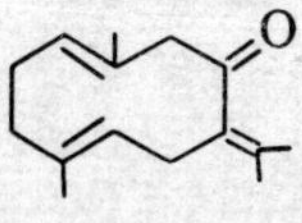   **Germacrone**

(c) Bicyclic Sesquiterpenoids:

The Cadinane Group:

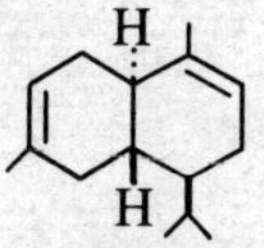

α-Cadinene

The Eudesmane Group:

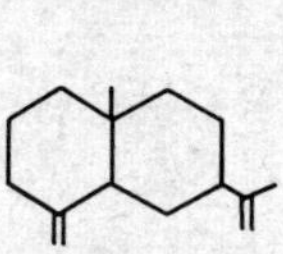

β-Selinene 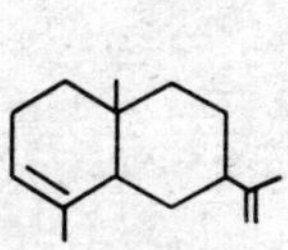   **α-Selinene** 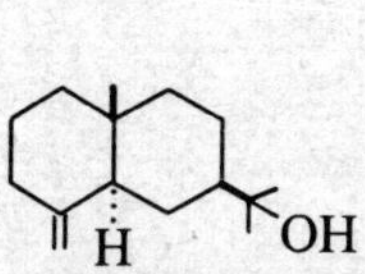   **(±)-β-Eudesmol** 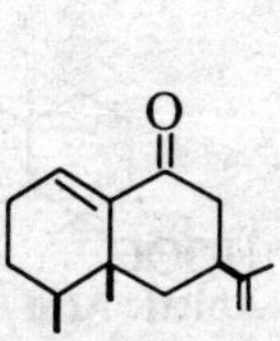   **Eremophilone**

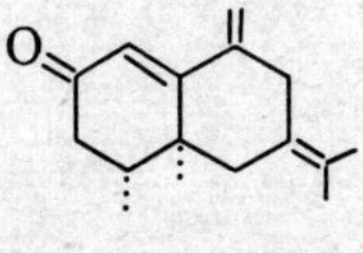

(±)-α-Vetivone 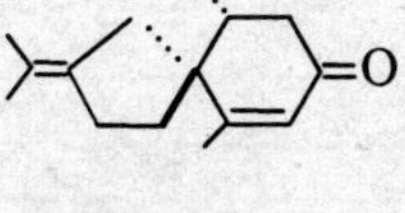   **(±)-β-Vetivone** 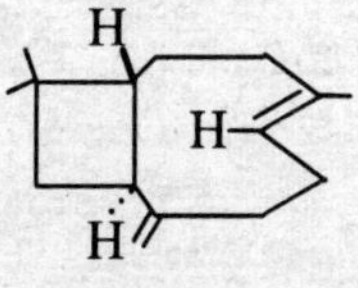   **Caryophyllene**

The Perhydroazulene Group:

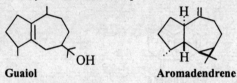

Guaiol   **Aromadendrene**

(d) Tricyclic Sesquiterpenoids:

The Cedrene Group:

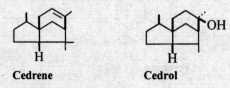

Cedrene   **Cedrol**

The Longifolene Group:

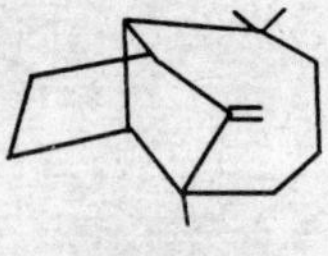

Longifolene

(III) *Diterpenoids*

(a) Acyclic Diterpenoids:

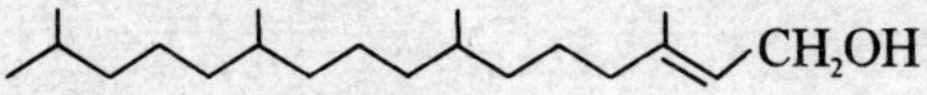

Phytol

(b) Monocyclic Diterpenoids:

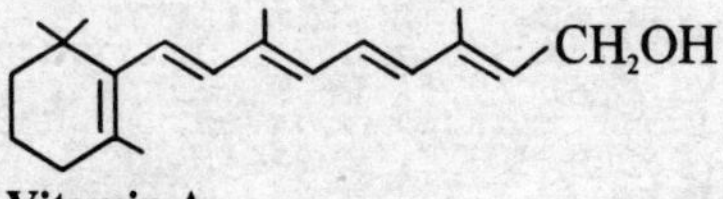

CH₂OH

Vitamin A₁   **Cembrene**

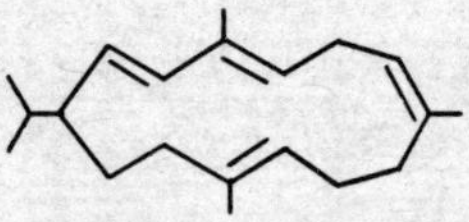

(c) Tricyclic Diterpenoids:

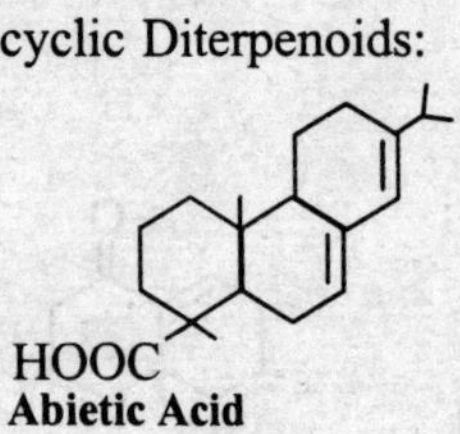

HOOC
Abietic Acid

(d) The Tetracyclic Diterpenoids:

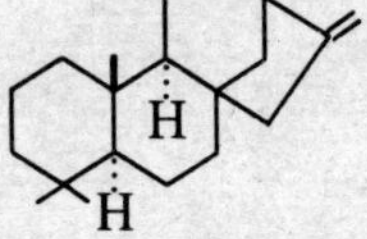

Phyllocladene

(IV) *Sesterterpenoids*

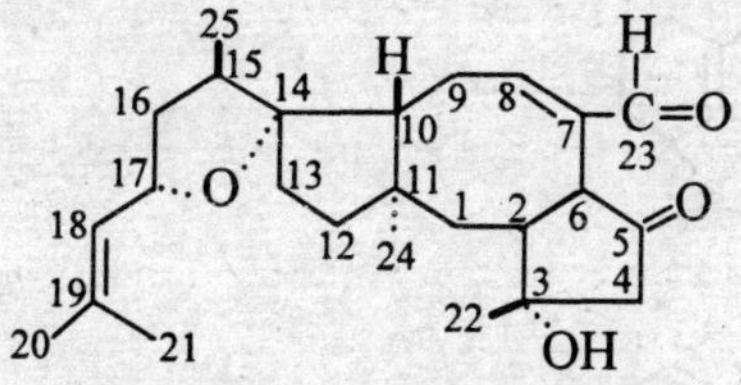

Ophiobolin

(V) *Triterpenoids*

(a) Acyclic Triterpenoids:

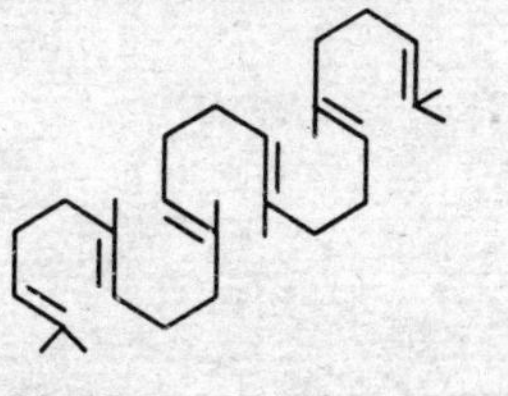

Squalene

(b) Tricyclic Triterpenoids:

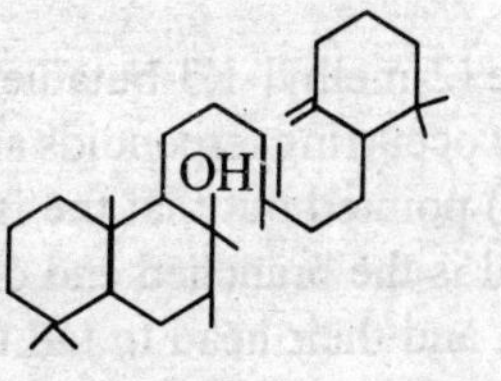

Ambrein

(c) Tetracyclic Triterpenoids:

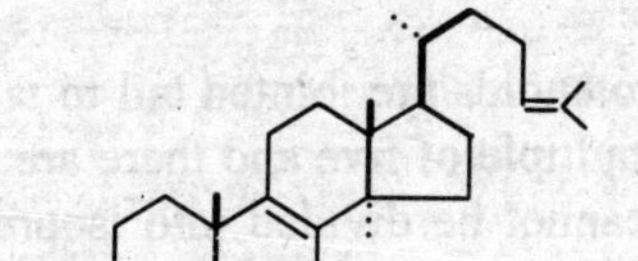

Lanosterol 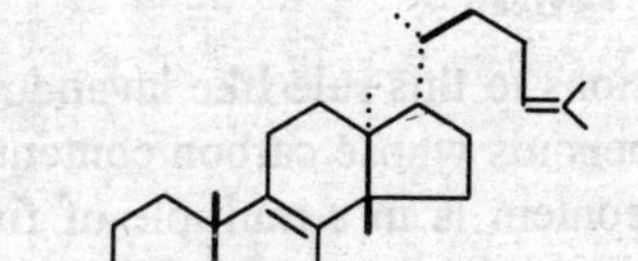   **Euphol**

(d) Pentacyclic Triterpenoids:

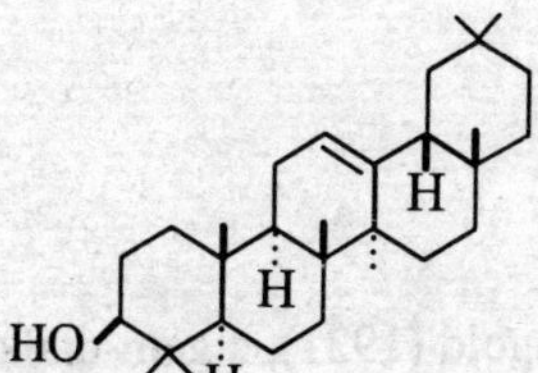

β-Amyrin 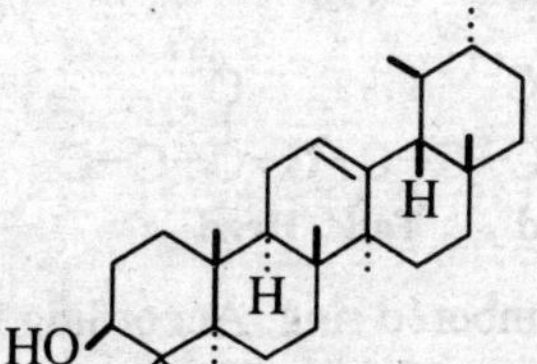   **α-Amyrin** 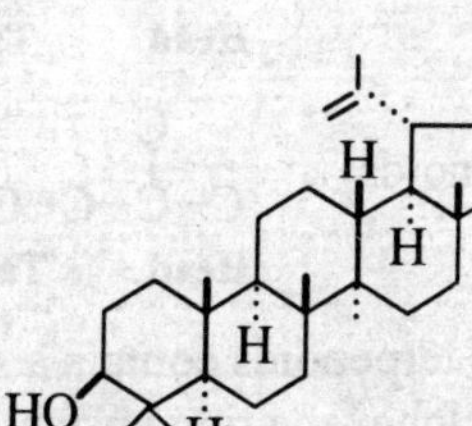   **Lupeol**

(VI) *Tetraterpenoids*

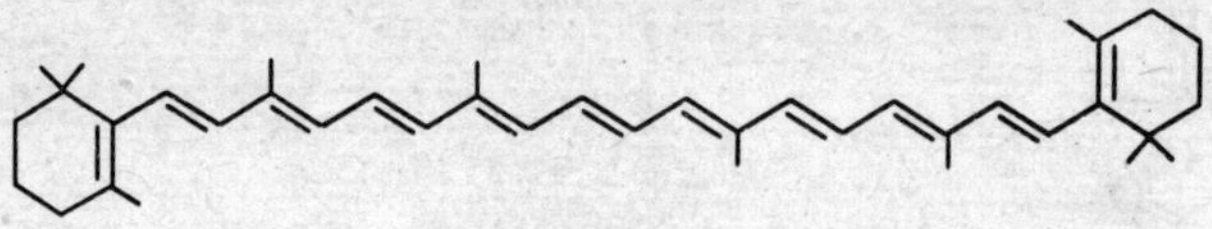

β-Carotene

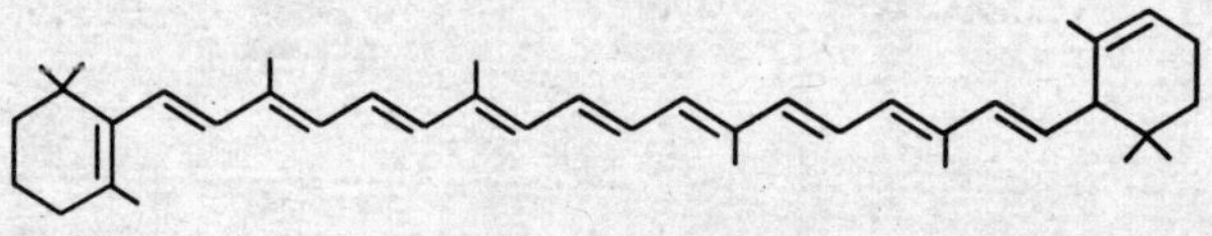

α-Carotene

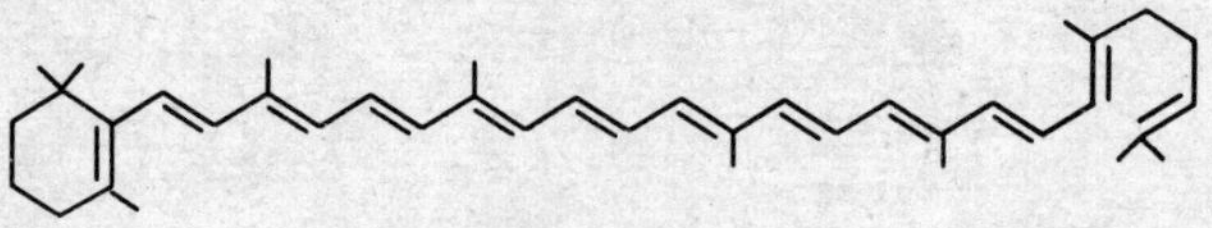

γ-Carotene

As seen besides the terpenoid hydrocarbons, the oxygenated derivatives of each class, mainly alcohols, aldehydes or ketones also occur in nature.

43.3 SPECIAL ISOPRENE RULE

All naturally occurring terpenoids on thermal decomposition give isoprene (2-methyl-1,3-butadiene) as one of the products. On this basis, Wallach (1887) pointed out that all naturally occurring terpenoids are built up of isoprene units. This is known as *Isoprene rule.* Further, Ingold (1925) pointed out that the isoprene in natural terpenoids is joined in 'head to tail' fashion (In isoprene, the head is the branched end of isoprene unit). Ingold referred to the divisibility of terpenoids into isoprene units, and their head to tail union as a *Special Isoprene Rule.* The rule, although useful, is not a fixed rule but is only a guiding principle.

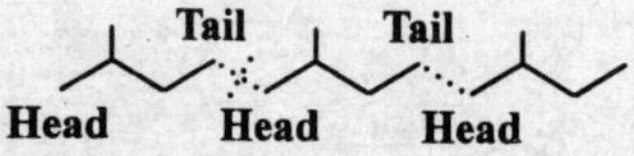

There are several exceptions to this rule like lavanduol, the carotenoids are jointed tail to tail at their centre. There are some terpenoids whose carbon content is not a multiple of five and there are instances where even if the carbon content is in a multiple of five, these cannot be divided into isoprene units.

43.4 CARBON SKELETON

The carbon skeleton of the open chain monoterpenoids and sesquiterpenoids may be represented as:

Monoterpenoids

```
   C              C
   |              |
C–C–C–C÷C–C–C–C
Head      Tail Head     Tail
```

Sesquiterpenoids

```
   C              C              C
   |              |              |
C–C–C–C÷C–C–C–C÷C–C–C–C
Head      Tail Head     Tail  Head
```

Monocyclic terpenoids contain a six membered ring. According to Ingold (1921), a gem-dialkyl group renders the cyclohexane ring unstable. Thus, in closing the open chain to a cyclohexane ring, the presence of dimethyl group limits the number of possible structures. The open chain monoterpenoid can thus give only one monocyclic monoterpenoid known as the p-cymene. Most of the natural monoterpenoids are derivatives of p-cymene which may be represented as follows:

Acyclic Structure   **p-Cymene Structure**

The Bicyclic monoterpenoids contain a six membered ring and another ring, which may be three, four, or five membered. According to Ingold (1921), the formation of cyclopropane and cyclobutane ring requires the introduction of a gem-dimethyl group to render them stable. Three bicyclic structure are possible and all are known.

43.5 ISOLATION

As terpenoids are widely distributed in nature, number of methods have been employed for their isolation.

   (a) Steam distillation

   (b) Isolation by extraction with volatile solvents.

   (c) Enfleurage (adsorption in purified fats)

   (d) Expression.

The most common method of isolation is by extraction with volatile solvents. The plant is macerated and extracted with petroleum ether (40-60°) and the solvent removed by distillation in vacuo to give the oil. Alternatively, the volatile oil may be obtained from the mascerated plant material by steam distillation. Steam distillation method cannot be used if the compound decomposes under these conditions.

The oil can also be isolated by enfleurage in which the fat is warmed (45-50°), flower petals spread on the surface of the fat until the latter is saturated. The fat is then refluxed with ethyl alcohol. The dissolved fat is removed by cooling (15-20°) and the residual alcoholic solution is distilled in vacuo to give the essential oil. The oil so obtained contains a number of terpenoids, which are separated by fractional distillation.

Recently, various types of chromatographic methods have been used for isolation and separation of terpenoids. The most common chromatographic method is by allowing the oil to pass through a column of suitable adsorbent (alumina, silica gel etc.). Various types of terpenoids are concentrated in different zones. The zones are cut and terpenoids are isolated by extraction with solvents. Gas chromatography is particularly useful for isolating pure configurational forms from synthetic mixtures. Another method used for the separation of terpenoids, especially for oxygenated derivatives is the counter current distribution.

By expression method only half the oil is recovered, the rest remains with the residue. This method is mostly used for obtaining citrus oil, lemon oil and grass oil.

43.6 DETERMINATION OF STRUCTURE

The essential requirement for determining the structure is the isolation of the compound in pure state. First of all its molecular formula is determined by elemental analysis and molecular weight by mass spectrometry. Its structure is then systematically determined as follows.

(i) *Function of Oxygen:* If the terpenoid contains oxygen, its nature is determined, i.e., whether it is present as hydroxy, aldehydic or ketonic function. The presence of hydroxy group is ascertained by acetylation (using acetyl chloride or acetic anhydride) or benzoylation (using benzoyl chloride). If hydroxy group is present their number is determined (by acetylation). The presence of aldehydic or ketonic group is ascertained by the formation of oxime, semicarbazone and phenylhydrazone.

(ii) *Presence of Olefinic Bonds:* It is ascertained by bromine. The number of double bonds are determined by analysis of the bromide, or by quantitative hydrogenation or by titration with monoperphthalic acid or perbenzoic acid. From the number of double bonds and the nature of the functional group, if any, the molecular formula of the parent hydrocarbon is derived, from which the number of rings present in the structure may be deduced.

(iii) *Number of C-methyl Groups:* It is estimated by Kuhn-Roth method.

(iv) *Presence of Isopropylidene Group:* It is detected and estimated by ozonolysis.

(v) *Dehydrogenation:* Dehydrogenation of cyclic terpenoids with sulphur, selenium, platinum or palladium converts them into parent or related aromatic compounds, usually p-cymene. This helps to deduce the carbon skeleton of the molecule.

(vi) *Oxidative Degradation:* The terpenoids are degraded into small fragments, which are isolated and identified. The usual reagents used for this purpose are ozone, potassium permanganate (acidic, neutral or alkaline), chromic acid, sodium hypobromite, osmium tetroxide, nitric acid, lead tetraacetate, peroxy acids and N-bromosuccinimide. Due to more information, it is now possible to use a reagent selectively for oxidising a particular group in the molecule. Oxidative degradation, thus is the most powerful tool in the structure elucidation of terpenoids.

(vii) *Refractive Index:* Measurement of refractive index gives the molecular refraction, which may be helpful to deduce the nature of the carbon skeleton, especially in sesquiterpenoids.

(viii) *Optical Exaltation:* Measurement of optical exaltation indicates the presence of double bonds in conjugation. The difference between observed and the calculated value of molecular refractivity is called optical exaltation or refractive exaltation. Zero value of optical exaltation indicates absence of conjugated system.

(ix) *Spectroscopic Methods*

(a) *Ultraviolet Spectroscopy:* It has been frequently used in terpenoid chemistry mainly to detect the conjugation. Thus, in simple acyclic dienes, λ_{max} is 217-227 nm (ε 15000-25000). In heteroannular dienes (semicyclic), i.e. the conjugated double bonds are not in the same ring, λ_{max} is 230-240 nm (ε 13000-20000). However, if the diene is homoannular, i.e. both double bonds are in the same ring,

λ_{max} is 256-265 nm (ε 2500-10000). In case, an α, β carbonyl system is present the λ_{max} is 220-250 nm (ε 10000-17500) along with a weak band at λ_{max} 315-330 nm (ε 15-100).

An empirical rule was devised by Woodward (1942) and later modified by Fieser (1948) for calculating the λ_{max} from the molecular structure of the compound. The method of calculation is as follows.

(i) *Woodward and Fieser rules for absorption of unconjugated polyenes.*

Parent homoannular dienes (basic value)	253 nm
Parent heteroannular and acyclic dienes (basic value)	214 nm
Increment for each double bond extending conjugation	30 nm
Increment for each C-substituent	5 nm
Increment for each exocyclic C=C in the system	5 nm
λ_{max} of the compound Total	

(ii) *Woodward and Fiesers rule for absorption of unsaturated ketones* of the type

$$-\overset{|}{\underset{\delta}{C}}=\overset{|}{\underset{\gamma}{C}}-\overset{|}{\underset{\beta}{C}}=\overset{|}{\underset{\alpha}{C}}-\overset{|}{C}=O$$

Parent system (one double bond) (basic value)	215 nm
For each C=C extended conjugation	30 nm
For each exocyclic C=C in the system	5 nm
For each C substituents at α	10 nm
at β	12 nm
at γ	18 nm
λ_{max} of the compound Total	

Following example of carvone explains the use of these rules.

λmax observed 235nm

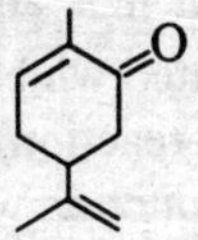

Calculated, Parent system		215 nm
C substituent	at α carbon	10 nm
	at β carbon	12 nm
λ_{max} Total		237 nm

(b) *Infrared Spectroscopy:* This technique is extremely valuable in detecting the presence of hydroxy group (approx. 3400 cm^{-1}) and an oxo group (saturated 1750-1700 cm^{-1}; α, β-unsaturated 1700-1660 cm^{-1}). The heteroannular dienes and unsubstituted α, β-unsaturated ketones (which cannot be distinguished by ultra violet spectroscopy—see Woodward Fieser rule) can be differentiated by their infrared spectra. The presence of isopropenyl group and cis-trans isomerism can also be inferred by this technique.

(c) *NMR Spectroscopy:* This technique is helpful to assign definite structures to a large number of terpenoids. It is extremely helpful to detect and identify double bonds, determine the number of rings present, nature of end groups and to ascertain the orientation of the methyl groups in the molecule.

(d) *Mass Spectrometry:* Using this technique it is possible to determine the molecular weight, molecular formula and also the relative positions of double bonds and the nature of various functional groups. This technique finds increasing use in the structure elucidation of terpenoids.

(x) *X-ray Analysis:* It is extremely useful in some cases to determine the structure and stereochemistry of terpenoids.

(xi) *Optical Rotation:* It is a valuable method for the structure determination of terpenoids. The ORD studies have been used for the assignment of absolute configuration.

The above methods normally gives a tentative structure of the terpenoid.Final confirmation of the structure is achieved by synthesis.

43.7   MONOTERPENOIDS

The monoterpenoids may be divided into three groups: acyclic, monocyclic and bicyclic. Further subdivisions are made on the basis of carbon skeleton.

43.7.1   The Acyclic Monoterpenoids

These may be subdivided into hydrocarbons, alcohols, aldehydes and ketones.

43.7.1.1   Hydrocarbons

The important member of this group is myrcene.

Myrcene: Myrcene is an acyclic monoterpenoid hydrocarbon and occurs in verbena and bay oils. Its structure is deduced as follows:

(i)   Its molecular formula as determined by analytical data is $C_{10}H_{16}$.

(ii)   It has three double bonds as on catalytic hydrogenation it takes up three molecules of hydrogen to give decane, $C_{10}H_{22}$. This also shows that it is an open chain compound.

(iii) Two of the double bonds are conjugated which is shown by the fact that myrcene forms addition product with maleic anhydride (Diels-Alder reaction). The presence of conjugation is supported by the fact that it shows optical exaltation and its reduction with sodium and alcohol to dihydro myrcene, $C_{10}H_{18}$ and is confirmed by its UV absorption (λ_{max} 224 nm).

(iv) Myrecene contains three double bonds, two of which are in conjugation. This has been confirmed by ozonolysis, which gives acetone, formaldehyde and a ketodialdehyde, $C_5H_6O_3$. The latter on oxidation with chromic acid gives succinic acid and carbon dioxide.

The above results can be explained by assigning the structure (I) for myrcene, [7-methyl-3-methylene octa-1,6-diene].

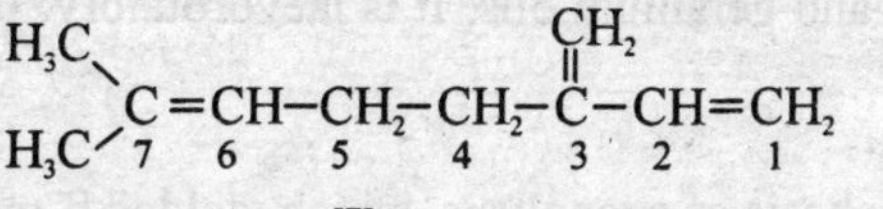

$$\underset{H_3C}{\overset{H_3C}{\diagdown}}\underset{7}{C}=\underset{6}{CH}-\underset{5}{CH_2}-\underset{4}{CH_2}-\underset{3}{\overset{\overset{CH_2}{\|}}{C}}-\underset{2}{CH}=\underset{1}{CH_2} \quad \text{or}$$

[I]   [Ia]

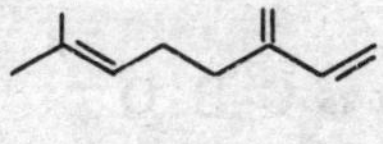

It is customary in terpenoid chemistry to use conventional formula rather than those of the type (I). This conventional formula in represented by lines. Thus structure (I) may be represented as (Ia), but in common practice in terpenoid chemistry the representation is made in the 'open' cyclohexane ring (ring fashion) as this representation clearly shows the relationship between various classes of terpenoids. Though myrcene may be represented by different ring structures for example II, III and IV, but the structure IV is recommended.

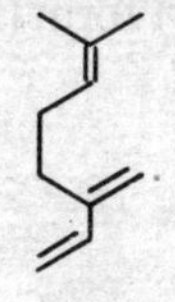

[II] 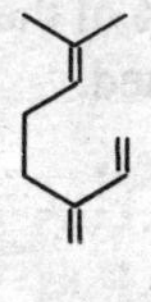   [III] 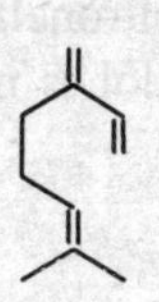   [IV]

On the basis of structure IV for myrcene, the products of ozonolysis and subsequent oxidation of the ketodialdehyde are represented as:

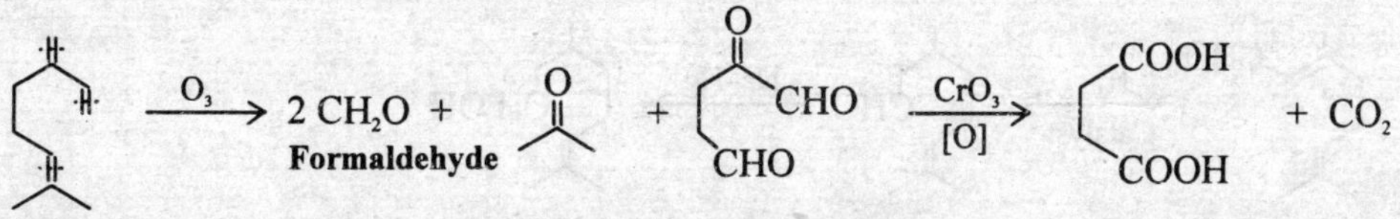

$$\text{Myrcene} \xrightarrow{O_3} 2\,CH_2O + \text{Acetone} + \text{Ketodialdehyde} \xrightarrow[\text{[O]}]{CrO_3} \text{Succinic acid} + CO_2$$

Formaldehyde

Myrcene   Acetone   Ketodialdehyde   Succinic acid

(v)   Final confirmation of this structure is achieved by its synthesis (Rusika, 1916)

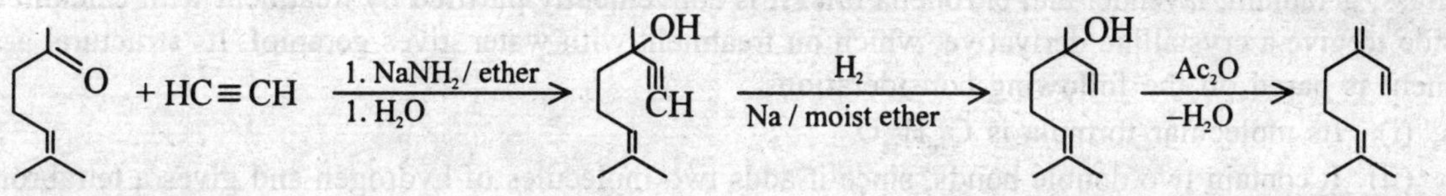

$$\text{Methylheptenone} + HC \equiv CH \xrightarrow[\text{1. H}_2\text{O}]{\text{1. NaNH}_2/\text{ether}} \xrightarrow[\text{Na / moist ether}]{\text{H}_2} \text{Linalool} \xrightarrow[\text{–H}_2\text{O}]{\text{Ac}_2\text{O}} \text{Myrcene}$$

Myrcene is used in the production of rubber like polymers. It is obtained on a large scale by the pyrolysis of β-pinene at 700° for 0.01 to 0.1 second.

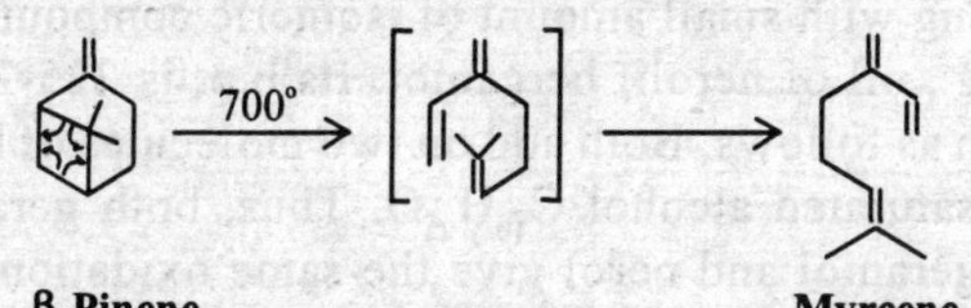

$$\beta\text{-Pinene} \xrightarrow{700°} [\quad] \longrightarrow \text{Myrcene}$$

β-Pinene   Myrcene
                                             [IV]

43.7.1.2 Alcohols
The well known members of this group are citronellol, geraniol, nerol and linalool.

Citronellol: It occurs naturally in Java citronella rose and geranium oils. It is laevorotatory. Its structure is determined as follows.

(i)   Its molecular formula is $C_{10}H_{20}O$.

(ii) Its structural assignment as (I) was made on the basis of ozonolysis, which yielded β-methyladipic acid and acetone as the main product.

| I | | Acetone | β-Methyladipic acid |

(iii) It was earlier believed that the natural citronellol contains 2-3% of the isopropenyl form (II) named rhodinol. But its presence in natural citronellol is no longer accepted.

II

Its structure has finally been confirmed by its synthesis from citral.

Citral   **(+)-Citronellol**

Geraniol: It is a sweet smelling liquid, b.p. 121°/17 mm (b.p. 229-230°/757 mm) present in rose, lemon grass, geranium, lavender and citronella oils. It is conveniently purified by treatment with calcium chloride to give a crystalline derivative, which on treatment with water gives geraniol. Its structural assignment is based on the following considerations.

(i)   Its molecular formula is $C_{10}H_{18}O$.

(ii) It contain two double bonds, since it adds two molecules of hydrogen and gives a tetrabromide.

(iii) It has a primary alcoholic group, since on oxidation it gives an aldehyde (citral-a), which on further oxidation gives an acid (geranic acid), both containing the same number of carbon atoms.

(iv) It has the same carbon frame work as in citral, which it gives on oxidation.

(v) Reduction of citral produces geraniol along with small amount of isomeric compound named nerol. Nerol occurs in various essential oils, e.g., oil of neroli, bergamot. Its b.p. is 225-226°. The structural identity of geraniol and nerol is shown as follows. Both add on two molecules of hydrogen on catalytic hydrogenation and give the same saturated alcohol $C_{10}H_{22}O$. Thus, both geraniol and nerol contain two double bonds. Further, both geraniol and nerol give the same oxidation product. Hence, geraniol and nerol are geometrical isomers. Geraniol has been assigned the trans-configuration and nerol the cis-configuration. This assignment is supported by the fact that cyclisation to α-terpineol by means of dilute sulphuric acid takes place about 9 times faster with nerol compared to geraniol. The fast rate with nerol is due to the proximity of the alcoholic group to the carbon atom (C*) which is involved in ring formation.

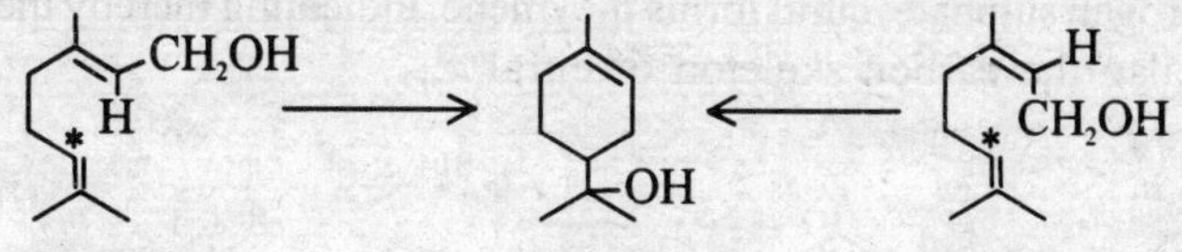

Geraniol
(trans or E) **α-Terpineol** **Nerol**
(cis or Z)

The mechanism of hydration of geraniol and nerol to α-terpineol may be shown as follows.

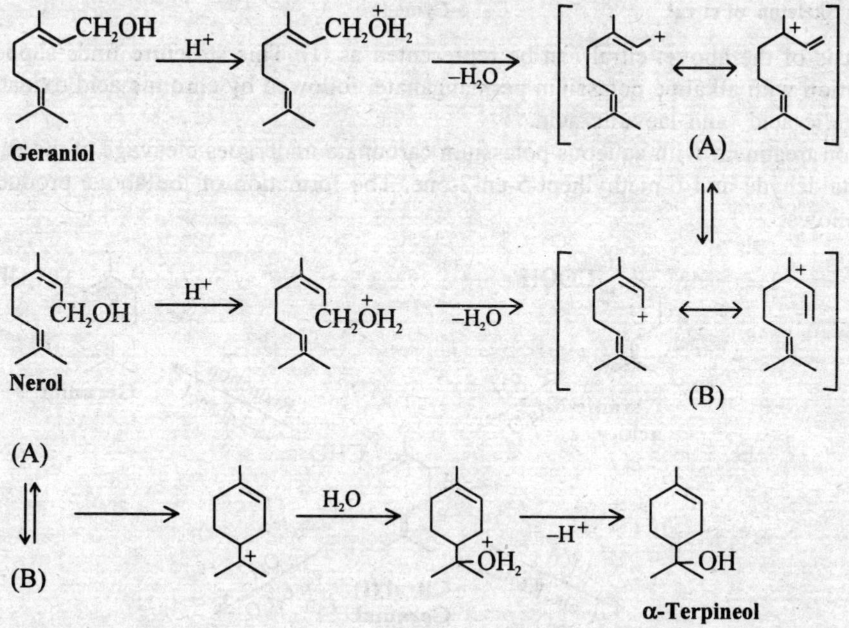

Geraniol (A)

Nerol (B)

(A)

(B) **α-Terpineol**

(vi) The assigned structures of geraniol and nerol have also been supported by NMR studies of trans and cis-methyl geranates, which can be reduced to geraniol and nerol, respectively.

43.7.1.3 Aldehydes and Ketones
The most important member of this group is citral.

Citral: Citral is the most important member of the aliphatic monoterpenoids and finds industrial applications. The structures of most of the other compounds in this group are based on citral. It occurs (60-80%) in lemon grass oil, from which it is obtained as its crystalline bisulphite derivative hydrolysis of which regenerates citral. Its structural assignment is made as follows.

(i) Its molecular formula is $C_{10}H_{16}O$.

(ii) It has two double bonds as shown by the formation of tetrabromide on bromination, $C_{10}H_{16}OBr_4$ and tetrahydrocitral $C_{10}H_{20}O$ on hydrogenation.

(iii) It contains an aldehyde group as shown by the formation of an oxime. Further on reduction (sodium-alcohol) it gives geraniol $C_{10}H_{18}O$, a primary alcohol and on oxidation gives geranic acid ($C_{10}H_{16}O_2$) both containing the same number of carbon atoms as in citral. The presence of α, β-unsaturated carbonyl system in citral is shown in its ultraviolet spectrum (λ_{max} 238 nm; ε 13500).

(iv) On heating with potassium hydrogen sulphate, citral forms p-cymene, indicating thereby the positions of methyl and isopropyl group and also the carbon skeleton of citral.

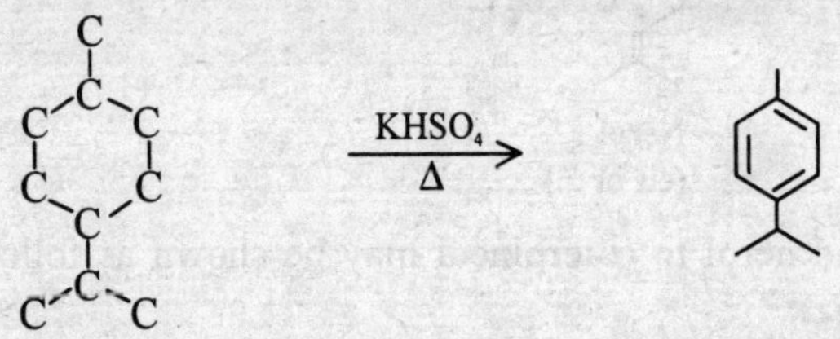

Carbon Skeleton of citral   **p-Cymene**

(v) On the basis of the above, citral can be represented as (I). This structure finds support by the fact that on oxidation with alkaline potassium permanganate, followed by chromic acid oxidation, citral gives acetone, oxalic acid and laevulic acid.

Further, citral on treatment with aqueous potassium carbonate undergoes cleavage at the α, β-double bond to give acetaldehyde and 6-methylhept-5-en-2-one. The formation of the above products can be represented as follows:

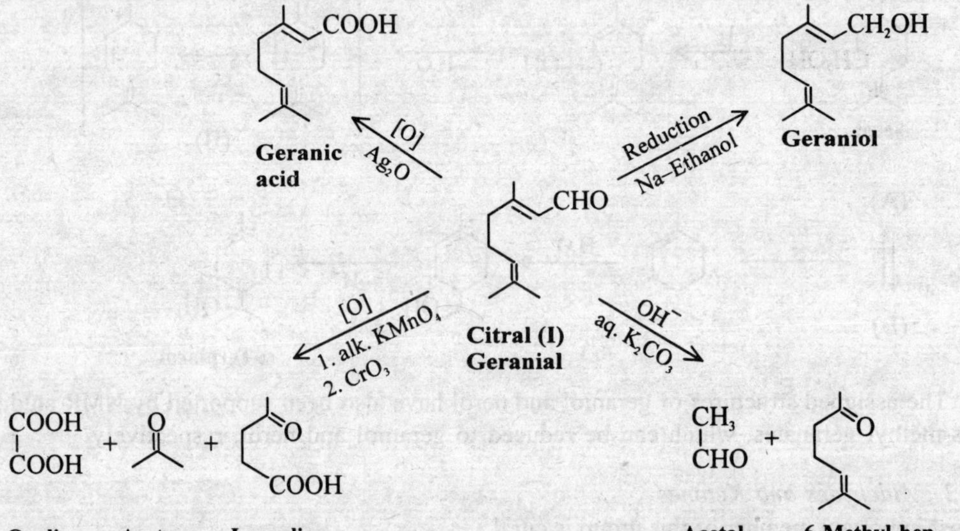

Geranic acid   **Geraniol**

Citral (I) Geranial

Oxalic acid   Acetone   Laevulic acid

Acetal-dehyde   6-Methyl-hep-5-ene-2-one

(vi) Structure (I) for citral is confirmed by its synthesis (Tiemann, 1898) from methylheptenone followed by its conversion into geranic ester and citral.

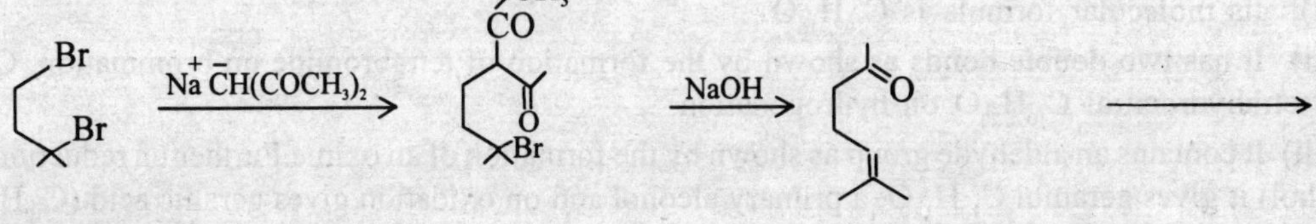

2,4-Dibromo-2-methylbutane   **6-Methylhept-5-ene-2-one**

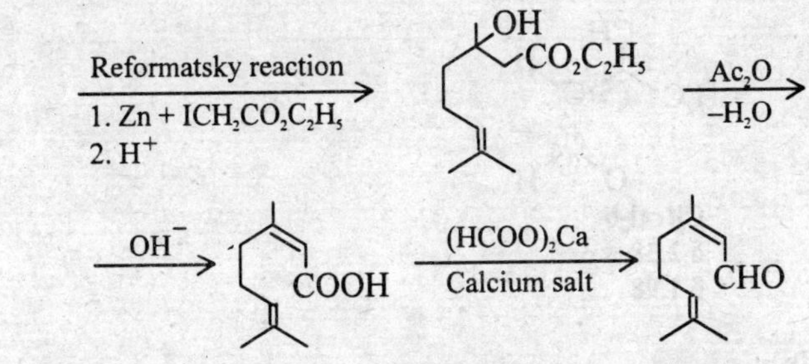

Another synthesis of citral (Arens and Dorp, 1948)

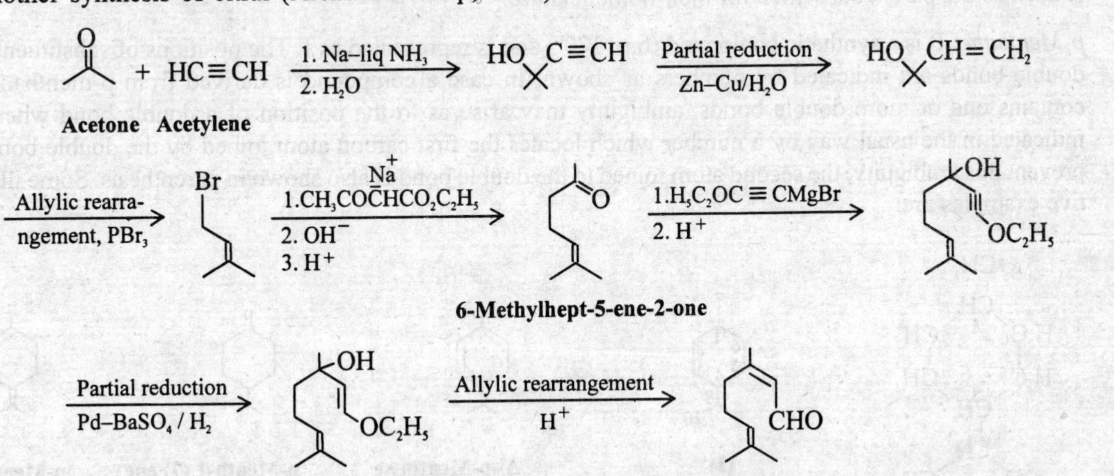

6-Methylhept-5-ene-2-one

Citral

(vi) *Geometrical Isomers of Citral:* A closer examination of the formula of citral suggests that two geometrical isomers are possible, the aldehyde group may be trans or cis with respect to methylene group of the main chain. Both isomers occur in natural citral. The presence of both isomers is confirmed by the formation of two semicarbazones and two alcohols (on reduction) viz., geraniol and nerol. The configuration of the two isomers has been determined from the ring closures of the corresponding alcohols (see geraniol). Thus, citral-a (also known as geranial) has b.p. 118-19°/20 mm and citral-b (also known as neral) has b.p. 117-18°/20 mm.

Trans (or E) isomer
citral - a
(Geranial)

Cis (or Z) isomer
citral - b
(Neral)

The above assignments have been confirmed by NMR of the two isomers. The δ values of CH_2 and CH_3 are different due to magnetic shielding effects of the carbonyl double bond.

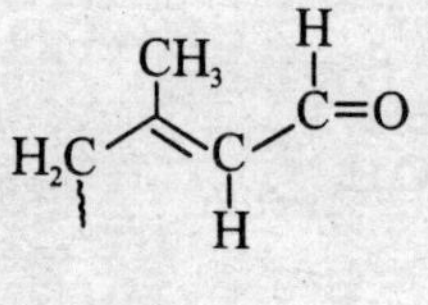

Citral-a
CH₂ δ 2.24
CH₃ δ 2.16

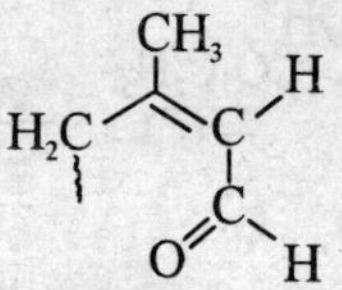

Citral-b
δ 2.58
δ 1.98

43.7.2   The Monocyclic Monoterpenoids

The monocyclic monoterpenoids have been derived from saturated hydrocarbon p-menthane, $C_{10}H_{20}$ which is used as the parent substance for their nomenclature.

p-Menthane: It is a synthetic compound (b.p. 170°) and is represented as I. The positions of substituents and double bonds are indicated by numbers as shown. In case a compound is derived from p-menthane and contains one or more double bonds, ambiguity may arise as to the position of a double bond when it is indicated in the usual way by a number which locates the first carbon atom joined by the double bond. To prevent this ambiguity, the second atom joined to the double bond is also shown in parenthesis. Some illustrative examples are:

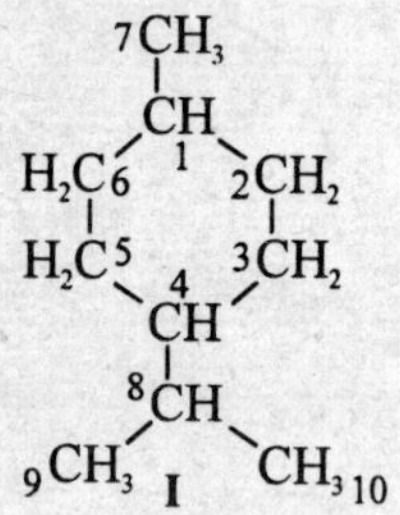

(I)

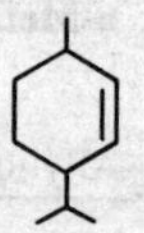

Δ²-p-Menthene
2-p-Menthene
p-Menth-2-ene

p-Menth-1 (7) ene

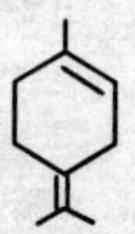

p-Mentha-1,
4 (8)- diene

The monocyclic monoterpenoids may be further subdivided into hydrocarbons and the oxygenated derivatives.

43.7.2.1   *Hydrocarbons*

The most important member of this group is limonene.

Limonene: It is the most important and most commonly occurring of the monoterpenoids. It is the main constituent of the terpenoid fraction of lemon and orange oils. Large quantities are produced in the *citrus fruit industry and in the oils of dill, caraway and bergamot* etc. It is found both in the (+) and (–) forms, and is also known as dipentene. It is used as a flavoring agent in beverages and foods. It is also used in medicines and in making synthetic resins, and high pressure lubricating oil additives. Its structure is assigned as follows:

(i)   Its molecule formula is $C_{10}H_{16}$.

(ii)   On dehydrogenation (heating with sulphur), it gives p-cymene. Hence, limonene has the p-cymene carbon skeleton.

(iii) It has two double bonds since it adds on four bromine atoms, and on catalytic reduction adds on two molecules of hydrogen to form p-menthane. Thus limonene is p-menthadiene.

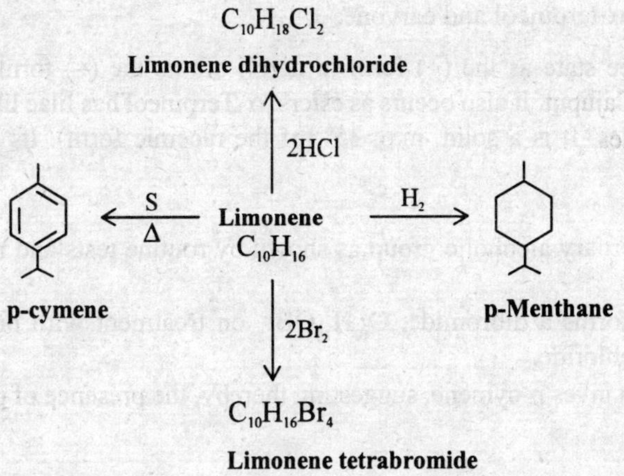

$$C_{10}H_{18}Cl_2$$
Limonene dihydrochloride

$$\uparrow \text{ 2HCl}$$

p-cymene $\xleftarrow[\Delta]{S}$ **Limonene** $C_{10}H_{16}$ $\xrightarrow{H_2}$ **p-Menthane**

$$\downarrow \text{ 2Br}_2$$

$$C_{10}H_{16}Br_4$$
Limonene tetrabromide

Further, it adds two molecules of hydrogen chloride to give dihydrochloride.

(iv) The positions of the two double bonds are established by its relation to α-terpineol. Dehydration of (+)- α-terpineol with potassium hydrogen sulphate gives (+)- limonene. Further, limonene (or dipentene) may be converted into α-terpineol on treatment with dilute sulphuric acid (hydration).

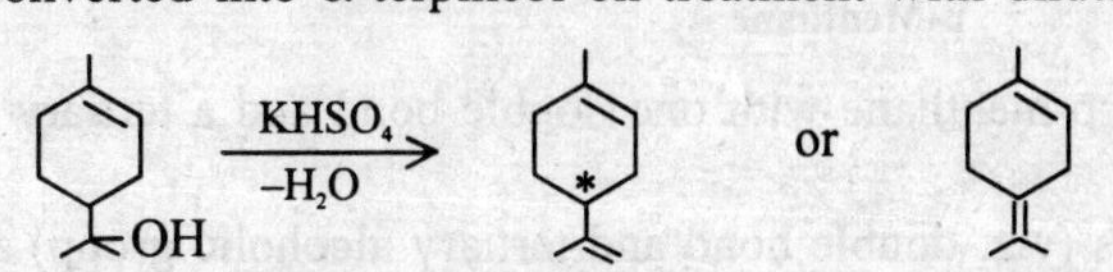

(+)-α-Terpineol $\xrightarrow[-H_2O]{KHSO_4}$ **I** or **II**

Thus, the carbon skeleton and the position of one double bond at C-1 in limonene are established. The other double bond may either be at position 8 (structure-I) or 4(8) (structure-II). Structure (I) has a chiral centre at C-4 (shown by asterisks) and hence exhibits optical activity, whereas, structure (II) being symmetric cannot be optically active. Therefore, limonene is represented by structure (I).

(v) The chemical proof for the structure of limonene is derived by its relation to carvone by the following reactions:

Limonene $\xrightarrow{NOCl}$ **Limonene nitrosochloride** $\xrightarrow[EtOH]{KOH}$ **Carvoxime**
(I)

The structure of carvoxime is known (Section 43.7.2.2). Therefore limonene must have the structure (I) i.e., the second double bond is present at C-8. This can be represented as

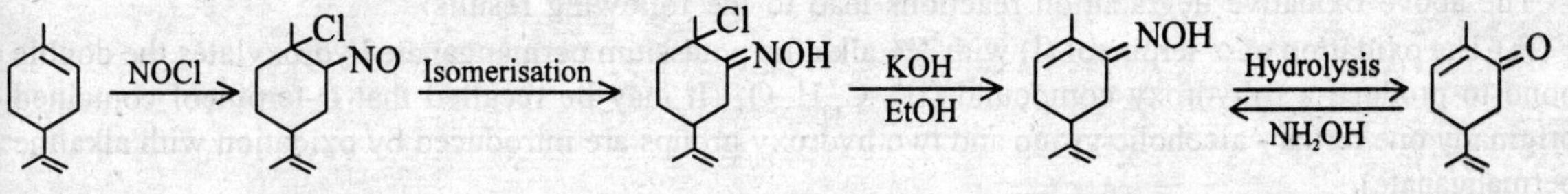

Limonene (I) $\xrightarrow{NOCl}$ **Limonene nitroso chloride** $\xrightarrow{Isomerisation}$ $\xrightarrow[EtOH]{KOH}$ **Carvoxime** $\underset{NH_2OH}{\overset{Hydrolysis}{\rightleftharpoons}}$ **Carvone**

Hence, limonene is 1,8(9)-menthadiene (I).

43.7.2.2   Oxygenated Derivatives

The important members of this group are α-terpineol and carvone.

α-Terpineol: It occurs naturally in the free state as the (+) form in neroli oil, as the (–) form in various camphor oils and the (±) form in the oil of Cajuput. It also occurs as esters. α-Terpineol has lilac like odor and is used in perfumes and cosmetic industries. It is a solid, m.p. 35° (of the racemic form). Its structure is deduced as follows:

(i)   Its molecular formula is $C_{10}H_{18}O$.

(ii)   The oxygen atom is present as a tertiary alcoholic group as shown by routine tests and reactions of α-terpineol.

(iii)   It has one double bond, since it forms a dibromide, $C_{10}H_{18}OBr_2$ on treatment with bromine and $C_{10}H_{18}O.NOCl$ on treatment with nitrosyl chloride.

(iv)   On dehydrogenation with sulphur it gives p-cymene, suggesting thereby, the presence of p-menthane carbon skeleton in α-terpineol.

$$\alpha\text{--Terpineol} \xrightarrow{\text{S}}$$

p-Cymene   **p-Menthane**

On the basis of the above, α-terpineol is p-menthane with one double bond and a tertiary alcoholic group.

(v)   The positions of the functional groups (viz. double bond and tertiary alcoholic group) are ascertained by oxidative degradation (Wallach, 1893-1895) of α-terpineol. The following scheme gives the results:

$$C_{10}H_{18}O \xrightarrow{\text{1\% Alk. KMnO}_4} C_{10}H_{20}O_3 \xrightarrow[\text{CrO}_3]{\text{[O]}} \text{[Ketohydroxy acid]}$$

α-Terpineol   Trihydroxy   (III)
(I)   compound
                       (II)

$$\longrightarrow C_{10}H_{16}O_3 \xrightarrow[\text{alk. KMnO}_4]{\text{Mild Oxid}^n} C_8H_{12}O_4 \xrightarrow{\text{KMnO}_4} C_7H_{10}O_4$$
                          warm

Keto lactone   Terpenylic acid   Terebic acid
(IV)   (V)   (VI)
                        +
                        Acetic acid

The above oxidative degradation reactions lead to the following results:

(a) The oxidation of α-terpineol (I) with 1% alkaline potassium permanganate hydroxylates the double bond to produce a trihydroxy compound (II), $C_{10}H_{20}O_3$ (It may be recalled that α-terpineol contained originally one tertiary alcoholic group and two hydroxy groups are introduced by oxidation with alkaline permanganate).

(b) The compound (II) on oxidation with chromic acid (chromium trioxide in acetic acid) gives a compound with molecular formula $C_{10}H_{16}O_3$(IV). Since IV contains the same number of carbon atoms as in I, the double bond must be present in the ring. If the double bond was present in the side chain, the formation of oxidation products (III and IV) would be accompanied by the loss of one carbon atom.

The compound IV is neutral (as it gives no reaction with sodium carbonate solution) and shown to ontain a ketonic group. However, IV on heating with excess of standard sodium hydroxide solution and then back titration, was found to consume an amount of alkali which corresponds to the presence of one carboxyl group. Thus, IV appears to be the lactone which is isolated and not the parent hydroxy acid (III). Thus, III may be γ-hydroxy acid, which lactonises spontaneously.

(c) The compound (IV) on warming with alkaline potassium permanganate gives acetic acid and a compound $C_8H_{12}O_4$(V). The formation of acetic acid suggests that IV is a methyl ketone i.e., CH_3CO-group is present. Thus, IV is a methyl ketone and a lactone and is known as homoterpenyl methyl ketone. Its structure has been confirmed by synthesis (Simmonsen,1932).

(d) The properties of terpenylic acid (V), show that it is the lactone of a monohydroxy dicarboxylic acid. Further oxidation of terpenylic acid, (V) gives terebic acid, $C_7H_{10}O_4$(VI), which is also the lactone of a monohydroxy dicarboxylic acid.

All the above reactions can be formulated as shown, by assuming (I) (p-menth-1-en-8-ol) as the structure of α-terpineol.

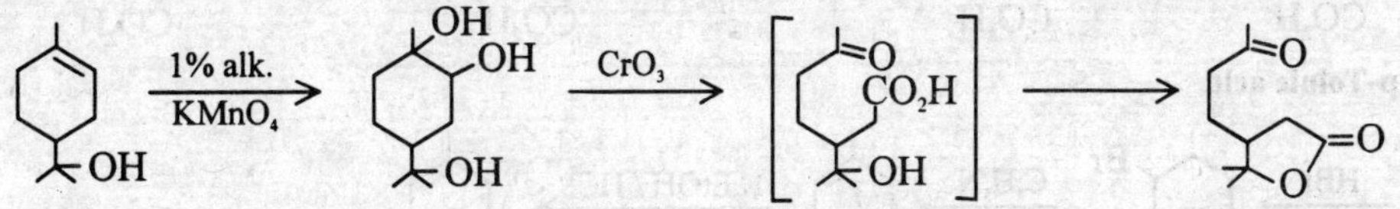

| α-Terpineol | Trihydroy compound | Ketohydroxy acid | Keto lactone |
| (I) | (II) | (III) | (IV) |

| | Acetic acid | Terpenylic acid (V) | Terebic acid (VI) |

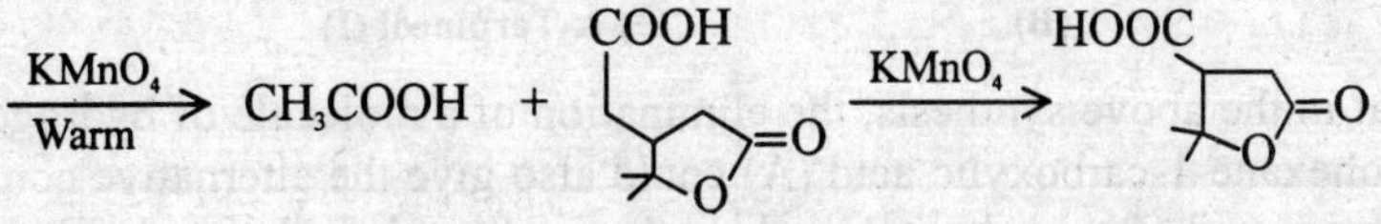

The structures of terpenylic acid (V) and terebic acid (VI) have been established by synthesis (Simonsen, 1907).

Synthesis of Terebic Acid (VI)

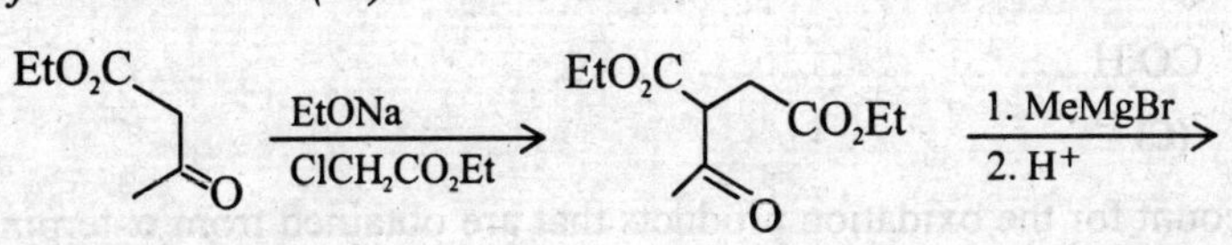

Ethyl acetoacetate

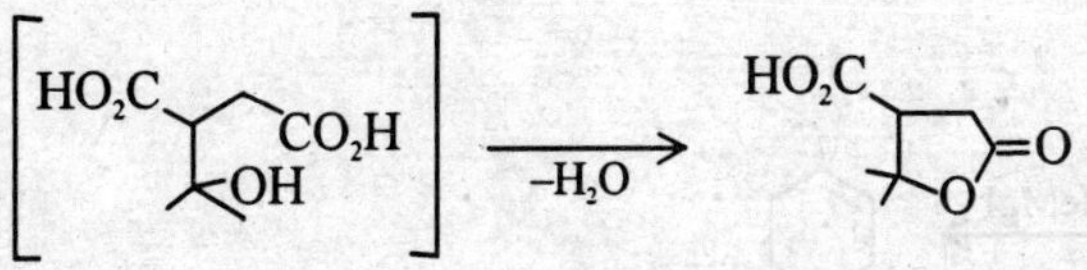

Terebic acid (VI)

Synthesis of Terpenylic Acid (V)

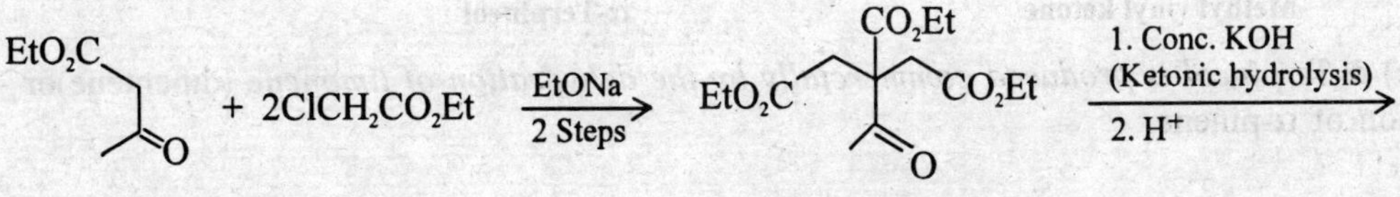

Ethyl acetoacetate   Ethyl-β-ketotricarballylate

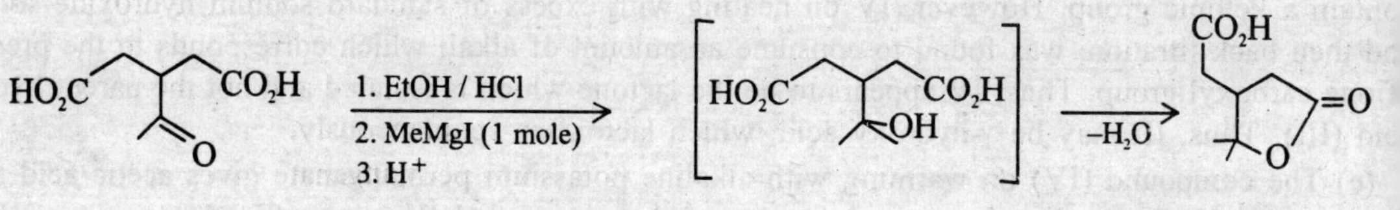

Terpenylic acid (V

(vi) The structure (I) for α-terpineol has been confirmed by its synthesis.

(a) *Perkin Jr, Meldrum and Fisher's Synthesis (1908):* For the synthesis of (+) and (−) terpineol the intermediate acid (B) was resolved with strychnine and each enantiomer treated as shown below

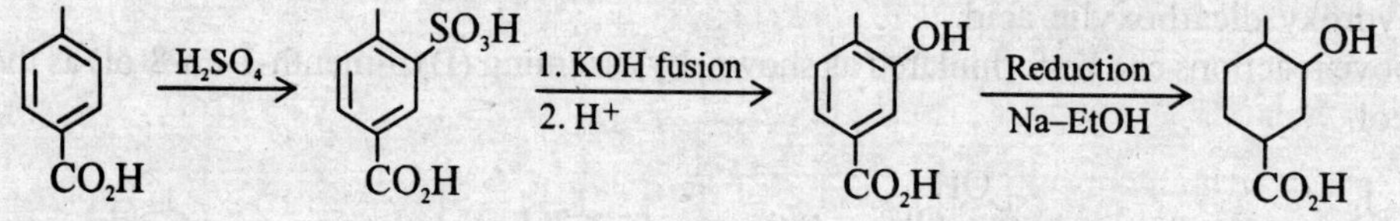

p-Toluic acid

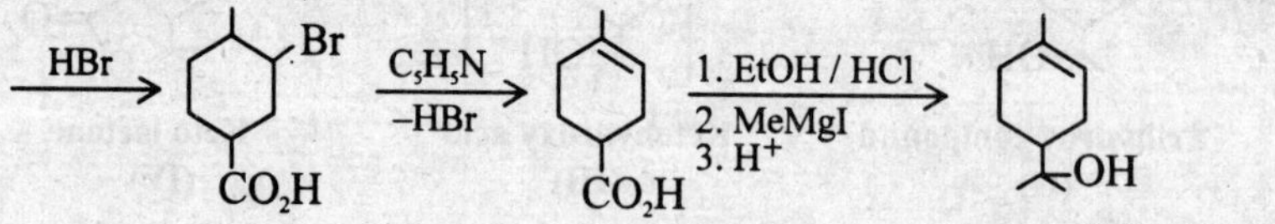

(A)   (B)   (±)-α-Terpineol (I)

It is of interest to note that in the above synthesis, the elimination of a molecule of hydrogen bromide from 3-bromo-4-methylcyclohexane-1-carboxylic acid (A) could also give the alternative compound (C) but that only the required compound (B) results is based on the analytical evidence for the position of this double bond.

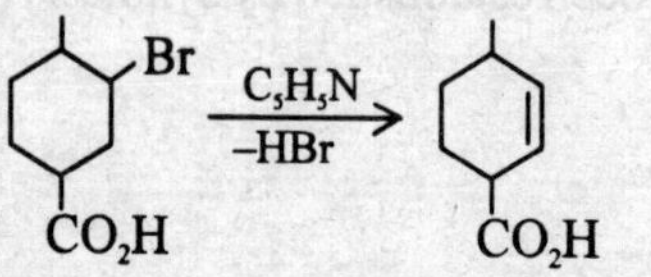

(A)   (C)

The structure (C) can not account for the oxidation products that are obtained from α-terpineol, where as the structure (B) does.

(b) *Alder and Vogt's Synthesis (1949)*

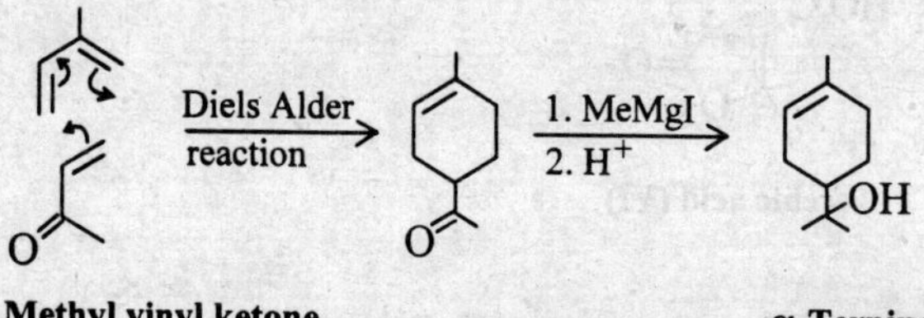

Methyl vinyl ketone   **α-Terpineol**

(c) *α-Terpineol is produced commercially by the dehydration of limonene* (dipentene)or by the hydration of α-pinene.

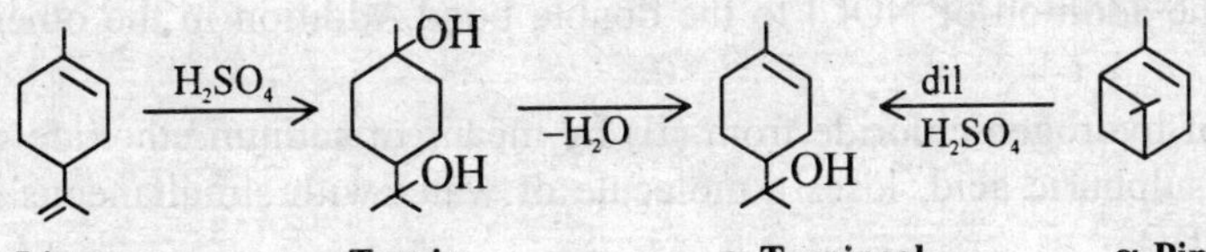

| Limonene (Dipentene) | Terpin | α-Terpineol | α-Pinene |

Two other terpineols viz., β and γ-terpineol are also known. Of these the γ-terpineol occurs naturally.

β-Terpineol   γ-Terpineol

Carvone: It occurs in both optically active and racemic forms in various oils, e.g. caraway, spearmint and dill oils. Its structure is based on following considerations:

(i) Its molecular formula is $C_{10}H_{14}O$.

(ii) It contains two double bonds, since it adds on four bromine atoms.

(iii) The presence of ketonic group is inferred by usual reactions.

(iv) The presence of two double bonds and a ketonic group implies that its parent saturated hydrocarbon is $C_{10}H_{20}$ (C_nH_{2n}) and so carvone is a monocyclic compound.

(v) On heating with phosphoric acid, carvone forms carvacrol (2-hydroxy-4-isopropyltoluene),indicating the presence of p-cymene structure and that the keto group is in the ring in the ortho position with respect to the methyl group.Thus the carbon skeleton of carvone may be represented as:

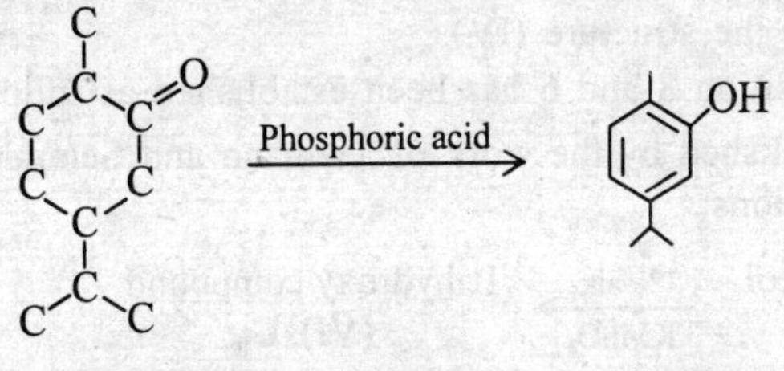

Carbon Skeleton of Carvone   Carvacrol

(vi) The structure of carvone is mainly based on the fact that carvone can be prepared from α-terpineol as follows:

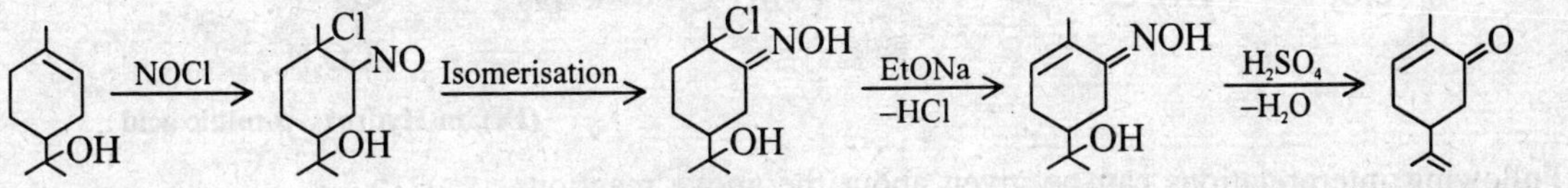

α-Terpineol   (I)   (II)   (III)   (IV) Carvone

The above sequence of reactions can be explained as follows:

(a) Addition of nitrosyl chloride to α-terpineol produces α-terpineol nitrosochloride (I);the addition follows the Markovnikov's rule, (chlorine being the negative part of NOCl).

(b) The nitrosochloride (I) rearranges spontaneously to the oximino compound (II). This rearrangement also proves the orientation of the addition of NOCl to the double bond.Addition in the other way will not give an oxime.

(c) Removal of a molecule of hydrogen chloride from (II) by means of sodium ethoxide gives (III), which on warming with dilute sulphuric acid, loses a molecule of water with simultaneous hydrolysis of the oxime to give carvone (IV).

On the basis of the above, Carvone is p-menth-6,8-dien-2-one (IV).

A closer examination, however, show that the carvone has the same carbon skeleton as α-terpineol, and also confirms the position of the keto group. They do not conclusively prove the positions of the double bonds.Instead of position 6 (in III), the double bond could be 1(7), and instead of position 8 (as in IV), the double bond could be 4(8).

(vii) The main problem is to ascertain the position of the two double bonds in the above carbon skeleton. This has been established by its relationship to limonene.

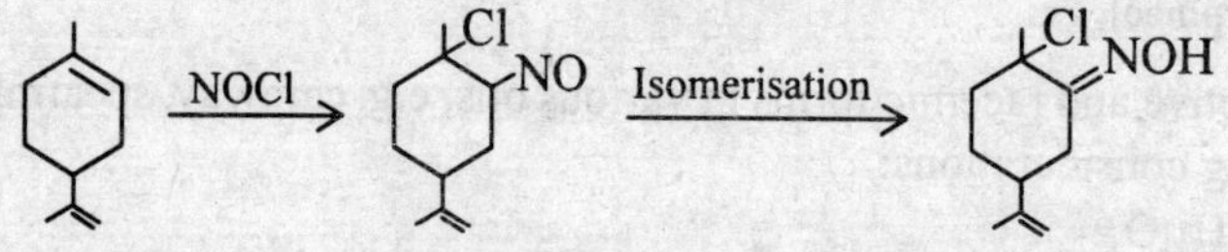

Limonene	Limonene nitrosochloride

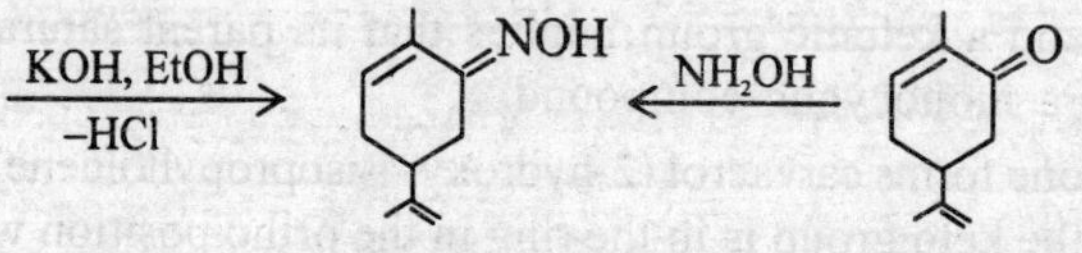

Carvoxime	(IV) Carvone

On the basis of the above carvone must have the structure (IV).

(viii) The position of the double bonds at position 8 and 6 has been established as follows:

The double bond in position 8 has been established by the work of Tiemann and Semmler (1895), who carried out the following sequence of reactions:

$$\text{Carvone (IV), } C_{10} \xrightarrow[(+4H)]{Na/EtOH} \text{Dihydrocarveol (V), } C_{10} \xrightarrow[KMnO_4]{1\% \text{ alk}} \text{Trihydroxy compound (VI), } C_{10}$$

$$\xrightarrow[CrO_3]{[O]} \text{Ketonic alcohol (VII), } C_9 \xrightarrow{NaOBr} \text{Hydroxy acid (VIII), } C_8 \xrightarrow[190°]{Br_2/H_2O}$$

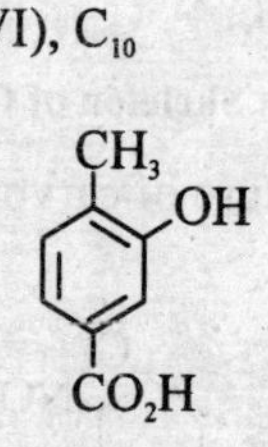

(IX), m-Hydroxy-p-toluic acid

Following interpretations can be given about the above reactions.

(a) Carvone (IV) on reduction with sodium and ethanol gives dihydrocarveol, $C_{10}H_{18}O(V)$, which is a secondary alcohol and has one double bond, i.e. one of the two double bonds and the keto group have been reduced.

(b) Alkaline potassium permanganate (1%) hydroxylates the double bond of dihydrocarveol (V), to produce a trihydroxy compound, $C_{10}H_{20}O_3(VI)$.

(c) Chromic acid oxidation of the trihydroxy compound (VI) results in cleavage of the glycol bond to give a ketonic alcohol, $C_9H_{16}O_2$(VII).

(d) Treatment of VII with sodium hypobromite losses one carbon atom to give a compound, $C_8H_{14}O_3$ (VIII), which is found to be a hydroxy monocarboxylic acid. Further, since one carbon atom is lost in the formation of VIII, its precursor (VII) must be a methyl ketone.

(e) Finally, dehydrogenation of VIII by heating with bromine water at 190° under pressure gave m-hydroxy-p-toluic acid (IX), a known compound.

The above reactions could be explained (Tiemann and Semmler) by assuming that one double bond in carvone is in 8 position.

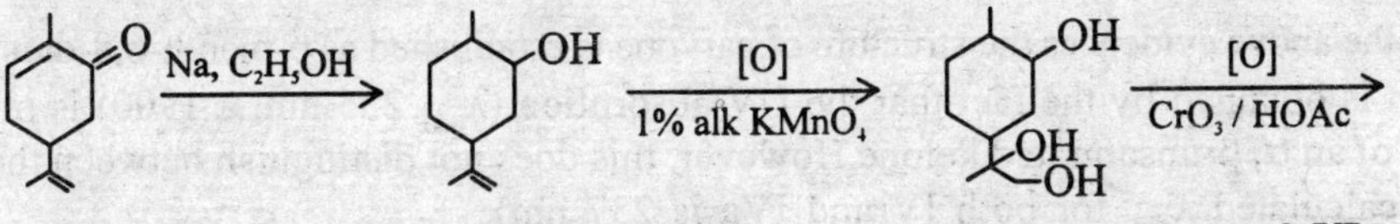

IV, Carrone   **Dihydrocarveol (V)**   **Trihydroxy compound (VI)**

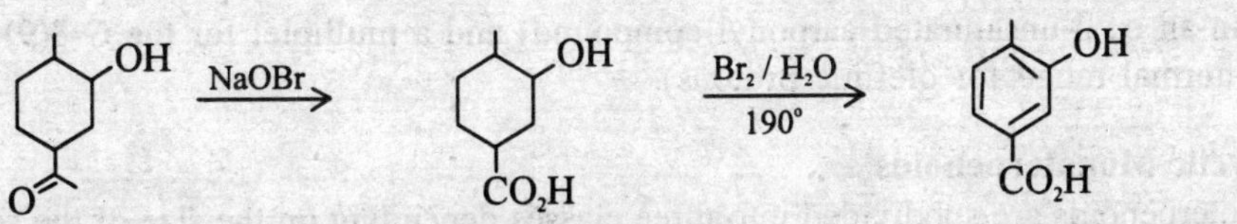

Ketonic alcohol (VII)   **Hydroxy acid (VIII)**   **m-Hydroxy-p-toluic acid (IX)**

In case, this double bond is in 4(8) position (IVa) then compound VIII and IX can not be obtained, since three carbon atoms would be lost during the oxidation.

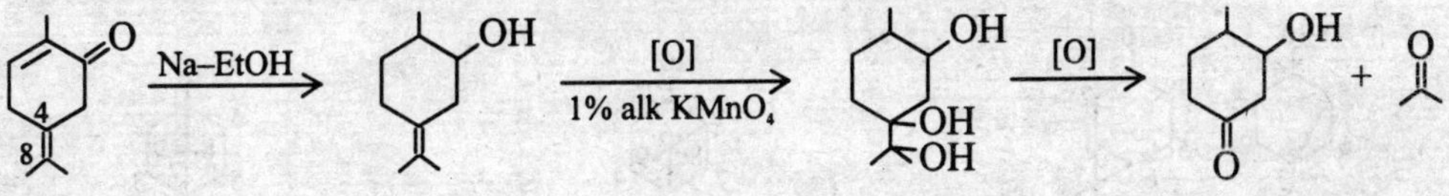

(IV a)

Further, the structure IV has a chiral centre, whereas IVa is symmetric and so cannot exhibit optical activity. Since carvone is known to occur in optically active forms, structure IV represents carvone.

The Double Bond in the 6-position

(a) The position 6 of the second double bond in carvone is established by the formation of isopropyl-succinic acid and pyruvic acid in its graded oxidation (Semmler, 1900) as shown below.

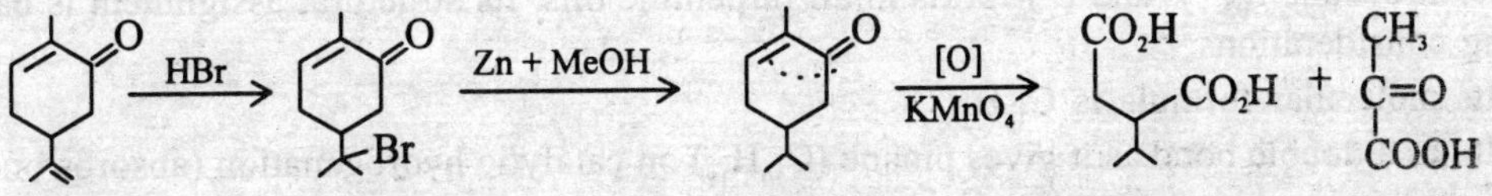

IV Carvone   **Carvone**   **Carvotan-**   **Isopropyl-**   **Pyruvic**
$C_{10}H_{14}O$ **hydrobromide**   **acetone**   **succinic acid**   **acid**
    $C_{10}H_{15}OBr$ $C_{10}H_{16}O$

The formation of these products can only be explained if there is a double bond at position 6 in the ring. In case the double bond is in the 1(7) position, formic acid (and not pyruvic acid) will be obtained.

(b) Further support for the double bond in 6-position is obtained (Simmonsen et al. 1922) by the oxidation of carvotanacetone (obtained above) with potassium permanganate to give 3-isopropylglutaric acid and acetic acid.

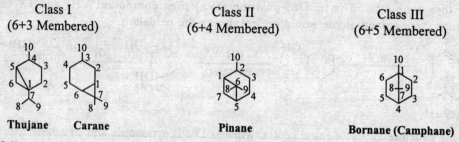

| Carvotan-
acetone | 3-Isopropyl-
glutaric acid | Acetic
acid |

On the basis of the above evidences the structure of carvone is established as p-menth-6,8-dien-2-one(IV).

Further support is obtained by the fact that the UV absorption (λ_{max} 235 nm; ε 1900) is in agreement with the structure of an α, β-unsaturated ketone. However, this does not distinguish between the structures IV and IVa (the calculated λ_{max} for both IV and IVa is 237 nm).

The NMR spectrum of carvone (IV) shows a multiplet at δ 6.75 for a C-6 proton (characteristic for α and β-protons in an α, β-unsaturated carbonyl compound) and a multiplet for the C-8(9) methylene group at δ 5.22 (normal range for olefinic protons).

43.7.3   The Bicyclic Monoterpenoids

The bicyclic monoterpenoids are subdivided into three classes depending on the size of the second ring. The first ring is six membered and the second ring may be three, four or five membered.

Class I (6+3 Membered)	Class II (6+4 Membered)	Class III (6+5 Membered)
Thujane Carane	Pinane	Bornane (Camphane)

In the following discussion, the structure of α-pinene (Pinane group) and camphor (Bornane group) have been discussed.

43.7.3.1   The Pinane Group

The most important terpenoid of this class is α-pinene.

α-Pinene: It occurs in (+)- and (–)-forms in all turpentine oils. Its structural assignment is based on the following considerations.

(i)   Its molecular formula is $C_{10}H_{16}$.

(ii)   It has a double bond as it gives pinane ($C_{10}H_{18}$) on catalytic hydrogenation (absorbs one molecule of hydrogen), and adds on two bromine atoms to form α-pinenedibromide, $C_{10}H_{16}Br_2$.

(iii)   The parent hydrocarbon $C_{10}H_{18}$ (obtained by hydrogenation), corresponds to the general formula C_nH_{2n-2} for bicyclic compounds. Thus α-pinene is a bicyclic compound.

(iv)   Treatment of α-pinene with ethanolic sulphuric acid gives α-terpineol, whose structure is well established.

$$C_{10}H_{18} \xrightarrow[\text{H}_2\text{O}]{\text{Ethanolic H}_2\text{SO}_4} \text{(\alpha-Terpineol structure)}$$

α-Pinene   α-Terpineol

The formation of α-terpineol leads to the following conclusions:

(a) α-Pinene contains a six membered ring and another ring.

(b) It establishes the position of the double bond.

(c) The gem dimethyl group $[-CH(CH_3)_2]$ is present in the second ring and not in the six membered ring.

(d) In the formation of α-terpineol one molecule of water is consumed and the hydroxy group appears on C-8 of α-terpineol. This indicates that the C-8 of α-terpineol is involved in the formation of the second ring of α-pinene. There are three possible points of attachment for this C-8 giving rise to two three membered (I and Ia) and one four membered rings (Ib).

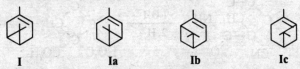

I   Ia   Ib   Ic

It is of interest to know that in fact there are four points of linkage for C-8; the three shown as I, Ia and Ib and the fourth being at the double bond as shown in (Ic). This last possibility (Ic) is rejected as it violates the Bredt's rule (1924), which states that a double bond cannot be formed by a carbon atom occupying the bridge head (of a bicyclic system). The reason for this is that such structures have a large amount of strain.

(v) The size of the second ring in α-pinene is shown to be a four membered one. This was estabilished by the following sequence of reactions (Baeyer,1896).

$$\alpha\text{-Pinene} \atop C_{10}H_{16} \;\; (I) \xrightarrow[\text{1\% alk KMnO}_4]{[O]} \;\; {\text{Pinene glycol} \atop C_{10}H_{16}O_2 \;\; (II)} \xrightarrow[\substack{\text{Warm alk}\\ \text{KMnO}_4}]{[O]} \;\; {\text{Pinonic acid} \atop C_{10}H_{18}O_2 \;\; (III)} \xrightarrow{\text{NaOBr}}$$

$$\xrightarrow{\quad} {\text{Pinic acid} \atop C_9H_{14}O_4, \;\text{(IV)}} \atop + \atop CHBr_3 \xrightarrow[\substack{\text{2. Ba(OH)}_2\\ \text{3. PbO}_2}]{\text{1. Br}_2} {\text{cis–Norpinic acid} \atop C_8H_{12}O_4 \;\text{(V)}}$$

(a) Hydroxylation of the double bond in α-pinene (I) with alkaline potassium permanganate gives pinene glycol, $C_{10}H_{18}O_2$ or $C_{10}H_{16}(OH)_2(II)$.

(b) Cleavage of the glycol bond in (II) gives pinonic acid, $C_{10}H_{16}O_3$ (III), which is a saturated keto-monocarboxylic acid.

(c) Formation of bromoform from III by oxidation with sodium hypobromite shows that III has an acetyl group $(-COCH_3)$. The pinic acid, $C_9H_{14}O_4$ (IV) formed along with bromoform is shown to be a saturated dicarboxylic acid.

It is now clear that α-pinene contains two methyl groups attached to the carbon atom in the second ring (structures I, Ia, Ib and Ic) and it is the other six membered ring (which contains a double bond) that has been opened up by the oxidation. The gem-dimethyl group, therefore, is carried by cis-norpinic acid (V), $(CH_3)_2C_4H_4(CO_2H)_2$.

Thus, the parent hydrocarbon of cis-norpinic acid (V) (considering methyl and carboxyl as substitutents) is C_4H_8, which corresponds to cyclobutane. Therefore cis-norpinic acid is dimethylcyclobutanedicarboxylic acid. Its structure has been confirmed by its synthesis (Kerr, 1929) from acetone, ethyl cyanoacetate and ammonia in alcoholic solution.

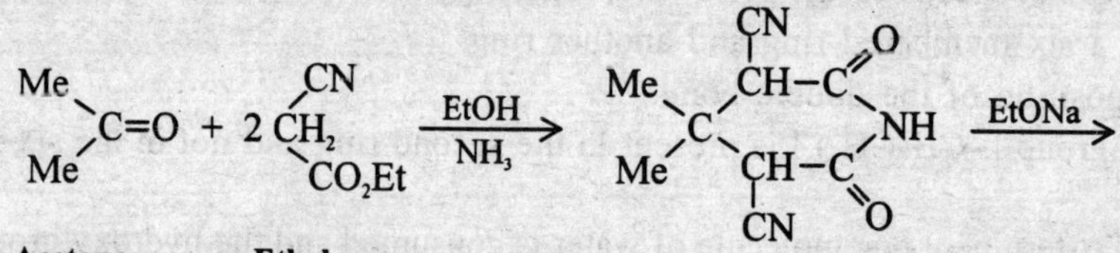

Acetone   Ethyl cyano-
              acetate

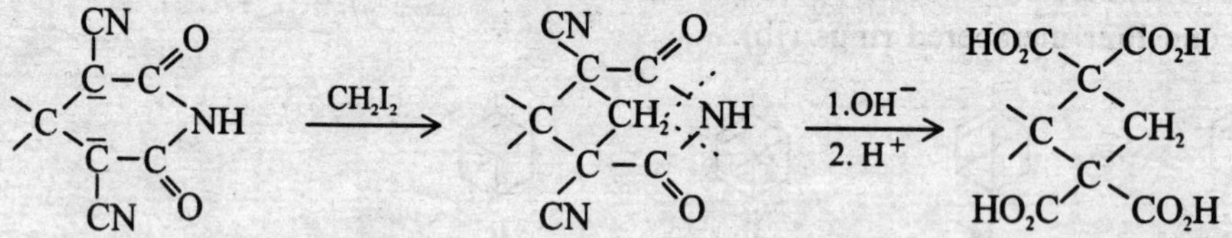

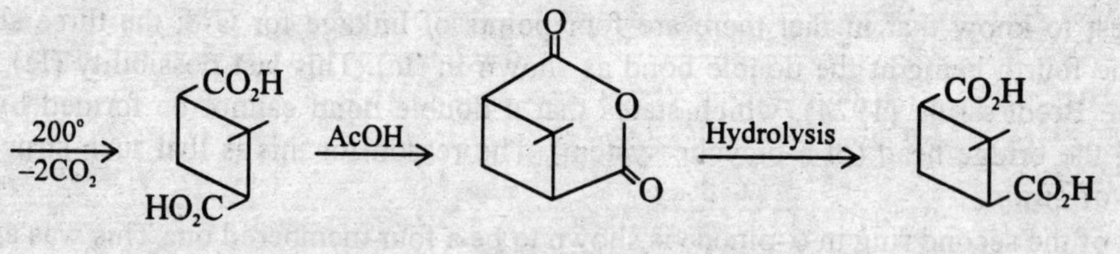

trans-Norpinic acid   cis-Anhydride   cis-Norpinic acid (V)

The formation of above oxidation products (I to V) [after establishment of the structure of cis-norpinic acid (V)] can be explained on the basis of the structure I for α-pinene as shown below:

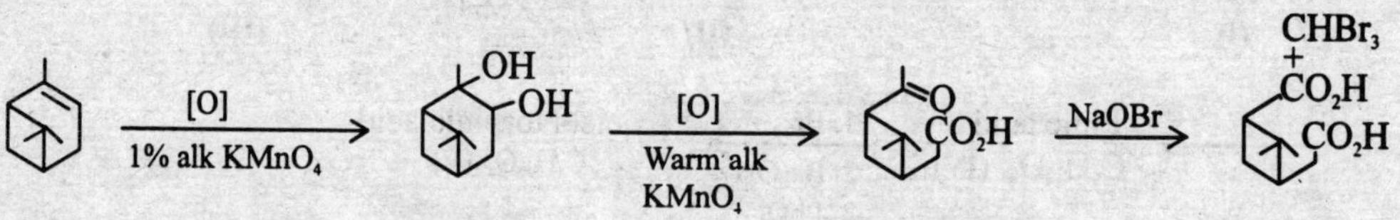

(I)   (II)   (III)   (IV)
α-Pinene   Pinene glycol   Pinonic acid   Pinic acid

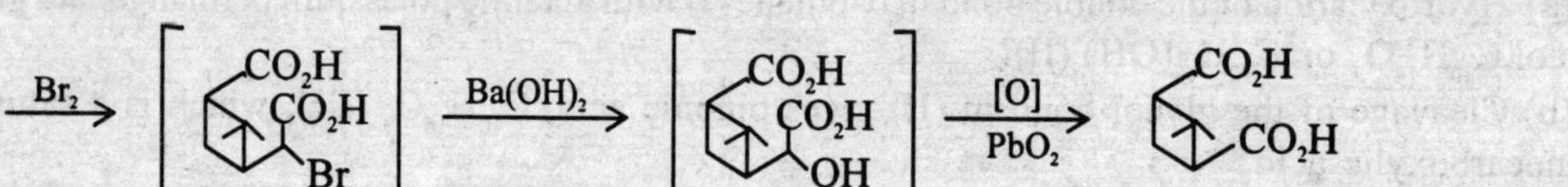

(V) cis-Norpinic acid

(vi) Final confirmation of structure (I) for α-pinene is done by its synthesis which is carried out in four steps.

Step-I: involves the synthesis of trans-norpinic acid, which has already been discussed.

Step-II: involves the synthesis of pinic acid (IV) from trans- norpinic acid (Guha et al 1937).

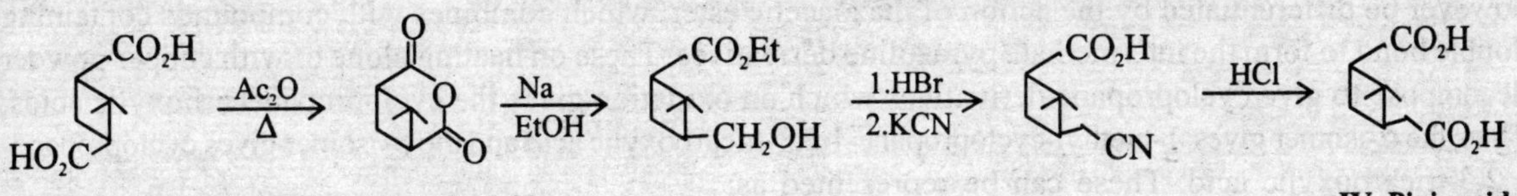

trans-Norpinic acid   **cis-Anhydride**   **IV, Pinic acid**

Step-III: involves the Synthesis of pinonic acid from synthetic pinic acid (Rao, 1943).

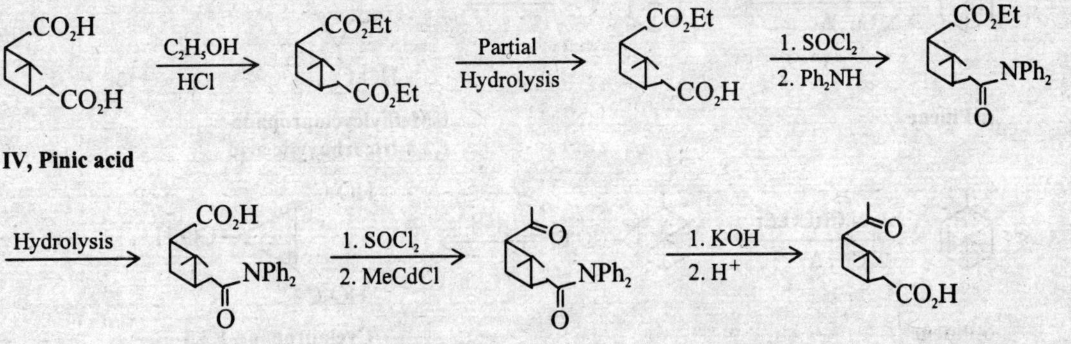

IV, Pinic acid

trans-Pinonic acid

Step-IV: involves the final conversion of trans-pinonic acid into α-pinene. This step was carried out by Ruzica et al. (1920-1924) using pinonic acid obtained by the oxidation of α-pinene.

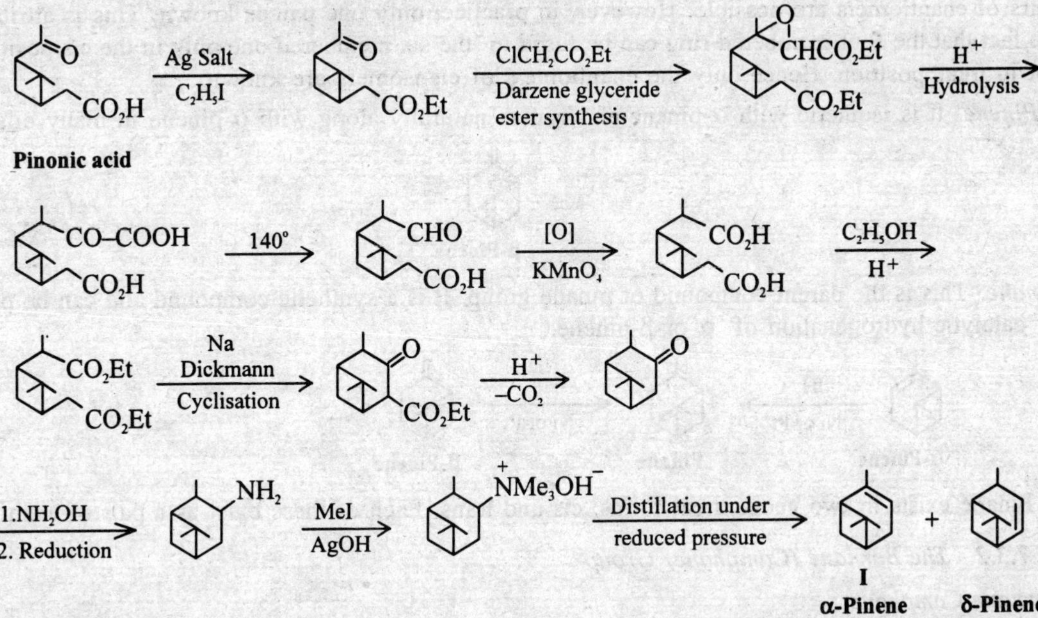

Pinonic acid

I

α-Pinene   **δ-Pinene**

In the above synthesis, the final step gives a mixture of two compounds, α- and δ-pinene. These were identified by preparation of nitrosochlorides; one of the nitrosochloride was identical with the one obtained from natural α-pinene. But this does not prove which is α- and which is δ-. The two isomers could, however be differentiated by the action of diazoacetic ester, which combines with compounds containing double bond to form the intermediate pyrazoline derivatives. These on heating alone or with copper powder decompose to give cyclopropane derivatives which on oxidation gives the cyclopropanecarboxylic acids. Thus, the α-isomer gives 1-methyl-cyclopropane-1,2,3-tricarboxylic acid and the δ- isomer gives cyclopropane-1,2,3-tricarboxylic acid. These can be represented as:

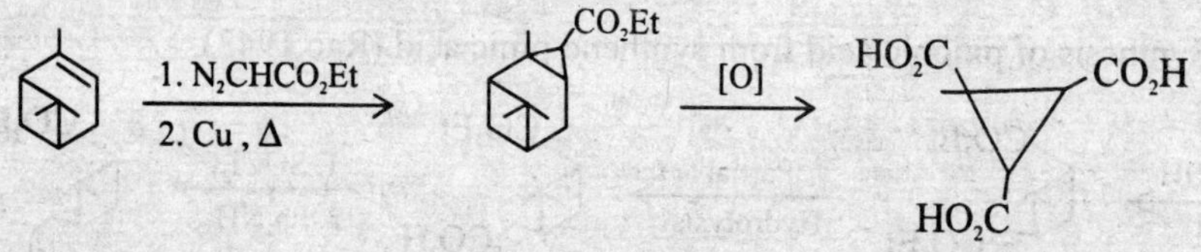

α-Pinene

1-Methylcyclopropane-
1,2,3-tricarboxylic acid

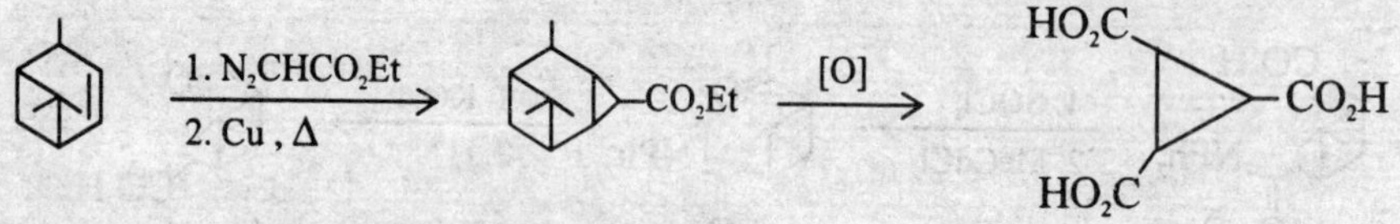

δ-Pinene

Cyclopropane-1,2,3-
tricarboxylic acid

The products obtained above are in accordance with the structures assigned to α and δ-pinene.

Stereoisomerism of α-pinene: The structure I for α-pinene contains two dissimilar chiral centres. So two pairs of enantiomers are possible. However, in practice, only one pair is known. This is attributed to the fact that the four membered ring can be fused to the six membered one only in the cis position and not in trans position. Hence only the enantiomers of cis-isomers are known.

β-Pinene: It is isomeric with α-pinene and occurs naturally along with α-pinene in many oils.

β-Pinene

Pinane: This is the parent compound of pinane group. It is a synthetic compound and can be prepared by catalytic hydrogenation of α or β-pinene.

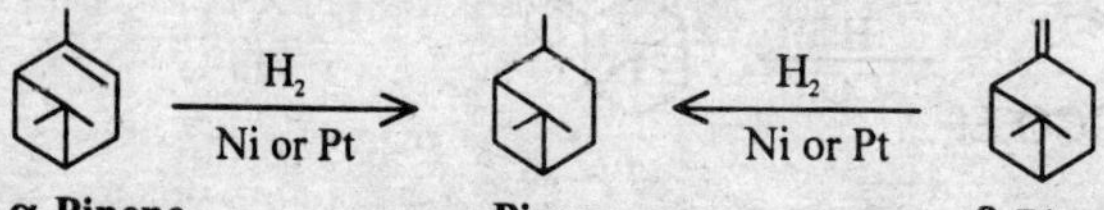

α-Pinene   Pinane   β-Pinene

Pinane exists in two geometrical forms, cis and trans. Each of these exist as a pair of enantiomers.

43.7.3.2   The Bornane (Camphane) Group

Bornane/Camphane

Bornane, $C_{10}H_{18}$, b.p. 156°, is an optically inactive synthetic compound. It is the hydrocarbon from which well known oxygenated bicyclic terpenoids are derived. It is prepared by the reduction of camphor.

(i)   Camphor on reduction forms a mixture of borneols. These are converted into bornyl iodides, which on reduction give bornane (Aschan, 1900).

Camphor   **Mixture of borneols**   **Mixture of bornyl iodide**   **Bornane**

(ii) Camphor is converted into bornane by the Wolff-Kishner reduction

Camphor   **Camphor hydrazone**   **Bornane**   + N₂

Camphor

It occurs in nature in the camphor trees of Formosa and Japan and is optically active. It occurs naturally in (+) and (−) forms and as a racemic mixture too. Camphor is a soft crystalline solid (m.p. 180°) with characteristic odor. It is used as an insect repellant, plasticizer for the preparation of cellulose and photographic films. It finds use in medicines since it has a high cryoscopic constant. It is used as a solvent in the Rast method for molecular weight determinations. Now a days most of the world requirements of camphor are met from the synthetic material obtained from α-pinene.

The structure determination of camphor involved much more efforts than for any other terpene. This can be judged from the fact that during the period 1820-1895, more than thirty structures assigned for camphor received consideration.

The special features concerned with the structure of camphor are given below:

(i) Its molecular formula is $C_{10}H_{16}O$.

(ii) The oxygen is present as a carbonyl function, since it forms an oxime, semicarbazone etc. The presence of carbonyl group is also supported by its reduction to alcohol, $C_{10}H_{18}O$ (This alcohol is a mixture of two isomeric alcohols called borneol and isoborneol, which can be oxidised back to camphor). Further, oxidation of camphor yields a dicarboxylic acid (having the same number of carbon atoms as in camphor) and not a monocarboxylic acid. Formation of dicarboxylic acid indicates that the carbonyl group is ketonic in nature (an aldehyde gives a monocarboxylic acid).

(iii) From the above facts, it is seen that the parent hydrocarbon of camphor is $C_{10}H_{18}$. This corresponds to the general formula C_nH_{2n-2}. Therefore camphor is a bicyclic compound.

(iv) Camphor contains a −CH₂CO− group, since it forms an oxime with nitrous acid and also a monobenzylidene derivative with benzaldehyde.

(v) Camphor on distillation with phosphorus pentoxide gives p-cymene, indicating that it contains a six membered ring substituted at 1,4 position. However, on distillation with iodine, it gives carvacrol. This indicates that the keto group in camphor is in ortho position with respect to methyl group.

Carvacrol   **p-cymene**

(vi) Though over thirty formulae were proposed for camphor, Bredt (1843) was the first to assign the correct formula. Bredt's structure is based on the fact that

(a) Oxidation of camphor with nitric acid gives camphoric acid, $C_{10}H_{16}O_4$ (Malaguti, 1837).

(b) Oxidation of camphoric acid (or camphor) with nitric acid gives camphoronic acid, $C_9H_{14}O_6$ (Bredt,1893).

$$\underset{C_{10}H_{16}O}{\text{Camphor}} \xrightarrow[HNO_3]{[O]} \underset{C_{10}H_{16}O_4}{\text{Camphoric acid}} \xrightarrow[HNO_3]{[O]} \underset{C_9H_{14}O_6}{\text{Camphoronic acid}}$$

Furthermore since camphoric acid has the same number of carbon atoms as camphor, the keto group must be in one of the rings of camphor.

To establish the structure of camphor, it is essential first to determine the structure of camphoric acid and camphoronic acid.

(vii) *Structure of Camphoronic Acid*

(a) Its molecular formula is $C_9H_{14}O_6$.

(b) It is shown to be a saturated tricarboxylic acid. Its formula can therefore be represented as $C_6H_{11}(COOH)_3$. Thus, its parent hydrocarbon is C_6H_{14} corresponding to the general formula C_nH_{2n+2} for open chain compound. Therefore, camphoronic acid is an acyclic compound.

(c) On distillation at atmospheric pressure, camphoronic acid yields isobutyric acid, trimethylsuccinic acid, carbon dioxide along with small amounts of some other products. The formation of the above products can be explained (Bredt,1893) if camphoronic acid is α,α,β-trimethylcarballylic acid (I). It is believed that the left hand side of the molecule (I) may break up as shown forming one molecule of carbon dioxide and two molecules of isobutyric acid (there is a shortage of two hydrogen atoms). However, breaking of the right hand side of the molecule forms one molecule of trimethylsuccinic acid, one molecule of carbon dioxide, one atom of carbon and two atoms of hydrogen, thereby making up the shortage of the left hand side of the molecule. This can be represented as below:

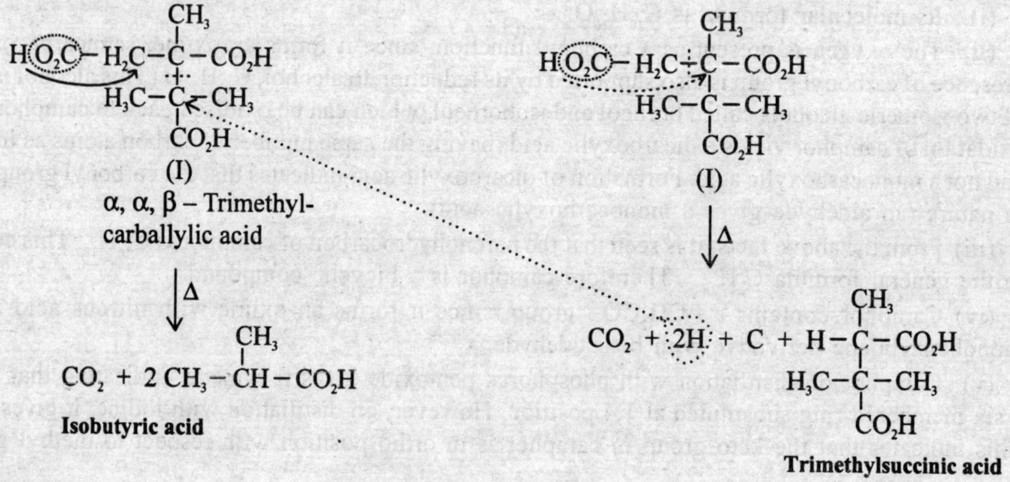

(d) The structure I for camphoronic acid is finally confirmed by its synthesis (Perkin and Thorpe 1897) from ethyl acetoacetate.

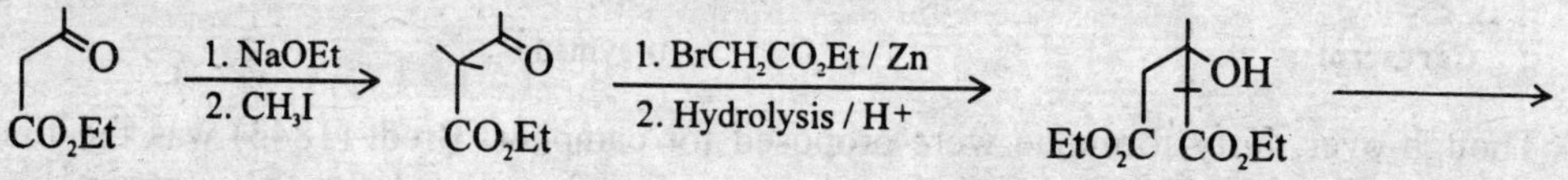

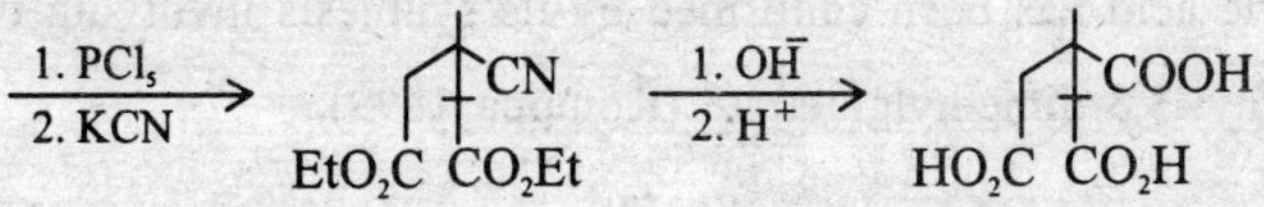

$$\xrightarrow[\text{2. KCN}]{\text{1. PCl}_5} \quad \overset{\text{CN}}{\underset{\text{EtO}_2\text{C} \ \ \text{CO}_2\text{Et}}{\big|}} \quad \xrightarrow[\text{2. H}^+]{\text{1. OH}^-} \quad \overset{\text{COOH}}{\underset{\text{HO}_2\text{C} \ \ \text{CO}_2\text{H}}{\big|}}$$

Camphoronic acid (I)

Thus, by the structure elucidation of camphoronic acid (I), nine out of ten carbon atoms of camphor have been accounted for.

(viii) *Structure of Camphoric Acid*

(a) Its molecular formula is $C_{10}H_{16}O_4$.

(b) It is shown to be a dicarboxylic acid and it does not contain carbon carbon double bond (as shown by its molecular refractivity). Out of the two carboxyl group in camphoric acid, one can easily be esterified to form monoester whereas the second carboxyl group is difficult to esterify. This suggests that the two carboxyl groups are differently situated viz. one is joined to a tertiary carbon and the second to a primary or a secondary carbon atom.

(c) Camphoric acid is formed from camphor involving the cleavage of the ring containing the keto group.Thus camphoric acid much be a monocyclic compound.

(d) Camphoric acid easily gives anhydride, which on bromination gives a monobromo derivative. This indicates that one of the carboxyl group is attached to a secondary carbon atom. Thus, the part structure of camphoric acid, its anhydride and the bromo derivative can be represented as follows.

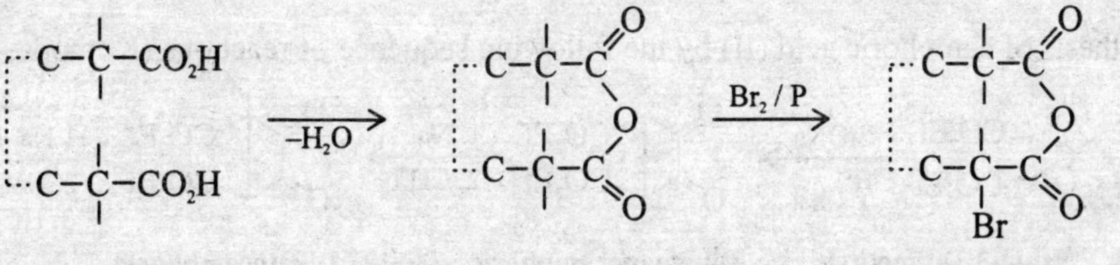

| Camphoric acid | Camphoric anhydride | Monobromo deriv of the anhydride |

(e) Camphoric acid contains three methyl group, two of which are gem dimethyl group. This is proved by the formula of camphoronic acid (I) which is obtained by the oxidation of camphoric acid (since I contains three methyl group, two of which are gem dimethyls). Camphoric acid ($C_{10}H_{16}O_4$), thus, can be represented as $(CH_3)_3C_5H_5(COOH)_2$. This shows that the parent hydrocarbon is C_5H_{10} and thus camphoric acid is a cyclopentanone derivative.

(f) On the basis of Blanc's rule, camphoric acid is a glutaric acid derivative, since on distillation with acetic anhydride it gives the anhydride.

(g) Although number of structures were proposed for camphoric acid, but Bredt, on the basis of the foregoing evidences proposed the structure (II) for camphoric acid.

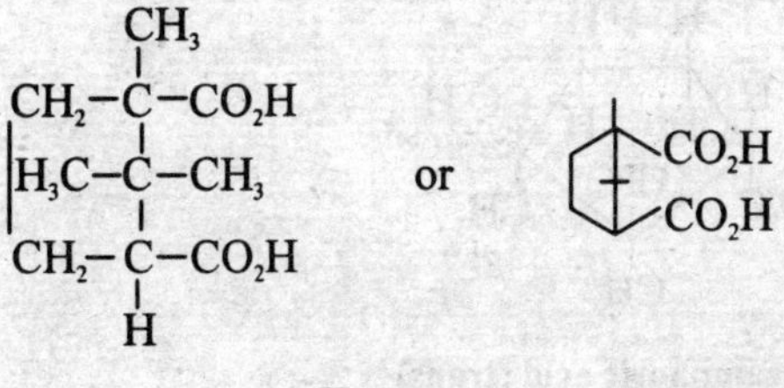

II

(h) The structure II for camphoric acid has been confirmed by its synthesis involving two steps.

Step-I: involves the synthesis of ethyl-3,3-dimethylglutarate (Komppa 1899).

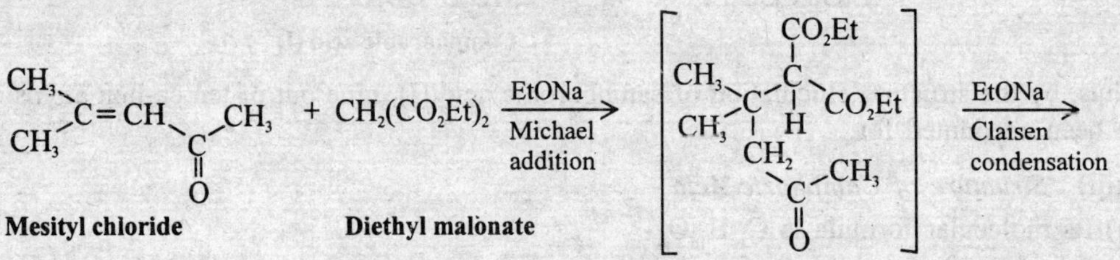

Mesityl chloride   Diethyl malonate

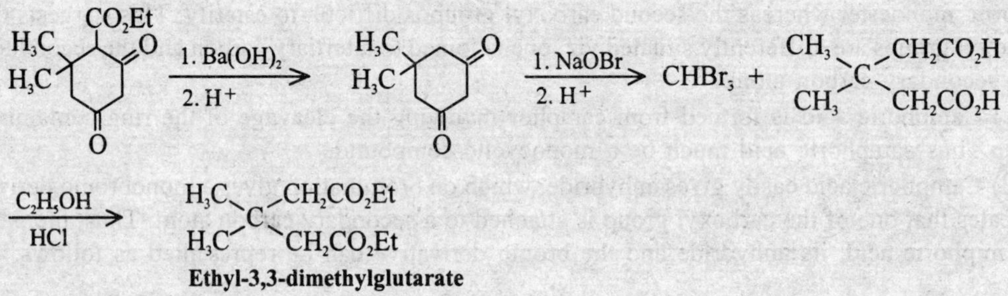

Ethyl-3,3-dimethylglutarate

Step-II: is the synthesis of camphoric acid (II) by the following sequence of reactions (Komppa, 1903):

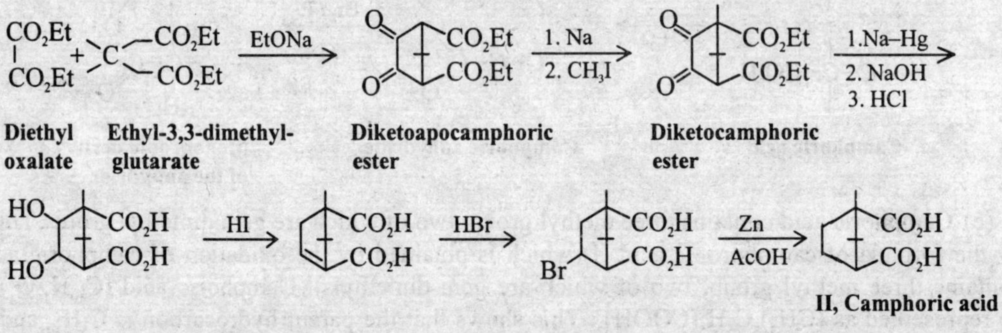

Diethyl   Ethyl-3,3-dimethyl-   Diketoapocamphoric   Diketocamphoric
oxalate   glutarate   ester   ester

II, Camphoric acid

Camphoric acid exists in two geometrical isomeric forms; cis and trans, neither of which have elements of symmetry thus four optically active forms are possible, all of which are known and correspond to (+) and (–) forms of camphoric acid and isocamphoric acid. Out of the two, camphoric acid forms an anhydride and isocamphoric acid does not. Thus, camphoric acid (m.p.187°) is the cis isomer and isocamphoric acid (m.p.171-72°) is the trans isomer.

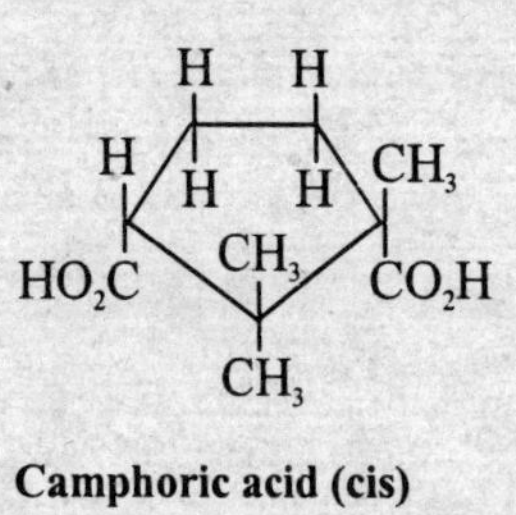

Camphoric acid (cis) 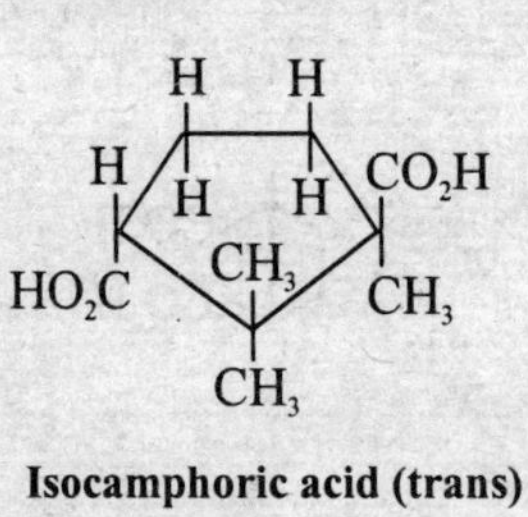   Isocamphoric acid (trans)

Having established the structures of camphoronic acid and camphoric acid we proceed to determine the structure of camphor.

Since the two carboxyl groups of camphoric acid are produced by the cleavage of –CH$_2$CO group in camphor, its structure may be represented as III (Bredt). On the basis of structure III for camphor, its oxidation into camphoric acid and comphoronic acid can be represented as follows:

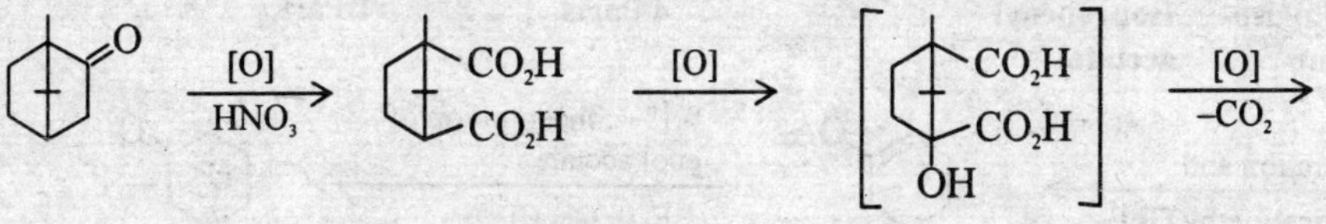

(III) Camphor　　　　**(II) Camphoric acid**

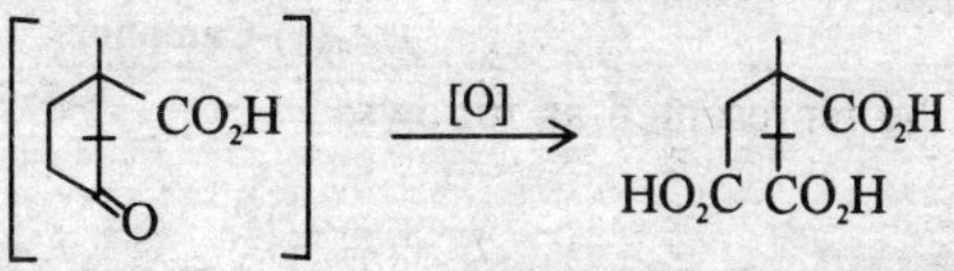

(I) Camphoronic acid

Thus the structure of camphor is established as III. It is appropriate to mention that another structure (IV) also accounts for the above conversions (Bredt). The main reason against the structure IV is that camphor gives carvacrol when distilled with iodine. The formation of carvacrol is possible from structure III but not from IV.

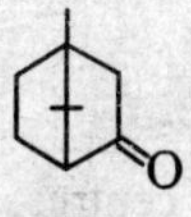

IV

Synthesis of Camphor: The structure of camphor as III has finally been confirmed by its synthesis (Haller, 1896) starting from camphoric acid. Since camphoric acid was later prepared by Komppa, this constitutes a total synthesis of camphor.

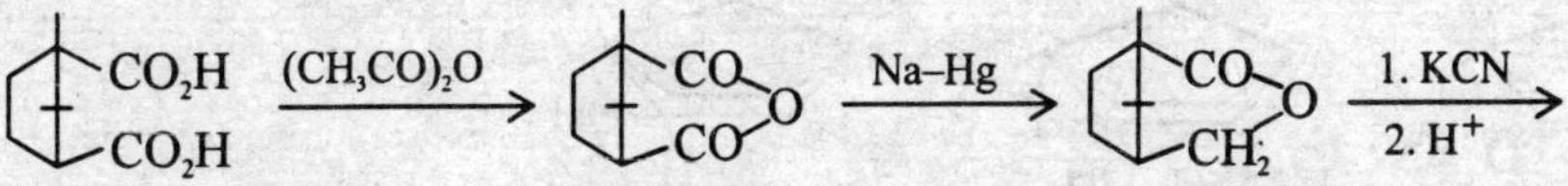

II, Camphoric acid　　　**Camphoric anhydride**　　**α-Campholide**

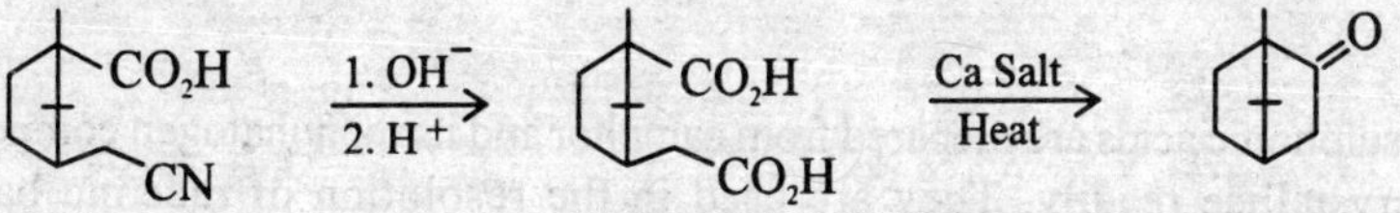

Homocamphoric acid　　　**III, Camphor**

It is not an unambiguous synthesis, because in the conversion of camphoric anhydride into campholide, the other keto group (near C–CH$_3$) might be reduced to form a campholide (structure V) which is actually β-campholide and gave compound IV on further treatments.

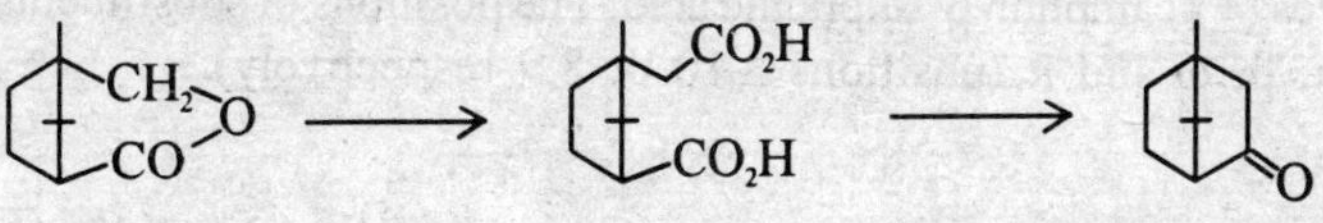

(V)　　　　　　　　　　　　　　**(IV)**

Biogenetic synthesis of camphor was accomplished from (±)-dihydrocarvone (T. Money et al., 1969).

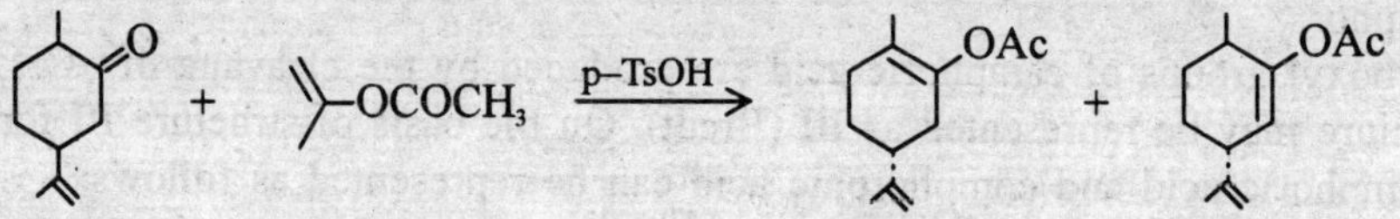

**(±)-Dihydro-   Isopropenyl   4 Parts   1 Part
carvone   acetate**

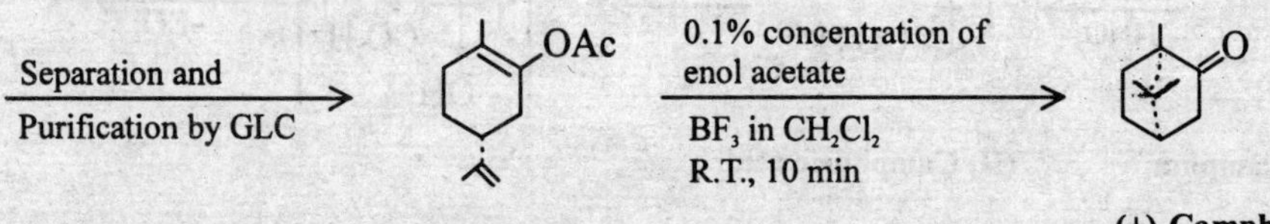

(+)-Camphor

Conversion of enol acetate into camphor may be represented as follows:

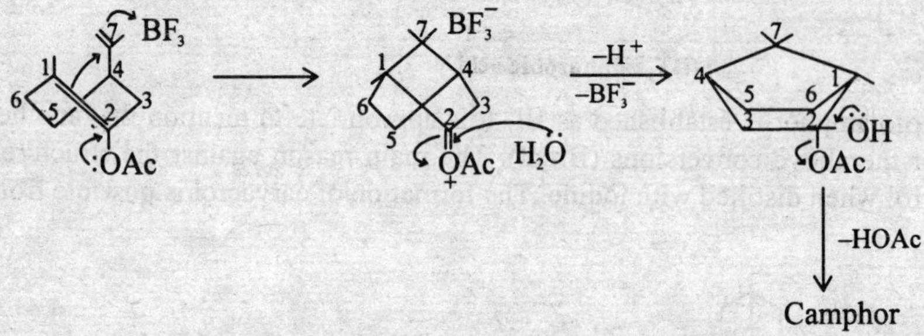

Camphor

Stereochemistry of Camphor: Camphor has two dissimilar chiral centres, but only one pair of enantiomers is known, as only cis fusion is possible. Trans fusion of the gem dimethyl methylene bridge to cyclohexane ring is not possible, as it will give too strained a molecule to exist (see also α-pinene).

Camphor and its derivatives exist in boat conformation.

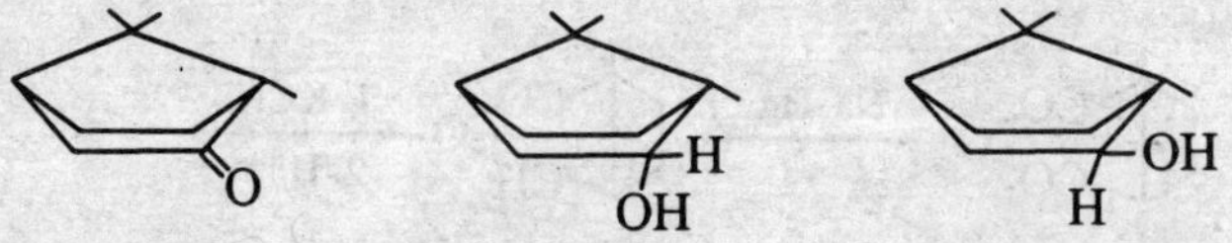

Camphor   Borneol   Isoborneol

Camphor Sulphonic Acids

A number of optically active sulphonic acids are prepared from camphor and its monohalogen compounds.These are strong acids and can crystallise readily. They are used in the resolution of racemic bases.

(+)-Camphor, on heating with bromine at 100° gives α-bromo-(+)-camphor, which on warming with sulphuric acid affords α-bromo-(+)-camphor-π-sulphonic acid. Its reduction gives (+)-camphor-π-sulphonic acid. (±)-camphor-π-sulphonic acid is obtained by the sulphonation of (+)-camphor with fuming sulphuric acid (under these conditions,(+)-camphor is racemised). However, sulphonation of (+)-camphor with sulphuric acid in acetic anhydride gives (+)-camphor-β-sulphonic acid. The positions of substituents in camphor are indicated by Greek letters α, β, ω and π (positions 3,10 or 8,9 respectively).

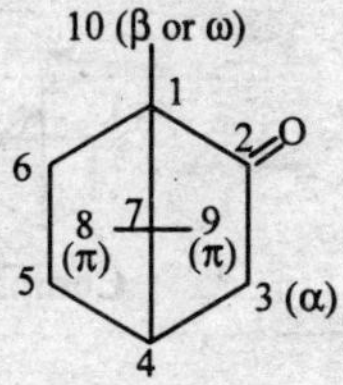

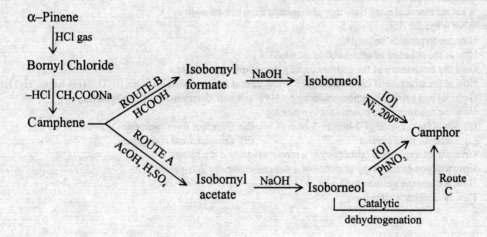

<table>
</table>

Position 3 = α

8 or 9 = π

10 = β or ω

A reaction of special interest of camphor is its fission on heating with potassium hydroxide. The fission is in accordance with the general rule for cyclic ketones, according to which fission occurs at the bond involving the least substituted carbon atom adjacent to the carbonyl group. Thus camphor gives campholic acid and isocampholic acid (Guerbert; 1912), the former being the major product.

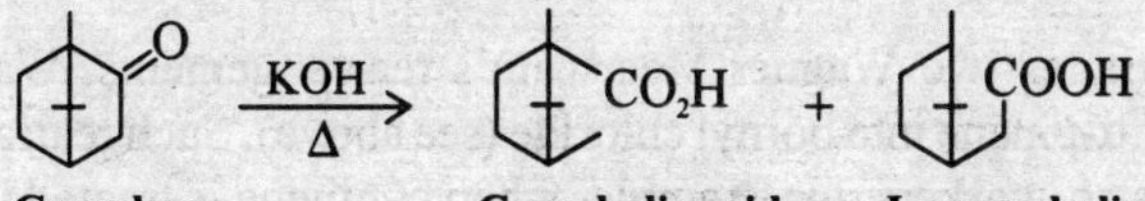

Camphor   **Campholic acid**   **Isocampholic acid**

Commercial Synthesis of Camphor

The natural source of camphor does not meet the demand of camphor. It is obtained on a large scale from α-pinene (which is obtained from turpentine oil using either of the three routes (A, B and C) shown below:

α–Pinene

| HCl gas

Bornyl Chloride

–HCl | CH₃COONa

Camphene

ROUTE B HCOOH → Isobornyl formate →NaOH→ Isoborneol

ROUTE A AcOH, H₂SO₄ → Isobornyl acetate →NaOH→ Isoborneol

Isoborneol — Catalytic dehydrogenation → Camphor (Route C)

Isoborneol [O] PhNO₂ → Camphor

Isoborneol [O] Ni, 200° → Camphor

The main route (A) is represented as follows:

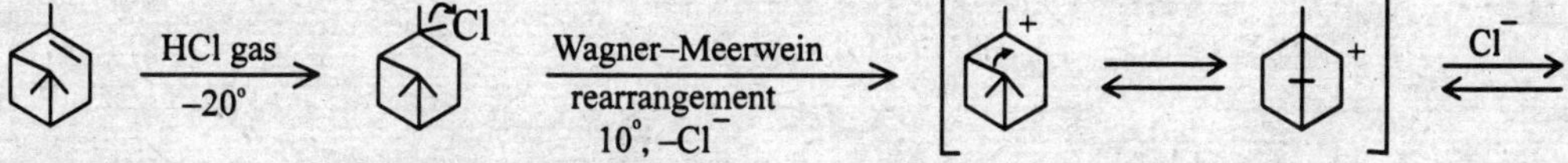

α-Pinene —HCl gas, −20°→ **Pinene hydrochloride** —Wagner–Meerwein rearrangement 10°, −Cl⁻→ [⇌] —Cl⁻→

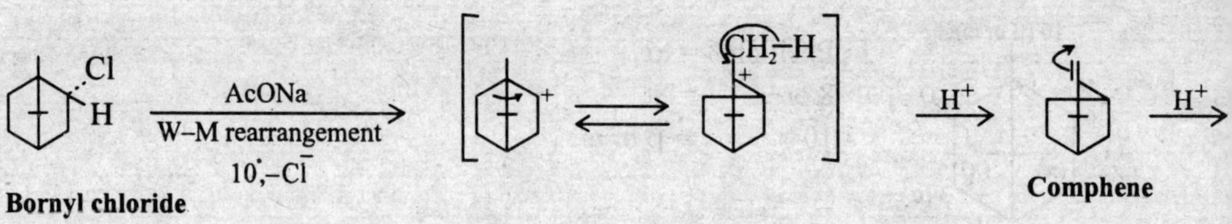

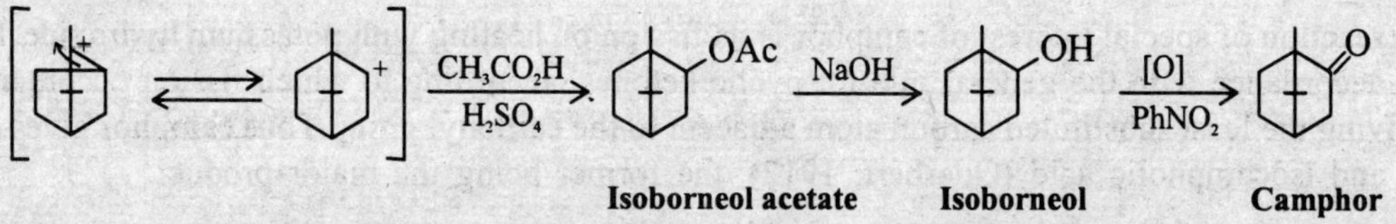

Bornyl chloride

AcONa / W–M rearrangement / 10°,–Cl⁻

Comphene

Isoborneol acetate   **Isoborneol**   **Camphor**

The above sequence of reactions constitutes two Wagner-Meerwein's rearrangements. Wagner first observed rearrangement in the conversion of α-pinene into bornyl chloride (see above). Such rearrangements were also investigated by Meerwein and his co-workers, for example, when α-pinene is treated in ethereal solution at – 20° with hydrogen chloride, the product is pinene hydrochloride, which is unstable and at 10° rearranges to bornyl chloride (Meerwein et al., 1922). Such rearrangements, which occurs with bicyclic monoterpenoids are known as Wagner-Meerwein rearrangements.

43.8   PROBLEMS

1. What are terpenes? How they are classified?
2. What are essential oils? How they are isolated from natural sources?
3. What is special isoprene rule?
4. How are terpenoids isolated?
5. How is the structure of terpenoid determined?
6. Give the mechanism of the conversion of geraniol into α-terpineol.
7. How is the structure of citral established? How does it react with: (i) NH_2OH, (ii) O_3 , (iii) Ag_2O, (iv) $KHSO_4$.
8. Geraniol and nerol are geometrical isomers. How will you determine their configuration?
9. How the following compounds are synthesised.
   (a) Citral from 2-methyl-2-hepten-2-one   (b) Camphor from α-pinene
   (c) Geraniol from citral-a   (d) Carvone from α-terpineol.
10. Give one example each (structure) of a mono, sesqui and diterpenes.
11. How will you establish the presence of a 4-membered ring in α-pinene?
12. How is α-pinene synthesised?
13. How is camphor synthesised?

44
Polymers

44.1 INTRODUCTION

Polymers are very large molecules which are formed by chemical combination of a large number of relatively small molecules called monomers. The polymers are of two types, the *natural polymers* like polysaccharides, proteins and the *synthetic polymers* like nylon, bakelite etc. The synthetic polymers have been successfully used in place of traditional materials such as wood, natural fibres and natural rubber. These are used for clothing (synthetic fibres), automobile tyres, tubes etc (synthetic rubber) and for making a variety of house hold goods telephone, television and radio cabinets, utensils, toys, chairs, balls etc.

44.2 CLASSIFICATION

Polymers are classified in different ways according to their structures, physical properties or the type of reactions by which they are prepared.

44.2.1 According to the Structure of the Polymer

On the basis of structure, the polymers may either be homopolymers or copolymers. In *homopolymers*, only one monomer unit is present and the arrangement of monomers may vary from polymer to polymer. Thus, a polymer may have a linear arrangement, side chain branching or different cross links. The nature and the type of polymer obtained depends on the type of monomer used and also on the method of polymerisation. In general, we may have the following three different types of polymers to describe the stereoisomers of substituted ethylenes having the general formula $R-CH=CH_2$.

(i) *Isotactic:* In this all the R-substituted carbon atoms have identical configuration.

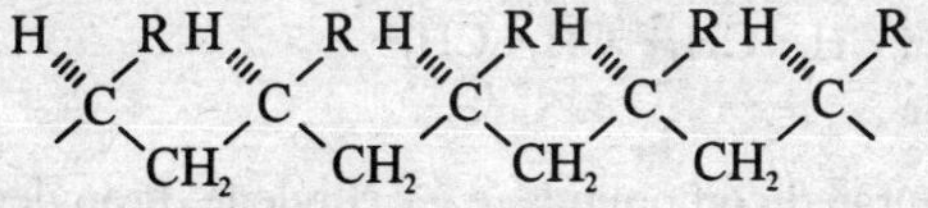

(ii) *Syndiotactic:* In this the configuration of R-substituted carbon atoms alternate in a regular fashion.

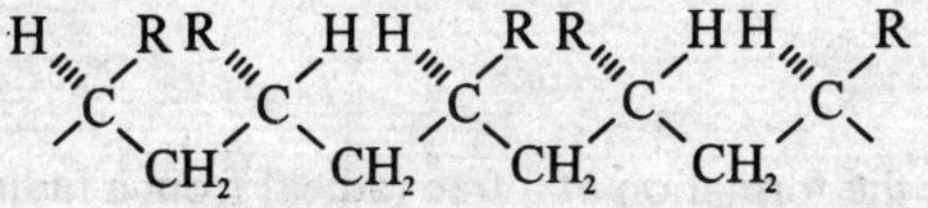

(iii) *Atactic:* In this there is a random stereochemical distribution.

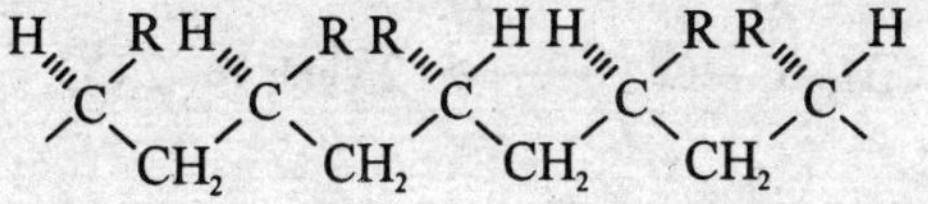

Copolymers, on the other hand, have two or more different monomer units. A copolymer may be of a random nature, i.e., it has a random arrangement of different monomers in the chain (atactic). Alternatively, it may have an orderly arrangement completely altered(syndiotactic). It can also be a block polymer. The three types of copolymers are represented below (the monomers used are A and B):

-A-A-B-A-A-B-A-A-B-B-A-A-B-

Random arrangement (Atactic)

-A-B-A-B-A-B-A-B-A-B-A-B-A-B-

Orderly arrangement (Syndiotactic)

-A-A-A-B-B-B-A-A-A-B-B-B-A-A-A-

Block Polymer

44.2.2   According to the Physical Properties of Polymers

On the basis of physical properties, following three types of polymers are recognised.

(i)   *Elastomers:* These can be easily stretched. However, some degree of rigidity in elastomers can be introduced by adding cross linking agents, e.g., vulcanised rubber.

(ii)   *Thermoplastics:* They become soft on heating and harden again on cooling. In other words, they undergo physical change on heating. Examples are polyethylene and polystyrene.

(iii)   *Thermosetting Plastics:* These undergo chemical changes on heating and on cooling set into a hard mass. The change is irreversible. Bakelite is generally used as a thermosetting polymer.

44.3   POLYMERISATION REACTIONS

It has already been stated that any reaction which converts a low molecular weight compound (monomer) into a high molecular weight compound (polymer) is called polymerisation reaction. Polymerisation reactions are of two types, viz. addition and condensation polymerisation.

44.3.1   Addition Polymerisations

In addition polymerisation, the monomers add to one another in such a way that the polymers contain all the atoms of the starting monomers, i.e., no elimination of water, alcohol or carbon dioxide etc takes place. For example, ethylene polymerises to polyethylene.

$$nCH_2=CH_2 \xrightarrow{\text{Polymerisation}} -CH_2-CH_2\{CH_2-CH_2\}_n CH_2-CH_2-$$
Ethylene   **Polyethylene**

The addition polymerisation is encountered in unsaturated compounds, e.g., ethylene, propylene, styrene, vinyl chloride etc. This polymerisation takes place by various mechanisms like free radical, cationic, anionic and co-ordination.

44.3.1.1   Free Radical Polymerisation

Initiation: It is the chain initiation step and involves the formation of a free radical from a radical initiator such as benzoyl peroxide, azobis-isobutyronitrile.

$$C_6H_5-\overset{O}{\underset{\|}{C}}-O-O-\overset{O}{\underset{\|}{C}}-C_6H_5 \xrightarrow{\Delta} 2\,C_6H_5-\overset{O}{\underset{\|}{C}}-O^{\cdot} \longrightarrow 2\,C_6H_5^{\cdot} + 2\,CO_2$$
Benzoyl peroxide

$$(CH_3)_2 - \underset{\underset{CN}{|}}{C} - N = N - \underset{\underset{CN}{|}}{C} - (CH_3)_2 \longrightarrow 2\,(CH_3)_2 - \overset{\cdot}{\underset{\underset{CN}{|}}{C}} + N_2$$

Azobis-isobutyronitrile

The radical so formed adds onto a molecule of the monomer producing another free radical.

$$\overset{\cdot}{R} + CH_2 = CH_2 \longrightarrow -R - CH_2 - \overset{\cdot}{C}H_2$$

$$[\overset{\cdot}{R} = C_6H_5 \quad \text{or} \quad (CH_3)_2\overset{\cdot}{\underset{\underset{CN}{|}}{C}}]$$

Propagation: The free radical $(R - CH_2 - \overset{\cdot}{C}H_2)$ formed above adds on to a monomer molecule forming another radical. The successive addition of monomers to these free radicals results in the propagation of the chain reaction.

$$R\,CH_2 - \overset{\cdot}{C}H_2 + CH_2 = CH_2 \longrightarrow RCH_2CH_2CH_2\overset{\cdot}{C}H_2$$

$$RCH_2CH_2CH_2\overset{\cdot}{C}H_2 + nCH_2 = CH_2 \xrightarrow{\text{n steps}} R \{CH_2 - CH_2\}_{\overline{n+1}} CH_2 - \overset{\cdot}{C}H_2$$

Termination: The termination of the long chain radicals usually occur by radical coupling or disproportionation reactions.

Radical Coupling:

$$2R\{CH_2 - CH_2\}_{\overline{n+1}} CH_2 - \overset{\cdot}{C}H_2 \longrightarrow$$

$$R\{CH_2 - CH_2\}_{\overline{n+1}} CH_2 - CH_2 - CH_2 - CH_2 \{CH_2 - CH_2\}_{\overline{n+1}} R$$

Disproportionation:

$$R\{CH_2 - CH_2\}_{\overline{n+1}} CH_2 - \overset{\cdot}{C}H_2 + \overset{\cdot}{C}H_2 - CH_2\{CH_2 - CH_2\}_{\overline{n+1}}R$$

$$\longrightarrow R\{CH_2 - CH_2\}_{\overline{n+1}} CH = CH_2 + CH_3 - CH_2\{CH_2 - CH_2\}_{\overline{n+1}}R$$

As seen, the initiator residues are incorporated into the polymer at one or both ends of the chain depending on the type of termination step.

Inhibitors or Retarders: Certain substances reduce the rate of polymerisation by terminating the reaction chains. Such substances are called inhibitors or retarders depending on whether they completely prevent the production of high polymers or are effective in reducing the rate of polymerisation. Phenols are generally used for such purposes. The phenolic hydrogen is easily abstracted by the radical to give a phenoxy radical which due to resonance stabilisation becomes unreactive and does not add to the monomer molecule.

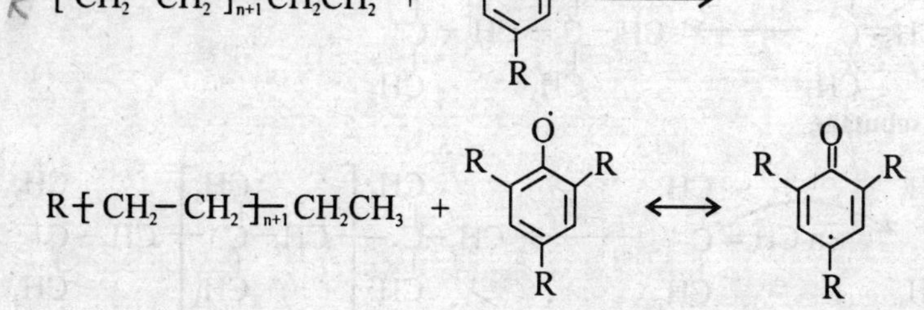

Phenoxy radical

For successful inhibition, the rate of reaction of a radical with the inhibitor should be much greater than the rate of its reaction with the monomer. Inhibitors, such as hydroquinone are generally used to stabilise monomers against spontaneous polymerisation when stored.

Copolymerisation: Polymers of a single monomer are called *homopolymers* and the process *homopolymerisation*. On the other hand, if two different kinds of monomers are polymerised, the polymer is called *copolymer* and the process *copolymerisation*. The later process gives polymers having desired properties. The composition of the copolymer is controlled by varying the composition of the two monomers and also by the reactivities of the monomers towards free radical addition. As an illustration, copolymerisation of styrene with methylmethacrylate gives the copolymer known as polystyrene co-methylmethacrylate, which is an industrially important process.

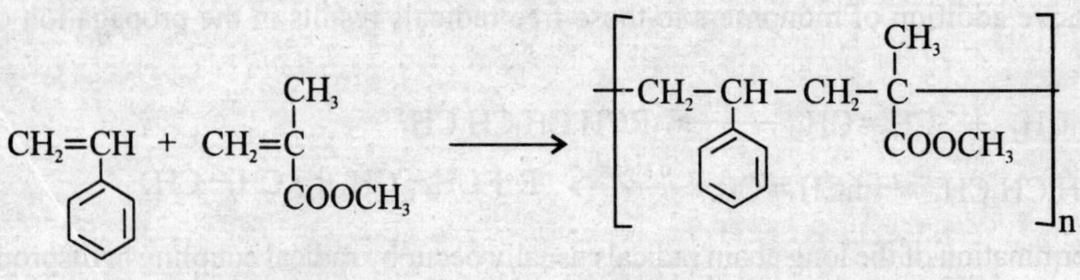

| Styrene | Methyl-methacrylate | Polystyrene co-methyl-methacrylate |

44.3.1.2 *Cationic Polymerisation*

Cationic polymerisation is initiated by catalysts which are electrophilic in nature, such as sulphuric acid, perchloric acid, halogen acids and Lewis acids viz. $AlCl_3$, BF_3 and $TiCl_4$.

The monomers readily accepts a proton from protic acids to form a carbocation. A second molecule of monomer donates an electron pair and forms a longer chain cation. The process continues till a high molecular weight cation is produced. The termination may occur by the loss of a proton.

In cationic polymerisations, the stability of the cation and the nature of the solvent used are important. Since isobutene gives a stable carbocation, it readily polymerises to give a high molecular weight polymer at low temperature. Various steps in the cationic polymerisation of isobutene are given below:

Initiation

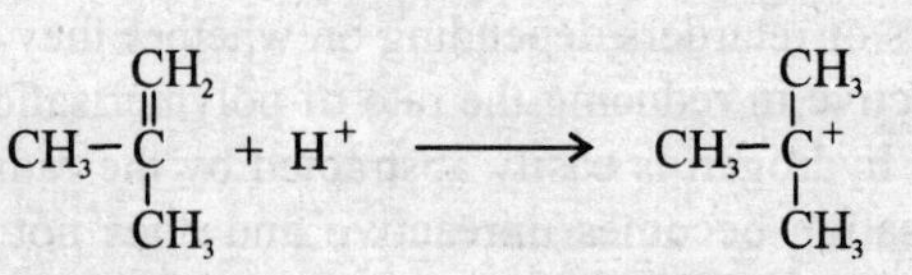

| Isobutene | | Carbocation |

Propagation

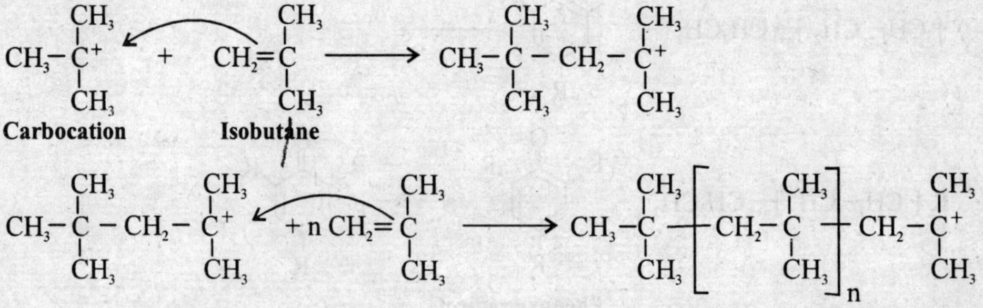

| Carbocation | Isobutane | | |

Termination

$$CH_3-\overset{\overset{\displaystyle CH_3}{|}}{\underset{\underset{\displaystyle CH_3}{|}}{C}}-\left[CH_2-\overset{\overset{\displaystyle CH_3}{|}}{\underset{\underset{\displaystyle CH_3}{|}}{C}}\right]_n CH_2-\overset{\overset{\displaystyle CH_3}{|}}{\underset{\underset{\displaystyle CH_3}{|}}{C^+}} \xrightarrow{-H^+} CH_3-\overset{\overset{\displaystyle CH_3}{|}}{\underset{\underset{\displaystyle CH_3}{|}}{C}}-\left[CH_2-\overset{\overset{\displaystyle CH_3}{|}}{\underset{\underset{\displaystyle CH_3}{|}}{C}}\right]_n CH_2-\overset{\overset{\displaystyle CH_3}{|}}{\underset{\underset{\displaystyle CH_2}{||}}{C}}$$

Lewis acids (like boron trifluoride) and water (co-catalyst) are also employed to initiate cationic polymerisations.

$$BF_3 + H_2O \rightleftharpoons [BF_3OH]^- H^+$$

$$[BF_3OH]^- H^+ + H_2\overset{\curvearrowleft}{C}=\overset{\overset{\displaystyle CH_3}{|}}{\underset{\underset{\displaystyle CH_3}{|}}{C}} \longrightarrow CH_3-\overset{\overset{\displaystyle CH_3}{|}}{\underset{\underset{\displaystyle CH_3}{|}}{C^+}}\,[BF_3OH]^-$$

The subsequent steps take place by addition of the carbocation to the new monomer unit.

$$[BF_3OH]^-\,CH_3-\overset{\overset{\displaystyle CH_3}{|}}{\underset{\underset{\displaystyle CH_3}{|}}{C^+}} + H_2\overset{\curvearrowleft}{C}=\overset{\overset{\displaystyle CH_3}{|}}{\underset{\underset{\displaystyle CH_3}{|}}{C}} \longrightarrow CH_3-\overset{\overset{\displaystyle CH_3}{|}}{\underset{\underset{\displaystyle CH_3}{|}}{C}}-CH_2-\overset{\overset{\displaystyle CH_3}{|}}{\underset{\underset{\displaystyle CH_3}{|}}{C^+}}\,[BF_3OH]^-$$

$$\downarrow$$

Polymer

The termination of the polymerisation takes place by proton transfer.

$$\left[\,CH_2-\overset{\overset{\displaystyle CH_3}{|}}{\underset{\underset{\displaystyle CH_3}{|}}{C^+}}\,[BF_3OH]^-\right]_n \longrightarrow \left[\,CH_2-\overset{\overset{\displaystyle CH_3}{|}}{\underset{\underset{\displaystyle CH_2}{||}}{C}}\right]_n + H^+[BF_3OH]^-$$

44.3.1.3 Anionic Polymerisation

In anionic polymerisation, the monomer possessing an electron withdrawing substituent such as vinyl, nitrile, phenyl or carboxyl is polymerised by nucleophiles.

$$Y^- + CH_2=\underset{\underset{\displaystyle X}{|}}{CH} \longrightarrow Y-CH_2-\underset{\underset{\displaystyle X}{|}}{\bar{C}H}$$

$$Y-CH_2-\underset{\underset{\displaystyle X}{|}}{\bar{C}H} + CH_2=\underset{\underset{\displaystyle X}{|}}{CH} \longrightarrow Y-CH_2-\underset{\underset{\displaystyle X}{|}}{CH}-CH_2-\underset{\underset{\displaystyle X}{|}}{\bar{C}H} \longrightarrow$$

The termination of the chain takes place by addition of a proton to the carbanion.

A number of reagent like n-butyllithium and lithium amide etc are used to bring about polymerisation.

$$nC_4H_9\overset{+}{Li} + \overset{\curvearrowleft}{C}H_2=\overset{\overset{\displaystyle CH_3}{\diagup}}{\underset{\underset{\displaystyle COOCH_3}{\diagdown}}{C}} \longrightarrow nC_4H_9-CH_2-\overset{\overset{\displaystyle CH_3}{|}}{\underset{\underset{\displaystyle COOCH_3}{|}}{C^-}}Li^+$$

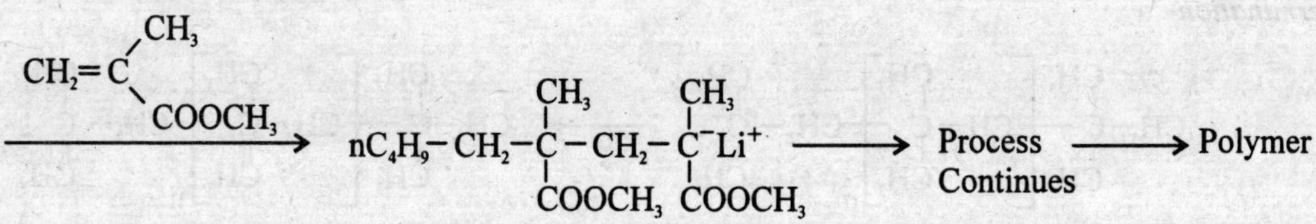

$$CH_2=C\begin{smallmatrix}CH_3\\\\COOCH_3\end{smallmatrix}$$

$$\xrightarrow{\quad\quad} nC_4H_9-CH_2-\underset{\underset{COOCH_3}{|}}{\overset{\overset{CH_3}{|}}{C}}-CH_2-\underset{\underset{COOCH_3}{|}}{\overset{\overset{CH_3}{|}}{C}}{}^-Li^+ \longrightarrow \begin{smallmatrix}Process\\Continues\end{smallmatrix} \longrightarrow Polymer$$

Anionic polymerisation using lithium amide is shown below:

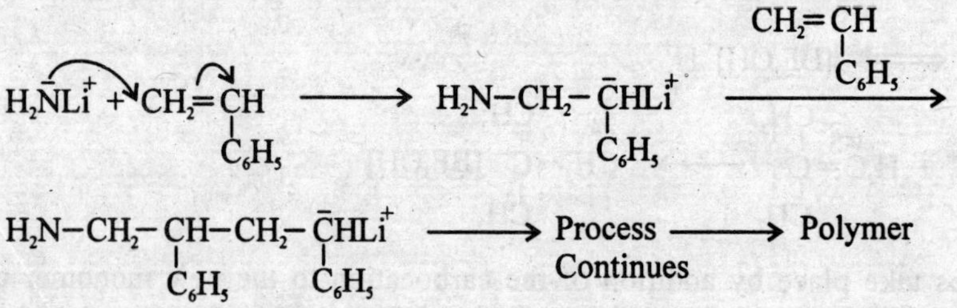

$$H_2\overset{\frown}{N}Li^+ + \overset{\curvearrowright}{CH_2}=\underset{\underset{C_6H_5}{|}}{CH} \longrightarrow H_2N-CH_2-\underset{\underset{C_6H_5}{|}}{\bar{C}H}Li^+ \xrightarrow{\underset{\underset{C_6H_5}{|}}{CH_2=CH}}$$

$$H_2N-CH_2-\underset{\underset{C_6H_5}{|}}{CH}-CH_2-\underset{\underset{C_6H_5}{|}}{\bar{C}H}Li^+ \longrightarrow \begin{smallmatrix}Process\\Continues\end{smallmatrix} \longrightarrow Polymer$$

The termination is effected by a hydride transfer.

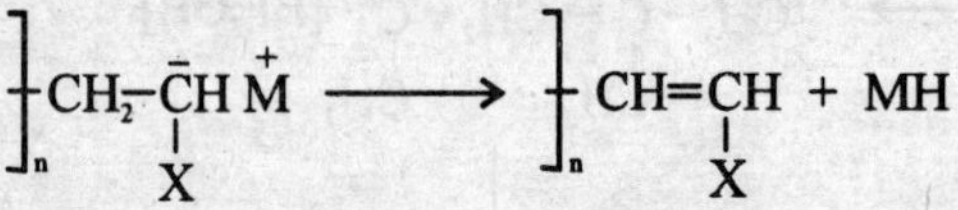

$$\left]{-}CH_2-\underset{\underset{X}{|}}{\bar{C}H}\ \overset{+}{M}\right]_n \longrightarrow \left]{-}CH=\underset{\underset{X}{|}}{CH}\right]_n + MH$$

Polymerisation of styrene can also be brought about via a radical anion obtained by the reaction of sodium metal with naphthalene. The styrene radical anion thus obtained dimerises to give a dianion which subsequently polymerises as shown below.

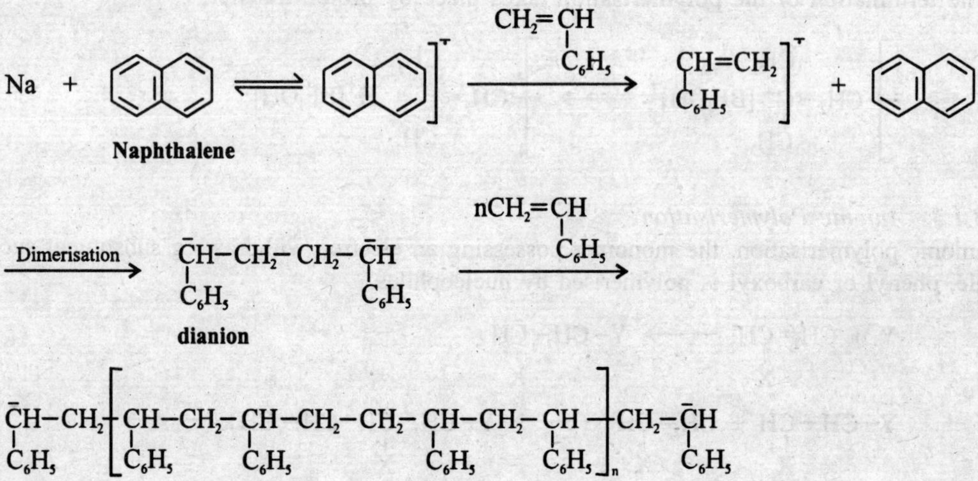

$$Na\ +\ \text{[naphthalene]} \rightleftharpoons \text{[naphthalene]}^{\overline{\bullet}} \xrightarrow{\underset{\underset{C_6H_5}{|}}{CH_2=CH}} \left[\underset{\underset{C_6H_5}{|}}{CH=CH_2}\right]^{\overline{\bullet}} + \text{[naphthalene]}$$

Naphthalene

$$\xrightarrow{Dimerisation} \underset{\underset{C_6H_5}{|}}{\bar{C}H}-CH_2-CH_2-\underset{\underset{C_6H_5}{|}}{\bar{C}H} \xrightarrow{\underset{\underset{C_6H_5}{|}}{nCH_2=CH}}$$

dianion

$$\underset{\underset{C_6H_5}{|}}{\bar{C}H}-CH_2\left[\underset{\underset{C_6H_5}{|}}{CH}-CH_2-\underset{\underset{C_6H_5}{|}}{CH}-CH_2-CH_2-\underset{\underset{C_6H_5}{|}}{CH}-CH_2-\underset{\underset{C_6H_5}{|}}{CH}\right]_n CH_2-\underset{\underset{C_6H_5}{|}}{\bar{C}H}$$

44.3.1.4   Ziegler-Natta Polymerisation

Prior to the discovery of Ziegler Natta catalyst, polymerisation of simple molecules like ethylene was carried out only at a very high temperature and pressure. With Z–N catalyst polymerisation, reactions occur easily under mild conditions of temperature and pressure. The polymers produced have more regular structure (isotectic and syndiotactic) and also have better physical properties.

It is also known as coordination polymerisation and is carried out in presence of complex solid catalysts (known as Ziegler-Natta catalysts) consisting of titanium tri- or tetrachloride and an alkyl aluminum compound. A commonly used catalyst of this type is a mixture of titanium tetrachloride and triethyl aluminum ($AlEt_3$-$TiCl_4$). The polymerisation is carried out in an inert solvent like heptane under mild conditions.

The mechanism of the polymerisation is not fully understood. It is believed that initial formation of titanium alkyl bonds takes place by the exchange of halogen atoms and the alkyl groups. The monomer is then inserted between the titanium and the alkyl group by coordination with titanium through π bond. The process is repeated and a straight chain polymer is obtained. Thus, with ethylene the polymerisation is represented as given below:

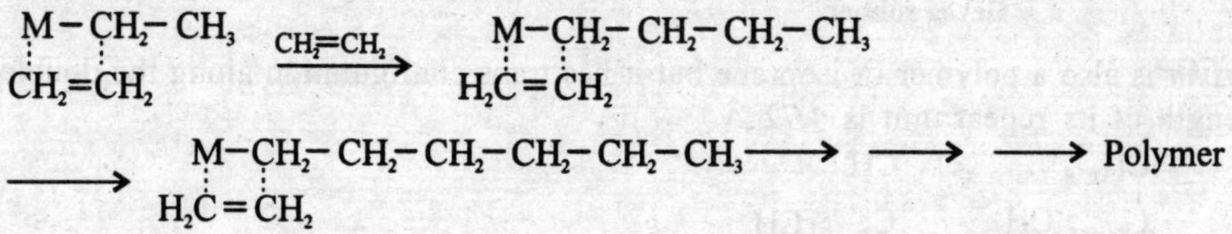

$$M-CH_2-CH_3 \quad \xrightarrow{CH_2=CH_2} \quad M-CH_2-CH_2-CH_2-CH_3$$
$$CH_2=CH_2 \qquad\qquad\qquad\qquad H_2C=CH_2$$

$$\xrightarrow{\qquad} \quad M-CH_2-CH_2-CH_2-CH_2-CH_2-CH_3 \longrightarrow \longrightarrow \longrightarrow Polymer$$
$$H_2C=CH_2$$

As already mentioned, in Ziegler-Natta polymerisations stereoregular polymers are obtained. Examples of this type are isotactic and syndiotactic.

44.3.2 Condensation or Step Polymerisation

Condensation polymerisation involves the reaction of polyfunctional molecules to give macro molecules with the loss of simple molecules like water etc. For example, a dicarboxylic acid like terephthalic condenses with a diol, (ethylene glycol) to give a polyester.

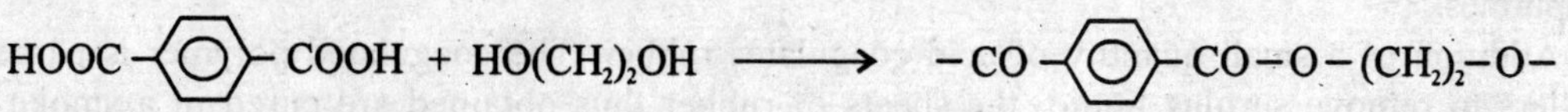

$$HOOC-\langle O \rangle-COOH + HO(CH_2)_2OH \longrightarrow -CO-\langle O \rangle-CO-O-(CH_2)_2-O-$$

When bifunctional molecules are involved, a linear polymer is produced e.g, polyester shown above. In case a polyfunctional compound is used, a cross linked polymer is the product.

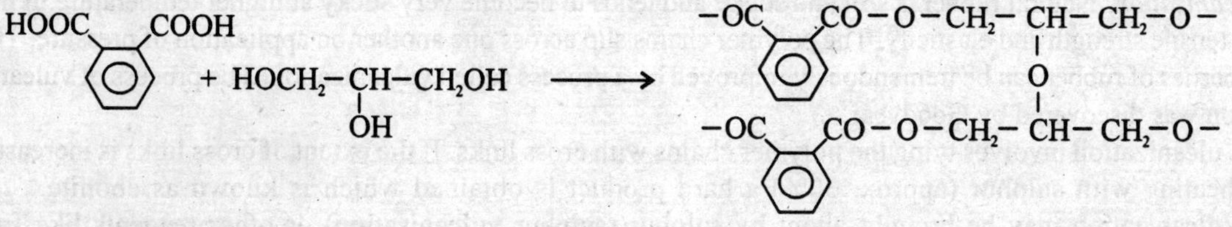

$$\begin{array}{c} HOOC \qquad COOH \\ \langle O \rangle \\ \end{array} + HOCH_2-CH-CH_2OH \longrightarrow$$
$$OH$$

$$\begin{array}{c} -OC \qquad CO-O-CH_2-CH-CH_2-O- \\ \langle O \rangle \qquad\qquad\qquad O \\ -OC \qquad CO-O-CH_2-CH-CH_2-O- \\ \langle O \rangle \end{array}$$

44.4 ELASTOMERS

44.4.1 Natural Rubber

Main source of natural rubber is the plant known as 'Hevea brasilliensis'. Rubber obtained from this plant is also known as Hevea rubber. Another form of natural rubber, known as Gutta Percha occurs in plants of Palaquium family. It is harder than Hevea rubber.

Structure of Natural Rubber

Destructive distillation of rubber produces mainly isoprene with a small amount of dipentene. Polymerisation of isoprene produced a rubber like product indicating that rubber is polyisoprene. The formation of levulinic aldehyde on ozonolysis of rubber indicates that polyisoprene units must be joined head to tail in rubber.

$$-\left(CH_2-\underset{\underset{CH_3}{|}}{C}=CH-CH_2\right)\!CH_2-\underset{\underset{CH_3}{|}}{C}=CH-CH_2\!-CH_2-\underset{\underset{CH_3}{|}}{C}=\cdots \xrightarrow{\text{Ozonolysis}} OHC-CH_2-CH_2-\underset{\underset{CH_3}{|}}{C}=O$$

$$\text{Levulinic aldehyde}$$

X-ray analysis indicates that rubber is cis-polyisoprene, the chain length of repeat unit is 8.10 Å.

Hevea rubber

Gutta percha is also a polymer of isoprene but it has trans configuration along the double bond and the chain length of its repeat unit is 4.72 Å.

Gutta percha

Isolation of Rubber

Rubber occurs as colloidal suspension (known as latex) in the bark of Hevea tree. Latex, collected by making an incision in the bark, is a milky fluid which contains 60% water, 35% rubber and 5% other impurities.

Addition of a small quantity of acid coagulates rubber. The coagulated polymer is passed through rollers to remove surplus liquid, the sheets of rubber thus obtained are cured in a smoke house.

A part of latex is creamed by centrifugation to remove excess of water and ammonia is added as an anticoagulant. It is used for making elastic threads and moulded articles.

Vulcanization: Natural rubber is soft and sticky and tends to become very sticky at higher temperature. It has low tensile strength and elasticity. The polymer chains slip across one another on application of pressure. The properties of rubber can be tremendously improved by a process called vulcanization. The process of vulcanization was discovered by Goodyear.

Vulcanization involves tying the polymer chains with cross links. If the extent of cross links is increased by heating with sulphur (approx. 30%) a hard product is obtained which is known as ebonite.

Vulcanization may be brought about by sulphur (sulphur vulcanisation) or other reagents like free radical generators, or metal oxides (non sulphur vulcanization).

Sulphur Vulcanization: It consists of heating rubber with 1-3% of sulphur, an accelerator, inert filler and an antioxidant. Sulphur helps to cross link the parallel polymer chains which may contain upto eight sulphur atoms. Some of the unsaturation is lost during vulcanization. It is believed to be due to addition of H_2S (to double bonds) obtained during vulcanization.

$$-X-X-X-X-X-X-$$
$$\underset{\displaystyle |}{S}\underset{\displaystyle |}{S}$$
$$-X-X-X-X-X-X-$$

(X is the Monomer unit)

To reduce the time required for combination of rubber and sulphur, accelerators are used. Accelerator are sulphur containing compounds that accelerate the process of vulcanization. Some of the commonly used accelerators are given below.

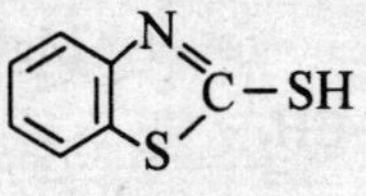

Mercaptobenzothiazole

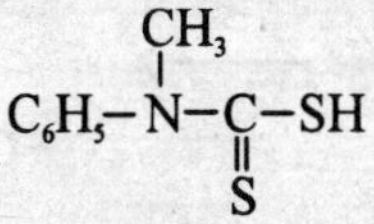

Phenylmethyldithiocarbamic acid

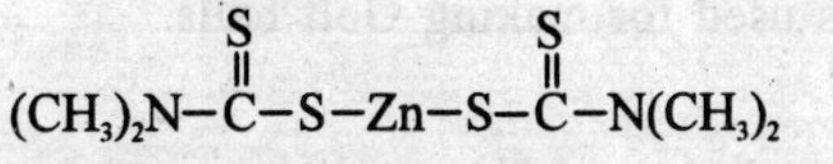

$(CH_3)_2N-\overset{\overset{S}{\|}}{C}-S-Zn-S-\overset{\overset{S}{\|}}{C}-N(CH_3)_2$
Zinc dimethyldithiocarbamate

The *inert fillers* increase the effectiveness of the accalerator. Zinc oxide or stearic acid is added as a filler to the light colored product and carbon black to dark colored product. Inert filler like carbon black also prevents the penetration of UV light and acts as a strengthening agent.

Natural rubber is very sensitive to oxidation by air and ozone. Oxidation results in reduction of the strength and elasticity of rubber. Clipping of the polymer chain occurs due to oxidation. This 'aging' of rubber can be delayed by the addition of compounds known as *antioxidants*.

Antioxidants are compounds which undergo easy oxidation and thus delay the oxidative degradation of rubber. Some of the commonly used antioxidants are given below.

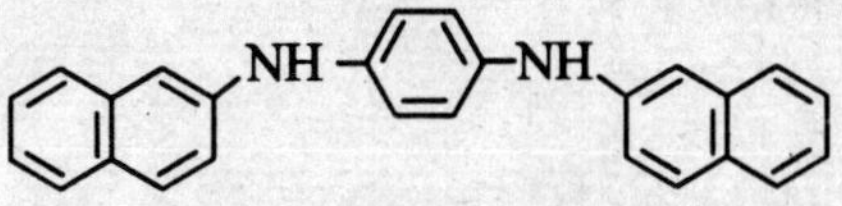

Di-β-naphthyl-p-phenylenediamine

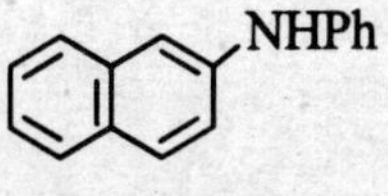

Phenyl-β-naphthylamine

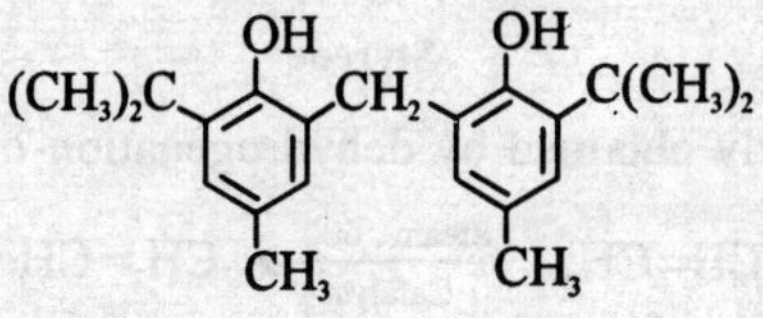

Oligomers are preferred over other simple molecules because the loss due to volatility of the former type of antioxidants is low.

Non Sulphur Vulcanization

(a) Free radical generators such as peroxides and azo compounds are used for this type of vulcanization. The mechanism of cross linking is believed to involve the following steps.

$$R-O-O-R \longrightarrow 2R-\overset{.}{O}$$

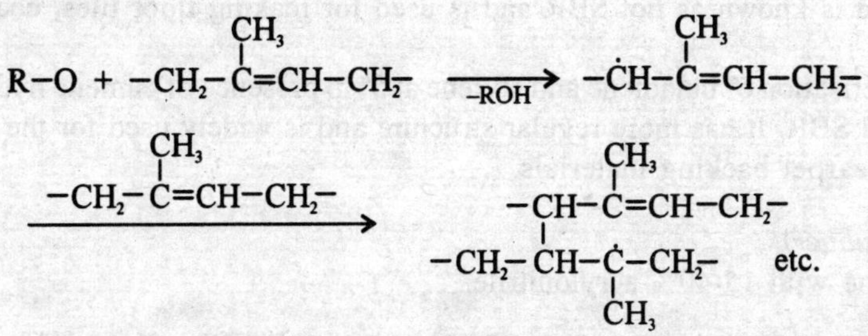

$$R-O + -CH_2-\overset{\overset{CH_3}{|}}{C}=CH-CH_2- \xrightarrow{-ROH} -\overset{.}{C}H-\overset{\overset{CH_3}{|}}{C}=CH-CH_2-$$

$$-CH_2-\overset{\overset{CH_3}{|}}{C}=CH-CH_2- \xrightarrow{\hspace{3cm}} \begin{array}{c} -\overset{\overset{CH_3}{|}}{C}H-\overset{\overset{}{}}{C}=CH-CH_2- \\ | \\ -CH_2-CH-\overset{\overset{CH_3}{|}}{\underset{\underset{CH_3}{|}}{C}}-CH_2- \quad \text{etc.} \end{array}$$

Termination occurs by the combination of rubber free radical with the radical of curing agent.

(b) Neoprene is vulcanized by using zinc or magnesium oxide and cross linking involves oxide links.

$$2 -CH_2- CH=C-CH_2- \underset{Cl}{|} \quad + \quad MgO \quad \longrightarrow \quad \begin{matrix} -CH_2-C=CH-CH_2 \cdots \\ | \\ O \\ | \\ -CH_2-C=CH-CH_2 \cdots \end{matrix} \quad + \quad MgCl_2$$

Natural rubber is still obtained in large quantities from Hevea brasilliensis. It is used for making a variety of rubber articles. Gutta percha, being hard is used for making Golf balls.

44.4.2   Synthetic Elastomers

Supply of rubber from natural sources was not sufficient to meet the ever increasing demand for rubber. During and after world wars, research efforts in Germany and later in Russia and the rest of the world led to the discovery of some important synthetic substitutes for natural rubber. Some of these are discussed in the following sections. Synthetic elastomers like natural rubber are subjected to vulcanization for improvement of their properties.

44.4.2.1   Styrene-Butadiene Rubber

It is also known as Buna-s or SBR or GSR (Government Styrene Rubber). An important synthetic elastomers is obtained by copolymerisation of styrene and butadiene. Styrene is obtained by dehydrogenation of ethylbenzene with super heated steam in presence of a catalyst.

Ethylbenzene $\xrightarrow[\text{Catalyst}]{\text{Steam , 600°}}$ Styrene $(+ H_2)$

Butadiene is similarly obtained by dehydrogenation of 1-butene.

$$CH_3- CH_2- CH=CH_2 \xrightarrow[\text{Catalyst}]{\text{Steam , 600°}} CH_2=CH-CH=CH_2 \quad (+ H_2)$$

1-Butene   Butadiene

Copolymerisation of styrene and butadiene in aqueous emulsion (sodium oleate as emulsifier) at 50° in presence of potassium persulphate and dodecylmercaptan produces SBR.

$$n \underset{C_6H_5}{\underset{|}{CH=CH_2}} + n\ CH_2=CH-CH=CH_2 \longrightarrow -(CH_2- CH=CH-CH_2- \underset{C_6H_5}{\underset{|}{CH-CH_2}})_n$$

Styrene   Butadiene   Buna-S

The polymer thus obtained is known as hot SBR and is used for making floor tiles, coated fabrics and shoe soles etc.

SBR obtained by copolymerisation of butadiene and styrene at 5° in presence of cumene hydroperoxide and glucose is known as cold SBR. It has more regular structure and is widely used for the production of car tyres, footwears and carpet backing materials.

44.4.2.2   Buna-N (Nitrile Rubber)

It is a copolymer of butadiene with 15-40% acrylonitrile.

Acrylonitrile is obtained by the following method.

(a) Addition of hydrogen cyanide to acetylene.

$$CH \equiv CH + HCN \longrightarrow CH_2 = CHCN$$

Acetylene   **Hydrogen**   **Acrylonitrile**
                cyanide

(b) Vapor phase ammonoxidation of propylene.

$$CH_2 - CH = CH_2 + NH_3 + 1/2 \ O_2 \xrightarrow{\text{Catalyst}} CH_2 = CH - CN + 3H_2O$$

Propylene   **Acrylonitrile**

(c) Acrylonitrile may also be prepared from ethylene by following sequence of reactions.

$$CH_2 = CH_2 \xrightarrow{\text{HOCl}} CH_2OH - CH_2Cl \xrightarrow{\text{KCN}} CH_2OH - CH_2CN \xrightarrow[-H_2O]{} CH_2 = CHCN$$

Acrylonitrile

Butadiene and acrylonitrile are copolymerised in an aqueous emulsion using soap as an emulsifier.

$$n \ CH_2 = CH - CH = CH_2 + n \ CH_2 = CHCN \longrightarrow \underset{\underset{CN}{|}}{(CH_2 - CH = CH - CH_2 - CH_2 - CH)_n}$$

Buna-N has outstanding resistance to heat, sunlight, oils and solvents and has high abrasion resistance. It is used for hose and tank lining.

44.4.2.3 Neoprene

Chloroprene required for the production of Neoprene is obtained as follows:

(a) From acetylene by its dimerisation followed by addition of hydrogen chloride.

$$2 \ CH \equiv CH \xrightarrow{Cu_2Cl_2} CH_2 = CH - C \equiv CH \xrightarrow{HCl} \underset{\underset{Cl}{|}}{CH_2 = CH - C = CH_2}$$

Acetylene   **Chloroprene**

(b) From butadiene by following sequence of reactions.

$$CH_2 = CH - CH = CH_2 + Cl_2 \xrightarrow[\text{Phase}]{\text{Vapor}} \underset{\underset{Cl}{|} \ \underset{Cl}{|}}{CH_2 - CH - CH = CH_2} + \underset{\underset{Cl}{|} \qquad \underset{Cl}{|}}{CH_2 - CH = CH - CH_2}$$

$$\xrightarrow[Cu_2Cl_2]{\text{Isomerisation}} \underset{\underset{Cl}{|} \ \underset{Cl}{|}}{CH_2 - CH - CH = CH_2} \xrightarrow[-HCl]{NH_4Cl, \ \Delta} \underset{\underset{Cl}{|}}{CH_2 = CH - C = CH_2}$$

Chloroprene is polymerised in aqueous emulsion at about 30° using soap as emulsifier and potassium persulphate as free radical catalyst.

$$\underset{\underset{Cl}{|}}{nCH_2 = CH - C = CH_2} \xrightarrow{\text{Polymerisation}} \underset{\underset{Cl}{|}}{(CH_2 - CH = C - CH_2)_n}$$

Chloroprene   **Neoprene (polychloroprene)**

Neoprene is resistant to attack by chemicals and solvents. It is used for protective clothing, linings, reaction vessels, hose covers, conveyor belts and floor tiles etc.

44.4.2.4   *Polyisoprene*

Isopropene required for polymerisation is obtained by the following reactions.

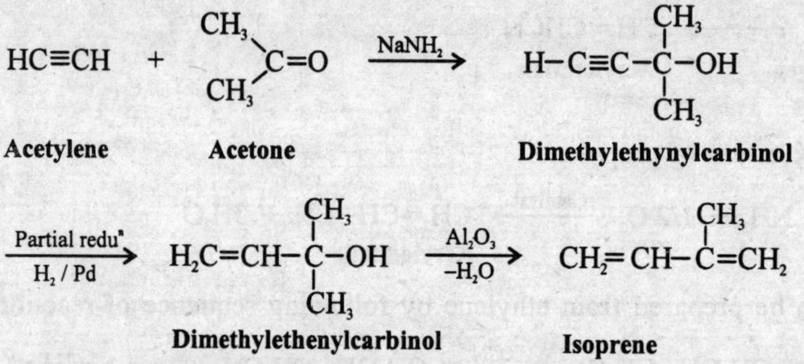

$$HC{\equiv}CH \;+\; \underset{CH_3}{\overset{CH_3}{\diagdown}}C{=}O \xrightarrow{\text{NaNH}_2} H{-}C{\equiv}C{-}\underset{CH_3}{\overset{CH_3}{\underset{|}{\overset{|}{C}}}}{-}OH$$

Acetylene   **Acetone**   **Dimethylethynylcarbinol**

$$\xrightarrow[\text{H}_2 / \text{Pd}]{\text{Partial redu}^n} H_2C{=}CH{-}\underset{CH_3}{\overset{CH_3}{\underset{|}{\overset{|}{C}}}}{-}OH \xrightarrow[-\text{H}_2\text{O}]{\text{Al}_2\text{O}_3} CH_2{=}CH{-}\underset{}{\overset{CH_3}{\underset{}{\overset{|}{C}}}}{=}CH_2$$

Dimethylethenylcarbinol   **Isoprene**

Polymerisation of isoprene prior to the discovery of Ziegler Natta catalyst did not give a product of great utility. However, the stereoregular polyisoprene obtained by use of Z–N catalyst resembles natural rubber, though it is somewhat inferior in properties to natural rubber. This is believed to be due to the absence of small amount of impurities present in the synthetic polymer.

44.4.2.5   *Butyl Rubber*

Polymerisation of isobutylene (obtained by dehydrogenation of isobutane with super heated steam at 600° and catalyst) produces a linear polymer which is not vulcanizable as it is a saturated polymer. Addition of about 3% of isoprene to isobutene during polymerisation results in the formation of butyl rubber which has some unsaturation functions along the chain and hence the product can be vulcanized. The copolymerisation is carried out in methyl chloride at about –90° using BF_3 or $AlCl_3$ as catalyst.

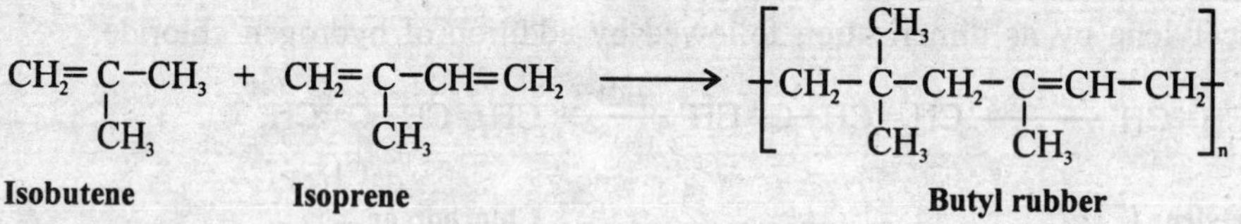

$$CH_2{=}\underset{CH_3}{\overset{|}{C}}{-}CH_3 \;+\; CH_2{=}\underset{CH_3}{\overset{|}{C}}{-}CH{=}CH_2 \longrightarrow \left[CH_2{-}\underset{CH_3}{\overset{CH_3}{\underset{|}{\overset{|}{C}}}}{-}CH_2{-}\underset{CH_3}{\overset{|}{C}}{=}CH{-}CH_2 \right]_n$$

Isobutene   **Isoprene**   **Butyl rubber**

Butyl rubber has very low permeability to air and is specially used for the production of tyre tubes.

44.4.2.6   *Polyurethane Foam*

Polyurethane foam is obtained by first preparing a prepolymer containing hydroxy end group (known as the diol) and then adding toluene diisocyanate.

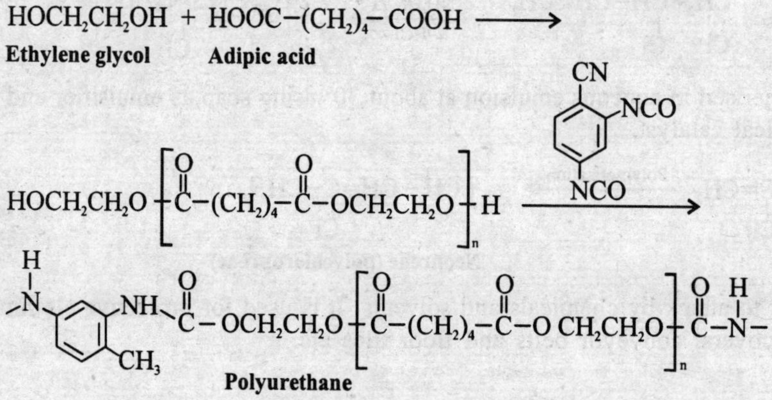

$$HOCH_2CH_2OH \;+\; HOOC{-}(CH_2)_4{-}COOH \longrightarrow$$

Ethylene glycol   **Adipic acid**

$$HOCH_2CH_2O\left[\overset{O}{\overset{\|}{C}}{-}(CH_2)_4{-}\overset{O}{\overset{\|}{C}}{-}OCH_2CH_2O \right]_n H \xrightarrow{}$$

$$\text{(toluene diisocyanate with } CN,\ NCO,\ NCO)$$

$$\overset{H}{\overset{|}{N}}{-}\ \ NH{-}\overset{O}{\overset{\|}{C}}{-}OCH_2CH_2O\left[\overset{O}{\overset{\|}{C}}{-}(CH_2)_4{-}\overset{O}{\overset{\|}{C}}{-}O\ CH_2CH_2O \right]_n \overset{O}{\overset{\|}{C}}{-}\overset{H}{\overset{|}{N}}{-}$$

Polyurethane

The foaming agent in the preparation of polyurethane is CO_2, which is made available insitu by the reaction of water with isocyanate group.

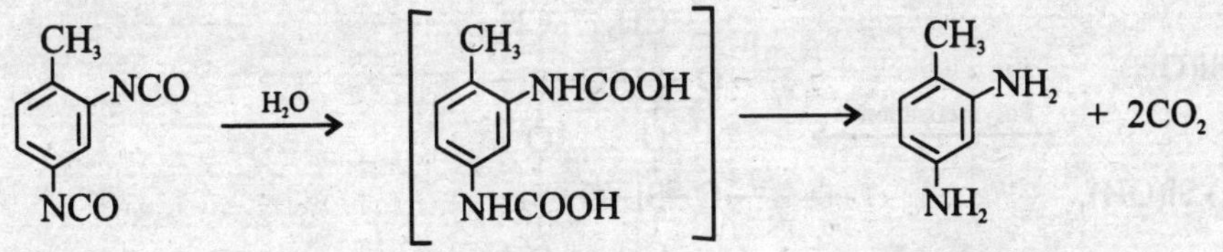

Toluene-2,4-diisocyanate   Carbamic acid   2,4-Diaminotoluene

The 2,4-diaminotoluene obtained above also reacts with isocyanate and forms urea linkages.

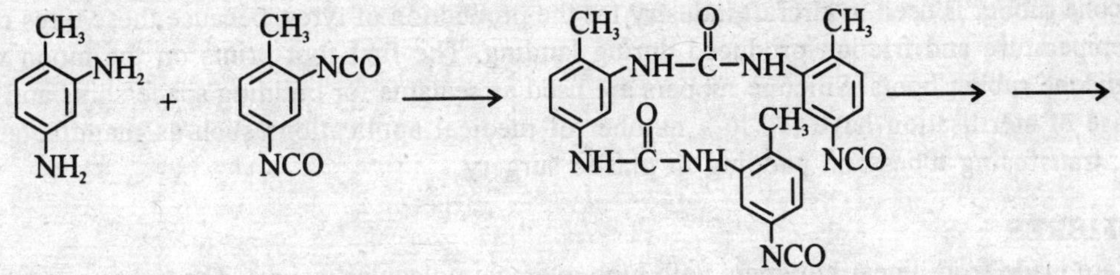

Great care has to be taken to maintain a required balance between the two reactions.
The toluene 2,4-diisocyanate required for the production of polymer is obtained as follows.

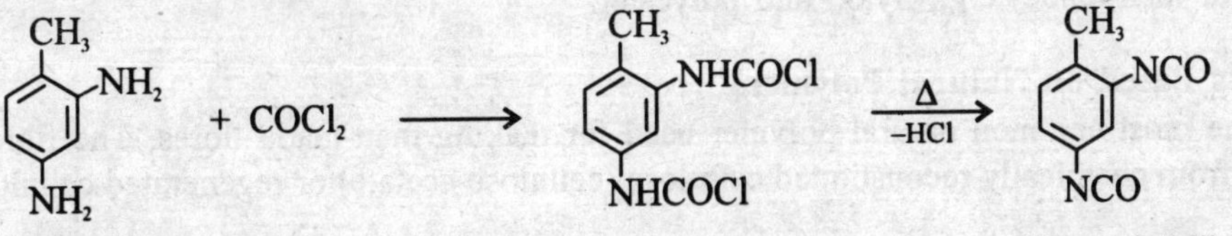

2,4-Diaminotoluene   Toluene-2,4-diisocyanate

The polyurethane foams are light and resistant to chemical and thermal degradation, used mainly in aircraft industry, manufacture of boats and for insulation etc.

44.4.2.7 Silicone Rubber

Silicone polymers are used as oils, gums and rubbers. the monomer is prepared by the following method.

$$Si(Cu) + 2CH_3Cl \longrightarrow (CH_3)_2SiCl_2 + Cu$$

Silicone copper
alloy

$$\downarrow H_2O$$

$$(CH_3)_2Si(OH)_2$$

The dihydroxysilane thus produced is polymerised to produce linear polymer known as silicone oil.

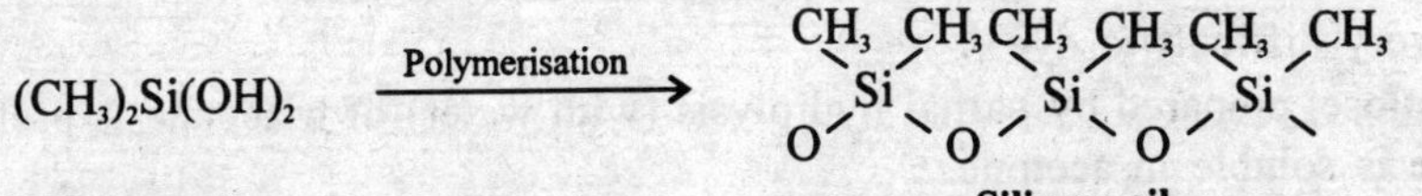

$$(CH_3)_2Si(OH)_2 \xrightarrow{\text{Polymerisation}}$$

Dihydroxysilane   Silicone oil

Silicone oil is used to give water repellent finish on textiles, leather and paper etc. If some $(CH_3)_3Si(OH)$ or $Si(OH)_4$ is used during polymerisation, cross linking takes place to produce silicone rubber.

$$CH_3Si(OH)_3$$
$$+$$
$$(CH_3)_2Si(OH)_2$$

$$\xrightarrow{\text{Polymerisation}}$$

$$
\begin{array}{cc}
CH_3 & CH_3 \\
| & | \\
-O-Si-O-Si- \\
| & | \\
O & O \\
| & | \\
-O-Si-O-Si- \\
| & | \\
CH_3 & CH_3
\end{array}
$$

The silicone rubber is flexible over a wide range of temperatures, highly resistant to chemical attack and weathering effects.

Silicone rubber is used in aircraft industry for the production of tyres because these tyres can tolerate high temperature and friction produced during landing. The first foot prints on the moon were made with silicone rubber boots. Silicone rubbers are used as sealants for building space ships and jet planes. The ease of sterilisation have led to a number of medical applications such as manufacture of heart valves, transfusing tubes and padding in plastic surgery.

44.5 FIBRES

Fibres are made from linear polymers with high average molecular weight. The polymeric material may be natural, e.g., silk and wool (polyamide) and cotton (polysaccharide) or the polymer may be obtained from nature but is treated chemically before use, e.g., rayon, cellulose acetate etc. Polymers may also be synthetic (or man-made) e.g., nylon and polyester.

44.5.1   Fibres Based on Natural Polymers

Cellulose is the most common natural polymer used for making man-made fibres. The fibres obtained may either be from chemically reconstituted cellulose (cellulose acetate) or regenerated cellulose (rayon).

44.5.1.1   Rayons

It is produced from wood pulp. For the production of rayon, the wood pulp is purified, bleached and treated with sodium hydroxide solution for several hours. The soda-cellulose thus obtained is expressed to remove excess alkali and allowed to 'age' in air.

Soda-cellulose is then churned with carbon disulphide to form sodium xanthole of cellulose, which when stirred with dilute caustic soda to form a honey like syrup known as 'viscose'. Fibres are spun from formed 'viscose' and coagulated with sulphuric acid. The yarn is then purified by washing and desulphurising with sodium sulphide solution and finally bleached with hydrogen peroxide, rewashed and dried.

44.5.1.2   Cellulose Acetate Fibres

Cellulose acetate is usually prepared from Linters (short cotton fibres). Acetylation of linters with acetic acid and acetic anhydride in presence of sulphuric acid produces the triacetate of cellulose, which may be converted into fibres. It is not used because of two main problems.

(i)   It is soluble only in chloroform which is expensive as well as toxic.

(ii)   The fibres are very difficult to dye.

The diacetate of cellulose, prepared by partial hydrolysis (with water) of triacetate is preferred over triacetate. The diacetate is soluble in acetone.

The cellulose acetate fibres are widely used for making dress material, under garments and swimming costumes etc.

44.5.2 Synthetic Fibres

44.5.2.1 Nylons: The Synthetic Polyamides

Synthetic polyamides are known as Nylons and are prepared as follows.

(i) By copolymerisation of a dibasic acid with a diamine.

(ii) By self polymerisation of an amino acid or a lactam.

To distinguish one nylon from the other, the number of carbon atoms in the monomer units is indicated. For example, Nylon 6 is the polyamide obtained from caprolactam (C_6-monomer), Nylon 66 is obtained by condensation polymerisation of a diamine containing 6 carbon atoms (hexamethylene diamine) and a dibasic acid containing 6 carbon atoms (adipic acid).

Nylon 66: Adipic acid required for the preparation of Nylon 66 can be prepared by the following methods.

(i) *From Cyclohexane:* Catalytic oxidation of cyclohexane with air under pressure gives a mixture of cyclohexanol and cyclohexanone. Subsequent oxidation of the mixture with nitric acid in presence of copper-vanadium catalyst, or with air in presence of acetic acid and catalyst gives adipic acid.

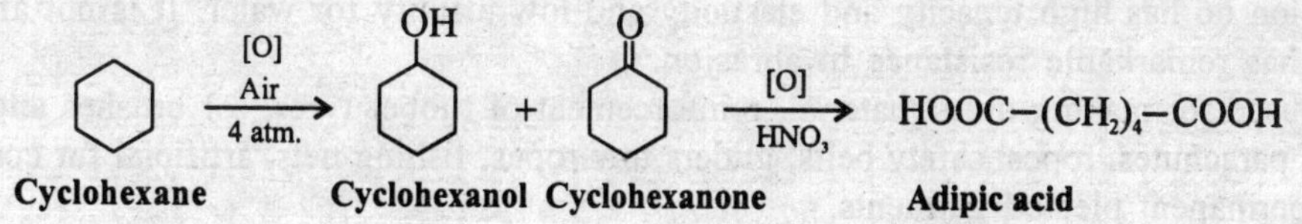

(ii) *From Butadiene:*

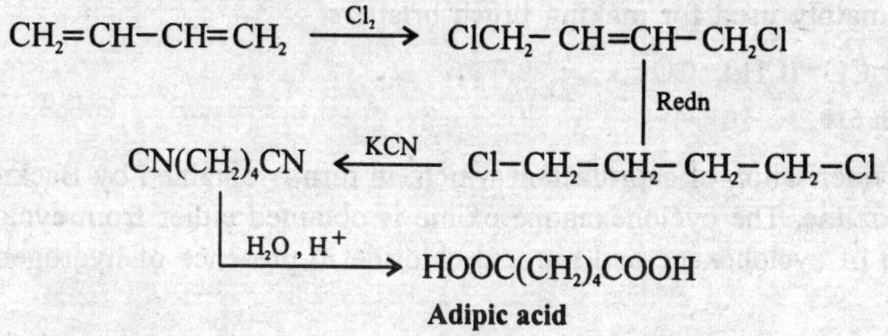

(iii) *From Furfural:*

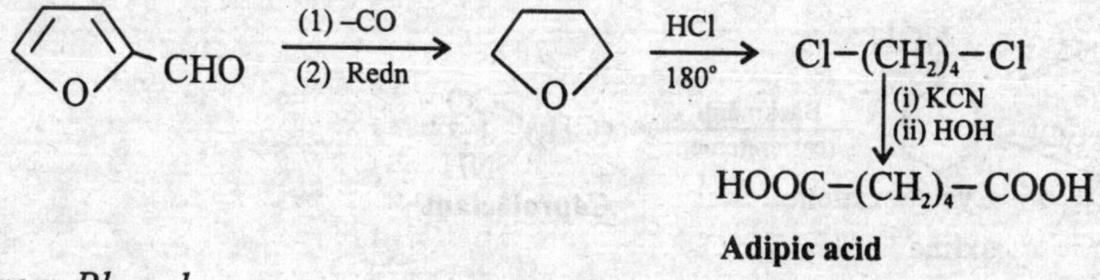

(iv) *From Phenol:*

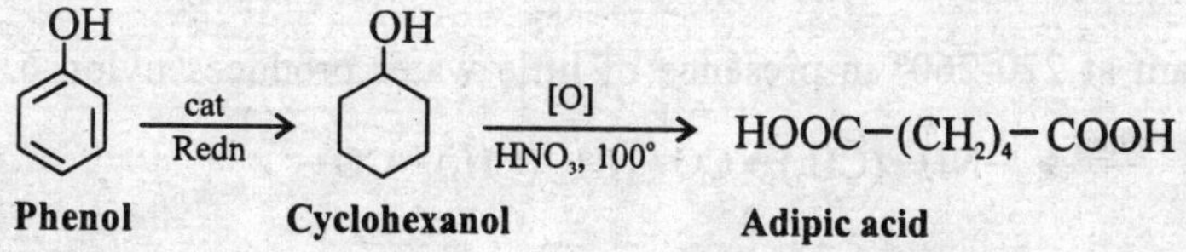

Hexamethylene diamine required for the preparation of Nylon 66 may be prepared from adiponitile by reduction.

$$NC-(CH_2)_4-CN \xrightarrow[130°]{H_2,\ Ni} H_2N-(CH_2)_6-NH_2$$

Adiponitrile   **Hexamethylene diamine**

Nylon 66 is precipitated by stirring together a solution of hexamethylene diamine and adipic acid. After drying, the polymer is melted and spun in an atmosphere of nitrogen to prevent decoloration due to air.

$$HOOC-(CH_2)_4-COOH + H_2N(CH_2)_6NH_2$$

$$\downarrow$$

$$\underset{O}{\overset{}{-\underset{\|}{C}}}-(CH_2)_4-\underset{O}{\overset{}{\underset{\|}{C}}}-\overset{H}{\underset{}{N}}-(CH_2)_6-\overset{H}{\underset{}{N}}-\underset{O}{\overset{}{\underset{\|}{C}}}-(CH_2)_4-\underset{O}{\overset{}{\underset{\|}{C}}}-\overset{H}{\underset{}{N}}-(CH_2)_6-\overset{H}{\underset{}{N}}-$$

Nylon 66

Properties: Nylon 66 has high tenacity and elasticity and low affinity for water. It is not affected by sea water and has remarkable resistance to abrasion.

It is widely used for making dress material, reinforcement of rubber tyres, for brushes and brooms, manufacture of parachutes, ropes, safety belts, gliders tow ropes, fishing nets, artificial fur coats, under garments and permanent pleated garments.

Nylon 610: It is obtained by the condensation polymerisation of hexamethylene diamine and sebacic acid (C_{10} dicarboxylic acid). It is mainly used for making brush bristlers.

$$-NH-(CH_2)_6-NH-CO-(CH_2)_8-CO-$$

Nylon 610

Nylon 6: It is obtained by polymerisation of caprolactam which, in turn is obtained by Backmann rearrangement of cyclohexanone oxime. The cyclohexanone oxime is obtained either from cyclohexanone or by photochemical reaction of cyclohexane and nitrocyl chloride in presence of hydrogen chloride.

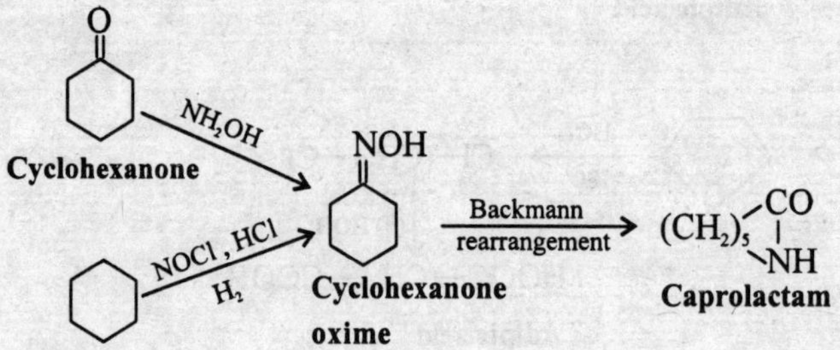

Cyclohexanone   **Cyclohexanone oxime**   **Caprolactam**

Cyclohexane

Polymerisation of caprolactam at 220-260° in presence of little water produces nylon 6.

$$(CH_2)_5\overset{CO}{\underset{NH}{|}} \xrightarrow{Polymerisation} -NH-(CH_2)_5-CO-NH-(CH_2)_5-CO-$$

Caprolactam   **Nylon 6**

Nylon 6 produces a softer and white fibre. It is used for making fabrics, fishing nets, parachutes, artificial fur and stokings etc.

44.5.2.2 Polyesters

Polyesters are obtained by condensation polymerisation of a dibasic acid and a diol. For example, terelene, kodel etc.

Terelene/Polyethylene Terephthalate/PET: It is also known as dacron. It is obtained by the condensation polymerisation of terephthalic acid or its ester and ethylene glycol under vacuum at elevated temperatures.

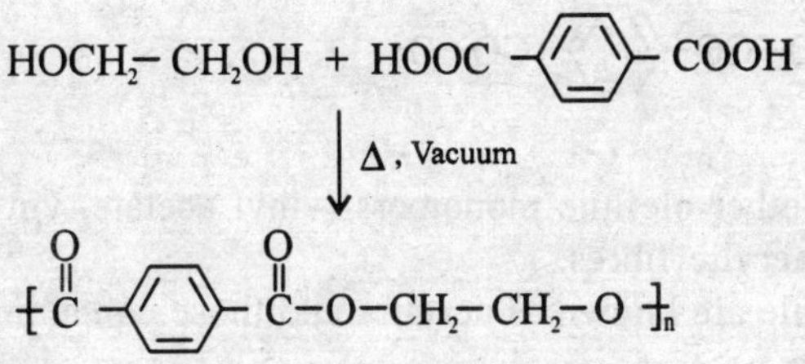

$$HOCH_2- CH_2OH + HOOC -\!\!\langle\bigcirc\rangle\!\!- COOH$$

$$\downarrow \Delta, \text{Vacuum}$$

$$\left[C -\!\!\langle\bigcirc\rangle\!\!- C-O-CH_2- CH_2-O \right]_n$$

Terelene

Glycol may be obtained from ethylene as follows:

$$CH_2\!\!=\!\!CH_2 \xrightarrow[\text{Aq. catalyst}]{\text{Air, }250°\text{–}325°} CH_2\!\!-\!\!CH_2 \xrightarrow[180°]{H_2O, H^+} \underset{\underset{OH\ \ OH}{|\ \ \ |}}{CH_2\!\!-\!\!CH_2}$$

Ethylene   **Ethylene oxide**   **Ethylene glycol**

Terephthalic acid is obtained by any of the methods given below.

(i) *Oxidation of p-Xylene:*

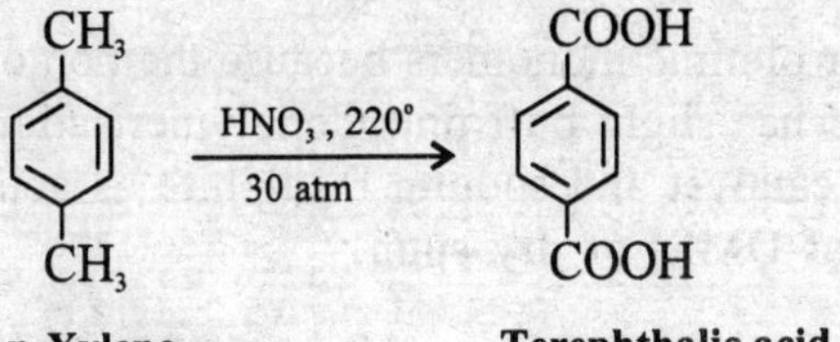

$$\underset{CH_3}{\overset{CH_3}{\langle\bigcirc\rangle}} \xrightarrow[30 \text{ atm}]{HNO_3,, 220°} \underset{COOH}{\overset{COOH}{\langle\bigcirc\rangle}}$$

p-Xylene   **Terephthalic acid**

(ii) *From Benzoic Acid:*

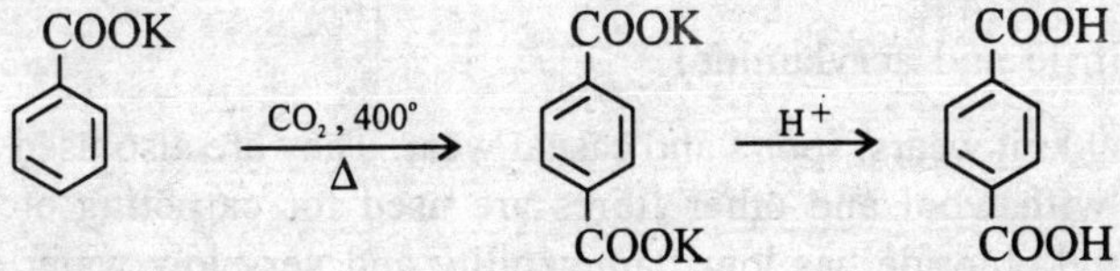

$$\underset{}{\overset{COOK}{\langle\bigcirc\rangle}} \xrightarrow[\Delta]{CO_2, 400°} \underset{COOK}{\overset{COOK}{\langle\bigcirc\rangle}} \xrightarrow{H^+} \underset{COOH}{\overset{COOH}{\langle\bigcirc\rangle}}$$

Potassium   **Terephthalic acid**
benzoate

Terelene can be heat set, and is used for making pleated skirts, neck ties etc. It is used for making curtains because it is resistant to degradation by sun light. It is also used for making filter sheets, conveyor and safety belts, fishing nets, sails and ropes etc.

It has also been processed into a highly tear resistant fibre called *Mylar,* used as backing for magnetic tapes which are used in tape recorders and computers.

Kodel: Kodel, a copolymer of terephthalic acid and cyclohexane dimethanol, possess good crease resistance, retention of shape and size during laundering and ironing and is used for making dress material.

$$HOH_2C-\hexagon-CH_2OH + HOOC-\hexagon-COOH$$

$$\downarrow \text{Polymerisation}$$

$$-OH_2C-\hexagon-CH_2-O-CO-\hexagon-CO-O-$$

Kodel

Polyacrylics: Copolymers of acrylonitrile with other olefinic monomers (vinyl acetate, vinyl chloride, styrene and acrylamide) are used for making acrylic fibres.

Those containing more than 80% of acrylonitile are known as acrylics and those containing less than 80% of acrylonitrile are known as modacrylics.

Acrylonitrile may be prepared from acetylene as shown.

$$CH\equiv CH + HCN \xrightarrow[85°]{CuCl} CH_2=CHCN$$

Acetylene   **Acrylonitrile**

It is also prepared from propylene by the following method.

$$CH_2=CH-CH_3 \xrightarrow{[O]} CH_2=CH-CHO \xrightarrow{NH_3} CH_2=CH-\underset{NH_2}{\overset{H}{\underset{|}{\overset{|}{C}}}}-OH \xrightarrow[(2)-H_2]{(1)-H_2O} CH_2=CH-CN$$

Propylene   **Acrylonitrile**

Acrylonitrile is generally co-polymerised with other olefinic monomers because the homopolymer of acrylonitrile is insoluble in most organic solvents and has slight buff tint. Copolymerisation is carried out in warm water using ammonium persulphate as catalyst and sodium bisulphite as activator. The resultant polymer is either wet spun (in a solution of DMF) or dry spun.

$$\underset{CH_2}{\overset{CN}{\underset{|}{CH}}}\underset{CH_2}{\overset{CN}{\underset{|}{CH}}}\underset{CH_2}{\overset{CONH_2}{\underset{|}{CH}}}CH_2$$

(a copolymer of acrylonitrile and acrylamide)

Acrylics are widely used for making knit wears, sports and casual wear. They are also used for making undergarments and dresses. Blends with wool and other fibres are used for carpeting etc.

Copolymer of acrylonitrile and vinyl chloride has low flammability and very low water absorption. It is resistant to attack by micro-organisms, insects and inorganic chemicals and is specially used for the manufacture of clothing for chemical workers, miners and military men.

44.6 PLASTICS

All plastics are man-made materials. The linear polymers are called thermoplastics, e.g, polyethylene and cross linked polymers are called thermoset plastics, e.g., bakelite.

44.6.1 Polyethylene

Initial attempts to polymerise ethylene proved difficult. Polymerisation could only be carried out at a very high pressure (2000 atmospheres) and temperature (170°). The product is now termed as a high pressure polyethylene. It has low density, high flexibility and is still used for production of squeeze bottles and cable insulation.

Polyethylene can now be prepared by the use of Ziegler Natta catalyst at a pressure of 70 atmospheres. It is known as low pressure polyethylene. It has higher degree of polymerisation, higher softening temperature, is more dense, crystalline and rigid as compared to high pressure polyethylene.

Expanded polyethylene has also been made by the use of blowing agents. It is used as an electrical insulator but is inferior to PUF and PVC foam in strength.

Polyethylene is water repellent, resistant to most reagents at room temperature. It is permeated by ethers and essential oils and hence these cannot be stored in polyethylene containers. It is oxidised when heated in air, is degraded by light. Addition of antioxidants and carbon black helps to prolong the life of polyethylene. Polyethylene films are used for packaging. It is extensively used for making house hold goods like kitchen bowls, baskets, bottles, waste bins etc.

44.6.2 Polypropylene

Use of Z–N type of catalyst produces stereoregular polypropylene. It is isotactic with some syndiotactic configuration. Isotactic polypropylene is without smell or color. It is harder and more rigid than polyethylene. It has a higher melting point (170°). The polymer is resistant to attack by inorganic and organic chemicals. Polypropylene is widely used for making washing machine parts, sterilisable medical and chemical equipments. It is also used for making washable wall paper, 'boil in the bag' food packs, book covering and packaging material.

44.6.3 Polyvinyl Chloride (PVC)

Vinyl chloride required for production of PVC is either made from acetylene or ethylene.

$$CH{\equiv}CH + HCl \xrightarrow[180°,\ 5\ atm]{HgCl_2} CH_2{=}CHCl$$

Acetylene **Vinyl chloride**

$$CH_2{=}CH_2 + Cl_2 \longrightarrow \underset{\underset{Cl}{|}\ \ \underset{Cl}{|}}{CH_2{-}CH_2} \xrightarrow[\substack{or\ heat\ with \\ NaOH\ solution}]{250°{-}500°} CH_2{=}CHCl$$

Ethylene **Vinyl chloride**

Vinyl chloride is polymerised either in water containing soap as an emulsifier and a persulphate initiator. Emulsion polymerisation produces hollow spheres of PVC and suspension polymerisation produces granular PVC. The emulsion polymer can take up a variety of additives such as plasticizers, fillers etc. Expanded PVC has also been prepared with the help of blowing agents.

$$CH_2{=}CHCl \xrightarrow{Polymerisation} {-}CH_2{-}\underset{\underset{Cl}{|}}{CH}{-}CH_2{-}\underset{\underset{Cl}{|}}{CH}{-}$$

Vinyl chloride **PVC**

PVC is a hard, rigid, tough polymer and is resistant to chemical attack. It is used in the manufacture of roofing sheets, road signs, tunnel lining, tank lining etc. Properties of PVC can be changed by co-polymerisation with other olefin monomers or by addition of large range of additives such as plasticizers, fillers, lubricants etc.

Addition of plasticizers turns tough and rigid PVC into soft and flexible product. Dibutylphthalate, dioctylphthalate and tricrysyl phosphate are commonly used as plasticizers. Plasticized PVC is used for making a variety of products like hand bags, wall coverings, packaging foils, long playing records, flexible tubing, protective clothing, ceiling lights, balls, toys, jars, food trays, cable covering, hose pipes etc.

44.6.4   Polystyrene

Polystyrene is obtained by polymerisation of styrene with heat, light or catalyst (free radical generator). Stereoregular polymers can be obtained by use of Ziegler Natta catalyst.

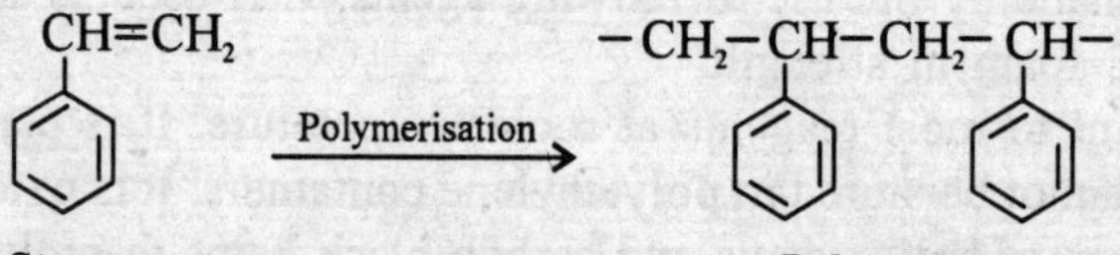

Styrene   Polystyrene

Polystyrene is a clear transparent resin having remarkable heat flow properties. It is highly resistant to attack by acids and other corrosive chemicals. It is an excellent electrical insulator. It, however, has poor impact strength and is affected by long exposure to sunlight and heat.

It is widely used for making house hold goods, plastic moulds, laminates, toys, lenses, battery boxes etc.

The expanded polystyrene is used in the manufacture of life jackets, boats and as a low density insulating and packaging material.

Properties of polystyrenes have been improved by copolymerisation with divinylbenzene, α-methylstyrene and vinyltoluenes.

44.6.5   Polyvinyl Acetate/PVA

Vinyl acetate is prepared by the addition of acetic acid to acetylene in presence of mercuric phosphate at 50°. Polymerisation of vinyl acetate by free radical initiation in water emulsion or suspension produces PVA.

$$CH\equiv CH + CH_3COOH \xrightarrow[50°]{\text{Mercuric phosphate}} CH_2 = CHCOOCH_3$$

Acetylene   **Vinyl acetate**

$\downarrow$ Polymerisation

$$-CH_2-CH-CH_2-CH- \\ \quad\quad\; | \quad\quad\quad\;\; | \\ \quad\quad OCOCH_3 \quad OCOCH_3$$

PVA

PVA is a colorless, nontoxic and odorless thermoplastic. It is mainly used in the adhesive industry. It is also used in chewing gums and in emulsion paints. It is also used for making wall tiles and artificial leather when mixed with fillers.

44.6.6   Teflon/Fluon

Polymerisation of tetrafluoroethylene by free radical initiators produces teflon. The polymerisation reaction is an explosively exothermic reaction.

$$CF_2{=}CF_2 \xrightarrow[\text{High Pressure}]{\text{Peroxide}} -CF_2{-}CF_2{-}CF_2{-}CF_2{-}$$

Tetrafluoro-
ethylene   **Teflon**

Tetrafluoroethylene is prepared from chloroform by the following method.

$$CHCl_3 + HF \xrightarrow{SbCl_3} CHF_2Cl + CHFCl_2 + CF_2Cl_2 + HCl$$

$$2CHF_2Cl \xrightarrow{650-800°} CF_2{=}CF_2 + 2HCl$$

Tetrafluoro-
ethylene

Teflon has high strength and remarkable resistance to chemical attack and temperature. It is non flammable, electrical insulator and is self lubricating. It is highly viscous and does not flow freely thereby making fabrication difficult.

It is widely used for providing non-stick, self lubricating surface. Joints in chemical apparatus have thin fluon layer to prevent locking of stoppers and joints. It is used for gaskets, protective lining in chemical plants, insulation of cables, transformers (used at high temperatures).

44.6.7 Phenol-Formaldehyde Resins

These are important group of polymers and are manufactured by the reaction of phenol and formaldehyde. The first step in their preparation is the addition of formaldehyde to phenol to give o- and/or p-hydroxybenzyl alcohol. This reaction is catalysed either by acids (hydrochloric acid, oxalic acid) or by bases (ammonia, sodium hydroxide etc.).

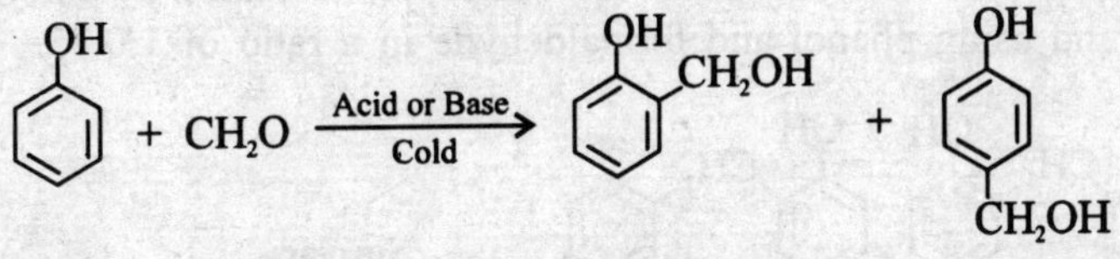

Phenol   Formaldehyde   O-Hydroxybenzyl   p-Hydroxybenzyl
**   alcohol   alcohol**

The mechanism involved in alkaline medium is given below:

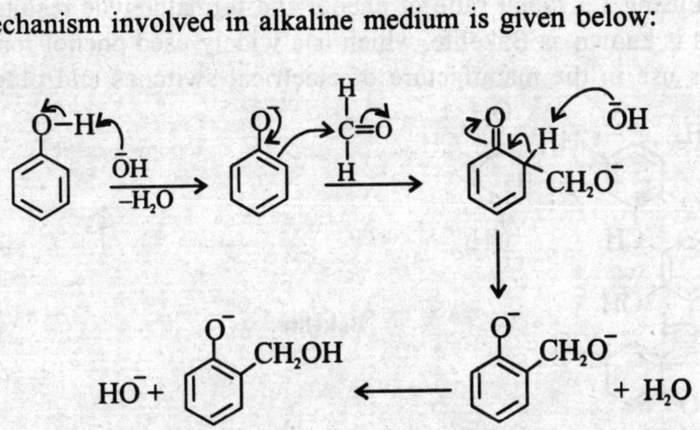

In acidic medium, the reaction follows the following route.

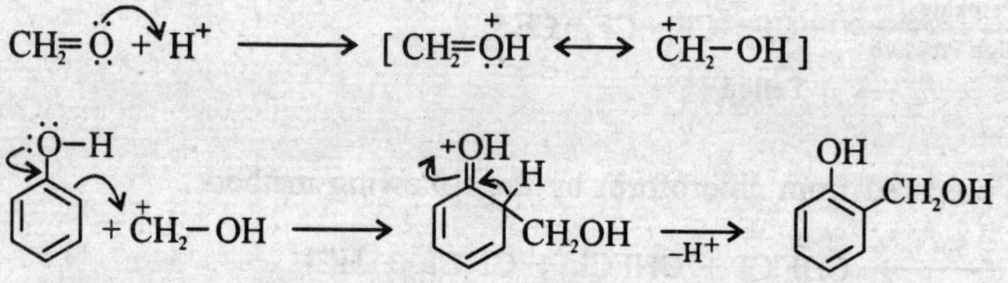

$$CH_2=\ddot{O} + H^+ \longrightarrow [CH_2=\overset{+}{O}H \longleftrightarrow \overset{+}{C}H_2-OH]$$

In the above mechanisms the condensation is shown at the o-position. It can be at p-position also which is more reactive. The reaction does not stop at the phenolic alcohol stage. It reacts further with formaldehyde to form bishydroxymethyl phenols and 4,4′-hydroxybisphenylmethane.

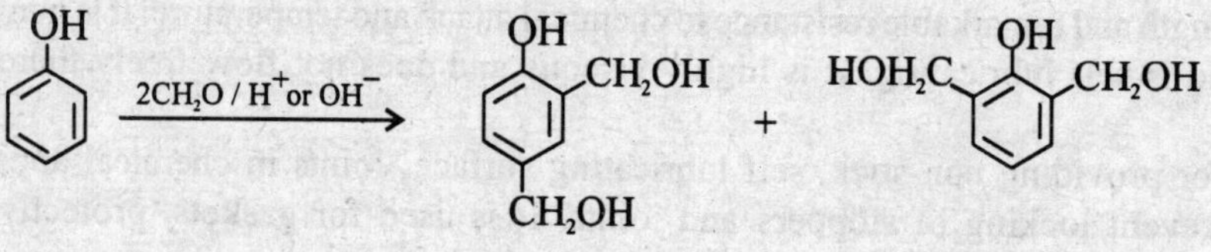

Bishydroxymethyl phenol

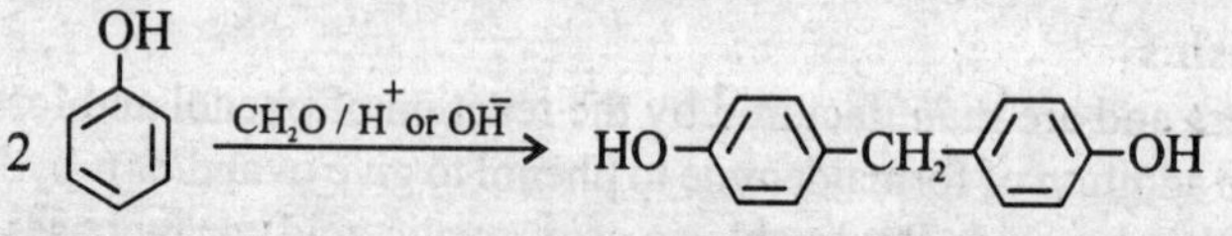

4,4′-Hydroxybisphenylmethane

Such condensations are repeated and the phenol formaldehyde resins are obtained.

The low molecular weight phenol-formaldehyde resins (Novolacs) are obtained by carrying out the condensation in acidic medium and using phenol and formaldehyde in a ratio of 1:0.75.

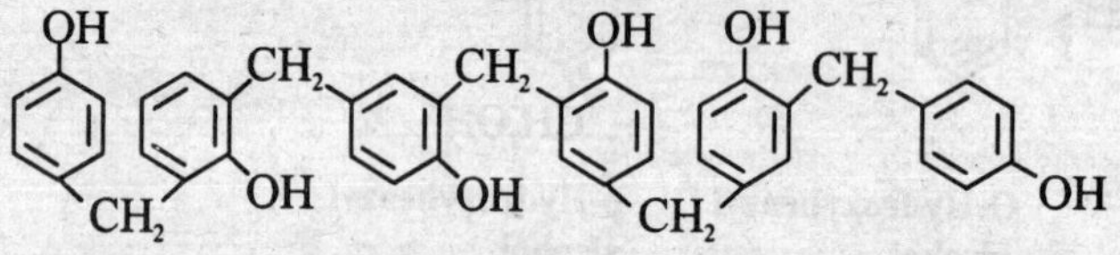

Novolac

On the other hand, condensation using 1:1 molar ratio of phenol and formaldehyde results in a three dimensional network polymer. This is known as Bakelite, which is a widely used phenol-formaldehyde resin. It is thermosetting and finds use in the manufacture of electrical switches and plugs.

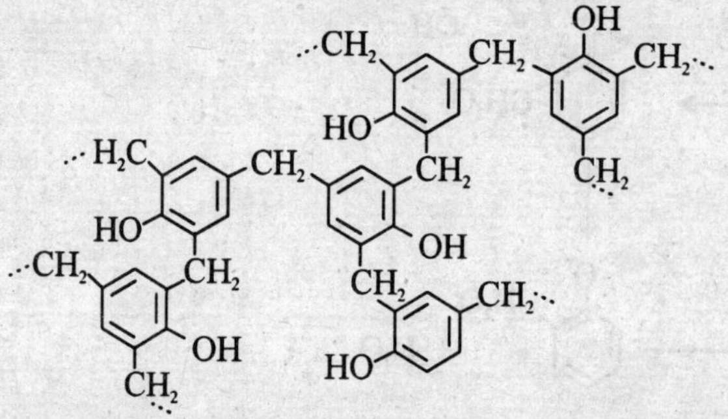

Bakelite

The starting material, phenol is obtained commercially by the following methods.

(i) *Cumene Phenol Process:* From benzene by Friedel Crafts alkylation followed by oxidative rearrangement.

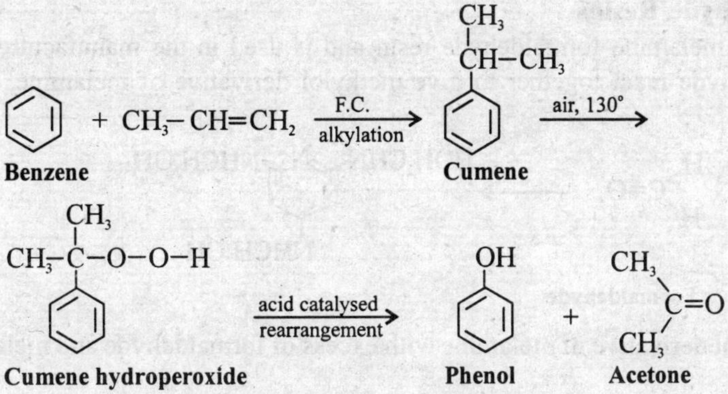

Benzene; Cumene; Cumene hydroperoxide; Phenol; Acetone

(ii) *From Chlorobenzene:*

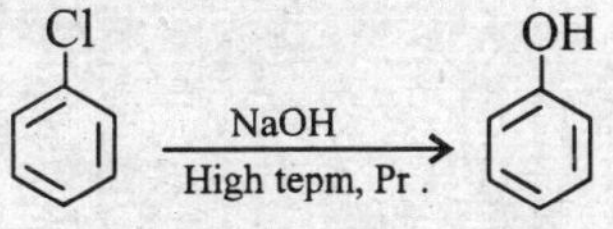

Chlorobenzene; Phenol

44.6.8 Urea Formaldehyde Resins

The urea formaldehyde resins are superior to phenol formaldehyde resins. They are colorless and do not get effected on exposure to light and are used in the manufacture of kitchen ware, laminates and as adhesives.

These resins are obtained from urea and formaldehyde. The reactions are analogous to phenol formaldehyde resin formation. Usual reaction gives low molecular weight branched products.

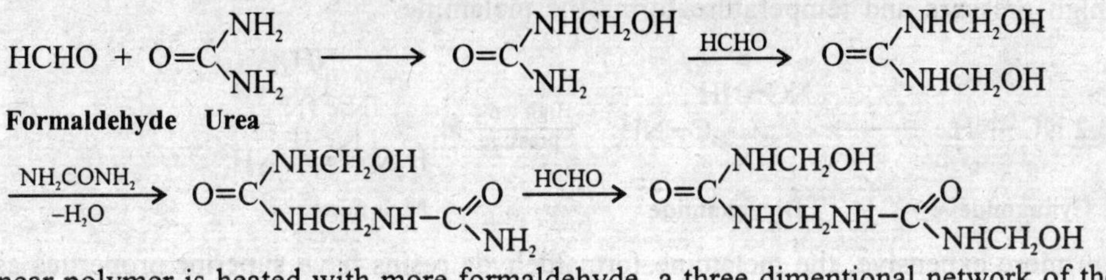

If the linear polymer is heated with more formaldehyde, a three dimentional network of the polymer is produced with the following structure.

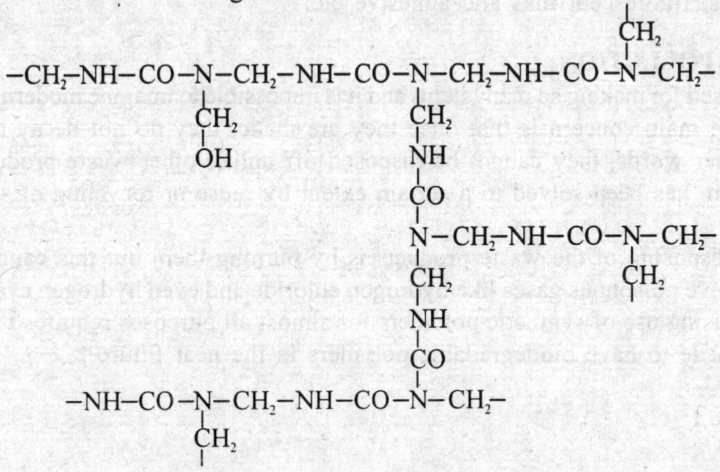

44.6.9   Melamine Formaldehyde Resins

Melmac is the most important melamine formaldehyde resin and is used in the manufacture of kitchen ware. Melamine and formaldehyde react together to give methylol derivative of melamine.

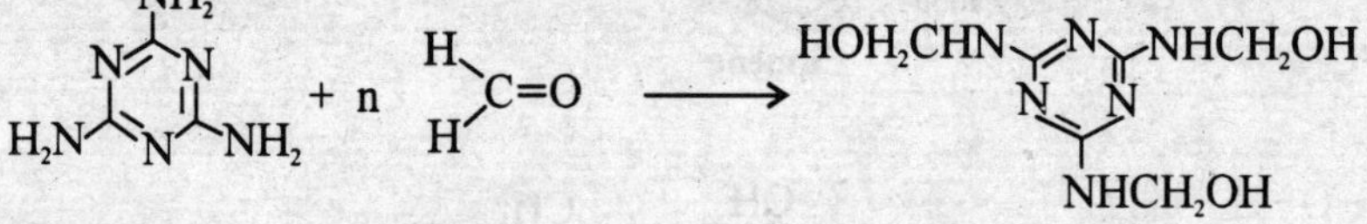

Melamine   Formaldehyde

Polycondensation of methylol derivative of melamine with excess of formaldehyde and melamine gives a cross linked risin.

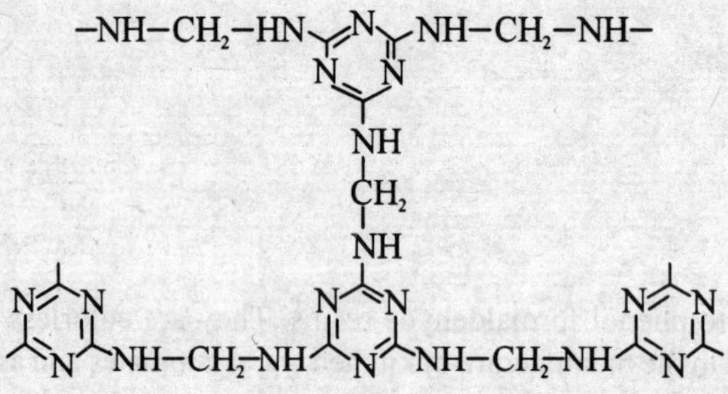

The required melamine is obtained from cyanamide. Dimerisation of cyanamide produces dicyanamide, which at high pressure and temperature forms the melamine.

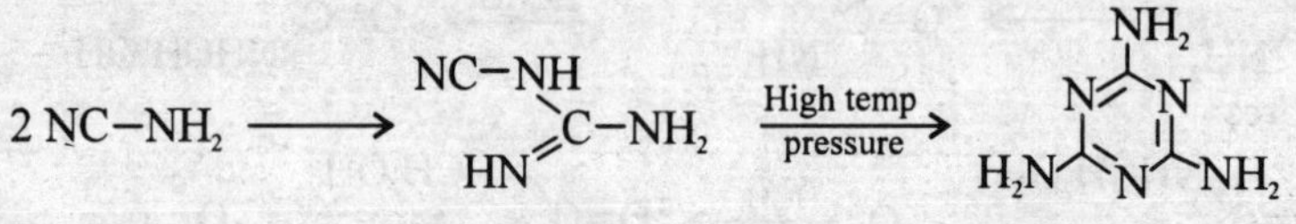

Cyanamide   Dicyanamide   Melamine

Although more expensive, the melamine formaldehyde resins have superior properties as compared to urea formaldehyde resins. These are largely used for the production of table ware, heavy duty electrical points, decorative laminates, fluorescent inks and adhesive etc.

44.7   ENVIRONMENTAL POLLUTION

The synthetic polymers are used for making so many items and it is impossible to imagine modern civilisation without them. However, the main concern is that once they are made, they do not decay and tend to remain for all times. In other words, they cannot be disposed off unlike other waste products, which are degradable. This problem has been solved to a certain extent by reuse or recycling of some of the polymers.

The normal method of disposing of the waste products is by burning them but this cannot be used for polymers, since they evolve poisonous gases like hydrogen chloride and even hydrogen cyanide (from polyacrylonitile). The increasing use of synthetic polymers for almost all purposes requires to be looked into. Attempts are being made to have biodegradable polymers in the near future.

44.8 PROBLEMS

1. What are polymers? How are they classified?
2. Discuss the mechanism of free radical polymerisation.
3. What are inhibitors or retarders? Explain their role in polymerisation.
4. What do you understand by the term copolymerisation? Explain with the help of a suitable example.
5. Give the mechanism of cationic and anionic polymerisation.
6. How is styrene polymerised?
7. Explain Ziegler-Natta polymerisation.
8. What do you understand by isotactic, syndiotactic and atactic polymers?
9. How the following obtained:
   Polyethylene, Polyvinyl chloride, Polyvinyl acetate, Acrylic fibres and Teflon.
10. What is condensation polymerisation? How is terylene and nylon obtained?
11. Explain the preparation and uses of different types of resins.
12. How is synthetic rubber obtained?
13. Discuss the structure of rubber.
14. Give the mechanism of vulcanization.

45
Dyes

45.1 INTRODUCTION

Dyes are colored organic compounds, which have the property of imparting their color to other materials such as textile fibres, plastics etc. The requirement from a substance to act as a dye is that, it must have a suitable color and is capable of being 'fixed' to the surface to be dyed. Also, the color of the dyed material should not be affected on prolonged exposure to light, water and soap, i.e. the dye must have fastness properties. It is to be noted that all colored compounds are not dyes.

The dyes were earlier obtained from vegetable or animal sources. For example, indigo (from *Indigo tinctoria*) and alizarin (from *Madder roots*) have been used as dyes since the advent of civilization. However, the modern dyes are mostly synthetic.

45.2 CLASSIFICATION

The dyes may be classified in a number of ways according to their origin, color, substance to which they are applied (eg. cotton, silk, wool etc.), methods of application and chemical structure. The present chapter discusses the two classes based on method of application and chemical structure.

45.2.1 Classification Based on Method of Application

(i) *Direct Dyes:* These are also known as substansive dyes and can be applied to the fibre directly from an aqueous solution. Such dyes are of two types.

(a) *Acid dyes:* These are sodium salts of sulphonic acids and nitro phenols and can easily dye wool and silk. The fibers are dyed by dipping in the solution of acid dyes after acidifying with dilute sulphuric acid or acetic acid. Examples of acid dyes are martius yellow (2,4-dinitrophenol) and metalin yellow.

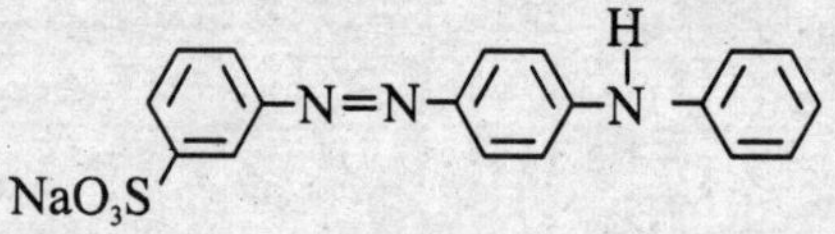

Metalin Yellow

(b) *Basic dyes:* These are the salts (hydrochloride or zinc chloride salts) of colored bases having basic groups (amino and imino) and are applied in water solution directly to animal fibre. Examples of this type are magenta, malachite green etc.

(ii) *Mordant Dyes:* These are also known as adjective dyes and do not dye a fibre directly. They require a mordant, which can be a metal hydroxide (for acidic dyes) and tannic acid or tannin (for a basic dye). The fibre to be dyed is first mordanted by dipping in the solution of the metallic salt or tannin and then dipped into the solution of the dye. By this process a colored lake is obtained which is insoluble and fast to washing. Examples of mordant dyes are alizarin and other anthraquinone dyes.

(iii) *Vat Dyes:* These cannot be applied directly since they are insoluble in water. They are, therefore, rendered soluble by reduction with sodium hydrosulphide to a colorless compounds known as leuco compounds. The fibre is dipped into the leuco compound solution and then exposed to air or chemical oxidation. By this process the leuco compound is oxidised back to the original insoluble dye which remains firmly impregnated in the fibre.

The reduction was originally carried out in large vats and hence the name. A common examples of vat dye is indigo.

(iv) *Disperse Dyes:* These are non-ionic dyes having low molecular weight and are used for dyeing fibres like nylon, terene etc. The dyes are generally applied in the form of a time dispersion in soap solution in presence of a stabilising agent. An example of this type of dye is 1-amino-4-hydroxyanthraquinone.

(v) *Ingrain Dyes:* These are produced in the fibre itself during the process of dyeing. For example, a piece of cloth to be dyed is soaked in an alkaline solution of β-naphthol and then dipped into a diazonium salt solution. Coupling takes place to produce an azo dye on the fibre itself.

45.2.2 Classification Based on Chemical Structure

The classificaton based on the method of applications is important to the dyer who is basically concerned with the reaction of the dyes towards the fibres being used. However, the chemists are more interested in the classification according to their chemical constitution.

The following table gives a brief account of different classes of dyes based on chemical structure with suitable examples. (Table 45.1)

Table 45.1: Some Important Dyes

Class; Basic Chemical Structure [chromophore]	Example	Remarks
1. Nitroso: o-Nitrosophenol or o-Nitrosonaphthol $[-N=O]$	Naphthol Green B	— Mordant dyes — Used only as lake of metal — Prepared by action of phenols or naphthols
2. Nitro: o- and p-Nitrophenols or o- and p-Nitroamines	Picric acid Martius yellow Naphthol yellow	— Prepared by the action of nitric acid on phenols, naphthols, amines.

Contd...

Class; Basic Chemical Structure [chromophore]	Example	Remarks
		— Most wide class of synthetic dyes
3. Azo dye: Aromatic azo compounds [—N=N—]		
(a) Monoazo [R—N=N—R']	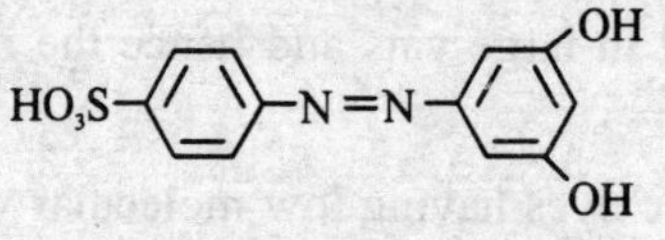 HO₃S—⟨⟩—N=N—⟨⟩ with OH, OH Resorcin yellow	— Prepared by action of nitrous acid on primary arylamine to give diazo compound which is coupled with aromatic amino or hydroxy compound. — Dyes sensitive to acids/bases, more commonly used as indicators.
	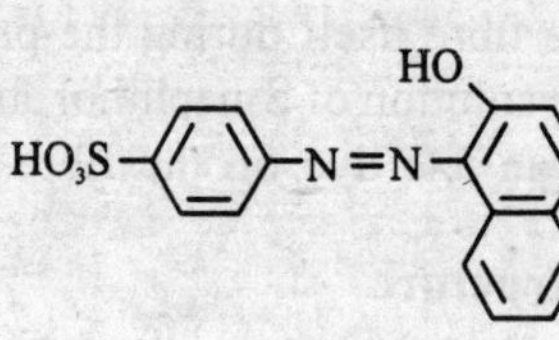 HO₃S—⟨⟩—N=N— naphthalene-HO Orange II	
	Other e.g., p-Aminoazobenzene, Butter yellow, Methyl orange, Methyl red	
(b) Disazo: [R—N=N—R'—N=N—R"]	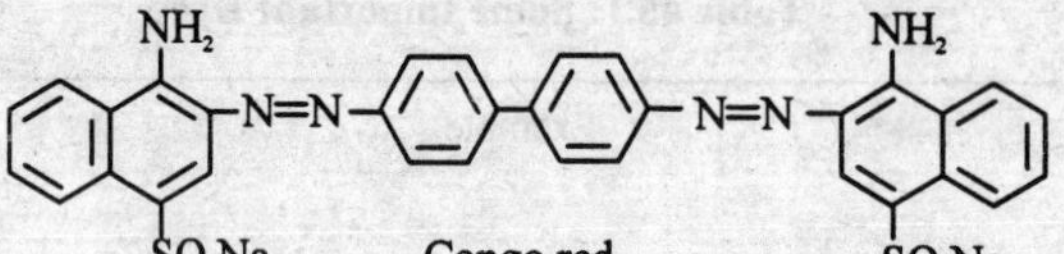 Congo red structure (NH₂, SO₃Na groups)	
	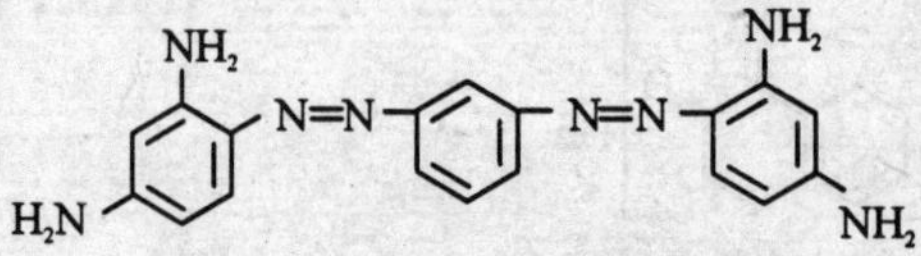 Bismark brown structure (NH₂, H₂N groups)	— Prepared by action of nitrous acid on primary diamine, followed by coupling of the resulting tetrazo compound with two moles of aromatic amino or hydroxy compounds.
(c) Trisazo: [R—N=N—R'— N=N—R"— N=N—R''']	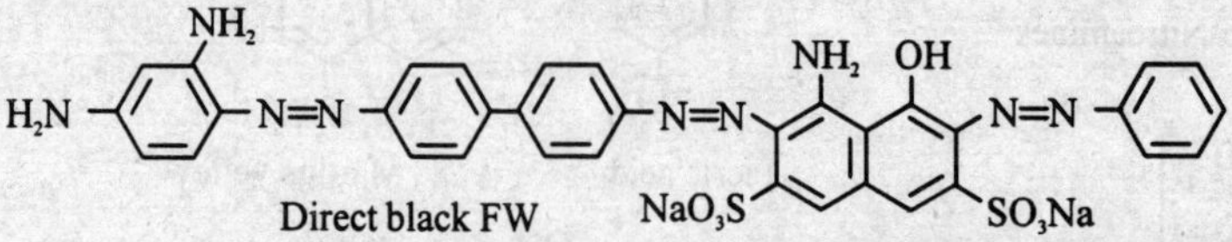 Direct black FW structure (NH₂, OH, NaO₃S, SO₃Na groups)	— Prepared from four intermediates by rediazotisations and coupling

Contd...

Class; Basic Chemical Structure [chromophore]	Example	Remarks
4. Diphenylmethane: (ketoneimine) $$\begin{bmatrix} -C- \\ \parallel \\ NH \end{bmatrix}$$	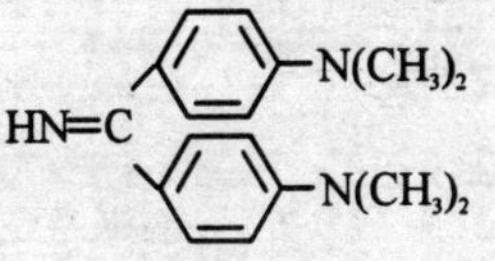 Auramine 0 (Basic yellow 2)	— From Michler's ketone ammonium and zinc chlorides
5. Triphenylmethane: $$\left[\begin{array}{c} \diagdown \\ \diagup C{=}Ar{=}NH \quad (or{=}O) \end{array} \right]$$ (a) Diamino:	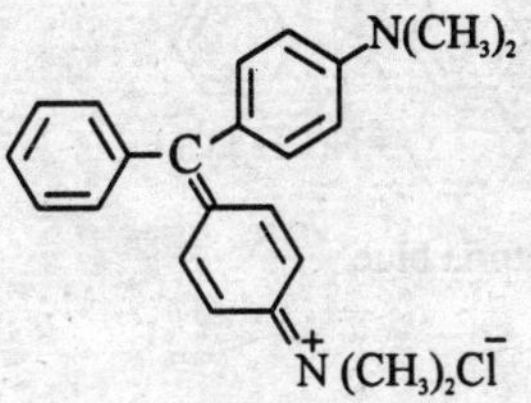 Malachite Green	— Prepared by condensation of aromatic aldehydes with aryl amines or phenols
(b) Triamino:	 Pararosaniline Rosaniline Crystal violet	
(c) Aminohydroxy:	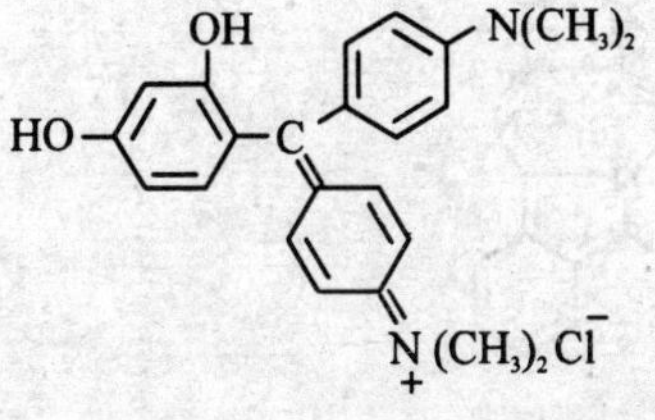 Resorcin violet	

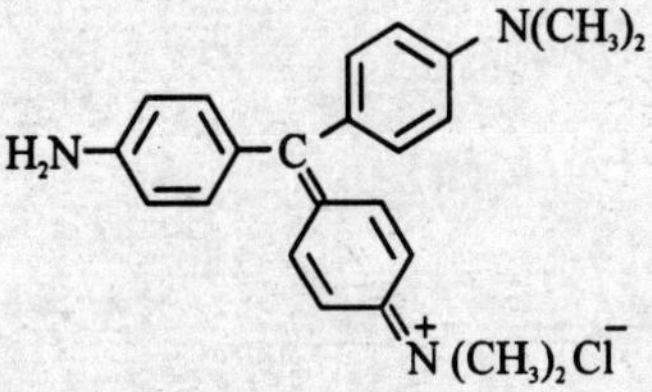

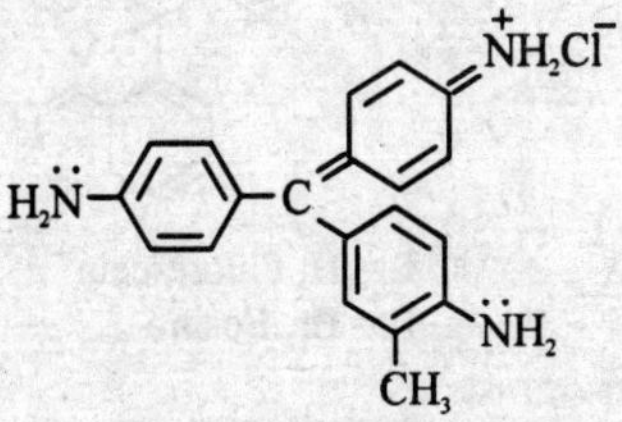

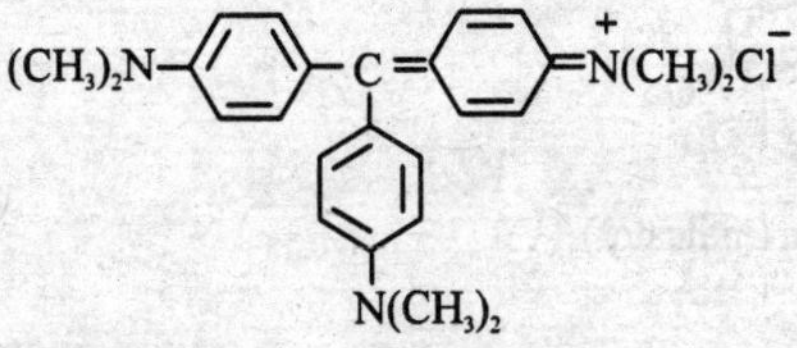

Contd...

Class; Basic Chemical Structure [chromophore]	Example	Remarks
(d) Hydroxy:	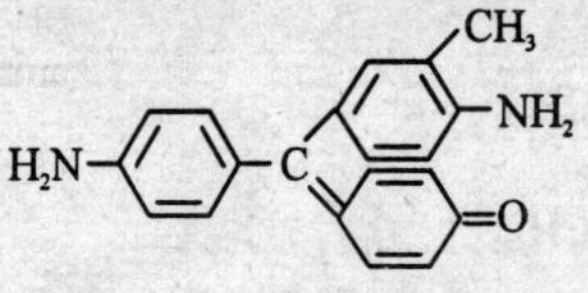 Aurin	
(e) Diphenylnaphthyl- methane:	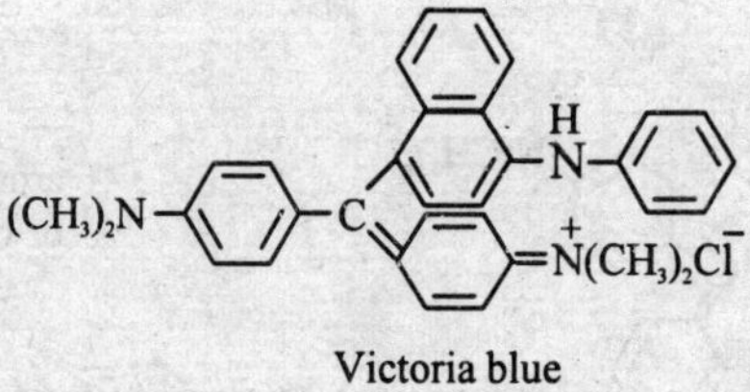 Victoria blue	
6. Phthalein dyes:	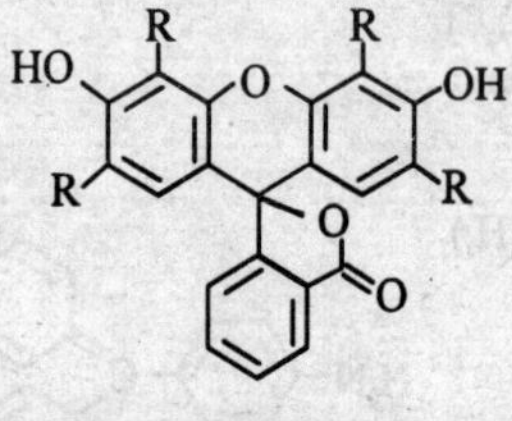 R = H, Fluorescein R = Br, Eosin Phenolphthalein (indicator)	— Obtained by condensing phenols and phthalic anhydride
7. Phthalocyanins:	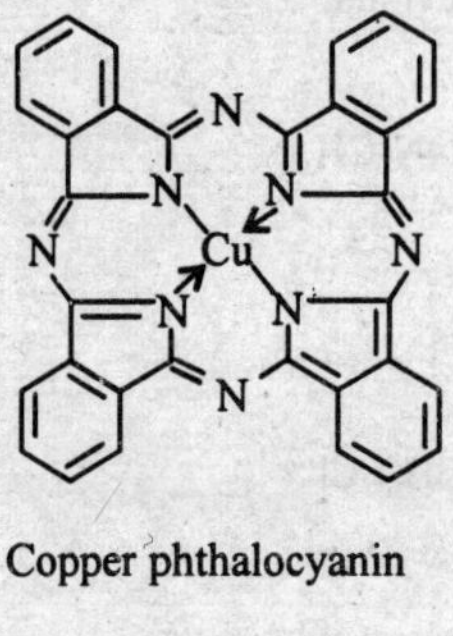 Copper phthalocyanin	— Used as organic dye as well as pigment

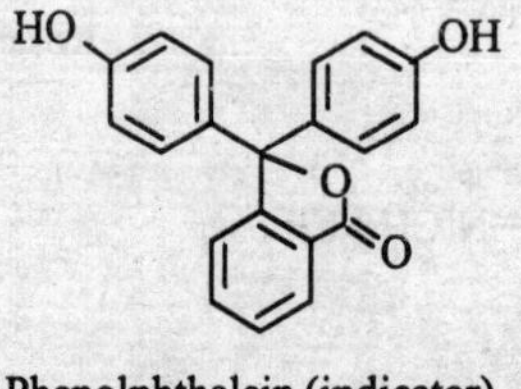

Contd...

Class; Basic Chemical Structure [chromophore]	Example	Remarks

— Vat dyes

8. Indigoid:

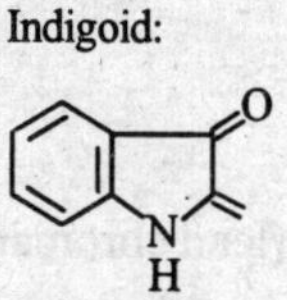

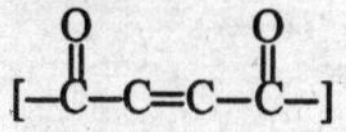

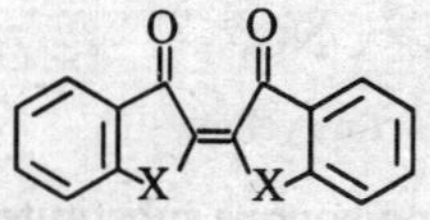

X=NH : Indigo (cis / trans)
X=S :Thioindigo (cis / trans)

(NH or S)

$[-\overset{O}{\underset{}{C}}-C=C-\overset{O}{\underset{}{C}}-]$

9. Anthraquinoid dyes:

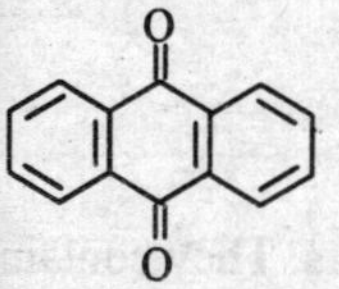

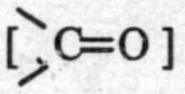

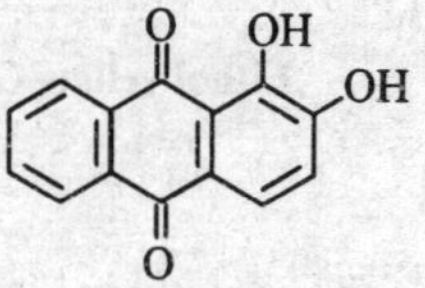

— Alizarin: Mordant dye, can be synthesised from anthraquinone

[$>C=O$]

Alizarin

45.2.3 Nitro and Nitroso Dyes

Nitroso Dyes: These are mordant dyes, used only as lakes of metal. Most of the nitroso dyes have one hydroxy group ortho to nitroso group. They are prepared by the action of nitrous acid on phenols or naphthols.

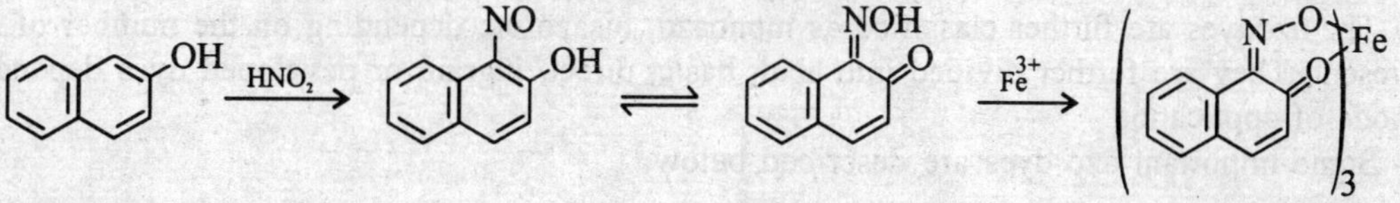

β-Naphthol   Naphthol green B

Nitro Dyes: These are prepared by action of nitric acid on phenols, naphthols and amines.

(a)   *Picric Acid:* It is prepared by nitration of phenol with fuming nitric acid in presence of concentrated sulphuric acid. It was used for dyeing wool and silk in bright yellow colour.

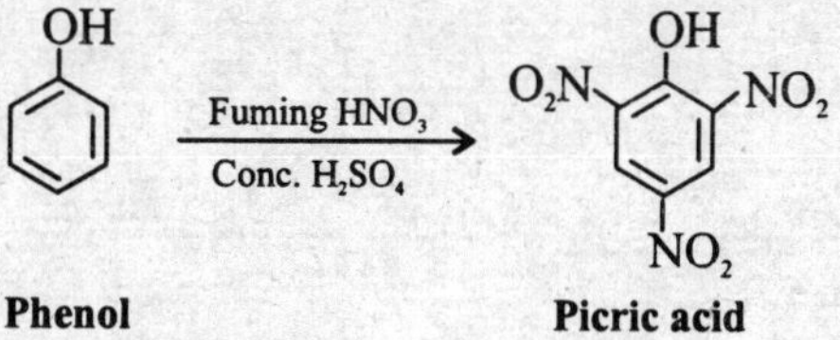

Phenol   Picric acid

(b)   *Naphthol Yellow:* It is obtained by nitrating 1-naphthol-2,4,7-trisulphonic acid.

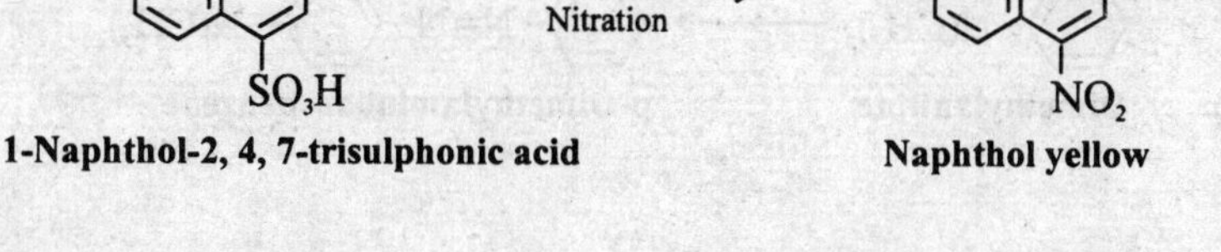

1-Naphthol-2, 4, 7-trisulphonic acid   Naphthol yellow

(c) *Martius Yellow:*It is obtained by nitrating 1-naphthol-2,4-disulphonic acid.

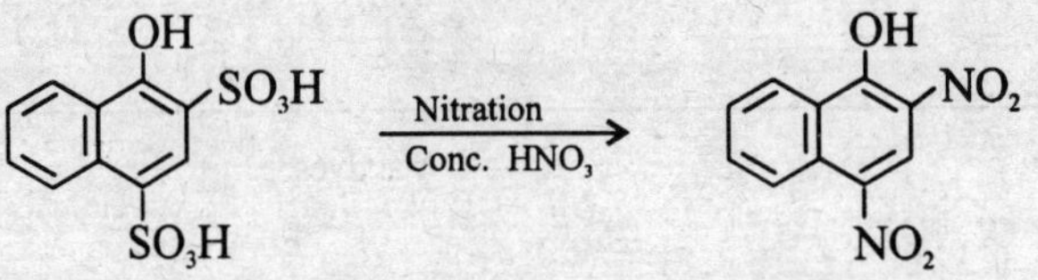

1-Naphthol-2, 4-disulphonic acid   **Martius yellow**

(d) *Lithol Yellow G:* It is a pigment and is a nonpoisonous substitute for chrome yellow (lead chromate). It is prepared by condensing 4-chloro-2-nitroaniline with formaldehyde.

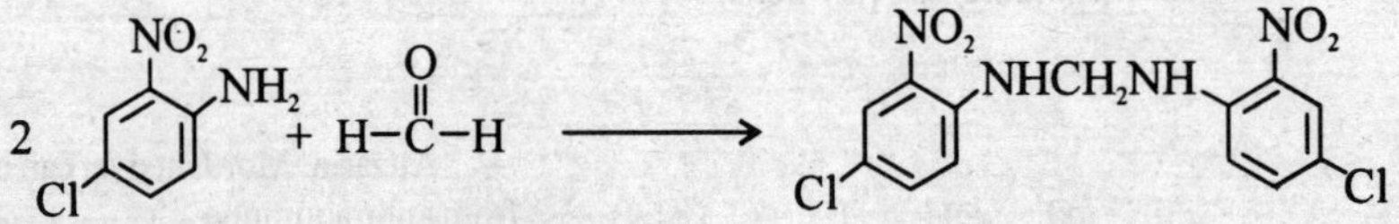

4-Chloro-2-   **Formaldehyde**   **Lithol yellow G**
nitroaniline

45.2.4   Azo Dyes

Azo dyes comprise the single largest group of dyes having almost a complete range of colors. They contain the azo group (−N=N−) and in addition contain sulphonic acid (which improves water solubility), hydroxy or amino groups. These are generally prepared by diazotising a primary amine. The coupling occurs at p-position to the hydroxy or amino group in benzene derivatives. In naphthalene series 1-substituted derivative couples at position 4, and if it is occupied then at position 2, but in 2-substituted derivatives coupling takes place only at position one.

The azodyes are further classified as monoazo, bisazo etc depending on the number of azo groups present. They are further divided into acid, basic, direct, ingrain or developed dyes depending on the mode of application.

Some important azo-dyes are described below:

1. *Aniline Yellow (p-Aminoazobenzene):* It is the simplest basic azo dye and is prepared by coupling benzenediazonium chloride with aniline.

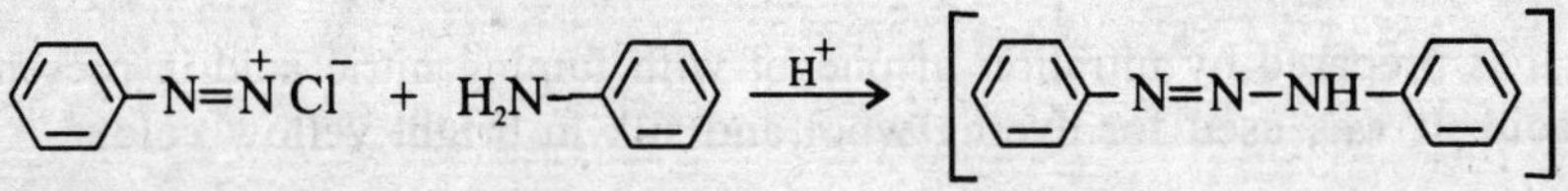

Benzenediazo-   **Aniline**
nium chloride

p-Aminoazobenzene

It is not used as a dye since it is sensitive to acids.

2. *Butter Yellow (p-Dimethylaminoazobenzene):* It is a basic azo dye having very little commercial value as a dye and is obtained by coupling benzenediazonium chloride with dimethylaniline.

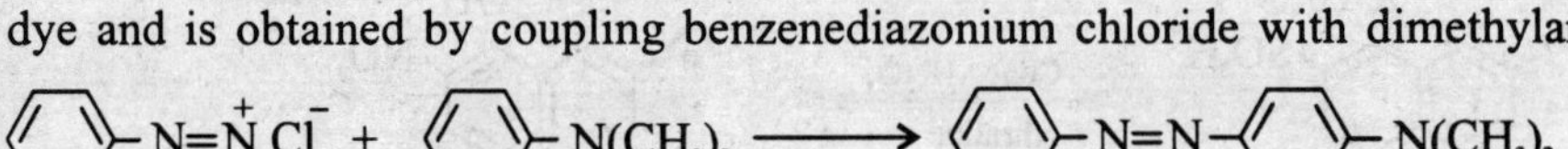

Benzenediazonium   **Dimethylaniline**   **p-Dimethylaminoazobenzene**
chloride

3. *Methyl Orange:* It is an acid dye and is used for dyeing wool and silk imparting them yellow color. However, the color is not fast to light or washing. It is prepared by coupling diazotised sulphanilic acid and dimethylaniline under acidic conditions followed by treatment with sodium hydroxide.

$$HO_3S-\langle\rangle-N{=}\overset{+}{N}\ \overset{-}{Cl}\ +\ \langle\rangle-N(CH_3)_2\ \xrightarrow[\text{2. NaOH}]{\text{1. }H^+}\ Na\overset{-}{O}_3S-\langle\rangle-N{=}N-\langle\rangle-N(CH_3)_2$$

| Diazotised sulphanilic acid | Dimethylaniline | | Methyl orange (sodium salt) |

It is not used as a dye but used as an indicator in acid alkali titrations. It gives yellow color with alkali and pink with acid. The change in color at the end point is due to change in structure of the ion.

$$\overset{-}{O}_3S-\langle\rangle-N{=}N-\langle\rangle-N(CH_3)_2\ \underset{H^+}{\overset{OH^-}{\rightleftharpoons}}\ \overset{-}{O}_3S-\langle\rangle-NH-N{=}\langle\rangle{=}\overset{+}{N}(CH_3)_2$$

Yellow in alk. solution
(pH $\geq$ 4.4)

p-Quinoid structure (Red)
(pH $\leq$ 3.1)

4. *Methyl Red:* It is used as an indicator in acid alkali titrations and is prepared by coupling diazotised o-aminobenzoic acid with dimethylaniline.

$$\langle\overset{\text{COOH}}{\rangle}-N{=}\overset{+}{N}\ \overset{-}{Cl}\ +\ \langle\rangle-N(CH_3)_2\ \xrightarrow{-HCl}\ \langle\overset{\text{COOH}}{\rangle}-N{=}N-\langle\rangle-N(CH_3)_2$$

| Diazotised | Dimethylaniline | | Methyl red |
| o-Aminobenzoic acid | | | |

5. *Orange II:* It is an acid monoazo dye and is used to dye wool, silk, nylon, leather etc. It is prepared by coupling diazotised sulphanilic acid with β-naphthol.

$$HO_3S-\langle\rangle-N{=}\overset{+}{N}\ \overset{-}{Cl}\ +\ \text{(β-Naphthol)}\ \xrightarrow[-HCl]{\text{NaOH}}\ Na\overset{-}{O}_3S-\langle\rangle-N{=}N-\text{(naphthol-OH)}$$

| Diazotised sulphanilic acid | β-Naphthol | | Orange II |

6. *Resorcin Yellow:* It is a golden yellow dye used for dyeing silk. It is prepared by coupling diazotised sulphanilic acid with resorcinol (1,3-dihydroxybenzene).

$$HO_3S-\langle\rangle-N{=}\overset{+}{N}\ \overset{-}{Cl}\ +\ \text{(Resorcinol)}\ \longrightarrow\ HO_3S-\langle\rangle-N{=}N-\text{(resorcinol)}$$

| Diazotised sulphanilic acid | Resorcinol | | Resorcin yellow |

7. *Congo Red:* It is a dark red disazo dye and is used as a direct dye for dyeing cotton. It is prepared by coupling tetrazotised benzidine with naphthionic acid (1-naphthylamine-4-sulphonic acid).

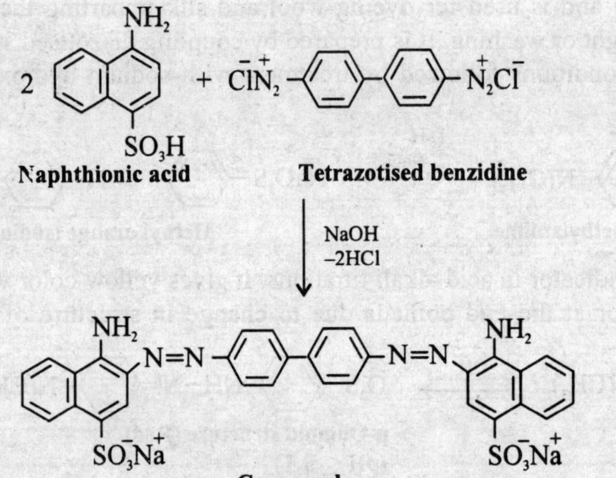

Naphthionic acid   **Tetrazotised benzidine**

$$\xrightarrow[\text{$-2HCl$}]{\text{NaOH}}$$

Congo red

Congo red is red in alkaline solution and its sodium salt dyes cotton red. It is very sensitive to acid, the color changes from red to blue in presence of inorganic acids. The dependence of color is attributed to the resonance shown below.

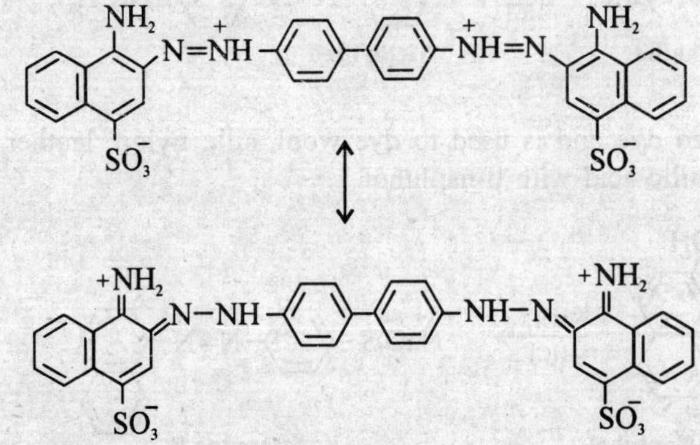

8. *Bismark Brown:* It is a basic disazo dye and is used in boot polishes and for wood polishing. It dyes wood and mordanted cotton. It is prepared by coupling tetrazotised-m-phenylenediamine with two molecules of m-phenylenediamine.

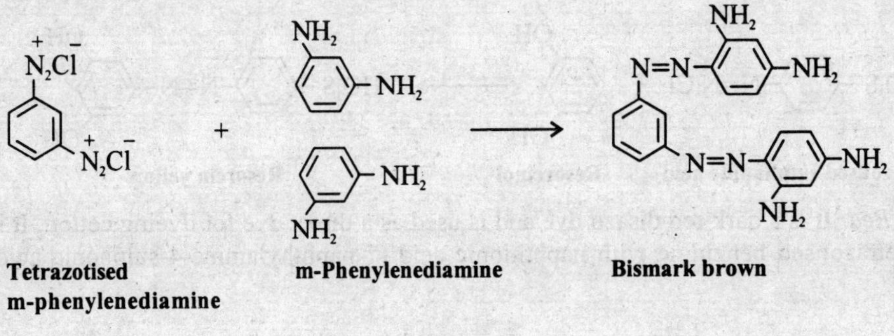

Tetrazotised   **m-Phenylenediamine**   **Bismark brown**
m-phenylenediamine

45.2.5 Triphenylmethane Dyes

These dyes have brilliant colors. However, the colors are not fast and they fade on exposure to light and washing. These are used for coloring paper and typewriter ribbons.

Triphenylmethane dyes (appropriately designated as triarylmethane dyes) are obtained by introducing $-NH_2$, $-NR_2$ or $-OH$ groups into the rings. The compounds so obtained are colorless leuco compounds, which on oxidation are converted into the corresponding tertiary alcohols. The color bases readily change from the colourless benzenoid form to the quinoid dyes in the presence of acid due to salt formation. The salts can be easily reconverted into the leuco base.

$$\text{Leuco base} \underset{\text{Reduction}}{\overset{\text{Oxidation}}{\rightleftarrows}} \text{Color base} \underset{\text{Alkali}}{\overset{\text{Acid}}{\rightleftarrows}} \text{Dye}$$
$$\text{(Colorless)} \qquad\qquad \text{(Colorless)} \qquad\qquad \text{(Colored)}$$

1. *Malachite Green:* It is used for dyeing wool and silk directly, and cotton mordanted with tannin. It is prepared by condensing dimethylaniline (2 mol) with benzaldehyde (1 mol) at 100° in presence of concentrated sulphuric acid. The formed leuco base is oxidised with lead dioxide in acetic acid solution containing hydrochloric acid. The resulting color base with exces of hydrochloric acid gives Malachite green.

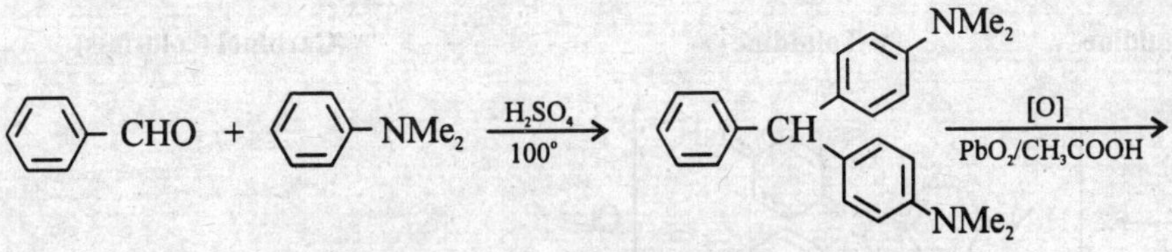

| Benzaldehyde | Dimethylaniline | Leuco base |

| Color Base | Malachite Green |

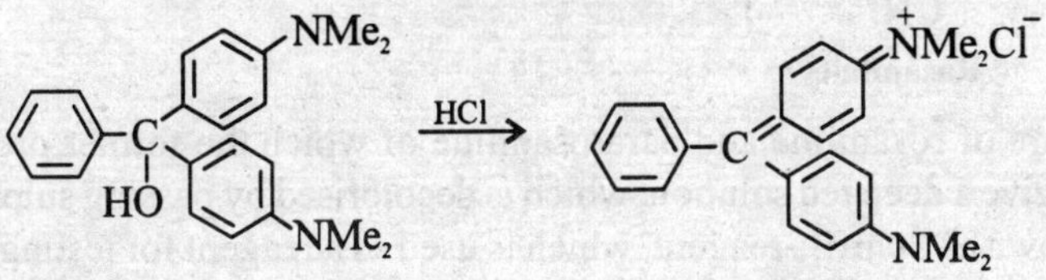

2. *Pararosaniline:* It dyes wool and silk directly producing a violet red color. Cotton is mordanted with tannin before dyeing. It is obtained by condening p-toluidine (1 mol) with aniline (2 mol) in presence of nitrobenzene, which serves both as a solvent and an oxidising agent. The carbinol formed on treatment with hydrochloric acid gives the dye.

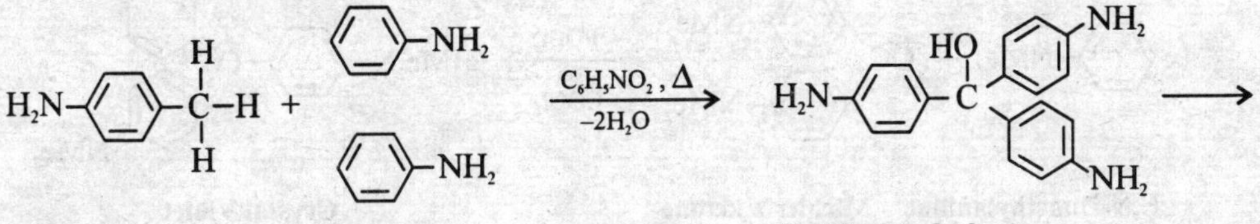

| p-Toluidine | Aniline | Carbinol (colorless) |

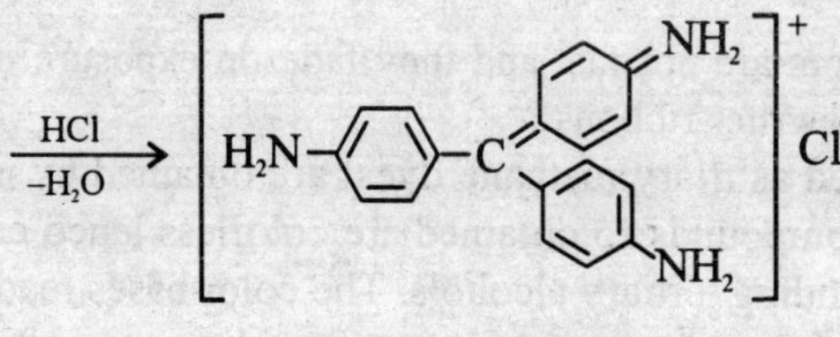

Pararosaniline

3. *Rosaniline:* It is used for dyeing cotton, silk and wool. It is prepared by condensing one mol each of aniline, p-toluidine and o-toluidine in nitrobenzene. The formed carbinol on treatment with hydrochloric acid gives the dye.

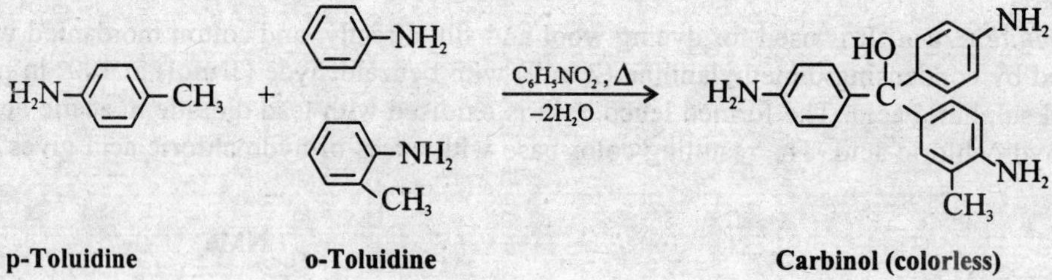

p-Toluidine   o-Toluidine   Carbinol (colorless)

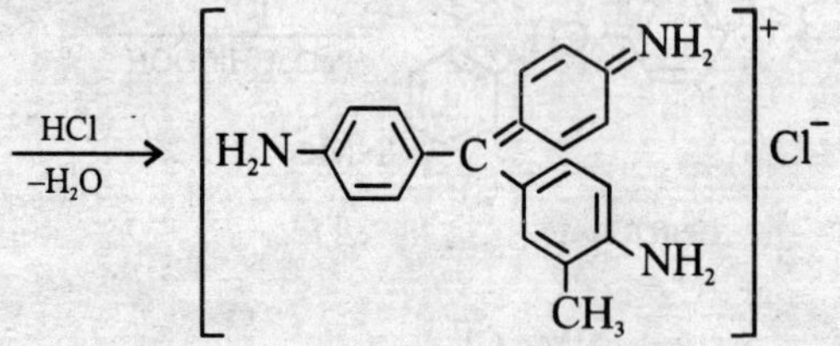

Rosaniline

The product obtained is a mixture of rosaniline and pararosaniline of which the former predominates. Rosaniline is soluble in water to give a deep red solution, which is decolorised by passing sulphur dioxide gas. The decolorised solution is known as Schiff's reagent, which is used as a reagent for testing aldehydes.

4. *Crystal Violet:* It is prepared by condensing Michler's ketone with N,N-dimethylaniline in presence of phosphorus oxychloride or phosgene. In case phosgene is used, crystal violet is directly prepared by heating phosgene and N,N-dimethylaniline. In the latter case Michler's ketone is first produced, which reacts further.

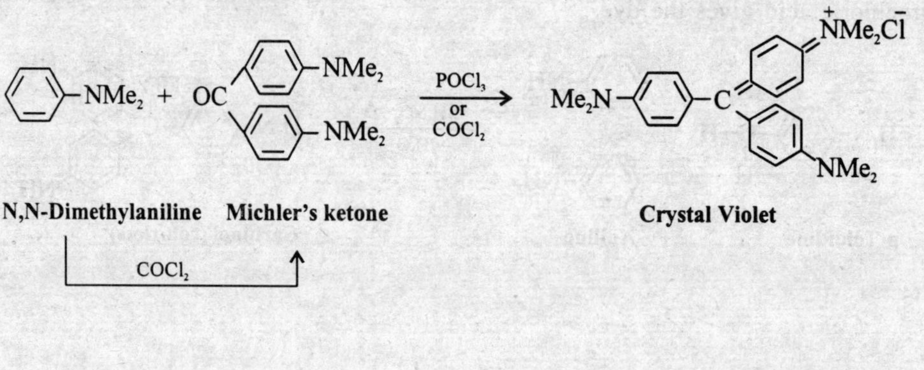

N,N-Dimethylaniline   Michler's ketone   Crystal Violet

It is soluble in water giving deep violet or purple color. A weakly acidic solution of crystal violet is purple. In strongly acidic solution the color is green and in still more strongly acid solution, the color is yellow.

Crystal violet is used in the manufacture of ink, stamping pads and typewriter ribbons. It is also used as an indicator for the determination of hydrogen ion concentration.

45.2.6 Phthalein Dyes

These are obtained by condensing phenols with phthalic anhydride in presence of dehydrating agents like concentrared sulphuric acid, fused zinc chloride or anhydrous oxalic acid. Some important dyes of this class are given below:

1. *Phenolphthalein:* It is prepared by heating phthalic anhydride (1 mol) with phenol (2 mol) in presence of concentrated sulphuric acid.

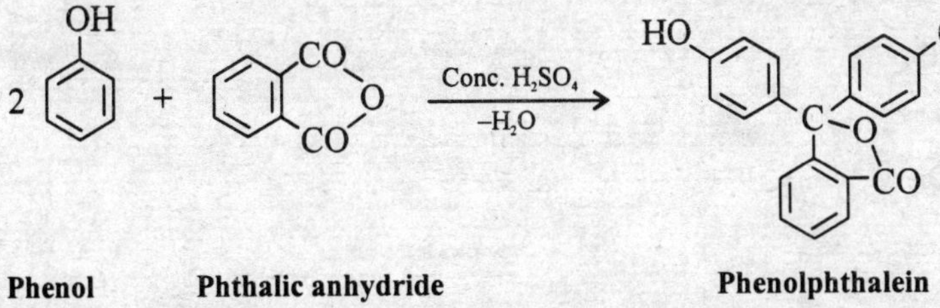

Phenol   **Phthalic anhydride**   **Phenolphthalein**

Phenolphthalein is a white crystalline solid and used as an indicator in acid-base titrations and not as a dye. With alkalis it gives deep red solution due to the formation of disodium salt(the ion is colored due to resonance). However, in presence of excess sodium hydroxide the solution becomes colourless due to formation of trisodium salt resulting in the loss of quinoid structure.

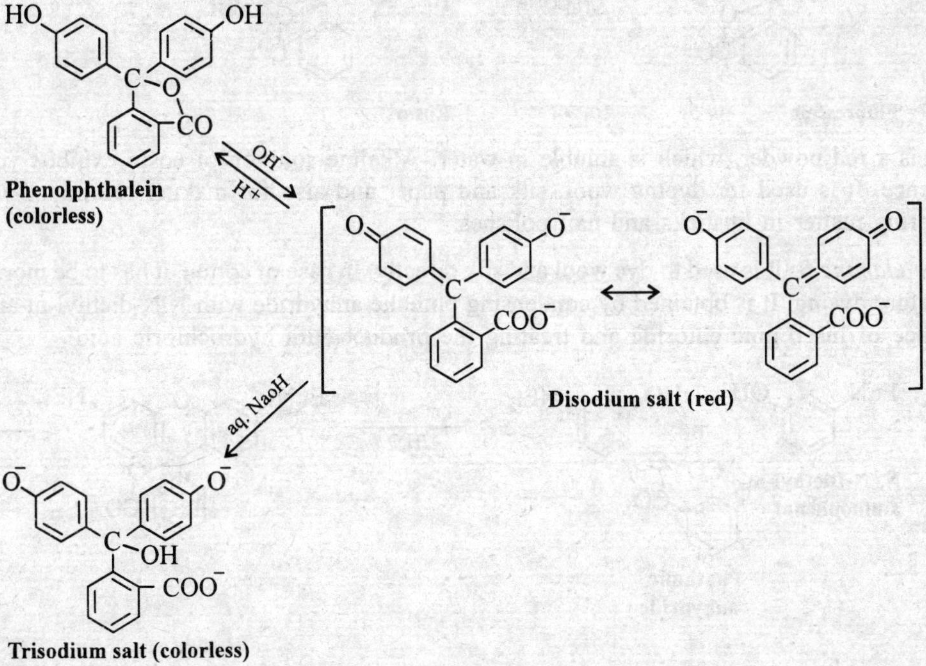

Phenolphthalein (colorless)

Disodium salt (red)

Trisodium salt (colorless)

2. *Fluorescein:* It is prepared by heating phthalic anhydride (1 mol) with resorcinol (2 mol) and fused zinc chloride at 190-200°.

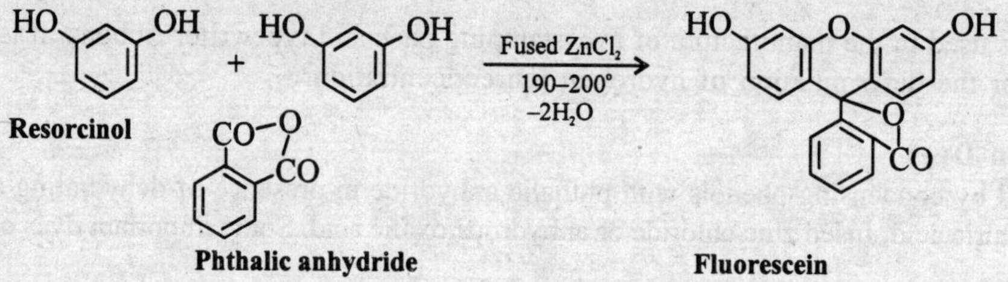

Resorcinol + Phthalic anhydride → (Fused ZnCl₂, 190–200°, –2H₂O) → Fluorescein

Fluorescein is not used as a dye. It is a red powder, insoluble in water and soluble in sodium hydroxide solution giving a strong yellow green flourescence. The sodium salt of fluorescein is called *uranine* and is used to dye wool and silk yellow.

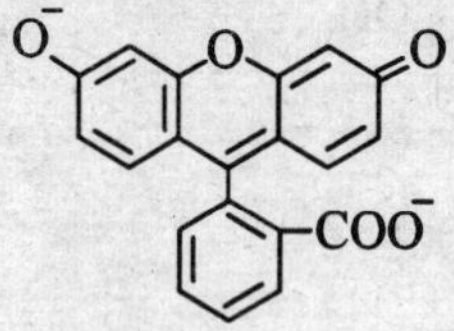

Uranine

3. *Eosin:* It is tetrabromofluorescein and is prepared by brominating fluorescein with bromine in glacial acetic acid.

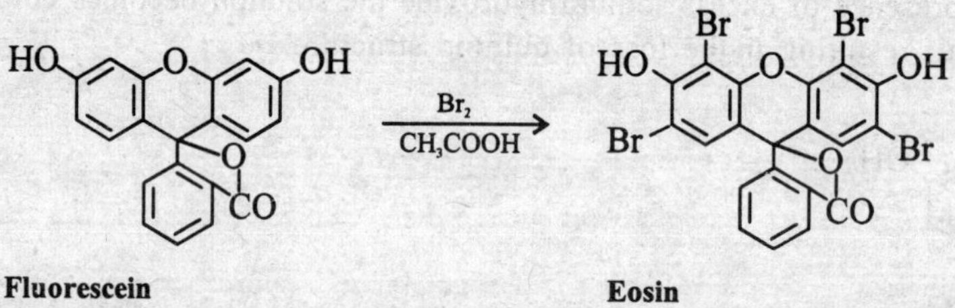

Fluorescein → (Br₂, CH₃COOH) → Eosin

Eosin is a red powder, which is soluble in water. Alkaline solution of eosin exhibits yellow green fluorescence. It is used for dyeing wool, silk and paper and also for making red ink. It is also used as a coloring matter in lipsticks and nail polishes.

4. *Rhodamine B:* It is used to dye wool and silk directly. In case of cotton, it has to be mordanted with tannin before dyeing. It is obtained by condensing phthalic anhydride with N,N-diethyl-m-aminophenol in presence of fused zinc chloride and treating the product with hydrochloric acid.

N, N-Diethyl-m aminophenol + Phthalic anhydride → (ZnCl₂, Δ, –2H₂O) →

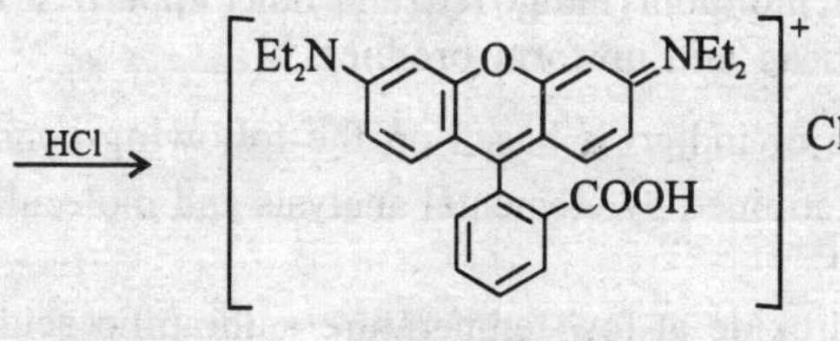

Rhodamine B

5. *Phthalocyanins:* The phthalocyanins are used as organic dyes and pigments. They are highly colored (green blue) and are fast to heat, light, acids and alkalis. They find application in the manufacture of paints, printing inks, plastics, fibres, rubber etc.

Phthalocyanins are prepared by passing ammonia into molten phthalic anhydride or phthalimide in presence of a metal salt. Alternatively, these can be prepared by heating o-cyanoarylamides or phthalonitriles with metals or metallic salts. The color of phthalocyanins depend on the metal (copper, magnesium, lead etc). The greenish shades are obtained by chlorination or bromination of phthalocyanins. The phthalocyanins are rendered water soluble by sulphonation and these soluble salts are used as direct dyes.

Copper Phthalocyanin: It is prepared by heating phthalonitrile with copper powder at 200°.

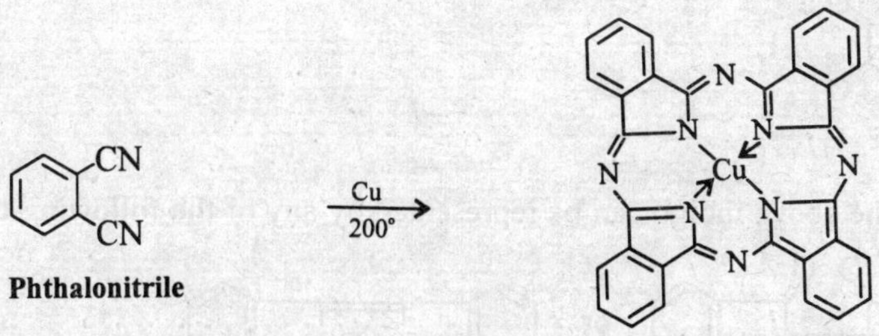

Phthalonitrile

Copper phthalocyanin

Copper phthalocyanin is rendered water soluble by sulphonation and is known as the *Alcian dye.*

45.2.7 Indigoid Dyes/Vat Dyes

Indigo: Indigo or indigotin is the oldest known organic dye, first prepared in India from the plants of Indigofera species about five thousand years ago. The leaves of the plants of indigofera group are soaked in water for several hours. The enzymes present in the plants cause fermentation and the indican (β-glucoside of indoxyl) present in the plant is converted into indoxyl and glucose. The indoxyl on exposure to air is oxidised to indigo.

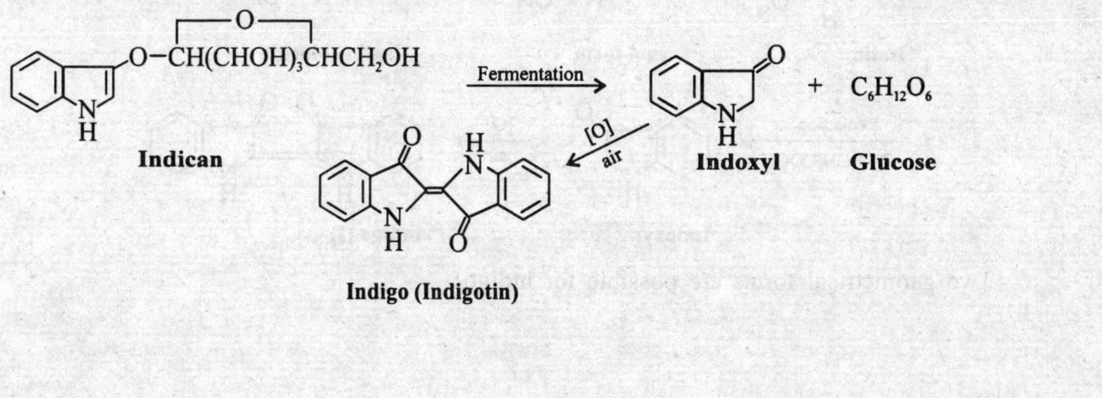

Indican

Indoxyl   **Glucose**

Indigo (Indigotin)

Natural indigo contains an isomer, indirubin (Indigo red) and other impurties. The indigo now a days use is synthetic as this gives pure, cheap and uniform product.

Structure of Indigo: The structure of indigo is based on the following consideration:

(i) Its molecular formula as determined by elemental analysis and molecular weight determination is $C_{16}H_{10}O_2N_2$.

(ii) On fusion with sodium hydroxide at low temperature anthranilic acid is obtained. This prove the presence of a benzene ring attached to one carbon atom and one nitrogen atom in orth position.

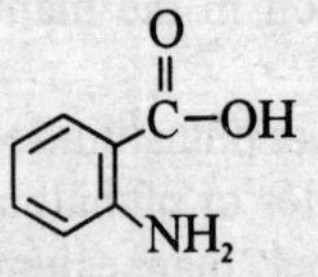

Anthranilic acid

(iii) Oxidation of indigo with nitric acid gives two molecules of isatin. This proves that indigo has tw identical units joined together.

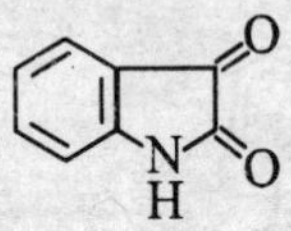

Isatin

On the basis of the above indigo can be represented by any of the following two structures (I or II).

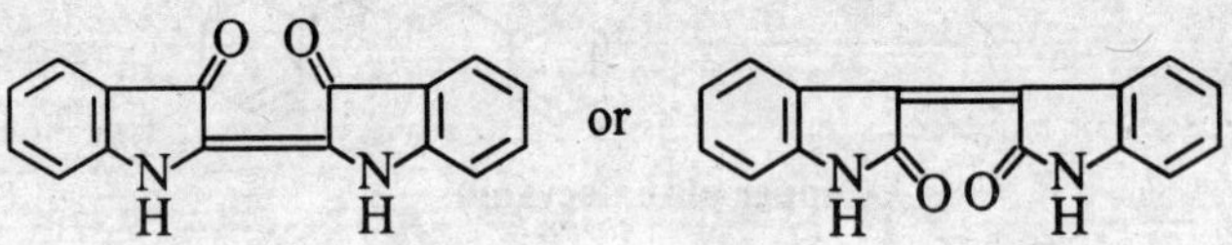

(iv) That the structure of indigo represented by I finds support from the fact that indoxyl on oxidatio by air yields indigo.

(v) The structure I has finally been confirmed by the following synthesis (Baeyer, 1872).

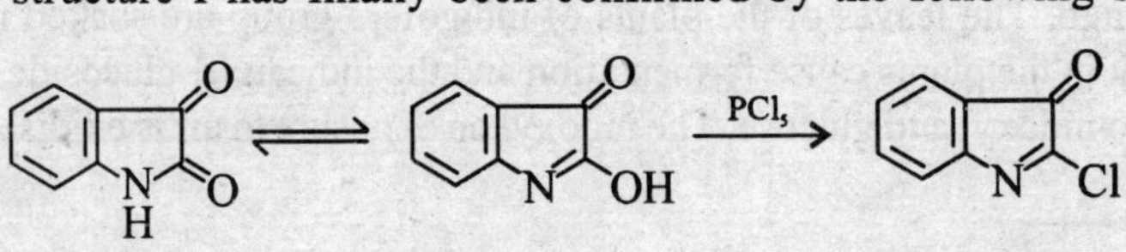

Isatin	enol form	Isatin chloride

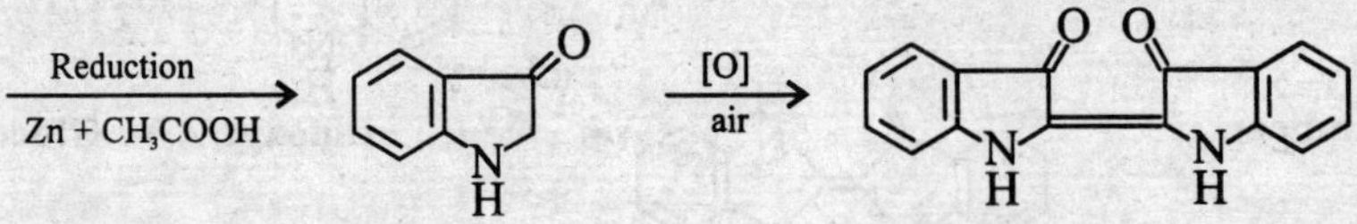

Indoxyl	Indigo (I)

Two geometrical forms are possible for indigo.

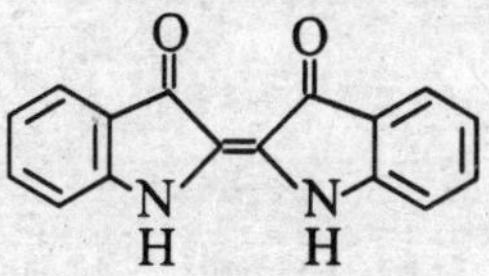

cis form 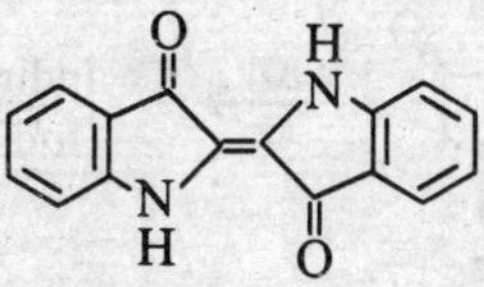   trans form

Though the derivatives of both the forms are known, the trans form is more stable. X-ray analysis shows that indigo exists mostly in the more stable trans form. The color of indigo is explained by its charged structure, which is stabilised by hydrogen bonding.

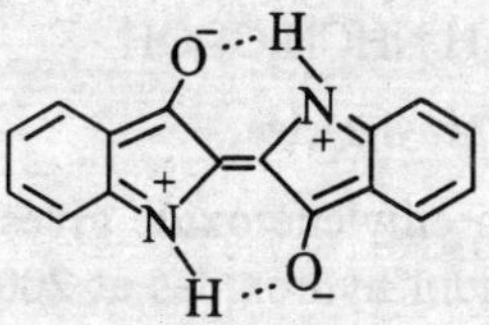

Charged structure

Synthesis of Indigo:

(i) *Heumann's Synthesis:* In this method anthranilic acid is heated with chloroacetic acid. The formed phenylglycine-o-carboxylic acid is heated with a mixture of potassium hydroxide and sodamide to give indoxylic acid, which on decarboxylation affords indoxyl. Final oxidation of indoxyl with air gives indigotin (indigo).

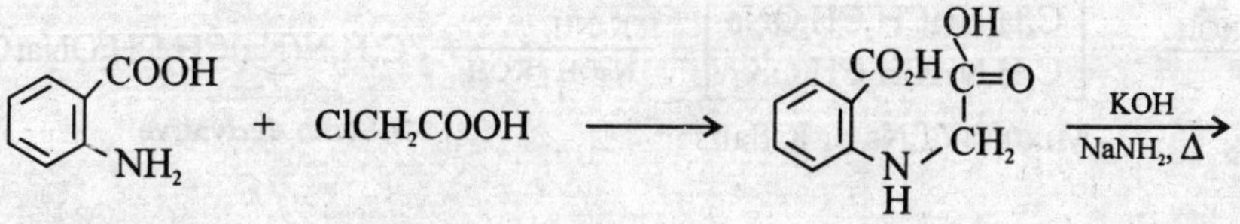

Anthranilic acid   Chloroacetic acid

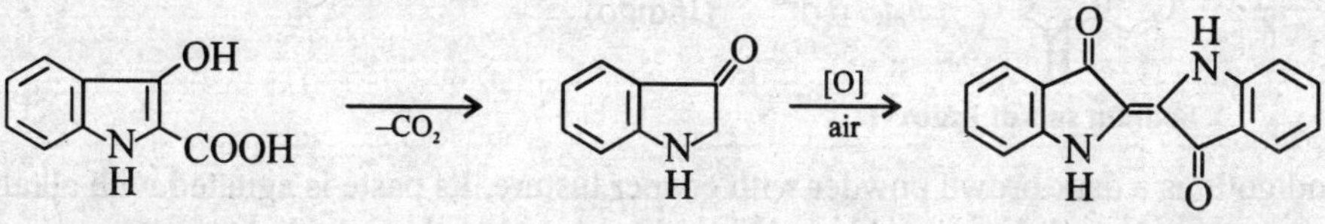

Indoxylic acid   Indoxyl   Indigotin (Indigo)

(ii) *From Phenylglycine:* Condensation of aniline with chloroacetic acid gave phenylglycine, which is converted into a mixture of sodium and potassium salts. The salts were subsequently fused with sodamide and a mixture of sodium and potassium hydroxide at 220-230°. The formed indoxyl is converted into indigotin by atomospheric oxygen.

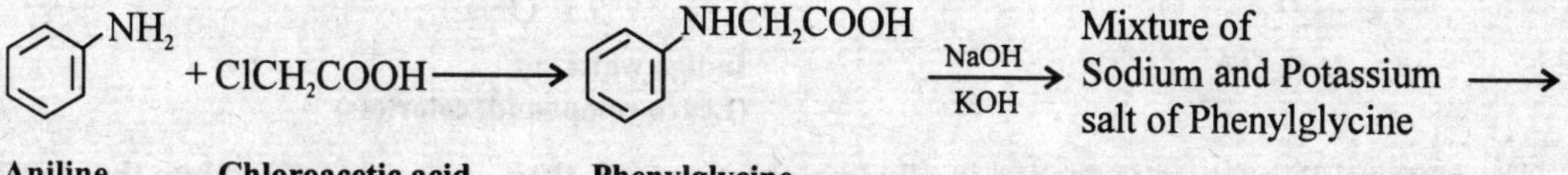

Aniline   Chloroacetic acid   Phenylglycine

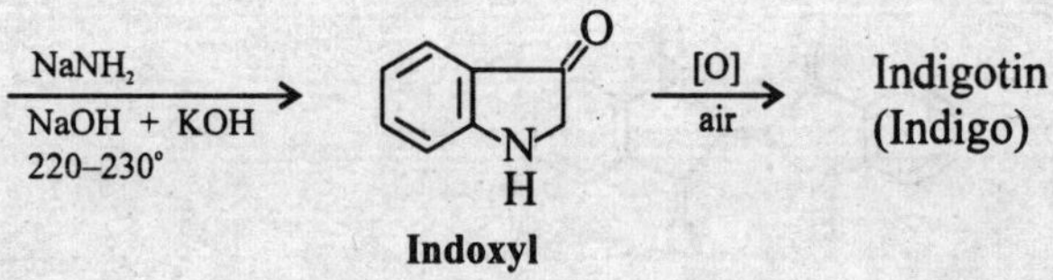

$$\xrightarrow[\substack{NaOH + KOH \\ 220–230°}]{NaNH_2}$$

Indoxyl

$\xrightarrow[air]{[O]}$ Indigotin (Indigo)

Indoxyl

The required phenylglycine is obtained conveniently from formaldehyde as follows:

$$CH_2O + NaHSO_3 \longrightarrow CH_2(OH)SO_3Na \xrightarrow[50–60°]{C_6H_5NH_2}$$

Formaldehyde　　　　　　　　　**Bisulphite addition product**

$$C_6H_5NHCH_2SO_3Na \xrightarrow[-Na_2SO_3]{NaCN} C_6H_5NHCH_2CN \xrightarrow[H_2O,\,70°]{Hydrolysis} C_6H_5NHCH_2COOH$$

Phenylglycine

(iii) *From Aniline and Ethylene Oxide:* The reaction of aniline with ethylene oxide gives N-phenyl-2-hydroxyethylamine. Its fusion with a mixture of sodium and potassium hydroxides at 200° gives the sodium and potassium alkoxide, which on heating with sodamide and a mixture of sodium and potassium hydroxide gave the corresponding N-sodio derivative. It dehydrogenates on heating to 300° and on cooling to 230° ring closure takes place giving the sodium salt of indoxyl. Its solution in water on exposure to air forms indigotin.

$$C_6H_5NH_2 + H_2C\overset{O}{-}CH_2 \longrightarrow C_6H_5NHCH_2CH_2OH$$

Aniline　　　　　　**Ethylene oxide**　　　**N-Phenyl-2-hydroxyethylamine**

$$\xrightarrow[220°]{NaOH / KOH} \begin{bmatrix} C_6H_5NHCH_2CH_2ONa \\ C_6H_5NHCH_2CH_2OK \end{bmatrix} \xrightarrow[NaOH / KOH]{NaNH_2} C_6H_5N(Na)CH_2CH_2ONa(OK)$$

Mixture of Na or K Salts　　　　　　　**N-Sodio derivative**

$$\xrightarrow[2.\,230°]{1.\,300°}$$

ONa

$\xrightarrow[air / H_2O]{[O]}$ Indigotin (Indigo)

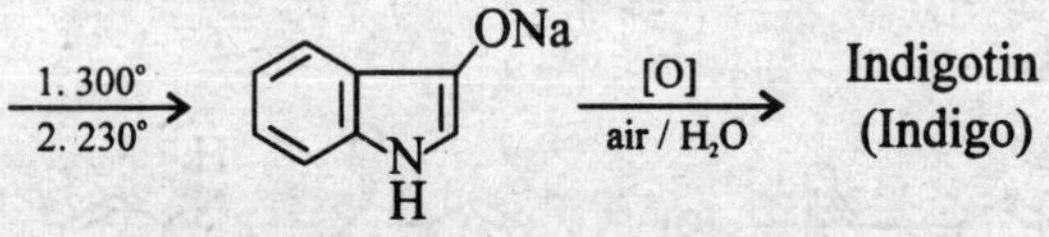

Sodium salt of Indoxyl

Vat Dyeing: Indigotin is a dark brown powder with copper lusture. Its paste is agitated with alkaline sodium hydrosulphite in large vats and the insoluble indigotin is reduced to the soluble leuco compound, indigotin white.

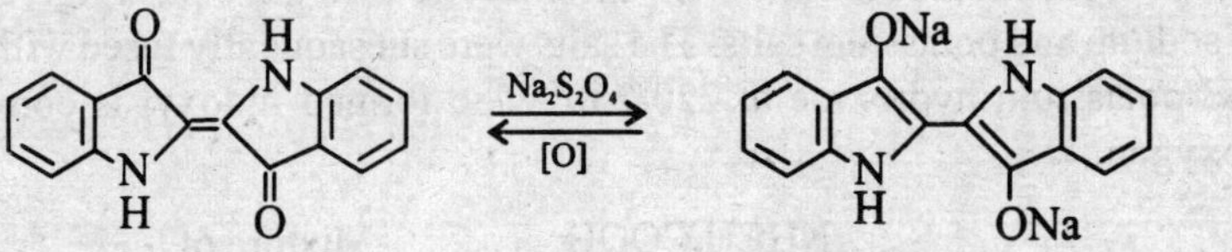

$$\underset{\textbf{Indigotin}}{\text{[structure]}} \underset{[O]}{\overset{Na_2S_2O_4}{\rightleftharpoons}} \underset{\substack{\textbf{Indigo white} \\ \textbf{(Leuco compound, colorless)}}}{\text{[structure]}}$$

The material to be dyed is soaked in alkaline solution and then exposed to air, when the original blue dye is regenerated on the cloth.

Since the dyeing with indigotin was carried out in wooden vats it is known as Vat dyeing.

Derivatives of Indigotin: Indigotin can be easily brominated or chlorinated to give 5,5,7,7-tetra substituted derivatives. The tetrabromo derivative, *Brilliant Indigo 2B* and tetrachloro derivative, *Brilliant Indigo B* are used commercially.

Tyrian Purple: The 6,6′-dibromoderivative of indigotin is known as Tyrian purple and was known in 1600 BC. It occurs naturally in the purple snail, *Murex brandaris*. It is not commercially availble, since much cheaper dyes of similar color are available.

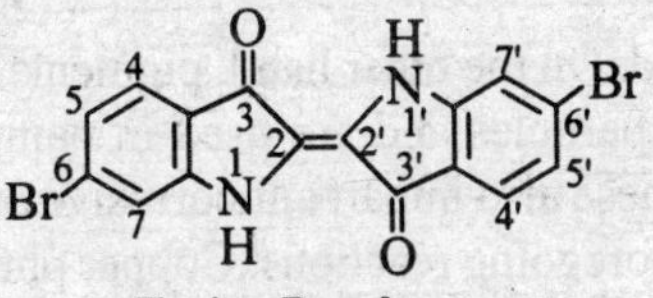

Tyrian Purple

Thioindigo: If the two -NH- groups of indigotin are replaced by sulphur, thioindigo is obtained. It is used to dye cotton, wool and polyester and is prepared from thiosalicylic acid (which is obtained from anthranilic acid) following the steps as in the case of indigotin.

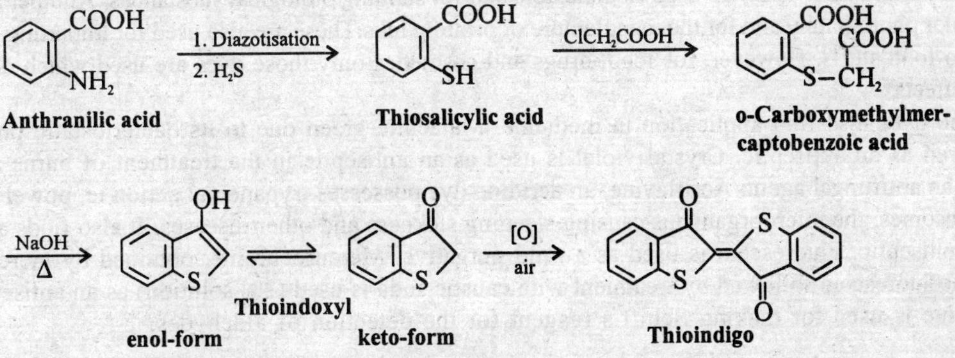

Anthranilic acid   **Thiosalicylic acid**   **o-Carboxymethylmer-**
                                                              captobenzoic acid

Thioindoxyl
enol-form   keto-form   **Thioindigo**

In thioindigo, though two geometrical isomers are possible, neither of these are stable due to very weak hydrogen bonding and in solution both forms exist in equilibrium.

45.2.8   Anthraquinoid Dyes

Alizarin: It is a mordant dye and the color of the lake depends on the metal used. Aluminum gives a red lake, iron (ferric) a violet lake and chromium a brown violet lake. Aluminum and iron lakes are used for dyeing cotton. Aluminum and chromium lakes are used for dyeing wool.

Alizarin is 1,2-dihydroxyanthraquinone and was first isolated from maddar roots. It is now manufactured from anthraquinone by sulphonation at high temperature and fusing the formed anthraquinone-2-sulphonic acid with sodium hydroxide and potassium chlorate at about 200° under pressure. Finally the disodium salt is acidified to yield alizarin.

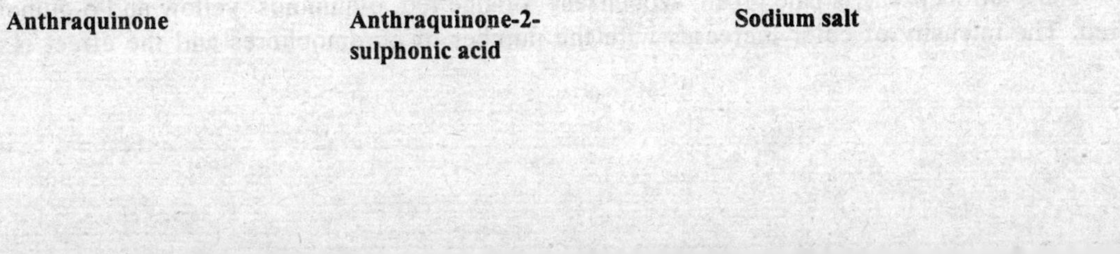

Anthraquinone   **Anthraquinone-2-**   **Sodium salt**
                           sulphonic acid

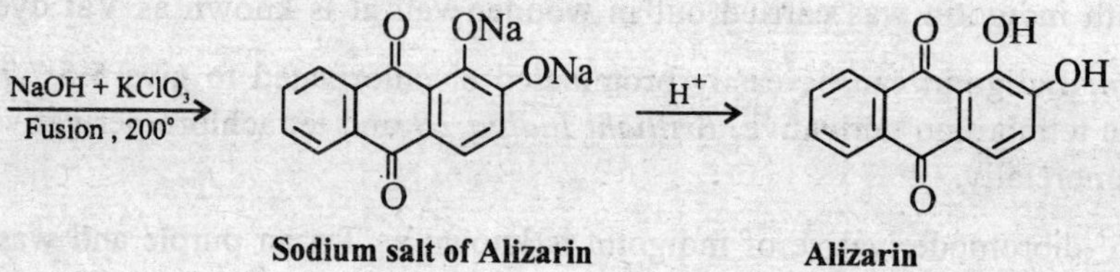

$$\xrightarrow[\text{Fusion , 200°}]{\text{NaOH + KClO}_3}$$

Sodium salt of Alizarin

$$\xrightarrow{\text{H}^+}$$

Alizarin

45.3 DYES Vs PIGMENTS

The dyes are actually absorbed by the material to be colored or dyed. On the other hand, pigments are applied to the surface. The pigments are insoluble powders of very fine particles and are used in paints, plastics, rubber, printing inks etc for imparting color and other properties. These also impart anticorrosive properties in paints. A number of examples of dyes have been mentioned in the foregoing reactions. Copper phthalocyanin is an example of an organic pigment.

45.4 USES OF DYES

Besides finding application in dyeing, large quantities are used in paper industry. They are used in plastic and metal industries also. They are used as indicators and for staining biological substances. Another application is in color photography and for the manufacture of printing inks.These are also used for imparting the desired color to food stuffs. However, for food, drugs and cosmetics only those dyes are used which do not have toxic effects.

Some dyes also find application in medicine. Malachite green due to its bacteriostatic properties is employed as an antiseptic. Crystal violet is used as an antiseptic in the treatment of burns and ulcers and as an antifungal agent. Acriflavine, an acridine dye possesses trypanocial action ie, power to destroy tryponosomes, the microorganisms causing sleeping sickness and other diseases. It also finds application as an antiseptic. Fluorescein is used as a mild purgative. Mercurochrome, obtained by mercuration of dibromofluorescein followed by treatment with caustic soda is used (2% solution) as an antiseptic. Pararosaniline is used for making Schiff's reagent for the detection of aldehydes.

45.5 COLOR AND CONSTITUTION

Some organic compounds are colored while others are not. What is the relationship between the color of the compound and its structure. Basically, two theories have been put forwards to explain this phenomenon.

(i) *Witt's Theory of Color:* According to the German chemist Otto Witt (1876), the color in an organic compound is associated with the unsaturated groups commonly known as *chromophores*. Some examples of chromophores are given below:

Nitro   **Azoxy**   **Nitroso**   **Azo**

p-Quinoid   **o-Quinoid**

Thus, nitrobenzene is pale green, azobenzene orange red, p-quinoids yellow and o-quinoids orange red. The intensity of color increases with the number of chromophores and the effect is still more when chromophores are conjugated. The compound containing the chromophoric group is called *Chromogen*.

Certain unsaturated groups produce color only when they are present in conjugation with chromophores. Such groups are carbonyl, $-C=C-$, azomethine etc. Thus, acetone is colorless whereas biacetyl

$$(CH_3-\overset{\overset{O}{\|}}{C}-\overset{\overset{O}{\|}}{C}-CH_3)$$ is yellow. However, acetylacetone $(CH_3-\overset{\overset{O}{\|}}{C}-CH_2-\overset{\overset{O}{\|}}{C}-CH_3)$ is colorless, since the

carbonyl groups are not conjugated. The groups which do not produce color themselves but are able to intensify color in a molecule along with a chromophore are called *auxochromes* by Witt. Some examples of auxochromes are $-OH$, $-NH_2$, $-NHR$ and $-NR_2$. Thus, nitrophenols (containing $-OH$ group as auxochrome) and nitroanilines (containing $-NH_2$ group as auxochrome) are more intensely colored than nitrobenzene. The auxochromes are salt forming groups. They deepen the color of the chromogen and their presence is essential to make the chromogen a dye. Following examples will clarify the points.

$$O_2N-\text{⟨benzene⟩}-N=N-\text{⟨benzene⟩}-R$$

R=H	: Red Orange
R=OH	: Deep Yellow
R=NH₂	: Red-Yellow
R=N(CH₃)₂	: Deep Red

Witt's theory is also known as the *Chromophore-auxochrome theory.*

Two more terms were subsequently introduced in the Witts theory. These are *bathochromic and hypsochromic groups.* These groups bring about deepening or lightening of the color of a dye respectively.

(ii) *Modern Theory of Color:* Though the Witts theory of color is useful, but for a clear understanding of the relationship between color and constitution we must consider the nature of light and its interaction with the matter. In other words, we must understand how color is produced.

When white light (750-400 nm) falls on the substance the light may be totally reflected (in this case the substance appears white) or totally absorbed (in this case the substance appears black). However, if a certain portion of the light is absorbed and the rest is reflected, the substance will have the color of the reflected light. If only a single band is absorbed, the substance has the complementary color of the absorbed band. If a substance absorbs all visible light except one band, which it reflects, the substance will have the color of that reflected band.

Thus, a substance can appear blue if it absorbs all the visible spectrum except blue. Obviously no dye gives a pure shade since it does not reflect only one band of wave length, e.g. malachite green reflects green light but also to a smaller extent red blue and violet. The following table gives the color absorbed at a particular wave length and the visible (complementary) color.

Wave length (nm)	Color absorbed	Visible (Complementary color)
400-435	Violet	Yellow-green
435-480	Blue	Yellow
480-490	Green-blue	Orange
490-500	Blue-green	Red
500-560	Green	Purple
560-580	Yellow-green	Violet
580-595	Yellow	Blue
595-605	Orange	Green-blue
605-750	Red	Blue-green

It has already been stated that bathochromic group bring about deepening of the color. In dye chemistry, the deepending of the color   generally means the following changes:

yellow - orange - red - purple - blue - green - black

This is the effective order of the visible complementary color of the absorbed band (see table above). Thus, we may say that the color of one compound is deeper than the other as the wavelength of maximum absorption of the former is longer that of the latter. Since the visible color is the complementary color of the absorbed band, bathochromic groups are said to have a *red shift* and hypsochromic groups a *blue shift*.

There are two approaches to the modern theory of color; the valence bond and the molecular orbital approach. The present discussion is restricted only to the valence bond approach.

Valence Bond Theory: According to valence bond theory, the electron pairs of a molecule in the ground state are in a state of oscillation. When a molecule is placed in the path of a beam of light, a photon of appropriate energy is absorbed since the amplitude of oscillation is increased. The result is that the molecule is in excited state. The wavelength of the photon absorbed depends on the energy difference between the excited and ground states, the smaller is the difference the longer is the wavelength.

As an illustration, let us consider the case ethylene. According to valence bond theory, ethylene is regarded as a resonance hybrid of I and II.

$$CH_2 = CH_2 \longleftrightarrow \overset{+}{C}H_2 - \overset{-}{C}H_2$$
(I)   **(II)**

The structure I represents the ground state predominantly and II represents the excited state. The energy of photon required to excite ethylene is very high (i.e. its wavelength is short) since the energy difference between the two structures I and II is very large.

The valence bond theory further states that:

(i)   Resonance among charged structures lowers the energies of both the ground and the excited states.

(ii)   Charged structures contribute more to the excited state than to the ground state.

(iii) The larger the number of electrons involved in resonance, the smaller is the energy difference between the ground and the excited state.

(iv) The more extended the conjugation in a molecule and the greater the contribution of charge structure, the longer is the wavelength of the photon required to excite the molecules. Thus, ethylene, 1,3-butadine and 1,3,5-hexatriene absorb at 175nm, 217nm and 258nm respectively and are colorless.

$CH_2 = CH_2$ $CH_2 = CH - CH = CH_2$ $CH_2 = CH - CH = CH - CH = CH_2$
Ethylene   **1,3-Butadine**   **1,3,5-Hexatriene**
λ_{max} 175 nm λ_{max} 217 nm λ_{max} 258 nm

However, with large conjugation the molecule is colored. Thus, β-carotene containing eleven conjugated double bonds is orange in color.

β-Carotene

In case of fused ring systems, as the number of fused rings increase, the position of absorption approaches the visible region and the compound gets color. Thus, benzene, naphthalene and anthracene are colorless. However, naphthacene is yellow and pentacene are blue. Graphite, which is considered as a combination of fused rings in all directions is black, i.e. it absorbs all colours almost completely.

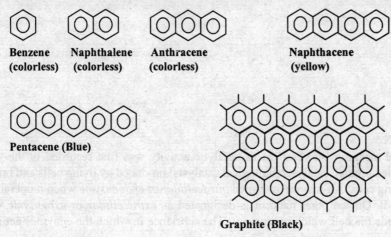

Benzene (colorless)   Naphthalene (colorless)   Anthracene (colorless)   Naphthacene (yellow)

Pentacene (Blue)

Graphite (Black)

45.6   PROBLEMS

1. What is a dye? What are the requirements from a compound to act as a dye?
2. How are dyes classified? Give two important examples of each type.
3. What are azodyes? How is methyl orange and orange II prepared?
4. Give the synthesis of malachite green, rosanilin and crystal violet.
5. How are phthalein dyes obtained? Fluorescein is an indicatior. How is it used as a dye?
6. What are vat dyes? How is the structure of indigo arrived at? Give one of its synthesis.
7. What are mordant dyes? Give one example. How is alizarin obtained synthetically?
8. What is difference between a dye and a pigment? Give some uses of dyes besides dyeing.
9. Give an account of color and constitution.
10. Give a brief account of the chemistry of triphenylmethane dyes.
11. Explain the terms: chromophores, auxochrome, chromogen, bathochromic shift and hypsochromic shift.
12. Write a note on valence bond approach to color.
13. Explain the use of phenolphthalein, methyl orange and Congo red as indicators.

46
Enzymes

46.1 INTRODUCTION

Enzyme is a Greek word meaning 'in yeast' as its catalytic activity was first recorded in the yeast cell juice. Enzymes are complex organic catalysts (biological catalysts) produced by living cells and bring about chemical reactions in living cells. An enzyme is termed *intracellular* or *endoenzyme* when it operates within the cell which produces it. On the other hand, it is designated as *extracellular* or *exoenzyme* when the site of its activity is outside the cell which produces it. The substance in which the enzymes act is termed as *substrate*.

The reactions in a biological system are in many respects similar to ordinary reactions in the laboratory. However, the basic difference between the two, is that the former takes place much faster than the latter and that these reactions occur at body temperature and in the physiological pH range (~ 7).

All enzyme are globular proteins. Some have been obtained in crystalline form.

46.2 CLASSIFICATION

There is no perfect method of nomenclature and classification of enzymes. Enzymes are generally named by adding the suffix *'ase'* to the substrate. For examples, *esterase* acts on esters (hydrolysis), *amylase* acts on starch (amylum), *urease* acts on urea, and *protease* acts on proteins. Some enzymes have retained their *trivial names* like emulsin, pepsin, trypsin etc. The names of some enzymes also denote the type of reaction with the substrate, for example, lactic acid dehydrogenase indicates that the enzyme catalyses the dehydrogenation of lactic acid.

The above nomenclature is still used. However, it has led to difficulties as more and more enzymes have been isolated. Due to these problems, the International Commission on Enzymes (1961) recommended a systematic method of nomenclature and classification of enzymes. According to this system, enzymes are divided into six main classes depending upon the nature of the reaction that it catalysed. Each class is further divided into several subclasses on the basis of the type of bond split, bond formed, functional group removed or transferred. The main classes and subclasses are indicated by index numbers. A third figure is also used for the subdivision of subclass(es). Finally, the fourth figure is used indicating the serial number of the specific enzyme within its own subclass. Thus, on the basis of the above, each enzyme is given a systematic code number known as Enzyme Commission (E.C.) number. For example, 1.1.1.1 is the enzyme code used for alcohol dehydrogenase. In this code, the first digit characterises the reaction type, the second subclass, the third sub-subclass and the fourth digit indicates the serial number of the enzyme of its sub-subclass.

The above system of nomenclature is complex, but is precise, descriptive and informative. Since these systematic names of enzymes are generally too long for ordinary use, the *trival names* are used. The trival names are derived from the names of the substrate, type of reaction catalysed and the suffix 'ase'. For example, alcohol dehydrogenase. Its systematic name is alcohol: NAD oxidoreductase and the Enzyme Commission (E.C.) has given 1.1.1.1. Following are given the major six classes of enzymes.

1. *Oxidoreductases:* The enzymes oxidoreductase catalyse oxidation-reduction reactions. This class include oxidases (direct oxidation with molecular oxygen) and dehydrogenases (which catalyse the removal of hydrogen from one substrate and pass it on to a second substrate).

2. *Transferases:* This group of enzymes catalyse the transfer of various functional groups from one substrate to another. For example, amino group (transamination), phosphate group (transphosphorylation) etc.

3. *Hydrolases:* These enzymes catalyse hydrolysis, i.e., direct addition of water molecule takes place across the bond, which is cleaved. For example, the hydrolysis of ester (esterases), ether (etherases), peptide (peptidases) glycosides (glycosidases) etc.

4. *Lyases:* These are of two types. The first type catalyses addition to double bond and the second type catalyses the elimination of groups to form the double bond.

5. *Isomerases:* This class includes enzymes which catalyse isomerisation, i.e., interconversion of optical, geometrical or position isomers. These enzymes are further subdivided depending on the reaction involved, e.g., recemases, intramolecular transferases etc.

6. *Ligases or Synthetases:* These enzymes catalyse the formation of a bond between two molecules. The process is accompanied by breaking down of a pyrophosphate bond of adenosine triphosphate (ATP) to adenosine diphosphate (ADP).

46.3 KINETICS OF ENZYME CATALYSED REACTIONS

Enzymes affect the rate of chemical reaction. The kinetics of enzyme catalysed reaction is the study of rate or velocity change from initial state to the final state of the reaction. Velocity is expressed in terms of change in the concentration of substrate or product per unit time. Fig. 46.1 shows the progress curves for an enzyme catalysed reaction.

The velocity of a reaction may be determined from the progress curve where product appearance is plotted against the time (Fig. 46.1). The slope of tangents to the curve yields the velocity at that time (V_o, the initial velocity and V_t, the velocity after time t). The figure also indicates that initial velocity increases with the increase in substrate concentration but reaches to a limit which is the characteristic of each enzyme.

46.3.1 Enzyme Unit

Several units are used to define velocity in terms of amount of enzyme present.

Enzyme Purity: It is expressed in terms of *specific activity* which is defined as micromoles of substrate converted to products per minute per milligram of enzyme protein. This indicates how fast 1 mg of an enzyme converts 1 μ mol of the substrate to product.

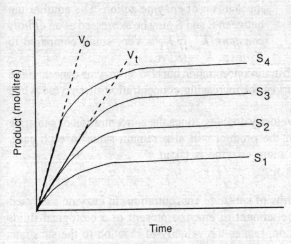

Fig. 46.1: An Enzyme Catalysed Reaction

Turnover Number: Turnover number expresses the rate of reaction in terms of moles of pure enzyme. When there are more than one catalytic centre per mole of enzyme, the term *Katal* is used. Katal is the number of micromoles of substrate converted to product per minute per micromole of enzyme active site. Katal clearly gives the comparison of relative catalytic activity between enzymes. For example, the enzyme catalase and Ó-amylase have enzyme activity as 5×10^6 and 2×10^4 Katals respectively, showing 2,500 times more activity of catalase compared to Ó-amylase.

Maximum Velocity: Maximum velocity, V_{max}, is the velocity obtained under the conditions of substrate saturation of enzyme in a given set of conditions for pH, temperature and ionic strength.

46.3.2   Enzyme and Substrate Interaction

We have already discussed that the initial velocity of an enzyme catalysed reaction depends on the concentration of the substrate (Fig. 46.1). As the concentration increases ($S_1 \rightarrow S_4$), the initial velocity increases till the enzyme is completely saturated with the substrate. The plot of initial velocity of product formation against the substrate concentration is obtained as a hyperbola (Fig. 46.2). This type of graph will be obtained for any process which involves interaction and binding of reactants at a specific but limiting number of

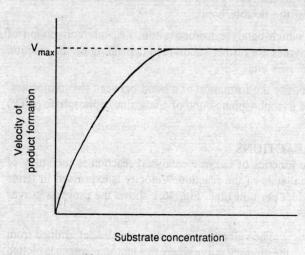

Fig. 46.2:   Velocity vs Substrate Plot for an Enzyme Catalysed Reaction

sites. The velocity of reaction reaches a limiting maximum (V_{max}) when all the available sites in enzyme are saturated.

In other words, we can represent the interaction of enzyme and substrate followed by product formation as given below:

$$E + S \underset{K_2}{\overset{K_1}{\rightleftharpoons}} ES \xrightarrow{K_3} E + P \quad .... \ 46.1$$

Where K_1 represents the rate constant for the formation of enzyme substrate (ES) complex, K_2 the rate constant for the disassociation of ES complex and K_3 the rate constant for the formation of product and enzyme. The above equation represents the general statement of the mechanism of enzyme action. The equilibrium between E and S may be expressed as an *affinity constant*, K_a. If K_3 is very small compared to K_2 then, $K_a = K_1/K_2$.

The velocity of enzyme action depends not only on substrate concentration but also on enzyme concentration as shown in Fig. 46.3, which shows the progress curves for increasing concentration of enzyme while the substrate concentration is kept constant.

As the enzyme concentration is doubled, the initial velocity doubles. Since the concentration of substrate remains same, the final equilibrium concentration of the product will also remain same in each case. However, equilibrium will be reached earlier where more enzyme is taken.

46.3.3   The Michaelis-Menten Equation

The velocity of a chemical reaction is expressed in terms of substrate concentration. In enzyme catalysed reactions, however, the velocity is correlated with the amount of enzyme present in a biological fluid. The rate equation, known as Michaelis-Menten equation, relates the velocity of reaction to the substrate concentration. Three basic assumptions are made to develop this equation.

1. ES complex is in steady state i.e., in the initial phase of reaction, concentration of ES complex remains constant.

2. Under saturation conditions all the enzyme is converted to ES complex.

3. If all the enzyme is in ES complex, rate of formation of product will be the maximum rate possible.

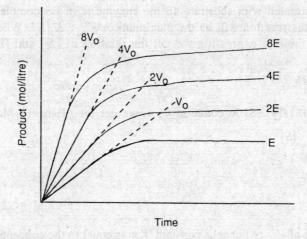

Fig. 46.3: Dependence of Enzyme Catalysed Reaction on Enzyme Concentration

In accordance with assumption 1 and equation 46.1 we can write the velocity of formation of ES and its breakdown as follows:

$$V_{formation} = K_1[S][E]$$
$$V_{breakdown} = K_2[ES] + K_3[ES]$$
$$= [ES](K_2 + K_3)$$

if rate of formation is equal to rate of breakdown,

$$K_1[S][E] = [ES](K_2 + K_3)$$

dividing both the sides by K_1, we get

$$[S][E] = [ES]\{(K_2 + K_3) / K_1\} \qquad \text{.... 46.2}$$

Defining ratio of rate constant $(K_2 + K_3)/K_1$ as Km, the Michaelis constant and substituting it in equation 46.2, we get:

$$[S][E] = [ES]Km \qquad \text{.... 46.3}$$

Since [E] is free enzyme and may be expressed in terms of its concentration as:

$$[E] = [E_t] - [ES]$$

Where E_t represents total enzyme. Substituting the value of [E] in the equation.

$$[S] ([E_t] - [ES]) = [ES]Km$$

dividing the equation by [S] and [ES] we get,

$$[E_t] - [ES] = \frac{[ES] Km}{[S]}$$

$$\frac{[E_t]}{[ES]} - 1 = \frac{Km}{[S]}$$

or,

$$\frac{[E_t]}{[ES]} = \frac{Km}{[S]} + 1 = \frac{Km + [S]}{[S]} \qquad \text{.... 46.4}$$

When the enzyme is saturated with substrate all the enzyme is in ES complex or $[E_t]$ = [ES], and the velocity observed for the reaction will be the maximum or $V_{max} = K_3[E_t]$. When $[E_t]$ is not equal to [ES], $V = K_3$ [ES]. From the two expression we get the relation of $[E_t]$ and [ES] as follows:

$$\frac{[E_t]}{[ES]} = \frac{V_{max}/K_3}{V/K_3} = \frac{V_{max}}{V}$$

Substituting the value of $[E_t]$/[ES] in equation 46.4, we get the Michaelis-Menten equation.

$$\frac{V_{max}}{V} = \frac{Km + [S]}{[S]}$$

or,

$$V = \frac{V_{max}\,[S]}{Km + S} \qquad\qquad ....\ 46.5$$

When V is half the value of V_{max}, Michaelis constant Km is equal to the concentration of the substrate.

$$\tfrac{1}{2}\,V_{max} = \frac{V_{max}\,[S]}{Km + S}$$

$$Km + [S] = 2[S] \text{ or } Km = [S]$$

The Michaelis-Menten constant is inversely proportional to the enzyme activity. A large value of Km indicates the requirement of high concentration of substrate as the enzyme has low affinity for the substrate.

In practice determination of Km from substrate saturation curve is not very accurate and it is determined by plotting the reciprocal of initial velocity vs the reciprocal of initial substrate concentration (Fig. 46.4). A straight line is obtained whose slope is Km/V_{max}, intercept on Y axis is $1/V_{max}$ and intercept on X axis is -1/Km.

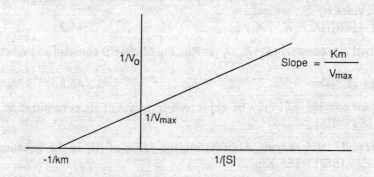

Fig. 46.4:   Lineweaver-Burk Plot

The Lineweaver Burk double reciprocal plot is also called the linear form of Michaelis-Menten equation and is used to determine the values of Km and V_{max}.

46.4   COFACTORS

Most of the enzymes require the presence of non-protein compounds in order to perform their catalytic action. These compounds are known as *Cofactors* or *activators* and belong to any of the following two groups.

(i) *Coenzymes*: It is well known that all enzymes (with no exception) are protein molecules. Some enzymes are simple proteins (i.e., their molecules contain only amino acids) and others are conjugated proteins. In conjugated proteins the non-protein part (non-amino acid part) is tightly bonded to protein molecule and is known as *prosthetic group*. The two fragments of same conjugated proteins (viz. protein and prosthetic group) can be separated by dialysis. Dialysis, however, results in the loss of catalytic activity of that enzyme, but the activity can be regained by mixing the two separated components. Thus, the dialyzable component is necessary for protein to act as an enzyme and this dialyzable material (prosthetic group) is termed coenzyme. The protein part of the conjugated protein is known as *apoenzyme* and the starting conjugated protein as *holoenzyme*.

$$\text{Conjugated protein} \rightleftharpoons \text{Protein} + \text{Prosthetic group}$$
(Haloenzyme)   **(Apoenzyme)**   **(Coenzyme)**

(ii) *Inorganic Ions*: Inorganic ions also act as cofactors. In some cases, the metal is tightly bonded to the enzyme which is referred to as *metalloenzyme*. In certain cases, the enzymes are 'metal activators'. The metal activators are univalent or bivalent metal cations like Na^+, K^+, Mg^{2+}, Zn^{2+}, Ca^{2+} etc.

The coenzymes (and prosthetic groups) act as carriers of specific atoms or functional groups. This is possible if these cofactors exist in two forms, one being converted into the other during a catalytic reaction and the latter is reconverted into the former by a coupled reaction.

It will be interesting to know some coenzymes which are nucleotides.

(a) *Nicotinamide-adenine Dinucleotide (NAD$^+$) (coenzyme I)*: It was earlier known as diphosphopyridine nucleotide (DPN). This coenzyme functions by accepting hydrogen atoms and electrons in presence of dehydrogenase and is itself reduced to *NADH*. Its structure is given below:

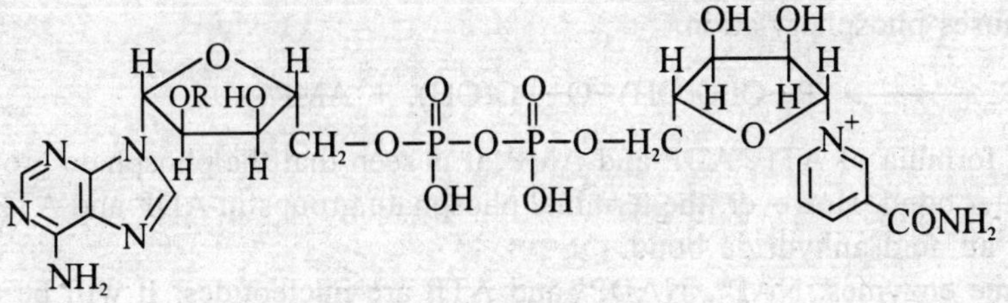

NAD^+ : R = H
$NADP^+$: R = PO_3H_2

In the above coenzyme, only the nicotinamide moiety is involved in H-transfer as shown below.

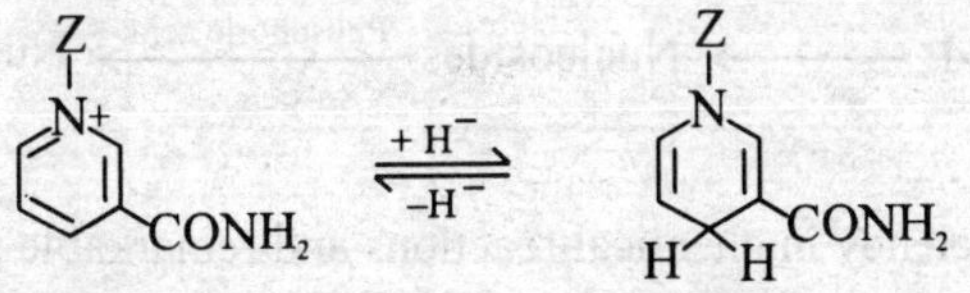

NAD$^+$ **NADH**

This coenzyme is involved in various degradative processes.

(b) *Nicotinamide-adenine Dinucleotide Phosphate (NADP) (coenzyme II)*: It was earlier known as triphosphopyridine nucleotide (TPN) and has the structure as shown above. Unlike NAD$^+$, this coenzyme (NADP$^+$) is involved in synthetic processes.

(c) *Adenosine Triphosphate (ATP):* It is involved in enzyme catalysed transphosphorylation reactions by transferring one phosphate group to the substrate and is itself converted into *adenosine diphosphate (ADP)*. The formed ADP can also transfer a phosphate group and is itself converted into *adenosine monophosphate (AMP)*

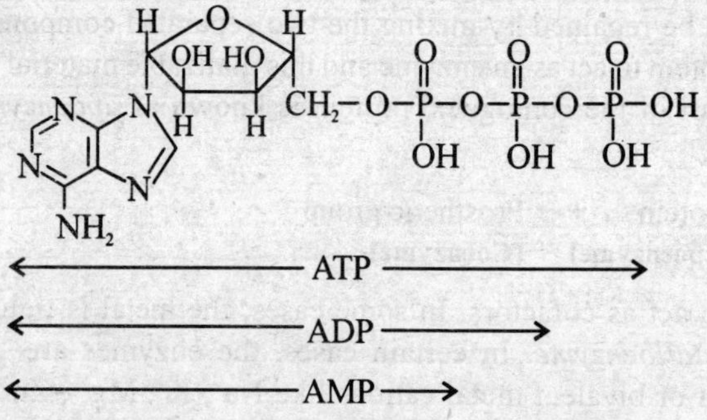

In the biosynthetic processes involved in transphosphorylation, the energy required to overcome the energy barrier is supplied by ATP in presence of a suitable enzyme.

$$R-OH + ATP \longrightarrow R-OPO(OH)_2 + ADP$$

The ADP in turn also behaves as a phosphorylating agent.

$$R-OH + ADP \longrightarrow R-OPO(OH)_2 + AMP$$

At times, ATP also causes phosphorylation.

$$R-OH + ATP \longrightarrow R-OPO(OH)-O-PO(OH)_2 + AMP$$

In the above structural formula of ATP, ADP and AMP, it is seen that the phosphate group in AMP is linked by the normal ester bond. However, the terminal phosphate groups in ADP and ATP are linked to a phosphate group by an acid anhydride bond.

As seen, the above three enzymes, NAD^+, $NADP^+$ and ATP are nucleotides. It will be appropriate to have the understanding of nucleotides (see Section 41.4). The nucleoproteins contain nucleic acid as the prosthetic group. The nucleic acids on hydrolysis yield nucleosides which are the combination of a base (purine or pyrimidine) and a sugar (ribose or deoxyribose). The combination of a nucleoside with phosphoric acid produces nucleotide, i.e., the nucleotides are nucleoside phosphates.

$$\text{Nucleoproteins} \longrightarrow \text{Nucleic acid} \longrightarrow \text{Nucleosides} \xrightarrow{\text{Phosphoric acid}} \text{Nucleotides}$$

46.5 PROPERTIES OF ENZYMES

Two important properties of enzymes are their efficiency in chemical reactions and remarkable specificity.

46.5.1 Enzyme Efficiency

A large number of rapid chemical conversions can be achieved by enzymes at comparatively low temperature. The efficiency of enzymes may be greater by as much as one hundred million times than the chemical catalysis.

46.5.2 Specificity of Enzymes

The most characteristic property of enzymes is their specificity of action. This specificity may be of three types:

(a) *Reaction Specificity:* An enzyme may catalyse a particular type of reaction. For example, esterase hydrolyses only esters. Such enzymes are called *reaction specific*. Some times, an enzyme may be specific to a particular compound or a class of compounds. Such enzymes are *substrate specific*. For example, urease hydrolyses only urea and phosphase hydrolyses only the phosphate esters.

(b) *Rate of Reactions:* The rate of a particular reaction depends on the enzyme use. Thus, esterases hydrolyse all esters, but different esters are hydrolysed at a different rate. For example, pepsin hydrolyses the peptide link but is most active for links in which the amino group is attached to an aromatic amino acid and carboxyl group is one of a dicarboxylic acid. In other words we say that the enzymes exhibit *kinetic specificity*.

(c) *Stereospecificity:* Stereospecificity is the most important property of enzymes. For example, maltase hydrolyses α-glycosides but not β-glycosides. On the other hand, emulsin hydrolyses the β-glycosides but not the α-glycosides.

Another illustration is the dehydrogenation of succinic acid to give only fumaric acid (and no maleic acid) whereas the chemical dehydrogenation gives a mixture of fumaric acid and maleic acid.

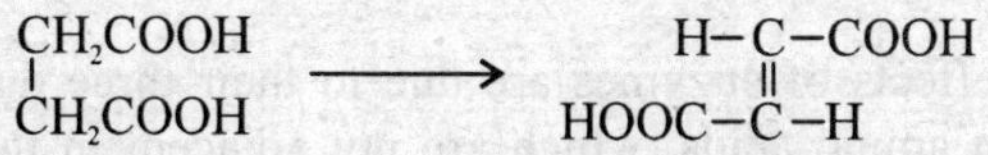

$$\begin{array}{ccc} \text{CH}_2\text{COOH} & & \text{H}-\text{C}-\text{COOH} \\ | & \longrightarrow & \| \\ \text{CH}_2\text{COOH} & & \text{HOOC}-\text{C}-\text{H} \\ \textbf{Succinic acid} & & \textbf{Fumaric acid} \end{array}$$

Number of examples are known in which a particular enzyme oxidises only one of the two optical isomers of D and L amino acids.

46.6 FACTORS INFLUENCING ENZYME ACTION

Following are given the factors that influence enzyme catalysed reactions.

(i) *Hydrogen Ion Concentration*: It has been shown that the enzyme activity is dependent on the pH of the solution. The enzymes are most effective over a narrow range of pH (between 5 and 9). The optimum pH is the characteristic for a particular enzyme and is determined experimentally. It has already been stated that at extreme pH, denaturation of proteins takes place. It is therefore reasonable to mention that the spatial arrangement of the molecular structure is responsible for enzymatic activity.

(ii) *Temperature*: It is well known that the enzymes can be denatured by heat. Thus, too high a temperature destroys the activity of the enzyme. Most of the enzymes have an optimum temperature range between 40-50°.

(iii) *Concentration of the Substrate and Enzyme*: The rate of enzyme catalysed reaction depends on the concentration of both the substrate and the enzyme. If the substrate is in excess, the rate is directly proportional to the concentration of the enzyme. However, if the enzyme concentration is kept constant, the rate increases as the substrate concentration increases.

46.7 MECHANISM OF ENZYME ACTION

A number of theories have been put forward to explain the mechanism of enzyme action. It is believed that the enzyme catalysed reactions proceed through a number of steps as shown below:

$$\text{E} + \text{S} \rightleftharpoons \text{ES} \rightleftharpoons \text{EP} \rightleftharpoons \text{E} + \text{P}$$

Where E is the enzyme (together with its cofactor), S is the substrate and P is the product obtained. The existence of the intermediates has been established by isolation (in several cases), spectroscopic studies and isotopic labeling studies.

In the formation of enzyme substrate complex (ES), the substrate (S) molecule attaches itself to certain specific points on the enzyme (E) molecules. These specific points on enzyme molecules are known as *active sites*. An enzyme may have one or more active sites. When all the active sites have been occupied, the enzyme becomes saturated and no further rate increase is possible.

The enzymes, like chemical catalysts lower the energy of activation (E_{act}) of the reactions which they catalyse. However, the enzymes are far more efficient than the chemical catalysts. This implies that the enzymes lower the energy of activation to a much greater extent. For example, in the decomposition of hydrogen peroxides:

$$H_2O_2 \xrightarrow{\text{Catalyst}} H_2O + 1/2\ O_2$$

E_{act} is 50.2 KJ/mol when platinum is used as catalyst, and E_{act} is only 2.5 KJ/mol when an enzyme is used in the above reaction.

The mechanism by which there is considerable rate accelerations is uncertain. The most important factor is the binding of the reaction molecules to the enzyme which results in increased concentration of the reactant molecules. Also binding causes the reactant molecules to be correctly oriented and so the transition state is readily reached.

It is now well established that the catalytic effects of enzymes are due to their three dimensional structure. X-ray studies have shown that certain amino acids, which are not adjacent in the primary structure, are brought together by folding, thereby producing an active site. The mode of folding depends on the sequence of the amino acids (primary structure). This is one factor that is responsible for the specificity of enzyme action. This implies that there is a steric relationship between enzyme and the substrate. On the basis of the above, Fischer (1894) proposed the 'Lock and Key' theory to explain enzyme specificity. According to this theory, the geometry of the enzyme (the Lock) is complementary to the substrate (the key). As a result the latter (substrate) fits into the former (enzyme) as a key fits into the lock. The stereospecificity of an enzyme can be explained on the basis of lock and key theory in the following way, the optically active compound is bonded to an enzyme through three points (assumption) which fit with either D- or L-enantiomer but not with both. For example, if D-enantiomer fits, the L-enantiomer will not and vice versa.

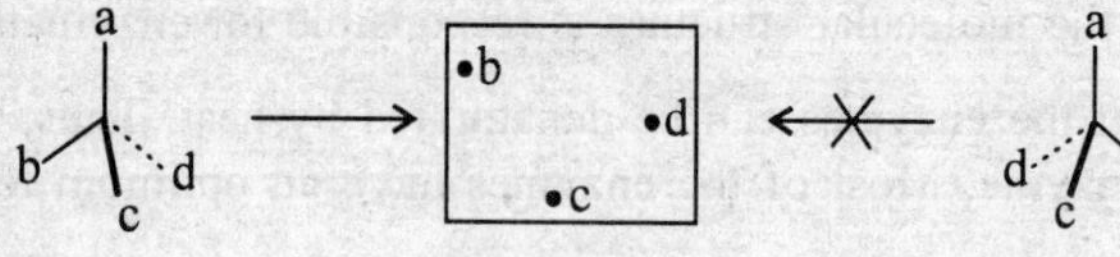

D-   Enzyme   L-

Similarly, the reduction of pyruvic acid to lactic acid can be explained. The pyruvic acid molecule fits into the enzyme in one way only and thus hydrogen transfer occurs to one face resulting in the formation of only one isomer of lactic acid.

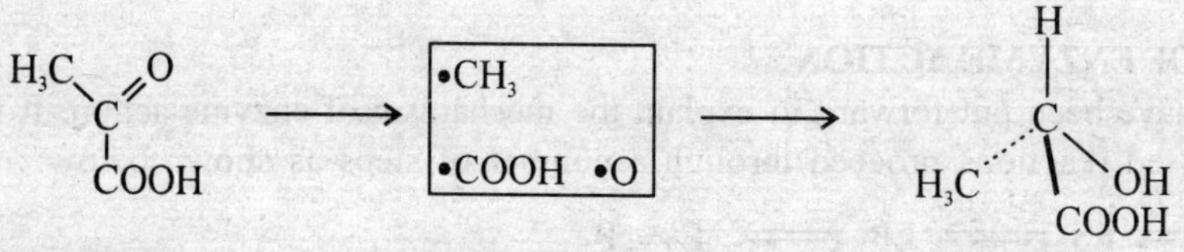

Pyruvic acid   Enzyme   Lactic acid

In the case of cofactor NAD^+ having a pair of enantiotropic faces accepts a hydride ion to give NADH containing enantiotropic hydrogen at position 4.

It has been shown that the NAD^+ - enzyme complex is usually stereospecific, only one hydrogen (Ha or Hb) reacts exclusively. Which hydrogen (Ha or Hb) from NADH is transferred depends on the nature of the enzyme.

46.8 ENZYME INHIBITION
It has already been stated that changes in temperature and pH inhibit enzyme activity. This type of inhibition is known as *non-specific inhibition*. However, compounds which inhibit the activity of only specific enzymes are called *specific inhibitors*. The specific inhibitions are of two types, reversible and irreversible.

46.8.1 Reversible Inhibition
Reversible inhibition may be competitive or non-competitive.

(A) *Competitive Inhibitor*: If there is a close resemblance between inhibiting substance and a particular substrate, the two compete with each other for the active site of the enzyme. The inhibitor may remain bound to the enzyme and prevents the substrate molecule from approaching the active sites. This type of inhibitor is called a *competitive inhibitor*. If the concentration of the inhibitor is large, it will block all active sites on the enzyme surface resulting in complete inhibition. However, in presence of higher substrate concentration, the inhibitor molecules are displaced from active sites and the inhibition is reversed. A typical example of competitive inhibition is the inhibition of enzyme activity of succinic acid dehydrogenase by malonic acid, which resembles succinic acid in configuration.

(B) *Non-Competitive Inhibition*: This type of inhibition cannot be reversed by increasing the concentration of the substrate. Here, the inhibitor binds at the active site on the enzyme other than the substrate binding site. Non competitive inhibitors can bind to free enzyme or to enzyme substrate complex or both.

Heavy metal ions such as Cu^{2+}, Hg^{2+} and Ag^+ or their derivatives act as non-competitive inhibitors of enzymes having -SH group at active sites. These inhibit the enzymes by forming mercaptides from -SH groups.

EDTA (ethylenediamine tetracetic acid) reversibly bind Mg^{2+} and other divalent cations and thus inhibits enzymes requiring these ions for activity.

46.8.2 Irreversible Inhibition
In irreversible inhibition, the inhibitor forms a stable enzyme-inhibitor complex via a covalent bond. If sufficient concentration of inhibitor is present, the catalytic effect of enzyme towards its normal substrate is lost.

For example alkylating agent like iodoacetamide has affinity for -SH group and react with -SH functional group of enzymes irreversibly.

$$Enzyme-SH + ICH_2CONH_2 \longrightarrow Enzyme-S-CH_2CONH_2 + HI$$

46.9　FERMENTATION

The decomposition of an organic substance into simpler compounds under the influence of micro-organisms (enzymes) is termed fermentation. The best known example of fermentation is the alcoholic fermentation of sugars by yeast. The sugars include glucose, fructose, mannose etc. The chemical changes involved in the process are given below.

$$C_6H_{12}O_6 \xrightarrow{\text{Yeast}} 2\ C_2H_5OH + 2CO_2 + \text{Energy}$$
Glucose　　　　　　　Ethyl alcohol

Besides ethyl alcohol, small amounts of acetaldehyde, glycerol, pyruvic acid, higher alcohols like amyl alcohol, succinic acid and fusel oil are also obtained as by products.

46.9.1　Mechanism

Fermentation is a complex process and involves a series of reactions, which are catalysed by specific enzyme-coenzyme combinations. The reactions are described below.

D-glucose is first converted into glucose-6-phosphate (by a hexakinase catalysed phosphorylation with ATP), which is isomerised into fructose-6-phosphate, (A) by phosphoglucoisomerase catalyst.

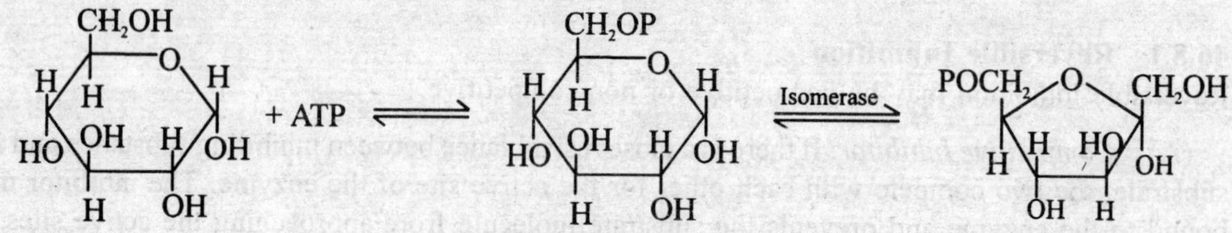

D-Glucose　　　　+ ATP　⇌　　Glucose-6-phosphate + ADP　⇌(Isomerase)　Fructose-6-phosphate (A)

The fructose-6-phosphate (A) is phosphorylated with phosphofructokinase to give fructose-1,6-diphosphate, which undergoes aldolase catalysed fragmentation forming a mixture of glyceraldehyde-3-phosphate and dihydroxyacetone phosphate.

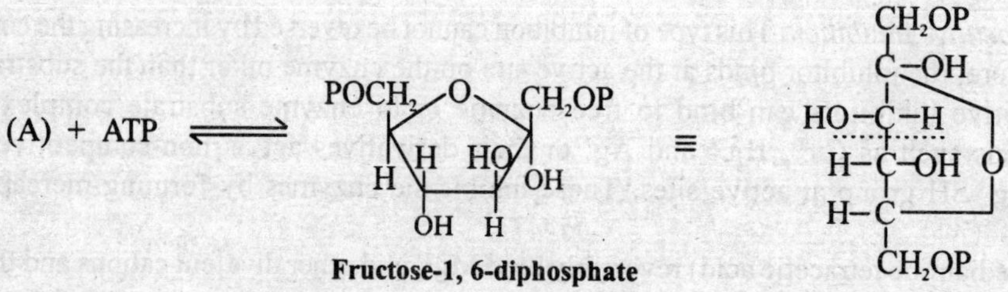

(A) + ATP　⇌　　Fructose-1, 6-diphosphate

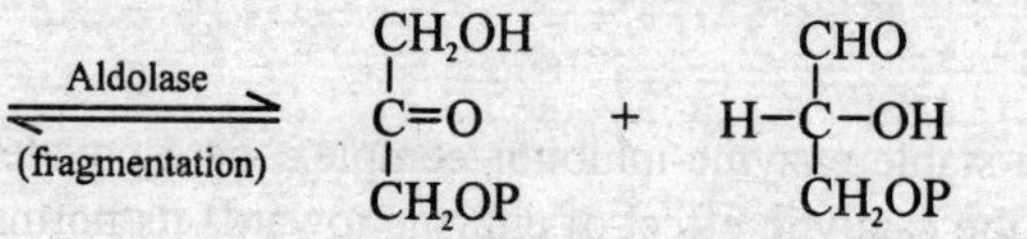

Aldolase (fragmentation) ⇌

CH₂OH | C=O | CH₂OP　Dihydroxy-acetone phosphate
+
CHO | H-C-OH | CH₂OP　3-Phosphogly-ceraldehyde (B)

Phosphate Isomerase

In the above sequence of reaction, only half the glucose is converted into the desired intermediate, i.e., glyceraldehyde-3-phosphate (B). The other byproduct, dihydroxyacetone phosphate is converted into the desired intermediate (B) by phosphate isomerase. The formed intermediate (B) is converted into 1,3-diphosphoglycerate (C) by NAD$^+$ in presence of inorganic phosphate. The reaction is catalysed by the enzyme glyceraldehyde-3-phosphatedehydrogenase. The 1,3-diphosphoglycerate (C) is converted into 3-phosphoglycerate (D).The transfer of one phosphate of (C) is effected by ADP in the presence of the enzyme phosphoglycerylkinase.

$$(B) \quad + NAD^+ + HPO_4^{2-} \rightleftharpoons$$

COOP
|
H——OH
|
CH$_2$OP

1,3-Diphospho-
glycerate
(C)

$\xrightarrow{ADP}$

COO$^-$
|
H——OH + ATP
|
CH$_2$OP

3-Phospho
glycerate
(D)

+ NADH + H$^+$

Intramolecular phosphorylation of 3-phosphoglycerate (D) in presence of phosphoglycerylmutase gives 2-phosphoglycerate (E), which on dehydration by the enzyme enolase gives 2-phosphoenylpyruvate (F).

$$(D) \xrightarrow[\text{Phosphorylation}]{\text{Intermolecular}}$$

COO$^-$
|
H——OP
|
CH$_2$OP
(E)

$\xrightarrow[-H_2O]{\text{Enolase}}$

COO$^-$
|
COP
‖
CH$_2$
(F)

3-Phosphoglycerate   **2-Phosphoglycerate**   **2-Phosphoenylpyruvate**

The final step is the phosphate transfer from 2-phosphoenylpyruvate (F) to ADP by pyruvickinase to form pyruvate (G) and ATP.

$$(F) \quad + \quad ADP \xrightarrow[\text{Kinase}]{\text{Pyruvic}}$$

COO$^-$
|
C=O
|
CH$_3$
(G)

$\xrightarrow[\substack{H^+ \\ \text{reduction}}]{\text{NaDH}}$

COO$^-$
|
CHOH + NAD$^+$
|
CH$_3$
(H)

Pyruvate   **Lactate**

The conversion of glucose (or glycogen) into pyruvic acid (referred above) is known as *Glycolysis*. All the steps discussed above can be summarised in the following equation.

$$C_6H_{12}O_6 + 2\text{Phosphate} + 2ADP \longrightarrow 2CH_3CHOHCOO^- + 2ATP + 2H_2O$$

As seen, the degradation of one mole of glucose produces 2 moles of lactate accompanied by phosphorylation of 2 moles of ADP to give 2 moles of energy rich ATP. After pyruvic acid is obtained, it can either be reduced to lactic acid (H). (Carbohydrate metabolism, see above) or can be converted into ethyl alcohol by the yeast via the formation of acetaldehyde (alcoholic fermentation).

COO$^-$
|
C=O
|
CH$_3$
(G) Pyruvate

$\xrightarrow[\substack{\text{Carboxylase} \\ -CO_2}]{\text{Yeast}}$

CHO
|
CH$_3$
Acetaldehyde

$\xrightarrow[\text{NADH}]{\text{Reduction}}$

CH$_2$OH
|
CH$_3$
Ethyl alcolol

46.10 KREB'S CYCLE

The oxidation of pyruvic acid to terminal products CO_2 and H_2O is accompanied by release of energy. Hans Kreb postulated that the above transformation is explained by a complicated sequence of reactions called citric acid cycle or *tricarboxylic acid cycle or Kreb's cycle*.

The Kreb's cycle is initiated by the transformation of pyruvate into acetyl-CoA which is catalysed by an enzyme system known as pyruvate dehydrogenase.

$$CH_3COCOO^- \xrightarrow[\text{dehydrogenase}]{\text{Pyruvate}} CH_3CO-SCoA$$

Pyruvate   **Acetyl-CoA**

$$CH_3COCOOH + CO_2$$

COCOOH
|
CH_2COOH
Oxalacetic acid

CH_3CO—SCoA   CoA—SH

CH_2COOH
|
C(OH)COOH
|
CH_2COOH
Citric acid

NADH

NAD$^+$

CH(OH)COOH
|
CH_2COOH
Malic acid

–H_2O

HC—COOH
||
C—COOH
|
CH_2COOH
cis-Aconitic acid

+ H_2O

HC—COOH
||
HOOC—CH
Fumaric acid

+ H_2O

HC(OH)COOH
|
CHCOOH
|
CH_2COOH
Isocitric acid

FADH_2

–2H

FAD

CH_2COOH
|
CH_2COOH
Succinic acid

NAD$^+$

NADH

COCOOH
|
CHCOOH
|
CH_2COOH
Oxalsuccinic acid

CoA—SH

GTP

GDP

Phosphate

CO—SCoA
|
CH_2
|
CH_2COOH
Succinyl–S–CoA

CoA—SH

2H   CO_2

–CO_2

COCOOH
|
CH_2
|
CH_2COOH
α–Ketoglutaric acid

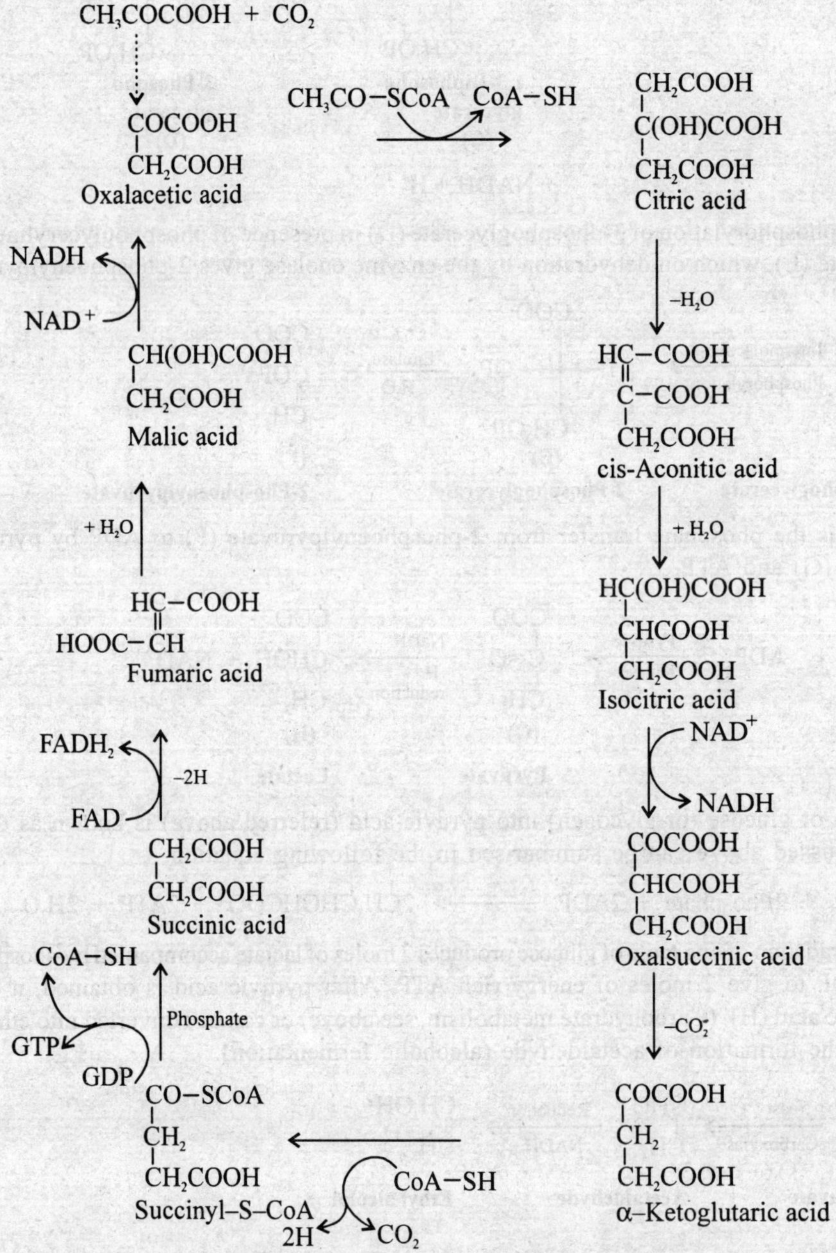

The acetyl-CoA condenses with oxalacetic acid (presumably derived from pyruvic acid and carbon dioxide) producing citric acid. The reaction is catalysed by the enzyme citrate synthase. The citric acid is isomerised by aconitase to isocitric acid via cis-aconitic acid. Subsequently, isocitric acid is oxidised to α-ketoglutaric acid and CO_2 via oxalsuccinic acid. The oxidation to α-ketoglutaric acid with α-ketoglutarate dehydrogenase in presence of CoA–SH gives succinyl-CoA, which on reaction with guanosine diphosphate (GDP) and phosphate gives succinic acid and guanosine triphosphate (GTP), the reaction being catalysed by succinicthiokinase. The succinic acid is oxidised to furmaric acid by a FAD-bound enzyme succinic dehydrogenase. The hydration of fumaric acid to L-malic acid is catalysed by fumarase. The tricarboxylic acid cycle is finally brought to completion by the oxidation of L-malic acid with NAD^+ linked L-maleatedehydrogenase producing oxalacetic acid, which starts a second cycle and so on.

As seen, the initial C–C bond formation between acetyl-CoA and oxalacetic acid initiates the Kreb's cycle. Further, in all the steps involved, there is an overall elimination of 2 molecules of carbon dioxide and about 14.3 Kcal/mol of energy is released.

46.11 PROBLEMS

1. What are enzymes. Define the term 'active site'.
2. Discuss the nomenclature and classification of enzymes.
3. What are cofactors? Discuss the different types of cofactors.
4. Give the structure of NAD^+, $NADP^+$ and ATP.
5. Give important properties of enzymes.
6. Discuss the factors that influence enzyme action.
7. Give mechanism of enzyme action. What do you understand by enzyme inhibition? Explain.
8. What is fermentation? Discuss mechanism of alcoholic fermentation.
9. Explain Kreb's cycle.

47

Drugs—Pharmaceutical Compounds

47.1 INTRODUCTION

A drug can be defined as a chemical substance used for the treatment or prevention of diseases in human beings or animals. Organic compounds, natural or synthetic, are the chief sources of drugs. Approximately 50% of the commonly used drugs come from chemical synthesis, where as 25% from higher flowering plants and about 10-12% from micro-organisms. Other sources include animals, minerals etc.

A drug may be a single chemical substance or a combination of two or more different substances.

47.2 CLASSIFICATION

The most common classification of drugs is based upon their mode of action. Following are given the major drug groups, their use and some common drugs of each type.

Drug Class and Uses	Common Drugs
Analgesics and Antipyretics for relieving pain and fever	Aspirin, Phenacetin, Paracetamol and Phenylbutazone
Gastro-Intestinal Agents for reducing excess stomach acids (antacids)	Sodium bicarbonate, Magnesium carbonate and Aluminum hydroxide
Sulpha Drugs as antibacterial agents	Sulphadiazine and Sulphaguanidine
Antibiotics as anti-infectious drugs	Penicillins, Ampicillin, Erythromycin, Streptomycin and Chloromycetin
Antimalarials	Pamaquine & Chloroquine
Antidepressants	Doxepin and Amitriptylin
Antihistaminics	Banadryl and Diphenylhydramine hydrochloride
Antiinflammatory Agents	Cortisone acetate and Prednisolone
Cardiovascular Drugs	Digitoxin
Diuretic Agents	Furosamide and Hydrochlorothiazide
CNS Stimulants: Sedatives and Hypnotics	Phenobarbital and Glutethimide
Tranquilizer	Methyldopa and Hydralazine
Antianxiety Agents	Meprobamate and Diazepam

In this chapter some important drugs belonging to the following classes will be discussed.

(a) Analgesics and antipyretics (b) Sulphonamides (sulpha drugs)
(c) Antimalarials (d) Antibiotics

47.3 ANALGESICS AND ANTIPYRETICS

Drugs that are used to decrease pain are known as analgesics. These act centrally to increase the capacity to tolerate pain without loss of consciousness. Drugs used to reduce body temperature or fever are referred to as antipyretics. Most of the analgesics have antipyretic effect and vice- versa. Some important examples are discussed in the following sections.

47.3.1 Aspirin/Acetylsalicylic Acid

It is the most common and widely used drug. It is used as an antipyretic, antiinflammatory and also as an analgesic drug in a variety of conditions ranging from headache, discomfort and fever associated with common cold, muscular pain and aches. It is recommended that aspirin can minimise the incidents of mycocardial infraction and also transient asthmatic attacks in small doses.

It is obtained by acetylation of salicylic acid with acetic anhydride in presence of one to two drops of concentrated sulphuric acid.

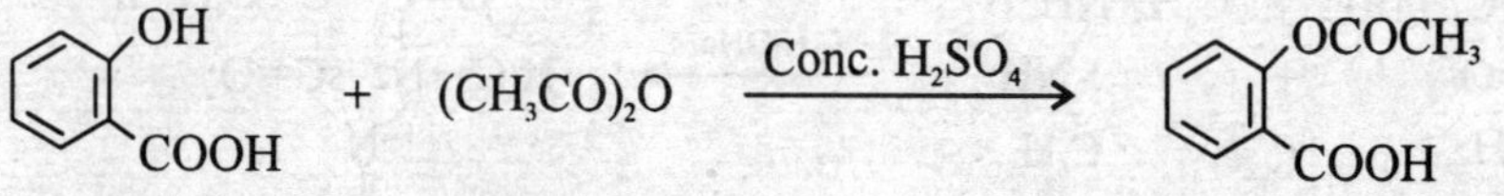

Salicylic acid **Acetic anhydride** **Aspirin**

Aspirin is also used for the treatment of rheumatism. The calcium salt of aspirin is known as soluble aspirin.

47.3.2 Phenacetin/p-Ethoxyacetanilide

It is an analgesic and antipyretic drug with effectiveness similar to that of aspirin. However, it is toxic and causes irreversible kidney damage and has been withdrawn in many countries. It is obtained from p-nitrophenol as shown below:

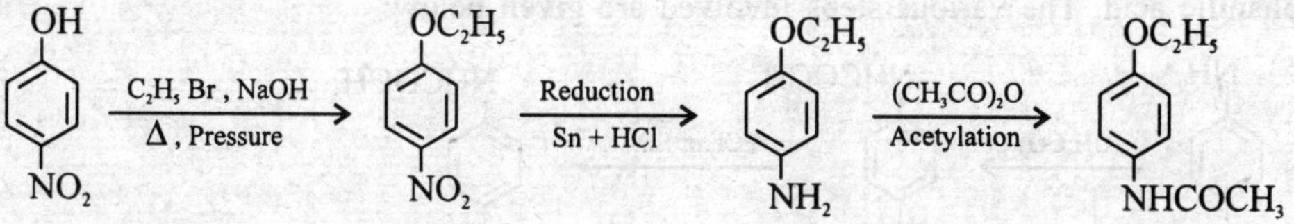

p-Nitrophenol **p-Nitrophenetol** **p-Phenetidine** **Phenacetin**

Phenacetin is also obtained from aniline by following sequence of reactions.

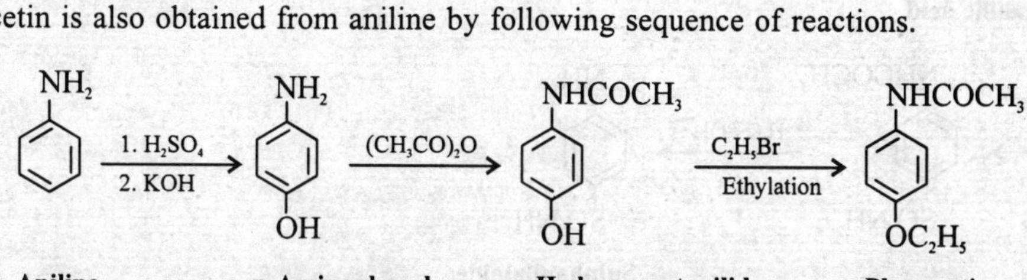

Aniline **p-Aminophenol** **p-Hydroxyacetanilide** **Phenacetin**

47.3.3 Paracetamol/p-Hydroxyacetanilide

It is used in broad spectrum of arthritic and rheumatic conditions linked with musculoskeletal pain, headache, neuralgias and dysmenorrhea and is particularly useful in aspirin sensitive patients. It is prepared from p-nitrophenol as shown below:

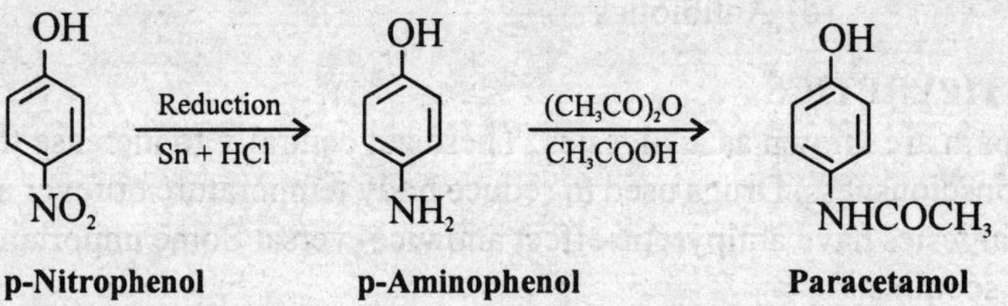

p-Nitrophenol   **p-Aminophenol**   **Paracetamol**

47.3.4 Phenylbutazone/4-Butyl-1,2-diphenyl-3,5-pyrazolinedione

It has antipyretic, analgesic and antiinflammatory actions. However, due to its toxicity, it is not used as a general antipyretic or analgesic. It is mostly used in the treatment or rheumatic disorders, such as rheumatoid arthritis, osteoarthrosis, where less toxic drugs have failed. It is prepared by the condensation of diethyl-n-butylmalonate with hydazobenzene in presence of sodium ethoxide.

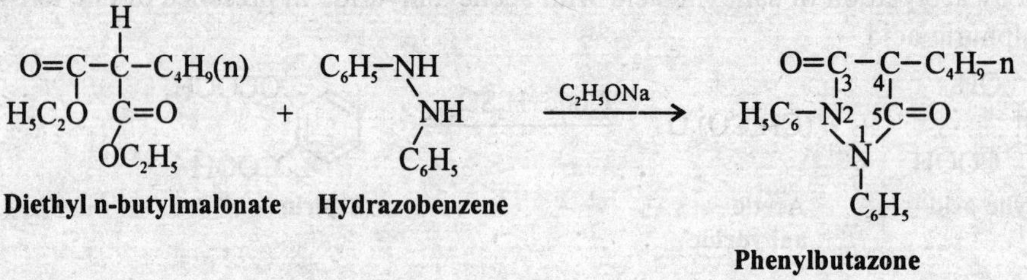

Diethyl n-butylmalonate   Hydrazobenzene

Phenylbutazone

47.4 SULPHA DRUGS/SULPHONAMIDES

Sulpha drugs were the first synthetic compounds found to be effective against pathogenic bacteria that cause pneumonia, tuberculosis etc. Some important sulpha drugs are described in following sections.

47.4.1 Sulphanilamide/p-Aminobenzenesulphonamide

It is the simplest and the first sulpha drug which was used for its bacteriostatic action. It is prepared from sulphanilic acid. The various steps involved are given below:

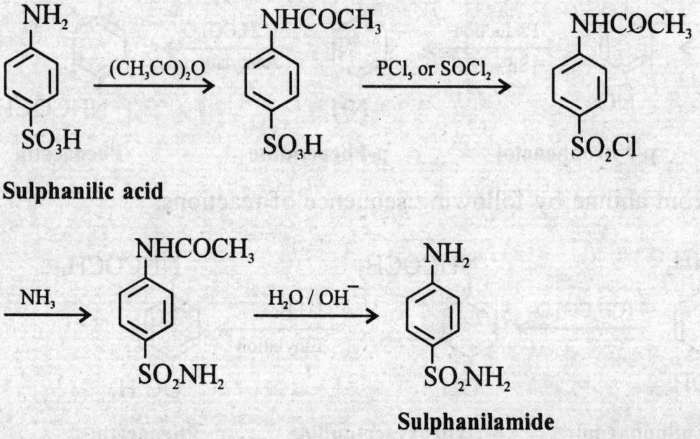

Sulphanilic acid

Sulphanilamide

Sulphanilamide is less potent compared to more complex sulpha drugs. Though it has been used in the treatment of a wide variety of infections, it is no longer the drugs of choice basically due to its side effects like cyanosis, nausea, vomiting, headache, dizziness and skin rashes. It is now banned for consumption by human beings, but is used for vaterinary purposes.

47.4.2 Sulphapyridine/2-Sulphanilamidopyridine

It was used in 1938 as the first specific drug for the treatment of pneumonia. It is obtained from sulphanilic acid as shown below:

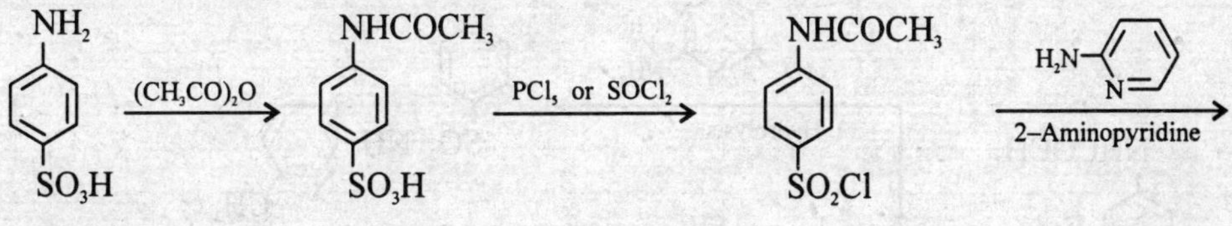

Sulphanilic acid

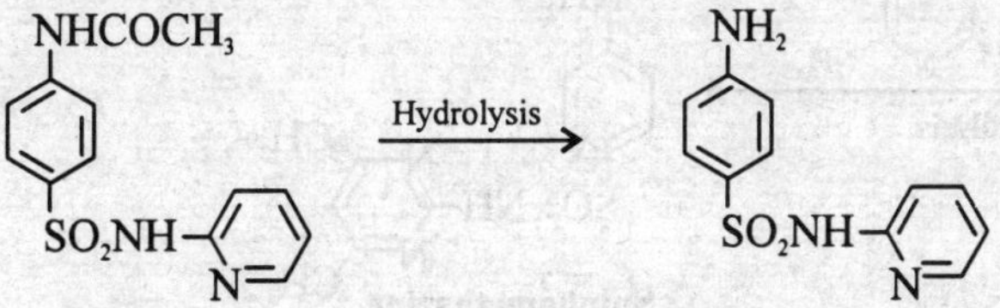

Sulphapyridine

Due to toxicity, sulphapyridine has been replaced by better sulpha drugs and antibiotics.

47.4.3 Sulphadiazine/2-Sulphanilamidopyrimidine

It is the most widely used sulpha drug for mild infections. It is obtained by the condensation of p-acetamido-benzenesulphonyl chloride (obtained as given earlier) with 2-aminopyrimidine in presence of mild alkali.

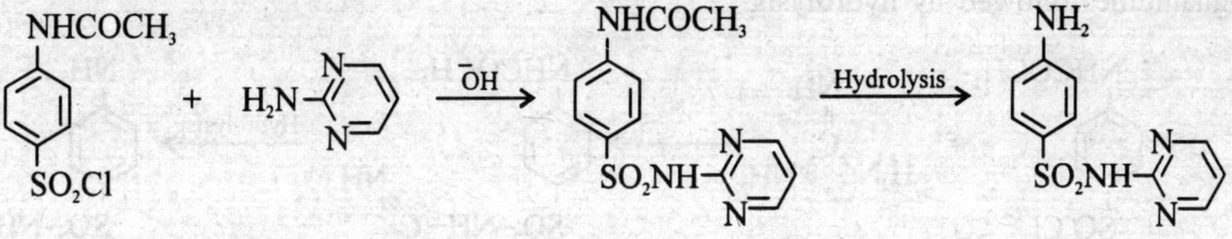

2-Aminopyri-midine

Sulphadiazine

The required 2-aminopyrimidine is obtained by following sequence of reactions.

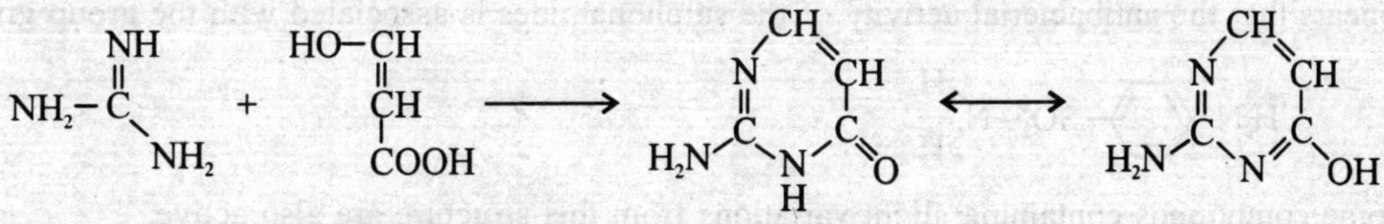

Guanidine   Formylacetic acid   Isocytosine

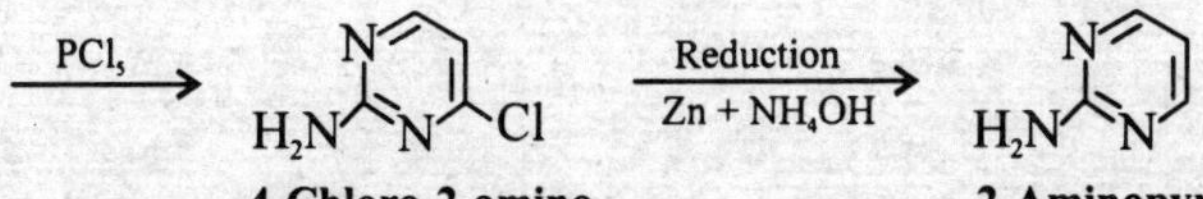

4-Chloro-2-amino-pyrimidine   2-Aminopyrimidine

Sulphadiazine is the drug of choice in a number of infections, viz., pneumococcal, meningococcal and H. influenzae. However, it shows some side effects like nausea, dizziness and skin rashes.

Two other derivatives of sulphadiazine which are in use are sulphamerazine and sulphamethazine. Both these are obtained in a similar manner as sulphadiazine using appropriate amino compounds.

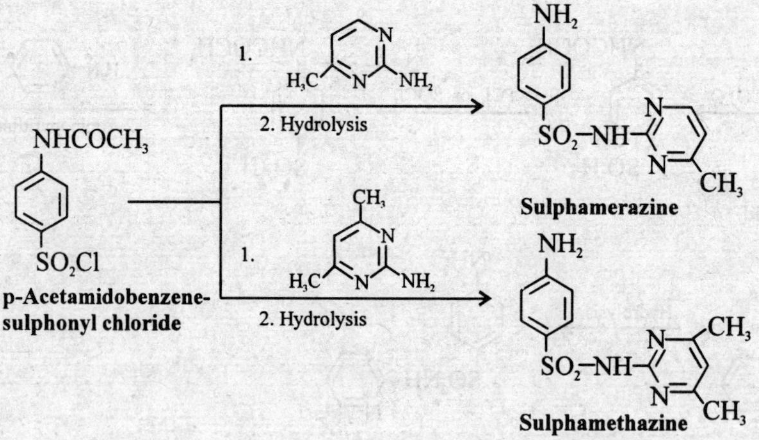

Sulphamerazine

Sulphamethazine

p-Acetamidobenzene-sulphonyl chloride

Both these drugs are useful against pneumococci infections.

47.4.4 Sulphaguanidine/2-Sulphanilamidoguanidine

It is only slightly absorbed in the intestinal tract and is therefore used in relatively larger doses in the treatment of bacillary dysentery. It is obtained by the condensation of p-acetamidobenzenesulphonyl chloride with guanidine followed by hydrolysis.

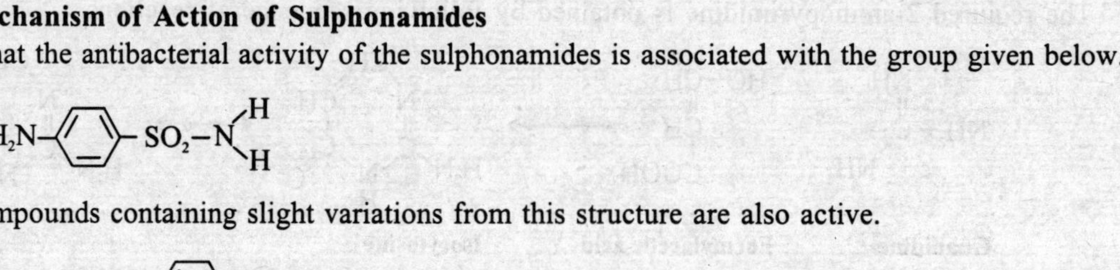

Guanidine

Sulphaguanidine

47.4.5 Mechanism of Action of Sulphonamides

It appears that the antibacterial activity of the sulphonamides is associated with the group given below.

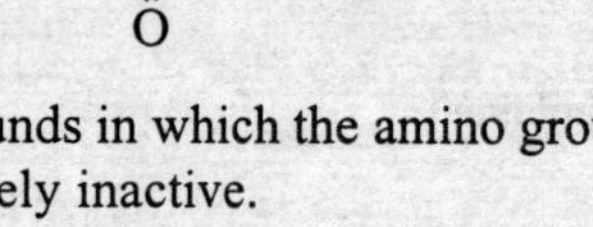

Some compounds containing slight variations from this structure are also active.

Compounds in which the amino group is ortho or meta to the sulphonamide group are either less active or completely inactive.

p-Aminobenzoic acid is an essential growth factor, for most bacteria are susceptible to the sulphonamides. The mechanism of action is that, owing to the similarity in structure of sulphonamide and p-aminobenzoic acid bacteria absorbs sulphonamides 'by mistake', and once it is absorbed, the bacteria cease to grow in numbers. In other words, the sulphonamides are not bactericidal but bacteriostatic.

47.5 ANTIMALARIALS

Malaria is one of the most prevalent of all human infections. It is most wide spread in tropical countries. The human beings are prone to four types of malaria caused by four species of plasmodium as given below:

Types of Malaria	Caused by
Malignant tertian malaria	Plasmodium falciparum
Benign tertian malaria	P.vivax
Quartan malaria	P.malarial
A rare type of malaria	P.ovale

Out of the four mentioned above, the first two are the most important.

Two types of antimalarial compounds are available for the treatment of malaria.

47.5.1 Naturally Occurring Antimalarials

The cinchona bark was used in the 16th century A.D. for the treatment of malaria. The roots of the Chinese plant Ch'ang Shan (Dichroa febrifuga) has been used as antimalarial in Chinese medicines from times immemorial. The cinchona bark is found to contain more than twenty five alkaloids, out of which only a few are clinically useful.

Cinchonine	R = H ; (+) 8R, 9S isomer
Cinchonidine	R = H ; (–) 8S, 9R isomer
Quinine	R = OCH_3 ; (–) 8S, 9R isomer
Quinidine	R = OCH_3 ; (+) 8R, 9S isomer

Out of these quinine is the most important anti-malarial.

47.5.2 Synthetic Antimalarial Drugs

Quinine is by no means an entirely satisfactory drug for the treatment of malaria. In chronic cases there is no effect of quinine. Therefore synthetic antimalarial drugs were developed. These may be classified into the following four groups:

(1) Quinoline derivatives   (2) Acridine derivatives
(3) Biguanidine derivatives   (4) Pyrimidine derivatives

Some typical examples are discussed in the following sections.

47.5.2.1 Quinoline Derivatives

These are further divided into 8-aminoquinoline derivatives (for example pamaquine, primaquine, pentaquine and isopentaquine) and 4-aminoquinoline derivatives (for example chloroquine and camaquine). Important quinoline compounds are discussed below:

(1) *Pamaquine/8-[(4-diethylamino-1-methylbutyl) amino]-6-methoxyquinoline:* It is the first synthetic antimalarial and is prepared by the condensation of 6-methoxy-8-aminoquinoline and 1-diethylamino-4-bromopentane.

$$CH_3O\text{-quinoline-}NH_2 \quad + \quad CH_3-\underset{\underset{Br}{|}}{CH}-(CH_2)_3N(C_2H_5)_2 \quad \longrightarrow \quad CH_3O\text{-quinoline-}NH-\underset{\underset{CH_3}{|}}{CH}-(CH_2)_3-N\underset{C_2H_5}{\overset{C_2H_5}{<}}$$

6-Methoxy-8-   **1-Diethylamino-**   **Pamaquine**
amino quinoline   **-4-bromopentane**

Pamaquine is gametocidal and in combination with quinine is used for the cure of vivax malaria.

(2) *Primaquine/8-[(4-amino-1-methylbutyl) amino]-6-methoxyquinoline:* It is less toxic and more effective than pamaquine and is prepared by condensing 6-methoxy-8-aminoquinoline with 4-bromopentylamine hydrobromide.

$$CH_3O\text{-quinoline-}NH_2 \quad + \quad CH_3\underset{\underset{Br}{|}}{CH}(CH_2)_3NH_2.HBr \quad \longrightarrow \quad CH_3O\text{-quinoline-}NH-\underset{\underset{CH_3}{|}}{CH}-(CH_2)_3-NH_2$$

6-Methoxy-8-   **4-Bromopentyl-**   **Primaquine**
aminoquinoline   **amine hydrobromide**

This drug is used for the radical cure and prevention of relapses of vivax malaria.

(3) *Chloroquine/4-[(4-diethylamino-1-methylbutyl) amino]-7-chloroquinoline:* It is a highly effective antimalarial and is most useful in P.vivax malaria as a suppressive agent. Some times chloroquine has been found to be very toxic causing respiratory problems and also cardiac arrest. It is prepared by the condensation of 4,7-dichloroquinoline and 1-diethylamino-4-aminopentane.

$$Cl\text{-quinoline-}Cl \quad + \quad CH_3-\underset{\underset{NH_2}{|}}{CH}-(CH_2)_3-N\underset{C_2H_5}{\overset{C_2H_5}{<}} \quad \longrightarrow \quad Cl\text{-quinoline-}NH-\underset{\underset{CH_3}{|}}{CH}-(CH_2)_3-N\underset{C_2H_5}{\overset{C_2H_5}{<}}$$

4, 7-Dichloro-   **1-Diethylamino-4-**   **Chloroquine**
quinoline   **aminopentane**

(4) *Camoquin:* It is three to four times more effective than quinine and is specially effective against vivax and falciparum malaria. It is prepared by the condensation of 4,7-dichloroquinoline with 4-amino-2-diethylaminomethylphenol.

$$Cl\text{-quinoline-}Cl \quad + \quad \text{(phenol)}\overset{OH}{\underset{NH_2}{}}CH_2N(C_2H_5)_2 \quad \longrightarrow \quad Cl\text{-quinoline-}NH\text{-(phenol)-}OH,\ CH_2-N\underset{C_2H_5}{\overset{C_2H_5}{<}}$$

4,7-Dichloro-   **4-Amino-2-diethyl-**   **Camoquin**
quinoline   **aminomethylphenol**

47.5.2.2 Acridine Derivatives
There is structural similarity between the acridine derivatives and chloroquine.

Mepacrine/Atebrin/3-chloro-9-[(4-diethylamino-1-methylbutyl) amino]-7-methoxyacridine: It is commonly used in the form of dihydrochloride or bis methylsulphonate salt for the rapid suppression and treatment of all types of malaria. However, on prolonged use it causes skin rash. It is synthesised by the condensation of 3,9-dichloro-7-methoxyacridine with 1-diethylamino-4-aminopentane.

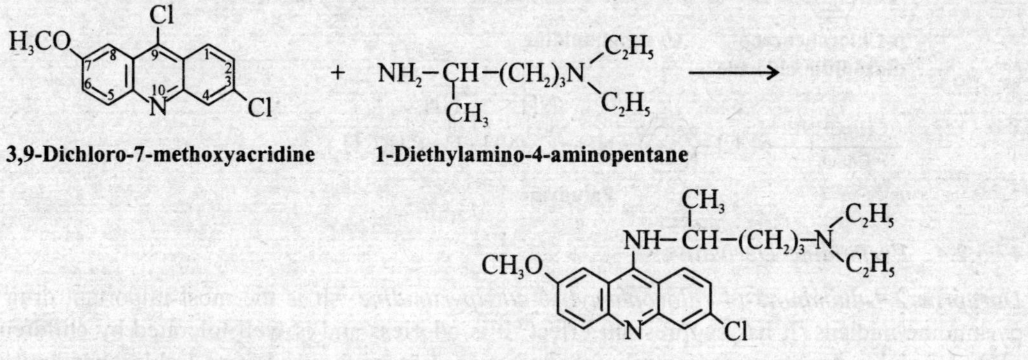

3,9-Dichloro-7-methoxyacridine   **1-Diethylamino-4-aminopentane**

Mepacrine

The substituted acridine and pentane are prepared as follows:

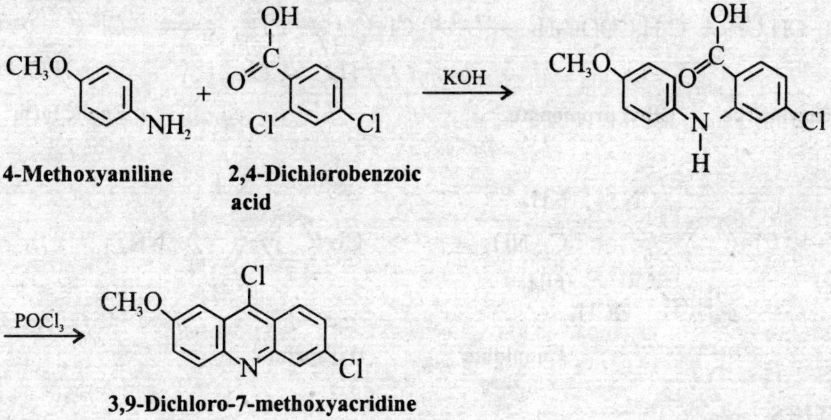

4-Methoxyaniline   **2,4-Dichlorobenzoic acid**

3,9-Dichloro-7-methoxyacridine

$$CH_3CO\overline{C}HCOOC_2H_5Na^+ + ClCH_2CH_2N(C_2H_5)_2 \longrightarrow CH_3-\overset{O}{\overset{\|}{C}}-\underset{\underset{COOC_2H_5}{|}}{CH}-CH_2-CH_2-N(C_2H_5)_2$$

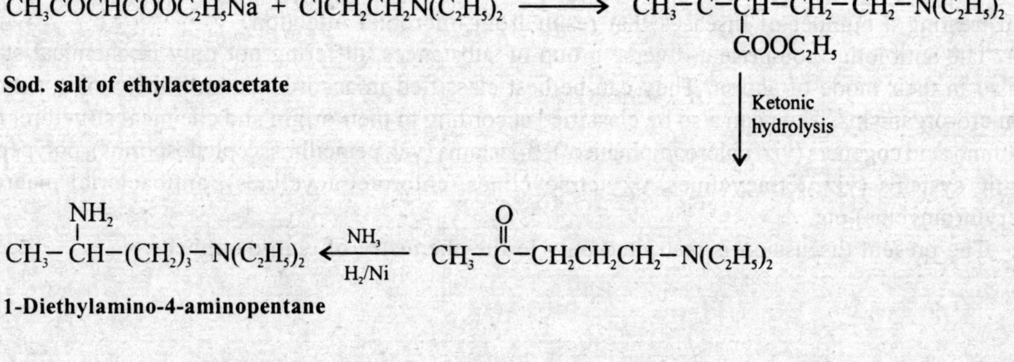

Sod. salt of ethylacetoacetate

Ketonic hydrolysis

$$\underset{\text{1-Diethylamino-4-aminopentane}}{CH_3-\overset{\overset{NH_2}{|}}{CH}-(CH_2)_3-N(C_2H_5)_2} \underset{H_2/Ni}{\overset{NH_3}{\longleftarrow}} CH_3-\overset{O}{\overset{\|}{C}}-CH_2CH_2CH_2-N(C_2H_5)_2$$

47.5.2.3 Biguanidine Derivatives

Paludrine: It is superior to mepacrine and chloroquine and is the best antimalarial drug known. It is prepared by coupling of p-chlorobenzenediazonium chloride with cyanoguanidine followed by treatment of the formed product with isopropylamine in presence of copper sulphate.

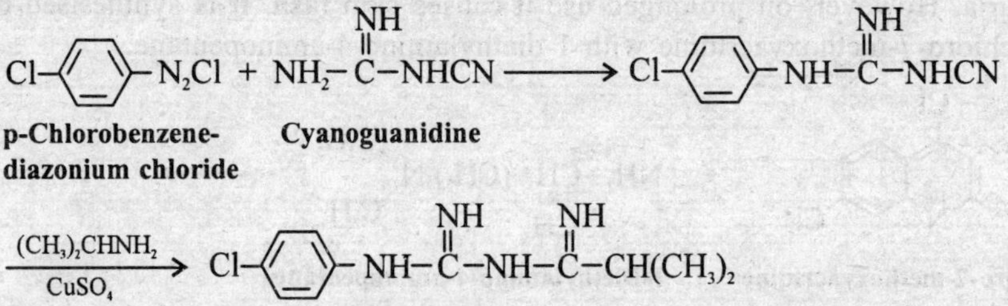

p-Chlorobenzene-   Cyanoguanidine
diazonium chloride

Paludrine

47.5.2.4 Pyrimidine Derivatives

Daraprim/2,4-diamino-5-(4-chlorophenyl)-6-ethylpyrimidine: It is the most important drug having the pyrimidine nucleus. It has suppressant effect. It is odorless and is well tolerated by children. Daraprim is very useful in transmission control of the infection. It is synthesised from 4-chlorobenzonitrile. Various steps involved are given below:

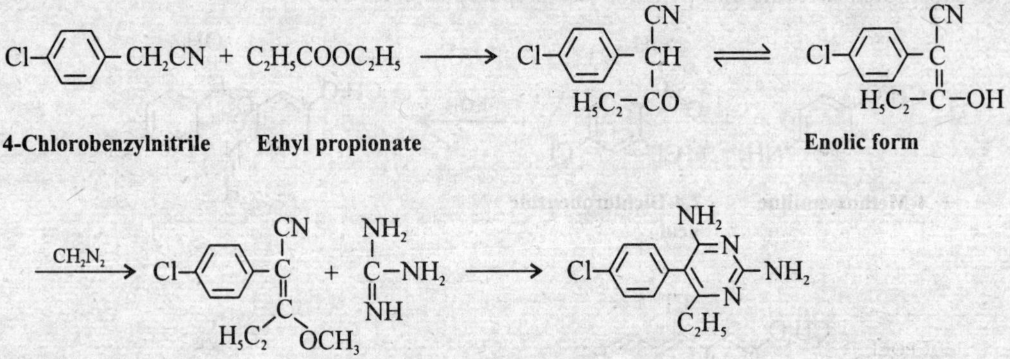

4-Chlorobenzylnitrile   Ethyl propionate   Enolic form

Guanidine   Daraprim

47.6 ANTIBIOTICS

An antibiotic is a chemical substance produced by micro-organisms (or derived from living organisms), which has the capacity to inhibit the growth of bacteria or even destroy them. It is of immense value in treating a number of diseases that result from microbial infection.

The antibiotics comprise a diverse group of substances, differing not only in chemical structure, but also in their mode of action. They can be best classified in accordance with their toxic action in target micro-organisms. They can also be classified according to their origin and chemical structure, for example amino acid cogeners (viz. chloroamphenicol), β-lactams (viz. penicillins, cephalosporins), polypeptides, fused ring systems (viz. tetracyclines, oxytetracyclines, chlorotetracyclines), antibacterial macrolides (viz. erythromycins) etc.

The present discussion is restricted only to the chemistry of chloroamphenicol.

47.6.1 Chloroamphenicol/Chloromycetin/D(−)Threo-2-dichloroacetamido-1-p-nitrophenylpropane-1,3-diol

It was isolated from the strains of Streptomyces venezuelae. It is a broad spectrum antibiotic and is active against gram-positive and gram-negative bacteria. It is an important drug for the treatment of typhoid fever and is also given in conjugation with penicillin, streptomycin and sulpha drugs in many conditions. Its structure (I) is based on the following evidences.

(i) Its molecular formula is $C_{11}H_{12}O_5N_2Cl_2$. It is laevorotatory. Both the chlorine are inert to silver ion.

(ii) It has a nitro group in the phenyl ring. This is inferred by the similarity of its ultra violet spectrum to that of nitrobenzene. The presence of nitro group is also confirmed by reduction of chloramphenicol with tin and hydrochloric acid, followed by diazotisation and coupling with β-naphthol to give an orange red precipitate.

(iii) Catalytic reduction of chloroamphenicol with palladium gives a product, whose absorption spectrum is similar to that of p-toluidine. Further, the solution contain ionic chlorine.

(iv) Hydrolysis of chloroamphenicol with acid or alkali gives dichloroacetic acid and an optically active base (II), $C_9H_{12}N_2O_4$. This base was shown to contain a primary amino group and gave back chloroamphenicol on treatment with methyl dichloroacetate. On this basis, it is assumed that the antibiotic must contain the group $NHCOCHCl_2$.

(v) Chloroamphenicol contains two hydroxy groups, since it gives a diacetate on treatment with acetic anhydride in pyridine. However, the base obtained from chloroamphenicol forms a triacetyl derivative on similar treatment. Also, the base consumes two molecules each of periodic acid to give one molecule of ammonia, formaldehyde, formic acid and p-nitrobenzaldehyde. The formation of these products can be accounted for if the base is given the structure 2-amino-1-p-nitrophenylpropane-1-3-diol (II). On this basis chloroamphenicol is given the structure (I).

I Chloroamphenicol
(λ_{max} 218 mµ ; $[\alpha]_d$ −25.2°)

II Base ($C_9H_{12}N_2O_4$)

$$NO_2-\langle\ \rangle-CHO + CH_2O + NH_3 + HCOOH$$

The structure of chloroamphenicol as (I) has also been confirmed by crystallographic studies (Dunitz, 1952). As seen chloroamphenicol has two chiral centres and thus there are two possible pairs of enantiomers. Of the four steroisomers only the D-threo form is active. Comparison of the properties of the base with those of norephedrine and norpseudo-ephedrine showed that the configuration of the base was similar to that of norpseudo-ephedrine. On this basis, chloroamphenicol is D(−)threo-2-dichloroacetamido-1-p-nitrophenylpropane-1,3-diol. It is noteworthy to see that chloramphenicol is the first natural compound found to contain a nitro group. Also the presence of the $CHCl_2$ group is rather unusual.

The structure of chloroamphenicol (I) has been confirmed by its synthesis (Long et. al., 1949).

p-Nitroacetophenone   **p-Nitrophenacyl bromide**

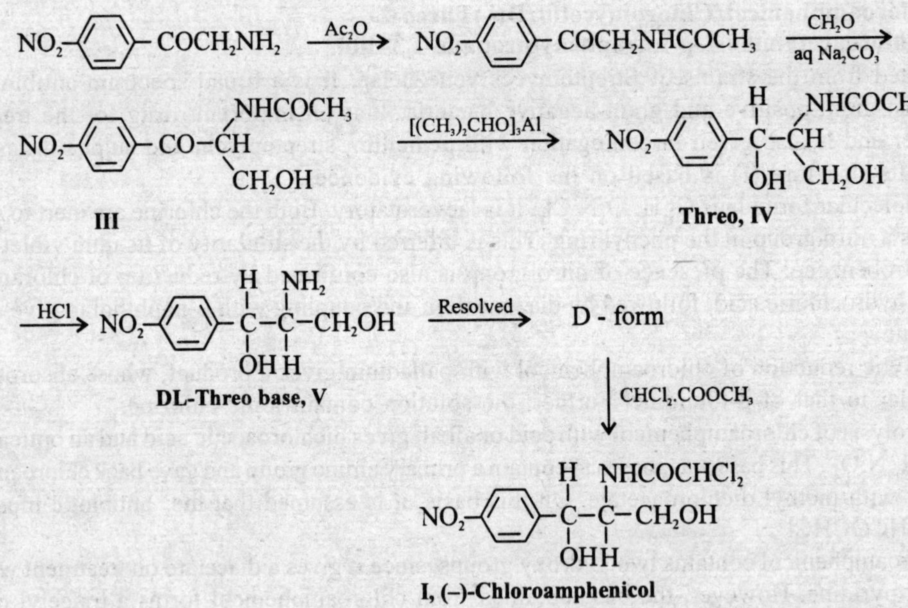

NO_2—⟨ ⟩—$COCH_2NH_2$ →(Ac₂O) NO_2—⟨ ⟩—$COCH_2NHCOCH_3$ →(CH₂O / aq Na₂CO₃)

NO_2—⟨ ⟩—CO–$\overset{NHCOCH_3}{\underset{CH_2OH}{CH}}$ →([(CH₃)₂CHO]₃Al) NO_2—⟨ ⟩—$\overset{H}{\underset{OH}{C}}$–$\overset{NHCOCH_3}{\underset{CH_2OH}{CH}}$

III   **Threo, IV**

→(HCl) NO_2—⟨ ⟩—$\overset{H}{\underset{OH}{C}}$–$\overset{NH_2}{\underset{H}{C}}$–$CH_2OH$ →(Resolved) **D - form**

DL-Threo base, V

(↓ CHCl₂.COOCH₃)

NO_2—⟨ ⟩—$\overset{H}{\underset{OH}{C}}$–$\overset{NHCOCHCl_2}{\underset{H}{C}}$–$CH_2OH$

I, (–)-Chloroamphenicol

In the above synthesis of chloramphenicol, reduction of (III) with aluminum isopropoxide gave predominantly the threo compound (IV) along with a small amount to the erythro isomer. The threo compound was separated from the erythro isomer by fractional crystallisation. Hydrolysis of threo compound (IV) gave threo base (V), which was resolved by means of (+)-camphorsulphonic acid into the D-form and finally converted into (–)-chloroamphenicol (I).

47.7   PROBLEMS

1. What is a drug? How drugs are classified?
2. Explain the term analgesics and antipyretic. How is aspirin and phenacetin obtained?
3. What is soluble aspirin? How is paracetamol and phenylbutazone obtained?
4. What are sulpha drugs? How is sulphanilamide and sulphadiazine obtained?
5. Give a method of preparation of sulphaguanidine.
6. What is an antimalarial? How is pamaquine, and primaquine obtained.
7. Give a synthesis of chloroquine and camaquine.
8. Define an antibiotic. How is the structure of chloroamphenicol established?
9. Describe a synthesis of chloroamphenicol.
10. Write the uses of the following:
    (a) Aspirin   (b) Phenylbutazone
    (c) Sulphaguanidine   (d) Chloroquine
    (e) Chloroamphenicol

48
Pesticides

48.1 INTRODUCTION
The chemicals used for the protection and control of pests are collectively known as pesticides. The adequate control of various insect and animal pests has become the main problem confronting masses now a days because a considerable percentage of crops and human lives are annually destroyed due to various vectors of infections diseases. Therefore, a continuous effort to improve the biological control measures is very necessary to combat the problems of pests.

48.2 CLASSIFICATION
The pesticides include wide variety of inorganic and organic compounds which are used against different pests. They may, therefore, be divided into various classes on the basis of either the nature of compound or nature of pest. Below is given various classes of pesticides and their important examples.

	Class	Examples
1.	Insecticides	
	(a) Naturally occurring insecticides	Pyrethrins, Rotenones etc.
	(b) Organochlorine insecticides	DDT, BHC
	(c) Carbamate insecticides	Carbaryl
	(d) Organophosphorus insecticides	Malathion, Parathion
	(e) Fumigants	Methyl bromide, Ethylene dibromide etc.
	(f) Nematicides	Furaden
	(g) Acaricides	Azobenzene
	(h) Phermones	Eugenol
	(i) Insect repellents	Dimethylphthalate
	(j) Biological methods	
	(k) Molluscides	Metaldehyde
	(l) Rodenticides	Sodium fluoroacetate
2.	Fungicides	Organosulphur compounds like Nabam, Zineb, Thiaram, Captam etc.
3.	Bactericides	Streptomycin in bacterial infections of maize, apples and tobacco
4.	Herbicides	Ferric sulphate, Copper sulphate, Sodium borate, 2,4-D etc.
5.	Plant growth regulators	1-Naphthylacetic acid (NAA), Ethanedial dioxime, Ancymidol etc.

The present discussion is restricted to insecticides only.

48.3   INSECTICIDES

48.3.1   Naturally Occurring Insecticides

Certain plants have developed protective mechanism like nicotine, pyrithrins, rotenoids etc. Natural, laevoro-tatory nicotine, principal extract from tobacco plants is used to control aphids or plant lice.

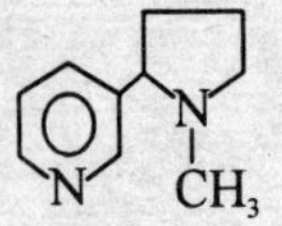

Nicotine 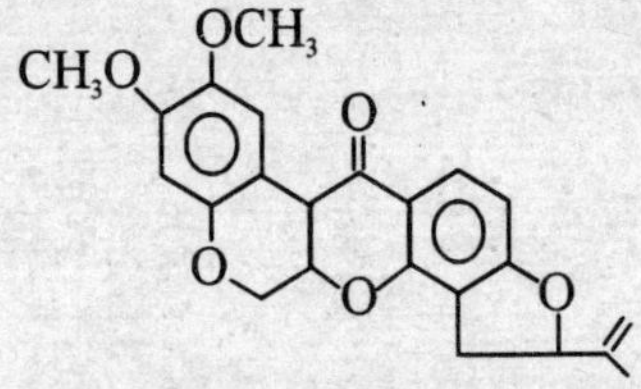   **Rotenone**

Rotenone, an important insecticide from the roots of Derris elliptica, is used against fishes, insects, aphids wasps and caterpillars.

A very useful contact insecticide, pyrethrum is extracted from the flower heads of Chrysanthemum cinerariae-folium which is a mixture of four insecticidal compounds Pyrethrin I, Pyrethrin II, Cinerin I and Cinerin II.

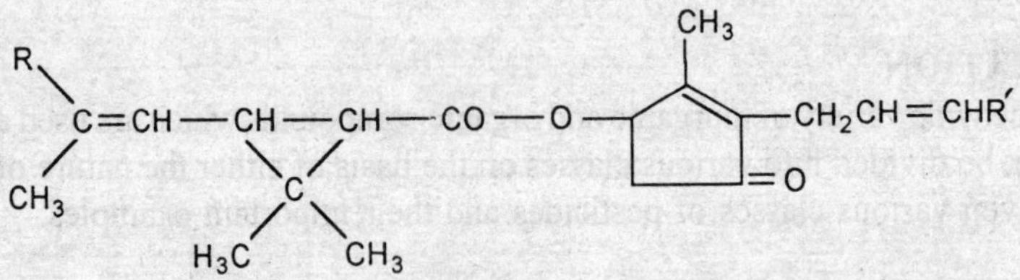

Pyrethrin I	R = CH_3	R' = $-CH=CH_2$
Pyrethrin II	R = CH_3COO	R' = $-CH=CH_2$
Cinerin I	R = CH_3	R' = $-CH_3$
Cinerin II	R = CH_3COO	R' = $-CH_3$

Pyrethrum owes its importance to the rapid excitations, convulsion, paralysis and finally death of the insect together with low mammalian toxicity. It is nonpersistent and does not leave any toxic residue. Its sprays are excellent home insecticides. It is used against houseflies, insects and rats.

Synergists: Pyrethrum, when mixed with another compound (which itself does not necessarily has any insecticidal activity), shows enhanced insecticidal activity. This phenomenon of potentialisation of insec-ticidal activity of pyrethrum is known as synergism and the compound added to it synergist. The commonly used synergists with pyrethrum are sesamin, piperonyl-butoxid etc.

Protein Insecticides: The protein toxins isolated from the cultures of *Bacillus* thuringiensis are specific stomach poisons for insects. They particularly affect the caterpillars and are toxic towards mammals. The toxins isolated are α, β, γ-exotoxin and δ-endotoxin of which δ-endotoxin is the most active one. The structure of these toxins has not been elucidated completely as yet.

Insecticides from Animals: A number of animals produce insecticidal compounds as defense poisons, but have not found any practical application so far. The defense poison of an Argentine ant, *Iridomyrmex humilis* is an effective insecticide. However, Nereistoxin (N,N-dimethyl-1,2-ditholan-4-amine) available from marine annelid worm (*Lumbriconeris heteropoda*) is used in Japan.

48.3.2 Organochlorine Insecticides
The important members of this class of insecticides have been discussed below:

48.3.2.1 DDT/1,1-(2,2,2-Trichloroethylidene)bis(4-chlorobenzene)
DDT, the most popular insecticide, was an excellent insecticide for many years. The properties associated with DDT are high insecticidal toxicity, low mammalian toxicity, wide spectrum, low price and simple in manufacturing and handling. It was effectively used for the control of malaria. The only drawback with organochlorine insecticides, in general, is their tendency to leave persistent residues which are accumulated in the environment and then in food chain. That is why the use of DDT has been restricted in most of the western industrialised countries.

DDT is synthesised by the condensation of chloral with chlorobenzene in presence of sulphuric acid or oleum in almost 100% yield.

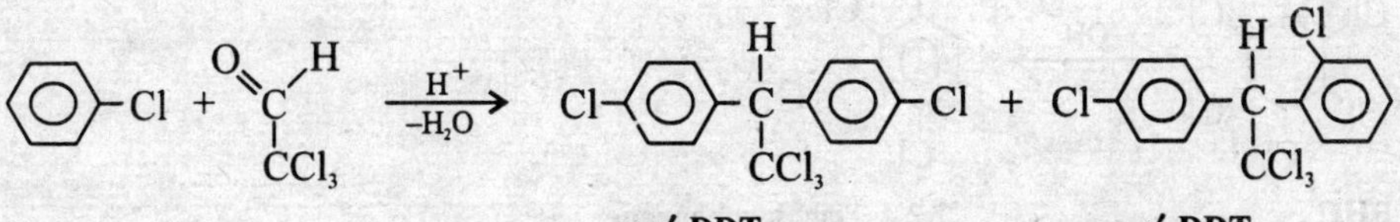

p,p'-DDT is produced in the above reaction along with a small amount of weakly insecticidal o,p'-DDT. It is stable to light, air and acids, insoluble in water but soluble in organic solvents.

DDT acts as a nerve poison and causes disturbance in the sodium balance of the nerve membranes. A number of organochlorine compounds having DDT like structure are popular insecticides.

	Compound	Structure
1.	Methoxychlor, DMDT	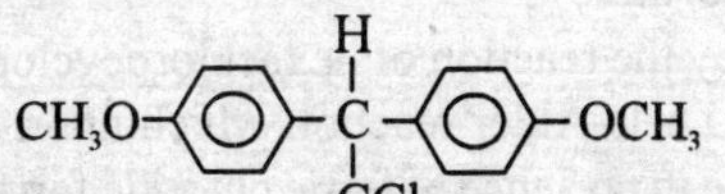
2.	DFDT: 1,1'-(2,2,2-Trichloroethylidene)-bis(4-fluorobenzene)	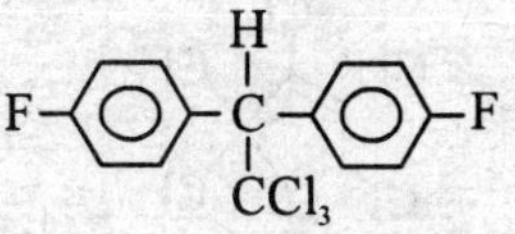
3.	TDE: 1,1'-(2,2-Dichloroethylidene)-bis(4-chlorobenzene)	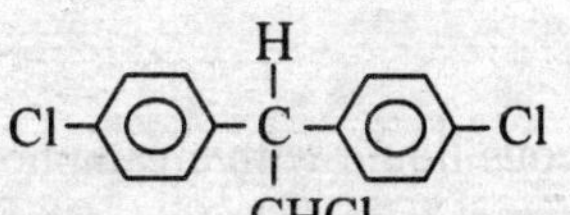

48.3.2.2   BHC/Hexachlorocyclohexane/Benzene hexachloride

It is a mixture of various stereoisomers of 1,2,3,4,5,6-hexachlorocyclohexane of which γ-1,2,3,4,5, 6-hexachlorocyclohexane is the active principal. It is commonly known as Gammexane or lindane.

BHC is prepared by the chlorination of benzene in presence of UV light. The isolation of γ-isomer, which is exclusively used now a days for pest control requires several concentration and purification steps.

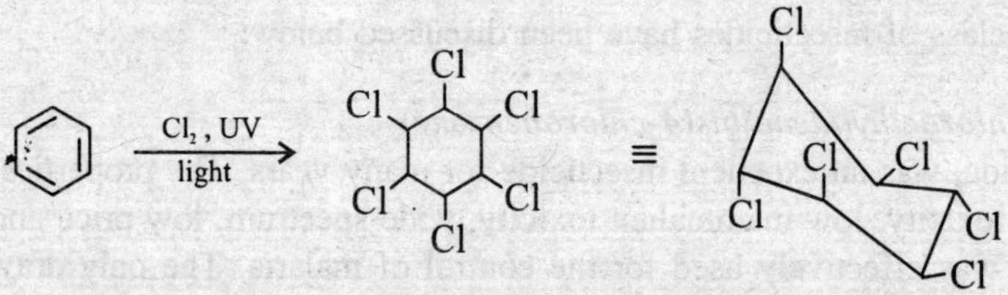

The concentration of γ-isomer or lindane in the raw product is 10-18% and the position of various chlorine atoms in cyclohexane ring is a,a,a,e,e,e where 'a' is axial and 'e' is equatorial. It is stable in acids but eliminates hydrogen chloride in alkaline solution to give 1,2,4-trichlorobenzene.

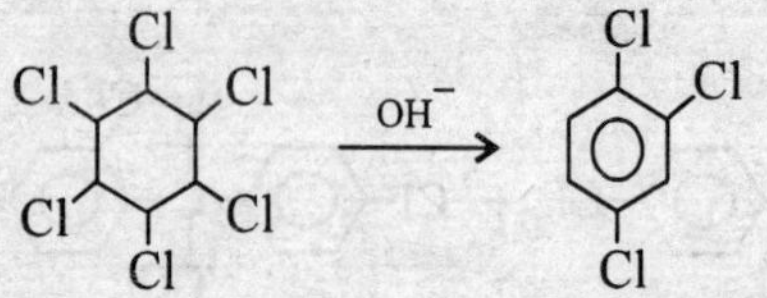

BHC   **1,2,4-Trichlorobenzene**

It is almost as active as DDT towards insects. It is a contact poison, stomach poison and respiratory poison. It is lethal to chewing and sucking insects but not to spider mites. It is also used to control soil pests like beetle larvae, wireworms, corn root worms, cabbage root fly etc. It is also used as a veterinary medicine for the control of ectoparasites such as ticks and mites.

It is accumulated in the body fat, milk and kidneys rapidly but is excreted relatively quickly (compared to DDT). Thus, the danger of accumulation is slight.

48.3.2.3   Chlordane

It is prepared by the reaction of hexachlorocyclopentadiene with cyclopentadiene. The adduct is formed as a result of endo Diels-Alder reaction which upon chlorination with either chlorine in carbon tetrachloride or thionyl chloride in presence of ferric chloride forms chlordane.

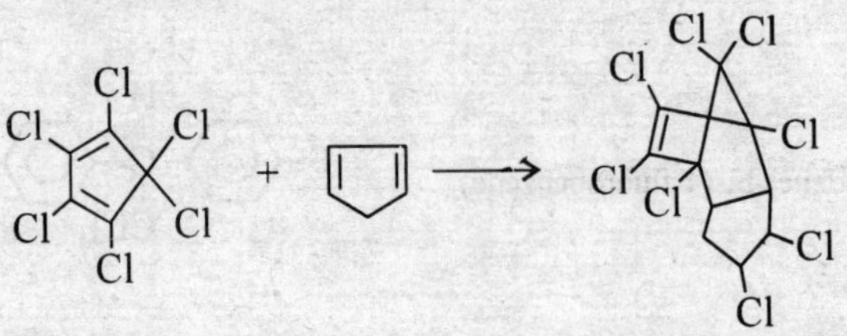

Chlordane

It is a brown, viscous liquid with a camphor like odor, insoluble in water but soluble in most organic solvents. It is stable towards acids but in presence of alkali, loses hydrogen chloride forming product with no insecticidal activity.

It is almost as effective an insecticide as DDT and BHC. It is a broad spectrum contact, stomach and respiratory poison, especially suitable for soil pests.

48.3.2.4 Aldrin

It is synthesised by the Diels-Alder reaction of hexachlorocyclopentadiene and bicyclo[2.2.1]-hepta-2,5-diene (norbornadiene) which involves refluxing of the two in toluene. Norbornadiene may be obtained in good yield by the reaction of cyclopentadiene and ethyne at 150-400° under 1-20 atm pressure.

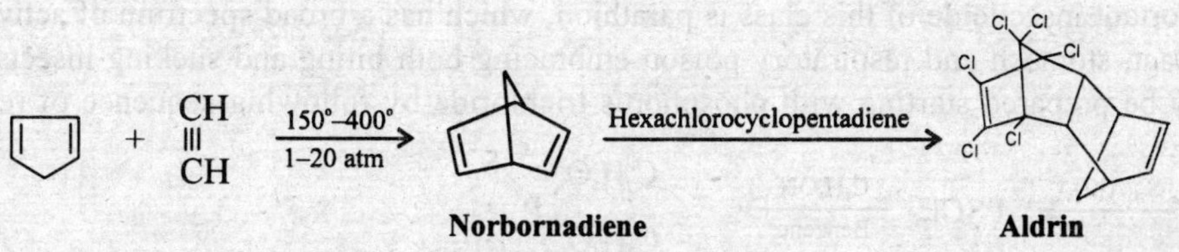

Norbornadiene   **Aldrin**

Aldrin is stable to dilute acid, alkali and high temperature. Concentrated acids and oxidising agents attack the double bond in the unchlorinated ring. However, the chlorinated part of the molecule is almost inert.

It is used as a soil insecticide in various fertiliser formulations. It is an effective contact, stomach and respiratory poison. It primarily acts as a nerve poison. It is used for the protection of crops of corn, potatoes, sugar beet, cane sugar and bananas.

48.3.2.5 Dieldrin

The epoxidation of aldrin forms dieldrin. The epoxidation may be done by hydrogen peroxide-acetic anhydride or peracetic acid or perbenzoic acid.

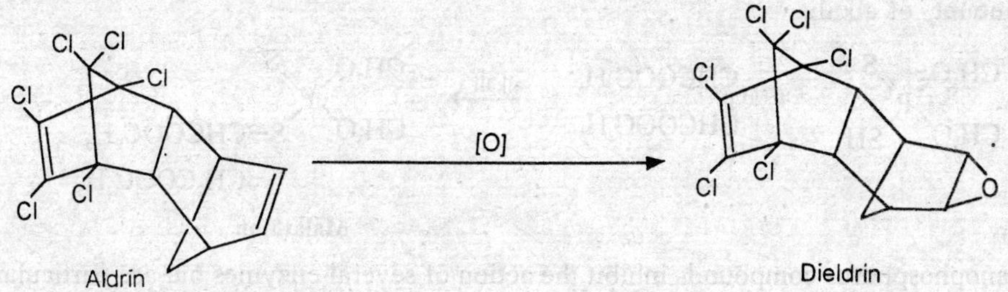

Aldrin   Dieldrin

Compared to aldrin, dieldrin has low volatility and chemical stability. It is a more potent insecticide than DDT. It is used in combination with other insecticides for the protection of forest corps. It is used to control crawling pests like crickets, cockroaches, mosquitoes and tse-tse flies.

Like other organochlorine insecticides, dieldrin is accumulated in lipids and fats via absorption through skin and causes liver damage.

48.3.3   Organophosphorus Insecticides

The basic structure of organophosphorus compounds having insecticidal activity is defined by Schrader's formula given below.

$$R^1 \diagdown \underset{R^2 \diagup}{P} \diagup O(S) \diagdown acyl$$

Where R^1 and R^2 may be alkoxy, alkyl or amino residues, 'acyl' represents organic and inorganic acids such as cyanate, thiocyanate or mercapto groups.

The most important insecticide of this class is parathion, which has a broad spectrum of activity. It is an effective contact, stomach and respiratory poison embracing both biting and sucking insects.

Parathion may be prepared starting with phosphorus trichloride by following sequence of reactions.

$$PCl_3 \xrightarrow{\text{S , AlCl}_3} PSCl_3 \xrightarrow[\text{Benzene}]{\text{C}_2\text{H}_5\text{OH}} \underset{C_2H_5O}{\overset{C_2H_5O}{\diagdown}} P \overset{S}{\underset{Cl}{\diagup}}$$

O,O-Diethylphosphorochlorothioate

$$\underset{C_2H_5O}{\overset{C_2H_5O}{\diagdown}} P \overset{S}{\underset{Cl}{\diagup}} + NaO\!-\!\bigcirc\!\!-\!NO_2 \longrightarrow NO_2\!-\!\bigcirc\!\!-\!O\,\overset{S}{\underset{OC_2H_5}{\diagup}}\!P\!\diagdown OC_2H_5$$

Sodium salt of p-nitrophenol   **Parathion**

The organophosphates have low mammalian toxicity and are easily biodegradable, thus, do not leave persistent residues like organochlorine insecticides.

Another organophosphate, malathion, is a good insecticide and acaricide together with low mammalian toxicity. It is also used for the eradication of anopheles mosquitoes from malaria zone.

It is prepared by the addition of O,O-dimethylphosphorodithioic acid to diethyl maleate in presence of catalytic amount of alkali.

$$\underset{CH_3O}{\overset{CH_3O}{\diagdown}} \overset{S}{\underset{SH}{P}} + \underset{CHCOOC_2H_5}{\overset{CHCOOC_2H_5}{\|}} \xrightarrow{\text{alkali}} \underset{CH_3O}{\overset{CH_3O}{\diagdown}} \overset{S}{\underset{S-CHCOOC_2H_5}{\underset{|}{P}}} \atop CH_2COOC_2H_5$$

Malathion

The organophosphorus compounds inhibit the action of several enzymes but are particularly active in-vivo against enzyme acetylcholinesterase. They, mimic natural substrate acetyl choline by binding itself to the esteratic site of acetylcholinesterase and very slowly hydrolysing the phosphorylated enzyme (I), in contrast to the hydrolysis of acetylated enzyme (II), thus blocking the effective hydrolysis of acetyl choline to choline.

$$(RO)_2\!-\!\overset{O,S}{\overset{\|}{P}}\!-\!Z + ECH_2OH \xrightarrow{-HZ} (RO)_2\!-\!\overset{O,S}{\overset{\|}{P}}\!-\!\underset{I}{OCH_2E}$$

$$\Big\downarrow \begin{array}{l}\text{Hydrolysis}\\ \text{(Slow)}\end{array}$$

$$(RO)_2\!-\!\overset{O,S}{\overset{\|}{P}}\!-\!OH + ECH_2OH$$

$$\underset{\textbf{Acetyl choline}}{(CH_3)_3\overset{+}{N}CH_2CH_2O\overset{\overset{O}{\|}}{C}CH_3} + \underset{\textbf{Enzyme}}{ECH_2OH} \rightleftharpoons \underset{\textbf{Choline}}{(CH_3)_3\overset{+}{N}CH_2CH_2OH} + \underset{\textbf{II}}{CH_3COOCH_2E}$$

$$\Big\downarrow \substack{\text{Hydrolysis}\\ \text{(fast)}}$$

$$\underset{\text{(Enzyme)}}{CH_3COOH + ECH_2OH}$$

Organosphosphorus compounds have good fungicidal properties also. They have low mammalian toxicity and are easily biodegradable, so leave no persistence residues like organochlorine insecticides.

48.3.4 Carbamate Insecticides

The insecticidal carbamates are represented by the general formula given below in which the parent compound (R′OH) is weakly acidic, R^2 is methyl and R^3 is usually hydrogen. They are systemic insecticides and are absorbed in the transport system of plant, i.e., xylem. It is therefore possible to control pests in roots as well as in shoots which was otherwise difficult. Some carbamates which are low in toxicity find applications as veterinary medicines as well as snail and bird repellents.

$$\underset{O=}{\overset{R^1O}{\diagdown}}C-N\underset{R^3}{\overset{R^2}{\diagup}}$$

Carbamate insecticides may be produced by any of the three routes given below. The substrate having a hydroxy group is treated with dimethylcarbamoyl chloride to give N,N-Dimethylcarbamates (Route I). On reaction with alkyl isocyanate, N-methylcarbamate is produced (Route II). The third route, however, involves the reaction of hydroxy compound with phosgene to form chloroformate which reacts with primary amine to give N-methylcarbamate.

$$ROH \longrightarrow$$

Route I
$$Cl-CO-N(CH_3)_2, -HCl \longrightarrow R-O-CO-N(CH_3)_2$$

Route II
$$CH_3-N=C=O, \text{ Catalyst} \longrightarrow R-O-CO-NHCH_3$$

Route III
$$CH_3NH_2, -HCl \uparrow$$

Route III
$$COCl_2, -HCl \longrightarrow R-O-CO-Cl$$

Carbamates, like organophosphates, owe their insecticidal properties to the inhibition of enzyme acetylcholinesterase (AchE) and prevention of effective nervous transmission across the synapse which is caused by the resultant accumulation of acetylcholine. The enzyme is poisoned by carbamoylation of active enzyme which is hydrolysed back to the active enzyme very slowly, thus blocking the hydrolysis of acetylcholine to choline.

$$\underset{\substack{\textbf{Enzyme}\\ \textbf{AchE}}}{ECH_2\overset{\cdot\cdot}{O}H} \quad \underset{\substack{\textbf{Carbamate}}}{\overset{RO}{\underset{CH_3HN}{\diagup}}C=O} \xrightarrow[-ROH]{} \underset{\substack{\textbf{Carbamoylated}\\ \textbf{enzyme}}}{ECH_2O-\overset{\overset{O}{\|}}{C}NHCH_3}$$

From toxicological point of view, carbamates require structural resemblance to the natural enzyme substrate acetylcholine so that it can compete strongly for the reactive site on acetylcholinesterase.

The resistance developed against insecticidal carbamates is mainly due to its detoxification via enzymatic hydrolysis. It has been found that carbaryl resistant houseflies have an abnormally high concentration of enzyme carbamate esterase which convert carbaryl into inactive α-naphthol.

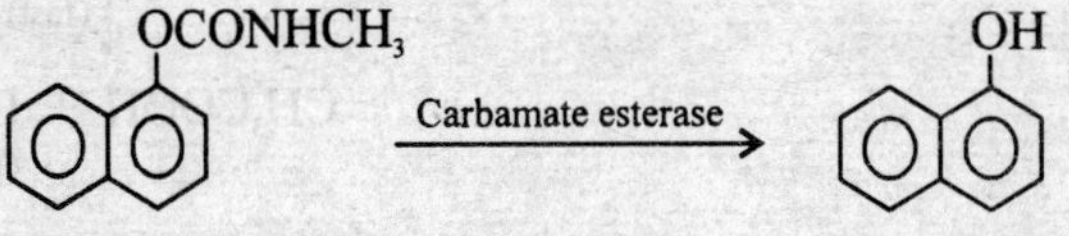

Carbaryl → Carbamate esterase → **α-Naphthol**

Some important carbamates and their uses have been summerised below.

Carbamate	Structure	Uses
1. Dimetan	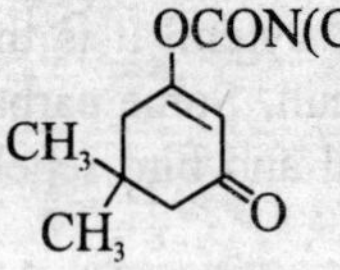	Contact insecticide with partial systemic activity.
2. Carbaryl	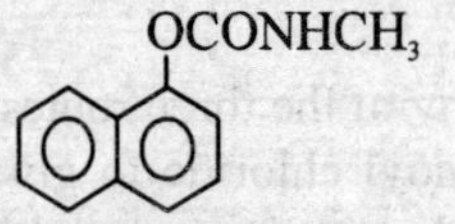	Broad spectrum, contact insecticide, nonsystemic, used to control cotton, fruit, vegetable etc.
3. Carbofuran	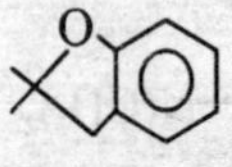	Broad spectrum insecticide, nematicide, miticide.

48.3.5 Fumigants

Fumigants are gaseous or easily vaporisable (low boiling) compounds that possess pesticidal activity. The insecticidal fumigants are associated with nematicidal, rodenticidal and sometimes fungicidal and bactericidal activities. They are used as soil insecticides, stored product protectants and for the control of household pests. They usually act on pests fast (time of contact is less) and do not alter the color, odor or taste of the stored product (have low water solubility and absorption). Some of the important compounds used as fumigants have been listed below.

1.	Methyl bromide	Fumigation of grains, fruits and vegetables
2.	Ethylene dichloride	Fumigation of stored grains
3.	Ethylene dibromide	Household fumigation
4.	Acrylonitrile	Fumigation of packed grain products and stored tobacco

5. Chloropicrin   For control of insects and rodents.

6. Sulphuryl fluoride   For fumigation of structures and wood products
                                from termites and beetles

48.3.6 Nematicides

Chemicals used to control nematodes are termed nematicides. They are pathogenic to plants, animals and human beings. In case of plants they can cause damage to the harvest qualitatively as well as quantitatively. In extreme cases the plants may die also. The important class of phytopathogenic nematodes include free living root nematodes, cyst nematodes, stem and leaf nematodes.

Nematicides include halogenated hydrocarbons, carbamates, organophosphates and fumigants. Some important nematicidal agents have been listed below:

1. Methyl bromide CH_3Br

2. Substituted benzyl halides

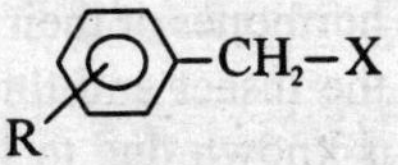

R= NO_2, NO, CN, SCN
X= Halogen

3. Dichlofenthion

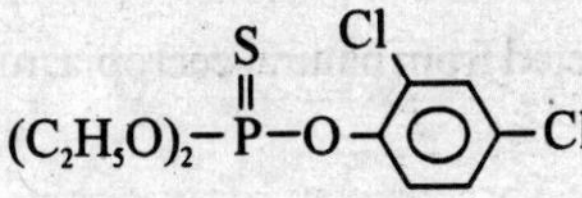

4. Furadan

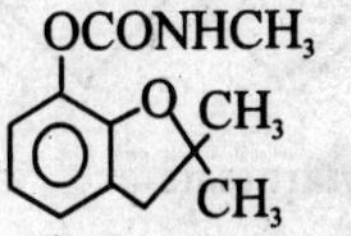

48.3.7 Acaricides

Acaricides are the compounds used for the control of mites. Mites that damage plants include spider mites(on fruits, roses, green house crops and vegetables), false spider mites (on green house crops), soft mites (on cotton and tea) and gall miles (on citrus plants).

A wide variety of organic compounds like nitrophenol derivatives, azo compounds, sulphide, sulphonates, diphenyl carbinols, fluorinated compounds, organophosphates, carbamates and heterocyclics are used as effective acaricides. Some important examples are given below:

1. Azobenzene

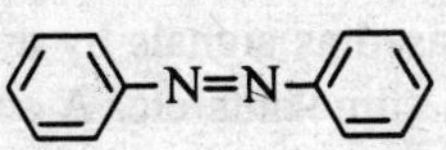

2. Fenson

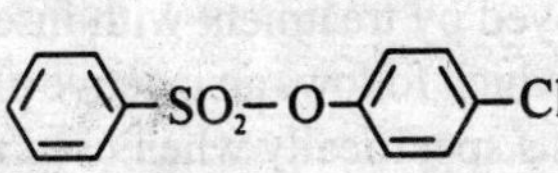

3. Dicofol

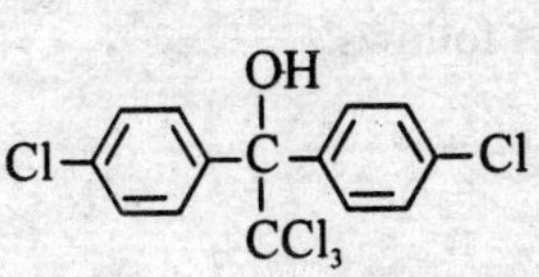

4.   Methiocarb

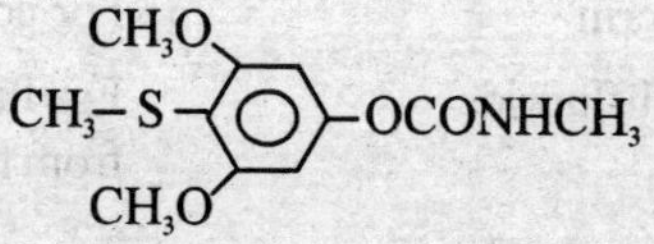

48.3.8 Hormones and Hormone Mimics

The crop protection by natural means include hormones and hormone mimics.

Invertebrates normally require three groups of hormones during their life cycle which regulate the growth, metamorphosis and reproduction. These hormones are Brain hormones, molting hormones and juvenile hormones. The brain hormone activates prothorax glands which release molting hormone, the ecdysone. The molting hormone initiates molting of larva or pupae. The juvenile hormone determines the nature of cuticle to be formed at each molting stage. If sufficiently high level of juvenile hormone is present along with ecdysone, the larva always molts to another larva.

A definite ratio of various hormones is present during the development of insects which may be disturbed by the loss or artificial supply of these hormones or their analogues. This may cause various morphogenetic disruptions resulting in the death of the insect and this is the basis of pest control by hormones.

No pesticide based on ecdysones is known due to its complicated chemical structure.

Juvenile hormones, however, have been used as effective insecticides. The change in juvenile hormone concentration during metamorphosis or in the last larval stage causes considerable irreversible disruptions to the insect development.

The juvenile hormone extracted from natural cecropia, methyl trans, trans-10,11-epoxyfarnesate has following structure.

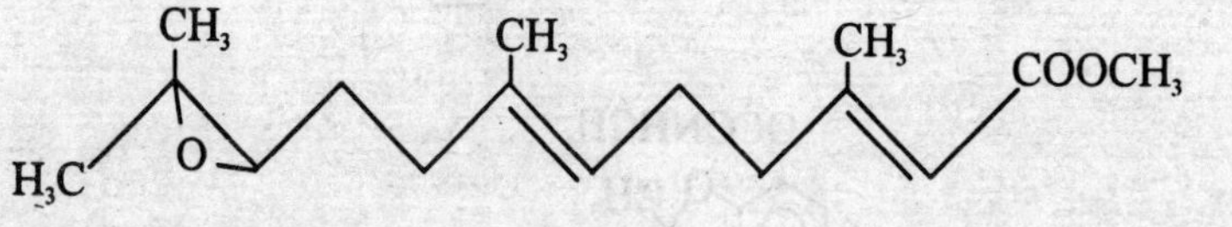

Methyl trans, trans-10,11-epoxyfarnesate

The synthetic compound having related structures show high biological activity. For example sesamax has excellent juvenile hormone activity.

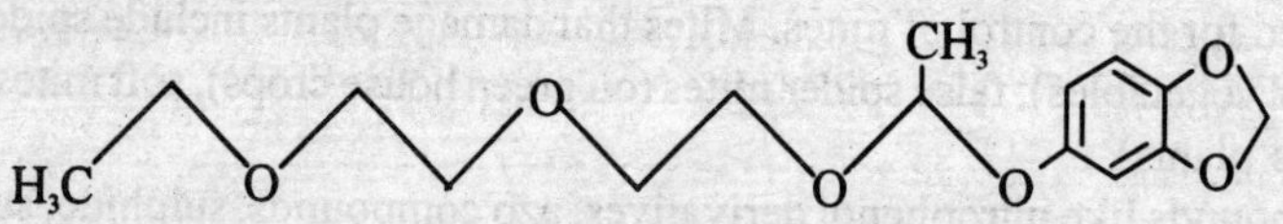

Sesamax

48.3.9 Phermones/Attractants

Phermones are the compounds released as signals by a particular organism as a result of certain reaction. Phermones serve as sex attractants, stimulants etc. A sex attractant may be used to attract the insect to a place where they can be destroyed by treatment with insecticides. Furthermore, the sex phermones can send insects to an hyperactive phase which follows an inactive phase due to exhaustion in which they do not copulate. The phermones can control pests specifically when used in appropriate amounts and also in pure state. Some common attractants are listed as follows:

1. Eugenol

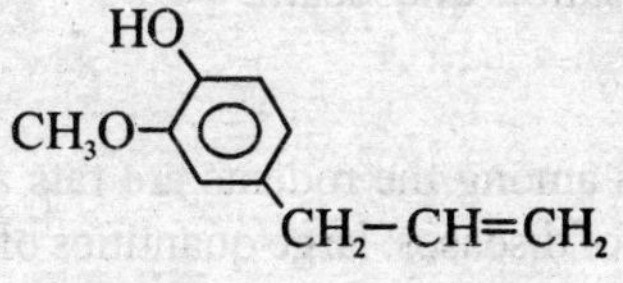

2. Muscalure or (Z)-9-Tricosene

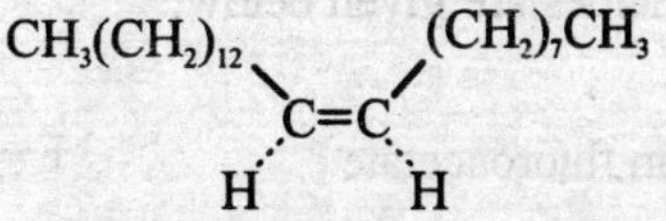

48.3.10 Insect Repellents

Insect repellents drive troublesome insects away rather than killing them. They are utilised most in human and veterinary sectors for the control of insects which are vectors of infectious diseases.

In the earlier times, plant extracts, essential oils, smoke etc. were used to repel insects. Now a days synthetic repellents are in use, for example, N,N-Dimethyl-3-methylbenzamide, Dimethyl phthalate (both used as mosquito repellent), and Di-n-propyl-2,5-pyridinedicarboxylate (used as housefly repellent).

48.3.11 Biological Methods

This involves other organisms to control or limit the population density of certain harmful animal or plant pest. The biological methods of pest control include.

1.Control with the aid of beneficial organisms which destroy or affect host. This includes pathogens, parasites etc.

2. Self destruction process in which sterilisation as a result of high energy radiation or chemical agent takes place. These sterilised insects are introduced for the control of screw-worm fly which affects cattle and goats and leads to complete eradication of this pest.

48.3.12 Molluscicides

Molluscs are crop pests and disease carriers. These are mainly slugs and snails which cause damage to the crops. They reproduce in large numbers and feed on plant parts, stored products etc. They also serve as intermediate host for certain developmental stages of insects which are animal and human parasites.

The commonly used molluscicides are quick lime (calcium oxide), calcium cyanamide used as sprays to control slugs. Copper (II) sulphate pentahydrate is used to control fresh water snail. Metaldehyde, prepared by acid catalysed tetramerisation of acetaldehyde in alcohol, is a commonly used moluscicide now a days.

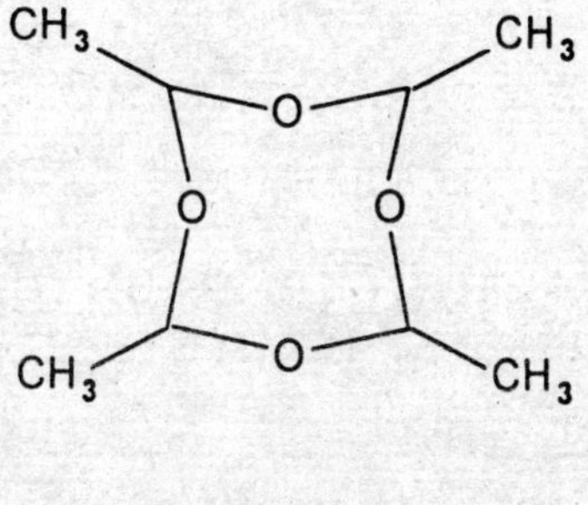

Metaldehyde

The monomer acetaldehyde and trimer paraldehyde have no muluscicidal activity. It is a contact poison, causes immobilisation and death.

48.3.13   Rodenticides

The most important pests among the rodents are rats and mice as they consume or destroy large quantities of food and act as vectors of diseases. large quantities of stored material is destroyed because of contamination with urine, faces, hair etc. Diseases like typhus and amoebic dysentery spread due to rodents. Some of the commonly used rodenticides are given below:

1.   Sodium fluoroacetate

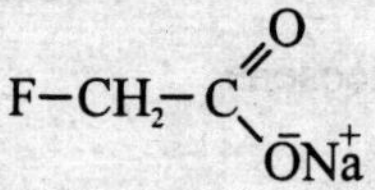

$$F-CH_2-C\begin{array}{c} \nearrow O \\ \searrow \bar{O}Na^+ \end{array}$$

2.   1-Naphthalenylthiourea

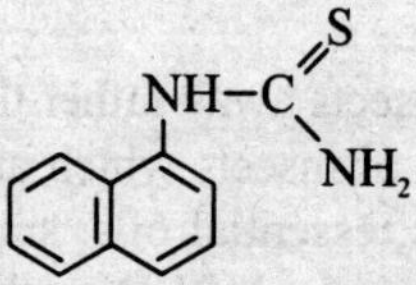

48.4   PROBLEMS

1. What are pesticides? How are they classified?
2. What are organochlorine insecticides? What are the disadvantages associated with them?
3. What are carbamate and organophosphate insecticides ? Discuss their mechanism of action.
4. Write a note on:
    - (a) Juvenile hormones
    - (b) Acaricides
    - (c) Molluscicides
    - (d) Insect repellents
    - (e) Attractants
5. Discuss the synthesis and uses of the following:
    - (a) Aldrin
    - (b) BHC
    - (c) Malathion
    - (d) Carbaryl
    - (e) Metaldehyde
6. What is acetyl cholinesterase? How is it affected by organophosphorus compounds?
7. What are contact and systemic insecticides? Give two examples of each.

Index